陕西省土地工程建设集团百篇优秀论文集

（2022）

《陕西省土地工程建设集团百篇优秀论文集》编委会　编

黄河水利出版社

·郑州·

图书在版编目(CIP)数据

陕西省土地工程建设集团百篇优秀论文集.2022/《陕西省土地工程建设集团百篇优秀论文集》编委会编.—郑州:黄河水利出版社,2023.9
ISBN 978-7-5509-3750-5

Ⅰ.①陕… Ⅱ.①陕… Ⅲ.①土地开发-中国-文集 Ⅳ.①F323.211-53

中国国家版本馆 CIP 数据核字(2023)第 192350 号

陕西省土地工程建设集团百篇优秀论文集(2022)
SHANXISHENG TUDI GONGCHENG JIANSHE JITUAN BAIPIAN YOUXIU LUNWENJI(2022)
《陕西省土地工程建设集团百篇优秀论文集》编委会

审　　稿　席红兵　13592608739

责任编辑　景泽龙　冯俊娜　　责任校对　杨秀英
封面设计　黄瑞宁　　责任监制　常红昕
出版发行　黄河水利出版社
地址:河南省郑州市顺河路 49 号　邮政编码:450003
网址:www. yrcp.com　E-mail:hhslcbs@126.com
发行部电话:0371-66020550
承印单位　河南瑞之光印刷股份有限公司
开　　本　890 mm×1 240 mm　1/16
印　　张　73.75
字　　数　2 180 千字
版次印次　2023 年 9 月第 1 版　　2023 年 9 月第 1 次印刷

定　　价　335.00 元

前　言

2022年，陕西省土地工程建设集团(以下简称“集团”)坚持以习近平新时代中国特色社会主义思想为指导，深入学习宣传贯彻党的二十大精神，紧密围绕粮食安全和乡村振兴等国家战略，坚持创新是第一动力，集团科研工作取得了长足进展，已成为国家在土地工程领域科技和实践的中坚力量。

这一年，正值集团成立30周年，集团继续坚持以高质量科技创新支撑高质量发展，收获了累累硕果。集团先后荣获省农业技术推广成果一等奖、省科学技术进步奖二等奖等省部级科技奖励9项，荣获中国创新方法大赛、省“三新三小”创新竞赛等奖励23项；建成全国最大的耕地土壤库，为全面摸清耕地土壤质量家底，加强耕地质量监测保护和管理提供了基础支撑；成为陕西省第三次全国土壤普查工作牵头单位和技术支持单位，负责统筹开展全省“三普”工作；聚焦土地工程基础研究及关键核心技术研发，获批省部级科研项目33项，国拨经费约2000万元；发表科技论文2031篇，其中SCI 81篇，单篇平均影响因子3.71，同比提升55%；持续夯实行业标准制定者地位，颁布实施地方标准1项，获批立项10项，年度获批数量创历史新高；出版专著2部，获批知识产权162项；聚焦国家战略需求，积极建言献策，联合完成决策咨询建议3份，其中1份被省委办公厅采纳；人才队伍建设卓有成效，1人获陕西省“特支计划”杰出人才，1人获陕西省青年科技新星。

为进一步推进土地工程学科建设，加强土地工程理论成果的交流推广和工程应用，本着突出集团主营业务、质量优先、兼顾全面的原则，经集团各部(室)、各单位推荐，编委会最终筛选出100篇编纂成集，内容涵盖土地整治工程、土地利用与保护、土地资源与土地信息、房地产及建筑工程、综合管理五大类，以期为从事土地工程及相关领域的科研人员及科研管理工作者提供参考，以便更好地服务土地工程科研工作发展，着力打造土地工程领域原创技术策源地。

论文集的成稿得益于很多人的付出和努力。感谢编委会对论文集的编撰给予的指导和支持，感谢对外合作与科技部人员在稿件整理、编撰过程中付出的大量心血，感谢各部(室)、各二级单位在论文集制作过程中给予的大力支持。本论文集得以出版，还得到了黄河水利出版社的热情支持，在此一并表示衷心的感谢!

编　者

2023年9月

目 录

1 土地整治工程

2　土地利用与保护

3 土地资源与土地信息

4　房地产及建筑工程

5　综合管理

1　土地整治工程

Adsorption of Toxic Tetracycline, Thiamphenicol and Sulfamethoxazole by a Granular Activated Carbon (GAC) under Different Conditions

Risheng Li[1,2], Wen Sun[1,2], Longfei Xia[1,2], Zia U[3], Xubo Sun[1,2], Zhao Wang[1,2], Yujie Wang[4] and Xu Deng[5]

(1. Shaanxi Provincial Land Engineering Construction Group Co., Ltd., Xi'an 710075, China; 2. Key Laboratory of Degraded and Unused Land Consolidation Engineering, the Ministry of Natural Resources, Xi'an 710075, China; 3. Department of Environmental Science and Engineering, School of Energy and Power Engineering, Xi'an Jiaotong University, Xi'an 710049, China; 4. Department of Chemistry, College of Resource and Environment, Baoshan University, Baoshan 678000, China; 5. Shaanxi University of Chinese Medicine, XiXian New Area, 712046, China)

【Abstract】 Activated carbon can be applied to the treatment of waste water loading with different types of pollutants. In this paper, a kind of activated carbon in granular form Granular Activated Carbon (GAC) was utilized to eliminate antibiotics from aqueous solution, in which Tetracycline (TC), Thiamphenicol (THI), and Sulfamethoxazole (SMZ) were selected as the testing pollutants. The specific surface area, total pore volume and micro pore volume of GAC were 1059.011 m^2/g, 0.625 cm^3/g, and 0.488 cm^3/g respectively. The sorption capacity of GAC towards TC, THI and SMZ were evaluated based on the adsorption kinetics and isotherm. It was found that the pseudo-second-order kinetic model described the sorption of TC, THI, and SMZ on GAC better than the pseudo-first-order kinetic model. According to Langmuir isotherm model, the maximum adsorption capacity of GAC towards TC, THI and SMZ were calculated to be 17.02, 30.40, and 26.77 mg/g respectively. Thermodynamic parameters of ΔG^0, ΔS^0, and ΔH^0 were obtained, indicating that all the sorptions were spontaneous and exothermic in nature. These results provided a knowledge basement on the utilization of activated carbon to the removal of TC, THI and SMZ from water.

【Keywords】 Adsorption; Antibiotics; Activated carbon; Water treatment

1 Introduction

Antibiotics can be used to kill or inhibit the growth of bacteria. Although antibiotics are used in humans and animals, roughly 80% of their total usage is on livestock and poultry for human consumption. Antibiotics are routinely added to the food and water of livestock to promote growth and improve feed-use efficiency. In addition, antibiotics are injected into animals when they are sick or at high risk of getting sick. According to recent sales data of world market, China is the biggest producer as well as user of antibiotics[1]. When the antibiotics are used, the organisms cannot absorb antibiotics fully so that they are released into the environment in an active form[2]. In 2013 in China, more than 50 $\times 10^3$ tons of antibiotics entered into aquifers as stated in a report. The aquifers include the outflow of sewage processing plant[3], drinking water, groundwater[4], rivers and lakes[5], and seawater. Antibiotics have different half-life, in which some are long lived[6] and their contagion rates in the environment have increased over the years. In a water-based environment, antibiotics are generally injurious, i. e., prevents the ability to breakdown micro-organisms deposit, destructs development of marine organisms as well as encourages maturation in bacterial drug-opposing genetic codes[7]. Various researches[8] have shown that any contact to antibiotics (μg/L-mg/L) creates negative effect and influence on the lives of water-based creatures, for example, the growth of their body and weight.

The residual of these antibiotics in water-based items comes into human body by the help of food chain and later mount up via biological enhancement[9]. Since the majority of antibiotics are cancerogenic, teratogenic, and mutagenic along with creating hormones related issues, so using antibiotics causes serious interference with anatomy of humans and immune system[7]. Antibiotics are obtaining the identification of rising environmental contaminants, are classified as fractious bio-accumulative substances[10-11], as well as are considered as harmful and toxic chemicals.

The outflow of Municipal treatment plants and pharmaceutical manufacturing plants are the basic sources of discharging antibiotics in water. According to Michael[12] and Rizzo[13], wastewater treatment plants in the cities are to be considered as the main source of releasing antibiotic in the environment. It is important to dispose residues of antibiotic before discharging wastewater in environment. There is an urgent need for case studies to provide cheap solution for the elimination of antibiotics. Sera Budi Verinda et al. used ozonation to remove ciprofloxacin in wastewater, and the removal rate reached 83.5%[14]; Shang, K.F. et al. indicates that the combination of DBD plasma and PMS/PDS is an efficient pretreatment method for bio-treatment of refractory SMX[15].

Adsorption process is[16] easy to plan and feasible to function.The adsorbent should be a biomass adsorbent such as agricultural waste, which is environmentally friendly and economical.[17]. Adsorption method is utilized to the removal of organic pollutants from contaminated waters over the surface of adsorbent[18]. Its application to the elimination of antibiotics for approximately 30 different compounds have been reported so far[19]. The performance of adsorption processes is largely influenced by the hydrogen bonding and electrostatic interactions.Different adsorbent materials are used for the elimination of antibiotics from the aqueous solution, for instance, clinoptilolite[20], soil[21], different kind of activated carbons[22-25], calcium phosphate materials, and core-shell magnetic nanoparticles[26]. Compared with traditional adsorption materials, activated carbon has excellent porosity, large specific surface area, low cost and environmental friendliness, and is reported to be an effective adsorbent for eliminating trace pollutants[27]. Granular activated carbon is divided into stereotyped and unshaped particles. It is mainly made of coconut shell, nut shell and coal, which is refined through a series of production processes. Its appearance is black amorphous particles; it has developed pore structure, good adsorption performance, high mechanical strength, low cost, etc. Therefore, granular activated carbon is widely used in drinking water, industrial water, wine, waste gas treatment, decolorization , desiccant, gas purification and other fields[28-29].

At present, there are very few related research on antibiotics in wastewater through activated carbon. In this paper, three different kinds of antibiotics including tetracycline, thiamphenicol and sulfamethoxazole were selected as the target pollutants and a kind of activated carbon in granular form was used as the adsorbent. The adsorption capacity and mechanism were studied. The purpose of this paper was to evaluate the removal efficiency of adsorption technology on the treatement of antibiotics-loading wastewater and to promote the application of activated carbon in such a field.

2 Materials and methods

2.1 GAC preparation

The Granular Activated Carbon used in these experiments was prepared by using corn stover that was collect from Shaanxi agriculture Technology Company (Xi'an, China). First this corn stover was put 1 month for air drying. After that it was chop into small pieces of length nearly 5 cm and positioned inside electrically operating container resistance furnace (LNB4-13Y; Haozhuang Co. Ltd., Shanghai, China). Under nitrogen atmosphere where pieces of cornstalk were heated at 500 ℃ for 2 h and then 700 ℃ for 2 h.

Physical activation management was done in an activation furnace (HHL-1; Huatong Co. Ltd., Henan, China) through superheated steam at very elevated temperatures (600 ℃, 2.0 MPa) for a duration of 2 hours to obtain the AC. The AC was also cleansed using 1.0M HCl-HF (1∶1, vol/vol) solution thrice and with ultrapure water a few times till the 7.0 pH value was obtained. After washing, drying of AC was then done at a temperature of 100 ℃ for a duration of 15 hours, and then crushed and pushed through a nylon mesh having a entrance of 5.0 mm. Finally, the AC's morphology was granular.

2.2 Chemicals

The physio-chemical characteristics and molecular structures of tetracycline (TC), thiamphenicol (THI), and sulfamethoxazole (SMZ) were given in Table 1. All the antibiotics have chromatographic clarity which was purchased from J&K scientific (Beijing). N, N-dimethylformamide with more than 99.9% purity was purchased by Sigma-Aldrich (Shanghai, China). Apparatus SPI-11-10T was used to prepare ultra-pure water (ULUPURE, Sichuan, China). The GAC sample used in this paper was supported by Fan et al.[30]

Table 1 Physicochemical properties of the three antibiotics

Property	Tetracycline (TC)	Thiamphenicol (THI)	Sulfamethoxazole (SMZ)
Molecular Formula & Chemical Structure	$C_{22}H_{24}N_2O_8$	$C_{12}H_{15}CL_2NO_5S$	$C_{10}H_{11}N_3O_3S$
Molar Mass	444.43	356.22	253.28
Solubility (25 ℃)	1700 mg/L	2270 mg/L	459 mg/L

2.3 Methodology for selecting antibiotics

Wastewater influents contain several types of antibiotics; however, due to the limited availability of information, only a few antibiotics will be selected for this work. The criteria for selecting antibiotic classes were defined by considering: (1) The relevance of antibiotic class to human medicine; (2) Usage amongst the different animal species; and (3) Their presence in wastewater treatment plants.

Based on the selection criteria mentioned above, the following antibiotics were selected for this work: (1) Tetracycline; (2) Thiamphenicol; (3) Sulfamethoxazole.

2.4 Sorption experiments

Stock solution preparation: 0.1000 gram of TC, THI, and SMZ were dissolved in a 50 mL volumetric flask with ultra-pure water, and then transfer into a 100 mL volumetric flask to obtain the stock solutions with final concentration of 1 g/L respectively. The testing solutions with different concentrations were obtained by diluting the stock solutions with ultrapure water.

Effect of GAC dosage: on the sorption of TC by the GAC, tests were conducted in a 100 mL beaker containing 50 mL TC testing solution with 25 mg/L. For GAC dosage, the amount of GAC ranged from 2 to 8 g/L, while the temperature was fixed at 25 ℃ and contact time was 100 min, the effect of GAC dosage was also conducted for the sorption of THI and SMZ.

Determination of the adsorption equilibrium time: to investigate the sorption of TC onto the GAC, batch experiments were carried out by using a 100-mL beaker containing 0.4000 g GAC and 50 mL TC testing solutions with various preliminary concentrations. Each beaker was put into a thermostatic reciprocating shaker (ZHWY-2102C) at 180 r/min and 25 ℃ in dark. The sample was withdrawn by a 5 mL syringe from one beaker after shaking for 5 mins, and passed through a 0.45 μm filter. The sampling times were 5, 10, 15, 20, 30, 40, 50, 60, 80, and 100 mins. The same steps were carried out to investigate the sorption of SMZ and THI on GAC, respectively. Filter liquor were used to determine the residual concentrations of the three antibiotics.

Adsorption isotherm: for measuring the adsorption isotherm, a 250 mL beaker containing 50 mL testing solution with different initial concentration of TC and 0.4000 g of GAC were aginated in dark under 180 rpm at 25 ℃. Samples were withdrawn after 100 minutes using a 5 mL pipette tip and filtered by using 0.45 μm filter membrane. This filter out solution of TC was used for remaining concentration analysis to found adsorption isotherm representation. For investigating adsorption isotherm of THI and SMZ, the same procedure followed.

Adsorption Thermodynamics: a 250 mL beaker containing 50 mL testing solution with 25 mg/L of TC and 0.4000 g of GAC were agitated in dark under 180 rpm at 15, 20 and 25 ℃. Samples were withdrawn after 100 minutes using a 5 mL pipette tip and filtered by using 0.45 μm filter membrane. This filter out solution of TC was used for remaining concentration analysis. For investigating adsorption isotherm of THI and SMZ, the same procedure followed.

Analysis: physisorption analyzer (ASAP-2020, Micrometrics, Beijing) was used to measure the prepared GAC characteristics, like surface area, pore volume, pore diameter at 77 K temperature. The Brunauer-Emmett-Teller (BET) method was used to measure surface area and the Barrett-Joyner-Halenda (BJH) method was used to measure pore size distribution. Scanning electron (TM-1000; Hitachi, Japan) was used to describe the GAC surface morphology and its porous structure. The zeta potential analyzer (Malvern Zetasizer Nano S90, Shanghai, China) was used to monitored GAC zeta potential value. The GAC was firstly convert it into powder form and then submerge it in NaCl solution (1 mmol = litter) for making a blend (0.1 gram = 1 L). For adjusting the pH value of solution from 3 to 9 HCl or NaOH was used. This solution was place in an ultrasonic treatment apparatus (25 ℃, 40 kHz) for 30 minutes. After the ultrasonic treatment, this solution was kept for 24 hours and then its zeta potential value was measured by using zeta potential analyzer. The Ultraviolet-visible spectrophotometer (model: SP-1915, Spectrum, Shanghai, China) was used to measure the antibiotics residual concentration. The calibration curves for TC, SMZ and THI were given as: $y=0.0324x+0.0307$, $y=0.0635x+0.0007$ and $y=0.1455x+0.0068$, respectively. For measuring the sorption capacity on GAC following formula can be used.

$$q_t = \frac{C_0 - C}{M} \times V \tag{1}$$

Where q_t = adsorption capacity at time t (mg/g); C_0 = adsorbate initial concentration in (mg/L); C = adsorbate residual concentration at time t (mg/L); V = volume of solution (L); M = mass of sorbent (g).

The removal efficiency at different initial concentration of antibiotics was calculated by using following formula:

$$R.E = \frac{C_0 - C_E}{C_0} \times 100\% \tag{2}$$

Where $R.E$ = removal efficiency (%); C_0 = initial concentration (mg/L); C_E = equilibrium concentration (mg/L).

2.5 Regeneration experiments

Saturated GAC(8.0000±0.0004)g adsorbing TC, THI and SMZ was placed in a quartz glass reactor, and N2 was used as a protective gas and placed in a microwave oven for irradiation, microwave power 730 W, microwave time 180 s[31], and carried out microwave regeneration test.

3 Results and discussion

3.1 Characterization of the granular activated carbon

Fig. 1(A) exhibits the outlook of GAC with an average length of 1-2 mm and diameter of 1 mm. Nitrogen adsorption/desorption at 77 K for the granular activated carbon was shown in Fig. 1(B), in which the GAC sample possessed a type I sorption isotherm. The details of BET are shown in Table 2: These results demonstrated that GAC was of high specific surface area and porous structure. In conclusion this type of GAC might be an ideal adsorbent for removing antibiotics from wastewater. In Fig. 1(A), the SEM image of granular activated carbon shows that external surface of GAC was multi-layer and rigid, which helped to increasing the specific surface area of granular activated carbon. The measurement of zeta potential is a technique for calculating the surface charge of activated carbon in a colloidal solution. The graph of activated carbon zeta potential in the solution was shown as a function of pH in Fig. 1(B). The graph significantly shows that the surface charge on activated carbon was linked with solution pH. The pH_{ZPC} zero-point charge of activated carbon is 4.0, indicating the charge on the activated carbon surface was positive when the pH of solution was less than pH_{ZPC} while negative when solution pH was greater than pH_{ZPC}. It can be seen from Fig. 1(D) that the infrared spectrum of GAC has a C-O characteristic absorption peak near 1000 cm^{-1}, a C=C characteristic absorption peak near 1600 cm^{-1}, and an O-H stretching vibration peak near 3200 cm^{-1}.

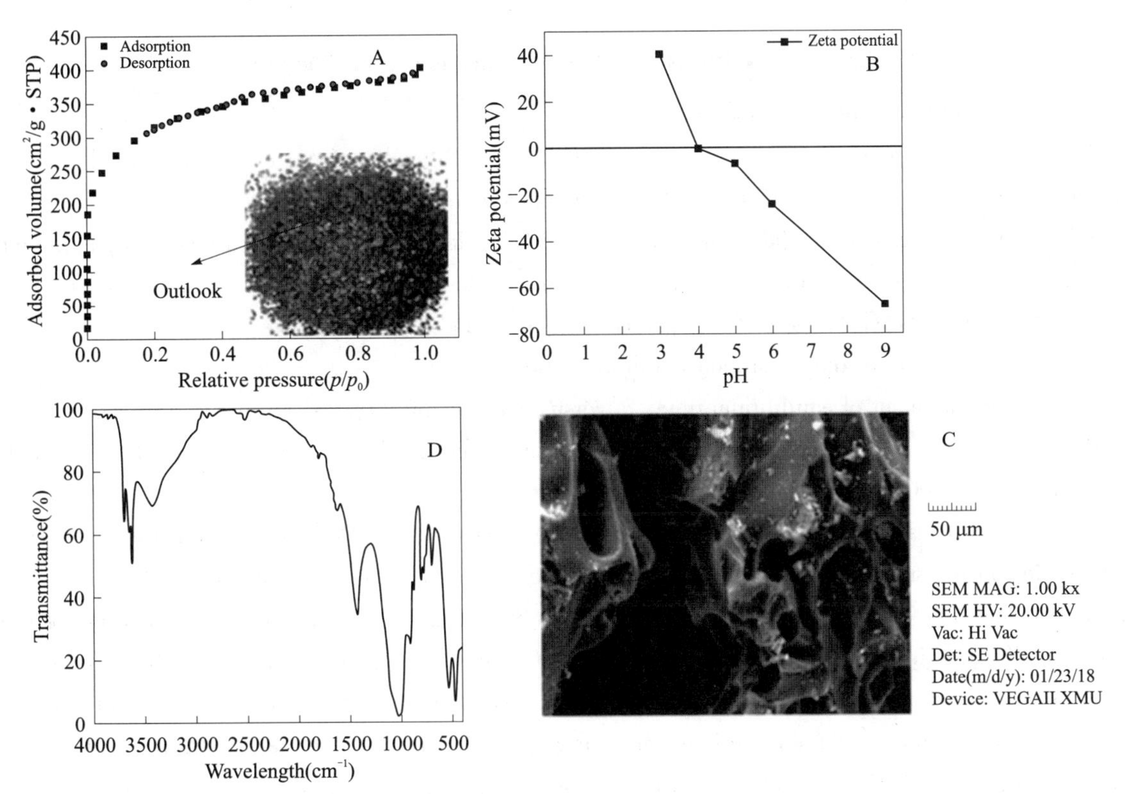

Fig. 1 A: Nitrogen (N_2) adsorption isotherm, B: Zeta potential of the GAC, C: SEM image, D: FTIR image

Table 2　Details result of the BET

BET	Specific Surface Area	Low Pressure (p/p_0<0.1) Adsorption Capacity	Hysteresis Loop (p/p_0 = 0.2)	Total Pore Volume	Micro Pore Volume
	1059.011 m^2/g	increased micropores	closed mesoporous	0.625 cm^3/g	0.488 cm^3/g

3.2　Effect of pH on GAC adsorption

Under the conditions of environmental conditions of 25 ℃, the antibiotic concentration of 25 mg/L, GAC dosage of 8 g/L, TC adsorption time of 100 min, and THI and SMZ adsorption time of 60 min: When the pH of the solution is 7, the effect of GAC on TC, THI, SMZ the best adsorption efficiencies are 91.76%, 96.34%, 94.23%, respectively.(Fig. 2)

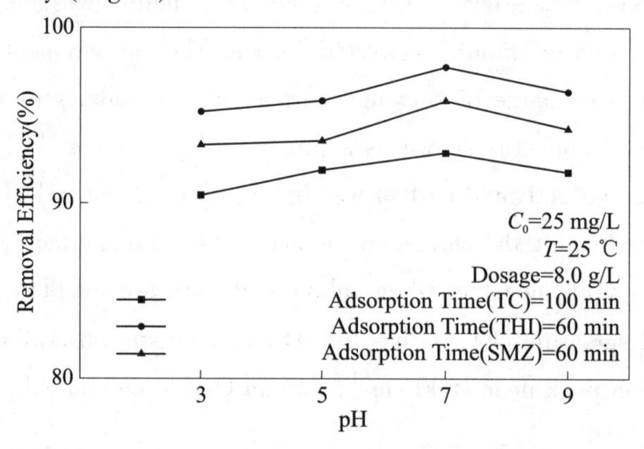

Fig. 2　The relationship between pH and adsorption efficiency

3.3　Effect of GAC dosage on the adsorption

Fig. 3 shows the removal efficiency versus GAC dosage. When the GAC dosage increased from 2 to 8 mg/g, all the removal efficiencies for the three antibiotics increased remarkably and then slightly from 6 to 10 mg/g. A possible reason was that with an increase in GAC dosage, the sorbent surface area, the number of sorption sites, and the contact area increased.[32-34] In this study, considering the removal efficiencies and economic benefits, a GAC dosage of 4 mg/g was selected as an optimum dosage.

3.4　Determination of the adsorption equilibrium time

For the measurement of equilibrium time, q_t versus contact time (t) was represented in Fig. 3. During the adsorption of TC, it was seen that the value of q_t increased quickly within the first 30 min and then slightly from 30 to 100 min for all the initial concentrations.

The possible cause was that during the initial stage of adsorption (0-30 min) there were enough adsorption sites on the surface of GAC. As the contact time prolonged, the adsorption sites provided by GAC became less and less, as a result, q_t increased slightly (30-100 min). In case of SMZ, the value of q_t increased significantly in the first 20 min and from 20 to 60 min the qt value increased slightly. Different from TC and SMT, the adsorption process of THI was faster, in which after 10 minutes no increase in the value of q_t was observed. Therefore, the equilibrium times for TC, SMT and THI were set at 100 min, 60 min and 10 min. In most cases, when the initial antibiotics concentration was low while the removal efficiency was high (Fig.4), the possible reason was that at low concentration there is much more adsorbing site were available for antibiotics molecule to absorbed on it.

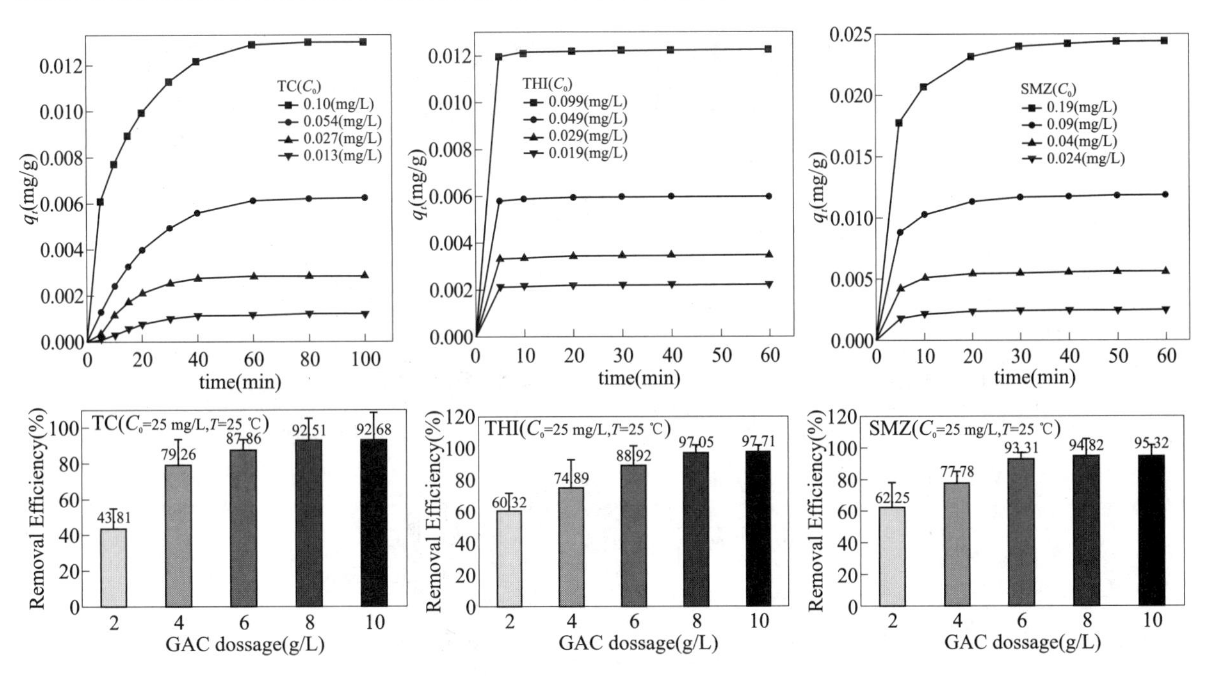

Fig. 3 Influence of contact time on the adsorption of the three antibiotics by GAC under different initial concentrations

3.5 Sorption kinetics

The pseudo-first-order and pseudo-second-order models, were used to describe all the data shown in Fig. 3 and the results were given in Table 3, in which Eqs. (3) and (4) represent the mathematical formula of the two models.

$$\ln(q_{e,exp} - q_t) = \ln(q_{e,cal}) - K_1 t \tag{3}$$

$$\frac{t}{q_t} = \frac{1}{K_2 q_{e,cal}^2} + \frac{1}{q_{e,cal}} t \tag{4}$$

Where $q_{e,exp}$ = adsorption amount in (mg/g) at equilibrium; q_t = adsorption amount in (mg/g) at time t; K_1 = rate constant of pseudo-first-order (min^{-1}); K_2 = rate constant of pseudo-second order [g/(mg · min)].

Fig. 4 show the linearized graph of t/q_t versus time of pseudo-second order kinetic. Table 3 contain all the parameters of these two kinetic models. The value of R^2 obtained from pseudo-second order model are given in Table 3, which are close to unity indicating that pseudo-second order kinetic model best fit the adsorption of antibiotic on GAC. Furthermore, the experimental value of $q_{e,exp}$ in (mg/g) agreed with calculated value of $q_{e,cal}$(mg/g). These results indicate that rate of adsorption on GAC is controlled by chemisorption, while valency forces with exchange or there might be a sharing of the electrons in-between these four antibiotics and the GAC, according to Table 3 primary initial adsorption rate $K^2 q_{e,cal}^2$ gradually increased with the increase in the concentration of primary four antibiotics, indicating the fact that the greater value of concentration enhanced driving forces that can basically help to overcome the barrier of mass transfer-resistance in-between the phases of solid and liquid.

Weber and Morris intra particle diffusion model was utilized to analyze the experimental data to determine the rate-limiting step during the adsorption process. All values were shown in Table 2.

$$q_t = K_{id} t^{1/2} + I \tag{5}$$

Where q_t = the removal amount at time t and reaction equilibrium (mg/g); K_{id} = the particle diffusion rate constant (mg/g · $min^{1/2}$); I = intercept, it gives the information about boundary layer effect. If the value of I is greater than the boundary layer has greater effect (mg/g).

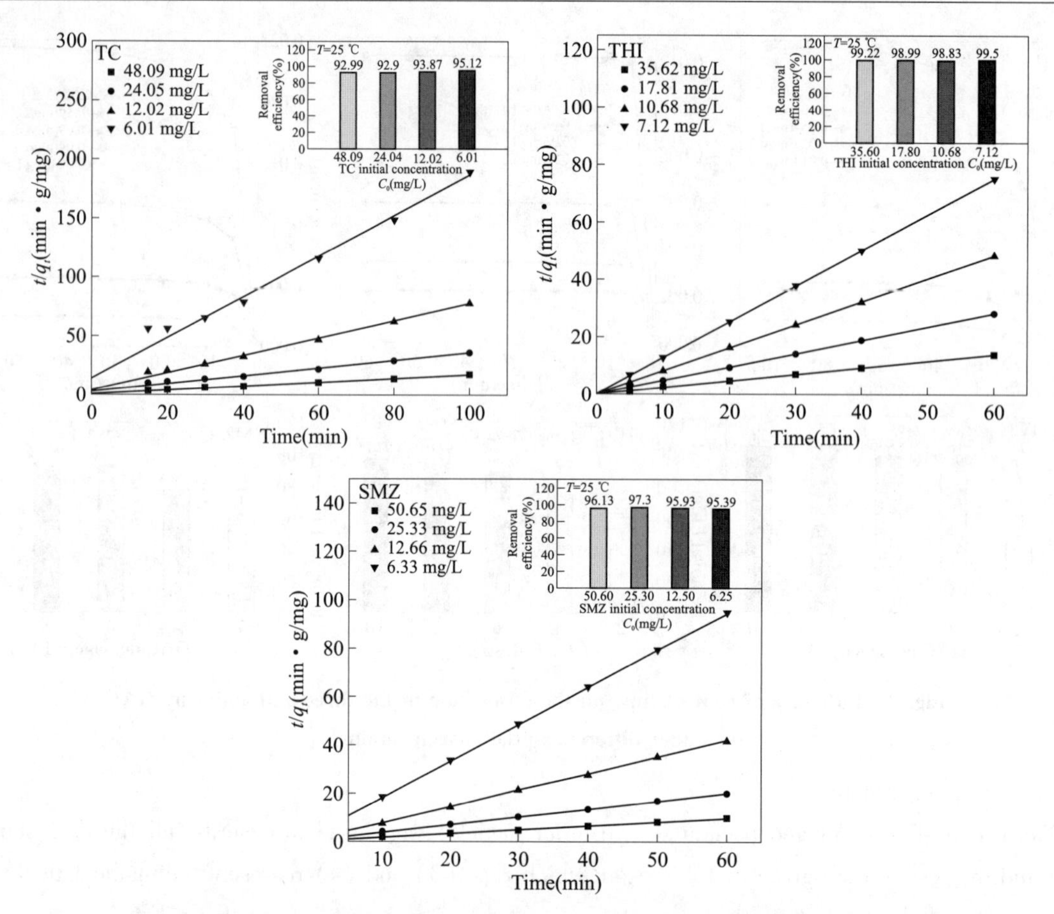

Fig. 4 The pseudo-second-order kinetics for the adsorption of the three antibiotics on the GAC at various initial concentrations

Table 3 Parameters of the pseudo-first-order and pseudo-second-order kinetic models for the sorption of the three antibiotics onto GAC at 25 ℃

Sorbate	C_0 (mg/L)	Pseudo-first-order model $q_{e,exp}$ (mg/g)	$q_{e,cal}$ (mg/g)	K_1 (min^{-1})	R^2	Pseudo-second-order model $q_{e,cal}$ (mg/g)	K_2 [g/(min · mg)]	$K_2q_{e,cal}^2$ [mg/(min · g)]	R^2
TC	6.01	0.52	2.332	0.023	0.77	0.49	0.704	0.169	0.68
	12.02	1.27	0.481	0.043	0.88	1.64	0.028	0.076	0.91
	24.04	2.78	2.773	0.053	0.98	3.46	0.014	0.167	0.98
	48.09	5.75	3.377	0.050	0.94	6.27	0.020	0.790	0.99
THI	7.12	0.79	15.7	0.109	0.73	0.80	5.14	3.306	1.00
	10.68	1.24	14.11	0.100	0.74	1.25	2.39	3.740	1.00
	17.81	2.13	12.23	0.091	0.76	2.14	2.65	12.158	1.00
	35.62	4.35	10.51	0.096	0.75	4.37	1.54	29.409	1.00
SMZ	6.33	0.63	4.57	0.021	0.78	0.65	0.811	0.349	0.99
	12.66	1.42	3.38	0.027	0.77	1.46	0.482	1.033	0.99
	25.32	3	1.11	0.038	0.84	3.10	0.185	1.781	0.99
	50.65	6.16	1.58	0.074	0.95	6.4	0.78	31.948	0.99

If the plot between q_t versus $t^{1/2}$ would be linear then intraparticle diffusion take place. When the plot also passes through origin ($I=0$) then the rate limiting is controlled by intraparticle diffusion. If the plots showed deviation from linearity, then it indicates the effect of boundary layer. Fig. 5 represents the plot of intraparticle for three antibiotics. This was observed that during whole time plots linear portion cannot pass through origin, indicating that both boundary layer and intraparticle diffusion take place during the adsorption of antibiotics on GAC.

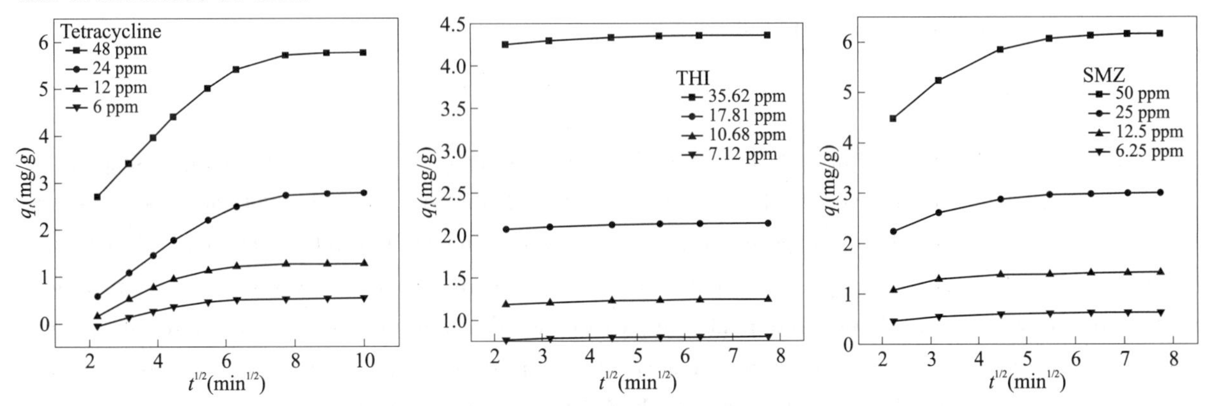

Fig. 5 Intraparticle diffusion plots for the antibiotics sorption on the GAC

In the Table 4, the rate constant and intercept values were shown. The intercepts C of the straight lines fitted by TC, THI, and SMZ are not 0, indicating that internal diffusion is not the only step controlling the removal of antibiotics by GAC, and the adsorption rate should be controlled by both external diffusion and intraparticle diffusion. According to Fig. 5, the plots of TC and SMZ have two different portions. Firstly, the steeper segment of plot depicts external surface adsorption. The second portion which is slowly adsorption showed intraparticle diffusion.

Table 4 Intraparticle diffusion kinetic model parameters at 25 ℃

Sorbate	C_0 (mg/L)	Intra-particle diffusion model					
		K_{id1}(mg/g · min$^{1/2}$)	I_1(mg/g)	R^2	K_{id2} (mg/g · min$^{1/2}$)	I_2(mg/g)	R^2
TC	6.01	0.13	-0.3	0.94	0.01	0.44	0.83
	12.02	0.25	-0.29	0.93	0.01	1.10	0.58
	24.04	0.47	-0.4	0.99	0.10	1.84	0.77
	48.09	0.66	1.31	0.98	0.13	4.58	0.77
THI	7.12	0.00	0.75	0.70	—	—	—
	10.68	0.01	1.17	0.85	—	—	—
	17.81	0.01	2.06	0.75	—	—	—
	35.62	0.01	4.23	0.79	—	—	—
SMZ	6.33	0.06	0.33	0.88	0.00	0.56	0.82
	12.66	0.13	0.82	0.76	0.01	1.32	0.86
	25.32	0.28	1.65	0.92	0.03	2.73	0.78
	50.65	0.6	3.21	0.94	0.09	5.49	0.76

Boyd kinetic model was used to further examine the kinetic data, in order to measure the actual slowest step, involved in the adsorption process.

$$F(t) = 1 - \left(\frac{6}{\pi}\right) \sum_{n=1}^{\infty} \left(\frac{1}{n^2}\right) \exp(-n^2 B_t) \tag{6}$$

Where $F(t) = q_t/q_e$ = ratio of the antibiotics adsorbed at time t and equilibrium; B_t = function of $F(t)$.

If the $F(t)$ value is higher than 0.85 then

$$B_t = 0.4977 - \ln(1 - F(t)) \tag{7}$$

If the $F(t)$ value is less than 0.85 then

$$B_t = \left(\sqrt{\pi} - \sqrt{\pi - \left(\frac{\pi^2 F(t)}{3}\right)}\right)^2 \tag{8}$$

As for the Boyd kinetic model, if the plot of B_t against t is linear and through the origin, it suggests the fact that intraparticle diffusion is controlling the process of mass transfer. When the plot can be seen as non-linear or linear but does not go through the origin, it means that the sorption rate is controlled through film diffusion. Fig.6 illustrates the Boyd plots for the three antibiotics on the GAC sample. The fact that Boyd plots were linear even through they were unable to pass through the origin was noticed. These results indicate the information regarding the fact that the major controlling process required for adsorption procedure was nonetheless diffusion with the layer at border. Table 5 lists the parameters of the Boyd kinetic model.

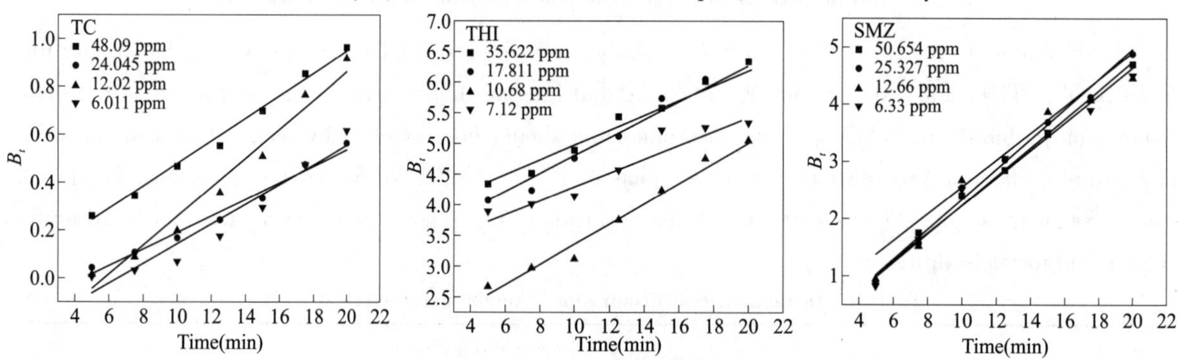

Fig. 6 Plots of Bod kinetic model

Table 5 The Boyd kinetic model parameters at 25 ℃

Sorbate	C_0(mg/L)	Boyd plot	
		Intercept	R^2
TC	6.01	-0.026	0.92
	12.02	-0.344	0.97
	24.04	-0.155	0.98
	48.09	0.008	0.99
THI	7.12	2.640	1.00
	10.68	1.700	0.97
	17.81	2.870	0.96
	35.62	3.220	0.94
SMZ	6.33	-0.168	0.97
	12.66	0.206	0.89
	25.32	-0.285	0.99
	50.65	-0.262	0.99

3.6 Sorption isotherm

Adsorption process can be defined as the process where mass transfer of adsorbate at the boundary layer in-between the solid adsorbent and liquid phase takes place. The adsorption isotherm is defined as the equilibrium relationship between solid adsorbent and adsorbate at constant temperature. For example: the ratio between amount of adsorbate absorbed on the solid and the remaining amount left in aqueous solution at equilibrium. There are several adsorption isotherms models like Langmuir, Freundlich, Temkin.

Experimental data can be fitted using these models to examine the suitability of model. The information obtained from these models can be used for designing the adsorption process. The parameters of adsorption isotherm normally estimated the sorption ability of several adsorbent for specific adsorbate with predetermine reaction condition. The performance of the sorption process depends not only on the rate at which mass transfer occurs but also on the sorbent-sorbate equilibrium concentration[35] Three different isotherm models, called the, Langmuir[36], Freundlich[37], and Temkin[21-22] were used. The mathematical formula of these isotherm models were shown as follows:

$$q_e = \frac{q_m K_L C_e}{1 + K_L C_e} \tag{9}$$

$$q_e = K_F C_e^{1/n} \tag{10}$$

$$q_e = B\ln(K_T C_e) \tag{11}$$

Where, K_L = constant from Langmuir (L/mg); C_e = adsorbate residual concentration (mg/L); q_e = The amount of adsorbate per unit mass of sorbent (mg/g); q_m = The maximum sorption capacity (mg/g); K_F(mg/L$^{(1-1/n)}$ · g) and n are constants of the Freundlich model; B is the Temkin constant of the sorption heat and K_T(1/mg) stands for the constant of Temkin isotherm.

Both Fig. 7 and Table 6 shows that Langmuir model is best fitted the experimental data than Freundlich and Temkin. The calculated values of q_{max}, from Langmuir indicates that the maximum degree of adsorption capacity of GAC sample to three antibiotics have been following the trend: THI > SMZ > TC.

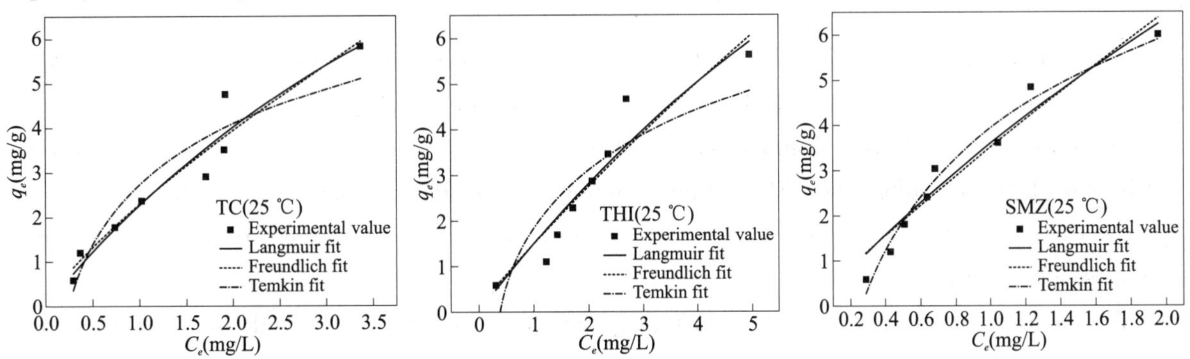

Fig. 7 Plots of adsorption isotherms for the sorption of antibiotics on the GAC

Table 6 Parameters of the isotherm models describing the sorption of antibiotics on GAC at 25 ℃

Sorbate	Langmuir			Freundlich			Temkin		
	q_m(mg/g)	K_L(L/mg)	R^2	K_F(mg/L$^{(1-1/n)}$.g)	n	R^2	B	K_T	R^2
TC	17.02	0.154	0.93	2.28	1.2	0.93	1.944	4.093	0.88
THI	30.42	0.530	0.92	12.17	1.1	0.91	—	—	—
SMZ	26.77	0.155	0.95	3.51	1.1	0.94	2.930	3.800	0.97

In 1974 Weber and Chakkravorti explained the Langmuir isotherm. The formula is given below.

$$R_L = \frac{1}{1 + K_L C_0} \tag{12}$$

Where, K_L = constant from Langmuir model (L/mg); C_0 = adsorbate initial concentration (mg/L); R_L is a separation factor, that explained the nature of adsorption. The explanation of this is given in Table 5 and Table 6. According to this formula, the value of R_L in all this experiment is $0<R_L<1$ which suggestion the favorable nature of adsorption of antibiotics.

Table 7 Separation factor

Value of R_L	Adsorption nature
$0<R_L<1$	Favorable
$R_L=0$	Irreversible
$R_L=1$	Linear
$R_L>1$	Unfavorable

3.7 Adsorption thermodynamics

The Gibbs free energy change (ΔG^0) is an indication of spontaneity of a chemical reaction and therefore is one of the most important criteria. It is calculated as follows:

$$\Delta G^0 = -RT\ln K \tag{13}$$

Where R is universal gas constant (8.314 J/(mol · K)), T is the absolute temperature in (K), and K is the thermodynamic equilibrium constant[38].

And

$$\Delta G^0 = \Delta H^0 - T\Delta S^0 \tag{14}$$

After combining equation 13 and 14 we get

$$\ln K_L = \frac{-\Delta H^0}{R} \times \frac{1}{T} + \frac{\Delta S^0}{R} \tag{15}$$

By constructing a plot of $\ln K_L$ versus $1/T$, from the intercept is calculated the change in entropy (ΔS^0) and by the slope, it is possible to calculate the change in enthalpy (ΔH^0)[39].

The value of K_L is obtained from Langmuir isotherm model, the value of K_L is in (L/mg) so first convert in into (L/g) by multiplying 1000 and then multiply it by antibiotic molecular mass then the value of K_L is become dimensionless, because for measuring correct value of thermodynamics parameters we need K in dimensionless unit[40,41].

The thermodynamics parameters ΔG^0, ΔH^0, and ΔS^0 are shown in Table 8, and also plot between $\ln K$ verses $1/T$ are shown in Fig. 8. The ΔG^0 negative values verified the process feasibility and the spontaneous nature of adsorption. As a rule of thumb, the decrease in the negative value of ΔG^0 with temperature increase indicates that the adsorption at higher temperature is more favorable. This may be possible because with the increase of temperature the mobility of adsorbate ion/molecule in the solution increases and that the adsorbate affinity to the adsorbent is high. On the contrary, an increase in the negative value of ΔG^0 with an increase in temperature implies that lower temperature facilitates adsorption.

The negative value of enthalpy ΔH^0 suggest that the sorption of three antibiotics is exothermic in nature. The positive value of ΔS^0 indicates a high randomness at the solid/liquid phase with some structural changes in the adsorbate and the adsorbent. The negative value of ΔS^0 suggests that the adsorption process is enthalpy driven. The negative value of entropy change (ΔS^0) also means that the disorder of the solid/liquid interface decreases during the adsorption process, resulting in the escape of adsorbed ion/molecules from the solid. Therefore, the amount of adsorbate adsorbed will decrease.

Table 8 Thermodynamics parameters for the adsorption of three antibiotics on GAC

Sorbate	T(℃)	ln$K$	ΔG^0(kJ/mol)	ΔH^0(kJ/mol)	ΔS^0[kJ/(mol·K)]
TC	15	11.522	-26.97		
	20	11.368	-27.67	-21.86	19.17
	25	11.215	-27.89		
THI	15	12.521	-29.98		
	20	12.412	-30.23	-33.58	-12.22
	25	12.049	-29.85		
SMZ	15	11.687	-27.98		
	20	11.132	-27.11	-81.39	-185.56
	25	10.545	-26.12		

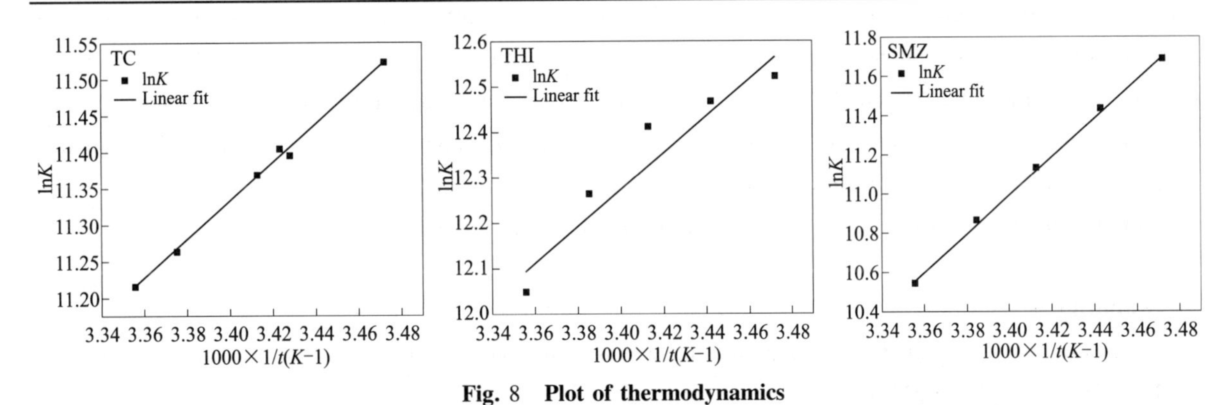

Fig. 8 Plot of thermodynamics

K is adsorption equilibrium constant of Langmuir isotherms. The value of K is in unit (L/mg). First it converts it into(L/g) and then multiplied by molecular formula of antibiotic, then its value become dimensionless. Because we required dimensionless value of K for measuring value of ΔG^0(Table 8).

3.8 Regeneration experiments

Through regeneration experiments, it was found that the adsorption capacity of granular activated carbon for TC, THI, and SMZ decreased with the increase in regeneration times. (Fig. 9) At an ambient temperature of 25 ℃, the initial concentration of antibiotics was 25 mg/L, and the dosage of GAC was 8 g/L, the adsorption of TC on GAC reached saturation after 60 minutes. The maximum adsorption efficiencies of initial GAC, 1 GAC regeneration, and 5 regeneration GAC were 92.54%, 85.73%, and 62.14%. Under the same conditions, the maximum adsorption efficiencies of initial GAC, 1 time regeneration GAC, and 5 times regeneration GAC for THI and SMZ were: 96.32%, 32.33%; 91.23%, 85.21%; 70.26%, 50.19%, respectively. Compared with other adsorbents, such as attapulgite, granular activated carbon has the characteristics of easy preparation and better regeneration.[42]

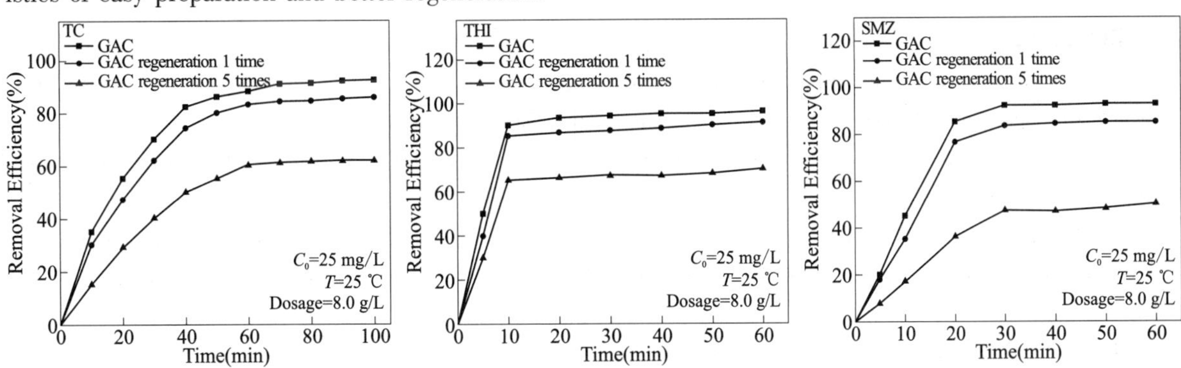

Fig. 9 Relationship between GAC regeneration adsorption efficiency and time

4 Conclusion

This research investigated the experimental results of granular activated carbon adsorbing different antibiotics that included tetracycline, thiamphenicol, and sulfamethoxazole, from aqueous solution. The BET experiment indicated the fact that GAC had high specific surface area approximated 1059 m^2/g and high pore volume 0.625 cm^3/g, meaning that it is very useful adsorbent for antibiotics removal. The equilibrium adsorption data of TC, THI, and SMZ were good expressed by Langmuir isotherm model, and maximum adsorption capacity were 17.02, 30.40, and 26.77 mg/g respectively. The kinetic data of sorption were well described by the pseudo-second-order model, indicating the sorption of the three antibiotics onto GAC involving valency forces through sharing or exchange of electrons between sorbent and sorbate. The Weber-Morris intraparticle diffusion model and Boyd kinetic model proved the main controlling step for the adsorption process was diffusion through the boundary layer. By using the adsorption equilibrium constant obtained from Langmuir isotherm, thermodynamic parameter ΔG^0, was calculated to tell the spontaneity of the adsorption reaction. The values of ΔH^0 and ΔS^0 were also obtained from a slope and intercept of the relationship between $\ln K$ and reaction temperature. Negative value of ΔG^0 and Negative value of ΔH^0 confirmed the spontaneous and exothermic nature of the adsorption process. In conclusion, GAC could be employed as an environmentally friendly adsorbent for the removal of antibiotics from water and wastewater.

Funding

This research was funded by the high precision determination of lead isotopes in environmental samples by MC-ICP-MS (DJNY2021-31)

Acknowledgments

This study was funded by Shaanxi Province science and technology resources open sharing platform project (Program No. 2021PT-03) and Key Laboratory of Degraded and Unused Land Consolidation Engineering, the Ministry of Natural Resources. The sponsors had no role in the design, execution, interpretation, or writing of the study. We are grateful to the anonymous reviewers, whose comments have helped to clarify and improve the text.

Conflicts of interest

The authors declare that they have no known competing financial interests or personal relationships that could have appeared to influence the work reported in this paper.

Data availability statement

The datasets used and/or analysed during the current study available from the corresponding author on reasonable request.

References

[1]Zhou, L. J. ; Ying, G. G. eds. Occurrence and fate of eleven classes of antibiotics in two typical wastewater treatment plants in South china. Science of the Total Environment 2013, 452-453, 365-376.

[2]Hirsch R.; Ternes, T. eds. Occurrence of antibiotics in the aquatic environment. Science of the Total Environment 1999, 225, 109-118.

[3]Zhang, Q.Q; Ying, G.G. eds. Comprehensive evaluation of antibiotics emission and fate in the river basins of China: source analysis, multimedia modeling, and linkage to bacterial resistance. Environmental science and technology 2015. 5, 1-39.

[4]Huang, C.H.; Renew, J. E. eds. Assessment of potential antibiotic contaminants in water and preliminary occurrence analysis. Journal of contemporary water research & education 2011, 10.1142/9789812799555_0004.

[5]Jiang, L.; Hu, X.L. eds. Occurrence, distribution and seasonal variation of antibiotics in the Huangpu River, Shanghai, China. Chemosphere 2011, 82 (6), 822-828.

[6]Daughton, C. G; Ternes, T. A. Pharmaceuticals and personal care products in the environment agents of subtle change. Environmental Health Perspectives 1999, 107, 907-938.

[7]Gao, P.; Mao, D. eds. Occurrence of sulfonamide and tetracycline resistant bacteria and resistance genes in aquaculture environment. Water Resources 2012, 46 (47), 2355-2364.

[8]Lai, H.T.; Hou, J.H. eds. Effects of chloramphenicol, florfenicol, and thiamphenicol on growth of algae Chlorella pyrenoidosa, Isochrysis galbana, and Tetraselmis chui. Ecotoxicology and Environmental Safety 2009, 72(72), 329-334.

[9]Pruden, A; Pei, R.T. eds. Antibiotic resistance genes as emerging contaminants: Studies in northern Colorado. Environmental science & technology 2006, 40 (23), 7445-7450.

[10]Chen, K; Zhou, J.L. Occurrence and behavior of antibiotics in water and sediments from the Huangpu River, Shanghai, China. Chemosphere 2014, 95, 604-612.

[11]Michael, I.; Rizzo, L. eds. Urban wastewater treatment plants as hotspots for the release of antibiotics in the environment. Water Resources 2013, 47(43), 957-995.

[12]Rizzo, L.; Manaia, C. eds. Urban wastewater treatment plants as hotspots for antibiotic resistant bacteria and genes spread into the environment. Science of the Total Environment 2013, 447, 345-360.

[13]Mehrjouei, M; Mueller, S. eds. Energy consumption of three different advanced oxidation methods for water treatment: a cost-effectiveness study. The Journal of Cleaner Production 2014, 65, 178-183.

[14]Sera B. V.;Muflihatul M.;Eko Y.eds. Degradation of ciprofloxacin in aqueous solution using ozone microbubbles: spectroscopic, kinetics, and antibacterial analysis.Heliyon 2022,8(8),e10137.

[15]Shang,K.F.;Rino,M.;Wang,N.eds.Degradation of sulfamethoxazole (SMX) by water falling film DBD Plasma/Persulfate: Reactive species identification and their role in SMX degradation. Chemical Engineering Journal 2022, 431 (1),133916.

[16]Ahmaruzzaman, M. Adsorption of phenolic compounds on low-cost adsorbents: a review Advances in Colloid and Interface Science. Process Biochemistry 2008, 143(141), 148-167.

[17]Han, R.; Ding, D. eds. Use of rice husk for the adsorption of congo red from aqueous solution in column mode. Bioresource Technology 2008, 99(98), 2938-2946.

[18]Lam, A.; Rivera, A. eds. Theoretical study of metronidazole adsorption on clinoptilolite. Microporous and Mesoporous Materials 2001, 49,157-162.

[19]Homem,V.;Santos, L. Degradation and removal methods of antibiotics from aqueous matrices-a review[J]. Journal of Environmental Management 2011,92(10), 2304-2347.

[20]Rabolle, M.; Spliid, N.H. Sorption and mobility of metronidazole, olaquindox, oxytetracycline and tylosin in soil. Chemosphere 2000, 40, 715-722.

[21]Ahmed, M. J. Microwave assisted preparation of microporous activated carbon from Siris seed pods for adsorption of metronidazole antibiotic. Chemical Engineering Journal 2013, 214, 310-318.

[22]Ahmed, M. J.; Theydan, S.K. Microporous activated carbon from Siris seed pods by microwave-induced KOH activation for metronidazole adsorption. Journal of Analytical and Applied Pyrolysis 2013a, 99, 101-109.

[23]Rivera-Utrilla, J.; Prados-Joya, G. eds. Removal of nitroimidazole antibiotics from aqueous solution by adsorption/bioadsorption on activated carbon. Journal of Hazardous Materials 2009, 170(1), 298-305.

[24]Ocampo-Pérez, R.; Orellana-Garcia, F. eds. Nitroimidazoles adsorption on activated carbon cloth from aqueous solution. Journal of Hazardous Materials 2013, 401, 116-124.

[25]Dan, C.; Jian, D. eds. Core-shell magnetic nanoparticles with surface-imprinted polymer coating as a new adsorbent for solid phase extraction of metronidazole. Analytical Methods 2013, 5, 722-728.

[26]He, J.; Dai, J. eds. Preparation of highly porous carbon from sustainable α-cellulose for superior removal performance of tetracycline and sulfamethazine from water. RSC Advances 2016, 6(33), 28023-28033.

[27]Fan, Y.; Zheng, C. eds. Preparation of Granular Activated Carbon and Its Mechanism in the Removal of Isoniazid, Sulfamethoxazole, Thiamphenicol, and Doxycycline from Aqueous Solution. Environ. Eng. Sci. 2019. 36, 1027-1040.

[28]He Y.; Cheng F.eds,The Affect Analysis of Microwave Regeneration on the Adsorption Properties of Granular Activated Carbon.Journal of Tianjin Institute of Urban Construction 2012.18(2):6.

[29] Ilavsk J.; BarlokovD.; Marton M. Removal of selected pesticides from water using granular activated carbon. IOP 2021.

[30]Zhu, Y. , Liu, L. eds.Experimental study of sewage plant tail water treatment by granular active carbon immobilized catalyst fenton-like. Shandong Chemical Industry 2018.

[31]Qin, Q. , Chen, Y. eds. Optimization of the modified components of mn-sn-ce/gac particle electrode by response surface method. Industrial Water Treatment. 2019, 1, 54-59.

[32]Gagliano E.;Falciglia P P.;Zaker Y.et al. Microwave regeneration of granular activated carbon saturated with PFAS. Water research 2021,15,198.

[33]Liu, P.; Wang, Q. eds. Sorption of sulfadiazine, norfloxacin, metronidazole, and tetracycline by grangranular activated carbon: Kinetics, mechanisms, and isotherms. Water air and soil pollution journal 2017, 228, 129 1027-1040.

[34]Ho, Y. S. Review of second-order models for adsorption systems. Journal of Hazardous Materials. 2006, B136, 681-689.

[35]El-Khaiary, I. M.; Malash, G.F. eds. On the use of linearized pseudo-second-order kinetic equations for modeling adsorption systems. Desalination. 2010, 257, 93-101.

[36]Mannarswamy, A.; Munson-Mcgee, S.H. eds. D-optimal experimental designs for Freundlich and Langmuir adsorption isotherms. Chemometrics and Intelligent Laboratory Systems 2009, 97, 146-151.

[37]Langmuir, I. The adsorption of gases on plane surfaces of glass,mica and platinum. Journal of American Chemical Society, 1918, 40, 1361-1403.

[38]Freundlich, H. Ueber die adsorption in Loesungen. Journal of Physical Chemistry 1906, 57, 385-470.

[39]Atkins, P.; Paula, J. Physical Chemistry. W. H. Freeman and Company, New York, 2010.

[40] Chang, R.; Thoman, J. W. Physical Chemistry for the Chemical Sciences. University Science Books, Canada, 2014.

[41]Çaliskan, E.; Göktürk, S. Adsorption characteristics of sulfamethoxazole and metronidazole on activated carbon. Separation Science and Technology 2010, 45, 244-255.

[42]Zhang S. Y.Adsorption and removal of sulfonamide antibiotics in water by granular activated carbon and modified attapulgite.Lanzhou Jiaotong University 2020.

本文曾发表于 2022 年《Molecules》第 27 卷

Soil Wind Erosion Resistance Analysis for Soft Rock and Sand Compound Soil: A Case Study for the Mu Us Sandy Land, China

Huanyuan Wang [1,2,3,4], Wei Tong[1,2,3,4], Jinbao Liu[1,2,3,4], Jichang Han [1,2,3,4] and Siqi Liu [1,2,3,4]

(1. Shaanxi Provincial Land Engineering Construction Group Co., Ltd., Xi' an 710075, PR China; 2. The Institute of Land Engineering and Technology, Shaanxi Provincial Land Engineering Construction Group Co., Ltd., Xi' an 710075, PR China; 3. Key Laboratory of Degraded and Unused Land Consolidation Engineering, Ministry of Natural Resources, Xi' an 710075, PR China; 4. Shaanxi Provincial Land Consolidation Engineering Technology Research Center, Xi' an 710075, PR China)

【**Abstract**】 Mixtures of soft rock and sand have been applied extensively in the Mu Us Sandy Land (also known as the Mu Us Desert) to limit the loss of top soil by wind erosion. In this study the efficacy of sand-fixing technology was investigated in a series of experiments. The sand-fixing effect of seven different mixtures (ratios) on soft rock and sand was evaluated in a wind tunnel. The results indicated that the Mu Us Sandy Land soils are susceptible to wind erosion, as the textural composition of sand is dominated by coarse particles. Mu Us Sandy Land soils dominated by silt and clay particle sizes are more resistant to wind erosion. Each soft rock and sand combination experiences severe wind erosion. However, wind erosion was significantly reduced when soft rock and sand were mixed. An increase in particle size was associated with an increase in the resistance of soft rock and sand mixtures to wind erosion. The ability to resist wind erosion was greatest when the ratio of soft rock and sand was between 1:2 and 1:5. This study provided data to support the approaches to sand-fixing commonly used at present in the Mu Us Sandy Land. The results of this study have important practical significance for the improvement of agricultural land potential in dry sandy areas.

【**Keywords**】 Mu Us Sandy Land; Sand; Soft rock; Wind erosion; Wind tunnel test

1 Introduction

The Mu Us Sandy Land (MUSL) is located in the semi-arid and arid climate zone of northern China, and it also known as the Mu Us desert. It lies within an ecotone for agriculture and animal husbandry in northern China. The area is subject to wind erosion and severe desertification. Traditional farming methods exacerbate the situation by overwintering and exposing sandy land in the process. In winter and spring, when it is dry and windy, wind erosion of the surface soil layer can be severe. Based on weather data from 1990 to 2014, the annual average daily high wind velocity (> 17 m/s) in MUSL occurs 10 to 40 days per annum, but not more than 95 days per year. Extended periods with continuous high winds, with a duration > a day account for 60% to 70% of the high wind days, while periods lasting 2 to 3 days account for 20% to 30%. Even longer periods with continuously high winds for 4 to 6 days account for approximately 5% of the high wind days. Dust storms occurred from 11 to 29 days per year. The frequent occurrence of wind and sand related environmental hazards has been associated with land degradation and hence had a serious impact on the economic and social development of the region. It seriously restricts the development of local resources and the sustained and stable development of society and local economy. In order to prevent and control wind and sand related hazards, a large number of agricultural, biological, and engineering related measures have been researched and implemented[1]. Commonly implemented examples include the planting of trees and grass, flood warping land, conservation tillage, soil replacement, and the application of chemical amendments[2]. However, many measures only aim to reduce or prevent wind erosion. Few of these interventions have been implemented in concert with utilization. Aspects of sandy land utilization that have been studied include the mixed utilization of sandy soil and loess[3], saline-alkali soil[4], coal gangue[5], and peat[6].

Soft rock is widely distributed in the MUSL[7]. It is hard when dry and expands rapidly under wet conditions, but it retains water well and can be used as a natural water retention agent. These properties of the soft rock complement the porosity of sand. The texture of sand is uniform and the structure is loose. The water and fertilizer leakage characteristics of the soft rock and sand have significant differences, although each of the two characteristics can compensate for the other's inherent defects. A compound soil that could provide plants with a favorable medium for growth was produced by mixing soft rock and sand[8-10], thereby mitigating the negative characteristics of each of the components. After years of experiment and practice[11], the Shaanxi provincial land engineering construction group applied soft rock and sand mixtures to land surfaces in several land renovation projects in MUSL, Yulin. By the end of 2014, the cumulative extent of treated soil was 4647 hm^2, of which 4400 hm^2 was arable land. In addition, the use of soft rock and sand compound soil technology has been extended to more than 13333 hm^2 in MUSL and surrounding areas. It has effectively increased the area of arable land, supplemented the index of construction land, improved the regional ecological environment, and explored a new way for sandy lands to be managed.

The recognition of the mechanism of soil wind erosion and the practice of sand control indicate that there is a close correlation between the occurrence of wind erosion and the grain size distribution of topsoil. The grain size of the topsoil is determined by soil formation[12]. Soil texture is an important determinant of the intensity of soil erosion, especially in the absence of vegetation. The physical and chemical properties of soil also directly affect the wind speed at which soil particles start to move. The essence of the of soft rock and sand mixture approach to soil improvement in sandy lands, is to improve the original textural characteristics (grain size) of the soil, with regard to susceptibility to erosion and application of fertilizer. This study has investigated the effect of the textural characteristics of soft rock and sand mixtures on sand-fixing (i.e. stabilization in relation to wind erosion). Experiments were carried out on different ratios of these mixtures in an indoor wind tunnel. The data produced in this study could be used to guide the development of methods to stabilize sandy lands and enhance the associated agricultural potential.

2 Materials and methods

2.1 Study area

The research area was located in the Yuyang district, Yulin city (Latitude, 38°27′53″N to 38°28′23″N; Longitude, 109°28′58″E to 109°30′10″E), on the southern margin of the Mu Us desert and on the middle reaches of Wuding river. It is a typical semi-arid continental monsoon climate in the temperate zone. The spatial and temporal distribution of precipitation is uneven, with a dry climate and abundant sunshine. The average annual temperature is 8.1 ℃, the accumulated temperature of ≥ 10 ℃ is 3307.5 and the duration is 168 days. The annual average frost-free period is 154 days. The average annual rainfall is approximately 413.9 mm, with 60%~90% falling from June to September. There were on average 2879 hours of sunshine per annum. According to the meteorological data for the years from 1990 to 2014 (Fig. 1), the average daily wind speed was approximately 9 m/s.

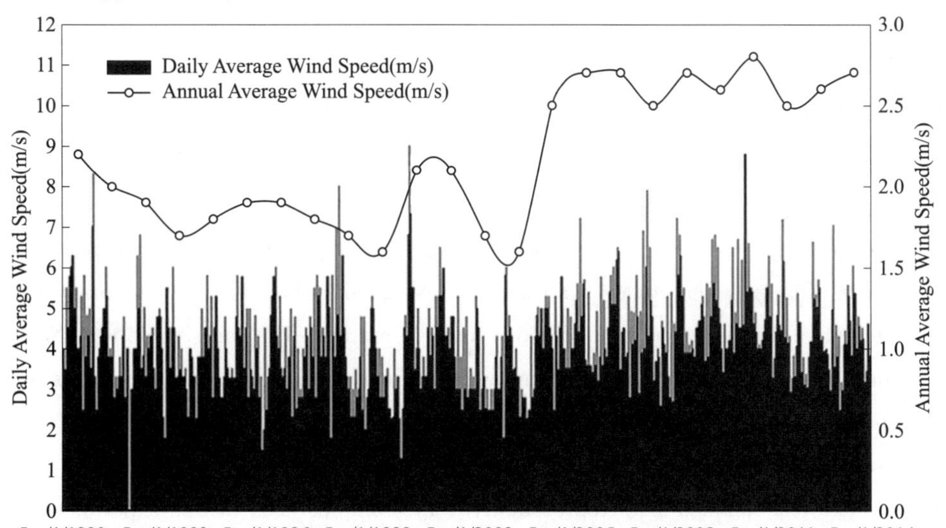

Fig. 1 Daily and annual average wind speed from 1990 to 2014 for the Yuyang district, Yulin, Shaanxi province, China

The soil types are mainly well-sorted aeolian sandy soils, with coarse sand grains, very few powder grains, a low particle surface activity, a low viscosity and strong looseness (Table 1). The nutrient content is low, with an average organic matter content of 3.32 g/kg and a total nitrogen content of 0.14 g/kg. In addition, a large percentage of the soft rock in the sandy land has a low level of diagenesis, has a low structural strength, is easily weathered, poorly cemented and poor in permeability. In contrast, due to the high content of sticky particles, it has a favorable water holding capacity and water retention capacity. The average content of organic matter is 6.20 g/kg, while the total nitrogen content is 0.13 g/kg. As a result, the nutrient content is relatively low.

Table 1 Physical and chemical properties of the tested soil

Sample	Particle(%)			Texture (USDA)	pH	TN ($g \cdot kg^{-1}$)	TP ($g \cdot kg^{-1}$)	TK ($g \cdot kg^{-1}$)	SOM ($g \cdot kg^{-1}$)
	Sand (0.05~2 mm)	Silt (0.002~0.05)	Clay (<0.002 mm)						
Sand	95.37	4.10	0.53	Sand	8.35	0.14	0.63	26.51	3.32
Soft rock	24.52	64.98	10.50	Silt loam	8.27	0.13	0.59	25.09	6.20

Note: USDA-United States Department of Agriculture, TN-total Nitrogen, TP-total phosphorus, TK-total potassium, SOM-soil organic matter.

2.2 Method

The experiment was carried out in the wind tunnel at the Institute of Soil and Water Conservation, Ministry of Water Resources, Chinese Academy of Sciences. The wind tunnel is 19.8 m long, 1.2 m high and 1.0 m wide[13]. It consists of five main parts, namely a power stage, an air regulation section, a rectifier section, a test section and a sand collecting section. The size of the sample cell used in the simulation experiment was 1.25 m × 1 m × 0.15 m (length × width × height), with the wind tunnel pushed from the test section during the test.

The wind speed is adjusted continuously by 0 to 20 m/s using a frequency converter that ranged from 0 to 50 Hz. Before the formal test, the wind speed in the wind tunnel was debugged uniformly, with each blowing simulation test set to blow for 10 minutes. The wind erosion amount of each blowing test was collected

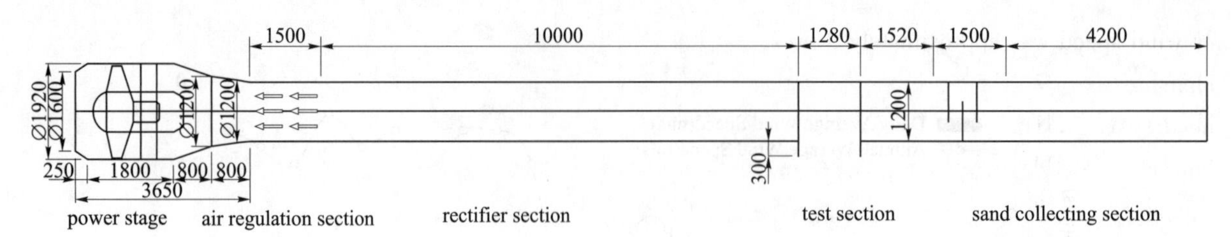

Fig. 2　Wind tunnel structure (Unit: mm)

by the tail sand-collecting device in the sand collecting section, as the total wind erosion volume of the secondary erosion simulation test. The eroded sand was then weighed by balance to an accuracy of 0.01 g. Each blowout test would make the surface of the soil more coarsely grained. The soil sample was therefore reloaded after each blowout. In order to ensure the uniform weight of each sample, each loading was weighed with a platform scale to ensure uniform bulk density.

2.3　Experimental design

In order to make test samples more representative, the samples used in this experiment were collected from Dajihan village, Xiaojihan town, Yuyang district, Yulin city. In total seven sets of ratios of soft rock to sand were considered, namely 1∶0, 5∶1, 2∶1, 1∶1, 1∶2, 1∶5 and 0∶1. Soil physical and chemical properties, such as water content, organic matter content, bulk density and particle size composition significantly affect wind erosion. In order to simply consider the effect of the proportions on wind erosion, two kinds of samples were first treated by air drying to ensure that their respective water content was less than 2% and that the organic matter content of the soft rock (0.78 g/kg) and sand (3.32 g/kg) was very low. The effects of water content and organic matter on wind erosion were therefore not considered any further in the study. Consequently, it was assumed that only the effect of particle size distribution on wind erosion was studied in the wind tunnel. The aeolian sandy soil was relatively homogeneous, with a particle size below 2 mm. In contrast, the particle size of the soft rock had a significant influence on the wind tunnel test. As a result, 2 mm, 8 mm and 20 mm were selected for assessment in the experiments. The soft rock with different diameters was mixed evenly with the sand for each treatment and subsequently replicated three times in the experiments. The sand grains in the MUSL are exposed to wind speeds of generally around 6 m/s, with a maximum daily mean wind speed of 9 m/s. Therefore, three wind speeds (viz. 7 m/s, 9 m/s and 11 m/s) were selected for investigation in this study.

3　Results

3.1　Grain size composition characteristics of different proportions of soft rock and soil

The particle size distributions of different proportions of soft rock and sand have been shown in Fig. 3. The grain size distribution of different proportions of soft rock and sand mixtures were different from the grain size distributions for aeolian sand and soft rock. The aeolian sand was well sorted with grain sizes concentrated in the sand grain segments (0.05～1 mm). In contrast, the grain size distributions of the soft rock were concentrated in the powder particle size and clay fraction (0.01～0.05 mm), with a size range larger than the size range of each soft rock and sand combination. The mixing of the two components thus changed the limitations of their grain size composition, with regard to susceptibility to wind erosion. The frequency distribution of their grain sizes can be divided into two distinct parts. In the finer fraction, with an increase in the proportion of soft rock in the mixture, the grain size of the smaller particle sizes increases gradually. In the coarser grain size fraction, the particle size gradually decreased, as the proportion of soft rock in the mixture increased. In general, as the proportion of soft rock in the mixture increased, the grain size distribution of the mixture changed in a finer direction. In terms of texture, the key grain size component (powder and clay) of the structure in the mixture is gradually improved, as is the coarse sand content. The soil texture changes from sandy loam to sandy soil and eventually to a loam soil. The mixture can therefore meet the need for crop root ventilation, improve water retention, provide a fertilizer protection effect and gradually improve the ability to resist both wind and water erosion. The addition of soft rock to sand is of great significance to strengthening the sand-fixing ability of soil in MUSL.

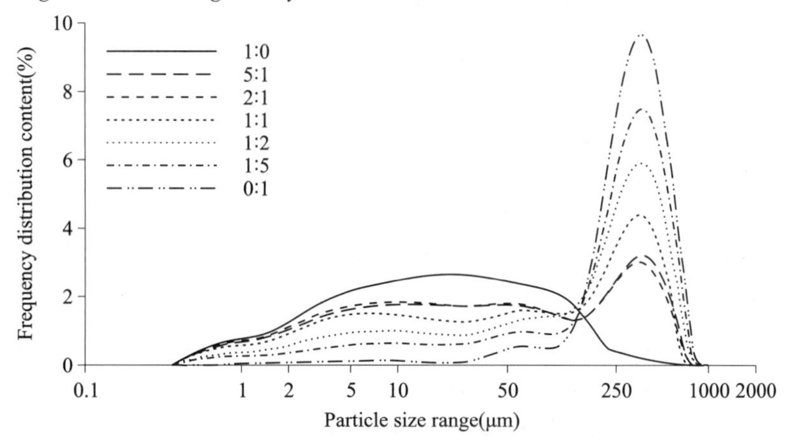

Fig. 3 **Grain size distribution of different proportions of soft rock and sand compound soil**

3.2 Wind erosion of different proportions of soft rock and soil

3.2.1 Influence of wind force on wind erosion of soft rock and sand compound soil

Fig. 4 shows the intensity of wind erosion in relation to wind velocity for seven ratios of soft rock to sand mixtures and three different soft rock grain size categories (viz. 2 mm, 8 mm and 20 mm). In general, the higher the wind velocity, the greater will be the intensity of wind erosion on a soil surface. One departure from this relationship was when the soft rock (2 mm) and sand mixture has a ratio of 1 :0 (i.e. erosion at 7 m/s > 9 m/s > 11 m/s). When soft rock size categories of 2 mm and 20 mm were used, there was a notable drop in sediment loss, when the soft rock was added to sand (i.e. ratio of 1 :5). This effect was largely lost when ratio of soft rock (2 mm) to sand mixture had a ratio of 5 :1. Perhaps counter to expectations the soft rock (2 mm) on its own was more susceptible to erosion at low wind velocities than higher ones. The 8 mm soft rock category also reduced wind erosion when used in a mixture, but the benefits were generally less than if the other two categories (i.e. 2 mm and 20 mm) were used. The reasons for the abnormal wind erosion of the soil mixture with a large particle size of 2 mm have been discussed further below. In terms of wind and sand dynamics, the finer particles are more easily entrained by the wind. Gillette[14] found that the particle size distribution of a land surface plays an important role in the onset velocity of wind erosion.

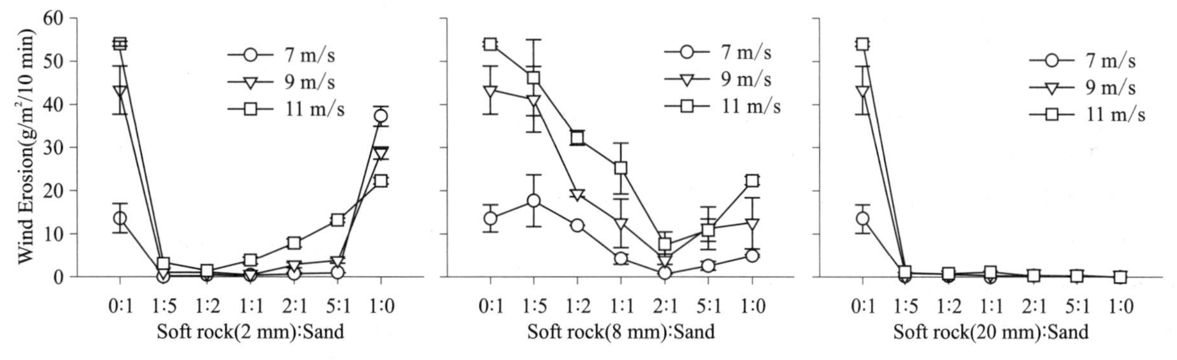

Fig. 4 **Wind erosion of different proportions of soft rock and sand compound soil for experiments carried out in a wind tunnel**

Table 2 Wind erosion of different proportions of soft rock and sand compound soil ($g/m^2/10$ min)

ratio	Soft rock: 2 mm			Soft rock: 8 mm			Soft rock: 20 mm		
	7 m/s	9 m/s	11 m/s	7 m/s	9 m/s	11 m/s	7 m/s	9 m/s	11 m/s
0:1	13.54±3.26	43.34±5.59	54.06±0.53	13.54±3.26	43.34±5.59	54.06±0.53	13.54±3.26	43.34±5.59	54.06±0.53
1:5	0.13±0.02	0.94±0.10	3.17±0.93	17.61±6.01	41.19±7.59	46.24±8.85	0.18±0.05	0.83±0.16	1.13±0.08
1:2	0.49±0.12	0.82±0.14	1.38±0.16	11.93±0.10	19.31±0.66	32.22±1.66	0.11±0.02	0.62±0.19	0.75±0.23
1:1	0.34±0.05	0.38±0.08	3.80±0.80	4.11±1.24	12.48±5.61	25.15±5.84	0.00±0.00	0.35±0.09	1.10±0.21
2:1	0.73±0.26	2.76±0.55	7.85±0.92	0.83±0.07	4.19±1.29	7.57±2.92	0.00±0.00	0.12±0.03	0.29±0.11
5:1	0.94±0.08	3.74±0.57	13.11±0.49	2.51±1.02	11.38±4.96	10.93±2.60	0.00±0.00	0.09±0.01	0.30±0.06
1:0	37.25±2.33	28.79±1.37	22.30±0.78	4.85±0.36	12.50±5.93	22.26±0.88	0.00±0.00	0.00±0.00	0.00±0.00

From the analysis above, we know that the grain size of the soft rock is mainly concentrated in the fine powder and clay segments. Fine granular soft rock is therefore easy to move with the wind. In the wind tunnel test, there is a distinct transition for the sandy soil at a wind speed of 6 m/s, as most of the particles of the sifted 2 mm sandstone are very fine. It has a distinct grain transition at a wind speed of approximately 4 m/s, with notable wind erosion evident. As the wind speed increased, the surface of the sample trough became rougher, as the fine particles are removed by the wind (Fig. 5). Even with notable erosion in the sample trough, there was no increase in the collection of tail sand in the collection plate. The main reason is that the wind speed at the outlet of the sand collection section is high, while the distance to the sand plate is small. High wind velocities at the outlet of the sand collection section were associated with a notable loss of fine sediment at the sand collection plate, which caused the wind erosion measurements to deviate greatly. Li and Shen[15] also pointed out that a sand sampler in the sand collecting section would lead to large deviation and low reliability of sand flow measurements.

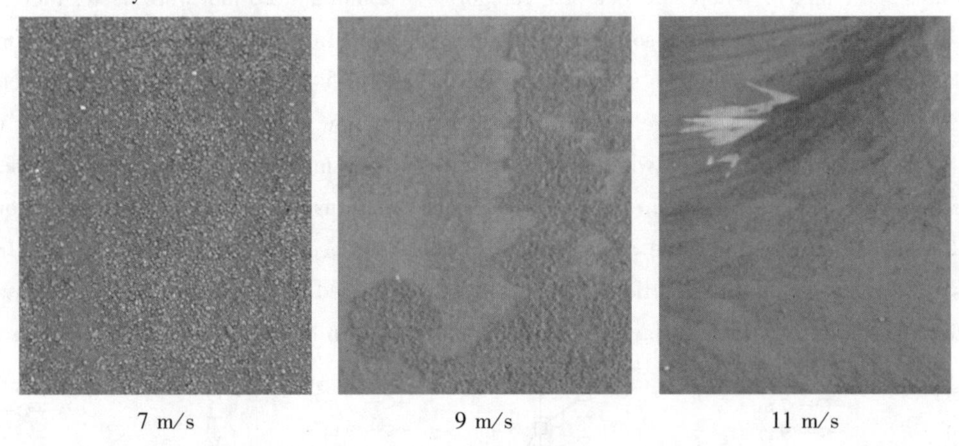

7 m/s　　9 m/s　　11 m/s

Fig. 5 Wind erosion of soil samples in the sample tank at different wind speeds

3.2.2 Effect of the particle size of soft rock on wind erosion of a mixed sediment sample

It can be seen in Fig. 4, that the particle sizes of the samples of the soft rock and the sandy soil have a significant effect on the amount of wind erosion. From the view of soil mineral particles, the particles larger than 2 mm are called gravel. Gravel cover is an effective sand fixing measure[16-18]. Therefore, the sample size of 8 mm and 20 mm is equivalent to the stabilization of gravel-covered sand. The amount of wind erosion at different soft rock and sand ratios with a grain size of 20 mm, under three kinds of wind speed is almost zero. This indicated that the particle size of the soft rock increases the surface roughness, absorbs and decomposes the surface wind movement and reduces the shear force[19-20] on the eroded bed surface. Moreover, the soft rock covered the surface layer of aeolian sand, which reduced the area of sand available for entrainment[16, 21-23] on the surface. Therefore, the soft rock is not blown away by the wind, and at the same time, it can protect the aeolian sand from wind erosion. In a land management context, the larger particle size soft rock (20 mm) and sand compound soil would be particularly useful for managing wind erosion in MUSL in winter and spring, if applied prior to the onset of these seasons. The use of a rotary tiller during crop planting would ensure the soils are evenly mixed, thereby improving water conservation and reducing the demand for fertilization of the tilled layer.

There are two explanations for the differences in the patterns of wind erosion recorded for soft rock with a particle size of 8 mm and soft rock with a particle size of 2 mm at three wind speeds. Firstly, the error in the collection of 2 mm tail sand is larger than the size of the particle size mentioned above, while the amount of wind erosion at the three wind speeds is underestimated at different soft rock (2 mm) and sand ratios. Secondly, although the sandstone with a particle size of 8 mm could be assigned to the gravel particle size class, it does not provide the same protective effect as a gravel surface cover. A possible reason for this could be the presence of grain sizes > 8 mm in the soft rock (8 mm) sample that have a low density and that are of poor quality. At the same wind speed, compared to the grain size, the sand particles may be more easily blown by wind erosion. Therefore, in practice, the crushed soil particle size should not be too fine, otherwise it may lead to more severe wind erosion than the original aeolian sand. When the soft rock and sand mixture is applied in agriculture, it will add the cumulative effect of other factors that increase resistance to wind erosion. These additional parameters include the organic matter content of the soil, the cementation of the clay particles and the cementation of the sand particles in the soft rock.

3.2.3 Effect of texture on wind erosion of soft rock and sand compound soil

The textural characteristics of soil are highly variable in nature. The texture and specific gravity of soil are important determinants of the susceptibility of a soil to wind erosion[24]. Different compound soils of soft rock and sand have different textural characteristics and hence variable moderating effects on wind erosion[25]. In order to illustrate the effect of particle size distribution on wind erosion, we have considered wind erosion at different wind speeds, when the soft rock particle size is 2 mm. Dong and Li's study[26] found that the main resistance to wind erosion in loose aeolian sand is the inertia of a grain and the strength of cohesion between grains. When particle size is < 0.09 mm, the inter particle strength of cohesion predominates, which means the wind speed required for entrainment to increase with the decrease in particle size. In contrast, when particle size is > 0.09 mm, inertia is the main determinant of the wind speed required for entrainment. It can be seen from Fig. 5 that the wind erosion of pure aeolian sandy soil is relatively serious under different wind speeds. More than 90% of the aeolian sandy soil has a grain size > 0.09 mm. In this size class, the water holding capacity of particles is poor and agglomeration is weak. As inertia is the dominant factor for the larger grain sizes (viz. > 0.09 mm), the resistance to wind erosion is relatively low. Although the content of soft clay particles is high, the cohesive strength they provide in this experiment is very low[27]. Although the wind erosion is less than that of the aeolian sandy soil, it still has a great influence on the results. After mixing the soft rock and sand, the grain size of the two components was modified and the resistance to wind erosion was enhanced.

3.2.4 The particle size distribution and erosion resistance of the tail sand

From the analysis of the particle size distribution of the tail sand in Fig. 6, it can be seen that the eroded particles are mainly concentrated in two grain size classes, namely 0.01 mm to 0.05 mm and 0.25 mm to 1 mm. The two size classes are the same size as the soft rock and aeolian soil, so the measures to mitigate wind erosion should focus on these two particle sizes. Neglecting the error associated with the collection of tail sand by the sand collector mentioned above, the erosion of the mixtures with ratios from 1:5 to 1:2 was significantly reduced, due to the single particle structure (i.e. limited adhesion) of the sandy soil. After adding soft rock, the clay particles in soft rock filled the gaps between the sand particles. The increase in the number of clay particles changed the wind erosion resistance force, which had been dominated by inertia up to that point, into a wind erosion resistance force progressively dominated by the forces of cohesion. The main factor determining resistance to wind erosion would then gradually change to dependence on the cohesive strength obtained from the fine-grained sediment weathered from the soft rock. As there is a high content of sand in this mixture, a small amount of sticky particles can efficiently fill the gaps between the sand particles, with indirect contact between the particles tighter.

After wind erosion of the loose particles between the surface particles, the specific gravity of the surface is increased, the roughness of the surface of the soil is increased and the fine particles are protected from further erosion. The greater the use of the mixture, the lower the volume of material that will be eroded[28]. From the observation of the wind erosion trough, it can be seen that the surface soil is only coarse grained at a wind speed of 7 m/s, the surface soil is coarse grained at a wind speed of 9 m/s, and the lower layer soil is still stronger in resisting wind erosion even if the particles are fine. Only at a wind speed of 11 m/s will both the top surface and the lower soil be severely eroded. The differences in wind erosion characteristics of mixtures in two ranges (viz. soft rock: sand, 1:1 to 5:1 vs 1:5 to 1:2), could be attributed to the infilling of interstitial spaces by sticky particles. The strength of cohesion between particles is less and relatively weak. This shows that in the case of dry soil, the high content of clay particles in the grain size of the soil may not enhance the cohesive strength and resistance to wind erosion, as the "pulling effect" of water molecules among the soil particles will be absent. Therefore, only when the particle size distribution is reasonable at all levels can the ability to resist wind erosion be enhanced.

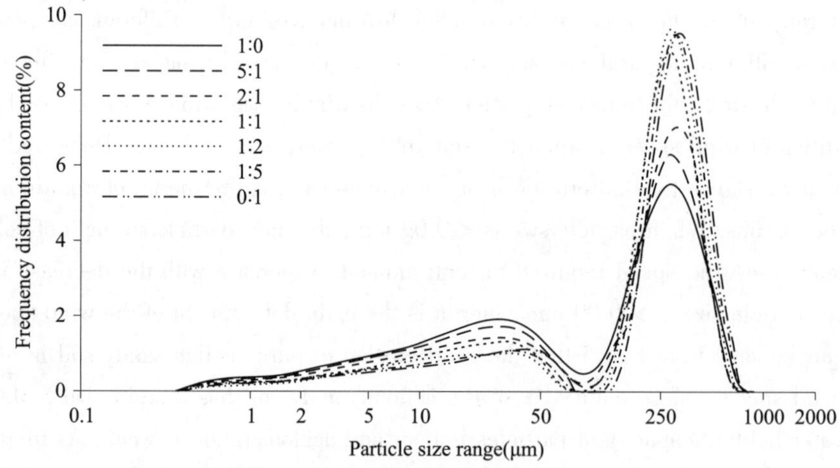

Fig. 6 Grain size distribution of tail sand of different proportions of soft rock and sand compound soil for experiments carried out in a wind tunnel

4 Discussion

As the main particle size of the soft rock was silt and clay, erosion of the rock by raindrop splash and irrigation, leads to the loss of clay and silt size particles which plug the gaps between sand grains[29-30]. As a result, a soil crust 2~6 mm thick can form on the surface of the compound soil. This crust can significantly increase the wind speed required for entrainment of grains from the land surface, and hence effectively prevent wind erosion. For uncultivated soil, this crust will gradually form soil biological crust with the improvement of ecological environment, which has a significant effect on wind erosion resistance[31]. For cultivated soil, the surface coverings during crop growth significantly improved the ability to resist wind erosion. After crop harvested, soil crusts can easily form on the surface of the compound soils. Therefore, sand fixation can be achieved in both the growing and fallow periods of crops[11]. The field tests showed that high wind velocities (Category 6, speed = 12 m/s) would not lead to the loss of particles from the surface of the compound soil. In addition, there is a strong water-retaining capacity of soft rock[29-30]. In winter, the compound soil of soft rock and sand will freeze and fix the wind-erosion particles from the surface layer, which can significantly reduce the wind erosion in windy weather in winter[31]. Therefore, from the perspective of the grain size of soft rock and sand compound soil compared to the sand soil, it had the ability to resist wind erosion with soil crusting, frozen cover. Soft rock and sand compound soil technology can guarantee the ability to resist wind erosion, the technology in the local promotion does not lead to more serious wind erosion and deterioration of the local ecological environment.

5 Conclusions

This study investigated the resistance of soft rock and sand mixtures to wind erosion. Seven mixtures (ratios) of soft rock and sand were used in the wind tunnel experiments. The results showed that in the absence of organic matter and water, particle size could determine the resistance of a soft rock and soil mixture to wind erosion. The grain size composition of the aeolian sandy soil in the Mu Us sandy land was dominated by the 0.05 mm and 1 mm size fractions. The texture was loose, with the interaction between particles weak. The soils were therefore susceptible to wind erosion. In the range of 0.01 mm to 0.05 mm, the particle size of the soft rock is high, and the cohesive force between particles is dominant. The soft rock is therefore much less susceptible to erosion than the sandy soils. The amount of wind erosion of single soft rock and sand is larger, while the wind erosion amount of these two is reduced significantly. Composite soils had a higher resistance wind erosion than sandy soils on their own. The ratios of soft rock to sand that provided the greatest level of protection from wind erosion ranged from 1∶5 to 1∶2. The results of our research can therefore be used to identify low cost, practical management measures, to protect sandy soils in ecologically vulnerable areas.

Acknowledgements

This work was supported by the Natural Science Basic Research Program of Shaanxi Province (2021JZ-57), the National Natural Science Foundation of China (No. 51679188), and Xi'an University of Technology Doctoral Dissertation Innovation Fund (310-252072018).

Author contributions

Huanyuan Wang and Wei Tong: Writing-original draft, methods, formal analysis;

Jinbao Liu and Jichang Han: Formal analysis, visualization, project administration;

Siqi Liu: Resources.

Conflict of interest

The authors state that they have no conflicts of interest.

Data availability statement

The datasets produced for this study are available from the corresponding author on reasonable request.

References

[1] Han Z W, Wang T, Dong Z B, et al. Main engineering measurements and mechanism of blown sand hazard control. Progress in Geography. 2004,23(1):13-21.

[2] Han L W, Li Q L, Shan X P, et al. Study on land desertification and desertification control measures. Soil and water conservation research. 2005,12 (5): 210-213.

[3] Zhou D W, Tian Y, Wang M L, et al. Study on saline-alkali land improvement and land reclamation technology by sand covering in Horqin Sandy Land-Songliao Plain crisscrocross Area [J]. Journal of natural resources. 2011,21 (6): 910-918.

[4] Zhang X W, Zhan Q. Development of soil improver for degraded land in mining area. Journal of liaoning technical university: natural science edition. 2010,29 (1) : 147 - 148.

[5] Wang R D, Wu X X. A new model for management of Mu us Sandy Land. Soil and water conservation research. 2009,16 (5): 176-180.

[6] Yu Z Y, Fan Z P, Chen F S, et al. A method of improving aeolian sand soil with peat and its application: China, 200810230100.8[P].2010-06-30.

[7] Wang Y C, Wu Y H, Li M, et al. Soil and water loss and its control ways in the sandstone region, Zhengzhou: The Yellow River Water Conservancy Press, 2007.

[8] Han J C, Xie J C, Zhang Y. Potential role of feldspathic sandstone as a natural water retaining agent in Mu Us sandy land northwest China. Chinese Geographical Science. 2012,22(5):550-555.

[9] Wang N, Xie J C, Han J C. A sand control and development model in sandy land based on mixed experiments of arsenic sandstone and sand:s case study in Mu Us sandy land in China. Chinese Geographical Science. 2013,23(6):700-707.

[10] Wang N, Xie J C, Han J C, et al. A comprehensive framework on land-water resources development in Mu Us Sandy Land. Land Use Policy, 2014,40:69-73.

[11] Wang H Y, Han J C, Tong W, et al. Analysis of water and nitrogen use efficiency for maize (Zea mays L.) grown on soft rock and sand compound soil. Journal of the Science of Food & Agriculture, 2016, 97(8): 2553-2560.

[12] Yu A Z, Huang G B. Effects of different tillage treatments on unerodible soil particles of wheat-field in spring in inland Irrigation region. Journal of Soil and Water Conservation. 2006,20(3):6-9.

[13] Liu Z D, Wang F, Li Gl. Influence of conical spoiler on airflow uniformity and stability in indoor wind tunnel. Journal of Northwest A&F University (Natural Science Edition). 2011, (11): 133-140.

[14] Gillette D. Define effective wind velocity by tunnel. Atmospheric Environment, 1978,12:2309-2313.

[15] Li X L, Shen X D. Experimental study on the distribution characteristics of the saltation particle of aeolian sediment in bare tillage. Transactions of the Chinese Society of Agricultural Engineering. 2006,22(5):74-77.

[16] Sun Y C, Ma M, Chen Z, et al. Wind tunnel simulation of impact of gravel coverage on soil erosion in arid farmland. Transactions of the Chinese Society of Agricultural Engineering. 2010,26(11):151-154.

[17] Wang Z Q, He Y F, Fu B F, et al. Wind tunnel study of wind erodibility of gravelly soil. Environmental Science and Technology. 2013,36(12M):103-105.

[18] Xue Y, Zhang W M, Wang T. Wind tunnel experiments on the effects of gravel protection and problems of field surveys. Acta Geographica Sinica. 2000,55(3):375-383.

[19] Neuman C M. Particle transport and adjustment of the boundary layer over rough surfaces with an unrestricted, upwind supply of sediment. Geomorphology. 1998,25:1-17.

[20] Raupach M R, Gillette D A, Leys J F. The effect of roughness elements on wind erosion threshold. Journal of Geophysical Research. 1993,98(D2):3023-3029.

[21] Jordan S B, Michael F J, Richard D C. Time series vegetation aerodynamic roughness fields estimated from moths observations. Agricultural and Forest Meteorology. 2005,135(8):252-268.

[22] Dong Z B, Gao S Y, Fryrear D W. Drag measurement of standing vegetation-clod cover surface. Journal of Soil and Water Conservation. 2000,14(2):7-11.

[23] Huang C H, Wang T, Zhang W M, et al. Contrastive study on sand drift movement over sandy and gravel Gobi desert in extremely arid regions. Arid Zone Research. 2007,24(4):556-562.

[24] Michels K, Sivakumar M V K, Allison B E. Wind erosion control using crop residue I. Effects on soil flux and soil properties. Field Crops Research. 1995,40:101-110.

[25] Stephane C A. Influence of soil texture on the binding energies of fine mineral dust particles potentially released by wind erosion. Geomorphology, 2008,93:157-167.

[26] Dong Z B, Li Z S. Wind erodibility of aeolian sand as influenced by grain-size parameters. Journal of Soil Erosion

and Soil and Water Conservation, 1998,4(4):1-5.

[27] Gu C Q, Sun Y. Discussion on the cohesion of soil changing with water content, cohesive soil content and dry density. Hydrogeology and Engineering Geology. 2005,1:34-36.

[28] Ha S. Effects of unerodible soil particles and tillage way on wind erosion on bashing highland. Journal of Desert Research. 1994,14(1):92-97.

[29] Wang H Y, Han J C, Luo L T, Tong W, Fu P, Zhang R Q. Regulatory Role of Sand in Soil Formation from Soft Rock and Sand. Chinese Journal of Soil Science. 2013, 45(2): 286-290.

[30] Han J C, Xie J C, Zhang Y. Potential role of soft rock as a natural water retaining agent in Mu Us Sandy Land, northwest China. Chinese Geographical Science. 2012,22:550-555.

[31] Luo L T, Wang H Y. Study on stability and sustainable utilization of soft rock and sand compound soil, 1st ed.; The Yellow River Water Conservancy Press, Henan, China, 2015: 153.

本文曾发表于2022年《Open Geosciences》第14卷

Soil Nutrient Content Analysis of Newly-Increased Farmland in the Process of Land Consolidation in Shaanxi Province, China

Wei Tong[1,2] and Jinbao Liu[1,2]

(1. Key Laboratory of Degraded and Unused Land Consolidation Engineering, the Ministry of Natural Resources of China, Xi'an 710075, China; 2. Shaanxi Provincial Land Consolidation Engineering Technology Research Center, Xi'an 710075, China)

【**Abstract**】The aim of this study is to analyze the characteristics of soil nutrient content of newly increased cultivated land in land consolidation, and identify the main factors affecting the productivity of newly increased cultivated land. The soil samples of 101 newly added cultivated land in northern, central and southern Shaanxi were collected, and their pH value, organic matter, total nitrogen, available phosphorus and available potassium contents were measured to analyze the soil nutrient status. The results indicated that: (1) The contents of soil organic matter, total nitrogen, available phosphorus and available potassium in the newly increased cultivated land were 0.21~23.63 g/kg, 0.20~8.27 g/kg, 2.34~34.16 mg/kg and 51~220 mg/kg, respectively. Compared with the nutrient grading standard of the second national soil survey, the content of soil nutrients in the newly increased cultivated land was relatively low, basically at the level of four to six. (2) The soil nutrient content of newly increased cultivated land in Shaanxi Province is greatly affected by the original land use types in northern Shaanxi and Guanzhong, and the soil nutrient difference in southern Shaanxi is mainly caused by the random changes among plots in the same region; The variance test showed that there were significant differences in soil pH organic matter and available potassium at 0.05 level, but no significant differences in soil total nitrogen and available phosphorus at 0.05 level. (3) It can be seen from the correlation analysis that among the nutrient indicators of the newly increased farmland in Shaanxi, pH organic matter has a significant negative correlation, and organic matter available potassium and available phosphorus available potassium have a very significant correlation. The correlation between soil nutrients in the newly increased farmland is weak, and the coordination between nutrients is poor. The low content of organic matter in soil is the main reason for low fertility. In the later stage of utilization, it is necessary to promote straw returning technology through the reasonable application of organic inorganic fertilizer and the application of soil testing formula fertilization technology, so as to finally realize the stable and high yield of newly added farmland.

【**Keywords**】Soil nutrients; Land consolidation; Newly-increased farmland

Introduction

China has more people and less land, and the contradiction between people and land is prominent. Food security is a major challenge facing China. In order to ensure the grain production capacity, in recent years, the state has vigorously implemented the policy of "balance between land occupation and land compensation", which means that the land shall be supplemented first and then occupied. Therefore, land consolidation has become an effective measure to increase the number of cultivated land and improve the productivity of cultivated land (Yang *et al.* 2008, Ma *et al.* 2010). Practice has proved that in the vast rural areas, land consolidation methods such as reclamation of abandoned homestead and upgrading of low standard land use can increase the area of effective cultivated land, improve the degree of intensive land use, and play a very important role in promoting the development of rural economy and urban-rural integration.

However, in the actual process of land consolidation, there are advantages and disadvantages, and paddy fields are used to supplement dry land, resulting in that the productivity of new cultivated land cannot reach the previous level. At present, after the completion of land consolidation, the "five supplies and one leveling" can be basically achieved (Long and Li 2006), and the soil nutrient status has become a key factor affecting agricultural production in the later period. Therefore, it is of great guiding significance to find out the soil nutrient status of fresh cultivated land for better playing the effect of land consolidation and evaluating the quality of new cultivated land and ecological environment (Zhao *et al.* 2015). Previously, there have been a lot of studies on soil fertility of existing basic farmland (Hou *et al.* 2007), but there are few studies on soil nutrient status of new cultivated land after land consolidation. Therefore, it is of great significance to study the soil fertility of the newly increased cultivated land after land consolidation and identify the short board of soil nutrients of the newly increased cultivated land in the region for improving the nutrient level of the newly increased cultivated land in the region and the classification and grading of agricultural land after the implementation of the project. The results can provide scientific basis for land management (Shi *et al.* 2015, Li *et al.* 2015).

In this study, typical land consolidation projects in 20 counties in Shaanxi Province in 2015 were selected to investigate the soil nutrients of the farmland transformed from non-agricultural land such as the original sandy wasteland, saline alkali land, waste grassland and bare rock gravel land. Based on the nutrient grading standard of the second national soil survey, the soil nutrient characteristics of the newly increased farmland were understood, and the soil nutrient conditions that restricted the productivity of the newly increased farmland were discussed, put forward targeted measures for soil fertility and improvement of new cultivated land, so that the new cultivated land can truly become a reliable guarantee for farmers to increase production and income.

Materials and methods

Shaanxi Province (31°42′~39°35′N, 105°29′~111°15′E) is located in the middle reaches of the Yellow River in China. With a large latitude span, the terrain is high from north to south and low in the middle, including mountains, plains, plateaus, basins and canyons. It is 870 km long from north to south, 200~500 km wide from east to west, and covers an area of 205800 km^2. It governs 10 cities and 1 district, including Xi'an City, Xianyang City, Baoji City, Weinan City, Yangling District, etc. From the north to the south, it is divided into three geomorphic regions: the Northern Shaanxi Plateau, the Guanzhong Plain, and the Qinling Bashan Mountains, which correspond to three regions, namely, the Southern Shaanxi and the Central Shaanxi and the Northern Shaanxi, respectively, accounting for 45%, 19% and 36% of the land area of the province. Due to the obvious influence of monsoon climate and continental climate, from south to north, it has the characteristics of northern subtropical humid climate, warm temperate semi humid climate and warm temperate and temperate semi-arid climate (Zhao *et al.* 2013). The average annual precipitation is 576.9 mm, the average annual temperature is 13 ℃, and the frost free period is about 218 days. In 2016, the total population of the province was 38.12 million, the annual GDP was 1916.539 billion yuan, the per capita GDP was 50395 yuan, the annual grain sown area was 3.0687 million hm^2, and the total grain output was 12.283 million tons.

The newly added cultivated land reclamation project in typical areas of Shaanxi Province was selected, involving 101 sample sites in 20 counties along the Great Wall, the Loess Plateau, Guanzhong Plain and Qinling Bashan Mountains (Fig. 1). The original land use types include sandy wasteland, saline alkaline land, gully land, grassland and bare rock gravel land (Table 1). On each type of land, according to the method of plum blossom sampling to collect soil samples, take soil samples of 0~30 cm soil layer on the top of the soil, about 500 g for each soil sample, bag and take it back to the laboratory for indoor analysis.

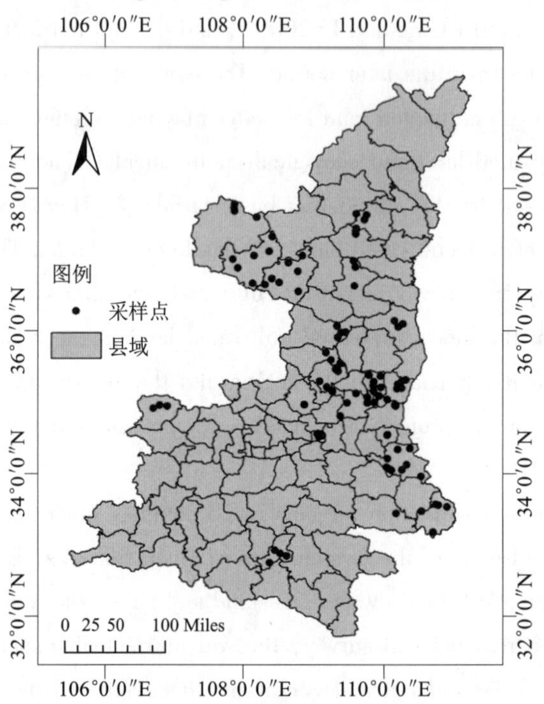

Fig. 1 Distribution map of sampling sites

Table 1 The former land-use patterns in newly-increased farmland

Sampling points (number of samples)	Land category	Sampling points (number of samples)	Land category
Dingbian County (6)	Saline and alkaline land, grassland	Chengcheng County (8)	Grassland
Zichang County (6)	Gully land	Dali County (5)	Saline and alkaline land, tidal flat land
Wuqi County (6)	Slope land	Heyang County (5)	Grassland, tidal flat land
Baota District (3)	Grassland	Pucheng County (6)	Saline and alkaline land, grass-land
Zhidan County (7)	Grassland	Huayin County (4)	Bare rock gravel, grassland
Yichuan County (3)	Grassland	Gaoling District (3)	Abandoned brick factory
Fuxian County (4)	Grassland	Longxian County (3)	Grassland
Huangling County (3)	Grassland	Luonan County (8)	Grassland
Yijun County (5)	Gully land	Shangnan County (6)	Tidal flat land
Yintai District (3)	Grassland	Hanyin County (4)	Slope land
Yaozhou District (3)	Rural wasteland		

Soil pH was measured by glass electrode method. Organic matter was determined by potassium dichromate oxidation and external heating method. Total nitrogen was determined by semi-micro Kjeldahl method. Available phosphorus was determined by sodium bicarbonate extraction-molybdenum-antimony resistance colorimetric method. The available potassium was determined by neutral ammonium acetate extraction and flame spectrophotometry. The same soil sample was measured three times in parallel with the same index, and its average value was obtained.

SPSS 19.0 was used for data entry, statistics and analysis.

Results and discussion

Fig. 2 for the spatial pattern of soil pH of newly increased cultivated land. As can be seen from Fig. 1, soil pH of newly cultivated land in Shaanxi Province ranges from 7.14 to 9.19, with an average value of 8.38. The coefficient of variation ranged from 0.33% to 23.57%, with an average value of 2.69%. The new cultivated land showed a weak alkalinity on the whole, and the soil pH variation was extremely mild, indicating that pH content was relatively stable in the soil samples of the new cultivated land. From the regional analysis, the pH of newly added farmland in northern Shaanxi is higher than that in Guanzhong and southern Shaanxi. The average pH of newly added farmland in Dingbian County, Wuqi County, Zhidan County, Yichuan County and Huangling County in Yulin City is more than 8.5, showing alkalinity. The pH of soil in Guanzhong and southern Shaanxi was neutral-weakly alkaline. The soil type, slope aspect and elevation are significantly related to the pH of the topsoil. The smaller the slope is, the higher the elevation is, and the greater the pH is. There are differences in the pH of different types of soil. It is recommended to prevent and control soil salinization in northern Shaanxi and Guanzhong, and soil acidification in southern Shaanxi, so as to promote sustainable agricultural development and ensure regional food security (Wang *et al.* 2021).

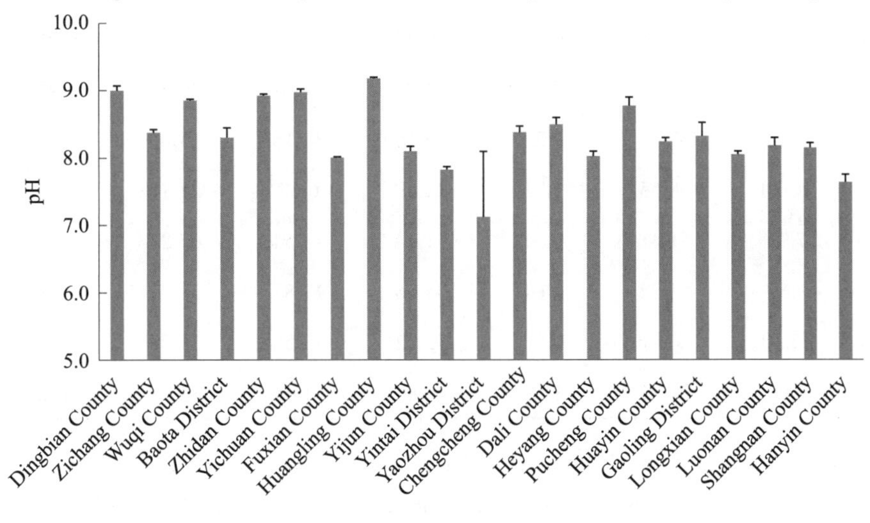

Fig. 2 The pH of newly increased farmland

After land consolidation in different areas of Shaanxi Province, the soil nutrient content of newly added cultivated land is shown in Tables 2 and 3.

Table 2 Statistical characteristics of soil nutrient content in newly-increased farmland in Shaanxi province

Sampling points	Organic matter (g/kg)		Total nitrogen (g/kg)		Available phosphorus (mg/kg)		Available potassium (mg/kg)	
	Av. value	Coefficient of variation (%)	Av. value	Coefficient of variation (%)	Av. value	Coefficient of variation (%)	Av. value	Coefficient of variation (%)
Dingbian County	2.13	9.01	0.70	23.11	2.34	14.00	60	12.87

Continued to Table 2

Sampling points	Organic matter (g/kg)		Total nitrogen (g/kg)		Available phosphorus (mg/kg)		Available potassium (mg/kg)	
	Av. value	Coefficient of variation (%)	Av. value	Coefficient of variation (%)	Av. value	Coefficient of variation (%)	Av. value	Coefficient of variation (%)
Zichang County	5.12	16.38	0.41	16.64	28.36	31.08	81	22.32
Wuqi County	2.98	52.65	0.62	114.97	8.44	29.04	70	8.33
Baota District	0.76	60.43	2.76	31.51	3.33	38.75	79	20.59
Zhidan County	0.21	23.34	0.98	86.13	5.00	83.30	64	15.37
Yichuan County	4.66	11.83	3.25	38.46	3.39	11.14	76	26.74
Fuxian County	3.88	14.90	1.25	30.73	9.18	13.98	58	10.05
Huangling County	7.87	23.57	0.85	16.47	4.80	24.17	105	16.83
Yijun County	5.33	30.06	1.08	18.77	5.53	14.76	89	22.89
Yintai District	4.67	10.09	8.27	20.80	3.87	27.54	126	16.23
Yaozhou District	13.10	19.13	0.91	31.28	10.83	54.78	143	26.38
Chengcheng County	6.57	30.49	1.03	20.91	3.60	25.48	51	27.74
Dali County	4.74	29.97	1.15	47.53	23.49	12.22	220	13.82
Heyang County	8.25	26.65	2.57	33.55	6.59	51.84	96	16.85
Pucheng County	23.6	7.04	0.20	56.72	11.25	22.75	155	25.08
Huayin County	5.72	14.91	2.33	109.67	25.87	26.34	242	34.72
Gaoling District	6.63	32.36	0.23	59.13	34.16	15.40	94	10.15
Longxian County	10.70	58.55	1.72	12.89	6.62	22.79	131	15.83
Luonan County	7.99	51.41	0.51	51.14	13.54	65.27	117	21.34
Shangnan County	9.53	80.22	1.64	24.86	17.87	51.74	70	15.54
Hanyin County	12.50	26.46	0.90	33.22	12.58	58.99	89	9.91
Average value	7.00	29.97	1.59	41.83	11.46	33.11	106	18.55

The soil organic matter content of the newly added cultivated land was 0.21~23.63 g/kg, with an average value of 7.00 g/kg, and the coefficient of variation was 7.04% to 80.22%, with an average value of 29.97%, which belonged to the medium variation. According to the nutrient classification standard of the second national soil survey, the content of soil organic matter in most of the newly increased cultivated land is at the level of grade 5 and 6, and the content of organic matter is generally low. Liu have shown that the content of soil organic matter in newly cultivated land is relatively low. The organic fertilizer combined with chemical fertilizer can effectively increase the content of soil organic matter in newly cultivated land, which can be 36.1% higher than that of ordinary fertilization. Optimizing fertilization is an effective measure to improve the stability of soil structure, fertilizer conservation characteristics and land productivity of newly renovated farmland(Liu *et al.* 2021). From the regional analysis, the content of soil organic matter in newly added farmland in Guanzhong was higher than that in northern and southern Shaanxi.

Table 3 The grades of soil nutrient of newly-increased farmland in Shaanxi province

Sampling points	Organic matter	Total nitrogen	Available phosphorus	Available potassium
Dingbian County	VI	V	VI	IV
Zichang County	VI	VI	II	IV
Wuqi County	VI	V	IV	IV
Baota District	VI	I	V	IV
Zhidan County	VI	IV	V	IV
Yichuan County	VI	I	V	IV
Fuxian County	VI	III	IV	IV
Huangling County	V	IV	V	III
Yijun County	VI	III	IV	IV
Yintai District	VI	I	V	III
Yaozhou District	V	IV	III	III
Chengcheng County	V	III	V	IV
Dali County	VI	III	II	I
Heyang County	V	I	IV	IV
Pucheng County	V	VI	III	II
Huayin County	VI	I	II	I
Gaoling District	V	VI	II	IV
Longxian County	V	I	IV	III
Luonan County	V	V	III	III
Shangnan County	V	I	III	IV
Hanyin County	IV	IV	III	IV

Note: The nutrient content grading standard refers to the nutrient grading standard of the second national soil survey.

The soil total nitrogen content of the newly added cultivated land was 0.20~8.27 g/kg, with an average value of 1.59 g/kg, and the coefficient of variation was 12.89% ~ 114.97%, with an average value of 41.83%, belonging to the medium variation. According to the nutrient classification standard of the second national soil survey, the soil total nitrogen content of newly cultivated land in most areas was at the level of 4~6, and a few areas such as Yijun County and Yintai District reached the level of 1, but the total nitrogen content was generally low. He showed that the total nitrogen mass ratio of soil also increased year by year with the increase of the implementation years of new cultivated land, and reached the maximum value of 0.92 g/kg 3 years after the implementation of the project. Compared with that before the project implementation, the total nitrogen mass ratio increased by 19.11%, 3.70% and 3.57% respectively at each stage of the project completion, with the largest increase after the project implementation. This is mainly due to the fact that after the implementation of the project, nitrogen will be supplemented through measures such as soil fertilization, and there will generally be a certain fallow period after the implementation of the project, so that soil nitrogen can be accumulated to a certain extent (He *et al.* 2022). Analysis shows that the change law of total nitrogen mass ratio is similar to that of organic matter. Regional analysis showed that the total nitrogen content of newly added farmland in Guanzhong was higher than that in northern and southern Shaanxi.

The available *P* content of newly added cultivated land ranged from 2.34 to 34.16 mg/kg, with an average of 11.46 mg/kg, and the coefficient of variation ranged from 11.14% to 83.30%, with an average of 33.11%, which belonged to the medium variation. According to the nutrient classification standard of the second national soil survey, the soil available *P* content of newly added cultivated land in most areas is at grade 4~grade 5 level, while the content of available *P* in Huayin City, Gaoling District and a few other areas reaches grade 2 level. Generally speaking, the content of available *P* is still low. Hu studied the soil nutrients of the new cultivated land in Heyang, Shaanxi Province, and found that the available phosphorus content in the soil was low, and the nutrient evaluation grade showed that the soil of the poor grade IV accounted for 83.4% (Hu *et al*. 2022), Song Jiajie showed that straw, fertilization and their interaction had a significant impact on the total phosphorus content in the soil layer of 20~40 cm ($P<0.05$). Compared with the treatment without fertilization, fertilization could significantly increase the total phosphorus content in the soil layer ($P<0.05$) (Song *et al*. 2022). From the regional analysis, the soil available *P* content of newly added farmland in southern Shaanxi was higher than that in northern Shaanxi and Guanzhong.

The soil available potassium content in the newly cultivated land ranges from 51 to 220 mg/kg, with an average value of 106 mg/kg, and the coefficient of variation ranges from 8.33% to 34.72%, with an average value of 18.55%, which belongs to the medium variation. According to the nutrient classification standard of the second national soil survey, the content of available potassium in soil of newly added cultivated land in most areas is at the third-fourth level. The content of available potassium in Dali County and Huayin City reaches the first level, and in general, the content of available potassium is at the medium level. Zhou researched on the newly added cultivated land of Jingtai County, Gansu Province's balance project of occupation and compensation of cultivated land shows that when the new cultivated land is 3 years old, the damage rate of soil structure is 2.05% lower than that when the new cultivated land is 1 year old, and the soil erosion resistance is significantly improved. The soil available potassium is 13.62% higher than that when the new cultivated land is 1 year old, respectively. The quality of the newly added cultivated land has the greatest impact on the grain production capacity, and the cultivation for 5 years can be 15.69% higher than that before the implementation(Zhou 2020). From the regional analysis, the content of soil available potassium in newly added cultivated land in Guanzhong was higher than that in northern and southern Shaanxi.

Soil fertility directly affects land productivity. In general, the soil nutrient content of new cultivated land in Guanzhong region is higher than that in northern and southern Shaanxi, mainly due to the difference of soil types of new cultivated land in various regions. Soil nutrients are greatly affected by terrain, soil type and human activities. Under different soil types, the nutrient content in lou soil is the highest, while that in black loessial soil is lower. The elevation is the dominant factor that causes the spatial variation of soil nutrients. The differences of agricultural production activities, such as fertilization, planting system and straw returning, also affect the spatial variability of soil nutrients to a large extent(Jia *et al*. 2022). The soil types in northern Shaanxi are mainly aeolian sandy soil and loess soil; The soil type in Guanzhong area is mainly lou soil; The soil type in southern Shaanxi is mainly paddy soil. Aeolian sandy soil and loess soil in northern Shaanxi, mountainous brown soil, mountainous yellow cinnamon soil and calcareous soil in southern Shaanxi have low nutrient content. The soil types with relatively high nutrient content are lou soil in Guanzhong Basin and paddy soil in Hanjiang Valley, and the land yield is relatively high.

In order to eliminate the influence of regional differences on soil nutrients, the variance analysis of soil nutrients between counties and different regions was carried out. It can be seen that the soil nutrient indexes of newly added cultivated land in Shaanxi Province were significantly different at the 0.05 level between counties in northern Shaanxi and Guanzhong, while the soil pH, total nitrogen and available potassium were significantly different at the 0.05 level between counties in southern Shaanxi Province, while the soil organic matter and available phosphorus were not significantly different at the 0.05 level(Table 4); There are significant differences in pH organic matter and available potassium of newly increased cultivated land between northern Shaanxi, Guanzhong and southern Shaanxi at 0.05 level, while there are no significant differences in total nitrogen and available phosphorus of soil at 0.05 level (Table 5). Zhao studied the spatial heterogeneity of soil nutrients in the transitional zone between Maowusu Desert and the Loess Plateau through semi variance analysis. The results showed that total nitrogen and available phosphorus had strong spatial correlation, and structural factors played a leading role in the variation; Available potassium has a medium intensity of spatial correlation, and structural factors and random factors play a leading role in the variationt(Zhao *et al*. 2016). The results showed that the soil nutrient difference of newly cultivated land in northern Shaanxi and Guanzhong was mainly affected by the land type in the original land consolidation area. The difference of soil nutrients in newly added cultivated land in southern Shaanxi was mainly caused by the random variation between plots in the same region.

Table 4　ANOVA of soil nutrients in newly increased farmland in rural areas of Shaanxi province

Sampling points	Index	Freedom	Sum of squares	Mean square	$F_{0.05}$	$P_r>F$
Northern Shaanxi	pH	7	5.101	0.729	60.900	<0.001
	Organic matter	7	183.301	26.186	31.510	<0.001
	Total nitrogen	7	27.698	3.957	9.860	<0.001
	Available phosphorus	7	2873.230	410.461	23.050	<0.001
	Available potassium	7	5903.652	843.379	5.350	0.0005
Guanzhong region	pH	9	6.908	0.768	3.760	0.0023
	Organic matter	9	1663.015	184.779	55.610	<0.001
	Total nitrogen	9	168.149	18.683	21.440	<0.001
	Available phosphorus	9	5233.262	581.474	44.980	<0.001
	Available potassium	9	146396.136	16266.237	13.960	0.0005
Southern Shaanxi	pH	2	0.882	0.441	6.890	0.0075
	Organic matter	2	53.590	26.795	0.910	0.4256
	Total nitrogen	2	4.430	2.215	21.110	<0.001
	Available phosphorus	2	89.147	44.573	0.590	0.5683
	Available potassium	2	7841.361	3920.681	11.290	0.0010

Table 5　ANOVA of soil nutrients in newly increased farmland in urban areas of Shaanxi province

Index	Freedom	Sum of squares	Mean square	$F_{0.05}$	$P_r>F$
pH	2	1.532	0.766	4.530	0.0256
Organic matter	2	142.768	71.384	3.450	0.0553
Total nitrogen	2	2.257	1.129	0.330	0.7243
Available phosphorus	2	155.677	77.839	0.850	0.4456
Available potassium	2	16776.318	8388.159	4.160	0.0338

The mean square comparison shows that the difference of soil nutrients of newly increased cultivated land in southern Shaanxi is mainly affected by the formation environment and the operation of consolidation technology, which is difficult to reflect the difference between the original land use types; However, the soil nutrient content of newly increased cultivated land in northern Shaanxi and Guanzhong regions is greatly affected by regions, forming significant differences. Fertilization methods should be properly adjusted according to land use types in different regions. Gao showed that part of soil nutrients mainly came from the decomposition of original minerals in soil parent materials, and different soil parent materials also determined the development of different soil types. In addition, the chemical reaction of soil water caused by artificial fertilization, crop consumption and irrigation and drainage in the later period will also have a significant impact on soil nutrients. And nutrient loss, which is one of the typical differences between cultivated soil and natural soil (Gao *et al.* 2021).

The content of soil nutrients is related to many factors, and each nutrient has a complex correlation. Organic matter, nitrogen, phosphorus and potassium are both independent and interrelated, which together guarantee the growth and development of crops. Understanding the relationship between various nutrient factors can give a more comprehensive understanding of the soil fertility structure, research on the consumption or accumulation of soil nutrients can reveal the cycle and balance of nutrients in the soil crop system. Long term application of organic and inorganic fertilizers can comprehensively improve the content of soil nitrogen, phosphorus, potassium, organic matter and CEC, and effectively regulate soil pH (Xu *et al.* 2006). See Table 6 for the correlation analysis of soil nutrients in the new cultivated land of land consolidation in Shaanxi Province.

Table 6 Correlation between available nutrients in newly increased farmland in Shaanxi province

Index	pH	Organic matter	Total nitrogen	Available phosphorus	Available potassium
pH	1				
Organic matter	−0.20*	1			
Total nitrogen	−0.19	−0.13	1		
Available phosphorus	−0.10	0.11	−0.18	1	
Available potassium	−0.09	0.31**	0.16	0.43**	1

Note: * Indicates significant correlation at 0.05 level; ** Indicates significant correlation at 0.01 level.

According to the test, among the nutrient indicators, pH organic matter has a significant negative correlation, and organic matter available potassium and available phosphorus available potassium have a very significant correlation. The correlation between soil nutrients in the newly added farmland in Shaanxi Province is weak, and the coordination between nutrients is poor. The high and stable yield of farmland requires sufficient and coordinated soil nutrients. Through fertilization and later cultivation, the soil nutrient content has been greatly improved, its spatial difference is weakened, and its homogenization is enhanced. The correlation between soil nutrient contents is weakened and the coordination is reduced, which is mainly due to the dependence on chemical fertilizer. In the later stage, the soil nutrient should be improved through field management and increased application of organic fertilizer (Zhang *et al.* 2022).

The content of organic matter in the soil of new cultivated land in Shaanxi Province is of medium variation, and the difference of organic matter content is obvious among different regions. The organic matter available potassium and available phosphorus available potassium have extremely significant correlation, with the correlation coefficients reaching 0.31 and 0.43 respectively. This is mainly because the new cultivated land is mainly fertilized by organic fertilizer, and the organic matter, N, P and other nutrients in the organic fertilizer are high. Increasing the application of organic fertilizer can improve the content of N, P, K and other nutrients in the soil. There is a significant negative correlation between soil pH and organic matter in the newly increased cultivated land, with a correlation coefficient of −0.20. Although the correlation with total nitrogen, available phosphorus and available potassium does not reach a significant level, there is still a certain negative correlation. Therefore, reducing soil acidity can increase the content of nutrients such as soil alkali hydrolyzable nitrogen and available potassium, and promote crop growth (Yang *et al.* 2009). Soil organic matter is an important source of plant nutrient elements. It plays a good role in improving soil structure, increasing soil microbial activity, enhancing soil enzyme activity, etc. By increasing the application of organic fertilizer, the quality fraction of soil organic matter is increased, so as to improve soil fertility (Xie *et al.* 2001).

Compared with the nutrient classification standard of the second national soil survey, it can be seen that after land consolidation in Shaanxi Province, the soil nutrient content of newly added cultivated land is not high. The soil nutrient content of new cultivated land in northern Shaanxi and Guanzhong was greatly affected by the land type of the original land consolidation area. The southern Shaanxi region is mainly caused by the random variation among the inner blocks in the same region. In Shaanxi Province, the correlation of soil nutrient contents was poor, except for pH-organic matter, organic matter-available potassium and available phosphorus-available potassium, the other nutrient contents showed no obvious correlation. According to existing research results, organic matter in soil nutrients is not only an important source of various nutrient elements in soil, especially N and P, but also conducive to improving soil fertility and buffering capacity, increasing soil nutrient availability, microbial activity and improving soil structure, thus improving the physical and chemical properties of soil (Huang *et al.* 2015). It can also be inferred that the low content of organic matter in newly added cultivated soil in land consolidation projects in Shaanxi Province is the main reason for low fertility, and the level of soil nutrients has become an important factor restricting the land production capacity and influencing the yield difference after land consolidation. Therefore, the following measures can be taken to improve the soil nutrient content of new cultivated land:

(1) As soil organic matter plays an important role in improving soil physical and chemical properties and adjusting soil fertility factors, maintaining soil organic matter balance and gradually increasing soil organic matter content should be taken as the central link of soil fertility in the future. In terms of soil fertility, the combination of organic and inorganic fertilizers should be advocated, which can expand soil total nitrogen content and effective nitrogen pool, and significantly improve soil nitrogen fertility level. At the same time, the reasonable application of organic and inorganic fertilizers is an effective measure to maintain the good basic physical and chemical properties of farmland soil, and a key soil fertility measure to maintain sustainable agricultural development.

(2) In terms of planting mode, it is recommended to plant annual green manure to press green, plant beans and other crops at the initial stage of reclamation, or plant rape and other crops (Ma *et al.* 2012); Through the combination of land cultivation and cultivation, crop rotation is implemented, straw returning measures are promoted, and more nitrogen and potassium fertilizers are applied to ensure the balanced supply of nutrients in the newly increased farmland, so as to achieve sustainable and stable yield and high yield of the developed farmland.

(3) Promote soil testing formula fertilization technology, find out soil fertility through soil testing, carry out accurate application of nitrogen, phosphorus, potassium, medium and trace elements in different growth periods according to crop growth rules, eliminate soil nutrient barriers, and improve fertilizer utilization efficiency, which can not only achieve the purpose of increasing and stabilizing yield, but also improve

plant agronomic properties and improve the quality of agricultural products (Luo *et al*. 2013, Xiang *et al*. 2006).

Acknowledgements

This work was financially supported by Shaanxi Land Engineering Construction Group fund (DJNY 2021-7).

References

Gao H R, Zhou Y, Liu J K and Wang L 2021. Spatial Correlation Between Soil Nutrients and Environmental Factors in Cultivated Land Based on Information Entropy. Water and Soil Conserv. Notifi. 41(6):226-236.

He Z J, He W, Li L R, Zhang J and Li H 2022. Analysis of factors influencing quality of newly increased cultivated land and grain productivity in arid highland area-taking occupation complementary balance project as an example. J. Drain. Irriga. Mach. Engin. (JDIME). 40(11):1151-1156,1166.

Hou L, Lei R D, Wang D X, Kang B W and Liu J J 2007. Study on soil nutrient and enzyme in hillclosed and afforested natural secondary Pinus tabulae form is forest land in Huanglong mountain. Journal of Northwest A & F university 35(2): 63-68.

Hu Y, Wang J and Li G 2022. Soil Nutrient Characteristics and Fertility Grade Evaluation of Newly-increased Farmland in Weibei Dryland: A Case Study of Heyang County. Chinese Agricul. Sci. Bull. 38(27): 94-100.

Huang D F, Wang L M, Li W H and Qiu X X 2014. Research progress on the effect and mechanism of fertilization measure on soil fertility. Chinese J. Eco-Agricul. 22(02): 127-135.

Jia L J, Yang L A, Feng Y T, Ji Y F and Li Y L 2022. Spatial variation of soil nutrients and its influencing factors in Baoji city. Resou. Environ. Arid Areas. 36(12): 135-143.

Li L, Shi Z B, Wang X R, Liu L and Zhao M 2015. Analysis of spatiotemporal variations in quality and productivity of newly-increased farmland by land on solidation: Taking Donghai County in Jiangsu Province as an example. Acta Agricul. Jiangxi. 27(05):95-99.

Liu Z, Zhang Y, Lei N, Zhang T Y, Xiong Y F, Zhang P P and Li Y N 2021. Effects of Optimized Fertilization Treatments on Soil Aggregate Characteristics and Organic Matter Content of Newly Reclaimed Cultivated Land in Loess Plateau. Water and Soil Conserv. Notifi. 41(5):99-106.

Long H L and Li X B 2006. Cultivated-land transition and land consolidation and reclamation in China: Research Progress and Frame. Progress in Geograp. Sci.25(5):67-76.

Luo X J, Feng S Y, Shi X P and Qu F T 2013. Farm households' adoption behavior of environment friendly technology and the evaluation of their environmental and economic effects in Taihu Basin-taking formula fertilization by soil testing technology as an example. J. Natural Resour. 28(11):1891-1902.

Ma C Q, Chen G X and Wang L X 2012. Analysis of soil nutrients content of newly-increased farmland in the process of land consolidation in Lintong district. Agricul. Res. Arid Areas. 30(02):47-51,61.

Ma C Q, Liu T M and Yang M H 2010. Analysis of soil nutrients of the newly-increased farmland in the process of land consolidation in Gaoling County. Journal of Northwest A&F University (Nat. Sci. Ed.). 38(5):175-180.

Shi C D, Han J C, Ma Z H, Zhang L and Zhang R Q 2015. Investigation of new farmland soil fertility after wild grassland reclamation in Weibei Plateau: A Case Study in Chengcheng County, Shaanxi Province. J. Anhui Agri. Sci. 43(18): 115-118.

Song J J, Xu X Y, Bai J Z, Yu Q, Cheng B H, Feng Y Z and Ren G X 2022. Effects of Straw Returning and Fertilizer Application on Soil Nutrients and Winter Wheat Yield. Environ. Sci. 43(9): 4839-4847.

Wang H, Cao J, Wu J H and Chen Y P 2021. Spatial and temporal variability in soil pH of Shaanxi Province over the last 40 years. Chinese J. Ecol. Agricul. 29(6):1117-1126.

Xiang X J, Wu Y M, Wang J P, Huang W H, Li J J, Gong D P and Wang D H 2006. Study and application of soil testing and fertilization technology. Hunan Agricul. Sci. 3:73-74,76.

Xie Z N, Xu W B, Zhuang Y M and Wang R J 2001. Correlation between the mass fractions of soil organic matter and availablenutrients in citrus and longan orchards. J. Fujian Agricul. and Forest. Univ. (Natural Science Edition). 30(1): 36-39.

Xu L, Zhang Y Z, Zeng X B, Zhou W J, Zhou Q and Xia H A 2006. Effects of Different Fertilizer's Application Systems on Soil Fertility and Rice Yield. J. Hunan Agricul. Univ. (Natural Sciences). 32(4): 362-367.

Yang J, Liu L, Sun C M and Liu X F 2008. Soil components and fertility improvement of added cultivated land. Transactions Chinese Soc. Agricul. Engineer. 24(7):102-105.

Yang J F, Sun Y, Wu H S and Zheng W Q 2009. Correlativity of Soil Nutrients in Pepper Garden. Chinese J. Tropical Agricul. 29(8):8-11.

Zhang B H, Zhang X L, Bai Z H, Zhang J P, Wang Z J and Dong J 2009. Study on Correlativity Change of Surface Soil Nutrients in Intensified Agricultural Regions. Shandong Agricul. Sci. 9: 67-69.

Zhao X, Han J C, Wang H Y, Zhang Y, Hao Q L, Sun Y Y and Zhang H O 2016. Soil nutrient spatial heterogeneity in the Mu Us Desert-Loess Plateau Transition Zone. Acta Ecologica Sinica. 36(22): 7446-7452.

Zhao Y H, Liu X J and Ao Y 2013. Analysis of cultivated land change, pressure index and its prediction in Shaanxi province. Transaction Chinese Soc. Agricul. Engineer. 29(11): 217-223.

Zhao Y T 2015. Spatial characteristics and changes of soil nutrients in cultivated land of Guanzhong region in Shaanxi Province based on GIS, Dissertation for Doctor Degree Northwest A&F University.

Zhou X H 2020. Quality Grades of Newly Cultivated Land and Factors Influencing Grain Productivity in Loess Tableland Area. Water Soil Conserv. Notifi. 40(4):237-243.

本文曾发表于2022年《Bangladesh J. Bot.》第51卷第4期

水稻种植对陕北盐碱地土壤理化性质的影响及耐盐碱水稻品种筛选

马增辉,曹　源,王启龙

(陕西省土地工程建设集团有限责任公司,西安 710075)

【摘要】为明确种植水稻对陕北盐碱地土壤理化性质的影响,筛选适宜陕北地区气候环境的耐盐渍水稻品种,进一步探索陕北盐碱地改良利用方式,以陕西省定边县堆子梁镇营盘梁村土地开发项目为例,开展大田试验,分别对种植不同水稻品种("宁粳 28""隆优 619""东稻 4 号")后的土壤及水稻产量展开研究,研究结果表明:水稻种植可以显著改善盐碱地表层(0~20 cm)土壤的理化性质,降低土壤酸碱度和含盐量,提高土壤养分含量,水稻种植后的表层土壤 pH 值和含盐量较整治前分别降低了 5.7%和 25.5%,全氮、有效磷、速效钾和有机质分别提高了 117.3%、45.9%、27.7%和 67.3%。另外,不同水稻品种对于盐碱地土壤酸碱度和含盐量基本无影响,但是对于表层土壤养分含量存在一定影响,水稻收获后,"东稻 4 号"土壤全氮、有效磷含量最高,"宁粳 28"有机质含量最高。从产量角度来看,种植"隆优 619"产量最高,分别较"宁粳 28""东稻 4 号"处理提高了 5.0%、42.8%,是本试验条件下产量最高的处理。

【关键词】水稻种植;盐碱地;理化性质;产量

Effects of Planting Rice on Soil Physical and Chemical Properties of Saline-alkali Land in Northern Shaanxi and Screening of Saline-alkali-tolerant Rice Varieties

Zenghui Ma, Yuan Cao, Qilong Wang

(Shaanxi Provincial Land Engineering Construction Group Co.Ltd.,Xi'an 710075)

【**Abstract**】In order to clarify the impact of rice planting on soil physical and chemical properties of saline alkali land in Northern Shaanxi, screen salt tolerant rice varieties suitable for climate environment in Northern Shaanxi, and further explore the improvement and utilization mode of saline alkali land in Northern Shaanxi, taking the land development project of yingpanliang village, duiziliang Town, Dingbian County, Shaanxi Province as an example, field experiments were carried out to plant different rice varieties ("Ningjing 28", "Longyou 619", "Dongdao 4"). The results show that rice planting can significantly improve the physical and chemical properties of soil in saline alkali surface layer (0~20 cm), reduce soil pH and salt content, and improve soil nutrient content. The pH value and salt content of topsoil after rice planting are reduced by 5.7% and 25.5% respectively compared with those before regulation Available potassium and organic matter increased by 117.3%, 45.9%, 27.7%

基金项目:陕西省重点科技创新团队计划项目(2016KCT-23);陕西省土地工程建设集团有限责任公司内部科研项目(DJNY 2019-1)。

and 67.3% respectively. In addition, different rice varieties have no effect on the pH and salt content of saline alkali soil, but have a certain effect on the nutrient content of surface soil. After rice harvest, the content of total nitrogen and available phosphorus in "Dongdao 4" soil is the highest, and the content of organic matter in "Ningjing 28" is the highest. From the perspective of yield, the yield of "Longyou 619" was the highest, which was 5.0% and 42.8% higher than that of "Ningjing 28" and "Dongdao 4" respectively.

【Keywords】 Rice planting; Saline alkali land; Physical and chemical properties; Yield

陕北地区盐碱地面积较大,是土地整治主要对象之一,截至目前,陕西省尚有约 1400 km^2 盐碱地尚未得到有效开发利用,土壤盐碱化给当地农业生产和粮食安全带来了巨大威胁[1]。不断加大盐碱地治理力度,对增加农民收入、保障重要农产品有效供给、保护生态安全等具有重要意义[2]。目前,盐碱地改良技术主要包括耕作改良技术[3]、地表覆盖技术[4]、化学改良技术[5]、水利改良技术[6]、工程改良技术[7]以及植物改良技术[8],其中植物改良主要通过植物根系生长,可改善土壤物理性状,根系分泌的有机酸可以中和土壤碱性,并可吸收、富集土壤盐分[9];减少土壤水分蒸发,从而防止土壤返盐[10]。在地下水位较高的盐碱地种植耐盐碱水稻,不仅可以提高土地利用率,还能起到保护环境的作用。

土壤养分是由土壤提供的植物必需的营养元素,是土壤肥力的重要物质基础,也是评价土壤肥力水平的重要内容之一。由于盐碱地土壤养分含量较低,不利于作物生长和利用。此外,土壤盐分和 pH 过高对于植物的伤害也是不容忽视的,土壤盐分过高会影响作物的光合作用和呼吸作用,会阻碍农作物蛋白质的合成,土壤的 pH 值过高会直接影响肥料的有效性,一些营养元素会变得难溶,有效性降低,从而不能被根系很好地吸收。因此,盐碱地的有效利用首先要解决土壤盐碱含量和土壤养分不足等问题。

陕西省定边县堆子梁地区盐碱地地下水位常年普遍较高,旱地作物无法生长,但其水热资源丰富[11],满足水稻生长基本条件,但陕北地区气候环境、土壤构成成分和盐碱度与其他区域不尽相同[12],因此明确水稻种植对于陕北盐碱地土壤理化性质影响,筛选适宜在陕北地区盐碱地种植的水稻品种,对于陕北地区盐碱地利用和生态环境改善有着重要意义。本研究通过在定边县堆子梁镇营盘梁村开展大田试验,研究水稻种植对于土壤理化性质影响和不同水稻品种生长状况,筛选合适的水稻品种,明确水稻种植对盐碱地的改良效果,从而为陕北盐碱地的改良利用提供理论与技术支撑,同时为其他地区盐碱地的改良利用提供借鉴。

1 材料与方法

1.1 研究区概况

项目区位于定边县堆子梁镇营盘梁村,地理坐标介于东经 108°15′00″~108°18′45″、北纬 37°37′30″~37°42′30″。项目区属大陆性半干旱季风气候,季节变化明显,四季冷暖干湿分明,大陆性特点显著。年平均气温 7.9 ℃,年均最高气温 37.7 ℃,最低气温 1.5 ℃。昼夜温差为 13.6 ℃,年平均日照时数 2700 h。多年平均无霜期 120~130 d。年平均降水量 320 mm,主要集中在 7~9 月内,占全年降水量的 61.34%,多年平均蒸发量 2522.8 mm。项目区位于定边县最低处,属于古河床上冲积母质发育的盐化草甸土壤,因为地势低洼,地下水出流不畅,受到地下水位埋藏浅、地下水矿化度高以及气候蒸发量大的共同作用,土壤表层积盐,盐渍化特征明显[12]。

1.2 试验设计

在营盘梁项目区设置田间小区试验。将试验小区划分为 8 个地块,并将每个地块四周起垄覆膜,做好地块间隔离措施,每个地块面积约 80 m^2,长 10 m,宽 8 m,根据种植水稻品种不同,试验设置 4 个处理,T1、T2、T3 处理分别种植"宁粳 28""隆优 619""东稻 4 号"3 个品种,CK 为空白对照,不种植水稻。试验开始前采集 0~100 cm 土层土壤样品,测定其基本理化性质,测定结果见表 1。

水稻于2019年6月初插秧,插秧密度为2万株/hm^2,10月初水稻成熟收获,试验小区采用统一的栽培管理措施,主要包括施肥、病虫害的防治和田间灌溉等,尿素、磷酸二胺、氯化钾的施肥量分别为334 kg/hm^2、196 kg/hm^2、96 kg/hm^2。试验田采用井水灌溉,每年的灌溉量为12000 m^3/hm^2,灌溉水的基本理化性质见表2。

表1 项目区盐碱土理化特性本底值

土层厚度/cm	pH	全氮/(g/kg)	有效磷/(mg/kg)	速效钾/(mg/kg)	有机质/(g/kg)	水溶性盐总量/(g/kg)
0~20	9.0	0.23	6.10	120	3.02	2.98
20~40	9.0	0.22	5.75	111	3.01	2.57
40~60	8.9	0.21	5.36	110	2.85	2.49
60~80	8.9	0.18	5.13	102	2.64	2.26
80~100	8.9	0.19	4.80	103	1.97	2.34

表2 灌溉水的基本理化性质

pH	全盐量/(mg/L)	矿化度/(mg/L)	总硬度/(mg/L)	钙离子/(mg/L)	镁离子/(mg/L)
8.1	0.23	748	186	369	74.1
钾离子/(mg/L)	钠离子/(mg/L)	硫酸根/(mg/L)	氯根/(mg/L)	碳酸根/(mg/L)	重碳酸根/(mg/L)
2.03	148	369	74.1	4.31	71.7

1.3 样品采集与分析

水稻收获后,在每个地块沿对角线随机选取3个点,用土钻按照0~20 cm、20~40 cm、40~60 cm、60~80 cm、80~100 cm逐层采集土壤剖面样品,将3个样点的同土层样品混合后按照四分法预留土样约1 kg。样品经风干研磨,过筛后备用。

采用质量法测定土壤中的水溶性总盐含量,电位法测定土壤样品pH(水和土壤比例为5:1),半微量凯氏法测定全氮,紫外可见分光光度计法测定土壤有效磷,乙酸铵提取-火焰光度计法测定土壤速效钾,重铬酸钾氧化外加热法测定有机质[13]。

1.4 数据分析方法

采用Microsoft Excel 2007分析软件处理试验数据,同时采用SPSS 11.5软件进行统计学分析,并对相关指标进行显著性分析,显著性水平为$P<0.05$,极显著性水平为$P<0.01$。

2 结果分析

2.1 水稻种植对土壤pH的影响

土壤的pH值会直接影响肥料的有效性,土壤的pH值过高或过低,一些营养元素会变得难溶,有效性降低,从而不能被根系很好地吸收[14]。水稻收获后土壤pH的剖面分布情况如图1所示。水稻种植前项目区土壤pH值较高,0~100 cm土体中,土壤pH平均值达到8.9,依据土壤酸碱度等级划分标准,土壤达到强碱性,严重影响农作物正常生长。种植水稻后,各处理表层土壤平均pH值为8.5,较水稻种植前土壤酸碱度得到了显著改善。种植水稻1年后,T1、T2、T3处理60~100 cm土层pH均为8.9,相较于空白对照CK处理,无显著变化,说明种植水稻1年,对于深层土壤(大于60 cm)pH影响不明显。另外,从图1中还可以看出,种植水稻的品种对于盐碱地土壤酸碱度影响不大,水稻收获后,T1、T2、T3处理0~100 cm土层pH平均值分别为8.7、8.8、8.7。

2.2 水稻种植对土壤水溶性盐总量的影响

土壤盐分过高对于植物的伤害是不容忽视的,土壤盐分过高会影响作物的光合作用和呼吸作用,会阻碍农作物蛋白质的合成[15]。通过对图2水稻种植对于0~100 cm土层土壤水溶性盐总量的影响

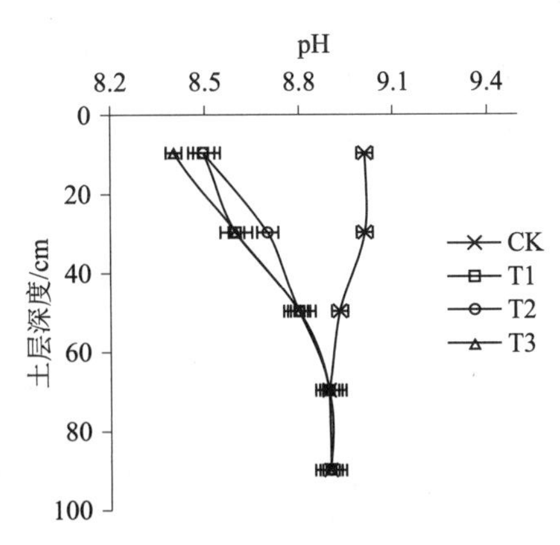

图1 水稻种植对0~100 cm土层土壤pH的影响

进行分析可知:水稻种植可以明显降低土壤含盐量,未种植水稻CK处理0~100 cm土层土壤平均含盐量为2.53 g/kg,土壤为重度盐化土,种植水稻后,土壤水溶性盐总量明显降低,平均值变为2.21 g/kg。另外,由图2还可以看出,种植水稻的品种对于盐碱地土壤含盐量基本无影响,T1、T2、T3的0~100 cm土层土壤水溶性盐总量分别为2.20 g/kg、2.21 g/kg和2.21 g/kg。

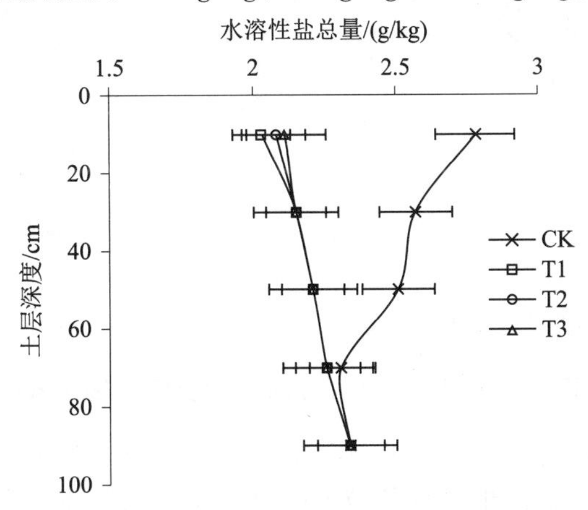

图2 水稻种植对0~100 cm土层土壤水溶性盐总量的影响

2.3 水稻种植对土壤养分的影响

土壤养分是指由土壤提供的植物必需的营养元素,它是土壤肥力的重要物质基础,也是评价土壤肥力水平的重要内容之一[16]。水稻种植对0~100 cm土层土壤养分的影响的剖面分布情况分别如图3所示:各个处理的全氮、有效磷、速效钾和有机质的含量均呈现出随土层深度的增加而降低的变化规律,CK处理0~100 cm土层的土壤全氮、有效磷、速效钾和有机质的平均含量较低,分别为0.19 g/kg、7.29 mg/kg、86.8 mg/kg、4.13 g/kg,根据我国土壤养分分级标准,其全氮和有机质含量为极缺乏,有效磷和速效钾为缺乏。种植水稻后,由于施肥等原因,土壤中养分含量得到了很大提高,种植水稻后,0~100 cm土层的土壤全氮、有效磷、速效钾和有机质含量分别平均达到0.34 g/kg、10.32 mg/kg、100 mg/kg、5.28 g/kg。另外,由图3可以看出,不同水稻品种对于深层土壤养分含量影响不明显,但是对于表层土壤养分含量也存在一定影响,T3处理表层土壤全氮、有效磷含量最高,分别达到0.57 g/kg、13.96 mg/kg,T1处理应该有机质含量最高,分别较T1和T2处理提高6.9%、7.2%,T1、T2、T3处理速效钾含量基本一致,分别是134 mg/kg、136 mg/kg、136 mg/kg。

2.4 水稻品种对于水稻产量影响

产量是土壤理化性质和水热条件的最终体现,通过对盐碱地进行作物种植可以显著提高盐碱地表层土壤的理化性质,进而对作物产量产生积极影响。表3为水稻品种对水稻产量的影响,通过分析可知,不同水稻品种对单穗穗粒数影响不显著($P>0.05$),对每平方米穗数影响显著($P<0.05$),不同水

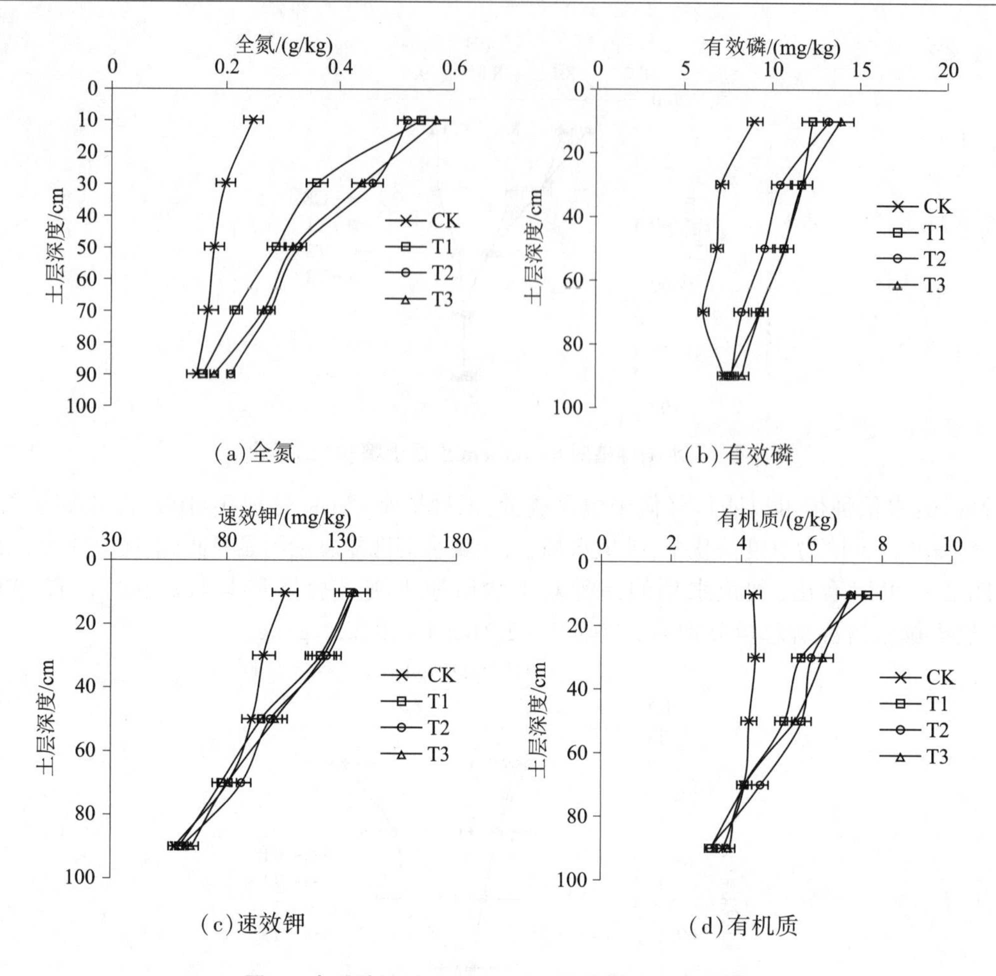

(a)全氮 (b)有效磷

(c)速效钾 (d)有机质

图 3 水稻种植对 0~100 cm 土层土壤养分的影响

稻品种对水稻百粒重和亩产量达到极显著水平($P<0.01$)。其中 T2 处理百粒重和亩产量均最大,分别达到 1.53 g 和 198.9 kg。T1 处理的单穗穗粒数和每平方米穗数均为最大,分别为 89 和 225,T3 处理的单穗穗粒数、每平方米穗数、百粒重以及亩产量均为最低。从产量角度来看,T2 处理分别较 T1、T3 处理提高了 4.96%、42.78%,是本试验条件下产量最高的处理。

表 3 水稻品种对水稻产量的影响

处理	单穗穗粒数	每平方米穗数	百粒重/g	亩产量/kg
T1	89±1.2 a	218±6.3 a	1.42±0.03 b	189.50±2.4 b
T2	87±2.1 b	214±5.8 b	1.53±0.02 a	198.90±1.2 a
T3	81±1.5 c	197±5.4 c	1.36±0.01 c	139.30±1.6 c
显著性 P 值	NS	*	* *	* *

3 讨论

在盐碱地内种植耐盐植物进行土壤改良,主要利用植物生长促进土壤积累有机质,改善土壤结构,降低地下水位,减少土壤中水分的蒸发,从而加速盐分淋洗、延缓或防止积盐返盐[17]。国内外学者对种植水稻改良盐碱地进行了大量研究和实践应用,特别是在新疆地区耐盐碱水稻广泛种植,并取得了很好的效果。本次试验结果表明种植水稻可以显著提高盐碱地表层(0~20 cm)土壤的理化性质,降低土壤酸碱度和含盐量,提高土壤养分含量,水稻种植后的表层土壤 pH 值和含盐量较整治前分别降低了 5.7% 和 25.5%,全氮、有效磷、速效钾和有机质分别提高了 117.3%、45.9%、27.7% 和 67.3%。程知言等[18]研究发现,耐盐水稻种植对滨海盐碱地改良具有良好作用,可有效控制耕作层

(0~20 cm)土壤盐度维持在较低的水平且保持稳定。另外,也有研究发现,种植水稻明显提高了稻田耕层 0~20 cm 土壤有机碳、有效磷、速效钾、壤全氮,不同种植方式下,土壤养分含量存在不同程度的差异[19]。这与本研究结果基本相同。

选育适合于陕北地区盐碱地种植的耐盐碱水稻品种,是开发利用陕北低洼盐碱地的有效途径之一[20]。不同水稻品种对于深层土壤养分含量影响不明显,但是对于表层土壤养分含量也存在一定影响,T3 处理表层土壤全氮、有效磷含量最高,分别达到 0.57 g/kg 、13.96 mg/kg,T1 处理应该有机质含量最高,分别较 T1 和 T2 处理提高 6.9%、7.2%,T1、T2、T3 处理速效钾含量基本一致,分别是 134 mg/kg、136 mg/kg、136 mg/kg。不同水稻品种对单穗穗粒数影响不显著($P>0.05$),对每平方米穗数影响显著($P<0.05$),不同水稻品种对水稻百粒重和亩产量达到极显著水平($P<0.01$),从产量角度来看,T2 处理分别较 T1、T3 处理提高了 5.0%、42.8%,是本试验条件下产量最高的处理。李全英等[21]认为盐碱地种植水稻并进行土壤改良的关键在于选育合适水稻品种,并针对内蒙古托克托县盐碱水土环境,开展了耐盐碱水稻品种筛选试验,试验结果表明,不同水稻品种的秧苗成活率存在显著差异,吉粳 113、吉粳 301、吉粳 515 的成活率降低到 50%以下,而"粳优 315""松粳 9""粳优 653"仍保持在 90%以上,表现出较强的耐盐性。李华等[22]研究发现,部分水稻品种虽有一定的耐盐性,但丰产性不佳,经济效益偏低,在盐碱地进行 6 个水稻品种的耐盐性适应性试验中,耐盐能力和适应性表现较强的是 9K-210 和 7K-339,表现较差的是南粳 9108;盐稻 12 号虽有一定的耐盐性,但丰产性不佳;宁粳 8 号和南粳 5055 可作为备选品种在盐碱地种植。本研究选取的适宜当地气候环境的 3 个稻品种,对于盐碱水土环境均有一定适应性,但是考虑经济效益及产量,建议选择 T2 处理"隆优 619"。

4　结论

水稻种植可以显著改善盐碱地表层(0~20 cm)土壤的理化性质,降低土壤酸碱度和含盐量,提高土壤养分含量,水稻种植后的表层土壤 pH 值和含盐量较整治前分别降低了 5.7%和 25.5%,全氮、有效磷、速效钾和有机质分别提高了 117.3%、45.9%、27.7%和 67.3%。另外,不同水稻品种对于盐碱地土壤酸碱度和含盐量基本无影响,但是对于表层土壤养分含量存在一定影响,水稻收获后,"东稻 4 号"土壤全氮、有效磷含量最高,"宁粳 28"有机质含量最高。从产量角度来看,种植"隆优 619"产量最高,分别较"宁粳 28""东稻 4 号"处理提高了 5.0%、42.8%,是本试验条件下产量最高的处理。

盐碱地的形成是一个长期而复杂的过程,盐碱地治理也是一个复杂的系统性工程[23]。通过低洼地区种植水稻进行盐碱地治理,对于地区盐碱地土壤理化性质改善存在一定局限性,建议耕作过程中采用综合措施进行改良,优先选用生理酸性有机肥和酸性调理剂进行改良,在补充土壤养分的同时中和土壤碱性。此外,尽可能选择盐碱含量小的灌溉水进行灌溉[24],避免由于灌溉导致盐碱累积,影响农作物生长。

参考文献

[1]李建国,濮励杰,朱明.土壤盐渍化研究现状及未来研究热点[J].地理学报,2012,67(9):1233-1245.

[2]刘长江, 李取生, 李秀军. 不同耕作方法对松嫩平原苏打盐碱化旱田改良利用效果试验[J]. 干旱地区农业研究, 2005, 23(5): 13-16.

[3]STAMFORD N P, SILVA A J N, FREITAS A D S, et al. Effect of sulphur inoculated with Thiobacillus on soil salinity and growth of tropical tree legumes[J]. Bioresource technology, 2002, 81(1): 53-59.

[4]李旭霖, 刘庆花, 柳新伟, 等. 不同改良剂对滨海盐碱地的改良效果[J]. 水土保持通报, 2015, 35(2): 219-224.

[5]侣小伟, 解建仓, 黄茹. 陕西卤泊滩盐碱地综合治理措施及效益分析[J]. 水土保持通报, 2009, 29(6): 177-181.

[6]BROWN T T, KOENIG R T, HUGGINS D R, et al. Lime effects on soil acidity, crop yield, and aluminum chemistry in direct-seeded cropping systems[J]. Soil Science Society of America Journal, 2008, 72(3): 634-640.

[7]秦都林，王双磊，刘艳慧. 滨海盐碱地棉花秸秆还田对土壤理化性质及棉花产量的影响[J]. 作物学报，2017，43(7)：1030-1042.

[8]李芙荣，杨劲松，吴亚坤，等. 不同秸秆埋深对苏北滩涂盐渍土水盐动态变化的影响[J]. 土壤，2013,45(6)：1101-1107.

[9]WANG R, SUN J, ZHAOHUA L U, et al. Effect of Soil Ameliorants on the Biochemical Properties of Coastal Saline-alkali Soil in the Yellow River Delta[J]. Acta Ecologica Sinica, 2017, 37(2): 425-431.

[10]HAJIBOLAND R, CHERAGHVAREH L, POSCHENRIEDER C. Improvement of Drought Tolerance in Tobacco (Nicotiana Rustica L.) Plants by Silicon[J]. Journal of Plant Nutrition, 2017, 40(12).:32-56.

[11]舒锟，曹源，王波，等.土壤调理剂对陕北盐碱地土体化学性质及水稻生长的影响[J].水土保持通报，2020，40(6)：175-180.

[12]王启龙.定边盐碱地治理改良技术研究与工程实践[J].南方农机，2019，50(21)：73，75.

[13]王海江，石建初，张花玲，等. 不同改良措施下新疆重度盐渍土壤盐分变化与脱盐效果[J].农业工程学报，2014，30(22)：102-111.

[14]倪海峰，朱尤东，刘树堂，等.保水剂及有机酸土壤调理剂对盐碱地的改良效果及小麦产量的影响[J].山东农业科学，2020，52(4)：121-125.

[15]王峰，姚宝林，孙三民，等. 矿化度对均质土壤毛管水上升特性与土壤盐分的影响[J]. 中国农村水利水电，2017(9)：27-31.

[16]郑普山，郝保平，冯悦晨，等. 不同盐碱地改良剂对土壤理化性质、紫花苜蓿生长及产量的影响[J]. 中国生态农业学报，2012，20(9)：1216-1221.

[17]李娟，韩霁昌，张扬，等. 盐碱地综合治理的工程模式[J]. 南水北调与水利科技，2016，14(3)：188-193.

[18]程知言，胡建，葛云，等.种植耐盐水稻盐碱地改良过程中的盐度变化趋势研究[J].矿产勘查，2020，11(12)：2592-2600.

[19]侯红燕，董晓亮，周红，等.滨海盐碱地不同氮肥用量对水稻干物质转运及稻米品质的影响[J].中国稻米，2021，27(1)：27-31.

[20]代金英，胡蕾，孙一标，等.播期对苏北盐碱地不同水稻品种生长发育进程及产量的影响[J].中国农学通报，2021，37(3)：1-6.

[21]李全英，李海波.内蒙古托克托县盐碱地适宜种植水稻品种筛选[J].中国稻米，2020，26(6)：109-111，113.

[22]李华，朱鹏飞，郁伟，等.几种不同水稻品种在盐碱地的适应性研究[J].中国稻米，2018，24(6)：110-111.

[23]李彦强，石称华，钱志红，等.不同土壤调理剂用量对油麦菜生长及土壤改良的效果试验[J].上海蔬菜，2020(2)：55-57，67.

[24]韩霁昌，解建仓，朱记伟，等.陕西卤泊滩盐碱地综合治理模式的研究[J].水利学报，2009，40(3)：372-377.

本文曾发表于2022年《中国稻米》第28卷第3期

陕西省靖边县农村土地整治项目与美丽乡村建设耦合关系研究

吴得峰

（陕西省土地工程建设集团有限责任公司西北分公司，陕西 榆林 719000）

【摘要】探索农村土地整治项目与美丽乡村建设耦合关系及变化机制，为美丽乡村建设背景下的土地整治项目实施方案及方向提供理论依据。以靖边县11个农村土地整治项目为例，首先建立起农村土地整治项目评价指标体系和美丽乡村建设评价体系，运用熵权法确定指标权重，用相对系数评价法对基础数据进行处理并计算出各自分值，然后引入sigmaplot10.0软件中的相关性分析曲线估计回归模型，探究农村土地整治项目与美丽乡村建设的耦合关系，最后通过相关系数R^2和耦合曲线走势分析土地整治项目与美丽乡村建设之间的相互关系。结果表明：(1)经济效益与村容村貌、生活水平、乡风文明和生产发展耦合关系大小顺序为：生活水平>生产发展>乡风文明>村容村貌。经济效益与村容村貌耦合关系不显著。(2)粮食产能与村容村貌、生活水平、乡风文明和生产发展耦合关系大小顺序为：生产发展>生活水平>乡风文明>村容村貌。粮食产能随着分值增加，村容村貌分值出现先增加后降低，最后增加的趋势，但波动范围较窄，介于50~70分。粮食产能与生活水平、乡风文明之间变化趋势均随着粮食产能分值不断增加，生活水平出现先降低后增加的趋势。(3)耕地等别与村容村貌、生活水平、乡风文明和生产发展耦合关系大小顺序为：生产发展>生活水平>乡风文明>村容村貌，耕地等别与村容村貌、乡风文明、生产发展之间均呈现先降低后增加，最后再降低的趋势，波动范围分别为50~70、20~80、30~80分。耕地等别与村容村貌耦合关系最低。通过耦合函数曲线和相关系数平方值关系分析可以得出，农村土地整治中，经济效益、粮食产能、耕地等别均与美丽乡村建设评价指标中生活水平、生产发展耦合关系好，与乡风文明关系次之，与村容村貌耦合关系有待进一步提高，为后续土地整治项目思路创新、制度创新提供新方案，最终实现两系统之间的互惠互利。

【关键词】土地整治；美丽乡村建设；耦合关系；靖边县；民众参与度

Coupling Relationships between Rural Land Consolidation Projects and Rural Beautification Construction at Jingbian County of Shaanxi Province

Defeng Wu

(Northwest Branch of Shaanxi Provincial Land Engineering Construction Group Co., Ltd., Yulin Shaanxi 719000, China)

【Abstract】To explore the coupling relationship and change mechanism between rural land consolidation projects and beautiful rural construction, and provide a theoretical basis for the implementation plan and direction of land consolidation projects in the context of beautiful rural construction. Taking 11 rural land consolidation projects

基金项目：陕西省土地工程建设集团有限责任公司内部科研项目“陕北风沙区土体有机重构技术措施及效应研究”(DJNY2021-4)。

in Jingbian County as an example, the evaluation index system of rural land consolidation projects and the evaluation system of beautiful rural construction were established. The entropy weight method was used to determine the index weight, and the relative coefficient evaluation method was used to process the basic data and calculate their respective scores. Then the correlation analysis curve in sigmaplot 10.0 software was introduced to estimate the regression model to explore the coupling relationship between rural land consolidation projects and beautiful rural construction. Finally, the correlation coefficient R^2 and the trend of the coupling curve were used to analyze the relationship between land consolidation projects and beautiful rural construction. The results showed that: (1) the order of coupling relationship between economic benefit and village appearance, living standard, rural civilization and production development was living standard > production development > rural civilization > village appearance. The coupling relationship between economic benefits and village appearance is not significant. (2) The order of coupling relationship between grain production capacity and village appearance, living standard, rural civilization and production development is production development > living standard > rural civilization > village appearance. Grain production capacity with the score increased, the village appearance score increased first and then decreased, and finally increased, but the fluctuation range is narrow, between 50~70. The change trend between grain production capacity and living standard and rural civilization is increasing with the increase of grain production capacity score, and the living standard decreases first and then increases. (3) The coupling relationship between cultivated land grade and village appearance, living standard, rural civilization and production development is in the order of production development > living standard > rural civilization > village appearance. The relationship between cultivated land grade and village appearance, rural civilization and production development shows a trend of first decreasing and then increasing, and finally decreasing. The fluctuation ranges are 50~70, 20~80 and 30~80 points, respectively. The coupling relationship between cultivated land grade and village appearance is the lowest. Through the analysis of the coupling function curve and the square value of the correlation coefficient, it can be concluded that in the rural land consolidation, the economic benefits, grain production capacity and cultivated land have a good coupling relationship with the living standard and production development in the evaluation index of beautiful rural construction, followed by the relationship with rural civilization, and the coupling relationship with the village appearance needs to be further improved. It provides a new scheme for the follow-up land consolidation project innovation and system innovation, and finally realizes the mutual benefit between the two systems.

【Keywords】 Rural land consolidation; Construction of Beautiful Village; Coupling Relationship; Jingbian County; Public participation

党在十八大报告中首次将生态文明建设提升到一个全新的高度,进行重点论述[1],同时首次将“美丽中国”作为未来生态文明建设的宏伟目标,对于推进我国现代化建设、实现中华民族伟大复兴意义重大[2-3]。美丽乡村建设主要措施包括完善基础设施建设,加大农村地区经济投入,促进农村发展、农业增效、提高农民收入,构建人与自然、物质与文化、生产与生活、传统与现代的融合体[4],使美丽乡村成为一种生产力,达到让广大农村地区群众有较高的幸福感和满意度的目的[5]。

农村土地整治,作为美丽乡村建设重要抓手与措施,日益受到当地政府的普遍关注[6-7]。土地整治工程主要通过增加耕地面积,完善电力水力等配套设施,增设防护林等对田水路林村进行综合整治,以达到提升耕地地力,改善生态环境的目的[8]。由此可见,美丽乡村建设与农村土地整治之间立足点、出发点相同,发展目标一致,存在一定的耦合关系[9-10]。但是,农村土地整治工程过分强调增加耕地面积,忽视了土地整治在美丽乡村建设过程中的应用,进而限制了土地整治与美丽乡村建设协同健康发展。因此,农村土地整治与美丽乡村建设耦合关系研究成为目前专家学者研究的热点与难点。张如林等[11]先后开展了大量的科学研究,也取得了丰硕的研究成果,但美丽乡村建设方面的研究主要集中在内涵、模式、路径、绩效等方面[12-13],土地整治主要是关于整治前后土壤养分变化、作物产量变化[14-15]以及测试方法的改进等方面[16],关于美丽乡村与土地综合整治协同互促关系的探讨,实现二者协调发展的相关机制、政策等方面研究鲜见报道。因此,在美丽乡村建设走向纵深发展的背景下,加强农村土地整治与美丽乡村建设的耦合关系研究,厘清两者之间的相互关系,并进行定量分析评价十分必要。

本文基于榆林市靖边县实施的农村土地整治项目和美丽乡村建设活动，通过分别建立农村土地整治与美丽乡村建设指标评价体系，分析两者之间的关系，厘清土地整治与美丽乡村建设的耦合关系，为优化基于美丽乡村建设背景下，农村土地整治模式与创新提供理论依据。

1 研究区域与方法

1.1 研究区概况

靖边地处东经 108°17′15″~109°20′15″，北纬 36°58′45″~38°03′15″，东西宽 91.3 km，南北长 116.2 km。全县海拔介于 1123~1823 m，地势南高北低，地貌类型属于北部风沙滩区，属半干旱大陆性季风气候，光照充足，温差大，气候干燥，通风条件好，雨热同季，四季明显。年平均降雨量 395.4 mm（348.3~431.3 mm），平均日照时数为 2768.2 h（2516.1~3037.7 h）。日照百分率年平均为 62%。多年平均太阳总辐射量为 137.19 kcal/cm^2，年平均气温 7.8 ℃，≥10 ℃的植物生长有效积温为 2800 ℃（2358.0~3356.2 ℃），年平均无霜期为 130 d（115~145 d），年冻土深度 106 cm，年降水量 395.4 mm。

靖边县通过采用大力实施“环境整治、民居改造、基础设施配套、公共服务提升、生态环境建设、产业升级”六大工程，着力构建布局合理、功能完善、质量提升的美丽乡村发展体系，美丽乡村建设各项工作有序推进，取得阶段性显著成效，靖边县农村土地整治项目规模面积 3666.4 hm^2，新增2736.4 hm^2。通过土地平整工程、灌溉排水工程、田间道路工程、农田防护与生态环境保持工程等四大工程，有效增加耕地面积，改善土地条件，对促进该县农业健康可持续发展创造有利条件。此外，该项目的实施能显著增加农民收入，社会、生态与经济效益显著，为美丽乡村建设作用突出。

1.2 研究思路

首先建立农村土地整治项目和美丽乡村建设指标评价体系，其次运用熵权法确定指标权重，用相对系数评价法对基础数据计算出各自分值，然后研究土地整治项目与美丽乡村建设相关性分析，选择一个相关性比较高的曲线配型，求值得出 R^2 数值[17]。最后通过相关系数 R^2 和耦合曲线走势分析土地整治项目与美丽乡村建设之间的相互关系。

1.3 评价指标体系的构建

1.3.1 农村土地整治项目指标体系构建

农村土地整治以增加有效耕地面积、提高耕地质量为目的，通过对项目区的田、水、路、林、村进行综合治理，配套农业基础设施建设，从而改善农业生产基本条件，提高土地利用率和产出率。本项目新增耕地来源主要对现有耕地进行提升改造，将其改造为稳产、高产的高标准基本农田，从而增加耕地数量，增加农民收入。土地整治项目实施内容由原来单一的仅仅增加耕地面积，变成通过田水路林等综合手段进行土地整治，新增耕地指标分析包括新增耕地来源分析、新增耕地潜力分析、新增耕地适宜性分析，水土资源平衡分析包括可供水量分析、需水量分析、水资源平衡分析及计算等。效益分析包含社会效益分析、生态效益分析和经济效益分析、耕地质量评价等[9,18]。本文根据项目实施实际情况，选取了项目经济效益、粮食产能、耕地等别等 3 个指标综合衡量农村土地整治项目。

1.3.2 美丽乡村建设评价体系构建

以美丽乡村建设为目标，通过设立村容村貌、生活水平、乡风文明、生产发展等 4 个方面进行综合衡量，参考相关文献中关于美丽乡村建设评价指标体系构建的相关内容，结合靖边县美丽乡村建设体系自身特点，选取指标层因子，运用熵权法先将各个指标的数据进行标准化处理，再根据信息熵的计算公式，计算出各个指标的信息熵，确定各指标相对权重，最后通过信息熵计算各指标的权重，进而构建美丽乡村指标体系[19]，见表 1。

表 1　美丽乡村建设评价指标体系

Table 1　Evaluation index system of beautiful rural construction

评价指标	评价内容	指标相对权重
村容村貌	道路清洁宽敞(+)	0.52
	村民穿戴得体精气神佳(+)	0.11
	林带分布合理(+)	0.37
生活水平	农民人均年收入(+)	0.49
	单位面积粮食产量(+)	0.47
	生态环境知识普及率(+)	0.04
乡风文明	文化娱乐消费比例(+)	0.31
	文体广场占地面积及参与度(+)	0.44
	环保意识及知识普及度(+)	0.25
生产发展	机械化耕作面积占比(+)	0.51
	种植用地土壤质量高低(+)	0.17
	水域覆盖面积及设施关乎率(+)	0.32

注:+、-分别表示指标的正负,正指标表示指标值越大越好,负指标表示指标值越小越好。

1.4　数据来源与处理

农村土地整治项目方面的数据主要来源于2018—2020年靖边县第一、二、三批次土地整治项目实施后实地观测及调查数据。研究区概况、各乡镇美丽乡村建设各项指标的参考数据主要来源于各个乡镇有关统计资料以及现场问卷调查。调查问卷主要针对同时设计土地整治项目和美丽乡村建设项目的村组,此次问卷调查活动2020年10月11日开始,10月14日结束,为期4 d。问卷共200份,回收有效问卷185份,回收率为92.5%,本次调研数据有效,同时对受访农户的基本信息也相应做了统计,包括性别、年龄、文化知识水平和所从事职业类型等,男女比例为52:48,平均年龄为43岁。为了能够测算出土地整治项目与美丽乡村建设之间的耦合关系,本文在参考已有研究的基础上,构建农村土地整治以及美丽乡村建设综合评价指数,其计算函数为[9]:

$$U = \sum U_{ij} \cdot b_{ij}$$

式中:U代表综合得分;U_{ij}代表第i个地区的第j个评价指标的权重;b_{ij}代表第i个地区的第j个评价指标的原始数值经过标准化处理后的数据。当指标是正指标时,其处理过程为:

$$b_{ij} = \frac{B_{ij} - \min B_j}{\max B_j - \min B_j} \times 100$$

当指标是负指标时,其处理过程为:

$$b_{ij} = \frac{\max B_{ij} - B_{ij}}{\max B_j - \min B_j} \times 100$$

式中:b_{ij}代表第i个地区的第j个评价指标的标准化评价分值,$0<b_{ij}\leqslant 1$;B_{ij}代表第i个地区的第j个评价指标的实际值;$\min B_j$代表第j个评价指标的最小数据值;$\max B_j$代表第j个评价指标的最大数据值。

根据以上构建的土地整治项目和美丽乡村建设评价体系、权重的确定及对基础数据的标准化处理,对靖边县龙洲镇、小河镇、海则滩镇、马连坑镇、掌高兔村、杨二村等11个村镇进行调查,农村土地整治项目与美丽乡村建设耦合关系评价结果如表2所示。

表 2 农村土地整治项目与美丽乡村建设指标评价分值

Table 2 Rural land improvement projects and beautiful rural construction index evaluation scores

序号	项目名称	农村土地整治项目指标评价分值			美丽乡村建设指标评价分值			
		经济效益	粮食产能	耕地等别	村容村貌	生活水平	乡风文明	生产发展
1	A 村土地整治项目	12.79	44.77	11.73	60.50	24.71	16.29	32.05
2	B 村土地整治项目	78.60	41.69	49.90	86.02	72.24	93.45	48.59
3	C 村土地整治项目	55.81	33.89	60.53	64.18	54.62	44.82	69.70
4	D 村土地整治项目	84.48	37.61	69.71	38.82	84.96	95.55	51.24
5	E 村土地整治项目	62.01	26.55	14.30	57.36	88.46	76.24	21.90
6	F 村土地整治项目	39.79	29.86	21.12	70.12	34.89	6.73	68.48
7	G 村土地整治项目	30.10	7.56	3.54	19.07	48.36	35.56	30.52
8	H 村土地整治项目	37.57	40.95	32.25	87.08	31.25	47.07	57.33
9	I 村土地整治项目	49.09	82.03	42.96	88.15	56.97	54.94	73.07
10	J 村土地整治项目	22.79	65.87	18.26	55.66	34.38	8.55	90.83
11	K 村土地整治项目	20.94	49.54	76.70	91.39	14.66	8.14	78.53

2 结果分析

2.1 经济效益与美丽乡村建设各指标间耦合关系分析

通过将数据导入 sigmaplot10.0,进行回归分析构建回归模型并从中找出农村土地整治项目与美丽乡村建设的最佳拟合曲线函数图(见图 1~图 3)[20],同时用模型中的三次函数 R^2 来判断二者之间耦合关系的高低,R^2 越大,说明耦合度越高,同时还可以根据曲线走势与散点的分布情况观测自变量(农村土地整治各项指标)与应变量(美丽乡村建设各项指标)在各分值区间的关系。

图 1 显示,农村土地整治项目经济效益与村容村貌、生活水平、乡风文明和生产发展耦合关系大小顺序为:生活水平($R^2=0.9466$)>生产发展($R^2=0.9463$)>乡风文明($R^2=0.7843$)>村容村貌($R^2=0.0989$)。经济效益与村容村貌耦合关系不显著,经济效益分值在 0~50 时,村容村貌几乎没有变化,但是当分值大于 50 时,村容村貌就会呈现先增加后降低的趋势,经济效益与乡风文明耦合关系较好,但是不显著,当经济效益分值在 0~20 时,乡风文明分值呈现下降趋势,这主要是因为土地整治项目实施给当地财政带来压力,使收入降低,但是当分值大于 20 时,乡风文明成上升趋势。经济效益与生活水平的耦合关系总体呈降低—增加—降低的趋势,当经济效益分值在 0~30 时,生活水平呈下降趋势,但当经济效益分值介于 30~70 时,生活水平呈增加趋势,但当分值大于 70 时,又出现下降趋势。经济效益与生产发展的关系为先降低后增加的趋势,当经济效益分值为 0~30 时,生产发展呈降低趋势,但是当分值大于 30 时,生产发展呈增加趋势。

2.2 粮食产能与美丽乡村建设各指标间耦合关系分析

图 2 显示,土地整治项目粮食产能与村容村貌、生活水平、乡风文明和生产发展耦合关系大小顺序为:生产发展($R^2=0.9546$)>生活水平($R^2=0.8986$)>乡风文明($R^2=0.7474$)>村容村貌($R^2=0.0504$)。粮食产能随着分值增加,村容村貌分值出现先增加后降低,最后增加的趋势,但波动范围较窄,介于 50~70。粮食产能与生活水平、乡风文明之间变化趋势均随着粮食产能分值不断增加,生活水平出现先降低后增加的趋势,当粮食产能分值为 80 时,出现降低趋势。粮食产能分值在 0~40 时,生产发展出现下降趋势,当分值大于 40 时,生产发展出现增加趋势。

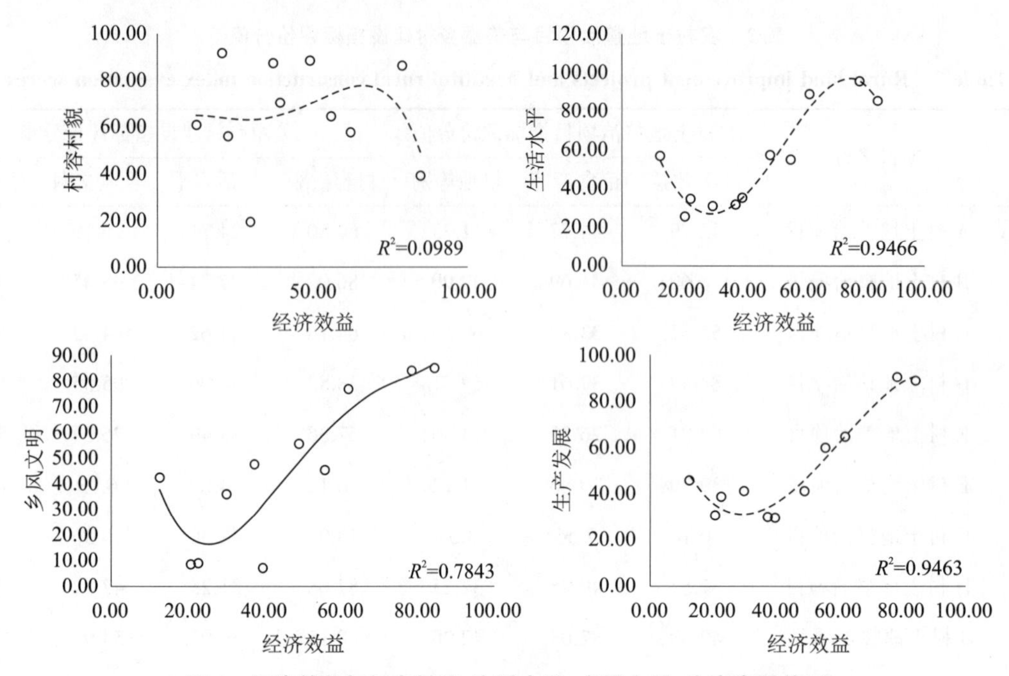

图 1　经济效益与村容村貌、生活水平、乡风文明、生产发展关系

Fig 1　The relationship between economic benefit and village appearance, living standard, rural civilization and production development

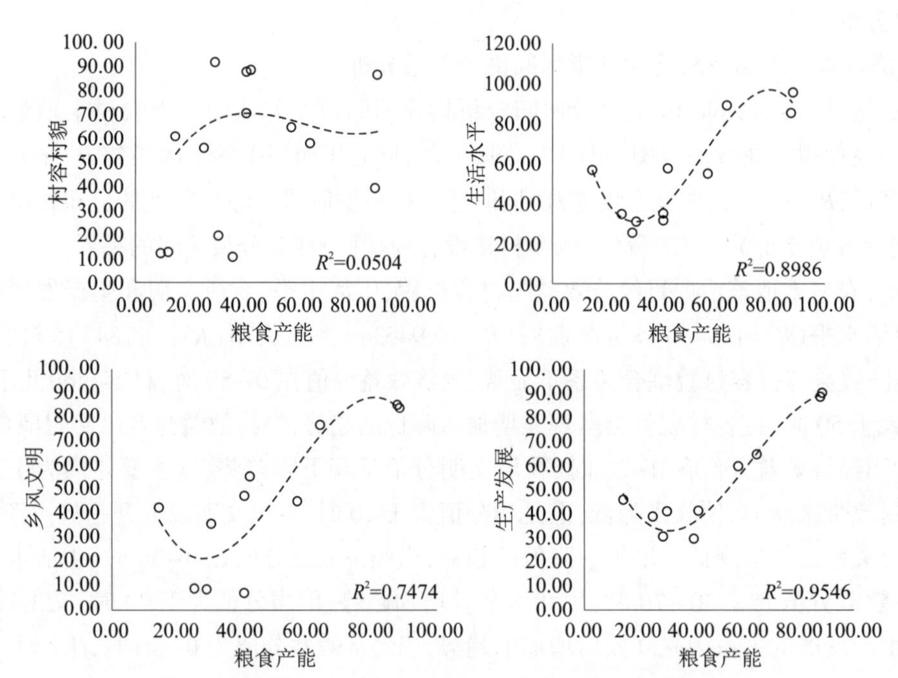

图 2　粮食产能与村容村貌、生活水平、乡风文明、生产发展关系

Fig 2　The relationship between grain production capacity and village appearance, living standard, rural civilization and production development

2.3　耕地等别与美丽乡村建设各指标间耦合关系分析

图 3 显示,土地整治项目耕地等别与村容村貌、生活水平、乡风文明和生产发展耦合关系大小顺序为:生产发展($R^2=0.9385$)>生活水平($R^2=0.7326$)>乡风文明($R^2=0.5855$)>村容村貌($R^2=0.0723$)。耕地等别与生活水平、乡风文明、生产发展之间均呈现先降低后增加,最后再降低的趋势,波动范围分别为 50~70、20~80、30~80。耕地等别与村容村貌耦合关系最低。

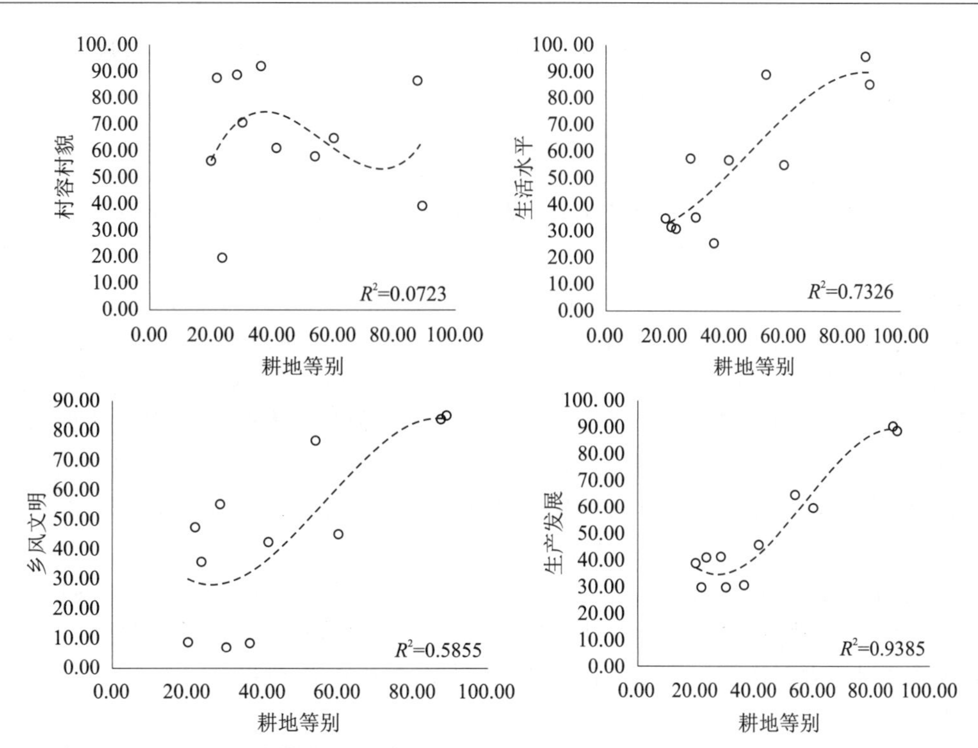

图 3 耕地等别与村容村貌、生活水平、乡风文明、生产发展关系

Fig 3 The relationship between cultivated land classification and village appearance, living standard, rural civilization and production development

综上所述，通过耦合函数曲线和相关系数平方值关系分析可以得出，农村土地整治中，经济效益、粮食产能、耕地等别均与美丽乡村建设评价指标中生活水平、生产发展耦合关系好，与乡风文明关系次之，与村容村貌耦合关系有待进一步提高。相关系数平方值如表 3 所示。

表 3 各模型相关系数平方值(R^2)

Table 3 The square value of the correlation coefficient of each model(R^2)

美丽乡村建设	生活水平	生产发展	乡风文明	村容村貌
经济效益	0.9466	0.9463	0.7843	0.0989
粮食产能	0.8986	0.9546	0.7474	0.0504
耕地等别	0.7326	0.9385	0.5855	0.0723

3 讨论与结论

3.1 讨论

(1)农村土地整治项目主要是采取土地平整工程、灌溉与排水工程、田间道路工程及农田防护与生态环境保持工程等四大工程，将生产条件差、产量低的地块进行集中整治，将其恢复至可耕种、高产稳产的状态，它与优化乡村环境的耦合关系并没有明显体现。研究结果表明，土地整治的经济效益与生活水平、生产发展耦合关系好，主要是因为土地整治项目实施以后，土地田块进行重新整合、平整，土壤质量得到显著提升，农民收益逐年增大，生活水平相应地就会提高。此外，农民增收会促进当地农业机械化发展进程，生产发展水平速度就会加快[21]。

(2)土地整治主要通过整合归并田块，进而提高田块的平整度，有利于农业机械化操作。其农田灌溉与排水工程、田间道路工程主要是通过改善农田水利、田间道路设施配套，进而创造出便利的适合于农业生产的便利条件，最终促进了农业机械化作业和规模化经营。农村土地整治项目的效果越好，农业机械化水平越高，生产发展的水平也越高。粮食产能与生活水平、乡风文明耦合关系次之，不仅促进了农业现代化的发展，增强了土壤的肥力，提高了粮食产能，也将会提高农业劳动生产率，节约生产成本,在新增耕地的同时也保证了其质量,提高了土地的产出率,增加了农民的收入,带动了当地的经济发展,物质水平的提高也会带动文化娱乐活动的发展,更有利于形成一个文明和谐的乡村环境。

(3)经济效益、粮食产量、耕地等别均与村容村貌耦合关系比较低,主要原因是土地整治是指采取一定的工程和生物措施,将被破坏和退化土地恢复到可利用的状态,它与优化村庄环境、建设美丽村庄理论上的耦合关系并没有明显体现。首先是因为土地整治资金占比很小,没有进行针对性的规划。土地整治的目标是形成耕地指标,进行耕地占补平衡指标交易,进而获得回款,产生经济效益,而美丽乡村建设,会无形中增加投资。目前土地整治主要还是看重“耕地数量的增加”,不重视对村庄周边林木等资源的保护。其次是缺乏整治监测和后续的管护措施,严重影响了土地整治对村庄周边生态环境及村容村貌的保护作用。加之当地村民对于美丽乡村建设重视程度不够、认识不足,故在村容村貌方面投资及保护力度很小,故经济效益与村容村貌耦合关系比较低。目前土地整治工作更加偏向于耕地数量的增加,对于耕地质量的提升不太理想,导致新增耕地的质量等级不高,而且缺乏相应的耕地基础配套设施,不能改善农民的生产条件。

目前各项土地整治的重点还是停留在农田土地整治上,以提高经济效益为目标,对项目建设环境尤其是村容村貌环境重视程度不够。在我国农村土地整治的实践中,伴随着大规模的农地整治活动的实施,对地表植被以及村庄整体景观生态建设方面考虑不充分,有待提升。

3.2 结论

(1)通过对农村土地整治项目和美丽乡村建设耦合函数曲线分析可以得到农村土地整治项目经济效益、粮食产能以及耕地等别均与生活水平、生产发展耦合关系好,与乡风文明耦合关系次之,与村容村貌关系最低。

(2)土地整治可以推动农业现代化、机械化发展,提高农业生产水平及效率,进而增加农民收入,带动农村经济发展。经济效益与村容村貌耦合关系比较低,在整治过程中,过分注重眼前经济效益,生态环境及环保意识比较低,在今后的土地整治过程中在注重经济效益的同时,应该加强社会效益、生态效益的提升。

参考文献

[1] 江东,林刚,付晶莹.“三生空间”统筹的科学基础与优化途径探析[J].自然资源学报,2021,36(5):1085-1101.

[2] 盖美,王秀琪.美丽中国建设时空演变及耦合研究[J].生态学报,2021,41(8):2931-2943.

[3] 邓倩.人与自然和谐共生:疫情防控实践的价值支撑[J].中学政治教学参考,2021(7):47-50.

[4] 燕连福,赵建斌,毛丽霞.习近平生态文明思想的核心内涵、建设指向和实现路径[J].西北农林科技大学学报(社会科学版),2021,21(1):1-9.

[5] 张浩泽.“土地整治+”与美丽乡村建设的契合性:以上海市“土地整治+”实践经验为依据[J].区域治理,2019(41):164-167.

[6] 李全宝.园地和残次林地开发关键问题研究:以江苏省新沂市为例[J].中国国土资源经济,2019,32(6):53-58.

[7] ALTES W K K, SANG BONG I M . Promoting Rural Development through the Use of Land Consolidation: The Case of Korea[J]. International Planning Studies, 2011, 16(2):151-167.

[8] 吴得峰.美丽乡村建设背景下残次林地土地开发项目经济效益分析:以靖边县掌高兔村项目为例[J].农业与技术,2021,41(1):156-159.

[9] 孙心如,周学武,王占岐.农村土地整治与生态文明建设耦合关系研究[J].水土保持研究,2017,24(2):267-271.

[10] 张晓燕.土地整治如何融入美丽乡村建设[J].中国土地,2015(7):21-23.

[11] 张如林,余建忠,蔡健,等.都市近郊区乡村振兴规划探索:全域土地综合整治背景下桐庐乡村振兴规划实践[J].城市规划,2020,44(S1):57-66.

[12] 李雪芬.美丽乡村建设背景下侗戏的传承与发展路径探究[J].四川戏剧,2021(4):128-130.

[13] 庞琳,张亦佳.基于美丽乡村背景的历史文化村镇“原乡”发展模式探讨:以河北省地区为例[J].农业经济,2021(3):54-55.

[14] 廖远立,钟传勇,宋育忠,等.五华县宜林荒山及残次林地植被与土壤调查[J].绿色科技,2016(11):115-116.

[15] Tang, et al. Analysis of cultivated land consolidation potential in China[J]. Transactions of the Chinese Society of Agricultural Engineering, 2012.

[16] 张会丽.贫瘠土地变良田[J].资源导刊,2019(7):38.

[17] 信桂新,杨朝现,杨庆媛,等.用熵权法和改进 TOPSIS 模型评价高标准基本农田建设后效应[J].农业工程学报,2017,33(1):238-249.

[18] 刘继志.天津市美丽乡村建设模式及效益评价体系构建[J].中国农业资源与区划,2019,40(10):256-261.

[19] 李炎,张金池,陈佩弦,等.农村土地整治与美丽乡村建设的耦合关系:以南京市 11 个项目区为例[J].水土保持通报,2019,39(2):317-324.

[20] John, et al. The environmental Kuznets curve: does one size fit all? [J] Ecological Economics, 1999.

[21] Ruether R, Ecofeminism R. Symbolic and Social Connections of the Oppression of Women and the Domination of Nature[J]. Feminist Theology, 1993, 3(9): 35-50.

本文曾发表于 2022 年《水土保持通报》第 42 卷第 2 期

Effects of Alternating Drying and Wetting on Soil Nitrogen and Phosphorus Loss in Root Zone of Dominant Plants in Phosphorus-Rich Mountainous Areas

Tao Wang, Runqing Tian, Dan Luo, Fangfang Liu and Jialei Wei

(Shaanxi Land Survey and Planning Institute, Shaanxi Provincial Land Engineering Construction Group Co., Ltd, Xi'an 710075, China)

【Abstract】 In the mountainous area of Yunnan-Guizhou Plateau, phosphorus and nitrogen are important factors that limit primary productivity and affect the distribution of plant communities. Alternation of wet and dry has a significant effect on the migration of various elements such as C, N, and P in soil. Taking the rhizosphere soil of the dominant plants in the phosphorus-rich mountainous area in the Dianchi Lake Basin as the research object. Six sets of different humidity treatment methods: blank constant humidity, blank dry and wet alternation, nitrogen constant humidity, nitrogen addition dry and wet alternation, carbon constant humidity, and carbon add dry and wet alternation were assessed. Responses of soluble total nitrogen and soluble total phosphorus to six treatments in plant root zone were investigated. The results showed that the contents of dissolved total nitrogen (DTN) and dissolved total phosphorus (DTP) in soil leachate and their changes of *Erianthus rufipilus* and *Eupatorium adenophorum* under different treatment conditions increased first and then decreased. In the carbon addition group, the DTN content of the soil leachate in the alternating dry and wet treatment was significantly higher than that in the constant humidity treatment, but there was no significant difference in the nitrogen addition group. The three factors of water treatment and sampling time have significant effects on the DTN content of soil leachate, and the interaction between the three also has a significant effect on it.

【Keywords】 Alternating wet and dry; Soil nutrient; DTN; DTP

Introduction

The Dianchi Lake Basin is one of the important phosphate rock industrial bases and agricultural areas in China. Phosphorus-rich areas account for about ten percent of the Dianchi Lake basin area (Zhang *et al.* 2007, Xue 2008). On the basis of this rich phosphorus, agricultural production still needs to apply a large amount of phosphorus fertilizer to the land for the growth of crops. The excess phosphorus that is not used by plants will enter surface water and groundwater through surface runoff, intrusion and leaching (leakage or underground runoff (Heskethn 2000). Coupled with the mining of phosphate rock in this area, this area has become the main source of non-point source pollution output in the Dianchi Lake Basin, resulting in eutrophication of the Dianchi Lake water body.

There is also a phosphorus-rich large area in the small watershed of Chaihe River in the south of Dianchi Lake Basin. However, due to the influence of human production and life, a large amount of phosphorus is exported from the terrestrial ecosystem in the phosphorus-rich areas in the soil of Dianchi Lake Basin, which poses a serious threat to the water security of Dianchi Lake. Restore the ecological environment of the mining area by establishing dominant plant communities and introducing species supplement (Hupfer *et al.* 2008, Thanh *et al.* 2005, Zhang *et al.* 2012). Forming a sustainable ecosystem requires not only achieving vegetation cover, but also restoring soil quality and function (Schnbrunneri *et al.* 2012). Climate change in the world is an indisputable fact in the scientific community. More and more regions experience long-term droughts after heavy rains, or areas with high precipitation are getting more and more arid, and they have to irrigate to maintain agricultural production. So the soil is more likely to experience multiple wet and dry alternations. Wet and dry alternation is one of the most extensive forms of stress experienced by soils (Gerstengarbe *et al.* 2008, Katterer *et al.* 1998, Kruse *et al.* 2004, Torbert *et al.* 1992, Schjonning *et al.* 1999), and it is a common phenomenon in nature. Low and random rainfall, warm and dry climate, etc. can cause dry and wet alternation, which will cause the soil to experience dry and wet alternation many times. Under the conditions of alternating dry and wet, the soil has undergone physical, chemical, and biological changes (Gordillo *et al.* 1997, Li *et al.* 2022, Hao *et al.* 2022) which are mainly manifested in the formation of soil structure, the decomposition and mineralization of organic matter, and the number and structure of soil (Tran *et al.* 2022). The content of mineral elements in soil varies from place to place. In the mountainous region of the Yunnan-Guizhou Plateau, phosphorus and nitrogen are considered to be important factors that limit primary productivity and affect the distribution of plant communities (Ullah *et al.* 2022, Ananna *et al.* 2022, Weber *et al.* 2021, Zhang *et al.* 2021). In recent years, by studying the effect of soil phosphorus migration and transformation under the condition of dry-wet alternation, the migration and transformation process of soil phosphorus under the condition of dry-wet alternation has been explored, and the biochemical cycle process of soil phosphorus has been recognized in the process of dry-wet alternation (Arruda *et al.* 2021, James *et al.* 2021, Albadri *et al.* 2021). The change of telecommunication has become a hot issue studied by many scholars at home and abroad (Thyagaraj *et al.* 2021). As core indicators of soil fertility, soil organic matter and soil nitrogen respond to major changes in soil. In recent years, with the further development of microbial denitrification theory and organic matter mineralization theory, many scholars have turned their attention to the impact of climate change on soil nitrogen and carbon cycle transformation (Dash *et al.* 2021, Watanabe *et al.* 2021, Zhou *et al.* 2020, Malumpong *et al.* 2021, Jin *et al.* 2021), making the dynamic change of soil organic carbon. Great progress has been made in research and the theory of soil nitrogen transformation. These research results have a very profound effect on reducing soil erosion, maintaining soil nutrients, reducing the impact of global climate change, rational fertilization management of farmland, and sustainable development of agriculture.

Materials and methods

The study area is located in Jinning County, Kunming City. The geographical position is 24°36′~37′N, 102°41′~42′E, and the altitude is 1936~2256 m. The landform is mountainous and semi-mountainous, and it is located from southeast to northwest. Chaihe is the small watershed of Dianchi Lake. This area belongs to the subtropical humid monsoon climate. The dry and wet seasons are distinct. The average annual temperature is 14.6 ℃ and the average annual rainfall is 925.4 mm. The types of soil are mostly mountain yellow-red soil and brown-red soil. The moisture content of the soil is 14.29%, the pH is 6.23, the total nitrogen of the soil is 0.78 mg/g, and the total phosphorus of the soil is 6.57 mg/g. The total potassium of the soil is 3.36 mg/g, and the organic matter is 3.21%. Details are shown in Table 1.

There are many early phosphate mining in this area, most of which are located in the 1/3 to 2/3 mountainside area. The soil in the area is very rich in phosphorus. With the early phosphate mining and other human production activities, the vegetation in the area has been severely destroyed, and a large amount of phosphorus has entered the lake with severe soil erosion. The plants selected in this experiment are the two most common dominant plants in the phosphorus-rich mountain areas of the Dianchi Lake Basin, *Erianthus rufipilus* and *Eupatorium adenophorum*. Soil sample collection: during the collection, three plants of uniform size are randomly selected for each plant, and the stripping method is used for collection. First, the dead branches and deciduous layers are removed to expose the base of the trunk of the dominant plants. Then, the soil in the area required for the experiment is taken with a soil knife. Then, the part of fibrous roots along the lateral roots is found, and the branch part is cut. Carefully take out the soil with fibrous roots and divide the samples. The soil falling after gently shaking is root zone soil (non rhizosphere soil). However, the soil still sticking to the root surface is rhizosphere soil, which is put into a self sealing bag together with the root and peeled off immediately after being taken back. The soil that sticks tightly can be gently knocked or carefully peeled off with a blade and taken back to the laboratory for enzyme activity analysis.

Table 1 Properties of soil and water

Indicator	soil	water
pH	6.23±1.21	7.21±1.34
Bulk density (g/cm)	96.41±3.27	—
Conductivity (mS/cm)	100.12±12.4	0.62±0.13
Organic matter content (%)	3.21	0.17
Soil moisture content (%)	14.29	—
Total potassiumcontent (mg/g) /(mg/L)	6.57±1.25	0.02
Total nitrogencontent (mg/g)/(mg/L)	0.78±0.12	0.24±0.02
Total potassiumcontent (mg/g)/(mg/L)	3.36±1.31	0.35±0.11
δ13c (‰)	−26.12	—
Sal (‰)	—	0.31±0.12

In this study, we used the alternate dry and wet condition to simulate the cultivation methods of the dry and rainy seasons in nature. The sieved soil was kept into a PVC tube with a diameter of 5 cm in the culture device, deionized water was added and covered with aluminum foil, and the silica gel was allowed to dry out at certain intervals. Six experimental groups were set up with different water treatment methods, namely blank with constant humidity (KA), blank with dry and wet alternate (KB), nitrogen added with constant humidity (NA), nitrogen added with alternate dry and wet (NB), carbon with constant humidity (CA), carbon with dry and wet alternately (CB). The six sets of experimental devices were all cultured in an artificial climate chamber (BIC-300 type) with constant temperature and humidity. The temperature was set to 25 ℃, the humidity was 50%, and the light time was 6:00 to 18:00. At the same time, the different experimental groups exchange positions were noted every day to ensure that the error caused by the artificial climate chamber interferes with the experimental data. The cultivation time was 60 days, which were divided into six batches of 2, 16, 29, 31, 46 and 60 days.

Using alkaline potassium persulfate digestion ultraviolet spectrophotometry, according to the national standard (GB 11894-89), the total nitrogen for water quality analysis was determined. Using ammonium molybdate spectrophotometric method, according to the national standard (GB 11893-89), the total phosphorus was determined. R software was used to carry out multivariate analysis of variance. OriginPro2021 software was used for drawing.

Results and discussion

The content and change of total soluble nitrogen in the soil leachate of sugarcane under different treatment conditions are shown in Fig. 1(a). It can be seen that the content change trend before 30 days roughly increased first and then decreased, and appeared after 30 days. The trend was basically to increase first and then decrease, and the changes in the carbon addition group were different.

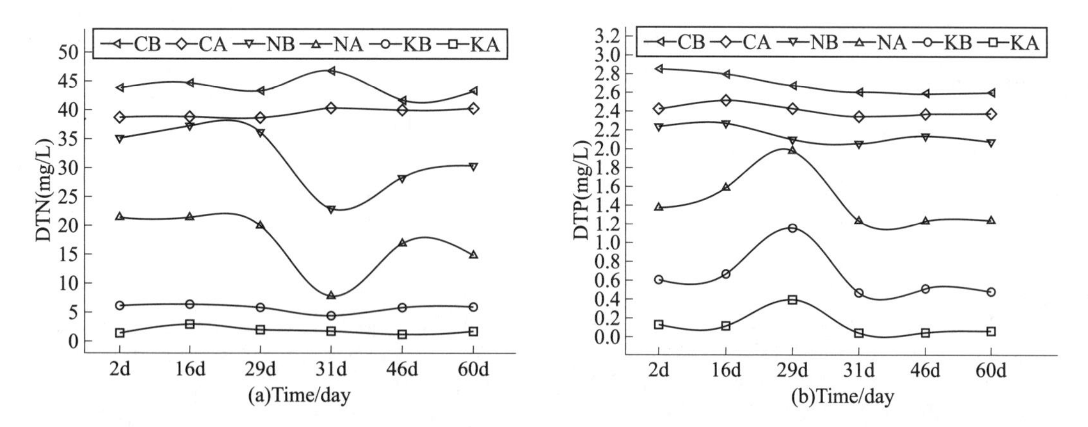

Fig. 1 The content and changes of DTN and DTP in soil leachate of *Erianthus rufipilus* under different treatment conditions

The content and change of soluble total phosphorus in the soil leachate of sugarcane under different treatment conditions are shown in Fig. 1(b). It can be seen that the content change showed an opposite trend, roughly increased first and then decrease before 30 days, and then regain the initial values after 30 days. The trend is also to increase first and then decrease.

The content and change of soluble total nitrogen in soil leachate of *B. adenophora* under different treatment conditions are shown in Fig. 2(a). There were a fluctuating tendency, where the general direction of the content change before 30 days was to increase first and then decrease, after 30 days. The changes of nitrogen addition group were different.

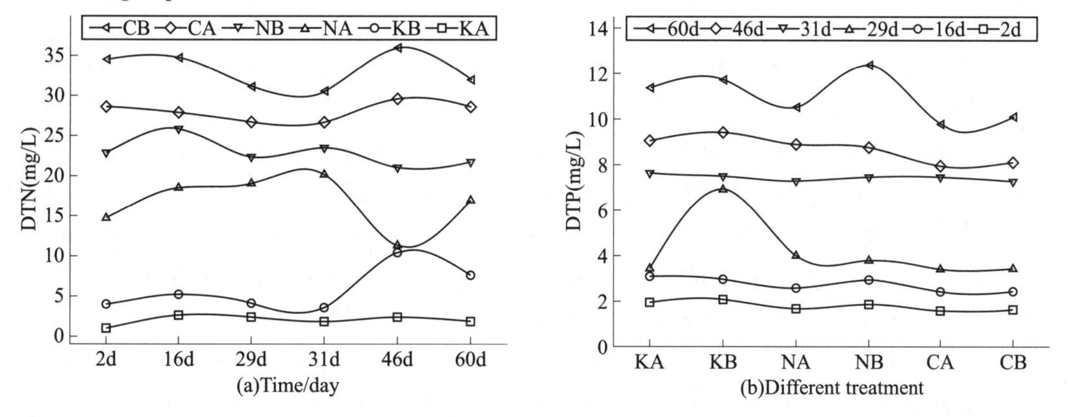

Fig. 2 The content and change of DTN/DTP in soil leachate of *Eupatorium adenophorum* under different treatment conditions

The content and change of total soluble phosphorus in the soil leachate of *B. adenophora* under different treatment conditions are shown in Fig. 2(b). It can be seen that the content change trend was roughly increased first and then decreased before 30 days, and appeared after 30 days. The basic trend was to decrease and then increase.

The effect of alternating wet and dry conditions on the total soluble nitrogen in the soil leachate of sugarcane grass under different treatment conditions is shown in Fig. 3(a). It can be seen that the highest content of soluble total nitrogen in the soil leachate is the nitrogen addition treatment group, and its DTN was significantly higher than other treatments, but in the nitrogen addition group, there was no significant difference in the DTN of the soil leachate between the dry and wet treatments and the constant humidity treatment; for the control group, the soil leachate DTN of the alternate dry and wet treatments and the constant humidity treatment also had no significant differences; but for carbon, in the addition group, the DTN content of soil leachate treated with alternating wet and dry treatments was significantly higher than that of constant humidity treatment.

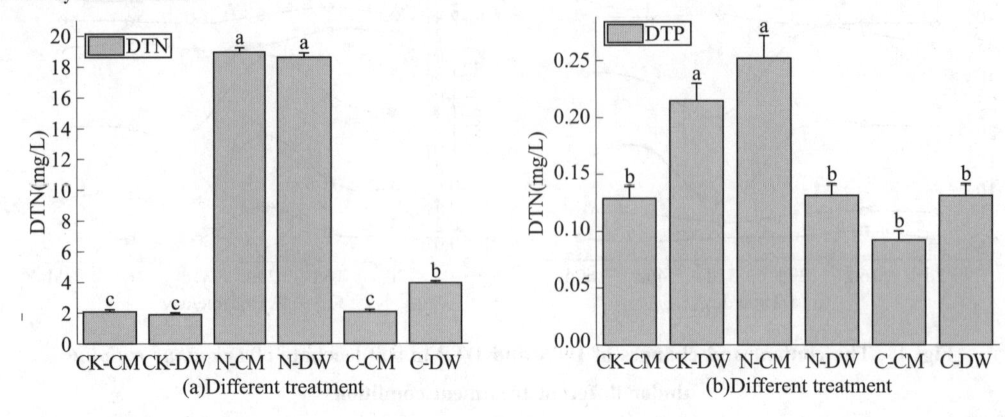

Fig. 3　The effect of alternation of drying and wetting under different treatment conditions on the DTN/DTP of the soil leachate of *Erianthus rufipilus*

The effect of alternating drying and wetting under different treatment conditions on the soluble total phosphorus in the soil leachate of *Saccharum vulgare* is shown in Fig. 3(b). It can be seen that the highest soluble total phosphorus content in the soil leachate is the constant humidity treatment of the nitrogen addition treatment group and the control group. There were significant differences between the two DTP treatments and other treatments in the dry wet alternate treatment; in the control group and the nitrogen group, the DTP of soil leachate under dry wet alternate treatment and constant humidity treatment was significantly different.

The effect of alternating wet and dry conditions on the total soluble nitrogen in soil leachate of *E. adenophorum* under different treatment conditions is shown in Fig. 4(a). It can be seen that the highest content of soluble total nitrogen in the soil leachate was the nitrogen addition treatment group, and the DTN is significant. It was higher than other treatments; and in the nitrogen addition group, there was a significant difference in the DTN of the soil leachate between the dry and wet treatment and the constant humidity treatment; but for the control group and the carbon addition group, the soil leachate DTN of the alternate dry and wet treatment and the constant humidity treatment also have no DTN significant differences.

The effect of alternating wet and dry conditions on the total soluble phosphorus of *E. adenophorum* soil leachate under different treatment conditions is shown in Fig. 4(b). It can be seen that the content of total soluble phosphorus in the soil leachate was the highest under the constant humidity treatment in the nitrogen addition group. It was the lowest in the carbon-added group under the constant humidity treatment condition; for the nitrogen-added group, the carbon-added group and the control group, although the contents were different, there was no significant difference in the statistical significance test.

The multi-factor analysis of variance affecting DTN and DTP of soil leachate is shown in the above table. It can be seen that the three factors of species, water treatment, and sampling time have significant effects on the DTN content of soil leachate, and the interaction between the three has a significant effect on the DTN content of soil leachate. It also has a significant impact. For soil leachate DTP content, species and water treatment have significant effects on it, but sampling time has no significant effect on the content, and the interaction between species and water treatment, and the interaction between species and sampling time on the DTP The content also has no significant effect; at the same time, the interaction of water treatment and sampling time and the interaction of the three factors have a significant impact.

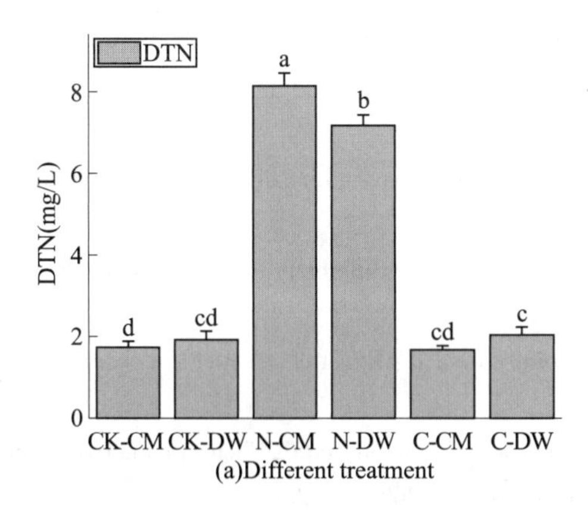

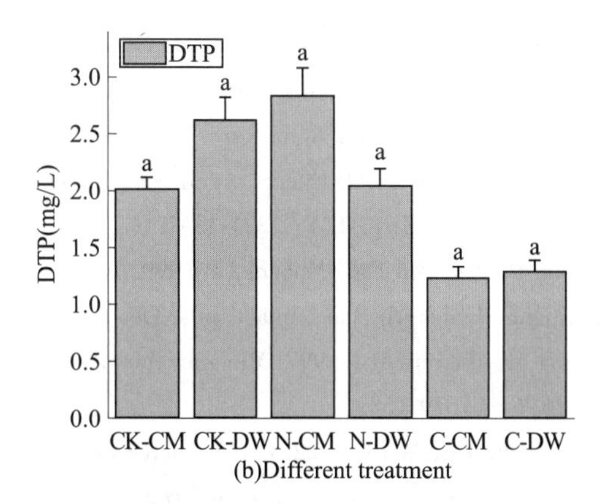

Fig. 4 The effect of alternating wet and dry conditions on the DTN/DTP of *Eupatorium adenophorum* soil leachate under different treatment conditions

Table 2 Multi-factor analysis of variance affecting soil leachate DTN and DTP

	DTN		DTP	
	F	*P*	F	*P*
Species	31.17	0.00	129.38	0.00
Moisture treatment	1087.24	0.00	3.00	0.01
Sampling time	11.93	0.00	1.76	0.13
Species * Moisture treatment	10.67	0.00	2.11	0.07
Species * sampling time	8.11	0.00	2.30	0.05
Moisture treatment * Sampling time	6.09	0.00	2.14	0.00
Species * Moisture treatment * Sampling time	8.17	0.00	2.13	0.00

Acknowledgments

Funding: this work by the Shaanxi Land Construction internal scientific research Project [DJNY2021-30, DJNY2022-16, DJNY2022-25, DJNY2022-32] and Shaanxi Province Enterprise Talent Lifting Plan Project [2021-1-2-2].

References

Zhang N M, Li C X, Li Y H 2007. Accumulation and releasing risk of phosphorus in soils in Dianchi watershed. Soils. 39: 665-667.

Xue B G 2008. Status and developing potentiality of the phosphorite resource around Dianchi Lake of Kunming. Geol. Chem. Miner. 30: 149-154.

Heskethn, Brookesp 2000. Development of an indicator for risk of phosphorus leaching. J. Environ. Qual. 29(1): 105-110.

Hupfer M, Lewandowski J 2008. Oxygen controls the phosphorus release from lake sediments-a long lasting paradigm in limnology. Int. Rev. Hydrobiol. 93(4-5):415-432.

Thanh N B, Marschnep P 2005. Effect of drying and rewetting on phosphorus transformations in red brown soils with different soil organic matter content. Soil Biol. Biochem. 37(8): 1573-1576.

Zhang B, Fang F, Guo J 2012. Phosphorus fractions and phosphate sorption-release characteristics relevant to the soil composition of water-level-fluctuating zone of Three Gorges Reservoir. Ecol. Eng. 40: 153-159.

Schnbrunneri M, Preiner S, Hein T 2012. Impact of drying and re-flooding of Sediment on phosphorus dynamics of river-floodplain systems. Sci. Total Environ. 432: 329-337.

Gerstengarbe F W, Werner P C 2008. Climate development in the last century-Global and regional. Int. J. Med. Microbiol. 298: 5-11.

Katterer T, Reichstein M, Anren O 1998. Temperature dependence of organic matter decomposition: A critical review using literature data analyzed with different models. Biol. Fertil. Soils. 27(3): 258-262.

Kruse J S, Kissel E, Cabrera M L 2004. Effects of drying and rewetting on carbon and nitrogen mineralization in soils and incorporated residues. Nutr. Cycl. Agroecosys. 69(3): 247-256.

Torbert H A, Wood C 1992. Effects of soil compaction and water-filled porespace on soil microbial activity and N losses. Commun. Soil Sci. Plan. 23(11-12): 1321-1331.

Schjonning P, Thonson I K, Moberg J P 1999. Turnover of organic matter in differently textured soils in Physical characteristics of structurally disturbed and intact soils. Geoderma. 89(3-4): 177-198.

Gordillo R M, Cabrera M L 1997. Mineralizable nitrogen in broiler litter, II: Effect of selected soil characteristics. J. Environ. Qual. 26(6): 1679-1689.

Li Y H, Zheng J L, Wu Q 2022. Zeolite increases paddy soil potassium fixation, partial factor productivity, and potassium balance under alternate wetting and drying irrigation. Agric. Water Manag. 260: 107294.

Hao M, Guo L J, Du X Z 2022. Integrated effects of microbial decomposing inoculant on greenhouse gas emissions, grain yield and economic profit from paddy fields under different water regimes. Sci. Total Environ. 805(01): 150295-150303.

Tran T H H, Kim S H, Jo H Y, Lee S 2022. Transient behavior of arsenic in vadose zone under alternating wet and dry conditions: A comparative soil column study. J Hazard Mater. 422(04): 126957.

Ullah K, Qureshi M I, Ahmad A 2022. Substitution potential of plastic fine aggregate in concrete for sustainable production. Structures. 35: 622-637.

Ananna F H, Amin M G M 2022. Groundwater contamination risks with manure-borne microorganisms under different land-application options. Water Sci. Engin. 14(4): 314-322.

Weber A M, Baxter B A 2021, McClung A. Arsenic speciation in rice bran: Agronomic practices, postharvest fermentation, and human health risk assessment across the lifespan. Environ. Pollut. 290: 117962.

Zhang Y C, Qu C K, Qi S H 2021. Spatial-temporal variations and transport process of polycyclic aromatic hydrocarbons in Poyang Lake: Implication for dry-wet cycle impacts. J Geochem Explor. 226:106738.

Arruda S, David L, Camelo D L 2021. Genesis of clay skins in tropical eutric soils: A case study from NE-Brazil. Catena. 202: 105236.

James J, Natesan A, Manohar A 2021. Potential of Portland pozzolana cement in the stabilization of an expansive soil subjected to alternate cycles of wetting and drying. Gradev. Mater. Konstr. Mater. Struct. 64(2): 81-91.

Xu X T, Shao L J, Huang J B 2021. Effect of wet-dry cycles on shear strength of residual soil. Soils Found. 61(3):782-797.

Albadri W M, Noor M, Jamaludin M D, Alhani I J 2021. The relationship between the shear strength and water retention curve of unsaturated sand at different hydraulic phases. Acta Geotech. 16(9): 2821-2835.

Thyagaraj T, Julina M 2021. Effect of physico-chemical interactions and cyclic wet-dry process on behaviour of compacted expansive soils. Indian Geotechnical Journal. 51(1): 225-235.

Dash S S, Sahoo B, Raghuwanshi N S 2021. How reliable are the evapotranspiration estimates by Soil and Water Assessment Tool (SWAT) and Variable Infiltration Capacity (VIC) models for catchment-scale drought assessment and irrigation planning. J Hydrol. 592: 125838.

Watanabe T, Katayanagi N, Agbisit R 2021. Influence of alternate wetting and drying water-saving irrigation practice on the dynamics of Gallionella-related iron-oxidizing bacterial community in paddy field soil. Soil Biol. Bioch. 152: 108064.

Zhou Y C, Hu B, Zhang W M 2020. Nitrous oxide emission from stormwater biofilters in alternating dry and wet weather. Enviro. Res. 191: 110137.

Malumpong C, Ruensuk N, Rossopa B 2021. Alternate Wetting and Drying (AWD) in Broadcast rice (Oryza sativa L.) management to maintain yield, conserve water, and reduce gas emissions in Thailand. Agric. Res. 10(1):116-130.

Jin W J, Cao W C, Liang F 2020. Water management impact on denitrifier community and denitrification in a soil at different of rice. Agric. Water Manage. 241: 106354.

Chu Y X, Xue K X, Feng G J 2022. Experimental study on mechanical properties of roots and stems of there herbaceous plants in Kunming. Forest Engineering. 38(1):15-26.

本文曾发表于2022年《Bangladesh J. Bot.》第51卷第3期

Impact of Co-contamination by PAHs and Heavy Metals on Micro-ecosystem in Bioretention Systems with Soil, Sand, and Water Treatment Residuals

Zhaoxin Zhang[a, b], Jiake Li[b], Huanyuan Wang[a], Yajiao Li[c], Xiaolong Duan[b]

(a. Institute of Land Engineering and Technology, Shaanxi Provincial Land Engineering Construction Group Co., Ltd., Xi'an 710075, China; b. State Key Laboratory of Eco-hydraulics in Northwest Arid Region of China, Xi'an University of Technology, Xi'an 710048, China; c. School of Architecture and Civil Engineering, Xi'an University of Science and Technology, Xi'an 710054, China)

【Abstract】 Rainfall runoff contained toxic pollutants such as polycyclic aromatic hydrocarbons (PAHs) and heavy metals. Bioretention systems can effectively remove PAHs and heavy metals from runoff. When PAHs and heavy metals accumulated in the media of bioretention systems, co-contamination occurred. To reveal the response processes of co-contamination in the media micro-ecosystem, bioretention systems were constructed and simulated rainfall tests were carried out to reveal the media micro-ecosystem evolution in the bioretention system under the co-contamination of PAHs and heavy metals. The simulated rainfall tests showed that the bioretention systems with different media had good purification effects with load reduction rates >90% (PAHs) and >40% (heavy metals). The microbial α-diversities in different media decreased by 6.67%~32.27%, and 9.21%~18.69% after the tests. The most dominant species of microorganisms in the media were *Proteobacteria* (all relative abundances >62%). As a typical PAHs-heavy metal degrading bacteria, the relative abundances of *Pseudomonas* increased significantly after the tests (both >9%). The co-contamination of PAHs and heavy metals inhibited all enzyme activities. The long-term operation of the bioretention system gradually increased the level of co-contamination in the media, and the stability of the media micro-ecosystem gradually deteriorated. The results of this paper can provide a theoretical basis and technical support for maintaining the operational efficiency of the bioretention system with modified-media and avoiding the risk of potential contamination.

【Keywords】 Bioretention system; Modified-media; Co-contamination; Micro-ecosystem; Ecological effect

1 Introduction

Urban expansion leads to the increase of urban impervious areas, and the resulting extreme rainfall events increase the risk of urban waterlogging. The resulting floods greatly threaten the urban environment and the safety of urban residents (Hou et al., 2019). Bioretention systems regulate stormwater runoff through the physical, chemical, and biological interactions of plant-media-microbial systems, which can effectively reduce loads of urban non-point source pollutants into surface/subsurface water bodies and avoid eutrophication. Bioretention systems have been widely used and promoted in many countries due to their low cost and environmental friendliness (Yang et al., 2022). The water volume reduction and water quality purification of the bioretention system on rainfall runoff are mainly responsible for the media (accounting for 70% to 80% of the total removal effects) (Tirpak et al., 2021). The purification capacities of the media restrict the removal effects of suspended solids, oily organic matter, heavy metals, nutrients, and other special pollutants (pathogenic microorganisms, organic micro-pollutants) in bioretention systems (Tian et al., 2019).

Many studies have used loamy sandy soil and sandy loam soil with high permeability as the best media types and carried out media improvement (Vijayaraghavan et al., 2021). Modifiers, such as water treatment residue (WTR), biochar, montmorillonite, sponge iron, and turf soil, are widely used to increase water holding and adsorption capacity, increase microbial abundance, and improve the plant growth environment (Xu et al., 2021). Especially for the resource reuse of solid waste, it not only realizes the recycling of waste but also achieves the effect of improving the pollutant purification capacity of the bioretention system (Biswal et al., 2022). WTR is rich in metal ions (Fe^{3+} and Al^{3+}), which can effectively adsorb phosphorus in the water. Xu et al. (2020) found that the addition of WTR effectively increased hydraulic retention time, enhance soil structural stability, and promote plant growth of bioretention systems.

The most important purpose of the current media improvement of bioretention systems is to improve the purification capacity of the media for runoff pollutants (Shrestha et al., 2018). Studies have shown that media adsorption was the most important way to remove runoff pollutants in bioretention systems (Tedoldi et al., 2016). These pollutants are mostly removed by the adsorption process of the media, and the accumulation of the pollutants in the media is not considered. With the improvement of media, pollutants were adsorbed by the media and gradually accumulated in the bioretention system (Zhang et al., 2021a). In bioretention systems, pollutants are usually adsorbed by the media and accumulate in the area of 0~30 cm, especially specific pollutants such as polycyclic aromatic hydrocarbons (PAHs) and heavy metals contained in the rainfall runoff. PAHs in rainfall runoff were mainly 4-ring PAHs (Fluoranthene and Pyrene) (Abdel-Shafy and Mansour, 2016), and heavy metals were copper (Cu), zinc (Zn), cadmium (Cd), and lead (Pb) (Li et al., 2021). Although the concentrations of these pollutants in the runoff are not large, the construction of the bioretention system makes a large amount of rainwater runoff be collected, resulting in large pollutants retained and accumulated in the facilities (Beryani et al., 2021). During the rainy season, continued rainfall results in the continuous collection of rainfall runoff by bioretention systems, and PAHs and heavy metals adsorbed in the media are unable to be efficiently biodegraded. The accumulation level will gradually accumulate to a higher pollution level. Studies have shown that the contents of PAHs in the bioretention system receiving road runoff were seriously exceeding the standard and had ecological and health risks (Zhang et al., 2022a). Heavy metals such as Cu, Pb, and Zn showed significant accumulation on the surface of the bioretention system (Zhang et al., 2021b). The increasing contents of heavy metals in media led to the saturation of the adsorption sites of the bioretention system (Yin et al., 2021).

The coexistence of organic pollutants and heavy metals in the soil formed co-contamination, while various direct or indirect interactions occurred between them, such as competition for adsorption sites, redox effects, and microbial stress (Liu et al., 2017). The co-contamination of PAHs and heavy metals were mainly due to the coordination of cation-π interactions (Ali et al., 2022). The co-contamination of PAHs-heavy metals changed soil properties, while the changes in physical/chemical properties easily led to changes in soil ecosystems (Wu et al., 2021). The accumulation of PAHs and heavy metals after the long-term operation of the bioretention system led to its pollution level reaching the same level as the actual polluted site (Chu et al., 2021). Even the PAHs and heavy metals accumulated in the media can be leached with surface runoff or flow into groundwater to cause pollution. For the present research, there are accumulations of PAHs and heavy metals in the bioretention system and the formation of co-contamination. The impacts of the co-contamination of PAHs and heavy metals on the operation efficiency and the stability of the ecosystem in bioretention systems were not discussed. It is necessary to evaluate the ecological stress of the co-contamination of PAHs and heavy metals on the micro-ecosystem in the bioretention system.

For soil pollution, the soil biological characteristics diagnosis method was used to evaluate the ecotoxicity and stress, which mainly includes: soil biomass, soil respiration, enzyme activity, nitrification potential, community structure and diversity, functional genes, and other indicators (Ali et al., 2022). To assess the changes of the micro-ecosystem in the bioretention system, the methods of soil can also be used. The health of microbial ecosystems in bioretention systems can be effectively assessed by studying microbial diversity and community structure (Chen et al., 2020). The analyses of the changes in the micro-ecosystem (microbial biomass, community structure, enzyme activity) during the operation, can reflect the operation effect of bioretention systems (Mehmood et al., 2021a). Especially for modified-media bioretention systems, the additions of exogenous modifiers have significant impacts on the indigenous microbial environment. When used sand, soil, and fly ash were as media in bioretention systems, 11 dominant bacterial groups were found and the diversity and metabolic effects of microorganisms were different (Zuo et al., 2019). Using wood chips as media modifiers in the bioretention system promoted the growth of denitrifying bacteria (*Dechloromonas*, *Acidoborax*, and *Pseudomonas*) and the process of denitrification (Liu et al., 2021). In bioretention systems with WTR as the modifier, Zhang et al. (2021a) found the adsorbed phosphorus in the media promoted the growth of phosphorus accumulating bacteria, like *Pseudomonas*.

Soil enzymes control various biochemical processes in the soil. The magnitudes of enzyme activities affect the intensity of biochemical reactions in soil, and its strength directly affects the material and energy conversion in the soil ecosystem (Jiang et al., 2019a). Monitoring enzyme activity as indicators can reveal the degree of pollution and remediation of heavy metals, pesticides, and petroleum hydrocarbons in the soil (Maurya et al., 2020). Since enzyme activity can effectively reflect the pollution degree of heavy metals and PAHs, the exploration of the change in enzyme activity can effectively reveal the accumulation level of pollutants in the media (Guo et al., 2019). Simultaneously, due to the increased operating time of the bioretention system, the activities of urease, dehydrogenase, and sucrase showed slow increase trends (Liu et al., 2020). The present studies mainly revealed the changes in enzyme activity in the bioretention system or only target the nitrogen cycle-related enzymes in the media, but do not establish relevance with enzyme activity and accumulation of pollutants especially the co-contamination of PAHs and heavy metals.

As PAHs and heavy metals and other highly toxic pollutants exist in rainfall runoff, it is easy to form a co-contamination of PAHs and heavy metals in bioretention systems after continuous rainfall events occur. The bioretention system has a better pollutant removal effect after the media improvement, but this also means that more pollutants will be absorbed and accumulated in the media. At present, there are some studies on the accumulation of pollutants after the media improvement of bioretention system, but they are mostly limited to the accumulation level of single pollutants. To reveal the impact of the co-contamination caused by PAHs and heavy metals accumulation in bioretention systems on the media micro-ecosystem, this study used different types of bioretention columns with soil, sand, and WTR as the main media: (i) to study the removal and accumulation level of PAHs and heavy metals by bioretention columns with different media types; (ii) to reveal the changes in the micro-ecosystem under the co-contamination of PAHs and heavy metals; (iii) to establish the relationship between environmental factors and dominant bacteria spices; (iv) to clarify the response process of media micro-ecosystem under co-contamination of PAHs and heavy metals.

2 Materials and methods

2.1 Device setting

The simulated rainfall tests were carried out at the sponge city test field of Xi'an University of Technology, Xi'an City, China. small-scale bioretention systems were established based on the review of relevant studies and previous research (Jiang et al., 2019a). A circular hollow PVC plastic tube with a height of 120 cm and an inner diameter of 40 cm constituted the bioretention system. The structure of each bioretention system from top to bottom was: ponding layer 15 cm, cover layer 5 cm, media layer 70 cm, and gravel drainage layer 15 cm. According to the climatic conditions in the semi-arid area, two plants, *Buxus sinica* and *Lolium perenne L.*, were planted in the bioretention systems (Zhang et al., 2021a). The media of the bioretention systems were configured by mixing soil, sand, and WTR in different proportions (by weight). The soil and sand used for the media were collected locally and WTR was taken from a water supply plant in Xi'an. Three kinds of bioretention systems with different media were constructed: (i) planting soil, PS; (ii) bioretention soil media, BSM; and (iii) BSM+WTR. PS was filled with 100 % soil, BSM was filled with 35% soil, 60% sand, and 5% organic matter, BSM+WTR was filled with 33.25% soil, 57% sand, 4.75% organic matter, and 5% WTR, while all percentages above were the mass ratio. Two controls for each media were set. Each media set two controls, and the media were evenly mixed. The media layer was divided into the upper and lower layer, wherein the upper layer was 0~35 cm below the cover layer, and the lower layer was 35~70 cm. Sampling holes are arranged in each of the upper and lower layers, at 5~20 cm and 45~60 cm below the cover layer, with a diameter of 2.5 cm. The structure of bioretention system is shown in Fig. 1(a), and the on-site photo is shown in Fig. 1(b).

2.2 Scheme design

After the construction of the bioretention systems, the simulated rainfall tests were carried out. The whole simulated rainfall tests are designed with 8 rainfall events, and the interval between each two rainfall events is guaranteed to be 5 d. For the water volume design of the simulated rainfall tests, a discharge ratio (20:1), 2 rainfall durations (120 min and 360 min), and 4 rainfall return periods (0.5 a, 2 a, 3 a, 5 a) were designed, combined with the rainstorm intensity in Xi'an (Equation (1)).

$$i = \frac{16.715(1 + 1.6158\lg P)}{(t + 16.813)^{0.9302}} \tag{1}$$

Where: i is the rainfall intensity (mm/min), P is the precipitation (mm), t is the rainfall duration (min).

The design water volume for each rainfall event is shown in Table 1. Considering the initial rain erosion effect in the actual rainfall process, the water inflow in the tests was divided into two stages: the first 5 min of each rainfall event was the initial stage (with high water inflow concentration), and then the later stage (with low water inflow concentration). Because the concentration of PAHs in the actual rainfall runoff was only concentrated in the early stage of rainfall, it was chosen to add naphthalene (NAP), fluoranthene (FLT), and pyrene (PYR) in the first 5 min of each test. The influent concentrations of PAHs and heavy metals such as copper (Cu), zinc (Zn), and cadmium (Cd) were determined by the corresponding pollutant concentrations in rainfall runoff from different underlying surfaces in Xi'an (Jiang et al., 2019a). The influent concentrations of heavy metals and PAHs are shown in Table 1.

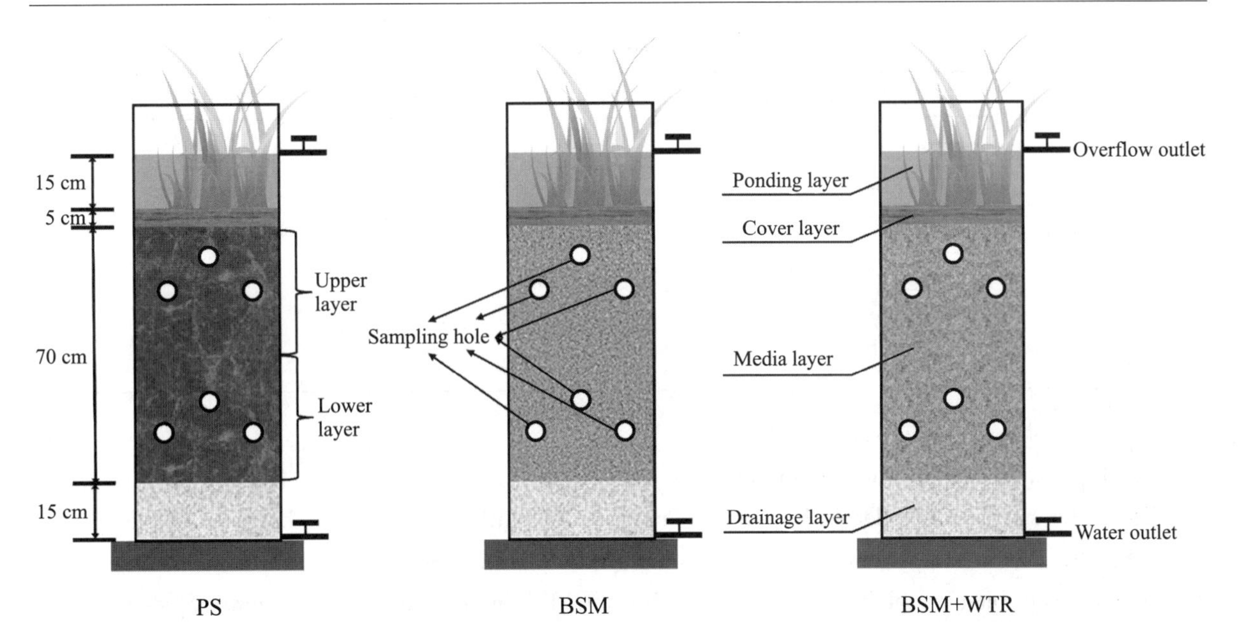

(a) The structure and characteristics of bioretention system

(b) The on-site photo

Fig. 1 The structure and on-site photo of devices

Table 1 Design of simulation rainfall tests scheme

	Rainfall tests number	Rainfall recurrence interval (a)	Duration (min)	Precipitation (mm)	Water volume (L)
Water quantity	1	0.5	120	13.41	33.70
	2	1		20.67	51.91
	3	2		27.92	70.13
	4	3		32.16	80.79
	5	0.5	360	15.68	39.39
	6	1		24.16	60.69
	7	2		32.64	81.99
	8	3		37.60	94.45

Continued to Table 1

	Pollutant (mg/L)		High-concentration	Low-concentration
Water quality	PAHs	NAP	1.4	—
		FLT	0.3	—
		PYR	0.3	—
	Heavy metals	Cu	1	0.3
		Zn	1.5	0.5
		Cd	0.5	0.1

2.3 Sampling and analysis method

2.3.1 Water sample collection and analysis

The water sample was collected by manual sampling. The outflow samples were collected in the first 2 h for 0, 5, 10, 20, 30, 60, 90 min, and 120 min. Then the samples were collected every hour until it was discharged. Each water sample was approximately 500 mL. The inflow and overflow water samples were also collected during the test. The concentration of NAP, FLT, PYR, Cu, Zn, and Cd were detected. PAHs in water samples were determined by gas chromatography-mass spectrometer (GCMS-TQ8040 NX, Shimadzu, Japan) (Agilent Technologies, 2019). The concentration of Cu, Zn, and Cd was determined by the flame atomic absorption method.

The calculation methods of volume reduction rate and load reduction rate are shown in Equation (2) ~ (5):

$$R_V = \frac{V_{in} - V_{out} - V_{over}}{V_{in}} \times 100\% \tag{2}$$

$$L \approx \sum_{0}^{n} C_t \cdot V_t \tag{3}$$

$$R_L = \frac{L_{in} - L_{out} - L_{over}}{L_{in}} \times 100\% \tag{4}$$

$$L_{retained} = L_{in} - L_{out} - L_{over} \tag{5}$$

Where: R_V is the volume reduction rate (%), R_L is the pollutant load reduction rate (%), V_{in}, V_{out}, and V_{over} are the inflow, outflow, and overflow water volume (m^3), C_t is the pollutant concentration at time t (mg/L), V_t is the volume at time t (L), L_{in}, L_{out}, and L_{over} are the pollutant load of inflow, outflow, and overflow (mg), $L_{retained}$ is the retained pollutant load by bioretention system (mg). The calculation results R_V, R_L, L_{in}, L_{out}, L_{over}, and $L_{retained}$ are shown in Table S1 and S2. The variability of pollutant load reduction by different bioretention systems was verified by the t-test.

2.3.2 Media sample collection and analysis

The media samples of different bioretention systems were collected before and after the tests. The samples which were collected from the sampling ports of 5~20 cm/40~60 cm were mixed and then can be regarded as the upper/lower layer. The contents of PAHs (NAP, FLT, PYR) and heavy metals (Cu, Zn, Cd) in the media were detected. The analysis methods were similar to the water sample analysis method, and only the pretreatment methods were different (Mehmood et al., 2021b).

The microbial ecosystem (micro-ecosystem) in media samples of bioretention systems was analyzed by high-throughput sequencing technology (Li et al., 2021). By extracting the DNA in the media samples, and then amplifying the 16s rDNA V3+V4 region, with the universal primer sequence 341F/806R (ACTCCTACGGGAGGCAGCAG/GGACTACHVGGGTWTCTAAT), the obtained molecular ecological information can be used to analyze the microbial ecosystem changes in the media samples. The analysis of the micro-ecosystem included microbial diversity (observed species, Shannon index), microbial community structure, and enzyme activities.

The Shannon index mainly reflects the microbial diversity of different samples under different sequencing quantities, and the calculation method is shown in Equation (6):

$$H_{\text{Shannon}} = -\sum_{i=1}^{S_{\text{obs}}} \frac{n_i}{N} \ln \frac{n_i}{N} \tag{6}$$

Where: H_{Shannon} is the Shannon index, S_{obs} is the actual number of OTUs measured, n_i is the number of OTUs containing i sequences, N is the number of all sequences.

The changes in the enzyme activities in the media were mainly determined by measuring four enzyme activities (dehydrogenases, urease, acid-phosphatase, and invertase). The determination methods of enzymatic activities were described by Zhang et al (2021a).

3 Results and discussion

3.1 Removal effects of PAHs and heavy metals

The load reduction rates of NAP, FLT, PYR, Cu, Zn, and Cd of bioretention systems with different media were calculated by Equation (3), and the results are shown in Fig. 2. The load reduction rates of the three PAH monomers all reached more than 90% ($p<0.05$), with average load reduction rates of 99.02% (planting soil, PS), 97.66% (BSM), and 97.06% (BSM+WTR). Especially for PS, the average load reduction rates for all PAH monomers were >99%. Organic matter in soil can effectively adsorb PAHs and control their environmental fates such as toxicity, migration, and transformation (Ukalska-Jaruga & Smreczak, 2020). Stable bioretention systems can effectively remove PAHs (Fairbairn et al., 2018), and the reduction effects of bioretention systems with different media on PAHs still showed differences. The infiltration rate of PS was the lowest, and PAHs had sufficient time to be absorbed by soil organic matter. PAHs in BSM and BSM+WTR were directly infiltrated with runoff, and the removal effects were lower than that of PS. Comparing the three PAH monomers, NAP had the best removal effects and was effectively removed by the bioretention system, while the load reduction effects of FLT fluctuated more than NAP. The low-ring PAHs in the soil were quickly adsorbed by soil organic matter and minerals (Boving & Neary, 2007), while the middle and high-ring PAHs have a poor adsorption effect after entering the soil (Ali et al., 2022). Since the removal effects of bioretention systems with different media were the same, the amounts of retained PAHs were the same (Table 2).

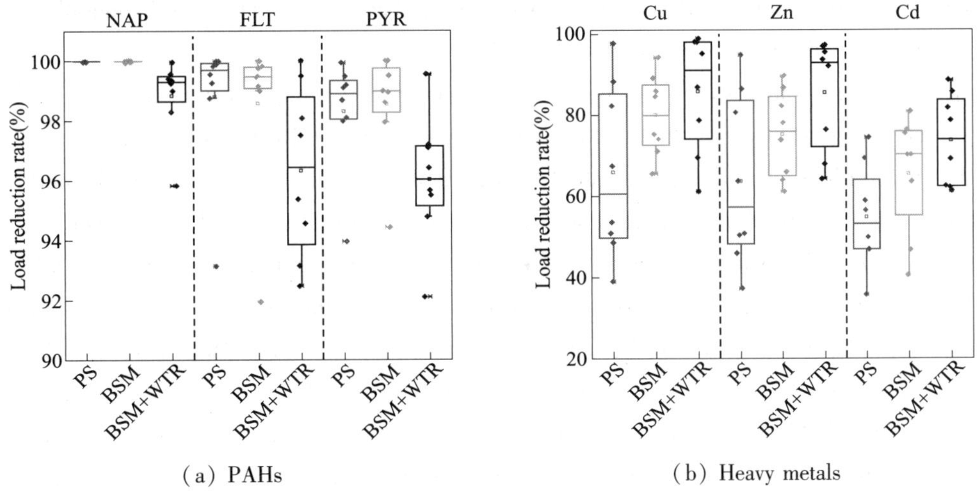

(a) PAHs (b) Heavy metals

Fig. 2 Load reduction rates for PAHs and heavy metals by bioretention systems

Table 2 The retained pollutant loads of the bioretention columns

Pollutant loads (mg)	PAHs			Heavy metals		
	NAP	FLT	PYR	Cu	Zn	Cd
PS	7.18	1.51	1.52	97.13	155.13	28.54
BSM	7.18	1.52	1.53	125.92	193.44	37.30
BSM+WTR	7.10	1.48	1.48	133.93	221.59	40.46

The removal effects of the bioretention systems on heavy metals fluctuated greatly, while the load reduction rates were all >40% and showed highly significant differences ($p<0.01$). There are significant differences in the load reduction effects of different media on Cu, Zn, and Cd, with the effect ranging from high to low: BSM+WTR>BSM≥ PS. The retained load was the total amount of pollutant load which was retained in the bioretention system. BSM+WTR has the best purification effect on Cu, Zn, and Cd, and its retained loads of Cu, Zn, and Cd were also the highest among the three bioretention systems. The efficiency of the bioretention system for heavy metal purification was stable whether the media were modified (Søberg et al., 2019). The removal of heavy metals mainly depends on organic matter or humus in media to reduce its bioavailability, while the removal effects of BSM and BSM+WTR were higher than that of PS due to the addition of organic matter (Jiang et al., 2019b).

3.2 Distribution of PAHs and heavy metals

The effects of rainfall tests on the contents of PAHs and heavy metals in the media are shown in Fig. 3. PAHs showed significant accumulation in bioretention systems. Before the rainfall tests, although PAHs were detected in the media (NAP<0.15 mg/kg, FLT<0.14 mg/kg, PYR<0.20 mg/kg), the distributions of PAHs were relatively uniform, and no significant stratification. There were significant increases in the contents of NAP, FLT, and PYR after the tests. The contents of NAP after the tests were 1.43~68.50 times than that before, while the contents of FLT and PYR were 16.08~175 times and 23.05~109.75 times. PAHs were easily absorbed by the media and accumulated in the bioretention systems (Tedoldi et al., 2016). The contents of PAHs in BSM were higher than in other media. PYR contents in BSM were 4.39~5.98 mg/kg, which were significantly higher than the others. After simulated rainfall tests, NAP, FLT, and PYR had uneven distributions in the media. The contents of NAP from the upper and lower layers in BSM+WTR after the tests showed differences, while the content of the upper layer was 1.51 mg/kg, and that of the lower layer was 0.04 mg/kg. PAHs showed obvious upper layer accumulation and generally accumulated in the upper layer 10~40 cm (Mehmood et al., 2021a).

After the tests, the contents of heavy metals in the three bioretention systems increased. The contents of Cu in the bioretention systems were higher than that before the tests. There are differences in the vertical distribution of Cu in different media. The contents of Cu in BSM+WTR were significantly higher than that in PS and BSM, while the upper layer was lower than the lower layer. BSM+WTR has better load reduction effects on the Cu (Fig. 2), Cu accumulated in the media. With the continuous scouring of rainfall, Cu in BSM+WTR continued to migrate downward, although no leaching phenomenon occurred, it resulted in higher Cu content in the lower layer than in the upper layer (Chu et al., 2021). The contents of Zn in PS were at high levels and there was a significant accumulation phenomenon. Due to the addition of modifiers (sand and WTR) in BSM and BSM+WTR, the Zn contents were higher than that of PS (all >63 mg/kg) after the construction of the bioretention systems. Then, the contents of Zn in these two bioretention systems showed significant decreasing trends after the tests. When the Zn contents in the media reached a certain level, the entry of the simulated rainfall led to the occurrence of Zn leaching (Mangangka et al., 2015). Although the contents of Cd varied to a certain extent in different bioretention systems, its contents fluctuated in a small range (always within the range of 0.13 ~ 0.18 mg/kg), so there were no accumulation phenomena of Cd. The heavy metals prepared in the tests were all dissolved heavy metals, it led to the strong downward migration of heavy metals with simulated rainfall (Søberg et al., 2019). Cu, Zn, and Cd in the bioretention systems did not show obvious stratification.

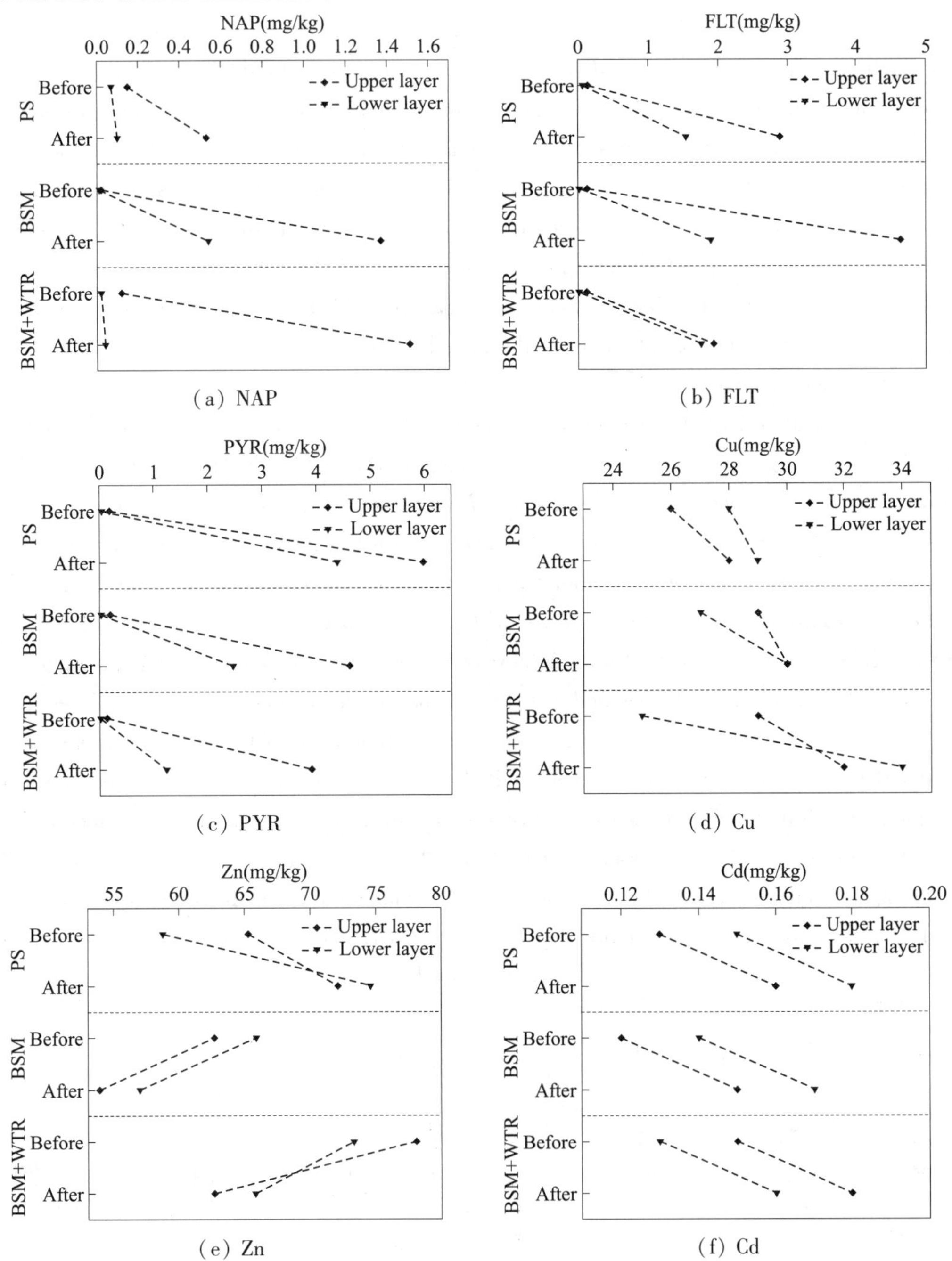

Fig. 3 The changes in the contents of PAHs and heavy metals in bioretention systems

3.3 Evolution of media micro-ecosystem

3.3.1 Microbial diversity

The analyses of observed species and Shannon indexes of the samples revealed the effects of runoff pollutants on the microbial diversity in the bioretention systems (Jiao et al., 2018). Fig. 4a shows the change in microbial diversity and indicates whether the differences in microbial diversity of the media were significant (Khudur et al., 2018), which was reflected in the drastic reduction of the α-diversity indexes. After the tests, the microbial α-diversities in different media decreased by 6.67%~32.27% (observed species) and 9.21%~18.69% (Shannon). The Shannon indexes of PS showed significant differences ($p<0.05$). The reduction degrees of microbial α-diversity indexes in PS were slight, which was closely related to the fact that the accumulation levels of heavy metals and PAHs in PS were lower than that of BSM and BSM+WTR. The difference in observed species of BSM was significant ($p<0.01$), while the difference in the Shannon index was significant ($p<0.001$). The α-diversity of microorganisms in BSM+WTR showed differences ($p<0.001$) before and after the tests. The pollutants in the rainfall runoff were efficiently absorbed by the bioretention systems, then these pollutants continued to accumulate in the media and formed the co-contamination (Mehmood et al., 2021b). The co-contamination of PAHs and heavy metals led to negative effects such as damage to microbial cells, and denaturation of DNA and protein structures and all of these effects led to dramatic reductions in microbial diversities (Rosner et al., 2021). The microbial diversities in media were negatively correlated with the degree of co-contamination of PAHs and heavy metals, and the accumulation of runoff pollutants caused great stress on the microbial ecosystem.

3.3.2 Dominant community

The co-contamination of PAHs and heavy metals also have significant effects on the media microbial community (Fig. 4b). At the phylum level, 10 dominant phyla in PS, BSM, and BSM+WTR were all found (*Proteobacteria*, *Actinobacteria*, *Acidobacteria*, *Firmicutes*, *Bacteroidetes*, *Chloroflexi*, *Gemmatimonadetes*, *Nitrospirae*, *Rokubacteria*, and *Verrucomicrobia*). Among them, *Proteobacteria* were the most dominant microorganisms, with relative abundances in all bioretention systems >60% after the tests. Through the simulated rainfall tests, the increases of *Proteobacteria* were 16.32% (PS), 12.42% (BSM), and 16.98% (BSM+WTR). *Proteobacteria* was always the most dominant microbial population in co-contaminated soil of PAHs and heavy metals and was proportional to the contents of PAHs (Kidd et al., 2021). The main reason is that *Proteobacteria* had a positive correlation with PAHs in soil (Kuppusamy et al., 2017). After the tests, the accumulation levels of PAHs in the bioretention systems were significantly increased compared with that before the tests, resulting in significant increases in the abundance of *Proteobacteria*. *Proteobacteria* are the main microorganisms in petroleum and heavy metal contaminated soils, with high growth capacity under heavy metal stress and direct participation in hydrocarbon degradation (Chai et al., 2022). The increases in relative abundances of *Proteobacteria* led to downward trends in the relative

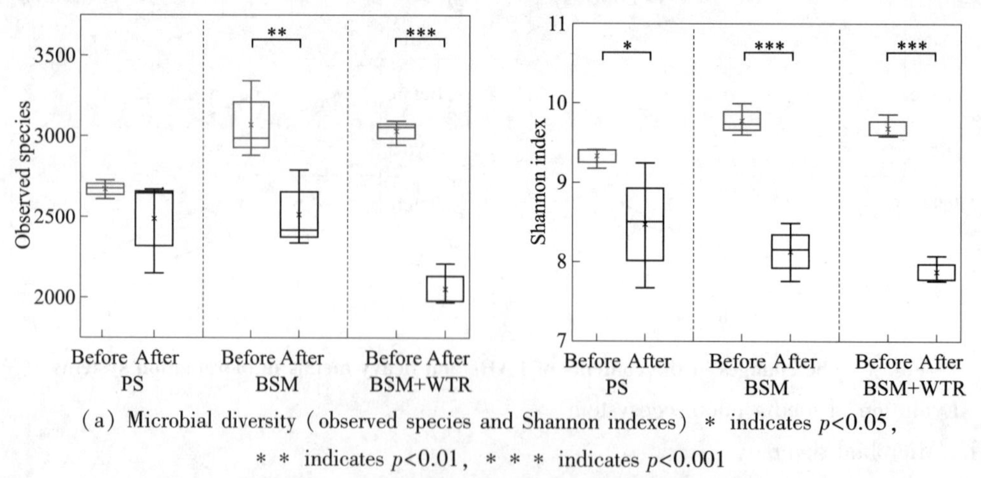

(a) Microbial diversity (observed species and Shannon indexes) * indicates $p<0.05$, ** indicates $p<0.01$, *** indicates $p<0.001$

Fig. 4 The evolution of micro-ecosystem under the stress of PAHs-heavy metal co-contamination in bioretention systems

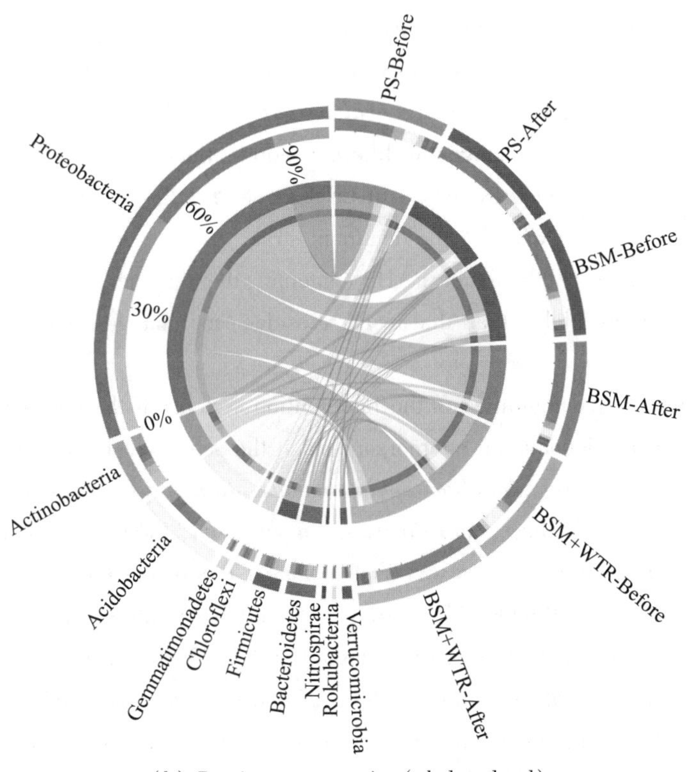

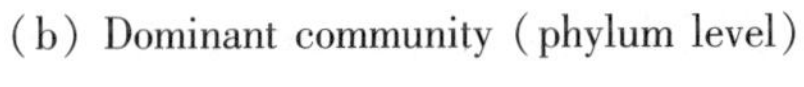
(b) Dominant community (phylum level)

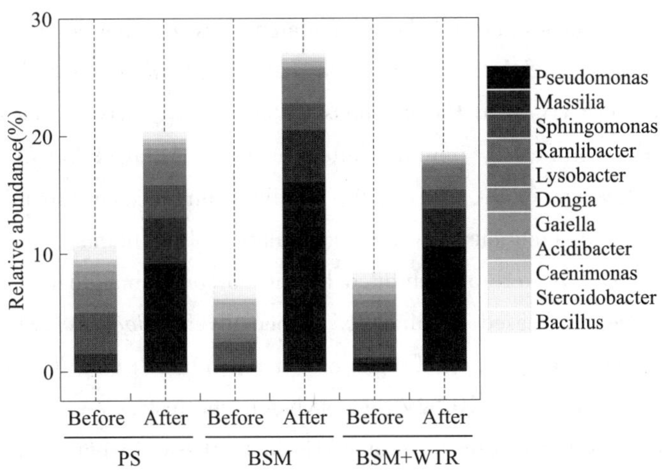

(c) Dominant community (genus level, with all relative abundances > 0.1%)

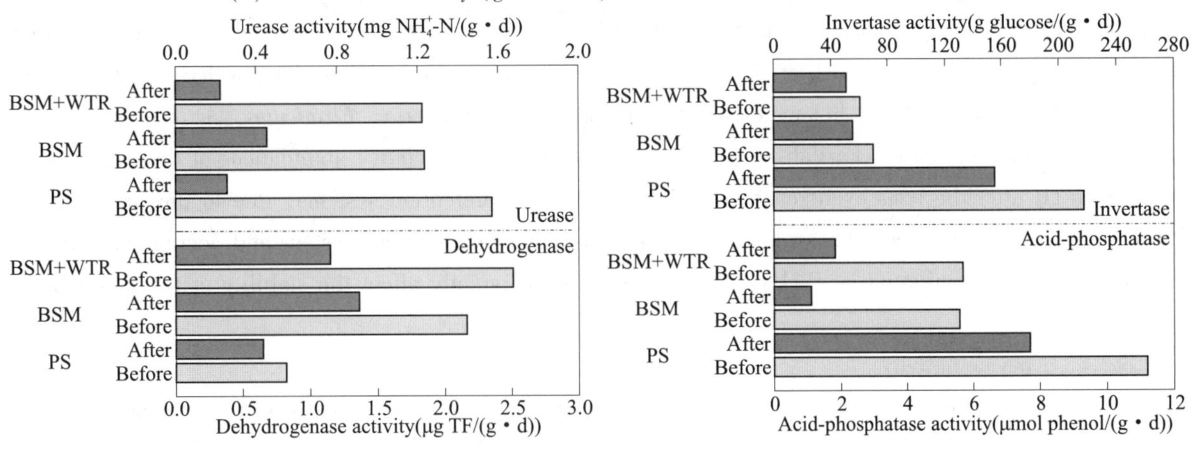

(d) Enzyme activity (dehydrogenase, urease, acid-phosphatase and invertase)

Continued to Fig. 4

abundances of most microorganisms except *Proteobacteria*. After the tests, the relative abundances of *Acidobacteria*, *Chloroflexi*, *Gemmatimonadetes*, *Nitrospirae*, *Rokubacteria*, and *Verrucomicrobia* were significantly lower than before the tests. In particular, *Acidobacteria* in BSM decreased by 11.37% (17.30% to 5.93%). *Actinobacteria* had different changes in different bioretention systems. It showed a decreasing trend in PS (from 9.10% to 7.71%), but it increased in BSM (from 5.21% to 5.41%), and BSM+WTR (from 6.56% to 11.03%). *Actinobacteria* was also the dominant strain in soil co-contamination of PAHs and heavy metals (Malicka et al., 2020). With higher PAHs contents, the relative abundances of *Actinobacteria* in BSM and BSM+WTR were also higher than that in PS. Besides, the changes of *Firmicutes* and *Bacteroidetes* were also different.

The changes in microbial communities in the bioretention systems at the genus level were analyzed, as shown in Fig. 4c. There were 11 dominant microorganisms at the genus level (with all relative abundances > 0.1%), which were *Pseudomonas*, *Massilia*, *Sphingomonas*, *Ramlibacter*, *Lysobacter*, *Dongia*, *Gaiella*, *Acidibacter*, *Caenimonas*, *Steroidobacter*, and *Bacillus*. Among them, the species with the largest relative abundance changes before and after the tests was *Pseudomonas*. Before the tests, the abundance of *Pseudomonas* was the highest at 0.74% (BSM+WTR). After the simulated rainfall tests, *Pseudomonas* became the most dominant specie, with the lowest relative abundance of 9.23% (PS) and the highest of 15.77% (BSM). *Pseudomonas*, as a typical degrading bacteria of petroleum hydrocarbons, can effectively use hydrocarbons and aromatic hydrocarbons as carbon sources and energy sources to grow and degrade PAHs in the soil, especially almost all low-ring PAHs (Rabodonirina et al., 2019). *Pseudomonas* was the dominant bacteria in the bioretention systems where PAHs accumulated. *Massilia* showed significant increases after the tests, especially the increase in BSM from 0.30% to 4.41%. *Massilia*, as a kind of PAH degradation ability, was more resistant to PAHs than *Pseudomonas* (Wang et al., 2016). The growth of *Sphingomonas* in PS and BSM+WTR was significantly inhibited after the tests. Although *Sphingomonas* has also shown PAH degradation ability (Jakoncic et al., 2007), the relative abundances of this microorganism were significantly reduced under eutrophic conditions after the accumulation of pollutants. The growth of *Gaiella*, *Acidibacter*, *Caenimonas*, and *Bacillus* were all inhibited by the co-contamination of PAHs and heavy metals, and their relative abundances decreased significantly. Especially for *Bacillus*, which played an important role in the transformation and decomposition of organic matter (Akpan et al., 2020), its relative abundance in BSM+WTR after the tests reduced to 10.52% of that before the tests.

In general, because the accumulation level of PAHs in BSM was higher than in others (Fig. 3), the microbial ecology in BSM changed significantly. The stress effects of PAHs in the bioretention systems on the microorganisms were significantly stronger than other pollutants, and the microorganisms were also the first organisms to respond to PAHs (Ali et al., 2022). Since most of the bacteria species that have been found to degrade and metabolize PAHs belong to the five phyla of *Proteobacteria*, *Firmicutes*, *Actinobacteria*, *Acidobacteria*, and *Chloroflexi* (Fazilah et al., 2020), the sum of the relative abundances of these five phyla in BSM+WTR after the tests even reached 89.09%. For the bioretention systems, due to the properties of PAHs and heavy metals (which can be easily removed by the bioretention system and accumulated in the media), the co-contamination of PAHs and heavy metals will greatly affect the stability of the bioretention system.

3.3.3 Enzyme activity

Soil enzyme activity is affected by accumulated pollutants in soil and effectively reflects the degree of biochemical reaction in the soil micro-ecosystem (Guo et al., 2019). As the co-contamination of PAHs-heavy metals also has more stresses on the microbial communities of the media, analyzing the changes in enzyme activities can better reveal the effect of co-contamination on the micro-ecosystem of the bioretention systems. The changes in enzyme activities are shown in Fig. 4d. Comparing the changes in enzyme activity, all four enzyme activities decreased to varying degrees after the tests. As the enzyme was considered an important indicator to characterize the co-contamination of PAHs and heavy metals (Mao et al., 2021), dehydrogenase activities were significantly suppressed by the co-contamination of PAHs and heavy metals. After the tests, the dehydrogenase activities of the bioretention systems were reduced by 20.69% (PS), 37.00% (BSM), and 54.04% (BSM+WTR). Due to the presence of the co-contamination, its inhibitory effects on dehydrogenase activity were significant (Cao et al., 2022). Similar to the change in dehydrogenase activity, the co-contamination of PAHs and heavy metals resulted in the decrease of urease, acid phosphatase, and invertase activities. Urease activity in soil was significantly negatively correlated with PAHs, and the activity decreased significantly with the increase in PAHs contents (Lipińska et al., 2015). After PAHs pollution, acid phosphatase and invertase in the soil will appear "low promotion and high inhibition" phenomenon. In the case of co-contamination of PAHs and heavy metals, the degree of inhibition of enzyme activity was much higher than that of heavy metals or PAHs (Cao et al., 2022). In the actual operating bioretention system, the PAHs and heavy metals in the runoff gradually accumulated in the media. The co-contamination of PAHs and heavy metals not only affected the microbial communities but also has significant inhibitory effects on enzymatic activities. The stability of the media micro-ecosystem was greatly threatened.

3.4 The relationship between media properties and micro-ecosystems

The accumulation of pollutants led to changes in soil/media physical/chemical properties, that affected microbial metabolism. By using redundancy analysis (RDA), the correlation between changes in media properties and the evolution of micro-ecosystems, caused by PAHs-heavy metal co-contamination in bioretention systems, can be revealed (Ahmad et al., 2016). As shown in Fig. 5, different media properties had different effects on microbial diversities, dominated species, and enzyme activities. The effects of the three PAH monomers on the microbial communities were the same. PAHs had positive correlations with *Pseudomonas*, *Massilia*, *and Ramlibacter*, and had negative correlations with other dominant species, especially *Bacillus*. The three PAH monomers were also negatively correlated with microbial diversities and enzyme activities, and the accumulation of PAHs inhibited microbial diversities and enzyme activities. The effects of changes in Cu and Cd contents on micro-ecosystems were similar to those of the three PAHs. The accumulation of Zn promoted microbial diversities and enzyme activities, but its effects were limited. The accumulation of rainfall runoff pollutants and the formation of co-contamination changed the properties of the media. In general, the co-contamination of PAHs and heavy metals led to changes in the media micro-ecosystem. The higher the degree of co-contamination, the greater the change in the micro-ecosystem, while the change was almost negative. These changes significantly affected the biomass of soil microorganisms, the composition of the microbial community, and the activity of microorganisms, and directly or indirectly affect the activity of enzymes (Jian et al., 2016).

3.5 Response processes of media micro-ecosystem to the co-contamination

In bioretention systems, the removal pathways of pollutants mainly include media adsorption, biological/chemical degradation, plant uptake, and volatilization. For PAHs and heavy metals, after they enter the bioretention systems, they were firstly adsorbed by the media, and then the soil-plant-microbe system thoroughly purified pollutants during the rainfall interval (Mehmood et al., 2021a). By comprehensively analyzing the changes in pollutant contents, microbial diversities and community structures, and enzyme activities, the response processes of the media micro-ecosystem to the co-contamination of PAHs and heavy metals were revealed (Fig. 6). The response processes were divided into four stages.

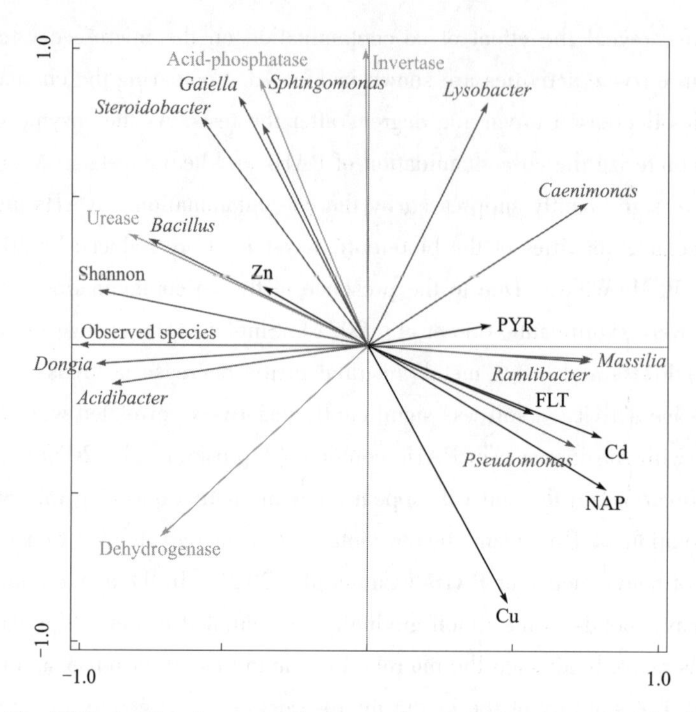

Fig. 5 The RDA result between media properties and micro-ecosystem in the media

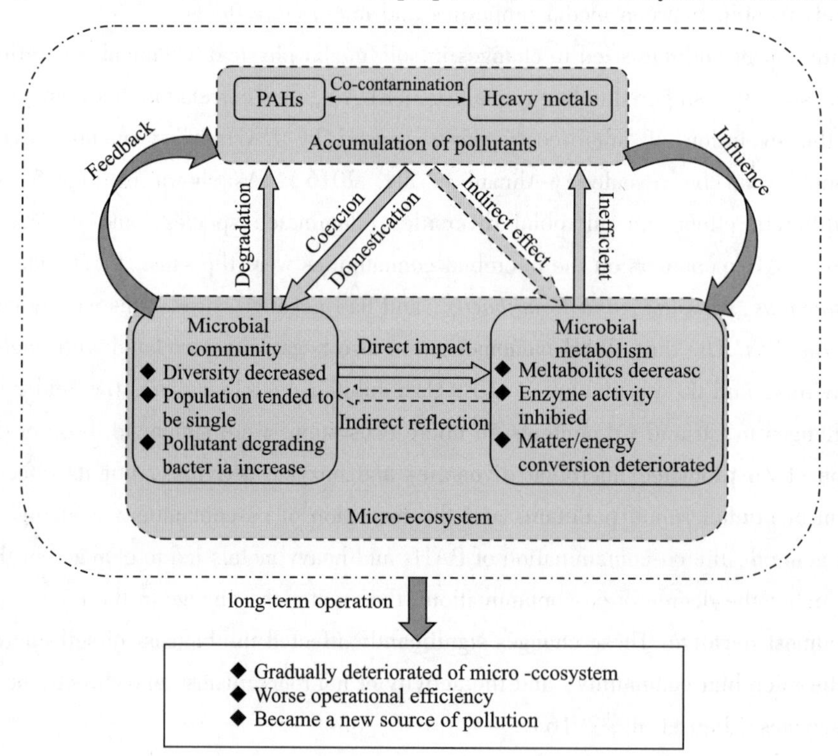

Fig. 6 Response mechanism of micro-ecosystem to PAHs-heavy metal co-contamination in bioretention system

Stage 1: Adsorption and accumulation of pollutants. When rainfall events occurred, rainfall runoff pooled into bioretention systems. The runoff pollutants were quickly trapped by the media and accumulated through the adsorption process. Since PAHs and heavy metals mainly existed in the runoff in the form of particles or colloids (De Buyck et al., 2021), PAHs and heavy metals were accumulated on the upper layers of the bioretention systems (Fig. 3). Stage 2: Adaptation of microbial community. The accumulation of pollutants first affected the micro-ecosystem in the media by changing the internal habitat and the microbial community. On the one hand, when pollutants continuedly accumulated, the relative abundances of microorganisms that can grow under low-nutrient conditions were significantly reduced, and microorganisms that could grow and reproduce quickly under eutrophic conditions increased greatly (Zhang et al., 2022b). Because the co-contamination of PAHs and heavy metals had certain stress effects on the microorganisms, microorganisms were domesticated to adapt to the accumulation of these pollutants (Ukalska-Jaruga & Smreczak, 2020). Microorganisms that were adapted to this condition were grown rapidly (such as *Pseudomonas* and *Massilia*), and that did not adapt to this environment were eliminated. These effects led to the distribution of microorganisms concentrated in one or a few categories. The relative abundances of *Proteobacteria* were all >60% in all bioretention systems, which also meant that the relative abundances of other microorganisms have dropped significantly (Zhang et al., 2021a). Stage 3: Change of microbial metabolism. The most intuitive impact of changes in the microbial communities was that the metabolism of microorganisms was affected, especially the enzyme activities (Liu et al., 2020). When the microbial communities changed under the stress of co-contamination by PAHs and heavy metals, their abilities to produce/secret enzymes were also affected. Due to the co-contamination of PAHs and heavy metals, the inhibitory effects on the enzyme activities of the media were significant. In the presence of PAHs and heavy metals, the activities of the four enzymes involved in the tests were significantly inhibited (the activities of the four enzymes after the tests were 16.45%~83.97% of those before the tests). Different types of enzymes played different roles in microbial metabolism and pollutant degradation, and the inhibition of enzyme activities affected the material and energy conversion processes in the media. Stage 4: Feedback of micro-ecosystem. After a certain period of adaptation of the microorganisms in the media of bioretention system, due to domestication, the microorganisms were concentrated on the species which can grow in the media environment. As the stress effect of the co-contamination on microbes was stronger (Liu et al., 2017), the microbial species tend to PAHs/heavy metals degrading bacteria. The growth of functional microorganisms involved in the carbon, nitrogen, and phosphorus cycle was inhibited. For example, the inhibition of dehydrogenase activity resulted in a reduction in the degradation activity of organic matter in the media (Guo et al., 2019). The simplification of the microbial population and the inhibition of enzyme activities gradually reduced the biodegradation efficiency of runoff pollutants, and the contents of the pollutants in the media increased with the operation time. The co-contamination of PAHs and heavy metals made the media micro-ecosystem gradually adapt to this environment, but also led to the gradual deterioration of the stability of the micro-ecosystem. Other runoff pollutants were not completely degraded, and the operating efficiency of the bioretention system was reduced. All of these evolved the bioretention systems into a new source of pollution and there are indeed potential contamination risks.

Faced with the potential risks of single contaminant or co-contamination in bioretention systems, future research needs to clarify the contamination processes, reveal the change mechanisms, and explore the pollution effects. The potential contamination risk in the bioretention system can be avoided through internal optimization (pretreatment, facility configuration, and plant design) and external intervention (media replacement, chemical, and bioremediation). The design parameters must be optimized at the design stage of the bioretention system to avoid potential contamination risks. Once the contamination risk occurs, the development of in-situ efficient pollutants decontamination technology is the key. Plants with high pollutants uptake ability (woody plants) can be planted to improve the uptake effect, and chemical inducers or exogenous microorganisms can be added to enhance the degradation effects. Balancing the pollutant adsorption performance and potential contamination risk of bioretention systems can facilitate the efficient application and promotion of bioretention systems.

4 Conclusions

The improvement of the media in the bioretention systems effectively improved the purification efficiency of runoff pollutants. After the bioretention system exhibits efficient contaminant removal performance, runoff pollutants will accumulate in the media. Due to the co-existence of PAHs and heavy metals in the media, the co-contamination of PAHs and heavy metals affected the stability of the media micro-ecosystem. The co-contamination had negative effects, including decreases in microbial diversities, the tendency of microbial communities to be single, and decreases in enzyme activities. In practical situations, the co-contamination in the bioretention systems was long-term, low-concentration, and high-frequency. If the bioretention systems with co-contamination operated for long times, the micro-ecosystem of the media was at poor levels, and the operating efficiency was also suppressed. The operating efficiency of bioretention systems was closely related to its media micro-ecosystem, and the stability of the micro-ecosystem was positively related to the degradation efficiency of accumulated pollutants. To ensure the operational efficiency of bioretention systems, some ideas can be used to keep the efficiency and extend the life span of bioretention systems with high pollutant accumulation levels: planting plants with high pollutants uptake ability (woody plants), and adding chemical inducers or exogenous microorganisms. In future research, it is necessary to ensure the adsorption capacity of the media and the purification capacity of the pollutants after excessive accumulation, so as to avoid the bioretention system becoming a new pollution source in the city.

CRediT authorship contribution statement

Zhaoxin Zhang: Validation, Formal analysis, Data curation, Writing-original draft. Jiake Li: Conceptualization, Methodology, Writing-original draft, Funding acquisition, Writing-review & editing. Huanyuan Wang: Writing-review & editing, Formal analysis. Yajiao Li: Writing-review & editing. Xiaolong Duan: Investigation, Formal analysis.

Declaration of competing interest

The authors declare that they have no known competing financial interests or personal relationships that could have influenced the work reported in this paper.

Acknowledgments

This work was financially supported by the National Natural Science Foundation of China (52070157 and 51879215).

References

Abdel-Shafy, H. I., Mansour, M. S. M. (2016). A review on polycyclic aromatic hydrocarbons: Source, environmental impact, effect on human health and remediation. *Egyptian Journal of Petroleum*, 25(1), 107-123.

Agilent Technologies (2019). Agilent bond elut specialty, disk and bulk solid phase extraction (SPE) selection guide.

Ahmad, M., Ok, Y. S., Rajapaksha, A. U., Lim, J. E., Kim, B.-Y., Ahn, J.-H., Lee, Y. H., Al-Wabel, M. I., Lee, S.E., & Lee, S. S. (2016). Lead and copper immobilization in a shooting range soil using soybean stover- and pine needle-derived biochars: Chemical, microbial and spectroscopic assessments. *Journal of Hazardous Materials*, 301, 179-186.

Akpan, S., Umana, S., & Etuk, S. (2020). Polycyclic aromatic hydrocarbon (PAH) degrading potential of vacteria isolated from Iko River sediment. *Microbiology Research Journal International*, 44-54.

Ali, M., Song, X., Ding, D., Wang, Q., Zhang, Z., & Tang, Z. (2022). Bioremediation of PAHs and heavy metals co-contaminated soils: Challenges and enhancement strategies. *Environmental Pollution*, 295, 118686.

Beryani, A., Goldstein, A., Al-Rubaei, A. M., Viklander, M., Hunt, W. F., & Blecken, G.-T. (2021). Survey of the operational status of twenty-six urban stormwater biofilter facilities in Sweden. *Journal of Environmental Management*, 297, 113375.

Biswal, B. K., Vijayaraghavan, K., Tsen-Tieng, D. L., & Balasubramanian, R. (2022). Biochar-based bioretention systems for removal of chemical and microbial pollutants from stormwater: A critical review. *Journal of Hazardous Materials*, 422, 126886.

Boving, T. B., & Neary, K. (2007). Attenuation of polycyclic aromatic hydrocarbons from urban stormwater runoff by wood filters. *Journal of Contaminant Hydrology*, 91(1-2), 43-57.

Cao, X., Cui, X., Xie, M., Zhao, R., Xu, L., Ni, S., & Cui, Z. (2022). Amendments and bioaugmentation enhanced phytoremediation and micro-ecology for PAHs and heavy metals co-contaminated soils. *Journal of Hazardous Materials*, 426, 128096.

Chai, G., Wang, D., Shan, J., Jiang, C., Yang, Z., Liu, E., Meng, H., Wang, H., Wang, Z., Qin, L., Xi, J., Ma, Y., Li, H., Qian, Y., Li, J., Lin, Y. (2022). Accumulation of high-molecular-weight polycyclic aromatic hydrocarbon impacted the performance and microbial ecology of bioretention systems. *Chemosphere*, 298, 134314.

Chen, Y., Shao, Z., Kong, Z., Gu, L., Fang, J., & Chai, H. (2020). Study of pyrite based autotrophic denitrification system for low-carbon source stormwater treatment. *Journal of Water Process Engineering*, 37, 101414.

Chu, Y., Yang, L., Wang, X., Wang, X., & Zhou, Y. (2021). Research on distribution characteristics, influencing factors, and maintenance effects of heavy metal accumulation in bioretention systems: Critical review. *Journal of Sustainable Water in the Built Environment*, 7(2), 03120001.

De Buyck, P.J., Van Hulle, S. W. H., Dumoulin, A., & Rousseau, D. P. L. (2021). Roof runoff contamination: a review on pollutant nature, material leaching and deposition. *Reviews in Environmental Science and Bio/Technology*, 20(2), 549-606.

Fairbairn, D. J., Elliott, S. M., Kiesling, R. L., Schoenfuss, H. L., Ferrey, M. L., & Westerhoff, B. M. (2018). Contaminants of emerging concern in urban stormwater: Spatiotemporal patterns and removal by iron-enhanced sand filters (IESFs). *Water Research*, 145, 332-345.

Fazilah, A., Ismail, N., & Darah, I. (2020). Biodegradation of PAH-polluted soil by indigenous bacteria. *IOP Conference Series: Earth and Environmental Science*, 494(1), 012002.

Guo, C., Li, J., Li, H., & Li, Y. (2019). Influences of stormwater concentration infiltration on soil nitrogen, phosphorus, TOC and their relations with enzyme activity in rain garden. *Chemosphere*, 233, 207-215.

Hou, J., Mao, H., Li, J., & Sun, S. (2019). Spatial simulation of the ecological processes of stormwater for sponge cities. *Journal of Environmental Management*, 232, 574-583.

Jakoncic, J., Jouanneau, Y., Meyer, C., & Stojanoff, V. (2007). The catalytic pocket of the ring-hydroxylating dioxygenase from Sphingomonas CHY-1. *Biochemical and Biophysical Research Communications*, 352(4), 861-866.

Jian, S., Li, J., Chen, J., Wang, G., Mayes, M. A., Dzantor, K. E., Hui, D., & Luo, Y. (2016). Soil extracellular enzyme activities, soil carbon and nitrogen storage under nitrogen fertilization: A meta-analysis. *Soil Biology and Biochemistry*, 101, 32-43.

Jiang, C., Li, J., Li, H., & Li, Y. (2019a). Remediation and accumulation characteristics of dissolved pollutants for stormwater in improved bioretention basins. *Science of The Total Environment*, 685, 763-771.

Jiang, C., Li, J., Li, H., & Li, Y. (2019b). An improved approach to design bioretention system media. *Ecological Engineering*, 136, 125-133.

Jiao, S., Chen, W., Wang, J., Du, N., Li, Q., & Wei, G. (2018). Soil microbiomes with distinct assemblies through vertical soil profiles drive the cycling of multiple nutrients in reforested ecosystems. *Microbiome*, 6(1), 146.

Khudur, L. S., Gleeson, D. B., Ryan, M. H., Shahsavari, E., Haleyur, N., Nugegoda, D., & Ball, A. S. (2018). Implications of co-contamination with aged heavy metals and total petroleum hydrocarbons on natural attenuation and ecotoxicity in Australian soils. *Environmental Pollution*, 243, 94-102.

Kidd, P. S., Álvarez, A., Álvarez-López, V., Cerdeira-Pérez, A., Rodríguez-Garrido, B., Prieto-Fernández, Á., & Chalot, M. (2021). Beneficial traits of root endophytes and rhizobacteria associated with plants growing in phytomanaged soils with mixed trace metal-polycyclic aromatic hydrocarbon contamination. *Chemosphere*, 277, 130272.

Kuppusamy, S., Thavamani, P., Singh, S., Naidu, R., & Megharaj, M. (2017). Polycyclic aromatic hydrocarbons (PAHs) degradation potential, surfactant production, metal resistance and enzymatic activity of two novel cellulose-degrading bacteria isolated from koala faeces. *Environmental Earth Sciences*, 76(1), 14.

Li, Y., Fu, H., Zhang, J., Zhang, Z., & Li, J. (2021). Study of pollutant accumulation characteristics and microbial community impact at three bioretention facilities. *Environmental Science and Pollution Research*, 28(32), 44389-44407.

Lipińska, A., Wyszkowska, J., & Kucharski, J. (2015). Diversity of organotrophic bacteria, activity of dehydrogenases and urease as well as seed germination and root growth Lepidium sativum, Sorghum saccharatum and Sinapis alba under the influence of polycyclic aromatic hydrocarbons. *Environmental Science and Pollution Research*, 22(23), 18519-18530.

Liu, C., Lu, J., Liu, J., Mehmood, T., & Chen, W. (2020). Effects of lead (Pb) in stormwater runoff on the microbial characteristics and organics removal in bioretention systems. *Chemosphere*, 253, 126721.

Liu, L., Wang, F., Xu, S., Sun, W., Wang, Y., & Ji, M. (2021). Woodchips bioretention column for stormwater treatment: Nitrogen removal performance, carbon source and microbial community analysis. *Chemosphere*, 285, 131519.

Liu, S.H., Zeng, G.M., Niu, Q.Y., Liu, Y., Zhou, L., Jiang, L.H., Tan, X., Xu, P., Zhang, C., & Cheng, M. (2017). Bioremediation mechanisms of combined pollution of PAHs and heavy metals by bacteria and fungi: A mini review. *Bioresource Technology*, 224, 25-33.

Malicka, M., Magurno, F., Piotrowska-Seget, Z., & Chmura, D. (2020). Arbuscular mycorrhizal and microbial profiles of an aged phenol-polynuclear aromatic hydrocarbon-contaminated soil. *Ecotoxicology and Environmental Safety*, 192, 110299.

Mangangka, I. R., Liu, A., Egodawatta, P., & Goonetilleke, A. (2015). Performance characterisation of a stormwater treatment bioretention basin. *Journal of Environmental Management*, 150, 173-178.

Mao, Y., Zhang, L., Wang, Y., Yang, L., Yin, Y., Su, X., Liu, Y., Pang, H., Xu, J., Hu, Y., & Shen, X. (2021). Effects of polycyclic aromatic hydrocarbons (PAHs) from different sources on soil enzymes and microorganisms of *Malus prunifolia* var. *Ringo*. *Archives of Agronomy and Soil Science*, 67(14), 2048-2062.

Maurya, S., Abraham, J. S., Somasundaram, S., Toteja, R., Gupta, R., & Makhija, S. (2020). Indicators for assessment of soil quality: A mini-review. *Environmental Monitoring and Assessment*, 192(9), 604.

Mehmood, T., Gaurav, G. K., Cheng, L., Klemeš, J. J., Usman, M., Bokhari, A., & Lu, J. (2021a). A review on plant-microbial interactions, functions, mechanisms and emerging trends in bioretention system to improve multi-contaminated stormwater treatment. *Journal of Environmental Management*, 294, 113108.

Mehmood, T., Lu, J., Liu, C., & Gaurav, G. K. (2021b). Organics removal and microbial interaction attributes of zeolite and ceramsite assisted bioretention system in copper-contaminated stormwater treatment. *Journal of Environmental Management*, 292, 112654.

Rabodonirina, S., Rasolomampianina, R., Krier, F., Drider, D., Merhaby, D., Net, S., & Ouddane, B. (2019). Degradation of fluorene and phenanthrene in PAHs-contaminated soil using Pseudomonas and Bacillus strains isolated from oil spill sites. *Journal of Environmental Management*, 232, 1-7.

Rosner, A., Armengaud, J., Ballarin, L., Barnay-Verdier, S., Cima, F., Coelho, A. V., Domart-Coulon, I., Drobne, D., Genevière, A.-M., Jemec Kokalj, A., Kotlarska, E., Lyons, D. M., Mass, T., Paz, G., Pazdro, K., Perić, L., Ramšak, A., Rakers, S., Rinkevich, B., … Cambier, S. (2021). Stem cells of aquatic invertebrates as an advanced tool for assessing ecotoxicological impacts. *Science of The Total Environment*, 771, 144565.

Shrestha, P., Hurley, S. E., & Wemple, B. C. (2018). Effects of different soil media, vegetation, and hydrologic treatments on nutrient and sediment removal in roadside bioretention systems. *Ecological Engineering*, 112, 116-131.

Søberg, L. C., Winston, R., Viklander, M., & Blecken, G.-T. (2019). Dissolved metal adsorption capacities and fractionation in filter materials for use in stormwater bioretention facilities. *Water Research*, 4, 100032.

Tedoldi, D., Chebbo, G., Pierlot, D., Kovacs, Y., & Gromaire, M.-C. (2016). Impact of runoff infiltration on contaminant accumulation and transport in the soil/filter media of Sustainable Urban Drainage Systems: A literature review. *Science of The Total Environment*, 569-570, 904-926.

Tian, J., Jin, J., Chiu, P. C., Cha, D. K., Guo, M., & Imhoff, P. T. (2019). A pilot-scale, bi-layer bioretention system with biochar and zero-valent iron for enhanced nitrate removal from stormwater. *Water Research*, 148, 378-387.

Tirpak, R. A., Afrooz, A. N., Winston, R. J., Valenca, R., Schiff, K., & Mohanty, S. K. (2021). Conventional and amended bioretention soil media for targeted pollutant treatment: A critical review to guide the state of the practice. *Water Re-*

search, 189, 116648.

Ukalska-Jaruga, A., & Smreczak, B. (2020). The Impact of Organic Matter on Polycyclic Aromatic Hydrocarbon (PAH) Availability and Persistence in Soils. *Molecules*, 25(11), 2470.

Vijayaraghavan, K., Biswal, B. K., Adam, M. G., Soh, S. H., Tsen-Tieng, D. L., Davis, A. P., Chew, S. H., Tan, P. Y., Babovic, V., & Balasubramanian, R. (2021). Bioretention systems for stormwater management: Recent advances and future prospects. *Journal of Environmental Management*, 292, 112766.

Wang, H., Lou, J., Gu, H., Luo, X., Yang, L., Wu, L., Liu, Y., Wu, J., & Xu, J. (2016). Efficient biodegradation of phenanthrene by a novel strain Massilia sp. WF1 isolated from a PAH-contaminated soil. *Environmental Science and Pollution Research*, 23(13), 13378-13388.

Wu, C., Li, F., Yi, S., & Ge, F. (2021). Genetically engineered microbial remediation of soils co-contaminated by heavy metals and polycyclic aromatic hydrocarbons: Advances and ecological risk assessment. *Journal of Environmental Management*, 296, 113185.

Xu, D., Lee, L. Y., Lim, F. Y., Lyu, Z., Zhu, H., Ong, S. L., & Hu, J. (2020). Water treatment residual: A critical review of its applications on pollutant removal from stormwater runoff and future perspectives. *Journal of Environmental Management*, 259, 109649.

Xu, Y., Liu, Y., Zhang, B., Bu, C., Wang, Y., Zhang, D., Xi, M., & Qin, Q. (2021). Enhanced removal of sulfamethoxazole and tetracycline in bioretention cells amended with activated carbon and zero-valent iron: System performance and microbial community. *Science of The Total Environment*, 797, 148992.

Yang, F., Fu, D., Zevenbergen, C., & Rene, E. R. (2022). A comprehensive review on the long-term performance of stormwater biofiltration systems (SBS): Operational challenges and future directions. *Journal of Environmental Management*, 302, 113956.

Yin, D., Chen, Y., Jia, H., Wang, Q., Chen, Z., Xu, C., Li, Q., Wang, W., Yang, Y., Fu, G., & Chen, A. S. (2021). Sponge city practice in China: A review of construction, assessment, operational and maintenance. *Journal of Cleaner Production*, 280, 124963.

Zhang, Z., Li, J., Li, Y., Wang, D., Zhang, J., & Zhao, L. (2021a). Assessment on the cumulative effect of pollutants and the evolution of micro-ecosystems in bioretention systems with different media. *Ecotoxicology and Environmental Safety*, 228, 112957.

Zhang, L., Lu, Q., Ding, Y., & Wu, J. (2021b). A procedure to design road bioretention soil media based on runoff reduction and pollutant removal performance. *Journal of Cleaner Production*, 287, 125524.

Zhang, Z., Li, J., Li, Y., Zhao, L., & Duan, X. (2022a). Accumulation of polycyclic aromatic hydrocarbons in the road green infrastructures of sponge city in Northwestern China: Distribution, risk assessments and microbial community impacts. *Journal of Cleaner Production*, 350, 131494.

Zhang, Z., Li, J., Jiang, C., Li, Y., & Zhang, J. (2022b). Impact of nutrient removal on microbial community in bioretention facilities with different underlying types/built times at field scale. *Ecological Engineering*, 176, 106542.

Zuo, X., Guo, Z., Wu, X., & Yu, J. (2019). Diversity and metabolism effects of microorganisms in bioretention systems with sand, soil and fly ash. *Science of The Total Environment*, 676, 447-454.

本文曾发表于2022年《Journal of Cleaner Production》第383卷第2023期

Advance in Remediated of Heavy Metals by Soil Microbial Fuel Cells: Mechanism and Application

Yingying Sun[1], Hui Wang[2,3], Xizi Long[4], Hui Xi[1], Peng Biao[1] and Wei Yang[1]

(1. Technology Innovation Center for Land Engineering and Human Settlements, Shaanxi Land Engineering Construction Group Co., Ltd. and Xi'an Jiaotong University, Xi'an, China; 2. State Key Laboratory of Eco-hydraulics in Northwest Arid Region, Xi'an University of Technology, Xi'an, China; 3. Department of Municipal and Environmental Engineering, School of Water Resources and Hydroelectric Engineering, Xi'an University of Technology, Xi'an, China; 4. International Center for Materials Nanoarchitectonics (WPI-MANA), National Institute for Materials Science, 1-1 Namiki, Tsukuba, Japan)

【Abstract】 In the past decade, studies on the remediation of heavy metals contaminated soil by microbial fuel cells (MFCs) have attracted broad attention because of the self-generated power and their multifield principles such as the extracellular electron transfer (EET) reduction, electromigration for heavy metals removal. However, given the bio electro-motive power from soil MFCs is weak and fluctuated during the remediation, we need to comprehensively understand the origination of driving force in MFC based on the analysis of the fundamental rationale of ion moving in cells and improve the performance *via* the appropriate configurations and operations. In this review, we first described the structures of soil MFCs for heavy metals remediation and compared the advantages of different types of configurations. Then, based on the theoretical models of heavy metal migration, enrichment, and reduction in soil MFCs, the optimization of soil MFCs including the length of the remediation area, soil conductivity, control of electrode reaction, and modification of electrodes were proposed. Accordingly, this review contributes to the application of bioelectrochemistry to efficiently remove heavy metals from soils.

【Keywords】 Soil microbial fuel cell; Heavy metal reduction; Mass transfer; Soil remediation; Redox reaction; Electric field intensity

Introduction

According to the Ministry of Natural Resources of China, there were 53598 mines by the end of 2020 (Ministry of Natural Resources, PRC, 2020b), while abandoned open-pit mines in key watersheds and regions, such as the Yellow River Basin and the Fenwei Plain, have significantly contributed to heavy metal pollution (Ministry of Natural Resources, PRC, 2020a). Metal (loid) exploitation and mining activities have led to increased toxicity of ecosystems and threaten human health when their residuals were released into the water, soil, air, or food chain (Liu et al., 2019). Over the last three decades, various *in-situ* and *ex-situ* soil remediation techniques have been developed to remediate heavy metal-contaminated soils, which can be grouped into soil washing, soil replacement, electrokinetic remediation, chemical fixation, chemical leaching, phytoremediation, and bioremediation (Liu et al., 2017; Ali et al., 2022). However, these physicochemical techniques are energy-intensive, cost-effective, and emit greenhouse gases into the atmosphere. Although the biological treatment techniques are economical and environmentally friend, they are greatly limited in the treatment of heavy metals with low bioavailability (Rajendran et al., 2022).

Based on the extracellular electron transfer (EET) process discovered in electroactive bacteria (EAB), such as the *Geobacter* sp., soil microbial fuel cells (MFCs) are increasingly developed and studied in recent years and have attracted significant attention as an environmentally sustainable bioelectrochemical technology (Li et al., 2017; Chen et al., 2021). Soil MFCs offer an alternative approach by providing electron donors/acceptors, thereby enhancing bioremediation processes, and migrating while reducing heavy metals in soil (Wang et al., 2020a; Zhang et al., 2020b). The principle of remediating heavy-metal-contaminated soils *via* MFCs lies in the fact that the microorganisms in the deeper subsoil (served as an anode in an anaerobic environment) can oxidize organic matter to generate electrons. It is followed by the electrons transfer to reach the aerobic surface layer of the soil (cathode) *via* an external circuit to the electron acceptor (oxygen or heavy metal). The redox reaction couples organic matter oxidation-oxygen/heavy metal reduction and accompanies the current generation. Simultaneously, heavy metals migrate from the subsoil to the surface layer under the electric field generated from the soil MFC (Wang H. et al., 2016). The remediation of heavy metals in soil by MFC mainly accomplished in two ways:

(1) reducing the bioavailability of heavy metals by electrical migration from soil (Wang H. et al., 2016; Zhang et al., 2020a; Hemdan et al., 2022). (2) bioelectrochemical reduction of heavy metal to the low valence, associated with the precipitation/ detoxication in the MFCs (Habibul et al., 2016b; Kabutey et al., 2019).

Nonetheless, heavy metals removal *via* soil MFCs is in its infancy and there are some issues to be studied urgently: (1) the effect of soil's physical and chemical properties on MFC's electricity generation and heavy metal removal; (2) the relationship between electricity generation and heavy metal migration, enrichment, and reduction; (3) optimizing the construction and components of soil MFCs. Hence, in the present review, the mechanisms of soil MFCs involved in the current generation and heavy metal remediation are discussed. In addition, theoretical models of heavy metal migration, enrichment, and reduction in soil MFCs are analyzed. It is highly expected that this review can provide useful information and suggestions to promote the practical and sustainable application of this technology.

The soil MFCs with different structural configurations and soils

The MFC established in the soil is developed based on the rationale of the fuel cell. The anode harvests electrons from the degraded organics, such as acetate, and amino acids in the soil by EAB. Then, the cathode receives electrons through an external circuit where oxygen is reduced to water, associated with the migration of protons (Dunaj et al., 2012). As shown in Fig. 1, various types of soil MFCs have been designed for heavy metal remediation. The single-chamber soil MFCs take the advantage of redox potential/oxygen gradients between the subsoil and the topsoil to set the anode and cathode, respectively, while the proton exchange membrane can be saved (Huang et al., 2020). Wang H. et al. (2016) established a single-chamber soil MFC to study the migration of copper. The six sections along the soil revealed an obvious accumulation of Cu^{2+} near the cathode from 150 to 250 mg/kg, indicating the feasibility of the single chamber soil MFC (Fig. 1A). However, heavy metal migration is not persistent because of the deficiency of carbon sources which leads to the power density weakening. Meanwhile, to replenish the carbon source, exterior organics such as straw, and sodium acetate were added to soil MFC. In addition, heavy metals cannot migrate from the soil to the cathode (such as activated carbon) because of heterogeneity, thereby minimizing the remediation effects (Wang H. et al., 2016). Hence, to enhance heavy metal removal, plant-microbial fuel cells (PMFCs), where plants have been placed in the topsoil and the root exudates supplied organics to EAB in the anode (Fig. 1B), have been developed and used for heavy metal contaminated waters and soils (Guan et al., 2019b; Li et al., 2020). Moreover, despite the cathodic bioelectrochemical reduction, the direct reduction by reducing microorganisms, plant uptake, and adsorption by electrodes, enriched the pathway and efficiency for heavy metal removal (Habibul et al., 2016b).

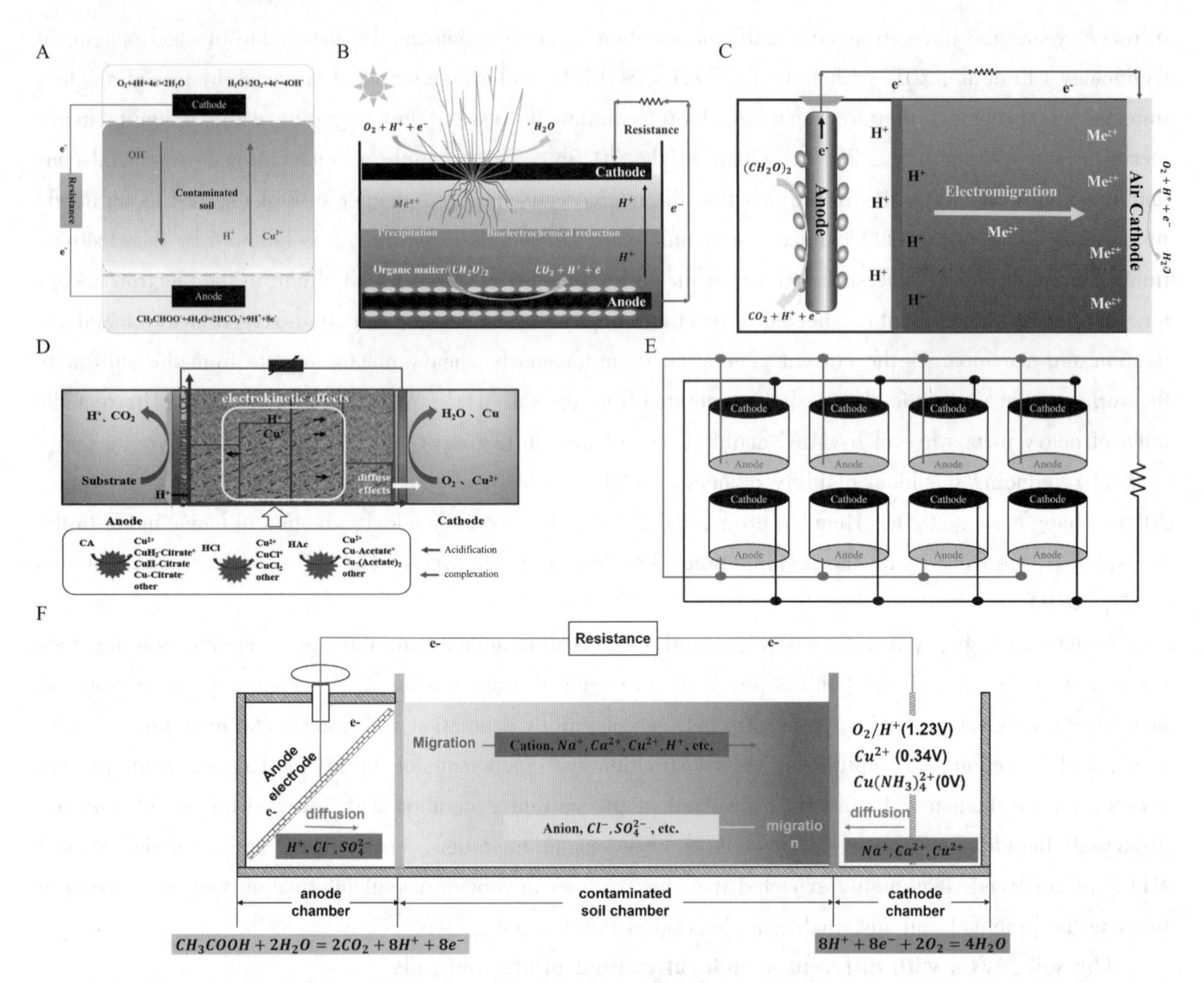

Fig. 1 Typical configurations of soil MFCs. (A) single-chamber (Wang H. et al., 2016), (B) plant (Guan et al., 2019b), (C) two-chamber (Habibul et al., 2016a), (D) three-chamber (Zhang et al., 2020a), (E) stack MFCs (Dziegielowski et al., 2021), and (F) Mass transfer of multi-ions in soil MFC induced by diffusion and migration. All panels are with the permission from publishers' copyright

Given that the EAB in single-chamber soil MFC would be inactive because of the toxicity of heavy metals and also would be difficult to be accumulated in soil located anode, the double-chamber air-cathode soil MFCs were constructed and were able to generate a stronger electrical field capable of powering electrokinetic remediation (Fig. 1C; Habibul et al., 2016a; Wang C. et al., 2016). Compared to the single-chamber soil MFC, the anode and cathode chambers here are separated by a proton/cation exchange membrane. The resultant removal efficiencies were 31.0% for Cd, 41.1% for Pb, and 99.1% for Cr, respectively (Habibul et al., 2016a; Wang C. et al., 2016). However, the heavy metals enriched near the cathode regions could not be processed further owing to the heterogeneity of soil and electrode, precipitation, metal species characteristics, etc. Thereby, to further enhance the cathode reduction, the three-chamber soil MFCs consisting of an anode, a contaminated soil chamber, and a cathode chamber for heavy metal removal were established (Fig. 1D; Wang et al., 2020a; Zhang et al., 2021b). This type of soil MFCs not only supports heavy metal migration from the soil toward the cathode but can also reduce the heavy metals in the cathode by adjusting the current or voltage generated. Regarding the use of soil MFCs as a power source and their efficient and stable performance over long periods, Dziegielowski et al. (2020, 2021) have developed and scaled a stack of soil MFCs to generate sufficient renewable energy for powering a water treatment electrochemical reactor, as a way of using soil MFCs as a power source (Fig. 1E).

Soil is a heterogeneous multiphase system, and its types affect the performance of soil MFCs. It was reported that the red soil generated a higher current than the fluvo-aquic soil while showing a higher Cr(VI) removal efficiency and cathode efficiency since there were more electron acceptors in red clay than in fluvo-aquic soil, such as Fe(III) (Wang C. et al., 2016). These electron acceptors would compete with the reduction of Cr(VI) while hindering the removal of Cr(VI) by MFC. However, more electrons will enhance the electricity-generating performance of MFC. And red soil contains more clay particles, which have stronger adsorption to Cr(VI), thus hindering the migration of Cr(VI) in the soil to the cathode (Wang C. et al., 2016). Moreover, Zhang et al. (2020b) believed that the physical and chemical properties of different soils would affect the removal of heavy metals in the soil MFCs. Decreasing soil pH, total organic carbon, and cation exchange capacity could promote the heavy metal of diffusion and electromigration. Higher soil electrical conductivity and pH could improve the electricity generation performance, which enhanced the electromigration of copper ions (Habibul et al., 2016a; Zhang et al., 2020b).

Ion migration is greatly inhibited when the soil moisture content is unsaturated; thus, regardless of the soil MFC type, considerable limitations exist when remediating contaminated soil (Popat and Torres, 2016; Kumar et al., 2017). As the anode and cathode pH significantly differ, protons generated in the former are transported more slowly to the latter when compared to the inverse route (Gil et al., 2003). This is because the cathode readily accepts electrons and is reduced to hydroxyl ions (De Schamphelaire et al., 2008). Consequently, heavy metals are precipitated from hydroxides on/near the cathode, restricting the migration of metal ions (Hicks and Tondorf, 1994). Thus, the construction of *in-situ* soil MFCs in contaminated soils remains a challenge. The low performance of soil MFCs for the removal of heavy metals is shown in Table 1. It is difficult to maintain sufficient soil organic matter content for sustaining the metabolism of electroactive microorganisms in a long-term operation. Whereas laboratory studies often add simple carbon sources to the soil for electricity production, and heavy metal removal and migration (Huang et al., 2020; Zhang et al., 2020b, 2021a); the addition of such carbon sources is not sustainable for *in situ* soil pollution remediation and engineering applications. Therefore, the configuration of soil MFC should be further studied along with the in-depth study of the EET mechanisms and its coupled redox reactions of heavy metals.

The model of migration, enrichment, and reduction process of heavy metals in soil MFCs

The underlying mechanisms of soil MFCs for heavy metal removal are based on electromigration and electroosmotic flow (Chen et al., 2015; Habibul et al., 2016a). It has been reported that Cu, Cd, and Pb gradually accumulate along the soil from the anode to the cathode owing to the established electric field (Figure 1F; Habibul et al., 2016a; Wang H. et al., 2016). However, even though the migration and movement of heavy metals have been observed in previous studies, experimental trials targeting effective improvement are still lacking, in part related to the ambiguity of rationale for the underlying process driving MFC cells. Conventionally, the process of electrokinetic remediation (EKR) of soil is regarded as an analogy to soil MFC. However, the differences underlying the externally supplied power of EKR and the self-constructed electric field inside the soil MFC cell are commonly neglected, where: (1) the EKR process was conducted under an electric field strength much higher than that of the soil MFC (Liu and Logan, 2004; Li et al., 2014); or (2) the distinct mechanisms are largely ignored as the redox reactions in EKR are the electrolysis of water, while the reactions in soil MFCs are established by the potential difference between the oxidation of organics and the reduction of oxygen/heavy metals (Logan et al., 2006).

The overall mass transfer of heavy-metal ions is driven by the electrochemical potential of the electric field. According to the bias of the potential and additional velocity of the solution, the flux of the ions is described by the Nernst-Planck equation (Bard and Faulkner, 2001):

$$J_i = -\frac{Z_i F}{RT} D_i C_i \nabla \Phi + C_i V - D_i \nabla C_i \tag{1}$$

where Z_i represents the valence of ions, F is the Faradaic constant (C/mol), R is the gas constant (J/K · mol), T is the temperature (K), D_i is the diffusivity of the ions (m^2/s), C_i is the concentration (mol), Φ is the electric field strength (V/m), and V is the fluid velocity (m^3/s). In general, heavy metals can migrate from the anode to the cathode *via* the electrical field capable of powering electrokinetic remediation. In addition, despite the cathodic bioelectrochemical reduction, the direct reduction by reducing microorganisms, plant uptake, and adsorption by electrodes, enriched the pathway and efficiency for heavy metal removal.

Optimizing soil MFC operation and configuration for heavy metal removal

For the convenience of the ensuing presentation, copper is taken as an example of the target pollutant (Fig. 1F). Electromigration is determined by the valence of ions as well as the external electric field and is regarded as the main driving force in EKR for heavy metal movement (Acar et al., 1995; Han et al., 2021). For soil MFCs, anode potential is primarily controlled by electroactive bacteria and the reduced species on the electrode. Accordingly, the anode potential can be estimated as $E_{anode} = -0.32$ V (for NAD^+/NADH redox pair), while the cathode potential as $E_{cathode} = 0.4$ V (for O_2 4 electrons reduction; Logan et al., 2006). Importantly, this voltage of −0.5 V is ≥200 times less than that used for the EKR process. Furthermore, because of the large internal resistance contributed by the long distance of the remediation area, the polarization further reduces the voltage (Wang H. et al., 2016). Nevertheless, Ca^{2+}, Na^+, and K^+ ions in the soil electrolyte/ buffer compete as electron carriers, thereby decreasing the transference number of heavy metals and undermining the electric migration capacity of heavy metals (Bard and Faulkner, 2001). As such, migration in the soil is severely impaired in MFCs.

As the electromigration and electroosmotic flow caused by strong voltage are predominant in the EKR process, corresponding discussions on the diffusion process are usually omitted. However, the importance of diffusion as the driving force for heavy metal movement must be emphasized in soil MFCs. Diffusion is caused by a concentration gradient, originating from the consumption of redox species on the electrode. In a soil MFC cathode, the competition between oxygen and Cu^{2+} reduction simultaneously dictates the priority of diffusion and migration. The reduction O_2 potential is higher than that of Cu^{2+}, preferentially favoring the O_2 reduction. Similarly, a comparatively large amount of O_2 over Cu^{2+} supports a slightly higher current reduction (i.e., the current output) for migration. In contrast, the weak Cu^{2+} reduction rate retards the formation of a concentration gradient for heavy metals, thereby impairing the diffusion process. Consequently, the control of soil MFCs for heavy metal removal requires further consideration, and such perspectives for system optimization are given below:

(1) The construction of the soil MFC should be designed depending on the length of the remediation area. Despite the intrinsic character of the redox reaction on the electrode, the electrical strength is directly related to the length of the cell. Zhang et al. (2021b) discovered that the internal resistance of soil decreased from 1,176 Ω (20 cm) to 583 Ω (5 cm; corresponding to 2.68 mV · cm^{-1} and 8.92 mV · cm^{-1}, respectively), while shorter lengths of soil MFC were associated with stronger total copper migration removal rates. After 63 days of remediation, the removal rates of acid-extractable copper in the soil were 42.50% and 12.40%, respectively. Considering that a higher internal resistance significantly affected cell polarization and deteriorated voltage output, the length of the soil MFC should be controlled at approximately 5 cm (Zhang et al., 2021b).

Table 1 Soil MFCs for the removal of heavy metals

Heavy metals	Configuration	MFC electrode	Power density	Carbon source/ electron donor	Removal/reducing	Driving force	References
1	Single-chamber	Granular activated carbon for anode and cathode	65.77 mW/m^2	Sodium acetate	Maximum 20%, 56 days	Electric migration	Wang H. et al., 2016
Cr		Carbon felts or graphite carbon felts for the anode and the cathode	469.21 mV	Root exudates	99%, 53 days	Electric migration, adsorption, and reduction	Guan et al., 2019b
Cr		Carbon felts or graphite carbon felts for the anode and the cathode	N/A	Root exudates	2.34-fold accumulated in cathode; 1.89-fold accumulated in plant root (near cathode) after 10 months	Electric migration, adsorption, and reduction	Guan et al., 2019a
Cd		Carbon felts for the anode and the cathode	22.93 mW/m^2	Sodium acetate	Maximum 30%, 50 days	Electric migration	Huang et al., 2020
Zn, Pb		Graphite felt pads for the anode and the cathode	25.7 mW/m^2	Straws	Maximum 30% (Pb) 15% (Zn), 50 days	Electric migration	Song et al., 2018
Cd, Cu, Cr, and Ni		Three carbon felt pads for the anode and the cathode	22.2±1.6 mW/m^2	Root exudates	35.1%, 32.8%, 56.9%, and 21.3% (Cd, Cu, Cr, and Ni in rice grains), 110 days	Electric migration	Gustave et al., 2020
As		Three carbon felt pads for the anode and the cathode	123.0 ± 2.2 mW/m^2	Organics in paddy soil	37.5% in pore water, 60 days	Electric migration	Gustave et al., 2019
As		Three carbon felt pads for the anode and the cathode	12.0 mW/m^2	Organics in paddy soil	47% at the anode, 50 days	Electric migration	Gustave et al., 2018
Cd, Cr	Double-chamber	Carbon brushes for the anode and carbon cloth for the cathode	48.8 mW/m^2	Sodium acetate	Maximum 7.6% (Cr) 12.1% (Cd), 50 days	Electric migration	Wang et al., 2020b
Pb, Cd		Graphite granules for the anode and carbon felt for the cathode	7.5 mW/m^2	Sodium acetate	Maximum 44% (Pb), 108 days; 31% (Cd), 143 days	Electric migration	Habibul et al., 2016a
Cr		Porous carbon felts for the anode and the cathode	200–300 mW/m^2	Sodium acetate	Maximum 35% (Cr), 16 days	Reduction, adsorption	Wang C. et al., 2016

Continued to Table 1

Heavy metals	Configuration	MFC electrode	Power density	Carbon source/ electron donor	Removal/reducing	Driving force	References
Zn, Cd	Three-chamber	Graphite for anode and Graphite mesh/Pt coated for cathode	0.4 mA/cm^2	Sodium acetate	25% (Zn), 18% (Cd), 78 days	Electric migration	Chen et al., 2015
Cu		Carbon felt for anode and stainless-steel mesh for cathode	222.72 mW/m^2	Sodium acetate	2.33-fold accumulated in soil, 100% removal in the cathode, 56 days	Electric migration, reduction	Wang et al., 2020a
Cu		Carbon felt for anode and stainless-steel plate for cathode	58.34 mW/m^2	Sodium acetate	41%, 74 days	Electric migration	Zhang et al., 2020a
Cu		Carbon felt for anode and stainless-steel plate for cathode	65.80 ± 1.29 mW/m^2	Sodium acetate	1.5-fold accumulated in soil, 100% removal in the cathode, 21 days	Electric migration, reduction	Zhang et al., 2021b
Cu		Carbon felt for dual anode and stainless-steel plate for cathode	42.48 mW/m^2	Sodium acetate	24.1%, 56 days	Electric migration,	Zhang et al., 2021a
Cu		Carbon felt for anode and stainless-steel plate for cathode	54 mW/m^2	Sodium acetate	19.3% ± 0.8%, 63 days	Electric migration,	Zhang et al., 2020b

(2) Soil conductivity should be controlled within an appropriate range to enhance MFC voltage output while ensuring the electromigration efficiency of heavy metals. Zhang et al. (2020a) compared the influence of the physical and chemical properties of soil on MFC power generation, observing that a higher soil conductivity promoted the current and decreased electrode polarization. However, the Phosphate-Buffered saline or electrolyte used in ordinary liquid MFCs is not suitable for soil addition. In fact, the widespread and abundant non-reaction ions in the MFC potential range (e.g., Ca^{2+}, Na^{+}, and K^{+}) function as supporting electrolytes in soil MFCs, severely minimizing the percentage of ion current from heavy metals and suppressing electromigration (Chen et al., 2021). To balance the conductivity of soil and the migration efficiency, desorption agents, such as small molecular organic acids (e.g., citric, tartaric, or acetic acid), inorganic acids (e.g., hydrochloric or nitric acid), and synthetic chelating agents (e.g., ethylenediaminetetraacetic acid), can effectively dissolve heavy metals in the acid extractable state, thereby increasing the conductivity of soil MFCs, and resulting in the higher power generation while simultaneously maintaining an increased migration rate of heavy metals (Zhang et al., 2020a).

(3) The electrode reaction rate should be controlled to overcome the competition between different electron acceptors. Clear competition between O_2 and Cu^{2+} occurs because the standard reduction potential (vs. SHE) of O_2/H_2O (1.229 V) is much higher than that of Cu^{2+}/Cu (0.337 V) or Cu with the anionic ligand $Cu(NH)^{2+}/Cu$ (0.0 V; Fig. 1F; Bard and Faulkner, 2001). Thermodynamically, O_2(1.229 V) was reduced on the electrode prior to Cu^{2+}. When the reaction rate (current) of the soil MFC cathode is slow, the electrons transferred to the cathode preferentially react only with the relatively abundant dissolved oxygen (not with Cu^{2+}). Conversely, when the reaction rate of the cathode is relatively fast, the cathode is controlled by the electrode and becomes diffusion-controlled. In this case, the concentration of oxygen and protons on the electrode surface is low and is difficult to replenish. The competition with Cu^{2+} is in turn mitigated, and significantly more Cu^{2+} is reduced to be associated with the higher current. Generally, to achieve a high removal efficiency of heavy metals in soil MFCs, the reaction rate of the electrode should be initially controlled at a low level (for example, loading a large external resistance) to rapidly establish a higher electric field strength for Cu^{2+} mitigation to the cathode. Then, a fast electrode reaction rate was applied to effectively reduce the heavy metals on the cathode, thereby accelerating Cu^{2+} reduction. In addition, multi-heavy metal ions still would be migrated or reduced since the electric field generated by soil MFC had no selectivity for the driving of charged ions (Zhang et al., 2022). Wang et al. (2020b) found that the interaction between negatively charged chromium and positively charged lead in the soil had no major effect on hindering migration. Moreover, the remediation performance of composite heavy metal contaminated soil was better than that of single heavy metal contaminated soil. It should be noted that some heavy metals with negative potential (e.g., Pb^{2+}, Cd^{2+}) could not be reduced unless performed in stacked microbial electrolysis cells with high series voltage (Zhang et al., 2015; Wang Q. et al., 2016).

(4) Designing the electrode in soil MFC to boost the energy conversion for heavy metal removal in soil MFC. Carbon materials are commonly selected to be the anode owing to their biocompatibility and chemical stability. For instance, after high-temperature pyrolysis, the biochar material develops cracks to form a pore structure, which greatly increases the specific surface area (Huggins et al., 2014). In addition, biochar owed good electrical conductivity and capacitance to accommodate electrons, which has greatly promoted the interspecific electron transfer (Sakhiya et al., 2020). Moreover, it was pseudo-discovered that doping metal oxides and conductive polymers with biochar can greatly improve the capacitance characteristics and lead to an increase in the current output of MFC (Thines et al., 2017; Norouzi et al., 2020). This pseudo-capaci-

tance increases the specific capacitance value of the electrode by a factor of 10~100 times compared to the ordinary electric double-layer capacitance, greatly improving the electron storage capacity of the interface (Liang et al., 2021). Meanwhile, these material modifications also introduce a large number of electrons transfer active sites (Feng et al., 2014). The high electrical conductivity, fast reversible redox ability of metal oxides, and abundant functional groups (such as amino and catechol functional groups) on the surface of conductive polymers 4~6 are favorable for the formation of active sites for electron transfer (Du et al., 2017).

Overall, three factors were observed when using soil MFCs to remediate heavy metal-contaminated soil. Firstly, improving soil conductivity and increasing the output voltage/current of soil MFCs will promote the migration of metal ions in the soil and the efficiency of cathode reduction; Secondly, more electrons from the anode can be used to reduce heavy metals by adjusting the cathode oxidation-reduction potential, thereby facilitating the reduction or morphological changes of heavy metals, reducing the concentration of heavy metals in the cathode, and increasing the transfer of heavy metals from the anode to cathode; and finally, the range of available remediation areas should be controlled because the electric field intensity generated by soil MFCs is much smaller than that generated by electric remediation.

Conclusion

While the bio-electrochemical method derived from soil MFCs has been developed in the last decades, the soil MFCs suffered from the constrained current, corresponding to the potential for heavy metal removal/reduction. Here, starting from the discussion of the configuration of MFCs, we collectively concluded the power generation and their removal of heavy metals in single-, double-, and triple-chamber soil MFCs. Meanwhile, by comparing the process of electrokinetic remediation (EKR) of soil, the migration, enrichment, and reduction process in soil MFCs were evaluated. Then we proposed the method to optimize soil MFCs operation/construction for heavy metal removal. Generally, our review concludes the perspective and challenge of the soil MFCs and would guide the improvement of soil MFC for heavy metal removal.

Author contributions

HW and XL contributed to the conception and design of the study. YS and PB wrote sections of the manuscript. HX and WY drew these pictures and checked the language. All authors contributed to manuscript revision, read, and approved the submitted version.

Funding

This work was financially supported by Technology Innovation Center for Land Engineering and Human Settlements, Shaanxi Land Engineering Construction Group Co., Ltd. and Xi'an Jiaotong University (2021WHZ0094), National Natural Science Foundation of China (42107030), Natural Science Basic Research Program of Shaanxi (2020JQ-617), and the postdoctoral program from Japan Society for the Promotion of Science (P20105).

Conflict of interest

YS, XH, BP, and YW were employed by the company Shaanxi Land Engineering Construction Group Co., Ltd.

The remaining authors declare that the research was conducted in the absence of any commercial or financial relationships that could be construed as a potential conflict of interest.

The authors declare that this study received funding from Shaanxi Land Engineering Construction Group Co., Ltd. The funder was not involved in the study design, collection, analysis, interpretation of data, the writing of this article or the decision to submit it for publication.

Publisher's note

All claims expressed in this article are solely those of the authors and do not necessarily represent those of their affiliated organizations, or those of the publisher, the editors and the reviewers. Any product that may be evaluated in this article, or claim that may be made by its manufacturer, is not guaranteed or endorsed by the publisher.

References

Acar, Y. B., Gale, R. J., Alshawabkeh, A. N., Marks, R. E., Puppala, S., Bricka, M., et al. (1995). Electrokinetic remediation: basics and technology status. *J. Hazard. Mater.* 40, 117-137.

Ali, M., Song, X., Ding, D., Wang, Q., Zhang, Z., and Tang, Z. (2022). Bioremediation of PAHs and heavy metals co-contaminated soils: challenges and enhancement strategies. *Environ. Pollut.* 295:118686.

Bard, A. J., and Faulkner, L. R. (2001). *Electrochemical Methods: Principles and Applications.* New York: Wiley, 386-428.

Chen, X., Li, X., Li, Y., Zhao, L., Sun, Y., Rushimisha, I. E., et al. (2021). Bioelectric field drives ion migration with the electricity generation and pollutant removal. *Environ. Technol. Innov.* 24:101901.

Chen, Z., Zhu, B.-K., Jia, W.-F., Liang, J.-H., and Sun, G.-X. (2015). Can electrokinetic removal of metals from contaminated paddy soils be powered by microbial fuel cells? *Environ. Technol. Innov.* 3, 63-67.

De Schamphelaire, L., Rabaey, K., Boeckx, P., Boon, N., and Verstraete, W. (2008). Outlook for benefits of sediment microbial fuel cells with two bio-electrodes. *J. Microbial. Biotechnol.* 1, 446-462.

Du, Q., An, J., Li, J., Zhou, L., Li, N., and Wang, X. (2017). Polydopamine as a new modification material to accelerate startup and promote anode performance in microbial fuel cells. *J. Power Sources* 343, 477-482.

Dunaj, S. J., Vallino, J. J., Hines, M. E., Gay, M., Kobyljanec, C., and Rooney-Varga, J. N. (2012). Relationships between soil organic matter, nutrients, bacterial community structure, and the performance of microbial fuel cells. *Environ. Sci. Technol.* 46, 1914-1922.

Dziegielowski, J., Metcalfe, B., and Di Lorenzo, M. (2021). Towards effective energy harvesting from stacks of soil microbial fuel cells. *J. Power Sources* 515:230591.

Dziegielowski, J., Metcalfe, B., Villegas-Guzman, P., Martínez-Huitle, C. A., Gorayeb, A., Wenk, J., et al. (2020). Development of a functional stack of soil microbial fuel cells to power a water treatment reactor: From the lab to field trials in north East Brazil. *Appl. Energy* 278:115680.

Feng, C., Lv, Z., Yang, X., and Wei, C. (2014). Anode modification with capacitive materials for a microbial fuel cell: an increase in transient power or stationary power. *Phys. Chem. Chem. Phys.* 16, 10464-10472.

Gil, G.-C., Chang, I.-S., Kim, B. H., Kim, M., Jang, J.-K., Park, H. S., et al. (2003). Operational parameters affecting the performannce of a mediator-less microbial fuel cell. *Biosens. Bioelectron.* 18, 327-334.

Guan, C.-Y., Hu, A., and Yu, C.-P. (2019a). Stratified chemical and microbial characteristics between anode and cathode after long-term operation of plant microbial fuel cells for remediation of metal contaminated soils. *Sci. Total Environ.* 670, 585-594.

Guan, C.-Y., Tseng, Y.-H., Tsang, D. C., Hu, A., and Yu, C.-P. (2019b). Wetland plant microbial fuel cells for remediation of hexavalent chromium contaminated soils and electricity production. *J. Hazard. Mater.* 365, 137-145.

Gustave, W., Yuan, Z.-F., Li, X., Ren, Y.-X., Feng, W.-J., Shen, H., et al. (2020). Mitigation effects of the microbial fuel cells on heavy metal accumulation in rice (Oryza sativa L.). *Environ. Pollut.* 260:113989.

Gustave, W., Yuan, Z.-F., Sekar, R., Chang, H.-C., Zhang, J., Wells, M., et al. (2018). Arsenic mitigation in paddy soils by using microbial fuel cells. *Environ. Pollut.* 238, 647-655.

Gustave, W., Yuan, Z.-F., Sekar, R., Ren, Y.-X., Liu, J.-Y., Zhang, J., et al. (2019). Soil organic matter amount determines the behavior of iron and arsenic in paddy soil with microbial fuel cells. *Chemosphere* 237:124459.

Habibul, N., Hu, Y., and Sheng, G.-P. (2016a). Microbial fuel cell driving electrokinetic remediation of toxic metal contaminated soils. *J. Hazard. Mater.* 318, 9-14.

Habibul, N., Hu, Y., Wang, Y. K., Chen, W., Yu, H. Q., and Sheng, G. P. (2016b). Bioelectrochemical chromium(VI) removal in plant-microbial fuel cells. *Environ. Sci. Technol.* 50, 3882-3889.

Han, D., Wu, X., Li, R., Tang, X., Xiao, S., and Scholz, M. (2021). Critical review of electro-kinetic remediation of contaminated soils and sediments: mechanisms, performances and technologies. *Water Air Soil Pollut.* 232, 1-29.

Hemdan, B., Garlapati, V. K., Sharma, S., Bhadra, S., Maddirala, S., Motru, V., et al. (2022). Bioelectrochemical systems-based metal recovery: resource, conservation and recycling of metallic industrial effluents. *Environ. Res.* 204:112346.

Hicks, R. E., and Tondorf, S. (1994). Electrorestoration of metal contaminated soils. *Environ. Sci. Technol.* 28, 2203-2210.

Huang, G., Zhang, Y., Tang, J., and Du, Y. (2020). Remediation of cd contaminated soil in microbial fuel cells: effects of cd concentration and electrode spacing. *J. Environ. Eng.* 146:04020050.

Huggins, T., Wang, H., Kearns, J., Jenkins, P., and Ren, Z. J. (2014). Biochar as a sustainable electrode material for electricity production in microbial fuel cells. *Bioresour. Technol.* 157, 114-119.

Kabutey, F. T., Antwi, P., Ding, J., Zhao, Q. L., and Quashie, F. K. (2019). Enhanced bioremediation of heavy metals and bioelectricity generation in a macrophyte-integrated cathode sediment microbial fuel cell (mSMFC). *Environ. Sci. Pollut. Res.* 26, 26829-26843.

Kumar, A., Hsu, L. H.-H., Kavanagh, P., Barriere, F., Lens, P. N. L., Lapinsonniere, L., et al. (2017). The ins and outs of microorganism-electrode electron transfer reactions. Nature reviews. *Chemistry* 1:0024.

Li, D., Tan, X.-Y., Wu, X.-D., Pan, C., and Xu, P. (2014). Effects of electrolyte characteristics on soil conductivity and current in electrokinetic remediation of lead-contaminated soil. *Sep. Purif. Technol.* 135, 14-21.

Li, X., Wang, X., Weng, L., Zhou, Q., and Li, Y. (2017). Microbial fuel cells for organic-contaminated soil remedial applications: A review. *Energ. Technol.* 5, 1156-1164.

Li, G.-X., Yang, H.-C., Guo, S., Qi, C.-F., Wu, K.-J., and Guo, F.-F. (2020). Remediation of chromium contaminated soil by microbial electrochemical technology. *Int. J. Electrochem. Sci.* 15, 6143-6154.

Liang, R., Du, Y., Xiao, P., Cheng, J., Yuan, S., Chen, Y., et al. (2021). Transition metal oxide electrode materials for supercapacitors: a review of recent developments. *Nanomaterials* 11:1248.

Liu, H., and Logan, B. E. (2004). Electricity generation using an air-cathode single chamber microbial fuel cell in the presence and absence of a proton exchange membrane. *Environ. Sci. Technol.* 38, 4040-4046.

Liu, J.-L., Yao, J., Lu, C., Li, H., Li, Z.-F., Duran, R., et al. (2019). Microbial activity and biodiversity responding to contamination of metal(loid) in heterogeneous nonferrous mining and smelting areas. *Chemosphere* 226, 659-667.

Liu, S.-H., Zeng, G.-M., Niu, Q.-Y., Liu, Y., Zhou, L., Jiang, L.-H., et al. (2017). Bioremediation mechanisms of combined pollution of PAHs and heavy metals by bacteria and fungi: A mini review. *Bioresour. Technol.* 224, 25-33.

Logan, B. E., Hamelers, B., Rozendal, R., Schröder, U., Keller, J., Freguia, S., et al. (2006). Microbial fuel cells: methodology and technology. *Environ. Sci. Technol.* 40, 5181-5192.

Ministry of Natural Resources, PRC (2020a). *China Mineral Resources* 2020. Beijing: Geological Publishing House.

Ministry of Natural Resources, PRC (2020b). *The report of mineral resources Saving and Comprehensive Utiliaztion in China* 2020.

Norouzi, O., Di Maria, F., and Dutta, A. (2020). Biochar-based composites as electrode active materials in hybrid supercapacitors with particular focus on surface topography and morphology. *J. Energy Storage* 29:101291.

Popat, S. C., and Torres, C. I. (2016). Critical transport rates that limit the performance of microbial electrochemistry technologies. *Bioresour. Technol.* 215, 265-273.

Rajendran, S., Priya, T. A. K., Khoo, K. S., Hoang, T. K. A., Ng, H.-S., Munawaroh, H. S. H., et al. (2022). A critical review on various remediation approaches for heavy metal contaminants removal from contaminated soils. *Chemosphere* 287:132369.

Sakhiya, A. K., Anand, A., and Kaushal, P. (2020). Production, activation, and applications of biochar in recent times. *Biochar* 2, 253-285.

Song, T. S., Zhang, J., Hou, S., Wang, H., Zhang, D., Li, S., et al. (2018). In situ electrokinetic remediation of toxic metal-contaminated soil driven by solid phase microbial fuel cells with a wheat straw addition. *J. Chem. Technol. Biotechnol.* 93, 2860-2867.

Thines, K. R., Abdullah, E. C., Mubarak, N. M., and Ruthiraan, M. (2017). In-situ polymerization of magnetic biochar-polypyrrole composite: a novel application in supercapacitor. *Biomass Bioenergy* 98, 95-111.

Wang, C., Deng, H., and Zhao, F. (2016). The remediation of chromium (VI)-contaminated soils using microbial fuel cells. *Soil Sediment Contam. Int. J.* 25, 1-12.

Wang, Q., Huang, L., Pan, Y., Zhou, P., Quan, X., Logan, B. E., et al. (2016). Cooperative cathode electrode and in situ deposited copper for subsequent enhanced cd(II) removal and hydrogen evolution in bioelectrochemical systems. *Bioresour. Technol.* 200, 565-571.

Wang, H., Long, X., Zhang, J., Cao, X., Liu, S., and Li, X. (2020a). Relationship between bioelectrochemical copper migration, reduction and electricity in a three-chamber microbial fuel cell. *Chemosphere* 241:125097.

Wang, H., Song, H., Yu, R., Cao, X., Fang, Z., and Li, X. (2016). New process for copper migration by bioelectricity generation in soil microbial fuel cells. *Environ. Sci. Pollut. Res.* 23, 13147-13154.

Wang, H., Zhang, H., Zhang, X., Li, Q., Cheng, C., Shen, H., et al. (2020b). Bioelectrochemical remediation of Cr (VI)/cd (II)-contaminated soil in bipolar membrane microbial fuel cells. *Environ. Res.* 186:109582.

Zhang, J., Cao, X., Wang, H., Long, X., and Li, X. (2020a). Simultaneous enhancement of heavy metal removal and electricity generation in soil microbial fuel cell. *Ecotoxicol. Environ. Saf.* 192:110314.

Zhang, J., Jiao, W., Huang, S., Wang, H., Cao, X., Li, X., et al. (2022). Application of microbial fuel cell technology to the remediation of compound heavy metal contamination in soil. *J. Environ. Manage.* 320:115670.

Zhang, J., Liu, Y., Sun, Y., Wang, H., Cao, X., and Li, X. (2020b). Effect of soil type on heavy metals removal in bioelectrochemical system. *Bioelectrochemistry* 136:107596.

Zhang, J., Sun, Y., Zhang, H., Cao, X., Wang, H., and Li, X. (2021a). Effects of cathode/ anode electron accumulation on soil microbial fuel cell power generation and heavy metal removal. *Environ. Res.* 198:111217.

Zhang, J., Wang, H., Zhou, X., Cao, X., and Li, X. (2021b). Simultaneous copper migration and removal from soil and water using a three-chamber microbial fuel cell. *Environ. Technol.* 42, 4519-4527.

Zhang, Y., Yu, L., Wu, D., Huang, L., Zhou, P., Quan, X., et al. (2015). Dependency of simultaneous Cr(VI), cu(II) and cd(II) reduction on the cathodes of microbial electrolysis cells self-driven by microbial fuel cells. *J. Power Sources* 273, 1103-1113.

本文曾发表于2022年《Frontiers in Microbiology》第13卷

Maltose-functionalized HILIC Stationary Phase Silica Gel Based on Self-assembled Oligopeptides and its Application for the Separation of Polar Compounds

Hailan Shi[1,2], Li Zhang[3]

(1. Shaanxi Provincial Land Engineering Construction Group Co. Ltd., Xi'an 710075, China; 2. Institute of Land Engineering and Technology, Shaanxi Provincial Land Engineering Construction Group Co. Ltd., Xi'an 710075, China; 3. School of Chemistry and Chemical Engineering, Shaanxi Normal University, Xi'an 710119, China)

【**Abstract**】 In this study, carbonyldiimidazole was used to bond maltose-modified oligopeptides (Ala-Glu-Ala-Glu-Ala-Lys-Ala-Lys) to the surface of silica spheres for hydrophilic interaction liquid chromatography (HILIC). Attenuated total reflectance-Fourier transform infrared spectroscopy, elemental analysis, X-ray photoelectron spectroscopy, thermogravimetric analysis, BET technique, and water contact angle measurement results confirmed the successful immobilization of the obtained material. Compared with the conventional method for preparing carbohydrate stationary phases, this method involves simpler steps and less time-consuming processes. The experimental results proved that the retention mechanism of the maltose-based HILIC column matched the typical HILIC retention mechanism. The column showed high separation efficiency and stability toward the separation of polar compounds such as amino acids, bases, nucleosides, water-soluble vitamins, and salicylic acid and its analogs. The column achieved high selectivity toward oligosaccharide separation. In addition, this efficient analysis demonstrates the applicability of the as-prepared material in the field of food inspection.

【**Keywords**】 HILIC; Surface modification; Polar compounds separation; Maltose-functionalized

Introduction

With the rapid development of glycoproteomics, metabolomics, and food and environment monitoring fields, highly polar and ionic compounds have become important research materials in biology and chemistry[1]. It is difficult to separate these compounds on reverse-phase liquid chromatography (RPLC), thereby limiting their applications. Although normal-phase liquid chromatography (NPLC) can separate these substances, this method is typically ineffective because of defects caused by the effects of the environment on peak trailing and retention time. Moreover, the mobile phase for NPLC is less soluble than the most hydrophilic samples[2]. HILIC can overcome the limitations of RPLC and serves as a good alternative to NPLC toward the effective separation of various highly polar compounds[3-8]. The use of water and water-soluble organic solvents as mobile phases can improve the dissolution of samples in the mobile phase[9] and allow suitable retention for various highly polar samples. In addition, a high proportion of the organic phase increases the flow rate, reduces the retention time, and effectively improves the detection sensitivity and speed of the instrument[10-12]; HILIC is also compatible with mass spectrometry. Thus, this analytical technique has received widespread attention. However, owing to the diverse and complex analytical requirements of various samples, the need for high chromatographic performance of the stationary phase is gradually increasing.

There is no standard research system for determining the retention mechanism of a new hydrophilic chromatographic stationary phase. Therefore, it is important to explore and develop high-water-affinity stationary-phase materials with novel structures and excellent performances.

Almost all HILIC stationary phases have been modified with high-polarity groups on the surface of a substrate. The polarity of the functional groups determines the separation performance of the stationary phases in hydrophilic and highly polar samples, which directly affects the application and development of HILIC[13-14]. There are many types of HILIC stationary phases with different functional groups, such as amino, poly(succinimide), cyano, glycol, zwitterion, and glycosyl groups. Carbohydrates possess abundant unique polyhydroxy groups and possess high polarity. New carbohydrate-based HILIC stationary phases have been recently explored and developed. Guo et al. used the click reaction to prepare a β-cyclodextrin-type HILIC stationary phase for the first time[15]. This motivated Liang's research group to use azido-alkynyl and sulfhydryl-alkenyl click reactions to prepare maltose-, glucose-, and β-cyclodextrin-modified HILIC stationary phases, which have a strong separation effect on small molecules, such as adenosine, amino acids, and sugars[16]. Huang et al. also used the click reaction to prepare chitosan-modified spherical-particle-based HILIC stationary phases to separate carbohydrates, nucleosides, and amino acid compounds; this column achieved a strong separation effect[17]. However, surface bonding in click chemistry requires the addition of azide or alkynyl groups on the sugar unit, which is difficult to achieve in polysaccharides because of their low solubility in organic solvents. The preparation process of this method is time consuming; therefore, a simpler and more effective method of polysaccharide immobilization is needed to prepare the HILIC stationary phase.

To this end, herein, carbonyldiimidazole (CDI) was used as a cross-linking agent, and AEK8-maltose was immobilized on the silica surface to prepare a maltose-based HILIC stationary phase (Scheme 1). This method for stationary phase preparation is simpler than those used for the development of conventional saccharide stationary phases and does not involve time-consuming processes. In the HILIC mode, the effects of different mobile phase compositions, salt concentrations, pH values, and temperatures on the solute retention behavior were investigated. The results confirmed that the retention mechanism of the maltose-based HILIC column matched that of the typical HILIC column. The column prepared using the assynthesized stationary phase exhibits high separation efficiency and stability for separating polar compounds, such as amino acids, bases, water-soluble vitamins, and organic acids, and achieves high selectivity toward oligosaccharide separation.

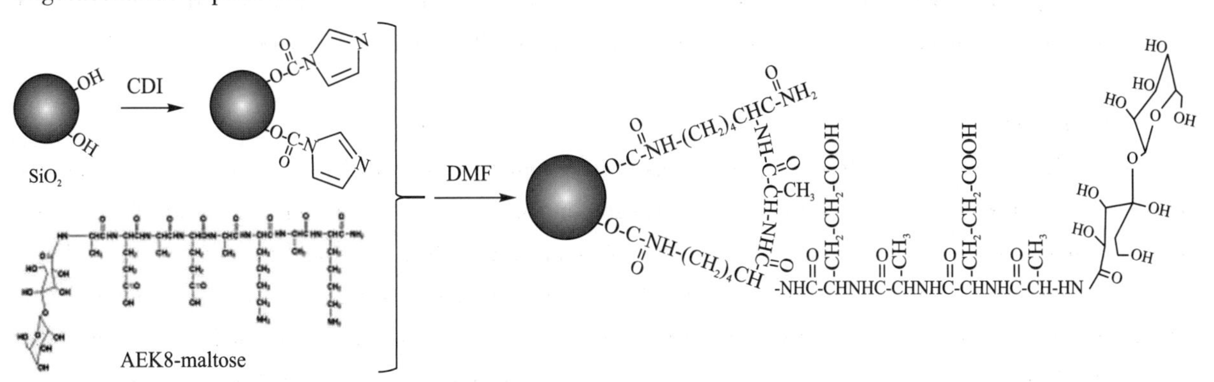

Scheme 1 Preparation procedure of the AEK8-maltose-functionalized SiO_2

Materials and methods

Materials and reagents

Silica gel, CDI, V_{PP}, V_{B2}, V_{B7}, V_{B1}, V_C, salicylamide, salicylic acid, trans-cinnamic acid, acetylsalicylic acid, salicylic acid, N,N-dimethylformamide (DMF), methanol, and acetonitrile were purchased from the Shanghai Chemical Reagents Company (China). 3,5-Dihydroxybenzoic acid, lysine, arginine, glutamic acid, aspartic acid, ribose, glucose, sucrose, melezitose, and trifluoroacetic acid were obtained from Sigma-Aldrich. Deionized water was purified using Milli-Q water prepared in the laboratory and all other chemicals were of analytical grade. All participants provided written informed consent.

Characterization techniques

Attenuated total reflectance-Fourier transform infrared (ATR-FTIR) spectroscopy was performed using a Tensor-27 infrared spectrometer (Bruker, Billerica, MA) with a wedged germanium crystal as an attenuated total reflectance accessory. Transmission electron microscopy (TEM) images were obtained using a JEM-2100 microscope (JEOL, Japan). Thermogravimetric analysis (TGA) was performed using a SACS Q600SDT thermogravimetric analyzer (USA). N2 adsorption surface areas were measured by the BET technique on an ASPS 2020 M analyzer (USA). X-ray photoelectron spectroscopy (XPS) was performed using an Axis Ultra X-ray photoelectron spectrometer (Kratos Analytical Ltd., Manchester, UK) with an Al X-ray source operating at 150 W (15 kV, 10 mA).

Synthesis of AEK8-maltose

The Fmoc-Ala-Glu(OtBu)-Ala-Glu(OtBu)-Ala-Lys(BOC)-MBHA resin was obtained by extending the peptide chain from the C-end to N-end according to the sequence AEAE-AKAK (Ala-Glu-Ala-Glu-Lys-Ala-Lys) on a Rink Amide-MBHA resin by solid-phase synthesis. Maltose and sodium cyanoborohydrin were added to the assynthesized oligopeptide and methanol was used as a solvent for the reflux reaction at 80 ℃, followed by suction filtration and washing. Then, a mixture of trifluoroacetic acid (TFA), triiso-propylsilane (TIS), and water was used as the cutting solution for the residue protective groups of amino acid in the oligopeptide. AEK8-maltose was separated from the resin at room temperature, and suction filtration was performed on it. The filtrate was precipitated using ice-cold ether. After centrifugation, the precipitate was dialyzed and freeze-dried to obtain maltose-modified oligopeptide (AEK8-maltose).

Preparation of maltose-type silica gel stationary phase

A total of 1.0 g of silica gel activated with 10% HCl and 1.17 g of CDI were mixed in 10 mL of a 70% methanol aqueous solution. The solution was refluxed for 5 h at 25 ℃, washed with water and methanol for several times, and dried at 80 ℃. Next, 1.0 g of the as-prepared CDI-activated silica gel was mixed with 6 mL of 15 mg/mL AEK8-maltose DMF solution at 40 ℃ for 5 h. The mixture was washed with DMF and methanol for several times and dried at 60 ℃ to obtain AEK8-maltose-modified silica gel (SiO_2-AEK8-maltose).

Column packing

SiO_2-AEK8-maltose was ultrasonically dispersed in chromatographic methanol, and the dispersion was transferred to a homogenization tube using chromatographic methanol as a replacement liquid. This system was maintained in a column packer at 35 MPa for 40 min, after which the nitrogen valve in the column packer was closed. When the pressure dropped to 0 MPa, the column was packed to obtain a 50 × 4.6 mm chromatographic column. The bare SiO_2 sphere column was prepared using the same method and specifications. After loading the column, it was activated with chromatographic methanol for 2 h prior to use.

Separation and application of maltose-based HILIC column

To explore the performance of the maltose-based HILIC stationary phase, experiments were performed to separate several common polar compounds, such as nucleosides, bases, amino acids, water-soluble vitamins, salicylic acid and its analogs, and oligosaccharides. The separation performance of the maltose-based HILIC column was compared with that of a self-made bare SiO_2 sphere chromatographic column under same conditions.

Reproducibility and stability test

Three batches of simult aneously prepared SiO_2-AEK8-maltose were loaded onto three chromatographic columns (50 mm × 4.6 mm). The samples of V_{PP}, V_{B2}, V_{B7}, V_{B1}, salicylamide, salicylic acid, and salicylide were analyzed under the mobile phase of 85% ACN-20 mM NH_4AcO at a pH 4.0. Then, the stability of the chromatographic behavior of the stationary phase was investigated and the same column was injected with all samples for 1, 5, 15, 30, 45, 60, and 75 days to analyze the changes in the retention behavior.

Results and discussion

Characterization of SiO_2-AEK8-maltose

The morphology of silica was analyzed by TEM. From the TEM images shown in Fig. 1a and b, no apparent changes in the silica spheres, before and after modification, can be observed. It may be because the functionalized AEK8-maltose molecule was too small to be observed. To investigate the success of AEK8-maltose modification on silica spheres, Fourier transform infrared spectroscopy (FT-IR) spectroscopy was performed for the functional groups present on the surface of SiO_2 and SiO_2-AEK8-maltose. As shown in Fig. 1c, the strong absorption peaks at 1100 and 805 cm^{-1} correspond to the symmetric and asymmetric stretching vibrations of Si-O-Si, respectively. After modifying the silica spheres with AEK8-maltose, the stretching vibration absorption peak of CH_2 appeared at 2900 cm^{-1}, and the absorption peaks of amide I and amide II appeared at 1640 and 1560 cm^{-1}, respectively. The characteristic absorption peak of sugar appeared at 3600 cm^{-1}. These results indicated the successful preparation of SiO_2-AEK8-maltose. Furthermore, the average pore size of the prepared stationary phase was approximately 12.4 nm, and the BET surface area was approximately 312 m^2/g (Fig. 1e). The TGA of bare SiO_2 spheres and SiO_2-AEK8-maltose was also analyzed (Fig. 1f). The endothermic mass loss, caused by the loss of water, was observed to be approximately 2.0% over the 0~100 ℃ range. At 200~600 ℃, a mass loss of 2% (from 12.39 to 12.14 g) was observed due to the breaking and dehydration of the silicon-oxygen bond in the silica gel. However, the mass of SiO_2-AEK8-maltose decreased from 9.54 to 9.34 g over 0~100 ℃ owing to the loss of water, resulting in a mass loss of 2.1%. At 200~600 ℃, the mass of SiO_2-AEK8-maltose decreased to 8.53 g, resulting in a mass loss of 8.4%. These observations preliminarily showed that the loss of AEK8-maltose on the silica surface was about 6.4%, indicating that the new stationary phase was modified with 6.4% wt. of AEK8-maltose to bare SiO_2 spheres.

To further confirm that AEK8-maltose was successfully modified on the surface of the silica spheres, the composition of SiO_2 and SiO_2-AEK8-maltose was analyzed by XPS. As shown in Fig. 1d, the XP spectra of both materials exhibit four peaks: O 1 s, C 1 s, Si 2 s, and Si 2p. The XP spectra of the AEK8-maltose-modified silica spheres had a new N 1 s peak, while the peak corresponding to N in the amino group of the peptide was only present, confirming that AEK8-maltose was successfully modified on the surface of the silica spheres. At the same time, the relative content of each element was calculated from the spec-trum; the results are summarized in Table 1. Elemental analysis (EA) was performed before and after modifying the silica spheres, as shown in Table 2. The contents of N, C, and H in the modified silica spheres were significantly increased, while AEK8-maltose contained only N (the trace amount of N in the bare SiO_2 spheres

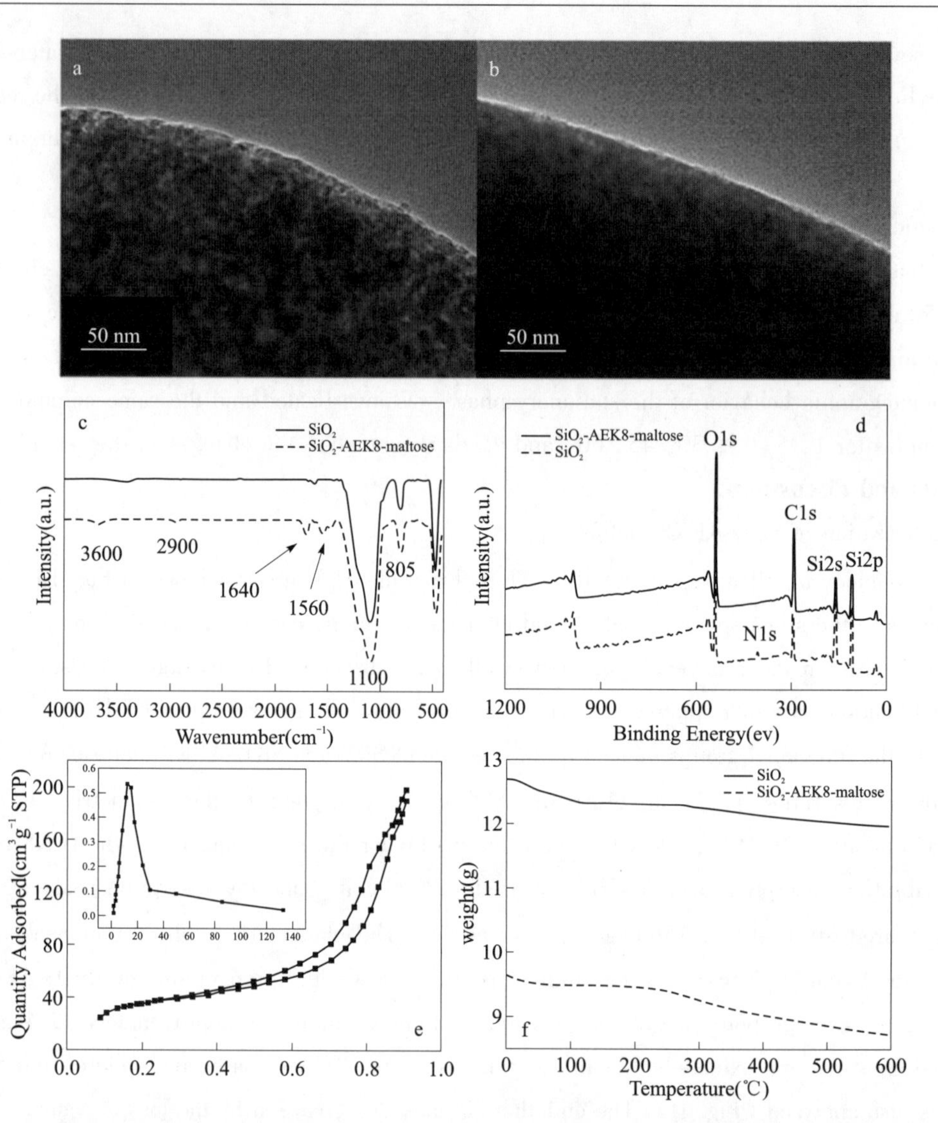

Fig. 1 TEM patterns of (a) SiO_2 and (b) SiO_2-AEK8-maltose. (c) FTIR spectra of SiO_2 and SiO_2-AEK8-maltose. (d) XP spectra of SiO_2 and SiO_2-AEK8-maltose. (e) N_2 adsorption/desorption isotherm of SiO_2-AEK8-maltose. (f) TGA curves of SiO_2 and SiO_2-AEK8-maltose

came from the air), indicating that AEK8-maltose and the silica spheres were successfully bonded. According to the calculation formula for surface bonding[18], the binding amount of AEK8-maltose on the surface of the maltose-based HILIC stationary phase was 0.806 μmol/m².

$$\text{coverage}(\mu\text{mol/m}^2)=\frac{\%X\times10^6}{(A_M)n100(1-\%X(M_W)/(A_M)n100)S}$$

Here, $\%X$ is the percentage increase in carbon or nitrogen in the bonded support determined by EA, A_M is the atomic mass of carbon or nitrogen, M_W is the molecular weight of the species bonded to the silica surface, n is the number of carbon or nitrogen atoms present in the bonded species, and S is the specific surface area of the silica support in meters squared per gram.

After coating SiO_2 and SiO_2-AEK8-maltose on the glass sheet, the water contact angle (WCA) of the coating was measured. As shown in Table 3, the measured contact angle of the SiO_2-AEK8-maltose coating was larger than that of the SiO_2 coating, indicating that AEK8-maltose was successfully modified on the surface of the silica spheres.

Table 1 Relative content of elements in SiO_2 and SiO_2-AEK8-maltose based on their XPS spectra

	C(%)	N(%)	O(%)	Si(%)
SiO_2	14.13	0	52.09	33.78
SiO_2-AEK8-maltose	28.66	2.44	41.89	27.01

Table 2 EA results of SiO_2 and SiO_2-AEK8-maltose

	N (%)	C (%)	H (%)
SiO_2	0.1	1.25	0.6
SiO_2-AEK8-maltose	1.6	6.39	1.486

Influence of water content in the mobile phase

The performance of HILIC is typically analyzed on the basis of the effect of the water content in the mobile phase on solute retention. If the increase in the water content weakens the retention, it is the HILIC mode. The organic phase of the HILIC mode mainly contains acetonitrile and methanol. Methanol is a polar protic solvent that can easily form hydrogen bonds with the stationary phase and competitively adsorbs the solutes, reducing solute retention. Although acetonitrile is more toxic, it does not react with the sample, and the column pressure of acetonitrile as the organic phase is the lowest, which prolongs the service life of the instrument and the chromatographic column. Therefore, acetonitrile was selected as the organic phase for this experiment. The effect of adjusting the proportion of acetonitrile (5%~96%) in the mobile phase on the retention factor (k) of salicylic acid and its analogs and water-soluble vitamins is shown in Fig. 2a. The overall retention behaviors of the two analytes were similar. In the ACN content range of 5%~70%, water-soluble vitamins, salicylic acid, and their analogs are almost not retained; in the high ACN range of the mobile phase, especially when ACN > 85%, k increases sharply. This shows that the maltose-based HILIC stationary phase had strong hydrophilicity, and the retention mechanism of the two analytes on the maltose-based HILIC column was mainly hydrophilic. Therefore, when the column was used to separate polar compounds, the ACN content in the mobile phase was selected within the range of 75%~100%.

Table 3 WCAs of the SiO_2 and SiO_2-AEK8-maltose coatings

Coating additives	WCAs, degrees[a]
SiO_2	3.0 ± 0.3
SiO_2-AEK8-maltose	8.0 ± 0.5

[a] Data are reported as the mean ± standard error (n = 6).

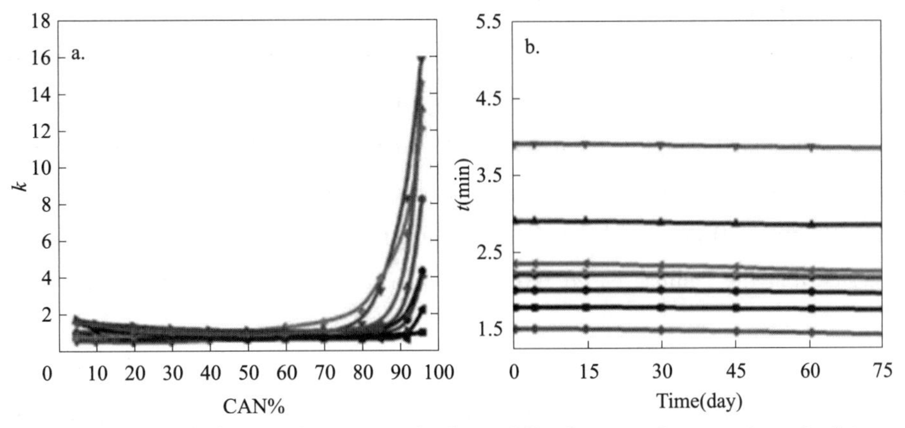

Fig. 2 Effect of (a) the ACN content in the mobile phase on the retention of solutes, and (b) column usage time on the solute retention time

Separation performance of the maltose-based HILIC column

To investigate the performance of the maltose-modified column toward the separation of polar compounds, experiments were performed to separate acids, bases, nucleosides, water-soluble vitamins, salicylic acid and its analogs, and oligosaccharides. This performance was compared with that of the bare SiO_2 sphere column.

Separation of amino acids

The experiment was performed under optimized chromatographic conditions. The mobile phase was 75% acetonitrile/20 mM KH_2PO_4, the pH was 3.50, and the UV detector wavelength was 190 nm. Using amino acids as solutes, the separation effects of the four amino acids on the maltose-based HILIC and bare SiO_2 sphere columns were compared under the same chromatographic conditions. As shown in Fig. 3a and b, the four analytes could be separated well within 5 min on the maltose-based HILIC column, and the peak shape is better than that obtained for the bare SiO_2 sphere column. The separation effect on the bare silica sphere column was weak, indicating that the maltose-based HILIC column can effectively separate amino acids.

Separation of bases and nucleosides

Naphthalene was used as a reference to separate the bases and nucleosides with gradient elution. The optimized chromatographic conditions were as follows: 0~5 min: 97% ACN, 5~15 min: 97%~80% ACN, 20 mM ammonium acetate, pH 4.0 gradient, and 0.6 mL/min flow rate. Under the same chromatogram conditions, the separation effects of these substances on maltose-based HILIC and bare SiO_2 sphere columns were compared. The results indicate that the maltose-based HILIC column allowed for a stronger separation effect than the bare SiO_2 sphere column, as represented in Fig. 3c and d, respectively.

Separation of water-soluble vitamins

To further investigate the separation effect of the maltose-based HILIC column, five water-soluble vitamins were separated. The optimized chromatographic conditions were as follows: 85% acetonitrile/20 mM ammonium acetate and 0.8% acetic acid, pH 4.0 mobile phase; 210 nm detection wavelength; and 0.6 mL/min flow rate. The separation effects of five water-soluble vitamins on the maltose-based HILIC and bare SiO_2 sphere columns under the same chromatographic conditions were compared. As shown in Fig. 3e and f, the five analytes achieved good separation within 8 min on the maltose-based HILIC column, while the bare SiO_2 sphere column could not separate them.

Separation of salicylic acid and its analogs

The optimized chromatographic conditions were as follows: 92% acetonitrile/20 mM ammonium acetate and 0.8% acetic acid, pH 4.0 mobile phase; 200 nm detection wavelength; and 0.6 mL/min flow rate. Salicylic acid and its analogs were used as analytes to investigate the maltose-based HILIC column. The separation effects of the same mixed sample on the maltose-based HILIC and bare SiO_2 sphere columns under the same separation conditions were compared. As shown in Fig. 4a and b, the six analytes achieved were more effectively separated within 10 min on the maltose-based HILIC column than on bare SiO_2 sphere column. This result suggested that the maltose-based HILIC column effectively separated salicylic acid and its analogs.

Separation of oligosaccharides

The maltose-based HILIC column was used to separate oligosaccharides. The optimized chromatographic conditions were as follows: 75% acetonitrile/water mobile phase; 35 ℃ column temperature; 350 kappa N_2

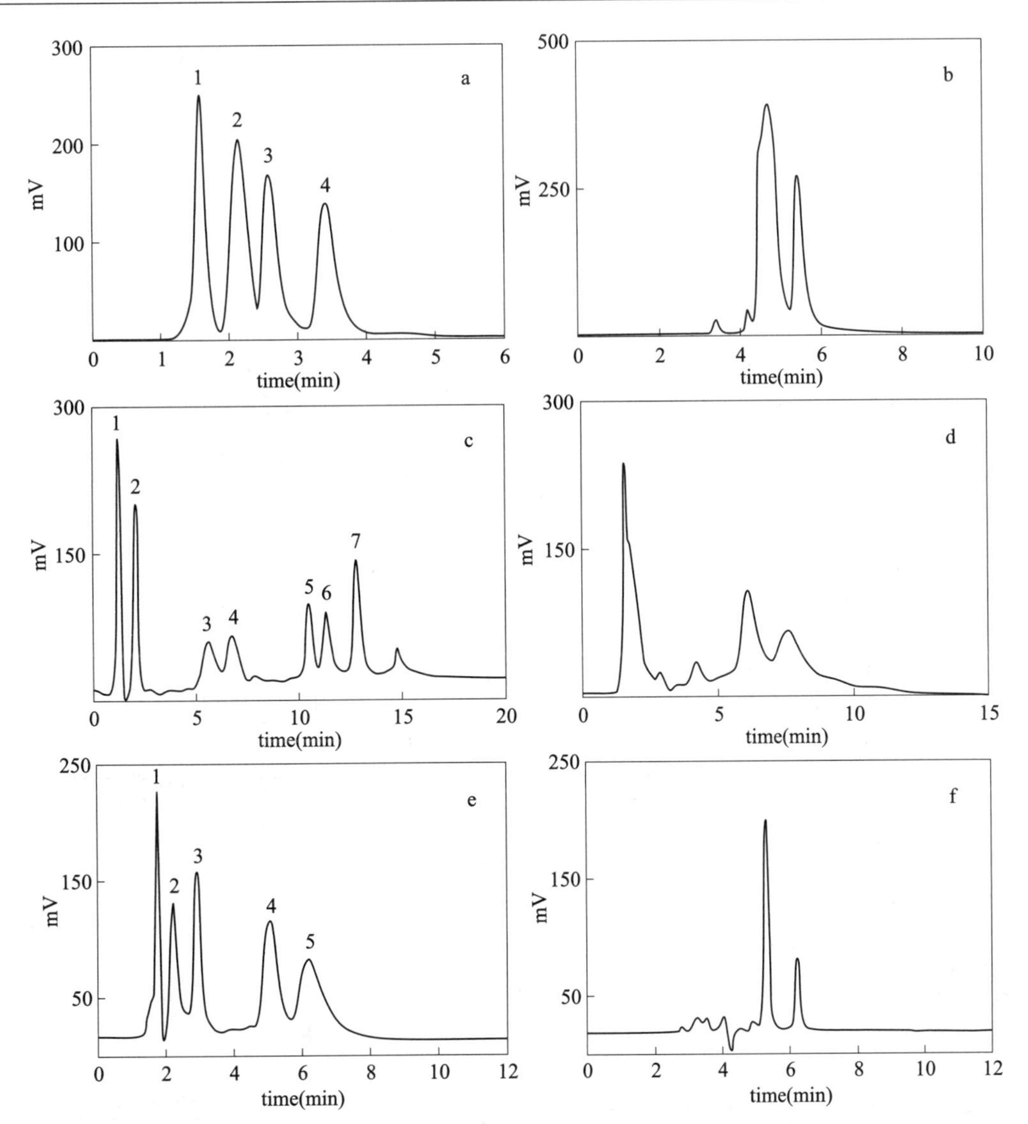

Fig. 3 Chromatograms of four amino acids on (a) maltose and (b) bare SiO_2 sphere columns (1—lysine, 2—arginine, 3—glutamic acid, 4— aspartic acid). Chromatograms of nucleic acid bases and nucleosides on (c) maltose and (d) bare SiO_2 sphere columns (1—naphthalene, 2—uracil, 3—adenosine, 4—adenine, 5—cytosine, 6—cytidine, 7— guanosine). Chromatograms of water-soluble vitamins on (e) maltose and (f) bare SiO_2 sphere columns (1—V_{PP}, 2—V_{B2}, 3—V_{B7}, 4—V_{B1}, 5—V_C)

pressure; 50 ℃ drift tube temperature; one filter; and 5 × 16 gain value. Under the same conditions, the separation effects of the same mixed sample on a maltose-based HILIC column and a bare SiO_2 sphere column were compared. As shown in Fig. 4c and d, six analytes achieved baseline separation within 4 min on the maltose-based HILIC column, while the silica sphere column could not separate the analytes, indicating that the maltose-based HILIC column was suitable for the separation of oligosaccharides.

Practical application

Owing to its high nitrogen content, melamine can be illegally added to infant milk formulas. In this study, the applicability of the maltose-based HILIC column was verified by analyzing melamine (Fig. 4e, f). With the mobile phase of 75/25 (v/v) ACN/water, flow rate of 1.0 mL/ min, and detection wavelength of 254 nm, the standard melamine retention was approximately 4 min (Fig. 4e, 1). The main components of infant milk were eluted within approximately 2 min (Fig. 4e, 2) causing no interference in the detection of melamine (Fig. 4f). This efficient analysis demonstrates the applicability of the asprepared material in the field of food inspection.

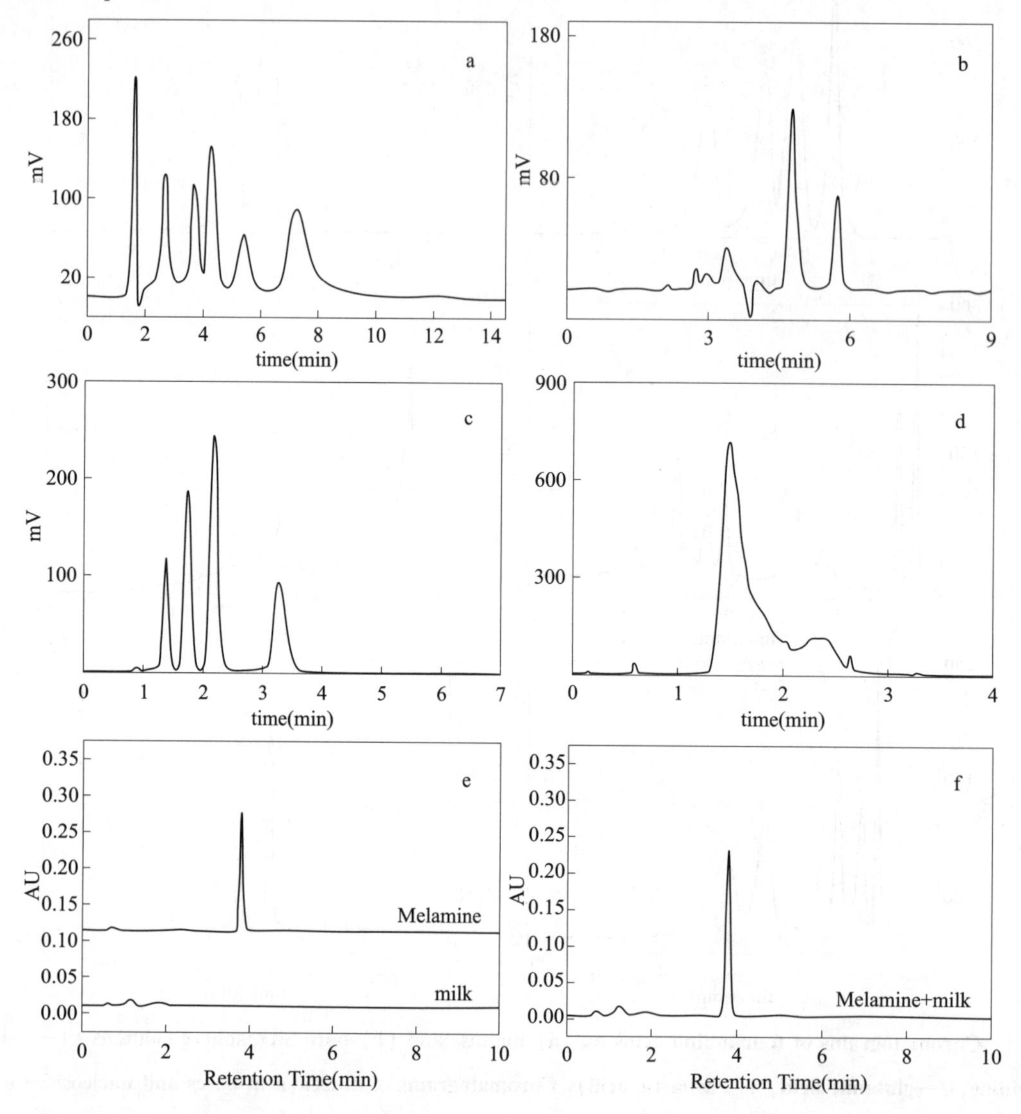

Fig. 4 Chromatograms of salicylic acid and its analogs on (a) maltose and (b) bare SiO_2 sphere columns (1—salicyla-mide, 2—salicylic acid, 3—trans-cinnamic acid, 4—acetyl-salicylic acid, 5—salicylic acid, 6—3,5-dihydroxybenzoic acid). Chromatograms of oligosac-charide on (c) maltose and (d) bare SiO_2 sphere columns (1— ribose, 2—glucose, 3—sucrose, 4—melezitose). (e) and (f) Practical application toward the separation of melamine in infant milk on maltose columns

Reproducibility and stability test

The reproducibility of different batches of the maltose-based HILIC stationary phase was investigated. Using the mobile phase of 85% ACN-20 mM NH_4AcO, pH 4.0, the retention times of water-soluble vitamins and salicylic acid compounds on the stationary phase of different batches were monitored. As shown in Table 4, the retention times of the four analytes were almost the same in these batches, indicating good inter-column reproducibility. In addition, the stability of the chromatographic behavior of the stationary phase was investigated. As shown in Fig. 2b, after 75 days of the continuous use of the same maltose-based HILIC column, the retention times of different analytes did not change, indicating that the maltose-based HILIC stationary phase had high stability.

Table 4 Solute retention time for different batches of columns

	t(min)						
	V_{PP}	V_{B2}	V_{B7}	V_{B1}	Salicylamide	Salicylic acid	Salicin
1	1.78	2.20	2.90	3.91	1.50	2.35	2.00
2	1.78	2.21	2.92	3.90	1.50	2.35	2.02
3	1.77	2.21	2.90	3.90	1.51	2.35	2.01

Conclusions

In this study, AEK8-maltose was bonded with silica spheres using CDI as a cross-linking agent to prepare a maltose-based HILIC stationary phase. The steps involved in this method are simple. In the HILIC mode, the effects of the compositions, salt concentrations, pH values, and temperatures of different mobile phases on the solute retention behavior were investigated. The experimental results indicated that the retention mechanism of the maltase-based HILIC stationary phase matched the typical HILIC retention mechanism. The maltose-based HILIC column showed a strong separation effect toward polar compounds such as amino acids, bases, and nucleosides, water-soluble vitamins, salicylic acid and its analogs, and oligosaccharides. In addition, the maltose-based HILIC column exhibited high reproducibility, high stability, and a long service life.

Funding This work was supported by the National Key R&D Program of China (2019YFB2103000), the Natural Science Foundation Project of Shaanxi Province (No. 2020JZ-24), and the Fundamental Research Funds for the Central Universities (GK201801006).

Declarations

Conflict of interest The authors declare no competing interests.

References

[1]Periat A, Krull IS, Guillarme D. Applications of hydrophilic interaction chromatography to amino acids, peptides, and proteins. J Sep Sci. 2015;38(3):357-367.

[2]Dell'aversano C, Hess P, Quilliam MA. Hydrophilic interaction liquid chromatography mass spectrometry for the analysis of paralytic shellfish poisoning (PSP) toxins. J Chromatogr A. 2005;1081(2):190-201.

[3]Paczkowska M, Mizera M, Tężyk A, Zalewski P, Dzitko J, Cielecka-Piontek J. Hydrophilic interaction chromatography (HILIC) for the determination of cetirizine dihydrochloride. Arab J Chem. 2019;12(8):4204-4211.

[4]Zborníková E, Knejzlík Z, Hauryliuk V, Krásný L, Rejman D. Analysis of nucleotide pools in bacteria using HPLC-MS in HILIC mode. Talanta. 2019;205:120161.

[5]Mathon C, Larive K. Separation of ten phosphorylated mono-and disaccharides using HILIC and ion-pairing interactions. Anal Chimica Acta. 2017;972:102-110.

[6]Shao W, Liu J, Yang K, Liang Y, Weng Y, Li S, Liang Z, Zhang L, Zhang Y. Hydrogen-bond interaction assisted branched copolymer HILIC material for separation and N-glycopeptides enrichment. Talanta. 2016;158:361-367.

[7]Rampler E, Schoeny H, Mitic M, Schwaiger M, Koellensperger G. Simultaneous non-polar and polar lipid analysis by on-line combination of HILIC, RP and high resolution MS. Analyst. 2018;143(5):1250-1258.

[8]Fan F, Wang L, Li Y, Wang X, Lu X, Guo Y. A novel process for the preparation of Cys-Si-NIPAM as a stationary phase of hydrophilic interaction liquid chromatography(HILIC). Talanta. 2020;218:121154.

[9]Pack BW, Risley DS. Evaluation of a monolithic silica column operated in the hydrophilic interaction chromatography mode with evaporative light scattering detection for the separation and detection of counter-ions. J Chromatogr A. 2005;1073(1-2):269-275.

[10]Tipke I, Bücker L, Middelstaedt J, Winterhalter P, Lubienski M, Beuerle T. HILIC HPLC-ESI-MS/MS identifica-

tion and quantification of the alkaloids from the genus Equisetum. Phytochem Anal. 2019;30(6):669-678.

[11]Tsochatzis E, Papageorgiou M, Kalogiannis S. Validation of a HILIC UHPLC-MS/MS method for amino acid profiling in triticum species wheat flours. Foods. 2019;8(10):514.

[12]Bento-Silva A, Gonçalves L, Mecha E, Pereira F, Patto M, Bronze R. An improved HILIC HPLC-MS/MS method for the determination of β-ODAP and its α isomer in Lathyrus sativus. Molecules. 2019;24(17):3043.

[13]Molnarova K, Kozlík P. Comparison of different HILIC stationary phases in the separation of hemopexin and immunoglobulin G glycopeptides and their isomers. Molecules. 2020;25(20):4655.

[14]Zhang J, Yang W, Li S, Yao S, Qi P, Yang Z, Feng Z, Hou J, Cai L, Yang M, Wu W, Guo D. An intelligentized strategy for endogenous small molecules characterization and quality evaluation of earthworm from two geographic origins by ultra-high performance HILIC/QTOF MS and Progenesis QI. Anal Bioanal Chem. 2016;408:3881-3890.

[15]Guo Z, Lei A, Liang X. Click chemistry: a new facile and efficient strategy for preparation of functionalized HPLC packings. Chem Commun. 2006;43:4512-4514.

[16]Guo Z, Jin Y, Liang T. Synthesis, chromatographic evaluation and hydrophilic interaction/reversed-phase mixed-mode behavior of a "Click β-cyclodextrin" stationary phase. J Chromatogr A. 2009;1216(2):257-263.

[17]Huang H, Jin Y, Xue M. A novel click chitooligosaccharide for hydrophilic interaction liquid chromatography. Chem Commun. 2009;45:6973-6975.

[18]Kibbey CE, Meyerhoff ME. Preparation and characterization of covalently bound tetraphenylporphyrin-silica gel stationary phases for reversed-phase and anion-exchange chromatography. Anal Chem. 1993;65(17):2189-2196.

本文曾发表于2022年《Analytical and Bioanalytical Chemistry》第414卷第13期

Effects of Several Organic Fertilizers on Heavy Metal Passivation in Cd-Contaminated Gray-Purple Soil

Luyao Wang[1,2], Siqi Liu[1], Jianfeng Li[1], Shunqi Li[3]

(1. Institute of Land Engineering and Technology, Shaanxi Provincial Land Engineering Construction Group Co., Ltd., Xi'an 710075, China; 2. Xi'an Jiao Tong University, School of Human Settlements and Civil Engineering, Xi'an 710075, China; 3. Zhejiang Dongda Environment Engineering Co., Ltd., Zhuji, 311800, China)

【**Abstract**】Soil heavy metal pollution has become one of the major ecological and environmental problems, and a serious threat to global food security. Organic fertilizer can not only improve soil quality and provide nutrients for plants, but also reduce the harm of heavy metal ions to a certain extent, which has become a current research hotspot in the field of heavy metal passivation. In this paper, a completely combined experimental design was used to compare the effects of five organic fertilizers (nutshell organic fertilizer (NOF), pig manure organic fertilizer (PMOF), sludge organic fertilizer (SOF), humus soil organic fertilizer (HSOF) and earthworm soil organic fertilizer (ESOF)) on available Cd in soil with different pollution levels at different dosages, and the passivation mechanism of soil Cd was preliminarily discussed. The results showed that all kinds of organic fertilizers were passivated by reducing the Cd availability, and their effects on the Cd availability of purple soil were closely related to the degree of soil pollution and the amount of organic fertilizers. The passivation effect of moderate Cd-contaminated soil was the best, which increased with the increase of organic fertilizer application rate, and the effects of NOF and SOF were the best. However, the passivation effect of organic fertilizers on soil Cd was the worst in mild Cd-contaminated soil, especially ESOF. The results of Cd morphological correlation analysis showed that Exe-Cd and FeMnOx-Cd in soil had significantly positive contributions to available Cd, while Res-Cd showed significantly negative contributions. And in moderate Cd contaminated soil, Exe-Cd content decreased by 7.12%~28.50%, while Res-Cd content increased by 19.74%~65.81%. In addition, the content of available Cd in soil decreased first and then increased with time after adding organic fertilizer, and reached the lowest value at 15 d and stabilized after 60 d. The conclusion of this paper can provide theoretical basis for rational use of organic fertilizer to reduce bioavailability of cadmium in Gray-Purple soil.

【**Keywords**】Organic fertilizer; Cd; Availability; Heavy metals speciation; Passivation

1 Introduction

Rapid industrial development and the extensive use of pesticides and mineral fertilizers have led to the widespread existence of heavy metals in agricultural soils. Rice is the largest food crop in China. Nearly two-thirds of the population live on rice. According to the survey, there are about 13000 hm^2 Cd-contaminated farmland soil in China, including 11 provinces and 25 regions (Guo, 1998; Hou, 2017). The results of the China soil pollution survey bulletin released in 2014 (CAEPI, 2014) showed that there was significant indigenous heavy metal pollution in cultivated land soil and even the whole soil in China. The heavy metal pollution of Pb, As, Ni and Cd was more serious. From the analysis of the distribution of soil pollution in China, the soil heavy metal pollution in southern China was more serious than that in northern China, and the large area of soil heavy metal content exceeded the standard in southwest and central south of China. The exceeding standard rate of Cd pollution was 7.0%, and the proportions of four different pollution levels from severe to slight pollution were 0.5%, 0.5%, 0.8% and 5.2%, respectively (Zhang, 2002). Cd is generally released into the environment through human activities such as mining and metal ore processing (Yang et al., 2018). It has the characteristics of strong toxicity, poor mobility, and easy accumulation in soil, and is toxic even at low exposure levels (Clemens et al., 2013). High concentration of Cd stress can induce many reactive oxygen species and malondialdehyde content in plant cells and cause the up-regulation of antioxidant enzyme gene expression. Cd stress can also destroy the integrity of photosynthetic system, inhibit photosynthesis, and reduce nutrient uptake by plants, thereby inhibiting photosynthesis and reducing the activities of several key enzymes, including superoxide dismutase (SOD), guaiacol peroxidase (POD), catalase (CAT), ascorbate peroxidase (APX) and glutathione peroxidase (GPX). (Araujo et al., 2017; Chandorkar et al., 2019; Zong et al., 2017). The transportation of Cd and other trace metals to human body through the food chain is also a great potential harm to human health (Rufus, 2015). Therefore, to eliminate the influence of Cd contaminated soil on the quality and safety of agricultural products, thereby reducing its harm to human beings is an important issue related to national economy and public' s health.

At present, the role of organic fertilizer in fertility improvement, soil biological activity improvement, nutrient recycling and sustainable agricultural development has been generally confirmed (Chen et al., 2005; Bulluck et al., 2002; Sharrm et al., 2005). The organic matter and beneficial microorganisms in organic fertilizer have strong adsorption and chelating effects on heavy metal ions. The bioavailability was reduced because of the reduction of contents of water-soluble heavy metals and exchangeable heavy metals in soil. Therefore, it showed very superior properties in improving farmland productivity and repairing heavy metals. (Andersson et al., 1991; Chen et al., 2003; Ansari and Mahmood, 2017). The mechanism of repairing heavy metal pollution in soil by organic fertilizer is due to its material composition and properties. Organic fertilizer contains a large number of low molecular simple organic acids and even polymer humus substances and other high molecular active substances, which have complex structures and can affect the conversion process, occurrence morphology and biological activity of heavy metals in soil through various mechanisms, including direct complexation/adsorption of heavy metal ions (Liu et al., 2008; Senesi et al., 1995), formation of organic-inorganic complexes with soil inorganic components (Matilainen et al., 2010), as electron shuttles which affects soil redox properties (Chen et al., 2011; Jiang et al., 2015), affecting soil pH and buffering performance (Österberg and Wei, 1999; Martina et al., 2009). There are great differences in the composition and properties of organic fertilizers due to the different sources of organic fertilizers, composting methods, and degree of maturity. Therefore, there are also significant differences in the regulation of soil heavy metal availability (Wang et al., 2010; Qin et al., 2014). Huang et al. (Huang et al., 2014) found that the application of cattle manure organic fertilizer could reduce exchangeable Cd and residual Cd by 28.84%~36.33% and 6.39%~19.29% respectively and increase carbonate-bound Cd and organic-bound Cd by 10.95%~75.27% and 44.91%~68.31% respectively, which proved that organic matter could reduce the bioavailability of Cd by changing the speciation of Cd in black soil, thereby inhibiting the absorption of Cd by plants. Ma et al. (Ma et al., 2015) showed that the application of bio-organic fertilizer with special functional microorganism and organic compound can reduce available Cd and Pb in acid paddy soil by 17.36% and 18.45%, respectively, and significantly reduce Cd and Pb contents in rice. Liu et al. (Liu et al., 2014) found that the three organic fertilizers can reduce the Cd content in various parts of wheat through the study of sheep manure, chicken manure and pig manure. The comprehensive effect is as follows: pig manure>sheep manure>chicken manure. Organic fertilizer reduces its bioavailability by reducing soil exchangeable and carbonate-bound Cd content, increasing iron-manganese oxidation, organic binding, and residual Cd content. However, some studies have found that the application of organic fertilizers has increased the availability of heavy metals in the soil and increased the accumulation of heavy metals in plants. Pan and Zhou (Pan and Zhou, 2007) found that after the application of straw organic fertilizer and pig manure organic fertilizer, the exchangeable Cd in the soil increased by 43.2% and 17.3% compared with the application of inorganic fertilizer, respectively. The Cd content in the wheat grain also exceeded the relevant Chinese standards by 60% and 70%, respectively. Shi et al. (Shi et al., 2009) found that after adding humic acid, the content of water-soluble Pb in soil increased, so it is necessary to reduce the harm of Pb through the adsorption of zeolite. Ushijima et al. (Ushijima et al., 2016) prepared organic materials by heat treatment of organic waste in the presence of Lewis acid catalyst. It was found that the content of Cd in alfalfa was related to the addition amount of organic materials. When the addition amount was 10%, humus may reduce the oxidative stress caused by Cd, thereby promoting the growth of alfalfa, and reducing its accumulation in plants. When the addition amount was 25%, the toxicity of Cd was dominant compared with humus fertilization, resulting in growth inhibition and higher Cd accumulation. Similarly, Wang et al. (Wang et al., 2014) also indicated that although organic materials could reduce the effectiveness of heavy metals in soil, they did not actually reduce the total amount of heavy metals in soil, and the dissolved organic matter and other factors generated by the decomposition of heavy metals in organic materials may lead to the aggravation of soil heavy metal pollution and soil acidification. Therefore, the use of organic materials to repair soil there are certain environmental risks, the effect of remediation depends on the type and composition of organic materials and soil environmental conditions. At present, the common sources of organic fertilizer in the world include livestock manure such as pigs, cattle and chickens, plant debris such as nutshell, fruit husk and rapeseed, organic sludge waste, and humus soil by various organic wastes. There are differences in the components of various types of organic fertilizers, and the influence and regulation performance on the form of heavy metals in the soil must be different. However, there is lack of systematic comparative study on the regulatory effect of different types of organic fertilizers and different degrees of heavy metals on the speciation effectiveness of heavy metals, which restricts the scientific application of organic fertilizers as soil heavy metal remediation materials, especially in the purple soil area of Chongqing where Cd pollution is more serious. Therefore, in this study, Cd, a heavy metal with great harm to public's health and environment and widespread pollution, was selected as the research object. Five kinds of common organic fertilizers were selected: nutshell organic fertilizer (NOF), pig manure organic fertilizer (PMOF), sludge organic fertilizer (SOF), humus soil organic fertilizer (HSOF) and earthworm soil organic fertilizer (ESOF). Through soil incubation experiment, the varying regularity of Cd availability and passivation effect of different types and different dosages of organic fertilizer on farmland soil with three different Cd pollution levels of the same soil type were studied, so as to provide scientific basis for the scientific application of organic fertilizer and soil pollution remediation.

2 Materials and methods

2.1 Materials

2.1.1 Organic fertilizer

In this study, five types of organic fertilizers with wide application and large differences in incubation were selected as follows: nutshell organic fertilizer (NOF), pig manure organic fertilizer (PMOF), sludge organic fertilizer (SOF), humus soil organic fertilizer (HSOF) and earthworm soil organic fertilizer (ESOF). The five organic fertilizers were all purchased from Hebei Shi Yuan Su Fertilizer Technology Co., Ltd., and the basic physical and chemical properties are shown in Table 1.

Table 1 Basic physicochemical properties of Manures

Organic fertilizer type	pH	DOC (g/kg)	TCd (mg/kg)	Organic matter (g/kg)	TN (g/kg)	TP (g/kg)	TK (g/kg)
NOF	8.08	66.1	0.31	300.41	11.54	10.21	7.96
PMOF	8.06	100	0.23	578.95	18.56	12.69	10.44
SOF	8.1	148	0.26	473.36	16.05	13.97	15.66
HSOF	8.09	251	0.18	733.69	8.42	10.32	8.29
ESOF	8.12	90.2	0.27	367.76	15.91	14.73	9.52

2.1.2 Soil collection, characterization

The soil in this research is purple-brown purplish soil (GPS, the same below) widely distributed in southwestern China, which was collected in the 0~20 cm layer from Fuling District, Chongqing, China. The gravel and plant residues were removed from the soil samples. Soil chemical parameters pH, DOC, TCd, OM, TN, TP, TK were determined, which is presented in Table 2.

Table 2 Basic physicochemical properties of soil sample

Soil	pH	DOC(g/kg)	TCd(mg/kg)	Organic matter(g/kg)	TN(g/kg)	TP(g/kg)	TK(g/kg)
GPS	8.15	1.59	0.02	23.42	1.24	0.73	0.81

2.2 Experimental methods

2.2.1 Preparation of Cd-contaminated soil

In order to accurately control the level of Cd pollution in soil, the artificial simulation of Cd-contaminated soil was carried out. After natural drying, the soil samples were milled and passed through 2 mm standard sieve. Refer to the secondary standard (0.3 mg/kg) in the Soil Environmental Quality Standard (GB 15618—1995), the air-dried and sieved soil was spiked with $CdCl_2$ to a final Cd concentration of 0, 1, 2, 5 mg/kg and wetted to 70% of its field capacity, which was recorded as Cd-0, Cd-1, Cd-2, Cd-5. The Cd-contaminated soil was incubated in dark at room temperature for three months for further use.

2.2.2 Organic fertilizer incubation experiment

The experiment was conducted as a completely randomized design in a 4×5×3 factorial scheme, with four levels of Cd contaminated soil (Cd-0, Cd-1, Cd-2, Cd-5), and five types of organic fertilizers (NOF, PMOF, SOF, HSOF, ESOF), and three rates of application (2 g/kg, 4 g/kg, 8 g/kg), with two replications. A control treatment without organic fertilizer was added, with three replications.

Five different types of organic fertilizers were incorporated with a 500 g Cd-contaminated air-dried soil in a 1000 mL plastic container with lids according to the three dosages, and homogeneously mixed. Deionized water was added to reach 70% field capacity. The aging process was simulated in the laboratory through an incubation experiment. Periodically during the incubation, deionized water was cautiously sprayed to bring samples back to 70% field capacity by weighting the pots. The experiment was kept under laboratory environment with constant temperature of 25 ℃. One soil sample of approximately 10 g was collected from each plastic container at 1, 7, 15, 30, 60, 90 d of incubation for the determination of Cd speciation. Repeated sampling 3 times for all measurements.

2.3 Analysis method

2.3.1 Basic physical and chemical properties of soil and organic fertilizer

The determination of soil pH, organic matter content, DOC and other basic physical and chemical properties refer to "Soil Agricultural Chemistry Analysis Method" (Lu, 2000). The total Cd content in the soil is digested with a mixture of aqua regia (HNO_3:HCl = 1:3): $HClO_4$ at a ratio of 2:1, and the Cd^{2+} in the solution after digestion is determined by flame atomic absorption spectrophotometry (TAS-900, Beijing Puxi).

2.3.2 Sequential extraction procedure

For the fractionation procedure, soil samples were collected from each experimental pot, air-dried and sieved to 2 mm sieve. Approximately 1 g of the Cd-contaminated soil treated with organic fertilizer was weighed into 50 mL polypropylene centrifuge tube to separate copper into six operationally defined fractions, according to Tessier et al. (Tessier et al., 1979), as follows:

Fraction 1: Soluble + exchangeable Cd (Exe-Cd), extracted with 16 mL of 0.1 M $MgCl_2$.

Fraction 2: Carbonates bound Cd (Carb-Cd), extracted with 16 mL of 1 M NaAOc, at pH 5.

Fraction 3: Fe and Mn oxide bound Cd (FeMnOx-Cd), extracted with 30 mL of 0.04 M NH_2OH. HCl (hydroxylamine hydrochloride) in acetic acid, at pH 3.

Fraction 4: Organic matter strongly bound Cd (OM-Cd), extracted with 10 mL of 30% H_2O_2 at pH 2, followed by the addition of 8 mL of 3.2 M NH_4AOc in 20% HNO_3.

Fraction 5: Residual fraction (Res-Cd), extracted with 8 mL of aqua regia (HNO_3:HCl = 1:3) and 2 mL $HClO_4$.

The available Cd content in soil was extracted and determined with 1 M CH_3COONH_4 solution. Analysis of Cd was performed by Atomic Absorption Spectrophotometry (TAS-900, Beijing Persee). The recovery rate Cd in all soil samples were higher than 90%.

2.3.3 Statistical analysis

Microsoft Excel 2007 and Origin Pro 8.5 for data recording and preliminary analysis. Analysis of variance (ANOVA) and Tukey's multiple range tests ($p<0.05$) were used to determine the statistical significance of the organic fertilizer treatment effects on Cd availability using SPSS 17.0 package.

3 Results

3.1 The effect of organic fertilizer on the availability of soil Cd

3.1.1 The effect of organic fertilizer types on available Cd in soil with different Cd pollution levels

After applying organic fertilizer to soils with different pollution levels for 90 days, the average content of available Cd in the soil is shown in Figure 1. It can be seen that the content of available Cd in moderate and severe Cd-contaminated soil can be reduced by organic fertilizer, and the effect of organic fertilizer on the available Cd content in the soil is related to the degree of soil pollution and the type of organic fertilizer. In moderate Cd-contaminated soil (Cd^{2+} = 2 mg/kg), all organic fertilizer treatments significantly reduced the availability of Cd in soil. Among them, NOF had the best inhibitory effect on the availability of Cd in soil, and its available Cd decreased by 27.15% compared with the control (0.55 mg/kg), and the other treatments decreased by 10.86%~16.55% compared with the control. There was no significant difference among the treatments. In severe Cd-contaminated soil (Cd^{2+} = 5 mg/kg), only SOF, HSOF and ESOF could significantly reduce available Cd content in the soil ($p<0.05$), which were 19.18%, 16.14% and 11.36% lower than the control (1.01 mg/kg), respectively. For mildly Cd-contaminated soil (Cd^{2+} = 1 mg/kg), the application of organic fertilizer increased the availability of Cd in soil. Compared with the control (0.24 mg/kg), the available Cd content in five organic fertilizer treatments increased by 1.79%~21.38%, respectively, and the ESOF treatment increased the most, reaching a significant level of $p<0.05$. Among them, SOF had the best effect.

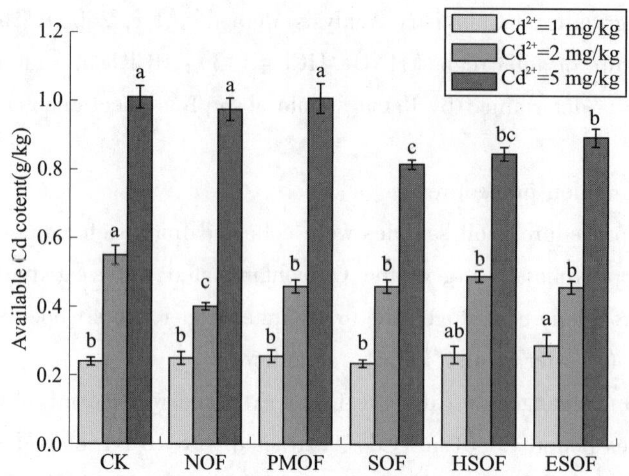

Fig. 1 The effect of organic fertilizer types on the contents of available Cd in soils with different Cd loads

(The letter on the bar shows the statistical differences between treatments at p<0.05)

3.1.2 The effect of organic fertilizer addition on available Cd in soil with different Cd pollution levels

Fig. 2 shows the variation of available Cd content in soil with organic fertilizer addition after 90 days of application of different types of organic fertilizers. It can be seen that for the same Cd pollution level of soil, with the increase of organic fertilizer dosage, the change trend of available Cd content in soil is basically similar, but the change trend of available Cd content in soil with different Cd pollution levels is different.

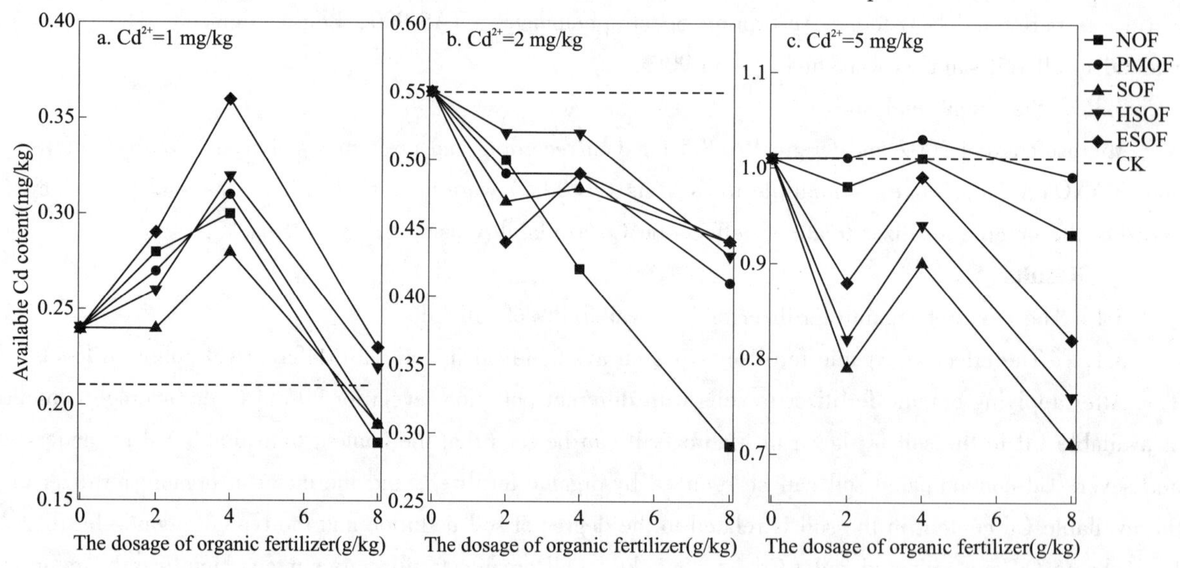

Fig. 2 Effects of organic fertilizer dosage on the contents of available Cd in soils with different Cd loads

In mild Cd-contaminated soil (Fig. 2a), the available Cd content decreased only when the organic fertilizer addition increased to 8 g/kg, and was lower than the control level. SOF had the best effect on reducing the available Cd content, which was reduced by 20.83% compared with the control. In moderate Cd-contaminated soil, it can be found from Fig. 2b that all kinds of organic fertilizers could significantly reduce soil available Cd content, and there was a significant negative correlation between available Cd content and organic fertilizer dosage. Among them, the effect of NM was the most obvious. When the addition amount was 8 g/kg, the soil available Cd content decreased by 47.27% compared with the control (0.55 mg/kg). In the severe Cd-contaminated soil (Fig. 2c), the soil available Cd content increased first and then decreased with the increase of organic fertilizer addition. Except that PM was useless at medium and low dosage, other organic fertilizers had inhibitory effects on soil Cd availability, especially at the highest dosage

(8 g/kg). The inhibitory order of various organic fertilizers on soil available Cd content was SOF>HSOF>ESOF>NOF>PMOF, and SM could reduce soil available Cd content by 29.70%.

3.1.3 Effect of organic fertilizer on the dynamic change of soil available Cd content

Taking moderate Cd-contaminated soil as an example, it can be observed from Fig.3 that after the addition of five organic fertilizers, the content of available Cd in soil decreased first and then increased with the increase of incubation time, i.e., it reached the minimum at 15~30 d, then began to rise, and finally stabilized after 60 d. The available Cd are all lower than the control, which indicating that all kinds of organic fertilizers had inhibitory effects on available Cd in soil at different times. When the addition amount of organic fertilizer was 2 g/kg, the soil available Cd content under the action of NOF, PMOF, SOF, HSOF and ESOF decreased by 8.60%, 11.07%, 15.18%, 6.13% and 19.30%, respectively, and ESOF had the strongest inhibitory effect on available Cd. When the addition of organic fertilizer was 4 g/kg and 8 g/kg, the soil available Cd content decreased by 24.24%, 11.07%, 12.71%, 5.30%, 11.07% and 48.12%, 25.89%, 19.30%, 20.95%, 20.13% under the action of five organic fertilizers, respectively. NOF had the strongest inhibitory effect on available Cd.

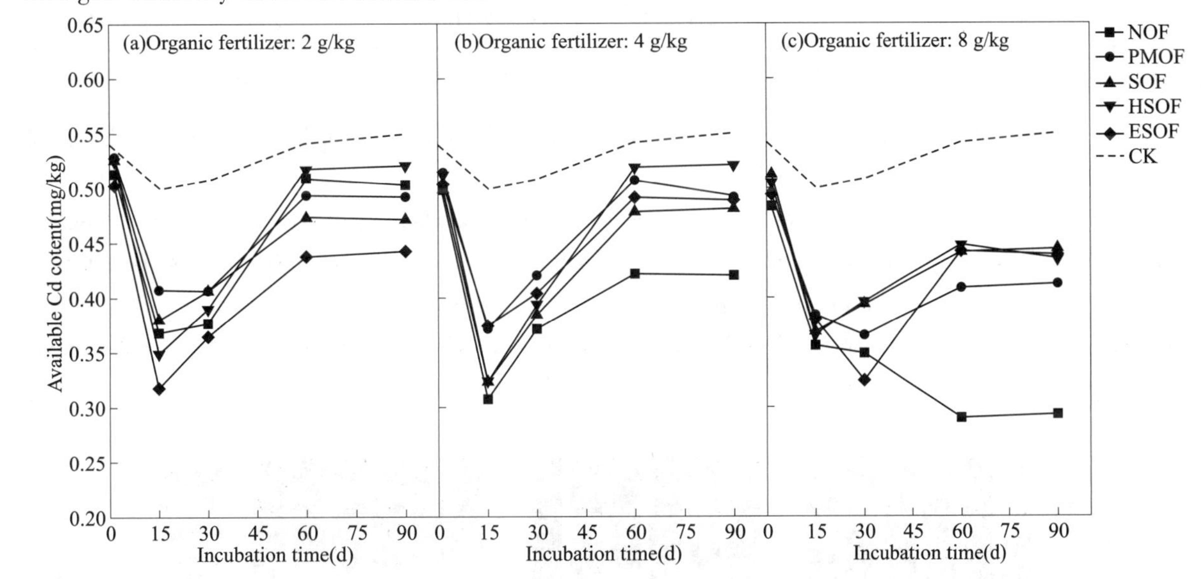

Fig. 3 Changes of available Cd content in soil treated with different dosages of organic fertilizers over time

3.2 Effects of organic fertilizer on speciation of Cd in soil

After 90 days of organic fertilizer application, the transformation of different speciation of Cd in soil tends to be balanced. The correlation analysis between available Cd and Cd speciation is shown in Table 3. It can be seen that the soil available Cd has a very significant positive correlation with soil Exe-Cd and Res-Cd, respectively, and a significant positive correlation with FeMnOx-Cd, indicating that the soil available Cd is mainly affected by the content of Exe-Cd, FeMnOx-Cd and Res-Cd.

Table 3 Analysis of the relationship between bioavailable Cd and Cd chemical speciation

Types of organic fertilizer	Exe-Cd	Carb-Cd	FeMnOx-Cd	OM-Cd	Res-Cd
NOF	0.413**	0.125	0.156*	0.151	−0.728**
PMOF	0.107**	0.058	0.088*	0.031	−0.528**
SOF	0.339**	0.211	0.171*	0.167	−0.447**
HSOF	0.201**	0.039	0.282*	0.071	−0.409**
ESOF	0.441**	0.142	0.115*	0.084	−0.702**

** means a significant correlation at the 0.01 level (two-sided), * means a significant correlation at the 0.05 level (two-sided).

In order to further determine the quantitative contribution of different speciation of Cd to soil available Cd, the optimal regression equation between soil available Cd and soil Cd speciation under different types of organic fertilizer was obtained by multiple regression analysis (Table 4). Exe-Cd and FeMnOx-Cd in soil have a significant positive contribution to available Cd, while Res-Cd has a significant negative correlation to soil available Cd, which indicated that the decrease of soil available Cd content was related to the decrease of soil Exe-Cd and FeMnOx-Cd content and the increase of Res-Cd content.

Table 4 Multiple regression analysis of available Cd and Cd speciation

Types of organic fertilizer	regression equation	R	R^2
NOF	$y=0.067+2.084x_1+1.471x_3-2.762x_5$	0.819	0.671
PMOF	$y=0.083+1.262x_1+1.253x_3-2.521x_5$	0.882	0.777
SOF	$y=0.064+2.031x_1+1.592x_3-1.767x_5$	0.798	0.636
HSOF	$y=0.059+1.801x_1+1.667x_3-1.581x_5$	0.846	0.716
ESOF	$y=0.091+2.177x_1+1.133x_3-2.715x_5$	0.843	0.711

y represents available Cd, x_1 represents soil Exe-Cd, x_3 represents soil FeMnOx-Cd, x_5 represents soil Res-Cd(mg/kg).

3.3 The effect of organic fertilizer on the Cd speciation distribution in soils with different levels of Cd pollution

Fig.4 shows the average distribution of Cd speciation in Cd-contaminated soil with different degrees after different organic fertilizer treatments. It can be seen from the figure that the difference of the effect of organic fertilizer on the distribution of Cd speciation in different degrees of Cd contaminated soil is mainly reflected in Exe-Cd, FeMnOx-Cd and Res-Cd, which is consistent with the correlation analysis in the previous section.

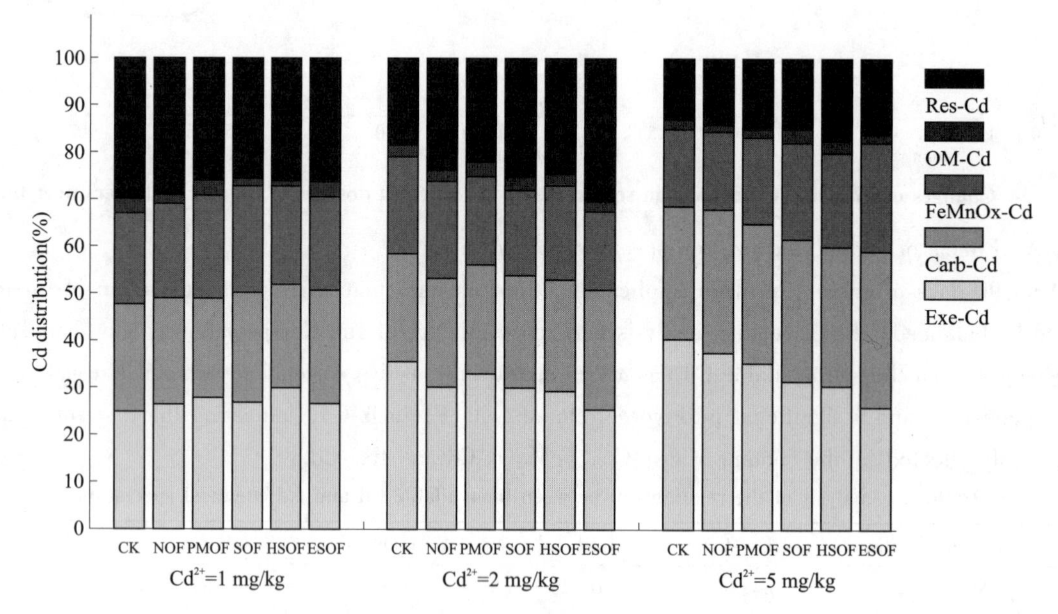

Fig. 4 Cd distribution among soil fractions after incubation with organic fertilizer from NOF, PMOF, SOF, HSOF, ESOF at different dosage of application

In mild Cd-contaminated soil, various organic fertilizers promoted the transformation of Cd to Exe-Cd. Compared with the control (24.87%), the content of Exe-Cd increased by 6.17% ~ 20.28%, and the increase of HSOF treatment was the largest. The content of FeMnOx-Cd also increased by 2.90% ~ 17.87%, and the increase of SOF treatment group was the largest. However, the Res-Cd content decreased by 3.19%~14.38%, with the largest decrease in PMOF treatment. In moderate Cd-contaminated soil, the application of organic fertilizer decreased the contents of Exe-Cd and FeMnOx-Cd, while the Res-Cd content increased significantly. Exe-Cd and FeMnOx-Cd were reduced by 7.12%~28.50% and 0.07%~19.47%, respectively. Res-Cd increased by 19.74%~65.81%, and ESOF treatment group had the largest decrease and increase. In severe Cd-contaminated soil, the content of Exe-Cd decreased significantly after adding organic fertilizer, while the contents of FeMnOx-Cd and Res-Cd increased slightly. Among them, Exe-Cd decreased by 7.25%~35.54%, ESOF decreased the most; the contents of FeMnOx-Cd and Res-Cd increased by 0.41%~42.41% and 10.28%~34.79%, respectively, and ESOF and HMOF showed the largest increase in these two forms, respectively.

3.4 The effect of organic fertilizer on the dynamic change of Cd speciation in soil

The dynamic changes of average distribution of three speciation of Cd in soils with different Cd pollution levels treated by organic fertilizer are shown in Figure 5. It can be seen that organic fertilizer does affect the speciation of Cd in soil, but the main influence is the distribution of Cd between different speciation, and the change trend of the same form of Cd is basically the same, and the speciation of Cd basically reach a stable state at 60 d.

In mild Cd-contaminated soil, the content of Exe-Cd increased sharply at the initial stage of incubation and reached the peak at 7 d, then continued to decline and stabilized after 60 d, which was finally higher than the control. The contents of FeMnOx-Cd and Res-Cd decreased first and then increased on the whole, but the content of FeMnOx-Cd increased compared with the control, while the content of Res-Cd decreased. In the medium Cd-contaminated soil, the change of Exe-Cd content showed the opposite trend with that in low-concentration Cd soil before 30 d of incubation, which decreased sharply at the beginning of incubation, then fluctuated and increased and stabilized after 60 d, and was finally higher than the control. FeMnOx-Cd and Res-Cd reached the highest values at 7 d, and then reached stability after 60 d. Finally, the content of FeMnOx-Cd decreased compared with the control, while the content of Res-Cd increased. In severe Cd-contaminated soil, the variation trend of each form was basically similar to that in moderate Cd-contaminated soil. Among them, the decrease and increase of Exe-Cd and FeMnOx-Cd contents in ESOF were the largest, respectively., but the increase of Res-Cd in HSOF was the largest.

4 **Discussion**

Purple soil is weathered from purple sedimentary rocks and is concentrated in hilly areas of Sichuan and Chongqing. According to its pH and calcium carbonate content, purple soil can be divided into three subgroups: acidic, neutral and calcareous purple soil (Yang et al., 2004). The test soil in this experiment was collected from Chongqing, belonging to calcareous purple soil. The heavy metal content of farmland soil in Chongqing was deeply affected by geochemical characteristics such as soil parent material, and was also strongly affected by human factors such as industry, mining and modern agriculture. Among them, Cd was the heavy metal with the highest proportion of exceeding the standard and the highest accumulation level in the region (He, 2004; Jia et al., 2018). The southwest region where Chongqing is located is a variety of cropping systems with rice and maize, and its grain yield also accounts for a considerable proportion in the country. Purple soil is one of the main agricultural soils in southern China, and Sichuan Basin is the most concentrated, accounting for more than 70% of the area (Yang et al., 1992). Therefore, heavy metal pollution in farmland, especially Cd pollution, will seriously endanger human health. The application of bio-organic fertilizer can not only change the physical and chemical properties of soil, improve soil fertility, but also improve the living environment of soil microorganisms, increase soil enzyme activity, and improve the

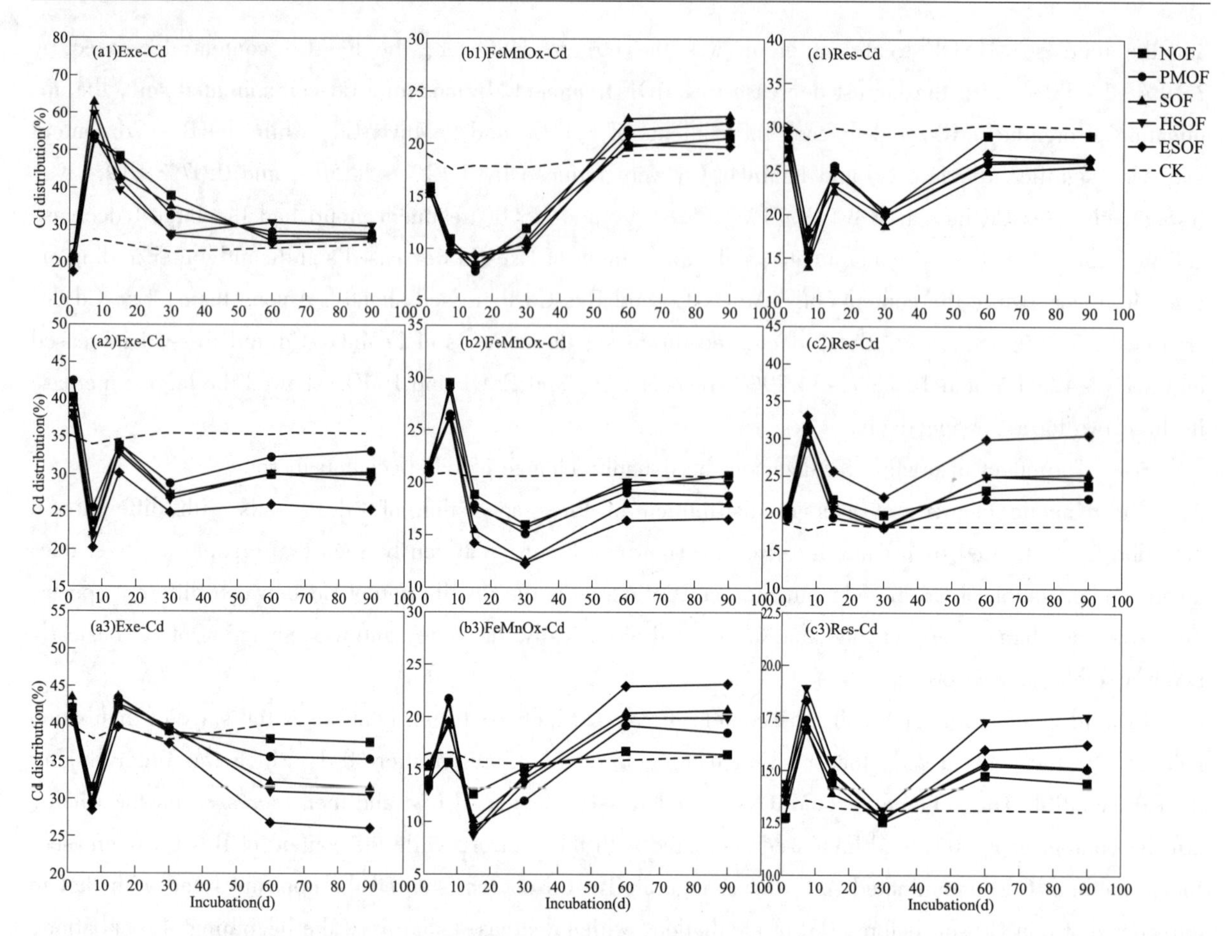

Fig. 5 Cd distribution among soil fractions after incubation with organic fertilizer from NOF, PMOF, SOF, HSOF, ESOF at different incubation time

heavy metal pollution of soil, which is crucial to the sustainable development of local agriculture (Yu and Zhao, 2013).

4.1 The change of Cd speciation in soil

The morphological characteristics of heavy metals in soil are important indicators to reveal the migration and transformation of heavy metals and bioavailability (Liu et al., 2020). Related studies have also shown that heavy metals in soil combine with different carriers into various forms, and different forms of heavy metals reflect different bioavailability (Huang et al., 2014; Liu et al., 2014; Kumar et al., 2021), which is further reflected in the change of soil available Cd content. According to the five-stage grouping method proposed by Tessier et al. (Tessier et al., 1979), heavy metals can be divided into exchangeable, carbonate-bound, Fe-Mn oxidized, organic-bound and residual states. Among them, the exchangeable and carbonate-bound heavy metals are easy to dissolve in the environment, and the exchangeable heavy metals are the most susceptible to biological utilization and have the strongest toxicity. Fe-Mn oxidized heavy metals can also be released when the redox potential changes, which is potentially effective for organisms. Residue heavy metals combined with sediments most firmly, it will not be absorbed, so its activity is minimal. The results of this study showed that organic fertilizer mainly affected the contents of Exe-Cd, FeMnOx-Cd and Res-Cd in soil, but had little effect on the contents of Carb-Cd and OM-Cd, and the effect was related to Cd content in soil. Liu et al. (Liu et al., 2020) found that bio-organic fertilizer increased the pH value of paddy soil, promoted the transformation of acid soluble Cd to oxidizable Cd, and reduced the bioavailability of Cd in soil. In this paper, the pH value of organic fertilizer was 8.06~8.12, and the pH value of gray-purple soil was 8.15, which were all weakly alkaline. The effect of organic fertilizer on soil pH value was minimal, so the addition of organic fertilizer almost did not affect the content of Carb-Cd. In heavy Cd-contaminated soil, organic fertilizer reduced the contents of Exe-Cd and FeMnOx-Cd, and increased the content of Res-Cd, which was similar to many research results. Xie et al. (Xie et al., 2018) showed that with the increase of humic acid application rate, the exchangeable Cd content in soil decreased significantly, and the residual Cd content increased by 53.6%~113.9%. However, when humic acid was added, the pH value of soil did not change significantly. The reason for the decrease of available Cd may be that humic acid could have a series of adsorption, complexation and chelating effects with Cd^{2+}. The combination of its products with soil clay particles strengthened the adsorption ability of soil clay particles to Cd^{2+}. At the same time, humic acid itself was adsorbed on the surface of soil colloid as an adsorbent, which increased the adsorption point of soil particles and further promoted the transformation of exchangeable Cd^{2+} to stable state. Liu et al. (Liu et al., 2014) found that the application of bioorganic fertilizer could promote the transformation of acid soluble Cd to reducible Cd by increasing soil organic matter content, CEC and pH, and showed that different organic fertilizers could reduce the content of available Cd in soil by increasing the content of soil organic matter. The mechanism was that organic fertilizer could directly increase the content of soil organic matter, and its various functional groups (-COOH, -OH, -CO, etc.) and organic matter with large specific surface area could form insoluble metal-organic complexes with heavy metals to enhance the adsorption of heavy metals (Zhou et al., 2018). Soil organic matter also has reducibility, Fe-Mn oxides of heavy metals can be released under reduction conditions, resulting in FeMnOx-Cd content increased. Moreover, organic matter can also improve soil structure, thereby indirectly changing the speciation distribution of Cd in soil. In addition, some studies have found that the change of soil microbial biomass and enzyme activity under organic fertilizer application is one of the important effective ways to affect the occurrence of Cd in soil (Huang et al., 2017). Xu et al. (Xu et al., 2015) found that heavy metals significantly affected MBC, MBN and various enzyme activities in soil. At the same time, the activities of UA, ACP and DH in soil were significantly correlated with the absorption of heavy metals by rice. Zhang et al. (Zhang et al., 2020) showed that the application of organic fertilizer to reduce the content of soil acid extractable Cd was mainly achieved by increasing the content of soil organic matter, soil dehydrogenase activity and microbial biomass carbon content. Therefore, it is suggested that in future studies, the influence mechanism of microbial biomass and enzyme activity in organic fertilizer on Cd forms in soil should be studied in depth. In this study, the ability of different types of organic fertilizers is different but does not change the evolution of specific speciation of Cd, which may be due to the morphological differentiation mechanism of various organic fertilizers on heavy metals in soil is basically similar.

4.2 The passivation effect of Cd in soil

The results showed that the passivation effect of soil Cd increased with the increase of organic fertilizer application rate, and in different Cd pollution levels of soil, the passivation effect of high organic fertilizer application amount (8 g/kg) in moderate Cd pollution soil was the best, while the passivation effect of medium organic fertilizer application amount (2 g/kg) in mild Cd pollution soil was the worst. Lu (Lu, 2003) studied the effect of bio-organic fertilizer on the bioavailability of Cd in soil. It was found that the higher the Cd content in soil, the more obvious the passivation effect of bio-organic fertilizer, which was similar to the results of this experiment. However, Liu et al. (Liu et al., 2009) studied the effects of different types of organic fertilizers and application rates on soil Hg pollution, migration and accumulation. The results showed that organic fertilizer had the best effect on low-level Hg-contaminated soil, which was contrary to the results of this experiment. Thus, although the effect of bio-organic fertilizer on the remediation of heavy metals in soil was studied, the results were different due to the different types of organic fertilizer, different treatment methods and experimental conditions and different research objects. For the five organic fertilizers used in this experiment, their own Cd content was 0.18~0.31 mg/kg, while the content of mild Cd contaminated soil was set to 1 mg/kg. The addition of organic fertilizer increased the Cd content in mild Cd contaminated soil by 18%~31%, which also greatly affected the final passivation effect of organic fertilizer. Organic fertilizer is rich in humus. Studies have shown that the chelation and adsorption of humus on metal ions coexist. When the concentration of metal ions is high, the exchange adsorption is dominant, and when the concentration is low, the chelation is dominant. The influence of the chelation on the migration of metal ions depends on whether the formed chelate is insoluble or soluble. When humus and metal ions form soluble chelate, it promotes the migration of heavy metals, and if insoluble chelate is formed, it hinders the migration of heavy metals. The chelate formed by humic acid and metal ions in humus composition is generally insoluble and can reduce the activity of heavy metals (Swift et al., 1992; Sahu and Banerjee, 1990), and the chelate formed by fulvic acid and heavy metal ions is soluble, thus promoting the mobility of heavy metal ions in soil. In this study, the passivation effect of high organic fertilizer application amount in heavily polluted soil was the best, and the passivation effect of medium and low biological organic fertilizer application amount in low polluted soil was the worst. The reason may be that when the Cd content in soil was high, various functional groups contained in organic fertilizer and organic matter with large specific surface area could form insoluble metal-organic complex by complexing with heavy metals, and the passivation effect was the best. When the Cd content was low, the Cd contained in organic fertilizer and the chelate formed by fulvic acid and Cd improved the availability of Cd in soil, resulting in the worst passivation effect.

5 Practical implications of this study

Heavy metal pollution, as a more serious type of soil pollution, has become increasingly serious, threatening almost every country, especially cadmium, which is particularly harmful to food safety and human health. Soil cadmium pollution control has become a difficult and hot spot in international research. Therefore, how to control cadmium pollution, and reduce its impact of polluted soil on the quality and safety of agricultural products, thereby reducing its harm to human beings, and achieving sustainable development of agriculture, is an important issue related to public ' s health. China is rich in organic fertilizer resources. It is convenient and economical to obtain materials, which not only plays an important role in improving soil heavy metal pollution, but also has great significance in improving land productivity. This is also incomparable to other inorganic improvement materials for controlling heavy metal pollution. Therefore, organic materials have broad application prospects and important significance in soil improvement. At the same time, in the future research, it is expected to study and launch organic fertilizer passivation formula suitable for different soil types and different heavy metal pollution.

6 Conclusion

This paper studied the regulation of different organic fertilizers on different degrees of Cd-contaminated soil, and provided theoretical support for the remediation of Cd-contaminated gray-purple soil in southwest China. The specific conclusions are as follows:

The effects of various organic fertilizers on Cd availability in purple soil are closely related to soil pollution and organic fertilizer dosage. In mild Cd contaminated soil, low dosage (<4 g/kg) of organic fertilizer promoted the activation of Cd, and only when the dosage increased to 8 g/kg, it showed inhibitory effect. In moderately Cd-contaminated soil, each organic fertilizer significantly inhibited the availability of soil Cd, and the inhibition effect increased with the increase of organic fertilizer dosage. In the severe Cd-contaminated soil, except that PMOF had no inhibitory effect on soil Cd availability at medium dosage, other organic fertilizers showed certain inhibitory effect at different application rates, and the effect was significant at low and high dosage. Therefore, it is highly recommended to use the five types of organic fertilizers mentioned in this paper to passivate moderately Cd contaminated soils.

The dynamic changes of soil available Cd content and morphological transformation were consistent after adding organic fertilizer, and both of them stabilized after 60 days of organic fertilizer application. Therefore, more definitive results are generally monitored after 60 days in practical applications. The availability of soil Cd is mainly restricted by Exe-Cd, FeMnOx-Cd and Res-Cd. Exe-Cd and FeMnOx-Cd have significant positive contributions to soil available Cd, and Res-Cd has significant negative contributions to soil available Cd.

The effect of organic fertilizer on Cd passivation in purple soil was significantly different in different polluted soils. In general, in moderate and severe Cd-contaminated soil, the effectiveness of organic fertilizers on soil Cd was significantly inhibited, among which NOF and SOF were the best, while in mild Cd-contaminated soil, the passivation effect of organic fertilizers on soil Cd was the worst, especially ESOF.

Acknowledgments

This research is supported by Natural Science Basic Research Program of Shaanxi (No.2020JQ-1002 and No.2021JQ-959), Innovation Capability Support Program of Shaanxi (No.2021KRM079), and the project of Shaanxi Provence Land Engineering Construction Group (Program No. DJNY2022-38).

Competing interests

The author(s) declare no competing interests.

References

Andersson, A., & Siman, G. (2009). Levels of cd and some other trace elements in soils and crops as influenced by lime and fertilizer level. Acta Agriculturae Scandinavica, 41(1), 3-11.

Ansari, R. A., & Mahmood, I. (2017). Optimization of organic and bio-organic fertilizers on soil properties and growth of pigeon pea. Scientia Horticulturae, 226, 1-9.

Araujo, D., Pereira, R., Almeida, D., et al. (2017) Photosynthetic, antioxidative, molecular and ultrastructural responses of young cacao plants to Cd toxicity in the soil. Ecotoxicology and Environmental Safety, 144, 148-157.

Borah, D. K., Rattan, R. K., & Banerjee, N. K. (1992). Effect of Soil Organic Matter on the Adsorption of Zn, Cu and Mn in Soils. Journal of the Indian Society Soil Science,40(2),277-282.

Bulluck, L. R., Brosius, M., Evanylo, G. K., et a1. (2002). Organic and synthetic fertility amendments influence soil microbial, physical and chemical properties on organic and conventional farms. Applied soil Ecology,19(2),147-160.

Chandorkar, V., Chimurkar, A., & Gomashe, A. (2019). Voltage production using metabolic activities of azatobacter species and other soil microbial flora in rice field using microbial fuel cell and microbial solar cell technology. Agricultural Science Digest-A Research Journal, 38(4),310-312.

Chen, A. L., Wang, K. R., & Xie, X. L. (2005). Efects of Fertilization Systems and Nutrient Recycling on Microbial Biom ass C, N and P in a Red-dish Paddy Soil. Journal of Agro-Environment Science,24(6), 1094-1099.

Chen, H., Chen, Y. C., & Yang, X. C. (2003). Regulation of Phyto-Availability of Hg, Cd, Pb in Soil by Limestone. Journal of Agro-Environment Science,22(5),549-552.

Chen, S. Y., Huang, S. W., Chiang, P. N., et al. (2011). Influence of chemical compositions and molecular weights of humic acids on Cr (VI) photo-reduction. Journal of Hazardous Materials,197,337-344.

Clemens, S., Aarts, M. G. M., Thomine, S., et al. (2013). Plant science: the key to preventing slow cadmium poisoning. Trends in Plant Science, 18(2),92-99.

Elloit, H. A., Liberati, M. R., & Huang, C. P. (1986) Competitive adsorption of heavy metals by soils[J]. Journal of Environmental Quality,15(3),214-219.

Guo, M. X., & Lin, Y. H. (1998). Using Micro-Ecosystem to Access the Bioactivity of Heavy Metals in Sediments. Acta Scientiae Circumstantie, 18(3),325-330.

Havelcová, M., Mizera, J., Sykorová, I. & Peka, M. (2009). Sorption of metal ions on lignite and the derived humic substances. Journal of Hazardous Materials, 161(1), 559-564.

He, F., Miao, J., & Wei, S. (2005). Speciation distribution and its chemical, biological availability of arsenic, lead added in acid, neutral and alkali purple soil. Transactions of The Chinese Society of Agricultural Engineering, S1,44-47.

He F. (2004). Correlation and Assessment of heavy metal contamination between agricultural soils and crops in Chongqing. Ph. D. Dissertation. Chongqing: Southwest Agricultural University.

Hou D., & Li F. (2017). Complexities surrounding China's soil action plan. Land Degradation & Development. 28(7), 2315-2320.

Huang, D., Liu, L., Zeng, G., Xu, P., & Wan, J. (2017). The effects of rice straw biochar on indigenous microbial community and enzymes activity in heavy metal-contaminated sediment. Chemosphere, 174, 545-553.

Huang, Q., Liu, B., Cai, H., & Bao, L. (2014). Effect of freeze-thaw cycles and organic fertilizer on the speciation of cadmium in black soils. Environmental Pollution & Control. 36(12), 38-42.

Jia, Z., Li, S., & Wang, L. (2018). Assessment of soil heavy metals for eco-environment and human health in a rapidly urbanization area of the upper yangtze basin. Scientific Reports, 8(1), 3256.

Jiang, T., Skyllberg, U., Wei, S., Wang D., Lu, S., & Jiang, Z., et al. (2015). Modeling of the structure-specific kinetics of abiotic, dark reduction of Hg (II) complexed by o/n and s functional groups in humic acids while accounting for time-dependent structural rearrangement. Geochimica Et Cosmochimica Acta, 154, 151-167.

Kumar, M., Bolan, N. S., Hoang, S. A., Sawarkar, A. D., & Rinklebe, J. (2021). Remediation of soils and sediments polluted with polycyclic aromatic hydrocarbons: to immobilize, mobilize, or degrade? Journal of Hazardous Materials, 420(4): 126534.

Liu, X. C., Gao, Y. M., Fan, Y. H., & Wang, B. S. (2008). Adsorption and desorption of heavy metal ions by organic fertilizers. Chinese Journal of Soil Science, 39(4), 942-945.

Liu, X. Z., Ma, Z. H., & Zhao, X. J. (2014). Effect of different organic manure on cadmium form of soil and resistance of wheat in cadmium contaminated soil. Journal of Soil and Water Conservation, 28(3):243-247.

Liu W., Chen X. M., Jing F., Hu S.M., Wen X., Li L. Q. (2020). Effects of applying bioorganic fertilizer on chemical form and transport characteristics of Cd in soil-rice system. Bulletin of Soil and Water Conservation, 40(1): 78-84.

Liu, W. B., Zhang, Q., & Zhang, C. Y. (2009). Effects of applying organic fertilizer and mercury added on Hg pollution in soils-winter wheat system. Journal of Agro-Environment Science, 28(5), 890-896.

Lu, K. G. (2003). The effect of bio-fertilizer on soil and absorption to cadmium and copper of Fuji apple root. MA. Thesis. Tai'an, Shandong: Shandong Agricultural University.

Lu, R. K. (2000). Agricultural chemical analysis of soil. Beijing: China Agricultural Science and Technology Press.

Ma, T. Z., Ma, Y. H., Fu, H. H., Wang, Q., Xu L. L., & Nie, J. R., et al. (2015). Remediation of biological organic fertilizer and biochar in paddy soil contaminated by Cd and Pb. Journal of Agricultural Resources & Environment, 32(1), 14-19.

Matilainen, A., Vepsalainen, M., & Sillanp, M. (2010). Natural organic matter removal by coagulation during drinking water treatment: a review. Advances in Colloid & Interface Science, 159(2), 189-197.

Österberg, R., Wei, S. Q., & Shirshova, L. (1999). Inert copper ion complexes formed by humic acids. Acta Chemica Scandinavica, 53(3), 172-180.

Pan, Y., & Zhou L.X. (2007). Influence of applying organic manures on the chemical form of cu and cd in the contaminated soil and on wheat uptake: field micro-plot trials. Journal of Nanjing Agricultural University, 30(2),142-146.

Qin, W. S., Wang, L., Tian, T., & Qiu, R. L. (2014). The effects of the structural characteristics of organic manure on cd morphology in contaminated soil. Sichuan Environment. 33(4), 24-28.

Ministry of Environmental Protection P.R.C., Ministry of Land and Resources P.R.C. (2014). Report on the national general survey of soil contamination[EB/OL]. https://www.mee.gov.cn/gkml/sthjbgw/ qt/201404/t20140417_270670_wh.htm

Rufus, L., & Chaney. (2015). How does contamination of rice soils with Cd and Zn cause high incidence of human cd disease in subsistence rice farmers. Current Pollution Reports, 1(1), 13-22.

Sahu, S., & Banerjee, D. K. (1990). Complexation properties of typical soil and peat humic acids with Copper (II) and Cadmium (II). International Journal of Environmental Analytical Chemistry, 42(1-4), 35-44.

Senesi, N., DOrazio, V., & Miano, T. M. (1995). Adsorption mechanisms of s-triazine and bipyridylium herbicides on humic acids from hop field soils. Geoderma, 66(3-4), 273-283.

Sharma, K. L., Mandal, U. K., Srinivas, K., Vittal, K., Mandal, B., & Grace, J. K., et al. (2005). Long-term soil management effects on crop yields and soil quality in a dryland alfisol. Soil and Tillage Research, 83(2), 246-259.

Shi, W. Y., Shao, H. B., Li, H., Shao, M. A., & Sheng, D. (2009). Co-remediation of the lead-polluted garden soil by exogenous natural zeolite and humic acids. Journal of Hazardous Materials, 167(1-3), 136-140.

Suksabye, P., Pimthong, A., Dhurakit, P., Mekvichitsaeng, P., & Thiravetyan, P. (2016). Effect of biochars and microorganisms on cadmium accumulation in rice grains grown in Cd-contaminated soil. Environmental Science and Pollution Research, 23(2), 962-973.

Swift, R. S., Rate, A. W., & Mclaren, R. G. (1995). Interactions of copper and cadmium with soil humic substances. In: Abstract, International Society of Soil Science Working Group M O, First Workshop, Edmonton, Alberta, Canada.

Tessier, A., Campbell, P., & Bisson, M. (1979). Sequential extraction procedure for the speciation of particulate trace metals. Analytical Chemistry, 51(7), 844-851.

Ushijima, K., Fukushima, M., Kanno, S., Kanno, I., & Ohnishi, M. (2016). Risks and benefits of compost-like materials prepared by the thermal treatment of raw scallop hepatopancreas for supplying cadmium and the growth of alfalfa (Medicago Sativa L.). Journal of Environmental Science & Health Part B, 51(3), 170-175.

Wang, M., Li, S. T., Ma, Y. B., Huang, S. M., Wang, B. R., & Zhu, P. (2014). Effect of long-term fertilization on heavy metal accumulation in soils and crops. Journal of Agro-Environment Science, 33(1):63-74.

Wang, Y. K., Zhang, H. C., Hao, X. Z., & Zhou, D. M. (2010). A review on application of organic materials to the remediation of heavy metal contaminated soils. Chinese Journal of Soil Science, 41(5), 1275-1280.

Xie, X. M., Fang, X. P., Liao, M., Huang, Y., & Huang X. H. (2018). Potential to ensure safe production from rice fields polluted with heavy cadmium by combining a rice variety with low cadmium accumulation, humic Acid, and sepiolite. Environmental Science, 39(9), 11:4348-4358.

Xu, C., Peng, C., Sun, L., Zhang, S., Huang, H., & Chen, Y., et al. (2015). Distinctive effects of TiO_2 and CuO nanoparticles on soil microbes and their community structures in flooded paddy soil. Soil Biology & Biochemistry, 86, 24-33.

Yang, W. J., Ding, K. B., Zhang, P., Qiu, H., Cloquet, C., Wen, H, J., et al. (2018). Cadmium stable isotope variation in a mountain area impacted by acid mine drainage. The Science of the total environment, 646, 696-703.

Yang, X. C., Mou, S. S., Tang, S. Y. (1992). The pollution and transference of heavy metals in purple soil area. Journal of Agro-environmental Science, 02, 61-65.

Yang X. L. (2004). Studies on surface charge properties of purple soil. Ph. D. Dissertation. Chongqing: Southwest Agricultural University.

Yu X. L., & Zhao M. J. (2013). Effect of bio-organic fertilizer application on nutrients in saline-alkaline soil. Journal of Jilin Agricultural University, 35(1), 50-54,57.

Zhang, J., Yang, W., Liao, B. H., & Wu H. (2020). Mechanisms and influence approaches od organic fertilizer on occurrence mode of Cd in acid paddy soil. Journal of Soil and Water Conservation, 34(1), 6: 365-370.

Zhang, J. M. (2002) The Pollution and Passivation Remediation of Soil As, Sb in Typical Antimony Mining Areas. MA. Thesis. Guizhou: Guizhou Education University.

Zhou, Y. J., Zhao, W., Luo, C., Xu, Y. J., & Wu W. D. (2018). Effects of organic manure on pb speciation in soil. Environmental Chemistry, 37(3), 534-543.

Zong, H., Liu, S., Xing, R., Chen, X., & Li, P. (2017). Protective effect of chitosan on photosynthesis and antioxidative defense system in edible rape (brassica rapa l.) in the presence of cadmium. Ecotoxicology and Environmental Safety, 138, 271-278.

本文曾发表于《Frontiers in Environmental Science》第10卷

The Response of Cd Chemical Fractions to Moisture Conditions and Incubation Time in Arable Land Soil

Nan Lu[1,2,3,4], Yang Wei[1,2,3,4], Zhaoxin Zhang[2,3,4], Yan Li[1,2,3,4], Gang Li[1,2,3,4] and Jichang Han[2]

(1. Shaanxi Land Engineering Construction Group Co., Ltd., Xi'an Jiaotong University, Technology Innovation Center for Land Engineering and Human Settlements, Xi'an 712000, China; 2. Key Laboratory of Degraded and Unused Land Consolidation Engineering, the Ministry of Natural Resources, Xi'an 710075, China; 3. Institute of Land Engineering and Technology, Shaanxi Provincial Land Engineering Construction Group Co., Ltd., Xi'an 710075, China; 4. Shaanxi Provincial Land Consolidation Engineering Technology Research Center, Xi'an 710075, China)

【Abstract】 Heavy metal pollution in soils is an issue of global concern, and many scholars have focused on Cadmium (Cd) because of its strong biological migration and toxicity. This study explored arable land soil, changes in external Cd contamination processes, its response to soil moisture conditions, and indoor simulation. After adding an external source of 5 mg/kg d.w., the distribution of soil Cd fractions content, EXC-Cd, CAB-Cd, FMO-Cd, OM-Cd, and RES-Cd, were continuously monitored under different water management regimes, and correlation analysis and regression equations were calculated. The results show that after external Cd entered arable land soils, the binging strength of pollutants and soil gradually increased with incubation time and the distribution of Cd chemical forms was more stable under different water management regimes. The over saturated water content promotes the transformation of EXC-Cd to other forms. The transformation of CAB-Cd fractions can be accelerated to other fractions by field capacity, and the active conversion period was 30~60 d. Not all Cd fractions correlated between each other under four water management regimes, but it seems that the reducibility of soil environment was more conducive to external Cd fixation and stability. The response surface design method (RSM) was used to establish quantitative regimes between Cd fractions with incubation time and soil moisture, and the soil moisture content and incubation time had an obvious effect on FMO-Cd content with $R^2=0.9542$.

【Keywords】 Cadmium; Bioavailability; Moisture conditions; Arable land

1 Introduction

Soil heavy metal pollution is a global environmental problem[1]. Cadmium (Cd) and its compounds have greater mobility and biological toxicity compared to other heavy metals[2-3]. It easily accumulates in the human body, and in Japan Itai-Itai Disease has appeared due to Cd poisoning[4]. The International Cancer Research Institution (IARC) classifies Cd as a human carcinogen; the EU lists Cd as a high-hazardous toxic substance and that is under regulation[5]. A National Communique on Soil Pollution Survey Bulletin issued in April 2014 found serious Cd pollution in the soil around non-ferrous metal mining areas. Areas exceeding the standard rate of pollution reached 7.0% in China[6]. Therefore, it is urgent to study the effects of soil Cd pollution and key influencing factors to ensure land safety.

The biological toxicity, migration characteristics and environmental effects of Cd in soil depend on not only speciation but also availability and mobility[7]. Soil is a complex multi-media multi-interface environment. After external Cd enters the soil, incorporation of Cd into the soil depends on the soil physical and chemical properties, human factors, mineral composition of the soil and microbial activity[8]. Thus, the Cd chemical fractions is an effective way to reflect its behavioral characteristics and soil environmental effects, which can aid in understanding dispersion and enrichment, and migration conversion regularity in soil.

Soil physical and chemical properties and environmental factors also have an important impact on the chemical fractions and bioavailability of Cd, and the migration transformation of water management controls is the focus of much research. Soil type, redox, organic matter content, etc. can affect the distribution of Cd fractions[9]. Water management includes agricultural measures that are low cost, green and environmentally friendly, easy to institute. Any modification of the soil water content will affect soil pH, Eh and sulfur oxidation-reduction state changes, and can also affect the distribution of soil microbial species and the decomposition of organic matter. In turn, the fractions and biological effectiveness of Cd is impacted[10]. Studies have shown that when soil water saturation increases, the acid paddy soil pH increases, the Eh is lowered, and the soil reducibility is enhanced. The extracted Cd^{2+} content is reduced because it forms precipitates with S^{2-}, and OH^-oriented resection ions[11-12]. The retention time of soil water content also plays a vital role in this process, and when the acid paddy soil is in an aerobic state, the bioavailable content of soil Cd will increase[13-14]. However, for different kinds of soil, an increase or decrease in Cd biological extractability may also occur under soil water saturation, and the conditions and influencing factors still need to be described.

In China, the land area contaminated by Cd is around 2×10^{13} m^2, of which 1.4% is arable land[15-16]. To explore the changes in external Cd chemical and the response of environmental factors, Cd in various forms was determined under different water management simulations such as maintaining field water capacity, alternating dry and wet, 65% field capacity, and long-term flooding. This study had the following goals: (1) to explore the effects of incubation time and field water capacity on the characteristics of external Cd, (2) to clarify changes in Cd chemical fractions with incubation time and water management, and (3) to identify the correlations and quantitative relationships between Cd chemical fractions and incubation time and water management. This research can provide information for developing arable land management practices and regulating Cd biotoxicity.

2 Materials and methods

2.1 Test soil

Testing was conducted in 2020 at the Key Laboratory of Degraded and Unused Land Consolidation Engineering, the Ministry of Natural Resources. The test soil was collected from 0~20 cm topsoil in the Tongguan mining area, Shaanxi Province (110°21′40″E, 34°30′16″N). The sundries in soil were first removed, and then soil samples were air dried and run through a 2.0 mm nylon sieve before analog tests. The field water capacity of test soil was measured according to the soil testing Part 22: Cutting ring methods for determination of field water-holding capacity in soil (NY/T 1121.22—2010)[17]. The basic physical and chemical characteristics of the test soil are shown in Table 1.

Table 1 Physical and chemical characteristics of test soil

pH	Conductivity ($dS \cdot m^{-1}$)	Organic matter ($g \cdot kg^{-1}$)	CEC ($cmol \cdot kg^{-1}$)	Total iron ($mg \cdot kg^{-1}$)	$CaCO_3$ (%)	Field water capacity(%)	Cd ($mg \cdot kg^{-1}$)
9.15	0.367	5.73	3.69	1.23	10.72	21	0.46

2.2 Materials and experimental design

According to the soil environmental quality standard for soil pollution risk control of agricultural lands (GB 15618—2018)[18], the control value of soil Cd pollution risk in agricultural lands (pH>7.5) was 4 mg/kg d.w., and 5 mg/kg d.w. was set as the target content of contaminated soil. Contaminated soil was prepared by spraying it with a $CdCl_2$ solution, and initial soil Cd content was considered as actual content. The test treatments were called GT1, GT2, GT3, and GT4, and the correspondence of codes are given in Table 2. Using a weighing method, the corresponding water management regimes were field capacity, alternating wet and dry, 65% of field capacity, and saturated water content. The sampling times were 0 d, 7 d, 15 d, 30 d, 60 d, 90 d, and 120 d, and the addition of Cd was made before the first sampling. To simulate changes in the formation and biotoxicity of external Cd in arable land soil, the distribution fractions characteristics of Cd, such as exchangeable form (EXC), carbonate form (CAB), Fe-Mn oxide form (FMO), organic form (OM), and residual fraction (RES), were dynamically monitored. The simulation experiments were carried out in 500 ml plastic beakers. Four hundred grams of test soil (with an accuracy of one hundredth) were placed into each beaker. Three repetitions were set for each treatment. The test design and measured Cd contents in polluted soil are shown in Table 2.

Table 2 Test design and measured Cd content in polluted soil (mean ±SE)

Treatment	Water management	Target content (mg/kg d.w.)	Measured content (mg/kg d.w.)	Sampling time (days)
GT1	Field capacity	5.0	5.63 ±0.296	0,7,15,30,60,90,120
GT2	Alternating wet and dry			
GT3	65% of field capacity			
GT4	Over saturated water content			

2.3 Indexes and methods for determination

2.3.1 Experimental methods

All experiments were carried out in a constant temperature incubator (DHP-9272, China) at 20 ℃± 0.5 ℃ during the 120 days. The experimental methods for the four different water management regimes were as follows:

(1) Field capacity. The water content in the basin was kept at field capacity by adding deionized water to supplement moisture due to evaporation loss every day.

(2) Alternating wet and dry. The initial water content was set to saturated, and when the soil moisture dried in 30 days, deionized water was added until saturated and cycled again.

(3) 65% of field capacity. The water content in the basin was kept to about 65% of the field capacity.

(4) Long-term flooding. The water content in the basin was kept at over saturated, and the water was about 5 cm above the soil surface.

2.3.2 Treatment and determination of soil samples

(1) Soil in the beaker was fully mixed before sampling, and 30.0 g of sample were taken each time. To prevent the sample changes from drying, fresh soil samples were used for continuous extraction. Soil water content was measured at the same time, and Cd chemical fractions contents were calculated based on dry weight. A modified five-step Tessier continuous extraction method was used to determine the Cd chemical

fractions contents[7,19].

(2) An acidometer (pHS-3C, China) was used to determine soil pH: water (v/v= 1:2.5). Microwave digestion was used for pretreatment, and then soil extracts were diluted after filtering (0.45 μm PTFE) before the concentrations of RES-Cd and total Cd determination were analyzed[20]. Cd chemical fractions concentrations in the soil extracts were determined using an AAS Zeenit 700P atomic absorption spectrometer. The accuracy of the analytical procedure adopted for atomic absorption spectrometer analysis was checked by running standard solutions every 20 samples. In addition, to control the experimental data error, the total number of chemical Cd forms was compared with the total measured Cd concentrations, and the deviation should not exceed 13%. Only excellent grade reagents were used for testing, with strict quality control[21].

2.4 Method for coefficient evaluation

The biological effective coefficient (BEC) was calculated using Eq.1:

$$\mathrm{BEC}(\%) = \frac{C_1 + C_2}{\sum_{i=1}^{k} C_i} \tag{1}$$

In the formula, k is the total number of steps extracted, $k=5$. i is the extraction order, 1 EXC-Cd, 2 CAB-Cd, 3 FMO-Cd, 4 OM-Cd, 5 RES-Cd. C_i is the extracted content of the i -th step (mg · kg^{-1}).

The reduced partition index (I_R)[22] was calculated using Eq.2 and describes the binding strength between heavy metals and soils. A higher strength coefficient value indicates a closer binging strength between heavy metals and soils.

$$I_R = \frac{\sum_{i=1}^{k} (F_i \times i^n)}{k^n} \tag{2}$$

In the formula, F_i is the ratio of the content of the i-th step and the total amount in the soil. n is an integer, with a value of 2.

The redistribution coefficient (U_{fi})[23] was calculated following Eq.3.

$$U_{fi} = F_i / F_{ci} \tag{3}$$

In the formula, F_{ci} is the ratio of the content of the i-th step and the total amount in soil before the addition of Cd.

The total redistribution coefficient (U_{ts}) is used to assess the distribution of various forms of heavy metal pollution in soil, compared to in uncontaminated soil. U_{ts} was calculated following Eq.4.

$$U_{ts} = \sum_{i=1}^{k} (F_i \times U_{fi}) \tag{4}$$

Because U_{ts} of clean soil is 1, a smaller difference between U_{ts} for contaminated soil and 1 means a more stable heavy metals distribution.

2.5 Statistical methods

Correlation analysis, ANOVA testing, and LSD multiple comparisons were performed at the 95% significance level using SPSS 22 (IBM SPSS Statistics, Version 22). One-way ANOVA was used to reveal the effects of different incubation times on the soil Cd fractions distribution and biological effective state changes in arable land soil. Least Significance Difference (LSD) testing was used to test the significance of each index across different treatment groups ($p < 0.05$). Pearson correlation analysis (two-tailed) was used to reveal the correlation relationship between Cd fractions and incubation time, and water management regimes.

All figures in this paper were created using ORIGIN PRO 2021 (OriginLab Corp., Northampton, USA). Response Surface Methodology (RSM) was used to establish the relationships between Cd chemical fractions, incubation time, and soil properties.

3 Results and discussion

3.1 The characteristics of changes with incubation time under GT1

3.1.1 Soil Cd fractions distribution and changes in arable land soil

The movement of Cd into the different soil fraction with incubation time in arable land soil under GT1 are shown in Fig. 1. After external Cd entered the soil, adsorption-desorption occurred between Cd and the soil particles. Further, the soil Cd fractions conversion, migration, and returning occurred in the soil environment. EXC-Cd can be present in a soil solution and hindered from conversion to other forms by soil particles[24]. Between 0 d and 7 d, the increase of EXC-Cd is approximately equal to the decrease of CAB-Cd, it seems that part of the CAB fraction moved to the EXC fraction. CAB fraction decreased slowly from 0 d to 30 d after which it decreased more drastically (at 60 d) and remained stable after. The big decrease seems to be related to the migration of part of this fraction into the FMO fraction (and slightly to EXC fraction). The percentage of the RES and OM fractions is very low, and the increment of the percentage is less than 1%.

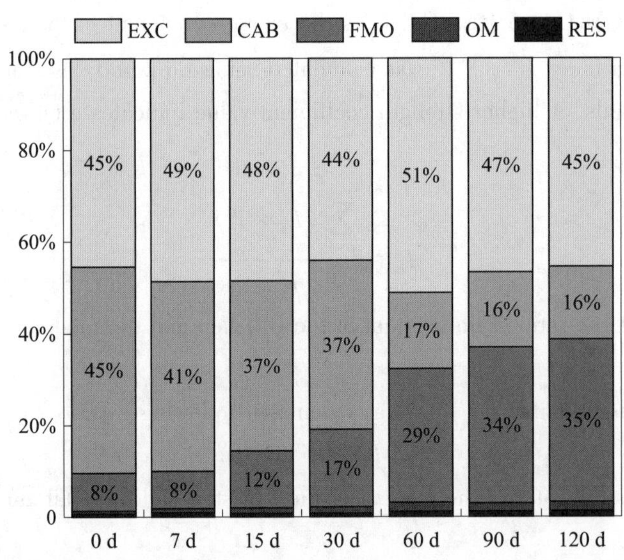

Fig. 1 The movement of Cd into the different soil fraction with incubation time in arable land soil under GT1

The Cd fractions change with incubation time in arable land soil under GT1 are shown in Fig. 2. After 30 days, the CAB-Cd conversion rate to FMO-Cd and OM-Cd was high, resulted in a large decrease in the CAB-Cd content. This finding was consistent with the experimental results of Lu et al.; that is, as the incubation time increased, Cd slowly converted from the exchangeable state to other forms[25]. The amount of CAB-Cd was always high during incubation due to the alkaline soil and higher soil calcium content[26].

3.1.2 Changes to the Cd biological effective coefficient in arable land under GT1

Heavy metal distribution characteristics can help evaluate their potential migration capacity and biological availability[27]. Bioavailability usually has a strong correlation with bioavailability, which is an important indicator of bioavailability. The binding strength of EXC-Cd, CAB-Cd and soil were weak; therefore, these two forms easily migrated with higher bioavailability and potential risk. The ratio of total EXC-Cd, CAB-Cd and total Cd content is commonly used to evaluate Cd availability. After external Cd entered the soil in the

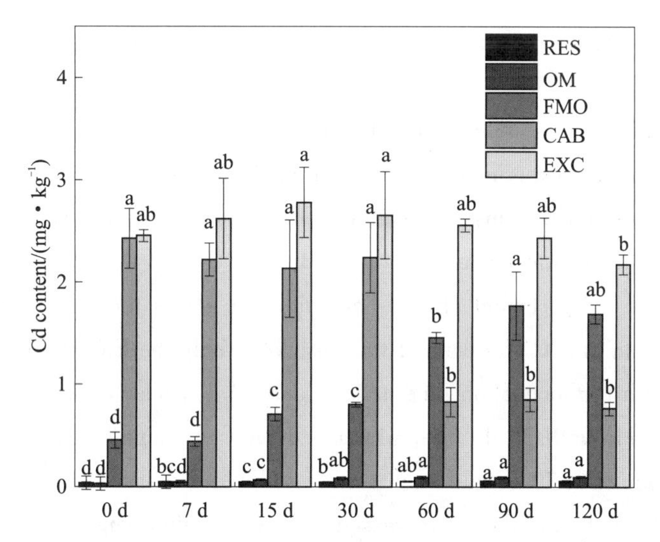

Fig. 2 The Cd fractions change with incubation time in arable land soil under GT1. Data shown are means ± standard error (n=3) , and different letters (a, b, c, d) present significant differences between time points of the same Cd fraction at p < 0.05

initial stage, it mainly existed as both EXC-Cd and CAB-Cd. Under GT1, the BEC remained stable from 0 d to 7 d, then decreased significantly from 7 d to 15 d, and remained stable until 30 d, after which it decreased significantly from 30 d to 120 d (Fig. 3). The biological effectiveness coefficients decreased from 90.1% to 61.5%, by nearly 30%. During the culturing time of 30~60 d, the bioeffective state declined by up to 16.2%. This result further confirms that the bioavailability decreases with the incubation proceed, and the pollutants exist in a more stable way[28]. The transformation of CAB-Cd to other forms especially FMO-Cd is the reason for the reduction in the bioavailability of Cd[29]. Obviously, Cd with lower bioavailability and mobility is more difficult to be exposed to the recipient population through the migration and transformation, with a lower environmental risk[30].

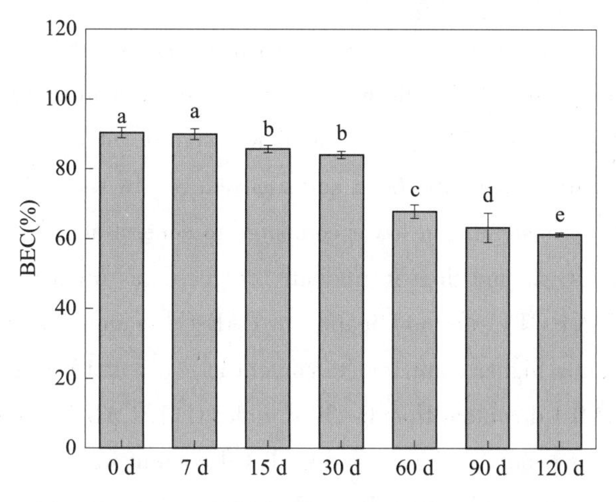

Fig. 3 Change in bioavailability coefficient of Cd with incubation time in arable land soil under GT1. Data shown are means ± standard error (n=3) , and different letters (a, b, c, d, e) present significant difference at p < 0.05

3.2 The characteristics of Cd fractions changes with incubation time

Water management regimes can change soil redox properties, affecting the physical chemistry of external heavy metals that further migrate and transform biological effectiveness[31-32]. The change characteristics of Cd fractions over different incubation times were analyzed by simulating different water management regimes. The formation of soil oxidative reduction properties was affected by the water management model.

The various forms, distributions, conversions, and migration were analyzed under different water management regimes.

Changes in Cd chemical forms under four different water management regimes and incubation times are shown in Fig. 4. The incubation experiment was carried out under water management regimes. Through continuous monitoring of the chemical forms of contaminated horizontal external Cd in arable land soil (up to 120 d), we found that different water management conditions resulted in different Cd chemical forms.

The EXC-Cd (Fig. 4 (a)) content changes are different across the four water management regimes. Under GT1, no modification in EXC-Cd content was observed, only the EXC-Cd content at 120 d decreased significantly compared with the incubation time at 15 d and 30 d. Under GT2, EXC-Cd increased from 0 d to 7 d, then remained stable until 30 d, after which it decreased significantly from 30 d to 60 d and was 23.74% less than the samples at 30 d, and then it increased and decreased again. Under GT3, the EXC-Cd content decreased from 0~7 d, and then it increased until 15 d, after which it decreased significantly from 15~60 d, and then EXC-Cd remained relatively stable and did not decrease with the extension of incubation time from 60~120 d. Under GT4, the EXC-Cd content decreased from 0~7 d, and then it remained relatively stable until 30 d, and decreased significantly from 30~60 d. It was relatively stable from 60~90 d, then further decreased from 90 d to 120 d. Except under GT1, EXC-Cd content decreased greatly from 30~60 d under the remaining water management regimes, which was the active period of EXC-Cd conversion to other forms. In order to see the effect of the different water régime on Cd fractions, the statistical comparison have been done between the treatments within each sampling time. Under GT2 and GT3, EXC-Cd content was always greater than under GT1 and GT4 from 7~120 d, and EXC-Cd content under GT1 was always the lowest from 0~30 d, and EXC-Cd content under GT4 was always the lowest during 60~120 d. It seems that the higher soil water content promotes the transformation of EXC-Cd to other forms. The change trend of EXC-Cd fraction during 60~120 d is consistent with Xu et al[33] that the continuous flooding has showed lowest content of soil EXC-Cd than other treatments.

For CAB-Cd (Fig. 4 (b)), Under GT1, the CAB-Cd content was relatively stable from 0~30 d. It was reduced by 63.06% during 30~60 d. Afterwards, the content remained relatively stable. Under GT2, CAB-Cd content at 7 d, 30 d, 60 d, 90 d, and 120 d was significant lower compared to content at 0 d. CAB-Cd content declined greatly during 30~60 d and was reduced by 44.79%. Under GT3, CAB-Cd content at 7 d, 30 d, and 120 d was significant lower compared to content at 0 d ($p < 0.05$). CAB-Cd content was relatively stable from 0~30 d, and then it gradually decreases as incubation time increased from 30~60 d and then stabilized. Under GT4, no modification in CAB-Cd content was observed, only the CAB-Cd content at 60 d was significantly higher compared to content at 7 d and 15 d. The statistical comparison result showed high levels of CAB-Cd content from 0~30 d under GT1. CAB-Cd content was always greater under GT4 than other water management regimes, and CAB-Cd content was the lowest from 60~120 d under GT1. CAB-Cd content decreased greatly from 30~60 d under the remaining water management regimes under GT1, which was the active period of CAB-Cd conversion to other fractions. Field water capacity or the increase of soil water content can accelerate the transformation of external Cd to more stable forms in the soils[30]. Results also suggested that the water regimes significantly determine the effective risk of extractable fractions[34].

Under the different treatments, FMO-Cd (Fig. 4 (c)) content increased rapidly during 15~90 d, and the maximum value was reached at 90 d or 120 d. Under GT1, the FMO-Cd content increased during 0~

90 d with a maximum value at 90 d. After that, the content was relatively stable, and it increased 2.91 times compared to the initial content. Under GT2, the FMO-Cd content reached a maximum at 120 d, and the content increased by 2.63 times compared to the initial content. Under GT3, the FMO-Cd content reached a maximum at 120 d, and the content increased by 3.35 times compared to the initial content. Under GT4, FMO-Cd content was significantly different ($p < 0.05$) compared with incubation time from 15~90 d, and was 4.55 times higher than the initial content. In both water management regimes for GT1 and GT4, FMO-Cd was converted more from other forms compared to GT2 and GT3, indicating that high soil water content or stronger reducibility is conducive to the transformation of FMO-Cd. The statistical comparison result showed that FMO-Cd content was always the lowest under GT4 than other water management regimes from 0~60 d, and then increases during 60~90 d, after that FMO-Cd content was relatively stable until 120 d. The result confirms that the contents of FMO fraction were higher in dry-wet cycling area[35]. There was high conversion capacity that resulted in a low existing state, so the reduction of oversaturation environment was more conducive to external Cd fixation and stability[36].

For OM-Cd and RES-Cd (Fig. 4 (d) and (e)), these two forms were the least proportional of the five, the sum of the two contents accounts for about 2.73%~3.91% of the total. Under GT1, OM-Cd fraction remained stable from 0~15 d, increased slowly from 15 d to 30 d after which it remained stable again. Under GT2, OM-Cd fraction remained stable from 0~30 d, increased rapidly until 60 d, after which it remained stable again, reached a maximum at 120 d, with a content increase of 2.00 times compared to the initial content. Under GT3, OM-Cd fraction remained stable from 0~15 d, increased until 30 d, after which it remained stable again, reached a maximum at 120 d, with a content increase of 1.77 times compared to the initial content. Under GT4, OM-Cd fraction remained stable from 0~15 d, increased until 30 d, after which it remained stable from 30~60 d, increased from 60~90 d, and then decreased slightly.

Under GT1, RES-Cd fraction increased from 0~7 d and decreased from 7~15 d, then increased again until 60 d, after which it remained stable, and reached the maximum value 90 d, with a content increase of 2.56 times compared to the initial content. Under GT2, RES-Cd fraction remained stable from 0~30 d, increased from 30~60 d, after which it remained stable again. Under GT3, RES-Cd fraction increased from 0~7 d and it remained stable until 120 d, only the RES-Cd content at 120 d reached significant levels compared with the incubation time at 30 d. Under GT4, RES-Cd fraction remained stable from 0~15 d, increased rapidly from 30~60 d, and then it remained stable until 120 d, the RES-Cd content at 30 d, 60 d, 90 d, and 120 d reached significant levels compared with the incubation time at 0 d, it was significantly higher compared to content at 0 d, and reached a maximum at 120 d, about 0.634 times the initial content.

The statistical comparison result between the treatments within each sampling time showed that OM-Cd fraction content was always the highest under GT1 from 7~60 d, and OM-Cd fraction content was always the highest under GT2 from 90~120 d. OM-Cd fraction content was the lowest under GT4 from 0~120 d. RES-Cd fraction content was always the highest under GT3 from 7~120 d. RES-Cd fraction content was always the lowest under GT2 from 0~30 d, and 90~120 d. RES-Cd content remained stable under different water management regimes at 60 d. It seems that the alternation of dry and wet promotes the formation of OM-Cd during 30~60 d, and 65% F.C. promotes the rapid formation of RES-Cd during 0~7 d and remain at a high level until 120 d.

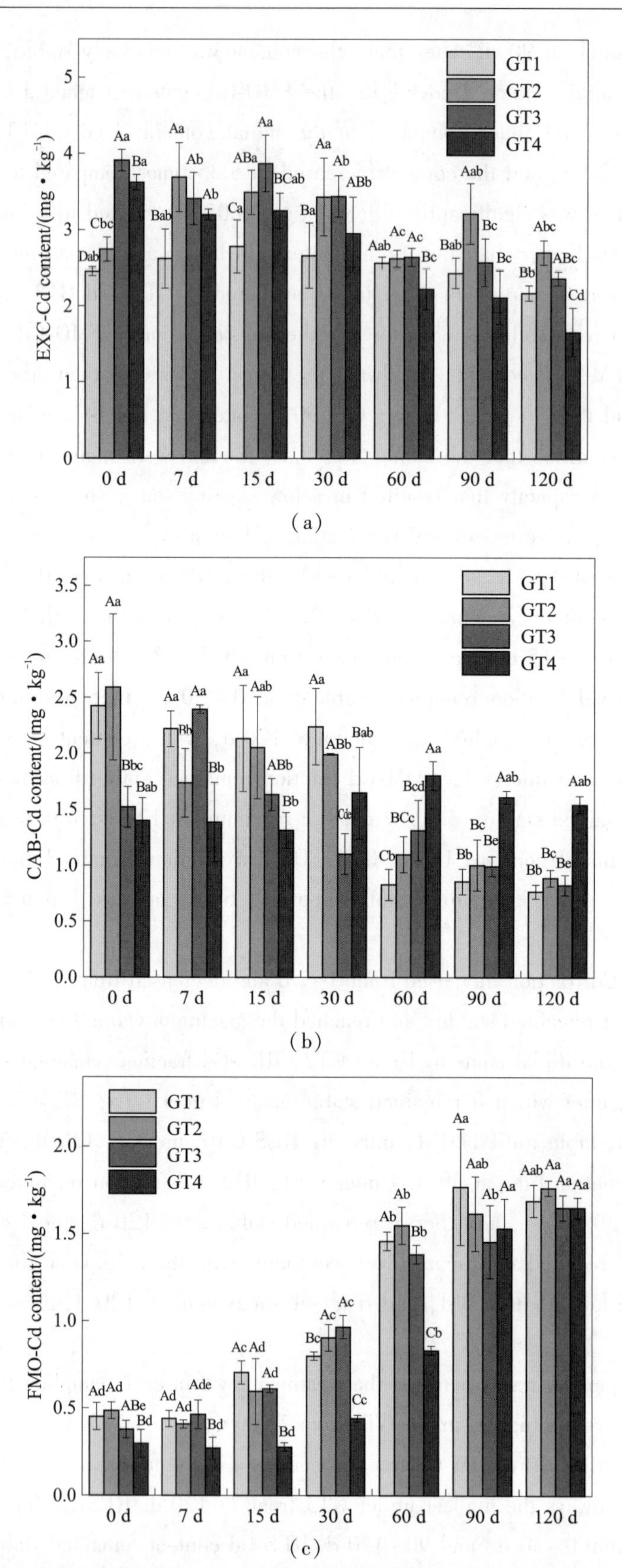

Fig. 4 Cd fractions changes with incubation time in arable land soil under different water management regimes. Data shown are means ± standard error ($n=3$), different letters (A, B, C, D) present significant differences between the treatments within the same sampling time at $p < 0.05$, and different letters (a, b, c, d) present significant differences between time points of the same treatment at $p < 0.05$

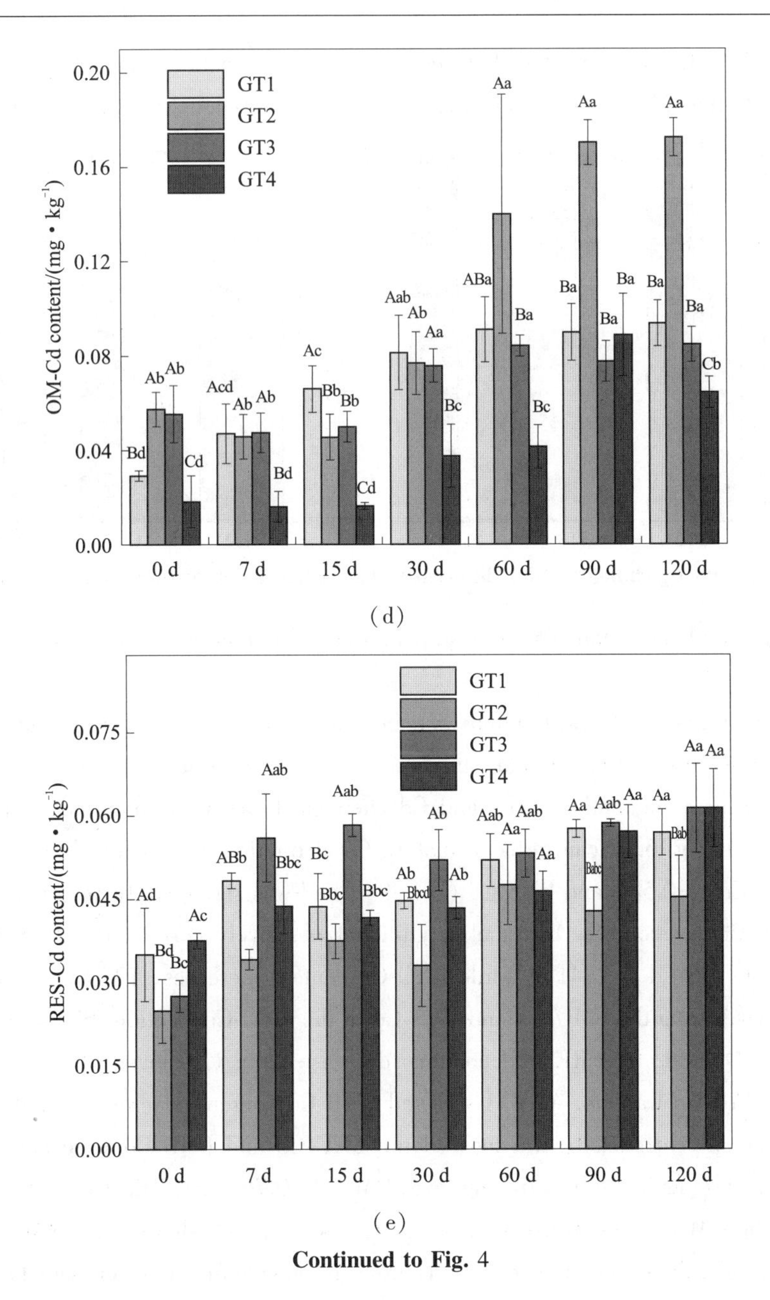

(d)

(e)

Continued to Fig. 4

3.3 Stability evaluation after entering arable land soil

After external Cd enters the soil, the binging strength of pollutants and soil gradually increased with incubation time and the distribution of Cd chemical forms was more stable under different water management regimes. The soil binding strength coefficient (I_R) and the total redistribution coefficient (U_{ts}) of external Cd of the initial and end stages under different water management regimes are shown in Fig. 5. Under GT1, I_R increased by 50.76% from the initial stage, and U_{ts} decreased by 9.41%. Under GT2, I_R increased by 45.00% from the initial stage, and U_{ts} decreased by 19.94%. Under GT3, I_R increased by 81.61% from the initial stage, and U_{ts} decreased by 11.74%. Under GT4, I_R and U_{ts} reached their maximum respective values of 103.29% and −34.86% compared with GT1, GT2, and GT3. When the soil was always in the reduction environment, the rate of external Cd transformation to forms that were difficult to extract was at a maximum. In addition to the GT4 treatment, the increases and decreases in I_R and U_{ts} were inconsistent. Under GT2 and GT1, the I_R increase was about the same, and was 60.77%~129.56% of the increase under GT3 and GT4. The reduction in U_{ts} was basically the same under GT1 and GT3 conditions, which was 69.90%~

270.37% of GT2 and GT4. This indicated that under the same incubation time, the chemical form stability conversion rate of external Cd was low, and the formal stability was longer than other treatments in GT1.

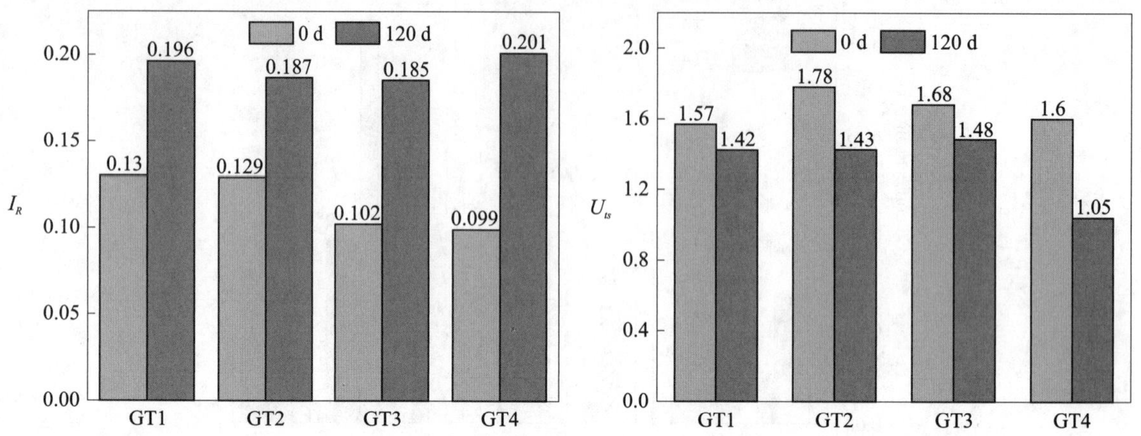

Fig. 5 I_R and U_{ts} changes under different water management regimes in arable land soil

3.4 Correlation and quantitative relationship between Cd fractions and incubation time, and water management model

3.4.1 Correlation analysis of Cd fractions and incubation time, and soil water content

Correlation analysis of quantitative data can show the relationship between factors and how strong it is. Data such as soil moisture, incubation time, and Cd chemical forms were analyzed using SPSS 22.0, and the results are shown in Fig. 6. It can be seen that EXC-Cd fraction was extremely significantly negatively correlated with FMO-Cd (−0.58) and RES-Cd (−0.39), and was significantly negatively related to OM-Cd (−0.24). The CAB-Cd fraction was extremely significant negatively correlated with FMO-Cd (−0.62), OM-Cd (−0.46) and RES-Cd (−0.41). While EXC-Cd content increases, FMO-Cd and RES-Cd contents decrease, and Cd moves from the FMO-Cd and RES-Cd to the EXC-Cd fraction. Similarly, CAB-Cd content increases, FMO-Cd, OM-Cd, and RES-Cd contents decrease, and Cd moves from the FMO-Cd, OM-Cd and RES-Cd to the CAB-Cd fraction. As EXC-Cd and CAB-Cd contents decreased, FMO-Cd, OM-Cd and RES-Cd contents gradually increased, and most of the EXC-Cd and CAB-Cd were converted to FMO-Cd. The FMO-Cd content was significantly correlated with OM-Cd (0.76) and RES-Cd (0.54). As the FMO-Cd content increased, the OM-Cd and RES-Cd contents increased, which showed that FMO-Cd played an important role for other forms in arable land soil[30]. There were no significant correlations between the Cd fractions content and soil moisture, and not all Cd forms correlated between each other. The EXC-Cd and CAB-Cd contents had extremely significant negative correlations with incubation time, and FMO-Cd, OM-Cd, and RES-Cd had extremely significant positive correlations with incubation time. As the incubation time increased, EXC-Cd and CAB-Cd fractions contents decreased, and FMO-Cd, OM-Cd, and RES-Cd increased, and the external Cd fractions tended to be more stable. The result was consistent with other studies, in which rapid reduction of bioavailable Cd was found within 10~20 days after adding heavy metals[37]. The results is also documented by Fan and Zhang[38-39] EXC-Cd and CAB-Cd fractions in soil showed a contrary pattern with RES-Cd and OM-Cd, respectively.

3.4.2 Quantitative relationship between formation and incubation time, and soil water content

The correlation analysis of Cd fractions content, soil moisture, and incubation time indicated complex relationships. To establish a quantitative relationship between the Cd chemical fractions and influencing factors, the response surface design method (RSM) was used to solve multivariate problems and establish quantitative regimes between factors and response values. This method is widely used in experimental design

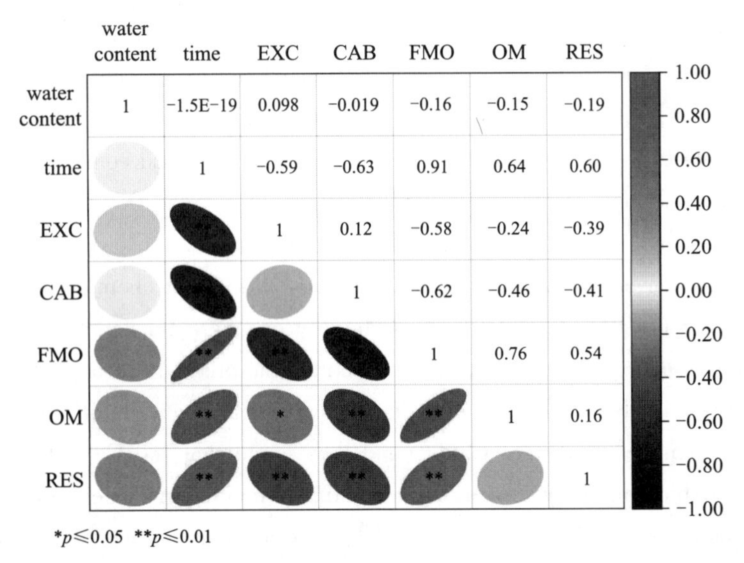

Fig. 6 Relationship between various fraction Cd, soil moisture, and incubation time

and data processing. In this study, the Cd fractions content data was used as a modeling sample under different processing conditions. The soil moisture content and incubation time are set as independent variables, and different Cd forms are set to establish a quantitative relationship model Cd fractions content, soil properties (soil moisture conditions), and environmental factor (incubation time), as shown in Table 3.

Table 3 Main factors and regression equations affecting Cd fractions content

Content	Factors(x_i)	Quantitative equation	C.V.(%)	Adjusted R^2
EXC-Cd (y_1)		$y_1 = 2.66112 + 3.21263x_2 - 3.47460 \times 10^{-3}x_1 - 0.013565x_2x_1 - 2.41585x_2^2 + 8.77084 \times 10^{-6}x_1^2$	9.31	0.7846
CAB-Cd (y_2)		$y_2 = 2.51371 + 0.35515x_2 - 0.026703x_1 + 0.023977x_2x_1 - 3.90473x_2^2 + 2.44129 \times 10^{-5}x_1^2 + 0.016047x_2^2x_1 - 1.98107 \times 10^{-4}x_2x_1^2 + 2.36993x_2^3 + 6.61833 \times 10^{-7}x_1^3$	18.82	0.7199
FMO-Cd (y_3)	incubation time (x_1), soil moisture (x_2)	$y_3 = 0.35994 + 0.34994x_2 + 0.016694x_1 - 0.017412x_2x_1 - 0.071505x_2^2 + 1.04641 \times 10^{-4}x_1^2 + 3.51351 \times 10^{-3}x_2^2x_1 + 1.18289 \times 10^{-4}x_2x_1^2 - 0.37843x_2^3 - 1.26830 \times 10^{-6}x_1^3$	12.15	0.9542
OM-Cd (y_4)		$y_4 = 0.033499 + 0.094948x_2 + 1.38432 \times 10^{-3}x_1 - 2.25142 \times 10^{-4}x_2x_1 - 0.12317x_2^2 - 5.57851 \times 10^{-6}x_1^2$	30.98	0.7029
RES-Cd (y_5)		$y_5 = 0.037624 - 0.084385x_2 + 5.73235 \times 10^{-4}x_1 - 3.53173 \times 10^{-4}x_2x_1 + 0.25973x_2^2 - 6.66872 \times 10^{-6}x_1^2 + 2.04227 \times 10^{-4}x_2^2x_1 + 1.59355 \times 10^{-6}x_2x_1^2 - 0.17599x_2^3 + 2.69750 \times 10^{-8}x_1^3$	14.47	0.5250

The model showed that Cd fractions content, soil moisture content, and incubation time had a good determination coefficient. There was a high degree of fit between FMO-Cd content, soil moisture content, and incubation time, with an R^2 of 0.9542. This indicated that the soil moisture content and incubation time had an obvious effect on FMO-Cd content. The EXC-Cd, CAB-Cd, and OM-Cd contents, soil moisture content, and incubation time also had high degrees of fit, with R^2 values of 0.7846, 0.7199, and 0.7029, respectively. The RES-Cd content, soil moisture content, and incubation time had a lower degree of fit with an R^2 of 0.5250. This was a very convenient way to calculate the contents of Cd chemical forms using quantitative models. The biological toxicity of Cd in soil and its stability can be estimated by monitoring soil moisture content and incubation time. Some scholars have also obtained the prediction models for the bioavailability of cadmium. Wen et al used a polynomial surface model to evaluate the bioavailability of Cd to rice (Oryza sativa L.) in the karst region, Southwestern China, and they found that soil Ca, pH, and total Cd are the controlling factors of Cd bioavailability[40]. Six bioaccessibility prediction models for Cd based on total Cd content have been constructed by ZHANG et al, and obtained a prediction model with higher correlation R^2 by introducing the Fe mineral content (R^2: 0.528~0.986)[41]. In the future research, on the basis of the existing prediction model, the influencing factors such as pH, Eh, and organic matter content will be introduced to further increase the correlation R^2, and the prediction models can help accurately assessing the ecological risks of Cd contaminated soil.

4 Conclusion

To explain Cd changes in soil and their response to soil moisture, indoor simulation experiments were carried out. Cd was added to arable land soil using external simulation, and the dynamic changes of Cd chemical forms were monitored under different water management regimes and incubation times. Correlations and quantitative relationships between Cd chemical forms, soil moisture, and incubation time were identified. The biologically effective coefficient gradually decreased and the refurbishment of Cd tended to stabilize under field capacity. The binging strength of external Cd and soil gradually became more integrated, and the Cd fractions distribution was close to stable under different water management regimes. The statistical comparison result between the different water regimes on Cd fractions within each sampling time showed that the over saturated water content promotes the transformation of EXC-Cd to other forms. The transformation of CAB-Cd fractions can be accelerated to other fractions by field capacity, and the active conversion period was 30~60 d. Correlation analysis of quantitative data showed that EXC-Cd fraction was extremely significantly negatively correlated with FMO-Cd and RES-Cd. The CAB-Cd fraction was extremely significant negatively correlated with FMO-Cd, OM-Cd, and RES-Cd. When the soil water regime was over saturated water content, EXC-Cd content decreased rapidly, FMO-Cd and RES-Cd contents increased, and Cd moves from the EXC-Cd fraction to the FMO-Cd and RES-Cd fraction. Similarly, when the soil water regime was field capacity, Cd moved from the CAB-Cd fraction to the fractions of FMO-Cd, OM-Cd and RES-Cd. The response surface design method (RSM) was used to establish quantitative regimes between Cd fractions with incubation time and soil moisture. It is theoretically and practically important to reveal the migration transformation and changes of Cd fractions in soil environments. In addition, research can also provide a basis for accurate development of soil pollution prevention and control measures.

Author contributions

Nan Lu, Yang Wei, and Zhaoxin Zhang conceived the idea. Nan Lu and Zhaoxin Zhang carried out the experiment, data analysis, and figures in this paper. Gang Li and Yan Li completed sample collection and analysis together. Nan Lu wrote the original draft. Jichang Han provided advice and edited drafts of the manuscript. All authors have read and agreed to the published version of the manuscript.

Funding

This research was funded by Technology Innovation Center for Land Engineering and Human Settlements, Shaanxi Land Engineering Construction Group Co., Ltd and Xi'an Jiaotong University (2021WHZ0094). Shaanxi Provincial Land Engineering Construction Group Internal Research Project (DJNY2021-20).

Conflicts of interest

The authors declare no conflict of interest.

References

[1]Sarwar, N.; Imran, M.; Shaheen, M.R.; Ishaque, W.; Kamran, M.A.; Matloob, A.; Rehim, A.; Hussain, S. Phytoremediation strategies for soils contaminated with heavy metals: Modifications and future perspectives. *Chemosphere* 2017,171,710-721.

[2]Song, W.; Chen, S.; Liu, J.; Chen, L.; Song, N.; Li, N.; Liu, B. Variation of Cd concentration in various rice cultivars and derivation of cadmium toxicity thresholds for paddy soil by species-sensitivity distribution. *Journal of Integrative Agriculture* 2015,14,1845-1854.

[3]Wu, Z. C.; Wang, F. H.; Liu, S.; Du, Y. Q.; Li, F. R.; Du, R. Y.; Wen, D.; Zhao, J. Comparative responses to silicon and selenium in relation to cadmium uptake, compartmentation in roots, and xylem transport in flowering Chinese cabbage (Brassica campestris L. ssp. chinensis var. utilis) under cadmium stress. *Environmental and Experimental Botany* 2016,131,173-180.

[4]Nogawa, K.; Suwazono, Y. Itai-Itai Disease. *Encyclopedia of Environmental Health* 2011,30(1),308-314.

[5]Clemens, S.; MA, J. F. Toxic heavy metal and metalloid accumulation in crop plants and food. *Annual Review of Plant Biology* 2016,67,489-512.

[6]Luo, Y.; Tu, C. The research and development of technology for contaminated site remediation. Twenty Years of Research and Development on Soil Pollution and Remediation in China; Springer, Singapore, 2018; Chapter 48, pp. 785-798.

[7]Tessier, A. P.; Campbell, P.G.C.; Bisson, M. X. Sequential extraction procedure for the speciation of particulate trace metals. *Analytical Chemistry* 1979,51,844-851.

[8]Cambier, P.; Michaud, A.; Paradelo, R.; Germain, M.; Mercier, V.; Guerin-Lebourg, A.; Revallier, A.; Houot, S. Trace metal availability in soil horizons amended with various urban waste composts during 17 years-monitoring and modelling. *Science of the Total Environment* 2019,651,2961-2974.

[9]Hussain B.; Ashraf M. N.; Shafeeq-ur-Rahman, Abbas, A.; Li, J. M.; Farooq, M. Cadmium stress in paddy fields: Effects of soil conditions and remediation strategies. *Science of the Total Environment* 2021,754,142188.

[10]Honma, T.; Ohba, H.; Kaneko-Kadokura, A.; Makino, T.; Nakamura, K.; Katou, H. Optimal soil Eh, pH, and water management for simultaneously minimizing arsenic and cadmium concentrations in rice grains. *Environmental Science & Technology* 2016,50,4178-4185.

[11]Ito, H.; Iimura, K. Absorption of cadmium by rice plants in response to change of oxidation-reduction conditions of soils. *Japanese Journal of Soil Science & Plant Nutrition* 1975,46,82-88.

[12]Fulda, B.; Voegelin, A.; Kretzschmar, R. Redox-controlled changes in cadmium solubility and solid-phase speciation in a paddy soil as affected by reducible sulfate and copper. *Environmental Science & Technology* 2013,47,12775-12783.

[13]Bingham, F. T.; Page, A. L.; Mahler, R. J.; Ganje, T. J. Cadmium availability to rice in sludge-amended soil under "Flood" and "Nonflood" culture. *Soil Science Society of America Journal* 1976,40,715-719.

[14]Nakamura, K.; Katou, H. Arsenic and cadmium solubilization and immobilization in paddy soils in response to alternate submergence and drainage. In Competitive Sorption and Transport of Heavy Metals in Soils and Geological Media; Selim, H. M., Ed.; CRC Press: Boca Raton, 2012; pp. 379-404.

[15]Liu, F.; Liu, X.; Ding, C.; Wu, L. The dynamic simulation of rice growth parameters under cadmium stress with the assimilation of multi-period spectral indices and crop model. *Field Crops Research* 2015,183,225-234.

[16]Xue, S.; Shi, L.; Wu, C.; Wu, H.; Qin, Y.; Pan, W.; Hartley, W.; Cui, M. Cadmium, lead, and arsenic

contamination in paddy soils of a mining area and their exposure effects on human HEPG2 and keratinocyte cell-lines. *Environmental Research* 2017,156,23-30.

[17]NY/T 1121.22-2010 *the field water capacity of test soil was measured according to the soil testing Part 22: Cutting ring methods for determination of field water-holding capacity in soil* Agricultural standard (Beijing: Ministry of rural agriculture)(in China).

[18]GB 15618-2018 *soil environmental quality risk control standard for soil contamination of agricultural land* National standard (Beijing: Ministry of ecological environment)(in China).

[19]Perin, G.; Crabole Dd A, L.; Lucchese, L.; Cirillo, R.; Orio, A. A. Heavy metal speciation in the sediments of Northern Adriatic sea. A new approach for environmental toxicity determination. Heavy Metals in the Environment, 1985, 2, 454-456.

[20]Lu, N.; Li, G.; Sun, Y.; Wei, Y.; He, L.; Li, Y. Phytoremediation potential of four native plants in soils contaminated with Lead in a mining area. *Land* 2021,10,1129.

[21]Lu, N.; LI, G.; Hav, J.C.; Wang, H.Y.; Wei, Y.; Sun, Y.Y. Investigation of lead and cadmium contamination in mine soil and metal accumulation in selected plants growing in a gold mining area. *Applied Ecology and Environmental Research* 2019,17,10587-10597.

[22]Han, F. X.; Banin, A.; Kingery, W. L.; Triplett, G. B.; Zhou, L. X.; Zheng, S. J.; Ding, W. X. New approach to studies of heavy metal redistribution in soil. *Advances in Environmental Research* 2003,8,113-120.

[23]Rezaei, M. J.; Farahbakhsh, M.; Shahbazi, K.; Marzi, M. Study of cadmium distribution coefficient in acidic and calcareous soils of Iran: Comparison between low and high concentrations. *Environmental Technology & Innovation* 2021, 22,101516.

[24]Rassaei, F.; Hoodaji, M.; Abtahi, S. A. Cadmium speciation as influenced by soil water content and zinc and the studies of kinetic modeling in two soils textural classes. *International Soil and Water Conservation Research* 2020,8,286-294.

[25]Lu, A.; Zhang, S.; Shan, X. Q. Time effect on the fractionation of heavy metals. *Geoderma* 2005,125,225-234.

[26]Rajaie, M.; Karimian, N.; Maftoun, M.; Yasrebi, J.; Assad. M. T. Chemical forms of cadmium in two calcareous soil textural classes as affected by application of cadmium-enriched compost and incubation time. *Geoderma* 2006, 136, 533-541.

[27]Safari Sinegani, A.A.; Jafari Monsef, M. Chemical speciation and bioavailability of cadmium in the temperate and semiarid soils treated with wheat residue. *Environmental Science & Pollution Research* 2016,23,9750-9758.

[28]Ma, P.; Tian, T.; Dai, Z.Y.; Shao, T.Y.; Zhang, W. M.; Liu, D. Assessment of Cd bioavailability using chemical extraction methods, DGT, and biological indicators in soils with different aging times. *Chemosphere* 2022,296,133931.

[29] Lu, H.; Qiao, D.; Han, Y.; Zhao, Y.; Wang, Y. Low Molecular Weight Organic Acids Increase Cd Accumulation in Sunflowers through Increasing Cd Bioavailability and Reducing Cd Toxicity to Plants. *Minerals* 2021,11,243.

[30]Liu, G.; Yu, Z.; Liu, X.; Xue, W.; Liu, Y. Aging Process of Cadmium, Copper, and Lead under Different Temperatures and Water Contents in Two Typical Soils of China. *Journal of Chemistry* 2020,1-10.

[31]Arao, T.; Kawasaki, A.; Baba, K.; Mori, S.; Matsumoto, S. Effects of water management on cadmium and arsenic accumulation and dimethylarsinic acid concentrations in Japanese rice. *Environmental Science & Technology* 2009, 43, 9361-9367.

[32]Hu, P.; Huang, J.; Ouyang, Y.; Wu, L.; Song, J.; Wang, S.; Li, Z.; Han, C.; Zhou, L.; Huang, Y.; Luo, Y.; Christie, P. Water management affects arsenic and cadmium accumulation in different rice cultivars. *Environmental Geochemistry & Health* 2013,35,767-778.

[33]Xu, W.; Hou, S.; Amankhan, M.; Chao, Y.; Dan, L.; Effect of water and fertilization management on Cd immobilization and bioavailability in Cd-polluted paddy soil. *Chemosphere*, 2021,276,130168.

[34]Tack, F. Watering regime influences Cd concentrations in cultivated spinach. *Journal of Environmental Management* 2017, 186(pt.2),201-206.

[35]Song, Z.; Dong, L.; Shan, B.; Tang, W. Assessment of potential bioavailability of heavy metals in the sediments of land-freshwater interfaces by diffusive gradients in thin films. *Chemosphere*,2018,191,218-225.

[36]Rizwan, M.; Ali, S.; Adrees, M.; Rizvi, H.; Zia-ur-Rehman, M.; Hannan, F.; Qayyum, M. F.; Hafeez, F.;

Ok, Y. S. Cadmium stress in rice: toxic effects, tolerance mechanisms, and management: a critical review. *Environmental Science & Pollution Research* 2016,23,17859-17879.

[37]Zheng, S. A.; Zheng, X. Q.; Chen, C. Transformation of metal speciation in purple soil as affected by waterlogging. *International Journal of Environmental Science and Technology* 2013, 10,351-358.

[38]Fan, S.; Zhang,Y.; The effect of different N rates on Cd speciation of Cd-contaminated soil with oilseed rape. *Journal of Plant Nutrition* 2017, 40,2680-2690.

[39]Yong, Y.; Xu, Y.; Huang, Q.; et al. Remediation effect of mercapto-palygorskite combined with manganese sulfate on cadmium contaminated alkaline soil and cadmium accumulation in pak choi (*Brassica chinensis* L.). *Science of The Total Environment* 2022, 813, 152636.

[40]Wen, Y.; Li, W.; Yang, Z.; Zhuo, X.; Ji, J. Evaluation of various approaches to predict cadmium bioavailability to rice grown in soils with high geochemical background in the karst region, Southwestern China. *Environmental Pollution* 2020, 258,113645.

[41]Zhang, J. W.; Tian, B.; Luo, J. J.; Wu, F.; Zhang, C.; Liu, Z.T.; Wang, X. N. Effect Factors and the Model Prediction of Soil Heavy Metal Bioaccessibility. *Environmental science* 2021,12-21.

本文曾发表于 2022 年《Sustainability》第 14 卷

Variation Characteristics of Particle Surface Electrochemical Properties during the Improvement of Reclaimed Soil from Hollow Village in Loess Area

Zhe Liu [1,2,3,4,5], Huanyuan Wang [1,3,4,5], Shiliu Cao [1], Zenghui Sun [1,3,4,5], Na Wang [1,3,4,5], Zhaoxin Zhang [1,3,4,5] and Yi Rong [1,3,4,5]

(1. Shaanxi Provincial Land Engineering Construction Group Co., Ltd., Xi'an 710075, China; 2. School of Human Settlements and Civil Engineering, Xi'an Jiaotong University, Xi'an 710049, China; 3. Institute of Land Engineering and Technology, Shaanxi Provincial Land Engineering Construction Group Co., Ltd., Xi'an 710075, China; 4. Key Laboratory of Degraded and Unused Land Consolidation Engineering, the Ministry of Natural Resources, Xi'an 710075, China; 5. Shaanxi Provincial Land Consolidation Engineering Technology Research Center, Xi'an 710075, China)

【Abstract】 Soil surface electrochemical properties, such as specific surface area and surface charge number, are important indexes to evaluate the agricultural soil quality change. However, there is not enough focus on the effect of different improved materials on the reclaimed soil surface electrochemical characteristics. Therefore, we selected maturing agent (TM), fly ash (TF), organic fertilizer (TO), maturing agent + organic fertilizer (TMO), fly ash + organic fertilizer (TFO), and no modified material (CK) treatment for 5 years of field location experiments to study the effects of different improved materials on the surface electrochemical properties of reclaimed soil from abandoned homestead. The results showed that, compared with CK treatment, the specific surface area, surface charge number, and surface charge density of reclaimed soil increased to 11.36~14.05 $m^2 \cdot g^{-1}$, 13.49~18.58 $cmol \cdot kg^{-1}$, and 1.14~1.76 cm^{-2} after five years of application of different improved materials, respectively, and the number of surface charge under TFO, TMO, and TO treatment increased by 28.9%, 25.2%, and 37.7% compared with CK, respectively. Meanwhile, the specific surface area increased significantly ($p < 0.05$), showing an order of TFO > TMO > TO > TF > TM > CK. The surface electric field strength can reach the order of 10^8 $V \cdot m^{-1}$. The statistical analysis results suggest that the contents of soil organic matter (SOM), silt, and clay were positively correlated with the soil surface electrochemical properties, which were the main factors for the changes of reclaimed soil surface electrochemical properties. Our research conclusion shows that in the process of reclamation of abandoned homestead in Loess Plateau, the application of different materials is helpful to improve the soil surface electrochemical properties, among which the organic-inorganic TFO treatment was a suitable improved material treatment for improving the surface electrochemical properties and fertility of reclaimed soil.

【Keywords】 Homestead reclamation soil; Modified materials; Physical properties; Soil organic matter; Particle surface electrochemical properties

1 Introduction

Due to the acceleration of urbanization and industrialization in Loess Plateau, coupled with the widespread problems of "building new houses without demolishing old ones" which constantly occupies high-quality cultivated land and expands to the peripheral area, a hollow village has been formed, which has resulted in the destruction and occupation of a large number of cultivated land resources and the idle waste of high-quality land resources, seriously threatening the protection and promotion of cultivated land quality in

Loess Plateau and becoming the main bottleneck affecting the regional food security and rural revitalization strategy[1-3]. Therefore, it is of great significance to carry out soil reclamation and quality improvement of abandoned homestead in hollow village for increasing the cultivated land resources in Loess Plateau, ensuring regional food security, and promoting rural revitalization[4-5]. However, the reclaimed soil of abandoned homestead comes from the old wall soil (raw soil) after the old houses are demolished, which loses the properties and functions of full-year cultivated soil. The low soil quality seriously limits the land productivity and utilization of reclaimed soil, and it is urgent to improve the reclaimed soil structure and fertility, and thus increase the productivity of reclaimed soil[6].

Previous studies have shown that the application of different soil improvement materials is an important way to improve the reclaimed soil quality in the abandoned homestead of hollow village in Loess Plateau, which can increase the soil nutrients and enzyme activities, improve the soil structure compactness, and enhance the water retention performance[7-8]. Organic fertilizer is rich in organic matter, nitrogen, phosphorus, potassium, and other nutrient elements, which can improve soil structure and pore condition, ripen soil, and boost soil fertilizer and productivity[9]. Fly ash is a solid waste residue discharged by coal-fired enterprises in the production process. With large specific surface area and strong adsorption capacity, it can promote the agglomeration and cementation of soil particles and enhance the ability of soil to retain water and fertilizer as a good soil improver and fertility filler[10]. Gao et al. showed that using ferrous sulfate as a soil amendment can improve soil physical and chemical properties, adjust soil pH, increase soil enzyme activity, loosen soil structure, and promote nutrient absorption of crops[11-12]. At present, there are many studies on the effects of different improved materials on soil properties and functions, mainly focusing on the changes of nutrient elements, aggregate stability, and water infiltration, but there are few studies on how to affect the electrochemical properties of reclaimed soil surface, with a lack of systematic understanding of the effects of soil surface electrochemical properties on soil water and fertilizer retention capacity and soil structure stability.

The surface electrochemical properties of soil particles include surface charge number, specific surface area, surface charge density, surface electric field strength, surface potential, and other parameters. The changes of these surface charge properties will cause changes in the interaction between soil particles, and have an important impact on many physical, chemical, and biological processes occurring on and around the soil particle surface[13-14]. Meanwhile, the electrochemical properties of the soil particle surface have a profound impact on the macro-processes of soil water and fertilizer retention, agglomerate breakage and stability, soil water movement, soil erosion, and agricultural non-point source pollutant migration[15-17]. The application of organic fertilizer, fly ash, and other improved materials is an important way to improve the quality of reclaimed soil in hollow village of Loess Plateau. The addition of these exogenous improved materials will inevitably have different effects on the electrochemical properties of reclaimed soil particles, which will affect the improvement effect of reclaimed soil fertility characteristics and structural stability[7,18]. Consequently, it is of great theoretical value and practical significance to study the effects of different improved materials on the electrochemical properties of reclaimed soil surface, and to accurately understand the variation characteristics of electrochemical properties of reclaimed soil surface, so as to thoroughly reveal the improvement mechanism of different materials on the fertility and structure of reclaimed soil in abandoned homestead. Therefore, this paper takes the reclaimed soil in hollow village of Loess Plateau as the research object, adopts the conjoint analysis of material surface properties to determine the surface electrochemical properties of soil particles treated with different improved materials, compares and studies the variation pattern of reclaimed soil surface electrochemical properties under the treatment of different improved materials, explores the correlation between the basic physical and chemical properties of reclaimed soil and the parameters of surface chemical properties after the application of improved materials, and discriminates the main factors for the change of electrochemical properties of soil surface during the improvement process so as to provide scientific basis for the improvement of soil structure and fertility in the hollow village.

2 Materials and Methods

2.1 Study Area

The long-term positioning test plot for reclaimed soil improvement in hollow village was set up in Fuping County Pilot Base (34°42′ N, 109°12′ E) in Weinan City, Shaanxi Province, and was completed on 15 June 2015. It is designed mainly for experimental research and engineering demonstration of key technologies for hollow village comprehensive improvement. The study area is located on the north side of Weibei Loess Plateau, with a climate type of warm temperate semi-humid continental monsoon climate zone, an average annual evaporation of 1154.2 mm, an average annual temperature of 13.3 ℃ and an average annual rainfall of 513.5 mm. The rainfall from June to September accounts for more than 60% of the annual rainfall. In the growing season of maize in 2020, the rainfall from June to September was 189.0, 133.4, 211.0, and 41.2 mm, respectively, and the monthly average temperature was 24.6, 25.4, 25.1, and 21.2 ℃, respectively (Fig. 1).

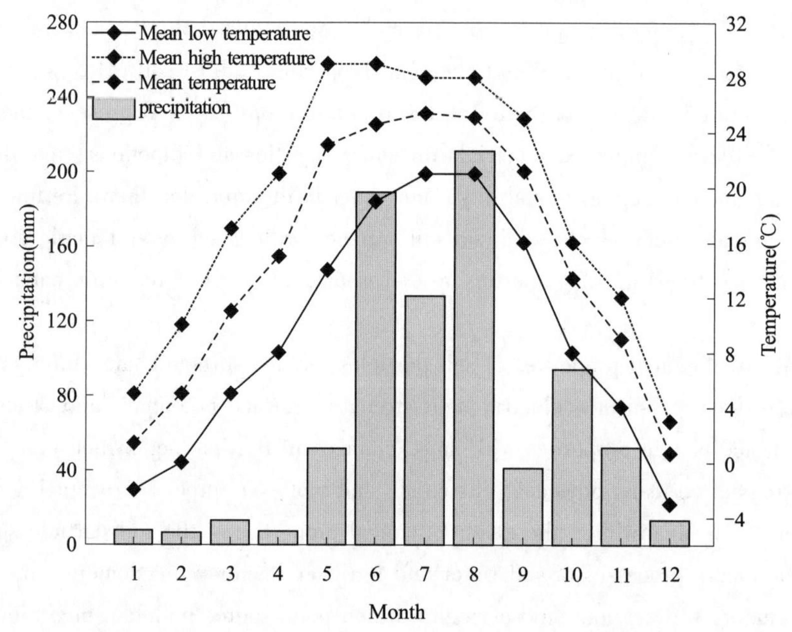

Fig. 1 Precipitation and temperature at the experimental site in Fuping County, Shaanxi Province, China in 2020

The reclaimed soil is from the backfilling of the old wall soil (raw soil) of the abandoned homestead land remediation project in Hollow Village. The backfill depth is 30 cm. After removing the gravel and other impurities, the reclaimed soil is cured and structurally improved by adding different improved materials to meet the requirement for the growth of food crops. The reclaimed soil is mainly developed from loess parent material. Before the experiment, the pH value of the surface soil was 8.5, and the soil texture was silty loam (USDA texture classification), with the organic matter content of 4.5 g · kg^{-1}, the total nitrogen content of 0.16 g · kg^{-1}, the available phosphorus content of 3.1 mg · kg^{-1}, the rapidly available potassium content of 61.4 mg · kg^{-1}, and the soil bulk density of 1.40 g · cm^{-3}. The soil quality is generally poor.

2.2 Experiment Design

The hollow village soil improvement experiment started in June 2015. Based on the survey and analysis of the characteristics of raw soil improvement materials and the review of published literature, this paper selected fly ash, organic fertilizer (decomposed chicken manure), and ferrous sulfate ($FeSO_4$) as the improvement materials of reclaimed soil to address the fertility and structural problems of reclaimed raw soil. The contents of environmental pollution indicators, As, Hg, Cu, Pb, Zn, and Cd, in fly ash were 13.59, 0.10, 91.6, 22.72, 57.81, and 0.06 mg · kg^{-1}, respectively, which was in line with the soil environmental quality assessment standard (GB 15618—2008, China)[19]. The experiment adopted a randomized block field experiment design with six treatments, maturing agent (TM), fly ash (TF), organic fertilizer (TO), maturing agent + organic fertilizer (TMO), fly ash + organic fertilizer (TFO), and no modified material (CK) treatment without the addition of improved materials. Each treatment has three replicates, totaling 18 experimental plots, with an 80 cm wide isolation belt between each group of treatments. The crop planting system is a two-year triple cropping system of winter wheat-summer maize rotation. The tested summer maize was sown in the first twenty days of June, with a sowing density of 6.5×10^4 plants per hectare, and harvested in the first ten days of October. The planting variety is Xianyu 958. Before sowing, compound fertilizer of 1500 kg · ha^{-1} was applied to all maize treatments, in which the contents of nitrogen, phosphorus, and potassium were 15%, 10%, and 20%, respectively. Then, the improved materials with different treatments were uniformly mixed into the reclaimed raw soil, and the soil improved materials were applied to each treatment in one-time application. The irrigation amount, fertilizer treatment, and other daily management indicators of the six treatments were consistent. See Table 1 for specific experimental treatments and the application amount of improved materials.

Table 1 Experimental treatments of reclamation soil improvement in hollow village

Number	Treatment	Improved Materials	Application Amount
1	CK	Control (no improved material)	0
2	TM	Maturing agent (ferrous sulfate)	0.6 t · ha^{-1}
3	TF	Fly ash	45 t · ha^{-1}
4	TO	Organic fertilizer (chicken manure)	30 t · ha^{-1}
5	TMO	Maturing agent + organic fertilizer	(0.6 + 30) t · ha^{-1}
6	TFO	Fly ash + organic fertilizer	(45 + 30) t · ha^{-1}

2.3 Soil Determination Indexes and Methods

After the summer maize harvest of this experiment in early October, 2020, 0 ~ 20 cm reclaimed soil surface samples were collected according to the experimental treatments, and three repeated soil samples were randomly taken from each treatment to analyze the effects of different improved materials treatments on the physical and chemical properties and surface electrochemical properties of reclaimed soil. After the soil sample is naturally air-dried indoors, impurities such as plant residues and gravel are removed, and the soil sample is ground through 0.25 mm, 1 mm, and 2 mm sieves. The basic physical and chemical properties of soil were determined by conventional classical analysis methods, including soil organic matter content by the $K_2Cr_2O_7$ heat capacity method[20], total nitrogen by the Kjeldahl method[21], available phosphorus by the sodium bicarbonate extraction-molybdenum antimony colorimetric method[22], rapidly available potassium by the ammonium acetate extraction-flame photometer method[23], and soil pH by the electrode method (water-soil mass ratio of 1:2.5). The bulk density and water content of soil are determined by the cutting ring method and drying method[24-25], and the soil particle sizes by MS3000 laser particle size analyzer. According to the American standard, the soil particles are divided into three grades: sand (2 ~ 0.05 mm), silt (0.05 ~ 0.002 mm), and clay (<0.002 mm).

2.4 Determination and Calculation of Soil Surface Electrochemical Properties

The soil surface electrochemical properties mainly include surface charge number, specific surface area, surface charge density, surface potential, and surface electric field strength. The soil surface electrochemical properties were determined by the conjoint determination method of material surface properties established by Li et al.[26]. Firstly, 200 g soil samples of 0.25 mm particles with different treatments were weighed and put into a 1000 mL beaker, and 600 mL of 0.5 mol · L^{-1} hydrochloric acid solution was slowly added. A glass rod was used to fully stir them until no obvious bubbles were generated when they were made standing. The supernatant was sucked with a straw, then the hydrochloric acid solution was added again. This process was repeated for many times until no carbon dioxide bubbles were produced in the soil samples, thus removing the calcium carbonate from the soil. Secondly, the soil without calcium carbonate was centrifuged to discard the supernatant, and 600 mL of 0.1 mol · L^{-1} HCl solution was added to shake for 5 hours before it was centrifuged to discard the supernatant. The hydrochloric acid solution with the same volume and concentration was continuously added, and the above operation was repeated for 3 times. After the last centrifugation, 600 mL of deionized water was added to oscillate until there was no Cl^- in the centrifuged solution, and then the soil sample was dried at 65 ℃ to obtain the hydrogen saturated soil sample. Finally, 5~10 g hydrogen saturated sample was weighed and put into a 100 mL centrifuge tube, and 40 mL of 0.0075 mol · L^{-1} NaOH and $Ca(OH)_2$ mixed solution was added to oscillate on a shaking table for 24 h (240 r · min^{-1}, temperature 25 ℃). Then, the pH of the mixed solution was adjusted to 6~8 with 1 mol · L^{-1} HCl. After shaking for 12 hours, the pH of the mixed solution was determined until reaching 7.0. The supernatant was obtained by centrifugation, and the concentrations of Ca^{2+} and Na^+ in the mixed solution were determined by atomic absorption spectrometer. The measurement of each treated soil sample was repeated three times. The measured data were brought into Formula (1) ~ (6)[26-28] to calculate the surface electrochemical properties, such as surface potential (φ_0, V · m^{-1}), surface charge number (σ_0, C · m^{-2}), surface electric field strength (E_0, V · m^{-1}), specific surface area (SSA, m^2 · g^{-1}), and surface charge density (SCN, cmol · kg^{-1}).

$$\varphi_0 = \frac{2RT}{(2\beta_{Ca} - \beta_{Na})F}\ln\frac{a^0_{Ca}N_{Ca}}{a^0_{Ca}N_{Ca}} \tag{1}$$

$$\sigma_0 = \mathrm{sgn}(\varphi_0)\sqrt{\frac{\varepsilon RT}{2\pi}[a^0_{Na}(e^{\frac{\beta_{Na}F\varphi_0}{RT}} - 1) + a^0_{Ca}(e^{\frac{2\beta_{Ca}F\varphi_0}{RT}} - 1) + (a^0_{Na} + 2a^0_{Ca})(e^{\frac{F\varphi_0}{RT}} - 1)]} \tag{2}$$

$$E_0 = \frac{4\pi}{\varepsilon}\sigma_0 \tag{3}$$

$$SSA = \frac{N_{Na}k}{ma^0_{Na}}e^{\frac{\beta_{Na}F\varphi_0}{2RT}} = \frac{N_{Ca}k}{ma^0_{Ca}}e^{\frac{2\beta_{Ca}F\varphi_0}{RT}} \times 10^{-2} \tag{4}$$

$$SCN = 10^5\frac{s\sigma_0}{F} \tag{5}$$

$$m = 0.5259\ln\left(\frac{c^0_{Na} + c^0_H}{c^0_{Ca}}\right) + 1.992 \tag{6}$$

Where φ_0(mV) is the surface potential; σ_0(C · m^{-2}) is the surface charge density; E_0(V · m^{-1}) is the surface electric field strength; SSA (m^2 · g^{-1}) is the specific surface area; SCN (cmol · kg^{-1}) is the surface charge number; R (J · K^{-1} · mol^{-1}) is the universal gas constant; T (K) is the absolute temperature; F (C · mol^{-1}) is the Faraday constant; Z is the charge of each ion species; β_{Na} and β_{Ca} are the corresponding modification factors of Z for Na^+ and Ca^{2+}, respectively; ε is the dielectric constant for water

$(8.9\times 10^{-9}\ C^2\cdot J^{-1}\cdot dm^{-1})$; $\beta_{Na} = 0.0213\ln(I^{0.5}) + 0.7669$, $\beta_{Ca} = -0.0213\ln(I^{0.5}) + 1.2331$; k (dm^{-1}) is the Debye-Hückel parameter; I $(mol\cdot L^{-1})$ is the ionic strength; $k = (4\pi F^2 \sum Zi^2\alpha i^0/\varepsilon RT)^{1/2}$; and $C^0_{Na}(mol\cdot L^{-1})$, and $C^0_{Ca}(mol\cdot L^{-1})$ are equilibrium Na^+ and Ca^{2+} concentrations in the bulk solution, respectively.

2.5 Statistical Analysis

The experimental data were organized and statistically analyzed by the software of Microsoft Excel 2013 and SPSS22.0, the mapping was made by Origin 2019, and redundancy analysis (RDA) was carried out using the software of Canoco 5. The least significant range (LSD) method was used to analyze the correlation between the physicochemical properties and particle surface electrochemical properties of reclaimed soil, with $p < 0.05$ indicating significant level and $p < 0.01$ indicating extremely significant level.

3 Results and Discussion

3.1 Physical Properties of Reclaimed Soil Treated with Different Improved Materials

After the application of different improved materials, the reclaimed soil physical properties of abandoned homestead changed significantly. Soil bulk density is influenced by soil particle composition, pore size, compactness, and other factors. The application of different improved materials makes soil bulk density decrease, which shows the order of TFO < TMO < TO < TF < TM < CK, with the lowest soil bulk density under the organic-inorganic combined treatment of TFO (Table 2). The results show that the application of different improved materials makes reclaimed soil loose, and the soil structure and permeability are gradually improved. With the long-term application of improved materials, the soil water content increased significantly ($p < 0.05$), especially under TFO and TMO treatments, under which the soil water content increased by 38.24% and 32.32%, respectively, compared with CK treatment, with the highest soil water content under TFO treatment. After more than five years of application of improved materials, the mean weight diameter (MWD) of reclaimed soil under organic-inorganic coupling TFO and TMO treatment is 2.50 times and 1.66 times that of the CK, respectively, and the MWD under TFO treatment increases to 0.80 mm. Organic fertilizer, fly ash, and other soil improvement materials can effectively improve the content of soil organic matter after being applied alone or in combination. The cementing materials formed in the process of transformation and decomposition contribute to the cementation and agglomeration of soil aggregates, which have significant influence on the aggregate size, distribution, and structural stability, and enhance the mutual adsorption and agglomeration ability of soil particles. Meanwhile, the application of the improved materials somewhat reduces the damage of soil structure caused by machinery and other artificial farming practices, and finally increases the agglomeration and stability of aggregate structure, thus promoting MWD value to increase[29-30]. Soil particle composition affects soil water and fertilizer status and is an important soil physical property. Long-term application of different improved materials reduces the content of soil sand, while increasing the content of clay and silt to a lesser extent. The reason for the improvement is as follows. The application of improved materials such as organic fertilizer, maturing agent, and fly ash could improve soil physical and chemical properties, increase soil enzyme activity, promote the formation of organic and inorganic colloids, stabilize the soil maturation environment, reduce the erosion of raindrops on the soil, and advance the growth of crop roots and the improvement of soil microbial activity. This increases the return amount of plant residues and roots into the soil, and the cementation substances such as polysaccharides and humus produced by decomposition, which enhance the adhesion, agglomeration, and cementation of soil particles, and increase the contents of clay and silt particles[11,31].

Table 2 Physical properties of reclaimed soil under the application of different modified materials

Treatment	Bulk Density (g · cm^{-3})	MWD (mm)	Water Content (%)	Particle Size Distribution (%) Clay (<0.002 mm)	Silt (0.002~0.05 mm)	Sand (0.05~2 mm)
CK	1.38 ± 0.02 a	0.32 ± 0.03 d	13.18 ± 0.32 d	10.00 ± 0.11 d	81.96 ± 0.13 d	8.04 ± 0.23 a
TO	1.22 ± 0.01 cd	0.45 ± 0.02 c	16.00 ± 0.10 b	11.58 ± 0.10 b	83.31 ± 0.32 b	5.11 ± 0.22 c
TF	1.25 ± 0.02 c	0.38 ± 0.02 c	14.99 ± 0.23 c	10.87 ± 0.09 c	82.39 ± 0.17 c	6.74 ± 0.09 b
TM	1.26 ± 0.01 b	0.32 ± 0.02 d	14.96 ± 0.13 c	10.08 ± 0.02 d	83.45 ± 0.27 b	6.47 ± 0.29 b
TFO	1.19 ± 0.00 d	0.80 ± 0.06 a	18.22 ± 0.15 a	12.66 ± 0.01 b	84.03 ± 0.15 a	3.31 ± 0.15 e
TMO	1.21 ± 0.01 d	0.53 ± 0.05 b	17.44 ± 0.66 a	11.56 ± 0.03 a	83.91 ± 0.11 a	4.53 ± 0.07 d

CK: no improved material; TO: organic fertilizer; TF: fly ash; TM: maturing agent (ferrous sulfate); TFO: fly ash + organic fertilizer; TMO: maturing agent + organic fertilizer; MWD: mean weight diameter. Different lowercase letters represent significant differences among different improved material treatments in the same index ($p < 0.05$).

3.2 Chemical Properties of Reclaimed Soil Treated with Different Improved Materials

The application of different improved materials had a significant impact on the reclaimed soil chemical properties (Table 3). The reclaimed soil was weakly alkaline, and with the application of different improved materials, the pH value of the reclaimed soil showed a significant decreasing trend (Table 3) ($p < 0.05$). The C, N, P, and K elements in soil are important nutrients that affected the crops normal growth and development, and played a key role in crop growth[32-33]. As the reclaimed soil of abandoned homestead is mainly the raw soil of loess parent material with poor fertility level, the application of different improved materials would gradually increase the soil nutrient content, and the soil organic matter (SOM), total nitrogen (TN), available phosphorus (AP), and available potassium (AK) contents in reclaimed soil showed a significant increasing trend ($p < 0.05$). Among them, the SOM content of reclaimed soil under TO, TF, TM, TFO, and TMO treatments increased by 91.2%, 84.7%, 11.8%, 132.3%, and 120.2%, respectively, compared with CK treatment, and the TN content increased by 17.3%, 9.6%, 5.8%, 42.3%, and 21.2%, respectively, compared with CK treatment, with the largest increase under TFO treatment. The content of AP and AK showed a similar trend to that of organic matter, with the highest content under TFO treatment and the lowest content under CK. Among them, organic-inorganic combined treatment of TFO and TMO had a better effect on improving the reclaimed soil nutrients. The results are consistent with those of Wei et al., who pointed out that the organic-inorganic combined treatments were more conducive to soil fertility and structure improvement[34-35]. The C/N in different improved materials, which is used to represent the balance of soil carbon and nitrogen nutrients, has an average value of 10.47, which is significantly higher than that in the control treatment. The application of improved materials increased the reclaimed soil C/N value.

Table 3 Chemical properties of reclaimed soil under the application of different modified materials

Treatment	SOM (g · kg^{-1})	TN (g · kg^{-1})	C/N	OP (mg · kg^{-1})	AK (mg · kg^{-1})	pH
CK	5.94 ± 0.17 c	0.52 ± 0.03 d	6.62 ± 0.31 c	14.29 ± 0.79 c	104.36 ± 8.38 c	8.66 ± 0.03 a
TO	11.36 ± 0.37 b	0.61 ± 0.01 b	10.81 ± 0.50 b	18.82 ± 1.49 a	114.72 ± 2.69 b	8.53 ± 0.05 b

Continued to Table 3

Treatment	SOM ($g \cdot kg^{-1}$)	TN ($g \cdot kg^{-1}$)	C/N	OP ($mg \cdot kg^{-1}$)	AK ($mg \cdot kg^{-1}$)	pH
TF	10.97 ± 0.56 b	0.57 ± 0.01 c	11.24 ± 0.68 b	16.43 ± 0.32 b	116.76 ± 2.24 b	8.52 ± 0.05 b
TM	6.64 ± 0.05 c	0.55 ± 0.02 cd	6.97 ± 0.29 c	15.44 ± 0.31 bc	111.30 ± 0.85 bc	8.38 ± 0.04 c
TFO	13.89 ± 0.78 a	0.74 ± 0.02 a	10.96 ± 0.76 b	18.70 ± 0.35 a	130.50 ± 2.62 a	8.27 ± 0.11 cd
TMO	13.08 ± 1.06 a	0.63 ± 0.02 b	12.36 ± 0.87 a	18.49 ± 0.22 a	123.99 ± 2.82 a	8.18 ± 0.07 d

CK: no improved material; TO: organic fertilize; TF: fly ash; TM: maturing agent (ferrous sulfate); TFO: fly ash + organic fertilizer; TMO: maturing agent + organic fertilizer; SOM: soil organic matter; TN: total nitrogen; AP: available phosphorus; AK: available potassium. Different lowercase letters represent significant differences among different improved material treatments in the same index ($p < 0.05$).

3.3 Ion Exchange Equilibrium Results and Surface Electrochemical Properties of Reclaimed Soil Treated with Different Improved Materials

According to the principle of conjoint determination of material surface parameters put forward by Li et al.[10], the calculation results of equilibrium activity (a^0_{Ca}, a^0_{Na}), equilibrium concentration (c^0_{Ca}, c^0_{Na}), and corresponding k, m, I, and diffusion layer ions (N_{Ca}, N_{Na}) of soil treated with six different improved materials are shown in Table 4. With the application of different improved materials, the adsorption capacity of Na^+ and Ca^{2+} ions increased continuously, especially under TFO, TMO, and TO treatments, and the adsorption capacity of Ca^{2+} increased by 13.7%, 12.8%, and 15.9%, respectively, compared with the CK treatment (Table 4).

Table 4 Calculation results of ion exchange equilibrium of reclaimed soil treated with different modified materials

Treatment	a^0_{Ca} ($mmol \cdot L^{-1}$)	a^0_{Na} ($mmol \cdot L^{-1}$)	c^0_{Ca} ($mmol \cdot L^{-1}$)	c^0_{Na} ($mmol \cdot L^{-1}$)	K (dm^{-1})	m	I	N_{Ca} (10^{-5} $mol \cdot g^{-1}$)	N_{Na} (10^{-5} $mol \cdot g^{-1}$)
CK	0.90 ± 0.11	6.29 ± 0.05	1.42 ± 0.19	7.04 ± 0.05	34,594,991	2.84 ± 0.07	0.011	5.33 ± 0.03	0.40 ± 0.03
TO	0.29 ± 0.04	6.03 ± 0.05	0.43 ± 0.05	6.63 ± 0.06	28,991,275	3.43 ± 0.06	0.008	6.18 ± 0.05	0.76 ± 0.05
TF	0.58 ± 0.01	6.32 ± 0.10	0.88 ± 0.02	7.02 ± 0.11	31,998,518	3.09 ± 0.01	0.010	5.79 ± 0.02	0.44 ± 0.06
TM	0.69 ± 0.02	6.32 ± 0.13	1.05 ± 0.03	7.03 ± 0.15	32,887,110	2.99 ± 0.02	0.010	5.64 ± 0.02	0.45 ± 0.09
TFO	0.38 ± 0.02	6.17 ± 0.05	0.57 ± 0.03	6.81 ± 0.06	30,071,527	3.30 ± 0.03	0.009	6.06 ± 0.03	0.60 ± 0.05
TMO	0.42 ± 0.02	6.22 ± 0.03	0.63 ± 0.03	6.88 ± 0.03	30,487,524	3.25 ± 0.02	0.009	6.01 ± 0.02	0.54 ± 0.03

CK: no improved material; TO: organic fertilize; TF: fly ash; TM: maturing agent (ferrous sulfate); TFO: fly ash + organic fertilizer; TMO: maturing agent + organic fertilizer; a^0_{Ca}: Ca^{2+} equilibrium activity in solution; a^0_{Na}: Na^+ equilibrium activity in solution; N_{Ca}: the adsorption capacity of Ca^{2+}; N_{Na}: the adsorption capacity of Na^+.

Surface potential (φ_0), surface charge density (σ_0), surface electric field strength (E_0), specific surface area (SSA), and surface charge number (SCN) are very important properties of soil colloid, which affect the physical, chemical, and biochemical processes in soil. Among them, the SCN is the key factor for crops to absorb nutrient elements, which determines the quantity of adsorbed ions in soil[36-37]. Compared with CK, different improved material treatments significantly increased the SCN ($p < 0.05$), and the SCN of reclaimed soil varied from 13.49 to 18.58 $cmol \cdot kg^{-1}$, among which those under TFO, TMO, and TO treatments increased by 28.9%, 25.2%, and 37.7%, respectively, compared with those under CK treatment (Table 5). From the adsorption amount of Na^+ and Ca^{2+} ions in Table 4, the adsorption amount of ions increased with the application of different improved materials. The huge specific surface area of soil colloid is an important place for soil adsorption reaction and ion exchange, which is closely related to soil's ability to maintain and supply nutrients and water necessary for crops[38-39]. After the application of different improved materials, the SSA of reclaimed soil increased significantly (Table 5), showing an order of TFO > TMO > TO > TF > TM > CK, with the specific surface area range increasing to 11.36~14.05 $m^2 \cdot g^{-1}$, and the largest SSA under the organic-inorganic TFO treatment. The reason may be that the application of different improved materials increased the organic matter and silt and clay particle contents in the soil, and the increase in the content of organic and inorganic colloids in the reclaimed soil further increased the specific surface area and SCN of the reclaimed soil. These results are similar to those of Yu et al., who found through indoor culture experiments that the number of inceptisols surface charge and specific surface area after straw application increased by 18.75% and 14.64%, respectively, compared with CK treatment[15].

The surface charge density of soil particles refers to the number of charges per unit area of soil particles, which affects the ion adsorption strength. The greater the charge density, the greater the ion adsorption capacity[40]. With the application of organic fertilizer, fly ash, and other improved materials, the σ_0 of reclaimed soil increased significantly ($p < 0.05$), ranging from 1.14 to 1.76 $C \cdot m^{-2}$, among which TO, TF, TM, TFO, and TMO treatments increased by 51.8%, 50.6%, 37.3%, 75.9%, and 73.4%, respectively, compared with CK treatment, with the highest σ_0 under TFO treatment. This shows that with the application of organic fertilizer, fly ash, and other improved materials, the amount of charges per unit area and the firmness of soil to retain nutrient ions increased. In addition, soil colloids with low surface charge density more easily form aggregates than those with high surface charge density, which implies that the surface charge properties of soil colloids play an important role in the formation process and structural stability of soil aggregates. After different improved materials are applied, the cementation and agglomeration ability of reclaimed soil will be changed to a certain extent with the change of surface charge density. As the effect of soil electric field measured by traditional methods is very small, the more accurate methods found that the surface electric field strength of reclaimed soil was as high as 10^8 orders of magnitude (Table 5). Under the treatment of returning different improved materials to the field, the E_0 of reclaimed soil showed a significant increasing trend, with the order of TFO > TMO > TO > TF > TM > CK, and the highest under the organic-inorganic TFO treatment. Such a strong surface electric field will inevitably affect the soil interface reaction, as well as the micro-process and macro-phenomenon in the soil[41-42].

Table 5 Soil surface electrochemical properties under the application of different modified materials

Treatment	SCN ($cmol \cdot kg^{-1}$)	SSA ($m^2 \cdot g^{-1}$)	σ_0 ($C \cdot m^{-2}$)	E_0 ($10^8\ V \cdot m^{-1}$)	φ_0 (mV)
CK	13.49 ± 0.12 f	10.22 ± 0.02 d	0.83 ± 0.03 d	11.67 ± 0.45 c	−122.35 ± 1.55 a
TO	18.58 ± 0.56 a	12.9 ± 1.94 bc	1.26 ± 0.06 b	17.80 ± 0.79 b	−142.04 ± 1.84 c
TF	15.36 ± 0.09 d	11.83 ± 0.28 bc	1.25 ± 0.03 b	17.66 ± 0.45 b	−139.79 ± 3.22 c
TM	14.54 ± 0.15 e	11.36 ± 0.31 cd	1.14 ± 0.07 c	16.35 ± 1.40 b	−134.23 ± 3.75 b
TFO	17.39 ± 0.26 b	14.05 ± 0.23 a	1.46 ± 0.05 a	20.94 ± 0.89 a	−141.21 ± 2.95 c
TMO	16.83 ± 0.20 c	13.25 ± 0.46 ab	1.44 ± 0.03 a	20.42 ± 0.42 a	−141.50 ± 2.83 c

CK: no improved material; TO: organic fertilize; TF: fly ash; TM: maturing agent (ferrous sulfate); TFO: fly ash + organic fertilizer; TMO: maturing agent + organic fertilizer; SCN: surface charge number; SSA: specific surface area; σ_0: surface charge density; E_0: surface electric field strength; φ_0: Surface potential. Different lowercase letters represent significant differences among different improved material treatments in the same index ($p < 0.05$).

3.4 Correlation Analysis between Surface Electrochemical Properties and Basic Physicochemical Properties of Reclaimed Soil

In order to further find out the correlation among the parameters of reclaimed soil treated with different improved materials, the correlation analysis of the relationship between soil physicochemical properties and surface electrochemical properties was carried out in this paper (Fig. 2). The contents of soil organic matter, clay particles, silt particles, C/N, and available phosphorus showed significantly positive correlation with the surface electrochemical properties (Fig. 2) ($p < 0.01$). Soil pH value had no significant correlation with specific surface area and surface charge quantity but had significant negative correlation with surface charge density and surface electric field strength ($p < 0.05$). Soil bulk density and sand content showed significantly negative correlation with surface electrochemical properties ($p < 0.05$), while total nitrogen and rapidly available potassium contents showed significantly positive correlation with surface charge density and specific surface area but had no significant relationship with surface charge quantity and surface electric field strength.

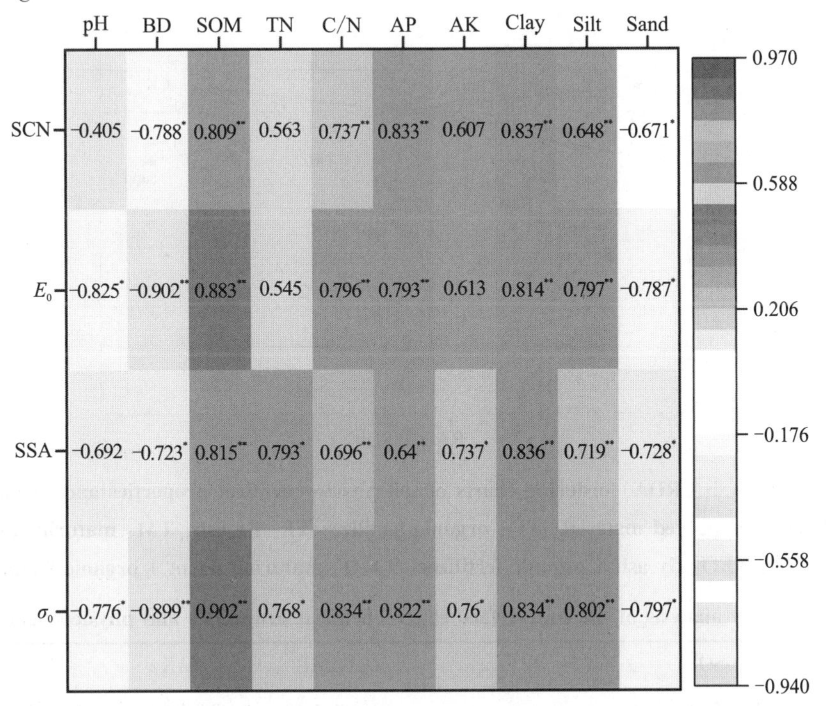

Fig. 2 Correlation analysis between soil surface electrochemical properties and basic physicochemical properties. SCN: surface charge number; SSA: specific surface area; σ_0: surface charge density; E_0: surface electric field strength; BD: soil bulk density; SOM: soil organic matter; TN: total nitrogen; AP: available phosphorus; AK: available potassium; * $p < 0.05$ and ** $p < 0.01$

The redundancy analysis (RDA) was made on the basic physical and chemical properties and the surface charge properties of soil treated with different organic improved materials (Fig. 3). In Fig. 3, the first axis and the second axis account for 88.59% and 7.72% of the total variation, respectively. The contents of organic matter and silt particles have strong correlation with the first sorting axis. There are differences in soil physical and chemical properties under different improved materials. The soil distribution under CK treatment is in the first quadrant, that under TFO and TO treatment is in the second quadrant, that under TMO and TF treatment is in the third quadrant, and that under TM treatment is in the fourth quadrant. Soil organic matter, silt, C/N and clay content are significantly positively correlated with surface charge quantity, specific surface area, surface electric field strength, and surface charge density, while pH value and sand content are negatively correlated with surface electrochemical properties. Soil organic matter and silt content ($F = 72.4$, $p = 0.002$; $F = 8.7$, $p = 0.01$) are the main factors that affect the reclaimed soil surface electrochemical properties, accounting for 69.4% and 8.3%, respectively (Table 6). Soil organic matter, silt, C/N, clay, available phosphorus, rapidly available potassium, and total nitrogen jointly account for 99% of the variation in surface electrochemical properties of the compound soil. The basic physical and chemical properties of reclaimed soil affect the surface electrochemical properties of soil in the order of soil organic matter, silt, C/N, clay, available phosphorus, rapidly available potassium, total nitrogen, pH, and sand.

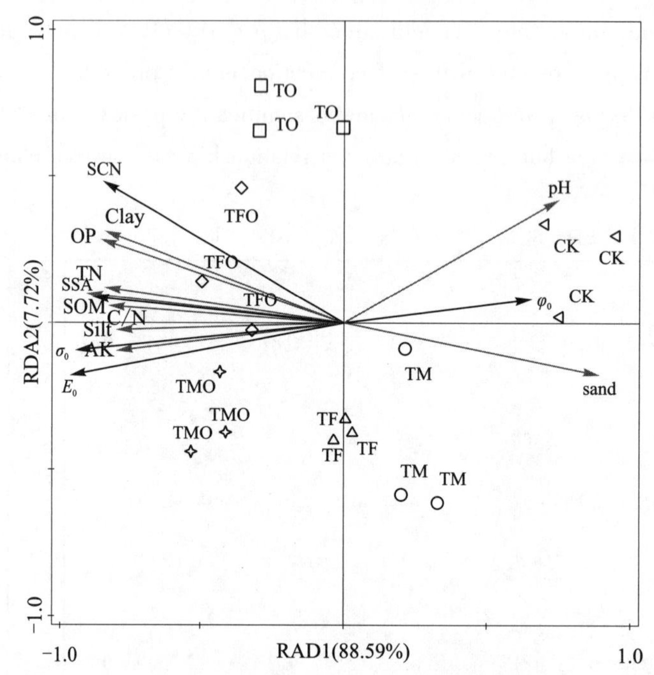

Fig. 3 Redundancy analysis (RDA) ordering charts of soil physicochemical properties and surface electrochemical properties. CK: no improved material; TO: organic fertilize; TF: fly ash; TM: maturing agent (ferrous sulfate); TFO: fly ash + organic fertilizer; TMO: maturing agent + organic fertilizer

Table 6 Redundancy analysis of the soil surface electrochemical properties and physicochemical properties

Soil Physicochemical Properties	SOM ($g \cdot kg^{-1}$)	Silt	C/N	Clay	AP ($mg \cdot kg^{-1}$)	AK ($mg \cdot kg^{-1}$)	TN ($g \cdot kg^{-1}$)	pH	Sand
Interpretation rate (%)	69.4	8.3	5.8	4.7	3.8	1.8	1.1	0.9	0.1
Contribution (%)	72.4	8.7	6.0	4.9	4.0	1.8	1.2	0.9	0.1
F	36.3	5.6	11.4	3.7	3.6	1.8	1.2	1.8	0.2
p	0.002	0.01	0.004	0.018	0.068	0.18	0.328	0.176	0.824

SOM: soil organic matter; AP: available phosphorus; AK: available potassium; TN: total nitrogen.

Correlation analysis (Table 6) and RDA analysis (Fig. 3) show that organic matter content and soil texture are the main factors affecting the surface electrochemical properties of reclaimed soil from abandoned homestead. As an important part of colloidal soil particles, organic matter can promote the cementation and aggregation of particles, enhance the stability of soil structure, change the soil colloidal state, and improve soil particle adsorption capacity[43]. Soil organic matter is generally negatively charged, and every 10 $g \cdot kg^{-1}$ increase in organic matter content can increase the amount of negative charge by 1 $cmol \cdot kg^{-1}$. In organic matter composition, the specific surface area of humus is about 800~900 $m^2 \cdot g^{-1}$, which is nearly 10 times that of common inorganic clay minerals. With the increase of organic matter content, the soil charge amount and specific surface area also gradually increase[44-45]. Therefore, the application of organic fertilizer, fly ash, and other exogenous improved materials increases the organic matter content and structural stability of reclaimed soil in the hollow village, promotes the formation of organic-inorganic complex in soil, and improves the ion adsorption capacity, thus increasing the surface charge quantity and specific surface area of soil particles. This is consistent with the research results of Yu et al., who showed that organic matter content has a significant positive correlation with soil specific surface area and cation exchange capacity. With the addition of straw, the organic matter content in inceptisols increased, and the soil surface charge quantity and specific surface area increased by 18.75% and 14.64%, respectively[36,46]. Clay and silt particles in soil are small and have a huge surface area, and their components are mainly layered aluminosilicate clay minerals and clay oxides. The isomorphic replacement of clay minerals and the dissociation of hydroxyl groups on the surface of clay oxides make the soil negatively charged, thus affecting the quantity and density of soil charges[47]. Through correlation analysis and RDA analysis, it was found that the content of clay particles in reclaimed soil has a highly significant positive correlation with the surface electrochemical properties. As fly ash is rich in clay particles and Al_2O_3, Fe_2O_3, and other oxides, it can significantly improve the mutual adsorption and agglomeration ability of soil particles. The humus and other cementing substances produced by the decomposition of organic fertilizer have high specific surface area and multi-level pores, which promote the adhesion, cementation, and agglomeration of soil particles. The improvement in clay and silt content directly increases the specific surface area and surface charge quantity of reclaimed soil[27,48]. This is consistent with the results of Hepper et al., who found that soil silt content has a significant positive correlation with specific surface area and cation exchange capacity, with silt content accounting for 70.0% of the variability of specific surface area and clay content accounting for 9.1%[49]. Therefore, soil clay and silt particles are closely related to the properties of soil surface charges. Higher content of soil clay and silt particles means larger specific surface area, more adsorption and exchange sites of particles, more negative soil charges, and greater amount of surface charges.

4 Conclusions

The application of different improved materials has a significant effect on the basic physical and chemical properties and surface electrochemical properties of reclaimed soil. Compared with the control treatment, the organic matter, total nitrogen, available phosphorus, rapidly available potassium, clay and silt particles, aggregate structural stability, and water content of reclaimed soil in hollow village increased after five years of application of different improved materials, while soil bulk density and pH value decreased. The application of different improved materials had a positive effect on the reclaimed soil surface electrochemical properties, such as specific surface area, surface charge quantity, and surface electric field strength. The results showed that the organic-inorganic combined treatment of TFO had the most significant effect on soil physical and chemical properties and surface electrochemical properties. Soil organic matter and silt particles contents showed a highly significant positive correlation with the surface electrochemical properties and were the main contributors to the variation in surface electrochemical properties of reclaimed soil, such as soil specific surface area and surface charge quantity, accounting for 69.4% and 8.3%, respectively. These findings provide important information for improving the reclaimed soil fertility and surface electrochemical properties in hollow village.

Author Contributions

Conceptualization, Z.L. and H.W.; methodology, H.W. and S.C.; software, Z.L.; formal analysis, Z. L.; data curation, Z.L. and S.C; writing—original draft preparation, Z.L.; writing—review and editing, Z. L. and S.C.; visualization, Z.L. and Z.S.; proofreading, Z.L. and Z.S.; supervision, N.W. and Z.Z.; project administration, H.W. and Y.R.; funding acquisition, H.W. and Y.R. All authors have read and agreed to the published version of the manuscript.

Funding

This research was supported by the Scientific Research Item of Shaanxi Provincial Land Engineering Construction Group (DJNY2022-15, DJNY2022-35, and DJTD-2022-5); The Fund for Less Developed Regions of the National Natural Science Foundation of China (No. 42167039).

Institutional Review Board Statement

Not applicable.

Informed Consent Statement

Not applicable.

Data Availability Statement

The data presented in this study are available on request from the corresponding author.

Acknowledgments

This work is grateful to the Institute of Land Engineering and Technology, Shanxi Provincial Land Engineering Construction Group, Xi'an, China. Special thanks go to the anonymous reviewers for their constructive comments in improving this manuscript.

Conflicts of Interest

The authors declare no conflict of interest.

References

[1] Liu, Y.S.; Li, J.T.; Yang, Y.Y. Strategic adjustment of land use policy under the economic transformation. *Land Use Policy* 2018, 74, 5-14.

[2] Liu, Y.S.; Liu, Y.; Cheng, Y.F.; Long, H.L. The process and driving forces of rural hollowing in China under rapid urbanization. *J. Geogr. Sci.* 2010, 20, 876-888.

[3] Liu, Y.S.; Long, H.L.; Chen, Y.F.; Wang, J.Y.; Li, Y.R.; Li, Y.H.; Yang, Y.Y.; Zhou, Y. Progress of research on urbanrural transformation and rural development in China in the past decade and future prospects. *J. Geogr. Sci.* 2016, 26, 1117-1132.

[4] Jiang, S.; Luo, P. A literature review on hollow villages in China. *China Popul. Res. Environ.* 2014, 24, 51-58.

[5] Liu, Y.S. Introduction to land use and rural sustainability in China. *Land Use Policy* 2018, 74, 1-4.

[6] Liu, Z.; Deng, L.; Zhou, W.; Chen, L.; Zou, G. Evaluation of effects of organic materials on soil fertilization of reclaimed homestead. *Soils* 2019, 51, 672-681. (In Chinese)

[7] Lei, N.; Han, J.; Mu, X.; Sun, Z.H.; Wang, H.Y. Effects of improved materials on reclamation of soil properties and crop yield in hollow villages in China. *J. Soils Sediments* 2019, 19, 2374-2380.

[8] Cox, D.; Bezdicek, D.; Fauci, M. Effects of compost, coal ash, and straw amendments on restoring the quality of eroded Palouse soil. *Biol. Fertil. Soils* 2001, 33, 365-372.

[9] Liu, Z.; Han, J.C.; Sun, Z.H.; Chen, T.Q.; Hou, Y.; Dong, Q.G.; He, J.; Lv, Y.Z. Long-term effects of different planting patterns on greenhouse soil micromorphological features in the North China Plain. *Sci. Rep.* 2019, 9, 2200.

[10] Ahmaruzzaman, A. A review on the utilization of fly ash. *Prog. Energ. Combust.* 2010, 36, 327-363.

[11] Liu, Z.; Zhang, Y.; Sun, Z.H.; Sun, Y.Y.; Wang, H.; Zhang, R.Q. Effects of the application of different improved materials on reclaimed soil structure and maize yield of Hollow Village in Loess Area. *Sci. Rep.* 2022, 12, 7431.

[12] Gao, Z.X.; Li, X.L.; Zhang, J.; Jin, L.Q.; Zhou, W. Effects of Different Fertilization Treatments on Soil Enzyme Activities in Coal Mining Residuals of Alpine Mining Area. *Acta Agrestia Sin.* 2021, 29, 1748-1756. (In Chinese)

[13] Li, S.; Li, H.; Xu, C.Y. Particle interaction forces induce soil particle transport during rainfall. *Soil. Sci. Soc. Am. J.* 2013, 77, 1563-1571.

[14] Hu, F.N.; Liu, J.F.; Xu, C.Y.; Wang, Z.L.; Liu, G.; Li, H.; Zhao, S.W. Soil internal forces initiate aggregate breakdown and splash erosion. *Geoderma* 2018, 320, 43-51.

[15] Yu, Z.H.; Li, H.; Liu, X.M.; Xu, C.Y.; Xiong, H.L. Influence of soil electric field on water movement in soil.

Soil Tillage Res. 2016, 155, 263-270.

[16]Hu, F.N.; Xu, C.Y.; Li, H.; Li, S.; Yu, Z.H.; Li, Y.; He, X.H. Particles interaction forces and their effects on soil aggregates breakdown. *Soil Tillage Res.* 2015, 147, 1-9.

[17]Yang, Z.H.; Hu, F.N.; Liu, J.F.; Xu, C.Y.; Ma, R.T.; Wang, Z.L.; Zhao, S.W. Water Infiltration in Soils Developed from Loess Affected by Surface Electric Field and Simulation. *Acta Pedol. Sin.* 2019, 56, 1359-1369. (In Chinese)

[18]Hou, J.; Li, H.; Zhu, H.L.; Wu L.S. Determination of clay surface potential: A more reliable approach. *Soil Sci. Soc. Am. J.* 2009, 73(5), 1658-1663.

[19]National Standards of the People's Republic of China, Environmental quality standards for soils, GB 15618—2008.

[20]Tiessen, H.; Moir, J.O. Total and Organic Carbon. In *Soil Sampling and Methods of Analysis*; Carter, M.R., Ed.; CRC Press: Boca Raton, FL, USA, 1993; Volume 38, pp. 187-199.

[21]Kalembasa, S.J.; Jenkinson, D.S. A comparative study of titrimetric and gravimetric methods for the determination of organic carbon in soil. *J. Sci. Food Agric.* 1973, 24, 1085-1090.

[22]Farrell, M.; Macdonald, L.M.; Baldock, J.A. Biochar differentially affects the cycling and partitioning of low molecular weight carbon in contrasting soils. *Soil Biol. Biochem.* 2015, 80, 79-88.

[23]Islam, K.; Singh, B.; McBratney, A. Simultaneous estimation of several soil properties by ultra-violet, visible, and near-infrared reflectance spectroscopy. *Soil Res.* 2003, 41, 1101.

[24]Rabot, E.; Wiesmeier, M.; Schlüter, S.; Vogel, H.J. Soil structure as an indicator of soil functions: A review. *Geoderma* 2018, 314, 122-137.

[25]Xue, B.; Huang, L.; Huang, Y.; Yin, Z.; Li, X.; Lu, J. Effects of organic carbon and iron oxides on soil aggregate stability under different tillage systems in a rice-rape cropping system. *Catena* 2019, 177, 1-12.

[26]Li, H.; Hou, J.; Liu, X.M.; Hou, J.; Liu, X.M.; Li, R.; Zhu, H.L.; Wu, L.S. Combined determination of specific surface area and surface charge properties of charged particles from a single experiment. *Soil. Sci. Soc. Am. J.* 2011, 75, 21-28.

[27]Liu, J.F.; Wang, Z.L.; Hu, F.N.; Xu, C.Y.; Ma, R.T.; Zhao, S.W. Soil organic matter and silt contents determine soil particle surface electrochemical properties across a long-term natural restoration grassland. *Catena* 2020, 190, 104526.

[28]Yu, Z.H.; Zheng, Y.Y.; Zhang, J.B.; Zhang, C.Z.; Ma, D.H.; Chen, L.; Cai, T.Y. Importance of soil interparticle forces and organic matter for aggregate stability in a temperate soil and a subtropical soil. *Geoderma* 2020, 362, 114088.

[29]Singh, J.S.; Pandey, V.C.; Singh, D.P. Coal fly ash and farmyard manure amendments in dry-land paddy agriculture field: Effect on N-dynamics and paddy productivity. *Appl. Soil. Ecol.* 2011, 47, 133-140.

[30]Zhang, S.X.; Li, Q.; Zhagn, X.P.; Wei, K.; Chen, L.J.; Liang, W.J. Effects of conservation tillage on soil aggregation and aggregate binding agents in black soil of Northeast China. *Soil Tillage Res.* 2012, 124, 196-202.

[31]Zhang, Z.Y.; Zhang, X.K.; Mahamood, M.; Zhang, S.Q.; Huang, S.M.; Liang, W.J. Effect of long-term combined application of organic and inorganic fertilizers on soil nematode communities within aggregates. *Sci. Rep.* 2016, 6, 31118.

[32]Schlesinger, W.H.; Andrews, J.A. Soil respiration and the global carbon cycle. *Biogeochemistry* 2000, 48, 7-20.

[33]Treseder, K.K.; Vitousek, P.M. Effects of soil nutrient availability on investment in acquisition of N and P in Hawaiian rain forests. *Ecology* 2001, 82, 946-954.

[34]Rautaray, S.K.; Ghosh, B.C.; Mittra, B.N. Effect of fly ash, organic wastes and chemical fertilizers on yield, nutrient uptake, heavy metal content and residual fertility in a rice-mustard cropping sequence under acid lateritic soils. *Biores. Technol.* 2003, 90, 275-283.

[35]Wei, W.; Yan, Y.; Cao, J.; Christie, P.; Zhang, F.; Fan, M. Effects of combined application of organic amendments and fertilizers on crop yield and soil organic matter: An integrated analysis of long-term experiments. *Agric. Ecosyst. Environ.* 2016, 225, 86-92.

[36]Yu, Z.H.; Zhang, J.B.; Zhang, C.Z.; Xin, X.L.; Li, H. The coupling effects of soil organic matter and particle interaction forces on soil aggregate stability. *Soil Tillage Res.* 2017, 174, 251-260.

[37]Rakhsh, F.; Golchin, A.; Agha, A.B.A. Effects of exchangeable cations, mineralogy and clay content on the min-

eralization of plant residue carbon. *Geoderma* 2017, 307, 150-158.

[38] Bayat, H.; Ebrahimi, E.; Ersahin, S.; Hepper, E. N.; Singh, D. N.; Amer, A. M.; Yukselen-Aksoy, Y. Analyzing the effect of various soil properties on the estimation of soil specific surface area by different methods. *Appl. Clay Sci.* 2015, 116, 129-140.

[39] Gruba, P.; Mulder, J. Tree species affect cation exchange capacity (CEC) and cation binding properties of organic matter in acid forest soils. *Sci. Total Environ.* 2015, 511, 655-662.

[40] Kalbitz, K.; Schwesig, D.; Rethemeyer, J.; Matzner, E. Stabilization of dissolved organic matter by sorption to the mineral soil. *Soil Biol. Biochem.* 2005, 37, 1319-1331.

[41] Xu, C.Y.; Yu, Z.H.; Li, H. The coupling effects of electric field and clay mineralogy on clay aggregate stability. *J. Soils Sediments* 2015, 15, 1159-1168.

[42] Liu, Y.L.; Li, H. Effect of electrostatic field originating from charged humic surface on the adsorption kinetics of Zn^{2+}. *J. Southwest Univ.* 2010, 32, 77-81. (In Chinese)

[43] Ma, R.T.; Hu, F.N.; Liu, J.F.; Zhao, S.W. Evaluating the effect of soil internal forces on the stability of natural soil aggregates during vegetation restoration . *J. Soils Sediments*. 2021, 21, 3034-3043.

[44] Ding, W.Q.; He, J.H.; Liu, X.M.; Hu, F.N.; Tian, R.; Li, H.; Zhu, H.L. Effect of organic matter on aggregation of soil colloidal particles in water bodies of Three Gorge Reservoir Region. *J. Soil Water Conserv.* 2017, 31, 166-171. (In Chinese)

[45] Zhao, Z.J.; Chang, E.; Lai, P.; Dong, Y.; Xu, R.K.; Fang, D.; Jiang, J. Evolution of soil surface charge in a chronosequence of paddy soil derived from Alfisol. *Soil Tillage Res.* 2019, 192, 144-150.

[46] Kweon, G.; Lund, E.; Maxton, C. Soil organic matter and cation-exchange capacity sensing with on-the-go electrical conductivity and optical sensors. *Geoderma* 2013, 199, 80-89.

[47] Wu, J.M.; Liu, Y.H.; Li, X.Y.; Ling, W.T.; Dong, Y.Y. Surface charge characteristics of soil colloids in China. *Acta Pedol. Sin.* 2002, 39, 177-183. (In Chinese)

[48] Yukselen-Aksoy, Y.; Kaya, A. Method dependency of relationships between specific surface area and soil physicochemical properties. *Appl. Clay Sci.* 2010, 50, 182-190.

[49] Hepper, E. N.; Buschiazzo, D. E.; Hevia, G. G.; Urioste, A.; Antón, L. Clay mineralogy, cation exchange capacity and specific surface area of loess soils with different volcanic ash contents. *Geoderma* 2006, 135, 216-223.

本文曾发表于2022年《Sustainability》第14卷

Stabilization of Soil Co-Contaminated with Mercury and Arsenic by Different Types of Biochar

Yang Wei[1, 2, 3, 4], Risheng Li[1, 2, 3, 4], Nan Lu[2, 3, 4] and Baoqiang Zhang [1, 2, 3, 4]

(1. Shaanxi Land Engineering Construction Group Co., Ltd., Xi'an Jiaotong University Technology Innovation Center for Land Engineering and Human Settlements, Xi'an 710075, China; 2. Land Engineering Technology Transformation Center, Shaanxi Provincial Land Engineering Construction Group Co., Ltd., Xi'an 710075, China; 3. Institute of Land Engineering and Technology, Shaanxi Provincial Land Engineering Construction Group Co., Ltd., Xi'an 710075, China; 4. Key Laboratory of Degraded and Unused Land Consolidation Engineering, Ministry of Natural Resources, Xi'an 710075, China)

【Abstract】 Mercury (Hg) and arsenic (As) are toxic and harmful heavy metals, with the exceedance rates of 1.6% and 2.7%, respectively, in soils across China. Compared with soils contaminated with Hg or As alone, co-contaminated soils pose complex environmental risks and are difficult to remediate. Biochar is widely used as a soil amendment to adsorb and immobilize pollutants such as heavy metals. However, only few studies have explored the efficiency of biochars produced from different crop straws to reduce the bioavailability of heavy metals in co-contaminated soils, and the effects on the soil biological properties are often overlooked. The aim of this study was to investigate changes in the physicochemical properties, enzyme activities, and heavy metal bioavailability of an industrial soil co-contaminated with Hg and As upon addition of different biochars from reed, cassava, and rice straws (REB, CAB, and RIB, respectively). The soil was amended with 1% biochar and planted with spinach in pots for 30 days. RIB was more effective than REB and CAB in increasing the soil pH, organic matter content, and cation exchange capacity. RIB and CAB exhibited similar positive effects on the soil dehydrogenase, catalase, invertase, and urease activities, which were higher than those of REB. The exchangeable fraction of both metals decreased upon biochar addition, and the residual fraction showed the opposite trend. All biochar amendments reduced the bioconcentration factors of heavy metals (especially Hg) in plants and decreased the metal bioavailability in soil. RIB is the optimal amendment for the stabilization of soil co-contaminated with Hg and As.

【Keywords】 Biochar amendment; Heavy metal co-contamination; Soil pollution; Enzyme activity

1 Introduction

In the wake of rapid economic development, heavy metal contamination caused by irrational exploitation, inappropriate environmental management and improper legal regulations has become a global environmental issue. For example, 1108~3784 t of mercury (Hg) and 28400~94000 t of arsenic (As) are released into soil every year in the globe owing to anthropogenic activity. Many heavy metal(loid)s, including Hg and As, are characterized by high toxicity, low biodegradability, and long environmental persistence[1]. They can accumulate in crop plants and then enter the human body through the food chain. All forms of Hg are toxic, with certain carcinogenicity, which can poison the nervous, reproductive immune, cardiovascular, and cerebrovascular systems[2]. In addition, long-term As exposure can lead to cancer of the skin or visceral organs[3], and oral administration of 0.1 g As_2O_3 can be fatal[4]. Compared with soils contaminated with Hg or As alone, co-contaminated soils pose complex environmental risks[3]. Therefore, the remediation of soil co-contaminated with Hg and As is of practical significance from a human health perspective.

According to the national soil pollution survey in China, the exceedance rates of Cd, Ni, As, Cu, Hg, Pb, Cr, and Zn are 7.0%, 4.8%, 2.7%, 2.1%, 1.6%, 1.5%, 1.1%, and 0.9%, respectively[5-6]. Although Cd is the most important heavy metal contaminant, soil pollution by As and Hg is also a non-negligible problem, especially in industrial wasteland. Wastewater, waste residue, and waste gas discharged from industrial production are the major sources of soil Hg and As[7]. In addition, the extensive use of Hg-and As-containing pesticides in agricultural production causes soil pollution[8]. Similar to Cd, Hg often exists as cations (e.g., Hg^{2+}) in the soil, whereas As mostly occurs as oxyanions (e.g., $H_2AsO_4^-$and $HAsO_3^{2-}$)[6]. Given the different behaviors of cations and oxyanions in the complex environmental matrix, it is difficult to remediate soils co-contaminated with Hg (or Cd) and As.

Numerous studies have been devoted to the removal or immobilization of heavy metals in soil using organic and inorganic additives, such as biochar (e.g.,[9]). As a major agricultural country, China contributes 15.0% to the global straw output[10]. But the straw utilization rate is only 80%, and incineration or disposal can easily lead to resource waste and environmental pollution[11]. The transformation of straw into biochar through pyrolysis is a sustainable technology for waste utilization and carbon (C) sequestration[12]. Compared with traditional physical and chemical techniques for soil remediation, biochar addition is pollution-free and cost-effective.

Biochar can be used as soil amendment to adsorb and immobilize heavy metals via different mechanisms[13]. First, biochar is characterized by high electronegativity and cation exchange capacity (CEC), with abundant functional groups (—OH, —COOH, —C=O—, and C=N) on the surface. It can interact directly with heavy metals via electrostatic attraction, ion exchange, complexation, and precipitation[14]. Second, biochar is rich in C and has a well-developed pore structure. It can indirectly affect heavy metal speciation by altering soil physicochemical properties, which, in turn, reduces the bioavailability of heavy metals[15].

The adsorption performance of biochars prepared from different feedstocks varies, due to differences in their elemental composition, functional groups, specific surface area, surface properties, and microscopic morphology[16]. Some researchers investigated the efficiency of biochars produced from cotton, corn, and rape straws for Hg adsorption and immobilization[17-19], whereas others applied biochars derived from wheat, peanut, and sugarcane straws to remove As from soil and water[20-22]. However, no previous studies have explored the use of biochars obtained from different crop straws to treat soil co-contaminated with Hg and As. The adsorption capacity and mechanisms of biochars on Hg-As in co-contaminated soil may be different from those on other heavy metals, such as Cd-Pb or Cd-Zn.

Biochar amendment may improve the soil biological properties[23], in addition to reducing the bioavailability of toxic heavy metals. In particular, the activity of soil enzymes has attracted attention for the remediation of heavy metal-contaminated soils[24]. Soil enzymes secreted by microorganisms play a critical role in nutrient cycling, redox reactions, and organic matter (OM) decomposition. Therefore, the soil enzyme activity reflects the intensity of biochemical reactions in soil and can be used to evaluate the quality of soils contaminated with heavy metals[24]. Biochar amendment can directly and indirectly affect the soil enzyme activity[25]. However, the relationships between specific physicochemical properties, enzyme activities, and metal bioavailability in biochar-amended soil contaminated with heavy metals have not been systematically studied.

Reed is a wetland plant with a wide range of sources and high levels of cellulose and hemicellulose. Reed straw-derived biochar is characterized by a loose internal structure, large specific surface area, and abundant active groups such as hydroxyl and carboxyl groups[26]. Thus, it may have a high ability to adsorb and stabilize heavy metals in soil. In addition, cassava and rice are common food crops, and their straw accounts for a large proportion of tropical agricultural waste[9]. Therefore, using reed, cassava, and rice straws to prepare biochar for the stabilization of heavy metal-contaminated soil can realize the comprehensive utilization of agricultural waste.

In the present study, we selected reed, cassava, and rice straw as feedstocks to prepare different types of biochars. A pot experiment was carried out to investigate biochar-induced changes in the physicochemical and biological properties as well as heavy metal bioavailability of soil co-contaminated with Hg and As. We also discussed the feasibility of using different biochars to stabilize heavy metal-co-contaminated soil. The present study can provide new insights into the efficiency of biochar for the green remediation of heavy metal-contaminated soil, and may represent a theoretical reference for the comprehensive utilization of agricultural waste.

2 Materials and methods

2.1 Materials

Soil co-contaminated with Hg and As was collected within a chemical plant that produced fumaric acid and its affiliated chemical products. There were coal yards, oil tank plants, and garbage dumps around the chemical plant, leading to substantial wastewater discharge and soil pollution. At the sampling site, weeds were removed from the surface and soil was collected the 0~50 cm depth using a shovel. The soil sample was transported to the laboratory and air-dried. After removing debris, the sample was ground and passed through a 2 mm sieve. The soil was classified as silty loam and its basic physicochemical properties are summarized in Table 1.

Table 1 Physicochemical properties of soil sample used in experiments

Particle size distribution (%)			Texture	pH	Organic matter ($g \cdot kg^{-1}$)	$CaCO_3$ (%)	Total Hg ($mg \cdot kg^{-1}$)	Total As ($mg \cdot kg^{-1}$)
Clay (0~2 μm)	Silt (2~50 μm)	Sand (50~2000 μm)						
11.29	71.16	17.55	Silty loam	8.96	8.73	8.95	8.74	106

Reed and rice straw samples were collected in Fuping County, Weinan City (109°11.78′E, 34°42.12′N) in Shaanxi Province, and the cassava straw sample was collected in Wuming County, Nanning City (107°58.26′E, 23°19.27′N) in Guangxi Province, China. Different types of biochars were produced by pyrolysis[27]. Briefly, the straw samples were washed with ultrapure water to remove surface dust and dirt, followed by oven-drying at 60 ℃ for > 2 days to remove moisture. The dry straw samples were milled to < 1 mm pieces. Each straw sample was loaded into a covered crucible and placed in a muffle furnace, then heated to 550 ℃ at a heating rate of 5 $℃ \cdot min^{-1}$ for 4 h. Afterward, the crucible was placed into a glass desiccator and cooled to room temperature. The biochar sample was then collected and ground to pass through a 0.149 mm sieve. The biochars obtained from reed, cassava, and rice straw samples are referred to as REB, CAB, and RIB, respectively.

2.2 Experimental design

A pot experiment was conducted in a greenhouse from May 13, 2021 to June 11, 2021. Before the experiment, equal amounts of soil samples were mixed with 1% REB, CAB, or RIB and filled into pots (20 cm diameter and 20 cm depth, 3.0 kg soil per pot). Basal fertilizer (N; 75 $mg \cdot pot^{-1}$) was applied before sowing. Spinach seeds (Chunqiu Seed Industry Co., Ltd., Shouguang, Shandong Province, China) were sown into the pots (40 seeds per pot) and covered with about 1 cm of soil. A blank control group without biochar amendment was included, and three replicates were used for each treatment group. After sowing, the pots were arranged in a randomized block design, with their positions changed regularly to maintain ventilation and light conditions. Water was added every other day to control the soil moisture at about 25%. Spinach plants were harvested 30 days after planting to determine their total Hg and As contents. Then, potted soil samples were collected to determine their physicochemical properties, enzyme activities, and heavy metal speciation.

Table 2 Experimental treatments used in the study

Treatment	Biochar type	Biochar dose ($g \cdot pot^{-1}$)	Plant
1	Reed straw biochar (REB)		
2	Cassava straw biochar (CAB)	30 (1%)	Spinach
3	Rice straw biochar (RIB)		
4	Control	0	

2.3 Physicochemical analysis

Air-dried soil subsamples were used for physicochemical analyses. In brief, mechanical composition tests were conducted using a MS2000 laser particle size analyzer (Malvern Instruments Ltd., Malvern, UK). pH and electrical conductivity (EC) measurements were made in 1:2.5 soil/deionized water slurries (w/w) using an S220-K acidometer and an S230-K conductivity meter (Mettler Toledo Instruments Ltd., Zurich, Switzerland), respectively. Total organic carbon (TOC) contents were determined by potassium dichromate oxidation with external heating and then converted to OM content using a conversion factor of 1.724[24]. Calcium carbonate ($CaCO_3$) content was determined by a titrimetric method[28]. CEC values were determined using the ammonium acetate method[29].

Biochar yields were calculated as the mass ratio of obtained biochar to crop straw used for pyrolysis. pH and EC measurements were conducted in 1:10 biochar/deionized water (w/w) slurries that had been agitated for 90 min at 20 ℃. Sulfur (S) content was analyzed using a Flash EA 1112 elemental analyzer (Thermo Fisher Scientific, Cleveland, OH, USA). Silica (SiO_2) content was determined by alkali fusion[30]. Scanning electron microscopy (SEM) images and pore sizes were obtained using an FEI Q45 scanning electron microscope (FEI Corp., Hillsboro, OR, USA). Specific surface areas were measured using a NOVA4200E analyzer (Quantachrome Instruments, Boynton Beach, FL, USA). Zeta potentials were measured using a Zeta potentiometer (DelsaMax Pro; Beckman Coulter, Brea, CA, USA).

2.4 Heavy metal analysis

An atomic fluorescence spectrometer (AFS-9760, Haiguang Instruments Corp., Beijing, China) was used to determine the total Hg and As contents in soil and plant samples according to the GB/T 22105.1—2008, GB/T 22105.2—2008 and GB 5009.17—2014, GB 5009.11—2014 Chinese standards, respectively.

The bioconcentration factors (BCFs) of Hg and As from soil to plants were calculated using Equation (1)[31]:

$$BCF = \frac{C_{pt}}{C_{so}} \tag{1}$$

where C_{pt} ($mg \cdot kg^{-1}$) is the Hg or As content in harvested plants and C_{so} ($mg \cdot kg^{-1}$) is the initial Hg or As content in soil samples.

The Tessier sequential extraction procedure was used to separate soil Hg and As into exchangeable (F1), carbonate-bound (F2), iron (Fe)/magnesium (Mn) oxide-bound (F3), organic-bound (F4), and residual (F5) fractions[32]. A higher proportion of F1 corresponds to more active and bioavailable heavy metals, whereas a higher F5 fraction denotes a lower bioavailability of heavy metals in the soil[33].

2.5 Enzyme activity assay

Fresh soil subsamples were used to measure enzyme activities. In brief, dehydrogenase activities were measured by the triphenyl tetrazolium chloride assay and expressed in μg · day^{-1} of triphenyl formamidine produced per gram of soil[34]. Catalase activities were measured by potassium permanganate titration and expressed in μmol · min^{-1} of potassium permanganate (0.1 mol · L^{-1}) consumed per gram of soil[35]. Invertase activities were assayed by 3,5-dinitrosalicylic acid colorimetry and expressed as in mg · day^{-1} of glucose produced per gram of soil[36]. Urease activities were assayed by sodium phenate-sodium hypochlorite colorimetry and expressed in mg · day^{-1} of NH_3-N produced per gram of soil[36].

2.6 Statistical analysis

Statistical analyses were performed using the SPSS 18.0 software (SPSS Inc., Chicago, IL, USA). One-way analysis of variance (ANOVA) was used to determine differences in sample means across different treatment groups.

3 Results and discussion

3.1 Comparison of biochars from different feedstocks

Table 3 summarizes the main properties of the three biochars, which showed significant variations with the feedstock type. Overall, the biochar samples exhibited considerable differences in terms of yield, pH, EC, pore size, specific surface area, S content, and SiO_2 content. The yield of RIB was remarkably higher than those of CAB and REB. All three biochars were alkaline, with a mean pH between 10.11 and 10.81. Similar pH values (10.76 and 10.13, respectively) were reported for biochars from canola and wheat straws (at 550 ℃)[37-38]. Biochars with alkaline pH could facilitate the treatment of heavy metal-contaminated soils[39]. The EC measures the content of dissolved salts and is a key indicator of biochar quality for use as amendment of heavy metal-contaminated soil[40]. In the present study, the EC of CAB was similar to that of RIB, whereas REB exhibited a considerably lower value. Previous studies have shown that sulfur and silicon can change the speciation of heavy metals and reduce their bioavailability to plants, thereby realizing the passivation and immobilization of heavy metals[41-42]. Biochar composition analysis revealed that RIB had the highest S and SiO_2 contents, while lower S and SiO_2 contents were observed in REB and CAB, respectively.

Table 3 Main properties of biochars from reed straw (REB), cassava straw (CAB), and rice straw (RIB)

Property	Biochar		
	REB	CAB	RIB
Yield (%)	29.24 ± 2.57 c	32.20 ± 1.35 b	45.71 ± 0.61 a
pH	10.42 ± 0.13 b	10.11 ± 0.09 c	10.81 ± 0.24 a
Electrical conductivity (μS · cm^{-1})	689.21 ± 1.36 b	1028.05 ± 2.73 a	1005.36 ± 2.59 a
S (%)	0.32±0.02 c	0.51±0.14 b	0.69±0.25 a
SiO_2(%)	4.02±0.91 b	3.58±0.24 b	11.08±0.37 a
Average pore size (μm)	10.36 ± 1.24 a	8.25 ± 1.69 b	3.09 ± 0.83 c
Specific surface area (m^2 · g^{-1})	10.61 ± 0.37 c	18.45 ± 0.27 b	23.05 ± 1.36 a
Zeta potential (mV)	−12.69±0.21 c	−14.18±0.44 b	−15.96±0.18 a

Data are shown as mean ± standard deviation. Different letters in the same column indicate significant differences between mean values corresponding to different treatment groups ($P < 0.05$).

The SEM images showed that REB, CAB, and RIB had porous structures (Fig. 1). In particular,

REB and CAB contained more macropores, while RIB had smaller pore sizes, with a more regular and compact pore arrangement. The mean pore size of the different biochars followed the order RIB < CAB < REB, while the specific surface area decreased in the order RIB > CAB > REB. The heavy metal adsorption capacity of biochar is strongly affected by its specific surface area and porosity; in particular, higher specific surface area and porosity values correspond to a stronger adsorption capacity for heavy metals[43]. Therefore, the present results indicate that, compared with REB and CAB, RIB had a greater ability to adsorb and immobilize heavy metals in soil.

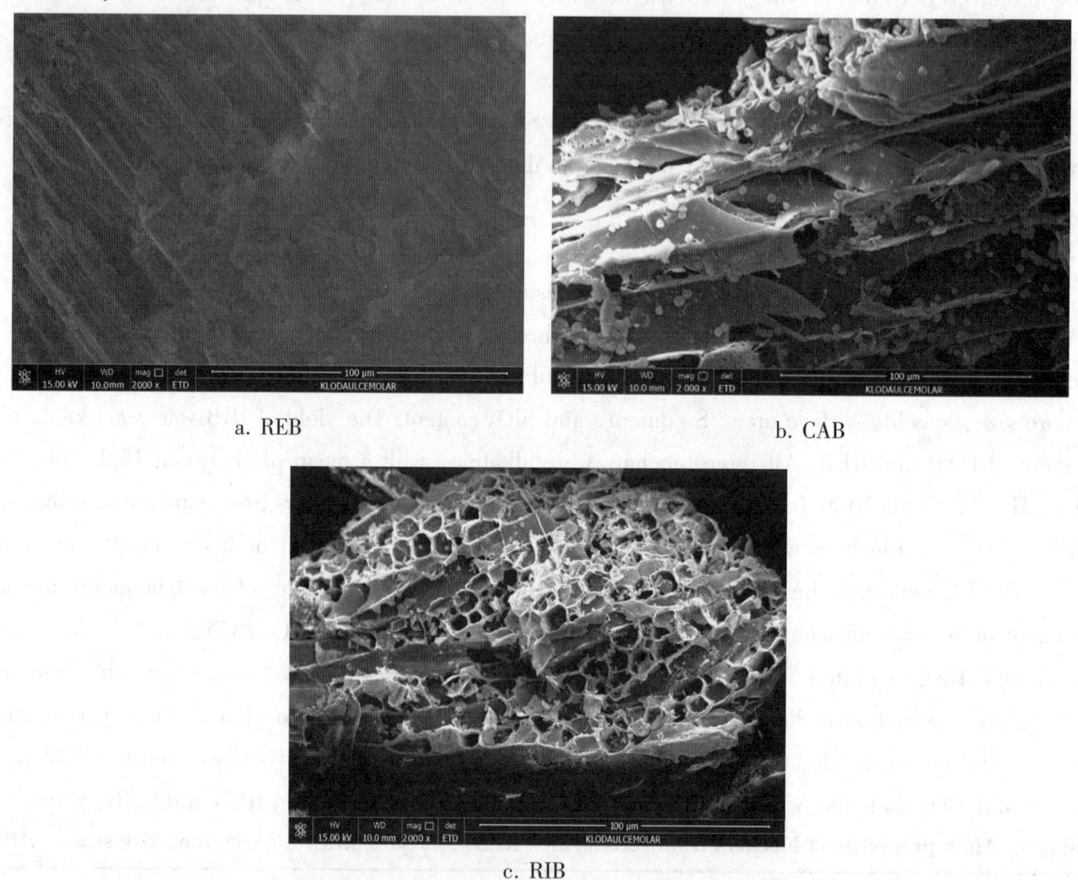

a. REB　　b. CAB　　c. RIB

Fig. 1　SEM images of biochar samples prepared by pyrolysis of different feedstocks. REB, reed straw biochar; CAB, cassava straw biochar; RIB, rice straw biochar

3.2　Effects of biochar amendments on soil physicochemical properties

All soil physicochemical properties analyzed in this study were significantly improved upon addition of different biochars (Fig. 2). Compared with the unamended control, the highest soil pH was observed for the RIB treatment, while the REB and CAB treatments led to smaller and similar increases in soil pH (Fig. 2a). Other studies also reported soil pH increases following application of wheat straw and rice husk biochars[44-45], although a decrease in soil pH was observed in some cases[46]. The soil EC showed the largest increase upon CAB and RIB addition (Fig. 2b), consistent with a previous study by Igalavithana et al. using pine cone biochar[47]. Compared with the unamended control, each biochar amendment significantly increased the soil OM, TOC, and CEC values, with the RIB treatment showing the strongest effect (Fig. 2c ~ e). Many studies also found that biochars produced from rice straw, wheat straw, and empty palm fruit bunches had a positive effect on soil OM[24], TOC[44, 48], and CEC[49], respectively.

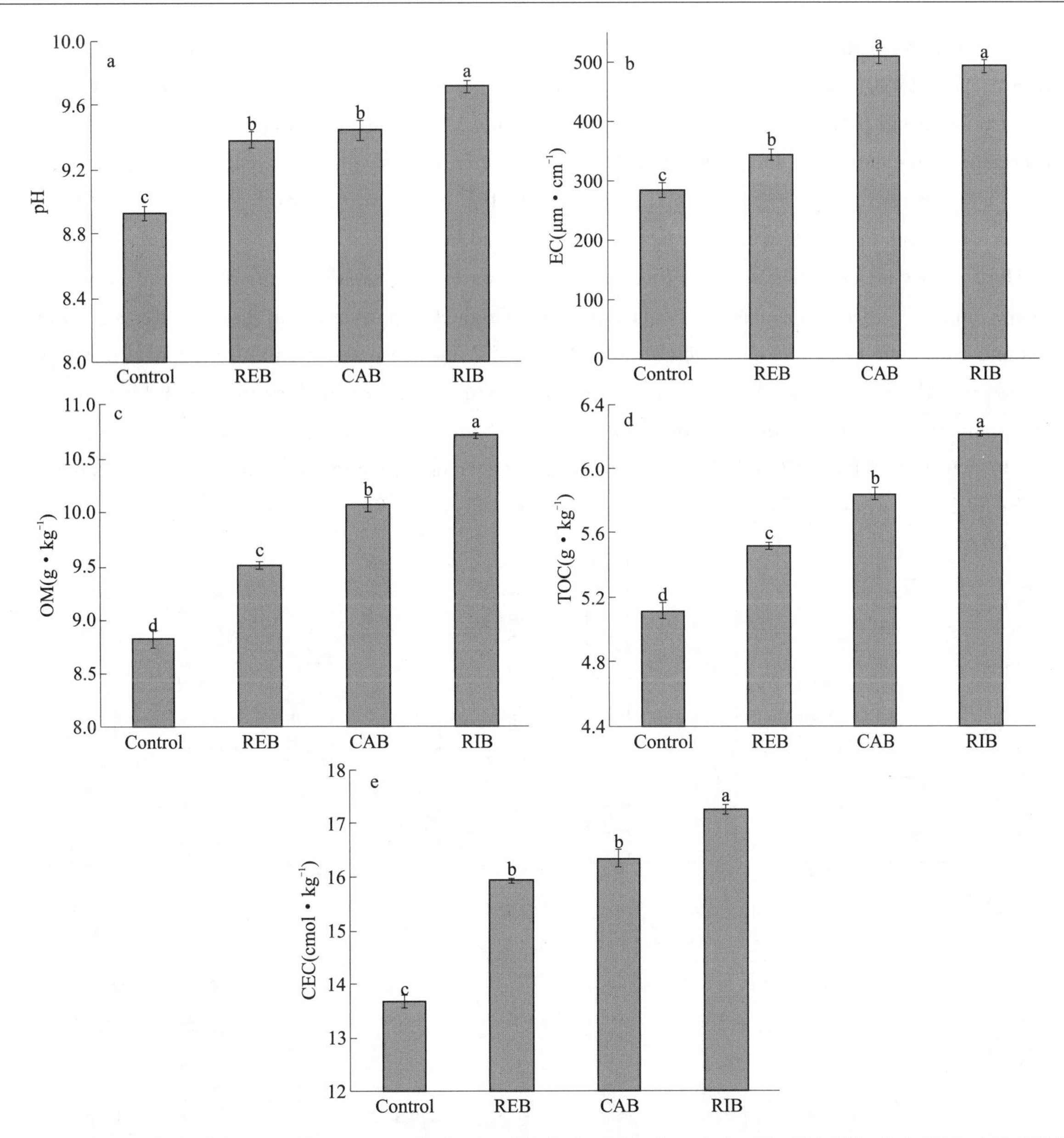

Fig. 2　Effects of biochar amendments on soil physicochemical properties: (a) pH, (b) EC, (c) OM, (d) TOC, (e) CEC. Error bars represent standard deviations of the means. Different letters above bars indicate significant differences between mean values of the treatment groups ($P < 0.05$). REB, reed straw biochar; CAB, cassava straw biochar; RIB, rice straw biochar

The improvements in soil physicochemical properties following biochar application may be a direct effect of the biochar components or result from a combination of different physicochemical factors. For example, the negatively charged functional groups, including phenolic, hydroxyl, and carboxyl groups on the surface of biochar can bind to the H^+ ions in the soil, causing an increase in its pH[50]. Biochar can also increase the soil NO_3-N content, which in turn leads to a pH increase[24]. NO_3-N may react with H^+ to suppress soil acidification. The marked EC increases observed for the three biochar treatments might be due to the higher EC of the biochars used in the experiment compared with that of the original soil. In addition, Tang et al. reported that biochar treatment can increase the available phosphorus (P) and potassium (K) contents in soil, leading to an increase in EC[24].

Biochar is a C-rich material with an organic C content of up to 90%, depending on the feedstock[51].

Biochar can increase the soil TOC content by adsorbing small organic molecules and then promoting their polymerization through surface catalysis[52]. Biochar can also efficiently increase the soil CEC, and a positive relationship between soil CEC and organic C content has been identified[53]. Liang et al. reported that organic C increases the soil CEC through a higher surface area with more cation adsorption sites, a higher charge density per unit surface area, or a combination of the two factors[54].

3.3 Changes in soil enzyme activities upon biochar addition

The soil enzyme activities serve as biological indicators for evaluating the quality of soil contaminated by heavy metals[20]. Dehydrogenase and catalase are oxidoreductase enzymes that directly alter ionic valence states and contribute to the detoxification of heavy metals[55]. Invertase and urease are involved in the transformation of soil nutrients such as C and N[55]. Compared with the unamended control, each biochar amendment enhanced the activities of dehydrogenase, catalase, invertase, and urease enzymes in soil samples (Fig. 3). Overall, RIB exhibited the greatest positive effect on soil enzyme activities (with increases of 29.2%~89.3%), while REB displayed the least positive effect (13.9%~27.7%).

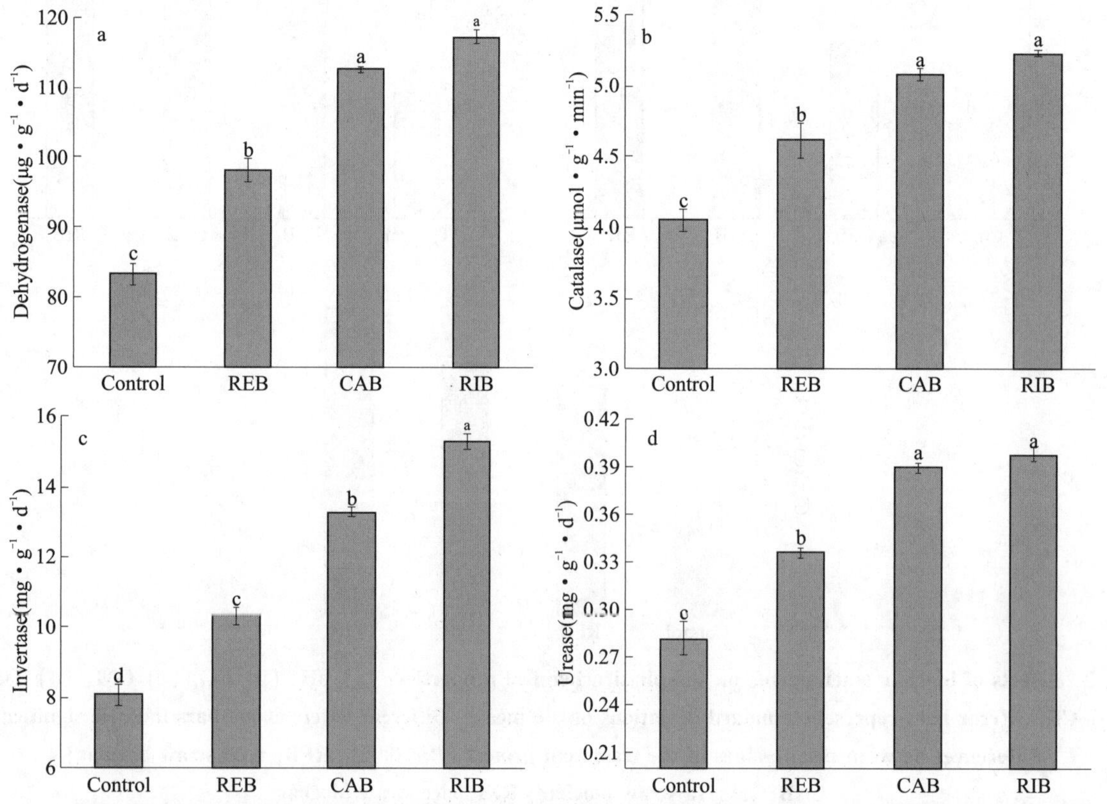

Fig. 3 Effect of biochar amendments on soil enzyme activities: (a) dehydrogenase, (b) catalase, (c) invertase, and (d) urease. Error bars represent standard deviations of the means. Different letters above bars indicate significant differences between mean values of the treatment groups ($P < 0.05$). REB, reed straw biochar; CAB, cassava straw biochar; RIB, rice straw biochar

Consistent with our results, many studies have concluded that biochar application can enhance the activity of soil enzymes such as dehydrogenase[44, 56], urease, and invertase[54]. The enhancement of soil enzyme activities by biochar amendment may be attributed to the stimulation of soil microorganisms that secrete enzymes. Gomez et al. showed that biochar addition increased microbial abundance and altered the community composition of a sandy loam soil[58]. When biochar was added to contaminated soil, the resistance of microbial groups to cadmium (Cd) and lead (Pb) tended to increase[27]. Biochar can promote the growth of microorganisms in soil by improving its quality, thus enhancing the soil enzyme activity[59]. Another factor that could influence the soil enzyme activity is that heavy metal stress levels in the soil may change following biochar addition[60].

Interestingly, in some cases biochar may inhibit the activity of soil enzymes such as dehydrogenase, catalase[24], invertase, and urease[25]. The enzymatic activities mainly depend on soil properties including OM content, pH value, and mineral nutrients[61]. Different biochar feedstocks, production methods, and amendment doses may result in distinct changes in soil enzyme activity[25, 62].

3.4 Bioavailability of heavy metals in biochar-amended soil

Soil heavy metals exist in different forms, with varied migration ability and biological toxicity[63]. Following the sequential extraction procedure, F5 (residual fraction) is generally considered an unavailable form, because it is fixed in the soil matrix and cannot be absorbed or utilized by plants. Conversely, F1 (exchangeable fraction) is considered a bioavailable form, whereas the other three fractions (F2-F4) are relatively difficult for plant uptake or utilization[64]. In the present study, Hg and As showed different distributions among the five fractions in soil samples with and without biochar amendments (Figs. 4, 5). In the unamended control, F4-Hg was the predominant Hg fraction (63.0%), followed by F1-Hg (19.0%), whereas F5-Hg, F2-Hg, and F3-Hg only accounted for small proportions of the total Hg (4.4% ~ 7.0%; Fig. 4a). In contrast, soil As mainly comprised F5-As (40.3%) and F1-As (33.2%), with small proportions of F4-As, F2-As, and F3-As (4.1% ~ 13.2%; Fig. 4b). The different distributions of Hg and As in the five fractions suggest a different bioavailability of these heavy metals in the original soil[65].

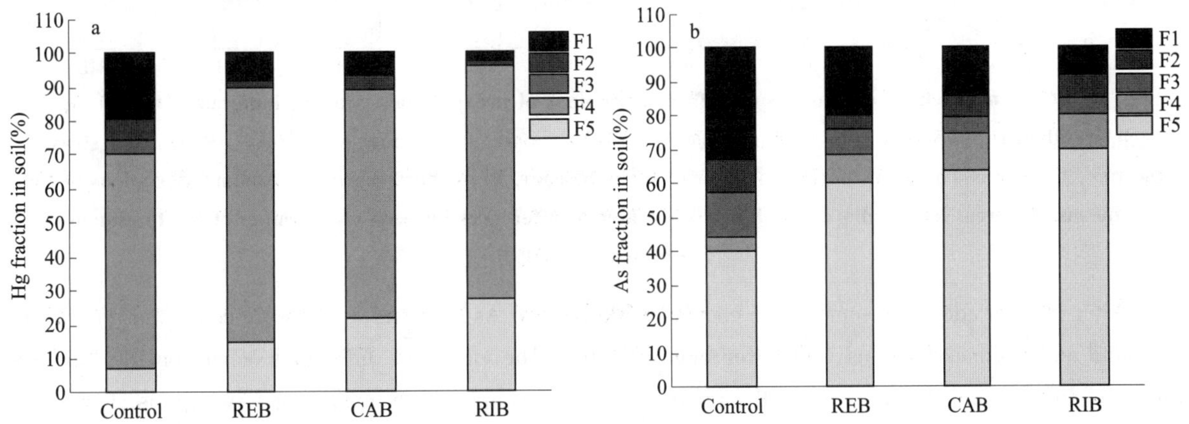

Fig. 4 Distribution of (a) Hg and (b) As heavy metals among five fractions in soil samples with different biochar amendments: F1, exchangeable fraction; F2, carbonate fraction; F3, Fe/Mn oxide-bound fraction; F4, organic-bound fraction; F5, residual fraction. REB, reed straw biochar; CAB, cassava straw biochar; RIB, rice straw biochar

In all treatments, F4-Hg was the highest among the five fractions, despite an increasing trend in the contribution of F5-Hg in soil with biochar amendments (Fig. 4a). A similar trend was observed for F5-As, which was the highest fraction in biochar-amended soil. For both heavy metals, the F1 fraction decreased significantly in all three biochar-treated soils compared with that in the unamended control. This indicated that REB, CAB, and RIB could effectively reduce the bioavailability of heavy metals in co-contaminated soil by facilitating the conversion of Hg and As from exchangeable to residual form. Among the biochar-amended soils, the largest decrease in F1-Hg and F1-As fractions was observed for the RIB treatment, along with the largest increase in F5-Hg and F5-As fractions. The increase in F5 suggests heavy metal passivation. F2 and F3 fractions of Hg and As showed no marked differences among all treatments. Compared with the unamended control, the F4 fraction moderately increased in all biochar-amended soils, but there was no significant difference among all the treatments. In summary, the results indicate that RIB exhibited the greatest effect on the stabilization and remediation of soil co-contaminated with Hg and As.

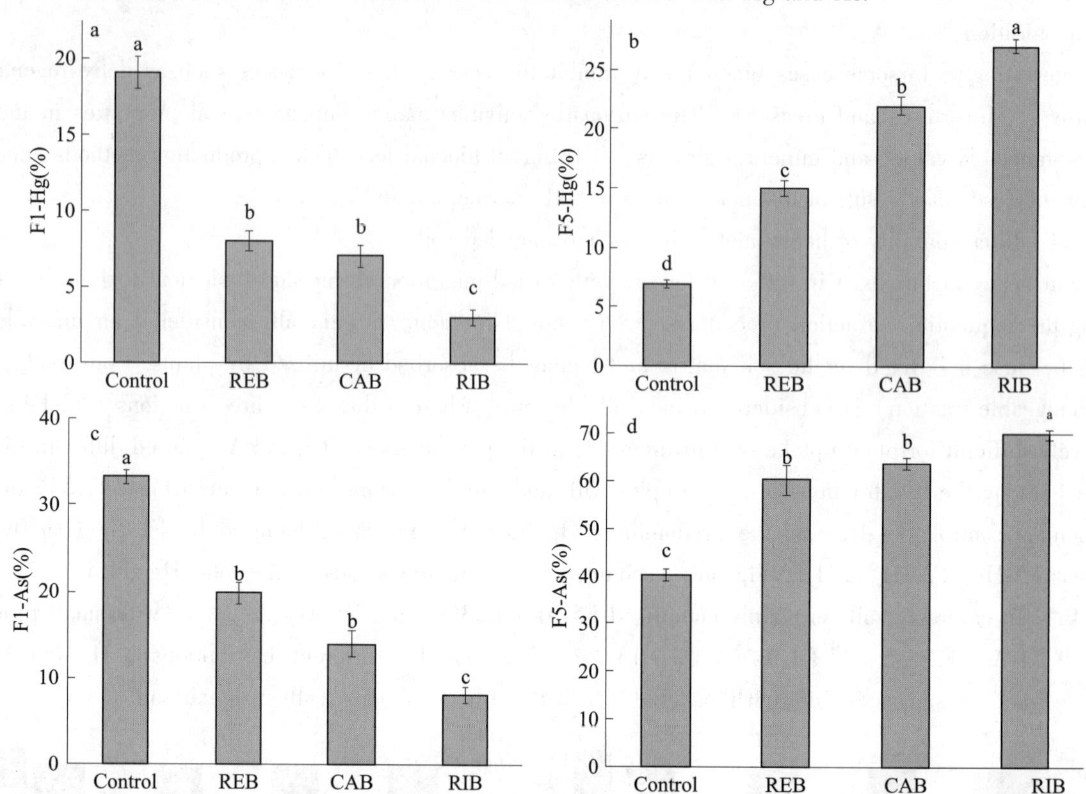

Fig. 5 Effects of biochar amendments on different fractions of heavy metals: (a) exchangeable Hg (F1-Hg), (b) residual Hg (F5-Hg), (c) exchangeable As (F1-As), and (d) residual As (F5-As). REB, reed straw biochar; CAB, cassava straw biochar; RIB, rice straw biochar. Error bars represent standard deviations of the means. Different letters above bars indicate significant differences between mean values of the treatment groups ($P < 0.05$)

After biochar addition, some of the exchangeable Hg and As fractions in soil were converted into organic-bound and residual fractions with lower bioavailability. The effects of different biochars on the bioavailability of soil Hg and As observed in the present study are supported by many previous reports. For example, Gamboa et al. found that, in a contaminated soil treated with biochar, the immobile Hg fraction increased by 76% compared with the control[66]. In addition, Gu et al. showed that the bioavailable fraction of As in soil exhibited a marked decrease of up to 18.0% after biochar treatment[67]. Yu et al. also indicated that the concentrations of bioavailable As species decreased upon biochar addition to contaminated soils[68].

The reduction of Hg and As bioavailability may be partly attributed to the direct effect of biochar on soil heavy metals. The three biochar materials had negative charges on their surface, with large specific surface area, many functional groups, and a high pH. All these factors can facilitate the immobilization of Hg cations through electrostatic interactions and chelation with functional groups on the biochar surface[57]. Gamboa et al. also indicated that highly porous biochars with abundant polar functional groups on the surface, along with high pH, EC, CEC, and ash percentage, can favor the adsorption and stabilization of Hg[66]. On the other hand, the adsorption of As oxyanions by biochar may be mainly related to its huge specific surface area and well-developed pore structure. After examining the microscopic morphology of the present systems, an obvious microporous structure could be observed in all three biochar materials after carbonization (Fig. 1). Overall, REB exhibited few isolated pores with relatively large size and the smallest specific surface area. However, RIB had a hierarchical pore structure with the smallest mean pore size and largest specific surface area, which may explain its strong ability to adsorb and immobilize both Hg and As in the co-contaminated soil.

Considering the elemental composition, all three biochars contained S and SiO_2 (Table 3). Hg in soil can adsorb to the surface of biochar and react with S groups (e.g., sulfate ester and sulfate radicals) loaded on biochar to form stable residual Hg-sulfide[42]. Compared with REB and CAB, RIB had a higher S content, which resulted in stronger immobilization of Hg. In addition, the water-soluble Si in Si-rich straw biochar plays a role in the adsorption and immobilization of heavy metals. Si can promote the conversion of exchangeable heavy metals with higher activity to the residue state[69]. Compared with reed and cassava, rice prefers Si and its straw is rich in phytolith ($SiO_2 \cdot H_2O$). Consequently, RIB had the highest SiO_2 content among the three biochars, which could enhance the passivation of heavy metals.

Furthermore, biochar-induced changes in soil properties (e.g., CEC and enzymatic activity) may indirectly reduce the bioavailability of soil heavy metals. Mohamed et al. indicated that the soil CEC is positively correlated with the residual fraction of heavy metals[70]. Liu et al. found a negative correlation between soil enzyme activity and the bioavailable content of heavy metals[71]. In this study, biochar amendments resulted in increased CEC and enhanced dehydrogenase, catalase, invertase, and urease activities in the soil. In turn, these changes effectively increased the F5-Hg and F5-As fractions, thereby reducing the bioavailability of heavy metals. In the present study, RIB resulted in the highest soil CEC and enzymatic activity, posing a strong effect on the immobilization of heavy metals.

Soil OM acts as an adsorbent for heavy metals, because it contains different functional groups (–OH and –COOH) that can easily bind metal ions to form strong complexes with low bioavailability[55, 72]. In biochar-amended soil samples, the increase of OM content could benefit plant growth and development, enhance plant stress resistance, and reduce plant uptake of available heavy metals. Since RIB contributed the most to soil OM content, this amendment could promote the transformation of heavy metals to the unusable residual form most prominently.

Generally, soil pH is considered one of the most important factors controlling heavy metal bioavailability[57, 73], because it controls the adsorption-desorption, dissolution-precipitation, and other processes of soil minerals. Biochar is an alkaline substance, and its application increases the soil pH, thereby affecting the adsorption and desorption of soil minerals. In turn, the increased pH promotes the transformation of active heavy metal species into stable residuals and thus stabilizes heavy metals such as Hg. Among the three biochars produced in the present study, RIB displayed the strongest effect on increasing the soil pH, which contributed to the effective reduction of Hg bioavailability. Unlike Hg, As is relatively stable in acid soil, and lowering the pH in alkaline soil environment is conducive to As stabilization, while the increase of pH can activate soil As to a certain extent[6]. Soil As exists mainly in the form of oxyanions, such as $H_2AsO_4^-$ and $HAsO_3^{2-}$. When pH increases, the OH^- concentration in the soil increases correspondingly. As OH^- competes with $H_2AsO_4^-$ and $HAsO_3^{2-}$ for the adsorption sites on biochar surface, the adsorption and stabilization of As in the soil may be decreased. However, in this study, the strong adsorption capability of biochars covered their mobilizing effect on As bioavailability, and significantly decreased the exchangeable As concentration (Fig. 5). Meanwhile, both effects of the increase in soil pH and adsorption of biochars significantly enhanced the stabilization of Hg^{2+} and reduced the bioavailability of Hg in soil. Moreover, the test soil is derived from loess parent material with high $CaCO_3$ content. The application of biochar could increase the contact area between Ca and As in the soil, promoting their reaction and co-precipitation, and consequently reducing the bioavailability of As.

3.5 Bioconcentration of heavy metals in plants after soil stabilization with biochar

The BCF indicates the ability of plants to accumulate heavy metals[74]. This factor can directly measure the difficulty in the migration and utilization of soil heavy metals, and indirectly reflect the efficiency of biochar for the stabilization of heavy metal-contaminated soils[75]. Irrespective of biochar amendments, the BCF of Hg (0.337) was markedly higher than that of As in spinach (0.095; Fig. 6), indicating a greater risk of Hg migration in the soil and subsequent accumulation in plants.

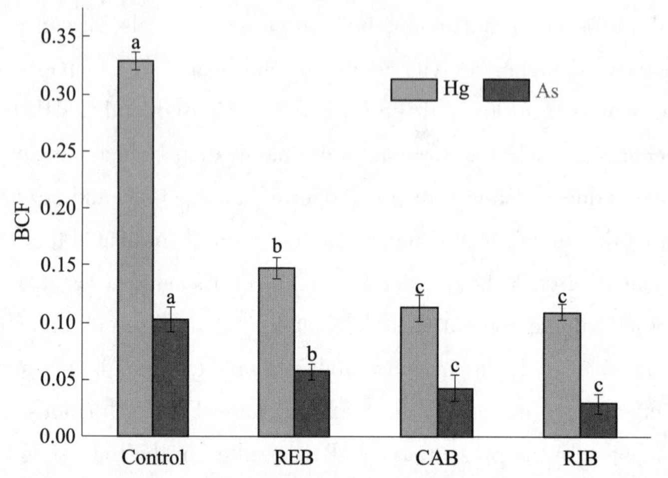

Fig. 6 Bioconcentration factors (BCFs) of heavy metals in spinach plants after biochar remediation. REB, reed straw biochar; CAB, cassava straw biochar; RIB, rice straw biochar. Error bars represent standard deviations of the means. Different letters above bars indicate significant differences between mean values of the treatment groups ($P < 0.05$)

For both Hg and As, the BCFs tended to decrease in the biochar-amended soil samples compared with the unamended control. These results suggest that the addition of biochars could enhance Hg and As adsorption by soil and effectively reduce heavy metal uptake by plant tissues. Yu et al. also reported that biochar addition could inhibit the accumulation of Hg in cabbage plants by preventing its migration[76]. Similar results were observed by Zama et al. showed that both silicon-modified and unmodified biochars were effective in minimizing the BCFs of As in spinach[77]. Among the different treatments, the BCFs of Hg and As in plants were the lowest in the RIB treatment and the highest in the REB treatment. These results indicate that the plant availability of Hg and As in soil decreased after amendment, and RIB had the best remediation effect on Hg and As, which was consistent with the results of the heavy metal speciation analysis.

4 Conclusions

The addition of different biochars from crop straw waste changed the physicochemical and biological properties of an industrial soil co-contaminated with Hg and As. All biochar amendments considerably improved the soil properties tested and led to enhanced soil enzyme activities. Moreover, the biochar amendments decreased the bioavailability of Hg and As, enhanced their adsorption on soil, and effectively reduced heavy metal uptake by plant tissues. The reduction of Hg and As bioavailability was attributed to the direct effect of biochar on soil heavy metals and the indirect effect of biochar-induced changes in soil properties. Compared with the biochars produced from reed and cassava straws, rice straw-derived biochar was the optimal amendment for the stabilization of co-contaminated soil, because it had the best pore and surface structures and exhibited the greatest positive effect on soil properties. Overall, the present results provide guidance for the use of biochars in the remediation of soil co-contaminated with Hg-As and possibly Cd-As. However, further studies are needed to determine the optimal dose of rice straw-derived biochar and the most suitable environmental conditions for its field application.

Author contributions: Conceptualization, Y.W. and R. L.; methodology, Y.W.; validation, Y.W. and N.L.; formal analysis, Y.W. and B.Z.; writing—original draft preparation, Y.W.; writing—review and editing, Y.W. and N.L. All authors have read and agreed to the published version of the manuscript.

Funding: This research was funded by the Technology Innovation Center for Land Engineering and Human Settlements, Shaanxi Land Engineering Construction Group Co., Ltd and Xi'an Jiaotong University, China, grant number 2021WHZ0094.

Institutional review board statement: Not applicable.

Informed consent statement: Not applicable.

Data availability statement: All data is presented in this study and thus contained within the article. There are no other available data in any publicly accessible repository.

Conflicts of interest: The authors declare no conflict of interest.

References

[1]Yan, A.; Wang, Y.; Tan, S. N.; Yusof, M. L. M.; Ghosh, S.; Chen, Z. Phytoremediation, A promising approach for revegetation of heavy metal-polluted land. *Front. Plant. Sci.* 2020, 11, 1-15.

[2]Feng, X. B.; Chen, J. B.; Fu, X. W.; Hu, H. Y.; Li, P.; Qiu, G. L.; Yan, H. Y.; Yin, X. F.; Zhang, H.; Zhu, W. Progress on environmental geochemistry of mercury. *Bull. Mineral. Petrol. Geochem.* 2013, 32(05), 503-530.

[3]Marziali, L.; Rosignoli, F.; Drago, A.; Pascariello, S.; Valsecchi, L.; Rossaro, B.; Guzzella, L. Toxicity risk assessment of Hg, DDT and As legacy pollution in sediments: A triad approach under low concentration conditions. *Sci. Total Environ.* 2017, 593-594, 809-821.

[4]Wright; David, A. *Environmental toxicology*. Cambridge University Press: USA, 2001.

[5]Esdaile, L. J.; Chalker, J. M. The mercury problem in artisanal and small-scale gold mining. *Chem. Eur. J.* 2018, 24, 6905-6916.

[6]Zhao, H.; Rong, Q.; Qin, X.; Nong, X.; Lu, D.; Zhang, C. Study on stabilization of arsenic in soil using FMBO incorporated with peat soil. *Environ. Sci. Technol.* 2021, 44(10), 53-59.

[7]Osterwalder, S.; Huang, J.; Shetaya, W.; Agnan, Y.; Frossard, A.; Frey, B.; Alewell, C.; Kretzschmar, R.; Biester, H.; Obrist, D. Mercury emission from industrially contaminated soils in relation to chemical, microbial, and meteorological factors. *Environ. Poll.* 2019, 250, 944-952.

[8]Chai, L.; Tang, J.; Liao, Y.; Yang, Z.; Liang, L.; Li, Q.; Wang, H.; Yang, W. Biosynthesis of schwertmannite by Acidithiobacillus ferrooxidans and its application in arsenic immobilization in the contaminated soil. *J. Soil. Sediment.* 2016, 16, 2430-2438.

[9]Lu, K.; Yang, X.; Gielen, G.; Bolan, N.; Ok, Y. S.; Niazi, N. K.; Xu, S.; Yuan, G.; Chen, X.; Zhang, X. Effect of bamboo and rice straw biochars on the mobility and redistribution of heavy metals (Cd, Cu, Pb and Zn) in contaminated soil. *J. Environ. Manage.* 2017, 186, 285-292.

[10]Wang, H.; Xu, J.; Sheng, L. Preparation of straw biochar and application of constructed wetland in China, A review. *J. Clean. Prod.* 2020, 273, 123131.

[11]Bai, B.; Zhao, H.; Zhang, S.; Zhang, X.; Yang, G. Forecasting of agricultural straw burning in the Northeastern China based on neural network. *China. Environ. Sci.* 2020, 40(12), 5205-5212.

[12]Shukla, P.; Giri, B. S.; Mishra, R. K.; Pandey, A.; Chaturvedi, P. Lignocellulosic biomass-based engineered biochar composites, A facile strategy for abatement of emerging pollutants and utilization in industrial applications. *Renew. Sust. Energ. Rev.* 2021, 152, 111643.

[13]Kamali, M.; Sweygers, N.; AL-Salem, S.; Appels, L.; Aminabhavi, T. M.; Dewil, R. Biochar for soil applications-sustainability aspects, challenges and future prospects. *Chem. Eng. J.* 2022, 428, 131189.

[14]Lian, F.; Xing, B. Black carbon (biochar) in water/soil environments: molecular structure, sorption, stability, and potential risk. *Environ. Sci. Technol.* 2017, 51 (23), 13517-13532.

[15]He, L.; Zhong, H.; Liu, G.; Dai, Z.; Brookes, P. C.; Xu, J. Remediation of heavy metal contaminated soils by

biochar: Mechanisms, potential risks and applications in China. *Environ. Poll.*, 2019, 252, 846-855.

[16]Zhan, G.; Chen, Z.; Tong, F.; Shen, H.; Gao, Y.; Liu, L.; Zhang, Z.; Lu, X. Effects of different biomass and pyrolysis technique on biochar characterization and immobilization of heavy metal in contaminated soil. *J. Ecol. Rural. Environ.* 2021, 37(1), 86-95.

[17]Hu, H.; Xi, B.; Tan, W. Effects of sulfur-rich biochar amendment on microbial methylation of Hg in rhizosphere paddy soil and methyl Hg accumulation in rice. *Environ. Pollut.* 2021, 286, 117290.

[18]Li, G.; Wang, S.; Wu, Q.; Wang, F.; Shen, B. Hg sorption study of halides modified biochars derived from cotton straw. *Chem. Eng. J.* 2016, 302, 305-313.

[19]Zhao, W.; Cui, Y.; Sun, X.; Wang, H.; Teng, X. Corn stover biochar increased edible safety of spinach by reducing the migration of Hg from soil to spinach. *Sci. Total. Environ.* 2020, 758, 143883.

[20]Kamran, M. A.; Bibi, S.; Chen, B. Preventative effect of crop straw-derived biochar on plant growth in an As polluted acidic ultisol. *Sci. Total Environ.* 2022, 812, 151469.

[21]Kumar, A.; Bhattacharya, T. Removal of As by wheat straw biochar from soil. *B. Environ. Contam. Tox.* 2022, 108, 415-422.

[22]Soares, M. B.; Santos, F. H.; Alleoni, L. R. F. Iron-modified biochar from sugarcane straw to remove As and lead from contaminated water. *Water. Air. Soil. Poll.* 2021, 232(9), 1-13.

[23]Garau, G.; Porceddu, A.; Sanna, M.; Silvetti, M.; Castaldi, P. Municipal solid wastes as a resource for environmental recovery, impact of water treatment residuals and compost on the microbial and biochemical features of As and trace metal-polluted soils. *Ecotoxicol. Environ. Saf.* 2019, 174, 445-454.

[24]Tang, J.; Zhang, L.; Zhang, J.; Ren, L.; Zhou, Y.; Zheng, Y.; Luo, L.; Yang, Y.; Huang, H.; Chen, A. Physicochemical features, metal availability and enzyme activity in heavy metal-polluted soil remediated by biochar and compost. *Sci. Total. Environ.* 2020, 701, 134751.

[25]Huang, D.; Liu, L.; Zeng, G.; Xu, P.; Huang, C.; Deng, L.; Wang, R.; Wan, J. The effects of rice straw biochar on indigenous microbial community and enzymes activity in heavy metal-contaminated sediment. *Chemosphere.* 2017, 174, 545-553.

[26]Song, Z.; Shi, X.; Liu, Z. Synthesis and characterization of reed-based biochar and its adsorption properties for Cu^{2+} and bisphenol A (BPA). *Environ. Chem.* 2020, 39(8), 2196-2205.

[27]Haddad, S. A.; Lemanowicz, J. Benefits of corn-cob biochar to the microbial and enzymatic activity of soybean plants grown in soils contaminated with heavy metals. *Energies.* 2021, 14, 5763.

[28]Bundy, L. G.; Bremner, J. M. A simple titrimetric method for determination of inorganic carbon in soils. *Soil Sci. Soc. Am. J.* 1972, 36(2), 273-275.

[29]Sumner, M.; Miller, W. *Cation exchange capacity and exchange coefficients. Methods of Soil Analysis Part 3-Chemical Methods*; (*methodsofsoilan*3). U.S. Environmental Protection Agency: Madison, USA, 1996; pp. 1201-1229.

[30]Elliott, C.; Snyder, G. H. Autoclave-induced digestion for the colorimetric determination of siliconin rice straw. *J. Agr. Food. Chem.* 1991, 39(6), 1118-1119.

[31]Zhuang, P.; Yang, Q.; Wang, H.; Shu, W. Phytoextraction of heavy metals by eight plant species in the field. *Water Air. Soil Poll.* 2007, 184, 235-242.

[32]Tessier, A.; Campbell, P. G. C.; Bisson, M. Sequential extraction procedure for the speciation of particulate trace metals. *Anal. Chem.* 1979, 51(7),844-850.

[33]O'Dell, R.; Silk, W.; Green, P.; Claassen, V. Compost amendment of Cu-Zn minespoil reduces toxic bioavailable heavy metal concentrations and promotes establishment and biomass production of *Bromus carinatus* (Hook and Arn.). *Environ. Poll.* 2007, 148(1), 115-124.

[34]Casida, L.E. Microbial metabolic activity in soil as measured by dehydrogenase determinations. *Appl. Environ. Microbiol.* 1977, 34(6), 630-636.

[35]Trasar-Cepeda, C.; Camiña, F.; Leirós, M. C.; Gil-Sotres, F. An improved method to measure catalase activity in soils. *Soil Biol. Biochem.* 1999, 31(3), 483-485.

[36]Guan, S. *Soil enzyme and its research methods*. China Agriculture Press: Beijing, China, 1986.

[37]Yang, C.; Lu, S. Straw and straw biochar differently affect phosphorus availability; enzyme activity and microbial functional genes in an Ultisol. *Sci. Total Environ.* 2022, 805, 150325.

[38]Yang, H. I.; Lou, K.; Rajapaksha, A. U.; Ok, Y. S.; Anyia, A. O.; Chang, S. X. Adsorption of ammonium in aqueous solutions by pine sawdust and wheat straw biochars. *Environ. Sci. Pollut. R.* 2018, 25, 25638-25647.

[39]Cárdenas-Aguiar, E.; Suárez, G.; Paz-Ferreiro, J.; Askeland, M. P. J.; Méndez, A.; Gascó, G. Remediation of mining soils by combining Brassica napus growth and amendment with chars from manure waste. *Chemosphere.* 2020, 261,127798.

[40]Kambo, H. S.; Dutta, A. A comparative review of biochar and hydrochar in terms of production, physico-chemical properties and applications. *Renew. Sust. Energ. Rev.* 2015, 45, 359-378.

[41]Li, L.; Zheng, C.; Fu, Y.; Wu, D.; Yang, X.; Shen, H. Silicate-mediated alleviation of Pb toxicity in banana grown in Pb-contaminated soil. *Bio. Trace Elem. Res.* 2012, 145(1), 101.

[42]Li, S.; Zheng, X.; Gong, J.; Xue, X.; Yang, H. Preparation of zinc chloride and sulfur modified cornstalk biochar and its stabilization effect on mercury contaminated soil. *Chin. J. Environ. Eng.* 2021,15(4), 1403-1408.

[43]Li, Q.; Liang, W.; Liu, F.; Wang, G.; Wan, J.; Zhang, W.; Peng, C.; Yang, J. Simultaneous immobilization of arsenic; lead and cadmium by magnesium-aluminum modified biochar in mining soil. *J. Environ. Manage.* 2022, 310, 114792.

[44]Chen, J.; Liu, X.; Zheng, J.; Zhang, B.; Lu, H.; Chi, Z.; Pan, G.; Li, L.; Zheng, J.; Zhang, X. Biochar soil amendment increased bacterial but decreased fungal gene abundance with shifts in community structure in a slightly acid rice paddy from Southwest China. *Appl. Soil Ecol.* 2013, 71, 33-44.

[45]Ibrahim, M.; Khan, S.; Hao, X.; Li, G. Biochar effects on metal bioaccumulation and arsenic speciation in alfalfa (*Medicago sativa* L.) grown in contaminated soil. *Int. J. Environ. Sci. Tech.* 2016, 13, 2467-2474.

[46]Zeng, G.; Wu, H.; Liang, J.; Guo, S.; Huang, L.; Xu, P.; Liu, Y.; Yuan, Y.; He, X.; He, Y. Efficiency of biochar and compost (or composting) combined amendments for reducing Cd, Cu, Zn and Pb bioavailability, mobility and ecological risk in wetland soil. *Rsc. Adv.*2015, 5, 34541-34548.

[47]Igalavithana, A. D.; Lee, S.; Lee, Y. H.; Tsang, D. C. W.; Rinklebe, J.; Kwon, E.; Ok, Y. S. Heavy metal immobilization and microbial community abundance by vegetable waste and pine cone biochar of agricultural soils. *Chemosphere.* 2017, 174, 593-603.

[48]Abujabhah, I. S.; Bound, S. A.; Doyle, R.; Bowman, J. P. Effects of biochar and compost amendments on soil physico-chemical properties and the total community within a temperate agricultural soil. *Appl. Soil Ecol.* 2016,98, 243-253.

[49]Cao, Y.; Gao, Y.; Qi, Y.; Li, J. Biochar-enhanced composts reduce the potential leaching of nutrients and heavy metals and suppress plant-parasitic nematodes in excessively fertilized cucumber soils. *Environ. Sci. Pollut. R.* 2018, 25(8), 7589-7599.

[50]Gul, S.; Whalen, J. K.; Thomas, B. W.; Sachdeva, V.; Deng, H. Physico-chemical properties and microbial responses in biochar-amended soils, mechanisms and future directions. *Agr. Ecosyst. Environ.* 2015, 206, 46-59.

[51]Beesley, L.; Moreno-Jiménez, E.; Gomez-Eyles, J. L. Effects of biochar and green waste compost amendments on mobility, bioavailability and toxicity of inorganic and organic contaminants in a multi-element polluted soil. *Environ. Pollut.* 2010, 158, 2282-2287.

[52]Song, D.; Xi, X.; Zheng, Q.; Liang, G.; Zhou, W.; Wang, X. Soil nutrient and microbial activity responses to two years after maize straw biochar application in a calcareous soil. *Ecotoxicol. Environ. Saf.* 2019, 180, 348-356.

[53]Machmuller, M. B.; Kramer, M. G.; Cyle, T. K.; Hill, N.; Hancock, D.; Thompson, A. Emerging land use practices rapidly increase soil organic matter. *Nat. Commun.* 2015, 6, 6995.

[54]Liang, B.; Lehmann, J.; Solomon, D.; Kinyangi, J.; Grossman, J.; O'Neill, B.; Skjemstad, J. O.; Thies, J.; Luizão, F. J.; Petersen, J.; Neves, E. G. Black carbon increases cation exchange capacity in soils. *Soil Sci. Soc. Am. J.* 2006, 70(5), 1719-1730.

[55]Yang, J.; Yang, F.; Yang, Y.; Xing, G.; Deng, C.; Shen, Y.; Luo, L.; Li, B.; Yuan, H. A proposal of "core enzyme" bioindicator in long-term Pb-Zn ore pollution areas based on topsoil property analysis. *Environ. Pollut.* 2016, 213, 760-769.

[56]Akmal, M.; Maqbool, Z.; Khan, K.S.; Hussain, Q.; Ijaz, S. S.; Iqbal, M.; Aziz, I.; Hussain, A.; Abbas, M. S.; Rafa, H. U. Integrated use of biochar and compost to improve soil microbial activity; nutrient availability, and plant growth in arid soil. *Arab. J. Geosci.* 2019, 12, 1-6.

[57]Jia, W.; Wang, B.; Wang, C.; Sun, H. Tourmaline and biochar for the remediation of acid soil polluted with heavy metals. *J. Environ. Chem. Eng.* 2017, 5, 2107-2114.

[58]Gomez, J. D.; Denef, K.; Stewart, C. E.; Zheng, J.; Cotrufo, M. F. Biochar addition rate influences soil microbial abundance and activity in temperate soils. *Eur. J. Soil Sci.* 2013, 65, 28-39.

[59]Wang, Y.; Shi, J.; Wang, H.; Lin, Q.; Chen, X.; Chen, Y. The influence of soil heavy metals pollution on soil microbial biomass, enzyme activity, and community composition near a copper smelter. *Ecotoxicol. Environ. Saf.* 2007, 67, 75-81.

[60]Abbas, T.; Rizwan, M.; Ali, S.; Adrees, M.; Zia-ur-Rehman, M.; Qayyum, M. F.; Ok, Y. S.; Murtaza, G. Effect of biochar on alleviation of cadmium toxicity in wheat (Triticum aestivum L.) grown on Cd-contaminated saline soil. *Environ. Sci. Poll. Res.* 2018, 25, 25668-25680.

[61]Wyszkowska, J.; Kucharski, J.; Kucharski, M. Activity of β-glucosidase; arylsulfatase and phosphatases in soil contaminated with copper. *J. Elem.* 2010, 15, 213-226.

[62]Bailey, V. L.; Fansler, S. J.; Smith, J. L.; Bolton, H. Reconciling apparent variability in effects of biochar amendment on soil enzyme activities by assay optimization. *Soil Biol. Biochem.* 2011, 43, 296-301.

[63]Zeng, X.; Xiao, Z.; Zhang, G.; Wang, A.; Li, Z.; Liu, Y.; Wang, H.; Zeng, Q.; Liang, Y.; Zou, D. Speciation and bioavailability of heavy metals in pyrolytic biochar of swine and goat manures. *J. Anal. Appl. Pyrol.* 2018, 132,82-93.

[64]Li, S.; Li, M.; Sun, H.; Li, H.; Ma, L. Lead bioavailability in different fractions of mining-and smelting-contaminated soils based on a sequential extraction and mouse kidney model. *Environ. Poll.* 2020, 262,114253.

[65]Li, J.; Li, K.; Cave, M.; Li, H. B.; Ma, L. Q. Lead bioaccessibility in 12 contaminated soils from China, correlation to lead relative bioavailability and lead in different fractions. *J. Hazard. Mater.* 2015, 295, 55-62.

[66]Gamboa-Herrera, J. A.; Ríos-Reyes, C. A.; Vargas-Fiallo, L. Y. Mercury speciation in mine tailings amended with biochar, Effects on mercury bioavailability; methylation potential and mobility. *Sci. Total Environ.* 2021, 760, 143959.

[67]Gu, Y.; Tan, X.; Cai, X.; Liu, S. Remediation of As and Cd contaminated sediment by biochars, Accompanied with the change of microbial community. *J. Environ. Chem. Eng.* 2022, 10, 106912.

[68]Yu, Z.; Qiu, W.; Wang, F.; Lei, M.; Wang, D.; Song, Z. Effects of manganese oxide-modified biochar composites on arsenic speciation and accumulation in an indica rice (*Oryza sativa* L.) cultivar. *Chemosphere.* 2017, 168, 341-349.

[69]Fei, Y. H.; Zhang, Z; Ye, Z.; Wu, Q.; Tang, Y.; Xiao, T. Roles of soluble minerals in Cd sorption onto rice straw biochar. *J. Environ. Sci.* 2022, 113, 64-71.

[70]Mohamed, B. A.; Ellis, N.; Kim, C. S.; Bi, X. The role of tailored biochar in increasing plant growth; and reducing bioavailability, phytotoxicity, and uptake of heavy metals in contaminated soil. *Environ. Poll.* 2017, 230, 329-338.

[71]Liu, B.; Huang, Q.; Su, Y.; Sun, L.; Wu, T.; Wang, G.; Kelly, R. M. Rice busk biochar treatment to cobalt-polluted fluvo-aquic soil, Speciation and enzyme activities. *Ecotoxicology*. 2019, 28 (10), 1220-1231.

[72]Chapman, E.; Dave, G.; Murimboh, J. D. A review of metal (Pb and Zn) sensitive and pH tolerant bioassay organisms for risk screening of metal contaminated acidic soils. *Environ. Pollut.* 2013, 179, 326-342.

[73]Karami, N.; Clemente, R.; Moreno-Jiménez, E.; Lepp, N.W.; Beesley, L. Efficiency of green waste compost and biochar soil amendments for reducinglead and copper mobility and uptake to ryegrass. *J. Hazard Mater.* 2011,191, 41-48.

[74]Bashir, S.; Hussain, Q.; Shaaban, M.; Hu, H. Efficiency and surface characterization of different plant derived biochar for cadmium (Cd) mobility, bioaccessibility and bioavailability to Chinese cabbage in highly contaminated soil. *Chemosphere.* 2018, 211, 632-639.

[75]Wang, S.; Wu, W.; Liu, F.; Liao, R.; Hu, Y. Accumulation of heavy metals in soil-crop systems, a review for wheat and corn. *Environ. Sci. Poll. Res. Int.* 2017, 24, 15209-15225.

[76]Yu, Y.; Yang, Y.; Zhang, C.; Yi, J.; An, S.; Wang, D. Effect of sewage sludge compost products application on total mercury and methylmercury in soil and plants. *Environ. Sci.* 2017, 38(1), 405-411.

[77]Zama, E. F.; Reid, B. J.; Sun, G. X.; Yuan, H. Y.; Li, X. M.; Zhu, Y. G. Silicon (Si) biochar for the mitigation of arsenic (As) bioaccumulation in spinach (*Spinacia oleracean*) and improvement in the plant growth. *J. Clean Prod.* 2018, 189, 386-395.

本文曾发表于2022年《Sustainability》第14卷

Experimental Study of Al-Modified Zeolite with Oxygen Nanobubbles in Repairing Black Odorous Sediments in River Channels

Chao Guo[1,2,3,4], Huanyuan Wang[1,2,3,4], Yulu Wei[1,2,3,4], Jiake Li[5], Biao Peng[1,2,3,4] and Shu Xiaoxiao[1,2,3,4]

(1. Shaanxi Provincial Land Engineering Construction Group Co., Ltd. Shaanxi Xi'an 710071, China; 2. Institute of Land Engineering and Technology, Shaanxi Provincial Land Engineering Construction Group Co., Ltd. Shaanxi Xi'an 710071, China; 3. Key Laboratory of Degraded and Unused Land Consolidation Engineering, the Ministry of Natural Resources., Shaanxi Xi'an 710071, China; 4. Shaanxi Provincial Land Consolidation Engineering Technology Research Center., Shaanxi Xi'an 710071, China; 5. State Key Laboratory Base of Eco-hydraulic Engineering in Arid Area, Xi'an University of Technology, Xi'an 710048, China)

【**Abstract**】 As an extreme phenomenon of water pollution, black odorous water not only causes ecological damage, but also severely restricts urban development. At present, the in-situ remediation technology of sediment from river channels is still undeveloped, and there are many bottlenecks in the key technologies of sediment pollution control and ecological restoration. In this study, three experimental tanks are used to explore the restoration effect of Al-modified zeolite with oxygen nanobubbles on black odorous sediment from Shichuan river. One of the tanks plants *Typha orientalis* and *Canna indica L.* (TC), the other plants the same plants and adds Al-modified zeolite with oxygen nanobubbles (TC+AMZON), and the last one is used as comparison test (CS). The results show that Nitrogen(N) and phosphorus(P) in the sediment are released violently into the surrounding water. However, TC+AMZON can effectively inhibit the release of P. The released amount of soluble reactive phosphorus (SRP) of pore water in the sediment reaches the maximum at 40 d, and it is 122.97% and 74.32% bigger in TC and CS than that in TC+AMZON. However, the released amount of total phosphorus (TP) reaches the maximum at 70 d, and it is 260.14% and 218.23% bigger in TC and CS than that in TC+AMZON, respectively. TC+AMZON can significantly increase the dissolved oxygen (DO) and oxidation-reduction potential(ORP) of pore water in the sediment in the early stage of the test. On the 0 d, the DO content in TC+AMZON reaches 10.6 mg/L, and it is 112.0% and 178.95% bigger than that in TC and CS, respectively. The change laws of ORP in the sediment is consistent with the DO. (3) TC+AMZON can significantly improve the transparency and reduce the content of $chlorophyll_a$ of the upper water, and can slightly reduce the content of N and Poverlying water. The transparency of the overlying water in TC+AMZON is increased by 130.76% and 58.73%, and the content of $chlorophyll_a$ is decreased by 55.6% and 50.0% compared with TC and CS.

【**Keywords**】 Al-modified Zeolite; Oxygen Nanoparticles; Black odorous sediments; Repairing

1 Introduction

With the development of industrial and agricultural production and the acceleration of urbanization, more and more rivers have been polluted[1]. Rivers become "black and stinky", and they have a serious impact on the lives of residents and the surrounding environment. As an extreme phenomenon of water pollution, black odorous water not only causes ecological damage, but also severely restricts urban development[2-3]. Remediation of black odorous water has become one of the most difficult environmental problems in water protection[4].

Many black odorous water remediation projects appear a strange phenomenon of "remediation every year, black odorous every year". It is mainly due to the fact that people usually focus on the process of remediation and short-term effects, but ignore the good water quality maintenance and long-term effects[5-6]. Therefore, how to choose the appropriate treatment technology is the main problem to be solved urgently[7]. In addition, preventing water deteriorate repeatedly and achieving the goal of long-term maintenance of good water quality are the key to completely eliminate the urban black odorous water after the remediation[8-9]. At this stage, the main technical bottlenecks are the control of the sediments pollution in the river channel and the reconstruction of healthy river ecosystems.

The primary task is to recognize the causes and take the effective control measures. The main causes of the urban rivers are the input of exogenous pollutants, such as industrial wastewater, domestic sewage, garbage, and non-point source pollution on both sides of the river. However, the released pollutants from river sediments accelerate the deterioration of water quality[10]. The environmental factors mainly include organic pollutants, nitrogen, phosphorus, iron, manganese, sulfide and other pollutants. The lack of oxygen and poor fluidity of water can further accelerate the black odorous. A lot of effective work on the control of exogenous pollutants have been done, and it can be achieved by improving the sewage pipeline network, increasing the efficiency of rain water and sewage diversion, and centralizing sewage treatment and discharging. For example, upgrading or shutting down the key sewage enterprises, improving the efficiency of garbage removal and transportation, etc. However, there are still many bottlenecks of the technology for the treatment of endogenous sediment pollution. Therefore, chemical inactivation using phosphorus (P) and nitrogen (N) binders has been increasingly used for the management of internal nutrients in the sediment of degraded bodies of water.[11-12]

Zeolite is an aluminosilicate mineral widely existing in nature. The main components are SiO_2, Al_2O_3 and iron oxides. Natural zeolite is one of the abundant reserves characterized by unique pore structure, large specific surface area, strong surface adsorption and ion exchange. Therefore, it is widely used in the adsorption of ammonia nitrogen and heavy metal ions[11]. However, the removal effect of natural zeolite on anions such as phosphate is poor, and it is affected by many factors such as temperature, particle size, etc.[12-13]. Modification of natural zeolite by different methods can effectively increase the removal efficiency of anions such as phosphate. Lin, J. *et al.*[14] showed that natural zeolite/hydrochloric acid modified zeolite and calcite composite can effectively reduce the release of N and P in sediments. The phosphorus passivation effect is increased by the dosage and the small particles, and the effect of adding zeolite is better than that of the calcite[15-16]. Gibbs, M. *et al.*[17-18] compared the effect of four passivators of modified zeolite Z2G1, phoslock, alum and allophane. The results show that Z2G1 can effectively inhibit the release of P in sediments, and heavy metal ions do not release into the surrounding water from sediments. In additon, most of the oxygen in the river sediment is consumed by reducing substances, and the amount of oxygen reaches the sediment-water interface is very small. Shi, M. *et al.*[19] showed that algae induced anoxia/hypoxia can be reduced or reversed after oxygen nanobubbles (Diameter is less than 1μm) are loaded the zeolite micropores and delivered to the anoxic sediment. The manipulation of microbial processes using the surface Oxygen nanobubbles potentially served as oxygen suppliers. Therefore, oxygen nanobubbles have the advantages of good stability and high oxygen mass transfer rate, which provide great potential for the research and development of precise oxygenation technology at the sediment-water interface of river sediments.

In this study, the Al-modified zeolite with oxygen nanobubbles is used to control the black odorous sediment pollutants in the river channel through three groups of pilot test, and aquatic plants are used for restoration. It will provide a new technology for the remediation of black odorous water in urban river channels.

2 Materials and methods

2.1 Materials

2.1.1 Al-modified zeolite with oxygen nanobubbles

Natural zeolite is modified using aluminum salt, and an efficient, low-risk, low-cost and in-situ passivation material for sediment pollution is developed. Namely Al-modified zeolite (natural zeolite 30% + aluminum salt 15% + Stone powder 55%). The main component of stone powder is $CaCO_3$, and it has a large specific surface area. After the zeolite is modified by aluminum salt, its Al^{3+} is hydrolyzed to form a positively charged $Al(OH)_3$ colloid. Oxygen nanobubbles are loaded zeolite micropores, and it could effectively increase the DO concentration at the sediment-water interface. Thus significantly inhibit the release of phosphorus from the sediment. Al-modified zeolite with oxygen nanobubbles oxidize the active Fe (Fe^{2+}) in the sediment to iron oxide (Fe^{3+}), and then form an iron oxide layer on the surface of the sediment, which can inhibit the release of Fe-P, and enhance the sediment P fixed capacity.

2.1.2 Aquatic plants

The aquatic plants of *Typha orientalis* and *Canna indica L* are used, and they have excellent pollutant adsorption ability. Physical-biological-microbial methods are used to optimize the structure of the local water ecosystem, and to improve the primary productivity of water and convert nutrients in water into plant tissues.

2.2 Test device

In March 2021, three test tanks with length×width×height = 2 m×1 m×1 m are made in Fuping pilot test base of Shaanxi Provincial Land Engineering Construction Group Co., Ltd. Black and odorous mud with a thickness about 10 cm is laid in the bottom, and the mud is taken from Shichuan River, Fuping County, Shaanxi Province. Then inject water about 20 cm, and the water needs to be placed for about 10 days to discharge the chlorine gas. A week later, one of the tanks plant aquatic plants such as *Typha orientalis* and *Canna indica L*, then add Al-modified zeolite with oxygen nanobubbles with a thickness of about 2 cm, and it is marked as TC+AMZON. The particle size of the zeolite is about 2~3 mm. The other tank is only planted *Typha orientalis* and *Canna indica L*, and it is marked as TC. The last one used as comparison test, and it is marked as CS. The test process is shown in Fig. 1. The test device is completed in May 2021.

2.3 Sample collection

On May 16th, 2021, Al-modified zeolite with oxygen nanobubbles is added into the device. On May 26th, the overlying water and sediment samples are collected. The water samples are taken from 5 cm below the water surface, and the sediment samples are taken within 2 cm below the sediment surface. After the samples are collected, the Al-modified zeolite is supplemented at the sampling point. Water and sediment samples are collected in three tanks every 10 days in the early stage, and they are collected every 15 or 20 days in the later period. The collected water and sediment samples are stored at −4 ℃, and the analysis is completed within one week. The test indicators and methods are shown in Table 1.

(a) Sediment drying (b) Sediment filling

(c) Scene drawing

Fig. 1 Test process Figures

2.4 Test methods

Table 1 Test indicators and methods

Indicators	Test methods
DO	Test in-siut
COD	Determination of the chemical oxygen demand-Dichrom method
Transparency	Methods of monitoring and analysis of water and wastewater[20]
NH_4^+-N	Air and exhaust gas-Determination of ammonia-Nessler' s reagent spetcrophotomatry
TN	Alkaline potassium persulfate digestion UV spectrophotometric method
TP	Ammonium molybdate spectrophotometric method
SRP	Analysis of water used in boiler and cooling system-Determination of phosphate
$Chlorophyll_a$	Determination of $chlorophyll_a$-spectrophotometric method
ORP	Test in-siut

3 Results and discussion

3.1 N and P of pore water in sediment

The contents of TP and SRP of pore water in the sediment in CS and TC show an increasing trend before 130 d, and they all maintain a high level of P content(Fig.2). It indicates that the P in the sediment releases into the surrounding water with the increase of temperature. The released amount of SRP reaches the maximum at 40 d, and it is 122.97% and 74.32% bigger in CS and TC than that in TC+AMZON at this

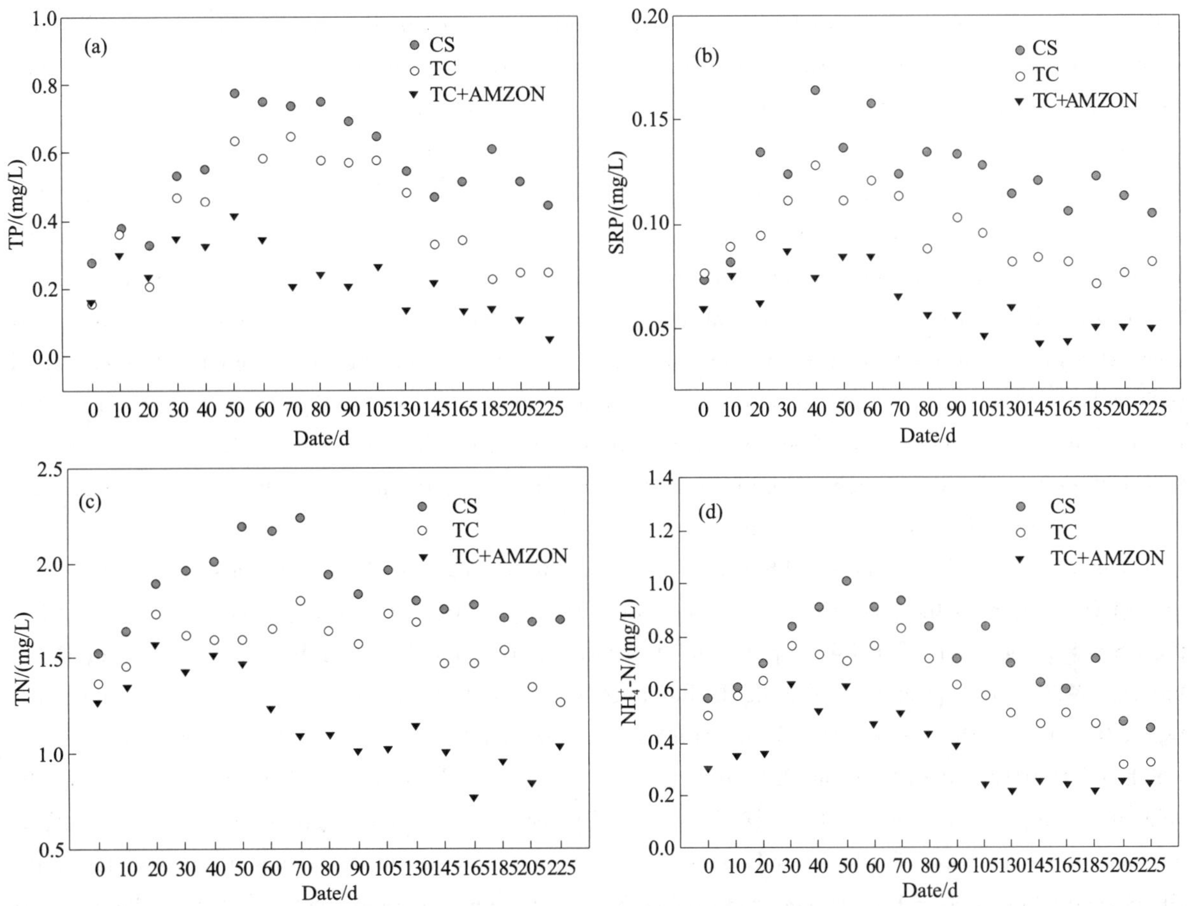

Fig. 2 N and P of pore water in sediment. (a) TP of pore water in sediment; (b) SRP of pore water in sediment; (c) TN of pore water in sediment; (d) NH_4^+-N of pore water in sediment

time, respectively. The released amount of TP reaches the maximum at 70 d, and it is 260.14% and 218.23% bigger in the CS and TC than that in TC+AMZON, respectively. The contents of TP and SRP in TC+AMZON increased slightly before 70 days, but both are less than those in CS and TC. The main reason is that during the anaerobic period in summer, Fe-P in the sediment is reduced to form a large amount of SRP, resulting in the strong release of P in the sediment[21]. However, the addition of the Al-modified zeolite with oxygen nanobubbles makes the sediment locally aerobic, and oxidizes Fe (Fe^{2+}) in the sediment to iron oxide (Fe^{3+}). Then it forms an iron oxide passivation layer, and Fe^{2+} is converted into Fe^{3+}. SRP/Fe-P is strictly fixed in the sediment, and effectively reduces the release of P. However, CS and TC is under the anaerobic conditions without the addition of Al-modified zeolite with oxygen nanobubbles, and it has a very high concentration gradient of Fe-P on the sediment surface, so it leads to a strong release of P in the sediment. When the Al-modified zeolite with oxygen nanobubbles in TC+AMZON is added to the surface of the sediment, the nanobubbles continuously releases oxygen and forms an oxide layer on the surface of the sediment, which not only reduces the concentration of P and Fe but also greatly reduces the concentration gradient of P on the sediment surface. Thereby the release of P in the sediment is inhibited. Fortunately, the oxygen nanobubble on Al-modified zeolite plays an important role. In addition, zeolite has a large specific surface area and strong adsorption[22-23]. The natural zeolite is rich in Al^{3+} after modification with aluminum salt, and the Al^{3+} is hydrolyzed to form $Al(OH)_3$ colloid, which plays an important role for the adsorption of PO_4^{3-} in the water[24-25]. It can reduce the release of P in sediment and improve the overlying water quality. Therefore, it achieves the dual purpose of controlling P pollution and preventing eutrophication of overlying water. At the same time, the available P of the pore water is absorbed by the aquatic plants, so the P in the sediment is effectively removed. The river sediment in-situ passivation with phytoremediation is a P pollutant treatment technology with high efficiency, low cost, environmental protection and aesthetics[26].

The TN and NH_4^+-N contents of the pore water in CS and TC sediments also show a trend of first increasing and then decreasing. The influence of TC+AMZON on TN and NH_4^+-N contents is great, and they are significantly lower than those in CS and TC. It shows that the sediment also releases the N to the surrounding water, and Al-modified zeolite with oxygen nanobubbles inhibits the N release. Studies have shown that nanobubbles have negative charges at the gas-liquid interface, which can interact with positively charged pollutants in water, and the free radicals and vibration waves generated can promote the removal of pollutants[27]. Therefore, the pollutants discharged from urban non-point source pollution to rivers and lakes can be repaired by in-situ Al-modified zeolite + phytoremediation technology in the sediment, which can achieve the effect of pollutant control and beautify the city. It is the most effective technical means at present.

3.2 DO and ORP of pore water in sediment

The addition of Al-modified zeolite with oxygen nanobubbles can significantly increase the dissolved oxygen(DO) content of the pore water in the sediment(Fig. 3). During the experimental period, the DO in the CS is kept between 2.9 and 4.3 mg/L, and it is varied from 5.3 to 7.5 mg/L in TC. However, the DO in the TC+AMZON is kept at 6.2~10.6 mg/L. Before 60 days, the DO content in TC+AMZON is significantly bigger than that in the CS and TC. On the 0th day, the DO content in TC+AMZON reaches 10.6 mg/L, which is 178.95% and 112.0% bigger than that in the CS and TC. With the experimental continual, the DO content of the pore water in the TC+AMZON is gradually decreased, which is mainly caused by the continuous oxygen consumption of reducing substances and microbial activities in the sediment. 60 days later, the DO concentration in TC+AMZON decreases below 5.5 mg/L, and it is consistent with that of the TC, and gradually becomes stable. The results show that the Al-modified zeolite with oxygen nanobubble has a good ability to increase the oxygen at the sediment-water interface. Therefore, the addition of Al-modified zeolite with oxygen nanobubbles is beneficial to increase the DO content in the pore water of the sediment.

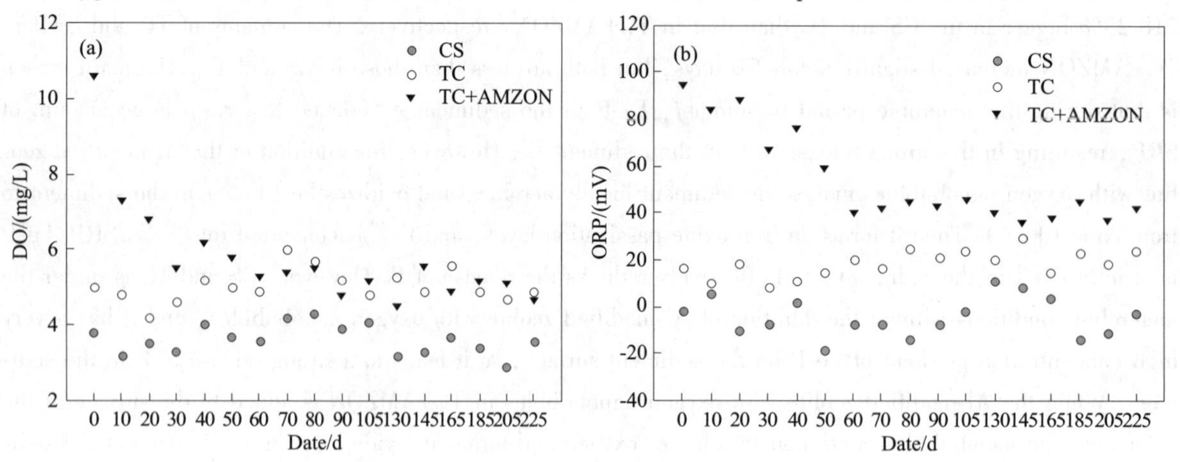

Fig. 3 DO and ORP of pore water in sediment. (a) DO of pore water in sediment; (b) ORP of pore water in sediment

The oxidation-reduction potential (ORP) on the sediment surface of the CS is varied from −35 to 20 mV, and it is kept between 5 and 35 mV in the TC, however the ORP in TC+AMZON is varied from 12 to 95 mV. Therefore, the ORP of pore water in the TC+AMZON is significantly increased. Moreover, the change process of ORP in the sediment is consistent with that of DO, and it indicates that the distribution of DO content in the sediment affects the level of ORP. Which is consistent with the research results of Shi M.

et al.[19]. The transformation and diffusion of most dissolved substances in sediments are affected by their ORP.

3.3 N, P and COD in overlying water

It can be seen from Fig. 4 that the TP content in the overlying water decreases with the experimental time. The TP content in the three tanks is in the order of CS > TC > TC+AMZON, and TP content in the TC and TC+AMZON is all smaller than that in the CS. It shows that the release of P in the sediment will not increase the P content in the overlying water. The main reason is that the Al-zeolite particles rapidly adsorb the dissolved active P in the sediment. Relevant studies have shown that aluminum ions of modified zeolites are hydrolyzed to form positively charged $Al(OH)_3$ colloids, which can strongly adsorb negatively charged bacteria and ions in water[28-29], such as PO_4^{3-}. At the same time, $Al(OH)_3$ colloids is the suspended matter with light weight and is not easy to precipitate[30-31]. But the $Al(OH)_3$ can increase its mass after adsorbing ions in water, and gradually settle to the bottom of the water, reducing the possibility of resuspension. In addition, the hydrolyzed product $Al(OH)_3$ provides bridging adsorption to adsorb suspended solids in water[32-33]. Al^{3+} is hydrolyzed to form a high molecular polymer with a linear structure, one side of the high molecular polymer can adsorb a particle far away, and the other side extends into the water to absorb another particle. The particle is bridged by the polymer adsorption that make the particles gradually become bigger and bigger[34]. With the increasing of the particle adsorbed by Al^{3+}, it gradually moves to the bottom of the water, and it can absorb the suspended particles in the water during the sinking process[35]. After the colloid settles to the surface of the sediment, a covering layer is formed on the surface of the sediment to prevent the pollutants in the sediment from being released to the overlying water.

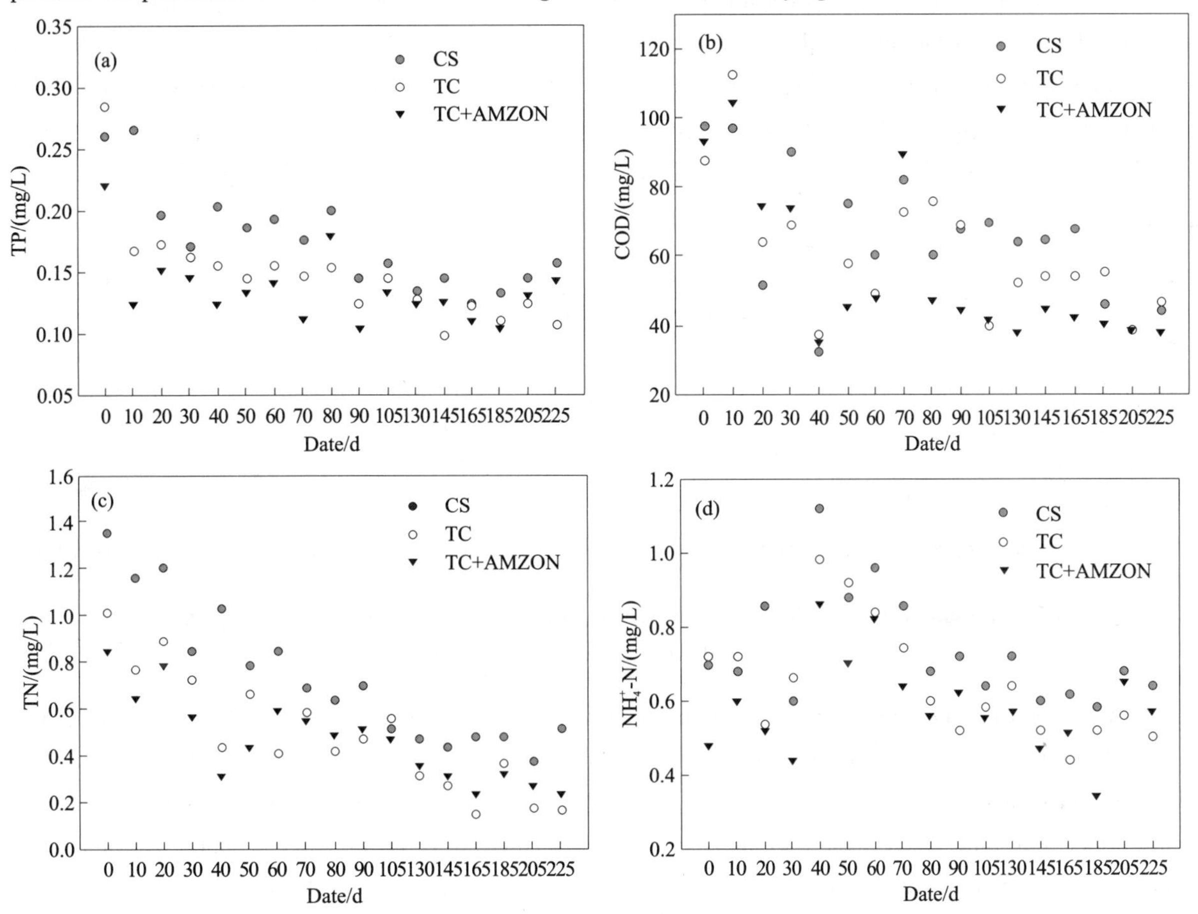

Fig. 4 N, P and COD in overlying water. (a) TP in overlying water; (b) COD in overlying water; (c) TN in overlying water; (d) NH_4^+-N in overlying water

The TN content in the overlying water shows a decreasing trend with the test time, and the NH_4^+-N shows a trend of increasing first and then decreasing gradually. The TN and NH_4^+-N contents in the TC and TC+AMZON is all smaller than that in the CS. Before 60 days, the contents of TN and NH_4^+-N in the TC+AMZON are slightly lower than those in TC, and the differences gradually decreased 60 days later. The main reason is that zeolite has an adsorption effect on NH_4^+-N, which reduces the N content in the overlying water. In addition, the absorption of available N by aquatic plants can also reduce the N content in water. Therefore, when using Al-modified zeolite to remediate pollutants in sediment, it is necessary to plant aquatic plants to absorb pollutants in sediment and water[36].

Al-modified zeolite has little effect on COD in the overlying water. In the early stage of the experiment, there is minor differences of COD in the overlying water among the three tanks. In the later stage, the COD content of the overlying water in the TC+AMZON and TC is slightly lower than those in CS. It is mainly because the adsorption of soluble organic pollutants by plants that reduce the COD content. Therefore, the application of Al-modified zeolite has no obvious improvement effect on COD in black odorous water.

3.4 Transparency and chlorophylla in overlying water

It can be seen from Fig. 5 that the transparency of the overlying water in the TC+AMZON and TC gradually increases with the experimental time, while it shows a trend of decreasing first, then remains stable in the later period in the CS. And the difference become bigger and bigger 60 days later. According to statistics, the transparency in the TC+AMZON remains at 17.6~30 cm 60 days later, and it varies from 15.5 to 22.4 cm in the TC. However, the transparency in the CS remains at 11.6~15.7 cm. Thus compared with the CS and TC, the transparency of the overlying water in the TC+AMZON is increased by 130.76% and 58.73%, respectively at the sampling same. Therefore, the application of Al-modified zeolite with oxygen nanobubbles has a significant effect on water purification.

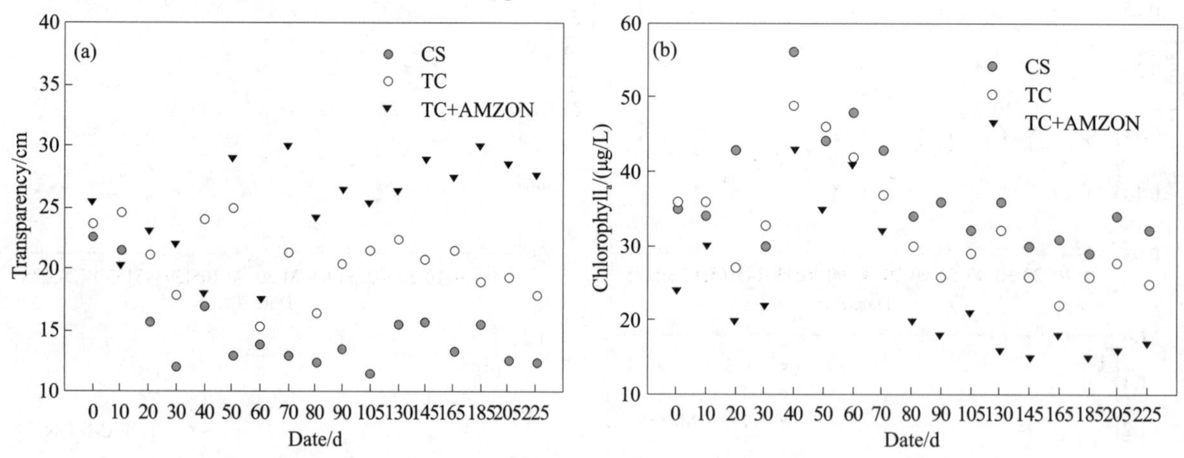

Fig. 5 Transparency and chlorophyll$_a$ in overlying water. (a) Transparency in overlying water; (b) Chlorophyll$_a$ in overlying water

The content of chlorophyll$_a$ in the overlying water generally increases at first and then decreased, and keeps stable finally. Tt begins to increase 30 days later, and reach the maximum value 50 days later. The main reason is that the experiment starts at the end of May. 30 days later, the temperature is high. There are more prokaryotic blue-green algae (cyanobacteria) and eukaryotic algae in the test tanks except for the green plants. The algae can synthesize some organic substances through photosynthesis, which converts light energy into chemical energy. Thereby increasing the content of chlorophyll$_a$ in water[37-38]. 80 days later, the content of chlorophyll$_a$ of the overlying water in the TC+AMZON is significantly less than those in TC and

CS. And the content of chlorophyll$_a$ in the TC+AMZON remaines between 15 and 21 μg/L, but it varies from 22 to 32 μg/L and from 29 to 36 μg/L in the TC and CS, respectively. The content of chlorophyll$_a$ in the TC+AMZON is decreased by 50.0% and 55.56% compared with TC and CS at the same sampling time. It shows that the content of chlorophyll$_a$ in water can be reduced by the addition of Al-modified zeolite with oxygen nanobubbles. 105 days later, the chlorophyll$_a$ content in the three tanks gradually becomes stable.

4 Conclusion

Al-modified zeolite with oxygen nanobubbles is used to repair the river sediments in this study, and it demonstrates that:

(1) Sediment in river channel releases P strongly to the surrounding water, but N has a certain release. However, the addition of Al-modified zeolite with oxygen nanobubbles can inhibit the release of P and N, and the effect is very obvious.

(2) The addition of Al-modified zeolite with oxygen nanobubbles can increase the DO and ORP content of pore water significantly in the early stage of the test. 60 days later, the DO content in TC+AMZON is reduced below 5.5 mg/L. The ORP variation of the pore water is consistent with the DO.

(3) P and N released from sediment does not increase their content of overlying water. There is minor difference of TN, TP and NH_4^+-N contents overlying waterin TC+AMZON and TC, and they are all smaller than those in CS. The addition of Al-modified zeolite with oxygen nanobubbles has small effect on COD of the overlying water. In the early stage of the experiment, the difference of COD content in the three tanks is small, however, it is less in the TC+AMZON and TC than that in the CS in later stage.

(4) The transparency of the overlying water is significantly improved by the addition of Al-modified zeolite with oxygen nanobubbles. 40 days later, the difference of the transparency in TC+AMZON is more and more obvious compared with the TC and CS. Moreover, the addition of Al-modified zeolite with oxygen nanobubbles can reduce the content of chlorophyll$_a$ of the overlying water. 80 days later, the chlorophyll$_a$ content overlying waterin the TC+AMZON is significantly less than that in the TC and CS.

Author contributions statement

Guo C., Wang H. Y. and Li J. K. conceived the experiments and analyzed the results, Wei Y. L. and Peng B. conducted the experiments, Shu X. X. analysed the date. All authors reviewed the manuscript.

Acknowledgments

This research is financially supported by the Key Research and Development Program of Shaanxi (2022ZDLSF-06-04 and 2020SF-420), the project of Shaanxi Provence Land Engineering Construction Group (DJNY2022-26) and the National Natural Science Foundation of China (No. 51879215).

Declaration of date

The authors declare that the datasets generated during and analyzed during the current study are not publicly available due to the reasons that the data is confidential, and it is the basis for further research, but are available from the corresponding author on reasonable request.

References

[1] Liu, C.; Shen, Q.; Zhou, Q.; Fan, C.; Shao, S. Precontrol of algae-induced black blooms through sediment dredging at appropriate depth in a typical eutrophic shallow lake. *Ecol. Eng.* 2015, 77, 139-145.

[2] Chai, X. L.; Wu, B. R.; Xu, Z. S.; Yang, N.; Song, L.; Mai, J.; Chen, Y.; Dai, X. Ecosystem activation system (EAS) technology for remediation of eutrophic freshwater. *Sci Rep-Uk* 2017, 7, 4818.

[3] Lalley, J.; Han, C.; Li, X.; Dionysiou, D.D.; Nadagouda, M. N. Phosphate adsorption using modified iron oxide-based sorbents in lake water: kinetics, equilibrium, and column tests. *Chem. Eng. J.* 2016, 284, 1386-1396.

[4]He, D. F.; Chen, R. R.; Zhu, E. H.; Chen, N.; Yang, B.; Shi, H. H.; Huang M. S. Toxicity bioassays for water from black-odor rivers in Wenzhou, China. *Environ. Sci. Pollut. R.* 2015, 22, 1731-1741.

[5]Sheng, Y.; Qu, Y.; Ding, C.; Yao, Q. A. combined application of different engineering and biological techniques to remediate a heavily polluted river. *Ecol. Eng.* 2013, 57, 1-7.

[6]Feng, Z. Y.; Fan, C. X.; Huang, W. Y.; Ding, S. Microorganisms and typical organic matter responsible for lacustrine "black bloom". *Sci. Total. Environ.* 2014, 470-471, 1-8.

[7]Wang, G. F.; Li, X. N.; Fang, Y.; Huang, R. Analysis on the formation condition of the alga-induced odorous black water agglomerate. *Saudi. J. Biol. Sci.* 2014, 21, 597-604.

[8]Suurnäkki, S.; Gomez-Saez, G. V.; Rantala-Ylinen, A;. Jokela, J.; Fewer, D. P.; Sivonen, K. Identification of geosmin and 2-methylisoborneol in cyanobacteria and molecular detection methods for the producers of these compounds. *Water Res.* 2015, 68, 56-66.

[9]Oh, H. S.; Lee, C. S.; Srivastava, A.; Oh,H. M.; Ahn, C. Y. Effects of environmental factors on cyanobacterial production of odorous compounds: geosmin and 2-methylisoborneol. *J. Microbiol Biotechnol* 2017, 27, 1316-1323.

[10]Sugiura, N.; Utsumi, M.; Wei, B.; Iwami, N.; Okano, K.; Kawauchi, Y.; Maekawa, T. Assessment for the complicated occurrence of nuisance odours from phytoplankton and environmental factors in a eutrophic lake. *Lakes & Reservoirs* 2004, 9,195-201.

[11]Yin, H. B.; Wang, J. F.; Zhang, R. Y.; Tang, W. Y. Performance of physical and chemical methods in the co-reduction of internal phosphorus and nitrogen loading from the sediment of a black odorous river. *Sci. Total. Environ* 2019,663, 68-77.

[12]Uzun, O.; Gokalp, Z.; Irik, H. A.; Varol,I. S.; Kanarya, F. O. Zeolite and pumice-amended mixtures to improve phosphorus removal efficiency of substrate materials from wastewaters. *J. Clean. Prod.* 2021, 317, 128444.

[13]Obiri-Nyarko, F.; Kwiatkowska-Malina, J.; Malina, M.; Wołowiec K. Assessment of zeolite and compost-zeolite mixture as permeable reactive materials for the removal of lead from a model acidic groundwater. *J. Contam. Hydrol.* 2020, 229,103597.

[14]Kostyniuk, A.; Bajec, D.; Likozar, B. Catalytic Hydrogenation, Hydrocracking and Isomerization Reactions of Biomass Tar Model Compound Mixture over Ni-modified Zeolite Catalysts in Packed Bed Reactor. *Renewable Energy* 2020, 167, 409-424.

[15]Lin, J. W.; Zhan, Y. H.; Zhu, Z. Evaluation of sediment capping with active barrier systems (ABS) using calcite/zeolite mixtures to simultaneously manage phosphorus and ammonium release. *Sci. Total. Environ.* 2011, 409, 638-646.

[16]Shahmansouri, A. A.; Bengar, H. A.; AzariJafari, H. Life cycle assessment of eco-friendly concrete mixtures incorporating natural zeolite in sulfate-aggressive environment. *CONSTR Build. Mater.* 2021, 268, 121136.

[17]Messaadi, C.; Ghrib, T.; Ghrib, M.; Al-Otaibi, A.L.; Glid, M.; Ezzaouïa, H. Investigation of the percentage and the compacting pressure effect on the structural, optical and thermal properties of alumina-zeolite mixture. *Results Phys.* 2018, 8, 422-428.

[18]Gibbs, M.; Ozkundakci, D. Effects of a modified zeolite on P and N processes and fluxes across the lake sediment-water interface using core incubations. *Hydrobiologia* 2011, 661, 21-35.

[19]Gibbs, M. M.; Hickey, C. W.; Ozkundakci, D. Sustainability assessment and comparison of efficacy of four P-inactivation a-gents for managing internal phosphorus loads in lakes: sediment incubations. *Hydrobiologia* 2011, 658, 253-275.

[20]Shi, W. Q.; Pan, G.; Chen Q. W.; Song, L. R.; Zhu, L.; Ji. X. N. Hypoxia Remediation and Methane Emission Manipulation Using Surface Oxygen Nanobubbles. *Environ. Sci. Technol.* 2018, 52, 8712-8717.

[21]Wei, F. S. Methods of monitoring and analysis of water and wastewater. *China Environmental Science Press* 2002, 231-232.

[22]Ding, S.; Sun, Q.; Xu, D.; Jia, F.; He, X.; Zhang C. High-resolution simultaneous measurements of dissolved reactive phosphorus and dissolved sulfide: the first observation of their simultaneous release in sediments. *Environ. Sci. Technol.* 2012, 46, 8297-8304.

[23]Schelske, C. L. Eutrophication: Focus on Phosphorus. *Science* 2009, 324,722.

[24]Gu, W.; Xie, Q.; Xing, M.; Wu, D. Enhanced adsorption of phosphate onto zinc ferrite by incorporating cerium.

Chemical Engineering Research and Design 2017, 117, 706-714.

[25]Xu, S. T.; Zhang, W. B.; Gao, L.; Wei, L. Q. Dynamic Changes of Phosphorus, Iron and Sulfur Concentrations in Water during the Decomposition of Green Tide Algae. *Ecology and Environmental Sciences* 2019, 28, 376-384.

[26]Zhu, G. R.; Cao, T.; Zhang, M.; Ni, L. Y.; Zhang, X. L. Fertile sediment and ammonium enrichment decrease the growth and biomechanical strength of submersed macrophyte Myriophyllum spicatum in an experiment. *Hydrobiologia* 2014, 727, 109-120.

[27]Chen, J. Z.; Meng, S. L.; Hu, G. D.; Qu, J. H.; Fan, L. M. Effect of ipomoea aquatic cultivation on artificial floating rafts on water quality of intensive aquaculture ponds. *Journal of Ecology and Rural Environment* 2010, 26, 155-159.

[28]Li, P.; Takahashi, M.; Chiba, K. Degradation of phenol by the collapse of microbubbles. *Chemosphere* 2009, 75, 1371-1375.

[29]Cai, L.; Zheng, S. W.; Shen, Y. J.; Zheng, G. D.; Liu, H. T.; Wu, Z. Y. Complete genome sequence provides insights into the biodrying-related microbial function of Bacillus thermoamylovorans isolated from sewage sludge biodrying material. *Bioresource Technolo.* 2018, 260, 141-149.

[30]Hsu, L. C.; Tzou, Y. M.; Chiang, P. N.; Fu, W. M.; Wang, M. K.; Teah, H. Y.; Liu, Y. T. Adsorption mechanisms of chromate and phosphate on hydrotalcite: A combination of macroscopic and spectroscopic studies. *Environmental Pollution* 2019, 247, 180-187.

[31]Ding, S.; Wang, Y.; Wang, D.; Li, Y. Y.; Gong, M. D.; Zhang, C. S. In situ, high-resolution evidence for iron-coupled mobilization of phosphorus in sediments. *Sci Rep-Uk* 2016, 6, 24341.

[32]Han, C.; Ding, S.; Yao, L.; Shen, Q. S.; Zhu, C. G.; Wang, Y.; Xu, D. Dynamics of phosphorus-iron-sulfur at the sediment-water interface influenced by algae blooms decomposition. *J. Hazard. Mater.* 2015, 300, 329-337.

[33]Rozan, T. F.; Taillefert, M.; Trouwborst, R. E.; Glazer, B.T.; Ma, S.; Herszage, J.; Valdes, L. M.; Iii, P. G. W. L. Iron-sulfurphosphorus cycling in the sediments of a shallow coastal bay: Implications for sediment nutrient release and benthic macroalgal blooms. *Limnol. Oceanogr* 2002, 47, 1346-1354.

[34]Xu, D.; Chen, Y.; Ding, S.; Sun, Q.; Wang, Y.; Zhang, C. S. Diffusive gradients in thin films technique equipped with a mixed binding gel for simultaneous measurements of dissolved reactive phosphorus and dissolved iron. *Environ. Sci. Technol.* 2013, 47, 10477-10484.

[35]Zhang, C. S.; Ding, S. M.; Xu, D.; Tang, Y.; Ming, H. W. Bioavailability assessment of phosphorus and metals in soils and sediments: a review of diffusive gradients in thin films (DGT). *Environmental Monitoring and Assessment* 2014, 186, 7367-7378.

[36]Chen, J. Z.; Meng, S. L.; Hu, G. D.; Qu, J. H.; Fan, L. M. Effect of ipomoea aquatic cultivation on artificial floating rafts on water quality of intensive aquaculture ponds. *Journal of Ecology and Rural Environment* 2010, 26, 155-159.

[37]Lu, H. B.; Wang, H. H.; Lu, S.Y.; Li, J X.; Wamg T. Response mechanism of typical wetland plants and removal of water pollutants under different levofloxacin concentration, *Ecol. Eng.* 2020, 158, 106023.

[38]Wang, Z.; Xu, Y.; Shao, J.; Wang, J.; Li, R.; Stal, L. J. Genes Associated with 2-Methylisoborneol Biosynthesis in Cyanobacteria: Isolation, Characterization, and Expression in Response to Light. *PLoS ONE* 2011, 6, 18665.

[39]He, D. F.; Chen, R. R.; Zhu, E. H.; Chen, N.; Yang, B.; Shi, H. H.; Huang M. S. Toxicity bioassays for water from black-odor rivers in Wenzhou, China. *Environ. Sci. Technol.* 2015, 22, 1731-1741.

本文曾发表于 2022 年《Water》第 14 卷

Effects of Soft Rock on Soil Properties and Bacterial Community in Mu us Sandy Land, China

Zhen Guo[1,2], Wei Hui[1], Juan Li[1,2], Chenxi Yang[1,2], Haiou Zhang[1,2] and Huanyuan Wang[1,2]

(1. Shaanxi Provincial Land Engineering Construction Group Co., Ltd., Shaanxi, Xi'an, China;
2. Institute of Land Engineering and Technology, Shaanxi Provincial Land Engineering Construction Group Co., Ltd., Shaanxi, Xi'an, China)

【**Abstract**】As a new material for the improvement of Mu Us Sandy Land, soft rock can be used for wind prevention and sand fixation, which is of great significance to ecological restoration and cultivated land replenishment in desert areas. Four treatments with the compound ratio of soft rock and sand of 0∶1 (CK), 1∶5 (P1), 1∶2 (P2) and 1∶1 (P3) were used as the research objects. Fluorescence quantitative PCR and high-throughput sequencing technology were used to analyze the structure and diversity of the bacterial community in the compound soil and its relationship with soil physical and chemical factors. The results showed that compared with the CK treatment, the contents of soil organic carbon (SOC), total nitrogen (TN) and NH_4^+-N increased significantly in P1 treatment, and the contents of available phosphorus (AP), available potassium (AK) and NO_3^--N increased significantly in P3 treatment. The bacterial gene copy number of P3 treatment was the largest, which was a significant increase of 182.05% compared with CK treatment. The three bacterial groups with the highest relative abundance at the Phylum level were *Actinobacteria*, *Proteobacteria* and *Chloroflexi*, which accounted for more than 70% of the total abundance. The bacterial α diversity showed a basically the same trend, the diversity and abundance index of P1 and P3 treatments were the highest, and the β diversity showed that the community structure of the two groups was similar. *Norank_f__Roseiflexacea*e and *Gaiella* (belonging to *Actinobacteria*) were significantly different in different compound ratios. NO_3^--N, NH_4^+-N and SOC were the main factors affecting bacterial community structure, and had a significant positive correlation with *Gaiella*. These species play an important role in stabilizing the soil structure of sandy land. Therefore, 1∶5 and 1∶1 compound soils were beneficial to the biological reproduction of sandy land and play an important role in biological sand fixation.

【**Keywords**】Soil properties; Community structure; Diversity; High-throughput sequencing; Mu Us Sandy Land

Introduction

Soil bacteria is one of the most diverse, abundant and functional groups in soil microorganisms (Steenwerth *et al.*, 2003). They are important drivers of biogeochemical cycles, participate in the transformation of soil nutrients, and are key organisms in the material cycle and energy flow of the ecosystem (Nacke *et al.*, 2011). Meanwhile, the structure and function of soil microbial community is also one of the important indicators reflecting the evolution of soil quality and fertility (Chen *et al.*, 2020). With more and more human disturbances to soil, such as changes in land regulation, fertilization methods and planting types, many studies have shown that these disturbances have a significant impact on the structure, diversity and even function of soil microbial community (Steenwerth *et al.*, 2003). However, the dominant environmental factors affecting soil microbial community structure have not been determined, especially in the Mu Us Sandy Land with development potential, the trend of microbial community change still needs to be further explored.

The Mu Us Sandy Land is located in a semi-arid egion of north China, which is a compound ecosystem area composed of grassland, forestry and agriculture (Zhang *et al.*, 2020). The ecological environment in Mu Us Sandy Land is fragile due to the lack of surface water resources, low vegetation coverage, vulnerability to human activities and serious soil wind erosion (Li *et al.*, 2017). A large number of domestic and overseas scholars have studied the impact of agricultural use patterns on the soil quality of farmland in sandy land based on different experimental areas, indicating that if there are reasonable land use methods and appropriate farming management measures, the regional habitat can be affected by increasing soil carbon and nitrogen storage (Di *et al.*, 2016). Conversely, over-use of land will reduce soil quality, leading to a decline in land productivity (Zhu *et al.*, 2020). Liu *et al.* (2010) found that conservation tillage and fine management of irrigated farmland were beneficial to soil environment improvement and ecosystem restoration in sandy land. Su *et al.* (2017) showed that after the desert sandy land was reclaimed into farmland, with the increase of the reclaiming period, although the soil fertility was significantly improved, the soil fertility in the area was still at a low level. He *et al.* (2020) used an engineering measure to improve the sandy land, indicating that soft rock is a loose rock widely distributed in the Mu Us Sandy Land, and its mixing with aeolian sandy soil can significantly improve the water and fertilizer retention capacity of the sandy land. It is also believed that the soft rock will become soft as mud when exposed to water, which can improve the chemical and physical characteristics of sandy soil and crop productivity, and increase the colloidal content of sandy land (Guo *et al.*, 2021). It can be seen that using soft rock to improve aeolian sandy soil in the Mu Us Sandy Land can not only realize the water and fertilizer retention capacity of the sandy land, but also increase the cultivated land area, promote the increase of crop yields, and solve the material demand for the improvement of the Mu Us Sandy Land and maintain the sustainable development of regional ecological environment (Zhang *et al.*, 2020; Han *et al.*, 2012).

Liu *et al.* (2019) believe that bacterial community composition in desert areas are greatly disturbed. Du *et al.* (2017) found that the number and diversity of microorganisms are the most abundant on the surface. The improvement of sandy land by soft rock is an engineering measure to organically reconstruct the sandy land, which has great disturbance. The study on the influence of soft rock on soil bacteria in sandy land is of great significance to reveal the response mechanism of underground microbial community to engineering measures and to study the improvement measures of soil quality in sandy land. However, previous studies on soft rock and sand compound soil mainly focused on the physical structure and chemical properties, while there are few reports on the differences in soil bacterial community structure and its driving factors during soil development. Therefore, this study took different proportions of soft rock and sand compound soil as the research object, using high-throughput sequencing technology to analyze the bacterial community structure and diversity of the 0~20 cm soil in the Mu Us Sandy Land. The aim was to clarify the differences in the bacterial community structure in the compound soil and the soil factors that regulate the bacterial community structure, and to clarify the mechanism of the change of bacterial diversity on soil improvement in desert areas.

Materials & methods

Overview of the test site

The soft rock and sand compound soil test field was located in the Mu Us Sandy Land (E109°28′58″~

109°30′10″, N38°27′53″~38°28′23″) in Yuyang District, Yulin City, which was located in northwest Shaanxi, China, and the altitude is between 1206~1215 m. The test area belongs to a typical mid-temperate semi-arid continental monsoon climate zone, with uneven distribution of precipitation in time and space, dry climate, long winter and short summer, four distinct seasons, and sufficient sunshine. The average annual temperature is 8.1 ℃, the average annual frost-free period is 154 days, the average annual precipitation is 413.9 mm, and 60.9% of the rainfall is concentrated in the three months from July to September. The annual average sunshine hours is 2879 hours, and the sunshine percentage is 65%. The soil type in the project area is mainly sandy soil.

Experiment design

The test field was to simulate the land condition of the mixed layer of soft rock and sand in the Mu Us Sandy Land. The experimental plot was to lay a mixture of soft rock and sand at 0~30 cm. The selected ratio of soft rock to sand (0:1(CK), 1:5(P1), 1:2(P2), 1:1(P3)) was repeated for three times in the experimental field with an area of 5 m × 12 m = 60 m^2. The field trial implements a potato cropping system once a year, planted in mid-April and harvested in mid-to-late September each year. Artificial planting mode is adopted throughout the year. During the farming years, the application of chemical fertilizers is used to promote the growth of crops and the accumulation of root exudates, and at the same time promote the metabolic activities of microorganisms, and promote the increase of the nutrient content of the compound soil of soft rock and sand. The test fertilizer types in the test field were urea, diammonium phosphate and potassium chloride. The fertilizer application amount was N 300 kg · ha^{-1}, P_2O_5 375 kg · ha^{-1} and K_2O 180 kg · ha^{-1} per year. All phosphate fertilizers and potash fertilizers are used as base fertilizers, and 50% of nitrogen fertilizers are used as base fertilizers. One to two days before planting, weigh the three kinds of fertilizers according to the required amount of each plot and mix them evenly, and sprinkle them evenly on the soil surface, and then properly rake the fertilizer to mix the topsoil. The remaining 50% of the nitrogen fertilizer is topdressed at the potato seedling stage and after flowering.

Soil sample collection

After the potatos were harvested in September 2021, soil samples of 0~20 cm soil layer were collected from each plot. Three mixed soil samples were collected in each plot, and each sample was uniformly collected and mixed by the five-point method. The collected soil samples were divided into two parts after removing animal and plant residues. One was naturally air-dried and screened with 1 mm and 0.149 mm for soil physical and chemical properties determination, and the other was stored in a refrigerator at −80 ℃ for microbial analysis.

Determination of soil physical and chemical properties

Soil organic carbon (SOC) was determined by potassium dichromate-concentrated sulfuric acid external heating method (Nelson and Sommers, 1996). Total nitrogen (TN) was determined by Kjeldahl digestion, Available phosphorus (AP) was determined by molybdate blue colorimetry, and Available potassium (AK) concentrations were measured using atomic absorption spectrometry (Dai *et al.*, 2017). NO_3^--N and NH_4^+-N were extracted at a ratio of 10 g fresh soil to 100 mL 2 M KCl. After shaking for 1 h, the extracts were filtered and analyzed by continuous flow analytical system (San++ System, Skalar, Holland) for NO_3^--N and NH_4^+-N (Di *et al.*, 2016). pH was measured using a pH meter (PHS-3E, INESA, China), and the soil-to-water ratio was 1:5 (Minasny *et al.*, 2011).

Soil DNA extraction and sequencing

The E.Z.N.A.® Soil DNA Kit (Omega, Inc., USA) was used to extract the total DNA of the soil sample, and then the concentration and purity of the DNA were determined by the NanoDrop 2000 spectrophotometer, and detected by 1% agarose gel electrophoresis. Using the total microbial DNA of each soil sample as a template, PCR amplification was carried out with bacterial V3-V4 region-specific primers 338F (5′-ACTCCTACGGGAGGCAGCAG-3′) and 806R (5′-GGACTACHVGGGTWTCTAAT-3′). The PCR products were recovered and purified by 2% agarose gel, eluted by Tris-HCl, and detected by 2% agarose electrophoresis. QuantiFluorTM-ST (Promega, USA) was used for quantitative detection. According to the standard operating procedures of the Illumina MiSeq platform (Illumina, SanDiego, USA), the purified amplified fragments were constructed to PE 2×300 library. Sequencing was performed using Illumina's Miseq PE300 platform (Chen *et al.*, 2020).

Fluorescence quantitative PCR amplification

Fluorescent quantitative PCR was performed using the same primers as the high-throughput sequencing above. The reaction system of 20 μL is: 10 μL 2 X ChamQ SYBR Color qPCR Master Mix, upstream and downstream primer (5 $\mu mol \cdot L^{-1}$) 0.8 μL each, 2 μL template, 0.4 μL 50 X ROX Reference Dye 1, 6 μL ddH_2O. The amplification program is: 95 ℃ pre-denaturation for 3 min; 95 ℃ denaturation for 5 s, 58 ℃ annealing for 30 s, and 72 ℃ extension for 1 min (Hassan *et al.*, 2021). ABI7300 fluorescence quantitative PCR instrument (Applied Biosystems, USA) was used for amplification. Set three replicates for each sample, and calculate the final gene abundance based on the soil dry weight.

Data processing and analysis

The experimental data is analyzed with SPSS 20.0 for variance analysis. Using QIIME (Version 1.9.1) to analyze the composition of the sample, obtain the data of the bacterial community composition and relative abundance of the sample at different taxonomic levels, and make the relative abundance maps of the species at the Phylum and Genus level. Use QIIME (Version 1.9.1) to analyze the dilution curve based on OTU and calculate the species diversity index. Use R language (Version 3.3.1) to draw the Principal Component Analysis (PCA) diagram of the soil bacterial community structure. Canoco was used to perform Redundancy Analysis (RDA) between bacterial community composition and environmental factors. The Spearman correlation coefficient is used to analyze the correlation between environmental factors and species, and the Heatmap is drawn with the aid of R software.

Results

Soil properties

The SOC content in the P1 and P2 treatments was higher, which was significantly increased by 46.07% ($n=12$, $df=3$, $P=0.0253$) and 43.46% ($n=12$, $df=3$, $P=0.0284$) compared with the CK treatment. Compared with CK, the TN content of P1 treatment was significantly increased by 112.50%, but there was no significant difference between P1, P2 and P3 treatments. The change trend of NH_4^+-N content was consistent with TN. The content of AP, AK and NO_3^--N was the highest in P3 treatment, and the NO_3^--N content increases with the increase of soft rock content. There was no significant difference in pH between treatments. The abundance of 16S rRNA genes of the four mixed soil bacteria was between $0.39\times10^9 \sim 1.10\times10^9$ copies g^{-1} dry soil. The P3 treatment had the largest bacterial gene copy number, which was significantly increased by 17.02%, 155.81% and 182.05% compared with P1, P2 and CK treatments, respectively. There was no significant difference between P2 and CK (Table 1).

Table 1　Soil properties under different compound ratio treatments

Treatments	SOC ($g \cdot kg^{-1}$)	TN ($g \cdot kg^{-1}$)	AP ($mg \cdot kg^{-1}$)	AK ($mg \cdot kg^{-1}$)	NO_3^--N ($mg \cdot kg^{-1}$)	NH_4^+-N ($mg \cdot kg^{-1}$)	pH	Genes copies number ($\times 10^9$)
CK	1.91±0.47 b	0.16±0.04 b	3.85±0.87 c	38.52±4.39 c	4.23±0.33 d	2.39±0.42 b	8.89±2.11 a	0.39±0.03 c
P1	2.79±0.21 a	0.34±0.06 a	7.57±2.14 b	63.47±6.38 b	13.08±1.96 c	3.77±1.43 a	8.78±0.91 a	0.94±0.05 b
P2	2.74±0.15 a	0.24±0.05 ab	5.53±1.52 c	44.26±4.56 c	17.44±4.05 b	3.44±0.88 a	8.87±1.74 a	0.43±0.06 c
P3	2.58±0.09 ab	0.27±0.07 ab	14.84±3.55 a	72.11±5.61 a	18.64±2.21 a	3.45±1.02 a	8.56±1.22 a	1.10±0.14 a

Note: CK, the volume ratio of soft rock to sand is 0∶1; P1, the volume ratio of soft rock to sand is 1∶5; P2, the volume ratio of soft rock to sand is 1∶2; P3, the volume ratio of soft rock to sand is 1∶1. SOC stands for soil organic carbon; TN stands for soil total nitrogen; AP stands for available phosphorus; AK stands for available potassium; NO_3^--N stands for nitrate nitrogen; NH_4^+-N stands for ammo-nium nitrogen. Lowercase letters indicate significant differences at the 5% level between different treatment.

Bacterial community composition

According to the Phylum classification level, the bacteria community composition of the compound soil under different soil layers was studied, and the results showed the abundance of the top 12 bacteria. Others classified the relative abundance less than 0.01 into one category (Fig. 1). The three dominant bacteria were *Actinobacteriota*, *Proteobacteria*, and *Chloroflexi*, respectively. However, *Cloacimonadota* unique bacteria Phylum appeared in P2 treatment. Compared with CK treatment, the relative abundance of *Actinobacteriota* in P3 treatment increased by 37.82%. Compared with CK treatment, the relative abundance of *Proteobacteria* in other treatments all showed a decreasing trend, with a larger decrease in P1. Compared with CK treatment, *Chloroflexi* abundance decreased by 20.56% under P2 treatment, and showed an increasing trend in P1 and P3 treatments.

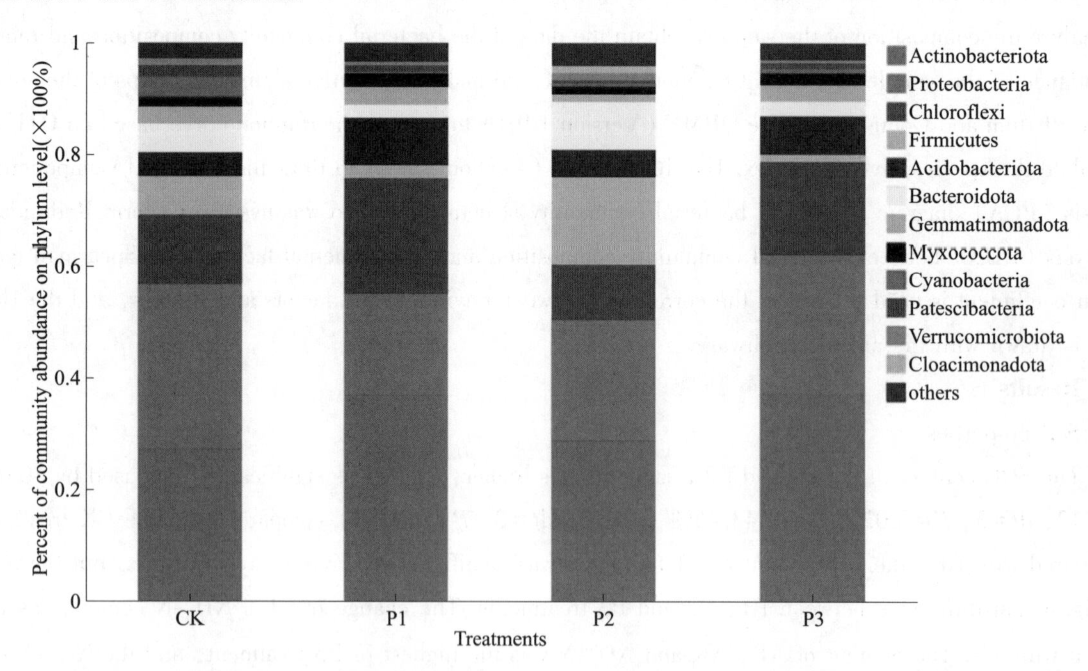

Fig. 1　Based on phylum level bacterial community composition. CK, the volume ratio of soft rock to sand is 0∶1; P1, the volume ratio of soft rock to sand is 1∶5; P2, the volume ratio of soft rock to sand is 1∶2; P3, the volume ratio of soft rock to sand is 1∶1

At the level of Genus classification, the differences of bacteria in different treatments increased (Fig. 2). The dominant bacteria of CK were *Arthrobacter* (6.33%), *norank_f__JG30-KF-CM45* (5.48%), and *Ly-*

sobacter (4.54%). The dominant bacteria of P1 were *Arthrobacter* (13.90%), *norank_f_JG30-KF-CM45* (4.83%), and *Blastococcus* (2.42%). The dominant bacteria of P2 were *Arthrobacter* (6.79%), *Pseudomonas* (6.23%), and *Rhodococcus* (4.26%). The dominant bacteria of P3 were *Arthrobacter* (11.01%), *norank_f_JG30-KF-CM45* (3.93%), and *Sphingomonas* (2.64%).

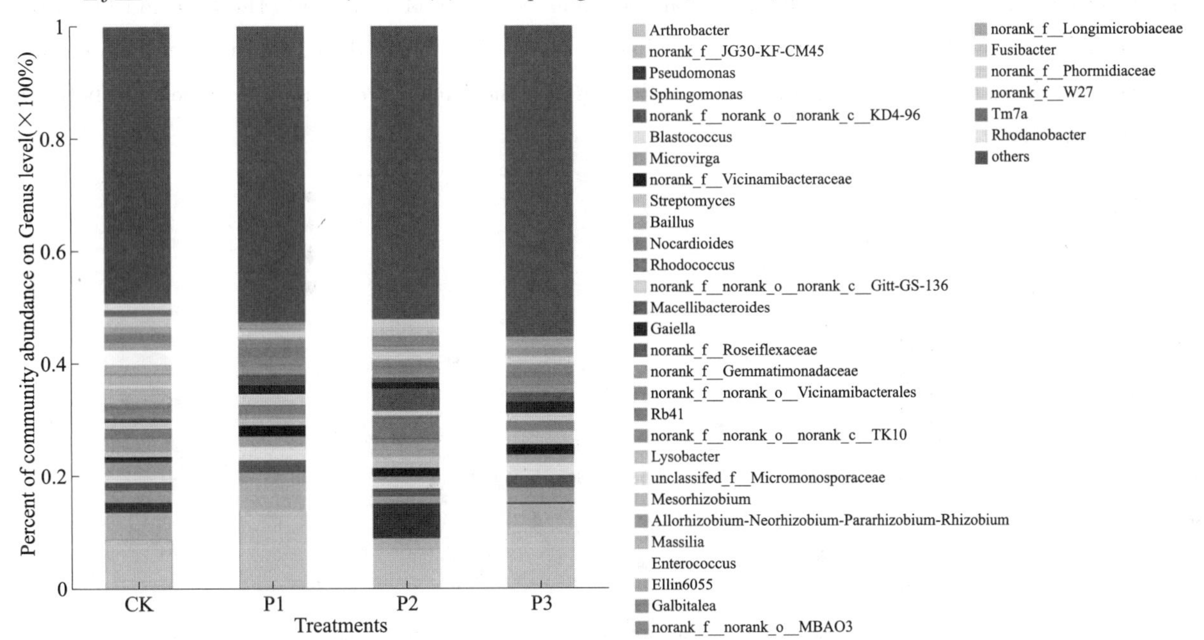

Fig. 2 Based on genus level bacterial community composition. CK, the volume ratio of soft rock to sand is 0∶1; P1, the volume ratio of soft rock to sand is 1∶5; P2, the volume ratio of soft rock to sand is 1∶2; P3, the volume ratio of soft rock to sand is 1∶1

Bacterial α diversity

Coverage refers to the sequencing accuracy of the sample library, and the higher the value, the higher the probability of the sequence in the sample being measured. The Coverage values in this study were all greater than 97%, indicating that the sequencing results were highly reliable and cover most of the sequencing information in the samples (Table 2). Chao and Ace indexes represent the abundance of bacterial communities, and the higher the value, the higher the abundance of community species. The results showed that the Chao index of P1 and P3 treatments was significantly higher than that of P2 and CK treatments. The change trend of Ace index was consistent with that of Chao index. Shannon index represented the diversity of bacterial community. The results showed that the addition of soft rock promoted the increase of bacterial diversity in sandy soil, but there was no significant difference between different treatments.

Table 2 Bacterial diversity of under different compound ratio treatments

Treatments	Reads	OTUs	Chao	Ace	Shannon	Coverage (%)
CK	46213	3125	2947.17±50.81 b	2916.35±49.85 b	5.67±0.60 a	98.07
P1	55139	3909	4274.92±40.56 a	4233.60±42.11 a	6.23±1.02 a	97.04
P2	60610	3776	2789.11±90.22 b	2821.63±88.21 b	5.89±0.98 a	98.40
P3	54746	3770	3937.10±83.41 a	3907.27±90.42 a	6.19±0.99 a	97.39

Note: CK, the volume ratio of soft rock to sand is 0∶1; P1, the volume ratio of soft rock to sand is 1∶5; P2, the volume ratio of soft rock to sand is 1∶2; P3, the volume ratio of soft rock to sand is 1∶1. Lowercase letters indicate significant differences between different compound ratios ($P<0.05$).

Bacterial community β diversity

PCA analysis results showed that CK was clearly distinguished from other processed samples on the PC1 axis, and other samples were located to the right of CK (Fig. 3). The results showed that the addition of soft rock has a certain impact on the bacterial community structure of sandy soil. The PC1 and PC2 axes explained 27.28% and 17.77% of the total variation, respectively. Among them, the distance between the P1 and P3 soil samples was relatively small, indicating that the bacterial community composition between them was similar.

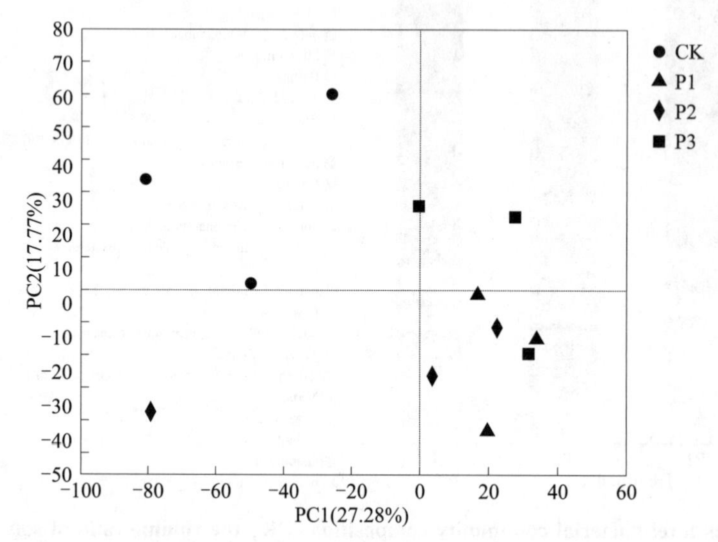

Fig. 3 The PCA analysis of bacterial community based on OTU level. CK, the volume ratio of soft rock to sand is 0:1; P1, the volume ratio of soft rock to sand is 1:5; P2, the volume ratio of soft rock to sand is 1:2; P3, the volume ratio of soft rock to sand is 1:1

Analysis of species differences between treatments

The top 15 species with relative abundance at the Genus level were selected for analysis of differences between treatments. The *norank_f_Roseiflexaceae* and *Gaiella* (belonging to *Actinobacteria* Phylum) showed significant differences between different compound proportions (Fig. 4). The *norank_f_Roseiflexaceae* had the largest abundance in P1 treatment, followed by P3 and P2 treatments, and CK treatment was the smallest. Compared with CK treatment, the relative abundance of *Gaiella* increased significantly with P3 treatment being the highest, followed by P1 treatment, and P2 treatment was the lowest.

The relationship between soil properties and bacterial communities

In the 0~20 cm soil layer, the interpretation degrees of the RDA1 axis and RDA2 axis was 75.15% and 14.96%, respectively, and the sum of the two axes was 90.11%. The degree of influence of various environmental factors on the composition of bacterial communities in soil samples was NN (NO_3^--N), AN (NH_4^+-N), and SOC with the greatest impact; AK, AP and TN had the second most impact; pH had the least impact. The bacterial community composition of P1 was positively correlated with other factors except pH (Fig. 5).

Heat map of the correlation between soil properties and bacterial communities

The top 15 relative abundance species at the Genus level were selected for correlation analysis with soil properties. The results showed that *Gaiella* (belonging to *Actinobacteria* Phylum) was significantly correlated with NN, SOC and AN, and *Microvirga* (belonging to *Proteobacteria* Phylum) was significantly correlated

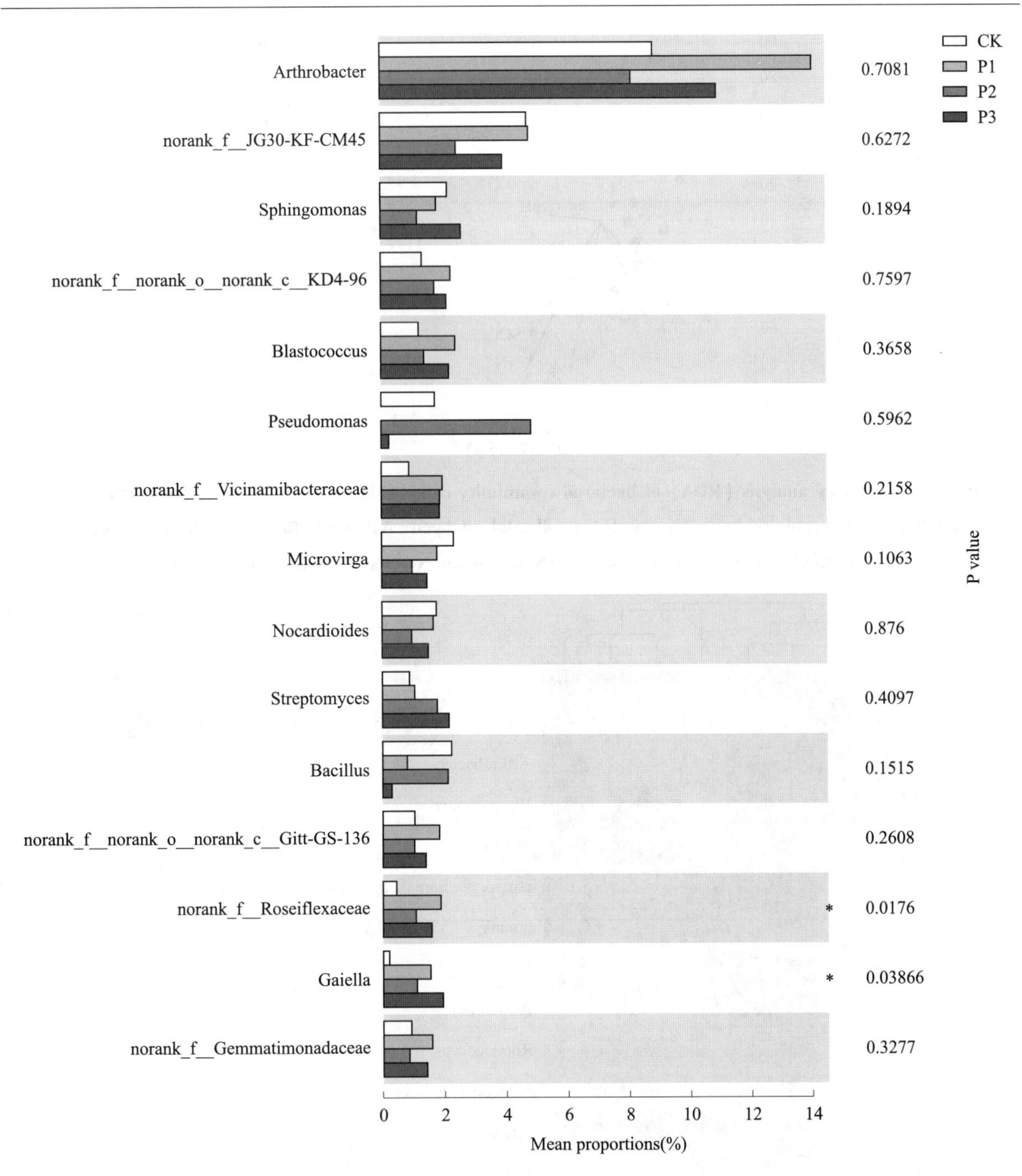

Fig. 4 Differences between treatments of bacteria at the Genus level. CK, the volume ratio of soft rock to sand is 0:1; P1, the volume ratio of soft rock to sand is 1:5; P2, the volume ratio of soft rock to sand is 1:2; P3, the volume ratio of soft rock to sand is 1:1. An asterisk (*) indicates signifificant differences at the 5% level between different treatment

with pH and NN (Fig. 6). There was no significant difference between other bacterial genera and soil properties.

Bacteria functional differences of compound soil

After soft rock improved the sandy land, the function of soil bacteria showed obvious differences (Fig. 7). The relative abundance of E, G, K, R and S was higher than 40%, and S>E>R>K>G. The relative abundance of S function was the highest, indicating that unknown functions still occupied most of the bacterial community in the mixed soil. Compared with CK treatment, E function increased significantly in

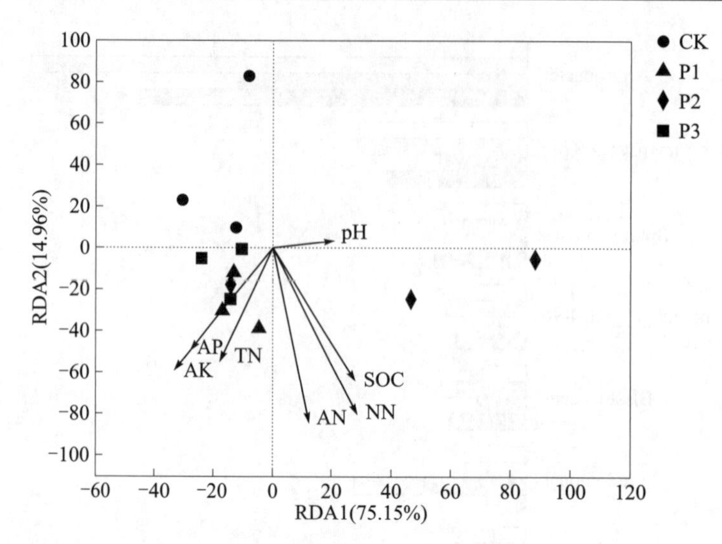

Fig. 5　Redundancy analysis (RDA) of bacterial community composition and soil chemical properties. SOC stands for soil organic carbon; TN stands for soil total nitrogen; AP stands for avail able phosphorus; AK stands for available potassium; NN stands for NO_3^--N; AN stands for NH_4^+-N

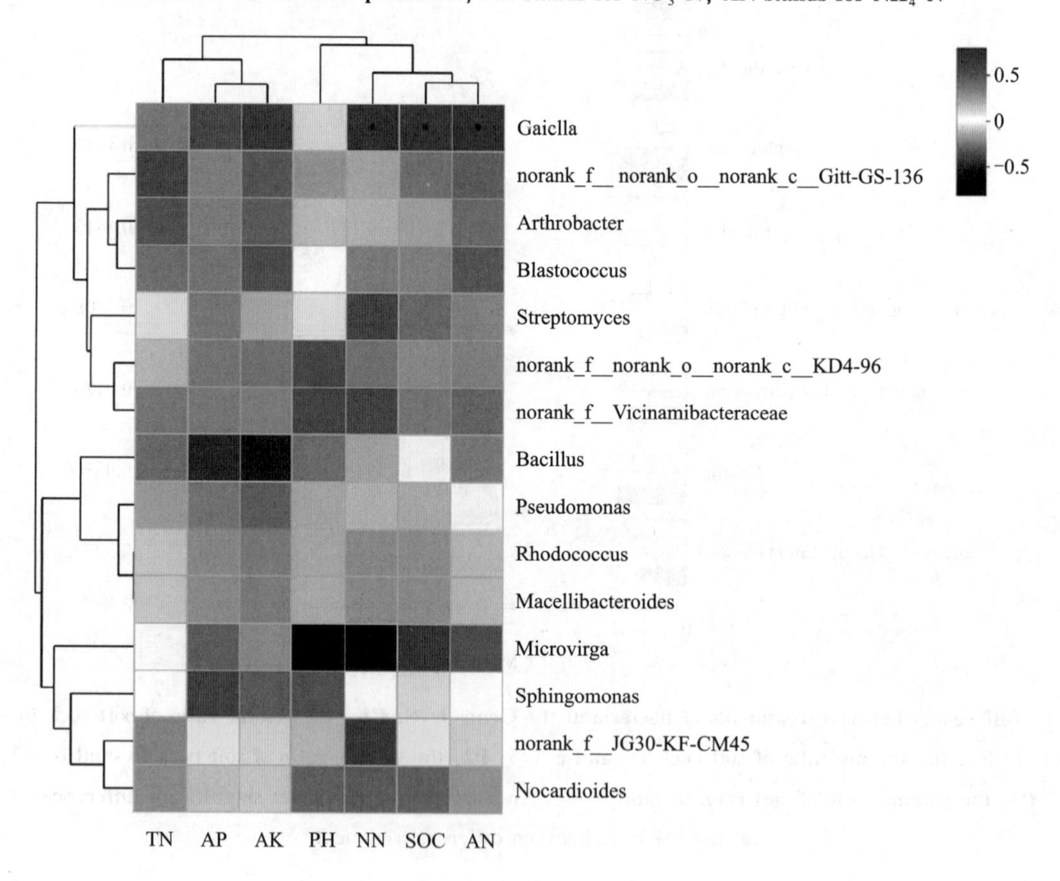

Fig. 6　A correlation heatmap of soil bacteria and soil chemical properties at the Genus level in different soil layers. SOC stands for soil organic carbon; TN stands for soil total nitrogen; AP stands for available phosphorus; AK stands for available potassium; NN stands for NO_3^--N; AN stands for NH_4^+-N. If the P value is less than 0.05, it is marked with an asterisk (＊)

P3 treatment, R function and G function increased significantly in P1 treatment, and K function increased significantly in P2 treatment. In P2 treatment, the functions of C, D, F, I, J, L, M, N, O, P, T, U, V and Z increased significantly compared with CK treatment, indicating that P2 treatment mainly enhanced the translation and transcription functions at the cellular level, thus promoting the utilization and release of nutrients.

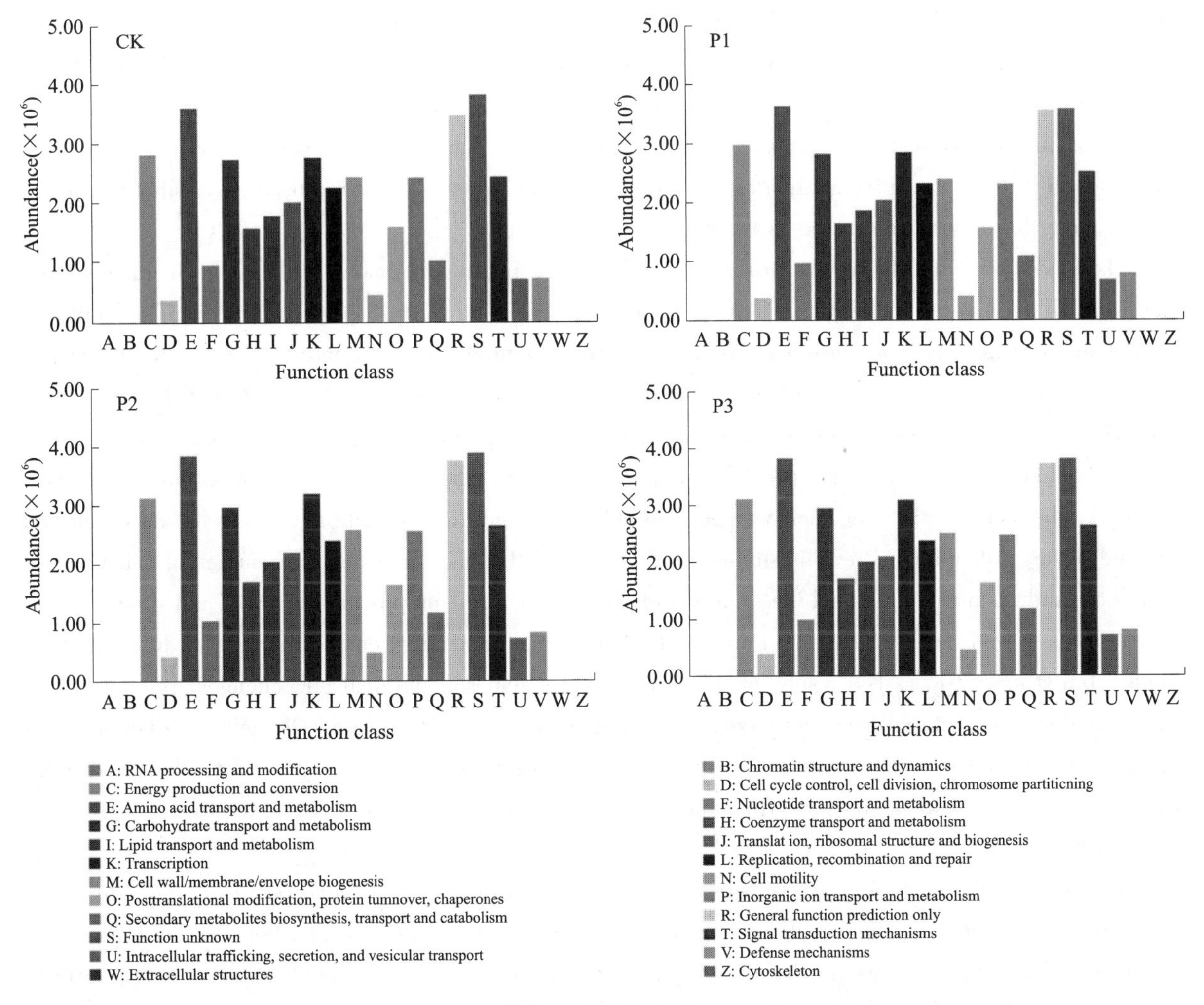

Fig. 7　The functional differences of bacterial communities in different composite treatments of soft rock and sand. CK, the volume ratio of soft rock to sand is 0∶1; P1, the volume ratio of soft rock to sand is 1∶5; P2, the volume ratio of soft rock to sand is 1∶2; P3, the volume ratio of soft rock to sand is 1∶1

Discussion

The results of Steven *et al.* (2013) showed that *Chloroflexi* in the soil has a greater abundance due to the role of the surface biological protection layer. Rao *et al.* (1990) pointed out that surface protection measures significantly affected the biological and physical properties of the top soil. The results of this study showed that the dominant bacterial community was different from the conclusion of Steven *et al.* (2013). The relative abundance of *Actinobacteriota*, *Proteobacteria* and *Chloroflexi* in the surface layer of compound soil was higher, and there were specific *Cloacimonadota* in P2 treatment. With the refinement of classification level, there were more endemic genera in the soil. The reason for this difference may be due to the difference in the nutrient content of the compound soil or the higher adaptability of endemic species to the new environment (Yu *et al.*, 2020), because the soft rock changed the microenvironment of the sandy land. The analysis results of soil bacteria α diversity and β diversity showed that the addition of different proportions of soft roc changed the richness index and diversity index of soil bacteria in the sandy soil, indicating that soft roc promoted the improvement of the biological characteristics of sandy soil. This was similar to the bacterial community structure in the Gurbantungut Desert (Liu *et al.*, 2019). This was because differences in soil nutrient content, pH and other environmental factors affect the distribution of soil microbial communities (Tiemann and Billings, 2012). Among the four treatments, the community structure of the P1 and P3 treatments was relatively similar, on the one hand because the α diversity of bacteria also changed, on the other hand, the abundance of the common species *Myxococcota* in the P1 and P3 treatments was higher than that of the P2 treatment. *Myxococcota* was a special species that can use live microbial cells or other biological macromolecules as food to obtain nutrients, and at the same time, it can respond to external nutritional thresholds to regulate the differentiation of vegetative cells into stress-resistant myxospores, so that the *Myxococcota* group has a good soil adaptability (Li *et al.*, 2019).

The study found that there was no significant difference in soil pH among the various treatments, and they were all alkaline, indicating that soft rock had little effect on the pH of sandy soil. The soil nutrient content was significantly different in each treatment. The content of SOC, TN and NH_4^+-N was the highest in the P1 treatment, and there was no significant difference between P1 and P2, which may be due to the point contact between the soft rock particles and the sandy soil particles (Han *et al.*, 2012). The contents of AP, AK and NO_3^--N were highest in P3 treatment, which may be because as the proportion of soft rock increases, the soil structure becomes compact and cohesive, and the available nutrients were absorbed and retained in large quantities. Previous studies have suggested that the main nutrient sources of soil bacteria were root exudates and litters, and the quality and quantity of nutrients provided by roots and litters for microorganisms were different, resulting in different soil bacterial community composition under different treatments (Dai *et al.*, 2017). The results of this study showed that soil physical and chemical properties had different effects on bacterial community composition, with NO_3^--N, NH_4^+-N and SOC being the largest. This was because the soil bacterial community can also choose specific environment, indicating that there were synergistic changes between soil properties and bacterial community in the process of improving sandy land with soft rock.

According to the analysis of differences between groups, it can be concluded that *norank_f_Roseiflexaceae* and *Gaiella* have significant differences in different treatments, and they all belong to the *Actinobacteria* phylum. Then through analysis, it can be concluded that the *Actinobacteria* phylum was mainly positively correlated with the content of NO_3^--N, SOC and NH_4^+-N in the compound soil. It can be seen that the *Actinobacteria* phylum was the first dominant group in the compound soil, and it has a relatively high relative abundance in each compound soil, and it was the main source of soil nutrient supply. This was related to the suitable growth of *Actinobacteria* in neutral and alkaline pH soils (Sun *et al.*, 2020). At the same time, *Actinobacteria*, as a common soil dweller, has a strong adhesion ability and can become a source of bacteria storage. The mucus secreted by *Actinobacteria* can bond sandy soil, and its filamentous bacteria are also conducive to the stability of soil structure, which can play a certain degree of sand fixation (Mummey *et al.*, 2006). In order to further understand the impact of changes in soil physical and chemical properties on the composition of soil bacterial communities in the process of sandy land improvement, the results of redundancy analysis showed that the soil NO_3^--N, NH_4^+-N, and SOC content have a greater impact on the bacterial community composition of the samples. This result also confirmed that in the process of sandy land improvement, soil properties and microbial communities have synergistic changes.

Conclusions

In the process of sandy land improvement, the addition of different proportions of soft rock changed soil physical and chemical properties and improved soil fertility. The soil bacterial community structure has also changed significantly. The dominant bacteria were *Actinobacteriota*, *Proteobacteria* and *Chloroflexi*, accounting for more than 70% of the total bacterial community. The richness, diversity index and gene copy number of soil bacteria were the highest in 1:5 and 1:1 compound soil, and the community structure between the two compound soils was relatively similar. Soil factors were the main factors driving the distribution of soil bacterial community. NO_3^--N, NH_4^+-N and SOC were the leading factors for the differentiation of bacterial community structure, and were highly correlated with *Actinobacteria*. At the same time, the 1:5 and 1:1 compound soils both showed strong carbohydrate transport and metabolism capabilities.

Acknowledgements

The authors thank the editor and anonymous reviewers for their constructive comments, which helped to improve the manuscript.

References

Chen J, Liu Y, Yang Y, Tang M, Wei Y. 2020. Bacterial community structure and gene function prediction in response to long-term running of dual graphene modified bioelectrode bioelectrochemical systems. *Bioresource Technology* 309:123398.

Dai X Q, Wang H M, Fu X L. 2017. Soil microbial community composition and its role in carbon mineralization in long-term fertilization paddy soils. *Sciene of the Total Environment* 580:556-563.

Dai Y T, Yan Z J, Xie J H, Wu H X, Xu L B, Hou X Y, Gao L, Cui Y W. 2017. Soil bacteria diversity in rhizosphere under two types of vegetation restoration based on high throughput sequencing. *Acta Pedologica Sinica* 54(3):735-748.

Di H J, Long X E, Shen J P, He J Z, Zhang L M. 2016. Contrasting response of two grassland soils to N addition and moisture levels: N_2O emission and functional gene abundance. *Journal of Soils and Sediments* 17(2):384-392.

Du C, Geng Z C, Wang Q, Zhang T T, He W X, Hou L, Wang Y L. 2017. Variations in bacterial and fungal communities through soil depth profiles in a Betula albosinensis forest. *Journal of Microbiology* 55(9):684-693.

Guo Z, Zhang H O, Wang H Y. 2021. Soft rock increases the colloid content and crop yield in sandy soil. *Agronomy Journal* 113(2):677-684.

Han J C, Xie J C, Zhang Y. 2012. Potential role of feldspathic sandstone as a natural water retaining agent in Mu Us Sandy Land, Northwest China. *Chinese Geographical Sciences* 22(5):550-555.

Hassan M A, Shirai Y, Husni M, Zainudin M, Mustapha N A. 2021. Effect of inorganic fertilizer application on soil microbial diversity in an oil palm plantation. *BioResources* 18(2):2203-2279.

He H H, Zhang Z K, Su R, Dong Z G, Zhen Q, Pang J Y, Lambers H. 2020. Amending aeolian sandy soil in the Mu Us Sandy Land of China with Pisha sandstone and increasing phosphorus supply were more effective than increasing water supply for improving plant growth and phosphorus and nitrogen nutrition of lucerne (Medicago sativa). *Crop & Pasture Science* 71(8):785-793.

Li Y R, Fan P C, Cao Z, Chen Y F, Liu Y S, Wang H Y, Liu H H, Ma F, Wan H. 2017. Sand-fixation effect and micro-mechanism of remixing soil by pisha sandstone and sand in the Mu Us Sandy Land, China. *Journal of Desert Research* 37(3):421-430.

Li Z K, Xia C Y, Wang Y X, Li X, Qiao Y, Li C Y, Zhou J, Zhang L, Ye X F, Huang Y, Cui Z L. 2019. Identification of an endo-chitinase from Corallococcus sp. EGB and evaluation its antifungal properties. *International Journal of Biological Macromolecules* 132:1235-1243.

Liu R T, Zhao H L. 2010. Effect of land use changes on soil properties in Horqin Sandy Land. *Ecology and Environmental Sciences* 19(9):2079-2084.

Liu Y B, Wang Z R, Zhao L N, Wang X, Liu L C, Hui R, Zhang W L, Zhang P, Song G, Sun J Y. 2019. Differences in bacterial community structure between three types of biological soil crusts and soil below crusts from the Gurbantunggut Desert, China. *European Journal of Soil Science* 70(3):630-643.

Minasny B, McBratney A B, Brough D M, Jacquier D. 2011. Models relating soil pH measurements in water and calcium chloride that incorporate electrolyte concentration. *European Journal of Soil Science* 62:728-732.

Mummey D, Holben W, Six J, Stahl P. 2006. Spatial stratification of soil bacterial populations in aggregates of diverse soils. *Microbial Ecology* 51(3):404-411.

Nacke H, Thürmer A, Wollherr A, Will C, Hodac L, Herold N, Schöning I, Schrumpf M, Daniel R, Gilbert J. 2011. Pyrosequencing-Based assessment of bacterial community structure along different management types in german forest and grassland soils. *Plos One* 6(2):e17000.

Nelson D W, Sommer L E. 1996. Total carbon, organic carbon and organic matter. In: Sparks DL, Page AL, Helmke PA, Loeppert RH, Soltanpour PN, Tabatabai MA, CT JH, Sumner ME, eds. Methods of soil analysis. Part 3-chemical methods. Madison: Soil Science Society of America.

Rao D L N, Burns R G. 1990. The effect of surface growth of blue-green algae and bryophytes on some microbiological, biochemical, and physical soil properties. *Biology and Fertility of Soils* 9:239-244.

Steven B, Gallegos-Graves L V, Belnap J, Kuske C R. 2013. Dryland soil microbial communities display spatial biogeographic patterns associated with soil depth and soil parent material. *FEMS Microbiology Ecology* 86(1):101-113.

Steenwerth K L, Jackson L E, Calderón F J, Stromberg M R, Scow K M. 2003. Soil microbial community composition and land use history in cultivated and grassland ecosystems of coastal california. *Soil Biology & Biochemistry* 34(11):1599-1611.

Sun P P, Qian C J, Yin X Y, Fan X K, Wang J, Yan X, Ma X F, Wang T. 2020. Effects of Artemisia vegetation built on soil bacteria in semi-arid sandy land. *Acta Ecologica Sinica* 40(16):5783-5792.

Su Y Z, Zhang K, Liu T N, Fan G P, Wang T. 2017. Changes in soil properties and accumulation of soil carbon after cultivation of desert sandy land in a marginal oasis in Hexi Corridor region, northwest China. *Scientia Agricultura Sinica* 50(9):1646-1654.

Tiemann L K, Billings S A. 2012. Tracking C and N flows through microbial biomass with increased soil moisture variability. *Soil Biology & Biochemistry* 49:11-22.

Yu H, Xue D M, Wang Y D, Zheng W, Zhang G L, Wang Z L. 2020. Molecular ecological network analysis of the response of soil microbial communities to depth gradients in farmland soils. *MicrobiologyOpen* 9(3):e983.

Zhang H O, Xie J C, Han J C, Nan H P, Guo Z. 2020. Response of fractal analysis to soil quality succession in long-term compound soil improvement of Mu Us Sandy Land, China. *Mathematical Problems in Engineering* 2020:5463107.

Zhu W, Gao Y, Zhang H, Liu L. 2020. Optimization of the land use pattern in horqin sandy land by using the clumondo model and bayesian belief network. *Sciene of the Total Environment* 739:139929.

本文曾发表于 2022 年《PeerJ》第 10 卷

Effects of Biochar on Soil Microbial Diversity and Community Structure in Clay Soil

Jing Zhang[1,2,3,4], Jianglong Shen[1,2,3,4]

(1. Shaanxi Provincial Land Engineering Construction Group Co., Ltd., Xi'an 710000, China; 2. Institute of Land Engineering and Technology, Shaanxi Provincial Land Engineering Construction Group Co., Ltd., Xi'an 710000, China; 3. Key Laboratory of Degraded and Unused Land Consolidation Engineering, the Ministry of Natural Resources, Xi'an 710000, China; 4. Shaanxi Provincial Land Consolidation Engineering Technology Research Center, Xi'an, China)

【**Abstract**】Purpose: We determined the microbial community diversity and structure in soil samples under different amounts of biochar added. Meanwhile, we also researched the relationships between soil microbial and soil physicochemical properties.

Method: In this study, a field experiment was set up, with a total of three experimental treatments: no biochar application, 10 t/m^3 biochar application and 20 t/m^3 application. High-throughput sequencing technologies were used for soil samples of different treatment groups to understand soil microbial diversity and community structure.

Results: We found that the soil physicochemical properties after biochar addition were better than those without biochar addition, and the alpha diversity was higher in biochar addition level of 20 t/m^3 than other processing groups. *Proteobacteria*, *Cyanobacteria*, and *Actinobacteria* were the dominant phyla of this study. The dominant genera were *Skermanella*, *Nostoc*, *Frankia*, and *Unclassified-p-protecbacteria*. At the gate level, *Actinobacteria* had significant differences among the three groups with different addition amounts. The microbial community structure was mainly influenced by soil porosity, soil moisture content, nitrogen fertilizer and potassium fertilizer other than soil phosphate fertilizer and organic matter.

Conclusions: The results suggested that changes under different amounts of biochar added generate changes in soil physicochemical properties and control the soil composition of microbial communities. This provides a new basis for soil improvement.

【**Keywords**】Biochar; MiSeq sequencing; Alpha diversity; Bacterial phyla; Bacterial community structure

Introduction

Soil is an important carbon "source" and "sink" in terrestrial ecosystems. Soil carbon pools are mainly divided into soil organic carbon pools and inorganic carbon pools (Atkinson et al. 2010). The main way to mitigate climate change in the short term is to increase the soil organic carbon pool and maintain the stability of the soil organic carbon pool (Liang et al. 2010). Agricultural land accounts for 35% to 37% of the global land area, and is the land most affected by human activities. The decline of organic carbon in farmland soil is the most serious degradation factor (Bronick et al. 2005). Therefore, the change of soil organic carbon in agricultural land has been widely concerned by scholars.

Biochar has high stability and cannot be decomposed well by soil microorganisms. The impact of biochar on soil microorganisms is mainly through changes to the soil environment (Wu et al. 2017).

Biochar has a wide range of carbonization raw materials and low price. As a renewable recycling resource, it play an important role to affect the change of soil organic carbon (Chen et al. 2013). Biochar has highly developed pore structure, huge specific surface area and strong ion adsorption and exchange capacity. This characteristic can change the indexes of soil surface area, porosity, aggregate and density (Steiner et al. 2010), provide niches for colonisation of soil microorganisms (Zackrisson and Wardle. 1996; Warnock et al. 2007; Richard et al. 2013), and affect soil aeration, water content, root movement, microbial habitat (Wildman and Derbyshire. 1991), and C and N cycling in the terrestrial ecosystem (Nguyen et al. 2017; Zhang et al. 2017a; Zhang et al. 2018).

Soil microorganism is an important participant in biochemical process. It promotes the microcirculation of vegetation soil ecosystem and has a significant effect on improving soil fertility. The diversity and community structure of soil microorganism can significantly affect soil quality, and are important factor to evaluate soil quality. The diversity and community structure of soil microorganism play an important role in influencing the soil fertility, soil health, ecosystem's function and productivity (Zou et al. 2017).

Clayey raw soil has the characteristics of poor permeability, small gap, poor ventilation and drainage performance, slow fertilizer release and nutrient transformation (Li et al. 2012; Pang et al. 2021), and cannot meet the nutrients and water required for crop growth. Under natural conditions, the natural maturation process of raw soil is slow, which seriously hinders the rapid development of agriculture. Therefore, it is a certain trend of current agricultural development to realize the rapid improvement of the quality of new cultivated land and degraded land through biochar.

At present, biochar has been widely studied to improve soil health (Keya. 2016; Yuan et al. 2018). It is mainly reflected in the effects of adding biochar on soil physical property (Wang et al. 2016; Stéphanie et al. 2005), chemical property (Wang et al. 2021; Zhang et al. 2018), soil microbial diversity (Gundale et al. 2007; Ahmad et al. 2014; Cheng et al. 2019; Ding et al. 2019). Grossman et al. (2010) study found that biochar in carbon-rich soils in the Amazon Basin can increase the number and diversity of soil bacterial communities. Khodadad et al. (2011) found that the relative abundance of actinomycetes and chlortetracyclines in soils supplemented with biochar increased significantly, indicating that inert biochar can affect bacterial community composition. Rondon et al. (2007) found that the application of biochar can significantly increase the biomass of fungi and gram-negative bacteria, and can promote the biological nitrogen fixation ability of rhizobia and improve the activity of soil nitrifying microbial flora. Numerous studies have shown that biochar addition has an effect on soil microorganisms. However, most studies focus on the effect of biochar addition on multi-year degraded soil and different soil types (Wang et al. 2013; Wang et al. 2016; Zhang et al. 2019).There are few studies on the effects of biochar application on soil microorganisms in clayey raw soil, and the optimal amount of biochar addition has not been determined. Clay soil has poor permeability, small voids, and low degree of maturity, which seriously affects soil quality and crop yield. Therefore, this paper adds biochar to clayey raw soil, and studies the sample plots with different gradient biochar addition, in order to achieve the following goals: (1) which biochar addition has the best effect on the improvement of clayey raw soil; (2) what is the mechanism or principle of the effect of different addition amounts on different microorganisms; and (3) which soil physicochemical properties have a significant impact on soil microorganisms.

Materials and methods

Experimental field

The experiment was carried out in Qinling field monitoring center station, which located in Shangwang village, Tangyu Town, Mei County, Baoji City, Shaanxi Province, China (33°59′~34°19′ N and 107°39′~108°00′ E). This area was characterized by a warm temperate semi humid continental climate, and its altitude was ranged from 442 to 3767 m. The mean annual precipitation was 609.5 mm, and the annual mean temperature was 12.9 ℃. The soil texture was clayey soil.

Experimental design and treatments

The raw material of biochar come from fruit tree residues (were manufactured by Shaanxi Yixin Bio-energy Technology Development Co., Ltd.). These biochar were dried in a continuous pyrolysis plant to $<5\%$ moisture content before carbonisation. The production process was slow pyrolysis, at a highest treatment temperature of 550 ℃, and a heating rate of 5 ~ 10 ℃/min(Zwieten et al. 2010). The feedstock was kept in the reactor for 30 min on average, then directly sieved (2 mm mesh). The properties of biochar were as follows: pH was 9.42; EC was 0.15 dS/m; the content of total C was 794 g/kg, the content of total N was 9.82 g/kg, the content of total H was 16.7 g/kg, and the organic carbon was 763 g/kg.

In September 2020, this experiment started to implement. This experiment adopted the method of field experiment. In this experiment, 9 test plots were set up, and the size of them was 1.5 m × 3 m. The biochar application amount was 0, 10, and 20 t/hm^2, and 3 treatments were set. The plot adopted the random block design, and each treatment was set for three repetitions. Sprinkled the biochar evenly on the soil surface, and mixed it with the plough layer soil (20 cm) by manual stirring, so that the color of the soil was uniform everywhere, and ridges were left to stand. The same N, P and K fertilization schemes were adopted in the experimental plots, which were basically consistent with the fertilization habits of local farmers, which were N: 150 kg / hm^2 respectively; P_2O_5:120 kg/hm^2; K_2O: 90 kg/hm^2. The crops planted in the experimental plot are the same as the local crops. Wheat is planted in winter and spring, and corn is planted in summer and autumn.

Sample collection and analysis

In June 2021, soil samples were collected. The plant residues and stones were moved away from the plots. Then samples were collected from three different regions of the plot by using a core sampler (20 mm internal diameter). The sampling depth was 20 cm. The soil samples were directly sieved (2 mm mesh), and subsamples were mixed to avoid heterogeneity and yield a soil sample for each plot. All soil samples were divided into two parts: one part was naturally air dried for the determination of soil physical and chemical properties, and the other part was frozen in refrigerator of − 20 ℃ for the extraction of soil macrogenomic DNA.

Chemical analysis

The soil moisture content (SMC) was measured by drying and weighing method (105℃ for 24 h). Soil porosity (SP) was determined by ring knife method. Ammonium nitrogen (AN) and nitrate nitrogen (NN) were extracted with 0.01mol/L calcium chloride and then determined by AA3 flow injection analyzer. Available phosphorus (AP) was extracted with 0.5 mol/L sodium bicarbonate (pH 8.5) and then determined by Smartchem 200 continuous flow injection analyzer. Available K (AK) was extracted with 1mol/L ammonium acetate (pH 7) and determined by flame photometer. Organic matter content (OMC) was determined by heating oxidation of potassium dichromate sulfuric acid and titration of ferrous sulfate. The required index measurement methods referred to Soil Agrochemical Analysis (Third Edition) written by Shidan Bao (Shidan. 2000). Each analysis was performed in three replicates, and the data were presented as the averages.

DNA extraction and high-throughput Miseq sequencing

The total genomic DNA in each soil sample was extracted using the MoBio Powersoil® DNA Isolation Kit (MoBio Laboratories, USA). This method performed equally well over a range of different soils (Wüst et al. 2016). The quality and concentration of DNA were verified by 1% agarose gel electrophoresis and a NanoDrop™ 1000 spectrophotometer (Thermo Scientific, USA).

The V3-V4 region of the bacterial 16S rRNA gene was amplified using the PCR primers 338F (5′-ACTCCTACGGGAGGCAGCAG-3′) and 806R (5′-GGACTACHVGGGTWTCTAAT-3′) and a sample tagging approach, the size of amplicon was 468bp (Caporaso et al. 2012). The formal PCR test used TransGen AP221-02: TransStart Fastpfu DNA Polymerase, 20 μl reaction system: 5×FastPfu Buffer 4 μl, 2.5 mM dNTPs 2 μl, Forward Primer (5 μM) 0.8 μl, Reverse Primer (5 μM) 0.8 μl, FastPfu Polymerase 0.4 μl, BSA 0.2 μl, Template DNA 10 ng, Supplement ddH_2O to 20 μl. The following thermal cycling scheme was used: 30 cycles of initial denaturation at 95 ℃ for 3 min, denaturation at 95 ℃ for 30 s, annealing at 55 ℃ for 30 s, and extension at 72 ℃ for 45 s, followed by a final extension at 72 ℃ for 10 min. Amplicons were extracted from 2% agarose gels, purified using an AxyPrep DNA Gel Extraction Kit (Axygen Biosciences, Union City, CA, USA) according to the manufacturer's instructions and quantified using a QuantiFluor™(Promega, USA). Purified amplicons were pooled in equimolar amounts and paired-end sequenced on an Illumina MiSeq platform (Majorbio, Shanghai) according to standard protocols.

Sequencing data processing

The data of each sample was distinguished according to the index sequence, and the extracted data was saved in fastq format. According to the overlap relationship between Paired-end reads, the paired reads were merged into a sequence by using Fastp and Flash software. At the same time, the quality of reads and the effect of merge were quality controlled and filtered. The samples were distinguished according to the barcode and primer sequence at the beginning and end of the sequence, the effective sequence was obtained, and the sequence direction was corrected. Using Uparse (version 7.0.1090) software, the biological information of OTU at 97% similar level was statistically analyzed. According to the Silva database (lease138), 97% OTU representative sequences with similar level were classified by RDP classifier Bayesian algorithm. Selected the OTU or other taxonomic level with 97% similarity, and used mother (version v.1.30.2) to calculate the alpha diversity index (Chao, ACE, Shannon, Smith-Wilson) under different random sampling.

Statistical analysis

Differences in the soil physicochemical properties at these plots were compared using one-way ANOVA with Tukey's test. Student test was used to analyze the differences between alpha diversity indexes. The species composition of different samples at the phylum level and genus level was analyzed by R. The evolutionary tree was constructed according to the maximum likelihood method, and then the distance matrix between samples was obtained by FastUniFrac. Finally, the sample distance Heatmap diagram was made in R (version 3.3.1) that was a programming language for statistical calculation and plotting. The beta diversity distance matrix was calculated with Qiime, and NMDS analysis was carried out with vegan packages of R. ANOSIM and PERMANOVA were calculated with Vegan package of R language. Kruskal Wallis H test was used to test the significant difference between groups at the phylum level, and the stats package of R was used to plot. Use LEfse software to carry out linear discriminant analysis (LDA) on samples according to different groups to find out the species that have significant differences in sample division. The relationships between soil physical and chemical properties and soil microbial diversity and community structure were determined using the RDA function in redundancy analysis (RDA) in the vegan package in R. The correlation Heatmap analysis was carried out with pheatmap package of R language to calculate the correlation coefficient between soil physical and chemical properties and selected species.

Results

Soil physical and chemical properties

With the increase of biochar added, there was no significant difference between ammonium nitrogen and available phosphorus. There was also no significant difference in soil water content between this field with addition amount of 10 t/hm^2(BS) and control group, but there was significant difference in soil water content between this field with addition amount of 20 t/hm^2(MCS) and control group. Porosity and nitrate nitrogen were significantly different among the three treatments, and showed a gradual increasing trend. Compared with the control group, there were significant differences between BS and MCS in available potassium and organic matter, and the content increased with the increase of dosage (Table 1).

Table 1 Soil physical and chemical properties between different amounts of biochar added

Treatments	SMC (%)	SP (%)	AN (mg/kg)	NN (mg/kg)	AP (mg/kg)	AK (mg/kg)	OMC (g/Kg)
MC	18±1.3a	47±00.3a	8.13±1.012	21.46±1.318a	21.50±5.200	131.67±3.786a	5.62±1.650a
BS	18±0.3a	51±00.1b	11.74±4.839	36.62±1.519b	25.87±3.092	291.00±84.894b	34.73±7.870b
MCS	20±0.5b	52±00.7c	10.15±2.121	42.04±2.090c	24.00±2.835	420.67±66.606b	45.33±10.957b

Notes: Different lower-case letters in the same column indicate a significant difference at the 0.05 level. MC, BS, and MCS represent the biochar addition amount of 0 t/m^3, 10 t/m^3, and 20 t/m^3 respectively. SMC: soil moisture content, SP: soil porosity, AN: ammonium nitrogen, NN: nitrate nitrogen, AP: available phosphorus, AK: available K, OMC: organic matter content.

The composition of the microbial community among different treatments was assessed by MiSeq sequencing, which produced 49122 to 56739 sequences with different numbers of phylogenetic Operational Taxonomic Units (OTUs). All rarefaction curves approached the saturation plateau, indicating that the data volume of sequenced reads was reasonable and that increasing the number of reads made only a small contribution to the total number of OTUs. However, there were significant differences in the rarefaction curves obtained from the samples, that the higher the amount of biochar addition, the higher richness (Fig. 1).

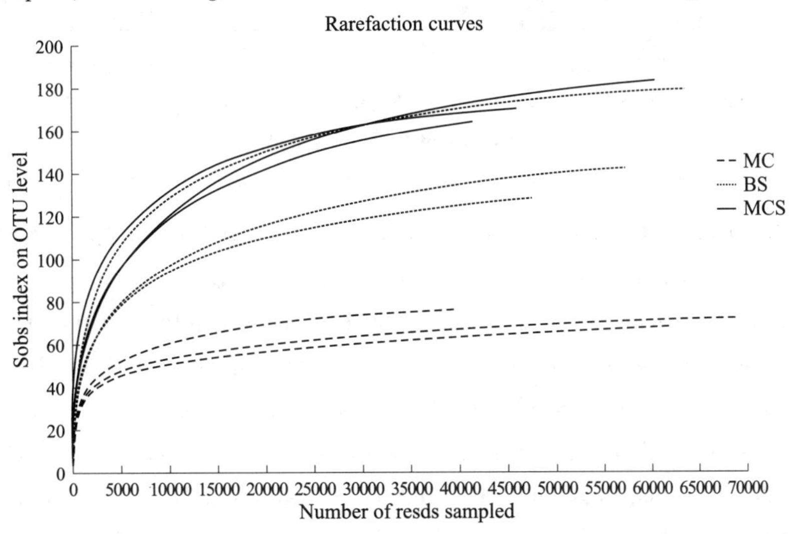

Fig. 1 Rarefaction curves between different amounts of biochar added

Notes: MC, BS, and MCS represent the biochar addition amount of 0 t/m^3, 10 t/m^3, and 20 t/m^3 respectively.

Effect of biochar addition on soil microbial community composition and overall diversity

The listed alpha diversity indices of soil bacterial were calculated based on the relative abundance of OTUs at 97% sequence similarity level and are shown in Fig. 2. Chao index and Ace index were used to describe community richness, Shannon index was used to describe community diversity, Smith-Wilson index was used to describe community evenness. BS and MCS had significantly higher Chao and Ace compared with MC, Smith-Wilson in BS and MCS also was significantly different but lower compared with MC, suggesting the biochar has been reported to be related to the richness and evenness. Shannon was not significantly different among these treatments, but Shannon in BS and MCS was higher than MC.

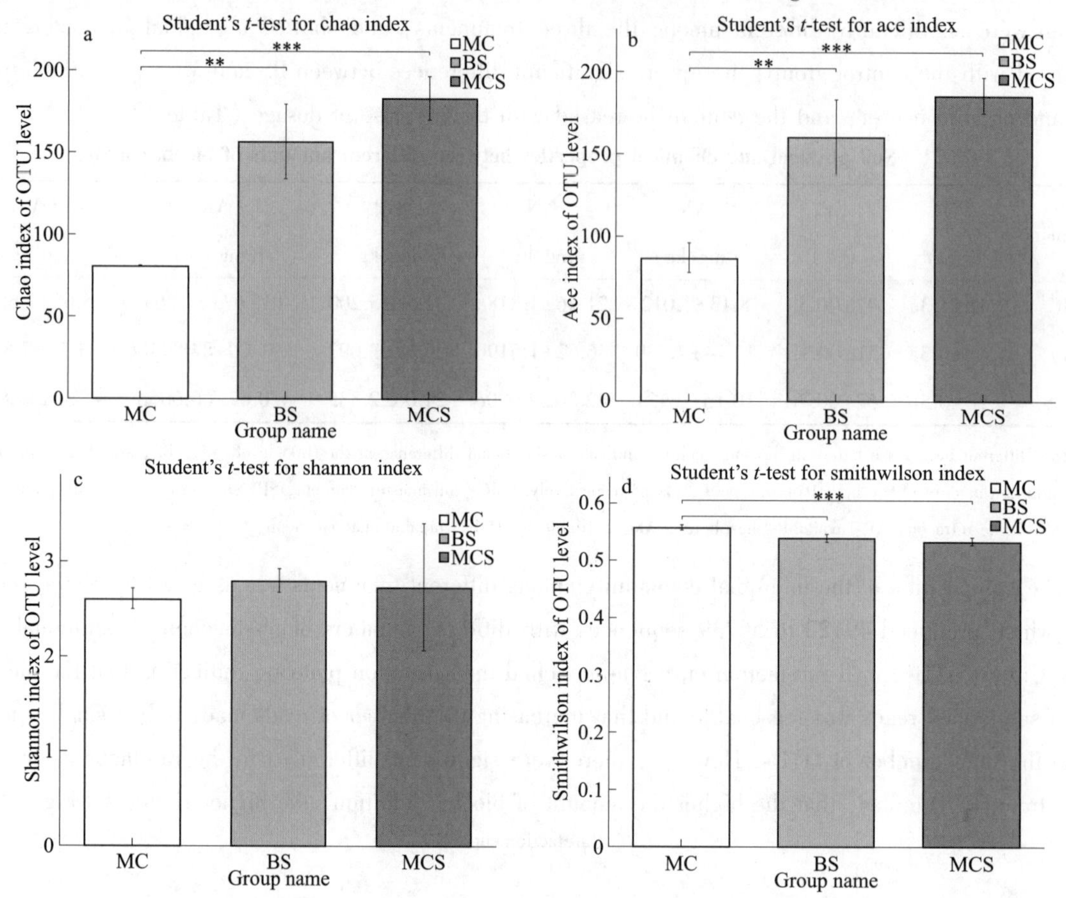

Fig. 2 Histogram of (a) Chao, (b) Ace, (c) Shannon, and (d) Smithwilson between different amounts of biochar added

Notes: MC, BS, and MCS are the same as in Fig. 1. * indicates $P \leq 0.05$, * * indicates $P \leq 0.01$, * * * indicates $P \leq 0.001$.

The relative abundances of major taxonomic groups have been showed in Fig. 3. OTUs were assigned into 6 bacterial phyla, 25 families, and 31 genera. The taxonomic classification of bacterial community composition showed that the dominant phyla, which accounted for more than 98% of the abundance of all species, were *Proteobacteria*, *Cyanobacteria* and *Actinobacteria*. All soils were dominated by the phylum *Proteobacteria*, accounting for 87.8% ~ 88.9% of all sequences among treatments. *Cyanobacteria* (1.1% ~ 12.1%) was the second most abundant phyla. It was worth noting that the content of *Actinobacteria* in MCS was significantly higher than that in BS and MC, and the content of *unclassified_k_norank_d_Bacteria* was higher in BS and MCS, which was almost absent in MC.

The dominant genera were *Skermanella*, *Nostoc*, *unclassified_p_Proteobacteria*, *unclassified_c_Alphaproteobacteria* and *Frankia*. *Skermanella* was the dominant genus accounting for 76.4% ~ 87.1%, and its content in MC was higher than BS and MCS. *Nostoc* (0.55% ~ 9.5%) was the second most abundant genera, which was almost absent in MCS. *Unclassified_p_Proteobacteria* and *unclassified_c_Alphaproteobacteria* appeared in BS and MCS. *Frankia* was unique in MCS.

Non-metric multidimensional scaling (NMDS) ordinations based on the Bray-Curtis similarity matrices was representative (Stress = 0.038 < 0.05) and indicated that experimental grouping was meaningful (ANOSIM, $p = 0.013 < 0.05$; PERMANOVA, $p = 0.004 < 0.01$). NMDS showed a clear separation of the bacterial

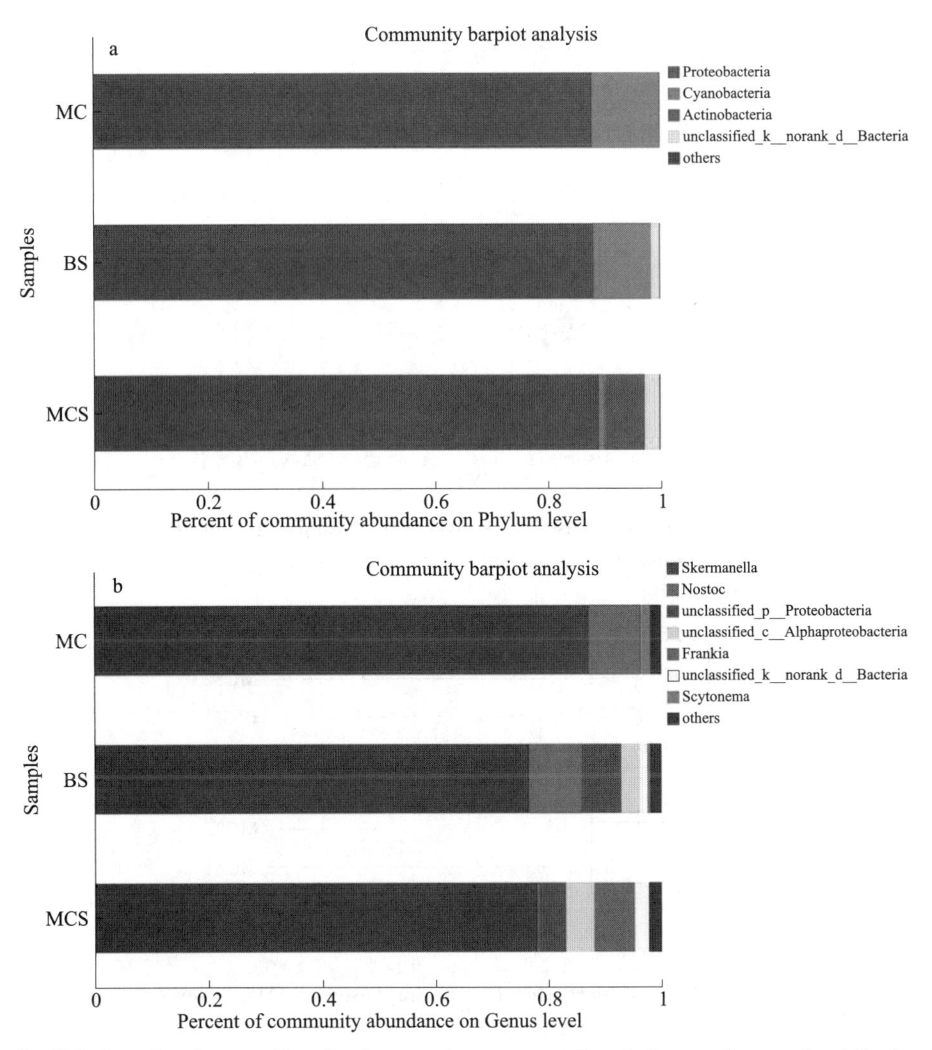

Fig. 3 Relative abundances of major taxonomic groups at the phylum and genus level for bacteria (a: phylum level; b: genus level)

Notes: MC, BS and MCS are the same as in Figure 1. Columns of different colors represent different species, and the length of the column represents the proportion of the species.

community structure in MCS from the other treatments, and MC was included in BS, which was more concentrated, the community structure of MC and BS was more similar (Fig. 4). The hierarchical clustering also indicated that MCS was separated from other treatments (Fig. 5). MC and BS were located in the same or similar branches, their community structure was basically similar; the close distance of MCS1, BS1 and BS3 indicates that they had similar effects on bacterial community structure.

Comparison of bacterial community structures in groups

The significance test of group differences at the gate level showed that *Actinobacteria* had significant differences among the three groups (Fig. 6, $p=0.035<0.05$), the average relative abundance was 7.55% in MCS. At the genus level, Frankia, *unclassified_p_Cyanobacteria*, *unclassified_o_Nostocales*, *unclassified_c_norank_p_Cyanobacteria* was significantly different among the three groups. *Frankia* had a higher average relative abundance in MCS, *unclassified_p_Cyanobacteria*, *unclassified_o_Nostocales*, and *unclassified_c_norank_p_Cyanobacteria* had a higher average relative abundance in MC. Lefse multistage species difference discriminant analysis showed that *F_unclassified_o_Nostocales*, *g_unclassified_o_Nostocales* is the marker of MC, *O_ Rhizobiales*, *p_Actinobacteria*, *c_Actinobacteria*, *o_Actinomycetales*, *f_Frankiaceae* is the marker of MCS (Fig. 7).

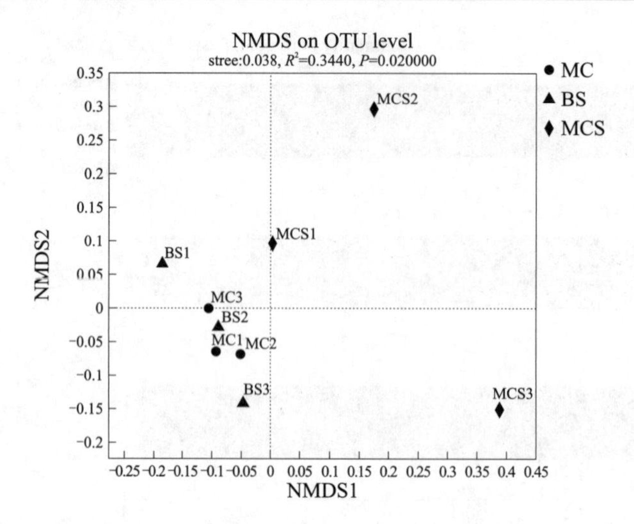

Fig. 4　NMDS plots based on Bray-Curtis dissimilarities of OTUs

Notes: MC, BS and MCS are the same as in Fig. 1.

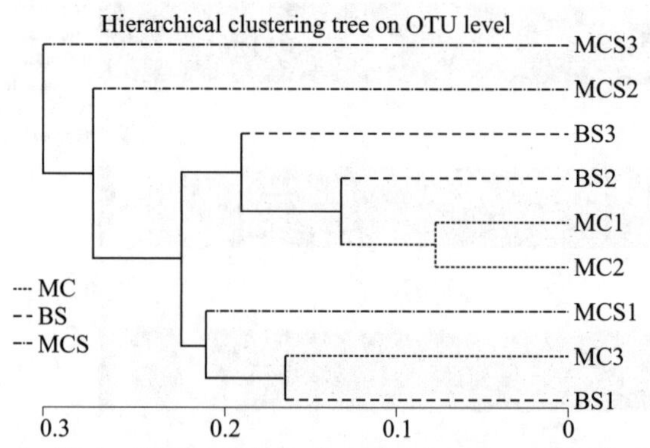

Fig. 5　Plots of hierarchical clustering based on OTU's level

Notes: MC, BS and MCS are the same as in Fig. 1.

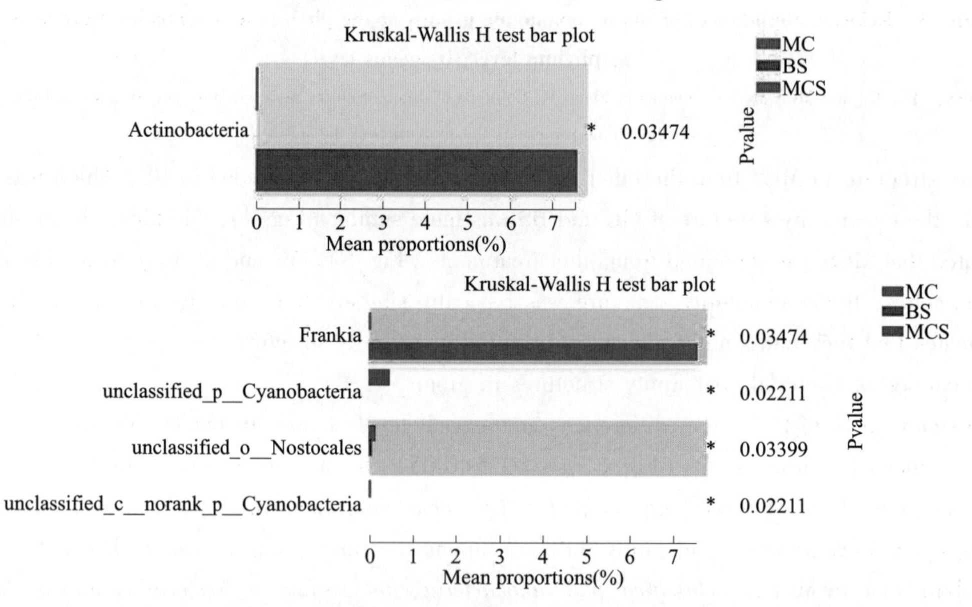

Fig. 6　Kruskal-Wallis H test plot at the phylum and genus level for bacteria

Notes: MC, BS and MCS are the same as in Fig. 1. Columns of different colors represent different groups. The rightmost is the p value and the corresponding symbol. * indicates $P \leq 0.05$, * * indicates $P \leq 0.01$, * * * indicates $P \leq 0.001$.

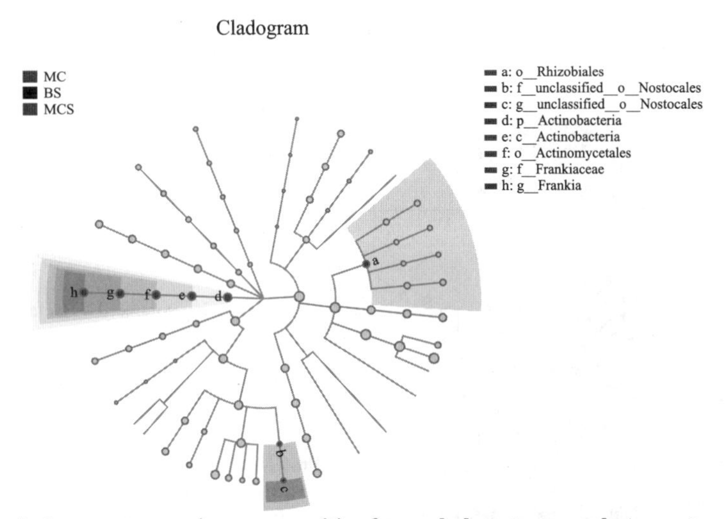

Fig. 7 Lefse cladogram comparing communities from phylum to genus between treatment groups

Notes: MC, BS and MCS are the same as in Fig. 1. Different color nodes represent microorganisms that are significantly enriched in the corresponding groups and have a significant impact on the differences between groups. Light yellow nodes indicate that there is no significant difference in different groups.

Correlations between soil microbial community composition and soil physical and chemical properties

Permanova analysis showed that there were significant positive correlation between SMC, AK and community structure. RDA results demonstrated that MCS were positively associated with SMC, SP, OMC, AK, NN and negatively associated with AP (Fig. 8). MC and BS had no significant effect on soil physical and chemical properties. Spearman correlation heatmap results were shown in Fig. 9. In addition to the negative correlation between AP and *Actinobacteria*, the influence of physical and chemical properties on Actinobacteria, *unclassified_d__Unclassified*, *unclassified_k__norank_d__Bacteria* and *Firmicutes* was positively correlated. Among them, *Actinobacteria* was significantly positively correlated with SP, SNC and AK, *Firmicutes* was significantly positively correlated with NN, and *unclassified_ k__ norank_ d__ Bacteria* was positively correlated with NN and sp. The influence of physical and chemical properties on *Proteobacteria* and *Cyanobacteria* was negatively correlated. Among them, *Cyanobacteria* was significantly negatively correlated with NN, SMC and AK. It is worth noting that NN has little effect on *Proteobacteria*, AN had little effect on *Firmicutes*, AP also had little effect on *Cyanobacteria*.

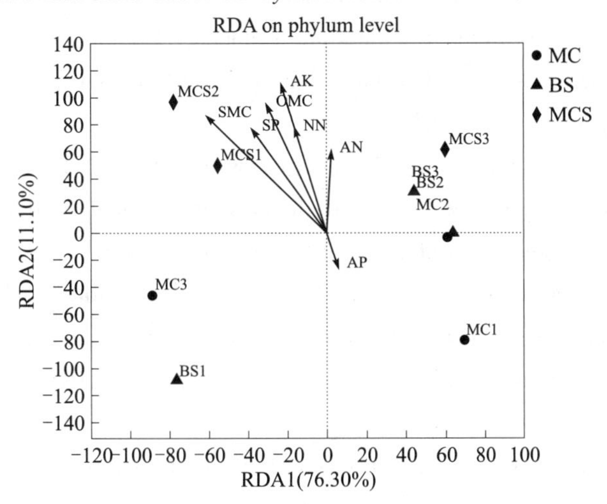

Fig. 8 RDA ordination plots showing the relationship between the bacterial community structure and physical and chemical properties

Notes: MC, BS and MCS are the same as in Fig. 1.

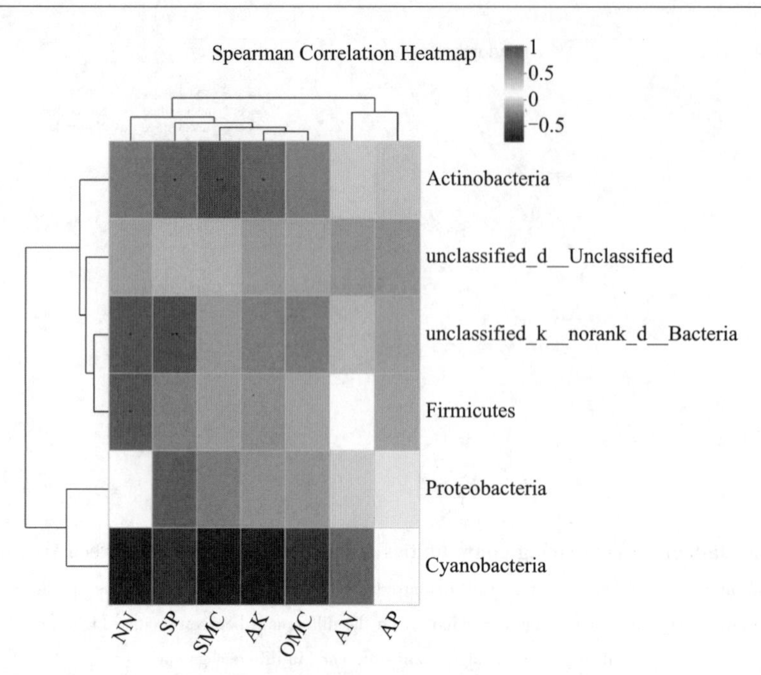

Fig. 9 Spearman correlation heatmap between bacteria at phylum level and physical and chemical properties

Notes: NN, SP, SMC, AK, OMC, AN and AP are the same as in Fig. 1. * indicates $P \leqslant 0.05$, * * indicates $P \leqslant 0.01$, * * * indicates $P \leqslant 0.001$.

Discussion

Biochar is mainly composed of carbon molecules. The addition of biochar can effectively change the physicochemical properties of soil. This study found that compared with the control, the treatment groups with different amounts of biochar were larger in the seven indexes of soil moisture content, soil porosity, ammonium nitrogen, nitrate nitrogen, available phosphorus, available potassium, and organic matter. With the increase of addition, soil moisture content, soil porosity, available potassium, nitrate nitrogen and organic matter also increased. The results of this study might support some result previously obtained. Yin et al. (2021) also found that the addition of biochar would change the physicochemical properties of soil and increase available phosphorus, total nitrogen, nitrate nitrogen, ammonium nitrogen and water content. Chen et al. (2018) found that the content of organic matter increased after the addition of biochar. Li et al. (2020) found that the soil porosity increased after the addition of biochar. This may be due to the porosity and composition of biochar, which increases the soil surface area, enhances the soil porosity and improves the micro ecological environment (Agusalim et al. 2010).

The biochar addition affected the physicochemical properties of soil, affected the living space of bacteria, and then affected the diversity of soil. This study found that the biochar addition increased the diversity of bacterial community and reduced the uniformity of bacterial community, and the species diversity showed an increasing trend with the increase of the amount of biochar, it is possible that the addition of biochar will change the soil microenvironment and cause the difference of bacterial community and biodiversity (Zhang et al. 2017), and this was consistent with many research (Nan et al. 2016; Wu et al. 2019; Hu et al. 2014; Thuy et al. 2014; Nguyen et al. 2018).

Studies have confirmed that biochar addition has an impact on microbial community composition (Hu et al. 2014). In this study, *Proteobacteria* and *Actinobacteria* are the dominant bacteria, this is consistent with the previous research results (Wu et al. 2019; Yin et al. 2021; Yao et al. 2017). Compared with the control, biochar addition significantly increased the relative abundance of *Actinobacteria* (Wu et al. 2019), it may be that after biochar was added, the soil nutrition was richer, and *Actinomycetes* was a eutrophic group, which can use the available carbon source to grow rapidly (Zeng et al. 2016), this showed that the addition of biochar to the soil makes *Actinomycetes* grow and reproduce better and had a significant impact on the structure of soil bacterial community, which was consistent with the previous research results (Zhang. 2014).

The porosity of biochar will create an aerobic environment, which was conducive to the growth and reproduction of soil microbial community. This study found that NN, SP, SMC and AK had significant effect on bacteria at the phylum level and were the main factors affecting the community structure. It is worth noting that the physical and chemical properties of soil had no significant effect on *Proteobacteria*, but *Actinobacteria* was positively correlated with SP, SMC and AK, *Cyanobacteria* was negatively correlated with NN, SMC and AK. This could be caused by *Proteobacteria* is the largest phylum in bacteria, with large intra phylum variability. *Proteobacteria* existed in large numbers in the study area, and the difference was not obvious. Therefore, the soil physicochemical properties had no significant effect on its abundance. The importance of soil physical and chemical properties in shaping microbial communities had been proved by several studies. As an important part of soil structure, porosity has a positive effect on the conduction of water and air in the soil (Luo et al., 2019), this is conducive to the growth and reproduction of aerobic bacteria. Deng et al. (2013) also found that soil porosity has an impact on soil microbial communities. Soil water content is one of the leading factors to maintain the life activities of soil microorganisms (Clark et al. 2009), and has a significant impact on soil microbial community, which was also confirmed by Li et al. (2020). Available potassium can be decomposed and utilized by microorganisms, and its content affected microbial diversity. The study by Wang et al. (2021) found that available potassium was negatively correlated with *Proteobacteria* and positively correlated with *Actinobacteria*, this was consistent with the research in this paper. Nitrate nitrogen is a kind of soil nitrogen fertilizer, and its nutrient content affects the abundance and diversity of soil microorganisms (Lan et al., 2017). Song et al. (2021) also found that nitrate nitrogen has a significant effect on microbial community structure.

Conclusions

Our study provides new basis for the rapid maturing technology of clayey raw soil using biochar. Our results indicated that the addition of biochar significantly improved the lack of fertility and low soil microbial diversity of clayey raw soil. The present study, using high-throughput sequencing technologies, provided a detailed picture of bacterial community variations on the phylum level among different biochar additions and showed the relationship between physical and chemical properties and soil microbial communities. Sequencing results and diversity indices indicated that the alpha diversity was higher in biochar addition level of 20 t/m^3 than other processing groups. The dominant phyla were *Proteobacteria*, *Cyanobacteria*, and *Actinobacteria*. At the gate level, *Actinobacteria* had significant differences among the three groups with different addition amounts, and the content was the highest in the treatment group with 20 t/m^3 addition amount. The microbial community structure was mainly influenced by soil porosity, soil moisture content, nitrogen fertilizer and potassium fertilizer other than soil phosphate fertilizer and organic matter. This experiment shows that high addition amount of biochar has better effect on soil improvement, but the range of biochar addition in this study is small, and it is necessary to continue to expand the range of biochar addition for further research.

Acknowledgements

This research was funded by Technology Innovation Center for Land Engineering and Human Settlements, Shaanxi Land Engineering Construction Group Co., Ltd and Xi'an Jiaotong University

(2021WHZ0089) and Shaanxi Provincial Land Engineering Construction Group (DJNY2022-53).

Authors' contributions

Corresponding author is the first author: guide the completion of this experiment and revise their papers. The second author: performed experiments and recorded data. Both authors read and approved the final manuscript.

Funding

This research was funded by Technology Innovation Center for Land Engineering and Human Settlements, Shaanxi Land Engineering Construction Group Co., Ltd and Xi'an Jiaotong University (2021WHZ0089) and Shaanxi Provincial Land Engineering Construction Group (DJNY2022-53).

Availability of data and materials

The data were obtained by the authors.

Declarations

Ethics approval and consent to participate.

The study did not violate ethics, and all participants agreed to publish the paper.

Consent for publication

Not applicable.

Competing interests

On behalf of all authors, the corresponding author states that there is no conflict of interest.

References

Agusalim M, Hadi U W, Syechfani M S (2010) Rice husk biochar for rice based cropping system in acid soil. the characteristics of rice husk biochar and its influence on the properties of acid sulfate soils and rice growth in West Kalimantan, Indonesia. Journal of Agricultural Science (1916~9752) 2(1):39-47.

Ahmad M, Rajapaksha A U, Lim J E et al (2014) Biochar as a sorbent for contaminant management in soil and water: A review. Chemosphere 99: 19-33.

Atkinson C J, Fitzgerald J D, Hipps N A (2010) Potential mechanisms for achieving agricultural benefits from biochar application to temperate soils: a review. Plant & Soil 337(1-2):1-18.

Bronick C J, Lal R (2005) Soil structure and management: a review. Geoderma 124:3-22.

Caporaso J G, Lauber C L, Walters W A et al (2012) Ultra-high-throughput microbial community analysis on the Illumina HiSeq and MiSeq platforms. Isme Journal Multidisciplinary Journal of Microbial Ecology 6(8):1621-1624.

Chen K, Xu X, Peng J, Feng X, Han X (2018) Effects of biochar and biochar-based fertilizer on soil microbial community structure. Scientia Agricultura Sinica 51:1920-1930.

Chen W F, Zhang W M, Meng J (2013) Advances and Prospects in Research of Biochar Utilization in Agriculture. Scientia Agricultura Sinica 46(16):3324-3333.

Cheng J, Lee X, Tang Y, Zhang Q (2019) Long-term effects of biochar amendment on rhizosphere and bulk soil microbial communities in a karst region, southwest China. Applied Soil Ecology 140:126-134.

Clark J S, Campbell J H, Grizzle H, Acosta-Martìnez V, Zak J C (2009) Soil microbial community response to drought and precipitation variability in the Chihuahuan Desert. Microbial Ecology 57(2):248-260.

Deng C, Dong B L, Qin J T et al (2013) Effects of long-term fertilization on soil property changes and soil microbial biomass. Soils 45(05):888-893.

Ding W S, Wei Z C, Meng L Q et al (2019) Effects of biochar on soil bacterial diversity in Chinese firplantations. Journal of Forest and Environment 39(06):584-592.

FastUniFrac. http://UniFrac.colorado.edu. Accessed 20 May 2022.

Grossman, Julie M et al. Amazonian Anthrosols Support Similar Microbial Communities that Differ Distinctly from Those Extant in Adjacent, Unmodified Soils of the Same Mineralogy. Microbial Ecology 60.1(2010):192-205.

Gundale M J, Deluca T H (2007) Charcoal effects on soil solution chemistry and growth of Koeleria macrantha in the ponderosa pine/Douglas-fir ecosystem. Biology & Fertility of Soils 43(3):303-311.

Hu L, Cao L, Zhang R (2014) Bacterial and fungal taxon changes in soil microbial community composition induced by short-term biochar amendment in red oxidized loam soil. World Journal of Microbiology & Biotechnology 30(3):1085-1092.

Keya Z. (2016) Effects of Soil Amendment Application on the Soil Quality and the Flue-Cured Tobacco Growth Based on Biochar. Master, Nanjing Agricultural University.

Khodadad, Christina L M., et al. Taxa-specific changes in soil microbial community composition induced by pyrogenic carbon amendments. Soil Biology and Biochemistry 43.2(2011):385-392.

Lan G, Li Y, Wu Z, Xie G (2017) Impact of tropical forest conversion on soil bacterial diversity in tropical region of China. EUR J SOIL BIOL 83:91-97.

Lefse Software. http://huttenhower.sph.harvard.edu/galaxy/root? tool_id=lefse_upload. Accessed 20 May 2022.

Li C Z, Zhang H, Yao W J et al (2020) Effects of biochar application combined with nitrogen fertilizer on soil physico-chemical property and winter wheat yield in the typical ancient region of Yellow River, China. Chinese Journal of Applied E-cology 31(10):3424-3432.

Li Q M, Zhang L L, Liu H M et al (2020) Effects of cover crop diversity on soil microbial community functions in a ki-wifruit orchard. JAES 39(02):351-359.

Li W B, Li X P, Li H Y, Ying Y et al (2012) Effects on micro-ecological Characteristics in clayey soil of tobacco area under different sand adding proportions. Journal of Northwest A & F University 40(11):85-90,96.

Liang B Q, Lehmann J, Sohi S P et a1 (2010) Black carbon affects the cycling of non-black carbon in soil. Organic Ge-ochemistry 41(2):206-213.

Luo J, Li L Z, Que Y X et al (2019) Effect of subsoiling depths on soil physical characters and sugarcane yield. Chinese J Appl Ecol 30(02):405-412.

Mother (version v.1.30.2). https://mothur.org/wiki/calculators. Accessed 20 May 2022.

Nan X, Tan G, Wang H, Gai X (2016) Effect of biochar additions to soil on nitrogen leaching, microbial biomass and bacterial community structure. European Journal of Soil Biology 74: 1-8.

Nguyen T, Wallace H M, Xu C Y et al (2017) Short-term effects of organo-mineral biochar and organic fertilisers on ni-trogen cycling, plant photosynthesis, and nitrogen use efficiency. Journal of Soils & Sediments,17(12):1-12.

Nguyen T T N, Wallace H M, Xu C Y et al (2018) The effects of short term, long term and reapplication of biochar on soil bacteria. Science of The Total Environment 636:142-151.

Pang C M, Guo X F, Pang Y N et al (2021) A dataset of grain yield and soil nutrient of wheat and maize in fluvo-aquic clayey soil region of Southwest Shandong Province based on the annual straw returning to the field from 2010 to 2020. China Scientific Data 6(04):187-195.

PiaK W, Nacke H, Kaiser K et al (2016) Estimates of Soil Bacterial Ribosome Content and Diversity Are Significantly Affected by the Nucleic Acid Extraction Method Employed. Applied and Environmental Microbiology 82(9).

Richard S, Quilliam, Helen C et al (2013) Life in the "charosphere"-Does biochar in agricultural soil provide a signifi-cant habitat for microorganisms? Soil Biology and Biochemistry 65(1):287-293.

Rondon, Marco A et al. "Biological nitrogen fixation by common beans (Phaseolus vulgaris L.) increases with bio-char additions." Biology & Fertility of Soils 43.6(2007):699-708.

Shidan Bao (2000). Soil Agrochemical Analysis (Third Edition). China Agricultural Press, Beijing.

Silva Database (lease138). http://www.arb-silva.de. Accessed 20 May 2022.

Song J S, Zhang X L, Kong F L et al (2021) Effects of biomass conditioner on soil nutrient and microbial community characteristics ofalpine desertified grassland in northwest Sichuan, China. Chinese J Appl Ecol 32(06):2217-2226.

Steiner C, Glaser B, Geraldes Teixeira W, Lehmann J, Blum W E H, Zech W (2010) Nitrogen retention and plant up-take on a highly weathered central Amazonian Ferralsol amended with compost and charcoal. Journal of Plant Nutrition and Soil Science 171(6):893-899.

Stéphanie T, Jean-François P, Ballof S (2005) Manioc peel and charcoal: a potential organic amendment for sustainable soil fertility in the tropics. Biology & Fertility of Soils 41(1):15-21.

Thuy T D, Corinne B, Yvan B, Thierry B, Thierry H T, Jean L J, Patrice L, Bo V N, Pascal J (2014) Influence of buffalo manure, compost, vermicompost and biochar amendments on bacterial and viral communities in soil and adjacent aquatic systems. Applied Soil Ecology 73(2):78-86.

Uparse Software (version 7.0.1090). http://drive5.com/uparse. Accessed 20 May 2022.

Wang J, Shi Y, Ziyuan L I et al (2016) Effects of Biochar Application on N 2 O Emission in Degraded Vegetable Soil and in Remediation Process of the Soil. Acta Pedologica Sinica 53(3):713-723.

Wang J, Xiong Z, Kuzyakov Y (2016) Biochar stability in soil: meta-analysis of decomposition and priming effects. GCB Bioenergy 8(3):512-523.

Wang L L, Cao Y G, Wang F et al (2021) Effect of Biochar on Reconstructed Soil Chemical Propertiesand Drought Resistanceof Medicagosativa. Research of Soiland Water Conservation 28(6): 105-114.

Wang X H, Guo G X, Zheng R L et al (2013) Effect of Biochar on Abundance of N-Related Functional Microbial Communities in Degraded Greenhouse Soil. Acta Pedologica Sinica, 50(3):624-631.

Wang Y, Sun C C, Zhou J H et al (2019) Effects of biochar addition on soil bacterial community in semi-arid region. China Environmental Science 39(5):2170-2179.

Wang Z Q, Zhang J X, Yang X L et al (2021) Characteristics of Soil Microbial Diversity in Different Patches of Alpine Meadow. Acta Agrestia Sinica 29(9):1916-1926.

Warnock D D, Lehmann J, Kuyper T W, Rillig M C (2007) Mycorrhizal responses to biochar in soil concepts and mechanisms. Plant & Soil 300:9-20.

Wildman J, Derbyshire F (1991) Origins and functions of macroporosity in activated carbons from coal and wood precursors. Fuel, 70(5):655-661.

Wu C, Shi L Z, Xue S G et al (2019) Effect of sulfur-iron modified biochar on the available cadmium and bacterial community structure in contaminated soils. Science of The Total Environment 647:1158-1168.

Wu S, He H, Inthapanya X et a1 (2017) Role of biochar on composting of organic wastes and remediation of contaminated soils-a review. Environmental Science & Pollution Research 24(20):16560-16577.

Wu Y G, Zhang G L, Lai X et al (2014) Effects of Biochar Applications on Bacterial Diversity in Fluvor-aquic Soil ofNorth China. Journal of Agro-Environment Science 33(5):965-971.

Wüst P K, Nacke H, Kaiser K et al (2016). Estimates of Soil Bacterial Ribosome Content and Diversity Are Significantly Affected by the Nucleic Acid Extraction Method Employed. Appl Environ Microbiol 82(9):2595-2607.

Yao Q, Liu J, Yu Z et al (2017) Changes of bacterial community compositions after three years of biochar application in a black soil of northeast China. APPL SOIL ECOL 113:11-21.

Yin Q Y, Li X, Wang D et al (2021) Effects of continuous application of biochar for 4 years on soil bacterial diversity and community structure. Journal of Henan Agricultural University 55(4):752-760,775.

Yuan J J, Tong Y A, Lu S H, et al (2018) Combined application of biochar and inorganic nitrogen influnces the microbial properties in soils of jujube orchard. Journal of Plant Nutrition and Fertilizers 24(4):1039-1046.

Zackrisson O, Wardle N (1996) Key ecological function of charcoal from wildfire in the Boreal forest. Oikos, 77(1): 10-19.

Zeng J, Liu X, Song L, Lin X, Chu H (2016) Nitrogen fertilization directly affects soil bacterial diversity and indirectly affects bacterial community composition. SBB 92:41-49.

Zhang G (2014) Effects of Biochar Applications on Bacterial Diversity in Fluvor-aquic Soil of North China. Journal of Agro-Environment Science 33(5):965-971.

Zhang J N, Zhou S, Li G N (2018) Improving the coastal mudflat soil chemical properties and rice growth using straw biochar. Journal of Agricultural Resources and Environment 35(6):492-499.

Zhang Q, Liu B J, Lu Y U, Wang R R, Li F M (2019) Effects of biochar amendment on carbon and nitrogen cycling in coastal saline soils: A review. Journal ofNatural Resources 34(12):2529-2543.

Zhang Y, Drigo B, Bai S H, Menke C, Zhang M, Xu Z (2018) Biochar addition induced the same plant responses as elevated CO2 in mine spoil. Environmental Science & Pollution Research 25:1460-1469.

Zhang Y J, Wu T, Zhao J et al (2017) Effect of biochar amendment on bacterial community structure and diversity in straw-amended soils. Acta Scientiae Circumstantiae 37(2):712-720.

Zhang Y L, Chen H, Bai S H, Menke C, Xu Z H (2017a) Interactive effects of biochar addition and elevated carbon dioxide concentration on soil carbon and nitrogen pools in mine spoil. Journal of Soils and Sediments 17(3):1-10.

Zou Q, An W H, Wu C et al (2017) Red mud-modified biochar reduces soil arsenic availability and changes bacterial composition. Environmental Chemistry Letters 16:615-622.

Zwieten L V, et al. "Influence of biochars on flux of N2O and CO2 from Ferrosol." Soil Research 48.11(2010):1043-1046.

本文曾发表于2022年《Annals of Microbiology》第72卷

Human Health Risk Assessment and Distribution of VOCs in a Chemical Site, Weinan, China

Yan Li[1,2,3], Bo Yan[1,2,3]

(1. Shaanxi Provincial Land Engineering Construction Group Co., Ltd., Xi'an 710075, China; 2. Institute of Land Engineering and Technology, Shaanxi Provincial Land Engineering Construction Group Co., Ltd., Xi'an 710021, China; 3. Technology Innovation Center for Land Engineering and Human Settlements, Shaanxi Land Engineering Construction Group Co., Ltd and Xi'an Jiaotong University, Xi'an 710075, China)

【Abstract】 The study assessed the volatile organic compound (VOCs) pollution characteristics in a chemical site in Weinan, China. The results indicated that chloroform, benzene, trichloroethylene, 1,2-dichloroethane, ethylbenzene, 1,2-dichloropropane and 1,2,3-trichloropropane were all exceeded the soil standard limit for soil contamination of development land (GB36600, PRC). Using pollution index, ambient severity, and correlation coefficient to revealed industrial production and relocation activities as sources of VOCs contamination in site. The carcinogenic risk assessed by human exposure to site VOCs through ingestion, respiration, exposure, etc. exceeded the potentially acceptable level (1.0×10^{-6}). 1,2,3-trichloropropane has the highest carcinogenic risk across all pathways, regions, and populations. The long-term exposure and emission of VOCs in the investigated sites could likely pose an adverse health risk to site staff and the surrounding sensitive groups. Therefore, it is necessary to carry out strict investigation and evaluation of the site, and timely repair and control to protect the water, soil and air environment and to avoid the long-term cumulative exposure risk to human health caused by VOCs emission.

【Keywords】 Carcinogenic and non-carcinogenic risk; Health risks; Chemical site; Volatile organic compounds

1 Introduction

The acceleration of urbanization has led to the relocation of existing urban industrial sites. Thus, many vacated industrial sites that need redevelopment also require remediation of contamination[1-2]. According to incomplete statistics, over 1 lakh high-pollution plants have been shut down or relocated in the past decade in China. Statistically, the main soil pollutants in China's contaminated sites are volatile organic compounds (VOCs) and semivolatile volatile organic compounds (SVOCs), which include benzene, chloroform, toluene, carbon tetrachloride, tetrachloroethylene, and dichloroethan. Moreover, most of them are cumulative, diverse, toxic and carcinogenic[3-4].

As the precursors of ozone and secondary organic aerosols[5], VOCs are ubiquitous in the atmosphere and the majority of them have been proved to be detrimental to uman health. VOCs in the environment were formed by natural and anthropogenic factors. As precursors of ozone and secondary organic aerosols, natural emissions contribute 91.9% of total volatile organic compounds at a global scale. But in human-inhabited areas, anthropogenic factors are comparable to natural factors[6-7], such as vehicles, Landfill, Liquefied Petroleum Gas/Natural Gas, Biomass/Biofuel Combustion, Industrial Production, Catering, Building Materials and Decoration[8-9]. A large number of VOCs potentially have carcinogenic effects on the human body, which is considered to be a more serious problem than other health-related effects. The main effects of VOCs on human health are usually related to the central nervous and hematopoietic systems[10-11]. Many studies have shown that VOCs polluted soil environment in industrial site soil enters the human settlement environment through precipitation, runoff and volatilization. In turn, it affects the central nervous system, blood, immune system and skin of the human body through diet, respiration, contact, etc[12-13]. Scholars have conducted research on the pollution status and sources of ambient VOCs in different regions. They found that sufficient sunlight and good air diffusion conditions in spring and summer are more conducive to photochemical reaction combustion and VOCs diffusion[14]. Therefore, the concentration in autumn and winter is higher than that in spring and summer[15]. Xiang et al[16] found that the concentration of VOCs in farmland around Shanghai industrial zone in China was higher than that in Beijing and Ningbo, China, but lower than that in Taiwan, China and Aliaga, Turkey. Hu et al[17] studied the emission characteristics of VOCs in different functional areas in Hefei, China, and found that the detected concentration of VOCs in traffic areas was the highest, followed by industrial areas, development areas and residential areas, indicating that vehicle exhaust is one of the main sources of high concentrations of VOCs in the air.

Several countries have conducted studies on VOCs contamination and potential hazards to human health in urban chemical sites soil or near-surface air[18-19]. The United States "Federal Positive Risk Assessment: Management Procedures" first proposed a four-step research system for health risk assessment in 1983. Subsequently, other countries had improved the assessment of legal and technical guidelines, and constantly deepening and perfecting the various types of potential pollutants, exposure pathways and risk assessment methods[20-21]. China's research on the health risk assessment of VOCs in the environment started relatively late, and in the early stage of the research, it mainly used foreign advanced assessment theories and methodologies for reference. Zhang et al[22] analyzed and evaluated the compositional changes and health risks of VOCs during the remediation process of a closed pesticide and chemical plant. Zhao et al[23-24] studied the most serious VOCs pollution during site excavation during the restoration of a site in Zhenjiang, which is closely related to the original production process and by-products. Through the research by many scholars on the types of pollutants, exposure routes, and risk assessment index systems in different environmental media, China has formed a relatively systematic assessment system, and standardized the technology, methods and content of site environmental risk assessment (Series standard HJ25). For example, Nie[25] used China HJ25.3 to evaluate the health risk status of a chemical plant in the south, and found that harmful VOCs in the workshop, raw materials and product areas entered the environment through respiration, skin contact and soil steaming heat, and had high cancer risk.

Numerous studies have shown that the characteristics and risks of VOCs caused by the remediation of contaminated sites have become the focus of environmental concerns[26]. Restoration of contaminated industrial sites, regardless of whether this includes on-site or off-site types of repair, generally requires digging of the polluted soil. During the excavation of contaminated soil, large amounts of VOCs may be released into the atmosphere, thereby exposing the operating staff to hazards. As an important source and sink of VOCs, it is of great gnificance to study the content, distribution, diffusion characteristics and risks of VOCs in soil[27]. However, in the early stage of investigation, the research on the distribution of VOCs in the site soil and health risk assessment is not enough. In order to explore the characteristics of site soil VOCs and the health risks for the surrounding sensitive bodies, and to provide scientific guidance for the restoration and health protection of similar polluted sites, this study was carried out in a chemical site, in Weinan City, China, 2018.

2 Materials and methods

2.1 Study area, sampling, and analysis

The study was based on a closed chemical plant (site) in Weinan, China, with an area of 92000 m^2[28-29]. The site was surrounded by Weihe River, Youhe River, sensitive residential areas, schools and major traffic ways. the site was a concentration area of small enterprises for the production such as pesticides, chemicals, and building materials. Finaly, the company's leading product was fumaric acid. The chemical site was mainly divided into the pesticide production area (P), the food additive production area (F), the benzene purification and storage area (B), product storage area (S), living area (L), office area (O), and wastewater treatment area(W) (Fig. 1). In the future, the site will be planned for residence and park.

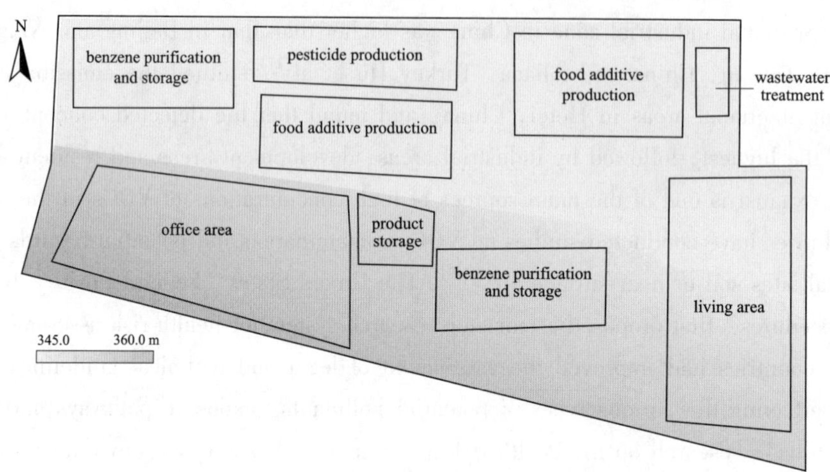

Fig. 1 The layout and location of the chemical site

The soil environment of the chemical site was investigated from August to October 2018. According to Chinese technical guidelines for monitoring during risk control and remediation of soil contamination of land for cinstruction (HJ25.2), and comprehensive zoning arrangement method and system arrangement method, a total of 36 soil sampling points were arranged, and the maximum sampling depth was 10 m. Soil samples were taken every 0.5 m at a depth of 0~3.0 m, and soil samples were taken every 1.0 m below 3.0 m. A total of 65 soil samples were collected. A drilling rig (QZ-50G produced by JIEKE, China) was used to drill the soils at different depth. All samples were stored in dedicated brown bottle to keep the soil fresh (at 4 ℃), and sent to the laboratory for analysis immediately. To check if the whole process of sample collection and analysis was contaminated, two blank samples (with 10 mL of collective modifier, and 2.0 g of quartz sand) was also conducted.

The VOCs of the chemical site soil were detected by the headspace-gas chromatography-mass spectrometry (GC-MS) method.The soil was extracted by shaking with methanol solution (150 times/min, 10 min), then quartz sand, 10 ml matrix modifier, 100 μL of methanol solution and 2.0 μL of internal standard (chlorobenzene) and substitutes were added to the extract then allowed to stand after shaking (150 times/min, 10 min). The extract was assayed on a VF-624MS column (60 m, 0.25 mm) on a GC-MS (Agilent 7890B-5977A, USA). The results showed that the detection components of the blanks of the whole process were lower than the detection limit(0.8~4 μg/kg). The samples with not less than 10% of the total were randomly selected for parallel analysis, and the calculated relative deviation was lower than 13.8% (no more than 10% of the total matrix spiked sample, the recovery of matrix peaked between 93.4% and 108.8%)[30].

2.2 Data analysis method

The *pollution index* (PI) was used to evaluate the degree of VOCs in chemical site, which also remains to be the first step to know VOCs pollution levels. The equation for calculating PI was the ratio of VOCs measured concentrations the soil divided by the value of environmental quality risk control standards for soil

(GB36600). According to the document, the variation in PI could be defined as follows: $PI \leqslant 1$ (non-pollution), $1 < PI \leqslant 2$ (minor pollution), $2 < PI \leqslant 3$ (light pollution), $3 < PI \leqslant 5$ (medium pollution), and $PI > 5$ (heavy pollution) [31].

As an important method for organic contaminants evaluation, *ambient severity* (AS) method is introduced into the VOCs evaluation soil of chemical site. The potential risk from VOCs was evaluated using the AS method, which could be calculated with the equation[32]:

$$AS_i = C_i / C_i^A \tag{1}$$

Where AS_i is the ambient severity for compound i in the soil; C_i is the measured concentration of compound i; and C_i^A is the target value for compound i, which can be defined as the risk screening values for soil environmental quality risk control standard four soil contamination of development land in China (GB 36600). It is assumed that the potential risk posed by each VOCs to humans has a linear relationship with its AS value[33], and that the total ambient severity (TAS) is the sum of the AS values for each compound, that is $TAS = \sum AS_i$ [34].

Human health risk assessment of contaminated sites is used as a means of pollution assessment[35-36]. Human health risk assessment is a quantitative method for quantifying the adverse effects of human exposure to VOCs from a contaminated environmental medium (e.g., soils and sediments). There are six pathways for adults and children to become exposed to soil VOCs: ingestion, inhalation, dermal adsorption, inhalation of gaseous pollutants from the surface and underlying soil in outdoor air, and inhalation of gaseous pollutants from the underlying soil in indoor air (HJ 25.3)[37]. Carcinogenic risk (CR) is defined as the probability of an individual developing any type of cancer throughout human lifetime due to exposure to carcinogenic. For VOCs, the sum of CR values of different organic compounds is called total carcinogenic risk (TCR). CR and TCR are calculated by the formula:

$$CR = ADI \times SF \tag{2}$$

$$TCR = CR_{ois} + CR_{dcs} + CR_{pis} + CR_{iovl} + CR_{iov2} + CR_{iivl} \tag{3}$$

where ADI is the average intake, SF is the carcinogenicity slope factor, other parameters are obtained from the study area and US EPA guidebook[38-39]. Risks $\leqslant 1.00 \times 10^{-6}$ are considered ignorable. Conversely, risks lying $> 1.00 \times 10^{-6}$ are generally considered unacceptable lifetime carcinogenic risk.

The Hazard Quotient (HQ) was calculated as the ratio of the ADI and the reference dose (R_{fD}) for a given contaminant (Equations (4))[40]. Used to characterize the level that a human exposed to a noncarcinogenic contaminant by a single route. The sum of the HQ values of all metals in the soil, called hazard index (HI), was used to assess the overall noncarcinogenic effects posed by multiple contaminants (Equations (5)).

HQ characterized the level which the human body is harmed by exposure to a non-carcinogenic pollutant through a single route.

$$HQ = ADI / R_{fD} \tag{4}$$

$$HI = HQ_{ois} + HQ_{dcs} + HQ_{pis} + HQ_{iovl} + HQ_{iov2} + HQ_{iivl} \tag{5}$$

If the HI value is <1, the exposed individual is unlikely to experience obvious adverse health effects; if the HI value is >1, there could be a risk of noncarcinogenic effects.

3 Result and discussion

3.1 Concentration, composition and distribution of VOCs

Seven VOCs were detected in soil samples from chemical sites: chloroform (CH), benzene(B), trichloroethylene (TH), 1,2-dichloroethane (1,2-DE), ethylbenzene (EB), 1,2-dichloropropane (DP-1,2) and 1,2,3-trichloropropane (1,2,3-TH), and had a relatively high detection rates (>23%) (Table 1). That was basically consistent with the research results in similar sites from Zhang and Kyab[41-42], which indicating that the VOCs in the environment are mainly benzene series and halogenated hydrocarbons. The total concentrations of the seven VOCs ranged from 0.19 to 50.98 mg/kg, with an average of 14.05 mg/kg. The EB concentrations in chemical site soil were never detected to 38.15 mg/kg. The mean of CH, B, EB and 1,2,3-TH were 0.42, 2.28, 12.74 and 0.47 mg/kg, respectively, and were all higher than the Chinese soil environmental quality risk control standards. In a whole, VOCs concentrations varied significantly across functional areas. The average concentrations of the VOCs in different functional areas in the order: P > B > F > W > other areas. Obviously, the production and raw material areas were higher than other areas[43].

Table 1 Descriptive statistics of VOCs in the soils of the chemical site (mg/kg)

VOCs	CH	1,2-DE	B	TH	1,2-DP	EB	1,2,3-TH	Σ
Chemical site (n = 37)								
Min	ND	ND	ND	ND	ND	ND	ND	0.19
Max	1.47	1.10	12.34	1.82	1.83	38.15	3.59	50.98
Mean	0.42	0.38	2.28	0.41	0.54	12.74	0.47	14.05
Detection rate (%)	43	29	43	23	20	85	57	—
SD	0.39	0.33	3.23	0.46	0.47	14.77	0.66	16.50
CV	1.07	1.17	0.71	0.90	1.17	0.86	0.70	0.85
Pesticide production area (n = 27)								
Min	0.07	ND	ND	0.08	0.02	0.35	0.02	0.07
Max	1.47	1.10	12.34	1.82	1.83	38.15	3.59	50.98
Mean	0.49	0.45	2.76	0.54	0.57	26.77	0.66	28.67
SD	0.38	0.31	3.40	0.47	0.47	9.39	0.73	12.25
CV	1.27	1.44	0.81	1.14	1.21	2.85	0.90	2.34
Food additive production area (n = 9)								
Min	ND	ND	0.12	0.05	ND	0.06	0.01	0.04
Max	ND	ND	0.37	0.07	0.18	2.10	0.14	6.49
Mean	—	—	0.25	0.06	0.18	1.39	0.04	1.95
SD	—	—	0.13	0.01	—	0.94	0.05	2.25
CV	—	—	1.96	7.54	—	1.48	0.84	0.87
Benzene purification and storage area (n = 19)								
Min	ND	ND	0.2	ND	ND	0.02	0.04	0.02
Max	0.01	0.01	5.21	ND	ND	3.01	0.05	7.25
Mean	—	—	1.49	—	—	0.86	0.04	2.21
SD	—	—	1.66	—	—	1.19	0.00	2.59
CV	—	—	0.90	—	—	0.72	9.81	0.85

Note: ND, not detected; SD, standard deviation; CV, coefficient of variation; —, no available data.

Fig.2 showed the composition of VOCs in the soils in different functional areas of the chemical site. The average concentrations of the seven VOCs at different depth soil decreased in the order: EB>>B>>1,2-DP> 1,2,3-TH> CH > TH > 1,2-DE, while B and EB also produced fumaric acid the main raw material[44-45]. Particularly, the content of EB in the soil of P area was up to 83% of the total VOCs, and the content of B in the soil of B area was up to 62%. In addition, 1,2,3-TH was detected in all functional areas of the site (except the L area), indicating that the production process of pesticides was backward at the beginning of the site's use, and the by-products were not properly disposed of, which eventually led to the spread of pollutants[46].

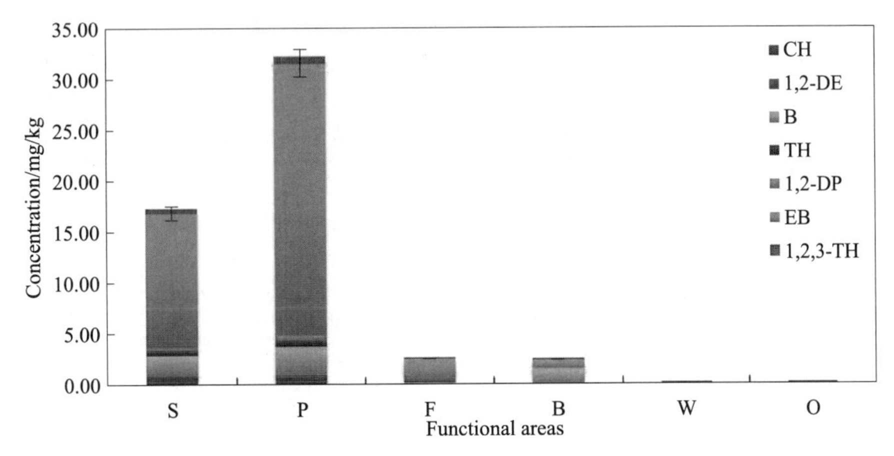

Fig. 2 VOCs concentration distributions in each functional area across chemical site

The vertical distribution of VOCs in soils was very different between functional areas[23,47]. The VOCs were mainly distributed in three production core functional areas such as P area, B area and F area, whereas only a few samples in O, L and S area detected B. The highest concentration of 1,2,3-TH was distributed at 1.0~2.0 m in the P area (Fig. 3 P(b)), and was also detected at 0~3.0 m in the adjacent F area (Fig. 3 F(b)). The CH and 1,2-DE were concentrated in the range of 0~10 m soil depth in the pesticide production area, and the maximum value occurred at the depth of 1.0~2.0 m. Similarly, the maximum value of 1,2-DP was observed at 3.0~4.0 m, and the B and EB are mainly distributed in the range of 0.5 to 4.0 m. The EB pollution extended to 10 m or even deeper soil layers in the P area, and the highest concentrations were found at 7.0 m depth. These results indicated that the VOCs in the soils of the chemical site were directly related to the production of pesticides and food additives, and were concentrated in the production area at both horizontally and vertically. At the same time, it was also the core areas of production and sewage disposal in the site. Leaks may have occurred during the dismantling of the sunken raw material tank, which extended to the deep soil along with the sewage facilities, and even had an impact on the groundwater environment[48].

3.2 Pollution characteristics of VOCs

According to results (Table 2), all of 65 samples of chemical site soil, 23% of CH, 12% of 1,2-DE, 18% of B, 37% of EB, 38% of 1,2,3-TH, and less than 10% TH and 1,2-DP respectively, have a PI > 1, suggesting contamination from these VOCs. Only in P area, there were about 41% of CH, 19% of 1,2-DE, 15% of TH, and 4% of 1,2-DP with its PI from 1 to 5, indicating that soil was somewhat contaminated by above index. Meanwhile, the high IP of EB was about 15% and about 74% from 1 to 5, and about 15% of B above 5. In B area, the PI of B was about 5% above 5 and 21% from 1 to 5, which was similar with the B result in the B area. This shows that both the production areas were under B contamination, and 14% of samples could be categorized as heavily contaminated by B[43]. In addition, the high IP of 1,2,3-TH was about 78% and 50% above 5 in both pesticide production and wastewater treatment areas, further explanation of these areas is heavily contaminated by 1,2,3-TH. The result was consistent with the site's utilization history, production process, raw material and by-product characteristics, etc.

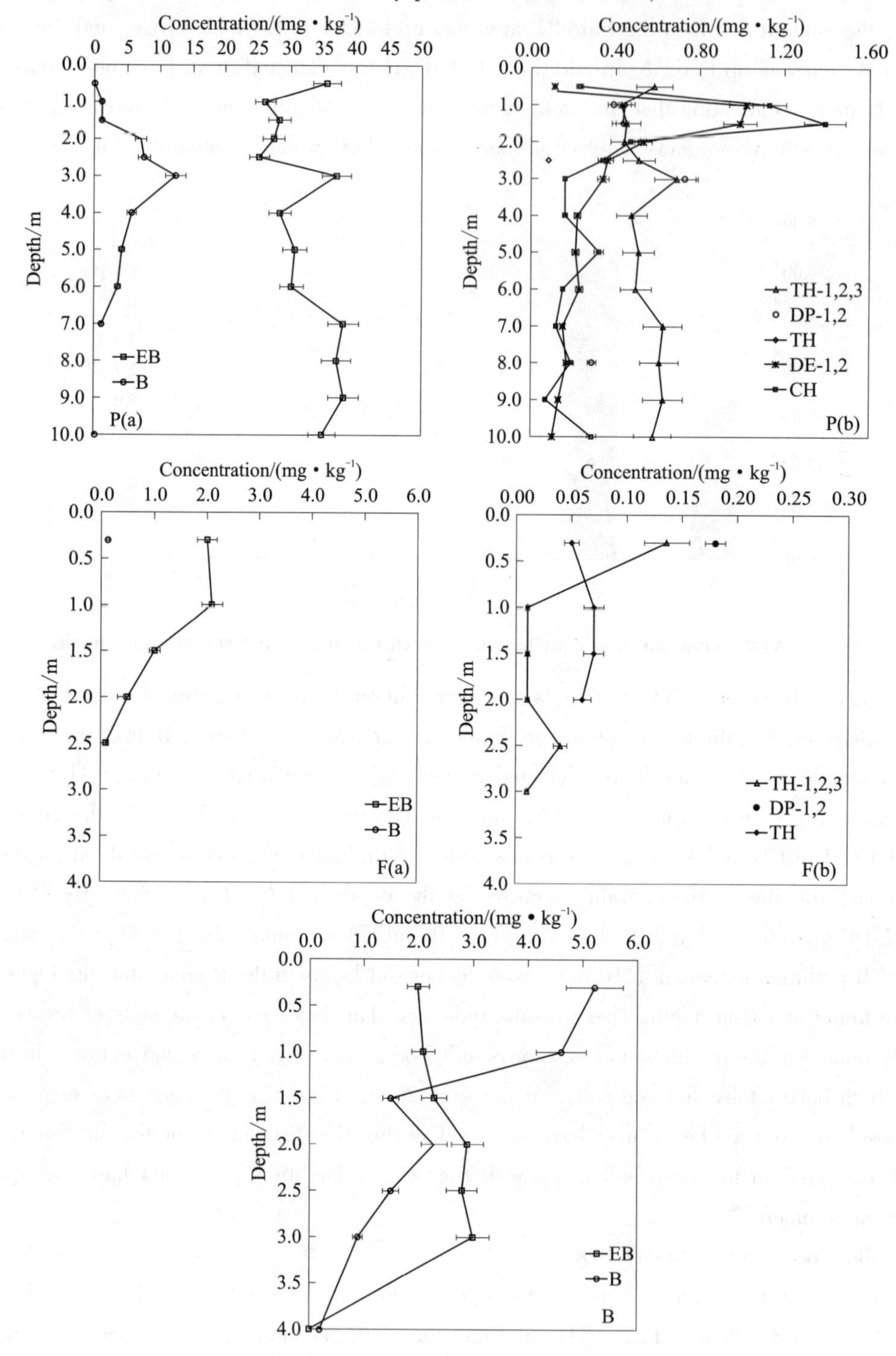

Fig. 3 Vertical distribution of VOCs at different depths of the chemical site soil

(P (a) and P (b) both for the pesticide production area, F (a) and F (b) both for the food additive production area, and B for the benzene purification and storage area)

In addition, the proportion of IP value of 1,2,3-TH higher than 5.0 in pesticide production area and sewage treatment area accounted for 78% and 50% respectively, which further proved the reason why the soil was seriously polluted by 1,2,3-TH.

Table 2 Class distribution of PI for VOCs in soil of chemical site

Functional areas	PI	Ratio (%)						
		CH	1,2-DE	B	TH	1,2-DP	EB	1,2,3-TH
Site	<1	77	88	82	96	98	63	62
	1~3	17	12	6	6	2	3	3
	3~5	6	0	5	0	0	28	0
	>5	0	0	8	0	0	6	34
P	<1	44	81	74	85	96	11	22
	1~3	41	19	4	15	4	7	0
	3~5	15	0	7	0	0	67	0
	>5	0	0	15	0	0	15	78
F	<1	100	100	100	100	100	100	96
	1~3	0	0	0	0	0	0	4
	3~5	0	0	0	0	0	0	0
	>5	0	0	0	0	0	0	0
B	<1	100	100	74	100	100	100	100
	1~3	0	0	16	0	0	0	0
	3~5	0	0	5	0	0	0	0
	>5	0	0	5	0	0	0	0
W	<1	100	100	100	100	100	100	0
	1~3	0	0	0	0	0	0	50
	3~5	0	0	0	0	0	0	0
	>5	0	0	0	0	0	0	50
Other areas	<1	100	100	100	100	100	100	100

The AS of chemical site soil VOCs was evaluated (Table 3), based on the detection rate and the results of the above organic pollution assessment. The AS for the VOCs were universal above 1.0, which indicates that the concentration of VOCs was higher than the target value for compound, and had a potential impact on the human health and environment. Obviously, highest TAS value was 83.6 in the pesticide production area, it was necessary to attach highly importance to the impact of its pollution on human health. Although the AS of other functional areas VOCs is less than 1.0, but the maximum TAS was above 1.0, it should also pay attention to its impact on human health. Out of the seven VOCs, the main impact posed came from 1,2,3-TH, B and CH, with the maximum AS for 71.8, 12.3 and 4.9.

Table 3 AS (>1.0) of VOCs in soil in the various functional areas of the chemical site

Functional area	Ratio (%)							TAS
	CH	1,2-DE	B	TH	1,2-DP	EB	1,2,3-TH	
Site	23	8	18	6	2	37	38	0.1~83.6
P	56	19	26	15	4	89	81	1.0~83.6
F	0	0	0	0	0	0	4	0.3~3.4

Continued to Table 3

Functional area	Ratio (%)							TAS
	CH	1,2-DE	B	TH	1,2-DP	EB	1,2,3-TH	
B	0	0	26	0	0	0	0	0.2~6.3
W	0	0	0	0	0	0	100	1.8~6.6
Other areas	0	0	0	0	0	0	0	0.1~0.8

To investigate the common characteristics of VOCs in the chemical, correlation analyses between VOCs were calculated. This analysis could effectively reveal the relationships among parameters and understand sources of chemical components. Correlations between the CH, 1,2-DE, B, TH, 1,2-DP, and 1,2,3-TH were significant at the $p < 0.01$ level as shown in Fig. 4(Table 4). This suggests that they had a common origin or similar chemical behavior. Obviously, the sources of soil pollution are possibly the production of pesticides and food additives, and the wastewater treatment. This result means that VOCs in the soil not only entered the atmospheric environment through emission, but also infiltrated into the deep soil and even polluted the groundwater environment[49-50].

Table 4 Correlation coefficients matrix of the VOCs in chemical site soils

Factor	CH	DE-1,2	B	TH	DP-1,2	EB	TH-1,2,3
CH	1						
DE-1,2	0.94**	1					
B	0.09	−0.19	1				
TH	0.05	0.13	0.97**	1			
DP-1,2	0.22	0.12	0.21	0.81**	1		
EB	0.27	0.14	0.38*	0.46	0.29	1	
TH-1,2,3	0.31	0.24	0.31	0.83**	0.84**	0.46**	1

Note: ** and * correlation is significant at the 0.01 and 0.05 level.

3.3 Health risk assessment of VOCs

The carcinogenic (CR) and noncarcinogenic risk value (HQ) of seven VOCs in chemical site soils due to six exposure pathways were shown in Table 5. The results imply that the CR and HQ of human exposure to CH, 1,2-DE, B, TH, 1,2-DP, EB, and 1,2,3-TH in chemical site soils were high and exceeded acceptable risk level, with all TCR values above 1.00×10^{-6}, and also exceeded the acceptable risk range of US EPA carcinogens (1.00×10^{-6} ~ 1.00×10^{-4}) and the 1.00×10^{-4} limit of Australia and the Netherlands[51-52]. The TCR for human exposed to CH, 1,2-DE, B, TH, 1,2-DP, EB, and 1,2,3-TH were 3.82×10^{-4}, 1.28×10^{-4}, 1.45×10^{-3}, 1.88×10^{-5}, 2.30×10^{-4}, 2.16×10^{-3} and 3.78×10^{-5} respectively. The CR of VOCs in the following order: $CR_{iivl} > CR_{ois} > CR_{iov2} > CR_{dcs} > CR_{pis} > CR_{iovl}$, suggesting that ingestion and inhalation were the main exposure pathway. Similarly, the HQ of different exposure pathway in the following order: $HQ_{iivl} > HQ_{iov2} > HQ_{pis} > HQ_{ois} > HQ_{dcs} > HQ_{iovl}$. All HI of VOCs above 1.0, and the HI of VOCs descended in the following order: 1,2,3-TH> B> 1,2-DP> TH> EB> CH> 1,2-DE[53].

Table 5 Health risks for each contaminant and exposure pathways

Factor	CH	1,2-DE	B	TH	1,2-DP	EB	1,2,3-TH
CR_{ois}	5.49×10^{-8}	4.96×10^{-8}	3.56×10^{-7}	5.28×10^{-28}	2.20×10^{-8}	6.13×10^{-7}	2.74×10^{-5}
CR_{dcs}	2.09×10^{-8}	1.89×10^{-8}	1.35×10^{-7}	2.01×10^{-28}	8.37×10^{-9}	2.33×10^{-7}	1.04×10^{-5}
CR_{pis}	6.93×10^{-7}	6.26×10^{-7}	8.58×10^{-7}	8.00×10^{-8}	1.04×10^{-7}	2.37×10^{-6}	—
CR_{iov1}	2.39×10^{-9}	2.16×10^{-9}	5.58×10^{-9}	2.76×10^{-10}	3.58×10^{-10}	8.17×10^{-9}	—
CR_{iov2}	1.40×10^{-6}	4.65×10^{-7}	5.31×10^{-6}	6.86×10^{-8}	8.41×10^{-7}	7.89×10^{-6}	—
CR_{iiv1}	3.80×10^{-4}	1.26×10^{-4}	1.44×10^{-3}	1.86×10^{-5}	2.29×10^{-4}	2.14×10^{-3}	—
HQ_{ois}	6.81×10^{-3}	6.15×10^{-3}	6.22×10^{-2}	8.82×10^{-2}	2.61×10^{-4}	2.14×10^{-2}	8.78×10^{-3}
HQ_{dcs}	2.23×10^{-3}	2.01×10^{-3}	2.04×10^{-2}	2.89×10^{-2}	8.55×10^{-5}	7.02×10^{-3}	2.87×10^{-3}
HQ_{pis}	5.83×10^{-3}	5.26×10^{-3}	6.95×10^{-2}	1.85×10^{-1}	4.93×10^{-2}	1.80×10^{-2}	9.82×10^{-1}
HQ_{iov1}	2.01×10^{-5}	1.81×10^{-5}	4.52×10^{-4}	6.37×10^{-4}	1.70×10^{-4}	6.19×10^{-5}	3.38×10^{-3}
HQ_{iov2}	1.18×10^{-2}	3.91×10^{-3}	4.30×10^{-1}	1.59×10^{-1}	3.99×10^{-1}	5.98×10^{-2}	1.88×10^{1}
HQ_{iiv1}	3.20×10^{0}	1.06×10^{0}	1.17×10^{2}	4.31×10^{1}	1.08×10^{2}	1.63×10^{1}	5.10×10^{3}
TCR	3.82×10^{-4}	1.28×10^{-4}	1.45×10^{-3}	1.88×10^{-5}	2.30×10^{-4}	2.16×10^{-3}	3.78×10^{-5}
HI	3.22×10^{0}	1.08×10^{0}	1.17×10^{2}	4.36×10^{1}	1.09×10^{2}	1.64×10^{1}	5.12×10^{3}

Human health risk assessments showed that the TCR and HI values were above 1.00×10^{-6} and 1.0, respectively, indicating an unacceptable threat of human health risk from the seven VOCs in the chemical site soil samples (Fig. 4). The TCR values for soil VOCs in the functional areas of chemical site were as follows: P > B > F > W > O > other areas, and HI values were as follows: P > F > B > W > O > other areas. The highest maximum TCR and HI value of VOCs for human was in the P area soils (4.60×10^{-3} and 5.42×10^{3}) and the lowest were in the O area soils (2.79×10^{-6} and 6.98×10^{-2}). Compared with different areas, both TCR and HI values for areas in study were all above threshold values (except HI for W and O area), reflecting that VOCs impact human health cannot be ignored. Thus, the sufficient attention should be paid to chemical site soil pollution before the development[54]. Moreover, highly toxic substances were often associated with high health risks[25].

It is noteworthy that the CR value of 1,2,3-TH contributed over 90% to TCR value in W and O area, and the B and EB also contributed at least 78% to the TCR value in the P, F, and B area. That reflecting 1,2,3-TH, B, and EB may pose a higher lifetime carcinogenic risk to human via exposure pathways compared with 1,2-DP, 1,2-DE, TH or CH. Similarly, 1,2,3-TH and B are the largest contributors to the HI value of the chemical site. It is noteworthy that the TCR and HI values of P area were significantly higher than other areas in chemical site, and all seven pollutants have a certain contribution (Fig. 4).

It was obvious that the TCR and HI produced by VOCs in the soil of each functional area through various exposure pathways were different (Fig. 5). The TCR values of W and O areas were due to the exposure pathways by ingestion (67% and 72%) and dermal adsorption (26% and 28%). Affected by heavy VOCs pollution in chemical site subsoil, VOCs contribute 75% or even 97% of lifetime carcinogenic risk through the inhalation of gaseous pollutants from the underlying soil in indoor air route compared to other exposure pathways. In addition, except for the above pathways, inhalation and inhalation of gaseous pollutants from the surface and underlying soil in outdoor air were also important pathways leading to noncarcinogenic

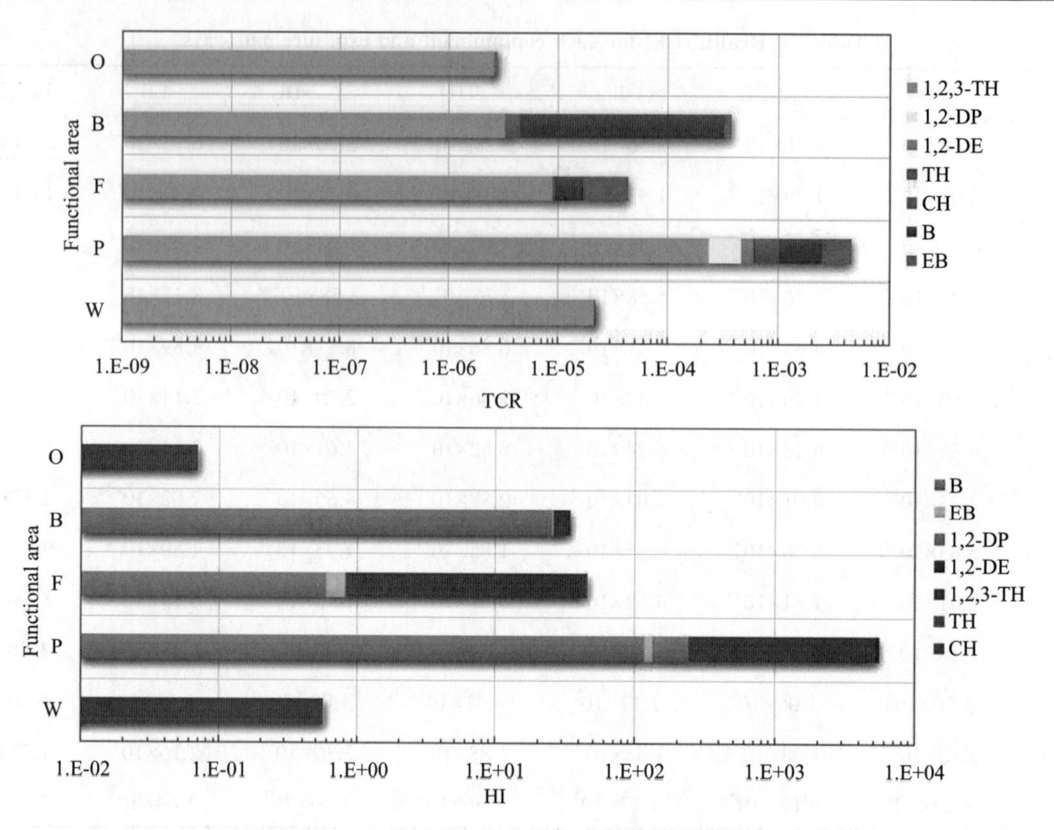

Fig. 4 The TCR and HI from VOCs in chemical site soils. Note that the horizontal axis was the logarithmic scale

risk[25,55]. Therefore, it was necessary to take measures to control the diffusion of VOCs in the soil of chemical plants into the air, and personnel should wear professional protective clothing when working. Comprehensive protection of personal safety and avoid direct contact with contaminated soil[56].

4 Conclusions

The site investigated poses was closely related to production activities. Seven VOCs were detected in the soil, and the phenomenon exceeded the standard limit of the soil. VOCs enter the human body through particulate matter via the respiratory system and skin contact in ways, thus posing a non-negligible health risk to the site population. Furthermore, this pathway was associated with unacceptable carcinogenic and noncarcinogenic risk. Therefore, it is strongly recommended that when investigating, excavating and repairing the site, the staff should wear protective clothing, wash their hands and bathe frequently to ensure personal safety. In addition, it is recommended that the government strengthen the construction of legislation and standards for the supervision of industrial site restoration and development, to ensure that industrial sites are first investigated and evaluated, then restored and managed, and finally green development. Always pay attention to site production safety and population health throughout the process.

Research funding

This research was funded by the Technology Innovation Center for Land Engineering and Human Settlements, Shaanxi Land Engineering Construction Group Co., Ltd and Xi'an Jiaotong University (2021WHZ0094), Shaanxi Province Enterprise Innovation Striving for the First Young Talents Support Program Project (2021-1-2), Shaanxi Provincial Land Engineering Construction Group Internal Research Project (DJNY2021-24), and Institute of Land Engineering and Technology, Shaanxi Provincial Land Engineering Construction Group Internal Pre-research Project (2020-NBYY-23). We are grateful for the reviews and support that this manuscript received from the reviewers and editor.

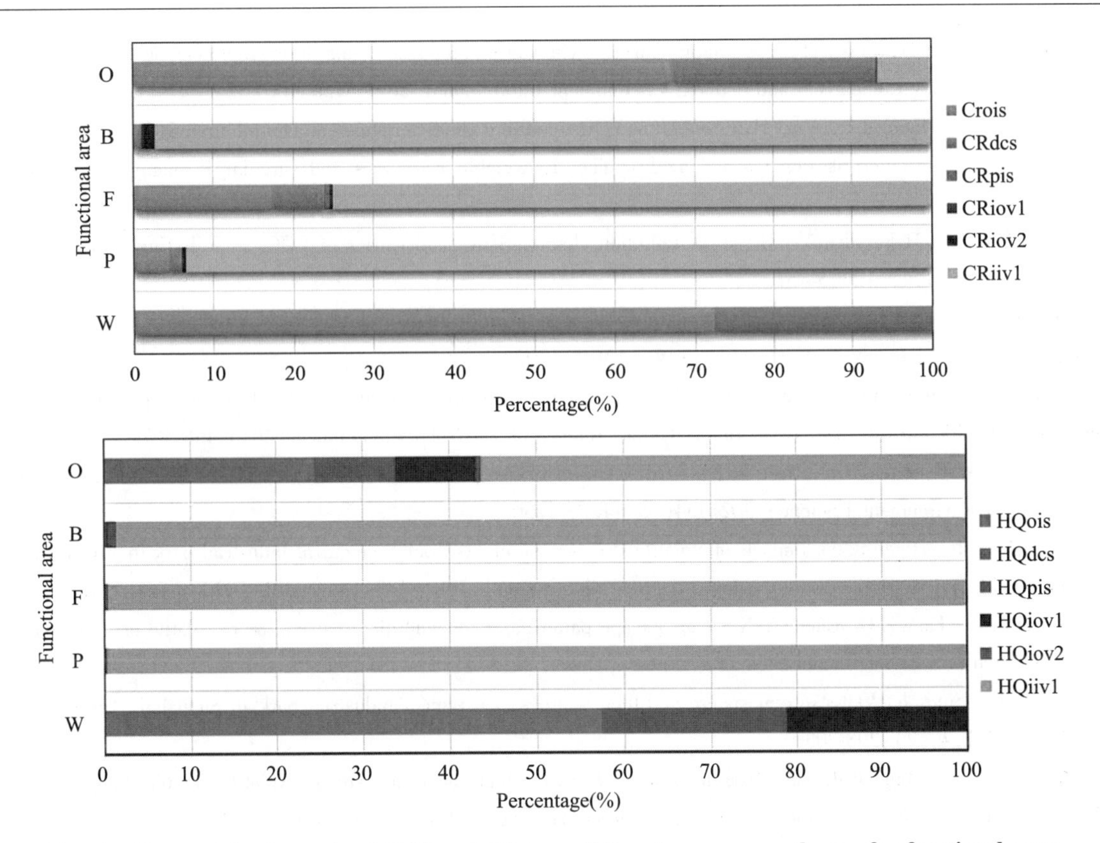

Fig. 5 The contribution rate of TCR and HI from different exposure pathways for functional areas

Author contribution

Yan Li proposed the framework of the study, performed all the statistical analysis, and drafted the manuscript. Bo Yan collected and analyzed the samples, interpreted the results and brought out environmental problem in the investigated site. All authors carried out the site investigation, revised and approved the final manuscript.

Conflicts of interests

The authors state no conflict of interest.

Data availability statement

All data generated or analysed during this study are included in this published article (and its supplementary information files).

References

[1] Yang L X. The World Health Organization (WHO) Environmental Health Criteria (EHC) programme and the practical Usefulness to China. Research of Environmental Science. 2021,34(12):3012-3028.

[2] Wang M E, Ding S K, Guo G L, et al. Advances in ecological risk assessment of soil in contaminated sites. Chinese Journal of Applied Ecology. 2020,31(11):3946-3958.

[3] Liu Y L. Monitoring methods and treatment technologies of volatile organic compounds. Arid Environ Monit. 2016,30(2):76-84.

[4] Lu X P, Guo H, Wang Y, et al. Hazardous volatile organic compounds in ambient air of China. Chemosphere. 2020, 246(C):125731.

[5] Kroll, J H, Seinfeld, J H. Chemistry of secondary organic aerosol: formation and evolution of low-volatility organics in the atmosphere. Atmos. Environ. 2008,42(16): 3593-3624.

[6] Ye L, Tai Q Q, Yu H M. Characteristics and Source Apportionment of Volatile Organic Compounds (VOCs) in the Automobile Industrial Park of Shanghai. Environmental Science. 2021, 42(2):624-633.

[7]Li B, Ho S S H, Li X, et al. A comprehensive review on anthropogenic volatile organic compounds (VOCs) emission estimates in China: Comparison and outlook. Environment International. 2021, 156:106710.

[8]Cheng Y J, Huang J L, Wu J H. Research of VOCs emission characteristics and control strategy for printing industry-a case study of printing enterprises in Guangdong Province. Recyclable Resources and Circular Economy. 2021,14(11): 37-41.

[9]Liu L H, Hu H F, Zhang C H, et al. Pollution characteristics and causes of VOCs in soil. Environmental Science and Management. 2021,46(12):54-57.

[10]Chen L, Miller S A, Ellis B R. Comparative Human Toxicity Impact of Electricity Produced from Shale Gas and Coal. Environmental Science & Technology. 2017,51, 13018-13027.

[11]Zheng J, Yu Y, Mo Z, et al. Industrial sector-based volatile organic compound (VOC) source profiles measured in manufacturing facilities in the Pearl River Delta, China. Science of The Total Environment. 2013,456-457.

[12]Zhang M M, Zhang C Y, Guo X X, et al. Refined risk assessment of soil benzene in unsaturated zone of coking site. Research of Environmental Science. 2021,34(05):1223-1230.

[13]Li X C. Investigating the Impacts of Volatile Organic Pollutants from X chemical-industrial park in Jilin Province on Population Health and its Hepatotoxicity Effect [doctoral dissertation]. Jilin: Jilin University; 2020. master' s thesis.

[14]He X W, Fang Z Q, Cheng Y X, et al. Escape pattern and concentration distribution of volatile organic compounds in the remediation process of contaminated sites. Environ Chem. 2015,34(2):284-292.

[15]Zhen J. Study on VOCs in atmosphere and their sources of a typical industrial park in Shanghai, China. Shanghai Normal Univ. 2017,46(2):298-303.

[16]Xiang L Y, Han D M. The characteristics and source analysis of atmospheric volatile organic compounds in the farmland surrounding by industrial zone in Shanghai, China. Time Agric Mach. 2016, 43(1):151-152.

[17]Hu R Y, Liu G J, Zhang H, et al. Levels, characteristics and health risk assessment of VOCs in different functional zones of Hefei City. Ecotoxicol Environ Saf. 2018,106:301-307.

[18]Shuai J, Kim S, Ryu H, et al. Health risk assessment of volatile organic compounds exposure near Daegu dyeing industrial complex in South Korea. International Journal of Environmental Research and Public Health. 2018,18,528-540.

[19]Nian S Y. Soil pollution investigation and risk assessment of three different industrial sites [dissertation]. Hefei: Anhui University of Science and Technology; 2020.

[20]Cachada A, Pato P, Rocha-Santos T, et al. Levels, sources and potential human health risks of organic pollutants in urban soils. Science of the Total Environment. 2012,430,184-192.

[21]Singkaew P, Kongtip P, Yoosook W, et al. Health risk assessment of volatile organic compounds in a highrisk group surrounding map taphut industrial estate, rayong province. Chotmaihet Thangphaet. 2013,96, 73-81.

[22]Zhang X F, Chen Q, Deng S P, et al. Advances in the study on secondary pollution of volatile organic compounds in remediation of contaminated site. Ecol Rural Environ. 2015,31(6):831-834.

[23]Zhao X, Ma H, Lu J, Yin T, Zhang Q, Dong X, et al. Characteristics and source apportionment of volatile organic compounds during the remediation of contaminated sites in Zhenjiang, China. Int J Environ Sci Technol. 2021;18:2271-2282.

[24]Yan Y Z, Xue N D, Zhou L L, et al. Distribution characteristics of HCHs and DDTs during excavation of a contaminated site. Res Environ Sci. 2014,27(6):642-648.

[25]Nie Y. Study on the status and health risk assessment of contaminated sites-a case in abandoned chemical factory [dissertation]. Beijing: North China Electric Power University; 2016.

[26]Ma Y, Dong B B, Du X M, et al. Secondary pollution and its prevention of VOC/SVOC-contaminated sites with ex situ remediation technologies. Environ Eng. 2017,35(04):174-178.

[27]Zhang M, Yshikawa M. An overview of remediation technologies for sites contaminated with volatile organic compounds. Geo Chicago. 2016, 273;295-301.

[28]Luo H Y, Zhou Y. Research and analysis of hydro geological conditions of Weinan urban area. Journal of Green Science and Technology. 2016, 6, 167-170.

[29]Shi W, Wang Y L. Evaluation and Analysis of Geological Environment Suitability for the Development of Underground Space in Weinan. Ground water. 2018,40(4):134-137.

[30]Wang L, Liu M, Tao W, et al. Pollution characteristics and health risk assessment of phthalate esters in urban soil in the typical semi-arid city of Xi'an, Northwest China. Chemosphere. 2018, 191, 467-476.

[31]Li X N, Cundy A B, Chen W P, et al. Systematic and bibliographic review of sustainability indicators for contaminated site remediation: Comparison between China and western nations. Environmental Research. 2021,111490.

[32]Liu F, Liu Y, Jiang D, et al. Health risk assessment of semi-volatile organic pollutants in Lhasa River China. Ecotoxicology. 2014, 23, 567-576.

[33]Dong W H, Lin X Y, Du S H, et al. Risk assessment of organic contamination in shallow groundwater around a leaching landfill site in Kaifeng, China. Environmental Earth Sciences. 2015,74, 2749-2756.

[34]Baasel W D. Economic methods for multipollutant analysis and evaluation. Dekker, U.S; 1985.

[35]Li W T,Li J J, Chen A,et al. Health risk assessment of a lubricant contaminated site. Asian Journal of Ecotoxicology. 2021,16(1) :137-146.

[36]Wang C, Li H L, Hu Q, et al. Analysis and prospects on soil environmental risk assessment technology in China. Asian Journal of Ecotoxicology. 2021,16(1):28-42.

[37]Jia H H, Gao S, Duan Y S, et al. Investigation of health risk assessment and odor pollution of volatile organic compounds from industrial activities in the Yangtze River Delta region, China. Ecotoxicology and Environmental Safety. 2021, 208:111474.

[38]US EPA. Risk Assessment Guidance for Superfund Volume I Human Health Evaluation Manual, Part A. Washington, DC, Office of Emergency and Remedial Response, US;1989.pp:1-24.

US EPA. Risk assessment guidance for superfund volume I human health evaluation manual, part A. Washington: DC, Office of Emergency and Remedial Response, US;1989. pp. 1-24.

[39]US EPA. Exposure Factors Handbook. Executive summary, Washington: DC, US; 2011.

[40]Zhang K, Liu S H, Wang S D, et al. Health risk assessment and distribution of VOCs during excavation processes for the remediation of contaminated sites. Human and Ecological Risk Assessment: An International Journal. 2019,25(8): 2073-2088.

[41]Zhang L, Zhu X Z, Wang Z R, et al. Improved speciation profiles and estimation methodology for VOCs emissions: A case study in two chemical plants in eastern China. Environmental pollution (Barking, Essex: 1987). 2021,291:118192.

[42]Kyab E, Chen W C, Song X D, et al. The identification, health risks and olfactory effects assessment of VOCs released from the wastewater storage tank in a pesticide plant - ScienceDirect. Ecotoxicology and Environmental Safety. 2019, 184:109665-109665.

[43]Sun Y Y, Li Y, Luo Y H. Investigation on the pollution of volatile organic pollutants in the air after the demolition of a chemical plant. Leather Manufacture and Environmental Technology.2021,2(20):76-77,79.

[44]Shi J H, Huang J, Cao H F, et al. Synthesis and research of diethyl fumarate. Textile Dyeing and Finishing Journal. 2020,42(11):20-22.

[45]Chen L. Comparison of Production Technology of Succinic Acid. Henan Chemical Industry. 2021, 38(4):15-16,20.

[46]Yu G H. Research on the utilization of industrial byproduct 1,2,3-trichloroproane[dissertation]. Henan: Zhengzhou University; 2007.

[47]Wunseon C, Sekyung J, Chul W L, et al. Assessment of Behavior for Volatile Organic Compounds and Trace Elements in the Changwon Industrial Complex. Journal of Korean Society for Atmospheric Environment. 2020,36(3):293-308.

[48]Wang Y K, Xu Y, Liu Y Y. Simulation and prediction analysis of heavy metals in groundwater of Weidong New City based on MODFLOW. Agriculture and Technology. 2019,39(14):47-51.

[49]Li L J, Wang H J, Ma J S. Pollution Characteristics and Health Risk Assessment of Volatile Organic Compounds in Groundwater in the Lower Liaohe River Plain, J. Rock and Mineral Analysis,2021,40(6): 930-943.

[50]Yu X, Ju Z Y, Lian H. Contamination Characteristics and Causes of Volatile Organic Compounds in the Groundwater at a Chemical Contaminated Site. Guangdong Chemical Industry. 2019,46(21):99-101,103.

[51]US EPA, Risk Assessment Guidance for Superfund (RAGS): Part D: Standardized planning, reporting, and review of superfund risk assessments, Washington, DC: Office of Emergency and Remedial Response. US EPA;001.

[52]Piet O, Lijzen J, Swartjes F, et al. Evaluation and revision of the CSOIL parameter set, proposed parameter set for

human exposure modelling and deriving Intervention Values for the first series of compounds, Bilthoven: National Institute for Public Health and the Environment,2001.

[53]Li W D, Zhang C Y, Guo X X, et al. Health risk assessment of trichloromethane at contaminated sites based on soil gas volatilization fluxes. Asian Journal of Ecotoxicology. 2021,16(1):87-96.

[54]Tan B, Wang T Y, Pang B, et al., Contamination Characteristics and Health Risk Assessment of Atmospheric Volatile Organic Compounds in Pesticide Factory, J. Environmental Science,2013,34(12): 4577-4584.

[55]Christina N, Lin F, Karoline K, et al. Sources of volatile organic compounds in suburban homes in Shanghai, China, and the impact of air filtration on compound concentrations. Chemosphere. 2019,231:256-268.

[56]Bari M A, Kindzierski B W. Ambient volatile organic compounds (VOCs) in Calgary, Alberta: Sources and screening health risk assessment. Science of the Total Environment. 2018,627-640.

本文曾发表于2022年《Open Chemistry》第20卷

Study of Solidification and Stabilization of Heavy Metals by Passivators in Heavy Metals Contaminated Soil

Shenglan Ye[1], Luyao Wang[1,2,3,4], Tiancheng Liu[1,2,3,4]

(1. Shaanxi Provincial Land Engineering Construction Group Co. Ltd.,Xi'an 710075, China; 2. Institute of Land Engineering and Technology, Shaanxi Provincial Land Engineering Construction Group Co. Ltd., Xi'an 710075, China; 3. Key Laboratory of Degraded and Unused Land Consolidation Engineering, Ministry of Natural Resources, Xi'an 710075, China;4. Shaanxi Key Laboratory of Land Consolidation, Xi'an 710075, China)

【**Abstract**】 In this study, the indoor constant temperature culture experiment was used to explore the mutual transformation of different forms of heavy metals. Appropriate types of passivating agents were screened and the optimal addition amount was determined, realize the solidification and stabilization of heavy metals. The results showed that the dissolved copper(Cu), zinc(Zn), and lead(Pb) content of the zeolite treatment decreased to the lowest. They were respectively 219, 819, and 40 g/kg, which were 31.2%, 6.5%, and 38.5% lower than no passivating agent added(CK); It gradually increased with the extension of time; 5% zeolite (Z4) treatment had the highest average content of Cu, Zn and Pb in the residue state, respectively 24 mg/kg, 48 mg/kg and 19 mg/kg; And at the end of the test, the residual zinc content of Z4 treatment reached 50 mg/kg, which was 72.4% higher than CK. Comprehensive analysis of the changes in the dissolved state of the four heavy metals in the soil shows that Cu, Zn, Cd and Pb treated with zeolite have the best effect, followed by sepiolite, and palygorskite last. Therefore, 5% zeolite can be used for the passivation restoration of heavy metals Cu, Zn, Cd and Pb in the soil.

【**Keywords**】 Heavy metals; Contaminated soils; Solidification and stabilization; Passivators; Constant temperature culture

1 Introduction

Soil is an important cornerstone for carrying the entire biological cycle, so the cleanliness and safety of the soil has an immeasurable impact on animals, plants, and humans. With the acceleration of industrialization and urbanization in today's society, human beings have neglected environmental protection while developing their economy rapidly. This makes soil pollution by heavy metals increasingly prominent, especially agricultural soils polluted by heavy metals. Due to its transport in the food chain, it poses a great threat to human health and ecological environment security[1-3]. Only when the soil is contaminated with heavy metals and accumulated to a certain degree can it be expressed through plant growth and human health. Therefore, once the soil is polluted by heavy metals, it will change the structure, properties and functions of the soil, and be transported to microorganisms and plants through the food chain, causing harm to human health[4-6]. At present, soil heavy metal pollution has become a global problem, especially in developing countries, and China is the largest developing country[7-8]. Therefore, it is imminent to study the large-scale remediation method of heavy metal contaminated soil.

Heavy metal pollution is an extremely severe type of pollution that is long-lasting, highly toxic and easy to migrate. Therefore, the cost, cycle and difficulty of repairing are extremely high[8-9]. Physical repair can take effect quickly, but it requires a huge cost. Phytoremediation has a long-term effect, but its repair time is longer[10]. Therefore, chemical passivation is the most practical method for soil contaminated by heavy metals. It is relatively fast and can ensure the safe production of crops [11-12]. Chemical passivation remediation refers to the addition of passivators to contaminated soil; passivators convert heavy metals from active forms to stable forms. This is beneficial to reduce the migration and bioavailability of heavy metals, and achieve a method of repairing soil contaminated by heavy metals [13-14]. This repair technology has the advantages of rapid repair and simple operation, so it is more suitable for large areas of farmland contaminated by heavy metals[15-16]. At present, the passivator materials that can be used mainly include various minerals, biochar and organic matter and so on. There are certain differences in their remediation results for different types of pollutants, soil types and pollution levels[17-18]. Inactivation agent repairing the soil can improve the physical and chemical properties of the soil and supplement nutrients. At the same time, through adsorption, precipitation, and complexation, heavy metals are changed from exchangeable state to organic combined state and residual state. This reduces the mobility and bio-combination of heavy metals[19].

There are many types of chemical passivators. The price of the passivator and the possibility of re-contamination should be fully considered. Minerals are cheap materials with the best environmental compatibility. In recent years, non-metallic minerals are often used as passivators for heavy metals in the soil. This research has achieved good results[20-21]. Studies have shown that the exchangeable cadmium content in vegetable soil can be reduced by 23.1%~41.2%. The maximum reduction of cadmium in the edible parts of lettuce, rapeseed and radish reached 51.8%, 47.0% and 24.9%, respectively [22]. Studies have shown that under the repairing effect of sepiolite, the absorption and accumulation of Cd in the edible part and roots of pakchoi can be significantly reduced.[23]. Some scholars have studied the control effects of different passivators on the soil-ryegrass system. It was found that passivators can change the chemical form and biological activity of the heavy metals, and reduce the migration and transformation of heavy metals into plants[24-25].

The remediation effect of passivators on heavy metals is restricted by factors such as climate, soil physical and chemical properties, water and fertilizer management, crop types and varieties. In actual research, passivation materials should be selected according to local conditions[26]. Therefore, it has two objects in this study, (i) to identify three mineral materials that were selected as passivation agents for remediation of heavy metal contaminated soil. Through constant temperature cultivation, the different forms of heavy metals in the soil at different cultivation times were studied. (ii) to discuss the best type and amount of passivator materials needed to change the chemical form and activity of soil heavy metals. It is conducive to the rational and effective use of heavy metals polluted soil under the condition of ensuring the quality of crops. Moreover, these objectives should be technically written.

2 Experimental materials and methods

2.1 Experimental materials

The test soil was the clay around Xi'an City, Shaanxi Province. The sampling depth was 0~20 cm. The soil samples were air-dried and passed through a 0.149 mm sieve for use. The contents of total copper, total cadmium, total zinc and total lead in the soil samples were 14, 0, 28 and 16 mg/kg, respectively. The passivating agent is sepiolite, zeolite and palygorskite. Heavy metal contaminated soils are expanded by 5 times in accordance with the second level of soil environmental quality standards (GB 15618—2018) and appropriate heavy metal salts are added to make the total copper, total cadmium, total zinc and total lead content in the soil reach 500, 1.5, 1250 and 600 mg/kg respectively.

Preparation of heavy metal compound contaminated soil: Prepared soil sample of 20 kg was added to a 4 L plastic pot. There are 5 plastic basins in total. The water content was controlled to 60% of the field water holding capacity. The soil bulk density is 1.21 g/cm^3. The soil samples were cultured in a constant temperature incubator at 25 ℃ for 7 days. According to the second level of soil environmental quality standards (GB

15618—2018), heavy metals in the form of sulfates are added to the soil. The total copper, total cadmium, total zinc and total lead content in the soil are 500, 1.5, 1250 and 1500 mg/kg, respectively. And add water according to 65% of the field water holding capacity. The soil is contaminated with heavy metals required for the test. The contaminated soil is then subjected to aging treatment. The soil moisture is adjusted by the weighing difference method every three days. The soil moisture content is maintained at 65%. After 30 days, a constant temperature incubation test was carried out.

2.2 Experimental method

The constant temperature incubation test was adopted. In the experimental setting, the ratio of passivation agent to air-dried soil sample was 1%, 2%, 3% and 5%. A mineral passivator was added to 150 g of heavy metal compound contaminated soil. The heavy metal compound contaminated soil after adding the passivator is thoroughly mixed. The mixed sample was placed in a 250 mL plastic bottle. The mouth of the bottle was sealed with plastic wrap, and several small holes were left in the middle of the plastic wrap. The treated soil samples were cultured in a constant temperature incubator at 25 ℃. During the cultivation process, a weighing method was used to supplement deionized water to 65% of the field water holding capacity. 10 g soil samples were collected at 20, 40, and 50 days of culture. After natural air drying, the soil sample was passed through a 0.149 mm sieve for later use. Palygorskite was recorded as P1~P4. The zeolite was recorded as Z1~Z4. Sepiolite was recorded as S1~S4. The control without passivation agent was recorded as CK. There were 13 treatments in total, with 3 repetitions for each treatment.

The content of different forms of copper, cadmium, zinc and lead in the soil was continuously extracted with 0.1 mol/L acetic acid, 0.1 mol/L hydroxylamine hydrochloride, and 8.8 mol/L H_2O_2. The leaching residue was digested with nitric acid-hydrofluoric acid-perchloric acid. Finally, the dissolved, reducible and oxidizable leachate and residue states were obtained. The content of Cd, Cu, Pb, and Zn in the extract and digestion solution was determined by atomic absorption spectrophotometry (Analytical Instruments AG in Jena, Germany. AAS ZEEnit700plus).

3 Results and discussion

3.1 Study the change trend of heavy metal forms in different culture time

At the beginning of the experiment, there was no significant difference in heavy metal content between different treatments. The content of dissolved heavy metals showed a downward trend with time. At the end of the experiment, the dissolved Cu, Zn, and Pb content of the zeolite treatment reached the lowest. They were 219, 819, and 40 mg/kg, which were 31.2%, 6.5%, and 38.5% lower than CK, respectively. The content of sepiolite in dissolved Cd is the lowest. It is 1.08 mg/kg. Followed by the content of palygorskite and zeolite is 1.09 mg/kg. Handing CK decreased by 10.7% and 9.9% respectively. The content of the reduced and oxidized states of the passivator treatment increases first and then decreases with the increase of time. The CK showed a continuous decline. The content of heavy metals in the residue of different treatments gradually increased with time; among them, the content of Cu, Zn, and Pb in the residue of zeolite reached the highest 48 mg/kg, 70 mg/kg, 41 mg/kg, which were 1.92, 2.41 and 1.86 times of CK respectively. However, the content of residual Cd was treated with sepiolite, which was 0.18 mg/kg, which was twice that of CK (Table 1). It shows that the form of heavy metals treated with passivating agent changes to a low-activity oxidizable state and a residue state. The content of heavy metals in the residues of different treatments gradually increased with time. This is consistent with the results of Gao Ruili[27]. Using montmorillonite to passivate and remediate heavy metal compound contaminated soil, the weakly acid-extracted Pb, Zn, and Cd in the soil were reduced by 12.0% ~ 15.9%, while the residue content increased by 62.5% ~ 110.1%. The results show that the bioavailability and mobility of these four heavy metals have been reduced. The main reason is that the addition of passivators can convert heavy metals in the soil from the dissolved state to the residue state, and the dissolved heavy metals are finally formed into a fixed residue form through the process of reduction and oxidation. Studies have shown that the types of passivators and soil pH, Eh have a close influence on the occurrence of heavy metals, coordination properties and charge adsorption. This can change the form of heavy metals in the soil and affect their migration and transformation in the soil environment[28-30]. Therefore, the dissolved heavy metals in the soil treated by the passivator quickly converted to the residue state, resulting in lower content than the control treatment.

Table 1 Changes in the forms content of four heavy metals treated (mg/kg) with different passivates over time

Time		Dissolved state				Reduced state				Oxidized state				Residual state			
		12.1	12.21	1.10	1.21	12.1	12.21	1.10	1.21	12.1	12.21	1.10	1.21	12.1	12.21	1.10	1.21
CK	Cu	367	357	346	321	214	106	67	58	38	31	27	14	15	20	23	25
	Zn	972	886	874	868	115	162	95	91	58	51	46	35	16	22	26	29
	Cd	1.35	1.31	1.26	1.21	0.10	0.11	0.11	1.09	0.02	0.04	0.02	0.01	0	0	0.06	0.08
	Pb	80	77	69	65	67	127	108	99	54	46	34	30	11	15	17	22
P	Cu	375	346	287	257	82	86	94	88	24	38	25	16	14	15	23	32
	Zn	973	923	865	841	126	134	130	122	41	57	40	27	28	32	43	48
	Cd	1.34	1.26	1.11	1.09	0.12	0.12	0.11	0.10	0.02	0.03	0.03	0.02	0.04	0.08	0.12	0.14
	Pb	84	69	66	63	81	113	109	103	56	69	30	18	14	14	22	36
Z	Cu	371	325	252	219	85	81	87	84	27	38	28	24	15	18	22	48
	Zn	989	912	850	812	117	153	145	132	45	51	43	37	32	38	49	70
	Cd	1.35	1.31	1.19	1.09	0.10	0.13	0.12	1.12	0.03	0.05	0.03	0.02	0.01	0.11	0.13	0.16
	Pb	84	77	61	40	84	101	96	93	61	73	40	19	16	17	25	41
S	Cu	372	318	347	265	81	86	82	880	17	26	28	15	17	19	20	35
	Zn	978	945	868	821	119	150	138	127	32	42	45	27	35	36	37	52
	Cd	1.31	1.29	1.16	1.08	0.10	0.12	0.10	0.10	0.01	0.03	0.02	0.01	0	0.12	0.15	0.18
	Pb	87	81	74	59	90	112	97	91	52	61	42	15	14	16	24	37

Note: The treatments in the table are CK-control without passivation agent; P-palygorskite; S-sepiolite; Z-zeolite, and the values of P, Z, and S represent the average heavy metal content of different proportions.

3.2 Study the change trend of dissolved heavy metal in different addition levels

Dissolved heavy metals are heavy metals that are exchange-adsorbed on soil clay minerals and other components, such as iron hydroxide, manganese hydroxide and humus. This form is most sensitive to changes in the soil environment. It has high activity and is easy to migrate and transform through leaching and leakage of irrigation, rainwater, etc. Eventually, it will seriously pollute the underlying soil, groundwater and growing plants. It can be seen from Fig. 1 that the content of dissolved Cu, Zn, Cd and Pb of CK is the largest. After adding different passivators, the content of dissolved heavy metals decreased to varying degrees. This is consistent with the results of existing studies. After adding passivation materials, the form of each heavy metal changes to an oxidizable state with low activity and a residue state. With the prolongation of the passivation time, the content of forms with high bioavailability gradually decreases. It shows that the passivation material added can effectively reduce the mobility and bioavailability of Pb, Cd, Zn, As[31-32]. In this study P3 has the lowest zinc content, which is 814 mg/kg, which is lower than other passivation treatments. When the dissolved heavy metals are treated with sepiolite, the effect of treatment with S3 (3% sepiolite) is the best, of which copper, zinc and lead are respectively 286, 856 and 60 mg/kg. For zeolite treatment, Z4 (zeolite 5%) performed best, and the dissolved heavy metals copper, cadmium, and lead are the lowest, which are 269, 0.94, and 55 mg/kg, respectively. Therefore, among the three passivating agents, palygorskite has the best passivation effect on copper, and 5% zeolite has the best comprehensive passivation effect on heavy metals.

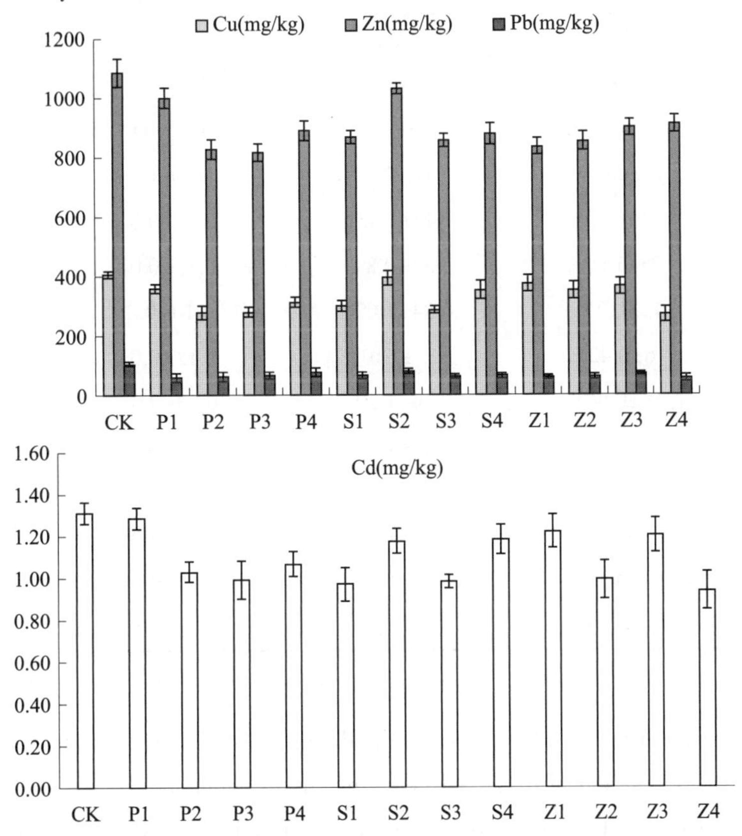

Fig. 1 Changes in dissolved heavy metal content in different treatments

Note: The treatments in the figure are CK-control without passivation agent; P1—Palygorskite 1%; P2—Palygorskite 2%; P3—Palygorskite 3%; P4—Palygorskite 5%; S1—Sepiolite 1%; S2—Sepiolite 2%; S3—Sepiolite 3%; S4—Sepiolite 5%; Z1—Zeolite 1%; Z2—Zeolite 2%; Z3—Zeolite 3%; Z4—Zeolite 5%

3.3 Study the change trend of residual heavy metal in different addition levels

Residual heavy metals are stable in crystal lattices such as quartz and clay minerals. Its nature is stable. It has little effect on the migration and bioavailability of heavy metals in the soil. Residual heavy metals in the soil will not be absorbed by plants, and they will not easily leak through rain or other means. It presents an extremely stable state. It can be seen from Table 2 that different passivation treatments significantly increased the content of residual heavy metals. The results are consistent with the conclusions of previous reports[32-33]. In this study, the heavy metal Cd is not easy to be solidified, and the residual Cd of S4, Z4 and S3 are 0.21 mg/kg, 0.18 mg/kg and 0.17 mg/kg, respectively. They are significantly different from CK ($P<0.05$). Among them, S4 and Z4 has the best fixation effect. This is consistent with the results of existing studies. The application of sepiolite in Cd-contaminated soil can significantly increase soil pH. It can promote the conversion of Cd in the soil from the exchange state with high activity to the residue state with low activity, thereby effectively reducing its bioavailability and migration ability[34-35]. The maximum Pb content in the residue of P2 treatment is 24 mg/kg. The residual Cu, Zn and Pb contents of Z4 treatment are the highest, which are 24 mg/kg, 48 mg/kg and 19 mg/kg. They are significantly higher than other treatments. Therefore, the effect of zeolite treatment is the best, followed by sepiolite. The main reason is that the application of zeolite powder can promote the formation of soil aggregates and increase soil organic matter[36]. In addition, zeolite powder belongs to clay minerals. Due to its large specific surface area and abundant negative charges on the surface, it has strong adsorption and ion exchange capabilities for metal ions[37].

Table 2 Effects of different passivators on the concentrations of residual heavy metals (mg/kg) in soil

Treatment	Cu	Zn	Cd	Pb
CK	$12±0.54^{d}$	$29±1.02^{ef}$	$0.08±0.01^{d}$	$8±0.22^{e}$
P1	$12±0.43^{d}$	$30±1.15^{e}$	$0.09±0.01^{d}$	$12±0.45^{d}$
P2	$17±0.64^{bc}$	$41±1.26^{bc}$	$0.10±0.01^{d}$	$16±0.41^{c}$
P3	$19±0.93^{b}$	$42±1.32^{bc}$	$0.11±0.01^{cd}$	$17±0.43^{c}$
P4	$25±1.00^{a}$	$43±0.89^{c}$	$0.15±0.01^{bc}$	$25±0.54^{b}$
Z1	$15±0.25^{c}$	$34±1.07^{d}$	$0.10±0.01^{d}$	$12±0.27^{d}$
Z2	$16±0.42^{bc}$	$33±1.01^{de}$	$0.12±0.01^{cd}$	$12±0.32^{d}$
Z3	$18±0.66^{b}$	$34±1.20^{d}$	$0.13±0.01^{cd}$	$24±0.66^{b}$
Z4	$24 ±1.21^{a}$	$48±1.52^{a}$	$0.18±0.01^{ab}$	$19±0.76^{c}$
S1	$12±0.27^{d}$	$27±0.68^{f}$	$0.13±0.01^{cd}$	$9±0.22^{de}$
S2	$16±0.48^{bc}$	$35±1.11^{d}$	$0.15±0.01^{bc}$	$11±0.38^{de}$
S3	$15±0.39^{c}$	$39±1.04^{c}$	$0.17±0.01^{ab}$	$17±0.47^{c}$
S4	$23±0.72^{a}$	$50±1.30^{a}$	$0.21±0.01^{a}$	$30±0.90^{a}$

Note: Different lowercase letters in the same column represent significant differences between treatments ($P<0.05$).

3.4 Changes in the content of heavy metals in polluted soil at different treatment times

3.4.1 Changes in the content of dissolved heavy metals in contaminated soil

It can be seen from Fig. 2 that the dissolved Cu, Zn, Cd and Pb in the soil showed a downward trend with the increase of the cultivation time under constant temperature culture conditions. The dissolved heavy metal content of each passivator treatment was basically less than the control CK. With the increase of the passivation dose in each treatment, the dissolved Cu, Zn, Cd and Pb content in the soil showed a downward trend. The research results of Ren LL [38] showed that mineral treatment significantly reduced the available Zn, Cd, Cu, and Pb content of heavy metals in the soil ($P<0.05$). The reduction rate was 99.1%, 91.4%, 85.6%, 46.1%. This is consistent with the results of this study. The Cu content of Z4 treatment decreased from 413 mg/kg to 299 mg/kg. The rate of decline is highest. It was 27.6%. Secondly, the copper content of P2 and S4 treatment decreased from 367 mg/kg and 377 mg/kg to 278 mg/kg and 286 mg/kg, respectively. The decline rates were 24.3% and 24.1%. The difference between the two is not significant. There was no significant difference in the dissolved Zn content of CK. The rate of decline is only 0.6%. The content of dissolved Zn in the soil treated with P4 was the lowest, only 814 mg/kg. And the rate of decline also reached the maximum, at 23.2%. Then it is Z4 processing. The dissolved Zn content decreased from 1077 mg/kg to 835 mg/kg. The decline rate reached 22.5%. And the difference between P4 and Z4 is not significant. In this study, only the dissolved Zn was reduced by 8.4% to 23.2%. This is consistent with the study by Liang[39]. The effective state of Zn in the soil was reduced by 23.48%~49.61% after the passivation agent treatment. The passivation rate in this study is low. Mainly due to passivation time is the shorter. Research by Zhao Jian[26] showed that the reduction range of soil available heavy metals increased with the extension of the test time. The overall dissolved content of Cd and Pb in the contaminated soil is CK> palygorskite> sepiolite> zeolite. Comprehensive analysis of the changes in the dissolved state of the four elements in the soil shows that Cu, Zn, Cd and Pb treated with zeolite have the best effect, followed by palygorskite, and sepiolite last.

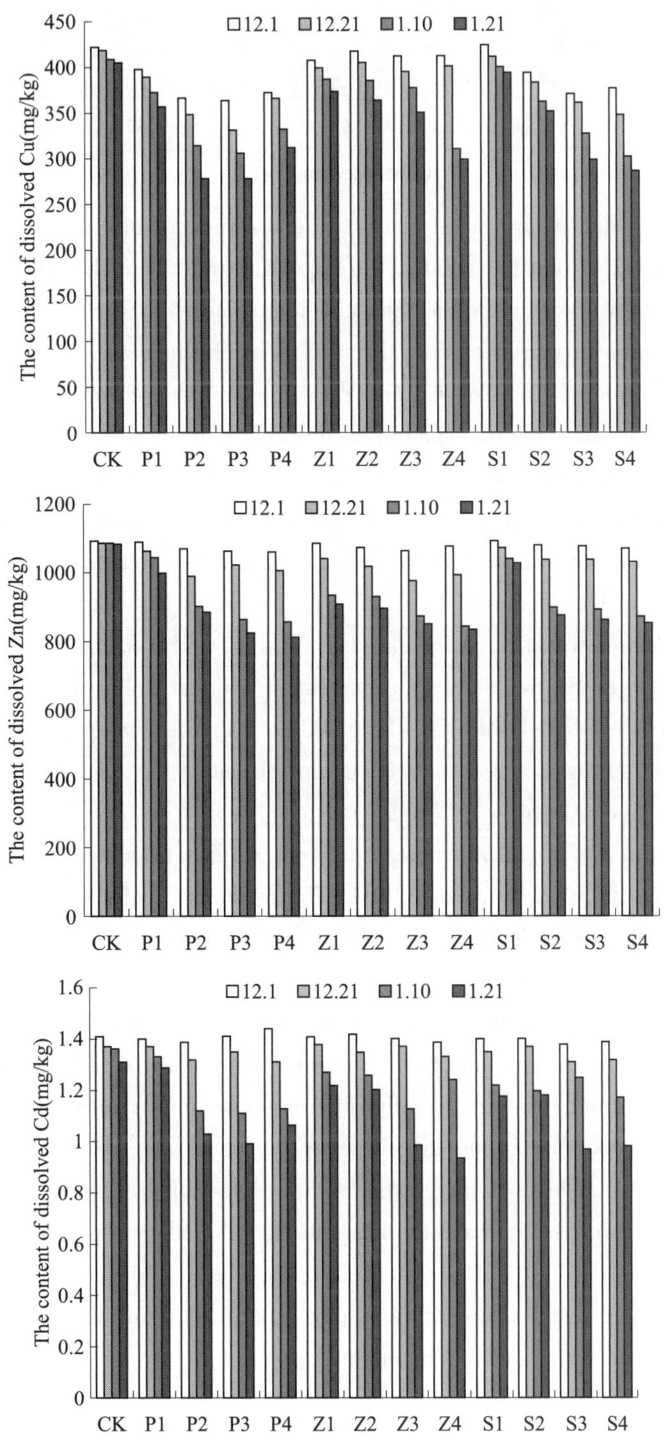

Fig. 2 Changes in the content of dissolved heavy metals in the soil at different incubation times

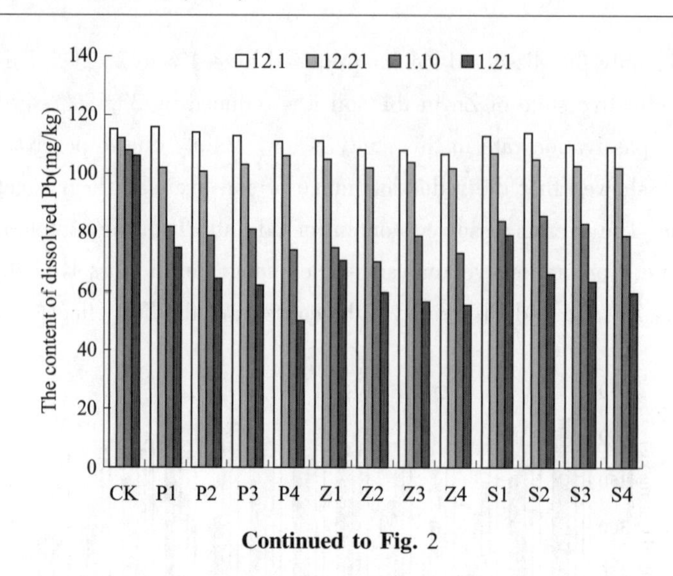

Continued to Fig. 2

3.4.2 Changes in residual heavy metal content in contaminated soil

It can be seen from Fig. 3 that the residual Cu, Zn, Cd and Pb in the soil showed an upward trend with the increase of the cultivation time under constant temperature culture conditions. The residual heavy metals content of each treatment is greater than that CK. With the increase of the passivation dose in each treatment, the content of residual Cu, Zn, Cd and Pb in the soil showed an upward trend. The overall Cu content in the soil residue is palygorskite>sepiolite>zeolite>CK. Among them, the residual Cu content of CK is 12 mg/kg. The residual Cu content of P4, S4 and Z4 treatments are 25 mg/kg, 24 mg/kg and 23 mg/kg. They are 2.1, 2.0 and 1.9 times of CK respectively. And it is significantly higher than other treatments, but the difference between the three is not significant. Residual Zn content changed from 23 mg/kg to 29 mg/kg. The fixation rate has increased by 2.1%. Z4 treatment has the highest residual zinc content, which is 50 mg/kg. Compared with CK, it increased by 72.4%, and it was significantly different from other zeolite treatments; followed by S4 treatment, the content of residual Zn in the soil was 48 mg/kg, which was an increase of 65.5% compared with CK. From P2 to P4, the content of residual Zn in the soil is 41 mg/kg, 42 mg/kg and 43 mg/kg, respectively. Compared with CK, it increased by 40.0%, 44.8% and 48.3% respectively. This is consistent with Wang Xiaoyu's research results[31]. With the increase of passivation time, the passivation rate of soil Zn increased. From 30 days to 60 days, the passivation rate of soil Zn increased from 32.2%~55.5% to 46.0%~63.9%. The residual Cd of Z4, S4 and S3 are 0.21 mg/kg, 0.18 mg/kg and 0.17 mg/kg, respectively. Compared with CK, it increases by 90.1%, 63.6% and 54.5%. This is consistent with the research results of Guo[40]. By applying passivating agent to the soil, the effective Cd content is significantly reduced, and the fixed content is increased. The content of residual Pb in the soil remains as zeolite>palygorskite> sepiolite>CK. The residual Pb content of Z4 reaches 30 mg/kg, which is 3.75 times higher than CK. And it is significantly different from other treatments; followed by P4 and Z3 treatments, which are 25 mg/kg and 24 mg/kg, respectively, which are 3.13 and 3.00 times higher than CK Comprehensive analysis of the changes in the content of the residues of the four elements in the soil shows that Cu, Zn, Cd and Pb treated with zeolite have the best effect, followed by sepiolite, and finally palygorskite.

4 Conclusion

In this experiment, the constant temperature culture experiment found that different kinds of passivators have a good fixation effect on heavy metal contaminated soil.

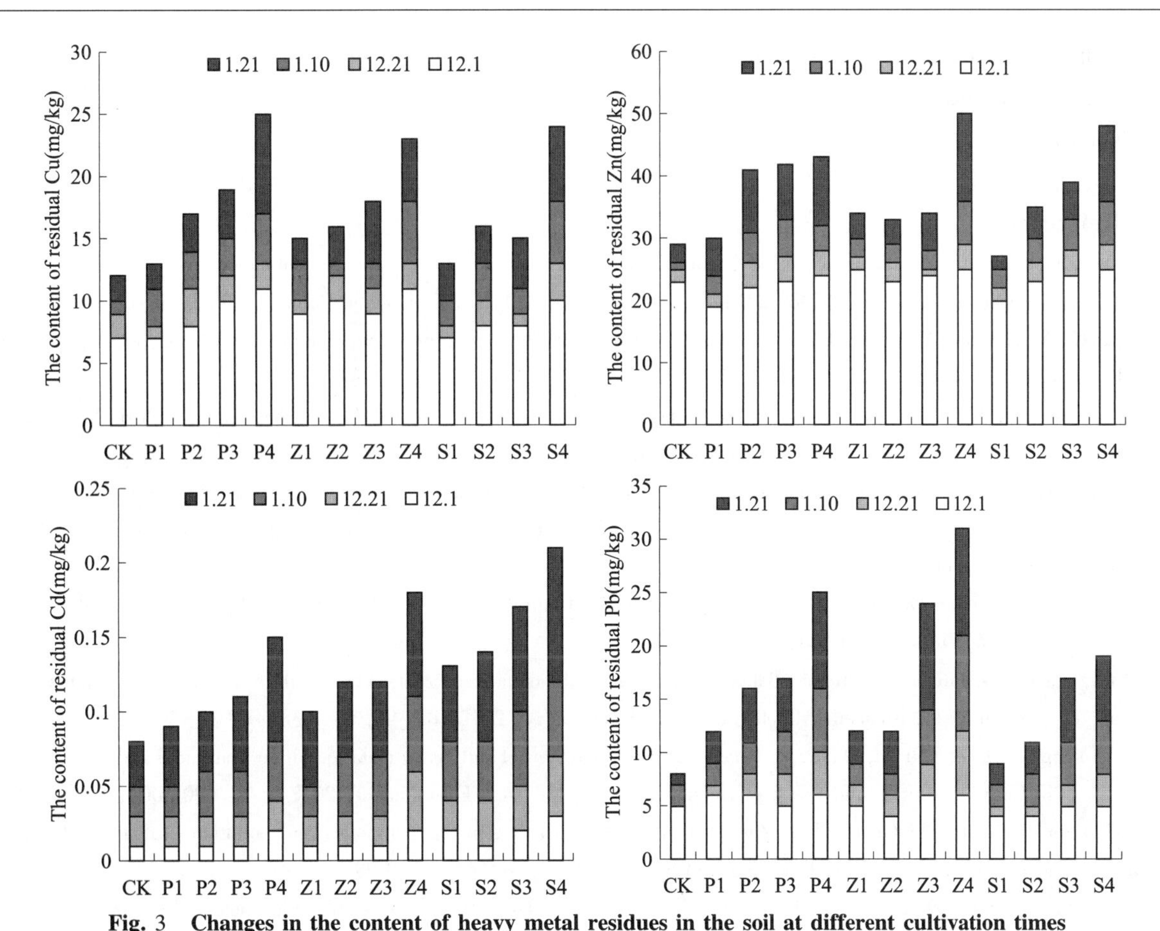

Fig. 3 Changes in the content of heavy metal residues in the soil at different cultivation times

Note: The bar chart in the figure uses December 1 as the base, showing the increase in the content of heavy metals in residues in the soil with different treatments over time.

(1) With different passivating agents, the dissolved Cu, Zn, and Pb content of the zeolite treatment has the highest decrease rate. The sepiolite treatment has the highest decrease rate of dissolved Cd. The content of heavy metals in the residue state of each treatment gradually increased with time.

(2) Different of passivators reduce the content of dissolved heavy metal and increase the content of residual heavy metal. Among the three passivating agents, 5% zeolite has the best passivation effect on heavy metals.

(3) With different amounts of passivation agents, the dissolved content of Cd and Pb in contaminated soil is CK> palygorskite> sepiolite> zeolite. The curing effect of Cu, Zn, Cd and Pb treated with zeolite is the best, followed by sepiolite, and finally palygorskite.

Comprehensive analysis shows that 5% zeolite has the best effect and can be used for passivation restoration of heavy metals Cu, Zn and Pb in the soil in the single passivation treatment. The low concentration of sepiolite can repair Cd better than other treatments.

Funding information: This research was supported byFund Project of Shaanxi Key Laboratory of Land Consolidation"AStudy onthe Influence of Loess Physical Properties on Heavy Metal Migration: A Case Study of Pb Pollution in Weidong New Town" (2019-JC06) and the project of Shaanxi Provence Land Engineering Construction Group (Program No. DJNY2021-23).

Author contributions: S.Y., L.W., T.L.—conception of the study; S.Y, L.W.—experiment; S.Y., L.W.—analysis and manuscript preparation; S.Y.—data analysis and writing the manuscript; T. L.—analysis with constructive discussions.

Conflict of interest: The authors declare they have no conflict of interests.

Ethical approval: The conducted research is not related to either human or animal use.

Data availability statement: All data generated or analyzed during this study are included in this published article.

References

[1]Li C, Sanchez G M, Wu Z F, et al. Spatiotemporal patterns and drivers of soil contamination with heavy metals during an intensive urbanization period (1989-2018) in southern China [J]. Environmental Pollution, 2020, 260.

[2]Hu Y A, Cheng H F, Shu T. The challenges and solutions for cadmium-contaminated rice in China: A critical review [J]. Environment International, 2016, 92-93, 515-532.

[3]Zhao F J, Ma Y B, Zhu Y G, et al. Soil Contamination in China: Current status and mitigation Strategies[J]. Environmental Science and Technology, 2015, 49(2), 750-759.

[4]Du H X, Nobuo H, Li F S. Responses of riverbed sediment bacteria to heavy metals: Integrated evaluation based on bacterial density, activity and community structure under well-controlled sequencing batch incubation conditions[J]. Water Research, 2018, 130, 115-126.

[5]Qu C S, Shi W, Guo J, et al. China' s Soil Pollution Control: Choices and Challenges[J]. Environmental Science and Technology,2016,50(24), 13181-13183.

[6]Solgi E, Esmaili-Sari A, Riyahi-Bakhtiari A, et al. Soil Contamination of metals in the three industrial estates, Arak, Iran[J]. Bulletin of Environmental Contamination and Toxicology, 2012, 88(4), 634-638.

[7]Yang Q., Li Z., Lu X., Duan Q., Huang L., Bi J. A review of soil heavy metal pollution from industrial and agricultural regions in China: Pollution and risk assessment. Science of the Total Environment, 2018, 642, 690-700.

[8]Chen Wenxuan, Li Qian, Wang Zhen, et al. Spatial distribution characteristics and pollution evaluation of heavy metals in farmland soils in China[J]. Environmental Science, 2020, 41 (6): 2822-2833.

[9]Zou T., Li T., Zhang X., Yu H., Huang H. Lead accumulation and Phyto stabilization potential of dominant plant species growing in a lead-zinc mine tailing. Environmental Earth Sciences, 2012, 65(3), 621-630.

[10]Yang GD, Zhang MZ, Feng T, et al. Research status and prospect of remediation technology for heavy metal polluted soil[J]. Modern Chemical Industry, 2020,42(12):50-54,58.

[11]Hu Hongqing, Huang Yizong, Huang Qiaoyun, et al. Research progress in chemical passivation remediation of heavy metal pollution in farmland soil[J]. Journal of Plant Nutrition and Fertilizer, 2017, 23(6): 1676-1685.

[12]Mench M, Renella G, Gelsomino A, et al. Biochemical parameters and bacterial species richness in soils contaminated by sludge borne metals and remediated with inorganic soil amendments[J]. Environmental Pollution, 2006, 144(1):24-31.

[13]Gao Binbin, Wang Xuan, Wang Jue, et al. Effects of chemical and clay mineral passivators on the transformation of phosphorus forms in cow manure straw compost[J]. Transactions of the Chinese Society of Agricultural Engineering, 2019, 35 (02): 242-249.

[14]Bolan N, Kunhikrishan A, Thangarajan R, et al. Remediation of heavy metals contaminated soils: To mobilize or to immobilize? [J]. Journal of Hazardous Materials, 2014, 266:141-166.

[15]Liang Yuan, Wang Xiaochun, Cao Xinde. Research progress on the remediation of heavy metal contaminated soil based on chemical passivation of phosphate, carbonate and silicate materials[J]. Environmental Chemistry, 2012, 31: 16-25.

[16]Li Yi, Qu Zhuangzhuang, Liu Yanjie, et al. Passivation effect of passivation agent on As in anaerobic fermentation of pig manure and process optimization[J]. Transactions of the Chinese Society of Agricultural Engineering, 2018, 34(12): 245-250.

[17]Chen Yiqun, Dong Yuanhua. Research and application progress of soil amendments[J]. Journal of Ecological Environment, 2008, 17(3): 1282-1289.

[18]Sun Jifeng, Wang Xu. Research and application progress of soil conditioners[J]. Soils and Fertilizers in China, 2013: 10-17.

[19]Chen Gongning. Remediation effect and mechanism of mineral passivator on heavy metal polluted red soil[M]. Guangzhou: South China University of Technology, 2017.

[20]Dai Rui, Zheng Shuilin, Jia Jianli, et al. Research progress of non-metallic mineral environmental materials[J].

China Non-metallic Mineral Industry Guide, 2009, (3): 3-10.

[21]Vondráková S, Hejcman M, SzákováJ. Effect of quick lime and dolomite application on mobility of elements (Cd, Zn, Pb, As, Fe, and Mn) in Contaminated soils. Pol J Environ Stud,2013, 22 (2): 577-589.

[22]Liang Xuefeng, Xu Yingming, Wang Lin, et al. Study on in-situ passivation remediation effect of natural clay combined with phosphate fertilizer on farmland soil cadmium and lead pollution[J]. Acta Scientiae Circumstantiae, 2011, 31(5): 1011-1018.

[23]Wang Yongxin, Sun Yuebing, Xu Yingming, et al. Application of chicken manure on the enhancement effect of sepiolite passivation and restoration of cadmium contaminated vegetable soil and soil enzyme activity[J]. Environmental Science, 2016, 35: 159-169.

[24]Xiao Liangliang, Ding Yuan. The regulation and mechanism of biochar matrix combined with medical stone on soil-ryegrass system[J]. Environmental Science, 2019, 40 (10): 4668-4677.

[25]Miao Xiurong, Lai Xuehui, Li Mengxi, et al. Effects of different passivators on available heavy metal contents in soil and their accumulation in pakchoi, Journal of Henan Agricultural Sciences, 2020,49(08):63-71.

[26]Zhao J. Effects of amendment materials on soil physicochemical properties, availability of heavy metals, and plant uptake in contaminated agricultural soils: A Meta-analysis [D].Zhe Jiang University,2020.

[27]Gao R L, Tang M, Fu Q L, et al. Fractions Transformation of Heavy Metals in Compound Contaminated Soil Treated with Biochar, Montmorillonite and Mixed Addition[J]. Environmental Science,2017,38(1):361-367.

[28]Jiang J,Xu R K,Jiang T Y,et al. Immobilization of Cu(II),Pb(II)and Cd(II)by the addition of rice straw derived biochar to a simulated polluted Ultisol[J].Journal of Hazardous Materials ,2012,229-230:145-150.

[29]Li F Y, Shen W Y, WU X, et al. Remediation of heavy metal contaminated soil by passivation of biochar complex minerals[J]. Chinese Journal of Soil Science, 2020, 51(1): 195-200.

[30]Wu P P, Li L J, Wang J J, et al. Effect of straw biochar on the transformation of heavy metals in polluted soil in mining area[J]. Journal of Ecology and Rural Environment, 2017, 33(5): 453-459.

[31]Wang Xiaoyu,Sun Lina, Lu Lianghe, et al. Study on the passivation effect of different passivators on Zn, Cd, Pb, As in soil[J]. Environmental protection and circular economy, 2021,51(5):52-58,84.

[32]Du Caiyan, Wang Panlei, Du Jianlei, et al. Influence of fixed addition of biochar, zeolite and bentonite on growth and Cd, Pb, Zn uptake by maize[J]. Ecology and Environmental Sciences, 2019,28(1): 190-198.

[33]Rees F, Germain C, Sterckeman T, et al., 2015. Plant growth and metal uptake by a non-hyperaccumulating species (Lolium perenne) and a Cd-Zn hyperaccumulator (Noccaea caerulescens) in contaminated soils amended with biochar [J]. Plant & Soil, 395(1-2):57-73.

[34]Bashir S, Ali U, Shaaban M, et al. Role of sepiolite for cadmium (Cd) polluted soil restoration and spinach growth in wastewater irrigated agricultural soil[J]. Journal of Environmental Management. 2020, 258: 110020.

[35]Wang Lin, Xu Yingming, Sun Guohong, et al. Effect and mechanism of immobilization of paddy soil contaminated by cadmium and lead using sepiolite and phosphate[J]. Ecology and Environment Sciences. 2012, 21(02): 314-320.

[36]Li H, Shi W Y, Shao H B, et al. The remediation of the lead-polluted garden soil by natural zeolite[J]. Journal of Hazardous Materials, 2009,169(1):1106-1111.

[37]Castaldi P, Santonal L, Enzo S, et al. Sorption processes and XRD analysis of a natural zeolite exchanged with Pb^{2+}, Cd^{2+}, and Zn^{2+}, cations[J]. Journal of Hazardous Materials, 2008,156(1-3):428-434.

[38]Ren L L, Cai Z P, Wang G N, et al. Effects of minerals with different immobilization mechanisms on heavy metals availability and soil microbial response[J]. Journal of Agro-Environment Science, 2021,40(7):1470-1480.

[39]Liang N, Jing T F, Li Z W, et al. Effects of different amendments on the availability of heavy metals and microbial communities in contaminated soils[J]. Asian Journal of Ecotoxicology,2021, 16(1):177-187 (in Chinese).

[40]Guo L M, Ai S Y, Tang, M D, et al. Effect of amendment on Cd uptake by Brassia chinensis in Cd-contaminated soils[J]. Chinese Journal of Eco-agriculture, 2010,18(3): 654-658.

本文曾发表于2022年《Open Chemistry》第20卷

Vertical Distribution of STN and STP in Watershed of Loess Hilly Region

Tingting Meng[1,2,3], Jinbao Liu[1,2,3], Huanyuan Wang[1,2,3], and Yichun Du[1]

(1. Shaanxi Provincial Land Engineering Construction Group Co. ,Ltd.,Xi' an 710075,China; 2. Institute of Shaanxi Land Engineering and Technology Co. , Ltd., Xi' an 710075,China; 3. Key Laboratory of Degraded and Unused Land Consolidation Engineering, Ministry of Land and Resources, Xi' an 710075, China)

【**Abstract**】 In order to explore the effects of land use change on the contents of total nitrogen and total phosphorus in deep soil, four land use types (cropland, grassland(7yr), grassland(30yr) and Jujube orchard) were selected from the Yuanzegou watershed in the loess hilly region of northern China. Soil samples at 0~10 m depth were collected to measure the contents of STN and STP, and their stocks were estimated. The results showed that the soil total nitrogen (STN) content showed a decreasing trend with the increase of soil depth, and the lowest STN content of grassland (7yr) was 0.09~0.17 g · kg^{-1}. The range of STN content in the watershed was 0.12~0.22 g · kg^{-1} and the coefficient of variation was 10.52%~25.90%, which belonged to medium variation. The soil total phosphorus (STP) content is stable regionally with the change of soil depth, and does not change much (except for grassland (30yr). STP content of the watershed is 0.81~1.05 g · kg^{-1} and the coefficient of variation is 9.37%~54.69%, which is a high variation. The change trend of STN and STP stocks is consistent with the nitrogen and phosphorus content. The results revealed the dynamic changes of STN and STP after land use change, and estimated the stocks of STN and STP in deep loess, which provided scientific basis for land and soil resource management and sustainable development of the project of returning farmland to forest or grassland in small watershed of loess hilly-gully region.

【**Keywords**】 Yuanzegou watershed; Nitrogen; Phosphorus; Deep soils; Land use

1 Introduction

Soil total nitrogen (STN) and phosphorus (STP) are two major elements influencing both plant growth and global biogeochemical cycles[1-2]. In terrestrial ecosystems, STN and STP play important roles by affecting soil properties[3], plant growth[4-5], and soil microbial activities[6]. In agricultural ecosystems, STN and STP are the major determinants and indicators of soil quality, which are closely related to soil productivity. The reduction of STN and STP levels can result in a decrease in soil nutrient supply, fertility, porosity, penetrability, and, consequently, in soil productivity[7]. In environmental science, nitrogen and phosphorus are the main non-point source pollutants of surface water and ground water. In addition, STN and STP are closely correlated to soil organic carbon (SOC) cycle[8], and they have dynamic effects on greenhouse gas emissions, which are linked to global climate change[9]. Thus, it is of great significance to scientifically evaluate the dynamic change of N and P for optimizing land management, ecological environment health and food security[10-11].

The Loess Plateau of North China is famous for its deep loess, unique landscapes and intense soil erosion[12]. The annual sediment input from the loess plateau of Shaanxi province to the Yellow River is 7.67 tons, accounting for 48% of the annual sediment transport volume of the Yellow River[13]. Excessive of soil and water losses on Loess Plateau cause serious nutrient losses, soil degradation and reduction of crop yields, which significantly restrict development of the regional economy and rise of the living standard for the farmers. In 1999 a large-scale ecological engineering program called "Grain for Green" was initiated to control serious soil erosion there by the central government of China[14]. Since then, the type of land use has changed.

The change of land use type is accompanied by the change of vegetation. Vegetation, as an important part of terrestrial ecosystem, is interdependent and mutually restricted between vegetation and soil in the ecosystem[15]. The direction and speed of vegetation occurrence, development and succession will be affected by the physical and chemical properties of soil. In turn, soil physical and chemical properties will change with the evolution of plant communities[16-17]. Human-induced land-use change has been identified as one of the major factors that affect soil C, N, and P cycles because it may alter plant species, land management practice, and soil microbial community structure[18-22].

There are many studies on the effects of land use change on soil nutrient distribution in loess hilly region. Wang et al.[2] and Liu et al.[23] used geostatistical methods to study the spatial heterogeneity of STN and STP in about 700 soil samples under 0~40 cm depth of different land use types across the entire Loess Plateau region of China. Wei et al[24] studied the distribution characteristics of STN and STP in the soil at a depth of 0~40 cm under three adjacent soil use types in the northern Loess Plateau. Xue et al[25] studied STN, nitrate nitrogen and ammonium nitrogen in 0~60 cm soil of different land use types in the hilly and gully region of The Loess Plateau in Ningxia. Qiao et al[26] studied the vertical distribution characteristics of STN and STP in 0~200 m deep soil from 703 soil samples in 5 regions of the Loess Plateau, but did not study the vertical distribution characteristics of STN and STP under different land uses. In summary, previous studies on STN and STP under different land uses on the Loess Plateau were mainly concentrated in the shallow soil layer, and the research on the vertical distribution of STN and STP in deep soil was limited to different regions of the Loess Plateau. Therefore, the research on the vertical distribution of STN and STP in deep soil under different land use types on the Loess Plateau is still relatively scarce.

Determining the vertical distributions of STN and STP as well as the factors that influence them under different land use can help evaluate the impact of patterns of land utilization conversion on soil N and P reserves[27-28]. In previous studies, researchers mainly focused on the effects of land-use change on dynamics of C, N, and P in topsoil ($\leqslant$30 cm) because this soil layer stores high levels of C, N, and P, which can be easily influenced by external disturbance[29-30]. However, due to limited annual precipitation (<600 mm) and thick loess soils on the loess plateau, perennial grass and forest species that have established during the last three decades in the region can extend their roots deeper than 10 m into soils, thereby modifying the water, carbon and nitrogen cycling[2].Thus, the objectives of this study were: (1) to investigate the vertical distributions of STN and STP with different land use types; (2) to analyze the factors that influence STN and STP; and (3) to evaluate the vertical distribution of the STN and STP stocks in deep soils.

2 Methods

2.1 Study area

The experimental area is located in the small watershed of Yuanzegou (37°150′N, 110°210′E) in the loess hilly region of northern China. The basin has deep steep gully, deep loess soil and serious erosion, covering an area of about 0.58 km^2, among which the gully area accounts for half of the total area of the sub-basin. The region has an arid and semi-arid climate, with an average annual precipitation of about 498 mm, 65% of which is mainly in autumn. This terrain is composed by complex features: gullied slopes of hills (20°<mostly with gradients<45°) in upper parts and deep, the loess soils are typical silt loams belonging to Inceptisols (United States Department of Agriculture), usually with >50% silt contents and <30% clay contents . Since the implementation of the conversion of farmland to forest, the main land use types in the basin include cropland, 7-year grassland, 30-year grassland and jujube orchard (as shown in Fig. 1).

Cropland　　Grassland(7yr)

Grassland(30yr)　　Jujube orchard

Fig. 1　Different land use types in the Yuanzegou watershed

2.2　Soil sampling and laboratory analysis

In August 2015, according to the area size and topographic factor (Table 1) of each land use type, 5, 4, 8 and 4 sample points were collected with soil drills on small watershed cropland, 7-year grassland, 30-year grassland and jujube forest, respectively, with a sampling depth of 0~10 m and 10 layers in total, according to the random sampling method. Some of the soil samples were collected and put into an aluminum box to determine the soil moisture content[31], and some were put into a self-sealed bag and taken back to the experiment for later use. The soil samples brought back to the laboratory were air-dried for 7 days and then passed through 2 mm and 0.25 mm sieve, respectively. The 2 mm soil sample was used to determine soil particle composition[32] and the 0.25 mm soil sample was used to determine soil organic carbon, STN and STP, The determination method of carbon, nitrogen and phosphorus was conventional experimental method[33]. Soil properties in watershed are shown in Table 2.

Table 1 Topographic factor for each land uses

Land uses	slope	altitude	aspect	Vegetation types
Cropland	23.86	987.82	248.90	*Setaria italic* & *Zea mays*
Grassland(7yr)	31.32	982.85	290.33	*Stipa bungeana* Trin & *Artemisia scoparia*
Grassland(30yr)	20.80	1061.62	229.79	*Tripolium*
Jujube orchard	28.55	1011.94	328.22	*Ziziphus jujube* Mill

Table 2 Soil properties with the depth of the soils under each land uses

Land uses	Soil depth (m)	Bulk density ($g \cdot cm^{-3}$)	Soil moisture (%)	SOC ($g \cdot kg^{-1}$)	Soil texture fractions		
					Clay(%)	Silt(%)	Sand(%)
Cropland	1	1.21	14.94	2.21	14.92	72.08	12.99
	2	1.28	13.48	1.77	15.58	72.20	12.23
	3	1.34	12.90	1.68	14.05	69.63	16.32
	4	1.31	13.80	1.55	14.77	72.70	12.53
	5	1.33	13.61	1.55	13.23	67.63	19.14
	6	1.22	13.71	1.50	18.24	61.65	20.11
	7	1.32	13.75	1.48	12.06	64.45	23.49
	8	1.28	14.10	1.40	15.73	68.74	15.54
	9	1.23	13.82	1.31	12.45	62.06	25.49
	10	1.33	13.93	1.23	11.27	58.80	29.92
Grassland(7yr)	1	1.30	11.91	1.20	15.05	69.29	15.66
	2	1.34	10.85	0.89	16.43	64.89	18.67
	3	1.32	12.23	1.52	15.96	63.33	20.70
	4	1.35	12.78	1.30	18.99	67.55	13.45
	5	1.20	11.29	0.85	15.96	64.83	19.21
	6	1.23	11.20	0.96	13.17	59.91	26.92
	7	1.33	11.84	1.33	15.92	66.44	17.64
	8	1.21	12.86	1.12	20.40	67.25	12.36
	9	1.22	12.24	1.03	18.71	65.08	16.21
	10	1.31	13.33	1.01	17.89	70.72	11.38
Grassland(30yr)	1	1.30	12.02	2.57	15.90	71.90	12.20
	2	1.38	13.29	2.23	15.95	69.75	11.12
	3	1.39	12.62	1.78	15.32	61.50	23.18
	4	1.40	12.11	1.69	14.59	70.87	14.55
	5	1.37	11.89	1.36	13.27	67.62	19.11
	6	1.40	12.81	1.44	13.26	68.00	18.74
	7	1.31	13.17	1.52	16.35	73.39	10.26
	8	1.32	13.30	1.46	12.16	42.94	44.90
	9	1.27	13.28	0.65	23.67	67.40	8.93
	10	1.32	12.72	0.76	15.89	61.53	22.59

Continued to Table 2

Land uses	Soil depth (m)	Bulk density (g · cm^{-3})	Soil moisture (%)	SOC (g · kg^{-1})	Soil texture fractions		
					Clay(%)	Silt(%)	Sand(%)
Jujube orchard	1	1.20	11.89	2.50	12.46	60.88	26.65
	2	1.19	11.18	2.36	13.70	66.80	13.26
	3	1.04	11.67	2.16	16.26	62.65	21.08
	4	1.22	10.98	1.80	15.56	69.45	14.98
	5	1.27	10.40	1.56	11.27	57.43	13.69
	6	1.23	11.23	1.33	13.22	67.18	19.59
	7	1.32	8.97	1.59	13.56	67.10	19.34
	8	1.25	10.22	0.73	5.65	28.89	11.40
	9	1.31	9.94	0.80	16.05	68.39	15.56
	10	1.25	9.54	0.53	19.51	69.44	11.05

2.3 Calculations

The stocks of STN and STP were calculated using the following equations[34].

$$STNS_i = D_i \cdot BD_i \cdot STNC_i \times \frac{1}{100}$$

$$STPS_i = D_i \cdot BD_i \cdot STPC_i \times \frac{1}{100}$$

where STNS and STPS are the stocks of STN and STP (kg · m^{-2}), respectively, i is the ith soil layer, D is the soil layer thickness (cm), BD is the bulk density (g · cm^{-3}), and STNC and STPC are the concentrations of STN and STP (g · kg^{-1}), respectively. while for lower layers, the soil bulk density was estimated using the following pedotransfer function(PTF) developed by Wang[35] from 1300 Loess Plateau datasets:

$$\begin{aligned} BD_i = {} & 1.8284 + 0.0429 \times \log_{10}(clay_i) + 0.0205 \times clay_i^{0.5} - 0.0125 \times \cos(clay_i) \\ & - 0.0061 \times silt_i + 0.0001 \times silt_i \times SG_i - 0.0098 \times SG_i - 0.0071 \times SOC_i \\ & - 0.0505 \times SOC_i^{0.5} + 0.0002 \times SOC_i^2 \end{aligned}$$

where clay and silt are contents at the ith depth, respectively; SG is slope gradient at the sample location; and SOC is contents at the ith depth.

2.4 Statistical analysis

Excel2020 and SPSS22.0 software were used for statistical analysis of the data. The CV was used to represent the overall variation of STN, STP, STNS and STPS. Pearson correlation coefficient indicates the strength of possible relationships between STN, STP, and other soil properties, and Origin2018 was used for mapping.

3 Results

3.1 Vertical distributions of STN and STP

Vertical distribution of STN and STP contents on the 0~10 m profile for different land use types in small watershed as shown Fig. 2 (a) and (b). The STN content of the four land use types decreased with the increase of soil depth. The STN content of cropland, grassland (7yr), grassland (30yr) and Jujube orchard are 0.12~0.21 g · kg^{-1}, 0.09~0.17 g · kg^{-1}, 0.09~0.26 g · kg^{-1}and 0.14~0.22 g · kg^{-1}. The STN content of grassland (7yr) was lower than that of the other 3 land use types (Fig. 2a). Except for grassland

(30yr), STP content tended to be stable with the increase of soil depth, but changed little, in cropland, grassland (7yr), and Jujube orchard are 0.96~1.21 g · kg^{-1}, 0.71~1.12 g · kg^{-1}, and 0.77~1.17 g · kg^{-1}. The STP content of grassland (30yr) fluctuated greatly, and the range was 0.65~1.08 g · kg^{-1}, especially in the 7 m soil layer, significantly decreased, and there was little difference in STP content among different land use types (Fig. 2b).

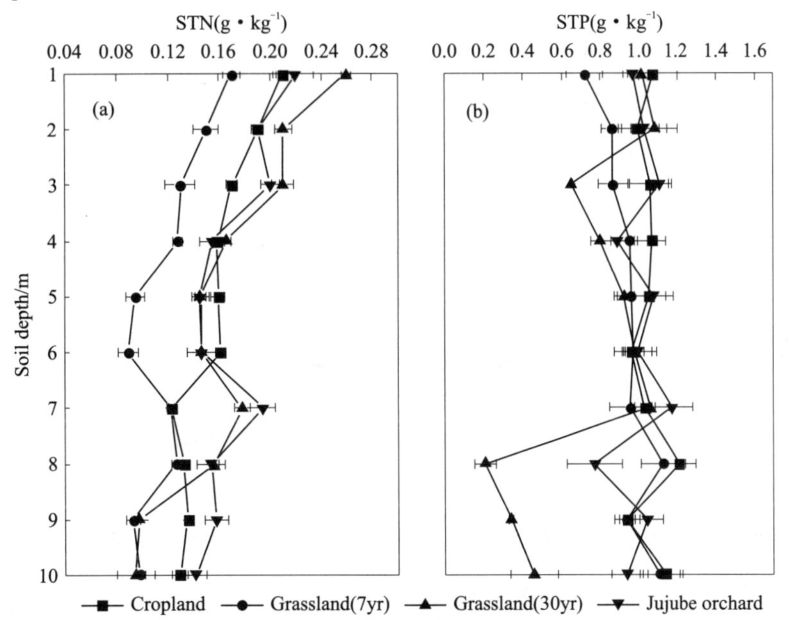

Fig. 2 STN and STP concentrations under different land uses in deep profiles (0~10 m).The error bar represent ±standard deviation

Distribution of STN content and its coefficient of variation under four land use patterns in Yuanzegou small watershed as shown Fig. 3 (a) and (b). The STN content of the four land use types ranged from 0.12 g · kg^{-1} to 0.22 g · kg^{-1} in the 0~10 m profile, with an average value of 0.15 g · kg^{-1}, and the variation coefficient ranged from 10.52% to 25.90%, showing moderate variation. STN content in the small watershed of Yuanzegou decreased with the increase of soil depth.

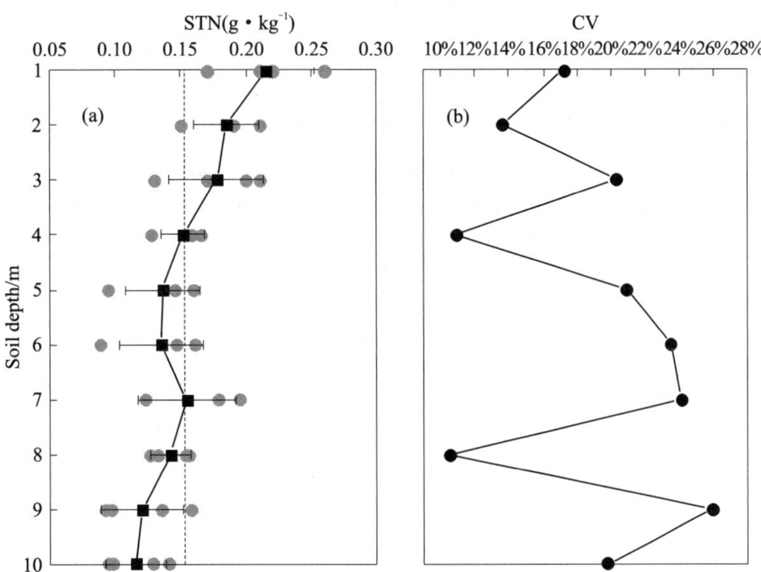

Fig. 3 STN content (gray circles) and means (blue circles) (a) and the coefficient of variation for STP content (b) at soil profile in four land uses on the watershed. The dashed red line is the mean of all the data for the entire 10 m profile. The error bars indicate the standard deviation

Distribution of STP content and its coefficient of variation under four land use patterns in Yuanzegou small watershed showed Fig. 4 (a) and (b). The STP content of the four land use types ranged from 0.81 g · kg^{-1} to 1.05 g · kg^{-1} in the 0~10 m profile, with an average value of 0.94 g · kg^{-1}, and the variation coefficient ranged from 9.37% to 54.69%, showing High variation. STP content in the small watershed of Yuanzegou increased firstly and then decreased with the increase of soil depth.

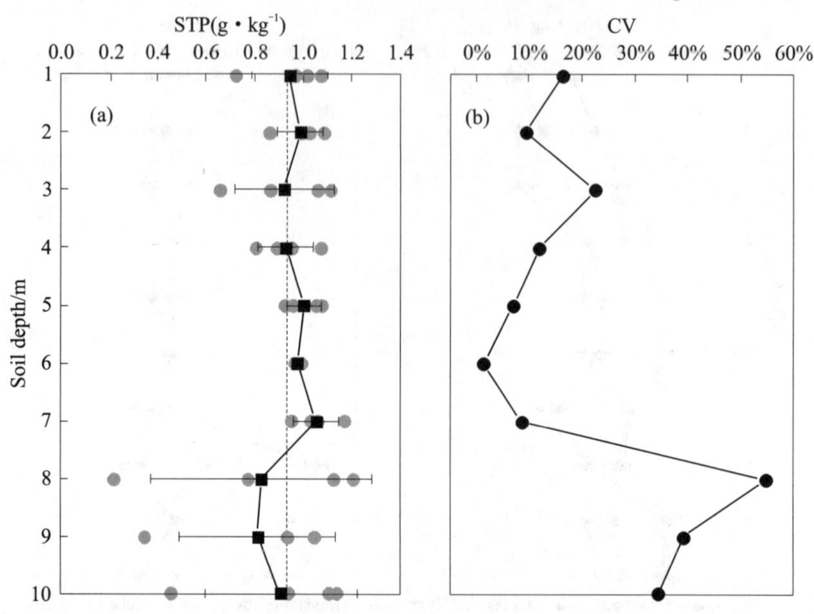

Fig. 4 STP content (gray circles) and means (blue circles) (a) and the coefficient of variation for STP content (b) at soil profile in four land uses on the watershed. The dashed red line is the mean of all the data for the entire 10 m profile. The error bars indicate the standard deviation

3.2 Correlation analysis

For the four land use types, the correlation between STN and STP contents in 0~10 m soil profiles and soil properties is shown in Table 3. Except for the jujube orchard, the contents of STN and STP were weakly correlated with bulk density and soil moisture. STN content was negatively correlated with sand content(r= −0.44, −0.47, −0.34, −0.48, $p<0.01$) and positively correlated with organic carbon content(r=0.47, 0.53, 0.51, 0.55, $p<0.01$) in four land use types STP content was negatively correlated with clay content (r= −0.41, −0.38, −0.37, −0.37, $p<0.01$) and weakly correlated with organic carbon content in the four land use types. There was not a significant correlation between STN and STP in four land use.

Table 3 Pearson's correlation coefficients between STN, STP and selected soil properties

Land uses	Variable	STN/(g · kg^{-1})	STP/(g · kg^{-1})	BD/(g · cm^{-3})	Sand/%	Silt/%	Clay/%	SM/%	SOC/(g · kg^{-1})
Cropland	STN	1	0.13	0.09	−0.44**	0.41**	0.38**	0.12	0.47**
	STP	0.13	1	−0.02	0.1	0.22*	−0.41**	0.08	0.20
Grassland(7yr)	STN	1	−0.05	0.11	−0.47**	0.21*	0.05	0.19	0.53**
	STP	−0.05	1	0.07	0.23*	0.39**	−0.38**	−0.11	0.07
Grassland(30yr)	STN	1	−0.07	0.11	−0.34**	0.45**	−0.08	−0.07	0.51**
	STP	−0.07	1	0.07	−0.08	−0.07	−0.37**	0.11	−0.04
Jujube orchard	STN	1	0.09	0.06	−0.48**	−0.28*	0.12	−0.20	0.55**
	STP	0.09	1	0.05	−0.14*	0.31*	−0.37**	−0.33*	0.12

Note: BD, Bulk density; SOC, soil organic carbon; SM, soil moisture.

3.3 Vertical distributions STNS and STPS

Vertical distribution of STN and STP stocks on the 0~10m profile for different land use types in small watershed as shown Fig. 5 (a) and (b). Same as STN, the STP stocks of the four land use types decreased with the increase of soil depth. The STN stocks of cropland, grassland (7yr), grassland (30yr) and Jujube orchard are 0.17~0.29 kg · m^{-2}, 0.13~0.24 kg · m^{-2}, 0.13~0.36 kg · m^{-2} and 0.20~0.31 kg · m^{-2}. The STN stocks of grassland (7yr) was lower than that of the other 3 land use types (Fig. 5a). Except for grassland (30yr), STP stocks tended to be stable with the increase of soil depth, but changed little, in cropland, grassland (7yr), and Jujube orchard are 1.35~1.58 kg · m^{-2}, 1.00~1.60 kg · m^{-2}, and 1.10~1.65 kg · m^{-2}. The STP content of grassland (30yr) fluctuated greatly, and the range was 0.03~1.50 kg · m^{-2}, especially in the 7 m soil layer, significantly decreased, and there was little difference in STP stocks among different land use types (Fig. 5b).

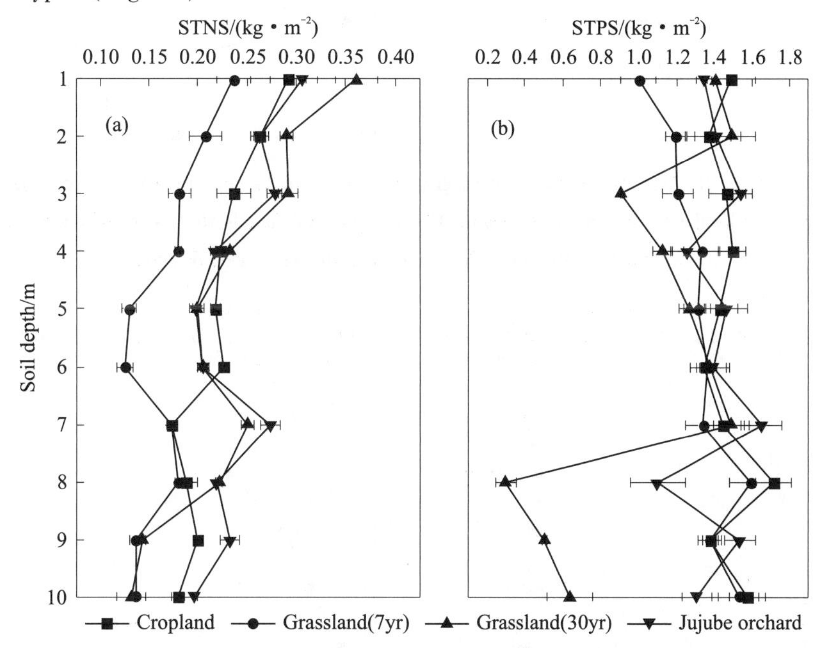

Fig. 5 STN and STP stocks under different land uses in deep profiles (0~10 m). The error bar represent ±standard deviation

Distribution of STN stocks and its coefficient of variation under four land use patterns in Yuanzegou small watershed showed Fig. 6 (a) and (b). The STN stocks of the four land use types ranged from 0.16 kg · m^{-2} to 0.26 kg · m^{-2} in the 0~10 m profile, with an average value of 0.21 kg · m^{-2}, and the variation coefficient ranged from 10.52% to 25.90%, showing moderate variation. The STN stocks in the small watershed of Yuanzegou decreased with the increase of soil depth.

Distribution of STP stocks and its coefficient of variation under four land use patterns in Yuanzegou small watershed showed Fig. 7 (a) and (b). The STP stocks of the four land use types ranged from 1.18 kg · m^{-2} to 1.37 kg · m^{-2} in the 0~10 m profile, with an average value of 1.31 kg · m^{-2}, and the variation coefficient ranged from 9.37% to 54.69%, showing high variation. The STP stocks in the small watershed of Yuanzegou increased firstly and then decreased with the increase of soil depth.

Discussion

Accordingly, in this study we found higher STN contents and stocks under the grassland (30yr) and jujube orchard than under cropland in the shallower layer. This result is consistent with other studies in the loess hilly region, where they found that cropland had lower STN content than forest land and grass-

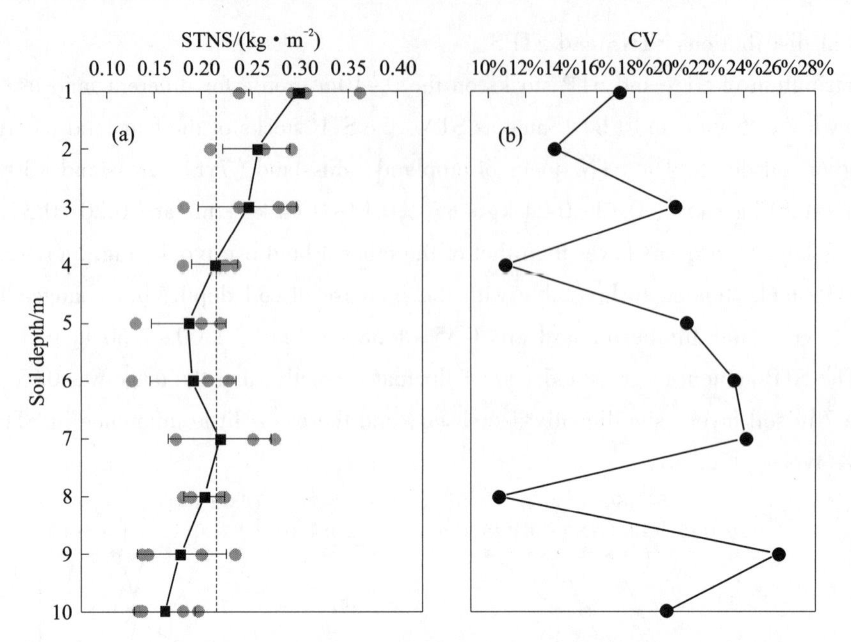

Fig. 6 STN stocks (gray circles) and means (blue circles) (a) and the coefficient of variation for STN stocks (b) at soil profile in four land uses on the watershed. The dashed red line is the mean of all the data for the entire 10 m profile. The error bars indicate the standard deviation

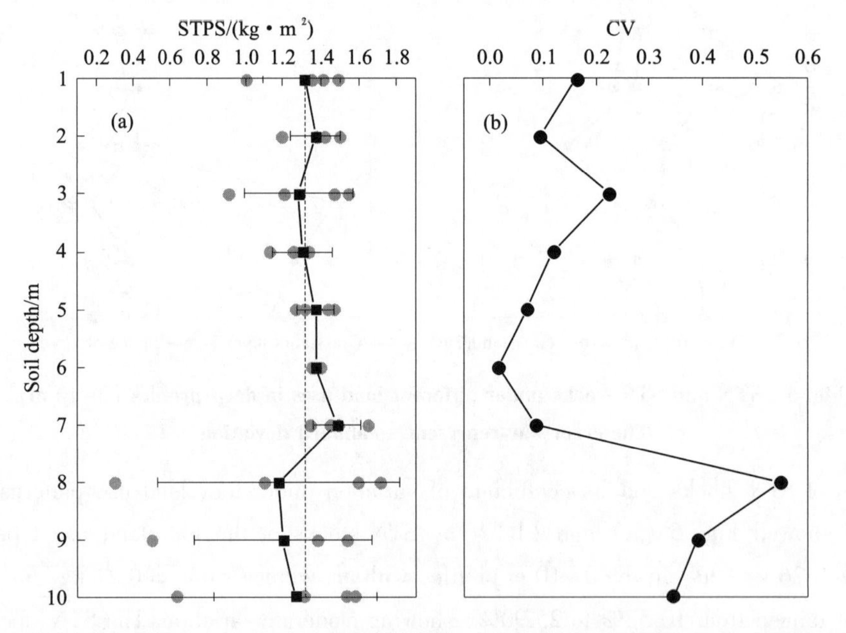

Fig. 7 STP stocks (gray circles) and means (blue circles) (a) and the coefficient of variation for STP stocks (b) at soil profil ein four land uses on the watershed. The dashed red line is the mean of all the data for the entire 10 m profile. The error bars indicate the standard deviation

land[36-37]. However, the STP content and stocks does not change much in the land use types .This result is similar to the Zaimes[38], that is, the change of land use type has little effect on STP. The main reason is that the soil erosion in a loess plateau is serious, STP content is generally low, and the phosphorus supply capacity is poor; the phosphorus is mainly related to the formation of the parent material.

Generally, the higher surface STN and STP under grassland can be attributed to greater above-ground and below-ground biomass, much lower soil erosion during heavy rainstorms[23, 39-40], slower mineralization of organic matter and better soil aggregation[41]. The main vegetation of the grassland (30yr) is Artemisia

Scoparia, which is a kind of herb of Artemisia, has less vertical root system, but the horizontal root system is more developed. The litter layer on the surface has a large water holding capacity, which can effectively absorb the water falling to the surface, delay the flow velocity of the surface runoff, and increase the infiltration time. In addition to the effective increase of infiltration, the underground root layer can effectively improve the soil's impact resistance and protect the surface soil nutrients. Therefore, STN and STP contents in the surface soil of the grassland (30yr) are higher. Although the above-ground biomass in jujube orchard can be greater than under cropland, inputs of litter (decayed leaves and branches) into surface soils can be negligible because of clean-cultivation soil management practices.

Cropland is generally considered to have a lower soil nutrient content than other natural land types (landless grasslands) because the biomass on the cropland is continuously removed and the disturbance of the land accelerates the decomposition and loss of nutrients[42-43]. However, in our study, STN and STP contents and stocks under cropland is higher than that of grassland (7yr), which is due to the cultivation of corn, millet, and cultivation during management. The organic fertilizer contains a large amount of organic matter, which is infiltrated into the soil with rainwater, so that STN and STP is replenished, and since the slope of the agricultural land is above 15 degrees, the soil is not disturbed by mechanical turning, and the soil nutrient loss is reduced. Grassland (7yr), because of the short time of abandoned land, the surface vegetation is sparse, and the erosion by rain is serious. In addition, at the early stage of vegetation restoration, vegetation growth needs to consume nutrients. Therefore, the contents and stocks of STN and STP in grassland (7yr) were lower than that of cropland.

In this study, the variation trend and coefficient of variation of STN storage and STP storage are consistent with the nitrogen and phosphorus content, mainly due to the small difference in soil bulk density in the loess hilly area. The STN content is greatly affected by the soil organic carbon content and sand content, which is consistent with Wang[2]. The STP content of the soil is greatly affected by the silt content, which is consistent with Qiao[26].

Conclusion

We investigated the vertical distribution of STN and STP at 0~10 m soil depth under four different land use types in the Yuanzegou small watershed in the loess hilly area. The contents of STN and STP in soil decreased with the increase of soil depth. There were significant differences in STN and STP content under different land use types. The stocks of STN and STP in deep soil is mainly affected by soil bulk density. The results provided scientific basis for land and soil resource management and sustainable development of the project of returning farmland to forest or grassland in small watershed of loess hilly-gully region.

Author contributions

Tingting Meng: Writing-original draft, Writing-review & editing, Methodology, Formal Analysis; Jinbao Liu: Writing-original draft, Formal Analysis, Visualization, Project administration. Huanyuan Wang and Yichun Du: Methodology, Analysis.

Acknowledgments

This work was jointly supported by the Scientific Research Item of Shaanxi Provincial Land Engineering Construction Group (DJNY2022-21).

References

[1] Bouwman A F, Beusen A H W, Billen G. Human alteration of the global nitrogen and phosphorus soil balances for the period 1970-2050. Global Biogeochemical Cycles. 2009; 23(4):1-16.

[2] Wang Y Q, Zhang X C, Huang C Q. Spatial variability of soil total nitrogen and soil total phosphorus under different

land uses in a small watershed on the Loess Plateau, China. Geoderma.2009;150(1-2):141-149.

[3] Hati K M, Swarup A, Mishra B, Manna M C, Waniari R H, Mandal K G, et al. Impact of long-term application of fertilizer, manure and lime under intensive cropping on physical properties and organic carbon content of an Alfisol. Geoderma. 2008;148:173-179.

[4] Norouzi M, Ayoubi S, Jalalian A, Khademi H, Dehghani A A. Predicting rainfed wheat quality and quantity by artificial neural network using terrain and soil characteristics. Acta Agriculturae Scandinavica Section B-Soil and Plant Science. 2010;60(4):341-352.

[5] Dokoohaki H, Gheysari M, Mehnatkesh A M, Ayoubi S. Applying the CSM-CERES-Wheat model for rainfed wheat with specified soil characteristic in undulating area in Iran. Archives of Agronomy and Soil Science. 2015;61(9):1231-1245.

[6] Tajik S, Ayoubi S, Lorenz N. Soil microbial communities affected by vegetation, topography and soil properties in a forest ecosystem. Applied Soil Ecology. 2020;149:103514.

[7] Huang B, Sun W X, Zhao Y C, Zhu J, Yang R Q, Zou Z, et al. Temporal and spatial variability of soil organic matter and total nitrogen in an agricultural ecosystem as affected by farming practices. Geoderma. 2007;139(3-4):336-345.

[8] Fu X L, Shao M A, Wei X R, Horton R. Soil organic carbon and total nitrogen as affected by vegetation types in Northern Loess Plateau of China. Geoderma. 2010;155(1-2):31-35.

[9] Jennings E, Allott N, Pierson D C, Schneiderman E M, Lenihan D, Samuelsson P, et al. Impacts of climate change on phosphorus loading from a grassland catchment: implications for future management. Water Res. 2009;43(17), 4316-4326.

[10] Lal R.Sequestering carbon in soils of agro-ecosystems. Food Policy.2011;36:S33-S39.

[11] Liu X, Ma J, Ma Z W, Li L H. Soil nutrient contents and stoichiometry as affected by land-use in an agro-pastoral region of northwest China. Catena. 2017;150:146-153.

[12] Chen L D, Wei W, Fu B J, Lu Y H. Soil and water conservation on the Loess Plateau in China: review and perspective. Prog Phys Geogr. 2007;31(4):389-403.

[13] Wang S A, Fu B J, Piao S L, Lu Y H, Ciais P, Feng X M, et.al. Reduced sediment transport in the Yellow River due to anthropogenic changes. Nat. Geosci. 2016;9(1):38-41.

[14] Liu Z P, Shao M A, Wang Y Q. Effect of environmental factors on regional soil organic carbon stocks across the Loess Plateau region, China. Agric Ecosyst Environ. 2011;142(3-4):184-194.

[15] Panico S C, Ceccherini M T, Memoli V, Maisto G, Pietramellara G, Barile R, et al. Effects of different vegetation types on burnt soil properties and microbial communities. International Journal of Wildland Fire. 2020;29(7):628-636.

[16] Fan Z Z, Lu S Y, Liu S, Li Z R, Hong J X, Zhou J X, et al. The effects of vegetation restoration strategies and seasons on soil enzyme activities in the Karst landscapes of Yunnan, southwest China. Journal of Forestry Research. 2019;31(5):1949-1957.

[17] Zhao X N, Wu P T, Gao X D, Persaud N. Soil quality indicators in relation to land use and topography in a small catchment on The Loess Plateau of China. Land Degradation and Development. 2015; 26(1):54-61.

[18] Khormali F, Ayoubi S, Foomani F K, Fatemi A, Hemmati K. Tea yield and soil properties as affected by slope position and aspect in Lahijan area, Iran. International Journal of Plant Production. 2007;1(1):95-111.

[19] Falahatkar S, Hosseini S M, Salman M A, Ayoubi S, Wang S Q. Soil organic carbon stock as affected by land use/cover changes in the humid region of northern Iran. Journal of Mountain Science. 2014;11(2): 507-518.

[20] Havaee S, Mosaddeghi M R, Ayoubi S. In situ surface shear strength as affected by soil characteristics and land use in calcareous soils of central Iran. Geoderma. 2015;237:137-148.

[21] Mokhtari K P, Ayoubi S, Mosaddeghi M R, Honarjoo N. Soil organic carbon pools in particle-size fractions as affected by slope gradient and land use change in hilly regions, western Iran. Journal of Mountain Science. 2012;9(1):87-95.

[22] Ayoubi S, Dehaghani S M. Identifying impacts of land use change on soil redistribution at different slope positions using magnetic susceptibility. Arabian Journal of Geosciences. 2020;13,(11):1-11.

[23] Liu Z P, Shao M A, Wang Y Q. Spatial patterns of soil total nitrogen and soil total phosphorus across the entire Loess Plateau region of China. Geoderma. 2013; 197:67-78.

[24] Wei X R, Shao M A, Fu X L, Horton R, Li Y, Zhang X C. Distribution of soil organic C, N and P in three adja-

cent land use patterns in the northern Loess Plateau, China. Biogeochemistry. 2009;96(1-3), 149-162.

[25] Xue Z J, Cheng M, An S S. Soil nitrogen distributions for different land uses and landscape positions in a small watershed on Loess Plateau, China. Ecological Engineering. 2013;60: 204-213.

[26] Qiao J B, Zhu Y J, Jia X X, Haung L M, Shao M A. Vertical distribution of soil total nitrogen and soil total phosphorus in the critical zone on the Loess Plateau, China. Catena. 2018;166:310-316.

[27] Wang T, Yang Y H, Ma W H. Storage, patterns and environmental controls of soil phosphorus in China. Acta Sci. Nat. Univ. Pekin, 2008;44 (6): 945-952. (in Chinese with English abstract)

[28] Roger A, Libohova Z, Rossier N, Joost S, Maltas A, Frossard E, et al. Spatial variability of soil phosphorus in the Fribourg canton, Switzerland. Geoderma. 2014;217:26-36.

[29] Cherubin M R, Franco A L C, Cerri C E P, Karle D L, Pavinato P S, Rodrigues M, et al. Phosphorus pools responses to land-use change for sugarcane expansion in weathered Brazilian soils. Geoderma. 2016;265:27-38.

[30] Spohn M, Novak T J, Incze J, Giani L. Dynamics of soil carbon, nitrogen, and phosphorus in calcareous soils after land-use abandonment-A chronosequence study. Plant and Soil. 2016;401(1-2):185-196.

[31] Gao X D, Wu P T, Zhao X N, Shi Y G, Wang J W, Zhang B Q. Soil moisture variability along transects over a well-developed gully in the Loess Plateau, China. Catena. 2011;87(3):357-367.

[32] Wang G L, Zhou S L, Zhao Q G. Volume fractal dimension of soil particles and its applications to land use. Acta Pedologica Sinca. 2005;42(4):545-550. (in Chinese with English abstract)

[33] Zhang Z S, Lu X G, Song X L, Guo Y, Xue Z S. Soil C, N and P stoichiometry of Deyeuxia angustifolia and Carex lasiocarpa wetlands in Sanjiang Plain, Northeast China. J Soils Sediments. 2012;12(9):1309-1315.

[34] Zhou Z Y, Sun O J, Huang J H, Li L H, Liu P, Han X G. Soil carbon and nitrogen stores and storage potential as affected by land-use in an agro-pastoral ecotone of northern China. Biogeochemistry. 2007;82(2):127-138.

[35] Wang Y Q, Shao M A, Liu Z P, Zhang C C. Prediction of bulk density of soils in the Loess Plateau region of China. Surveys in Geophysics. 2014;35(2):395-413.

[36] Chen L D , Gong J , Fu B J , Huang Z L, Huang Y L,Gui L D. Effect of land use conversion on soil organic carbon sequestration in the loess hilly area, loess plateau of China. Ecological Research. 2007; 22(4):641-648.

[37] Wang Y Q, Shao M A, Zhu Y J, Liu Z P. Impacts of land use and plant characteristics on dried soil layers in different climatic regions on the Loess Plateau of China. Agricultural and Forest Meteorology. 2011;151(4):437-448.

[38] Zaimes G N, Schultz R C, Isenhart T M. Total phosphorus concentrations and compaction in riparian areas under different riparian land-uses of Iowa. Agriculture ecosystems & environment. 2008;127(1-2):22-30.

[39] Fang X, Xue Z J, Li B C, An S S. Soil organic carbon distribution in relation to land use and its storage in a small watershed of the Loess Plateau, China. Catena. 2012, 88(1):6-13.

[40] Ayoubi S, Emami N, Ghaffari N, Honarjoo N, Sahrawat K L. Pasture degradation effects on soil quality indicators at different hillslope positions in a semiarid region of western Iran. Environmental earth sciences. 2014;71(1):375-381.

[41] Six J, Elliott E T, Paustian K, Doran J W. Aggregation and soil organic matter accumulation in cultivated and native grassland soils. Soil Science Society of America Journal. 1998; 62(5):1367-1377.

[42] Chen L D , Huang Z L , Gong J , Fu B J, Huang Y L. The effect of land cover/vegetation on soil water dynamic in the hilly area of the loess plateau, China. Catena. 2007;70(2):200-208.

[43] Deng L, Liu G B, Shangguan Z P. Land-use conversion and changing soil carbon stocks in China's "Grain-for-Green"Program: a synthesis. Global Change Biology. 2014;20(11):3544-3556.

本文曾发表于2022年《Open Geosciences》第14卷

Effect of Brackish Water Irrigation on the Movement of Water and Salt in Salinized Soil

Panpan Zhang [1,2,3,4], Jianglong Shen[1,2,3,4]

(1. Key Laboratory of Degraded and Unused Land Consolidation Engineering, Ministry of Natural Resources, Xi'an, 710075, China; 2. Shaanxi Provincial Land Consolidation Engineering Technology Research Center, Xi'an, 710075, China; 3. Land Engineering Technology Innovation Center, Ministry of Natural Resources, Xi'an, 710075, China; 4. Institute of Land Engineering and Technology, Shaanxi Provincial Land Engineering Construction Group Co., Ltd, Xi'an, 710075, China)

【**Abstract**】In China, fresh water resources are scarce, brackish water resources are abundant. Reasonable utilization of brackish water is one of the important measures to alleviate the contradiction of water shortage. In order to study the effect of brackish water irrigation on water and salt transport in saline-alkali soils, one-dimensional brackish water infiltration experiments of soil columns was conducted. The influence of brackish water with different salinity on water and salt transport in salinized soil was compared and analyzed. The results showed that under brackish, the Kostiakov model could better simulate the change of soil infiltration rate with time. the soil infiltration capacity had a positive response to the salinity of irrigation water. There was a good linear relationship between cumulative infiltration and the wetting front distance. Under different salinity conditions, the depth of soil desalination, Na^+ and Cl^- removal is different, which is inversely proportional to the degree of salinity; with the salinity of irrigation water increases, the water, salt content and the concentration of Na^+ and Cl^- increased gradually, but the difference in the desalination zone was not obvious. Therefore, brackish water irrigation has a certain effect on the distributions of water and salt in saline soil.

【**Keywords**】Brackish water; Water and salt movement; Mineralization degree; Salinized soil

1 Introduction

China is a country with a severe shortage of water resources, and the distribution of time and space is uneven. Agricultural water use is difficult to be generally satisfied, especially in the arid and semi-arid areas of northwest China, where precipitation is scarce and the average annual precipitation is only 300 mm[1], which seriously restricts the sustainable development of agriculture. China is rich in brackish water resources and can exploit 13 billion m^3 annually[2]. As an alternative resource, it can be rationally developed and utilized to increase irrigation water source, alleviate the pressure of insufficient freshwater resources, improve irrigation guarantee rate and ensure crop yield, which has important practical significance.

The requirements for safe and sustainable development of brackish water irrigation are high. Appropriate salt tolerant crops, fine irrigation levels and good soil characteristics are needed. Reasonable brackish water irrigation can ensure crop yield and improve product quality[3-8]. There was no significant difference in the yield of fresh water and brackish water (3~5 g/L) irrigated during the jointing stage of winter wheat in North China for four consecutive years, compared with dry farming, the yield increased by an average of 31.6 %, and the water use efficiency could be improved[3]. Using brackish water to plant tomatoes in coastal saline-alkali land, it was found that brackish water irrigation with $EC_i \leq 4.7$ dS/m could produce tomatoes with good quality under the condition of ensuring the balance of yield and soil salinity[4]. However, long-term brackish water irrigation may still aggravate soil salinization, thereby affecting crop growth. The technologies of saline-fresh water rotation irrigation[3,4,9], plastic film mulching and soil improvement are important means to improve the efficiency of brackish water irrigation. The harm of salt to crops and its effect on yield are less than that of water. It is suggested that the soil salt content and soil relative solution concentration under saline water irrigation should not exceed the salt tolerance limit of crops. Appropriate use of saline water irrigation can ensure agricultural production[10]. The mulching technology can reduce water evaporation, inhibit salt accumulation and effectively regulate soil water and salt distribution[11,12]. Using biochar, gypsum, straw, earthworm casts and other improved materials[13-18] combined with brackish water irrigation reference literature, the soil saline-alkali obstacle reduction and crop yield and quality improvement were studied, and the improvement effect of salinized soil was remarkable.

There are many models for rational use of brackish water at home and abroad. However, due to regional climate differences, different physicochemical properties of brackish water, different salt tolerance of crops, and uneven planting structure and economic development, the research results have their own characteristics. Although some mechanisms of brackish water utilization are revealed according to a certain water-salt migration model There are many models for rational use of brackish water at home and abroad[19,20]. However, due to regional climate differences, different physicochemical properties of brackish water, different salt tolerance of crops, and uneven planting structure and economic development, the research results have their own characteristics. Although some mechanisms of brackish water utilization are revealed according to a certain water-salt migration model[21-25]. For example, in the eastern coastal areas of China, the vertical distribution of soil water content and salt content was significantly affected by different flow rates and opening rates of brackish water film furrow irrigation, which generally increased the water content of crop root layer and reduced the soil salinity in 0 ~ 40 cm soil layer[12]. Under the condition of brackish water irrigation in Ningxia, the effects of sand gradation and sand thickness on soil water and salt migration and dynamic distribution were studied. It was found that the effect of sand gradation on soil water and salt distribution was not as large as that of sand thickness, and the water holding capacity of sand layer increased with the increase of sand thickness[19]. Some scholars also studied the soil water and salt migration characteristics of water-repellent soil under different salinity of brackish water, and found that the cumulative infiltration amount after brackish water infiltration had a good linear relationship with the advancing distance of wetting front. The content of water and salt in the same section of water-repellent soil was lower than that of non-water-repellent soil, and a certain amount of water repellency was generated after brackish water infiltration[25]. However, the characteristics of the study area are obvious, and it is difficult to replicate and promote in a large area. The utilization mechanism has not been fully revealed, and the corresponding water and salt migration model has not been fully established. Especially in areas with serious shortage of fresh water resources, how to control the process of soil water and salt migration, prevent soil secondary salinization and ensure the safe and sustainable development of brackish water irrigation needs further study. Therefore, this paper analyzes the characteristics of soil water and salt migration under different salinity brackish water irrigation conditions by indoor soil column one-dimensional infiltration test, and further reveals the law of water and salt migration in the study area, which provides a theoretical reference for the rational use of brackish water in saline areas where freshwater resources are lacking.

2 Methods

The test soil samples belonged to Salinized Soil and were taken from Manas County, Xinjiang Uygur

Autonomous Region in 2020. Remove debris from the soil, dry it naturally, grind and sieve it for later use. The physical properties of soil samples are shown in Table 1. According to the international classification standard, the soil texture is silty loam.

Table 1 Partial physical properties of soil in the study area

Physical properties	Texture(%)			Electrical conductivity of extract $EC_{1:5}$(μS/cm)	Volume weight (g/cm³)
	Clay particles	Powder particle	Sand particles		
Value	12.01	66.98	21.01	1113.50	1.45

The one dimensional soil column infiltration test is carried out indoors. The experimental device is shown in Fig. 1. The infiltration test device is mainly composed of a Markov bottle water supply device, an infiltration soil column and a fixed bracket group. The infiltration soil column is made of plexiglass (Φ8 cm, height 60 cm), with holes perforated at the bottom. Use a Markov bottle to control the constant infiltration head (2 cm) and automatically supply water. In order to reduce the influence of evaporation during the infiltration process, the soil column is covered with a film, and small holes are evenly opened in the film.

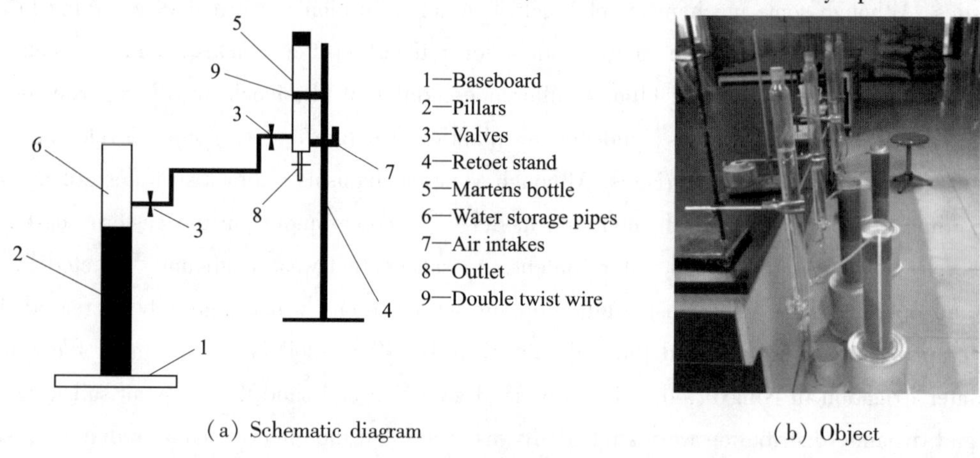

(a) Schematic diagram (b) Object

Fig. 1 Experimental installation

The salinized soil is packed into soil columns in layers according to the designed soil bulk density (1.45 g/cm³), each layer is 5 cm, and the soil height is 40 cm. In order to compare the influence of brackish water with different salinity on soil infiltration, the experiment set 3 different salinity (1.7, 3.4 and 5.1 g/L), and distilled water infiltration as a control, a total of 4 treatments, each three soil columns were installed in each treatment. At the end of the test, a soil column with the same infiltration time was selected for sampling and analysis for each treatment. The brackish water is made up of NaCl particles and distilled water. When the infiltration depth reaches 30 cm, the test is stopped and samples are taken for analysis.

During the test, a stopwatch was used to record the time, record the water level change of the Markov bottle, and observe the change process of the soil column wetting front. After the test, take soil samples every 3 cm in the vertical direction, use the drying method to determine the soil mass moisture content, use the conductivity meter method to determine the conductivity of the soil extract EC1:5, and use the atomic absorption spectrophotometer to determine the Na^+ content. Titrimetric determination of Cl^- content.

3 Results

3.1 The influence of irrigation water salinity on soil infiltration performance

3.1.1 The impact of brackish water irrigation with different salinity on soil cumulative infiltration

The cumulative infiltration of soil is one of the important indicators of soil infiltration performance. The difference of irrigation salinity has a certain influence on the accumulated infiltration of soil. In order to analyze and compare the influence of different irrigation salinity on the cumulative infiltration of soil, the soil cumulative infiltration duration curve during irrigation with 0, 1.7, 3.4, 5.1 g/L salinity irrigation water is plotted on the same graph, as shown in Fig. 2.

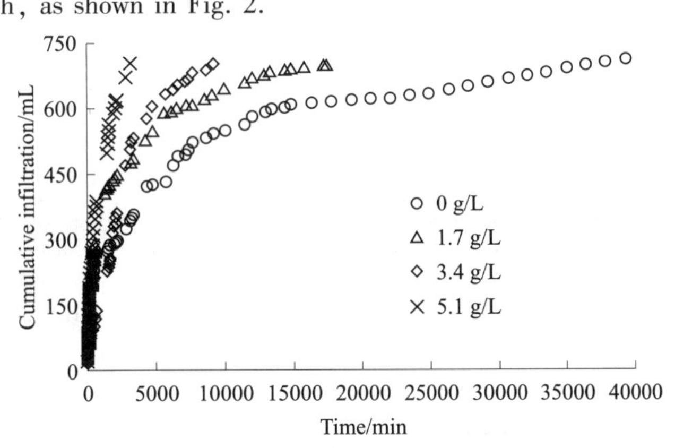

Fig. 2 Variations of cumulative infiltration versus time

It can be seen from Fig. 2 that when the infiltration depth is constant (30 cm), the infiltration duration of 5.1 g/L salinity irrigation water is the shortest, followed by 3.4 g/L salinity, and 0 g/L salinity is the longest. It shows that the higher the degree of mineralization of irrigation water, the more conducive to the infiltration of water. In the same time, with the increase in the salinity of irrigation water, the cumulative infiltration of the soil becomes larger. This is because the salt concentration is high, the ion concentration is relatively high, and the soil colloidal gel capacity is also strong, which promotes the improvement of soil water conductivity, strong infiltration capacity.

3.1.2 Index system

There are many soil water infiltration models, and the commonly used ones are Philip model, Green-Ampt model, Kostiakov two-parameter model and Kostiakov three-parameter model. Yue Haijing[26], Liu Chuncheng[27] showed that the three prediction models (Philip model, Kostiakov two-parameter model and Kostiakov three-parameter model) are all feasible, and the Kostiakov two-parameter model is higher in terms of prediction accuracy. Therefore, this paper selects the Kostiakov two-parameter model to simulate the change of soil infiltration rate. The Kostiakov model is:

$$i_t = \alpha t^{-k} \tag{1}$$

In the formula, i_t is the time infiltration rate (mm/min); t is the infiltration duration (min); α and k are the infiltration parameters.

The change curve of soil infiltration rate under different irrigation water salinity is shown in Fig. 3, and the related parameters of soil infiltration rate fitted by Kostiakov model are shown in Table 2.

It can be seen from Fig. 3 that in the early stage of infiltration, the soil infiltration rate of each treatment was relatively large. With the extension of the infiltration duration, the soil infiltration rate gradually decreased and stabilized. With the increase of the salinity of irrigation water, the infiltration duration will gradually decrease at the same infiltration depth. For example, the infiltration duration of 0 g/L salinity treatment is 655 h, and the salinity increases to 5.1 g/L. The infiltration duration is reduced to 52.5 h. The greater the salinity of the irrigation water, the higher the infiltration rate that will eventually stabilize. It can be seen that brackish water is conducive to the infiltration of soil moisture.

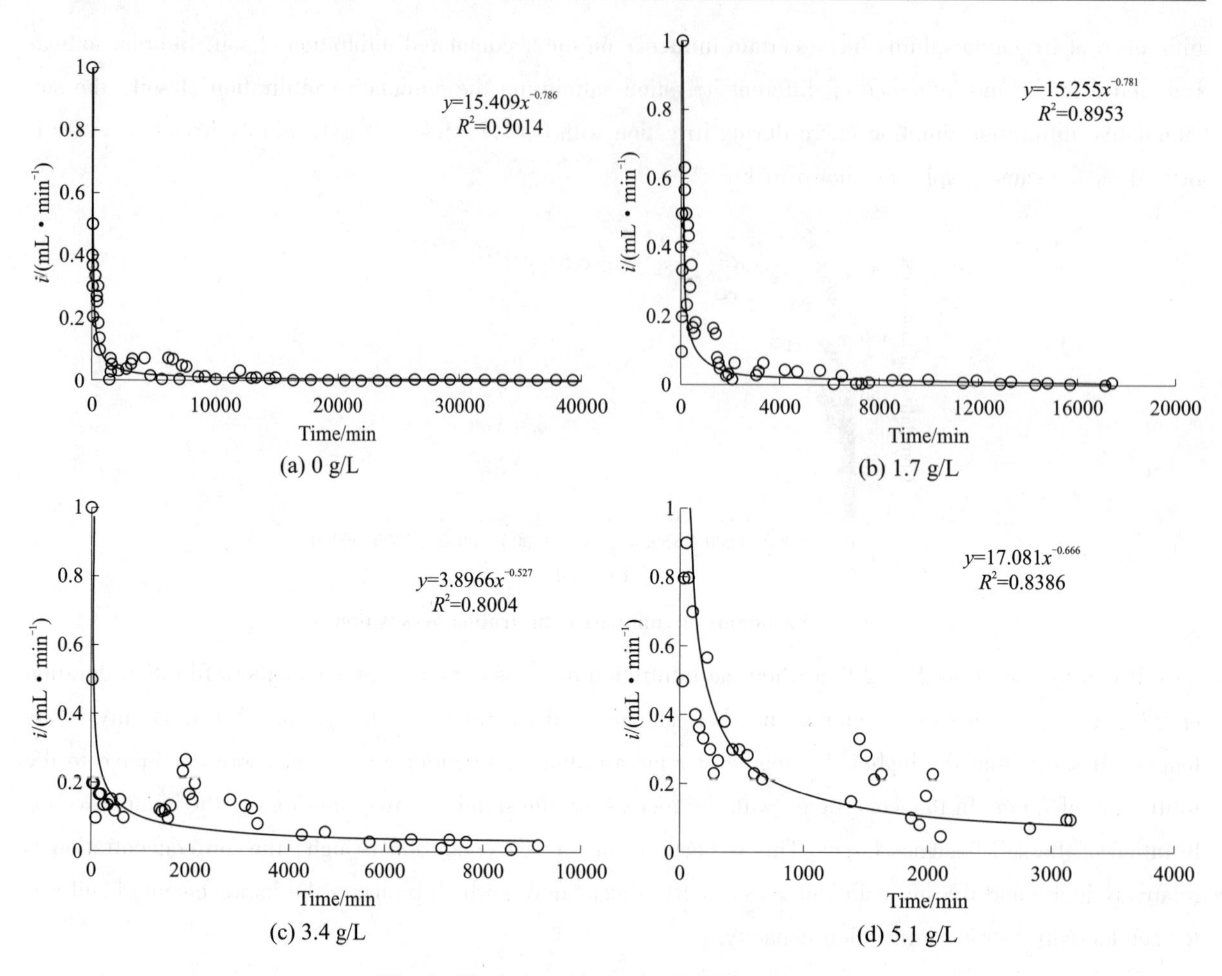

Fig. 3 Variation curve of soil infiltration rate

Table 2 Parameters fitted by Kostiakov equation

Salinity/(g/L)	α	k	R^2
0	15.409	0.786	0.9014
1.7	15.255	0.781	0.8953
3.4	3.8966	0.527	0.8004
5.1	17.081	0.666	0.9004

It is easy to see from Table 2 that the simulation effect of Kostiakov model is better, and R^2 is above 0.8. Under the test conditions, the model infiltration parameter α value decreases with the increase of the salinity, but when the salinity exceeds 3.4 g/L, the α value will increase. the model infiltration parameter k value and the infiltration parameter α have the same trend of change.

3.2 Relationship between wet front advance distance and cumulative infiltration under brackish water irrigation

For the same duration, there is a certain quantitative linear relationship between the wet front advancement distance (Z_f) and the cumulative infiltration amount (I). The expression can be written as $I=n \cdot Z_f$, where n is the fitting parameter. This relationship was used to fit the cumulative infiltration and the advancing distance of the wet front under the condition of brackish water with different salinity. The results are shown in Fig. 4.

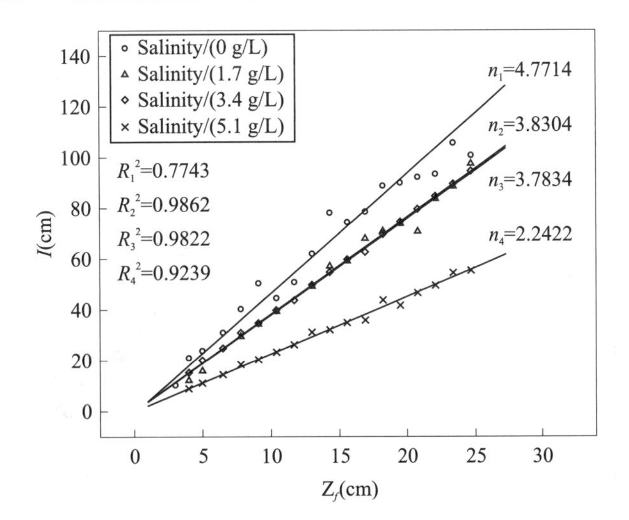

Fig. 4 Fitting of cumulative infiltration amount and advancing distance of wetting front

It can be seen from Fig. 4 that the fitting R^2 of the cumulative infiltration volume and the advancing distance of the wet front is above 0.77, indicating that the two have a good linear relationship. The value of n decreases with the increase of the salinity of irrigation water. This may be because: the wetting depth of the control plan is 33 cm. The higher the salinity of irrigation water, the faster the infiltration rate and the shorter the infiltration duration. At the infiltration depth, the cumulative infiltration volume tends to decrease (Fig. 3), indicating that brackish water is beneficial to the infiltration of soil moisture, but is not conducive to the storage of soil moisture.

3.3 The impact of brackish water irrigation on soil water and salt transport

3.3.1 Distribution of soil water and salt in profile

Under the condition of fresh water irrigation, the soil salinity moves to the lower layer with the water, which has the effect of leaching the salinity downward. When using brackish water with different salinity for irrigation, because the irrigation water itself has a certain amount of salt, the salt separators in the irrigation water and the salt in the soil may undergo a certain reorganization, and a certain physical and chemical reaction occurs, and its salt leaching effect it may be different for freshwater irrigation. After irrigating with brackish water of different salinity, the distribution of soil moisture and salt in different soil layers is shown in Fig. 5.

It can be seen from Fig. 5a that after the infiltration of irrigation water with different salinity, the soil moisture content of each soil layer has little difference on the soil surface, and the values are relatively close. In the depth range of 0~3 cm, the soil moisture content decreases sharply. In the depth range greater than 3~21 cm, the soil moisture content changes little. In the depth range of 21~30 cm, the soil moisture content decreases again. In the same soil layer, the soil moisture content increases with the increase in the salinity of irrigation water, especially in the depth range of 3~27 cm, which is consistent with the change trend of the cumulative infiltration of the soil in the same duration.

It can be seen from Fig. 5b that in the same soil layer, the 0~21 cm soil layer generally shows that the salt content of the soil layer increases with the increase in the salinity of the irrigation water. In the 21~30 cm soil layer, the soil salinity is still the highest after 5 g/L brackish water irrigation, while the regularity of other salinity levels is not obvious. In addition, after 3.4 and 5.1 g/L brackish water irrigation, the desalination depth is very close, about 15~16 cm, and after 0 and 1.7 g/L irrigation water (fresh water), the desalination depth is about 17 ~ 18 cm. In the desalination zone, the salt content between different treatments is much lower than the initial salt content, and the difference between treatments is small, and

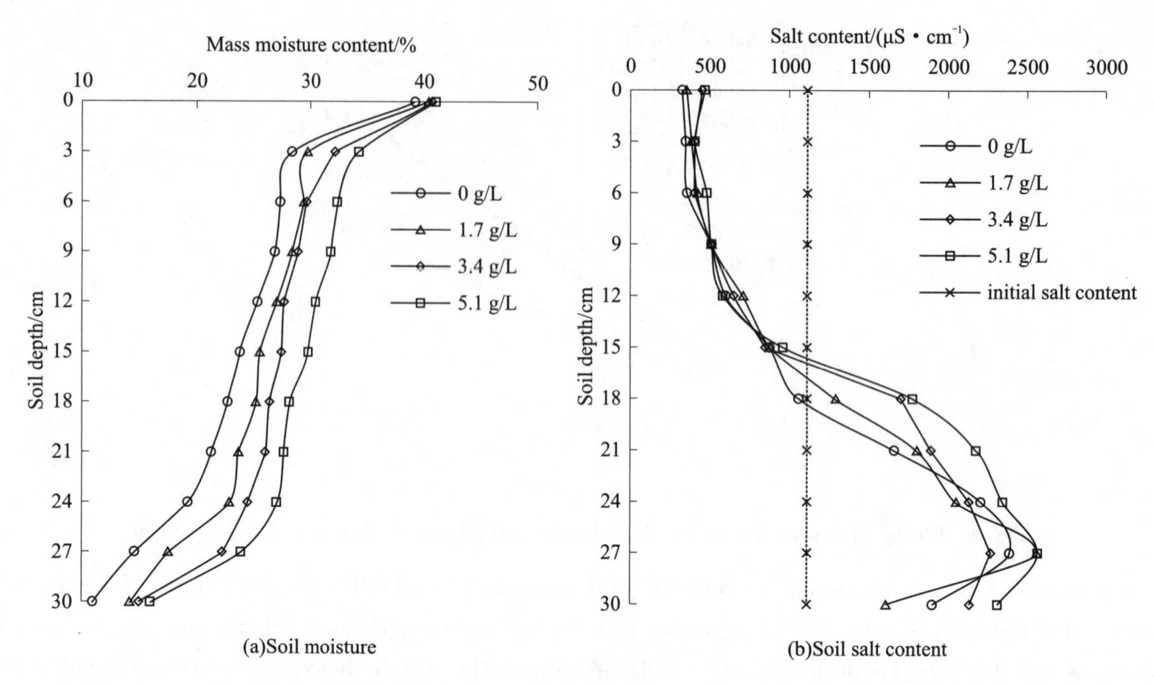

Fig. 5 Distributions of soil water content and soil salt content in profiles

below the desalination depth, with the in-crease of salinity in general, the salt content of soil and irrigation water changes with water infiltration and migration, resulting in a gradual increase in soil salt content.

3.3.2 Distribution of Na^+ and Cl^- in soil profile

The harm of salt damage to plants is mainly manifested in: (1) Osmotic stress, that is, too much soluble salt in the soil reduces the soil water potential, which makes it difficult for plants to absorb water. In severe cases, it even causes water infiltration in plant tissues, resulting in physiological drought. (2) The photosynthesis decreased, and the activities of PEP carboxylase and RuBP carboxylase decreased due to excessive salt. (3) Ion imbalance, too much of a certain ion in the soil often eliminates the absorption of other ions by plants. For example, when wheat grows in an environment with excessive Na^+, its body lacks K^+, which hinders soil colloidal ion exchange, reduces soil water vapor permeability, and endangers crop growth. In addition, plants have low demand for Cl^-, and the content of Cl^- in the soil generally exceeds the amount required for plant growth. Therefore, it is of great significance to analyze the migration of Na^+ and Cl^-. After irrigation of different salinity irrigation water, the distribution of Na^+ and Cl^- in the soil profile is shown in Fig. 4.

Fig. 6a shows that the Na^+ content of the surface soil is very low due to the effect of leaching. As the depth of the soil layer increases, the Na^+ content of the soil layer increases slightly, and there was basically no difference in the content of Na^+ in the profile soil after irrigated with brackish water of different salinities, but when the depth of the soil layer reached 15 cm, with the increase of soil depth, the content of Na^+ increases rapidly, and the content of Na^+ in the soil profile is significantly different after irrigation with brackish water of different salinity. In the same soil layer, in general, the Na^+ content increases as the salinity of irrigation water increases. After irrigating with brackish water of different salinity, there is a depth of "de-Na^+" (15 ~ 16 cm), and the higher the salinity, the lower the depth, but the difference is not obvious. When the salinity is 0 g/L, the areas of desalination and salt accumulation are basically the same, and with the increase of salinity, the area of salt accumulation is larger than the area of de-salination area. This is because there is a certain amount of Na^+ in brackish water of different salinity, and the initial value of irrigation water salinity and soil Na^+ content affects the Na^+ content of the soil profile after irrigation.

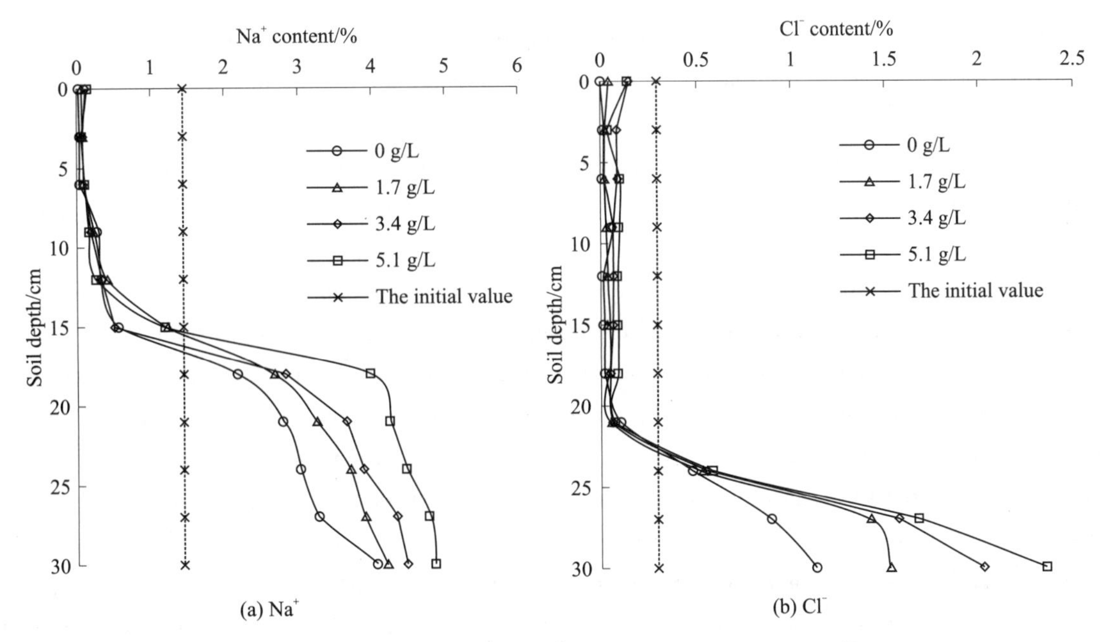

Fig. 6 Distributions of soil Na^+ and Cl^- content in the profiles

Fig. 6b shows that the Cl^- content of the surface soil is very low due to the effect of leaching. As the depth of the soil layer increases, the Cl^- content of the soil layer in-creases slightly, with a smaller amplitude and different salinity. The Cl^- content of soil has a small difference, but compared with Na^+, the difference is slightly larger. However, when the depth of the soil layer exceeds 24cm, the Cl^- content increases rapidly as the depth of the soil layer increases, and the Cl^- content of the soil profile is significantly different after irrigation with brackish water of different salinity. In the same soil layer, in general, the Cl^- content increases with the increase of the salinity of irrigation water. After irrigating with brackish water with different salinity, there is a depth of "Cl^- removal" (22~23 cm), and the difference is not obvious. When the degree of salinity is 0 g/L, the areas of desalination and salt accumulation are basically the same. With the increase of salinity, the area of salt accumulation is larger than that of desalination, this is because there is a certain amount of Cl^- in brackish water with different salinity, and the initial value of irrigation water salinity and soil Cl^- content together affect the Cl^- content of the soil profile after irrigation.

Salt in soil moves with soil moisture. Different initial values of soil ion content and unique water distribution characteristics under experimental conditions will inevitably affect the distribution of different salt ions. The distribution characteristics of different salt ions in soil are mainly related to the concentration and charge number of ions : Na^+ is easily adsorbed by soil colloids, while Cl^- is difficult to be adsorbed by negatively charged soil colloids[28], resulting in the difference of Na^+ and Cl^- content in each layer of soil.

4 Discussion

Due to brackish water contains a large number of chemical elements, the influence of these elements on soil infiltration characteristics is relatively clear, but some mechanisms of action have not yet been revealed, such as what kind of trace elements are contained in brackish water and how these trace elements interact with the soil. The mechanism of action and other aspects has not yet been fully understood. At the same time, the chemical composition of brackish water varies from region to region. In order to facilitate research and the popularization and application of research results, salinity is usually used to comprehensively reflect the salt segregant content of brackish water. A large number of studies have shown that as the salinity of brackish water increases, the infiltration capacity gradually increases. When the salinity reaches 3~5 g /L, the infiltration capacity reaches the maximum, and then as the salinity increases, the infiltration capacity It gradually weakens, so the soil infiltration capacity and brackish water salinity show a parabolic change process. The test results of Ma Donghao et al. showed that although the salinity of brackish water affects the infiltration capacity of soil, the overall performance is similar to that of freshwater infiltration, that is, the cumulative infiltration volume, wetting front and infiltration time all have a power function relationship. The infiltration volume has a linear relationship with the wetting front, but the cumulative infiltration volume and the wetting front under the condition of brackish water infiltration are larger than those of fresh water[23]. Shi Xiaonan et al. calculated the soil hydraulic parameter values under different infiltration conditions with different salinity using the infiltration model, and the results showed that the increase in salinity effectively improved the diffusion rate and saturated hydraulic conductivity of the soil[24]. The increase in salinity is conducive to enhancing the flocculation of the soil, increasing the effective pores of the soil, thereby improving the water conductivity of the soil. Generally, it is believed that the Na ion concentration in the soil solution is too high, causing the soil aggregate structure to disperse and expand, resulting in connectivity, the pores become smaller and blocked, reducing the water conductivity of the soil. The pores become smaller and blocked, reducing the soil's water conductivity. This research is consistent with the above research results. There are not many studies on Na^+ and Cl^- content in about desalting area, so this study has some innovative and instructive significance for fresh water shortage area.

Under the condition that the accuracy requirements are met, the linear model is simple and easy to implement compared with other prediction models such as nonlinear models. The use of linear models can greatly simplify the workload and is more conducive to carrying out or guiding agricultural production activities. Kostiakov two-parameter and three-parameter infiltration model has the advantages of simple form, simple calculation and less constraint conditions, which has become a recognized and widely used empirical model at home and abroad. The physical relationship between Philip infiltration model and soil physical and chemical properties is clear and widely used. Therefore, it is feasible to establish Kostiakov two-parameter infiltration model, three-parameter infiltration model and Philip infiltration model parameter prediction model by using the basic physical and chemical parameters of soil easily obtained and multiple linear regression method[26,27]. The average error and relative error of the three models were compared. The accuracy of each parameter prediction model of the two-parameter infiltration model was higher than that of the three-parameter infiltration model and the Philip infiltration model, and the relative error of the cumulative infiltration amount at a given time was also smaller than that of the two infiltration models. Therefore, Kostiakov two-parameter infiltration model was recommended to predict the soil moisture infiltration process. Although the Kostiakov two-parameter infiltration model can be established by linear regression method to study soil water infiltration, the accuracy of the established linear model is low due to the nonlinear relationship between the parameters and the physical and chemical properties, which needs further study to improve the prediction accuracy. In addition, these infiltration models do not reflect the impact of land use and topography changes on infiltration, ignoring the impact of spatial variability. Future research should be strengthened. At the same time, we should further study how to extend the single point infiltration model to a larger area, and apply high-tech and means (such as RS, RIS, data assimilation technology) to soil infiltration research.

This experiment mainly considered the water and salt migration law of sa-line-alkaline soil under the condition of brackish water irrigation with four different salinities. On the one hand, due to lack of experience, repeated soil pillars were not all sampled, which made it impossible to analyze the significance of differences. Moreover, the soil type is single, and other soil types need to be further compared and analyzed in the later stage. On the other hand, this experiment is a short-term continuous infiltration experiment. The transport of soil water and salt under long-term brackish water irrigation needs further study. In the later period, we will simulate 3~5 years of long-term brackish water irrigation and treat the soil water and study the law of salt migration. Finally, this experiment focuses on the mechanism. According to the experimental results, in the actual planting process, According to the root distribution of the crop and the salt tolerance threshold, in order to increase the leaching depth, the salinity of the irrigation water can be controlled or the irrigation amount can be increased to ensure the normal growth and development of the crop.

5 Conclusions

(1) Under the condition of brackish water irrigation, the Kostiakov model could better simulate the change of soil infiltration rate with time; When the infiltration depth was fixed (30 cm), the infiltration duration of 5.1 g/L salinity irrigation water was the shortest, 3.4 g/L salinity was the second, and 0 g/L salinity was the longest. The soil infiltration capacity had a positive response to the salinity of irrigation water, there was a good linear relationship between soil cumulative infiltration and wetting front advancing distance, but the fitting parameter n decreased with the increase of irrigation water salinity, which was not conducive to the storage of soil moisture.

(2) Under the condition of brackish water irrigation, the depth of soil desalination was about 15 to 18 cm, the depth of removing Na^+ was 15 to16 cm, and the depth of removing Cl^- was 22 to 23 cm, all which was inversely proportional to the level of salinity; As the salinity increasing of irrigation water, the contents of soil moisture, salt, Na^+ and Cl^- in each soil layer gradually increased in the salt accumulation area, but the difference was not obvious in the desalination area.

Author Contributions: Methodology and software, P.Z. and J.S.. formal analysis, resources and data curation, investigation, writing—original draft preparation, P.Z.. writing—review and editing, J.S.. supervision and project administration, P.Z. and J.S.. All authors have read and agreed to the published version of the manuscript.

Acknowledgements: This research was funded by Technology Innovation Center for Land Engineering and Human Settlements, Shaanxi Land Engineering Construction Group Co., Ltd., and Xi'an Jiaotong University (2021WHZ00891).

References

[1] Wang C H, Zhang S N, Li KC, Zhang F M, Yang K. Change Characteristics of Precipitation in Northwest China from 1961 to 2018. Chinese Journal of Atmospheric Sciences. 2021;45(04):713-724.

[2] Liu J, Gao Z. Advances in study and practice of brackish water irrigation in China. Water Resources and Hydropower Engineering. 2012;43(1): 101-104.

[3] Gao C S, Shao L W, Yan Z Z, Li L, Chen S Y, Zhang X Y. Annual soil salt balance and crop performance under brackish water irrigation during the winter wheat season. Chinese Journal of Eco-Agriculture. 2021; 29(5): 809-820.

[4] Li D, Wan S Q, Kang Y H, Li X B. Effects of Water-salt Regulation on Tomato Growth and Quality under Drip Irrigation with Brackish Water in Coastal Saline-alkali Soil. Journal of Irrigation and Drainage. 2020; 39(7):39-50.

[5] Li X B, KangY H. Water-salt control and response of Chinese rose (Rosa chinensis) root on coastal saline soil using drip irrigation with brackish water. Transactions of the CSAE. 2019; 35(11):112-121.

[6] Guo X H, Bi Y J, Sun X H, Ma J J, Kong X Y. Prediction model of soil water and salt transport on yield of summer squash under mulch drip irrigation with brackish water. Transactions of the CSAE. 2019;35(8):167-175.

[7] Zhang J, Cao C, Feng D, Sun J, Li K, Liu H. Effects of different planting patterns on cotton yield and soil water-salt under brackish water irrigation before sowing. Transactions of the Chinese Society for Agricultural Machinery. 2013;44(2): 97-102.

[8] Wang Z, He Y, Jin M, Wang B. Optimization of mulched drip-irrigation with brackish water for cotton using soil-

water-salt numerical simulation. Transactions of the CSAE. 2012;28 (17): 63-70.

[9] Huang M Y, Zhang Z Y, Xu H, Zhai Y M, Wang C, Zhu C L. Effects of Cycle Irrigation with Brackish and Fresh Water and Biochar on Water and Salt Transports of Coastal Saline Soil. Transactions of the Chinese Society for Agricultural Machinery. 2021;52(01);238-247.

[10] Zhang Z, Zhang Z, Lu P, Feng G, Qi W. Soil water-salt dynamics and maize growth as affected by cutting length of topsoil incorporation straw under brackish water irrigation. Agronomy. 2020;10(2),246.

[11] Li X W, Jin M G, Yuan J J, Huang J O. Evaluation of soil salts leaching in cotton field after mulched drip irrigation with brackish water by freshwater flooding. Journal of Hydraulic Engineering.

[12] Zhang Z, Feng G, Ma H, Chen Y, Si H. Analysis of soil water-salt distribution and irrigation performance under brackish water film hole furrow irrigation. Transactions of the Chinese Society for Agricultural Machinery. 2013;44 (11): 112-116.

[13]Zheng F L, Yao R J, Yang J S, Wang X P, Chen Q, Li H Q. The Effects of Soil Amendment with Different Materials on Soil Salt Distribution and Its Ion Composition under Brackish-water Drip Irrigation. Journal of Irrigation and Drainage. 2020;39(8):60-71.

[14]Lu P, Zhang Z, Sheng Z, Huang M, Zhang Z. Effect of surface straw incorporation rate on water-salt balance and maize yield in soil subject to secondary salinization with brackish water irrigation. Agronomy. 2019;9 (7),341.

[15]Zamparas M, Kyriakopoulos G L, Kapsalis V C, Drosos M, Kalavrouziotis I K. Application of novel composite materials as sediment capping agents: Column experiments and modelling. Desalin Water Treat. 2019;170:111-118.

[16]Zamparas M, Kyriakopoulos G L, Drosos M, Kapsalis V C, Kalavrouziotis I K. Novel Composite Materials for Lake Restoration: A New Approach Impacting on Ecology and Circular Economy. Sustainability; 2020, 12 (8), 3397.

[17]Yla C, Jiao W B, D M. Effects of earthworm casts on water and salt movement in typical Loess Plateau soils under brackish water irrigation. Agr Water Manage. 2021;252,106930.

[18]Zamparas M, Kyriakopoulos G L, Drosos M, Kapsalis V C. Phosphate and ammonium removal from wastewaters using natural-based innovative bentonites impacting on resource recovery and circular economy. Molecules. 2021;26 (21),6684.

[19]Tan J L, Wang X N, Jin H J, Du F F, Tian J C. The Effects of Grading and Thickness of Gravel Mulching on Water and Salt Movement in Soil under Brackish Water Irrigation. Journal of Irrigation and Drainage. 2020; 39(9): 7-13.

[20] Dong L Y, Wang W H. Study of movement of water and salts in dry red soil under drip irrigation with brackish water. Agricultural Research in the Arid Areas. 2019; 37 (2):37-43.

[21] Mu X, Yang J, Zhao D, Feng, Y, Xing H, Pang G. Effect of Brackish Water Irrigation on soil Water-salt Distribution. IOP Conference Series: Earth and Environmental Science. 2018; 208 (1): 012062.

[22] Zhang K, Niu W, Wang Y, Xue W, Zhang, Z. Characteristics of water and salt movement in soil under moistube-irrigation with brackish water. Transactions of the Chinese Society for Agricultural Machinery. 2017; 48 (1):175-182.

[23] Ma D H. Research on the characteristics of soil water and salt transport. Master thesis, Xi'an University of Technology, Xi'an, 2005.

[24] Shi X N, Wang Q J,Su Y. Effect of slight saltwater quality on the characteristics of soil water and salt transference. Arid Land Geography. 2005;28(4): 516-520.

[25] Liu CC, Li Y, Guo LJ, Guan BY, Liao YQ, Wang J. Effect of brackish water irrigation on water and salt movement in repellent soils. Transactions of the CSAE, 2011; 27(8): 39-45.

[26] Yue H J, Fan G S. Comparison of Multivariate Linear Prediction Model Precision of Soil Infiltration Model Parameters. Journal of Irrigation and Drainage. 2016;35(10): 73-77,96.

[27] Liu C C, Li Y, Ren X, Ma X Y. Applicability of four infiltration models to infiltration characteristics of water repellent soils. Transactions of the CSAE. 2011;27(5): 62-67.

[28] Wang D, Kang Y H, Wan S Q. Distribution characteristics of different salt ions in soil under drip irrigation with saline water. Trans actions of the CSAE. 2007; 23(2): 83-87.

本文曾发表于2022年《Open Geosciences》第14卷

Effect of Soil Layer Thickness on Organic Carbon Mineralization in Improved Sandy Land

Zhen Guo[1,2,3,4], Juan Li[1,2,3,4], Haiou Zhang[1,2,3,4], Huanyuan Wang[1,2,3,4], Jian Wang[1,2,3,4], Tingting Cao[1,2,3,4]

(1.Shaanxi Provincial Land Engineering Construction Group Co., Ltd., Xi'an 710075, China; 2.Institute of Land Engineering and Technology, Shaanxi Provincial Land Engineering Construction Group Co., Ltd., Xi'an 710021, China; 3.Key Laboratory of Degraded and Unused Land Consolidation Engineering, the Ministry of Natural and Resources of China, Xi'an 710021, China; 4.Shaanxi Provincial Land Consolidation Engineering Technology Research Center, Xi'an 710075, China)

【**Abstract**】 Soft rock and sand are interspersed in the Mu Us Sandy Land, and the land degradation is severe. Using soft rock and sand as resources on the spot is of great significance to promote the environmental construction and human settlement structure of the area. In this paper, the soft rock and sand are mixed according to the volume ratio of 0 : 1 (CK), 1 : 5 (S1), 1 : 2 (S2) and 1 : 1 (S3) to construct a soil plow layer. And based on the 0~30 cm soil layer, the mineralization of soil organic carbon was studied. The results show that as the proportion of soft rock increases, the organic carbon content of the 10~20 cm soil layer gradually decreases. With the increase of the soil layer, the soil organic carbon mineralization rate and cumulative mineralization amount are the highest in the 10~20 cm soil layer, followed by 0~10 cm and 20~30 cm. With the extension of the cultivation time, the mineralization intensity showed a trend of first decreasing, then increasing and finally decreasing steadily. At the end of the cultivation, the organic carbon mineralization rate of the compound soil was between 3.10~10.47mg/(kg · d^{-1}). The cumulative mineralization rate of soil organic carbon in each soil layer has different performances. The average value is the lowest at 0~10 cm, which is 20.54%, and its value increases with the increase of arsenic content. The change trend of the potential mineralizable organic carbon content is consistent with the cumulative mineralization amount. The mineralization amount treated with S1 and S2 is relatively small, and the half-turnover period is opposite to the change trend of the mineralization rate constant. The difference was significant only in the 0~10 cm soil layer in the half turnover period, and the treatment of adding arsenic sandstone was significantly higher than that of CK treatment. In conclusion, the organic carbon in the 0~10 cm soil layer is relatively stable and still in the process of soil formation and development, and the carbon sequestration ability is stronger when the ratio of arsenic sandstone to sand reclamation is between 1 : 5 and 1 : 2.

【**Keywords**】 Mineralization parameters; Soft rock; Aeolian sandy soil; Soil layer thickness; Land engineering

Introduction

In terrestrial ecosystems, the decomposition of soil organic carbon is an important part of the carbon cycle process, and plays an important role in maintaining atmospheric balance and stable development of soil carbon pools[1]. Soil organic carbon is twice the atmospheric carbon and three times the total carbon of the earth's vegetation, and 80% of the total carbon involved in the earth's terrestrial carbon cycle exists in the soil in the form of soil organic carbon[2-3]. The Mu Us Sandy Land is rich in light and heat resources, which can be used as a granary reserve resource with development potential[4]. It covers a large area, and small changes can lead to significant changes in atmospheric CO_2 concentration[5]. In recent years, the dynamic change of soil carbon in Mu Us Sandy Land may be affected by the change of land use type and the effect of land consolidation and restoration[6]. Through the implementation of soil organic reconstruction technology in Mu Us Sandy Land, organic carbon storage can be increased and carbon sink effect can be achieved[7]. Therefore, it is of great practical significance and development prospect to study the carbon pool effect of Mu Us Sandy Land in the practice of "carbon neutrality" concept.

In the context of global climate change, plant roots grow into deeper soil, and the production of root exudates and the input of plant and animal residues will change the carbon input of deep soil, thus triggering the carbon mineralization decomposition[8]. It is believed that deep soil is a major organic carbon pool in terrestrial ecosystems, containing more than half of soil carbon[9]. Small changes in the deep soil organic carbon pool may have important effects on the overall soil carbon dynamics and ultimately affect the global carbon cycle[10]. Studies have shown that about 60% of soil organic carbon storage within the depth of 1 m is stored in the depth of 20~100 cm, and the biological activity and availability of soil organic carbon in deep layer is lower than that in surface layer, but its stability is higher than that in surface layer, which is also one of the reasons for the slow decomposition of soil organic carbon in deep layer[11]. Guan et al. [12] studied the mineralization of red soil in the subtropical Masson pine forest, and showed that the mineralization rate of soil organic carbon decreased with the increase of soil depth in the early stage of mineralization, and the mineralization rate of soil organic carbon in the deep layer was higher than that in the surface layer in the later stage of mineralization. It can be seen that soil layer thickness is an important factor affecting the emission of organic carbon mineralization, and the intensity of organic carbon mineralization in soil below the surface layer has a significant impact on the effect of atmospheric circulation[9]. Along with the downward growth of roots and the addition of exudates and other foreign substances, the original soil will have a certain stimulating effect. Huang et al.[13] pointed out in their study on the mineralization effect of black soil that there was no significant difference in the mineralization rate between the surface soil and the lower soil unit organic carbon in the absence of exogenous additives, and glucose stimulated more mineralization release of the surface soil organic carbon, while the mineralization decomposition of the lower soil organic carbon decreased. However, Paterson et al.[14] found in their study that the mineralization of soil organic carbon in upper and lower layers had a consistent response to the outside world. Therefore, without the addition of external carbon sources, the intensity of soil mineralization on the vertical scale needs to be further clarified with the gradual growth of crop roots to the lower layer, especially in sandy land with greater productivity potential and carbon sink effect.

The Mu Us Sandy Land is one of the earliest, most comprehensive and most systematic ecological restoration areas in China because of its good hydrothermal conditions and active cultural and economic activities[15]. Han et al. [16] innovatively discovered the complementarity of the two material structures through the study of soft rock and sand in the Mu Us Sandy land, obtained the combination formula of soft rock and sand suitable for the growth needs of different crops, and successfully realized the combination of sand and soft rock. However, facing the development of global economy and the implementation of national carbon neutralization strategy, it is urgent to clarify the degree of carbon cycling effect in Mu Us Sandy Land. Therefore, this study focused on the Mu Us Sandy Land at the edge of monsoon in China, focused on soil organic carbon mineralization at different soil depths, and combined with first-order dynamic equation to determine the mineralization effect and carbon sequestration degree. It is expected to provide effective scientific basis for perfecting the theory of soil organic reconstruction, reasonably estimating the soil carbon pool in sandy land, and practicing the concept of carbon peak and carbon neutralization.

Materials and methods

Study site. The long-term fertilization experiment of soft rock and sandy soil is located in Fuping County, Shaanxi Province (109°11′E, 34°42′N), which belongs to semi-arid climate of continental monsoon warm zone, with annual total radiation of 5187.4 MJ/m^2 and annual average sunshine duration of 2389.6 h. The terrain of the land slopes from northwest to southeast, the north is high and the south is low. The altitude is 375.8~1420.7 m, the average annual precipitation is 527.2 mm (1990~1995), the average annual temperature is 13.1 ℃, the inter-annual variation of precipitation is large and the coefficient of variation can reach more than 20%[17].

Experiment design. According to the site conditions, the test plots are laid out in a "one" shape from south to north. A new batch of test plots will be built in 2016. The test consists of 4 treatments, each with 3 replicates, and a total of 12 test plots. The experimental setup took the pure sand treatment as the control (CK), and the three grades of soft rock were added at 16.67% (S1), 25.0% (S2) and 50% (S3), respectively. In the experimental field, winter wheat and summer corn were planted in June and harvested in October each year, while winter wheat was planted in November and harvested in May the following year. The types of fertilizers are urea (about 46.4% nitrogen content), diammonium phosphate (16% nitrogen content, 44% available phosphorus content, calculated as P_2O_5) and potassium sulfate (52% K_2O content).

Sample Collection. After the harvest of summer maize in October 2020, samples of mixed soil layers of 0~30 cm (0~10 cm, 10~20 cm, 20~30 cm) were collected at 10 cm intervals in each plot. Each plot uses the "S" shape sampling method to collect 7 sample points and mix them into a mixed sample, and then pass through a 2 mm sieve to remove animal and plant residues and large gravels for use. A part was stored in a 4 ℃ refrigerator for organic carbon mineralization culture experiments. The other part was air-dried under natural indoor conditions, and then passed through a 0.149 mm soil sample sieve for grinding for the determination of soil organic carbon. Soil organic carbon was measured by potassium dichromate-concentrated sulfuric acid external heating method[18]。

Laboratory mineralization incubation experiment. Organic carbon mineralization was tested by alkaline absorption method[19]. Weigh 30 g of soil stored at 4 ℃ and place it in a 50 mL beaker. Deionized water is used to adjust the water content to about 65%. At this point, the beaker is weighed at a constant value. Place the lye and beaker with soil in a thermos, seal and incubate in the dark. The lye is a solution of 0.1 mol/L NaOH. The lye solution was taken out and titrated with dilute acid on the 1st, 3rd, 5th, 7th, 9th, 16th, 23rd, 30th, 37th, 44th, 51st and 58th day of culture. Dilute acid is 0.1 mol/L HCl solution. Before each titration, add 2 mL of 1 mol/L $BaCl_2$ solution and 2 drops of phenolphthalein indicator to the alkali solution. After the dilute acid titration is completed, replace with a new lye solution. At this time, the beaker containing the soil was taken out and weighed, and deionized water was added to maintain a constant weight, and the culture was continued.

Data processing. The soil organic carbon mineralization rate was calculated as described by Wang et al[20]. After the indoor cultivation was completed, the first-order kinetic equation in the Origin 2017 software was used to fit the soil organic carbon accumulation mineralization parameters, and the potential mineralized organic carbon and turnover rate constants were obtained. With soil depth and compound ratio as two factors in this study, Duncan's multiple comparisons and ANOVA were performed on soil organic carbon and mineralization parameter data using Microsoft Excel 2010 and SPSS 20.0.

Results and discussion

Organic carbon content of compound soil. After 4 years of planting, the soil organic carbon content in

0~10 cm, 10~20 cm and 20~30 cm layers ranged from 2.04~2.65 g/kg, 1.45~2.70 g/kg and 1.49~2.03 g/kg, respectively, and the average content decreased gradually with the deepening of soil layer. The organic carbon content in 0~10 cm soil layer was significantly higher than that in 20~30 cm soil layer (Fig.1). This is due to the high temperature and moderate humidity of the surface soil of the composite soil of soft rock and sand, which is conducive to the decomposition of crop litter and the continuous accumulation of organic carbon in the surface soil. Pure sand is laid in 30~70 cm soil layer, and organic carbon in 20~30 cm soil layer may also be transported downward with rainfall or irrigation[21]. In the 0~10 cm soil layer and the 20~30 cm soil layer, there was no significant difference in the organic carbon content under all treatments ($P>0.05$), while in the 10~20 cm soil layer, the organic carbon content of the CK treatment was significantly higher compared with other compound ratio treatments ($P<0.05$), there was no significant difference among other treatments. This is because the pores between the soft rock particles and the sand particles become larger after the soft rock and sand are compounded. The higher the content of soft rock, the more large pores, and the soil organic carbon migrates downward under the action of gravity potential and matrix potential[22]. The results showed that with the increase of the content of soft rock, the content of soil organic carbon in each soil layer was redistributed, which promoted the accumulation of soil organic carbon in the 0~10 cm surface layer. Four years after planting, the compound ratio and soil layer had a significant impact on the soil organic carbon content in the sandy land. At this time, the compound soil was still in the soil development stage, and many properties were still unstable.

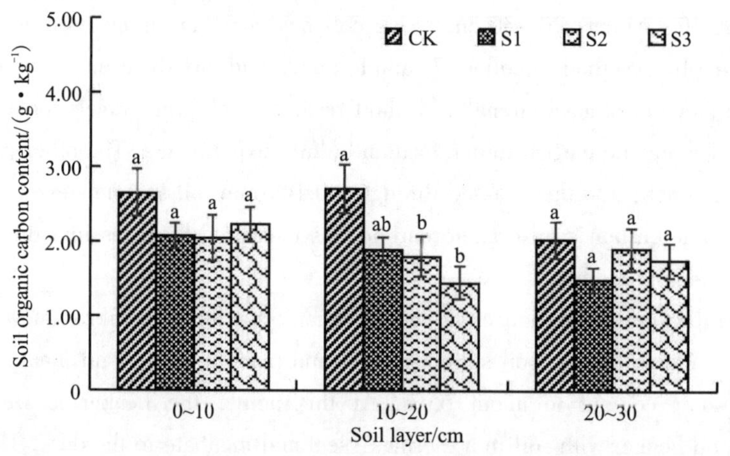

Fig.1 Organic carbon content of soft rock and sand compound soil planted for 4 years. Different letters on the column indicate significant differences among different treatments under the same soil layer ($P<0.05$)

Organic carbon mineralization rate of compound soil. Under different soil layer thicknesses, the mineralization rate of organic carbon in the soft rock-sand composite soil showed obvious periodic changes with the extension of the culture time. During the first 5 days of indoor culture, the mineralization rate of organic carbon in the compound soil was unstable, showing a trend of decreasing and then increasing by high-speed mineralization. This stage can be called the stress change stage (Fig.2). Mainly, the compound soil is maintained to a certain water content in the early stage of cultivation, which stimulates the activity of microorganisms[23]. Then, with the stabilization of the compound soil culture environment and the recovery of soil microbial activity, the organic carbon mineralization rate increased to the maximum on the 5th day of culture, and the change range of the organic carbon mineralization rate was 13.79~43.56 mg/(kg · d^{-1}) (0~10 cm), 27.59~42.83 mg/(kg · d^{-1}) (10~20 cm) and 18.88~41.02 mg/(kg · d^{-1}) (20~30 cm). Another reason for the peak on the 5th day of cultivation may also be due to the high content of easily mineralized organic carbon components in the soil, along with the decomposition of light group organic carbon in the soil, a large amount of nutrients for microbial metabolism are produced, thus leads to an increase in the activity of soil microorganisms and an increase in the rate of mineralization of organic carbon[24]. After that, the mineralization rate of soil organic carbon decreased slowly and gradually became stable. This stage can be called the slow release stage. At the end of the incubatioon, the mineralization rates of organic carbon in the 0~10 cm soil layers S1, S2 and S3 were 6.87 mg/(kg · d^{-1}), 8.13 mg/(kg · d^{-1}), 8.22 mg/(kg · d^{-1}), which was significantly higher than CK treatment by 1.48 mg/(kg · d^{-1}). The average rate of organic carbon mineralization in the 20~30 cm soil layer is basically the same as that in the 0~10 cm soil layer, but both are less than those in the 10~20 cm soil layer. It can be seen that within the same cultivation time period, the organic carbon mineralization in the 10~20 cm soil layer is faster, and the corresponding turnover time is shorter.

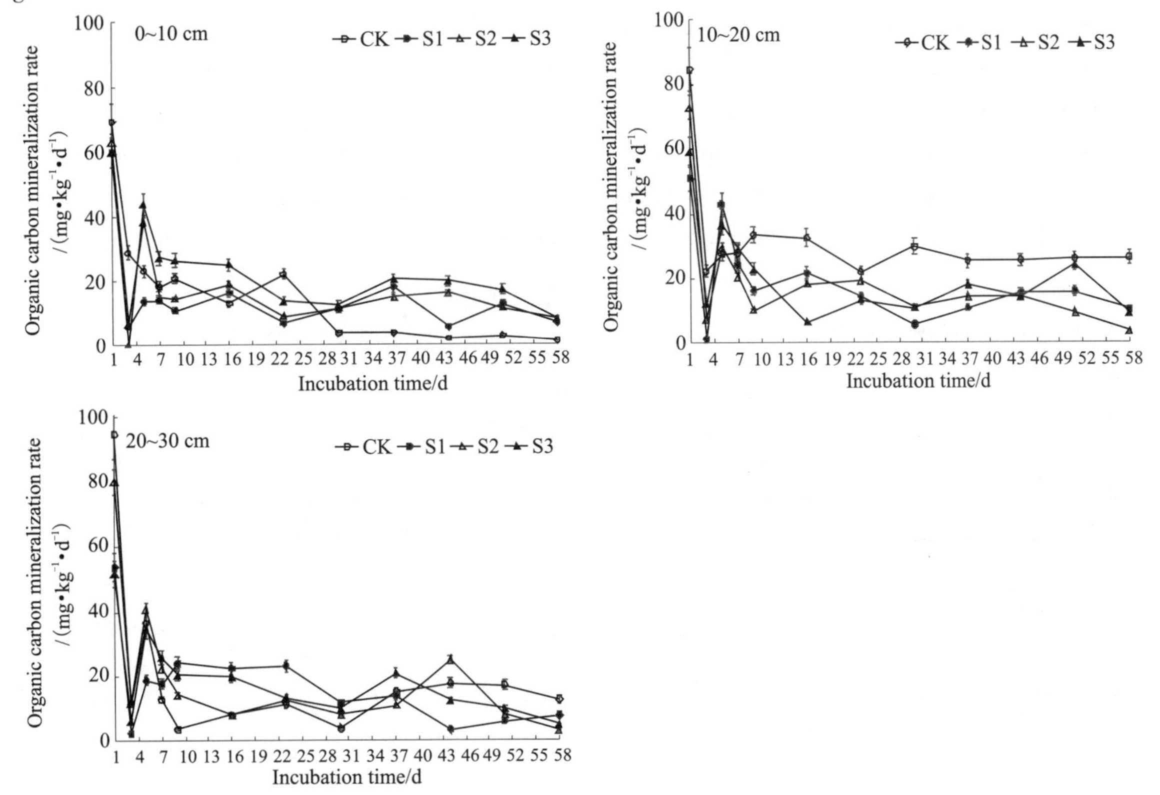

Fig.2 Organic carbon mineralization rate of soft rock and sand compound soil planted for 4 years

Cumulative mineralization of organic carbon in compound soil. The cumulative mineralization of soil organic carbon in each soil layer of soft rock and sand composite soil increased exponentially with the cultivation time, and the average value was the highest in the 10~20 cm soil layer (491.84.55 mg/kg), followed by the 0~10 cm soil layer. The soil layer (480.65 mg/kg), the 20~30 cm soil layer was the lowest (467.77 mg/kg) (Fig.3). Possibly due to the lower nitrogen content in the middle layer of soil, according to the hypothesis of "microbial nitrogen utilization", the lower nitrogen availability in the soil is conducive to the decomposition of stubborn organic matter by soil microorganisms to obtain nitrogen sources, so the lower effective nitrogen in the 10~20 cm nitrogen content may enhance microbial mineralization of organic carbon[25]. The 0~10 cm top-soil is rich in readily available substrates (such as litter, rhizosphere sediments, and roots) and is suitable for microbial growth[11], therefore, soil microorganisms did not mineralize more soil organic carbon in the topsoil. The accumulated mineralization amount of organic carbon was different between treatments with different compound ratios in the same soil layer. In the 0~10 cm soil layer, the accumulated mineralization amount of organic carbon in the S3 treatment was the largest, which was 595.17

mg/kg, followed by the S2 and S1 treatments. Among them, S3 treatment was significantly higher than S1 and CK, and there was no significant difference between S1 and CK treatments.

In the 10~20 cm soil layer, the CK treatment had the largest cumulative mineralization of organic carbon, which was 874.68 mg/kg, which was significantly higher than other treatments

with an increase of 65.72%~86.53%. And there is no significant difference among S1, S2 and S3 treatments, which is consistent with the change trend of organic carbon content. In the 20~30 cm soil layer, there was no significant difference in the accumulated mineralization of organic carbon among all treatments, and the average value was between 449.09 mg/kg and 486.68 mg/kg. The most important factor affecting the gradient of organic carbon accumulation and mineralization is the difference and redistribution of carbon content. Zha et al.[26] also pointed out that soil organic carbon is the main driving factor of soil fertility and has a significant impact on the mineralization process. In this study, the rate of organic carbon mineralization (Fig. 2) and the cumulative amount of mineralization (Fig. 3) were larger in the S3 treatment, followed by the S2 treatment, and the S1 treatment was the smallest. It may be that the proportion of soil nutrient content and active organic carbon in the S3 treatment is higher. Combining all soil layers, the average value of accumulated mineralization of soil organic carbon was the largest in CK treatment, which was 567.29 mg/kg, followed by S3, S2, and S1 treatments.

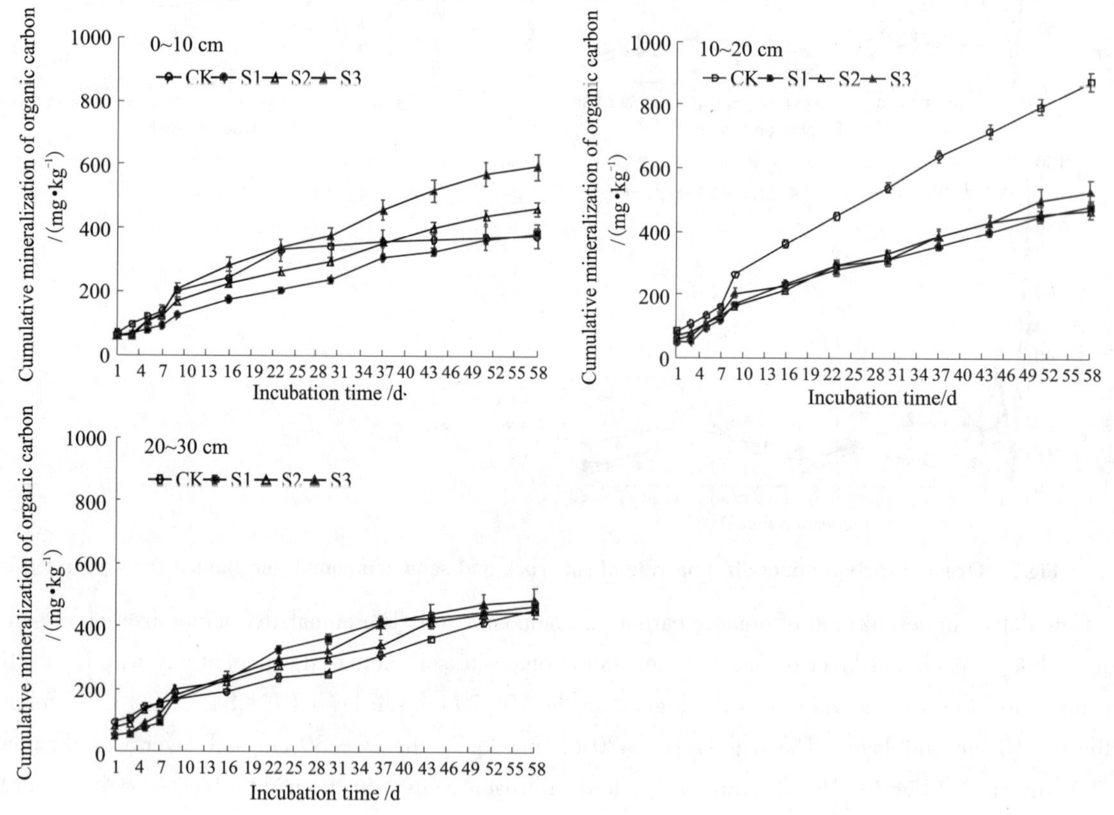

Fig.3 Organic carbon cumulative mineralization of soft rock and sand compound soil planted for 4 years

Cumulative mineralization rate of organic carbon in compound soil. There are certain differences in the cumulative mineralization rate of soil organic carbon in different soil layers. Among them, the cumulative mineralization rate of organic carbon in the 0~10 cm soil layer has a similar trend of change with the cumulative mineralization amount, both of which are higher in the S3 treatment and the lowest in the CK treatment (Fig.4). It can be seen that the addition of soft rock to the sandy soil can improve the soil structure and nutrients. The cumulative mineralization rate of organic carbon in the 10~20 cm soil layer was 1.46

times that of the 0~10 cm soil layer, indicating that the soil organic carbon in the 10~20 cm soil layer was unstable and had a high degree of mineralization. In the 10~20 cm soil layer, the cumulative mineralization rate of organic carbon in the CK and S3 treatments was significantly higher than that in the S1 and S2 treatments, and there was no significant difference between the S1 and S2 treatments. In the 20~30 cm soil layer, there was no significant difference in the cumulative mineralization rate of organic carbon in the CK, S2 and S3 treatments, but it was significantly reduced by 1.13~1.41 times compared with the S1 treatment. It can be seen that the cumulative mineralization rate of organic carbon in the S1 treatment and the S2 treatment is lower, which is beneficial to the carbon pool storage, and the organic carbon in the 0~10 cm soil layer is relatively stable. This may be due to the fact that dissolved organic carbon in the underlying soil is easy to be decomposed and utilized by microorganisms, and the depolymerization and dissolution of organic carbon are the prerequisite for its mineralization. It is generally believed that organic carbon must be converted into soluble organic carbon in soil solution before it can be converted into CO_2[27].

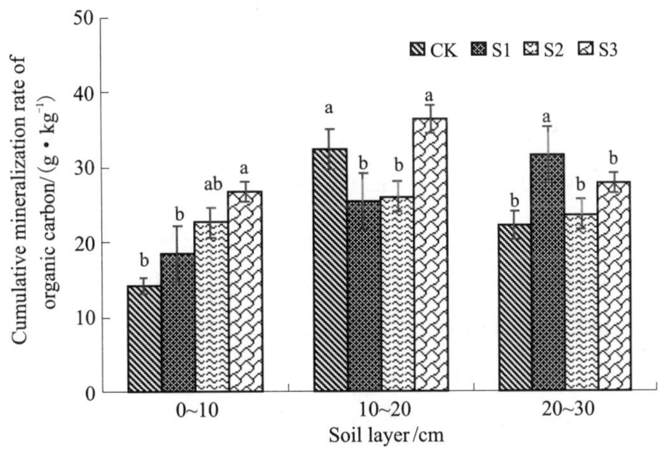

Fig.4 Organic carbon cumulative mineralization rate of soft rock and sand compound soil planted for 4 years

Fitting parameters of the kinetic equation of organic carbon mineralization in compound soil. According to the CO_2-C release measured at each cultivation stage in the constant temperature room for 58 days, the first-order kinetic equation was carried out for the cumulative mineralization of organic carbon in the soil layers of 0~10 cm, 10~20 cm and 20~30 cm with different compound ratios ($C_t = C_0(1-e^{-kt})$). The fitting effect is good (R^2>0.96), and the potential mineralizable organic carbon (C_0), mineralization rate Changshu (k) and semi-turnover period ($T_{1/2}$) are obtained (Table 1). Among all soil layers, the C_0 value of the 10~20 cm soil layer was larger, and the difference between the 0~10 cm and 20~30 cm soil layers was consistent with the change law of the cumulative mineralization. The addition of arsenic sandstone reduced the mineralization intensity of the 10~20 cm soil layer, but promoted it in the 0~10 cm and 20~30 cm soil layers. The change trend of C_0/SOC is consistent with the cumulative mineralization rate. The 0~10 cm soil layer is the smallest, which is 28.26%, the second is 20~30 cm, which is 32.92%, and the 10~20 cm soil layer is the largest, which is 40.33%. The changing trends of k and $T_{1/2}$ are opposite. The comprehensive analysis showed that the $T_{1/2}$ of the 0~10 cm soil layer showed a significant increasing trend compared with the CK treatment, and the $T_{1/2}$ of the 10~30 cm soil layer had no significant difference between the treatments.

In the 0~10 cm soil layer, since it is in direct contact with the outside atmosphere, the addition of soft rock improves the cementation, permeability and structure of the sandy soil, and the availability of substrates such as litter and rhizosphere sediments is improved. Natural addition promotes the increase of substrates available to microorganisms, and the increase of biomass also promotes the reproduction of microorganisms. As the soil environment changes, soil microbial diversity and community composition also change. For example, the proportion of fungi with higher carbon utilization efficiency increases, and due to the increase of other substrates available in the soil, the proportion of respiration carbon allocated during the decomposition of organic carbon will decrease, which is conducive to the maintenance of soil organic carbon[28]. The high mineralization intensity of the 10~20 cm soil layer may be due to the adsorption and colloidal properties of the arsenic sandstone itself, which can adsorb the soluble carbon in the soil solution. This increases the decomposition release of organic carbon by microorganisms, thereby reducing the ability and quantity of available carbon to continue downward migration in the soil profile[29]. In conclusion, the mineralization of the 0~10 cm and 20~30 cm soil layers is small, and the carbon sequestration ability of the S1 and S2 treatments is stronger.

Table 1 Kinetic parameters of organic carbon mineralization of soft rock and sand compound soil planted for 4 years

Soil layer (cm)	Compound ratio	C_0(mg · kg^{-1})	k($\times 10^{-3}$ d^{-1})	$T_{1/2}$(d)	C_0/SOC (%)	R^2
0~10	CK	390.34±11.49 cC	115.03±9.63 aA	6.03±2.01 bB	14.73	0.9839
	S1	576.43±10.65 bC	35.45±9.86 cB	19.55±3.56 aA	27.80	0.9705
	S2	608.47±26.54 bA	44.54±9.29 bB	15.56±2.41 aA	29.82	0.9753
	S3	906.07±21.47 aA	35.14±5.32 cB	19.73±3.82 aA	40.70	0.9922
10~20	CK	1088.30±97.15 aA	54.31±8.39 aB	12.76±3.73 aA	40.26	0.9856
	S1	624.62±59.10 cA	44.38±6.95 bAB	15.62±2.89 aA	33.04	0.9871
	S2	614.50±65.18 cA	48.06±8.76 abB	14.42±2.55 aA	34.08	0.9786
	S3	782.32±65.16 bB	34.75±10.37 cB	19.95±4.12 aA	53.95	0.9680
20~30	CK	515.85±75.13 cB	52.11±2.63 bB	13.30±2.93 aA	25.45	0.9616
	S1	612.96±40.69 bB	49.93±5.81 bcA	13.88±3.24 aA	41.21	0.9920
	S2	498.88±58.34 cB	66.34±8.15 aA	10.45±2.17 aA	26.22	0.9719
	S3	680.12±51.15 aC	42.35±5.16 cA	16.37±2.35 aA	38.77	0.9929

Mean ± standard deviation, lowercase letters indicate significant differences among different treatments under the same soil layer, and uppercase letters indicate significant differences among different soil layers under the same treatment ($P<0.05$).

Conclusions

The change trend of the organic carbon mineralization rate curve and the cumulative mineralization amount curve of the different soil layers of the soft rock and sand composite soil is basically the same, and they all show the characteristics of first decreasing, then increasing and finally stabilizing at a certain level. However, with the deepening of the soil layer, the mineralization rate and cumulative mineralization of the composite soil gradually increased, with the most prominent being 10~20 cm. These characteristics may be affected by the vertical input of organic carbon, soil structure and soil active organic carbon composition. Along with the organic carbon mineralization process of the composite soil, with the increase of soil layer, the mineralization rate and mineralization amount of soil organic carbon increased, and the cumulative mineralization rate and C_0/SOC value of the 0~10 cm soil layer were small. Compared with other treatments, the loss of carbon can be reduced when the ratio of soft rock and sand is 1 : 5 and 1 : 2. The organic carbon mineralization rate of the 0~10 cm composite soil layer is slow and the accumulated mineralization is small, which may become a carbon sink or a carbon source, because the composite soil is a long-term development process. Therefore, in the study of global carbon cycle, deep soil organic carbon and surface soil organic carbon should be placed in the same important position for research.

Acknowledgements

This study is financially supported by Natural Science Basic Research Program of Shaanxi (2021JZ-57), Shaanxi Province Innovative Talents Program-Youth Science and Technology Rising Star Project (2021KJXX-88), Technology innovation Center for Land Engineering and Human Settlements, Shaanxi Land Engineering Construction Group Co., Ltd. and Xi'an Jiaotong University (2021WHZ0087).

Conflicts of interest

The authors declare no conflict of interest.

References

[1] Song, K.C., Wang, G.H., Xu, D.M., Wang, X. (2021) Soil organic carbon mineralization and temperature sensitivity in desert steppe with different enclosure years. Journal of Ecological Environment. 30(3), 453-459.

[2] Körschens, M. (2010) Der organische kohlenstoff im boden (corg)-bedeutung, bestimmung, bewertung soil organic carbon (corg)-importance, determination, evaluation. Archives of Agronomy and Soil Science. 56, 375-392.

[3] Ganjegunte, G.K., Vance, G.F., Preston, C.M., Schuma, G.E., Welker, J.M. (2005) Soil organic carbon composition in a northern mixed-grass prairie. Soil Science Society of America Journal. 69, 1746-1756.

[4] Li, Y.F., Xie, T., Shi, W.L., Li, X.J. (2021). Effects of vegetation restoration on soil organic carbon mineralization in the southeastern edge of Tengger Desert. Journal of Lanzhou University (Natural Science Edition). 57(1), 14-23.

[5] Feng, K., Wang, T., Liu, S.L. (2021) Path analysis model to identify and analyse the causes of aeolian desertification in Mu Us Sandy Land, China. Ecological Indicators. 124, 107386.

[6] Sun, Z.H., Han, J.C., Wang, H.Y., Zhang, R.Q., Hu, Y. (2021) Use and economic benefit of soft rock as an amendment for sandy soil in Mu Us Sandy Land, China. Arid Land Research and Management. 35(1), 15-31.

[7] Liu, Y., Li, C.F., Dang, X.H., Bai, H.J., Li, K.W., Yan, Y. (2020) Research progress on composite soil technology of arsenic sandstone and aeolian sandy soil in Mu Us Sandy Land. Inner Mongolia Forestry Science and Technology. 46(1), 43-46.

[8] Lal, R., Follett, R.F., Stewart, B.A., Kimble, J.M. (2007) Soil carbon sequestration to mitigate climate change and advance food security. Soil Science. 172, 943-956.

[9] Fontaine, S., Barot, S., Barré, P., Bdioui, N., Mary, B., Rumpel, C. (2007) Stability of organic carbon in deep soil layers controlled by fresh carbon supply. Nature. 450, 277-280.

[10] Wang, Q.K., Wang, Y.P., Wang, S.L., He, T., Liu, L. (2014) Fresh carbon and nitrogen inputs alter organic carbon mineralization and microbial community in forest deep soil layers. Soil Biology and Biochemistry. 72, 145-151.

[11] Wang, S., Huang, M., Shao, X., Mickler, R.A., Li, K., Ji, J. (2004) Vertical distribution of soil organic carbon in China. Environmental Management. 33, S200-S209.

[12] Guan, Y.Y., Zhang, X.Y., Wen, X.F. (2017) Effects of glucose addition on carbon mineralization and temperature sensitivity of red soil at different depths. Soil Bulletin. 48(5), 1132-1140.

[13] Huang, S.S., Huo, C.F., Xie, H.T., Wang, P., Cheng, W.X. (2019) Mineralization rate and excitation effect of organic carbon in surface and sublayer no-till black soil. Chinese Journal of Applied Ecology. 30(6), 1877-1884.

[14] Paterson, E., Sim, A. (2013) Soil-specific response functions of organic matter mineralization to the availability of labile carbon. Global Change Biology. 19, 1562-1571.

[15] He, J., Shi, X.Y., Fu, Y.J., Yuan, Y. (2020) Evaluation and simulation of the impact of land use change on ecosystem services trade-offs in ecological restoration areas, china. Land Use Policy. 99, 105020.

[16] Han, J.C., Liu, Y.S., Luo, L.T. (2012) Research on the core technology of rapid composite soil formation of arsenic sandstone and sand in the Mu Us Sandy Land. China Land Science. 26(8), 87-94.

[17] Guo, Z., Shi, C.D. (2021) Prediction of bacterial community structure and function in soils with different com-

pounding ratios. Environmental Science and Technology. 44(1), 69-76.

[18] Smethurst, P.J. (2000) Soil solution and other soil analyses as indicators of nutrient supply: a review. Forest Ecology & Management. 138(1-3), 397-411.

[19] Ribeiro, H.M., Fangueiro, D., Alves, F., Vasconcelos, E., Coutinho, J., Bol, R., Cabral, F. (2010) Carbon-mineralization kinetics in an organically managed Cambic Arenosol amended with organic fertilizers. Journal of Plant Nutrition and Soil Science. 173, 39-45.

[20] Wang, Q., Wang, S., He, T., Liu, L., Wu, J. (2014) Response of organic carbon mineralization and microbial community to leaf litter and nutrient additions in subtropical forest soils. Soil Biology and Biochemistry. 71, 13-20.

[21] Shi, C.D., Wang, S.G., Li, J., Hua, D.W. (2020) Research on aggregate changes of soft rock and sand compound soil after cultivation. People's Yellow River. 42(S1), 65-66,68.

[22] Cao, T.T., Sun, Y.Y., Wang, H.Y., Du, Y.C. (2019) Characteristics of soil carbon distribution in different years of land consolidation. Soil and Water Conservation Research. 26(2), 86-92.

[23] Luo, L.J., Zhang, L.H., Lu, M.H. (2020) Effects of different water treatment and litter decomposition on soil CO_2 release in Minjiang estuary wetland. Wetland Science and Management. 16(4), 35-40.

[24] Liu, D.Y., Song, C.C. (2008) Effects of phosphorus enrichment on mineralization of organic carbon and contents of dissolved carbon in a freshwater marsh soil. China Environmental Science. 28, 769-774.

[25] Mason-Jones, K., Schmucker, N., Kuzyakov, Y. (2018) Contrasting effects of organic and mineral nitrogen challenge the N-mining hypothesis for soil organic matter priming. Soil Biology and Biochemistry. 124, 38-46.

[26] Zha, Y., Wu, X.P., He, X.H., Zhang, H.M., Gong, F.F., Cai, D.X., Zhu, P., Gao, H.J. (2014) Basic soil productivity of spring maize in black soil under long-term fertilization based on DSSAT model. Journal of Integrative Agriculture. 13, 577-587.

[27] Tang, S.R., Cheng, W.G., Hu, R.G., Nakajima, M, Guigue, J., Kimani, S.M., Sato, S., Tawaraya, K., Xu, X. (2017) Decomposition of soil organic carbon influenced by soil temperature and moisture in Andisol and Inceptisol paddy soils in a cold temperate region of Japan. Journal of Soils and Sediments. 17, 1843-1851.

[28] Zhu, M.T., Liu, X.X., Wang, J.M., Liu, Z.W., Zheng, J.F., Bian, R.J., Wang, G.M., Zhang, X.H., Li, L.Q., Pan, G.X. (2020) Effects of biochar on microbial diversity of paddy soil aggregates. Chinese Journal of Ecology. 40(5), 1-12.

[29] Li, Y., Wei, Z.C., Li, H.T., Qiu, Y.X., Zhou, C.F., Ma, X.Q. (2017) Effects of biochar on soil carbon and nitrogen mineralization in Chinese fir plantations. Journal of Agricultural and Environmental Sciences. 36(2), 314-321.

本文曾发表于2022年《Fresenius Environmental Bulletin》第31卷第6A期

Effects of Organic Fertilizers on Growth Characteristics and Fruit Quality in Pear-jujube in the Loess Plateau

Shenglan Ye [1,2,3], Biao Peng [1,2,3], Tiancheng Liu [2,3]

(1.Institute of Land Engineering and Technology, Shaanxi Provincial Land Engineering Construction Group Co., Ltd., Xi'an 710075, China;2.Shaanxi Provincial Land Engineering Construction Group Co., Ltd., Xi'an 710075, China;3.Key Laboratory of Degraded and Unused Land Consolidation Engineering, Ministry of Natural Resources, Xi'an 710075, China)

【Abstract】 The ecological environment of the hilly and gully area of the Loess Plateau in northern Shaanxi is fragile and the soil fertility is low. As a result, the yield and quality of Pear-jujube which constitute one of the dominant economic forests in this region, have been severely restricted. At present, the scientific application of fertilizer is important for comprehensively improving the quality of fruit trees, and for devising the optimal management of fruit trees. In particular, the application of organic fertilizers plays an important role in improving soil and improving fruit quality. In this experiment, a field study was conducted to understand the effects of different organic fertilizer applications on physiological growth, photosynthetic characteristics, reproductive growth and nutritional quality of Pear-jujube in the Loess Plateau. The results showed that organic fertilizer significantly promoted the physiological growth of Pear-jujube. The Pear-jujube bearing branch and leaf area under the soybean cake fertilizer (SC) treatment were 20.17 cm and 1246 mm^2/leaf, respectively, which are increased by 34% and 44.46% compared with the no fertilizer treatment, which was a control check (CK). The total chlorophyll content of fertilization treatment was significantly higher than that of CK ($P < 0.05$). The maximum of chlorophyll content was 10.90 mg/dm^2 under the biogas fertilizer (BM). The content of LAI was in the order BM > conventional fertilizer (CF) > sheep manure (SM) > SC > CK. The changing trend of gap fraction was opposite to that of LAI, and the density of light was consistent with that of LAI. The density of light BM was the largest, reached 38.06 mol/(m^2 · d), which was 15.13% higher than that of CK. Organic fertilizer significantly improved the net photosynthetic rate (Pn) and water use efficiency (WUEp) of Pear-jujube. The WUEp of SC was up to 3.30%. Organic fertilizer significantly promoted the reproductive growth and improved the nutritional quality of Pear-jujube. The yield under the SC was 19177 kg/hm^2, increased by 138.5% compared with that of CK. The fruit water content (FWC), total soluble solids (TSS), solid-acid ratio (TSS/TA), Vc and total flavonoids content improved, and the maxima of FWC, TSS, TSS/TA, Vc and total flavonoids content under the SC treatment were 86.30%, 18.48%, 40.17, 46.18 mg/kg and 14.35 mg/kg, respectively, which were significantly different from those of CK ($P < 0.05$). Organic fertilization effectively promotes the growth, development, yield and fruit quality of Pear-jujube in the Loess Plateau and the effect of the soybean cake fertilizer is the most significant.

【Keywords】 Pear-jujube;Organic fertilizer; Physiological growth; Water use efficiency; Fruit quality

1 Introduction

The ecological environment in the hilly and gully regions of the Loess Plateau in northern Shaanxi is fragile. Soil fertility is low. To protect the environment and make use of land resources, a large number of economic forests are planted in this area to reconvert farmlands to forests. This can effectively green the barren hills, conserve water and soil, and improve the ecological environment. This can also increase the economic income of farmers. Pear-jujube is a rare and precious variety of jujube. It has rich nutritional value. At present, it has become one of the leading plant varieties in the economic forest industry in northern Shaanxi[1]. The long-term and sustainable development of the Pear-jujube Forest is inseparable from a good water-and-fertilizer environment. However, there is a problem with the long-term application of quick-acting chemical fertilizers and pesticides in the production of Pear-jujube. This will not only cause serious environmental pollution but also affect the nutritional quality of Pear-jujube[2-3]. Chen et al.[4] studied the effect of different nitrogen, phosphorus and potassium fertilizers on the yield and fruit quality of jujube, and concluded that the highest yield fertilizer ratio was 1 : 0.84 : 0.11. And it is proposed that under the same fertilizer base conditions, the total acidity of red dates is positively correlated with the nitrogen fertilizer application amount, and the reducing sugar and vitamin C content and nitrogen and phosphorus fertilizer application amounts show a trend that is first increasing and then decreasing. Excessive application or partial application of chemical fertilizer is to exchange high input for high yield; although the yield problem was solved, it also caused a series of environmental effects such as groundwater pollution and serious soil degradation[5-6]. Moreover, excessive application of chemical fertilizers will also have adverse effects on the metabolism of organic compounds in plants[7]. Because some fruit farmers only pay attention to the application of chemical fertilizers with less or no use of organic fertilizers, problems such as reduced soil organic matter, acidification, nutrient imbalance, and serious diseases in orchards are caused[8]. The application of those highly toxic and high residue pesticides increase the toxic components of grains, vegetables and fruits. Food safety is not guaranteed, thus endangering human health. In recent years, with the accompanying increase in the number of chemical fertilizers and crop yields, the number of organic fertilizers has also increased[9].

At present, the fertilization system and technology in organic planting in China are not mature and perfect. Therefore, studies on how to reasonably use field management measures in organic planting, and appropriately regulates the types of fertilizers are needed. Scientific fertilization and soil fertilization are very necessary. The application of organic fertilizer can not only improve soil fertility[10-11], improve soil physical and chemical properties[12-13], and enhance soil water storage capacity[14-15] but also effectively promote the vegetative growth and reproductive growth of plants, thereby improving the quality of plants[16-19]. The experimental results of Yu et al.[20] showed that organic fertilizer can promote rice to reduce ineffective tillers, increase panicle rate, increase effective panicle number and late green leaf area, increase leaf-grain ratio, and improve rice seed setting rate and grain filling fullness. This increases the yield effect of increased panicle weight. The application of different organic fertilizers can significantly reduce the nitrate content of peppers[21]. Wang et al.[22] found that the yield of jujubes with medium and high dosages of organic fertilizers without biochar addition was the highest, reaching 39.49 and 40.16 kg per plant. The application of a high proportion of organic fertilizer and biochar could not significantly improve the photosynthetic capacity of jujube; on the contrary, the treatment of a low application amount of organic fertilizer without biochar could significantly improve the photosynthetic capacity of jujube. Biochar has high adsorption. Excessive application of biochar increased soil porosity, thereby reducing soil effective porosity. The movement of nutrients and water to the root system of crops is affected. It is not conducive to the absorption of nutrients and water, thereby reducing the photosynthetic capacity of jujube. This shows that the combined application of organic fertilizer and biochar is not the best fertilization mode for jujube in the short term. Zaituniguli et al.[23] researched that long-term fertilization can increase the chlorophyll content in crop leaves. At the grain filling stage, the SPAD value of the NP treatment was the lowest, which was 37.07; the SPAD value of the NPK + organic fertilizer treatment was the highest, reaching 44.62, and the difference was significant ($P<$

0.05). The biological yield of NPK + organic fertilizer treatment was the highest, equaling 94.81 t/hm^2, which was 97.95% higher than that of CK. Therefore, research on the application of organic fertilizer and photosynthesis is key in exploring the yield and quality of Pear-jujube.

With the rapid growth of the industrial economy in the Loess Plateau region of northern Shaanxi and the acceleration of the construction of energy and chemical bases, the layout, production structure and product structure of the agricultural industry can be further optimized. Taking advantage of the huge reserves of farmyard manure in the Loess Plateau, it can supply and develop characteristic organic agriculture and become an effective way to increase farmers' income. At present, there are few research reports on the effects of applying organic fertilizers on the photosynthetic physiological characteristics and water use of Pear-jujube in the hilly and gully regions of the Loess Plateau. For this reason, this study takes Pear-jujube in the loess hilly area as the research object, and applies different organic fertilizers during the flowering and fruit setting period. We also analyze the comprehensive effects of organic fertilizer on the vegetative growth, reproductive growth, photosynthetic characteristics and water use of Pear-jujube. The research further explores the organic fertilizer source suitable for the growth and development of Pear-jujube. This can provide a theoretical basis and technical support for the production of organic Pear-jujube.

2 Materials and methods

2.1 Overview of the test area

The experiment was carried out at the Pear-jujube micro-irrigation technology experimental demonstration base (37.78′N, 110.23′E, 870 m above sea level) in Mizhi County, Yulin City, Shaanxi Province from 2018 to 2019. This area is a typical hilly and gully area of the Loess Plateau, with a mid-temperate semi-arid climate, with an average annual temperature of 8.5 ℃ and an average annual rainfall of 451.6 mm, mainly from July to September. The main soil in the test area is Lossiah soil. The physical and chemical properties of the soil layer at 0~20 cm before fertilization are as follows: the available nitrogen, phosphorus, and potassium contents are 34.73, 2.90, and 101.9 mg/kg, respectively, and the organic matter content is 2.1 g/kg and pH 8.6, bulk density of 1.21 g/cm^3 and the soil particle composition is as shown in Table 1. The entire slope of the orchard has a height difference of approximately 100 m. The terrain is complex, and the slope directions are different. During the trial period, the drip irrigation quota was 135 m^3/hm^2, and irrigation was carried out on May 15th and June 10th, respectively.

Table 1 Soil particle composition

Layers (cm)	Sand (%) (1~0.05mm)	Silt particle (%) (0.05~0.001mm)	Clay (%) (<0.001mm)	Physical clay (%) (<0.01mm)
0~20	28.6	67.9	2.01	17.4
20~40	28.1	68.0	2.38	16.3
40~60	32.2	65.3	2.31	15.8

2.2 Experimental design

We selected the 7-year-old mountain dwarf and densely planted Pear-jujube (*Zizyphus jujuba* Mill.cv.) with a uniform tree body and good growth. We set up five fertilization treatments under drip irrigation conditions, namely: no fertilizer (CK); conventional fertilizer[24] (CF), urea (containing N 46%) 0.48 kg/plant, superphosphate (containing P_2O_5 12%) 1.35 kg/plant, potassium sulfate (containing 50% K_2O) 0.61 kg/plant; fermented and decomposed soybean cake fertilizer (SC), 5 kg/plant; decomposed sheep manure (SM), 15 kg/plant; and biogas fertilizer (BM), 100 kg/plant. The application rate is calculated based on the content of each nutrient in the organic fertilizer and based on the consistent application of nitrogen, phosphorus, and potassium. A single plant is regarded as one treatment, and a total of five repetitions are set. The test area was 300 m^2, and the planting density was 2 m×3 m. The soybean cake fertilizer used in the test is the oil residue obtained after oil extraction. It is sealed and fermented for 35 days under the condition of relative humidity of 70%. It was then used as a decomposed soybean cake fertilizer. The mixture of biogas slurry and biogas residue, decomposed sheep manure is a mixture of crushed straw and sheep manure buried in a water pit and subjected to microbial anaerobic fermentation. The nutrient content of each organic fertilizer is shown in Table 2.

Table 2 Nutrient content of the experimental organic fertilizers

Organic fertilizer	Moisture content(%)	Organic matter(%)	Total N(%)	Total P(%)	Total K(%)
Soybean cake fertilizer	22.74	7.69	4.13	0.91	1.57
Sheep manure fertilizer	8.23	24.72	0.95	0.73	2.25
Biogas manure fertilizer	99.43	0.17	0.21	0.096	0.11

2.3 Measurement indicators and methods

Organic fertilizer was applied in early April 2019. On May 31, June 6, June 14, June 22, and July 1, 2019, 12 jujube hangings were selected from the upper parts of the plant in four directions: east, west, south and north. The jujube hangings are in good shape and uniform in growth. The leaf area and leaf relative water content of the 5th leaf from the top were determined. In mid-June, the leaf chlorophyll content, photosynthetic index, canopy index and soil water content (SWC) were measured, and the number of flowering plants was investigated. In mid-July the number of fruits and was counted and the fruit set rate was calculated. On October 8, the ripe fruits were collected from each tree to determine the quality.

The length of the jujube hanging was measured with steel tape; 12 jujube hangings were fixedly selected in different directions on the upper part of each plant, and the length was measured regularly with a steel tape measure.

The leaf area was measured with a portable leaf area meter model AM300(ADC-UK). The leaf area was periodically measured on the fifth leaf of the fixed jujube.

The relative water content (RWC) of the leaf was measured using the saturation weighing method. The fresh weight (W_f) of the leaves was first weighed. Then the leaves were immersed in distilled water for 6~8 h, so that the leaves were saturated with water. After the leaves were taken out, surface water was wiped with absorbent paper, and immediately put into a weighing bottle of known weight to weigh. Then the leaf was immersed into distilled water for a while, taken out and dried outside, and then weighed until the weight no longer increases. This is the saturated water weight (W_t) of the leaf. The samples were then dried at 70 ℃ at constant weight to obtain tissue dry weight (W_d). The calculation formula is $RWC = (W_f - W_d)/(W_t - W_d) \times 100\%$.

The chlorophyll content of leaves was measured using the acetone method, and the content of chlorophyll a, b and the total amount of chlorophyll were calculated[25]. Fresh leaves with thick veins were cut into pieces, and extracted with 80% acetone in the dark. The extract was taken and measured for colorimetry using a 722 type visible light spectrophotometer at the absorbance wavelengths of 663, 645 and 652nm, and 80% acetone was used as a reference.

The photosynthetic indexes were measured using the LI-6400XT photosynthesis instrument (Li-COR, USA). The red and blue light sources that come with the leaf chamber were used. The intensity of photosynthetically active radiation was set to 2500 μmol/(m^2 · s) , with an open gas path. The temperature was 25 ℃. The CO_2 concentration was 400 μmol/mol. P_n, C_i, G_s and T_r of leaves were determined after Pear-jujube matured, with five replicates per plant.

WUEp was calculated using P_n and T_r when the photosynthetically active radiation was 2500 μmol/(m^2 · s). The formula was WUEp = P_n/T_r.

The measured P_n, G_s, C_i and T_r were compared according to the values when the photosynthetically active radiation was 2500 μmol/(m^2 · s).

Canopy indicators were determinedusing WinsCanopy2005a canopy analysis. Image acquisition was performed in the field. Then, canopy analysis and supporting analysis software was used to process the images. The leaf area index (LAI), forest gap fraction (GFR), total radiant flux above the canopy and total radiant flux below the canopy were measured, and the transmittance and the density of light were calculated.

Transmittance = total average density of photosynthetically effective radiation below the canopy / total average density of photosynthetically effective radiation above the canopy;

The density of light = the average density of total photosynthetically effective radiation above the canopy-the average density of total photosynthetically effective radiation below the canopy[26].

SWC was measured using conventional soil drilling and drying method. The sampling location was 30 cm away from the tree body, the measuring depth was 0~100 cm soil layer, and the sampling interval was 20 cm.

Determination of fruit quality: The water content of the fruit was measured using the normal pressure heating and drying method[27]; the total soluble solids (TSS) were measured using the 2WAJ-Abbe refractometer; the titratable acid (TA) was measured using the 0.1 mol/L NaOH standard solution Titration method[28]; reduced vitamin C was titrated with 2,6-dichloroindophenol[29]; and total flavonoids were determined using $NaNO_2$-$Al(NO_3)_3$ spectrophotometry[30,31].

The average value of each treatment was taken as the final result. The experimental data was organized in Microsoft Excel 2003. Data analysis of variance was performed using DPS 7.05 software to test significant differences between data processing.

3 Results and analysis

3.1 Effect of different organic fertilizers on the growth of Pear-jujube

3.1.1 Effect of different organic fertilizers on the bearing branch length of Pear-jujube

Jujube-bearing branch has the dual role of fruiting and photosynthesis[32-33]. It can be seen from Fig.1 that different organic fertilizer treatments have a significant impact on the growth of jujube-bearing branches. Among them, the longest jujube-bearing branch in the SC treatment is 20.17 cm, which is significantly higher than that in CK and CF; the jujube-bearing branch length in the SC, SM and BM treatment are increased by 34%, 23% and 25% compared with that in CK, and the difference is significant ($P<0.05$), compared with CF treatment, the increase is 22%, 12%, and 14%, respectively. Starting from May 31st, the length of the jujube-bearing branch in each treatment increased significantly. By the end of June, the length of the jujube-bearing branch in the CK and CF treatments grew slowly, while the length of the jujube-bearing branch in the SC, SM and BM treatments increased by 3.76 cm, 2.53 cm, and 2.96 cm respectively.

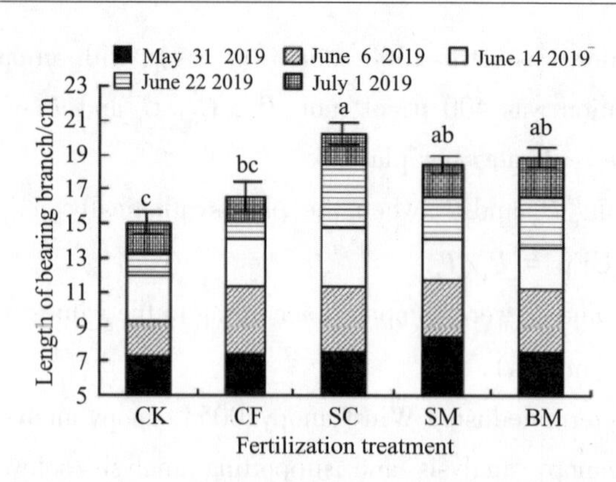

Fig.1 Effects of different fertilizer treatments on bearing a branch of Pear-jujube

Note: At different levels, each population mean follows a normal distribution with the same variance. The least significant difference (LSD) method was selected for analysis of variance in DPS software. Different lowercase letters indicate significant difference between treatments ($P<0.05$).

3.1.2 Effect of different organic fertilizers on the leaf area of Pear-jujube

Fig.2 shows that the increase in leaf area of each fertilization treatment is significantly different from that of the CK ($P<0.05$). The leaf area of the SC treatment is the largest, with a leaf area of 1246 mm^2/leaf. The leaf area of the SC and BM treatments are significantly different from that of the CK, CF and SM treatments ($P<0.05$); Compared with that of CK, the leaf area of CF, SC, SM and BM treatments increased by 18.34%, 44.46%, 26.67%, and 41.65%. Compared with CF treatment, SC, SM, and BM organic fertilizer treatments increased by 22.08%, 7.04%, and 19.70%, respectively. Soybean cake fertilizer (SC) and biogas fertilizer (BM) saw the most significant increases.

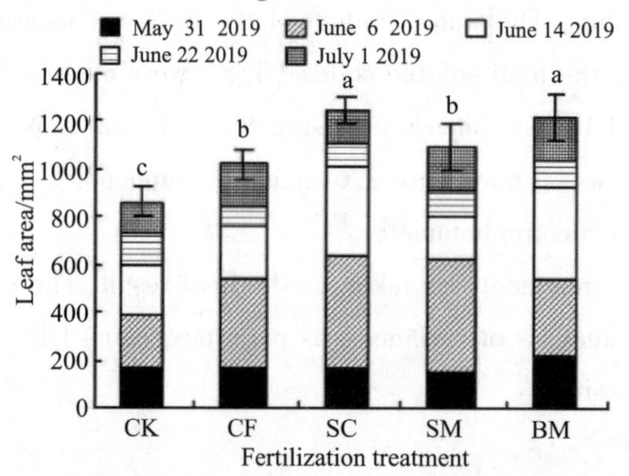

Fig.2 Effects of different fertilizer treatments on leaf area of Pear-jujube

Note: At different levels, each population mean follows a normal distribution with the same variance. The least significant difference (LSD) method was selected for analysis of variance in DPS software. Different lowercase letters indicate a significant difference between treatments ($P<0.05$).

3.2 Effect of organic fertilizers on the photosynthetic effect of Pear-jujube during the flowering and fruit setting period

3.2.1 Effect of organic fertilizers on the chlorophyll content of Pear-jujube during flowering and fruit setting

The content of chlorophyll affects the absorption and conversion of light energy. The change in the ratio of chlorophyll a and chlorophyll b (chl a/b) can reflect the strength of leaf photosynthetic activity and the amount of light energy used by plants[34]. The photosynthetic performance of Pear-jujube during the flowering and fruit setting period directly affects the supply of nutrients required for the flowering and fruit setting of Pear-jujube, which is of great significance to the final yield. It can be seen from Table 3 that four treatments have different effects on the chlorophyll content of Shandi Pear-jujube during the flowering and fruit setting period. The total chlorophyll content is significantly higher than that of CK. CF, SC, SM and

BM increased by 22.86%, 26.73%, 39.31% and 43.01%, respectively, comparing with CK. The content of chlorophyll a is in the order SM > BM > SC > CF > CK. It is found that organic fertilizer can significantly improve the chlorophyll a content. The changing trend of chlorophyll b content is consistent with the total content. The analysis found that the chl a/b content of CK is the highest. The chl a/b content of BM, SC and CF are significantly less than CK. It may be due to fertilizing that the synthesis of chlorophyll b was greater than that of chlorophyll a. This can significantly increase the absorption of the plants to blue-green light, enhance the photosynthetic activity of the blade. Among them, the application of biogas fertilizer has the most significant effect on the chlorophyll content, and the total chlorophyll content is 10. 90 mg/dm^2, while chl a/b is only 3.27.

Table 3 Effects of different fertilizer treatments on chlorophyll contents of Pear-jujube

Treatment	Chl a(mg/dm^2)	Chl b(mg/dm^2)	TC(mg/dm^2)	Chl a/b
CK	6.08±0.04 c	1.54±0.06 c	7.62±0.10 d	3.97±0.13 a
CF	6.97±0.09 b	2.10±0.02 b	9.36±0.34 c	3.61±0.01 b
SC	7.76±0.19 a	2.11±0.15 b	9.66±0.04 bc	3.60±0.30 b
SM	8.38±0.21 a	2.33±0.01 ab	10.61±0.34 ab	3.82±0.05 a
BM	8.34±0.61 a	2.55±0.17 a	10.90±0.78 a	3.27±0.03 c

Note: TC—total chlorophyll; The data are the mean ± standard deviation of the parameters in this group. At different levels, each population mean follows a normal distribution with the same variance. The least significant difference (LSD) method was selected for analysis of variance in DPS software. Different lowercase letters in the same column indicate significant difference between treatments ($P < 0.05$).

3.2.2 Effects of organic fertilizers on the canopy structure and canopy optical properties of Pear-jujube during flowering and fruit setting

It can be seen from Fig.3 that LAI of Pear-jujube after fertilization is significantly higher than that of CK. Compared with CK, CF, SC, SM, and BM increases by 15. 75%, 12. 46%, 15. 62%, and 24.50%, respectively. The LAI of BM is up to 2.17, and it is significantly different from other treatments ($P < 0.05$). The changing trend of gap fraction in different treatments is opposite to that of LAI. Canopy photosynthetically active radiation is the most important indicator for evaluating canopy transmittance and light interception ability. The density of light of different treatments is consistent with LAI, which is expressed as BM>CF>SM>SC >CK. Among them, the density of light of BM is the largest. It reaches 38.06 mol/(m^2 · d). CF, SC, SM and BM respectively increase by 11.54%, 8.09%, 7.96% and 15. 13% compared with CK, and the difference is significant. The canopy transmittance of jujube is BM<SM< SC<CF<CK, which are reduced by 8. 64%, 14. 78%, 21.62% and 34.16% compared with CK, respectively. The organic fertilizer treatments are significantly different from CK ($P<0.05$).

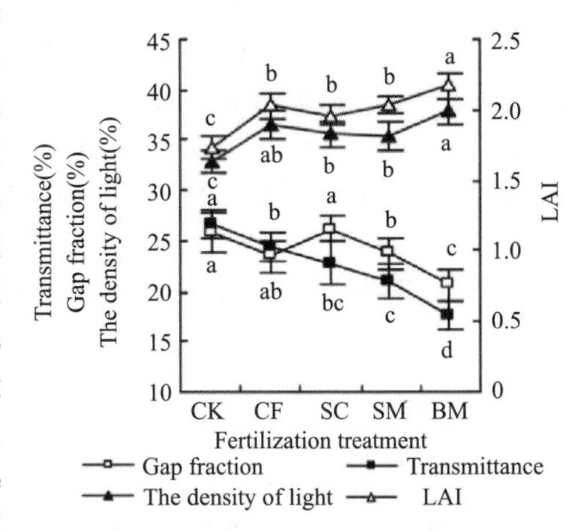

Fig.3 Effects of different fertilizer treatments on canopy characteristics of Pear-jujube

Note: At different levels, each population mean follows a normal distribution with the same variance. The least significant difference (LSD) method was selected for analysis of variance in DPS software. Different lowercase letters indicate a significant difference between treatments ($P<0.05$).

3.2.3 Effects of organic fertilizers on photosynthetic characteristics and leaf water use efficiency of Pear-jujube during flowering and fruit setting

The organic fertilizer significantly increases the Pn and Gs of Pear-jujube. The Pn and Gs in each treatment are in the order BM>SM>SC>CF>CK. The Pn and Gs of BM are the highest, which are 22.38 μmol/(m^2 · s) and 0.5014 mmol/(m^2 · s), respectively. After fertilization, the Ci is less than that of CK. It shows that fertilization effectively reduce Ci. Among them, the effect of BM which is only 240.8 μmol/mol is the most obvious. Transpiration is the same as photosynthesis; they are regulated by many factors. Stomatal transpiration is the main method of transpiration[35-36]. The Tr of different treatments is expressed in the order BM>CK >CF>SM>SC. The highest Tr of BM reaches 8.66 μmol/moL. It may be related to higher LAI, and the instantaneous water use efficiency of SC is highest, which reaches 3.30%. The WUEp of CF, SC, SM and BM treatments increase by 22.4%, 64.2%, 44.3% and 30.8%, respectively, compared with that of CK. It reaches a significant difference level (P < 0.05). The instantaneous water use efficiency of SC, SM and BM is 0.84, 0.44 and 0.17 percentage points higher than that of CF, respectively (Table 4).

Table 4　Effects of different organic fertilizers on photosynthetic characters of Pear-jujube

Treatment	Pn [μmol/(m^2 · s)]	Gs [mmol/(m^2 · s)]	Ci (μmol/mol)	Tr (μmol/mol)	WUEp(%)
CK	18.30±1.87 c	0.4011±0.0304 c	270.2±3.16 a	8.54±0.31 a	2.01±0.12 d
CF	20.80±1.61 b	0.4547±0.0456 b	255.3±4.67 bc	8.30±0.45 a	2.46±0.17 c
SC	22.19±0.90 a	0.4912±0.0320 a	252.0±4.11 c	6.57±0.33 c	3.30±0.26 a
SM	22.37±0.49 a	0.4991±0.0352 a	255.9±3.57 b	7.67±1.06 b	2.90±0.19 b
BM	22.38±0.97 a	0.5014±0.0329 a	240.8±3.74 d	8.66±0.20 a	2.63±0.13 bc

Note: Pn—Net photosynthetic rate, Gs—Stomatal Conductance, Ci　Intercellular CO_2 concentration, Tr—transpiration rate, WUEp—water use efficiency. The data are the mean ± standard deviation of the parameters in this group. At different levels, each population mean follows a normal distribution with the same variance. The least significant difference (LSD) method was selected for analysis of variance in DPS software. Different lowercase letters in the same column indicate significant difference between treatments (P<0.05).

3.3　Effect of organic fertilizers on water utilization during flowering and fruit setting period of Pear-jujube

3.3.1　Effect of organic fertilizers on SWC during flowering and fruit setting of Shandi Pear-jujube

The SWC of each treatment increases rapidly with the increase of soil depth and then tends to be gentle. Among them, the SWC of SC and SM is significantly higher than that of BM, CF and CK. This is because the application of organic fertilizer can make the soil absorb a lot of water and prevent the infiltration of water. The average SWC of each treatment in the 0~80 cm soil layer is SC>SM>BM>CF>CK. Compared with CK (9.37%), the SC, SM, BM, and CF increased by 3.69%, 3.18%, 1.11% and 0.40% points, respectively. Organic fertilizer is beneficial to increase the water content of the soil. Among them, soybean cake fertilizer (SC) has the largest increase, which is significantly different from CK (P < 0.05) (Fig.4).

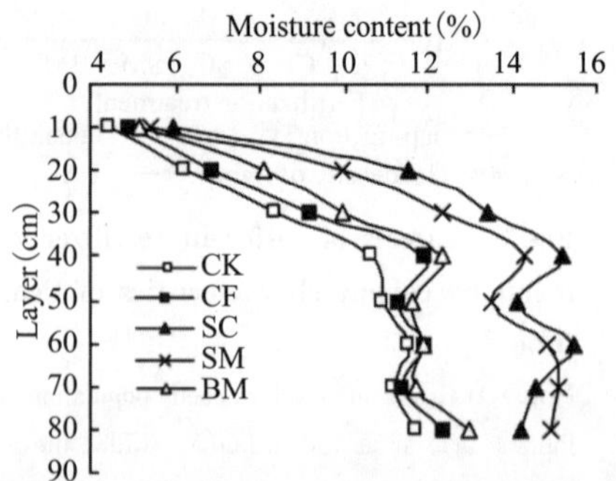

Fig.4　Effects of different fertilizer treatments on soil water content

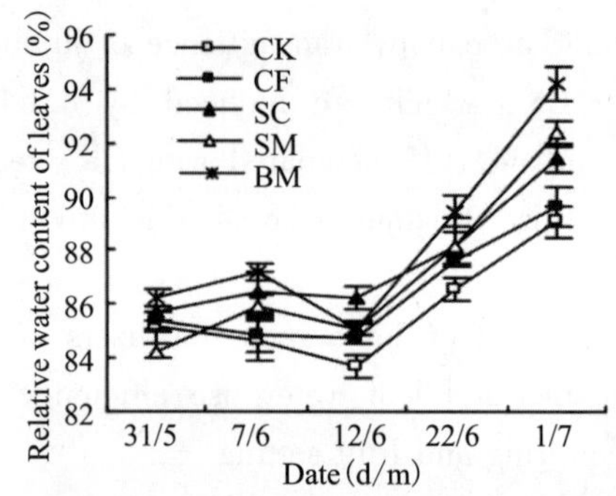

Fig.5　Effects of different fertilizer treatments on relative leaf water content of Pear-jujube

3.3.2 Effect of organic fertilizers on relative water content during flowering and fruit setting of Pear-jujube

The flowering period requires sufficient humidity, so the high relative water content (RWC) helps to set up flowers and improve the photosynthesis of Pear-jujube. It can be seen from Fig.5 that the trend of leaves RWC changes of Pear-jujube in each treatment is a slow rise. It decreases slightly on June 12. This may be related to the large demand for water of jujube trees at this time, in which the application of organic fertilizer can improve the RWC of Pear-jujube leaves, thus supplying the water required for flowering. By July 1, the RWC of the leaves reaches the highest. The RWC for each treatment is in the order BM > SM > SC > CF > CK. The RWC of BM reaches 94.20%, which is significantly different from CK (P<0.05). It can be seen that the application of organic fertilizer can improve the RWC of leaves. The RWC of BM, SM and SC increase by 5%, 3% and 2%, respectively, compared with that of CF.

3.4 Effect of different organic fertilizers on the reproductive growth of mountain Pear-jujube

It can be seen from Table 5 that the total flowers, fruit number and yield per plant after fertilization are all greater than those of CK, and SC treatment has the most significant effect on reproductive growth (P<0.05). Compared with CK, the total flower number of CF, SC, SM and BM increases by 42.59%, 82.68%, 14.48% and 40.39%, respectively, and the fruit yield per plant increased by 95.47%, 157.27%, 68.79% and 148.48%, respectively. It shows that different fertilization treatments significantly increase the fruit setting rate of Pear-jujube. The highest fruit setting rate of BM is 4.57%. The fruit setting rate and yield of each fertilization treatment are significantly higher than CK. Different organic fertilizers have varying degrees of influence on the reproductive growth of Pear-jujube. In combination with the above indicators, SC and BM have the most significant effects.

Table 5 Effects of different fertilizer treatments on reproductive growth of Pear-jujube

Treatment	Flower No. (No./plant)	Fruit No. (No./plant)	Fruit setting rate (%)	Yield (kg/hm^2)
CK	14790±1061 c	383±19 c	2.54±0.20 c	8040±568 d
CF	21564±1660 b	748±81 b	3.50±0.65 b	11797±741 c
SC	27629±4552 a	985±121 a	3.58±0.20 b	19177±836 a
SM	18246±2044 bc	646±51 b	3.58±0.58 b	12025±891 c
BM	21232±5072 bc	951±132 a	4.57±0.61 a	14142±812 b

Note: The data are the mean ± standard deviation of the parameters in this group. At different levels, each population mean follows a normal distribution with the same variance. The LSD least significant difference method was selected for analysis of variance in DPS software. Different lowercase letters in the same column indicate significant difference between treatments (P<0.05).

3.5 Effect of different organic fertilizers on the nutritional quality of Pear-jujube

Table 6 shows that the fruit water content of SC and BM increases by 9.15 and 8.26 percentage points, respectively, compared with that of CK, and it reaches a significant difference (P<0.05). TSS, solid-to-acid ratio (TSS/TA), Vc and total flavonoid content of organic fertilizer treatments are higher than those of CK. The TSS of all treatments except SM is significantly different than that of CK (P < 0.05). The titratable acid (TA) content is in the order CK>SM>BM>SC, and the lowest TA of SC is 0.23% points lower than that of CK. The Vc content of SC is the highest, reaching 46.18 mg/kg. It is an increase of 78.92% compared with CK. The Vc content of BM and SM also increase by 59.59% and 43.36%, respectively, compared with those of CK. The total flavonoid content of each fertilization treatment is SC> SM>BM>CK. The total flavonoid content of SC reaches 14.35 mg/kg, which is 24.57% higher than that of CK. The total flavonoid content of SM and BM increase by 17.01% and 9.2%, respectively, compared with that of CK. Moreover, each treatment is significantly different from CK (P<0.05). There is no significant difference in the total flavonoids content between SM and BM.

Table 6 Effects of different fertilizer treatments on nutrient quality of Pear-jujube

Treatment	FWC (%)	TSS (%)	TA (%)	TSS/TA	Vc (mg/kg)	Total flavones (mg/kg)
CK	77.15 b	15.45 c	0.69 a	22.39 c	25.81 c	11.52 c
SC	86.30 a	18.48 a	0.46 c	40.17 a	46.18 a	14.35 a
SM	76.32 b	16.63 bc	0.53 b	31.38 b	37.00 b	13.48 b
BM	85.41 a	17.05 b	0.51 b	33.43 b	41.19 b	12.58 b

Note: FWC—Fruit water content; TSS—Total Soluble Solid; TA—Titratable Acid; At different levels, each population mean follows a normal distribution with the same variance. The least significant difference (LSD) method was selected for analysis of variance in DPS software. Different lowercase letters in the same column indicate significant difference between treatments ($P<0.05$).

4 Discussions

The flowering and fruit setting periods of Pear-jujube are lasts for a long time. This overlapping period shows the coexistence of vegetative growth and reproductive growth. It is more sensitive to nutrition and water conditions. Adequacy of its nutrient supply is directly related to the growth and yield of jujube. So scientific fertilization and soil cultivation fat are very necessary. Organic fertilizer can activate the nutrients in the substrate, improve the physical and chemical properties of the soil, promote the absorption of nutrients by plants, increase the nutrient content[37], provide the nutrients needed for dry matter accumulation, and promote vegetative growth and reproductive growth[13,38].

A study by Xue et al.[39] has shown that the proportion of leaves and short branches in the seaweed fertilizer treatment was the highest (59.53%), while the proportion of long branches and leggy branches was the lowest (33.23%), and the leaf area reached 35.91 cm^2 which was significantly higher than that of the control. The results of this study showed that the new shoots and leaf areas of SC, SM, and BM were significantly different from those of CK ($P < 0.05$) after the application of organic fertilizers, which increased by 34%, 23%, 25% and 44.46%, 26.67%, and 41.65% compared with that of CK.

Chlorophyll is the main pigment for plant photosynthesis, and it plays a central role in light absorption in photosynthesis[40]. A study by Wang et al.[41] has shown that after applying organic fertilizer, the chlorophyll content of wheat leaves is the highest, and SPAD value may reach up to 60.1, which is 58% higher than that of CK. A study by Li[42] has shown that organic fertilizers from different plant sources increased the Pn of crisp jujube leaves to varying degrees and adjusted the transpiration rate of the leaves. This is consistent with the results of this research. The total chlorophyll content in the flowering and fruit setting period was in the order BM > SM > SC > CF > CK. The Pn indicates the accumulation of photosynthetic products in plants, and increasing the photosynthetic efficiency can produce more organic matter, which is beneficial to plant growth and development. The changing trend of Pn and Gs was consistent with the change in chlorophyll content. This may be because different organic fertilizers improve soil fertility[43], which in turn enhances the photosynthesis of leaves and makes them accumulate more photosynthetic products. The intercellular CO_2 concentration was opposite to the changing trend of leaf photosynthetic rate and stomatal conductance. It may be that non-stomata factors reduce the utilization of CO_2 and cause the accumulation of CO_2[44-46]. Therefore, the application of organic fertilizer significantly increases the chlorophyll content of the Pear-jujube during the flowering and fruit setting period, enhanced the intensity of photosynthesis, caused a decrease in the intercellular CO_2 concentration, and accelerated the synthesis and accumulation of photosynthetic products.

A study by Chen et al.[47] has shown that organic fertilization can significantly increase the rhizosphere moisture content and reduce the loss rate of soil water. In this study, the rhizosphere moisture content with organic fertilizer was significantly increased, especially in the depth of 40～80 cm in the soil layer of the widest distribution of absorbing roots. The SWC of SC and SM was significantly higher than that of CK and CF. This may be related to the strong water retention effect of organic fertilizers. At the same time, it also increased the relative water content of the leaves at the flowering stage. The relative water content of the leaves of BM was the highest (94.20%). Compared with CF, it increased by 5%. Organic fertilizer significantly improveds the water use efficiency of Pear-jujube. This may be due to the application of organic fertilizer being able to improve the total soil porosity and nutrient status[48-49], which is beneficial to plant growth and water use. Although BM has the highest relative water content, its transpiration rate is also large, which reduces its water use efficiency. The SWC of SM and SC are both high, but the SC has the highest water use efficiency, therefore, the application of SC can significantly improve the water utilization of Pear-jujube, and effectively retain water; this can meet the water demand of Pear-jujube during the flowering period.

The application of organic fertilizer improves the vegetative growth and water utilization of Pear-jujube. This is conducive to the accumulation of dry matter during the flowering period, and provides sufficient nutrition for reproductive growth to improve quality. In the organic fertilizer treatment, the fruit setting rate of the application of biogas fertilizer is the highest. This may be because the Pear-jujube itself has less flowering and requires relatively fewer nutrients. However, the Pear-jujube treated by the soybean cake fertilizer has too much flower volume, and in the case that the number of fruits did not reach a significant difference, the fruit setting rate was relatively low. The yield of Pear-jujube after applying organic fertilizer was significantly different from that of CK ($P < 0.05$). Among them, soybean cake fertilizer had the best effect, which is138.52% higher than CK. Not only the yield affects the sales of Pear-jujube, but also the water content and TSS/TA of the fruit directly affect the sales of Pear-jujube. A study by Zang et al.[43] has showns that the application of organic fertilizer can increase the content of soluble solids and Vc in jujube fruit, and significantly reduce the content of titratable acid. After applying organic fertilizer, the water content of the fruit was significantly increased, and the difference between SC and CK treatments was significant ($P<0.05$). TSS, TSS/TA, Vc and total flavonoids content were all higher than those of CK, and titratable acids were all lower than those of CK.Zhu[50] found that the application of organic fertilizer can increase TSS and TSS/TA of citrus. Luo et al.[51] found that organic fertilizer can increase the Vc content of Feicheng peach fruits and reduce the titratable acid content. Organic fertilizer could continuously and slowly release fertilizer efficiency, which was conducive to synchronizing with the physiological needs of crops. It could effectively promote the coordinated and balanced nutrient metabolism of crops to ensure high yield and high quality of crops[52].

Fertilizer application can be based on leaf synergistic efficiency and fruit quality as reference indicators[53]. Based on the above results, applying organic fertilizers can promote vegetative growth, reproductive growth, quality improvement and water utilization of jujube trees. Among them, the effect of SC fertilizer is relatively best, which can provide a theoretical basis and reference for scientific fertilization for the production of organic Pear-jujube in the hilly area of the Loess Plateau. However, further research is needed to determine the appropriate application rate of organic fertilizer, comprehensive effect of organic fertilizer application, and cumulative effect of long-term application of organic fertilizer and its impact on soil fertility.

5 Conclusions

(1) The application of soybean cake fertilizer (SC) had the most significant effect on the improvement of jujube hanging and leaf area; the application of organic fertilizer increased the LAI, canopy light interception density average, Pn and Gs, where the BM processed reach a maximum. The SWC of SC was significantly higher than that of CK and CF.

(2) The application of organic fertilizer can significantly increase the fruit yield of Pear-jujube, and the yield of SC treatment is the highest.

(3) Each organic fertilizer treatment significantly improved the fruit nutritional quality. TSS, TSS/TA, Vc and total flavonoid content of each treatment were higher than those of CK. The SC treatment had the highest water content, and the titratable value of SC acid content is minimal.

In the semi-arid area of northern Shaanxi, water and fertilizer are the main factors affecting Pear-jujube. From the perspective of fertilizer, the application of organic fertilizer in the loess hilly area of northern Shaanxi can effectively promote the growth and development of dwarf densely planted Pear-jujube, increase yield, and significantly improve fruit quality, among which cake fertilizer has the most significant effect. Compared with conventional fertilization, applying 5 kg of soybean cake fertilizer per plant can supply Pear-jujube trees all year round and need complete fertilizer. Therefore, the fertilizer should be applied at one time when the base is applied, and the required dosage should be reasonable, and the organic fertilizer must be fully decomposed.

Data availability

All data generated or analyzed during this study are included in this published article.

Acknowledgments

This research is supported by Natural Science Basic Research Program of Shaanxi (Program No. 2020JQ-1002), the Technology Innovation Center for Land Engineering and Human Settlements, Shaanxi Land Engineering Construction Group Co. Ltd and Xi' an Jiaotong University (Program No. 2021WHZ0093), Innovation Team for Soil and Water Quality Improvement in New Urbanization Construction (Program No.DJTD-2022-4) and the project of Shaanxi Provence Land Engineering Construction Group (Program No.DJNY2021-17 and Program No.DJNY2022-27).

Author contributions

Conceptualization: Shenglan Ye, Tiancheng Liu. Data curation: Shenglan Ye. Formal analysis: Biao Peng. Funding acquisition: Shenglan Ye, Biao Peng, Tiancheng Liu. Investigation: Shenglan Ye, Tiancheng Liu. Methodology: Shenglan Ye, Biao Peng. Project administration: Tiancheng Liu. Resources: Shenglan Ye, Tiancheng Liu. Supervision: Tiancheng Liu. Writing-original draft: Shenglan Ye. Writing-review & editing: Biao Peng, Tiancheng Liu.

Competing interests

The authors declare no competing interests.

References

[1] Su J.J. Research on fruit quality traits of fresh jujube in northern Shaanxi. Northwest, A&F University, (2019)

[2] Ramos C, Agut A, Lidon A L. Nitrate leaching in important crops of the valencian community region, Environ. Poll., 118, 215-223, (2002)

[3] Xiao Y., Li Y.W., Zhu L.Z. The empirical research of farmer' s fertilization behavior and its influencing factors based on the structural equation model, Soils and Fertilizers Sciences in China, (04), 167-174, (2017)

[4] Chen B.L., Sheng J.D., Li J.G., Wang Z. Effects of nitrogen, phosphorus and potassium fertilizers on yield and quality of jujube[J]. Northern Horticulture, (3), 1-3, (2011)

[5] Chen S. M., Lin H. P. Organic Agriculture and Its Source Products, Beijing China Environmental Science Press, (2004)

[6] Ramos C., Agut A., Lidon A.L. Nitrate leatching in import crops of the valencian community region(Spain), Environmental Pollution, 18,215-223, (2002)

[7] Zhu J.H., Li J.L., Li X.L., Zhang F.S. Effects of Several Compound Fertilizers Application on Soil Environmental Quality in Vegetable Protected Areas Agricultural Environmental Protection, (1), 5-8,(2002)

[8] Chai Z.P., Jiang P.A., Wang X.M., Chen B.L., Sun X. The investigation research of fertilization about several main characteristic fruit trees in Xinjiang, Chinese Agricultural Science Bulletin, (11): 231-234, (2008)

[9] Lin B. Fertilizers and Pollution-Free Agriculture Beijing: China Agricultural Press, (2003)

[10] Ning C.C., Wang J.W., Cai K.Z. The Effects of Organic Fertilizers on Soil Fertility and Soil Environmental Quali-

ty: A Review, Ecology and Environmental Sciences, 25, 175-181, (2016)

[11] LI Q., Pei H. D., Ma Z. M., Luo J. J., Lin Y.H. Effects of potassium fertilizer and organic fertilizer on rhizosphere soil enzyme activity, nutrient content and bulb yield of Lily, Soils and Fertilizers Sciences in China, (1), 91-99,(2020)

[12] Li Y.B., Li P., Wang S. H., Xu L.Y., Deng J. J., Jiao J. G. Effects of organic fertilizer application on crop yield and soil properties in rice-wheat rota-tion system: A meta-analysis, Chinese Journal of Applied Ecology, 32, 3231-3239,(2021)

[13] Zhou Y., Li Y. M., Fan M. P., Wang Z. L., Xu Z., Zhang D., Zhao J. X. Effects of different base fertilizer treatments on mountain red earth soil nutrition, enzyme activity, and crop yield, Chinese Journal of Applied & Environmental Biology, 26, 603-611,(2020)

[14] Wenhai Mi,Lianghuan Wu,Philip C. Brookes,Yanling Liu,Xuan Zhang,Xin Yang. Changes in soil organic carbon fractions under integrated management systems in a low-productivity paddy soil given different organic amendments and chemical fertilizers, Soil & Tillage Research. (2016)

[15]Sebastiana M, Juan C, Ruiz P. Chemical and biochemical properties in a silty loam soil under conventional and organic management, Soil&Till. Res., 60, 162-170, (2009)

[16] Xu Z.T., Lu Q., Wu Y.L., Zhang H.J., Xiao R., Deng H.L., Li F.Q., Wang Z.Y. Effects of Partial Replacement of Chemical Fertilizers with Different Organic Fertilizers on Growth Dynamics and Grain Yield of Seed Corn in Hexi Oasis, AGRICULTURAL ENGINEERING, 11, 129-134,(2021)

[17] Hou X.P., An T.T., Zhou Y.N., Li C.H., Zhang X.L. Effect of Adding Organic Fertilizer on Summer Maize Production and Soil Properties, Journal of Maize Sciences, 26, 127-133,(2018)

[18] Yu H.L., Xu B.B., Xu G.Y., Shao W., Gao D.T., Si P. Effects of bio-organic fertilizers on apple seedling growth, physiological characteristics and functional diversity of soil microorganisms, China Agricultural Science Bulletin, 2022, 38, 32-38 (2022)

[19] Li J.L., Liang Y.Y., Liu W.J., Yang Q., Xu W.X., Tang S.R., Wang J.J. Effects of replacing chemical nitrogen fertilizers with organic fertilizers on the growth of rubber seedlings and soil environment, Chinese Journal of Applied Ecology, 32, 1-10,(2021)

[20] Yu S.M., Jin Q. Y., Zhu L.F., Ouyang Y. N., Xu D. H. The Effects of Nutri SmartTM Eco-fertilizer on the Yields and Cash Income in Rice, CHINA RICE, (4)M,72-74,(2007)

[21] Yao X.W., Liang Y.L., Zeng R., Wu X. Eeffects of different organic fertilizers on the yield and quality of pepper, Journal of Northwest A&F University, (10),157-162, (2011)

[22] Wang C.X., Yue X.W., Shi L.T., Li K., Li X.Y., Fang H.D., Pan Z.X. Responses of photosynthesis and yield of jujube to different proportions of organic fertilizer and biochar in Yuanmou dry-hot valley, Chinese Journal of Tropical Crops, 43, 128-136,(2022)

[23]Zaituniguli K., Tuerxun T., Tu Z.D., Aikebaier Y.H. Effects of Different Fertilizers on Photosynthetic Characteristics and Yield of Sorghum Leaves in Arid Areas, Acta Agriculturae Boreali-Sinica, 36, 127-134,(2021)

[24]Yang Y. Study on effect of fertilization on Jujube growth and yield in northern shanxi under different water manegement, Shaanxi: Ms thesis, Institute of Soil and Water Conservation, Chinese Academy of Science and Ministry of Water Resources (2011)

[25]Shu Z., Zhang X.S., Chen J. The simplification of chlorophyll content measurement, Plant Physiol. Comm., 46, 399-402 (2010)

[26]Gao D.T., Han M.Y., Li B.Z. Zhang L.S., Bai R. The characteristic of light distribution in apple tree canopy using WinsCanopy2004a, Acta Agric. Bor-Occid. Sin., 15, 166-170,(2006)

[27]Shi D.L. Dtermination and comparison on the dry matter and water contentin fruit of jujube varieties,Animal Husbandry Feed Sci., 2009, 30, 17-18,(2009)

[28]Zhao S.J., Shi G.A., Dong X.C. Plant physiology experimentguide(1st Edn.), Beijing: China Agricultural Science and Technology Press, 84 (2002)

[29]Gao J.F. Plant physiology experimentguid, Beijing: Higher Education Press (2006)

[30]Li M.F., Xi F., Li Q.L., Zhu H.H. The extraction of flavonoids from red date and its analytical method,Acta Agric. Univ. Jiangxiensis, (6), 1156-1159,(2009)

[31]Huo W.L, Liu B.Y, Cao Y.P. Study on extraction and auti-oxidation of the total flavones fromzizyphus jujube in north of Shaanxi,Food Sci. Technol., (10), 45-47,(2006)

[32]Yang Y.R, Zhao J., Liu M.J. Advances in research on bearing branch of Chinese jujube, Acta Agric. Boreali-

Sin., (S2), 52-57 (2007)

[33] Wang S., Yan C. Comparison of the Accumulation Ability of Photosynthetic Product between Two Types of Bearing Shoots of Southern China Fresh-Eat Jujube, Scientia Silvae Sinicae, (6), 90-97,(2014)

[34]Hui H.X., Xu X., Li Q.R. Effects of NaCl stress on betaine, chlorophyll fluorescence andchloroplast pigment of leaves of Lycium Barbarum L.,Agric. Res. Arid Areas, 22, 109-114,(2004)

[35]Pang G.B., Xu Z.H., Wang H.X., Zhang L.Z., Wang T.Y. Effect of Irrigation with Slight Saline Water on Photosynthesis Characteristics and Yield of Winter Wheat,Journal of Irrigation and Drainage, 37, 35-41,(2018)

[36] Zuo Y.F., He K.N., Chai S.X., Yu G.F., Li Y.H., Lin S., Chen Q., Wang Q.L. Simulation of daily transpiration of Picea crassifolia in growing season based on Penman-Monteith equation, Acta Ecologica Sinica, 41, 3656-3668,(2021)

[37]Xiao Q.,Wang Q.Q., Wu L., Cai A.D., Wang C.J., Zhang W.J., Xu M.G.Fertilization impacts on soil microbial communities and enzyme activities across China' s croplands: a meta-analysis, Plant Nutrition and Fertilizer Science 24, 1598-1609,(2018)

[38]Yuan Q., Feng B., Zhong Y.H., Wang J., Xu X.J., Ma H.B. Partial Replacement of Chemical Fertilizer with Organic Fertilizer:Effects on Eggplant Yield,Quality and Soil Fertility, Chinese Agricultural Science Bulletin, 37, 59-63 (2021)

[39]Xue X.M., Nie P.X., Han X.P., Chen R., Wang Ji.Z. Effects of Bio-organic Fertilizer on Tree, Leaf, Yield and Quality of Red Fuji Apple in Early Fruiting Stage, Tianjin Agricultural Sciences, 24, 51-53,65,(2018)

[40] Yao H.S., Zhang Y.L., Yi X.P., Xu J., Luo Y., Luo H.H., Zhang W.F. Study on differences in comparative canopy structure characteristics and photosynthetic carbon assimilation of field-grown pima cotton (*Gossypium barbadense*) and upland cotton (G. *hirsutum*), Scientia Agricultura Sinica, (2):251-261,(2015)

[41]Wang J.L. The Effect on long-term fertilization to chlorophyll content of winterwheat (Triticum aestivum) and summer corn (Zea mays), Chin. Agric. Sci. Bull., 26, 182-184, (2010)

[42]Li Y.L., He X., Zhang L. Response of leaf photosynthesis and fruit quality to different organic fertilizer ratios Ziziphus jujuba 'Zhongqiu Sucui', Journal of Central South University of Forestry & Technology, 41, 45-51,(2021)

[43] Zang X.P., Ma W.H., Zhou Z.X., Wei C.B., Zhong S., Zuo X.D., Liu Y.X. Effects of Different Organic Fertilizers on Zizyphus mauritiana Lam. Yield, Quality and Soil Fertility, Chinese Journal of Soil Science,(6):1445-1449,(2014)

[44]Su Y., Huang S.L. Effects of Bio-organic Fertilizer on Flue-cured Tobacco Photosynthetic Characteristics and Rhizosphere Soil Microorganism, Journal of Agricultural Science and Technology, 24, 164-171,(2022)

[45]Winter K.,Schramm M.J. Analysis of stomatal and nonstomatal components in the environmental control of CO_2 exchanges in leaves of Welwitschia mirabilis, Plant Physiol., 82, 173-178, (1986)

[46]Chen G.Y., Chen J., Xu D.Q. Thinking about the relationship between net photosynthetic rate and intercellular CO_2 concentration,Plant Physiol. Comm., 46,64-66 (2010)

[47]Chen Q., Xie J.H., Li L.L., Wang L.L., Zhou Y.J., Li J.R., Xie L.H., Wang J.B. Effects of different proportions of organic fertilizer substitutes for chemical fertilizer on growth characteristics and water use efficiency of maize, Agricultural Research in the Arid Areas, 39, 162-170,(2021)

[48]Tang J.W., Xu J.K., Wen Y.C., Tian C.Y., Lin Z.A., Zhao B.Q. Effects of organic fertilizer and inorganic fertilizer on the wheat yields and soil nutrients under long-term fertilization,Plant Nutrition and Fertilizer Science, 25, 1827-1834,(2019)

[49]Angers D., Chantigny M., MacDonald J.D., Rochette P., Côté D. Differential retention of carbon, nitrogen and phosphorus in grassland soil profiles with long-term manure application, Nutr. Cycl. in Agroecol. Syst., 86, 225-229,(2010)

[50]Zhu C.B, Wu S.H., Zhang X.Y., Zhou D.P., Fan J.Q., Jiang Z.F. Effects of organic fertilizer application on soil fertility and leaf nutrients and fruit quality of citrus, Acta Agric. Shanghai, 8, 65-68,(2012)

[51]Luo H., Li M., Hu D.G., Song H.R., Hao Y.J., Zhang L.Z. Effects of organic fertilization on fruit yield and quality of Feicheng peach ,Plant Nutr. Fert. Sci., 18, 955-964 (2012)

[52]Wang Y.P., Liu Y.H., Ruan R.S., Yang Li., Li J.H., Liu C.M., Wan Y.Q. Study on the effect oforganic manure on improving the quality of farm products, Chin. Agric. Sci. Bull., 7, 51-56 (2011)

[53]Zhang C.X., Yang T.Y, Luo J, Wang X.L. Jiang A.L., Ye Z.W. Effects of different fertilizers and application methods on photosynthetic characteristics and quality of Kyoho grape, Southwest China Journal of Agricultural Sciences, (2): 440-443 (2010)

本文曾发表于2022年《Nature Portfolio》第12卷

Novel Fabrication of Hydrophobic/Oleophilic Human Hair Fiber for Efficient Oil/Water Separation Through One-Pot Dip-Coating Synthesis Route

Chenxi Yang[1,2,3,4], Jian Wang [1,2,3,4], Haiou Zhang[1,2,3,4], Tingting Cao[1,2,3,4]
Hang Zhou[1,2,3,4], Jiawei Wang[1,2,3,4], Bo Bai[5,6]

(1.Institute of Land Engineering and Technology, Shaanxi Provincial Land Engineering Construction Group Co., Ltd., Xi'an 710075, China;2.Shaanxi Provincial Land Engineering Construction Group Co., Ltd., Xi'an 710075, China;3.Key Laboratory of Degraded and Unused Land Consolidation Engineering, the Ministry of Natural Resources, Ltd., Xi'an 710075, China;4.Shaanxi Provincial Land Consolidation Engineering Technology Research Center, Xi'an 710075, China;5.Key Laboratory of Subsurface Hydrology and Ecological Effects in Arid Region of the Ministry of Education, Chang'an University, Xi'an 710054, China;6.School of Water and Environment, Chang'an University, Xi'an 710054, China)

【**Abstract**】 Frequent oil spill accidents and industrial wastewater discharge has always been one of the most severe worldwide environmental problems. To cope with this problem, many fluorine-containing and high-cost materials with superwettability have been extensively applied for oil-water separation, which hinders its large-scale application. In this work, a novel human hair fiber (HHF)-polymerized octadecylsiloxane (PODS) fiber was fabricated with a facile one-pot dip-coating synthesis approach, inspired by the self-assembly performance and hydrophobicity of OTS modification. The benefits of prominent hydrophobic/lipophilic behavior lie in the low surface energy, and a rough PODS coating was rationally adhered on the surface of HHF. Driven solely by gravity and capillary force, the HHF-PODS showed excellent oil/water separation efficiency (>99.0%) for a wide range of heavy and light oil/water mixtures. In addition, HHF-PODS demonstrated durability toward different harsh environments like alkaline, acid, and salty solutions.

【**Keywords**】 Oil/water separation; Human hair fiber; Modification

1 Introduction

In recent years, the frequent occurrence of discharge of industrial oily wastewater and oil spills has long been focused on the risk to environmental protection and human health[9, 12, 18, 37]. To solve these oil-leakage problems and protect the limited water resources, many advanced materials[24, 48], equipment[28] and technologies for sustainable and efficient water purification have gained rapidly increasing research interests[6, 39]. With the rapid development of materials and interfacial science over the past decade, absorption and separation strategies become the most promising strategy for the removal of oil contaminations owing to high efficiency, low cost, and environmentally friendly advantages[5, 46]. What's more, absorption and separation strategies can transform oil into a solid or semisolid phase for further utilization[19]. Benefiting from these merits, preparation advanced absorption and separation materials are of great significance to solve the problem of oil spills and discharge of industrial oily wastewater. In general, adsorption and separation materials require intriguing wettability, i. e., hydrophobic/lipophilic or hydrophilic/oleophobic[14]. Specifically, these special wettable materials can be used to absorb or separate oil-water mixture according to its special affinity for liquid. From this point of view, the development of advanced oil/water separation materials with environmental compatibility, high hydrophobicity, low cost, good reusability, and efficiency are significant for the problems of oil spills and discharge of industrial oily wastewater.

Recently, there have been many reports showing the potential to prepare hydrophobic/oleophilic functional materials for oil-water separation[2, 7, 52]. For instance, Oscar et al[1] successfully fabricated the functionalized three-dimensional graphene sponges from graphene oxide solutions with different concentrations and trichloro (1H, 1H, 2H, 2H-perfluorooctyl) silane. These advanced 3D graphene sponges showed excellent absorption efficiency of hexane solution. Eunjoo et al[16] fabricated a membrane using polyamide 6 (PA6) as the substrate, and trichloro (1H, 1H, 2H, 2H-perfluorooctyl) silane (F-POSS) was used to modify the substrate of PA6 with a fluorinated silane (F-silane) to endow a primary hydrophobic coating on the surface of PA6. This obtained membrane showed good separation and purification performance. However, it has been verified that these absorption and separation materials still have limitations for commercialization due to the fluorine in the molecules[31]. Although the fluorine family has been extensively utilized as oil/water separation and absorption materials for decades, environmental and health risks have recently attracted attention due to its persistent and bioaccumulative characteristics[17, 54]. Moreover, these special wettability materials involve complicated chemical/instrumentation reaction processes, high cost of the substrate, toxic chemicals, and environmental incompatibility. Therefore, new oil/water separation materials with facile fabrication processes, good environmental compatibility, good reusability, and fluorine-free properties are significant for the development of the advanced oil/water separation materials.

Human hair fiber (HHF), an important solid waste, mainly covalently linked by 18-methyldocosanoic acid (18-MEA) and its protein components[3, 43]. Notably, the tightly cross-linked outer layers provide good stability and a strategy of protection against mechanical collision, which meet the complex environmental challenge. Moreover, water can easily pass through the endocuticle, the low cross-link the density of endocuticle combined with its hydrophilic state makes it highly water swellable[43]. Relying on these effects, HHF is a kind of natural nanocomposite absorption substrate with good economic performance. Moreover, HHF, as a natural polymer, will not have a negative effect on water when it is used as an absorbent and oil/water separation material[4]. Herein HHF, a cheap solid waste, was rationally selected as a substrate to fabricate the advanced oil/water separation material by attaching hydrophobic coating onto the HHF surface. The core of this innovation lies in the selection of HHF as the substrate, and the special wettability of the rough hydrophobic coating is endowed by a simple dip-coating and self-assembly method, thereby promoting oil/water separation capacity. Moreover, the as-obtained modified HHF could be conducted to selectively absorb several oils or organic solvents while repelling water completely. The HHF-PODS separate oils of different densities by a simple homemade oil/water mixture separation testing system, and all separation efficiencies were higher than 99.0% for different density oils/water mixtures. Overall, the results of this study provide a measure to prepare hydrophobic/oleophilic solid waste for application in the oil/water separation field.

2 Experimental

2.1 Material

Human hair fiber (HHF) was obtained from a local barber shop (Shaanxi, China) (The length is 5 ~ 20 cm, the diameter is 70 ~ 80 μm). Engine oil and castor oil were obtained from a local shop (Shaanxi, China). Ethanol and methylbenzene were supplied by Tianjin Chemical Reagent Factory (Tianjin, China). Octadecyltrichlorosilane (OTS) was furnished by Aladdin (China). All chemicals were used directly.

2.2 Preparation of HHF-PODS

The raw HHF was cleaned in ethanol and deionized water and dried in an oven at 60 ℃. Then, OTS-toluene solution was prepared by adding 0.5 mL OTS into 20 mL toluene solution. The HHFs were immersed in OTS-toluene solution for 15 minutes. Then, the HHF was removed and dried (humidity 50%, temperature 15 ℃). The dried HHF were cured in an oven at ambient temperature to obtain the modified HHF-PODS.

2.3 Characterization

The functional groups were confirmed using a PerkinElmer FTIR System 2000 via KBr pellet. The morphology of HHF and HHF-PODS was carried out with field-emission scanning electron microscopy (Hitachi S-4800, Japan) using an In-Lens detector operated at accelerating voltages of 5 and 20 kV. Elemental analysis was performed using an energy dispersive spectroscopy (EDS) detector equipped with an SEM. Contact angles were observed by a Krüss CCA200 contact angle goniometer at ambient temperature and the volumes of probing liquids in the measurements were approximately 5.0 μL. The values were averages from goniometers at least three different positions for HHF and HHF-PODS.

2.4 Measurements of oil absorption and oil/water separation capacity

Toelucidate the maximal oil absorption capacity of hydrophobic/oleophilic HHF-PODS for different oils, absorption tests of oil absorption were carried out in toluene, petroleum ether, machine oil, castor oil, and the final maximal absorption capacity was calculated by an average value of 3 times.

$$Q=\frac{G_2-(G_0+G_1)}{G_1} \tag{1}$$

where Q (g/g) represents the oil absorbency, defined as grams of oil per gram of the modified HHF-PODS, G_1(g) and G_2(g) are the weights of samples before and after oil absorption, and G_0(g) are the weights of absorption bag.

The HHF-PODS was placed at the bifurcation and completely covered the hole. Then, the oil/water mixture (1 : 1, V/V) was poured into the top of the container, and the oil and water were driven by gravity and capillarity. The oil in the oil-water mixture and the separated oil are V_0 and V_1, respectively. The separation efficiency is defined as Q (%) and the separation efficiency is calculated using the following formula:

$$Q=\frac{V_1}{V_0}\times 100\% \tag{2}$$

The flux of the immiscible oil/water mixtures separation procedure was elucidated by Flux = V/S_t, where V is the volume of the permeated oil (mL), S is the contact area (cm^2) of HHF-PODS, and t is the separation time (min)[50].

3 Results and discussion

3.1 Formation of hydrophobic/lipophilic HHF-PODS

HHF-PODS was prepared by the self-assembly of PODS onto the HHF skeleton through a facile dip-coating method. The proposed synthesis procedure and reaction mechanism are elucidated in the scheme.

On the whole, the formation of a PODS hydrophobic coating consists of a two-step process: hydrolysis and condensation[30, 35]. First, OTS was injected into the toluene solution. Trace amounts of water in toluene and water in air hydrolyze the OTS to generate free–OH groups. Specifically, the water required for the hydrolysis of OTS to form free–OH groups generally come from the solution and air. In consequence, humidity is a significant factor for OTS hydrolysis[15]. McGovern[32] and Lieberman[42] showed that the optimum amount of water required to prepare a densely packed monolayer was between that of the spiked solvent (0.3 mg/100 mL) and that of the anhydrous substrate (approximately 0 mg/100 mL). From this point of view, the 0.03% water content of analytical pure toluene can fully hydrolyze OTS and the influence of humidity on the formation of the PODS self-assembled film is negligible. Second, the OTS after hydrolysis is cross-linked by intermolecular polycondensation of−Si−OH groups on neighboring molecules, forming PODS in the later high-coverage stages[20, 33]. Third, when the hydrolyzed OTS physisorbed on the substrate surface, some PODS molecules organized themselves into crystalline or near crystalline order due to different temperatures, and the formation characteristics of coating depended on the deposition temperature[34]. Relying on these effects, the PODS coating will be highly ordered and tightly packed at low temperatures ($T <$ 16 ℃). Therefore, the lower 15 ℃ was chosen as the reaction temperature in this experiment.

Predicting the structure and composition of the obtained HHF-PODS, this material might be very suitable for the preparation of superior absorbent and oil/water separation materials for oil spillage. First, the HHF substrate has good absorption ability for liquids due to the low cross-link density of endocuticles and high swellability. Moreover, central to the functional groups and structure of proteins and amino acids, which exhibit both inertness and selective reactivity, provide brand-newed insight into the chemical modifications of HHF. Moreover, HHF, as a natural polymer waste, will not have a negative effect on water when it is used as an oil/water absorbent. Second, the HHF substrate after curing at ambient temperature yielded a hydrophobic PODS nanocoating, and the prominent hydrophobic/lipophilic behavior from the low surface energy and rough PODS coating was rationally adhered on the surface of the HHF. Such unique hydrophobicity inevitably provides advantages for oil/water separation. From these points of view, the formed HHF-PODS composite materials, combined with the synergistically high absorptivity of HHF and the separation ability of hydrophobicity/lipophilicity, may shine new light for solving industrial oily wastewater, oil spills and waste utilization.

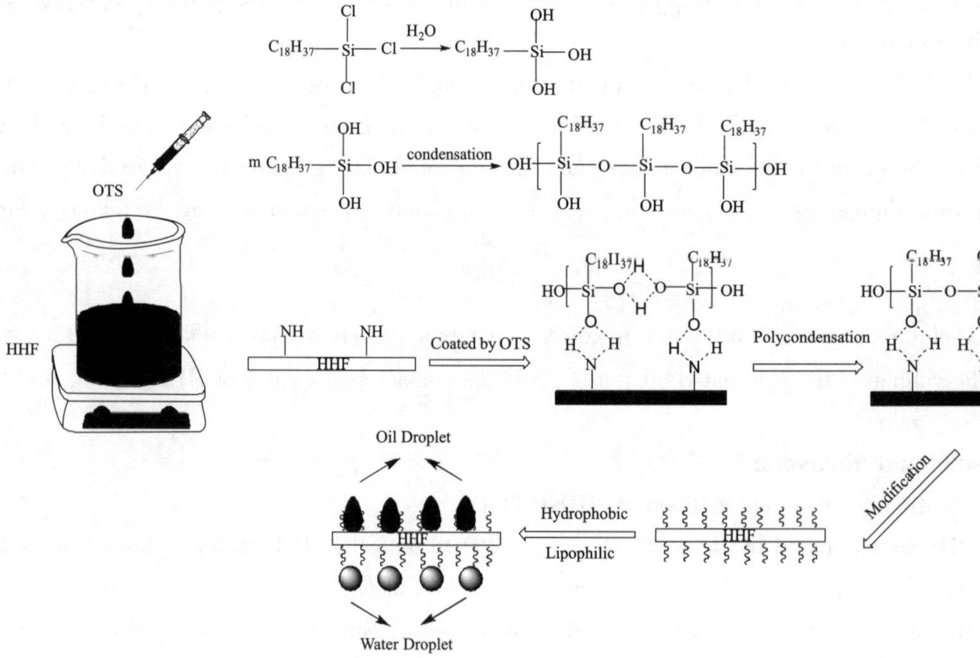

Scheme 1 Schematic illustration of the formation mechanism of the PODS and modified human hair fiber

To validate the successful synthesis of HHF-PODS, the FTIR spectra of HHF and HHF-PODS were registered. The FTIR results are filed in Fig.1. In Fig.1a, the spectrum of HHF exhibited prominent peaks at 3460 cm^{-1}, which were assigned to the stretching vibrations of −OH and −NH. The intramolecular hydrogen bonds of amino acids and the hydrogen bonds formed with hydroxyl groups cause such strong absorption peaks. Peaks at 2924 cm^{-1} were indicative of C−H bending of the methylene group. What's more, two small peaks at 1712 cm^{-1} and 1465 cm^{-1} were attributed to stretching vibrations of C=O and −NH bonds in amino acids[41]. The above peaks were characteristic absorption peaks in HHF. Moreover, compared with the FTIR spectrum of HHF (Fig.1a), HHF-PODS (Fig.1b) had new characteristic absorption peaks after dip-coating modification. The peaks at approximately 2926 and 2879 cm^{-1} may be attributed to the typical stretching vibration of $-CH_2$ in the PODS coating[8]. Moreover, the HHF-PODS has weak characteristic absorption peaks at 1115 cm and 1040 cm^{-1}, which are attributed to the Si−O−Si asymmetric stretching mode confirming the formation of long chain linear polysiloxane[53]. The peaks at 895 cm^{-1} were attributed to the incomplete polycondensation of ODS. Relying on the weak peak at 895 cm^{-1}, it can be verified that almost all ODS molecule cross-linked by Si−O−Si condensation with each other, demonstrating the existence of PODS within the structure of HHF-PODS[49]. Ultimately, it can be concluded that the PODS self-assembled layer successfully adhered to the surface of the HHF.

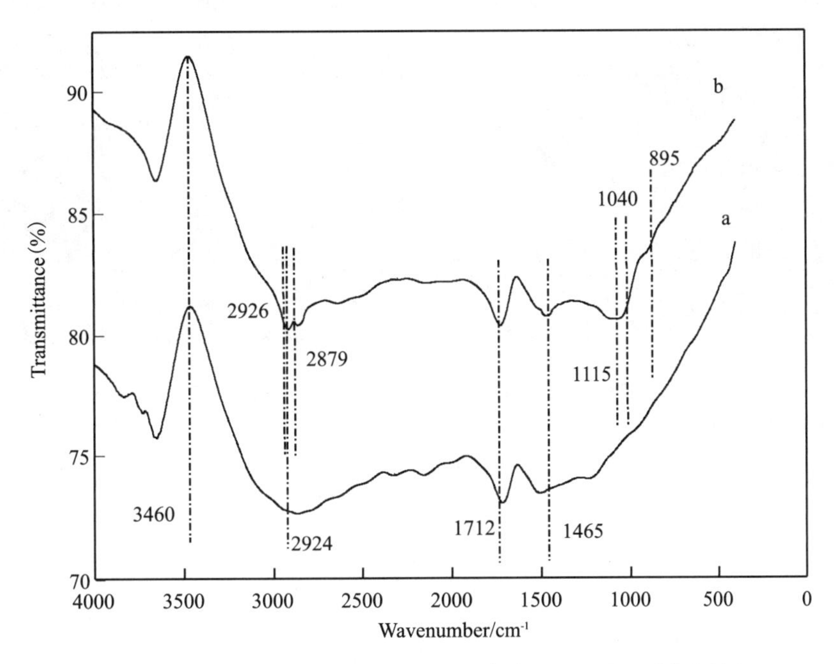

Fig.1 FTIR spectra analysis of HHF (a) and HHF-PODS (b)

The surface characteristics of the obtained materials were further registered in SEM images. Fig.2 shows the surface morphology of the HHF (a) and HHF-PODS (b).

Fig.2a shows an intrinsic HHF surface structure, which offered highly accessible rough structure. Specifically, the hair cells are fused densely and oriented parallel to the axis of the hair fiber. Each hair cell is packed with fine, axially oriented filaments (microfibrils and macrofibrils)[43-45]. The different packing dispositions of macrofibrils make the surface of the hair rough[43]. However, HHF surface is still smooth in large scale. Consequently, this smooth structure endows HHF with good hydrophilicity. Unlike the relatively smooth surface of HHF, HHF-PODS showed a rough surface in the image. Such morphology variation of HHF-PODS provides solid evidence that the PODS were successfully adhered onto the HHF surface.

Fig.2(c, d) shows the corresponding mapping images of the HHF and HHF-PODS, respectively. From Fig.2c, an overlay of element EDX maps verified the clear distribution of the elements. The surface EDS mapping of HHF shows a large amount of O and S elements, which is determined by the composition of HHF. Furthermore, Au on the surface of the HHF was introduced by ion sputtering before SEM characterization. After modification with OTS, the Si signal remained strongly enhanced. Moreover, Cl on the surface of HHF-PODS is that some OTS are not completely hydrolyzed, which leads to the introduction of Cl into the surface of HHF-PODS. Elemental mapping analysis verified that we successfully synthesized the HHF-PODS by dip-coating.

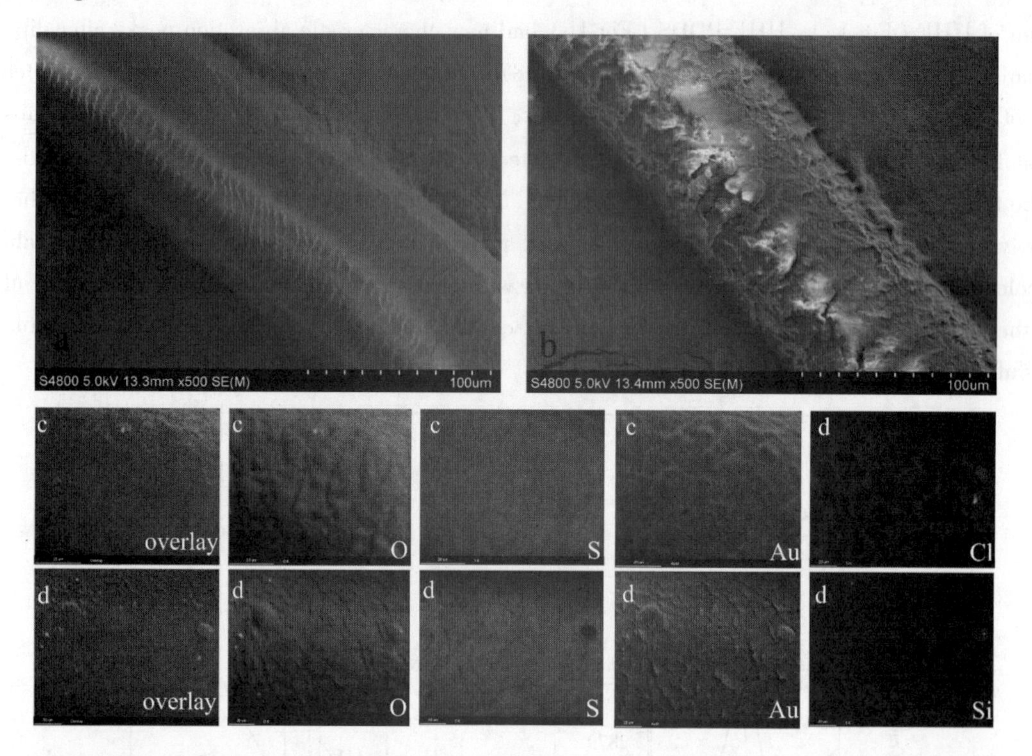

Fig.2 SEM photos of HHF (a) and HHF-PODS (b). Elemental mapping images of HHF (O, S, Au) (c) and HHF-PODS (O, S, Au, Si, Cl) (d)

3.2 Wettability of materials

The wetting behavior of material surfaces is dependent on the chemical composition and the morphological structure of surfaces. In general, the Young formula was used to elucidate the wettability of the smooth surface[11, 27]:

$$\cos\theta=\frac{\sigma_{sv}-\sigma_{sl}}{\sigma_{lv}} \tag{3}$$

where σ_{sv}, $\sigma_{lv,}$ and σ_{sl} are the interfacial tensions of the solid-vapor, liquid-vapor, and solid-liquid phases, respectively. Nevertheless, the general separation material surface is not absolutely smooth. Therefore, the Wenzel formula was used to determine the contact angle w of liquids on a relatively rough surface by Eq. (4):

$$\cos\theta_w = r\cos\theta \tag{4}$$

where r represents the roughness coefficient, which is the ratio of the actual area of a rough surface to the geometrically projected area, and r is usually greater than 1.

When water does not impregnate the rough surface, the formation of air pockets between solid-liquid interface transpires, which further enhances the contact angle of the hierarchical surface, and then the contact angle of such a surface is given by the Cassie-Baxter equation[13]:

$$\cos\theta_w = rf\cos\theta-1+f \tag{5}$$

where f is defined as the fraction of the solid surface contacting the different liquids. The reflections from air bubbles trapped under the water droplet validate the Cassie-Baxter wetting state [36, 53]. (Fig.S2)

Specifically, the hydrophobicity can be improved through the introduction of the microstructures [10, 29]. This phenomenon may be attributed to the introduction of air pockets between the solid surface and the liquid, thus improving the hydrophobicity [38]. Moreover, a suitable rough structure can synergistically turn a relatively smooth oleophilic solid surface into a more oleophilic surface[27]. Hereby, studying the wetting behavior of the sample is very significant for the oil/water separation problem. In this study, the wettability of HHF and HHF-PODS was filed in Fig. 3 by measuring the contact angle on the obtained material surface. As depicted in Fig.3, the WCA and OCA of HHF are 10° and 27°, respectively, which indicate that HHF is an amphiphilic substrate. This result also verifies that HHF has unique properties; that is, low cross-linking density combined with its hydrophilic behavior makes it highly water swellable. Obviously, the WCA and OCA of HHF-PODS are 132° and 8°, respectively, indicating that the self-assembled hydrophobic coating was rationally attached to the surface of HHF. Meanwhile, this wetting behavior verifies the significance of surface roughness and low surface energy coating for the fabrication of hydrophobic/oleophilic materials.

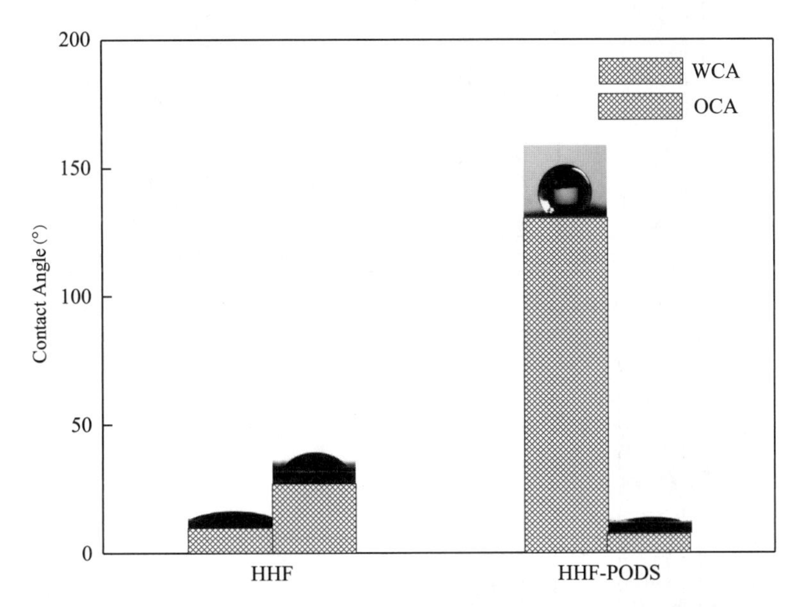

Fig.3 The different contact angles of water and oil on the HHF and HHF-PODS surfaces

3.3 Oil absorption capacity

The absorption capacities of the raw HHF and as-prepared HHF-PODS for different organic solvents and oil/water mixtures are shown in Fig.4. In Fig.4a, the absorption capacities of HHF-PODS for castor oil, motor oil, toluene, and petroleum ether are 11. 5, 6. 1, 3. 1, and 2. 0 g/g, respectively. Nevertheless, the absorption capacity of HHF is only 6.5, 5.2, 2.0, and 1.6 g/g. Obviously, this shows that the modification of HHF improves the oil absorption ability of the HHF substrate. This phenomenon indicates that the as-prepared HHF-PODS improves the oil absorption capacity. In theory, the reason for this phenomenon is that HHF-PODS synergistically reduces the surface energy and increases the surface roughness of the HHF. In consequence, the modification improves the oil affinity. Additionally, HHF and HHF-PODS have the largest absorption capacity for castor oil. The possible reason for this phenomenon was that the high viscosity oil can adhere to the substrate surface more easily. Sticking to this principle, HHF and HHF-PODS showed good absorption properties for castor oil.

In view of practical applications, the ability to absorb oil in oil/water mixtures is an important indicator for evaluating the performance of adsorbents. To study the oil absorption performance of the as-obtained materials in oil-water mixtures, HHF and HHF-PODS were put into the oil-water mixture to recover the oil, and the results are exhibited in Fig.4b. Fig.4b shows that the maximum absorption capacity of HHF-PODS for castor oil, motor oil, toluene, and petroleum ether is 10.9, 5.9, 3.0, and 2.0, respectively. Comparing the results with the absorption capacity in organic solvents, the oil absorption has a slight decrease. As anticipated, the possible reason for the slight decrease in oil absorption capacity is that water occupies the surface absorption site of HHF.

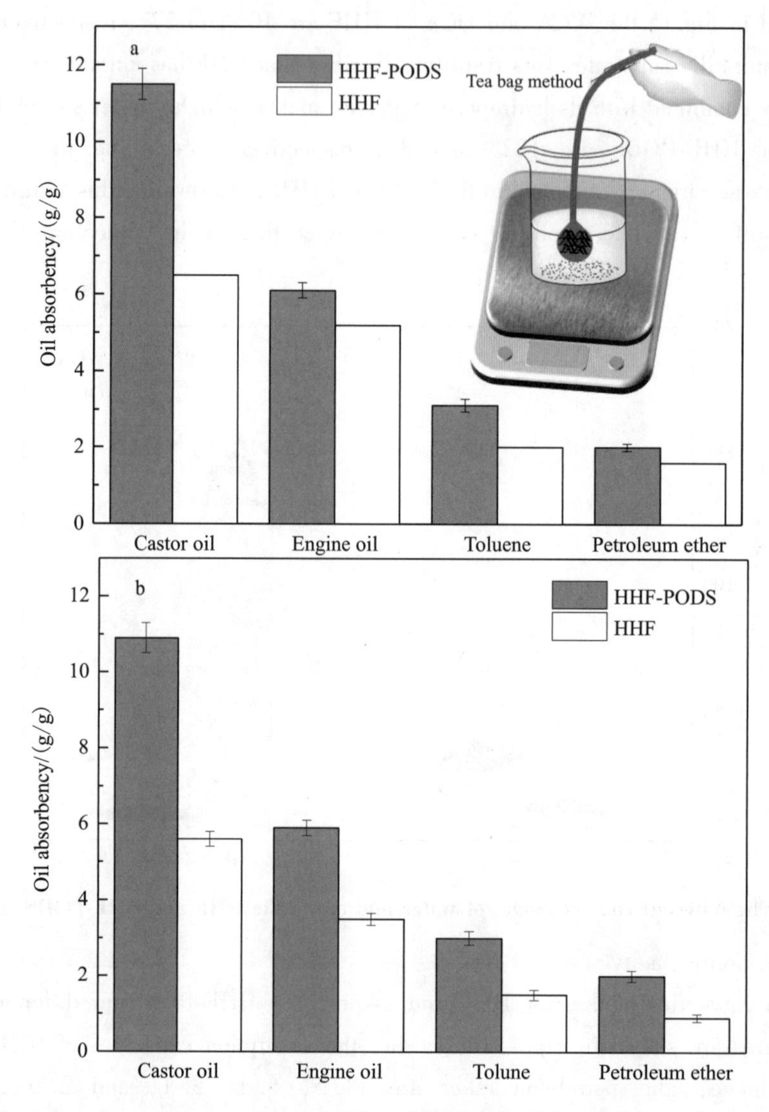

Fig.4 Maximum absorption capacity of HHF and HHF-PODS for different oils in the pure oil system (a) and the oils floating on the water surface system (b)

3.4 Application in water/oil separation

The separation of oil from an oil/water mixture was conducted by putting hydrophobic/oleophilic HHF-PODS into an oil/water mixture. To intuitively verify the oil/water mixture ability of the as-prepared HHF-PODS, a homemade oil/water mixture separation testing system was carried out to assess the oil/water mixture separation performance of the as-obtained HHF-PODS, and the oil-water separation schematic diagram is shown in Fig.5a. The oil/water mixture is poured into a tube. The oil spread on the HHF due to the capillary effect and drain vertically through the HHF into the beaker by gravity. Meanwhile, the water droplets are spherical on the surface of HHF-PODS due to the hydrophobicity of HHF-PODS and then flow horizontally through the tube into another beaker. Simultaneously, the homemade oil-water separation system can be conducted continuously, which verifies that HHF-PODS maintains hydrophobicity/lipophilicity even when wetted by organic solvents [47]. From Fig.5b, the quantitative separation efficiency was also studied, and can be calculated by the ratio between the weight of oil initially added to the oil/water mixture and after separation. All oil/water separation efficiencies were higher than 99.0% through calculation.

As HHF-PODS is compactable in continuous oil water separation experiments, the fluxes were tested after compacting the HHF-PODS until stabilization. The various immiscible oil/water mixtures were used to assess the oil water separation capacity as shown in Fig.5b. The different oil fluxes of HHF-PODS are around 30 $mL/cm^2 \times min$ driven by gravity. Moreover, the HHF-PODS exhibited good separation efficiency without external pressure. It is easy to know that the HHF-PODS can separate oil and water quickly and effectively with high flux. From Table 1, the oil-water separation performances of HHF-PODS is comparable or even superior to the materials in reference. This oil water separation performance owing to the wettability of HHF-PODS.

Table 1 Comparison of the wettability and oil-water separation performances for the HHF-PODS with the sample in references

Samples	Wettability	Oil water separation efficiency (%)	Ref
Waste brick grain (WBG)	Underoil WCA: 138.3°	99.1	[25]
Potato residue coated-mesh (PRCM)	WCA > 150° Underwater with OCA: 152° ± 1.3°	96.5	[22]
Waste from pulp modified by silane	WCA: 156°	99	[51]
Porous waste epoxy resin (PEP)	WCA: 120°	99.99	[40]

3.5 Durability of the HHF-PODS toward different environments

In general, the stability of the obtained material surface wettability toward different environments is valuable for investigating in addition to the oil/water mixture separation efficiency [23, 55]. To further verify the durability of the HHF-PODS toward different harsh environments, such as alkaline, acidic, and salt solutions. The obtained HHF-PODS was immersed into the abovementioned solutions for 4 h, and the microtopography and WCAs of the HHF-PODS are shown in Fig.6 (pH 2: 123°, Fig.6a; 3.5 wt% NaCl: 131°, Fig.6b; pH 10: 121°, Fig.6c, respectively). Interestingly, although the WCA of the as-prepared HHF-PODS under acidic and alkaline conditions was less than that at 3.5 wt% NaCl (neutral pH), it still shows hydrophobicity (greater than 120°). Moreover, the microtopography remained almost constant. This phenomenon could infer that during the harsh environmental process, the rough PODS coating firmly adhered to the HHF surface. Consequently, the HHF-PODS showed excellent resistance to extreme conditions.

Furthermore, the efficiency of separating the light oil (toluene)/water mixture and heavy oil (CCl_4)/water was still satisfactory after 10 cycles. The efficiencies were all higher than 99.0% for both toluene/water mixtures and CCl_4/water mixtures. These separation results are well in line with the SEM analysis and wetting behavior and can help better understand the oil/water separation and durability of HHF-PODS. In consequence, this HHF-PODS can withstand harsh environments, maintain separation capacity for both heavy oil/water and light oil/water mixtures, and perform with good durability.

Karl Fischer method is usually used to evaluate water content in various organic solvents[21]. The water content in the CCl_4 after separation was tested by injecting liquid into a titrator for Karl Fischer titration analysis. The water content of CCl_4 after HHF-PODS separation is ~9600 ppm (i.e., ~0.96 wt% water). The total organic carbon (TOC) of water after HHF-PODS separation light oil-water mixture (toluene/water) were analyzed by Beiya TOC analyzer (HTY-DI1000C)[26]. The result showed that TOC could be maintained around 9800 ppm. The Karl Fischer and TOC results are well in line with the oil/water separation performance of HHF-PODS.

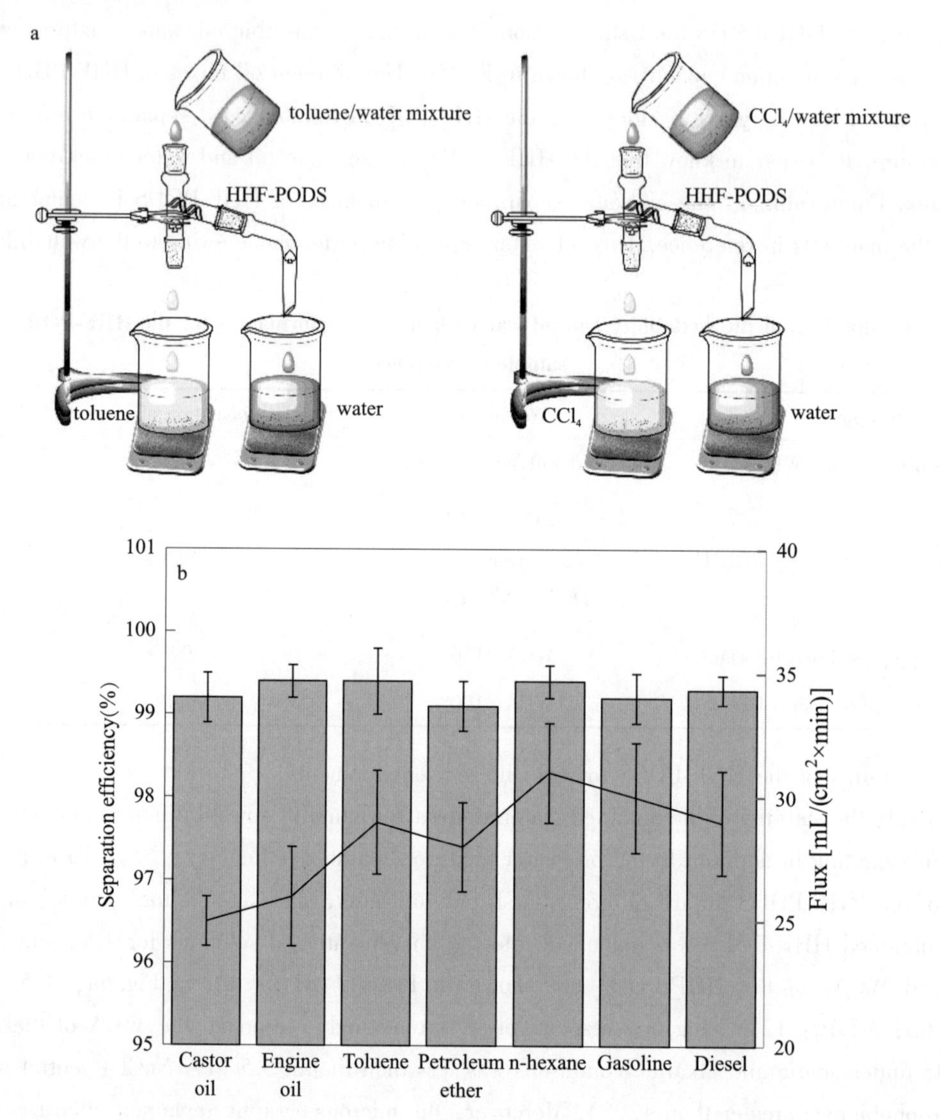

Fig.5 (a) Schematic diagram of the homemade oil/water mixture separation system. (b) The separation efficiency and flux of HHF-PODS for different organic solvents/water mixtures

The abrasion could destroy the surface topography and wettability of HHF-PODS. To evaluate the durability of HHF-PODS, here, the linear abrasion was used to evaluate the durability. Hereby, as showed in Fig.6f, HHF-PODS was placed on sandpaper (2000 Cw), and a 100 g weight was placed and moved 10 cm with a speed of 2 cm/s[50]. Clearly, HHF-PODS could sustain hydrophobicity and good oil-water separation performance after linear abrasion repeatedly, clarifying the excellent adhesion of the rough PODS coating on the HHF. The evidence of the linear abrasion experiment states clearly that the HHF-PODS is mechanically durable.

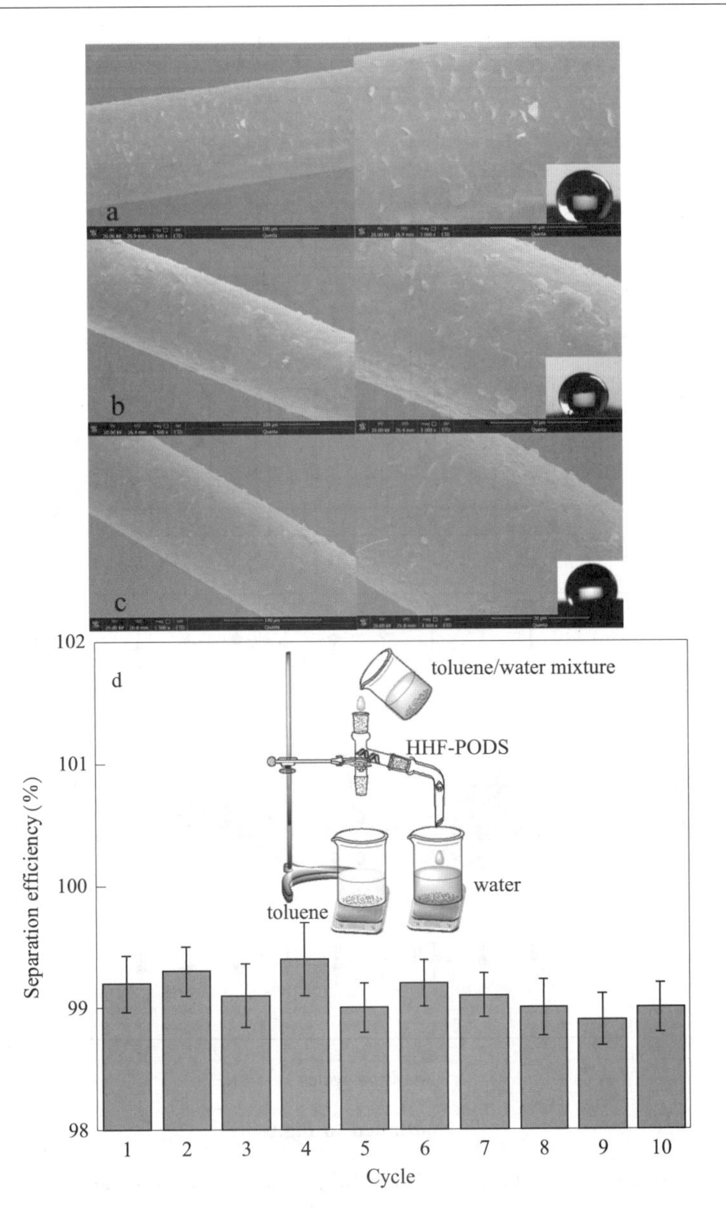

Fig.6 After immersion in pH = 2, 7 (3.5 wt% NaCl solution) and 10 for 4 h, the morphology of HHF-PODS was almost unchanged, and a dense coating of PODS on the HHF-PODS surface was preserved, while intact fibrosis also verified the excellent resistance (a, b, c). Reusability experiments of HHF-PODS for separating oil/water mixtures. The light oil (toluene)/water mixture was separated 10 times (d). The heavy oil (CCl_4)/water mixture was separated 10 times (e), with an oil/water mixture separation efficiency higher than 99.0%. The WCA of HHF-PODS and separation ability after the 10 successive reusability tests with sand paper (f)

4 Conclusion

In conclusion, we synthesized HHF-PODS with a facile dip-coating method, inspired by the self-assembly performance and hydrophobicity of OTS modification. The core of our innovation lies in the fact that the remarkable hydrophobic coating rationally surrounded onto the surface of HHF. Notably, the concept was embodied by a rational integration of waste utilization and special wettability, which exhibited extraordinary hydrophobicity and oil/water separation capacity. Moreover, the HHF-PODS can separate oils of different densities by a simple homemade oil/water mixture separation testing system. The results show that the efficiency was satisfactory and all higher than 99.0% for different density oils/water mixtures. In addition,

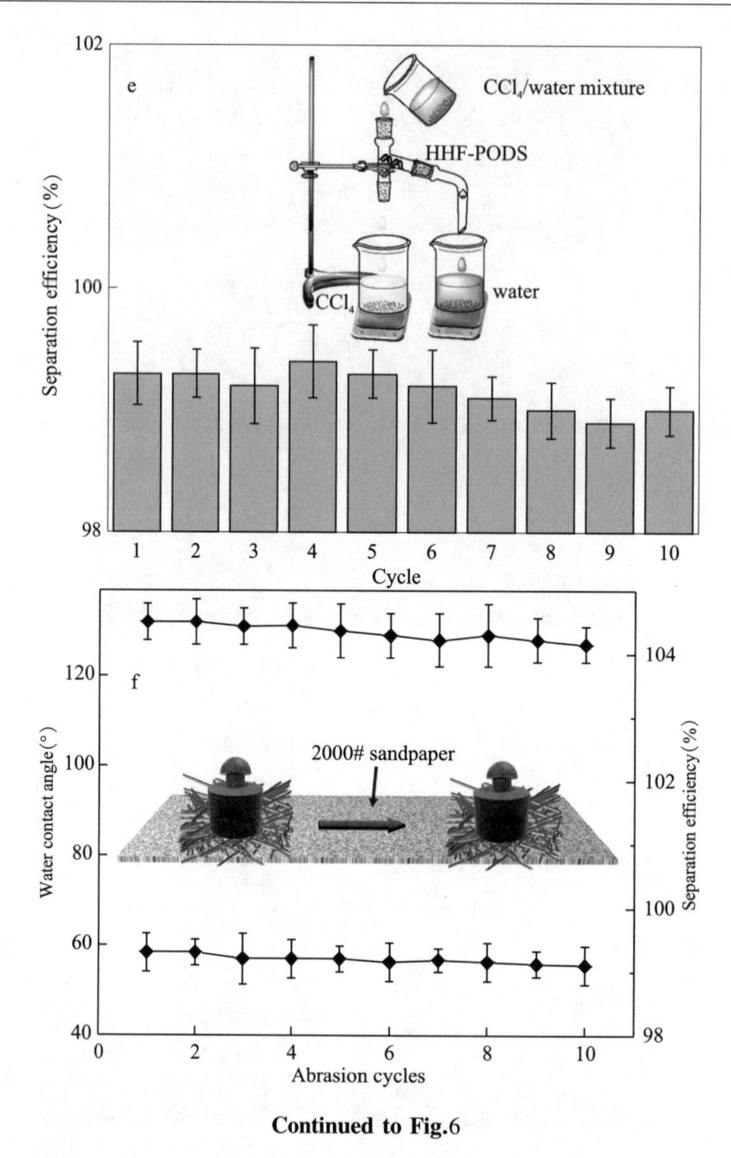

Continued to Fig.6

HHF-PODS exhibited a prominent resistance and durability. Benefiting from the merits of superior resistance and durability, this HHF-PODS can withstand harsh environments, maintain outstanding separation ability for both heavy oil/water and light oil/water mixtures, and perform with good durability. Owing to the merits of waste utilization, good hydrophobicity, durability toward different harsh environments, and reusability, the combination of waste material and special wettability may provide a promising avenue for the replacement of traditional oil/water separation materials.

Conflicts of interest

The authors declare that they have no known competing financial interests or personal relationships that could have appeared to influence the work reported in this paper.

Author contributions

Yang Chenxi and Bai Bo contributed to the experimental design and manuscript writing. Wang Jian, Zhang Haiou, Zhou hang, Wang Jiawei and Cao Tingting contributed to the experimental process and data analysis.

Acknowledgements

The authors are grateful for financial supported by the Fundamental Research Funds for the Central Universities, CHD (300102292504). This work was also supported by the Key Research & Developement

Plan of Shaanxi Province (Program No. 2022NY-082), Shaanxi Provincial Land Engineering Construction Group internal research project (DJNY2021-22).

References

[1]Bagoole O, Rahman M M, Shah S, et al. (2018) Functionalized three-dimensional graphene sponges for highly efficient crude and diesel oil adsorption. Environmental Science and Pollution Research 25:23091-23105.

[2]Bauza M, Turnes Palomino G, Palomino Cabello C. (2021) MIL-100(Fe)-derived carbon sponge as high-performance material for oil/water separation. Separation and Purification Technology 257:117951.

[3]Bhushan B. (2008) Nanoscale characterization of human hair and hair conditioners. Progress in Materials Science 53:585-710.

[4]Buffoli B, Rinaldi F, Labanca M, et al. (2014) The human hair: from anatomy to physiology. International Journal of Dermatology 53:331-341.

[5]Chang J, Shi Y, Wu M, et al. (2018) Solar-assisted fast cleanup of heavy oil spills using a photothermal sponge. Journal of Materials Chemistry A 6:9192-9199.

[6]Chen J, Zhang W, Wan Z, et al. (2019) Oil spills from global tankers: Status review and future governance. Journal of Cleaner Production 227:20-32.

[7]Chen T, Li M, Zhou L, et al. (2020) Bio-Inspired Biomass-Derived Carbon Aerogels with Superior Mechanical Property for Oil-Water Separation. ACS Sustainable Chemistry & Engineering 8:6458-6465.

[8]Chenxi Y, Juan L, Jian W, et al. (2020) Dip-coated rapeseed meal composite as a green carrier for light-induced controlled release of pesticide. New Journal of Chemistry 44:15983-15993.

[9]Colvin K A, Lewis C, Galloway T S. (2020) Current issues confounding the rapid toxicological assessment of oil spills. Chemosphere 245:125585.

[10]Feng L, Zhang Y, Cao Y, et al. (2011) The effect of surface microstructures and surface compositions on the wettabilities of flower petals. Soft Matter 7:2977-2980.

[11]Ge J, Zhao H Y, Zhu H W, et al. (2016) Advanced Sorbents for Oil-Spill Cleanup: Recent Advances and Future Perspectives. Advanced Materials 28:10459-10490.

[12]Gong C, Lao J, Wang B, et al. (2020) Fast and all-weather cleanup of viscous crude-oil spills with Ti3C2TX MXene wrapped sponge. Journal of Materials Chemistry A 8:20162-20167.

[13]Gupta P, Kandasubramanian B. (2017) Directional Fluid Gating by Janus Membranes with Heterogeneous Wetting Properties for Selective Oil-Water Separation. ACS Applied Materials & Interfaces 9:19102-19113.

[14]Hoang A T, Nižeti ć S, Duong X Q, et al. (2021) Advanced super-hydrophobic polymer-based porous absorbents for the treatment of oil-polluted water. Chemosphere 277:130274.

[15]Kelkar S S, Chiavetta D, Wolden C A. (2013) Formation of octadecyltrichlorosilane (OTS) self-assembled monolayers on amorphous alumina. Applied Surface Science 282:291-296.

[16]Koh E, Lee Y T. (2021) Preparation of an omniphobic nanofiber membrane by the self-assembly of hydrophobic nanoparticles for membrane distillation. Separation and Purification Technology 259:118134.

[17]Krafft M P, Riess J G. (2015) Selected physicochemical aspects of poly-and perfluoroalkylated substances relevant to performance, environment and sustainability—Part one. Chemosphere 129:4-19.

[18]Kukkar D, Rani A, Kumar V, et al. (2020) Recent advances in carbon nanotube sponge-based sorption technologies for mitigation of marine oil spills. Journal of Colloid and Interface Science 570:411-422.

[19]Kumar A, Petrič M, Kričej B, et al. (2015) Liquefied-Wood-Based Polyurethane-Nanosilica Hybrid Coatings and Hydrophobization by Self-Assembled Monolayers of Orthotrichlorosilane (OTS). ACS Sustainable Chemistry & Engineering 3:2533-2541.

[20]Kumar A, Ryparovú P, Hosseinpourpia R, et al. (2019) Hydrophobicity and resistance against microorganisms of heat and chemically crosslinked poly(vinyl alcohol) nanofibrous membranes. Chemical Engineering Journal 360:788-796.

[21]Kwon G, Kota A K, Li Y, et al. (2012) On-Demand Separation of Oil-Water Mixtures. Advanced Materials 24:3666-3671.

[22]Li J, Li D, Yang Y,et al. (2016) A prewetting induced underwater superoleophobic or underoil (super) hydrophobic waste potato residue-coated mesh for selective efficient oil/water separation. Green Chemistry 18:541-549.

[23]Li J, Zhao Z, Shen Y,et al. (2017) Fabrication of Attapulgite Coated Membranes for Effective Separation of Oil-in-Water Emulsion in Highly Acidic, Alkaline, and Concentrated Salty Environments. Advanced Materials Interfaces 4:1700364.

[24]Li R, Li J, Rao L,et al. (2021) Inkjet printing of dopamine followed by UV light irradiation to modify mussel-inspired PVDF membrane for efficient oil-water separation. Journal of Membrane Science 619:118790.

[25]Li Z, Wu J, Yue X,et al. (2020) Study on the application of waste bricks in emulsified oil-water separation. Journal of Cleaner Production 251:119609.

[26]Liao Y, Tian M, Wang R. (2017) A high-performance and robust membrane with switchable super-wettability for oil/water separation under ultralow pressure. Journal of Membrane Science 543:123-132.

[27]Liu F, Ma M, Zang D,et al. (2014) Fabrication of superhydrophobic/superoleophilic cotton for application in the field of water/oil separation. Carbohydrate Polymers 103:480-487.

[28]Liu L, Zhao L, Yang X,et al. (2019) Innovative design and study of an oil-water coupling separation magnetic hydrocyclone. Separation and Purification Technology 213:389-400.

[29]Lm S J, Kim D, Kim Y,et al. (2020) Hydrophobicity Evolution on Rough Surfaces. Langmuir 36:689-696.

[30]Lobato N C C, Ferreira A D M, Weidler P G,et al. (2020) Microstructure and chemical stability analysis of magnetic core coated with SILICA and functionalized with silane OTS. Applied Surface Science 505:144565.

[31]Ma W, Zhang M, Liu Z,et al. (2019) Fabrication of highly durable and robust superhydrophobic-superoleophilic nanofibrous membranes based on a fluorine-free system for efficient oil/water separation. Journal of Membrane Science 570-571:303-313.

[32]Mcgovern M E, Kallury K M R, Thompson M. (1994) Role of Solvent on the Silanization of Glass with Octadecyltrichlorosilane. Langmuir 10:3607-3614.

[33]Parikh A N, Schivley M A, Koo E, et al. (1997) n-Alkylsiloxanes: From Single Monolayers to Layered Crystals. The Formation of Crystalline Polymers from the Hydrolysis of n-Octadecyltrichlorosilane. Journal of the American Chemical Society 119:3135-3143.

[34]Poda A, Anderson A, Ashurst W R. (2010) Self-assembled octadecyltrichlorosilane monolayer formation on a highly hydrated silica film. Applied Surface Science 256:6805-6813.

[35]Puri S, Thaokar R M. (2019) Study of dependence of elasticity on the microstructure of microcapsules using electro-deformation technique. Colloids and Surfaces A: Physicochemical and Engineering Aspects 569:179-189.

[36]Qiu S, Li Y, Li G,et al. (2019) Robust Superhydrophobic Sepiolite-Coated Polyurethane Sponge for Highly Efficient and Recyclable Oil Absorption. ACS Sustainable Chemistry & Engineering 7:5560-5567.

[37]Ruan X, Xu T, Chen D,et al. (2020) Superhydrophobic paper with mussel-inspired polydimethylsiloxane-silica nanoparticle coatings for effective oil/water separation. RSC Advances 10:8008-8015.

[38]Saleh T A, Baig N. (2019) Efficient chemical etching procedure for the generation of superhydrophobic surfaces for separation of oil from water. Progress in Organic Coatings 133:27-32.

[39]Singh H, Bhardwaj N, Arya S K,et al. (2020) Environmental impacts of oil spills and their remediation by magnetic nanomaterials. Environmental Nanotechnology, Monitoring & Management 14:100305.

[40]Tian F, Yang Y, Wang X L,et al. (2019) From waste epoxy resins to efficient oil/water separation materials via a microwave assisted pore-forming strategy. Materials Horizons 6:1733-1739.

[41]Wang M, Zhao T, Wang G,et al. (2014) Blend films of human hair and cellulose prepared from an ionic liquid. Textile Research Journal 84:1315-1324.

[42]Wang Y, Lieberman M. (2003) Growth of Ultrasmooth Octadecyltrichlorosilane Self-Assembled Monolayers on SiO_2. Langmuir 19:1159-1167.

[43]Wolfram L J. (2003) Human hair: A unique physicochemical composite. Journal of the American Academy of Dermatology 48:S106-S114.

[44]Wortmann F J, Popescu C, Sendelbach G. (2008) Effects of reduction on the denaturation kinetics of human

hair. Biopolymers 89:600-605.

[45]Wortmann F J, Stapels M, Chandra L. (2009) Humidity-dependent bending recovery and relaxation of human hair. Journal of Applied Polymer Science 113:3336-3344.

[46]Wu M B, Huang S, Liu T Y, et al. (2021) Compressible Carbon Sponges from Delignified Wood for Fast Cleanup and Enhanced Recovery of Crude Oil Spills by Joule Heat and Photothermal Effect. Advanced Functional Materials 31:2006806.

[47]Xue C H, Ji P T, Zhang P, et al. (2013) Fabrication of superhydrophobic and superoleophilic textiles for oil-water separation. Applied Surface Science 284:464-471.

[48]Yan S, Li Y, Xie F, et al. (2020) Environmentally Safe and Porous MS@ TiO2@ PPy Monoliths with Superior Visible-Light Photocatalytic Properties for Rapid Oil-Water Separation and Water Purification. ACS Sustainable Chemistry & Engineering 8:5347-5359.

[49]Yang C, Bai B, He Y, et al. (2018) Novel Fabrication of Solar Light-Heated Sponge through Polypyrrole Modification Method and Their Applications for Fast Cleanup of Viscous Oil Spills. Industrial & Engineering Chemistry Research 57: 4955-4966.

[50]Yang C, Wang J, Li J, et al. (2021) Facile fabrication of durable mesh with reversible photo-responsive wettability for smart oil/water separation. Progress in Organic Coatings 160:106520.

[51]Yu M, Mishra D, Cui Z, et al. (2021) Recycling papermill waste lignin into recyclable and flowerlike composites for effective oil/water separation. Composites Part B: Engineering 216:108884.

[52]Zhan Y, He S, Hu J, et al. (2020) Robust super-hydrophobic/super-oleophilic sandwich-like UIO-66-F4@ rGO composites for efficient and multitasking oil/water separation applications. Journal of Hazardous Materials 388:121752.

[53]Zhang L, Zhou A G, Sun B R, et al. (2021) Functional and versatile superhydrophobic coatings via stoichiometric silanization. Nature Communications 12:982.

[54]Zhao J, Zhu W, Wang X, et al. (2020) Fluorine-Free Waterborne Coating for Environmentally Friendly, Robustly Water-Resistant, and Highly Breathable Fibrous Textiles. ACS Nano 14:1045-1054.

[55]Zhao T, Zhang D, Yu C, et al. (2016) Facile Fabrication of a Polyethylene Mesh for Oil/Water Separation in a Complex Environment. ACS Applied Materials & Interfaces 8:24186-24191.

本文曾发表于 2022 年《Nature Portfolio》第 12 卷

治沟造地背景下延安市生境质量时空演变特征

王　晶[1,2],胡　一[1,3],李　鹏[1],白清俊[2]

(1. 陕西省土地工程建设集团有限责任公司,自然资源部退化及未利用土地整治工程重点实验室,西安 710075;2.西安理工大学水利水电学院,西安 710048;3.西安交通大学人居环境与建筑工程学院,西安 710054)

【摘要】研究治沟造地工程背景下延安市土地利用及生境质量的变化,对黄土高原生态保护和高质量发展具有重要意义。本文基于延安市遥感影像数据,运用 ArcGIS 和 InVEST 模型定量分析了 2010—2018 年延安市土地利用和生境质量的时空变化特征,并结合各县(区)治沟造地工程实施情况,剖析其与区域生境质量的关系。结果表明:(1)延安市主要土地利用类型为草地、林地和耕地,2010—2018 年土地利用类型变化主要为三者互相转化;(2)延安市生境质量处于中等水平,高和较高等级生境质量面积占 50%以上,呈现东南和西南高、北部较低的分布格局,2010—2018 年间,延安市生境质量指数下降了 1.4%,宝塔区中部由于城市扩张指数下降最为显著;(3)延安市生境质量具有一定的空间聚集性,"H-H"集聚区主要位于西南部林区,"L-L"集聚区主要位于西北部的吴起县以及东部的延长、宜川两县,2010—2018 年间生境质量的空间集聚程度有分散的趋势,中部呈现较大的点状退化;(4)治沟造地工程对延安市生境质量有一定的负面影响。通过因地制宜的生态保护模式,延安市在维持生境质量较为稳定的同时,保证了耕地红线和粮食安全,实现了高质量发展和生态保护之间的平衡关系。

【关键词】生境质量;InVEST 模型;土地利用;延安市;治沟造地

Spatial and Temporal Variations of Habitat Quality Under the Background of Gully Control and Land Consolidation in Yan'an, China

Jing Wang[1,2], Yi Hu[1,3], Peng Li[1], Qingjun Bai[2]

(1.Shaanxi Provincial Land Engineering Construction Group Co. Ltd., Key Laboratory of Degraded and Unused Land Consolidation Engineering, Ministry of Natural Resources, Xi'an, 710075, China;2.Institute of Water Resources and Hydro-Electric Engineering, Xi'an University of Technology, Xi'an, 710048, China;3.School of Human Settlements and Civil Engineering, Xi'an Jiaotong University, Xi'an, 710054, China)

【Abstract】 Studying variations in land use and habitat quality in Yan'an City under the background of the gully control and land consolidation project (GCLC) has significant implications for ecological protection and quality development of the Loess Plateau. Therefore, this study quantitatively analyzed the temporal and spatial variations of land use and habitat quality in Yan'an City from 2010 to 2018 using the ArcGIS and InVEST models. The relationship between the status of implementation of the GCLC project in various districts in Yan'an City and regional habitat quality was also analyzed. The results showed that: (1) The land use types in Yan'an City mainly

基金项目:国家重点研发计划课题(2017YFC0504705),陕西省土地工程建设集团内部科研项目(DJNY2021-21)。

consisted of grassland, forest land, and cultivated land. From 2010 to 2018, the land use types remained relatively stable with limited interchangeable transformation among cultivated land, forest land, and grassland. (2) The habitat quality of Yan'an City was at the medium level, and the area of higher habitat quality levels accounted for more than 50% of the total area. The distribution pattern of habitat quality was higher in the southeast and southwest, and lower in the north. During the study period, the habitat quality index of Yan'an City decreased by 1.4%, and this trend was the most evident in central Baota District due to urbanization. (3) The habitat quality of Yan'an City had a certain degree of spatial aggregation. The "H-H" cluster is mainly located in the southwest forest region, while the "L-L" cluster is mainly located in Wuqi County in the northwest and Yanchang and Yichuan counties in the east. From 2010 to 2018, the spatial agglomeration of habitat quality tended to be scattered, and a large point degradation was observed in the central part. (4) The GCLC project had negative impact on the habitat quality of Yan'an City to a certain degree. Yan'an City maintained a relatively stable habitat quality and fulfilled the red line for cultivated land and food security concomitantly through ecological protection modes adapted to local conditions, resulting in a balance between quality development and ecological protection.

【Keywords】Habitat quality; InVEST model; Land use; Yan'an city; Gully control and land consolidation

黄土高原位于黄河流域中部,是中国古代文化的摇篮,同时也是“人-地”关系矛盾最突出的生态环境脆弱区,在黄河流域生态保护和高质量发展战略中具有十分重要的地位。延安市是黄土高原最具代表性的城市,作为21世纪中国规模最大的生态环境建设项目“退耕还林工程”的重点区和示范区,延安市在退耕还林工程期间,生态环境显著提升[1-2],但是随着工程深入推进,产生了耕地面积减少、粮食安全危机等问题[3-4],且随着城市发展和红色旅游的兴起,人口增长和城镇化发展增强了对可利用土地的需求。因此,针对黄土高原丘陵沟壑区特殊地貌,为增加耕地面积,保障粮食安全,延安市启动了集坝系建设、旧坝修复、盐碱地改造、荒沟闲置土地开发利用和生态建设为一体的治沟造地工程[1]。

生境质量是指生态系统提供适宜物种生存和持续发展的能力,是衡量区域生物多样性和生态系统服务价值的重要指标[5]。土地利用类型在空间上的配置不同,其生境质量也有所差异。人类活动能够改变土地资源及其利用方式,土地利用变化对生境斑块之间物质流、能量流的循环过程产生影响,从而影响区域的生境质量、分布格局和演变特征[6-7]。随着“3S”技术的发展,相比较传统实地调查采样评估生境质量的方法,模型由于空间分析功能、数据获取、评估精度、动态性等方面的优势,越来越广泛地应用于生境评价,如InVEST(Integrated Valuation of Ecosystem Services and Tradeoffs)模型[8]、SoLVES(Social Values for Ecosystem Services)模型[9]、ARIES(Artificial Intelligence for Ecosystem Services)模型[10]等。其中,InVEST模型生境质量模块(Habitat Quality)中基于生境胁迫评估生境质量的方法在生境质量动态评估和时空变化分析方面应用更加广泛[6]。国外学者多使用模型评估某种野生动物的生境质量[11-12];国内学者的相关研究集中在流域[13-14]、省[15-16]、城市[17]、区域[18-19]尺度的土地利用和生境质量时空变化,以及人类活动对生境质量的影响[20]。

有学者指出,延安的造地运动会破坏水系和生态系统[21],治沟造地工程引起的地下水位变化还会导致土壤盐渍化问题[22];但也有研究表明,治沟造地项目能显著提升区域植被覆盖率,改善生态环境[23],促进区域生态系统良性演化[24]。目前前人对治沟造地工程生态环境影响的研究基本围绕项目开展,工程实施前后区域生境质量是如何变化的?治沟造地与生境质量变化是否存在相关性?这些问题尚不明确。因此,本文以延安市为研究对象,基于遥感影像数据,运用ArcGIS和InVEST模型定量分析了2010—2018年治沟造地前后延安市土地利用和生境质量的时空演变,剖析治沟造地与区域生境质量变化的关系,为治沟造地生态效益评价、黄土高原生态保护和高质量发展提供科学依据和决策支撑。

1　研究区概况

延安市(35°20′39″~37°53′31″ N,107°38′59″~110°34′46″ E)位于黄土高原腹地,黄河中游,东西

宽 256 km,南北长 236 km,总面积 37037 km²。属半湿润半干旱大陆性季风气候,冬季寒冷干燥,夏季炎热多雨,多年平均降水量为 562.1 mm,7—9 月降雨相对集中,且多暴雨,易诱发滑坡、水土流失等灾害。全年平均气温 9 ℃,平均无霜期 179 d,农作物一年一熟有余,两熟不足。延安市属黄土丘陵沟壑区,地势西北高、东南低,北部以黄土梁峁、沟壑为主,占全市总面积的 72%;南部以黄土塬沟壑为主,占总面积的 19%;石质山地占总面积的 9%。“干流深切,支沟密布”是延安市河流水系分布的主要特征,境内 1000 m 以上沟道 20889 条。区内地形破碎,沟壑纵横,冲沟下切强烈,地形坡度较大,水土流失严重。延安市辖 1 区 12 县,16 个街道办事处、84 个镇、12 个乡,总人口 219 万人。2020 年,延安市生产总值(GDP)1601.48 亿元,占陕西省的 6.1%。

延安市治沟造地工程于 2010 年在宝塔区和子长、延川两县先行试点,2012 年 9 月被列入国家土地整治重大项目给予支持,2013 年 11 月正式批复[25]。项目涉及全市 13 个县(区),共 197 个子项目,集中实施在 2013—2017 年,完成建设规模 36986.22 hm²,新增耕地 7848.72 hm²,建成高标准农田 28233.29 hm²,完成投资 36.64 亿元(见表 1)。同时,2010 年延安市退耕还林工程已进入成果巩固阶段,全市森林覆盖率达 36.6%,区域生态环境得到很大恢复。2013 年,延安市启动新一轮退耕还林工程,计划逐步将 25°以上坡耕地全部退耕还林,实现陡坡全绿化、林草全覆盖。

表 1 延安市各县(区)治沟造地工程实施规模

Table 1 Implementation scale of GCLC project in different district of Yan' an city

序号 Number	县(区) District	项目个数 Projects number	建设规模 Construction scale/hm²	新增耕地 Newly increased cultivated land/hm²	高标准农田建设规模 High standard farmland/hm²	投资金额 Investment amount/ten thousand yuan
1	安塞县	20	2405.95	631.72	1910.32	21229.54
2	宝塔区	34	7734.74	1190.80	6183.23	80892.24
3	富县	16	3070.66	428.47	2460.37	28438.50
4	甘泉县	19	3278.88	529.10	2625.58	30824.60
5	黄龙县	7	820.93	157.38	707.20	7518.75
6	洛川县	13	2191.04	616.59	1782.51	22917.90
7	吴起县	10	2159.59	838.22	1244.78	19880.75
8	延川县	14	3561.59	801.65	2795.02	38754.82
9	延长县	16	3403.09	831.10	2698.84	34277.77
10	志丹县	3	330.95	95.94	134.13	3375.71
11	子长县	29	4919.54	817.47	3214.24	50370.23
12	宜川县	8	1873.35	611.62	1423.52	16289.26
13	黄陵县	8	1235.91	297.66	1053.55	11637.61
	合计	197	36986.22	7847.72	28233.29	366407.68

2 数据来源与研究方法

2.1 数据来源

延安市 2010 年、2018 年两期土地利用现状遥感监测数据(分辨率均为 30 m)来源于中国科学院资源环境科学数据中心(http://www.resdc.cn)。其中,2010 年数据由 Landsat TM 遥感影像数据解译而成,2018 年数据用 Landsat 8 遥感影像数据解译而成。按照刘纪远[26]建立的 LUCC 分类系统,将土地利用类型分为 6 个一级分类和 18 个二级类型。治沟造地工程规模数据来自延安市治沟造地领导小组办公室。

2.2 研究方法

2.2.1 生境质量

采用 InVEST 模型 Habitat Quality 模块对延安市治沟造地前后生境质量进行评估。该模型利用每种地类威胁的相对影响、生境类型对每一种威胁的相对敏感性,生境栅格与威胁源之间的距离,用生境质量作为连接生物多样性与不同土地覆被类型之间的指标。首先计算生境退化度[8]:

$$D_{xj} = \sum_{r=1}^{R} \sum_{y=1}^{Yr} (W_r / \sum_{r=1}^{R} W_r) r_y \, i_{rxy} \, \beta_x \, S_{jr} \tag{1}$$

$$i_{rxy} = 1 - (\frac{d_{xy}}{d_{r\max}}) \text{ (线性衰减)} \tag{2}$$

$$i_{rxy} = \exp (\frac{-2.99 \, d_{xy}}{d_{r\max}}) \text{ (指数衰减)} \tag{3}$$

式中:D_{xj}为生境类型 j 中 x 栅格的生境退化度;R 为威胁因子个数;Y_r 为威胁因子的栅格个数;W_r 为威胁因子 r 的权重;r_y为栅格 y 的威胁值;i_{rxy}为栅格 y 的威胁因子 r_y对栅格 x 的胁迫水平;β_x 为威胁因子对栅格 x 的可达性(本文不考虑法律保护程度因子,因此将β_x 设为 1);S_{jr}为生境类型 j 对威胁因子 r 的敏感度;d_{xy}为栅格 x(生境)与栅格 y(威胁因子)的直线距离;$d_{r\max}$为胁迫因子 r 的最大胁迫距离。生境退化度介于 0~1 之间,值越大退化程度越明显。

生境质量计算公式:

$$Q_{xj} = H_j \left[1 - \left(\frac{D_{xj}^z}{D_{xj}^z + k^z} \right) \right] \tag{4}$$

式中:Q_{xj}为生境类型 j 中 x 栅格的生境质量指数;D_{xj}为生境类型 j 中栅格 x 所受胁迫水平;k 为半饱和系数,一般设置为生境退化程度最大值的 1/2,z 为模型默认参数,值为 2.5;H_j为生境类型 j 的生境适宜度,取值在 0~1 之间;生境质量值在 0~1 之间,值越高生境质量越好。

根据模型参数值设置原则[5]、相关文献[15-18]以及熟悉该区域专家的建议,将受人类活动干扰大的城镇用地、农村居民点、耕地、工矿交通用地及裸地定为威胁因子,生境质量威胁因子影响距离、权重以及各土地利用类型对生境威胁因子的敏感度具体数据见表 2、表 3。

表 2 威胁因子影响距离与权重

Table 2 Distance and weight of threat factors

威胁因子 Threat factor	最大影响距离 Maximum influence distance/km	权重 Weight	空间衰退类型 Spatial decay type
城镇用地 Urban land	5	0.8	指数
农村居民点 Rural settlements	3	0.6	指数
水田 Paddy field	1.5	0.3	线性
旱地 Non-irrigated arable land	1	0.5	线性
工矿交通用地 Industrial, mining and transportation land	3	0.7	指数
裸地 Bare land	1	0.2	线性

表3 不同土地利用类型生境适宜度及其对威胁因子敏感性

Table 3 Habitat suitability of different land use types and their sensitivity to threat factors

地类代码 Land use code	土地利用类型 Land use type	生境适宜度 Habitat suitability	城镇用地 Urban land	农村居民点 Rural settlements	水田 Paddy field	旱地 Non-irrigated arable land	工矿交通用地 Industrial, mining and transportation land	裸地 Bare land
11	水田	0.3	0.8	0.9	0.4	0.3	0.5	0.2
12	旱地	0.2	0.7	0.8	0.3	0.3	0.4	0.2
21	有林地	1.0	0.8	0.7	0.4	0.4	0.7	0.3
22	灌木林地	0.8	0.7	0.6	0.4	0.4	0.6	0.3
23	疏林地	0.7	0.6	0.6	0.3	0.3	0.6	0.2
24	其他林地	0.7	0.6	0.6	0.3	0.3	0.6	0.2
31	高覆盖草地	0.8	0.7	0.7	0.4	0.3	0.5	0.2
32	中覆盖草地	0.7	0.6	0.6	0.3	0.2	0.4	0.1
33	低覆盖草地	0.6	0.6	0.6	0.3	0.2	0.4	0.1
41	河渠	0.9	0.9	0.8	0.6	0.6	0.6	0.3
42	湖泊	0.9	0.9	0.8	0.6	0.6	0.6	0.3
44	水库	0.8	0.8	0.7	0.6	0.5	0.5	0.2
46	滩地	0.6	0.8	0.6	0.5	0.5	0.5	0.2
51	城镇用地	0	0	0	0	0	0	0
52	农村居民点	0	0	0	0	0	0	0
53	其他建设用地	0	0	0	0	0	0	0
63	盐碱地	0	0	0	0	0	0	0
65	裸地	0	0	0	0	0	0	0

2.2.2 空间自相关分析

空间自相关是对空间临近的区域单元属性值相似程度和空间关联模式的定量描述,用于揭示相邻地域间某种地理现象在空间上的分布特征以及变量间的聚集程度,分为全局空间自相关和局部空间自相关。本文用这两种分析方法进一步探讨延安市生境质量在空间上的分布规律。

全局空间自相关分析可以判别区域整体生态质量的空间关联情况,用于描述生境质量在整个区域上有无集聚效应,本文利用 Moran's I 指数来探究其空间相关性,其计算公式为[27]:

$$I = \frac{n}{\sum_{i=1}^{n}\sum_{j=1}^{n} W_{ij}} \times \frac{\sum_{i=1}^{n}\sum_{j=1}^{n} W_{ij}(x_i - \bar{x})(x_j - \bar{x})}{\sum_{i=1}^{n}(x_i - \bar{x})^2} \tag{5}$$

式中:x_i和 x_j 为协调效率的观测值;W_{ij} 为 i 和 j 的权重连接矩阵;Moran's I 的值介于 −1 到 1 之间,大于 0 表示空间单元属性呈正相关,反之表示负相关,值越大表示相关程度越高;趋于 0,则表明空间单元随机分布。

局部空间自相关可以识别局部相邻区域间要素和属性的关联模式,本文采用 Anselin[28] 提出的空间关系局域指标 LISA(Local Indicators of Spatial Associations)来识别延安市每个栅格的生境质量

与其周围栅格生境质量的空间差异程度及其显著性水平,揭示生境质量在空间上同一属性值的集聚和离散特征。其计算公式为:

$$I_i = z_i \sum_{j}^{n} w_{ij} z_j \tag{6}$$

式中:z_i、z_j分别为区域 i、j 上观测值的标准化。

LISA 集聚分析将生境质量的空间格局分为 5 种类型,即 H-H(高高集聚,生境质量高值聚集)、H-L(高低聚集,生境质量高值被周围低值包围)、L-H(低高聚集,生境质量低值被周围高值包围)、L-L(低低集聚,生境质量低值聚集)和 NS(不显著,不存在显著的空间集聚现象)。其中,H-H 和 L-L 为正相关类型,H-L 和 L-H 为负相关类型。

3 结果与分析

3.1 延安市土地利用动态变化特征

由表 4 可以看出,草地、林地和耕地是延安市土地利用面积最大的类型,比例分别约为 47%、26% 和 25%,三者面积占延安市总面积的 99%以上。其中林地主要分布在延安市西南和东南两侧黄陵、黄龙、宜川、富县几个县(区),耕地在子长、宝塔、洛川三县分布较为集中,草地同时分布在全市各个县(区)(见图 1)。二级分类中,面积最大的是草地中的中覆盖草地、耕地中的旱地和林地中的疏林地。

表 4 延安市 2010—2018 年土地利用类型变化

Table 4 Land use changes of Yan' an city from 2010 to 2018

土地利用类型 Land use type		2010		2018		2010—2018	
		面积 Area/km²	比例 Proportion/%	面积 Area/km²	比例 Proportion/%	面积变化量 Area change/hm²	变化比例 Change proportion /%
耕地 Cultivated land	水田	10.22	0.03	10.27	0.03	0.05	0.49
	旱地	9346.14	25.24	9333.85	25.20	-12.29	-0.13
	耕地合计	9356.36	25.27	9344.12	25.23	-12.24	-0.13
林地 Forest	有林地	1699.52	4.59	1698.76	4.59	-0.75	-0.04
	疏林地	5656.56	15.28	5795.78	15.65	139.22	2.46
	灌木林	1740.14	4.70	1738.46	4.69	-1.68	-0.10
	其他林地	694.27	1.87	693.42	1.87	-0.86	-0.12
	林地合计	9790.49	26.44	9926.42	26.80	135.93	1.39
草地 Grassland	高覆盖草地	2255.66	6.09	2109.74	5.70	-145.92	-6.47
	中覆盖草地	11472.85	30.98	11444.95	30.90	-27.90	-0.24
	低覆盖草地	3748.13	10.12	3732.59	10.08	-15.54	-0.41
	草地合计	17476.64	47.20	17287.28	46.68	-189.36	-1.08
水域 Water	河渠	73.04	0.20	75.67	0.20	2.63	3.60
	湖泊	2.31	0.01	1.28	0.00	-1.03	-44.59
	水库	25.18	0.07	33.14	0.09	7.96	31.61
	滩地	58.21	0.16	56.23	0.15	-1.99	-3.42
	水域合计	158.74	0.43	166.31	0.45	7.58	4.78

续表 4

Contined to Table 4

土地利用类型 Land use type		2010 面积 Area/ km^2	2010 比例 Proportion/ %	2018 面积 Area/ km^2	2018 比例 Proportion/ %	2010—2018 面积变化量 Area change/ hm^2	2010—2018 变化比例 Change proportion /%
建设用地 Construction land	城镇用地	46.39	0.13	54.82	0.15	8.43	18.17
	农村居民点	161.25	0.44	165.59	0.45	4.34	2.69
	其他建设用地	25.90	0.07	59.74	0.16	33.84	130.66
	建设用地合计	233.55	0.63	280.15	0.76	46.61	19.96
未利用地 Unuse land	盐碱地	0.05	0.00	0.05	0.00	0.00	0.00
	裸地	14.27	0.04	31.21	0.08	16.94	118.71
	未利用地合计	14.32	0.04	31.26	0.08	16.94	118.30

从 2010—2018 年延安市土地利用类型面积变化来看,主要为耕地、林地、草地三者互相转化。林地、建设用地、未利用地及水域面积均有增加,其中疏林地面积增加最大,为 139.22 km^2;草地、耕地面积有所减少,其中高覆盖草地面积下降最多,为 145.92 km^2。从土地利用变化比例来看,研究时段未利用地面积增加幅度最大,为 118.30%,其次为建设用地 19.96%(见表 4)。研究期间,延安市各类用地面积维持在相对稳定的状态。随着经济的快速增长,建设用地有了较大幅度的增加,但占延安市面积比例不到 1%,土地开发强度控制在合理范围内。

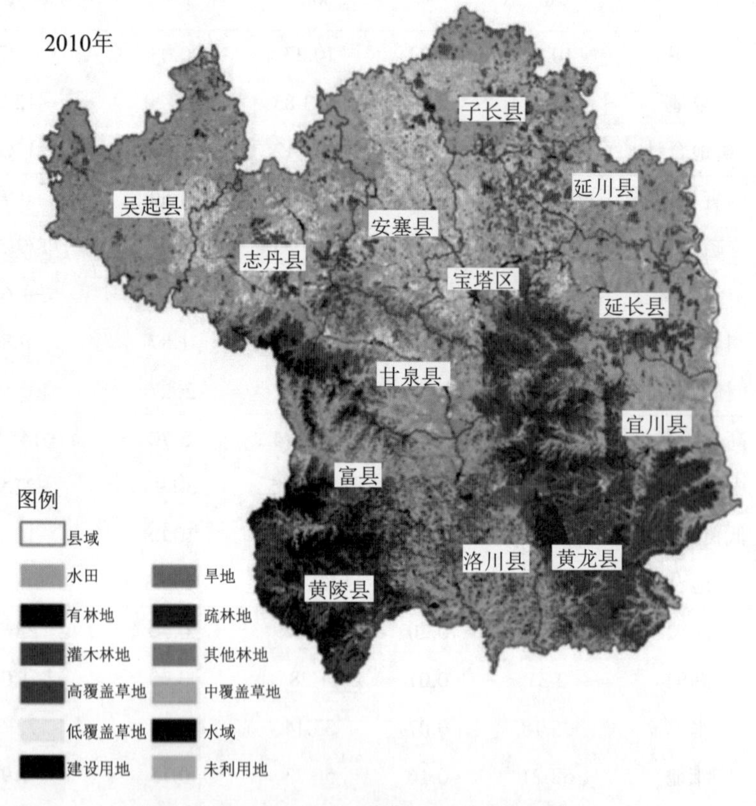

图 1　2010 年和 2018 年延安市土地利用类型

Fig.1　Land use of Yan' an city in 2010 and 2018

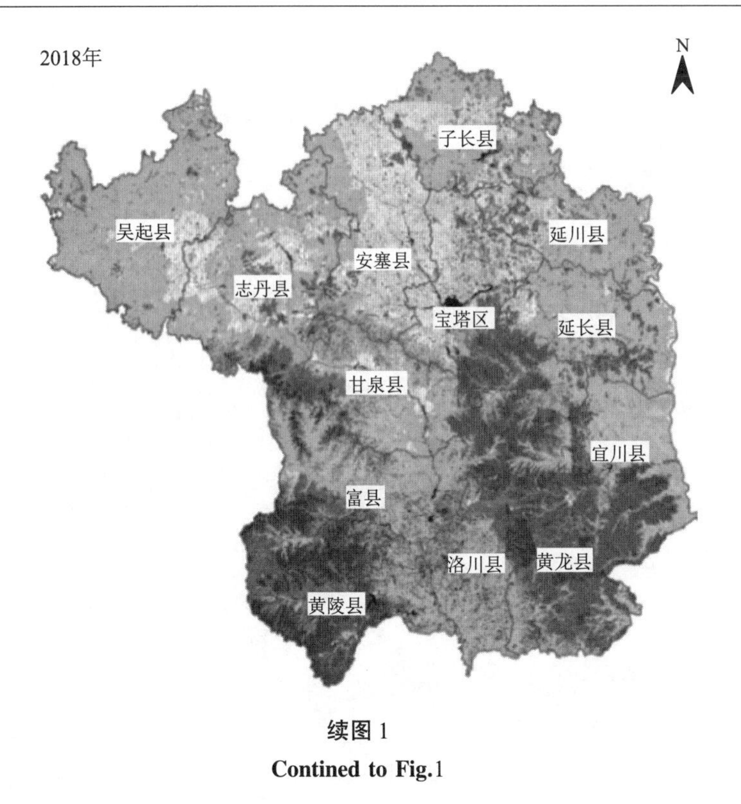

续图 1

Contined to Fig.1

3.2　延安市生境质量时空变化特征

3.2.1　时间变化

基于 InVEST 模型的生境质量模块得到延安市 2010 年和 2018 年的生境质量空间分布图,利用 ArcGIS 的等间法将研究结果划分为低(0,0.2)、较低(0.2,0.4)、中等(0.4,0.6)、较高(0.6,0.8)、高(0.8,1)5 个等级(见图 2),各生境质量等级的土地面积见表 5。

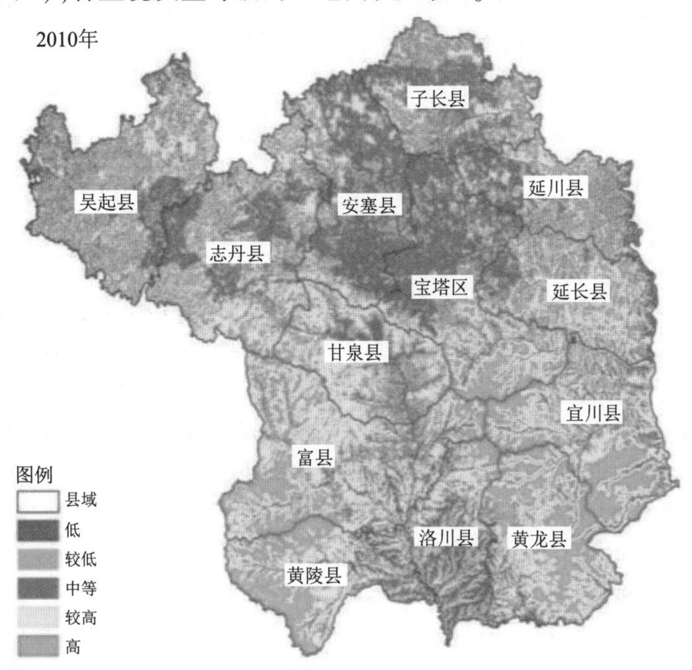

图 2　2010 **年和** 2018 **年延安市生境质量空间分布**

Fig.2　**Spatial distribution of habitat quality of Yan' an city in** 2010 **and** 2018

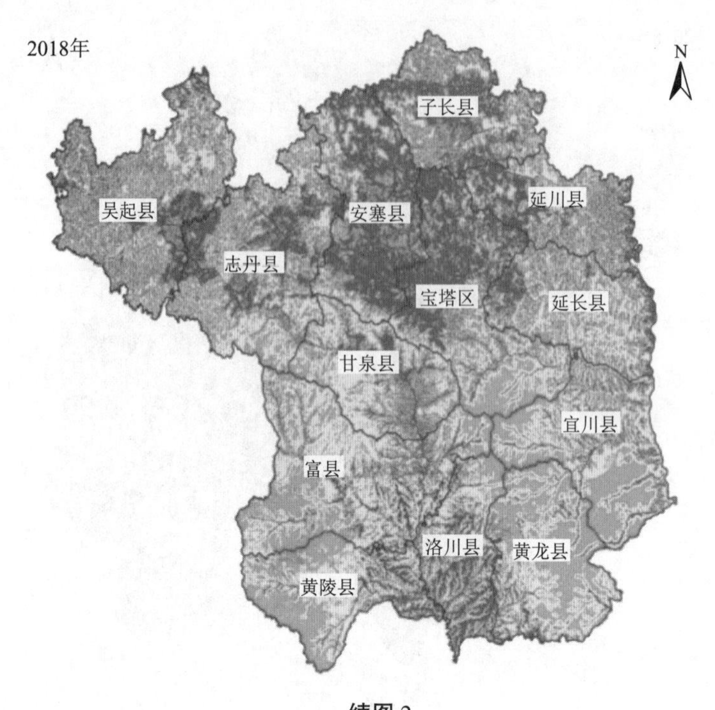

续图 2

Contined to Fig.2

表 5 2010 年和 2018 年延安市不同等级生境质量面积

Table 5 Different grades of habitat quality area of Yan'an city from 2010 and 2018

生境质量等级 Habitat quality grade	2010 年		2018 年	
	面积 Area/km²	比例 Proportion/%	面积 Area/km²	比例 Proportion/%
低 Low(0~0.2)	9604.01	25.93	9648.84	26.05
较低 Relatively low(0.2~0.4)	10.21	0.03	10.31	0.03
中等 Medium(0.4~0.6)	3728.81	10.07	3711.58	10.02
较高 Relatively high(0.6~0.8)	19139.57	51.68	19107.01	51.59
高 High(0.8~1)	4550.40	12.29	4555.27	12.30

从时间尺度上看,2010 年和 2018 年延安市生境质量平均值分别为 0.5914 和 0.5906,生境情况处于中等水平,8 年间生境质量呈现下降的趋势,降幅为 1.4%。生境质量指数的标准差从 0.2511 上升至 0.2521,表明生境质量在空间上差异在扩大。由表 5 可以看出,较高等级生境质量面积最大,占到研究区面积的一半以上,其次是低等级生境质量,占比 26%左右,高等和中等两个级别各占约 12%和 10%,较低等级生境质量占比小于 1%。2010—2018 年间,较高、中等和低等级生境质量面积有较小程度的变化。低等级生境质量面积趋于上升,增长了 0.12%,较高和中等等级的生境质量面积趋于下降,分别下降 0.05%和 0.09%。总体来讲,由于组成延安市主要土地利用类型的草地、林地和耕地的比例较为稳定,全市整体生境质量变化较小。

3.2.2 空间变化

从空间分布上看,延安市生境质量表现出明显的空间分异特征,生境质量分布与土地利用类型分布大体一致,林地分布区域的生境质量较高,耕地和建设用地集中分布区域生境质量较低。为进一步探究延安市不同区域生境质量变化,根据延安市行政区划,将不同等级生境质量面积按照不同县(区)统计(见表 6)。生境质量较高的县(区)为黄陵、黄龙、宜川、富县,这几个县分布在延安市东南和西南部，均为陕西省五大林区，雨水相对充沛，自然条件较好，林草覆盖率高，生境适宜性高，且有众多自然保护区，是全省重要的生态保护和水源涵养区，因此生境质量普遍较高。生境质量较低的县(区)为子长县、安塞县、延川县、吴起县和洛川县。子长、安塞、延川、吴起四县分布于延安市北部，该区域为高原温带半干旱区，降雨量小，自然环境及植被覆盖率相对较差，以中低覆盖度草地及耕地分布较广，为延安市生境质量主要的低值区；而洛川虽然自然条件较好，但海拔较低的河谷耕种条件好，生境质量更容易受到人类干扰，因此生境质量也较差。

表 6 2010—2018 年延安市各县(区)生境质量变化

Table 6 Habitat quality changes of different districts of Yan'an city from 2010 to 2018

序号 Number	县(区) District	生境质量平均值 Mean habitat quality			生境质量等级变化 Changes in habitat quality grades				
		2010 年	2018 年	2010—2018 年	低	较低	中等	较高	高
1	安塞县	0.5099	0.5097	−0.0002	0.0326	0.0000	−0.0812	0.0812	−0.0326
2	宝塔区	0.5600	0.5565	−0.0035	0.4355	0.0000	−0.3390	−0.1142	0.0176
3	富县	0.6738	0.6723	−0.0015	0.2709	−0.0001	0.0128	−0.2869	0.0033
4	甘泉县	0.5901	0.5898	−0.0003	0.0558	0.0001	0.0017	−0.0637	0.0060
5	黄陵县	0.7315	0.7297	−0.0018	0.1815	0.0009	−0.0763	−0.1692	0.0632
6	黄龙县	0.7302	0.7295	−0.0007	0.1072	0.0010	0.0064	−0.1126	−0.0021
7	洛川县	0.5290	0.5277	−0.0013	0.1590	0.0000	0.0204	−0.1689	−0.0105
8	吴起县	0.5210	0.5205	−0.0005	0.0733	0.0000	−0.0188	−0.0532	−0.0013
9	延长县	0.5712	0.5715	0.0003	−0.0170	0.0000	−0.0113	0.0070	0.0213
10	延川县	0.5118	0.5116	−0.0002	0.0315	0.0026	−0.0316	−0.0739	0.0714
11	宜川县	0.6588	0.6589	0.0001	0.0131	0.0000	−0.0025	−0.0501	0.0395
12	志丹县	0.5682	0.5678	−0.0004	0.0690	0.0000	−0.0239	−0.0595	0.0144
13	子长县	0.5002	0.5001	−0.0001	−0.0028	0.0000	0.0156	−0.0188	0.0060

2010—2018 年间延安市生境质量空间分布格局较为类似，除延长和宜川两县外，其他县(区)生境质量都有降低的趋势。生境质量降低最显著区域为宝塔区，与城市扩张区域基本重合，因为城镇化程度加深和旅游业的发展，宝塔区城市中心建设用地逐渐增多，一部分中等质量和较高质量的生境区域转化为低质量，所以生境质量指数降低。黄陵、富县、洛川和黄龙几县也有相当幅度的下降，这一区域主要分布在延安市南部，基本由于较高质量的生境面积减少，低质量的生境面积增加，导致整体生境质量下降。其余县域生境质量相对稳定，变化幅度较小。

3.3 空间自相关分析

利用 ArcGIS 软件计算研究区生境质量的 Moran's I 指数，延安市 2010 年和 2018 年 Moran's I 指数分别为 0.2719($P=0,Z=93$)和 0.2410($P=0,Z=92$)，表明延安市生境质量具有一定空间正相关性，在空间上存在聚集特征，即生境质量较高的区域倾向于与其他生境质量较高的区域毗邻，而生境质量较低的区域倾向于与其他生境质量较低的区域毗邻。2018 年 Moran's I 指数比 2010 年有所降低，表明生境质量的空间集聚程度有分散的趋势。

根据生境质量 LISA 集聚分析(见图 3)，延安市"H-H"集聚区域面积占全市面积的 10.86%，连片分布于延安西南部的林区，该区域生境质量普遍较好，因此出现高值聚集，其他呈不连续的片、点状分布于全市；"L-L"集聚区面积占 5.93%，主要呈密集点状分布于西北部生境质量低值区的吴起县以及东部的延长、延川两县；"H-L"聚集区占比为 5.89%，大部分位于志丹、安塞、子长三县；"L-H"聚集区占比较小，为 4.85%，未出现明显的集中分布。从时间变化来看，2010—2018 年，"H-H"集聚区增加 0.77%，在黄龙县北部聚集范围增加，而在西南部的黄陵县、富县聚集面积减小；"L-L"集聚区增加

0.72%,主要增加在东部延川、延长、宜川几县;“H-L”聚集区减少了0.9%,主要在西南部黄陵县、富县交界处减少;“L-H”聚集区研究期间上升了1.32%,集中出现在宝塔区城镇附近,这一区域因城市建设、资源开发等人类活动对生境的威胁,导致生境呈点状退化。

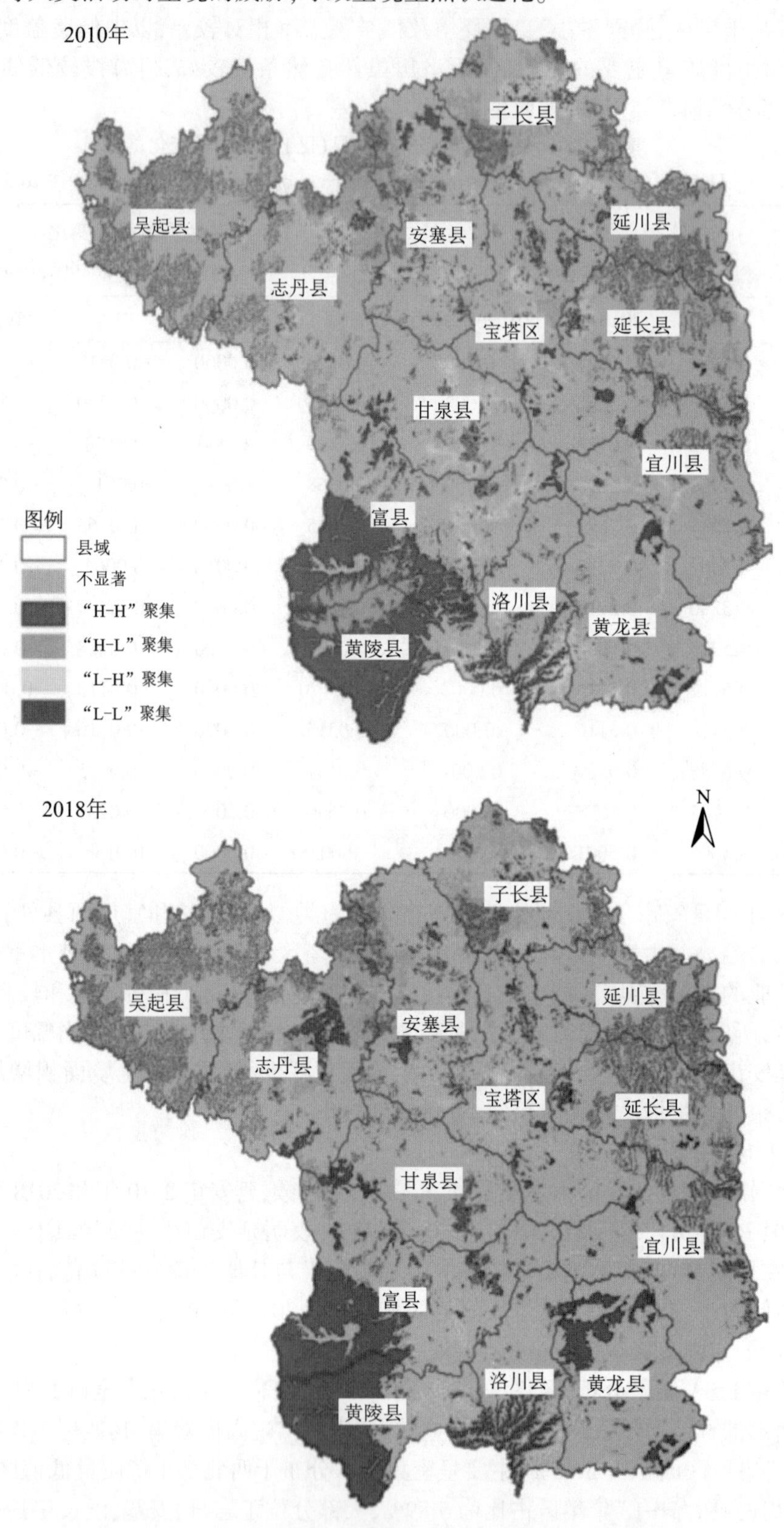

图3 2010年和2018年延安市生境质量LISA集聚图

Fig.3 LISA agglomeration of habitat quality of Yan' an city in 2010 and 2018

3.4 生境质量与治沟造地指标的相关分析

基于皮尔逊相关系数,本文采用 SPSS18.0 软件,对延安市 13 个县(区)治沟造地工程实施规模数据(见表 1)与生境质量变化值(见表 6)进行相关性分析。结果表明(见表 7),2010—2018 年,治沟造地工程对延安市不同县(区)生境质量变化的影响存在差异,其中建设规模、高标准农田建设规模与投资金额与生境质量变化呈一定的负相关关系($P<0.1$),项目个数和新增耕地面积与生境质量变化无显著相关性。相关分析表明治沟造地工程对生境质量有一定的负面影响。随着工程实施人为扰动和土地利用方式的变化,不可避免引起生境质量的下降。但是整体来看,延安市生境质量变化幅度不大。

表 7 治沟造地工程实施指标与生境质量变化相关性

Table 7 Correlation between implementation index of GCLC project and habitat quality change

治沟造地工程实施指标 Implementation index of GCLC project	项目个数 Projects number	建设规模 Construction scale/hm^2	新增耕地 Newly increased cultivated land /hm^2	高标准农田建设规模 High standard farmland/hm^2	投资金额 Investment amount/ ten thousand yuan
相关系数 Correlation coefficient	-0.350	-0.481*	-0.308	-0.518*	-0.483*

注:* 表示在 0.1 水平上显著相关。

从延安市治沟造地规模和生境质量分布中可以看出,延安市由于各县(区)生境特征不同,在生态保护、粮食安全及经济发展中承担着不同的角色,治沟造地工程的实施重点及其对生境质量变化的影响也有差异。宝塔区是延安市市政府所在区域,承担着城市政治和经济发展的任务,快速城镇化使得该区域大量耕地被建设用地占用,需要实施治沟造地集约发展农业,提高经济和人口承载能力,因此宝塔区是延安市实施治沟造地面积最大、新增耕地最多、建设高标准农田最多的区域,工程实施加之城市建设,使宝塔区成为生境质量下降最为显著的区域;东部子长、延川、延长几县地势较为低缓,有较为适宜发展农业的地形基础,但目前未发挥出粮食生产的潜力,因此实施治沟造地面积较大,且以改善土地质量、提高农产品产量为目标实施,新增高标准农田面积较大,单位投资较多,但不会大规模改变土地利用类型,因此生境质量变化较小;延安市南部,尤其是黄陵、黄龙两县是重要的生态功能地区,其主要任务是保护生态,涵养水源,创造生态产品和价值,这一区域以集中连片林地为主,治沟造地实施虽然规模不大,新增耕地少,但景观的破碎化仍引起了生境质量一定程度的下降;北部的吴起、志丹、安塞几县生态环境本底条件不好,实施治沟造地的主要目标在于有效提高农业基础设施水平,全力保障和改善民生,因此生境质量仅小幅下降。

4 讨论

4.1 延安市土地利用和生境质量变化

与退耕还林时期相比[29],8 年来,延安市土地利用类型维持在较稳定的动态变化中,生境质量虽然有下降的趋势,但幅度不大。2010 年起,退耕还林工程进入成果巩固阶段,实施力度小于上一时期,同时治沟造地工程开始大规模开展,使得这一时期的农业用地和生态用地做到了较好的协调,在一定程度上保障了延安各种用地资源的平衡。这与侯孟阳等[30]研究结果中预测的延安市耕地与草地林地达到平衡状态的结果一致。这一时期,延安市 GDP 从 2010 年的 885.42 亿元增加到 2018 年的 1558.91 亿元,经济社会快速发展,尤其是随着红色旅游的兴起,第三产业产值增长了 176.8%[31],而建设用地扩张在城市发展中往往必不可免,相比较黄土高原其他区域[32-33],延安市的城市扩张保持在可控范围内,没有出现大面积的生境质量退化。

土地整治等人类活动通过对土地资源及其利用方式进行再组织和再分配,作用于区域生态系统,进而改变区域生境质量及其分布格局和功能[6]。虽然治沟造地工程实施过程中对土地的扰动及生境破碎化对项目区的生境质量产生了负面影响,但通过田块整合、水资源调控、改善农田基础设施和农业生产条件、合理调整种植结构等措施,科学优化土地利用格局和结构,减弱了这种负面影响。

钟丽娜等[6]、单薇等[34]对土地整治后的生境质量研究也表明,通过对土壤的改良和景观格局的优化,最终改善了项目区整体的生境质量,在整治完成3~5年后,生境质量开始逐渐好转。

4.2　治沟造地背景下延安市生态与经济的协同发展

黄土高原在过去相当长的时间里是中国生态环境最为脆弱的区域之一,当时区域发展的重点是遏制水土流失、改善生态环境。通过十多年的生态治理,生态环境显著提高[35],但耕地面积以平均每年10%的速度持续减少,同时食物生产能力下降了16%[36],粮食生产供应不足极有可能威胁到退耕还林的成果。随着社会经济的发展和人口的增长,人地矛盾不断增大。因此,区域发展不能单单追求生境质量的提高,必须同时权衡生态保护、粮食安全和经济发展多重目标。

经济效益是土地生态质量的重要组成,经济增长能够推动产业发展及升级,进而提高地区的生态效率,有利于保护地区的生态环境[37]。如果单纯在生态保护建设中做出经济发展的牺牲,也会引发产业结构不合理、生态效率低下。因此,延安市根据不同区域的生境特征与差异,以及角色、功能和定位,有针对性地进行开发与保护,通过对土地利用格局的改变,合理规划土地利用,防止经济发展带来的环境恶化,在保证生境质量较为稳定的同时,有效地缓解了区域内耕地面积大幅下降、食物生产能力不足的现状,保证了耕地红线和粮食安全,平衡了生态保护与农民生计的冲突,实现了生态安全与粮食安全、高质量发展和生态保护之间的平衡关系。未来应该进一步因地制宜制定相应的优化策略,通过绿色产业体系发展经济,加强生态红线区域的环境保护,防止经济发展引起的环境恶化,促进地区生态、经济、社会的最优权衡。

4.3　展望

本研究基于InVEST模型对延安市生境质量时空演变进行分析,既能得出定量分析结果,也能将生境质量结果可视化分析,同时将延安市主要的土地利用类型进行了二级共18个分类,详细地计算了生境质量指数,有利于得出更加细化的生境质量变化,为黄土高原生态保护和高质量发展制定科学政策提供依据。研究的不足及未来需要改进的方面主要有:(1)仅分析评价了治沟造地工程与工程背景下延安市各县(区)生境质量变化之间的相关关系,缺少治沟造地区域的空间表达,下一步,在获取治沟造地项目空间信息的基础上,利用空间分析手段得出治沟造地与生境质量的关系,为科学评价治沟造地生态效益提供更有效的理论依据;(2)未考虑土地质量提升对区域生境质量的影响。下一步,进一步结合实际调研数据,综合考虑耕地质量,深入分析治沟造地对项目区生境质量及其他生态系统服务功能如水源涵养、土壤保持等的影响,为这一区域其他生态建设项目的实施提供借鉴和参考。

5　结论

(1)延安市的土地利用类型主要为草地、林地和耕地,三者面积占延安市总面积的99%以上,二级分类中面积最大的是草地中的中覆盖草地、耕地中的旱地和林地中的疏林地。2010—2018年间延安市土地利用变化处于较稳定状态,主要为耕地、林地、草地三者互相转化。

(2)延安市2010年和2018年生境质量平均值分别为0.5914和0.5906,处于中等水平,较高等级生境质量面积占50%以上,呈现东南和西南高、北部较低的分布格局;2010—2018年间,延安市生境质量下降了1.4%,无剧烈变化,宝塔区生境质量由于城市扩张降低最为显著。

(3)延安市生境质量具有一定的空间聚集性,"H-H"集聚区主要位于西南部林区,"L-L"集聚区主要位于西北部的吴起县以及东部的延长、宜川两县;2010—2018年间生境质量的空间集聚程度有分散的趋势,中部呈现较大的点状退化。

(4)治沟造地工程对延安市生境质量有一定的负面影响,由于各县(区)的生境特征不同,治沟造地工程的实施重点及其对生境质量变化的影响有差异。

参考文献

[1]贺春雄. 延安治沟造地工程的现状、特点及作用. 地球环境学报, 2015, 6(4): 255-260.

[2]韩磊, 火红, 刘钊, 赵永华, 朱会利, 陈芮, 赵子林. 基于地形梯度的黄河流域中段植被覆盖时空分异特征: 以延安市为例. 应用生态学报, 2021, 32(5): 1581-1592.

[3]何立恒, 贾子瑞, 王志杰. 延安市土地利用/土地覆被格局变化特征. 南京林业大学学报: 自然科学版, 2015, 39(6): 173-176.

[4]周卫健, 安芷生. 实施与"退耕还林还草"并重的"治沟造地"重大方针的建议. 中国科学报, 2014-09-19.

[5]Song Y N, Wang M, Sun X F, Fan Z M. Quantitative assessment of the habitat quality dynamics in Yellow River Basin, China. Environmental Monitoring and Assessment, 2021, 193(9): 614.

[6]钟莉娜, 王军. 基于InVEST模型评估土地整治对生境质量的影响. 农业工程学报, 2017, 33(1): 250-255.

[7]Haddad N M, Brudvig L A, Clobert J, Davies K F, Gonzalez A, Holt R D, Lovejoy T E, Sexton J O, Austin M P, Collins C D, Cook W M, Damschen E I, Ewers R M, Foster B L, Jenkins C N, King A J, Laurance W F, Levey D J, Margules C R, Melbourne B A, Nicholls A O, Orrock J L, Song D X, Townshend J R. Habitat fragmentation and its lasting impact on Earth's ecosystems. Science Advances, 2015, 1(2): e1500052.

[8]Sharp R, Tallis H T, Ricketts T, Guerry A D, Wood S A, Chapin-Kramer R, Nelson E, Ennaanay D, Wolny S, Olwero N, Vigerstol K, Pennington D, Mendoza G, Aukema J, Foster J, Forrest J, Cameron D, Arkema K, Lonsdorf E, Kennedy C, Verutes G, Kim C K, Guannel G, Papenfus M, Toft J, Marsik M, Bernhardt J, Griffin R, Gowinski K, Chaumont N, Perelman A, Lacayo M, Mandle L, Hamel P, Vogl A L, Rogers L, Bierbower W. InVEST 3.2.0 User's Guide. The Natural Capital Project, Stanford University, University of Minnesota, The Nature Conservancy, and World Wildlife Fund, 2015.

[9]Sherrouse B C, Semmens D J. Social Values for Ecosystem Services, Version 3.0 (SolVES 3.0): Documentation and User Manual. Reston, VA: U. S. Geological Survey, 2015.

[10]Bagstad K J, Villa F, Johnson G W, Voigt B. ARIES-Artificial Intelligence for Ecosystem Services: A Guide to Models and Data, Version 1.0 Beta. Bilbao: The ARIES Consortium, 2011.

[11]Goertz J W. The influence of habitat quality upon density of cotton rat populations. Ecological Monographs, 1964, 34(4): 359-381.

[12]Smith J A M, Reitsma L R, Marra P P. Moisture as a determinant of habitat quality for a nonbreeding Neotropical migratory songbird. Ecology, 2010, 91(10): 2874-2882.

[13]王军, 严有龙, 王金满, 应凌霄, 唐倩. 闽江流域生境质量时空演变特征与预测研究. 生态学报, 2021, 41(14): 5837-5848.

[14]李子, 张艳芳. 基于InVEST模型的渭河流域干支流生态系统服务时空演变特征分析. 水土保持学报, 2021, 35(4): 178-185.

[15]李胜鹏, 柳建玲, 林津, 范胜龙. 基于1980—2018年土地利用变化的福建省生境质量时空演变. 应用生态学报, 2020, 31(12): 4080-4090.

[16]赵晓冏, 王建, 苏军德, 孙巍, 晋王强. 基于In VEST模型和莫兰指数的甘肃省生境质量与退化度评估. 农业工程学报, 2020, 36(18): 301-308.

[17]冯琰玮, 甄江红, 马晨阳. 呼和浩特市生境质量对城市用地扩展的时空响应. 干旱区地理, 2020, 43(4): 1014-1022.

[18]黄木易, 岳文泽, 冯少茹, 张嘉晖. 基于In VEST模型的皖西大别山区生境质量时空演化及景观格局分析. 生态学报, 2020, 40(9): 2895-2906.

[19]石小伟, 冯广京, 苏培添, 何改丽, 邹逸江, 王小锋. 大都市郊区土地利用时空演变特征与生境质量评价. 农业工程学报, 2021, 37(4): 275-284.

[20]周婷, 陈万旭, 李江风, 梁加乐. 1995—2015年神农架林区人类活动与生境质量的空间关系研究. 生态学报, 2021, 41(15): 6134-6145.

[21]Li P Y, Qian H, Wu J H. Environment: accelerate research on land creation. Nature, 2014, 510(7503): 29-31.

[22]Jin Z, Guo L, Wang Y Q, Yu Y L, Lin H, Chen Y P, Chu G C, Zhang J, Zhang N P. Valley reshaping and dam-

ming induce water table rise and soil salinization on the Chinese Loess Plateau. Geoderma, 2019, 339: 115-125.

[23]He M N, Wang Y Q, Tong Y P, Zhao Y L, Qiang X K, Song Y G, Wang L, Song Y, Wang G D, He C X. Evaluation of the environmental effects of intensive land consolidation: a field-based case study of the Chinese Loess Plateau. Land Use Policy, 2020, 94: 104523.

[24]李裕瑞, 李怡, 范朋灿, 刘彦随. 黄土丘陵沟壑区沟道土地整治对乡村人地系统的影响. 农业工程学报, 2019, 35(5): 241-250.

[25]张信宝, 金钊. 延安治沟造地是黄土高原淤地坝建设的继承与发展. 地球环境学报, 2015, 6(4): 261-264.

[26]刘纪远. 中国资源环境遥感宏观调查与动态研究. 北京: 中国科学技术出版社, 1996.

[27]Moran P A P. Notes on continuous stochastic phenomena. Biometrika, 1950, 37(1/2): 17-23.

[28]Anselin L. Local indicators of spatial association-LISA. Geographical Analysis, 1995, 27: 93-115.

[29]韩磊, 朱会利, 刘钊. 延安市退耕还林前后土地利用动态变化分析. 西北师范大学学报: 自然科学版, 2017, 53(5): 101-108.

[30]侯孟阳, 姚顺波, 邓元杰, 丁振民, 鲁亚楠, 郑雪, 李雅男. 格网尺度下延安市生态服务价值时空演变格局与分异特征:基于退耕还林工程的实施背景. 自然资源学报, 2019, 34(3): 539-552.

[31]陕西省统计局, 国家统计局陕西调查总队. 陕西统计年鉴2010—2018. 北京: 中国统计出版社, 2010-2018.

[32]周亮, 唐建军, 刘兴科, 党雪薇, 慕号伟. 黄土高原人口密集区城镇扩张对生境质量的影响:以兰州、西安-咸阳及太原为例. 应用生态学报, 2021, 32(1): 261-270.

[33]刘春芳, 王川. 基于土地利用变化的黄土丘陵区生境质量时空演变特征:以榆中县为例. 生态学报, 2018, 38(20): 7300-7311.

[34]单薇, 金晓斌, 孟宪素, 杨晓艳, 徐志刚, 顾铮鸣, 周寅康. 基于多源遥感数据的土地整治生态环境质量动态监测. 农业工程学报, 2019, 35(1): 234-242.

[35]李蕴琪, 韩磊, 朱会利, 赵永华, 刘钊, 陈芮. 基于土地利用的延安市退耕还林前后生态服务价值变化. 西北林学院学报, 2020, 35(1): 203-211.

[36]谢怡凡, 姚顺波, 邓元杰, 贾磊, 李园园, 高晴. 延安市退耕还林(草)工程对生境质量时空格局的影响. 中国生态农业学报(中英文), 2020, 28(4): 575-586.

[37]任保平, 杜宇翔. 黄河流域经济增长-产业发展-生态环境的耦合协同关系. 中国人口·资源与环境, 2021, 31(2): 119-129.

本文曾发表于2022年《生态学报》第42卷第23期

砒砂岩与沙复配土壤组成变化及玉米产量可持续性分析

张海欧[1,2,3,4]，曹婷婷[2,3,4]，杨晨曦[2,3]

（1.陕西省土地工程建设集团有限责任公司，西安 710075；2.陕西地建土地工程技术研究院有限责任公司，西安 710021；3.自然资源部退化及未利用土地整治工程重点实验室，西安 710021；4.陕西省土地整治工程技术研究中心，西安 710021）

【摘要】砒砂岩与沙复配成土技术在毛乌素沙地土地整治中已广泛应用，然而，探索适合当地主要粮食作物高产稳产的砒砂岩与风沙土的适宜混合比例，提高复配土壤的农业适应性有待进一步研究。利用2010—2018年田间定位试验数据，分析玉米种植模式下不同比例复配土壤粉粒和黏粒含量及空间迁移规律，探索玉米产量可持续性及稳定性。结果表明：(1) 随着试验的开展，0～30 cm土层中不同比例复配土壤粉粒、黏粒含量均向下层土体运移，运移速率为1∶5>1∶1>1∶2，下层土壤中的粉粒和黏粒含量增加，使得土体剖面耕层增加至30～40 cm土层厚度。(2) 玉米产量在1∶1，1∶2，1∶5复配土壤上差异较大，与2010年相比较，到2018年产量分别提高了43.9%、105.9%、58.5%，并且1∶2复配土壤上玉米多年平均产量超过了15000 kg/hm^2，与当地高产田产量持平。(3) 1∶2复配土壤对玉米增产的效果最优，SYI值最大，C_V值最小，即产量稳定性和可持续性最好。因此，确定1∶2复配比例最适合玉米生长，其朝着有利于玉米生长发育并获得高产的方向发展。

【关键词】复配土壤；粉粒；黏粒；迁移规律；产量可持续性

Analysis of Soil Particle Size Change and Corn Yield Sustainability in Soft Rock and Sand Compound Soil

Haiou Zhang[1,2,3,4], Tingting Cao[2,3,4], Chenxi Yang[2,3]

(1.Shaanxi Provincial Land Engineering Construction Group Co., Ltd., Xi'an 710075, China; 2.Institute of Land Engineering and Technology, Shaanxi Provincial Land Engineering Construction Group Co., Ltd., Xi'an 710021, China; 3.Key Laboratory of Degraded and Unused Land Consolidation Engineering, the Ministry of Natural Resources, Xi'an 710021, China; 4.Shaanxi Provincial Land Consolidation Engineering Technology Research Center, Xi'an 710021, China)

【Abstract】The technology of soft rock and sand compound soil has been widely used in the land remediation of Mu Us sandy land. However, exploring the suitable mixing ratio of soft rock and aeolian sandy soil that is suitable for the high and stable yield of the main local economic crops, and improving the agricultural adaptability of the compound soil needs further research. This paper uses the field positioning test data from 2010 to 2018 to analyze the silt and clay content and spatial migration rules of the compound soil at different proportions under the corn planting mode, and explore the sustainability and stability of corn yield. The results showed that: (1) with the development of the test, the compound soil silt and clay contents in the 0～30 cm soil layer migrated to the lower soil

资助项目：陕西省土地整治重点实验室开放基金“基于土地整治工程新增耕地的土壤有机碳稳定性研究”(2019-JC07)。

at a rate of 1 : 5>1 : 1> 1 : 2. The silt and clay contents in the lower soil layer increased, making the surface layer of the soil profile increase to the thickness of 30~40 cm soil layer. (2) Corn yield was significantly different in 1 : 1, 1 : 2 and 1 : 5 compound soil. Compared with 2010, the annual output of corn increased by 43.9%, 105.9% and 58.5% respectively by 2018. Moreover, the annual average yield of corn in 1 : 2 compound soil exceeded 15000 kg/hm^2 for many years, which was equal to the yield of local high-yielding fields. (3) The 1 : 2 compound soil has the best effect on increasing corn yield, with the maximum SYI value and the minimum C_V value, that is, the yield stability and sustainability were the best. Therefore, the 1 : 2 compound ratio was determined to be the most suitable for corn growth, which was in favor of corn growth and development and obtained high yield.

【Keywords】 Compound soil; Silt; Clay; Migration law; Yield sustainability

毛乌素沙地以光照资源丰沛,光合效率高,成为玉米等作物主要生产基地,春玉米正是榆林当地的主要农作物之一,其经济效益巨大。然而,沙地受风蚀堆积作用,地貌多为起伏不平的沙丘,颗粒呈无团聚的分散状态,抗风蚀性差,土体营养物质贫瘠,不利于植物生长,只在“肥力岛”处滋生有零星的小灌丛,植被极为稀疏,无经济价值。Han 等[1]研究发现砒沙岩具有天然保水剂的作用,揭示了砒砂岩的保水机制,提出了规模化使用砒沙岩改良毛乌素沙地的可行性,即将一定量的砒砂岩和沙两种物质经科学配比,使其成为满足作物生长的新增耕地的耕作土壤层。风沙土中添加砒砂岩能够改善土壤质地[2],提升黏粒含量和有机质含量[3],降低氮素的淋溶和损失[4],提高土壤结构稳定性和肥力[5-6]。然而,探索适合当地主要粮食作物高产稳产的砒砂岩与风沙土的适宜混合比例,提高复配土壤的农业适应性有待进一步研究。本文利用 2010—2018 年砒砂岩与沙体积比为 1 : 1、1 : 2、1 : 5 复配土壤的田间定位试验数据,分析随着玉米种植年限的增加不同比例复配土壤粉粒和黏粒含量及空间迁移规律,探索玉米产量可持续性及稳定性,以期为砒砂岩与沙复配土壤—植被系统的良性循环发展提供理论基础。

1 试验区概况与方法

1.1 试验区概况

研究团队于 2010 年在毛乌素沙地榆林市榆阳区建立了砒砂岩与沙复配成土野外科学观测试验小区,开展长期监测。研究区环境条件具有典型的代表性,该地区气温年际变化较大,冬季(1 月)平均温度在-9.5~-12 ℃,夏季(7 月)平均温度在(24±2)℃;降水时空分布不均匀,秋季(尤其 8 月)降水几乎占全年降水量的 60%~75%,并且年际间降水量也呈现出显著差异,即湿润年是干旱年降水量的 2~4 倍。此外,结合该地区光照条件充足、地下水埋藏较浅等特点,具备生产出高产量玉米的环境条件。研究区域以风沙土为主,结构疏松,持水能力差,蒸散量大,导致经常缺水。当地松软易风化的砒砂岩,结构强度低,透水性差,但具有较好的持水能力和保水能力,并且当地地下水能够为植物生长提供水分,因此将砒砂岩与风沙土按照一定比例混合形成本试验的复配土。

1.2 试验小区设计

自 2010 年砒砂岩与沙复配成土整治示范工程项目完成起,建立了复配土长期定位试验小区(长 5 m×宽 12 m)。在当地原始沙地表层按照试验需求,仅将 0~30 cm 土层分别按砒砂岩与风沙土体积比 1 : 1、1 : 2、1 : 5 进行复配后,通过机械翻耙,使其充分混合,每种比例设置 3 个重复试验小区,共计 9 个试验小区。种植当地主要粮食作物春玉米(榆丹 9 号),每年播种时间为 5 月中旬,9 月下旬进入收获期,种植制度为一年一季。播种前 1~2 d 施入复合肥(90 kg N /hm^2,40 kg P/hm^2,75 kg K/hm^2),在玉米拔节期以 187 kg N/ hm^2 追施尿素 1 次。灌溉的时间和量,根据天气干旱、作物生长需要,以 60 cm 土层内保持田间最大持水量的 75%~80%为宜。为了避免由于种植不同玉米品种而对试验造成的影响,按照试验周期内所种植的玉米品种,将试验分为 2 个时间段,2010—2014 年和 2015—2018 年种植玉米品种分别为榆丹 9 号、先玉 335。试验于每年玉米收获后采用“S”形采样方法,分别采集各处理下 0~30 cm 土样,进行土壤物理和养分指标测定及分析。

土壤颗粒组成测定采用马尔文激光粒度分析仪 Mastersizer 2000(英国),其粒度分级采用1951年美国农业部(USDA)制定的分级标准。每年按小区收获计算玉米实收产量。

1.3　产量稳定性及可持续性指数计算

1.3.1　产量稳定性计算

采用变异系数(C_V)表示产量稳定性,衡量随着年际变化同一品种作物平均产量间的变异程度,其值越小,表明产量稳定性越高[7]。

$$C_V = \sigma/\overline{Y}$$

式中:$\overline{Y}$为平均产量;σ为标准差。

1.3.2　产量可持续性指数计算

产量可持续性指数(SYI)作为衡量土地可持续生产力的指标,其值越大,说明可持续性越好[8-9]。

$$SYI = (\overline{Y} - \sigma)/Y_{max}$$

式中:Y_{max}为试验点最高产量。

2　结果与分析

2.1　复配土壤粉粒和黏粒迁移规律分析

土壤是由不同粒径的黏粒、粉粒和砂粒等颗粒组成,合理的机械组成是土壤良好发育的基础,其影响着植物根系的生长[10-11]。不同种植年限1∶1、1∶2、1∶5复配土体剖面粉粒和黏粒迁移规律分别如图1~图3所示,试验小区0~30 cm土层为砒砂岩与沙复配土壤,相比较于30 cm以下原始沙地土层中粉粒、黏粒含量高。随着玉米种植年限的增加,3种比例复配土壤表层中粉粒和黏粒均向下层土体运移,使下层土壤中的粉粒和黏粒含量增加。2010—2013年表层土壤中粉黏粒含量呈下降趋势,30 cm以下土层其含量增加,2014年之后表层土壤粉粒、黏粒含量整体呈稳定状态,30 cm以下土层粉粒、黏粒含量增加速率减缓。其中,1∶1复配土壤10~20 cm土层中粉粒含量积累最多,30~40 cm土层中粉粒含量次之,黏粒含量在0~15 cm土层中积累量最多。1∶2复配土壤中10~20 cm土层中粉粒和黏粒含量均积累最多。1∶5复配土壤中0~10 cm土层中粉粒含量最大,10~20 cm土层中黏粒含量最大。

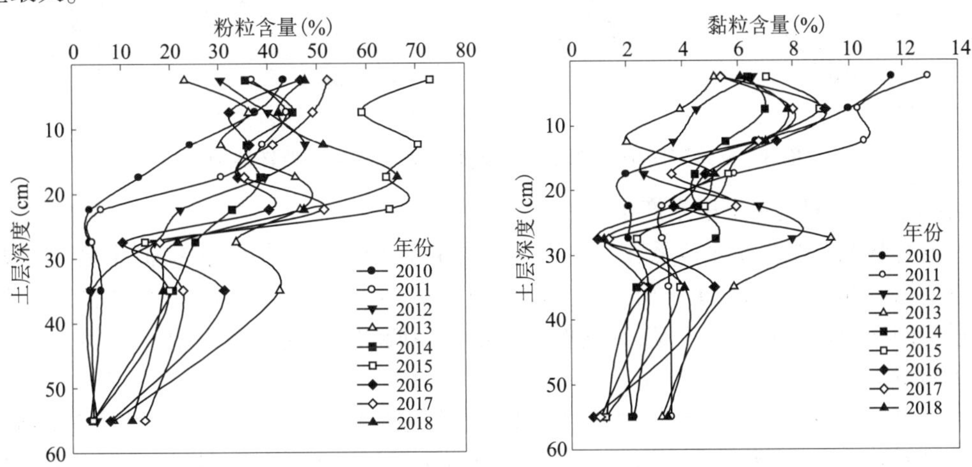

图1　玉米不同种植年限1∶1复配土壤黏粒和粉粒迁移规律

随着试验的开展,表层土壤粉粒、黏粒向下运移的速率1∶5>1∶1>1∶2,这是由于风沙土中随着砒砂岩含量的增加,粉粒、黏粒含量逐渐增加,1∶5复配土壤中砂粒含量多,属于砂质土壤,颗粒粗糙,组成的粒间大孔隙数量多。2014年之后,1∶2复配土壤表层粉粒、黏粒积累速率大于1∶5、1∶1,并且其向下运移速率小于1∶1、1∶5,这可能是作物种类、种植年限及混合比例等因素之间的交互作用影响了不同比例复配土壤的理化性状,使得玉米种植模式下1∶2复配土壤稳定性较好。随着上层土壤中粉粒、黏粒的运移,2014年之后30~40 cm土层粉粒、黏粒含量累积量不断增加,2016—

2018 年 1∶1、1∶2、1∶5 复配土壤粉粒含量平均值分别为 24.09%、8.75%、53.23%,分别是 2010 年的 3.10 倍、2.89 倍和 6.89 倍,黏粒含量平均值为 3.94%、1.87%、2.49%,与 2010 年相比较分别提高了 4.19 倍、5.04 倍和 1.47 倍,40 cm 以下土层中粉粒含量在 6%以下、黏粒含量在 2%以下,仍然很少,没有达到作物生长需求。因此,随着作物种植年限的增加和粉粒、黏粒的向下运移,使得土体剖面的耕层增加至 30~40 cm 土层厚度。

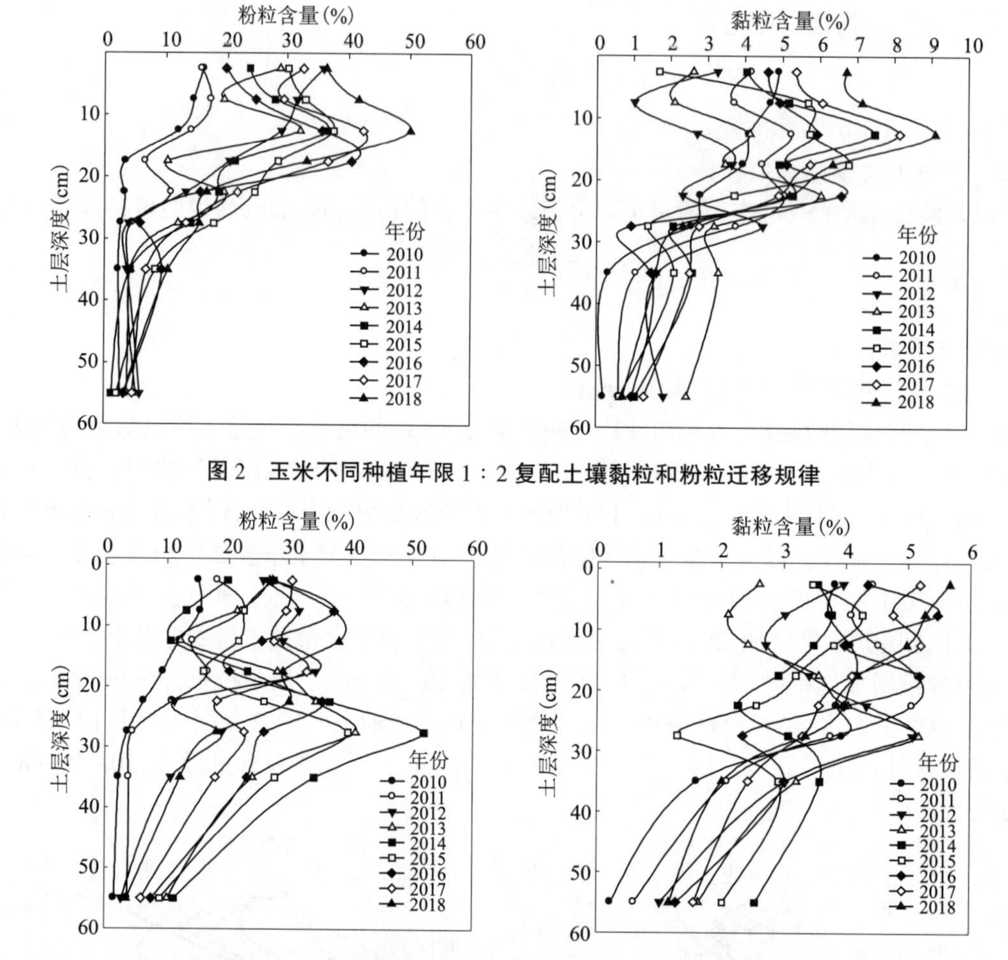

图 2　玉米不同种植年限 1∶2 复配土壤黏粒和粉粒迁移规律

图 3　玉米不同种植年限 1∶5 复配土壤黏粒和粉粒迁移规律

2.2　不同比例复配土壤上玉米产量的变化特征

土壤的发育状况及肥力水平高低在一定程度上决定了土地生产力的好坏,而土壤生产力作为土壤质量的重要组成部分,其高低水平衡量了土壤质量的好坏[12]。土壤肥力、气候条件及人为管理措施等多因素共同作用影响着作物产量,因此作物产量的高低和变化趋势可以作为土壤可持续性分析的一个重要衡量指标[13]。

从图 4 可以看出,3 种比例复配土壤随着玉米种植年限的增加产量整体呈现出上升趋势,不同复配比例间产量差异较大,说明混合比例对玉米产量有着重要的影响,即不同比例复配土壤结构和肥力特征不同。与 2010 年相比较,到 2018 年 1∶1、1∶2、1∶5 复配土壤玉米产量分别提高了 43.9%、105.9%、58.5%,2015 年之后 3 种比例复配土壤玉米产量增加速率显著,大小顺序为 1∶2>1∶5>1∶1,尤其是 1∶2 复配土壤 2015 年之后产量大于或等于当地高产田玉米产量,1∶5 复配土壤玉米产量接近当地高产田产量。这是由于 2015 年之后,砒砂岩与沙复配土壤结构及肥力状况达到相对稳定的有机-无机复合状态,并且 1∶2 复配土壤产量增加速率最大,产量最高。

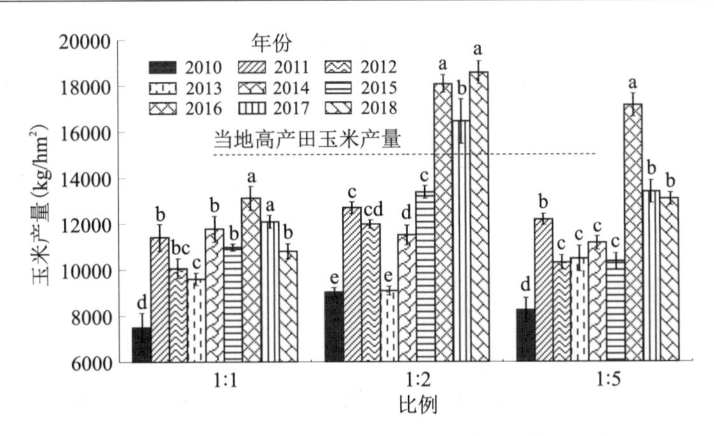

图 4 3 种比例复配土壤上玉米产量随时间的变化

不同比例复配土壤上玉米产量随时间变化的拟合方程见图 5,方程的斜率即为玉米产量年变化速率[g/(kg・a)]。随着玉米种植季数的增加,不同比例复配土壤产量整体呈现出上升的趋势,其中 1∶2 复配土壤上玉米产量增加速率极显著($p<0.001$),每年达到 1249 kg/hm^2;其次是 1∶5 复配土壤上玉米产量增加速率较显著($p<0.005$),每年为 792.7 kg/hm^2;1∶1 复配土壤上玉米产量与种植年限相关性较小,每年增加速率为 546.8 kg/hm^2。因此,由于复配土壤质地、种植年限、作物生长特性等之间的交互作用,1∶2 复配土壤对玉米产量提升更加显著。

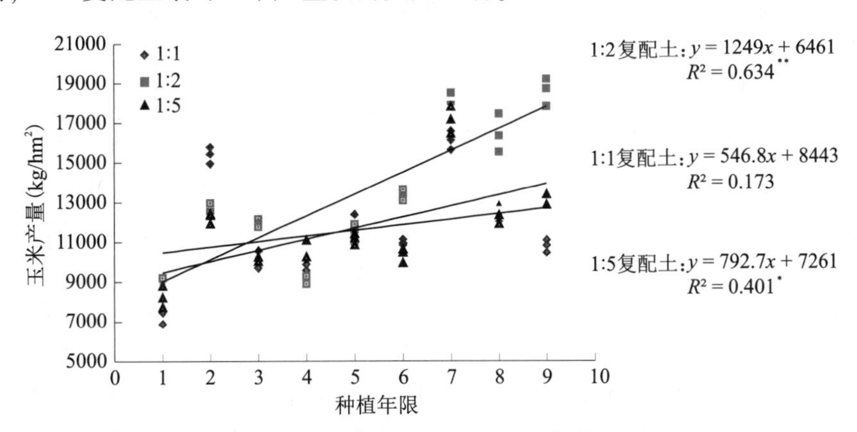

图 5 不同比例复配土壤上玉米产量随时间的拟合方程

注:* * 年变化率极显著($p<0.001$),* 年变化率显著($p<0.005$)。

2.3 玉米产量可持续性及稳定性分析

由表 1 可知,不同比例复配土壤上玉米产量可持续性指数(SYI)不同,1∶2 复配土壤能显著提高玉米 SYI 值。不同比例复配土壤上玉米 SYI 值大小顺序为:2010—2014 年:1∶5>1∶2>1∶1;2015—2018 年:1∶2>1∶5>1∶1,两个时间段中 1∶2、1∶5 复配土壤产量可持续性较好。在 2015 年以后,3 种比例复配土壤 SYI 均大于 0.6,产量可持续性均较好,说明 2015 年之后复配土壤向良好的土壤状态发育,尤其 1∶2 复配土壤 SYI 值高达 0.771,产量可持续性最好。

不同比例复配土壤上玉米生产的 C_V 值越小,表示稳定性越高。不同比例复配土壤上玉米产量的 C_V 值不同,2010—2018 年 1∶1 复配土壤 C_V 值较大,1∶5 复配土壤 C_V 值次之,1∶2 复配土壤 C_V 值最小,因此 1∶2 复配土壤上玉米产量稳定性最高。2015—2018 年玉米平均产量大小顺序为:1∶2>1∶5>1∶1,并且 1∶2 复配土壤上玉米多年平均产量超过了 15000 kg/hm^2,与当地高产田产量持平。综上表明,1∶2 复配土壤上玉米产量最高,产量稳定性和可持续性最好,说明砒砂岩与沙混合比例为 1∶2 时最适合玉米生长。

表1 不同比例复配土壤上玉米产量可持续性指数和变异系数

砒砂岩:沙	2010—2014年(榆丹9号)			2015—2018年(先玉335)		
	产量平均值(kg/hm²)	C_V(%)	SYI	产量平均值(kg/hm²)	C_V(%)	SYI
1:1	10864	25.5	0.526	12493	19.1	0.635
1:2	10956	14.7	0.729	16617	13.2	0.771
1:5	10474	13.2	0.746	13236	18.7	0.646

3 讨论

砒砂岩与沙复配成土,首要是进行风沙土颗粒组成优化,引入砒砂岩中粉粒、黏粒,改良沙地的不良粒径组成。然而这种新造复配土壤属于非均质土壤,粒径组成状态是否稳定,在利用期间受耕作、灌溉以及其农艺措施作用的影响较大,进而影响土壤结构的发育。高亚军等[14]研究结果表明,由于土壤颗粒组成在剖面中的垂直分异及其在土体中的含量不同,土壤颗粒组成的变化与黏粒的形成过程就是土壤的发育过程,黏粒丰富的土壤中,胶体含量越多,吸附土壤养分的能力越强。植被演替对土壤结构和质量的变化也具有重要的影响,植被对土壤条件的响应是生物界最常见的自然现象之一[15]。砒砂岩与沙复配成土技术就是重构沙地土体构造和根区土层的颗粒,然而在新造土壤利用过程中,其土层中黏粒、粉粒含量的变化,对土壤的质量演替过程的研究有着重要的作用。研究发现,玉米的种植促进了土壤细颗粒和养分在垂直剖面上的分化,使得耕层厚度增加。

土壤颗粒组成在土体剖面中的垂直分布及含量直接影响着土壤的水、肥、气、热等特性[16],因此可以说土壤的形成就是黏粒的形成与机械组成的变化[17-18]。复配土壤上作物种植不仅能有效地阻止沙地细颗粒物质的流失,还能促进细颗粒物质的沉积,增加细颗粒物质含量。2011—2013年试验区表层(0~30 cm)土层3种比例复配土壤粉粒、黏粒含量呈下降趋势,2014年之后整体呈稳定状态,在30~40 cm土层形成了一层相对致密的黏化层,黏化层的形成可有效防止水肥的渗漏,有利于提高复配土土体的保水保肥性。这主要是由于种植初期,新造复配土壤处于发育初期,结构不稳定,土壤结构松散,土体空隙较大,具有很强的滑动性,粉粒、黏粒易被雨水冲刷向下层迁移。2014年之后玉米根系不断产生的胶结物质、根系分泌物对土壤颗粒的黏结,加强了土壤颗粒之间的联结形成土壤微团聚体,从而降低了粉粒、黏粒的向下迁移。同时,表层耕作土壤中枯枝落叶的分解和种植过程中有机肥的施用,使表层土壤质地和养分改善,土壤稳定性提高,新造复配土壤结构不断向良好的状态发育。

作物产量的高低和变化趋势在一定程度上能够确切反映土壤生产力水平[19-21]。为了获得玉米的高产,首先要选择适合玉米生长的土壤。研究发现,2015年以后,随着土壤结构发育状态相对稳定,3种比例复配土壤SYI较大,C_V值减小,尤其1:2复配土壤对玉米增产的效果最优,SYI值明显高于1:1、1:5,C_V值小于1:1、1:5,即产量稳定性和可持续性最好,说明1:2复配土壤受耕作管理措施、气象因素及环境条件影响较小,土壤结构发育及肥力水平较高。这主要是由于随着种植年限的增加,以及玉米生长对土质的需求和改善作用,不同质地类型土壤结构发育不同,然而土壤结构和肥力质量是衡量土壤生产力的综合指标,砒砂岩与沙混合比例不同,复配土壤具有不同的特性,各种作物属性不同,对土壤的需求也不同。这与张卫华等[22]、Sun等[23]研究结果一致,其研究显示玉米种植下砒砂岩和风沙土按1:2复配后土壤养分累积和水分的利用效率较好。Dorraji等[24]研究显示土壤改良剂对沙质土壤中玉米产量具有显著影响,说明应用砒砂岩改良风沙土为毛乌素沙地的治理提供了切实可行的措施,不仅可以增加沙区农业用地面积,还可促进当地的农业经济可持续发展。

4 结论

复配土耕种年限和植被生长的综合作用,改善了新造复配土壤粉粒、黏粒含量的空间分布特征,使得土体剖面的耕层增加至30~40 cm土层厚度,土壤颗粒级配趋于合理化,提高了土壤结构稳定性。1∶2 复配土壤对玉米增产的效果最优,SYI 值最大,C_V 值最小,即产量稳定性和可持续性最好。因此,1∶2 复配土壤理化性状改良效果最好,使其朝着有利于玉米生长发育并获得高产的方向发展,确定 1∶2 复配比例最适合玉米生长,提高了农业适应性,该复配模式可作为研究区砒砂岩改良风沙土复配比例的最佳选择。本研究为促进砒砂岩与沙复配土壤—植被系统的良性循环和稳定性的提高提供理论依据。

参考文献

[1] Han J C, Xie J C, Zhang Y. Potential role of feldspathic sandstone as a natural water retaining agent in Mu Us Sandy Land, northwest China[J]. Chinese Geographical Science, 2012,22(5):550-555.

[2]张露,韩霁昌,马增辉,等.砒砂岩与沙复配“土壤”的质地性状[J].西北农业学报,2014,2014,23(4):166-172.

[3]柴苗苗,韩霁昌,罗林涛,等.砒砂岩与沙混合比例及作物种植季数对复配土壤性质和作物产量的影响[J].西北农林科技大学学报,2013,41(10):179-184,192.

[4]罗林涛,程杰,王欢元,等.玉米种植模式下砒砂岩与沙复配土氮素淋失特征[J].水土保持学报,2013,27(4):58-66.

[5]李娟,韩霁昌,李晓明.砒砂岩与沙复配成土对小麦光合生理和产量的影响[J].麦类作物学报,2014,34(2):203-209.

[6]魏彬萌,王益权,李忠徽.烟秆生物炭对砒砂岩与沙复配土壤理化性状及玉米生长的影响[J].水土保持学报,2018,32(2):219-222.

[7]张雅蓉,李渝,刘彦伶,等.长期施肥对黄壤有机碳平衡及玉米产量的影响[J].土壤学报,2016,53(5):179-189.

[8] Manna M C, Swarup A, Wanjari R H, et al. Long-term effect of fertilizer and manure application on soil organic carbon storage, soil quality and yield sustainability under sub-humid and semi-arid tropical India[J]. Field Crops Research, 2005,93(2-3):264-280.

[9]李忠芳,唐政,李继光,等.长期施肥对辽西褐土区土壤有机碳含量和玉米产量的影响[J].土壤与作物,2013,2(4):150-156.

[10] Behzad G, Hugh D. Fractal dimension of soil fragment mass-size distribution: A critical analysis[J]. Geoderma, 2015,245-246:98-103.

[11] Meysam M, Mahmoud S, Mohammad H M, et al. Characterizing spatial variability of soil textural fractions and fractal parameters derived from particle size distributions[J]. Pedosphere, 2019,29(2):224-234.

[12]徐明岗,梁国庆,张夫道,等.中国土壤肥力演变[M].北京:中国农业科学技术出版社,2006.

[13]左小安,赵学勇,赵哈林.沙地退化植被恢复过程中灌木发育对草本植物和土壤的影响[J].生态环境学报,2009,18(2):643-647.

[14]高亚军.陕北农牧交错带土地荒漠化演化机制及土壤质量评价研究[D].陕西杨凌:西北农林科技大学,2003.

[15]刘勇.黄土高原植被演替过程中植被与土壤养分、水分关系研究进展[J].吉林农业科学,2010,35(5):25-27.

[16]苏志珠,刘蓉,梁爱民,等.晋西北沙化土地土壤机械组成与有机质的初步研究[J].水土保持研究,2018,25(6):61-67.

[17] Abbas A, Mohammad-Reza N, Hassan R, et al. Fractal dimension of soil aggregates as an index of soil erodibility[J]. Journal of Hydrology, 2011,400(3):305-311.

[18]华瑞,徐学选,张少妮,等.不同退耕年限林草地土壤颗粒分形特征研究[J].水土保持学报,2016,30(4):206-209.

[19]李忠芳,徐明岗,张会民,等.长期施肥和不同生态条件下我国作物产量可持续性特征[J].应用生态学报,2010,21(5):1264-1269.

[20]罗倩,张珍明,向准,等.不同种植年限鸟王茶产地土壤物理性质及生长特征[J].西南农业学报,2017,30(12):2746-2750.

[21]刘彦伶,李渝,张雅蓉,等.长期氮磷钾肥配施对贵州黄壤玉米产量和土壤养分可持续性的影响[J].应用生态学报,2017,28(11):3581-3588.

[22]张卫华,韩霁昌,王欢元,等.砒砂岩对毛乌素沙地风成沙的改良应用研究[J].干旱区资源与环境,2015,29

(10):122-127.

[23] Sun Z H, Han J C. Effect of soft rock amendment on soil hydraulic parameters and crop performance in Mu Us Sandy Land, China[J]. Field Crops Research, 2018,222:85-93.

[24] Dorraji S S, Golchin A, Ahmadi S. The effects of hydrophilic polymer and soil salinity on corn growth in sandy and loamy soils[J]. Clean-Soil Air Water, 2015,38(7):584-591.

本文曾发表于2022年《水土保持研究》第29卷第1期

砒砂岩改良风沙土后玉米不同种植年限下土壤团聚体变化特征

张海欧[1,2,3,4],师晨迪[1,2,3,4],李 娟[1,2,3,4]

(1.陕西省土地工程建设集团有限责任公司,西安 710075;
2.陕西地建土地工程技术研究院有限责任公司,西安 710021;
3.自然资源部退化及未利用土地整治工程重点实验室,西安 710021;
4.陕西省土地整治工程技术研究中心,西安 710021)

【摘要】毛乌素沙地砒砂岩与沙复配土壤属于人为重构土壤,水稳性团聚体作为自然土壤形成和发育的最重要标志之一,其成土母质与后期发育土壤具有本质性区别。本文基于2010—2018年1∶1、1∶2、1∶5砒砂岩与沙复配土壤田间小区试验,分析玉米不同种植年限下三种比例复配土壤水稳性团聚体的组成、分布以及平均重量直径等指标的变化特征。结果表明,随着玉米种植年限的增加,三种比例复配土壤<0.25 mm粒径团聚体含量逐渐减少,>0.25 mm粒径团聚体呈持续稳定增加趋势,并且以0.25~0.5mm粒径团聚体占比最大,说明改变了团聚化物质基础;种植9年后,1∶1、1∶2和1∶5复配土壤水稳性团聚体平均重量直径(MWD)值分别增加了1.13倍、1.85倍和1.58倍,并且玉米种植模式下1∶2复配土壤>0.25 mm粒径团聚体含量和MWD值增加速率较快,土壤团聚化作用和稳定性较高。综上,砒砂岩与沙混合比例和玉米种植年限二者之间的交互作用对复配土壤水稳性团聚体形成及稳定性具有显著的影响,随着耕种年限的增加不同比例复配土壤团聚化作用增强,土壤结构发育不断完善,本研究对抑制毛乌素沙地土地风蚀沙化及水土流失具有重要作用。

【关键词】风沙土;砒砂岩;水稳性团聚体;平均重量直径;稳定性

Changes of Soil Aggregates in Different Planting Years of Corn after Aeolian Sandstone Improved by Soft Rock

Haiou Zhang[1,2,3,4], Chendi Shi[1,2,3,4], Juan Li[1,2,3,4]

(1.Shaanxi Provincial Land Engineering Construction Group Co., Ltd., Xi'an 710075; 2. Institute of Land Engineering and Technology, Shaanxi Provincial Land Engineering Construction Group Co., Ltd., Xi'an 710021; 3.Key Laboratory of Degraded and Unused Land Consolidation Engineering, the Ministry of Natural Resources, Xi'an 710021; 4. Shaanxi Provincial Land Consolidation Engineering Technology Research Center, Xi'an 710021, China)

【Abstract】 The mixed soil of soft rock and sand in the Mu Us sandy land belongs to artificial new soil. As one of the most important indicators of the formation and development of natural soil, water-stable aggregates have essential differences between the parent material of soil formation and later soil development. Therefore, the study on the change characteristics of soil aggregates in the reconstituted soil of soft rock and sand after long-term crop planting aims to provide theoretical basis for understanding the structural development and sustainable utilization of

基金项目:陕西省土地整治重点实验室开放基金(2019-JC07);陕西省土地工程建设集团科研项目(DJNY2020-13,DJNY2020-14)。

the compound soil. Based on the field experiment with the volume ratio of soft rock to sand of 1 ∶ 1, 1 ∶ 2, and 1 ∶ 5 from 2010 to 2018, analyze the composition, distribution and average weight diameter of the water-stable aggregates of the three proportions of compound soil under different years of corn planting. With the increase of corn planting years, the content of aggregates in the three proportions of compound soils with a particle size of <0.25 mm gradually decreased, and aggregates with a particle size of >0.25 mm showed a continuous and stable increase trend, and the proportion of aggregates with a particle size of 0.25~0.5 mm was the largest, Indicating that the material basis of agglomeration has been changed; After 9 years of planting, the average weight diameter (MWD) values of the 1 ∶ 1, 1 ∶ 2 and 1 ∶ 5 compound soil water-stable aggregates increased by 1.13 times, 1.85 times and 1.58 times, respectively, and the 1 ∶ 2 compound soil >0.25 mm particle size aggregates content and MWD value increased at a faster rate, with higher soil agglomeration and stability in corn planting mode. The interaction between the mixing ratio of soft rock and sand and the planting years of corn has a significant impact on the formation and stability of water-stable aggregates in the compound soil. With the increase of corn planting years, the compounding soil agglomeration has enhanced, and the soil structure has been continuously improved.

【**Keywords**】Sand soil; Soft rock; Water stable aggregates; Average weight diameter; Stability

土壤团聚体的数量、粒径分布和稳定性等是土壤结构良好发育的重要指标[1],尤其是水稳性团聚体含量对土壤的物理、化学、生物性质有着显著的影响[2-3],其主要作用是保水保肥、协调土壤的水肥气热适宜作物生长,其是农作物生长、发育、高产和稳产的基础条件[4-5]。土壤团聚体组成和稳定性是用来衡量土壤发育程度和可持续利用的重要评价指标[6-7]。平均重量直径(MWD)是评价土壤团聚体特征及稳定性的常用方法,其值越大,表示土壤团聚体的平均粒径团聚度越高,结构越稳定,发育状况越好[8]。

毛乌素沙地区内存在础砂岩与沙两种不同的物质,导致土地风蚀沙化及水土流失严重。Han等[9]发现二者之间具有互补特性,将二者复配形成重构土壤。础砂岩与沙复配“土壤”具有农业生产的基本理化性质,农业利用过程中,既存在促使土壤团聚化发生的物质积累及动力学过程,又同时存在着分散的风险与机制。摄晓燕等[10]指出,础砂岩的加入可有效提高风沙土保水保肥能力,降低风沙土的养分流失及水分入渗率。Sun 等[11]通过水分模型分析指出,础砂岩的添加使得风沙土的水力学参数得到有效改善,能够显著增加作物的产量。李裕瑞等[12]研究指出固沙导向下础砂岩与沙最优复配比例为1∶1~1∶5,其保水和持水性能最强。前人研究多基于复配土壤养分和水分方面,而关于长时间序列的土壤结构研究未见报道。分析础砂岩与沙复配土壤水稳性团聚体分布及稳定性特征,有助于了解新造复配土壤发育状况及结构变化趋势[13]。为此,需要开展复配土团聚化过程、状态的系列研究,以评价复配的科学价值及效应。本研究于2010—2018年在1∶1、1∶2、1∶5础砂岩与沙复配土壤田间试验小区开展研究,分析不同种植年限下三种比例复配土壤水稳性团聚体的组成、分布以及平均重量直径,来表征复配土壤水稳性团聚体的稳定性及土壤结构发育状况,以期为毛乌素沙地使用础砂岩改良风沙土的技术提供科学支撑。

1 试验设计与研究方法

1.1 研究区概况

毛乌素沙地面积422万 hm^2,是我国四大沙地之一,区内土地风蚀沙化严重;同时该地区分布着约167万 hm^2 的础砂岩,水土流失严重,是黄河粗砂主要来源[9]。试验区位于毛乌素沙地榆林市榆阳区,该地区一年内冬夏季节气温相差较大,冬季(1月)平均温度在-9.5~-12 ℃,并且早晚温差较大,夏季(7月)平均温度在(24±2)℃;年内降雨几乎集中在秋季(尤其8月),降水量占全年降水量的60%~75%,并且年际间降水量也呈现出显著差异,即湿润年是干旱年降水量的2~4倍。区域光照充足,地下水埋藏较浅并且丰富,能够满足当地主要农作物玉米的灌溉及生长。试验区主要分布着风沙土和础砂岩,二者呈相间分布,沙土颗粒呈无团聚的分散状态,结构性差,土质疏松,表层土壤蒸发量较高,土壤持水保水性差。研究区础砂岩储量非常丰富,与沙相间分布,运距短,可以满足促沙成土对材料“量”的需求。同时,砒砂岩具有丰富的次生黏土矿物,能够促使土壤团聚化,持水性较好,其也满足了促沙成土对材料“质”的需求[10-11]。

1.2　试验小区设计

砒砂岩与沙复配成土野外科学观测试验小区于2010年在毛乌素沙地榆林市榆阳区建立,设置砒砂岩与沙比例为1∶1、1∶2、1∶5三种比例下各三个重复试验,每个小区长12 m×宽5 m,共计9个试验小区,开展复配土质量长期监测试验。由于所种植玉米为浅根系植物,其根系生长所需要的营养区在0~30 cm土层厚度,30 cm以下的土层是作物根系生长的固定区,因此在研究区风沙土表层(0~30 cm)按照体积比为1∶1、1∶2、1∶5的砒砂岩与沙土的比例覆盖复配土,并使用机械耙翻措施,将两种材料混合充分。试验小区种植当地主要农业作物春玉米,于每年5月初进行播种,播种前1~2天施入复合肥(90 kg N /hm^2,40 kg P/hm^2,75 kg K/hm^2),在玉米拔节期以187 kg N/hm^2追施尿素1次。

1.3　试验方法

试验于每年10月初玉米收获后采用“S”形采样方法[14],使用铝盒采集各比例下表层20 cm原状土,用于测定土壤机械稳定性团聚体和水稳定性团聚体的含量。机械稳定性大团聚体组成用干筛法测定[15],水稳定性团聚体组成用湿筛法测定[16],利用团聚体分析仪测定>2 mm、2~1 mm、1~0.5 mm、0.5~0.25 mm、<0.25 mm各粒径下水稳性团聚体含量。每小区选取上中下3个样点,每个样点选取2 m×2 m大小,测量玉米实收产量,折算为标准单位。

1.4　土壤团聚体稳定性的指标计算

水稳性团聚体是判定土壤结构和质量好坏的重要指标,采用>0.25 mm粒级的水稳性团聚体含量(WSAC)来衡量团聚体质量[17],计算公式如下:

$$\mathrm{WSAC} = \sum_{i=1}^{n} (W_i) \tag{1}$$

平均重量直径(MWD)是评价土壤团聚体稳定性常用方法,其表征土壤团聚体大小分布状况,MWD值越大,表示复配土壤团聚体的平均粒径团聚度越高,土壤结构越稳定,并且发育状况越好[17]。计算公式如下:

$$\mathrm{MWD} = \sum_{i=1}^{n} (\overline{X_i} \cdot W_i) \tag{2}$$

式中:W_i代表第i粒级团聚体含量,g;X_i代表第i粒级范围内团聚体MWD,mm。

2　结果与分析

2.1　复配土壤水稳性团聚体年际变化特征

土壤水稳性团聚体的粒径组成和含量多少是表征土壤结构发育状况的重要指标[4,18]。从玉米不同种植年限下不同比例复配土壤水稳性团聚体组成状况来看(见表1),种植前<0.25 mm水稳性团聚体含量占比最大,变化范围为84.2%~86.3%。随玉米种植年限的增加,水稳性团聚体分布发生显著变化,1∶1、1∶2、1∶5复配土壤均呈现出<0.25 mm水稳性团聚体含量显著下降,与种植前相比较,种植9年后,分别降低了53.9%、67.9%、60.7%,相应地,1∶1、1∶2、1∶5复配土壤水稳性团聚体0.5~0.25 mm粒径含量分别增加了2.69倍、3.96倍、3.34倍,1~0.5 mm粒径水稳性团聚体含量分别增加了2.56倍、2.63倍、1.86倍,2~1 mm粒径水稳性团聚体含量分别增加了3.43倍、6.05倍、8.25倍,>2 mm水稳性团聚体含量分别增加了3.93倍、8.18倍、8.75倍,说明土壤结构不断向着团聚化的方向发展。

进一步比较不同比例复配土壤团聚体含量的变异系数,明显可以看出,三种比例复配土壤中,粒径>2 mm水稳性团聚体变异系数显著高于其他粒级,说明大团聚体稳定性较差,其受外力破坏作用的风险较高,容易受成土母质、利用方式、气候条件、人为耕作措施等因素的影响。其中1∶1复配土壤不同粒级水稳性团聚体含量变异系数整体呈现小于1∶2和1∶5,说明1∶1复配土壤的团聚体随时间变化浮动不大,而1∶2和1∶5复配土壤随玉米种植年限的增加团聚体分布变化较大,种植年限对二者的影响较大。

表1 玉米不同种植年限下不同比例复配土壤水稳性团聚体组成

Table 1 The composition of compound soil water-stable aggregates under different planting years of corn

砒砂岩∶沙	年限(年)	水稳性团聚体百分比含量(%)				
		>2 mm	2~1 mm	1~0.5 mm	0.5~0.25 mm	<0.25 mm
1∶1	0	1.4±0.07d	2.1±0.11d	2.7±0.08e	9.6±0.42e	84.2±9.71a
	1	1.9±0.12d	3.3±0.12d	3.4±0.27d	14.1±0.54d	77.3±7.94ba
	3	3.2±0.30c	4.7±0.23cd	6.5±0.34c	18.9±1.26c	66.7±5.35b
	5	4.8±0.44b	6.4±0.49b	9.9±0.52b	26.1±2.73b	52.8±7.48b
	7	6.1±0.61a	8.2±0.47a	10.4±1.29a	32.6±1.96a	42.7±5.72c
	9	6.9±0.35a	9.3±0.61a	9.6±0.77a	35.4±3.19a	38.8±3.55c
变异系数(%)		55.5	49.6	48.2	45.2	30.8
1∶2	0	1.1±0.15d	1.9±0.05e	3.2±0.27e	7.5±1.07e	86.3±7.07a
	1	1.7±0.31d	3.5±0.17d	3.8±0.31e	12.1±0.63d	78.9±6.74a
	3	3.9±0.47c	5.2±0.39c	4.3±0.44de	17.1±1.82c	69.5±10.16b
	5	5.4±0.43c	8.1±0.42b	7.4±0.62c	23.6±1.41b	55.5±2.75c
	7	8.5±0.56b	12.6±0.72a	10.3±1.35b	36.8±3.17a	31.8±4.49d
	9	10.1±0.37a	13.4±1.68a	11.6±1.16a	37.2±2.35a	27.7±2.87e
变异系数(%)		70.8	64.0	52.8	55.9	41.9
1∶5	0	0.8±0.08d	1.2±0.07e	4.3±0.63cd	8.1±2.51f	85.6±5.26a
	1	1.4±0.13d	2.9±0.25d	4.7±0.45c	11.2±1.29e	79.8±6.87b
	3	2.8±0.19c	4.4±0.61c	3.3±0.59d	18.0±0.83d	71.5±5.09c
	5	3.7±0.22c	6.3±0.36b	5.2±0.37c	25.4±1.75c	59.4±4.72d
	7	5.9±0.47b	10.7±0.82a	7.5±0.94b	31.4±2.24b	44.5±2.45e
	9	7.8±0.39a	11.1±0.79a	12.3±1.15a	35.2±3.48a	33.6±3.58f
变异系数(%)		72.0	66.9	52.9	50.8	32.7

2.2 不同比例复配土壤>0.25 mm水稳性团聚体分布

种植前粒径>0.25 mm水稳性团聚体含量大小为:1∶1>1∶5>1∶2(见图1),无显著性差异,变化幅度为13.7%~15.8%。随着种植年限的增加,三种比例复配土壤中粒径>0.25 mm水稳性团聚体含量总体呈增加趋势,种植9年后,1∶1、1∶2和1∶5复配土壤粒径>0.25 mm水稳性团聚体含量相比种植前增加了2.87倍、4.28倍、3.61倍,其中1∶2复配土壤此粒径范围团聚体百分比高达72.3%,1∶5复配土壤此粒径范围团聚体百分比高达66.4%,1∶1复配土壤占比为61.2%。这说明玉米种植在一定程度上改善了复配土壤团聚体的质量,尤其是1∶2、1∶5复配土壤团聚化作用明显,并且不同比例复配土壤结构随着种植年限的增加结构发育趋势良好。

三种比例复配土壤中0.25~2 mm粒径水稳性团聚体的年际变化特征见图2。种植前0.25~2 mm粒径水稳性团聚体含量大小为:1∶1>1∶5>1∶2,变化幅度为14.4%~12.6%,各比例之间无显著性差异($P>0.05$)。随着试验开展,三种比例复配土壤中0.25~2 mm粒径水稳性团聚体均呈不断增加趋势,土壤团聚体组成由以<0.25 mm粒径为主逐渐转变为以0.25~2 mm的细大团聚体为主,与种植前

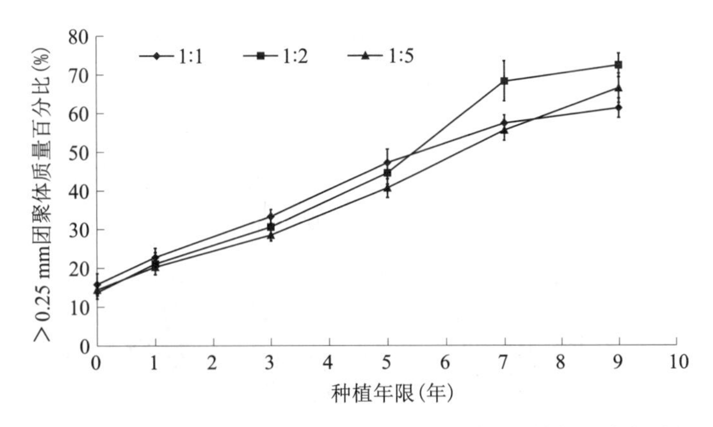

图1　不同比例复配土壤中>0.25 mm水稳性团聚体年际变化特征

Fig.1　Interannual variation of >0.25 mm water-stable aggregates in compound soils of different proportions

相比较,种植9年后1∶1、1∶2和1∶5复配土壤0.25~2 mm粒径水稳性团聚体含量分别增加了2.77倍、3.94倍、3.31倍,占比为54.3%~62.2%。这表明,玉米生长、耕种年限增强了复配土壤团聚化作用,使微小团聚体(<0.25 mm)向细大团聚体(0.25~2 mm)不断转化。因此,随着玉米种植年限的增加,主要改变了复配土壤的团聚化物质基础,为进一步的团聚奠定了基础。

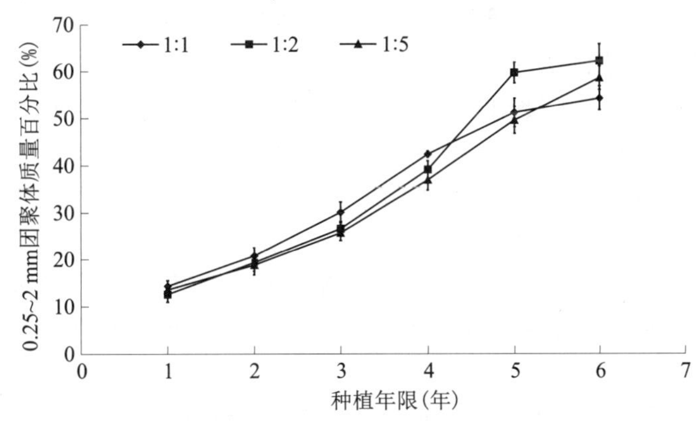

图2　不同比例复配土壤中0.25~2 mm水稳性团聚体年际变化特征

Fig.2　Interannual variation of 0.25~2 mm water-stable aggregates in compound soils of different proportions

2.3　不同比例复配土壤团聚体稳定性

随着种植年限的增加,三种比例复配土壤水稳性团聚体MWD值均表现为不断增加的趋势(见图3)。1∶1复配土壤种植前与种植1年后差异不显著($P>0.05$),种植3年、5年、7年MWD值显著提高($P<0.05$),种植7年后逐渐趋于稳定,不同年际间MWD值差异不显著($P>0.05$)。1∶2复配土壤水稳性团聚体MWD值种植前与种植1年后差异不显著,种植3年后MWD值呈现显著增加趋势,并且3年、5年、7年、9年之间MWD值具有显著差异($P<0.05$)。1∶5复配土壤种植前、1年、3年、5年间MWD值差异不显著($P>0.05$),种植5年后MWD值显著增加,并且不同年际间具有显著差异($P<0.05$)。与种植前相比较,种植9年后1∶1、1∶2和1∶5复配土壤MWD值增加了1.13倍、1.85倍和1.58倍,其中1∶2和1∶5复配土壤增加速率显著高于1∶1复配比例。

3　讨论

土壤团聚体的组成、数量、分布及稳定性是评价土壤结构的重要因素[19-20]。复配比例(砒砂岩含量)是风沙土团聚化的物质基础及条件,玉米种植年限是团聚化的动力学过程及影响因素,二者都是影响砒砂岩与沙复配土壤团聚体含量的重要因子,研究不同种植年限不同比例复配土壤团聚体的组成及稳定性是探索新造复配土壤发育过程中土壤结构和功能变化的基本要素。作物种植前,≥0.25 mm水稳性团聚体含量较低,是由于复配土组分只是原始砒砂岩和沙,属于成土母质,此时团聚体间以砒

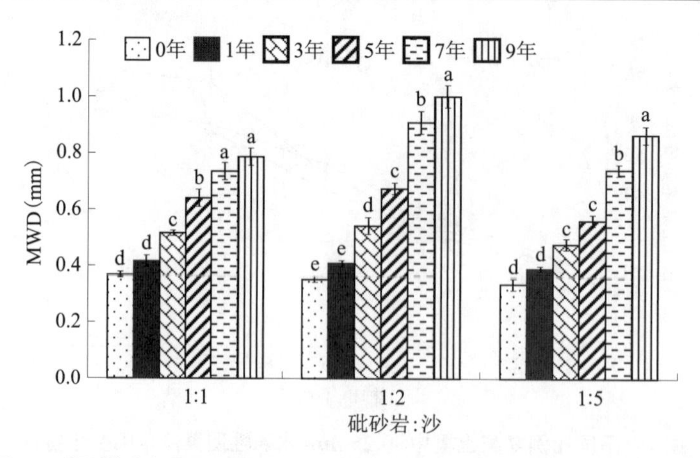

图 3 不同耕种年限不同比例复配土壤水稳性团聚体的 MWD 变化特征

Fig.3 MWD variation characteristics of compound soil water-stable aggregates in different tillage years

砂岩自身的无机胶结为主,砒砂岩富含碳酸盐矿物,能够促进砒砂岩与沙之间无机胶结形成团聚体。

砒砂岩与沙复配成土的核心目标之一是从物质上确定颗粒团聚化,减弱原风沙土的分散性,达到复配土的状态稳定。一般土壤团粒结构体是指>0.25 mm 粒径的团聚体,为土壤中最优的结构体,其数量与成土母质、土地利用方式、土壤肥力状况、土壤管理措施等密切相关。本研究发现作物种植年限对新造复配土壤团聚体具有显著影响,种植 9 年后,0.25~0.5 mm 粒径水稳性团聚体含量占比最大,其为>0.25 mm 粒径团聚体的主要组成部分,主要改变了复配土壤团聚化的物质基础,为以后更进一步团聚奠定了基础。这主要是由于有机肥的施用以及植物根系分泌物和土壤中微生物代谢产物不断增加,促进了土壤颗粒间的有机胶连作用,团聚作用增强,促进了<0.25 mm 粒径转变为 0.25~2 mm 粒径范围的团聚体,并且 1∶2 复配土壤粒径>0.25 mm 团聚体含量显著高于 1∶1 和 1∶5 复配比例,这主要是由玉米种植年限和砒砂岩与沙混合比例的交互作用对复配土壤水稳性团聚体的影响。因此,随着种植年限的增加,复配土团聚化作用明显,且有更进一步团聚的态势,土壤结构发育状态较好,不需要多年后再次进行复配。

MWD 是评价土壤团聚体稳定性的重要指标[21-22],随着种植年限的增加,三种比例复配土壤水稳性团聚体 MWD 均表现为逐渐增加的趋势,与种植前相比较,种植 9 年后 1∶1、1∶2 和 1∶5 复配土壤 MWD 值分别增加了 1.13 倍、1.85 倍和 1.58 倍,其中 1∶2 和 1∶5 复配土壤增加速率显著高于 1∶1 复配比例。这主要是由于成土母质、耕种年限和玉米种植三者间的综合作用影响着复配土壤团聚体的形成和稳定性。随着复配土壤中作物种植年限的增加,作物根系和其分泌物能固结团聚土粒,形成稳定的团粒结构。此外,由相关研究[11,23,24]可知,玉米生长特性及砒砂岩与沙混合比例二者因素的交互作用使 1∶2 复配土壤的黏粒、有机质含量高于 1∶1 和 1∶5 复配比例,而土壤有机质和黏粒含量直接影响到土壤颗粒的凝聚和胶结,土壤黏粒、有机质含量越高,土壤团聚化作用越强,土壤稳定性越高[25]。

4 结论

砒砂岩与沙混合比例、玉米种植年限二者的交互作用对复配土壤团聚体的形成和稳定性具有重要影响。随着玉米种植年限的增加,三种比例复配土壤>0.25 mm 粒径团聚体含量和 MWD 值呈持续增加趋势,不同比例复配土壤 MWD 值总体为:1∶2>1∶5>1∶1,其中 1∶2 和 1∶5 复配土壤增加速率显著高于 1∶1 复配比例,并且种植 9 年后,>0.25 mm 粒径团聚体的主要组成部分为 0.25~0.5 mm 粒径含量占比最大。在相同的农艺管理措施下,随着种植年限的增加,三种比例复配土壤不断熟化,团聚化作用不断增加,提升了土壤团聚体的稳定性能,土壤结构发育不断完善,实现了从“土”到“壤”质的提升过程。

参考文献

[1]黄昌勇.土壤学[M].北京:中国农业出版社,2000:71-102.

[2]卢金伟,李占斌.土壤团聚体研究进展[J].水土保持研究,2002,9(1):81-85.

[3]瞿晴,徐红伟,吴旋,等.黄土高原不同植被带人工刺槐林土壤团聚体稳定性及其化学计量特征[J].环境科学,2019,40(6):2904-2911.

[4]王丽,李军,李娟,等.轮耕与施肥对渭北旱作玉米田土壤团聚体和有机碳含量的影响[J].应用生态学报,2014,25(3):759-768.

[5]张世祺,王沛裴,王昌全,等.不同植烟年限对土壤团聚体稳定性的影响及其相关因素分析[J].土壤,2017,49(6):1229-1236.

[6]徐红伟,吴阳,乔磊磊,等.不同植被带生态恢复过程土壤团聚体及其稳定性:以黄土高原为例[J].中国环境科学,2018,38(6):2223-2232.

[7]刘敏英.植茶年限对土壤团聚体组成及其有机碳组分影响[D].成都:四川农业大学,2012.

[8]胡尧,李懿,侯雨乐.不同土地利用方式对岷江流域土壤团聚体稳定性及有机碳的影响[J].水土保持研究,2018,25(4):22-29.

[9]HAN J C, XIE J C, ZHANG Y. Potential role of feldspathic sandstone as a natural water retaining agent in Mu Us Sandy Land, northwest China[J]. Chinese Geographical Science, 2012, 22(5):550-555.

[10]摄晓燕,张兴昌,魏孝荣.适量砒砂岩改良风沙土的吸水和保水特性[J].农业工程学报,2014,30(14):115-123.

[11]SUN Z H, HAN J C. Effect of soft rock amendment on soil hydraulic parameters and crop performance in Mu Us Sandy Land, China [J]. Field Crops Research, 2018, 222:85-93.

[12]李裕瑞,范朋灿,曹智,等.毛乌素沙地砒砂岩与沙复配农田的固沙效应及其微观机理[J].中国沙漠,2017(3):421-430.

[13]孙涛,刘艺杉,孙崇玉,等.石灰岩山地植被恢复对土壤水稳性团聚体的影响[J].草业科学,2018,35(6):1361-1367.

[14]鲍士旦.土壤农化分析[M].北京:中国农业出版社,2000:30-172.

[15]郭军玲,王虹艳,卢升高.亚热带土壤团聚体测定方法的比较研究[J].土壤通报,2010,41(3):542-546.

[16]赵玉明,高晓飞,刘瑛娜,等.不同水稳性团聚体测定方法的对比研究[J].水土保持通报,2013,33(2):138-143.

[17]祁迎春,王益权,刘军,等.不同土地利用方式土壤团聚体组成及几种团聚体稳定性指标的比较[J].农业工程学报,2011,27(1):340-347.

[18]蔡立群,齐鹏,张仁陟.保护性耕作对麦-豆轮作条件下土壤团聚体组成及有机碳含量的影响[J].水土保持学报,2008,22(2):41-145.

[19]华瑞,徐学选,张少妮,等.不同退耕年限林草地土壤颗粒分形特征研究[J].水土保持学报,2016,30(4):206-209.

[20]文海燕,傅华,赵哈林.退化沙质草地开垦和围封过程中的土壤颗粒分形特征[J].应用生态学报,2006,17(1):155-159.

[21]苏志珠,刘蓉,梁爱民,等.晋西北沙化土地土壤机械组成与有机质的初步研究[J].水土保持研究,2018,25(6):61-67.

[22]刘文利,吴景贵,傅民杰,等.种植年限对果园土壤团聚体分布与稳定性的影响[J].水土保持学报,2014,28(1):129-133.

[23]罗林涛,程杰,王欢元,等.玉米种植模式下砒砂岩与沙复配土氮素淋失特征[J].水土保持学报,2013,27(4):58-66.

[24]WANG H Y, HAN J C, TONG W , et al. Analysis of water and nitrogen use efficiency for soft rock and sand compound soil [J]. Journal of the Science of Food and Agriculture, 2017,97(8):2553-2560.

[25]谷忠元,康黎,罗梦娟,等.湘东地区典型土壤团聚体稳定性的影响因素[J].水土保持通报,2018,38(5):58-63.

本文曾发表于2022年《干旱区资源与环境》第36卷第2期

5 种矿区土著植物对铅污染土壤的修复潜力研究

卢　楠[1,2,3,4,5],魏　祥[1,2,3,4,5],李　燕[1,2,3,4,5]

(1.陕西省土地工程建设集团有限责任公司,西安 710075;
2.陕西地建-西安交大土地工程与人居环境技术创新中心,西安 712000;
3.自然资源部退化及未利用土地整治工程重点实验室,西安 710075;
4.陕西省土地整治工程技术研究中心,西安 710075;
5.陕西地建土地工程技术研究院有限责任公司,西安 710075)

【摘要】为评估我国西北某矿区茵陈蒿、蒲公英、苜蓿、大叶苦菜和车前草等 5 种土著作物对含铅土壤的修复效果,采用盆栽试验的手段,设置了 4 种不同铅含量水平(0、2‰、3‰和 5‰,质量分数),监测了作物不同组织器官和种植前后土壤中 Pb 含量,以及根际与非根际土壤微生物生物量碳含量、过氧化氢酶活性等指标,对修复效果进行综合评价。结果表明:茵陈蒿和车前草作为土壤重金属铅污染修复植物的潜力较大。土壤铅含量水平不同,适种作物有所不同。多种铅含量水平下,茵陈蒿和车前草均适合种植,且对土壤铅去除率在 12%~32%。除苜蓿和车前草外,植物对铅的累积量随初始土壤重金属铅含量的增大而增大,车前草根系和茵陈蒿茎叶对铅的最高累积量分别为 3.62‰和 0.72‰,显著高于其他作物。此外,作物种类、原始土壤 Pb 污染程度等因素对根际、非根际微生物生物量碳含量和过氧化氢酶活性都有一定影响。根际土壤微生物生物量碳含量高于非根际土壤微生物生物量碳含量 2.37%~13.89%。土壤 Pb 抑制了根际与非根际土壤过氧化氢酶活性,使其活性低于对照组 0.44%~22.3%,根际土壤过氧化氢酶活性比非根际过氧化氢酶活性高 0.89%~8.09%。

【关键词】铅污染土壤;植物修复;累积;过氧化氢酶;微生物生物量碳

Phytoremediation Potential of Five Native Plants in Soils Cintaminated With Lead in a Mining Area

Nan Lu, Yang Wei, Yan Li

【Abstract】 Metal pollution in soils is an issue of global concern, and lead (Pb) pollution is considered to be the most serious type. The arid and semi-arid areas of Northwest China are rich in Pb ore resources. In this study, four native crops [wormwood (*Artemisia capillaris*), dandelion (*Taraxacum mongolicum*), alfalfa (*Medicago sativa*), sauce (*Lxeris chinensis*), and plantain (*Plantago asiatica* L.)] that grow naturally around tailings slag in a mining area in Northwest China were selected to screen their ecological restoration impacts on Pb-contaminated soil. In pot experiments, different metal lead pollution gradients (0, 2‰, 3‰, and 5‰, *w/w*) were set, the changes of soil Pb content in different tissues and organs before and after planting were analyzed. The results showed that a certain degree of soil lead pollution could promote plant growth, the accumulation of Pb in crops increased as soil Pb content increased. Under the same treatment conditions, the accumulation of heavy

基金项目:陕西地建-西安交大土地工程与人居环境技术创新中心开放基金项目(2021WH20094);陕西省土地整治重点实验室开放基金项目(2019-JC04);陕西省土地工程建设集团内部科研项目(DJNY2021-20)。

metal Pb in the underground plantain components and the aboveground dandelion components was more advantageous than other plants. Crop types and the initial soil Pb content have an effect on soil microbial biomass carbon and catalase activity. Microbial biomass carbon content in rhizosphere soil is above 2% higher than in bulk soil. The catalase activity in rhizosphere soil is about 0.89% to 8.07% higher than the content in bulk soil. It was found that wormwood and plantain have great potential as remediation plants for soil metal lead pollution in the arid and semi-arid lead bearing mining areas of Northwest China.

【**Keywords**】 Lead contaminated soil; Phytoremediation; Accumulation; Catalase activity; Microbial biomass carbon

0 引言

土壤重金属污染是全球关注的环境问题之一,而 Pb 污染被认为是最严重的金属污染,是造成环境污染的主要原因[1-2]。Pb 是一种对人体危害极大的有毒重金属,也是第二大有害重金属[3],根据 2014 年 4 月发布的全国土壤污染状况调查结果,有色金属矿区周边土壤 Pb 污染较为严重,点位超标率达 1.5%[5]。陕西省地处西北内陆腹地,属干旱半干旱区,秦岭山区矿产资源丰富,其中 Pb 矿产储量居西部 12 省前列。工业、矿业开采和冶炼等活动导致工矿区及其周边土壤 Pb 含量日益累积,亟须采取有力措施,解决土壤 Pb 污染问题。

植物修复技术具有稳定地表土层结构,可减少因采矿活动造成的地质灾害,改善土壤养分,降低污染物含量等优势[6],且绿色环保,成本低廉,实现污染修复与生态恢复的同时进行,因而,国内外应用植物修复技术修复工矿区重金属污染较为广泛。国外对重金属 Pb 污染的植物修复研究开展得比较早,主要集中于筛选 Pb 超富集植物,发现了包括遏蓝菜属、蓝云英属等 500 多种 Pb 超富集植物;研究了植物对重金属 Pb 的累积效应,探索植物生长代谢与污染胁迫间的交互影响,揭示了不同的植物组织器官对 Pb 累积效应存在差异。有研究者发现 *Thlaspi rotundifolium* (L.)地上部分可吸收高达 8 500 mg(Pb)/kg(干重)[7],生长于营养液中的 *Brassica juncea* (L.)茎的 Pb 含量可以达到 1.5%[8]。通过评估铅锌冶炼厂周边多种植物不同组织中 Pb 含量发现,万寿菊和鬼针草在叶片中的累积量较高,黑穗草则在根系中累积量较高,分别表现出具有较大的修复潜力和稳定重金属的潜力[9-10]。在土壤生态系统中,酶参与有机质的矿化分解,其质量和活性大小表征了土壤肥力的高低以及碳和养分循环速率的快慢,在一定程度上能反映出土壤生态系统物质循环速率、方向及土壤质量的演变规律[11-12]。

茵陈蒿、蒲公英、苜蓿、大叶苦菜和车前草等 5 种多年生草本植物是西北干旱半干旱区常见种,具有耐旱、耐寒、耐贫瘠,且可作药用的特点,与引进物种相比,具有较高成活率和生物量,为评价几种植物对土壤中 Pb 的耐受及修复潜力,通过模拟矿区土壤 Pb 污染特点,围绕以下内容开展研究:①厘清土壤-植物间 Pb 的再分配关系;②植物不同组织器官对 Pb 的富集规律;③根际与非根际微生物生物量碳、过氧化氢酶活性随土壤 Pb 含量的变化情况。通过综合比较,总结筛选出具有 Pb 修复潜力的土著植物,为 Pb 污染矿区废弃地植物修复和土壤环境质量评价提供理论依据。

1 材料与方法

1.1 研究区描述

试验于 2019—2020 年在自然资源部退化及未利用土地整治工程重点实验室温室进行。试验所在地富平县地处暖温带半干旱气候区,属显著大陆性季风气候,年平均气温 13.3 ℃,年平均地面温度 15.6 ℃,年均日照时长 2352.3 h,年平均降水量 513.5 mm,太阳总辐射 550×10^7 J/m^2,年平均风速 2.4 m/s,全年无霜期 223 d。

1.2 试验材料

1.2.1 供试土壤

供试土壤采自矿区农田(110°21′40″E, 34°30′16″N)0~20 cm 表层土壤,除去杂物,风干过 5.0 mm 尼龙筛后用于盆栽试验。供试土壤基本理化性质见表 1。

表1 供试土壤的物理化学性质

Table 1 Physical and chemical characteristics of test soil

pH	电导率(dS/m)	粒径组成(%)			质地(USDA)
		黏粒(<0.002 mm)	粉粒(0.05~0.002 mm)	砂粒(0.05~2 mm)	
9.15	0.367	5.47	78.78	15.75	粉壤土

速效钾(mg/kg)	有效磷(mg/kg)	全氮(g/kg)	有机质(g/kg)	CEC(cmol/kg)	全量铁(mg/kg)	$CaCO_3$(%)
85.70	3.2	0.39	5.73	3.69	1.23	10.72

1.2.2 供试植物

选用茵陈蒿(*Artemisia capillaries*)、蒲公英(*Taraxacum Officinale* L.)、苜蓿(*Medicago sativa*)、大叶苦菜(*Lxeris chinensis*)和车前草(*Plantagoasiatica* L.)等5种西北地区常见多年生药用及饲草植物作为研究对象。种子来源于上一年秋天在矿区自行收集。

1.3 试验设计

根据GB 15618—2018《土壤环境质量 农用地土壤污染风险管控标准》,Pb的农用地土壤污染风险管制值(pH>7.5)为1000 mg/kg,标准值的2~3倍为轻度-中度污染,5倍及以上为重度污染,采用喷施$Pb(NO_3)_2$溶液的形式制备不同程度Pb污染土壤。

试验处理分别设置为CK、T1、T2和T3,对应污染土壤目标Pb含量分别为0、2‰、3‰和5‰(*w*/*w*)。制备完成后平衡1个月,测定土壤Pb含量,作为实测含量,见表2。向稳定后的各梯度Pb浓度土壤中施入尿素、磷酸二氢钾和硫酸钾粉末等基肥,使土壤氮含量225 mg/kg、磷含量65 mg/kg、钾含量227 mg/kg。将各处理土壤定量称取2.5 kg移入盆钵,筛选颗粒饱满且大小均匀、无虫蛀的茵陈蒿、蒲公英、苜蓿、大叶苦菜和车前草等5种植物种子,分别以字母A、T、M、L和P代表,在室温下直接播入,待出苗后,及时间苗,保持每盆种植数量一致。采用称重法,保持盆中含水量约为田间持水量的65%。每个处理设3个重复,并于播种90 d后收获植物。试验设计及污染土壤实测Pb含量见表2。

表2 试验设计及污染土壤实测Pb含量(平均值±标准差)

Table 2 Test design and measured Pb content in polluted soil (mean±SE) ‰

处理	目标含量	实测含量	种植作物
CK	0	0.06547 ±0.00640	A, T, M, L, P
T1	2	2.23942 ±0.08999	
T2	3	3.25626 ±0.14722	
T3	5	5.10648 ±0.10070	

1.4 测定指标与方法

1.4.1 土壤样品处理及指标测定

(1)土壤样品处理:植物收获时,取根际(植物根部的少量土壤)与非根际土壤鲜样,用于测定土壤微生物生物量碳、过氧化氢酶活性;土壤样品风干后,用木棒粉碎和研磨,筛分(<0.149 mm)并储存,用于测定土壤Pb含量。

(2)土壤样品测定:准确称取并记录已处理好的土壤样品重量(精度万分之一),采用微波消解,消解液过0.45 μm聚四氟乙烯滤膜,稀释并测量。样品总Pb含量采用原子吸收光谱仪(AAS Zeenit 700P)进行测定。每运行20个样品通过测定标准溶液校正仪器的准确性[13]。土壤微生物生物量碳含量采用氯仿熏蒸-K_2SO_4直接浸提法测定[14]；土壤过氧化氢酶活性采用紫外分光光度法，以每 20 min 内每 1 g 土壤分解的过氧化氢的微克数(μg)表示[15]。

1.4.2 植物样品处理及指标测定

(1)植物生长指标：收获后，将鲜样剪分为地上部分和地下部分，并于 105 ℃杀青 15 min，75 ℃烘干至恒重，采用电子天平称重(精确至 0.001 g)。

(2)植物不同组织中 Pb 含量的测定：采用玛瑙研钵研磨后，过 60 目筛，准确称取 0.1000 g，采用硝酸-盐酸体系进行消解，消解液测定方法同土壤样品。

1.5 数据处理

植物转移系数（Transfer coefficient, TF）、污染土壤 Pb 的去除率(η,%)以及 Pb 富集系数(Plant bioaccumulation factor, BCF)等参数计算分别参见式(1)~式(3)。

$$TF=\frac{C_s}{C_r} \tag{1}$$

$$\eta=\frac{C_{so}-C'_{so}}{C_{so}} \tag{2}$$

$$BCF=\frac{C_{pt}}{C_{so}} \tag{3}$$

式中：C_s和 C_r分别是植物茎叶和根系中的 Pb 含量，mg/kg；C_{so}和 C'_{so}分别是作物种植前后土壤 Pb 含量，mg/kg；C_{pt}是收获作物组织中的 Pb 含量，mg/kg。

1.6 数据统计分析与画图

采用单因素方差分析(One-way ANOVA)，以揭示不同浓度重金属铅胁迫对茵陈蒿、苜蓿、蒲公英、大叶苦菜、车前草的生长及对铅吸收积累的影响，并用 Duncan 检验法检验每个指标在不同处理组(p<0.05)的差异显著性。采用 Origin8.5 作图。

2 结果和分析

2.1 植物组织器官中 Pb 的累积与分布

土壤 Pb 不同污染水平下，各植物组织器官中 Pb 的累积量如图 1 所示。

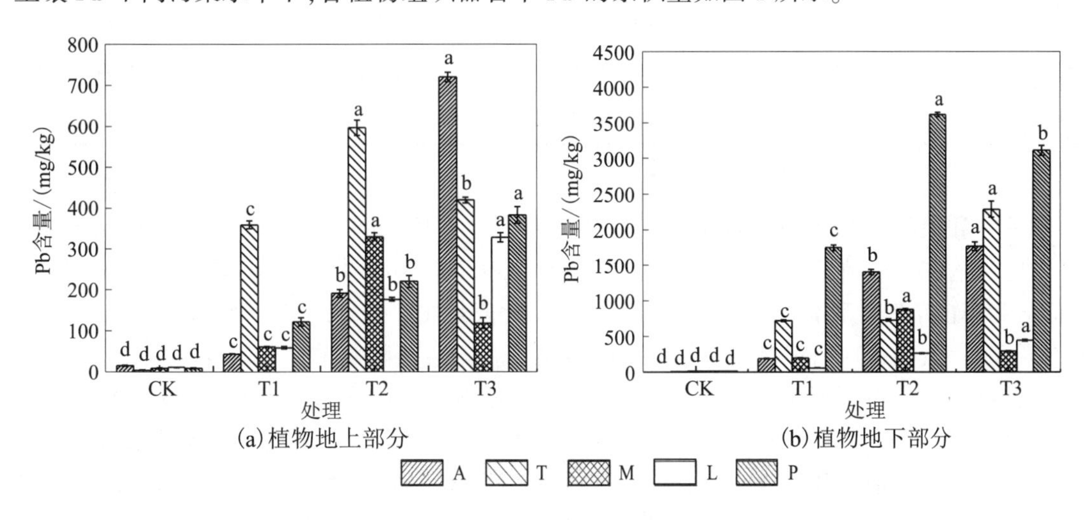

图 1 不同处理条件下植物中 Pb 含量变化情况

Fig.1 Pb concentration in the organs of the native plants species at different treatments, different letters (a, b, c, d) represent significant differences at p < 0.05.

不同处理条件下，各植物组织中 Pb 含量与对照处理相比，均达显著水平(p<0.05)，表明土壤 Pb 污染对西北矿区 5 种常见植物 Pb 含量存在不同程度的影响。不同植物对污染土壤中 Pb 元素的吸收和累积特征有所不同，且随着土壤 Pb 含量的升高，植物不同组织器官对 Pb 的累积量差异较大。

在污染土壤目标 Pb 含量不高于 5‰时,茵陈蒿 Pb 含量随土壤 Pb 含量的升高而增加($p<0.05$),茎叶和根系中 Pb 含量最高分别达 720 mg/kg 和 1768 mg/kg,分别是对照处理茎叶和根系 Pb 含量的 48.3 倍、491 倍。大叶苦菜的 Pb 含量随土壤 Pb 含量的升高而增加($p<0.05$),茎叶和根系中 Pb 含量最高分别为 340 mg/kg 和 456 mg/kg,分别是对照处理茎叶和根系 Pb 含量的 30.4 倍、48.4 倍。在污染土壤目标 Pb 含量不高于 3‰时,苜蓿 Pb 含量随土壤 Pb 含量的升高而增加($p<0.05$),茎叶和根系部分 Pb 含量最高分别达 330 mg/kg 和 876 mg/kg,分别为对照处理茎叶和根系 Pb 含量的 39.3 倍、78.9 倍。也有一些植物地上部分和地下部分对 Pb 的最大累积量,出现在污染土壤目标 Pb 含量不同的处理中,如污染土壤目标 Pb 含量不高于 5‰时,蒲公英根系和车前草茎叶 Pb 含量随土壤 Pb 含量的升高而增加($p<0.05$),Pb 的最大累积量分别达 2284 mg/kg 和 383 mg/kg,分别是对照处理根系和茎叶 Pb 含量的 439.2 倍、47.9 倍。污染土壤目标 Pb 含量不高于 3‰时,蒲公英茎叶和车前草根系 Pb 含量随土壤 Pb 含量的升高而增加($p<0.05$),Pb 的最大累积量分别达 596 mg/kg 和 3617 mg/kg,分别是对照处理根系和茎叶 Pb 含量的 161.1 倍、267.9 倍。对比 5 种植物不同组织中 Pb 含量,从整体情况来看,根系含量普遍高于茎叶。这与目前大多数报道结论一致,可能原因在于 Pb 进入土壤后主要以硫化物和氧化物的二价态形式存在,形成的不溶性沉淀物,通常较难移动,不利于植物吸收[16],但仍能被富集。Pb 在根系主要以磷酸铅和碳酸铅等沉淀形式存在,根系吸收的 Pb 向地上部运输困难[17]。本实验的污染土壤为中性-弱碱性土,这也进一步增加了 Pb 离子活化的难度。

利用式(1)计算茵陈蒿、蒲公英、苜蓿、大叶苦菜和车前草等 5 种植物对重金属 Pb 的转移系数,结果见表 3。

表 3　不同污染水平对植物中 Pb 转移系数的影响

Table 3　Effect of Pb pollution level on lead translocation factor

作物	处理			
	CK	T1	T2	T3
茵陈蒿	4.194±0.426 a	0.228±0.014 b	0.136±0.010 b	0.408±0.016 b
蒲公英	0.711±0.050 b	0.499±0.010 c	0.819±0.020 a	0.184±0.007 d
苜蓿	0.759±0.080 a	0.310±0.016 c	0.377±0.009 bc	0.407±0.045 b
大叶苦菜	1.119±0.025 a	1.008±0.053 b	0.662±0.011 d	0.728±0.029 c
车前草	0.593±0.088 a	0.092±0.004 b	0.069±0.005 b	0.061±0.004 b

注:不同小写字母代表经 LSD 检验的差异显著性($p < 0.05$)。

植物茎叶和根系对 Pb 元素的吸收和分布累积特征也不尽相同。转移系数(TF)能反映出 Pb 在植物体内的运输和分配情况[18]。由土壤-植物体系中 Pb 的转移系数(TF)可知(见表 3),茵陈蒿、蒲公英、苜蓿和车前草等 4 种植物的 TF 均低于 1,仅大叶苦菜在部分处理条件下 TF 略高于 1。相同土壤 Pb 含量条件下,在矿区自然状态生长的大叶苦菜、蒲公英、茵陈蒿和苜蓿转移系数高于车前草,有一定运输 Pb 的能力,特别是大叶苦菜,在土壤 Pb 质量浓度处于中度污染程度以下时,TF 达 1.008,能够将相当量的 Pb 运输至茎叶中去,便于采用收获或刈割的方式,将 Pb 从污染土壤中去除,对矿区 Pb 污染土壤具有一定的修复潜力,可大大提高矿区污染土地的修复效率。在不同土壤 Pb 质量浓度含量条件下,茵陈蒿、苜蓿、大叶苦菜和车前草转移系数均低于对照,这表明 Pb 由根部转移至茎叶分布阻力较大。车前草对 Pb 的转移系数显著低于其余 4 种作物,尤其是车前草的转移系数在污染水平升至轻度时即降至较为稳定的状态,表明该植物可将 Pb 元素固定于根系部分,阻控 Pb 向植物体的地上部分转移,减少 Pb 对植物体的毒害作用。

此外,植物的生长代谢与铅污染之间具有复杂的交互影响机制。Gupta 等[19]在研究植物对 Pb 的吸收和转移过程中发现,植物体可依靠隔离外排、诱导螯合、酶促反应等机制,有效避免因 Pb 累积引发的毒性效应，提高自身耐受能力。Malar 等[10]探明了植物可通过调控光合色素、丙二醛含量和抗氧化酶水平等对 Pb 产生耐受，沙参能够将 Pb 储存于根部而不影响生长。Liu 等[20]在研究毛竹幼苗对 Pb 的耐受性及修复潜力中发现，在受到一定程度 Pb 胁迫时，植物生长受到一定程度的抑制，但其在根系形成了较多内含物，耐受能力和生物量并未受到显著影响。

2.2 植物种植对土壤中 Pb 的修复效果

植物生长对土壤重金属 Pb 的去除率变化的影响见图 2。由图 2 可知，种植作物种类、初始土壤 Pb 含量等因素对土壤中 Pb 含量的减少都有一定影响。总体而言，5 种植物的种植对土壤 Pb 含量的降低均有一定的作用。不同土壤 Pb 污染水平下，植物种植 90 天后，对土壤 Pb 的去除率［式(2)］排序有所不同。当土壤 Pb 处于轻度及以下污染水平时，蒲公英>茵陈蒿>车前草>大叶苦菜>苜蓿。当土壤 Pb 处于轻度-中度污染水平时，茵陈蒿>车前草>苜蓿>大叶苦菜>蒲公英。当土壤 Pb 处于中度-重度污染水平时，蒲公英>苜蓿>车前草>茵陈蒿>大叶苦菜。当土壤 Pb 处于重度及以上污染水平时，茵陈蒿>车前草>大叶苦菜>苜蓿>蒲公英。对不同植物对土壤 Pb 的去除率求取平均值进行排序发现，茵陈蒿和车前草对土壤 Pb 的去除率高于苜蓿、大叶苦菜和蒲公英，去除率分别为 26.25%、19.72%、18.63%、17.43%和 13.01%。随着土壤 Pb 污染程度的不断提高，车前草对土壤 Pb 的去除率不断降低，由 24.03%降低至最终的 12.82%。茵陈蒿对土壤 Pb 的去除率与污染程度无显著关系，但去除率在土壤 Pb 为重度污染程度时达最大，去除率达 31.89%。蒲公英和苜蓿对土壤 Pb 的最大去除率均出现在土壤 Pb 处于轻度-中度污染水平，大叶苦菜对土壤 Pb 的最大去除率出现在土壤 Pb 处于轻度及以下污染水平。

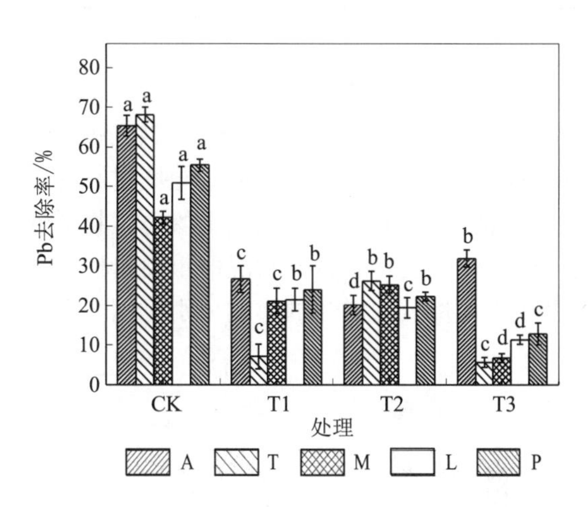

图 2 不同处理条件下土壤 Pb 去除率的变化情况

Fig.2 The removal rate of Pb in soil at different treatments, different letters (a, b, c, d) represent significant differences at $p < 0.05$

土壤 Pb 污染水平对土壤-植物体系中 Pb 的富集系数的影响见表 4。由表 4 可知，植物 BCF 对 Pb 污染胁迫的响应存在差异，且普遍存在地下部分 BCF 高于地上部分 BCF 的现象。除在土壤 Pb 处于轻度及以下污染水平时，蒲公英地上部分 BCF、茵陈蒿地下部分 BCF 与对照处理无显著差异性之外，其余不同 Pb 污染水平胁迫下，植物 BCF 与对照处理均存在显著差异($p<0.05$)。茵陈蒿、苜蓿和车前草地上部分 BCF 随着 Pb 污染程度的提高出现下降，其中茵陈蒿地上部分 BCF 在无污染情况下数值达最大。仅车前草地下部分 BCF 在土壤 Pb 轻度-中度污染水平下大于 1，其余植物各器官富集系数均小于 1。在不同土壤 Pb 含量胁迫处理下，蒲公英地下部分 BCF 随着土壤 Pb 污染程度的提高逐渐增大($p<0.05$)，且在土壤 Pb 污染程度不高于重度时数值达最大 0.474。车前草根系 BCF 在一定范围内，随着土壤 Pb 污染程度的提高逐渐增大，在土壤 Pb 污染程度不高于中度时达最大值1.430，随着土壤 Pb 含量的进一步提高，车前草地下部分 BCF 出现下降，说明土壤 Pb 含量在一定范围内，可促

进车前草根系对重金属 Pb 的富集,进一步证明了根系是植物直接吸收、转移营养元素和 Pb 的主要组织器官。

表 4　土壤 Pb 污染水平对土壤-植物体系中 Pb 富集系数的影响

Table 4　Effect of Pb pollution level on lead bioaccumulation factor in soil-plant system

BCF	作物	处理			
		CK	T1	T2	T3
根系	茵陈蒿	0.160±0.040 a	0.121±0.008 a	0.539±0.016 b	0.509±0.022 b
	蒲公英	0.248±0.010 d	0.349±0.007 b	0.303±0.010 c	0.474±0.019 a
	苜蓿	0.293±0.030 b	0.114±0.006 c	0.360±0.014 a	0.061±0.002 d
	大叶苦菜	0.294±0.027 c	0.122±0.016 d	0.423±0.062 b	0.786±0.078 a
	车前草	0.462±0.020 d	1.062±0.007 b	1.430±0.022 a	0.931±0.022 c
茎叶	茵陈蒿	0.662±0.111 a	0.028±0.001 c	0.074±0.006 c	0.207±0.003 b
	蒲公英	0.177±0.012 b	0.174±0.001 b	0.248±0.004 a	0.087±0.002 c
	苜蓿	0.222±0.019 a	0.036±0.001 c	0.136±0.007 b	0.025±0.003 c
	大叶苦菜	0.328±0.024 b	0.124±0.020 c	0.279±0.037 b	0.572±0.058 a
	车前草	0.274±0.039 a	0.074±0.006 b	0.087±0.006 b	0.086±0.007 b

注:不同小写字母代表经 LSD 检验的差异显著性($p < 0.05$)。

2.3　根际和非根际土壤微生物生物量及酶活性

Pb 污染土壤易造成植被稀少、营养元素缺乏,种植土著植物,改善土壤环境,可对土壤环境变化敏感的土壤微生物生物量和酶活性产生直接或间接影响。土壤过氧化氢酶是生物防御体系的关键酶之一,且活性与土壤呼吸及微生物活动密切相关,可表征土壤总生物活性[21],土壤微生物生物量碳和过氧化氢酶活性是监测土壤环境有效性变化和营养状况的重要指标[22]。重金属 Pb 污染土壤中植物生长对根际和非根际土壤微生物生物量及酶活性的影响分别见图 3 和图 4。

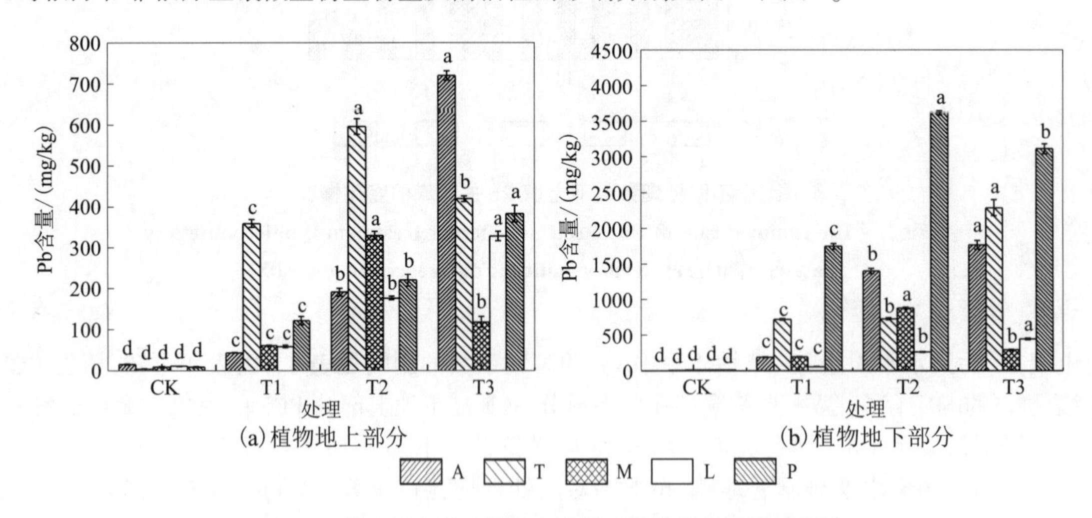

图 3　不同处理条件下根际与非根际微生物生物量碳含量

Fig.3　The content of microbial biomass carbon in rhizosphere soil and bulk soil at different treatments, different letters (a, b, c) represent significant differences at $p < 0.05$

由图 3 可知,作物种类、原始土壤 Pb 污染程度等因素对根际、非根际微生物生物量碳含量都有一定影响。整体上,根际土壤微生物生物量碳含量高于非根际土壤微生物生物量碳含量 2.37%~13.89%。不同种类植物根际与非根际土壤微生物生物量碳含量对土壤 Pb 含量变化的响应不同。种植车前草的根际与非根际土壤微生物生物量碳含量与污染土壤 Pb 含量差异不显著($p>0.05$)，种植茵陈蒿非根际土壤微生物生物量碳含量与污染土壤 Pb 含量同样差异不显著($p>0.05$)，即车前草的根际、茵陈蒿与车前草的非根际土壤微生物生物量碳含量不随土壤 Pb 含量的变化而变化。蒲公英和大叶苦菜的根际与非根际土壤微生物生物量碳含量在土壤目标 Pb 含量为 2‰时与对照处理相比差异显著($p<0.05$)；当土壤目标 Pb 含量高于 2‰时，蒲公英和大叶苦菜的根际与非根际土壤微生物生物量碳含量保持相对稳定，低于对照组 11.2%～19.6%，不再随污染水平的提高大幅变化。当土壤目标 Pb 含量在 2‰～3‰时，蒲公英的非根际土壤微生物生物量碳含量保持相对稳定；当土壤目标 Pb 含量高于 3‰时，土壤微生物生物量碳含量随土壤 Pb 含量的升高出现降低，低于对照组 35.6%。

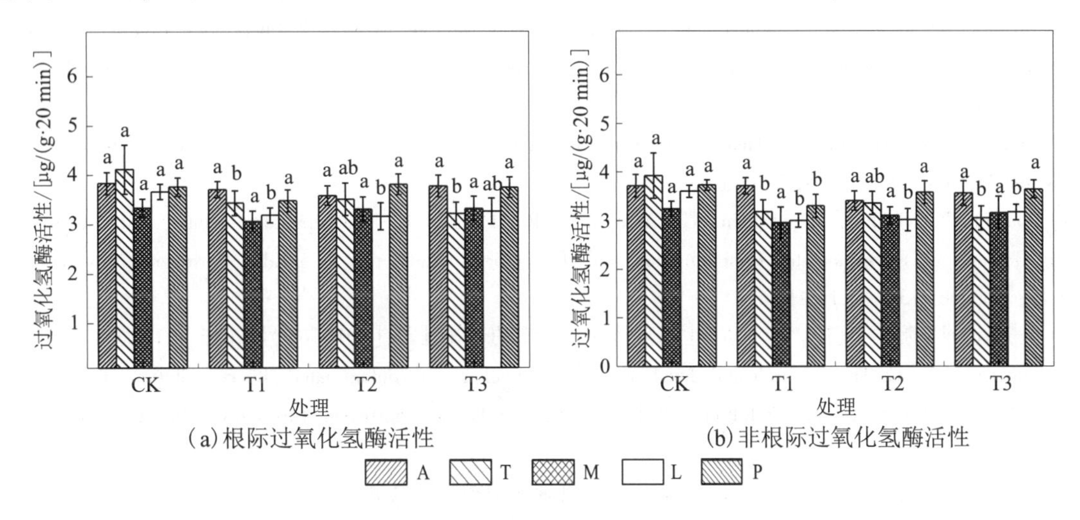

图 4　不同处理条件下根际与非根际过氧化氢酶活性

Fig.4　The content of catalase in rhizosphere soil and bulk soil at different treatments

由图 4 可知，土壤无污染且种植相同作物条件下，根际与非根际土壤过氧化氢酶活性均高于含 Pb 土壤过氧化氢酶活性。不同种类植物根际与非根际土壤过氧化氢酶活性对土壤 Pb 含量变化的响应不同。种植茵陈蒿、苜蓿和车前草的根际土壤过氧化氢酶活性与污染土壤 Pb 含量差异不显著($p>0.05$)，种植茵陈蒿、苜蓿的根际土壤过氧化氢酶活性与污染土壤 Pb 含量差异不显著($p>0.05$)，即茵陈蒿、苜蓿的根际与非根际土壤过氧化氢酶活性和车前草的根际土壤过氧化氢酶活性不随土壤 Pb 含量的变化而变化。蒲公英和大叶苦菜根际与非根际土壤过氧化氢酶活性在土壤目标 Pb 含量为 2‰时与对照处理相比差异显著($p<0.05$)，当土壤目标 Pb 含量高于 2‰之后，土壤过氧化氢酶活性保持相对稳定，不再随污染水平的提高大幅变化，根际土壤过氧化氢酶活性低于对照组 10.8%～21.8%，非根际土壤过氧化氢酶活性低于对照组 12.0%～22.3%。车前草的非根际土壤过氧化氢酶活性在土壤目标 Pb 含量为 2‰时与对照处理相比差异显著($p<0.05$)，其余处理与对照处理差异不显著，即在土壤目标 Pb 含量不高于 2‰时，车前草的非根际土壤过氧化氢酶活性与对照处理相比出现了降低，低于对照处理非根际土壤过氧化氢酶活性 11.5%。

3　结果和分析

通过模拟土壤 Pb 不同含量状况，采用盆栽试验种植了茵陈蒿、蒲公英、苜蓿、大叶苦菜和车前草等 5 种作物，监测了作物种植前后土壤 Pb 含量、植物茎叶及根系中 Pb 累积量等指标，探索了不同植物对土壤 Pb 的耐受阈值、土壤-植物间 Pb 的再分配关系及植物体内不同部位的富集规律，筛选出了针对不同梯度 Pb 含量土壤适宜种植作物。当土壤 Pb 含量不高于 2‰时，适宜种植茵陈蒿和车前草，土壤 Pb 去除率不低于 24%，车前草、蒲公英根系 Pb 的累积量分别为 1.75‰和 0.72‰；当土壤 Pb 含量不高于 3‰时，适宜种植蒲公英、苜蓿、车前草、茵陈蒿等作物，土壤 Pb 去除率高于 20%，车前草和茵陈蒿根系中 Pb 的累积量分别为 3.62‰和 1.40‰；当土壤 Pb 含量不高于 5‰时，种植车前草、蒲公英和茵陈蒿等作物，土壤 Pb 去除率高于 11%，3 者根系中 Pb 的累积量为 1.76‰～3.11‰。植物对

Pb 的累积量随初始土壤重金属 Pb 含量的增大而增大,车前草根系 Pb 累积量最高达 3.11‰,茵陈蒿茎叶中 Pb 累积量最高为 0.72‰,显著高于其他植物。此外,作物的种类、初始土壤 Pb 含量对土壤微生物生物量碳和过氧化氢酶活性具有不同程度的影响,根际土壤微生物生物量碳含量高于非根际含量 2%以上。根际与非根际过氧化氢酶活性相差不大,根际较非根际含量高 0. 89%~8.09%。经综合比较,茵陈蒿和车前草具有较大的修复潜力,可作为西北干旱、半干旱含铅矿区土壤重金属铅污染修复植物。同时,研究成果或可用于公路沿线受车辆燃油污染带的种植修复,应用前景较广泛。

参考文献

[1] Sarwar, N.; Imran, M.; Shaheen, M.R.; Ishaque, W.; Kamran, M.A.; Matloob, A.; Rehim, A.; Hussain, S. Phytoremediation strategies for soils contaminated with heavy metals: Modifications and future perspectives. Chemosphere 2017, 171, 710-721.

[2] Kumar, G.H.; Kumari, J.P. Heavy metal lead influative toxicity and its assessment in phytoremediating plants—A review. Water Air Soil Pollut. 2015, 226, 324.

[3] Pourrut, B.; Shahid, M.; Dumat, C.; Winterton, P.; Pinelli, E. Lead uptake, toxicity, and detoxification in plants. In Reviews of Environmental Contamination and Toxicology; Whitacre, D., Ed.; Springer: New York, NY, USA, 2011; Volume 213, pp. 113-136.

[4] 国务院.国务院关于引发"十三五"生态环境保护规划的通知,国发〔2016〕65 号,2016 年 12 月 5 日.

[5] Luo, Y.; Tu, C. The research and development of technology for contaminated site remediation. Twenty Years of Research and Development on Soil Pollution and Remediation in China; Springer, Singapore, 2018; Chapter 48, pp. 785-798.

[6] 吴仁杰, 陈银萍, 曹雯婕,等. 营养元素与螯合剂强化植物修复重金属污染土壤研究进展[J]. 中国土壤与肥料, 2021,5:328-337.

[7] Reeves, R.D.; Brooks, R.R. Hyperaccumulation of lead and zinc by two metallophytes from mining areas of Central Europe. Environ. Pollut. Ser. A Ecol. Biol. 1983, 31, 277-285.

[8] Kumar, P.N.; Dushenkov, V.; Motto, H.; Raskin, I. Phytoextraction: The use of plants to remove heavy metals from soils. Environ. Sci. Technol. 1995, 29, 1232-1238.

[9] Salazar, M.J.; Pignata, M.L. Lead accumulation in plants grown in polluted soils. Screening of native species for phytoremediation. J. Geochem. Explor. 2014, 137, 29-36.

[10] Malar, S.; Manikandan, R.; Favas, P.J.C.; Sahi, S.V.; Venkatachalam, P. Effect of lead on phytotoxicity, growth, biochemical alterations and its role on genomic template stability in Sesbania grandiflora: A potential plant for phytoremediation. Ecotoxicol. Environ. Saf. 2014, 108, 249-257.

[11] 马志良,赵文强,刘美.高寒灌丛生长季根际和非根际土壤多酚氧化酶和过氧化氢酶活性对增温的响应[J].应用生态学报,2019,30(11): 3681-3688.

[12] 王孝涛,李淑芹,许景钢,裴占江. 生物肥对大豆根际过氧化氢酶和脉酶活性的影响[J].东北农业大学学报,2012,43(5): 96-99.

[13] Lu, N.; LI, G.; Hav, J.C.; Wang, H.Y.; Wei, Y.; Sun, Y.Y. Investigation of lead and cadmium contamination in mine soil and metal accumulation in selected plants growing in a gold mining area. Appl. Ecol. Environ. Res. 2019, 17, 10587-10597.

[14] 孙凯,刘娟,凌婉婷.土壤微生物量测定方法及其利弊分析.土壤通报,2013, 44(4):1010-1016.

[15] 杨兰芳,曾巧,李海波,等.紫外分光光度法测定土壤过氧化氢酶活性[J].土壤通报,2011,42(10):207-210.

[16] Pitchell, J.; Kuroiwa, K.; Sawyerr, H.T. Distribution of Pb, Cd, and Ba in soils and plants of two contaminated soils. Environ. Pollut. 1999, 110, 171-178.

[17] Romeh, A.A.; Khamis, M.A.; Metwally, S.M. Potential of Plantago major L. for phytoremediation of Lead-contaminated soil and water. Water Air Soil Pollut. 2016, 227, 9.

[18] Buscaroli, A. An overview of indexes to evaluate terrestrial plants for phytoremediation purposes. Ecol. Indic. 2017, 82, 367-380.

[19] Gupta, D.K.; Huang, H.G.; Corpas, F.J. Lead tolerance in plants: Strategies for phytoremediation. Environ. Sci. Pollut. Res. 2013, 20, 2150-2161.

[20] Liu, D.; Li, S.; Islam, E.; Chen, J.R.; Wu, J.S.; Ye, Z.Q.; Peng, D.L.; Yan, W.B.; Lu, K.P. Lead accumulation and tolerance of Moso bamboo (Phyllostachys pubescens) seedlings: Applications of phytoremediation. J. Zhejiang Univ.-Sci. B 2015, 16, 123-130.

[21] 张双,肖听,白兴雷,贾红霞. Cu 和 Cd 胁迫对小麦生长过程中土壤过氧化氢酶活性的影响[J].安徽农业科学, 2009, 37 (20): 9422-9424.

[22] 罗明霞,胡宗达,刘兴良,等. 川西亚高山不同林龄紫果云杉人工林土壤微生物生物量及酶活性[J].生态学报, 2021, 41(14).

本文曾发表于 2022 年《环境工程》第 40 卷第 11 期

城市绿地不同管理方式土壤重金属污染及生态风险评价

孟婷婷[1,2,4],刘金宝[1,2,3,4],董　浩[1,2,4],王　博[1,2,4],张国剑[1]

(1.陕西省土地工程建设集团有限责任公司,西安 710075;2.陕西地建土地工程技术研究院有限责任公司,西安 710075;
3.西安理工大学,西安 710048;4.自然资源部退化及未利用土地整治工程重点实验室,西安 710075)

【摘要】土壤重金属污染是威胁城市生态环境的关键问题,为探究城市绿地不同管理措施下土壤修复效果,以城市不同管理措施绿地土壤为研究对象,测定复垦土地不同管理措施土壤中的 Cr、Ni、Cu、Zn、As、Cd、Pb 含量,并通过单因子污染指数法、内梅罗综合污染指数、地累积指数法、潜在生态风险评价法定量分析了不同管理方式下土壤重金属污染及生态风险程度。结果表明:①不同管理方式土壤 Cr、Ni、Cu、Zn、As、Cd、Pb 含量均高于陕西省背景值,不同管理方式土壤重金属含量大小依次为:自然生长>灌溉>施肥。②各重金属含量随土壤深度增加的变化趋势为逐渐减小或先减小后增加,且在 20~40 cm 深度取得最小值。③Cd 是造成该地区生态风险的主要元素,为轻污染,且是污染程度最高的元素。灌溉和施肥措施对土壤重金属污染修复效果优于自然生长,合理布设城市绿地管理措施有利于城市绿地土壤重金属污染修复、城市生态环境质量改善。

【关键词】城市绿地;土壤修复;重金属污染;生态风险;管理方式

Soil Heavy Metal Pollution and Ecological Risk Assessment under Different Management Methods of Urban Green Land

Tingting Meng[1,2,4],Jinbao Liu[1,2,3,4],Hao Dong[1,2,4],Bo Wang[1,2,4],Guojian Zhang[1]

(1.Shaanxi Provincial Land Engineering Construction Group Co. ,Ltd. ,Xi' an 710075,China;
2.Institute of Shaanxi Land Engineering and Technology Co. , Ltd. , Xi' an 710075,China;
3.Xi' an University of Technology , Xi' an 710048,China;4.Key Laboratory of Degraded and
Unused Land Consolidation Engineering,Ministry of Land and Resources, Xi' an 710075, China)

【**Abstract**】Treatment of heavy metal pollution in urban ecological environment become one of the urgent environmental problems. In order to explore the difference of remediation effect of heavy metal pollution in urban green space under different management measures, experimental plots with different management methods were established as the research object, and soil samples were collected and determined the contents of Cr, Ni, Cu, Zn, As, Cd and Pb in soil . The quantitative analysis of soil heavy metal pollution and ecological risk under differ-

基金项目:陕西省科技成果转移与推广计划(2021CGBX-03);长安大学重点科研平台开放基金(300102351501)。

ent management methods by combining the single factor pollution index method, the Nemerow comprehensive pollution index, the geo-accumulation index method and the potential ecological risk assessment method The results showed: (1) The contents of Cr, Ni, Cu, Zn, As, Cd and Pb in soil of different management methods were higher than the background value of Shaanxi Province. The contents of heavy metals in soil of different management methods were natural growth > irrigation > fertilization. (2) The variation trend of heavy metal content with the increase of soil depth was gradually decreasing or first decreasing and then increasing, and the minimum value was obtained at the depth of the 20 ~ 40 cm. (3) The Cd was the main element causing ecological risk in the region, which is light pollution and the highest pollution level. Irrigation and fertilization measures have better remediation effect on soil heavy metal pollution than natural growth. Reasonable layout of urban green space management measures is conducive to the remediation of soil heavy metal pollution in urban green space and the improvement of urban ecological environment quality.

【Keywords】 Urban green land; Soil remediation; Heavy metal pollution; Ecological risk; Regulative measure

0　引言

土壤重金属污染是指人类活动将重金属引入土壤中致使有害元素含量超过背景值,并造成土壤污染的现象[1]。相关研究表明,受重金属污染,我国有约16.1%土壤样本超国家二级质量标准[2-3]。重金属元素通过扬尘沉降、暴雨径流、污水灌溉等方式进入生态系统循环后,不仅影响土地质量、粮食作物生长质量[4-6],还会扰乱土壤结构,改变土壤性质,破坏土壤环境稳定性[7-8]。重金属元素具有难降解、活动性强、累积性强等特点,且在极低的浓度下就可能对动物和人体健康产生严重影响,土壤重金属污染问题已经成为城市化和工农业发展过程中不可忽视的环境问题[2,9]。目前,国内外学者们对于土壤重金属污染的研究主要集中于利用单因子指数法、地质累积指数法、富集系数法、潜在生态风险评价法等方法[10],描述区域重金属污染现状及其空间分布特征。作为城市生态系统的重要组成部分[11-13],城市绿地不仅可以美化环境、净化空气,还可以修复受污染的土壤。植物修复是一种绿色清洁技术,植物修复(phytoremediation)一般采用的方法是利用超富集植物超强的重金属富集能力,吸收土壤重金属后及时收获植株,从而降低土壤中的重金属含量[14-16]。在人类农业活动和工矿业等活动的影响下,造成城市绿地土壤的重金属污染问题将给城市环境和人类健康产生极大影响[1, 17-18]。近年来,对于城市重金属污染的元素赋存形态[19]、来源解析、重金属的空间分布特征[20]等方面污染国内学者展开了大量研究[18,21],但对于不同管理措施下的土壤重金属污染修复效果研究尚不多见。鉴于此,本研究通过构建试验小区,以不同管理措施下的城市绿地土壤为研究对象,定量分析不同管理措施下土壤修复效果差异,旨在探明城市绿地重金属污染现状和生态风险程度,以期为城市绿地土壤重金属污染防治和修复提供依据。

1　材料与方法

1.1　研究区概况

试验小区位于陕西省咸阳市渭城区兰池大道附近(108°50′38″E,34°23′26″N),地处关中中部渭河北岸,属暖温带大陆性半干旱季风性气候,四季冷暖分明,平均气温14.7 ℃,平均风速2.1 m/s,空气相对湿度63%,无霜期228 d,年均降水量443 mm。试验区为村庄经撤、扩、并后的废弃宅基地,造林面积为3.07 hm^2;2018年4月,试验区经推平、清表、翻耕后扦插栽植速生景观杨,栽植行距和株距分别为1.0 m、0.5 m。

1.2　试验设计

2018年5月,选择杨树长势基本一致的地块划定试验小区,如图1所示。每个试验区面积60 m^2(15 m×4 m),共有9个试验区。试验设置灌溉处理,施用农家肥+灌溉(简称"肥灌")处理,自然生长处理。3个处理在同方位内,每个试验小区长、宽间隔为3 m。灌溉方式为漫灌,单个试验小区的灌溉量为4 m^3,施肥方式为将农家肥作为基肥在每年的第一个月施入,每穴8 kg,后期采用追肥方式,每穴施肥量为尿素0.5 kg+农家肥2 kg,施肥后灌溉,试验周期与灌溉施肥频次如表1所示。

图 1　试验小区

Fig.1　Experimental plot

表 1　试验周期与灌溉施肥频次

Table 1　Test period and frequency of irrigation and fertilization

试验年限	处理	肥灌频次			
		3 月	5 月	8 月	11 月
2018	灌溉	—	√	√	√
	肥灌	—	√	√	√
	自然生长	—	—	—	—
2019	灌溉	√	√	√	√
	肥灌	√	√	√	√
	自然生长	—	—	—	—
2020	灌溉	√	√	√	√
	肥灌	√	√	√	√
	自然生长	—	—	—	—

1.3　样本采集

于 2020 年 9 月使用环刀对三种管理方式杨树林地土壤进行样品采集,采集方法为五点取样法,采样点及试验样地分布如图 2 所示,按照不同深度 0~20 cm、20~40 cm、40~60 cm、60~80 cm、80~100 cm 进行采样,将同一管理方式相同深度土壤混合装于自封袋,3 种处理下共采取 45 个样点,共 225 个土壤样本。去除石砾、植物根系等后,将样本带回实验室放置阴凉处自然风干。

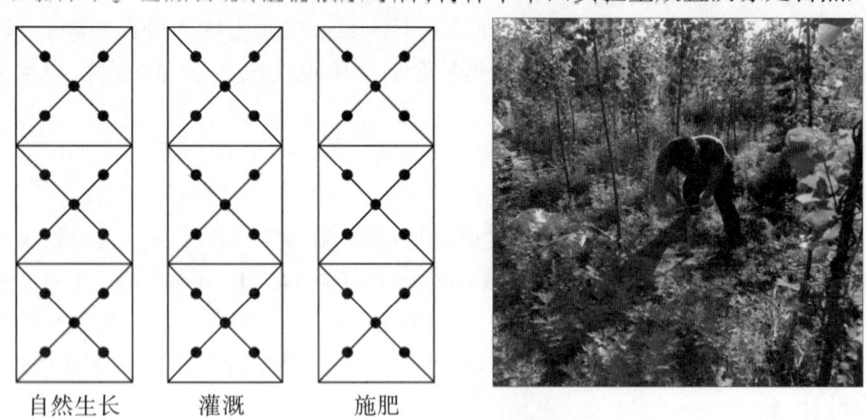

图 2　采样点分布和采样图

Fig.2　Sampling point distribution

1.4　指标测定

采用王水-微波消解作为土壤前处理方法,利用电感耦合等离子体质谱(ICP-MS)测定土壤的重金属元素[22]。

1.5 计算方法

分别采用单因子指数法和内梅罗综合污染指数法评价试验小区不同管理措施下土壤重金属污染程度，单因子指数法和内梅罗综合污染指数法计算公式及其分级标准分别如式(1)、式(2)及表2所示[10]：

$$P_i = \frac{C^i}{C_s^i} \tag{1}$$

$$P_n = \sqrt{\frac{P^2_{i\max} + P^2_{iave}}{2}} \tag{2}$$

式中：P_i为重金属i的单因子污染指数，无量纲；C^i为重金属含量实测值，mg/kg；C_s^i为重金属含量土壤风险筛选值，mg/kg，本研究取区域背景值，参照陕西省土壤元素背景值，分别取为Cr(62.5 mg/kg)、Ni(28.8 mg/kg)、Cu(21.4 mg/kg)、Zn(69.4 mg/kg)、As(11.0 mg/kg)、Cd(0.094 mg/kg)、Pb(21.4 mg/kg)；P_n为重金属综合污染指数，无量纲；$P_{i\max}$为单因子污染指数最大值；P_{iave}为单因子污染指数平均值。

采用地质累计指数法评价土壤中重金属污染水平[10]，计算公式如下：

$$I_{geo} = \log_2\left(\frac{C^i}{k \times S_i}\right) \tag{3}$$

式中：I_{geo}为重金属i的地质累计指数，无量纲；C^i为重金属含量实测值，mg/kg；S_i为重金属背景值，mg/kg；k为修正系数，取1.5，地质累计指数等级划分如表2所示。

重金属潜在生态风险指数计算公式如下[10]：

$$\mathrm{RI} = \sum_{i=1}^{n} E_r^i = \sum_{i=1}^{n} (T_r^i \times C_f^i) = \sum_{i=1}^{n} \left(T_r^i \times \frac{C^i}{C_n^i}\right) \tag{4}$$

式中：C^i为土壤重金属含量实测值，mg/kg；C_n^i为区域重金属背景值，mg/kg；E_r^i为重金属i潜在生态风险指数；T_r^i为重金属i毒性响应系数，参照已有成果，分别取Cr、Ni、Cu、Zn、As、Cd、Pb毒性响应系数为2、5、5、1、10、30、5；RI为综合潜在生态风险指数，生态风险指数分级情况如表2所示[23]。

表2 污染等级、地质累计指数分级、生态风险程度划分

Table 2 Classification of pollution level, geological cumulative index and ecological risk degree

等级	I_{geo}	污染程度	P_i	P_N	污染程度	E_r^i	RI	危害程度
0	<0	无污染	<0.7	<0.7	无污染	<40	<150	轻微危害
1	0~1	轻污染	0.7~1 .0	0.7~1 .0	警戒值	40~80	150~300	中等危害
2	1~2	中污染	1 .0~2.0	1 .0~2.0	轻度污染	80~160	300~600	强危害
3	2~3	中污染-重污染	2.0~3.0	2.0~3.0	中度污染	160~320	600~1200	很强危害
4	3~4	重污染	>3.0	>3.0	重度污染	≥320	≥1200	极强危害
5	4~5	重-极重污染						
6	>5	极重污染						

1.6 数据处理

采用Office Excel 2016进行参数计算，绘制重金属含量分布表、计算结果表，Spss20进行重金属含量相关性分析，使用ArcGIS10.2进行试验小区位置图绘制。

2 结果与分析

2.1 不同管理措施下土壤重金属含量分布特征

不同管理方式土壤各重金属元素含量及pH分布状况如表3所示，试验小区土壤pH总体处于

表 3　重金属含量和 pH 分布状况

Table 3　Distribution of heavy metal content and pH

处理	深度(cm)	Cr(mg/kg)	Ni(mg/kg)	Cu(mg/kg)	Zn(mg/kg)	As(mg/kg)	Cd(mg/kg)	Pb(mg/kg)	pH
自然生长	0~20	73.44±4.05Ac	36.87±1.98Ac	37.29±8.99Ac	96.52±7.1Ab	13.26±0.14Ab	0.27±0.02Aa	27.77±1.49Aa	8.65±0.11Ac
	20~40	70.04±0.46Ac	34.36±1.16Ac	36.9±7.99Ac	78.78±3.1Ac	12.9±0.05Ab	0.18±0Ab	24.24±1.12Ac	8.7±0.02Ac
	40~60	82.09±11.44Ab	42.17±5.67Ab	42.37±2.75Ab	89.66±8.06Ab	16.09±1.74Aa	0.19±0.01Ab	26.09±0.93Ab	8.68±0.06Ac
	60~80	88.39±7.63Aa	48.02±4.89Aa	46.66±10.9Ab	96.13±3.52Ab	15.88±1.07Aa	0.2±0Ab	26.09±1.02Ab	8.87±0.01Ab
	80~100	89.58±1.73Aa	48.67±2.4Aa	56.23±20.16Aa	101.85±0.47Aa	16.57±2.3Aa	0.19±0.01Ab	27.24±0.41Aa	8.98±0.28Aa
	0~100	80.71±8.75Ab	42.02±6.43Ab	43.89±7.98Ab	92.59±8.85Ab	14.94±1.72Aa	0.21±0.04Ab	26.29±1.36Ab	8.69±0.14Ac
灌溉	0~20	64.7±8.31Bc	31.97±4.95Bc	29.11±2.58Bc	84.78±9.51Bc	13.05±0.16Ab	0.23±0.05Bd	34.82±11.46Ac	8.67±0.23Bc
	20~40	68.45±2.71Bc	33.58±0.05Bc	30.96±0.4Bc	75.19±1.96Bd	13.01±0.21Ab	0.17±0.02Bc	23.24±0.3Ab	8.66±0.07Bc
	40~60	81.04±12.92Bb	40.85±7.54Bb	37.93±9.03Bc	86.02±13.2Bc	15.87±2.05Ab	0.17±0.01Bb	25.36±1.96Ab	8.65±0.01Bc
	60~80	89.44±6.14Bb	47.56±5.54Bb	39.13±0.25Bb	92.83±1.14Bb	17.21±0.82Ab	0.19±0Ba	26.04±0.94Ab	8.9±0.05Bb
	80~100	93.81±4.27Ba	50.07±0.42Ba	43.71±2.45Ba	107.15±7.02Ba	17.31±3.34Aa	0.19±0.01Ba	27.81±0.4Aa	9.18±0Ba
	0~100	74.87±11.83Bc	37.7±6.73Bb	34.04±6.36Bc	86.58±11.58Bc	15.16±2.52Ab	0.19±0.02Ba	26.58±4.69Ab	8.65±0.15Bc
施肥	0~20	62.4±5.05Bc	30.35±2.66Bc	28.27±1.39Cc	82.08±5.7Cc	12.74±0.28Bc	0.22±0.04Bb	34.35±12.13Bb	8.61±0.43Ab
	20~40	66.58±0.08Bc	33.03±0.82Bc	29.02±2.34Cc	73.41±0.55Cc	13.28±0.18Bc	0.16±0.01Bc	22.21±1.15Bb	8.55±0.08Aa
	40~60	72.93±1.45Bb	36.56±1.48Bb	32.15±0.85Cb	82.79±8.63Ca	14.61±0.27Bb	0.17±0.01Ba	24.01±0.05Bb	8.6±0.08Ab
	60~80	80.11±7.05Ba	41.33±3.28Ba	37.26±2.89Ca	90.25±2.51Cb	16.21±2.24Ba	0.18±0.01Ba	25.27±2.04Ba	8.79±0.21Ab
	80~100	92.32±6.38Ba	47.22±3.6Ba	43.5±2.75Ca	104.34±11Ca	18.97±0.99Ba	0.19±0.01Ba	27.05±1.47Ba	8.9±0.4Ab
	0~100	73.34±12.51Bb	36.7±8.31Bb	30.31±7.57Cc	82.62±18.8Ca	13.93±2.42Bc	0.18±0.01Ba	23.72±1.83Bb	8.78±0.1Ab

注：大写字母表示不同管理措施之间差异显著($P<0.05$)，小写字母表示不同深度之间差异显著($P<0.05$)。

8.55~9.98 范围内,属碱性土。不同管理方式下土壤 pH 均值大小关系为:灌溉(8.65)<自然生长(8.69)<施肥(8.78)。自然生长条件土壤重金属含量均值分别为:Cr(80.71 mg/kg)、Ni(42.02 mg/kg)、Cu(43.89 mg/kg)、Zn(92.59 mg/kg)、As(14.94 mg/kg)、Cd(0.21 mg/kg)、Pb(26.29 mg/kg)。与自然生长相比,灌溉条件下土壤各重金属元素含量均值分别减少了 7.23%、10.28%、22.44%、6.49%、-1.49%、9.78%、-1.11%,施肥条件下分别减少了 9.13%、12.65%、30.95%、10.76%、6.78%、14.85%、9.75%。不同管理方式土壤重金属含量随土壤深度增加变化趋势分别为:施肥方式下,重金属元素含量均随土壤深度增加先降低后升高,且在 20~40 cm 深度为最小值,在 80~100 cm 深度为最大值。自然生长方式下,土壤重金属含量均随土壤深度增加先降低后升高,且在 20~40 cm 深度为最小值,Cr、Ni、Cu、Zn、As 含量在 80~100 cm 深度为最大值,Cd、Pb 在 0~20 cm 深度为最小值。灌溉方式下,随土壤深度增加,Zn、Cd、Pb 含量先降低后升高,Cr、Ni、Cu、As 含量逐渐升高。

2.2 不同管理措施下土壤重金属污染程度分析

结合表 2 与单因子污染指数和内梅罗综合污染指数结果(见表 4)可知,自然生长条件下,土壤重金属单因子污染指数处于 1.13(Pb)~2.92(Cd)范围内,Cu、Pb 单因子污染指数分别为 2.05、2.19,属中度污染,Cr、Ni、Zn、As、Cd 单因子污染指数均处于 1.0~2.0 范围内,属轻度污染。灌溉条件与施肥条件下,各重金属元素单因子污染指数分别处于 1.00(Cr)~2.37(Cd)、0.92(Cr、Ni)~1.93(Cd)范围内,污染程度均为轻度污染。相比同种管理方式下其他重金属元素,Cd 元素单因子污染指数均为最高,Cd 元素单因子污染指数大小依次为自然生长(2.19)>灌溉(1.98)>施肥(1.86)。内梅罗综合污染指数表明,三种管理方式重金属综合污染状况均为中度污染,自然生长内梅罗指数最高,为 1.98,灌溉次之,为 1.73,施肥最小,为 1.62。

表 4 单因子污染指数及内梅罗综合污染指数

Table 4 P_i and P_n distribution

	自然生长				灌溉				施肥			
	最大值	最小值	平均值	C_V	最大值	最小值	平均值	C_V	最大值	最小值	平均值	C_V
Cr	1.43	1.15	1.29	0.11	1.48	1.00	1.20	0.16	1.39	0.92	1.17	0.17
Ni	1.69	1.19	1.46	0.15	1.64	1.05	1.31	0.18	1.55	0.92	1.27	0.23
Cu	2.63	1.72	2.05	0.18	2.03	1.32	1.59	0.19	1.82	0.96	1.42	0.25
Zn	1.47	1.14	1.33	0.10	1.50	1.06	1.25	0.13	1.42	0.84	1.19	0.23
As	1.51	1.17	1.36	0.12	1.72	1.16	1.38	0.17	1.49	0.99	1.27	0.17
Cd	2.92	1.92	2.19	0.19	2.37	1.74	1.98	0.12	1.93	1.71	1.86	0.05
Pb	1.30	1.13	1.23	0.05	1.61	1.04	1.24	0.18	1.21	1.00	1.11	0.08
P_n	2.35	1.66	1.98	0.15	1.94	1.51	1.73	0.10	1.74	1.42	1.62	0.09

土壤重金属地质累计指数结果如表 5 所示,结合地质累计指数等级划分标准(见表 2)可知,Cu 在自然生长和灌溉土壤中为轻污染程度,大小关系为自然生长(0.43)>灌溉(0.07),在施肥土壤中为无污染。三种管理方式下土壤重金属 Cd 均为轻污染程度,大小依次为自然生长(0.53)>灌溉(0.39)>施肥(0.31)。其他重金属元素地质累计指数均值均小于 0,均为无污染状态。由此表明,Cu、Cd 在土壤重金属污染中起主要作用。

表 5 地质累计指数

Table 5 I_{geo} numerical results

	自然生长				灌溉				施肥			
	最大值	最小值	平均值	C_V	最大值	最小值	平均值	C_V	最大值	最小值	平均值	C_V
Cr	−0.07	−0.42	−0.22	−0.71	−0.02	−0.59	−0.34	−0.66	−0.11	−0.71	−0.37	−0.68
Ni	0.17	−0.33	−0.05	−4.16	0.13	−0.51	−0.21	−1.18	0.04	−0.71	−0.27	−1.28
Cu	0.81	0.20	0.43	0.58	0.44	−0.18	0.07	3.98	0.28	−0.64	−0.12	−3.11
Zn	−0.03	−0.40	−0.17	−0.82	0.00	−0.50	−0.28	−0.68	−0.08	−0.84	−0.37	−0.96
As	0.01	−0.36	−0.15	−1.13	0.20	−0.37	−0.14	−1.69	−0.01	−0.59	−0.26	−0.98
Cd	0.96	0.35	0.53	0.47	0.66	0.21	0.39	0.44	0.37	0.19	0.31	0.25
Pb	−0.21	−0.41	−0.29	−0.26	0.10	−0.53	−0.29	−0.83	−0.31	−0.58	−0.44	−0.25

2.3 土壤重金属生态风险程度分析

单因子潜在生态风险指数及综合潜在生态风险指数计算结果如表 6 所示,结合潜在生态风险程度划分(见表 2)可知,自然生长、灌溉、施肥方式下不同重金属潜在生态风险大小顺序均为 Cd>As>Cu>Ni>Pb>Cr>Zn,其中 As、Cu、Ni、Pb、Cr、Zn 潜在生态风险指数均小于 40,为轻微危害程度,Cd 在三种管理方式下均为中等危害,大小顺序为自然生长(65.70)>灌溉(59.27)>施肥(55.94)。不同管理方式土壤综合潜在生态风险指数大小依次为自然生长(106.89)>灌溉(97.41)>施肥(91.14),分布范围分别为 92.88~125.04、85.09~106.01、78.34~99.47,均为轻微危害程度。

表 6 潜在生态风险指数

Table 6 Ecological risk index

	自然生长				灌溉				施肥			
	最大值	最小值	平均值	C_V	最大值	最小值	平均值	C_V	最大值	最小值	平均值	C_V
Cr	2.87	2.24	2.58	0.11	2.95	2.00	2.40	0.16	2.78	1.84	2.35	0.17
Ni	8.45	5.97	7.29	0.15	8.20	5.27	6.54	0.18	7.73	4.60	6.37	0.23
Cu	13.14	8.62	10.25	0.18	10.16	6.61	7.95	0.19	9.08	4.80	7.08	0.25
Zn	1.47	1.14	1.33	0.10	1.50	1.06	1.25	0.13	1.42	0.84	1.19	0.23
As	15.07	11.73	13.58	0.12	17.25	11.58	13.78	0.17	14.85	9.95	12.66	0.17
Cd	87.65	57.52	65.70	0.19	71.01	52.12	59.27	0.12	58.01	51.29	55.94	0.05
Pb	6.49	5.66	6.14	0.05	8.03	5.19	6.21	0.18	6.03	5.02	5.54	0.08
P_n	125.04	92.88	106.89	0.11	106.01	85.09	97.41	0.09	99.47	78.34	91.14	0.10

2.4 土壤重金属相关分析

重金属间的相关程度可以反映污染源属性,重金属元素含量相关分析(见表 7)表明,Cr、Ni、Cu、Zn、As 五种元素含量之间均呈显著相关,相关系数均大于 0.772($P<0.01$)。Cd 与 Pb 含量呈显著相关,相关系数为 0.675($P<0.01$),与其余 5 种重金属含量无显著相关,表明 Cd、Pb 与其他 5 种重金属元素的来源、迁移途径不同或不存在复合污染现象,Cr、Ni、Cu、Zn、As 可能具有共同的污染源,且与 Cd、Pb 污染源可能相同。这可能与试验小区建立前后农业活动和道路交通有关。土壤中重金属元素主要来源于成土母质分化或人类活动,研究表明 Cr、Ni、As 主要来源于地质环境背景,Cd、Pb、As 主要与化肥、农药、燃料使用等工农业活动有关[2, 13]。

表 7 重金属元素含量相关分析

Table 7 Correlation analysis of heavy metal elements

元素	Cr	Ni	Cu	Zn	As	Cd	Pb
Cr	1						
Ni	0.991**	1					
Cu	0.877**	0.895**	1				
Zn	0.879**	0.877**	0.785**	1			
As	0.948**	0.925**	0.772**	0.815**	1		
Cd	0.123	0.134	0.274	0.439	0.029	1	
Pb	0.191	0.209	0.319	0.441	0.236	0.675**	1

注:$n=225$, ** 表示在 $P<0.01$ 水平上差异显著,双侧。

3 讨论

本研究表明,与自然生长方式相比,施肥对土壤重金属污染减轻效果最为显著,灌溉方式下除 As 和 Pb 含量高于自然生长,其余重金属含量均低于自然生长,且随土壤深度增加,不同管理方式土壤重金属含量变化趋势为逐渐减小或先减小后增加并在 20~40 cm 深度取得最小值。这与已有研究结果一致,窦韦强等[2]研究表明随土壤深度增加,重金属含量变化趋势为逐渐降低或先降低后增加的趋势。王国贤等[24]也在研究中发现部分重金属在土壤中垂直迁移过程中存在逐渐减小的趋势。Ma 等[25]在研究中也得出了类似的结论。这可能是由于在植物根系和水分双重影响造成的,相关研究表明,在雨水冲刷或人工灌水的影响下,重金属元素会随水分下渗过程在土壤中发生迁移沉降现象[26],而植物对受重金属元素污染土壤的修复作用主要发生位置在于植物根部[14],通过根部对重金属的富集、吸收、向地上部分的转移、植物组织与重金属的螯合等过程,达到降低土壤中重金属含量的目的[24]。且植物根系分泌物不仅会调节根际环境的理化性质,如土壤性状、pH、电位条件等,还对土壤中多种微生物有促生效应,部分微生物及其代谢产物可以改变土壤中重金属的赋存形态,更利于植物吸收[9, 15],而试验小区土壤中 20~40 cm 深度植物根系分布最为密集,富集吸收效应导致 20~40 cm 深度重金属含量较低。

单因子污染指数分析表明,不同管理方式下土壤污染程度最高的重金属是 Cd,且自然生长(2.19)>灌溉(1.98)>施肥(1.86)。研究表明,城市土壤中的 Cd 主要来源于汽车尾气的排放、石油泄漏、橡胶轮胎的磨损等[27],试验小区处于城市道路旁边,大量 Cd 累积,加之不同管理措施的影响,导致自然生长条件下重金属元素富集现象最为严重。在进行城市土壤修复治理时,应注重对重金属 Cd 的处理,可采取限制车辆通行等措施来降低城市绿地土壤的 Cd 输入。地质累计指数及潜在生态风险指数分析表明,自然生长和灌溉条件下土壤中 Cu 为轻污染,三种管理措施下土壤重金属 Cd 均为轻污染状态,中等危害程度,其余重金属为无污染。这不仅与土壤中 Cd 的含量较高有关,还与 Cd 元素的毒性系数远高于其他重金属有关[27-28]。

内梅罗综合污染指数及潜在生态风险指数分析表明,该地区不同管理方式土壤均为中度污染状态,且污染程度及潜在风险程度大小均依次为:自然生长>灌溉>施肥。自然生长和灌溉条件下,Cd 和 Pd 元素在土壤表层(0~20 cm)含量最高,施肥条件更利于 Cd 和 Pd 元素迁移过程的发生。这可能是由于施肥改变了土壤化学性质,研究表明土壤理化性质会对重金属迁移产生较大影响,随着 pH 升高土壤表面的负电荷增加,增加与有机酸的配合概率,进而导致土壤对重金属吸附能力减弱,增强了重金属在土壤中的迁移性[2-3, 9],最终形成土壤重金属污染修复能力大小依次为:施肥>灌溉>自然生长。

4 结论

本研究通过建立试验小区,分析比较了陕西省咸阳市渭城区不同管理措施绿地土壤 Cr、Ni、Cu、

Zn、As、Cd、Pb 七种重金属污染状况和生态风险程度,得到了如下结论:

(1)随土壤深度增加,各重金属含量变化趋势为逐渐减小或先减小后增加,并在 20~40 cm 深度取得最小值。

(2)土壤中 Cd 是造成土壤重金属污染和生态风险的主要来源,为轻污染状态,且处于中等危害程度。

(3)城市绿地管理措施对重金属污染治理效果存在差异,施肥条件下土壤重金属污染修复效果最好,灌溉条件次之,合理布设城市绿地施肥、灌溉条件有助于城市环境重金属污染治理。

参考文献

[1]万凯,袁飞,李光顺,等.恩施州矿山周边耕地土壤重金属污染特征及评价[J].资源环境与工程,2020,34(S1):28-32.

[2]窦韦强,安毅,秦莉,等.农田土壤重金属垂直分布迁移特征及生态风险评价[J].环境工程,2021,39(2):166-172.

[3]于亚军,贺泽好.三种复垦类型煤矸山土壤重金属有效态含量及其影响因素分析[J].环境工程,2018,36(5):189-192.

[4] KUPK D, KANIA M, PIETRZYKOWSKI M, et al. Multiple Factors Influence the Accumulation of Heavy Metals (Cu, Pb, Ni, Zn) in Forest Soils in the Vicinity of Roadways[J]. Water, Air, & Soil Pollution. 2021, 232(5): 141-148.

[5]张传严,席北斗,张强,等.堆肥在土壤修复与质量提升的应用现状与展望[J].环境工程:1-15[2021-11-15].

[6]孙厚云,卫晓锋,孙晓明,等.钒钛磁铁矿尾矿库复垦土地及周边土壤-玉米重金属迁移富集特征[J].环境科学,2021,42(3):1166-1176.

[7] CHEN X Y, LI F, ZHANG J D, et al. Heavy metal pollution characteristics and health risk evaluation of soil around a tungsten-molybdenum mine in Luoyang, China[J]. Environmental Earth Sciences, 2021, 80(7).

[8]李晓强,董炜华,宋扬.路域农田大型土壤动物对公路运营过程重金属累积的响应[J].地理研究,2020,39(12):2842-2854.

[9]刘伟,张永波,贾亚敏.重金属污染农田植物修复及强化措施研究进展[J].环境工程,2019,37(5):29-33.

[10]范拴喜,甘卓亭,李美娟,等.土壤重金属污染评价方法进展[J].中国农学通报,2010,26(17):310-315.

[11]郭志娟,周亚龙,王乔林,等.雄安新区土壤重金属污染特征及健康风险[J].中国环境科学,2021,41(1):431-441.

[12]姚文文,陈文德,黄钟宣,等.重庆市主城区土壤重金属形态特征及风险评价[J].西南农业学报,2021,34(1):159-164.

[13]梁青芳,杨宁宁,高煜,等.宝鸡市不同功能区灰尘重金属污染特征与评价[J].环境工程,2019,37(10):216-221.

[14]刘睿,聂庆娟,王晗.木本园林植物对土壤重金属的富集及修复效应研究进展[J].北方园艺,2021(8):117-124.

[15]金明兰,王悦宏,姚峻程,等.植物对重金属污染土壤的生态修复[J].科学技术与工程,2020,20(32):13493-13496.

[16]赵晓光,张亦扬,杜华栋.陕北矿区不同土地类型下土壤重金属污染评价[J].环境工程,2019,37(9):188-193.

[17] ZHANG H L, ZHANG M, WU Y T, et al. Risk sources quantitative appointment of ecological environment and human health in farmland soils: a case study on Jiuyuan District in China[J]. Environmental Geochemistry and Health, 2021, 43(11): 4789-4803.

[18] 张广映,吴琳娜,欧阳坤长,等.都柳江上游沿岸喀斯特地区土壤重金属污染特征及风险评价[J/OL].中国岩溶:1-11[2021-11-15].

[19]王丹丹,林静雯,丁海涛,等.牛粪生物炭对重金属镉污染土壤的钝化修复研究[J].环境工程,2016,34(12):183-187.

[20] HAQUE E, THORNE P S, NGHIEM A A, et al. Lead (Pb) concentrations and speciation in residential soils from an urban community impacted by multiple legacy sources[J]. Journal of Hazardous Materials, 2021, 416: 125886.

[21]赵超,赵杰,苏胜男,等.济南市城市绿地土壤重金属污染及生态风险评价[J].能源环境保护,2021,35(1):60-66.

[22]王昭.ICP-MS 法测定土壤 6 种重金属元素[J].西部大开发(土地开发工程研究),2019,4(4):22-26.

[23]王昌宇,张素荣,刘继红,等.雄安新区某金属冶炼区土壤重金属污染程度及风险评价[J].中国地质,2021:1-16.

[24]王国贤,陈宝林,任桂萍,等.内蒙古东部污灌区土壤重金属迁移规律的研究[J].农业环境科学学报,2007(S1):30-32.

[25]MA W F, LIU F, CHENG X, et al. Environmental evaluation of the application of compost sewage sludge to landscaping as soil amendments: a field experiment on the grassland soils in Beijing[J]. Desalination and Water Treatment, 2015,54(4/5): 1118-1126.

[26]杨宾,罗会龙,刘士清,等.淹水对土壤重金属浸出行为的影响及机制[J].环境工程学报,2019,13(4):936-943.

[27]董燕,孙璐,李海涛,等.雄安新区土壤重金属和砷元素空间分布特征及源解析[J].水文地质工程地质,2021,48(3):172-181.

[28]陈锐,杜双杰,徐伟,等.南京城郊某典型退耕农用地土壤重金属含量特征与污染评价分析[J/OL].环境工程:1-16[2021-11-16].

本文曾发表于 2022 年《环境工程》第 40 卷第 12 期

施用不同改良材料对黄土区空心村复垦土壤结构和有机质含量的影响

刘　哲[1,2,3,4]，王欢元[1,2,3,4]，孙婴婴[1,2,3,4]，孙增慧[1,2,3,4]，张瑞庆[1,2,3,4]

(1.陕西省土地工程建设集团有限责任公司，陕西 西安 710075；2.自然资源部退化及未利用土地整治工程重点实验室，陕西 西安 710021；3.陕西地建土地工程技术研究院有限责任公司，陕西 西安 710021；4.陕西省土地整治工程技术研究中心，陕西 西安 710075)

【摘要】目的：旨在解决黄土区空心村废弃宅基地复垦土壤结构不良、肥力低下的严重问题，为空心村复垦土壤的质量提升提供理论依据。方法：选取6种不同改良材料处理进行了5年的田间定位试验，分析了对照(CK)、熟化剂(TM)、粉煤灰(TF)、有机肥(TO)、熟化剂+粉煤灰(TMF)、熟化剂+有机肥(TMO)、粉煤灰+有机肥(TFO)处理对复垦土壤结构稳定性和有机质含量的改善效应。结果与对照处理相比，长期施用不同改良材料后空心村复垦土壤有机质含量明显增加，促使水稳性微团聚体(<0.25 mm)胶结形成水稳性大团聚体(>0.25 mm)，水稳性大团聚体含量均呈增加趋势。在0~15 cm土层，TM、TF、TO、TMF、TMO、TFO处理下水稳性大团聚体含量分别比CK处理增加了328.2%、130.0%、87.8%、81.1%、36.7%、12.2%。TF、TO、TMF、TMO、TFO处理显著增加了土壤团聚体平均重量直径(MWD)、几何平均直径(GMD)值，降低了不稳定团粒指数(E_{LT})、分形维数(D)($p<0.05$)。土壤有机质含量与MWD、GMD、>2 mm粒级水稳性团聚体含量呈极显著正相关关系，与E_{LT}、D<0.25 mm粒级水稳性微团聚含量呈极显著负相关。其中，粉煤灰+有机肥处理下空心村复垦土壤的水稳性大团聚体和有机质含量最高，土壤结构稳定性提升效果最好。结论：在黄土区空心村废弃宅基地复垦土壤整治中，粉煤灰+有机肥的有机无机结合处理是黄土区空心村复垦土壤结构改善和肥力提升适宜的改良措施。

【关键词】空心村；宅基地复垦土壤；改良材料；土壤水稳性团聚体；团聚体稳定性；土壤有机质

Effects of Different Modified Materials on Reclaimed Soil Structure and Organic Matter Content of Hollow Village in Loess Region

Zhe Liu[1,2,3,4], Huanyuan Wang[1,2,3,4], Yingying Sun[1,2,3,4], Zenghui Sun[1,2,3,4], Ruiqing Zhang[1,2,3,4]

(1.Shaanxi Provincial Land Engineering Construction Group Co., Ltd., Xi'an, Shaanxi 710075, China; 2.Key Laboratory of Degraded and Unused Land Consolidation Engineering, the ministry of Natural Resources, Xi'an, Shaanxi 710021, China; 3.Institute of Land Engineering and Technology, Shaanxi Provincial Land Engineering Construction Group Co., Ltd., Xi'an, Shaanxi 710021, China; 4.Shaanxi Provincial Land Consolidation Engineering Technology Research Center, Xi'an, Shaanxi 710075, China)

【Abstract】 Objective: The aim is to address the poor structure and low fertility of the reclaimed soil of aban-

资助项目：陕西省土地工程建设集团内部科研项目“不同有机物料处理下土壤团聚体稳定性的内力作用机制”(DJNY2020-25)；陕西省土地工程建设集团内部科研项目(DJNY2022-15，DJNY2022-35，DJTD-2022-5)；陕西地建-西安交大土地工程与人居环境技术创新中心开放基金资助项目(2021WHZ0092)。

doned homestead in Hollow Village of the loess area, providing a theoretical basis for improving the quality of reclaimed soils. Methods: Six modified materials were selected with different treatments for a five-year field located experiment, and the effects of modified materials, with different treatments of control (CK), maturing agent (TM), fly ash (TF), organic fertilizer (TO), maturing agent + fly ash (TMF), maturing agent + organic fertilizer (TMO), and fly ash + organic fertilizer (TFO) on the structural stability and soil organic matter (SOM) content in reclaimed soil were analyzed. Results: Compared with CK, the SOM content has been significantly increased after the application of different modified materials, which promoted the cementation and aggregation of water-stable microaggregates (<0.25 mm) into water-stable macroaggregates (>0.25 mm) which showed an increasing trend. In the 0~15 cm soil layer, the contents of water-stable macroaggregates under TM, TF, TO, TMF, TMO, and TFO treatments increased by 328.2%, 130.0%, 87.8%, 81.1%, 36.7%, and 12.2% than CK, respectively. Moreover, through the TF, TO, TMF, TMO, TFO treatments, the mean weight diameter (MWD) and geometric mean diameter (GMD) of soil aggregates were significantly increased, while the unstable-aggregate index (E_{LT}) and fractal dimension (D) were decreased ($p<0.05$). The SOM content was significantly positively correlated with MWD, GMD, and >2 mm water-stable aggregates, and remarkably negatively correlated with the E_{LT}, D, and water-stable microaggregates. Especially, the reclaimed soil under TFO had the highest content of SOM and water-stable macroaggregates. Conclusion: The combined treatment of organic fertilizer and fly ash was an appropriate measure for the improvement of the reclaimed soil structure and fertility in Hollow Village of the loess region.

【**Keywords**】Hollow village; Homestead reclamation soil; Modified materials; Soil water-stability aggregate; Aggregate stability; Soil organic matter

黄土塬区面临着生态条件脆弱、耕地面积减少、耕地后备资源匮乏、人地矛盾加剧的现实问题,伴随着城镇化、工业化进程的快速推进实施,农村青壮年劳动力大量迁移城市,造成农村废弃空心化现象更加严重,使得大量的废弃宅基地资源闲置浪费[1-3]。而且农村宅基地建设普遍存在"建新不拆旧"、不断占用优质耕地向外围扩展、农村人口转移减少而宅基地规模却扩大的不良态势,形成了村庄新房建设规模不断扩大、村内宅基地废弃闲置的农村空心化的普遍现象,造成了大量耕地资源的破坏占用以及优质土地资源的闲置浪费,严重威胁着耕地资源的保护与区域粮食安全,成为限制美丽乡村建设和统筹城乡高质量发展面临的主要瓶颈[4-6]。对此,结合黄土区人地矛盾不断加剧,当地对于废弃宅基地进行土地复垦的需求增大,深入推进空心村土地综合整治,开展废弃宅基地复垦对于新时期城乡一体化高质量发展、实行耕地"占补平衡"、保障国家粮食安全、缓解区域人地矛盾具有迫切的战略意义[6-8]。由于空心村复垦土壤主要来源于废弃宅基地的老墙土,而老墙土受到自然和人为因素的共同影响,多为常年没有耕种的生土,缺乏耕作土壤自然的功能与属性,其物理结构不良,土壤肥力低下严重,极大地限制了空心村复垦土壤产能和健康可持续发展,亟待改善宅基地复垦土壤的结构性能与肥力特性,提高复垦土壤的产能[9-10]。

土壤团聚体的数量和结构不仅是影响土壤侵蚀、板结、紧实程度等结构状况的重要指标,而且在供储土壤养分、调节土壤持水能力、维持土地生产力方面发挥着关键作用,是能很好地评价土壤肥力特性和环境质量变化的关键指标[11-13]。相比于非水稳性团聚体,水稳性团聚体在维持土壤结构稳定性、抗侵蚀性、土壤养分保持方面发挥着更为重要的作用[14-15]。研究发现,有机质含量的提升与水稳性团聚体数量和稳定性的提升有着直接的关系。因此,施用土壤改良材料可有效地增加土壤的有机质含量,促进土壤团聚体的胶结团聚与结构稳定性的提升,改善复垦土壤肥力和抗侵蚀能力[16-17]。粉煤灰是燃煤电厂排出的粉状固体残渣,具有发达的比表面积和多级孔隙,含有丰富的黏土颗粒和 Al_2O_3、Fe_2O_3 等氧化物,可显著提高土壤颗粒间的相互吸附和团聚能力,增强土壤保水保肥能力[18-19]。硫酸亚铁熟化剂在改善土壤结构的同时,降低土壤 pH 值,也可用作肥料,对植物的吸收起着重要的促进作用[20]。有机肥富含有机质和多种营养物质,可有效改善土壤肥力和结构,提高土壤生产力,保持土壤健康和作物产量的可持续性[21-22]。因此,通过应用不同的土壤改良材料对空心村复垦土壤进行熟化和改良,增强复垦土壤结构稳定性和肥力特性,使其迅速恢复耕作土壤的功能和性质具有现实意义和紧迫性。然而,目前关于空心村整治主要集中在整治潜力评估、演变规律、区域宏观整治模式、调控政策等方式上,这些对空心村废弃宅基地整治有着重要意义[23-25],但空心村复垦土壤作为区域补充耕地资源的关键举措,对于分析不同土壤改良材料对复垦土壤结构改善和稳定性提升的研究较少。为此,本文以黄土塬区空心村复垦土壤为研究对象,对比研究不同改良材料处理对复垦土壤结构及有机质含量的影响,评估不同改良材料施入后复垦土壤有机质含量与水稳性团聚体各指标间的相关关系,以期筛选出适宜空心村土壤改良的适宜材料,为改善空心村复垦土壤结构,提高复垦土壤肥力和抗侵蚀性提供科学参考。

1 材料与方法

1.1 空心村土地复垦方式及试验区概况

空心村复垦土壤改良长期定位试验小区依托自然资源部退化及未利用土地整治工程重点实验室,布设于陕西省富平县中试基地(34°42′N,109°12′E),该研究区属于渭北黄土塬区,气候类型为暖温带半湿润大陆性季风气候,年均气温为 12.2 ℃,年平均降水量为 533.2 mm,年光能辐射总量为 135.44 kcal/cm^2,四季分明。试验小区建于 2015 年 6 月 15 日,空心村废弃宅基地复垦土壤为土地整治项目的老墙土进行回填,回填深度为 30 cm,剔除宅基地复垦老墙土中的瓦砾、石块等杂质后,通过添加土壤改良材料对复垦土壤进行熟化和结构改良,以满足小麦、玉米等粮食作物的正常生长。复垦土壤主要为黄土母质发育而成,试验开始前 0~30 cm 复垦土壤基础理化指标如下:pH 值(1∶2.5 水土质量比)为 8.5,土壤质地为粉壤土(美国制),其中黏粒(<0.002 mm)含量为 10.15%,粉粒(0.05~0.002 mm)含量为 77.82%,砂粒(0.05~2 mm)含量为 12.65%,有机质含量为 4.4 g/kg,土壤全氮含量为 0.15 g/kg,有效磷含量为 70.4 mg/kg,速效钾含量为 61.3 mg/kg。土壤容重为 1.40 g/cm^3,>2 mm、1~2 mm、0.5~1 mm、0.25~0.5 mm 和<0.25 mm 粒级水稳性团聚体的比例分别为 0.67%、0.84%、1.67%、3.55%、93.27%。综上所述,空心村废弃宅基地复垦土壤以黄土母质的老墙土为主,土壤肥力和结构较差。

1.2 试验设计

空心村复垦土壤长期定位田间改良试验开始于 2015 年 6 月,依据还田改良材料的特性调查和已发表文献分析,针对复垦土壤存在的问题,本文采用粉煤灰、有机肥(腐熟鸡粪)、和硫酸亚铁($FeSO_4$)作为复垦土壤改良材料,粉煤灰中环境污染指标 As、Hg、Cu、Pb、Zn、Cd 的含量分别为 13.59 mg/kg、0.10 mg/kg、91.6 mg/kg、22.72 mg/kg、57.81 mg/kg、0.06 mg/kg,符合土壤环境质量评价标准 GB 15618—2018。本试验采用随机区组田间试验设计,共设置 7 种处理,分别为熟化剂(TM)、粉煤灰(TF)、有机肥(TO)、熟化剂+粉煤灰(TMF)、熟化剂+有机肥(TMO)、粉煤灰+有机肥(TFO)、无改良材料添加对照(CK)处理,每种处理设置 3 个重复,每个处理小区为 2 m×2 m 的正方形田块,共计 21 个小区,每个小区中间设有 80 cm 宽的隔离带。作物种植制度为冬小麦-夏玉米轮作的两年三熟制度,供试冬小麦在 10 月中旬播种,播种量为 220.5 kg/hm^2,第二年 5 月下旬收获,品种为长武 134。夏玉米 6 月中上旬播种,播种密度为每公顷 26.5 万株,10 月上旬收获,品种为先玉 958。采用人工播种的方式,播种前所有处理撒施复合肥 1 500 kg/hm^2,复合肥中氮磷钾含量分别为 15%、10%、20%,然后将 6 种不同处理的改良材料均匀地撒施在土壤表面,通过人为耕作的方式将不同的改良材料均匀拌入到 0~30 cm 的空心村复垦土壤中,每种试验处理一次性施入土壤改良材料,7 种处理化肥用量、浇水量、病虫害防治等其他管理措施与水平保持一致,具体土壤改良材料处理及用量见表 1。

表 1 空心村复垦土壤改良试验设计

序号	处理	改良材料	施用量/(t/hm²)
1	CK	对照(无添加)	0
2	TM	熟化剂(硫酸亚铁)	0.6
3	TF	粉煤灰	45
4	TO	有机肥(腐熟鸡粪)	30
5	TMF	熟化剂+粉煤灰	(0.6+45)
6	TMO	熟化剂+有机肥	(0.6+30)
7	TFO	粉煤灰+有机肥	(45+30)

1.3 测定指标及计算方法

供试土样于 2020 年 5 月下旬冬小麦收获后,分 0~15 cm 和 15~30 cm 两个土层分别采集测定土壤团聚体的原状土样和测定土壤有机质含量的混合土样,每个处理分 3 个重复进行取样,土壤团聚体采用不锈钢铝盒进行保存,运输过程中尽量避免对土壤团聚体结构的影响。带回实验室自然风干后,原状土样按照土样自然裂痕轻轻掰成 10 mm 左右的小土块,剔除石块等杂质后备用,进行土壤水稳性团聚体、有机质等指标的检测。土壤容重采用环刀法测定,土壤含水量采用烘干法测定,土壤有机质含量采用重铬酸钾-外加热法测定[26]。0~15 cm 和 15~30 cm 两个土层土壤水稳性团聚体分布与大小采用湿筛法测定[27],团聚体结构稳定性指标平均重量直径(MWD)、几何平均直径(GMD)、土壤不稳定团粒指数(E_{LT})和分形维数(D)详细计算见式(1)~式(4) 和相关的参考文献计算分析[28-30]。

$$\mathrm{MWD}=\frac{\sum_{i=1}^{n}(\bar{x}_i w_i)}{\sum_{i=1}^{n} w_i} \tag{1}$$

$$\mathrm{GMD}=\exp\left(\frac{\sum_{i=1}^{n} w_i \ln \bar{x}_i}{\sum_{i=1}^{n} w_i}\right) \tag{2}$$

$$E_{LT}=\frac{M_T-R_{0.25}}{M_T}\times 100\% \tag{3}$$

$$\frac{M(r<\bar{x}_i)}{M_T}=\left(\frac{\bar{x}_i}{x_{max}}\right)^{3-D} \tag{4}$$

1.4 数据统计分析

通过 SigmaPlot12.5 软件进行图形绘制,采用 Microsoft Excel 2010 和 SPSS 22.0 软件对所有试验数据进行统计和方差分析,采用 LSD 法进行不同处理间多重比较($p<0.05$)。

2 结果与分析

2.1 改良材料对复垦土壤水稳性团聚体分布的影响

水稳性团聚体是评价土壤结构稳定性和抗侵蚀能力的重要指标,其数量和分布状况能很好地反映土壤结构和土壤质量的变化[31]。与对照处理相比,长期施用不同改良材料处理后 0~15 cm 和 15~30 cm 土层复垦土壤水稳性团聚体分布状况发生了显著变化(图 1、图 2)($p<0.05$)。在 0~15 cm 深度土层,各改良材料对复垦土壤的水稳性大团聚体(>0.25 mm)含量呈增加趋势。其中,除熟化剂(TM)处理外,粉煤灰(TF)、有机肥(TO)、熟化剂+粉煤灰(TMF)、熟化剂+有机肥(TMO)、粉煤灰+有机肥(TFO)相比于对照(CK)处理均显著增加了>2 mm、1~2 mm、0.5~1 mm 粒级水稳性团聚体的含量,尤其是>2 mm 粒级水稳性团聚体的数量增加最为明显(见图 1)。TF、TO、TMF、TMO、TFO 处理下>2 mm 粒级水稳性团聚体分别比 CK 处理增加了 88.1%、194.5%、203.7%、376.2%、781.7%,水稳性大团聚体含量呈现出 TFO(35.8%) > TMO(20.7%) > TO(16.9%) >TMF(16.3%) > TF 处理(12.3%) > TM 处理(10.1%) > CK 处理(9.0%)的大小顺序,水稳性大团聚体分别比对照处理增加了 328.2%、130.0%、87.8%、81.1%、36.7%、12.2%,增幅最大为 328.2%。总体来看,6 种不同改良材料处理均增加了空心村复垦土壤的水稳性大团聚体的含量,促进了水稳性微团聚体(<0.25 mm)向水稳性大团聚体(>0.25 mm)的团聚胶结,其中 TFO 处理对于增加的水稳性大团聚体的效果最好, TMO、TO 和 TMF 处理的效果次之,TF 和 TM 处理的效果不是很显著。

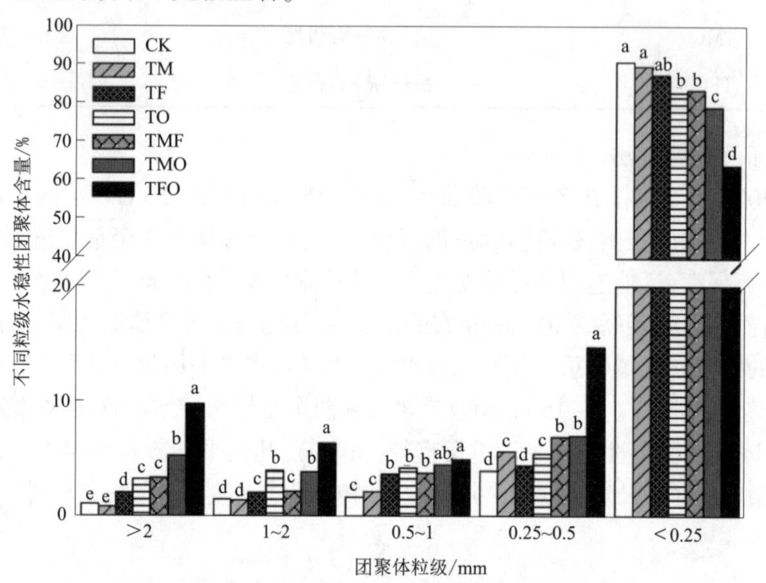

图 1　长期施用不同改良材料处理下 0~15 cm 土层各粒级水稳性团聚体百分含量

注:小写字母表示不同改良材料处理下差异显著性($p<0.05$),下同。

在 15~30 cm 深度土层,与 CK 处理相比,水稳性团聚体变化呈现出与 0~15 cm 土层相似的趋势,TF、TO、TMF、TMO、TFO 处理均显著增加了>2 mm、1~2 mm 粒级水稳性团聚体的含量,降低了水稳性微团聚体的含量(见图 2)($p<0.05$)。其中 TF、TO、TMF、TMO、TFO 处理后>2 mm 粒级水稳性团

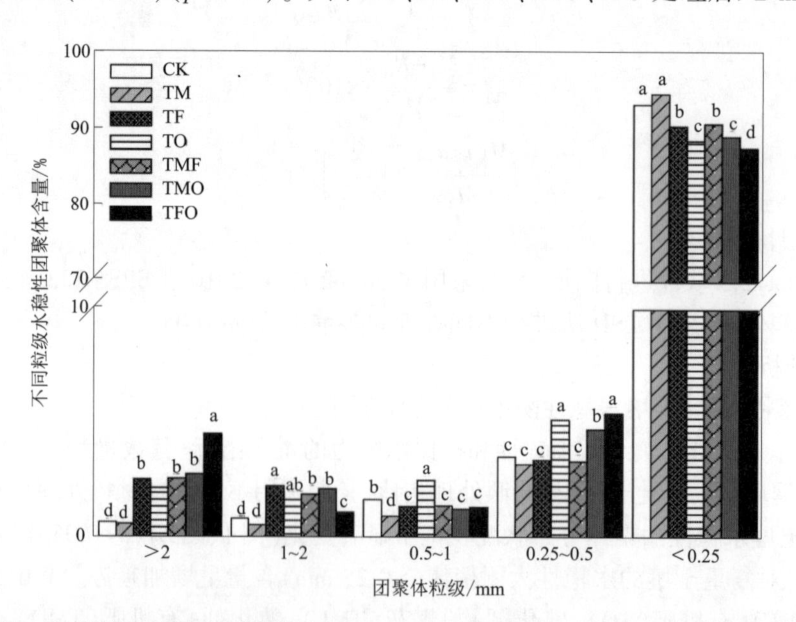

图 2　长期施用不同改良材料处理下 15~30 cm 土层各粒级水稳性团聚体百分含量

聚体分别比 CK 处理增加了 130.3%、94.5%、133.9%、151.4%、309.2%，TFO 处理对于水稳性大团聚数量的增加最为显著。相比于 0~15 cm 土层，随着土层深度的增加，15~30 cm 土层水稳性大团聚体数量呈现出逐步降低的趋势。

2.2 改良材料对复垦土壤水稳性团聚体结构的影响

平均重量直径(MWD)、几何平均直径(GMD)、不稳定团粒指数(E_{LT})、分形维数(D)是反映土壤团聚体结构几何形状和结构稳定性的重要指标，研究表明 MWD、GMD 值越高，E_{LT}、D 值越小，团聚体的结构稳定性越好，土壤结构越好[32-33]。与 CK 处理相比，施用不同改良材料处理后，TF、TO、TMF、TMO、TFO 处理下 MWD、GMD 值呈显著增加的趋势，D、E_{LT}值呈显著减小的趋势($p<0.05$)，TM 处理对团聚体稳定性指标影响不显著(见表 2)。在 0~15 cm 深度土层，TF、TO、TMF、TMO、TFO 处理下 MWD 值分别比 CK 处理增加了 6.19%、27.66%、22.16%、49.71%、125.96%，GMD 值分别比 CK 处理增加了 4.09%、12.46%、9.34%、19.82%、49.15%，E_{LT}值分别比 CK 处理减小了 1.35%~29.5%，D 值分别比 CK 处理减小了 0.76%~4.35%，TF、TO、TMF、TMO、TFO 处理均不同程度地改善了团聚体的团聚能力，增强了复垦土壤的结构稳定性和抗侵蚀的能力，其中 TFO 处理对团聚体结构稳定性的改善效果最佳。在 15~30 cm 深度土层，与 CK 处理相比，TF、TO、TMF、TMO、TFO 处理下 MWD、GMD、D、E_{LT}值也呈现出显著的改善效应($p<0.05$)。通过团聚体稳定性指标数据来看，相比于 0~15 cm 土层，15~30 cm 深度土层水稳性团聚体结构稳定性呈现出减小的趋势，这可能与 0~15 cm 土层土壤有机质含量较高有关。

表 2 施用不同有改良材料对水稳性团聚体稳定性指标的影响

土层	处理	MWD/mm	GMD/mm	E_{LT}/%	D
0~15 cm	对照(CK)	0.35d	0.28c	91.01a	2.97a
	熟化剂(TM)	0.32d	0.28c	89.87ab	2.99a
	粉煤灰(TF)	0.38c	0.29bc	87.69b	2.97a
	有机肥(TO)	0.45c	0.31b	83.08b	2.95b
	熟化剂+粉煤灰(TMF)	0.43c	0.31b	83.72b	2.95b
	熟化剂+有机肥(TMO)	0.53b	0.34b	79.31b	2.92b
	粉煤灰+有机肥(TFO)	0.80a	0.42a	64.17c	2.84c
15~30 cm	对照(CK)	0.30c	0.27c	93.28a	2.99a
	熟化剂(TM)	0.29c	0.26c	94.66a	2.99a
	粉煤灰(TF)	0.39b	0.29b	90.43b	2.96b
	有机肥(TO)	0.37b	0.29b	88.50bc	2.97b
	熟化剂+粉煤灰(TMF)	0.39b	0.29b	90.77b	2.96b
	熟化剂+有机肥(TMO)	0.39b	0.29b	89.07b	2.96b
	粉煤灰+有机肥(TFO)	0.48a	0.30a	87.58c	2.93c

2.3 改良材料对复垦土壤有机质含量的影响

施用不同改良材料后空心村复垦土壤 0~15 cm 和 15~30 cm 土层土壤有机质含量整体均呈现出增加的趋势(见图 3)。在 0~15 cm 土层，TM、TF、TO、TMF、TMO、TFO 处理下土壤有机质含量分别比 CK 增加了 9.6%、79.0%、90.0%、61.4%、120.1%、131.7%，表明不同改良材料对复垦土壤有机质含量均具有重要的提升作用。不同改良材料处理对复垦土壤 0~15 cm 土层有机质含量的提升呈现出 TFO> TMO > TO> TF > TMF > TM >CK 的大小顺序，添加有机肥处理的 TO、TMO、TFO 处理能够显著提升复垦土壤的有机质含量($p<0.05$)，其中 TFO 处理最有利于土壤有机质含量的提升。在 15~30

cm 深度土层中,显著性分析结果表明,TO、TMF、TMO、TFO 处理均显著提高了土壤有机质含量($p<0.05$),TM、TF 处理对土壤有机质含量的提升差异不显著,其中 TFO 处理对复垦土壤有机质含量的提升效果最为明显。

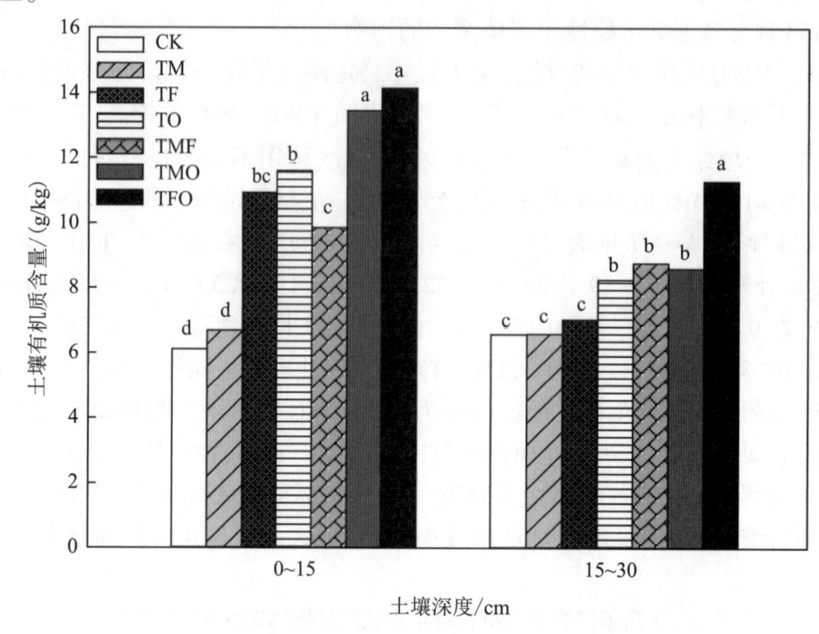

图 3　施用不同改良物料对土壤有机质含量的影响

2.4　改良材料对复垦土壤容重和含水量的影响

施用不同改良材料后复垦土壤 0~15 cm 和 15~30 cm 土层土壤容重显著降低($p<0.05$)(见图 4a)。在 0~15 cm 土层,TM、TF、TO、TMF、TMO、TFO 处理下土壤容重较 CK 分别降低了 5.71%、7.74%、8.57%、8.69%、8.79%和 9.13%,说明不同改良材料对复垦土壤容重都有一定的降低作用。但由于改良材料特性不同,对复垦土的疏松效果不同,复垦土壤的容重表现为 TFO > TMO >TO > TF > TMF > TM > CK 的大小趋势。有机-无机改良材料组合可有效降低复垦土壤容重,TFO 处理下的容重最小为 1.19 g/cm^3。在 15~30 cm 土层,通过方差分析,不同改良材料对复垦土壤容重的影响呈现出与 0~15 cm 土层容重相似的下降趋势。

施用不同改良材料后 0~15 cm 和 15~30 cm 土层复垦土壤含水量显著增加($p<0.05$),且两个土层含水量的变化规律基本相似,大小顺序为 TFO>TMO>TMF>TO>TF≈TM>CK(见图 4b)。在 0~15 cm 土层,TM、TF、TO、TMF、TMO、TFO 处理下土壤含水量较 CK 分别增加 13.5%、13.8%、21.4%、21.9%、32.4%和 38.3%。TMO 和 TFO 对复垦土壤含水量的提高效果较好,土壤质量含水量分别为 17.4%和 18.2%。综上所述,与 CK 相比,不同改良材料施用在提高土壤有机质含量、促进团聚体形成和稳定的同时,增加了水分滞留和传输,有利于保持更多的水分,其中,有机无机改良材料组合的 TFO 和 TMO 处理有利于保持更多的土壤水分。

2.5　土壤有机质及水稳性团聚体各参数间相关性分析

为进一步探明 6 种不同改良材料处理后复垦土壤各参数之间的相关性,本文对有机质及不同粒级水稳性团聚体各参数之间的相关关系进行了回归分析。从表 3 可知,土壤有机质含量与 MWD、GMD、>2 mm 粒级水稳性团聚体含量呈极显著正相关关系,与 E_{LT}、D、<0.25 mm 粒级水稳性微团聚含量呈极显著负相关关系,说明土壤有机质是影响土壤水稳性团聚体形成及其结构稳定性的重要因素,有机质含量越高,越能促进水稳性大团聚体的形成,土壤的结构稳定性越强。>2 mm、1~2 mm、0.5~1 mm 粒级水稳性团聚体含量与 MWD、GMD 值呈显著正相关,与 E_{LT}、D 值呈极显著负相关关系;<0.25 mm 水稳性团聚体与 MWD、GMD 值呈极显著负相关,与 D 值呈显著正相关关系;这些说明较大粒级水稳性团聚体含量的增加,有助于促进土壤团聚体结构稳定性的提升。综上表明,TM、TF、TO、

TMF、TMO、TFO 改良材料处理在促进空心村复垦土壤有机质含量增加的同时，能够有效地促进水稳性大团聚体的团聚形成与结构稳定性的提升，尤其是 TFO 处理是改善空心村复垦土壤结构特性、培肥地力和增强抗侵蚀性的最佳处理。

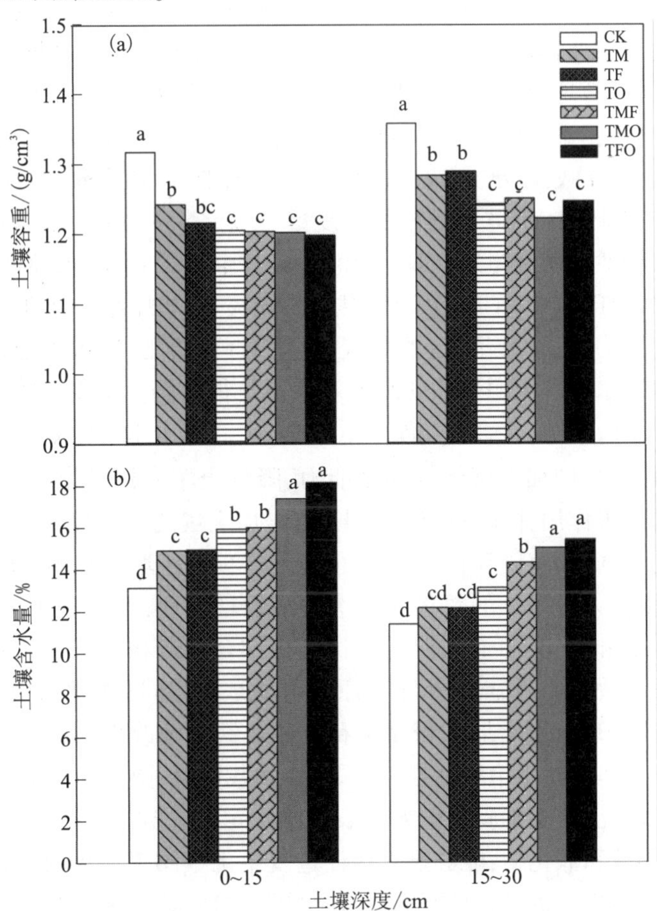

图 4　施用不同改良物料对土壤容重和含水量的影响

表 3　土壤有机质及水稳性团聚体各参数间相关性分析

指标	SOM/(g/kg)	MWD/(mm)	GMD/mm	E_{LT}/%	D	粒级/mm				
						>2	1~2	0.5~1	0.25~0.5	<0.25
SOM	1									
MWD	0.7177**	1								
GMD	0.6960**	0.9798**	1							
E_{LT}	−0.6948**	−0.9364**	−0.9814**	1						
D	−0.7003**	−0.9926**	−0.9540**	0.9001**	1					
>2	0.7316**	0.9846**	0.9457**	−0.8948**	−0.9935**	1				
1~2	0.5949*	0.8140**	0.8835**	−0.8881**	−0.7639**	0.7596**	1			
0.5~1	0.5450	0.8522*	0.8980*	−0.9256**	−0.8190**	0.8080**	0.6602**	1		
0.25~0.5	0.5521	0.4727	0.5708	−0.6744**	−0.4167	0.4234	0.7138*	0.5190	1	
<0.25	−0.6948**	−0.9364**	−0.9814**	1.0000	0.9001**	−0.8948**	−0.8881**	0.6744**	−0.9256**	1

注：* 显著性水平 $p<0.05$；** 显著性水平 $p<0.01$。

3 讨论

3.1 长期施用改良材料对空心村复垦土壤水稳性团聚体分布及稳定性影响

土壤有机质、氧化物和黏粒是土壤团聚体形成与稳定的主要胶结物质,对土壤团聚体的数量、大小分布及结构稳定性起着重要作用[34]。有机肥、熟化剂(硫酸亚铁)、粉煤灰等土壤改良材料单独或者结合还田施用后,能够有效地改善土壤有机质的含量,其转化分解过程中形成的胶结物质有助于土壤团聚体的胶结团聚,对于团聚体的大小、分布状况及结构稳定性有明显的影响[12,35]。本研究中,相比于 CK 处理,除熟化剂(TM)处理外,粉煤灰(TF)、有机肥(TO)、熟化剂+粉煤灰(TMF)、熟化剂+有机肥(TMO)、粉煤灰+有机肥(TFO)均显著增加了复垦土壤>2 mm、1~2 mm 粒级水稳性团聚体的含量,降低了<0.25 mm 水稳性微团聚体的含量($p<0.05$),其中>0.25 mm 水稳性大团聚体含量呈现出 TFO> TMO > TO >TMF > TF> TM > CK 的大小顺序,粉煤灰+有机肥结合处理对空心村复垦土壤水稳性大团聚体含量的增加最为明显,熟化剂+粉煤灰、熟化剂+有机肥、有机肥处理的改善效果次之,单施熟化剂和粉煤灰处理对水稳性大团聚体的改善效果不是很显著。董少文等研究表明,由于粉煤灰具有发达的比表面积和多级孔隙,富含黏粒及 Al_2O_3、Fe_2O_3 等氧化物,施入土壤后能明显增强土壤颗粒间的相互吸附与团聚能力,增加土壤保水保肥能力,降低土壤容重[36-37]。硫酸亚铁熟化剂对北方土壤的碱性具有一定的改善作用,同时可以疏松土壤,增加土壤的保水性,改善土壤结构[38-39]。有机肥输入土壤腐解产生的多糖、腐殖质等胶结物质,能进一步促进土壤颗粒的团聚和微团聚体向大团聚体的堆叠胶结,增加水稳性大团聚体的数量和土壤的结构稳定性,增加土壤持水性能[40-41]。本研究结果显示,粉煤灰+有机肥和熟化剂+有机肥的有机无机结合处理相比于单施有机肥处理,对复垦土壤土壤有机质含量的改善效果更明显,分析主要原因可能是在相同有机肥施用量的情况下,粉煤灰+有机肥和熟化剂+有机肥处理分别增加了粉煤灰和硫酸亚铁土壤改良剂的施用,有机无机的组合施用使得复垦土壤颗粒相互作用时土壤酶活性、根系等作物残留量、微生物数量、矿物种类、孔隙结构等因素相互影响,使得在改善土壤有机质含量方面表现出一定的差异[39,42-44]。由于粉煤灰发达的比表面积和多级孔隙,富含黏粒和氧化物,能够显著地促进复垦土壤团聚体形成与稳定,增加土壤的蓄水保墒能力;加之硫酸亚铁熟化剂也具有良好的改善土壤理化性状、增加土壤酶活性的作用,从而促进了作物根系的生长和土壤微生物活性的提升,进而增加了由于土壤中微生物活性提高而形成腐殖物和作物根系等残留物、有机肥分解产生如多糖、蛋白质、木质素等不同种类有机质的数量,因此粉煤灰+有机肥和熟化剂+有机肥的有机无机结合处理提升复垦土壤有机质的程度高于单施有机肥处理,这与高志香等和 Singh 等的研究结果相似,他们的研究表明,粉煤灰、硫酸亚铁土壤无机改良剂与有机肥配合施用下土壤有机质含量要明显高于单施有机物料处理[39,42-43]。同样,这也与本研究中的团聚体稳定性指标相似,TF、TO、TMF、TMO、TFO 处理下 MWD、GMD 值整体呈现出显著增加的趋势,D、E_{LT}值呈现出显著减小的趋势($p<0.05$),其中 6 种改良材料处理中 TFO 和 TMO 处理下的团聚体稳定性指标最好,其中粉煤灰+有机肥(TFO)的结合处理对空心村复垦土壤团聚体结构稳定性的改善效果最好,是提高复垦土壤团聚体数量和结构稳定性的最佳途径。此外,粉煤灰+有机肥组合处理还显著降低了复垦土壤容重,增加了土壤水分入渗,提高了土壤的蓄水保墒能力,Singh 等和魏俊岭等的研究结果也表明粉煤灰与有机物料或者配方施肥处组合施用,可以有效地降低土壤容重,增加了水分稳定入渗速率和持水能力[42,44]。

3.2 改良材料施用对复垦土壤有机质及各参数间相关性分析

不同改良物料还田施用是提高复垦土壤有机质的重要来源,有机肥、熟化剂(硫酸亚铁)、粉煤灰等土壤改良材料单独或者结合还田施用后,能够提高土壤的肥力水平,促进作物生物量的提高,增大植物残体和根系进入土壤的归还量,从而提高了土壤有机质含量[33]。本研究中,施用不同改良材料后复垦土壤 0~15 cm 和 15~30 cm 土层土壤有机质含量均呈现出整体增加的趋势。在 0~15 cm 深度土层中,TM、TF、TO、TMF、TMO、TFO 处理下土壤有机质含量分别比 CK 处理增加了 9.6%、79.0%、

90.0%、61.4%、120.1%、131.7%，表明不同改良材料对复垦土壤有机质含量的提升均具有重要作用，但由于不同改良材料自身的理化特性不同以及对复垦土壤的微生物活性的影响过程差异，导致对于复垦土壤有机质含量的改善效应不同。其中，粉煤灰+有机肥（TFO）的提升效果最佳，熟化剂+有机肥（TMO）、有机肥（TO）处理次之，这与粉煤灰和有机肥的养分含量及结构状况有着直接的关系，粉煤灰发达的比表面积和多级孔隙使其具有很好的保肥特性，加之有机肥本身就富含有机物质和多种营养元素，所以粉煤灰+有机肥（TFO）的有机无机结合对有机质的改善效果最为显著。张旭辉等[45]的研究表明，有机质的累积量与0.25～2 mm粒级团聚体的含量有着密切的关系，团聚体的粒级越大，有机碳的含量越高。粉煤灰+有机肥（TFO）处理显著地提高了0.25～2 mm粒级团聚体的含量，这对于土壤有机质含量的提升也有着重要的促进作用。这与雷娜等[20]的研究结果相似，雷娜等研究表明腐熟的鸡粪有机肥和无机改良材料粉煤灰的复合还田能显著地提高土壤肥力水平。空心村复垦土壤各参数之间的相关性分析结果表明，土壤有机质含量与MWD、GMD、>2 mm粒级水稳性团聚体含量呈极显著正相关关系，与E_{LT}、D<0.25 mm粒级水稳性微团聚含量呈极显著负相关关系，>2 mm、1～2 mm、0.5～1 mm粒级水稳性团聚体含量与团聚体稳定性指标MWD、GMD值呈显著相关关系，说明土壤有机质含量影响着土壤团聚体的胶结团聚与结构稳定性[27,47]，外源改良材料的还田施用促进了空心村复垦土壤有机质含量的提高，为土壤中小颗粒的团聚胶结创造了良好的条件，有助于水稳性微团聚体通过有机质的黏合团聚形成水稳性大团聚体，进而增强了土壤的结构稳定性和抗侵蚀性。

4　结论

通过综合对比分析，施用不同改良材料5年后空心村复垦土壤有机质含量均呈现增加趋势，其中粉煤灰+有机肥的结合处理对0～15 cm土层有机质的提升效果最好，增幅达131.7%。随着有机质含量的提升，不同改良材料处理促进了水稳性微团聚体向水稳性大团聚体（>0.25 mm）的胶结团聚，0～15 cm土层水稳性大团聚体含量呈现出TFO > TMO > TO>TMF> TF > TM > CK处理的大小顺序，粉煤灰+有机肥处理的增幅最大为328.2%。团聚体结构稳定性指标MWD、GMD、D、E_{LT}值整体得到改善，复垦土壤水稳性大团聚体数量和结构稳定性得到提高。6种不同改良材料处理中，粉煤灰+有机肥的有机无机结合处理是空心村复垦土壤最适宜的改良材料，二者结合使用可以有效地增强空心村复垦土壤结构稳定性和抗侵蚀性，提高土壤肥力。

参考文献

[1] 刘彦随，刘玉.中国农村空心化问题研究的进展与展望[J].地理研究，2010，29(1)：35-42.

[2] Liu Yansui, Li Jintao, Yang Yuanyuan. Strategic adjustment of land use policy under the economic transformation [J]. Land Use Policy, 2018,74:5-14.

[3] Huang Fang, Wang Ping. Vegetation change of ecotone in west of Northeast China plain using time-series remote sensing data [J]. Chinese Geographical Science, 2010,20(2):167-175.

[4] 刘彦随，刘玉，翟荣新.中国农村空心化的地理学研究与整治实践[J].地理学报，2009，64(10)：1193-1202.

[5] 胡智超，彭建，杜悦悦，等.基于供给侧结构性改革的空心村综合整治研究[J].地理学报，2016，71(12)：2119-2128.

[6] 李娟，韩霁昌，陈超，等.山地丘陵区不同土地利用方式对空心村整治还田土壤团聚体特征的影响[J].水土保持研究，2017，24(4)：174-181.

[7] 姜绍静，罗泮.空心村问题研究进展与成果综述[J].中国人口·资源与环境，2014，24(6)：51-58.

[8] Liu Yansui. Introduction to land use and rural sustainability in China [J]. Land Use Policy, 2018,74:1-4.

[9] 黄耀华，王侃，苏婷婷，等.重庆农村土墙型复垦宅基地土壤肥力特征及改造利用研究[J].西南大学学报(自然科学版)，2015，37(1)：33-39.

[10] 刘子骁，邓良基，周伟，等.有机物料对宅基地复垦土壤培肥效果评价[J].土壤，2019，51(4)：672-681.

[11] Verchot L V, Dutaur L, Shepherd K D, et al. Organic matter stabilization in soil aggregates: Understanding the bi-

ogeochemical mechanisms that determine the fate of carbon inputs in soils [J]. Geoderma, 2011,161(3/4):182-193.

[12] 王晓娟,贾志宽,梁连友,等.旱地施有机肥对土壤有机质和水稳性团聚体的影响[J].应用生态学报,2012,23(1):159-165.

[13] 刘哲,韩霁昌,孙增慧,等.$\delta^{13}C$ 法研究砂姜黑土添加秸秆后团聚体有机碳变化规律[J].农业工程学报,2017,33(14):179-187.

[14] Blanco-Moure N, Moret-Fernúndez D, López M. Dynamics of aggregate destabilization by water in soils under long-term conservation tillage in semiarid Spain [J]. Catena, 2012, 99:34-41.

[15] 王勇,姬强,刘帅,等.耕作措施对土壤水稳性团聚体及有机碳分布的影响[J].农业环境科学学报,2012,31(7):1365-1373.

[16] Bronick C J, Lal R. Soil structure and management: A review [J]. Geoderma, 2005,124(1/2):3-22.

[17] 刘恩科,赵秉强,梅旭荣,等.不同施肥处理对土壤水稳定性团聚体及有机碳分布的影响[J].生态学报,2010,30(4):1035-1041.

[18] 孙红娟,曾丽,彭同江.粉煤灰高值化利用研究现状与进展[J].材料导报,2021,35(3):3010-3015.

[19] Parab N, Sinha S, Mishra S. Coal fly ash amendment in acidic field: Effect on soil microbial activity and onion yield [J]. Applied Soil Ecology, 2015,96:211-216.

[20] 雷娜,陈田庆,董起广,等.空心村整治还田材料土壤培肥效果分析[J].水土保持学报,2018,32(4):222-226.

[21] 邱吟霜,王西娜,李培富,等.不同种类有机肥及用量对当季旱地土壤肥力和玉米产量的影响[J].中国土壤与肥料,2019(6):182-189.

[22] Liu Zhe, Han Jichang, Sun Zenghui, et al. Long-term effects of different planting patterns on greenhouse soil micromorphological features in the North China Plain [J]. Scientific Reports, 2019,9:2200.

[23] 尹娟,邱道持,潘娟.基于 XF-GF 空间合成技术的农村居民点复垦潜力分析:以潼南县双江镇双林村为例[J].农机化研究,2011,33(11):5-9.

[24] 崔宝敏.天津市"以宅基地换房"的农村集体建设用地流转新模式[J].中国土地科学,2010,24(5):37-40.

[25] 冯巍仑,李裕瑞,刘彦随.基于微观视角的农村转型发展评价及对策探析:以河南省获嘉县楼村为例[J].地域研究与开发,2018,37(2):133-137.

[26] 鲍士旦.土壤农化分析[M].3 版.北京:中国农业出版社,2000.

[27] Six J, Bossuyt H, Degryze S, et al. A history of research on the link between (micro)aggregates, soil biota, and soil organic matter dynamics [J]. Soil and Tillage Research, 2004,79(1):7-31.

[28] 尚杰,耿增超,赵军,等.生物炭对塿土水热特性及团聚体稳定性的影响[J].应用生态学报,2015,26(7):1969-1976.

[29] Liu Zhe , Sun Zenghui, Wang Huanyuan, et al. Effects of straw decomposition on aggregates composition and aggregate-associated organic carbon in different mineral types of soil[J]. Applied Ecology And Environmental Research,2020,18(5):6511-6528.

[30] 周虎,吕贻忠,杨志臣,等.保护性耕作对华北平原土壤团聚体特征的影响[J].中国农业科学,2007,40(9):1973-1979.

[31] Wang Y, Ji Q, Liu S., Sun, H, et al. Effects of tillage practices on water-stable aggregation and aggregate-associated organic C in soils [J]. Journal of Agro-Environment Science, 2012,31:1365-1373.

[32] 刘哲,孙增慧,吕贻忠.长期不同施肥方式对华北地区温室和农田土壤团聚体形成特征的影响[J].中国生态农业学报,2017,25(8):1119-1128.

[33] Wang F, Ya Tong, Js Z, et al. Effects of various organic materials on soil aggregate stability and soil microbiological properties on the Loess Plateau of China [J]. Plant,Soil and Environment, 2013,59(4):162-168.

[34] 刘满强,胡锋,陈小云.土壤有机碳稳定机制研究进展[J].生态学报,2007,27(6):2642-2650.

[35] Lei Na, Han Jichang, Mu Xingmin, et al. Effects of improved materials on reclamation of soil properties and crop yield in hollow villages in China [J]. Journal of Soils and Sediments, 2019,19(5):2374-2380.

[36] 董少文,马淑花,初茉,等.粉煤灰基土壤调理剂作用下盐碱土壤微观结构变化规律[J].过程工程学报,2022,22(3):357-365.

[37] 米美霞,陈玉鹏,武小钢,等.粉煤灰和蚯蚓粪施用对土壤蒸发的影响[J].节水灌溉,2021(11):25-31.

[38] 张露,魏静.不同还田材料对空心村整治后土壤肥力的影响[J].水土保持通报,2018,38(3):74-78.

[39] 高志香,李希来,张静,等.不同施肥处理对高寒矿区渣山改良土酶活性和理化性质的影响[J].草地学报,2021,29(8):1748-1756.

[40] 姜灿烂,何园球,刘晓利,等.长期施用有机肥对旱地红壤团聚体结构与稳定性的影响[J].土壤学报,2010,47(4):715-722.

[41] 槐圣昌,刘玲玲,汝甲荣,等.增施有机肥改善黑土物理特性与促进玉米根系生长的效果[J].中国土壤与肥料,2020(2):40-46.

[42] Singh J S, Pandey V C, Singh D P. Coal fly ash and farmyard manureamendments in dry-land paddy agriculture field: effect on N-dynamics and paddy productivity[J].Applied Soil Ecology, 2011, 47(2):133-140.

[43] Rautaray S K, Ghosh B C, Mittra B N. Effect of fly ash, organic wastes and chemical fertilizers on yield, nutrient uptake, heavy metal content and residual fertility in a rice-mustard cropping sequence under acid lateritic soils [J]. Bioresource Technology, 2003,90(3):275-283.

[44] 魏俊岭,金友前,郜红建,等.施肥措施对砂姜黑土水分入渗性能的影响[J].中国生态农业学报,2014,22(8):965-971.

[45] 张旭辉,李恋卿,潘根兴.不同轮作制度对淮北白浆土团聚体及其有机碳的积累与分布的影响[J].生态学杂志,2001,20(2):16-19.

[46] 任顺荣,邵玉翠,杨军.宅基地复垦土壤培肥效果研究[J].水土保持学报,2012,26(3):78-81.

[47] Chivenge P, Vanlauwe B, Gentile R, et al. Organic resource quality influences short-term aggregate dynamics and soil organic carbon and nitrogen accumulation [J]. Soil Biology and Biochemistry, 2011,43(3):657-666.

本文曾发表于2022年《水土保持通报》第42卷第5期

城市雨水花园集中入渗对土壤氮、磷及重金属的影响

郭　超[1,2,3]，谢　潇[1,2,3]，李家科[4]

(1.陕西省土地工程建设集团 自然资源部退化及未利用土地整治工程重点实验室，陕西 西安 710075；2.陕西省土地工程建设集团 陕西省土地整治工程技术研究中心，陕西 西安 710075；3.陕西省土地工程建设集团 自然资源部土地工程技术创新中心，陕西 西安 710075；4.西安理工大学水利水电学院，陕西 西安 710048)

【摘要】目的：研究城市雨水花园集中入渗对土壤污染的影响，为城市雨水径流集中入渗工程技术的合理配置与推广应用提供科学依据。方法：以西安理工大学校园内运行 8~9 年的 2 个雨水花园(RD_1：受纳屋面径流雨水；RD_2：受纳路面和屋面径流的混合雨水)为研究对象。2016—2018 年对雨水花园 RD_1 和 RD_2 监测了 16~18 场降雨事件，确定其径流中化学需氧量(COD)、总悬浮物(TSS)、N、P、重金属的 EMC 浓度和雨水花园单位面积受纳污染物负荷量。2017 年 4 月至 2019 年 2 月，采集 7 次花园内不同土层深度处的土样，测定土壤中 NH_4-N、NO_3-N、TN、TP、总有机碳(TOC)及 Cu、Zn、Cd 含量，明确其在土壤垂向上的分布规律。结果：RD_1 径流中 COD、TSS、NH_4-N、TN、TP 的 EMC 浓度均大于 RD_2，而重金属小于 RD_2。土壤 NH_4-N 和 TN 含量随土层深度增加逐渐减小；NO_3-N 和 TP 含量随土层深度增加逐渐增大，50 cm 以下不同深度土壤 NO_3-N 和 TP 含量大多大于 0~50 cm 土层。TOC 含量随土层深度逐渐减小。0~30 cm 土壤重金属含量较高，Cu 和 Zn 主要以铁-锰氧化物结合态和残渣态形式存在，而 Cd 主要以可交换态和碳酸盐结合态形式存在，雨水花园土壤重金属 Cu、Zn、Cd 与 TOC 具有较好的拟合关系($R^2>0.8$)。结论：雨水花园集中入渗对土壤氮、磷、重金属有一定的影响，NO_3-N 和 TP 发生了淋溶，土壤 NH_4-N 和 TN 主要富集在 0~50 cm 土层，重金属主要富集在 0~30 cm。

【关键词】雨水花园；集中入渗；土壤 N，P，TOC；重金属

Influence of Concentrated Infiltration on Soil N, P and Heavy Metals in Urban Rain Garden

Chao Guo[1,2,3], Xiao Xie[1,2,3], Jiake Li[4]

(1. Shaanxi Provincial Land Engineering Construction Group, Key Laboratory of Degraded and Unused Land Consolidation Engineering, Ministry of Natural Resources, Xi'an, Shaanxi 710075, China; 2. Shaanxi Provincial Land Engineering Construction Group, Shaanxi Provincial Land Consolidation Engineering Technology Research Center, Xi'an, Shaanxi 710075, China; 3. Shaanxi Provincial Land Engineering Construction Group, Land Engineering Technology Innovation Center, Ministry of Natural Resources, Xi'an, Shaanxi 710075, China; 4.Xi'an University of Technology, Xi'an, Shaanxi 710048, China)

【Abstract】 Objective: The influence of concentrated infiltration of rainfall runoff on soil of rain gardens was studied in order to provide scientific reference for the reasonable configuration and application of the concentrated infiltration technology of urban rainfall runoff. Methods: Two rain gardens (RD_1: accept roof rainfall runoff; RD_2: accept roof and road rainfall runoff) run for 8~9 years on the campus of Xi'an University of Technology, were used as the research objects. From 2016 to 2018, 16~18 rainfall events were monitored to determine the event

资助项目：陕西省重点研发项目“城市内河黑臭水体原位综合治理技术开发与应用示范”(2020SF-420)；陕西省土地工程建设集团内部科研项目(DJNY2020-27、DJNY2021-17)。

mean concentration (EMC) and the pollutant load of chemical oxygen demand (COD), total suspended solids (TSS), N, P, heavy metals in the rainfall runoff of the two rain gardens. From April 2017 to February 2019, a total of 7 soil samples at different soil depths in the two gardens were collected to determine NH_4-N, NO_3-N, TN, TP, total organic carbon (TOC) and heavy metals of Cu, Zn, and Cd and clarify the vertical distribution of N, P, TOC and heavy metal content in the soil of rain garden. Results: The EMC concentrations of COD, TSS, NH_4-N, TN, and TP in RD_1 were all bigger than those in RD_2, but opposite results were obtained about heavy metals. The contents of NH_4-N and TN showed a gradually decreasing trend with the depth of the soil layer. But the contents of NO_3-N and TP in the soil gradually increased with the depth of the soil layer, and they were all bigger below 50 cm than those of the upper layer (0~50 cm). The TOC content in the soil of the rain garden showed a gradually decreasing trend with the depth of the soil layer. The contents of heavy metals were big at the soil layer of 0~30 cm. Cu and Zn mainly existed in the form of iron-manganese oxide combined state and residue state, but Cd mainly existed in the form of exchangeable and carbonate bound forms. The contents of Cu, Zn and Cd in the rain garden soil had the good relationship with TOC (R^2>0.8). Conclusion: The concentrated infiltration of rain gardens had certain influence on soil N, P and heavy metals. NO_3-N and TP leaching occured in soil. The contents of NH_4-N and TN in rain garden soil mostly concentrated within 50 cm, however, the heavy metals mostly concentrated within 30 cm.

【**Keywords**】Rain garden; Concentrated infiltration; Soil N, P, total organic carbon (TOC); Heavy metals

传统城市化带来了城市内涝、水环境污染与地下水缺乏补给等严峻问题,人们通过反思与研究,提出了许多应对之策,如美国的低影响开发(LID)、英国的可持续城市排水系统(SUDS)、澳大利亚的水敏感城市设计(WSUD)等[1],均取得了良好效果。我国根据自身的水文、地理条件,提出了建设自然积存、自然渗透、自然净化的“海绵城市”的要求。在我国的海绵城市建设中,雨水花园、生物滞留设施等雨水径流集中入渗设施应用较多。这类措施主要利用入渗性能和污染物吸附性能较好的改良材料作为填料,使降雨径流携带的污染物经填料吸附、过滤、离子交换、微生物降解等物理、化学、生物作用得到自然净化,然后下渗补给地下水或收集再次利用,具有减缓城市内涝、净化水质与涵养地下水资源的综合功能。由于城市雨水径流存在严重的面源污染,且集中入渗的水量负荷与污染负荷强度大,长期集中入渗会造成填料堵塞、污染物吸附饱和等众多问题,降低设施运行效率,缩短设施运行寿命,这些都是众多专家非常关注的科学问题[2]。

目前,国内外研究大多针对雨水花园、生物滞留设施等 LID 设施本身的结构、效果、影响机制与模拟研究展开[3-5]。如王璐等[6]将雨水花园与工程隔盐技术相结合,通过室内模拟和正交试验方法,提出适于上海滨海盐碱地区的 3 种雨水花园结构模式。Gurung 等[7]研究表明,雨水花园对降雨径流中氮、磷、重金属具有较好的去除效果。罗鸣等[8]基于下凹绿地和雨水花园构建了 SWMM 模型,模拟 1 年一遇、2 年一遇、5 年一遇重现期暴雨时的出水口径流过程及各污染物负荷总量,表明下凹绿地和雨水花园对径流和水质均有一定的削减作用,且雨水花园作用效果更显著。对低影响开发(LID)模式下雨水径流集中入渗对土壤的影响开展了一些研究,如 Kim 等[9]研究了雨水花园中重金属在土壤中的迁移特性,证实了其会对土壤产生污染;Gunawardena 等[10]论述了城市雨水径流中常用的雨水入渗技术和城市雨水中主要的污染物,并对这些污染物入渗后的去向进行了调查; Tedoldi 等[11]认为 SUDS 的普遍应用会对土壤造成影响,并通过文献回顾分析了解决问题的方法。Tedoldi 等[11]和 Xie 等[12]研究了流体中携带的污染物在土壤环境中的迁移行为和分布规律。但是,近年来,雨水花园集中入渗雨水径流对土壤影响过程的量化分析缺乏系统研究,影响了这类措施的合理使用。

本文对运行 7~8 年的雨水花园土壤进行多次取样分析,重点研究黄土地区雨水径流集中入渗对土壤氮、磷、重金属含量的影响过程,为黄土地区雨水径流集中入渗措施的合理配置与推广应用提供科学依据和理论支撑,促进中国海绵城市的健康发展。

1 研究区域及方法

1.1 研究区域

试验雨水花园位于陕西省西安市西安理工大学校园内,西安市位于黄河流域中部的陕西关中盆地,属于大陆性季风气候,平均气温 13.3 ℃,冷热干湿,四季分明,冬季干冷,春季干燥,夏季湿热,秋季多雨,属半湿润气候区。根据西安市 1951—2008 年(58 年)的降雨统计资料,多年平均降雨量 580.2 mm,2017 年研究区降水总量为 642.9 mm,降雨量年际变化相差较大,最大年降雨量达 903.2 mm(1983 年)。西安是缺水城市,人均水资源占有量仅为全国的 1/6,且地下水超采严重。西安地区的黄土层极为丰厚,黄土入渗性能非常好,且为入渗、滞留自然降雨径流提供了天然的屏障。故以 LID 设施为单个海绵体,用于排除和蓄存雨水径流是海绵城市建设过程中重要的措施,为缓解城市内涝灾害和补充地下水资源匮缺提供良好的天然条件。本研究涉及两个雨水花园 RD_1 和 RD_2。

雨水花园 RD_1 建成于 2011 年,用于处理办公楼屋面雨水径流,屋面面积即汇水面积约为 605 m^2,雨水花园面积为 30.24 m^2,汇流比为 20 : 1。上部蓄水层 15 cm,下部结构层填充西安市本地黄土 20 cm,渗透系数大约为 2.346 m/d,结构层以下为原状土层(未受扰动),花园底部不做防渗处理,雨水径流经土壤基质直接入渗补给地下水。入流口安装 45°三角堰,溢流口安装 30°三角堰。雨水花园 RD_1 东侧 30 m 处有一垃圾场,主要用于收集校园内生活垃圾以及枯枝落叶等,垃圾场底部为水泥混凝土硬质地面,垃圾渗滤液不会发生垂向或水平渗漏,主要是清理不及时带来的少量垃圾碎屑沉积物随径流雨水冲刷而进入花园内。

雨水花园 RD_2 建成于 2012 年,大致为椭圆形,长轴为 6 m,短轴为 2 m,雨水花园面积约为 9.42 m^2,主要汇集路面和屋面的混合雨水,汇水面积约为 140 m^2,汇流比为 15 : 1。花园中间用隔板分割为两个面积相同的雨水花园,其中一侧做防渗处理,底部埋设穿孔排水管,排水管周围填筑 15 cm 砾石反滤层;另一侧不做防渗处理,雨水径流可直接入渗补给地下水。在花园入流口安装两个 45°三角堰,出水口和溢流口安装 30°三角堰。上部蓄水层 50 cm,下部填充西安市本地黄土 60 cm,底层为原状土层(未受扰动)。雨水花园结构见图 1。

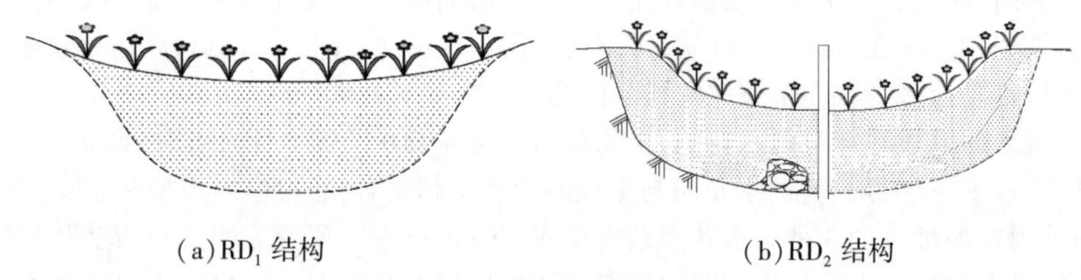

(a) RD_1 结构　　(b) RD_2 结构

图 1　雨水花园结构

1.2 试验方法

每次降雨开始时在入流、出流、溢流三角堰处记录各场次降雨的入流、出流瞬时流量,降雨初期一般每隔 5 min 采集一次进、出水水样,在降雨持续 20 min 后可按照每隔 10 min 采集 1 次进、出水样,降雨持续 1 h 后按照每隔 20 min 或 30 min 采集 1 次进、出水样,一般降雨水样采集数量大于 8 个,现场采集的水样及时放入 -4 ℃冰箱内,水样分析指标包括 TSS、COD、NH_4-N、NO_3-N、TN 和 TP,水质指标一般在 3~5 d 内分析完毕。

2017 年 4 月至 2019 年 2 月在两个雨水花园 RD_1 和 RD_2(入渗一侧)内共采集了 7 次土样(距离花园 RD_1 西侧 5 m 处的绿化地中采集一对照土样 CK),土壤含水率保持在 11.3%~30.1%,pH 值保持在 7.55~7.76,呈微碱性,主要测定土壤 N、P、TOC、重金属含量。为了保持取样的一致性,一般定于降雨后 2~3 d 内进行土样采集,采用 3 点(花园 RD_1 为同心圆上 3 点,花园 RD_2 为直线上 3 点)取样混合法采集土样。雨水花园 RD_1 土层深度较深,分 5 层进行土样采集,分别为 0~10 cm、20~30 cm、40~50 cm、70~80 cm、90~100 cm;雨水花园 RD_2 土层较浅(60 cm),分 3 层进行土样采集,分别为 0~10

cm、20~30 cm、40~50 cm;对照组 CK 分 4 层采集,分别为 0~10 cm、20~30 cm、40~50 cm、70~80 cm。样品采集后分成两部分,一部分立即进行含水率测定,另一部分晾在阴凉处,在自然状态下进行通风阴干,剔除其中的草根等杂物,将其充分混匀,采用四分法取少量阴干土壤,过 2 mm 筛,测定 NH_4-N、NO_3-N,另一部分过 1 mm 筛测定 TN、TP,重金属 Cu、Zn、Cd 超出最小检测线值记为 0。经处理后的土样放入-20 ℃冰箱内待测,两周内分析完毕。7 次土样采集前降雨情况见表 1,水样和土样分析方法见表 2。重金属 Cu、Zn、Cd 的形态采用经典的 Tessier 等连续五级提取法[13],包括可交换态(S_1)、碳酸盐结合态(S_2)、铁-锰氧化物结合态(S_3)、有机结合态(S_4)和残渣态(S_5)。

表 1 7 次土样采集前降雨和入渗状况

降雨时间/年-月-日	土样采集时间/年-月-日	降雨类型	降雨时长	降雨量/mm	RD_1 入渗量/m^3	RD_2 入渗量/m^3
2017-04-26	2017-04-27	中雨	9 h 20 min	5.2	2.99	0.70
2017-07-06	2017-07-07	大雨	16 h	28.1	16.15	3.81
2017-10-11	2017-10-14	暴雨	11 h 10 min	31.0	17.82	4.20
2018-01-21	2018-01-22	小雨	7 h 40 min	5.0	2.87	0.68
2018-05-06	2018-05-07	中雨	10 h 10 min	13.4	7.70	1.81
2018-09-28	2018-09-30	大雨	9 h 30 min	18.8	10.81	2.55
2019-02-20	2019-02-21	中雨	8 h 50 min	8.9	5.12	1.21

表 2 水样和土样分析方法

测试指标		检测方法	检测仪器
雨水花园入流水样	COD	重铬酸钾快速消解分光光度法	紫外分光光度计
	TSS	称重法	—
	NH_4-N	纳氏试剂比色法	流动分析仪
	NO_3-N	紫外分光光度法	流动分析仪
	TN	碱性过硫酸钾消解分光光度法	紫外分光光度计
	TP	过硫酸钾消解钼酸铵分光光度法	紫外分光光度计
	Cu	原子吸收分光光度法	原子吸收分光光度计
	Zn		
	Cd		
雨水花园土样	TOC	容量法	滴定管
	NH_4-N	LY/T 1228—2015	紫外分光光度计
	NO_3-N	LY/T 1228—2015	紫外分光光度计
	TN	LY/T 1228—2015	半自动定氮仪
	TP	LY/T 1228—2015	紫外分光光度计
	Cu	土壤元素近代分析法	原子吸收分光光度计
	Zn		
	Cd		

2 结果与分析

2.1 雨水花园入流污染物浓度和负荷分析

2016—2018 年对雨水花园 RD_1 和 RD_2 监测了 16~18 场降雨事件,主要采集雨水花园入流水样,测定其径流中污染物含量,其径流中各污染物 EMC 浓度和雨水花园单位面积受纳污染负荷量见表 3。

表 3 雨水花园入流 EMC 浓度和单位面积受纳的污染负荷量(2016—2018 年)

指标	RD_1		RD_2	
	平均 EMC 浓度/($mg \cdot L^{-1}$)	单位面积受纳污染负荷量/($g \cdot m^{-2}$)	平均 EMC 浓度/($mg \cdot L^{-1}$)	单位面积受纳污染负荷量/($g \cdot m^{-2}$)
COD	103.91	1179	57.78	483
TSS	93.55	1062	65.04	544
NH_4-N	2.45	28	1.14	10
NO_3-N	1.11	13	1.57	12
TN	4.24	48	3.09	27
TP	0.32	4	0.23	2
Cu	0.044	0.662	0.106	1.201
Zn	0.281	4.212	0.630	7.102
Cd	0.022	0.032	0.031	0.035

由表 3 可以看出,雨水花园 RD_1 入流中 COD、TSS、NH_4-N、TN、TP 等污染指标的平均 EMC 浓度较 RD_2 大,但 NO_3-N 和重金属 Cu、Zn、Cd 含量均小于 RD_2。其中,RD_1 中 COD 的 EMC 浓度是 RD_2 的 1.80 倍,单位面积受纳污染负荷量是 RD_2 的 2.44 倍,说明 RD_1 径流受有机物污染较为严重,降雨径流进入雨水花园后将破坏土体平衡,COD 的降解将会消耗土体中氧气,成为厌氧状态,影响生物生存[14]。雨水花园 RD_1 东侧 30 m 处有一校园垃圾场,垃圾场三面设有高 2 m 的水泥围墙(敞开一面背离雨水花园),受校园垃圾日常清扫、垃圾装卸等活动的影响,雨水花园 RD_1 汇水渠中堆积的粉尘较多,夏季降雨突然来不及清扫时,粉尘随径流会进入 RD_1 中,导致雨水花园 RD_1 径流中 TSS、N、P 等污染物含量较高。

雨水花园 RD_2 中重金属含量较高,其中重金属 Cu、Zn 的 EMC 含量分别是 RD_1 的 2.41 倍、2.24 倍,单位面积受纳污染负荷量分别是 RD_1 的 1.81 倍、1.69 倍,说明重金属含量主要来自路面径流。主要是由于城市道路行驶车辆较多,汽车尾气排放、油脂泄漏、车胎磨损将导致路面沉积物中重金属 Zn 含量升高。韦毓韬等[15]研究表明,道路雨水径流中 Cu、Zn、Cd、Pb 等重金属主要受交通因素的影响。

2.2 土壤 N、P、TOC 含量垂向分布

雨水花园 0~100 cm 不同土层深度 N、P 含量见图 2,具体分析如下:

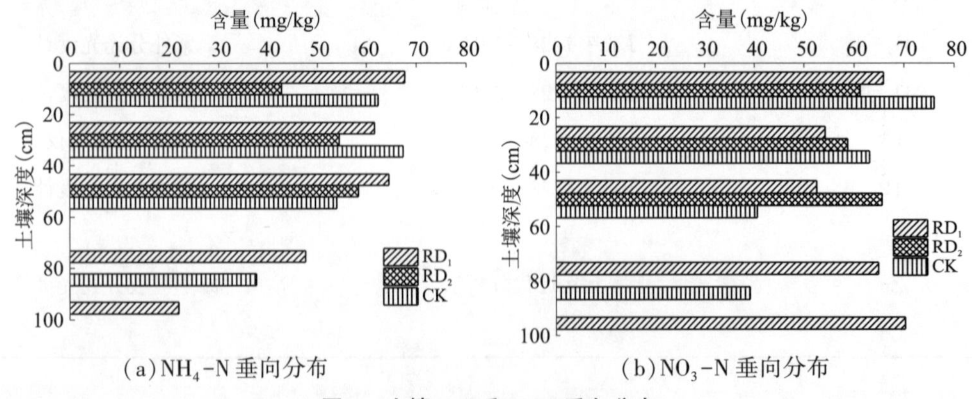

(a) NH_4-N 垂向分布　　(b) NO_3-N 垂向分布

图 2 土壤 N、P 和 TOC 垂向分布

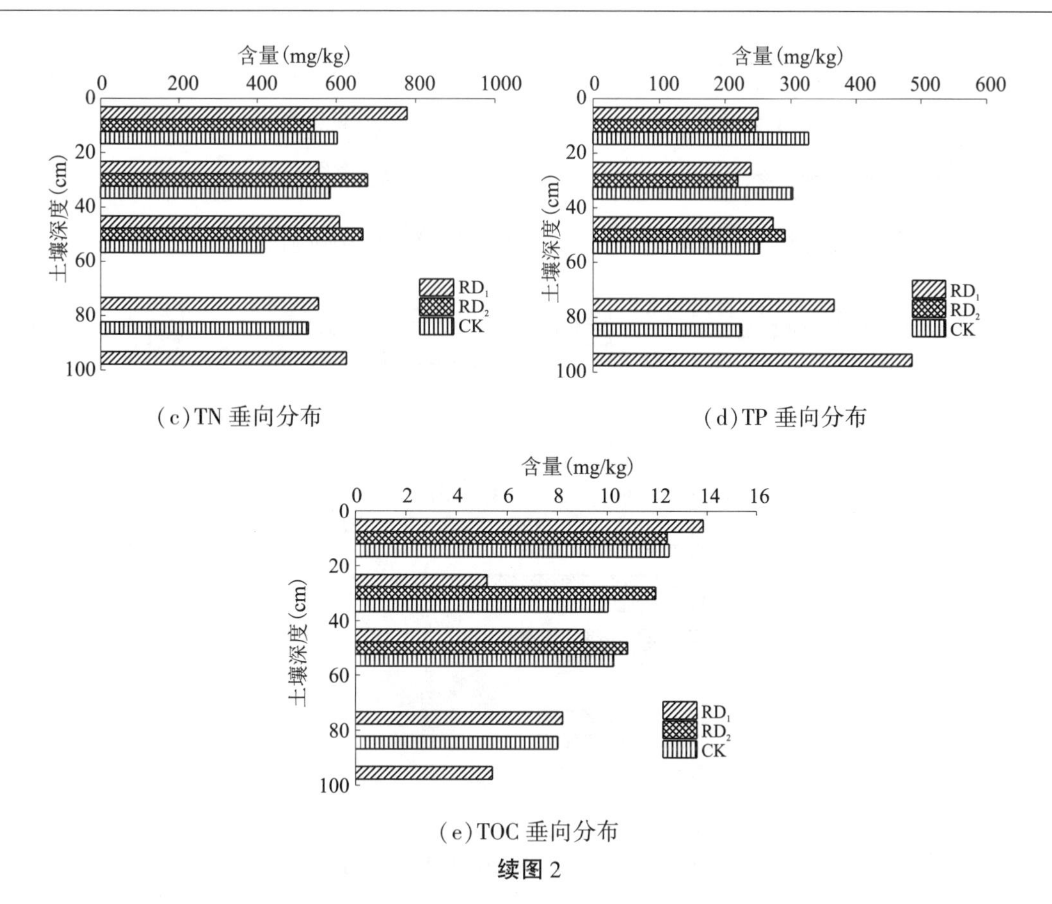

(c)TN 垂向分布　　　　(d)TP 垂向分布

(e)TOC 垂向分布

续图 2

由图 2 可以看出,雨水花园和 CK 土壤中 NH_4-N 和 TOC 含量随土层深度增加呈逐渐递减的趋势,上层土壤中 NH_4-N 和 TOC 含量明显高于下层土壤。其中 RD_1 土壤中 NH_4-N 含量由表层下 10 cm 的 67.69 mg/kg 降低至表层下 100 cm 的 21.75 mg/kg,降低了 211.18%。这主要是由于降雨径流中 NH_4-N 主要以颗粒态形式存在,进入雨水花园后被拦截在表层土壤。此外,NH_4-N 主要带正电荷,而土壤大多带负电荷,雨水径流中 NH_4-N 易吸附在土壤胶体表面,在降雨冲刷作用下不易随水分向下迁移。RD_1 土壤中 TOC 含量由表层下 10 cm 的 13.83 g/kg 降低至表层下 100 cm 的 5.42 g/kg,降低了 155.38%。随取样深度增加,雨水花园和 CK 土壤中 NO_3-N 含量呈增加趋势,上层土壤中 NO_3-N 含量小于下层土壤。RD_1 土壤中 NO_3-N 含量由表层下 10 cm 的 65.84 mg/kg 增加至表层下 100 cm 的 70.23 mg/kg,下层土壤 NO_3-N 含量比上层增加了 6.68%。这就说明降雨径流中 NO_3-N 进入土壤后发生了淋溶现象,NO_3-N 随水分入渗易向下迁移。王禄等[16]通过人工滤柱试验发现,NH_4-N 主要集中于表层下 50 cm 土壤,并随滤池深度增加呈逐渐降低的趋势,而 NO_3-N 随水流被快速排除试验系统。有关研究表明[17],硝酸盐作为阴离子不被土壤介质所吸附,在土壤/水系统中通常具有很强的移动性。与 NH_4-N 含量的变化规律一致,雨水花园土壤中 TN 含量随土层深度增加呈逐渐减小的趋势。雨水花园土壤中 TP 含量随土层深度增加逐渐增大,而 CK 土壤中 TP 含量随深度增加呈减小趋势。RD_1 不同层深度 TP 含量分别为 250.56 mg/kg、239.93 mg/kg、273.09 mg/kg、366.49 mg/kg、485.83 mg/kg,表层下 100 cm 土壤 TP 含量较表层下 10 cm 增加了 93.90%。说明雨水花园土壤中可溶性磷随水分入渗发生了明显的迁移作用,具有显著的深层富集现象。因此,雨水花园、生物滞留设施不宜用于地下水位较高的地区,若确实需要应用,需做好前处理,避免硝酸盐和磷酸盐的淋溶而发生次生灾害。

2.3　土壤重金属垂向分布

将雨水花园 2017 年 4 月至 2019 年 2 月采集的 7 次土样中 3 种重金属含量分别取均值,得到雨水花园土壤重金属含量在垂向上的分布规律(见图 3)。

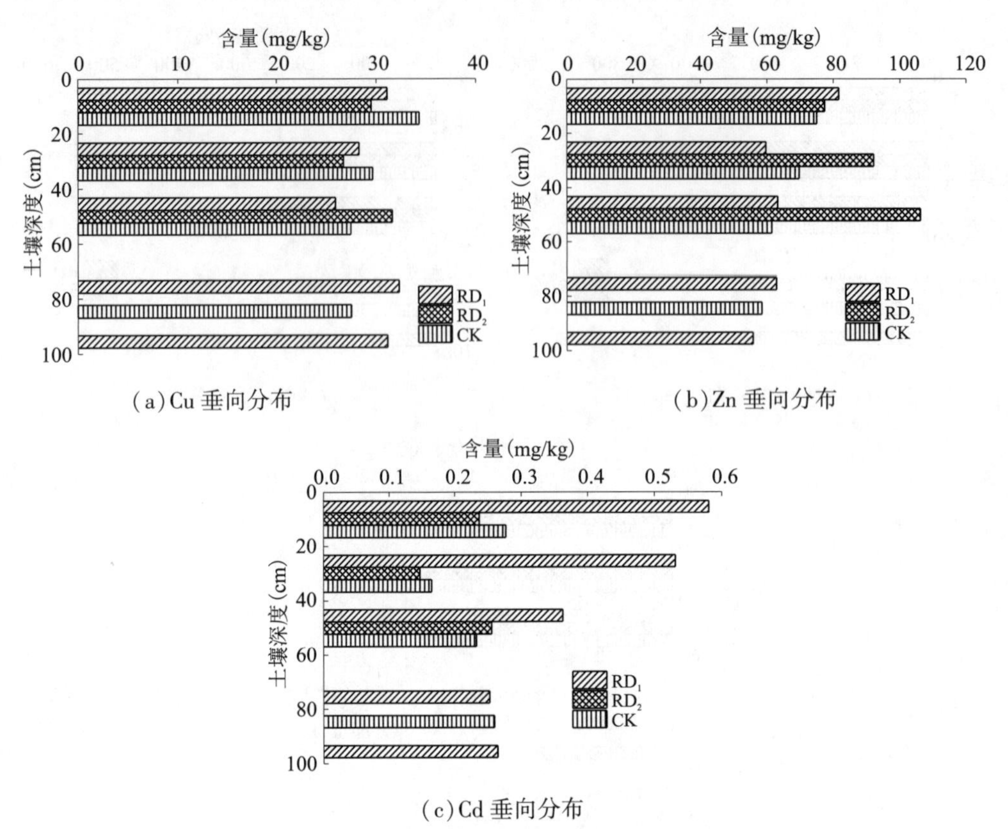

(a) Cu 垂向分布

(b) Zn 垂向分布

(c) Cd 垂向分布

图 3　雨水花园土壤重金属垂向分布

由图 3 可以看出,雨水花园和 CK 土壤重金属 Zn 和 Cd 大多富集在土壤表层下 30 cm 范围内,并且土壤 Cd 含量随土层深度增加呈逐渐减小的趋势。土壤中重金属 Cu 的分布较均匀,RD_1 的 90~100 cm 和 RD_2 的 40~50 cm 土层中 Cu 含量略大于其他各层,而 CK 表层土壤重金属 Cu 含量略大于其他各层。研究表明,城市道路径流雨水中重金属 Zn 主要来源于车辆轮胎磨损、发动机润滑油遗漏;重金属 Cu 主要来源于金属电镀、车辆轴承及制动部件磨损;重金属 Cd 主要来源于车辆轮胎磨损[18]。董微砾[19]研究新疆乌鲁木齐市城市道路雨水中重金属 Cu 平均值为 209.7~237.4 μg/L,Zn 平均值为 530.4~745.5 μg/L,Cd 平均值为 0.7~3.3 μg/L。Xie 等[12]通过大量研究认为 Zn 在市区的主要来源是车辆的轮胎磨损、制动磨损、大气干湿沉积和石油、天然气的燃烧作用。

降雨径流导致了雨水花园土壤中重金属含量的增加,0~50 cm 土层,雨水花园 RD_1 和 RD_2 土壤 Zn 含量分别为 81.58 mg/kg、70.66 mg/kg、59.73 mg/kg、56.93 mg/kg、58.48 mg/kg 和 77.36 mg/kg、84.74 mg/kg、92.13 mg/kg、99.18 mg/kg、106.23 mg/kg,可以看出,除表层土壤(0~10 cm)外,其余各层土壤,雨水 RD_2 中 Zn 含量分别是 RD_1 的 1.20 倍、1.54 倍、1.74 倍、1.68 倍,表明雨水花园土壤 Zn 含量受降雨汇水面类型的影响较大,汇水面为道路的雨水花园,其土壤中重金属 Zn 含量明显高于汇水面为屋面的雨水花园。由于 RD_2 主要接纳路面径流,故 RD_2 入流水质中 Zn 浓度远高于 RD_1。

不同汇水面对雨水花园土壤中重金属 Cd 的影响也不同,0~50 cm 各层土壤,雨水花园 RD_1 土壤重金属 Cd 含量分别为 0.581 mg/kg、0.531 mg/kg、0.363 mg/kg,RD_2 土壤重金属 Cd 含量分别为 0.237 mg/kg、0.147 mg/kg、0.256 mg/kg,而 CK 土壤分别为 0.277 mg/kg、0.165 mg/kg、0.233 mg/kg。可以看出,雨水花园 RD_2 和 CK 土壤中 Cd 含量差异较小,但 RD_1 土壤 Cd 含量远大于 RD_2 和 CK,分别是 RD_2 和 CK 的 2.45 倍、3.62 倍、1.42 倍、2.09 倍、3.22 倍、1.56 倍。分析其原因,主要是受 RD_1 东侧的垃圾场影响。雨水花园 RD_1 主要汇集屋面径流,降雨径流经坡屋面自由落水进入地面汇水沟后集中进入雨水花园,而垃圾场西侧正好为雨水花园部分地面汇水沟,每天校园内清理垃圾时,有大量粉尘落入东侧地面汇水沟和雨水花园 RD_1,粉尘中可能含有重金属 Cd,在降雨冲刷作用下进入了雨水花园，导致雨水花园 RD_1 中 Cd 含量远大于 RD_2。

2.4 土壤重金属与 TOC 的关系

由图 4 可以看出，土壤中重金属 Cu 和 Zn 与土壤有机碳（TOC）呈直线关系，R^2 分别达 0.840、0.861，重金属 Cd 与 TOC 呈二次曲线关系，R^2 达 0.802。TOC 是通过微生物作用所形成的腐殖质、动植物残体和微生物体的合称，土壤 TOC 是雨水花园径流雨水中重金属的重要载体，TOC 含量较高的土壤可有效削减径流雨水中的重金属，从而减少重金属对土壤中动、植物的危害，并且可防止重金属向下迁移，影响地下水水质。王腾云等[20]通过福建沿海地区土壤—稻谷重金属含量关系与影响因素研究，发现富含有机碳的土壤条件有利于阻断稻谷对土壤 Cd 的吸收，降低土壤 Cd 污染的生态风险。谢娜等[21]通过研究不同土地利用方式土壤有机碳变化特征及与重金属的相关性，发现典型土壤重金属 Hg、Cd、Cr、As、Pb 与 SOC 储量之间存在显著相关性，园地和建设用地最高 $r=0.99$，林地和耕地次之，分别为 $r=0.87$ 和 $r=0.86$。可见，土壤中 TOC 与土壤重金属具有密切相关性。

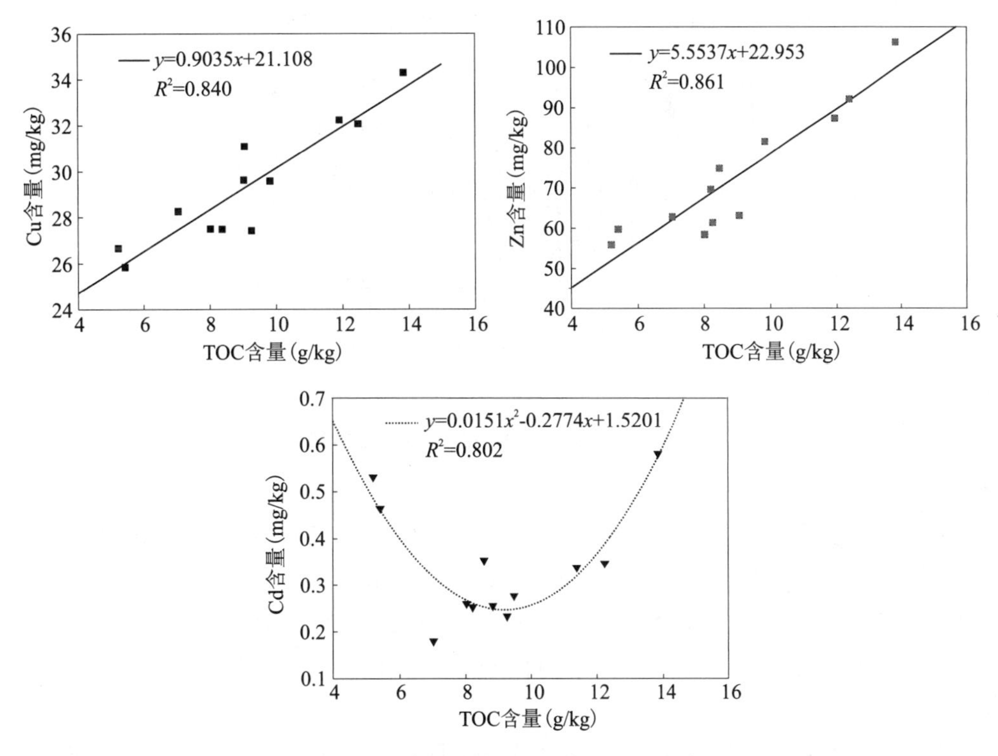

图 4 雨水花园土壤重金属与 TOC 的关系

2.5 土壤重金属形态分布

可交换态和碳酸盐结合态两种形态的重金属不大稳定，随水分入渗易向下迁移，铁-锰氧化物结合态、有机结合态和残渣态在土壤中的存在形态较稳定。重金属 Cu、Zn、Cd 的各形态含量如图 5 所示。

由图 5 可以看出，雨水花园和 CK 土壤中重金属 Cu 的 5 种形态关系为：$S_5>S_4>S_3>S_2>S_1$，而 Zn 的 5 种形态关系为：$S_5>S_3>S_2>S_4>S_1$，故雨水花园土壤中重金属 Cu 和 Zn 主要以铁-锰氧化物结合态和残渣态形式存在，存在形态相对稳定，随水分入渗不易发生向下迁移。雨水花园和 CK 土壤中重金属 Cd 的 5 种存在形态与 Cu 和 Zn 明显不同，其关系为 $S_1>S_2>S_3>S_4>S_5$。由此可见，雨水花园土壤中 Cd 主要以可交换态和碳酸盐结合态形式存在，这就说明 Cd 在土壤中很不稳定，随水分入渗有向下迁移的风险，易进入地下水。

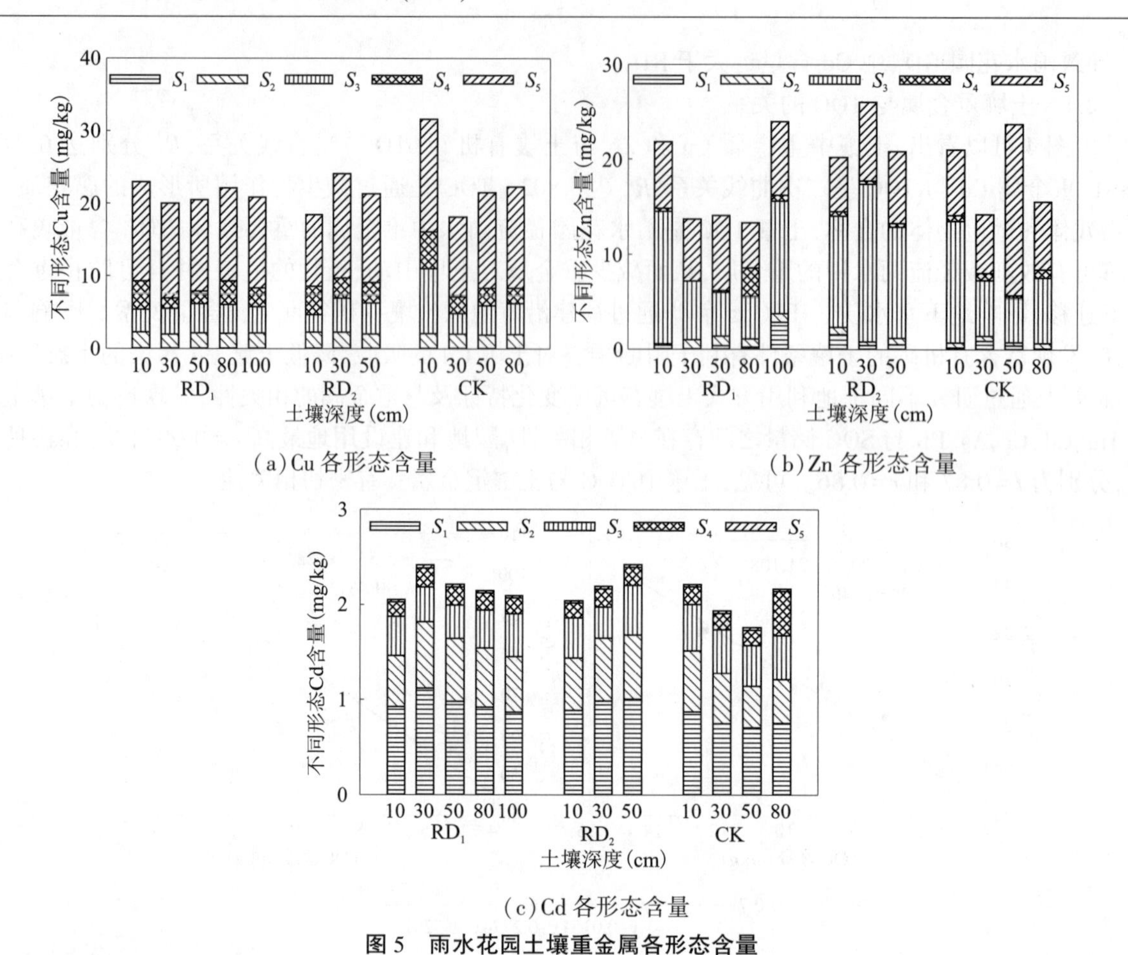

(a) Cu 各形态含量

(b) Zn 各形态含量

(c) Cd 各形态含量

图 5 雨水花园土壤重金属各形态含量

3 结论

(1) 雨水花园汇水面类型影响降雨径流中污染物浓度,雨水花园 RD_1 入流中 COD、TSS、N、P 等污染物含量较高,而雨水花园 RD_2 入流中重金属含量远高于 RD_1。

(2) 受降雨径流集中入渗污染影响,雨水花园 0~50 cm 土壤 NH_4-N 和 TN 含量较高。土壤 NH_4-N、TN 和 TOC 随土层深度增加逐渐减小,而 NO_3-N 和 TP 随土层深度增加逐渐增大,发生了明显的淋溶。

(3) 雨水花园土壤重金属大多集中在 0~30 cm 范围。Cu 和 Zn 主要以铁-锰氧化物结合态和残渣态形式存在,Cd 主要以可交换态和碳酸盐结合态形式存在,随水分入渗易向下迁移。雨水花园土壤重金属与 TOC 具有较好的拟合关系(R^2>0.8)。

参考文献

[1]Fletcher T D, Shuster W, Hunt W F, et al. SUDS, LID, BMPs, WSUD and more-The evolution and application of terminology surrounding urban drainage [J]. Urban Water Journal, 2015,12(7):525-542.

[2] Mehring A S, Hatt B E, Kraikittikun D, et al. Soil invertebrates in Australian rain gardens and their potential roles in storage and processing of nitrogen [J]. Ecological Engineering, 2016,97:138-143.

[3] Jia Zhonghua, Tang Shuangcheng, Luo Wan, et al. Small scale green infrastructure design to meet different urban hydrological criteria [J]. Journal of Environmental Management, 2016,117:92-100.

[4]郭效琛,杜鹏飞,辛克刚,等.基于监测与模拟的海绵城市典型项目效果评估[J].中国给水排水,2019,35(11):130-134.

[5]李家科,张兆鑫,蒋春博,等.海绵城市生物滞留设施关键技术研究进展[J].水资源保护,2020,36(1):1-8.

[6]王璐,于冰沁,陈嫣,等.上海滨海盐碱地区雨水花园适应性结构设计:以临港海绵城市建设示范区为例[J].上

海交通大学学报(农业科学版),2019,37(4):29-36.

[7] Gurung S B, Geronimo F K, Hong J, et al. Application of indices to evaluate LID facilities for sediment and heavy metal removal [J]. Chemosphere, 2018,206:693-700.

[8]罗鸣,叶兴成,王飞,等.下凹绿地与雨水花园在小区尺度应用中的对比[J].节水灌溉,2018(2):117-121.

[9] Kim B S M, Salaroli A B, Ferreira P A D L, et al. Spatial distribution and enrichment assessment of heavy metals in surface sediments from Baixada Santista, Southeastern Brazil [J]. Marine Pollution Bulletin, 2016,103(1/2):333-338.

[10]Gunawardena J, Egodawatta P, Ayoko G A, et al. Atmospheric deposition as a source of heavy metals in urban stormwater [J]. Atmospheric Environment, 2013,68:235-242.

[11] Tedoldi D, Chebbo G, Pierlot D, et al. Impact of runoff infiltration on contaminant accumulation and transport in the soil/filter media of Sustainable Urban Drainage Systems: A literature review[J]. Science of the Total Environment, 2016,569/570:904-926.

[12] Xie Yunfeng, Chen Tongbin, Lei Mei, et al. Spatial distribution of soil heavy metal pollution estimated by different interpolation methods:accuracy and uncertainty analysis [J]. Chemosphere, 2011,82(3):468-476.

[13] Tessier A, Campbell P G C, Bisson M. Sequential extraction procedure for the speciation of particulate trace metals [J]. Analytical Chemistry, 1979,51(7):844-851.

[14]蔡俊驰,李嘉,任婷婷,等.水质因子对白鹤滩藻类生长影响模拟试验研究[J].四川大学学报(工程科学版),2014,46(S1):37-41.

[15]韦毓韬,姜应和,张校源,等.雨水径流中重金属污染现状及其相关性分析[J].环境保护科学,2018,44(5):68-72.

[16]王禄,喻志平,赵智杰.人工快速渗滤系统氨氮去除机理[J].中国环境科学,2006,26(4):500-504.

[17] Li Liqing, Davis A P. Urban Storm-water runoff nitrogen composition and fate in bioretention systems[J]. Environment Science & Technology, 2015,48(6):3403-3410.

[18]申丽勤,车伍,李海燕,等.我国城市道路雨水径流污染状况及控制措施[J].中国给水排水,2009,25(4):23-28.

[19]董微砾.快速城市化地区雨水径流重金属污染特征研究:以乌鲁木齐市为例[D].乌鲁木齐:新疆师范大学,2014.

[20]王腾云,周国华,孙彬彬,等.福建沿海地区土壤—稻谷重金属含量关系与影响因素研究[J].岩矿测试,2016,35(3):295-301.

[21]谢娜,冯备战,李春亮.不同土地利用方式土壤有机碳变化特征及与重金属的相关性分析[J].中国农学通报,2019,35(26):115-120.

本文曾发表于2022年《水土保持通报》第42卷第1期

黄土丘陵沟壑区沟道造地土壤水分时空变异特征

王　晶[1,2],白清俊[1],王欢元[2],雷　娜[2],何　靖[2]

(1.西安理工大学水利水电学院,陕西 西安 710048;2.陕西省土地工程建设集团有限责任公司 陕西地建土地工程技术研究院有限责任公司 自然资源部退化及未利用土地整治工程重点实验室 陕西省土地整治工程技术研究中心 自然资源部土地工程技术创新中心,陕西 西安 710075)

【摘要】基于 Van Genuchten 模型测定土壤水动力学参数,采用定点监测方法,在沟道中不同位置以及对照坡面进行土壤水分观测,分析了延安市典型治沟造地项目沟道造地土壤水分的时空变异特征,阐明治沟造地工程对沟道土壤水分的影响。结果表明:(1) 土壤水力学参数在沟道土层深度为 40 cm 附近发生了显著改变,0~40 cm 土层土壤容重 1.12~1.25 g/cm^3,导水率达到 40 mm/min 以上,入渗能力强,同时饱和含水率较大,40 cm 以下土层土壤容重在 1.5 g/cm^{3}左右,导水率在 1.25~ 1.41 mm/min,入渗速率明显减小。(2) 沟道土壤水分显著大于对照坡面,其季节变化稍滞后于降水的季节变化,整个生长季在 15.76%~21.91%波动,高出对照坡面 5%左右,垂直分布随土层深度的增加而增加,表层最低,为 15.07%,160 cm 土层最高,为 22.84%,深层土壤含水量优势更加显著;沟道土壤水分变异系数在 0.131~ 0.234,相比较坡面,沟道表层以下土壤水分存在较强的空间异质性,100 cm 以下土层变异系数在 0.2 左右,土壤水分变化活跃。(3) 沟口土壤含水量显著高于沟头,土层深度 40 cm 以下的土壤含水量长期达到甚至超过田间持水量。通过治沟造地工程水分综合调控体系,沟道能够为作物生长提供水分充足的生境条件,提高水资源利用效率的同时改善农业生态环境。

【关键词】治沟造地;沟道;土壤水分;时空变异;黄土丘陵沟壑区

Spatial and Temporal Variations of Soil Moisture of Gully Cultivated Land in the Loess Hilly and Gully Region

Jing Wang[1,2], Qingjun Bai[1], Huanyuan Wang[2], Na Lei[2], Jing He[2]

(1.Institute of Water Resources and Hydro-Electric Engineering, Xi'an University of Technology, Xi'an 710048, China; 2.Shaanxi Provincial Land Engineering Construction Group Co. Ltd., Institute of Land Engineering and Technology, Shaanxi Provincial Land Engineering Construction Group Co. Ltd., Key Laboratory of Degraded and Unused Land Consolidation Engineering, Ministry of Natural and Resources, Shaanxi Provincial Land Consolidation Engineering Technology Research Center, Land Engineering Technology Innovation Center, Ministry of Natural Resources, Xi'an 710075, China)

【Abstract】The characteristics of spatial and temporal variation of soil moisture in gully were discussed in order to provide scientific basis for the rational allocation, efficient utilization of water resources and agricultural

基金项目:国家重点研发计划课题(2017YFC0504705);陕西省土地工程建设集团内部科研项目(DJNY2021-21)。

management in the Loess Plateau. Based on the Van Genuchten model, the soil hydrodynamic parameters were measured. The fixed-point monitoring method for observing soil moisture at different positions in gully and on control slope was used to analyze the temporal and spatial variation characteristics of soil moisture in the gully control and land consolidation project in Yan' an City, and clarify the impact of the project on the soil moisture in the gully. The results show that: (1) The soil hydraulic parameters change significantly at the depth of around 40 cm of gully. The soil bulk density of layer 0~40 cm in the gully is 1.12~1.25 g/cm^3, the hydraulic conductivity is more than 40 mm/min, show the characteristics of stronger infiltration capacity and higher saturated moisture content. While the soil bulk density of layer below 40 cm is about 1.5 g/cm^3, the hydraulic conductivity is 1.25~1.41 mm/min, and the infiltration rate is significantly reduced. (2) The soil water content of the gully is significantly greater than that of the control slope. Its seasonal variation characteristics are slightly lagging behind the falling seasonal, and it is fluctuate from 15.76% to 21.91% and is about 5% higher than control slope in the whole growing season. The vertical distribution increases with the increase of soil depth, the lowest is 15.07% in the surface layer and the highest is 22.84% in 160 cm.The advantage of deep soil water content is significant. The variation coefficient of soil moisture in gully is between 0.131 and 0.234. Compared with slopes, the soil moisture below the surface of gully has strong spatial heterogeneity, and changes actively below 100 cm with the variation coefficient is about 0.2. (3) The soil water content of the exit of the gully is significantly higher than that of the head of the gully, and the soil water content below 40 cm at the entrance reaches or even exceeds the field water capacity for a long time. Through the comprehensive water control system in gully control and land consolidation project, the gully cultivated land can provide sufficient water conditions for crop growth, improving the utilization efficiency of water resources while the agricultural ecological environment.

【**Keywords**】The loess hilly and gully region; Gully control and land consolidation; Gully; Soil moisture; Spatial and temporal variation

土壤水分是黄土高原地区植物生长的限制因子,也是这一地区农业耕作和植被恢复的关键[1]。黄土高原地形复杂,土壤疏松,水力侵蚀作用形成各种侵蚀沟。沟道由于地势较低,是流域水分的聚集区,土壤水分显著高于坡面[2]。治沟造地是延安市针对黄土高原丘陵沟壑区特殊地貌,集坝系建设、旧坝修复、盐碱地改造、荒沟闲置土地开发利用和生态建设为一体的沟道治理新模式[3]。治沟造地工程中,土地平整工程的实施改变了地表坡度及植被覆盖,进而影响地表径流的汇流、入渗和蒸发等过程[4],农田水利工程的建设更是人为改变了区域的水资源配置[5],引起土壤水分的变化。

治沟造地工程实施后,一些学者对沟道水文变化做了研究。在治沟造地对径流的影响方面,娄现勇等[6]研究沟道整治工程产汇流发现,治沟造地有利于降水就地入渗,沟道整治比例越大,沟道地表径流平均流速越小;孙彭成[7]研究表明,沟道土地整治能减少地表径流中面源污染物的输出;Kang等[8]对延河流域的研究表明,治沟造地工程具有良好的减流减沙效益,对流量和输沙的影响主要集中在汛期;在治沟造地对水分循环的影响方面,雷娜等[9]、Yin 等[10]、Jin 等[11]的研究均表明,由于农田水利工程影响了土壤水分的自然补给和排出,对沟道地下水位抬升起到促进作用;Zhao 等[12-13]通过研究 5 m 深度的土壤水分,得出与未进行治沟造地流域相比,治沟造地流域相邻季节之间土壤出水量的动态变化更小,沟道通过增加降水的渗透来存储大量土壤水分,充当土壤水库来缓解缺水问题。目前研究多集中在治沟造地后水文环境的变化,对土壤水分的时空变化未见深入研究。因此,本文以延安市宝塔区南泥湾镇九龙泉治沟造地项目沟道土壤水分为研究对象,分析其水力学参数、时空变化特征及沟道不同位置土壤水分差异,讨论治沟造地工程对沟道土壤水分的影响,从而为提高黄土高原地区沟道土壤水分利用效率,加强农业耕作管理及促进区域高质量发展提供科学依据。

1 材料与方法

1.1 研究区概况

延安市宝塔区(36°11′~37°02′N,109°14′~110°07′E)位于黄土高原中部,南北长 96 km,东西宽 76 km。属半湿润半干旱大陆性季风气候,夏季炎热多雨,冬季寒冷干燥;多年平均降水量为 562.1 mm,70%的降水量集中在 6—9 月。全年平均气温 9 ℃,无霜期 179 d。研究区属于黄土丘陵沟壑区,土壤类型为黄绵土。区内地形破碎,沟壑纵横,冲沟下切强烈,坡度较大,水土流失严重。九龙泉沟位于宝塔区南部的梁峁丘陵沟壑区,属河谷川、台地貌,东北—西南走向,沟道长约 9.5 km,河道海拔自北向南由 1170 m 降至 1093 m,平均比降 0.78%,河谷宽一般在 250~500 m,流域面积 62.63 km^2。流域坡面植被以草地和天然灌木为主。沟道土地多开垦为农田,种植前为旱地,水资源比较匮乏,无灌溉条件,粮食单产水平偏低。

治沟造地工程于 2013 年初开工建设,主体工程于 2014 年初竣工,总建设规模 360.91 hm^2,总投资 4411.65 万元。工程利用"截水沟蓄水排渍一体"的沟道水分调控技术,通过截水沟、灌排渠相互连通形成的水分综合调控体系,达到截分流结合、聚贮流结合、供节水结合的目的。雨季发生强降雨时,水库拦蓄和存储降水,超过蓄水量时利用排洪沟泄洪,同时通过截水沟将田块内多余积水引出;旱季沟道内田块缺水时,利用排洪沟和截水沟内水位的水势差,将水库中的蓄水引入截水沟和灌排两用渠,最终进入田块,实现了"旱时灌、涝时排"。

1.2 研究方法

1.2.1 土壤水动力学参数测定

在项目区沟道中游,选取典型沟道造地田块,挖取 150 cm 深土壤剖面,用环刀采集 0~120 cm 深度原装土壤样品,20 cm 为一层,测定其土壤容重和土壤水分特征曲线。土壤容重采用烘干法测定,土壤水分特征曲线用离心机测定,质地采用马尔文激光粒径分析仪(MS2000,英国)进行测定[14]。基于 Van Genuchten 模型利用 RETC 软件拟合土壤水分特征曲线,利用水分特征曲线推导不同深度处的土壤饱和导水率、残余含水率、田间持水量等土壤水动力学参数[15-16]。

1.2.2 土壤水分定点监测

在沟道中游,按照距坡顶距离,分别选取坡面和沟道不同空间位置的 15 个观测点,其中 1~9 分布于坡面上、中、下位置; 10 ~12 分布于沟道已耕种农田, 13~15 分布于沟底未扰动区域(见图 1),坡面的坡向为半阳坡,每个观测点垂直埋设长度为 2 m 的土壤水分观测管,进行长期十壤水分监测,监测时段为 2017 年 5—11 月,每月测定 1 次,观测深度为 0~160 cm。采用 TRIME-HD 高精度 TDR(Time Domain Reflect) ,从地表开始向下,按照 10 cm 间距逐层测定,每层重复测定 3 次,每次探针方向旋转 120°,取平均值作为该测点该层次的土壤含水量,所测土壤含水量为体积含水量。

根据沟道地形地貌条件,沿沟道走向在沟头和沟口共设置 5 个固定观测点,所有观测点均为农田。垂直埋设长度为 2 m 的土壤水分观测管进行监测。监测时间为 2019 年,每月测定一次,测定方法同上。

1.2.3 数据处理

研究采用经典统计分析、地统计分析相结合的方法,采用 Excel 进行统计分析;沟道不同位置土壤水分等值线图采用 Sigmaplot 12.0 绘制。

2 结果与分析

2.1 沟道造地土壤水分参数特征

从表 1 可以看出,沟道内土层深度 0~20 cm 容重最小,为 1.12 g/cm^3,20~40 cm 为 1.25 g/cm^3,40 cm 以下容重在 1.5 g/cm^3 左右。0~40 cm 为耕作层,受农业活动的影响,其土壤相对疏松,容重较小,而深度 40 cm 以下土层为耕作形成的犁底层,容重较大。

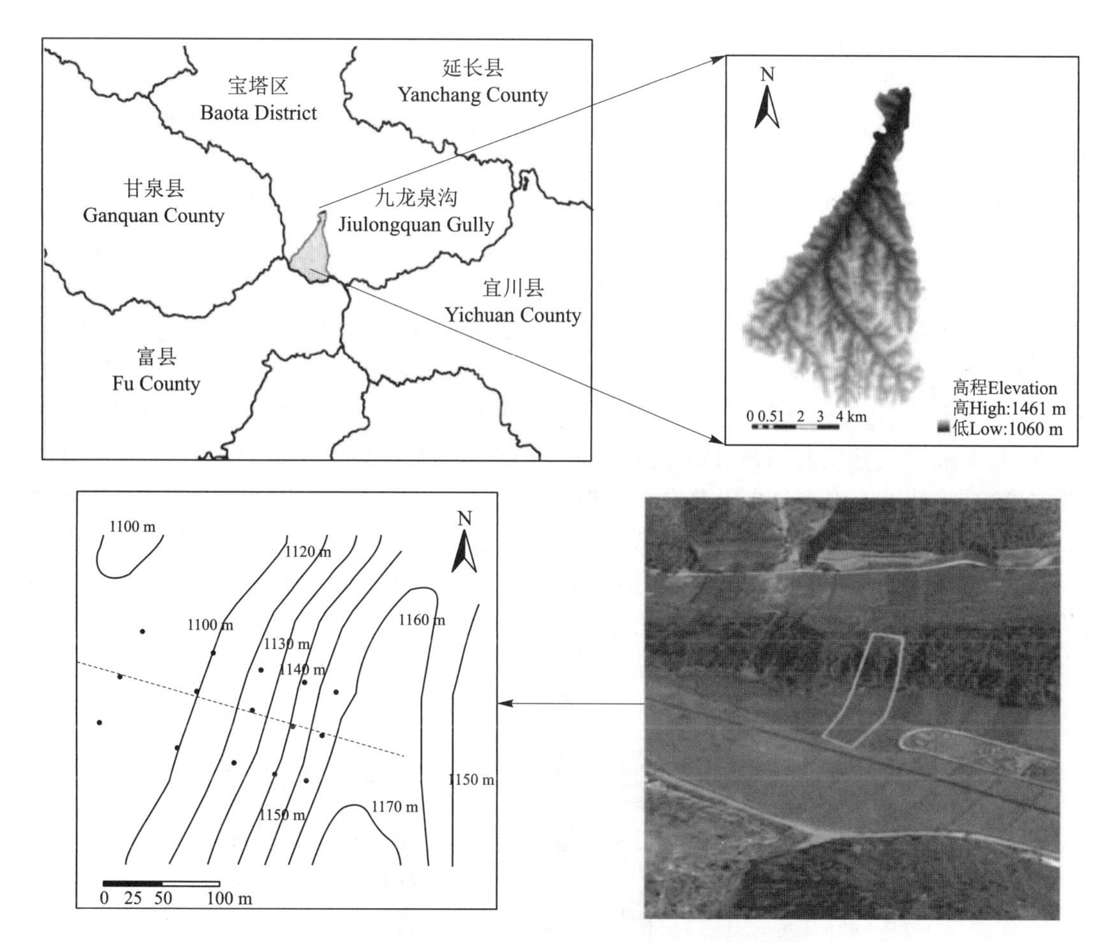

图 1 研究区位置及土壤水分定位监测点示意图

Fig.1 Location of study area and monitoring points for positioning of soil moisture

从土壤水分特征来看,表层(0~40 cm)土壤导水率较大,可达到 40 mm/min 以上,入渗能力强,同时土壤饱和含水率较大;下层(>40 cm)土壤导水率很小,仅为 1.25~1.41 mm/min,土壤饱和含水率较小。田间持水量、残余含水率呈现出表层低、下层高的特点。

土壤水力学参数反映了不同深度土壤的水力学特征。土壤性质在土层深度为 40 cm 处发生了显著改变,在土壤水分运移到该深度时,其运动状态极易发生改变,入渗速率将会明显减小,当土壤含水量积累达到田间持水水平、接近饱和时很有可能发生侧向运移,从而使侧向入渗强度增大,上层土壤水分的下行入渗补给减少。

表 1 沟道造地土壤水力学参数

Table 1 Soil hydraulic parameters of gully cultivated land

土层深度/cm Soil depth	容重/(g/cm^3) Bulk density	饱和含水率/% Saturated moisture content	田间持水量/% Field capacity	残余含水率/% Residual moisture content	导水率/(mm/min) Hydraulic conductivity
0~20	1.12	47.8	21.5	5.89	40.8
20~40	1.25	46.2	21.7	4.64	41.9
40~60	1.49	41.5	26.8	7.82	1.41
60~80	1.52	43.0	26.2	7.59	1.28
80~100	1.47	41.7	27.3	7.62	1.25
100~120	1.48	44.3	28.1	7.48	1.37

2.2　沟道土壤水分的时空变化特征

2.2.1　土壤水分的季节变化

取沟道和对照坡面采样点各层次的平均值作为每月土壤含水量值,得出土壤水分季节变化(见图2)。

沟道土壤含水量在15.76%～21.91%波动,5—6月在16%上下,7月开始显著增加,8月达到最高(21.91%),9月回落至18%,10—11月有所上升,在19%左右(见图2)。降水是黄土高原土壤水分最重要的补给源,因此土壤水分的时间变化特征主要取决于年内降水的季节性变化[17]。沟道土壤水分的季节变化稍滞后于降水的季节变化。生长前期降水量较小,气温回升,日照逐渐强烈,加之土壤蒸发和植被蒸腾作用,土壤水分得到补给较少。7—8月进入雨季,土壤水分得到大量降雨补充,虽然作物进入快速生长期,强烈日照蒸腾使耗水量增大,但补充的水分仍然多于消耗的水分,土壤含水量增加;9月降雨较少,蒸发和蒸腾作用仍然较强,土壤水分有明显的回落过程;10月气温逐渐降低,日照强度减弱,补充土壤水分多于蒸发和蒸腾的消耗量,所以10—11月土壤含水量再次增加。

由于地形差异以及沟道内的农田水利措施,坡面和沟道对降雨的再分配和入渗规律的影响各不相同。坡面土壤水分变化规律与沟道相似,但沟道在整个生长季土壤含水量都高出对照坡面5%左右,说明沟道由于汇流作用,有利于水分的储存。

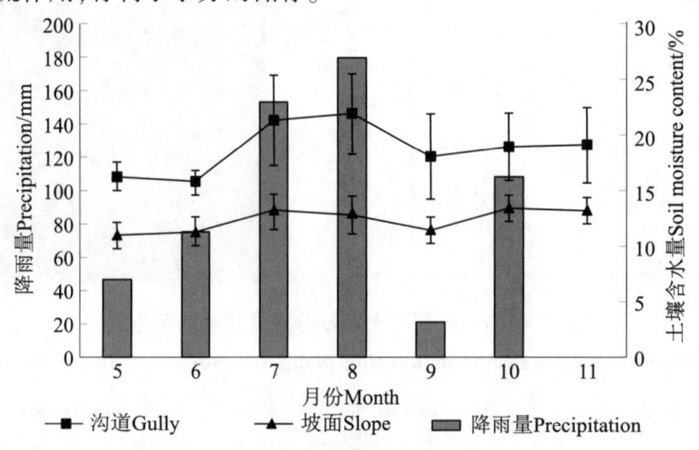

图2　沟道与坡面土壤水分季节变化

Fig.2　Seasonal variation of soil moisture in gully and slope

2.2.2　土壤水分的垂直分布

受降水入渗、植物蒸散、根系吸收以及地下水补充这几个过程影响,土壤剖面各层次土壤含水量存在差异。在地表以下160 cm土层中,沟道和坡面的土壤水分含量基本呈现出随土层深度的增加而增加的趋势(见图3)。

沟道表层土壤含水量为15.07%,20～60 cm土层土壤含水量随土层深度增加逐渐增加,60 cm土层达到峰值(19.19%)。坡面径流由于重力作用向沟底汇聚,同时因为沟底地势较平,土壤导水率在40～60 cm土层急剧减小,水分可以有效保存,因此土壤含水量较高。60～80 cm土层土壤含水量随土层深度的增加缓慢下降,土壤容重高于表层土壤,水分向下入渗缓慢,不易得到降水和地下水补给,土壤含水量有下降趋势。80～160 cm土层土壤水分含量呈现增加趋势,在160 cm土层土壤含水量达到最大值22.84%。一方面,沟道土壤水分在下渗过程中损失较小,水分能一直下渗到深层,具有较好的蓄水保水能力;另一方面,由于地下水埋深浅,土壤水分能够得到沟道中浅层地下水的向上补给。

坡面土壤含水量随土层深度的变化处于较为平稳的状态。由于坡面降雨大部分随地表径流流失,土壤水分入渗少,同时半阳坡蒸发损失较大,因此表层土壤含水量较低,在10.0%左右,最高在160 cm土层达到13.31%。与坡面相比,沟道较高的土壤含水量对于作物的生长极为有利,同时100 cm以下土层较高的土壤含水量能够为作物提供深层次的水分补充。

2.2.3　土壤水分变异特征

根据监测数据，计算出沟道和坡面不同层次土壤水分时间变化的标准差和变异系数（见表2）。沟道土壤水分变异系数呈现随土层深度的增加，先增大，随后减小，最后再增大的趋势。坡面则随着土层深度的增加逐渐减小。

坡面表层（0～20 cm）土壤水分变异系数较高，为0.215。土壤表层受降水、温度、风力等气象因子的作用，发生大量的物质和能量交换，因此土壤水分变化较为剧烈。随后土壤水分变异系数逐渐变小，50 cm以下土层在0.1以下，基本进入一个较为稳定的状态，直至160 cm土层变异系数降至最低。土壤深层风力、温度等外界因素作用减弱，降雨的补偿滞后，土质较上层紧实，土壤水分入渗阻力较大，补给也相对缓慢，因此变异系数较小。

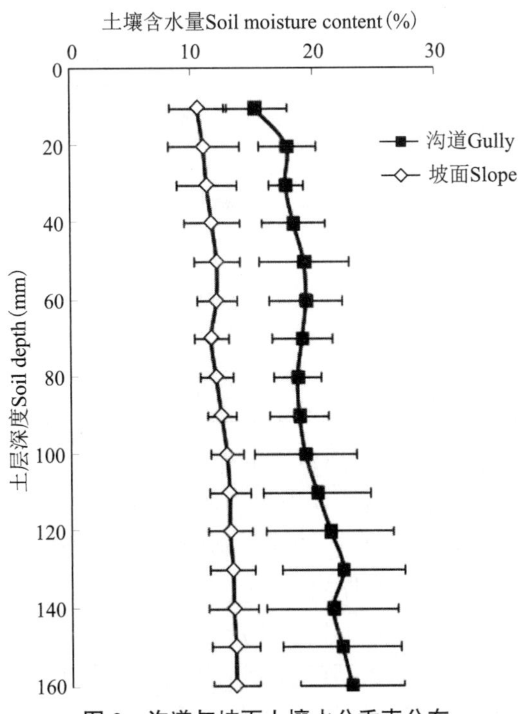

图3　沟道与坡面土壤水分垂直分布

Fig.3　Vertical distribution of soil moisture in gully and slope

与坡面相比，沟道表层以下土壤水分存在较强的空间异质性。沟道变异系数在0.131～0.234，除0～20 cm土层土壤含水量变异系数小于坡面外，其他各个层次均大于坡面。沟道由于地形的原因，不仅受降雨、入渗、产流、蒸散等因素的影响，还有坡面降水的汇集，加之沟道水利工程的影响，土壤水分动态变化比坡面更加强烈，因此变异系数更大。100 cm以下土层土壤水分含量变异系数再次增大，说明这个土层的土壤水分含量变化活跃，经常处于水分流失和补充的不稳定状态。

表2　沟道与坡面土壤水分变异系数

Table 2　Variation coefficient of soil moisture in gully and slope

土层深度/cm Soil depth	沟道 Gully		坡面 Slope	
	标准差 Standard deviation	变异系数 Coefficient of variation	标准差 Standard deviation	变异系数 Coefficient of variation
10	2.253	0.148	2.215	0.215
20	2.606	0.150	1.924	0.179
30	2.730	0.158	1.528	0.139
40	4.236	0.234	1.327	0.116
50	4.078	0.218	1.135	0.096
60	3.956	0.210	1.001	0.085
70	3.639	0.195	1.101	0.097
80	2.943	0.160	1.132	0.096
90	2.415	0.131	1.042	0.086
100	2.672	0.141	1.012	0.080
110	3.970	0.195	1.073	0.084
120	4.069	0.189	1.030	0.080

续表 2

Continued to Table 2

土层深度/cm Soil depth	沟道 Gully		坡面 Slope	
	标准差 Standard deviation	变异系数 Coefficient of variation	标准差 Standard deviation	变异系数 Coefficient of variation
130	5.047	0.218	0.966	0.074
140	5.281	0.222	0.993	0.076
150	4.275	0.184	0.857	0.065
160	5.129	0.205	0.622	0.047

2.3　沟道不同位置土壤水分差异

沟道自沟头到沟口具有明显的坡降。通过对沟道不同位置土壤水分进行测定,分析土壤水分含量的时间变化特征,绘制沟头和沟口各深度下土壤相对含水量(土壤含水量与土壤田间持水量的比值)在1—12月期间的等值线(见图4)。从图4可以看出,沟口的土壤相对含水量显著高于沟头,沟头7—12月60 cm以下土层土壤含水量达到并超过田间持水量,5月、6月干旱时100 cm土层土壤含水量均达不到田间持水量。沟口处雨季20 cm以下土层即可达到田间持水量,全年40 cm以下土层长期处于或超过田间持水量水平。由于坡降以及土壤厚度较小[9]的原因,沟口比沟头位置更易储存水分,也更易获得地下水的补给,因此土壤含水量更大。

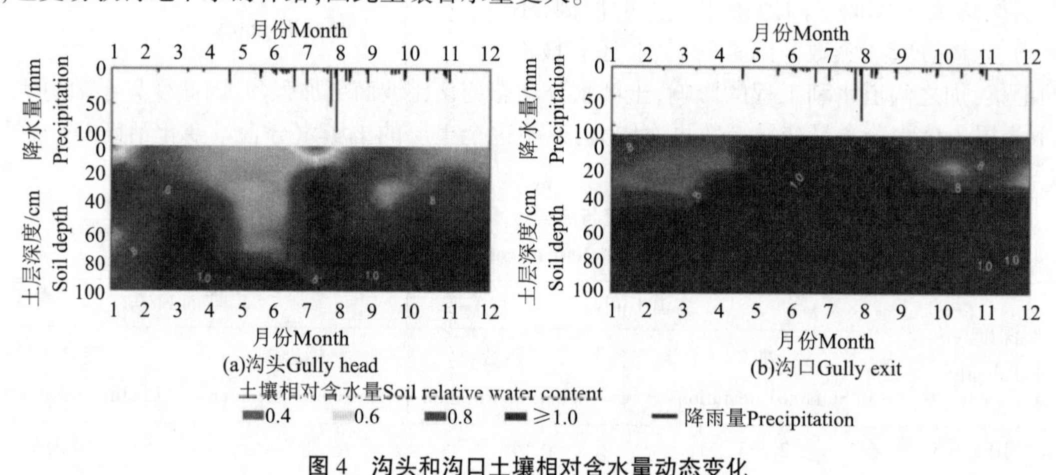

图4　沟头和沟口土壤相对含水量动态变化

Fig.4　Dynamic change of soil relative water content at gully head and gully exit

3　讨论

土地整治项目会压实土壤,研究表明整治后土壤容重增大,孔隙度减小,土壤含水量低,紧实度极高,渗水率极低[18]。但是治沟造地项目在经过一段时间的耕作后,表层土壤容重较小,导水率较大,入渗能力强,保证了土壤水分的储存和下渗。于洋等[19]对黄土丘陵区整地后的土壤水分含量研究表明,对比坡面,工程措施长期开展后表层土壤水分明显改善。

在延安地区,水资源问题是限制农业耕作和生态农业可持续发展的最突出因素,而提高深层土壤水分利用率是解决干旱缺水的有效措施[20]。根据以往研究,土地利用方式、坡度及其交互作用对土壤水分有显著影响[21]。沟道本身就有水分的聚集作用,一方面获得上部坡面的径流补充汇集大量降水,另一方面是降雨的直接补给,治沟造地工程充分利用了沟道良好的水分条件,同时改变了沟道中的微地形地貌,使耕地更加平整,增加了水分的存留时间,因此土壤含水量显著优于对照坡面,且在100 cm以下土层仍然保持较高的水平,即使在干旱季节,作物的根系层仍能得到较深层次的水分补给,有为作物提供大量水分供给的潜力。

研究发现,黄土高原地区土壤水分变异系数一般随土层深度增加而逐渐减小,100 cm 以下土层基本为稳定层[22-23]。本研究中坡面的变化规律如此,但是沟道土壤水分存在较强的空间异质性,变异系数高于坡面,这与 Gao 等[2, 24]的研究结果一致。治沟造地工程通过灌排工程的调控措施,实现对沟道地表水、土壤水和地下水资源的调控与利用,使土壤水分变化更加活跃,尤其是地下水参与土壤水分的活动更加频繁。通过截水沟、灌排渠相互连通形成的水分综合调控体系,雨季减弱水流的侵蚀力,减小水土流失,降低滑坡、崩塌等自然灾害的发生,保护农田不被洪水冲毁; 旱季通过灌溉缓解旱情,实现了从田面表层到土壤深层、从旱季到雨季对沟道水资源的时空调控,提高深层土壤水分的利用效率,增加作物产量,保证粮食安全。

黄土高原切沟中,通常沟口土壤含水量明显高于沟头,且地下水更为丰富[25-26],部分地区由于特殊的地质地形环境,河谷新造耕地的地下水位抬升,沟道中水分过剩,甚至出现了明显的盐渍化问题[27]。本研究也发现,沟道下游,尤其是沟口位置,40 cm 以下土层土壤水分即超过田间持水量。由于地下水位高,降水后地下水位进一步上升,土壤内长期存在积水,易造成农田盐渍化且不利于作物根系生长。治沟造地工程应注意利用农田水利措施,及时将田块内多余的土壤水分排出,降低地下水位,减小淹水对作物生长的影响,防范盐渍化风险。同时根据沟道造地土壤水分的空间分布和变化特征,可在沟道不同位置有区别地配置作物类型,比如上游种植玉米或小麦,下游种植水稻,实现黄土高原沟道旱区农业的水田化,不仅提高沟道水资源利用效率,且形成黄土高原沟道旱—水结合的生态农业景观。

4　结论

(1)土壤水力学参数反映沟道土壤性质在土层深度为 40 cm 附近发生了显著改变,0~40 cm 土层土壤容重较小,导水率较大,入渗能力强,同时饱和含水率较大,田间持水量、残余含水率小; 40 cm 以下土层容重较大,导水率较小,入渗速率明显减小,土壤饱和含水率较小,田间持水量、残余含水率较大。

(2)沟道土壤水分显著大于对照坡面,其季节变化稍滞后于降雨的季节变化,整个生长季土壤含水量在 15.76%~21.91%波动,高出对照坡面 5%左右;垂直分布随土层深度的增加而增加,深层土壤含水量优势更加显著; 沟道土壤水分变异系数在 0.131~0.234,相比较坡面,沟道表层以下土壤水分存在较强的空间异质性, 100 cm 以下土层土壤水分变化活跃。

(3)沟口土壤含水量显著高于沟头,沟口土层深度 40 cm 以下土壤含水量长期处于或超过田间持水量,雨季在土层深度为 20 cm 以下即可达到田间持水量。治沟造地工程应注意利用农田水利措施降低地下水位,防范盐渍化风险。

参考文献

[1]胡伟, 邵明安, 王全九, 等. 黄土高原退耕坡地土壤水分空间变异性研究[J]. 水科学进展, 2006, 17(1): 74-81.

HU W, SHAO M A, WANG Q J, et al. Study on spatial variability of soil moisture on the recultivated slope-land on the Loess Plateau[J]. Advances in Water Science, 2006, 17(1): 74-81.

[2]高晓东, 吴普特, 张宝庆, 等. 黄土丘陵区小流域土壤有效水空间变异及其季节性特征[J]. 土壤学报, 2015, 52(1):57-67.

GAO X D, WU P T, ZHANG B Q, et al. Spatial variability of available soil moisture and its seasonality in a small watershed in the hilly region of the Loess Plateau[J]. Acta Pedologica Sinica, 2015, 52(1):57-67.

[3]贺春雄. 延安在治沟造地基础上如何发展现代农业[J]. 延安大学学报(社会科学版), 2013, 35(3):60-63.

HE C X. How does Yan'an develop modern agriculture on the basis of gully control and land consolidation[J]. Journal of Yan'an University (Social Science Edition), 2013, 35(3):60-63.

[4]高海东. 黄土高原丘陵沟壑区沟道治理工程的生态水文效应研究[D].杨凌:中国科学院大学(中国科学院教育部水土保持与生态环境研究中心), 2013.

GAO H D. Hydro-ecological impact of the gully erosion control works in Loess Hilly-gully Region[D]. Beijing: University of Chinese Academy of Sciences (Research Center of Soil and Water Conservation and Ecological Environment, Ministry of Education), 2013.

[5]张正峰, 赵伟. 土地整理的生态环境效应分析[J]. 农业工程学报, 2007, 23(8):281-285.

ZHANG Z F, ZHAO W. Effects of land consolidation on ecological environment[J]. Transactions of the Chinese Society of Agricultural Engineering, 2007, 23(8): 281-285.

[6]娄现勇, 高建恩, 韩赛奇, 等. 黄土丘陵沟壑区沟道土地整治工程对流域产汇流的影响[J]. 水电能源科学, 2016, 34(10):23-27.

LOU X Y, GAO J E, HAN S Q, et al. Influence of land consolidation engineering of gully channel on watershed runoff yield and concentration in loess hilly and gully region[J]. Water Resources and Power, 2016, 34(10):23-27.

[7]孙彭成. 典型沟道土地整治工程对降水转化影响的模拟试验研究[D]. 杨凌:西北农林科技大学, 2017.

SUN P C. Simulation experiment on the influence of typical gully land consolidation project on precipitation conversion [D].Yangling:Northwest A&F University, 2017.

[8]KANG Y C, GAO J E, SHAO H, et al. Evaluating the flow and sediment effects of gully land consolidation on the Loess Plateau, China [J]. Journal of Hydrology, 2021,600:126535.

[9]雷娜, 韩霁昌, 高红贝,等. 延安治沟造地工程水资源调控与利用分析[J]. 中国农村水利水电, 2017(5): 26-30.

LEI N, HAN J C, GAO H B, et al. An analysis of regulation and utilization of water resources of gully control and land reclamation in yan'an[J]. China Rural Water and Hydropower, 2017(5):26-30.

[10]YIN X X, CHEN L W, HE J D, et al. Characteristics of groundwater flow field after land creation engineering in the hilly and gully area of the Loess Plateau[J]. Arabian Journal of Geosciences, 2016, 9(14):646.

[11]JIN Z, GUO L, WANG Y Q, et al. Valley reshaping and damming induce water table rise and soil salinization on the Chinese Loess Plateau [J]. Geoderma, 2019, 339: 115-125.

[12]ZHAO Y L, WANG Y Q, WANG L, et al. Exploring the role of land restoration in the spatial patterns of deep soil water at watershed scales[J]. Catena, 2019, 172:387-396.

[13]ZHAO Y L, WANG Y Q, SUN H, et al. Intensive land restoration profoundly alters the spatial and seasonal patterns of deep soil water storage at watershed scales[J]. Agriculture, Ecosystems and Environment, 2019, 280:129-141.

[14]鲁如坤. 土壤农业化学分析方法[M].北京:中国农业科技出版社, 1999.

LU R K. Methods for Soil agrochemical Analysis [M]. Beijing: China Agricultural Science and Technology Press, 1999.

[15]VAN GENUCHTEN M T, LEIJ F J,YATES S R, et al. The RETC code for quantifying the hydraulic functions of unsaturated soils: project summary[J]. Robert S. Kerr Environmental Research Laboratory, Research and Development, U.S. Environmental Protection Agency, 1992.

[16]王小华, 贾克力, 刘景辉, 等. Van Genuchten 模型在土壤水分特征曲线拟合分析中的应用[J]. 干旱地区农业研究, 2009, 27(2):179-188.

WANG X H, JIA K L, LIU J H, et al. Application of Van Genuchten model to analysis of soil moisture characteristics curve[J]. Agricultural Research in the Arid Areas, 2009, 27(2):179-188.

[17]张北赢, 徐学选, 刘文兆, 等.黄土丘陵沟壑区不同降水年型下土壤水分动态[J]. 应用生态学报, 2008, 19(6):1234-1240.

ZHANG B Y, XU X X, LIU W Z, et al. Dynamic changes of soil moisture in loess hilly and gully region under effects of different yearly precipitation patterns[J].Chinese Journal of Applied Ecology, 2008, 19 (6):1234-1240.

[18]孟会生, 王静, 郭建奎,等. 黄土区土地整理压实土壤物理性状的初步研究[J]. 中国农学通报, 2009, 25(24):549-552.

MENG H S, WANG J, GUO J K, et al. Preliminary study on physical properties of soil compact in land consolidation in Loess Plateau[J]. Chinese Agricultural Science Bulletin, 2009,25(24):549-552.

[19]于洋, 卫伟, 陈利顶, 等. 黄土丘陵区小流域典型造林整地工程土壤水分特征曲线模拟[J]. 生态学报, 2018, 38(18):6511-6520.

YU Y, WEI W, CHEN L D, et al. Simulation of a soil water retention curve of typical soil and water conservation engineering measures in the Loess hilly watershed[J]. Acta Ecologica Sinica, 2018, 38(18): 6511-6520.

[20]王晶，朱清科，赵荟,等. 陕北黄土区阳坡微地形土壤水分特征研究[J].水土保持通报，2011,31(4):16-21.

WANG J, ZHU Q K, ZHAO H, et al. Soil moisture characteristics of micro-topography in south slope of loess region in northern Shaanxi Province[J]. Bulletin of Soil and Water Conservation, 2011, 31(4):16-21.

[21]邱德勋，赵佰礼，尹殿胜,等. 黄土丘陵沟壑区土壤水分垂直变异及影响因素[J]. 中国水土保持科学，2021,19(3):72-80.

QIU D X, ZHAO B L, YIN D S, et al. Vertical variation of soil moisture in the loess hilly and gully region and its influence factors [J]. Science of Soil and Water Conservation, 2021,19(3):72-80.

[22]马婧怡，贾宁凤，程曼. 黄土丘陵区不同土地利用方式下土壤水分变化特征[J]. 生态学报，2018，38(10): 3471-3481.

MA J Y, JIA N F, CHENG M. Water characteristics of soil under different land-use types in the Loess Plateau region [J]. Acta Ecologica Sinica, 2018, 38(10):3471-3481.

[23]张静,王力,韩雪,等. 黄土塬区农田蒸散的变化特征及主控因素[J]. 土壤学报，2016,53(6): 1421-1432.

ZHANG J, WANG L, HAN X, et al. Evapotranspiration of farmland on loess tableland and its major influencing factors [J]. Acta Pedologica Sinica, 2016,53(6): 1421-1432.

[24]GAO X D, WU P T, ZHAO X N, et al. Soil moisture variability along transects over a well-developed gully in the Loess Plateau, China[J]. Catena, 2011, 87: 357-367.

[25]陈明玉，邵明安，李同川,等. 黄土高原典型切沟土壤水分时空分布特征及其影响因素[J].土壤学报，2021, 58(2):381-390.

CHEN M Y, SHAO M A, Li T C,et al. Characteristics and influencing factors of spatiotemporal distribution of soil moisture in typical gully of the Loess Plateau[J]. Acta Pedologica Sinica, 2021, 58(2): 381-390.

[26]袁鸿猷，樊军，金沐,等. 黄土高原淤地坝土壤水分和浅层地下水时空分布特征解析[J]. 灌溉排水学报，2020, 39(10):50-56.

YUAN H Y, FAN J, JIN M, et al. Spatiotemporal distribution of soil water and shallow groundwater in check dams in the Loess Plateau of China[J]. Journal of Irrigation and Drainage, 2020, 39(10): 50-56.

[27]陈淑敏，金钊，张晶,等. 陕北不同沟道土地盐碱化现状及影响因素[J]. 地球环境学报，2020，11 (1): 81-89.

CHEN S M, JIN Z, ZHANG J, et al. Situation and impact factors of soil salinization in different dammed farmlands in the valley area of the Northern Shaanxi Province[J]. Journal of Earth Environment, 2020, 11(1):81-89.

本文曾发表于2022年《干旱地区农业研究》第40卷第4期

不同水稻品种在陕北盐碱地的适宜性

何振嘉[1,2],王启龙[1],罗林涛[1],杜宜春[1,2]

(1.陕西省土地工程建设集团有限责任公司,陕西 西安 710075;
2.中陕高标准农田建设集团有限公司,陕西 杨凌 712100)

【摘要】陕北定边县部分盐碱地培育和筛选适宜盐碱地种植的水稻品种,对于盐碱地的利用和生态环境的改善有着重要意义。在定边县堆子梁镇营盘梁村开展大田试验,设置4个处理T1、T2、T3和T4,分别为隆优619、宁靖28、东稻4号、水稻FL478,均采用井水作为灌溉水源,以研究不同品种水稻种植对盐碱土壤理化性质、养分、土壤离子变化、水稻生长状况以及产量的影响。结果表明:与种植水稻前相比,各处理土壤pH值和全盐含量均有不同程度降低($P<0.05$,$P<0.01$)。种植水稻后,各处理促进了土壤黏粒含量的降低、粉粒含量的提高,而砂粒含量均有所降低。各处理土壤有机质含量和全氮含量均有不同程度提高($P<0.01$),但有效磷和速效钾含量显著降低($P<0.05$)。处理T1、T2、T3和T4的土壤耕作层有机质含量较种植水稻前分别提高219.87%、398.34%、218.91%和277.53%;全氮含量分别提高150.00%、300.00%、160.71%和29.23%。各处理Ca^{2+}、Mg^{2+}、Cl^-、SO_4^{2-}、HCO_3^-含量分别较本底值降低46.76%、63.78%、33.91%、63.78%、43.18%,种植水稻对土壤脱盐效果显著。各处理与水稻株高生长和每平方米穗数之间具有统计学意义($P<0.05$);与盐碱地种植水稻百粒质量和产量具有统计学意义($P<0.01$),处理T1分别较T2、T3和T4提高了4.96%、42.78%和161.64%。种植隆优619、宁靖28水稻产量较高,是较为适宜在陕北盐碱地种植的水稻品种,可在陕北地区进行推广。

【关键词】盐碱地;水稻种植;理化性质;土壤离子;产量

Suitability of Different Rice Varieties in Saline-alkali Land in Northern Shaanxi

Zhenjia He[1,2], Qilong Wang[1], Lintao Luo[1], Yichun Du[1,2]

(1.Shaanxi Provincial Land Engineering Construction Group Co. Ltd., Xi' an, Shaanxi 710075, China;
2. China Shaanxi High-standard Farmland Construction Group Co., Ltd, Yangling, Shaanxi 712100, China)

【Abstract】 The cultivation and screening of rice varieties suitable for cultivation in saline-alkali land in Dingbian County, northern Shaanxi are of great significance for the utilization of saline-alkali land and the improvement of the ecological environment. A field experiment was carried out in Yingpanliang Village, Duiziliang Town, Dingbian County, and four treatments T1, T2, T3 and T4 were set up, namely Longyou 619, Ningjing 28, Dongdao No. 4, and rice FL478, all using well water as irrigation Water sources were used to study the effects of different rice varieties on the physical and chemical properties, nutrients, soil ion changes, rice growth status and yield of saline-alkali soils. The results show that: Both pH value and total salt content decreased to different

基金项目:陕西省重点科技创新团队计划项目(2016KCT-23);陕西省土地工程建设集团有限责任公司内部科研项目(DJNY2019-1、DJNY2022-53、DJNY2022-54);陕西省土地工程建设集团有限责任公司创新团队项目(DJTD-2022-5)。

degrees ($P<0.05$, $P<0.01$). After planting rice, each treatment promoted the reduction of clay content, the increase of powder content, and the decrease of sand content. The soil organic matter content and total nitrogen content of each treatment increased to different degrees ($P<0.01$), but the content of available phosphorus and available potassium decreased significantly ($P<0.05$). T1, T2, T3, and T4 treatments increased the soil organic matter content in the tillage layer by 219.87%, 398.34%, 218.91%, and 277.53%, respectively, and the total nitrogen content increased by 150.0%, 300.0%, 160.71%, and 277.53% respectively 29.23%. The contents of Ca^{2+}, Mg^{2+}, Cl^-, SO_4^{2-} and HCO_3^- in each treatment were 46.76%, 63.78%, 33.91%, 63.78% and 43.18% lower than the background value, respectively. Planting rice has a significant effect on soil desalination. The relationship between each treatment and rice plant height growth and number of ears per m^2 was statistically significant ($P<0.05$); it was statistically significant with the 100-grain mass and per mu yield of rice planted in saline-alkali land ($P<0.01$). Treatment T1 Compared with T2, T3 and T4, they were improved by 4.96%, 42.78% and 161.64% respectively. Longyou 619 and Ningjing 28 have higher rice yields and are more suitable rice varieties for planting in saline-alkali land in northern Shaanxi, and can be promoted in northern Shaanxi.

【**Keywords**】Saline-alkali land; Rice cultivation; Physical and chemical properties; Soil ion; Yield

随着经济发展水平不断提高,对土地资源的需求量也日渐加大,优质耕地数量减少、现有耕地质量降低等因素更加凸显了耕地资源的稀缺性。截至目前,陕西省尚有约 1400 km^2 盐碱地尚未得到有效开发利用,土壤盐碱化给当地农业生产和粮食安全带来了巨大威胁。目前,盐碱改良技术主要包括耕作改良技术、地表覆盖技术、化学改良技术、水利改良技术、工程改良技术以及植物改良技术,其中植物改良主要为植物根系生长,可改善土壤物理性状,根系分泌的有机酸可以中和土壤碱性,并可吸收、富集土壤盐分,减少土壤水分蒸发,从而防止土壤返盐[1]。在地下水位较高的盐碱地种植耐盐碱水稻不仅可以提高土地利用率,还能起到保护环境的作用。"海水稻"作为抗逆性作物的一种,可以生长在滩涂和盐碱地,可以将盐碱水净化,同时保持土壤,增加土壤的有机质,具有极高的生态价值和社会价值[2]。

大量研究表明[3-5],通过对盐碱地进行作物种植可以显著改善盐碱地表层土壤的理化性质。定边县部分盐碱地地下水位常年普遍较高,旱地作物无法生长,但其水热资源丰富,满足水稻生长基本条件;水稻既是禾本科植物的模式物种,也是一种对盐中度敏感的作物,可作为盐碱地开发利用的先锋植物,但陕北地区气候环境、土壤构成成分和盐碱度与其他区域并不完全相同[6],因而培育和筛选适宜在陕北地区盐碱地种植的水稻品种,对于陕北地区盐碱地利用和生态环境改善具有重要意义。文中通过在定边县堆子梁镇营盘梁村开展大田试验,研究不同品种水稻的生长状况,以筛选适宜的水稻品种,探明水稻种植对盐碱地的改良效果,从而为陕北盐碱地的改良利用提供理论与技术支撑。

1　材料与方法

1.1　试验区概况

试验于 2018 年 4 月至 2019 年 11 月在陕西省榆林市定边县堆子梁镇营盘梁村进行,地理坐标位于 108°15′46″E、37°38′29″N。项目区地处陕北黄土高原和鄂尔多斯高原外延交界,为毛乌素沙地边缘的风沙草滩区;气候条件为典型的温带干旱大陆性季风气候,年平均降水量为 316 mm,年平均蒸发量为 2 850 mm,年平均气温为 7.9 ℃,1 月和 7 月平均气温分别为-8.8 ℃和 22.3 ℃,日照 2 728 h,无霜期为 125 d,年平均冻土深度为 98.9 cm,试验地土壤质地主要为砂壤土和壤砂土,为中度盐碱化地块。土壤 pH 平均值为 9.3,有机质质量比平均值为 3.48 g/kg,全氮平均值为 0.27 g/kg,电导率平均值为 32.13 mS/cm;其土壤表层 0~5 cm 含盐质量比为 16 g/kg,5~10 cm 土壤含盐质量比为 5.08 g/kg,10~50 cm 含盐质量比为 3.85 g/kg,50~100 cm 土壤含盐质量比低于 2 g/kg,全盐量在耕作层(0~50 cm)平均值为 5.19 g/kg(中重度盐碱地)。地下水位平均埋深为 1.0~1.2 m,矿化度平均达到 3.8 g/L,土壤表层积盐层 5~8 mm。主要离子中,Ca^{2+}、Mg^{2+}、Cl^-、CO_3^{2-}、SO_4^{2-}、HCO_3^-的质量比分别为 0.103 g/kg、0.078 g/kg、0.072 g/kg、0.021 g/kg、0.192 g/kg、0.541 g/kg。总体上,有机质含量低,土壤肥力不足,且易受风蚀和旱灾危害。

1.2 试验方案及处理

在项目区内分别选择盐碱程度具有代表性的 1 个地块设置田间试验(试验布置如图 1 所示)。试验地块面积占地约 7 200 m^2(60 m×120 m),试验地四周起垄后,划分为 4 个大田块,采用随机区组排列,5 月 15 日播种,5 月 27 日移栽,4 次重复完全随机设计,株行距为 15 cm×30 cm,每穴插 3~5 株苗。将小田块地表进行人工平整,在翻耕种植前及时采集土壤样品。由于陕北地区气候环境、土壤构成成分和盐碱度与南方及陕西关中地区并不相同,因而在耐盐水稻品种筛选时初步参考同一纬度、气候环境和种植条件等相似的榆林、宁夏地区目前已进行种植的水稻品种;试验共设置 4 个处理,分别为 T1(隆优 619:LY)、T2(宁靖 28:NJ)、T3(东稻 4 号:DR)以及 T4(水稻 FL478:FL);灌溉水源均为井水灌溉,灌水量为 3 800 m^3/hm^2。试验施肥采用复合肥,选择尿素(N:46%)、过磷酸钙(P_2O_5:20%)、硫酸钾(K_2O:50%),施肥量为 135 kg/hm^2。试验田的播种育苗、插秧及灌溉等田间具体情况参考当地生产实际进行。基础样地土壤理化特性本底值见表 1,表中物理量分别为电导率 EC、质量比 ω、质量分数 δ。

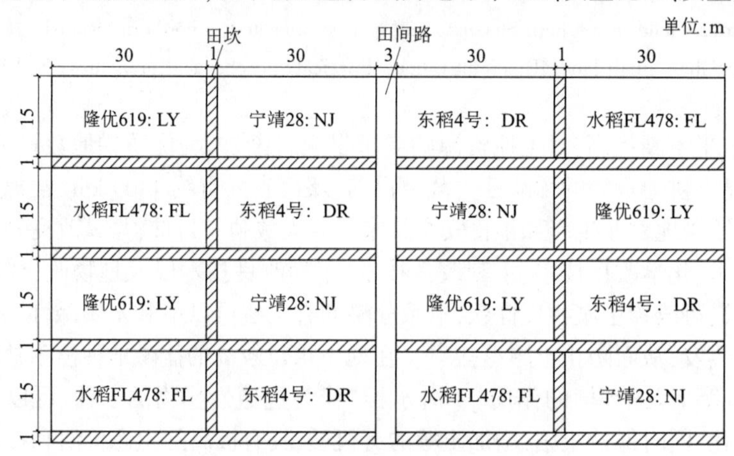

图 1 小区平面布置示意图

Fig.1 Schematic diagram of plot layout

表 1 样地土理化特性本底值

Table 1 Background values of physical and chemical properties of sample soil

处理	pH	EC/($mS\cdot cm^{-1}$)	ω/(g/kg)					δ/%		
			全盐量	有机质	全氮	有效磷	速效钾	黏粒	粉粒	砂粒
T1	9.12	32.6	2.8	3.12	0.30	1.70×10^{-2}	0.551	25.97	73.84	0.19
T2	9.38	33.7	2.9	3.01	0.23	1.01×10^{-2}	0.537	25.73	73.79	0.48
T3	9.22	30.3	2.5	4.23	0.28	1.24×10^{-2}	0.552	25.69	73.60	0.71
T4	9.52	31.9	2.2	3.56	0.26	1.17×10^{-2}	0.550	26.28	72.66	1.06

1.3 测定项目与方法

测定作物生育期内水样 pH 值、水样电导率以及水样全盐含量。同时在作物关键生育期分层采集土壤样品,通过室内试验检测土壤水分、pH、电导率、含盐量以及土壤养分的变化情况,在水稻收获后检测土壤中 Ca^{2+}、Mg^{2+}、Cl^-、CO_3^{2-}、SO_4^{2-}以及 HCO_3^-的含量,采样深度分别设为 0~20 cm、20~40 cm、40~60 cm。在作物关键生育期(苗期、分蘖期、拔节期等)选取长势比较有代表性的 3~5 株植株,定期利用卷尺测定株高,采集有代表性的 3~5 株植株样品,按营养器官分别进行烘干称重,通过室内测定植株样品中的全盐含量。在作物收获后,每个小区选取有代表性的 5 蔸水稻,带回实验室进行考种,计算有效穗数、每穗粒数和百粒质量,剩余部分各小区实收核产。水稻收割后,机械粉碎稻田秸秆,深翻还田。pH 采用 pH 计测定(1∶5 土水比),电导率采用电导率仪测定,全盐量采用残渣烘干法测定。有机质质量比采用 TOC 仪测定；全氮质量比采用凯氏定氮法测定；速效钾质量比采用 NH_4OAc浸提，火焰光度计测定；速效磷采用 $NaHCO_3$ 浸提，分光光度计比色测定。Ca^{2+} 和 Mg^{2+} 采用 EDTA 滴定法测定，Cl^- 采用硝酸银滴定法测定，SO_4^{2-} 采用 EDTA 间接络合滴定法测定，CO_3^{2-} 和 HCO_3^- 采用双指示剂-中和滴定法测定。作物产量在各小区收获后取 1 m^2 样品测定、计算产量。

1.4 数据分析方法

采用 Microsoft Excel 2007 分析软件处理试验数据，同时采用 SPSS 11.5 软件进行统计学分析，并对相关指标进行显著性分析，显著性水平为 $P<0.05$，极显著性水平为 $P<0.01$。

2 结果分析

2.1 种植水稻对盐碱土物理性质的影响

表 2 为盐碱土种植水稻对土壤物理性质的影响，表中物理量分别为土层厚度 h、全盐质量比 ω_s。分析可知，项目区属风沙草滩区，土壤盐碱化程度较高，与种植水稻前相比，各处理土壤 pH 值均有不同程度降低，其中处理 T1 最低，T4 最高；除 T3 外，其他处理均表现出表层土壤 pH 值较低、深层土壤 pH 值较高的分布规律，种植水稻对土壤 pH 值的改善具有显著性($P<0.05$)。土壤盐碱程度过高，会严重阻碍作物对土壤养分的吸收效果，因此对盐碱土壤改良的首要目的为降低土壤中含盐量。种植水稻能极显著降低土壤含盐量($P<0.01$)，这主要是由于经过井水灌溉，对土壤中全盐含量进行了淋洗，水稻种植后，降低了地表蒸发。由于水稻生长过程中能分泌一定量的有机酸，可在一定程度上疏松土壤，降低土壤碱性，同时水稻生长可以吸收、富集土壤中多余的盐分，进一步降低土壤含盐量。

由于盐碱土结构性差，土壤剖面以活性毛管孔隙为主，缺少较为粗大的通气孔隙，故土壤水气不够协调。而水稻根系与盐碱土的相互作用对于促进土壤团聚体形成以及改善土壤结构具有显著影响，增加了土壤孔隙度，阻碍了毛细作用，抑制了土壤盐分上升的通道，在一定程度上减少了返盐，因此导致土壤全盐含量平均值显著降低。

处理 T1、T2 和 T4 的土壤含盐量在底层出现了一定量的涨幅，这主要是由于盐碱土壤的含盐量受地下水埋深、灌溉以及降雨影响较大。项目区地下水位以上土壤层为沙壤质到轻壤质土，土壤剖面属于均质构型，土壤毛细连通性好，毛管水活动强烈，毛管水上升高度可以达到该区域，在地下水位以上的土层处在毛管上升水的控制区域，虽然属于包气带的概念范畴，但土层内有水无气，土壤含水量过高，且随着地下水位变化以及田间灌溉行为的继续，使得土壤含盐量略有提高。

整体上，种植水稻后，各处理促进了黏粒含量的降低、粉粒含量的提高，而砂粒含量均有所降低。这主要是由于在地下水位的变化和农业耕作条件下，增加了土壤孔隙度，促进黏粒通过水分动向下层运动，在表层土层形成适宜作物生长的稳定耕作层，并在耕作层下仍保有稳定的犁底层结构进而保水保肥。

2.2 盐碱土种植水稻对土壤养分的影响

表 3 为盐碱土种植水稻对土壤化学性质的影响。由表 3 可见，在种植水稻后，各处理土壤有机质含量和全氮含量均有不同程度提高($P<0.01$)。种植水稻后，处理 T1、T2、T3 和 T4 的土壤耕作层有机质质量比平均值分别为 9.98 g/kg、15.00 g/kg、13.49 g/kg 和 13.44 g/kg，较种植水稻前本底值分别提高 219.87%、398.34%、218.91%和 277.53%。土壤全氮含量整体随土层深度增加而降低，各处理全氮质量比平均值分别较种植水稻前本底值提高 150.00%、300.00%、160.71%和 29.23%。土壤有机质含量的高低对土壤肥力的提高和土壤保墒保肥效果的提升以及土壤通气性的增加具有重要影响，可以作为土壤肥力的综合反映。有效磷和速效钾含量则显著降低($P<0.05$)。土壤有效磷含量整体表现为随土层深度增加呈先降低后小幅增加的趋势，但整体分布情况较为均衡，其质量比分别较种植水稻前降低 52.17%、29.70%、34.92%和 19.66%。除处理 T4 外，其他处理的土壤速效钾含量均随土层深度增加而降低，土壤速效钾质量比分别较种植前降低 45.07%、31.09%、42.69%和 80.30%，这可能与水稻种植过程中施肥量太少以及部分养分淋失所导致。因此，在种植水稻时须采取一定措施改良土壤稳定性，确保肥效能最大化利用。综合观察，处理 T2 的土壤有机质含量、全氮含量以及速效钾含量均显

著高于其他处理,保肥效果最好。

表2 盐碱土种植水稻对土壤物理性质的影响

Table 2 Effects of planting rice on saline-alkali soil on soil physical properties

处理	h/cm	pH	ω_s/(g/kg)	δ/%		
				黏粒	粉粒	砂粒
T1	0~20	8.77±1.05c	0.93±0.22b	22.24±0.54b	77.68±3.25b	0.08±0.07a
	20~40	8.83±1.11a	0.79±0.18c	23.95±0.56a	76.05±3.65c	0.00±0.00b
	40~60	9.08±1.08b	1.07±0.23a	21.95±0.48c	78.05±3.17a	0.00±0.00b
T2	0~20	8.83±1.13b	1.06±0.25b	22.41±0.45a	77.55±3.22a	0.04±0.01c
	20~40	8.88±1.20b	0.97±0.19c	19.64±0.41b	77.51±3.14a	2.85±2.02a
	40~60	9.24±1.14a	1.12±0.26a	22.16±0.43a	77.44±3.25a	0.40±0.02b
T3	0~20	8.90±1.23a	1.05±0.17b	18.67±0.48b	80.65±3.45a	0.68±0.14a
	20~40	8.86±1.15a	1.14±0.24a	22.93±0.53a	77.07±3.52b	0.00±0.00b
	40~60	8.83±1.11a	1.08±0.20c	22.41±0.47a	77.59±3.34b	0.00±0.00b
T4	0~20	8.51±1.12b	1.21±0.23a	24.17±0.45c	75.02±2.07b	0.81±1.87a
	20~40	8.53±1.17b	1.17±0.21b	25.52±0.47a	74.15±2.18a	0.33±2.08b
	40~60	9.09±1.21a	1.26±0.25a	25.21±0.44b	73.98±2.32a	0.81±1.98b
显著性 P 值		*	* *	NS	NS	*

注:同列不同小写字母表示不同处理间差异在 $P<0.05$ 水平下具有统计学意义;NS 表示差异不具有统计学意义,* 表示具有统计学意义($P<0.05$),* * 表示具有统计学意义($P<0.01$)。下同。

表3 盐碱土种植水稻对土壤化学性质的影响

Table 3 Effects of rice planting on saline soil on soil nutrients

处理	h/cm	ω/(g/kg)		ω/(mg/kg)	
		有机质	全氮	有效磷	速效钾
T1	0~20	10.90±1.87a	0.80±0.23a	8.3±1.56b	355±23.26a
	20~40	10.40±1.69ab	0.78±0.19ab	8.5±1.74a	272±14.63b
	40~60	8.66±1.35b	0.68±0.15b	7.6±1.56c	281±13.68b
T2	0~20	15.20±2.18b	0.94±0.22ab	5.3±1.19b	394±25.69a
	20~40	18.40±2.63a	1.03±0.24a	7.9±1.62ab	381±23.54ab
	40~60	11.40±1.95c	0.78±0.20b	8.1±1.54a	335±22.89b
T3	0~20	12.90±2.08b	0.84±0.25a	11.2±2.02a	365±23.63a
	20~40	9.48±1.65c	0.70±0.21b	6.2±1.26b	311±19.85b
	40~60	18.10±2.58a	0.66±0.18c	6.8±1.30b	273±15.37c
T4	0~20	13.06±1.75b	0.35±0.12ab	9.6±1.68b	93±6.52c
	20~40	14.27±1.69a	0.37±0.11a	9.9±1.69a	107±6.35b
	40~60	12.98±1.58c	0.29±0.09b	8.7±1.58c	125±6.89a
显著性 P 值		* *	* *	*	*

2.3 盐碱土种植水稻对土壤离子的影响

表4为盐碱土种植水稻对土壤离子的影响情况。各处理对土壤Ca^{2+}、Mg^{2+}、Cl^-以及HCO_3^-的影响具有统计学意义($P<0.05$),对SO_4^{2-}影响具有统计学意义($P<0.01$),而对CO_3^{2-}影响不具有统计学意义($P>0.05$)。其中,处理T4的Ca^{2+}和Cl^-质量比均达到最高,平均值分别为0.079 g/kg和0.183 g/kg,分别较本底值降低23.30%和113.86%;处理T3的均最低,分别为0.043 g/kg和0.041 g/kg,分别较本底值降低58.90%和25.21%。处理T1的Mg^{2+}质量比最高,为0.037 g/kg,较本底值降低52.56%;处理T2的最低,为0.018 g/kg,较本底值降低77.35%。SO_4^{2-}和HCO_3^-的质量比均为处理T3的最大,分别为0.114 g/kg和0.332 g/kg,分别较本底值降低40.45%和41.84%;处理T1的SO_4^{2-}质量比最低,为0.045 g/kg,较本底值降低76.39%;处理T4的HCO_3^-质量比最低,为0.257 g/kg,较本底值降低52.43%。各处理之间CO_3^{2-}含量无显著变化,较本底值降低19.05%。

由于经过水稻种植,在水稻叶片的覆盖作用下降低了部分土壤水分蒸发,抑制了土壤盐分表聚现象的发生,各处理Ca^{2+}、Mg^{2+}、Cl^-、SO_4^{2-}和HCO_3^-的质量比平均值较本底值分别降低46.76%、63.78%、33.91%、63.78%和43.18%,可见种植水稻对土壤脱盐效果显著。

根据试验效果,处理T1对土壤SO_4^{2-}的吸收富集效果最佳,T2对土壤Mg^{2+}的吸收富集效果最佳,T3对土壤Ca^{2+}和Cl^-的吸收富集效果最佳,T4对土壤HCO_3^-的吸收富集效果最佳。另外,在水稻生长过程中,主要通过根系吸收作用对土壤中各类盐分离子进行了有效富集,因此土壤中各类盐分离子的含量得到有效降低。水稻收获后,各处理作物体内全盐质量比分别为0.26 g/kg、0.31 g/kg、0.22 g/kg和0.18 g/kg,分别占土壤本底值的9.29%、10.69%、8.80%和8.18%,处理T2表现出更好的脱盐效果。

表4 盐碱土种植水稻对土壤离子的影响

Table 4 Effects of rice planting on saline-alkali soil on soil ions

处理	h/cm	ω/(g/kg)					
		Ca^{2+}	Mg^{2+}	Cl^-	SO_4^{2-}	CO_3^{2-}	HCO_3^-
T1	0~20	0.078±0.02a	0.054±0.01a	0.054±0.02b	0.101±0.03a	—	0.412±0.09a
	20~40	0.036±0.01c	0.020±0.00b	0.069±0.03a	0.032±0.01b	—	0.309±0.07b
	40~60	0.051±0.03b	—	0.052±0.01b	0.003±0.00c	0.017±0.00a	0.257±0.05c
T2	0~20	0.052±0.02a	0.016±0.00ab	0.048±0.02b	0.059±0.02b	0.017±0.00a	0.292±0.06b
	20~40	0.049±0.02b	0.020±0.00a	0.052±0.02b	0.108±0.04a	0.017±0.00a	0.292±0.05b
	40~60	0.028±0.01c	0.017±0.00ab	0.071±0.03a	0.095±0.03ab	0.017±0.00a	0.360±0.08a
T3	0~20	0.044±0.02ab	0.017±0.00b	0.038±0.01b	0.110±0.03b	—	0.309±0.07c
	20~40	0.048±0.03a	0.024±0.00a	0.034±0.01b	0.144±0.03a	0.017±0.00a	0.326±0.07b
	40~60	0.035±0.02b	0.024±0.00a	0.052±0.02a	0.089±0.02c	0.017±0.00a	0.360±0.08a
T4	0~20	0.069±0.03b	0.031±0.01b	0.038±0.01b	0.028±0.00b	—	0.206±0.05c
	20~40	0.100±0.03a	0.031±0.00b	0.014±0.00c	0.157±0.03a	—	0.309±0.07a
	40~60	0.068±0.02b	0.048±0.01a	0.049±0.08a	0.155±0.03a	0.017±0.00a	0.257±0.05b
显著性P值		*	*	*	**	—	*

2.4 盐碱土种植对水稻生长和产量的影响

通过对盐碱地进行作物种植可以显著改善盐碱地表层土壤的理化性质,进而对作物产量产生积极影响。表5为盐碱土种植对水稻生长和产量的影响,表中物理量为株高h_p、分蘖a、每平方米穗数b、单穗穗粒数N_g、百粒质量m、产量Y。分析可知,各处理对水稻株高生长和每平方米穗数影响具有统计学意义($P<0.05$),对水稻分蘖和单穗穗粒数影响不具有统计学意义($P>0.05$),对盐碱地种植水稻百粒质量和产量影响具有统计学意义($P<0.01$)。其中处理 T1 的株高、分蘖数、百粒质量和产量均最大,分别达到 74.27 cm、17.84 个、1.53 g 和 2893.5 kg/hm^2;处理 T3 的株高最低,T4 的分蘖数最少。而单穗穗粒数和每平方米穗数均为处理 T2 的最大,分别为 89 粒和 225 穗/m^2;单穗穗粒数、每平方米穗数、百粒质量以及产量均为处理 T4 的最低。从产量角度看,处理 T1 分别较 T2、T3 和 T4 提高了 4.96%、42.78%和 161.64%,是本试验条件下产量最高的处理。

表 5 盐碱土种植对水稻生长和产量的影响

Table 5 Effects of saline-alkali soil cultivation on rice growth and yield

处理	h_p/cm	a/个	N_g/粒	b/(穗/m^2)	m/g	Y/(kg/hm^2)
T1	74.27±2.6a	17.84±1.5a	87±3.2b	214±8.7b	1.53±0.02a	2 983.50±36.0a
T2	72.35±3.6ab	16.92±1.3b	89±2.8a	225±7.8a	1.42±0.03b	2 842.50±37.5b
T3	69.22±2.8c	16.34±1.2c	78±2.6c	197±6.5c	1.36±0.02c	2 089.50±39.0c
T4	70.25±3.1bc	13.8±1.1d	65±2.6d	145±5.6d	1.21±0.01d	1 140.30±31.5d
显著性 P 值	*	—	—	*	* *	* *

3 讨论

在盐碱地内种植耐盐植物进行土壤改良,主要利用植物生长促进土壤积累有机质,改善土壤结构,降低地下水位,减少土壤中水分的蒸发,从而加速盐分淋洗、延缓或防止积盐返盐。陈淑娟等[7]研究了宁夏地区盐分胁迫条件下种植水稻对土壤 pH 的影响,结果表明各种类型土壤 pH 值均在水稻收获后显著降低,与本研究结果一致。高彦花等[8]研究了耐盐植物对海滨地区盐碱地的改良效果,研究表明,与对照相比,耐盐植物种植后土壤相比盐分质量分数显著降低,文中研究发现种植耐盐水稻能极显著降低土壤含盐量($P<0.01$)。董起广等[9]在沿黄地区盐碱地水稻种植中研究结果表明,土壤的机械组成并没有因种植水稻发生明显改变,而文中研究各处理促进了黏粒含量的降低、粉粒含量的提高,而砂粒含量均有所降低。这可能是由于文中研究持续时间较长,且经过了水稻秸秆深翻还田等田间管理措施,导致黏粒含量有了一定程度的降低。何海锋等[10]在宁夏盐碱地种植柳枝稷,研究其对盐碱地土壤养分的影响,结果表明,盐碱地种植柳枝稷 5 年后,耕作层土壤养分均有显著提高,文中研究表明,在种植水稻后,各处理土壤耕作层有机质含量有不同程度的提高($P<0.01$)。李玉等[11]研究发现,有机肥施用对滨海盐碱地土壤全氮含量影响不显著,文中研究表明,盐碱地种植水稻对全氮含量提升效果显著($P<0.01$)。这可能是试验地区土壤条件和施用化肥种类不同所导致。而有效磷和速效钾含量显著降低($P<0.05$),这可能是水稻种植过程中施肥量太少,且部分养分淋失,以及种植年限较短等原因所导致。

王旭等[12]研究了盐地碱蓬对重度盐碱地土壤盐分的影响,结果表明,土壤含盐量随种植年限增加而下降,根区土壤中 Cl^- 含量随种植年限增加而显著降低;Ca^{2+} 和 SO_4^{2-} 不易随水移动,淋洗程度低;HCO_3^- 和 Mg^{2+} 淋洗效果随年限延长而增加;经过 3 年种植后土壤中 Cl^- 在表层盐分组成中的比例下降,Ca^{2+} 比例上升。由于试验条件及种植作物种类有所区别,文中研究各处理对土壤 Ca^{2+}、Mg^{2+}、Cl^- 以及 HCO_3^- 的影响具有统计学意义($P<0.05$),对 SO_4^{2-} 影响具有统计学意义($P<0.01$),而对 CO_3^{2-} 影响不具有统计学意义($P>0.05$)。各处理 Ca^{2+}、Mg^{2+}、Cl^-、SO_4^{2-} 和 HCO_3^- 质量比分别较本底值降低 46.76%、63.78%、33.91%、63.78%以及 43.18%,种植水稻对土壤脱盐效果显著,这与 Khadim 等[13]的研究结果一致。水稻收获后各处理的作物体内全盐质量比分别占土壤本底值的 9.29%、10.69%、8.80%和 8.18%,处理 T2 表现出更好的脱盐效果。

各处理对水稻株高生长和每平方米穗数影响具有统计学意义($P<0.05$),对水稻分蘖和单穗穗粒数影响不具有统计学意义($P>0.05$),对盐碱地种植水稻百粒质量和产量影响具有统计学意义($P<$

0.01)，处理 T1 分别较 T2、T3 和 T4 提高了 4.96%、42.78%和 161.64%。种植年限对水稻产量及其构成因素均有显著影响，这与罗成科等[14]在宁夏盐碱地种植水稻得到的研究结果一致。

4　结论

在文中试验条件下，处理 T2(宁靖 28)的保肥和脱盐效果最好，T1(隆优 619)的产量最高，是较为适宜在陕北盐碱地种植的水稻品种。根据研究情况，需要适当增施肥料以促进土壤有效磷和速效钾提高。该研究结果对于陕北地区盐碱地利用和生态环境改善有着重要意义。

参考文献

[1]史文娟，杨军强，马媛.旱区盐碱地盐生植物改良研究动态与分析[J].水资源与水工程学报，2015，26(5)：229-234.

SHI W J，YANG J Q，MA Y. Review on saline-alkali soil improvement with planting halophyte method in arid region [J].Journal of Water Resources and Water，2015，26(5)：229-234.

[2] Brown T T，Koenig R T，Huggins D R，et al. Lime effects on soil acidity，crop yield，and aluminum chemistry in direct-seeded cropping systems[J]. Soil Science Society of America Journal，2008，72(3)：634-640.

[3] 李帅，杨敏，曹惠翔，等.连年种植菊芋对滨海盐碱地的生态修复效果与机制[J].南京农业大学学报，2021，44(6)：1107-1116.

LI S，YANG M，CAO H X，et al. Ecological restoration effect and mechanism of continuous-year cultivation of Jerusalem artichoke on coastal saline-alkali land[J].Journal of Nanjing Agricultural University，2021，44(6)：1107-1116.

[4] 郭耀东，程曼，赵秀峰，等.轮作绿肥对盐碱地土壤性质、后作青贮玉米产量及品质的影响[J].中国生态农业学报，2018，26(6)：856-864.

GUO Y D，CHEN M，ZHAO X F，et al. Effects of green manure rotation on soil properties and yield and quality of silage maize in saline-alkali soils[J].Chinese Journal of Eco-Agriculture，2018，26(6)：856-864.

[5] 颜安，吴勇，徐金虹，等.有机肥氮替代化肥氮和土壤改良剂对盐碱地棉花产量和土壤养分的影响[J].中国土壤与肥料，2021(6)：72-77.

YAN A，WU Y，XU J H，et al. Efetes of geen manure roato on sol pre and ved nd qull of lage maein saline-alkali soils [J].Soil and Fertilizer Sciences in China，2021(6)：72-77.

[6] 郑普山，郝保平，冯悦晨，等.不同盐碱地改良剂对土壤理化性质、紫花苜蓿生长及产量的影响[J].中国生态农业学报，2012，20(9)：1216-1221.

ZHENG PuShan，HAO BaoPing，FENG YueChen，et al.Effects of different saline-alkali land amendments on soil physico-chemical properties and alfalfa growth and yield[J].Chinese Journal of Eco-Agriculture，2012，20(9)：1216-1221.

[7] 陈淑娟，何文寿，王晓军，等.不同措施对碱化土壤培肥及水稻植株养分吸收的影响[J].干旱地区农业研究，2011，29(6)：115-118，129.

CHEN Shujuan，HE Wenshou，WANG Xiaojun，et al. The influence of different fertilizering measures on soil fertility and nutrient uptaking of rice plant in alkaline soil[J]. Agricultural Research In The Arid Areas，2012，20(9)：1216-1221.

[8] 高彦花，张华新，杨秀艳，等.耐盐碱植物对滨海盐碱地的改良效果[J].东北林业大学学报，2011，39(8)：43-46.

GAO Yanhua，ZHANG Huaxin，YANG Xiuyan，et al. Ameliorative Effect of Saline-Alkali Tolerant Plants in Coastal Saline-Alkali Land[J].Journal of Northeast Forestry University，2011，39(8)：43-46.

[9] 董起广，何振嘉，高红贝，等.沿黄地区盐碱地种植水稻土壤理化性质的比较[J].植物资源与环境学报，2017，26(2)：110-112.

DONG Qiguang，HE Zhenjia，GAO Hongbei，et al.Comparison on soil physicochemical properties of saline and alkaline land planted with Oryza sativa in the area along the Yellow River[J].Journal of Plant Resources and Environment，2017，26(2)：110-112.

[10] 何海锋，吴娜，刘吉利，等.柳枝稷种植年限对盐碱土壤理化性质的影响[J].生态环境学报，2020，29(2)：285-292.

HE Haifeng，WU Na，LIU Jili，et al.Effects of Planting Years of Panicum virgatum on Soil Physical and Chemical Properties[J].Ecology and Environment Sciences,2020, 29(2):285-292.

[11] 李玉,田宪艺,王振林,等.有机肥替代部分化肥对滨海盐碱地土壤改良和小麦产量的影响[J].土壤,2019, 51(6):1173-1182.

LI Yu,TIAN Xianyi,WANG Zhenlin, et al. Effects of Substitution of Partial Chemical Fertilizers with Organic Fertilizers on Soil Improvement and Wheat Yield in Coastal Saline and Alkaline Land[J].Soils,2019, 51(6):1173-1182.

[12] 王旭,田长彦,赵振勇,等.滴灌条件下盐地碱蓬(Suaeda salsa)种植年限对盐碱地土壤盐分离子分布的影响[J].干旱区地理,2020, 43(1):211-217.

WANG Xu,TIAN Changyan,ZHAO Zhenyong, et al. Effects of different planting years of Suaeda salsa on the soil ions distribution in saline-sodic soil under drip irrigation[J].Arid Land Geography,2020, 43(1):211-217.

[13]Khadim F K , Su H , Xu L , et al. Soil Salinity Mapping in Everglades National Park Using Remote Sensing Techniques and Vegetation Salt Tolerance[J]. Physics and Chemistry of the Earth, 2019,1(4):31-35.

[14] 罗成科,田蕾,毕江涛,等.种稻年限对盐碱土微量元素及水稻产量和品质的影响[J].生态环境学报,2019, 28(8):1577-1584.

LUO Chengke,TIAN Lei,BI Jiangtao,et al.Effects of Rice Planting Years on Saline-alkali Soil Trace Elements,Rice Yield and Quality[J].Ecology and Environment Sciences,2019, 28(8):1577-1584.

本文曾发表于2022年《排灌机械工程学报》第40卷第8期

Dynamic Changes of Soil Moisture in Hilly and Gully Regions of Loess Plateau

Jing He[1,2,3], Shaodong Qu[4]

(1.Shaanxi Provincial Land Engineering Construction Group, Key Laboratory of Degraded and Unused Land Consolidation Engineering, Ministry of Natural Resources, Xi'an, China; 2.Shaanxi Provincial Land Engineering Construction Group, Shaanxi Provincial Land Consolidation Engineering Technology Research Center, Xi'an, China; 3.Shaanxi Provincial Land Engineering Construction Group, Land Engineering Technology Innovation Center, Ministry of Natural Resources, Xi'an, China; 4.Shaanxi Provincial Land Consolidation Engineering, Xi'an, China)

【Abstract】 The interflow is extremely difficult come into being in Loess Plateau, and our group bring the gully control and land reclamation projects into force, the interflow has been found. In order to study interflow in gully control and land reclamation projects, TRIME-TDR sensor has been used to monitoring soil moisture changes with depth and time in Nanniwan, Yan'an, to analysis the dynamic change of soil moisture. The results show that with the height decrease the soil moisture increases as the same depth of soil layer; soil moisture content in hillside slope toe is higher than that which are on the hillside; as well as, the cultivated land has highest and more stable soil moisture content; soil moisture changes over time for seasonal precipitation variability. And there are some conclusions: 1. With the elevation higher the soil moisture content is lower, on the contrary, the elevation lower the soil moisture content higher; 2. soil moisture from up to down in the vertical direction : the soil moisture content decreases, and then increases, is different from the existing growth and reduce type; 3. soil moisture content changing in the time scale is depending on seasonal rainfall and the local crop planting situation; 4. the interflow may exists in this region which inferred with soil moisture variation tendency at 40 to 60 cm depth. This article as a preliminary research result of gully control and land reclamation on soil moisture in hilly and gully regions of Loess Plateau.

【Keywords】 Soil moisture; TRIME-TDR; Dynamic change

Introduction

Soil moisture is one of the important characteristics of soil[1], and it has long been one of the important research contents of experts and scholars. In terms of spatial variation of soil moisture, according to different soil profile water content, in the vertical direction according to the fluctuation amplitude of water content is divided into active layer, sub-active layer and relatively stable layer[2], in the study of soil moisture in loess hilly area, Yang Kaibao used a similar layered method. There are also experts and scholars according to the use of vegetation roots on soil moisture stratification research[3], Ruan Chengjiang and other soil moisture stratification for Hippophae rhamnoides forest root weak use layer, root use layer, supplementary regulation layer and weak regulation layer four layers[4]; Zou et al[5] divided the soil moisture of poplar sand mixed forest and poplar pure forest into two layers : the weak utilization layer of forest root soil moisture and the utilization layer of forest root. The vertical changes of soil moisture in the Loess Plateau are mainly growth type[6], and there is also decrease type[7]. In the horizontal direction, there are a lot of research work at this stage, but due to the complexity of spatial scale changes and soil spatial variability, there is no relatively perfect research results. The change of soil moisture on time scale also has a lot of research results, mainly from seasonal change[10] and inter annual change[11]. This study is mainly based on the discovery of soil middle flow phenomenon in the construction of the project in this area, so as to apply TRIME-TDR to real-time monitoring and regular observation records of soil moisture and temperature in typical plots, and to study the dynamic change process of soil moisture in gully construction on the Loess Plateau, aiming to provide reference for subsequent research.

Materials and Methods

Overview of test area. The test area is located in Nanniwan Town, Baota District, located in the upper reaches of Fenchuan River and Jiulongquanchuan Jiulongquangou land consolidation area. The geographical coordinates are between 36°14′40″~36°19′25″N and 109°35′50″~109°39′50″E, belonging to the temperate monsoon climate zone. The annual average temperature is 9 ℃, the annual average frost-free period is 179 days, and the annual average precipitation is 573 mm. Geology is part of the Ordos platform, in the Mesozoic bedrock and Cenozoic red soil layer formed by the terrain covered with deep aeolian loess. The loess cover depth is 50~150 m, and the soil thickness in Chuandao area is 0.4~5.0 m. The soil in the test area is mainly gray cinnamon soil, and the vegetation is mainly planted in plantations, including Robinia pseudoacacia, Toona sinensis, Pine cypress, apricot, jujube, etc.

Method. In the project area, the typical trenching field is selected as the research object. Based on the field area, 18 experimental observation plots are established (Fig.1), and each plot area is about 50 m^2. In each plot, 1 m depth TRIME soil moisture observation tube was installed, and the portable soil surface moisture observation instrument and TRIEM-TDR observation probe were used to regularly observe the soil moisture stratification in the depth range of 0~1 m. The observation layer was 10 cm once. Microsoft Excel 2013 software was used to process and graph the data.

Results and Discussion

Vertical monitoring results. The monitoring results of soil moisture is the average value of long-term monitoring results, in order to eliminate the instrument error and human monitoring error in the monitoring process, as far as possible to truly reflect the dynamic change process of soil moisture in the monitoring area. There are many experimental data. Due to the length problem, only a few experimental results are listed. Fig.2 is the same slope, different elevation point soil moisture monitoring results, Fig.3 is different slope, the same elevation soil moisture monitoring results, because 13,14,15 points are located at the intersection of slope foot and farming land, 16,17,18 points are located in the slope farmland, so the monitoring results are given, respectively, see Fig.3 (b), (c).

It can be seen from Fig.2 that the soil moisture content at the measurement point with higher elevation is lower than that at the measurement site with low elevation, and the dynamic changes of soil moisture at different elevation points are inconsistent, and some fluctuations are large, such as 14 measurement points, soil moisture above 40 cm depth. The content continues to decrease until the 40 cm measuring point, and then gradually increases, and the increase is greater; the depth of the soil layer continues from 60 cm downwards, and the soil moisture at the same elevation point has a small fluctuation range and is basically stable. As the elevation gradually decreases, the soil moisture content gradually increases. The soil moisture content on the hillside is relatively small, and the soil moisture content in the farmland is the largest. The 17th measurement points are located in farmland, and the soil moisture content from the high level of the surface layer to the lowest level at a depth of 40 cm, after which the soil moisture content begins to increase again, until it is basically stable.

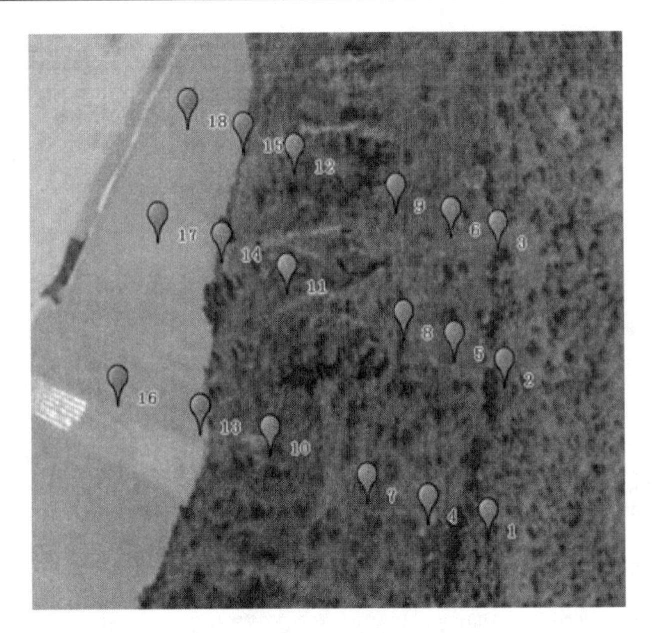

Fig.1 TRIME pipe layout

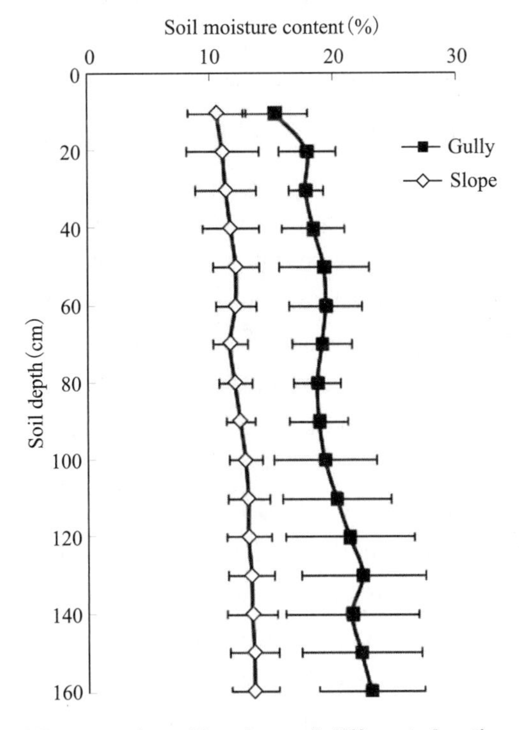

Fig.2 The same slope direction and different elevation points

Fig.3(a) describes the dynamic change process of soil moisture in different slope directions, the same elevation, and different regions. The overall trend of soil moisture is that the shallow soil moisture content is higher than the deeper soil layer, but as the depth of the soil layer increases, the soil moisture content gradually increases, and reaches a relatively stable soil moisture value with a small fluctuation range. During the dynamic change of the soil moisture content of point 1 measuring point, the fluctuation range is small and the moisture content is relatively stable. This is due to the fact that there are more trees in the forest area and water conservation reduces water loss; the 2 and 3 measuring points have higher soil moisture surface layer content. The moisture content decreases between 10 cm and 30 cm, and then gradually increases to a relatively stable level.

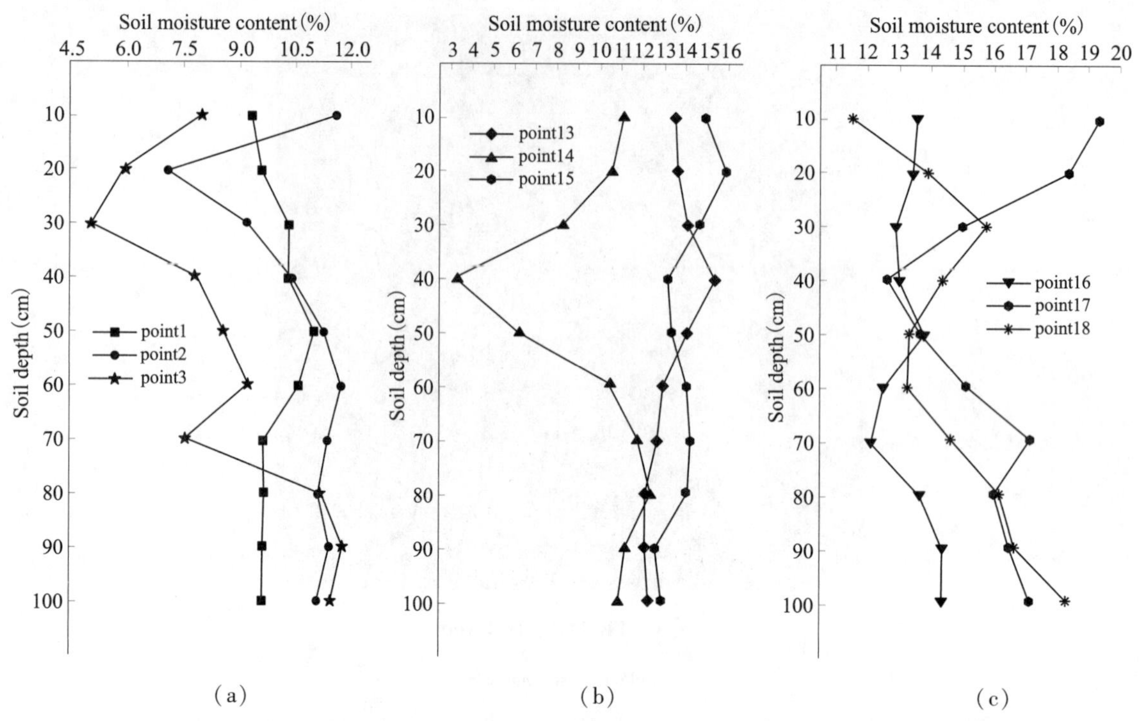

Fig.3 Different slope direction and the same elevation points

The change of soil moisture in Fig.3 (b), (c) is basically the same as the change of (a), so there's no repeat.

There are generally two situations in the change trend of soil moisture content from top to bottom, one is increasing and the other is decreasing[1]. This result is inconsistent with the existing research results, because the soil moisture content depends on the local time and space conditions[1] and the types and growth of plants in the monitoring area. The soil moisture content changes drastically in the range of 0~40 cm, which is due to the influence of the weak utilization layer of forest root soil moisture on the soil moisture. Below 40 cm belongs to the forest root utilization layer, which has little impact on soil moisture content, and the fluctuation range of soil moisture is small[3].

There are two obvious changes in soil moisture at different depths underground. There are two points: 1) the fine root area of the plant is located in the depth area of the soil layer, which absorbs a large amount of soil moisture in the area, resulting in the water content of the area compared to the shallow soil and the deeper layer. Soil moisture content is low; while surface soil moisture is affected by precipitation and tree canopy (which affects precipitation interception and evaporation)[8], its content is relatively high. 2) there is a soil flow phenomenon, although the loess texture is not prone to soil flow phenomenon, but the Loess Plateau. The underground environment is different in different regions, especially in mountainous areas, where the soil structure has not been studied in detail, and a dense and impervious layer may appear in the deep soil layer, which will lead to the phenomenon of soil flow. And when the TRIME pipe was pre-embedded in the early stage of the experiment, the expected burying depth was 2 m, but after the depth of 1 m, the gravel layer appeared in the ground. The TRIME pipe was extremely difficult to bury, and it was not easy to continue burying when it continued to deepen about 50 cm.

Results of soil moisture change over time. Fig.4 shows the change process of soil moisture with time at different depths of point 1, 14 and 17. Soil moisture at six different time points was collected during the experiment.

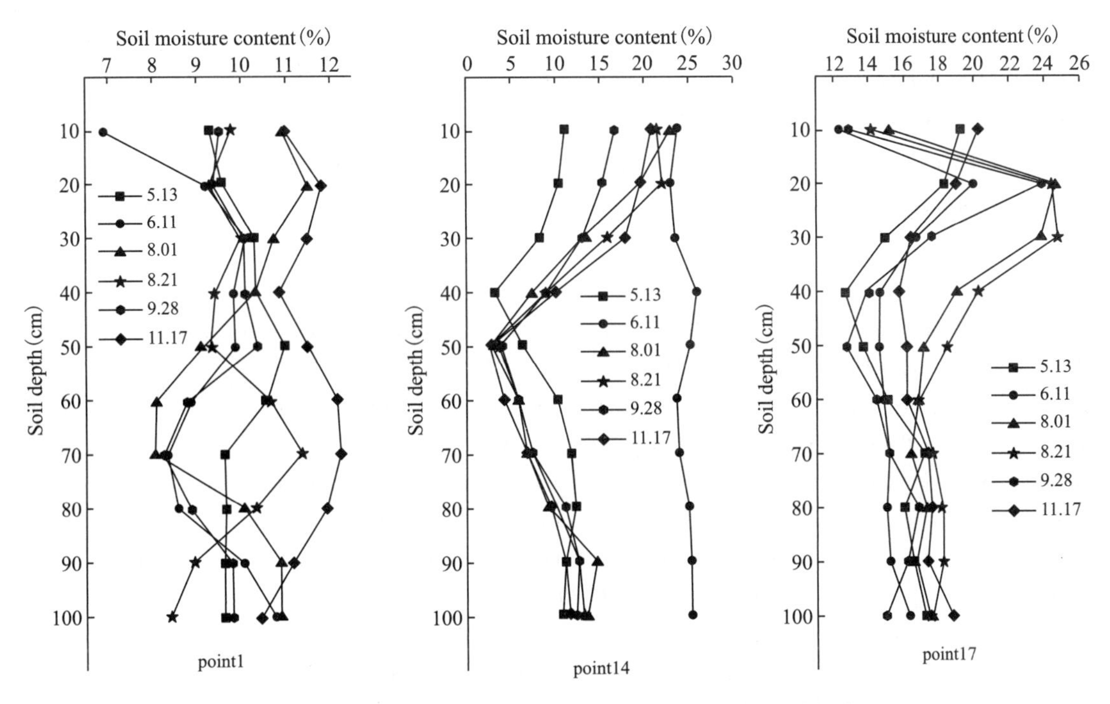

Fig.4 The dynamic process of soil moisture changes with time

It can be seen from Fig. 4 that the changes in soil moisture over time are seasonal precipitation changes, and it is difficult to quantitatively analyze the changes on the time scale. Existing research only conducts targeted mechanism analysis from the two aspects of rainy and dry seasons. The measurement point 1 is located on the hillside, and its soil moisture changes little on the time scale, mainly because the forest trees are perennial plants, which consume soil moisture to varying degrees throughout the year and have the function of conserving water. The soil moisture can be maintained at a relatively stable level throughout the year. Point 14 measuring point is located at the intersection of the slope foot and the farmland. The soil moisture at different soil depths shows an increasing type on the time scale, but it shows a decreasing type at 40~60 cm, and the soil moisture content at each time period is lower than the rest of the monitoring layer; at this measurement point, the monitoring data in June is much higher than the monitoring data in the time period, which is due to the influence of precipitation in the rainy season. The point 17 measurement point is located in the farmland. The soil moisture performance is relatively stable on the time scale. Only in August and September, the fluctuation range is relatively large. This is due to the rainy season with a lot of precipitation, and the crop is in the fruiting and mature stage, and the water consumption is reduced; 40~60 cm soil moisture content is lower than the rest of the time period, crop growth is less affected, and there may be soil flow. Changes in soil moisture on the time scale are mainly affected by the local climate seasonal precipitation[1], and crop types also have a certain impact. The soil moisture content in November is relatively stable, because this time period is in the period of slow moisture loss in late autumn and winter (late September to late March of the following year)[9], and the water consumption can be ignored.

Conclusion

The vertical change of soil moisture from top to bottom is neither a growth type nor a decrease type, but a new change pattern: from high to low, from low to high. The higher the elevation, the lower the soil moisture content; conversely, the lower the elevation, the higher the soil moisture content. The fluctuations of soil moisture on the time scale are presented as seasonal precipitation changes. The soil moisture content is larger in the rainy season and smaller in the dry season. The soil moisture content is very low between the depth of 40~60 cm in the soil layer. Under the conditions of excluding the influence of seasonal rainfall and water consumption for crop growth, there may be soil flow in the soil, which requires further study.

Acknowledgements

Funding: This research was funded by intra-scientific research project of Shaanxi Provincial Land Engineering Construction Group Co., Ltd., and grant number DJNY2022-53 and DJNY2022-54.

References

[1]He, Q.H., He, Y.H., Bao, W.K. (2003) Research on dynamics of soil moisture in arid and semiarid mountainous areas. Journal of Mountain Science. 21(2), 149-156.

[2]Wang, M.B., Li, H.J. (1995) Quantitative study on the soil water dynamics of various forest plantations in the Loess Plateau region in Northwestern Shanxi. Acta Ecologica Sinica. 15(2), 178-184.

[3]Yang, K.B., Li, J.L., Guo, P.C., Zhang, G.Y. (1999) Law of soil moisture change in terrance land section in Loess hilly region. Journal of Soil Erosion and Soil and Water Conservation. 5(2), 64-69.

[4]Ruan, C.J., Li, D.Q. (1999) Soil moisture and its influence on seaduckthorn growth in semi-arid Loess Hilly region. Bulletin of Soil and Water Conservation. 19(5), 27-30.

[5]Zou, G.X., Li, T.J., Li, X.H., Guo, L.Z. (2000) Study on regularity of soil moisture of poplar mixture seabuckthon in gentle slope of North-west Liaoning. Journal of Soil and Water Conservation. 14(5), 55-57.

[6]Qiu, Y., Fu, B.J., Wang, J., Chen, L.D. (2001) Spatial variability of soil moisture content and its relation to environmental indices in a semi-arid gully catchment of the Loess Plateau, China. Journal of Arid Environments. 49, 723-750.

[7]Yang, X.M. (2001) Study on the characteristics of water environmental in shrubbery land of Loess Plateau. Arid Zone Research. 18(1), 8-13.

[8]Li, L.S., Zhao, X.N., Gao, X.D., Wu, P.T., Li, H.C., Ling, Q., Sun, W.H. (2016) Soil water dynamic of rain-fed jujube (Ziziphus jujube) with stand age on Loess Plateau. Transactions of the Chinese Society of Agricultural Engineering (Transactions of the CSAE). 32(14), 145-152. (in Chinese with English Abstract)

[9]Lu, Z.F., Zhang, X.C., Su, M., Lin, H.P. (1995) Studies on the dynamic changes of soil moisture and the benefits of soil and water conservation from growing grasses on the Loess Plateau. Journal of Arid Land Resources and Environment. 9(1), 40-49.

[10]Sheng, C.L., Liu, L.H., Liu, W.Y. (2000) Biomass and dynamics of soil environment during the early stage of vegetation restoration in a degraded dry-hot mountain area of Nanjian, Yunnan. Acta Phytoecologica Sinica. 24(5), 575-580.

[11]Li, H.J., Wang, M.B., Chai, B.F. (1998) Study on characteristics of soil water of planted forest and its relation to precipitation in Northwestern Shanxi. Journal of Soil Erosion and Soil and Water Conservation. 4(4), 60-65.

本文曾发表于2022年《Fresenius Environmental Bulletin》第31卷第4期

2　土地利用与保护

Assessment of the Vertical Characteristics and Contamination Levels of Toxic Metals in Sediment Cores from Typical Chinese Intertidal Zones

Haihai Zhuang[a], Yantao Hu[a], Chaofan Yan[a], Zenghui Ma[a], Maosheng Gao[b], Jia Zhang[c,d]

(a.Shaanxi Provincial Land Engineering Construction Group CO. Ltd., Xi'an 710075, China;
b.Qingdao Institute of Marine Geology, China Geological Survey, Qingdao 266071, China;
c.Krirk University, Bangkok 10220, Thailand; d.Xi'an International Technician College, Xi'an 710075, China)

[Abstract] The heavy metal (As, Cd, Cr, Cu, Hg, Pb, and Zn) contents of 24 sediment cores were analyzed and obtained from 12 typical Chinese intertidal zones. The results revealed a gradual improvement in the environmental quality of the intertidal zone. Enrichment factor and geoaccumulation index analyses demonstrated a generally good environmental quality of intertidal sediment, with some areas of serious contamination, such as the Xiamen Jiulong Estuary, the Yangtze River Estuary, and the Pearl River Delta. Relative to the guidelines for sediment quality, the studied intertidal zones were moderately impacted, with a risk of biotoxic impacts. This research can reveal the status of toxic metal pollution in the intertidal zone of China and provide a reference for coastal area development.

[Keywords] Sediment; Intertidal zone; Toxic metal; Assessment of contamination

1 Introduction

Coastal regions represent important ecological-social-economic natural areas with compound functions. The intertidal zone is the area that is most directly affected by both the ocean and the land and is a significant "sink" and secondary "source" of toxic metals in aquatic ecosystems; intertidal sediments play a significant role in the transport, storage and research of toxic metals (Hartmann et al., 2005). Once this area is polluted, the related biological systems are also severely damaged.

Toxic metals represent pollutants characterized by accumulation in the environment, long-lasting residues, difficult elimination, and resistance to degradation from microbes (Kang et al., 2017; Viers et al., 2009). Toxic metals are distributed and build up in organisms via the ecosystem food cycle and can result in the disruption of metabolic and physiological processes (Harikrishnan et al., 2017; Sakan et al., 2009). The majority of toxic metals entering surface water bodies sink to the bottom sediment. Therefore, analyses of sediments can provide an indication of the accumulation of toxic metals in water bodies (Liu et al., 2013). Therefore, toxic metal pollution in intertidal zone sediments is one of the focuses of domestic and foreign research (Punniyakotty and Ponnusamy, 2018; Sun et al., 2015). Additionally, offshore sediments provide a historical record of the source, migration, and transformation of various contaminants associated with anthropogenic activities and provide an optimal tool to trace the history of marine pollution (Li et al., 2011; Wang et al., 2005).

In recent years, many scholars have completed much research on the contamination by and distribution of toxic metals in China's offshore sediments, but the relevant research is mainly concentrated in estuaries, bays and local coasts (Xu et al., 2009; Yu et al., 2011; He et al., 2021), and few studies have concentrated on the entire coastal zone in China. The present study analyzed the contents of toxic metals and their vertical profiles, drivers, and concentrations in sediment cores taken from twelve typical Chinese intertidal zones. The results of the present study can help improve knowledge of the levels and spatial distributions of toxic metals and their historical trends in China, thereby facilitating the sustainable development of marine resources.

2 Study area

The coastline of mainland China starts from the Yalu River Estuary in the north and reaches the Beilun River Estuary in the south, with a length of more than 18000 kilometers. China is one of the countries with the longest coastline in the world and the most abundant coastline types in the world (Liu et al., 2021). The present study selected twelve typical intertidal zones in China for sampling sediument based on category, level of development, spatial area, and effects of anthropogenic activities on sediment quality.

From north to south, the sampling points were located in Liaohe River Delta (Typical areas affected by human activities, such as oilfield development and sea reclamation), Tianjin Hangu intertidal zone (Located in the largest bay in China and a typical silt-silt intertidal zone), Yellow River Delta (The world's highest sediment content and China's second longest river estuary), Qingdao Jiaozhou Bay (Closed silty intertidal zone with frequent industrial and agricultural activities), Jiangsu Yancheng Shoal (China's largest plain, silt tidal flat and national nature reserve), Yangtze River Delta (China's longest river estuary and typical estuarine silty intertidal zone), Zhejiang Hangzhou Bay (The most typical strong tidal bay-type silty intertidal zone in China), Fujian Minjiang Estuary (Diverse types of intertidal zones), Xiamen Jiulong Estuary (Intertidal zone interacting with river runoff and ocean currents), Pearl River Delta (The most industrialized region in China), Guangxi Yingluo Bay (One of the famous wetland ecological environments in China) and Hainan Dongzhai Port (Typical mangrove reserve in China).

Two different capital letters will be used in the article to indicate the collection location of sediment cores, as shown in Table 1.

Table 1 Specific coordinates and typicality of sampling points

Sampling sites	Cores	East longitude	North latitude
LH (Liaohe River Delta)	LH1	122°08′21.21″	40°36′34.37″
	LH2	122°09′42.74″	40°35′31.92″
HG (Tianjin Hangu intertidal zone)	HG1	117°57′04.03″	39°12′50.44″
	HG2	118°03′27.42″	39°12′49.55″
DY (Yellow River Delta)	DY1	118°55′47.48″	37°25′21.87″
	DY2	118°56′56.65″	37°23′47.37″
QD (Qingdao Jiaozhou Bay)	QD1	120°10′19.71″	36°12′10.48″
	QD2	120°09′07.18″	36°10′44.58″
YC (Jiangsu Yancheng Shoal)	YC1	120°46′23.78″	33°17′0.41″
	YC2	120°46′57.52″	33°15′15.30″
DT (Yangtze River Delta)	DT1	121°58′57.54″	31°29′44.10″
	DT2	121°58′59.34″	31°29′42″
CX (Zhejiang Hangzhou Bay)	CX1	121°24′28.24″	30°18′59.89″
	CX2	121°24′57.33″	30°18′47.52″

Continued to Table 1

Sampling sites	Cores	East longitude	North latitude
FZ (Fujian Minjiang Estuary)	FZ1	119°38′3.85″	26°01′50.85″
	FZ2	119°37′53.33″	26°1′54.86″
JL (Xiamen Jiulong Estuary)	JL1	117°57′0.16″	24°24′18.20″
	JL2	117°56′29.15″	24°24′38.96″
ZJ (Pearl River Delta)	ZJ1	113°39′10.04″	22°26′21.92″
	ZJ2	113°39′27.15″	22°26′5.45″
YL (Guangxi Yingluo Bay)	YL1	109°45′26.34″	21°29′1.30″
	YL2	109°45′28.40″	21°28′53.43″
DZ (Hainan Dongzhai Port)	DZ1	110°36′23.34″	20°00′13.16″
	DZ2	110°36′10.16″	20°00′21.10″

3 Materials and methods

Sampling of sediment cores extended between September and December 2014, with 24 cores obtained (two cores per site). All sediment cores were collected by driving plastic, fiber-reinforced graduated tubes (inner diameter and wall thickness of 100 mm and 5 mm, respectively) into the sediment, which protects the relative positions between sediments. The sample survey used a handheld global positioning system (GPS) to georeference sites. The instrument was a Yli X28/S7, which had an error of less than 5 m.

After preserving half of each core, the other half was subdivided into 2 cm intervals in the laboratory. The analysis of each subsample batch used whole-process blanks to exclude the risk of contamination. Measurements of the contents of toxic metals utilized parallel samples and reference material for offshore sediment (GBW07314) for quality control. Inductively coupled plasma-mass spectrometry (ICP-MS) was used to measure the contents of toxic metals using X-Series 2 equipment. The laboratory analyses were performed in the Marine Geological Experimental Testing Centre, Ministry of Land and Resources. The measurement error of the above analyses was regulated to within ±5%. Statistical analyses were conducted in Microsoft Excel 2013, and the results were plotted in Grapher 10.0 and Coreldraw X6.

The enrichment factor (EF) and the geoaccumulation index (I_{geo}) were used to calculate the pollution status of the sediments.

Al was utilized within normalization for calculating EF:

$$EF = \frac{C_S(Me)/C_S(Al)}{C_B(Me)/C_B(Al)}$$

where $C_S(Me)$ and $C_S(Al)$ represent the contents of target toxic metals and Al of sediment samples, respectively, and $C_B(Me)$ and $C_B(Al)$ represent toxic metal background contents and Al of sediment in the study area, respectively. The background contents of toxic metals showed high regional variation. Therefore, the application of the background values of toxic metals for one region to another within the assessment of the local sedimentary environment can produce inaccurate results. Therefore, as demonstrated in Table 2, the current study applied the results of sediment quality surveys for a range of Chinese shallow seas by Chi and Yan (2007) as background values of toxic metals (Chi and Yan, 2007).

Sutherland (2000) categorized the toxic metal sediment enrichment levels into five classes according to the EF of an element: no or slight enrichment ($EF<2$), moderate enrichment ($2 \leq EF<5$), significant enrichment ($5 \leq EF<20$), high enrichment ($20 \leq EF<40$) and extremely high enrichment ($EF \geq 40$) (Sutherland, 2000).

Table 2 Background values of sediments from different locations

Sites	Cu (mg/kg)	Pb (mg/kg)	Zn (mg/kg)	Cr (mg/kg)	Cd (mg/kg)	As (mg/kg)	Hg (mg/kg)	Al_2O_3 (mg/kg)
Bohai Bay	22	20	64	57	0.09	22	0.036	12.02
Yellow Sea	18	22	67	64	0.088	18	0.024	11.74
East China Sea	14	21	66	61	0.068	8	0.025	10.02
South China Sea	13	19	61	53	0.053	7.2	0.027	9.91

I_{geo} is a quantitative index proposed by Muller (1969) to characterize toxic metal contamination (Zhang et al., 2007). The associated equation is as follows:

$$I_{geo} = \log_2\left(\frac{C_n}{1.5B_n}\right)$$

where C_n represents the content of sediment metal n, B_n represents metal n's background content, and the constant of 1.5 was applied to represent variations in background data attributable to variations in lithology. The values of the geoaccumulation index can be classified into seven classes according to Muller: Class 0 ($I_{geo}<0$), uncontaminated; Class Ⅰ ($0 \leq I_{geo}<1$), uncontaminated to moderately contaminated; Class Ⅱ ($1 \leq I_{geo}<2$), moderately contaminated; Class Ⅲ ($2 \leq I_{geo}<3$), moderately contaminated to strongly contaminated; Class Ⅳ ($3 \leq I_{geo}<4$), strongly contaminated; Class Ⅴ ($4 \leq I_{geo}<5$), strongly contaminated to extremely contaminated; and Class Ⅵ ($I_{geo}>5$), extremely contaminated.

The present study used sediment quality guidelines (SQG_s) to examine the effects of contaminants in sediments on the ecology (Macdonald et al., 1996). The SQG_s include a range of contaminants and have adopted the most up-to-date baseline values, thereby producing highly confident results. Past studies initially applied SQG_s to examine single toxic metals. Long et al. (1998) generated a multitoxic metal ecological risk index based on SQG_s($SQG-Q$) (Zhuang et al., 2020):

$$\text{PEL-}Q_i = \frac{C_i}{\text{PEL}_i}$$

$$\text{SQG-}Q = \frac{\sum_{i=1}^{n}(\text{PEL}-Q_i)}{n}$$

In the formula, PEL-Q_i represents the probable effect level (PEL) for the i-th toxic metal; C_i is the measured content of the i-th toxic metal; PEL_i is the PEL for the i-th toxic metal; and n represents the number of toxic metals. The PEL*s* of Cu, Pb, Zn, Cr, Cd, and Hg were 108.00, 112.00, 271.00, 160.0, 4.21, 41.6, and 0.7 mg/kg, respectively. Damaging biotoxic impacts seldom occur when toxic metal contents are below the PEL. The calculated SQG-Q indicated that levels of ecological risk due to sediment toxic metals can be categorized as follows: no damaging biotoxic impacts (SQG-$Q \leq 0.1$); moderate effects with potential damaging biotoxic impacts ($0.1<$SQG-$Q \leq 1$); and strong effects with extremely strong damaging biotoxic impacts (SQG-$Q>1.0$).

4 Results

Fig.1 illustrates the variability in toxic metal contents with depth in sedimentary cores taken from typical Chinese intertidal zones. Each panel in Fig.1 has a bottom and top X-axis, with the top axis representing Cu, Pb, Zn, Cr, and As and the bottom axis representing Cd and Hg.

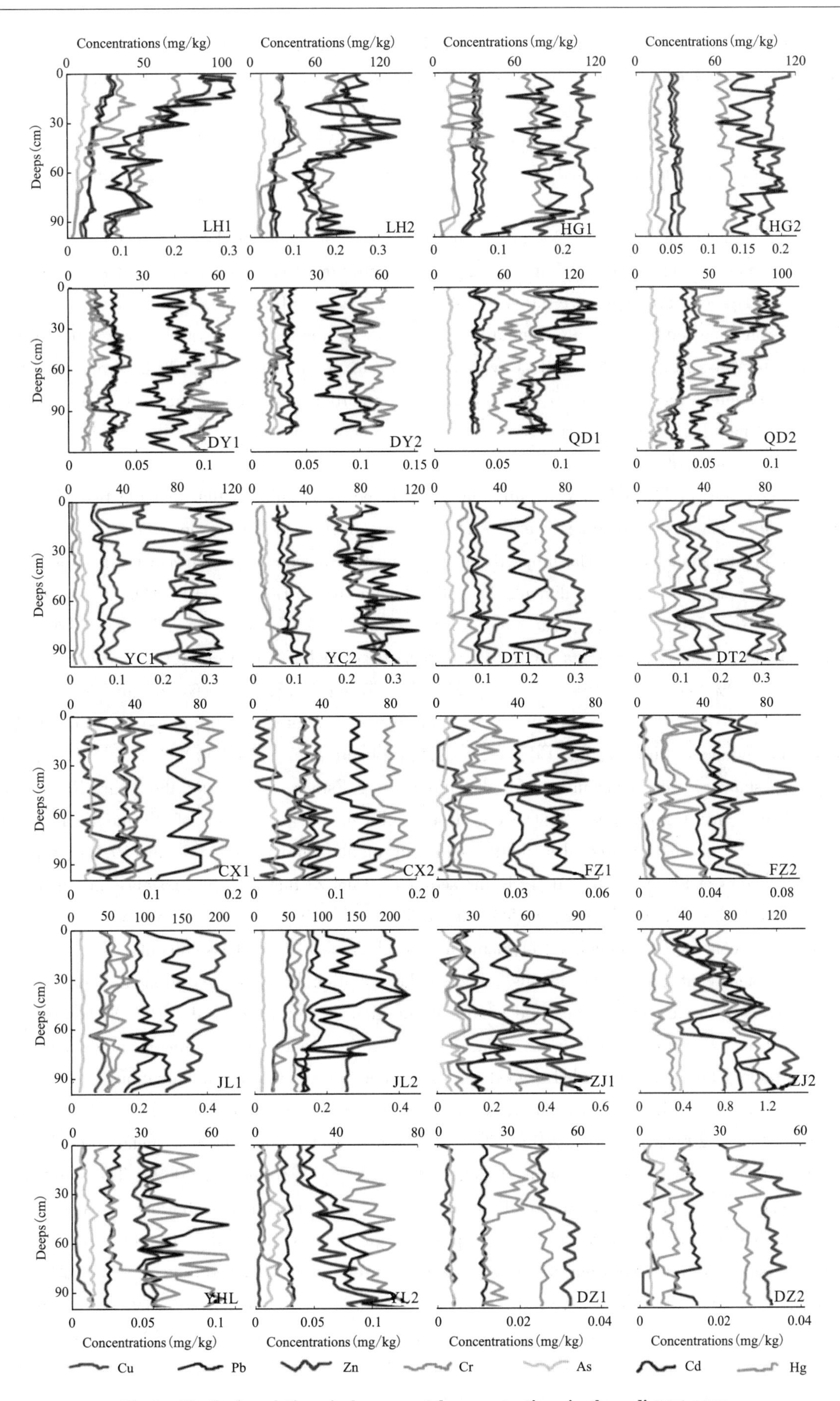

Fig.1 Vertical variations in heavy metal concentrations in the sediment cores

In China's marine sediment quality code (GB 18668—2002), the first-class marine sediment quality standard was applicable to marine fishery waters, marine nature reserves, human-contact sediments, etc.; this level was the most stringent standard (Zhao et al., 2019). According to this standard, the highest contents of Cu, Pb, Zn, Cr, Cd, Hg, and as were 35, 60, 150, 80, 0.5, 0.20, and 20 mg/kg, respectively. In this investigation, Cu concentrations ranged between 1.0~84.2 mg/kg (mean of 26.2 mg/kg), Pb concentrations were 10.0~133.0 mg/kg (mean of 33.9 mg/kg), Zn concentrations were 1.0~228.0 mg/kg (mean of 74.3 mg/kg), Cr concentrations were 7.3~112.7 mg/kg (mean of 63.4 mg/kg), Cd concentrations were 0.040~1.280 mg/kg (mean of 0.195 mg/kg), As concentrations were 0.6~36.8 mg/kg (mean of 10.7 mg/kg), and Hg concentrations were 0.004~0.340 mg/kg (mean of 0.053 mg/kg) (Table 3), which met the first-class marine sediment quality standard overall, showing that China's marine environment was generally good. In May 2021, the Ministry of Environmental Protection released the 2020 Environmental Quality Bulletin in Offshore Areas of China. In this report, the proportion of Cu concentrations in the coastal sea sediments that met the first-class marine sediment quality standard was 89.2%, and the remaining toxic metals were more than 95.0%.

The proportions of Cu and Cr concentrations in 961 sediment samples that met the first-class marine sediment quality standard were 68.9% and 73.9%, respectively, and the proportions of all other toxic metals were greater than 90.0%. The proportions of Cu and Cr concentrations that met the first-class marine sediment quality standard in the sediments from the Liaohe River Delta (LH), Hangu Intertidal Zone (HG), Dagu River Estuary (QD), Yancheng Coastal Zone (YC), Yangtze River Estuary (DT), Hangzhou Bay (CX), Jiulong River Estuary (JL), and Pearl River Delta (ZJ) were 100% and 82.6%, 62.1% and 91.9%, 43.9% and 38.6%, 64.7% and 21.2%, 50.0% and 76.5%, 19.1% and 36.5%, 22.1% and 92.6%, and 42.6% and 69.1%, respectively, and those from other sampling points were 100%. In addition, the sediments with Pb, Zn, Cd, As, and Hg concentrations exceeding the first-class marine sediment quality standard were concentrated in the Jiulongjiang Estuary (JL) and the Pearl River Estuary (ZJ), and the remaining sampling points were close to 100%. The proportions of Pb and Zn in the sediments from the JL that met the first-class marine sediment quality standard were 0% and 19.1%, respectively. The proportions of Pb, Cd, As and Hg in the sediments from ZJ that met the first-class marine sediment quality standard were 67.6%, 55.9%, 51.5% and 76.5%, respectively. Therefore, we believe that Cu and Cr are the most concerning contaminant elements in typical Chinese intertidal sediments.

The sediment contents of toxic metals reflected the pollution level of the sediment, and its vertical changes were used to characterize localized historical pollution (Pennington et al., 1973). The majority of sites selected in the present study were adjacent to estuaries, with most sediment produced by river basins. Therefore, these sites provided a good representation of toxic metal contamination across the entire basin. The present study identified sources of toxic metal contamination based on their horizontal distributions (Gao, 2009; Yu et al., 2011). The offshore sediment contents of toxic metals provide an indication of regionalized environmental quality (Mckee et al., 2004; Viers et al., 2009). The sediment contents of toxic metals in typical intertidal zones had various vertical distribution types in different regions, which were roughly divided into three types. For the first characteristic type, the toxic metal concentrations tended to increase from the bottom to the surface, including sediment cores LH1, LH2, QD1, QD2, JL1, and JL2. There were two reasons for this characteristic. The first reason was that the smaller the sediment size was, the better the toxic metal adsorption (Thuy et al., 2000; Zhuang et al., 2018). The weak hydrodynamic effect in these areas resulted in small grain sizes for surface sediments and large grain sizes for bottom sediments. The second reason was that the sampling sites were close to emerging

Table 3 Statistical characteristics of heavy metal concentrations in the sediment cores

Sampling points		Concentration	Cu (mg/kg)	Pb (mg/kg)	Zn (mg/kg)	Cr (mg/kg)	Cd (mg/kg)	As (mg/kg)	Hg (mg/kg)	Al_2O_3 (mg/kg)
Bohai	LH1	Mean	17.1	18.4	52.8	51.2	0.151	8.5	0.061	11.2
Sea	(n=43)	Range	8.3~29.2	10.0~31.7	24.3~106.8	28.7~73.1	0.067~0.309	3.2~14.4	0.008~0.126	8.3~13.0
	LH2	Mean	25.6	26.7	75.4	73.3	0.201	10	0.072	12.2
	(n=43)	Range	16.6~42.5	19.9~35.7	44.5~134.8	53.6~87.7	0.099~0.350	5.8~13.5	0.022~0.129	8.9~15.2
	HG1	Mean	34.3	30.1	106.9	74.9	0.168	13.3	0.04	15.7
	(n=43)	Range	23.1~39.7	21.1~34.3	73.8~119.5	55.2~97.9	0.075~0.216	6.1~16.1	0.011~0.093	13.3~16.8
	HG2	Mean	31.8	28	102.3	69	0.159	12.2	0.032	15.5
	(n=43)	Range	27.5~36.7	25.1~33.3	92.4~116.2	60.4~79.7	0.115~0.208	8.6~15.8	0.026~0.045	13.6~16.9
	DY1	Mean	16.7	18	54.7	57.1	0.077	8.8	0.017	8.7
	(n=60)	Range	8.0~25.0	14.8~21.5	43.7~68.2	47.1~66.5	0.051~0.100	7.5~10.0	0.010~0.030	7.8~9.7
	DY2	Mean	13.7	18	50.3	55.8	0.081	10.2	0.014	8.3
	(n=54)	Range	8.1~21.4	15.0~21.0	44.5~60.8	40.4~67.1	0.059~0.110	8.0~11.7	0.006~0.028	7.8~9.3
Yellow	QD1	Mean	41.8	35.7	101.9	85.4	0.094	13	0.06	13.9
Sea	(n=54)	Range	32.1~55.6	30.2~43.6	81.3~128.0	76.4~99.8	0.057~0.130	11.2~15.7	0.044~0.080	12.5~15.0
	QD2	Mean	31.6	30.9	82.6	77.6	0.069	11.9	0.043	13.3
	(n=60)	Range	18.2~42.7	24.5~36.5	56.4~104.0	59.8~94.2	0.040~0.110	9.0~15.2	0.012~0.080	11.9~14.4
	YC1	Mean	33.6	22.5	75	91.3	0.299	8.7	0.011	12.9
	(n=43)	Range	20.4~59.4	16.5~27.3	47.9~96.5	74.8~112.7	0.222~0.360	4.9~14.2	0.004~0.023	10.3~14.5
	YC2	Mean	31.8	24.3	76.3	83.3	0.263	10.7	0.024	12.7
	(n=42)	Range	15.2~44.3	18.3~36.4	56.2~98.3	62.7~109.2	0.186~0.360	8.1~16.1	0.009~0.070	11.2~14.4
East	DT1	Mean	31.4	25.1	83.8	68.5	0.202	9.2	0.063	10.4
Sea	(n=34)	Range	23.0~44.4	20.2~32.9	70.1~98.4	60.5~77.9	0.120~0.290	7.0~13.3	0.034~0.090	9.5~11.9
	DT2	Mean	40.7	32.1	80.4	79.3	0.249	12.4	0.086	11.9
	(n=34)	Range	23.8~53.9	21.7~40.3	61.5~95.2	63.0~89.4	0.170~0.330	7.5~17.0	0.040~0.120	10.0~13.0

Continued to Table 3

Sampling points		Concentration	Cu (mg/kg)	Pb (mg/kg)	Zn (mg/kg)	Cr (mg/kg)	Cd (mg/kg)	As (mg/kg)	Hg (mg/kg)	Al_2O_3 (mg/kg)
	CX1	Mean	40.7	32.8	16.2	84.8	0.14	12.9	0.071	12.9
	(n=34)	Range	22.3~52.1	21.3~41.5	5.5~33.6	71.2~97.0	0.105~0.180	8.7~16.2	0.054~0.090	10.1~14.6
	CX2	Mean	38.1	30.4	12.5	80.9	0.131	11.7	0.061	12.5
	(n=34)	Range	29.6~49.2	24.1~39.6	1.0~28.0	69.9~94.0	0.100~1.160	8.9~14.4	0.042~0.080	11.1~14.0
	FZ1	Mean	8.3	42	55.2	14.6	0.048	3.9	0.016	7.9
	(n=34)	Range	1.0~24.8	33.6~66.0	37.3~74.0	7.3~30.4	0.040~0.061	0.6~7.6	0.008~0.030	6.3~11.0
	FZ2	Mean	6.7	40.7	65.9	17.5	0.046	4.3	0.023	8
	(n=34)	Range	1.0~16.5	35.9~52.0	53.3~100.0	12.3~29.4	0.040~0.061	2.6~11.4	0.011~0.043	6.4~11.0
	JL1	Mean	42.2	93.1	180.5	58	0.284	14.1	0.124	17.9
	(n=34)	Range	24.8~54.9	70.1~121.0	129.0~216.0	34.6~81.6	0.150~0.400	9.9~17.0	0.060~0.200	16.5~19.7
	JL2	Mean	42.8	37.9	182	70.1	0.236	12.4	0.098	18.3
	(n=34)	Range	25.7~58.3	66.9~133.0	134.0~228.0	56.0~82.9	0.110~0.430	10.4~15.9	0.048~0.150	17.5~20.0
South	ZJ1	Mean	29.6	34.2	73.4	58.5	0.357	17.7	0.077	8.7
Sea	(n=34)	Range	12.3~51.9	21.2~57.9	42.2~94.8	42.6~75.0	0.200~0.498	12.0~25.0	0.038~0.120	5.1~12.5
	ZJ2	Mean	57	73.6	90.7	86.6	0.917	26.6	0.208	13.6
	(n=34)	Range	30.9~84.2	42.3~102.0	21.2~139.0	59.0~106.0	0.330~1.280	13.6~36.8	0.089~0.340	7.0~18.0
	YL1	Mean	3.2	16.3	31.7	45.7	0.055	7.2	0.031	5.3
	(n=34)	Range	1.0~8.7	11.9~20.6	26.0~37.8	30.2~67.2	0.040~0.096	3.6~10.8	0.014~0.073	3.6~8.2
	YL2	Mean	2.2	15.7	33.7	54.8	0.075	7.7	0.016	4.7
	(n=34)	Range	1.1~4.5	11.8~19.1	20.9~59.8	34.0~74.3	0.040~0.130	3.5~14.4	0.005~0.032	1.9~7.0
	DZ1	Mean	5.1	19.7	51.8	43.6	—	5.9	0.015	7.1
	(n=34)	Range	1.1~7.8	17.4~21.6	42.0~60.9	32.9~50.0	—	4.4~7.1	0.001~0.022	5.9~8.5
	DZ2	Mean	4.1	18.4	47.1	39	—	4.6	0.009	5.7
	(n=34)	Range	1.0~11.0	12.4~23.2	30.0~60.1	32.9~46.5	—	2.8~9.0	0.004~0.013	2.2~7.9

cities, and the poorly regulated economic development model was accompanied by pollutants. For the second characteristic type, the toxic metal contents oscillated from the bottom to the surface, including sediment cores CX1, CX2, DT1, DT2, DY1, DY2, FZ1, FZ2, HG1, HG2, and YC1. This characteristic was mostly due to the direct impacts of human activities, such as reclamation projects and port construction, on sediments. For the third characteristic type, the concentrations of toxic metals had a downward trend from the bottom to the surface, including sedimentary cores YC2, ZJ1, ZJ2, YL1, YL2, DZ1, and DZ2. The reason for this result was the continuous improvement of the ecological environment. In general, the distributions of toxic metals in the offshore sediments of the Bohai Sea and the Yellow Sea belonged to the first type, the East China Sea offshore sediments belonged to the second type, and the South China Sea offshore sediments belonged to the third type.

Cu, Pb, Zn, Cr and Hg average contents showed similar spatial variation trends. The left vertical axis of the Fig.2 shows the content of Cu, Pb, Zn, Cr and As, and the right vertical axis shows the content of Cd and Hg. Areas with high contents of these toxic metals were in the East China Sea or Yellow Sea, while the areas with the lowest values were found in the South China Sea. Regions with elevated contents of As and Cd were in the South China Sea, followed by the Yellow Sea, with the lowest contents in the Bohai Sea. In general, the sediments in the South China Sea were relatively clean, but the sediment contents of toxic metals in some areas (ZJ) increased the average values of the South China Sea.

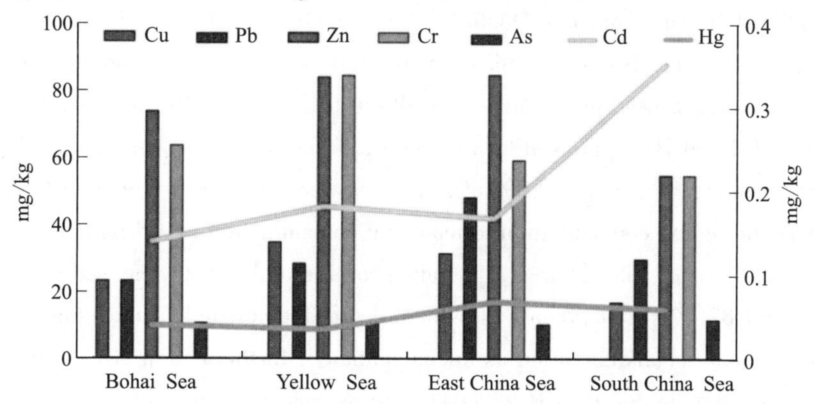

Fig.2 Heavy metal concentrations in different sea areas

5 Discussion

5.1 Heavy metal enrichment analysis

The sediment chemical composition and particle dimension characteristics regulate the contents of trace elements (Thuy et al., 2000). The enrichment factor (EF) is used to remove the effects of the above two aspects, thereby providing an indication of sediment trace element enrichment (Loska et al., 2003).

Clearly, the EF values of toxic metals in most intertidal sediments were lower than 2 (Fig.3). Compared with the sediments in the other regions, those in southeast coastal regions were significantly disturbed by human, which is consistent with the prosperity of China's economic development. Among them, the most seriously disturbed region was the Pearl River Delta (ZJ), where the EF values of Cu, Pb, As, and Hg in the sediments all exceeded 2, indicating moderate enrichment, while Cd reached significant enrichment. Similarly, studies have shown that the mean mass ratios of Cu, Cd, and Pb elements in the soils of the Pearl River Delta estuary are higher than those in Quaternary sediments (Fu et al., 2019). The average mass ratios for this site were followed by the Jiulong River Estuary (JL) and the Yangtze River Delta (DT), both of which had moderate enrichment for 3 elements. In addition, Pb in the Minjiang Estuary (FZ) reached significant enrichment. According to the 2020 Environmental Quality Bulletin in Offshore Areas of China, China's ocean dumping volume has increased annually, and dumping activities have mainly occurred in the coastal waters of Guangdong and the Yangtze River Estuary, which supports our conclusions.

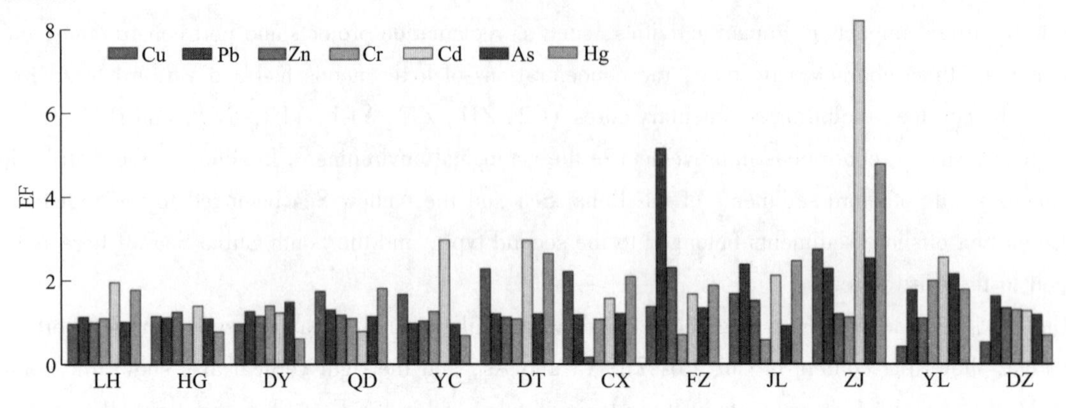

Fig.3 Enrichment factor (EF) values of heavy metals

5.2 Geoaccumulation index analysis

The sediment cores were categorized into two categories according to toxic metal contents: I_{geo} values < 0, i.e., uncontaminated, and I_{geo} values > 0, i.e., contaminated.

According to the 2020 Environmental Quality Bulletin in Offshore Areas of China, most of the main rivers and bays were in a subhealthy state. Some intertidal zones are slightly polluted by individual toxic metals, and a few intertidal zones are polluted by multiple toxic metals. In the Liaohe Estuary (LH), the average I_{geo} values of Cd and Hg in the sediments were 0.25 and 0.02, respectively, and other toxic metals were lower than 0. In total, 14.0% and 9.3% of sediments were moderately contaminated, and 50.0% and 39.5% of sediments showed no contamination to moderate contamination by Cd and Hg, respectively. In the Hangu intertidal (HG) zone, the average I_{geo} values of Zn and Cd in the sediments were 0.12 and 0.26, with 88.4% and 82.2% of sediments classified as uncontaminated to moderately contaminated, respectively, and other toxic metals classified as uncontaminated. In the Yellow River Delta (DY), all sediments were uncontaminated. In Jiaozhou Bay (QD), the average I_{geo} values of Cu, Pb and Hg in the sediments were 0.40, 0.001 and 0.39, respectively, and other toxic metals were lower than 0. In total, 1.8%, 26.3% and 7.9% of the sediments were moderately contaminated, and 86.8%, 25.4% and 75.4% of the sediments were uncontaminated to moderately contaminated with Cu, Pb and Hg, respectively. In Yancheng Shoal (YC), the average I_{geo} values of Cu and Cd in the sediments were 0.24 and 1.07, respectively, and other toxic metals were lower than 0. In total, 1.2% and 62.4% of sediments were moderately contaminated, and 37.6% and 76.5% of sediments showed no contamination to moderate contamination by Cd and Cu, respectively. The average I_{geo} values of Cu, Cd, and Hg in sediments of the Yangtze River Delta (DT) were 0.75, 1.11 and 0.94, respectively, and those of other toxic metals were lower than 0. Twenty-five percent, 62.7%, and 46.3% of sediments were moderately contaminated, and 74.6%, 37.3%, and 52.2% showed no contamination to moderate contamination by Cu, Cd and Hg, respectively.

In Hangzhou Bay (CX), the average I_{geo} values of Cu, Cd, As and Hg in the sediments were 0.89, 0.40, 0.02 and 0.80, respectively, and other toxic metals were lower than 0. A total of 29.4% and 22.1% of sediments were moderately contaminated with Cu and Hg, and 70.6%, 98.5%, 58.8% and 77.9% of sediments showed no contamination to moderate contamination by Cu, Cd, As and Hg, respectively. In the Minjiang Estuary (FZ), only the I_{geo} value of Pb in the sediments was higher than 0, with 2.9% and 97.1% of sediments classified as moderately contaminated and uncontaminated to moderately contaminated, respectively. In the Jiulong River Estuary (JL), the average I_{geo} values of sediment Cu, Pb, Zn, Cd, As and Hg were 0.98, 1.51, 0.85, 1.28, 0.13 and 1.51, respectively, and the average I_{geo} value of Cr was lower than zero. A total of 1.5%, 1.5%, and 7.4% of sediments showed moderate to strong contamination by Pb, Cd and Hg, respectively; 60.3%, 98.5%, 33.8%, 73.5% and 77.9% of sediments showed moderate contamination by Cu, Pb, Zn, Cd and Hg, respectively; and 39.7%, 66.2%, 25.0%, 76.5% and 14.7% of sediments showed no contamination to moderate contamination by Cu, Zn, Cd, As and Hg, respectively. In the Pearl River Delta (ZJ), the average I_{geo} values of Cu, Pb, Cd, As and Hg in the sediments were 0.89, 0.62, 2.42, 1.28, 0.81 and 1.65, respectively, and the mean I_{geo} values of Zn and Cr were below zero. There were 36.8% and 16.2% of sediments that were strongly contaminated with Cd and Hg, respectively; 1.5%, 32.4% and 14.7% of sediments showed moderate to strong contamination by Cu, Cd and Hg, respectively; 45.6%, 30.9%, 29.4%, 38.2% and 47.1% of sediments were moderately contaminated with Cu, Pb, Zn, Cd, As and Hg, respectively; and 42.6%, 45.6%, 1.5%, 61.8% and 22.1% of sediments showed no contamination to moderate contamination by Cu, Pb, Zn, Cd, As and Hg, respectively. In Yingluo Bay (YL), the average I_{geo} values of all toxic metals were less than 0, but some sediments showed no contamination to moderate contamination by Cd, As and Hg. In Dongzhai Port (DZ), all sediments were uncontaminated.

In summary, we infer that the sediments in China's intertidal zones are generally clean. The Jiulong River Estuary(JL) and the Pearl River Estuary(ZJ) were polluted by a variety of heavy metals, the most prominent elements were Cd and Hg, followed by Pb and Cu.

5.3 Sediment quality guideline analysis

The toxic metal contents of the sediments of twelve typical Chinese intertidal zones were shown to have moderate effects with potential damaging biotoxic impacts. Compared with other regions, the potential ecological hazards of sediments in the Pearl River Delta and Jiulong River Estuary were relatively high. The difference was that in the Liaohe Estuary (LH), Minjiang Estuary (FZ), Yingluo Bay (YL) and Dongzhai Port (DZ), 5.71%, 8.82%, 23.5% and 14.7% of the sediments, respectively, had no effect and no damaging biotoxic impacts (Table 4), while the remaining sediments showed moderate impact with potential damaging biotoxic impacts.

The ranking of the twelve intertidal zones in terms of potential ecological risk due to toxic metals was as follows: Xiamen Jiulong Estuary (JL) > Pearl River Delta (ZJ) > Qingdao Jiaozhou Bay (QD) > Tianjin Hangu intertidal zone (HG) > Yangtze River Delta (DT) > Jiangsu Yancheng Shoal (YC) > Zhejiang Hangzhou Bay (CX) > Liaohe River Delta (LH) > Yellow River Delta (DY) > Fujian Minjiang Estuary (FZ) > Guangxi Yingluo Bay (YL) > Hainan Dongzhai Port (DZ).

Table 4 Statistical characteristics of SQG-Q values in the sediment cores.

Areas	LH	HG	DY	QD	YC	DT	CX	FZ	JL	ZJ	YL	DZ
Mean	0.20	0.26	0.16	0.27	0.23	0.25	0.23	0.13	0.40	0.36	0.12	0.11
Range	0.08~0.33	0.18~0.29	0.14~0.19	0.18~0.36	0.17~0.31	0.19~0.33	0.19~0.30	0.09~0.21	0.30~0.52	0.20~0.61	0.09~0.16	0.08~0.14

6 Conclusions

The present study characterized the vertical distributions and contaminations of sediment toxic metals in typical Chinese intertidal zones. The average contents of Cu, Pb, Zn, Cr, Cd, As and Hg were 26.2, 33.9, 74.3, 63.4, 0.195, 10.7 and 0.053 mg/kg, respectively. The study shows that there were three types of vertical distribution of metal content in typical intertidal sediments in China. The first characteristic type was the toxic metal concentrations tended to increase from the bottom to the surface mainly including the Bohai Sea and the Yellow Sea coast sediment. The second characteristic type was the toxic metal contents oscillated from the bottom to the surface mainly including the East China Sea coast sediments. The third characteristic type was the concentrations of toxic metals had a downward trend from the bottom to the surface mainly including the South China Sea coast sediments.

Sediment quality analysis based on the geoaccumulation index and sediment quality guidelines indicated a generally good quality sedimentary environment in the Chinese intertidal zone, with only partial contamination. Compared with other regions, sediments in southeast coastal areas of China were polluted by some heavy metals, mainly Cd and Hg, followed Cu and Pb.

References

Chi, Q.H., Yan, M.C., 2007. Handbook of elemental abundance for applied geochemistry. Geological Publishing House Beijing China. 2007.96-97 (in Chinese). ISBN 978-7-116-05536-0.

Fu S.Q., Wei Z.Q., Yuan S.X., Xiong H.X., Guo Y.H., 2019. Heavy metal pollution features and potential ecological risk assessment in sediments and soils of the Pearl river estuary. Journal of Safety and Environment 19, 600-606 (in Chinese with English abstracts).

Gao, S., 2009. Modeling the preservation potential of tidal flat sedimentary records, jiangsu coast, eastern china. Continental Shelf Research 29, 1927-1936.

Harikrishnan, N., Ravisankar, R., Chandrasekaran, A., Suresh, G. M., Kanagasabapathy, K. V., Prasad, M. V. R., Satapathy, K.K., 2017. Assessment of Heavy Metal Contamination in Marine Sediments of East Coast of Tamil Nadu Affected by Different Pollution Sources. Marine Pollution Bulletin 121, 418-424.

Hartmann, P.C., Quinn, J.G., Cairns, R.W., King, J.W., 2005. Depositional history of organic contaminants in Narragansett Bay, Rhode Island, USA. Marine Pollution Bulletin 50, 388-395.

He, Y., Hong, X., Bi, X.Y., Chen, C.F., Lu, Q., Pan, R.X., Tian, Y., Chen, B., 2021. Characteristics and sources of heavy metal pollution in water environment of Jiuzhou River basin. Environmental Chemistry 1, 240-253 (in Chinese with English abstracts).

Kang, X., Song, J., Yuan, H., Duan, L., Li, X., Li, N., Liang, X., Qu, B., 2017. Speciation of heavy metals in different grain sizes of Jiaozhou Bay sediments: Bioavailability, ecological risk assessment and source analysis on a centennial timescale. Ecotoxicology and Environmental Safety 143, 296-306.

Long, E.R., Field, L.J., Macdonald, D.D., 1998. Predicting toxicity in marine sediments with numerical sediment quality guidelines. Environmental Toxicology and Chemistry 17, 714-727.

Loska, K., Danuta, W., Korus, I., 2003. Metal contamination of farming soils affected by industry. Environment International 30, 159-165.

Li, F.Y., Li, X.G., Qi, J., Song, J.M., 2011. Accumulation of heavy metals in the core sediments from the Jiaozhouwan Bay during last hundred years and its environmental significance. Journal of Marine Sciences 29, 35-45 (in Chinese with English abstracts).

Liu L., Wang H.J., Yue Q., 2021. Current situation and management countermeasures of coastline protection and utilization in China. Marine Environmental Science 33, 723-731 (in Chinese with English abstracts).

Liu, Q., Liang, L., Wang, F., Y., Liu, F., 2013. Characteristics of heavy metals pollution in sediments of the hydrofluctuation belt in the Liao River of Liaoning Province, Northeast China. China Environmental Science 33, 2220-2227 (in Chinese with English abstracts).

Macdonald, D.D., Carr, R.S., Calder, F.D., Long, E.R., Ingersoll, C.G., 1996. Development and evaluation of sediment quality guidelines for Florida coastal waters. Ecotoxicology 5, 253-278.

Mckee, B.A., Aller, R.C., Allison, M.A., Bianchi, T.S., Kineke, G.C., 2004. Transport and transformation of dissolved and particulate materials on continental margins influenced by major rivers: benthic boundary layer and seabed processes. Continental Shelf Research 24, 899-926.

Muller, G., 1969. Index of geoaccumulation in sediments of the rhine river. Geo Journal 2(3), 109-118.

Pennington, W.,Tutin, T.G.,Cambray, R.S., Fisher, E.M., 1973. Observations on lake sediments using fallout 137cs as a tracer. Nature 242, 324-326.

Punniyakotti, J., Ponnusamy, V., 2018. Environmental radiation and potential ecological risk levels in the intertidal zone of southern region of Tamil Nadu coast (HBRAs), India. Marine Pollution Bulletin 127, 377-386.

Sakan, S.M., Dordevic, D.S., Manojlovic, D.D., Predrag, P.S., 2009. Assessment of heavy metal pollutans accumulation in the Tisza river sediments. Journal of Environmental Management 90, 3382-3390.

Sutherland, R. A., 2000. Bed sediment-associated trace metals in an urban stream, oahu, hawaii. Environmental Geology39(6), 611-627.

Sun, Z.G., Mou, X.J., Tong, C., Wang, C.Y., Xie, Z.L., Song, H.L., Sun, W.G., Lv, Y.C., 2015. Spatial variations and bioaccumulation of heavy metals in intertidal zone of the Yellow River estuary, China. Catena 126, 43-52.

Thuy, H.T., Tobschall, H.J., An, P.V., 2000. Trace element distributions in aquatic sediments of Danang-Hoian area. Vietnam. Environmental Geology 39, 733-740.

Viers, J., Dupré, B., Gaillardet, J., 2009. Chemical composition of suspended sediments in World Rivers: new insights from a new database. Science of the Total Environment 407, 853-868.

Wang, X., Sato, T., Xing, B., Tao, S., 2005. Health risks of heavy metals to the general public in Tianjin, China via consumption of vegetables and fish. Science of the Total Environment 350, 28-37.

Xu, B., Yang, X., Gu, Z., Zhang, Y., Chen, Y., Lv, Y. , 2009. The trend and extent of heavy metal accumulation over last one hundred years in the Liaodong Bay, China. Chemosphere 75, 442-446.

Yu, W.J., Zou, X.Q., Zhu, D.K., 2011. Caofeidian Laolongkou modern sedimentary environments and heavy metal pollution. China Environmental Science 31, 1366-1376 (in Chinese with English abstracts).

Zhao, M.W., Wang, E.K., Xia, P., Feng, A.P., Chi, Y., Sun, Y.G., 2019. Distribution and pollution assessment of heavy metals in the intertidal zone environments of typical sea areas in China. Marine Pollution Bulletin 138, 397-406.

Zhang, L.P., Ye, X., Feng, H., Jing, Y.H., Ouyang, T., Yu, X.T., Liang, R.Y., Gao, C.T., Chen, Q.P., 2007. Heavey metal contamination in western Xiamen Bay sediments and its vicinity, China. Marine Pollution Bulletin 54, 974-982.

Zhuang, H.H., Gao, M.S., Yan, C.F., Cao, Y., Zhang, J., 2020. Vertical profiles and contamination assessments of heavy metals in sediment cores from typical intertidal zones in northern China. Marine Pollution Bulletin. 159, 111442.

Zhuang, H.H., Xu, S.H., Gao, M.S., Zhang, J., Hou, G.H., Liu, S., Huang, X.Y., 2018. Heavy metal pollution characteristics in the modern sedimentary environment of northern Jiaozhou Bay, China. Bulletin of Environmental Contamination and Toxicology 101, 473-478.

本文曾发表于2022年《Marine Pollution Bulletin》第185卷

Contamination and Health Risk Assessment of Heavy Metals form a Typical Pb-Zn Smelter in Northwest China

Yantao Hu, Defeng Wu, Jinbao Liu, Chaofan Yan, Jiangfeng Sun

(Shaanxi Provincial Land Engineering Construction Group Co. Ltd., Xi'an, Shaanxi, 710075, China)

【**Abstract**】Soil contamination by heavy metals due to metal smelting activities poses a serious threat to the ecological environment and to human health, as it is considered to be one of the most significant sources of soil pollution. The objective of this study was to analyze the pollution status and human health risks caused by heavy metals emitted from metal smelting activities of a Pb-Zn smelter. Contamination levels were evaluated using the potential ecological risk index (RI). Human health risks were assessed using the health risk assessment model developed by the US EPA. The results showed that the soils are seriously polluted, and migrated down the soil vertical profile. The index of RI indicated a very high potential ecological risk overall in the entire study area, especially, Cd. The health risk analysis showed that adults and children are exposed to significant noncarcinogenic health risks, and there are higher noncarcinogenic health risks for children than for adults. Additionally, the carcinogenic risks of Cr were higher than those of Cd for the two population groups, and children were more susceptible than adults, which should receive greater attention. These results are useful for management, prevention, control and remediation of heavy-metal contamination. Meanwhile, this research provides methods, experiences, and reference to other study of similar heavy-metal soil pollution.

【**Keywords**】Heavy metal; Ecological risk; Health risk; Non-carcinogenic risks; Soil pollution

1 Introduction

With the rapid development of industrialization and urbanization in developing countries, large amounts of heavy metals (HMs) produced by anthropological activities enter into the environmental medium, which becomes polluted or causes adverse ecological effects when it exceeds the load of the environmental medium (Gao et al. 2014; Salmanighabeshi et al. 2015; Agomuo et al. 2017; Li et al. 2018). Currently, soil contamination by HMs has more invisibility and great harmfulness and is regarded as the most adverse environmental issue in the universe, not only because of its acute and chronic toxicity to plants, animals, microorganisms, and the ecosystem but also because of its environmental persistence, bioaccumulation, non-degradable and slow removal process (Islam et al. 2015; Islam Md et al. 2018; Nkansah et al. 2017; Moghtaderi et al. 2018). Numerous previous studies have shown that HM pollution in soil has been both serious and widespread in many areas in China, which has become a severe obstacle for regional economic and social development and human health (Li et al. 2014; Li et al. 2016; Padoan et al. 2017; Wu et al. 2018; Steffan et al. 2018; Xu et al. 2018; Yang et al. 2018). According to the State Environmental Protection Administration, China faces serious soil HM pollution; approximately 10 million m^2 of arable land has been polluted, and 12 million tons of grains have been contaminated by HMs in the soil in China (Teng et al. 2010; CSC 2012; Chen et al. 2015; Li et al. 2018). The HM pollution in China has drawn worldwide attention. Many investigations have confirmed that mining activities (including excavating, crushing, grinding, separation, smelting, refining and tailings) are the primary source of HMs in the environment, which pose the greatest potential risk to human health and the environment (Ramana et al. 2012 and 2013; Ettler et al. 2014; Li et al. 2015; El Azhari et al.

2017; Shen et al. 2017; Ahirwar et al. 2018; Gu et al. 2018; Lee et al. 2018; Zhu et al. 2018). In many pollution sources and paths, activities associated with mining, including industrial mining, metal flotation, smelting and processing, artisanal gold mining, and uranium mining, have been regarded as four of the world's ten pollution problems (Ericson et al. 2008; Csavina et al. 2012). All mining exploitation, including mining, crushing, grinding, screening, smelting, refining, casting, metal processing and tailings management, and even including the transportation of ore, produce large quantities of dust and aerosols with high levels of heavy metals, which are released into the air and deposited as dust. Atmospheric particles discharged into the air by mining activities are as an important component of air pollution, and even affect the entire biosphere, including atmosphere, hydrosphere, and pedosphere. Mineral dust is one of the primary contributors of atmospheric aerosol. Some research has suggested that dust and aerosols from mining activities are normally associated with significantly elevated levels of one or more of these contaminants, including Pb, Cr, Hg and As (Meza-Figueroa et al. 2009; Brotons et al. 2010; Corriveau et al. 2011). A great deal of dust loaded high levels of heavy metals can be released into the air and deposited on the surface of the soil as dust as a result of mining activities, including mining, crushing, grinding, screening, smelting, refining and tailings management, and enter into soil via deposition and precipitation (Csavina et al. 2012; Li et al. 2015).In particular, the smelting of metal ores is consider as one of the most serious sources in all HM pollution sources. The atmospheric particulates with high levels of metal and metalloid not only can cause a substantial impact on the environment, but also pose a threat to human health, and the magnitude of this impacts and harms is dependent both on the pollution concentration and the size of the dust particles (Nriagu and Pacyna 1988; Li et al. 2015; Li et al. 2016; Li et al. 2018(a); Li et al. 2018). The smelting of ore concentrates powder causes large quantities of Pb, Zn, Cd, Cu, As and Hg, and other elements to be released into the environment, which can cause bioaccumulation and biomagnifications in the ecosystem (Shang et al. 2017). Pb, Cd, Cu, Cr and As have been identified as poisonous and harmful heavy metallic elements to human health by the World Health Organization (WHO) (Song et al. 2015). In addition, a latest assessment report for the global health impacts of contaminants indicated Pb, Cr, As, Hg, pesticides, and radionuclides as the six most toxic pollutants that threaten human health (McCartor and Becker 2010). Many investigations have indicated that there is a relationship between mortality and living near mining and smelting areas (Hawkesworth et al. 2013; Song et al. 2013; Song et al. 2015). The dust and aerosol particles from mining activities may carry highly toxic metallic and nonmetallic elements, including the neurotoxic elements such as Pb and As, which are easy to accumulate in sediment and vegetation. There are three main size ranges in atmospheric dust and aerosol, including ultrafine, accumulative and coarse, and all of these types of patterns are closely related to mining-related emission (Kříbek et al. 2010; Csavina et al. 2012). Among them, ultrafine particles are mainly generated from hot vapors in the smelting furnace, which diffuse quickly into the air, and they would collide and coagulate into larger particles at residence times in the air of minutes to hours, form accumulative particles. The accumulative particles are too large to diffuse or coagulate in a short time, but they are too small to settle by gravity, so they remain at an average residence time of 8~10 days in the air. However, coarse dust are mainly generated by crushing and grinding of ore and wind erosion of mine tailings, which settle rapidly into soil and water in minutes to hours. Researches have also confirmed that the particle sizes of dust and aerosols can affect the deposition efficiency (Krombach et al. 1997; Park and Wexler 2008; Valiulis et al. 2008; Csavina et al. 2012). Besides, epidemiological studies showed that ultrafine dust may has much effect on the health (Shaheen et al. 2005; Moreno et al. 2006; Querol et al. 2006; Csavina et al. 2006). Moreover, heavy metal elements in soil and atmospheric particulates easily enter into the human body by inhalation, ingestion and dermal contact, and might lead to poisoning or even death if people excessively intake of these elements, especially in children (Lu et al. 2009; Ali Ubaid et al. 2017; Doabi et al. 2018; Li et al. 2018(a); Steffan et al. 2018). In recent years, the problem of HM pollution have become increasingly serious, and protecting environment from pollution and ensuring people to keep healthy have become a problem needed to address urgently (Duan et al. 2016; Akopyan et al. 2018; Li 2018). Although some studies have analyzed and assessed the pollution levels, spatial distribution state, potential risks, and health risks of heavy metals from mining and smelting area soil, the regions of heavy metals contamination from mining activities have received relatively limited attention.

It is therefore imperative to continually monitor and assess the levels of HM contamination from mining activities, and study the pollution characteristics of HMs, evaluate their pollution degrees and assess the health risks to protect the soil environment and human health and to formulating pollution prevention strategies.

Baoji is rich in mineral resources of many varieties and is main a base of lead-zinc minerals in China. In the course of the exploitation of metal ore, the environment could been vulnerable to pollute in these areas and its surroundings. Emissions of heavy metals can pollute atmosphere, soils, surface water, groundwater, and food crops, even which can threaten the health to residents near mining areas. Feng County (33°34′50″~34°18′13″N, 106°24′19″~107°10′26″E) is located to the southwest of Baoji City in the Shaanxi Province of China. Feng County is much enriched in lead-zinc (Pb-Zn) mineral resources and deposits probably reached 4.5 million tons, as one of the four biggest Pb-Zn mineral bases in China (Shen et al. 2017; Fan et al. 2019). One of the largest Pb-Zn smelters in Baoji is located in Feng County. The Pb-Zn smelter lies in a canyon area, which is dominated by mountains. The refining dusts and exhaust gases are difficult to diffuse, and those refining dusts contain toxic and harmful heavy metals such as Pb, Zn, and Cd. Long-term mining activities have caused serious pollution of this area, and ever the accidents of excessive amounts lead in the blood occurred in 2012 (Shen et al. 2017; Fan et al. 2019). Shen et al. (2017) studied the physicochemical parameters of soil, spatial-temporal distributions of HMs and potential ecological risks in this smelter area three years ago. Even now, the Pb-Zn smelting activities are ongoing. Although the smelting process has been considerably improved and the metalliferous dust emission significantly decreased, the soil has still been contaminated in recent years. The soil contaminated by Pb-Zn smelting activities still needs to be further investigated, and this information is very important to control and manage the contaminated lands and to provide a theoretical basis for management, prevention, control and remediation of heavy-metal contamination in the future.

In order to supplement predecessor's research for the soil contamination nearby Pb-Zn smelting, obtain the most new information of the pollution status, the objectives of this study were to: (1) quantify the concentrations of metals (Zn, Pd, Cd, Cr, Cu and Mn) in the soil; (2) evaluate the enrichment degree of the metals (Zn, Pd, Cd, Cr, Cu and Mn); (3) assess the ecological risk of HMs; and (4) evaluate the health risks from exposure to HMs in the study area.

2 Materials and methods

2.1 Study area

The Pb-Zn smelter located in the northwest part of Feng County at longitude 106°32′2.69″E and latitude 33°56′43.02″N is approximately 3 km north of the county (Fig.1). The Pb-Zn smelter started to be built in 2000 and was started in 2001 by the Dongling Group subsidiary. The smelter mainly engaged in nonferrous metal smelting, sulfuric acid production, coking production, calcine and other zinc byproducts, with annual output of 6.0×10^4 tons zinc and 1.2×10^5 tons sulfuric acid in recent years. There was a village to the north at about 350 m; however, most of the inhabitants have long since been evacuated, leaving only a few people. The north soil near the smelter was once used for agriculture; however, this area was planted with poplar forests to currently suppress smelter dusts. The west is near the Hong Tang Shuang Road. The XiaoRui River flows through the west of the Pb-Zn smelter from north to south, which flows into the Jialing River. The Pb-Zn smelter is located at the bottom of a canyon, and mountains are to its east. The smelter is still currently in production. This area lies among the mountainous and mild climate, with an average annual temperature of about 11.4 ℃. The mean annual precipitation is approximately 613.2 mm. The annual dominant wind direction is east winds and southwest winds, with an annual mean wind speed of 0.7 m/s.

2.2 Soil sampling

Sampling was conducted from April to May 2017. Altogether 138 soil samples were collected using a stainless-steel drill from the soil around the Pb-Zn smelter in Feng County, including 46 surface soil (0~20 cm) and 92 related vertical profile soil (0~60 cm, with one soil sample was extracted per 20 cm). In order to make the taking of samples homogenous and representative, we collected 3 samples from each sampling site, and mixed together as one sample to provide the individual composite samples for the study. All the samples were placed in cloth bags respectively, and properly labeled and recorded, then transporting to our laboratory.

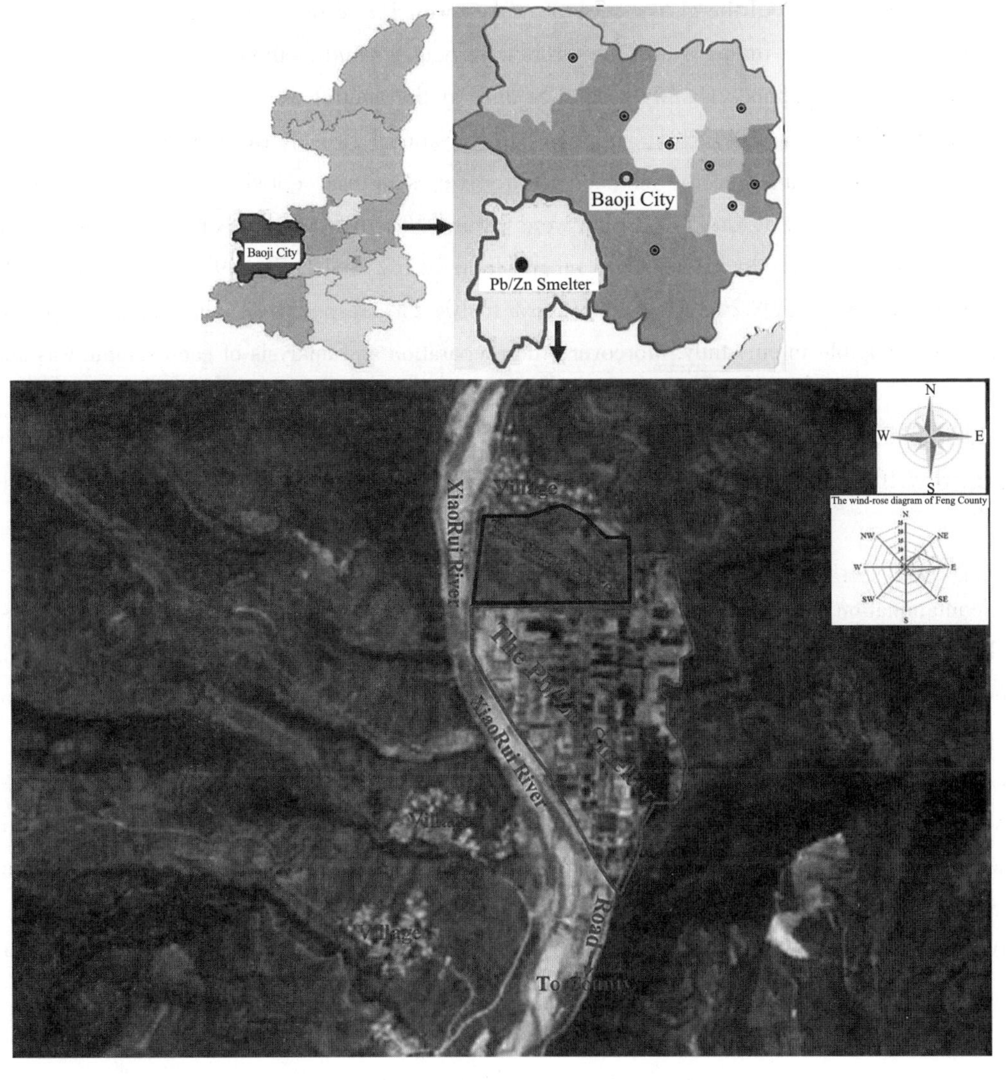

Fig. 1 The location of the studied regional

2.3 Experimental procedure

The collected soil samples were naturally dried to a constant weight in a dark place in the laboratory with indoor ventilation at room temperature. They were constantly crushed with hand in the natural drying process, and pick out stones, plant root residues and tree leaves. All the samples were crushed into power with a round wooden, and passed through a 0.15 mm (100 mesh) nylon sieve, storing in labeled cloth bags for analysis.

After a pretreatment, all soil samples were digested with $HCl-HNO_3-HF-HClO_4$ (volume ratio of 2:1:1:1) wet digestion. The detailed extraction procedure was as described in study of Fan et al. (2019). Finally, all extracting samples were filtered using a 0.45 μm pore size cellulose acetate filter, and the filtrate was collected into acid-washed polyethylene sample bottles for HM analysis. The content of Pb, Zn, Cd, Cr, Cu and Mn was determined using an air-acetylene flame atomic absorption spectrophotometer (SHIMADZU AA-6800).

2.4 Quality control

In this study, all reagents were guaranteed reagent (GR) grade, and all the chemical solution was prepared with ultra-pure water. All the glassware were soaked with 1% nitric acid for 24 h, then washed with ultra-pure water and dried in a drying oven. The errors from reagents and methods were reduced with analysis of replicates and the method blanks. Quality assurance and quality control were controlled using certified reference materials for the soils (GBW08301, supplied by the National Research Center for CRMs in China) that were used to verify the accuracy of the method. The recoveries were accepted when the determined standard concentrations for Pb, Cd, Zn, Cu, Cr and Mn were within 95%~105% of the certified limits. The recoveries of Pb, Cd, Zn, Cu, Cr and Mn in our study ranged from 98.6% to 103.7%, 97.5% to 103.2%, 96.2% to 105.9%, 99.3% to 104.9%, 98.2 to 104% and 95.3% to 104.2%, respectively. Therefore, the errors from instruments were negligible in our study. Moreover, the preparation and analysis of each sample was analyzed in triplicate.

To ensure the sensitivity and stability of analytical instruments, a standard reference solution was analyzed after every 10 samples. The $mg \cdot kg^{-1}$ in this study means the contents of HMs in per kilogram of dry soil.

2.5 Assessment methods for HM contamination

Soil contamination degree is usually assessed by comparing the measured values of pollution status for HMs with the geochemical background values. Currently, there are various indices for evaluating the pollution extent of HMs. In order to understand the level of pollution of HMs and the toxic effect of HM pollutants for the environment, the potential ecological risk (PER) index were used in this study.

On the basis of sedimentology, Håkanson (1980) established the potential ecological risk index (RI), which was introduced to evaluate the contamination level of analyzed HMs in sediments. RI is the total potential ecological risks of all HMs, representing the sensitivity of biology community to toxic substances and illustrating the potential ecological risk caused by the contaminants (Yi et al. 2011; Bahloul et al. 2018; Barkett et al. 2018; Izah et al. 2018; Li et al. 2018(b)). Håkanson (1980) established three Eqs. (1-3) to calculate RI. E_r^i(Eq. 2) is calculated based on the contamination factor (C_f^i) of the element (Eq. 3).

$$RI = \sum_{i=1}^{n} E_r^i \tag{1}$$

$$E_r^i = T_r^i \times C_f^i \tag{2}$$

$$C_f^i = \frac{C_s^i}{C_n^i} \tag{3}$$

Where n is the number of studied HMs; i is the ith studied element; RI is the potential ecological risk index of the HM; E_r^i is the potential risk factor for the individual HM; T_r^i is the toxic-response factor of an HM, which was given by Håkanson (1980) (i.e., Pb, Cu, and Ni=5, Zn=1, Cd=30, Cr=2, and As=10), accounting for the toxic requirement and the sensitivity requirement, reflecting the toxicity level and environmental sensitivity of the HM; C_f^i is the contamination factor; C_s^i is the actual concentration of the HM in the soil, mg · kg^{-1}; C_n^i is the mean background concentration of studied element, mg · kg^{-1}. The soil background values of elements used were those reported by Xue (1986) and the limiting value of II level standard of State Environment Standard (GB 156182—1995) (Table 2). According to Håkanson (1980), the E_r^i and RI can be classified into five categories, and classified as: low risk(E_r^i <40, RI<150); moderate risk(40≤ E_r^i <80, 150 ≤ RI < 300); considerable risk (80 ≤ E_r^i < 160, 300 ≤ RI < 600); high risk (160≤ E_r^i <320, 600≤RI<1200); very high risk(E_r^i ≥320, RI≥1200).

2.6 Health risk assessment of soil HMs

Human health risk assessment is to estimate the probability of adverse health effects in humans who may be exposed to harmful and toxic substances in contaminated environment (Li et al. 2014; Fan & Wang 2017; Li et al. 2018; Fan et al. 2019). Human health risk from direct exposure to the HM contaminated soil should not be ignored. In general, humans are three main pathways to expose in soil contaminated with HMs, including ingestion, inhalation and dermal contact (Fan & Wang 2017; Jiang et al. 2017; Li et al. 2018). Ingestion through the mouth is the highest of all exposure pathways caused by soil pollution. In order to systematically understand the adverse effects caused by soil contamination with HMs and to protect human health, we have the necessity to carry out human risk evaluations of soil contaminated with HMs. The steps of a health risk assessment are as follows: risk identification, dose-response estimation, exposure assessment, noncarcinogenic risk assessment and carcinogenic risk assessment. Seven HMs of Cd, Cr, As, Pb, Cu, Zn, and Ni were preferentially considered in the health risk assessment, mainly because these heavy metals are relatively strong toxicity to humans, and there are detailed and published dose-response relationships (Ordóñez et al. 2011; Jiang et al. 2017). The model used for human health risk assessment was originally formulated and recommended by the United States Environmental Protection Agency, and published the assessment guidelines and Exposure Factors Handbook of the US Environmental Protection Agency (USEPA 1986, 1989, 2001, 2002, 2003, 2004, 2011; Hadzi et al. 2018). In this study, the health risk assessment model recommended by the USEPA was used to evaluate the health risk from soil contaminated with HMs. In consideration of behavioral and physiological differences, the health risk assessment was divided into two groups of children and adults in this study.

In this study, the risk assessment to human health from the exposure of pollution was characterized using exposure assessment, noncarcinogenic risk and carcinogenic risk.

2.6.1 Exposed dose model

Human health exposure risk has close relation with exposure frequency, exposure time, exposure does, and exposure path. The purpose of exposure assessment is qualitative and quantitative to determine exposure risk from soil contaminated with HMs.

Dose-response assessment is to quantitatively evaluate the relationship between the exposure level of harmful factors and the incidence of health hazard effects on exposed humans, with the foundation for the quantification of the health risk assessment (Li et al. 2018). Different dose response may be due to the toxicity degrees of different elements and total intake of toxicity elements. Moreover, the behavioral and physiological effects of different people are different for different dose responses. Thus, this study divided the affected populations into children and adults, and respectively evaluates their health risk.

The risk exposure pathways caused by HM contaminated soils may occur in three main pathways: (a) direct ingestion of soil particles, termed ingestion; (b) inhalation of suspended particles through the mouth and nose, termed inhalation; and (c) dermal absorption of toxic elements from particles adhered to exposed skin (Ordóñez et al. 2011; Li et al. 2014). The research results of Ordóñez et al. (2011) showed that direct ingestion of soil particles is the most common risk exposure pathway for Pb, Cd, Zn, Cu, Cr, Ni and As for the mercury mining areas of Northern Spain. According to the human health risk evaluation manual (Part A) and supplemental guidance for dermal risk assessment (Part E) (USEPA 1989 and 2004), the average daily dose (ADD) of HMs via each pathway can be calculated as follows (Li et al. 2015; Han et al. 2017; Moghtaderi et al. 2018):

$$ADD_{ing}=C\times\frac{IR\times EF\times ED}{BW\times AT} \tag{4}$$

$$ADD_{inh}=C\times\frac{IR\times EF\times ED}{BW\times AT\times PEF} \tag{5}$$

$$ADD_{dermal}=C\times\frac{SA\times AF\times ABS\times EF\times ED}{BW\times AT} \tag{6}$$

$$ADD=ADD_{ing}+ADD_{inh}+ADD_{dermal} \tag{7}$$

where ADD_{ing}, ADD_{inh}, and ADD_{dermal} are the average daily intake doses of HMs from soil via ingestion, inhalation, and derma, respectively, with units of $mg\cdot kg^{-1}\cdot d^{-1}$; C is the measured concentration of HM in the soil, with units of $mg\cdot kg^{-1}$; ADD is the sum of the average daily intake soil doses via the three pathways; IR is the ingestion rate from soil contaminated by HMs, with units of $mg\cdot day^{-1}$; EF is the exposure frequency, with units of $days\cdot year^{-1}$; ED is exposure duration, with units of years; BW is the body weight of the exposed individual, with units of kg; AT is the average contact time, with units of day; SA is the individual exposed skin surface area, with units of $cm^2\cdot day^{-1}$; AF is the skin adherence factor, with units of $mg\cdot cm^{-2}\cdot day^{-1}$; PEF is the particle emission factor, with units of $m^3\cdot kg^{-1}$; and ABS is the dermal absorption factor, unitless.Table 1 shows the various parameter values for the two calculation formulas.

Table 1 Exposure dose of health risk assessment models

Factor	Values (children)	Values(adults)	Reference
EF	350 days · year^{-1}	350 days · year^{-1}	Environmental site assessment guideline (2009)
IR	200 mg · day^{-1}	100 mg · day^{-1}	USEPA, 2011
PEF	1.36×10^{9}	1.36×10^{9}	USEPA, 2011
ED	24 years	6 years	USEPA, 2011
SA	2800 cm^{2} · day^{-1}	5700 cm^{2} · day^{-1}	Environmental site assessment guideline (2009)
AF	0.2 mg · cm^{-2}	0.07 mg · cm^{-2}	USEPA, 2004
BW	15 kg	70 kg	Environmental site assessment guideline (2009)
AT	ED×365 (Noncarcinogenic); 70×365 (Carcinogenic)		USEPA, 1989
ABS	0.001	0.001	Chabukdhara and Nema, 2013

2.6.2 Noncarcinogenic risk model

According to the health risk evaluation model recommended by the USEPA, the human health risk from the HMs was classified into noncarcinogenic risk and carcinogenic risk. In this study, hazard quotient (HQ) was used for evaluation the noncarcinogenic risks caused by the contaminated soil with HMs. The values of hazard index (HI) equal to the sum of all HQs from the three main exposure pathways, with meaning the total potential noncarcinogenic risks of all the elements studied. HQ and HI were used to estimate the noncarcinogenic risk. The noncarcinogenic risks of the HMs are given as Formulas (8) and (9).

$$HQ_{ij}=\frac{ADD_{ij}}{RfD_j} \tag{8}$$

$$HI=\sum_{i=1}^{n}\sum_{j=1}^{3}HQ_{ij}=\sum_{i=1}^{n}\sum_{j=1}^{3}\frac{ADD_{ij}}{RfD_j} \tag{9}$$

where ADD_{ij} is daily intake of a certain toxic metal (i) through an exposure pathway (j); HQ_{ij} is the noncarcinogenic risk that estimates the risk level for the single element (i) in an exposure pathway (j), which equal to divide the average daily dose by a specific reference dose (RfD_j); RfD_j indicates the exposed populations intake the toxic elements maximum levels that didn't cause adverse reactions via an exposure pathway (j) in unit weight and unit time, with units of mg · kg^{-1} · day^{-1}, the values of RfD_j in this study are as follows: RfD_{ing}, Pb = 3.50×10^{-3}, Zn = 3.00×10^{-1}, Cd = 1.00×10^{-3}, Cr = 3.00×10^{-3}, Cu = 4.00×10^{-2}; RfD_{inh}, Pb = 3.52×10^{-3}, Zn = 3.00×10^{-1}, Cd = 1.00×10^{-3}, Cr = 2.86×10^{-5}, Cu = 4.02×10^{-2}; RfD_{dermal}, Pb = 5.25×10^{-4}, Zn = 6.00×10^{-2}, Cd = 1.00×10^{-5}, Cr = 6.0×10^{-5}, Cu = 1.20×10^{-2} (USEPA 1989, 2004; Bai et al. 2017; Li et al. 2018(b); Moghtaderi et al. 2018). HI represents the total noncarcinogenic risk from the three exposure pathways of all individual toxic metal; and i represent the different contaminants. Generally, HQ or HI < 1 means that there is no possibility of adverse health effects for exposed populations, whereas a HQ or HI > 1 may be possible adverse health effects (USEPA 1989).

2.6.3 Carcinogenic risk model

The cancer risks were used to signify the carcinogenic effects. The carcinogen risk (RI) reflects the caused cancer probability of the populations exposed to the potential carcinogen within the entire lifetime. In assessment models of RI, the values of RI represent a level of cancer risk, which are equal to the exposure doses of each exposure pathway are multiplied by the slope coefficient (SF). The SF shows the maximal probability of the carcinogenic effect for the human body upon exposure to a certain dose of pollutant, with units of mg · kg^{-1} · day^{-1} (USEPA 2002). According to the USEPA, Cd, Cr, Co and Ni are considered carcinogens only via inhalation, therefore, we only consider the carcinogenic risk of Cr and Cd in this study, and the SF values of the studied metals are SF_{inh}-Cd = 6.30 and SF_{inh}-Cr = 42.00. The carcinogenic risk levels are divided into five categories. RI values below 10^{-6} show there are no significant health effects, and this is also set as the maximum limit of the acceptable risk level for carcinogens by the USEPA. Then, $1\times10^{-6}\sim1\times10^{-5}$ indicates low risk, $1\times10^{-5}\sim1\times10^{-4}$ indicates medium risk, $1\times10^{-4}\sim1\times10^{-3}$ indicates high risk, and $>10^{-3}$ indicates very high risk and is perceived as being concerning and needs an effective method for reducing the exposure and resulting risk (Rapant et al. 2011; Li and Ji 2017; Han et al. 2017; Tepanosyan et al. 2017). The following formulas (10-11) are used to calculated the carcinogenic risk of Cr and Cd (USEPA 1989).

$$RI_{ij}=ADD_{ij}\times SF_{ij} \tag{10}$$

$$RI=\sum_{i=1}^{n}ADD_{ij}\times SF_{ij} \tag{11}$$

where RI_{ij} is the carcinogen risk of an i metal via an exposure pathway (j), SF_{ij} is the slope coefficient for a single element (i) through an exposure pathway (j), and RI is total carcinogen risk.

3 Results and discussion

3.1 Contamination level of HMs in soil

The concentrations of HMs in the 138 soil samples and the background values for the local soil are summarized in Table 2.

Table 2 Descriptive statistics of HMs content in soils

Depth/cm	Parameters	Pb	Zn	Cd	Cr	Cu	Mn
	Max/(mg · kg^{-1})	551.90	12505.80	161.53	69.98	61.75	338.89
0~20	Min/(mg · kg^{-1})	78.10	964.63	19.40	16.18	23.73	281.17
	Mean/(mg · kg^{-1})	225.42	4004.94	65.15	34.69	44.08	313.86
	Max/(mg · kg^{-1})	89.45	1160.30	17.48	69.28	52.13	327.31
20~40	Min/(mg · kg^{-1})	28.88	125.10	1.83	11.38	22.88	256.02
	Mean/(mg · kg^{-1})	48.18	409.70	6.38	34.84	31.90	285.41
	Max/(mg · kg^{-1})	62.23	484.73	7.25	53.50	47.10	343.77
40~60	Min/(mg · kg^{-1})	21.20	74.68	1.55	22.65	20.08	276.53
	Mean/(mg · kg^{-1})	38.83	215.27	3.47	36.02	28.96	299.13
Background values of Shaanxi		16.30	65.80	0.12	65.70	23.50	557.00
Grade Ⅱ standards		350	300	0.60	250	100	—

Note: Background value, based on a report on heavy metal content by Xue (1985) in agricultural soils of Guanzhong area, Shannxi Province, China; Grade Ⅱ standards = the Grade Ⅱ environmental quality standard for soils in China (GB 15618—1995).

As shown in Table 2, the results showed that the contents of heavy metals in soils were distinct changes. The range of concentration change of Zn, Pb, Cd, Cr, Cu and Mn in the soils of 0 ~ 20 cm was 964.63 ~ 12505.80, 78.10 ~ 551.90, 19.40 ~ 161.53, 16.18 ~ 69.98, 23.73 ~ 61.75 and 281.17 ~ 338.89 mg · kg^{-1}, respectively, and the mean concentrations were 4004.94, 225.42, 65.15, 34.69, 44.08 and 313.86 mg · kg^{-1}, respectively. The ranges of Zn, Pb, Cd, Cr, Cu and Mn in the soils of 20 ~ 40 cm were 25.10 ~ 1160.30, 28.88 ~ 89.45, 1.83 ~ 17.48, 11.38 ~ 69.28, 22.88 ~ 52.13 and 256.02 ~ 327.31 mg · kg^{-1}, respectively, and the mean concentrations were 409.70, 48.18, 6.38, 34.84, 31.90 and 285.41 mg · kg^{-1}, respectively. They were 74.68 ~ 484.73, 21.20 ~ 62.23, 1.55 ~ 7.25, 22.65 ~ 3.50, 20.08 ~ 47.10 and 276.53 ~ 343.77 mg · kg^{-1} in the soils of 20 ~ 40 cm, and the mean concentrations were 215.27, 38.3, 3.47, 36.02, 28.96 and 299.13 mg · kg^{-1}, respectively.

From this, we can see the mean concentrations of Cd, Zn, Pb and Cu were observably higher than the background values of Shaanxi Province, especially for Cd, Zn and Pb, at 0 ~ 20 cm, 20 ~ 40 cm and 40 ~ 60 cm. Furthermore, the mean concentrations, including Cd and Zn at 0 ~ 20 cm and 20 ~ 40 cm as well as Cd at 40 ~ 60 cm, far exceeded the soil environmental standard of National Second Grade (Pb ≤ 350, Zn ≤ 300, Cd ≤ 0.60, Cr ≤ 250, Cu ≤ 100) (GB 15618—1995), especially at 0 ~ 20 cm, and other elements did not exceed the soil environmental standard of National Second Grade for each layer. Mn did not exceed the soil environmental standard of National Second Grade and the background values of Shaanxi Province, largely because the average concentration range of Mn in soil of the world varies from 270 mg · kg^{-1} (in Podzoles) to 525 mg · kg^{-1} (in Cambisols) (Demková et al. 2017). The Cd concentration is very low in natural soil, and being often below 0.1 mg · kg^{-1} throughout the world (Baize and Sterckeman et al. 2001; Demková et al. 2017). The background value for Shaanxi Province is below 0.12 mg · kg^{-1}. The concentration of Cd and Zn exceeded all the low exceeded all the value of environmental standards the soil environmental standard of National Second Grade and the background values of Shaanxi Province at all sampling sites in our study area. The concentration of Cr in soil is generally lower in China, whereas the concentration of Cd in soil has been at high values in most cities of China (Wei and Yang 2010; Liu et al. 2018). But, Cd is one of the most toxic HMs, which can cause negative damange to human health and to the biodiversity and activity of soil microbial communities (Li et al. 2017; Demková et al. 2017; Fan et al. 2019).

3.2 Ecological risk assessment

For ecological risk assessment, we first calculated the monomial potential ecological risk index (E_r^i), which is the individual ecological risk factor associated with the contribution of HMs. On the basis of the E_r^i calculation, we calculated the potential ecological risk (RI). The calculation formula of RI synthetically considers HM toxicity, transfer and transformation of HMs within study areas, sensitivity to HM pollution, and differences in regional background values of HMs to remove the influence of regional differences and sources. The calculation results for E_r^i and RI are shown in Fig.2. Based on the above results, the contents of Mn in the soils was low and did not exceed the soil environmental standard of National Second Grade and the background values of Shaanxi Province, meaning no pollution and has thus been chosen as a background element in many studies. Meanwhile Mn will no longer be discussed with regard to the ecological risk assessment and health risk assessment.

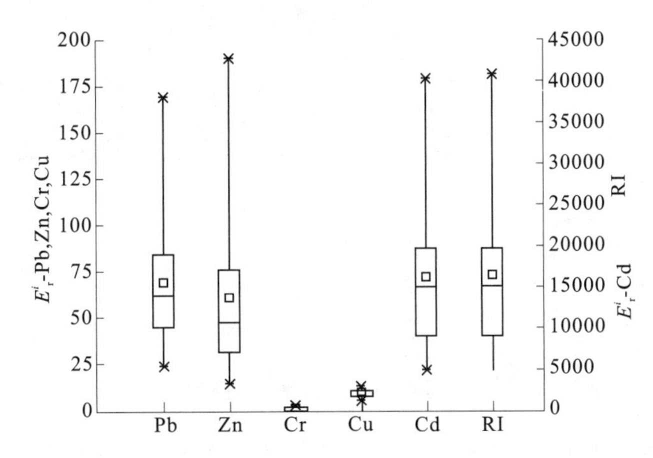

Fig. 2 Spatial distribution of ecological risk for HMs in soil near Pb-Zn smelter

Comparing the monomial potential ecological risk index (E_r^i) (Fig.2) with its grade classification, the E_r^i values for Cu and Cr were less than 40, showing a low potential ecological risk overall and that they hardly posed threats in the study area. However, among the five HMs, Cd presented the highest ecological risk as a result of its high toxicity factor, which was in the range of 4850.0~40381.25 with a mean value of 16286.81, mainly originated from the smelting activities of the Pb-Zn smelter. Suresh et al. (2012) also thought that nonferrous metal mining, refining and manufacture are the main anthropogenic sources of Cd in the environment. In addition, the E_r^i values of Zn and Pb were in the ranges of 14.66~190.06 and 23.96~169.29, respectively, between low risk and high risk. Overall, the individual potential risk for the average E_r^i for the HMs is Cd > Zn ≥ Pb > Cu > Cr. Additionally, the calculated RI values ranged from 4902.29 to 40753.80 with an average value of 16427.25, which showed a very high potential ecological risk primarily caused by Cd, Zn and Pb. In particular, there is a risk from Cd because of its high ecological toxicity. Therefore, this may require further attention when considering environmental remediation activities.

Furthermore, the spatial distribution of ecological risk for heavy metals is shown in Fig. 3. The spatial distribution characteristics of E_r^i and RI for Pb, Zn, Cd and Cu were consistent, which showed a high ecological risk overall in the entire study area and the highest near the smelter chimneys in the southeast and downwind of the smelter in the north. This suggests that the enrichment of metal concentrations even poses some threat to the ecological environment, which may be caused by smelting activities. Additionally, Cr was a low ecological risk overall in the entire study area, the spatial distribution pattern of Cr was different from the other tested metals, and the hot-spot areas of Cr were in the southeast part of the study region.

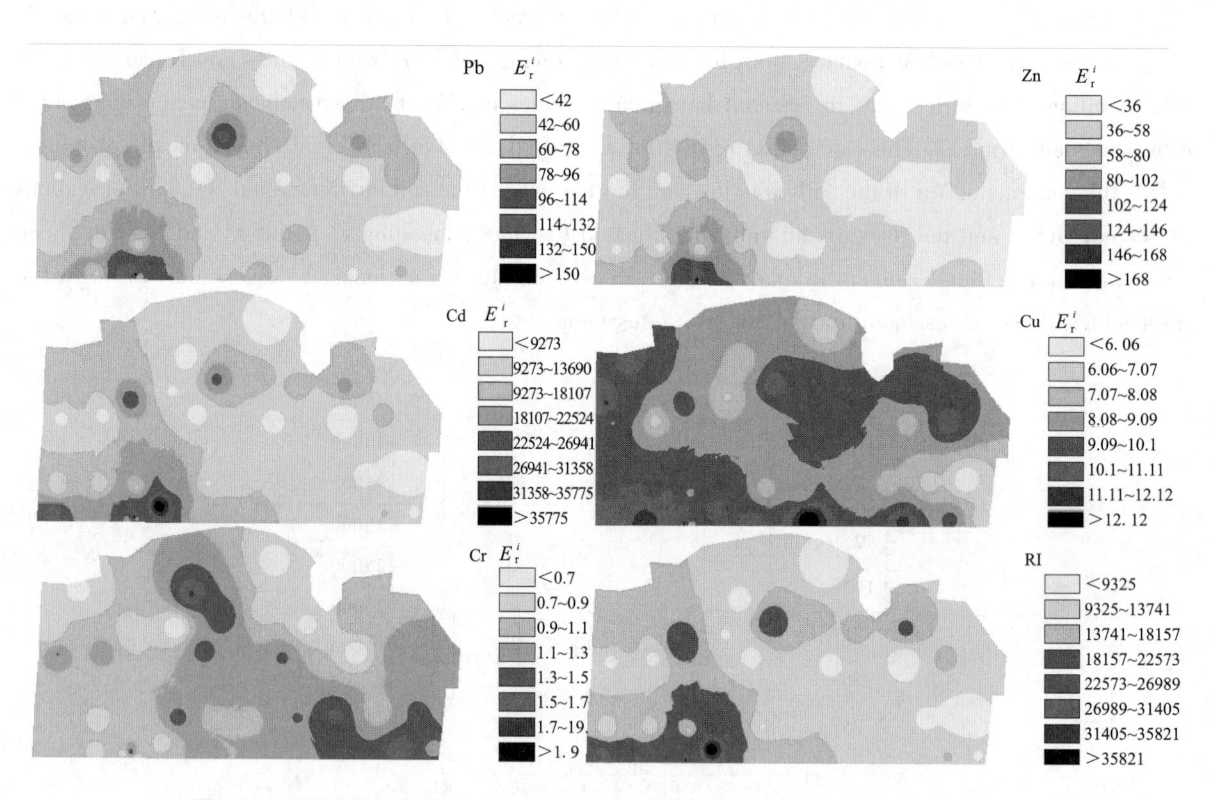

Fig. 3 Spatial distribution of ecological risk for HMs in soil near Pb-Zn smelter

3.3 Health risk assessment

3.3.1 Dose-response assessment

The average daily dose (ADD) of heavy metals via several pathways for children and adults from the soil near the Pb-Zn smelter are listed in Table 3.

Table 3 Daily dose of soil HMs in three models

Elements	Statistical metrics	Children			Adults		
		ADD_{ing}	ADD_{inh}	ADD_{dermal}	ADD_{ing}	ADD_{inh}	ADD_{dermal}
Pb	Max	7.06×10^{0}	5.19×10^{-9}	1.98×10^{1}	7.56×10^{-1}	5.56×10^{-10}	7.54×10^{-1}
	Min	9.99×10^{-1}	7.34×10^{-10}	2.80×10^{0}	1.07×10^{-1}	7.87×10^{-11}	1.07×10^{-1}
	Mean	2.88×10^{0}	2.12×10^{-9}	8.07×10^{0}	3.09×10^{-1}	2.27×10^{-10}	3.08×10^{-1}
Zn	Max	1.60×10^{2}	1.18×10^{-7}	4.48×10^{2}	7.99×10^{1}	5.88×10^{-8}	1.71×10^{1}
	Min	1.23×10^{1}	9.07×10^{-9}	3.45×10^{1}	6.17×10^{0}	4.53×10^{-9}	1.32×10^{0}
	Mean	5.12×10^{1}	3.77×10^{-8}	1.43×10^{2}	2.56×10^{1}	1.88×10^{-8}	5.47×10^{0}
Cd	Max	7.08×10^{-1}	5.21×10^{-10}	1.98×10^{1}	1.90×10^{-2}	1.39×10^{-11}	7.57×10^{-2}
	Min	8.50×10^{-2}	6.25×10^{-11}	2.38×10^{-1}	2.28×10^{-3}	1.67×10^{-12}	9.09×10^{-3}
	Mean	2.86×10^{-1}	2.10×10^{-10}	8.00×10^{-1}	7.65×10^{-3}	5.62×10^{-12}	3.05×10^{-2}
Cr	Max	3.07×10^{-1}	2.26×10^{-10}	8.59×10^{-1}	8.22×10^{-3}	6.04×10^{-12}	3.28×10^{-2}
	Min	7.09×10^{-2}	5.21×10^{-11}	1.99×10^{-1}	1.90×10^{-3}	1.40×10^{-12}	7.58×10^{-3}
	Mean	1.52×10^{-1}	1.12×10^{-10}	4.26×10^{-1}	4.07×10^{-3}	3.00×10^{-12}	1.63×10^{-2}
Cu	Max	7.89×10^{-1}	5.81×10^{-10}	2.21×10^{0}	8.46×10^{-2}	6.22×10^{-11}	8.44×10^{-2}
	Min	3.03×10^{-1}	2.23×10^{-10}	8.49×10^{-1}	3.25×10^{-2}	2.39×10^{-11}	3.24×10^{-2}
	Mean	5.64×10^{-1}	4.14×10^{-10}	1.58×10^{0}	6.04×10^{-2}	4.44×10^{-11}	6.02×10^{-2}

As outlined in Table 3, the average daily exposure intake of Pb, Zn, Cd, Cr and Cu in topsoil near the Pb-Zn smelter was as follows: for children, the exposure dose ($ADD_{ing}/ADD_{inh}/ADD_{dermal}$) ranges for Pb, Zn, Cd, Cr and Cu were $9.99\times10^{-1}\sim7.06\times10^{0}/7.34\times10^{-10}\sim5.19\times10^{-9}/2.80\times10^{0}\sim1.98\times10^{1}$, $1.23\times10^{1}\sim1.60\times10^{2}/9.07\times10^{-9}\sim1.18\times10^{-7}/3.45\times10^{1}\sim4.48\times10^{2}$, $8.50\times10^{-2}\sim7.08\times10^{-1}/6.25\times10^{-11}\sim5.21\times10^{-10}/2.38\times10^{-1}\sim1.98\times10^{1}$, $7.09\times10^{-2}\sim3.07\times10^{-1}/5.21\times10^{-11}\sim2.26\times10^{-10}/1.99\times10^{-1}\sim8.59\times10^{-1}$, and $3.03\times10^{-1}\sim7.89\times10^{-1}/2.23\times10^{-10}\sim5.81\times10^{-10}/8.49\times10^{-1}\sim2.21\times10^{0}$, respectively; for adults, the ranges were $1.07\times10^{-1}\sim7.56\times10^{-1}/7.87\times10^{-11}\sim5.56\times10^{-10}/1.07\times10^{-1}\sim7.54\times10^{-1}$, $6.17\times10^{0}\sim7.99\times10^{1}/4.53\times10^{-9}\sim5.88\times10^{-8}/1.32\times10^{0}\sim1.71\times10^{1}$, $2.28\times10^{-3}\sim1.90\times10^{-2}/1.67\times10^{-12}\sim1.39\times10^{-11}/9.09\times10^{-3}\sim7.57\times10^{-2}$, $1.90\times10^{-3}\sim8.22\times10^{-3}/1.40\times10^{-12}\sim6.04\times10^{-12}/7.58\times10^{-3}\sim3.28\times10^{-2}$, and $3.25\times10^{-2}\sim8.46\times10^{-2}/2.39\times10^{-11}\sim6.22\times10^{-11}/3.24\times10^{-2}\sim8.44\times10^{-2}$, respectively. Thus, the average daily dose of HMs for the children was significantly higher than that of adults for all five metals, which was similar to the study by Xiao et al. (2017). Thus, considerable attention should be paid to the risk exposure for children in daily life. Additionally, the ADDs for different exposure routes for children and adults were different: the ADD of Pb, Zn, Cd, Cr and Cu for children decreased in the order of dermal contact > ingestion > inhalation with dermal contact and ingestion playing the most important roles for children; however, the average daily intake of Pb, Zn and Cu for adults decreased in the order of ingestion > dermal contact > inhalation, and Cd and Cr decreased in the order of dermal contact > ingestion > inhalation. This result is in accordance with the true circumstances. For children, dermal contact is the main exposure pathway for Pb, Zn, Cd, Cr and Cu. In contrast, for adults, ingestion is the main exposure pathway for Pb, Zn and Cu; however, dermal contact is a more common exposure pathway for Cd and Cr. Analogously, Li et al. (2014) also deemed that dermal absorption is the main exposure pathway for Cd and Cr, whereas ingestion is a more common exposure pathway for Pb and Zn. Furthermore, the average daily intake of each toxic metal for children and adults via the three exposure routes followed the descending order of Zn > Pb > Cu > Cd > Cr.

3.3.2 Noncarcinogenic risk assessment

Too much exposure to elevated heavy metals has noncarcinogenic effects on human health. The HQ values for different population groups vary, as shown in Table 4.

Table 4 Hazard quotients of soil HMs for children and adults

Metals	Statistical metrics	Children				Adults			
		HQ_{ing}	HQ_{inh}	HQ_{dermal}	HQ	HQ_{ing}	HQ_{inh}	HQ_{dermal}	HQ
Pb	Max	2.02×10^{-3}	1.47×10^{-12}	3.76×10^{-4}	2.39×10^{-3}	2.16×10^{-4}	1.58×10^{-13}	1.44×10^{-5}	2.30×10^{-4}
	Min	2.85×10^{-4}	2.09×10^{-13}	5.33×10^{-5}	3.39×10^{-4}	3.06×10^{-5}	2.23×10^{-14}	2.03×10^{-6}	3.26×10^{-5}
	Mean	8.23×10^{-4}	6.02×10^{-13}	1.54×10^{-14}	9.77×10^{-4}	8.82×10^{-5}	6.45×10^{-14}	5.87×10^{-6}	9.41×10^{-5}
Zn	Max	5.33×10^{2}	3.92×10^{-7}	7.46×10^{3}	7.99×10^{3}	2.66×10^{2}	1.96×10^{-7}	2.85×10^{2}	5.51×10^{2}
	Min	4.11×10^{1}	3.02×10^{-8}	5.76×10^{2}	6.17×10^{2}	2.06×10^{1}	1.51×10^{-8}	2.20×10^{1}	4.25×10^{1}
	Mean	1.71×10^{2}	1.26×10^{-7}	2.39×10^{3}	2.56×10^{3}	8.53×10^{1}	6.28×10^{-8}	9.12×10^{1}	1.77×10^{2}
Cd	Max	7.08×10^{-4}	5.21×10^{-13}	1.98×10^{-5}	7.28×10^{-4}	1.90×10^{-5}	1.39×10^{-14}	7.57×10^{-7}	1.97×10^{-5}
	Min	8.50×10^{-5}	6.25×10^{-14}	2.38×10^{-6}	8.74×10^{-5}	2.28×10^{-6}	1.67×10^{-15}	9.09×10^{-8}	2.37×10^{-6}
	Mean	2.86×10^{-4}	2.10×10^{-13}	8.00×10^{-6}	2.94×10^{-4}	7.65×10^{-6}	5.62×10^{-15}	3.05×10^{-7}	7.95×10^{-6}
Cr	Max	1.02×10^{-4}	7.91×10^{-16}	1.43×10^{-6}	1.04×10^{-4}	2.74×10^{-6}	2.12×10^{-17}	5.46×10^{-8}	2.79×10^{-6}
	Min	2.36×10^{-5}	1.83×10^{-16}	3.31×10^{-7}	2.40×10^{-5}	6.33×10^{-7}	4.9×10^{-18}	1.26×10^{-8}	6.46×10^{-7}
	Mean	5.07×10^{-5}	3.92×10^{-16}	7.10×10^{-7}	5.14×10^{-5}	1.36×10^{-6}	1.05×10^{-17}	2.71×10^{-8}	1.38×10^{-6}
Cu	Max	1.97×10^{-3}	1.44×10^{-12}	1.84×10^{-2}	2.04×10^{-2}	2.11×10^{-4}	1.55×10^{-13}	7.03×10^{-4}	9.15×10^{-4}
	Min	7.58×10^{-4}	5.55×10^{-13}	7.08×10^{-3}	7.84×10^{-3}	8.13×10^{-5}	5.94×10^{-14}	2.70×10^{-4}	3.51×10^{-4}
	Mean	1.41×10^{-3}	1.03×10^{-12}	1.31×10^{-2}	1.46×10^{-2}	1.51×10^{-4}	1.1×10^{-13}	5.02×10^{-4}	6.53×10^{-4}

As shown in Table 4, the HQ values for Pb, Cd, Cr and Cu for adults and children via the different pathways were less than 1, and the total HQ values via the three pathways were less than 1. The HQ values for Zn via ingestion (HQ_{ing}) and dermal contact (HQ_{dermal}) and the total HQ values were greater than 1. These results illustrated that Cd, Pb, Cu and Cr had no possibility of adverse health effects for exposed populations (adults and children). However, Zn showed possible adverse health effects, and for children, it was greater than adults; thus, the risk of noncarcinogenic exposure for children cannot be ignored; however, it should not be exaggerated. In addition, according to the HQ values, it was obvious that children tended to have a higher probability than adults, indicating that children are more susceptible to environmental contaminants, which may be due to the behavioral and physiological characteristics of children. The HQ values of different heavy metals for children and adults was in the order of Zn> Cu > Pb > Cd> Cr. The HQ values for the three exposure pathways for children decreased in the following order: for Pb, ingestion > inhalation > dermal contact; for Zn and Cu, dermal contact > ingestion > inhalation; for Cd and Cr, ingestion > dermal contact > inhalation. For adults, the HQ values decreased in the following order: for Zn and Cu, dermal contact > ingestion > inhalation; for Pb, Cd and Cr, ingestion > dermal contact > inhalation. These results are likely due to the fact that children are more likely to contact heavy metals via inadvertent ingestion, such as via pica behavior, hand or finger sucking, or outdoor play activities (Mielke et al. 1999; Karim and Qureshi 2014; Han et al. 2018).

According to the results of the noncarcinogenic risk assessment, the hazard indices for exposed populations are shown in Fig. 4 and Fig. 5.

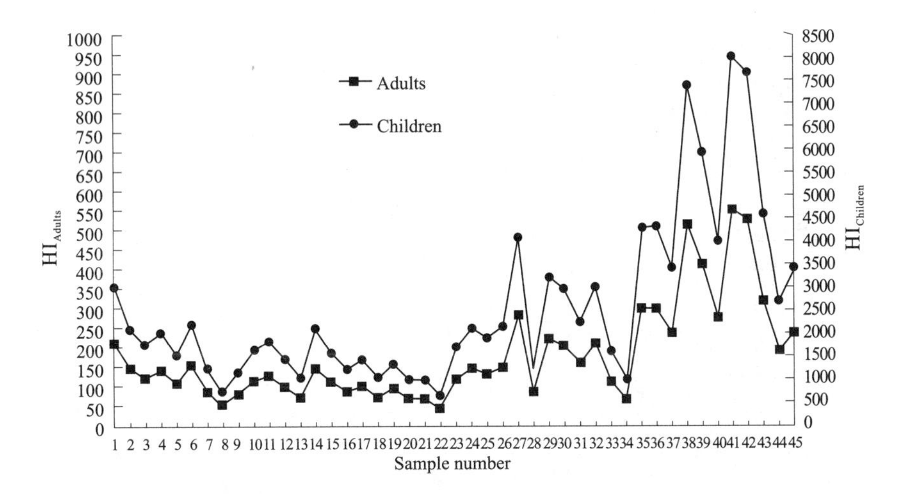

Fig. 4 The distribution map of hazard index (HI) in adults and children

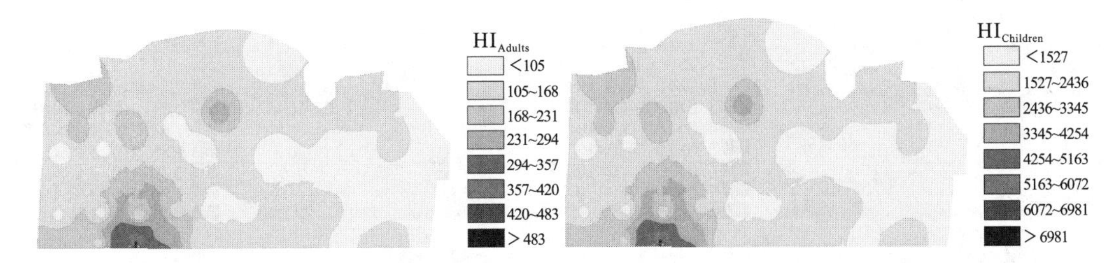

Fig. 5 The distribution pattern of hazard index (HI) for adults and children

As shown in Fig. 4, the calculated HI values for children ranged from 616.67 to 7994.60, and the average was 256.25; for adults, the calculated HI values ranged from 42.52 to 551.29, and the average was 176.55. Obviously (Fig. 5), for adults and children, the HI values for the five metals from all soil samples far exceeded the safe levels (=1), suggesting that adults and children are exposed to significant noncarcinogenic health risks, which should be addressed and studied in more detail. Additionally, children have higher health risks that are noncarcinogenic compared with adults based on their higher calculated HI values, indicating that children are exposed to a significant noncarcinogenic risk due to their behavioral and physiological characteristics, especially hand-to-mouth transfer of soil. Similar results have also been observed in other studies (Li et al. 2014; Tepanosyan et al. 2017; Xiao et al. 2017; Han et al. 2018). The accumulation of Zn is the main cause of the noncarcinogenic risk based on their high HQ values, and excessive intake of Zn leads to chronic diseases that affect the healing of wounds, the immune system response, the ability to taste and smell and stunted growth (Steffan et al. 2018). Thus, the risks for people, and especially children, from exposure to multiple metals in the soil from the Pb/Zn smelter require considerable attention. Zn should be regarded as a priority control pollutant, although the results may not reveal that people actually experience adverse health effects.

3.3.3 Carcinogenic risk assessment

Although the five metals in this study have chronic noncarcinogenic health risks, only two metals (Cd and Cr) have a carcinogenic risk, and the carcinogenic risks for Cd and Cr were considered only via inhalation, as shown in Fig. 6 and Fig. 7.

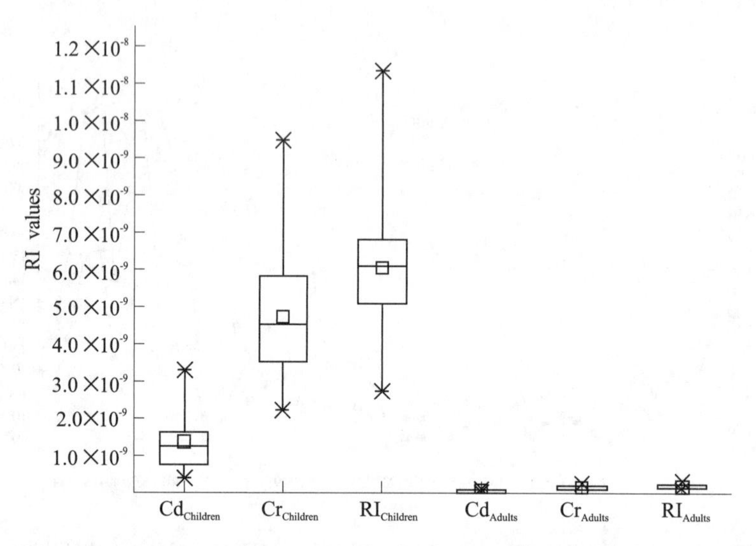

Fig.6 Boxplots of carcinogenic risks of Cd and Cr for children and adults

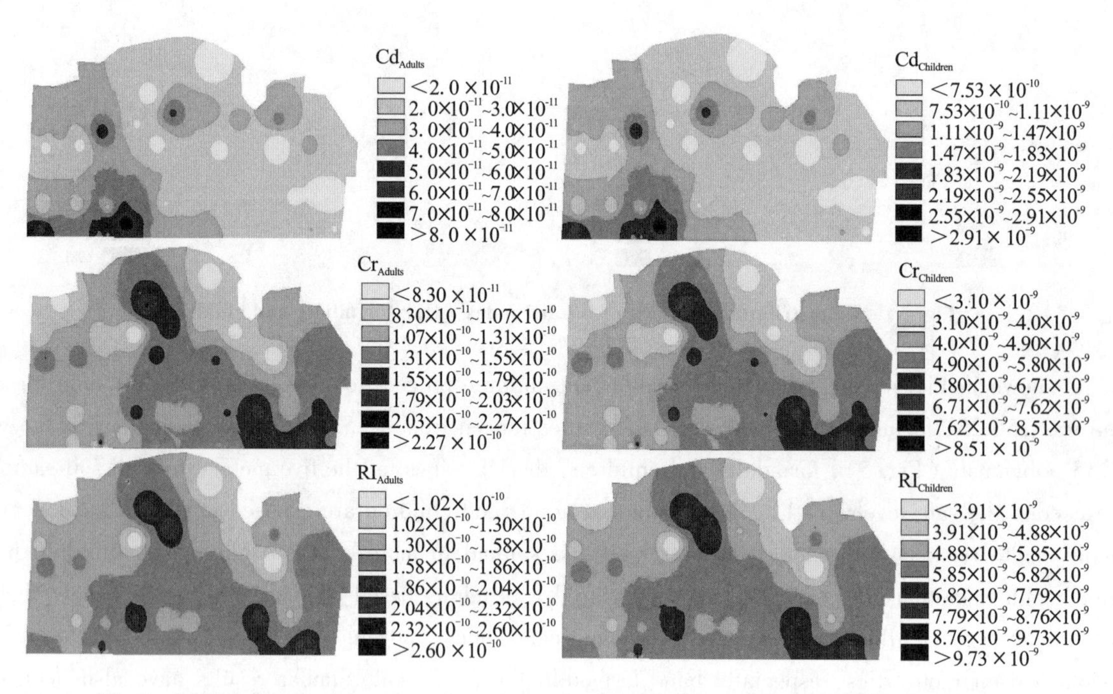

Fig.7 Spatial distribution maps of carcinogenic risks of Cd and Cr for children and adults

For the carcinogenic risk (Fig. 6), the single carcinogenic risk values for Cd and Cr for children were in the ranges of $3.94\times10^{-10}\sim3.28\times10^{-9}$ and $2.19\times10^{-9}\sim9.47\times10^{-9}$ with means of 1.32×10^{-9} and 4.07×10^{-9}, respectively; the single carcinogenic risk values for Cd and Cr for adults were in the ranges of $1.06\times10^{-11}\sim8.79\times10^{-11}$ and $5.87\times10^{-11}\sim2.54\times10^{-10}$ with means of 3.54×10^{-11} and 1.26×10^{-10}, respectively. The total carcinogen risk values (RI) for children and adults were in the ranges of $2.74\times10^{-9}\sim1.13\times10^{-8}$

and $7.33\times10^{-11}\sim3.04\times10^{-10}$, respectively. These results show that all the carcinogenic risk values for the two population groups were less than 10^{-6} overall in the entire study area (Fig.7), which is not considered to pose significant health effects. Thus, children and adults faced an acceptable carcinogenic risk. In addition, the carcinogenic risk levels for children were higher than those for adults, and the carcinogenic risks for the two population groups all showed that Cr posed a higher risk than Cd. Overall, the cancer risk for all HMs in this study were within the acceptable range, implying negligible carcinogenic risk; however, more attention needs to be given to this health issue.

4 Conclusions

A total of 138 samples were collected from near the Pb-Zn smelter in 2017. The concentrations of six potentially toxic HMs elements (Pb, Zn, Cd, Cu, Cr and Mn) in the soil near the Pb-Zn smelter were determined by using an air-acetylene flame atomic absorption spectrophotometer (SHIMADZU AA-6800). The pollution characteristics of the HMs were statistically analyzed by using a mathematical statistics method. The pollution levels were assessed using the potential ecological risk index (RI). The health risk upon exposure to soil HMs was assessed for children and adults using the health risk assessment model developed by the USEPA.

The results showed the following. (1) The mean concentrations of Pb, Zn, Cd and Cu, excluding Mn and Cr, were significantly higher than the background values of Shaanxi Province at 0~20 cm, 20~40 cm and 40~60 cm; the mean concentrations for Cd and Zn at 0~20 cm and 20~40 cm, as well as for Cd at 40~60 cm, far exceeded the soil environmental standard of National Second Grade, indicating that Zn, Cd, Pb and Cu pollution in soil around the smelter pollution is very serious, especially in the topsoil (0~20 cm). (2) The ecological risk assessment indicated the following. For Cu and Cr, there was an overall low potential ecological risk, whereas for Cd, the ecological risk was the highest. For Zn and Pb, it was between low risk and high risk. The RI results exhibited a very high potential ecological risk, mainly caused by Cd, Zn and Pb, especially Cd. The spatial distribution of E_r^i for Pb, Zn, Cd and Cu and the RI showed a high ecological risk overall for the entire study area, which was the highest near the smelter chimneys in the southeast and downwind of the smelter in the north. Cr was lowest overall ecological risk in the entire study area. (3) The health risk analysis showed that dermal contact was the dominant exposure pathway for Zn, Pb, Cd, Cr and Cu for children, and for adults, ingestion was the main exposure pathway for Zn, Pb and Cu. However, dermal contact is a more common exposure pathway for Cd and Cr, and for children and adults, the average daily intake of each toxic metal via the three exposure routes followed the descending order of Zn > Pb > Cu > Cd > Cr. The HQ values showed that Pb, Cd, Cr and Cu exhibited no possibility of adverse health effects for the exposed populations (adults and children), but Zn exhibited possible adverse health effects. The HI values for Pb, Zn, Cd, Cr and Cu from all soil samples far exceeded the safe levels (=1), suggesting that adults and children are exposed to significant noncarcinogenic health risks, and children are under higher noncarcinogenic health risks than adults. For the carcinogenic risk, the single carcinogenic risks from Cd and Cr for children were higher than for adults, and the carcinogenic risks from Cr for the two population groups were higher than from Cd; however, there was no significant carcinogenic risk for the two population groups. Although the carcinogenic risk from Cd and Cr fell within acceptable values, children were more susceptible than adults and suffered higher noncarcinogenic risks from exposure to metals in the soil, which should receive greater attention.

Funding

The study was financially supported by Natural Science Foundation of Shaanxi Province (2020SF-438); and Shaanxi Provincial Key Discipline of Geography.

References

Agomuo E. N., Amadi P. U., (2017), Accumulation and toxicological risk assessments of heavy metals of top soils from markets in Owerri, Imo state, Nigeria, Environmental Nanotechnology Monitoring and Management, 8, 121-126.

Ahirwar N. K., Gupta G., Singh R., Singh V., Isolation, (2018), Assessment of Present Heavy Metals in Industrial Affected Soil Area of Mandideep, Madhya Pradesh, India, International Journal of Current Microbiology and Applied Sciences, 7(1), 3572-3582.

Akopyan K., Petrosyan V., Grigoryan R., Melkom Melkomian D., (2018), Assessment of residential soil contamination with arsenic and lead in mining and smelting towns of northern Armenia, Journal of Geochemical Exploration, 184, 97-109.

Ali M. U., Liu G., Yousaf B., Abbas Q., Ullah H., Munir M. A. M., Fu B., (2017), Pollution characteristics and human health risks of potentially (eco) toxic elements (PTEs) in road dust from metropolitan area of Hefei, China, Chemosphere, 181, 111-121.

Bahloul M., Baati H., Amdouni R., Azri C., (2018), Assessment of heavy metals contamination and their potential toxicity in the surface sediments of Sfax Solar Saltern, Tunisia, Environmental Earth Sciences, 77(1).

Bai H., Hu B., Wang C., Bao S., Sai G., Xu X., Zhang S., Li Y., (2017), Assessment of Radioactive Materials and Heavy Metals in the Surface Soil around the Bayanwula Prospective Uranium Mining Area in China, International Journal of Environmental Research and Public Health, 14(3), 300.

Baize D., Sterckeman T., (2001), Of the necessity of knowledge of the natural pedo-geochemical background content in the evaluation of the contamination of soils by trace elements, Science of the Total Environment, 264(1-2), 127-139.

Barkett M. O., Akün E., (2018), Heavy metal contents of contaminated soils and ecological risk assessment in abandoned copper mine harbor in Yedidalga, Northern Cyprus, Environmental Earth Sciences, 77(10).

Brotons J. M., Díaz A. R., Sarría F. A., Serrato F. B., (2010), Wind erosion on mining waste in southeast Spain, Land Degradation and Development, 21(2), 196-209.

Chabukdhara M., Nema A. K., (2013), Heavy metals assessment in urban soil around industrial clusters in Ghaziabad, India: Probabilistic health risk approach, Ecotoxicology and Environmental Safety, 87, 57-64.

Chen H., Teng Y., Lu S., Wang Y., Wang J., (2015), Contamination features and health risk of soil heavy metals in China, Science of the Total Environment, 512-513, 143-153.

Corriveau M. C., Jamieson H. E., Parsons M. B., Campbell J. L., Lanzirotti A., (2011), Direct characterization of airborne particles associated with arsenic-rich mine tailings: Particle size, mineralogy and texture, Applied Geochemistry, 26(9-10), 1639-1648.

Csavina J., Field J., Taylor M. P., Gao S., Landázuri A., Betterton E. A., Sáez A. E., (2012), A review on the importance of metals and metalloids in atmospheric dust and aerosol from mining operations, Science of the Total Environment, 433, 58-73.

CSC(China State Council), Chinese gov't vows to curb soil pollution, (2012).

Demková L., Árvay J., Bobul'ská L., Tomáš J., Stanovič R., Lošák T., Harangozo L., Vollmannová A., Bystrická J., Musilová J., Jobbágy J., (2017), Accumulation and environmental risk assessment of heavy metals in soil and plants of four different ecosystems in a former polymetallic ores mining and smelting area (Slovakia), Journal of Environmental Science and Health Part A, 52(5), 479-490.

Doabi S. A., Karami M., Afyuni M., Yeganeh M., (2018), Pollution and health risk assessment of heavy metals in agricultural soil, atmospheric dust and major food crops in Kermanshah province, Iran, Ecotoxicology and Environmental Safety, 163, 153-164.

Duan Q., Lee J., Liu Y., Chen H., Hu H., (2016), Distribution of Heavy Metal Pollution in Surface Soil Samples in China: A Graphical Review, Bulletin of Environmental Contamination and Toxicology, 97(3), 303-309.

El Azhari A., Rhoujjati A., El Hachimi M. L., Ambrosi J., (2017), Pollution and ecological risk assessment of heavy metals in the soil-plant system and the sediment-water column around a former Pb/Zn-mining area in NE Morocco, Ecotoxicology and Environmental Safety, 144, 464-474.

Environmental site assessment guideline, (2009), DB11/T 656-2009. (In Chinese).

Ericson B., Hanrahan D., Kong V., (2008), The world's worst pollution problems: top ten of the toxic twenty, New

York: Blacksmith Institute.

Ettler V., Konečný L., Kovářová L., Mihaljevič M., Šebek O., Kříbek B, Majer V, Veselovský F, Penížek V., Vaněk A., Nyambe I., (2014), Surprisingly contrasting metal distribution and fractionation patterns in copper smelter-affected tropical soils in forested and grassland areas (Mufulira, Zambian Copperbelt), Science of The Total Environment, 473-474, 117-124.

Fan S., (2014), Assessment of heavy metal pollution in stream sediments for the Baoji City section of the Weihe River in Northwest China, Water Science and Technology, 70(7), 1279-1284.

Fan S., Wang X., (2017), Analysis and assessment of heavy metals pollution in soils around a Pb and Zn smelter in Baoji City, Northwest China, Human and Ecological Risk Assessment, 23(5), 1099-1120.

Fan S., Wang X., Lei J., Ran Q., Zhou J., (2019), Spatial distribution and source identification of heavy metals in a typical pb/zn smelter in an arid area of northwest china, Human and Ecological Risk Assessment, (1), 1-27.

Gao X., Zhou F., Chen C. T. A., (2014), Pollution status of the Bohai Sea: An overview of the environmental quality assessment related trace metals, Environment International, 62, 12-30.

Gu J. D., (2018), Mining, pollution and site remediation, international biodeterioration and biodegradation, 128, 1-2.

Hadzi G., Y., Essumang D. K., Ayoko G. A., (2018), Assessment of contamination and health risk of heavy metals in selected water bodies around gold mining areas in Ghana, Environmental Monitoring and Assessment, 190(7), 460.

Håkanson L., (1980), An ecological risk index for aquatic pollution control a sedimentological approach, Water Research, 14(8), 975-1001.

Han W., Gao G., Geng J., Li Y., Wang Y., (2018), Ecological and health risks assessment and spatial distribution of residual heavy metals in the soil of an e-waste circular economy park in Tianjin, China, Chemosphere, 197, 325-335.

Han X., Lu X., Qinggeletu., Wu Y., (2017), Health Risks and Contamination Levels of Heavy Metals in Dusts from Parks and Squares of an Industrial City in Semi-Arid Area of China, International Journal of Environmental Research and Public Health, 14(8), 886.

Han Z., Guo X., Zhang B., Liao J., Nie L., (2018), Blood lead levels of children in urban and suburban areas in China (1997-2015): Temporal and spatial variations and influencing factors, Science of The Total Environment, 625, 1659-1666.

Hawkesworth S., Wagatsuma Y., Kippler M., Fulford A. J., Arifeen S. E., Persson L. A., Moore E. S., Vahter M., (2012), Early exposure to toxic metals has a limited effect on blood pressure or kidney function in later childhood, rural Bangladesh, International Journal of Epidemiology, 42(1), 176-185.

Islam Md S., Kormoker T., Ali M. M., Proshad R., (2018), Ecological risk analysis of heavy metals toxicity from agricultural soils in the industrial areas of Tangail District, Bangladesh, SF Journal of Environmental and Earth Science, 1 (2), 1022.

Islam S., Ahmed K., Habibullah-Al-Mamun., Masunaga S., (2015), Potential ecological risk of hazardous elements in different land-use urban soils of Bangladesh, Science of the Total Environment, 512-513, 94-102.

Izah S., C., Bassey S. E., Ohimain E. I., (2018), Ecological risk assessment of heavy metals in cassava mill effluents contaminated soil in a rural community in the Niger Delta Region of Nigeria, Molecular Soil Biology, 9(1), 1-11.

Jiang Y., Chao S., Liu J., Yang Y., Chen Y., Zhang A., Cao H., (2017), Source apportionment and health risk assessment of heavy metals in soil for a township in Jiangsu Province, China, Chemosphere, 168, 1658-1668.

Ju X. T., Kou C. L., Christie P., Dou Z. X., Zhang F. S., (2007), Changes in the soil environment from excessive application of fertilizers and manures to two contrasting intensive cropping systems on the North China Plain, Environmental Pollution, 145(2), 497-506.

Karim Z., Qureshi B., A., (2014), Health Risk Assessment of Heavy Metals in Urban Soil of Karachi, Pakistan, Human and Ecological Risk Assessment, 20(3), 658-667.

Kim B. S. M., Angeli J. L. F., Ferreira P. A. L., de Mahiques M. M., Figueira R. C. L., (2018), Critical evaluation of different methods to calculate the Geoaccumulation Index for environmental studies: A new approach for Baixada Santista-Southeastern Brazil, Marine Pollution Bulletin, 127, 548-552.

Kříbek B., Majer V., Veselovský F., Nyambe I., (2010), Discrimination of lithogenic and anthropogenic sources of metals and sulphur in soils of the central-northern part of the Zambian Copperbelt Mining District: A topsoil vs. subsurface soil concept, Journal of Geochemical Exploration, 104(3), 69-86.

Krombach F., Münzing S., Allmeling A. M., Gerlach J. T., Behr J., Dörger M., (1997), Cell size of alveolar macrophages: an interspecies comparison, Environmental Health Perspectives, 105(suppl 5), 1261-1263.

Lee S. W., Cho H. G., Kim S. O., (2018), Comparisons of human risk assessment models for heavy metal contamination within abandoned metal mine areas in korea, Environmental Geochemistry and Health, (1), 1-25.

Li B., Wang Y., Jiang Y., Li G., Cui J., Wang Y., Zhang H., Wang S. C., Xu A., Wang R., (2016), The accumulation and health risk of heavy metals in vegetables around a zinc smelter in northeastern China, Environmental ence and Pollution Research, 23(24), 25114-25126.

Li F., (2018), Heavy Metal in Urban Soil: Health Risk Assessment and Management, Heavy Metals, doi:10.5772/intechopen.73256

Li F., Cai Y., Zhang J., (2018(a)), Spatial Characteristics, Health Risk Assessment and Sustainable Management of Heavy Metals and Metalloids in Soils from Central China, Sustainability, 10(2), 91.

Li F., Wang T., Xiao M. S., Cai Y., Zhuang Z. Y., (2018(b)), Ecological risk assessment and carcinogen health risk assessment of arsenic in soils from part area of the Daye City, China, IOP Conference Series: Earth and Environmental Science, 108, 042048.

Li F., Zhang J., Huang J., Huang D., Yang J., Song Y., Zeng G., (2016), Heavy metals in road dust from Xiandao District, Changsha City, China: characteristics, health risk assessment, and integrated source identification, Environmental Science and Pollution Research, 23(13), 13100-13113.

Li H., Ji H., (2017), Chemical speciation, vertical profile and human health risk assessment of heavy metals in soils from coal-mine brownfield, Beijing, China, Journal of Geochemical Exploration, 183,22-32.

Li K., Liang T., Wang L., Yang Z., (2015), Contamination and health risk assessment of heavy metals in road dust in Bayan Obo Mining Region in Inner Mongolia, North China, Journal Geographical Sciences, 25(12),1439-1451.

Li P., Lin C., Cheng H., Duan X., Lei K., (2015), Contamination and health risks of soil heavy metals around a lead/zinc smelter in southwestern China, Ecotoxicology and Environmental Safety, 113, 391-399.

Li X., Li Z., Lin C J., Bi X., Liu J., Feng X., Zhang H., Chen J., Wu T., (2018), Health risks of heavy metal exposure through vegetable consumption near a large-scale Pb/Zn smelter in central China, Ecotoxicology and Environmental Safety, 161, 99-110.

Li Z., Ma Z., van der Kuijp T. J., Yuan Z., Huang L., (2014), A review of soil heavy metal pollution from mines in China: Pollution and health risk assessment, Science of The Total Environment, 468-469, 843-853.

Liu H., Wang H., Zhang Y., Yuan J., Peng Y., Li X., Shi Y., He K., Zhang Q., (2018), Risk assessment, spatial distribution, and source apportionment of heavy metals in Chinese surface soils from a typically tobacco cultivated area, Environmental Science and Pollution Research, 25(17), 16852-16863.

Looi L. J., Aris A. Z., Yusoff F. M., Isa N. M., Haris H., (2018), Application of enrichment factor, geoaccumulation index, and ecological risk index in assessing the elemental pollution status of surface sediments, Chemistry and Ecology.

Lu X., Li L. Y., Wang L., Lei K., Huang J., Zhai Y., (2009), Contamination assessment of mercury and arsenic in roadway dust from Baoji, China, Atmospheric Environment, 43(15), 2489-2496.

Meza-Figueroa D., Maier R. M., de la O-Villanueva M., Gómez-Alvarez A., Moreno-Zazueta A., Rivera J., Campillo A., Grandlic C. J., Anaya R., Palafox-Reyes J., (2009), The impact of unconfined mine tailings in residential areas from a mining town in a semi-arid environment: Nacozari, Sonora, Mexico, Chemosphere, 77(1),140-147.

Mielke H. W., Gonzales C. R., Smith M. K., Mielke P. W., (1999), The Urban Environment and Children's Health: Soils as an Integrator of Lead, Zinc, and Cadmium in New Orleans, Louisiana, U.S.A., Environmental Research, 81(2), 117-129.

Moghtaderi T., Mahmoudi S., Shakeri A., Masihabadi M. H., (2018), Heavy metals contamination and human health risk assessment in soils of an industrial area, Bandar Abbas-South Central Iran, Human and Ecological Risk Assessment, 24(4),1058-1073.

Moreno T., Querol X., Alastuey A., Viana M., Salvador P., Sánchez de la Campa A., Artiñano B., Jesús de la Rosa., Gibbons W., (2006), Variations in atmospheric PM trace metal content in Spanish towns: Illustrating the chemical complexity of the inorganic urban aerosol cocktail, Journal of Crystal Growth, 40(35), 6791-6803.

Müller G., (1969), Index of geoaccumulation in sediments of the Phine River, Geological Journals, 2,109-118.

Nkansah M. A., Darko G., Dodd M., Opoku F., Bentum Essuman T., Antwi-Boasiako J., (2017), Assessment of pollution levels, potential ecological risk and human health risk of heavy metals/metalloids in dust around fuel filling stations from the Kumasi Metropolis, Ghana, Cogent Environmental Science, 3(1).

Nriagu J. O., Pacyna J. M., (1988), Quantitative assessment of worldwide contamination of air, water and soils by trace metals, Nature, 333(6169), 134-139.

Ordóñez A., Álvarez R., Charlesworth S., De Miguel E., Loredo J., (2011), Risk assessment of soils contaminated by mercury mining, Northern Spain, Journal of Environmental Monitoring, 13(1), 128-136.

Padoan E., Romè C., Ajmone-Marsan F., (2017), Bioaccessibility and size distribution of metals in road dust and roadside soils along a peri-urban transect, Science of the Total Environment, 601-602:89-98.

Park S. S., Wexler A. S., (2008), Size-dependent deposition of particles in the human lung at steady-state breathing, Journal of aerosol science, 39(3), 266-276.

Querol X., Viana M., Alastuey A., Amato F., Moreno T., Castillo S., Pey J., de la Rosa J., Sánchez de la Campa A., Artíñano B., Salvador P., García Dos Santos S., Fernández-Patier R., Moreno-Grau S., Negral L., Minguillón M. C., Monfort E, Gil J. I., Inza A., Ortega L. A., Santamaría J. M., Zabalza J., (2007), Source origin of trace elements in PM from regional background, urban and industrial sites of Spain, Atmospheric environment, 41(34), 7219-7231.

Ramana S., Biswas A. K., Ajay Singh A. B., Ahirwar N. K., (2012), Phytoremediation of Chromium by Tuberose, National Academy Science Letters, 35(2), 71-73.

Rapant S., Fajcíková K., Khun M., Cvecková V., (2010), Application of health risk assessment method for geological environment at national and regional scales, Environmental Earth Sciences,64(2), 513-521.

Salmanighabeshi S., Palomo-Marín M. R., Bernalte E., Rueda-Holgado F., Miró-Rodríguez C., Fadic-Ruiz X., Vidal-Cortez V., Cereceda-Balic F., Pinilla-Gil E., (2015), Long-term assessment of ecological risk from deposition of elemental pollutants in the vicinity of the industrial area of Puchuncaví-Ventanas, central Chile, Science of the Total Environment, 527-528, 335-343.

Shaheen N., Shah M. H., Jaffar M., (2005), A Study of Airborne Selected Metals and Particle Size Distribution in Relation to Climatic Variables and their Source Identification, Water, Air, and Soil Pollution, 164(1-4), 275-294.

Shang S., Zhong W., Wei Z., Zhu C., Ye S., Tang X., Chen Y., Tian L., Chen B., (2017), Heavy metals in surface sediments of lakes in Guangzhou public parks in China and their relations with anthropogenic activities and urbanization, Human and Ecological Risk Assessment, 23(8), 2002-2016.

Shen F., Liao R., Ali A., Mahar A., Guo D., Li R. H., Sun X. N., Awasthi M. K., Wang Q., Zhang Z., (2017), Spatial distribution and risk assessment of heavy metals in soil near a Pb/Zn smelter in Feng County, China, Ecotoxicology and Environmental Safety, 139, 254-262.

Song D., Jiang D., Wang Y., Chen W., Huang Y., Zhuang D., (2013), Study on Association between Spatial Distribution of Metal Mines and Disease Mortality: A Case Study in Suxian District, South China, International Journal of Environmental Research and Public Health, 10(10), 5163-5177.

Song D., Zhuang D., Jiang D., Fu J., Wang Q., (2015), Integrated Health Risk Assessment of Heavy Metals in Suxian County, South China, International Journal of Environmental Research and Public Health, 12(7), 7100-7117.

Steffan J. J., Brevik E. C., Burgess L. C., Cerdà A., (2017), The effect of soil on human health: an overview, European Journal of Soil Science, 69(1), 159-171.

Suresh G., Sutharsan P., Ramasamy V., Venkatachalapathy R., (2012), Assessment of spatial distribution and potential ecological risk of the heavy metals in relation to granulometric contents of Veeranam lake sediments, India, Ecotoxicology and Environmental Safety, 84, 117-124.

Teng Y., Ni S., Wang J., Zuo R., Yang J., (2010), A geochemical survey of trace elements in agricultural and non-agricultural topsoil in Dexing area, China, Journal of Geochemical Exploration, 104(3), 118-127.

Tepanosyan G., Sahakyan L., Belyaeva O., Maghakyan N., Saghatelyan A., (2017), Human health risk assessment and riskiest heavy metal origin identification in urban soils of Yerevan, Armenia, Chemosphere, 184, 1230-1240.

USEPA, (1986), Superfund Public Health Evaluation Manual, 540, pp. 1-86, Office of Emergency and Remedial Response, 20460 EPA, Washington, DC.

USEPA, (1989), Risk Assessment Guidance for Superfund. Human Health Evaluation Manual (Part A), Vol.I; EPA/540/1-89/002, Office of Superfund Remediation and Technology Innovation, U.S. Environmental Protection Agency, Washington, DC.

USEPA, (2001), Baseline Human Health Risk Assessment, Vasquez Boulevard and I-70 superfund site Denver, Denver, CO, US Environmental Protection Agency.

USEPA, (2002), Supplemental guidance for developing soil screening levels for superfund sites; Office of Emergency and Remedial Response: Washington, DC, Soild Waste and Emergency Response.

USEPA, (2003), Example exposure scenarios National Center for Environmental Assessment, Washington, DC, EPA/600/R-03/036, National Information Service.

USEPA, (2004), Risk Assessment Guidance for Superfund Volume I: Human Health Evaluation Manual (Part E, Supplemental Guidance for Dermal Risk Assessment), Washington: Office of Superfund Remediation and Technology Innovation, US Environmental Protection Agency, D5-D7.

USEPA, (2011), Exposure factors handbook. National Center for Environmental Assessment, Washington, DC: United State Environmental Protection Agency.

Valiulis D., Jonas Šakalys, Kristina Plauškaitė, (2008), Heavy metal penetration into the human respiratory tract in Vilnius, lithuanian journal of physics, Lithuanian Journal of Physics, 48(4), 349-355.

Wei B., Yang L., (2010), A review of heavy metal contaminations in urban soils, urban road dusts and agricultural soils from China, Microchemical Journal, 94(2), 99-107.

World Health Organization (WHO), (2001), Codex Maximum Level for Cadmium in Cereals. Pulses and Legumes; WHO: Geneva, Switzerland, CAC/GL39.

Wu J., Lu J., Li L., Min X., Luo Y., (2018), Pollution, ecological-health risks, and sources of heavy metals in soil of the northeastern Qinghai-Tibet Plateau, Chemosphere, 201, 234-242.

Xiao R., Wang S., Li R., Wang J. J., Zhang Z., (2017), Soil heavy metal contamination and health risks associated with artisanal gold mining in Tongguan, Shaanxi, China, Ecotoxicology and Environmental Safety, 141, 17-24.

Xu Y., Dai S., Meng K., Wang Y., Ren W., Zhao L., Christie P., Teng Y., (2018), Occurrence and risk assessment of potentially toxic elements and typical organic pollutants in contaminated rural soils, Science of the Total Environment, 630, 618-629.

Xue C., Xiao L., Wu Q., Li D., Li H., Wang R., (1986), Studies of background values of ten chemical elements in major agricultural soils in Shaanxi Province, J Northwest A & F University(Natural Science Edition), 14(3), 30-40.

Yang Q., Li Z., Lu X., Duan Q., Huang L., Bi J., (2018), A review of soil heavy metal pollution from industrial and agricultural regions in China: Pollution and risk assessment, Science of the Total Environment, 642, 690-700.

Yi Y., Yang Z., Zhang S., (2011), Ecological risk assessment of heavy metals in sediment and human health risk assessment of heavy metals in fishes in the middle and lower reaches of the Yangtze River basin, Environmental Pollution, 159(10), 2575-2585.

Zhu D., Wei Y., Zhao Y., Wang Q., Han J., (2018), Heavy Metal Pollution and Ecological Risk Assessment of the Agriculture Soil in Xunyang Mining Area, Shaanxi Province, Northwestern China, Bulletin of Environmental Contamination and Toxicology, 101(2), 178-184.

本文曾发表于2022年《Bangladesh J. Bot.》第51卷第4期

延河流域耕层土壤养分空间变异及与地形因子的相关性研究

罗 丹[1],毛忠安[2],张庭瑜[2,3,4,5],常庆瑞[6]

(1.陕西地建土地勘测规划设计院有限责任公司,西安 710075;2.陕西省土地工程建设集团有限责任公司,西安 710075;3.陕西地建土地工程技术研究院有限责任公司,西安 710075;4.自然资源部退化及未利用土地整治重点实验室,西安 710075;5.陕西省土地整治工程技术研究中心,西安 710075;6.西北农林科技大学资源环境学院,陕西杨凌 712100)

【摘要】了解耕层土壤养分空间变异规律及其与地形因子的相关关系,以期为土壤养分有效利用和农业管理提供理论依据。以延河流域为研究对象,利用地统计学与 GIS 相关方法,研究土壤空间分布及变异规律,并基于 DEM 数据提取相关地形因子,分析流域尺度下地形因子对土壤养分空间分布的影响。研究区耕层全氮和有机质处于较低水平,速效钾和有效磷处于中等水平,pH 呈弱碱性,空间变异大小依次为速效钾>有效磷>全氮>有机质>pH。构建的土壤养分指标半方差函数决定系数均大于 0.6;土壤 pH 空间分布最为破碎;土壤速效钾和 pH 块金系数小于 25%,主要为结构性因素影响;土壤全氮、有效磷、有机质具有中等水平的块金效应,受结构性因素和随机因素共同作用;其中有效磷块金系数最大,人类活动对其空间变异影响更大。全氮与有机质分布规律由西北向东南先递增后递减,速效钾呈现由西北向东南波浪形递减的分布格局,有效磷整体分布趋势表现为由中游向上下游降低,土壤 pH 沿某一方向变化的趋势不强。结论:土壤养分空间变异情况与地形因子的相关关系相一致。在地势较低且平坦的区域内,更有利于农业生产,土壤养分较为充足;在高海拔、地形起伏较大区域养分含量相对较低。

【关键词】地统计学;土壤养分;空间变异;地形因子;延河流域

Spatial Variability of Topsoil Nutrients and its Correlation with Topographic Factors in Yanhe River Basin

Dan Luo[1], Zhongan Mao[2], Tingyu Zhang[2,3,4,5], Qingrui Chang[6]

(1.Land Surveying, Planning and Design Institute of Shaanxi Provincial Land Engineering Construction Group Co. Ltd., Xi'an 710075; 2.Shaanxi Province Land Engineering Construction Group Co. Ltd., Xi'an 710075;
3.Institute of Land Engineering and Technology, Shaanxi Provincial Land Engineering Construction Group Co. Ltd. , Xi'an 710075;
4.Key Laboratory of Degraded and Unused Land Consolidation Engineering, the Ministry of Natural Resources, Xi'an 710075;
5.Shaanxi Provincial Land Consolidation Engineering Technology Research Center, Xi'an 710075;
6.College of Nature Resources and Environment, Northwest A&F University, Yangling, Shaanxi 712100)

【Abstract】To provide a basis for the effective utilization of soil nutrients and agricultural management, the spatial variation rule of topsoil nutrients and its correlation with topographic factors were studied. The authors took Yanhe

基金项目:中央高校基本科研业务费资助项目"基于深度学习的灾损土地语义空间分布规律研究及分析"(300102351502);陕西省耕地地力调查与质量评价项目;陕西省土地工程建设集团科研项目"基于 DEM 的陕北高原地貌形态空间格局研究"(DJNY2019-8);陕西省土地工程建设集团有限责任公司科研项目"基于深度学习网络融合多源遥感数据的废弃地块识别研究"(DJNY2021-10)。

River basin as the research object, the spatial distribution and variation of soil were studied by Geostatistics and GIS. Based on DEM data, relevant topographic factors were extracted to analyze the influence of topographic factors on soil nutrient spatial distribution in watershed scale. The results showed that the total nitrogen and organic matter in the surface layer of the study area were at a low level, the available potassium and available phosphorus were at a medium level, and the pH was weakly alkaline. The spatial variability was in the following order: available potassium > available phosphorus > total nitrogen > organic matter >pH. The determination coefficients of the semi-variance function on soil nutrients were all more than 0.6. The spatial distribution of soil pH was the most fragmented. Which mainly influenced by structural factors, the nugget coefficients of soil available potassium and pH were less than 25%. With the influenced by structural and random factors, soil total nitrogen, available phosphorus and organic matter had medium level of nugget effect. And the nugget coefficient of available phosphorus was the highest, indicated that human activities has a greater influence on its spatial variation. The distribution of total nitrogen and organic matter increased first and then decreased from northwest to southeast. The distribution pattern of available potassium showed a wave decreasing from northwest to southeast. The overall distribution trend of available phosphorus was reduced from middle to the upstream and downstream. The trend of pH change along a certain direction had no obvious changes. The correlation between soil nutrient spatial variation and topographic factors was consistent. In the region of low and flat terrain, the soil nutrients were more abundant which conducive to agricultural production. And the nutrient content was relatively low in the areas with large relief at high altitude.

【**Keywords**】Geostatistics; Topsoil nutrients; Spatial variability; Topographic factors; Yanhe River basin

0 引言

土壤养分是农作物摄取养分的重要来源之一,是土地生产力的基础,也是土壤质量评价的重要指标[1]。土壤养分空间变异是指在土壤质地相同的区域内,同一时刻不同空间位置上也具有明显差异[2]。土壤空间变异是普遍存在的,分析掌握变异规律是土地可持续利用的重要组成部分[3]。随着地统计学和“3S”技术的普遍应用,土壤养分空间变异研究取得了很大进展。其中郭旭东等[4]基于GIS和地统计学对河北省遵化市5种养分要素空间变异规律进行了研究,研究结果可用于有效指导农业生产;秦松等[5]通过对土壤空间变异与地形因子的相关关系的研究,得出地形因素与养分分布存在显著相关性的结论;高洁等[6]对阜康市1982年、2010年、2018年3个时期土壤有机质时空变化规律及其影响因素进行分析,评价了该地区的土壤肥力情况。延河流域是典型的黄土高原丘陵沟壑区,生态环境脆弱,加之人为因素干扰,受到土地利用和地形因素的显著影响,土壤属性空间变异程度较大[7]。研究土壤养分空间变异特征对地区合理施肥和防治水土流失有着十分关键的作用。已有学者对延河流域水土流失、景观格局、植被恢复等方面进行研究,在协调空间布局和改善生态治理方面贡献了理论支持[8-10]。对于流域内土壤空间变异规律及其影响因素还未有研究进行深入探讨,特别是对于黄土高原地区,土壤肥力丰缺是影响农作物产量的重要因素。本文以延河流域为研究对象,利用地统计学与GIS相关方法,研究了该地区土壤空间分布及变异规律,并基于DEM数据提取相关地形因子,分析流域尺度下地形因子对土壤养分空间分布的影响,以期为土壤养分有效利用和农业管理提供理论依据。

1 材料与方法

1.1 研究区概况

延河是陕北第二条大水系,全长286.9 km,源起榆林市靖边县天赐湾乡周山白于山附近,由西北向东南,流经延安市志丹县、安塞县、宝塔区,于延长县南河沟凉水岸附近汇入黄河,流域面积7725 km^2。流域以北为清涧河流域,西南临北洛河流域,南与云岩河流域相接[11],地势西北高、东南低,海拔在500~1800 m(见图1),平均海拔1212.34 m。延河分为上、中、下游三部分,属于中游发达的河流,地貌形态特征较为分明。从源头到安塞县的化子坪为上游,地貌形态为黄土峁梁丘陵沟壑区,山大沟深,坡陡谷窄,植被稀少,地形破碎,侵蚀强烈,水土流失严重;从化子坪到宝塔区的甘谷驿为中游,地貌形态为黄土峁状丘陵沟壑区,河谷宽阔,两岸阶地宽广平坦,支流众多,河流侵蚀没有上游严重;从甘谷驿到河口为下游,地貌形态为破碎塬区,河流下切基岩,形成落差较大的陡峭峡谷,水土流失量低于上游。延河主要支流有杏子河、平桥川、西川河、南川河、蟠龙川等[12],集中在中上游。受区位影响和地形制约,流域具有典型的大陆季风气候特征,平均温度为8.8~10.2 ℃,夏秋季多雨,冬春季干旱少雨,年降水量为400~550 mm,从上游到下游径流量逐渐减少,季节性明显,夏季汛期径流量明显增加,冬季常有枯水。由此导致自然景观和农业生产结构也具有明显的过渡性特征,以干旱草本植物为主,间有块状灌木林地,少量乔木林地。

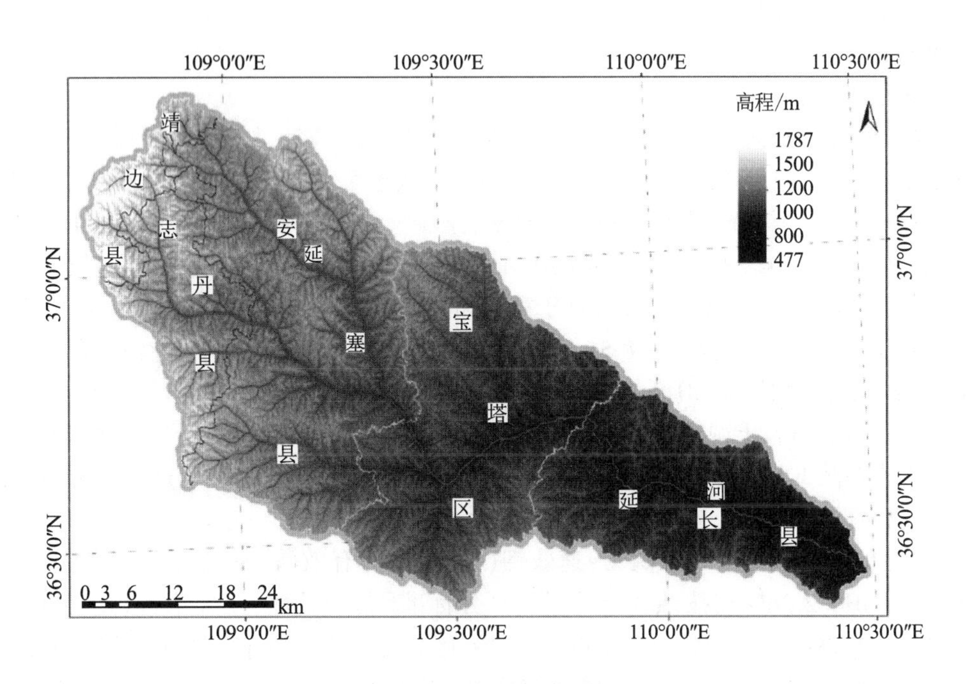

图1 延河流域DEM图

1.2 数据来源及处理

土壤样品于2016年作物收获后、施肥前采集,遵循科学性、均衡性、代表性原则[13],选择不同土壤类型、不同地貌形态的研究区耕层土壤作为样本,用GPS定位,同时记录灌溉条件、肥料施用等情况。土壤有机质、全氮、速效钾、速效磷、pH采用常规测定方法[14]。运用3倍标准差方法剔除异常值[15],为保证样本量,将剔除的异常值用最大值或最小值代替。

本文使用的ASTER GDEM数据来源于地理空间数据云,空间分辨率为30 m×30 m。在ArcGIS10.3支持下,从DEM中提取流域范围,通过Spatial Analyst模块Hydrological Analyst工具,Fill填充洼地—Flow direction分析流向—Flow accumulation计算栅格水流量—Stream link提取栅格河网—Stream to feature栅格河网矢量化—Watershed生成流域,得到研究区DEM影像。基于研究区DEM运用Spatial Analyst模块表面分析中Slope、Aspect工具,生成坡度、坡向图,运用邻域分析工具及地图代数中栅格计算器工具求得地形起伏度,对高程、坡度、坡向、地形起伏度进行重分类,得到延河流域地形因子分级及面积统计数据。

1.3 研究方法

半方差函数是地统计学中研究土壤养分变异性的关键函数。土壤在空间上应该是连续变异的,采用相关函数对土壤养分进行拟合,生成曲线方程即为半方差函数理论模型。采用GS+软件计算半方差函数及模型并分析,用以对土壤养分含量空间连续变异进行描述。计算公式如式(1)所示。

$$\gamma(h)=\frac{1}{2N(h)}\sum_{i=1}^{N(h)}[Z(x_i)-Z(x_i+h)]^2 \tag{1}$$

式中:$\gamma(h)$为间距为h时的半方差;h为采样点间隔距离;$N(h)$为采样点间距为h的所有样本点对

数和；$Z(x_i)$ 和 $Z(x_i+h)$ 分别为 x_i 和 x_i+h 处实测值。

半方差函数值随着样点间距的增加而增大,并在一定的间距(变程)升大到一个基本稳定的常数(基台),基台代表了空间总变异,变程代表了空间变异尺度范围;本该通过坐标原点,在位置趋于零时而并不为零时的非零值即块金方差,代表了随机变异。

利用 GS+软件进行数据正态性检验、趋势分析和半方差函数构建;借助 ArcGIS10.3 地统计学模块克里金插值进行空间内插得到延河流域土壤养分空间分布趋势图;利用 SPSS 软件对土壤养分与地形因子进行相关性分析等操作。

2 结果与分析

2.1 土壤养分总体统计特征

将处理异常值后的延河流域土壤养分数据进行基本特征统计,结果见表 1。土壤全氮含量为 0.27~0.77 g/kg,平均含量为 0.52 g/kg,根据陕西省第二次土壤普查养分分级标准[16],全氮集中分布在第 6 级别(0.5~0.75 g/kg),属于较低含量水平;土壤速效钾含量为 50~168 mg/kg,平均含量为 105.56 mg/kg,处于标准的第 3、4 和 5 级别(70~150 mg/kg),属于中等含量水平;土壤有效磷含量为 4.5~13.6 mg/kg,平均含量为 9.07 mg/kg,处于第 5 和 6 级别(5~15 mg/kg),属于中等含量水平;有机质含量在 4.9~12.2 g/kg,平均含量为 8.55 g/kg,集中分布在第 7 级别(8~10 g/kg),属于较低含量水平;土壤 pH 7.2~9.3,平均值为 pH 8.19,集中在第 4 级别(pH 8.1~8.5),整体呈弱碱性。对不同尺度区域背景值进行对比分析,土壤养分平均含量与全省土壤养分背景值有较大差异,各指标明显低于全省平均含量;而在与陕北地区土壤养分平均含量对比中,各指标呈基本持平,这也符合较大地域差异影响土壤养分含量的规律。

各土壤养分的变异系数大小依次为:速效钾(19.34%)>有效磷(16.27%)>全氮(16.09%)>有机质(13.98%)>pH (2.36%),按照变异等级的划分标准[弱变异性,变异系数(C_v)<10%;中等变异性,10%≤C_v≤100%;强变异性,C_v>100%],研究区各土壤养分除 pH 外在空间上都存在中等程度的变异性。土壤 pH 在土壤性质中变异性较低,属于比较稳定的因素,不易受施肥、灌溉等人为因素影响[8]。经过 Box-Cox 变换后,各养分指标符合或基本符合正态分布,达到空间克里金插值的前提条件。

表 1 土壤养分统计特征

土壤养分	样本数	最大值	最小值	平均值	标准差	变异系数	BV01	BV02
全氮 TN/(g/kg)	7488	0.77	0.27	0.52	0.08	16.09	0.90	0.56
速效钾 AK/(mg/kg)	8167	168.00	50.00	105.56	20.42	19.34	146.27	119.22
有效磷 AP/(mg/kg)	8153	13.60	4.50	9.07	1.48	16.27	17.45	9.92
有机质 OM/(g/kg)	8169	12.20	4.90	8.55	1.19	13.98	14.44	8.58
pH	8208	9.30	7.20	8.19	0.19	2.36	7.75	8.01

注:BV01 为陕西省土壤养分平均含量背景值[17],BV02 为陕北地区土壤养分平均含量背景值[17]。

2.2 土壤养分空间变异特征

利用 GS+软件求取土壤养分半方差函数,以决定系数接近于 1、残差趋向于 0 选取理论模型,土壤全氮最优拟合模型为球状模型,速效钾符合高斯模型,其他 3 种指标最优模型为指数模型,半方差函数决定系数均大于 0.6,表明半方差函数拟合精度较高(见表 2、图 2)。在变程范围内,距离越近的点相关性越大,而距离大于变程则不具备相关性,pH 的变程相对其他土壤养分较小,在空间上表现得更离散,实地分布更加破碎。所有指标块金值均大于 0,表示养分指标存在人为等随机性变异;均不同程度地具有一定的基台值,说明存在实验误差和短距离变异,土壤有效磷的基台值最大,表明空间总变异最大;块金与基台两者的比值为块金系数,比例大小被称为块金效应,表示变量的空间相关性程度,块金系数小于25%表明空间相关性强,块金系数在25%~75%表明空间相关性中等,块金系数大于75%表明空间相关性非常弱[18]。气候、地形、土壤类型等结构性自然因素会导致土壤养分空间相关性增强,而施肥、种植制度、耕作措施等人为随机性因素会致使其空间相关性减弱。延河流域土壤速效钾和pH块金系数小于25%,具有强烈的空间相关性,说明受到随机因素的影响较小;土壤全氮、有效磷、有机质块金系数均位于中等空间相关性区间内,说明同时受到结构性因素和随机因素共同影响;有效磷块金系数最大,为50%,表明其受到人类活动影响更大。

表2 土壤养分最优半方差拟合模型及相关参数

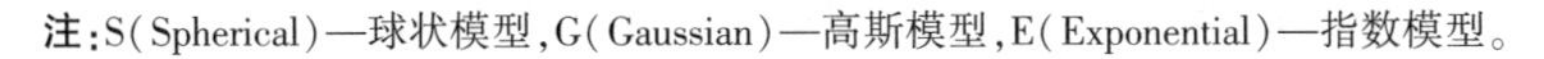

土壤养分	理论模型	变程	块金值	基台值	块金系数	R^2
全氮 TN	S	45800	0.001	0.004	0.416	0.726
速效钾 AK	G	191565	0.016	0.116	0.141	0.967
有效磷 AP	E	45600	1.095	2.191	0.500	0.897
有机质 OM	E	35700	0.020	0.044	0.461	0.839
pH	E	6900	0.0004	0.0004	0.121	0.611

注:S(Spherical)—球状模型,G(Gaussian)—高斯模型,E(Exponential)—指数模型。

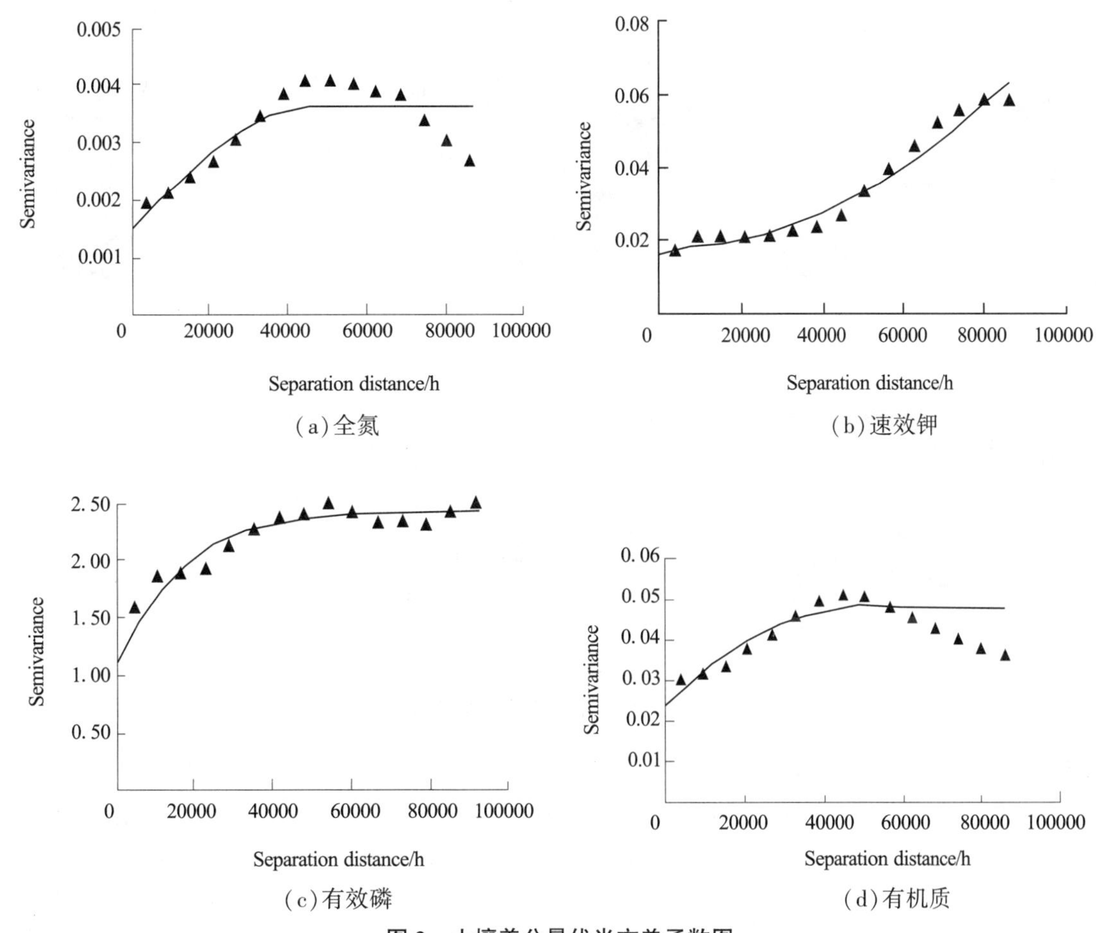

图2 土壤养分最优半方差函数图

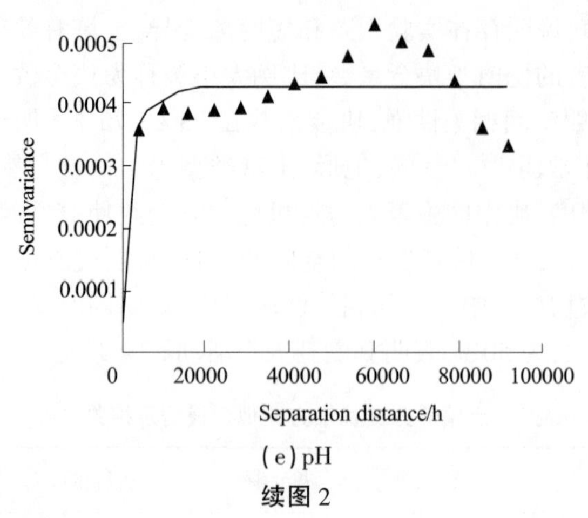

(e) pH

续图 2

2.3　土壤养分空间分布特征

基于 Kriging 插值获得研究区土壤养分含量分布图(见图 3)。插值结果对采样点进行了一定的延展,通过分级仍可表现出养分的空间分布差异和规律。延河流域全氮与有机质分布规律较为相似,低值位于西北部海拔较高处和东南部海拔较低处,高值位于流域中部宝塔区附近,呈现出由西北向东南先递增后递减的规律。这与延河流域泥沙量存在明显地区差异相关,从上游到中游含沙量和输沙模数是逐渐减少的,从中游到下游逐渐增多,延河上、下游水土流失严重,中游相对较轻,有机质和全氮在中游含量相对较高。速效钾与全氮、有机质分布规律不同,低值分布在西北高海拔地区,高值分布在东南部延长县附近,呈现由西北向东南波浪形递减的分布格局。有效磷除少量高值位于西北部靖边县附近外,整体分布趋势表现为由中游宝塔区向两端减少。土壤 pH 沿某一方向变化的趋势不强,各个级别呈块状分布。

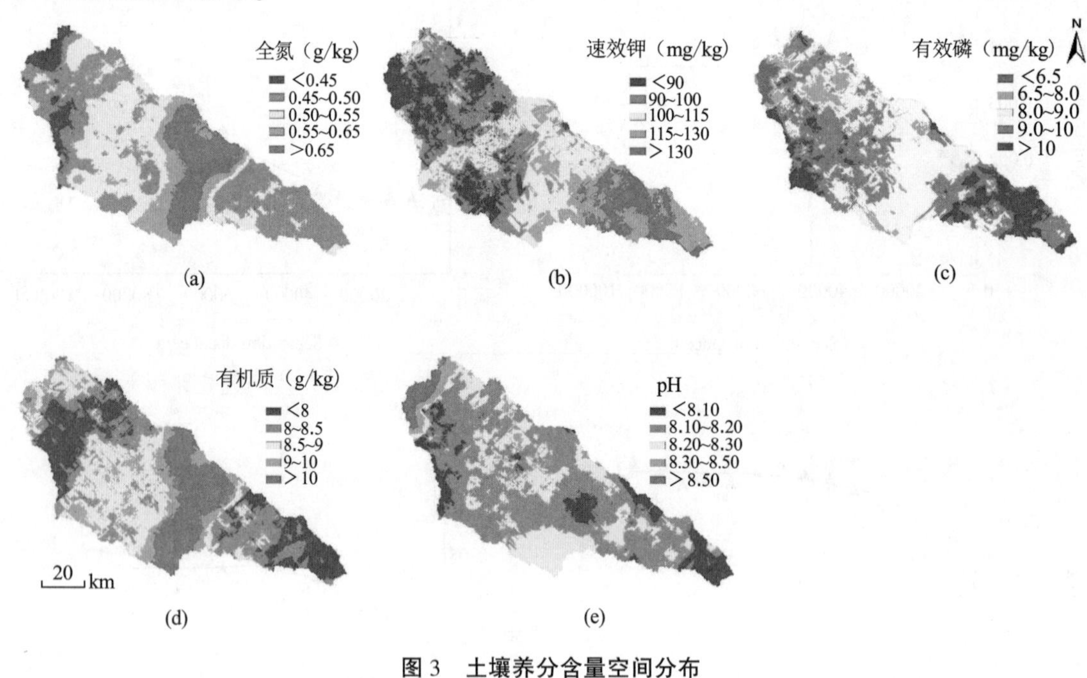

图 3　土壤养分含量空间分布

2.4　土壤养分与地形因子的关系

地形条件通过水热条件和成土物质再分配,造成土壤养分空间差异[19]。通过 ArcGIS 地统计模块,计算耕地地形因子和土壤养分平均值,利用 SPSS 软件计算其相关系数,结果见表 3,各土壤养分指标与地形因子之间均存在不同程度的相关。其中,全氮、速效钾、有效磷与所有地形因子均呈现负相关,全氮与高程、地形起伏度、坡向达到极显著水平;速效钾除坡向外,与其他地形因子均达到极显著相关;有效磷与高程、坡度呈极显著负相关;有机质与高程、地形起伏度呈极显著负相关,与坡向呈极显著正相关;pH 除与高程呈极显著正相关外,与其他地形因子无显著相关关系,说明选取的地形因子对土壤 pH 变化解释能力不强。

表 3 土壤养分与地形因子的相关性分析

土壤养分	高程	坡度	地形起伏度	坡向
全氮 TN	-0.385**	0.035	-0.099**	-0.148**
速效钾 AK	-0.617**	-0.092**	-0.114**	-0.065
有效磷 AP	-0.186**	-0.077**	-0.037	-0.058
有机质 OM	-0.283**	0.036	-0.105**	0.108**
pH	0.287**	0.012	0.029	-0.050

注:**表示相关性达到极显著水平。

对研究区各地形因子进行分级,分别统计不同地形因子条件下土壤养分分布特征,结果见表 4。海拔变化会引起温度、湿度变化,从而影响土壤中营养元素的积累、迁移和分解,导致土壤养分有所差异[20-21]。研究区耕地大部分位于海拔 1000~1400 m,占耕地总面积的 78.07%。随着海拔的升高,全氮、速效钾、有效磷、有机质呈现降低趋势,全氮、有机质在 1200~1400 m 海拔范围内变化幅度较大,速效钾在 1000~1200 m 变化幅度较大,有效磷在 1400~1600 m 变化幅度较大;土壤 pH 呈升高趋势,但变化幅度较大。坡度表示了该局部地表坡面的倾斜程度,坡度的大小直接影响地表物质流动与能量转换的强度与规模[22],坡度变化越大,养分越不容易积累。研究区耕地坡度主要分布在 6°~25°,占土地总面积的 96.90%。随着坡度的增加,速效钾、有效磷呈减小趋势;全氮、有机质在不同地形因子级别下变化不大,在 6°~25°分布较离散。土壤养分空间分布具有海拔梯度性[23],地形起伏度是描述区域地形特征的宏观指标,尤其是对于丘陵沟壑区等切割性较强的地区,地形起伏度变化越大,地形地貌越复杂多样[24]。研究区地形起伏度主要分布在 70~150 m,占总面积的 91.85%,其中 110~150 m 所占比例为 64.50%。地形起伏度小于 30 m 区域内,耕地土壤养分含量高且分布集中;随着地形起伏度升高,全氮、速效钾、有效磷、有机质呈下降趋势,全氮变化幅度逐渐增加。坡向是决定地表面局部地面接收阳光和重新分配太阳辐射量的重要地形因子之一,直接造成局部地区气候特征的差异[25],从而导致土壤养分差异。将坡向划分为平地、阴坡、半阳坡、阳坡、半阴坡,研究区各坡向面积相差不大,分布较为均衡。各方位的地面光热资源分配及地表径流也表现较为均衡,除平地外,不同坡向土壤养分变化规律不明显,变化幅度也基本相当,平地上各营养指标变化幅度低于有坡向地区。地形因子对土壤养分的影响与土壤养分空间分布格局基本一致。在地势较低且平坦的区域内,温度湿度较高,土壤养分较为充足;在高海拔、地形起伏较大区域土壤贫瘠,养分含量低。

表 4 不同地形因子分级下土壤养分含量分布特征统计

地形因子		全氮 TN/(g/kg)	速效钾 AK/(mg/kg)	有效磷 AP/(mg/kg)	有机质 OM/(g/kg)	pH
高程	<1000 m	0.59±0.12	123.81±36.56	10.03±1.88	9.24±1.81	8.11±0.23
	1000~1200 m	0.62±0.16	118.65±37.96	9.54±2.44	9.20±2.03	8.13±0.27
	1200~1400 m	0.59±0.18	111.40±33.51	8.90±2.47	8.87±2.14	8.24±0.21
	1400~1600 m	0.45±0.14	94.04±30.60	8.03±3.68	8.55±1.94	8.39±0.38
	>1 600 m	0.44±0.11	86.87±28.96	8.07±3.73	8.93±1.44	8.39±0.36

续表 4

地形因子		全氮 TN/(g/kg)	速效钾 AK/(mg/kg)	有效磷 AP/(mg/kg)	有机质 OM/(g/kg)	pH
坡度	0°~2°	0.55±0.14	117.99±29.50	9.95±1.65	8.97±1.61	8.21±0.17
	2°~6°	0.54±0.21	114.69±43.64	8.26±3.34	8.80±2.19	8.33±0.40
	6°~15°	0.55±0.23	109.55±50.82	8.16±3.81	8.93±2.31	8.32±0.45
	15°~25°	0.54±0.23	106.32±48.41	8.06±3.82	8.92±2.19	8.31±0.44
	>25°	0.54±0.19	105.82±37.58	8.04±2.84	8.93±2.08	8.34±0.34
地形起伏度	0~30 m	0.59±0.10	124.66±9.07	9.79±0.49	9.81±0.77	8.17±0.08
	30~70 m	0.55±0.15	121.65±36.68	9.56±2.04	8.79±2.18	8.17±0.21
	70~110 m	0.54±0.23	109.14±51.23	8.16±3.81	8.93±2.31	8.32±0.43
	110~150 m	0.54±0.23	108.74±47.30	8.03±3.82	9.01±2.22	8.32±0.45
	>150 m	0.54±0.23	107.31±40.97	7.96±3.41	8.89±2.15	8.36±0.39
坡向	平地	0.47±0.11	118.50±33.50	9.85±1.95	8.30±2.00	8.30±0.50
	阴坡	0.54±0.23	108.50±58.50	9.05±4.55	8.50±3.60	8.45±0.85
	半阴坡	0.52±0.24	107.00±57.00	9.05±4.55	8.50±3.60	8.50±0.70
	阳坡	0.53±0.25	108.50±58.50	9.00±4.50	8.55±3.55	8.25±1.05
	半阳坡	0.52±0.25	108.50±58.50	8.95±4.45	8.60±3.60	8.55±0.75

3 结论

延河流域土壤全氮、有机质平均含量处于较低水平,土壤速效钾、有效磷属于中等水平,土壤 pH 整体呈弱碱性。各土壤养分空间变异大小依次为速效钾>有效磷>全氮>有机质>pH,除 pH 外,在空间上都存在中等程度变异。通过构建半方差函数得到养分指标理论模型和相应参数,模型决定系数均大于 0.6,拟合精度较高。土壤 pH 空间分布最为破碎;所有指标均存在人为等随机性变异;土壤有效磷空间总变异最大;土壤速效钾和 pH 块金系数较小,主要影响因素为气候、地形地貌、成土母质等结构性因素;土壤全氮、有效磷、有机质具有中等水平的块金效应,受结构性因素和随机因素共同作用;其中有效磷块金系数最大,受到人类活动等随机因素影响更大。

土壤全氮与有机质空间分布为西北和东南低、中部高,整体呈现由西北向东南先递增后递减的分布规律,与高程、地形起伏度呈极显著负相关;速效钾含量西北低、东南高,呈现由西北向东南波浪形递减的分布格局,与高程、坡度、地形起伏度呈极显著负相关;有效磷整体分布趋势表现为由中游向上下游降低,与高程、坡度呈极显著负相关;土壤 pH 沿某一方向变化的趋势不强,除与高程呈极显著正相关外,与其他地形因子无显著相关关系。对研究区地形因子进行分级,分析不同高程、坡度、地形起伏度、坡向下土壤养分差异,结果表明土壤养分变异情况与地形因子的相关性计算结果相一致,地形因子对土壤养分的影响与土壤养分空间分布格局基本一致。在地势较低且平坦的区域内,更有利于农业生产,土壤养分较为充足;在高海拔、地形起伏较大区域养分含量相对较低。

4 讨论

地形条件影响土壤和环境间物质能量交换,是造成土壤养分空间变异的主要因素之一[26]。海拔除了影响温、光、水、热条件,还与土层厚度及成土母质相关,土壤养分随着海拔的升高均呈降低趋势。坡度和地形起伏度越大,实地地势越陡峭,养分流失更严重;而地势平缓的地区易于保水保肥,土壤养分更丰富。光照在一定程度上影响养分的分解速率,所以坡向也会导致养分差异。延河流域上、下游水土流失与中游相比更加严重,除暴雨因素外,与上游坡大沟深的丘陵沟壑地形,土壤抗蚀力差,下游农业行动频繁,又缺乏植被保护也有一定关系。而中游植被较好,特别是延安以南地区生态环境修复、植树造林等因素影响,中游水土流失低于上、下游。全氮与有机质空间结构较为相似,氮素容易流失淋溶[27],在上、中、下游表现出较强的空间变异性。在东南海拔较低区域农业生产中施用钾肥较多,植物吸收量低或钾肥随径流流向下游造成速效钾含量较高。植物对磷肥吸收利用较差,中下游有效磷随泥沙流失造成含量降低,而中游含量较高。

土壤养分空间变异是普遍存在的,主要由自然因素和人为因素引发。前人研究表明,土壤养分空间变异主要取决于气候、地形、土壤类型等结构性自然因素。此外,施肥、种植制度、耕作措施等人为随机性因素也会对土壤养分空间变异产生较大影响。本文仅对土壤空间变异特征及地形因素的影响进行了分析,在今后的研究中仍需对于土壤类型等自然因素、土壤养分间相互影响以及人为因素在农业生产过程中的影响进行深入研究,以期为地区土壤养分管理、合理配方施肥提供依据。

参考文献

[1]刘占锋,傅伯杰,刘国华,等.土壤质量与土壤质量指标及其评价[J].生态学报,2006,26(3):901-913.

[2]杨玉玲,文启凯,田长彦,等.土壤空间变异研究现状及展望[J].干旱区研究,2001,18(2):50-55.

[3]Florinsky I V, Eilers R G, Manning G R, et al. Prediction of soil properties by digital terrain modeling[J]. Environ. model.soft,2002,17(3):295-311.

[4]郭旭东,傅伯杰,马克明,等.基于 GIS 和地统计学的土壤养分空间变异特征研究:以河北省遵化市为例[J].应用生态学报,2000(4):557-563.

[5]秦松,樊燕,刘洪斌,等.地形因子与土壤养分空间分布的相关性研究[J].水土保持研究,2008,18(1):46-49,52.

[6]高洁,武红旗,李新梅,等.近 40 年阜康市耕层土壤养分变化特征[J].西南农业学报,2020,33(10):2294-2302.

[7]连纲,郭旭东,傅伯杰,等.黄土高原小流域土壤养分空间变异特征及预测[J].生态学报,2008,45(3):946-954.

[8]夏岩,张姝琪,高文冰,等.黄土高原变绿对黄河中游延河流域径流演变的影响估算[J].地球科学与环境学报,2020,42(6):849-860.

[9]杨殊桐.黄土高原典型流域植被恢复对生态系统服务功能权衡协同关系的影响[D].西安:西安理工大学,2020: 1-5.

[10]王天宇,惠怡安,师莹,等.人为干扰度视角下黄土丘陵沟壑区延河流域景观格局演变预测及优化:以陕西省延长县为例[J].水土保持通报,2020,40(6):1-8.

[11]李晶,周自翔.延河流域景观格局与生态水文过程分析[J].地理学报,2014,69(7):933-944.

[12]刘胤汉.陕西省延安地区地理志[M].西安:陕西人民出版社,1983:731-738.

[13]李松.汉中市土壤养分空间变异及耕地地力评价研究[D].杨凌:西北农林科技大学,2016:45-46.

[14]鲍士旦.土壤农化分析[M].北京:中国农业出版社,2000:25.

[15]杨静涵,刘梦云,张杰,等.黄土高原沟壑区小流域土壤养分空间变异特征及其影响因素[J].自然资源学报,2020,35(3):743-754.

[16]郭兆元.陕西土壤[M].北京:科学出版社,1992:156-159.

[17]于洋.陕西省耕地土壤养分状况及耕地地力评价研究[D].杨凌:西北农林科技大学,2016:36-46.

[18]郭旭东,傅伯杰,陈利顶,等.河北省遵化平原土壤养分的时空变异特征:变异函数与 Kriging 插值分析[J].地理学报,2000,55(5):555-566.

[19]Miller P M, Singer M J, Nielsen D R. Spatial variability of wheat yield and soil properties on complex hills[J]. Soil Science Society of America Journal,1988,52:1133-1141.

[20]张忠华,胡刚,祝介东,等.喀斯特森林土壤养分的空间异质性及其对树种分布的影响[J].植物生态学报,2011,35(10):1038-1049.

[21]何志祥,朱凡.雪峰山不同海拔梯度土壤养分和微生物空间分布研究[J].中国农学通报,2011,27(31):73-78.

[22]罗丹,王涛,常庆瑞.县域农村居民点适宜性评价:以陕西省陇县为例[J].中国农学通报,2019,35(14):157-164.

[23]张素梅,王宗明,张柏,等.利用地形和遥感数据预测土壤养分空间分布[J].农业工程学报,2010,26(5):188-194.

[24]宋丰骥.黄土丘陵沟壑区土壤养分空间变异及其丰缺评价研究[D].杨凌:西北农林科技大学,2012:36.

[25]杨昕,张茜.基于数字高程模型的太阳辐射模拟[C]//武汉:中国地理学会2003年学术年会文集,2003:1.

[26]杨建虎,常鸿莉,魏琪.黄土高原小流域土壤养分空间特征及其与地形因子的相关性[J].西北农林科技大学学报(自然科学版),2014,42(12):85-90.

[27]陈皓,章申.黄土地区氮磷流失的模拟研究[J].地理科学,1991,11(2):142-148.

本文曾发表于2022年《中国农学通报》第38卷第12期

Utilizing the GOA-RF Hybrid Model, Predicting the CPT-based Pile Set-up Parameters

Zhilong Zhao[1], Simin Chen[1], Dengke Zhang[1], Bin Peng[1], Xuyang Li[1] and Qian Zheng[2]

(1.Shaanxi Construction of Land Comprehensive Development Co. Ltd, Xi'an Shaanxi, 710000, China;
2.Faculty of Civil Engineering, UAE Branch, Islamic Azad University, Dubai, 502321, UAE)

[Abstract] The undrained shear strength of soil is considered one of the engineering parameters of utmost significance in geotechnical design methods. In-situ experiments like cone penetration tests (CPT) have been used in the last several years to estimate the undrained shear strength depending on the characteristics of the soil. Nevertheless, the majority ofthese techniques rely on correlation presumptions, which may lead to uneven accuracy. This research's general aim is to extend a new united soft computing model, which is a combination of random forest (RF) with grasshopper optimization algorithm (GOA) to the pile set-up parameters' better approximation from CPT , based on two different types of data as inputs. Data type 1 contains pile parameters, and data type 2 consists of soil properties. The contribution of this article is that hybrid GOA-RF for the first time, was suggested to forecast the pile set-up parameter from CPT . In order to do this, CPT data and related bore log data were gathered from 70 various locations across Louisiana. With an R^2 greater than 0.9098, which denotes the permissible relationship between measured and anticipated values, the results demonstrated that both models perform well in forecasting the set-up parameter. It is comprehensible that, in the training and testing step, the model with data type 2 has finer capability than the model using data type 1, with R^2 and RMSE are 0.9272 and 0.0305 for the training step and 0.9182 and 0.0415 for the testing step. All in all, the models' results depict that the A parameter could be forecasted with adequate precision from the CPT data with the usage of hybrid GOA-RF models. However, the RF model with soil features as input parameters results in a finer commentary of pile set-up parameters.

[Keywords] Cone penetration test; Grasshopper optimization algorithm; Pile parameters; Pile set-up parameter *A*; Random forest model; Soil properties

1 Introduction

Development in the field of geotechnical Engineering is closely related to laboratory studies (Esmaeili-Falak *et al.* 2020, 2018). Cohesive soil and cohesion-less soil in both of them found that the pile resistance increments by spending time. In the process of driving, the soil around a pile is significantly deformed and perturbed radially (Li and Li 2021). The vastness of such a disrupted area is generally commensurate with the dislocated soil amount (Poorjafar *et al.* 2021). Additionally, the remarkable extension of excess pore water pressure (EPWP) reduces the soil's effective shear power (Esmaeili Falak and Sarkhani Benemaran 2022). The pile's resistance decreases as a consequence. After the pile drive is complete, the disturbed soil's prolonged excess pore water pressure gradually decomposes over time, followed by subsequent reconsolidation and remolding. As a result, the soil's power increases, which results in increased pile resistance. The term "pile set-up" refers to the enhancement of a pile's resistance that is time-dependent (Archer and Kimes 2008, Bullock 2008). Some factors, such as soil features, pile specs, and other factors, determine the extent of pile setup and the amount of time required (Esmaeili-Falak 2013, Maghsoodi *et al.* 2013). On the opposite, sometimes for very over-consolidated soils, negative pore water pressure can be extended where the circumambient soil experiences a reduction in the general strength causes to a smaller pile resistance, which is said as "Relaxation" (Richardson 2011).

1.1 Mechanism

A remolded region has developed all over the pile as it is being put up in the soil; this area allows the pile to move with the least amount of resistance, just like the soil volume does. A transition region with different soil features occurs after the remolded area. In the remolded region, there is significant extra pore water pressure (Bond and Jardine 1991, Haque 2015, Lee *et al.* 2011, Lee *et al.* 2010, Steward and Wang 2011). The extra PWP dissipates over time, reconsolidating the remoulded region. The first supporter of the increased pile resistance or setup is this reassembling of the remoulded region (Paikowsky *et al.* 1994). The pile setup does not end to increment even later than the whole waste of the extended EPWP that happens owing to the soil aging (Bergahl 1981, Komurka *et al.* 2003, Schmertmann, 1991). Three phases made up the pile setup process (Komurka *et al.* 2003):

Step 1: EPWP waste level is non-linear with the time method;

Step 2: EPWP waste level is linear with the time method;

Step 3: Soil aging.

However, due to the uneven treatment of soil, soils at different depths along the pile shaft may go through different stages of building up at the prescribed period.

1.2 Empirical models

Numerical methods have also been used to evaluate pile setup in addition to experimental and analytical approaches (Elias 2008, Ng *et al.* 2013). To estimate the set-up value, an empirical technique was developed (Wang and Reese 1989). The formula was based on a DLT analysis result of clayey soil taken from Shanghai, Pei, China. Based on this relationship, the pile power after the time t (day of setting up) is able to be approximately calculated as follows.

$$\frac{R_t}{R_{t0}}=0.263[1+\lg t]R_{\max} \tag{1}$$

In this equation, t shows time spacing after setting up and R_t shows the primary pile power.

The sensitivity (S_t) connection to pile set-up was firstly offered by analyzing 70 cohesive soils of test piles on 20 different structure operations in the littoral zone of East China (Guang-Yu 1988). He mentioned that incremented sensitivity (S_t) raises the pile resistance retake. The outcomes appraisal was found in this essay to which of *SLT* to show the formula below.

$$\frac{R_t}{R_{14}}=0.375\ S_t+1 \tag{2}$$

However, Denver and Skoy expanded the frequently utilized formula among the empirical formulas (Skov and Denver 1988). They introduce a semi-logarithmic technique based on data from four German and Danish file records. Based on soil's three kinds (chalk, clay, and sand), the formula below was determined.

$$\frac{R_t t}{R_{t_0}}=1+A\ \lg\frac{t}{t_0} \tag{3}$$

Where R_t shows the power of pile at a time (t), R_{t_0} shows the strength of the pile at the elementary time (t_0), t shows passed time from the finishing time of driving, t_0 shows reference time, and A shows the set-up parameter.

In the beginning time, t_0, is often defined as the time after the drive has finished at that logarithmically linear pile setup rate. It is hard to evaluate since it is dependent on the size of the piles and the kind of soil. Typically, apile's larger diameter results in a larger t_0. (Camp Ⅲ and Parmar, 1999). Knowing the ultimate pile resistance at different close intervals of time is required for accurate measurement of t_0, that is exceedingly challenging to perform in reality. Hence, to get t_0, the researchers used a back calculation, made an assumption based on stored data, or used an empirical formula. For PSC and H-piles, t_0 was obtained as two days (Camp Ⅲ and Parmar 1999). However, they said that a day is a reasonable quantity. In non-cohesion soil, a similar $t_0 = 1$ day was obtained for PSC piles (Axelsson1998). Furthermore, $t_0 = 1$ or 2 days were used (Svinkin *et al.* 1994). Numerous studies have also shown that $t_0 = 1$ day is the ideal (Paul Joseph Bullock 1999, McVay *et al.* 1999). The pile set-up variable A used in Eq. (3) linked to elements such as pile material, pile size, pile type, pile resistance, and soil type (Camp Ⅲ and Parmar 1999, Nafiul Haque *et al.* 2014 m Svinkin *et al.* 1994).$A = 0.2$ and 0.6 was suggested for clay and sand (Rikard Skov and Denver, 1988). Clay had a counting t_0 value of one day, whereas sand had a value of 0.5 days. However, that suggested value is independent of both the depth and excess PWP waste (Paul Joseph Bullock 1999, McVay *et al.* 1999). In order to get the A variable, the scholars had to do a back calculation or make an assumption using an empirical formula or field data. The range among 0.27~0.75 (Chow *et al.* 1998), among 0.2~0.8 (Axelsson 1998), and average value of 0.20 (Komurka *et al.* 2003) was suggested for A by utilizing fully detailed literature research.

1.3 Developed prediction models

Today, more than ever, consideration is being given to the application of artificial intelligence in the area of civil engineering (Liu *et al.* 2021, Luat *et al.* 2020, Sarkhani Benemaran *et al.* 2020, Xiang *et al.* 2021, Yang *et al.* 2022, Yuan *et al.* 2022, Zhu *et al.* 2022), especially on the field of geotechnical engineering (Johari, *et al.* 2011, Johari *et al.* 2011, Johari *et al.* 2016). Therefore, the constitutive model of di materials may be approximated at a significantly reduced time and cost by applying non-linear artificial neural regressions (Sarkhani Benemaran *et al.* 2022). Additionally, a number of studies have documented the use of artificially based techniques in a variety of engineering disciplines, such as admixtures in concrete (Benemaran and Esmaeili-Falak 2020), chloride diffusion in cement mortar (Hoang *et al.* 2017), triaxial compressive strength and Young's modulus of frozen sand (Mahzad Esmaeili-Falak *et al.* 2019), so on. In the case of soil attributes adapted from CPT, several artificial neural network (*ANN*) techniques are established according to the descriptions, such as investigating *ANN* to evaluate the undrained shear power of soil (Abu-Farsakh and Mojumder 2020), and final pile valence and pile set-up variable A (Mojumder 2020). Other hybrid techniques, such as the random forest model with optimization methods and Harris hawk optimization (HHO) for synchronized energy management (Abdelsalam *et al.* 2021), creep index forecast according to PSO and RF (Zhang *et al.* 2020), landslide susceptibility mapping based on RF hyperparameter optimization utilizing bayes method (Sun *et al.* 2020), and grasshopper optimization algorithm (GOA)-based proceed forward to categorize epileptic *EEG* signals (Singh *et al.* 2019), landslide susceptibility mapping (Sun *et al.* 2021, Sun *et al.* 2021, Zhou *et al.* 2021), durability evaluation of GFRP rebars (Iqbal *et al.* 2021), for heart diseases diagnosis (Asadi *et al.* 2021), prediction of proton-exchange membrane fuel cell (Huo *et al.* 2021), UHPFRC under ductility requirements for seismic retrofitting applications (Abellán-García and Guzmán-Guzmán 2021), smart meter data classification (Zakariazadeh 2022), rapid chloride permeability of SCC (Ge *et al.* 2022), so on, have also been effectively and realistically deployed.

1.4 Main objective of study

This research's main target is to extend a new united soft computing model, for example GOA-RF, which is a combination of random forest (RF) with grasshopper optimization algorithm (GOA) to

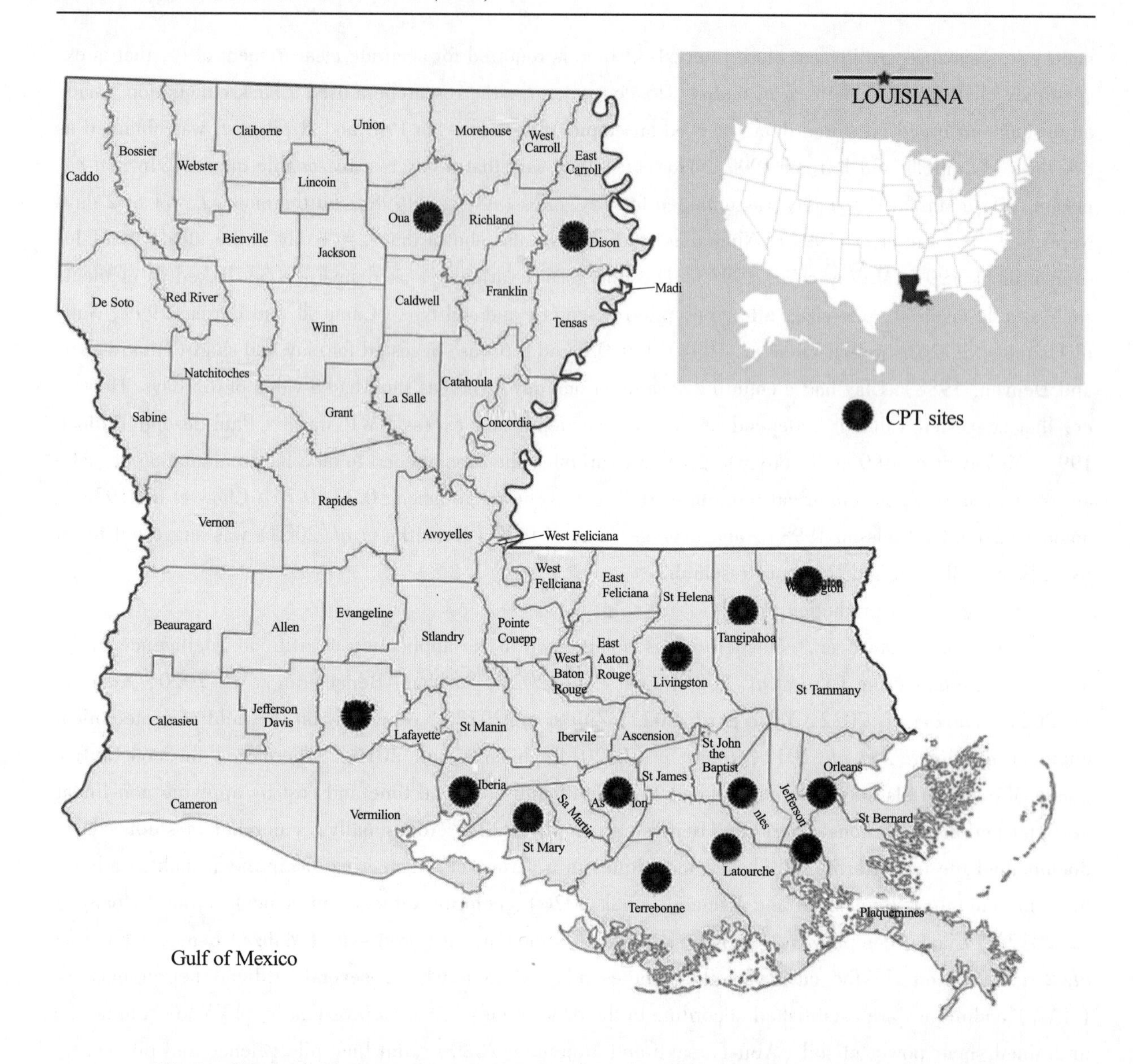

Fig.1 Cone penetration test locations in Louisiana (Abu-Farsakh and Mojumder 2020)

the pile set-up variables A's greater approximation from cone penetration test (CPT), according to two different types of data as inputs. Data type 1 contains mean emended cone tip resistance (q_t), mean skin friction (f_s), and effective overburden pressure (σ_{vo}), and data type 2 consists of plasticity index (PI), undrained shear strength of soil (S_u), and over consolidation ratio (OCR). The unique aspect of this work is that hybrid GOA-RF was initially suggested as a way to forecast pile set variables from CPT. In order to do this, CPT data and related bore log data were gathered from 70 various locations across Louisiana. Statistic evaluation variety is taken into account while evaluating the effectiveness of the constructed models.

2 Methodology

2.1 CPT and bore log database

CPT data and accompanying bore log data were gathered from 70 various locations in Louisiana in order to estimate the pile set-up variable from the CPT, as displayed in Fig. 1 (Abu-Farsakh and Mojumder 2020, Mojumder 2020). These studies performed the primary empirical experiments on samples of soil like grain size distribution (ASTM D422-63, 2017), water content (ASTM D4643-17, 2017), Atterberg limits (ASTM D4318-00, 2017), and unit weight (ASTM D7263-21, 2021) and in-situ CPT testing at aforesaid locations in Louisiana.

Undistributed Shelby tube samples modified from different borehole depths at every location were part of the empirical experiment plan. Triaxial unconsolidated undrained experiments were conducted in accordance with ASTM-D 2850-03a to assess the soil's undrained shear power (ASTM D2850-03, 2017). CPT (ASTM D3441-16, 2018) and RCPT were primarily used in the in-situ testing, that were conducted all around the boreholes that were bored using 10 cm^2 and 15 cm^2 piezocone penetrometers. If the RCPT also assessed excess pore pressure behind the cone (u_2) in alongside the previously specified characteristics, the cone penetration experiment might evaluate cone tip resistance (q_c) and cone sleeve friction (f_s). At constant distances of 2 cm in-depth, recordings were taken while the cone was being drilled in at a constant penetration level of 2 cm/sec .

The effectiveness of the model creation is significantly impacted by the parameters used as model inputs. While a parameter's big number increases the size of the network, that might cause a drop in the net work's efficiency due to slowed processing speed, the right selection of the variables can significantly improve the model's efficiency. In the current research, parameters of two various kinds were chosen as inputs to examine the different parameters' impact on the forecasting of A , in accordance with the literature. Selected variables for data type 1 were mean modified cone tip resistance (q_t), mean skin friction (f_s), and efficient overburden pressure (σ_{vo}), and for data type 2 were plasticity index (PI), undrained soil's shear power (S_u), and over consolidation ratio (OCR) for train models to identify the network that produces the greatest outcomes. In this case, the pile set-up variable (A) was the result, with values for data types 1 and 2 ranging from 0.07 to 0.53 and from 0.12 to 0.53, respectively. The power of the different layers was assessed based on the strain gauge measurements of the load diffusion during the static load experiments. The set-up level variable A was then calculated from Eq. (3) since the resistance of the pile (R_t) at a given time (t) and the resistance of the pile (R_{t0}) at the starting time (t_0) had already been measured (Skov and Denver, 1988). In this case, variable A heavily depends on soil characteristics. CPT is a strong, effective instrument that has been used for this purpose throughout time to gauge the soil's power. Following the dataset collection, inputs were modified to identify the finest model that produces the finest forecast for evaluating the A .

The inexpensive independent data collection is split into two subcategories during the split step: training data (used to develop) and testing (used to determine when to halt the training). Different dataset ratios among these subgroups were used (Hammerstrom 1993, Looney 1996, Shahin *et al.* 2004, Stone 1974). In the current research, the data gathered from locations in Louisiana were accidentally split into 25% for testing and 75% for training. All of the records' available patterns are included in the training data since it was given in this manner. The test data will check the model's extrapolation capacity rather than its interpolation ability when the points with excessive value are excluded from the training set, that results in having modes that are not relevant. Table 1 provides a statistics breakdown of the input and output parameters utilized in the model's development. Also, the histogram figures of inputs and output for data types 1 and 2, along with their normal distribution, are plotted in Figs. 2 and 3.

Table 1　The statistical values of the inputs and outputs from the train and test data

Category	Property	Data type 1				Data type 2			
		Inputs			Output	Inputs			Output
		q^t	f_s	G_{vO}	A	PI	S_u	OCR	A
		tsf^*	tsf	tsf	–	%	tsf	–	–
Train data	Min	1.52	0.04	0.55	0.10	4	0.07	1	0.12
	Max	108.05	2.09	10.27	0.53	84	1.57	3.05	0.53
	St. deviation	21.65	0.47	2.06	0.13	19.5719	0.434	0.689	0.1101
	Median	15.44	0.49	2.62	0.31	37	0.475	1	0.305
	Average	22.01	0.68	3.21	0.30	40.5192	0.6658	1.5019	0.3042
	Skewness	2.72	0.82	1.23	−0.02	0.338	0.5091	1.0448	0.06
Testdata	Min	3.32	0.10	1.09	0.07	4	0.07	1	0.12
	Max	80.24	1.54	9.15	0.42	84	1.45	3.05	0.53
	St. deviation	18.75	0.41	2.47	0.11	20.4054	0.3896	0.7392	0.1214
	Median	14.73	0.45	4.42	0.26	28	1.11	1.77	0.18
	Average	21.04	0.53	4.42	0.26	33.1765	0.9382	1.8124	0.2424
	Skewness	2.00	1.25	0.16	−0.02	1.1445	−0.7432	0.3483	1.1112

Note: * tsf = Tons Force per Square Foot.

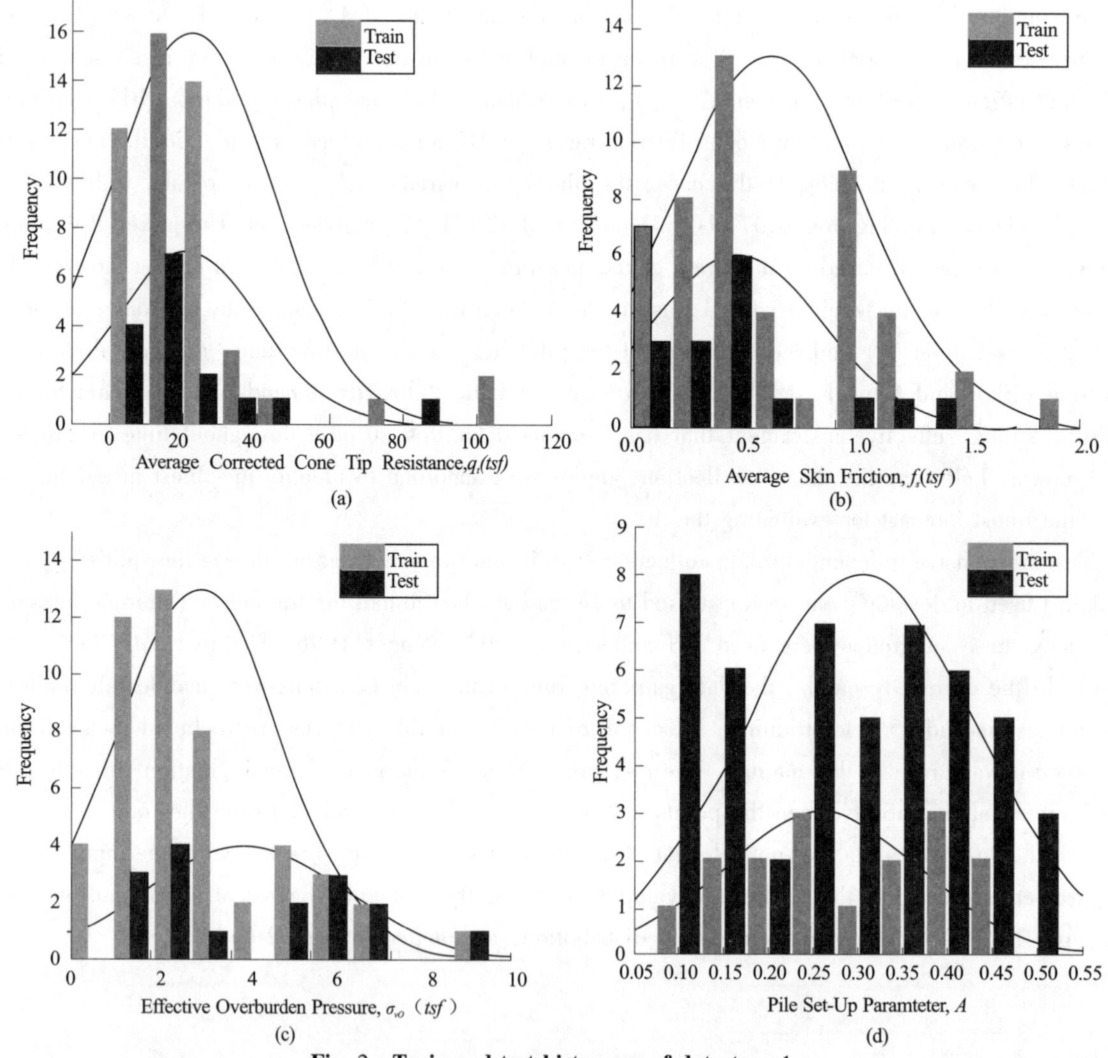

Fig. 2　Train and test histogram of data type 1;

(a) q_t, (b) f_s, (c) σ_{vO} and (d) A (with their normal distribution)

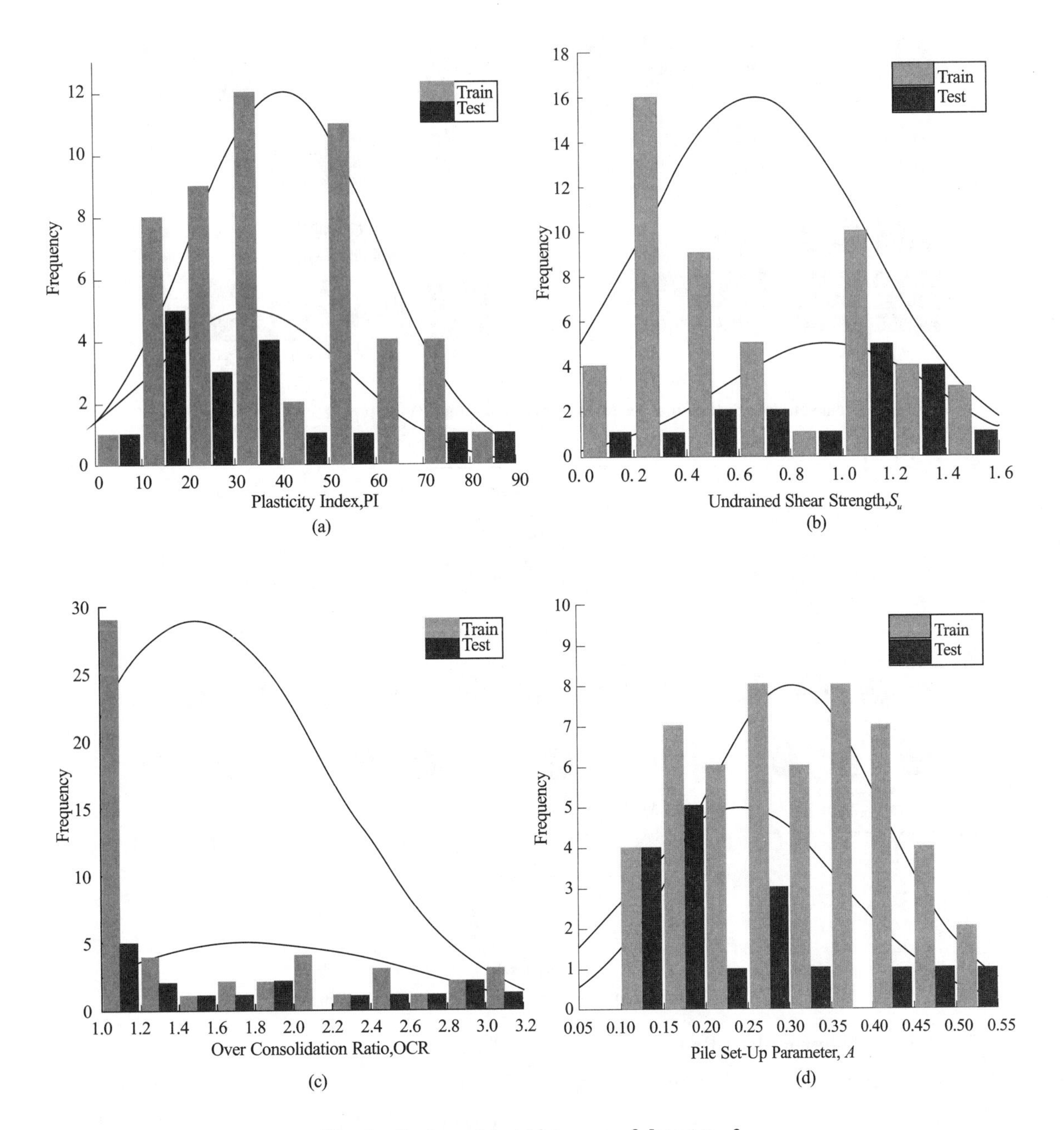

Fig. 3 Train and test histogram of data type 2;
(a) PI, (b) S_u, (c) OCR and (d) A (with their normal distribution)

Then, the relation coefficients among the variables under consideration were calculated and compiled in Figs. 4(a) and 4(b) for data types 1 and 2, see. A large positive or negative correlation coefficient between the parameters may indicate that the procedures have a low yield and that it is difficult to discern how the explanatory factors affect the result. As it can be obtained from Fig. 4(a) for data type 1, there are moderate positive correlation between f_s and q_t. A and f_s also have a moderately negative correlation. Additionally, there is no remarkable linear link among σ_{vo} or any of the parameters. Turning to data type 2 (Fig. 4(b)), there are moderate correlation between OCR and PI, as well as between OCR and S_u. Moreover, an almost high correlation is depicted between A with PI and S_u.

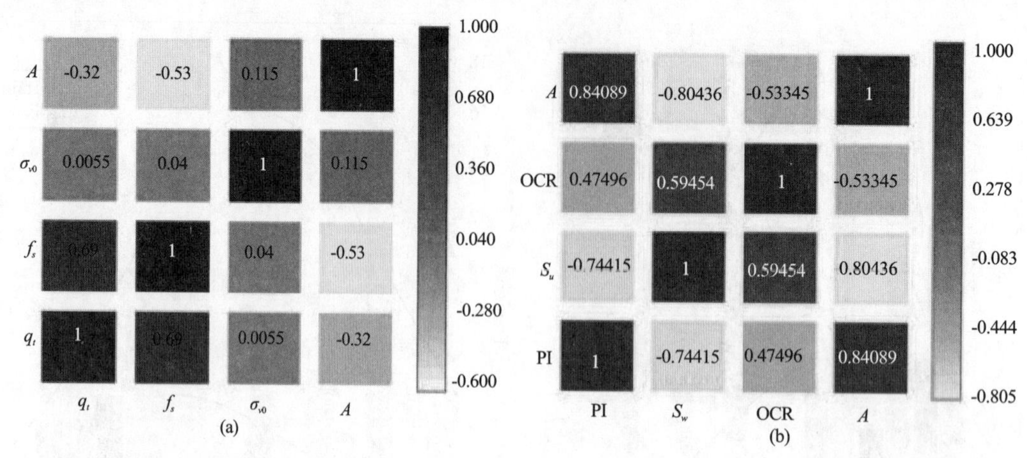

Fig. 4 Pearson correlation matrix of the variables; (a) Data type 1 and (b) Data type 2

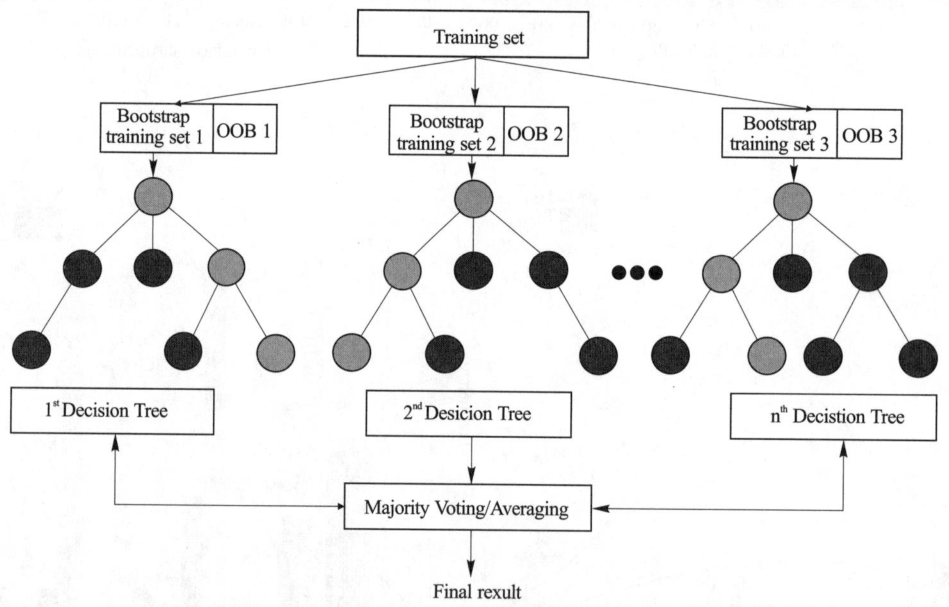

Fig. 5 Manufacturing of a random forest's Flowchart

2.2 Random Forest model (RF)

Breiman suggested RF model, which is recognized as a robust method to gain the answer for unsupervised learning, regression, and categorization issues (Breiman 2001, Liaw and Wiener 2002). RF with passable performance outcomes has been extended in several scopes (Nhu *et al.* 2020, Shozib *et al.* 2021, Wang *et al.* 2021). Compared to different techniques like great accuracy with accurate data information containing tiny parameters and calibrating, RF contains several benefits (Hong *et al.* 2016, Stumpf and Kerle 2011). The bagging technique is used to voluntarily choose the parameters nominees within the entire data collection in order to calibrate models in the categorization issues case (Chen *et al.* 2014). In the finding procedure, an Out-Of-Bag sample and two types of faults called decrement in precision and reduction in Gini (Eq. (4)) were calculated due to these faults might be accepted to level and selected parameters (Archer and Kimes 2008, Biau *et al.* 2008). Considering every variables, the function specifies the fault for the model when the variable values are exchanged throughout the *OOB* computation (Trigila *et al.* 2015). The manufacturing of a random forest procedure is presented in Fig. 5.

$$\mathrm{GI}(t_{X(xi)}) = 1 - \sum_{j=1}^{m} f(t_{X(xi),j})^2 \tag{4}$$

In general, the seek procedure of nature-inspired optimization algorithms contains two phases: (1) exploration and (2) exploitation. The exploration phase requires suddenly altering location. In the time of exploitation, the insects usually have local movements. Grasshoppers carry out exploration and exploitation as matures and nymphs. These both natural treatment procedures of grasshoppers were studied and designed to propose the "grasshopper optimization algorithm" (Mafarja *et al.* 2018, Mirjalili *et al.* 2018, Saremi *et al.* 2017, Talaat *et al.* 2020, Topaz *et al.* 2008). Therefore, the GOQ was expanded by imitating the grasshopper's treatment. The mathematical pattern of a grasshopper swarm could be demonstrated as Eq. (5).

$$X_i = S_i + G_i + A_i \tag{5}$$

where X_i is the location, S_i shows power of social interplay, G_i present gravity power of the ith grasshopper, and A_i is the evolution of air. In the search area, the search factor is dispensed accidentally. To gain accidental treatment, Eq. (5) is rectified as Eq. (6).

$$X_i = r_1 S_i + r_2 G_i + r_3 A_i \tag{6}$$

In this equation, r_1, r_2, and r_3 are accidental numbers between [0,1]. The force of social interplay of the ith grasshopper could be computed by

$$X_i = \sum_{\substack{j=1 \\ j \neq i}}^{N} S_s(d_{ij}) \widehat{d_{ij}} \tag{7}$$

where d_{ij} is the interval among the ith and the jth grasshopper; N is the whole number of grasshoppers; $\widehat{d_{ij}}$ is monad vector from i^{th} to j^{th} grasshopper; S_s is the rigidity of the social powers subordinate.

Particularly, S_s is computed by Eq. (8).

$$S_s(r) = fe^{\frac{-r}{l}} - e^{-r} \tag{8}$$

Here f show the strength of attraction, and I is the attractive length scale. The suggested value for I is 1.5 and for f is 0.5 (Khan *et al.* 2013, Talaat *et al.* 2020). The gravity power of the ith grasshopper, G_i is computed by

$$G_i = -g\hat{e}_g \tag{9}$$

where $\hat{e}_g$ is monad vector into the middle of the Earth, and g is the gravitational constant. Wind evolution, A_i is able to computed by

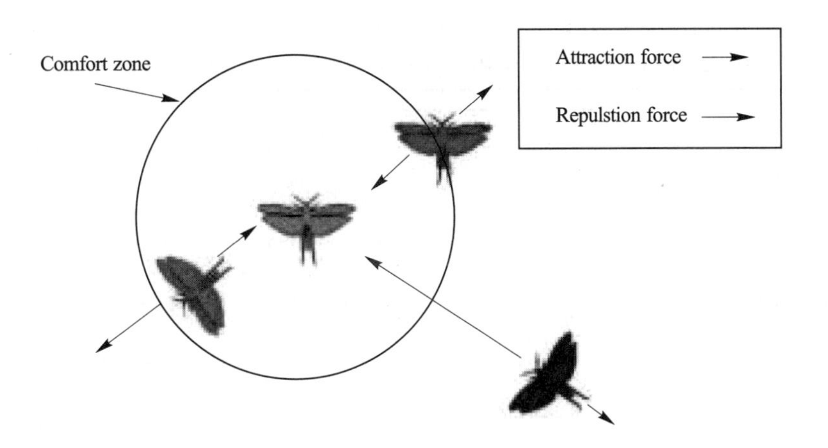

Fig. 6 Primitive corrective patterns between individuals in a swarm of grasshoppers (**Saremi *et al.*** 2017)

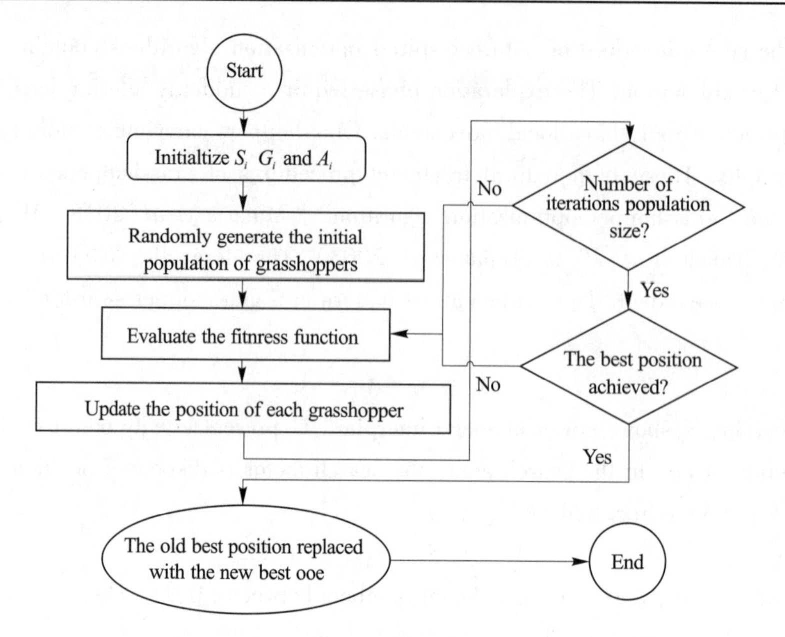

Fig. 7 The flowchart of GOQ

$$A_i = u\,\widehat{e_w} \tag{10}$$

where $\widehat{e_w}$ is the unit vector is the wind direction, and u present the drift constant. Eq. (11) could be gained by replacing the value of A_i, G_i, and S_i in Eq. (5).

$$X_i = \sum_{\substack{j=1 \\ j\neq 1}}^{N} S(|x_j - x_i|)\frac{x_j - x_i}{d_{ij}} - G\hat{e}_g + u\,\widehat{e_w} \tag{11}$$

The past mathematical method is not able to be utilized straightly for fixing optimization issues due to a grasshopper's swarm does not converge to a determined aim. Eq. (11) is modified as Eq. (12) to gain the optimal solvation.

$$X_i^d = c\left(\sum_{\substack{j=1 \\ j\neq 1}}^{N} c\frac{ub_d - lb_d}{2}s|x_j^d - x_i^d|\frac{x_j - x_i}{d_{ij}}\right) + \widehat{T_d} \tag{12}$$

where u_{bd} is lower limitation in the dth dimension; l_{bd} is upper limitation in the dth dimension; $\widehat{T_d}$ is value of the aim in the dth dimension; c is a coefficient which reduces in ratio to the number of iterations.

Table 2 Lists the primary RF hyperparameters and describes each one

Hyperparameters	Explanation
Max_Depth	The highest depth of DTs
Min_Samples_Split	The lower number of samples for the split
Min_Samples_Leaf	The lower number of samples at the leaf node
Max_DT	The lower number of RT models in the ensemble
Max_Features	The number of features considered during the selection of the best splitting

The coefficient c is able to be computed as bellows

$$c = c_{\max} - k\frac{c_{\max} - c_{\min}}{k_{\max}} \tag{13}$$

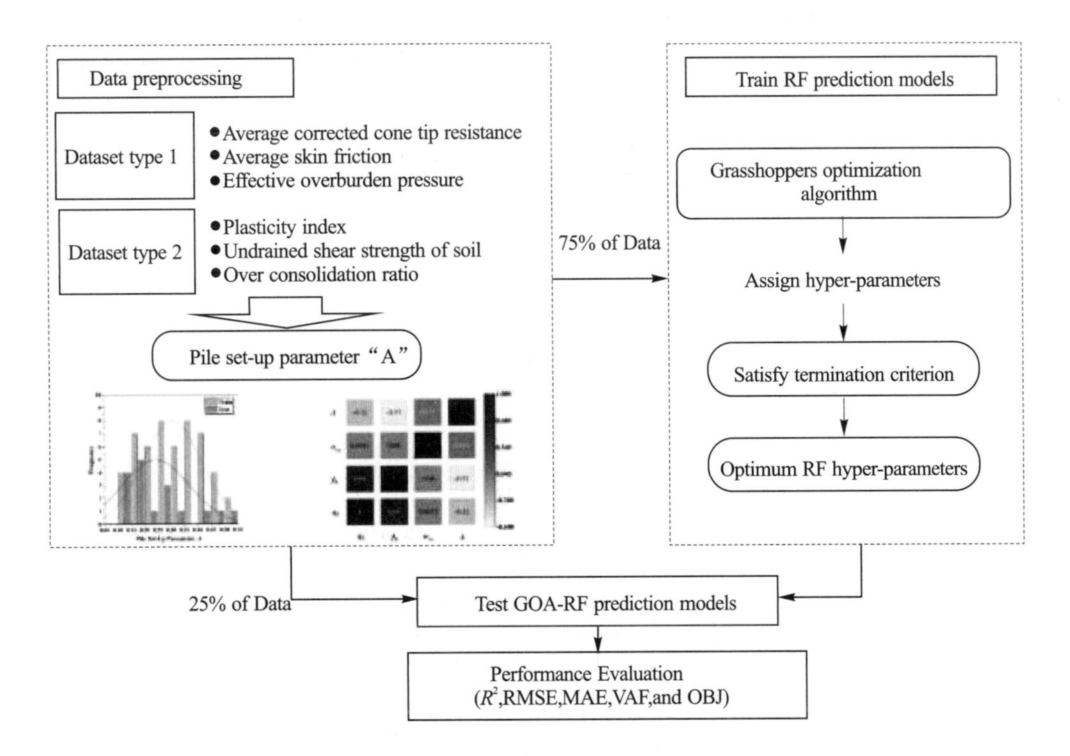

Fig. 8 Procedure for developing GOQ-RF models

In Eq. (13), c_{max} show maximum value of the reducing coefficient and c_{min} depict minimum value of the reducing coefficient, k_{max} is maximum iteration, and k specify the present iteration. The primitive corrective patterns between individuals in a swarm of grasshoppers and the flowchart of the GOA are presented in Figs. 6 and 7, respectively.

2.3 Models' framework

In order to model the non-linear correlation among the dependent parameters and independent parameters, the RF categorization and regression technique is used in the current research. In order to gain a high-performance model, Applying the GOA method adjusted the RF model's five significant hyperparameters that play an axial part in the learning procedure (Qi *et al.* 2018). Table 2 adjusted the necessary hyperparameters and summarized their description. To reduce the random separating's randomness, five RF models were created regarding every parameters' set. In the optimization methods, the trial-experiment procedure determined the entire variable tuning. The training efficiency from 10-fold C_V was used as the GOA's fitness function throughout the variables adjusting procedure. Within the optimization techniques, a particle represented every group of hyperparameters. Locations modified to get the maximum suitable value as well as the taken into consideration hyperparameters were improved in accordance by repeating the optimization phases. Finally, in the testing phase, the random forest model with the ideal hyperparameters was verified. Fig. 8 provides the process for creating GOA-RF models for data types 1 and 2.

2.4 RF models assessment

Two RF models optimized with GOA were developed in order to forecast the pile set-up variable A in the present study. Training and testing data efficiency has been used to assess the accuracy of mentioned approach's predictions. In the present paper, six model performance statistical criteria were used to evaluate the accuracy and precision of the approaches, such as coefficient of determination (R^2), root mean square error (RMSE), mean absolute error (MAE), the variance accounted factor (VAF), and OBJ . The lower

RMSE , MAE , PI and OBJ values and higher VAF are more reliable statistical effects. The higher R^2 values represent the greater fit between the analytical and predicted values. These evaluations are described as follows

$$R^2 = \left(\frac{\sum_{n=1}^{N} (x_N - \bar{x})(y_N - \bar{y})}{\sqrt{[\sum_{n=1}^{N} (x_N - \bar{x})^2][\sum_{n=1}^{N} (y_N - \bar{y})^2]}} \right)^2 \tag{14}$$

$$\text{RMSE} = \sqrt{\frac{1}{N}\sum_{n=1}^{N} (y_n - x_n)^2} \tag{15}$$

$$\text{MAE} = \frac{1}{N}\sum_{n=1}^{N} | y_n - x_n | \tag{16}$$

$$\text{VAF} = \left(1 - \frac{\text{var}(x_p - y_p)}{\text{var}(x_p)}\right) \times 100 \tag{17}$$

$$\text{OBJ} = \frac{n_{\text{train}}}{N} \cdot \left(\frac{\text{RMSE}_{\text{train}} + \text{MAE}_{\text{train}}}{R^2_{\text{train}} + 1}\right) + \frac{n_{\text{test}}}{N} \cdot \left(\frac{\text{RMSE}_{\text{test}} + \text{MAE}_{\text{test}}}{R^2_{\text{test}} + 1}\right) \tag{18}$$

where, y_p, x_p, $\bar{x}$, and $\bar{y}$ represent the predicted values of the Nth pattern, the target values of the Nth pattern, the averages of the target values, and the averages of the predicted values, respectively.

3 Results and discussion

3.1 Hyperparameters tuning

The presented Fig. 9 determines the RMSE's lowest value specified by GOA with repetitions for two developed models of RF . It is evident from curves that the RMSE values declined continuously within the optimization methods' iteration. RMSE reduction curve with GOA for data type 1 was the worst one to stand at about 0.0382 at about 288 iterations. While, for data type 2, RMSE could obtain a lower value at about 0.0305. It could be concluded that data type 2 with the consideration of soil properties (PI , S_u, and OCR) is more appropriate than type 1 for predicting A parameter, but other aspects need to be scrutinized to prove this matter. As a result, it was determined that the RF hyperparameters shown in Table 3 were the best values for the GOA-RF models that had been constructed.

3.2 Predictive capability of the models

Here are the outcomes of the hybrid RF models that were created to forecast the pile set-up variable A . In the current research, the non-linear correlation among the dependent parameters and independent parameters is modeled using the RF technique. In order to gain a high-performance model, Applying the GOA (GOA-RF) method adjusted the RF model's five significant hyperparameters that play an axial part in the learning procedure. In this research, the data gathered from locations in Louisiana were split at random into 25% for testing and 75% for training. A foresaid hybrid models were trained to utilize two different types of properties, first using pile parameters (q_t, f_s, and Gvo) and second with soil properties (PI , S_u, and OCR). In the training and testing phases, Figs. 10 and 11 provide the comparison of the measured and forecasted values by GOA-RF for models with data types 1 and 2.

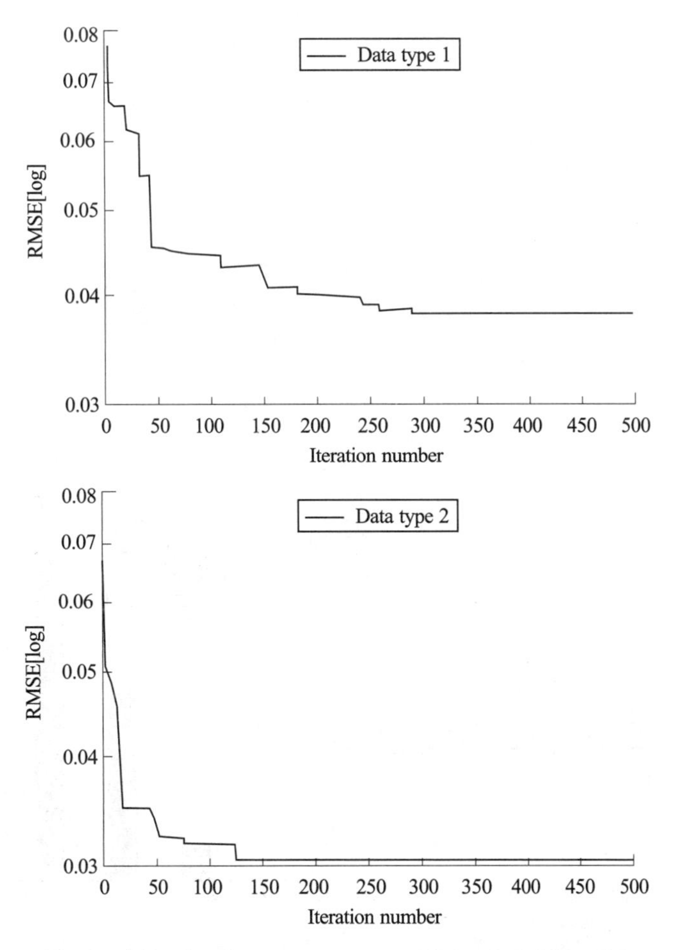

Fig.9 Adjusting hyperparameters utilizing hybrid models

Table 3 General hyperparameters' value implicated in GOA-RF for data type 1 and 2

Hyperparameters	Developed models with GOA-RF	
	Data type 1	Data type 2
Max_Depth	15	12
Min_Samples_Split	5	3
Min_Samples_Leaf	1	1
N_estimator	845	689
Max_Features	0.834	0.957

Concerning the statistical assessment criteria, five performance indices (R^2, RMSE, MAE, VAF, and OBJ) were taken into consideration; the findings are compiled in Table 4 and allow for a thorough comparison of the utility of the constructed hybrid algorithms. Outcomes indicate that both models forecast the set-up variable (A) with reasonable accuracy, with an R^2 value of more than 0.9098, which indicates a reasonable degree of relationship among measured and forecasted values.

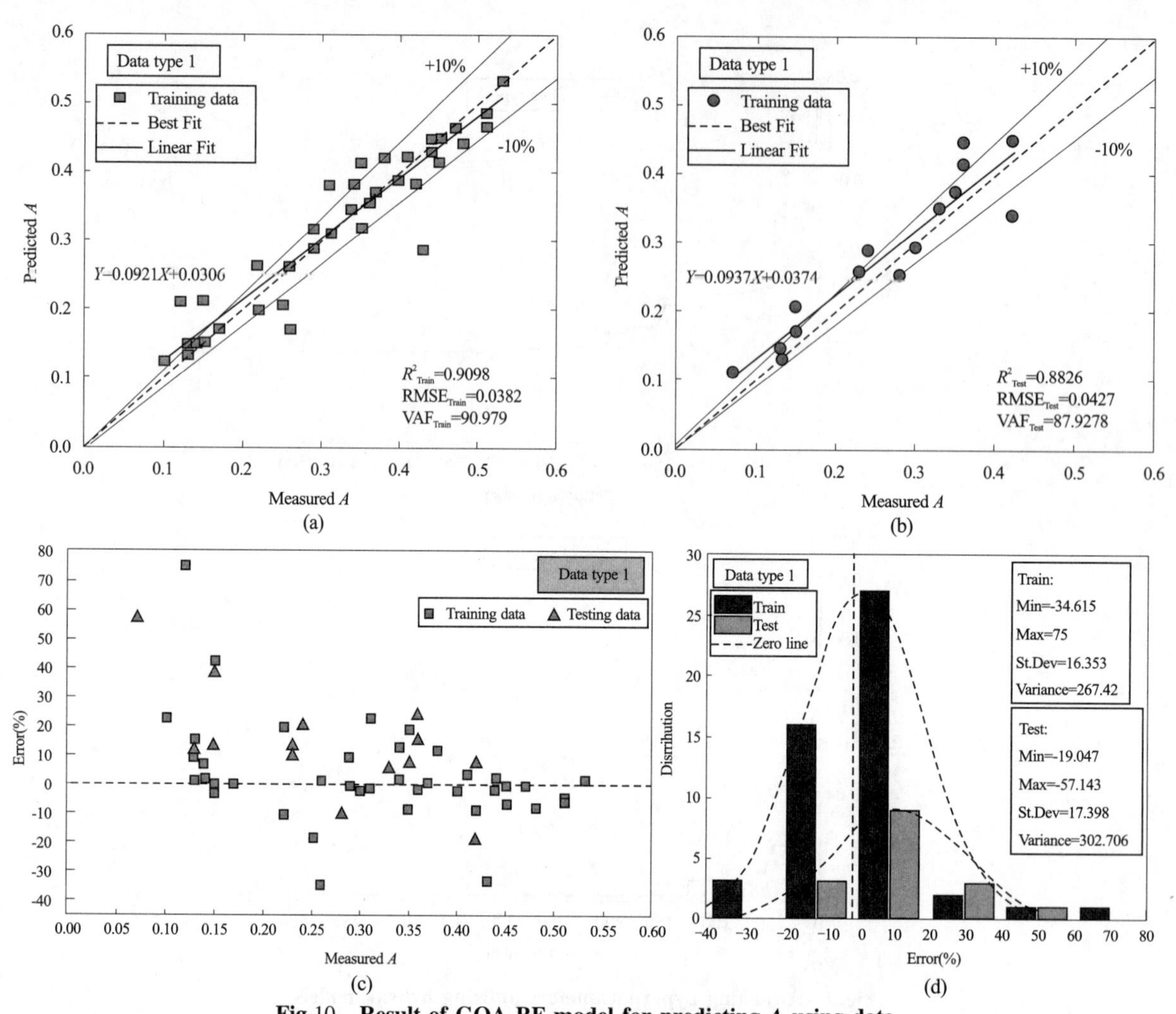

Fig.10 Result of GOA-RF model for predicting A using data type 1: (a) and (b) Forecasted against measured; (c) Measured against Fault and (d) Fault's diffusion

Table 4 Hybrid GOA-RF models' performance appraisal for predicting A

Models	GOA-RF								
	Training phase				Testing phase				
	R^2	RMSE	MAE	VAF (%)	R^2	RMSE	MAE	VAF (%)	OBJ
Data type 1	0.9098	0.0382	0.0233	90.9795	0.8826	0.0427	0.0354	87.9278	1.1208
Data type 2	0.9272*	0.0305	0.0231	92.35	0.9182	0.0415	0.0312	88.7095	1.0453

It is comprehensible that, in the validating and learning step, the model with data type 2 (input parameters were PI, S_u, and OCR) has finer expertise than the model using data type 1 (q_t, f_s, and σ_{vo} as inputs), with R^2 and RMSE values presented as 0.9272 and 0.0305 for the training step and 0.9182 and 0.0415 for testing data. According MAE and VAF values, predicting A with data type 2 results in better values than type 1. Moreover, other performance indexes can consider other indexes (R^2, RMSE, and MAE) simultaneously is OBJ, in such a way that the lower OBJ value is the most massive model precision. When this index is taken into consideration, the GOA-RF model with data type 2 has a lower value, that results in it outperforming the model with soil variables in comparison to the model with pile variables for forecasting A variable, and is thus chosen as the suggested model. Overall, the models' findings show that using hybrid

GOA-RF models, the pile set-up variable may be forecasted from the CPT data with sufficient accuracy. A variable's commentary is improved by RF models that use soil features as input variables.

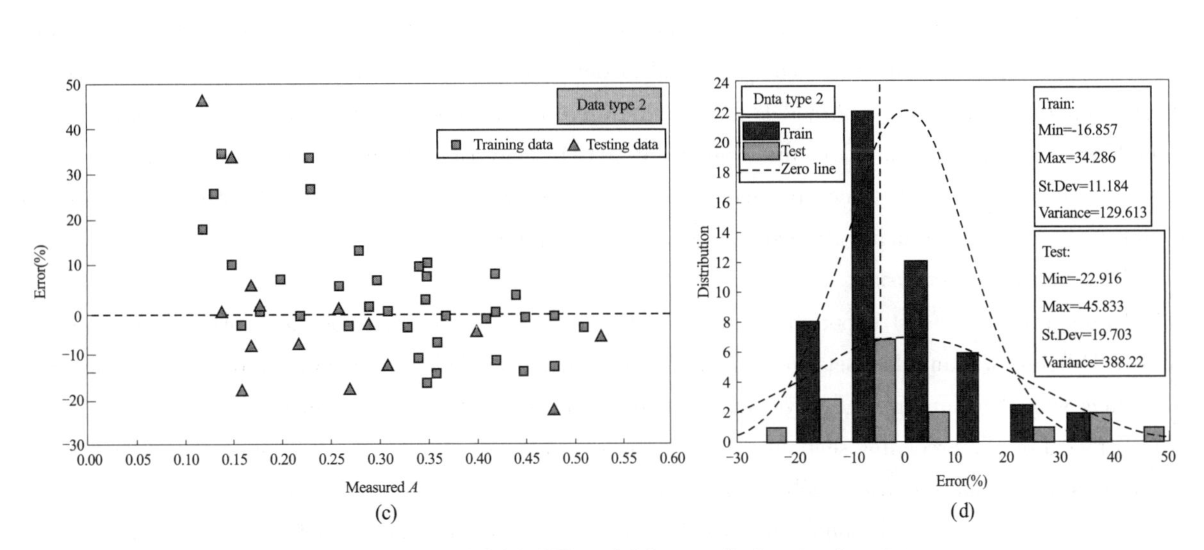

Fig. 11　Result of GOA-RF model for predicting A using data type 2: (a) and (b) Forecasted against measured, (c) Measured against Fault and (d) Fault's diffusion

The results of the GOA-RF models for data types 1 and 2 are shown in the provided Figs. 10 and 11, which reflect the forecasted, measured, and fault percentage of the A variable as well as the faults' diffusion. The scatter plot of measured A values versus forecasted values is shown in Figs. 10(a)、(b) and 11(a)、(b), and it demonstrates the allowable relationship among measured and forecasted values. Additionally, Figs. 10(c)、(d) and 11(c)、(d) show that the model's accuracy increases as the faults' diffusion around zero increases. Therefore, the suggested GOA-RF technique results in a higher number of instances around zero, with a diffusion curve resembling a Gaussian bell; this means that the model utilizing soil characteristics has a greater fault diffusion than the model utilizing pile variables in forecasting A variable

3.3　Comparison of developed models with ANN

The outcomes of both constructed models were contrasted with findings that had been suggested in the research to have comparison (Mojumder 2020). Table 5 compares the R^2 and RMSE values, in which for data type 1 (containing pile parameters), developed model yield much better performance than ANN, with a rise of about 0.14 in R^2 and a decline of about 23% for RMSE. Regarding the model with data type 2 (soil properties), slight R^2 improvement has resulted in comparison with ANN . Same as the results of literature, it is concluded that applying soil properties (PI , S_u, and OCR) yield a better interpretation of A parameter.

Table 5 Comparison of created systems

Model	Citation	Inputs	R^2	RMSE
ANN (3-4-1)	(Mojumder, 2020)	Data type 1	0.76	0.05
GOA-RF	This study		0.9098	0.0382
ANN (3-4-1)	(Mojumder, 2020)	Data type 2	0.92	0.03
GOA-RF	This study		0.9275	0.0305

3.4 Limitations and future recommendations

In order to examine the pile set-up parameters from the CPT , which are influenced by a number of factors, this research applies artificial intelligence. The type and number of parameters are very important and helpful when building the framework of algorithms. These limitations on future works might well be eased by increasing the volume of data from different projects, and it will also become more obvious how adaptable the models developed to evaluate the pile set-up parameters are. The idea of using new optimization techniques to improve the output of computational models and artificial intelligence leads to the creation of a unique generation or direction of models that may be used in a variety of settings. The structure of the fundamental RF model was optimized in this work using a GOA approach. Each optimization approach contains distinct, useful elements that need parametric analysis in order to attain the ideal state. Therefore, this constraint might be lifted by using alternative optimization techniques.

3.5 Sensitivity analysis

To determine the most important input variables, a sensitivity analysis of the hybrid system was performed. Different training data sets were created by concurrently eliminating a single input, and the test data set, along with the training set, provided the quantities of two statistical performance criteria, RMSE, and MAE . Based on the preceding sections, the data set is separated into 75% for training and 25% for testing. The results depict that removing each parameter causes to decrease in the performance of the model. The results are in Table 6, which is shown that the S_u is the most influential parameter for predicting the A parameter increased RMSE and MAE from 0.0305 to 0.0396, and 0.0231 to 0.0309 in the training phase, and from 0.0415 to 0.0502, and from 0.0312 to 0.0379 in the testing stage, respectively. The lowest effect belonged to OCR with respect to the model with all parameters. It is worth noting that reducing input parameters could only cause a little performance loss change for the model, but in the current research, since the analysis was founded on experimental data, omitting factors may reduce the model's generalizability. Given that the multicollinearity issue has no substantial influence on model fit and often does not impress much on predictions, the current research does not recommend eliminating any variables.

Table 6 sensitivity analysis by GOA-RF model

Inputs	Removed parameter	RMSE		MAE	
		Train	Test	Train	Test
	—	0.0305	0.0415	0.0231	0.0312
PI	PI	0.0375	0.0484	0.0282	0.0364
S_u	S_u	0.0396	0.0502	0.0309	0.0379
OCR	OCR	0.0342	0.0463	0.025	0.0349

4 Conclusions

The main objective of this research is to create a brand-new integrated soft computing model called GOA-RF , that combines the grasshopper optimization algorithm (GOA) and random forest (RF) to better estimate the pile set-up variables A from cone penetration test (CPT) using two various kinds of data as inputs. Data type 1 contains mean emended cone tip resistance (q_t), mean skin friction (f_s), and efficient overburden pressure (σ_{vo}), and data type 2 consists of plasticity index (PI), undrained soil's shear power (S_u), and over consolidation ratio (OCR). In order to do this, CPT data and related bore log data were gathered from 70 various locations across Louisiana.

· With an R^2 greater than 0.9098, which denotes the permissible relationship among measured and forecasted values, the outcomes demonstrate that both models operate well in forecasting the set-up variable.

· It is comprehensible that in both the validating step and learning step, the model with data type 2 has finer expertise than the model using data type 1, with R^2 and RMSE value is 0.927 2 and 0.030 5 for the training step and 0.9182 and 0.0415 for testing data. By taking into account the OBJ index, the GOA-RF model using data type 2 has a smaller value, leading to performing better the model with soil parameters (PI , S_u, and OCR) with respect to model with pile parameters (q_t, f_s, and σ_{vo}) for predicting A parameter, as a result, is determined as the suggested model.

· All in all, the models' outcomes depict that the pile set-up variable could be forecasted with adequate precision from the CPT data with the hybrid GOA-RF models' usage. However, RF models with soil characteristics (PI , S_u, and OCR) as input parameters result in a better commentary of A parameter.

· The type and number of parameters are very important and helpful when building the framework of algorithms. These limitations on future works might well be eased by increasing the volume of data from different projects, and it will also become more obvious how adaptable the models developed to evaluate the pile set-up parameters are. The structure of the fundamental RF model was optimized in this work using a GOA approach. Each optimization approach contains distinct, useful elements that need parametric analysis in order to attain the ideal state. Therefore, this constraint might be lifted by using alternative optimization techniques.

References

Abdelsalam, M., Diab, H.Y. and El-Bary, A.A. (2021), "A metaheuristic harris hawk optimization approach for coordinated control of energy management in distributed generation based microgrids", *Appl. Sci.*, 11(9), 4085.

Abellán-García, J. andGuzmán-Guzmán, J.S. (2021), "Random forest-based optimization of UHPFRC under ductility requirements for seismic retrofitting applications", *Constr. Build.Mater.*, 285, 122869.

Abu-Farsakh, M.Y. and Mojumder, M.A.H. (2020), "Exploring artificial neural network to evaluate the undrained shear strength of soil from cone penetration test data", *T. Res. Record*, 2674(4), 11-22.

Archer, K.J. and Kimes, R.V. (2008), "Empirical characterization of random forest variable importance measures", *Comput. Stat. Data Anal.*, 52(4), 2249-2260.

Asadi, S., Roshan, S. and Kattan, M.W. (2021), "Random forest swarm optimization-based for heart diseases diagnosis", *J. Biomed. Inform.*, 115, 103690.

ASTM D2850-03. (2017), "Standard test method for unconsolidated-undrained triaxial compression test on cohesive soils".

ASTM D3441-16. (2018), "Standard test method for mechanical cone penetration testing of soils".

ASTM D422-63. (2017), "Standard test method for particle-size analysis of soils".

ASTM D4318-00. (2017), "Standard test methods for liquid limit, plastic limit, and plasticity index of soils".

ASTM D4643-17. (2017), "Standard test method for determination of water content of soil and rock by microwave oven heating".

ASTM D7263-21. (2021), "Standard test methods for laboratory determination of density and unit weight of soil specimens".

Axelsson, G. (1998), "Long-term set-up of driven piles in non-cohesive soils evaluated from dynamic tests on penetration rods", *Geotech. Site Character.*, 895-900.

Benemaran, R.S. and Esmaeili-Falak, M. (2020), "Optimization of cost and mechanical properties of concrete with admixtures using MARS and PSO", *Comput. Concrete*, 26(4), 309-316.

Bergahl, U. (1981), "Load tests on friction piles in clay", *Proceedings of the 10th Int. Conf. on SMFE.*

Biau, G., Devroye, L. and Lugosi, G. (2008), "Consistency of random forests and other averaging classifiers", *J. Machine Learning Res.h*, 9(9).

Bond, A.J. and Jardine, R.J. (1991), "Effects of installing displacement piles in a high OCR clay," *Geotechnique*, 41(3), 341-363.

Breiman, L. (2001), "Random forests", *Machine Learning*, 45(1), 5-32.

Bullock, Paul J. (2008), "The easy button for driven pile setup: dynamic testing", *From Research to Practice in Geotechnical Engineering*, 471-488.

Bullock, Paul Joseph. (1999), "Pile friction freeze: A field and laboratory study," University of Florida.

Camp III, W.M. and Parmar, H.S. (1999), "Characterization of pile capacity with time in the Cooper Marl: study of applicability of a past approach to predict long-term pile capacity", *T. Res. Record*, 1663(1), 16-24.

Chen, W., Wang, Y., Cao, G., Chen, G. and Gu, Q. (2014), "A random forest model based classification scheme for neonatal amplitude-integrated EEG", *Biomed. Eng. Online*, 13(2), 1-13.

Chow, F.C., Jardine, R.J., Brucy, F. and Nauroy, J.F. (1998), "Effects of time on capacity of pipe piles in dense marine sand", *J. Geotech. Geoenviron. Eng.*, 124(3), 254-264. https://doi.org/10.1061/(ASCE)1090-0241(1998)124:3(254).

Elias, M.B. (2008), "Numerical simulation of pile installation and setup", 70(1).

Esmaeili-Falak, M. (2013), "Two-dimensional finite element analysis of influence of plasticity on the seismic soil-micropiles-structure interaction", *Tech. J. Eng. Appl. Sci.*, 3(13), 1301-1305.

Esmaeili-Falak, M, Katebi, H. and Javadi, A.A. (2018), "Experimental study of the mechanical behavior of frozen soils-a case study of Tabriz subway", *Periodica Polytechnica Civil Eng.*, 62(1), 117-125.

Esmaeili-Falak, M., Katebi, H. and Javadi, A.A. (2020), "Effect of freezing on stress-strain characteristics of granular and cohesive soils", *J. Cold Regions Eng.*, 34(2), 05020001.

Esmaeili-Falak, Mahzad, Katebi, H., Vadiati, M. and Adamowski, J. (2019), "Predicting triaxial compressive strength and Young's modulus of frozen sand using artificial intelligence methods", *J. Cold Regions Eng.*, 33(3), 4019007.

Esmaeili Falak, M. and Sarkhani Benemaran, R. (2022), "Investigating the stress-strain behavior of frozen clay using triaxial test", *J. Struct. Constr. Eng.*.

Ge, D.M., Zhao, L.C. and Esmaeili-Falak, M. (2022), "Estimation of rapid chloride permeability of SCC using hyperparameters optimized random forest models", *J. Sustain. Cement-Based Mater.*, 1-19.

Guang-Yu, Z. (1988), "Wave equation applications for piles in soft ground", *Proceedings of the 3rd International Conference on the Application ofStress-Wave Theory to Piles. Canada: Ottawa.*

Hammerstrom, D. (1993), "Neural networks at work", *IEEE Spectrum*, 30(6), 26-32.

Haque, M.N., Abu-Farsakh, M.Y., Chen, Q. and Zhang, Z. (2014), "Case study on instrumenting and testing full-scale test piles for evaluating setup phenomenon", *T. Res.Record*, 2462(1), 37-47.

Haque, M.N. (2015), "Field instrumentation and testing to study set-up phenomenon of driven piles and its implementation in LRFD design methodology."

Hoang, N.D., Chen, C.T. and Liao, K.W. (2017), "Prediction of chloride diffusion in cement mortar using multi-gene genetic programming and multivariate adaptive regression splines", *Measurement*, 112, 141-149.

Hong, H., Pourghasemi, H.R. and Pourtaghi, Z.S. (2016), "Landslide susceptibility assessment in Lianhua County (China): a comparison between a random forest data mining technique and bivariate and multivariate statistical models", *Geomorphology*, 259, 105-118.

Huo, W., Li, W., Zhang, Z., Sun, C., Zhou, F. and Gong, G. (2021), "Performance prediction of proton-exchange membrane fuel cell based on convolutional neural network and random forest feature selection", *Energ. Convers. Manag.*, 243, 114367.

Iqbal, M., Zhang, D. and Jalal, F.E. (2021), "Durability evaluation of GFRP rebars in harsh alkaline environment using optimized tree-based random forest model", *J. Ocean Eng. Sci.*

Johari, A, Habibagahi, G. and Ghahramani, A. (2011), "Prediction of SWCC using artificial intelligent systems: A comparative study", *Scientia Iranica*, 18(5), 1002-1008.

Johari, A, Javadi, A.A. and Habibagahi, G. (2011), "Modelling the mechanical behaviour of unsaturated soils using a genetic algorithm-based neural network", *Comput. Geotech.*, 38(1), 2-13.

Johari, Ali, Javadi, A.A. and Najafi, H. (2016), "A genetic-based model to predict maximum lateral displacement of retaining wall in granular soil", *Scientia Iranica*, 23(1), 54-65.

Khan, M.M., Ahmad, A.M., Khan, G.M. and Miller, J.F. (2013), "Fast learning neural networks using cartesian genetic programming", *Neurocomput.*, 121, 274-289.

Komurka, V.E., Wagner, A.B. and Edil, T.B. (2003), *Estimating soil/pile set-up*. Citeseer.

Lee, J., Prezzi, M. and Salgado, R. (2011), "Experimental investigation of the combined load response of model piles driven in sand", *Geotech. Test. J.*, 34(6), 653-667.

Lee, W., Kim, D., Salgado, R. andZaheer, M. (2010), "Setup of driven piles in layered soil", *Soils Found.*, 50(5), 585-598.

Liaw, A. and Wiener, M. (2002), "Classification and regression by random forest", *R News*, 2(3), 18-22.

Liu, J., Jiang, Y., Zhang, Y. and Sakaguchi, O. (2021), "Influence of different combinations of measurement while drilling parameters by artificial neural network on estimation of tunnel support patterns", *Geomech. Eng.*, 25(6), 439-453.

Looney, C.G. (1996), "Advances in feedforward neural networks: demystifying knowledge acquiring black boxes", *IEEE T. Knowledge Data Eng.*, 8(2), 211-226.

Luat, N.V., Lee, K. and Thai, D.K. (2020), "Application of artificial neural networks in settlement prediction of shallow foundations on sandy soils", *Geomech. Eng.*, 20(5), 385-397.

Mafarja, M., Aljarah, I., Heidari, A.A., Hammouri, A.I., Faris, H., Ala'M, A.Z. and Mirjalili, S. (2018), "Evolutionary population dynamics and grasshopper optimization approaches for feature selection problems", *Knowledge-Based Syst.*, 145, 25-45.

Maghsoodi, V., Atermoghaddam, F. and Esmaeili-Falak, M. (2013), "Parametric and two dimensional study of seismic behavior of micro pile group in sandy soil", *Intl. Res. J. Appl. Basic. Sci.*, 6(7), 901-909.

McVay, M.C., Schmertmann, J., Townsend, F. and Bullock, P. (1999), "Pile friction freeze: a field investigation study", *Research Report No. WPI* 0510632.

Mirjalili, S.Z., Mirjalili, S., Saremi, S., Faris, H. and Aljarah, I. (2018), "Grasshopper optimization algorithm for multi-objective optimization problems", *Appl. Intell.*, 48(4), 805-820.

Mohammad, L.N., Raghavendra, A., Medeiros, M., Hassan, M. and King, W. "Bill" (2018), "Louisiana transportation research center", *Louisiana State University*, 70808(225).

Mojumder, M.A.H. (2020), "Evaluation of undrained shear strength of soil, ultimate pile capacity and pile set-up parameter from Cone Penetration Test (CPT) using Artificial Neural Network (ANN)", LSU Master's Theses. 5145.

Ng, K.W., Suleiman, M.T. and Sritharan, S. (2013), "Pile setup in cohesive soil. II: Analytical quantifications and design recommendations", *J. Geotech. Geoenviron. Eng.*, 139(2), 210-222.

Nhu, V.H., Hoang, N.D., Duong, V.B., Vu, H.D. and Bui, D.T. (2020), "A hybrid computational intelligence approach for predicting soil shear strength for urban housing construction: a case study at Vinhomes Imperia project, Hai Phong city (Vietnam)", *Eng. with Comput.*, 36(2), 603-616.

Paikowsky, S.G., Regan, J.E. and McDonnell, J.J. (1994), *A simplified field method for capacity evaluation of driven piles final report.*

Poorjafar, A., Esmaeili-Falak, M. and Katebi, H. (2021), "Pile-soil interaction determined by laterally loaded fixed head pile group", *Geomech. Eng.*, 26(1), 13-25.

Qi, C., Chen, Q., Fourie, A. and Zhang, Q. (2018), "An intelligent modelling framework for mechanical properties of cemented paste backfill", *Miner. Eng.*, 123, 16-27.

Richardson, B.D. (2011), "A case study on pile relaxation in dilative silts," University of Rhode Island".

Saremi, S., Mirjalili, S. and Lewis, A. (2017), "Grasshopper optimisation algorithm: theory and application", *Adv. Eng. Softw.*, 105, 30-47.

Sarkhani Benemaran, R., Esmaeili-Falak, M. and Javadi, A. (2022), "Predicting resilient modulus of flexible pavement foundation using extreme gradient boosting based optimised models", *Int. J. Pavement Eng.* 1-20.

Sarkhani Benemaran, R., Esmaeili-Falak, M. and Katebi, H. (2020), "Physical and numerical modelling of pile-stabilized saturated layered slopes", *Proceedings of the Institution of Civil Engineers - Geotechnical Engineering*, 1-50.

Schmertmann, J.H. (1991), "The mechanical aging of soils", *J. Geotech. Eng.*, 117(9), 1288-1330.

Shahin, M.A., Maier, H.R. and Jaksa, M.B. (2004), "Data division for developing neural networks applied to geotechnical engineering", *J. Comput. Civil Eng.*, 18(2), 105-114.

Shozib, I. A., Ahmad, A., Rahaman, M. S. A., majdi Abdul-Rani, A., Alam, M. A., Beheshti, M. and Taufiqurrahman, I. (2021), "Modelling and optimization of microhardness of electroless Ni−P−TiO_2 composite coating based on machine learning approaches and RSM", *J. Mater. Res. Technol.*, 12, 1010-1025.

Singh, G., Singh, B. and Kaur, M. (2019), "Grasshopper optimization algorithm−based approach for the optimization of ensemble classifier and feature selection to classify epileptic EEG signals", *Med. Biol. Eng. Comput.*, 57(6), 1323-1339.

Skov, R. and Denver, H. (1988), "Time-Dependence of bearing capacity of piles", *Proceedings of the 3rd Int. Conf. App. Stress-Wave Theory to Piles.*

Skov, R. and Denver, H. (1988), "Time-dependence of bearing capacity of piles", *Proceedings of the 3rd International Conference on the Application of Stress-Wave Theory to Piles. Ottawa.*

Steward, E.J. and Wang, X. (2011), "Predicting pile setup (freeze): a new approach considering soil aging and pore pressure dissipation", In *Geo-Frontiers* 2011: *Advances in Geotechnical Engineering.*

Stone, M. (1974), "Cross - validatory choice and assessment of statistical predictions", *J. Roy. Stat. Soc.: Series B (Methodological)*,36(2),111-133.

Stumpf, A. and Kerle, N. (2011), "Object-oriented mapping of landslides using random forests", *Remote Sens. Environ.*, 115(10), 2564-2577.

Sun, D., Shi, S., Wen, H., Xu, J., Zhou, X. and Wu, J. (2021), "A hybrid optimization method of factor screening predicated on GeoDetector and random forest for landslide susceptibility mapping," *Geomorphology*, 379, 107623.

Sun, D., Wen, H., Wang, D. and Xu, J. (2020), "A random forest *CC*model of landslide susceptibility mapping based onhyperparameter optimization using Bayes algorithm",*Geomorphology*, 362,107201.

Sun, D., Xu, J., Wen, H. and Wang, D. (2021), "Assessment of landslide susceptibility mapping based on Bayesian hyperparameter optimization: A comparison between logistic regression and random forest", *Eng. Geol.*, 281, 105972.

Svinkin, M.R., Morgano, C.M. and Morvant, M. (1994), "Pile capacity as a function of time in clayey and sandy soils",*Proceedings of the Deep Foundations Institute Fifth International Conference and Exhibition on Piling and Deep Foundations.*

Talaat, M., Hatata, A.Y., Alsayyari, A.S. and Alblawi, A. (2020),"A smart load management system based on the grasshopper optimization algorithm using the under-frequency load shedding approach", *Energy*, 190, 116423.

Topaz, C.M., Bernoff, A.J., Logan, S. and Toolson, W. (2008), "A model for rolling swarms of locusts", *Eur. Phys. J. Spec. Topics*,157(1), 93-109.

Trigila, A., Iadanza, C., Esposito, C. andScarascia-Mugnozza, G.(2015), "Comparison of logistic regression and random forests techniques for shallow landslide susceptibility assessment in Giampilieri (NE Sicily, Italy)", *Geomorphology*, 249, 119-136.

Wang, J., Fa, Y., Tian, Y. and Yu, X. (2021), "A machine-learning approach to predict creep properties of Cr - Mo steel with timetemperature parameters", *J. Mater. Res. Technol.*, 13, 635-650.

Wang, S.T. and Reese, L.C. (1989), "Predictions of response of piles to axial loading", *Predicted and Observed Axial Behavior ofPiles: Results ofa Pile Prediction Symposium*, 173-187.

Xiang, G., Yin, D., Cao, C. and Yuan, L. (2021), "Application of artificial neural network for prediction of flow ability of soft soil subjected to vibrations", *Geomech. Eng.*, 25(5), 395-403.

Yang, C., Feng, H. and Esmaeili-Falak, M. (2022), "Predicting the compressive strength of modified recycled aggregate concrete", *Structural Concrete*.Yuan, J., Zhao, M. and Esmaeili - Falak, M. (2022), "A comparative study on predicting the rapid chloride permeability of selfcompacting concrete using meta - heuristic algorithm and artificial intelligence techniques", *Struct. Concrete*, 23(2), 753-774.

Zakariazadeh, A. (2022), "Smart meter data classification using optimized random forest algorithm", *ISA Transactions*, 126, 361-369.

Zhang, P., Yin, Z.Y., Jin, Y.F. and Chan, T.H.T. (2020), "A novel hybrid surrogate intelligent model for creep index prediction based on particle swarm optimization and random forest", *Eng.Geol.*, 265, 105328.

Zhou, X., Wen, H., Zhang, Y., Xu, J. and Zhang, W. (2021), "Landslide susceptibility mapping using hybrid random forest with GeoDetector and RFE for factor optimization", *Geosci.Front.*, 12(5), 101211.

Zhu, W., Huang, L., Mao, L. and Esmaeili - Falak, M. (2022), "Predicting the uniaxial compressive strength of oil palm shell lightweight aggregate concrete using artificial intelligencebased algorithms", *Struct.Concrete*.

本文曾发表于2022年《Geomechanics and Engineering》第31卷第2期

Characteristics of Rural Sewage Discharge and a Case Study on Optimal Operation of Rural Sewage Treatment Plant in Shaanxi, China

Yi Rong[a, c], Yang Zhang[a, c], Zenghui Sun[a, c], Zhe Liu[a, c], Xin Jin[b], Pengkang Jin[b]

(a.Institute of Land Engineering and Technology, Shaanxi Provincial Land Engineering Construction Group Co., Ltd., Xi'an, Shaanxi 710021, China; b.School of Human Settlements and Civil Engineering, Xi'an Jiaotong University, Xi'an, Shaanxi 710049, China; c.Shaanxi Provincial Land Engineering Construction Group Co., Ltd., Xi'an, Shaanxi 710075, China)

【Abstract】 Rural wastewater treatment has become an important issue in China. This study investigated 76 rural wastewater treatment plants (WWTPs) in Shaanxi over a one-year period. We found that although rural sewage is suitable for biological treatment (BOD_5/COD = 0.38~0.41), it has an obvious intermittent flow cutoff (5~19 h/d) characteristic, resulting in low sewage loading and poor pollutant removal in rural WWTPs. In order to improve operation efficiency and increase economic benefits of the existing rural WWTPs, a pilot-scale A^2/O bioreactor was established, and the effect of rural sewage discontinuity (cutoff duration 10 h/d) on A^2/O process was studied. An intermittent aeration and dissolved oxygen (DO) regulation operation strategy was developed. The results demonstrated that ammonia oxidizing bacteria (AOB), nitrite oxidizing bacteria (NOB), and polyphosphate accumulating bacteria (PAOs) were all enriched when $T_{aeration}:T_{stop}=5$ h:5 h and DO = 2.0~3.0 mg/L. During the subsequent process of reducing DO concentration to 0.5~1.0 mg/L, AOBs (2.43%) and PAOs (1.91%) were enriched and NOBs (1.61%) were panned. The transformation of functional bacteria in the system stimulated partial nitrification and denitrification phosphorus removal (DPR) (nitrite accumulation rate = 46.56%±6.88%, DPR rate = 43.15%±3.54%), which improved the nutrient removal rates. Finally, the optimal operation strategy was applied to a typical town-level WWTP in Ankang, China. The removal rates of NH_4^+-N, TN, and TP increased from 85.85%±1.31%, 59.77%±2.86%, and 23.91%±1.77% to 94.17%±2.02%, 65.32%±4.75% and 61.05%±2.34%, respectively, and the operating cost was reduced by 118600 CNY (17541 USD) per year.

【Keywords】 Rural sewage; Case study; A^2/O process; Nutrient removal; Optimal operation

1 Introduction

In the last decade, China's economy has developed at an astonishing rate, dramatically improving living conditions in rural districts. At the same time, the amount of domestic sewage produced in rural districts has increased significantly. According to a survey by the Ministry of Ecology and Environment of the People's Republic of China[1], rural districts in China were responsible for almost 40% of China's major water pollutants in 2020, including 51% of the chemical oxygen demand (COD), 30% of the total nitrogen (TN), and 39% of the total phosphorus (TP). Less than 35% of rural areas had wastewater treatment systems by the end of 2020[2], therefore a large amount of untreated or insufficiently treated wastewater is discharged from rural areas each year. This not only results in serious pollution, but may also threaten safe drinking water for rural inhabitants, raising concern from both academics and government. With increasing attention being paid by the Chinese government to the improvement of rural living environments, the effective treatment of domestic sewage in rural areas has become a pressing task[3-4].

In recent years, with the support of various domestic policies, the construction of rural sewage treatment facilities in China has accelerated[5]. However, applying urban wastewater treatment plant (WWTP) design in rural areas has caused a large number of existing rural WWTPs to be unsuited to rural sewage. This results in significant government investment producing limited improvements in rural sewage treatment.

Compared with urban areas, rural residents in China use less water and live at lower population densities, and a large amount of rural domestic sewage evaporates or seeps into the ground during the collection process[6]. Additionally, the living habits and customs of rural residents causes the volume of sewage discharge to fluctuate. Typically, peak flows occur during the morning, midday, and evening meal times, and almost no sewage is generated during the rest of the day[7-8]. These discontinuities in flow cause real problems that cannot be ignored in the processing of rural sewage. In addition, urbanization and industrialization in China has resulted in a large number of rural young adults flooding into the cities, where they live and work except during the Spring Festival or other traditional holidays. As a result, the countryside population is predominantly elderly and children, and the actual population of these "hollowing villages" in China is far less than the registered population. Because the design scale of village WWTPs is often based on the registered population, a large number of existing rural WWTPs are often in a state of insufficient influent and cannot operate efficiently and stably.

At present, anaerobic/anoxic/aerobic (A^2/O) process is the most common process in rural WWTPs in China[3]. However, village WWTPs with A^2/O process are often not capable of long-term stable operation, particularly with discontinuous or insufficient influent. In order to develop wastewater treatment technologies that can operate well under conditions typical of rural sewage in China, basic research in Chinese rural areas is necessary. To date, few such studies have been undertaken, especially on water quality and sewage discharge quantities and rates. In addition, the effects of rural sewage discontinuity or insufficient flow on biological nutrients removal systems (BNR) is not clear. Therefore, it is of great significance to study the quality and characteristics of rural sewage discharge, and its influence on BNR, in order to improve the operation of existing and future rural WWTPs.

The 76 village WWTPs in the northern plateau, central plain, and southern mountainous areas of Shaanxi Province were investigated for one year in this study, in order to solve the problems of low pollutant removal efficiency and economic benefit caused by insufficient and discontinuous influent. The quality and quantity of influent and the pollutant removal efficiency at each WWTP were analyzed, and the effects of rural sewage discontinuity on A^2/O process was investigated. Based on the above research, an optimal operation strategy based on dissolved oxygen (DO) concentration regulation was proposed, and the mechanisms of nitrogen (N) and phosphorus (P) removal were identified. Finally, taking a typical town-level WWTP in southern Shaanxi as the research case, the optimal operation strategy was applied.

2 Materials and methods

2.1 Basic information of rural WWTPs investigated

The rural WWTPs used in this study were selected in order to be representative of the different topographic characteristics of Shaanxi Province, i.e., 15 in the Loess Plateau region of northern Shaanxi Province; 16 in the plain region of central Shaanxi Province; and 45 in the mountainous region of southern Shaanxi Province. The treatment processes used by these 76 rural sewage treatment stations included A^2/O (27), membrane bioreactor (37), constructed wetlands (7) and oxidation pond (4), and every village

WWTP chosen for the survey had been in operation for at least one year. The survey took place from June 2019 to May 2020. Basic information describing all WWTPs is presented in the Supplementary Material (Table S1 and Fig. S1).

2.2 Monitoring method and frequency

The evaluations of rural WWTPs were performed in two parts, influent quantity analysis at each site, and water quality analysis conducted in the laboratory. Influent quantity analysis was performed using regular meter readings by field operation and maintenance personnel. Samples of the influent to each treatment station were taken for the analysis of water quality. The samples were collected three times on a sampling day (i.e., in the morning, noon and evening) and then mixed together. In each month, the sampling was conducted on three separate days. There was no rain for 3 days before the sampling. Samples were transported to the laboratory in refrigerated storage at 4 ℃ and were an analyzed within 2 days. In the laboratory, COD was measured using Hach chemical reagents. Five-day biochemical oxygen demand (BOD_5), mixed liquor suspended solids (MLSS), NH_4^+-N, TN and TP were determined using standard Chinese State Environmental Protection Administration methods[9].Each sample was filtered through 0.45 μm pore size filters prior to testing. Statistical analyses were performed using Microsoft Excel and the software package OriginPro 9.5 (OriginLab Corporation, Northampton, MA, USA).

2.3 Bioreactor operation

A pilot-scale bioreactor (Fig. S2) with A^2/O process was established. The bioreactor was made of square steel plate with the thickness of 10 mm, and it had an approximate working volume of 10 m^3 ($L \times W \times H$ = 3.75 m × 1.50 m × 2.00 m). Baffles were used to divide the plant into 6 compartments, and the last three compartments were equipped with independently adjustable, microporous aerator pipes on the bottom. The main reaction areas of the reactor include anaerobic, anoxic, aerobic and precipitation zones, where $V_{anaerobic}:V_{anoxic}:V_{aerobic} = 1:2:3$. The influent and aeration times of the bioreactor were controlled by time relays, and the operation was maintained at ambient temperature (18±2)℃. The sludge retention time (SRT), hydraulic retention time (HRT), internal and external recycle are 15 d, 14 h, 200% and 100%, respectively. The seed sludge was activated sludge taken from an aerobic zone of a town-level WWTP in Ankang, China. The experiments were conducted after 30 d of seeding when the bioreactor reached a steady state as indicated by the relatively constant MLSS and mixed liquor volatile suspended solids (MLVSS) concentrations along with the stable nutrient removal performance. The raw wastewater of the bioreactor was taken from the regulating tank of a town-level WWTP in Ankang, China (as shown in Table S2). The following two pilot experiments were conducted in this study: impacts of rural sewage discontinuity on A^2/O process and optimization operation of A^2/O process based on intermittent aeration and DO control. The purpose of the "impacts of rural sewage discontinuity on A^2/O process" experiment was to study the effect mechanism of rural sewage discontinuity on A^2/O process, so as to provide theoretical basis for reducing consumption and increasing efficiency of rural WWTPs. The operation parameters of the bioreactor applied in this experiment are described in Table 1.

The purposes of the "optimization operation of A^2/O process based on intermittent aeration and DO control" experiment were to investigate the optimal operation mode of A^2/O process under discontinuity influent condition, and the mechanism of advanced N and P removal under this operation mode. The operation parameters of the bioreactor applied in this experiment are described in Table 2.

Table 1 The operation parameters of the bioreactor in the "impacts of rural sewage discontinuity on A^2/O process" experiment

Phase	Name	Time(d)	Influent cutoff time	Influent cutoff duration (h/d)	DO in aerobic zones (mg/L)
1	Acclimation stage	0~29	—	0	2.0~3.0
2	Continuous influent operation stage	30~59	—	0	2.0~3.0
3	Discontinuous influent operation stage	60~89	10 pm to 8 am	10	2.0~3.0

Table 2 The operation parameters of the bioreactor in the "optimization operation of A^2/O process based on intermittent aeration and DO control" experiment

Phase	Name	Time (d)	Influent cutoff time	Influent cutoff duration/(h/d)	$T_{aeration}$: T_{stop}	DO in aerobic zones (mg/L)
1	Acclimation stage	0~29	—	0	—	2.0~3.0
2	Intermittent aeration stage	30~59	10 pm to 8 am	10	0h:10h	2.0~3.0
3	Intermittent aeration stage	60-89	10 pm to 8 am	10	2h:8h	2.0~3.0
4	Intermittent aeration stage	90~119	10 pm to 8 am	10	5h:5h	2.0~3.0
5	Intermittent aeration stage	120~149	10 pm to 8 am	10	8h:2h	2.0~3.0
6	Micro-oxygen operation stage	150~179	10 pm to 8 am	10	5h:5h	2.0~3.0
7	Micro-oxygen operation stage	180~209	10 pm to 8 am	10	5h:5h	1.0~2.0
8	Micro-oxygen operation stage	210~239	10 pm to 8 am	10	5h:5h	0.5~1.0

2.4 Analytical methods

In the pilot experiments and case study, the concentrations of COD, NH_4^+-N, NO_2^--N, NO_3^--N, TN, TP, PO_4^{3-}-P and MLVSS were determined following standard Chinese State Environmental Protection Administration methods[9]. The temperature and DO concentrations were monitored using WTW340i probes (WTW Company, Germany). The concentration of poly-b-hydroxybutyrate (PHB) was detected as detailed in a previous study[10].

2.5 Batch tests and calculations

The activity of ammonia oxidizing bacteria (AOB) (characterized by the specific rate of ammonia oxidation, SAOR) and nitrite oxidizing bacteria (NOB) (characterized by the specific rate of nitrite oxidation, SNOR) were detected by batch tests each 10 d, and the activity of denitrifying phosphorus-accumulating organisms (DPAOs) was detected by batch tests in the last 3 d of each phase during the "optimization operation of A^2/O process based on intermittent aeration and DO control" experiment. The simultaneous nitrification and denitrification (SND) efficiency is defined as the loss of N in the aerobic zone, and nitrite accumulation rate (NAR) is defined as the proportion of nitrite at the end of the aerobic zone. The SAOR and SNOR were defined as indicators of the activity of AOB and NOB, respectively[11]. The detail experimental workflow of microbial activity batch tests and computational formulas can be seen in supporting information.

2.6 High-throughput sequencing analysis of activated sludge

Sludge samples were taken from activated sludge in the last 3 days of each phase during the pilot experiments. Three sludge samples were taken for each phase, and immediately after sampling, the supernatant was separated and removed by centrifuge, and stored at −20 ℃ for future use. After the end of pilot experiments, all sludge samples were sent to Shanghai Personal Biotechnology Co., Ltd (Shanghai, China) for high-throughput sequencing. The methods and process of DNA extraction, PCR amplification and sequence analysis can be seen in a previous study[12].

2.7 Description of the town-level WWTP

A case study was conducted at a town-level WWTP in Ankang, China. The WWTP was built in 2014 to use A^2/O process, with a design scale of 1000 m^3/d. The volume ratio of anaerobic, anoxic, and aerobic tank was 1 : 2 : 3, and the total construction area was 1989 m^2. There were three air compressors in the WWTP, two of which are running while the third was a standby. The power of each air compressor was 7.5 kW, and the SRT, HRT, internal recycle and external recycle are 15 d, 14 h, 200% and 100%, respectively. Discharge water complied with the Class 1A of GB 18918—2002[13]. The process flow chart of the WWTP is shown in Fig. S3. The actual wastewater treated by the WWTP was 400~500 t/d, and the sewage load was < 50%. The WWTP operated without influent for approximately 10 h/d. In addition, the concentration of pollutants in the influent was low and the carbon source content was insufficient. A summary of the influent characteristics is listed in Table S2. The operation parameters of the WWTP applied in the case study are described in Table 3.

Table 3 The operation parameters of the WWTP in the case study

Phase	Name	Time (month)	Influent cutoff time	Amount of glucose added (kg/d)	Influent cutoff duration (h/d)	$T_{aeration}$: T_{stop}	DO in aerobic zones (mg/L)
1	Original operation stage	Jan to Mar	10 pm to 8 am	50	10	24 h : 0 h	>3.0
2	Acclimation stage	Apr		50		24 h : 0 h	2.0~3.0
3	Co-cultivation stage	May		0		5 h : 5 h	2.0~3.0
4	Screening stage	Jun		0		5 h : 5 h	0.5~2.0
5	Stable operation stage	Jul to Dec		0		5 h : 5 h	0.5~1.0

3 Results and discussion

3.1 Discharge characteristics of rural wastewater

3.1.1 Characteristics of rural wastewater quality

Concentrations of pollutants in the influents to the 76 rural sewage treatment stations, observed from June 2019 to May 2020, are shown in Fig. 1. The average concentrations of COD, BOD_5, NH_4^+-N, TN, TP, and SS, decreased from northern, to central, and then to southern Shaanxi. The BOD_5/COD ratios of rural sewage in northern, central, and southern Shaanxi were 0.40±0.065, 0.38±0.047 and 0.41±0.030, respectively, suggesting that rural sewage in Shaanxi has biochemical characteristics appropriate for biochemical wastewater treatment[14]. However, the COD/TN ratios of rural sewage in northern, central, and southern Shaanxi were 5.42±0.73, 5.24±0.77, and 4.23±0.81, respectively. The COD/TN and COD/TP ratios in all three regions were lower than the theoretical values required for nutrient (N and P) removal in BNR[15]. The sewage discharged from villages and towns in Shaanxi Province is an insufficient carbon source, and this has a significant negative impact on the effectiveness of BNR.

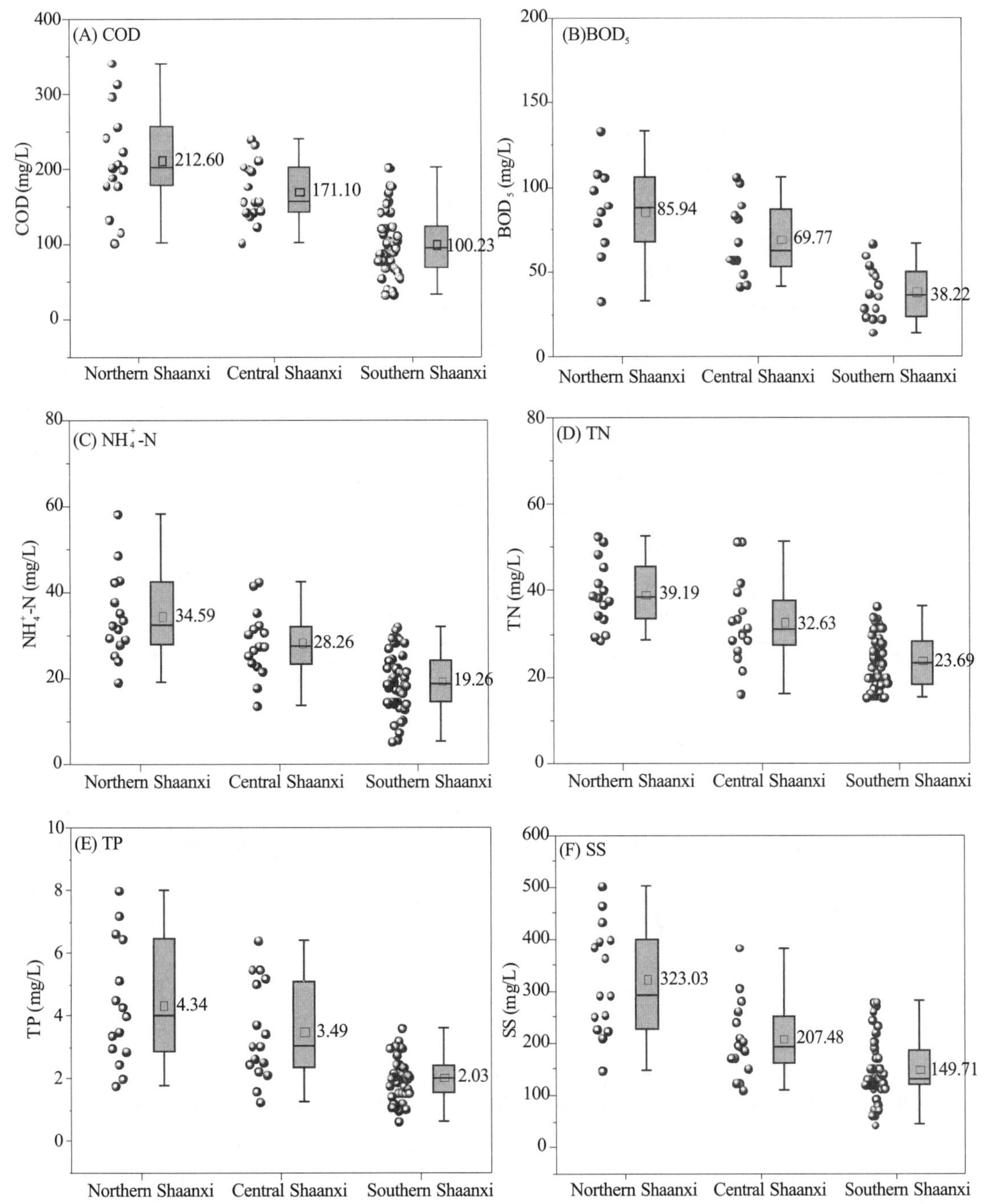

Fig.1 Characterization of water quality within rural wastewater for three different regions in Shaanxi Province, China, observed from June 2019 to May 2020

Because the construction mode of urban WWTPs was copied in earlier rural WWTP designs, there may be cases in which the design of influent pollutant concentration was too high, resulting in excessive treatment and energy consumption losses after completion. The concentration of pollutants in the influent of the rural WWTPs in this survey was measured (Table 4), and the values of COD, BOD_5, NH_4^+-N, TN, TP, and SS of the influent ranged from 100~213, 38~86, 19~35, 24~39, 2~5, and 150~323 mg/L, respectively.

Table 4　The concentration of influent pollutants in the rural WWTPs in Shaanxi Province, China, as observed from June 2019 to May 2020

Item	Northern Shaanxi (mg/L)	Central Shaanxi (mg/L)	Southern Shaanxi (mg/L)
COD	212.60	171.10	100.23
BOD_5	85.94	69.77	38.22
NH_4^+-N	34.59	28.26	19.26
TN	39.19	32.63	23.69
TP	4.34	3.49	2.03
SS	323.03	207.48	149.71

3.1.2　Fluctuation and discontinuity characteristics

Rural sewage systems in China often experience difficulties with flow due to low water consumption of rural residences and their dispersed nature. Additionally, the living habits and customs of rural residents in China resulted in significant rural sewage discharge fluctuations during the daytime, and almost no sewage flow at night.

The daily fluctuation of rural wastewater for different regions is shown in Fig. 2. The daily inflow quantity of the village WWTPs (20~200 t/d) had a daily fluctuation rhythm (Fig. 2A), with peak inflows concentrated at three time periods: 07:00 to 09:00 in the morning; 11:00 to 13:00 at midday, and 17:00 to 19:00 in the afternoon (all times are local times). The average daily influent loads of village WWTPs in northern, central, and southern Shaanxi were 36%, 26% and 22%, respectively. The daily inflow quantity of the town WWTPs (240~2000 t/d) also showed daily fluctuation with peak inflows concentrated in three time periods: 08:00 to 11:00, 13:00 to 16:00, and 19:00 to 21:00 (Fig. 2B). The average daily influent quantity load of town WWTPs in northern, central, and southern Shaanxi were 50%, 45%, and 36%, respectively. The peak inflows of town WWTPs occur later than those of village WWTPs, most likely due to sewage transportation distances in towns compared to those in villages.

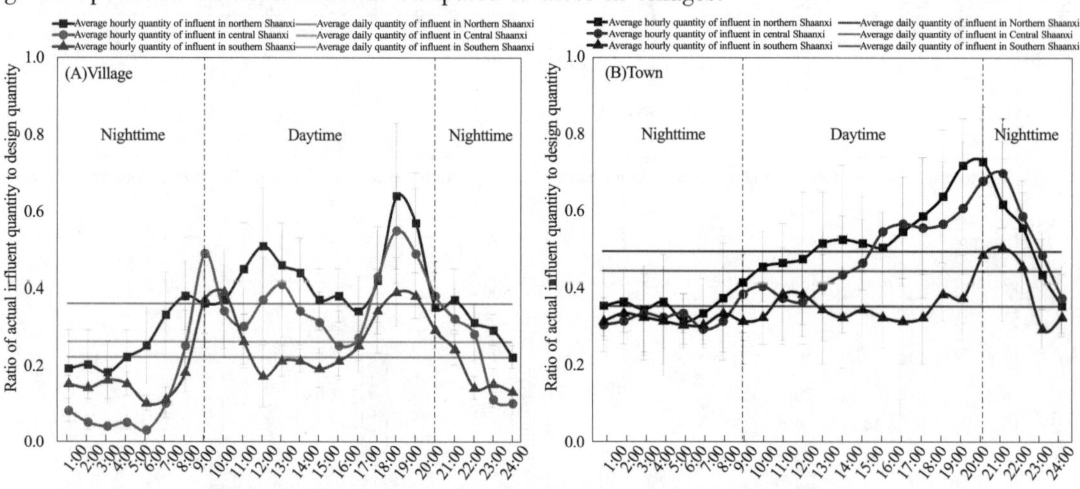

Fig.2　Daily fluctuation characterization of rural wastewater for three different regions of Shaanxi province, China, as observed from June 2019 to May 2020

The annual fluctuation of rural wastewater for different regions is shown in Fig. 3. The monthly average influent amounts of sewage treated by village (Fig. 3A) and town (Fig. 3B) WWTPs fluctuated with the seasons, and the average monthly influent of village and town WWTPs in northern, central, and southern Shaanxi was higher in summer and autumn, and lower in winter and spring. Additionally, the influent quantity increased significantly during the Spring Festival (February). The main reason for this is that many young people return to their native villages and towns from cities during the Spring Festival, temporarily increasing the rural population and resulting in a corresponding surge in sewage discharge.

Except for the high average monthly influent load during the Spring Festival, the average monthly influent load of the village and town WWTPs did not exceed 50% during summer and autumn (June to November), thus the WWTPs are overscaled. As a result, village and town WWTPs were not economically efficient.

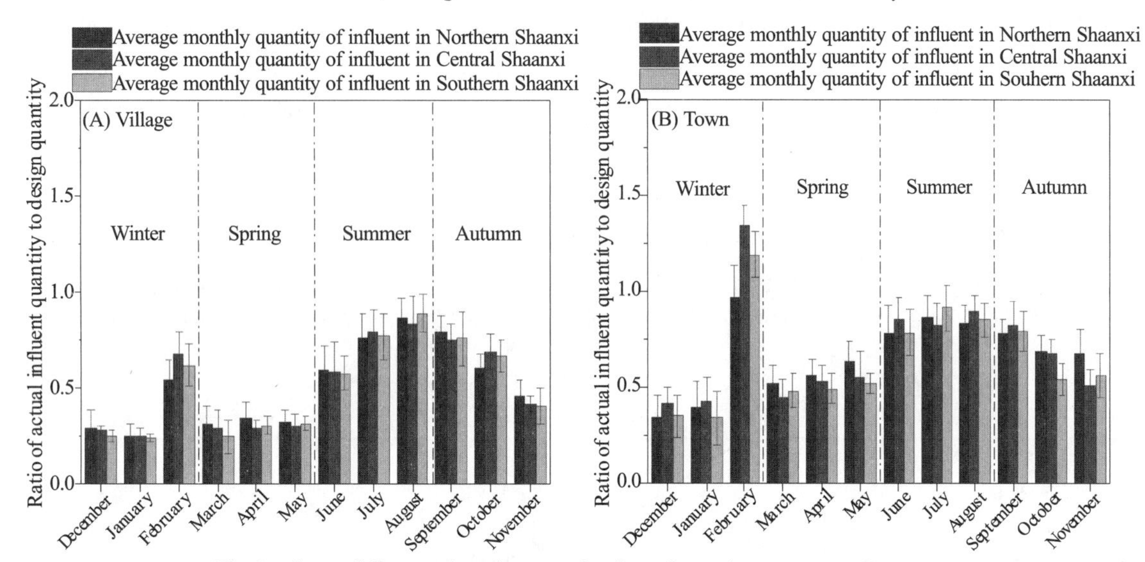

Fig.3 Annual fluctuation characterization of rural wastewater for three different regions of Shaanxi province, China, as observed from June 2019 to May 2020

Because of the flow characteristics described above, sewage-lifting pump flow is not met for many treatment stations most of the time, resulting in no wastewater treatment. This is described further in Fig. 4, which shows the cutoff duration of rural wastewater for the different regions of Shaanxi province. The average cutoff durations of village WWTPs in northern, central, and southern Shaanxi were 11, 16, and 19 h/d, respectively. The corresponding average cutoff durations for town WWTPs in the three regions were 5, 9, and 12 h/d, respectively. WWTP cutoff primarily occurred at night (8pm to 8am) for both village and town WWTPs. Most rural WWTPs experienced intermittent flow of influent, which is a significant obstacle to the stable operation of BNR.

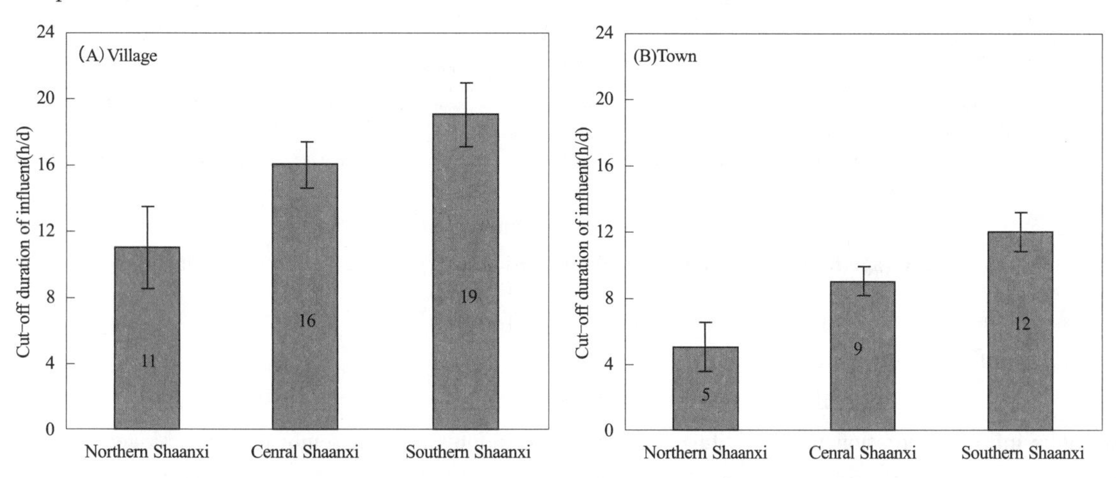

Fig.4 Cutoff durations of rural wastewater treatment stations in villages and towns for three different regions of Shaanxi province, China, as observed from June 2019 to May 2020

3.1.3 Correlation between pollutant removal rate and discontinuity of influent

The relationship between pollutant removal efficiency and cutoff duration is shown in Fig. 5. The removal rates of COD, NH_4^+-N, TN, and TP at the 76 rural WWTPs were negatively correlated with cutoff duration, i.e., the removal rates of all four pollutant indices decreased with increased cutoff duration.

The direct effect of discontinuous influent flow on BNR is an insufficient supply of substrate for the survival of the WWTP microorganism community, which results in starvation of the functional bacteria. Normally, under such conditions, bacteria adjust their metabolic processes to reduce their need for energy, making them less active. In addition, as a separate mechanism, bacteria may also begin programmed cell death (PCD), a genetically programmed process of cell self-destruction that occurs only under extreme conditions of starvation. The main purpose of this process is to maintain the activity of part of the bacterial population through intracellular material and to avoid losing the advantage in competition with other microbial populations[16].

For a BNR system, the most direct macro influence brought by either the reduction in activity of functional bacteria or cell death is the reduction of the system's pollutant removal efficiency. This is the primary internal reason that WWTPs using this process cannot operate normally when sewage inflow is intermittent.

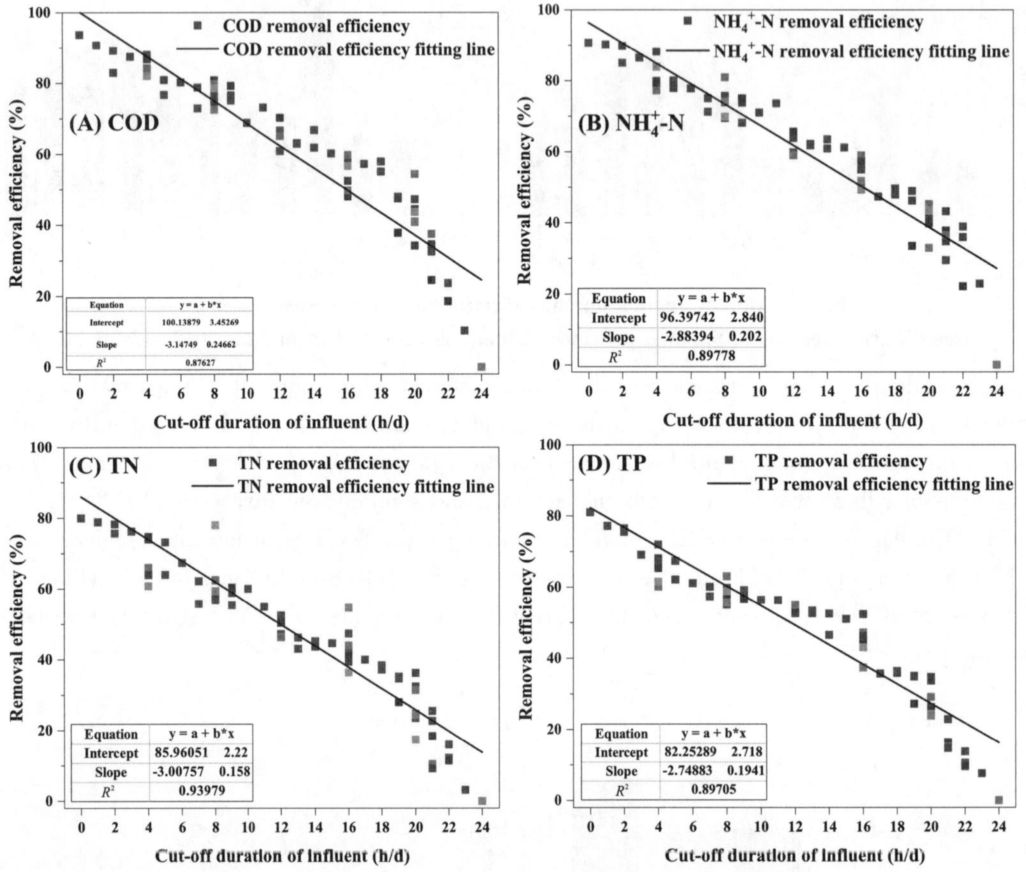

Fig.5 Relationship between pollutant removal efficiency of rural sewage treatment stations and the variation of treatment station cutoff duration

3.2 Impacts of rural sewage discontinuity on A^2/O process

3.2.1 Influence on pollutant removal efficiency

The characteristics of pollutant removal in the pilot-scale A^2/O system are shown in Fig. 6. During the continuous influent operation period (Stage 2), highly efficient nutrient removal was achieved with COD, NH_4^+-N, TN, and TP average effluent concentrations of (32.72±3.39) mg/L, (0.46±0.27) mg/L, (11.39±1.84) mg/L, and (0.62±0.17) mg/L, respectively. The average removal rates were (70.95±6.20)%, (97.70±1.83)%, (57.51±6.99)% and (73.16±4.56)%, respectively. However, the average effluent concentrations of COD, NH_4^+-N, TN, and TP during the discontinuous influent operation period (Stage 3) were (32.27±3.49) mg/L, (6.00±3.46) mg/L, (16.77±4.37) mg/L, and (1.67±0.45) mg/L, with COD, NH_4^+-N, TN, and TP average removal rates of (72.39±5.36)%, (74.86±12.80)%, (40.88±7.17)%, and (27.97±12.42)%, respectively. In addition to COD, the removal efficiencies of NH_4^+-N, TN, and TP were inhibited by the 10 h/d cutoff condition.

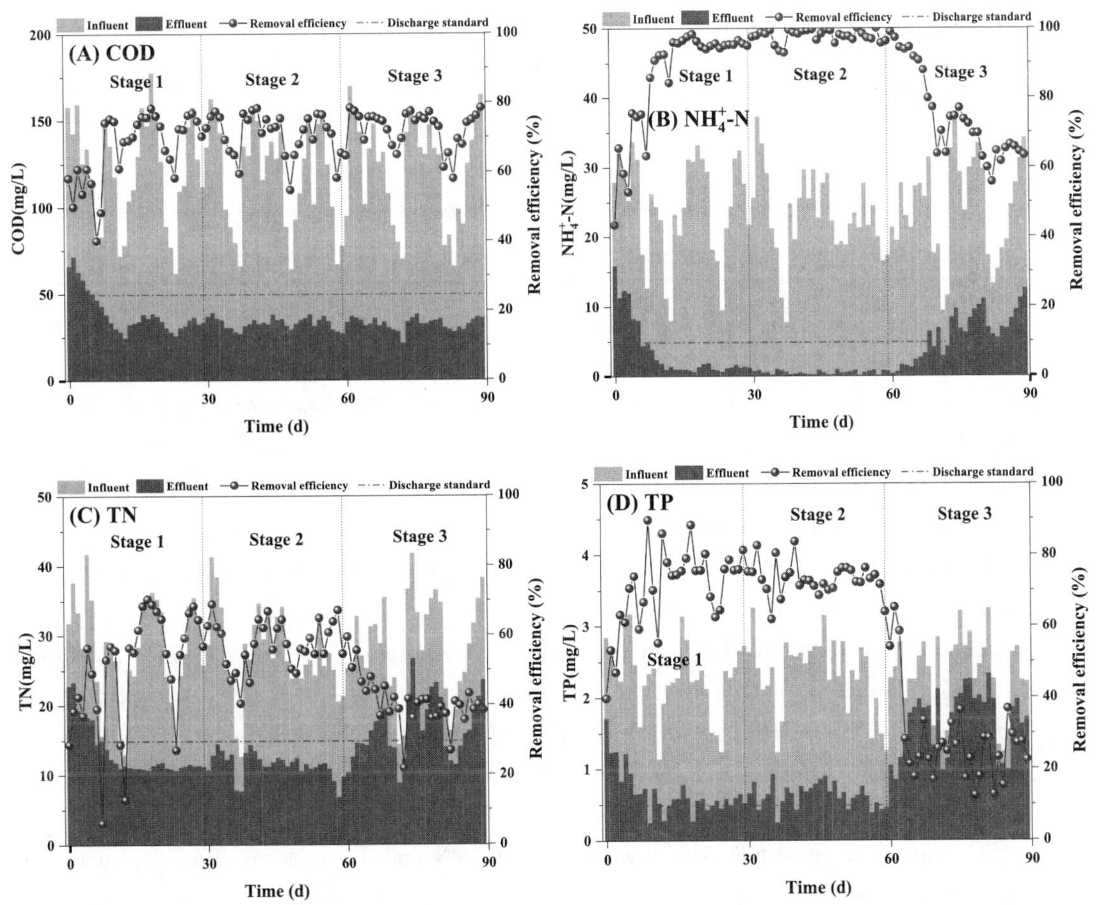

Fig.6 Removal performance of COD (A), NH_4^+-N (B), TN (C), and TP (D) in each stage

3.2.2 Influence on activity and community structure of functional bacteria

The activity of the functional bacteria is shown in Fig. S4. The activity of nitrifying bacteria decreased rapidly during Stage 3, and had decreased by 48.93% at the end of that stage. Although the activity of denitrifying and phosphorus-accumulating bacteria (PAOs) decreased during Stage 3, the decrease was small relative to that of nitrifying bacteria. The activity of denitrifying bacteria had decreased by 21.23% by the end of Stage 3. The specific release phosphorus rate and specific uptake phosphorus rate decreased by 32.40% and 34.68% by the end of Stage 3. These results demonstrated that functional bacteria activity decreased due to the starvation condition caused by insufficient substrate. In the presence of external substrates (influent time), heterotrophic bacteria are able to store large amounts of substrates, such as PHA and glycogen, in the form of intracellular polymers, and then consume them in endogenous processes to generate energy for cell maintenance and growth[17-18]. In addition, heterotrophic bacteria can oxidize cellular tissues such as proteins and RNA in order to avoid cell death during prolonged starvation[19-20]. Unlike heterotrophic bacteria, autotrophic bacteria, and particularly AOB, have not been reported to store large amounts of energy substrate. No significant tissue degradation, such as RNA for AOB, was observed under starvation conditions[21]. Therefore, denitrobacteria and PAOs are more adaptable to discontinuous influent conditions than are nitrifying bacteria.

In order to study the influence of different influent modes on the microbial community in this system, DNA was extracted from the activated sludge in the aerobic zone at the end of each stage, and microbial diversity was analyzed using the PacBio platform and the 16S rRNA sequencing method (Table S3). The species richness and diversity indices of the sequenced samples under 97% similarity are shown in Table 5. The Chao1, ACE, Shannon, and Simpson indices decreased under discontinuous influent condition (S3), indicating that intermittent cutoff of rural sewage can reduce the richness and diversity of microbial populations in A^2/O systems.

Table 5 The species richness and diversity indicators of activated sludge samples

No.	ACE	Chao1	Simpson	Shannon	Coverage
S1	596.3655	613.7619	0.9917	7.7916	0.9987
S2	599.1167	597.9091	0.9889	7.8125	0.9972
S3	410.3342	419.0532	0.6699	4.0124	0.9989

The relative abundance of bacteria with different functions at the genus level was analyzed. The microbial population structure consisted of 313 genera, and only the relative abundances of common functional bacteria with obvious changes in this experiment are shown (Fig. 7). The PAOs included *Acinetobacter* (2.15%~ 4.65%), *Candidatus_Accumulibacter* (0.51%~0.91%), *Tetrasphaera* (0.22%~0.49%), and *Dechloromonas* (3.75%~6.18%). The relative abundances of these genera decreased under discontinuous influent conditions (S3), and were consistent with variation in TP removal efficiency in the system. This indicated that the growth and reproduction of these bacteria were greatly affected by sewage discontinuity, and that they important in TP removal in the A^2/O system.

The nitrification and denitrification bacteria included *Nitrosospira* (0.83%~1.11%), *Nitrosomonas* (0.73%~2.85%), *Nitrobacter* (1.73%~2.59%), *Devosia* (0.09%~0.31%), *Nitrospira* (2.33%~4.47%), *Dokdonella* (0%~0.31%), *Paracoccus* (0.48%~0.81%), *Denitratisoma* (0.31%~0.53%), *Thauera* (0.92%~1.78%), *Haliangium* (0.14%~0.18%), *Hyphomicrobium* (0.10%~0.16%) and *Zoogloea* (0.37%~0.51%). The relative abundances of these 12 genera decreased under discontinuous influent conditions (S3), and were consistent with variation in NH_4^+-N and TN removal efficiency in the system. This indicated that the growth and reproduction of these 12 genera were strongly affected by sewage discontinuity, and that these genera were important in the removal of NH_4^+-N and TN in the A^2/O system.

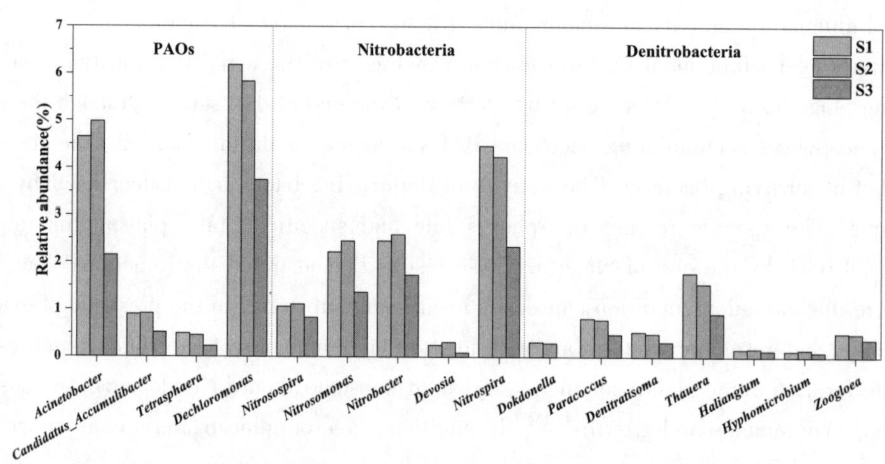

Fig.7 Relative abundances of microorganisms at genus level in samples at different stages

3.2.3 Effect mechanism and operation strategy

The mechanism through which rural sewage discontinuity effects A^2/O process and operation strategy can be seen in Fig. 8. The above results show that the activity of nitrifying and denitrifying bacteria and PAOs in the system were severely inhibited under discontinuous influent conditions. The discontinuity of rural sewage caused system operation to alternate between famine and feast. During the famine period (without influent), the activity of nitrifying bacteria was severely inhibited due to the absence of substrate (N source). Denitrifying bacteria and PAOs can maintain cellular activity to some extent through endogenous respiration[22]. The decrease in the activity of nitrifying bacteria lead to increases in DO concentration in the aerobic zone of the system and in the anaerobic and anoxic zones through internal/external recycle, shifting the system into an oxygen-rich operation state (Fig. S5). Studies have shown that high DO concentration enhances microbial metabolic activities, resulting in insufficient nutritional supply and self-oxidation[23]. Under the effects of oxygen-rich operation and insufficient carbon source, the activity of denitrifying bacteria and PAOs will be further inhibited, and the system will collapse.

During the famine period, external carbon sources such as acetate, glucose, methanol, and ethanol, could be utilized as carbon additives to improve N and P removal in A^2/O processes[24]. However, the high purchase cost (up to 70% of the total operation and maintenance costs of WWTPs) limits their use[25]. Studies have shown that an intermittent aeration operation strategy can significantly increase the relative abundance of nitrifying bacteria and improve the nitrification performance of activated sludge systems when influent substrate is insufficient[26-27]. In addition, under the condition of low DO, the community structure of nitrifying bacteria may change, but the abundance is enriched, and the nitrification of activated sludge system can also increase when substrate is insufficient[28]. However, whether the operation strategy of intermittent aeration and DO regulation can improve the pollutant removal efficiency of A^2/O system with discontinuous flow of rural sewage is not clear. Therefore, this study intends to test intermittent aeration and DO regulation to improve the pollutant removal efficiency of A^2/O system under starvation conditions.

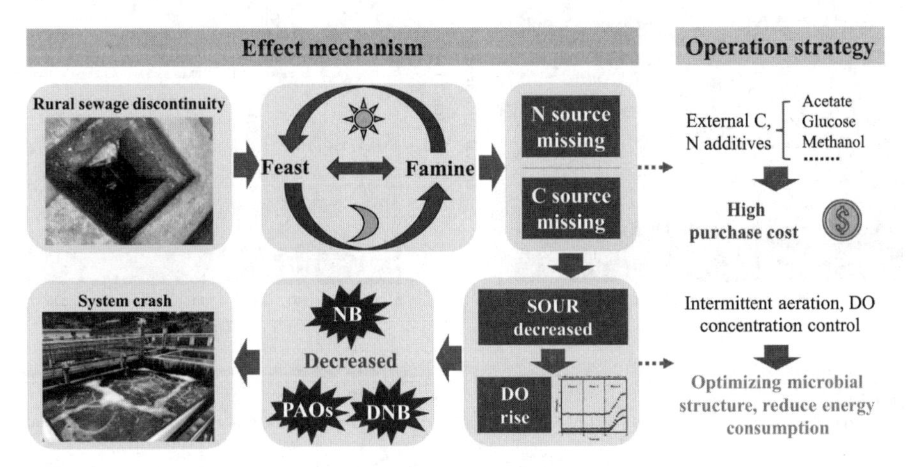

Fig.8 Effect mechanism of rural sewage discontinuity on A^2/O process and operation strategy

3.3 Optimal operation of A^2/O process based on intermittent aeration and DO control

3.3.1 Effects on pollutant removal efficiency

The nutrient removal performance of the pilot-scale A^2/O system treating rural sewage under different aeration modes and DO concentrations was investigated over a 240 d operational period. Throughout the test period, the average effluent concentration was (34.28±7.39) mg/L and the average removal efficiency of COD was and 76.79%±5.86% (Fig. 9A). This demonstrated that the aeration mode and DO concentration had no distinct effect on COD removal.

The removal efficiency of NH_4^+-N was strongly affected by aeration mode (Fig. 9B). The nitrification of the system deteriorated when aeration stopped or was excessive (phases 2 and 5) during famine period, resulting in effluent NH_4^+-N concentration exceeding the discharge standard. The system showed better nitrification performance when $T_{aeration}:T_{stop}$ = 5 h:5 h (phase 4), and the average effluent NH_4^+-N concentration and removal rate were (1.32±0.83) mg/L and 94.78%±2.66%, respectively. When aeration was $T_{aeration}:T_{stop}$ = 5 h:5 h, the gradual reduction of DO concentration did not influence the nitrification performance of the system, and the system maintained an ideal NH_4^+-N removal rate in phases 7 and 8.

The removal efficiency of TN was strongly affected by aeration mode and DO concentration (Fig. 9C). The aeration time in the famine period had no distinct effect on TN removal during phases 2~4. However, denitrification of the system deteriorated due to excessive aeration (phase 5) during the famine period, resulting in effluent TN concentration exceeding the discharge standard. In the $T_{aeration}:T_{stop}$ = 5 h:5 h aeration mode, the gradual reduction of DO concentration improved the denitrification performance of the system, and the system had optimal TN removal performance during phase 8.

The removal efficiency of TP was strongly affected by aeration mode and DO concentration (Fig. 9D). Increased aeration time improved TP removal efficiency during phases 2~4. However, the TP removal performance of the system decreased due to excessive aeration (phase 5) during the famine period. In the $T_{aeration}:T_{stop}$ = 5 h:5 h aeration mode, the reduction of DO concentration improved the TP removal performance of the system, and TP removal performance was optimal during phase 8. Pollutant removal was most efficient during phase 8 ($T_{aeration}:T_{stop}$ = 5 h:5 h, DO = 0.5~1.0 mg/L), during which the average removal efficiencies of NH_4^+-N, TN, and TP were (95.52±1.44)%, (74.15±2.96)%, and (72.27±3.35)%, and the average effluent concentrations were (1.13±0.38) mg/L, (8.45±1.39) mg/L, and (0.75±0.20) mg/L, respectively.

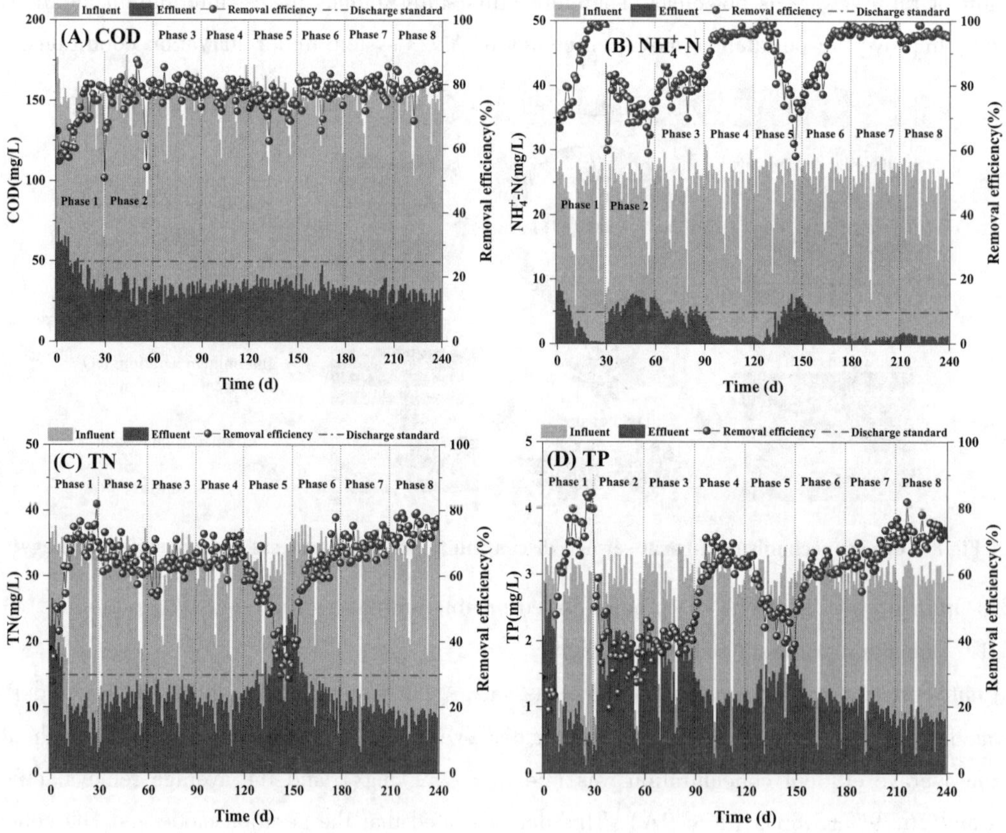

Fig. 9 Removal performance of COD (A), NH_4^+-N (B), TN (C), and TP (D) in each stage

3.3.2 N removal mechanism of A^2/O process under intermittent aeration and micro-oxygen conditions

Studies have shown that when AOB and NOB are disturbed by hypoxia under the condition of insufficient substrate, NOB will be eliminated, and when aerobic conditions were restored again, AOB activity recovered faster than NOB, leading to the accumulation of NO_2^--N and achieving a stable partial nitrification pathway[29-30]. The key to achieving a partial nitrification pathway is to eliminate NOB while enhancing AOB activity[31]. A batch test was conducted to investigate the activities of AOB and NOB during each stage (Fig. 10A). After the suspension of aeration during the famine period in phase 2, the system could not provide sufficient electron acceptors for the metabolism of AOB and NOB, therefore the activities of both AOB and NOB were inhibited. When the aeration duration was increased, AOB and NOB recovered (phases 3 and 4). However, excessive aeration during the famine period inhibited AOB and NOB activity, which was lowest during phase 5. AOB and NOB recovered during phase 6 ($T_{aeration}:T_{stop}$ = 5 h:5 h). In the process of reducing the DO concentration during phases 7 and 8, the activity of AOB increased (0.21 g(N)/g(VSS) · d), while the activity of NOB decreased (0.18 g(N)/g(VSS) · d), resulting in NO_2^--N accumulation at the end of phase 8. The average NO_2^--N concentration during phase 8 was (4.10 ±0.47) mg/L with NAR of (46.56±6.88)% (Fig. 10B). In this study, there were two main reasons for AOB enrichment and NOB elimination in the system: first, intermittent aeration prolonged the anoxic period, during which NOB activity decreased rapidly, but it recovered slowly during the aerobic period[32-33], resulting in a gradual decline. Second, AOB has a stronger affinity for DO than NOB, that is, AOB can better adapt to the low-DO environment[34], so the activity of AOB increased while DO concentration decreased and enriched AOB.

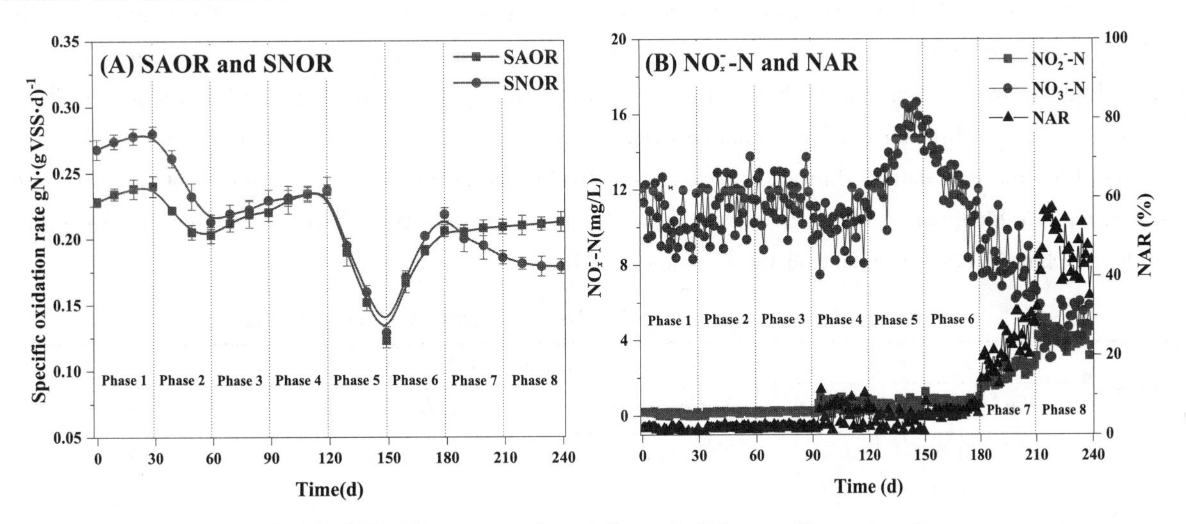

Fig.10 NO_x^--N concentration at the end of the aerobic zone and NARs in each stage (A), SAOR and SNOR in each stage (B)

To further investigate the microbial community structure of AOB and NOB, the genus-level 16S rRNA gene sequencing results for activated sludge at the end of each stage of the system were sorted. The AOB in the system consisted of *Nitrosomonas* and *Nitrosospira* and the NOB consisted of *Nitrobacter*, *Devosia* and *Nitrospira* (Table 6). At the end of the acclimation stage (phase 1), the dominant bacterial community of nitrifying bacteria in the system was NOB (4.37%). Based on the changes in the concentration of N along the reactor (Fig. S6A), the N removal approach in the system at this time was conventional[35]. However, after intermittent aeration and micro-oxygen optimization, the dominant nitrifying bacteria in phase 8 changed from NOB (1.61%) to AOB (2.43%), and the N removal pathway in the system was transformed from conventional N removal to partial nitrification (Fig. S6B)[12].

3.3.3 Premoval mechanism of A^2/O process under intermittent aeration and micro-oxygen conditions

A batch test was conducted to investigate the activity of DPAOs in each stage. The DPAOs in anoxic zones show low levels of activity from phases 1 to 5, with an average DPAOs/PAOs ratio of (13.00± 1.58)%, consistent with the average denitrification phosphorus-removing (DPR) rate of (5.40 ± 0.88)% (Fig. S7). The activity of the DPAOs increased beginning in phase 6, and peaked at the end of phase

Table 6 Relative abundance of nitrifying bacteria at the genus-level in each stage

Phase	Relative abundance of AOB (%)			Relative abundance of NOB (%)			
	Nitrosomonas	*Nitrosospira*	Total	*Nitrobacter*	*Devosia*	*Nitrospira*	Total
1	1.54	1.17	2.71	2.69	0.25	1.43	4.37
2	1.09	1.02	2.11	1.97	0.18	1.35	3.50
3	1.19	1.09	2.28	2.04	0.17	1.36	3.57
4	1.29	1.09	2.38	2.05	0.19	1.35	3.59
5	1.01	0.72	1.73	1.62	0.09	1.12	2.83
6	1.22	1.00	2.22	1.81	0.13	1.20	3.14
7	1.12	1.18	2.30	1.21	0	1.09	2.30
8	1.19	1.24	2.43	0.92	0	0.69	1.61

8. The average DPAOs/PAOs ratio in phase 8 was (49.00±2.42)%, with an average DPR rate of (43.15±3.54)%. This indicated that $T_{aeration}:T_{stop}$ = 5 h:5 h and DO = 0.5~1.0 mg/L is beneficial to the enrichment of DPAOs. To further investigate the microbial community structure of PAOs, the genus-level 16S rRNA gene sequencing results for activated sludge at the end of each stage were sorted. Three generally recognized genera were detected in each sample: *Ca_Accumulibacter*, *Dechloromonas*, and *Tetrasphaera* (Table 7). The relative abundances of *Dechloromonas* and *Tetrasphaera* during phase 8 increased by 48.78% and 81.58% from the abundances during phase 1, respectively, while the relative abundance of *Ca. Accumulibacter* in phase 8 decreased by 19.23% from phase 1. The changes in the relative abundances of these three PAOs may have a crucial impact on the P removal pathways in the system.

Table 7 Relative abundance of PAOs at the genus-level in each stage

Phase	Relative abundance of PAOs (%)			
	Ca_Accumulibacter	*Dechloromonas*	*Tetrasphaera*	Total
1	1.04	0.82	0.38	2.24
2	0.97	0.84	0.42	2.23
3	0.94	0.91	0.49	2.34
4	1.02	0.95	0.53	2.50
5	0.44	0.41	0.25	1.10
6	0.96	0.97	0.55	2.48
7	0.91	1.11	0.64	2.66
8	0.84	1.22	0.69	2.75

The variations in PO_4^{3-}-P, NO_x^--N, and PHB concentrations along the bioreactor from days 27 to 29 (phase 1) and days 257 to 259 (phase 7) were analyzed in order to investigate the pathway of phosphorus removal. The concentration of PO_4^{3-}-P released in the anaerobic zone in phase 1 was 7.32 mg/L, and the subsequent PO_4^{3-}-P uptakes in the anoxic and aerobic zones were 0.61 and 7.87 mg/L, respectively (Fig. 11A). The ratio of PO_4^{3-}-P uptake in the anoxic to aerobic zones was 7.75%, indicating that most of the phosphorus was removed in the aerobic zones during phase 1, which is a typical conventional enhanced biological phosphorus removal pathway[36-37]. After optimal operation ($T_{aeration}:T_{stop}$ = 5 h:5 h and DO = 0.5~1.0 mg/L) in phase 8, the concentration of PO_4^{3-}-P released in the anaerobic zone was 13.14 mg/L, and the subsequent PO_4^{3-}-P uptakes in the anoxic and aerobic zones were 5.17 and 7.06 mg/L, respectively (Fig. 11B). The ratio of PO_4^{3-}-P uptake in the anoxic to aerobic zones was 73.23%, indicating that the P removal pathway changed from EBPR during phase 1 to DPR combined with EBPR during phase 8. In this study, DPR occurs primarily because reducing the concentration of DO in the aerobic zone not only reduces the oxygen carried in returned sludge, but also reduces the NO_3^--N with increased denitrification. This not only ensures a pure anaerobic environment in the anaerobic zone but also stops the inhibitory effect of NO_3^--N on the anaerobic phosphorus release of PAOs. Therefore, PAOs can synthesize sufficient internal carbon source PHB (9.66±0.94) mmol/L during the anaerobic phosphorus release process, creating the necessary conditions for anoxic DPR. In addition, the accumulation of NO_2^--N at the end of the aerobic zone in the system causes a large amount of NO_2^--N to be carried with internal recycling to the anoxic zone, which provides DPAOs with another electron acceptor (NO_2^--N) required for anoxic DPR, further stimulating anoxic DPR in the system.

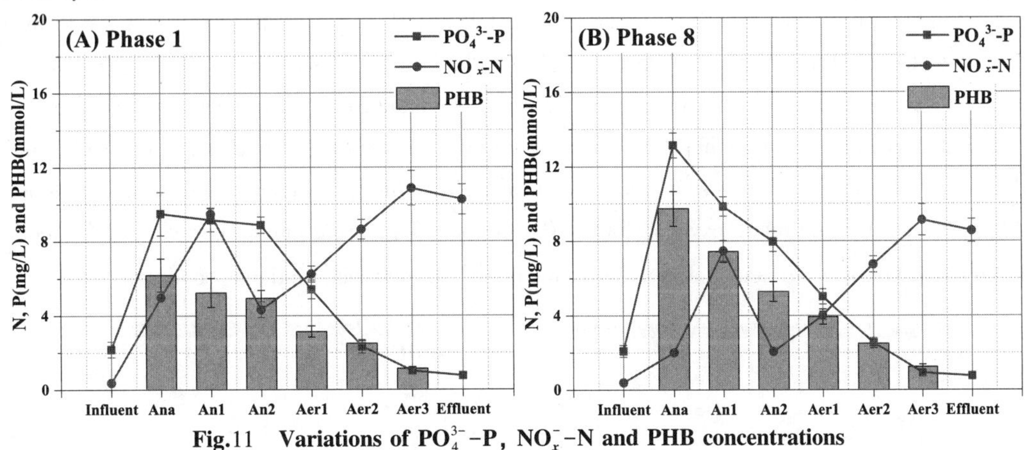

Fig.11 Variations of PO_4^{3-}-P, NO_x^--N and PHB concentrations along the bioreactor from days 27 to 29 (phase 1) and days 237 to 239 (phase 8)

3.3.4 Optimal operation strategy based on intermittent aeration and DO control

The dominant bacteria species in the pilot-scale A^2/O reactor changed as a result of changes in aeration mode and DO regulation. During the famine period, the operation mode $T_{aeration}:T_{stop}=1:1$ resulted in AOB (dominated by *Nitrosomonas* and *Nitrosospira*) and NOB (dominated by *Nitrobacter*, *Devosia*, and *Nitrospira*) accumulating together. Meanwhile, PAOs mainly composed of *Ca_Accumulibacter*, *Dechloromonas*, and *Tetrasphaera* can also achieve growth in this operating mode. However, in the process of further reducing DO concentration in the reactor, AOB are enriched due to their strong adaptability to the micro-oxygen environment. NOB were eliminated due to their lack of adaptability. In addition, PAOs (mainly *Dechloromonas* and *Tetrasphaera*) grew in the micro-oxygen environment.

In summary, the optimal operation strategy based on intermittent aeration and DO control can be divided into three operational stages. The first stage is the intermittent aeration co-cultivation stage, that is, the aeration mode is set as $T_{aeration}:T_{stop}=1:1$ and DO = 2.0~3.0 mg/L, so that AOB, NOB, and PAOs in the aerobic zone of the system can be enriched together. This prepares the bacterial community reserve for the next stage of screening. The second stage is the microbial community screening stage of micro-oxygen operation. Under the condition of $T_{aeration}:T_{stop}=1:1$ in the famine period, the DO concentration is controlled within the range of 0.5~1.0 mg/L, so that the environment in the aerobic zone is more conducive to the growth and metabolism of AOB. Under this operating condition, AOB and DPAOs increase, and NOB are eliminated after 2~3 SRT. The third stage is stable operation, that is, maintain $T_{aeration}:T_{stop}=1:1$ and DO = 0.5~1.0 mg/L in the famine period in order to make the A^2/O system provide efficient and stable pollutant removal under intermittent influent conditions.

3.4 Case study

3.4.1 Water quality change in the WWTP under optimal operation

The optimal operation strategy was applied to a town-level WWTP in Shaanxi, China, and the water quality change before and after application is shown in Fig. 12. The average effluent concentration and removal efficiency of COD throughout the test period were (35.30±5.98) mg/L and (76.39±1.66) %, respectively (Fig. 12A). The average effluent COD concentration of the WWTP consistently met the Class 1A standards (below 50 mg/L) of GB 18918—2002, which further verified that aeration mode and DO concentration had no distinct effect on COD removal in the A^2/O process.

Compared with the original operation during phase 1, NH_4^+-N, TN, and TP removal during the stable operation stage in phase 5 was significantly improved (Fig. 12B~D). The average effluent concentrations of NH_4^+-N, TN, and TP in phase 5 were (1.56±0.74) mg/L, (9.93±0.95) mg/L, and (0.86±0.05) mg/L, with average removal rates of (94.17±2.02)%, (65.32±4.75)%, and (61.05±2.34)%, respectively. The average removal rates of NH_4^+-N, TN, and TP in phase 5 increased by 8.32%, 5.55%, and 37.14%, respectively, relative to phase 1, and the effluent concentration of each pollutant could stably meet Class 1A standards (TP could stably meet Class 1B, below 1.0 mg/L). The results further indicated that the optimal operation of intermittent aeration ($T_{aeration}:T_{stop}$ = 5 h:5 h) and micro-oxygen (DO = 0.5~1.0 mg/L) during famine periods could effectively improve the pollutant removal efficiency of A^2/O systems under discontinuous influent conditions.

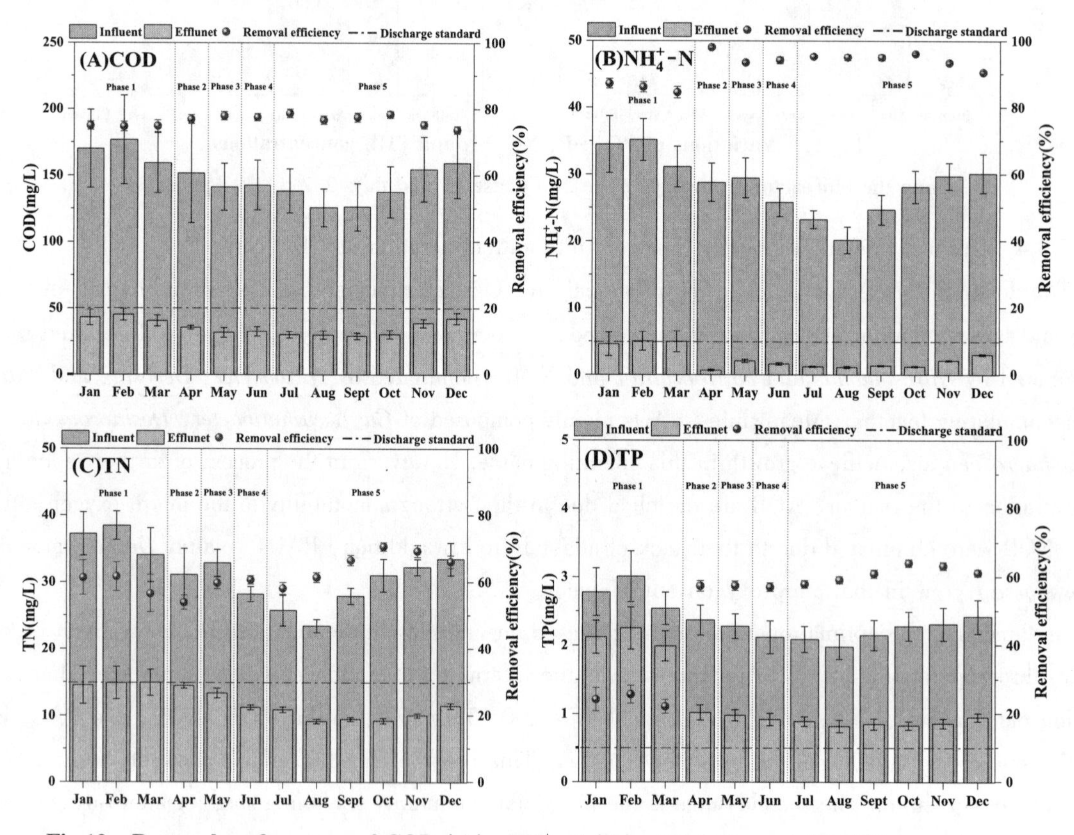

Fig.12 Removal performance of COD (A), NH_4^+-N (B), TN (C), and TP (D) in each stage

3.4.2 Mass balance calculations and economic benefit

Mass balance analysis of the town-level WWTP was conducted to reveal the nutrient removal pathway (Fig. 13 and Table S4). During stable operation (phase 5), the amounts of N and P removed via microbial metabolism were 72.00% and 68.54% of the total removal amount, respectively, significantly higher than the 40.85% and 25.98% removal rates observed in the original operation stage (phase 1). This is mainly because the removal pathways of N and P in the system changed during phase 5, and reduced the carbon requirement for N and P removal in the system. Therefore, under the same influent conditions, the denitrification and DPR processes in the anoxic zone of the system were enhanced to 46.65% (including 14.09% of N loss by DPR) and 29.75%, respectively. Meanwhile, the decreased DO concentration created a micro-oxygen environment which is conducive to the occurrence of SND in the aerobic zone[38]. In phase 5, 13.90% of N was removed by SND, further improving N removal efficiency. In addition, 5.81% of N was removed by other pathways in phase 5, which was significantly higher than in phase 1 (2.63%), possibly due to anaerobic ammonium oxidation in the anoxic zone[39].

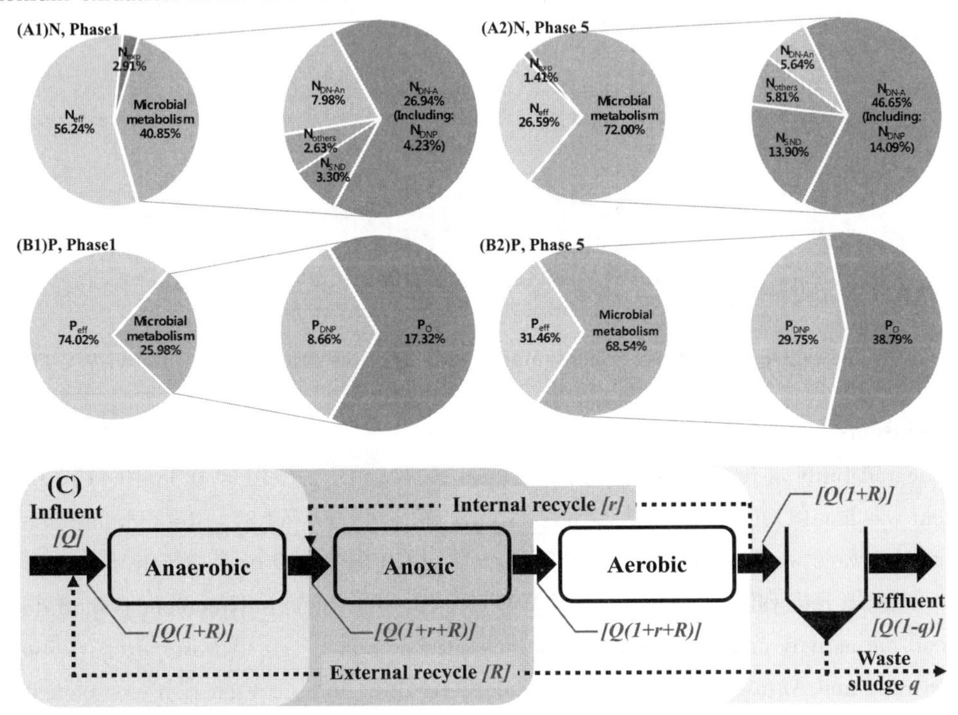

Fig.13 Mass balance calculations for COD, N and P in the town-level WWTP

N_{eff} and P_{eff}: TN and TP in the effluent; N_{exp}: TN in the waste sludge; $N_{DN\text{-}An}$: N loss by denitrification in anaerobic zone; $N_{DN\text{-}A}$: N loss by denitrification in anoxic zone; N_{DNP} and P_{DNP}: N and P loss by DPR in anoxic zone; N_{SND}: N loss by simultaneous nitrification and denitrification; N_{others}: N loss by other pathway; P_0: P uptake by EBPR in oxic zone; Q: average influent flowrate (m^3/d); q: waste sludge ratio (%); R: external recycle ratio (%); r: internal recycle ratio (%).

The sewage treatment power consumption and additives cost of the town-level WWTP can be seen in Fig. 14. The economic benefits of intermittent aeration and micro-oxygen operation come primarily from reducing the cost of electricity and additives. During the original operation stage from January to March, two of the three air compressors in the WWTP were kept running continuously, and the power consumption per ton of wastewater was (0.72±0.01) kW/h. In terms of aeration alone, the electricity cost per ton of wastewater was (0.58±0.01) CNY (the local electricity cost was 0.8 CNY · $(kW \cdot h)^{-1}$). In addition, in order to maintain microbial activity, 50 kg of glucose must be added daily. The cost of this additive per ton of wastewater was 0.25 CNY (the cost of glucose for industrial is 2500 CNY/ton). During the stable operation stage from July to December, the duration of air compressor operation was reduced to 17 h/d, and only one air compressor was needed. Therefore, at this stage, the power consumption per ton of wastewater reduced to (0.23±0.02) kW/h, and the electricity cost per ton of wastewater reduced to (0.18±0.01) CNY. In addition, no additives were used in the stable operation stage, thus saving this part of the cost. Compared with the original operation stage, the electricity consumption and additive cost saved during stable operation were 0.40 and 0.25 CNY/ton respectively, and the operating cost can be reduced by 118600 CNY/a (17541 USD/a), with significant improvement in economic benefits.

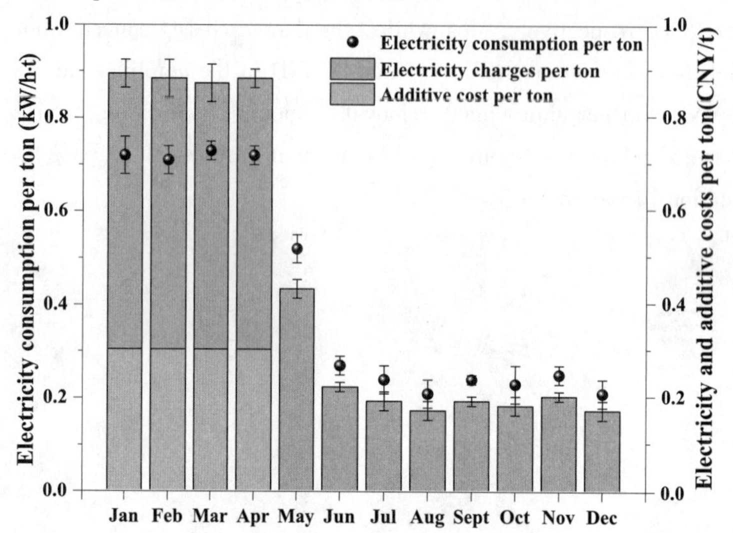

Fig.14 Sewage treatment power consumption and additives cost of the town-level WWTP

4 Conclusions

The biodegradability of rural sewage in Shaanxi was good (BOD_5/COD = 0.38~0.41), but the source carbon content was low (COD/TN = 4.23~5.42, COD/TP = 48.99~49.37). In addition, the discontinuous flow of rural sewage was significant, and influent was cutoff for 5~19 h/d, which was an important factor affecting pollutant removal efficiency in rural WWTPs. The starvation environment (cutoff duration = 10 h/d) inhibited the activity and relative abundance of nitrobacteria (e.g., *Nitrosospira*, *Nitrosomonas*, *Nitrobacter*, *Devosia*, and *Nitrospira*) in the aerobic zone, resulting in an oxygen-rich environment in the system at the same aeration rate. This further inhibited the activity and reduced the relative abundance of denitrobacteria (e.g., *Dokdonella*, *Paracoccus*, *Denitratisoma*, *Thauera*, *Haliangium*, *Hyphomicrobium*, and *Zoogloea*) and PAOs (e.g., *Acinetobacter*, *Candidatus_Accumulibacter*, *Tetrasphaera*, and *Dechloromonas*), resulting effluent concentrations of NH_4^+-N (6.00±3.46) mg/L, TN (16.77±4.37) mg/L, and TP (1.67±0.45) mg/L exceeding the Class 1A discharge standard of GB 18918—2002. The optimal operation strategy of $T_{aeration}:T_{stop}$ = 1:1 and DO = 0.5~1.0 mg/L can enrich AOB (*Nitrosomonas* and *Nitrosospira*) and DPAOs (*Dechloromonas* and *Tetrasphaera*), stimulating partial nitrification and DPR (NAR = (46.56±6.88)%, DPR rate = (43.15±3.54)%. Therefore, nutrient removal rates were improved. The optimal removal rates of NH_4^+-N, TN, and TP were (95.52±1.44)%, (74.15±2.96)%, and (72.27±3.35)% with effluent concentrations of (1.13±0.38) mg/L, (8.45±1.39) mg/L, and (0.75±0.20) mg/L. After the optimized operation strategy was applied to a town-level sewage treatment plant in Ankang, Shaanxi, China, the removal efficiency of NH_4^+-N, TN, and TP increased from (85.85±1.31)%, (59.77±2.86)%, and (23.91±1.77)% to (94.17±2.02)%, (65.32±4.75)%, and (61.05±2.34)%, respectively, and the operating cost per ton of wastewater was decreased by 0.65 CNY. This approach can reduce operation cost by 118600 CNY (17541 USD) per year, reducing consumption and increasing efficiency.

Declaration of competing interest: The authors declare that they have no known competing financial interests or personal relationships that could have appeared to influence the work reported in this paper.

Acknowledgements: This study was supported by the Key Research and Development Project of Shaanxi Province (2021ZDLSF05-06), the National Natural Science Foundation of China (52070151, 52170052).

References

[1] PRC M E E. The Bulletin of the Second National Pollution Source Investigations. (2020).

[2] Y. Xu, H. Li, Y. Li, X. Zheng, C. Zhang, Y. Gao, P. Chen, Q. Li, L. Tan, Systematically assess the advancing and limiting factors of using the multi-soil-layering system for treating rural sewage in China: From the economic, social, and environmental perspectives, J. Environ. Manage. 312 (15) (2022).

[3] P.P. Cheng, Q. Jin, H. Jiang, M. Hua, Z. Ye, Efficiency assessment of rural domestic sewage treatment facilities by a slacked-based DEA model, J. Cleaner Prod. 267 (2020), 122111.

[4] Y.D. Xie, Q.H. Zhang, M. Dzakpasu, Y.C. Zheng, X.C. Wang, Towards the formulation of rural sewage discharge standards in china, Sci. Total. Environ. 759 (12) (2020), 143533.

[5] Y. Ma, Y. Zhai, X. Zheng, S. He, M. Zhao, Rural domestic wastewater treatment in constructed ditch wetlands: Effects of influent flow ratio distribution, J. Cleaner Prod. 225 (2019) 350-358.

[6] F. Chen, Q. Yao, The Development of Rural Domestic Wastewater Treatment in China, Adv. Mat. Res. 1073-1076 (2015) 829-832.

[7] B. Xi, X. Li, J. Gao, Y. Zhao, H. Liu, X. Xia, T. Yang, L. Zhang, X. Jia, Review of challenges and strategies for balanced urban-rural environmental protection in China, Front. Env. Sci. Eng. 9 (3) (2015) 371-384.

[8] Q. Liao, S. You, M. Chen, M. Yang, The Application of Combined Sewage Treatment Technology in Rural Polluted Water Prevention and Control, Appl. Mech. Mater. 2972 (2014) 782-785.

[9] Chinese N E P A. Water and wastewater monitoring methods 4th ed. Beijing: Chinese Environmental Science Publishing House. (2002).

[10] T. Ren, Y.L. Chi, Y. Wang, X. Shi, P.K. Jin, Diversified metabolism makes novel Thauera strain highly competitive in low carbon wastewater treatment, Water Res. 206 (2021), 117742.

[11] T, Zhang, B. Wang, X.Y. Li, Q. Zhang, L. Wu, H. Yang, Y.Z. Peng, Achieving partial nitrification in a continuous post-denitrification reactor treating low C/N sewage, Chem. Eng. J. 188 (09) (2017) 330-337.

[12] Y. Rong, X.C. Liu, L.J. Wen, X. Jin, X. Shi, P.K. Jin, Advanced nutrient removal in a continuous A^2/O process based on partial nitrification-anammox and denitrifying phosphorus removal, J. Water Process. Eng. 36 (2020), 101245.

[13] T.L. Zheng, R. Xiong, W.K. Li, W.J. Wu, Y.Q. Ma, P.Y. Li, X.S. Guo, An enhanced rural anoxic/oxic biological contact oxidation process with air-lift reflux technique to strengthen total nitrogen removal and reduce sludge generation, J. Cleaner Prod. 348 (2022), 131371.

[14] R.C. Testolin, L. Mater, E. Sanches-Simões, A. Conti-Lampert, A. Corrê, M. Groth, M. Oliveira-Carneiro, C.M. Radetski, Comparison of the mineralization and biodegradation efficiency of the Fenton reaction and Ozone in the treatment of crude petroleum-contaminated water, J. Environ. Chem. Eng. 8 (5) (2020), 104265.

[15] D. Chen, Z. Wang, M. Zhang, X. Wang, S. Lu, Effect of increasing salinity and low C/N ratio on the performance and microbial community of a sequencing batch reactor, Environ. Technol. 42 (8) (2019) 1213-1224.

[16] X. Hao, Q. Wang, X. Zhang, Y. Cao, C.M. Loosdrecht, Experimental evaluation of decrease in bacterial activity due to cell death and activity decay in activated sludge, Water Res. 43 (14) (2009) 3604-3612.

[17] E. Van den Eynde, L. Vriens, M. Wynants, H. Verachtert, Transient behaviour and time aspects of intermittently and continuously fed bacterial cultures with regard to filamentous bulking of activated sludge, Appl. Microbiol. Biotechnol. 19 (1984) 44-52.

[18] M.C.M.V. Loosdrecht, M.A. Pot, J.J. Heijnen, Importance of bacterial storage polymers in bioprocesses, Water Sci. Technol. 35 (1) (1997) 41-47.

[19] J.G. Kramer, F.L. Singleton, Variations in rRNA content of marine Vibrio spp. during starvation-survival and recovery, Appl. Environ. Microbiol. 58 (1) (1992) 201-207.

[20] P. Roslev, G.M. King, Aerobic and anaerobic starvation metabolism in methanotrophic bacteria. Appl. Environ. Microbiol. 61 (4) (1995) 1563-1570.

[21] A. Bollmann, I. Schmidt, A.M. Saunders, M.H. Nicolaisen, Influence of starvation on potential ammonia-oxidizing activity and amoA mRNA levels of Nitrosospira briensis, Appl. Environ. Microbiol. 71 (3) (2005) 1276-1282.

[22] X. Hao, Q. Wang, J. Zhu, C.M. Loosdrecht, Microbiological endogenous processes in biological wastewater treatment systems, Critical Reviews in Environ. Sci. Technol. 40 (2010) 239-265.

[23] C.M. Fitzgerald, P. Camejo, J.Z. Oshlag, D.R. Noguera, Ammonia-oxidizing microbial communities in reactors with efficient nitrification at low-dissolved oxygen, Water Res. 70 (2015) 38-51.

[24] Y.Q. Gao, Y.Z. Peng, J.Y. Zhang, S.Y. Wang, J.H. Guo, L. Ye, Biological sludge reduction and enhanced nutrient removal in a pilot-scale system with 2-step sludge alkaline fermentation and A^2/O process, Bioresour. Technol. 102 (5) (2011) 4091-4097.

[25] P. Elefsiniotis, D. Li, The effect of temperature and carbon source on denitrification using volatile fatty acids, Biochem. Eng. J. 28 (2) (2006) 148-155.

[26] Y.Y Miao, L. Zhang, D.H. Yu, J.H. Zhang, W.K. Zhang, G.C. Ma, X.C. Zhao, Y.Z. Peng, Application of intermittent aeration in nitrogen removal process: development, advantages and mechanisms, Chem. Eng. J. (2021), 133184.

[27] J.S. Huang, L.J. Xu, Y.Y. Guo, D.S. Liu, Intermittent aeration improving activated granular sludge granulation for nitrogen and phosphorus removal from domestic wastewater, Bioresour. Technol. 15 (2021), 100739.

[28] G. Liu, J. Wang, Long-Term Low DO Enriches and Shifts Nitrifier Community in Activated Sludge, Environ. Sci. Technol. 47 (10) (2013) 5109-5117.

[29] E.M. Gilbert, S. Agrawal, F. Brunner, Response of different nitrospira species to anoxic periods depends on operational DO, Environ. Sci. Technol. 48 (5) (2014) 2934.

[30] S.J. Ge, Y.Z. Peng, S. Qiu, A. Zhu, N.Q. Ren, Complete nitrogen removal from municipal wastewater via partial nitrification by appropriately alternating anoxic/aerobic conditions in a continuous plug-flow step feed process, Water Res. 55 (2014) 95-105.

[31] C. Hellinga, A.A.J.C. Schellen, J.W. Mulder, M.C.M. van Loosdrecht, J.J. Heijnen, The Sharon process: an innovative method for nitrogen removal from ammonium-rich wastewater, Water Sci. Technol. 37 (1998) 135-142.

[32] N.M. Cruz, K. Bournazou, H. Hooshiar, G. Arellano-Garcia, G. Arellano-Garcia, Model based optimization of the intermittent aeration profile for SBRs under partial nitrification, Water Res. 47 (10) (2013) 3399-3410.

[33] R. Yu, K. Chandran, Strategies of nitrosomonas europaea 19718 to counter low dissolved oxygen and high nitrite concentrations, BMC Microbiology, 10 (1) (2010) 70.

[34] A. Guisasola, I. Jubany, J.A. Baeza, J. Carrera, J. Lafuente, Respirometric estimation of the oxygen affinity constants for biological ammonium and nitrite oxidation, J. Chem. Technol. Biot. 80 (4) (2010) 388-396.

[35] J. Qian, M. Zhang, Y. Wu, J. Niu, C. Xing, H. Yao, S. Hu, X. Pei, A feasibility study on biological nitrogen removal (BNR) via integrated thiosulfate-driven denitratation with anammox, Chemosphere, 208 (2018) 793-799.

[36] T. Mino, M. Loosdrecht, J.J. Heijnen. Review Paper: microbiology and biochemistry of the enhanced biological phosphate removal process, Water Res. 32 (11) (1998) 3193-3207.

[37] W.H. Zhao, X.J. Bi, Y.Z. Peng, M. Bai, Research advances of the phosphorusaccumulating organisms of Candidatus Accumulibacter, Dechloromonas and Tetrasphaera: Metabolic mechanisms, applications and influencing factors, Chemosphere, (2022).

[38] Y. Hyungseok, K.-H. Ahn, H.-J. Lee, K.-H. Lee, Y.-J. Kwak, K.-G. Song, Nitrogen removal from synthetic wastewater by simultaneous nitrification and denitrification (SND) via nitrite in an intermittently-aerated reactor, Water Res. 33 (1) (1999) 145-154.

[39] J.W. Li, Y.Z. Peng, L. Zhang, J.J. Liu, X.D. Wang, R.T. Gao, L. Pang, Y.X. Zhou, Quantify the contribution of anammox for enhanced nitrogen removal through metagenomic analysis and mass balance in an anoxic moving bed biofilm reactor, Water Res. 160 (2019) 178-187.

本文曾发表于2022年《Chemical Engineering Journal》第453卷

Components of Respiration and Their Temperature Sensitivity in Four Reconstructed Soils

Na Lei[1,2,3], Huanyuan Wang[1,3], Yang Zhang[1,3], Tianqing Chen[1,3]

(1.Shaanxi Provincial Land Engineering Construction Group Co., Ltd., Xi'an, China;2.Institute of Soil and Water Conservation, Northwest A&F University, Yangling Shaanxi, China;3.Institute of Land Engineering and Technology, Shaanxi Provincial Land Engineering Construction Group Co., Ltd., Xi'an, China)

【Abstract】Seasonal changes characteristics in the respiration of four reconstructed soil masses in a barren gravel land were monitored. The results showed that: (1) Respiration and heterotrophic respiration of the four reconstructed soils with added meteorite, shale, sand increased gradually with increasing soil temperatures, reaching its maximum in summer and decreasing to its minimum in winter. The average annual respiration of reconstructed soil with sand was 4.87 μmol · m^{-2} · s^{-1}, which was significantly higher than the other reconstructed soils ($p<0.05$). (2) Themaximum and minimum values of autotrophic respiration for the four reconstructed soils appeared in August 2018 and January 2018, respectively. The proportion of autotrophic respiration to total respiration was 12.5% ~ 38.0%,9.5%~42.0%, 7.7%~41.2%, and 5.0%~39.3% for the soils with reconstituted meteorite, shale, sand, and soft rock, respectively. (3) The relationship between respiration and the temperature of reconstructed soils can be represented by an exponential function. The 90% to 93% changes in reconstructed soils respiration were caused by soil temperature. The temperature sensitivity (Q_{10}) of reconstituted soil with added sand was significantly higher than that of the other three reconstituted soils.

【Keywords】 Seasonal changes; Reconstructed soils; Soil respiration; Soil temperature; Temperature sensitivity

Introduction

Soil respiration is the primary process whereby terrestrial ecosystems release CO_2 into the atmosphere[1-2], with the annual release of CO_2 via this route being more than 10 times that released by the combustion of fossil fuels[3]. The temperature sensitivity of soil respiration is considered the main factor affecting the response of terrestrial ecosystems to global warming and also determines the feedback of soil respiration to atmospheric CO_2 concentrations[4]. In the context of continuous global warming, research on the temperature sensitivity of soil respiration has been a constant focus of scholars[5-6], with the mainstream consensus being that soil respiration is particularly sensitive to variations in temperature[7-8]. However, although temperature and moisture are considered the main factors influencing soil respiration[9], in reality, the rate of soil respiration is a compound effect, reflecting the mutual influence of multiple factors, including temperature, humidity, and organic carbon content, which accordingly contribute to the complexity of the responses of soil respiration to changes in temperature[10]. Moreover, these responses to temperature change are characterized by spatio-temporal variability[11], which inevitably exacerbates the complexity of research on the temperature sensitivity of soil respiration.

Although soil respiration comprises both autotrophic and heterotrophic respiration, the contributions of these components to temperature sensitivity remain unclear. Thus, to gain a sufficient understanding of the responses of soil respiration to changing temperature, it is necessary to accurately determine the proportionate contributions of autotrophic and heterotrophic respiration to total soil respiration. Given the differing biological and ecological processes involved in the different components of soil respiration, their responses to temperature change will similarly differ[12], and consequently, dividing soil respiration into different components is considered key to understanding the mechanisms underlying the response soil respiration to temperature change[13].

The findings of research conducted to date on the temperature sensitivity of soil respiration and its components have tended to be somewhat inconsistent, with some authors contending that the temperature sensitivity of autotrophic respiration is greater than that of heterotrophic respiration[14-15], whereas others have indicated that heterotrophic respiration makes a greater contribution in this regard[16-17]. Consequently, the precise mechanisms underlying temperature sensitivity have yet to be sufficiently determined.

Soil reconstruction is a process whereby humans, on the premise of respecting the laws of nature, adopt engineering methods, such as replacement, compounding, increase and decrease, and other technical means, to reconstruct soil structures and improve the quality of the land environment. In the reconstruction process, soils that are considered difficult to use or are unusable, such as those from degraded, contaminated, or inefficiently utilized sires, are transformed into soils that are conducive to the survival and reproduction of living organisms. For example, in an area of coal mining subsidence, mechanical rolling and disturbance caused by construction has been found to alter the original structure and profile of the soil, which in turn has modified important environmental factors affecting the rate of soil respiration rate, thereby resulting in a reduction in the soil respiration Q_{10} value[18]. Furthermore, the findings of a study that examined four types of newly structured soil (sandy loess, sandy loess + weathered coal, sandy loess + weathered coal + soft rock, and sandy loess + soft rock) revealed that weathered coal promoted respiration within the newly structured soils, improved the carbon release rate, and altered the diurnal pattern of soil respiration[19]. Compared with natural soils, differences in those in coal gangue filling and reconstruction areas have been found to lead to certain distinctions between the soil respiration processes. During reconstruction, it was established that differences in the thickness of the upper layer of the coal gangue influenced soil surface respiration to varying extents, with the soil carbon sequestration capacity of the 60~100 cm layer being notably most robust, thereby indicating that soils of these depths would be a more suitable thickness of covering soil[20].

In this study, we examined the properties four reconstructed soils supplemented with meteorite, shale, sand, or soft rock, respectively. Using a soil carbon flux measurement system, our aim was to determine the seasonal changes in soil respiration and its components in these four types of reconstructed soils, along with the temperature sensitivity of soil respiration. We also sought to clarify the respiratory processes and dynamic change mechanisms of soils reconstituted with different materials. This, we hoped, would enable us to gain a more complete understanding of the potential contribution of reconstituted soil respiration in land remediation, to promote the further development of the carbon cycle theory, and to provide a theoretical basis for accurately assessing regional CO_2 emissions and thus formulating appropriate CO_2 emission reduction measures. On the basis of our findings, we propose the use of soil respiration to characterize the environmental friendliness of reconstituted soils, which we anticipate will provide a reference for guiding the future selection of suitable materials for soil reconstitution.

Materials and methods

Overview of test plots

The test plot is located in Shangwang Village, Tangyu Town, Meixian County, Baoji City, Shaanxi Province (107°53′50″E, 34°8′33″N), and a demonstration area for the barren gravel land remediation project. The total area is 8.00 hm^2, and the newly added cultivated land is 6.80 hm^2. Four materials of soft rock, sand, shale, and meteorite were selected, crushed through a 10 mm sieve, disinfected, sterilized, and mixed with the constructed soil source to form a mixed layer (30 cm) of meteorite, shale, sand and meteorite, and soil. Lou soil, which was the local common soil type, was used for construction. Finally, four reconstituted soils were formed, i.e., gravel + meteorite + lou, gravel + shale + lou, gravel + sand + lou, and gravel + soft rock + lou soil types (hereinafter referred to as meteorite, shale, sand, and soft rock reconstituted soil masses) long-term positioning test[24] (Fig.1). The dosage of meteorite, shale, sand, and soft rock was 1×10^{-3} m^3/m^2. The dimensions of all test plots were 20 m× 30 m.

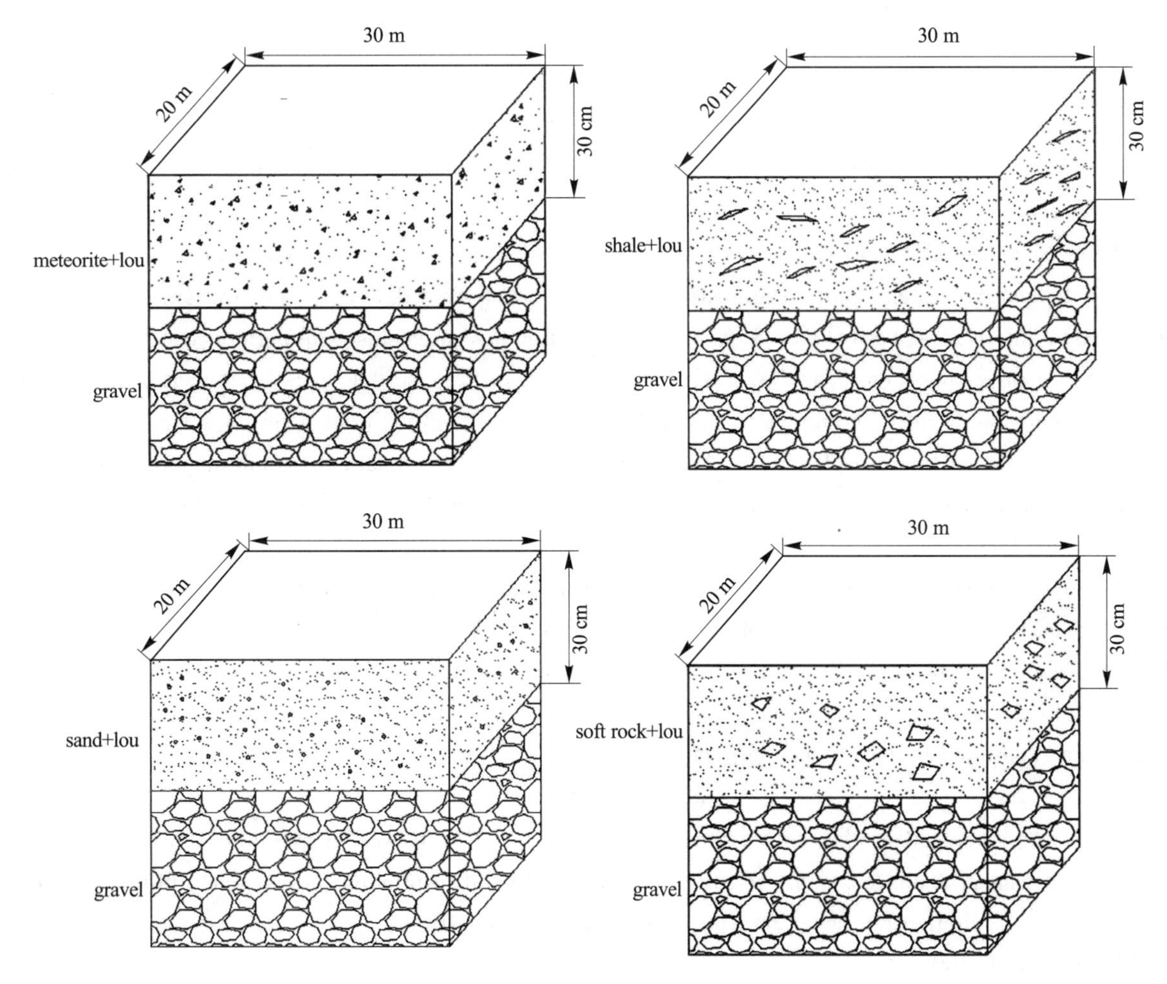

Fig.1 Stereograms of the test sample

Three soil respiration rings (inner diameter 10 cm) were buried in each of the four test plots, ensuring that the tops of the rings were 2 cm above the ground. At the same time, three small rectangular plots(2 m× 2 m)were randomly set up as root exclusion treatment plots. A soil respiration ring of the same specification was buried in each of root exclusion treatment plots. A small trench with a depth of 40 cm was excavated around the root exclusion treatment plots. The excavated ditches were partitioned with as oards, and the soil was backfilled according to the profile level. The vegetation on the ground was cut off in soil respiration rings, ensuring that no vegetation grew in soil respiration rings during the observation period[24]. The physical and chemical properties of the test plots are shown in Table 1.

Table 1 Basic physical and chemical properties of four reconstructed soil at 0~20 cm depth

Detection Indicator		Reconstituted soil mass types			
		Meteorite	Shale	Sand	Soft rock
pH		8.55	8.49	8.51	8.49
Organic carbon ($g \cdot kg^{-1}$)		3.41	3.75	3.7	4.77
Total nitrogen ($g \cdot kg^{-1}$)		0.56	0.36	0.44	0.48
Available phosphorus ($mg \cdot kg^{-1}$)		12.93	26.33	27.27	21.7
Available potassium ($mg \cdot kg^{-1}$)		136.96	130.15	115.54	111.65
Size grading	<0.002 mm	16.47	16.88	15.17	17.85
	0.002~0.05 mm	79.87	76.09	79.99	79.22
	>0.05 mm	6.04	7.03	4.84	2.93

Research methods

From November 2017 to October 2018, all the soil respiration rings of four test plots were measured on the three typical days each month. The measurement time per typical day was from 9:30 am to 11:00 am, and the time interval was basically 6~8 days. Soil respiration measurements were performed using a soil carbon flux measurement system (LI-8100, LI-COR Biosciences, Lincoln, NE, USA) to measure soil carbon flux, soil temperature at 5 cm and water content at 10 cm. Each soil respiration ring was measured 3 times and the measurement time was 4 min[24]. Autotrophic respiration was obtained by subtraction, that is, soil autotrophic respiration should be the difference between soil total respiration sinus heterotrophic respirations.

Data analyses

One-way ANOVA was used to analyze differences in soil respiration of the four reconstructed soils. All statistical tests were carried out using SPSS software (version 16.0; SPSS Inc., Chicago, IL, USA). Nonlinear regression was used to assess the relationship between soil respiration and hydrothermal influence factors of the four reconstructed soils, and Q_{10} was estimated. The relationship between soil respiration and soil temperature was fitted by an exponential model (Eq.1):

$$R_S = ae^{bT}, \quad Q_{10} = e^{10b} \tag{1}$$

Where R_S is the soil respiration rate ($\mu mol \cdot m^{-2} \cdot s^{-1}$); T is the soil temperature (℃); a and b are the model parameters, and Q_{10} is the sensitivity coefficient of soil respiration, which refers to the change in entropy of soil respiration rate when the soil temperature rises by 10 ℃.

Results and analysis

Respiration and heterotrophic respiration of reconstructed soils

We found that the total and heterotrophic respiration of the four assessed reconstructed soils, supplemented with meteorite, shale, sand, and soft rock, respectively, exhibited the same seasonal trends with respect to soil temperature. Specifically, both total and heterotrophic respiration increased gradually in response to increasing soil temperatures, with the trend being highest in summer and lowest in winter. Throughout the entire year, seasonal changes in the total respiration of the meteorite, shale, sand, and soft rock amended soils ranged from 0.16 to 7.97, 0.21 to 9.69, 0.26 to 10.87, and 0.20 to 8.71 $\mu mol \cdot m^{-2} \cdot s^{-1}$, respectively, whereas the corresponding rates of heterotrophic respiration varied from 0.14 to 4.94, 0.19 to5.62, 0.24 to 6.39, and 0.19 to 5.42 $\mu mol \cdot m^{-2} \cdot s^{-1}$, respectively (Fig.2). Among the four reconstructed soils, the annual average respiration rate of the soil reconstructed with sand (4.87 $\mu mol \cdot m^{-2} \cdot s^{-1}$) was found to be significantly higher than that of the other three assessed soils ($p>0.05$), whereas differences among these three soils were shown to be non-significant.

Autotrophic respiration rate of reconstructed soils and its relationship with heterotrophic respiration

The autotrophic respiration of the four reconstructed soils was found to show clear seasonal dynamic changes, with the highest and lowest rates being observed in August 2018 and January 2018, respectively. Among the four reconstructed soils supplemented with meteorite, shale, sand, and soft rock, the highest autotrophic respiration rates were 3.03, 4.07, 3.29, and 5.62 μmol · m^{-2} · s^{-1}, respectively, whereas the corresponding minimum values were 0.02, 0.02, 0.01, and 0.19 μmol · m^{-2} · s^{-1}(Fig.2). Apart from the summer months, during which we detected significant differences in autotrophic respiration in the meteorite and sand-supplemented soils ($p>0.05$), there were no significant differences in the autotrophic respiration of the four reconstructed soils in other seasons ($p>0.05$). Comparisons of autotrophic and heterotrophic respiration revealed significant differences among the four reconstructed soils in January ($p>0.05$). During winter, there were significant differences in the autotrophic and heterotrophic respiration of the soft rock-supplemented soil ($p > 0.05$), although during the remaining months, we detected no significant differences in the four reconstructed soils with respect to the two types of respiration ($p > 0.05$) (Fig. 3). Moreover, we established that throughout the entire year, the annual average soil autotrophic respiration rates of the reconstructed soil masses supplemented with meteorite, shale, sand, and soft rock accounted for 12.5%~38.0%, 9.5%~42.0%, 7.7%~41.2%, and 5.0%~39.3% of the total soil respiration, respectively (Table 2).

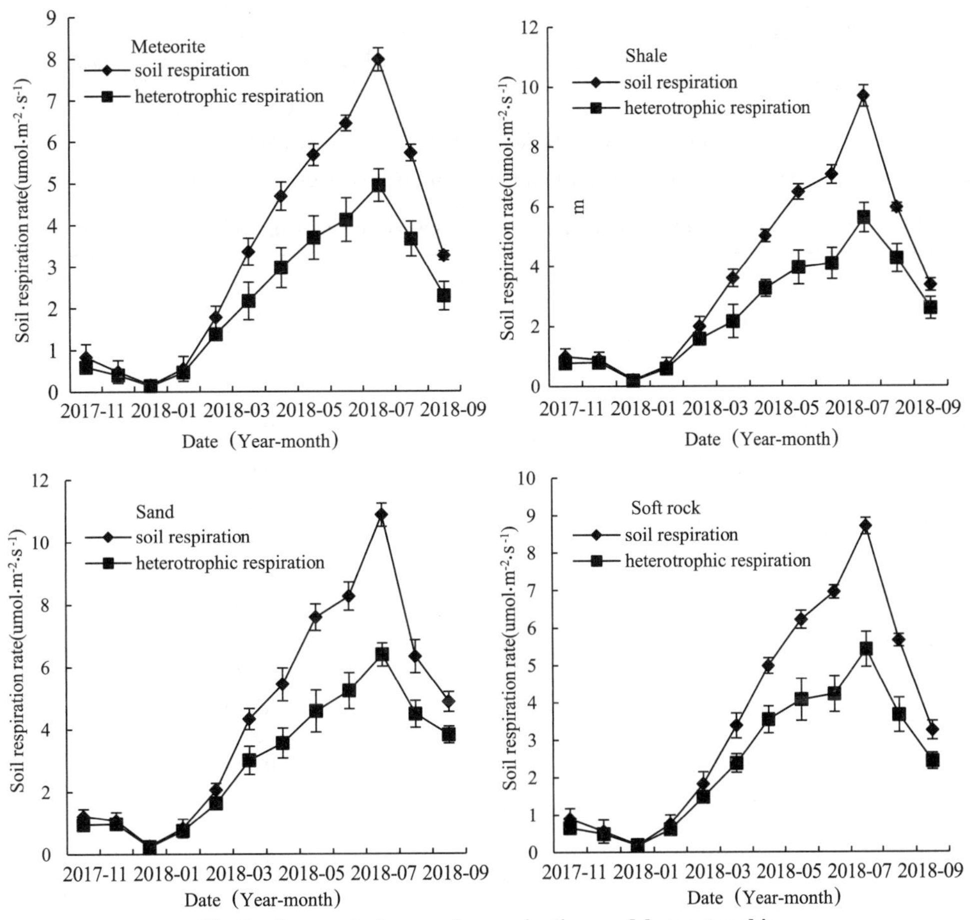

Fig.2 Seasonal changes in respiration and heterotrophic respiration of the four reconstructed soils (mean±standard error)

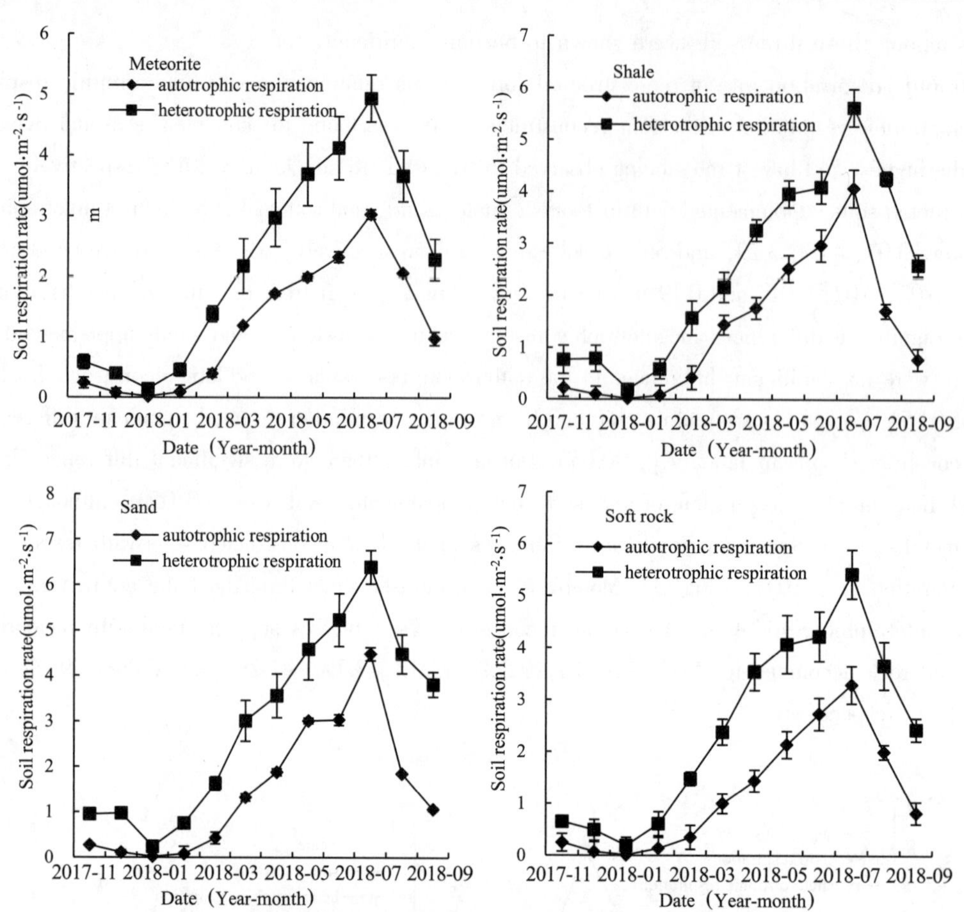

Fig.3 Autotrophic respiration rate of reconstructed soils and its relationship with heterotrophic respiration (mean±standard error)

Table 2 The ratio of autotrophic respiration to total respiration in reconstructed soils (%)

Year-month	Meteorite	Shale	Sand	Soft rock
2017-11	0.289	0.222	0.221	0.278
2017-12	0.167	0.112	0.102	0.125
2018-01	0.125	0.095	0.077	0.050
2018-02	0.164	0.119	0.096	0.187
2018-03	0.225	0.211	0.209	0.191
2018-04	0.354	0.400	0.305	0.296
2018-05	0.367	0.351	0.346	0.289
2018-06	0.350	0.389	0.397	0.345
2018-07	0.360	0.421	0.367	0.393
2018-08	0.380	0.420	0.412	0.378
2018-09	0.360	0.286	0.292	0.353
2018-10	0.301	0.231	0.218	0.252

The relationships among soil respiration, soil temperature, and water content

Among the four reconstructed soils, we detected a very significant correlation between the rate of soil respiration and soil temperature ($p>0.01$), whereas in contrast, there was no obvious correlation between respiration and soil water content. Furthermore, in these soils, the components of soil respiration were found to be significantly correlated with soil temperature ($p>0.01$) (Table 3), which was identified as the main factor affecting soil respiration. From the perspective of seasonal change, the relationships between soil respiration and soil temperature in the four reconstructed soils can be characterized by exponential functions (Table 4). Among the soils supplemented with meteorite, shale, sand, and soft rock, approximately 90%~93% of the changes in soil respiration rate could be attributed changes in soil temperature, with respective Q_{10} values of 3.23, 3.66, 3.89, and 3.50, respectively. Of these values, the Q_{10} of soil reconstructed with sand was found to be significantly higher than that of the other three soils, among which we detected no significant differences.

Table 3 Correlation between soil respiration and soil temperature (*T*), water content (*W*)

Respiration component	Meteorite		Shale		sand		soft rock	
	T	*W*	*T*	*W*	*T*	*W*	*T*	*W*
Total Respiration	0.952**	0.416	0.942**	0.566	0.937**	0.363	0.955**	0.487
heterotrophic respiration	0.960**	0.429	0.942**	0.571	0.944**	0.380	0.971**	0.493
autotrophic respiration	0.934**	0.491	0.913**	0.541	0.899**	0.323	0.914**	0.468

Note: **, $p<0.01$; *, $p<0.05$.

Table 4 Relationship between annual soil respiration (*R*) rate and temperature (*T*)

Reconstituted soil mass types	Relationship	Model types	R^2	Q_{10}
Meteorite	*R*&*T*	$R=0.3021e^{0.1343T}$	0.90	3.23a
Shale	*R*&*T*	$R=0.3593e^{0.1242T}$	0.92	3.66a
Sand	R_S&*T*	$R=0.3838e^{0.1182T}$	0.92	3.89b
Soft rock	*R*&*T*	$R=0.3194e^{0.1227T}$	0.93	3.50a

Discussion

The variation law of soil respiration rate and its components in different reconstructed soils

Among the four reconstructed soils assessed in this study, we detected clear variations in the rates of soil respiration and its components, with high and low values be detected in summer and winter, respectively. These observations are consistent with those previously reported for the seasonal characteristics of reconstructed soil respiration in areas with coal mining subsidence[18], which have been shown to be determined by temperature and soil moisture conditions[25]. With respect to the different components of soil respiration, organic carbon is primarily released into the atmosphere via heterotrophic respiration, thereby contributing to ecosystem carbon cycling, and thus influencing global climate change[26]. In the present study, we identified consistent trends in the heterotrophic and total respiration of the four reconstructed soils, all of which showed a single-peak curve. These findings are similar to those obtained for the soil respiration characteristics of newly constructed soils in sandy loess dumps located in the Shanxi, Shaanxi, and Mongolian mining areas of China[19]. With respect to autotrophic respiration in the four reconstructed soils, we found that the proportional contribution of this component to the total respiration ranged from 5.0% to 42.0%. Comparatively, a previous study has reported percentage autotrophic respirations ranging from 13% to 94%[27], and in a further study, the proportion of autotrophic respiration in the soil of a cold zone was found to be 50%~93%, whereas that in a temperate zone was 33%~62%[28]. These findings would accordingly tend to indicate that the proportion of autotrophic respiration within soils can be influenced to varying extents by vegetation, time, temperature, and methods of measurement[29]. Given its sources, autotrophic respiration shows clear changes in response to changes in climate, time of day, and season, and predictably, the contribution of autotrophic respiration to total soil respiration will typically be higher during the growing season, and relatively low during the time of year when growth ceases or is substantially reduced[27].

The relationship between soil respiration and hydrothermal factors

Although water content and temperature have been established to be the main environmental factors influencing soil respiration in Chinese farmland ecosystems, this respiration and its components are characterized by differential responses to variations in temperature and water content[30]. The findings of numerous studies have indicated that soil temperature is the main factor influencing soil respiration, which is clearly reflected in the observed seasonal changes[31-32]. Consistent with the opinion of a majority of scholars[33-34], we detected an exponential correlation between soil respiration and soil temperature among the four assessed reconstructed soils. In contrast, in a study examining the CO_2 flux of reconstituted soil under different ecological restoration modes (vegetation type and covering soil thickness) in the Huainan mining area, the authors concluded that the relationship between respiration and soil water volume can be represented by a quadratic function[35]. Moreover, correlation analyses revealed a non-significant association, with corresponding R^2 values of between 0.08 and 0.44, which is broadly consistent with our finding for the four reconstructed soils examined in the present study (R^2 values of between 0.363 and 0.487). Compared with temperature, observed differences in the influence of soil moisture on soil respiration tend to be a little more complex. For example, differences in the total annual precipitation and soil structure of different study sites area may contribute to modifying the relationships between soil moisture and respiration. Furthermore, it can be envisaged that there exists a threshold determining the influence of soil moisture on soil respiration, and that the effect is manifested only when this threshold is exceeded.

The temperature sensitivity of soil respiration varies depending on soil and climatic conditions[36], and can serve as an important indicator in quantifying and predicting the responses of ecosystems and global carbon cycles to climate change. In China, it has been established that soil respiration Q_{10} values range between 1.09 and 6.27, with an average value of 2.26[37], and that among different ecosystem types, values follow the order, forest (2.35) > farmland (2.18) > grassland (2.03)[37]. In the present study, we found that respiration in the four reconstructed soils was particularly sensitive to changes in temperature, with corresponding changes in Q_{10} values of between 3.23 and 3.89. The temperature sensitivity of the reconstructed soils was found to be more pronounced than that of farmland ecosystems, particularly after manual intervention. Our findings tend to indicate that the physical and chemical properties of the reconstructed soils and the ecological environment in the study area have, to varying extents, contributed to modifying the gaseous and material circulation processes. In particular, changes in the underlying soil surface have led to changes in soil temperature and moisture, which in turn have influenced respiration within the reconstructed soils.

Conclusion

(1) Respiration within the reconstructed soils and the corresponding carbon emissions were found to be dependent on the materials used to supplement these soils. Our findings indicate that soils reconstructed with meteorite would be beneficial with respect to protection of the ecological environment, whereas soil reconstructed with sand would be unsuitable in this regard.

(2) Soil heterotrophic respiration (soil microbial and animal respiration) can be used to represent total soil respiration. In future studies, it will be necessary to examine the contributions of microbial and animal respiration in reconstructed soils to facilitate the development of a better soil mass structure that is ecologically and organically beneficial.

(3) When governments implement land remediation plans, if budgets permit, they should prioritize amendment using materials that contribute to environmental protection. Furthermore, carbon dioxide emissions from reconstructed soils should be taken into consideration, thereby enabling the formulation of effective regional measures that are deemed ecologically sound.

Acknowledgements

This study was supported by Natural Science Basic Research Program of Shaanxi (Project No. 2021JQ-961); the Scientific Research Item of Shaanxi Provincial Land Engineering Construction Group (DJNY2021-29).

Author contributions

L.N. wrote the main manuscript, L.N. and W.H.Y conceived the study, experimentation, Z.Y. revised manuscript and figures and C.T.Q analyzed the data.

References

[1] Wang, C. & Yang, J. Rhizospheric and heterotrophic components of soil respiration in six Chinese temperate forests. *Global Change Biol*. 13, 123-131(2007).

[2] Zhao, X., Li, L., Xie, Z. & Li, P. Effects of nitrogen deposition and plant litter alteration on soil respiration in a semiarid grassland. *Sci Total Environ*. 740, 1-10(2020).

[3] Jia, X., Shao, M. & Wei, X. Responses of soil respiration to N addition, burning and clipping in temperate semiarid grassland in northern China. *Agr Forest Meteorol*. 166, 32-40(2012).

[4] Meyer, N., Meyer, H. & Welp, G. Soil respiration and its temperature sensitivity (Q_{10}): Rapid acquisition using mid-infrared spectroscopy. *Geoderma*. 323, 31-40(2018).

[5] Gao, Q. *et al*. Effects of litter manipulation on soil respiration under short-term nitrogen addition in a subtropical evergreen forest. *Forest Ecol Manag*. 429, 77-83(2018).

[6] Wang, Z. *et al*. Soil respiration response to alterations in precipitation and nitrogen addition in a desert steppe in northern China. *Sci Total Environ*. 688, 231-242(2019).

[7] Luo, J. *et al*. Temporal-spatial variation and controls of soil respiration in different primary succession stages on glacier forehead in Gongga Mountain, China. *Plos One*. 7, 1-9(2012).

[8] Tong, X. *et al*. Ecosystem carbon exchange over a warm-temperate mixed plantation in the lithoid hilly area of the North China. *Atmos Environ*. 49, 257-267(2012).

[9] Hursh, A., *et al*. The sensitivity of soil respiration to soil temperature, moisture, and carbon supply at the global scale. *Glob. Change Biol*. 23, 2090-2103 (2017).

[10] Huang, S.D. *et al*. Autotrophic and heterotrophic soil respiration responds asymmetrically to drought in a subtropical forest in the southeast China. *Soil Biol Biochem*. 123, 242-249(2018).

[11] Zeng, X., Song, Y., Zhang, W. & He, S. Spatio-temporal variation of soil respiration and its driving factors in semi-arid regions of north China. *Chin. Geogr. SCI*. 28, 12-24 (2018).

[12] Li, X., *et al*. Contribution of root respiration to total soil respiration in a semi-arid grassland on the Loess Plateau, China. *Sci Total Environ*. 627, 1209-1217(2018).

[13] Luo, Y. & Zhou, X. Soil Respiration and the Environment. 3-4, (Elsevier, 2006).

[14] Bhupinderpal, S. *et al*. Tree root and soil heterotrophic respiration as revealed by girdling of boreal Scots pine forest: extending observations beyond the first year. *Plant Cell Environ*. 26, 1287-1296(2003).

[15] Lavigne, M. *et al*. Soil respiration responses to temperature are controlled more by roots than by decomposition in balsam fir ecosystems. *Can J Forest Res*. 33, 1744-1753(2003).

[16] Rey, A. *et al*. Annual variation in soil respiration and its components in a coppice oak forest in Central Italy. *Global Change Biol*. 8, 851-866(2002).

[17] Hartley, I., Heinemeyer, A., Evans, S. & Ineson, P. The effect of soil warming on bulk soil vs. rhizosphere respiration. *Global Change Biol*. 13, 2654-2667(2007).

[18]Zheng, Y., Zhang, Z., Hu, Y., Yao, D. & Chen, X. Seasonal variation of soil respiration and its environmental effect factors on refactoring soil in coal mine reclamation area. *Journal of China Coal Society*. 39, 2300-2306(2014).

[19]Ren, Z. *et al*. Effect of weathered coal on soil respiration of reconstructed soils on mining area's earth disposal sites in Shanxi-Shaanxi-Inner Monglia adjacent area. *Transactions of the CSAE*. 31, 230-237(2015).

[20]Wang, F. Effect of coversoil thickness on reconstruction soil respiration characteristics in coal mining areas-A case from Panji mining area in Huainan, China. *Huainan: Anhui University of Science and Technology*. 1, 59-60(2017).

[21]Sun, Z. H., Han J. C. & Wang, H. Y. Soft rock for improving crop yield in sandy soil of Mu Us sandy land, China. *Arid Land Res Manag*. 33, 136-154(2019).

[22]Sun, Z. H. & Han J. C. Effect of soft rock amendment on soil hydraulic parameters and crop performance in Mu Us sandy land, China. *Field Crop Res*. 222, 85-93(2018).

[23]Liu Y. S., Yang Y. Y., Li Y. Y. & Li J. T. Conversion from rural settlements and arable land under rapid urbanization in Beijing during 1985-2010. *J Rural Stud*. 51, 141-150 (2017).

[24]Lei, N. & Han, J. C. Effect of precipitation on soil respiration of different reconstructed soils. *SCI REP-UK*. 10, 7328(2020).

[25]Jin, Z., Qi, Y., Yun, S. & Domroes, M. Seasonal patterns of soil respiration in three types of communities along grass-desert shrub transition in Inner Mongolia, China. *Adv atmos Sci*. 26, 503-512(2009).

[26]Wang, X. *et al*. Soil respiration under climate warming: differential response of heterotrophic and autotrophic respiration. *Global Change Biol*. 20, 3229-3237(2014).

[27]Zhao, C., Zhao, Z., Yilihamu, Hong, Z. & Jun, L. Contribution of root and rhizosphere respiration of Haloxylon ammodendron to seasonal variation of soil respiration in the Central Asian desert. *Quatern Int*. 244, 304-309 (2011).

[28]Hanson, P., Edwards, N., Garten, C. & Andrews, J. Separating root and soil microbial contributions to soil respiration: A review of methods and observations. *Biogeochemistry*. 48, 115-146(2000).

[29]Liu, H.&Li, F. Effects of shoot excision on in situ soil and root respiration of wheat and soybean under drought stress. *Plant Growth Regul*. 50, 1-9(2006).

[30]Han, X., Zhou, G. & Xu, Z. Research and prospects for soil respiration of farmland ecosystems in China. *J Plant Ecol*. 32, 719-733 (2008).

[31]Tong, D., Xiao, H., Li, Z., Nie, X & Huang, J. Stand ages adjust fluctuating patterns of soil respiration and decrease temperature sensitivity after revegetation. *Soil Sci Soc Am J*. 84, 760-774 (2020).

[32]Gromova, M., Matvienko, A., Makarov, M., Cheng, C. & Menyailo, O. Temperature Sensitivity (Q_{10}) of soil basal respiration as a function of available carbon substrate, temperature, and moisture. *Eurasian Soil ence*. 53, 377-382 (2020).

[33]Meyer, N., Welp, G. & Amelung, W. The temperature sensitivity (Q_{10}) of soil respiration: Controlling factors and spatial prediction at regional scale based on environmental soil classes. *Global Biogeochem Cy*. 32, 204-210(2018).

[34]Tang, X., Shao, H. & Liang, H. Soil respiration and net ecosystem production in relation to intensive management in moso bamboo forests. *Catena*. 137, 219-228(2016).

[35]Zhou, Y., Wang, F., Chen, X., Chen, M & Liu, B. Effects of ecological restoration patterns on Diurnal Variation of CO_2 flux from rehabilitated soil of coal mining areas in Huainan City. *Bulletin of Soil and Water Conservation*. 36, 40-46 (2016).

[36]Lellei, K. *et al*. Temperature dependence of soil respiration modulated by thresholds in soil water availability across European shrub land ecosystems. *Ecosystems*. 19, 1460-1477(2016).

[37]Zhan, X., Yu, G., Zheng, Z. & Wang, Q. Carbon emission andspatial pattern of soil respiration of terrestrial ecosystems in China: Based on geostatistic estimation of flux measurement. *Advancesin Earth Science*. 31,97-108(2012).

本文曾发表于2022年《Nature Portfolio》第12卷

Effects of the Application of Different Improved Materials on Reclaimed Soil Structure and Maize Yield of Hollow Village in Loess Area

Zhe Liu[1,2,3,4,5], Yang Zhang[1,2,3,4], Zenghui Sun[1,2,3,4],
Yingying Sun[1,2,3,4], Huanyuan Wang[1,2,3,4], Ruiqing Zhang[1,2,3,4]

(1.Shaanxi Provincial Land Engineering Construction Group Co.,Ltd., Xi'an 710075, China;2.Key Laboratory of Degraded and Unused Land Consolidation Engineering, the Ministry of Natural Resources, Xi'an 710021, China;3.Institute of Land Engineering and Technology,Shaanxi Provincial Land Engineering Construction Group Co.,Ltd., Xi'an 710021, China; 4.Shaanxi Provincial Land Consolidation Engineering Technology Research Center, Xi'an 710075, China;5.School of Human Settlements and Civil Engineering, Xi'an Jiaotong University, Xi'an 710049, China)

【Abstract】 In order to solve the soil problem of poor structure and low fertility after the abandoned homestead reclamation of Hollow Village in Loess Area, and to improve the quality of the reclaimed soil in Hollow Village, a five-year field experiment was conducted here. The field experiment included seven treatments: no modified material (CK), maturing agent (TM), fly ash (TF), organic fertilizer (TO), maturing agent + fly ash (TMF), maturing agent + organic fertilizer (TMO) and fly ash + organic fertilizer (TFO), and we study the effects of different improved materials on soil properties and crop yield. The results showed that: the content of soil organic matter(SOM) and total nitrogen increased significantly after the application of different improved materials, which promoted the cementation and aggregation of water-stable microaggregates (< 0.25 mm), and the water-stable macroaggregates showed an increasing trend. In the 0~15 cm soil layer, the proportion of water-stable macroaggregates under TM, TF, TO, TMF, TMO, and TFO treatment increased by 328.2%, 130.0%, 87.8%, 81.1%, 36.7%, and 12.2% compared with CK, respectively. Meanwhile, TF, TO, TMF, TMO, TFO treatments significantly increased the mean weight diameter (MWD) and geometric mean diameter (GMD) values, reduced soil bulk density, the stable aggregate index (E_{LT}) and fractal dimension (D) values ($P<0.05$), and the stability of soil structure and the capacity of soil moisture retention has been significantly improved. The SOM content has a very significant positive correlation with MWD, GMD, and >2 mm water-stable aggregates, and a significant negative correlation with the E_{LT}, D, and water-stable microaggregates. Specially, the organic-inorganic coupling treatment of TFO had the highest content of SOM, soil moisture content, water-stable macroaggregates and maize yield, which was the most appropriate amendment for improving the reclaimed soil structure and fertility of Hollow Village in Loess Area.

Introduction

At present, the loess plateau region is facing the realistic problems of fragile ecological conditions, decreasing arable land area, lack of arable land reserve resources, and intensifying human-land conflicts. With the rapid promotion and implementation of urbanization and industrialization, a large number of young and strong rural laborers migrate to cities, resulting in the continuous deterioration of abandoned and hollowed rural areas[1-2]. Therefore, a great deal of abandoned housing resources is left idle and wasted[3]. Moreover, the construction of rural housing land generally has the undesirable trend of "building the new but not demolishing the old", constantly occupying high-quality arable land to expand to the periphery, and expanding the scale of housing land while the rural population are transferred and decreased, resulting in the widespread phenomenon of rural hollowing with that the scale of new house construction is constantly expanded while the housing lands in the village are abandoned and left idle, which has caused the destruction and occupation of a large number of arable land resources and the idleness and waste of high-quality land resources. It seriously threatens the protection of arable land resources and regional food security, and has become a major bottleneck limiting the construction of beautiful villages and the coordinated high-quality development of urban and rural areas[4-6]. In view of the above problems, combined with the increasing contradiction between human and land in the loess plateau region and the realistic situation of increasing local demand for reclaiming the abandoned housing land, it is of urgent strategic significance to deeply promote comprehensive land improvement in hollow villages and carry out reclamation of abandoned housing land for developing the urban-rural integration based on high quality, implementing "requisition-compensation balance" of arable land, guaranteeing national food security, and alleviating the regional contradiction between human and land[6-8] However, hollow village reclaimed soil mainly comes from the old wall soil on the abandoned housing land, which is mostly raw soil that has not been cultivated for years due to the combined influence of natural and human factors, and has lost its natural functions and properties. Its physical structure and soil fertility are seriously damaged, which greatly limits the land productivity and the healthy and sustainable development of new arable land in hollow villages. It is urgent to improve the structural properties and fertility features of the reclaimed soil on the housing land and to improve the productivity of the reclaimed soil[9-10].

The number and distribution ratio of soil aggregates are not only important indicators of soil erosion, hardening, compaction and other structural conditions, but also play a key role in supplying and storing soil nutrients, regulating soil water holding capacity and maintaining land productivity, and they are key indicators that can well evaluate the changes in soil fertility feature and environmental quality1[11-13]. Compared with non-water-stable aggregates, water-stable aggregates play a more important role in maintaining soil structural stability and erosion resistance, and retaining soil nutrients[14-15]. It was found that the increase of organic matter content is directly related to the increase of the number and stability of water-stable aggregates, and the application of different improved materials can effectively increase the organic matter content of the soil, promote the cemented agglomeration and structural stability of soil aggregates, and improve the fertility and erosion resistance of reclaimed soil[16-17]. Fly ash is the powdery solid residue discharged from coal-fired power plant after the combustion of pulverized coal, which had well-developed specific surface area and multi-level pores and contained abundant clay particles and oxides such as Al_2O_3 and Fe_2O_3, it could significantly improve mutual adsorption and aggregation ability between soil particles, and enhance the soil abilities of water and fertility retention[18-19].The maturing agent of ferrous sulfate can improve soil structure and reduce the pH value of soil while loosening the soil, it can also be used as fertilizer, which plays an important role in the absorption of plants[20]. Organic fertilizers are rich in organic matter and various nutrients, which can effectively improve soil fertility and structure, increase soil productivity and maintain the sustainability of crop yield[21-22].Therefore, it is of practical importance and urgency to enhance the structural stability and fertility feature of the reclaimed soil by applying different soil improved materials to mature and improve the reclaimed soil of hollow villages, so that the reclaimed soil can quickly recover its original functions and properties. However, the current approaches to remediation of hollow villages are mainly focused on the assessment of remediation potential, evolution laws, regional macro-remediation models, and regulation policies, which further affirm the significance of the remediation of abandoned housing land in hollow villages[23-25].

The reclaimed soil in hollow villages is taken as a key initiative to the regional supplementation to arable land resources; however, little research is conducted to compare and analyze the role of different improved materials in the structure and fertility improvement of reclaimed soil, thus it is difficult to distinguish the influences of different improved materials on the number of aggregates, structural stability, nutrient content and crop yield of reclaimed soils. Therefore, with the research on reclaimed soil in hollow villages of the loess plateau region, this paper compares the influences of different improved materials on the water-stable aggregates, structural stability, nutrient content and maize yield in reclaimed soil, and evaluates the correlation between the soil organic matter and the water-stable aggregates stability indexes of reclaimed soil after the application of different improved materials, expecting to screen out an appropriate amendment material for hollow village remediation, and provide a theoretical basis for enhancing the structure and fertility of hollow village reclaimed soil.

Results and analysis

Effects of application of different improved materials on soil properties in reclaimed soil

Soil organic matter(SOM) and total nitrogen(TN). After the application of different improved materials, the content of SOM and TN in both 0~15 cm and 15~30 cm soil layers of the hollow village reclaimed soil showed an overall increasing trend (Fig.1). In the 0~15 cm soil layer, the organic matter content increased by 9.6%, 79.0%, 90.0%, 61.4%, 120.1%, and 131.7% respectively under TM, TF, TO, TMF, TMO and TFO treatments compared with CK treatment, indicating that different improved materials all played important roles in improving the organic matter content of reclaimed soil(Fig.1a). The improvement of organic matter content in the 0~15 cm soil layer of the reclaimed soil after the treatments of different improved materials showed the order of TFO>TMO>TO>TF>TMF>TM>CK, and TO, TMO and TFO treatments with organic fertilizer addition could significantly improve the organic matter content of the reclaimed soil ($P<0.05$), among which TFO treatment was the most effective on improvement of the organic matter content. In the 15~30 cm soil layer, the results of significance analysis showed that TO, TMF, TMO and TFO treatments all significantly increased the organic matter content ($P<0.05$), while TM and TF treatments had no significant difference in improving the soil organic matter content, with TFO treatment having the most significant effect.

Compared with CK, the concentration of TN in the two soil layers had similar increasing trends to SOM after the application of different improved materials (Fig.1b). In the 0~15 cm soil layer, TM, TF, TO, TMF, TMO and TFO increased by 14.29%, 16.33%, 26.53%, 20.41%, 28.57%, and 51.02%, respectively. In the 15~30 cm soil layer, the concentration of TN also showed an increasing trend in varying degrees.

Size distribution of water-stable aggregates. Water-stable aggregates are important indicators for evaluating the structural stability and erosion resistance of soil, and their quantity and distribution can well reflect the changes in soil structure and quality[26]. Compared with CK, significant changes in the distribution of water-stable aggregates were shown in the reclaimed soils in the 0~15 cm and 15~30 cm layers after the long-term application of different improved materials (Fig.2 and 3) ($P<0.05$). In the 0~15 cm layer, after the long-term application of different improved materials in hollow village reclaimed soil, the proportion of water-stable macroaggregates (particle size>0.25 mm) showed an increasing trend on the whole, and the content of water-stable microaggregates (particle size <0.25 mm) showed a decreasing trend. In particular, it showed that except for the treatment of maturing agent (TM), the proportion of water-stable aggregates (with particle size >2 mm, 1~2 mm, and 0.5~1 mm) under the TF, TO, TMF, TMO and TFO were significantly increased compared with CK, especially that of particle size >2 mm (Fig.2). The proportion of > 2 mm water-stable aggregates was increased by 88.1%, 194.5%, 203.7%, 376.2%, and 781.7% respectively under TF, TO, TMF, TMO and TFO compared with CK. The proportion of water-stable macroaggregates showed the order of TFO (35.8%)> TMO (20.7%)> TO (16.9%)> TMF (16.3%)> TF (12.3%)> TM (10.1%)> CK (9.0%), and the water-stable macroaggregates were increased by 328.2%, 130.0%,87.8%,81.1%, 36.7%, and 12.2% respectively compared with CK, with the maximum increase of 328.2%.In general, all six different amendment material treatments increased the proportion of water-stable macroaggregates in reclaimed soil and promoted the aggregation and cementation of water-stable microaggregates (<0.25 mm) to water-stable macroaggregates (>0.25 mm). And the TFO showed the best effect on the increase of water-stable macroaggregates, followed by TMO, TO, and TMF, and TF and TM treatments were not very effective.

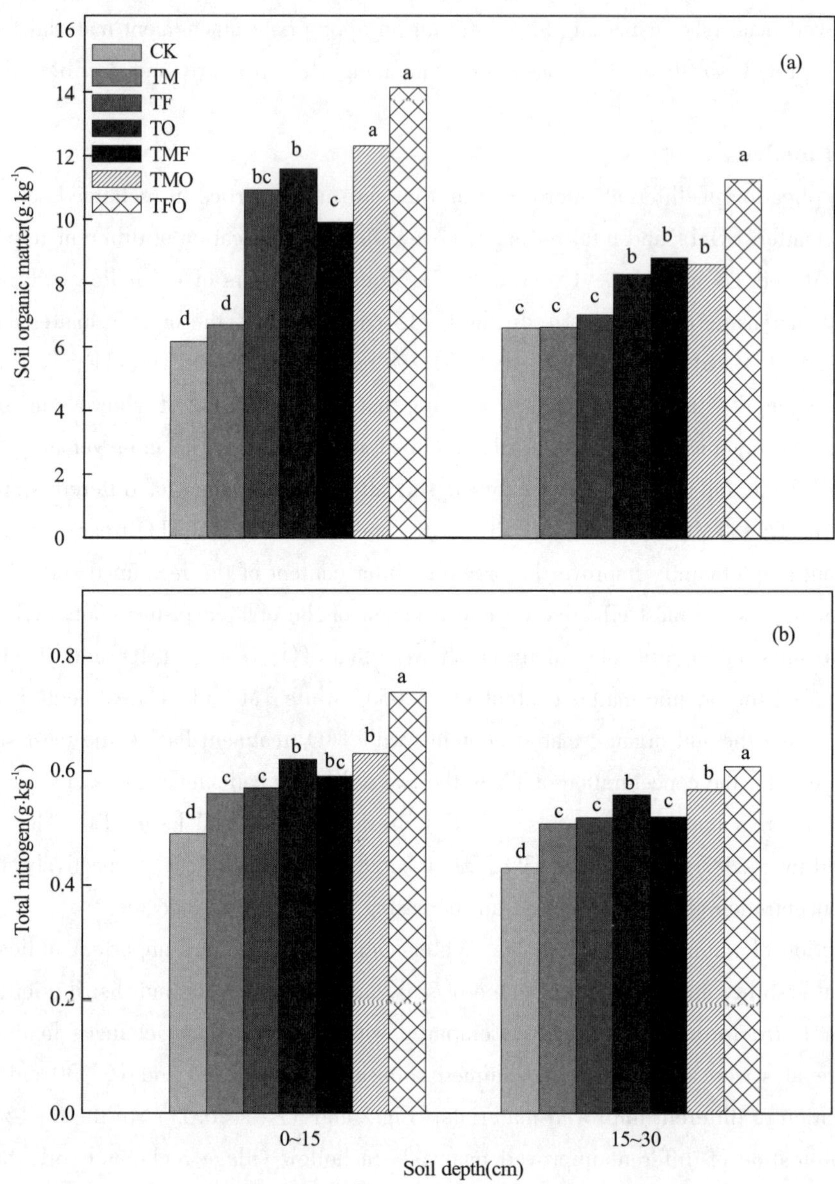

Fig.1 Effects of long-term application of different improved materials on SOM and TN

CK: no modified material; TM: maturing agent (ferrous sulfate); TF: fly ash; TO: organic fertilize; TMF: maturing agent + fly ash, TMO: maturing agent + organic fertilizer; TFO: fly ash + organic fertilizer; SOM, soil organic matter; TN, total nitrogen. Different lowercase letters represent significant differences between different improved material treatments in the same soil layer.

In the 15~30 cm soil layer, the change of water-stable aggregates showed a similar trend to that in the 0~15 cm soil layer compared with CK treatment. TF, TO, TMF, TMO, and TFO treatments significantly all increased the contents of >2 mm and 1~2 mm water-stable aggregates, and decreased the content of water-stable microaggregates ($P<0.05$) (Fig.3). In particular, TF, TO, TMF, TMO, and TFO treatments increased the proportion of >2 mm water-stable aggregates by 130.3%, 94.5%, 133.9%, 151.4%, and 309.2% respectively compared with CK, and TFO treatment showed the most significant effect on the increase of the number of water-stable macroaggregates. Compared with the 0 ~ 15 cm layer, the proportion of water-stable macroaggregates in the 15 ~ 30 cm layer showed a gradual decrease with the increase of soil depth.

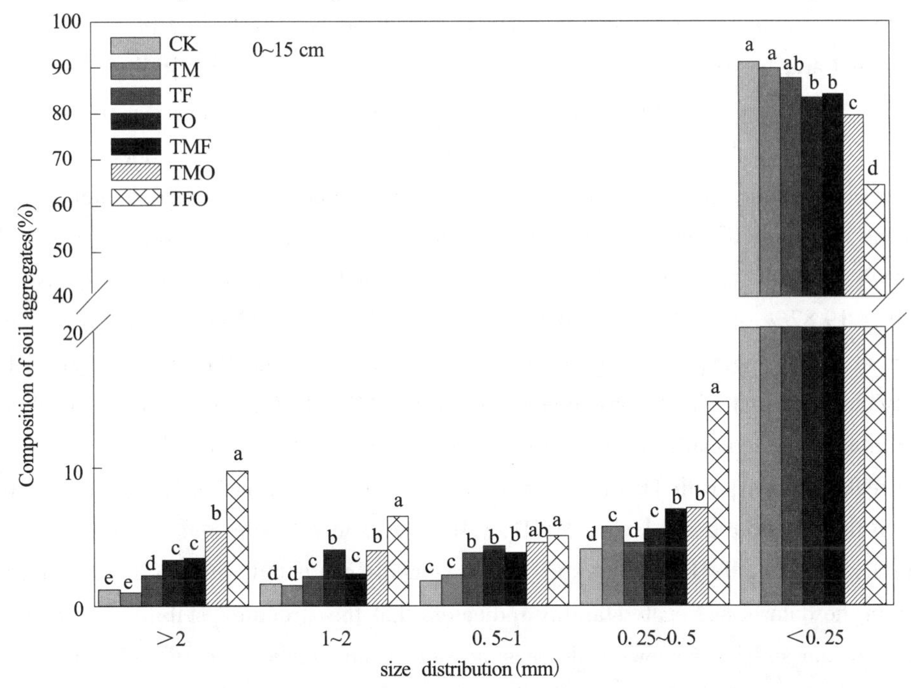

Fig.2 Percentage (%) of soil water-stable aggregates under long-term application of different improved materials at 0 ~ 15 cm depth

CK: no modified material; TM: maturing agent (ferrous sulfate); TF: fly ash; TO: organic fertilize; TMF: maturing agent + fly ash, TMO: maturing agent + organic fertilizer; TFO: fly ash + organic fertilizer. Different lowercase letters represent significant differences between different improved material treatments the same particle-size aggregates.

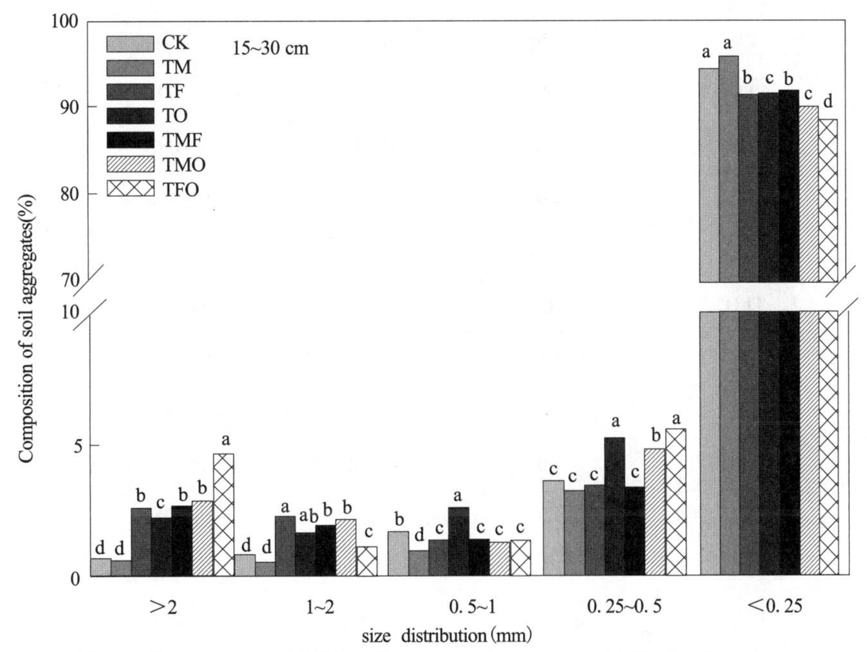

Fig.3 Percentage of (%) soil water-stable aggregates under long-term application of different improved materials at 15 ~ 30 cm depth

CK: no modified material; TM: maturing agent (ferrous sulfate); TF: fly ash; TO: organic fertilize; TMF: maturing agent + fly ash, TMO: maturing agent + organic fertilizer; TFO: fly ash + organic fertilizer. Different lowercase letters represent significant differences between different improved material treatments the same particle-size aggregates.

Water-stable aggregates structure stability. The mean weight diameter (MWD), geometric mean diameter (GMD), unstable aggregate index (E_{LT}), and fractal dimension (D) are important indicators reflecting the structural geometry and stability of soil aggregates, and it has been indicated in research that the higher the MWD and GMD and the smaller the E_{LT} and D, the better the structural stability of the aggregates and the soil structure[27-28]. Compared with CK treatment, the MWD and GMD showed a trend of significant increase while the D and E_{LT} showed a trend of significant decrease ($P<0.05$) under TF, TO, TMF, TMO and TFO treatments after the application of different improved materials , and TM treatment had no significant effect on the indicators of aggregate stability (Table 1). In the 0~15 cm soil layer, the MWD is increased by 6.19%, 27.66%, 22.16%, 49.71%, 125.96% and the GMD value is increased by 4.09%, 12.46%, 9.34%, 19.82%, 49.15% respectively under TF, TO, TMF, TMO and TFO treatments compared with CK treatment, while the E_{LT} is decreased by 1.35% to 29.5%, and the D is decreased by 0.76% to 4.35% respectively compared with CK treatment. TF, TO, TMF, TMO, and TFO treatments all improved the aggregation ability of aggregates to different degrees and enhanced the structural stability and erosion resistance of the reclaimed soil, with TFO treatment having the best effect on improving the structural stability of aggregates. In the 15~30 cm soil layer, MWD, GMD, D, and E_{LT} values also show significant improvement effects under TF, TO, TMF, TMO, and TFO treatments compared with CK treatment ($P<0.05$). It can be seen from the data of aggregate stability indicators that the structural stability of water-stable aggregates in the 15~30 cm soil layer shows a decreasing trend compared with the 0~15 cm soil layer, which may be related to the higher organic content in the 0~15 cm soil layer.

Table 1 Effects of long-term application of different improved materials on water-stable aggregate stability indexes

Soil layer	Treatments	MWD(mm)	GMD(mm)	E_{LT}(%)	D
0~15 cm	CK	0.35d	0.28c	91.01a	2.97a
	TM	0.32d	0.28c	89.87ab	2.99a
	TF	0.38c	0.29bc	87.69b	2.97a
	TO	0.45c	0.31b	83.08b	2.95b
	TMF	0.43c	0.31b	83.72b	2.95b
	TMO	0.53b	0.34b	79.31b	2.92b
	TFO	0.80a	0.42a	64.17c	2.84c
15~30 cm	CK	0.30c	0.27c	93.28a	2.99a
	TM	0.29c	0.26c	94.66a	2.99a
	TF	0.39b	0.29b	90.43b	2.96b
	TO	0.37b	0.29b	88.50bc	2.97b
	TMF	0.39b	0.29b	90.77b	2.96b
	TMO	0.39b	0.29b	89.07b	2.96b
	TFO	0.48a	0.30a	87.58c	2.93a

GMD, geometric mean diameter; MWD, mean weight diameter; E_{LT}, unstable aggregate index; D, fractal dimension. Different lowercase letters represent significant differences between different improved material treatments in the same aggregate stability index.

Soil bulk density and soil moisture content. Soil bulk density (BD) is one of the important indicators reflecting soil quality, and the BD at the 0~15 cm and 15~30 cm soil layers of the reclaimed soil decreased significantly after the application of different improved materials(P<0.05) (Fig.4a). In the 0~15 cm soil layer, the BD under TM, TF, TO, TMF, TMO, and TFO treatments was decreased by 5.71%,7.74%, 8.57%, 8.69%, 8.79%, and 9.13% compared with CK, which indicated that the application of different improved materials all have a certain reducing effect on the BD. However, the loosening effect on the reclaimed soil was different due to the different characteristics of the improved materials, and the BD of reclaimed soil showed the order of TFO > TMO > TO > TF > TMF > TM > CK. The combination of organic-inorganic improved materials can effectively reduce the BD of reclaimed soil, and the BD under TFO treatment was the smallest at 1.19 $g \cdot cm^{-3}$. In the 15~30 cm soil layer, through variance analysis, the effect of different improved materials on the BD showed a similar decreasing trend to BD In the 0~15 cm soil layer.

The soil moisture content(SMC) of the reclaimed soil in the 0~15 cm and 15~30 cm soil layers increased significantly after the application of different improved materials (P<0.05), and the variation law of SMC in the two soil layers was basically similar, showing the order of TFO>TMO>TMF>TO>TF≈TM>CK (Fig.4b). In the 0~15 cm soil layer, the SMC under TM, TF, TO, TMF, TMO, and TFO treatments was increased by13.5%、13.8%、21.4%、21.9%、32.4% and 38.3% compared with CK. The TMO and TFO showed the best increment effect on the SMC of reclaimed soil, and the mass water content was 17.4% and 18.2% respectively. In conclusion, compared with CK, these improved materials increased the SOM content and porosity, promoted the formation and stability of aggregates, and increased the retention and transmission of water, which was helpful to maintain more water. Among them, the coupling treatment of organic and inorganic improved materials can hold more soil moisture, and the most obvious increase was observed under TFO and TMO.

Correlation analysis between parameters of soil organic matter and stability parameters of water-stable aggregates

To further explore the correlation between the parameters of the reclaimed soil after the application of six different improved materials, a regression analysis was conducted in this paper on the correlation between the parameters of organic matter and water-stable aggregates with different particle sizes. From Table 2, it could be seen that the organic matter content had a highly significant positive correlation with MWD, GMD and the content of >2 mm water-stable aggregates, and a highly significant negative correlation with E_{LT}, D and the content of water-stable microaggregates(<0.25 mm), indicating that soil organic matter was an important factor affecting the formation of water-stable aggregates and their structural stability, and higher organic matter content would promote the formation of macro water-stable aggregates and improve the structural stability of soil. The content of water-stable aggregates (particle size >2 mm, 1~2 mm, and 0.5~1 mm) had a significant positive correlation with MWD and GMD values and a highly significant negative correlation with E_{LT} and D values; water-stable microaggregates (<0.25 mm) had a highly significant negative correlation with MWD and GMD values and a significant positive correlation with D, indicating that the increase in the content of water-stable aggregates with larger particle size helped to promote the structural stability of soil aggregates. In summary, it showed that TM, TF, TO, TMF, TMO, and TFO treatments of

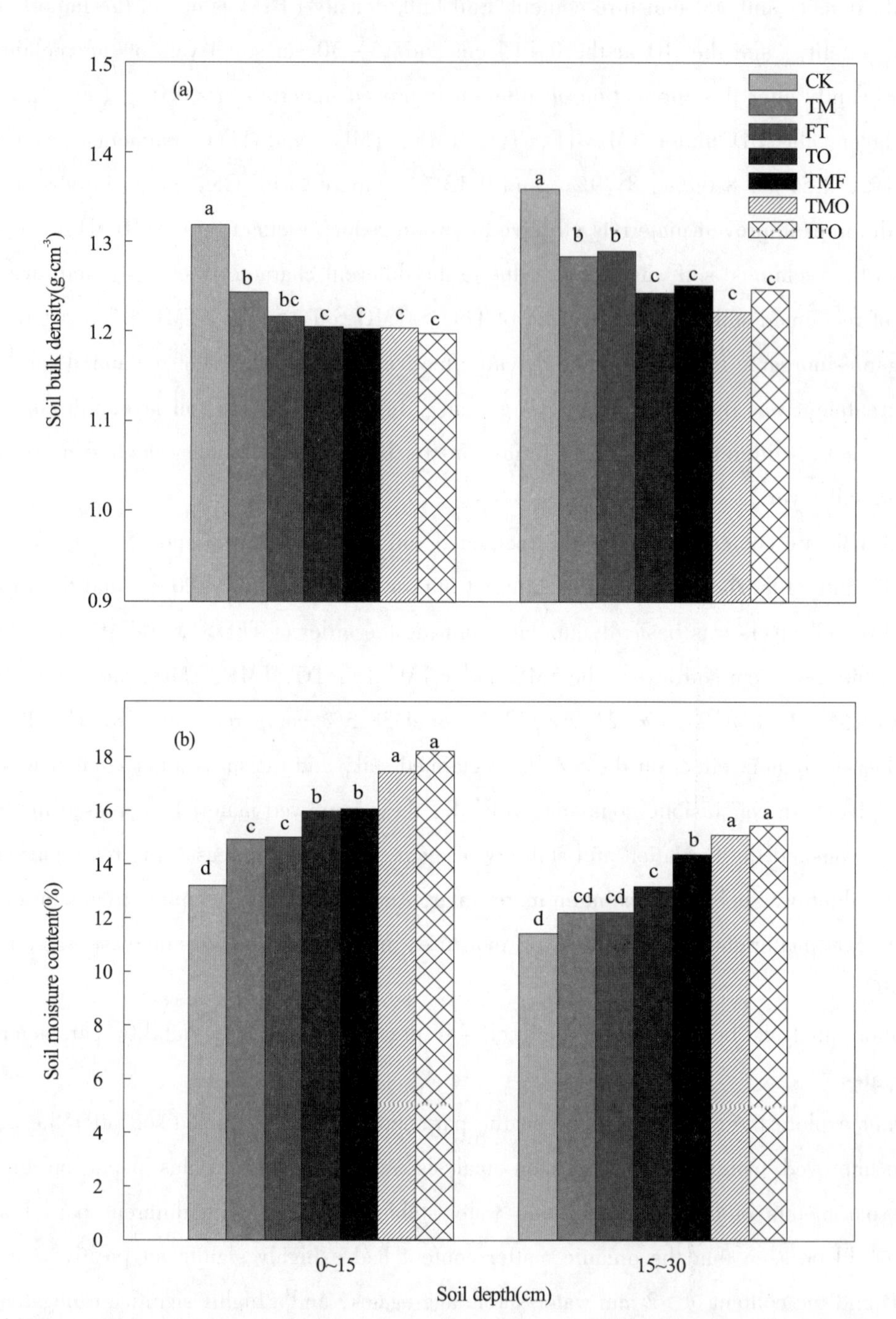

Fig.4 Effects of long-term application of different improved materials on BD and SMC

CK, no modified material; TM, maturing agent (ferrous sulfate); TF, fly ash; TO, organic fertilize; TMF: maturing agent + fly ash, TMO, maturing agent + organic fertilizer; TFO, fly ash + organic fertilizer; BD, soil bulk density; SMC, soil moisture content. Different lowercase letters represent significant differences between different improved material treatments in the same soil layer.

improved materials can effectively promote the formation of macro water-stable aggregates and improve the structural stability while promoting the increase of organic matter content in the hollow village reclaimed soil. Especially, TFO treatment is the best to improve the structural properties of hollow village reclaimed soil, enrich the soil fertility, and enhance the erosion resistance.

Table 2 Correlation analysis between SOM and parameters of water-stable aggregation

Index	SOM (g · kg^{-1})	WMD (mm)	GMD (mm)	E_{LT} (%)	D	Size (mm)				
						>2	1~2	0.5~1	0.25~0.5	<0.25
SOM	1									
WMD	0.7177**	1								
GMD	0.6960**	0.9798**	1							
E_{LT}	-0.6948**	-0.9364**	-0.9814**	1						
D	-0.7003**	-0.9926**	-0.9540**	0.9001**	1					
>2	0.7316**	0.9846**	0.9457**	-0.8948**	-0.9935**	1				
1~2	0.5949*	0.8140**	0.8835**	-0.8881**	-0.7639**	0.7596**	1			
0.5~1	0.5450	0.8522*	0.8980*	-0.9256**	-0.8190**	0.8080**	0.6602**	1		
0.25~0.5	0.5521	0.4727	0.5708	-0.6744**	-0.4167	0.4234	0.7138*	0.5190	1	
<0.25	-0.6948**	-0.9364**	-0.9814**	1.0000	0.9001**	-0.8948**	-0.8881**	0.6744**	-0.9256**	1

SOM, soil organic matter; GMD, geometric mean diameter; MWD, mean weight diameter; E_{LT}, unstable aggregate index; D, fractal dimension. * means significant correlation at 0.05 level; ** means highly significant correlation at 0.01 level.

Effects of application of different improved materials on maize yield

Different improved materials showed a different impact on maize yield (Table 3). The order of maize yield was TFO>TMO>TO>TMF>TF>TM>CK, and different improved materials all significantly increased maize yield than CK ($P<0.05$). The order of average kernels per ear and 100-kernel weight showed a similar increasing trend with maize yield, highest in TFO, following by TMO and TO. Compared with CK, the 100-grain weight increased by 2.0%, 3.9%, 8.1%, 4.8%, 4.9%, and 12.5% respectively under TM, TF, TO, TMF, TMO and TFO treatments, and the maize yield increased by 10.1%, 18.2%, 34.1%, 24.9%, 38.8%, and 53.4%, respectively. The maize yield under TFO was the highest, up to 11 558.79 kg · hm^{-2}. In summary, it showed that organic improved materials had stronger ability to increase maize yield than inorganic improved materials. The organic-inorganic coupling treatment of TFO and TMO had the best effect on improving the 100-grain weight and maize yield of the reclaimed soil. The possible reason was that the combination of organic and inorganic constituents can effectively increase the contents of soil organic matter and total nitrogen, promote the formation and cementation of aggregates, increase the retention and transmission of water, and improve the structural stability of hollow village reclaimed soil, which were confirmed by the results of the previous effects on SOM, total nitrogen, soil moisture content, aggregate proportion and aggregates structure stability index.

Table 3 Maize yield under different improved material treatments

Treatments	Row number	Kernels/row	Kernels/spike	100-kernel weight (g)	Theoretical yield(kg · hm^{-2})
CK	14.67±1.15b	34.33±1.53c	500.22±25.73e	27.38±0.38d	7532.48f
TM	14.67±0.58b	35.67±0.58bc	516.44±15.79de	27.93±0.25cd	8293.37e
TF	15.00±1.00ab	36.67±1.00bc	544.67±10.67cd	28.44±0.98cd	8906.10de
TO	15.33±1.15ab	38.67±1.15a	593.78±16.80b	29.59±0.80b	10102.46bc
TMF	15.33±1.15ab	37.33±1.52ab	570.44±8.46bc	28.69±0.15bc	9411.57bcd
TMO	15.67±0.58ab	38.67±1.15a	606.22±9.79b	28.73±0.99bc	10451.27b
TFO	16.67±1.54a	39.33±1.15a	652.44±13.81a	30.81±0.16a	11558.79a

CK: no modified material; TM: maturing agent (ferrous sulfate); TF: fly ash; TO: organic fertilize; TMF: maturing agent + fly ash, TMO: maturing agent + organic fertilizer; TFO: fly ash + organic fertilizer; BD, soil bulk density; SMC, soil moisture content. Different lowercase letters represent significant differences between different improved material treatments in the same indicator.

Discussion

The effects of long-term application of different improved materials on the water-stable aggregates and crop yield in reclaimed soil of hollow village.

Soil organic matter, oxides and clay particles are the main cementing substances for the formation and stability of soil aggregates, and play important roles in the number, size distribution, and structural stability of soil aggregates[29-30]. Improved materials such as organic fertilizer, maturing agent (ferrous sulfate) and fly ash can effectively improve the content of soil organic matter when applied alone or returned to the field after coupling. The cementing substances formed during conversion and decomposition contribute to the cementation and aggregation of soil aggregates and have significant effects on the size, distribution, and structural stability of the aggregates[12,31]. In this study, compared with CK treatment, except for the treatment with Maturing Agent (TM), Fly Ash (TF), Organic Fertilizer (TO), Maturing Agent + Fly Ash (TMF), Maturing Agent + Organic Fertilizer (TMO) and Fly Ash + Organic Fertilizer (TFO) all significantly increased the content of water-stable aggregates (particle size >2 mm and 1~2 mm) in the reclaimed soil and decreased the content of water-stable microaggregates (particle size<0.25 mm) ($P<0.05$). In particular, the content of water-stable macroaggregates (particle size>0.25 mm) showed the order of TFO>TMO>TO>TMF>TF>TM>CK. The coupling treatment of Fly Ash + Organic Fertilizer (TFO) increased the content of macro water-stable aggregates in hollow village reclaimed soil most significantly, with Maturing Agent + Fly Ash (TMF), Maturing Agent + Organic Fertilizer (TMO) and Organic Fertilizer (TO) treatments followed, and the Maturing Agent (TM) and Fly Ash (TF) treatment are not very effective in improving the macro water-stable aggregates.

The Study by Blissett *et al* showed that, as fly ash had well-developed specific surface area and multi-level pores and contained abundant clay particles and oxides such as Al_2O_3 and Fe_2O_3, it could significantly improve the mutual adsorption and aggregation ability between soil particles after being applied to soil[32]. The maturing agent of ferrous sulfate could effectively improve soil structure and reduce the pH value of soil while loosening the soil[33]. The cementing substances such as polysaccharide and humus generated by the decomposition after applying organic fertilizer to soil could further promote the aggregation of soil particles and the stacking and cementation of micro aggregates to macroaggregates, increasing the number of water-stable macroaggregates and soil structural stability[34]. Therefore, the organic/inorganic coupling treatment of fly ash+organic fertilizer can effectively increase the organic matter content of soil. In addition, fly ash has well-developed specific surface area and multi-level pores, and contains abundant clay particles and oxides, therefore it can significantly promote the formation and stabilization of the aggregates in reclaimed soil. This is similar to the stability indicators of aggregates in this study that MWD and GMD values showed a significant increase under TF, TO, TMF, TMO and TFO treatments while D and E_{LT} values showed a significant decrease ($P<0.05$). Among six amendment material treatments, the indicator of aggregate stability under TFO treatment is the best, so the coupling treatment of fly ash+organic fertilizer (TFO) has the best effect on improving the structural stability of hollow village reclaimed soil aggregates and is the best way to increase the number and enhance the structural stability of reclaimed soil aggregates. Moreover, the coupling treatment of TFO can also significantly reduce soil bulk density and increase total nitrogen, and soil moisture content than other treatments, which was helpful to promote the growth and yield of maize crops. The research result is similar to the present research results of Chang *et al* and Wei *et al*, who indicated that the combination of organic and inorganic modified materials can effectively improve soil quality and crop yield than either amendment alone[35-36].

Correlation analysis between the organic matter and other parameters in hollow village reclaimed soil after the application of different improved materials.

The application of different improved materials is the important source of the enhancement of organic matter and total nitrogen in reclaimed soil. Improved materials of soil such as organic fertilizers, maturing agents (ferrous sulfate), and fly ash, when applied alone or returned to the field after coupling, can enhance the fertility of the soil, promote the increase of crop biomass, and increase the amount of plant residues and roots returned into the soil, thus increasing the soil organic content[31]. In this study, after the application of different improved materials, the contents of organic matter and total nitrogen in both 0~15 cm and 15~30 cm soil layers of the reclaimed soil showed a trend of overall increase. In the 0~15 cm soil layer, the organic matter content increased by 9.6%, 79.0%, 90.0%, 61.4%, 120.1%, and 131.7% respectively under TM, TF, TO, TMF, TMO, and TFO treatments compared with CK treatment, indicating that different improved materials all played important roles in improving the organic matter content of reclaimed soil. However, the improvement effect on the contents of organic matter and total nitrogen in reclaimed soil differed due to the different physicochemical properties of different improved materials and the difference in the process of influencing the microbial activity in reclaimed soil. In particular, fly ash+organic fertilizer (TFO) had the best improvement effect, followed by maturing agent+organic fertilizer (TMO) and organic fertilizer (TO), which was directly related to the nutrient content and structural condition of fly ash and organic fertilizer. The well-developed specific surface area and multi-level pores of fly ash endow it with the property of good fertilizer retention. In addition, organic fertilizers are rich in organic substances and various nutrients, so the organic/inorganic coupling of fly ash+organic fertilizer (TFO) has the most significant effect on the improvement of organic matter. The research of Zhang Xuhui et al and Six et al[37-38] showed that the accumulation of organic matter is closely related to the content of 0.25~2 mm aggregates, and the larger the particle size, the higher the content of organic carbon. TFO treatment significantly increased the content of 0.25~2 mm aggregates, which also had an important contribution to the improvement of soil organic content. This was similar to the findings of Leina et al[31,39], the research of Leina et al. showed that the compound returning of the organic fertilizer of rotted chicken manure and the inorganic improved material of fly ash could significantly increase organic matter content.

The results of the correlation analysis between the parameters of hollow village reclaimed soil showed that the soil organic content had a highly significant positive correlation with MWD, GMD, and the content of water-stable aggregates (particle size >2 mm), and a highly significant negative correlation with E_{LT}, D, and the content of micro water-stable aggregates (particle size <0.25 mm). The content of water-stable aggregates (particle size>2 mm, 1~2 mm, and 0.5~1 mm) was significantly correlated with the values of MWD and GMD, indicators of aggregate stability, indicating that the content of soil organic matter affected the cementation, aggregation, and structural stability of soil aggregates[40-41]. The returning and application of exogenous improved materials promoted the increase of organic matter content in hollow village reclaimed soil, which created good conditions for the cementation and aggregation of small and medium-sized soil particles and helped micro water-stable aggregates form macro water-stable aggregates through the adhesion and aggregation of organic matter, thus enhancing the structural stability and erosion resistance of soil.

Conclusion

After the long-term application of different improved materials, the content of soil organic matter and total nitrogen in hollow village reclaimed soil showed an increasing trend, among which the coupling treatment of fly ash+organic fertilizer had the best effect in enhancing the organic matter and total nitrogen in the 0~15 cm soil layer, with an increase of 131.7% and 51.02%. Meanwhile, the various improved materials significantly increased soil moisture content, and reduced soil bulk density. With the increase of organic content, different amendment materials promoted the cementation and aggregation of water-stable microaggregates to water-stable macroaggregates (>0.25 mm). The content of water-stable macroaggregates in 0~15 cm soil layer showed the order of TFO>TMO>TO>TMF>TF>TM>CK, and the largest increase was shown in fly ash+organic fertilizer treatment, being 328.2%. The values of MWD, GMD, D, and E_{LT}, indicators of aggregate structural stability were improved on the whole, and the number and structural stability of water-stable macroaggregates in reclaimed soil were enhanced. Among the six different improved material treatments, the organic/inorganic coupling treatment of fly ash+organic fertilizer was the most suitable for the improvement of hollow village reclaimed soil, and the coupling of them can effectively increase soil organic matter, total nitrogen, soil moisture content and water-stable macroaggregates, and enhance the structural stability and erosion resistance of the hollow village reclaimed soil, therefore improve the soil fertility.

Materials and methods

Experimental site. Based on the Key Laboratory of Degraded and Unused Land Remediation Project, Ministry of Natural Resources, the long-term positioning test plot of reclaimed soil improvement of hollow villages was set up in the pilot base of Fuping County, Shaanxi Province , China(34°42′N, 109°12′E) (Fig.5). The study area was in the Weibei loess plateau region, and the test plot was built on June 15, 2015. The climate in this area is characterized by a continental temperate, semi-arid, and semi humid climate. The annual average temperature, annual average rainfall, and average sunshine hours are 13.4 ℃, 533.2 mm and 2389.6 h, respectively. The reclamation method is to reclaim and re-organize the old wall soil of the abandoned and idle housing land through engineering techniques such as house demolition, digging and filling, land leveling, and organic reconstruction of the tillage layer. In order to simulate the condition of reclaimed soil remediation and returning of abandoned housing land in hollow villages, the old wall soil of the reclaimed land remediation project of hollow villages in Chengcheng County was used for off-site backfilling, with the backfilling depth of 30 cm. The part 30cm below the soil layer was undisturbed soil. After removal of impurities such as rubbles and stones in the old wall soil reclaimed from the housing land, the reclaimed soil was matured and structurally improved by soil fertilization and adding improved materials, so as to meet the normal growth of wheat, corn and other crops.

The basic physicochemical indicators of the reclaimed soil at the depth of 0 to 30 cm before the start of the experiment were as follows: the pH value (water-soil mass ratio 1:2.5) was 8.5, the soil bulk density was 1.40 $g \cdot cm^{-3}$, the soil texture was silt loam (US Soil Taxonomy) with 10.15% clay (particle size <0.002 mm), 77.82% silt (particle size of 0.05~0.002 mm), 12.65% sand (0.05~2 mm), 4.4 $g \cdot kg^{-1}$ organic matter, 0.15 $g \cdot kg^{-1}$ total nitrogen, 61.3 $mg \cdot kg^{-1}$ available potassium, and 70.4 $mg \cdot kg^{-1}$ effective phosphorus. The proportion of water-stable aggregates of >2, 1~2, 0.5~1, 0.25~0.5 and <0.25 size are 0.67%, 0.84%, 1.67%, 3.55%, 93.27%, respectively. In summary, the reclaimed soil of the housing land in hollow villages was mainly developed from loess parent material, and the soil fertility and structure were comparatively poor.

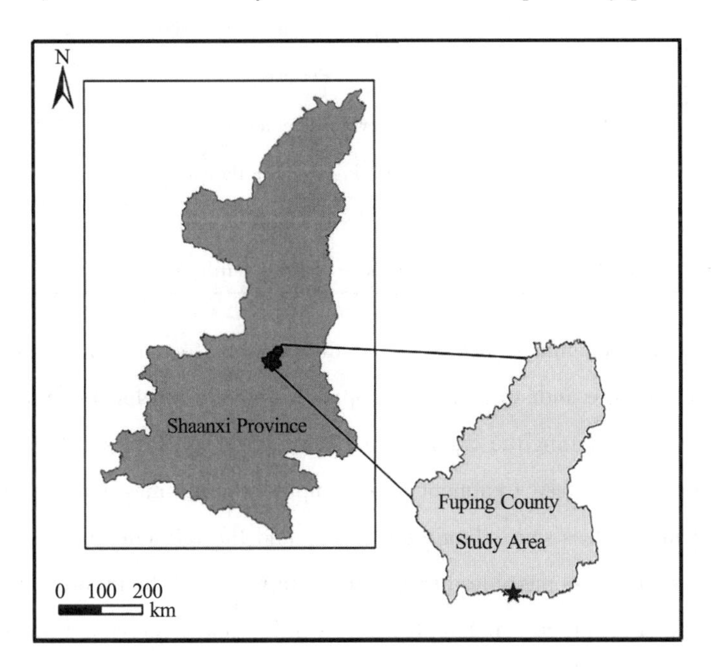

Fig.5 Location map of Long-term field experiment area. The map was produced using ESRI ArcGIS software (version 10.3)

Experimental design. The long-term localized field improvement test of reclaimed soil in hollow villages began in June 2015. Based on the investigation and analysis of the returned improved materials, fly ash, organic fertilizer (well-rotted chicken manure), and ferrous sulfate ($FeSO_4$) were selected as the improved materials for reclaimed soil (Table 4). The test was designed as a randomized group field test with seven treatments: maturing agent (TM), fly ash (TF), organic fertilizer (TO), maturing agent + fly ash (TMF), maturing agent + organic fertilizer (TMO), fly ash + organic fertilizer (TFO), and control (CK) treatment with no improved material added. Each treatment was repeated three times, and each treatment plot was a square field of 2 m×2 m. There were a total of 21 treatment plots, with an isolation zone in the middle of each treatment plot. The cropping system was a two-year triple cropping one with the rotation of winter wheat-summer maize, and the winter wheat for the test was seeded in mid-October with a seeding rate of 220.5 $kg \cdot hm^{-2}$ and harvested in late May of the following year. And the variety was Changwu 134. The summer maize was seeded in mid to early June with a density of 6.5×10^4 plants per hectare and harvested in early October. And the variety was Xianyu 335. The seeds were sown artificially, and 1 500 kg · ha^{-1} compound fertilizer was applied before seeding. Then, six improved materials with different treatments were evenly spread on the soil surface and mixed into the hollow village reclaimed soil at the depth of 0 to 30 cm by artificial tillage, each experimental plot only used improved materials once, and other management measures and levels such as watering amount, chemical fertilizers consumption, pest control, etc. were kept consistent. See Table 1 for the specific treatment and amount of different improved materials.

Table 4 Experimental design of reclamation soil improvement in hollow village

Number	Treatment	Improved materials	Application amount
1	CK	Control (no modified material)	0
2	TM	Maturing agent (ferrous sulfate)	0.6 t · hm^{-2}
3	TF	Fly ash	45 t · hm^{-2}
4	TO	Organic fertilizer(chicken manure)	30 t · hm^{-2}
5	TMF	Maturing agent + fly ash	(0.6+45) t · hm^{-2}
6	TMO	Maturing agent + organic fertilizer	(0.6+30) t · hm^{-2}
7	TFO	Fly ash + organic fertilizer	(22.5+15) t · hm^{-2}

Sampling and soil physicochemical analysis. The soil samples were tested in late September 2020 after the winter wheat harvest, and the undisturbed soil samples were collected for determining soil aggregates and the mixed soil samples for determining the organic content from the soil layers at the depth of 0 to 15 cm and 15 cm to 30 cm respectively. Every treatment was sampled three times repeatedly. The aggregate soil samples were preserved in stainless steel aluminum boxes, and the influence on the structure of soil aggregates should be avoided as much as possible during transportation. After being brought back to the laboratory for natural air-drying, the undisturbed soil samples were gently broken into small pieces of about 10 mm according to the natural cracks, and impurities such as stones were removed for the testing of water-stable aggregates, organic matter and other indicators. The content of soil organic matter and total nitrogen was determined by the potassium dichromate—external heating method and Kjeldahl method[42-43], soil bulk density and water content were determined by using cutting ring method and drying method at 105 ℃[44], and the water-stable aggregate distribution and size in two soil layers at the depth of 0~15 cm and 15~30 cm were determined by the wet-sieve method[45]. See Equations (1) to (4) and relevant references for the indicators of aggregate structural stability such as detailed calculation of the Mean Weight Diameter (MWD) and Geometric Mean Diameter (GMD), Unstable Aggregate Index (E_{LT}), and Fractal Dimension (D)[45-47]. Maize yield was determined by random sampling method. Ten representative plants were randomly selected for each treatment at maize maturity, and the effective panicle number and number of kernels per spike was counted by hand. The maize kernels were at 105 ℃ dry for 48 hours to ensure that the moisture content was controlled below 15%, 100-kernel weight of maize was measured by electronic balance, and finally the theoretical yield of maize was calculated.

$$\mathrm{MWD} = \frac{\sum_{i=1}^{n} (\overline{x}_i w_i)}{\sum_{i=1}^{n} w_i} \tag{1}$$

$$\mathrm{GMD} = \exp\left(\frac{\sum_{i=1}^{n} w_i \ln \overline{x}_i}{\sum_{i=1}^{n} w_i}\right) \tag{2}$$

$$E_{\mathrm{LT}} = \frac{M_{\mathrm{T}} - R_{0.25}}{M_{\mathrm{T}}} \times 100\% \tag{3}$$

$$\frac{M(r < \overline{x}_i)}{M_{\mathrm{T}}} = (\frac{\overline{x}_i}{x_{\max}})^{3-D} \tag{4}$$

where n denotes the number of aggregate size fractions, $\overline{x}_i$ is the mean diameter of aggregates retained in the ith sieve, W_i is the weight of aggregates retained in the ith sieve, $M(r \leqslant x_i)$ is the weight of aggregates with a fraction diameter less than or equal to x_i, M_{T} is the gross weight of aggregates, and $R_{0.25}$ is macroaggregates with diameters of >0.25 mm.

Statistical analysis. Data statistics and analysis were performed by using Microsoft Excel 2010 (Microsoft, Inc., red-mond WA, USA, https://www.microsof.com/zh-cn/download/ofce.aspx) and SPSS25.0 (SPSS sofware,25.0, SPSS Institute Ltd, Chicago, USA, https://www.ibm.com/cn-zh/analytics/spss-statistics-software). Differences among treatments were evaluated using one-way analysis of variance (ANOVA), and the least significant range (LSD) method was used for multiple comparisons ($P<0.05$).

Data availability. All data generated or analysed during this study are included in this published article.

Statement. "Xianyu 335", the maize(Zea mays L.) cultivar that we used in the present experiment, complied with institutional, national, international guidelines. We complied with the IUCN Policy Statement on Research Involving Species at risk of extinction and the Convention on International Trade in Endangered Species of Wild Fauna and Flora.

Acknowledgments

This research was supported by the Scientific Research Item of Shaanxi Provincial Land Engineering Construction Group(DJNY2020-25), Shaanxi Province Innovative Talent Promotion Plan (Youth Science and Technology Rising Star Project) (2020KJXX-051). The authors gratefully acknowledge the Institute of Land Engineering and Technology, Shanxi Provincial Land Engineering Construction Group, Xi'an, China.

Author contributions

Z.L., Y.Z. and Z.S. wrote the main manuscript text. Z.L., H.W. and R.Z. designed the experiment. Z. L. performed the experiments and collected the data. Z.L. and Y.Z. prepared the Fig.1-5. Y.S. and R.Z. contributed to the statistical analysis. Z.L. and Z.S. provided English language editing. All the authors revised the manuscript.

References

[1]Liu, Y. & Liu Y. Progress and prospect on the study of rural hollowing in China. *Geographical Research*. 29: 35-42 (2010).

[2]Huang, F. & Wang, P. Vegetation change of ecotone in west of northeast china plain using time-series remote sensing data. *Chinese Geographical Science*. 20, 167-175 (2010).

[3]Liu, Y. S., Li, J. T. & Yang, Y. Y. Strategic adjustment of land use policy under the economic transformation. *Land*

Use Policy, 74, 5-14 (2018).

[4]Liu, Y. S., Liu, Y., Cheng, Y. F. & Long, H. L. The process and driving forces of rural hollowing in China under rapid urbanization. *Journal of Geographical Sciences*, 20, 876-888 (2010).

[5]Hu, Z., Peng, J., Du, Y., Song, Z., Liu, Y. & Wang Y. Reconstructing hollow villages in the view of structural reform of the supply side. *Acta Geographica Sinica*. 71, 2119-2128 (2016).

[6]Liu, Y. S. et al. Progress of research on urbanrural transformation and rural development in China in the past decade and future prospects. *Journal of Geographical Sciences*. 26, 1117-1132 (2016).

[7]Jiang, S. & Luo, P. A literature review on hollow villages in china. *China Population Resources and Environment*. 24, 51-58 (2014).

[8]Liu, Y.S. Introduction to land use and rural sustainability in China. *Land Use Policy*. 74, 1-4 (2018).

[9]Huang, Y., Wang, K., Su, T. & Yang, J. Study on fertility characteristics and reforming of reclaimed soil for earth-walled rural homestead in Chongqing. *Journal of Southwest University (Natural Science)*. 37, 33-39 (2015).

[10]Liu, Z., Deng, L., Zhou, W., Chen, L. & Zou, G. Evaluation of effects of organic materials on soil fertilization of reclaimed homestead. *Soils*. 51, 672-681 (2019).

[11]Wang, D., Fonte, S., Parikh, S., Six, J. & Scow, K. Biochar additions can enhance soil structure and the physical stabilization of c in aggregates. *Geoderma* 303, 110-117 (2017).

[12]Zhang, S. X. et al. Effects of conservation tillage on soil aggregation and aggregate binding agents in black soil of Northeast China. *Soil Till. Res*. 124, 196-202 (2012).

[13]Song, Z. W. et al. Effects of long-term fertilization on soil organic carbon content and aggregate composition under continuous maize cropping in Northeast China. *J. Agri. Sci*. 153, 236-244 (2015).

[14]Blanco-Moure, N., Moret-Fernández, D. & López, M. Dynamics of aggregate destabilization by water in soils under long-term conservation tillage in semiarid Spain. *Catena* 99, 34-41(2012).

[15]Wang, Y., Ji, Q., Liu, S., Sun, H. & Wang, X. Effects of tillage practices on water-stable aggregation and aggregate-associated organic c in soils. *Journal of Agro-Environment Science*. 31, 1365-1373 (2012).

[16]Bronick, C. & Lal, R. Soil structure and management: a review. *Geoderma* 124, 3-22 (2005).

[17]Liu, E., Zhao, B., Mei, X., Hwat, B., Li, X. & Li, J. Distribution of water-stable aggregates and organic carbon of arable soils affected by different fertilizer application. *Acta Ecologica Sinica*. 30, 1035-1041(2010).

[18]Ahmaruzzaman, A. A review on the utilization of fly ash. *Progress in Energy and Combustion Science*, 36, 327-363 (2010).

[19]Parab, N., Sinha, S. & Mishra, S. Coal fly ash amendment in acidic field: effect on soil microbial activity and onion yield. *Appl Soil Ecol*, 96, 211-216 (2015).

[20]Lei, N., Chen, T., Dong, Q. & Luo, L. Effects of the returning materials on soil fertility of reclaimed hollow village. *Journal of Soil and Water Conservation*. 32, 222-226 (2018).

[21]Xin, X. L., Zhang, J. B., Zhu, A. N. & Zhang, C. Z. Effects of long-term (23 years) mineral fertilizer and compost application on physical properties of fluvo-aquic soil in the North China Plain. *Soil Tillage Res*. 156, 166-172 (2016).

[22]Liu, Z. *et al.* Long-term effects of different planting patterns on greenhouse soil micromorphological features in the North China Plain. *Sci. Rep-UK* 6, 31118 (2019).

[23]Yin, J., Qiu, D. & Pan, J. Analysis on land reclamation potential of rural residential area based on xf-gf space synthesis recognition model: a case study of Shuanglin village of Shuangjiang town of Tongnan city. *Journal of Agricultural Mechanization Research*, 33, 5-9 (2011).

[24]Cui, B. Exchange isolated farmers' residential land with well-planned house: a new transfer mode of rural collective construction land in Tianjin city. *China Land Science*, 24, 37-40,46 (2010).

[25]Feng, W., Li, Y. & Liu, Y. Development evaluation of rural transformation and its countermeasure based on micro-perspective: a case study of Loucun village of Huojia county in Henan Province. *Areal Research and Development*. 37, 133-137 (2018).

[26]Verchot, L., Dutaur, L., Shepherd, K. & Albrecht, A. Organic matter stabilization in soil aggregates: understanding the biogeochemical mechanisms that determine the fate of carbon inputs in soils. *Geoderma* 161, 182-193 (2011).

[27]Wang, F., YA, T., JS, Z., PC, G., & JN, C. Effects of various organic materials on soil aggregate stability and

soil microbiological properties on the Loess Plateau of China. *Plant. Soil. Environ.* 59, 162-168 (2013).

[28]Tripathi, R. et al. Soil aggregation and distribution of carbon and nitrogen in different fractions after 41 years long-term fertilizer experiment in tropical rice-rice system. *Geoderma* 213, 280-286 (2014).

[29]Wang, F., Tong, Y., Zhang, J., Gao, P. & Coffie, J. Effects of various organic materials on soil aggregate stability and soil microbiological properties on the Loess Plateau of China. Plant, *Soil and Environment*. 59, 162-168 (2013).

[30]Bottinelli. *et al.* Tillage and fertilization practices affect soil aggregate stability in a Humic Cambisol of Northwest France. *Soil Tillage Res.* 170, 14-17 (2017).

[31]Lei, N., Han, J., Mu, X., Sun, Z. & Wang, H. Effects of improved materials on reclamation of soil properties and crop yield in hollow villages in China. *J Soils Sediments*. 19, 2374-2380 (2019).

[32]Blissett RS. & Rowson NA. A review of the multi-component utilisation of coal fly ash. *Fuel* 97, 1-23 (2012).

[33]Zhang, L. & Wei, J. Effects of different field-returning materials on soil fertility after remediation in vacant village. *Bulletin of Soil and Water Conservation*. 38, 74-78 (2018).

[34]Mikha, M. M., Hergert, G. W., Benjamin, J. G., Jabro, J. D. & Nielsen, R. A. Long-term manure impacts on soil aggregates and aggregate-associated carbon and nitrogen. *Soil Sci. Soc. Am. J.* 79, 626-636 (2015).

[35]Chang, P. Y. The effect of fly ash and biogas residue on corn growth and soil. Dissertation, *Shanxi Agricultural University*, 2013.

[36]Wei, W., Yan, Y., Cao, J., Christie, P., Zhang, F. & Fan, M. Effects of combined application of organic amendments and fertilizers on crop yield and soil organic matter: An integrated analysis of long-term experiments[J]. *Agriculture, Ecosystems & Environment*, 225, 86-92 (2016).

[37]Zhang, X., Li, L. & Pan, G. Effect of different crop rotation systems on the aggregates and their SOC accumulation in paludalfs in north huai region, China. *Chinese Journal of Ecology*. 2, 16-19 (2001).

[38]Six, J., Bossuyt, H., Degryze, S. & Denef, K. A history of research on the link between (micro)aggregates, soil biota, and soil organic matter dynamics. *Soil Tillage Res.* 79, 7-31 (2004).

[39]Ren, S., Shao, Y. & Yang, J. Study on effects of fertilizations on homestead reclamation soil. *Journal of Soil and Water Conservation*. 26, 78-81 (2012).

[40]Chivenge, P., Vanlauwe, B., Gentile, R. & Six, J. Organic resource quality influences short-term aggregate dynamics and soil organic carbon and nitrogen accumulation. *Soil Biol. Biochem*. 43, 657-666 (2011).

[41]Yu, H. Y. *et al.* Accumulation of organic C components in soil and aggregates. *Sci. Rep-UK* 5, 13804 (2015).

[42]Tiessen, H. & Moir J. O. Total and Organic Carbon. In: Carter, M.R. (Ed.), Soil Sampling and Methods of *Analysis. J. Environ. Qual*. 38, 187-199 (1993).

[43]Bremner, J. M. Determination of nitrogen in soil by the Kjeldahl method. *The Journal of Agricultural Science*. 55, 11-33 (1960).

[44]Rabot, E., Wiesmeier, M., Schlüter, S. & Vogel, H. J. Soil structure as an indicator of soil functions: A review. *Geoderma* 314, 122-137 (2018).

[45]Zhou, H., Li, B. G. & Lu, Y. Z. Micromorphological analysis of soil structure under no tillage management in the black soil zone of northeast china. *J. Mt. Sci.* 6, 173-180 (2009).

[46]Meng, Q. *et al.* Distribution of carbon and nitrogen in water-stable aggregates and soil stability under long-term manure application in solonetzic soils of the Songnen plain, northeast China. *J. Soil Sediment*. 14, 1041-1049 (2014).

[47]Zhu, G., Shangguan, Z. & Deng, L. Soil aggregate stability and aggregate-associated carbon and nitrogen in natural restoration grassland and Chinese red pine plantation on the Loess Plateau. *Catena* 149, 253-260 (2017).

本文曾发表于2022年《Nature Portfolio》第12卷

Spatial Distribution Characteristics and Evaluation of Soil Pollution in Coal Mine Areas in Loess Plateau of Northern Shaanxi

Na Wang[1,2,3,4], Yuhu Luo[1,2,3,4], Zhe Liu[1,2,3,4], Yingying Sun[1,2,3,4]

(1.Institute of Land Engineering and Technology, Shaanxi Provincial Land Engineering Construction Group Co., Ltd., Xi'an 710075, China; 2.Shaanxi Provincial Land Engineering Construction Group Co., Ltd., Xi'an 710075, China; 3.Key Laboratory of Degraded and Unused Land Consolidation Engineering, the Ministry of Natural Resources, Xi'an 710075, China; 4.Shaanxi Provincial Land Consolidation Engineering Technology Research Center, Xi'an 710075, China)

【**Abstract**】The ecological environment in Loess Plateau of Northern Shaanxi is fragile, so the soil pollution caused by the exploitation of coal resources cannot be ignored. With Shigetai Coal Mine in Loess Plateau of Northern Shaanxi as the object of study for field survey and sampling, the content of heavy metals in soil is analyzed, the environmental pollution in the research area is evaluated by the single factor pollution index method, comprehensive pollution index method and potential ecological risk index method, and the spatial distribution characteristics of heavy metals are discussed by the geostatistics method. According to the study results, the average contents of heavy metals Hg, Cd, Pb and Cr are 2.03, 1.36, 1.11 and 1.23 times of the soil background values in Shaanxi Province respectively and the average contents of other heavy metals are lower than the soil background values in Shaanxi Province; Hg and Cd show moderate variation while As, Pb, Cr, Zn, Ni and Cu show strong variation; the skewness coefficients and kurtosis coefficient of Cd, As and Cu in the soil within the research area are relatively high, and these elements are accumulated in large amounts. Single factor pollution index (P_i) and potential ecological risk index (E) indicate that heavy metal Hg is the main pollution factor and mainly distributed in the east and north of the research area. The comprehensive index of potential ecological risk (RI) of the research area is 1 336.49, showing an extremely high ecological risk, and the distribution characteristics of potential ecological risk are consistent with that of potential ecological risk index (E) of Hg. The results of ecological risk warning show that Hg is in a slight warning status, while Cd, Pb and Cr are in a warning status. The areas with high ecological risk warning values are mainly distributed in the east and north, and the whole research area shows relatively obvious zonal distribution law. The soil is disturbed greatly during the coal mining, so the ecological governance of the mine area shall adapt to the local natural conditions and regional environmental characteristics and follow the principle of "adjusting governance measures based on specific local conditions and classifications". An environmentally sustainable governance manner shall be adopted to realize the protection of the ecological environment and high-quality development of coal resources.

Introduction

In recent years, the cultivated land area has been reduced and the natural environment has been seriously polluted due to the industrial revolution, rapid population growth, rural-urban migration and unplanned and uncontrolled urbanization[1]. With the high toxicity, long-term retention, persistent bioavailability and recalcitrance, pollutants such as heavy metals and metalloids (mercury, lead, zinc, copper, cadmium, chromium, nickel, arsenic) are considered to be the main harmful trace elements[2-4]. The sources of heavy metals in soil mainly include natural factors and human factors. The natural factor mainly refers to the rock weathering in the process of soil formation, and many studies showed that the concentration of the heavy metals formed by the natural factor is usually harmless to the ecological environment[5]. Human activities, such as exploitation of mineral resources, metal processing and smelting, chemical production, factory discharge and sewage irrigation, have been proved to be the main sources of heavy metal pollution[6-8]. According to statistics, the land area polluted by heavy metals in China is more than 10 million hectares, seriously threatening the agricultural development and quality and safety of ecological environment in China[9-10]. In addition, excessive heavy metals may enter the human body through the food chain, causing a variety of human diseases[11]. For example, excessive lead in the human body will damage their nervous and immune systems, and long-term exposure to cadmium will lead to potential risks of lung cancer and fracture, long-term intake of high concentrations of copper and zinc will damage human pancreas, liver and kidney and lower the HDL cholesterol level[12], long-term consumption of crops and vegetables grown on the soil polluted by the arsenic will increase the risks of skin cancer, bladder cancer and lung cancer[13-14]. Therefore, in order to protect the ecosystem and human health, it is essential to evaluate the distribution, sources and potential environmental risks of heavy metals in soil.

Mining is one of the most polluted activities in the world. It is well-known that mining activities will lead to the loss of surrounding biodiversity[15-16]. In particular, mining activities will cause a wide range of landform disturbances, such as the destruction of geological continuity on the surface, soil pollution, hydrological impact on runoff capacity, and changes in landscape morphology. Excavation and overburden deposition aggravate the change of surface topography and cause the degeneration of the aquifers[17]. After the founding of new China, in order to vigorously develop the economy, China adopted an extensive economic development model, that is, treatment after pollution, which led to serious environmental problems. The open-pit coal mining is one of these models. No effective preventive and protective measures have been taken in the process of coal mining, resulting in serious environmental problems around the mine area. Heavy metal is one of the main pollutants in the coal mine area. According to statistics, about 3.2 million hectares of land in China were affected by mining activities prior to 1996, including cultivated land, woodland and pasture[18]. Many previous studies have shown that heavy metals are released into farmland soil, surface water and plant leaves around the mine area during coal mining and processing, endangering human health through the food chain[19-20]. In the process of coal mining, such as ore concentration, mineral extraction, topsoil stripping, tailings accumulation, wastewater treatment and transportation, a large amount of heavy metal dust and dusty wastewater are generated around the mine area[21]. After being mixed with soil, these refractory heavy metals migrate to the water environment through surface runoff, which amplifies their toxicity in the ecological environment[22]. In addition, long-term overexploitation causes land subsidence, soil erosion and deterioration of groundwater quality, seriously disturbing the local natural environment[23]. It is well-known that coal gangue and fly ash contain a variety of toxic heavy metal elements, which are released into the environment through coal transportation, smoke and dust discharging from coal-fired power plants, sewage discharging from mine areas and other coal-related industry activities. In addition, the long-term accumulation of coal gangue and fly ash is another way to release toxic elements into the soil[24]. *Masto et al* (2015)[25]. and other[26] researchers showed that the coal dust contains heavy metals such as iron, zinc, manganese, copper, lead, chromium, nickel, strontium, zirconium and arsenic, which are seeped into soil and even groundwater through surface runoff. According to statistics, the average concentrations of arsenic in bituminous coal and lignite in the world are 9.0 mg/kg and 7.4 mg/kg, respectively[27]. *Chen et al*(2013)[28] found that the contents of arsenic, selenium, mercury and antimony discharged into the atmosphere from coal-fired power plants are 236, 637, 172 and 33 tons, respectively.

With rich coal, natural gas and oil resources, Northern Shaanxi is an important energy and chemical base in Shaanxi Province. Its ecosystem is very fragile and sensitive to environmental impact[29]. The geological tectonic unit of the mine area belongs to Ordos platform slope of North China platform and the north-central part of northern Shaanxi platform depression. No magmatic rock formation, magmatism and volcanic eruption occur and earthquakes are rare in this area. It borders the sandy grassland on the southern margin of Maowusu Desert in its north, and borders the hinterland of Loess Plateau in its south, with crisscrossed gullies and crisscross hills and ridges. At present, the study on heavy metals in mining industry mainly focuses on the evaluation, nature, mechanism, ecological improvement and biological effects of heavy metal pollution[30]. There are only sporadic research reports on Shenmu Mine Area, a super-large energy base located in water-wind erosion area and few reports on the spatial distribution and potential pollution evaluation of heavy metals in soil in this area. Geostatistics is used to predict the extent of soil and groundwater pollution as well as to calculate the risk in active or abandoned mining, waste disposal and urban sites, by accounting for the spatial distribution and uncertainty of the estimates. It facilitates quantification of the spatial characteristics of soil parameters and enables spatial interpolation[31-33].

Therefore, with Shigetai Coal Mine in Shenmu as the research area, the pollution level of toxic substances is evaluated and their spatial distribution characteristics are discussed by using the single factor pollution index method, Nemerow pollution index method and potential ecological evaluation method based on GIS technology and geostatistical theory, in order to provide scientific support and basis for environmental management and standardization.

Materials and methods

Overview of the research area

The research area is located in Shigetai Coal Mine in the north of Loess Plateau of Northern Shaanxi on the southern margin of Maowusu Desert (Fig.1). The coal field is bordered by Ulanmulun River in the west, Halagou coal field in the south, Battuta coal field in the north, Qigaigou and Shaanxi-Inner Mongolia border in the east. The geographical coordinates of the area are: 110°09′41″E ~ 110°18′35″E and 39°17′02″N − 39°35′16″N. The research area is dominated by a semi-arid continental climate in the mid-temperate zone, with cold winter and hot summer. The temperature difference between day and night is large, with the maximum value of about 20 ℃. The maximum frozen soil depth is 146 cm and the maximum snow depth is 12 cm from November to March of the following year. The monsoon period lasts from the beginning of January to the beginning of May and the prevailing wind direction is northwest wind. The annual average wind speed is 2.5 m/s and the maximum wind speed is 25 m/s. The annual average temperature is 8.5 ℃, the extreme maximum temperature is 38.9 ℃, the extreme minimum temperature is 28.5 ℃, and the annual average precipitation is 434.10 mm, which mainly occurs from July to September. The average annual evaporation is 1907.2 ~ 2122.7 mm, which is about 4 ~ 5 times of the precipitation.

Shigetai Coal Mine, one of the main production mines of Shendong Coal Group, was officially put into operation on January 15, 2006, with a geological reserve of 893 million tons and a recoverable reserve of 657 million tons. Characterized by low ash content, low sulfur content, low phosphorus content, medium to high calorific value in quality, the coal mined belongs to long flame coal with high volatile component and non-caking coal, which is the high-quality steam coal, chemical and metallurgical coal.

Sample collection

The sampling time is June 2020 and the soil sampling depth is 0 ~ 30 cm. One sample of about 0.5 kg is collected at the sampling point first, and then the 0.5 kg soil sample is collected at the other two sampling points within 2 m near this point respectively. Mix three soil samples evenly, take 500 g with quartering method to form one sample, and put it into a cloth bag. Put the collected samples into a cloth bag first to drain the most of the water before putting them into a polyethylene bag. Dry the sample in the air, and take some samples dried to be screened by 0.149 mm sieve, and then test Hg, Cd, As, Pb, Cr, Zn, Ni and Cu. See Fig.1 for the location of the research area and the distribution of sampling points.

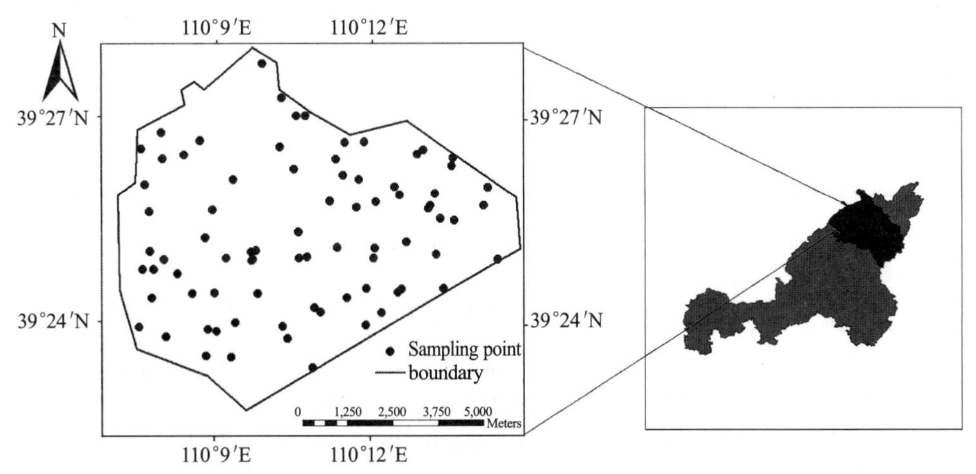

Fig.1 Location of the research area and the distribution of sampling points

Test method

See Table 1 for test indexes of heavy metals. The parallel test is carried out on each sample for three times. During the test, the quality shall be controlled against the national standard soil reference substance (GSS-12) and duplicate samples. The test results show that the recovery rates of all elements are within the allowable range.

Table 1 Executive standard for test of heavy metal contents

Element	Test basis	Test limit(mg/kg)	Test equipment
Hg	GB/T 22105.1—2008	0.002	BAF-2000 Atomic Fluorescence Spectrophotometer
Cd	GB/T 17141—1997	0.01	SOLAAR M6 Atomic Absorption Spectrometer
As	GB/T 22105.2—2008	0.01	BAF-2000 Atomic Fluorescence Spectrophotometer
Pb	GB/T 17141—1997	0.10	SOLAAR M6 Atomic Absorption Spectrometer
Cr	HJ 350—2007	0.40	ICP-5000 Inductively Coupled Plasma Emission Spectrometer
Zn	HJ 350—2007	0.10	ICP-5000 Inductively Coupled Plasma Emission Spectrometer
Ni	HJ 491—2019	3.0	SOLAAR M6 Atomic Absorption Spectrometer
Cu	HJ 491—2019	1.0	SOLAAR M6 Atomic Absorption Spectrometer

Evaluation methods

Nemerow pollution index method. With the soil background value in Shaanxi Province as a reference, the soil pollution status is evaluated by using the single factor pollution index method (P_i) and Nemerow pollution index method (NPI). The calculation equation is:

$$P_i = \frac{C_i}{C_n^i} \quad (1)$$

$$\mathrm{NPI} = \sqrt{\frac{P_{iave}^2 + P_{imax}^2}{2}} \quad (2)$$

Where, C_i is the measured content of heavy metal i (mg/kg); C_n^i is the background value of heavy metal i (mg/kg); P_i is the single factor pollution index of heavy metal i; NPI is Nemerow pollution index, P_{iave} is the average value of each single factor pollution index of heavy metals, and P_{imax} is the maximum value of single factor pollution index of heavy metals. See Table 2 for pollution index classification standard.

Table 2 Classification standard for single factor pollution index and nemerow pollution index

Pollution grade	P_i	Pollution level	NPI	Pollution level
1	$P_i \leqslant 1$	Without pollution	NPI≤0.7	Clean/safe
2	$1< P_i \leqslant 2$	Mild pollution	0.7<NPI≤1	Almost clean/close to the warning line
3	$2< P_i \leqslant 3$	Moderate pollution	1<NPI≤2	Mild pollution
4	$P_i >3$	Heavy pollution	2<NPI≤3	Moderate pollution
5	—	—	NPI>3	Heavy pollution

Potential ecological risk index method. Potential ecological risk index method can reflect the pollution level of a single heavy metal element and the comprehensive effect of all heavy metal elements from the perspective of the biological toxicity of heavy metals. The calculation equation is:

$$\mathrm{RI}_j = \sum_{i=1}^{n} E_j^i = \sum_{i=1}^{n} T_i \times C_j^i = \sum_{i=1}^{n} T_i \times \frac{c_j^i}{c_r^i} \tag{3}$$

Where: RI_j is the comprehensive potential ecological risk index of multiple heavy metals from the sampling point j, E_j^i is the single potential ecological risk index of heavy metal i from the sampling point j, C_j^i is the pollution index of heavy metal i from the sampling point j, c_j^i is the measured concentration of heavy metal i from the sampling point j, c_r^i is the reference value of heavy metal i (the background value of soil environment in Shaanxi is taken as the reference value in this study), and T_i is the toxicity coefficient of heavy metal i[34-35]. See Table 3 for c_r^i and T_i.

Table 3 Environmental background values and toxic-response parameters of heavy metals in the soil

Element	Hg	Cd	As	Pb	Cr	Zn	Ni	Cu
c_r^i	0.063	0.76	11.1	21.4	62.5	69.4	28.8	21.4
T_i	40	30	10	5	2	1	5	5

Ecological risk warning index method. In this research, the warning for the ecological risks of heavy metal pollution in open-pit coal mining areas is evaluated based on the ecological risk warning index (I_{ER}) proposed by Rapant and Kordik[36]. The calculation equation is:

$$I_{\mathrm{ER}} = \sum_{i=1}^{n} I_{\mathrm{ER}j} = \sum_{i=1}^{n} (P_i - 1) \tag{4}$$

Where: I_{ER} is the ecological risk warning index, $I_{\mathrm{ER}j}$ is the ecological risk index of the ith heavy metal, and P_i is the pollution index of the heavy metal i. See Table 4 for classification of pollution risk levels of RI and I_{ER}.

Table 4 Classification of risk degree of RI and I_{ER}

Class	RI		I_{ER}	
Ⅰ	$E \leqslant 40$; RI≤150	Low	$I_{\mathrm{ER}} \leqslant 0$	No
Ⅱ	$40< E \leqslant 80$; 150<RI≤300	Medium	$0< I_{\mathrm{ER}} \leqslant 1$	Early
Ⅲ	$80< E \leqslant 160$; 300<RI≤600	High	$1< I_{\mathrm{ER}} \leqslant 3$	Low
Ⅳ	$160< E \leqslant 320$; 600<RI≤1200	Very high	$3< I_{\mathrm{ER}} \leqslant 5$	Medium
Ⅴ	$E >320$; RI>1200	Extremely high	$I_{\mathrm{ER}} >5$	High

Geostatistics method

Geostatistics is an effective method to study the spatial distribution structure characteristics of regional variables. Its basic tool is a semivariate function that can be estimated by the following formula[37]:

$$\gamma(h)=\frac{1}{2N(h)}\sum_{i=1}^{N(h)}[Z(x_i)-Z(x_i+h)]^2 \tag{5}$$

Where, $\gamma(h)$ is the variation function, $Z(x)$ is the value of the regionalised variable at the sampling point x, $N(h)$ is the number of pairs with interval h and h is the interval, which is called the lag distance. Variograms can reflect and describe many properties of regionalised variables, and it is an important tool to analyze their spatial variation.

Data analysis

The data is summarized and processed by SPSS10.0 software, and the spatial distribution of single factor pollution index (P_i), Nemerow pollution index (NPI), potential ecological risk index (E), RI and ecological risk warning index (IER) is determined by ArcGIS 13.0 Kriging interpolation method.

Results and discussion

Analysis of contents of heavy metals in wasteland soil

The test results show(Table 5) that the contents of Hg, Cd, As, Pb, Cr, Zn, Ni and Cu in the surface soil within Shigetai Coal Mine vary from 0.043 to 0.255, 0.44 to 2.23, 2.66 to 18.40, 11.80 to 42.80, 40.50 to 118.60, 18.90 to 70.10, 4.31 to 28.10, 4.96 to 46.25 mg/kg, respectively; the average contents of Hg, Cd, As, Pb, Cr, Zn, Ni and Cu are 0.128, 1.03, 4.73, 23.08, 76.22, 46.94, 16.11 and 12.10 mg/kg, respectively. The average contents of Hg, Cd, Pb and Cr in soil within the research area are 2.03, 1.36, 1.11 and 1.23 times of the soil background values in Shaanxi Province, respectively. The average contents of As, Zn and Cu are lower than the soil background value in Shaanxi Province, but the maximum contents of these three elements are 1.65, 1.01 and 2.16 times of the soil background values in Shaanxi Province, respectively. It is reported that the average concentration of lead in agricultural soil affected by coal mines is relatively high (433 mg · kg^{-1})[38]. Lead is usually related to minerals in coal and occurs mainly in the form of sulfide such as PbS and PbSe[39]. In addition, aluminosilicate and carbonate also contain lead[40]. Chromium is a non-volatile element, which is related to aluminosilicate minerals[41]. In the mining process, chromium may be accumulated in coal, gangue or other tailings, and then enter the soil or water body through rain leaching[42].

The coefficient of variation (C_V) of Hg and Cd contents in soil within the research area is 0.050 and 0.37, respectively, with moderate variation, indicating that the content of these two heavy metals is less affected by the external factors; the coefficient of variation (C_V) of As, Pb, Cr, Zn, Ni and Cu contents is 2.81, 7.46, 18.00, 13.51, 5.44 and 5.64, respectively, with strong variation (C_V>0.50)[43], indicating that the content of these eight heavy metals may be affected by some local pollution sources. The skewness coefficient (SK) ranges from −3 to 3, and the larger its absolute value, the greater its skewness. When SK>0, it is positive skewness; when SK<0, it is negative skewness. Kurtosis coefficient is the characteristic value representing the peak value of probability density distribution curve at the average value[44]. The skewness coefficient and kurtosis coefficient of Cd, As and Cu elements in the soil within the research area are relatively high, indicating that these three elements are accumulated in the soil within the research area in large amounts.

Table 5　Statistics of contents of heavy metals in wasteland soil (n=79)

Parameter	Hg	Cd	As	Pb	Cr	Zn	Ni	Cu
Minimum(mg/kg)	0.043	0.44	2.66	11.80	40.50	18.90	4.31	4.96
Maximum(mg/kg)	0.255	2.23	18.40	42.80	118.60	70.10	28.10	46.25
Average(mg/kg)	0.128	1.03	4.73	23.80	76.22	46.94	16.11	12.10
Coefficient of variation(C_V)	0.050	0.37	2.81	7.46	18.00	13.51	5.44	5.64
Skewness coefficient	0.389	0.36	0.60	0.31	0.24	0.29	0.34	0.47
Kurtosis coefficient	0.808	1.91	1.39	0.76	0.36	−0.36	−0.01	2.90
Background value of Shaanxi soil(mg/kg)	0.063	0.76	11.1	21.4	62.5	69.4	28.8	21.4

Characteristics of heavy metal pollution in the research area

It can be seen from Table 6 that the average single factor pollution index (P_i) of heavy metals in the surface soil within the research area is Hg(2.03), Cr(1.22), Cd(1.14), Pb(1.11), Zn(0.68), As (0.59), Cu(0.57) and Ni(0.56) in descending order.

The average value of each element in the research area is at the mild pollution level or above, Hg is at a moderate pollution level, Cd, Pb and Cr are at the mild pollution level, and other heavy metals are at the clean level. Among them, the sampling points with the single pollution index of element Hg at moderate pollution level account for 21.52% of the total number of sampling points, and the sampling points with the single pollution index of Cd, Pb and Cr elements at the mild pollution level account for 43.04%, 55.70% and 77.22% of the total number of sampling points, respectively. NPI of eight heavy metal elements in the surface soil within the research area is between 0.80 and 3.20, with an average value of 1.64, so it is at the mild pollution level. The results show that the heavy metal elements in the surface soil within the research area are affected by human activities, and Hg is the most important pollution factor. This is consistent with the research result obtained by *Li et al* (2018)[45].

Table 6　Ecological risk of heavy metals in the surface soil within the research area

Parameter	P_i								NPI
	Hg	Cd	As	Pb	Cr	Zn	Ni	Cu	
Minimum	0.68	0.57	0.24	0.55	0.65	0.27	0.15	0.23	0.80
Maximum	4.05	2.94	1.66	2.00	1.92	1.01	0.98	2.16	3.20
Average	2.03	1.14	0.59	1.11	1.22	0.68	0.56	0.57	1.64
Pollution class	Ⅲ	Ⅱ	Ⅰ	Ⅱ	Ⅱ	Ⅰ	Ⅰ	Ⅰ	Ⅲ

Spatial distribution pattern of heavy metal pollution in soil

The spatial distribution pattern of P_i and NPI values of eight heavy metals in the surface soil of Shigetai Coal Mine is drawn by ArcGIS 13.0 software based on GIS technology and geostatistical analysis (Fig.2).

As can be seen from Fig.2, the heavy metal Hg in the surface soil within the research area ranks top in the pollution degree and area. The heavy metal Hg is at the moderate pollution level in the east, south and north of the research area, while it is at the mild pollution level in the west and middle of the research area. The heavy metal Cr in the research area ranks second in the pollution area. The element Cr is at the clean level in the middle of the research area, but at the mild pollution level in other areas, with small clean area. The heavy metal Pb in the research area ranks third in the pollution area. The element Pb is at the clean level in the south and middle (with small clean area) of the research area, but at the mild pollution level in other areas, with large area of mild pollution. The heavy metal Cd in the research area ranks fourth in the pollution area. The element Cd is at the clean level in the northwest (with small clean area) and south of the research area, but at the mild pollution level in other areas. Other heavy metal elements in the research area are at the clean level. Relevant researches show that the heavy metal pollution in soil of the coal mine is mainly caused by various mining activities[46]. Compared with natural soils, the elevated concentrations of heavy metals in the mining-affected soils were also reported elsewhere, e.g., Bangladesh[38] and India[47].

According to the study results of *Sun and Li et al* (2013)[48], in the process of coal mining, a large amount of coal gangue and fly ash is produced. During the process of rain leaching, many heavy metals, such as Pb, Hg, Crand Cd are released from coal gangue and fly ash. This is the most important sources of Pb, Hg, Cd, and Cr. In addition, traffic activities such as the wear of motor vehicle brake blocks and other parts, exhaust emissions, etc. are also one of the sources of heavy metal (Hg) pollution in soil[49]. Combined with the actual situation of the research area, the research area is adjacent to the Shaanxi-Inner Mongolia border in the east, so the exhaust emissions of motor vehicles may also be another cause of Hg pollution in the research area. Based on the spatial distribution pattern of NPI of heavy metal in the research area, its distribution status is basically consistent with the spatial distribution of P_i of heavy metal Hg.

Evaluation of potential ecological risks of heavy metal pollution in soil

By taking the soil background value of Shaanxi Province as the reference value, the single potential ecological risk index (E) and RI of heavy metals at each sampling point in Shigetai Coal Mine are calculated, and the ecological risk is evaluated according to the classification standard of potential ecological risks. It can be seen from Table 7 that the average value of E of the heavy metal in the surface soil within the research area is Hg(81.01), Cd(34.13), As(5.90), Pb(5.56), Cu(2.83), Ni(2.80), Cr(2.44) and Zn(0.68) in descending order. Except the heavy metal Hg, the average value of potential ecological risk index of other heavy metals in the surface soil within the research area is less than 40, so it is at a mild pollution level. The potential ecological risk of the element Hg ranges from 27.30 to 161.90, with an average value of 81.01, so it is in a relatively high ecological risk. It can be seen that Hg is the most important ecological risk factor in the research area. RI in the research area ranges from 53.44 to 6400.00 and the average value of comprehensive potential ecological risk index is 1336.49, so it is in an extremely high ecological risk.

Table 7 Evaluation of ecological risk of heavy metals in the surface soil within the research area

Parameter	E								RI
	Hg	Cd	As	Pb	Cr	Zn	Ni	Cu	
Minimum	27.30	17.21	2.40	2.76	1.30	0.27	0.75	1.16	53.44
Maximum	161.90	88.11	16.58	10.00	3.80	1.01	4.88	10.81	6400.00
Average	81.01	34.13	5.90	5.56	2.44	0.68	2.80	2.83	1336.49
Pollution class	High	Low	Low	Low	Low	Low	Low	Low	Extremely high

Note: E stands for the individual potential ecological risk index.

Spatial distribution pattern of potential ecological risks of heavy metal pollution in soil

The spatial distribution pattern of E and RI of heavy metals in the surface soil of Shigetai Coal Mine is as shown in Fig.3. It can be seen from Fig.3 that the spatial distribution of E value of the element Hg is basically consistent with the spatial distribution of P_i, the distribution area of Hg is relatively large, and the potential ecological risk index is in the moderate risk, indicating that Hg is the main ecological risk factor in this mine area.

The spatial distribution of E values of other heavy metals Cd, As, Pb, Cr, Zn, Ni and Cu is quite different. Among them, Ni, Zn and Pb are mainly distributed in the east and midwest of the mine area, Cr and As are mainly distributed in the east and southeast of the mine area, the heavy metal Cu is only distributed in the south of the mine area, and the heavy metal Cd is mainly distributed in the northwest and in the east (small area) of the mine area.

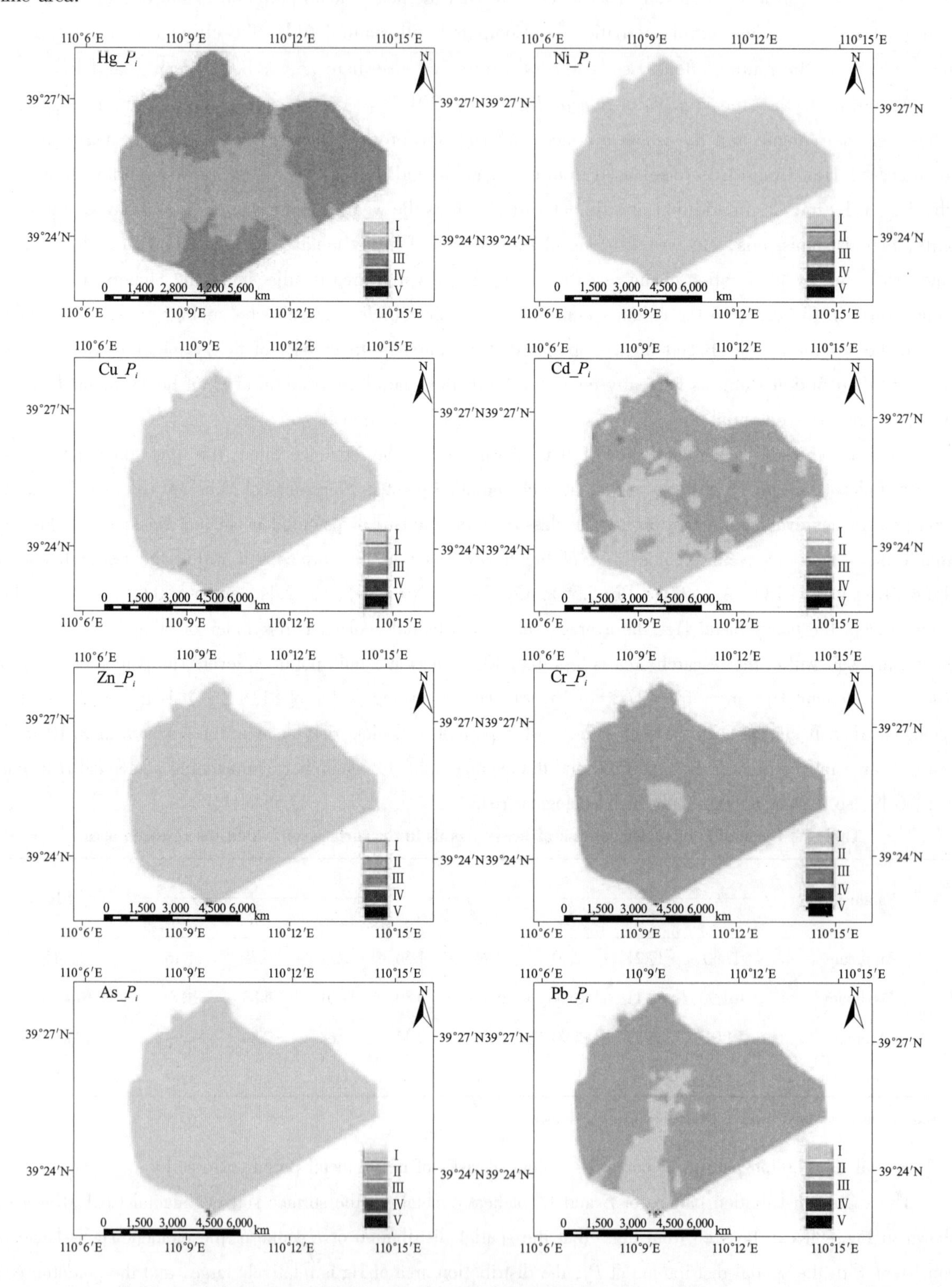

Fig.2 Spatial distribution of P_i and NPI of heavy metal in the study area

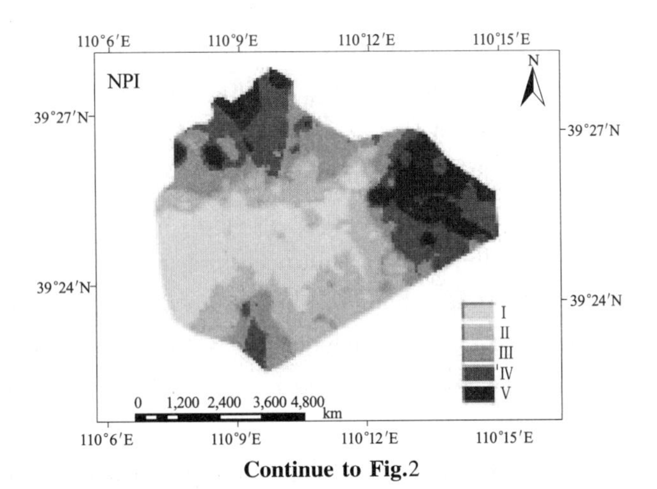

Continue to Fig.2

As can be seen from Fig.4, the ecological risk index is high in the east and north of the research area from the spatial distribution pattern of RI in the research area, showing extremely strong risks, while it is in the slight risk in the middle and west, and moderate risk in other areas. On the whole, the spatial distribution pattern of RI values of heavy metals in the research area shows an obvious horizontal zonal distribution pattern, which is basically consistent with the distribution pattern of NPI.

Ecological risk warning of heavy metal pollution in soil

Based on the ecological risk evaluation, the ecological risk warning evaluation emphasizes the research on possible risk warning in ecosystem. The warnings for the ecological hazards caused by surface soil heavy metal pollution in the Shigetai Coal Mine in northern Shaanxi are evaluated based on the classification standard of ecological risks proposed by Rapant and Kordik[36]. It can be seen from Table 8 that the average values of I_{ER} of eight elements including Hg, Cd, As, Pb, Cr, Zn, Ni and Cu are 1.03, 0.14, −0.41, 0.11, 0.22, −0.32, −0.44 and −0.43, respectively. Among them, Hg is in the slight warning status, Cd, Pb and Cr are in the warning status, and As, Zn, Ni and Cu are in no warning status. I_{ER} values in the research area range from −34.80 to 81.00 and the average value is −1.13, so it is in no warning status.

Based on the spatial distribution of I_{ER} values (Fig.5), the northern and eastern areas in the research area show serious warning, the transition area between the east and the north shows moderate warning to slight warning, the western and central areas show no warning, and other areas show warning to slight warning. The areas with high ecological risk warning values are mainly distributed in the east and north, with relatively serious pollution, showing relatively obvious zonal distribution law.

Table 8 **Warning of ecological risks of heavy metals in surface soil within the research area**

Parameter	I_{ERi}								I_{ER}
	Hg	Cd	As	Pb	Cr	Zn	Ni	Cu	
Minimum	−0.32	−0.43	−0.76	−0.45	−0.35	−0.73	−0.85	−0.77	−34.80
Maximum	3.05	1.94	0.66	1.00	0.90	0.01	−0.02	1.16	81.00
Average	1.03	0.14	−0.41	0.11	0.22	−0.32	−0.44	−0.43	−1.13
Warning class	Low	Early	No	Early	Early	No	No	No	No

Conclusion

(1) The contents of Hg, Cd, As, Pb, Cr, Zn, Ni and Cu in the soil within the research area vary from 0.043 to 0.255, 0.44 to 2.23, 2.66 to 18.40, 11.80 to 42.80, 40.50 to 118.60, 18.90 to 70.10, 4.31 to 28.10, 4.96 to 46.25 mg/kg, respectively. Among them, the average contents of heavy metals Hg, Cd, Pb and Cr are 2.03,

1.36,1.11 and 1.23 times of the soil background values in Shaanxi Province respectively and the average contents of other heavy metals are lower than the soil background values in Shaanxi Province; based on the coefficient of variation, Hg and Cd show moderate variation while As, Pb, Cr, Zn, Ni and Cu show strong variation; the skewness coefficient and kurtosis coefficient of Cd, As and Cu in the soil within the research area are relatively high, and these elements are accumulated in large amounts.

(2) The analysis of single factor pollution of heavy metals in the soil within the research area shows that the heavy metal Hg pollutes the soil seriously, so it is at the moderate pollution level, and it is mainly distributed in the east and north of the research area as well as a small area in the south, while three elements including Cd, Pb and Cr are at the mild pollution level. The analysis of Nemerow pollution index method shows that Nemerow pollution index in the research area reaches level Ⅲ due to mining activities, and Hg is the most important pollution factor.

(3) The average value of potential ecological risk index of heavy metals in soil within the research area is Hg(81.01), Cd(34.13), As(5.90), Pb(5.56), Cu(2.83), Ni(2.80), Cr(2.44) and Zn(0.68), respectively. The potential ecological risk of the heavy metal Hg ranges from 27.30 to 161.90, with an average value of 81.01, so it is in a relatively high ecological risk. Other heavy metals are in the low risk.

(4) From the perspective of spatial distribution, the eastern and northern ares in the research area are in the high risk, the central and western areas are in the low risk, and the other areas are in the medium risk; the areas with high ecological risk warning values are mainly distributed in the east and north, with relatively serious pollution, and the whole research area shows relatively obvious zonal distribution law. Due to the long-term human activities, the spatial heterogeneity of heavy metal pollution is obviously enhanced, and the potential risks may be increased beyond our expectations. Therefore, the ecological risks and human health risks of heavy metal pollution in this area shall be comprehensively evaluated in the future study, so as to provide a basis for the prevention and control strategies of heavy metal pollution in coal mine areas.

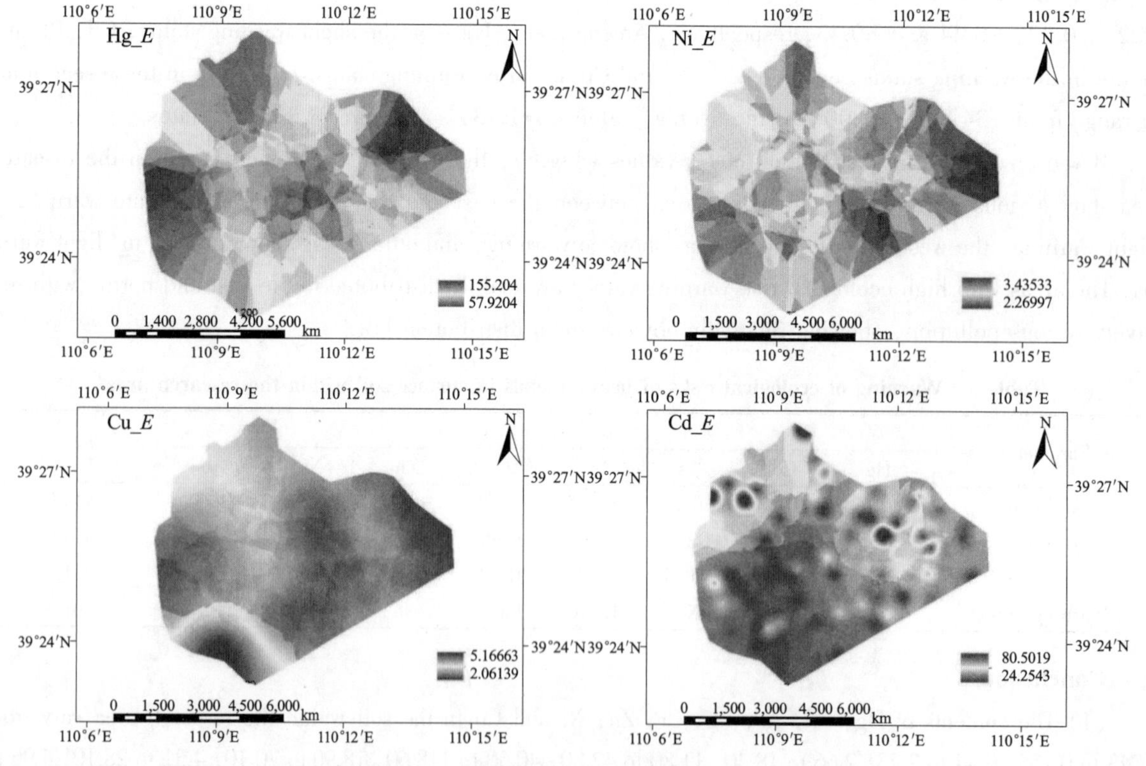

Fig.3 Spatial distribution of single ecological risk index (E) of heavy metals

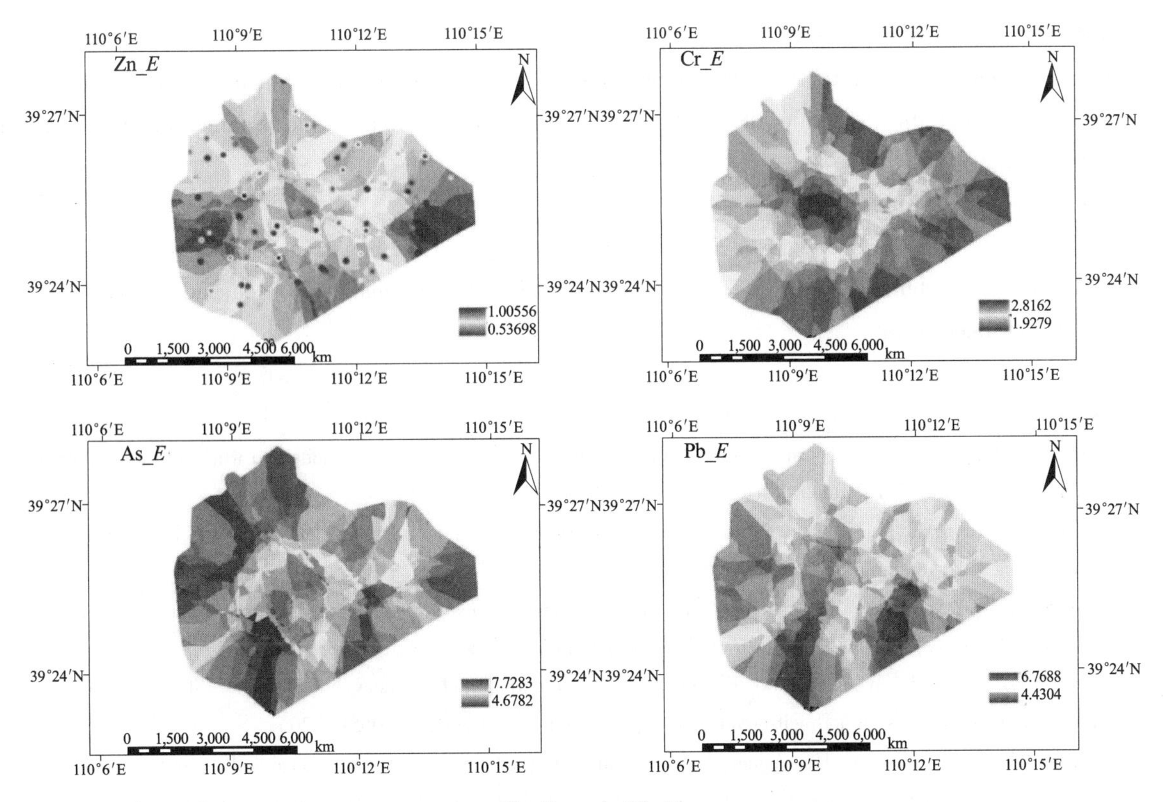

Continue to Fig.3

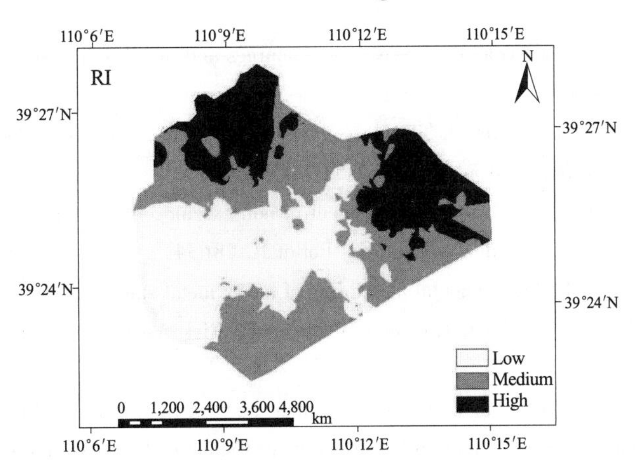

Fig.4 Spatial distribution of RI of heavy matals

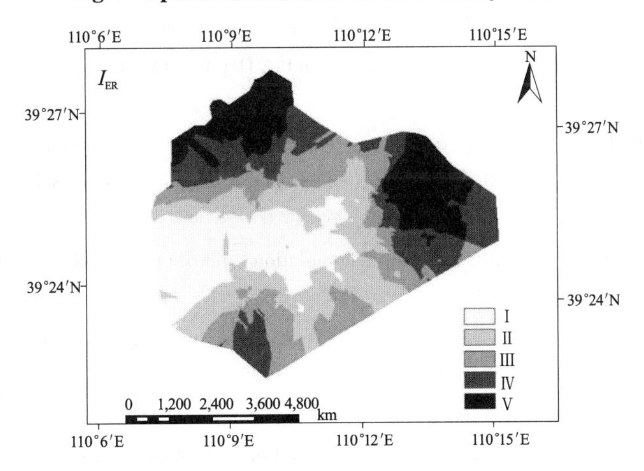

Fig.5 Ecological risk warning assessment of heavy metals in the study area

Acknowledgments

This research was supported by Natural Science Basic Research Plan in Shaanxi Province of China (2021JQ－960) and the Scientific Research Item of Shaanxi Provincial Land Engineering Construction Group(DJNY2022-20)

Author contributions

N.W. designed the study and wrote the paper. Y. L performed the experiments and collected the data. Z. L. contributed the statistical analyses. Y.S. reviewed and edited the manuscript All authors read and approved the final manuscript.

References

[1]Ozturk, A., & Arici, O. K. Carcinogenic-potential ecological risk assessment of soils and wheat in the eastern region of Konya (Turkey).Environ Sci Pollut R.28(12), 15471-15484(2021).

[2]Dai, L., Wang, L., Li, L.et al. Multivariate geostatistical analysis and source identification of heavy metals in the sediment of Poyang Lake in China.Sci Total Environ.621, 1433-1444(2018).

[3]Mazurek, R., Kowalska, J., Ga siorek, M.et al. Assessment of heavy metals contamination in surface layers of Roztocze National Park forest soils (SE Poland) by indices of pollution. Chemosphere.168, 839-850(2017).

[4]Wang, Z., Xiao, J., Wang, L.et al. Elucidating the differentiation of soil heavy metals under different land uses with geographically weighted regression and self-organizing map. Environ Pollut.260, 114065(2020).

[5]Javed, M. T., Saleem, M. H., Aslam, S. et al. Elucidating silicon-mediated distinct morpho-physio-biochemical attributes and organic acid exudation patterns of cadmium stressed Ajwain (Trachyspermum ammi L.). Plant Physiol Bioch.157, 23-37(2020).

[6]Sun, J., Yu, R., Hu, G.et al. Tracing of heavy metal sources and mobility in a soil depth profile via isotopic variation of Pb and Sr. Catena.171, 440-449(2018).

[7]Wen, H., Zhang, Y., Cloquet, C.et al. Tracing sources of pollution in soils from the Jinding Pb-Zn mining district in China using cadmium and lead isotopes. Appl Geochem. 52, 147-154(2015).

[8]Chakraborty, B., Bera, B., Roy, S.et al. Assessment of non-carcinogenic health risk of heavy metal pollution: evidences from coal mining region of eastern India. Environ Sci Pollut R. 28(34), 47275-47293(2021).

[9]Abdelhafez, A. A., & Li, J. Environmental monitoring of heavy metal status and human health risk assessment in the agricultural soils of the Jinxi River area, China. Human and Ecological Risk Assessment: An International Journal. 21(4), 952-971(2015).

[10]Zhou, M., Liao, B., Shu, W.et al. Pollution assessment and potential sources of heavy metals in agricultural soils around four Pb/Zn mines of Shaoguan city, China. Soil and Sediment Contamination: An International Journal. 24(1), 76-89 (2015).

[11]Wei, L., Wang, K., Noguera, D. R.et al.Transformation and speciation of typical heavy metals in soil aquifer treatment system during long time recharging with secondary effluent: Depth distribution and combination. Chemosphere. 165, 100-109(2016).

[12]Shi, G. L., Lou, L. Q., Zhang, S.et al. Arsenic, copper, and zinc contamination in soil and wheat during coal mining, with assessment of health risks for the inhabitants of Huaibei, China. Environ Sci Pollut R. 20(12), 8435-8445 (2013).

[13]Zhao, F. J., Stroud, J. L., Eagling, T.et al. Accumulation, distribution, and speciation of arsenic in wheat grain. Environ Sci Technol. 44(14), 5464-5468(2010).

[14]Melo, É. E., Guilherme, L. R., Nascimento, C. W.et al. Availability and accumulation of arsenic in oilseeds grown in contaminated soils. Water Air Soil Poll. 223(1), 233-240(2012).

[15]Horasan, B. Y., Ozturk, A., & Tugay, O. Nb-Sr-Pb isotope analysis in soils of abandoned mercury quarry in northwest Black Sea (Turkey), soil and plant geochemistry, evaluation of ecological risk and its ımpact on human health. Environ Earth Sci. 80(15), 1-19(2021).

[16] Mammola, S., Cardoso, P., Culver, D. C. et al. Scientists' warning on the conservation of subterranean ecosystems. BioScience. 69(8), 641-650(2019).

[17] Vatalis, K. I., & Kaliampakos, D. C. An overall index of environmental quality in coal mining areas and energy facilities. Environ Manage. 38(6), 1031-1045(2006).

[18] Li, M. S. Ecological restoration of mineland with particular reference to the metalliferous mine wasteland in China: a review of research and practice. Sci Total Environ. 357(1-3), 38-53(2006).

[19] Chen, Y., Zhao, H. X., Xie, Z. H.et al. Heavy Metal Pollution Characteristics in the Kaili Coal Mining Region, Guizhou Province, China. Journal of Residuals Science & Technology. 12, S123-S131 (2015).

[20] Tozsin, G. Hazardous elements in soil and coal from the Oltu coal mine district, Turkey. Int J Coal Geol. 131, 1-6 (2014).

[21] Qiao, M., Cai, C., Huang, Y.et al. Characterization of soil heavy metal contamination and potential health risk in metropolitan region of northern China. Environ Monit Assess. 172(1), 353-365(2011).

[22] Islam, A. R. M. T., Hasanuzzaman, M., Islam, H. T.et al. Quantifying source apportionment, co-occurrence, and ecotoxicological risk of metals from upstream, lower midstream, and downstream river segments, Bangladesh. Environ Toxicol Chem. 39(10), 2041-2054(2020).

[23] Chen, Y., Yuan, L., & Xu, C. Accumulation behavior of toxic elements in the soil and plant from Xinzhuangzi reclaimed mining areas, China. Environ Earth Sci. 76(5), 1-8(2017).

[24] Candeias, C., Da Silva, E. F., Salgueiro, A. R.et al. Assessment of soil contamination by potentially toxic elements in the Aljustrel mining area in order to implement soil reclamation strategies. Land Degrad Dev. 22(6), 565-585 (2011).

[25] Masto, R. E., Sheik, S., Nehru, G.et al. Assessment of environmental soil quality around Sonepur Bazari mine of Raniganj coalfield. India. Solid Earth. 6(3), 811-821(2015).

[26] De, S., & Mitra, A. K. Mobilization of heavy metals from mine spoils in a part of Raniganj coalfield, India: Causes and effects. Environmental Geosciences. 11(2), 65-76(2004).

[27] Yudovich, Y. E., & Ketris, M. P. Arsenic in coal: a review. Int J Coal Geol. 61(3-4), 141-196(2005).

[28] Chen, J., Liu, G., Kang, Y.et al. Atmospheric emissions of F, As, Se, Hg, and Sb from coal-fired power and heat generation in China. Chemosphere. 90(6), 1925-1932(2013).

[29] Wu, C., Deng, H., Shu, S. The study on the land development process in the border area between Shaanxi and Inner Mongolia. Geographical Research.33(8):1579-1592(2014).(in Chinese)

[30] Solgun, E., Horasan, B. Y., & Ozturk, A. Heavy metal accumulation and potential ecological risk assessment in sediments from the southwestern Konya district (Turkey). Arab J Geosci. 14(8), 1-15(2021).

[31] Modis, K., & Komnitsas, K. Optimum sampling density for the prediction of acid mine drainage in an underground sulphide mine. Mine Water Environ. 26(4), 237-242(2007).

[32] Modis, K., Vatalis, K., Papantonopoulos, G.et al. Uncertainty management of a hydrogeological data set in a greek lignite basin, using BME. Stoch Env Res Risk A. 24(1), 47-56(2010).

[33] Modis, K., & Vatalis, K. I. Assessing the risk of soil pollution around an industrialized mining region using a geostatistical approach. Soil and Sediment Contamination: An International Journal. 23(1), 63-75(2014).

[34] Suresh, G., Sutharsan, P., Ramasamy, V.et al. Assessment of spatial distribution and potential ecological risk of the heavy metals in relation to granulometric contents of Veeranam lake sediments, India. Ecotox Environ Safe.84, 117-124 (2012).

[35] Shen, F., Liao, R., Ali, A.et al. Spatial distribution and risk assessment of heavy metals in soil near a Pb/Zn smelter in Feng County, China. Ecotox Environ Safe. 139, 254-262(2017).

[36] Rapant, S., Kordík, J. An environmental risk assessment map of the Slovak Republic: application of data from geochemical atlases. Environ Geol. 44(4), 400-407 (2003).

[37] Zhan, J., He, Y., Zhao, G.et al. Quantitative Evaluation of the Spatial Variation of Surface Soil Properties in a Typical Alluvial Plain of the Lower Yellow River Using Classical Statistics, Geostatistics and Single Fractal and Multifractal Methods. Applied Sciences. 10(17), 5796(2020).

[38] Bhuiyan, M. A., Parvez, L., Islam, M. A.et al. Heavy metal pollution of coal mine-affected agricultural soils in the northern part of Bangladesh. J Hazard Mater. 173(1-3), 384-392(2010).

[39] Hower, J. C., & Robertson, J. D. Clausthalite in coal. Int J Coal Geol. 53(4), 219-225(2003).

[40] Wang, J., Sharma, A., & Tomita, A. Determination of the modes of occurrence of trace elements in coal by leaching coal and coal ashes. Energ Fuel. 17(1), 29-37(2003).

[41] Zhou, C., Liu, G., Wu, S.et al. The environmental characteristics of usage of coal gangue in bricking-making: a case study at Huainan, China. Chemosphere. 95, 274-280(2014).

[42] You, M., Huang, Y., Lu, J., & Li, C. Characterization of heavy metals in soil near coal mines and a power plant in Huainan, China. Anal Lett. 48(4), 726-737(2015).

[43] Eziz, M., Mamut, A., Mohammad, A.et al.Assessment of heavy metal pollution and its potential ecological risks of farmland soils of oasis in Bosten Lake Basin. Acta Geographica Sinica. 72(9), 1680-1694(2017).

[44] Dong, Z., Wang, L., & Tian, F. The pollution status and spatial distribution of Cu and Pb in soil in Zhenjiang Region. Arid Environmental Monitoring. 28(4), 149-153(2014).

[45] Li, F., Li, X., Hou, L., & Shao, A. Impact of the coal mining on the spatial distribution of potentially toxic metals in farmland tillage soil. Scientific reports. 8(1), 1-10(2018).

[46] Liu, X., Bai, Z., Shi, H.et al. Heavy metal pollution of soils from coal mines in China. Nat Hazards. 99(2), 1163-1177(2019).

[47] Mishra, V. K., Upadhyaya, A. R., Pandey, S. K.et al. Heavy metal pollution induced due to coal mining effluent on surrounding aquatic ecosystem and its management through naturally occurring aquatic macrophytes. Bioresource Techno. 99(5), 930-936. (2008).

[48] Sun, X. B., & Li, Y. C. The spatial distribution of soil heavy metals and variation characteristics of Datong abandoned coal mine area in Huainan City. Scientia Geographica Sinica.13(10), 1238-1244(2013).

[49] Apeagyei, E., Bank, M. S., Spengler, J., D. Distribution of heavy metals in road dust along an urban-rural gradient in Massachusetts. Atmos Environ. 45(13): 2310-2323(2011).

本文曾发表于2022年《Nature Portfolio》第12卷

Ecological Risk Evaluation and Source Identification of Heavy Metal Pollution in Urban Village Soil Based on XRF Technique

Siqi Liu[1,2,3], Biao Peng[2,4], Jianfeng Li[2,3]

(1.Institute of Land Engineering and Technology, Shaanxi Provincial Land Engineering Construction Group Co., Ltd., Xi'an 710075, China;2.Shaanxi Provincial Land Engineering Construction Group Co., Ltd., Xi'an 710075, China; 3.Key Laboratory of Degraded and Unused Land Consolidation Engineering, The Ministry of Natural Resources, Xi'an 710075, China;4.School of Land Engineering, Chang'an University, Xi'an 710000, China)

【**Abstract**】The rapid urbanization in China has resulted in significant differences between urban and rural areas. The emergence of urban villages is inevitable in this context, for which complex problems regarding land use, industrial management and ecological environment have arisen. This study performed a case study on a typical urban village, by assessing heavy metal pollution and ecological risk in soil. It detected a total of 80 basic units through portable X-ray fluorescence (XRF) instrument. A total of 25 high-risk contaminated points were selected, sampled and analyzed in laboratory as confirmation. The results showed the mean concentrations of Pb, Cu, Zn and Ni in soil were significantly higher than background values. Pb, Zn and Ni showed obvious pairwise correlation, and the high-value zones could be attributed to automobile traffic and industrial activities. In addition, the pollution problem is complicated by a combination of agricultural activities, the absence of clear division between different functional zones, as well as a general lack of environmental awareness. All of these lead to increased ecological risk and are a serios threaten to public health.

【**Keywords**】Soil pollution; Heavy metal; Potential ecological risk; Urban village; Industrial land

1 Introduction

Land use change is an essential part of the earth's environmental change, and the most direct manifestation of the impact of human activities on the earth's surface system.[1-2]. China has experienced an extremely rapid urbanization in the past four decades. From 1981 to 2019, China's urbanization rate increased from 20.16% to 60.60%, with an average annual increase of 1.06%[3]. During this period, the area of urban construction land in China soared from 6 720.0 km^2 to 58 307.7 km^2, an increase of 767.67%[4]. The rapid growth of construction land mainly comes from non-agricultural transformation of agricultural land[5-6]. One of the most prominent features of urban expansion was a growing number of industrial lands. The rapid growth of industry had promoted China's modernization and urbanization. However, China's industrial development had been in an extensive mode in the past, which mainly manifested in low land utilization efficiency and serious environmental pollution[7]. Therefore, it intensified the contradiction between supply and demand of urban land, which seriously destroyed the balance of the man-land interrelationship[8-10]. In 2019, China's industrial and manufacturing land accounted for 19.69% of the total areas of urban construction land, while the proportion in developed countries was often less than 10%[11-12]. At present, China is at an important stage of industrial development and transformation, including industrial structure upgrading, land intensification and green sustainable development. In terms of environmental sustainability, the government has implemented a series of measures, including closing scattered industrial factories in downtown and moving industrial sites to suburban areas, to keep ecological environmentally friendly and urban living environment healthy. In addition, rationally controlling the scale and layout of industrial land is an important way to improve urban eco-efficiency[13-14].

Urban soil contamination mainly comes from industrial production and related activities. The raw materials and compounds of Cd, Cr, Cu, Ni, Pb, Zn, Hg and As are widely utilized in human daily life and production, as well as the most polluted[15-16]. From 2005 to 2013, China carried out the first national soil survey, in which the over-standard rates of pollution points of heavily polluted enterprise land, industrial wasteland and industrial park were 36.3%, 34.9% and 29.4%, respectively[17]. The problem of heavy metal pollution in soil was very prominent. For single heavy metal element, the over-standard rates of Cd, Ni, As and Cu reached 7.0%, 4.8%, 2.7% and 2.1%, respectively, which posed a great threat to human health[18]. Soil is not only the carrier of human activities, but also the final receptor of various pollution emissions[19]. Previous studies showed that industrial manufacturing and fossil fuel combustion would lead to the concentrations of heavy metals in urban soil significantly higher than their natural background values[20-22]. Remediation of heavy metal pollution in soil is complicated owning to its persistence, bio-accumulation and low microbial degradability properties. Its migration and transformation are affected by soil types, heavy metal ions and their occurrence forms[23-24]. Soil pollution of heavy metals endangers human health through oral intake, respiratory inhalation and skin contact[25-26].

There have existed obviously regional differences in the process of urbanization in China. In general, the growth rate of urbanization decreases progressively from east to west in turn[27]. With the in-depth development of urbanization, urban expansion and urban renewal have become the two dominant modes of urban land change[28]. It is an important issue in coordinating urban development and environmental sustainability. A study of livable cities in China released by the Chinese Academy of Sciences noted that the average environmental health score of all 40 cities participating in the survey was only 58.24, lower than 60. Environmental health problem has become the main limitation of construction of livable cities in China. Furthermore, this study pointed out that the environmental health problem in Xi'an was serious, which directly affected its livability[29]. Over a long period in the past, Xi'an's industrial development had the characteristics of high energy consumption, heavy pollution and high emission. With the implementation of green sustainable strategy, people have been focusing on soil health and living environment issues.

Many researchers have conducted qualitative and quantitative studies on heavy metal pollution in soil, analyzing the spatial distribution rule and assessing the pollution degree[30-32]. However, the city is quite large in relation to its land area. There are a very limited number of sampling points and their layout lacks exact data support in most studies. Urban village is a transitional zone between urban and rural areas. There have existed huge urban-rural differences in aspects of land use and landscape pattern. Furthermore, there are dense population, poor environment and inadequate infrastructure[33-34]. Many researchers have pointed out that the urban village issue is the key to urban high-quality development[35-36]. However, seldom have researchers focused on soil environment in urban village and the internal relations among land use, industry and heavy metal concentration in soil. Through developing an integrated approach, this research aimed at (1) finding out high concentration zones of Cd, Cr, Cu, Ni, Pb, Zn, Hg and As in research areas based on the XRF technique, (2) assessing the pollution and ecological risk, (3) analyzing its distribution rule and pollution sources, and (4) putting forward some advices based on health and sustainability. This research may provide an efficient and reliable method for large-scale urban soil investigation. Its results may offer a scientific basis for comprehensive environmental management and are of great significance for urban sustainable development.

2 Materials and Methods

2.1 Research Area and Detection Points

As one of the nine national central cities, Xi'an is located in the northwest of China. It is an important industrial base of China, and its main industries include automobile manufacturing, pharmaceutical manufacturing and electronic equipment manufacturing. In addition, Xi'an has been experiencing rapid urban sprawl during the last decade. A large amount of farmland has been transformed into urban construction land. Under the sharp conflict of urban-rural industry, population and land, many urban villages have formed as a result. In total, there were 156 urban villages in Xi'an until 2018, most of which were located in suburban areas[37]. As shown in Fig.1, the industrial land of Xi'an is mainly distributed in suburban areas as well, especially the western suburbs. Therefore, the research area is in the western suburbs of Xi'an, which is of obvious urban-rural dual structure and characteristics. The research area is a typical urban village, locating in Chang'an district at latitude 34°13′09″~34°13′40″ N, longitude 108°48′03″~108°48′40″ E. The total land area is approximately 0.78 km^2. It is dominated by industrial land and residential land, with a small amount of open space and farmland. The industries in this urban village mainly involves electronic equipment manufacturing, construction materials and automobile maintenance.

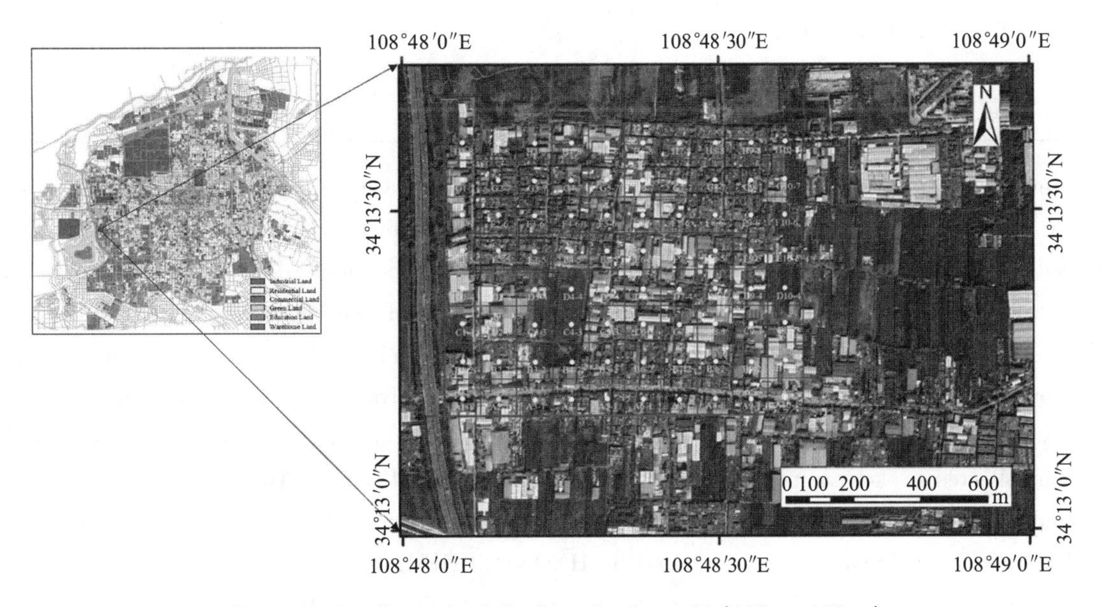

Fig.1 Research area and the investigation grid (100 m×100 m)

Through the recognition of remote sensing (RS) image, it indicated that the research area was a similar 1000 m × 800 m (L×W) rectangle. According to Technical Guidelines for Investigation on Soil Contamination of Land for Construction (HJ 25.1—2019) and The Technical Specification for Soil Environmental Monitoring (HJ/T 166—2004), the research area was transformed into a 10 × 8 basic grid and each unit was a 100 m × 100 m rectangle[38-39]. The grid was unified with the geographical coordinates by geographic information system (GIS) and global positioning system (GPS) technologies (Fig. 1). The geometric center of each rectangle unit was determined as a quick detection point with X-ray fluorescence (XRF) instrument. Consequently, in total there were 80 quick detection points.

2.2 Analytic Methods

At present XRF has been widely used to confirm the type and quantity of elements in substances. Some researchers applied this technology to the detection of heavy metals in soil, showing the characteristics of fast, accurate, portable and no damage to samples[40]. The sources causing heavy metal pollution in urban soil are complex. Theoretically, the more detection points, the more accurate it can reflect the overall soil pollution situation of the site. Traditional soil sampling and laboratory analysis need to consume a lot of manpower and material resources. However, XRF has obvious advantages in this aspect, which can realize large-scale full coverage and efficient investigation of heavy metals in soil.

The sundries on soil surface were cleaned by a plastic shovel (non-metallic). Heavy metal concentrations in topsoil were quantified in situ with field portable XRF instrument (Sci-Aps, 200) set to Geo-Env soil mode. In this soil mode, three beams under different voltage can detect different elements (Table 1). This research focused on the concentrations of Cd, Cr, Cu, Ni, Pb, Zn, Hg and As. Therefore, beam 1 and 3 were selected, while beam 2 was skipped. Each measurement was collected for a total of 30 s, which was very quick and well directed. The calibration verification, determination of instrument precision, accuracy and limit of detection (LOD) fulfills the requirements of United States Environmental Protection Agency (EPA) Method 6200[41].

Table 1 Detectable elements in Geo-Env soil mode of SciAps X-200

Mode	Detectable elements(Beam 1)	Detectable elements(Beam 2)	Detectable elements(Beam 3)
Soil	Ti, V, Cr, Mn, Fe, Co, Ni, Cu, Zn, As, Se, Sr, Rb, Zr, Mo, W, Tl, Hg, Pb, Bi	Mg, Al, Si, P, S	Ag, Cd, Sn, Sb, Ba

Through soil analysis with XRF, the concentrations of Cd, Cr, Cu, Ni, Pb, Zn, Hg and As in soil of each unit were quantified. EPA Method 6200 recommends a minimum 5% of all samples tested by XRF be confirmed by laboratory analysis. Considering the difference of pollution degree of each heavy metal element, 5%~10% of quick detection results of different heavy metal were regarded as high-risk contaminated, respectively. Finally, high-risk contaminated points of above eight heavy metals were integrated. These points would be confirmed with soil sampling and laboratory analysis. This research focused on urban topsoil, which is significantly affected by human activities[42]. Three topsoil subsamples, at a depth of 0-20 cm, were collected at each sampling point and mixed into a final soil sample. These soil samples were naturally dried and then filtrated through 100-mesh nylon sieve. Using a total-digestion EPA analytical reference method 3050B[43], soil samples were digested by HNO_3-HCl-HF-H_2O_2 and HNO_3-HCl, respectively. The concentrations of Cd, Cr, Cu, Ni, Pb, Zn, As were quantitatively analyzed by ICP-MS (Agilent, 7700e). Concentrations of Hg were quantitatively analyzed by Atomic Fluorescence Spectrometer (JINSUOKUN, SK-2003AZ). This laboratory analysis adopted national standard materials for soil composition analysis (GSS-8) for analytic quality control (QC).

2.3 Comprehensive Pollution Index (CPI) Method

Comprehensive pollution index (CPI) method is a multi-factor environmental quality assessment method highlighting the maximum value. It is one of the important methods to evaluate pollution in soil, water and sediment[44]. Based on single pollution index (SPI), CPI can comprehensively reflect the different effects of various heavy metal pollutants on soil. The CPI is computed as follows:

$$P_i = C_i / S_i \quad (1)$$

$$P_c = \sqrt{\frac{P^2_{imax} + P^2_{iave}}{2}} \quad (2)$$

where C_i is the measured concentrations of a specific heavy metal, S_i is the evaluation criteria of heavy metal pollution. In this research, the criteria mainly use Risk Control Standard for Soil Contamination of Development Land (GB 36600—2018) and Risk Control Standard for Soil Contamination of Agricultural Land (GB 15618—2018) for reference[45-46]. P_i and P_c represent SPI and CPI, respectively. P_{imax} is the maximum pollution index of a specific heavy metal and P_{iave} is the average pollution index of all heavy metal factors. P_c is classified into five categories that $P_c \leq 0.7$ (Level I and clean), $0.7 < P_c \leq 1$ (Level II and low polluted), $1 < P_c \leq 2$ (Level Ⅲ and moderately polluted), $2 < P_c \leq 3$ (Level IV and highly polluted), $3 < P_c$ (Level V and significantly highly polluted).

2.4 Geo-Accumulation Index (I_{geo}) Method

Geo-accumulation index (I_{geo}) method is an important quantitative evaluation method of heavy metal pollution. It is widely used in the quantitative evaluation of heavy metal pollution of soil and sediment. This method takes into account the difference of background values caused by natural geological processes and the influence of heavy metal pollution caused by human activities[47]. I_{geo} is an important parameter, which can evaluate the impact of human activities on soil pollution. The I_{geo} is computed as follows:

$$\left[\frac{C_n}{1.5 \times BV_n}\right] \tag{3}$$

where C_n is the measured concentration of a specific heavy metal in soil, and BV_n is the geochemical background value of a specific heavy metal in soil. According to the value of I_{geo}, pollution grades are classified into seven categories: $I_{geo} < 0$ (not polluted), $0 \leq I_{geo} < 1$ (not polluted to moderately polluted), $1 \leq I_{geo} < 2$ (moderately polluted), $2 \leq I_{geo} < 3$ (moderately polluted to heavily polluted), $3 \leq I_{geo} < 4$ (heavily polluted), $4 \leq I_{geo} < 5$ (heavily polluted to extremely polluted), and $5 \leq I_{geo}$ (extremely polluted).

2.5 Potential Ecological Risk Index (RI) Method

Potential ecological risk index (RI) method was established by Hakanson for evaluating heavy metal pollution and ecological risk based on the principles of sedimentology[48]. This method takes toxic effects of different heavy metals on the environment into consideration, associating heavy metal pollution and its ecological effect. *RI* can comprehensively reflect the potential ecological risk level of heavy metals on soil environment. The calculation formula is as follows:

$$RI = \sum_{i=1}^{n} E_r^i = \sum_{i=1}^{n} T_r^i \times \frac{C_s^i}{C_n^i} \tag{4}$$

where C_s^i is the measured concentrations of a specific heavy metal in soil, C_n^i is the evaluation criteria. This research used soil background values of Shaanxi province for criteria reference[49]. T_r^i is the specific biological toxic response factor for a heavy metal element. E_r^i is the potential ecological risk index of single heavy metal. RI is calculated as the sum of potential ecological risk of all researched heavy metals. C_n^i and T_r^i are shown in Table 2. With regard to potential ecological risk caused by single heavy metal, E_r^i is classified into five categories: $E_r^i < 40$ (low level), $40 \leq E_r^i < 80$ (moderate level), $80 \leq E_r^i < 160$ (considerable level), $160 \leq E_r^i < 320$ (high level), and $320 \leq E_r^i$ (significantly high level). As for overall potential ecological risk, RI is classified into four categories: $RI < 150$ (low level), $150 \leq RI < 300$ (moderate level), $300 \leq RI < 600$ (high level), and $600 \leq RI$ (significantly high level).

Table 2 The evaluation criteria (C_n^i) and toxic response factor (T_r^i) of heavy metals in soil

Heavy Metals	Cd	Cr	Cu	Ni	Pb	Zn	Hg	As
C_n^i ($mg \cdot kg^{-1}$)	0.76	62.5	21.4	28.8	21.4	69.4	0.063	11.1
T_r^i	30	2	5	5	5	1	40	10

2.6 Methods of Sources Identification

This research analyzed the numerical relation between XRF and corresponding laboratory analysis results, especially in high-risk areas. The XRF results of 80 sampling points were corrected through regression fit lines. Then, principal component analysis (PCA) and K-Means Clustering (KMC) analysis were adopted to conduct pollution sources identification in IBM SPSS Statistics 26. According to previous research[50-51], the principal components (PCs) were selected when initial contribution rate was larger than 80.00%. In addition, the pairwise correlation between different heavy metals were analyzed and the similarity of pollution sources was inferred and identified.

3 Results

3.1 Soil Analyses with XRF

In total, there were 80 quick detection points analyzed by XRF. Cd and Hg were not detected. It indicated that the concentrations of Cd and Hg in soil did not reach LOD of the XRF instrument. This research analyzed and visualized the concentrations of heavy metals by Inverse Distance Weighted (IDW) in Arcgis 10.3 (ESRI). The results intuitively reflected their spatial distribution and the variation trends (Fig.2). The mean concentrations of Cr, Cu, Ni, Pb, Zn and As were 53.37 mg · kg^{-1}, 51.27 mg · kg^{-1}, 53.05 mg · kg^{-1}, 69.43 mg · kg^{-1}, 165.91 mg · kg^{-1} and 10.94 mg · kg^{-1}, respectively. Pb, Cu, Zn and Ni were 3.24, 2.40, 2.39 and 1.84 times as large, respectively, as the background values in Shaanxi province. The concentrations of Pb at point 2-2 (*X-Y*) were 512.72 mg · kg^{-1}, exceeding 400 mg · kg^{-1}(risk screening values for soil contamination of development land in GB 36600—2018). There were 13 points whose concentrations of Pb were larger than 120 mg · kg^{-1}(risk screening values for soil contamination of agricultural land in GB 15618—2018), accounting for 16.25%. It was found that large value points were mostly distributed in the south of the research area, close to the traffic road. The maximum values of Zn were 2114.93 mg · kg^{-1}. Moreover, its coefficients of variation (C_V) and standard deviation (SD) were the largest, reaching 173.71% and 288.19, respectively. It indicated the spatial distribution of Zn varied greatly and it was probably affected by land use type, industry and human activity. The concentration of Zn in eight detection points was more than 250 mg · kg^{-1}(evaluation criteria), accounting for 10%.

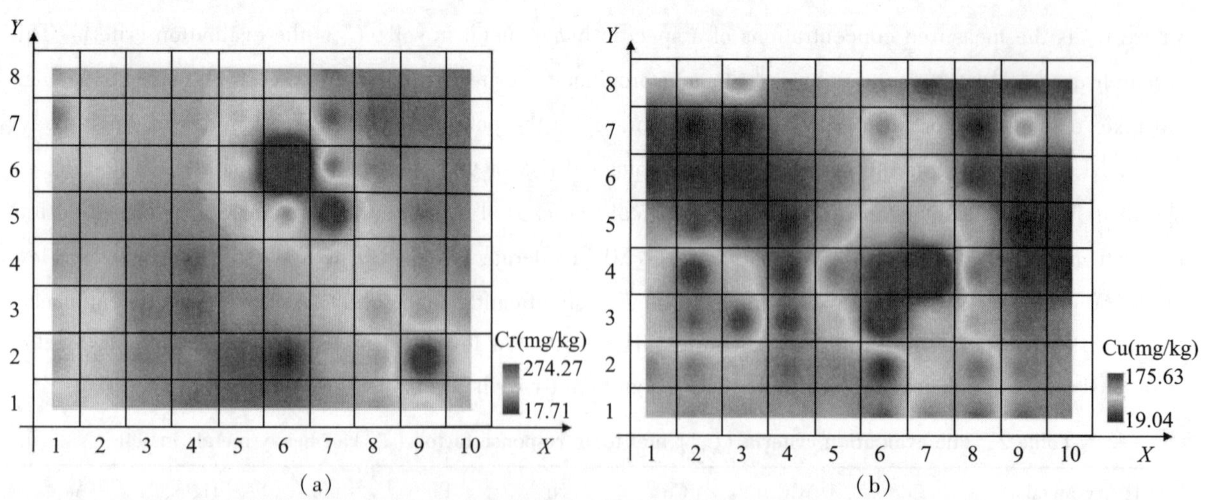

Fig.2 The concentrations distribution of Cr (a), Cu (b), Ni (c), Pb (d), Zn (e), As (f) by IDW. The concentrations of Cd and Hg were not detected (<LOD)

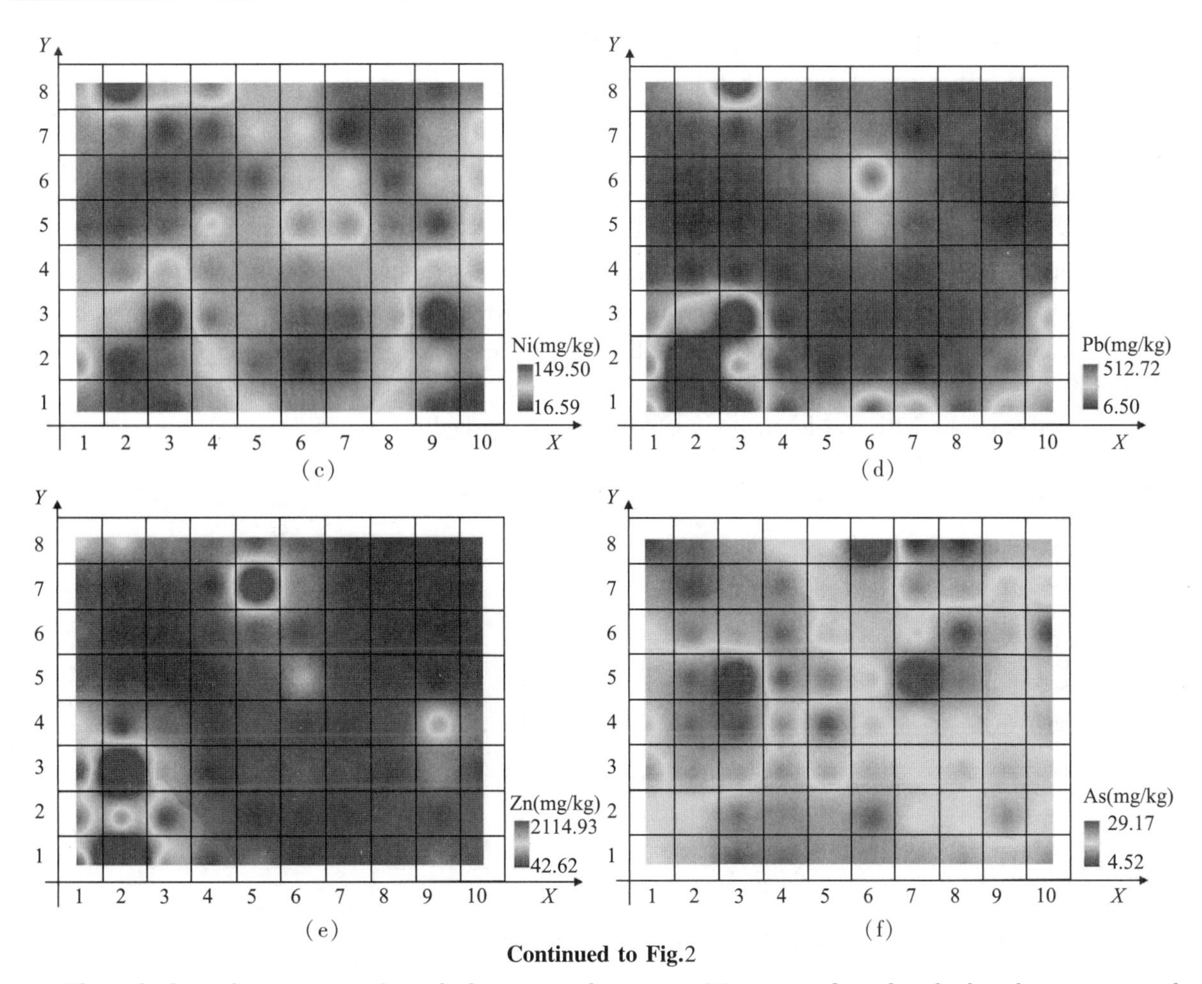

Continued to Fig.2

Through the risk screening of quick detection, there were 25 points selected as high-risk contaminated. Based on the identification of RS image and on-site survey, the land use types in this research mainly included residential land, industrial land and open space. Additionally, residential land and open space were divided into three clusters, respectively (Fig.3). A total of 84% of the high-risk contaminated points were located in industrial areas.

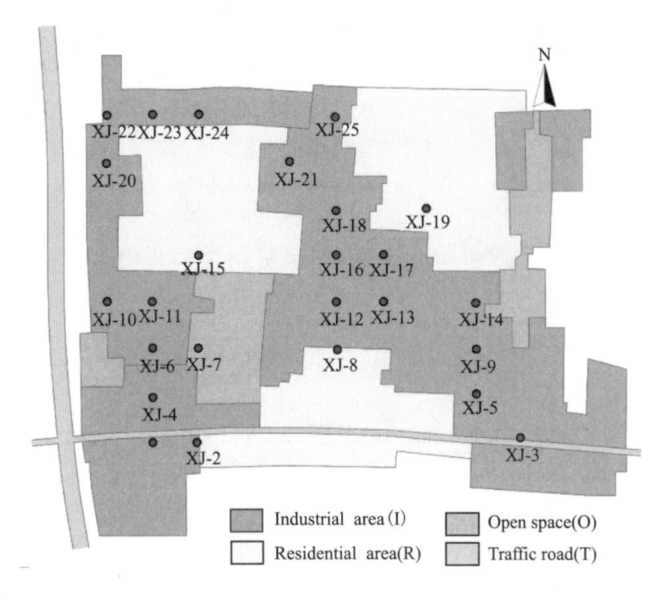

Fig.3 Location of sampling points in different functional areas

3.2 Confirmation Laboratory Analyses

In total, 25 high-risk contaminated points were sampled. The mean absolute percent differences between XRF and laboratory analysis results of Cr, Cu, Ni, Pb, Zn and As were 7.94%, 8.98%, 8.41%, 10.28%, 11.17% and 19.49%, respectively. Generally, the correlation coefficients (Pearson's r) between these two arrays were all greater than 93.26%. Results from ICP-MS and Atomic Fluorescence Spectrometer corroborated the findings achieved from XRF quick detection. As shown in Fig.4, satisfying correlations ($R^2>0.9$) were observed between XRF and laboratory analysis concentrations of Cr, Cu, Ni, Pb, and Zn. Among them, the correlations of Zn and Pb were most significant. In addition, XRF concentrations were slightly higher than the results from laboratory analysis in general. Especially for As, the average concentration of laboratory analysis was 10.99% less than that for XRF. Through linear regression fit, the mathematical relations between the concentrations measured by XRF and laboratory analysis were established.

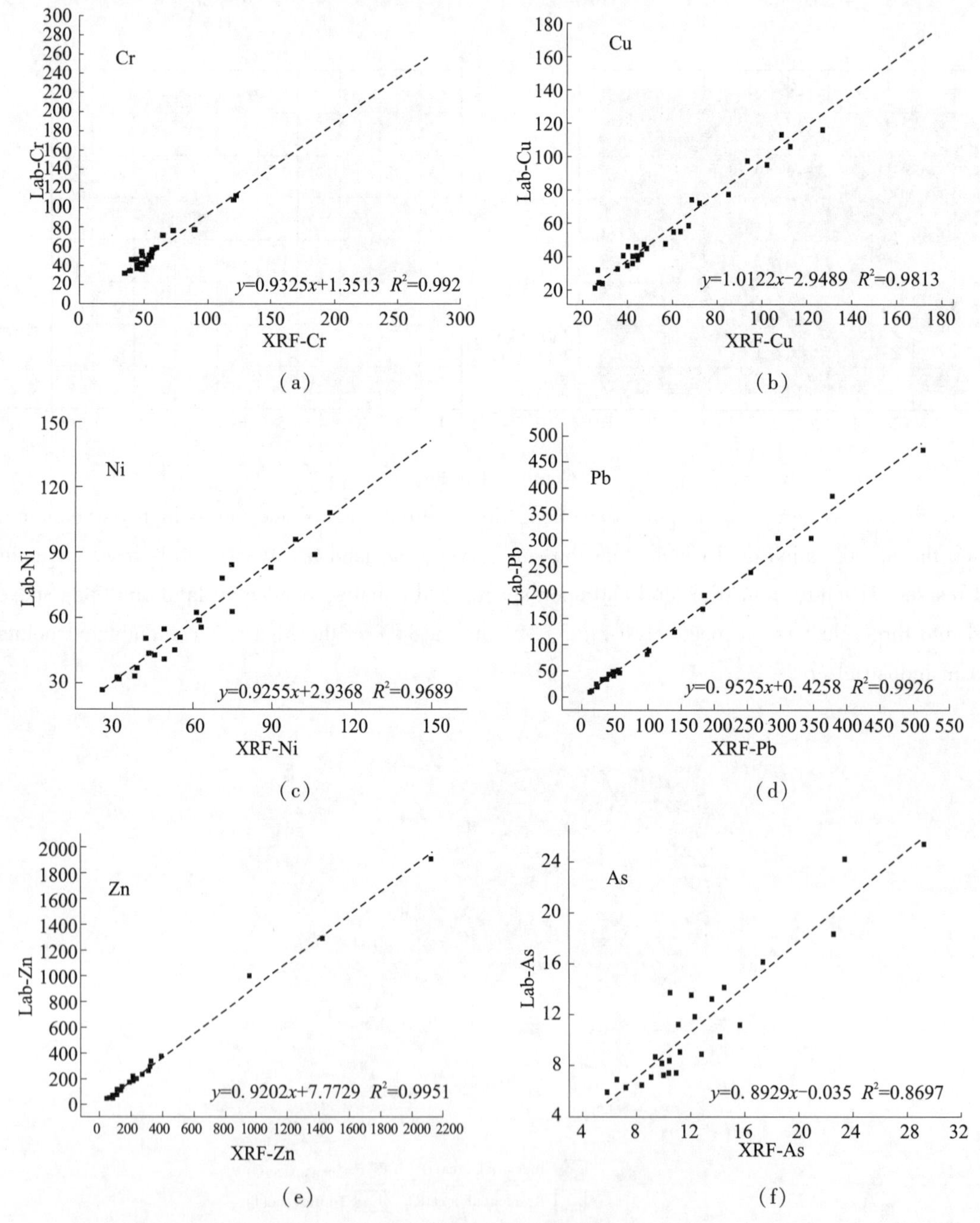

Fig.4 Regression fit lines between XRF(x) and laboratory analysis (y) of heavy metal concentrations ($mg \cdot kg^{-1}$) of Cr (a), Cu (b), Ni (c), Pb (d), Zn (e), As (f) in soil samples

The statistics of heavy metal concentrations at 25 sampling points are shown in Table 3. The mean concentrations in soil of Cd, Cr, Cu, Ni, Pb, Zn, Hg and As were 0.38 mg · kg^{-1}, 63.90 mg · kg^{-1}, 63.81 mg · kg^{-1}, 67.25 mg · kg^{-1}, 116.95 mg · kg^{-1}, 309.98 mg · kg^{-1}, 0.19 mg · kg^{-1} and 11.30 mg · kg^{-1}, respectively. Except for Cd, the mean concentrations of other seven heavy metals exceeded the background values of Shaanxi province. In particular, the mean concentrations of Pb and Zn were 5.47 and 4.47 times larger, respectively, than background values. It indicated that specific human activities in urban village resulted in the significant increase in concentrations of Pb and Zn. The C_V of Pb and Zn were larger than 100%, reaching 109.91% and 142.21%. It reflected the concentrations of Pb and Zn in different sampling points were divergent, and the overall data dispersion was high. By comparison, the C_V of As was the lowest, which was 46.30%. It showed the numerical variation of As at different points was relatively small.

As shown in Fig.5, it was found that all the heavy metal elements analyzed in this research had obvious outlier, except Ni. The background value of Cd was remarkably greater than the upper limit of the box plot. It showed that the overall concentration values of Cd were low, and most values were significantly lower than the background value. The box' s height of Cr was low. The mean concentrations of Cr, approaching to background value, were slightly higher than the upper interquartile of the box plot. It indicated that the numerical distribution was very concentrated, except the outlier, and these values were relatively low. Additionally, the skewness of Cr was 3.70, which was the largest. It indicated the right-skewness distribution of Cr was significant. Meanwhile, only 24% of all sampling points exceeded background value regarding Cr. This situation reflected the dispersion of Cr, especially in relatively high-value zone, was obvious. Similarly, the skewness of Zn was 2.73 (Skew > 0). However, its lower interquartile was greater than background value. It not only showed most of the Zn values exceeded background value, but also the data dispersion was significant. The increased concentrations of specific heavy metal elements in soil were probably affected by regional human activities.

Table 3 Descriptive statistics of heavy metal concentrations at 25 sampling points

Heavy Metals	Cd	Cr	Cu	Ni	Pb	Zn	Hg	As
Mean(mg · kg^{-1})	0.38	63.90	63.81	67.25	116.95	309.98	0.19	11.30
Median(mg · kg^{-1})	0.27	49.89	48.07	55.93	49.72	173.84	0.17	9.17
SD	0.29	45.20	37.53	33.43	128.54	442.37	0.13	5.23
Min(mg · kg^{-1})	0.11	35.26	21.64	27.38	10.37	51.89	0.07	6.03
Max(mg · kg^{-1})	1.22	259.32	182.81	143.90	475.15	1908.13	0.59	25.44
C_V(%)	75.45	70.74	58.82	49.71	109.91	142.71	67.01	46.30

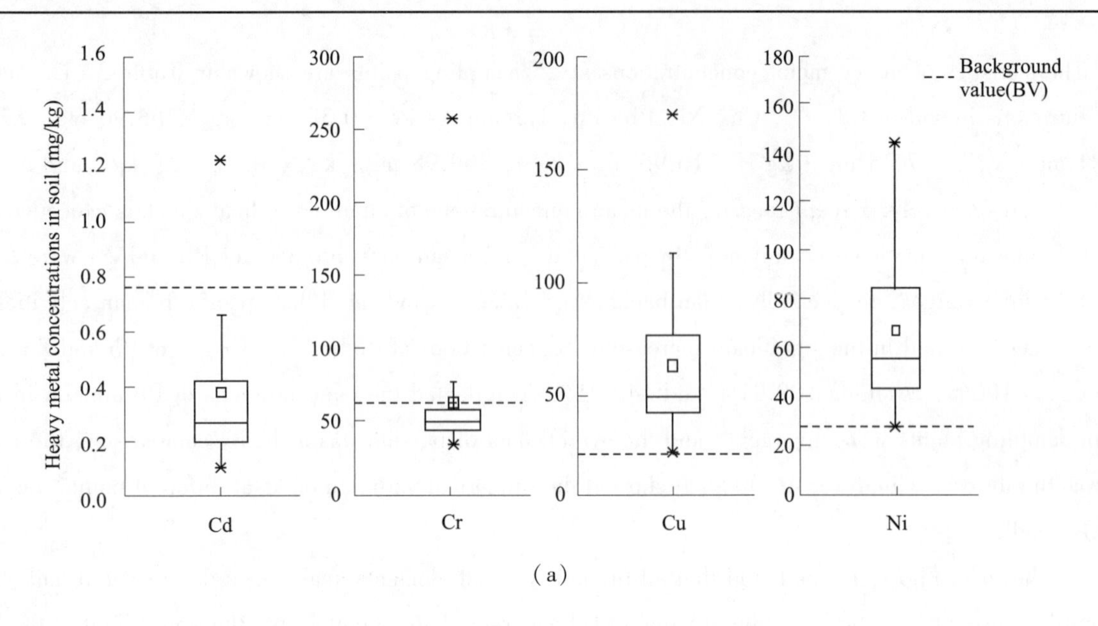

(a)

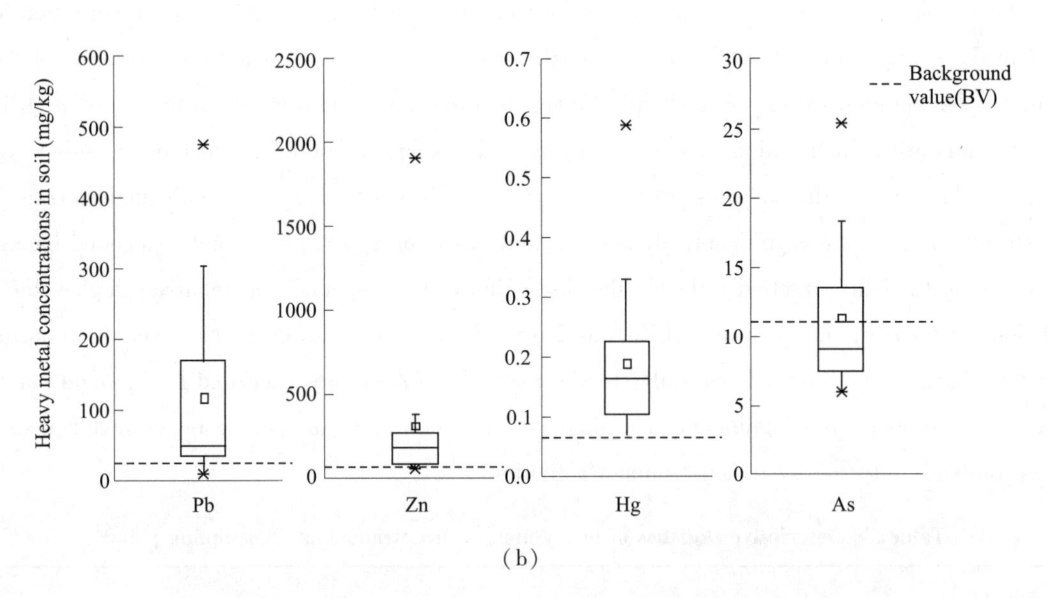

(b)

Fig.5 Box plots of Cd, Cr, Cu, Ni and the background values (a); Box plots of Pb, Zn, Hg, As and the background values (b)

From the calculation results of SPI, the proportions of Cd, Zn and Cr, whose P_i was larger than 1, were 44%, 28% and 24%, respectively. The results of CPI are shown in Fig.6. According to the evaluation criteria in this research, the following results are noted: a total of 14 samples were clean (Level I), accounting for 56% of the total; four samples were low polluted (Level II), accounting for 16% of the total; one sample was moderately polluted (Level III), accounting for 4% of the total; six samples were significantly highly polluted (Level V), accounting for 24% of the total. It manifested that the difference in CPI between points was huge. The maximum was 15.45 at point 1, and the minimum was 0.21 at point 19. All the significantly highly polluted points were located in industrial areas, which showed obvious agglomeration characteristics.

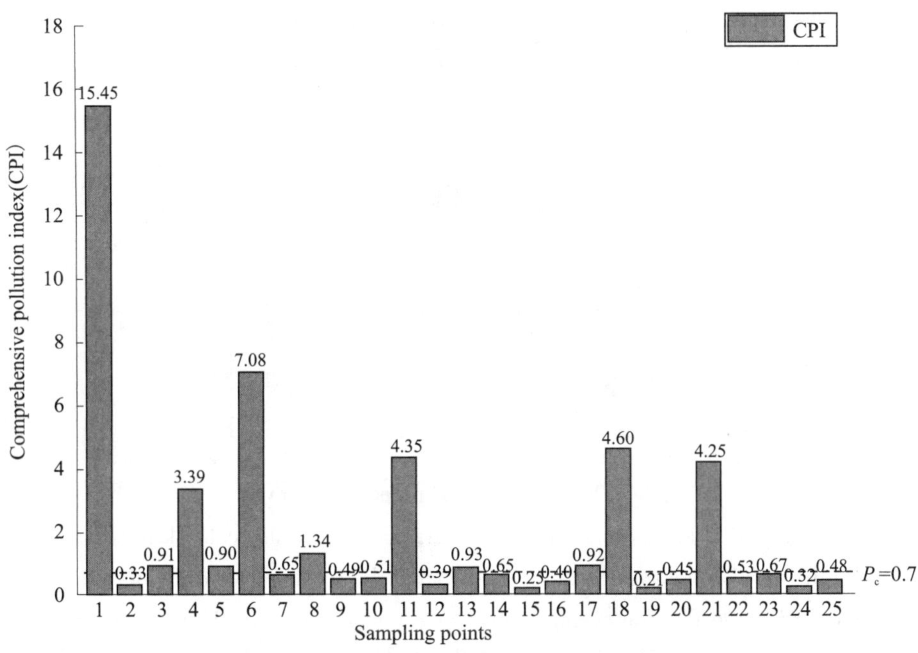

Fig.6 The results of CPI at 25 sampling points

The statistical results of I_{geo} are shown in Fig.7. The mean I_{geo} values were arranged in the order of Pb(1.09) > Cu(0.78) > Zn(0.77) > Hg(0.76) > Ni(0.48) > As(−0.68) > Cr(−0.73) > Cd(−1.89). A large majority of I_{geo}(Cd), I_{geo}(Cr) and I_{geo}(As) were not polluted, accounting for 96%, 88% and 88%, respectively (Fig.8). For I_{geo}(Cd) and I_{geo}(As) in particular, the values were all lower than 1. These heavy metals concentrations were relatively low, which were close to natural background values. However, Cu, Pb, Zn and Hg exhibited signs of heavy pollution. In total, 28% of all sampling points were above moderate pollution regarding I_{geo}(Pb). I_{geo}(Pb) at point 4 reached 3.89 among them. The maximum I_{geo}(Zn) existed at point 1, which was 4.20 (heavily polluted to extremely polluted). Furthermore, the I_{geo} values of Pb and Zn fluctuated clearly. It indicated that the concentrations of Pb and Zn were high in general, and several samples were seriously contaminated.

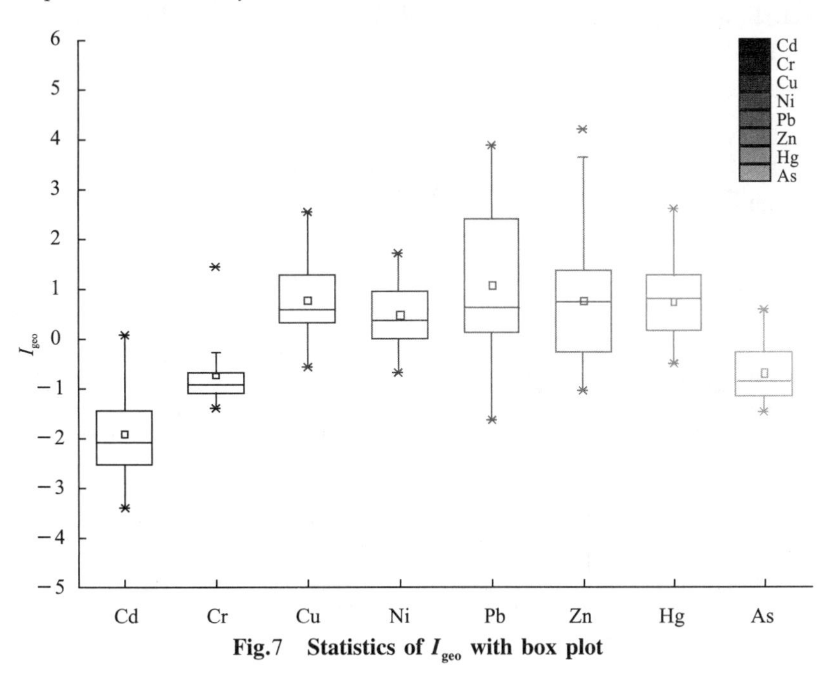

Fig.7 Statistics of I_{geo} with box plot

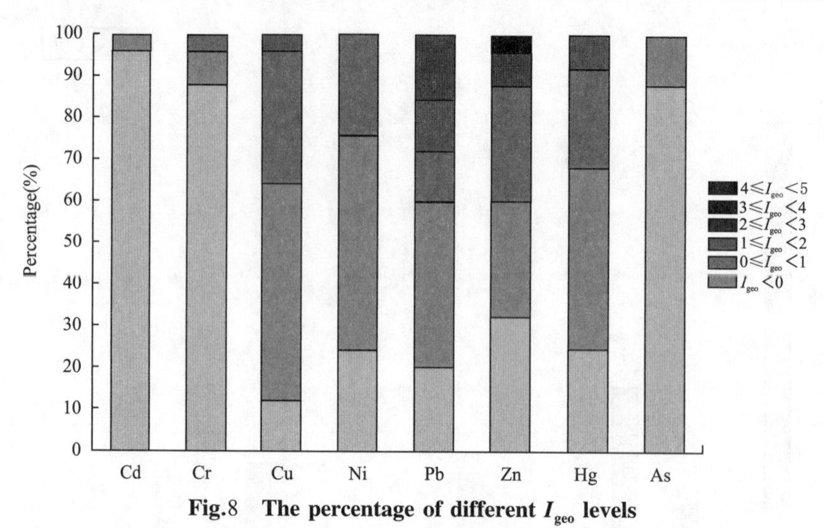

Fig.8 The percentage of different I_{geo} levels

RI values were calculated at all 25 sampling points. The potential ecological risk index of single heavy metal was calculated. From the view of biological toxicity assessment, the results showed the overall potential ecological risks were low except for Hg and Pb. The average values of E_r(Hg) were 120.30, which was wholly at considerable level. The maximum values of E_r(Hg) were 373.97 at point 1, reaching significantly high level. Although the average values of E_r(Pb) were 27.33, less than 40 (low level), 24% of total samples exceeded 40 with its maximum E_r(Pb) 111.02 at point 4. It showed that there were considerable ecological risks in the specific region. The results of SPI and I_{geo} reflected the concentrations of Zn were high. However, since T_r of Zn was 1, the E_r(Zn) was only 4.46. It had no potential risks to the ecological environment.

The results of RI were shown in Fig.9. Generally, nine samples were at low risk level, accounting for 36%. In total, 14 samples were at moderate risk level, accounting for 56%. Point 1 and point 9 were at high risk level, which were 557.05 and 366.76, respectively. The large values of E_r(Hg) was the main factor causing high ecological risk. From the perspective of spatial distribution and land use, the ecological risk of industrial land was dramatically higher than that of residential land. The RI at point 8, 15 and 19 were located in residential areas, which were 146.83, 83.59 and 102.01, respectively. They were all less than 150 and at low risk level. In addition, point 7 was sampled at open space. Its RI value was 217.04, which was at moderate risk level.

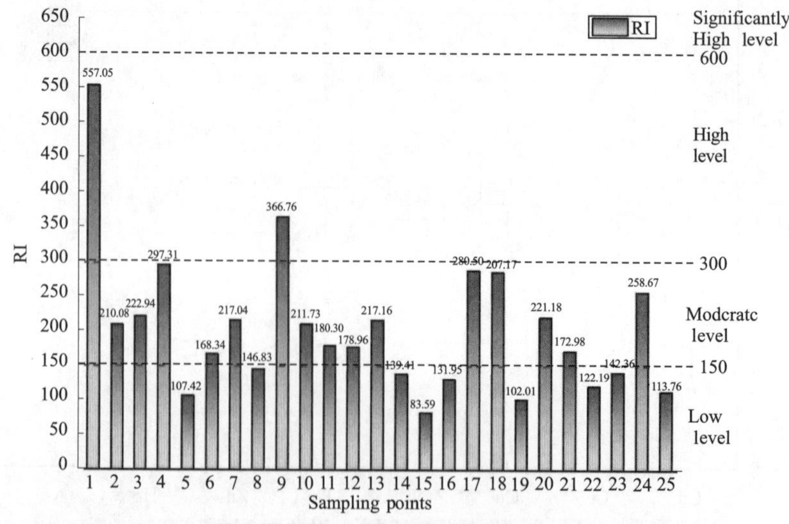

Fig.9 RI and comprehensive potential ecological risk level of each point

3.3 Source Identification

According to Table 4, Pb-Ni and Pb-Zn showed highly significant positive correlation at $p < 0.01$ level, the correlation coefficients reaching 0.557 and 0.417, respectively. Ni-Zn showed significant positive correlation at $p < 0.05$, whose correlation coefficient was 0.254. Therefore, there were notable pairwise correlation among Pb, Zn and Ni, indicating these probably resulted from similar pollution sources. Based on KMC results (Table 5), 80 sampling points were classified into four clusters. Cluster 1 included 73 samples, which was the largest and accounted for 91.25% in total. It showed that the heavy metal concentrations in most research areas were in a stable and normal condition. In general, these samples were all lower than the criteria of risk control. Clusters 2 and 3 indicated the variation of Pb, Zn and Ni were notably similar and accordant. In addition, it was consistent with the results of Pearson correlation analysis.

Table 4 Pearson correlation coefficients among heavy metal concentrations in soil

Heavy metals	Cr	Cu	Ni	Pb	Zn	As
Cr	1					
Cu	0.097	1				
Ni	0.102	0.148	1			
Pb	0.160	0.132	0.557 **	1		
Zn	0.047	0.104	0.254 *	0.417 **	1	
As	0.196	0.171	−0.008	−0.047	0.033	1

** Correlation is highly significant at $p < 0.01$ (two-tailed); * Correlation is significant at $p < 0.05$ (two-tailed); number of samples ($n = 80$).

Table 5 KMC of heavy metal concentrations at 80 sampling points

Heavy metal	Cluster			
	1	2	3	4
Cr	51.19	57.18	45.23	57.43
Cu	47.49	43.67	67.00	68.68
Ni	49.25	85.45	92.04	56.97
Pb	47.10	359.23	335.82	91.79
Zn	108.09	1953.93	200.18	1095.01
As	9.73	8.85	8.96	11.65
Cluster quantity	73	1	4	2

Based on Pearson correlation analysis, this research conducted PCA and results were shown in Tables 6 and 7. The Kaiser-Meyer-Olkin (KMO) measure of sampling adequacy was 0.598 (KMO > 0.5), and the significance of Bartlett's test of sphericity was 0.000 (Sig. < 0.001). This research extracted four PCs, whose initial eigenvalues were larger than 0.8. These four PCs cumulative contribution rates were over 80.00%, reaching 81.67%. The contribution rates of PC1 were 32.29%. PC1 showed high positive loadings in Ni, Pb and Zn, reaching 0.763, 0.838 and 0.640, respectively. Combined with correlation analysis, it was confirmed that Ni, Pb and Zn originated from similar pollution sources. These heavy metal materials are widely used in petroleum, automobile, printing and food processing industries. Therefore, this result was in line with the industrial characteristics of the research areas. The contribution rates of PC2 were 20.80%, and it showed high positive loadings in Cr and As. The research areas were located in the suburbs of the city, and agricultural and industrial activities were particularly active. Pesticides, industrial sewage and dust contain these metallic elements, leading to the accumulation of heavy metal concentrations in soil.

Table 6 PCA and total variance explained of heavy metal concentrations

Component	Initial eigenvalues			Extracted sums of squared loadings		
	Total	% of Variance	Cumulative %	Total	% of Variance	Cumulative %
1	1.938	32.293	32.293	1.938	32.293	32.293
2	1.248	20.801	53.094	1.248	20.801	53.094
3	0.908	15.129	68.223	0.908	15.129	68.223
4	0.807	13.443	81.667	0.807	13.443	81.667
5	0.701	11.688	93.355			
6	0.399	6.645	100.000			

Table 7 Initial factor loading matrix of heavy metal concentrations (PCs loadings > 0.5 are shown in bold)

Heavy metals	Component			
	1	2	3	4
Cr	0.318	0.554	−0.655	−0.224
Cu	0.362	0.464	0.673	−0.368
Ni	0.763	−0.184	−0.032	−0.253
Pb	0.838	−0.223	−0.103	−0.046
Zn	0.640	−0.159	0.110	0.611
As	0.107	0.786	0.050	0.426

4 Discussion

4.1 Method of Investigation

Previous studies have pointed out the advantages of the XRF technique that is non-destructive, more efficient and low cost, compared with conventional sampling and laboratory-based measurement[52-53]. This research adopted methods combined XRF quick detection with confirmation laboratory analysis. It greatly reduced the amount of sampling points and improved efficiency, while ensuring the accuracy of detection results. It proved the advantages and worth of the XRF technique in soil environment investigation. Additionally, this research proposed to carry out laboratory-based analysis for high-value zones. The regression fit lines were used for numerical correction. This method can increase the accuracy and reliability of results. In this study, the results from XRF and laboratory analysis coincided. The mean absolute percent differences in between were less than 12%, except As, reaching 19.49%. Radu et al. researched the heavy metal concentrations at silver mines and showed an excellent correlation between portable XRF instruments and laboratory-based methods[54]. Moreover, the study found that concentrations measured by laboratory analysis were commonly lower than XRF's results. Meanwhile, this was consistent with the results of Caporale et al.[55], who pointed out the main cause is the incomplete chemical dissolution in the laboratory. Generally, the XRF technique can be used as a replacement for conventional laboratory-based method when the calibration is adequate[56].

4.2 Multi-Sources of Pollution

In this research, the measured values of Pb, Cu, Zn and Ni were significantly higher than the regional background values. The high-value zones of Pb, Zn and Ni were distributed around the south traffic road. Previous study showed these existed large ecological and health risk in suburb areas, and traffic emission is the main pollution source[57]. Through on-site survey, there were many garages, metal recycling and processing factories on both sides of the road. The traffic flow was large, particularly the trucks (Fig.10). Walraven et al. found that Pb pollution in roadside soils was mainly derived from gasoline, which was generally polluted at 0~15 cm depth[58]. There was a building materials production and logistics base in the southwest of research area. The maximum values of I_{geo}(Pb) and I_{geo}(Zn) were found in this area. This was consistent with the results of Jeong et al., which found that non-exhaust traffic emissions, e.g., particles from brake pads and tires, contain heavy metal elements. They were demonstrated to be important pollution sources causing the increase in concentrations of Pb and Zn[59]. Zn is one of the essential elements for human body, animals, and plant growth. However, excessive concentrations of Zn will hinder the growth of animals and plants, reducing biological diversity[60-61]. The result from this research indicated the concentration of Zn were relatively high, and yet E_r(Zn) was low. This was principally because the biological toxicity of Zn is low. T_r^i of Zn is only 1, which is the lowest numerically among above eight heavy metals.

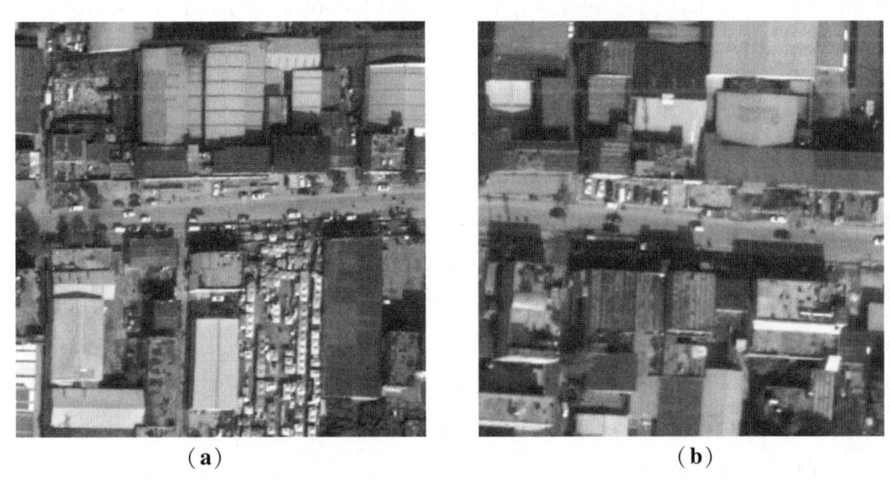

(a) (b)

Fig.10 Auto related industries around the south traffic road image by Google Earth: freight transport and parking (a); automobile maintenance (b)

The concentrations of Cd, Cr, Cu, Hg and As in soil had been proved to be related to industrial activities, which was consistent with the results of Wang et al.[62]. Hu et al. showed that industrial activities have significant correlation with heavy metal pollution[63]. In this research, 84% of the high-risk points were located in industrial areas. The sampling points located in residential areas were all at low risk level. The average SPI and I_{geo} of Hg were only 0.38 and 0.76, respectively; however, the heavy metal element Hg has strong bioaccumulation. It can cause great harm to the whole ecosystem and has great chemical toxicity. Therefore, E_r(Hg) was generally the highest among eight heavy metals. A mixed use of farmland and construction land is a significant feature of urban village located in suburban areas (Fig.11). Industrial land will increase heavy metal pollution risk of surrounding agricultural land[64]. The chemical raw materials, chemical reaction equipment, industrial dust and waste gas are potential pollution resources, resulting in the increase in heavy metals concentrations in soil. In general, industrial sources and traffic sources are the main factors leading to the increase in heavy metal concentrations in soil. However, as the opinion of Huang et al., agricultural land pollution and agricultural pollutions source should be brought to the attention at peri-urban area[65].

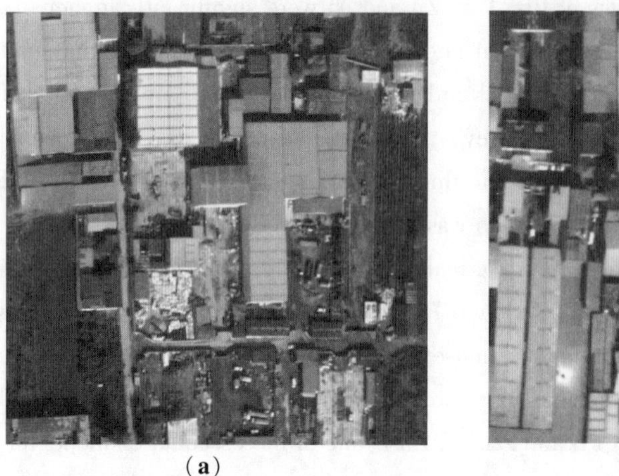

(a) (b)

Fig.11 Mixed-use of farmland and construction land image by Google Earth (a,b)

4.3 Sustainability of Urban Village

In China, small-scale enterprises such as equipment manufacturing, materials processing and printing are common in urban village. The production equipment and technical process are relatively primitive in urban village[66]. It probably causes direct harm to the ecological environment. Urban village basically involved industrial land, residential land and open space in this research. Different land types were mixed and intertwined. Through on-site survey, it was found that the workers' dormitory was close to the industrial plant, and the restaurant was adjacent to the garage. This chaotic layout posed a great threat to the ecological environment and human health. There were many uses for open space, including stacking sundries, dumping domestic waste and planting crops, which represented potential pollution risk. In China, urban village is the inevitable product of urbanization[67]. However, with the improvement of urban environment and high-quality industrial transformation, the living environment of urban villages have continuously improved. According to the list published by the Xi'an government from 2017 to 2019, many enterprises in urban villages had been urged to rectify or forced to close regarding environmental pollution issue. Therefore, strict zoning system should be implemented in urban village, which can reduce pollution risk to the environment.

5 Conclusions

In this research, the method combining in situ XRF detection and laboratory-based analysis was proposed. It turned out to be reliable and cost efficient. The measured concentrations of Pb and Zn were significantly higher than background values. Industrial land was at high risk for heavy metal pollution. There existed obvious pairwise correlation among Pb, Zn and Ni, and these heavy metal accumulations originated from similar automobile traffic and industries activities. Industrial sources and traffic sources were regarded as main factors causing heavy metal accumulation in soil. In addition, agricultural sources, e.g., chemical fertilizers and pesticides, increased the risk of pollution in urban village. In recent years, soil safety has been paid close attention. This research provided a fast, efficient and accurate method for urban large-scale investigation of the soil environment.

In China, urban village is an important part of the city. This research explored the relation among land, industry and heavy metal pollution. The result was of great significance to environmental governance as well as urban planning.

Author Contributions: Conceptualization, S.L.; methodology, S.L.; software, J.L.; validation, S.L., B.P. and J.L.; formal analysis, S.L.; investigation, J.L.; resources, B.P.; data curation, B.P.; writing—original draft preparation, S.L.; writing—review and editing, B.P.; visualization, J.L.; supervision, B.P.; project administration, S.L.; funding acquisition, S.L. All authors have read and agreed to the published version of the manuscript.

Funding: This research was funded by Shaanxi Provincial Department of Science and Technology through Innovation Capability Support Program (Grant No. 2021KRM079).

Institutional Review Board Statement: Not applicable.

Informed Consent Statement: Not applicable.

Data Availability Statement: The data presented in this study are available on request from the corresponding author.

Conflicts of Interest: The authors declare no conflict of interest.

References

[1]Lawler, J.J.; Lewis, D.J.; Nelson, E.; Plantinga, A.J.; Polasky, S.; Withey, J.C.; Helmers, D.P.; Martinuzzi, S.; Pennington, D.; Radeloff, V.C. Projected land-use change impacts on ecosystem services in the United States. *Proc. Natl. Acad. Sci. USA* 2014, 111, 7492-7497.

[2]Xie, H.L.; He, Y.F.; Choi, Y.; Chen, Q.R.; Cheng, H. Warning of negative effects of land-use changes on ecological security based on GIS. *Sci. Total Environ.* 2020, 704, 135427.

[3]National Bureau of Statistics of China. *China Statistical Yearbook*; China Statistics Press: Beijing, China, 2020; pp. 32-49.

[4]Ministry of Housing and Urban-Rural Development of the People's Republic of China (MOHURD). China Construction Statistical Yearbook. Beijing, 2020.

[5]Ning, J.; Liu, J.Y.; Kuang, W.H.; Xu, X.; Zhang, S.; Yan, C. Spatial-temporal patterns and characteristics of land-use change in China during 2010-2015. *J. Geogr. Sci.* 2018, 28, 547-562.

[6]Fang, L.; Tian, C.H. Construction land quotas as a tool for managing urban expansion. *Landsc. Urban Plan.* 2020, 195, 103727.

[7]Li, Q.; Zeng, F.; Liu, S.; Yang, M.; Xu, F. The effects of China's sustainable development policy for resource-based cities on local industrial transformation. *Resour. Policy* 2021, 71, 101940.

[8]Hu, Q.; Huang, H.P.; Kung, C.C. Ecological impact assessment of land use in eco-industrial park based on life cycle assessment: A case study of Nanchang High-tech development zone in China. *J. Clean. Prod.* 2021, 300, 126816.

[9]Zhang, J.F.; Zhang, D.X.; Huang, L.Y.; Wen, H.Z. Zhao, G.C.; Zhan, D.S. Spatial distribution and influential factors of industrial land productivity in China's rapid urbanization. *J. Clean. Prod.* 2019, 234, 1287-1295.

[10]Zhang, Z.F.; Liu, J.; Gu, X.K. Reduction of industrial land beyond Urban Development Boundary in Shanghai: Differences in policy responses and impact on towns and villages. *Land Use Policy* 2019, 82, 620-630.

[11]Bertaud, A.; Renaud, B. Socialist Cities without Land Markets. *J. Urban Econ.* 1997, 41, 137-151.

[12]Gao, W.; Ma, K.X.; Liu, H.M. Policy evolution of the economical and intensive utilization of industrial land in China since 1978. *China Land Sci.* 2013, 27, 37-43.

[13]Zhao, X.; Shang, Y.P.; Song, M.L. Industrial structure distortion and urban ecological efficiency from the perspective of green entrepreneurial ecosystems. *Soc.-Econ. Plan. Sci.* 2020, 72, 100757.

[14]Yin, G.Y.; Lin, Z.L.; Jiang, X.L.; Qiu, M.L.; Sun, J. How do the industrial land use intensity and dominant industries guide the urban land use? Evidences from 19 industrial land categories in ten cities of China. *Sustain. Cities Soc.* 2020, 53, 101978.

[15]Peng, C.; He, Y.L.; Guo, C.H.; Xiao, X.Y.; Zhang, Y. Characteristics and risk assessment of heavy metals in urban soils of major cities in China. *Environ. Sci.* 2021, 43, 1-10.

[16]Yuan, X.H.; Xue, N.D.; Han, Z.G. A meta-analysis of heavy metals pollution in farmland and urban soils in China over the past 20 years. *J. Environ. Sci.* 2021, 101, 217-226.

[17]Sun, L.; Guo, D.K.; Liu, K.; Meng, H.; Zheng, Y.J.; Yuan, F.Q.; Zhu, G.H. Levels, sources, and spatial

distribution of heavy metals in soils from a typical coal industrial city of Tangshan, China. *CATENA* 2019, 175, 101-109.

[18] Ministry of Ecology and Environmental of P. R. China (MEE); Ministry of Natural Resources of P. R. China (MNR). The Communique for Soil Contamination Status Survey in China. Beijing, 2014.

[19] Chen, T.B.; Zheng, Y.M.; Lei, M.; Huang, Z.C.; Wu, H.T. Chen, H.; Fan, K.K.; Yu, K.; Wu, X.; Tian, Q. Z. Assessment of heavy metal pollution in surface soils of urban parks in Beijing, China. *Chemosphere* 2005, 60, 542-551.

[20] Guo, G.H.; Wu, F.C.; Xie, F.Z.; Zhang, R.Q. Spatial distribution and pollution assessment of heavy metals in urban soils from southwest China. *J. Environ. Sci.* 2012, 24, 410-418.

[21] Taati, A.; Salehi, M.H.; Mohammadi, J.; Mohajer, R.; Díezc, S. Pollution assessment and spatial distribution of trace elements in soils of Arak industrial area, Iran: Implications for human health. *Environ. Res.* 2020, 187, 109577.

[22] Nikolaeva, O.; Tikhonov, V.; Vecherskii, M.; Kostina, N.; Fedoseeva, E.; Astaikina, A. Ecotoxicological effects of traffic-related pollutants in roadside soils of Moscow. *Ecotoxicol. Environ. Saf.* 2019, 172, 538-546.

[23] Briffa, J.; Sinagra, E.; Blundell, R. Heavy metal pollution in the environment and their toxicological effects on humans. *Heliyon* 2020, 6, e04691.

[24] Sharma, N.; Sodhi, K.K.; Kumar, M.; Singh, D.K. Heavy metal pollution: Insights into chromium eco-toxicity and recent advancement in its remediation. *Environ. Nanotechnol. Monit. Manag.* 2021, 15, 100388.

[25] Bo, L.J.; Wang, D.J.; Li, T.L.; Li, Y.; Zhang, G.; Wang, C.; Zhang, S.Q. Accumulation and risk assessment of heavy metals in water, sediments, and aquatic organisms in rural rivers in the Taihu Lake region, China. *Environ. Sci. Pollut. Res.* 2015, 22, 6721-6731.

[26] Tepanosyan, G.; Maghakyan, N.; Sahakyan, L.; Saghatelyan, A. Heavy metals pollution levels and children health risk assessment of Yerevan kindergartens soils. *Ecotoxicol. Environ. Saf.* 2017, 142, 257-265.

[27] Zhou, L.; Che, L.; Sun, D.Q. The coupling coordination development between urbanization and economic growth and its influencing factors in China. *Econ. Geogr.* 2019, 39, 97-107.

[28] Pan, W.; Du, J. Towards sustainable urban transition: A critical review of strategies and policies of urban village renewal in Shenzhen, China. *Land Use Policy* 2021, 111, 105744.

[29] Zhang, W.Z.; Yu, J.H.; Zhang, D.S.; Ma, R.F. *A Study of Livable Cities in China*; Science Press: Beijing, China, 2016; pp. 3-11.

[30] Zhang, H.L.; Walker, T.R.; Davis, E.; Ma, G.F. Ecological risk assessment of metals in small craft harbour sediments in Nova Scotia, Canada. *Mar. Pollut. Bull.* 2019, 146, 466-475.

[31] He, Y.S. Pollution characteristics and ecological risk assessment of heavy metals in Haikou urban soils. *Chin. J. Ecol.* 2014, 33, 421-428.

[32] Varol, M.; Sünbül, M.R.; Aytop, H.; Yılmaz, C.H. Environmental, ecological and health risks of trace elements, and their sources in soils of Harran Plain, Turkey. *Chemosphere* 2020, 245, 125592.

[33] Wu, Y.Z.; Sun, X.F.; Sun, L.H.; Choguill, C.L. Optimizing the governance model of urban villages based on integration of inclusiveness and urban service boundary (USB): A Chinese case study. *Cities* 2020, 96, 102427.

[34] Tan, Y.Z.; He, J.; Han, H.Y.; Zhang, W.W. Evaluating residents' satisfaction with market-oriented urban village transformation: A case study of Yangji Village in Guangzhou, China. *Cities* 2019, 95, 102394.

[35] Li, J.; Sun, S.; Li, J. The dawn of vulnerable groups: The inclusive reconstruction mode and strategies for urban villages in China. *Habitat Int.* 2021, 110, 102347.

[36] Gu, Z.; Zhang, X. Framing social sustainability and justice claims in urban regeneration: A comparative analysis of two cases in Guangzhou. *Land Use Policy* 2021, 102, 105224.

[37] Yu, X.; Wang, M.Y.; Dong, X.; Chen, X.; Lu, J.X. A study on migration tendency of floating population in urban villages: Taking the city of Xi'an as an example. *Mod. Urban Res.* 2021, 8, 10-16.

[38] Ministry of Ecology and Environmental of P. R. China (MEE). Technical Guidelines for Investigation on Soil Contamination of Land for Construction, 2019. Available online.

[39] Ministry of Ecology and Environmental of P. R. China (MEE). The Technical Specification for Soil Environmental Monitoring. 2004.

[40] Brent, R.N.; Wines, H.; Luther, J.; Irving, N.; Collins, J.; Drake, B.L. Validation of handheld X-ray fluores-

cence for in situ measurement of mercury in soils. *J. Environ. Chem. Eng*. 2017, 5, 768-776.

[41]United States Environmental Protection Agency (U.S. EPA). SW-846 Test Method 6200: Field Portable X-Ray Fluorescence Spectrometry for the Determination of Elemental Concentrations in Soil and Sediment. 2007.

[42]United States Environmental Protection Agency (U.S. EPA). SW-846 Test Method 3050B: Acid Digestion of Sediments, Sludges, and Soils. 1996.

[43]Xu, G.L.; Wen, Y.; Cai, S.Y.: Luo, X.F. Review for the effects of urban topsoil on the ecological health. *Geogr. Res*. 2019, 38, 2941-2956.

[44]Xue, Z.B.; Li, L.; Zhang, S.K.; Dong, J. Comparative study between Nemerow Index method and Compound Index method for the risk assessment of soil heavy metal pollution. *Sci. Soil Water Conserv*. 2018, 16, 119-125.

[45]Ministry of Ecology and Environmental of P. R. China (MEE). Soil Environment Quality Risk Control Standard for Soil Contamination of Development Land. 2018.

[46]Ministry of Ecology and Environmental of P. R. China (MEE). Soil Environment Quality Risk Control Standard for Soil Contamination of Agricultural Land. 2018.

[47]Muller, G. Index of geoaccumulation in sediments of the Rhine River. *GeoJournal* 1969, 2, 109-118.

[48]Hakanson, L. An ecological risk index for aquatic pollution control. A sedimentological approach. *Water Res*. 1980, 14, 975-1001.

[49]Ministry of Ecology and Environment of the People's Republic of China; China National Environmental Monitoring Centre. *Background Values of Soil Elements in China*; China Environmental Science Press: Beijing, China, 1990; pp. 87-90.

[50]Feng, Q.W.; Wang, B.; Ma, X.J.; Jiang, Z.H.; Chen, M. Pollution Characteristics and Source Analysis of Heavy Metal in Soils of Typical Lead-Zinc Mining Areas in Northwest Guizhou, China. *Bull. Mineral. Petrol. Geochem*. 2020, 39, 863-870.

[51]Deng, M.H.; Zhu, Y.W.; Shao, K.; Zhang, Q.; Ye, G.H.; Shen, J. Metals source apportionment in farmland soil and the prediction of metal transfer in the soil-rice-human chain. *J. Environ. Manag*. 2020, 260, 110092.

[52]Peralta, E.; Pérez, G.; Ojeda, G.; Alcañiz, J.M.; Valiente, M.; López-Mesas, M.; Súnchez-Martín, M.-J. Heavy metal availability assessment using portable X-ray fluorescence and single extraction procedures on former vineyard polluted soils. *Sci. Total Environ*. 2020, 726, 138670.

[53]Parsons, C.; Margui Grabulosa, E.; Pili, E.; Floor, G.H.; Roman-Ross, G.; Charlet, L. Quantification of trace arsenic in soils by field-portable X-ray fluorescence spectrometry: Considerations for sample preparation and measurement conditions. *J. Hazard. Mater*. 2013, 262, 1213-1222.

[54]Radu, T.; Diamond, D. Comparison of soil pollution concentrations determined using AAS and portable XRF techniques. *J. Hazard. Mater*. 2009, 171, 1168-1171.

[55]Caporale, A.G.; Adamo, P.; Capozzi, F.; Langella, G.; Terribile, F.; Vingiani, S. Monitoring metal pollution in soils using portable-XRF and conventional laboratory-based techniques: Evaluation of the performance and limitations according to metal properties and sources. *Sci. Total Environ*. 2018, 643, 516-526.

[56]Messager, M.L.; Davies, I.P.; Levin, P.S. Development and validation of in-situ and laboratory X-ray fluorescence (XRF) spectroscopy methods for moss biomonitoring of metal pollution. *MethodsX* 2021, 8, 101319.

[57]Heidari, M.; Darijani, T.; Alipour, V. Heavy metal pollution of road dust in a city and its highly polluted suburb; quantitative source apportionment and source-specific ecological and health risk assessment. *Chemosphere* 2021, 273, 129656.

[58]Walraven, N.; van Os, B.J.H.; Klaver, G.T.; Middelburg, J.J.; Davies, G.R. The lead (Pb) isotope signature, behaviour and fate of traffic-related lead pollution in roadside soils in The Netherlands. *Sci. Total Environ*. 2014, 472, 888-900.

[59]Jeong, H.; Ryu, J.S.; Ra, k. Characteristics of potentially toxic elements and multi-isotope signatures (Cu, Zn, Pb) in non-exhaust traffic emission sources. *Environ. Pollut*. 2022, 292, 118339.

[60]Kang, M.J.; Kwon, Y.K.; Yu, S.Y.; Lee, P.K.; Park, H.S.; Song, N. Assessment of Zn pollution sources and apportionment in agricultural soils impacted by a Zn smelter in South Korea. *J. Hazard. Mater*. 2019, 364, 475-487.

[61]Shao, Y.Y.; Yan, T.; Wang, K.; Huang, S.M.; Yuan, W.Z.; Qin, F.G.F. Soil heavy metal lead pollution and its stabilization remediation technology. *Energy Rep*. 2020, 6, 122-127.

[62]Wang, Q.; Hao, D.; Wang, F.; Wang, H.; Huang, X.; Li, F.; Li, C.; Yu, H. Development of a new framework to estimate the environmental risk of heavy metal(loid)s focusing on the spatial heterogeneity of the industrial layout. *Environ. Int.* 2021, 147, 106315.

[63]Xu, X.; Hu, X.; Wang, T.; Sun, M.; Wang, L.; Zhang, L. Non-inverted U-shaped challenges to regional sustainability: The health risk of soil heavy metals in coastal China. *J. Clean. Prod.* 2021, 279, 123746.

[64]Ji, W.; Yang, T.; Ma, S.; Ni, W. Heavy Metal Pollution of Soils in the Site of a Retired Paint and Ink Factory. *Energy Procedia* 2012, 16, 21-26.

[65]Huang, Y.; Chen, Q.; Deng, M.; Japenga, J.; Li, T.; Yang, X.; He, Z. Heavy metal pollution and health risk assessment of agricultural soils in a typical peri-urban area in southeast China. *J. Environ. Manag.* 2018, 207, 159-168.

[66]Li, Q.; Xu, X.D. Research hotspots and trend of city village renovation. *Urban Probl.* 2018, 8, 22-30.

[67]Lin. X.B.; Ma, X.G.; Li, G.C. Formation and governance of informality in urban village under the rapid urbanization process. *Econ. Geogr.* 2014, 34, 162-168.

本文曾发表于2022年《Sustainability》第14卷

Heavy Metal Pollution and Environmental Risks in the Water of Rongna River Caused by Natural AMD Around Tiegelongnan Copper Deposit, Northern Tibet, China

Yuhu Luo[1,2,3,4], Jiaoping Rao[5,6], Qinxian Jia[7]

(1.Institute of Land Engineering & Technology, Shaanxi Provincial Land Engineering Construction Group Co., Ltd., Xi'an, China; 2.Shaanxi Provincial Land Engineering Construction Group Co., Ltd., Xi'an, China; 3.Key Laboratory of Degraded and Unused Land Consolidation Engineering, the Ministry of Natural Resources, Xi'an, China; 4.Shaanxi Provincial Land Consolidation Engineering Technology Research Center, Xi'an, China; 5.China University of Geosciences (Beijing), Beijing, China; 6.Key Laboratory of Metallageny and Mineral Assessment, Ministry of Natural resources, Institute of Mineral Resources, Chinese Academy of Geological Sciences, Beijing, China; 7.Key Laboratory of Saline Lake Resources and Environments, Ministry of Natural Resources, Institute of Mineral Resources, Chinese Academy of Geological Sciences, Beijing, China)

【Abstract】 Acid mine drainage (AMD) is one of the biggest environmental challenges associated with in the mining process. Most of the current research on AMD focuses on developed deposits, whereas there is almost no research on naturally-produced AMD from undeveloped deposits. In this study, river water and AMD were collected to analyze the distribution characteristics of heavy metals and the phytoplankton community. In addition, the environmental risks of heavy metals were evaluated by single-factor pollution index, Nemerow pollution index and health risk assessment model. The results show that the pH of the Rongna River water ranged from 6.52 to 8.46, and the average concentrations of Mn and Ni were 867.37 and 28.44 μg/L, respectively, which exceed the corresponding Grade Ⅲ Environmental Quality Standard of Surface Water. The results of the environmental health risk assessment show that the river section of the Rongna River was seriously polluted by the heavy metal Mn after AMD confluence, and the health risk assessment indicates that oral ingestion of Mn posed a potential non-carcinogenic risk to children and adults. A total of 35 phytoplankton species were found in the Rongna River. The phytoplankton biomass was negatively correlated with the concentration of major heavy metals, indicating that the heavy metal concentration exceeded the tolerance limit of phytoplankton, thereby affecting their normal growth. Finally, statistical analysis shows that Cu, Zn, Ni, Mn and Cd in the Rongna River were mainly derived from AMD.

【Keywords】 Rongna River; Heavy metals; AMD; Environmental risk assessment; Phytoplankton; Northern Tibet

Introduction

The exploitation of mineral resources provides human beings with a large amount of resources and energy, but also causes heavy metal pollution in water[1-2]. Previous studies have shown that rivers flowing through mining areas are more susceptible to heavy metal pollution[3-4]. Due to both man-made and natural factors, some deposits rich in sulfides (pyrite, chalcopyrite, galena, etc.) are exposed to air, and through the action of extended periods of rain and weathering, a large amount of acid mine drainage (AMD) rich in heavy metal ions is formed[5-6]. When this AMD enters a river, it causes great harm to the ecological environment of the given water body[7-8]. After entering the environment, some toxic heavy metals are not only non-biodegradable, but they also accumulate in the environment[9]. Furthermore, some dissolved heavy metals are very easily used by aquatic organisms[10], and may also enter the human body through drinking water, skin absorption and biological chains, ultimately endangering human health[11-12].

Remediating heavy metal pollution of rivers caused by mineral mining is often costly, time-consuming and labor-intensive, and immediate results are difficult to achieve[13]. Even many years after mining has ceased, the impact of heavy metals on the environment still exists[14-15]. For example, although the abandoned lead-zinc mine in northern Idaho in the western United States has been closed for 75 years, it still has a significant impact on the river ecosystem of the region[16]. Moreover, acidic wastewater enhances the solubility of heavy metals, which allows them to migrate long distances, thereby causing harm to rivers, nearby soil and even groundwater[17]. Operations at China's Dexing copper mine have caused serious river pollution around the mining area, which has spread to farmland soil through the use of river water for irrigation[18]. China's Dabaoshan iron polymetallic mine has formed a large amount of tailings and accumulated substantial waste rock, which are quickly oxidized after being in contact with air, resulting in acidic wastewater. At the same time, a large amount of toxic and harmful heavy metal ions are released, causing serious pollution of the Hengshi River within the mining area. The pollution has spread to the downstream town of Xinjiang, causing the death of a large number of fish and shrimp in the river[19]. The highly toxic, large-scale pollution caused by the acid wastewater from the mine has caused devastating harm to the ecological environment of the area.

The Tiegelongnan copper deposit is located in the hinterland of the northern Tibetan Plateau, which represents a fragile ecological environment. The copper mine belongs to a super-large high-sulfur, porphyry-type epithermal copper deposit with a preliminary estimated copper ore reserve of more than 11 million tons[20]. At present, the deposit has not been mined, but part of the ore body is exposed to air and easily oxidizes to form acidic wastewater.. The Rongna River, originating from a mountain spring far away from the Tiegelongnan copper deposit, is about 30 kilometers long. When the Rongna River flows through the Tiegelongnan copper deposit, AMD (Fig.1B) that is naturally formed in the middle of the ore body flows into the river. As a result, a large amount of mineral extracts are carried into the Rongna River, causing serious harm to the river's ecological environment. It can be seen from field observations that before pollution from the deposit flows in, the vegetation on both sides of the river bank is luxuriant (Fig.1A). However, the vegetation disappears from areas located after the AMD flows into the river (Fig.1C) and many yellow bubbles appear in the water, indicating that the influx of AMD causes serious water pollution in the Rongna River. In this study, the concentration of heavy metals, pH, and phytoplankton distribution characteristics in river water were determined to assess environmental risk by single-factor pollution index, Nemerow pollution index and human health risk assessment model. The purpose of this study was to (1) investigate the concentrations, spatial distributions and sources of Cu, Pb, Zn, As, Mn, Cd, Cr, Ni, and Hg in the water of the Rongna River; (2) evaluate the environmental risk of heavy metal pollution in the river water, and evaluate the carcinogenic and non-carcinogenic risks caused by heavy metals; and (3) analyze the influence of heavy metal pollution on the distribution characteristics of phytoplankton. In view of the distinct ecological environment in northern Tibetan and the existence of naturally occurring AMD, the results of this study can provide a reference for investigating heavy metal pollution in rivers under special geographical environment and conditions.

Fig.1 Photographs of the Rongna River. Panel A represents the uncontaminated upstream section of the Rongna River; panel B represents the AMD from the deposit; panel C represents the river section after the AMD flows into it; and panel D represents the river section far away from the deposit area

Materials and methods

Study area

The Tiegelongnan copper deposit is located in the Wuma Township, Gaize County, Northern Tibet. It lies between 83°23′ E~83°27′ E longitude and 32°47′ N~32°50′ N latitude, at an altitude of 4800~5100 m. The study area belongs to a plateau subtropical semi-arid monsoon climate. The annual average temperature is −0.1 ℃ to −2.5 ℃, with a large temperature difference between day and night. The annual rainfall is 308.3 mm, and the rainy season is concentrated from July to August. The Tiegelongnan copper deposit is a large-scale polymetallic sulfide deposit with an average Cu grade of 0.64% with copper resources exceeding 11 million tons. The metal minerals within the deposit include chalcopyrite, pyrite, bornite, magnetite, iron ore, sphalerite, blue chalcocite and malachite.[20].

Experimental reagents and characterization of materials

Suprapur nitric acid, 4% formaldehyde solution, Lugol's solution, standard solution from Center of National Standard Reference Material of China (GSB04-1767—2004), 0.45μm glass fiber filter membrane, No. 25 plankton net (200 mesh), polyethylene sampling bottle, portable multi-parametric meter (HI9828 HANNA Italy).

Sample collection and chemical analysis

Seven sampling points (R1-R7) were chosen from the uncontaminated upper reaches of the Rongna River to the end of the river, and three sampling points (S1-S3) were set up in the AMD section. At the same time, water samples from the Bolong River (BL1-BL3), away from the mining area, were collected as a control (Fig.2). Water samples for heavy metals and phytoplankton analysis were collected at each point, and a portable multi-parameter meter (HI9828, HANNA, Italy) was used to determine the pH of the water on site.

The samples used for heavy metal determination were filtered through 0.45 μm glass fiber filter membranes immediately after collection to remove large suspended solids. Next, HNO_3 was added to ensure the pH was less than 2, and then the samples were kept sealed at 4 ℃. After being transported to the laboratory, Cu, Pb, Zn, Mn, Cd, Cr and Ni in the water samples were measured by inductively coupled plasma mass spectrometer system (ICP-MAS, PE300D), while As and Hg were measured by atomic fluorescence spectrometry (AFS-9760). The precision and accuracy for the analysis of heavy metals in water were validated using standard reference materials from the Center of National Standard Reference Material of China (GSB04-1767—2004). The recovery rates of heavy metal contents in the standard reference materials were between 90% and 110%. For quantitative analysis of phytoplankton, No. 25 plankton net was used to collect samples under the water surface. The collected phytoplankton-containing water samples were allowed to stand for 24 hours, and then the supernatant was carefully drawn with a pipette and concentrated to 50 mL, and 4% formaldehyde was added as a fixative. For quantification of phytoplankton, 1 L of water was collected at each sampling point, and 10 mL of Lugol's solution was added for fixation. The water samples were returned to the laboratory and then concentrated to 100 mL. Species identification and cell counts were performed under a microscope (Olympus CX21) at 400 times magnification[21-22].

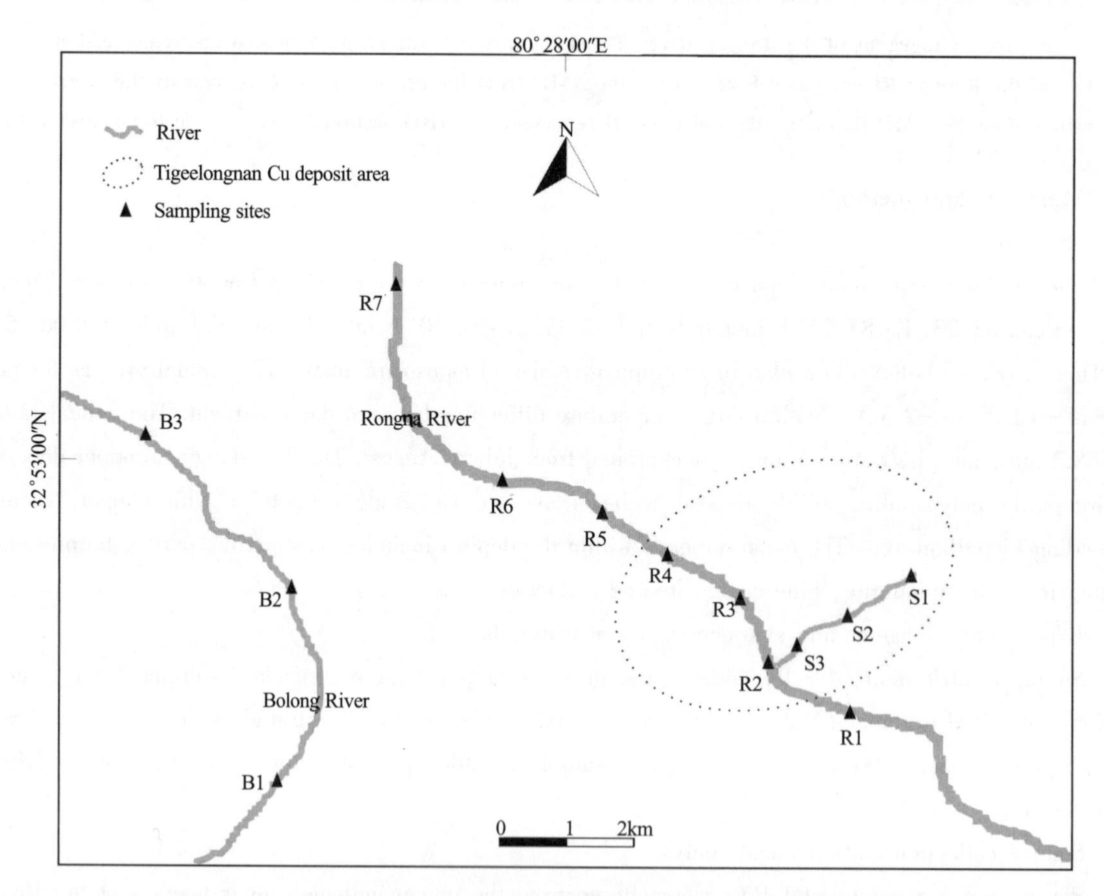

Fig.2 Location of the study area and the distribution of sampling points.
The location coordinate map of the study area was drawn according to the USGS National Map Viewer

Environmental risk assessment

Pollution index

The single-factor pollution index and Nemerow pollution index are often used to evaluate the pollution status of heavy metals in water bodies[23-24]. The single-factor evaluation method, which evaluates the pollution level of a single heavy metal in river water, was calculated as shown in Eq. (1)[25]:

$$P_i = C_i/B_i \tag{1}$$

where, P_i represents the single-factor pollution index of element i; C_i represents the actual concentration of element i (μg/L); and B_i represents the evaluation standard of element i. In this study, the surface water environmental quality standard of the National Environmental Protection Agency of China (NEPAC) was used as the evaluation standard (GB 3838—2002)[26].

The Nemerow pollution index not only reflects the pollution degree of single-factor heavy metals, but also describes the comprehensive pollution of multiple heavy metals. Additionally, it highlights the impact and effect of the pollutant with the largest pollution index on environmental quality. This index is currently used most frequently because it is a comprehensive method that evaluates the environmental quality of water bodies. The Nemerow pollution index was calculated as shown in Eq. (2)[27].

$$P_n = \sqrt{0.5 \times [\max(P_i)^2 + \text{ave}(P_i)^2]} \tag{2}$$

where, C_i represents the measured concentration of heavy metal i; $\max(P_i)$ represents the maximum value of the heavy metal single-factor pollution index; and $\text{ave}(P_i)$ represents the average value of the single-factor pollution index for each heavy metal. The pollution level classifications for the single-factor pollution index (P_i) and the Nemerow pollution index (P_n) are shown in Table 1[28-29].

Table 1 Classification of pollution levels for P_i and P_n

P_i	Pollution level	P_n	Pollution degree
$P_i<1$	Unpolluted	$P_n \leqslant 0.7$	Safe
$1 \leqslant P_i<2$	Slightly polluted	$0.7<P_n \leqslant 1$	Precaution
$2 \leqslant P_i<3$	Moderately polluted	$1<P_n \leqslant 2$	Slight pollution
$3 \leqslant P_i<5$	Highly polluted	$2<P_n \leqslant 3$	Moderate pollution
$P_i \geqslant 5$	Very highly polluted	$P_n>3$	Heavy pollution

Health risk assessment

Surface water heavy metal elements can enter the human body through daily drinking water intake, skin absorption and respiration. For humans, intake and skin absorption are the two main exposure pathways for aquatic heavy metals[30-31]. The daily dose of water intake and skin absorption, and the carcinogenic risk and non-carcinogenic risk of heavy metals to the human body were determined according to relevant documents from the US EPA[32], using the following equations:

$$\text{ADD}_{\text{ingestion}} = \frac{C_i \times \text{IR} \times \text{EF} \times \text{ED}}{\text{BW} \times \text{AT}} \tag{3}$$

$$\text{ADD}_{\text{dermal}} = \frac{C_j \times \text{SA} \times K_p \times \text{ET} \times \text{EF} \times \text{ED}}{\text{BW} \times \text{AT}} \times 10^{-3} \tag{4}$$

$$\text{HI} = \Sigma\,\text{HQ} = \Sigma \frac{\text{ADD}}{\text{RfD}} \tag{5}$$

$$TCR = \sum CR = \sum ADD \times CSF \tag{6}$$

Among them: $ADD_{ingestion}$ and ADD_{dermal} are the daily doses for drinking water intake and skin absorption, respectively; HQ is the risk quotient; HI is the risk index; CR is the carcinogenic risk; TCR is the total carcinogenic risk of all exposed metals; C_i is the concentration of heavy metals in the water body; IR is the average daily drinking water intake; EF is the exposure frequency; ED is the exposure time; BW is the average body weight; AT is the average exposure time; SA is the skin exposure area; SL is the skin adhesion factor; ET is the exposure time; and K_p is the permeability coefficient of heavy metals in water. The specific exposure parameters are shown in Table 2[33]. Table 3 shows the RfD, CSF and K_p values of the heavy metals[34]. HQ and HI are used to describe the non-carcinogenic risk of heavy metals. When HQ or HI<1, there is no non-carcinogenic health risk; otherwise, there is a potential non-carcinogenic health risk, with larger values representing higher risk. CR and TCR are used to describe the carcinogenic risk of heavy metals. When $CR<10^{-6}$, there is no carcinogenic risk; when CR is between $10^{-6} \sim 10^{-4}$, the risk is acceptable; and when $CR>10^{-4}$, the heavy metals in the water body are likely to cause cancer risk to the human body.

Table 2 Exposure parameters for the health risk assessment models

Parameters	Unit	Value	
		Child	Adult
IR	$L \cdot d^{-1}$	0.64	2
EF	$d \cdot year^{-1}$	350	350
ED	years	6	30
BW	kg	15	70
AT	d	2190 (For non-carcinogens) 25550 (For carcinogens)	10950 (For non-carcinogens) 25550 (For carcinogens)
SA	cm^2	6600	18000
ET	h	1	0.58

Table 3 RfD, CSF and K_p of heavy metals

Element	$RfD_{ingestion}$ [μg/(kg·day)]	RfD_{dermal} [μg/(kg·day)]	$CSF_{ingestion}$ [mg/(kg·day)]	CSF_{dermal} [mg/(kg·day)]	K_p (cm/h)
Cu	40	8			0.001
Pb	1.4	0.42			0.0001
Zn	300	60			0.0006
As	0.3	0.285	1.5	3.66	0.001
Mn	24	0.96			0.001
Cd	0.5	0.025			0.001
Cr	3	0.075			0.002
Ni	20	0.8			0.0002
Hg	0.3	0.021			0.001

Statistical analysis

Correlation analysis, cluster analysis and principal component analysis (PCA) can effectively reflect the source of heavy metals[35-36]. In order to understand the heavy metal sources in the Rongna River, this study used IBM SPSS Statistics 24 software to conduct Pearson correlation analysis (two-tailed), cluster analysis and PCA on the heavy metals and pH in the water at the sampling points. For the PCA, the principal component was calculated based on the correlation matrix, VARIMAX was used to normalize the rotation, and the principal component was extracted only when the eigenvalue was greater than or equal to 1.

Results and discussion

Distribution characteristics of heavy metals in the water of Rongna River

The heavy metal content and pH characteristics of the Rongna River water are shown in Table 4. The pH of the water body ranged from 6.52 to 8.46, with an average value of 7.26, which meets the corresponding Grade Ⅲ national surface water standard[26]. The pH range of AMD was 2.86~3.06, with an average value of 2.98, which is much lower than the corresponding Grade Ⅲ national surface water standard and denotes serious acidification. Under the action of humans, as well as some natural destructive forces, the original stable protective layer on the metal sulfide deposits and surrounding rocks can be destroyed, which exposes them to the atmospheric oxygen-containing environment, resulting in a large amount of AMD[6]. Acidic water increases the solubility of heavy metals, which further increases their diffusion capacity[17]. Therefore, the acidic water produced in the mining area may be the main reason that the Rongna River is polluted by heavy metals.

Table 4 Characteristics of heavy metal concentrations in the water

Parameters	Rongna River		AMD		Bolong River		Grade Ⅲ
	Range	Mean	Range	Mean	Range	Mean	
Cu/(μg/L)	1.89~806.00	280.98±259.04	1890.00~2272.00	2072.67±156.4	5.42~7.37	6.50±0.81	1000
Pb/(μg/L)	0.49~2.41	1.32±0.63	1.21~2.17	1.65±0.4	0.34~0.64	0.52±0.13	50
Zn/(μg/L)	13.00~415.00	178.66±135.39	1321.00~1515.00	1404.67±81.41	39.80~56.70	46.73±7.23	1000
As/(μg/L)	0.01~3.83	0.88±1.27	0.07~0.64	0.29±0.25	1.09~4.96	2.43±1.79	50
Mn/(μg/L)	43.10~2041.00	867.37±678.69	7488.00~8072.00	7835±250.78	37.80~51.90	46.47±6.19	100
Cd/(μg/L)	0.12~0.64	0.43±0.16	0.84~1.54	1.2±0.29	0.11~0.54	0.35±0.18	5
Cr/(μg/L)	1.56~6.37	4.24±1.58	3.13~7.24	4.87±1.73	1.74~3.08	2.32±0.56	50
Ni/(μg/L)	7.45~60.10	28.44±16.8	167.00~189.00	178.67±9.03	10.7~18.22	13.84±3.19	20
Hg/(μg/L)	0.001~0.012	0.0053±0.0042	0.001~0.005	0.0023±0.0019	0.010~0.011	0.0097±0.0012	0.1
pH	6.52~8.46	7.26±0.67	2.86~3.06	2.98±0.08	7.95~8.06	7.95±0.09	6~9

Concentrations of the heavy metals Cu, Pb, Zn, As, Cd, Cr, Ni and Hg in the water of Rongna River were all within the limits of the Grade Ⅲ national surface water environmental quality standard, whereas both Mn and Ni exceeded the Grade Ⅲ standard by 8.67 and 1.42 times, respectively. The concentrations of Cu, Zn, Mn and Ni in the AMD exceed the Grade Ⅲ standard by 2.07, 1.40, 78.35 and 8.93 times, respectively. In the river sections before and after the AMD inflow point, the spatial distribution of heavy metal concentrations in the Rongna River changed substantially (Fig.3). The concentration of heavy metals in the upper reaches of the Rongna River (R1) was similar to that of the Bolong River, which was used as the control because it is far away from the mining area. This shows that without the influx of AMD, the natural weathering of rocks may not cause serious heavy metal pollution to the river. The heavy metal concentrations for Cu, Zn, Mn and Ni in the AMD were 318.71, 30.06, 168.62 and 12.91 times higher, respectively, than those in the Bolong River. After the AMD entered the Rongna River (R2-R6), the heavy metals in the water were significantly greater than the Bolong River. The concentrations of Cu, Zn, Mn, Pb, Cr and Ni in the water from sites R2-R6 were 60.36, 5.23, 25.65, 3.00, 2.15 and 2.60 times higher, respectively, than those in the Bolong River. The Cu, Zn, and Mn concentrations in the polluted reaches of the Rongna River were greater larger than those of the Heihe River (Table 5), which is distributed in the mining area but has no AMD discharge[37]; Conversely, the content of heavy metals in Rongna River was less than that in the Gyamaxung-chu River, which is distributed in the mining area and polluted by AMD[38]. Comparing the rivers in two different mining areas shows that AMD is the main factor causing pollution of the rivers in the mining area. At the end of the Rongna River (R7), the heavy metal concentrations in the water were close to those in the Bolong River. Furthermore, compared with the source of some rivers in northern Tibet and the Lhasa River distributed around a city, the concentration of heavy metals at the end of the Rongna River was close to that of the Yellow River, Buha River, Shule River and Lhasa River[39-40]. This indicates that after the long-distance self-purification of the river, the heavy metal concentrations in the river returned to normal levels.

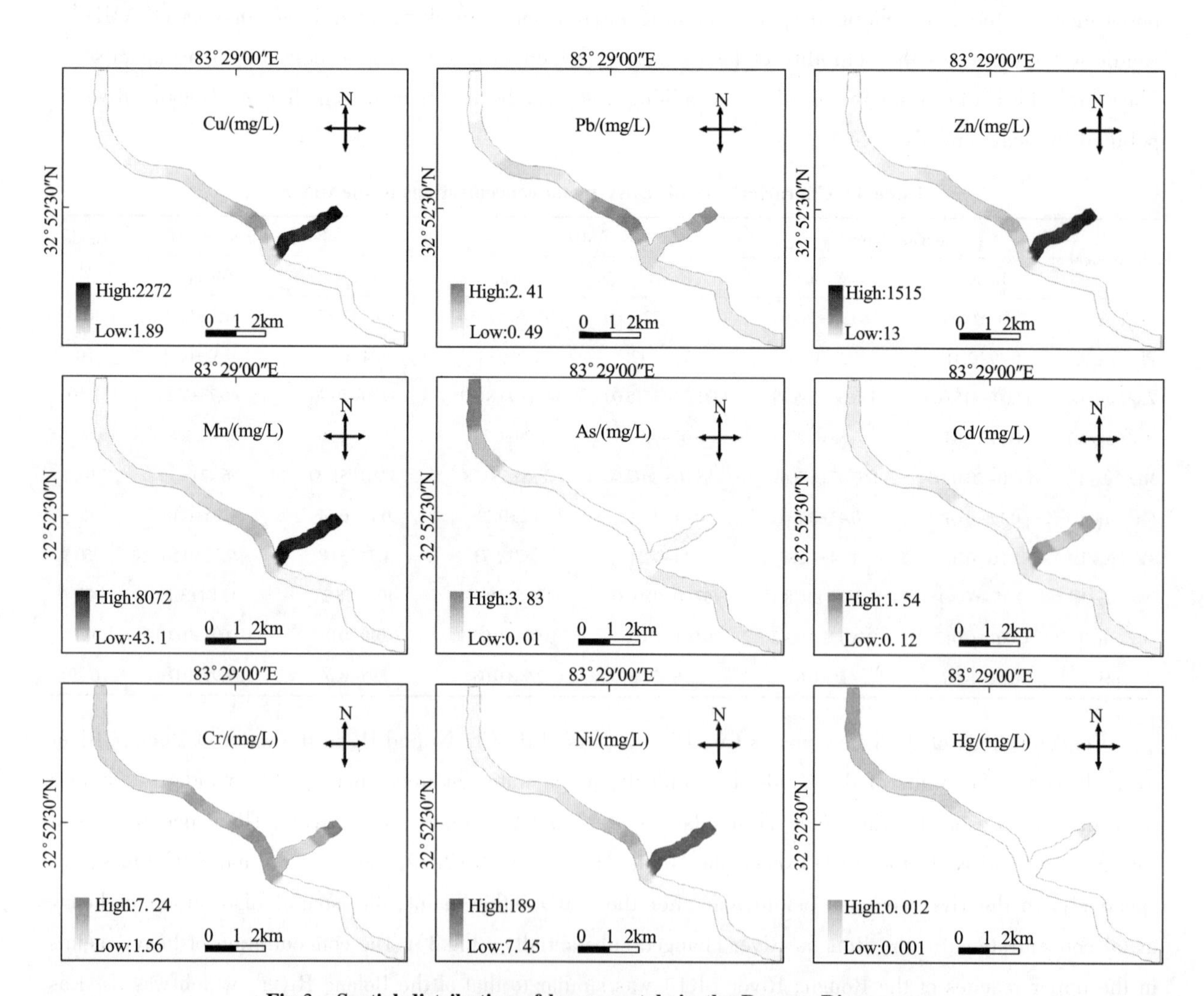

Fig.3 Spatial distribution of heavy metals in the Rongna River water.

The location coordinate map of the study area was drawn according to the USGS National Map Viewer

Table 5 Comparison of heavy metal contents in other rivers in Tibet

Location	Cu	Pb	Zn	As	Mn	Cd	Cr	Ni	Hg	pH
Yellow River	1	0.1	4.4	1.2	3.3	<0.007	1.8	—	—	8.0~8.6
Buha He	1.4	<0.05	3.7	0.9	3.8	N.D.	2	—	—	8.5
Shule He	0.8	0.1	1.5	1.4	6.3	<0.004	2	—	—	8.3~8.8
Lhasa River	2.863	0.056	0.829	3.071	6.237	0.042	3.156	—	0.005	8.8
Heihe River	6.02	6.14	66.70	2.68	46.75	0.65	6.57	20.37	—	8.6~8.7
Gyamaxung-chu	5800	695.67	2454.67	—	1061	2.87	2.30	23.20	—	—

Note: The heavy metal units in the table are all μg/L.

Environmental risk assessment of heavy metals

Single-factor evaluation method and Nemerow pollution index. Fig. 4 shows the results of the single-factor pollution index (P_i) and Nemerow pollution index (P_n) at different points along the Rongna River. The P_i of all heavy metals in the unpolluted section of R1, the upstream section of the Rongna River, was less than 1, which represents no pollution. In the AMD (S1-S3), the single-factor pollution index of Mn and Ni was greater than 5, which represents extremely serious pollution, while the average P_i value of Cu was 2.07 (moderate pollution). The average P_i of Zn was 1.40 (slight pollution), while the P_i of other heavy metals was less than 1 (no pollution). Where AMD flows into the river section (R2-R6), the average P_i of Mn was 11.92 (extremely serious pollution), and the average P_i of Ni was 1.81 (light pollution). The P_i of all heavy metals at the end of the Rongna River (R7) was less than 1 (no pollution). The characteristics of P_i correspond to the spatial distribution characteristics of heavy metals in the Rongna River that were presented in the previous section (3.1). This indicates that the heavy metal pollution in the Rongna River water body is mainly affected by deposit-associated AMD.

The P_n in both the AMD (S1-S3) and the main reach of Rongna River (R2-R6) were greater than 3, indicating heavy pollution. This shows that AMD caused most of the Rongna River to be polluted. The upper (R1) and lower (R2) reaches of Rongna River and the Bolong River (control) were all less than 0.7, which indicates a clean state. The P_n of the R2-R6 reach showed a trend of first increasing and then decreasing as the distance of the river increased. This indicates that the heavy metals are gradually diluted by the clean river water and, along with self-purification of the river itself, the water basically returns to an unpolluted state by the time it reaches the end of the river.

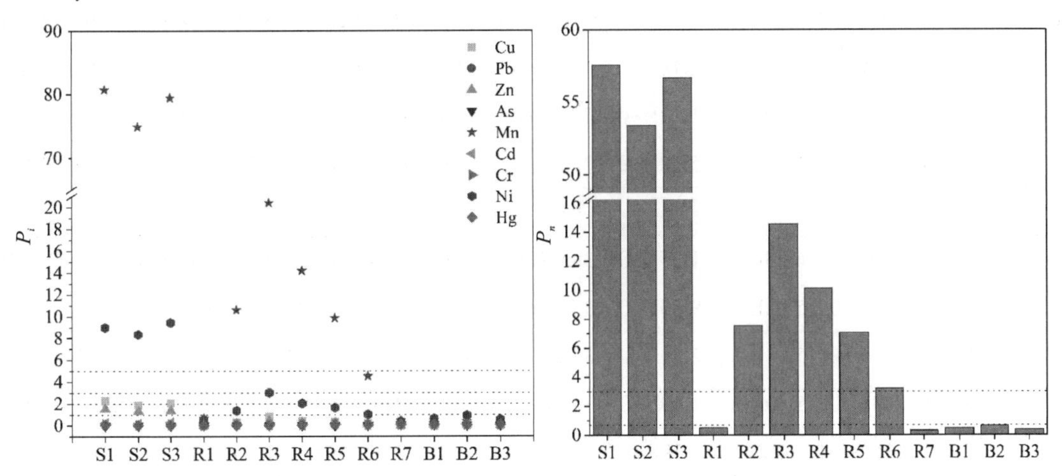

Fig.4 P_i and P_n of heavy metals in the Rongna River

Health risk assessment. Table 6 and Table 7 show the calculated non-carcinogenic and carcinogenic risk coefficients HQ and CR for children and adults based on the model provided by the US EPA. These values can be used to judge the potential carcinogenic and non-carcinogenic risks of polluted water to the exposed population. It can be seen from Fig.5 that in the AMD (S1-S3), the $HQ_{ingestion}$, HQ_{dermal} and HI values of the element Mn for children and adults are all greater than 1, indicating that ingestion or skin exposure to Mn will cause a certain non-carcinogenic risk to humans. The $HQ_{ingestion}$ and HI values of Cu for children and adults were greater than 1, while HQ_{dermal} was less than 1, indicating that water intake of Cu will cause potential non-carcinogenic health risks to children and adults, whereas skin contact with Cu will not cause potential non-carcinogenic risks. At sites R2-R5, the $HQ_{ingestion}$ and HI values of Mn to children and adults were greater than 1, while HQ_{dermal} was less than 1, indicating that oral ingestion of Mn poses a potential non-carcinogenic risk to children and adults, while skin contact does not cause potential non-carcinogenic risks. At sites R1, R6 and R7, the HI for adults and children was less than 1, indicating that exposure at these sites would not pose a potential non-carcinogenic risk to children and adults. The $HQ_{ingestion}$ of all heavy metal elements in the river water was greater than HQ_{dermal}. This shows that oral intake is the main exposure mode for heavy metal non-carcinogenic risk, which agrees with the results of a previous study[41]. For both HQ and HI, the values for children were significantly higher than those for adults, indicating that children have a higher non-carcinogenic risk under the same environmental conditions. Previous studies have also come to this conclusion[30-31].

Table 6　Hazard quotient (HQ) and cancer risk (CR) of heavy metals from Rongna River for children

Site	Cu	Pb	Zn	As	Mn	Cd	Cr	Ni	Hg	As
					$HQ_{ingestion}$					$CR_{ingestion}$
S1	2.32E+00	6.34E-02	2.07E-01	9.55E-03	1.38E+01	9.90E-02	9.87E-02	3.68E-01	6.82E-04	3.68E-07
S2	1.93E+00	4.59E-02	1.80E-01	8.73E-02	1.28E+01	6.87E-02	4.27E-02	3.42E-01	1.36E-04	3.37E-06
S3	2.10E+00	3.54E-02	1.88E-01	2.18E-02	1.35E+01	1.26E-01	5.80E-02	3.87E-01	1.36E-04	8.42E-07
R1	1.93E-03	2.78E-02	1.77E-03	1.54E-01	1.17E-01	3.60E-02	2.13E-02	2.27E-02	1.36E-04	5.94E-06
R2	2.76E-01	3.59E-02	2.70E-02	2.18E-02	1.81E+00	4.42E-02	7.45E-02	5.56E-02	1.36E-04	8.42E-07
R3	8.24E-01	7.04E-02	5.66E-02	4.09E-03	3.48E+00	3.11E-02	8.69E-02	1.23E-01	1.36E-04	1.58E-07
R4	4.07E-01	5.96E-02	3.50E-02	1.36E-03	2.42E+00	9.82E-03	5.92E-02	8.26E-02	6.82E-04	5.26E-08
R5	3.57E-01	2.54E-02	3.44E-02	2.18E-02	1.68E+00	5.24E-02	7.69E-02	6.69E-02	1.23E-03	8.42E-07
R6	1.43E-01	3.65E-02	1.39E-02	1.12E-01	7.71E-01	2.78E-02	4.38E-02	4.11E-02	1.09E-03	4.31E-06
R7	1.99E-03	1.43E-02	1.85E-03	5.22E-01	7.35E-02	4.50E-02	4.24E-02	1.52E-02	1.64E-03	2.01E-05
					HQ_{dermal}					CR_{dermal}
S1	1.21E-01	2.20E-04	6.45E-03	1.05E-04	3.58E+00	2.06E-02	8.22E-02	1.92E-02	1.01E-04	9.35E-09
S2	1.01E-01	1.59E-04	5.62E-03	9.56E-04	3.32E+00	1.43E-02	3.55E-02	1.78E-02	2.03E-05	8.55E-08
S3	1.09E-01	1.23E-04	5.87E-03	2.39E-04	3.52E+00	2.62E-02	4.83E-02	2.01E-02	2.03E-05	2.14E-08
R1	1.01E-04	9.63E-05	5.53E-05	1.69E-03	3.04E-02	7.49E-03	1.77E-02	1.18E-03	2.03E-05	1.51E-07
R2	1.44E-02	1.25E-04	8.43E-04	2.39E-04	4.70E-01	9.20E-03	6.20E-02	2.90E-03	2.03E-05	2.14E-08

Continued to Table 6

Site	Cu	Pb	Zn	As	Mn	Cd	Cr	Ni	Hg	As
	HQ ingestion									CR ingestion
R3	4.29E-02	2.44E-04	1.77E-03	4.48E-05	9.05E-01	6.47E-03	7.23E-02	6.40E-03	2.03E-05	4.01E-09
R4	2.12E-02	2.07E-04	1.09E-03	1.49E-05	6.30E-01	2.04E-03	4.93E-02	4.30E-03	1.01E-04	1.34E-09
R5	1.86E-02	8.82E-05	1.07E-03	2.39E-04	4.38E-01	1.09E-02	6.40E-02	3.48E-03	1.82E-04	2.14E-08
R6	7.45E-03	1.27E-04	4.34E-04	1.22E-03	2.00E-01	5.79E-03	3.64E-02	2.14E-03	1.62E-04	1.10E-07
R7	1.04E-04	4.97E-05	5.79E-05	5.72E-03	1.91E-02	9.37E-03	3.53E-02	7.93E-04	2.43E-04	5.12E-07

Table 7 Hazard quotient (HQ) and cancer risk (CR) of heavy metals from Rongna River for adults

Site	Cu	Pb	Zn	As	Mn	Cd	Cr	Ni	Hg	As
	HQ ingestion									CR ingestion
S1	1.56E+00	4.25E-02	1.38E-01	6.39E-03	9.21E+00	6.63E-02	6.61E-02	2.47E-01	4.57E-04	1.23E-06
S2	1.29E+00	3.07E-02	1.21E-01	5.84E-02	8.55E+00	4.60E-02	2.86E-02	2.29E-01	9.13E-05	1.13E-05
S3	1.41E+00	2.37E-02	1.26E-01	1.46E-02	9.07E+00	8.44E-02	3.88E-02	2.59E-01	9.13E-05	2.82E-06
R1	1.29E-03	1.86E-02	1.19E-03	1.03E-01	7.82E-02	2.41E-02	1.42E-02	1.52E-02	9.13E-05	1.99E-05
R2	1.85E-01	2.41E-02	1.81E-02	1.46E-02	1.21E+00	2.96E-02	4.99E-02	3.73E-02	9.13E-05	2.82E-06
R3	5.52E-01	4.72E-02	3.79E-02	2.74E-03	2.33E+00	2.08E-02	5.82E-02	8.23E-02	9.13E-05	5.28E-07
R4	2.73E-01	3.99E-02	2.35E-02	9.13E-04	1.62E+00	6.58E-03	3.96E-02	5.53E-02	4.57E-04	1.76E-07
R5	2.39E-01	1.70E-02	2.30E-02	1.46E-02	1.13E+00	3.51E-02	5.15E-02	4.48E-02	8.22E-04	2.82E-06
R6	9.59E-02	2.45E-02	9.32E-03	7.49E-02	5.16E-01	1.86E-02	2.93E-02	2.75E-02	7.31E-04	1.44E-05
R7	1.34E-03	9.59E-03	1.24E-03	3.50E-01	4.92E-02	3.01E-02	2.84E-02	1.02E-02	1.10E-03	6.75E-05
	HQ dermal									CR dermal
S1	4.06E-02	7.39E-05	2.17E-03	3.51E-05	1.20E+00	6.92E-03	2.76E-02	6.44E-03	3.41E-05	1.57E-08
S2	3.38E-02	5.35E-05	1.89E-03	3.21E-04	1.12E+00	4.81E-03	1.19E-02	5.97E-03	6.81E-06	1.44E-07
S3	3.68E-02	4.12E-05	1.97E-03	8.03E-05	1.18E+00	8.81E-03	1.62E-02	6.76E-03	6.81E-06	3.59E-08
R1	3.38E-05	3.23E-05	1.86E-05	5.67E-04	1.02E-02	2.52E-03	5.95E-03	3.97E-04	6.81E-06	2.53E-07
R2	4.83E-03	4.19E-05	2.83E-04	8.03E-05	1.58E-01	3.09E-03	2.08E-02	9.72E-04	6.81E-06	3.59E-08
R3	1.44E-02	8.21E-05	5.94E-04	1.51E-05	3.04E-01	2.17E-03	2.43E-02	2.15E-03	6.81E-06	6.73E-09
R4	7.11E-03	6.95E-05	3.68E-04	5.02E-06	2.12E-01	6.86E-04	1.66E-02	1.44E-03	3.41E-05	2.24E-09
R5	6.24E-03	2.96E-05	3.60E-04	8.03E-05	1.47E-01	3.66E-03	2.15E-02	1.17E-03	6.13E-05	3.59E-08
R6	2.50E-03	4.26E-05	1.46E-04	4.11E-04	6.73E-02	1.94E-03	1.22E-02	7.19E-04	5.45E-05	1.84E-07
R7	3.49E-05	1.67E-05	1.94E-05	1.92E-03	6.42E-03	3.15E-03	1.19E-02	2.66E-04	8.17E-05	8.59E-07

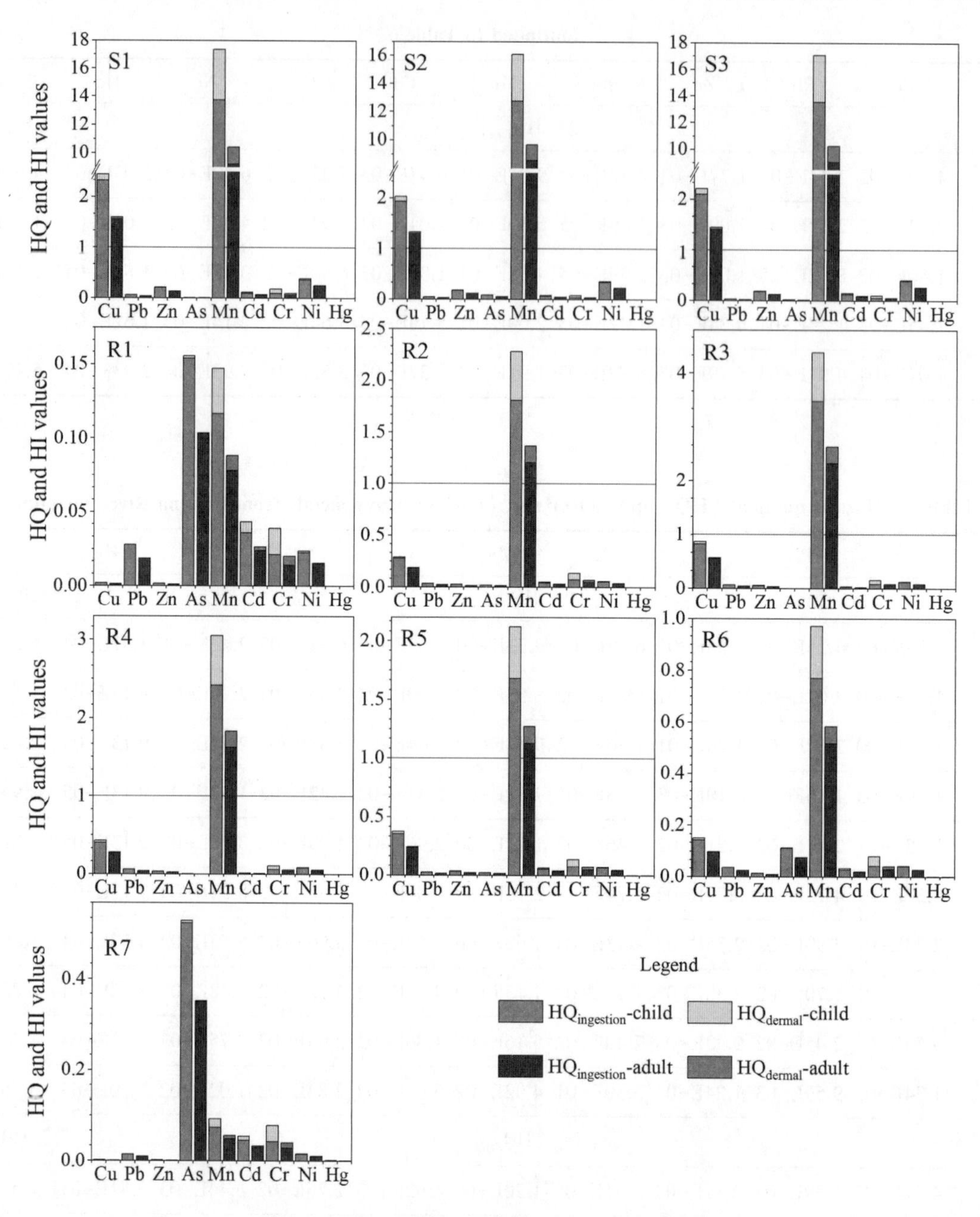

Fig.5 HQ and HI of the Rongna River water

The US EPA classifies As as a carcinogen that is harmful to humans. This study calculated the carcinogenic risk of As to children and adults using equations (3)~(6). It can be seen from the results (Fig. 6) that the carcinogenic risk through ingestion and skin exposure was between 1.34×10^{-9}~6.75×10^{-5}, which is less than 10^{-4}, the acceptable carcinogenic risk stated by the US EPA. At the same time, the carcinogenic risk for adults via the oral intake and skin exposure routes was higher than that of children, which may be attributed to the larger amount of water consumed by adults and their larger skin area.

The impact of heavy metal pollution on the distribution of phytoplankton

As the most important phytoplankton in aquatic ecosystems, algae are widely distributed in rivers, lakes and seas[42]. Algae are not only closely related to their living environment, but they also play an important role in processes related to material circulation, energy conversion and information transmission. Different types of algae have different sensitivities to changes in the aquatic environment, and their species

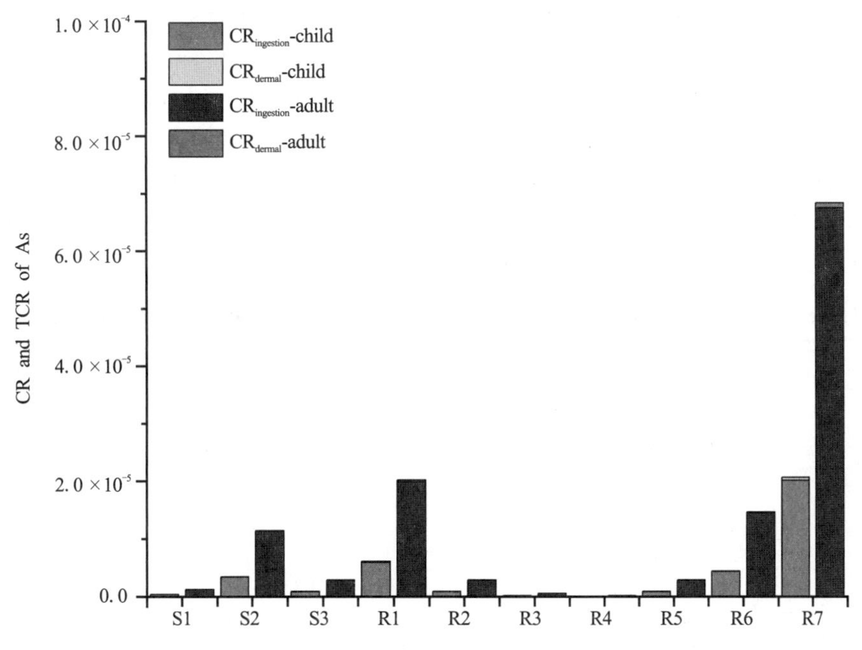

Fig.6 CR and TCR of As in the Rongna River water

composition, community structure and biomass are closely related to the environment factors in which they live[43-44]. Moreover, they can be used as indicators of water quality changes and a means to monitor and evaluate the health of the aquatic environment because some algae have a fast growth cycle and the community structure is very sensitive to environmental changes[45].

Table 8 shows the distribution characteristics of algae at the different water sites. A total of 30 algal species were detected in the study area, namely 5 species of Cyanophyta, 23 species of Diatoms and 2 species of Chlorophyta. Among these 30 phytoplankton species, *Pseudoanabaena* sp. exhibited the highest average density in the entire river at 20.14×10^4 cell/L and a maximum dominance of 0.645. *Melosira* sp. (3.068×10^4 cell/L and 0.098 dominance) and *Pinnularia* sp. (2.044×10^4 cell/L and 0.065 dominance) were the next most predominant. The density of phytoplankton in the AMD was 0.1×10^4 cell/L. The density of photoplankton increased from R1 to R6, with R6 registering 110.85×10^4 cell/L.

Due to the different water pollution levels at the different sites of the Rongna River, the corresponding phytoplankton species composition and population abundance varied greatly. A total of 11 algal species were found in the upstream clean river section (R1), specifically 10 species of Diatoma and 1 species of Chlorophyta. *Navicula bacilloides Hust* and *Ulothrix* sp. were only found in the clean river water in the upper reaches of the river. *Diatoma vulgare* appeared in both the upper and lower reaches of the Rongna River. Therefore, these three types of algae can be used as indicator species for clean water bodies.The diversity of algae was low in the AMD, due to the existence of a high concentration of heavy metal elements, especially Cu, which is toxic to algae[46]. In fact, only *Chroococcus* sp. was observed in the AMD. This indicates that *Chroococcus* sp. can survive in AMD, has strong adaptability to acid and heavy metal pollution, and can be used as a typical indicator alga for AMD. Algae at sites R2-R6 included *Pseudoanabaena* sp., *Melosira* sp., *Aphanizomenon* sp., *Anabaena* sp., *Pinnularia* sp., *Navicula* sp., *Navicula* sp., *Oscillatoria* sp. and *Cymbella* sp. At the end of the Rongna River (R7), the dominant community members were *Plumbonia*, *Navicula* and *Melosira*, with dominance of 0.388, 0.239 and 0.164, respectively. The comparison found that cyanobacteria only appeared in the AMD and polluted sections of the Rongna River, indicating that cyanobacteria can adapt to heavy metal-polluted environments and can be used as an indicator of heavy metal pollution in river water. This is related to the strong tolerance of cyanobacteria to heavy metals[47]. Pearson correlation coefficients between the density of plankton and heavy metal concentrations are shown in Table 9. It can be seen that the pollutant elements Cu, Zn, Mn, Cd, and Ni are negatively correlated with phytoplankton biomass, indicating that the concentration of these heavy metals exceeds the tolerance range of algae and affects their normal growth. The cyanobacteria and Hg were positively correlated at the level of 0.05, which is similar to previous studies, indicating that Hg has a certain promoting effect on the growth of algae within a safe concentration range[48]. Chlorophyta were negatively correlated with Hg concentrations, which may be because the detected Chlorophyta are more sensitive to Hg.

Table 8 Distribution characteristics of photoplankton in the study area ($\times 10^4$ cell/L)

Category	Species	R1	R2	R3	R4	R5	R6	R7	S1-S3
Cyanophyta	*Chroococcus* sp.								0.1
	Pseudoanabaena sp.				19.4	55.6	86.15		
	Oscillatoria sp.				4.4		5.15		
	Anabaena sp.				1.7				
	Aphanizomenon sp.				4	8.25			
Diatoms	Melosira sp.			10.35	1.8	1.55	10.3	0.55	
	Tabellaria sp.				0.15	0.4	0.1	0.05	
	Diatoma vulgare	0.1					0.1	0.05	
	Fragilaria sp.			1.15	0.55	0.7	0.6		
	Fragilaria intermedia			0.35	0.15	0.55	0.6		
	Synedra sp.		0.2	0.05					
	Eunotia sp.	0.05		0.15					
	Navicula sp.	0.4	1.4	1	1.15	2.4	1.9	0.8	
	Navicula bacilloides Hust	0.05							
	Pinnularia sp.	0.4	1.45	1.7	2.8	4	4.7	1.3	
	Pinnularia viridis		0.1						
	Cymbella sp.	0.1	0.8	0.45	0.5	0.8	0.8	0.2	
	Cymbella aspera			0.05					
	Cymbella cistula	0.1	0.5	0.45	0.1	0.15		0.15	
	Cymbella turgidula			0.05					
	Gomphonema sp.	0.05		0.05	0.05		0.1	0.1	
	Gomphonema constrictum					0.05			
	Achnanthes sp.		0.05						
	Hantzschia sp.					0.05			
	Nitzschia sp.	0.1	0.5	0.15	0.55	0.55	0.2	0.15	
	Nitzschia linearis		0.05						
	Cymatopleura sp.						0.15		
	Surirella sp.	0.05	0.05	0.05					
Chlorophyta	*Crucigenia quadrata*		0.05	0.1	0.05				
	Ulothrix sp.	0.55							

Table 9 Pearson correlation coefficients between the density of photoplankton and heavy metal concentration

Category	Cu	Pb	Zn	As	Mn	Cd	Cr	Ni	Hg
Cyanophyta	−0.535	−0.633	−0.497	0.786	−0.512	−0.327	−0.756	−0.508	0.853*
Diatoms	−0.630	−0.187	−0.673	0.616	−0.687	−0.635	−0.605	−0.651	0.240
Chlorophyta	−0.166	0.567	−0.251	−0.482	−0.239	−0.480	0.176	−0.213	−.823*

* Correlation is significant at the 0.05 level (2-tailed).

Source analysis of heavy metals

Correlation analysis is an important basis for determining the source of heavy metal elements. A significant correlation between the metal elements means that the elements are homologous or have some relevance[49]. Correlation analysis of the contents of 9 heavy metal elements in the water at each sampling point (Fig.7) showed that Cu, Zn, Ni, Mn and Cd exhibited extremely significant positive correlations ($P<0.01$), indicating that Cu, Zn, Ni, Mn and Cd have homology. The pH and the heavy metals Cu, Zn, Ni, Mn and Cd showed a very significant negative correlation ($P<0.01$). The lower the pH, the higher the content of heavy metals in the water, indicating that the heavy metals Cu, Zn, Ni, Mn, Cd and Mn in the water of the Rongna River are related to AMD. AMD increases the solubility of heavy metals, causing a large amount of heavy metal ions to enter the river with the AMD. Cluster analysis is used to group heavy metals with homologous characteristics to determine their source. In the water of the Ronna River, three distinct clusters were identified (Fig.7). Mn, Ni, Zn, Cu and Cd, which had higher content, were classified into the same cluster, and these elements may come from AMD. The content of Pb and Cr were slightly higher than that of the Bolong River, and they were grouped into the same cluster. The source of these two heavy metals may be related to the oxidation of sulfide mines. The contents of As and Hg were similar to those of the Bolong River and were classified into the same cluster, which indicates that these two heavy metals may originate from lithogenic sources.

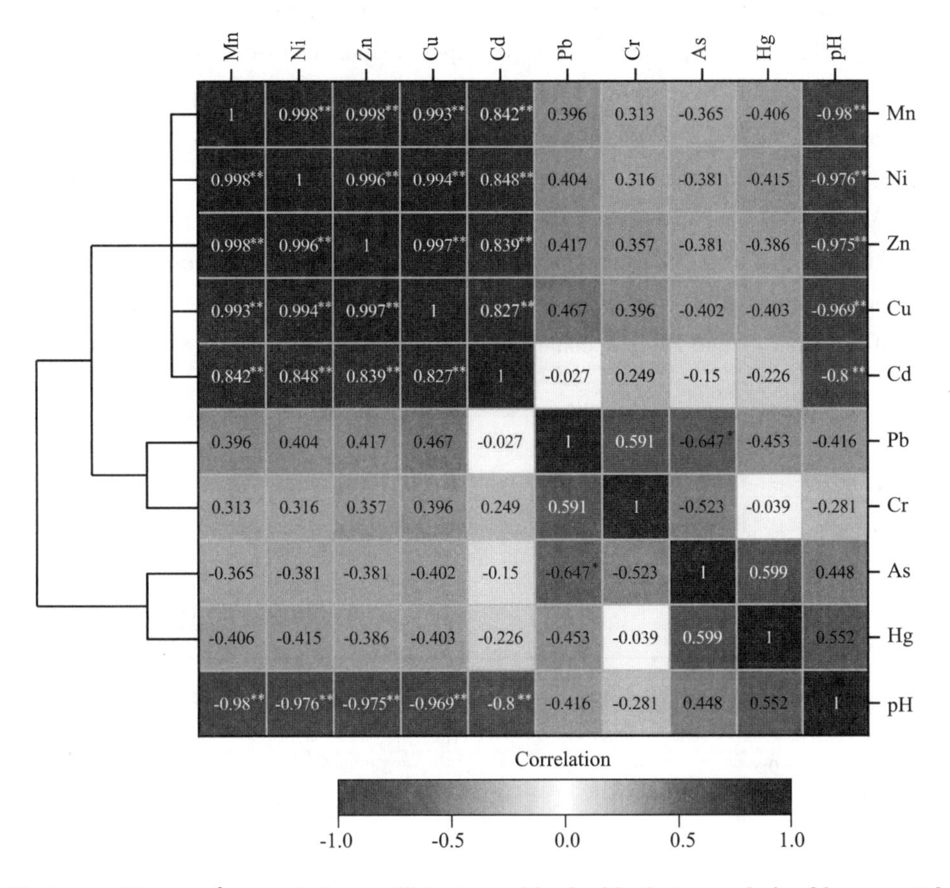

	Mn	Ni	Zn	Cu	Cd	Pb	Cr	As	Hg	pH
Mn	1	0.998**	0.998**	0.993**	0.842**	0.396	0.313	-0.365	-0.406	-0.98**
Ni	0.998**	1	0.996**	0.994**	0.848**	0.404	0.316	-0.381	-0.415	-0.976**
Zn	0.998**	0.996**	1	0.997**	0.839**	0.417	0.357	-0.381	-0.386	-0.975**
Cu	0.993**	0.994**	0.997**	1	0.827**	0.467	0.396	-0.402	-0.403	-0.969**
Cd	0.842**	0.848**	0.839**	0.827**	1	-0.027	0.249	-0.15	-0.226	-0.8 **
Pb	0.396	0.404	0.417	0.467	-0.027	1	0.591	-0.647*	-0.453	-0.416
Cr	0.313	0.316	0.357	0.396	0.249	0.591	1	-0.523	-0.039	-0.281
As	-0.365	-0.381	-0.381	-0.402	-0.15	-0.647*	-0.523	1	0.599	0.448
Hg	-0.406	-0.415	-0.386	-0.403	-0.226	-0.453	-0.039	0.599	1	0.552
pH	-0.98**	-0.976**	-0.975**	-0.969**	-0.8 **	-0.416	-0.281	0.448	0.552	1

Fig.7 Heatmap of Pearson's correlation coefficients combined with cluster analysis of heavy metals and pH

* * Correlation is significant at the 0.01 level. * Correlation is significant at the 0.05 level

Table 10　PCA results of heavy metals in the Rongna River

Component	Initial eigenvalues			Rotation sums of squared loadings			Element	Principal component	
	Total	% of variance	Cumulative %	Total	% of variance	Cumulative %		1	2
1	5.52	61.30	61.30	5.52	61.30	61.30	Cu	0.98	0.16
2	1.80	19.98	81.27	1.80	19.98	81.27	Pb	0.55	−0.71
3	0.98	10.90	92.17				Zn	0.97	0.21
4	0.45	4.96	97.13				As	−0.56	0.69
5	0.22	2.45	99.58				Mn	0.97	0.23
6	0.03	0.39	99.97				Cd	0.79	0.51
7	0.002	0.021	99.99				Cr	0.48	−0.48
8	0.001	0.012	100.00				Ni	0.97	0.22
9	0	0	100.00				Hg	−0.52	0.38

PCA can effectively determine the source of heavy metal pollution[50]. Table 10 shows the PCA results for the heavy metals. Two principal components with a rotation value greater than 1 were extracted, and the cumulative contribution rate of the two principal components reached 81.27%, which can explain most of the information for the heavy metal elements. The principal component loading diagram (Fig.8) shows that the contribution rate of principal component 1 reached 61.30%, and mainly represented Cu, Zn, Ni, Mn and Cd. This agrees with the result of the correlation analysis and cluster analysis, further suggesting that Cu, Zn, Ni, Mn and Cd may come from the same source. Moreover, the main pollution in the water was also Cu, Zn, Ni, Mn and Cd, indicating that these heavy metals are mainly derived from AMD. Principal component 2 explained 1.9% of the total variance. The loadings of Hg and As were 69% and 51%, respectively, and the concentrations of Hg, As, Cr and Pb were similar to those of the control river Bolong River. This indicates that Hg, As, Cr and Pb may come from a background source related to rock weathering in the environment.

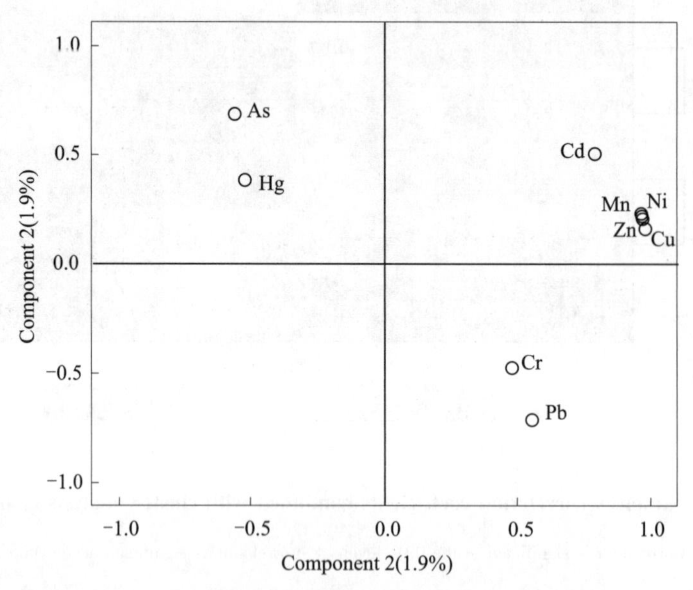

Fig.8　Score plot for the principal component analysis of heavy metals in the Rongna River

Conclusions

In this study, the concentration characteristics of heavy metals and the distribution of phytoplankton in water from the Rongna River within the unmined Tiegelongnan deposit area were investigated, and a risk assessment of heavy metal pollution in the river water was carried out. The results show that naturally occurring AMD caused serious heavy metal pollution in the Rongna River. Cu, Zn, Mn and Ni in the AMD exceeded the Grade Ⅲ national surface water environmental quality standard. After the AMD flows into the river, Mn and Ni exceeded the Grade Ⅲ national surface water environmental quality standard, while the concentrations of Cu, Zn, Mn, Pb, Cr and Ni were 60.36, 5.23, 25.65, 3.00, 2.15 and 2.60 times higher, respectively, than those in the water of the control river Bolong. The results of the heavy metal pollution evaluation indicate that the Rongna River is heavily polluted by heavy metals, with Mn posing a certain non-carcinogenic risk to humans. Thus, the water is no longer suitable for drinking or bathing. There were 30 species of algae detected in the Rongna River. The phytoplankton biomass was negatively correlated with the concentration of major heavy metals, indicating that the heavy metal concentration exceeded the tolerance limit of phytoplankton and thus affected their normal growth. Statistical analysis shows that the heavy metals Cu, Zn, Ni, Mn and Cd are mainly sourced from AMD, while Hg, As, Cr and Pb may come from rock weathering.

Author contributions

Investigation: Yuhu Luo, Qinxina Jia.

Methodology: Yuhu Luo, Jiaoping Rao, Qinxian Jia.

Software: Yuhu Luo, Jiaoping Rao.

Writing—original draft: Yuhu Luo.

Writing—review and editing: Jiaoping Rao, Qinxian Jia.

References

[1] Yaraghi N, Ronkanen A- K, Haghighi A T, Aminikhah M, Kujala K, Klove B. Impacts of gold mine effluent on water quality in a pristine sub-Arctic river. Journal of Hydrology. 2020;589.

[2] Chen M, Li F, Tao M, Hu L, Shi Y, Liu Y. Distribution and ecological risks of heavy metals in river sediments and overlying water in typical mining areas of China. Mar Pollut Bull. 2019;146.

[3] Luis A T, Antonio Grande J, Duraes N, Miguel Davila J, Santisteban M, Almeida S F P, et al. Biogeochemical characterization of surface waters in the Aljustrel mining area (South Portugal). Environ Geochem Health. 2019;41(5).

[4] Ruiz Canovas C, Riera J, Carrero S, Olias M. Dissolved and particulate metal fluxes in an AMD-affected stream under different hydrological conditions: The Odiel River (SW Spain). CATENA. 2018;165.

[5] Xiao H Y, Zhou W B, Zeng F P, Wu D S. Water chemistry and heavy metal distribution in an AMD highly contaminated river. Environmental Earth Sciences. 2010;59(5).

[6] Duruibe J O, Ogwuegbu M O C, Egwurugwu J N. Heavy metal pollution and human biotoxic effects. International Journal of the Physical Sciences. 2007;2(5).

[7] Affandi F A, Ishak M Y. Impacts of suspended sediment and metal pollution from mining activities on riverine fish populationa review. Environ Sci Pollut Res Int. 2019;26(17).

[8] Byrne P, Wood P J, Reid I. The Impairment of River Systems by Metal Mine Contamination: A Review Including Remediation Options. Crit Rev Environ Sci Technol. 2012;42(19).

[9] Zeng X, Liu Y, You S, Zeng G, Tan X, Hu X, et al. Spatial distribution, health risk assessment and statistical source identification of the trace elements in surface water from the Xiangjiang River, China. Environ Sci Pollut Res Int. 2015;22(12).

[10] Eggleton J, Thomas K V. A review of factors affecting the release and bioavailability of contaminants during sediment disturbance events. Environ Int. 2004;30(7).

[11]Long J, Luo K. Elements in surface and well water from the central North China Plain: Enrichment patterns, origins, and health risk assessment. Environ Pollut. 2020;258.

[12]Turdi M, Yang L. Trace Elements Contamination and Human Health Risk Assessment in Drinking Water from the Agricultural and Pastoral Areas of Bay County, Xinjiang, China. Int J Environ Res Public Health. 2016;13(10).

[13]Bird G. The influence of the scale of mining activity and mine site remediation on the contamination legacy of historical metal mining activity. Environ Sci Pollut Res Int. 2016;23(23).

[14]Jung M C. Contamination by Cd, Cu, Pb, and Zn in mine wastes from abandoned metal mines classified as mineralization types in Korea. Environ Geochem Health. 2008;30(3).

[15]Sarmiento A M, Miguel Nieto J, Olias M, Canovas C R. Hydrochemical characteristics and seasonal influence on the pollution by acid mine drainage in the Odiel river Basin (SW Spain). Appl Geochem. 2009;24(4).

[16]Lefcort H, Vancura J, Lider E L. 75 years after mining ends stream insect diversity is still affected by heavy metals. Ecotoxicology. 2010;19(8):1416-25.

[17]Hammarstrom J M, Seal R R, Meier A L, Jackson J C. Weathering of sulfidic shale and copper mine waste: secondary minerals and metal cycling in Great Smoky Mountains National Park, Tennessee, and North Carolina, USA. Environ Geol. 2003;45(1):35-57.

[18]Liu G, Tao L, Liu X, Hou J, Wang A, Li R. Heavy metal speciation and pollution of agricultural soils along Jishui River in non-ferrous metal mine area in Jiangxi Province, China. J Geochem Explor. 2013;132.

[19]Fu S M, Zhou Y Z, Zhang C B, Yang X Q, Ding J, Zhao Y Y. Environmental response to manganese contamination of Dabaoshan mine in environmental system of lower reaches. Acta Scientiarum Naturalium Universitatis Sunyatseni. 2007; 46(2).

[20]Tang J, Song Y, Wang Q, Lin B, Yang C, Guo N, et al. Geological Characteristics and Exploration Model of the Tiegelongnan Cu (Au-Ag) Deposit: The First Ten Million Tons Metal Resources of a Porphyry-epithermal Deposit in Tibet. Acta Geoscientia Sinica. 2016;37(6).

[21]Hu H, Wei Y. Chinese freshwater algae:systems,ecology and classification. Beijing: Science Press; 2006.

[22] Chinese Academy of Sciences Qinghai-Tibet Plateau Comprehensive Scientific Expedition Team Team. Tibetan algae. Beijing: Science Press; 1992.

[23]Mishra S, Kumar A, Shukla P. Estimation of heavy metal contamination in the Hindon River, India: an environmetric approach. Applied Water Science. 2021;11(1).

[24]Wu J, Lu J, Zhang C, Zhang Y, Lin Y, Xu J. Pollution, sources, and risks of heavy metals in coastal waters of China. Human and Ecological Risk Assessment. 2020;26(8).

[25]Hakanson L. AN ECOLOGICAL RISK INDEX FOR AQUATIC POLLUTION-CONTROL-A SEDIMENTOLOGICAL APPROACH. Water Res. 1980;14(8):975-1001.

[26]NEPAC. Environmental Quality Standards for Surface Water. 2002.

[27]Chen H, Teng Y, Lu S, Wang Y, Wang J. Contamination features and health risk of soil heavy metals in China. Sci Total Environ. 2015;512.

[28]Zhang C, Qiao Q, Piper J D A, Huang B. Assessment of heavy metal pollution from a Fe-smelting plant in urban river sediments using environmental magnetic and geochemical methods. Environ Pollut. 2011;159(10).

[29]Cheng H, Li M, Zhao C, Li K, Peng M, Qin A, et al. Overview of trace metals in the urban soil of 31 metropolises in China. J Geochem Explor. 2014;139:31-52.

[30]Gao B, Gao L, Gao J, Xu D, Wang Q, Sun K. Simultaneous evaluations of occurrence and probabilistic human health risk associated with trace elements in typical drinking water sources from major river basins in China. Sci Total Environ. 2019;666.

[31]Xiao J, Wang L, Deng L, Jin Z. Characteristics, sources, water quality and health risk assessment of trace elements in river water and well water in the Chinese Loess Plateau. Sci Total Environ. 2019;650.

[32]USEPA. Risk Assessment Guidance for Superfund Volume I: Human Health Evaluation Manual (Part E, Supplemental Guidance for Dermal Risk Assessment). Washington DC, USA: Office of Superfund Remediation and Technology Innovation U.S. Environmental Protection Agency; 2004.

[33]Tong S, Li H, Tudi M, Yuan X, Yang L. Comparison of characteristics, water quality and health risk assessment of trace elements in surface water and groundwater in China. Ecotoxicol Environ Saf. 2021;219.

[34]Githaiga K B, Njuguna S M, Gituru R W, Yan X. Water quality assessment, multivariate analysis and human health risks of heavy metals in eight major lakes in Kenya. J Environ Manag. 2021;297.

[35]Wang J, Liu G, Liu H, Lam P K S. Multivariate statistical evaluation of dissolved trace elements and a water quality assessment in the middle reaches of Huaihe River, Anhui, China. Sci Total Environ. 2017;583.

[36]Zeng J, Han G, Yang K. Assessment and sources of heavy metals in suspended particulate matter in a tropical catchment, northeast Thailand. Journal of Cleaner Production. 2020;265.

[37]Wei W, Ma R, Sun Z, Zhou A, Bu J, Long X, et al. Effects of Mining Activities on the Release of Heavy Metals (HMs) in a Typical Mountain Headwater Region, the Qinghai-Tibet Plateau in China. Int J Environ Res Public Health. 2018;15(9).

[38]Huang X, Sillanpaa M, Gjessing E T, Peraniemi S, Vogt RD. Environmental impact of mining activities on the surface water quality in Tibet: Gyama valley. Sci Total Environ. 2010;408(19).

[39]Qu B, Zhang Y, Kang S, Sillanpaa M. Water quality in the Tibetan Plateau: Major ions and trace elements in rivers of the"Water Tower of Asia". Sci Total Environ. 2019;649.

[40]Mao G, Zhao Y, Zhang F, Liu J, Huang X. Spatiotemporal variability of heavy metals and identification of potential source tracers in the surface water of the Lhasa River basin. Environ Sci Pollut Res Int. 2019;26(8).

[41]Zhang Y, Li F, Li J, Liu Q, Tu C, Suzuki Y, et al. Spatial Distribution, Potential Sources, and Risk Assessment of Trace Metals of Groundwater in the North China Plain. Human and Ecological Risk Assessment. 2015;21(3).

[42]Field C B, Behrenfeld M J, Randerson J T, Falkowski P. Primary production of the biosphere: Integrating terrestrial and oceanic components. Science. 1998;281(5374).

[43]Beardall J, Young E, Roberts S. Approaches for determining phytoplankton nutrient limitation. Aquatic Sciences. 2001;63(1):44-69.

[44]Nweze N O. Seasonal variations in phytoplankton populations in Ogelube Lake, a small natural West African Lake. Lakes Reserv Res Manag. 2006;11(2):63-72.

[45]Bellinger E G, Sigee D C, Bellinger E G, Sigee D C. Freshwater Algae: Identification, Enumeration and Use as Bioindicators, 2nd Edition2015.

[46]Verma S K, Singh H N. EVIDENCE FOR ENERGY-DEPENDENT COPPER EFFLUX AS A MECHANISM OF CU2+ RESISTANCE IN THE CYANOBACTERIUM NOSTOC-CALCICOLA. FEMS Microbiol Lett. 1991;84(3).

[47]Fiore M F, Trevors J T. CELL COMPOSITION AND METAL TOLERANCE IN CYANOBACTERIA. Biometals. 1994;7(2):83-103.

[48]Jie QIN, Wen Z, Peng Z. The environment mercury pollution toxicity effect to the alga and their influencing factors. J Biol. 2011;28(3).

[49]Gailey F A, Lloyd O L. GRASS AND SURFACE SOILS AS MONITORS OF ATMOSPHERIC METAL POLLUTION IN CENTRAL SCOTLAND. Water Air Soil Pollut. 1985;24(1):1-18.

[50]Gao S, Wang Z, Wu Q, Zeng J. Multivariate statistical evaluation of dissolved heavy metals and a water quality assessment in the Lake Aha watershed, Southwest China. Peerj. 2020;8.

本文曾发表于2022年《Plos One》第17卷第4期

Effects of Oil Pollution on Soil Microbial Diversity in the Loess Hilly Areas, China

Lei Shi[1,2], Zhongzheng Liu[1], Liangyan Yang[1], Wangtao Fan[1]

1.Shaanxi Provincial Land Engineering Construction Group Co., Ltd, Xi'an, 710075, China;
2.School of Human Settlements and Civil Engineering, Xi'an Jiaotong University, Xi'an 710075, China)

【Abstract】 Purpose: Data support and theoretical basis for bioremediation and treatment of petroleum-contaminated soils in the Loess hills of Yan'an, northern Shaanxi. Methods: The evolutionary characteristics of soil microbial diversity and community structure under different levels of oil pollution were studied by field sampling, indoor simulation experiments, and analyzed through assays, using the mine soils from Yan'an, Shaanxi Province, as the research object. Results: Compared with clean soil, the microbial species in contaminated soil were significantly reduced, the dominant flora changed, and the flora capable of degrading petroleum pollutants increased significantly. The soil microbial diversity and community structure differed, although not significantly, between different pollution levels, but significantly from clean soil. In the uncontaminated soil (CK), the dominant soil microbial genera were mainly *Pantoea*, *Sphingomonas*, *Thiothrix*, and *Nocardioides*. The abundance of *Pseudomonas*, *Pedobacter*, *Massilia*, *Nocardioides* and *Acinetobacter* in the soil increased after oil contamination, while *Thiothrix*, *Sphingomonas* and *Gemmatimonas* decreased significantly. Conclusions: After the soil was contaminated with petroleum, the microbial species in the soil decreased significantly, the dominant genera in the soil changed, and the relative abundance of bacteria groups capable of degrading petroleum pollutants increased. The genera that can degrade petroleum pollutants in the petroleum-contaminated soil in the study area mainly include *Pseudomonas*, *Acinetobacter*, *Pedobacter*, *Acinetobacter and Nocardioides*, *etc.*, which provide a scientific basis for exploring It provides a scientific basis for exploring remediation methods suitable for petroleum-contaminated soil in this region.

【Keywords】 Loess hills; Oil pollution; Microbial diversity; Microbial community structure

Introduction

The loess hills area in northern Shaanxi, China, is an important oil and gas reserve and development area. Oil well leaks, crude oil pipeline, and storage tank leaks during the development of oil and gas fields cause petroleum-based products to fall on the ground soil, causing serious damage to the local ecological environment, in turn, affecting the ecological balance and leads to gradual environmental degradation (Zhen et al. 2015; Wang et al. 2017). Petroleum hydrocarbons are hard-to-degrade organic pollutants that can seriously affect soil productivity and cause different degrees of harm to microorganisms (Mambwe et al. 2021), plants, and animals in the ecosphere around the pollution source (Wang et al. 2016). The entry of petroleum-based substances into the soil has caused a dramatic increase in soil organic carbon content, resulting in an imbalance between soil C content and N content (He et al. 2021; Zhang et al.2018; Wen et al. 2017). This imbalance destroys the habitat for soil microbial life and changes the number and diversity of soil microbial populations (Zheng et al. 2021; Dong 2020). Microorganisms play a very important role in maintaining soil ecosystem functions; they can effectively decompose plant and animal residues, promote soil C, N, S, P, and other nutrient cycle (Pan et al. 2012; Li et al. 2017; Mu et al. 1994), regulate soil material and energy cycle, etc. (Iyobosa et al. 2021; Kong et al. 2021; Wang 2021). Therefore, it was important to investigate the current status of petroleum-contaminated soil in northern Shaanxi and analyze the effects of petroleum pollution on the diversity of soil microbial populations to explore remediation methods suitable for petroleum-contaminated soil in the region (Ajona et al. 2021; Fan et al. 2015).

The current research on microorganisms of petroleum-contaminated soil has mainly focused on the remediation effect of soil microorganisms (screening of efficient degradation bacteria, degradation characteristics, etc.) (Ueno et al. 2010), while the structural diversity of indigenous microbial communities of petroleum-contaminated soil in Loess hilly areas has been less studied (Margesin et al. 2007). In this study, we investigated the soil microbial diversity under the effects of different levels of oil pollution and analyzed the response of soil microbial community structure to oil pollution through field sampling (Yang et al. 2013), indoor simulation tests, detection, and analysis, to provide a scientific basis for bioremediation and treatment of oil-contaminated soil in the Loess hills of northern Shaanxi Province. This study aims to provide a scientific basis for the bioremediation and treatment of oil-contaminated soil in the Loess hills (Wang et al. 2020; Li et al. 2021).

Materials and methods

Sample collection

The test soil for this study was collected from a typical oil well area in Baota District, Yan'an City, Shaanxi Province, China. The soil samples were collected in the 0~30 cm range by the S-spotting method in the vicinity of oil wells, selected from areas not contaminated with oil, following the principles of random, equal, and multi-point mixing. The samples were then passed through a 2-mm sieve to remove impurities and preserved at low temperature for indoor simulation tests (Li et al. 2009).

Indoor simulation test

Crude oil was weighed according to the ratio of oil content of 15 g/kg (C15), 30 g/kg (C30) and 45 g/kg (C45), added to clean soil, and mixed thoroughly to prepare different oil contamination gradient soil (Li et al. 2011). After one week of static aging, indoor simulated incubation tests were carried out, i.e., incubation under constant conditions (ambient temperature controlled at (25±2)℃ and soil moisture content maintained at 15%) for 21 days, with three replicates for each treatment (Table 1). After incubation, soil samples were collected from four different locations in each container using the multi-point mixed soil sample collection method, and then mixed thoroughly to form a mixed sample of about 20 g for determining the microbiological properties of the soil (Liu et al. 2012; Liu et al. 2020).

Table 1 Experimental treatments

Oil pollution levels (g/kg)	Sample number		
0 (CK)	CK 1	CK 2	CK 3
15 (C15)	C 1-1	C 1-2	C 1-3
30 (C30)	C 2-1	C 2-2	C 2-3
45 (C45)	C 3-1	C 3-2	C 3-3

Testing indicators and measurement methods

Soil microbial diversity was determined by DNA extraction and high-throughput sequencing, and the genomic DNA was extracted by CTAB or SDS method, followed by agarose gel electrophoresis to check the purity and concentration of DNA. The PCR was performed using specific primers with Barcode, Phusion High-Fidelity PCR Master Mix with GC Buffer from New England Biolabs, and high efficiency high fidelity enzymes to ensure amplification efficiency and accuracy. The primers corresponding to the region for the identification of bacterial diversity were 16S V4 region primers (515F and 806R) (Zhang et al. 2022).

The PCR products were detected by electrophoresis using a 2% agarose gel; the PCR products that passed the test were purified by magnetic beads, quantified by enzyme labeling, mixed in equal amounts according to the concentration of PCR products, mixed thoroughly and then detected by electrophoresis using a 2% agarose gel, and the target bands were recovered using a gel recovery kit provided by Qiagen (Fig.1). The library was constructed using TruSeq DNA PCR-Free Sample Preparation Kit. The constructed library was quantified by Qubit and Q-PCR, and after the library was qualified, the library was sequenced using NovaSeq6000 (Lusk et al. 2021).

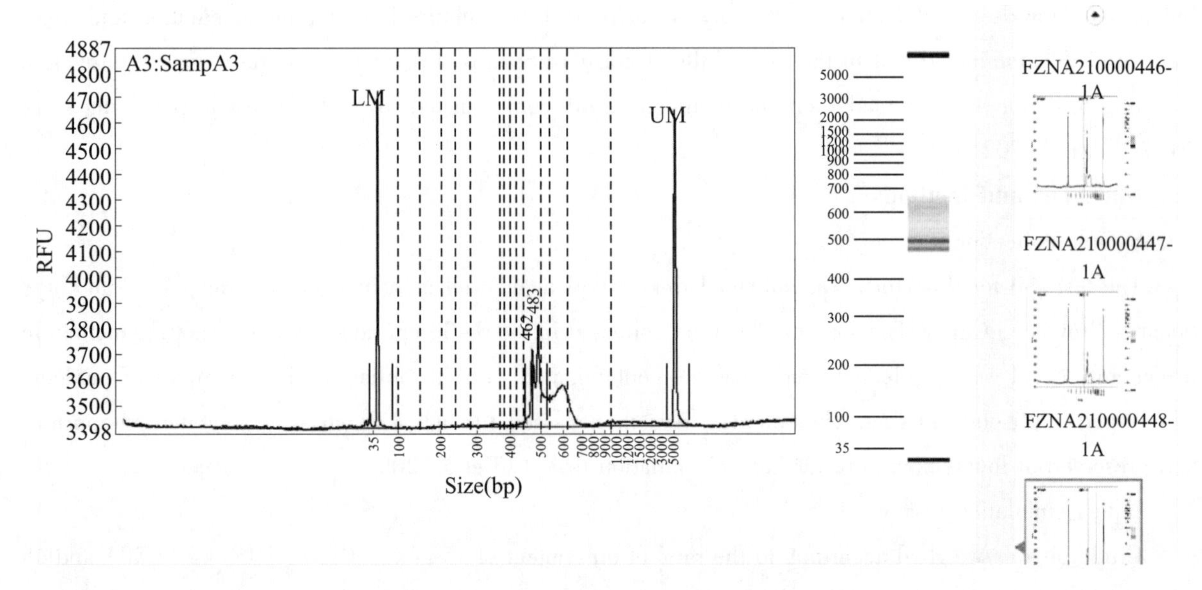

Fig.1 Amplification product result graph

Statistical analysis of data

According to Barcode sequence and PCR amplification primer sequence to split each sample data from the downstream data, truncated the Barcode and primer sequence and used FLASH (V1.2.7) to splice the reads of each sample (Liu et al. 2001), the spliced sequence was the original Tags data (Raw Tags); the spliced Raw Tags, need to go through strictly. The spliced Raw Tags need to be filtered strictly to get the high quality Tags data (Clean Tags). Referring to the quality control process of Tags in Qiime (V1.9.1), the following operations were performed: a) Tags interception: Raw Tags were truncated from the first low-quality site where the number of consecutive low-quality bases (default quality threshold ≤19) reached a set length (default length value is 3) (Ali et al. 2020); b) Tags length filtering: Tags were intercepted to obtain the Tags The Tags obtained after the above processing need to be processed to remove the chimeric sequences, and the Tags sequences were compared with the species annotation database to detect the chimeric sequences, and finally removed the chimeric sequences to obtain the final effective data (Effective Tags) (Wagh et al. 2001).

Results and discussion

Dilution curve

For the dilution curve, a certain number of sequences were randomly selected from the sample, and the Alpha diversity index of these sequences corresponding to the sample was counted. The horizontal coordinate represents the number of sequencing strips randomly selected from the sample, and the vertical coordinate represents the number of operational taxonomic units (OTUs) that can be constructed based on this number of sequencing strips to plot the curve, and the adequacy of the current sequencing data volume is judged according to whether the curve reaches a flatness (Ma et al. 2020). According to Fig.2, the microbial OTU in petroleum-contaminated soils was significantly reduced compared to the uncontaminated controls, i.e., the overall number of microbial species decreased. The number of soil microbial OTUs at different oil contamination concentrations was in the order of C15> C45> C30, where the difference between C30, and C45 was small, indicating that the number of soil microbial species first decreased and then leveled off as the contamination level increased (Chen et al. 2020).

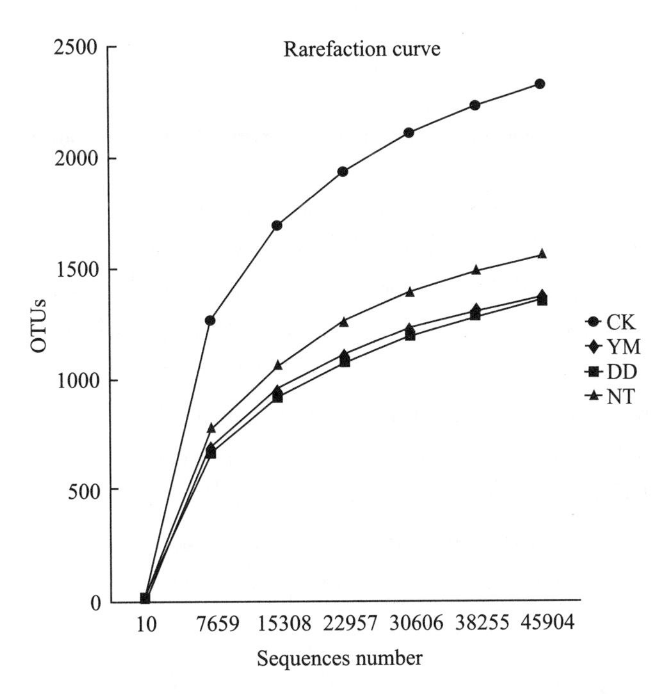

Fig.2 Dilution curves of soil microorganisms under different oil pollution concentrations

Microbial alpha (α)-diversity-analysis for variability between groups

Alpha diversity(α-diversity) refers to the species diversity within a community or habitat, mainly focusing on the species diversity within the community. The common method is to calculate Chao, Shannon, Simpson and coverage 4 indexes based on OTU results for biodiversity analysis(Liu YP, et al. 2021). In this study, the Shannon and Simpson indices were selected to characterize the α-diversity of soil contaminated with different additives, and to examine the differences between treatments, the inter-group difference test of indices was used. The results showed that the microbial α-diversity of petroleum-contaminated soils (C15, C30, C45) was significantly lower than the control group with no petroleum contamination, but the differences in soil microbial α-diversity between varying concentrations of contamination did not reach a statistically significant level (Fig.3).

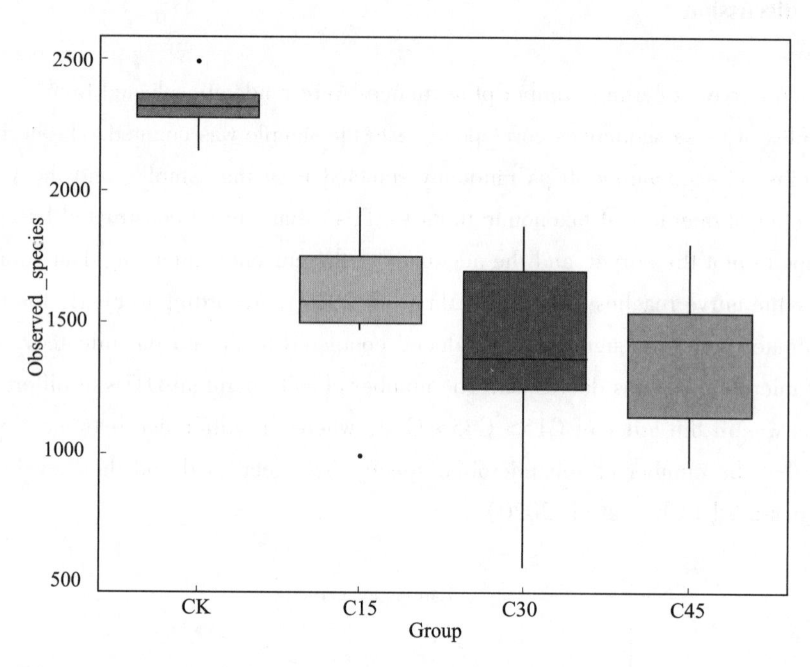

Fig.3 Soil microbial α-diversity under different oil pollution concentrations

Microbial species composition analysis through Venn diagrams

Venn diagrams were obtained by comparing OTUs between samples or between subgroups, which can visualize the similarity and overlap of OTU composition of environmental samples (Wang et al. 2019). The results showed that each soil sample produced a total of 14090 OTUs, of which 1791 were shared, accounting for 12.71% of the total OTUs (Fig. 4). the number of OTUs in the C15, C30 and C45 treatments were 3719, 3194 and 3345, respectively, which were lower than that of CK (3832). The number of shared OUTs between contaminated and clean soils showed a decreasing trend as the oil contamination gradient increased. The highest overlap between different pollution gradients was found in the C15 and C45 treatments, which had a high similarity of bacterial community structure. In addition, the different oil contamination treatments did not significantly affect the number of soil-specific OTUs.

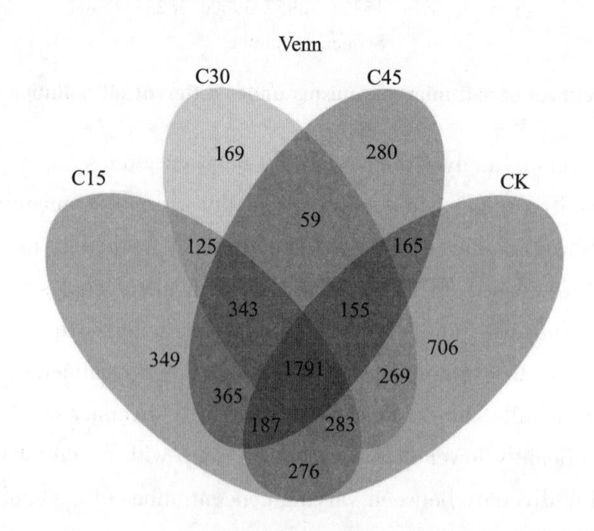

Fig.4 Venn diagram of soil microbial OTU numbers at different oil pollution concentrations concentrations

Microbial β-diversity analysis for variability between groups

To study the similarity or difference relationship of different sample community structures, cluster analysis was performed on the sample community distance matrix to build a sample hierarchical clustering tree. This method effectively identifies the "major" elements and structures in the data, removes noise and redundancy, reduces the dimensionality of the original complex data, and reveals the simple structure hidden behind the complex data (Zhang et al. 2017).

Fig.5 shows that the different shape legends in the Fig.represent the control and three soil samples with different concentrations of oil contaminants. The differences in the microbial community composition of CK, C15, C30, and C45 were found by principal component analysis. The microbial community composition of the contaminated soil differed, although not significantly, between the different contamination treatments and differed significantly From CK. the PC1 and PC2 axes explain 20.03% and 9.23% of the results, respectively.

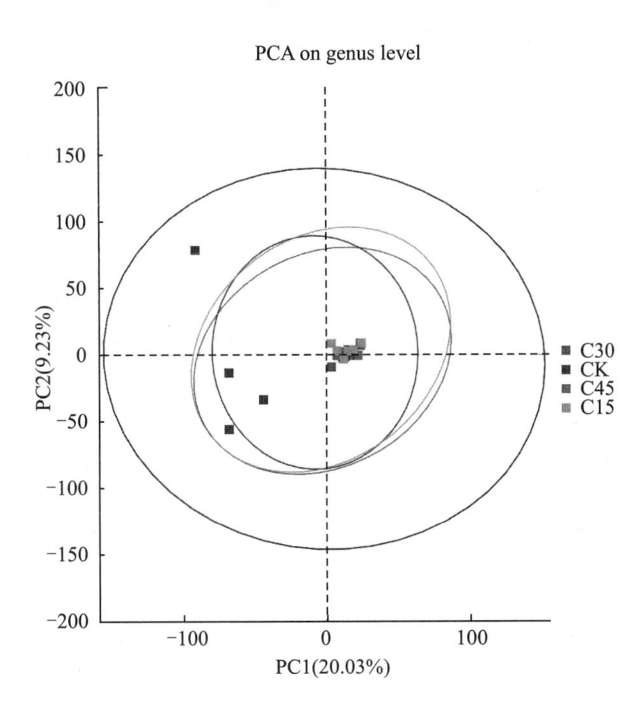

Fig.5 Principal component analysis of soil microorganisms under different oil pollution concentrations

Differential analysis of microbial community composition

The histogram presents the community composition and species abundance at different taxonomic levels (Zhu et al. 2020). In this study, community composition and species abundance analysis were conducted at the genus level. In the absence of oil contamination (CK), the dominant genera included *Pantoea*, *Streptomyces*, *Alkanindiges*, *Massilia*, etc. After oil contamination, the dominant genera of the microbial community changed significantly, with *Pantoea* being dominant. After D2 treatment, the abundance of *Streptomyces* increased, and the difference between C30 and C45 decreased (Fig.6). In addition, the abundance of *Nocardioides* and *Acinetobacter* increased with increasing oil pollution concentration, while *Thiothrix*, *Sphinggomonas*, and *Gemmatimonas* decreased significantly.

Conclusions

In this study, we used high-throughput sequencing technology to analyze soil microbial diversity under different gradients of petroleum pollution concentration and sequenced 16S amplicons by the soil, and the Il-

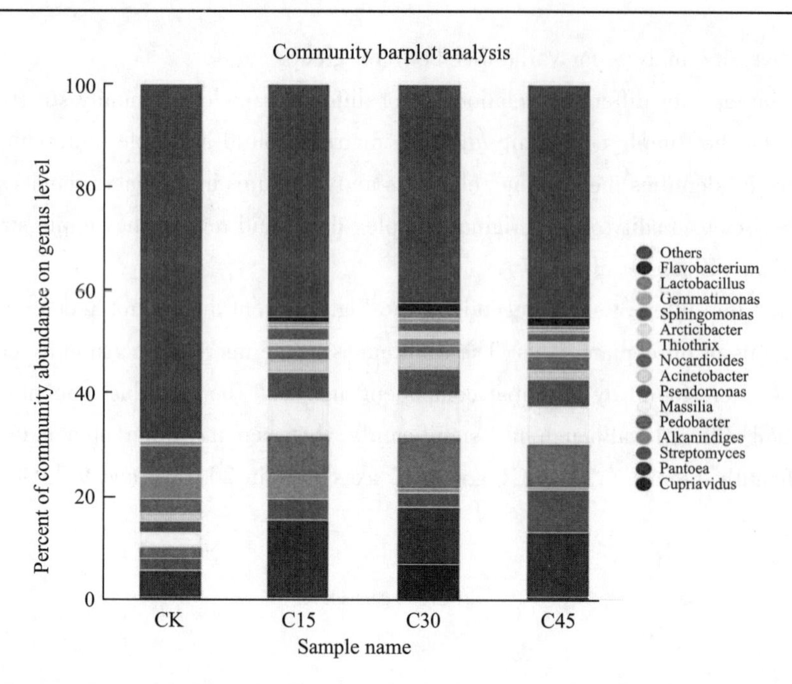

Fig.6 Community composition characteristics of soil microorganisms at different genus levels

luminaMiSeq platform. The results showed that the microbial species in contaminated soil were more abundant than in clean soil, the main dominant bacterial groups were changed, and significantly more bacteria could degrade petroleum pollutants. At the same time, while the soil microbial diversity and community structure differed, it was not significant between different contamination levels and differed significantly from that of clean soil, consistent with the results of previous studies (Wu et al. 2020; Xu et al. 2021; Sun et al. 2021; Zhao et al. 2020).

At the genus level, *Pantoea*, *Sphingomonas*, *Thiothrix*, and *Nocardioides* were dominant in uncontaminated soil in high abundance. The intake of petroleum hydrocarbons resulted in a decrease in the abundance of *Thiothrix*, *Sphingomonas*, *Gemmatimonas*, *Pseudomonas*, *Acinetobacter*, *Pedobacter*, and the abundance of *Pseudomonas*, *Acinetobacter*, *Pedobacter*, and *Massilia* increased significantly. This may be because petroleum contaminants caused changes in water content, total organic carbon, total petroleum hydrocarbons, total carbon, total sulfur, and pH values, which are the limiting factors for the growth of the dominant genera (Liang et al. 2016), thus changing the relative abundance of *Sphingomonas*, *Thiothrix*, and *Pseudomonas*. The oleophilic bacteria genera in the contaminated soil in the typical oil extraction area in the Loess hilly area mainly include *Pseudomonas*, *Acinetobacter*, *Pedobacter*, *Acinetobacter*, *Nocardioides*, and *Alkanindiges* among others (Xu et al. 2021; Zhen et al. 2015; Zhen et al. 2021).

These data can help us to better understand the evolutionary characteristics of various indigenous microorganisms in oil-contaminated soils in the region and provide theoretical reference and data support for remediation and treatment technologies of oil-contaminated soils in the region. In the future, several kinds of oleophilic indigenous bacteria need to be screened and domesticated from the contaminated soil in the study area to make bacteriological agents and conduct indoor simulation experiments to investigate and explore the applicability of indigenous oleophilic microorganisms in the bioremediation of petroleum-contaminated soil.

Acknowledgments

This work was supported by the projects of "Technology Innovation Leading Program of Shaanxi (Program No. 2021CGBX-03)" and Shaanxi Provincial Land Engineering Construction Group (Program No. DJNY2021-33).

Ethical statements

I certify that this manuscript is original, has not been published anywhere, and will not be submitted elsewhere for publication while being considered by the Annals of Microbiology. The study is not split up into several parts to increase the number of submissions and submitted to various journals or to one journal over time. No data have been fabricated or manipulated (including images) to support the conclusions. No data, text, or theories by others are presented as if they were our own.

The submission has been received explicitly from all co-authors, and authors whose names appear on the submission have contributed sufficiently to the scientific work and therefore share collective responsibility and accountability for the results.

Authors' contributions

LS: guided the completion of this study, completed the first draft paper and made revisions. WF: guided the experimental methods and thesis writing. ZL: performed experiments, recorded data, and data analysis. LY: performed data analysis. All authors read and approved the final manuscript.

Funding

This work was supported by the projects of "Technology Innovation Leading Program of Shaanxi (Program No. 2021CGBX-03)", and Shaanxi Provincial Land Engineering Construction Group (Program No. DJNY2021-33, DJNY2022-37).

Availability of data and materials

The data were obtained by the authors.

Declarations

Ethics approval and consent to participate: The study did not violate ethics, and all participants agreed to publish the paper.

Consent for publication: Not applicable.

Competing interests: The authors declare that they have no competing interests.

References

Ajona M, Vasanthi P. (2021) Bio-remediation of crude oil contaminated soil using recombinant native microbial strain. Environmental Technology & Innovation, 23(11): 101635.

Ali F, Nadeem M A, Habyarimana E, et al. (2020) Molecular characterization of genetic diversity and similarity centers of safflower accessions with ISSR markers. Brazilian Journal of Botany, 43(1): 1-13.

Chen F F, Cong X, Xiang J Z, et al. (2020) Application of Illumina MiSeq high-throughput sequencing technology to analyze the inter-rooted soil microbial diversity of Corydalis sativus. Hubei Agricultural Science, 59(17): 6.

Dong Z J. (2020) Study of soil contamination law by seepage of an oil pipeline in loess area. Engineering Survey, 48(5): 6.

Fan G P, Zhu H Y, Hao X Z, et al. (2015) Electrokinetic remediation of an electroplating contaminated soil with different enhancing electrolytes. China Environmental Science, 35(5): 1458-1465.

He Y J, Zhou K P, Rao Y X, et al. (2021) Environmental risks of antibiotics in soil and bioremediation technology of contaminated soil. Journal of Biological Engineering, 37(10): 18.

Iyobosa E, Zhu, S F., Ning H J, et al. (2021) Development of a robust bacterial consortium for petroleum hydrocarbon degradation. Fresenius Environmental Bulletin, 30(3): 2356-2367.

Kong L L, Wang X Y, Gu M G, et al. (2021) Analysis of factors influencing enhanced biodegradation of petroleum hydrocarbons by biochar. Environmental Science and Technology, 44(3): 8.

Li J, Cao X T, Sui H, et al. (2017) Current status and prospect of research on remediation technology of petroleum-contaminated soil. Journal of Petroleum (Petroleum Processing), 033(005): 811-833.

Li J L, Liu G G, Cao S X, et al. (2011) Optimization of RAPD-PCR reaction system for partridge tea using one-way and orthogonal design. Journal of Agronomy, 1(4): 8.

Li Q, He L S, Wang Y F, et al. (2021) Soil contamination characteristics and distribution of smelting industry sites in China. Journal of Ecology and Environment, 30(3): 10.

Li S L, Liu G B, Xu M X. (2009) Characteristics of petroleum pollution in soils and surface waters of oil fields in northern Shaanxi. Soil and Water Conservation Research, 16(5): 4.

Liang J F, Yang J K, Yang Y, et al. (2016) Relationship between bacterial community structure and environmental factors in oil-contaminated soils of the Karamay oilfield. Journal of Microbiology, 56(8): 13.

Liu J M, Peng J P. (2001) One-way analysis of variance for biological experimental data. Journal of Zoology, 36(6): 4.

Liu L, Wang D S, Zeng Q S, et al. (2012) Homogeneous design experiments for microwave remediation of crude oil-contaminated soil. Journal of Environmental Engineering, 6(6): 5.

Liu M Y, Ma J H, Li Y, et al. (2020) Effects of 16S rRNA gene high variation regions V4 and V3-V4 and sequencing depth on the analysis of bacterial flora in oil reservoirs. Microbiology Bulletin, 2: 10.

Liu Y P, Mao Y F, Hu Y L, et al. (2021) Effects of grass cultivation on soil microbial diversity, enzyme activity and carbon fraction in apple orchards. Journal of Plant Nutrition and Fertilizer, 27(10): 14.

Lusk R, Stene E, Banaei-Kashani F, et al. (2021) Aptardi predicts polyadenylation sites in sample-specific transcriptomes using high-throughput RNA sequencing and DNA sequence. Nature Communications, 12(1): 1652.

Ma R, Jiang M, Li H, et al. (2020) Analysis of microbial diversity in northeastern bean paste based on high-throughput sequencing technology. Food Industry Science and Technology, 41(12): 6.

Mambwe M, Kalebaila K K, Johnson T. (2021) Remediation technologies for oil contaminated soil. Global Journal of Environmental Science and Management, 1: 1-20.

Margesin R, Mmerle M, Tscherko D. (2007) Microbial activity and community composition during bioremediation of diesel-oil-contaminated soil: effects of hydrocarbon concentration, fertilizers, and incubation time. Microbial Ecology, 53(2): 259-269.

Mu C R. (1994) The impact of oil development on the ecological environment of the Loess Plateau. Geographical Studies, 13(4): 9.

Pan F, Chen L H, Fa S J, et al. (2012) A study on the transport performance of the petroleum contaminants in soil of the Longdong loess plateau. Journal of Environmental Science, 32(2): 410-418.

Sun J, Wang N, Chen H K, et al. (2021) Microbial community distribution characteristics of petroleum-contaminated soil. Journal of Petroleum (Petroleum Processing), 37(2): 12.

Sun Y R, Tan L, Chen Y. (2018) Soil contamination in typical oilfield areas in arid regions. Investigation and analysis of soil petroleum pollution in typical oilfield areas in arid regions. Environment and Development, 30(11): 2.

Wagh G A, Tiwari P D. (2020) On the diversity and abundance of avian species from grassland and wetland areas of an industrial zone of tropical Maharashtra. Bioscience Biotechnology Research Communications, 13(2): 770-780.

Wang D, Ma B, Gao H, et al. (2020) Study on the remediation characteristics of petroleum-contaminated soil by microbial enhancement. Journal of Agricultural Environmental Science, 39(12): 8.

Wang J, Qi Y C, Feng Q, et al. (2016) Relationship between the restoration of oil-contaminated soil vegetation and soil CAT activity on the Loess Plateau. Shaanxi Agricultural Science, 62(6): 6.

Wang J C, Jing M B, Duan C Y, et al. (2017) Effects of petroleum hydrocarbon pollution on the biological and non-biological properties of soils on the Loess Plateau of Longdong. Soil and Water Conservation Bulletin, 37(1): 8.

Wang J D, Qu C T, Song S F. (2021) Temperature-induced changes in the proteome of Pseudomonas aeruginosa during petroleum hydrocarbon degradation. Archives of Microbiology, 1: 1-11.

Wang R, Zhu J, Jin T, et al. (2019) Characterization of soil ammonia-oxidizing microbial abundance and community structure in rice fields under rice and shrimp crop model. Journal of Plant Nutrition and Fertilizer, 25(11): 13.

Wen G Y, Zhao Q Z, Ma H G, et al. (2017) Research and application of high-efficiency and low-cost bioremediation technology for petroleum-contaminated soil. China Safety Production Science and Technology, S1: 5.

Wu H, Song A, Zheng J, et al. (2020) Progress of functional genes for microbial degradation of petroleum hydrocarbons. Microbiology Bulletin, 47(10): 14.

Xu Y R, Wu M L, Wang L, et al. (2021) Changes in microbial populations and influencing factors of petroleum-contaminated soils in northern Shaanxi. China Environmental Science, 41(9): 11.

Yang M Q, Li L M, Li C, et al. (2013) Microbial community structure and distribution characteristics of petroleum-contaminated soil. Environmental Science, 34(2): 6.

Zhang L, Zhao Q, Wu W N, et al. (2018) Current status and prospect of bioremediation technology for petroleum-contaminated soil. Modern Chemical Industry, 38(1): 5.

Zhang J, Xu M. (2022) Characteristics of soil bacterial community diversity in different vegetation types in loess hilly areas. Journal of Ecology and Rural Environment, 38(2): 225-235.

Zhang G Q, Zhao P, Dong Y X, et al. (2017) High-throughput sequencing analysis of the effect of environmental fertilizer enhancers on the change of soil fungal diversity in potato inter-rhizosphere. Microbiology Bulletin, 44(11): 8.

Zhao M Y, Wang X, Li F M, Guo S H, et al. (2020) Soil oil pollution and its microbial community characteristics in Liaohe oilfield. Journal of Applied Ecology, 31(12): 10.

Zhen L, Gu J, Hu T, et al. (2015) Microbial community structure and metabolic characteristics of petroleum-contaminated soils on the Loess Plateau. Journal of Ecology, 35(17): 8.

Zheng Y M, He X S, Shan G C. (2021) Effects of oil pollution on bacterial communities in sites and their feedback mechanisms. Environmental Science Research, 34(4): 9.

Zhu S S, Xia B, Hao W L, et al. (2020) Functional diversity of soil microbial communities on eroded slopes in loess areas. China Environmental Science, 40(9): 7.

本文曾发表于2022年《Scientific Reports》第12卷

Effect of Plant Waste Addition as Exogenous Nutrients on Microbia Remediation of Petroleum Contaminated Soil

Lei Shi[1,2], Zhongzheng Liu[1], Liangyan Yang[1], Wangtao Fan[1]

(1.Shaanxi Provincial Land Engineering Construction Group Co., Ltd, Xi'an, 710075, China;
2.School of Human Settlements and Civil Engineering, Xi'an Jiaotong University, Xi'an 710075, China)

【**Abstract**】Purpose: This study investigates the feasibility of bio-enhanced microbial remediation of petroleum-contaminated soil, and analyzes the effect of different plant wastes as exogenous stimulants on microbial remediation of petroleum-contaminated soil and the effect on soil microbial community structure, in order to guide the remediation of soil in long-term petroleum-contaminated areas with nutrient-poor soils. Methods: The study was conducted in a representative oil extraction area in the Loess Hills, a typical ecologically fragile area in China. Through indoor simulated addition tests, combined with the determination of soil chemical and microbiological properties, the degradation efficiency of petroleum pollutants and the response characteristics of soil microbial community structure to the addition of different plant wastes in the area were comprehensively analyzed to obtain the optimal exogenous additive and explore the strengthening mechanism of plant wastes on microbial remediation of petroleum-contaminated soil. Results: Compared with the naturally decaying petroleum-contaminated soil, the addition of plant waste increased the degradation rate of petroleum pollutants, that is, it strengthened the degradation power of indigenous degrading bacteria on petroleum pollutants, among which the highest degradation rate of petroleum pollutants was achieved when the exogenous additive was soybean straw; compared with the naturally decaying petroleum-contaminated soil, the addition of soybean straw and dead and fallen leaves of lemon mallow made the microbial species in the contaminated soil significantly reduced and the main dominant flora changed, but the flora capable of degrading petroleum pollutants increased significantly; the addition of exogenous nutrients had significant effects on soil microbial diversity and community structure. Conclusions: Soybean straw can be added to the contaminated soil as the optimal exogenous organic nutrient system, which improves the physicochemical properties of the soil and gives a good living environment for indigenous microorganisms with the function of degrading petroleum pollutants, thus activating the indigenous degrading bacteria in the petroleum-contaminated soil and accelerating their growth and proliferation and new city metabolic activities, laying a foundation for further obtaining efficient, environmentally friendly and low-cost microbial enhanced remediation technology solutions. The foundation for further acquisition of efficient, environmentally friendly and low-cost microbial enhanced remediation technology solutions. It is important for improving soil remediation in areas with long-term oil contamination and nutrient-poor soils.

【**Keywords**】Loess hilly area; Petroleum hydrocarbons; Plant waste; Microbial remediation

1 Introduction

Oil is one of the main energy sources for social and economic development nowadays, and is inseparable from people's production and life (Wang P, et al.2018). However, in the process of oil extraction, storage and transportation and use, oil spills and leaks and other accidents have led to increasingly serious oil pollution (Blumer M, et al.1973). The entry of petroleum hydrocarbon compounds into the soil affects the soil structural composition, biochemical cycles and soil microbial community diversity, causing soil pollution, which in turn affects the ecological balance and leads to gradual environmental degradation (Rous

JD, et al.1994;Zheng J, et al. 2021). The Loess Plateau area in northern Shaanxi Province, China is an important oil and gas reserve and development area in China, and the relevant survey shows that the area of moderately contaminated soil is about 708.16 million m^2(Ding Q, et al.2020). Therefore, it is urgent to investigate and analyze the current situation of petroleum-contaminated soil in northern Shaanxi Province and explore the remediation methods suitable for petroleum-contaminated soil in the area (Lei Q, et al. 2019; Wu ML, et al.2018; He LJ, et al. 2004).

At present, the remediation methods of petroleum-contaminated soil mainly include physical, chemical and biological methods (Shi X, et al.2009; Zhong L, et al.2021). Compared with the traditional physical and chemical methods, bioremediation is widely recommended because of its high efficiency, low cost and low pollution (Zhen LS, et al. 2021; Li L, et al.2020; Yang T, et al.2021). Bioremediation is a treatment technology that stabilizes the contaminants by degrading, absorbing or converting them into harmless forms such as CO_2 and H_2O, mainly by remediating the soil organisms (Youngsook O, et al.2001). In the process of soil petroleum pollutant treatment, in order to improve the bioremediation efficiency and remediation effect, many researchers also target soil environmental factors such as soil physicochemical properties (Long F, et al.2020; Gao CT, et al.2019), pollution status and environmental conditions, and achieve the purpose of soil remediation by accelerating microbial metabolic activities through the use of fertilizers or other additives into oil-contaminated soil (Zhang LJ, et al.2021; Xu YR, et al.2021).

At present, most of the domestic and foreign research on oil-contaminated soil remediation technology focuses on how to improve the biochemical properties of soil after oil-contaminated soil (Zhang P, et al.2018; Ames R, et al.1987; Eivazi F, et al. 2018), and there is less research on improving oil-contaminated soil by using natural organic matter of agricultural waste (Struecker, et al.2015; Deng H, et al.2008; Cheng GS, et al. 1987). Some researchers have studied the effect of different additives on the remediation of petroleum-contaminated soil by adding agricultural wastes such as hay, straw and leaves to the petroleum-contaminated soil (Wu PY, et al.2019; Ueno A, et al.2010; Larkin RP, et al. 2010). Therefore, in this study, to address the current situation of petroleum contamination in the soil of the petroleum extraction area of the Loess Plateau Land in northern Shaanxi Province, three common agricultural wastes in northern Shaanxi Province, such as soybean straw, corn cobs and lemon leaves, were used as additives to study the effect on petroleum hydrocarbon contaminated soil and the improvement effect on it (Wichern F, et al.2020; Macias-Benitez S, et al. 2020; Ibekwe AM, et al. 2010). Through field sampling, indoor experiments, testing and analytical measurements, we investigated the effects of different types of nutrient additives on soil microbial diversity, elucidated the structural characteristics of petroleum-contaminated soil microbial communities in response to different nutrient additives, and comprehensively analyzed the response mechanism of soil-microbial ecosystem to nutrient additives in this area, with a view to providing scientific basis for bioremediation and treatment of petroleum-contaminated soils on the Loess Plateau (Song X, et al.2019).

2 Materials and methods

2.1 Sample collection

The soil tested in this study was collected from a typical Loess hilly area in China, namely Baota District, Yan'an City, Shaanxi Province, China. It is located in the middle reaches of the Yellow River and the central and southern part of the Loess Plateau. It belongs to the warm temperate semi-humid and drought-prone climate zone. Monsoon circulation, the annual average temperature is about 8.7 ℃, and the annual average rainfall is about 500 mm. The main soil in the sample collection area is the calcareous beginning soil (Loess soil) developed from the Loess parent material, which is yellow, has no bedding, silty texture, loose, large voids, most of the soil nutrients are at a low level, and the soil is weakly alkaline as a whole(Wang BR, et al.2008). The uncontaminated area was selected as the sample plot. Following the principles of randomness, equal amount and multi-point mixing, soil samples of 0~50 cm were collected by the S-spot method. The relevant geographic location information of the sample collection points is shown in Table 1 (Chen XP, et al.2008).

Table 1 Geographical information of sample collection sites

Sample number	Slope(°)	Elevation (m)	Longitude	Latitude
YN1	22	1104	109°58′98″	36°81′02″
YN2	23.5	1156	109°57′32″	36°25′28″
YN1	26	957	109°39′02″	36°55′03″
YN1	25	1213	109°59′02″	36°80′15″

2.2 Indoor simulation test

Weigh crude oil with an oil content of 15 g/kg and dissolve it in petroleum ether, add it to clean soil, mix well, and after standing for one week, carry out an indoor simulated addition culture test, that is, according to the mass percentage of 30% (w/w) Addition amount The crushed and decomposed soybean straw (DD), corncob (YM) and caragana litter (NT) were added to the prepared oil-contaminated soil without exogenous addition treatment. (CK) is a blank control, cultivated for 120 days under constant conditions (environmental temperature controlled at (25±2)℃, soil moisture content maintained at 15%), with regular stirring and ventilation, a total of 4 treatments, each treatment setting 3 a parallel experiment. The specific experimental design is shown in Table 1. After the cultivation, soil samples were collected from eight different locations in each container, and after thorough mixing, a mixed sample of about 200 g was formed, which was used to determine the total petroleum hydrocarbon content and microbial diversity in the soil (Huang TL, et al.2008). The specific restoration protocols are shown in Table 2.

Table 2 Test program

Additives	Sample no.		
CK	CK-1	CK-2	CK-3
Soybean straw (DD)	DD-1	DD-2	DD3
Corn cob (YM)	YM-1	YM-2	YM-3
Lime leaf drop (NT)	NT-1	NT-2	NT-3

2.3 Testing indicators and measurement methods

The total amount of petroleum hydrocarbons in soil was determined by infrared spectroscopy, that is, an appropriate amount of soil samples were added to carbon tetrachloride and left to stand overnight.

Soil microbial diversity was determined by DNA extraction and high-throughput sequencing. The genomic DNA of the sample was extracted by CTAB or SDS method, and then the purity and concentration of the DNA were detected by agarose gel electrophoresis. Using the diluted genomic DNA as a template, according to the selection of the sequencing region, use specific primers with Barcode, Phusion® High-Fidelity PCR Master Mix with GC Buffer from New England Biolabs, and high-efficiency and high-fidelity enzymes to carry out PCR to ensure amplification Efficiency and accuracy. The corresponding regions of primers to identify bacterial diversity are 16S V4 primers (515F and 806R). The PCR products were detected by electrophoresis on agarose gel with a concentration of 2%; the qualified PCR products were purified by magnetic beads, quantified by enzyme labeling, and the samples were mixed in equal amounts according to the concentration of PCR products, and 2% agar was used after thorough mixing (Lusk R, et al.2021; Liu MY, et al.2020). The PCR products were detected by glycogel electrophoresis, and the target bands were recovered using the gel recovery kit provided by Qiagen. The TruSeq® DNA PCR-Free Sample Preparation Kit was used for library construction. The constructed library was quantified by Qubit and Q-PCR. After the library was qualified, NovaSeq6000 was used for on-machine sequencing (Li JL, et al.2011).

2.4 Data statistics and analysis

One-way ANOVA was performed in SPSS 21.0, and Duncan's method was used for multiple comparisons among different treatments ($P<0.05$). Software R3.3.2 was used for data analysis and graph drawing. Shannon and Simpson indices were used to estimate species diversity, Chao 1 and ACE were used to calculate and estimate species richness, and Good's coverage reflected the coverage of community microorganisms by sequencing data. The UPGMA (Unweighted Pair-Group Method with Arithmetic means) cluster analysis was performed by considering both species diversity and abundance, and the overall differences in microbial community structure were evaluated based on Bray-Curtis distances principal coordinate analysis (PCoA) (Liu JM, et al.2001; Ali F, et al.2020; Wagh GA, et al.2001).

3 Results and discussion

3.1 Effects of different exogenous additions on the degradation rate of petroleum pollutants

After 120 days of cultivation, the total amount of petroleum hydrocarbons in the oil-contaminated soil under the action of different exogenous additives was measured, and the degradation rate was calculated from this [degradation rate = (initial oil content-final oil content) / (initial oil content × 100)]. From Table 3, it can be seen that the content of petroleum pollutants in soil decreased to different degrees. The degradation rates of petroleum pollutants in soil under CK, YM, DD and NT treatments reached 42.46%, 58.63%, 74.06%, and 62.13%, respectively. , compared with the blank control group, the degradation rates of petroleum pollutants in the soil under the addition of exogenous nutrients were improved, and the degradation rate under the DD treatment was the highest, reaching 74.06%. It can be seen that adding corncob, soybean straw and Caragana caragana litter can improve the degradation efficiency of petroleum pollutants in soil, and soybean straw has the best enhancement effect on pollutant degradation.

Table 3 Degradation rate of petroleum pollutants under different treatment of exogenous substances

No.	Treatment	Petroleum hydrocarbon content before strengthening (g/kg)	Petroleum hydrocarbon content after strengthening (g/kg)	Degradationrate (%)
CK	Natura lattenuation	15	8.63	42.46
YM	Addition of Corncob	15	6.21	58.63
DD	Addition of Soybeanstraw	15	3.89	74.06
NT	Addition of litter Leaves of Caraganakorshinskii	15	5.68	62.13

3.2 Effects of different pollution concentrations on soil microbial diversity

3.2.1 Dilution curve

The dilution curve is to randomly select a certain number of sequences from the sample, and count the Alpha diversity index of the samples corresponding to these sequences. Draw a curve, and judge whether the amount of sequencing data is sufficient according to whether the curve is flat. It can be seen from Fig.1 that the dilution curve tends to be flat, indicating that the detection ratio of microbial communities in environmental samples is close to saturation, that is, the current sequencing volume can cover most of the species in the samples. In addition, the number of soil microbial OTUs under different treatments was YM>CK>DD>NT. It can be seen that the number of soil microbial species decreased significantly under DD and NT treatments, but the difference between the two was small; while the number of microbial species was slightly higher under YM treatment in CK (Chen FF, et al.2020).

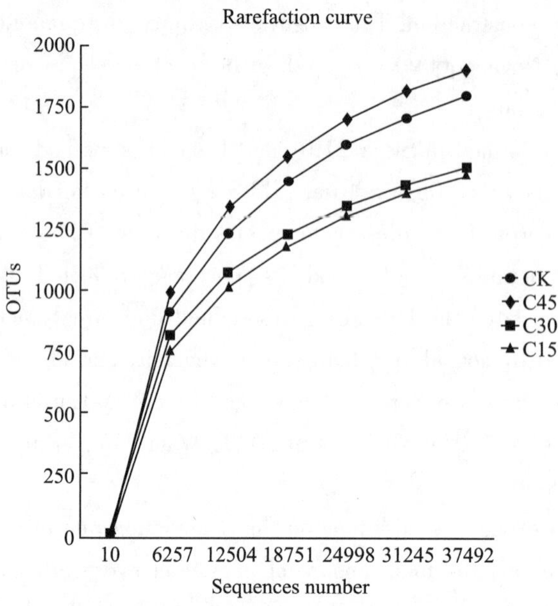

Fig.1 Microbial dilution curves of petroleum-contaminated soils under different types of agricultural waste addition treatments

3.2.2 Species Venn diagram analysis

Venn diagrams can be used to count the number of common and unique species (OTUs) in multiple (group) samples, visually display the similarity and specificity of OTU composition in environmental samples, and perform bioinformatics statistics on OTUs classified by 97% similarity (Wang R, et al.2019). Analysis, the results showed that each soil sample produced a total of 10526 OTUs, of which the total number of OTUs was 1261, accounting for 11.98% of the total number of OTUs (Fig.2). The number of OTUs processed by DD and NT were 2367 and 2456, which were lower than CK (2785), and the number of OTUs processed by YM was 2918, slightly higher than that of CK. Compared with CK, the number of unique OUTs in YM, DD and NT treatments were all reduced, and the reduction value of DD and NT was greater than 200. It can be seen that different exogenous nutrient addition treatments had a significant effect on the number of unique OTUs in soil. In addition, among the different treatments, the DD and NT treatments had the highest degree of overlap, and their bacterial community structures had a high similarity.

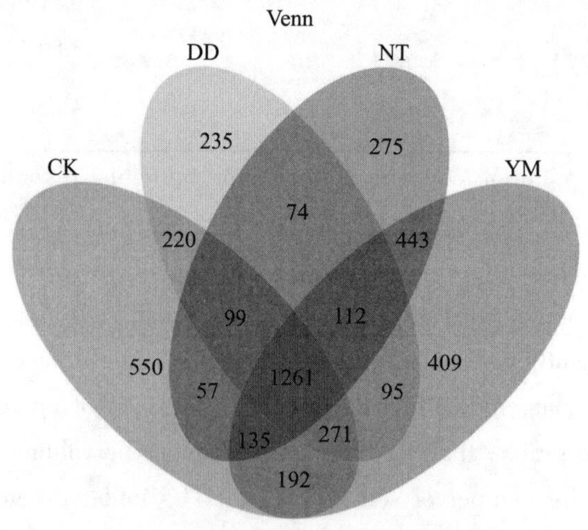

Fig.2 Venn diagram of soil microbial OTU counts under different types of farm waste addition treatments

3.3 Effects of different exogenous treatments on soil microbial community structure

3.3.1 β diversity-analysis of variability between groups

In order to study the similarity or difference relationship of different sample community structures, clustering analysis can be performed on the sample community distance matrix, and a sample hierarchical clustering tree can be constructed. The different shape legends in Fig.3 represent the control and soil samples treated with three different types of plant waste as exogenous nutrients, respectively. The differences in the microbial community composition of CK, YM, DD and NT were analyzed by PCA. The results showed that the soil microbial community composition was different under different exogenous treatments. The difference between DD, NT and CK was significant, but the difference between YM and CK was not significant. The PC1 axis and PC2 axis explained 12.56% and 17.19% of the results, respectively (Zhang GQ, et al. 2017).

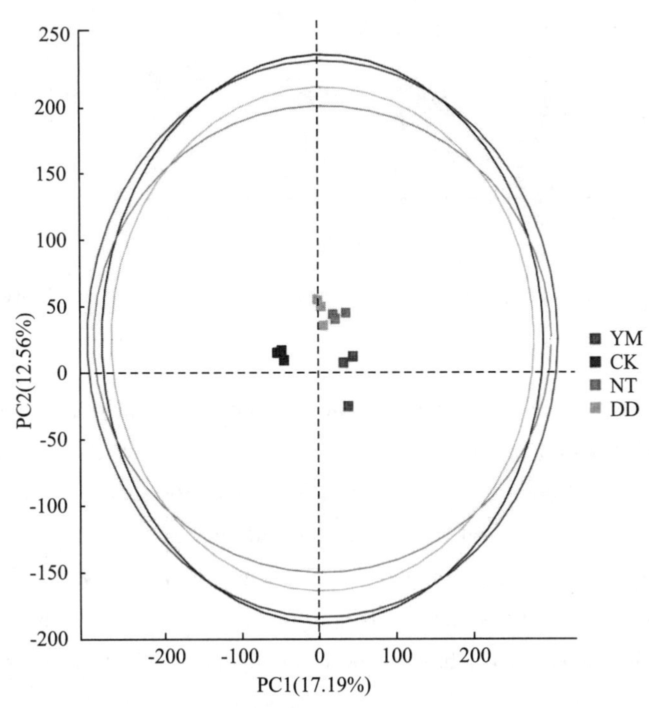

Fig.3 PCA analysis of soil microorganisms under different exogenous treatments

3.3.2 Analysis of the variability of community composition

The histogram can present the community composition and species abundance at different taxonomic levels (Zhu SS, et al.2020). In this study, community composition and species abundance were analyzed based on the genus level. The dominant genera in petroleum-contaminated soil without the addition of exogenous substances (CK) mainly included *Nocardioides*, *Nocardia*, *Rhodococcus*, *Sphingomonas* and *Marmoricola*. The abundance of *Sphingomonas*, *Thermomonas*, *TM7a* and *Lysobacter* increased significantly under YM treatment. *Nocardioides*, *Nocardia*, *Rhodococcus*, *Mycobacterium* and *Marmoricola* significantly decreased in abundance under DD treatment, while *Sphingomonas*, *Brevundimonas Sphingomonas*, *Thermomonas*, *TM7a*, *Lysobacter*, *Bacteroides*, *Pseudoxanthomonas*, *Sphingopyxis*, *Bacteroides*, *Flavobacterium* were increased in abundance; *Cupriavidus*, *TM7a* and *Brevundimonas* were significantly increased in abundance under NT treatment, and *Pseudomonas and Bacteroides* were increased in abundance but not significantly (Fig.4).

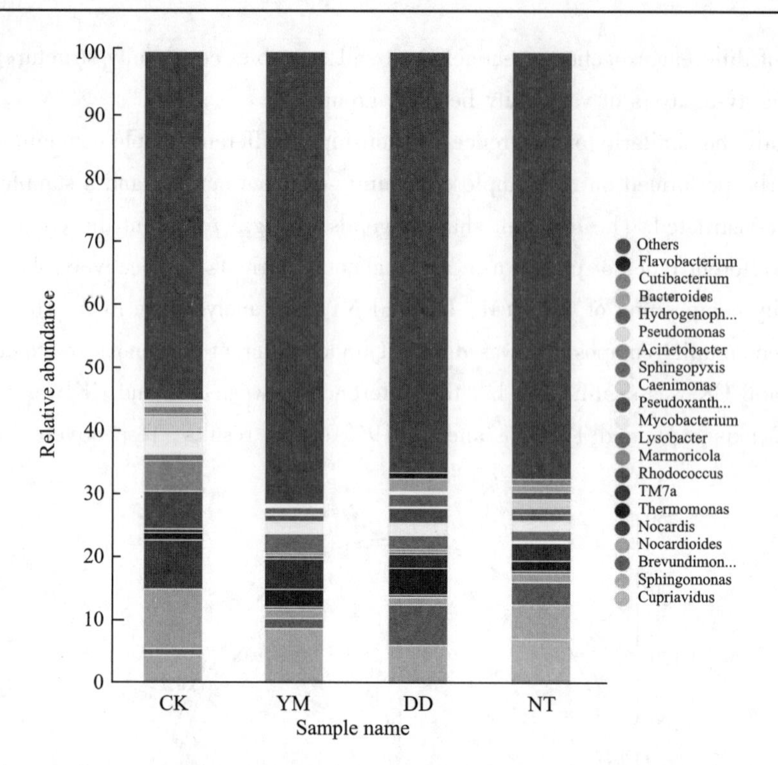

Fig.4 Community composition characteristics of soil microorganisms at different genus levels

4 Conclusions

Using plant wastes as exogenous nutrients to biostimulate the remediation of contaminated soil will change the soil micro-ecological environment. In this study, by measuring the total amount of petroleum hydrocarbons in the soil under the addition of different exogenous substances, it can be seen that compared with the oil-contaminated soil with natural decay, the degradation rate of oil pollutants was improved after adding plant waste, that is to say, it strengthened the The degradation ability of indigenous degrading bacteria to petroleum pollutants, in which the exogenous additive is soybean straw, the degradation rate of petroleum pollutants reaches the highest, so it has the best enhancement effect on the degradation of petroleum pollutants by indigenous microorganisms, it may be because the carbon-nitrogen ratio of soybean straw is 20.4:1, and the phosphorus content is 0.07%, which is relatively high, which can acidify the soil, greatly promote soil respiration, and significantly increase invertase and hydrogen peroxide. Enzyme, alkaline phosphatase and urease activity (Li A, et al.2018).

Through high-throughput sequencing of soil microorganisms, the results show that, compared with the naturally attenuated oil-contaminated soil, the addition of soybean straw and Caragana litter leaves significantly reduced the types of microorganisms in the contaminated soil, but the main dominant flora changed, but the bacteria that can degrade petroleum pollutants increased significantly. The addition of exogenous substances has a significant impact on soil microbial diversity and community structure, which may be due to the application of organic nutrient systems to improve the physical and chemical properties of soil, giving indigenous microorganisms with the function of degrading petroleum pollutants. A good living environment, thereby activating the indigenous degrading bacteria and accelerating their growth, proliferation and metabolic activities.

In this study, the methods of bio-enhanced microbial remediation of oil-contaminated soil were studied, and the strengthening effect of different plant wastes as exogenous stimuli on microbial remediation of oil-contaminated soil and the impact on soil microbial community structure were analyzed. The low-cost microbial enhanced remediation technology scheme provides theoretical and data support, and is of great significance for improving soil remediation in areas that have been polluted by oil for a long time and have poor soil nutrients. Since this study is limited to the laboratory, further research and exploration are still needed on the enhancement effect of the in-situ remediation technology for oil-contaminated soils.

Acknowledgements

This work was supported from the projects of "Technology Innovation Leading Program of Shaanxi (Program No. 2021CGBX-03)", and Shaanxi Provincial Land Engineering Construction Group (Program No. DJNY2021-33,DJNY2022-37).

Authors' contributions

LS: guided the completion of this experiment, completed the first draft paper, and made revisions. WF: guided the experimental methods and thesis writing. ZL: performed experiments, recorded data, and data analysis. LY: data analysis. All authors read and approved the final manuscript.

Funding

This work was supported from the projects of "Technology Innovation Leading Program of Shaanxi (Program No. 2021CGBX-03)", and Shaanxi Provincial Land Engineering Construction Group (Program No. DJNY2021-33, DJNY2022-37).

Availability of data and materials

The data were obtained by the authors.

Declarations

Ethics approval and consent to participate: I certify that this manuscript is original, has not been published, and will not be submitted elsewhere for publication while being considered by Annals of Microbiology. The study is not split up into several parts to increase the quantity of submissions and submitted to various journals or to one journal over time. No data have been fabricated or manipulated (including images) to support your conclusions. No data, text, or theories by others are presented as if they were our own.

The submission has been received explicitly from all co-authors, and authors whose names appear on the submission have contributed sufficiently to the scientific work and therefore share collective responsibility and accountability for the results.

The study did not violate ethics, and all participants agreed to publish the paper.

Consent for publication: Not applicable.

Competing interests: The authors declare that they have no competing interests.

References

Wang P, Tang S. (2018) Research on remediation technology of contaminated soil in oil production and transportation. Chemical Engineer, 32(11):56-59.

Blumer M, Ehrhardt M, Jones J H. (1973) The environmental fate of stranded crude oil. Deep Sea Research & Oceanographic Abstracts, 20(3): 239-259.

Lei Q, Weng D, Luo J, et al. (2019) Progress and development direction of petroleum oil and gas extraction engineering technology in China. Petroleum Exploration and Development, 46(1):7.

Lusk R, Stene E, Banaei-Kashani F, et al. (2021) Aptardi predicts polyadenylation sites in sample-specific transcriptomes using high-throughput RNA sequencing and DNA sequence. Nature Communications, 12(1):1652.

Rous J D,Sabatini D A, Suflita J M, et al. (1994) Influence of surfactants on microbial degradation of organic compounds. Critical Reviews in Environmental Science and Technology, 24: 325-370.

Ding Q, Liu H. (2020) Research progress of microbial method for treating petroleum-contaminated soil. Industrial Catalysis, 28(9):5.

Zheng J, Fu Y, Song Q, et al. (2021) Progress of microbial enhanced remediation of petroleum-contaminated soil. Journal of Biological Engineering, 37(10):14.

Liu M Y, Ma J H, Li Y, et al. (2020) Effects of 16S rRNA gene high variation regions V4 and V3-V4 and sequencing depth on the analysis of bacterial flora in oil reservoirs. Microbiology Bulletin, 2:10.

Shi X. (2009) Research on the damage of petroleum pollutants on soil and environmental remediation technology. Science and Technology Information, (20):452-455.

Wang B R, Yang J J, An S S, et al. (2018) Effects of vegetation and topography features on ecological stoichiometry of soil and soil microbial biomass in the Hilly-Gully region of the Loess plateau, China, 29(1):247-259.

Li J L, Liu G G, Cao S X, et al. (2011) Optimization of RAPD-PCR reaction system for partridge tea using one-way and orthogonal design. Journal of Agronomy, 1(4):8.

Liu J M, Peng J P. (2001) One-way analysis of variance for biological experimental data. Journal of Zoology, 36(6):4.

Youngsook O, Dooseup S, Sangjin K. (2001) Effects of nutrients on crude oil biodegradation in the upper intertidal zone. Marin Pollution Bulletin, 42(12): 1367-1372.

Wagh G A, Tiwari P D. (2020) On the diversity and abundance of avian species from grassland and wetland areas of an industrial zone of tropical Maharashtra. Bioscience Biotechnology Research Communications, 13(2):770-780.

Sun Y, Wei W, Zhao M. (2009) Microbial remediation technology for petroleum hydrocarbon contaminated soil. Journal of Northeast Forestry University, 10:4.

Ali F, Nadeem M A, Habyarimana E, et al. (2020) Molecular characterization of genetic diversity and similarity centers of safflower accessions with ISSR markers. Brazilian Journal of Botany, 43(1):1-13.

Long F, Guo B, Wang X, Wang X. (2020) Research progress of microwave for petroleum-contaminated soil remediation. Modern Chemical Industry, 1:5.

Gao C T, Han Z T, Chen C Z, et al. (2019) Remediation of crude oil-contaminated soil by combined oxidation-microbial degradation with hydrogen peroxide and sodium persulfate activation. Chemical Environmental Protection, 39(1):6.

Chen F F, Cong X, Xiang J Z, et al. (2020) Application of Illumina MiSeq high-throughput sequencing technology to analyze the inter-rooted soil microbial diversity of Corydalis sativus. Hubei Agricultural Science, 59(17):6.

Wang R, Zhu J, Jin T, et al. (2019) Characterization of soil ammonia-oxidizing microbial abundance and community structure in rice fields under rice and shrimp crop model. Journal of Plant Nutrition and Fertilizer, 25(11):13.

Zhang L J, Li K R, Zhang X Y. (2021) Uptake and accumulation of petroleum pollutants by different plants in the Loess Plateau of northern Shaanxi. Journal of Northwest Agriculture and Forestry University of Science and Technology: Natural Science Edition, 8:7.

Xu Y R, Wu M L, Wang L, et al. (2021) Changes in microbial populations and influencing factors of petroleum-contaminated soils in northern Shaanxi. China Environmental Science, 41(9):11.

Zhen L S, Gu , Hu T, et al. (2015) Microbial community structure and metabolic characteristics of petroleum-contaminated soils on the Loess Plateau. Journal of Ecology, 35(17):8.

Zhang G Q, Zhao P, Dong YX, et al. (2017) High-throughput sequencing analysis of the effect of environmental fertilizer enhancers on the change of soil fungal diversity in potato inter-rhizosphere. Microbiology Bulletin, 44(11):8.

Zhang P, Ren C, Sun H, et al. (2018) Sorption, desorption and degradation of neonicotinoids in four agricultural soils and their effects on soil microorganisms. The Science of the Total Environment, 615:59-69.

Ames R, Bethlenfalvay G. (1987) Mycorrhizal fungi and the integration of plant and soil nutrient dynamics. Journal of Plant Nutrition, 10(9):1313-1321.

Eivazi F, Afrasiabi Z, Jose E. (2018) Effects of Silver Nanoparticles on the Activities of Soil Enzymes Involved in Carbon and Nutrient Cycling. Pedosphere, 28(2):209-214.

Struecker, Juliane, Joergensen, et al. (2015) Microorganisms and their substrate utilization patterns in topsoil and subsoil layers of two silt barns, differing in soil organic C accumulation due to colluvial processes. Soil Biology & Biochemistry, 91:310-317.

Deng H, Li S, Huang Y Z, et al. (2008) Progress in soil microbial pollution-induced community tolerance[J]. Journal of Ecotoxicology, 3(5):428-437.

Cheng G S. (1987) A new trend in soil microbiology-soil contamination microbiology. Journal of Agricultural Environmental Science, 02:24-25.

Wu P Y, Chao Q F, Zhao Y G, et al. (2019) Microbial community structure and its metabolic characteristics of petroleum-contaminated soil in Karamay. Guangxi Agricultural Bioscience, 5(5):5.

Ueno A, Ito Y, Yamamoto Y, et al. (2010) Bacterial community changes in diesel oil contaminated soil microcosms biostimulated with Luria-Bertani medium or bioaugmented with a petroleum-degrading bacterium, Pseudomonas aeruginosa strain WatG. Journal of Basic Microbiology, 46(4):310-317.

Larkin R P, Griffin T S, Honeycutt CW. (2010) Rotation and Cover Crop Effects on Soilborne Potato Diseases, Tuber Yield, and Soil Microbial Communities. Plant Disease, 94(12):1491-1502.

Wichern F, Islam R, Hemkemeyer M, et al. (2020) Organic amendments alleviate salinity effects on soil microorganisms and mineralisation processes in aerobic and anaerobic paddy rice soils. Frontiers in Sustainable Food Systems, 4(30):1-14.

Macias-Benitez S, Garcia-Martinez AM, Jimenez PC, et al. (2020) Rhizospheric organic acids as biostimulants: monitoring feedbacks on soil microorganisms and biochemical properties. Frontiers in Plant Science, 11:633.

Ibekwe A M, Poss J A, Grattan S R, et al. (2010) Bacterial diversity in cucumber (Cucumis sativus) rhizosphere in response to salinity, soil pH, and boron. Soil Biology & Biochemistry, 42(4):567-575.

Zhong L, Qing J W, Chen H Y, et al. (2021) Advances in microbial remediation of petroleum hydrocarbon soil contamination. Journal of Biological Engineering, 37(10):17.

Zhu S S, Xia B, Hao W L, et al. (2020) Functional diversity of soil microbial communities on eroded slopes in loess areas. China Environmental Science, 40(9):7.

Wu M L, Qi Y Y, Zhu C C, et al. (2018) Effect of composting on the removal of petroleum hydrocarbons from soil and microbial communities. China Environmental Science, 38(8):7.

He L J, Li P J, Wei D Z, et al. (2004) Nutrient balance and degradation mechanism of petroleum hydrocarbon microbial degradation. Environmental Science, 25(1):4.

Li L, Dong W T, Zhang X, et al. (2020) Research progress of remediation technology for petroleum-contaminated soil. Sichuan Environment, 39(4):6.

Yang T, Guo B, Yang Y, et al. (2021) Experimental study on cleaning process and bioremediation of high concentration petroleum contaminated soil. Contemporary Chemical Industry, 50(9):5.

Song X, Wang J C, Jing M B, et al. (2019) Integrated response of alfalfa to ecological remediation of oil-contaminated soil sites in Longdong Loess Plateau. Grassland Science, 36(7):11.

Chen X P, Yi X Y, Tao X I, et al. (2008) Screening and characterization of pyrene-degrading microbial bacteria for efficient degradation of pyrene in petroleum-contaminated soil. Journal of Environmental Engineering, 2(3):413-417.

Huang T L, Tang Z X, Xu J L, et al. (2008) Indoor simulation study on bioremediation of petroleum-contaminated soils in loess areas. Journal of Agricultural and Environmental Sciences, 27(6):5.

Li A, Wang T, Wang A, et al. (2018) Microbial remediation promotion technology for petroleum hydrocarbon contaminated soil. Chemical Environmental Protection, 38(3):4.

本文曾发表于 2022 年《Annals of Microbiology》第 72 卷

Spatial and Temporal Variations of Vegetation Coverage and Their Driving Factors Following Gully Control and Land Consolidation in Loess Plateau, China

Jing Wang[1,2,3], Yi Hu[1,2,4], Liangyan Yang[1,2], and Qingjun Bai[3]

(1.Key Laboratory of Degraded and Unused Land Consolidation Engineering, Ministry of Natural Resources, Xi' an 710075, China; 2.Institute of Land Engineering and Technology, Shaanxi Provincial Land Engineering Construction Group Co., Ltd., Xi' an 710075, China; 3.Institute of Water Resources and Hydro-Electric Engineering, Xi' an University of Technology, Xi' an 710048, China; 4.School of Human Settlements and Civil Engineering, Xi' an Jiaotong University, Xi' an 710054, China)

【Abstract】 Comprehensive management of the ecological environment and sustainable ecological development, such as the Gully Control and Land Consolidation (GCLC) project may affect surface vegatation. The normalized difference vegetation index (NDVI) is a sensitive indicator of vegetation dynamics, but in-depth study that continually monitors spatial and temporal variation of regional vegetation before and after the implementation of the GCLC project is still scarce. To address this issue, we analyzed the the spatial and temporal variations of Landsat surface reflectance derived NDVI data in the Jiulongquan watershed in Yan' an City, China, from 2010 to 2019, and examined the main driving factors for the variation. Results showed high overall vegetation coverage in evaluated watershed. The NDVI was spatially varied and tended to be low in the gully area and high on the slope. From 2010 to 2019, the NDVI values exhibited an increasing trend and the most evident changes concentrated in the gully areas. The changes of NDVI were mainly driven by human activities rather than the evaluated climatic factors. This work indicates that the GCLC project had positive effects on ecological and agricultural environment at a regional scale.

【Keywords】 NDVI; Correlation analysis; Trend analysis; Climatic factors; Human activities

1 Introduction

Vegetation is a vital constituent of the ecosystem and serves as the hub for mass circulation and energy exchange directly linking the atmosphere, soil, water, and human activity[1-2]. The Normalized Difference Vegetation Index (NDVI) is a measure of the red: near-infrared reflectance ratio and is usually derived from remote-sensing (satellite) data . Since the NDVI is very sensitive to biophysical characteristics of the vegetation such as the growth status, biomass, and photosynthetic intensity, the relationship between the NDVI and vegetation productivity has been well documented . This has enabled the NDVI one of the most widely applied tools for monitoring vegetation dynamics at varied spatiotemporal scales[3-4] and one of the most feasible indicators of eco-environment quality. Development of vegetation and its response to environmental change (e.g., global warming) have been evaluated by many scholars using the NDVI data at regional and global scales in recent years[5-7].

Previous studies suggested that spatiotemporal variation of the NDVI is mainly affected by climatic factors and human activities. The change of the NDVI over longer time periods appears to be driven by long-term climate change, but the change in the shorter time scales is related to human activities[8-9]. Precipitation and temperature are two dominant climatic factors that influence vegetation growth[10]. For example, global warming was reported to enhance plant growth in northern mid-latitudes and high latitudes, but had a negative impact on vegetation in arid and semi-arid areas due to increased evapotranspiration[11,12]. On the other hand, rainfall generally has a positive correlation with the NDVI[13].

Human activities may also have positive and negative impacts on vegetation simultaneously. In forestry ecological engineering projects, afforestation as well as soil and water conservation practices could promote the increment of vegetation coverage[14,15]. In contrast, human activities such as overgrazing, deforestation or urbanization could directly or indirectly destroy the surface vegetation[16].

The *Gully Control and Land Consolidation* (GCLC) project is another major project of ecological environment management in the Loess Plateau following the *Grain for Green* project. The implementation of a series projects directly worked on and thus changed the surface vegetation. The GCLC project started in 2011, and some researchers have quantitatively studied the vegetation changes in the area since then. For example, He et al.[17] and Du et al.[18] both showed an increasing trend of vegetation coverage in the watershed following the GCLC project.

However, these studies simply compared the vegetation coverage before and after the GCLC project. In-depth study that continually monitors the spatial and temporal variation of regional vegetation in the area is still scarce, and driving factors for such variation have not been determined. To quantitively analyze the impact of the GCLC project on vegetation in the area over time series, the Jiulongquan project in the Nanniwan Town, Yan'an City, China was chosen as the study area. Based on the Landsat Surface Reflectance-derived NDVI and meteorological data, we studied the characteristics and trend of the spatiotemporal variation of vegetation in the project area from 2010 to 2019 and analyzed the main driving factors for such variation using trend analysis and correlation analysis. This study aimed to provide a scientific basis for subsequent land engineering and management in this area as well as places with similar geological and climatic conditions.

2 Overview of the study area

The Baota District of Yan'an City (36.18°~37.03°N and 109.23°~110.12° E) is characterized by a semi-humid and semi-arid continental monsoon climate with hot and rainy summers and cold and dry winters. The average annual precipitation is 562.1 mm and has an uneven temporal distribution with 70% concentrated in the summer (June to September). The annual average temperature is 9 ℃ and the average frost-free period is 179 days.

The Jiulongquan watershed is located in the Fenchuan River basin and belongs to the hilly and gully region of the south Baota District (Fig.1). The terrain mainly comprises of river valley and terrace following a southwest-northeast direction. The terrain is fragmented by many crisscrossed gullies that are deep with sharp slopes, and the soil erosion is severe. Soil type in the area is calcic Cambisol as defined by the Food and Agriculture Organization (FAO) of the United Nations. The river channel is about 9.5 km long. The elevation of the river channel drops from 1170 m in the north to 1093 m in the south, with an average gradient of 0.78%. The valley is generally 250~500 m wide, and total area of the watershed is 62.63 km^2.

The vegetation on slopes in the watershed was dominated by grassland and natural shrubs prior to the GCLC projects. Most of the gully land has been reclaimed into rainfed farmland with limited water availability and consequently, the grain productivity is low. The project started at the beginning of 2013 with a total construction area of 360.91 hm^2 and a total investment of CNY 44.12 million. The main project was completed in 2014 and farming began since then.

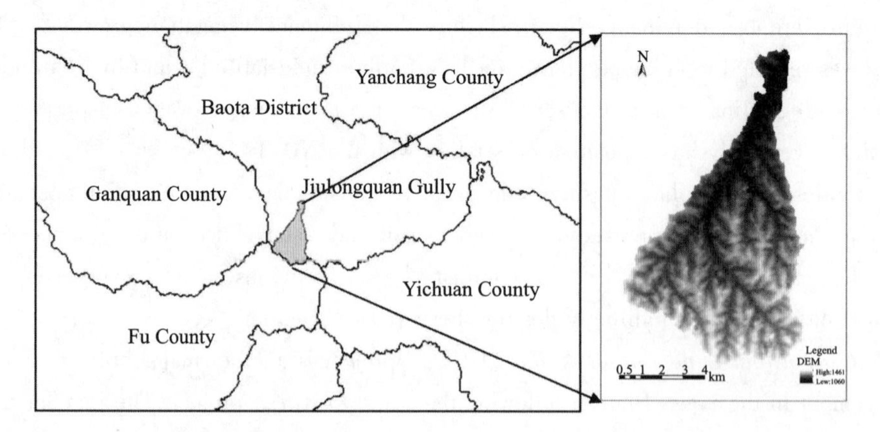

Fig.1 Location and Digital Elevation Model (DEM) of the Jiulongquan watershed

3 Data acquisition, calibration, and analysis

3.1 Source of data

3.1.1 Remote sensing images

The Landsat data (path 127, row 35 with a spatial resolution of 30 m) managed by the United States Geological Survey (USGS) were obtained from the Geospatial Data Cloud website (http://www.gscloud.cn). Landsat images from June to August in each year between 2010 and 2019 when the vegetation was vigorously growing and the study area was as cloud-free as possible (e.g., cloudiness<15%), were selected to study the changes of vegetation coverage in the watershed before and after the GCLC project. The data information and key parameters are shown in Table 1. Striped data missing occurred for some images in 2012 and 2018 due to malfunction of theairborne scan line corrector (SLC) of the Landsat 7 satellite. Therefore, the missing stripes on such images were repaired by pixel spatial interpolation with the gap-fill (triangulation) function of the ENVI 5.3 software[19]. Spatial interpolation is one of the most common techniques to reconstruct single remote sensing image[20].

Table 1 Research data information

Satellite	Date of acquisition	Sensor typea	Cloudiness/%
Landsat 5	2010-06-17	TM	2.48
Landsat 7	2010-08-28	ETM+	2.98
Landsat 7	2011-06-28	ETM+	0
Landsat 8	2013-06-25	OLI	0.34
Landsat 7	2013-08-04	ETM+	2.74
Landsat 8	2014-07-14	OLI	0.30
Landsat 8	2015-07-01	OLI	0.07
Landsat 7	2015-07-25	ETM+	0.29
Landsat 8	2016-06-17	OLI	0.17
Landsat 8	2016-07-03	OLI	11.27
Landsat 7	2016-07-27	ETM+	0.15
Landsat 8	2017-06-20	OLI	0.82
Landsat 8	2019-08-13	OLI	6.39
Landsat 8	2019-08-29	OLI	0.65

The images were radiometric calibrated and atmospheric corrected in the ENVI 5.3 software and clipped according to the vector boundary of the study area. To eliminate the interference due to cloud, atmospheric effect, and solar altitude angle, the universally adopted maximum value composite (MVC) method[20-21] was used to obtain the maximized annual NDVI data, which reflects the vegetation coverage in the best season of vegetation growth. Eight images were obtained from 2010 to 2019.

3.1.2 Meteorological data

Monthly precipitation and temperature data from January 2010 to December 2019 from two weather stations in Yan'an City were obtained from the National Meteorological Information Center of China Meteorological Administration (CMA Meteorological Data Centre, http://data.cma.cn). Annual mean temperature and precipitation in the study area were calculated using the spatial interpolation method as implemented by the ANUSPLIN 4.2 software (The Australian National University, Canberra, Australia). The grid size was set as 30 m×30 m.

3.2 Data analysis

3.2.1 Trend analysis

To quantitatively analyze the change of vegetation coverage during the study period, unitary linear regression was employed. Trends of each pixel NDVI in the study area from 2010 to 2019 were calculated using the following formula[22]:

$$\text{Slope} = \frac{n \times \sum_{i=1}^{n} (i \times \text{NDVI}_i) - \sum_{i=1}^{n} i \sum_{i=1}^{n} \text{NDVI}_i}{n \times \sum_{i=1}^{n} i^2 - (\sum_{i=1}^{n} i)^2} \tag{1}$$

Where slope represents the slope of the NDVI regression equation which is indicative of the rate of change; i represents the year number; n represents the total number of years during the study period; NDVI_i represents the NDVI value of the ith year. Slopes >0 and <0 represent an increasing and decreasing trend of the vegetation coverage, respectively. According to the calculations, the change rates of slope were divided into seven grades including severely degraded, moderately degraded, slightly degraded, relatively stable, slightly improved, moderately improved, and significantly improved[23].

3.2.2 Correlation analysis

Partial correlation coefficient was calculated to characterize the degree of correlation between climatic factors (i.e., annual precipitation and mean temperature) and the NDVI. The linear correlation coefficient was first calculated with the following formula[24]:

$$R_{xy} = \frac{\sum_{i=1}^{n} (x_i - \bar{x})(y_i - \bar{y})}{\sqrt{\sum_{i=1}^{n} (x_i - \bar{x})^2 \sum_{i=1}^{n} (y_i - \bar{y})^2}} \tag{2}$$

Where x is the NDVI value, $\bar{x}$ is the averaged NDVI value of the corresponding pixel from 2010 to 2019; y is the climatic factor, namely annual precipitation and mean temperature; $\bar{y}$ is the multi-year average of the two climatic factors; i is the year number; n is the total number of years.

Based on the calculated linear correlation coefficient, the partial correlation coefficient was calculated:

$$R_{xy,z} = \frac{R_{xy} - R_{xz} R_{yz}}{\sqrt{1 - R_{xz}^2} \sqrt{1 - R_{yz}^2}} \tag{3}$$

Where $R_{xy,z}$ is the partial correlation coefficients between the dependent variable x and the independent variable y after the independent variable z is fixed; R_{xy}、R_{xz}、R_{yz} are the linear correlation coefficients of the corresponding variables. The partial correlation coefficient was statistically analyzed by t-test using the following equation:

$$t = \frac{R_{xy,z}}{\sqrt{1 - R_{xy,z}^2}} \sqrt{n - m - 1} \tag{4}$$

Where $R_{xz,y}$ is partial correlation coefficients; n is the total number of samples; m is the number of independent variables.

Climatic factors are usually interdependent, and their impacts on vegetation are not opposed. Therefore, multiple correlation analysis was adopted to study the combined effects of the two climatic factors on the NDVI[25]. The calculation formula of multiple correlation is as follows:

$$R_{x,yz} = \sqrt{1 - (1 - R_{xy}^2)(1 - R_{xz,y}^2)} \tag{5}$$

Where $R_{x,yz}$ represents the multiple correlation coefficient between the dependent variable x and the independent variables y and z; R_{xy} represents the linear correlation coefficient between x and y; $R_{xz,y}$ represents the partial correlation coefficient between the dependent variable x and the independent variable z after fixing the independent variable y. The F test was used to test the statistical significance of the multiple correlation coefficient, and the calculation formula is as follows:

$$F = \frac{R_{x,yz}^2}{1 - R_{x,yz}^2} \times \frac{n - k - 1}{k} \tag{6}$$

Where $R_{x,yz}$ is the multiple correlation coefficients; n is the number of samples; k is the number of independent variables.

4 Results

4.1 Spatial characteristics of vegetation

To study the spatial distribution of the NDVI in the Jiulongquan watershed, the averaged NDVI value of pixels in each image from 2010 to 2019 was calculated and its spatial distribution map was composed (Fig.2). Together with the Digital Elevation Model (DEM) map of the project area (Fig.1), data showed that the NDVI tended to be low in the gully and high on the slope. The area with the lowest elevation was the main gully and its tributaries, and the area with low NDVI was mainly distributed as strips along the gully bottom of the watershed. The patches with high NDVI were widely distributed on the slopes of the whole watershed, with a maximum value of about 0.9.

Fig.2 Spatial distribution of averaged NDVI in the Jiulongquan watershed from 2010 to 2019

4.2 Spatial and temporal variation of vegetation

4.2.1 Inter-annual variation

The averaged NDVI value of each image from 2010 to 2019 represented the optimal status of vegetation coverage in the study area and was illustrated in Fig.3. In general, vegetation coverage in the watershed was in good conditions as indicated by the >0.75 averaged NDVI each year. During the study period, the averaged NDVI value showed an increasing trend ($p > 0.05$) from 0.788 in 2010 to 0.824 in 2019 with an annual growth rate of 0.004.

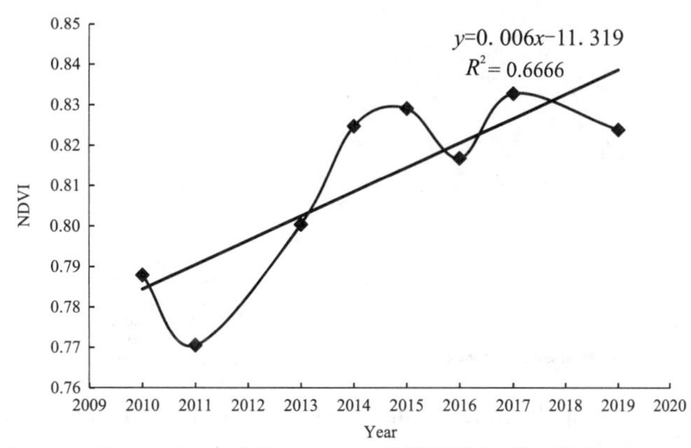

Fig.3 Interannual variation of the averaged NDVI in the Jiulongquan watershed

4.2.2 Trend analysis

To better understand the magnitude of variation and spatial distribution of vegetation coverage in the Jiulongquan watershed, unitary linear regression analysis was carried out on each pixel NDVI at each sampling time. The slopes of the regreassion lines varied between −0.13 and 0.11 during 2010−2019. Classification based on calculated slopes indicated that the improved area was 34.17 km^2, accounting for 54.55%of the study area, in which the slightly improved area was the largest, accounting for 50.64% of the study area. The degraded area was relatively small, accounting for only 3.39% (2.12 km^2). The other 42.05%was relatively stable (Table 2).

Table 2 Statistics of NDVI change in Jiulongquan watershed from 2010 to 2019

Slope	Classification	Area/km^2	Percentage/%
Slope≤−0.05	Severely degraded	0.04	0.07
−0.05 <Slope ≤−0.02	Moderately degraded	0.46	0.74
−0.02<Slope ≤ −0.005	Slightly degraded	1.62	2.58
−0.005<Slope ≤0.005	Relatively stable	26.34	42.05
0.005<Slope ≤0.02	Slightly improved	31.71	50.64
0.02<Slope ≤0.05	Moderately improved	2.39	3.81
Slope>0.05	Significantly improved	0.07	0.10

The areas with the most significantly changed slopes of NDVI regression lines overlapped with those having the greatest degradation or improvement and were mainly concentrated as strips in the upper and middle part of the gully (Fig.4). Despite the evident changes in the gully, a majority of the watershed remained relatively unchanged or slightly improved.

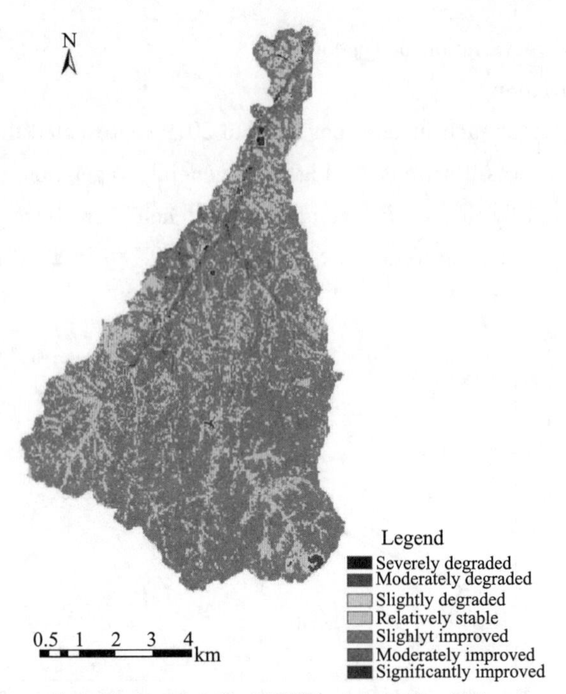

Fig.4 Slope grading diagram of the NDVI value in the Jiulongquan watershed

4.3 Analysis of driving factors

4.3.1 Correlation between the vegetation index and climatic factors

Annual precipitation and mean temperature were selected as representative climatic factors for the following analysis. The partial correlation coefficients between the NDVI and annual precipitation was between −0.95 and 0.96 during the study period. The area with positive and negative correlation accounted for 64.98 and 35.02% of the total study area, respectively. However, only 1.3% of the total area had statistically significant partial coeeficients(Fig.5a).

As shown in Fig.5b, the partial correlation coefficient between the NDVI and mean temperature was between −0.98 and 0.98. A negative correlation existed in 88.43% of the watershed, of which 56.29% reached statistical significance and was mainly distributed on the slopes of the middle and lower reaches of the watershed. In contrast, the area with positive correlation accounted for 11.57% of the total area and was mainly distributed in the gully, of which only 1.38% had significantly positive partial correlation coefficients.

To study the compounding influence of climatic factors on vegetation coverage, multiple correlation among annual precipitation, mean temperature, and the NDVI of each pixel in the Jiulongquan watershed from 2010 to 2019 was performed. The multiple correlation coefficient greatly varied ranging between 0.002 and 0.98. The area with statistically significant multiple correlation coefficients accounted for 13.6% of the total area and was mainly distributed on the downstream slope of the watershed (Fig.5c).

4.3.2 Driving factors partition of vegetation index changes

To further reveal the driving factors of the spatiotemporal changes of vegetation coverage, the driving factors partition method employed by Chen et al.[26] was adopted. Climate background and characteristics of the Loess Plateau were also considered.

According to the grid statistics of the partition results, change in about 14.05% of the total area was driven by climatic factors (Fig.6). In detail, the area driven by precipitation and both precipitation and temperature accounted for 0.38% and 0.02% of the total area, respectively. Such area was scattered across

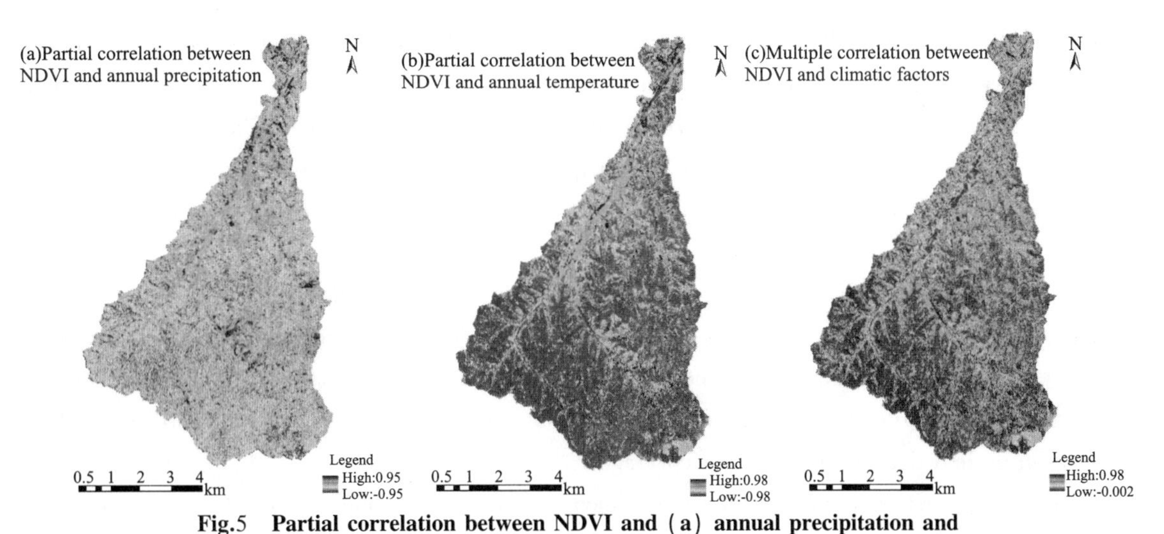

Fig.5 Partial correlation between NDVI and (a) annual precipitation and (b) mean temperature and multiple correlation, (c) in the Jiulongquan watershed from 2010 to 2019

the study area. The area driven by temperature itself accounted for about 13.65% of the study area and was mainly distributed on the southwest downstream slope. The variation of the NDVI in 85.92% of the Jiulongquan watershed were driven by non-climate factors.

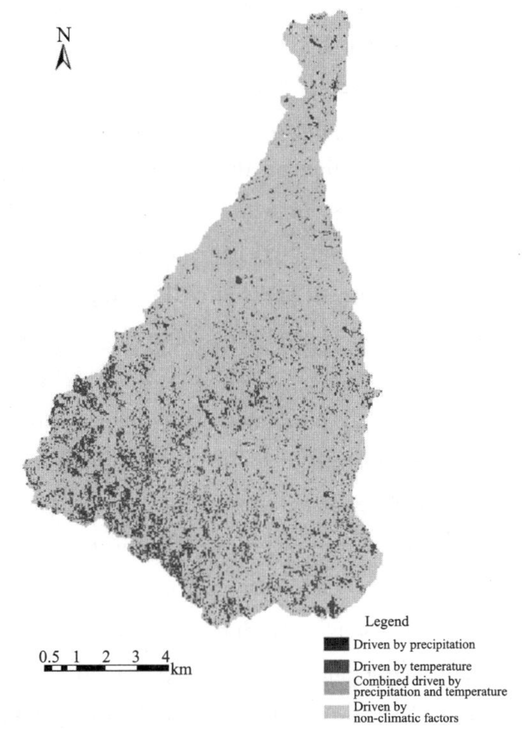

Fig.6 Factor driving zoning of NDVI change in the Jiulongquan watershed

5 Discussion

5.1 Distribution and change of vegetation

The low NDVI in the gully bottom of the study area is probably due to prenneial crop cultivation and the disturbance of the GCLC project which had a negative impact on surface vegetation. In contrast, the high NDVI on the slopes may be explained by the vigorous growth of tree and shrubs as a result of the *Grain for Green* projects which was impelemented prior to the GCLC projects.

Our study showed that following the implementation of the Jiulongquan GCLC project, the vegetation coverage of the watershed tended to improve to a limited extent. The improved area was much larger than the degraded area inconsistent with other land consolidation projects. For example, Zhong and Wang[27] and Shan et al.[28] both showed that the regional habitat quality tended to decrease within 3~5 year after land consolidation. This is probably because the objectives of different land consolidation projects highly varied. The implementation of a series of land consolidation projects in those studies had a large-scale disturbance on surface soil and thus usually destroys the surface vegetation of gullies as the goal was to create more cultivated land. The monocultural cropping system is another factor leading to the reduction and degradation of the species and quantity of primary and secondary vegetation in the project area[29]. If such goal was excessively pursued, the area of other ecological land types, such as forestland and grassland, will inevitably decrease, resulting in a decreased vegetation coverage[30-31].

In contrast, the GCLC project in this study aimed to restore, add, and improve the quality of the cultivated land in the gully. As a result, the disturbance by construction was concentrated in the gully and thus the primary vegetation in other areas was not destroyed on a large scale. Although the increase of cultivated land would occupate some grassland and woodland[32], the project greatly restored the vegetation damage caused by the disturbance through a systematic construction of the farmland forest network. Some of the shrub land were converted to more ecologically suitable native forest due to biological measures also promoted the vegetation restoration[33]. On the other hand, the fragile agricultural ecological environment in and around the project area was improved by hydraulic engineering, which effectively reduced the occurrence of drought and flooding events, prevented soil and water erosion, and thus promoted the protection of vegetation in the whole basin. After the project, the improvement in soil quality and grain yield of the cultivated land effectively reduced the reclamation on the slopes and consolidated the achievements of the *Grain for Green* project. Such results are consistent with Du et al.[18]. That study performed in the Gutun Watershed in Yan 'an city showed that the mean NDVI in the watershed increased by about 54% owing to the combined effects of the *Grain for Green* project and the GCLC project. Taken together, the GCLC project has an overall positive effect on the change of vegetation coverage in this area.

5.2 Driving factors

Our work found that NDVI was positively correlated with precipitation and negatively correlated with temperature in evaluated watershed. Soil moisture is one of the dominant factors limiting plant growth in the Loess Plateau. Climate change, particularly global warming may increase evapotranspiration. Uneven spatiotemporal distribution of precipitation subsequently aggravates the water shortage, leading to an limited regional vegetation growth. In contrast, the increase of rainfall is conducive to the improvement of soil water status and promotes vegetation growth consistent with Sun and Wang[34-35]. However, such correlation was statistically singnificant in only <15% of the total study area, suggeting that the evaluated climatic factor may not be the main driving factors for the development of NDVI in the area.

Our Further partition analysis of the driving factors showed that development of the NDVI during 2010–2019 in >85% of the total area was driven by non-climatic factors such as human activities. The improvement of the NDVI may be attributed to the implementation of the GCLC project. Many studies investigating the change of vegetation coverage in the Loess Plateau also showed that human activities are the main attibutes of vegetation change in this area after 2000[36]. Among all the human activities, ecological engineering, particularly the *Grain for Green* project has largely promoted the increase of vegetation coverage[37-38]. The GCLC project had little disturbance to the region except for the gullies. Meanwhile, the GCLC project changed the factors affecting vegetation growth such as hydrological environment[39] and soil erosion process[40-41], extending the influence to the whole basin through measures such as field integration, water resource regulation, improvement of agricultural infrastructure and production conditions, and reasonable adjustment of the cropping systems. Li et al.[32] showed that the GCLC project could contribute to the improvement of land use structure and help to optimize the landscape pattern, thus affecting the changes of geographical environment. Such results suggest that the overall goal of coordinated development of cultivated land quantity, quality, and ecology was achieved.

In the current study we performed a quantitative analysis on the change of regional vegetation coverage and its driving factors before and after the GCLC project on the project scale over time series. In the future, residual analysis method[42], regression model method[43] and other methods may be used to further analyze the change of vegetation coverage caused by specific human activities in detail.

6 Conclusion

The vegetation coverage of the Jiulongquan watershed was generally high. The spatial distribution of the NDVI tended to be low in the gully area and high on the slope. From 2010 to 2019, vegetation coverage and ecological environment of the watershed tended to improve following the GCLC project as indicated by the increased NDVI values, with a annual growth rate of 0. 004. The significant NDVI changes were concentrated in the gullies, while a majority of the watershed was relatively stable or slightly improved, indicating that the engineering disturbance had little negative impact on the whole watershed. The change of NDVI was mainly driven by human activities rather than the climatic factors including annual precipitation and mean temperature.

These results indicated that the GCLC project has significant ecological implications in the Loess Plateau. Comprehensive management and utilization of the gullies could be more effective in increasing cultivated land resources, improving the quality of cultivated land, and ensuring food security than traditional land consolidation projects. The GCLC projects could also consolidate the achievements of the Grain for Green project on slopes of watershed, and thus have positive effects on the local ecological and agricultural environment. Our work could be cited as a reference for other comprehensive ecological control projects in the Loess Plateau.

Funding information: This work was supported by the National Key Research and Development Project of China (Grant No. 2017YFC0504705), the Natural Science Basic Research Project of Shaanxi Province (Grant No. 2022JM-168), and the Internal Research Project of Shaanxi Provincial Land Engineering Construction Group (Grant No. DJNY2021-21).

Author contributions: JW performed the study, analyzed the data, and drafted the originalmanuscript; YH reviewed and edited the manuscript; LY helped with the visualization of the data; QB managed the entire research project and offered supervision.

Conflict of interest: Authors state no conflict of interest.

References

[1]Parmesan C, Yohe G. A globally coherent fingerprint of climate change impacts across natural systems. Nature. 2003; 421:37-42.

[2]Piao S L, Fang J Y. Seasonal changes in vegetation activity in response to climate changes in China between 1982 to 1999. Acta Geographica Sinica. 2003; 58(1):119-125.

[3]Jong R D, Bruin S D, Wit A D, Schaepman M E, Dent D L. Analysis of monotonic greening and browning trends from global NDVI timeseries. Remote Sens Environ. 2011;115(2):692-702.

[4]Begue A,Vintrou E, Ruelland D, Claden M, Dessay N. Can a 25-year trend in Soudano-Sahelian vegetation dynamics be interpreted in terms of land use change? A remote sensing approach. Sci Tech. 2011;21(2):413-420.

[5]Mu S J, Li J L, Chen Y Z, Gang C C, Zhou W, Ju W M. Spatial differences of variations of vegetation coverage in Inner Mongolia during 2001-2010. Acta Geographica Sinica. 2012;67(9):1255-1268.

[6]Pettorelli N, Vik JO, Mysterud A, Gaillard JM, Tucker CJ, Stenseth NC. Using the satellite-derived NDVI to assess ecological responses to environmental change. Trends Ecol Evol. 2005;20(9):503-510.

[7]Tucker C J, Pinzon J E, Brown M E, Slayback D A, Pak E W, Mahoney R, et al. An extended AVHRR 8 km NDVI dataset compatible with MODIS and SPOT vegetation NDVI data[J]. Int J Remote Sens. 2005;26(20): 4485-4498.

[8]Neigh C S R, Tucker C J, Townshend J R G. North American vegetation dynamics observed with multi-resolution satellite data. Remote Sensing of Environment Remote Sens Environ. 2008;112(4): 1749-1772.

[9]Levin N. Human factors explain the majority of MODIS-derived trends in vegetation cover in Israel: a densely populated country in the eastern Mediterranean. Rer Environ Change. 2016; 16(4):1197-1211.

[10]Sun J, Cheng G W, Li WP, Sha Y K, Yang Y C. On the variation of NDVI with the principal climatic elements in the Tibetan Plateau. Remote Sens-Basel. 2013;5:1894-1911.

[11]Nemani R R, Keeling C D, Hashimoto H, Jolly W M, Piper S C, Tucker CJ, et al. Climate-driven increases in global terrestrial net primary production from 1982 to 1999. Science. 2003;300:1560-1563.

[12]Zhao X, Tan K, Zhao S, Fang J.Changing climate affects vegetation growth in the arid region of the northwest China. J Arid Environ. 2011;75(10):946-952.

[13] Liu X F, Zhu X F, Pan Y Z, Li S S, Ma Y Q, Nie J. Vegetation dynamics in Qinling-Daba Mountains in relation to climate factors between 2000 and 2014. J Geogr Sci. 2016; 26:45-58.

[14]Zhao L H, Wang P, Ouyang X Z, Wu Z W.An analysis of the spatio-temporal variation in fractional vegetation cover and its relationship with non-climate factors in Nanchang City, China. Acta Ecologica Sinica. 2016; 36(12):3723-3733.

[15]Guo J K, Wu X Q, Dong G H, Li Y S, Wu R. Vegetation coverage change and relative effects of driving factors based on MODIS /NDVI in the Tarim River Basin. Arid Zone Research. 2017; 34(3):621-629.

[16]Wei Y Q, Lu H Y, Wang J N, Sun J, Wang X F. Responses of vegetation zones, in the Qinghai-Tibetan Plateau, to climate change and anthropogenic influences over the last 35 years. Pratatical Science. 2019; 36(4):1163-1176.

[17]He M N, Wang Y Q, Tong Y P, Zhao Y L, Qiang X K, Song Y G, et al. Evaluation of the environmental effects of intensive land consolidation: A field-based case study of the Chinese Loess Plateau. Land Use Policy. 2020; 94:104523.

[18]Du P C, Xu Q, Zhao K Y, Guo P, Peng S Q, Guo C. Impacts of gully reclamation project on cropland distribution and vegetation restoration in North Shaanxi Province-A case study at Gutun Watershed of Yan' an City. Bulletin of Soil and Water Conservation. 2019; 39(6): 1-8.

[19]Zhang C , Li W , Travis D . Gaps-fill of SLC-off Landsat ETM+ satellite image using a geostatistical approach. Int J Remote Sens. 2007; 28(22): 5103-5122.

[20]Li S, X u L, Jing Y H, Yin H, Li X H, Guan X B. High-quality vegetation index product generation: a review of NDVI time series reconstruction techniques[J]. Int J Appl Earth Obs. 2021; 105: 102640.

[21]Stow D, Petersen A, Hope A, Engstrom R, Coulter L. Greenness trends of Arctic tundra vegetation in the 1990s: comparison of two NDVI data sets from NOAA AVHRR systems. Int J Remote Sens.2007; 28(21):4807-4822.

[22]Stow D, Daeschner S, Hope A, Douglas D, Petersen A, Myneni R, et al. Variability of the seasonally integrated normalized difference vegetation index across the north slope of Alaska in the 1990s. Int J Remote Sens. 2003; 24(5): 1111-1117.

[23]Jin K, Wang F, Han K J, Shi S Y, Ding W B. Contribution of climatic change and human activities to vegetation NDVI change over China during 1982-2015. Acta Geographica Sinica. 2020; 75(5): 961-974.

[24]Gutman G, Ignatov A. The derivation of the green vegetation fraction from NOAA/AVHRR data for use in numerical weather prediction models. Int J Remote Sens. 1998;19(8):1533-1543.

[25]Liu Y L, Lei H M. Responses of natural vegetation dynamics to climate drivers in China from 1982 to 2011. Remote Sens-Basel. 2015;7(8):10243-10268.

[26]Chen Y H, Li X B, Shi P J. Variation in NDVI driven by climate factors across China 1983-1992. Acta Plant Ecology. 2001; 25(6): 716-720.

[27]Zhong L N, Wang J. Evaluation on effect of land consolidation on habitat quality based on InVEST model. Transac-

tions of the Chinese Society of Agricultural Engineering. 2017; 33(1):250-255.

[28]Shan W, Jin X B, Meng X S, Yang X Y, Xu Z G, Gu Z M, et al. Dynamical monitoring of ecological environment quality of land consolidation based on multi-source remote sensing data. Transactions of the Chinese Society of Agricultural Engineering. 2019; 35(1): 234-242.

[29]Yamaguchi H, Umemoto S, Maenaka H. Floral composition of the vegetation on levees of traditional and reconstructed paddies in Sakai city, Japan. Weed Res. 1998; 43(3):249-257.

[30]Crecente R, Alvarez C, Fra U. Economic, social and environmental impact of land consolidation in Galicia. Land Use Policy. 2002;19(2):135-147.

[31]Yang B, Wang Z Q, Hu X D. Performance assessment and impact factors analysis of land consolidation project based on improved extension matter-element model. China Land Science. 2018; 32(7):66-73.

[32]Han X L, Lv P Y, Zhao S, Sun Y, Yan S Y, Wang M H, Han X N,Wang X R.The effect of the Gully Land Consolidation Project on soil erosion and crop production on a typical watershed in the Loess Plateau. Land. 2018; 7(4):113.

[33]Li Y R, Li Y, Fan P C, Long H L. Impacts of land consolidation on rural human-environment system in typical watershed of the Loess Plateau and implications for rural development policy. Land Use Policy. 2019; 86:339-350.

[34]Sun W Y, Song X Y, Mu X M, Gao P, Wang F, Zhao GJ. Spatiotemporal vegetation cover variations associated with climate change and ecological restoration in the Loess Plateau. Agr Forest Meteorol. 2015; 209-210(1): 87-99.

[35] Wang H , Liu G H , Li Z S , Ye X , Wang M, Gong L. Driving force and changing trends of vegetation phenology in the Loess Plateau of China from 2000 to 2010. J Mt Sci. 2016;13:844-856.

[36]Qu L L , Huang Y X , Yang L F , Li Y R. Vegetation restoration in response to climatic and anthropogenic changes in the Loess Plateau, China. Chinese Geogr Sci. 2020;30(1):89-100.

[37]Zheng K, Wei J Z, Pei J Y, Cheng H, Zhang X L, Huang F Q,et al. Impacts of climate change and human activities on grassland vegetation variation in the Chinese Loess Plateau[J]. Sci Total Environ.2019;660(10):236-244.

[38]Dong Y, Yin D Q, Li Y, Yan Q L, Wang S H. Spatio-temporal patterns of vegetation change and driving forces in the Loess Plateau. Journal of China Agricultural University. 2020; 25(8):120-131.

[39]Lou X Y, Gao J E, Han S Q, Guo Z H, Yin Y. Influence of land consolidation engineering of gully channel on watershed runoff yield and concentration in loess hilly and gully region. Water Resources and Power. 2016; 34(10):23-27.

[40] Ji Q Q,Gao Z, Li X Y, Gao J E, Zhang G G, Ahmad R, Liu G, Zhang Y Y, Li W Z, Zhou F F, Liu S X. Erosion transportation processes as influenced by Gully Land Consolidation Projects in highly managed small watersheds in the Loess Hilly-Gully Region, China .Water. 2021;13(11):1540.

[41]Liu Y S, Guo Y J, Li Y R, Li Y H. Gis-based effect assessment of soil erosion before and after gully land consolidation: a case study of wangjiagou project region, loess plateau. Chinese Geogra Sci, 2015;25(2), 137-146.

[42]Evans J, Geerken R. Discrimination between climate and human induced dryland degradation. J Arid Environ. 2004; 57:535-554.

[43]Mueller T, Dressler G, Tucker C J, Pinzon J E, Leimgruber P, Dubayah R O, et al.Human land-use practices lead to global long-term increases in photosynthetic capacity. Remote Sens-Basel. 2014; 6(6): 5717-5731.

本文曾发表于2022年《Open Geosciences》第14卷

Research on Remediation of Soil Petroleum Pollution by Fenton-like Catalyst Carried by Permutite

Yuhu Luo[1,2,3,4], Nan Lu[1,2,3,4], Yang Wei[1,2,3,4]

(1.Institute of Land Engineering and Technology, Shaanxi Provincial Land Engineering Construction Group Co., Ltd., Xi'an 710075, China;2.Shaanxi Provincial Land Engineering Construction Group Co., Ltd., Xi'an 710075, China;3.Key Laboratory of Degraded and Unused Land Consolidation Engineering, the Ministry of Natural Resources, Xi'an 710075, China;4.Shaanxi Provincial Land Consolidation Engineering Technology Research Center, Xi'an 710075, China)

【Abstract】The supported Fenton-like catalyst was prepared by the impregnation method with permutite as the carrier, and the optimal preparation conditions of the catalyst, the use conditions and catalytic mechanism of the catalyst to degrade oil-contaminated soil were studied. The results show that when the mass ratio of $FeSO_4 \cdot 7H_2O$ to permutite is 1∶1, the prepared catalyst has the best degradation effect. When the catalyst was used to degrade 10 g of 50 $g \cdot kg^{-1}$ heavy petroleum-contaminated soil, the degradation efficiency reached the highest 76.79% when the catalyst dosage was 2 g and the oxidant dosage was 6 mL. After 24 hours, the reaction of the catalyst to degrade oil-contaminated soil basically ends, and the degradation efficiency reaches the highest. The analysis results of petroleum hydrocarbon components before and after the reaction showed that the prepared catalyst had the highest degradation efficiency for saturated hydrocarbons in petroleum hydrocarbons, followed by aromatic hydrocarbons, and finally for colloid and bitumen. The prepared Fenton-like catalyst was characterized by scanning electron microscope and X-ray energy dispersive spectrometer. The results showed that Fe was successfully and uniformly supported on the surface or void of permutite in a dispersed state by this method, which solved the limitation of the catalyst by pH conditions, effectively improve the catalytic efficiency of the catalyst.

【Keywords】Fenton-like; Soil petroleum pollution; Oxidative degradation

Introduction

As an important chemical raw material, petroleum plays a vital role in agriculture, heavy industry, transportation, and national defense construction. However, the abnormal leakage of petroleum exploration, exploitation, processing, storage, and transportation often causes serious soil pollution problems, and the contaminants then enter the human body through the food chain or drinking water, resulting in a series of economic, environmental and social problems[1]. At present, the remediation methods of soil oil pollution mainly include physical remediation, chemical remediation and bioremediation[2-4], among which chemical remediation technology is widely used due to its advantages of strong adaptability, high remediation efficiency and short remediation period[5-7].

Fenton technology is a common means of chemical repair. It is a reaction system composed of Fe^{2+} and H_2O_2, which can generate strong oxidizing hydroxyl radicals ($\cdot OH$), and its standard redox potential is 2.8 V, second only to fluorine. Therefore, it can oxidize most of the organic pollutants into organic compounds with simple molecular structure, and even into carbon dioxide and water, and finally achieve the purpose of remediating soil organic pollution[8]. With the development of this technology, it has been widely used in the remediation of organic pollutants such as petroleum hydrocarbons, printing and dyeing wastes, and pesticides[9-11]. However, the traditional homogeneous Fenton reagent needs to reach a high degradation efficiency under the strong acid condition of pH 3. Moreover, the catalyst Fe^{2+} is easy to form precipitation during the reaction process, which reduces the catalytic efficiency. Therefore, in order to overcome the defects in the homogeneous Fenton reaction system, scholars replaced Fe^{2+} with an iron-containing solid catalyst to form a heterogeneous Fenton-like reagent to catalyze the decomposition of hydrogen peroxide to generate ·OH, thereby achieving efficient degradation of organic pollutants. For example, the use of the soil's own iron oxides can catalyze the degradation of organic pollution. Common iron oxides in the soil include goethite (α-FeOOH), hematite (α-Fe_2O_3), magnetite (Fe_3O_4), water Iron ore (α-$Fe_{10}O_{15} \cdot 9H_2O$), etc.[12-13]. These iron-containing minerals in the soil matrix can act as catalysts, catalyzing the decomposition of hydrogen peroxide into hydroxyl radicals, and then oxidatively degrade the organic matter in the soil. However, not all soils have sufficient iron oxides to meet the requirements for catalytic degradation of pollutants. Therefore, people began to load Fe on the surface of solid materials to form solid catalysts by artificial means to meet the needs of degrading pollutants. Du Ting et al. loaded Fe on the surface of activated carbon to degrade COD and phenol in wastewater. The results showed that compared with the traditional Fenton method, the degradation efficiency of COD and phenol was significantly improved by this method[14]. Lu Tianyu et al. used attapulgite to load Fe to degrade phenol wastewater, and the removal efficiency of phenol reached 98% after 2.5 hours of reaction[15]. Xiang Chunyan et al. supported Fe oxide on diatomite to prepare a supported heterogeneous Fenton catalyst, and the degradation efficiency of the modified catalyst for phenol reached more than 99%[16].

The previous research results show that the supported Fenton catalyst has a good catalytic effect, solves the limitation of the reaction system by pH conditions, and achieves a high degradation effect. However, most of these studies were used to degrade organic pollutants in water, and few studies were used to degrade organic pollutants in soil. Therefore, in this study, a low-cost mineral material artificial zeolite with developed specific surface area, which will not cause secondary pollution after being applied into the soil, was selected as the carrier material. Fe is supported on the surface of artificial zeolite by impregnation method to prepare a supported Fenton-like catalyst, and the preparation regulations of catalysts, the application conditions and reaction mechanism of degrading oil-contaminated soils are studied, in order to improve the theoretical basis and data support for the remediation of oil-contaminated soils.

Materials and methods

Materials and instruments. Test materials: artificial zeolite (60~80 mesh), $FeSO_4 \cdot 7H_2O$ (AR), hydrogen peroxide (30% AR), crude oil, non-polluting soil.

Main instruments: OIL-8 infrared oil measuring instrument, cyclotron oscillator, JSM-6390A scanning electron microscope, X INCA-350 ray energy spectrometer, digital blast drying oven, constant temperature water bath, mechanical stirrer, gyratory shaking stirrer.

Preparation of catalysts. The artificial zeolite used in the experiment is 60~80 mesh uniform particles, and the catalyst is prepared by impregnation method. Specifically, according to the parameter settings in Table 1, 10 g, 20 g and 30 g of $FeSO_4 \cdot 7H_2O$ were respectively taken in a beaker, 200 mL of distilled water was added, and then 20 g of 60~80 mesh artificial zeolite were added respectively. Next, place the beaker on a 90 ℃ constant temperature water bath for four hours with mechanical stirring at a speed of 250 rmp to remove the remaining liquid, and then use distilled water to wash the remaining substances on the material until no iron ions are dissolved. Finally, the catalysts were dried in an oven at 110 ℃ to complete the preparation of catalysts P1, P2 and P3.

Table 1 Preparation parameters of catalysts

Catalyst type	Zeolite(g)	$FeSO_4 \cdot 7H_2O$(g)	Method
P1	20	10	Impregnation and heating evaporation for 4 hours
P2	20	20	
P3	20	30	

Oil pollution degradation test. Take the uncontaminated soil in Guanzhong area of Shaanxi as the test soil, remove the 0~20 cm topsoil that is easily affected by human activities, take 20~50 cm bottom soil, air dry, remove large particles such as gravel and plant debris, and grind it to 2 mm mesh screen. Oilfield crude oil was used as the pollution source, and petroleum ether was used as the solvent to prepare heavily oil-contaminated soil with a pollution concentration of 50 $g \cdot kg^{-1}$. The prepared soil was aged for two weeks in a fume hood, and stored in the dark for later use. The texture and basic physicochemical properties of the tested soils are shown in Table 2.

Table 2 Test soil texture and basic physical and chemical properties

Items	Particle size composition(%)			Texture (USDA)
	Cosmid (<0.002 mm)	Powder (0.05~0.002 mm)	Grit (0.05~2 mm)	
Average value	19.60	79.61	0.69	Silty loam
Items	pH	Conductivity ($\mu S \cdot cm^{-1}$)	Organic matter content ($g \cdot kg^{-1}$)	Total nitrogen content ($g \cdot kg^{-1}$)
Average value	8.15	125.8	3.14	0.19

Then, the prepared Fenton-like catalysts (P1, P2, P3) were used for the degradation test of oil-contaminated soil. During the test, the room temperature was kept at 25 ℃, and 10 g of oil-contaminated soil was taken into a stoppered conical flask each time, and three catalysts P1, P2, and P3 were added respectively, and the dosage of oxidant is 4 mL, 5 mL, 6 mL, 7 mL, and 8 mL, respectively. During the test, the soil-water ratio is controlled to be 1:3 (g:mL). After the test conditions are met, seal the conical flask, then oscillated at 250 $r \cdot min^{-1}$ for 30 min on a gyratory shaker, and then allowed the conical flask to stand for 24 h to allow the reaction to complete. Finally, the amount of petroleum hydrocarbons in the soil was determined by infrared spectrophotometry, and the petroleum-contaminated soil without reagents was used as the control, and the degradation efficiency of petroleum hydrocarbons in the soil was calculated by comparing the changes of petroleum hydrocarbons in the soil before and after the reaction. The components of petroleum hydrocarbons in the control group and the reaction group were determined, and the degradation effects of petroleum hydrocarbons of different components were analyzed. At the same time, during the reaction, samples were collected every 6 hours to determine the degradation efficiency of oil-contaminated soil in different time periods.

Characterization of the catalyst. Scanning Electron Microscopy (SEM) was used to observe the surface morphology of the catalyst, and X-ray photoelectron spectroscopy was used to determine its valence and elemental composition. The catalytic mechanism of the prepared catalysts was studied by analysis.

Test method. The content of petroleum hydrocarbons in the soil is based on " Determination of Petroleum in Soil by Infrared Spectrophotometry" (HJ 1051—2019), extracted with tetrachloroethylene, and measured with an infrared oil measuring instrument; the microstructure of the catalyst was observed with a scanning electron microscope (SEM); The composition of catalyst elements was analyzed by X-ray energy dispersive spectrometer (EDS); the separation and determination of petroleum hydrocarbon components in the soil were based on "Analysis of Soluble Organic Matter and Crude Oil Group Components in Rocks" (SY/T 5119—2016); soil pH was determined Refer to "Determination of Soil pH" (NY/T 1377—2007), and the determination of electrical conductivity refers to "Electrode Method for Determination of Soil Conductivity HJ 802—2016"; the determination of organic matter refers to "Soil Testing Part 6: Determination of Soil Organic Matter" (NY /T 1121.6—2006); for the determination of total nitrogen, refer to "Determination of Soil Total Nitrogen (Semi-Micro Kelvin Method)" (NY/T 53—1987).

Results and discussion

Research on catalytic reaction conditions. According to the conditions of 2.3, the prepared zeolite-supported iron-based Fenton catalyst was used to degrade petroleum-contaminated soil. The test results are shown in Fig.3. It can be seen that with the increase of the oxidant dosage, the degradation efficiency of petroleum hydrocarbons in the soil shows an obvious trend of first increasing and then decreasing. It shows that when using Fenton's reagent to degrade soil organic pollution, the dosage of hydrogen peroxide and the ratio of hydrogen peroxide to catalyst must be well controlled. When the dosage of hydrogen peroxide is too small, it is difficult to fully degrade, and when the dosage is too large, side reactions will occur, resulting in the self-decomposition of hydrogen peroxide, which reduces the catalytic efficiency.

It can be seen from Fig.1(A) that when the dosage of catalyst is 1 g, the dosage of hydrogen peroxide is 7 mL, and the degradation efficiency of P2 catalyst is the highest, which is 60.21%; Fig.1(B) shows that when the dosage of catalyst is 2 g , the dosage of hydrogen peroxide is 6 mL, the degradation efficiency of P2 catalyst is the highest 76.79%; Fig.1(C) shows that when the dosage of catalyst is 3 g, the dosage of hydrogen peroxide is 6 mL, and the degradation efficiency of P2 catalyst is the highest 65.37%.

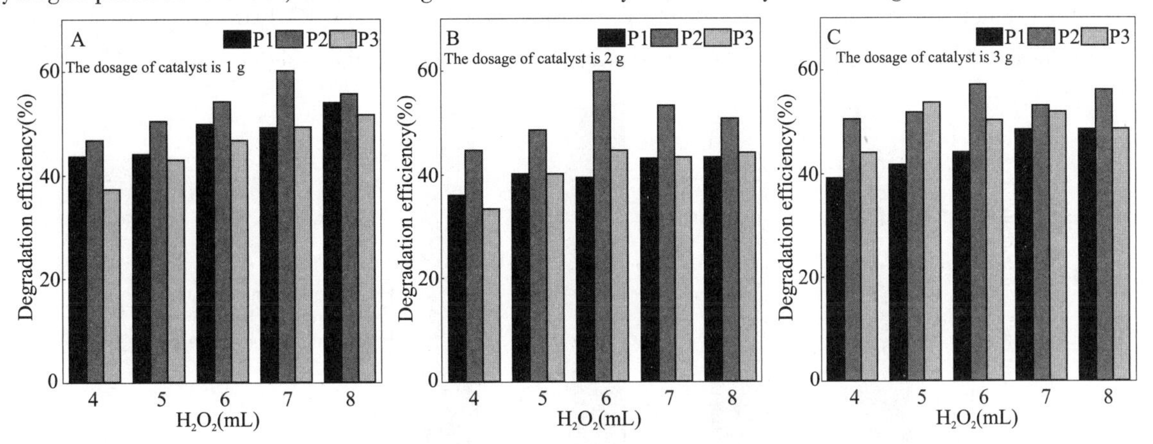

Fig.1 Experiment on the degradation of soil oil pollution by catalyst

The comprehensive comparison shows that the catalyst P2 degradation efficiency is the highest when the dosage of P2 catalyst is 2 g and the dosage of hydrogen peroxide is 6 mL. When using this Fenton-like reagent to degrade soil oil pollution, the dosage of hydrogen peroxide and the ratio of hydrogen peroxide to catalyst must be well controlled.

Response time. In order to understand the degradation efficiency of Fenton-like catalysts on oil-contaminated soil under different reaction times, the dosage of hydrogen peroxide was 6 mL, and the dosage of three types of catalysts P1, P2, and P3 were all 2 g, and the soil was sampled every 6 hours for determination the degradation efficiency of petroleum hydrocarbons in soil. The test results are shown in Fig.2. It can be seen that from 0 to 18 hours, under the action of the three catalysts, the degradation efficiency of petroleum hydrocarbons increases significantly. From 18 to 24 hours, the rate of increase in degradation efficiency slows down. From 24 to 30 hours, the degradation efficiency was almost unchanged, which indicated that the oxidative degradation reaction of petroleum hydrocarbons in the soil in the Fenton-like reagent was complete after 24 hours.

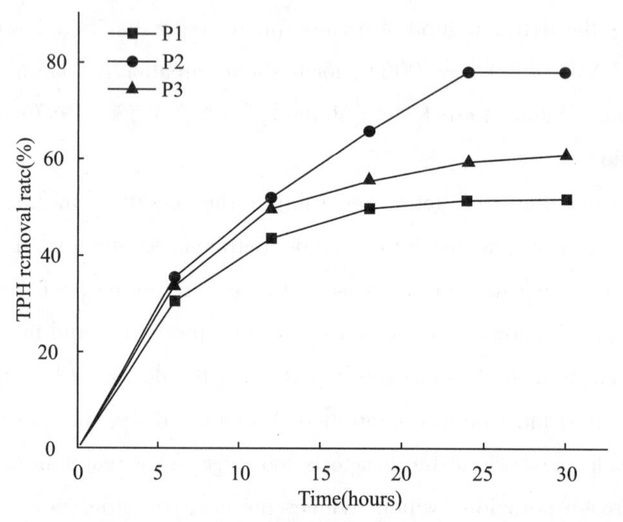

Fig.2 Degradation efficiency of petroleum hydrocarbons with different reaction times

Petroleum hydrocarbon components. Petroleum is a mixture of various components, including saturated hydrocarbons, aromatic hydrocarbons, colloids and asphaltenes[17]. Each component has different structures and has different degrees of oxidative degradation. By analyzing the changes of petroleum components, the degradation mechanism of Fenton-like catalysts on petroleum pollution was further explored. In Fig.3, CK is the control without adding catalyst and oxidant. The dosage of P1, P2, and P3 hydrogen peroxide is all 6 mL, the dosage of the corresponding three types of catalysts is 2 g, and the reaction time is 24 hours.

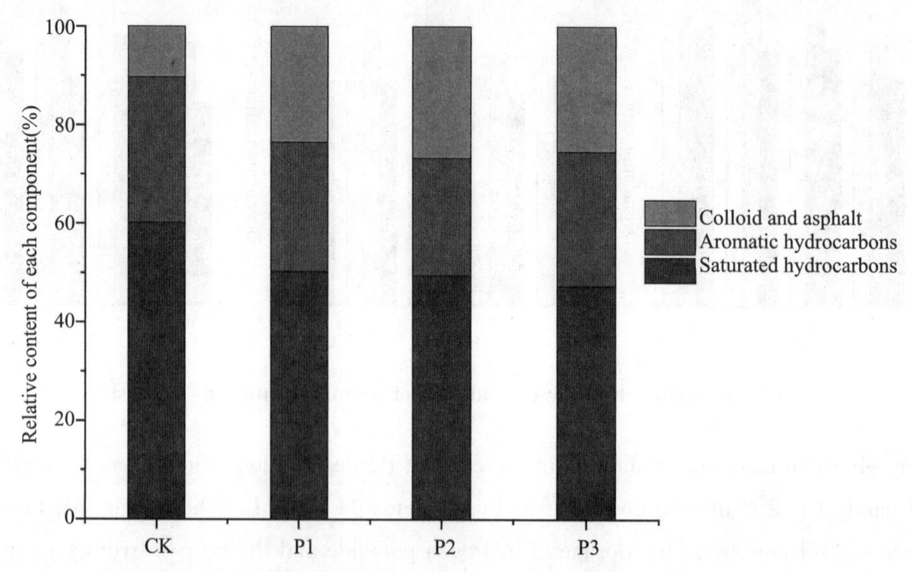

Fig.3 Variation characteristics of petroleum hydrocarbon components

It can be seen from Fig.3 that the relative content of saturated hydrocarbons in P1, P2, and P3 of petroleum hydrocarbons decreased by 9.9%, 10.7%, and 12.8%, respectively, and the relative contents of aromatic hydrocarbons decreased by 3.2%, 5.6%, and 2.2%, respectively. The relative content of asphalt and asphalt increased by 13.1%, 16.4% and 15.1%, respectively. This result shows that the prepared catalyst has the highest degradation efficiency for saturated hydrocarbons in petroleum hydrocarbons, followed by aromatic hydrocarbons, and finally colloid and asphalt.

Analysis of catalyst morphology. Fig.4 is the scanning electron microscope (SEM) images of the permutite and the iron-supported catalyst before and after loading, respectively. The surface of the permutite has a well-developed concave-convex pore structure with fewer surface particles. The surface of the permutite loaded with iron appears dense and fine particulate matter, which is wrapped on the surface of the permutite. These particles should be Fe loaded in the surface and voids of the permutite in the form of oxides, forming an iron-supported artificial zeolite catalyst.

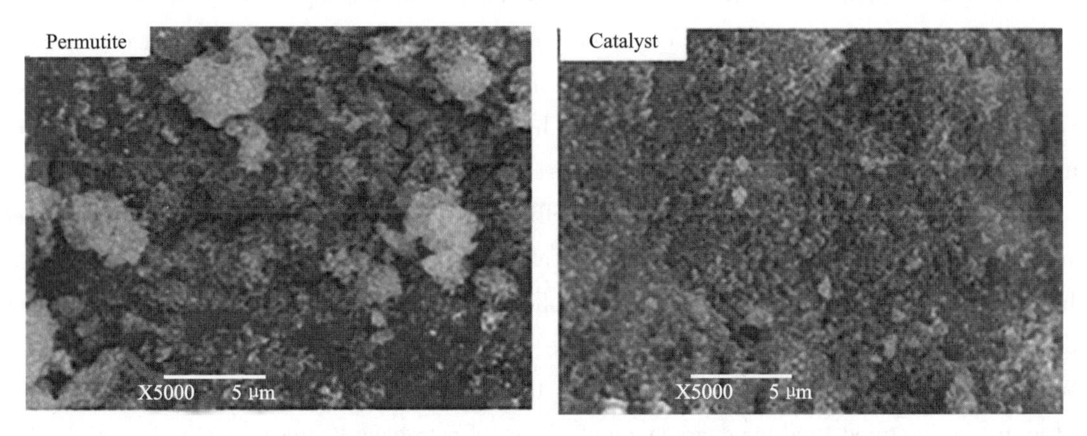

Fig.4 Surface structure of permutite and Fenton-like catalyst

Analysis of chemical composition of catalysts. Fig.5 is the energy dispersive spectrometer (EDS) of permutite and catalyst. The element composition of catalyst obtained by energy spectrum analysis is shown in Table 3. The main elements in permutite include Si, C, Al, Na, O, etc. The catalyst has Fe element, and its mass fraction reaches 15.1%, indicating that Fe was successfully loaded on the surface of permutite after impregnation loading.

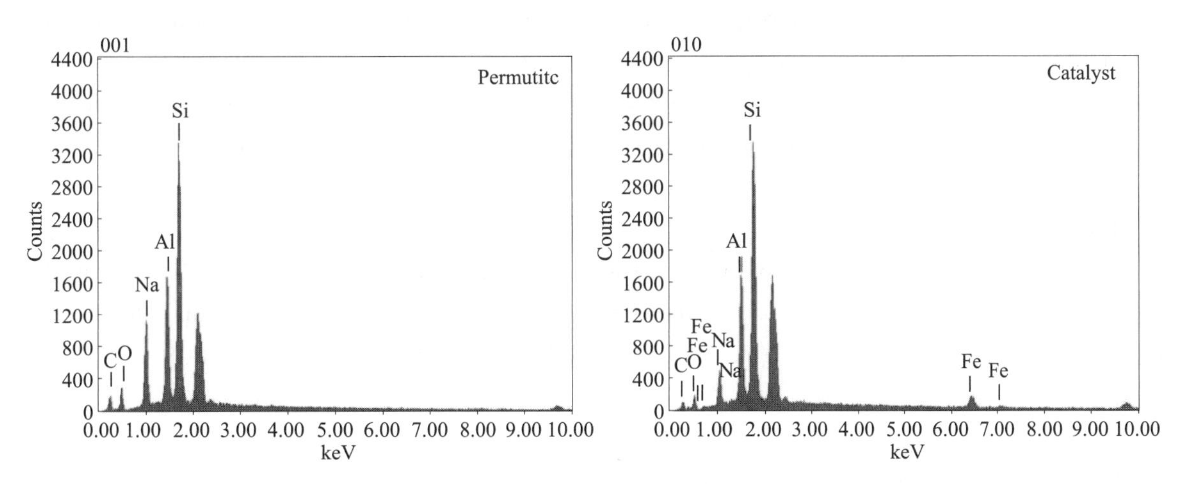

Fig.5 Energy spectrum of permutite and Fenton-like catalyst

Table 3 Elemental composition of permutite and Fenton-like catalyst

Element	Permutite		Catalyst	
	100% quality	Error/%	100% quality	Error/%
C	20.14	0.54	11.17	0.74
O	9.71	0.46	5.64	0.54
Na	12.2	0.23	6.09	0.38
Al	16.94	0.25	17.9	0.35
Si	41.02	0.32	44.1	0.44
Fe	—	—	15.1	2.45

Conclusion and outlook

The supported Fenton-like catalyst was prepared by impregnation method with permutite as the carrier. When the mass ratio of $FeSO_4 \cdot 7H_2O$ to permutite was 1∶1, the catalyst P2 had the best degradation efficiency. The catalyst can degrade heavily petroleum-contaminated soil under the original soil pH condition. When the dosage of catalyst is 2 g and the dosage of hydrogen peroxide is 6 mL, the degradation efficiency of petroleum hydrocarbons in soil can reach the highest 76.79%. When the prepared catalyst was used to degrade petroleum-contaminated soil, the degradation efficiency of petroleum hydrocarbons showed a trend of first increasing and then decreasing with the increase of the dosage of oxidant hydrogen peroxide. This shows that the appropriate dosage of oxidant must be controlled when using this type of Fenton reagent to ensure high degradation efficiency.

With the increase of the catalytic reaction time, the degradation of petroleum hydrocarbons in the soil showed an increasing trend, and the degradation efficiency of petroleum hydrocarbons in the soil reached the highest after 24 hours, and the degradation reaction basically ended. The prepared catalyst has the highest degradation efficiency for saturated hydrocarbons in petroleum hydrocarbons, followed by aromatic hydrocarbons, and finally colloid and asphalt.

The prepared catalysts were characterized by SEM and EDS, indicating that Fe was successfully loaded on the surface or in the voids of permutite in the form of oxides by impregnation method, forming an effective component that plays a catalytic role in the catalyst.

The catalyst prepared by the impregnation method successfully solved the limitation of the pH condition of the traditional Fenton reagent, and could achieve a higher degradation efficiency under the original soil pH condition. However, the catalyst is difficult to achieve secondary utilization, resulting in high cost. Therefore, the preparation of a recyclable Fenton-like catalyst is the future development direction of degrading oil-contaminated soil.

Ackowledgements

Funding. The Project Supported by Natural Science Basic Research Plan in Shaanxi Province of China (2021JQ-958), Enterprise top innovative young talents support plan (2021-1-2-6) and the Research Project of Shaanxi Provincial Land Engineering Construction Group in China (DJNY2021-19).

References

[1]Wang, S., Wang, D., Yu, Z., Dong, X., Liu, S., Cui, H., Sun, B. (2021) Advances in research on petroleum biodegradability in soil. Environmental Science-Processes and Impacts. 23(1), 9-27.

[2]Kulik, N., Goi, A., Trapido, M., Tuhkanen, T. (2006) Degradation of polycyclic aromatic hydrocarbons by combined chemical pre-oxidation and bioremediation in creosote contaminated soil. Journal of Environmental Management. 78(4), 382-391.

[3]Usman, M., Faure, P., Hanna, K., Abdelmoula, M., Ruby, C. (2012) Application of magnetite catalyzed chemical oxidation (Fenton-like and persulfate) for the remediation of oil hydrocarbon contamination. Fuel. 96(1), 270-276.

[4]Venny, V., Gan, S., Ng, H.K. (2012) Modified Fenton oxidation of polycyclic aromatic hydrocarbon (PAH)-contaminated soils and the potential of bioremediation as post-treatment. Science of the Total Environment. 419, 240-249.

[5]Yap, C.L., Gan, S., Ng, H.K. (2011) Fenton based remediation of polycyclic aromatic hydrocarbons-contaminated soils. Chemosphere. 83(11), 1414-1430.

[6]Ferrarese, E., Andreottola, G., Oprea, I.A. (2008) Remediation of PAH-contaminated sediments by chemical oxidation. Journal of Hazardous Materials. 152(1), 128-139.

[7]Yang, Y., Zhang, Y., Li, S., Zhang, X., Ying, P., Wang, X. (2019) A Case Study on Design and Application of In-Situ Chemical Oxidation High Pressure Injection Remediation. Environment Engineering. 37(8), 185-189.

[8]Fenton, H.J.H. (1894) LXXIII:Oxidation of tartaric acid in presence of iron. J. Chem. Soc. Trans. 65, 899-910.

[9]Ouriache, H., Arrar, J., Namane, A., Bentahar, F. (2019) Treatment of petroleum hydrocarbons contaminated soil by Fenton like oxidation. Chemosphere. 232, 377-386.

[10]Unal, B.O., Bilici, Z., Ugur, N., Isik, Z., Harputlu, E., Dizge, N., Ocakoglu, K. (2019) Adsorption and Fenton oxidation of azo dyes by magnetite nanoparticles deposited on a glass substrate. Journal of Water Process Engineering. 32, 100897.

[11]Ming, C., Tian, K., Lu, Z., Ake Kouassi Marius, H. (2014) Pretreatment of cartap pesticide wastewater by Fenton reagent. Chinese Journal of Environmental Engineering. 8(12), 5135-5140.

[12]Kanel, S.R., Neppolian, B., Choi, H., Yang, J.W. (2003) Heterogeneous catalytic oxidation of phenanthrene by hydrogen peroxide in soil slurry: Kinetics, mechanism, and implication. Soil and Sediment Contamination. 12(1), 101-117.

[13]Watts, R.J., Stanton, P.C., Howsawkeng, J., Teel, A.L. (2002) Mineralization of a sorbed polycyclic aromatic hydrocarbon in two soils using catalyzed hydrogen peroxide. Water Research. 36(17), 4283-4292.

[14]Du, T., Wang, S., Chen, X. (2012) Treatment of Phenol Wastewater with the Improved Fenton Method. Enuivonmental Science and Technology. 35(2), 159-161,166.

[15]Lu, T., Zhou, P., Zhang, N., Zhang, L., Sheng, Y. (2018) Research on the treatment of phenol wastewater by attapulgite heterogeneous Fenton catalysts. Environmental Pollution and Control. 40(2), 155-160.

[16]Xiang, C., Pu, S., Wang, M., Hu, X. (2017) Preparation of Supported-Diatomite Catalyst of Heterogeneous Fenton and Its Capability of Degrading Phenol. Environment Engineering. 35(2), 5-9,14.

[17] Frankenberger, J. W. T. (1997) Factors Affecting the Fate of Urea Peroxide Added to Soil. Bulletin of Environmental Contamination and Toxicology. 59(1), 50-57.

本文曾发表于 2022 年《Fresenius Environmental Bulletin》第 31 卷第 10 期

北方农牧交错带花生种植模式对荒漠化的影响

齐　丽[1,2],何振嘉[1]

(1.陕西省土地工程建设集团有限责任公司, 西安 710075;
2.陕西地建土地工程技术研究院有限责任公司,西安 710075)

【摘要】人类活动作为影响土地荒漠化的重要因素越来越被学者们重视,寻求合适的土地利用模式对于抑制土地荒漠化具有重要意义。因此,本文以彰武县北部主要花生种植区的不同土地利用种植模式为研究对象,针对其2020年多时相遥感影像的物候特征和作物光谱特征进行相关分析,在此基础上现场采集土壤样本,应用化学实验方法,结合SPSS等软件构建荒漠化程度评价指标体系,运用累积曲线分级法将土地荒漠化程度划分为潜在荒漠化、轻度荒漠化、中度荒漠化、重度荒漠化四个级别,与课题组2015年该地区的荒漠化数据库进行荒漠化程度对比检验。结果表明:农林复合模式、沙平地防护林-花生模式、砂质丘陵防护林-花生模式、沙平地-花生模式、砂质丘陵-花生模式五种模式对减轻荒漠化的能力依次降低。农林复合模式、沙平地防护林-花生模式、砂质丘陵防护林-花生模式均有使荒漠化逆向演替的效果,但演替程度逐渐降低,其中砂质丘陵防护林-花生模式有一定的局限性;沙平地-花生、砂质丘陵-花生模式均有使荒漠化正向演替的效果,演替程度逐渐加剧。

【关键词】农牧交错带;花生;种植模式;荒漠化

Effects of Peanut Planting Modes on Desertification in the North Agro-pastoral Zone

Li Qi[1,2],Zhenjia He[1]

(1. Shaanxi Provincial Land Engineering Construction Group Co., Ltd., Xi'an 710075, China;
2.Institute of Land Engineering and Technology, Shaanxi Provincial Land Engineering Construction Group Co., Ltd., Xi'an 710075, China)

【Abstract】As an important factor affecting land desertification, human activities have been paid more and more attention by scholars. The search for a suitable land use model is of great significance to inhibit land desertification. Therefore, this article takes the different land use and planting patterns in the main peanut planting areas in the northern part of Zhangwu County as the research object, and conducts correlation analysis on the phenological characteristics and crop spectral characteristics of its multi-temporal remote sensing images in 2020. On this basis, soil samples are collected on site. Applying chemical experimental methods, combined with software such as SPSS to

基金项目:国家重点研发计划项目(2021YFD1900700);陕西省土地工程建设集团有限责任公司内部创新团队项目(DJTD-2022-5);陕西土地工程建设集团有限责任公司内部科研项目(DJNY2022-23,KJNY2022-54)。

construct a desertification degree evaluation index system, using the cumulative curve classification method to divide the degree of land desertification into four levels: potential desertification, mild desertification, moderate desertification, and severe desertification. The desertification database of the region conducts a comparative test of the degree of desertification.The results showed that the five models of agroforestry, sand flat shelter forest-peanut model, sandy hill shelter forest-peanut model, sand flat land-peanut model, sandy hill-peanut model reduced the ability to reduce desertification in order. The agroforestry model, the sand flat shelterbelt-peanut model, and the sandy hill shelterbelt-peanut model all have the effect of reverse succession of desertification, but the degree of succession is gradually reduced. Among them, the sandy hill shelterbelt-peanut model has certain limitations; sand Both the flat land-peanut and sandy hill-peanut model have the effect of positive succession of desertification, and the degree of succession is gradually increasing. The conclusions of this study will provide a scientific and reasonable basis for the efficient and sustainable use of resources and the environment in the peanut planting area in the north of Zhangwu County.

【**Keywords**】Farming-pastoral transition zone; Peanut; Planting pattern; Desertification

荒漠化是指人类活动以及气候变异等多种因素导致的土地退化,在全世界广泛发生,也是目前最严重的社会经济问题与环境问题之一[1]。土地荒漠化会影响作物产量和土地可持续性,甚至导致人类生存环境的恶化[2]。就我国地理环境而言,农牧交错带是遏止荒漠化、沙化东移南下的最后一道屏障,特别是位于长城一线以北、草原东侧和南侧的北方农牧交错带是我国重要的生态治理区域和北方地区重要的生态防线[3-4]。但是伴随着人为干预的不断增多,人类对土地进行不合理的开垦以及放牧,致使农牧交错带不仅没有起到生态屏障的作用,反而使其成了生态脆弱带[5]。基于当前情况,农牧交错带荒漠化问题已经对我国社会经济发展造成了一定的阻碍,同时,此问题也引起了政府和科学界的普遍关注。

我国北方农牧交错带位于半干旱和干旱气候的交汇处,作为京津冀地区重要的水源涵养带与中东部地区重要的生态安全屏障,对维护国家的生态安全具有重要的意义[6]。而辽北农牧交错带作为北方农牧交错带的重要组成部分,位于东北平原下沉区,是防止科尔沁沙地向南入侵华北平原、向东南入侵东北平原最重要的生态屏障[2]。该地区因为受降水量少、蒸发量大、气候干旱以及人为干扰等影响,区域内的植被、地貌和水文条件等环境因素都发生了一系列的变化,随之而来的土地荒漠化也日渐严重,并且呈南侵及东南侵的态势。研究表明,高植被覆盖率能有效抑制土地沙漠化[7-8]。而研究区具有悠久的花生种植历史,虽然多年来获得了较高的经济收益,但是由于其种植模式陈旧,致使地表的土壤流失、土表粗化、地力急剧下降。因此,探讨适合研究区的花生种植模式,对于土地的可持续利用以及防止土地沙漠化、保护区域生态具有重要意义。本研究选取辽北农牧交错带典型区域彰武县作为研究区,以多时相遥感作物识别理论和土地荒漠化评价理论为基础,应用 RS 和 GIS 等相关软件,选取土地利用覆被状况和土壤质量为评价指标,对不同种植模式下的土地进行了荒漠化评价,探索了不同种植模式与荒漠化演替方向的影响,分析了不同种植模式下的优劣,旨在为规范和改善不合理种植模式提供参考。

1　材料与方法

1.1　研究区概况

彰武县隶属于辽宁省阜新市,地处辽宁省西北部,科尔沁沙地南部,东连康平、法库两县,南接新民市,西隔绕阳河与阜新蒙古族自治县相邻,北依内蒙古自治区通辽市的库伦旗和科尔沁左翼后旗。本研究中涉及研究区域为彰武县北部章古台镇、大冷乡、四合城乡、阿尔乡,耕地总面积为 30037.93 hm^2,其中花生种植面积为 13835.95 hm^2。研究区位于北纬 42°07′~42°51′、东经 121°53′~122°58′,地势北高南低,东西丘陵,北部沙荒,中、南部为平原。彰武县属于温和半湿润的季风大陆性气候,四季分明,雨热同季,光照充足。彰武是国家商品粮生产基地,截至 2010 年,粮豆作物 6 万 hm^2,经济作物 4.2 万 hm^2。

1.2　数据来源

应用的数据主要为彰武县 2015 年、2020 年 Landsat5TM 数据、彰武县统计年限数据、彰武县 1∶1 万土地利用现状图和 1∶1万地形图等,其中遥感数据主要是利用 3 和 4 波段数据,数据主要来源于彰武县自然资源局与统计局。

1.3　数据预处理

1.3.1　研究区域的剪裁

首先,将 3、4 波段影像图导入到 ENVI4.5 中。其次,选择 Overlay 菜单中的 Vectors 选项,将研究区域的边界叠加到影像上。然后,在 Vector Parameters 对话框中的 File 菜单中,打开 Export Active Layer to ROIs 选项,选择 Convert all vectors to one ROI。最后,在 Tools 菜单中选择 Region of Interest 中的 ROI Tool。在其对话框中选择 File 中的 Subset Data via ROIs 选项,并将 Background Value 设为 0,然后保存在一个新的文件夹中。另一期数据采用同样方法处理。

1.3.2　花生种植面积遥感数据的提取与处理

以 2015 年、2020 年 Landsat 5TM 遥感影像为基本数据源,除 TM 6 波段外,其余波段空间分辨率均为 30 m,为 p120r31,基本无云覆盖,已经过辐射校正和几何校正,数据质量较好,以研究区行政区划图、1∶50000 数字高程模型(DEM)、1∶10000 土地利用现状图、野外调查数据以及自然、社会、经济方面的文字资料为辅助资料,在 ArcGIS 9.3 和 ENVI 4.7 等遥感影像处理软件的支持下,对 TM 影像进行多波段融合。

在遥感数据信息提取的基础上,根据研究需要以及彰武县实际情况,对遥感影像初步判读种植模式进行修订,以花生种植区周边的土地利用覆被现状及地形地貌特征为划分依据,选择 5 种花生种植模式,分别为沙平地花生种植模式以阿尔乡为代表,砂质丘陵花生种植模式以大冷乡、阿尔乡为代表,沙平地防护林花生种植模式以章古台镇、四合城乡、大冷乡为代表,砂质丘陵防护林花生种植模式以大冷乡、阿尔乡为代表,农林复合种植模式以章古台镇、大冷乡为代表。

1.4　土壤理化性质的测定

测定土壤机械组成、有机质、全氮、HA/FA(胡敏酸与富里酸的比),分别采用比重法、重铬酸钾容量法、凯氏定氮法进行测定[9]。

1.5　荒漠化评价指标体系的建立

采用专家打分法[10]并结合研究区的实际情况,初步选出遥感、立地条件、理化性状和土地利用指数 4 大部分指标,如表 1 所示。

指标计算公式:

$$\mathrm{NDVI}=\frac{\mathrm{Nir-Red}}{\mathrm{Nir+Red}} \tag{1}$$

式中:NDVI 为植被覆盖指数;Nir、Red 分别为 Landsat-5TM 遥感影像的 TM4 和 TM3 波段亮度值。

利用 Liang[11]建立的 Landsat-TM 数据的反演模型(2)估算研究区地表反照率:

$$\mathrm{Albedo} = 0.356\rho_{\mathrm{TM}_1} + 0.130\rho_{\mathrm{TM}_3} + 0.373\rho_{\mathrm{TM}_4} + 0.085\rho_{\mathrm{TM}_5} + 0.072\rho_{\mathrm{TM}_7} - 0.0018 \tag{2}$$

式中:Albedo 为地表反照率;ρ 为不同波段地面相对反射率。

本研究采用层次分析法确定评价因子权重,通过建立层次结构与构造判断矩阵确定权重,如表 2 所示。

表 1　彰武县北部荒漠化评价指标体系

Table 1　Northern ZhangWu desertification evaluation index system

目标层 Target layer(G)	准则层 Criterion layer(B)	指标层 Index layer(C)	数据来源 Data sources
彰武县北部荒漠化评价指标体系 Evaluation index system of desertification in the north of Zhangwu County	遥感 Remote sensing (B1)	地表反照率 Surface albedo(%) (C1)	遥感调查、模型计算 Remote sensing survey, model calculation
		植被覆盖指数 Normalized Difference Vegetation Index(%)(C2)	遥感调查、模型计算 Remote sensing survey, model calculation
	立地条件 Site conditions (B2)	土壤砂黏比 Soil sand viscosity ratio (%) (C3)	土壤采样化验 Soil sampling test
		坡度 Slope (°) (C4)	DEM 数据库 DEM database
	理化性状 Physical and chemical properties (B3)	全氮 Total nitrogen (g/kg) (C5)	土壤采样化验 Soil sampling test
		胡敏酸/富里酸 Humic acid/Fulvic acid (%)(C6)	土壤采样化验 Soil sampling test
		土壤有机质 Soil organic matter (g/kg) (C7)	土壤采样化验 Soil sampling test
	土地利用指数 Land use index (B4)	林木平均半径 Average radius of forest (m)(C8)	实地调查 Field investigation
		农田防护林完备程度 Completeness of farmland shelterbelts(%) (C9)	实地调查 Field investigation

表 2　荒漠化评价因子组合权重

Table 2　Desertification evaluation factor combination weight

层次 Level G	遥感 Remote sensing B1	理化性状 Physical and chemical properties B2	立地条件 Site conditions B3	土地利用指数 Land use index B4	组合权重 Combination weight
C1	0.2500				0.0151
C2	0.7500				0.0453
C5		0.0669			0.0109
C6		0.2200			0.0357
C7		0.7132			0.1157
C3			0.1667		0.0480
C4			0.8333		0.2399
C8				0.2500	0.1224
C9				0.7500	0.3671

对于各评价因子的隶属函数的确定,采用戒上型函数和概念型函数2种类型函数。

戒上型函数模型:

$$y_i = \begin{cases} 0 & u_i \leqslant u_t \\ 1/[1 + a_i(u_i - c_i)^2], u_i < c_i & (i = 1,2,\cdots,m) \\ 1 & u_i \geqslant c_i \end{cases} \quad (3)$$

式中:y_i为第i个因素评语;u_i为样品观测值;c_i为标准指标值;a_i为系数;u_t为指标下限值。

对于概念型评价因子的隶属度由专家打分法[10]得出,按照指标对于荒漠化影响程度特点直接进行赋值(见表3)。

表3 概念型评价因子隶属度

Table 3 The membership degree of generalizing evaluation factors

评价因子 Evaluation factor	分级 Grading	隶属度 Membership
林木平均半径 Average radius of forest	0	0.1
	≤50	0.4
	50~100	0.7
	≥100	0.9
坡度 slope	≤2°	0.9
	2°~6°	0.5
	>6°	0.1
农田防护林完备程度 Completeness of farmland shelterbelt	无 no	0.1
	中等 medium	0.4
	良好 good	0.7
	优秀 excellent	0.9

在此基础上,进行综合评价的计算,但是,在评价因子中,某些评价因子对荒漠化程度存在着明显的限制性,即某些评价因子存在着极限值,当这些因子的值变化超过极限值时,会使荒漠化程度加剧。本研究借鉴已有的相关研究资料,采用特尔斐法[12]确定并经实地验证,最终确定了研究区荒漠化程度极限因子及其极限值(坡度大于等于25°或者小于2°)。采用修正的加权指数和法计算各评价单元的综合指数以确定用地适宜性等级,计算公式为:

$$S_j = \begin{cases} \sum_{i=1}^{n} (W_i \cdot P_{ij}) & (\text{评价因子值未超过极限值}) \\ 0 & (\text{任意评价因子值超过其极限值}) \end{cases}$$

式中:S_j为综合指数;W_i为评价因子的权重;P_{ij}为评价因子的隶属度。

1.6 荒漠化等级确定

本研究运用累积曲线分级法划分研究区荒漠化程度[13],将彰武县北部研究区荒漠化程度分为4个级别,分别是为无荒漠化(综合分数>0.79)、轻度荒漠化(0.63~0.79)、中度荒漠化(0.40~0.63)、重度荒漠化(综合分数<0.4)。

2 结果与分析

2.1 不同种植模式下土壤机械组成

为了方便表达,将农林复合、沙平地防护林-花生、砂质丘陵防护林-花生、沙平地-花生、砂质丘陵-花生5种模式分别用Ⅰ、Ⅱ、Ⅲ、Ⅳ和Ⅴ来表示。从不同种植模式剖面质地的变化(见图1)可以看出,5种植模式在剖面的每个深度层次上砂粒、黏粒、粉粒的含量依次减少,并且砂粒含量比其他粒级含量大很多,这说明这5种种植模式均有不同程度的沙化。农林复合模式、沙平地防护林-花生两种种植模式的砂粒随着深度的增加而增加,即20~40 cm砂粒含量在各种植模式下都是最大,砂质丘陵防护林花生模式随着深度的增加变化幅度不明显,沙平地-花生模式则随着深度的增加砂粒含量呈先升高后降低的趋势,砂质丘陵-花生模式则随着深度的增加砂粒含量逐渐降低。粉粒、黏粒都有减少的趋势,其中农林复合模式黏粒随深度的增加变化不明显,有防护林的耕地次之,没有防护林的会随深度的增加粉粒、黏粒有增加的趋势。这说明农林复合模式、沙平地防护林模式均有不同程度防止颗粒粗化的能力。

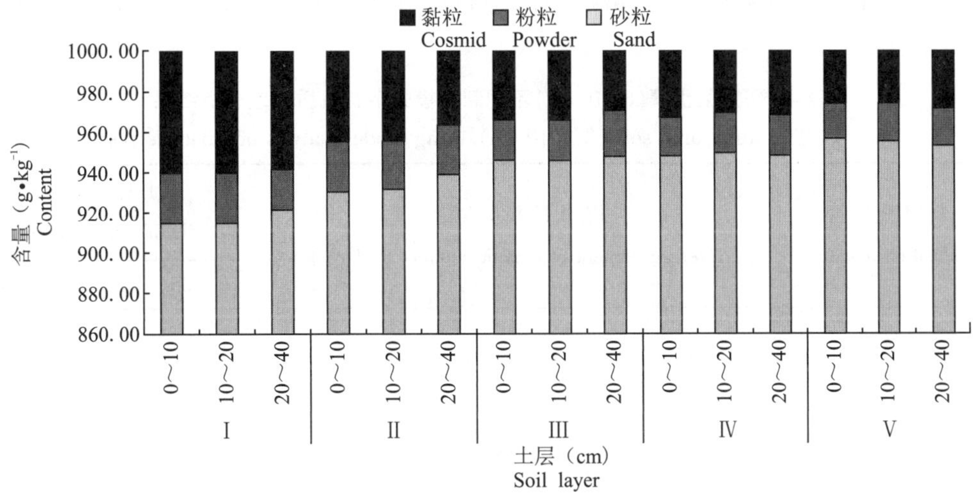

图1 不同种植模式下土壤剖面机械组成

Fig.1 Mechanical composition of soil profile under different planting modes

土壤砂黏比(见图2)在0~10 cm、10~20 cm处农林复合、沙平地防护林-花生、砂质丘陵防护林-花生、沙平地-花生、砂质丘陵-花生5种模式依次递增,而在沙平地出现先增后减的趋势,可能是由于在秋起花生后,裸露的地表得到了一定的保护,这说明在过去的某一时间曾有过沙化侵袭,现在有所恢复。农林复合、沙平地防护林-花生、砂质丘陵防护林-花生3种种植模式在耕层深度上随着深度的增加砂黏比升高,说明现在的荒漠化程度和过去的某一时期相比有所减轻。而砂质丘陵花生耕层深度上随着深度的增加砂黏比降低,说明现在的荒漠化程度和过去的某一时期相比有所增加。

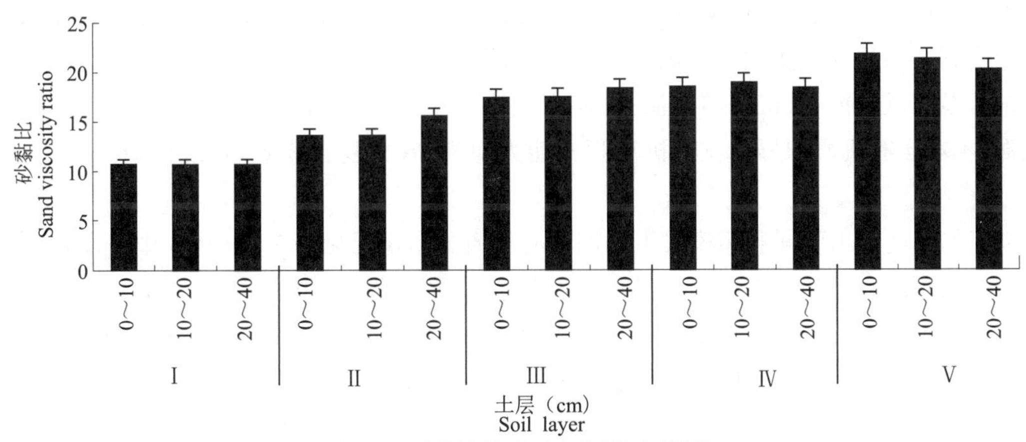

图2 不同种植模式下土壤剖面砂黏比

Fig.2 Soil profile sand viscosity ratio under different planting modes

2.2　不同种植模式下有机质含量比较

5种不同利用模式采样点表层风沙土有机质含量变化在1.76~21.34 $g \cdot kg^{-1}$,平均为6.58 $g \cdot kg^{-1}$;有机质含量低于6.58 $g \cdot kg^{-1}$的土壤约占整个样品数的59%。对表层土壤(0~20 cm)不同种植方式的平均有机质含量进行方差分析,见表4。从表4中可以看出,5种不同模式下土壤有机质含量差异显著,说明不同种植模式下荒漠化对土壤有机质的影响程度也不同,其中耕地有机质平均含量低于林地,这说明固定风沙土林地被开辟为农田以后,其土壤有机质受人为活动的影响越来越大,因为农业生产将地上部的生物量进行收获,仅有地下部的根系作为有机物质的主要来源,减少了有机物的输入。同时,作为耕种土壤,不断的耕作活动,加速了原有有机质的矿化。这样有机质含量就会不断降低。与草地相比较而言,要高于草地的有机质平均值。这说明该地区草地土壤正处于退化阶段,有机质含量降低。而当地采取的高留茬、施用有机肥、种植防护林等措施对耕地有机质的积累起到了一定的作用。

表4　研究区土壤(0~10 cm)不同利用模式下有机质方差分析结果

Table 4　The study area soil (0 ~ 10 cm) using mode analysis of variance results

利用模式 Utilization mode	有机质平均含量 Average content of organic matter($g \cdot kg^{-1}$)	差异显著性 Significant difference	
		$\alpha=0.05$	$\alpha=0.01$
农林复合 Agroforestry	8.49	a	A
沙平地防护林-花生 Peanut in shapingdi shelterbelt	7.34	a	AB
砂质丘陵防护林-花生 Peanut in sandy hilly shelterbelt	5.10	b	BC
沙平地-花生 Shaping ground peanuts	3.36	b	BC
砂质丘陵-花生 Sandy hilly peanuts	2.56	b	C

土壤表层有机质含量总体变化趋势(见表4)为农林复合模式、沙平地防护林-花生模式、砂质丘陵防护林-花生模式、沙平地-花生模式、砂质丘陵-花生模式依次降低,各利用模式土壤有机质的含量均受到不同程度荒漠化的影响,含量普遍很低。这些土壤大部分处于风沙区,受风蚀影响,表层的细颗粒被风吹蚀,有机质随风流失,使有机质含量下降;且表层常处于通气良好状态,好气分解较强,不利于有机质积累。就有机质积累能力而言,农林复合模式最强,砂质丘陵-花生模式最差。

从剖面的角度得出不同利用模式下的剖面有机质分布情况见图3。从图3中可以看出,5种利用模式中,农林复合模式、沙平地防护林-花生模式、砂质丘陵防护林-花生模式的有机质含量在各自的剖面上均符合随着深度的增加而减少的规律。3种利用模式土表层(0~10 cm)有机质含量都高于下部层次,表层0~10 cm有机质含量变化是农林复合模式>沙平地防护林-花生模式>砂质丘陵防护林-花生模式。而沙平地-花生模式和砂质丘陵-花生模式的有机质含量在各自剖面大体符合随深度增加而增加的趋势,其中沙平地-花生模式出现先增后减的趋势,可能是由于在秋起花生后,裸露的地表得到了一定的保护,这说明在过去的某一时间曾有过沙化侵袭,现在有所恢复。

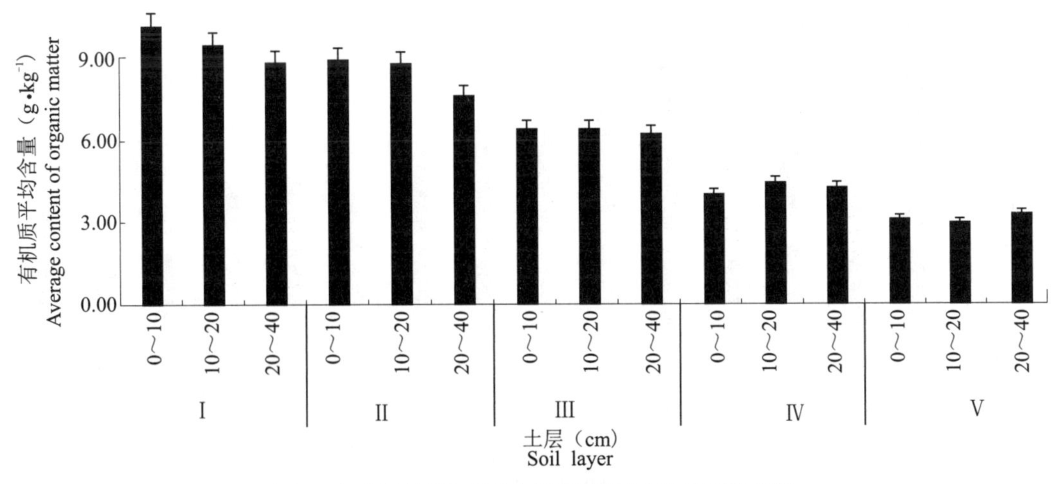

图3 不同利用模式下土壤剖面有机质的变化情况

Fig.3 Changes of soil profile organic matter under different utilization modes

2.3 不同种植模式下全氮含量

研究区表层土壤全氮含量变化范围在 0.62~1.59 g · kg^{-1},其平均值为 0.94 g · kg^{-1}。根据各种利用模式的平均值与其总平均值的差异程度可以看出,土壤氮素与有机质有相似的分布。砂质丘陵花生模式风沙土表层的全氮含量处于 5 种模式中的最低区,平均为 0.69 g · kg^{-1},农林复合模式、沙平地防护林-花生模式、砂质丘陵-花生模式、沙平地-花生模式表层全氮含量分别为 1.02 g · kg^{-1}、0.98 g · kg^{-1}、0.8 g · kg^{-1}、0.74 g · kg^{-1}、0.69 g · kg^{-1}。对不同利用模式下全氮平均含量的方差分析结果见表 5,从表 5 中可以看出,风沙土平均全氮含量以农林复合模式、沙平地防护林-花生模式、砂质丘陵-花生模式、沙平地-花生模式、砂质丘陵-花生模式的顺序依次降低,农林复合与砂质丘陵之间差异极显著。由于氮素在土壤中极易挥发、流失,人为耕作中即使增加氮肥的施用量,土壤中氮素含量增加也很不明显。

表5 研究区土壤(0~10 cm)不同利用模式下全氮含量方差分析结果

Table 5 Soil (0 ~ 10 cm) in the study area using the variance analysis results under the mode of total nitrogen content

利用模式 Utilization mode	全氮平均含量 Average total nitrogen content(g · kg^{-1})	差异显著性 Significant difference	
		$\alpha=0.05$	$\alpha=0.01$
农林复合 Agroforestry	1.02	a	A
沙平地防护林-花生 Peanut in shapingdi shelterbelt	0.98	ab	A
砂质丘陵防护林-花生 Peanut in sandy hilly shelterbelt	0.88	abc	AB
沙平地-花生 Shaping ground peanuts	0.74	bc	AB
砂质丘陵-花生 Sandy hilly peanuts	0.69	d	B

不同种植模式的剖面土壤全氮含量都有随着深度增加而减少的趋势(见图 4),但是在 0~10 cm 处明显大于其他层次。其主要原因是随土壤垂直深度的增加,生物量的积累和有机质的分解强度是决定土壤氮素含量的主要因素。其中农林复合模式和沙平地防护林模式随深度变化明显,而其他 3 种变化较为平缓。这是因为耕地受人为因素的影响。全氮变化相对较稳定,随着深度的增加也有降低的趋势,3 种种植模式差异不大。

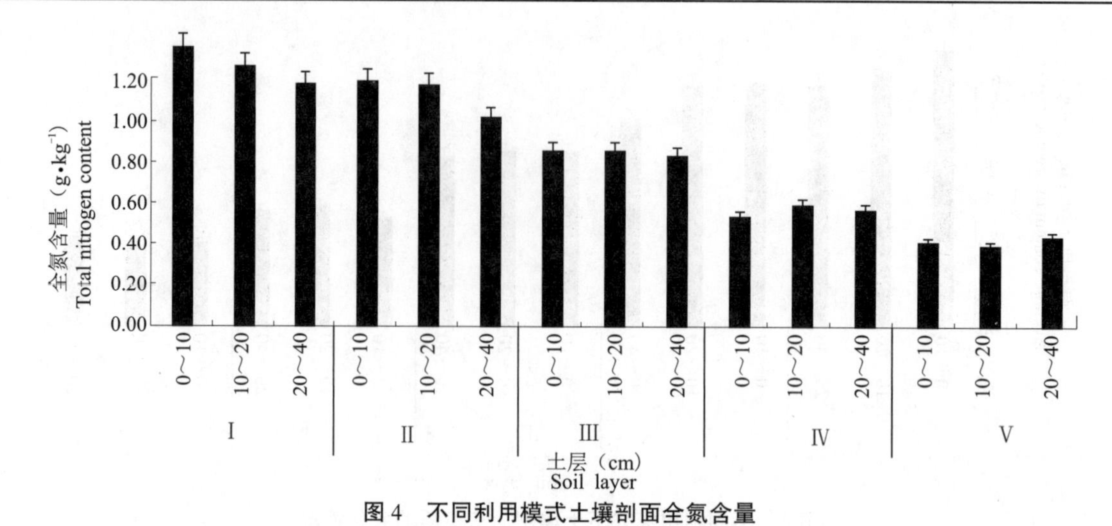

图 4 不同利用模式土壤剖面全氮含量

Fig.4 Total nitrogen content of soil profile in different utilization modes

2.4 不同种植模式下 HA/FA 比较

胡敏酸与富里酸的比例被用来衡量腐殖质品质,胡敏酸比例越高,腐殖质的活性越强,品质越好。研究区表层土壤 HA/FA 变化范围在 0.26~0.90。根据各种利用模式的平均值与其总平均值的差异程度可以看出,土壤 HA/FA 与有机质有相似的分布。砂质丘陵-花生模式风沙土表层的 HA/FA 处于 5 种模式中最低的最低区,平均为 0.33,农林复合模式、沙平地防护林-花生模式、砂质丘陵防护林-花生模式、沙平地-花生模式表层 HA/FA 分别为 0.59、0.54、0.46、0.39。对不同利用模式下的 HA/FA 方差分析结果见表 6,从表 6 可以看出,风沙土平均 HA/FA 以农林复合模式、沙平地防护林-花生模式、砂质丘陵-花生模式、沙平地-花生模式、砂质丘陵-花生模式的顺序依次降低,因此腐化程度也依次降低,农林复合与砂质丘陵之间差异极显著。

表 6 研究区土壤(0~10cm)不同利用模式下 HA/FA 方差分析结果

Table 6 The study area soil (0 ~ 10 cm) using different mode of HA/FA variance analysis results

利用模式 Utilization mode	HA/FA	差异显著性 Significant difference	
		$\alpha=0.05$	$\alpha=0.01$
农林复合 Agroforestry	0.59	a	A
沙平地防护林-花生 Peanut in shapingdi shelterbelt	0.54	ab	A
砂质丘陵防护林-花生 Peanut in sandy hilly shelterbelt	0.46	abc	AB
沙平地-花生 Shaping ground peanuts	0.39	bc	AB
砂质丘陵-花生 Sandy hilly peanuts	0.33	c	B

土壤腐殖质具有复杂的复合结构,其影响土壤的物理性质、化学性质和生物学特性,土壤腐殖质的形成是积累的过程,很大程度上可以影响土壤的肥力,是农作物品质和产量的保证。研究区内不同利用种植模式 HF/FA 的比例有着不同的规律,如图 5 所示。表层 0~10 cm 土壤 HA/FA 变化是农林复合模式>沙平地防护林-花生模式>砂质丘陵防护林-花生模式。而沙平地-花生模式和砂质丘陵-花生模式的 HA/FA 在各自剖面大体符合随深度增加而增加的趋势。

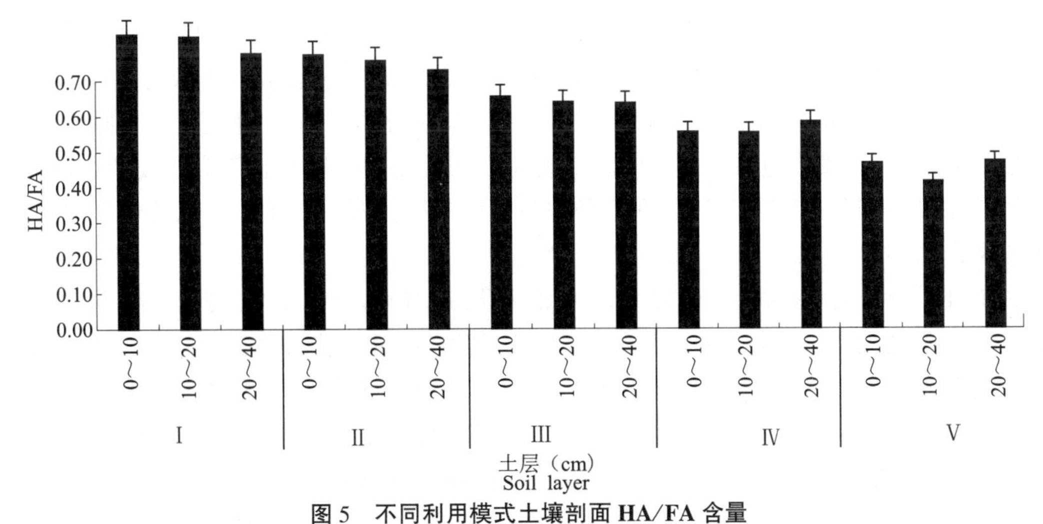

图 5　不同利用模式土壤剖面 HA/FA 含量

Fig.5　HA/FA content in soil profile of different utilization modes

2.5　不同种植模式下荒漠化程度分析

由分析得知：在 2020 年农林复合模式下，无荒漠化>轻度荒漠化>中度荒漠化，不存在重度荒漠化，而 2015 年农林复合模式相同点位荒漠化信息提取中显示（见图 6），无荒漠化样本数为 4 个，轻度荒漠化为 2 个。中度荒漠化为 4 个，重度荒漠化为 2 个。对比可以看出，中度荒漠化和重度荒漠化均有所减少，而潜在荒漠化和轻度荒漠化均增加，其中潜在荒漠化增加最多，这可以说明在这 5 年过程中荒漠化土地发生了较大程度的逆向演替，土壤理化条件得到了显著改善。

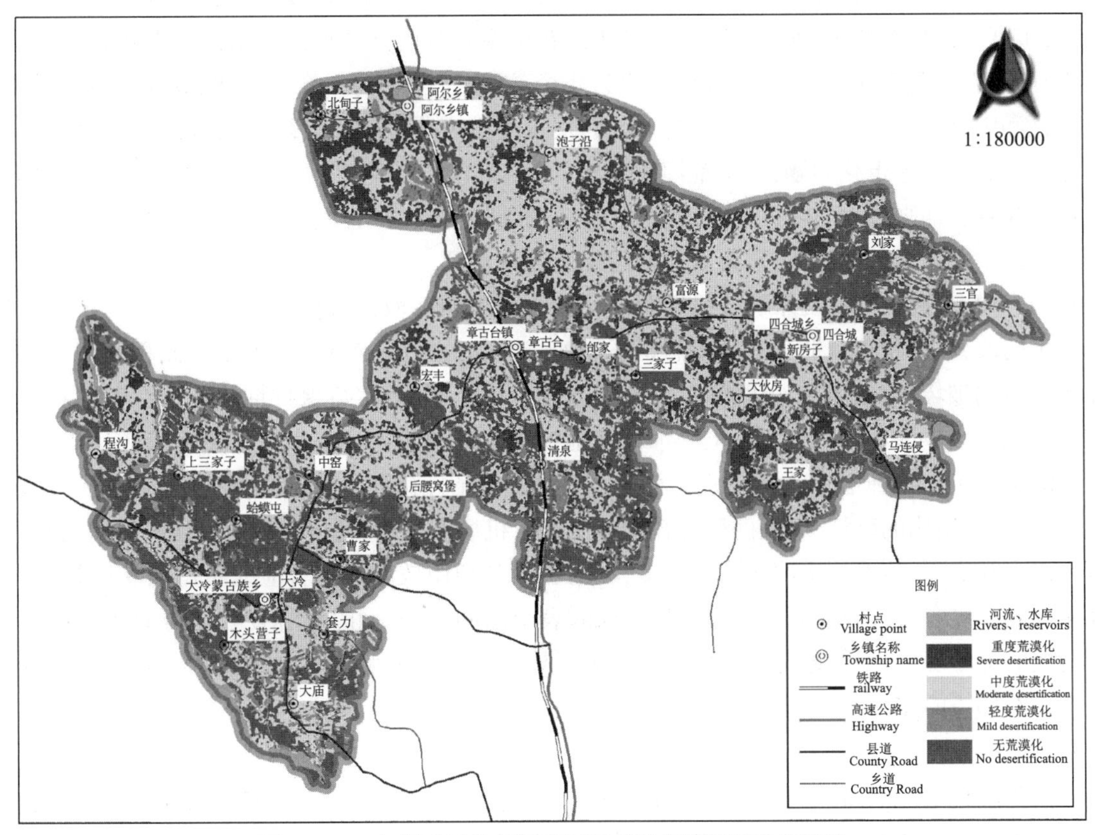

图 6　2015 年彰武县北部研究区遥感信息提取荒漠化等级

Fig.6　Northern Zhangwu remote sensing information extraction of desertification in the study area in 2015 levels

在2020年沙平地防护林-花生模式下,轻度荒漠化>中度荒漠化>无荒漠化,不存在重度荒漠化,而2015年沙平地防护林-花生模式相同点位荒漠化信息提取中,无荒漠化样本数为3个,轻度荒漠化为2个,中度荒漠化为10个,重度荒漠化为7个。对比2020年可以看出,重度荒漠化减少最多,中度荒漠化也向轻度荒漠化发生逆向演替,而无荒漠化减少可能是由于防护林不完备或者在风口处将林地开垦为耕地,使得部分土地发生退化,但总体来说,沙平地防护林-花生模式荒漠化为逆向演替,但程度较农林复合模式要低。

在2020年砂质丘陵防护林-花生模式下,无荒漠化点位样本数为0个,轻度荒漠化为2个,中度荒漠化为9个,重度荒漠化数为0个,而2015年砂质丘陵防护林-花生模式相同点位荒漠化信息提取中,无荒漠化样本数为2个,轻度荒漠化为0个,中度荒漠化为5个,重度荒漠化为4个。对比2020年可以看出,重度荒漠化减少最多,荒漠化发生逆向演替,而无荒漠化减少说明砂质丘陵可以在一定程度上加重荒漠化的正向演替,但总体来说,砂质丘陵防护林-花生模式可在重度或中度荒漠化土地上抑制荒漠化的正向演替,在无荒漠化土地和轻度荒漠化土地上会是荒漠化正向演替,这也说明了砂质丘陵会在一定程度上加速土地荒漠化的正向演替,因此砂质丘陵防护林-花生模式虽然可以起到防风固沙使荒漠化逆向演替的作用,但是有一定的局限性。

在2020年沙平地-花生模式下,无荒漠化点位样本数为0个,轻度荒漠化为0个,中度荒漠化为8个,重度荒漠化数为4个,而2015年沙平地-花生模式相同点位荒漠化信息提取中,无荒漠化样本数为0个,轻度荒漠化为2个,中度荒漠化为6个,重度荒漠化为4个。对比2020年可以看出,轻度荒漠化向中度荒漠化发展,重度荒漠化保持不变,这说明沙平地-花生模式存在一定土地退化,发生荒漠化的正向演替。

在2020年砂质丘陵-花生模式下,无荒漠化点位样本数为0个,轻度荒漠化为0个,中度荒漠化为0个,重度荒漠化数9个,而2015年砂质丘陵-花生模式相同点位荒漠化信息提取中,无荒漠化样本数为0个,轻度荒漠化为3个,中度荒漠化为4个,重度荒漠化为2个。对比2020年可以看出,轻度荒漠化和中度荒漠化均向重度荒漠化方向演替,重度荒漠化个数显著增加,这说明砂质丘陵-花生模式存在较大的土地退化问题,发生荒漠化的正向演替,比较其他4种模式,砂质丘陵-花生模式荒漠化正向演替更加强烈,而且程度大、速度快,造成局部地区土壤条件恶化,因此砂质丘陵-花生模式是5种模式中最不合理的模式,需要有关部门加大管理力度。

总的来说,彰武县北部历来属于风蚀荒漠化较为严重的区域,经过长期的造林活动,土壤条件得到了一定的改善,但是群众的生态环境意识还需加强,同时在固定沙丘上,由于大面积开垦导致部分天然植被遭到破坏,存在潜在沙化的问题,应适当减少耕地面积,实行保护性耕作。在秋起花生后增加秸秆覆盖还田,可以对花生产量、土壤水分利用效率的影响达到显著水平。覆盖量较适中,秸秆覆盖还田既增加了来年种植时土壤有机质含量,又能挡风遮阴,有效抑制风蚀,使表土层不易被大风吹走,减少对土壤结构的破坏,腐烂的秸秆在增加土壤有机质的同时,还可以使土壤结构变得疏松,提高降水入渗率和渗深,从而让土壤接收雨水多,提高了水分利用效率。再者应完备网格状防护林建设,积极采取农林复合模式以保证秋起花生后对裸露表土层的保护,同时农林复合也可起到抑制土地风蚀所造成的地势变化,保持地势平坦。政府大力倡导构建有利于生态环境的农林复合模式,以起到保护土壤生态环境的作用。

3 讨论

防治荒漠化是人类面临的共同挑战,需要国际社会携手应对[14]。本研究应用RS和GIS技术,在对遥感影像进行处理的基础上,运用实验手段分析不同种植模式下土壤机械组成、土壤有机质含量、土壤HA/FA以及土壤全氮含量的变化趋势,通过构建土地荒漠化评价指标体系,探索不同种植模式与荒漠化演替方向的影响,分析不同种植模式下的优劣,以此来对不合理种植模式进行规范和改善。本研究由差异性分析得知:5 种不同模式下土壤有机质含量差异显著,风沙土林地被开辟为农田以后,其土壤有机质受人为活动的影响越来越大,因为农业生产将地上部的生物量进行收获,仅有地下部的根系作为有机物质的主要来源,减少了有机物的输入。同样地,针对土壤氮素与 HA/FA 进行差异性分析可以看出,其与土壤有机质有相似的分布。同时,结果显示:农林复合模式、沙平地防护林-花生模式、砂质丘陵防护林-花生模式、沙平地-花生模式、砂质丘陵-花生模式 5 种模式对减轻荒漠化的能力依次降低。农林复合模式、沙平地防护林-花生模式、砂质丘陵防护林-花生模式均有使荒漠化逆向演替的效果,但演替程度逐渐降低,其中砂质丘陵防护林-花生模式有一定的局限性;沙平地-花生、砂质丘陵-花生模式均有使荒漠化正向演替的效果,演替程度逐渐加剧。当前学术界对于该方向的研究还鲜有报道,但是不少学者针对土地荒漠化及其驱动因素进行了研究。张博[15]研究指出,人类活动是影响土壤荒漠化的重要因素,不同的人类活动方式对于土壤荒漠化起到不同方向的影响;Ma 等[16]根据收集到的 50 年的数据与资料,分析得出,人类活动在影响民勤县的土地荒漠化中占据主要地位;张东杰[17]以共和盆地作为研究区,通过对 50 年来数据的分析得出人类活动是影响土地荒漠化的主要因素这一结论;章予舒等[18]使用灰色关联度方法对土地荒漠化因素进行分析,得出的结论是:安西县土地荒漠化的最大影响因素是过度放牧,其次是人口数量的变化以及农业活动造成的土地开垦。同样地,奥布力·塔力普等[19]针对我国南疆地区土地荒漠化因素进行了分析,并指出造成土地荒漠化的因素分为自然因素和人为因素;何鹏杰等[20]对河西地区土地荒漠化影响因素进行分析,其结果指出:人类活动是荒漠化一个重要影响因素,其中人类活动包括垦荒、放牧等活动,并指出调整产业发展模式,使产业结构合理化、高级化,是解决荒漠化的有效途径。这都与本文的研究结果一致,即不同的种植模式对于土壤荒漠化的影响是不一样的,既有可能有正向影响,也有可能有负向影响。不同的是,目前各位学者对于驱动因素的分析大多是针对其驱动因素与土地荒漠化的相关关系进行分析,但很少去具体地分析各种因素是如何影响土地荒漠化的,或者说难以确定何种模式对于土地荒漠化有更好的抑制和改善效果。因此,确定具体的影响关系,从而建立科学合理的种植模式,对于土地荒漠化的治理具有重要意义。

参考文献

[1]廖兴亮,张腾,徐艳.半干旱区荒漠化演变趋势研究:以科尔沁左翼后旗为例[J].中国农业资源与区划,2020,41(4):299-307.

LIAO X L, ZHANG T, XU Y. Research on the evolution trend of desertification in semi-arid area:a case study of Kezuohouqi county[J]. Chin. J. Agric. Resour. Region. Plan., 2020, 41(4): 299-307.

[2]贾树海,王潇雪,杨亮.辽北农牧交错带土地荒漠化及景观格局变化研究[J].中国水土保持,2014(5):51-55,73.

JIA S H,WANG X X,YANG L. Soil and water conservation in China [J]. Chin. Soil.Water Conser.,2014(5):51-55,73.

[3]宋乃平,卞莹莹,王磊,等.农牧交错带农牧复合系统的可持续机制[J].生态学报,2020,40(21):7931-7940.

SONG N P, BIAN Y Y, WANG L,et al. Sustainable mechanism of agro-pastoral complex system in agro-pastoral ecotone [J]. Acta Ecolo.Sin.,2020,40(21):7931-7940.

[4]王涛.浑善达克沙区土地沙漠化过程及其生态环境效应[D].北京:北京林业大学,2020.

WANG T. The aeolian desertification dynamics and related ecological environment effect in the Otindag Sandy Land [D]. Beijing: Beijing Forest. Unive., 2020.

[5]马明德,杨美玲.基于 STIRPAT 模型的农牧交错带草地面积变化影响因素分析:以宁夏回族自治区盐池县为例[J].中国农业资源与区划,2018,39(3):48-54.

MA M D , Yang M L. The influence factors of grassland area change in farming-pastoral transitional zone by stirpat model:a case of Yanchi county of Ningxia Hui autonomous region in China [J]. Chin. J. Agric. Resour. Region. Plan., 2018,39(3):48-54.

[6]李旭亮,杨礼箫,田伟,等. 中国北方农牧交错带土地利用/覆盖变化研究综述[J].应用生态学报,2018, 29(10): 331-339.

LI X L , Yang L X , Tian W , *et al*. Land use and land cover change in agro-pastoral ecotone in Northern China:A review[J]. Chin.J.Appli.Eco., 2018,29(10): 331-339.

[7]郭泽呈,魏伟,石培基,等.中国西北干旱区土地沙漠化敏感性时空格局[J].地理学报,2020,75(9):1948-1965.

GUO Z C, WEI W, SHI P J, et al. Spatiotemporal changes of land desertification sensitivity in the arid region of Northwest China[J]. Acta Geogr. Sin.,2020,75(9):1948-1965.

[8]崔向慧,却晓娥,杨柳.土地退化和荒漠化防治领域国际标准化现状与思考[J].中国水土保持科学(中英文),2020,18(6):147-152.

CUI X H, Que X E, Yang L. Current status and thoughts on international standardization in the field of mitigating land degradation and combating desertification[J]. Sci.Soil.Water Conser.Chin., 2020,18(6):147-152.

[9]黄昌勇.土壤学[M].中国农业出版社,2000:15-18,31.

[10]何超,李萌,李婷婷,等.多目标综合评价中四种确定权重方法的比较与分析[J].湖北大学学报(自然科学版), 2016,38(2):172-178.

HE C, LI M, LI T T, et al. Comparison and analysis of the four methods of determining weights in multi-objective comprehensive evaluation[J]. J. Hubei Univer., 2016,38(2):172-178.

[11]黄妙芬,邢旭峰,王培娟,等.利用 LANDSAT/TM 热红外通道反演地表温度的三种方法比较[J].干旱区地理,2006(1):132-137.

HUANG M F , Xing X , Wang P J , et al. Comparison between three different methods of retrieving surface temperature from Landsat TM thermal infrared band[J]. Ari.Land Geogr., 2006(1):132-137.

[12]李雪. 辽宁地区耕地质量等别评价系统开发研究[D].辽宁工程技术大学,2017.

LI X. Research on the development of evaluation system of cultivated land quality in Liaoning [D]. Liaoning Tec. Unive., 2017.

[13] 殷贺,李正国,王仰麟,等.基于时间序列植被特征的内蒙古荒漠化评价[J].地理学报,2011,66(5):653-661.

Yin He, Li Zhengguo, et al. Assessment of Desertification Using Time Series Analysis of Hyper-temporal Vegetation Indicator in Inner Mongolia[J]. Acta Geogr.Sin., 2011, 66(5): 653-661.

[14] 新华社. 习近平致信祝贺联合国防治荒漠化公约第十三次缔约方大会高级别会议召开[J].国土绿化,2017(9) : 6.

[15] 张博. 1999—2018 年青海省土地荒漠化遥感监测及其驱动力分析[D].中国地质大学 北京,2020.

ZHANG B. Remote sensing monitoring and driving force analysis of land desertification in Qinghai Province from 1999 to 2018 [D]. Beijing: Chin.Unive.Geosci., 2020.

[16] Ma Y, Fan S, Zhou L, et al. The temporal change of driving factors during the course of land desertification in arid region of North China: the case of Minqin County[J]. Envi.Geolo.,2007, 51(6): 999-1008.

[17] 张东杰.共和盆地近 50 年来草地荒漠化驱动因素定量研究[J].水土保持研究,2010,17(4):166-169.

ZHANG D J. Quantitative Research of Driving Factors on Grassland Desertification over Last 50 Years in Gonghe Basin [J]. Res.Soil.Water.Conse., 2010,17(4):166-169.

[18] 章予舒,王立新,张红旗,等.疏勒河流域土地利用变化驱动因素分析:以安西县为例[J].地理科学进展,2003(3):170-178.

ZHANG Y S , WANG L X, ZHANG H Q , et al. An Analysis on Land Use Changes and Their Driving Factors in Shule River: an Example From Anxi County[J]. Progr.Geogr., 2003(3):170-178.

[19] 奥布力·塔力普,阿里木江·卡斯木.南疆地区经济发展对荒漠化程度的影响研究[J].冰川冻土,2017,39

(1):220-228.

Obul Talip, Alimujiang Kasimu. Study of the social economic development impacting the desertification in southern Xinjiang[J]. J. Glaci. Geocry.,2017,39(1):220-228.

[20] 何鹏杰,张恒嘉,王玉才,等.河西地区临泽县土地荒漠化影响因素分析[J].环境工程,2016,34(S1):1111-1116.

HE P J, ZHANG H J, WANG Y C, et al. Analysis of influencing factors of land desertification of HeXi corridor of linzecountyy[J]. Envir.Engin., 2016,34(S1):1111-1116.

本文曾发表于2022年《中国农业科技导报》第24卷第9期

不同施氮水平下AM真菌对高粱生物量及氮磷吸收的交互效应

王　健[1,2],张海欧[1,2],杨晨曦[1,2],李　娟[1,2]

(1.自然资源部退化及未利用土地整治重点实验室,陕西 西安 710075;
2.陕西地建土地工程技术研究院有限责任公司,陕西 西安 710075)

【摘要】目的:氮沉降是影响陆地生态系统稳定的主要胁迫之一。丛枝菌根(Arbuscular mycorrhizal, AM)真菌是自然界广泛分布的一类真菌,对植物生长及抗逆性具有重要的调节作用,然而针对氮沉降对菌根共生体的研究鲜有报道。研究氮沉降背景下AM真菌对植物的生长影响,旨在为全球气候变化背景下AM真菌生理生态学的研究提供一定的科学依据。方法:本研究利用盆栽试验对高粱幼苗(*Sorghum hicolor* (L.) Mocrnch)接种AM(*Glomus mosseae*)菌剂,模拟氮沉降,添加不同水平(0 kg·N·hm^{-2}·a^{-1}、10 kg·N·hm^{-2}·a^{-1},20 kg·N·hm^{-2}·a^{-1},30 kg·N·hm^{-2}·a^{-1},分别指N0、N1、N2、N3)的NH_4NO_3,生长16周后进行菌根侵染率、植株生物量及组织氮磷含量的测定。结果:①接种AM真菌显著提高了高粱根系的菌根侵染率(all $P < 0.001$),且随着氮添加浓度的增加,菌根侵染率逐渐降低。②在未施氮处理(N0)中,接种AM真菌显著促进了高粱地上生物量及总生物量(all $P < 0.05$, T-test),而在高浓度氮添加(N3)下,接种AM真菌显著抑制了高粱地上生物量及总生物量(all $P < 0.05$, T-test)。③在未施氮处理(N0)中,接种AM真菌显著促进了高粱组织的氮、磷含量及组织氮磷比(all $P < 0.05$, T-test),而在N2和N3氮水平下,接种AM真菌显著抑制了高粱组织的氮、磷含量及组织氮磷比(all $P < 0.05$, T-test),除磷在N2水平无显著差异外($P > 0.05$, T-test)。④高粱的菌根生长效应(MGR)、菌根氮吸收效应(MNR)及菌根磷吸收效应(MPR)均随着氮梯度的增加逐渐由正效应转为负效应。结论:AM真菌接种和模拟氮添加对高粱生物量及组织氮磷吸收存在显著的交互效应。

【关键词】AM真菌;氮沉降;高粱;生物量;氮磷吸收;菌根效应

Interactive Effects of AM Fungi and on the Biomass and Nitrogen and Phosphorus Uptake of *Sorghum hicolor* (L.) Mocrnch under Simulated Nitrogen Deposition

Jian Wang[1,2], Haiou Zhang[1,2], Chenxi Yang[1,2], Juan Li[1,2]

(1. Key Laboratory of Degraded and Unused Land Consolidation Engineering, the Ministry of Nature and Resources, Xi'an, Shaanxi 710075, China;2. Institute of Land Engineering and Technology, Shaanxi Provincial Land Engineering Construction Group Co., Ltd., Xi'an, Shaanxi 710075, China)

【Abstract】Objective: Nitrogen (N) deposition is one of the main stresses affecting the stability of terrestrial ecosystems. Arbuscular mycorrhizal (AM) fungi, a type of fungi widely distributed in natural ecosystems, which play an important regulatory effects on plant growth and stress resistance. However, there were few reports on the

项目基金:陕西省创新人才推进计划——青年科技新星项目(2021KJXX-88);陕西省科协企业创新争先青年人才托举计划项目(2021-1-2)。

research of N deposition on mycorrhizal symbionts. Studying the effects of AM fungi on plant growth under the background of N deposition to provide a scientific basis for the study of AM fungi's physiology and ecology under the background of global climate change. Method: In order to assess the effect of AM fungi on the growth of host plant under N deposition, a greenhouse pot experiment was conducted to survey the effects of AM fungi (*Glomus mosseae*) on the growth of *Sorghum hicolor* (L.) Mocrnch under different N addition levels (0 kg · N · hm^{-2} · a^{-1}, 10 kg · N · hm^{-2} · a^{-1}, 20 kg · N · hm^{-2} · a^{-1}, 30 kg · N · hm^{-2} · a^{-1}, respectively refer to N0、N1、N2、N3). After 16 weeks cultivation, mycorrhizal colonization, plant biomass and N and phosphorus (P) content were measured. Result: The results showed that: ①Inoculating AM fungi significantly increased the mycorrhizal colonization of sorghum roots (all $P < 0.001$), besides, the mycorrhizal colonization gradually decreased as the increased N addition gradient. ②In the N free treatment (N0), inoculating AM fungi significantly improved the above-ground biomass and total biomass of sorghum (all $P < 0.05$, *T*-test), while inoculating AM fungi obviously inhibited the aboveground biomass and total biomass of sorghum (all $P < 0.05$, *T*-test) under high-dose N (N3). ③In the N free treatment (N0), inoculating AM fungi significantly enhanced the N & P content of sorghum tissues and the ratio of N to P (all $P < 0.05$, *T*-test), surprisedly, the result is just the opposite at N2 and N3 treatment (all $P < 0.05$, *T*-test), except that there was no significant difference was observed on P content between inoculation and inoculation free at N2 treatment ($P > 0.05$, T-test). ④The mycorrhizal growth effect (MGR), mycorrhizal N-uptake effect (MNR) and mycorrhizal P-uptake effect (MPR) of sorghum gradually changed from positive effect to negative effect as the N gradient increased. Conclusion: Inoculating AM fungi and simulated N addition have significant interaction effects on sorghum biomass and tissue N & P uptake.

【**Key words**】AM fungi; Nitrogen deposition; *Sorghum hicolor* (L.) Mocrnch; Biomass; N and P uptake; Mycorrhizal response

0 引言

【研究意义】自 20 世纪中叶以来，化石燃料的大量燃烧、化肥的过量使用以及过度放牧等人类活动导致大气中的活性氮氧化物含量激增，大气氮素沉降也呈迅猛增加的趋势[1]。研究表明，在 2010 年我国华北平原部分经济发达地区氮沉降速率已接近 30 kg · N · hm^{-2} · a^{-1} [2]。人为干扰下的大气氮素沉降已成为全球氮素生物化学循环的一个重要组成部分，大气氮沉降速率的急剧增加将严重影响生态系统的生产力和稳定性[3]。由于自然生态系统生产力和生物多样性对外源氮的敏感性，氮沉降对植物的影响受到越来越多学者的关注[4-6]。AM(Arbuscular Mycorrhizal)真菌是土壤里广泛分布的一类有益微生物，能与绝大多数的陆生植物形成菌根互惠共生体[7]。AM 真菌通过其密集的根外菌丝网络帮助宿主植物汲取土壤中的水分和无机营养，来换取宿主植物光合作用产物用于自身的生长繁殖。因此，大力研究和预测大气氮沉降速率增加背景下，AM 真菌对作物，尤其是对庄稼作物的影响，对于农业生态系统的施肥管理具有重要的理论和实践意义。

【前人研究进展】研究表明，AM 真菌能显著提高宿主植物对氮、磷的吸收，进而提高其生物量[8]。此外，氮沉降引起的土壤氮含量增加会直接影响 AM 真菌的多样性和功能，进而影响植物的生长[9]。与此同时，由氮沉降引起的植物变化也将对共生的 AM 真菌产生间接的影响[10]。事实上，生态系统中多数植物是以植物-菌根共生体的形式存在[11]。

【本研究切入点】一年生草本植物高粱(*Sorghum hicolor* (L.) Mocrnch)因其侵染率高、生物量大、根系发达、生长速度快，常被作为 AM 真菌功能研究的模式宿主之一。

【拟解决的关键问题】本试验以高粱为研究对象，通过温室盆栽模拟实验探究 AM 真菌和氮添加对高粱生物量及氮磷吸收的影响，旨在为全球气候变化背景下 AM 真菌生理生态学的研究提供一定的科学依据，为我国高粱的高产种植以及微生物肥料的制备提供实践指导。

1 材料与方法

1.1 试验材料

● 供试菌种为摩西球囊霉(*Glomus mosseae*)，以高粱为宿主植物扩繁得到，密度为 110 个孢子/g 菌剂，含植物根断、菌丝等。菌种来自兰州大学细胞活动与逆境适应教育部重点实验室。

• 供试高粱品种为京杂抗四杂交高粱(*Sorghum hicolor* (L.) Mocrnch),来自陕西省农业科学院。

• 供试盆栽基质为河沙与沸石混合物(河沙:沸石,$V/V=1:1$),沸石和河沙购自北票市天翊沸石矿业有限公司,河沙粒径1 cm,沸石粒径2 cm。混合均匀的盆栽基质,121 ℃灭活2 h,隔天再灭一次。

1.2 试验设计

试验采用完全随机设计,设置2个接种处理和4个施氮处理。2个接种处理为接种AM真菌菌剂(+AMF)和接种灭活AM菌剂(- AMF);4个施氮水平(以NH_4NO_3形式添加)分别为0 mg N/kg、200 mg N/kg、400 mg N/kg、500 mg N/kg,相当于0 kg·N·hm^{-2}·a^{-1}、10 kg·N·hm^{-2}·a^{-1}、20 kg·N·hm^{-2}·a^{-1}、30 kg·N·hm^{-2}·a^{-1}氮沉降水平,分别用N0、N1、N2和N3表示;试验共8个处理,每个处理重复6次,总计48盆。

1.3 试验过程与方法

试验于2020年5月15日至2020年9月15日在陕西地建土地工程技术研究院温室内进行。高粱种子用10%的NaClO表面消毒后,于无菌沙中催芽。待种子出芽后,移苗于培养盆[37.5 cm(直径)×34.5 cm(高)]内(基质约需10 kg/盆),采用hoagland营养液按需给予水肥。2周后,幼苗定苗至每盆5株,并在+AMF处理组幼苗根系穴施AM菌剂15 g(约1500个孢子),在-AMF处理组幼苗根系穴施AM菌剂15 g灭活AM菌剂。将定量的NH_4NO_3溶于水中,分4次(2周一次)加入不同施氮处理盆中。温室光强为120 μmol photons·m^{-2}·s^{-1},昼夜交替时间为15 h/9 h,昼夜交替温差为23 ℃/16 ℃,生长周期为4个月,每隔1周随机移动花盆的位置[12]。待植物生活史完成后,分别收集植物地上茎叶和地下根系,取部分鲜细根系用于侵染率的测定,剩余样品用于生物量的测定,烘干的植物组织粉碎后用于氮、磷含量测定。

1.4 测定指标与方法

1.4.1 菌根侵染率测定

取鲜根样约1 g装入试管中,加入10% KOH溶液没过根样,于水浴锅(70~80 ℃)中碱化处理15~30 min,用自来水冲洗3遍;加入2% HCl溶液没过根样,于水浴锅中酸化处理30 min,用自来水冲洗3遍;加入0.05% 台盼蓝(1:1:1的乳酸甘油水($V:V:V$))染色10 min,用自来水洗3遍;染色后的根样剪成1 cm左右的根断,均匀地铺在涂有PVLG的载玻片上,盖上盖玻片,小心地压片至皮层被压开。制备好的装片置于显微镜下(200×)观察并计数[13]。

菌根侵染率:
$$RLC\% = \frac{T-N}{T} \times 100\% \tag{1}$$

丛枝侵染率:
$$AC\% = \frac{A}{T} \times 100\% \tag{2}$$

泡囊侵染率:
$$VC\% = \frac{V}{T} \times 100\% \tag{3}$$

式中:A为每个样品中观察到的有丛枝结构的视野数;V为每个样品中观察到的有泡囊结构的视野数;N为每个样品中没观察到的任何菌根侵染的视野数;T为每个样品中观察到的总视野数。

1.4.2 生物量及氮磷含量测定

将高粱收获后的地上部分、洗净后的地下部分分别装信封后,放在恒温干燥箱(60 ℃)中,48 h烘干至恒重后,称其生物量。烘干后的植物组织经球磨仪粉碎,过100目筛,制成供试样品。取0.2 g供试样品,精准置于消煮管底部,加1 g催化剂(K_2SO_4:$CuSO_4$:硒粉 = 100:10:1),加入5 mL浓硫酸静置碳化30 min,后置于消煮炉上,420 ℃消煮约40 min,直至消煮液呈现透明的青绿色。待冷却后,取下消煮管,将消煮液转移到容量瓶内,用蒸馏水定容至50 mL,然后上流动注射分析仪测定植物组织氮磷含量(Quik Chem 8500)[14]。

1.5 菌根效应计算

菌根生长效应(Mycorrhizal Growth Response, MGR)、菌根氮吸收效应(Mycorrhizal N-uptake Response, MNR)和菌根磷吸收效应(Mycorrhizal P-uptake Response, MPR)被用来评估不同氮梯度下

AM 真菌对植物的影响作用[15],下面用 MGR 为例来进行说明。

$$\left.\begin{aligned} MGR &= (1-\frac{NM_{mean}}{AM}) \times 100 \qquad NM_{mean} < AM \\ MGR &= \frac{AM}{NM_{mean}}-1) \times 100 \qquad NM_{mean} > AM \end{aligned}\right\} \tag{4}$$

式中:NM_{mean}为每一个氮水平下-AMF 处理组高粱生物量的平均值;AM 为每一个氮水平下+AMF 处理组高粱的总生物量;同样地,MNR 和 MPR 也按式(4)来计算。

1.6 数据处理与分析

采用 Microsoft Excel 2013 软件进行数据整理,原始数据经经验符合正态分布后进行下一步数据分析,由 R 语言(R version 3.3.2)进行统计分析和图表绘制。施氮和接种 AM 真菌处理对侵染率变量、植物生物量以及植物组织氮、磷含量的影响利用双因素方差分析进行分析。采用 Duncan 多重检验比较不同处理下各因变量之间的差异($P \leq 0.05$),采用 T-test 比较各因变量在+AMF 处理和-AMF 处理之间的差异($P \leq 0.05$)。此外,利用线性回归来分析不同氮水平下菌根效应的变化规律。

2 结果与分析

2.1 氮添加及接种 AM 真菌对高粱菌根侵染率的影响

由表 1 可以看出,摩西球囊霉可与高粱宿主形成良好的共生。不接种 AM 真菌处理(-AMF)组的高粱根系均没有被 AM 真菌侵染;而接种 AM 真菌(+AMF)显著促进了高粱根系的菌根侵染率(F = 3101.01, P < 0.001)、丛枝侵染率(F = 235.49, P < 0.001)和泡囊侵染率(F =259.13, P < 0.001)。此外,菌根侵染率(F = 44.57, P < 0.001)和泡囊侵染率(F = 19.24, P < 0.001)受氮添加影响显著,表明在不同氮水平处理下,AM 真菌对高粱的侵染能力存在明显差异;随着氮施肥梯度增加,菌根侵染率和泡囊侵染率随之增加,可见高氮水平可以抑制 AM 真菌对宿主的侵染。

表 1 不同处理下高粱菌根侵染状况

Table 1 Colonization of sorghum plants inoculated with AM fungi at different N levels

氮梯度	AMF(+/-)	RLC (%)	AC (%)	VC (%)
N0	+	56.66±3.14 a	13.17±1.21	23.73±1.98 a
N1	+	43.02±1.78 b	11.44±1.25	24.01±1.32 a
N2	+	32.57±3.03 c	9.78±1.29	17.78±0.57 b
N3	+	33.32±1.24 c	10.10±1.21	19.44±2.27 b
N0	-	0±0	0±0	0±0
N1	-	0±0	0±0	0±0
N2	-	0±0	0±0	0±0
N3	-	0±0	0±0	0±0
处理效应				
氮添加	F-value	44.57	1.47	19.24
	P-value	< 0.001	0.237	< 0.001
AM 真菌接种	F-value	3101.01	235.49	259.13
	P-value	< 0.001	< 0.001	< 0.001

注:表中数据为均值±标准误差;不同字母代表组间差异显著(P < 0.05);ns, P > 0.05; *, P < 0.05; **, P < 0.01; ***, P < 0.001。

Note: Data are mean±standard error; different letters represent significant differences between groups (P < 0.05); ns, P > 0.05; *, P < 0.05; **, P < 0.01; ***, P < 0.001.

2.2 氮添加及接种 AM 真菌对高粱生物量及根冠比的影响

由图 1 可以看出,施氮显著促进了高粱的地上生物量(F = 64.47, P < 0.001)和总生物量(F = 18.45, P < 0.001),降低了高粱根冠比(F = 24.72, P < 0.001),但对地下生物量(P > 0.05)影响没达到统计学上的显著水平。地上生物量和总生物量随着氮施肥梯度的增加而增加,根冠比则随着氮施肥梯度的增加而降低。接种 AM 真菌对高粱地上生物量、地下生物量、总生物量以及根冠比无显著影响(all P > 0.05),表明在各个氮处理水平下接种 AM 真菌对高粱的生长影响不大。由 T-test 结果分析可知,在未施氮处理(N0)组,接种 AM 真菌显著促进了高粱的地上生物量和总生物量(all P < 0.05, T-test);在高水平氮添加下(N3),接种 AM 真菌显著抑制了高粱的地上生物量和总生物量(all P < 0.05, T-test);在 N2 和 N3 处理下,接种 AM 真菌显著增加了高粱的根冠比(all P < 0.05, T-test)。此外,由图 1(d)可以看出,接种 AM 真菌和氮添加处理对高粱的地上生物量(F = 27.08, P < 0.001)、地下生物量(F = 3.01, P < 0.05)、总生物量(F = 12.01, P < 0.001)以及高粱根冠比(F = 3.88, P < 0.05)存在显著的交互作用。

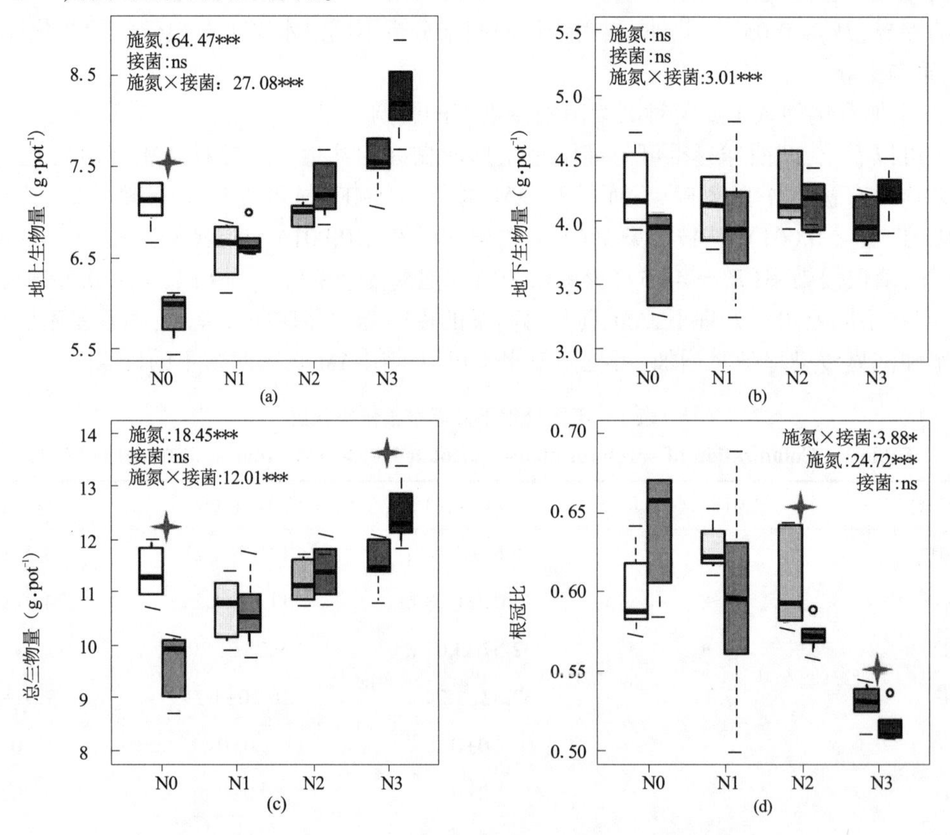

图 1 不同氮水平下接种 AM 真菌对高粱生物量的影响

Fig.1 Effects of inoculation with AM fungi on biomass of sorghum at different N levels

注: * 表示在同一个氮梯度下接种和不接种 AM 真菌之间差异达到显著水平(P ≤ 0.05, T-test)。The red stars indicate significant differences between the AM and AM free treatment at a same N addition level according to T-test (P ≤ 0.05).

2.3 氮添加及接种 AM 真菌对高粱氮磷吸收的影响

施氮显著增加了高粱组织的全氮含量(F = 11.38, P < 0.001),接种 AM 真菌对高粱组织的全氮含量在各个氮水平梯度下无显著差异(P > 0.05),施氮和接种对高粱组织的全氮含量存在显著交互作用(F = 19.70, P < 0.001)。接种 AM 真菌显著增加了高粱植株的全氮含量(P < 0.05, T-test)在未施氮处理组,而在高水平氮添加下(N2 和 N3),接种 AM 真菌显著降低了高粱植株的全氮含量(P < 0.05, T-test)(见图 2a)。由图 2b 可以看出,高粱组织的全磷含量(F = 3.18, P < 0.05)受氮添加处理的影响达到了显著水平,而受 AM 真菌接种的影响较弱。此外,施氮和接种 AM 真菌对高粱组织的

全磷含量存在显著交互作用($F = 9.22$, $P < 0.001$),在未施氮处理下,接种 AM 真菌显著促进了高粱植株的全磷含量($P < 0.05$, T-test),而在高水平氮添加下(N3),接种 AM 真菌对植株全磷含量则表现出显著的抑制作用($P < 0.05$, T-test)。施氮显著促进了高粱植株的氮磷比($F = 32.82$, $P < 0.001$),接种 AM 真菌对高粱植株的氮磷比在各氮水平梯度下无显著影响($P > 0.05$),但在未施肥处理组,接种 AM 真菌显著提高了高粱的氮磷比。此外,接种 AM 真菌和氮添加对高粱氮磷比的影响也不存在显著的交互作用($P > 0.05$)(见图 2c)。

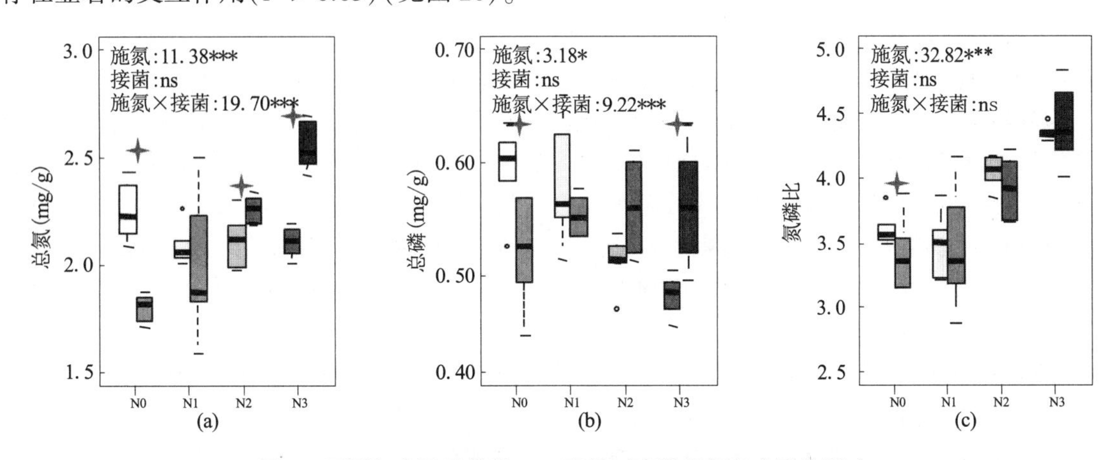

图 2 不同氮水平下接种 AM 真菌对高粱组织氮含量的影响

Fig.2 Effects of inoculation with AM fungi on tissue N concentrations of sorghum at different N levels

注:* 表示在同一个氮梯度下接种和不接种 AM 真菌之间差异达到显著水平($P \leq 0.05$, T-test)。The red stars indicate significant differences between the AM and AM free treatment at a same N addition level according to T-test ($P \leq 0.05$).

2.4 氮添加及接种 AM 真菌对高粱生长的交互效应

由图 3a 可知,施氮显著影响了高粱的菌根生长效应,在 N0 和 N3 处理组,AM 真菌对高粱植株的促生效应(菌根生长效应,MGR)的影响均达到了显著水平($P < 0.05$, T-test),菌根生长效应随氮施肥梯度的增加而逐渐降低,呈强烈的线性关系($R^2 = 0.81$, $P < 0.001$)。菌根氮吸收效应(MNR)($R^2 = 0.93$, $P < 0.001$)和菌根磷吸收效应(MPR)($R^2 = 0.88$, $P < 0.001$)与氮添加梯度呈强烈的线性关系,均随氮添加浓度的增加而降低,且在 N0、N2 和 N3 均达到显著水平。

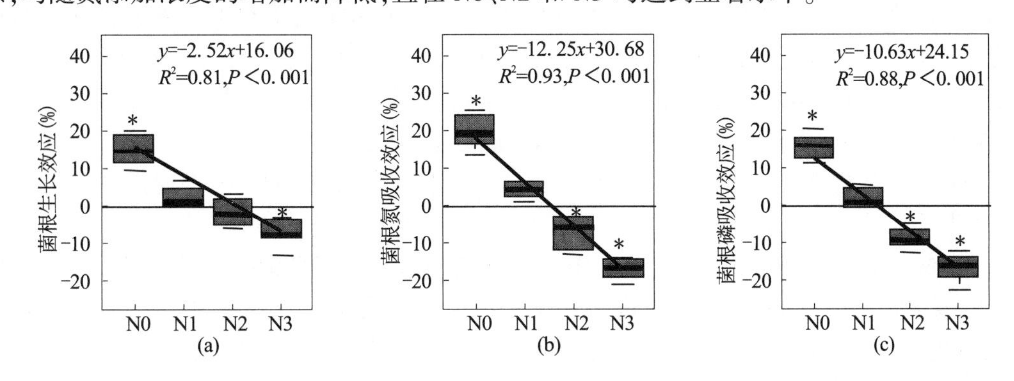

图 3 不同氮水平下高粱的菌根生长效应(a)、菌根氮吸收效应(b)和菌根磷吸收效应(c)

Fig.3 Mycorrhizal growth response (MGR), nitrogen uptake response (MNR) and phosphorus uptake response (MPR) (a~c) of *Sorghum bicolor* (L.) Moench along an N gradient

注:* 表示菌根效应差异达到显著水平($P \leq 0.05$, T-test),实线代表菌根效应原始变量和氮添加梯度之间线性回归拟合直线($P \leq 0.05$)。An asterisk indicates a significant difference between the mycorrhizal response and zero according to t-tests: *, $P \leq 0.05$. The solid lines represent the fitted linear or quadratic regressions ($P \leq 0.05$ are shown) between the raw data of mycorrhizal response variables and the level of N addition.

3 讨论与结论

菌根侵染率是AM真菌与宿主植物建立共生成功与否的重要标志,并在一定程度上决定着宿主植物的生长和抗逆能力[16]。泡囊和丛枝是AM真菌的根内菌丝特化的根内结构,其数量决定着菌根共生体内物质交换的频率和强度[17]。菌根侵染率既受宿主植物限制,也受土壤速效氮、磷含量制约[18]。目前,土壤速效磷含量对菌根侵染率的影响已被深入研究[19-20],相对而言,氮添加处理对菌根侵染率的影响研究较少。已有研究结果表明,土壤中速效氮含量过高会降低AM真菌对宿主植物的侵染[21],而在氮素缺乏的地区,施氮则有利于AM真菌对宿主植物的侵染[22]。本研究中,在未施氮处理组,AM真菌的菌根侵染率和泡囊侵染率均最高,且随着氮水平的不断增加,AM真菌的菌根侵染率和泡囊侵染率逐渐降低,Jiang等[23]在模拟氮沉降对青藏高原优势种垂穗披碱草的生长效应时,也发现氮添加逐渐降低了AM真菌的菌根侵染率。张彩丽[24]从超微结构上得出,氮添加对植物根系造成伤害,导致细胞线粒体肿胀,细胞质分解等现象。此外,本研究中发现丛枝受氮添加的影响不大,这可能与丛枝的特性有关,不同于泡囊和菌丝,丛枝在植物根系皮层细胞内的寿命较短,一般新形成的丛枝1~2周就会被植物根系细胞的内分泌酶消化掉,目前理论尚不能解释为什么丛枝的寿命如此之短。

氮、磷是植物生长不可或缺的营养元素,参与了植物体内许多重要化合物的组成及代谢途径[25]。前人的研究表明,接种AM真菌可以促进宿主植物对土壤氮、磷的吸收,尤其在土壤养分匮乏的地区[26-28]。本研究在未施肥处理组(N0)中,接种AM真菌显著提高了高粱的地上生物量及组织氮、磷含量,这与徐如玉等[27]的研究一致。AM真菌对宿主植物的促生作用主要得益于AM真菌的密集的根外菌丝和无隔的菌丝内壁。通过AM真菌菌丝的磷转运要比通过植物根系的磷转运效率高10倍左右,植物体内约90%的磷和25%的氮是通过AM真菌的菌丝获得的[29]。此外,AM真菌提高植物营养吸收的原因还在于土壤里密集的菌丝吸收面积,尤其土壤中的速效磷容易受土壤胶体吸附,扩散系数低,随着土壤中植物根系对磷的吸收,很快就会形成一个磷匮乏区,而AM真菌的根外菌丝网可以到达植物根系到达不了的地方,扩大了植物根系的吸收面积,大大增加了植物对磷的吸收[30]。

AM真菌对宿主植物的促生效应与土壤环境条件密切相关,在一定土壤养分范围内,菌根的促生效应随土壤养分的减少而增强,原因在于土壤养分较低时,植物生长受限,光合作用产物向根系及AM真菌分配减少,此时,适当施肥便能增加AM真菌侵染,从而增加宿主植物对土壤速效养分的吸收,促进其生长发育,大部分的研究证明了这一点[31-33]。此外,AM真菌和土壤养分在促进植物生长方面存在交互作用,一方面,植物的生长需要充足的氮源;另一方面,过量的氮素又会抑制AM真菌和宿主植物的共生,进而影响菌根的有益效应[34]。本文通过测试高粱的生物量及组织氮磷含量,发现随着氮添加梯度的增加,菌根对植物的有益效应由正效应(共生关系)逐渐变为负效应(寄生关系),这一点和Jiang等[23]的研究结果一致。主要是由于高水平氮添加情况下,植物通过根系可以吸收氮素满足植物生长需求,而不需要额外负担真菌的营养分配,高氮水平下的低菌根侵染率可以很好地说明这一点。本试验表明,在低水平氮添加下,接种AM真菌可以提高高粱的氮、磷含量及生物量,而在高氮条件下,接种AM真菌高粱组织氮、磷含量及生物量不升反降。因此,要充分发挥AM真菌的菌根效应,需首先了解当地的土壤养分条件,尤其在全球氮沉降背景加剧的情况下。此外,氮沉降导致高粱生物量降低,一定程度上也是因为高氮水平下,AM真菌和宿主植物处于一种寄生关系,AM真菌在不需要消耗能量的情况下,参与宿主植物的光合产物分配,最后以根外菌丝、孢子及其次级代谢产物(球囊霉素)的形式存在。研究表明,AM真菌孢子以及球囊霉素是土壤碳库的重要组成部分[35],换句话说,AM真菌在土壤高氮水平下,一定程度上减缓了土壤碳的释放。

综上所述,接种AM真菌显著增加了高粱的菌根侵染率,随着氮浓度的增加,高粱菌根侵染率和泡囊侵染率逐渐降低。本研究发现接种AM真菌和模拟氮沉降对高粱的菌根生长效应(MGR)、菌根氮吸收效应(MNR)及菌根磷吸收效应(MPR)存在显著的交互效应。未施氮处理(N0)时,接种AM真菌显著促进了高粱的MGR、MNR以及MPR;但随着氮梯度的增加,高粱的MGR、MNR以及MPR逐渐由正效应转为负效应，当模拟氮沉降浓度达到 20 kg · N · hm^{-2} · a^{-1}(N2)时，接种 AM 真菌显著抑制了高粱的 MNR 以及 MPR；当模拟氮沉降浓度继续增加到 30 kg · N · hm^{-2} · a^{-1}(N3)时，高粱的 MGR、MNR 以及 MPR 均受 AM 真菌抑制。此外，本试验选用的是单一的 AM 菌种，不同的 AM 菌种，尤其是混合菌种对氮的耐受性不尽一致，且本研究所采用的盆栽试验与外界大田所处的氮沉降仍会有一定差异，故而仍需进一步研究。

参考文献

[1]许稳. 中国大气活性氮干湿沉降与大气污染减排效应研究[D]. 北京：中国农业大学，2016.

Xu W. Studies on dry and wet deposition of atmospheric reactive nitrogen and air pollution control effects in China [D]. Beijing: China Agricultural University, 2016. (in Chinese)

[2]郑丹楠，王雪松，谢绍东，等. 2010 年中国大气氮沉降特征分析[J]. 中国环境科学，2014，34(5)：1089-1097.

Zhen D N, Wang X S, Xie S D, et al. Simulation of atmospheric nitrogen deposition in China in 2010 [J]. China Environmental Science, 2014, 34(5): 1089-1097. (in Chinese)

[3]徐丽，杨雁茹，张军辉，等. 模拟氮沉降增加对中国陆地生态系统土壤呼吸 Q_{10} 的影响[J]. 生态学杂志，2019，38(5)：1560-1569.

Xu L, Yang Y R, Zhang J H, et al. Effects of simulated N deposition on Q_{10} of soil respiration in Chinese terrestrial ecosystems [J]. Journal of Ecology, 2019, 38(5): 1560-1569. (in Chinese)

[4]王洪义，常继方，王正文. 退化草地恢复过程中群落物种多样性及生产力对氮磷养分的响应[J]. 中国农业科学，2020(13)：2604-2613.

Wang H Y, Zhang J F, Wang Z W. Responses of community species diversity and productivity to nitrogen and phosphorus addition during restoration of degraded grassland [J]. China Agricultural Sciences, 2020(13): 2604-2613. (in Chinese)

[5]蒯晓妍，邢鹏飞，张晓琳，等. 短期不同水平氮添加对农牧交错带草地植物群落多样性和生产力的影响[J]. 中国草地学报，2019，41(5)：104-110.

Kuai X Y, Xing P F, Zhang X L, et al. Effects of Short-term Nitrogen Addition on Plant Community Diversity and Productivity of Grassland in Agro-pastoral Ecotone [J]. Chinese Journal of Grassland, 2019, 41(5): 104-110. (in Chinese)

[6]王玉冰，孙毅寒，丁威，等. 长期氮添加对典型草原植物多样性与初级生产力的影响及途径[J]. 植物生态学报，2020，44(1)：22-32.

Wang Y B, Xun Y S, Ding W, et al. Effects and pathways of long-term nitrogen addition on plant diversity and primary productivity in a typical steppe [J]. Acta Plant Ecology, 2020, 44(1): 22-32. (in Chinese)

[7]Smith S E and Read D J 2008. Mycorrhizal Symbiosis [M]. third ed. Academic Press, London. pp.13.

[8]Selosse M A, Rousset F. The plant-fungal marketplace [M]. Science. 2011, 333, 828-829.

[9]Han Y, Feng J, Han M, Zhu, B. Responses of arbuscular mycorrhizal fungi to nitrogen addition: A meta-analysis [J]. Global Change Biology, 2020, 26(12): 7229-7241.

[10]Chen Y L, Xu Z W, Xu T L, et al. Nitrogen deposition and precipitation induced phylogenetic clustering of arbuscular mycorrhizal fungal communities [J]. Soil Biology and Biochemistry, 2017, 115: 233-242.

[11]Soudzilovskaia N A, Vaessen S, Barcelo M, et al. Fungal Root: global online database of plant mycorrhizal associations [J]. New Phytologist, 2020, 227(3): 955-966.

[12]Bahadur A, Jin Z C, Long X L, et al. Arbuscular mycorrhizal fungi alter plant interspecific interaction under nitrogen fertilization [J]. European Journal of Soil Biology, 2019, 93: 103094.

[13]McGonigle T, Miller M, Evans D, et al. A new method which gives an objective measure of colonization of roots by vesiculararbuscular mycorrhizal fungi [J]. New phytologist. 1990, 115, 495-501.

[14]鲁如坤. 土壤农业化学分析力法[M]. 北京：中国农业科技出版社，2000.

[15]Veiga R S L, Jansa J, Frossard E, et al. Can arbuscular mycorrhizal fungi reduce the growth of agricultural weeds? PLoS ONE. 2011, 6, e27825.

[16]王晓英，王冬梅，陈保冬，等. 丛枝菌根真菌群落对白三叶草生长的影响[J]. 生态学报，2010，30(6)：1456-1462.

Wang X Y, Wang D M, Chen B D, et al. Growth response of white clover to inoculation with different Arbuscular mycorrhizal fungi communities [J]. Acta Ecologica Sinica, 2010, 30(6): 1456-1462. (in Chinese)

[17]田蜜, 陈应龙, 李敏, 等. 丛枝菌根结构与功能研究进展[J]. 应用生态学报, 2013(8): 285-292.

Tian M, Chen Y L, Li M, et al. Structure and function of arbuscular mycorrhiza: A review[J]. Journal of Applied Ecology, 2013(8): 285-292. (in Chinese)

[18]王振楠, 杨美玲, 刘鸯, 等. 丛枝菌根真菌对红花生长及根际土壤微环境的影响[J]. 江苏农业学报, 2016, 32(4): 904-909.

Wang Z N, Yang M Liu Y. et al. Effects of arbuscular mycorrhization on the growth of safflower and the microenvironment of rhizosphere soil [J]. Journal of Jiangsu Agricultural Sciences, 2016, 32(4): 904-909. (in Chinese)

[19]Jaitieng S, Sinma K, Rungcharoenthong P, et al. Arbuscular mycorrhiza fungi applications and rock phosphate fertilizers enhance available phosphorus in soil and promote plant immunity in robusta coffee [J]. Soil Science and Plant Nutrition, 2021, 67(1): 97-101.

[20]Yang G, Liu N, Lu W, et al. The interaction between arbuscular mycorrhizal fungi and soil phosphorus availability influences plant community productivity and ecosystem stability [J]. Journal of Ecology, 2014, 102(4): 1072-1082.

[21]王淼焱, 徐倩, 刘润进. 长期定位施肥土壤中 AM 真菌对寄主植物的侵染状况[J]. 菌物学报, 2006(1): 131-137.

Wang M Y, Xu Q, Liu R J. Colonization status of arbuscular mycorrhizal fungi on host plants grown in long-term fixed fertilization field [J]. Mycosystema, 2006(1): 131-137. (in Chinese)

[22]Van Diepen L T A, Lilleskov E A, Pregitzer K S. Simulated nitrogen deposition affects community structure of arbuscular mycorrhizal fungi in northern hardwood forests [J]. Molecular Ecology, 2011, 20(4): 799-811.

[23]Jiang, S, Liu, Y, Luo, J, et al. Dynamics of arbuscular mycorrhizal fungal community structure and functioning along a nitrogen enrichment gradient in an alpine meadow ecosystem[J]. New Phytologist, 2018, 220(4), 1222-1235.

[24]张彩月. AM 真菌和施氮量对五味子生长和化学成分的交互效应[D]. 石家庄: 河北大学, 2006.

Zhang C Y. Effects of AM fungi on the growth and chemical composition of *Schisandra Chinesis* under different nitrogen levels [D]. Shijiazhuang: Hebei University, 2006. (in Chinese)

[25]李国军. 大量营养元素对玉米苗期铬吸收及生理特性的影响研究[D]. 太原: 山西大学, 2010.

Li G J. Colonization status of arbuscular mycorrhizal fungi on host plants grown in long-term fixed fertilization field [D]. Taiyuan: Shanxi University, 2010. (in Chinese)

[26]甄莉娜, 王润梅, 杨俊霞, 等. 丛枝菌根真菌与氮肥对羊草生长的影响[J]. 中国草地学报, 2018, 40(3): 51-56.

Zhen L N, Wang R M, Yang J X, et al. Effects of Arbuscular Mycorrhizal Fungi and Nitrogen Fertilizer on the Growth of *Leymus chinensis* [J]. Chinese Journal of Grassland, 2018, 40(3): 51-56. (in Chinese)

[27]徐如玉, 左明雪, 袁银龙, 等. 增施摩西管柄囊霉对甜玉米氮肥增效及土壤丛枝菌根真菌多样性的影响[J]. 福建农业学报, 2020, 197(4): 25-37.

Xu R Y, Zuo M X, Yuan Y L, et al. Effects of *Funneliformis mosseae* Application on Nitrogen Utilization by Sweet Corn and AM Fungi Diversity in soil [J]. Fujian Journal of Agricultural Sciences, 2020, 197(4): 25-37. (in Chinese)

[28]贾艳艳, 杨文飞, 杜小凤, 等. 接种 AM 真菌和施氮对还田稻秆氮素释放和小麦产量的影响[J]. 江西农业学报, 2020(3): 8-13.

Jia Y Y, Yang W F, Du X F, et al. Effects of AM Fungi Inoculation and Nitrogen Application on Rice-straw Nitrogen Release and Wheat Yield [J]. Acta Agriculturae Jiangxi, 2020(3): 8-13. (in Chinese)

[29]蔺吉祥, 杨雨衡, 王英男, 等. 氮沉降对植物-丛枝菌根共生体影响的研究进展[J]. 草原与草坪, 2015(3): 88-94.

Ling J X, Yang Y H, Wang Y N, et al. Research progress on effects of nitrogen deposition on symbiont of plant-Arbuscular mycorrhizal [J]. Grassland and Turf, 2015(3): 88-94. (in Chinese)

[30]George E, Marschner H, Jakobsen I. Role of arbuscular mycorrhizal fungi in uptake of phosphorus and nitrogen from soil [J]. Critical Reviews in Biotechnology, 1995, 15(3-4): 257-270.

[31]Johnson, N C, Rowland, D L, Corkidi, L, et al. Plant winners and losers during grassland N-eutrophication differ

in biomass allocation and mycorrhizas [J]. Ecology. 2008, 89, 2868-2878.

[32]Johnson, N C, Rowland, D L, Corkidi, L, et al. Nitrogen enrichment alters mycorrhizal allocation at five mesic to semiarid grasslands [J]. Ecology. 2003, 84(7), 1895-1908.

[33]Frater P N, Borer E T, Fay P A, et al. Nutrients and environment influence arbuscular mycorrhizal colonization both independently and interactively in Schizachyrium scoparium [J]. Plant and Soil, 2018, 425(1): 493-506.

[34]王红新，李富平，国巧真，等. AM 真菌生长发育影响因素及其对植物作用的研究[J]. 中国土壤与肥料，2006 (1): 52-56.

Wang H X, Li F P, Guo Q Z, et al. The growth influence factor of AM and the function of it for the plants [J]. China Soil and Fertilizer, 2006 (1): 52-56. (in Chinese)

[35]黄彬彬，邢亚娟，闫国永，等. 兴安落叶松林球囊霉素相关土壤蛋白含量对年际间模拟氮沉降的响应[J]. 生态环境学报，2019，28(3):22-30.

Huang B B, Xing Y J, Yan G Y, et al. Response of GRSP Content to Interannual Simulated Nitrogen Deposition in *Larix gmelinii* Forest in Greater Khingan Mountains [J]. Ecology and Environmental Sciences, 2019, 28(3):22-30. (in Chinese)

本文曾发表于2022年《福建农业学报》第37卷第7期

土地综合整治项目实施效益评价

——基于 AHP-PCE 模型

慕哲哲[1, 2, 3]

(1. 自然资源部退化及未利用土地整治工程重点实验室,西安,710021;
2. 陕西省土地工程建设集团有限责任公司,西安,710075;
3. 陕西地建土地工程技术研究院有限责任公司,西安,710021)

【摘要】为研究土地综合整治工程实施前后项目区总体建设效益情况,明晰土地综合整治工程建设过程中存在的问题及可提升潜力指标,以法库县龙家村土地整治项目为例,从经济效益、社会效益、生态效益和景观效益 4 个方面选取 14 个评价指标,采用多层次模糊综合评价法测算各效益分值及综合分值,并划分评价等级。研究结果表明:项目区土地综合整治工程综合效益分值为 82.10,综合评价等级为良好。经济效益、社会效益、生态效益和景观效益表现为“一优三良”。其中,社会效益分值为 86.15,评价等级为优秀;经济效益、生态效益、景观效益分值分别为 80.74、82.12 和 79.41,评价等级均为良好。经济效益、社会效益、生态效益和景观效益均有所提升,但经济效益和景观效益与项目预期建设目标存在差距,其中新增耕地指标、项目区总产值指标及景观丰富度指标有较大提升潜力。

【关键词】土地整治;效益评价;层次分析法;模糊综合评价

Implementation Benefit Evaluation of Comprehensive Land Consolidation Project: Based on AHP-PCE Model

Zhezhe Mu[1,2,3]

(1. Key Laboratory of Degraded and Unused Land Consolidation Engineering, Ministry of Natural Resources, Xi' an 710021, China; 2.Shaanxi Provincial Land Engineering Construction Group Co., Ltd., Xi' an 710075, China; 3. Institute of Land Engineering and Technology, Shaanxi Provincial Land Engineering Construction Group Co., Ltd., Xi' an 710021, China)

【Abstract】 In order to study the overall construction benefit of the project area before and after the implementation of the comprehensive land consolidation project, and clarify the problems existing in the construction process of the comprehensive land consolidation project and the potential indicators for improvement, this study takes the land consolidation project of Longjia Village in Faku County as an example. 14 evaluation indexes are selected from four aspects of economic benefit, social benefit, ecological benefit and landscape benefit. The multi-

基金项目:陕西省土地工程建设集团内部项目(DJNY2021-25、DJNY2022-55)。

level fuzzy comprehensive evaluation method is used to calculate the benefit scores and comprehensive scores, and the evaluation grades are divided. The results show that : (1) The comprehensive benefit score of land comprehensive improvement project in the project area is 82.10, and the comprehensive evaluation level is good. (2) Economic benefits, social benefits, ecological benefits and landscape benefits are 'one excellent three good'. Among them, the social benefit score is 86.15, the evaluation grade is excellent ; the scores of economic benefit, ecological benefit and landscape benefit were 80.74, 82.12 and 79.41, respectively, and the evaluation levels were good. (3) Economic benefit, social benefit, ecological benefit and landscape benefit have been improved, but there is a gap between economic benefit and landscape benefit and the expected construction goal of the project.Among them, the new cultivated land index, the total output value index of the project area and the landscape richness index have great potential for improvement.

【Keywords】 Land consolidation; Benefit evaluation; Analytic hierarchy process; Fuzzy comprehensive evaluation

0 引言

人多地少、耕地资源分布不均、后备资源相对匮乏是我国耕地资源的典型特征。随着社会经济发展,工业化、城市化快速推进,耕地数量减少、质量下降,农业生产和粮食安全受到威胁。为了保护耕地资源,保证我国十几亿人口的吃饭问题,国家投入大量资金,在农村开展土地综合整治工作,并取得了一定的成效。然而,目前农村土地综合整治多数是以完成建设目标为主,工程质量较低,资金使用缺少管护,新增耕地未能达标[1-2]。为了使土地综合整治工作健康持续推进,使土地综合整治发挥最大效益,这就需要对农村土地综合整治项目的效益进行评价。研究和分析项目实施前后存在的问题,剖析出现问题的原因以及关键之处,为后续土地综合整治工作提供经验借鉴,并指明方向[3-4]。

为此,我国学者就土地综合整治的发展历程、概念内涵、效益评价、指标体系构建及评价方法等多方面开展研究。夏方舟等[5]梳理和总结了我国近 40 年国土综合整治的演进与变化,提出新时期国土综合整治内涵是以提高国土利用效率和效益、保障国土资源永续利用、改善生态景观环境为主要目的。程传胜[6]结合残次林地土地综合整治工程施工情况,采用适宜的耕地质量评价体系,开展土地综合整治前后质量评价。吴家龙等[7]以广东省为例,基于问题导向和目标导向,采用专家咨询等方法,构建了涵盖自然资源、经济和社会三个维度的土地综合调查评估指标体系。梁敏[8]从社会结构、人口素质和社会进步等方面构建了土地整治效益评价体系,以此来判断项目实施是否达到预期效果。范业婷等[9]通过解析全域土地综合整治的内涵特征并揭示其助推乡村重构的机制,采用案例分析法,提出全域土地综合整治试点实践需立足乡村重构的现实需求。丁继辉等[10]从生产、生活、生态的角度出发,提出土地综合整治的目标及评价指标体系,采用多层次模糊综合评价模型评价土地整治的综合效益。在评价方法上,我国学者大多是通过构建数学模型,定性和定量相结合,综合评价土地综合整治效益。刘姝驿等[11]利用层次分析法及模糊综合评价法对重庆市 26 个村农村土地综合整治项目进行效益评价。潘珍妮等[12]通过建立模糊物元模型,进而评价土地综合整治效益。程文仕等[13]借助 RAGA-PPC 模型对甘肃省庆阳市典型土地整治项目进行综合效益评价。

总的来说,农村土地综合整治效益评价多集中在经济、社会和生态等方面的单项评价,也不乏对经济、社会效益与生态效益相结合的综合效益的评价。然而,将景观效益作为一个评价层面,与经济、社会和生态相融合,构建综合评价指标体系的研究相对较少。在目前生态文明的宏观背景下,农村土地综合整治的景观建设也不容忽视。因此,本研究以法库县龙家村土地综合整治项目为例,从经济、社会、生态和景观 4 个层面构建符合地区特征的评价指标体系,采用多层次模糊综合评价法对该项目进行评价,分析其存在的问题和潜在的提升空间,研究结果能够为类似土地整治项目开展提供思路和方向,为项目的科学决策提供参考和依据。

1 研究区概况与数据来源

土地综合整治项目区位于法库县东部柏家沟镇,建设地点在龙家村(123°32′42″~123°36′50″E,42°30′21″~42°35′20″N)。项目区位于低丘平原区,总体地势北高南低,平均海拔 110 m。项目区现状耕地为 1 293.35 hm^2、园地 4.12 hm^2、林地 73.11 hm^2、其他草地 14.35 hm^2。项目区总面积 1 563.22 hm^2,建设规模为 1 329.74 hm^2。研究数据主要来源于项目的可行性研究报告、项目结题报告、项目预算书、土地利用现状图、《法库县统计年鉴》和《法库县土地利用总体规划(2006—2020 年)》等。

2 土地综合整治效益评价研究方法

2.1 研究思路

首先,收集和整理研究区土地整治项目建设总面积、项目投资、新增耕地面积等基础数据。其次,分析和整理已有学者对土地整治效益评价的研究成果,确定本研究中使用层析分析法和多层次模糊综合评价的方法。最后,对评价结果进行分析,并提出有关政策建议。

2.2 研究方法

2.2.1 构建评价指标体系并确定权重

土地整治效益评价体系的构建需要尽可能全面地涵盖耕地作用的各个方面,既应体现经济效益,又能反映对农业生产和生态环境条件的综合改善程度。同时,评价指标的选取需考虑系统性、典型性以及可量化的原则[14-16]。在参考王栋[17]、刘永利[18]、周婷[19]、徐玲玲[20]等学者研究成果的基础上,结合项目区自然、经济和社会条件,本研究从经济效益、社会效益、生态效益和景观效益 4 个方面分别选取人均年纯收入变化率、路网密度变化率、防护林网面积变化率、田块平整度指数等 14 个评价指标,构建了符合项目区实际情况的土地整治综合效益评价指标体系(见表 1)。

指标权重的确定采用层次分析法(AHP)。首先,建立"目标层—准则层"和"准则层—指标层"两级判矩阵。其次,进行标准化处理、特征向量及最大特征根的求解、一致性偏离度(CR)的检验等数学计算。最后,得出各指标的相对权重和组合权重[21-23],测算模型如下,计算结果见表 1。

判断矩阵模型

$$a_{ij} = \frac{1}{a_{ji}} \tag{1}$$

式中:a_{ij}为要素 i 与要素 j 重要性比较结果。

一致性检验模型

$$\mathrm{CI} = \frac{\lambda - n}{n - 1} \tag{2}$$

$$\mathrm{RI} = \frac{\mathrm{CI}_1 + \mathrm{CI}_2 + \cdots + \mathrm{CI}_n}{n} \tag{3}$$

$$\mathrm{CR} = \frac{\mathrm{CI}}{\mathrm{RI}} \tag{4}$$

表 1 土地综合整治项目实施效益评价指标体系及权重

Table 1 Evaluation index system and weight of the implementation benefit of the comprehensive land improvement project

目标层	准则层（权重）	指标层	测算方法	数据来源/处理方法	相对权重/组合权重
土地综合整治项目实施效益	经济效益（0.488）	人均年纯收入变化率	（整治后人均年纯收入-整治前人均年纯收入）÷整治前人均年纯收入	调研数据/统计分析	0.252/0.123
		项目区总产值变化率	（整治后项目区总产值-整治前项目区总产值）÷整治前项目区总产值	调研数据/统计分析	0.482/0.235
		新增耕地率	新增耕地面积÷项目建设规模	项目设计/统计分析	0.266/0.130
	社会效益（0.212）	路网密度变化率	（整治后路网密度-整治前路网密度）÷整治前路网密度	地理空间数据/GIS分析	0.127/0.027
		人均耕地面积变化率	（整治后人均耕地面积-整治前人均耕地面积）÷整治前人均耕地面积	项目设计/统计分析	0.234/0.050
		居民满意度	项目区满意土地整治人数÷项目区总人数	调研数据/统计分析	0.520/0.110
		灌排设施变化率	（整治后灌排设施总长度-整治前灌排设施总长度）÷整治前灌排设施总长度	项目设计/统计分析	0.119/0.025
	生态效益（0.225）	防护林网面积变化率	（整治后防护林网面积-整治前防护林网面积）÷整治前防护林网面积	地理空间数据/GIS分析	0.321/0.072
		人均绿地面积变化率	（整治后人均绿地面积-整治前人均绿地面积）÷整治前人均绿地面积	调研数据、项目设计/统计分析	0.218/0.049
		耕地灌溉保证率指数	（整治后满足灌溉耕地面积-整治前满足灌溉耕地面积）÷整治前满足灌溉耕地面积	项目设计/统计分析	0.259/0.058
		土地垦殖率	项目耕地面积÷项目区建设规模	项目设计/统计分析	0.202/0.045
	景观效益（0.075）	田块平整度指数	（整治前田面坡度-整治后田面坡度）÷整治前田面坡度	地理空间数据/GIS分析	0.386/0.029
		景观破碎度指数变化率	（整治前斑块个数-整治后斑块个数）÷整治前斑块个数	地理空间数据/GIS、Fragstats 分析	0.289/0.022
		景观丰富度指数变化率	（整治后景观丰富度指数-整治前景观丰富度指数）÷整治前景观丰富度指数	地理空间数据/GIS、Fragstats 分析	0.325/0.024

式中：CI 为一致性指标（CI=0，有完全的一致性；CI 接近于 0，有满意的一致性，CI 越大，不一致越严重）；λ 为判断矩阵特征根；n 为阶数；RI 为随机一致性指标，当 $n=14$ 时，查表可知 RI=1.59；CR 为检验系数，一般认为 CR<0.1，该判断矩阵通过一致性检验。

2.2.2 多层次模糊综合评价模型构建

（1）各评价因子隶属度矩阵构建。根据统计整理的基础数据，利用 CAD、ArcGIS 和 Excel 软件进行标准化处理，运用特尔斐法与模糊统计法，邀请专家教授对各个评价指标完成情况进行打分，确定各评价指标的隶属度[24-27]，见表 2。

表2 各评价因子隶属度

Table 2 Membership degree of each evaluation factor

评价指标	标准化结果/%	隶属度			
		V_1	V_2	V_3	V_4
人均年纯收入变化率	46.30	0.2	0.4	0.3	0.1
项目区总产值变化率	20.40	0.2	0.5	0.3	0.0
新增耕地率	0.03	0.1	0.3	0.4	0.2
路网密度变化率	3.70	0.2	0.5	0.1	0.2
人均耕地面积变化率	16.50	0.2	0.3	0.3	0.2
居民满意度	96.80	0.5	0.4	0.1	0.0
灌排设施变化率	75.60	0.4	0.2	0.4	0.0
防护林网面积变化率	56.20	0.1	0.5	0.3	0.1
人均绿地面积变化率	2.72	0.2	0.3	0.3	0.2
耕地灌溉保证率指数	75.60	0.2	0.5	0.1	0.2
土地垦殖率	97.30	0.5	0.4	0.1	0.0
田块平整度指数	34.60	0.2	0.3	0.4	0.1
景观破碎度指数变化率	86.50	0.3	0.2	0.5	0.0
景观丰富度指数变化率	47.20	0.2	0.2	0.4	0.2

注: V 为各评价指标隶属度。其中,V_1 为优秀,V_2 为良好,V_3 为合格,V_4 为不合格。

(2) 模型构建及等级划分。多层次模糊综合评价模型,即对某一土地综合整治项目分不同的层级分别进行模型构建,其评价过程由下而上逐级进行。根据范业婷等[9]、丁继辉等[10]、刘永利[18]等学者对土地整治项目效益评价的研究思路和研究方法,确定本研究土地整治多层次模糊综合评价计算模型。

$$B_i = R_i \times V_i = [r_{11}, r_{12}, \cdots, r_{1n}] \times \begin{bmatrix} v_{11} & v_{12} & \cdots & v_{1n} \\ v_{21} & v_{22} & \cdots & v_{2n} \\ \vdots & \vdots & & \vdots \\ v_{m1} & v_{m2} & \cdots & v_{mn} \end{bmatrix} \tag{5}$$

式中:B_i 为经济、社会、生态和景观效益评价结果矩阵;R_i 为各评价指标权重集矩阵;V_i 为各评价指标隶属度矩阵;r_{1n} 为各评价指标相对权重;v_{mn} 为各评价指标隶属度。

为了使计算结果具有可比性,结合研究区实际情况,将各准则层计算结果划分为 4 个等级。结合模糊综合评价计算方法,引入学生成绩评定标准,进一步计算经济、社会、生态效益和景观效益分值及综合效益分值[28-30]。即优秀($Q \geqslant 85$)、良好($70 \leqslant Q<85$)、合格($60 \leqslant Q<70$)和不合格($Q<60$)。

$$Q_i = B_i \times T = B_i \times [100, 85, 70, 60]^{\mathrm{T}} \tag{6}$$

式中:Q_i 为经济、社会、生态和景观效益测算分值及综合分值;T 为学生成绩评定标准函数矩阵。

3 评价结果与分析

3.1 评价结果分值测算

依据多层次模糊综合评价测算模型(5)和模型(6),分别测算研究区土地综合整治项目实施经济效益、社会效益、生态效益、景观效益和综合效益分值,其测算结果见表 3。

表 3　各效益指标测算结果

Table. 3　Calculation results of benefit indicators

效益指标	评价隶属度				效益等级（分值）
	V_1	V_2	V_3	V_4	
经济效益	0.1734	0.4216	0.3266	0.0784	良好（Q_1=80.74）
社会效益	0.3798	0.3655	0.1825	0.0722	优秀（Q_2=86.15）
生态效益	0.2285	0.4362	0.2078	0.1275	良好（Q_3=82.12）
景观效益	0.2289	0.2386	0.4289	0.1036	良好（Q_4=79.41）
综合效益	0.2337	0.3993	0.2770	0.0900	良好（Q=82.10）

3.2　评价结果分析

经济效益评价过程中选取人均年纯收入变化率、项目区总产值变化率和新增耕地率 3 个指标。根据测算结果及各评价因子隶属度矩阵分析得知，经济效益评价结果 V_1 占 17.34%，V_2 占 42.16%，V_3 占 32.66%，V_4 占 7.84%。结合最大隶属原则，确定经济效益整体评价等级为良好（$70 \leqslant Q_1 < 85$）。说明通过土地综合整治，项目区农民收入和农业产值均有所提升，经济效益较为显著，达到了较为理想的效果。但项目区整体新增耕地较少，主要是因为本项目工程侧重耕地基础设施建设和产能提升。

社会效益评价过程中选取路网密度变化率、人均耕地面积变化率、居民满意度和灌排设施变化率 4 个指标。根据测算结果及各评价因子隶属度矩阵分析得知，社会效益评价结果 V_1 占 37.98%，V_2 占 36.55%，V_3 占 18.25%，V_4 占 7.22%。结合最大隶属原则，确定社会效益整体评价等级为优秀（$Q_2 \geqslant 85$）。说明项目区土地综合整治社会效益显著，道路、灌排等基础设施建设较为完善，耕地质量有所提升，居民对项目区建设较为满意。

生态效益评价过程中选取防护林网面积变化率、人均绿地面积变化率、耕地灌溉保证率指数和土地垦殖率 4 个指标。根据测算结果及各评价因子隶属度矩阵分析得知，生态效益评价结果 V_1 占 22.85%，V_2 占 43.62%，V_3 占 20.78%，V_4 占 12.75%。结合最大隶属原则，确定生态效益整体评价等级为良好（$70 \leqslant Q_3 < 85$）。说明项目区土地综合整治对生态建设和保护较为重视，通过增加防护林网面积、绿地面积及抽水、截水、蓄水等设施，防止耕地受到风蚀、水蚀等影响，保证耕地灌溉，提升耕地产能，保护生态环境。

景观效益评价过程中选取田块平整度指数、景观破碎度指数变化率和景观丰富度指数变化率 3 个指标。根据测算结果及各评价因子隶属度矩阵分析得知，景观效益评价结果 V_1 占 22.89%，V_2 占 23.86%，V_3 占 42.89%，V_4 占 10.36%。结合最大隶属原则，确定景观效益整体评价等级为良好（$70 \leqslant Q_4 < 85$）。结合研究区实际情况说明，通过田、水、路、林、渠等土地综合整治工程建设，项目区整体景观效益增加，耕地利用由“零散”逐渐变为“集中连片”，田块规整度提高，有利于耕地资源节约、集约、高效利用。

通过综合计算得知，项目区综合效益评价结果 V_1 占 23.37%，V_2 占 39.93%，V_3 占 27.70%，V_4 占 9.00%。结合最大隶属原则，确定综合效益整体评价等级为良好（$70 \leqslant Q < 85$）。总的来说，通过土地综合整治，改善了项目区农业生产条件，提高了耕地质量和农业产值，项目区经济效益、社会效益、生态效益和景观效益有所提升。但横向比较可以得出，社会效益评价分值最高，景观效益、经济效益评价分值较低，各效益之间发展不平衡。这也说明，该项目区景观效益和经济效益有较大的提升空间，建议在后期“旱改水”“高标准农田建设”等工程项目中，注重经济、社会、生态和景观的有机统一，确保

各效益之间平衡发展。

4 结论与讨论

4.1 结论

本研究从4个方面选取14个评价因子构建了土地综合整治项目实施效益评价体系,基于AHP-PCE评价模型测算了研究区土地整治综合效益分值,并划分了评价等级,得到以下结论:

(1)项目区土地综合整治中经济、社会、生态和景观层面指标效益分值表现为“一优三良”,土地整治综合效益分值为82.10,综合评价等级为良好。其中,经济效益分值为80.74,综合评价等级为良好;社会效益分值为86.15,综合评价等级为优秀;生态效益分值为82.12,综合评价等级为良好;景观效益分值为79.41,综合评价等级为良好。

(2)项目区整体经济效益提升,人均年纯收入较建设前提升46.30%,项目区总产值较建设前提升20.04%,新增耕地率达到0.03%。项目区整体社会效益提升,居民满意度较建设前提升96.80%、灌排设施较建设前提升75.6%,农业生产便捷程度提高,居民满意度增加。项目区整体生态效益提升,耕地灌溉保证率和土地垦殖率变化明显,较建设前分别提升75.60%和97.30%。项目区景观水平进一步优化。通过工程建设,田块更为集中、规整,景观破碎度降低,丰富度指数提升,景观效益显著。

(3)项目区土地综合整治经济效益、社会效益、生态效益和景观效益均有所提升,但较建设前相比,景观效益和经济效益提升不明显,各效益间发展不平衡,与项目预期建设目标存在一定差距。

4.2 讨论

土地综合整治是保证我国粮食安全和农业生产的重要手段,土地综合整治效益评价是评定土地整治目标实施程度的有效方法,其评价指标的选取是保证研究科学性和合理性的关键所在。本研究从土地综合整治项目入手,建立了符合项目区的土地综合整治效益评价指标体系,在大尺度范围内应根据项目所在区实际情况,科学合理选取代表性指标,因地制宜建立适宜的评价指标体系。目前,我国土地综合整治效益评价正处于发展阶段,随着时间和空间的变化,不同区域、不同项目综合效益评价指标的选取是动态变化的。因此,全面的、科学的、完整的土地综合整治效益评价框架理论及指标体系是我们目前研究的关键所在。

参考文献

[1]刘楚杰. 土地整治项目后评价研究:以宁乡县两个村为例[D]. 长沙:湖南农业大学, 2017.

Liu Chujie. Study on the post evaluation of land consolidation project: a case study of two villages in Ningxiang county [D]. Changsha: Hunan Agricultural University, 2017.

[2]杨俊. 土地整治项目实施后效益评价研究:以湖北省长阳土家族自治县土地整治项目为例[D]. 武汉:中国地质大学, 2012.

Yang Jun. Study on post benefit evaluation of land consolidation project implementation: a case study of land consolidation project in Changyang Tujia autonomous county, Hubei province [D]. Wuhan: China University of Geosciences, 2012.

[3]杨俊, 王占岐, 金贵, 等. 基于AHP与模糊综合评价的土地整治项目实施后效益评价[J]. 长江流域资源与环境, 2013, 22(8):1036-1042.

Yang Jun, Wang Zhanqi, JinGui, et al. Post: benefit evaluation of land consolidation project implementation based on AHP and FUZZY comprehensive evaluation [J]. Resources and Environment in the Yangtze Basin, 2013, 22(8): 1036-1042.

[4]李鸣慧, 芦艳艳, 刘桢. 基于河南省土地整治规划的土地整治效益评价研究:2016—2020年[J]. 农业与技术, 2019, 39(3):172-174.

[5]夏方舟，杨雨濛，严金明. 中国国土综合整治近40年内涵研究综述：阶段演进与发展变化[J]. 中国土地科学，2018，32(5)：78-85.

Xia Fangzhou, Yang Yumeng, Yan Jinming. The connotation research review on integrated territory consolidation of China in recent four decades: staged evolution and developmental transformation [J]. China Land Science, 2018, 32(5): 78-85.

[6]程传胜. 残次林地土地综合整治耕地质量等级评价研究：以高圈村等5村土地开发项目为例[J]. 国土资源情报，2021(12)：18-22.

Cheng Chuansheng. Study on the cultivated land quality grade evaluation of comprehensive management of substandard forestland: a case study of the land development project of 5 villages [J]. Land and Resources Information, 2021(12): 18-22.

[7]吴家龙，苏少青，杨远光，等. 全域土地综合整治调查评估指标体系构建：以广东省为例[J]. 中国国土资源经济，2022，35(2)：77-82,89.

Wu Jialong, Su Shaoqing, Yang Yuanguang, et al. Construction of the index system of comprehensive land consolidation for investigation and evaluation in the whole region: a case study of Guangdong province [J]. Natural Resource Economics of China Economics, 2022, 35(2): 77-82,89.

[8]梁敏. 农村土地整治社会效益评价研究[D]. 南京：南京农业大学，2011.

Liang Min. Study on social benefits evaluation of rural land reclamation [D]. Nanjing: Nanjing Agricultural University, 2011.

[9]范业婷，金晓斌，张晓琳，等. 乡村重构视角下全域土地综合整治的机制解析与案例研究[J]. 中国土地科学，2021，35(4)：109-118.

Fan Yeting, JinXiaobin, Zhang Xiaolin, et al. Mechanism analysis and case study of comprehensive land consolidation from the perspective of rural restructuring [J]. China Land Science, 2021, 35(4): 109-118.

[10]丁继辉，朱永增，张梦婷，等. 基于"三生"视角的土地整治综合效益评价[J]. 人民黄河，2020，42(10)：86-91.

Ding Jihui, Zhu Yongzeng, Zhang Mengting, et al. Comprehensive benefit evaluation of land consolidation based on perspective of production, living and ecology [J]. Yellow River, 2020, 42(10): 86-91.

[11]刘姝驿，杨庆媛，何春燕，等. 基于层次分析法(AHP)和模糊综合评价法的土地整治效益评价：重庆市3个区县26个村农村土地整治的实证[J]. 中国农学通报，2013，29(26)：54-60.

Liu Shuyi, Yang Qingyuan, He Chunyan, et al. The benefit assessment of land consolidation based on analytical hierarchy process (AHP) and Fuzzy synthetic evaluation: case of 26 villages in Chongqing [J]. Chinese Agricultural Science Bulletin, 2013, 29(26): 54-60.

[12]潘珍妮，刘应宗，高红江. 基于模糊物元的农村土地综合整治效益评价[J]. 电子科技大学学报(社科版)，2012，14(3)：66-69,75.

Pan Zhenni, Liu Yingzong, Gao Hongjiang. Benefit evaluation of rural land comprehensive consolidation based on Fuzzy Matter-Element [J]. Journal of University of Electronic Science and Technology of China (Social Sciences Edition), 2012, 14 (3): 66-69,75.

[13]程文仕，乔蕻强，刘志，等. 基于RAGA-PPC模型的土地整治综合效益评价：以甘肃省庆阳市15个土地整治项目为例[J].水土保持通报，2016，36(4)：257-261,268.

Cheng Wenshi, QiaoHongqiang, Liu Zhi, et al. Assessment of comprehensive benefits from i and remediation based on RAGA-PPC model: a case study of 15 projects in Qingyang city of Gansu province [J]. Bulletin of Soil and Water Conservation, 2016, 36(4): 257-261,268.

[14]张桂川. 农村土地综合整治效益评价指标体系探究[J]. 农村经济与科技，2017，28(6)：16-17.

[15]原伟鹏，刘新平，马耘秀，等. 基于多指标综合评价的农村土地整治项目效益评价[J]. 南方农村，2016，32(4)：35-39.

[16]周婷，胡庆国. 土地综合整治项目效益评价指标体系研究[J]. 科技信息，2013(3)：165-166.

Zhou Ting, Hu Qingguo. Land comprehensive project performance evaluation index system research[J]. Science & Technology Information, 2013(3): 165-166.

[17]王栋. 基于层次分析与模糊综合评价的土地整治项目实施后效益评价:以安定区苏家岔流域基本农田整理项目为例[J]. 安徽农业科学, 2014, 42(27): 9566-9569.

Wang Dong. The expost benefit evaluation of land consolidation project based on AHP and Fuzzy comprehensive evaluation: a case of basic farmland consolidation project in Sujiacha river basin, Anding zone [J]. Journal of Anhui Agricultural Sciences, 2014, 42(27): 9566-9569.

[18]刘永利. 保定市土地综合整治效益研究[D]. 保定:河北农业大学, 2013.

Liu Yongli. Benefit evaluation research of land comprehensive improvement of Baoding [D]. Baoding: Hebei Agricultural University, 2013.

[19]周婷. 土地综合整治项目效益评价研究[D].长沙:长沙理工大学, 2013.

Zhou Ting. Land comprehensive improvement project benefit evaluation research [D]. Changsha: Changsha University of Science & Technology, 2013.

[20]徐玲玲. 土地整治效益评价方法研究:以洛川县为例[D]. 西安:长安大学, 2012.

Xu Lingling. The research of method on the benefit of land reclamation on Luo chuan county [D]. Xi'an: Chang'an University, 2012.

[21]齐梅, 王燕. 土地整理综合效益评价:以山东省章丘市为例[J]. 资源与产业, 2008(4):62-66.

Qi Mei, Wang Yan. A case study of Zhangqiu city: assessment of comprehensive benefit of land consolidation [J]. Resources & Industries, 2008(4): 62-66.

[22]吕添贵, 吴次芳, 李冠, 等. 基于耦合模型的赣江水源保护区城市发展与水资源环境关系研究[J]. 水土保持研究, 2014, 21(6):271-277.

Lv Tiangui, Wu Cifang, Li Guan, et al. Study on the relationship between the urban development and the environment of water resources in Ganjiang river based on coupling model [J]. Research on Soil and Water Conservation, 2014, 21(6): 271-277.

[23]朱建军. 层次分析法的若干问题研究及应用[D].沈阳:东北大学, 2005.

Zhu Jianjun. Research on some problems of the analytic hierarchy process and its application [D]. Shenyang: Northeastern University, 2005.

[24]敖佳, 张凤荣, 李何超, 等. 川西平原全域土地综合整治前后耕地变化及其效益评价[J]. 中国农业大学学报, 2020, 25(8):108-119.

Ao Jia, Zhang Fengrong, Li Hechao, et al. Changes and benefit evaluations of cultivated land before and after comprehensive land consolidation in West Sichuan Plain [J]. Journal of China Agricultural University, 2020, 25(8): 108-119.

[25]曾璇, 胡笑涛. 土地整理项目综合效益评价研究:以靖边县土地整理项目为例[J]. 农业与技术, 2020, 40(15):165-166.

Zeng Xuan, Hu Xiaotao. Study on comprehensive benefit evaluation of land consolidation project: taking Jingbian county land consolidation project as an example [J]. Agriculture and Technology, 2020, 40(15): 165-166.

[26]丁继辉, 朱永增, 张梦婷, 等. 基于"三生"视角的土地整治综合效益评价[J]. 人民黄河, 2020, 42(10): 86-91.

Ding Jihui, Zhu Yongzeng, Zhang Mengting, et al. Comprehensive benefit evaluation of land consolidation based on perspective of production, living and ecology [J]. Yellow River, 2020, 42(10): 86-91.

[27]曾涛, 吕婧, 史佳良, 等. 基于多层 AHP-FCE 评价模型的土地整治重大工程效益评价研究[J]. 江西农业大学学报, 2017, 39(6):1234-1243.

Zeng Tao, Lu Jing, Shi Jialiang, et al. Benefit analysis and evaluation of the key land consolidation and readjustment projects [J]. Acta Agriculturae Universitatis Jiangxiensis, 2017, 39(6): 1234-1243.

[28]刘志铭，王瑷玲. 基于模糊综合评价的土地整治效益研究:以诸城市黑龙沟等村土地整治项目为例[J]. 山东农业大学学报(社会科学版)，2020，22(2):77-82,148.

Liu Zhiming, Wang Ailing. Land consolidation efficiency evaluation based on Fuzzy comprehensive method: a case study of villages such as Hei Long-gou in Zhucheng City,Shandong Province [J]. Journal of Shandong Agricultural University (Social Science Edition), 2020, 22(2): 77-82,148.

[29]刘楚杰，喻瑶，李帅. 农村土地整治项目综合效益后评价研究:以湖南省宁乡县2个村为例[J]. 山西农业科学，2017，45(9):1543-1548.

Liu Chujie, Yu Yao, Li Shuai. Study on Post-evaluation of comprehensive benefit in rural land renovation project:take two villages in Ningxiang county of Hunan province as an example [J]. Journal of Shanxi Agricultural Sciences, 2017, 45(9): 1543-1548.

[30]鲁胜晗，朱成立，周建新，等. 生态景观视角下土地整治的生态效益评价[J]. 水土保持研究，2020，27(5): 311-317.

Lu Shenghan, Zhu Chengli, Zhou Jianxin, et al. Evaluation on ecological benefit of land remediation from the perspective of ecological and landscape [J]. Research on Soil and Water Conservation, 2020, 27(5): 311-317.

本文曾发表于2022年《中国农机化学报》第43卷第12期

生物有机肥施用量对土壤有机碳组分及酶活性的影响

孟婷婷[1,2,3], 杨亮彦[1,2,3], 孔　辉[1,2,3],刘金宝[1,2,3]

(1.陕西省土地工程建设集团有限责任公司,陕西 西安 710075;
2.陕西地建土地工程技术研究院有限责任公司,陕西 西安 710075;
3.自然资源部退化及未利用土地整治工程重点实验室,陕西 西安 710075)

【摘要】以𪣻土为试材,采用室内盆栽培养方法,设置 0(T0)、10 g ·kg^{-1}(T1)、20 g ·kg^{-1}(T2)、30 g ·kg^{-1}(T3)、40 g ·kg^{-1}(T4)5 个生物有机肥施用水平,研究了生物有机肥施用量对土壤有机碳(SOC)、溶解性有机碳(DOC)、易氧化性有机碳(ROC)、微生物生物量碳(MBC)含量及过氧化氢酶、蔗糖酶活性的影响,以期为陕西关中𪣻土区生物有机肥的施用提供参考依据。结果表明:生物有机肥可显著增加 SOC、DOC、ROC、MBC 及过氧化氢酶、蔗糖酶的含量。在整个培养期内,SOC 含量随着时间的延长而降低,DOC、ROC、MBC 及蔗糖酶和过氧化氢酶活性随着时间的延长而增加。Pearson 相关分析表明,除蔗糖酶外,有机碳组分间及有机碳组分与过氧化氢酶有极显著正相关关系。生物有机肥的施用提高了𪣻土有机碳活性组分含量和酶活性,但各组分的最大值对施用量的反映不同。

【关键词】生物有机肥;有机碳组分;酶活性;施用量;Pearson 相关分析

Effect of Application Amount of Bio-organic Fertilizer on Soil Organic Carbon Components and Enzyme Activities

Tingting Meng[1,2,3], Liangyan Yang[1,2,3], Hui Kong[1,2,3],Jinbao Liu[1,2,3]

(1.Shaanxi Provincial Land Engineering Construction Group Co. ,Ltd. ,Xi' an,Shaanxi 710075;
2.Institute of Shaanxi Land Engineering and Technology Co. , Ltd. , Xi' an,Shaanxi 710075;
3. Key Laboratory of Degraded and Unused Land Consolidation Engineering,
Ministry of Land and Resources, Xi' an,Shaanxi 710075)

【Abstract】Taking lou soil as the test material, the indoor potted cultivation method was used, 0 (T0), 10 g ·kg^{-1} (T1), 20 g ·kg^{-1}(T2), 30 g ·kg^{-1}(T3), 40 g ·kg^{-1}(T4), 5 biological organic fertilizer fertilization level were set up, the biological organic fertilizer seems to organic carbon (SOC), dissolved organic carbon (DOC), radily oxidizable organic carbon (ROC), microbial biomass carbon (MBC) contents and the influence of catalase, sucrase activity were studied ,in order to provide evidence for the application of biological organic fertilizer in the agricultural area of China. The results showed that the contents of SOC, DOC, ROC, MBC, catalase and sucrase could be significantly increased by bio-organic fertilizer. With the extension of culture time, SOC content decreased, while the content of other carbon components and enzyme activity increased. Pearson correlation

基金项目:陕西省科技成果转移与推广计划(2021CGBX-03);长安大学重点科研平台开放基金(300102351501)。

analysis showed that, except for sucrase, there were significant positive correlations between organic carbon components and catalase. The application of biological organic fertilizer increased the content of active organic carbon components and enzyme activities in soil. However, the maximum values of each component varied in response to the application rate.

【**Key words**】Bio-organic fertilizer; Organic carbon composition; Enzyme activity; Application Amount; Pearson correlation analysis

化肥的过量施用带来的环境污染、土壤退化等问题已严重影响到我国农业的高质量和可持续发展，是我国农业发展亟待解决的现实问题。从已取得的成绩来看，各种有机肥配施减少化肥的危害，有机肥逐步替代化肥在这一过程中发挥至关重要的作用。

有机肥是以动物粪便及植物枯落物为底物进行发酵生产出的一种肥料。而生物有机肥是在此基础上添加某些特定有益微生物而衍生出的一种新型有机肥。有机肥自身含有丰富的有机质和养分，合理施用有机肥可以显著改善贫瘠土壤的理化及生物性状[1-2]，从而解决贫瘠土壤肥效差、漏水漏肥的问题，提高土壤肥力，最终实现农业增产[3-4]。与普通有机肥相比，生物有机肥中还含有丰富的酵母菌、乳酸菌、纤维素分解菌和固氮菌等有益微生物和功能菌，这些微生物的存在能有效改善根系土壤环境，对土壤的改良效果更显著[5-6]。

土壤有机碳(SOC)表征土壤肥力大小和质量[7]。根据有机碳的存在方式和稳定性，一般可将有机碳分为活性有机碳和稳定性有机碳。土壤活性有机碳以溶解性有机碳(DOC)、易氧化性有机碳(ROC)和微生物量碳(MBC)为主。土壤中移动快、稳定性差、易氧化、易矿化，并对植物和土壤微生物活性较高的活性有机碳常用来作为评价土壤质量的指标[8]。多数研究表明，施用有机肥在提高土壤有机碳及其组分方面有积极作用[9-10]，Yousefi 等[11]的研究表明，随着有机肥用量增加，土壤团聚体稳定性、有机碳、水溶性有机碳(DOC)和弱酸水解化合物的含量进一步增加。Li 等[12]的研究结果表明，有机肥施用增加了有机碳和活性组分(DOC 和 LFOC)。土壤有机碳活性组分对外界环境变化十分敏感，可及时反映施肥方式及施肥量对土壤有机碳的动态变化[13]。此外，有机碳组分之间的转化与土壤中各种酶有着密切的关系。土壤酶活性反映了土壤中进行各种生物化学过程的动力和强度，许多研究均表明土壤酶活性可以作为土壤肥力、土壤质量和微生物活性的重要指标[14-15]。

塿土作为陕西关中地区主要生产土壤之一，占陕西省耕地面积的 34.1%[16]，面积较少，人地矛盾突出。为增加塿土有机质含量及提高土壤肥力，实现作物增产增效，常常采用有机肥与无机肥配合、土壤改良剂与有机无机肥料配施的方法。但是，随着不同有机肥的施入，塿土有机碳组分与酶活性的变化趋势还不清晰。因此，该研究以关中地区塿土为试验对象，通过盆栽培养模拟试验，在塿土中施入不同质量生物有机肥，培养一定周期，研究分析塿土中有机碳组分和土壤酶活性变化，以期为合理管理塿土、维持和提高其可持续生产能力提供科学依据。

1　材料与方法

1.1　试验材料

试验采用陕西富平褚塬村(北纬 34°42′，东经 108°11′)农田塿土作为供试土壤，其基本理化性质见表 1。装土容器采用口径为 17.5 cm、高度为 13.2 cm 的花盆，生物有机肥采用河北德沃多肥料有限公司生产的生物有机肥(884)，有机质含量≥40%；微生物活菌≥1 亿/g，养分含量见表 1。试验在秦岭野外监测中心站(北纬 34°08′，东经 107°53′)温室大棚内进行。

1.2　试验方法

试验采用室内盆栽培养方法，设置 0(T0)、10 g $\cdot$ kg^{-1}(T1)、20 g $\cdot$ kg^{-1}(T2)、30 g $\cdot$ kg^{-1}(T3)、40 g $\cdot$ kg^{-1}(T4)5 个生物有机肥施用水平，每个处理 3 次重复，共设置 15 个盆栽试验。每个花盆装土+有机肥(过 2 mm 筛)500 g，将有机肥按设定量与土混合均匀加入花盆中，加水至田间持水量的 70%，使土壤充分湿润，置于温室内培养，培养期间采用称重法将土壤水分保持到田间持水量的 70%。

表 1　供试塿土和生物有机肥基本理化性质

Table 1　Experiment of soil and biological organic fertilizer basic principles

试验材料 Experiment material	pH	有机质 Soil organic matter /($g \cdot kg^{-1}$)	全氮 Soil total nitrogen /($g \cdot kg^{-1}$)	全磷 Soil total phosphorus /($g \cdot kg^{-1}$)	有效磷 Soil available phosphorus /($mg \cdot kg^{-1}$)	速效钾 Soil available potassium /($mg \cdot kg^{-1}$)
塿土 Lou soil	7.3	16.21	0.51	0.74	12.3	124
生物有机肥 Bio-organic fertilizer	7.5	8.56	0.17	0.15	5.6	64

1.3　项目测定

在培养的第 90、120、150、180 天,用直径 2 cm 的土钻在每个花盆内,均匀采取 3 个土样,混合后作为一个土壤样品,自然风干后,研磨过 2 mm 和 0.149 mm 筛。土壤有机碳(SOC)含量采用高锰酸钾外加热法[17]测定,溶解性有机碳(DOC)含量采用总有机碳分析仪[17]测定,易氧化性有机碳(ROC)含量采用 K_2MnO_4氧化法-比色法[17]测定,微生物生物量碳(MBC)含量采用氯仿熏蒸-K_2SO_4溶液浸提法[17]测定,蔗糖酶活性采用硫代硫酸钠滴定法[18]测定,过氧化氢酶活性采用高锰酸钾滴定法[18]测定。

1.4　数据分析

采用 Excel 2010 软件进行数据处理,采用 SPSS 22.0 软件进行单因素方差分析和 Person 相关分析,并用 LSD 法检验差异显著性($P<0.05$),采用 Origin 2018 软件绘图。

2　结果与分析

2.1　生物有机肥施用量对活性有机碳组分的影响

由图 1a、b 可知,生物有机肥不同施用量下,SOC 和 DOC 含量表现为 T4>T3>T2>T1>T0,即施用生物有机肥均可提高 SOC、DOC 含量,并且随着生物有机肥施用量的增加而增加,SOC 含量除 150 d 外,DOC 含量除 90 d 和 180 d 外,其他各处理间均有显著性差异($P<0.05$)。培养至 180 d 时,与 T0 处理相比,T1~T4 处理,SOC 和 DOC 含量分别增加了 42%~154%、38%~173%。T4 处理时,SOC 和 DOC 含量最高,分别为 25.05 $g \cdot kg^{-1}$和 74.89 $mg \cdot kg^{-1}$。随着培养时间的延长,SOC 含量呈降低趋势,各处理与培养初期(90 d)相比,SOC 含量降低了 6%~17%,T4 处理下,培养初期(90 d)与其他培养时间段有显著性差异,其他处理下,各培养时间段差异不显著。DOC 含量随着培养时间的延长呈增加趋势,各处理与培养初期(90 d)相比,DOC 含量增加 3%~13%,T1、T2 处理时,DOC 含量在各培养时间段有显著性差异($P<0.05$),但在 T3、T4 处理时,仅在培养前期(90 d)和后面各培养时间段有显著性差异。

由图 1c、d 可以看出,生物有机肥不同施用量下,ROC 和 MBC 含量表现为 T3>T4>T2>T1>T0,即施用生物有机肥可同时增加 ROC 和 DOC 含量,且随着生物有机肥用量的增加,ROC 和 MBC 含量呈先增加后降低的趋势,超过 T3 处理时,ROC 和 MBC 含量下降,但大于 CK,不同处理间在各培养阶段均有显著性差异($P<0.05$)。培养至 180 d 时,与 T0 处理相比,T1~T4 处理,ROC 和 MBC 含量分别增加了 38%~173%、4%~9%。随着培养时间的延长,ROC 和 MBC 含量增加,与培养初期(90 d)相比,ROC 和 MBC 含量分别增加了 15%~64%、3%~25%,且各处理在培养末期(180 d)与培养初期(90 d)有显著性差异,ROC 含量在 150 d 时,含量最大,MBC 含量在 180 d 时,含量最大。

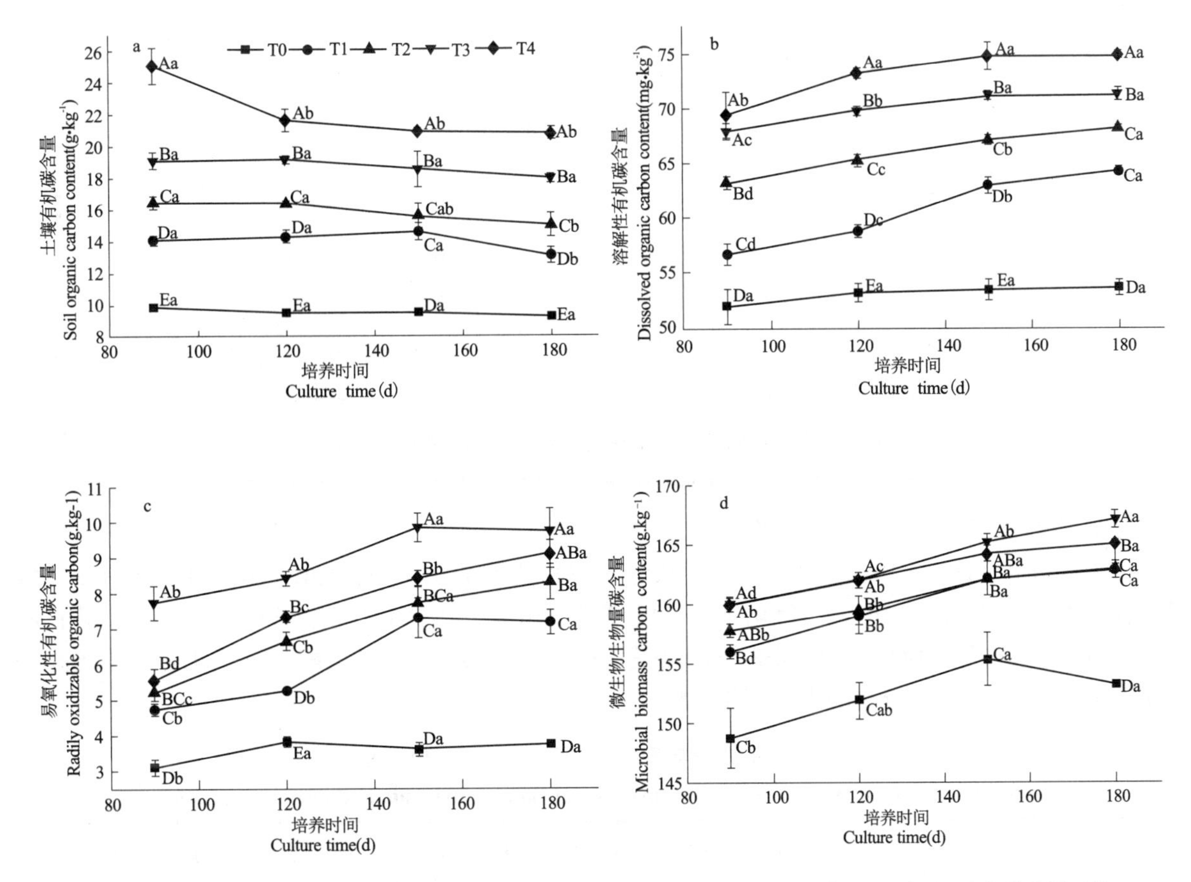

注：小写字母表示同一处理下各培养时间之间的差异显著，大写字母表示同一培养时间段各处理之间的差异显著。

Note: Lowercase letters indicate the significant differences among the incubation times under the same treatment, and the uppercase letters indicate the significant differences among the treatments in the same incubation period.

图1 生物有机肥施用量对SOC、DOC、ROC和MBC的影响

Fig. 1 Effects of application amount of bio-organic fertilizer on SOC, DOC, ROC and MBC

2.2 生物有机肥施用量对酶活性的影响

由图2a可知，生物有机肥不同施用量下，过氧化氢酶活性的大小表现为T2>T1>T4>T3>T0，即施用生物有机肥可增加过氧化氢酶的活性，但是，过氧化氢酶活性随着生物有机肥施用量的增加，呈先增加后降低的趋势，T2处理时，含量最高为6.89 mg ·g^{-1}，超过T2处理时，过氧化氢酶活性下降，但大于CK，不同处理间在各培养阶段均有显著性差异（$P<0.05$）。培养至180 d时，与T0处理相比，T1～T4处理的过氧化氢酶活性增加了1%～12%。随着培养时间的延长，过氧化氢酶活性呈增加趋势，T3、T4处理时，各培养时间段间有显著性差异（$P<0.05$），与培养初期（90 d）相比，过氧化氢酶活性增加了4%～10%。

由图2b可以看出，生物有机肥不同施用量下，蔗糖酶活性的大小表现为T2>T1>T4>T3>T0，即施用生物有机肥可增加蔗糖酶的活性，与过氧化氢酶一样，蔗糖酶活性也随着生物有机肥用量的增加，呈先增加后降低的趋势，T2处理时，含量最高为0.97 mg ·g^{-1}，超过T2处理时，蔗糖酶活性下降，但大于CK，各培养时间段T0、T1、T2、T4间蔗糖酶活性有显著性差异（$P<0.05$）。培养至180 d时，与T0处理相比，T1～T4处理的蔗糖酶活性分别增加了1%～23%。各培养时间段，T3、T4处理间差异不显著，培养前120 d，T1、T2、T4处理间有显著性差异。随着培养时间的延长，蔗糖酶活性增加，与培养初期（90 d）相比，蔗糖酶活性增加了1%～31%，T1、T2处理时，培养前期与后期蔗糖酶活性有显著性差异。

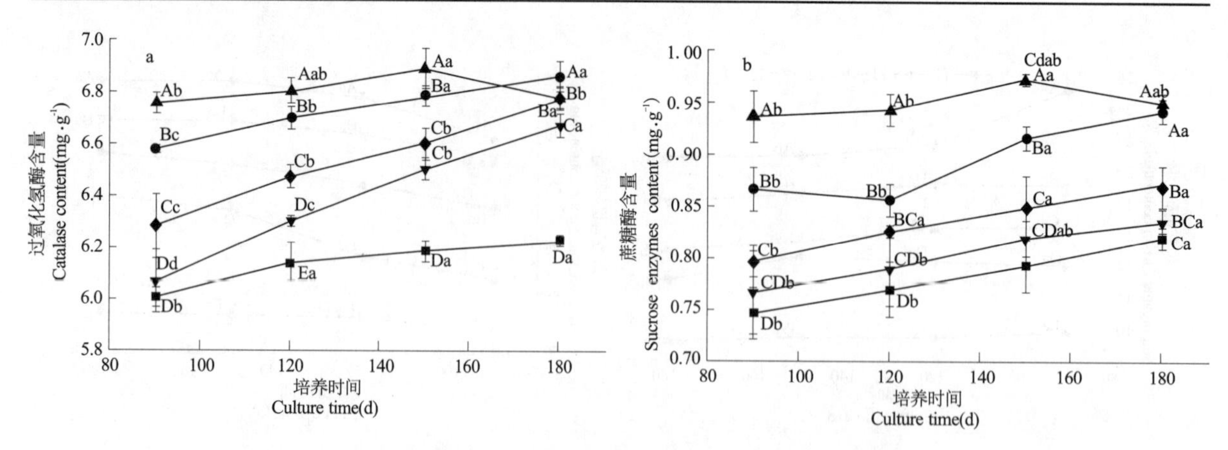

图 2　生物有机肥施用量对过氧化氢酶和蔗糖酶的影响

Fig.2　Effects of application amount of bioorganic fertilizer on catalase and sucrase

2.3　有机碳组分与酶活性的相关关系

由表 2 可知,SOC 与 DOC、ROC、MBC 呈极显著正相关关系;DOC 与 ROC、MBC、过氧化氢酶呈极显著正相关关系;ROC 与 MBC、过氧化氢酶呈极显著正相关关系;MBC 与过氧化氢酶呈极显著正相关关系;过氧化氢酶与 DOC、ROC、MBC、蔗糖酶呈极显著正相关关系;蔗糖酶与过氧化氢酶呈极显著正相关关系,与有机碳组分之间无显著相关性。

表 2　有机碳组分与酶活性之间的 Pearson 相关关系

Table 2　Pearson correlation between organic carbon components and enzyme activities

指标 Inders	土壤有机碳 SOC	溶解性有机碳 DOC	易氧化性有机碳 ROC	微生物生物量碳 MBC	过氧化氢酶 CAT	蔗糖酶 Sucrase
土壤有机碳 SOC	1					
溶解性有机碳 DOC	0.884**	1				
易氧化性有机碳 ROC	0.645**	0.886**	1			
微生物生物量碳 MBC	0.670**	0.874**	0.915**	1		
过氧化氢酶 CAT	0.205	0.415**	0.480**	0.586**	1	
蔗糖酶 Sucrase	−0.132	0.023	0.109	0.193	0.771**	1

注:**0.01 水平下的显著性(双尾)。

Note: ** Significant correlations at the 0.01 probability level (two-tailed).

3　讨论

该研究结果表明,施用生物有机肥处理的土壤有机碳活性组分(SOC、DOC、ROC 和 MBC)含量均得到明显提高,这与张鹏鹏等[19]的研究结果一致,即生物有机肥将外源有机碳带入土壤,促进了土壤有机碳各组分的提高,同时,生物有机肥作为一种功能菌肥料,含有多种有益微生物,可提高土壤中微生物种群和数量。各处理下,SOC 含量随着时间的延长呈降低趋势,DOC、ROC 和 MBC 含量随着培养时间的延长呈增加趋势,其原因是,随着时间的延长,微生物肥料被分解,惰性碳被分解成活性碳,所以有机碳活性组分增加,而有机碳总量减少。土壤有机碳含量受气候影响很大,尤其是土壤的水热状况和物理、化学过程的性质与强度,通常来讲,作为一个大地区,其土壤有机质含量应该是相对稳定的。但是作为小范围,有机肥资源充足,且有机肥的有机质含量较高的情况下,土壤施用的有机肥越多,土壤有机碳含量提高得也越高,因此在小范围的盆栽土壤上,生物有机肥施用量越多,SOC 和 DOC 含量也越高。T3 处理时,ROC 和 MBC 含量最高,这表明采用 T3 处理的生物有机肥用量,可以更好地改善土壤肥力状况。

该研究中,各处理下过氧化氢酶和蔗糖酶活性均高于对照,说明施用生物有机肥能够显著提高土壤酶活性,这与前人的研究结果一致[20-21]。其原因是生物有机肥是以农家肥为主的一种腐熟高效肥料,施入土壤后,不仅能改善土壤理化性状,而且能使土壤中的碳氮比适宜,并促进作物和土壤微生物的生长,从而增加土壤酶的活性[22],而当有机肥用量超过处理 T2 水平时,土壤中营养元素配比失衡,导致土壤离子之间产生拮抗作用,因此又表现出蔗糖酶、过氧化氢酶的降低[23]。

土壤有机碳组分相互转化过程本身主要是由微生物主导的生物和生物化学反应过程,微生物分泌的酶能够催化有机碳分解和合成反应[24]。因此,研究有机碳组分与酶活性的相关关系可间接了解土壤有机碳与微生物两者之间的相互关系。该研究中,SOC、DOC、ROC、MBC、过氧化氢酶之间有极显著正相关关系,蔗糖酶除与过氧化氢酶有显著正相关关系外,与其他皆无显著关系,这与曲成闯等[24]的研究结果相一致。

4 结论

生物有机肥的施用显著提高了塿土有机碳活性组分含量和土壤酶活性,30 g $\cdot$ kg^{-1}施用量时,塿土中 ROC 和 MBC 含量最高,土壤酶活性较高。在陕西关中地区塿土实际生产中可适当配施一定量的生物有机肥,提高塿土的肥力和生产力。该研究是室内模拟培养试验,土壤环境范围小,且肥料充足,与实际大田土壤环境有一定的差别,因此在实际生产中生物有机肥最佳施用量可能会与该结果有一定的差距。

参考文献

[1] 宁川川, 王建武, 蔡昆争. 有机肥对土壤肥力和土壤环境质量的影响研究进展[J]. 生态环境学报, 2016, 25(1): 175-181.

[2] 王利辉. 不同来源有机肥及其配合施用对土壤性质的影响[D]. 长春: 吉林农业大学, 2007.

[3] 周伟红. 有机肥对土壤培肥和作物产量的影响[D]. 长沙: 湖南农业大学, 2007.

[4] 张运龙. 有机肥施用对冬小麦-夏玉米产量和土壤肥力的影响[D]. 北京: 中国农业大学, 2017.

[5] 沈德龙. 我国生物有机肥的发展现状及展望[J]. 中国土壤与肥料, 2007(6): 1-5.

[6] 何蔚娟. 生物有机肥料生产问题研究[J]. 陕西农业科学, 2018, 64(6): 90-92.

[7] HE Y T, ZHANG W J, XU M G, et al. Long-term combined chemical and manure fertilizations increase soil organic carbon and total nitrogen in aggregate fractions at three typical cropland soils in China[J]. Science of The Total Environment. 2015, 532: 635-644.

[8] DUMALE J W, MIYAZAKI T, NISHIMURAT, et al. CO_2 evolution and short-term carbon turnover in stable soil organic carbon from soils applied with fresh organic matter[J]. Geophysical Research Letters, 2009(36): 143-153.

[9] 李娟, 葛磊, 曹婷婷, 等.有机肥施用量和耕作方式对旱地土壤水分利用效率及作物生产力的影响[J]. 水土保持学报, 2019, 33(2): 121-127.

[10] 张莉, 李玉义, 逄焕成, 等. 玉米秸秆颗粒还田对土壤有机碳含量和作物产量的影响[J]. 农业资源与环境学报, 2019, 36(2): 160-168.

[11] YOUSEFI M, HAJABBASI M, SHARIATMADARI H. Cropping system effects on carbohydrate content and water-stable aggregates in a calcareous soil of Central Iran[J]. Soil and Tillage Research, 2008, 1(101): 57-61.

[12] LI J, WU X, GEBREMIKAEL M T, et al. Response of soil organic carbon fractions, microbial community composition and carbon mineralization to high-input fertilizer practices under an intensive agricultural system.[J]. PloS One, 2018, 4(13): e195144.

[13] 寇智瑞, 周鑫斌, 徐宸, 等. 有机无机肥配施对黄壤烟田有机碳组分的影响[J]. 土壤, 2020, 52(1): 195-201.

[14] 田小明, 李俊华, 危常州, 等. 不同生物有机肥用量对土壤活性有机质和酶活性的影响[J]. 中国土壤与肥料,2012(1): 26-32.

[15] 路磊, 李忠佩, 车玉萍. 不同施肥处理对黄泥土微生物生物量碳氮和酶活性的影响[J]. 土壤,2006(3): 309-314.

[16] 谢钧宇，彭博，王仁杰，等.长期不同施肥对塿土大团聚体中有机碳组分特征的影响[J]. 植物营养与肥料学报，2019，25(7)：1073-1083.

[17]鲍士旦.土壤农化分析[M].北京：中国农业出版社，1982.

[18]关松荫.土壤酶及其研究法[M].北京：农业出版社，1986.

[19]张鹏鹏，刘彦杰，濮晓珍，等.秸秆管理和施肥方式对绿洲棉田土壤有机碳库的影响[J].应用生态学报，2016，27(11)：3529-3538.

[20]王冬梅，王春枝，韩晓日，等.长期施肥对棕壤主要酶活性的影响[J].土壤通报，2006，37(2)：263-267.

[21]BASTIDA F, KANDELER E, HERNANDEZ T, et al. Long-tern effect of municipal solid waste amendment on microbial abundance and humus-associated enzymed activities under semiarid conditions[J]. Microb. Ecol，2008，55：651-661.

[22]施娴，刘艳红，张德刚，等.猪粪与化肥配施对植烟土壤酶活性和微生物生物量动态变化的影响[J].土壤，2015，47（5）：899-903.

[23]吴平江，夏叶，薛勇，等.生物有机肥对绿洲温室黄瓜产量、品质及土壤酶活性的影响[J].中国水土保持，2019(4)：53-56.

[24]曲成闯，陈效民，张志龙，等.施用生物有机肥对黄瓜连作土壤有机碳库和酶活性的持续影响[J].应用生态学报，2019，30(9)：3147-3154.

本文曾发表于2022年《北方园艺》第17期

MulTiple Linear Regression and Correlation Analysis of Yield of Summer Maize Varieties in Shaanxi Province, China

Qili Hao

(Shangluo Branch of Shaanxi Land Engineering
Construction Group Co., Ltd.,Shaanxi Xi'an 710075, China)

【**Abstract**】The main factors affecting the yield of mid-ripening maize varieties were clarified to provide reference for breeding and production of summer maize varieties in Guanzhong Plain, Shaanxi Province. According to the relevant agronomic trait data of 11 new maize varieties planted in the Guanzhong Plain summer maize regional trial introduced in Shaanxi Province in 2019, analysis of the main characters of the new maize varieties were carried out using the multiple linear regression method. It can provide reference for the selection of maize agronomic traits. The agronomic traits that were significantly related to yield of summer maize varieties in Shaanxi Province were 100-grain weight, number of grains in row, seed rate, empty stalk rate and ear length, etc. When selecting maize varieties, under the premise of ensuring proper 100-grain weight and ear length, one should pay attention to the selection of varieties with high seed rate and low empty stalk rate. In the breeding of high-yielding maize varieties, lodging resistance should be enhanced, the rate of empty stalks should be reduced, the rate of seed production should be increased, 100-grain weight and ear diameter should be appropriately increased, and ear height should be reduced.

【**Keywords**】 Maize varieties; Yield; Multiple linear regression; Correlation analysis

Introduction

With the rapid development of science, technology and economy, the demand for corn in China is also increasing year by year (Liu *et al.* 2017, Axel *et al.* 2021, Singh 2021). With the increasing uncertainty of international factors such as geopolitics, countries and people are paying more and more attention to food security. The conflict between the two major grain-producing countries, Russia and Ukraine, has further increased the country's sense of worry about grain production (Xiang *et al.* 2020). Yield remains an important goal pursued by corn breeders, national and individual producers at this stage. Shaanxi Province is the main summer corn producing area of China and a commercial grain production base (Ali *et al.* 2019). The selection of suitable crop varieties is of great significance for the promotion of crop planting, the increase of grain production and the increase of farmers' income (Qiu *et al.* 2015, Nafziger and Srinivasan 2021, Shahhosseini 2021). Shaanxi Province has a vast territory. Beishan and Qinling Mountains divide Shaanxi Province into three natural regions: the northern Shaanxi Plateau, the central Guanzhong Plain, and the southern Qinba Mountains (Nascimento 2020, Yue *et al.* 2021). The Guanzhong Plain has been known as a granary since ancient times (He and Zhou 2016, Forte *et al.* 2017, Ruis 2021, Simão and Johnston 2021, Lu *et al.* 2022). In the process of maize planting, it is necessary to study the main agronomic traits and their correlation with yield, to clarify the main factors affecting yield, and to provide a basis for selecting superior hybrids (Fernando 2020, Hou *et al.* 2020, Tao 2020). Analysis of the correlation between the main agronomic traits and yield of 11 maize varieties introduced in Guanzhong Plain, Shaanxi Province in 2019, would provide a reference for the selection of maize agronomic traits, and promote suitable planting varieties, increase crop yield, provide strong support to increase farmers' income.

Materials and methods

The experiment was set up in Ducun Town, Fuping County, Weinan City. The area has a semi-arid continental climate, and the rainfall distribution is extremely uneven throughout the year, mostly concentrated in July to September, accounting for 49% of the annual rainfall, and other seasons are relatively dry. The dry and wet seasons are distinct, and the dry season is longer than the wet season. Especially in spring, it is windy and rainy, and the evaporation is large. The annual evaporation is 1000~1300 mm, which is 2.0 to 2.3 times of the rainfall. The maximum evaporation (189.5 mm) was in June, the smallest (44. 9 mm) was in December. The frost-free period is 225 days, and the annual average temperature is 13. 4 ℃, the highest temperature in summer is 41.8 ℃, the lowest temperature in winter − 22 ~ −10 ℃, the regional conditions are suitable for the cultivation of crops such as corn and wheat.

The experimental materials 11 corn varieties are suitable for the Guanzhong summer planting area in the sixth batch of application for introduction of corn varieties in Shaanxi Province in 2019: AnNong 591, NonGDan 476, JinAo 608, WoFeng 9, ZhengDan 5176, RuiFeng 168, ZhengDa 1473, BoFa 707, XiWang 3088, XianYu 1653, XianYu 1568.

Corn planting density was 60000~67500 plants/hm^2, randomly block arranged, in 5 rows of blocks, 6 m row length, and the middle 3 rows of actual receipts in each plot were used to calculate the yield. The experiment was repeated three times. The eight main agronomic traits selected for the study were ear length (cm), ear diameter (cm), bald tip length (cm), number of grains per row (grain/row), weight of 100-grains (g), and empty culm rate (%), inversion rate (%) and seeding rate (%).

According to the requirements of multiple linear regression theory, the yield of 11 corn hybrid combinations and 8 related agronomic traits were subjected to regression analysis, and *t*-test was used for significance test. Data analysis were performed using the R statistical programming language.

Results and discussion

Since the dimensions of the inspection indicators are inconsistent, it is necessary to initialize the original data to make it dimensionless, so as to facilitate the comparison of data between different dimensions. The Z-Score standardization method was used to standardize the data. Applicable to large samples of more than 30, the mean of the standardized indicators is 0, and the variance is 1. The formula is as follows:

$$y_{ij} = \frac{x_{ij} - \overline{x}_j}{S_j} \quad (i = 1,2,\cdots,n;j = 1,2,\cdots,m) \tag{1}$$

Among them, $\overline{x}_j$ is the average value of index j, S_j is the variance of index j.

In order to investigate the dependence between the yield and agronomic traits of different corn hybrid varieties, it is necessary to establish a multiple linear regression model.

$$Y=\beta_0+\beta_1X_1+\beta_2X_2+\beta_3X_3+\cdots+\beta_pX_p+\varepsilon \tag{2}$$

Among them, β_0 is the regression constant, β_1, β_2, β_3, $\cdots$, β_p are the overall regression parameters. When $p=1$, formula (2) is called a univariate linear regression model, and when $p\geqslant 2$, it is called a multiple linear regression model. ε is a random error and obeys the distribution of $\varepsilon \sim N(0,\sigma^2)$.

The estimation method of parameter β adopts the least square estimation method, and its objective function is to minimize:

$$Q(\beta)=\sum_{i=1}^{n}\| y_i - x_i\beta \| \tag{3}$$

Among them, Y stands for yield, X_1 stands for 100-grain weight, X_2 stands for the number of grains in row, X_3 stands for seed rate, X_4 stands for empty stalk rate, X_5 stands for ear length, X_6 stand for ear tip-barrenness, X_7 stands for ear diameter, and X_8 stands for fold rate. When getting n sets of observation data (X_1, X_2, X_3, X_4, X_5, X_6, X_7, X_8, y_i) ($i=1,2,3,\cdots,n$), The linear regression model can be expressed as

$$\begin{cases} y_1=\beta_0+\beta_1X_{11}+\beta_2X_{12}+\beta_3X_{13}+\cdots+\beta_8X_{18}+\varepsilon_1 \\ y_2=\beta_0+\beta_1X_{21}+\beta_2X_{22}+\beta_3X_{23}+\cdots+\beta_8X_{28}+\varepsilon_2 \\ \cdots\cdots \\ y_n=\beta_0+\beta_1X_{n1}+\beta_2X_{n2}+\beta_3X_{n3}+\cdots+\beta_8X_{n8}+\varepsilon_n \end{cases} \tag{4}$$

It is written in matrix form as $Y=X\beta+\varepsilon$, Usually X is called the design matrix, β is the regression coefficient vector.

From the established multiple regression model and the regression coefficients that have been obtained, the fitting test of the entire regression equation was carried out, and the R^2 test was used. Goodness of fit was calculated based on decomposing the square of the total deviation. The formula for calculating the sum of squares of the total deviation is:

$$SST=SSE+SSR \tag{5}$$

Among them, SSE is the residual sum of squares, SSR is the regression sum of squares, and SST is the total squared deviation. Calculated as follows:

$$SSE=\sum_{i=1}^{n}(y_i-\hat{y}_i)^2 \tag{6}$$

$$SSR=\sum_{i=1}^{n}(\hat{y}_i-\bar{y})^2 \tag{7}$$

$$SST=\sum_{i=1}^{n}(y_i-\bar{y})^2 \tag{8}$$

Among them, $\bar{y}$ is the mean value of the observed value of the sample, $\hat{y}$ is the estimated value, and the coefficient of determination R^2 is the ratio of the sum of squares of the regression to the sum of squares of the total deviation. The calculation formula is:

$$R^2=\frac{SSR}{SST}=\frac{\sum_{i=1}^{n}(\hat{y}_i-\bar{y})^2}{\sum_{i=1}^{n}(y_i-\bar{y})^2} \tag{9}$$

R^2 reflects the goodness of fit of the regression line to the data, and its value is between [0,1]. The closer R^2 is to 1, the better the fitting result of the regression equation, and the closer R^2 is to 0, the worse the fitting of the regression equation.

The t-test is to test whether each regression coefficient in the regression model is significant, so that only those factors that have a significant impact on the dependent variable are retained in the model. When testing, first calculate the statistic t_1, and then look up the t distribution table according to the given significant level α and the degrees of freedom n-k-1 to obtain the critical value t_α or $t_{\alpha/2}$, if $t > t - \alpha$ or $t_{\alpha/2}$, then the regression coefficient. There is a significant difference between β_i and 0, on the contrary, there is no significant difference. The calculation formula is:

$$t = \frac{\bar{x} - \mu}{\sigma_x / \sqrt{n - 1}} \tag{10}$$

where $\bar{x}$ is the sample mean, μ is the population mean, σ_x is the sample standard deviation and n is the sample size.

Using the data in Table 1 to solve the reference factors, the results in Table 2 can be obtained.

Table 1 Normalization results of raw data

Tested varieties	Yield	Ear length	Ear thickness	Ear tip-barrenness	Number of grains	100-grain weight	Empty stalk rate	Inversion rate	Seed yield
AnNong 591	0.042	1.222	2.083	0.498	0.909	0.392	0.278	6.354	−0.280
NongDan 476	0.044	0.047	3.819	−2.488	−0.909	0.283	−1.667	−4.696	−1.212
JinAo 608	0.026	0.047	5.556	−2.488	0	−0.215	0	−0.552	1.119
WoFeng 9	−0.032	0.987	2.083	−2.488	0	−0.121	3.889	−3.315	−1.678
ZhengDa 5176	−0.004	2.162	−8.333	−1.493	0.909	−0.261	−0.833	3.591	−0.435
RuiFeng 168	0.034	0.517	−1.389	6.468	0.909	−0.775	−2.222	−1.934	0.031
ZhengDa 1473	0.063	1.222	3.819	3.483	0	−0.246	−0.278	−3.315	0.653
BoFa 707	−0.028	−3.008	−3.125	−4.478	−0.909	−0.012	0.278	3.591	2.517
XiWang 3088	−0.028	−1.128	0.347	1.493	−0.455	0.283	−1.111	3.591	−1.523
XianYu 1653	0.026	0.047	5.556	−2.488	0	−0.215	0	−0.552	1.119
XianYu 1568	0.037	0.041	2.612	2.221	−1.234	−0.351	0.575	−0.513	2.312

Table 2 Model summary

Model	R	R square	Adjusted R square	Std. error of the estimate	Change statistics					Durbin-Watson
					R-squared change	F value change	df_1	df_2	Significant F value change	
1	0.992	0.984	0.946	0.0182	0.984	15.461	8	2	0.0013	2.138

Table 2 showed how well the model fits. As can be seen from the table, the complex correlation coefficient of the model is 0.992, the determination coefficient is 0.984, the adjusted coefficient of determination was 0.946 and the standard error of the estimated value was 0.0182, and the Durbin-Watson test statistic was 2.138. When DW≈2, the residuals were independent.

Table 3 showed ANOVA results for the model. It was observed that the value of the F statistic corresponding to the model was 15.462, and the p value was 0.0013. When the significance level is 0.05, it can be considered that there is a linear relationship between crop yield and different traits.

Table 3 ANOVA

Model	Sum of squares	df	Mean square	F	Sig.
Regression	0.011	8	0.001	15.462	0.0013
Residual	0.000	2	0.000		
Total	0.011	10			

Table 4 Parameter estimation results of multiple linear regression

Regression coefficients	Parameter estimates	Standard error	t	p-value
β_0	33.675	3.8724	3.527	0.0487
β_1	0.1891	0.0137	3.172	<0.001
β_2	0.0321	0.0021	1.026	<0.001
β_3	0.0328	0.0073	2.186	<0.001
β_4	0.2157	0.0342	1.823	<0.001
β_5	0.1274	0.0225	3.524	<0.001
β_6	0.3251	0.0673	−3.434	<0.001
β_7	0.1325	0.0101	−3.052	0.0321
β_8	0.0132	0.0031	4.871	0.0211

According to the p value results presented in the table, it can be found that the parameters β_0, β_7, β_8 were significant at the 95% confidence level. Parameters β_1, β_2, β_3, β_4, β_5, β_6 are significant at the 99 per cent confidence level. The results showed that among the 8 main agronomic traits selected, factors such as 100-grain weight, number of grains per row, seed rate, empty stalk rate, ear length and other factors had a significant impact on yield, and were closely related to yield. Through the analysis, the multiple linear regression model about the yield was obtained.

$$Y = 0.1891X_1 + 0.0321X_2 + 0.0328X_3 + 0.2157X_4 + 0.1274X_5 + 0.3251X_6 + 0.1325X_7 + 0.0132X_8 + 33.675 \quad (11)$$

By calculating the correlation coefficient between yield and 8 main agronomic traits, it was found that the correlation between different traits and yield was quite different. The correlation between 100-kernel weight and yield exceeded 0.9, the strongest correlation. The number of grains per row and seedling emergence rate in these two traits have a strong correlation with yield, ranging from 0.8 to 0.9, while the correlation between empty culm rate, ear length, bald length, ear diameter and fold rate and yield was less than or equal to close to 0.8, indicating relatively weak correlations between these traits and yield.

Table 5 Correlation coefficients between yield and different traits of maize varieties

Tested varieties	Yield	Ear length	Ear thickness	Ear tip-barrenness	Number of grains	100-grain weight	Empty stalk rate	Inversion rate	Seed yield
AnNong 591	0.78	0.872	0.902	0.728	0.923	0.947	0.798	0.729	0.88
ongDan 476	1	0.825	0.622	0.614	0.945	0.909	0.867	0.769	1
JinAo 608	0.994	0.83	0.624	0.795	0.946	0.995	0.879	0.793	0.996
WoFeng 9	0.804	0.864	0.63	0.793	098	0.915	0.86	0.717	0.803
ZhengDa 5176	0.658	0.834	0.736	0.621	0.942	0.935	0.837	0.607	0.858
RuiFeng 168	0.895	0.847	0.693	0.726	0.838	0.949	0.78	0.508	0.897
ZhengDa 1473	0.783	0.828	0.55	0.786	0.932	0.925	0.793	0.777	0.881
BoFa 707	0.792	0.818	0.733	0.608	0.931	0.924	0.835	0.736	0.892
XiWang 3088	0.792	0.818	0.733	0.708	0.931	0.924	0.835	0.736	0.892
XianYu 1653	0.794	0.817	0.766	0.688	0.901	0.915	0.821	0.752	0.819
XianYu 1568	0.837	0.804	0.742	0.701	0924	0.901	0.875	0.713	0.872

Eleven varieties of summer corn were selected in the Guanzhong Plain of Shaanxi Province in 2019 to conduct multiple linear regression analysis of main agronomic traits. Results showed that the agronomic traits closely related to yield of summer maize varieties in Shaanxi Province were 100-grain weight, number of grains per row, seed rate, empty stalk rate and ear length, etc that are the main influencing factors of hybrid breeding. It showed that when selecting corn varieties, on the premise of ensuring proper 100-grain weight and ear length, one should pay attention to selecting varieties with high seed rate and low empty stalk rate. In the selection of high-yielding maize varieties, lodging resistance should be enhanced, the rate of empty stalks should be reduced, the rate of seed production should be increased, the 100-grain weight and ear diameter should be appropriately increased, and the ear height should be reduced.

Acknowledgements

This study is financially supported by Shaanxi Province Natural Science Basic Research Program (2022JQ-457); Basic scientific research funds of central universities (300102351502); The Internal Scientific Research Projects of Shaanxi Provincial Land Engineering Construction Group (DJNY2021-30, DJNY2022-25).

References

Ali S, Xu Y Y, Ma X C, Ahmad I, Manzoor J Q M, Akmal M, Hussain Z, Arif M and Cai T 2019. Deficit irrigation strategies to improve winter wheat productivity and regulating root growth under different planting patterns. Agr. Water Manage. 219(1): 1-11.

Axel H, Giuliana D, Francisco G A and Fernandez F A 2021. Sustainable production of microalgae in raceways: Nutrients and water management as key factors influencing environmental impacts. Cleaner Product. 287(1): 125005.

Fernando S G 2020. Investigation of azospirillum brasilense inoculation and silicon application on corn yield responses. Soil Sci. Plant Nutri. 102(1): 1-13.

Forte A, Fiorentino N, Fagnano M and Fierro A 2017. Mitigation impact of minimum tillage on CO_2 and N_2O emissions from a Mediterranean maize cropped soil under low-water input management. Soil Tillage Res. 166(1): 167-178.

He H J, Zhou G S 2016. Climate-associated distribution of summer maize in China from 1961 to 2010. Agricul. Ecosyst. Environ. 232(1): 326-335.

Hou H J, Han Z D, Yang Y Q, Abudu S and Li Z C 2020. Soil CO_2 emissions from summer maize fields under deficit irrigation. Environ. Sci. Pollu. Res. 27(1): 4442-4449.

Liu Y, Qin Y, Ge Q, Dai J and Chen Q 2017. Reponses and sensitivities of maize phenology to climate change from 1981 to 2009 in Henan Province, China. Geograph. Sci. 27: 1072-1084.

Lu J S, Ma L H, Hu T T, Geng C M and Yan S C 2022. Deficit drip irrigation based on crop evapotranspiration and precipitation forecast improves water-use efficiency and grain yield of summer maize. Sci. Food Agricul. 102(2):653-663.

Nascimento L O 2020. Effect of reduced spacing on relationship of physiological, morphological and productive traits of corn yield. Australian J. Crop Sci. 14(7): 1202-1208.

Nafziger E D, Srinivasan V 2021. Nitrogen deficiency and corn yield with delayed N application. Agron. J. 113(4): 3665-3674.

Qiu Q Y, Wu L F, Ouyang Z, Li B B, Xu Y Y, Wu S S and Gregorich E G 2015. Effects of plant-derived dissolved organic matter (DOM) on soil CO_2 and N_2O emissions and soil carbon and nitrogen sequestrations. Appl. Soil Ecol. 96(1): 122-130.

Ruis S J 2021. Corn residue baling and grazing impacts on corn yield under irrigated conservation tillage systems. Agron. J. 113(3): 2387-2397.

Shahhosseini M 2021. Corn yield prediction with ensemble CNN-DNN. Front. Plant Sci. 12(1): 709008.

Simão L M and Johnston A M 2021. Winter wheat residue impact on soil water storage and subsequent corn yield. Agron. J. 113(1): 276-286.

Singh R 2021. Effects of broiler litter application rate and time on corn yield and environmental nitrogen loss. Agron. J. 114(1): 415-426.

Tao L. 2020. DeepCropNet: a deep spatial-temporal learning framework for county-level corn yield estimation. Environ. Res. Lett. 15(3): 034016.

Xiang K Y, Yi L, Robert H and Hao F 2020. Similarity and difference of potential evapotranspiration and reference crop evapotranspiration-a review. Agricul. Water Manage. 232(1): 106043.

Yue H W, Jiang X V, Wei J W, Xie J L, Chen S P, Peng H C and Bu J Z 2021. A study on genotype × environment interactions for the multiple traits of maize hybrids in China. Agron. J. 113(6): 4889-4899.

本文曾发表于2022年《Bangladesh J. Bot.》第51卷第3期

关中地区土地生态安全调整策略研究

朱 坤

(中陕高标准农田建设集团有限公司,陕西 咸阳 712000)

【摘要】近年来,随着国家经济的快速发展,对土地资源的需求量也与日俱增,而在土地生态安全管理上仍旧存在着问题,以致土地资源利用率难以提高。常见问题:草地退化严重、土壤污染问题,而加强对土地生态安全管理,目的就是促进国家可持续发展。本文就关中地区土地生态安全管理存在问题展开分析,并提出几点有效的调整策略,希望以此来帮助土地快速恢复并提高其利用率,这对于国家经济快速增长有着帮助作用。

【关键词】关中地区;土地生态;利用率

0 引言

目前,我国土地生态环境正面临着较大的威胁,而土地资源是国家经济发展不可缺少的重要资源,相关部门应通过科学合理管理提高资源利用率。但是仍旧存在着问题,使得土地破坏较为严重。故而笔者根据实际情况,对目前关中地区土地生态安全管理进行研究,并提出具体的应对策略,希望以下观点可以为相关人员提供参考,进而解决土地生态安全管理上的问题,并为我国的长远发展奠定土地资源基础,这对于人类发展也至关重要。

1 关中地区土地生态安全管理存在问题

1.1 草地退化严重

如今,土地生态安全管理已经成为国家比较关注的一个话题,因为国家的发展离不开土地资源的支持,而随着时间的推移,土地资源数量也在逐渐减少,同时土地生态安全管理过程中也存在着一些问题,其中关中地区草地退化现象比较严重,由于耕地和林地面积减少,将会导致水土流失问题的出现。其中水土流失会对周边的居民生活带来一定的影响,因为水土流失有着一定的危险性,随着水土流失情况的不断加重,将会威胁到周边的农业,使得农业生产产量下降[1]。

另外,我国城镇化建设的不断推进,房屋建筑对耕地的占用也变得越来越多,这也会使土地生态资源变得越来越少,而我国林业面积相对于其他国家来讲比较少,一旦因耕地与林地面积减少,会使得草地退化严重,从而引发水土流失问题。

1.2 土壤污染问题

我国地大物博,随着时间的推移,土地资源也变得越来越少,其中关中地区的土地生态安全管理效率比较低,这也是由于内部存在着问题。其中土壤污染问题愈发突出,而致使土壤污染的原因有很多。其一,工业时代的到来,企业生产规模逐步扩大,在日常生产过程中所排放的污水量也变得越来越多,倘若没有按照要求去排放污水,就会对周边的水资源造成污染,也会间接影响周边居民的生活。而工业废水中含有有毒有害的气体,一旦没有妥善处理,就会对大气环境、水体以及土壤造成污染[2]。其二,随着人们生活条件的改善,污水量也变得越来越多,倘若这些污水没有达到排放标准就随意排放,也会引起周边土壤污染,这都是目前我国土地生态安全管理所面临的一个问题,迫切需要去处理。

1.3 土地利用率低

土地生态安全问题对于人类的发展也有着影响,土地遭到破坏将会威胁人类的生存。其中土地生态安全管理中所面临的问题越来越多,关中地区土地资源利用率相对来讲比较低,此类问题出现的原因,有以下几点:第一,土地资源规划不合理,相关部门并没有对现有的土地资源进行合理的分配,这不仅亵渎了土地资源的利用价值,也会使得土地生态安全管理效率低下。如今,土地生态安全一直是我国最关注的话题,其与人类的发展也有着直接关系,但是因有关部门在土地开发和建设过程中过于注重外表的华丽,忽视了对土地资源的科学合理规划,使得土地资源利用率较低[3]。第二,因相关机制缺失,也会导致土地破坏问题没有得到及时的处理,再加上土地生态安全相关法律法规不够健全,针对那些对土地资源产生破坏的行为举止,也没有加大惩处力度,这样都不利于我国未来的发展。

2 关中地区土地生态安全调整策略研究

2.1 树立保护土地生态安全意识

关中地区土地生态安全管理中存在问题,所以为了解决问题,我们需要做出以下的调整:第一,应树立保护土地生态安全意识,这就需要加大管理力度,以便于为后续土地生态安全管理工作提供便利。现如今,土地生态安全成为人们所探讨的焦点,因为人类的发展也离不开土地资源,所以我们每一个人都应树立正确的土地生态安全保护意识,这不仅能够提高土地资源利用率,也能够为我国长远发展奠定良好的基础。第二,相关部门也要加大宣传,让更多的人参与到土地生态安全管理行列中,同时也可以鼓励一些企业参与其中,通过不断宣传,增强各行各业土地生态安全意识,通过制订详细的方案,对现有土地生态资源进行合理规划。第三,要处理好人与土地之间的关系,通过合理规划,既可以实现提高土地资源使用效率的目的,又能够促进国家更好的发展。我国地大物博,为了解决以往土地生态安全管理上的问题,加大宣传力度是非常有必要的,进而呼吁更多的人参与到其中,从而为人类的可持续发展创造良好的条件,切实将土地生态安全管理工作落到实处。

2.2 严格控制对土地生态的规划

我国土地生态安全管理存在不足,所以关中地区相关部门还需要及时做出调整,这就需要严格控制对土地生态的规划,这是保障土地生态系统完整性和平衡性的必然前提,需要相关部门给予高度的重视。土地生态安全,近几年来成为人们所探讨的焦点,它也是一个全新的主题,但是以往由于诸多因素的影响,使得我国的土地生态安全问题较为严峻,所以进行土地开发时,有关部门应严格审核开发力度和开发强度,只有通过对其有效规划,才能够实现提高管理水平的目的[4]。同时对于那些无关紧要的规划也要严厉禁止,这是从源头上避免问题产生的重要内容。另外,为了保障土地生态平衡,政府部门也需要发挥职能作用,比如:通过派遣相关工作人员,进一步对土地规划和开发进行严格的监督和管理,针对其存在的不足之处能够及时提出解决方案,这有助于提高土地生态资源质量。

2.3 加强对土壤污染的管控力度

关中地区要想解决以往土地生态安全管理中的问题,还需要加强对土壤污染方面的管理力度。而土壤之所以出现问题,一部分原因来自工业,另一部分原因来自人们生活。如今,伴随着工业时代的到来,工业废水的排放会产生很多有毒有害的气体,一旦长期排放,就会对周边的土壤、空气及水体造成污染,也会间接影响到周边居民生活以及农业生产。所以,我们还需要通过加大对土壤污染的管理力度,进而改善土壤环境,确保农作物的正常生长。因为人类的生存也离不开农作物的支持,所以注重对土壤污染的管理是非常重要的[5]。另外,政府也要加大对工业废水排放治理力度,针对随意排放污水的企业应加大惩治力度,进一步给其他工厂警示,让他们能够自觉做好污水处理。比如:可以安装无毒处理装置,如此一来,确保污水排放达到国家标准,这不仅可以避免土壤污染,又能够保障土地生态的平衡。

2.4 健全土地生态安全相关法律

关中地区要想保障土地生态安全管理工作顺利开展,还需要健全相关法律法规。因为土地生态管理离不开法律制度的支持。当前时代背景之下,城市化建设不断推进,同时土地生态资源也遭到了严重的破坏,这个问题若不及时处理,将会阻碍国家经济发展。所以,我们还需要结合实际健全土地生态安全相关法律,通过利用法律,对肆意破坏土地生态安全的人员进行约束,有关部门也要加大惩治力度,让他们能够认识到自身行为举止的错误性,并自觉做到遵纪守法。现如今,土地生态安全一直是国家所关注的问题,不及时处理,对于国家的发展也会造成制约,所以为了促进国家发展,更加需要利用法律武器来对相关人员进行约束,这不仅可以发挥法律法规的作用和价值,又能为我国发展奠定土地资源基础。另外,为了保障土地生态安全管理效率得到提高,关中地区相关部门也要做好本职工作,利用现有法律法规增强对破坏土地生态安全人员的惩治力度,这也有助于提高土地资源的使用效率。

2.5　加强关中地区土地生态恢复

现今阶段,人们面临着土地生态问题,比如:草地退化、土壤污染,这些问题的存在将影响国家的发展。首先,关中地区有关部门需要加强土地生态恢复,而土地自身就有着一定的恢复能力,当然,这也需要经过漫长的时间。所以,为了更好地帮助土地进行生态恢复,相关部门还需要投入更多的精力。比如:利用恢复生态学中的相关理论并与实践相结合,这不仅可以逐步提高土地生态恢复能力,也能为人类生存保驾护航。现如今,土地资源日渐匮乏,这也是城市化建设的不断推进所致,使得用地面积越来越大,而我国的土地生态安全问题也随之而来,故而土地生态恢复工作势在必行,这可以避免土地沙漠化、土地盐碱化等问题的出现。除此之外,相关部门也可以积极发展生态工程技术,这也是一种帮助土地恢复原有面貌的方式,同时也可以通过退耕还林的方式来提高土地生态恢复能力,从而解决我国土地生态安全问题,这也有助于改善当地人们的生活环境。

2.6　建立完善土地生态补偿机制

要想满足土地生态安全管理工作需求,关中地区相关部门还需要健全法律法规,这也显得尤为重要。首先,国家政府部门应发挥职能作用,通过建立完善的土地生态补偿机制,避免土地破坏不可逆等问题的出现。与此同时,针对已经出现的土地生态安全问题,有关部门也要及时采取应对措施去处理。这也可以避免水土流失、土地沙漠化等问题。当然,国家也要提供充足的资金、技术手段,以便于为土地生态安全管理工作提供更多支持,政府也应为其提供资金支持,这都有助于改善土地的实际情况。近几年来,随着时代的发展,城市化建设也逐渐推进,同时占地面积越来越大,而在施工过程中也会对周边的生态环境造成破坏,使得土地生态安全问题愈发严重,这将威胁到人们的生存,所以建立完善的生态补偿机制刻不容缓。另外,为了使土地生态可以得到有效改善,相关部门也可以通过发起集体投资及个人投资等活动来获得更多的资金,随后将这部分资金投入环境保护上,其中人类的生活离不开生态环境,倘若生态环境遭到破坏,将难以提高生活质量,故而建立完善的生态补偿机制很关键,最终实现提高土地生态安全管理水平的目的。

3　结语

综上所述,现今阶段,关中地区土地生态安全管理工作中面临一些挑战,比如:土壤污染问题、土地利用率较低,这些问题的存在将会制约地区经济发展。所以,为了解决问题,我们应树立保护土地生态安全意识并严格对土地进行规划,同时也要加强对土壤污染的管控力度且健全相关法律法规,在必要的前提下还需要建立完善的土地生态补偿机制,更要加强土地生态恢复工作,进而保障土地生态安全,促进人类可持续发展。

参考文献

[1]王一山,张飞,陈瑞,齐亚霄,刘长江. 乌鲁木齐市土地生态安全综合评价[J]. 干旱区地理,2021,44(2):427-440.

[2]许思维,徐育红,赵小汎. 辽宁省土地生态安全评价及生态管理[J]. 安徽农业科学,2017,45(10):200-202,248.

[3]陆威,赵源,冯薪霖,朱艳婷,常毅. 土地资源生态安全研究综述[J]. 中国农学通报,2016,32(32):88-93.

[4]石美珍. 我国土地生态的安全管理问题分析[J]. 北京农业,2015(3):194.

[5]袁丽娟. 土地资源生态安全评价研究综述[J]. 西部资源,2013(3):179-181.

[6]薛亮,任志远.基于格网 GIS 的关中地区生态安全研究[J]. 地理科学,2011(9):64.

本文曾发表于 2022 年《中文科技期刊数据库(全文版)自然科学》第 6 卷

旱塬区新增耕地质量和粮食产能影响因素分析
——以占补平衡项目为例

何振嘉[1],贺　伟[2],李刘荣[3],张　俊[1,4],李　河[1]

(1.陕西省土地工程建设集团有限责任公司,陕西 西安 710075;2.西安理工大学水利水电学院,
陕西 西安 710048;3.中交一公局西北工程有限公司,陕西 西安 710199;
4.陕西地建关天投资建设有限公司,陕西 宝鸡 721000)

【摘要】为揭示渭北旱塬区耕地占补平衡项目对粮食产能的影响,以冬小麦和春玉米为量化标准,采用Logistic回归模型,对典型的渭北旱塬区千阳县2017年占补平衡项目新增耕地粮食产能影响因素进行研究;同时,对项目实施前后新增耕地土壤理化性质变化进行分析。结果表明:通过占补平衡项目实施使新增耕地等级提升了1~2个等级,且新增耕地中水浇地面积占比越大,新增耕地等别越高,单位面积上水浇地较旱地的粮食产能提高率增加25%。随着项目实施年限增加,土壤理化性质不断得到改善,总孔隙度、大团聚体、水稳性团聚体和微团聚体均显著变大($P<0.05$),土壤结构破坏率显著减小($P<0.05$),抗侵蚀能力显著提高($P<0.05$)。土壤有机质质量比显著提高($P<0.05$),土壤养分状况不断改善。新增耕地的面积、等别、基础设施、单位面积投资、年限等因素均与粮食产能之间呈现不同程度的正相关关系;各因素对新增耕地粮食产能影响程度,按因素排序由大到小表现为等别、面积、基础设施、年限、投资。

【关键词】渭北旱塬;新增耕地;占补平衡项目;土壤理化性质;粮食产能

Analysis of Factors Influencing Quality of Newly Increased Cultivated Land and Grain Productivity in Arid Highland Area
—Taking Occupation Complementary Balance Project as an Example

Zhenjia He[1], Wei He[2], Liurong Li[3], Jun Zhang[1,4], He Li[1]

(1. Shaanxi Provincial Land Engineering Construction Group Co., Ltd., Xi'an, Shaanxi 710075, China;
2.School of Water Resources and Hydropower, Xi'an University of Technology, Xi'an, Shaanxi 710048;
3.CCCC First Highway Northwest Engineering Co., Ltd., Xi'an, Shaanxi 710199, China;
4. Shaanxi Dijian Guantian Investment Construction Co., Ltd., Baoji, Shaanxi 721000, China)

【Abstract】In order to reveal the impact of the arable land complementary balance project on the grain production capacity of the Weibei platform area, this paper takes winter wheat and spring corn as the quantitative standards, and adopts the Logistic regression model to study the factors influencing grain productivity of newly added farmland for typical arable land complementary balance project in Weibei platform area Qianyang County in 2017. And at the same time, the changes in the physical and chemical properties of the newly added arable land are analyzed before and after

基金项目:国家重点研发计划资助项目(2021YFD1900700);陕西省土地工程建设集团有限责任公司内部创新团队项目(DJTD-2022-5);陕西省土地工程建设集团有限责任公司内部科研项目(DJNY2022-11,DJNY2022-15,DJNY2022-54)。

the implementation of the project. The results show that the implementation of complementary balance project increased the level of new cultivated land by 1 to 2 levels, and the larger the proportion of irrigated land in the newly increased cultivated land, the higher the level of newly added cultivated land. The increase rate of grain production capacity per unit area of irrigated land is 25% higher than that of dry land. With the increase in the implementation years of the complementary balance project, the physical and chemical properties of soil have been continuously improved, the total soil porosity, large aggregates, water-stable aggregates and micro-aggregates all increase significantly ($P<0.05$), the damage rate of soil structure was significantly reduced ($P<0.05$), and the erosion resistance was significantly improved ($P<0.05$). The quality ratio of soil organic matter increased significantly ($P<0.05$), and the status of soil nutrients continued to improve. Factors such as the area, the level, infrastructure, the investment per mu and the number of years of newly increased cultivated land, which all show positive correlations with grain production capacity. The degree of impact on grain production capacity of newly increased cultivated land is shown in the order of factors from large to small: the level, land area, infrastructure, years, investment per mu.

【Keywords】Weibei dryland; Newly increased cultivated land; Complementary balance project; Soil physical and chemical properties; Grain productivity

耕地是保障粮食生产和粮食安全的重要资源,而保证耕地数量不减少和耕地质量不降低是中央农村工作会议提出的明确要求。根据中国耕地质量评定结果,耕地普遍存在以下问题:耕地等别较低、粮食产能较低、土壤有机质不足、基础地力差,且受到不同程度的农业污染、工业污染、土壤污染以及地下水污染等,总体耕地质量较差。在国家山水林田湖草一体化背景下,生态型土地整治项目已成为解决荒地、坡耕地水土流失和土壤侵蚀等现象的有效方式,对于耕地保水保墒和增加作物产量、整合零散破碎的土地资源以及保障中国粮食安全具有重要的推进作用。

占补平衡项目通过机械施工整平未利用土地,有利于消除地形坡度、降低地表径流;同时,配合田间管理和保护性耕种等农艺措施,能有效涵养水土,特别是有利于土壤养分、有机质含量提高。此外,通过相应的水土保持措施,提高耕地抗侵蚀程度,是保障中国粮食安全的一项重要手段[1-2]。通过占补平衡项目增加耕地数量、提高耕地地力水平,已初步取得成效,以粮食产能为核心的耕地占补平衡新模式也逐渐被推广。土地整治对新增耕地质量具有显著影响。谢向向等[3]研究了土地整治项目对粮食产能的贡献情况,结果表明,粮食产能随着新增耕地面积增加而增加,但受投资强度影响,低投入产生的新增耕地质量较差,会对粮食产能产生负面影响。芦艳艳[4]等研究表明,通过土地整治项目实施,改善了基础设施条件,在对表层土壤质地、有机质含量以及灌溉保证率提升的基础上,有效提高了土地自然等别和利用等别。陈正发等[5]针对云南坡耕地障碍因子进行综合治理,重点对田面坡度、土壤有机质、灌溉保证率、有效磷、速效钾、pH 值进行针对性改善,有效提升了耕地质量。赵海乐等[6]对内蒙古河套-土默川平原区耕地整治区域进行划定,优先开展对土地质量提升潜力较高并且集中连片的地块进行整治。邵雅静等[7]对宝鸡地区耕地质量障碍因子进行分析,由土层深度、有机质含量和灌溉保证率等因素限制耕地质量提升的土地,超过了宝鸡市耕地总面积的 66.11%。

新增耕地粮食产能影响因素较多,目前已有较多研究从耕地数量变化[8]、耕地资源变化[9]以及耕地质量变化[10]对粮食影响方面进行了研究,也提出了通过科技投入、生产条件改善和资源配置优化等措施[11]提升耕地质量的方法和路径,但其研究多集中于中国东北、华北等粮食生产大省以及平原地区,对于旱塬沟壑区新增耕地对粮食产能影响因素相关研究很少。因此,开展旱塬区粮食产能影响因素研究具有重大意义。文中以典型的旱塬沟壑区千阳县 2017 年耕地占补平衡项目为例,利用计量经济学 Logistic 回归模型,从构成粮食产能的主要因素出发,分析不同因素对粮食产能的影响效果;同时,对项目实施前后新增耕地土壤理化性质变化进行分析,为渭北旱塬区耕地保护和粮食产能的提高提供一定借鉴。

1 试验区概况

千阳县位于陕西省宝鸡市西北部，地形地貌为典型的渭北黄土旱塬，地理坐标介于106°56′15″~107°22′31″E、34°34′34″~34°56′56″N，海拔710.0~1 545.5 m。千阳县气候为温带大陆性季风区半湿润气候，年均降水量627.4 mm。耕地总面积236.52 km^2，人均耕地面积0.02 hm^2，土壤以黄性土、红土、紫色土和黑垆土的覆盖面积为大，分别占耕地面积的41.2%、12.8%、10.8%和24.7%。土层厚度为10~20 m，土壤pH值为7~8，0~30 cm耕作层平均土壤容重为1.39 g/cm^3；平均有机质、土壤氮素、有效磷和速效钾质量比分别为14.02 g/kg、0.68 g/kg、13.87 mg/kg和165.42 mg/kg；土壤疏松，透水、透气性好，蓄水保墒能力强，主要种植小麦、玉米等。2017年千阳县实施并竣工的耕地占补平衡土地整治项目共3个，实施其他草地开发总面积109.74 hm^2，实现新增耕地103.73 hm^2，其中水浇地70.48 hm^2，旱地33.25 hm^2，新增耕地等别为10~11等。

2 数据来源与研究方法

2.1 数据来源及参数确定

研究中所使用的主要数据来源：农用地分等数据来源于1:10000陕西省县级耕地质量等别年度更新数据（耕地质量利用等数据）[12]、千阳县自然资源局提供的基于《农用地质量分等规程》（GB/T 28407—2012）和《陕西省高标准农田新增耕地和新增产能工作技术指南》（试行）[13]的耕地质量等别评定报告[14]以及2017年项目耕地土壤检测报告[15]。技术依据：《陕西省耕地质量等别年度更新评价县级技术手册》[16]，千阳县2017年耕地质量等别数据库[12]，千阳县2017年土地利用现状图和影像图[12]。依据《耕地质量等级》（GBT 33469—2016），千阳县属黄土高原区—渭北陇东黄土旱塬区，其标准耕作制度为春玉米和冬小麦，二年三熟。中春玉米光温生产潜力指数和气候生产潜力指数分别为2346和1968，冬小麦光温生产潜力指数和气候生产潜力指数分别1077和784，指定作物产量比系数为1。选取地形坡度、土壤侵蚀度、有效土层厚度、表层土壤质地、土壤有机质含量以及灌溉保证率作为分等因素。

粮食产能计算公式[12]为

$$Y=S_n(D-K_n)\times 15\times 100 \tag{1}$$

式中：Y为粮食产能，kg；S_n为新增耕地面积；D为产能计算常数，$D\leq 16$（当产能为0时，$D=16$）；K_n为新增耕地平均质量等别。

新增耕地平均质量等别计算公式为

$$K_n=\sum_{i=1}^{n}(K_i\cdot S_i)/\sum_{k=1}^{n}S_i \tag{2}$$

式中：K_n为项目区内耕地平均质量等别；n为项目区内核定单元总个数；K_i为第i个核定单元的耕地质量利用等别；S_i为第i个核定单元的耕地面积。

2.2 样品采集和数据计算

在千阳县2017年3个占补平衡项目（分别记为A1、A2、A3）范围内进行土样采集；分别采集项目实施前、后以及项目实施1年后和3年后的土壤数据。利用直径6 cm的土钻按对角线法设在面积为1 m×1 m的样方内分层取土样3个，取土样深度为0~30 cm，每层为10 cm，试验做3次重复。土壤粒度及微团聚体采用MS3000型粒度分析仪测定。土壤氮素含量采用UDK129型凯氏定氮仪测定；有机质含量采用DU-30G油锅法测定；有效磷含量采用752N紫外可见分光光度计测定；速效钾含量采用M420火焰光度计测定[17]。

土壤分散系数计算公式[18]为

$$M=\frac{a}{b}\times 100\% \tag{3}$$

式中：M为土壤分散系数（%）；a为土壤微团聚体分析黏粒含量；b为土壤机械分析黏粒含量。

土壤结构系数计算公式[18]为

$$N = (1 - M) \times 100\% \tag{4}$$

式中:N 为结构系数(%);M 为土壤分散系数(%)。

土壤结构破坏率计算公式[18]为

$$P = \frac{m_f - m_w}{m_f} \times 100\% \tag{5}$$

式中:P 为结构破坏率(%);m_f为>0.25 mm 粒度的力稳性团聚体质量,g;m_w为>0.25 mm 粒度的水稳性团聚体质量,g。

采用环刀法测定土壤容重及孔隙度,土壤总孔隙度计算公式[19]为

$$Q = \frac{d_s - r_s}{d_s} \times 100\% \tag{6}$$

式中:Q 为总孔隙度(%);d_s为土粒密度,g/cm^3;r_s为土壤容重,g/cm^3。

水稳性团聚体按照湿筛法测定,采用 2000 μm、250 μm 和 53 μm 的分样筛,分别获得>2000 μm、250~2000 μm、53~250 μm 和<53 μm 的土壤团聚体和黏粉粒含量。称取土样 50 g,在室温下将土样放在 2000 μm 的套筛上,使桶内的去离子水量刚好淹没土样,保持浸泡 5 min;利用自动筛分仪上下振动筛分,每次振动幅度为 3 cm,频率为 30 次/min,上下振荡 5 min。分别将 2000 μm、250 μm 和 53 μm筛子上的部分用去离子水冲洗到已烘干称重的铝盒中;筛分 2000 μm、250~2000 μm、53~250 μm土壤团聚体时,去掉水表面漂浮的植物残体等,于 60 ℃下烘干至恒重;<53 μm 部分则在 4000 r/min下使用离心机离心 5 min 后收集,将样品烘干后称重,记录各粒级团聚体质量,再分别计算各粒级土壤团聚体质量所占比例。其中:>2000 μm 和 250~2000 μm 为水稳性大团聚体,53~250 μm 和<53 μm 为水稳性微团聚体。

2.3 研究方法

为综合考虑耕地占补平衡对粮食产能的影响,以千阳县主要作物冬小麦和夏玉米作为量化标准,评价千阳县耕地占补平衡新增耕地的粮食产能。新增耕地平均耕地等别采用项目区所在千阳县 2017 年度耕地质量等别年度更新评价相关参数,以新增耕地面积、新增耕地等别、新增耕地基础设施(含道路条件、灌排设施等)、新增耕地单位面积投资、新增耕地年限等为新增耕地粮食产能影响因素。

利用 Logistic 回归模型分析各因素对粮食产能的影响,计算式为

$$\ln \frac{p(y=1)}{1 - p(y=1)} = \alpha + \sum_{i=1}^{k} \beta_i x_i \tag{7}$$

因变量为新增耕地产能 Y;自变量即影响因素:新增耕地面积 X_1、新增耕地等别 X_2、新增耕地基础设施 X_3、新增耕地单位面积投资 X_4、新增耕地年限 X_5。各变量含义及赋值情况见表 1,预期影响"+"表示因变量和自变量间存在正向关系。

表 1 变量含义及赋值情况

Table 1 Variable meaning and assignment

变量	含义及赋值	预期影响
因变量 Y:新增耕地产能	1 为低产能,2 为中产能,3 为高产能	
自变量 X_1:新增耕地面积		+
自变量 X_2:新增耕地等别	9,10,11 级	+
自变量 X_3:新增耕地基础设施(道路条件、灌排设施等)	不完善为 1,基本完善为 2,完善为 3	+
自变量 X_4:新增耕地单位面积投资	元/hm^2	+
自变量 X_5:新增耕地年限	a	+

采用 Excel 2007 处理数据,对相关试验数据进行 Pearson 相关检验,当显著性低于 0.05 时,认为存在显著相关关系。

3 结果分析

3.1 千阳县 2017 年新增耕地等别和粮食产能计算

表 2 为 2017 年千阳县内新增耕地粮食产能变化,2017 年实施的 3 个耕地占补平衡项目实现的耕地面积分别为:A1 项目(63.8306 hm^2,其中水浇地 60.9761 hm^2,旱地 2.8545 hm^2)、A2 项目(水浇地 9.5095 hm^2)以及 A3 项目(旱地 33.2538 hm^2)。分析可知,新增耕地等别和面积对粮食产能具有较大影响。A1、A2 及 A3 项目的粮食产能分别为 570.19 t、85.58 t 以及 249.40 t,可以看出粮食产能随新增耕地面积增大而增大,但受新增耕地地类影响较大。水浇地项目的粮食产能显著高于旱地项目的粮食产能,A1 项目中,新增水浇地面积占比为新增耕地的 95.53%,而创造的粮食产能占比为96.25%。由项目 A2 和 A3 比较分析可知,新增水浇地单位面积粮食产能为 9.0 t/hm^2,而新增旱地单位面积上粮食产能为 7.5 t/hm^2,单位面积上水浇地较旱地的粮食产能增加率提高 25%。这是由于通过占补平衡项目的实施,增加了资金和技术投入,改善了原有土地土壤、水源以及生产条件,促进了新增耕地等别的提高,新增耕地土壤地力条件不断加强,进而促进了新增耕地产能的提升。

表 2 2017 年千阳县占补平衡项目粮食产能计算

Table 2 Calculation of grain production capacity of Qianyang County's 2017 complementary balance project

A1				A2				A3				合计			
耕地面积(hm^2)		平均耕地等别	粮食产能(t)	耕地面积(hm^2)		平均耕地等别	粮食产能(t)	耕地面积(hm^2)		平均耕地等别	粮食产能(t)	耕地面积(hm^2)		平均耕地等别	粮食产能(t)
63.8306		10	570.19	9.5095		10	85.58	33.2538		11	249.40	106.5939		11	905.17
水浇地	60.9761	10	548.78	水浇地	9.5095	10	85.58	水浇地	0	10	0	水浇地	70.4856	10	634.36
旱地	2.8545	11	21.41	旱地	0	11	0	旱地	33.2538	11	249.40	旱地	36.1083	11	270.81

3.2 占补平衡项目对新增耕地土壤理化性质的影响

项目实施完成后项目区种植冬小麦和夏玉米,两年三熟制。表 3 为占补平衡项目实施不同年限后,新增耕地土壤物理性质变化情况。分析可知,土壤总孔隙度、大团聚体、水稳性团聚体和微团聚体均随着项目实施年限增加而显著变大($P<0.05$),土壤结构破坏率显著降低($P<0.05$),抗侵蚀能力得到显著提高($P<0.05$)。土壤总孔隙度表征土壤疏松程度,可为微生物提供良好的生产环境,有利于促进作物根系生长发育。

由表 3 可以看出,在项目实施后,土壤总孔隙度首先出现了一定程度的下降,降幅为 1.56%,而在项目实施 1 年和 3 年后,出现显著提高($P<0.05$),分别较项目实施前提高 8.19%和 9.36%。这主要是由于项目实施过程中对地块进行了机械压实,休耕时期在降雨作用下增强了土壤紧实度。而随着耕种和土壤翻耕等农业措施的进行,导致耕作层土壤质地疏松,总孔隙度变大。土壤团聚体是构成土壤结构的基础,其数量和稳定性也是土壤熟化程度的重要表现之一。分析可知,土壤大团聚体与项目实施年限呈显著正相关关系($P<0.05$),而微团聚体与项目实施年限呈显著负相关关系($P<0.05$),且均为项目实施后变化幅度最大,项目实施 1 年和 3 年后变化幅度较低。经过长期耕种、施肥等措施,水稳性团聚体随项目实施年限的增加显著增加($P<0.05$),与项目实施前相比,各年度增幅分别达到 0.48%、5.708%和 7.18%,说明土壤养分状况也不断得到改善。土壤结构系数是指参与团聚化的黏粒与黏粒总量的比值,结构系数越大,微团聚体的水稳性越强,土壤结构的养分状况也越好。分析可知,在项目实施前后,土壤结构系数表现为先降低后增加规律,在项目实施 3 年后达到最大,为 19.13;土

壤分散系数则表现为相反的变化规律。结合土壤结构破坏率观察,土壤耕作年限越久、土壤熟化程度越高,土壤结构破坏率也越低,这主要由于土壤翻耕和水肥施入使得土壤有机质和养分不断积累,促进了土壤颗粒团聚作用。与项目实施前相比,土壤结构破坏率仅在项目实施后出现了一定程度的增大,在项目实施1年和3年后显著降低($P<0.05$),分别较项目实施前降低0.89%和1.44%,土壤抗侵蚀能力得到显著改善。

表3 占补平衡项目对新增耕地土壤物理性质的影响

Table 3 Impact of complementary balance project on physical properties of new cultivated land

项目实施情况	总孔隙度/%	大团聚体/%	水稳性团聚体/%	微团聚体/%	土壤分散系数/%	土壤结构系数/%	土壤结构破坏率/%
项目实施前	45.43±1.32b	80.43±2.74b	12.58±1.05c	3.40±0.03a	81.25±0.55b	18.75±0.74b	76.55±3.62a
项目实施后1年内	44.72±1.14b	82.25±2.54a	12.64±1.12c	3.36±0.01a	82.07±0.64a	17.92±0.70c	76.58±3.56a
项目实施1年后	49.15±1.04a	82.38±2.36a	13.36±1.07b	3.24±0.02b	81.17±0.69b	18.83±0.75b	75.87±3.20b
项目实施3年后	49.68±1.14a	82.47±2.36a	14.32±1.18a	3.15±0.02c	80.87±0.38c	19.13±0.77a	75.45±3.15b

注:同一列中所带字母不同,表示具有统计学意义($P<0.05$),下同。

3.3 占补平衡项目对新增耕地土壤养分的影响

项目实施后,通过旋耕施加生物有机肥进行土壤培肥,施肥量为3000 kg/hm^2。表4为项目实施后不同年份新增耕地土壤养分变化情况。分析可知,经过占补平衡项目实施,土壤养分含量整体呈良性变化,均在项目实施3年后达到较高水平。土壤有机质质量比在项目实施后达到显著提高($P<0.05$),较项目实施前提高10.20%,这主要是由于项目实施完成后,会通过客土培肥等措施增大有机质质量比,而随着耕种、施肥行为的开展和持续进行,土壤熟化程度不断提升,导致有机质质量比逐渐增大,在实施3年后达到最大值15.78 g/kg,但增加幅度逐渐趋于缓和。全氮质量比是土壤氮素供应能力的重要体现,随着项目实施年限增加,土壤全氮质量比也逐年增加,并于项目实施3年后达到最大值0.92 g/kg。与项目实施前相比,全氮质量比在项目完成各阶段增幅分别为19.11%、3.70%和3.57%,项目实施完成后增量最大,这主要是由于项目实施完成后,会通过客土培肥等措施补充氮素,且项目实施后普遍会存在一定的休耕期,使得土壤氮素得到一定积累,分析可知,全氮质量比变化规律与有机质变化规律相似。土壤有效磷质量比在项目实施后出现一定程度的降低,较项目实施前降低4.47%,而实施1年和3年后得到逐步提升,分别较项目实施前提高10.45%和13.05%,这主要是由于项目实施中所采用的培肥材料一般为尿素和生物有机肥,其中所含磷肥较低,且随着土壤有机质质量比增加,加剧了土壤微生物活动强度,肥料中的部分磷肥被吸附固定。另外,由于项目区土质为黄绵土,土壤中较高的碳酸钙含量也会对施入磷肥产生一定的固定作用,而随着耕种和土壤翻耕的田间管理措施的介入,土壤中有效磷质量比逐渐得到显著提高($P<0.05$)。整体上,土壤速效钾质量比变化在项目实施后有所提高,较项目实施前提高5.59%($P<0.05$),但随着实施年限提升,速效钾质量比虽略有提高,但均处于相对稳定的状态,占补平衡项目的实施可有效改善土壤养分状况。

表4 占补平衡项目对新增耕地土壤养分的影响

Table 4 Impact of complementary balance project on soil nutrients of new cultivated land

项目实施情况	有机质/(g/kg)	全氮 /(g/kg)	有效磷/(mg/kg)	速效钾 /(mg/kg)
项目实施前	14.02±0.17c	0.68±0.05c	13.87±0.15b	165.42±2.85b
项目实施后1年内	15.45±0.23b	0.81±0.08b	13.25±0.19c	174.67±3.24a
项目实施1年后	15.43±0.22b	0.84±0.09b	15.32±0.21a	172.44±3.13a
项目实施3年后	15.78±0.22a	0.87±0.11a	15.68±0.21a	177.87±3.05a

3.4 新增耕地粮食产能影响因素

根据构成影响粮食产能的重要因素,对新增耕地面积、新增耕地单位面积投资、新增耕地年限、新增耕地等别以及新增耕地基础设施等进行 Logistic 回归模型验证。表 5 为不同因素对新增耕地粮食产能的影响情况,表中 Wald 值表示卡方值,OR 值表示优势比。通过计算,平行线检验中卡方为29.6,显著性为 0.253,大于 0.05,说明 Logistic 回归模型可进行各因素对粮食产能的影响模拟,通过似然比检验,模型具有较高的精度。

表 5 新增耕地粮食产能影响因素分析

Table 5 Analysis of factors affecting grain production capacity of new cultivated land

变量	变量及哑变量	回归系数	Wald 值	OR 值	显著性
新增耕地面积	hm^2	0.134	2.158	—	0.013*
新增耕地单位面积投资	元/hm^2	0.183	3.047	—	0.045*
新增耕地年限	年/a	0.016	0.012	—	0.033*
新增耕地等别	11 级=1	−2.135	15.470	0.127	0.000**
	10 级=2	−1.734	14.940	0.169	0.000**
	9 级=3	—	—	—	—
新增耕地基础设施（道路条件、灌排设施等）	不完善=1	−1.897	12.876	0.197	0.027*
	基本完善=2	−1.034	10.230	0.278	0.042*
	完善=3	—	—	—	—

注: *表示差异具有统计学意义($P<0.05$),**表示差异具有统计学意义($P<0.01$)。

新增耕地面积与粮食产能之间呈现显著的正相关关系($P<0.05$),显著性水平为 0.013,影响系数为 0.134,说明新增耕地面积越大所创造的粮食产能也越高。但值得注意的是,单位耕地面积上的粮食产量是有限的,由于粮食产量受耕地面积或实际播种面积影响较大,因此在实际粮食产能的估算中要做好后期移交管护工作,确保新增耕地不出现撂荒等现象,切实保证新增耕地粮食产能提高。新增耕地单位面积投资额的高低综合体现了耕地开发难度和投入程度,一般而言,单位面积投资的提高主要是由于提高耕地基础地力(例如土壤翻耕和土壤培肥等)、配套完善田间灌排系统,增大水浇地开发面积,因此有利于产能提高。从分析结果观察,粮食产能随新增耕地单位面积投资提高而增大,影响系数为 0.183,高于新增耕地面积影响系数,且具有统计学意义($P<0.05$)。新增耕地年限与粮食产能成正向影响($P<0.05$),这是由于新增耕地土壤熟化程度不高,且土壤中有机质和其他营养元素缺乏,但随着年限增加,土壤地力条件得到改善,会逐渐促进粮食产能提高。

耕地等别是气候、地貌、土壤、植被、水文等以及与耕地利用有关社会经济条件的综合反映,对新增耕地粮食产能的影响具有统计学意义($P<0.01$)。以耕地等别 9 级为参照,可以看出,10 级=2 的优势比 OR 值为 0.169,11 级=1 的优势比 OR 值为 0.127,呈显著降低趋势,说明新增耕地等别越高,粮食产能也越大。由于耕地等别能综合反映宏观尺度上的耕地质量,可直接与粮食产能挂钩,因此现有自然资源系统对于粮食产能的计算均以单位面积上的新增耕地等别为依据,估算精度也较高。

新增耕地基础设施完善情况是提高粮食产能的有力保障。由于千阳县地处旱塬丘陵地带,可开发后备资源分布环境较差,水源保障率低,交通不便,尤其是田间道路基础较差,不仅难以满足耕作,更易增加各种耕种材料运输成本,极大地阻碍了机械化进程和当地群众的耕作意愿。因此,新增耕地基础设施的配套完善也是促进粮食产能提高的重要影响因素。从结果可见,基础设施对粮食产能的影响系数为负值,新增耕地基础设施与粮食产能产生负向影响,且随着基础设施完善程度的提高,其影响系数逐渐增大,说明道路条件和灌排设施完善程度与新增耕地粮食产能呈显著正相关关系($P<$

0.05)。综合分析,各因素对新增耕地粮食产能影响程度由大到小依次是等别、面积、基础设施、年限和单位面积投资。

4 讨论

粮食产能高低是新增耕地质量各因素综合作用的结果,由于粮食产能是单产与其对应耕地面积的乘积,因此各新增耕地面积的大小对粮食产能大小有着重要的影响。在各外部影响因素相同的情况下,新增耕地粮食产能随新增耕地面积增加而增大,但由于不同占补平衡项目实施区域存在的问题以及针对不同情况进行土地整治的模式有所不同,导致各区域提高新增耕地粮食产能的途径也有所不同。根据项目实施区域的土壤理化特性、土壤剖面构型、地形坡度、水源情况以及道路通达情况等条件的不同,有效提高耕地粮食产能的途径主要有通过改造、消除耕作中的限制因素以提高耕地生产能力,通过完善基础设施提高耕地生产能力,以及通过增加对新增耕地的投入引导农户对耕地的投入、经营以提高耕地生产能力。恶劣的土壤条件是限制作物增产、提高生产能力最重要的因素,例如有效耕作土层较薄、土壤有机质含量不足、氮磷钾等营养成分较少,以及土壤存在侵蚀度较高和盐碱化等情况,均会对粮食产能的提高产生不利影响,因此要根据不同区域的土壤情况,进行针对性的土壤翻耕、培肥等措施以提高其土地生产能力。新增耕地农田水利灌溉排水设施对新增耕地粮食产能的提高具有显著影响,水利设施完善可提高农田灌溉保证率,既有利于增加水浇地面积,又能通过配套的田间水利工程以及水土保持措施提高新增耕地抗旱、固沙以及减少水土流失的能力,极大地降低了新增耕地减产风险。而道路通达情况有利于促进农业机械集约化、规模化生产,也便于提高土地的经营管理和集中流转。

占补平衡项目实施前后,会有大量的资金和技术等投入到新增耕地建设中。而随着新增耕地单位面积投资的提高,原有的土壤条件、水源条件以及其他生产条件均能得到很大的改善,形成了良好的耕地基础,并通过向完成项目的镇、村移交等工作,明确权属和后期管护责任,促进了土地所有人对新增耕地的重视程度,能有效影响其增加对耕地的投入和集约经营。此外,通过土地整治能显著提高耕地质量,且随着土地整治年限的延长,土壤理化特性和生物特性均会显著改善,土壤抗蚀性能也可得到显著提高。耕地自然属性的重点在于土壤质量,而社会经济属性包括耕地的管护质量和土地利用质量。耕地自然属性受作物种植区域水文气候条件、地形地貌条件、土壤理化性质和土壤养分等影响,但更重要的是,农业生产投入、耕地利用方式、田间灌排设施、道路通达程度、农业科技投入以及后期管护等措施,均会对耕地的社会经济属性产生重要影响,进而对粮食产能的提升产生显著推进作用。中国现有耕地资源约 1.28 亿 hm^2,但其中仍有 70%以上属于中低产田,为耕地资源保护和粮食安全保障带来了巨大压力。耕地质量对粮食产量的提高具有决定性影响,因此要在符合实际情况的前提下,通过土壤翻耕、培肥等基础地力提升措施改善土壤条件,通过提高单位面积投资,配套完善田间水利设施等方式,尽可能通过水利设施建设提高灌溉保证率,提高粮食产能。

耕地等别虽对粮食产能的影响最大,但由于单位耕地面积上的粮食产量是有限的,因此实际耕作面积是制约粮食产能提高的一项重要因素。另外,新增耕地在建设完成初期,理论产量和实际产量均会处于相对较低水平,是否能随着耕作年限的延长得到有效保护和利用将是提高粮食产能的关键,因此要在做好工程措施提高耕地等别的基础上,切实落实好项目后期管护工作,并加大政策扶持,避免出现新增耕地撂荒现象,提高新增耕地利用效率。

5 结论

以典型的黄土旱塬丘陵沟壑区千阳县 2017 年占补平衡项目为例,以冬小麦和夏玉米为量化标准,对新增耕地粮食产能的影响因素进行了分析,并利用 Logistic 回归模型分析了新增耕地各因素对粮食产能的影响,得到以下主要结论:通过占补平衡项目实施可使耕地等别提升 1~2 等,且新增耕地中的水浇地面积占比越大,新增耕地等别越高;水浇地项目的粮食产能显著高于旱地项目的粮食产能,单位面积上水浇地较旱地的粮食产能增加率提高 25%。土壤总孔隙度、大团聚体、水稳性团聚体和微团聚体均随着项目实施年限增加而显著变大($P<0.05$),土壤结构破坏率显著减小($P<0.05$),抗侵蚀能力得到显著提高($P<0.05$)。土壤耕作年限越久、土壤熟化程度越高,土壤结构破坏率也越低,土壤抗侵蚀能力得到有效改善。通过项目实施,土壤有机质和土壤养分含量较项目实施前不断改善。整体而言,文中试验条件下,对新增耕地粮食产能影响程度按因素排序由大到小表现为新增耕地等别、新增耕地面积、新增耕地基础设施、新增耕地年限、新增耕地单位面积投资。

参考文献

[1] Zhang J H, Su Z A, Liu G C. Effects of terracing and agroforestry on soil and water loss in hilly areas of the Sichuan Basin[J], China. Journal of Mountain Science, 2008,5(3): 241-248.

[2] Vancampenhout K, Nyssen J, Gebremichael D, et al. Stone bunds and soil conservation in the northern Ethiopian highlands: Impacts on soil fertility and crop yield [J]. Soil and Tillage Research, 2006, 90(1/2): 1-15.

[3] 谢向向,汪晗,张安录,等.土地整治对中国粮食产出稳定性的贡献[J].中国土地科学,2018,32(2):55-62.

XIE Xiangxiang, WANG Han, ZHANG Anlu, et al. Land Consolidation Contribution to Grain Production Stability in China[J].China Land Science,2018,32(2):55-62.

[4] 芦艳艳,樊雷,刘桢.基于土地利用系数修正的土地整治重大项目区耕地质量等别评价——以河南省延津县为例[J].地域研究与开发,2018,37(5):147-151.

LU Yanyan, FAN Lei, LIU Zhen. Cultivated Land Quality Evaluation of Land Consolidation Major Project Based on Land Use Coefficient Correction Method: Major Project of Yanjin County, Henan Province[J].Areal Research and Development, 2018,37(5):147-151.

[5] 陈正发,龚爱民,张刘东,等.基于质量评价的省域尺度坡耕地质量调控体系构建[J].农业工程学报, 2021,37(20):136-145.

CHEN Zhengfa, GONG Aimin, ZHANG Liudong, et al. Construction of the quality regulation system for provincial scale slope farmland based on quality evaluation[J].Transactions of the Chinese Society of Agricultural Engineering,2021,37(20): 136-145.

[6] 赵海乐,徐艳,张国梁,等.基于限制因子改良与耕地质量潜力耦合的耕地整治分区[J].农业工程学报,2020, 36(21):272-282,324.

ZHAO Haile, XU Yan, ZHANG Guoliang, et al. Farmland consolidation zoning based on coupling of improved limiting factors and farmland quality potential[J].Transactions of the Chinese Society of Agricultural Engineering,2020,36(21):272-282,324.

[7] 邵雅静,员学锋.基于限制因子分析的耕地质量提升潜力分区[J].生态学杂志, 2019,38(8):2442-2449.

SHAO Yajing, YUAN Xuefeng. Zoning of cultivated land quality improvement potential based on limiting factor analysis [J].Chinese Journal of Ecology,2019,38(8):2442-2449.

[8] 王静怡,李晓明.近20年中国耕地数量变化趋势及其驱动因子分析[J].中国农业资源与区划,2019,40(8): 171-176.

WANG Jingyi, LI Xiaoming, et al. Research on the change trend of farmland quantity in china for recent 20 years and its driving factors[J].Chinese Journal of Agricultural Resources and Regional Planning,2019,40(8):171-176.

[9] 韦宇婵,张丽琴.鄂豫地区耕地资源变化时空特征及其影响因素[J].水土保持通报,2019,39(2):293-300.

WEI Yuchan, ZHANG Liqin. Spatial Characteristics of Cultivated Land Changes and Influence Factors in Hubei and Henan Provinces[J].Bulletin of Soil and Water Conservation,2019,39(2):293-300.

[10] 祝锦霞,徐保根.基于变化向量的耕地利用方式变化下耕地质量评价[J].农业工程学报,2020,36(2): 292-300.

ZHU Jinxia, XU Baogen. Evaluation of cultivated land quality under changed cultivated land use pattern based on change vector analysis[J].Transactions of the Chinese Society of Agricultural Engineering,2020,36(2):292-300.

[11] 杨建锋,马军成,王令超.基于多光谱遥感的耕地等别识别评价因素研究[J].农业工程学报,2012, 28(17): 230-236.

YANG Jianfeng, MA Juncheng, WANG Lingchao. Evaluation factors for cultivated land grade identification based on multi-spectral remote sensing[J].Transactions of the Chinese Society of Agricultural Engineering,2012, 28(17):230-236.

[12] 何振嘉,范王涛,杜宜春.占补平衡项目对千阳县新增耕地等别和粮食产能的影响[J].中国农机化学报,2021,42(2):209-216.

HE Zhenjia, FAN Wangtao, DU Yichun. Effects of the complementary balance project on the newly added arable land and grain production capacit in Qianyang County[J].Journal of Chinese Agricultural Mechanization,2021,42(2):209-216.

[13]陕西省农业农村厅,陕西省自然资源厅. 陕西省高标准农田项目新增耕地和新增产能工作技术指南(试行)[R]. 西安: 陕西省土地工程建设集团有限责任公司, 2021.

[14]千阳县国土资源局. 千阳县草碧镇仰塬村等3镇6村土地开发项目新增耕地质量等级评定报告[R]. 西安: 陕西德诺地产评估事务有限责任公司, 2017.

[15]千阳县国土资源局. 千阳县草碧镇仰塬村等3镇6村土地开发项目土样检测报告[R]. 西安: 国土资源部退化及未利用土地整治工程重点实验室, 2017.

[16]陕西省国土资源厅. 陕西省耕地质量等别年度更新评价县级技术手册[R]. 西安: 陕西省土地工程建设集团有限责任公司, 2014.

[17] 陈立华,张欢,姚宇阗,等.盐地碱蓬覆被对滨海滩涂土壤理化性质的影响[J].植物资源与环境学报,2021,30(2):19-27.

CHEN Lihua, ZHANG Huan, YAO Yutian, et al. Effects of Suaeda salsa covering on soil physicochemical properties in coastal beach[J].Journal of Plant Resources and Environment,2021,30(2):19-27.

[18] 周欣花.黄土丘陵沟壑区坡改梯土壤质量效应研究[J].人民长江,2020,51(5):74-78.

ZHOU Xinhua. Effect of slop to terrace projects on soil quality in loess hilly and gully region[J].Yangtze River,2020,51(5):74-78.

[19] 鲁如坤.土壤农业化学分析方法[M].北京:中国农业科技出版社,2000.

本文曾发表于2022年《排灌机械工程学报》第40卷第11期

碳中和背景下矿区生态修复减排增汇实现对策

何振嘉[1,2],罗林涛[1],杜宜春[1,2],缑丽娜[1]

(1.陕西省土地工程建设集团有限责任公司,陕西 西安 710075;
2. 中陕高标准农田建设集团有限公司,陕西 杨凌 712100)

【摘要】中国是矿产资源消耗大国,在推动工业化和城市化的进程中,能源消耗所产生的温室气体已造成严重的生态问题。我国于 2021 年对国际社会做出承诺,将于 2060 年实现"碳中和",以高质量发展改变高耗能的粗放发展模式。我国能源消耗的主要来源是煤炭资源,在全球"碳中和"背景下,实现矿区"碳中和"已成为实现我国社会主义新时代生态文明建设整体布局的一项重要途径。本文在分析了矿区开采引发的生态系统碳排放问题基础上,揭示了我国目前在矿区生态修复中存在的不足,并针对性地提出"树立固碳增汇意识,深化固碳机制研究""节约集约用地,降低碳排放当量""优化复垦技术,提升碳固存能力""创新矿区土地综合利用模式"等矿区"碳中和"的实现对策,为新时代"碳中和"背景下矿区生态修复指明方向,为我国实现"碳达峰、碳中和"提供必要借鉴。

【关键词】碳中和;矿区生态修复;减排增汇;碳固存;土地复垦

Countermeasures to Realize Ecological Restoration and Emission Reduction and Increase of Sinks in Mining Areas Under the Background of Carbon Neutrality

Zhenjia He[1,2], Lintao Luo[1], Yichun Du[1,2], Lina Gou[1]

(1.Shaanxi Provincial Land Engineering Construction Group Co., Ltd., Xi 'an, Shaanxi 710075, China;
2. China Shaanxi High-standard Farmland Construction Group Co., Ltd., Yangling, Shaanxi 712100, China)

【Abstract】 China is a big consuming country of mineral resources. In the process of ensuring industrialization and urbanization, greenhouse gases generated by energy consumption have caused serious ecological problems. China made a commitment to the international community in 2021 to achieve carbon neutrality in 2060 and change the extensive development model of high energy consumption with high-quality development. The main source of energy consumption in China is coal resources. In the context of global "carbon neutrality", achieving carbon neutrality in mining areas has become an important way for China to realize the overall layout of ecological civilization construction in the new era of socialism in China. Based on the analysis of the ecological system carbon emissions caused by mining area mining, this paper deeply analyzes the current status of ecological restoration in China's mining areas, and specifically proposes "establishing the awareness of carbon sequestration and increasing sinks, deepening carbon se-

基金项目:国家重点研发计划资助项目(2021YFD1900700);陕西省土地工程建设集团有限责任公司创新团队项目(DJTD2022-5);陕西省土地工程建设集团有限责任公司内部科研项目(DJNY2022-23、DJNY2022-54)。

questration mechanism research", "saving and intensive research". Land use, reducing carbon emission equivalent", "optimizing restoration technologies, enhancing carbon sequestration capacity" and "accelerating industrial integration, innovating restoration models" and other countermeasures to achieve carbon neutrality in mining areas, are for the ecological restoration of mining areas in the context of carbon neutrality in the new era Point out the direction in order to provide the necessary reference for China to achieve "carbon peak and carbon neutrality".

【Keywords】Carbon neutrality; Ecological restoration in mining areas; Reduce emissions and increase foreign exchange ; Carbon sequestration; Land reclamation

引言

煤炭、石油和天然气等化石燃料使用以及土地利用变化是影响碳排放和造成全球气候变暖的重要因素[1]。气候变化不只会对全球范围内的生态安全产生威胁,也不利于人类社会经济可持续发展。工业文明时期,人类活动对化石能源具有强烈的依赖性,但由于化石燃料的固有属性和无节制消耗,大量 CO_2 等温室气体排放引发的气候变化问题已成为人类面临的巨大威胁和挑战[2]。从 1992 年起,《联合国气候变化框架公约》就对温室气体排放进行了相关规定,随后,《巴黎协定》《欧洲绿色协议》《气候变化法案》《欧盟气候法》等先后对实现"碳达峰、碳中和"提出了具体的内容和明确的时间节点,实现温室气体净零排放,统筹做好节能减排、固碳增汇已成为世界各国共同应对全球气候变化问题的共识[3-5]。我国已进入新的发展时期,为推进社会主义生态文明建设,实现美丽中国建设的远大愿景,2020 年 9 月,习近平总书记在第七十五届联合国大会一般性辩论上宣布,力争在 2030 年前二氧化碳排放达到峰值,努力争取 2060 年前实现碳中和;2020 年 12 月,习近平总书记在气候雄心峰会上宣布到 2030 年中国单位国内生产总值二氧化碳排放将比 2005 年下降 65%以上,非化石能源占一次能源消费比重将达到 25%左右;同年,我国"十四五"规划和 2035 年远景目标纲要提出将碳排放达峰后稳中有降列入中国 2035 年远景目标,"碳达峰、碳中和"已成为我国社会主义新时代生态文明建设的重要内容[6-7]。

我国是一个发展中国家,也是矿产资源生产和消耗大国,目前仍处于工业化、城市化进程快速发展的关键时期,不可避免地会产生巨大的资源消耗。而能源消耗的主体仍是煤炭资源,仅 2020 年,我国煤炭资源利用所产生的 CO_2 接近世界总 CO_2 排放量的 20%[8-9]。从碳达峰到力争实现碳中和目标,仅有 30 年时间,在"碳中和"总量和强度的双重压力下,矿产资源开发和利用产生的温室气体排放已成为影响和实现我国碳中和目标的重要因素。能源需求的增长和碳排放下降的约束将使我国完成碳中和目标更具挑战[10-11]。另外,伴随着大规模、高强度的各类矿产资源开采和利用,虽然保障了经济稳定持续发展,但同时也引发了生态环境破坏、土地资源损毁及矿区碳平衡破坏等问题。自 1987 年至 2020 年,仅因煤炭资源开采损毁的土地资源就超过 180 万 hm^2,不仅造成了极大的土地资源浪费,也降低了生态系统碳固存的能力和效果[12]。随着全球气候日益恶化,矿产资源开发引发的"碳中和"问题已成为我国现阶段以及未来较长一段时期必须面临及重视的问题。

1 矿区资源开发引发的碳排放问题及矿区修复进展

矿产资源开发会对区域内碳源生态系统产生一定影响,通过矿产资源开发,大型机械施工和开采行为会直接扰动区域内土地利用结构,加剧对土壤理化特性的破坏及生态环境损害,造成原有耕地、草地、林地等一系列高碳汇型土地资源的固碳能力减弱或丧失。矿产资源开发的强度和规模直接影响矿区固碳能力,矿产资源开发前期,区域内生态系统较为稳定,大气、水域、植被和土壤碳库接近于平衡状态,且在生态系统循环作用下,有利于降低和固定系统内产生的 CO_2 排放。碳排放量随着采矿强度、开采规模的增大而增大,在矿产资源开采过程中,大型机械、车辆使用均会产生极大的燃料消耗,会直接造成矿区内碳排放量加剧。同时,开采过程中产生的 CH_4 溢散、煤炭与煤矸石氧化产生的温室气体溢散也会造成碳排放量加剧。新建的矿山用地和工业用地均会通过资源开采、运输、利用、排放等过程消耗能源，造成土体、水源、大气污染和 CO_2 大量排放，严重危害“三生空间”系统。在矿产资源利用过程中，也会持续不断产生直接和间接碳排放，如煤矸石堆积自燃产生的 CO 和 CO_2 等温室气体。此外，矿产资源开采后，由于土地结构受到破坏，损毁土地的土壤抗侵蚀度、土壤渗透能力以及土壤养分等均受到一定破坏，矿区碳固定能力不断降低[13]，且由于矿区所处特殊的地理位置，在资源开发利用的整体过程中，由点向面、向网的扩散模式对于生态系统恢复和碳循环系统的持续稳定也会造成深远的影响，严重危害矿区辐射范围内土壤、大气以及植被的碳储存和碳交换[14]。

我国矿区面临重大的生态环境问题，为加快矿区推进地质环境恢复和综合治理工作，2001 年以来，我国先后出台了《矿山地质环境保护规定》和《矿山地质环境保护与治理规划》等政策，加快废弃矿山摸底调查和专项治理进度。随着生态文明建设步伐的稳步推进，国家对环境保护的要求也提到了新高度，废弃矿区造成的严重危害已成为阻碍新时代生态文明建设的巨大阻碍。国务院于 2016 年出台《土壤污染防治行动计划》（国发〔2016〕31 号）；2018 年 6 月，中共中央、国务院印发《关于全面加强生态环境保护坚决打好污染防治攻坚战的实施意见》（中发〔2018〕17 号）等众多政策规划制定了生态保护红线并提出“绿水青山就是金山银山”理论，对矿区生态修复工作提出了十分必要的指导建议和有效的修复方法，同时也对废弃矿山生态修复工作提出了更高的要求。自然资源部发布《关于探索利用市场化方式推进矿山生态修复的意见》（自然资规〔2019〕6 号）将矿区生态功能修复和后续资源开发利用、产业发展统筹考虑，明确了一系列激励政策，有力推进了矿山生态修复工作的步伐。截至目前，全国矿区修复治理面积已超过 92 万 hm^2，治理率超过 28.75%[15]。

2　矿区生态修复存在的不足

2.1　修复目标单一，固碳增汇理念不足

结合目前废弃矿山修复情况来看，我国矿山修复关注点主要在于生态系统功能的恢复和土地利用结构的优化，对减少碳排放、促进碳中和方面关注较少。其修复目标主要是以最小的经济代价快速实现矿区污损土地修复和生态复绿，在有条件建设区域将其复垦为耕地或还林还草，矿区固碳增汇理念不足，缺乏必要的实践探索，且目前应用于固碳增汇的技术含量较低，尤其缺乏必要的水、土资源检测等基础性调研，整体修复效果不佳。2013 年 5 月，习近平总书记指出，推进生态文明建设，必须全面贯彻落实党的十八大精神，树立尊重自然、顺应自然、保护自然的生态文明理念，坚持节约资源和保护环境的基本国策，坚持节约优先、保护优先、自然恢复为主的方针，着力树立生态观念、完善生态制度、维护生态安全、优化生态环境，形成节约资源和保护环境的空间格局、产业结构、生产方式、生活方式。2018 年 5 月，习近平总书记在全国生态环境保护大会上的讲话中提出，山水林田湖草是生命共同体，要统筹兼顾、整体施策、多措并举，全方位、全地域、全过程开展生态文明建设。各项政策和精神与实现碳中和理念高度契合，但结合目前矿区生态修复和固碳增汇方面的表现来看，生态功能优先和固碳增汇的修复理念尚未得到充分展现，部分用于实现矿区生态功能和固碳增汇的各类举措尚处在初级阶段。

2.2　修复方法单一，固碳增汇水平较低

矿区生态修复工作是一项系统、复杂的综合性工程，新形势下由于缺乏整体规划、周密调查和必要的科技投入，且对节能减排、固碳增汇方面考虑不足，现有的矿区修复手段和方法较为单一，主要通过工程措施进行地质灾害治理，简单地与土地整治模式相结合进行简单粗放的复垦。我国淮南矿区对采煤塌陷区进行了简单的土地复垦，但并未对原矿区内污染物进行去除，复垦土壤中 Mn、Pb 含量仍严重超标，危害土壤安全和人类健康。我国山东邹城东滩矿区经过单一土地复垦，对土壤 pH、有机碳、微生物碳等产生了显著影响，但降低了固碳相关菌门的丰度，不利于土壤碳汇水平的提高。土壤碳库是陆地生态系统中的核心组成部分，也是解决碳汇问题的关键，是削减碳排放、缓解全球气候变化的重要路径。土壤环境问题是矿区修复的重点，由于矿区会产生大量温室气体、重金属及其他有毒有害物质，且污染周期长、危害大，在此基础上直接进行简单的土地复垦势必会严重影响土地质量，降低土壤固碳能力，也

不利于耕地产能提升。更严重的是矿区各类有毒有害物质会在土壤中积聚和迁移,进一步污染地表和地下水资源,对陆地生态系统碳库造成极大损害。针对矿区土壤碳库固存问题,目前普遍采用物理置换、植物固定、生物调理等方式改善土壤条件,进一步提升土壤固碳能力,但由于修复成本较大、周期较长而导致效果并不理想。此外,单一的土壤修复方式难以满足固碳增汇的目标,且经济效益显著、可行性强的修复技术尚未得到充分开发,很多固碳增汇技术需要进一步的完善和研发。

2.3 修复模式单一,固碳增汇成效不显著

矿区生态系统修复工作的主要载体在矿区自身,但同时也与社会经济系统之间有极深的耦合关系。现有的矿区生态修复工作整体统筹规划以及对山水林田湖草生命共同体的内在机制和规律认识不足,在规划设计阶段就未能将修复区域作为一个完整的生态系统进行综合考虑,仅通过完成每一项具体的单项工程来实现修复,模式较为单一。虽在短期内可取得明显修复效果,但由于缺乏长期有效的监管机制,修复后的矿区资源无法得到充分利用和相应管护,生态系统在一段时期后将再次发生损毁或破坏,造成“生态-经济-社会”系统不稳定,无法达到预期的生态系统服务价值,同时也不利于陆地生态碳库的持续稳定。我国榆神府矿区针对因资源开发而受到破坏的矿区进行生态恢复和景观修复,虽一定程度上实现了矿区的复垦复绿,但整体规划中未能充分考虑后续资金投入和相关产业植入,固碳水平和综合效益并不显著。此外,现有的矿区生态修复未能充分考虑废弃矿区土地利用现状和开发潜力等因素,修复后缺乏必要的土地管理措施和林网工程建设,在一定程度上减弱了土壤和植被碳汇效应。同时,大部分矿区也未能因地制宜结合自身特色推动产业经济发展,修复后的矿区难以形成长期稳定且能够自维持的生态系统。整体来看,矿区生态修复和固碳增汇成效不显著。

3 基于碳中和背景矿区固碳增汇实现对策

3.1 树立固碳增汇意识,深化固碳机制研究

“碳达峰、碳中和”是习近平总书记立足于社会主义生态文明建设、建设美丽中国提出的一项气候变化国家战略,也是影响我国未来较长一段时间内经济发展的重要战略。我国是目前全世界最大的碳排放国,在切实推进我国经济持续稳定增长的同时,碳排放引发的问题也成为阻碍我国生态环境改善的重要因素。习近平总书记提出的“碳达峰、碳中和”目标,为我国绿色低碳的矿区生态修复工作指明了前进方向,也为全球气候治理向前迈进注入了新动能,顺应时代发展需求,也是保障我国经济可持续发展的重要举措。我国目前矿区生态修复工作取得了一定成效,但基于碳中和背景的矿区固碳增汇工作尚存在不足,矿区应充分借助国家战略提供的政策指引,牢固树立固碳增汇意识,切实将碳中和理念深度融入矿区生态修复工作中,将技术诱导与自然恢复相结合,注重本土植物和生态重建。同时,要针对不同土地利用结构进一步研究矿区生态系统中碳源构成,碳源转化方式及其影响因素、作用机制、过程机制和固碳效应,提高林地、草地、滩涂地等高碳汇型土地资源保有率。结合矿区资源开发模式、产业结构及生态过程等方面的需求,重点对矿区修复过程中土壤、植被、生物等方面固碳机制进行充分研究,更好地指导工程或其他相关措施实现碳中和,持续提升固碳增汇能效。

3.2 节约集约用地,降低碳排放当量

在碳中和背景下,矿区生态修复已成为集矿山生态修复、提高土地利用率兼顾提升矿区碳汇能力的重要方式。矿产资源开发利用的基础载体为土地资源,矿区土地的节约集约利用有利于缓解矿区范围内碳排放量和排放强度。相关研究表明,通过矿区生态修复可显著增加土壤有机碳库容量和CO_2吸收量,通过矿区复垦并种植水稻,将CO_2吸收量由复垦前的 1.1×10^6 kg 提高到 2.67×10^6 kg,增幅达 142.73%,是实现碳中和十分有利的一项举措[16-17]。矿区资源开发利用的整体过程中,会伴随大量温室气体排放,要降低矿区碳排放,实现碳中和,必须以低碳发展为目标导向,要切实以国土空间规划和用途管制为手段,合理选定矿区范围和可开发范围,优化土地利用结构、规模和布局,提前规划好矿区资源开采、运输、利用、排放以及土地资源复垦再利用全过程的实施方案,重点对场地选址、采区空间优化以及排土场空间优化等方面进行减排设计,通过提高运输效率和缩短运距等方式实现矿区

降低碳排放强度、提高碳汇水平和固碳效率，做到“少占地、少损毁”以及“多造地、快复垦”。我国山西省朔州市平朔矿区土地损毁碳排放量高达591.21万t，在煤炭开采过程中通过缩短岩土剥离过程中的运输距离，优化开采施工作业场地等措施，显著减缓了矿区土地损毁碳排放量。将来，排土场微地形优化、适宜物种的筛选以及优化植被配置模式也是推动矿区生态修复和系统碳库恢复的重要研究方向。因此，在规划设计前期，要切实做好土地资源利用调查，避免或尽可能减少林地、天然牧草地等高碳汇型土地资源的开发和扰动，同时要从有利于节能、减排、降耗等角度出发进行规划设计，实现矿区平面布局和空间布局最优规划。此外，要充分结合土地利用结构、矿区开发规模和矿产资源分布特征等制订相应的复垦方案，采用“边采边复”的方式及时修复治理受到轻微破坏的生态环境，确保“土壤-植被”碳库的稳定性，最大限度地节约集约利用矿区土地资源，助力矿业生产低碳循环与绿色发展。

3.3　优化复垦技术，提升碳固存能力

实现矿区碳中和的重点在于确保矿区修复后土地再利用程度的提高，土地复垦后土壤固碳能力提升和生态环境的改善是生态修复效果最佳的体现方式。已有研究表明，通过实施土地复垦，可较好地恢复矿产资源开发过程中产生的碳损失，显著提升土壤固碳能力。我国晋东南黄土区煤矿区，煤炭资源开采导致地面沉陷后地表出现4~5 m马鞍状落差，原有水浇地遭到破坏，耕地转换为旱薄地，土地生产力严重下降。通过工程措施进行表土剥离后，通过有机无机肥配施，促进了土壤中大粒径团聚体的形成，提高了大团聚体对有机碳的富集能力，有助于土壤固碳。因而，新时期矿区生态修复可通过优化复垦技术来实现固碳能力提升。

首先，要统筹规划，全面制订经济合理、安全高效、切实可行的矿区土地复垦方案，精细化做好矿区范围内地质调查和土壤分析、水样检测等基础工作。针对存在粉尘、重金属及其他有毒害物质地区，要做好土壤污染修复专项方案，利用土体有机重构、生物修复、农业生态修复及改性生物炭材料等技术切实推动污损土地修复，不断改善矿区土壤质量和植被覆盖率，最大程度提升土壤和植被固碳能力。此外，不断优化复垦技术，针对不同区域、不同产业、不同作物对耕地产能提升的需求，以土壤质量提升、产能提高和固碳量有效增加为目的，综合考虑待复垦土地损毁类型和程度、自然适宜性和生产潜力以及工程地质条件和应用机械的可能性，结合社会环境条件和经济因素，采用高标准建设，开展绿色清洁有机肥制备理论、制备技术和施用方法研究，同时优化配置水资源，重视土壤培肥等农艺措施，通过提升复垦土地等别来推动生态修复效果的提升，实现耕地质量和碳固定量的双提升。最后，要重视生态型修复措施的应用，统筹考虑废弃矿区山水林田湖草综合体构建，在不具备复垦条件的地区充分利用固废复绿技术和无土喷播技术，按“宜林则林，宜草则草”原则进行修复，全方位拓展固碳渠道，同时要加强废弃矿区资源再利用，最大程度地改善矿区生态环境[18-20]。

3.4　创新矿区土地综合利用模式

矿区生态修复的主体是土地，实现矿区碳中和的主要方式是将其复垦为耕地，部分地区也依据“宜林则林、宜草则草”的原则因地制宜开展修复工作。在“碳达峰、碳中和”背景下，矿区生态修复工作的核心已发展为山水林田湖草绿色理念的综合体现，而现有的矿区土地复垦模式主要采用单一的土地平整和客土回填等工程措施，固碳增汇能力较弱，因而实现碳中和远大目标也需向多元化发展模式转变。矿区的经济和社会价值包括综合开发关键矿产，同时降低运营风险并促进矿山修复[21]，要充分考虑废弃矿区土地利用现状和开发潜力等因素，结合生态功能修复和后续资源开发利用、产业发展等需求，充分尊重土地权利人的意见，合理确定矿区内各类空间用地的规模、结构、布局和时序，大力发展和施行“矿区修复+产业融合”模式推动生态修复、经济增长和产业升级，降低人类活动对表层土壤的干扰，综合提升植被和土壤碳汇功能，最终实现碳中和目标。

结合景观生态学，着力构建矿区绿色生态廊道、丰富矿区绿地系统、加强矿区植物配置规划，增强碳汇能力，如上海“深坑酒店”项目、陕西潼关小秦岭金矿国家矿山公园、黄石国家矿山公园、湖北应

城国家矿山公园、威海华夏城矿坑修复项目等,通过矿区创面危险消除、削坡平整土地施工以及生态环境恢复等工程措施,将矿区土地资源优化再利用,不仅完成了对矿区生态修复的目标和任务,提升了矿区土壤碳固存效果,同时也结合自然和人文景观,因地制宜盘活土地资源利用,创造了新的收益获取和增长点,而取得的收益可回馈于矿区生态环境管护,将生态优势转化为经济优势的延伸型补偿,顺应了新时代生态文明建设和"绿水青山就是金山银山"的时代要求。

4 结语

碳中和是顺应时代发展和符合人类社会发展需求的大趋势。为深入落实《中共中央 国务院关于全面加强生态环境保护坚决打好污染防治攻坚战的意见》(中发〔2018〕17 号)、响应党的十九大报告提出的"山水林田湖生命共同体"和习近平总书记"绿水青山就是金山银山"理论,实现 2060 年碳中和愿景目标,加快矿区生态修复和碳中和进展具有重大的战略意义。本文从矿区资源开采引发的生态系统碳排放问题出发,揭示了我国矿区生态修复过程中存在的一些不足,并针对性地提出了矿区节能减排、固碳增汇的实现对策,以期为我国矿区生态修复、实现碳中和提供必要借鉴。

参考文献

[1] 彭英健,姚有利,董川龙. 煤中低分子化合物的甲烷溶解能力研究[J]. 矿产综合利用, 2019(4):139-144.

PENG Y J, YAO Y L, DONG L C. Study on Gas Dissolution of Low Molecular Compounds in Coal[J]. Multipurpose Utilization of Mineral Resources, 2019(4):139-144.

[2] 刘玉芹,马鸿文,邓鹏,等. 热还原制备金属镁的反应热力学与工艺过程评价[J]. 矿产综合利用,2013(3):39-44.

LIU Y Q, MA H W, DENG P, et al. Thermodynamic Analysis and Processing Evaluation of Magnesium Preparation by Thermal Reduction Method[J]. Multipurpose Utilization of Mineral Resources, 2013(3):39-44.

[3] 杨光,张淑会,杨艳双. 烧结烟气中气态污染物的减排技术现状及展望[J]. 矿产综合利用, 2021(1):45-56.

YANG G, ZHANG S H, YANG Y S. Current Status and Prospects of Emission Reduction Technology for Gaseous Pollutants in Sintering Flue Gas[J]. Multipurpose Utilization of Mineral Resources, 2021 (1):45-56.

[4] 樊大磊,李富兵,王宗礼,等.碳达峰、碳中和目标下中国能源矿产发展现状及前景展望[J].中国矿业, 2021, 30(6):1-8.

FAN D L, LI F B, WANG Z L, et al. Development status and prospects of China's energy minerals under the target of carbon peak and carbon neutral[J]. China Mining Magazine, 2021,30(6):1-8.

[5] 张雅欣,罗荟霖,王灿,等.碳中和行动的国际趋势分析[J].气候变化研究进展,2021, 17(1):88-97.

ZHANG Y X, LUO H L, WANG C. Progress and trends of global carbon neutrality pledges[J]. Climate Change Research, 2021, 17(1):88-97.

[6] 白暴力,程艳敏,白瑞雪.新时代中国特色社会主义生态经济理论及其实践指引:绿色低碳发展助力我国"碳达峰、碳中和"战略实施[J].河北经贸大学学报, 2021, 42(4):26-36.

BAI B L, CHENG Y M, BAI R C. The Ecological Economic Theory on Socialism with Chinese Characteristics for a New Era and its Practical Guidelines:Green and low-carbon development helps the implementation of China's strategy of"peak carbon dioxide emissions,carbon neutrality"[J].Journal of Hebei University of Economics and Trade, 2021, 42(4):26-36.

[7] 张宁,赵玉.中国能顺利实现碳达峰和碳中和吗? 基于效率与减排成本视角的城市层面分析[J]. 兰州大学学报(社会科学版), 2021,49(4):13-22.

ZHANG N, ZHAO Y. Can China Achieve Peak Carbon Emissions and Carbon Neutrality:An Analysis Based on Efficiency and Emission Reduction Cost at the City Level[J]. Journal of Lanzhou University (Social Sciences), 2021,49(4):13-22.

[8] 欧阳志远,史作廷,石敏俊,等. "碳达峰碳中和":挑战与对策[J].河北经贸大学学报, 2021,42(5):1-11.

OUYANG Z Y, SHI Z T, SHI M J, et al. Challenges and Countermeasures of "Carbon Peak and Carbon Neutrality"[J]. Journal of Hebei University of Economics and Trade, 2021,42(5):1-11.

[9] WANG Qiang, LI Rongrong. Journey to burning half of global coal: trajectory and drivers of China's coal use[J].

Renewable & Sustainable Energy Reviews, 2016,58:341-346.

[10] 冯爱青,岳溪柳,巢清尘,等.中国气候变化风险与碳达峰、碳中和目标下的绿色保险应对[J].环境保护, 2021, 49(8): 20-24.

FENG A Q, YUE X L, CHAO Q C, et al. Climate Change Risk and Green Insurance Response Under the Goal of Carbon Peak and Carbon Neutral in China[J]. Environmental Protection, 2021, 49(8): 20-24.

[11] 王江,唐艺芸.碳中和愿景下地方率先达峰的多维困境及其纾解[J].环境保护, 2021, 49(15): 31-36.

WANG J, TANG Y Y. The Multi-dimensional Dilemma and its Solution of the Local Governments Taking the Lead in Peaking Carbon Dioxide Emissions Under the Vision of Carbon-neutrality[J]. Environmental Protection, 2021, 49(15): 31-36.

[12] 杨博宇,白中科.碳中和背景下煤矿区土地生态系统碳源/汇研究进展及其减排对策[J].中国矿业, 2021, 30(5): 1-9.

YANG B Y, BAI Z K. Research advances and emission reduction measures in carbon source and sink of land ecosystems in coal mining area under the carbon neutrality[J]. China Mining Magazine, 2021, 30(5): 1-9.

[13] REYNOLDS Brandon, REDDY K J. Infiltration rates in re-claimed surface coal mines[J].Water Air Soil Pollut, 2012,223 : 5941-5958.

[14] 白中科,周伟,王金满,等.再论矿区生态系统恢复重建[J].中国土地科学,2018,32(11):1-9.

BAI Z K, ZHOU W, WANG J M, et al. Rethink on Ecosystem Restoration and Rehabilitation of Mining Areas[J]. China Land Science, 2018,32(11):1-9.

[15] 于辉胜.区域性废弃矿山生态修复实践与思考[J].中国国土资源经济,2021, 34(6): 84-89.

YU H S. Practice and Thinking of Eco-Restoration of Regional Abandoned Mines[J]. Natural Resource Economics of China, 2021, 34(6): 84-89.

[16] RAJ K SHRESTHA, RATTAN LAL. Ecosys tem carbon budgeting and soil carbon sequestration in reclaimed mine soil[J]. Environ Int, 2006, 32 (6):781-796.

[17] 张黎明,张绍良,侯湖平,等.矿区土地复垦碳减排效果测度模型与实证分析[J].中国矿业,2015,24(11): 65-70.

ZHANG L M, ZHANG S L, HOU H P, et al. Evaluation model and empirical study of carbon emission reduction effect from mining land reclamation[J]. China Mining Magazine, 2015,24(11) :65-70.

[18] 阎赞,王想,徐名特,等. 尾矿资源化研究在铅锌尾矿中的应用[J]. 矿产综合利用, 2017 (1): 1-5.

YAN Z, WANG X, XU M, et al. Utilization Situation and Development Trend of Lead and Zinc Tailing Resources[J]. Multipurpose Utilization of Mineral Resources, 2017 (1): 1-5.

[19] 刘虹利,张均,王永卿,等. 磷矿固体废弃物资源化利用问题及建议[J]. 矿产综合利用, 2017 (1): 6-11.

LIU H L, ZHANG J, WANG Y Q, et al. Problems and Proposals of Solid Waste Utilization of Phosphate[J]. Multipurpose Utilization of Mineral Resources, 2017 (1): 6-11.

[20] 张金山,孙春宝,董红娟,等. 内蒙古大青山煤矸石资源化综合利用探讨[J]. 矿产综合利用, 2017 (2): 8-11.

ZHANG J S, SUN C B, DONG H J, et al. Exploration of Comprehensive Utilization of Coal Gangue Resource in Daqing Mountain[J]. Multipurpose Utilization of Mineral Resources, 2017 (2): 8-11.

[21] 吴西顺,张炜,杨添天,等. 地质冶金学发展及对建设智慧矿山的意义[J]. 矿产综合利用, 2021(5): 67-75.

WU X S, ZHANG W, YANG T T, et al. Development of Geometallurgy and its Significance to the Construction of Wisdom Mines[J]. Multipurpose Utilization of Mineral Resources, 2021(5): 67-75.

本文曾发表于 2022 年《矿产综合利用》第 2 期

A Study of the Differences in Heavy Metal Distributions in Different Types of Farmland in a Mining Area

Yangjie Lu[1,2,3,4], Yan Xu[1,2,3,4], Zhen Guo[1,2,3,4], Chendi Shi[1,2,3,4], ,Jian Wang[1,2,3,4]

(1. Key Laboratory of Degraded and Unused Land Consolidation Engineering, the Ministry of Land and Resources of China, Shaanxi, Xi'an, China; 2.Shaanxi Provincial Land Consolidation Engineering Technology Research Center, Shaanxi, Xi'an, China; 3.Shaanxi Provincial Land Engineering Construction Group Co., Ltd., Shaanxi, Xi'an, China; 4.Institute of Land Engineering and Technology, Shaanxi Provincial Land Engineering Construction Group Co., Ltd., Shaanxi, Xi'an, China)

【**Abstract**】To investigate the effects of different types of land use and soil depths on the distributions of heavy metals in the soil in mining areas, heavy metals in different soil layers of five types of agricultural land in the Tongguan gold mining area were studied. The results revealed that the land use type had a greater impact than soil layers on the distribution of heavy metals in the soil. Among the five types of agricultural land examined, the risk values were only exceeded for the heavy metals lead(Pb) and mercury(Hg) in the pepper field, indicating combined pollution of Pb and Hg. Furthermore, some of the heavy metals, such as Pb, zinc(Zn), cadmium(Cd) and Hg, were highly significantly and positively correlated with each other. The pepper field should be monitored to prevent pollution from other heavy metals.

【**Key Words**】Heavy metal pollution; Farmland; Pollution evaluation; Ecological risk

Introduction

Soil is the natural environment on which humans depend and an important resource for carrying out activities related to agricultural production (Zhang 2018). At present, global food, resource and environmental issues are closely related to soil pollution. With the rapid development of economic globalization, pollution containing heavy metals can enter the soil through different channels (Xiao et al. 2018), which results in heavy metal pollution in the environment. Many countries are facing the problem of heavy metal pollution in the soil, which seriously hinders agricultural production (Zhou et al. 2019). Heavy metal pollution in the soil can cause a decline in the yield and quality of crops and may endanger human health through the food chain. It may also lead to further deterioration of the air quality and aquatic environments, which has caused widespread concern in many countries around the world (Chen et al. 1999; Guo 2018; Zhou 1999).

The term heavy metal refers to metal elements that have a density of 5.0 $g \cdot cm^{-3}$ or greater (Song et al. 2018). Despite the fact that arsenic(As) and selenium(Se) are nonmetallic elements, their toxicity and chemical properties are similar to those of heavy metals, and therefore, these two elements are commonly included in the range of heavy metal contaminants (Jiang 2018). The heavy metals involved in environmental pollution mainly include mercury(Hg), lead(Pb), chromiun(Cr) and metalloid As, which exhibit significant biological toxicity, as well as a number of other toxic heavy metals, such as zinc(Zn), copper(Cu), cobalt(Co), nickel(Ni), cadmium(Cd), tin(Sn), and vanadium(V) (Han 2018).

Since soil is an open, dynamic system, exogenous heavy metals are inevitably involved in the process of material and energy exchange with the surrounding environment (Zhang et al. 2018). Heavy metal pollution in the soil, which results in the accumulation of foreign heavy metals in the surface layer of the soil, mainly originates from industrial waste; the diffusion, sedimentation and accumulation of heavy metals in waste gas (Chen 1996); the irrigation of farmland with heavy metal wastewater; and the extensive application of pesticides containing heavy metals and phosphate fertilizers (Sun 2018). At present, the area of cultivated land that is polluted by three types of industrial waste in China is as high as 1×10^7 hm^2, of which more than 3.3×10^6 hm^2 involves farmland that uses sewage irrigation (Liu and Zhang 2019). As a result, China's annual production of rice is reduced by more than 10 million tons due to heavy metal pollution (Fan et al. 2013). In addition, the amount of food contaminated by heavy metals is as high as 12 million tons per year, resulting in a total economic loss of at least 20 billion yuan (Liu et al. 2011). The heavy metal pollution in the soil is increasing annually (Liu et al.2007; Liao 2005).

Soil heavy metal pollution has become a research hot-spot for scholars. Chon and Lee (2004) investigated the contents of heavy metals in minerals and surrounding soils in several abandoned mining areas in South Korea. The study showed that the content of As and Cd in the soil was higher than other heavy metals according to the Soil Pollution Index (PI), and the study also evaluated the risk of contaminated soil on human's health. Fernández et al. (2009) investigated the contaminated soil around abandoned mining area in southwestern Spain and discussed the migration and distribution of heavy metals in the soil. Kim et al. (2002) investigated and analyzed the content of heavy metals in minerals, soil and crops in a mining area in southwest Korea. The analysis results showed that the soil was heavily polluted by heavy metals. Some scholars have also studied the heavy metal pollution in Tongguan County, Shaanxi Province. Xu et al. (2007) evaluated the heavy metal pollution of farmland soil in Tongguan gold mining area. Xie Juan et al. (2008) studied the pollution of farmland soil in the Shuangqiao River Basin of Xiaoqinling Mountain. Dai (2004) estimated the total amount of Hg produced in the past two decades in Tongguan and the whole country during gold metallurgy, but the research mainly studied the heavy metal pollution evaluation, and there are few studies on the pollution status of different types of agricultural land and the correlation of heavy metals.

Taking different types of agricultural land in the mining area as the research object, this study is to provide a scientific basis for risk assessment and treatment direction of farmland after heavy metal treatment in the mining area by monitoring soil pH, nutrient content and heavy metal content. Therefore, objectives of the current study were to: (1) study the effects of agricultural land types and soil depth on the distribution of heavy metals; (2) analyze whether there is a correlation between different heavy metals; (3) evaluate the heavy metal pollution status of farmland. We hypothesized that (1) both agricultural land types and soil depth will have an influence on the distribution of heavy metals, and the influence of agricultural land types is more obvious; (2) the farmland in the mining area as a whole is at a low ecological risk level, and there is a correlation between heavy metals.

Methods and materials

Overview of the study area

Tongguan County is located at the eastern end of the Guanzhong Plain in Shaanxi Province, at the junction of Shanxi, Shaanxi and Henan. This region is characterized by a warm, temperate, continental, monsoon, arid climate with four distinct seasons, including long winter and summer seasons and short spring and autumn seasons. In addition, the solar energy resources in this area are sufficient, the temperature and precipitation are low, and there are large spatial and temporal differences in environmental conditions. Due to the disorderly development of mineral resources in the past, the supervision of slag discharge is not strict, and a large number of illegal, small-scale and large-scale mining activities are difficult to prohibit. A large amount of the waste left after mining is randomly stored without any strict treatment, polluting the soil and surrounding environment. Tailing piles are often found in front of the houses of many local residents. Due to the lack of awareness of environmental protection, local residents have not stopped farming on heavily polluted land. Heavy metals, such as those accumulated in the soil, can enter the human body through the food chain, causing serious harm to the health of the local residents. The study area belonged to the tailings stacking area. Different crops were planted after uniform covering and leveling. The soil heavy metal pollution in the treated study area was regarded as a unified level, and the distribution was different after absorption and enrichment of different crops.

Test materials

On March 27, 2018, field investigations were conducted in Anle town, Tongguan County (110°09′32″~110°25′27″E, 34°23′33″~34°39′01″N). Surface soils from nearby contaminated farmland were sampled, ensuring that each sampling point was representative, while detailed information, such as the land and crop types, was recorded. To ensure that the samples were representative, a random sampling method was used (Lei et al. 1996).

There are five different types of land use selected in this study, namely wild grass field, wheat field, rapeseed field, pepper field and pomegranate field. During sampling, GPRS is used for 3D positioning during sampling and the information such as vegetation type was recorded (Table 1). To ensure that the samples were representative, the research adopted random point distribution method for sampling. Three points were randomly selected from each sampling area, and then conducted stratified collection of fresh soil at layers of 0~10 cm, 10~20 cm, 20~40 cm, 40~60 cm, 60~80 cm, and 80~100 cm. There are 75 soil samples in total.

Stainless steel soil drill was used for soil sample collection. Weeds, branches, rocks and other sundries on the surface of the sampling area were cleaned before sample collection. After the collection, bamboo slices were used to take out the soil that was not in contact with the sampler, and rocks, plant and animal residues were picked out and put the soil sample into the sample bag. Soil samples were air-dried and then sifted by a plastic sieve (100 mesh) before use.

Table 1 Title of sampling points

Sampling area	Type of agricultural land	Latitude(N)	Longitude(E)	Altitude (m)
1	Wild grass	34°29′32″	110°14′37″	740
2	Wheat field	34°29′37″	110°14′54″	750
3	Rapeseed field	34°29′37″	110°14′54″	760
4	Pepper field	34°29′38″	110°14′53″	760
5	Pomegranate field	34°32′35″	110°15′2″	570

Test methods

Environmental science mainly studies Hg, Cd, Pb, Cr and As with significant biotoxicity and Cu, Zn and Ni with toxicity (Zhao 2006). So, the test indicators selected the necessary test items for the risk screening value of agricultural land soil pollution: Cr, Ni, Cu, Zn, Cd, Pb, As and Hg (GB 15618—2018).

Determination of heavy metals Cr, Ni, Cu, Zn, Pb and Cd: weighed 0.1g of the soil sample (accurate to 0.0002g), added a small amount of water to moisten, then added 6 mL HNO_3, 2 mL H_2O_2 and 2 mL HF for digestion, after the digestion, diluted the volume to 50 mL, used ICP-MS (7700e, Agilent, USA) to measure (HJ 803—2016).

Determination of heavy metals Hg and As: weighed 0.1 ~ 1 g of the soil sample (accurate to 0.0002 g), added a small amount of water to moisten, added 10 mL of aqua regia and digested in a boiling water bath, and then diluted to 50 mL. Took a certain amount of digestion solution into a 50 mL colorimetric tube, added 3 mL HCl, 5 mL CH_4N_2S, 5 mL ascorbic acid solution to make the volume to 50 mL, took the supernatant liquid and measured it with an atomic fluorescence spectrometer (SK-2003AZ, Haiguang, China) (GB/T 22105—2008).

This study measured soil pH and nutrient indicators that may affect the content of heavy metals (Table 3). Soil pH, potassium, phosphorus, nitrogen and organic matter content were determined as described in *Soil Agrochemical Analysis* (Bao 1981).

Certified reference materials used to ensure quality of data were GSS-32 for metals, HTSB-2 for total nitrogen, organic matter, available potassium and available phosphorus, and ASA-9 for pH.

Data processing

The average and standard deviation of the data were calculated using Excel 2013 software. The data analysis was performed using SPSS 21.0 software, and a linear correlation analysis was performed.

Results and discussion

Quality assurance

From the analysis of the reference materials, the efficiency of the methods used for the determination of heavy metals, soil pH, potassium, phosphorus, nitrogen and organic matter were found to be satisfactory (Table 2).

Table 2 Recovery test results

Item	Reference material	Recovery(%)	RSD(%)
Cr	GSS-32	102.53	3.80
Ni		97.30	5.41
Cu		103.85	7.69
Zn		101.56	7.81
Cd		103.03	10.61
Pb		103.85	7.69
As		98.43	5.51
Hg		97.30	5.41
pH	ASA-9	100.35	0.82
Available potassium	HTSB-2	101.42	2.85
Available phosphorus		95.24	9.52
Total nitrogen		101.52	3.03
Organic matter		103.03	8.08

Distribution of heavy metals in the agricultural land in the mining area

Soil nutrients and pH are listed in the table below (Table 3).

Table 3 Soil pH and nutrient content

Item	pH	Available potassium (mg · kg^{-1})	Available phosphorus (mg · kg^{-1})	Total nitrogen (mg · kg^{-1})	Organic matter (g · kg^{-1})
Content	8.21±0.13	70.58 ±13.32	10.02 ±2.56	0.72±0.17	1.19±0.38

The contents of the heavy metal Cr in the soil ranged from 30.06 ~ 71.71 mg · kg^{-1} in different land use types in the farmland area (Fig.1). Among the different land use types, the Cr content in the soil of the pepper field exhibited large fluctuations, ranging from 30.00 ~ 72.00 mg · kg^{-1}. The Cr content in other land use types was in the range of 30.00 ~ 40.00 mg · kg^{-1}. This value changed little with the increase in soil depth, and there was no obvious trend. The analysis of variance demonstrated that the type of land had a significant effect on the concentration of the heavy metal Cr, while the soil layer did not have a significant effect on the Cr content (Table 4).

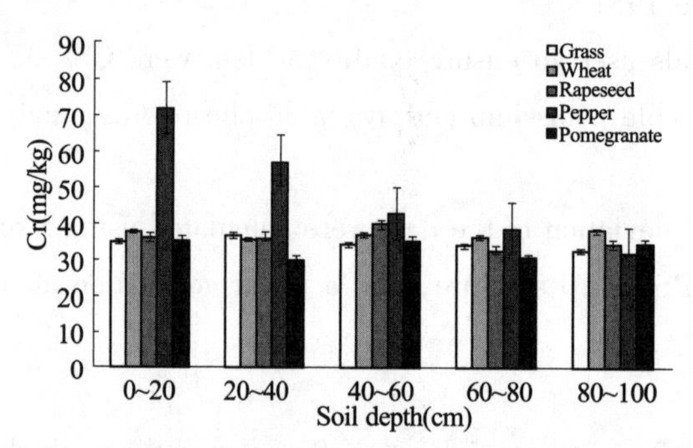

Fig.1 Cr

Table 4 P value of heavy metals in soil layer and different land use types

Item	Cr	Ni	Cu	Zn	Cd	Pb	As	Hg
Soil layer	0.320	0.155	0.191	0.023	0.016	0.114	0.693	0.255
Land use type	0.027	0.029	0.075	0.028	0.037	0.091	0.005	0.095

The Ni content in the agricultural land in the mining area ranged from 22.36 mg · kg^{-1} to 41.89 mg · kg^{-1} (Fig.2). The Ni content in the soil of the pepper field decreased with increasing soil depth and was highly concentrated in the 0 ~ 20 cm soil layer. The Ni content in the soil fluctuated within the range of 25.00 ~ 30.00 mg · kg^{-1}, with little change. The analysis of variance showed that the land use type had significant effects on the content of Ni, while changes in the content according to the soil layer were not significant (Table 4).

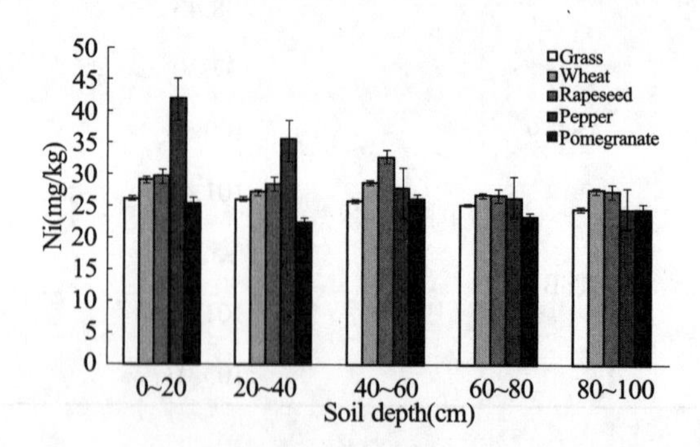

Fig.2 Ni

The content of the heavy metal Cu in the agricultural land of the mining area was between 14.81 and 82.87 mg · kg^{-1} (Fig.3). The Cu was mainly distributed in the top 0～40 cm of the soil in the pepper field, and in the other land use types, its distribution changed slightly. Except in the pomegranate field, the Cu content in the soil generally showed a decreasing trend with increasing soil depth. Table 1 shows that the influence of the soil layer and agricultural land type on the Cu content was not significant. Fig.3 shows that the Cu content was very high in the pepper field compared with that in the rest of the land use types. Thus, the land use type did affect the Cu content (Table 4).

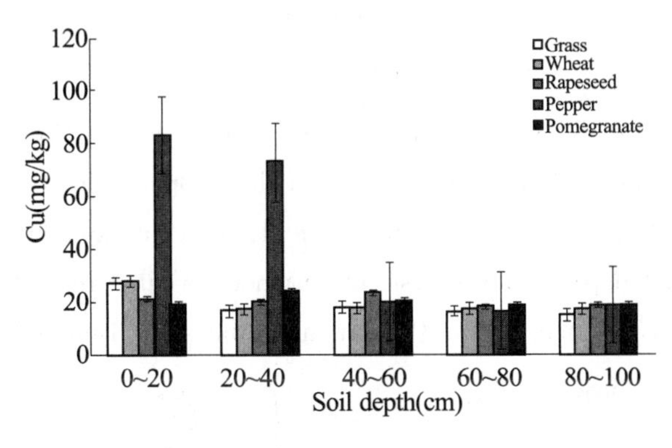

Fig.3 Cu

The soil content of the heavy metal Zn in the agricultural land of the mining area ranged from 40.25～98.22 mg · kg^{-1} (Fig.4). The distribution of Zn in different types of agricultural land was quite different. In the grassland, wheat field and pepper field, the Zn was concentrated in the upper layer of the soil and decreased with increasing soil depth. In the rapeseed and pomegranate fields, the Zn was mainly distributed in the middle soil layers, with an initial increase and then a decrease in Zn content with increasing soil depth. According to the results of the analysis of variance, both soil layer changes and agricultural land use types had significant effects on the Zn content in the soil (Table 4).

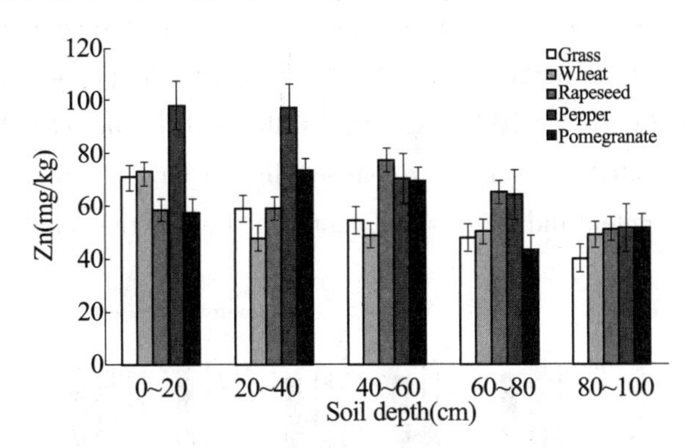

Fig.4 Zn

The soil content of the heavy metal Cd in the agricultural land of the mining area ranged from 0.08～0.56 mg · kg^{-1} (Fig.5). The Cd was mainly distributed in the 0～20 cm soil layer of the grassland and wheat field and in the 0～60 cm soil layer of the pepper field; the Cd content in the other land use types was low. According to the results of the analysis of variance, both soil layer changes and agricultural land use types had significant effects on the content of Cd in the soil (Table 4).

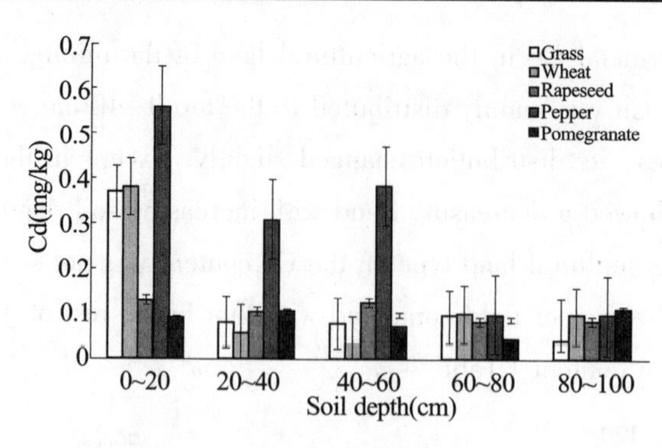

Fig.5 Cd

As shown in Fig.6, the content of Pb in the soil ranged from 11.65~206.01 mg · kg^{-1}, and this heavy metal was mainly distributed in the 0~20 cm soil layer of the pepper field. Compared with that of Cr, the fluctuation of Pb was larger. Except in the pomegranate field, the content of Pb in the soil showed a decreasing trend with increasing soil depth. The analysis of variance revealed that changes in the soil layer and land use types had no significant effect on the Pb content in the soil (Table 4).

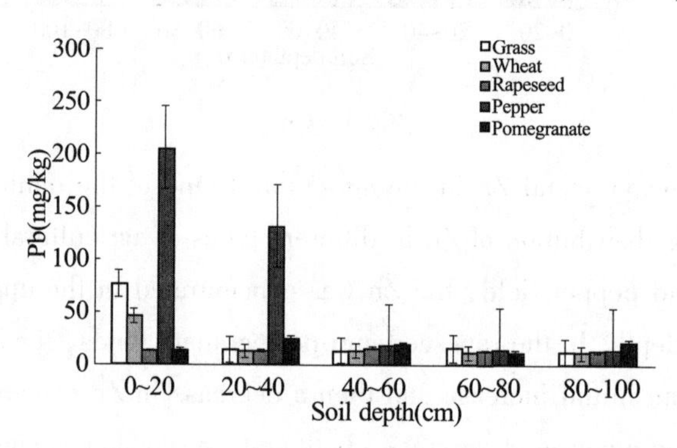

Fig.6 Pb

As shown in Fig.7, the content of the heavy metal As in the farmland ranged from 4.18~11.52 mg · kg^{-1}, with the exception of the pepper field. The As content in the pepper field increased with increasing soil depth, whereas it undulated by 10.00 mg · kg^{-1} with depth in the other land use types. The results of the analysis of variance revealed that soil layer changes did not have a significant effect on the content of As in the soil; however, its content did differ significantly in the different land use types (Table 4).

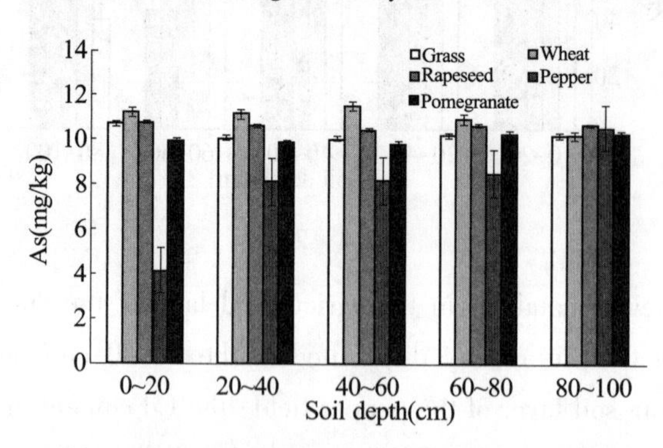

Fig.7 As

As shown in Fig.8, the content of the heavy metal Hg in the agricultural land in the mining area ranged from 0.01~5.89 mg · kg^{-1}. Except in the rapeseed field, in which there was a high Hg content in the 80~100 cm soil layer, the Hg was mainly observed in the 0~20 cm and 20~40 cm soil layers and showed a decreasing trend with increasing soil depth. The analysis of variance showed that the effects of the land use type and the soil layer depth on the distribution of Hg were not significant (Table 4).

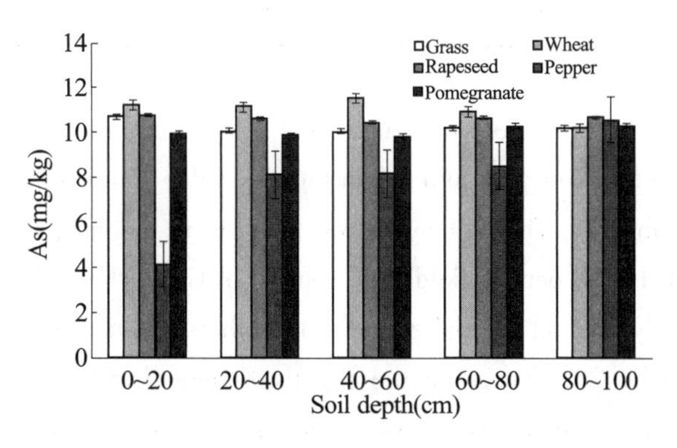

Fig.8 Hg

Correlations between the heavy metals in the farmland

As shown in Table 5, there was a significant positive correlation between the contents of Cr and Cu in the agricultural land in the mining area, while there was a highly significant and negative correlation between Cu and As. In addition, Pb had a highly significant and positive correlation with Zn, Cd and Hg. The negative correlation between As and Pb as well as the correlation between Cd and Hg also reached a significant level. Among the heavy metals in the agricultural land in the mining area, there was not a significant correlation between Ni and the other heavy metals. Combined with the analysis of sampling sites, the sampling area is in the mineral resource development zone. The source of heavy metal pollution is mainly ore smelting. In Mao's research, the ore and wall rock of the gold mine contain Hg, Pb, Zn, etc., and mineral smelting was also the maiWn source of heavy metal Cb, which is basically consistent with the correlation between heavy metals (Mao et al. 2000).

Table 5 Correlation coefficients between the heavy metals in the agricultural land in the mining area

	Cr	Ni	Cu	Zn	Cd	Pb	As	Hg
Cr	1							
Ni	0.929	1						
Cu	0.923**	0.849	1					
Zn	0.757	0.736	0.818	1				
Cd	0.743	0.666	0.744	0.751	1			
Pb	0.902	0.808	0.967	0.782**	0.816**	1		
As	−0.850	−0.669	−0.780**	−0.670	−0.671	−0.790*	1	
Hg	0.827	0.798	0.892	0.701	0.680*	0.876**	−0.655	1

Note: * $P<0.05$; ** $P<0.01$.

Table 6 Soil heavy metal risk screening value of agricultural land ($mg \cdot kg^{-1}$)

Heavy metals	Cr	Ni	Cu	Zn	Cd	Pb	As	Hg
Screening values	250	190	100	300	0.6	170	25	3.4

Comparison of the heavy metal content and risk value in the farmland

The content of Cr in the farmland was lower than the risk value, indicating that there was no risk of Cr pollution in the farmland (Table 7). The Ni content was also low in the soil; therefore, there was no risk of Ni pollution in the farmland. In the pepper field, the content of Cu in the soil reached 83% of the risk value, while it was approximately 25% of the risk value in the other land use types, indicating that there is no risk of heavy metal pollution by Cu. Except for the slightly higher distribution in the pepper field, the Zn content in the soils of the other land use types was similar, and the risk value was similar to that of Cu. However, the Cd content in the soil in the pepper field reached 93% of the risk value, while that in the grassland and wheat field was also more than 60% of the risk value. Despite the lack of risk of pollution by Cd, the concentration of this metal should be monitored. Furthermore, the Pb in the soil of the pepper field exceeded the risk value by 21%; thus, there was a risk of Pb pollution. The Pb content in the other land use types was less than 50% of the risk value, indicating that there was no pollution risk. The content of the heavy metal As in the different land use types in the mining area was similar, ranging between 40% and 50% of the risk value, suggesting no risk of As contamination. Finally, the content of the heavy metal Hg in the soil of the pepper field exceeded the risk value by 73%, indicating that there is a risk of Hg pollution. The Hg content in the other land use types varied greatly. For example, the Hg content in the pomegranate field only reached 3% of the risk value, while the Hg content in the rapeseed field reached 98% of the risk value. However, the contents of Hg in these fields did not exceed the risk value, thus, there was no risk of Hg pollution. In this study, the heavy metal Hg had a higher risk of pollution, followed by Pb and Cd, which is consistent with the study by Xu et al. (2008)

Table 7 Comparison of heavy metals and risk values of the agricultural land in the mining area(%)

Land use type	Cr	Ni	Cu	Zn	Cd	Pb	As	Hg
Wild grass	0.15±0.02	13.79±0.78	26.83±2.01	23.61±0.41	61.67±2.33	45.21±1.55	42.88±0.21	64.71±1.78
Wheat field	0.15±0.01	15.33±0.53	27.63±0.72	24.16±0.39	63.33±1.98	26.99±0.85	46.08±0.18	26.47±0.88
Rapeseed field	0.05±0.00	17.26±0.66	23.27±0.55	25.75±0.54	21.67±0.69	9.21±0.03	43.16±0.76	97.94±1.63
Pepper field	0.22±0.03	22.05±1.20	82.87±1.49	32.74±0.63	93.33±2.57	121.18±4.38	42.40±0.85	173.24±8.97
Pomegranate field	0.04±0.00	13.76±0.22	23.85±0.26	24.26±0.45	18.33±0.34	15.12±0.52	41.32±0.47	3.24±0.12

Note: The screening values represent the soil pollution risk for nonfield agricultural land at a risk value of pH>7.5 (GB 15618—2018).

Risk assessment of the heavy metals in the farmland

A single-factor evaluation can reflect different degrees of environmental pollution by different pollutants. Generally, this method is based on soil environmental quality standards (Li et al. 2009). In this study, the risk screening value of the soil environmental quality standards was used as the critical value in the heavy metal evaluations. The risk screening value was calculated as follows:

$$P_i = C_i / S_i \quad (1)$$

In equation 1, P_i is the pollution index of the i-th heavy metal element; C_i is the measured content of the i-th heavy metal element; and S_i is the critical value of the evaluation standard for a given heavy metal element.

The Nemerow composite index is one of the most commonly used methods to calculate the comprehensive pollution index locally and regionally (Li 2012). It considers the impact of high concentrations of pollution and the combined pollution from multiple heavy metals on soil environmental quality. Considering the polluted area as a whole, an evaluation of the pollution status was performed (Wang 2010) using the following formula:

$$P_n = \sqrt{\frac{P_{imax}^2 + P_{iave}^2}{2}} \tag{2}$$

In equation 2, P_n is the integrated Nemerow index of pollution; P_{imax} is the maximum pollution index of each pollution element; and P_{iave} is the arithmetic average of the pollution index of each pollution element.

According to the single-factor pollution index, the heavy metal contamination in the 1 m soil layer of the agricultural land in the Tongguan gold mining area was relatively light, with an overall value of $P_i<1$, which indicates no contamination. Only the P_i of Hg in the pepper field was close to 1, suggesting that pollution is possible. According to the values of the Nemerow pollution index for these six heavy metal elements, the pollution in the pepper field in the agricultural land near the Tongguan gold mine reached the warning level, indicating that there is a potential pollution risk. The values of the integrated pollution indexes of these six heavy metal elements from high to low were as follows: pepper > wheat > rape seed> grass > pomegranate. Therefore, we should focus on how to reduce heavy metal pollution in pepper fields (Table 8 and Table 9).

Table 8 Classification of the soil environmental quality assessment

Level	Single-factor pollution index		Nemerow composite index	
	P_i	Pollution assessment	P_n	Pollution assessment
Ⅰ	$P_i \leqslant 1$	Pollution free	$P_n \leqslant 0.7$	Clean (safe)
Ⅱ	$1<P_i \leqslant 2$	Slight pollution	$0.7<P_n \leqslant 1.0$	Clean (warning limit)
Ⅲ	$2<P_i \leqslant 3$	Light pollution	$1<P_n \leqslant 2.0$	Light pollution
Ⅳ	$3<P_i \leqslant 5$	Moderately polluted	$2<P_n \leqslant 3.0$	Moderately polluted
Ⅴ	$P_i>5$	Severe pollution	$P_n>3$	Severe pollution

Table 9 Results of the evaluation of the heavy metals in the agricultural land

Land use type	Single-factor pollution index (P_i)					Nemerow composite index
	Cd	Pb	As	Hg	Average value	(P_n)
Grass	0.30	0.21	0.41	0.24	0.29	0.36
Wheat	0.30	0.14	0.44	0.14	0.26	0.36
Rapeseed	0.18	0.08	0.43	0.36	0.26	0.36
Pepper	0.55	0.58	0.29	0.91	0.58	0.76
Pomegranate	0.17	0.11	0.40	0.01	0.17	0.31

Hakanson established a set of methods for evaluating the potential ecological hazards of heavy metals based on their properties (Hakanson 1980). These methods, which can objectively and comprehensively evaluate soil heavy metal pollution, take into account the soil heavy metal content, the ecological and environmental effects of the heavy metals, and the toxicological characteristics of the heavy metals (Lu and Li 2017). The potential ecological hazards of heavy metals are calculated as follows:

$$E_i = T_i \times \frac{C_i}{S_i} \tag{3}$$

$$\mathrm{RI} = \sum_{i=1}^{m} E_i \tag{4}$$

In equations 3 and 4, C_i and Si are the same as previously described; E_i is the potential ecological risk factor for a single heavy metal element; RI is the comprehensive potential ecological hazard index of the heavy metal elements; and T_i is the toxicity response coefficient of a single heavy metal element according to related research (Xue et al. 1986). The toxicity response coefficients of Cd, Pb, As, and Hg are 30, 5, 10, and 40, respectively.

Scholars have classified the individual ecological risks and comprehensive potential ecological risks of heavy metals. The classification results are shown in Table 10. Table 11 shows the index values of the potential ecological risk in the study area. The results showed that the single-factor ecological risk index values of Cd, Pb, As, and Hg were 9.00, 1.12, 3.94, and 13.30, respectively. These heavy metals all had low ecological risk levels, and from high to low, their index values were as follows: Hg > Cd > As > Pb. The average value of the comprehensive ecological risk index in this region was 27.34, and the maximum value was 58.70. Therefore, it was concluded that the agricultural land as a whole had a low ecological risk.

Table 10 Potential ecological risk level of the heavy metals

Individual ecological risk index		Comprehensive ecological risk index	
Level	Points	Level	Points
Low ecological risk	<40	Low ecological risk	<150
Medium ecological risk	40~80	Medium ecological risk	150~300
Higher ecological risk	80~160	High ecological risk	300~600
High ecological risk	160~320	Very high ecological risk	>600
Very high ecological risk	>320		

Table 11 Prediction of the potential ecological risk of the heavy metals in the soil

Land use type	Single-factor ecological risk (*E*)				Comprehensive ecological risk
	Cd	Pb	As	Hg	(RI)
Grass	9.00	1.05	4.10	9.60	23.75
Wheat	9.00	0.70	4.40	5.60	19.70
Rapeseed	5.40	0.40	4.30	14.40	24.50
Pepper	16.50	2.90	2.90	36.40	58.70
Pomegranate	5.10	0.55	4.00	0.40	10.05
Average value	9.00	1.12	3.94	13.30	27.34

Discussion and conclusion

Land use type has significant effects on the contents of heavy metals, such as Cr, Ni, Zn, Cd and As, in the soil. In this study, the land use type had a significant effect on the As in the soil, while the soil layer affected only the contents of Zn and Cd in the soil. Zn and Cd were greatly affected by both the soil depth and land use type. Cr, Ni and As were greatly affected by land use type, while Cu, Pb and Hg were affected by soil depth and land use type to a lesser extent. Studies have shown that changes in land use type have a greater impact than soil depth on heavy metals, which may be related to the migration of heavy metals and root adsorption by plants (Duo ang Wang 2014; Zheng 2010; Song et al. 1999).

In this study, according to the single-factor pollution index, the heavy metal contamination in the 1 m soil layer of the agricultural land in Tongguan gold mining area was not serious, with an overall value of $P_i<1$, which at unpolluted level. Only the P_i of mercury in the pepper field was close to 1, suggesting that is dangerous to pollution. According to the values of the Nemerow pollution index for these six heavy metal elements, the pollution in the agricultural land of the pepper field in Tongguan gold mine reached the warning level, indicating that there is a potential pollution risk. Judging by the comprehensive ecological risk index, it was concluded that the agricultural land as a whole had a low ecological risk.

But from the perspective of the risk screening value of heavy metals in agricultural land alone, the heavy metal content in the soils of the wheat, rapeseed, grass and pomegranate fields did not exceed the risk screening value, which has no risk of heavy metal pollution. However, the maximum values of Pb and Hg in the pepper field were 206.01 mg · kg^{-1} and 5.89 mg · kg^{-1}, 1.21 and 1.73 times greater than the risk screening values, which were Pb-Hg complex contamination. This is the same as the results obtained by Lin et al. (2020) and Zhang et al. (2019). In addition, there is a significant positive correlation among the heavy metals Hg, Pb, and Cd in the farmland of the mining area. The heavy metal Cd of pepper field reached 93.33% of the risk screening value. Therefore, while pepper field was focused on dealing with heavy metal Hg and Pb pollution, it should also prevent heavy metal Cd pollution. At the same time, the heavy metal Hg in the rapeseed field is also close to the risk screening value, so the rapeseed field also needs to prevent heavy metal Hg pollution.

In general, both different types of agricultural land(different crops) and soil depth will affect the distribution of heavy metals in the agricultural land of mining areas, but different crops had a greater impact on them. Heavy metals Hg and Pb related to gold smelting in some areas (pepper field) were beyond the risk value and have significant correlation. However, the agricultural land in the mining area was at a low ecological risk level. At present, it is necessary to treat the heavy metal Hg and Pb in pepper land. At the same time, heavy metal Cd Hg pollution should be prevented in pepper field due to the significant positive correlation among the heavy metals Cd, Hg, and Pb, the rapeseed field also need to prevent heavy metal Hg pollution.

Acknowledgments

We thank the research staff for their contributions to this work. This study was funded by Shaanxi Provincial Land Engineering Construction Group Co., Ltd. (Grant No: DJNY2020-17).

References

Bao S D (2000) Soil Agrochemical Analysis. 3rd Edition. China Agriculture Press.

Chen H M, Zheng C R, Tu C, Zhu Y G (1999) Status and Prevention Countermeasures of Heavy Metal Pollution in Soils in China. AMBIO-Journal of Human Environment 28: 130-134,207.

Chen H M (1996) Pollution of Heavy Metals in Soil-Plant System. Science Press, Beijing.

Chon H T, Lee J S (2004) Heavy metal contamination and human risk assessment around some abandoned Au-Ag and base metal mine sites in Korea. Seoul: School of Civil, Urban and Geosystem Engineering Seoul Mational University 151-744.

Dai Q J (2004) Environmental Geochemistry of Mercury in China's Amalgamation Gold Mining Area: A Case Study of Tongguan, Shaanxi.

Duo J W, Wang W (2014) Study on the effect of root exudates on the adsorption of heavy metals Pb~($^{2+}$). Science and Technology Innovation Review 11: 5-6,8.

Fan T, Ye W L, Chen H Y, et al.(2013) Review on contamination and remediation technology of heavy metal in agricultural soil. Ecology and Environmental Sciences 22:1727-1736.

Fernández-Caliani J C, Barba-Brioso C, González I (2009) Heavy metal pollution in soils around the abandoned mine sites of the Iberian Pyrite Belt (Southwest Spain). Water, air, and soil pollution 200(1-4): 211-226.

GB/T 22105—2008 Soil quality: Determination of total mercury, total arsenic and total lead.

GB 15618—2018 Soil Environmental Quality Standard.

Guo X F(2008) Effects of biochar and AM fungi on soil nutrient and growth of Wangjiangnan under heavy metal pollution. Acta Prairie Sinica 11: 150-161.

Hakanson L(1980) An ecological risk index for aquatic pollution control.a sedimentological approach. Water Research 14: 975-1001.

Han X Y(2018) Study on the migration of heavy metal pollutants in soil. China Resources Comprehensive Utilization 36: 145-146,150.

Jiang X(2018) Study on the migration and enrichment of heavy metals in shallow soils. Proceedings of the 2018 National Academic Annual Conference of Environmental Engineering. Editorial Department of Environmental Engineering 6.

Kim J Y, Kim K W, Lee J U(2002) Assessment of As and heavy metal contamination in the vicinity of Duckum Au-Ag mine, Korea. Environmental Geochemistry and Health 24(3): 213-225.

Lei Z D, Yang S X, Luo Y(1996) Research on the method of monitoring the situation of public opinion in the field. Irrigation and Drainage 03: 9-15.

Li C K, Wang Li D, Li Y, Zhou X S(2009) Research progress on evaluation methods of soil heavy metal pollution. Rock and Soil Mechanics 30: 155-159.

Li G F(2012). Comparison of Different Pollution Evaluation Methods for Heavy Metals in Soil: A Case Study of the Drainage Ditches of the Gold Mine in Tongguan County. Guizhou Agricultural Sciences 40: 222-225.

Liao G L(2005) Study on migration and pollution evaluation of heavy metals in typical non-ferrous metal mines. Central South University.

Lin Y, Liang W J, Jiao Y, et al. (2020) Ecological health risk assessment of heavy metals in farmland soil around gold mining area in Tongguan County, Shaanxi. Geology of China 1-21.

Liu J L, Zhang J T(2019) Current status and treatment of heavy metal pollution in soil. Shandong Industrial Technology 07: 229.

Liu J, Teng Y G, Cui Y F, Wang J S (2007) A Review of Ecological Risk Assessment Methods for Soil Heavy Metal Pollution. Environmental Monitoring Management and Technology 03: 6-11.

Liu Y, Yue L L, Li J C(2011) Heavy metal pollution in soils of Taiyuan City and its potential ecological risk assessment. Journal of Environmental Science 31: 1285-1293.

Lu N, Li G(2017) Evaluation of Heavy Metal Pollution in Soil Around Tailings Slag in Gold Mining Area. China Mining Industry 26: 81-87.

Mao J W, Gu S Y, Zhang Q H(2000) Heavy metal pollution in the process of gold mining.Guizhou Environmental Protection Technology (03):17-22,28.

Song S Q, Wu H, Huang S Y(1999) Study on the Migration and Transformation Law of Heavy Metals in Soil-Crop System. Journal of Guangxi Teachers College(Natural Science Edition) 04: 87-92.

Song W H, Yao J, Wang W F (2018) Study on Source, Distribution and Risk Assessment of Heavy Metal Pollution in Soil. Resource Conservation and Environmental Protection 10: 78-79.

Sun X Y(2018) Study on the availability of heavy metal plants in lead-zinc mining areas. China University of Geosciences

(Beijing).

Wang X W (2010) Discussion on the relationship between farmland soil heavy metal pollution and crops in gold mining area. Chang' an University.

Xiao R, Nie Y J, Cao Q F, Qi C H, Zhao J, Li Q(2018) Characteristics and treatment of moderate and mild heavy metal contaminated farmland. Chinese Agricultural Science Bulletin 33: 101-106.

Xie J, Xu Y N, Qian H, He K(2008) Analysis and Evaluation of Heavy Metals in Farmland Soils in Shuangqiao River Basin. Gold 29(3): 46-50.

Xu Y N, Ke H L, Zhao A N, Liu R P, Zhang J H (2007) Evaluation of soil heavy metal pollution in a gold mining area in Xiaoqinling. Soil Science Bulletin 04: 732-736.

Xu Y N, Zhang J H, Zhao A N(2008) Evaluation of potential ecological hazards of heavy metal pollution in farmland soil of a gold mining area in Xiaoqinling. Geological Bulletin 27(8): 1273-1275.

Xue C Z, Xiao L, Wu Q F, Li D Y, Wang K X, Li H G(1986) Study on the background values of ten elements in the main agricultural soils of Shaanxi Province. Journal of Northwest A & F University (Natural Science Edition) 03: 30-53.

Zhang J, Hu F J, Lu CB, Liu Z W, Yang X Y, Yan Y Q (2018) Suggestions on the study of soil heavy metal pollution control in rare earth mining areas. Applied Chemicals 47: 1254-1257,1262.

Zhang J L, Lei J X, Zhao X J , An X L, Bai Y S(2019) Analysis and Evaluation of Heavy Metal Pollution in Farmland of Tongguan County, Shaanxi Province.Northwest Journal of Agriculture 28(2):247-252.

Zhang S(2018) Harm of heavy metal pollution in soil and its control measures. World Nonferrous Metals 18: 285-286.

Zheng S A (2010) Study on the transformation and migration characteristics of heavy metals in typical farmland soils in China. Zhejiang University.

Zhou Y C, Sun H L, Chen X G, Zhou L, Wu S S(2019) Heavy metal pollution characteristics and ecological risk assessment of surface soil in Yining City, Oasis City. Arid Area Resources and Environment 02: 127-133.

Zhou Z Y (1999) Heavy Metal Pollution and Control in Chinese Vegetables. Researches on Resource Ecology Environment 10: 21-27.

本文曾发表于2022年《Bulletin of Environmental Contamination and Toxicology》第109卷第5期

The Influence of Cognitive Level on the Guaranteed Behavioral Response of Landless Farmers in the Context of Rural Revitalization—An Empirical Study Based on PLS-SEM

Yangjie Lu[1,2], Hao Dong[1,2], Huanyuan Wang[1,2]

(1.Institute of Land Engineering and Technology, Shaanxi Provincial Land Engineering Construction Group Co., China; 2.Shaanxi Provincial Land Engineering Construction Group Co., Ltd., China)

【Abstract】 With the continuous acceleration of urbanization and agricultural modernization in China, the trend of concentration of rural land transfer is irreversible. For landless farmers, the absence of land guaranteed function inevitably gives rise to the substitution effect of other guaranteed methods. And the subjective preferences exhibited by farmers in making guaranteed behavior decisions can be quantitatively described as guaranteed behavioral responses. Based on the analytical framework of distributed cognitive theory, this paper adopts the validated factor analysis method of structural equation modeling to quantitatively study the cognitive basis and behavioral responses of landless farmers' guaranteed behavior by combining the survey data of rural households in typical rural areas of Wuhan urban area. The study shows that the guaranteed behavioral responses of landless farmers are significantly influenced by the cognitive level. "Locality power", "cultural power", and "personal power" are the main, important, and effective cognitive levels that influence farmers' guaranteed behavioral responses, respectively. Policy-based protection occupies a dominant position in the rural social guaranteed system, savings-based protection still plays an important function in rural areas, and market-based protection has greater development potential.

【Keywords】 Cognitive hierarchy; Landless farmers; Guaranteed behavioral response; Distributed cognitive theory; Structural equation modeling

1 Introduction

China is currently in the stage of rapid industrialization and urbanization, and a large amount of agricultural land in good locations around cities has been expropriated. It is projected that the urbanization rate in China will exceed 75% in 2030, with an average annual increase of 1%~2%, and the new urban population will be about 17 million per year, while the number of landless farmers will exceed 78 million (Bao and Peng, 2016), and landless farmers have become a huge interest group at this stage. In reality, local governments and developers often argue and contradict with farmers over land acquisition compensation and resettlement measures, and many farmers do not want their land to be expropriated, resulting in conflicts of wills and more serious consequences of violent conflicts.

Since the "globalization of land expansion" in 2000, conflicts about landless peasants have been happening continuously and have become more and more intense. Therefore, scholars have begun to focus on the conflict behavior of landless peasants, and have conducted many useful explorations from different perspectives. Bao et al. (2021) point out that in the past 10 years, the attitude of land-lost farmers towards land acquisition has changed greatly, from the initial compromise and acceptance to resistance and rights protection. He believes that China's social transformation and urbanization process awaken the land-lost farmers' awareness of land protection, and then affect land-lost farmers' land acquisition conflict behavior (Li et al., 2020; Su et al., 2020). Liu et al. (2018) analyzed the formation mechanism of land-lost farmers' right protection behavior in land expropriation from the micro level, and applied Logistic model to analyze the influencing factors of land-lost farmers' right protection behavior. Mishra and Mishra (2017) believes that economic interest factors (per capita income level of family, Engel's coefficient, compensation ratio of land expropriation loss) have a greater impact. Through the game analysis of the behaviors of local government, central government and peasant households in land expropriation conflicts, Cao and Zhang (2018) found that peasant households' rights protection behavior depends on the cost of rights protection, compensation standard and land expropriation behavior of local governments. Bao et al. (2020) use the theoretical framework of social action to study the action strategies of land-lost farmers in the land expropriation environment, and believes that the changes of land-lost farmers' rights protection methods and behavioral strategies at different stages are to maximize their own interests. Based on Korf's scenario model and in combination with India's macro-social and economic environment, Ghelichi et al. (2018) proposed the context-scenario model and applied it for the first time to study farmers' participation in land acquisition conflicts in India, explored the influencing factors and action paths of farmers' conflict behaviors, and concluded that the relationship between cadres and groups in rural society should be mainly managed. By studying the living conditions of peasants in Southeast Asia, Printsmann et al. (2022) proposed the concepts of "survival theory" and "moral economy", arguing that when peasants' survival morality and social justice are violated, they will have a strong will to resist and even resort to desperate measures. Other scholars, from the perspective of social problems in rural China, argue that there is a significant correlation between the occurrence of mass incidents in rural China and the rate of urbanization, and that the rent-seeking behavior of local governments and the lack of legitimate organizations representing farmers' rights and interests have led landless farmers to resort to non-institutionalized violence to defend their rights. Through a review of the existing literature, it can be found that previous studies on landless farmers' conflict behavior have mainly focused on the choice of conflict behavior, the context of behavior generation and causes, while not enough attention has been paid to the psychological behavior of individuals or groups in the process of land acquisition. In general, the literature on conflict willingness in China is scarce, and most of it remains in the qualitative research stage, and no scholars have conducted a comprehensive and systematic study on landless farmers' conflict willingness.

The research methods are mostly limited to statistical analysis and quantitative analysis, and the research conclusions are not the same. On the basis of summarizing the existing literature, this paper aims to explore the cognitive logic of the guaranteed behavior decision-making of land-lost farmers under the current situation, and further study the cognitive basis behind the guaranteed behavior of farmers by combining the micro survey data of farmers in Xi'an urban circle and the confirmatory factor analysis method of structural equation model. It also provides policy suggestions for improving risk management and guaranteed level of land-lost farmers and promoting rational improvement of rural social guaranteed system.

2 Theoretical analysis and research hypotheses

Psychology has a long history of research on individual cognitive activity, and Hatch's research team pioneered the concentric circles model of distributed cognition (see Fig.1) back in 1993 (Rogers and Ellis, 1994), which posits that individual cognition is influenced by a combination of individual human, regional, and cultural power (Liu et al., 2008). Distributed Cognition Theory (DCT) was thus born. As a new perspective to observe the complete process of cognitive activities, distributed cognition no longer emphasizes the influence of individual characteristics on cognitive activities unilaterally, but takes the cognitive level of individuals' processing of environmental information as the basic unit of research (Roessler et al., 2022). Relevant empirical studies have proved that distributed cognition theory has strong explanatory power for individual cognitive activities in complex environments, and its analytical framework is applicable to the study of farmers' behavior (Yakhlef, 2008).

The guaranteed behavioral response (GBR) studied in this paper is a customized abstract concept that connotes the subjective preferences that farmers exhibit when making guaranteed behavioral decisions (Lyver et al., 2019). According to the basic framework of distributed cognition (Belland, 2011), the cognitive environment of farmers can be subdivided into three cognitive levels: individual power (IP), regional power (RP) and cultural power (CP) for the specific problem of GBR (Lu et al., 2020), and there is a theoretical influence path of "cognitive level → behavioral response" (see Fig.1). Based on this, the hypothesis to be tested in this paper is as follows.

IP is the basis and core of distributed cognition, located in the innermost circle of the concentric circle model (Yen and Tsao, 2020), which emphasizes the characteristics and subjective motivation of the cognitive subject (Stiller and Schworm, 2019). The individual characteristics that may have significant influence include gender, age, social class status and education level of farmers, while the effective manifestation of their subjective motivation is their part-time work ability (Jordakieva et al., 2020). Theoretically, the cognitive level of individuals is proportional to their growing experience, and their behavioral decision-making process will tend to be rational with the accumulation of experience.

Hypothesis (H1). IP is positively related to GBR.

RP is the key of distributed cognition, which is located in the middle layer of concentric circle model and emphasizes the interaction between cognitive subject and cognitive environment (Valles-Colomer et al., 2019; Meyer et al.). When choosing the guaranteed mode, farmers will make behavioral decisions to a large extent depending on certain family roles, so the family is the main cognitive environment for individual farmers at this time (Meert et al., 2005). For this special group of land-lost farmers, the family environmental factors that may have a significant impact include family income, livelihood resilience, land transfer ratio, life attitude and family happiness (Fang et al., 2016). The above indicators can measure the quality of family life of farmers to a certain extent, and theoretically farmers are more inclined to guarantee high-quality family life.

Hypothesis (H2). RP is positively related to GBR.

CP is an abstract event that can have an indirect influence on specific cognitive activities (Gardner et al., 2000). It is located in the outermost layer of concentric circle model, and its influence on cognitive activities can not be ignored (Jang and Kim, 2019). Farmers mainly live in traditional villages inhabited by acquaintances (Zhao and Zou, 2017), and their social relationship network is relatively simple. Their choice of guaranteed mode is mainly influenced by herd psychology, policy publicity and policy satisfaction (Du et al., 2018). Theoretically, the direction of public opinion and public policy is largely the same, that is, to improve the level of rural social guaranteed and farmer household guaranteed is a basic consensus.

Hypothesis (H3). CP is positively related to GBR.

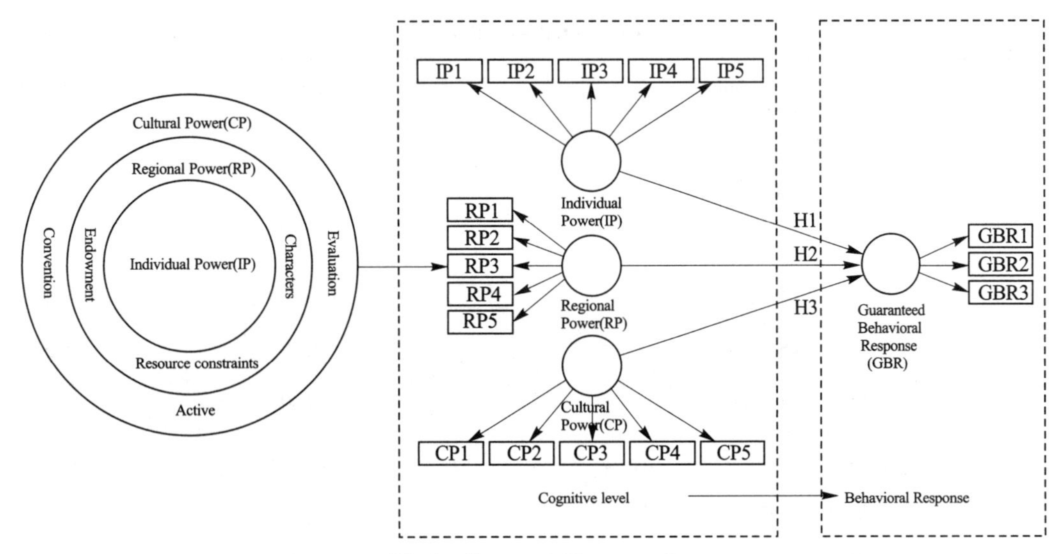

Fig.1 Research Framework

3 Materials and methods

3.1 Sampling and data collection

The questionnaire was answered by the most widely used questionnaire survey website "Questionnaire Star" in mainland China, and sampling was completed through the Ministry of Agriculture and Rural Affairs and the Shaanxi Provincial State-owned Assets Supervision and Administration Commission. Xi'an urban area is located in central Shaanxi Province, which is an important urban agglomeration in central China and a typical sample area for studying the "three rural issues". The data used in this study were obtained from a survey conducted by the research team in May-June 2022 on farmland transfer households in a typical rural area of Xi'an city circle. The sample areas include but are not limited to Xi'an (Xizhangpo and Podi villages), Xianyang (Bailizhuang, Baitu and Yuanjia villages), Yan'an (Haojiahe and Shi'er villages), Yulin (Jinjiisha and Miaowan villages), Weinan (Beizhuang and Yohong villages), and Baoji (Jianhe and Wangjiazhuang villages). The sample sampling method was Accidental Sampling, and the survey was conducted by one-on-one semi-structured interviews with farmers. 450 questionnaires were distributed, of which 285 were valid, with a valid rate of 63.3%.

Table 1 Scale development and questionnaire design

Characteristics	Samples	Percent (%)
Education Level		
Illiterate	202	70.9
Primary School	15	5.3
Junior High School	12	4.2
High School	23	8.1
Bachelors	22	7.7
Master	9	3.1
PhD	2	0.7
Social Identity		
Ordinary Villagers	270	94.7

Continued to Table 1

Characteristics	Samples	Percent (%)
Village officials	15	5.3
age		
< 35 years	32	11.2
35~44 years	43	15.1
45~54 years	100	35.1
55~64 years	100	35.1
> 65 years	10	3.5
Degree of part-time work		
Professional farmers	95	33.3
Professional-part-time farmers	85	29.8
Part-time farmers	70	24.6
Part-time-non farmers	25	8.8
Non-farmers	10	3.5

3.2 Measures

The article contains four variables: IP, RP, CP, GBR. The Likert 7-level scale is used to measure the above variables, and the measurement range is from "very dissatisfied" to "very satisfied" corresponding to the numbers "1" to "7". Based on the above analysis and theoretical hypothesis, this paper designed a scale for measuring farmer household guaranteed behavior based on distributed cognitive theory (see Table 1). The scale contains 4 subject variables and 18 observation indicators. The influence paths among the 4 subject variables are shown in Fig.1, which constitutes the cognitive logic of farmer household guaranteed behavior. The 18 observation indicators are divided into 4 groups, and the specific meanings of 4 main variables are measured respectively, so as to meet the quantitative research needs from "abstract concept" to "concrete reality". On the basis of scale development and combined with the results of semi-structured interviews with farmers in the pre-survey stage, 18 questionnaire items were designed in this study, and the index values were represented in the form of ordered categorical variables of Likert scale. The specific scale development and questionnaire design results are shown in Table 1.

3.3 Data analysis technique

Potential biases were considered in the survey, protocol design, and data analysis. Several approaches (e.g. direct contact by phone and assurance to share the results) were adopted to ensure the highest response rate and avoid a non-response bias (Frohlich and Westbrook, 2002). We used a partial least squares structural equation modelling (PLS-SEM) technique to analyse the data. This technique has been adopted because this process gives better results in the analysis of this type of exploratory study. This process can also analyse those data that are not normally distributed (Hair et al., 2012). This technique does not impose any sample restriction to conduct the survey. This process involves quantification of responses on a specific scale.

Table 2 Descriptive characteristics of the sample

Variable	Observation indicators	Measurement item	Sources
Individual Power (IP)	IP1	Gender of survey respondents	(Sirmon et al., 2007)
	IP2	Age of survey respondents	
	IP3	Social class status of survey respondents	
	IP4	Education level of survey respondents	
	IP5	Extent of part-time employment of survey respondents	
Regional Power (RP)	RP1	Average annual gross income of surveyed households in the past three years	(Jehn, 1997)
	RP2	How the employment status of survey respondents' household labor force would change if existing livelihoods were unsustainable	
	RP3	Proportion of survey respondents' household land transfer area	
	RP4	Survey respondents' projections of household living conditions in the next 5 years	
	RP5	Survey respondents' evaluation of family happiness at this stage	
Cultural Power (CP)	CP1	The extent to which survey respondents accept "advice from acquaintances" when making decisions about safeguarding behaviors	(Lee et al., 2010)
	CP2	Receptiveness of survey respondents to "policy advocacy" when making decisions about safeguarding behaviors	
	CP3	Satisfaction level of survey respondents with agricultural subsidy policies	
	CP4	Satisfaction of survey respondents with rural pension insurance policies	
	CP5	Satisfaction of survey respondents with rural medical insurance policies	
Guaranteed Behavioral Response (GBR)	GBR1	Survey respondents' agreement with strict implementation of regular household savings plan	(Cohen and Levinthal, 1990)
	GBR2	Survey respondents' recognition of active participation in rural medical and pension insurance	
	GBR3	Survey respondents' agreement with buying insurance products provided by commercial organizations	

4 Analysis result

4.1 Measurement model assessment

The results showed that the measurement model satisfies all general requirements (see Table 3). First, all the standardized factor loadings of all the first-order and second-order constructs are above the minimum value of 0.808 (Fornell and Larcker, 1981). Second, the Cronbach's alpha scores ranged between 0.707 and 0.934 while the composite reliability scores ranged between 0.866 and 0.930 which are above the recommended value of 0.70 indicating adequate construct validity. In addition, all the constructs have an AVE value above 0.50, suggesting that latent variables achieved convergent validity. Finally, this study follows three approaches to assess the discriminant validity i.e, (1) Fornell-Larcker criterion, (2) cross loading, and (3) the heterotrait-monotrait ratio of correlations (HTMT).

Table 3 Reliability and validity

Variable	Item	Cross loadings	Convergent validity Composite reliability	AVE	Cronbach's alpha	Multicollinearity VIF
Individual Power (IP)	IP1	0.776	0.907	0662	0.873	2.033
	IP2	0.844				2.505
	IP3	0.846				2.247
	IP4	0.788				1.860
	IP5	0.811				1.795
Regional Power (RP)	RP1	0.763	0.882	0.599	0.833	1.622
	RP2	0.812				1.808
	RP3	0.749				1.660
	RP4	0.783				1.781
	RP5	0.762				1.665
Cultural Power (CP)	CP1	0.745	0.866	0.564	0.808	1.470
	CP2	0.771				1.613
	CP3	0.761				1.580
	CP4	0.770				1.736
	CP5	0.707				1.559
Guaranteed Behavioral Response (GBR)	GBR1	0.860	0.930	0.815	0.886	2.100
	GBR2	0.934				3.598
	GBR3	0.912				2.955

The correlation matrix in Table 4 shows that for each pair of constructs, the AVE square root of each construct is higher than the absolute value of their correlation (Fornell and Larcker, 1981). The results of cross loading show that all items are loaded higher on their respective constructs than on the other constructs and the cross-loading differences are much higher than the suggested threshold of 0.1. In all cases the HTMT values are below the threshold of 0.85. These results conenterpriseed that the discriminant validity is present in this study. The statistical values of each goodness of fit index met the threshold conditions, and the PLS-SEM had a good fitting effect on sample data, and the model passed the robustness test.

Table 4 Discriminant validity-Fornell-Larcker criterion and Heterotrait-Monotrait ratio

Variables	Mean	S.D	1	2	3	4
Individual Power (IP)	4.69	1.12	0.813	0.786	0.595	0.733
Regional Power (RP)	4.964	1.22	0.672 * *	0.774	0.619	0.695
Cultural Power (CP)	5.012	1.20	0.510 * *	0.510 * *	0.751	0.567
Guaranteed Behavioral Response (GBR)	5.06	1.05	0.657 * *	0.601 * *	0.486 * *	0.903

Note: Significant level: $p<0.10$; * $p<0.05$; * * $p<0.01$; * * * $p<0.001$, Bold diagonal entries are square root of AVEs, Heterotrait-Montrait ratios (HTMT) (Underlined) are below 0.85.

4.2 Structural model assessment

This study followed Hair et al.(Hair et al., 2012) to estimate the structural model. First, the results show minimal collinearity in the structural model as all VIF values are far below the common cutoff threshold of 5 Hair et al (Hair et al., 2012). Second, following the rules of thumb, the R^2 values of GBR (0.493) exceed the minimum value of 0.10 recommended by Hair which is a satisfactory level of predictability as shown in Table 5. Similarly, results from blindfolding with an omission distance of 7 yield Q^2 values well above zero (Table 5). This supporting the model's predictive relevance in terms of out-of-sample prediction. Further analysis of the composite-based standardized root mean square residual (SRMR) yields a value of 0.063, which conenterprises the overall fit of PLS path model (Henseler et al., 2014). Applying the bootstrapping procedure (5000 bootstrap samples; no sign changes) provides the p-values as well as the corresponding 95% bias-corrected and accelerated (BCa) bootstrap confidence intervals (Table 5). The empirical results support the vast majority of hypothesized path model relationships among the constructs.

Table 5 Significant testing results of the structural model path coefficients

	Path coefficient	t-value	p-value	95% BCa confidence interval	Conclusion
Individual Power (IP)→ Guaranteed Behavioral Response (GBR)	0.416	6.993	0.000	(0.294,0.529)	H1 supported
Regional Power (RP)→ Guaranteed Behavioral Response (GBR)	0.246	2.937	0.003	(0.066,0.392)	H2 supported
Cultural Power (CP)→ Guaranteed Behavioral Response (GBR)	0.149	2.988	0.003	(0.058,0.253)	H3 supported

SRMR composite model = 0.063.

$R^2_{GBR}=0.493$; $Q^2_{ICMS}=0.391$.

Based on the above results, the analysis is as follows:

First, the standardized path coefficients of IP, RP and CP to GBR were significant at 0.001, 0.05 and 0.05 levels, respectively. Theoretical hypotheses H1, H2 and H3 were all effectively verified, indicating that the cognitive basis of farmers' choice of security mode conforms to the distributed cognitive framework, and farmers' choice behavior of security mode is influenced by three cognitive levels: IP, RP and CP.

Second, the standardized path coefficient of RP→GBR is 0.246. RP is the main cognitive level that affects farmers' security behavior response. Factor loading coefficients of RP1, RP2, RP3, RP4 and RP5 of RP observation indexes were 0.763, 0.812, 0.749, 0.783 and 0.762, respectively. It shows that the increase of household income can effectively promote the response of peasant households' security behavior, and the area proportion of land transfer and their yearning for a better future life, while the happiness of peasant households' life and the re-employment ability of family members also affect peasant households' security behavior to a certain extent.

Thirdly, the standardized path coefficient of CP → GBR was 0.586, which was relatively small among the three cognitive levels of distributed cognition, indicating that CP was an important cognitive level affecting farmers' security behavior. The factor loading coefficients of CP1, CP2, CP3, CP4, CP5 were 0.745, 0.771, 0.761, 0.770 and 0.707, respectively. It shows that farmers' satisfaction with endowment insurance policy has a significant impact on their security behavior, followed by their satisfaction with agricultural subsidies and medical insurance policy, and the promotion effect of policy publicity on farmers' security behavior needs to be improved.

Fourthly, the standardized path coefficient of IP→GBR is 0.416, which is largest among the three cognitive levels of distributed cognition, indicating that IP is an effective cognitive level affecting farmers' security behavior. Factor load coefficients of IP1, IP2, IP3, IP4 and IP5 are -0.776, 0.844, 0.846, 0.788 and 0.811, respectively, indicating that with the improvement of education level, farmers are more inclined to take appropriate security behaviors. The higher the level of non-agricultural livelihood of farmers, the higher the corresponding degree of their security behavior will be. It should be noted that the effect of gender and age on IP is negative, indicating that for land-lost farmers, female group and elderly farmers have certain limitations on their cognition of security behavior.

Fifthly, the factor loads of GBR1, GBR2 and GBR3 on GBR are 0.860, 0.934 and 0.912, respectively, and reach the significance level of 0.01, indicating that among the three security modes, land-lost farmers are more inclined to choose policy-based security, followed by traditional savings security, and their acceptance of market-based security needs to be improved.

5 Conclusions and policy recommendations

5.1 Conclusion

First, landless farmers' choices of guaranteed methods follow the basic framework of distributed cognition, and their behavioral decision-making mechanisms are influenced by a combination of cognitive levels such as individual power, territorial power, and cultural power. Among them, regional power is the main cognitive level, cultural power is an important cognitive level, and personal power is an effective cognitive level.

Second, the increase of household income can significantly improve the level of farmers' guaranteed behavior, and promoting farmers' income increase is the core of improving the level of rural social guaranteed; the function of land guaranteed can be replaced by other guaranteed methods to a certain extent, and farmers have the potential incentive to withdraw from land and choose other guaranteed methods; the atmosphere of farmers' family life and household labor endowment can promote their guaranteed behavior to a certain extent.

Third, farmers' guaranteed behavior is largely influenced by the policy, and the performance evaluation of policy implementation (satisfaction) has a greater effect on farmers' guaranteed behavior than policy publicity, suggesting that landless farmers' behavioral decisions are more likely to be performance-oriented than opinion-oriented, and the effect of policy implementation promotes farmers' adverse choice of guaranteed methods.

Fourth, the improvement of farmers' individual quality (including literacy and part-time work ability) can promote their guaranteed behavior to a certain extent, but the "accumulation of experience" as they grow older may inhibit them from taking effective guaranteed measures. In addition, the survey found that there is gender discrimination in the decision-making process of landless farmers' households, and the guaranteed needs of female groups are difficult to be met effectively.

Fifth, there is a clear preference in the choice of protection methods among landless farmers, with government-led policy-based protection being the mainstay of the rural social guaranteed system, while farmers still rely to some extent on traditional savings methods, and the acceptance of market-based insurance products and financial services in rural areas still needs to be further enhanced.

5.2 Policy recommendations

First, regulate land transfer and improve the level of protection. With the agglomeration effect of urban development becoming more and more prominent, a large number of rural laborers are moving to the cities, the rural areas in the suburbs tend to decline, and the abandonment and abandonment of arable land are serious, and the small farmer economy is in trouble. In this context, China has tried to solve the real problems of abandonment and fragmentation of arable land by implementing land management rights transfer. Scholars have pointed out that the standardized implementation of land transfer policy can effectively improve farmers' welfare and promote farmers' household income, but at the present stage, land transfer in China still suffers from the "double-low dilemma" of low level and low efficiency (Ma et al., 2020). The results of the analysis of farmers' perception of "territorial power" show that landless farmers have a tendency to "exchange land for guaranteed", and the transformation of family livelihood through land transfer can effectively promote the response of farmers' guaranteed behavior. Based on this, the government should accelerate the establishment of a rural land property rights trading platform, standardize the flow process, expand the scope of the flow, and introduce a market-based price competition mechanism to ensure the reasonable realization of farmers' land property rights, so as to further improve the quality and level of farmers' guaranteed behavior.

Second, strengthen policy guidance and establish a feedback mechanism. China is at a critical stage of transition from traditional smallholder economy to agricultural modernization, and the livelihood environment of rural households is subject to exogenous shocks from institutional changes, making them typically "risk averse" (Wilson et al., 2018). Studies show that government subsidies, insurance and financial policies through fiscal transfers are still the main supply of social guaranteed services in rural areas, and rural residents are largely path-dependent on them. As the demand side of risk protection, the farming community has the initiative and necessity to supervise government actions. However, the current rural social guaranteed system has not yet established a reliable policy feedback mechanism. Government departments focus unilaterally on the "top-down" system design, ignoring the possible intersection of various risk factors and uncertainties, and the lack of "bottom-up" complaint channels for farmers' demands for protection. The analysis of farmers' perceptions of "cultural power" shows that farmers' subjective evaluations of the effectiveness of policy implementation have a more significant impact on their response to protection behavior than the government's policy promotion efforts. If this cognitive "anchoring effect" persists for a long time, it will definitely have a negative impact on the government's credibility. On the one hand, the government should further strengthen its policy propaganda work and improve the efficiency of policy guidance and policy implementation; on the other hand, relevant departments should establish a feedback mechanism for grassroots farmers on various protection policies and transform farmers' subjective evaluations into objective evaluations of policy performance, so as to ensure the realism and effectiveness of various protection policies.

Third, be vigilant about the fragmentation of farm households and focus on vulnerable groups. With the continuous refinement of social division of labor, many aspects of agricultural production have been replaced by specialized social service institutions, and the group of farmers has transformed from traditional "agricultural production labor" to new "agricultural business decision makers". The survey found that the typical rural areas in the sample regions are now in a state of coexistence of multiple agricultural business entities, with pure farmers, semi-part-time farmers and non-farmers all accounting for a certain proportion, and the farming groups showing a heterogeneous and differentiated development. The results of the analysis of farmers' perception of "personal manpower" show that the higher the level of education and part-time employment of farmers, the higher the level of their guaranteed needs; as they grow older, their motivation to adopt effective guaranteed behavior decreases significantly; at the same time, female farmers recognize guaranteed behavior less than male farmers. The above analysis shows that the divergence of farmers' guaranteed behavior mainly comes from the difference of education and part-time employment. The government should be alert to the possible class division and class conflict of farmers caused by this divergence trend, especially for the disadvantaged groups of farmers (such as women and the elderly), and should give some attention and policy favor to them, so as to maintain social equity and justice.

Author contributions

Methodology and software, H.D. and H.Y.W.; formal analysis, H.D. and Y.J.L.; resources and data curation, H.D. and Y.J.L.; investigation, H.D.; writing-original draft preparation, Y.J.L.; writing-review and editing, H.D. and Y.J.L.; supervision and project administration, H.Y.W.; All authors have read and agreed to the published version of the manuscript.

Ethics statement

No animal studies are presented in this manuscript.

No human studies are presented in this manuscript.

No potentially identifiable human images or data is presented in this study.

Conflicts of interest

Author Hao Dong was employed by Shaanxi Provincial Land Engineering Construction Group Co., Ltd. The remaining authors declare that the research was conducted in the absence of any commercial or financial relationships that could be construed as a potential conflict of interest.

Reference

Akinwumi, S., and Dada, J. (2020). Access to Finance, Indigenous Technology and Food Security in Nigeria: Case Study of Ondo Central Senatorial District. *Economics and Culture* 17, 75-87.

Bao, H., Dong, H., Jia, J., Peng, Y., and Li, Q. (2020). Impacts of land expropriation on the entrepreneurial decision-making behavior of land-lost peasants: An agent-based simulation. *Habitat International* 95, 102096.

Bao, H., Han, L., Wu, H., and Zeng, X. (2021). What affects the "house-for-pension" scheme consumption behavior of land-lost farmers in China? *Habitat International* 116, 102415.

Bao, H., and Peng, Y. (2016). Effect of land expropriation on land-lost farmers' entrepreneurial action: A case study of Zhejiang Province. *Habitat International* 53, 342-349.

Belland, B. R. (2011). Distributed Cognition as a Lens to Understand the Effects of Scaffolds: The Role of Transfer of Responsibility. *Educ Psychol Rev* 23, 577-600.

Cao, Y., and Zhang, X. (2018). Are they satisfied with land taking? Aspects on procedural fairness, monetary compensation and behavioral simulation in China's land expropriation story. *Land Use Policy* 74, 166-178.

Chen, K., Long, H., Liao, L., Tu, S., and Li, T. (2020). Land use transitions and urban-rural integrated development: Theoretical framework and China's evidence. *Land Use Policy* 92, 104465.

Cohen, W. M., and Levinthal, D. A. (1990). Absorptive Capacity: A New Perspective on Learning and Innovation. *Administrative Science Quarterly* 35, 128.

Du, H., Liu, D., Sovacool, B. K., Wang, Y., Ma, S., and Li, R. Y. M. (2018). Who buys New Energy Vehicles in China? Assessing social-psychological predictors of purchasing awareness, intention, and policy. *Transportation Research Part F: Traffic Psychology and Behaviour* 58, 56-69.

Fang, Y., Shi, K., and Niu, C. (2016). A comparison of the means and ends of rural construction land consolidation: Case studies of villagers' attitudes and behaviours in Changchun City, Jilin province, China. *Journal of Rural Studies* 47, 459-473.

Fornell, C., and Larcker, D. F. (1981). Evaluating Structural Equation Models with Unobservable Variables and Measurement Error. *Journal of Marketing Research* 18, 39.

Frohlich, M. T., and Westbrook, R. (2002). Demand chain management in manufacturing and services: web-based integration, drivers and performance. *Journal of Operations Management* 20, 729-745.

Gardner, W. L., Pickett, C. L., and Brewer, M. B. (2000). Social Exclusion and Selective Memory: How the Need to belong Influences Memory for Social Events. *Pers Soc Psychol Bull* 26, 486-496.

Ge, D., Wang, Z., Tu, S., Long, H., Yan, H., Sun, D., et al. (2019). Coupling analysis of greenhouse-led farmland transition and rural transformation development in China's traditional farming area: A case of Qingzhou City. *Land Use Policy* 86, 113-125.

Ghelichi, Z., Saidi-Mehrabad, M., and Pishvaee, M. S. (2018). A stochastic programming approach toward optimal design and planning of an integrated green biodiesel supply chain network under uncertainty: A case study. *Energy* 156, 661-687.

Hair, J. F., Sarstedt, M., Ringle, C. M., and Mena, J. A. (2012). An assessment of the use of partial least squares structural equation modeling in marketing research. *J. of the Acad. Mark. Sci.* 40, 414-433.

Henseler, J., Dijkstra, T. K., Sarstedt, M., Ringle, C. M., Diamantopoulos, A., Straub, D. W., et al. (2014). Common Beliefs and Reality About PLS: Comments on Rönkkö and Evermann (2013). *Organizational Research Methods* 17, 182-209.

Jang, K. M., and Kim, Y. (2019). Crowd-sourced cognitive mapping: A new way of displaying people's cognitive perception of urban space. *PLOS ONE* 14, e0218590.

Jehn, K. (1997). A Qualitative Analysis of Conflict Types and Dimensions in Organizational Groups. *Administrative Science Quarterly* 42.

Jordakieva, G., Grabovac, I., Steiner, M., Winnicki, W., Zitta, S., Stefanac, S., et al. (2020). Employment Status and Associations with Workability, Quality of Life and Mental Health after Kidney Transplantation in Austria. *International Journal of Environmental Research and Public Health* 17, 1254.

Lee, S., Park, G., Yoon, B., and Park, J. (2010). Open innovation in SMEs—An intermediated network model. *Research Policy* 39, 290-300.

Li, L., Tan, R., and Wu, C. (2020). Reconstruction of China's Farmland Rights System Based on the "Trifurcation of Land Rights" Reform. *Land* 9, 51.

Liu, B., Tian, C., Li, Y., Song, H., and Ma, Z. (2018). Research on the effects of urbanization on carbon emissions efficiency of urban agglomerations in China. *Journal of Cleaner Production* 197, 1374-1381.

Liu, Z., Nersessian, N., and Stasko, J. (2008). Distributed Cognition as a Theoretical Framework for Information Visualization. *IEEE Transactions on Visualization and Computer Graphics* 14, 1173-1180.

Lu, H., Hu, L., Zheng, W., Yao, S., and Qian, L. (2020). Impact of household land endowment and environmental cognition on the willingness to implement straw incorporation in China. *Journal of Cleaner Production* 262, 121479.

Lyver, P. O., Ruru, J., Scott, N., Tylianakis, J. M., Arnold, J., Malinen, S. K., et al. (2019). Building biocultural approaches into Aotearoa-New Zealand's conservation future. *null* 49, 394-411.

Ma, T., Sun, S., Fu, G., Hall, J. W., Ni, Y., He, L., et al. (2020). Pollution exacerbates China's water scarcity and its regional inequality. *Nat Commun* 11, 1-9.

Meert, H., Van Huylenbroeck, G., Vernimmen, T., Bourgeois, M., and van Hecke, E. (2005). Farm household survival strategies and diversification on marginal farms. *Journal of Rural Studies* 21, 81-97.

Meyer, K., Garzón, B., Lövdén, M., and Hildebrandt, A. Are global and specific interindividual differences in cortical thickness associated with facets of cognitive abilities, including face cognition? *Royal Society Open Science* 6, 180857.

Mishra, S. K., and Mishra, P. (2017). Determinants of households' resistance against land acquisition for mining: Experiences at Talcher coalfields in India. *Land Use Policy* 66, 10-17.

Morris, C., and Potter, C. (1995). Recruiting the new conservationists: Farmers' adoption of agri-environmental schemes in the U.K. *Journal of Rural Studies* 11, 51-63.

Printsmann, A., Nugin, R., and Palang, H. (2022). Intricacies of Moral Geographies of Land Restitution in Estonia. *Land* 11, 235.

Roessler, M., Schneckenberg, D., and Velamuri, V. K. (2022). Situated Entrepreneurial Cognition in Corporate Incubators and Accelerators: The Business Model as a Boundary Object. *IEEE Transactions on Engineering Management* 69, 1696-1711.

Rogers, Y., and Ellis, J. (1994). Distributed Cognition: An Alternative Framework for Analysing and Explaining Collaborative Working. *Journal of Information Technology* 9, 119-128.

Sirmon, D. G., Hitt, M. A., and Ireland, R. D. (2007). Managing Firm Resources in Dynamic Environments to Create Value: Looking Inside the Black Box. *AMR* 32, 273-292.

Stiller, K. D., and Schworm, S. (2019). Game-Based Learning of the Structure and Functioning of Body Cells in a Foreign Language: Effects on Motivation, Cognitive Load, and Performance. *Frontiers in Education* 4.

Su, Y., Qian, K., Lin, L., Wang, K., Guan, T., and Gan, M. (2020). Identifying the driving forces of non-grain production expansion in rural China and its implications for policies on cultivated land protection. *Land Use Policy* 92, 104435.

Tu, S., Long, H., Zhang, Y., Ge, D., and Qu, Y. (2018). Rural restructuring at village level under rapid urbanization in metropolitan suburbs of China and its implications for innovations in land use policy. *Habitat International* 77, 143-152.

Valles-Colomer, M., Falony, G., Darzi, Y., Tigchelaar, E. F., Wang, J., Tito, R. Y., et al. (2019). The neuroactive potential of the human gut microbiota in quality of life and depression. *Nat Microbiol* 4, 623-632.

Wang, M., Li, M., Jin, B., Yao, L., and Ji, H. (2021). Does Livelihood Capital Influence the Livelihood Strategy of Herdsmen? Evidence from Western China. *Land* 10, 763.

Wilson, G. A., Hu, Z., and Rahman, S. (2018). Community resilience in rural China: The case of Hu Village, Sichuan Province. *Journal of Rural Studies* 60, 130-140.

Xu, X., Zhang, L., Chen, L., and Liu, C. (2020). The Role of Soil N_2O Emissions in Agricultural Green Total Factor Productivity: An Empirical Study from China around 2006 when Agricultural Tax Was Abolished. *Agriculture* 10, 150.

Yakhlef, A. (2008). Towards a post-human distributed cognition environment. *Knowl Manage Res Pract* 6, 287-297.

Yen, G.-F., and Tsao, H.-C. (2020). Reexamining Consumers' Cognition and Evaluation of Corporate Social Responsibility via a DANP and IPA Method. *Sustainability* 12, 529.

Zhang, Y., Halder, P., Zhang, X., and Qu, M. (2020). Analyzing the deviation between farmers' Land transfer intention and behavior in China's impoverished mountainous Area: A Logistic-ISM model approach. *Land Use Policy* 94, 104534.

Zhao, W., and Zou, Y. (2017). Un-gating the gated community: The spatial restructuring of a resettlement neighborhood in Nanjing. *Cities* 62, 78-87.

本文曾发表于2022年《Frontiers in Psychology》第10卷

Novel Analytical Expressions for Determining van der Waals Interaction Between a Particle and Air-water Interface: Unexpected Stronger van der Waals Force than Capillary Force

Yichun Du[a,b], Scott A. Bradford[c], Chongyang Shen[b], Tiantian Li[d], Xiaoyuan Bi[b], Dong Liu[b], Yuanfang Huang[b]

(a.Key Laboratory of Degraded and Unused Land Consolidation Engineering, Ministry of Natural Resources of the People's Republic of China, Xi'an, Shaanxi 710075, China; b.Department of Soil and Water Sciences, China Agricultural University, Beijing 100193, China; c.USDA, ARS, SAWS Unit, Davis, CA 95616, United States; d.School of Environmental Engineering, Henan University of Technology, Zhengzhou, Henan 450001, China)

Graphicalabstract

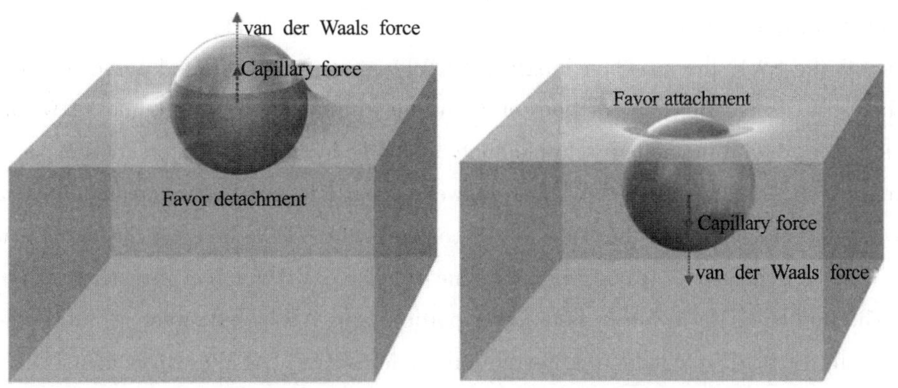

Dominance of van der Waals force over capillary force for a colloid at air-water interface

【Abstract】 Hypothesis: Analytical expressions for calculating Hamaker constant (HC) and van der Waals (VDW) energy/force for interaction of a particle with a solid water interface has been reported for over eighty years. This work further developed novel analytical expressions and numerical approaches for determining HC and VDW interaction energy/force for the particle approaching and penetrating air-water interface (AWI), respectively.

Methods: The expressions of HC and VDW interaction energy/force before penetrating were developed through analysis of the variation in free energy of the interaction system with bringing the particle from infinity to the vicinity of the AWI. The surface element integration (SEI) technique was modified to cal-culate VDW energy/force after penetrating.

Findings: We explain why repulsive VDW energy exists inhibiting the particle from approaching the AWI. We found very significant VDW repulsion for a particle at a concave AWI after penetration, which can even exceed the capillary force and cause strong retention in water films on a solid surface and at air-water-solid interface line. The methods and findings of this work are critical to quantification and understanding of a variety of engineered processes such as particle manipulation (e.g., bubble flotation, Pickering emulsion, and particle laden interfaces).

【Keywords】 Particle; Air-water interface; van der Waals force; Hamaker constant

1 Introduction

Investigation of colloidal particle transport in porous media is crucial to various environmental concerns and engineered applications[1]. For example, biocolloids (e.g., viruses and bacteria) are ubiquitous in soil porous media. Understanding mechanisms governing biocolloid transport in soil is critical to accurately determining the potential for biocolloid contamination of groundwater[2]. In addition, engineered nanomaterials such as nanoscale zerovalent iron (nZVI) have been frequently used to treat contaminants in wastewater[3,4]. Knowledge of nanomaterial transport in porous media is critical to extending their application to insitu soil remediation because the remediation efficiency is highly related to their ability to reach polluted sites[5,6].

Transport of colloidal particles in porous media are mainly controlled by deposition and straining[7-8]. The deposition and straining are governed by interaction of a particle with only one and multiple solid-water interfaces (SWIs) in saturated porous media, respectively[1,9,10]. Attachment of a particle on a SWI has received numerous attention[1]. The particle-SWI interaction is controlled by the Derjaguin-Landau-Verwey-Overbeek (DLVO) interaction forces composed of van der Waals (vdW) and double layer (DL) interactions as well as shortrange repulsions[11]. Analytical expressions have been developed for calculating the VDW or DL energy between two parallel flat surfaces per unit area (or between an area element and an infinite planar surface)[12-13]. Based on these expressions, the interaction energies/forces between a particle and SWI can be determined by using the integration methods such as Derjaguin approach and surface element integration (SEI) techniques[14-15].

In addition to the SWI, the air-water interface (AWI) exists in unsaturated porous media[2,16,17]. Similar to the particle-SWI interaction, interactions of a particle with an AWI also include VDW and DL interactions[18]. In addition, due to the superhy-drophobicity of AWI, hydrophobic interaction exists between a hydrophobic particle and the AWI. The VDW force between a particle and SWI is commonly attractive, which is responsible for the attachment on the SWI[19]. In contrast, the VDW force for the particle with AWI has been elucidated to be repulsive[10,16,18]. Similarly, the repulsive VDW force also exists between a bubble and SWI[20]. The repulsive VDW forces between a particle and AWI or between a bubble and SWI have been experimentally verified by conducting atomic force microscopy (AFM) experiments[21-22]. In fact, the VDW force that can be repulsive has been illustrated since the landmark work of Hamaker[12]. This study indicated that the VDW interaction may be a repulsion if the interacting bodies in a fluid are of different composition. Using the Liftshitz theory, the VDW force was mathematically revealed to be repulsive for the particle-AWI and bubble-SWI interactions, as indicated by the negative values of Hamaker constant[23,19]. However, the mechanisms causing the VDW repulsion for the particle-AWI and bubble-SWI interactions remain unclear to date[10,20].

Existing literature[20,24] used the same methods to calculate the VDW forces between particles and AWIs as those for the particle-SWI interaction such as the Hamaker expressions, Derjaguin approach and SEI techniques. However, the values of Hamaker constants for the particle-AWI interaction commonly use measurements such as atomic force microscopy (AFM) experiments or estimates from Liftshitz theory[23,25]. Due to the complex of the mathematic equations of Liftshitz theory, approximations are commonly necessary to obtain the values of Hamaker constants[19,23]. There are still no analytical expressions available for estimating Hamaker constants for the particle-AWI interaction. In addition, existing studies only estimated VDW force between the particle and AWI before the particle attaches to the AWI. Once the particle adheres to and penetrates the AWI, meniscus normally form due to gravity (particle weight and buoyance force) and/or wetting properties of the particle surface[18,26]. The VDW force/energy for a particle at the curved AWI meniscus has not been evaluated to date.

It should be noted that when the particle penetrates the AWI, it will also experience a capillary force due to Laplace pressure or the surface tension forces[18]. The capillary force is regarded to be the main reason of film straining of particles and retention at the air-water-solid (AWS) interface line[17,27-30]. Calculation of the capillary force demands accurate determination of the geometry of the AWI meniscus around the particle and collector, which is controlled by the force field within the water film. Therefore, accurate calculation of the VDW force between the particle and AWI is critical to determination of the geometry of the AWI meniscus and accordingly the capillary force. Notably, quantification of the VDW force is also essential to elucidating the relative contribution of capillary and VDW forces in processes such as particle retention via film straining and the AWS interface line and self-assembly.

The analytical expressions for determining Hamaker constant and VDW force/energy for particle-SWI interactions has been reported for over eighty years by the landmark work of Hamaker[12]. This study developed novel analytical expressions to calculate the Hamaker constant and accordingly VDW force/energy for the particle-AWI interactions. We also presented a numerical approach by modifying the SEI technique to calculate the VDW interaction force/energy for the particle penetrating the AWI. We were the first to theoretically explain why the VDW energy of particle-AWI or bubble-SWI interaction is repulsive, and the observed discrepancies of favorable and unfavorable attachment of particles at AWIs[31-34]. We also revealed that the VDW repulsion for particle penetrating the AWI was significantly larger than before the penetration and can even exceed the capillary force. Various processes such as retention and release of a particle at the AWS interface line or in water films and self-assembly were attributed to only the capillary force[35-37]. Our finding, however, is in contrast to the common belief, which clearly demonstrated that the VDW interaction played a significant role in these processes, which can even exceed the capillary force.

2 Theory

2.1 VDW force and energy

The AWI can be taken as a bubble collector in solution with infinitely large radius. To determine the VDW energy (U) between a particle and the AWI or bubble collector (depicted with subscript P and C, respectively) in water (W), we have to consider not only the particle and the bubble collector, but also a particle and a collector with the same sizes but composed of water[10,12] (see Fig. 1). When the particle mobilizes from infinity to the vicinity of the bubble collector, the energy of the particle will change from U_P to $U_P + U_{PC} - U_{PW}$, where U_P is the VDW energy for the particle in water at infinity, U_{PC} is the VDW energy for particle-bubble collector interaction in vacuum, and U_{PW} is the VDW energy for particle-water collector in a vacuum. Note that a water particle is also moved towards infinity at the same time. The corresponding energy will change from $U_W + U_{WC} - U_{WW}$ to U_W, where U_W is the VDW energy for the water particle in water at infinity, U_{WC} is the water particle-bubble collector VDW energy in a vacuum, and U_{WW} is the water particle-water collector VDW energy in a vacuum. The energy variation because of change in separation distance between the particle and the bubble collector (i.e., the VDW energy) in water U_{VDW} is thus written as:

$$U_{VDW} = U_{PC} + U_{WW} - U_{PW} - U_{WC} \tag{1}$$

The values of U_{PC} and U_{WC} are much smaller than those of U_{WW} and U_{PW} because the VDW energy is proportional to the atomic densities of the interacting bodies[12] and the density of air is extremely small compared to that of particle or water. Consequently, Equation (1) can be recast as:

$$U_{VDW} = U_{WW} - U_{PW} \tag{2}$$

The values of U_{PW} and U_{WW} change with the particle-bubble collector separation distance (θ). The expressions for determining the U_{WW} and U_{PW} have the form of $-A_{WW}f(H)$ and $-A_{PW}f(H)$, respectively, where $f(H)$ is a function that is only related to the geometry of an interaction configuration (e.g., particle

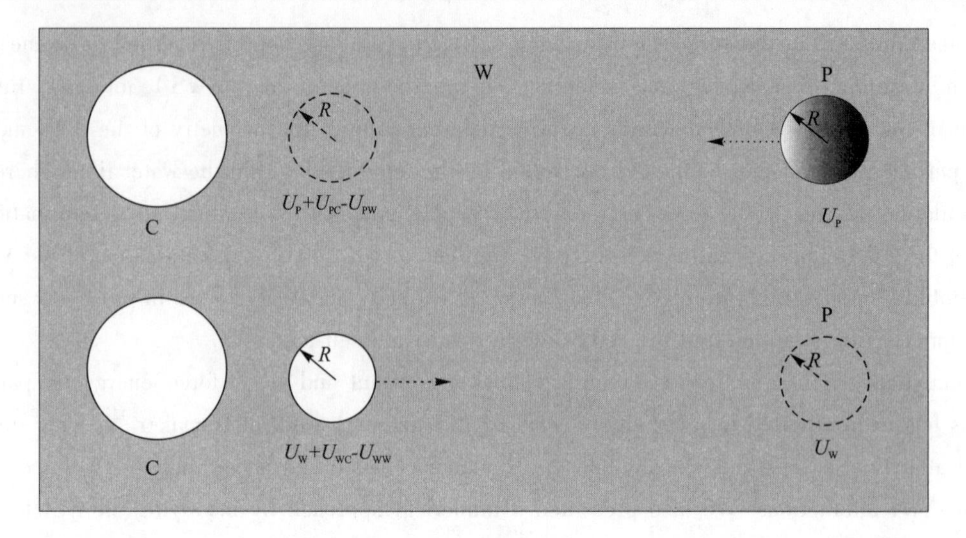

Fig. 1 Illustration of change in free energy of the VDW system when the particle and AWI is brought together from infinity. The particle, water, and bubble collector are denoted as *P*, *W*, and *C*, respectively. U_p is the VDW energy for the particle in water at infinity, U_{pc} is the VDW energy between the particle and the bubble collector in vacuum, U_{pw} is the VDW energy between the particle and water collector (with the same size of the bubble collector) in a vacuum, U_w is the VDW energy for the water particle in water at infinity, U_{wc} is the VDW energy between the water particle (with the same size of the particle) and the bubble collector in a vacuum, and U_{ww} is the VDW energy between the water particle and the water collector in a vacuum

size and shape, separation distance (H), A_{WW} is the water-water Hamaker constant, and A_{PW} is the particle-water Hamaker constant. Equation (2) can then be rewritten as

$$U_{VDW} = -A_{WW}f(H) + A_{PW}f(H) = -(A_{WW} - A_{PW})f(H) \tag{3}$$

The Hamaker constant for the particle-AWI interaction is thus $A_{WW}-A_{PW}$. Note that the expressions of $f(H)$ are the same for the Hamaker approach and Liftshitz method for simple interaction configurations such as sphere-flat surface interaction. For the spherical particle-flat AWI interaction, Equation (3) can be writtenas

$$U_{VDW} = -\frac{A_{WW}-A_{PW}}{6}\left[\frac{R}{H}+\frac{R}{H+2R}+\ln\left(\frac{H}{H+2R}\right)\right] \tag{4}$$

Where R is radius of the spherical particle. The equation used to determine the spherical particle-AWI VDW force (F_{VDW}) is

$$F_{VDW} = \frac{(A_{WW} - A_{PW})R}{3}\left[\frac{1}{H} - \frac{R}{H^2} - \frac{1}{H + 2R} - \frac{R}{(H + 2R)^2}\right] \tag{5}$$

Meniscus AWI may form (see Fig. 2) once the particle pene-trates through the AWI. The SEI technique[14] was modified to determine the VDW force for the particle-meniscus AWI interaction depicted by the Cartesian coordinate system. The x-y plane is parallel to the flat AWI (i.e., before deformation due to the particle penerating) and passes through the bottommost point of the particle. The z axis passes through the particle center and points towards the AWI. The particle surface was first discretized into small element areas dS, and the AWI was correspondingly discretized into elemental areas dS' related to the dS. The total VDW energy (U_{VDW}) was obtained by summing the differential VDW energy (E_{VDW}) between each pair of dS and dS':

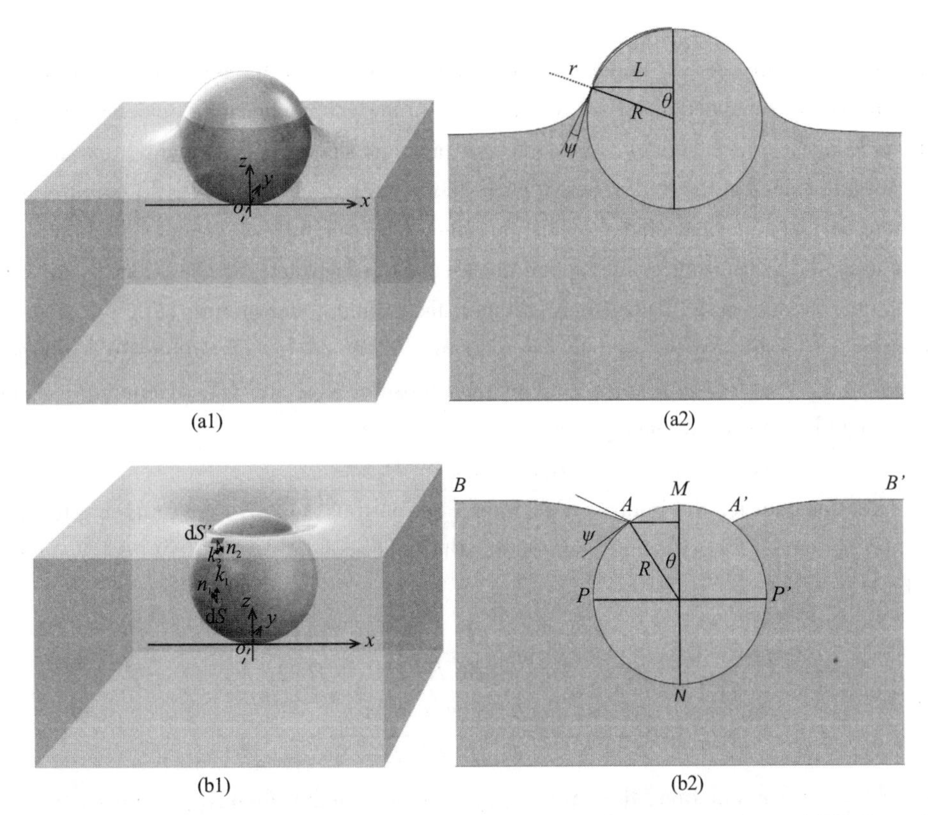

Fig. 2　Interaction of a spherical particle with a (a) concave or (b) convex meniscus AWI. Parts (a2) and (b2) are cross section images of parts (a1) and (b1) passing through the particle center, respectively. d*S* and d*S'* are differential area elements on particle and AWI surfaces, respectively, n_1 and n_2 are unit outwards normal to the particle surface and AWI, respectively, k_1 and k_2 are unit vectors directed towards the positive and negative *z*-axis, respectively, *R* and *r* are radii of the particle and AWI, respectively, *L* is the radius of the circle due to contact of the particle surface with AWI, θ is the angle between *z* axis and the line through the particle and AWI arc centers, *w* is contact angle of the water surface with the particle.

$$U_{\mathrm{VDW}}(H)=\int_{s}(n_1\cdot k_1)(n_2\cdot k_2)E_{\mathrm{VDW}}(h)\,\mathrm{d}S$$
$$=\int_{s}(n_2\cdot k_2)\frac{n_1\cdot k_1}{|n_1\cdot k_1|}E_{\mathrm{VDW}}(h)\,\mathrm{d}A \tag{6}$$

where dA is the projected area of dS on the x-y plane, $\theta$is distance between dS and dS', n_1 and n_2 are unit outward normal to the element dS and dS', respectively, k_1 and k_2 are unit vectors directed along the positive and negative z axis, respectively. The expression of E_{VDW} can be derived from Equation (4), which is

$$E_{\mathrm{VDW}}=-\frac{A_{\mathrm{WW}}-A_{\mathrm{PW}}}{12\pi h^2} \tag{7}$$

The Equation (6) can also be used to determine the total VDW force for the particle-meniscus AWI interaction (U_{VDW}) using the differential VDW force (P_{VDW}) instead of the differential VDW energy. The expression used to calculate the value of P_{VDW} can be derived from Equation (7), which is

$$P_{\mathrm{VDW}}=\frac{A_{\mathrm{WW}}-A_{\mathrm{PW}}}{6\pi h^3} \tag{8}$$

The shape of the AWI may be determined using the Young-Laplace equation of capillarity by accounting for the balance of the forces including gravity, buoyancy, and interfacial forces such as VDW force[38]. However, the VDW interaction between the particle and AWI, as mentioned previously, is still not unclear.

Our work is exactly an effort to understand the VDW force/energy for this interaction configuration. The shape of AWI was simply simulated as an arc with a radius of r which connects the undeformed AWI and a water film on the particle surface (see Fig. 2), as adopted in previous studies[17,39]. Notably, the aforementioned approach can be used to calculate the VDW energy/force between a particle and AWI with an arbitrary shape. Matlab programs were developed for calculation of the particle-AWI VDW forces/energies based on a code presented in our previous work[40]. Dimensionless interaction energies are present later in the paper by dividing the calculated energies by the product of k (Boltzmann constant) and absolute temperature (T).

The Equation (1) can also be used to calculate the VDW energy of bubble-SWI interaction. In this case, the values of U_{PC} and U_{PW} are much smaller than those of U_{WW} and U_{WC}. Hence, the expression used to calculate the bubble-SWI VDW energy is

$$U_{UDW}=U_{WW}-U_{WC} \tag{9}$$

The Hamaker constant for the bubble−SWI interaction is thus $A_{WW}-A_{WC}$by replacing the Hamaker constant in Equation(4) and (5) with this expression, the spherical bubble-flat SWI VDW energy and force can be calculated, namely,

$$U_{VDW}=-\frac{A_{WW}-A_{WC}}{6}\left[\frac{R}{H}+\frac{R}{H+2R}+\ln\left(\frac{H}{H+2R}\right)\right] \tag{10}$$

$$F_{VDW}=\frac{(A_{WW}-A_{WC})R}{3}\left[\frac{1}{H}-\frac{R}{H^2}-\frac{1}{H+2R}-\frac{R}{(H+2R)^2}\right] \tag{11}$$

For the bubble-AWI interaction, the value of A_{WC} is much smaller than A_{WW}, hence, the Hamaker constant for interaction of a bubble with AWI is A_{WW}(i.e., $A_{WW}-A_{WC}\approx A_{WW}$). The Equations (10) and (11) can be used to calculate the VDW energy and force for spherical bubble-flat AWI interaction by using A_{WW} to replace $A_{WW}-A_{WC}$.

2.2 DL, hydrophobic, and capillary interactions

Because the DL interaction is a surface force, the expression used to calculate the DL energy for the particle-AWI interaction is the same as that of the particle-SWI interaction. The equation of differential DL energy (E_{DL}) between a particle and AWI under constant surface potential condition is[41]

$$E_{DL}=\frac{\varepsilon\kappa}{2}(\psi_1^2+\psi_2^2)\left[1-\coth(\kappa h)+\frac{2\psi_1\psi_2}{(\psi_1^2+\psi_2^2)}\mathrm{cosech}(\kappa h)\right] \tag{12}$$

where ε is the dielectric permittivity of the medium, z is ion valence, e is electron charge, ψ_1 and ψ_2 are surface potentials of the particle and AWI, respectively, κ is the inverse Debye screening length.

Donaldson et al.[42] developed a general expression for calculating Hydrophobic (HR) force, which is

$$F_{HR}=2\frac{\gamma D}{H_0}\exp\left(-\frac{H}{H_0}\right) \tag{13}$$

where F_{HR} is the HR force per unit area, γ is the interfacial tension, D is the hydro parameter (taken as 1 when both surfaces are hydrophobic), and H_0 is the decay length. The expression used to calculate hydrophobic interaction energy between a particle and AWI(U_{HR}) can be derived by using $F_{HR}=-\frac{dE_{HR}}{dH}$ and the Derjaguin approach $U_{HR}=2\pi R\int_0^\infty E_{HR}(H)dH$.The expression of hydrophobic interaction energy is written as

$$U_{HR}=-4\pi\gamma DRH_0\exp\left(-\frac{H}{H_0}\right) \tag{14}$$

The values of γ and θ_0 were taken as 50 mJ/m^2 and 0.3 nm, respectively[36].

The capillary force F_C that acts on the particle at concave AWI (Fig. 2a) can be calculated using the follow expression[43]

$$F_C = \pi R^3 \rho g \left\{ \frac{D\beta + R\beta[1+\cos(\theta)]}{R\beta} \sin^2(\pi-\theta) + \frac{2}{(R\beta)^2} \sin(\pi-\theta+\psi)\sin(\pi-\theta) \right\} \quad (15)$$

Where ρ is the density of water, g is acceleration due to gravity, D is thickness of the free water film, $\beta = \sqrt{(\Delta\rho g)/\sigma}$ is the capillary number, $\Delta\rho$ is the difference between water and air density, σ is surface tension of water, θ is the angle between z axis and the line through centers of the particle and AWI arc, and ψ is contact angle of the water surface with the particle. Note that the direction of the capillary force for the particle-concave AWI interaction configuration is along the negative z-direction of the coordinate system. However, if the water film thickness exceeds the particle diameter (Fig. 2b), the direction of the capillary force will be along the positive z-direction. In this case, the maximum capillary force is given by[18,44]

$$F_C = 2\pi R\sigma \sin^2\left(\frac{\psi}{2}\right) \quad (16)$$

3 Results and discussion

3.1 Hamaker constant

The Hamaker constant for the particle-AWI interaction system is A_{WW}-A_{PW}. This expression and Equation (2) or (3) were obtained through exact analysis of the change in free energy of the VDW system by bringing the particle from infinity to the vicinity of AWI without assumptions, which thus can accurately calculate the Hamaker constant for this interaction system. Notably, the value of A_{PW} is calculated as $\sqrt{A_{PP}}\sqrt{A_{WW}}$, where A_{PP} is the Hamaker constant for particle-particle system in vacuum.

Table 1 Hamaker constant for various materials in combinations with water and air
The value of A_{WW} is taken as 3.7×10^{-20} J for calculating the values of Hamaker constant for particle-water-air

Materials	$A_{PP}(10^{-20}$ J)	Hamaker constant for particle-water-air(10^{-20} J)	
		This work	Hough and White (1980)
Quartz	8.83	-2.02	-1.83
Silica	6.55	-1.22	-1.03
Calcite	10.1	-2.41	-2.26
Polystyrene	6.58	-1.23	-1.06
Teflon	2.75	0.51	0.69
Carbon	8.2	-1.81	—
lron	21	-5.11	—

Table 1 presents calculated values of Hamaker constant for the particle composed of various materials interacting with the AWI (i.e., particle-water-air system). The values of A_{PP} and A_{WW} from Hough and White[19] were adopted for these calculations. The Hamaker constant values for particle-AWI systems are generally negative, indicating that the VDW forces/energies between these materials and AWI are repulsive. This is because the value of $A_{WW}-A_{PW}$ is equal to $\sqrt{A_{WW}}(\sqrt{A_{WW}}-\sqrt{A_{PP}})$, and the value of A_{WW} is generally smaller than A_{PP}(due to larger density of a solid material compared to that of water). An exception is that the valueof A_{PP} for Teflon is smaller than A_{WW}, causing a positive value of Hamaker constant for the Teflon-AWI system and an attractive VDW energy/force between the Teflon particle and the AWI. Conversely, the smaller A_{PP} than A_{WW} causes a negative value of Hamaker constant and attractive VDW force between the Teflon and SWI[45-46]. This is in contrast to the common attractive VDW energy/force for the particle-SWI interaction. Table 1 shows that calculated values of Hamaker constant for the particle-AWI systems via our method are comparable to those of Hough and White[23] which used Lifshitz theory for calculations.

As mentioned previously, the Hamaker constant for the bubble SWI interaction is $A_{WW}-A_{WC}[=\sqrt{A_{WW}}(\sqrt{A_{WW}}-\sqrt{A_{CC}})]$ (A_{CC} is Hamaker constant for the solid collector-solid collector system in vacuum). Therefore, the value of the Hamaker constant for the bubble-SWI interaction is also negative and the VDW energy is repulsive because the value of A_{WW} is commonly smaller than A_{CC}(due to larger density of a solid collector material compared to that of water). The Hamaker constant for interaction of a bubble with the AWI or another bubble is A_{WW}, as mentioned previously. This indicates that the bubble-AWI or bubble-bubble interaction is attractive, which is the reason of the favorable coalescence of a bubble with the AWI or another bubble[47,48].

3.2 Interaction energies for a particle before contacting the AWI

Fig. 3a presents calculated VDW interaction energies between a polystyrene latex sphere with different diameters and the AWI as function of separation distances. The value of Hamaker constant of the polystyrene-AWI system was taken as -1.23×10^{-20} J (see Table 1). For a given particle size, the repulsive VDW energy increases with decreasing separation distance. The repulsive VDW energy increases almost linearly with particle size. A linear relationship also exists between the repulsive VDW energy and Hamaker constant as shown by Equation (4). Therefore, particles with larger sizes and composed of materials with larger Hamaker constants experience greater repulsive VDW energies due to the AWI. Similarly, larger bubbles experience greater repulsion when they approach a SWI. In contrast, greater VDW attraction exists between larger bubbles in solution. Therefore, nanobubbles have greater stability in solution due to small VDW attraction between them and high Brownian diffusion effect[49-51].

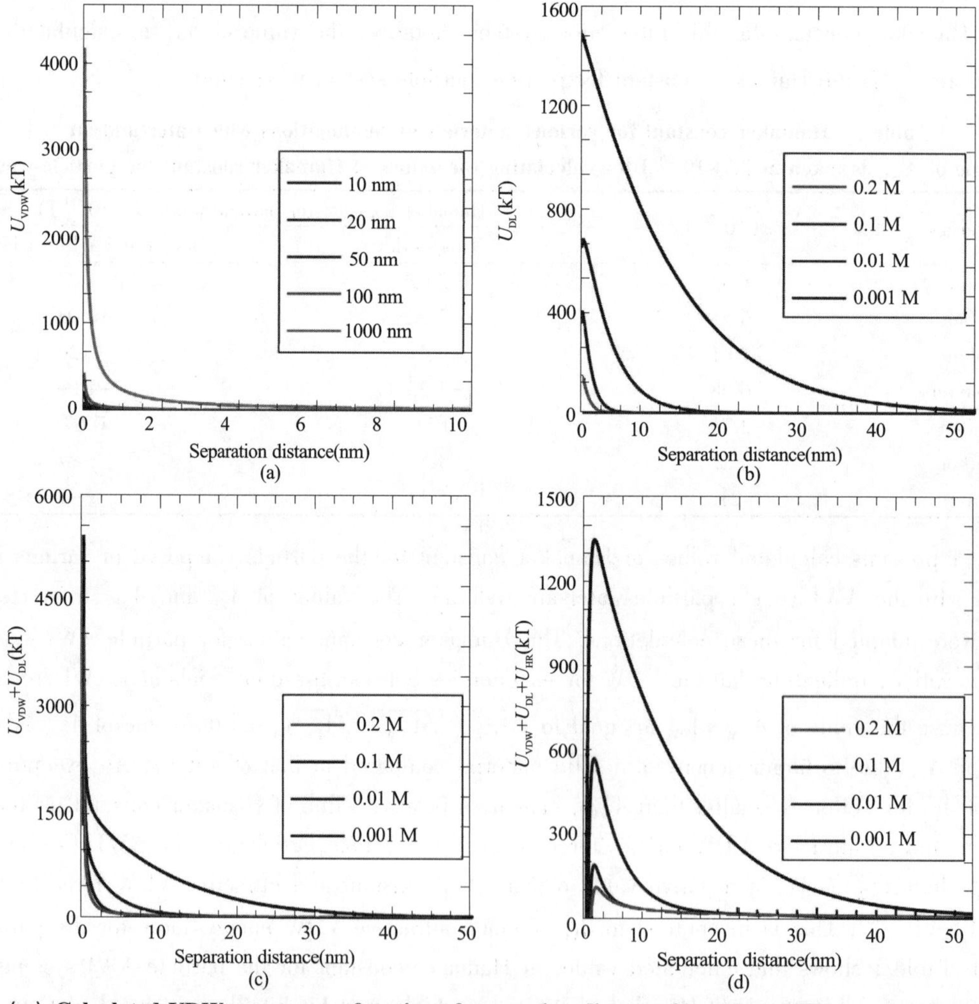

Fig. 3 (a) Calculated VDW energy as a function of separation distance from the AWI for a negatively charged polystyrene latex sphere with different diameters, (b) DL energy, (c) Sum of VDW and DL energy, and (d) sum of VDW, DL, and HR energy for the interaction between a 1 μm polystyrene latex sphere as a function of separation distances from the AWI under different IS conditions

Fig. 3b presents the sum of VDW and DL energies between a negatively charged 1 lm polystyrene latex sphere and the AWI at different separation distances in NaCl at different solution ionic strengths (ISs). The values of zeta potentials for the latex particle were taken as 47 mV, 41 mV, 38 mV, and 25 mV at ISs of 0.001 M, 0.01 M, 0.1 M, and 0.2 M, respectively[52]. The values of zeta potentials for the AWI were −43 mV, −27 mV, −20 mV, and 12 mV at 0.001 M, 0.01 M, 0.1 M, and 0.2 M, respectively[53]. The negative charge of AWI was believed to be resulted from an excess of OH^- due to preferential orientation of water molecules at the interface, but the exact mechanisms are still not clear to date[18]. Reduction of IS increases the DL repulsive energy, causing increased total repulsion acting on the latex particle at low ISs due to the AWI (e.g., ≤0.01 M). Fig. 3c shows that if the attractive hydrophobic interaction is involved, the particle-AWI interaction becomes attractive at small separation distances due to the domi-nance of hydrophobic attraction over repulsive VDW and DL energies, resulting in a deep primary minimum. However, at larger separation distances, a repulsive energy barrier exists at all solution ISs. This is because the decay of hydrophobic force with separation distance is much rapider than the VDW and DL forces[54]. Interestingly, although the DL repulsion is minor at the largest IS (0.2 M), a significant energy barrier (>100 kT) still exists in the energy profile because of the VDW repulsion. This energy barrier is much larger with regard to the average kinetic energy of a colloidal particle in solution (i.e., 1.5 kT). Therefore, the particle cannot overcome the large energy barrier via Brownian diffusion and enter the primary minimum for attachment. Our results theoretically explain why the AWI is not favorable for particle (either hydrophobic or hydrophilic) attachment when the particle surface and AWI are like charged as observed in the literature[27,55,56].

Fig. 4 shows the sum of VDW and DL energies between a positively charged polystyrene latex sphere with diameters of 1 μm or 100 nm and the AWI as a function of separation distance in NaCl at different solution ISs. The values of zeta potentials for the positively charged latex particle were taken as 22 mV, 20 mV, 10 mV, and 5 mV at ISs of 0.001 M, 0.01 M, 0.1 M, and 0.2 M, respectively[57]. The attractive DL energy dominates over the repulsive VDW energy at all separation distances at the lowest IS (i.e., 0.001 M). Therefore, the positively charged latex particles are favorable to be attached at AWI at this IS. At 0.01 M, the competition between VDW repulsion and DL attraction results in short-range repulsion, a primary minimum, and a repulsive energy barrier (11 kT and 0.5 kT for the 1 μm and 100 nm particles, respectively) in the interaction energy profiles. The nanoparticle can readily attach at the primary minima thorough overcoming the very small energy barrier via Brownian diffusion, whereas the attachment cannot occur for the microparticle. At ISs of 0.1 and 0.2 M, the VDW repulsion exceeds the DL attraction at all separation distances and no energy mini-mum exists in the profile. Therefore, both particles cannot be attached at the AWI at the two high ISs. However, including the hydrophobic interaction for a hydrophobic particle can cause deep primary minima where the nanoparticle may be attached by over-coming the relatively small energy barriers at the two high ISs (see Fig. 4).

The above calculations demonstrate that negatively charged latex particles cannot attach at the AWI even if they are hydropho-bic. Positively charged microparticles (hydrophilic or hydrophobic) can attach at the AWI only at very low solution ISs. Positively charged nanoparticles can attach at the AWI at medium ISs if they are hydrophilic, and the attachment can occur even at high solution ISs when they are hydrophobic. The elucidated dependence of particle retention at AWI on particle size, DL interaction, and hydrophobicity theoretically explain the observed favorable and unfavorable retention at AWI in the experimental literature[31-34,58-61].

While the nZVIs has held promise for *in-situ* soil remediation[62], the mobility of nZVIs is very limited in granular media such as soil[63-65]. The retention of hydrophilic nZVIs has been traditionally attributed to attachment on the SWI. Our calculations, however, showed that nZVIs could also be retained at the AWI at very low ISs. For example, Fig. 5 shows that the competition between the attractive DL energy and VDW repulsion causes a remarkable primary minimum (10.1 kT) and a very small energy barrier (1.6 kT) for the

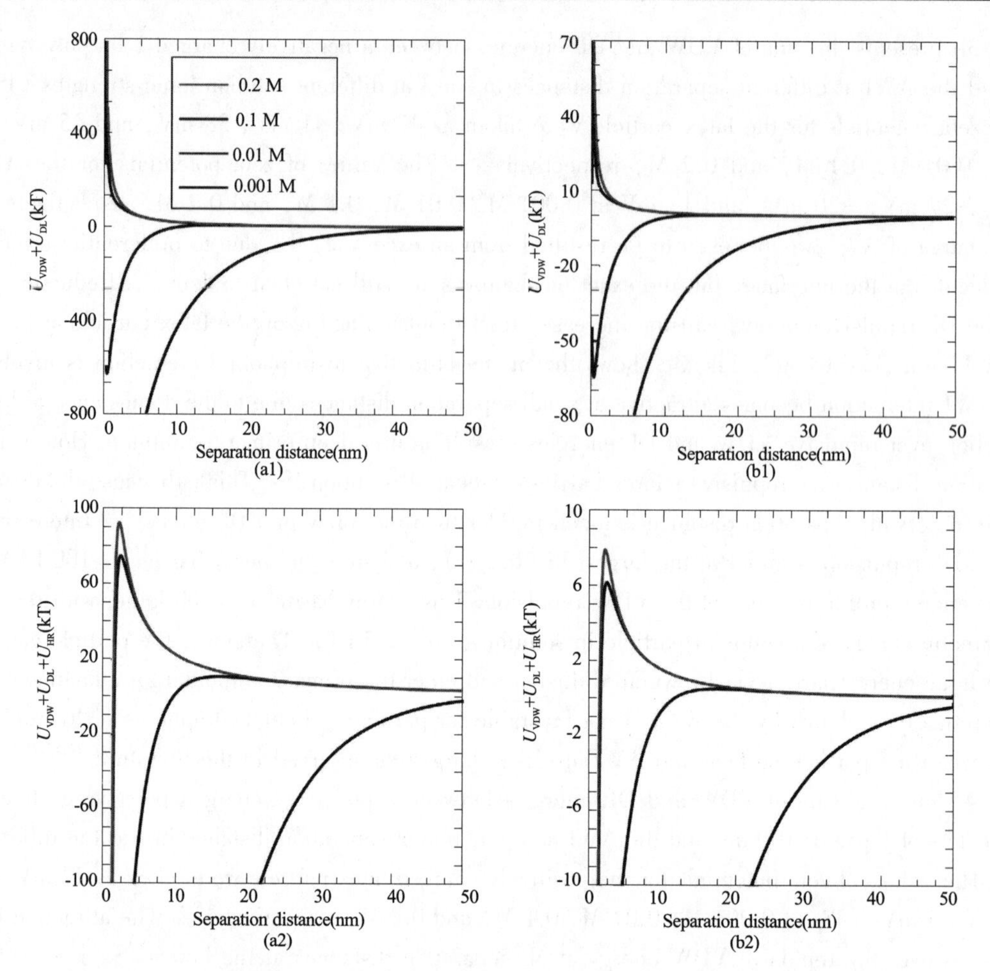

Fig. 4 (a) Calculated sum of (1) VDW and DL energy or (2) sum of VDW, DL, and HR energy between a (a) 1 μm or (b) 100 nm positively charged polystyrene latex sphere and AWI as a function of separation distances for different IS conditions

100 nm nZVI at 0.001 M. The attachment of the nZVI on the AWI via primary-minimum association could occur via overcoming this small energy barrier. If the nZVIs are oxidized, the zeta potentials of the nZVIs surfaces become positive and the DL attraction (e.g., for magnetite nanoparticles) is further increased. Therefore, the oxidized particles could be attached on the AWI at higher ISs.

3.3 Interaction energy for a particle penetrating through AWI

Fig. 6 presents calculated VDW energy as a function of θ (i.e., the angle between z axis and the line through the particle and AWI arc centers in Fig. 2) for a 1 μm particle interacting with a concave AWI with radius of 100 nm (triangle) or 10 nm (circle). We changed the value of θ to examine the variation of the position of the contact point between the AWI and particle surface on the VDW energy. The calculated VDW energy is solely related to the force acting along the z direction. There is no VDW force in the x or y direction due to the symmetry of the interaction configuration of the particle with the AWI. Positive and negative values mean that the particle experiences repulsion and attraction from the AWI in z direction, respectively. The particle experiences repulsive VDW energy for the range of θ values (between $\frac{1}{18}\pi$ and $\frac{4}{9}\pi$) considered in Fig.6. Comparison of Fig. 6 with Fig. 3 show that for small values of θ, the VDW repulsion of a particle at the AWI are significantly larger than those before the particle penetrates the AWI. For a given particle size, the VDW repulsion decreases with increasing value of θ. This is because the area elements of the AWI impose more repulsion on the particle in the horizontal direction at a larger value of θ, which are canceled by each other due to the symmetry of the interaction configuration. Particularly, the VDW energy becomes zero at $\theta=\frac{1}{2}\pi$. Due to a similar reason, the VDW repulsion increases with decreasing radius of the AWI for a given value of θ. For small values of θ, the VDW repulsion is much larger than the repulsive energy barrier between a 1 lm particle and SWI (normally smaller than several thousands of κTs)[52]. Therefore, when the particle is captured in a water film adjacent to a solid surface under unfavorable conditions, the VDW repulsion from the AWI and the repulsive energy from the SWI act on the particle in opposite directions (e.g., when the SWI is located under the particle) and the VDW repulsion can readily overcome the energy barrier between the particle and the SWI, resulting in strong attachment[29,30,66].

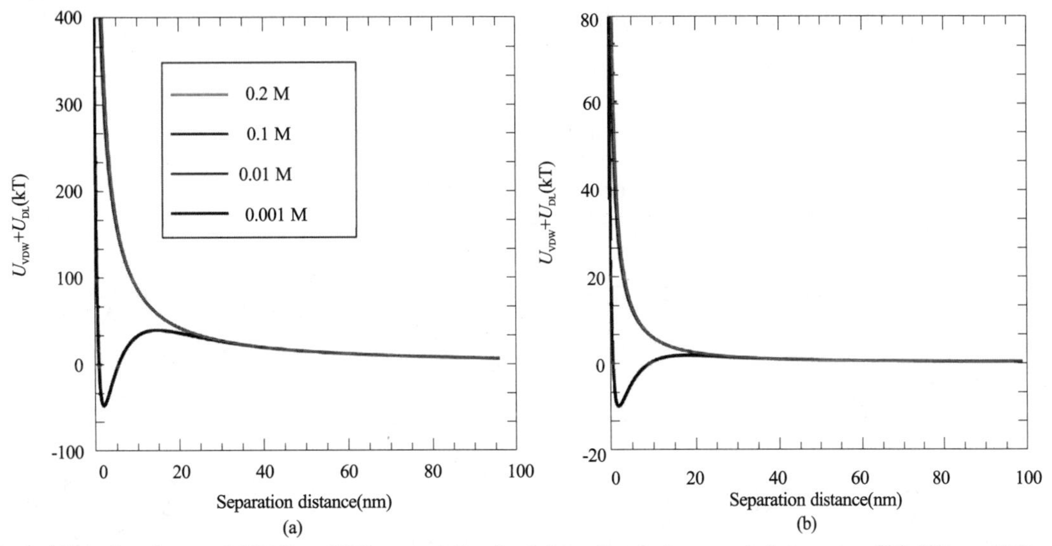

Fig. 5 (a) Calculated sum of VDW and DL energy for the interaction between a (a) 1 μm or (b) 100 nm ZVI particle and AWI at different separation distances with different ISs. The zeta potentials of ZVI particles were taken as 0

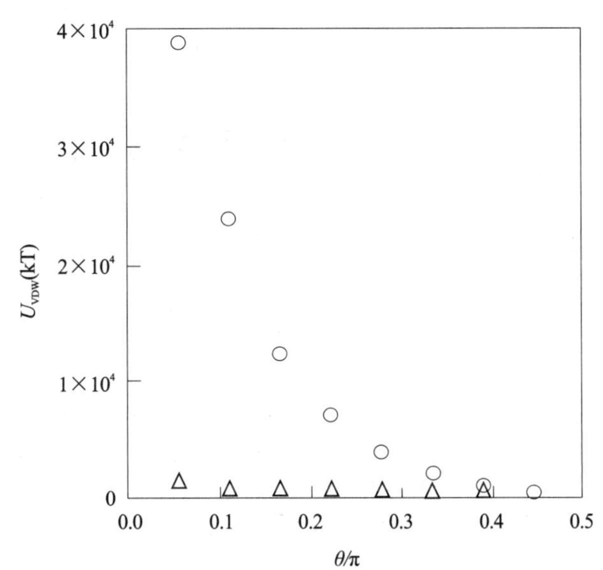

Fig. 6 Calculated VDW energy as a function of θ for a 1 μm particle interacting with a concave AWI with radius of 100 nm (triangle) or 10 nm (circle)

It is interesting to note that the DL energy exhibits a nonmonotonic variation with θ when the particle is at a concave AWI (see Fig. 7a). This is because the DL interaction energy does not interacting with a surface. Specifically, a maximum value of DL energy is located at a small separation distance. The DL energy decreases with further decreasing or increasing the separation distance. The separation distance corresponding to the maximum

DL energy changes at different solution ISs, causing a nonmonotonic variation of the DL energy with IS for some values of θ. For a particle in a water film adjacent to the SWI, inclusion of DL repulsion between the particle and concave AWI can facilitate attachment of the particle on the SWI because the VDW and DL repulsions from the concave AWI (Fig. 7b) are significantly larger than the the energy barrier between the particle and the SWI reported in the literature[1,7,8,19,52]. However, if hydrophobic interaction is accounted for, the total interaction becomes attractive at small values of θ (Fig. 7c). The attractive interaction energy could reach several hundred thousand kTs, demonstrating that the particle is tightly captured by the AWI. At large values of θ, the repulsive VDW and DL energies exceed the hydrophobic attraction. In these cases, the AWI imposes repulsion on the particle in the z direction. This is because the hydrophobic interaction decays much more rapidly (i.e., exponentially) that the VDW and DL forces, as mentioned previously.

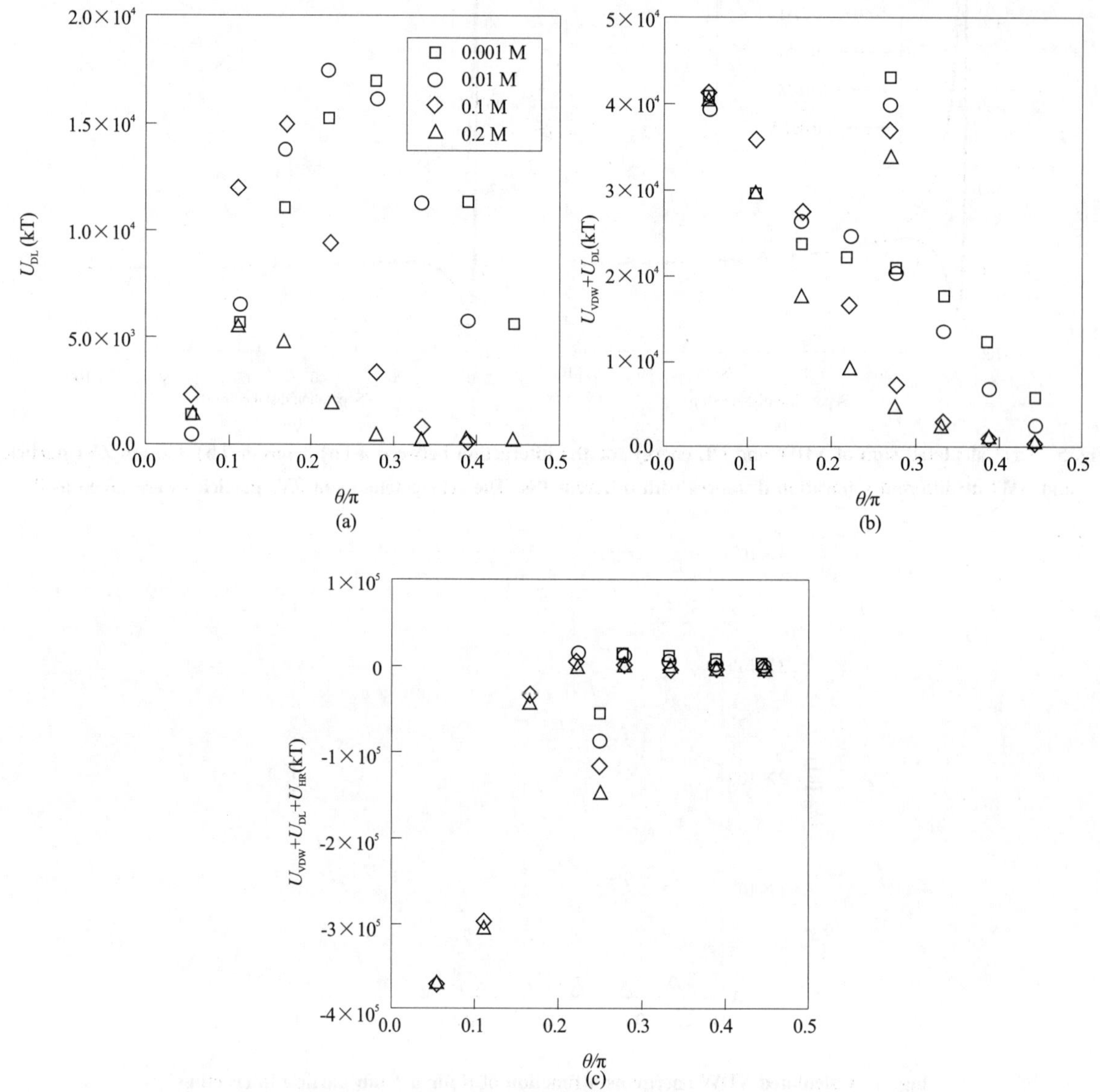

Fig. 7 Calculated (a) DL energy, (b) sum of VDW and DL energies, and (c) sum of VDW, DL and HR energies as a function of θ for a 1 μm particle interacting with a concave AWI with radius of 10 nm at different solution ISs

For the interaction of the particle with a convex AWI in Fig. 2b, the AWI of AB or $A'B'$ not only exerts a VDW repulsion on the SWI of APN or $A'P'N'$ in water, but also has a VDW attraction with the particle surface of *AM* or *MA′* in air. Fig. 8 presents calculated VDW energy as a function of θ for a 1 μm particle interacting with a convex AWI with radius of 2 μm or 5 μm. For the VDW interaction of the AWI of *AB* or *A′B′* with *AM* or *MA′* of the particle surface in air in Fig. 2, the value of A_{WP} was taken as 4.58 10^{-20} J. The angle between the line through contact point *A* and the center of the convex AWI and the axis in Fig. 2 was fixed to be p/6. The calculations show that only when the contact point *A* between the convex AWI and particle surface is very close to the apex point *M* of the particle (i.e., when most fraction of the particle is immersed in the water), the VDW repulsion from the AWI of *AB* and *A′B′* slightly exceeds the VDW attraction between the AWI and the particle surface of *AM* and *MA′*. Therefore, the particle experiences a small VDW repulsion due to the convex AWI. When the fraction of the particle immersed in the water decreases, the VDW attraction between the AWI and the particle surface of *AM* and *MA′* exceeds the VDW repulsion from the AWI of *AB* and *A′B′*. Therefore, the particle experiences a VDW force in the upward direction due to the convex AWI. This upward VDW force can facilitate the release of colloids from a collector surface in unsaturated porous media during imbibition[67-71]. The release of colloids due to imbibition has been traditionally attributed solely to capillary force. The VDW energy that acts on the particle in the upward direction becomes more significant when the contact point *A* is closer to the bottom of the particle (see Fig. 8). These results offer a plausible explanation for the observation that drainage to a lower water content and subsequent imbibition caused significant release of particles in unsaturated porous media[71].

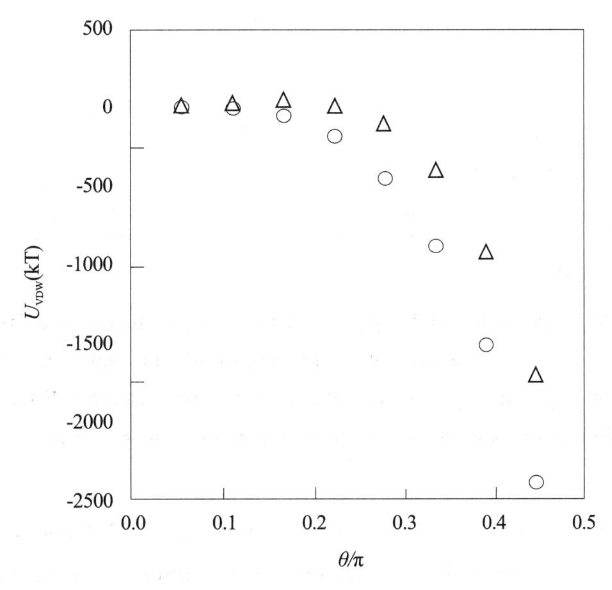

Fig. 8 Calculated VDW energy as a function of θ for a 1 μm particle interacting with a convex AWI with radius of 2 μm (blue triangle) or 5 μm (red circle). For the VDW interaction of the AWI of *AB* or *A′B′* with *AM* or *MA′* of the particle surface in air in Fig. 2, the value of A_{WP} was taken as 4.58×10^{-20} J. The angle between the line through point *A* and the center of the convex AWI and the x axis was fixed to be $\pi/6$

3.4 Comparison of VDW force with capillary force

Fig. 9 compared VDW and capillary forces at different values of θ for a 1 μm particle interacting with a concave AWI with radii of 100 nm. Our interface geometry calculations show that the VDW force exceeds the capillary force when the contact point of AWI with the particle surface is located near the apex point of the particle. Note that DLVO forces have been traditionally believed to be several orders of magnitude smaller than the capillary force. The VDW force decreases with increasing θ when the value of θ is larger than p/4, and the capillary force becomes dominant. As mentioned previously, the VDW force for the particle at the convex AWI will act in the positive z-direction (see Fig. 2). A change of the VDW force to the positive z-direction can cause release of particles in unsaturated porous media. For example, nanoscale protruding asperities on the SWI have been shown to reduce attraction or even cause repulsion between particles and the SWI[1,72]. VDW repulsion from particles at a concave AWI can push particles toward nanoscale protruding asperities (NPAs) on the SWI and cause attachment when the water film thickness decreases. However, if the shape of the AWI changes from concave to convex due to imbibition, the VDW force will operate in the opposite direction, facilitating the release of particles from NPAs. Notably, if the particles at the concave AWI are firmly attached on the SWI beneath the water film (e.g., at deep primary energy minima), the capillary force may be necessary for particle detachment. The capillary force is significantly larger than the VDW force for a particle at the convex AWI (see Fig. 9b).

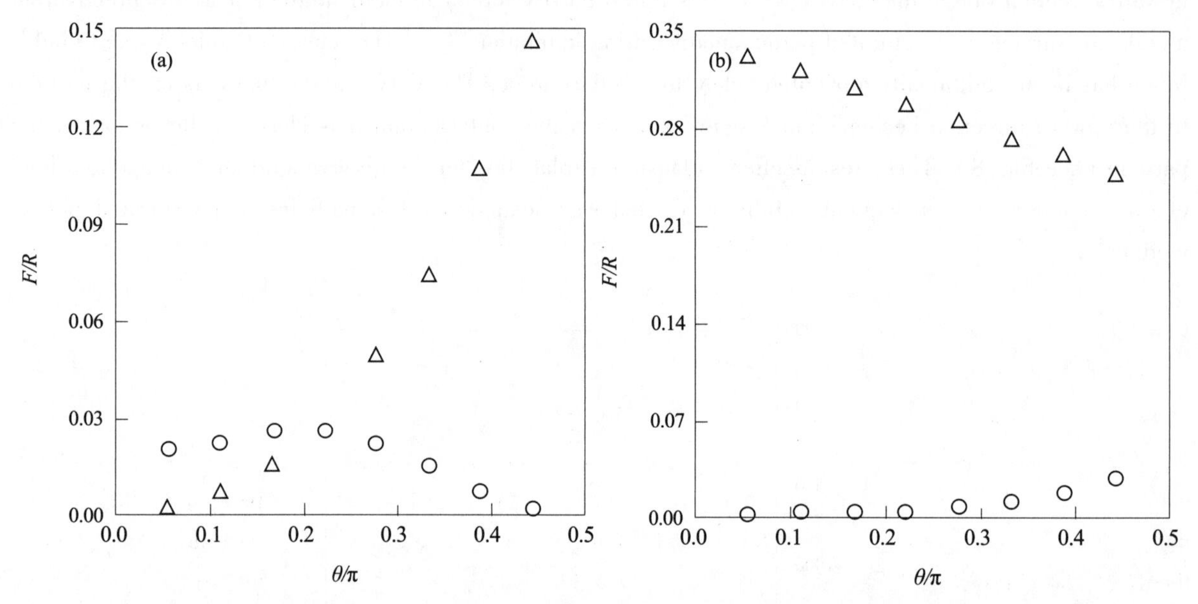

Fig. 9 Calculated VDW (red circle) and capillary (blue triangle) forces as a function of θ for a 1 μm particle interacting with a concave and convex AWI with radii of 100 nm and 2 μm, respectively. Note that the VDW and capillary forces of (a) and (b) are along the negative and positive z-direction of Fig. 2, respectively. All simulation conditions were the same as those of Fig. 6 and Fig. 8

3.5 Implication

The repulsive VDW and/or DL energy between a particle and a AWI not only facilitates attachment of the particle in a water film on the SWI, but also favors the retention of particles at the AWS interface line that is formed due to the intersection of the AWI and the SWI. The retention of colloids at the AWS interface line has been widely reported in the literature[73-76]. Whereas particle retention at the AWS interface line has been traditionally attributed only to capillary force, our calculations demonstrated that particles are favorably retained at this location even if the capillary force is absent. For example, the strong repulsion acting on a particle from the concave AWI can exceed the energy barrier between the particle and SWI, causing attachment at the SWI. This VDW repulsive can even exceed the capillary force. These theoretical calculations can explain experimental observations such as trapping of colloids in the wedge formed by the SWI and the AWI without being attached at the AWI (i.e., in the absence of capillary force)[77]. However, when the shape of the AWI changes from concave to convex, the VDW force that acts on the particle will be directed in the opposite direction, which favors particle detachment from the SWI. It is worthwhile mentioning that other colloid processes such as self-assembly are also frequently attributed to only capillary force[35-37], our work revealed that the VDW force also played a significant role in these processes.

Our work has important implication to optimal deliver of nanomaterials such as nZVI for *in-situ* soil remediation. Our calculations suggest that injection of nanomaterial suspensions with high solution ISs can enhance transport of the nanomaterials by reducing attachment at the AWI. Note that the attachment of positively charged nanomaterials or nZVI at the SWI is also reduced at high ISs because the DL attraction between the particles and collector surfaces (prevalently with negative charges in subsurface environments) are decreased. However, if bubbles are used as vehicles for delivery of nanoparticles[78-80], preparing the suspensions with both bubbles and nanoparticles using solutions with low ISs can favor aggregation of the bubbles with the nanoparticles.

Our theoretical calculations are also critical to accurately manipulate colloidal particles (e.g., biocolloids and surfactant micelles) using the AWI such as detachment from SWI, bubble flotation, Pickering emulsions, particle laden interfaces, and fabrication of new nanoand microstructured functional materials[29,81-87]. For example, the results in Figs. 5, 6, and 9 show that the interaction energy/force between a particle and concave or convex AWI changes significantly with the contact position of the AWI with the particle surface. The desired particle attachment or detachment can be achieved through accurate control of the thickness of the water film (determining the contact position) based on the theoretical calculations. Such accurate control of AWI curvature for manipulation of particles shows its promise as a novel green technique for separating and recycling colloids and nanoparticles in addition to the magnetic separation and antisolvents (e.g., CO_2)[88].

4 Conclusions

The landmark work of Hamaker[12] presented analytical expressions for determining the VDW force/energy for particle-SWI interaction for over eighty years. Existing literature[18,20,22,25,27,41,60 69,89] used the same expressions to calculate the VDW force/energy for the particle-AWI interaction before penetration but the values of Hamaker constant were taken as the measurements such as AFM experiments or estimates from Lift-shitz theory for the particle-AWI interaction. The validity of this method, however, is unclear to date. Moreover, the VDW force/energy for the particle-AWI interaction after the penetration has not been examined.

This work developed novel analytical expressions and numerical approaches to determine the Hamaker constant and VDW force/energy for the particle-AWI interaction before and after penetration. Our calculations were the first to theoretically explain why the VDW energy for particle-AWI or bubble-SWI interaction before penetration is repulsive. Our theoretical calculations also reasonably explain the observed discrepancies of favorable and unfavorable retention of particles at AWIs in the experimental literature[31-34,58-61]. Surprisingly, we found that strong VDW interaction energy existed after the particle penetrated the AWI, which could even exceed the capillary force and cause very stable retention of the particle at the AWS interface line and in thin water films next to the SWI. The result challenges the common belief that the colloidal processes such as stable attachment at the aforementioned locations and self-assembly were solely due to capillary force[17,27,41,89]. The methods and findings in this work are critical to quantifying and understanding numerous engineered applications such as *in-situ* soil remediation and particle manipulation (e.g., bubble flotation, Pickering emulsions, and particle laden interfaces)[81-88]. Further theoretical work is under way to reveal the couple of AWI shape and forces acting on the particle at the AWI including the VDW interaction using the Young-Laplace equation of capillarity.

Credit authorship contribution statement

Yichun Du: Writing-original draft, Investigation, Software. Scott A. Bradford: Conceptualization,

Writing-review & editing, Validation. Chongyang Shen: Methodology, Writing-original draft. Tiantian Li: Writing-original draft. Xiaoyuan Bi: Investigation. Dong Liu: Writing-original draft. Yuanfang Huang: Conceptualization, Software, Validation.

Declaration of competing interest

The authors declare that they have no known competing financial interests or personal relationships that could have appeared to influence the work reported in this paper.

Acknowledgments

We acknowledge the National Natural Science Foundation of China (no. 41922047, 41671222), and the 2115 Talent Development Program of China Agricultural University (1191-00109011) for financial support.

References

[1] C. Shen, Y. Jin, J. Zhuang, T. Li, B. Xing, Role and importance of surface heterogeneities in transport of particles in saturated porous media, Crit. Rev. Environ. Sci. Technol. 50 (3) (2020) 244-329.

[2] S. Sasidharan, S.A. Bradford, J. Šimů nek, S. Torkzaban, Minimizing virus transport in porous media by optimizing solid phase inactivation, J. Environ. Qual. 47 (5) (2018) 1058-1067.

[3] P. Opr čkal, A. Mladenovič, J. Vidmar, A. Mauko Pranji ć, R. Mila čič, J. Š čančar, Critical evaluation of the use of different nanoscale zero-valent iron particles for the treatment of effluent water from a small biological wastewater treatment plant, Chem. Eng. J. 321 (2017) 20-30.

[4] Y. Monga, P. Kumar, R.K. Sharma, J. Filip, R.S. Varma, R. Zbo řil, M.B. Gawande, Sustainable synthesis of nanoscale zerovalent iron particles for environmental remediation, ChemSusChem. 13 (13) (2020) 3288-3305.

[5] T. Phenrat, H.-J. Kim, F. Fagerlund, T. Illangasekare, G.V. Lowry, Empirical correlations to estimate agglomerate size and deposition during injection of a polyelectrolyte-modified Fe^{0} nanoparticle at high particle concentration in saturated sand, J. Contam. Hydrol. 118 (3-4) (2010) 152-164.

[6] D. O'Carroll, B. Sleep, M. Krol, H. Boparai, C. Kocur, Nanoscale zero valent iron and bimetallic particles for contaminated site remediation, Adv. Water Resour. 51 (2013) 104-122.

[7] J.N. Ryan, M. Elimelech, Colloid mobilization and transport in groundwater, Colloids Surf. A 107 (1996) 1-6.

[8] I.L. Molnar, W.P. Johnson, J.I. Gerhard, C.S. Willson, D.M. O'Carroll, Predicting colloid transport through saturated porous media: A critical review, Water Resour. Res. 51 (9) (2015) 6804-6845.

[9] S.A. Bradford, J. Simunek, M. Bettahar, M.T. van Genuchten, S.R. Yates, Significance of straining in colloid deposition: Evidence and implications: Significance of straining in colloid deposition, Water Resour. Res. 42 (12) (2006).

[10] C. Shen, S.A. Bradford, M. Flury, Y. Huang, Z. Wang, B. Li, DLVO interaction energies for hollow particles: The filling matters, Langmuir 34 (2018) 12764-12775.

[11] E. J. W. Verwey, J. T. G. Overbeek, Theory of the stability of lyophobic colloids, Elsevier, Amsterdam, Netherland, 1948.

[12] H.C. Hamaker, The London-van der Waals attraction between spherical particles, Physica. 4 (10) (1937) 1058-1072.

[13] R. Hogg, T.W. Healy, D.W. Fuerstenau, Mutual coagulation of colloidal dispersions, Trans. Faraday Soc. 62 (1966) 1638-1651.

[14] S. Bhattacharjee, M. Elimelech, Surface element integration: A novel technique for evaluation of DLVO interaction between a particle and a flat plate, J. Colloid Interface Sci. 193 (1997) 273-285.

[15] S. Lin, M.R. Wiesner, Exact analytical expressions for the potential of electrical double layer interactions for a sphere-plate system, Langmuir 26 (22) (2010) 16638-16641.

[16] J. Wan, J.L. Wilson, Visualization of the role of the gas-water interface on the fate and transport of colloids in porous media, Water Resour. Res. 30 (1) (1994) 11-13.

[17] B. Gao, T.S. Steenhuis, Y. Zevi, V.L. Morales, J.L. Nieber, B.K. Richards, J.F. Mccarthy, J.-Y. Parlange,

Capillary retention of colloids in unsaturated porous media, Water Resour. Res. 44 (2008) W04504.

[18] M. Flury, S. Aramrak, Role of air-water interfaces in colloid transport in porous media: A review, Water Resour. Res. 53 (7) (2017) 5247-5275.

[19] J.N. Israelachvili, Intermolecular and Surface Forces, Academic Press, San Diego, USA, 2011.

[20] S. Hamamoto, T. Sugimoto, T. Takemura, T. Nishimura, S.A. Bradford, Nanobubble retention in saturated porous media under repulsive van der Waals and electrostatic conditions, Langmuir 35 (21) (2019) 6853-6860.

[21] W.A. Ducker, Z. Xu, J.N. Israelachvili, Measurements of hydrophobic and DLVO forces in bubble-surface interactions in aqueous solutions, Langmuir 10 (9) (1994) 3279-3289.

[22] R.F. Tabor, R. Manica, D.Y.C. Chan, F. Grieser, R.R. Dagastine, Repulsive van der Waals forces in soft matter: Why bubbles do not stick to walls, Phys. Rev. Lett. 106 (2011) 064501.

[23] D.B. Hough, L.R. White, The calculation of Hamaker constants from Liftshitz theory with applications to wetting phenomena, Adv. Colloid Interface Sci. 14 (1) (1980) 3-11.

[24] A. Khaled Abdella Ahmed, C. Sun, L. Hua, Z. Zhang, Y. Zhang, T. Marhaba, W. Zhang, Colloidal properties of air, oxygen, and nitrogen nanobubbles in water: Effects of ionic strength, natural organic matters, and surfactants, Environ. Eng. Sci. 35 (7) (2018) 720-727.

[25] C. Yang, T. Dabros, D. Li, J. Czarnecki, J.H. Masliyah, Analysis of fine bubble attachment onto a solid surface within the framework of classical DLVO theory, J. Colloid Interface Sci. 219 (1) (1999) 69-70.

[26] P.A. Kralchevsky, K. Nagayama, Capillary interactions between particles bound to interfaces, liquid films and biomembranes, Adv. Colloid Interface Sci. 85 (2-3) (2000) 145-192.

[27] V. Lazouskaya, L.-P. Wang, H. Gao, X. Shi, K. Czymmek, Y. Jin, Pore-scale investigation of colloid retention and mobilization in the presence of a moving air-water interface, Vadose Zone J. 10 (4) (2011) 1250-1260.

[28] A. Dathe, Y. Zevi, B.K. Richards, B. Gao, J.-Y. Parlange, T.S. Steenhuis, Functional models for colloid retention in porous media at the triple line, Environ. Sci. Pollut. Res. 21 (15) (2014) 9067-9080.

[29] Y.E. Yu, S. Khodaparast, H.A. Stone, Separation of particles by size from a suspension using the motion of a confined bubble, Appl. Phys. Lett. 112 (18) (2018) 181604.

[30] T. Wu, Z. Yang, R. Hu, Y.-F. Chen, H. Zhong, L. Yang, W. Jin, Film entrainment and microplastic particles retention during gas invasion in suspension-filled microchannels, Water Res. 194 (2021) 116919.

[31] Wan, T.K. Tokunaga, Partitioning of clay colloids at air-water interfaces, J. Colloid Interface Sci. 247 (1) (2002) 54-61.

[32] J.T. Crist, J.F. McCarthy, Y. Zevi, P. Baveye, J.A. Throop, T.S. Steenhuis, Pore-scale visualization of colloid transport and retention in partly saturated porous media, Vadose Zone J. 3 (2) (2004) 444-450.

[33] T. Knappenberger, M. Flury, E.D. Mattson, J.B. Harsh, Does water content or flow fate control colloid transport in unsaturated porous media-, Environ Sci. Technol. 48 (2014) 3791-3799.

[34] W. Zhang, S. Wu, Y. Qin, S. Li, L. Lei, S. Sun, Y. Yang, Deposition and mobilization of viruses in unsaturated porous media: Roles of different interfaces and straining, Environ. Pollut. 270 (2021) 116072.

[35] P.W.K. Rothemund, Using lateral capillary forces to compute by self-assembly, PNAS 97 (3) (2000) 984-989.

[36] A. Huerre, M. De Corato, V. Garbin, Dynamic capillary assembly of colloids at interfaces with 10,000g accelarations, Nat. Commun. 9 (2018) 3630.

[37] G. Soligno, M. Dijkstra, R. van Roij, Self-assembly of cubic colloidal particles at fluid-fluid interfaces by hexapolar capillary interactions, Soft Matter 14 (1) (2018) 42-50.

[38] K. Hinsch, Holographic study of liquid surface deformations produced by floating particles, J. Colloid Interface Sci. 92 (1) (1983) 243-255.

[39] P.A. Kralchevsky, N.D. Denkov, K.D. Danov, Particles with an undulated contact line at a fluid interface: Interaction between capillary quadrupoles and rheology of particulate monolayers, Langmuir 17 (2001) 7694-7705.

[40] C. Shen, H. Wang, V. Lazouskaya, Y. Du, W. Lu, J. Wu, H. Zhang, Y. Huang, Cotransport of bismerthiazol and montmorillonite colloids in saturated porous media, J. Contam. Hydrol. 177-178 (2015) 18-19.

[41] P. Sharma, M. Flury, J. Zhou, Detachment of colloids from a solid surface by a moving air-water interface, J.

Colloid Interface Sci. 326 (1) (2008) 143-150.

[42] S.H. Donaldson, A. Røyne, K. Kristiansen, M.V. Rapp, S. Das, M.A. Gebbie, D.W. Lee, P. Stock, M. Valtiner, J. Israelachvili, Developing a general interaction potential for hydrophobic and hydrophilic interactions, Langmuir 31 (7) (2015) 2051-2064.

[43] L. Zhang, L. Ren, S. Hartland, More convenient and suitable methods for sphere tensiometry, J. Colloid Interface Sci. 180 (2) (1996) 493-503.

[44] O. Pitois, X. Chatcau, Small particle at a fluid interface: Effect of contact angle hysteresis on force and work of detachment, Langmuir 18 (2002) 9751-9756.

[45] S. Lee, W.M. Sigmund, AFM study of repulsive van der Waals forces between Teflon AFTM thin film and silica or alumina, Colloids Surf. A 204 (2002) 43-50.

[46] C.D. Bohling, W.M. Sigmund, Predicting and measuring repulsive van der Waals forces for a Teflon AF^{TM}-solvent-α-alumina system, Colloids Surf. A 462 (2014) 137-146.

[47] I.U. Vakarelski, R. Manica, X. Tang, S.J. O'Shea, G.W. Stevens, F. Grieser, R.R. Dagastine, D.Y.C. Chan, Dynamic interactions between microbubbles in water, PNAS 107 (25) (2010) 11177-11182.

[48] S. Orvalho, M.C. Ruzicka, G. Olivieri, A. Marzocchella, Bubble coalescence: Effect of bubble approach velocity and liquid viscosity, Chem. Eng. Sci. 134 (2015) 205-216.

[49] J.N. Meegoda, S. Aluthgun Hewage, J.H. Batagoda, Stability of nanobubbles, Environ. Eng. Sci. 35 (11) (2018) 1216-1227.

[50] N. Nirmalkar, A.W. Pacek, M. Barigou, Interpreting the interfacial and colloidal stability of bulk nanobubbles, Soft Matter 14 (47) (2018) 9643-9656.

[51] B.H. Tan, H. An, C.-D. Ohl, Stability, dynamics, and tolerance to undersaturation of surface nanobubbles, Phys. Rev. Lett. 122 (2019) 134502.

[52] C. Shen, B. Li, Y. Huang, Y. Jin, Kinetics of coupled primary-and secondary-minimum deposition of colloids under unfavorable chemical conditions, Environ. Sci. Technol. 41 (20) (2007) 6976-6982.

[53] C. Yang, T. Dabros, D. Li, J. Czarnecki, J.H. Masliyah, Measurement of the zeta potential of gas bubbles in aqueous solutions by microelectrophoresis method, J. Colloid Interface Sci. 243 (1) (2001) 128-135.

[54] D. Grasso *, K. Subramaniam, M. Butkus, K. Strevett, J. Bergendahl, A review of non-DLVO interactions in environmental colloidal systems, Rev. Environ. Sci. Biotechnol. 1 (1) (2002) 17-18.

[55] W. Zhang, V.L. Morales, M.E. Cakmak, A.E. Salvucci, L.D. Geohring, A.G. Hay, J.-Y. Parlange, T.S. Steenhuis, Colloid transport and retention in unsaturated porous media: Effect of colloid input concentration, Environ. Sci. Technol. 44 (2010) 4965-4972.

[56] C. Wang, R. Wang, Z. Huo, E.n. Xie, H.E. Dahlke, Colloid transport through soil and other porous media under transient flow conditions-a review, Wiley Interdiscip. Rev. Water 7 (4) (2020).

[57] M. Elimelech, Kinetics of capture of colloidal particles in packed beds under attractive double layer interactions, J. Colloid Interface Sci. 146 (2) (1991) 337-352.

[58] A. Schäer, H. Harms, A.J.B. Zehnder, Bacterial accumulation at the air-water interface, Environ. Sci. Technol. 32 (23) (1998) 3704-3712.

[59] V.I. Syngouna, C.V. Chrysikopoulos, Experimental investigation of virus and clay particles cotransport in partially saturated columns packed with glass beads, J. Colloid Interface Sci. 440 (2015) 140-150.

[60] H. Bai, N. Cochet, A. Pauss, E. Lamy, DLVO, hydrophobic, capillary and hydrodynamic forces acting on bacteria at solid-air-water interfaces: Their relative impact on bacteria deposition mechanisms in unsaturated porous media, Colloids Surf. B 150 (2017) 41-49.

[61] D. Predelus, L. Lassabatere, C. Louis, H. Gehan, T. Brichart, T. Winiarski, R. Angulo-Jaramillo, Nanoparticle transport in water-unsaturated porous media: Effects of solution ionic strength and flow velocity, J. Nanopart. Res. 19 (2017) 104.

[62] Y. Liu, T. Wu, J.C. White, D. Lin, A new strategy using nanoscale zero-valent iron to simultaneously promote remediation and safe crop production in contaminated soil, Nat. Nanotechnol. 16 (2) (2021) 197-205.

[63] N. Saleh, H.-J. Kim, T. Phenrat, K. Matyjaszewski, R.D. Tilton, G.V. Lowry, Ionic strength and composition affect the mobility of surface-modified Fe^0 nanoparticles in water-saturated sand columns, Environ. Sci. Technol. 42 (9) (2008) 3349-3355.

[64] R.L. Johnson, G.O. Johnson, J.T. Nurmi, P.G. Tratnyek, Natural organic matter enhanced mobility of nano zerovalent iron, Environ. Sci. Technol. 43 (14) (2009) 5455-5460.

[65] Y. Hu, M. Zhang, X. Li, Improved longevity of nanoscale zero-valent iron with a magnesium hydroxide coating shell for the removal of Cr(VI) in sand columns, Environ. Int. 133 (2019) 105249.

[66] J. Wan, T.K. Tokunaga, Film straining of colloids in unsaturated porous media: Conceptual model and experimental testing, Environ. Sci. Technol. 31 (1997) 2413-2420.

[67] T. Cheng, J.E. Saiers, Mobilization and transport of in situ colloids during drainage and imbibition of partially saturated sediments, Water Resour. Res. 45 (2009) W08414.

[68] S. Aramrak, M. Flury, J.B. Harsh, Detachment of deposited colloids by advancing and receding air-water interfaces, Langmuir 27 (16) (2011) 9985-9993.

[69] V. Lazouskaya, L.-P. Wang, D. Or, G. Wang, J.L. Caplan, Y. Jin, Colloid mobilization by fluid displacement fronts in channels, J. Colloid Interface Sci. 406 (2013) 44-50.

[70] Q. Zhang, S.M. Hassanizadeh, N.K. Karadimitriou, A. Raoof, B. Liu, P.J. Kleingeld, A.Imhof, Retention and remobilization of colloids during steady-state and transient two-phase flow, Water Resour. Res. 49 (12) (2013) 8005-8016.

[71] S.A. Bradford, Y. Wang, S. Torkzaban, J. Šm ů nek, Modeling the release of E. coli D21g with transients in water content, Water Resour. Res. 51 (5) (2015) 3303-3316.

[72] T. Li, C. Shen, S. Wu, C. Jin, S.A. Bradford, Synergies of surface roughness and hydration on colloid detachment in saturated porous media: Column and atomic force microscopy studies, Water Res. 183 (2020) 116068.

[73] S.S. Thompson, M.V. Yates, Bacteriophage inactivation at the air-water-solid interface in dynamic batch systems, Appl. Environ. Microbiol. 65 (3) (1999) 1186-1190.

[74] Y. Zevi, A. Dathe, J.F. McCarthy, B.K. Richards, T.S. Steenhuis, Distribution of colloid particles onto interfaces in partially saturated sand, Environ. Sci. Technol. 39 (18) (2005) 7055-7064.

[75] Y. Zevi, A. Dathe, B. Gao, W. Zhang, B.K. Rchards, T.S. Steenhuis, Transport and retention of colloidal particles in partially saturated porous media: Effect of ionic strength, Water Resour. Res. 45 (2009) W12403.

[76] C. Douarche, J.-L. Sikorav, A. Goldar, Aggregation and adsorption at the air-water interface of bacteriophage / X174 single-stranded DNA, Biophys. J. 94 (1) (2008) 134-146.

[77] G. Chen, M. Flury, Retention of mineral colloids in unsaturated porous media as related to their surface properties, Colloids Surf. A 256 (2-3) (2005) 207-216.

[78] X. Shen, L. Zhao, Y. Ding, B.o. Liu, H. Zeng, L. Zhong, X. Li, Foam, a promising vehicle to deliver nanoparticles for vadose zone remediation, J. Hazard. Mater. 186 (2-3) (2011) 1773-1780.

[79] L.B. Mullin, L.C. Phillips, P.A. Dayton, Nanoparticle delivery enhancement with acoustically activated microbubbles, IEEE Trans. Ultrason. Ferroelectr. Freq. Control 60 (1) (2013).

[80] V. Prigiobbe, A.J. Worthen, K.P. Johnston, C. Huh, S.L. Bryant, Transport of nanoparticle-stabilized CO_2-foam in porous media, Transport Porous Med. 111 (2016) 265-285.

[81] W. G. Pitt, M. O. McBride, A. J. Barton, R. D. Sagers, Air-water interface displaces adsorbed bacteria, Biomaterials 14 (8) (1993) 605-608.

[82] M. Cavallaro, L. Botto, E.P. Lewandowski, M. Wang, K.J. Stebe, Curvature-driven capillary migration and assembly of rod-like particles, PNAS 108 (52) (2011) 20923-20928.

[83] E.M. Furst, Directing colloidal assembly at fluid interfaces, PNAS 108 (52) (2011) 20853-20854.

[84] J. Eastoe, S. Nave, A. Downer, A. Paul, A. Rankin, K. Tribe, J. Penfold, Adsorption of ionic surfactants at the air-solution interface, Langmuir 16 (10) (2000) 4511-4518.

[85] D.S. Valkovska, G.C. Shearman, C.D. Bain, R.C. Darton, J. Eastoe, Adsorption of ionic surfactants at an expanding air-water interface, Langmuir 20 (11) (2004) 4436-4445.

[86] A. Mendoza, E. Guzman, F. Martinez-Pedrero, H. Ritacco, R.G. Rubio, F. Ortega,V.M. Starov, R. Miller, Par-

ticle laden fluid interfaces: Dynamics and interfacial rheology, Adv. Colloid Interface Sci. 206 (2014) 393-419.

[87] J. Zhang, J. Hwang, M. Antonietti, B.V.K.J. Schmidt, Water-in-water pickering emulsion stabilized by polydopamine particles and crosss-linking, Biomacromolecules 20 (2019) 204-211.

[88] O. Myakonkaya, Z. Hu, M.F. Nazar, J. Eastoe, Recycling functional colloids and nanoparticles, Chem. Eur. J. 16 (39) (2010) 11784-11790.

[89] V. Lazouskaya, Y. Jin, D. Or, Interfacial interactions and colloid retention under steady flows in a capillary channel, J. Colloid Interface Sci. 303 (1) (2006) 171-184.

本文曾发表于2022年《Journal of Colloid and Interface Science》第601卷

3　土地资源与土地信息

陕南秦巴山区不同类型梯田侵蚀效应试验设计探究

李 鹏

（陕西省土地工程建设集团有限责任公司延安分公司，陕西 延安 716000）

【摘要】陕南秦巴山区生态环境问题突出，很多地区地形破碎复杂，土壤土层较薄，土壤侵蚀、土地退化和水土流失问题突出，严重制约着当地的农业生产和农民脱贫致富。目前，在秦巴山区坡地应用推广坡改梯工程是治理水土流失、减少土壤侵蚀最为有效的工程措施。据了解，不同类型的梯田受土壤侵蚀的程度相同，影响因素也较多。为了更好地揭示不同类型不同坡度梯田田面下土壤侵蚀及养分迁移变化，拟开展小区模拟试验进行探究。本文以商洛市商州区为例，主要通过科学的试验设计为后期深入研究降水侵蚀对梯田土壤养分及产沙量的影响提供一套技术方案。

【关键词】秦巴山区；梯田；土壤侵蚀；土壤养分；试验设计

秦巴山区泛指秦岭和巴山山区，即位于长江最大支流——汉水上游的秦岭大巴山及其毗邻地区，涉及甘肃、四川、陕西、重庆、河南、湖北六省市，其主体位于陕南地区。秦巴山区由诸多的山间谷地和小盆地相连接，其中以汉中盆地、西乡盆地、安康盆地、汉阴盆地、商丹盆地和洛南盆地最为著称。秦巴山地是我国南北过渡带的主体和青藏高原向东部平原重要的大尺度生态廊道，也是气候变化的敏感区和生态环境的脆弱区，地势自西向东逐渐降低，地形地貌复杂多样，地形起伏较大。该区域地处我国北亚热带和暖温带过渡区，南北两侧气候差异较大，气温和降水由南到北呈递减趋势，区内年均温 12~16 ℃，年降水量 450~1300 mm。

陕南秦巴山区耕地的主要类型为坡耕地，据统计，地处秦巴山区腹地的汉中市坡耕地面积高达 1.26 万 hm^2，占全市耕地面积的 63%[1]。陕南秦巴山区的流域面积仅为长江流域的 4%，而土壤侵蚀量高达 1.175 亿 t，占长江总输沙量的 12%[2]。坡耕地是陕南秦巴山区水土流失最主要的源头，耕地流失的细颗粒泥沙成为该流域内的河流泥沙重要组成部分。而坡改梯又是坡耕地水土流失治理典型的工程举措。因此，采用坡改梯工程，对于保护培育陕南秦巴山区耕地资源，减少水土流失，提高土地生产力，维护社会稳定和生态安全，促进经济和社会发展具有重要意义。

1 土壤侵蚀危害及研究进展

陕南秦巴山区存在的主要问题：生态问题突出，生态类型多样、本底脆弱，很多地区地形破碎复杂[3]；土壤侵蚀、土地退化和水土流失问题严重[4]；应对干旱和极端暴雨灾害的能力低下，对气候变化和人类干扰敏感；有效耕地面积少、粮食安全问题突出[5]。其中商洛市商州区处于我国秦淮南北自然分界线上，地势西高东低，南北高中间低，山峦叠嶂，犹如掌状由西北向东南倾斜，最高处秦王山海拔2087 m，最低处腰市镇南湾海拔 543 m。地貌类型复杂多样，山地面积占 84%，海拔 543~2087 m。全区年平均侵蚀模数 1977 t/km^2，水土流失面积占总面积的 45.5%，降水量大，暴雨频繁，地质灾害频发，生态环境极其脆弱（图 1）。

基金项目：2020 年度陕西省土地工程建设集团有限责任公司内部科研项目（DJNY2020-32）。

图1 陕南秦巴山区地貌

已有研究表明,坡耕地的水土流失随坡度增高、坡长增加而增强。降水量较大时,水流在下坡位汇集,汇流动能较大,下坡位土壤易被水冲走,最后在土壤凹处淤积。上坡位土壤在雨强较大时,土壤受雨滴溅蚀扰动,耕层土壤极易离散,顺水流方向向下移动,动能明显高于上溅土粒,因此土壤侵蚀严重。降雨强度较大的情况下,极易在过道、田埂下部、土壤抗水蚀性较差的薄弱环节形成冲刷沟,进一步使坡耕地犁底层土壤部分冲刷,这样会使土壤养分含量较高的表层土壤大量流失,以化肥、有机肥等为主的人为养分归还得不到及时补充,又遭破坏,尤其在降雨较大的年份和暴雨频发的雨季,这种情况将愈发严重。由于土壤团粒结构(有机成分)减少,干旱年份或冷热交替的冬春季节土壤保水性能差,土壤饱和含水量较低,加重旱情或降水量减少,又使坡耕地地力下降,耕地产量明显减低。

梯田是指在坡地上沿等高线修建的、断面呈阶梯状的田块,一般以梯面为水平或向内、或向外倾斜而分为水平、顺坡、反坡梯田3种形式。在坡地上修筑梯田,仍是我国最为重要也最为常见的水土保持工程措施,梯田能有效地消除或减缓地面坡度,截短径流流线,削减径流冲刷动力,强化降水就地入渗与拦蓄、保持水土,改善坡耕地生产条件,为作物稳产高产和生态环境建设创造条件。梯田是防治水土流失的得力措施,可以变跑水、跑土、跑肥的“三跑田”为保水、保土、保肥的“三保田”,这是其水土保持效益;梯田又具有保收、增收等经济效益,是农业可持续发展的有力保障,可提高粮食产量、促进生产力的发展[6-8]。针对陕南秦巴山区特殊的地质地貌和坡耕地居多的现状,当地政府已在商洛市商州区开展了大规模的坡改梯田土地整治工程(图2)。

图2 项目区坡改梯田土地整治工程

梯田修筑工程实践中,梯田田面设计及田坎的防护是确保梯田整体稳定的关键因素[9]。试验区选在商洛市商州区腰市镇,属于秦巴山区低山丘陵地带,区内土层薄、石砾多,薄层土覆盖于半风化砂泥岩之上[10-11]。土壤以始成褐土为主,黏粒含量高,多具有膨胀性。东部山顶有部分紫色土类属砂壤质地,保水保肥性能差,且下部半风化砂泥岩遇水易崩解成碎散体,透水能力强,地质条件不利于梯田修筑的稳定性[12]。此外,根据气象资料,项目区属于暖温带半湿润季风山地气候[13],多年平均降水量 740 mm,每逢夏季,短时间暴雨,造成山地径流严重,洪水泛滥,易因田面积水过多下渗而造成梯田垮塌,水土大量流失[14]。目前对黄土高原坡改梯(水平梯田)后土壤水分和养分的变化研究,水平梯田田坎坡度及田面宽度研究较多,而选择在陕南秦巴山区研究不同类型梯田不同坡度田面侵蚀效应的较少。

若修筑的梯田田面可及时有效排走多余的水分,进而有效减少梯田水土及养分流失,减轻田坎负担,则可确保梯田田面及田坎整体的稳定性。鉴于此,本试验拟计划在陕南秦巴山区开展小区试验,构筑不同类型梯田,分析不同类型梯田不同坡度田面水土流失及养分迁移变化,探寻最适合陕南秦巴山区的坡改梯工程的梯田类型设计,为大面积推广保水保肥的坡改梯工程提供参考依据。

2　研究内容及目标

2.1　研究内容

(1)探明不同类型梯田不同坡度下土壤侵蚀量变化。

通过测定自然降雨条件下,不同类型(顺坡、反坡和水平)梯田不同坡度下(0°、5°和 10°),土壤质地、径流量、产沙量、降雨量等关键指标,评价不同类型梯田不同坡度下土壤侵蚀量的变化。

(2)探明雨强与不同类型、不同坡度梯田坡面径流量及产沙量的关系。

(3)探明不同类型梯田不同坡度下土壤及坡面径流的养分迁移变化。

通过测定自然降雨条件下,不同类型(顺坡、反坡和水平)梯田不同坡度下(0°、5°和 10°),土壤中有机质、氮、磷、钾等土壤养分前后变化的情况,分析其养分迁移变化规律。

2.2　研究目标

本项目拟计划在陕南秦巴山区开展小区试验,构筑不同类型梯田坡面,揭示不同类型、不同坡度田面下土壤侵蚀量及养分迁移变化,探究雨强与其径流量及产沙量的关系。通过对比分析,寻找一个最佳的梯田坡面设计条件,为坡改梯田田面设计提供应用依据。通过研究主要达到以下目标:

(1)探明不同类型梯田不同坡度条件下土壤侵蚀量变化;

(2)探明雨强与不同类型不同坡度梯田坡面径流量及产沙量的关系;

(3)探明不同类型梯田不同坡度条件下土壤及坡面径流的养分迁移变化。

2.3　拟解决的关键科学问题

通过对比分析不同类型不同坡度梯田田面下土壤侵蚀及养分迁移变化,探究雨强与其径流量及产沙量的关系,为陕南秦巴山区坡改梯田田面设计提供工程应用依据。

3　拟采取的研究方案及可行性分析

3.1　研究方案

3.1.1　试验设计

试验采用田间小区试验,项目区地处商州区腰市镇,地理坐标介于东经 109°56′26″~109°57′28″,北纬 34°02′31″~34°03′44″。在同一坡面上分别设置标准水平梯田、顺坡梯田(田面坡度 5°和 10°)和反坡梯田(田面坡度 5°和 10°)共 5 个径流试验小区(表 1),小区坡面投影到水平面的宽度为 2 m,坡长为 8 m,有效土层厚度为 50 cm。坡面无任何植被,为无人为扰动的裸坡地。在每个梯田田面坡底放置 A、B 两个径流桶。按当地可能发生的最大暴雨、径流量设计,每个桶直径为 80 cm,桶高 1 m。其中 A 桶有 1/4 可流入 B 桶。桶壁装有水尺,可直接读数计算地表径流量。

试验用土取至当地 0~50 cm 深的土壤,按自然状态下土壤的密度进行压实或者原状土装填。土壤中的根系、碎石块不做处理,保持原状。

表 1　不同类型梯田设计

处理编号	梯田类型
CK	水平梯田
T5	顺坡梯田,田面坡度 5°
T10	顺坡梯田,田面坡度 10°
S5	反坡梯田,田面坡度 5°
S10	反坡梯田,田面坡度 10°

自然气候条件下,定期采集各试验小区田面坡顶(距小区坡头边沿 1 m,横向位于正中间)、坡中(位于小区正中心)、坡底(距小区坡底边沿 1 m,横向位于正中间)0～30 cm 土壤样品以及径流泥沙,进行质地、有机质及养分含量等测定,并分析坡面径流量、养分含量与土壤和泥沙质地的相关性,探析不同田面坡度水土流失及养分迁移变化。梯田设计示意图见图 3。

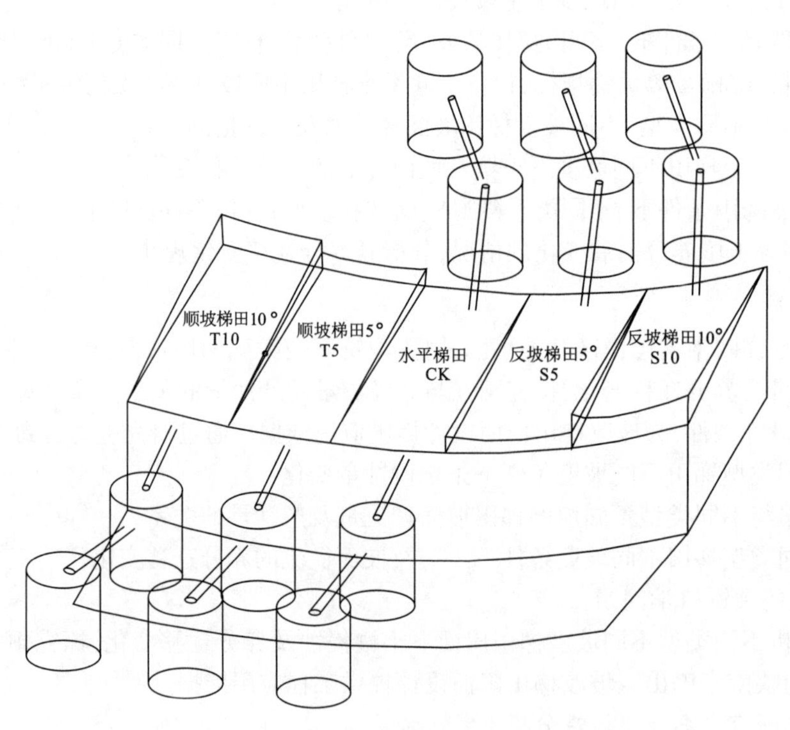

图 3　梯田设计示意图

3.1.2　试验指标

土样采集:每个小区从坡头向坡底方向垂直向下取点。定期采集各试验小区田面坡顶(距小区坡头边沿 1 m,横向位于正中间)、坡中(位于小区正中心)、坡底(距小区坡底边沿 1 m,横向位于正中间)0～10 cm、10～20 cm、20～30 cm 土壤样品,采用四分法进行土样采集,通风阴干,剔除杂物并磨细过筛待测。

水样采集及产流量及产沙量的测定:每次降雨之后测定径流桶中流失的泥沙总量,并将桶中水土搅拌均匀后采集 500 mL 水样,静置过滤,烘干称重,测定泥沙含量,含沙量与水的体积乘积则为产沙量;然后将径流桶中液体清理干净,下次待测。

土壤物理指标测定:用环刀法环刀分层取样,测定土壤容重;激光粒度分析法测定土壤质地;烘干法测定土壤含水量。

土壤化学性质的测定:有机质采用重铬酸钾氧化-外加热法测定;全氮采用凯氏定氮法;全磷采用 NaOH 熔融-钼锑抗比色法;有效磷采用 $NaHCO_3$浸提比色法;全钾采用 NaOH 熔融-火焰光度法;速效钾采用醋酸铵浸提-火焰光度法。

降雨量等气象数据通过试验现场设置雨量计,每次降雨后计取降雨数据。

3.1.3 技术路线

技术路线见图 4。

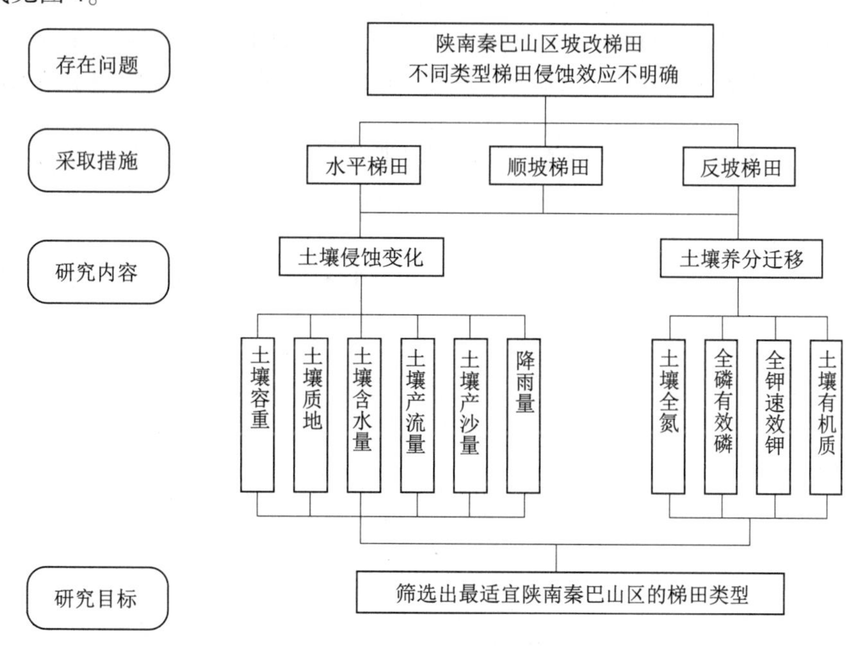

图 4 技术路线

3.2 可行性分析

目前课题组已在商洛市商州区金凤山区梯田中开展了植物根系护坡效应研究,多次前往项目区采样调研,对项目区自然地理气候条件及坡改梯田项目进展情况等比较熟悉,基本摸清了该地区土壤背景状况及坡改梯田工程实践当中遇到的工程实际问题。项目区已有修筑好的水平梯田,为本试验开展提供了良好基础。试验设计简单、可操作性强,检测的土壤养分、土壤理化性状及侵蚀量等指标都是常规测定指标,测定方法较成熟可靠,检测仪器较普遍,数据的准确性有所保障。另外,课题组成员长期从事土地工程技术研究、土地工程相关试验经验丰富。最重要的是当地大量的坡改梯工程或多或少存在不同问题,政府和当地村民多重视工程实践,不注重土地工程科学试验,不能将科学试验与工程实践有机结合起来,为开展该试验提供可行性和应用推广的广阔空间。

通过对比分析不同类型不同坡度梯田田面下土壤侵蚀及养分迁移变化,为陕南秦巴山区坡改梯田田面设计提供工程应用依据,进而为开展丘陵山区土地整治研究,有效增加耕地面积,防止水土流失,减少地质灾害,增加作物产量,促进当地农民脱贫增收和经济社会发展,改善当地生态环境提供有力支持。

参考文献

[1]涂洋,王向东,刘卉芳,等.秦巴山区坡耕地“药用植物篱+土坎”整治模式生态效益研究[J].科学技术与工程,2013,13(11):2938-2942.

[2]孙国梅,况明生,曲华.陕西秦巴山区地质灾害研究[J].水土保持研究,2005,12(5):240-243.

[3]程圣东,杭朋磊,李鹏,等.陕南土石山区坡改梯对坡面稳定性的影响[J].水土保持研究,2018,25(5):157-161.

[4]张晓佳.秦岭北麓“坡改梯”农田土壤肥力状况及水土保持效应研究[D]. 杨凌:西北农林科技大学,2014.

[5]徐勇,党丽娟,汤青,等.黄土丘陵区坡改梯生态经济耦合效应[J].生态学报,2015, 35(4):1258-1266.

[6]薛蓮,刘国彬,张超,等.黄土高原丘陵区坡改梯后的土壤质量效应[J].农业工程学报,2011,27(4):310-315.

[7]张靖宇,杨洁,王昭艳,等.红壤丘陵区不同类型梯田产流产沙特征研究[J].人民长江,2010,41(14):99-103.

[8]赵万广,高玉凤,王锦志,等.黄土高原土石山区坡改梯后土壤水分及养分特征[J].中国水土保持,2018(4):50-54.

[9]马良瑞,梅再美.梯田断面设计与优化研究[J].贵州科学,2012,30(2):45-48.

[10]赵文亮,任云平.浅议商洛地区工程地质环境与水土流失[J].陕西水利,2012,24(4):34-39.

[11]吴珍,王新军,张小虎,等.商洛山区土壤状况与适种中药材研究[J].水土保持通报,2005,25(5):62-68.

[12]殷志有,周丹,陈明彬.商洛山地地质灾害分布及诱发成因研讨[J].陕西气象,2005(1):19-23.

[13]李刘荣,谢军政.生态护坡在陕南秦巴山区土地整治中的应用研究[J].土地开发工程研究(西部大开发),2018,12(3):35-44.

[14]何凡,王向东,尹婧.秦巴山区坡耕地整治模式及水土保持效益分析[J].南水北调与水利科技,2010,8(5):125-128.

本文曾发表于2022年《科学与技术》第18期

耕地质量等别评价方法及相关建议

——以陕西扶风县为例

张晶晶

（陕西省土地工程建设集团有限责任公司，陕西 西安 710075）

【摘要】耕地等别评定是对耕地质量进行的综合评价工作，耕地质量分等成果在土地整治项目实施的多个阶段都有重要的参考意义。本文以扶风县为例展开耕地等别评定研究，对耕地等别评定方法进行详细说明，应用相关公式对扶风县 2018 年耕地质量进行综合评定，同时针对现行耕地等别评定过程中存在的一些问题提出了针对性的建议，以期为我国耕地等别评定工作提供一定参考。

【关键词】耕地等别；粮食产能；评定方法；相关建议

Cultivated Land Quality Grade Evaluation Method and Related Suggestions

——Take Fufeng County as an Example

Jingjing Zhang

(Shaanxi Provincial Land Engineering Construction Group Co. Ltd., Xi'an, Shaanxi 710075)

【Abstract】Cultivated land grade evaluation is a comprehensive evaluation of cultivated land quality, and the results of cultivated land quality classification have important reference significance in various stages of land remediation project implementation. In this paper, the evaluation of cultivated land classification in Fufeng County is studied, and the evaluation method of cultivated land classification is described in detail. At the same time, the relevant formulas are used to comprehensively evaluate the quality of cultivated land in Fufeng County in 2018. The problem puts forward targeted suggestions, in order to provide some reference for the evaluation of cultivated land classification in my country.

【Keywords】Cultivated land classification; Grain production capacity; Evaluation method; Related suggestions

以农业为主导产业的现状决定了我国在未来相当长的时间内，耕地仍是拉动经济的重要基础和农民最基本的生活保障[1]。因此，进行土地开发整理，增加耕地面积，改善耕地质量，进而改善农民的基本生产条件是一项长期的工作。土地整治项目是一种通过工程措施提升耕地质量和改善耕作环境的有效举措，通过对项目区统一规划，合理布局，增加有效耕地面积，提高耕地质量，实现区域耕地总量动态平衡，增强农业发展后劲，改善农民生产、生活条件，增加农民经济收入，促进农业经济的可持续发展，进而为解决好“三农”问题、加快社会主义新农村建设步伐提供有力支持[2-4]。耕地质量等别评价是自然资源部门依据耕地质量好坏而进行的评价性工作，耕地等别越好，则其粮食产能也越高。因而，以提升粮食产能为目的的耕地等别评定在整个土地整治项目中则显得尤为重要[5-7]。本文以扶风县为例展开耕地等别评定研究，以期为我国耕地等别评定工作提供一定参考。

1　研究区基本情况

扶风县地处陕西省中部偏西，位于西安、宝鸡、咸阳三市交界处，隶属宝鸡市。地理坐标居于东经 107°45′28″～108°45′2″，北纬 34°12′54″～34°37′56″之间。研究区地处中国东部暖温带区的西缘,属半湿润大陆性季风气候。气候特点:四季分明,春、秋季短;日照充足,光能、热量资源丰富;夏温高却没有酷暑,冬温低但不严寒;降水集中,雨热同期,但秋霜严重。扶风县全县日照平均时数为2134.3 h,年有效辐射为56.66 kcl/cm^2,年平均气温为12.4 ℃,无霜期历时209 d。扶风县受大陆性季风气候影响较大,热量条件好,但降水量不足,且年际差异大,时空分布不均。干湿季节分明,一般降水不能满足农作物生长需要。水资源总量49.86亿 m^3,其中46.6亿 m^3 为过境流量,人均水资源量323 m^3,耕地亩均水资源量453 m^3。地下水可开采量1.47亿 m^3,占可开发利用量的44.6%。宝鸡峡、冯家山两大水利工程可供本县水量1.83亿 m^3,占可开发利用量的56.4%。研究区生物资源以农作物为主,三大类作物共37种,其中粮食作物11种,冬小麦、玉米是主栽作物;经济作物22种,主栽作物为油菜、棉花、瓜类;饲料和绿肥作物4种,为苜蓿、毛苕子、草木栖、沙打旺。全县林木共有19科69种,以乔木为主。截至2018年年底,扶风县完成地区生产总值139.87亿元,比上年增长8.6%。其中,第一产业完成增加值20.82亿元,比上年增长2.8%;第二产业完成增加值78.79亿元,比上年增长11.2%;第三产业完成增加值40.26亿元,比上年增长7.5%。2018年年末,全县三次产业结构为14.9∶56.3∶28.8。按常住人口计算,全县人均生产总值33055元。非公有制经济实现增加值78.9亿元,占地区生产总值的比重为56.42%。

2 土地利用现状

根据2018年土地利用现状变更调查,扶风县土地总面积为70528.69 hm^2。其中耕地38197.59 hm^2,占土地总面积的54.16%,耕地中不涉及水田,水浇地35099.06 hm^2,占耕地总面积的91.89%,旱地3098.53 hm^2,占耕地总面积的8.11%;园地5410.92 hm^2,占土地总面积的7.67%;林地13198.00 hm^2,占土地总面积的18.71%;草地1485.45 hm^2,占土地总面积的2.11%;城镇村及工矿用地8400.65 hm^2,占土地总面积的11.91%;交通运输用地1929.79 hm^2,占土地总面积的2.74%;水域及水利设施用地1329.44 hm^2,占土地总面积的1.88%;其他土地576.85 hm^2,占土地总面积的0.82%。扶风县2018年度土地利用现状面积汇总表,如表1所示。

表1 扶风县2018年度土地利用现状面积汇总表

地类名称		面积/hm^2	占比/%
耕地(01)	水浇地(012)	35099.06	49.77
	旱地(013)	3098.53	4.39
	小计	38197.59	54.16
园地(02)	果园(021)	5405.63	7.66
	其他园地(023)	5.29	0.01
	小计	5410.92	7.67
林地(03)	有林地(031)	10951.73	15.53
	灌木林地(032)	1640.43	2.33
	其他林地(033)	605.84	0.86
	小计	13198.00	18.71

续表 1

地类名称		面积/hm²	占比/%
草地（04）	天然牧草地(041)	107.78	0.15
	其他草地(043)	1377.67	1.95
	小计	1485.45	2.11
城镇村及工矿用地（20）	建制镇(202)	2107.09	2.99
	村庄(203)	5486.32	7.78
	采矿用地(204)	433.35	0.61
	风景名胜及特殊用地(205)	373.89	0.53
	小计	8400.65	11.91
交通运输用地（10）	铁路用地(101)	63.08	0.09
	公路用地(102)	556.89	0.79
	农村道路(104)	1309.43	1.86
	管道运输用地(107)	0.39	0.00
	小计	1929.79	2.74
水域及水利设施用地（11）	河流水面(111)	172.86	0.25
	水库水面(113)	195.02	0.28
	坑塘水面(114)	93.38	0.13
	内陆滩涂(116)	183.95	0.26
	沟渠(117)	629.11	0.89
	水工建筑用地(118)	55.12	0.08
	小计	1329.44	1.88
其他土地（12）	设施农用地(122)	200.30	0.28
	田坎(123)	375.56	0.53
	裸地(127)	0.99	0.00
	小计	576.85	0.82
合计		70528.69	100.00

3 评价方法

3.1 分等因素指标区的确定

分等因素指标区是依据主导因素原则和区域分异原则划分的区域，是区域内决定农用地自然质量的各种因素的组合。本次耕地质量等别年度更新沿用 2010 年度耕地质量等级补充完善工作采用的分等因素指标区，即在国家层面和省级范围内划分三个层次，根据划分结果，扶风县所在区域的国家标准耕作制度一级区为黄土高原区，国家标准耕作制度二级区为汾渭谷地，陕西省分等因素指标区为关中渭河平原区。

3.2 基准作物和指定作物的确定

1）基准作物。基准作物是理论标准粮的折算基准，是某个地区比较普遍的主要粮食作物。陕西省农用地分等工作全省统一选取冬小麦为基准作物，本次耕地质量等别年度更新评价工作仍沿用冬小麦为基准作物。

2)指定作物。指定作物是指行政区所属耕作区标准耕作制度中所涉及的农作物。根据扶风县的标准耕作制度以及多年统计年鉴中农作物播种面积占总播种面积的比例,并结合耕地质量等级补充完善工作时确定的指定作物,由相关专家论证后调整确定。扶风县耕地质量等别年度更新评价工作指定作物与2017年度耕地质量等别更新评价工作确定的指定作物一致,为冬小麦和夏玉米。

3.3　标准耕作制度的确定

标准耕作制度是在当前社会经济水平、生产条件和技术水平下,有利于生产或最大限度发挥当地土地生产潜力,有较大发展前景,不造成生态破坏,能满足社会需求,并已为当地普遍采用的农作方式。扶风县标准耕作制度为“冬小麦-夏玉米”,复种类型为“一年二熟”。

3.4　计算公式

1)耕地等别计算公式

①耕地自然质量分计算公式:

$$C_{Lij} = \sum_{k=1}^{m} w_k \cdot f_{ijk}/100 \tag{1}$$

式中,C_{Lij}——第 i 个分等单元的第 j 种指定作物的农用地自然质量分;

w_k——第 k 个分等因素的权重;

f_{ijk}——第 i 个分等单元内第 j 种指定作物第 k 个分等因素的指标分值,取值为0~100;

m——分等因素指标数量。

②耕地自然质量等指数计算公式:

$$R_{ij} = \alpha_{ij} \times C_{Lij} \times \beta_j \tag{2}$$

式中,R_{ij}——第 i 个分等单元第 j 种指定作物的自然等指数;

C_{Lij}——第 i 个分等单元第 j 种指定作物的农用地自然质量分,由公式(1)计算可得;

α_{ij}——第 i 个分等单元第 j 种作物的生产潜力指数,耕地有灌溉条件时采用光温生产潜力指数,耕地无灌溉条件时采用气候生产潜力指数;

β_j——第 j 种作物的产量比系数。

国家级自然质量等=ROUND((16-(R_{ij}+200.01)/400),0)

2)土地利用系数确定。耕地利用等指数分指定作物计算:

$$Y_{ij} = R_{ij} \times K_{Lj} \tag{3}$$

式中,Y_{ij}——第 i 个分等单元第 j 种指定作物的利用等指数;

R_{ij}——第 i 个分等单元第 j 种指定作物的自然等指数,由公式(2)计算可得;

K_{Lj}——评价单元所在等值区第 j 种指定作物土地利用系数。

按照关中地区的标准耕作制度,其熟制为“二年三熟”,耕地利用等指数:

$$Y_i = \sum Y_{ij}/2 \tag{4}$$

式中,Y_i——第 i 个分等评价单元耕地利用等指数;

Y_{ij}——第 i 个分等评价单元内第 j 种指定作物的利用等指数,由公式(3)计算可得。

国家级利用等=ROUND((16-(Y_{ij}+100.01)/200),0)

3)计算土地经济系数。耕地经济等指数分指定作物计算:

$$G_{ij} = Y_{ij} \times K_{Cj} \tag{5}$$

式中,G_{ij}——第 i 个分等评价单元第 j 种指定作物的经济等指数;

Y_{ij}——第 i 个分等评价单元内第 j 种指定作物的耕地利用等指数,由公式(3)计算可得;

K_{Cj}——评价单元所在等值区的第 j 种指定作物的土地经济系数。

按照关中地区的标准耕作制度,其熟制为“二年三熟”,耕地经济等指数:

$$G_i = \sum G_{ij}/2 \tag{6}$$

式中，G_i——第 i 个分等评价单元耕地经济等指数；

G_{ij}——第 i 个分等评价单元内第 j 种经济作物经济等指数，由公式(5)计算可得。

国家级经济等=ROUND((16-(G_i+100.01)/200),0)

4 数据来源

收集扶风县2017年度的县级耕地质量等别评价成果，包括文字成果、数据库成果和数据表格成果，扶风县经部确认的2018年土地利用变更调查成果，收集和整理扶风县2018年度内验收的各级各类土地整治项目和农业综合开发、农田水利建设项目的可研、设计和竣工验收资料，以及项目耕地质量等别评定资料。重点对土地整治和农业综合开发项目范围内耕地有效土层厚度、表层土壤质地等分等因素属性值进行调查，进一步完善耕地分等因素图。收集2018年度内由农业、水利等其他部门组织完成的中低产田改造、农业综合开发、农田水利建设等项目的设计、验收资料。收集扶风县2018年统计年鉴及国民经济和社会发展统计年报，得到扶风县2018年经济资料及农作物产量。另外，还需要收集自然条件资料，主要包括以下几个方面。

1)地形地貌：由2018年土地利用现状变更调查的耕地坡度成果获取。

2)气候：收集扶风县气象站气象统计数据，重点抽取各月平均温度、≥0 ℃积温、≥10 ℃积温、降水量、蒸发量、无霜期、灾害气候等资料。

3)水文：收集扶风县农业生产灌溉水源类型(地表水、地下水)、水量、灌溉保证率、水质以及水资源分布图等资料。

4)土壤：收集第二次土壤普查资料集及图件，主要包括有效土层厚度、表层土壤质地、土壤盐渍化程度、土壤有机质含量、土壤 pH、土壤剖面构型、土壤侵蚀状况、土壤砾石含量、地表岩石露头度、土壤成分化验分析报告以及土壤图等资料。

5 评价结果与分析

2018年度内，扶风县耕地总面积38197.59 hm^2，其中水浇地35099.06 hm^2，旱地3098.53 hm^2。全县耕地国家自然等为7~11等，其中7等地23430.92 hm^2，8等地11160.88 hm^2，9等地505.87 hm^2，10等地949.48 hm^2，11等地2150.43 hm^2；耕地国家利用等为7~12等，其中7等地27230.29 hm^2，8等地3140.41 hm^2，9等地4448.95 hm^2，10等地962.72 hm^2，11等地2402.75 hm^2，12等地12.47 hm^2；耕地国家经济等为7~12等，其中7等地0.77 hm^2，8等地5635.09 hm^2，9等地23547.55 hm^2，10等地4569.13 hm^2，11等地2098.51 hm^2，12等地2346.55 hm^2。扶风县2018年度耕地国家利用等各地类等别分布情况，如表2所示。

表2 扶风县2018年度耕地国家利用等分地类等别面积汇总表 单位：hm^2

地类名称	7等	8等	9等	10等	11等	12等	合计
水浇地	27227.83	3136.24	4428.54	295.02	11.43	0.00	35099.06
旱地	2.45	4.17	20.41	667.70	2391.33	12.47	3098.53
总计	27230.29	3140.41	4448.95	962.72	2402.75	12.47	38197.59

由表2可知，2018年度全县耕地利用等中水浇地为7~11等，7等地所占面积最多，为27227.83 hm^2，11等地所占面积最少，为11.43 hm^2；旱地为7~12等，11等地所占面积最多，为2391.33 hm^2，7等地所占面积最少，为2.45 hm^2。

由表3可知，2018年度扶风县耕地全县各个乡镇均有分布，全县耕地利用等分布情况为：城关镇耕地所占面积最多，为6499.54 hm^2，分布在7~11等，各等别中7等地所占面积最多，为4066.61 hm^2，11等地所占面积最少，为54.88 hm^2，城关镇平均利用等为7.6392等，略高于全县平均等；法门镇耕地所占面积次之,为6213.62 hm²,分布在7~12等,各等别中7等地所占面积最多,为4139.63 hm²,12等地所占面积最少,为0.23 hm²,法门镇平均利用等为7.9599等,略低于全县平均等;南阳镇耕地所占面积最少,为1848.21 hm²,分布在7~12等,各等别中9等地所占面积最多,为666.07 hm²,12等地所占面积最少,为11.47 hm²,南阳镇平均利用等为9.5181等,远低于全县平均等。全县大部分乡镇的平均利用等都高于全县平均等,其中绛帐镇、召公镇、上宋乡、段家镇等乡镇的平均利用等较高,仅法门镇、太白乡、南阳镇、天度镇4乡镇平均利用等低于全县平均等,其中南阳镇、天度镇的平均等别较低。总体来看,南阳镇、天度镇耕地平均利用等与其他各乡镇差别较大,除南阳镇、天度镇之外的乡镇平均利用等差别不大。

表3　扶风县2018年度耕地国家利用等分乡(镇)等别面积汇总表　　单位:hm²

乡镇名	7等	8等	9等	10等	11等	12等	合计	平均利用等
城关镇	4066.61	1030.65	1137.70	209.70	54.88	0.00	6499.54	7.6392
绛帐镇	3304.96	22.86	12.86	38.03	0.00	0.00	3378.71	7.0481
杏林镇	1763.09	380.91	291.13	13.35	4.32	0.00	2452.80	7.4161
召公镇	3643.01	251.43	21.71	46.92	4.97	0.00	3968.03	7.1148
法门镇	4139.63	341.50	579.75	147.87	1004.63	0.23	6213.62	7.9599
上宋乡	2269.44	45.08	0.24	21.58	0.00	0.00	2336.35	7.0472
段家镇	2372.73	117.35	0.17	18.36	0.00	0.00	2508.61	7.0689
午井镇	3771.62	266.73	120.71	194.44	30.04	0.00	4383.53	7.2764
太白乡	1542.01	46.66	780.15	26.72	9.53	0.00	2405.08	7.7173
南阳镇	155.42	245.00	666.07	61.31	708.93	11.47	1848.21	9.5181
天度镇	201.76	392.25	838.45	184.44	585.45	0.77	2203.12	9.2550
总计	27230.29	3140.41	4448.95	962.72	2402.75	12.47	38197.59	7.6440

6　相关建议

2018年度扶风县实施了3634.25 hm²质量建设耕地,质量建设后平均利用等提高了0.0792等,等别提高不显著。一方面,土地整治、高标准基本农田建设等项目围绕土地平整、灌溉与排水、农田道路、农田防护与生态环境保持四类工程进行农田建设[8-10],全省耕地分等各指标区采用的分等因素包括有效土层厚度、表层土壤质地、土壤盐渍化程度、土壤有机质含量、排水条件、地形坡度、灌溉保证率、灌溉水源、土壤侵蚀程度、地表岩石露头度、土壤剖面构型和pH。因此,一些工程建设内容不能够通过现有分等指标因素体现出其建设成效,如农田道路等,因此项目实施后耕地质量等别变化不能完全反映出。但现行分等因素指标体系在自然资源部启动下一轮耕地质量等别全面更新之前不能随意调整,以保持现阶段等别成果的持续性与可比性。另一方面,目前采用的耕地质量等别主要为国家级等别,划分等别时等指数的跨度区间较大,除"旱改水"外的其他单项工程建设对分等因素的影响一般较小,等指数提高但等别不变的现象普遍存在。因此,需要紧跟自然资源部的指导思想,尽快建立"即用即评"的耕地质量定级制度,全面服务于土地整治项目耕地质量评价。

参考文献

[1] 周伟,石吉金,苏子龙,等. 耕地生态保护与补偿的国际经验启示——基于欧盟共同农业政策[J].中国国土资源经济, 2021, 34(8):37-43.

[2] 李华英.论农村耕地地力保护补贴工作的落实——以邹圩镇耕地地力保护补贴为例[J].农村科学实验,2020(1):102-103.

[3] 钟海辉,钟雪芹.做好耕地地力保护补贴发放工作的基层实践[J].当代农村财经,2018(10):30-31.

[4] 汤怀志,桑玲玲,郧文聚.我国耕地占补平衡政策实施困境及科技创新方向[J].中国科学院院刊,2020(5):637~644.

[5] 李俊颖,吴克宁,宋文.基于作物营养当量修正产量比系数法的农用地质量分等[J].农业工程学报,2019(9):238-245.

[6] 周子健,吴克宁,马建辉,等.耕地质量等级监测中县域土地利用系数更新方法研究——以北京市大兴区为例[J].中国农业资源与区划,2013(3):66-72.

[7] 向武,周卫军,肖彦资,等.县域耕地地力与农用地自然质量等级差异及关联性研究[J].中国生态农业学报,2014,22(7):821-827.

[8] 吴龙驰.土地整治项目对台塬区耕地质量的影响研究[J].南方农机,2022,53(4):72-74.

[9] 陈丽丽.关于耕地质量等别年度更新评价工作的思考[J].山西农经,2021(21):28-29,81.

[10] 李志芳,沈新磊,王锐.漯河市耕地质量等别划分与评价[J].中国农学通报,2021,37(9):79-84.

本文曾发表于2022年《南方农机》第8期

基于 DEM 的陕北黄土高原水文相关地形因子的分析研究

王媛媛

(陕西地建土地勘测规划设计院有限责任公司,陕西 西安 710075)

【摘要】水文要素是研究黄土高原水土流失的重要因子,为揭示陕北黄土高原水文地形要素之间的相关规律。基于 90 m SRTM(Shuttle Radar Topography Mission) DEM (Digital Elevation Model)数据,提取陕北黄土高原相关水文地形因子,包括坡度坡长因子、地形湿度指数及沟壑密度,并对三类水文因子进行了相关性分析。地形湿度指数与坡度值采用指数拟合效果最好,R^2值为 0.8168,坡度坡长指数与坡度值采用线性拟合效果最好,R^2值为 0.993,沟壑密度与坡度值采用多项式拟合效果最好,R^2值为 0.0604。地形湿度指数与坡度坡长因子采用指数拟合效果最好,R^2值为 0.8385,坡度坡长指数与沟壑密度采用二次多项式拟合效果最好,R^2值为 0.2077,地形湿度指数与沟壑密度采用多项式拟合效果最好,R^2值为 0.2801。坡度坡长因子、地形湿度指数、沟壑密度均在一定程度反映了出实际地貌。坡度坡长指数与坡度存在线性显著相关关系。坡度越大地形湿度指数越小。沟壑密度随汇流阈值设置的增大而减小。

【关键词】水文要素;水土流失;陕北黄土高原;地形测绘;坡度坡长因子;沟壑密度;地形湿度指数

Analysis of Terrain Factors Related to Hydrology in the Loess Plateau of Northern Shaanxi Based on DEM

Yuanyuan Wang

(Land Surveying, Planning and Design Institute of Shaanxi Provincial Land Engineering Construction Group, Xi'an 710075, Shaanxi, China)

【Abstract】Hydrological element is considered as one of the significant factors in the study of soil erosion in the Loess Plateau area. In order to reveal the correlation rule between each hydrographic and hydrological element of the Loess Plateau area in northern Shaanxi. The relevant hydrographic and hydrological factors were extracted based on the SRTM DEM with 90 m resolution, including the slope, slope length, topographic humidity index, and gully density. Moreover, the three types of hydrological factors was analyzed. The topographic humidity index and the slope value show the best index fitting performance, the R^2 of which is 0.8168. On the other hand, the slope length index and the slope value show the best quadratic polynomial fitting performance, the R^2 of which is 0.993. The ravine density and the slope value show the best polynomial fitting performance, the R^2 of which is 0.0604. The topographic humidity index and the slope and slope length factor show the best index fitting performance, the R^2 of which is 0.8385. The slope and slope length index and the gully density show the best quadratic polynomial fitting performance, the R^2 of which is 0.2077. The topographic humidity index and the gully density show the best polynomial fitting performance, the R^2 of which is 0.2801. In conclusion, the condition of actual landform is reflected by slope, length factor, terrain humidity index, and gully density. In addition, the significant linear correlation relationship existed between the slope length index and the slope. The slope is inversely proportional to the terrain humidity index. Similarly, the gully density is inversely proportional to threshold of conflux value.

【Keywords】Hydrological element; Soil erosion; Loess plateau in Northern Shaanxi; Topographic mapping; Slope length and slope gradient factor; Gully density; Topographic wetness index

0 引言

黄土高原是土壤发育脆弱区,土壤侵蚀严重,前人针对黄土高原的水文情况主要做了两方面研究。一类是通过 DEM 针对河网水系的研究。李景星采用数字化等高线生成 DEM,提取了陕北黄土区部分县区的河网水系并分析其分布及分级状况[1]。祝士杰基于 DEM 研究了黄土高原多流域面积高程积分谱系[2]。管伟瑾利用遥感影像水体指数法和 DEM 水文分析两种方法提取了黄河兰州段河流信息[3]。颜明以 DEM 为基础提取黄河流域河网并揭示其空间分布与归一化植被指数(Normalized Difference Vegetation Index, NDVI)的关系[4]。赵卫东基于 DEM 研究黄土高原小流域的汇流累积量的变化模式[5]。另一类,对黄土高原水文环境进行研究分析。王文静采用水土流失方程对陕北黄土高原水土流失敏感性进行综合评价[6]。张宝庆研究了黄土高原退耕还林实施后大规模植被恢复对黄土高原生态水文过程的影响[7]。蒋凯鑫采用多种方法对黄土高原典型流域无定河的水沙变化进行归因对比分析[8]。

同时,对于利用遥感数据提取各类水文因子国内学者进行了多方面研究。针对坡度坡长因子方面,主要的研究有 DEM 栅格单元大小对汇水区及坡度坡长因子的影响[9]及 Landsat TM 影像提取的植被覆盖度与高程、坡度坡长因子的相关性分析[10]。针对地形湿度指数方面,主要研究有单流向法地提取的形湿度指数在不同地形区间和不同尺度下的差异[11]及 DEM 栅格单元异质性对地形湿度指数提取的影响[12]。针对沟壑密度方面,主要的研究有山坡地形曲率分布特征及其水文影响[13]、集水面积阈值与沟谷密度的关系[14]及评价 1″分辨率的 DEM 在流域分析中的适用性[15]。

前人的研究主要集中在一是利用卫星遥感影像提取研究河网水系以及生态水文过程影响的模拟,对水文因子间相关性分析较少。本文在总结前人研究的基础上,以地形水文相关因子为切入点,以陕北黄土高原为研究区,以空间分辨率 90 m 的 SRTM DEM 为数据源,采用 ArcGIS10 空间分析提取多种水文相关地形因子,包括坡度坡长因子、地形湿度指数和沟壑密度,并以县域为研究单元对地形因子进行相关分析,揭示其内部规律,为宏观层面地形水文研究提供科学依据。

1 区域概况与研究方法

1.1 研究区概况

研究区共涉及陕北黄土高原榆林、延安两个市中 25 个县区,经度 107.2°~111.3°E,纬度 35.3°~39.6°N,属于半干旱区,夏季降水集中,植被稀疏,水土流失严重。

水分对植被生长具有重要意义,准确表达土壤侵蚀严重区水文分布情况尤为迫切,本文使用多种水文模型对陕北黄土高原 25 个县区水文状况进行模拟,阐释其分布规律。

本文所使用的 DEM 数据来源于中国科学院计算机网络信息中心地理空间数据云平台(http://www.gscloud.cn)。

1.2 坡度坡长因子

坡度坡长因子是土壤侵蚀模型(Revised Universal Soil Loss Equation, RUSLE)中重要地形因素,由坡长因子(L)和坡度(S)共同决定,具体见公式(1):

$$LS = L \cdot S \tag{1}$$

坡长通常指在地面上某点沿水流方向到其流向起点间的最大地面距离在水平面上的投影长度。坡长因子是水土保持上的重要因子,坡长越长,汇水量越大,土壤侵蚀力越强,发生水土流失的可能性越大。坡长依据公式(2)、(3)、(4)计算得来:

$$L = (\lambda/22.13)^{\alpha} \tag{2}$$

$$\alpha = \beta/(\beta + 1) \tag{3}$$

$$\beta = (\sin\theta/0.0896)/[3.0(\sin\theta)^{0.8} + 0.56] \tag{4}$$

式中,LS 为坡度坡长因子;L 为指坡长因子;S 为坡度;λ 为单元水平坡长;α 为坡长指数;θ 为坡度值。λ 系数来源于单元水流方向值,若方向值为 1、4、16、2、64,则格网间距离为 1,否则为 $\sqrt{2}$,$\lambda=\lambda$ 系数×cell size。

土壤侵蚀因子中的坡度因子不是简单的坡度,它是根据坡度大小及地区不同形成的经验公式,本文采用张文杰等[16]在纸坊沟流域中的研究公式(5):

$$S = \begin{cases} 10.8\sin\theta + 0.03, \theta < 9\% \\ 16.8\sin\theta - 0.50, 9\% \leqslant \theta < 14\% \\ 21.9\sin\theta - 0.96, \theta \geqslant 14\% \end{cases} \tag{5}$$

式中,S 为坡度因子;θ 为由 DEM 提供的坡度。

1.3 地形湿度指数

地形湿度指数(Topographic Wetness Index, TWI)是一种基于 DEM 对径流路径长度、产流面积等的定量描述,在一定程度上反映区域持水能力的空间分布特征,它是单位汇水面积与坡度的函数。它能一定程度上表明地形变化对土壤径流的影响,在径流模拟中广泛应用[17]公式(6):

$$w = \ln(\text{Area}_{汇水}/\tan(\text{Slope})) \tag{6}$$

式中:w 为地形湿度指数;$\text{Area}_{汇水}$ 为栅格单元的汇流面积,由汇流累计量与栅格格网尺寸计算得来;Slope 为弧度坡度值,本文采用单流向法计算湿度指数,即假设格网流向是单一的。一般提取过程:①DEM洼地填充;②生成水流方向矩阵;③生成汇流累积量矩阵;④计算单位格网的汇流面积;⑤计算坡度的正切值;⑥计算地形湿度指数。

1.4 沟壑密度

沟壑密度,指单位面积内沟壑的总长度,单位一般以 km/km^{2}表示,具体计算方式见公式(7)。

$$D=\text{Length}_{沟谷}/\text{Area}_{研究区} \tag{7}$$

式中,D 为沟壑密度;$\text{Length}_{沟谷}$ 为流域内沟壑总长度(单位为 km);$\text{Area}_{研究区}$ 为研究区面积(单位为 km^2)。一般提取过程:依据汇流累积栅格矩阵设置阈值为(2500、5000、7500、10000)提取沟谷河网,生成河流链接,栅格河网转矢量河网,提取沟谷网络及出水口,计算各研究区面积及沟壑密度。

2 结果与分析

2.1 坡度坡长因子统计分析

坡长因子是依据栅格格网大小结合坡度相关经验指数得来,依据公式本研究汇总底数一般为 4.067或 5.751,由于坡度不同导致坡长指数 α 不同,α 变化区间[0,0.737 5],坡长因子最终结果在[1,3.626]之间。坡度因子根据坡度不同依据经验公式分级计算,坡度以弧度为单位,分为小于 9%、在9%~14%之间及大于 14%,坡度因子取值区间[0.03,14.6147]。

如图 1 所示,定边县、靖边县坡度坡长因子南北差异明显,榆阳区、神木县东西差异明显。各区县坡度坡长均值值域区间分别为:榆阳区和神木县为(1,4),横山县、定边县、靖边县值域区间为(5,6),府谷县、佳县、米脂县的值域区间(6,7),绥德县、吴堡县、洛川县、子洲县、清涧县值域区间(8,10),甘泉县、富县值域区间(10,11),宝塔区、子长县、黄陵县、黄龙县、延川县、黄龙县、延长县、安塞县、宜川县、志丹县等 9 个县区值域区间(11,12),吴起县坡度坡长均值大于 12。

2.2 地形湿度指数统计分析

地形湿度指数是格网单元汇流面积与坡度值的函数,与坡度成反比,与单位汇流面积成正比。

图 2 显示了陕北黄土高原地形湿度指数,地形湿度指数的均值从值上分成四组,其中子长县、志丹县、延川县、延长县、清涧县、宜川县、宝塔区、吴起县、安塞县地形湿度指数均值大于 4.8 小于 5,

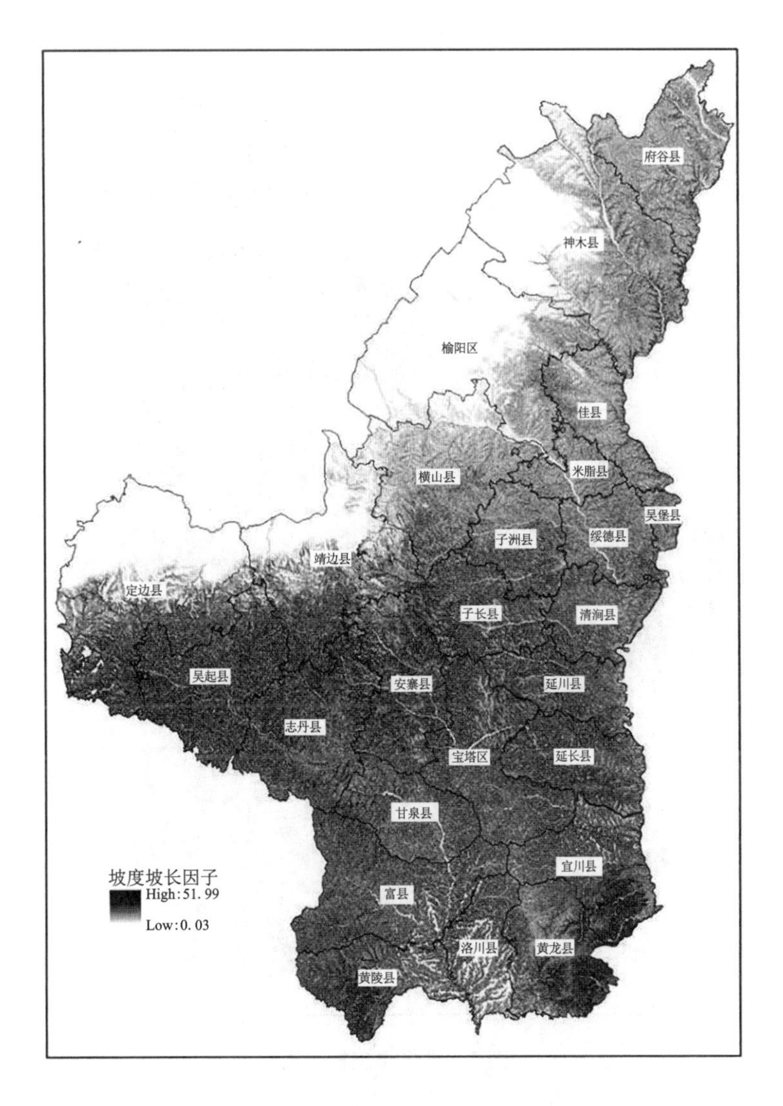

图 1　坡度坡长因子图

米脂县、黄陵县、佳县、府谷县、洛川县、横山县地形湿度指数均值大于 5 小于 6,靖边县、神木县、定边县地形湿度指数均值大于 6 小于 7,榆阳区地形湿度指数大于 7。

2.3　沟谷河网分流域统计分析

如表 1 所示,研究区沟壑密度与汇流累积量阈值设置有关,由于回流累计量阈值增大,河网总长度减小,因此沟壑密度随着阈值增大而减小。不同汇流阈值下,各县区沟壑密度的相对大小排列顺序差异明显,其中阈值 2500 和阈值 5000 县域沟壑密度大小排序相对接近,阈值 7500 和阈值 10000 县域沟壑密度大小排序相对接近。汇流阈值 2500 时各县区沟壑密度均值在0.120到 0.201 之间,汇流阈值 5000 时各县区沟壑密度均值在 0.075 到 0.134 之间,汇流阈值 5000 时各县区沟壑密度均值在 0.039 到 0.116 之间,汇流阈值 10000 时各县区沟壑密度均值在 0.024 到 0.108 之间。

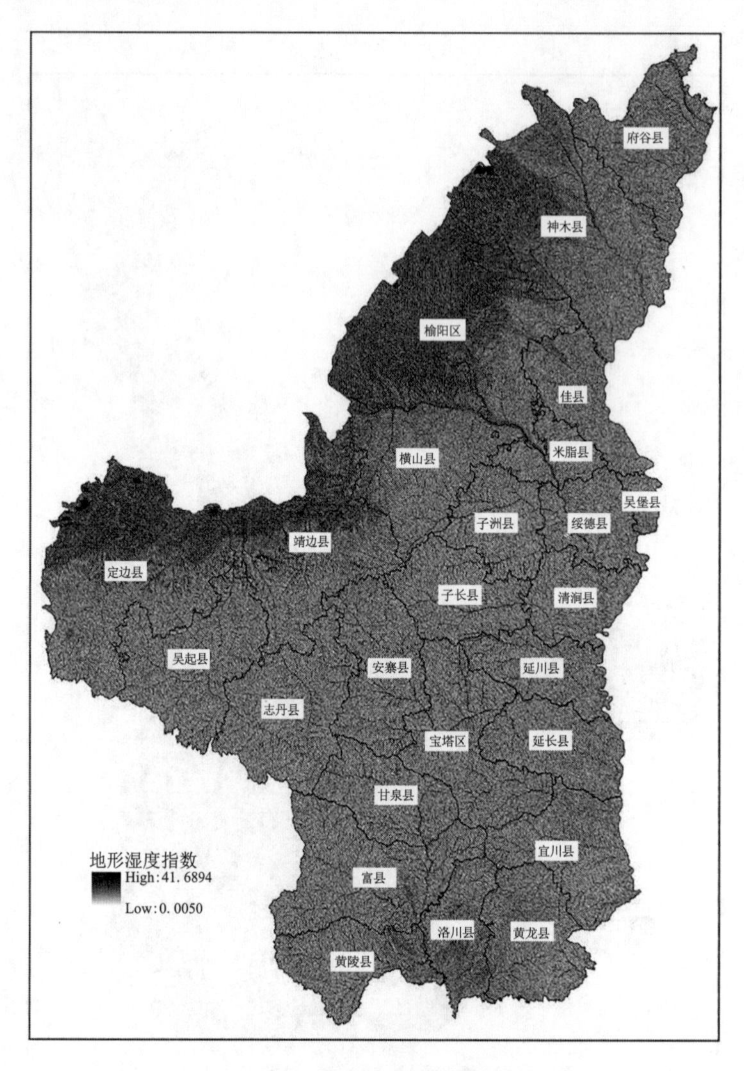

图 2 地形湿度指数图

表 1 陕北黄土高原各县区不同汇流阈值下的沟壑密度表 单位:km/km²

延安市各县区	阈值 2500	阈值 5000	阈值 7500	阈值 10000	榆林市各县区	阈值 2500	阈值 5000	阈值 7500	阈值 10000
安塞县	0.157	0.112	0.090	0.079	定边县	0.155	0.104	0.078	0.067
宝塔区	0.159	0.113	0.094	0.081	府谷县	0.131	0.094	0.082	0.069
富县	0.166	0.121	0.099	0.081	横山县	0.160	0.119	0.099	0.084
甘泉县	0.167	0.118	0.087	0.074	佳县	0.156	0.126	0.105	0.090
黄陵县	0.184	0.133	0.111	0.094	靖边县	0.154	0.106	0.087	0.074
黄龙县	0.145	0.097	0.075	0.063	米脂县	0.120	0.075	0.067	0.055
洛川县	0.188	0.134	0.110	0.099	清涧县	0.167	0.115	0.088	0.081
吴起县	0.160	0.122	0.100	0.090	神木县	0.162	0.116	0.098	0.083
延川县	0.161	0.123	0.104	0.095	绥德县	0.156	0.111	0.094	0.083
延长县	0.151	0.104	0.088	0.076	吴堡县	0.137	0.075	0.039	0.024
宜川县	0.162	0.124	0.116	0.108	榆阳区	0.201	0.120	0.085	0.070
志丹县	0.152	0.103	0.083	0.074	子洲县	0.156	0.113	0.097	0.089
子长县	0.133	0.090	0.071	0.056					

图 3 为不同汇流阈值下陕北黄土高原各县域的沟谷河网图,阈值大的河网线会覆盖阈值小的河网线的主干部分。同一地区内部由于地形地貌差异,沟谷河网分布也会出现明显不同。而沟谷密度不能反映实际水流量,仅从地貌上反映了沟谷大致分布情况。如在 7500 和 10000 阈值下沟谷密度最小的吴堡县,沟谷分布量西北部多于东南部,东南部地处黄河沿岸,地势平而水流量大,西北部土壤侵蚀严重为典型黄土沟壑地貌。榆林市榆阳区的沟谷密度也具有典型性,该地处于鄂尔多斯台地东部,境内以明长城为界,沿北为占总面积 75%的风沙草滩区,沿南属丘陵沟壑区。榆阳区的沟谷密度在阈值 2500 时沟壑密度最大,阈值 5000、7500 及 10000 下排列处于较中间位置。

2.4 坡度坡长因子、地形湿度指数、沟壑密度与坡度相关分析

以下分析采用 2500 阈值下求取的沟壑密度同其他各项指数进行拟合。

图 4 中的 A、B、C 分别显示了地形湿度指数、坡度坡长指数、沟壑密度与坡度之间的相关关系。对于不同因子分别采用指数模型、线性模型及多项式等多种拟合方式,然后保留拟合效果最好的结果做相关性比较。其中地形湿度指数与坡度值采用指数拟合效果最好,R^2值为 0.8168,坡度坡长指数与坡度值采用线性拟合效果最好,R^2值为 0.9930,沟壑密度与坡度值采用多项式拟合效果最好,R^2值为 0.0604。

图 3 沟谷河网图

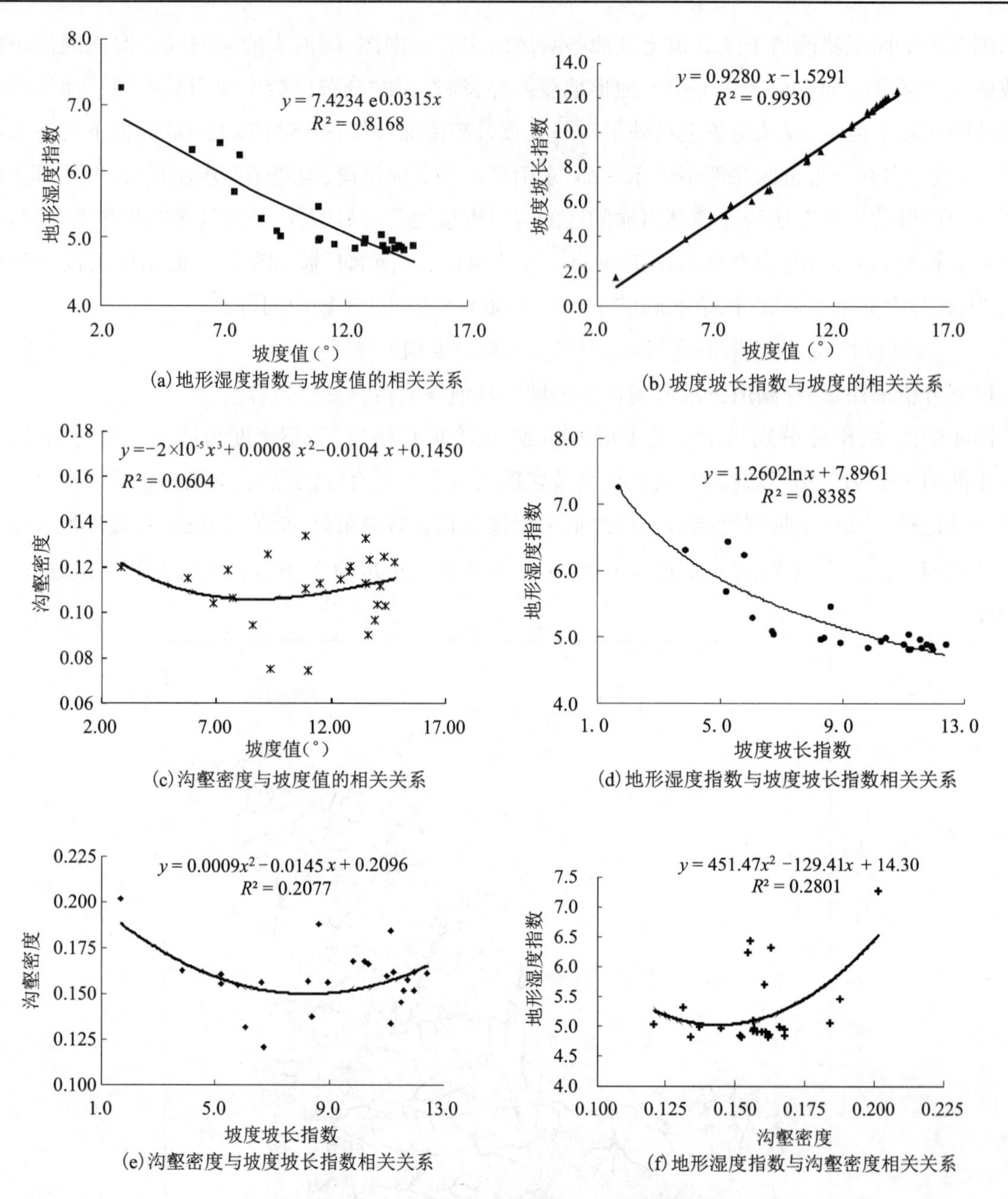

(a)地形湿度指数与坡度值的相关关系　(b)坡度坡长指数与坡度的相关关系

(c)沟壑密度与坡度值的相关关系　(d)地形湿度指数与坡度坡长指数相关关系

(e)沟壑密度与坡度坡长指数相关关系　(f)地形湿度指数与沟壑密度相关关系

图 4　与水文相关指数的各类指数相关性分析图

图 4 中的 D、E、F 分别显示了地形湿度指数、坡度坡长指数、沟壑密度之间的相关关系，各水文因子采用指数模型和多项式两种拟合方式，然后保留拟合效果最好的结果做相关性比较。其中地形湿度指数与坡度坡长因子采用指数拟合效果最好，R^2 值为 0.8385，坡度坡长指数与沟壑密度采用二次多项式拟合效果最好，R^2 值为 0.2077，地形湿度指数与沟壑密度采用多项式拟合效果最好，R^2 值为 0.2801。

坡度坡长指数与坡度存在线性显著相关关系，地形湿度指数与坡度值、坡度坡长指数均存在高度相关关系，沟壑密度与坡度坡长指数、地形湿度指数、坡度值相关性极弱。

3　讨论

3.1　坡度坡长因子的相关阐述

坡度坡长因子是综合坡度、单元格网大小的经验指数，一般需采用与本地区高度相似的地区的各项系数值，本研究中采用同位于陕北黄土高原的纸坊沟流域中的研究经验数据[16]。从结果来看：定边县北部、靖边县西北部、榆阳区及神木县西北部位于风沙草滩区,地势较平,坡度坡长指数值比其他地区小,吴起县、志丹县位于黄土高原丘陵沟壑区,坡度坡长指数均值高于其他县区。符素华等[18]研究中的坡长坡度因子计算工具中的主要步骤与本研究中的处理方式一致。Hickey 等和 Van Remortel[19-20]等在 Arc/Info 平台上编制出基于 DEM 格网累积的坡长自动计算方法。Liu 等[21-22]的相关研究成果表明变化相对有限的条件下,不同算法间坡度因子差异较小,无论变化是否稳定,不同算法间的坡长因子相差较大。

3.2 地形湿度指数的相关阐述

汇流累积量以洼地填充后方向矩阵为基础,流出栅格和流向栅格一致,格网单元汇流值增加。薛丰昌等[9]的研究表明栅格格网变大,汇水区的面积减小,此结果与本研究的不同阈值下汇水面积量的变化趋势一致。地形湿度指数和沟壑线均以栅格单元的汇流累积量为基础提取,地形湿度指数同时综合了坡度,坡度越大地形湿度指数越小。该结果与于海洋等[15]在鹤壁汤河流域的研究结果一致。同时刘金涛等[13]的研究表明凹形山坡、收敛形山坡对应的土壤含水量、湿度指数更高,而凸形和发散形山坡对应的土壤含水量及湿度指数较低,本研究中地形湿度指数与坡度相关程度呈现类似规律。多项研究表明[12,14]基于高程的栅格指数与地形湿度指数偏差之间均存在显著的负相关性,本研究前期相关数据也表明地形湿度指数在同坡度情况下随高程的降低而增大。

3.3 沟壑密度的相关阐述

不同汇流阈值下,各县区沟壑密度的相对大小排列顺序差异明显,其中阈值 2500 和阈值 5000 县域沟壑密度大小排序相对接近,阈值 7500 和阈值 10000 县域沟壑密度大小排序相对接近。沟壑密度与河网总长度成正比,与研究区面积成反比。吴秉等[23]通过研究沟道覆盖区栅格数与汇流累积栅格二值化阈值间的关系,表明不同用途或尺度的研究应选用合适的阈值。同一地区内部由于地形地貌差异,沟谷河网分布也会出现明显不同。而沟壑密度不能反映实际水流量,仅从地貌上反映了沟谷大致分布情况。汇流累积量阈值设置得越大,河网总长度越小,沟壑密度越小,此结果与杨华容等[24]的相关研究结果一致。

各类水文因子相关性分析情况如下:地形湿度指数与坡度值采用指数拟合效果最好,R^2 值为 0.8168,坡度坡长指数与坡度值采用线性拟合效果最好,R^2 值为 0.993,沟壑密度与坡度值采用多项式拟合效果最好,R^2 值为 0.0604。地形湿度指数与坡度坡长因子采用指数拟合效果最好,R^2 值为 0.8385,坡度坡长指数与沟壑密度采用二次多项式拟合效果最好,R^2 值为 0.2077,地形湿度指数与沟壑密度采用多项式拟合效果最好,R^2 值为 0.2801。各项水文指数的相关性分析可为陕北黄土高原相关水文模拟分析提供可靠的误差矫正参数。

3.4 本研究需改进及展望

本研究仅从以各阈值下河流长度提取计算沟壑密度,而未采用河网分级形式定量分析陕北黄土高原的水文情况,下步研究应从河网分级结合河流长度处综合分析。地形湿度指数是单位汇水面积和坡度因子的复合函数,受到尺度效应的影响。龚秒的研究证明地形湿度指数与 DEM 分辨率存在着明显的线性相关[25]。胡璐锦的研究表明 DEM 分辨率越低坡长取值越大[26]。于海洋的研究中表述 1″ DEM 在微地貌以及高坡度地形区存在失真相关[15]。本研究仅从用 90 m 分辨率的 DEM,具有一定的局限性,下一步将从更大尺度上讨论各项水文因子与生态环境的关系。胡刚等[27]分析了以 20 m 和 20.13 m 为基准坡长对不同土壤侵蚀模型的影响,得出结论:在坡度较小时侵蚀模型间差距不大,随着坡度增大侵蚀模型差异明显。本研究仅以 20.13 m 为基准坡长,下步应对照多基准坡长分析比较。王洪明等[28]研究了小流域尺度土壤水分与地形湿度指数的相关性,本研究仅对 DEM 提取的各类水文指数的相关性进行了分析,而未提取实际的土壤水分来验证,研究下步将在大尺度小流域研究区内进行实地验证。陈见影等[29]的研究结果表明不同地类空间分布与地形因子坡度、高程、沟壑密度关系密切,并呈现一定的规律。本研究仅从各水文地形因子相关程度上进行研究,而未涉及地类分布,下步将会把地类纳入研究内容。

4 结论

坡度坡长因子、地形湿度指数、沟壑密度均在一定程度上反映了实际地貌。坡度越大地形湿度指数越小。沟壑密度随汇流阈值设置的增大而减小。坡度坡长指数与坡度存在线性显著相关关系,地形湿度指数与坡度值、坡度坡长指数均存在高度相关关系,沟壑密度与坡度坡长指数、地形湿度指数、坡度值相关性极弱。

参考文献

[1]李景星. 基于 DEM 的黄土区河网水系特征研究[D].杨凌:西北农林科技大学,2012.

[2]祝士杰. 基于 DEM 的黄土高原流域面积高程积分谱系研究[D].南京:南京师范大学, 2013.

[3]管伟瑾, 曹泊, 王晓艳,等. 河流信息提取方法比较[J]. 人民黄河, 2017(2):51-55.

[4]颜明, 许炯心, 贺莉,等. 黄河流域河网密度的空间特征及其影响因素[J]. 水土保持研究, 2018, 25(2):288-292,299.

[5]赵卫东, 杨文韬, 龚俊豪,等. 基于DEM 的黄土高原小流域主沟道汇流累积量沿程变化模式[J]. 合肥工业大学学报:自然科学版, 2018,41(1):112-118.

[6]王文静. 陕北黄土高原地区水土流失敏感性评价[D].西安:西北大学, 2016.

[7]张宝庆,邵蕊,吴普特,等.大规模植被恢复对黄土高原生态水文过程的影响[J].应用基础与工程科学学报,2020,28(3):594-606.

[8]蒋凯鑫,于坤霞,曹文洪,等.黄土高原典型流域水沙变化归因对比分析[J].农业工程学报,2020,36(4):143-149.

[9]薛丰昌, 唐步兴, 黄敏敏. DEM 栅格单元大小对汇水区提取的影响研究[J]. 科技通报, 2019(3):18-25.

[10]汤巧英, 戚德辉, 宋立旺, 等. 基于 GIS 和 RS 的延河流域植被覆盖度与地形因子的相关性研究[J]. 水土保持研究, 2017, 24(4): 198-203.

[11]王海力,韩光中,谢贤建.单流向法地形湿度指数尺度效应的不同地形区差异分析[J]. 地理与地理信息科学, 2016, 32(4):23-29.

[12]马建超,林广发,陈友飞,等. DEM 栅格单元异质性对地形湿度指数提取的影响分析[J].地球信息科学学报,2011,13(2):157-163.

[13]刘金涛, 冯德锃, 陈喜, 等. 山坡地形曲率分布特征及其水文效应分析——真实流域的野外实验及相关分析研究[J]. 水科学进展, 2011, 22(1):1-6.

[14]荀娇娇, 罗明良, 王飞. 影响黄土高原集水面积阈值的地形因子主成分分析[J].武汉大学学报:信息科学版, 2017, 42(5):704-710.

[15]于海洋, 罗玲, 马慧慧, 等. SRTM (1″) DEM 在流域水文分析中的适用性研究[J].国土资源遥感, 2017, 29(2):138-143.

[16]张文杰 ,程维明 ,李宝林, 等. 黄土高原丘陵沟壑区切沟侵蚀与地形关系分析——以纸坊沟流域为例[J]. 地球信息科学学报, 2014, 16(1):87-94.

[17]张娜.生态学中的尺度问题:内涵与分析方法[J].生态学报, 2006, 26(7):2340-2355.

[18]符素华, 刘宝元, 周贵云, 等. 坡长坡度因子计算工具[J]. 中国水土保持科学, 2015, 13(5):105-110.

[19]Hickey R, Smith A, Jankowski P. Slope length calculations from a DEM within ARC/INFO GRID[J]. Computers, Environment and Urban Systems, 1994, 18(5):365-380.

[20]Van Remortel R D, Maichle R W, Hickey R J. Computing the LS factor for the Revised Universal Soil Loss Equation through array-based slope processing of digital elevation data using a C++ executable[J]. Computers & Geosciences, 2004, 30(9-10):1043-1053.

[21]Liu B Y, Nearing M A, Shi P J. et al. Slope length effects on soil loss for steep slopes[J]. Soil Society of American Journal,2000, 64(5): 1759-1763.

[22]Liu B Y, Nearing M A, Risse L M. Slope gradient effects on soil loss for steep slopes[J]. Transactions of the ASAE, 1994,37(5): 1835-1840.

[23]吴秉, 侯雷, 宋敏敏, 等. 基于汇流累积计算的沟壑密度分析方法[J]. 水土保持研究, 2017(3):39-44.

[24]杨华容，文路军，彭文甫，等. 基于 DEM 和 GIS 的流域水文信息提取——以巴中市为例[J]. 人民长江，2016，47(8)：34-38，25.

[25]龚秒. 基于 DEM 的地形湿度指数不确定性研究[D].南京：南京师范大学，2015.

[26]胡璐锦，王亮，陶坤旺. DEM 内插算法对区域坡长提取的影响分析——以陕北黄土丘陵区为例[J]. 干旱区资源与环境，2013，27(10)：169-175.

[27]胡刚，宋慧，刘宝元，等. 黑土区基准坡长和 LS 算法对地形因子的影响[J]. 农业工程学报，2015，31(3)：166-173.

[28]王洪明，杨勤科，姚志宏. 小流域尺度土壤水分与地形湿度指数的相关性分析[J]. 水土保持通报，2009(4)：112-115.

[29]陈见影，孙虎，常占怀. 渭北旱塬小流域土地利用空间分布与地形因子的关系[J]. 水土保持通报，2014，34(2)：163-167.

本文曾发表于 2022 年《农学学报》第 12 卷第 5 期

Understanding Farmers' Eco-friendly Fertilization Technology Adoption Behavior Using an Integrated S-O-R model: The Case of Soil Testing and Formulated Fertilization Technology in Shaanxi, China

Hao Dong[1,2,3], Bo Wang[1,2], Jichang Han[2], Lintao Luo[2], Huanyuan Wang[1,2], Zenghui Sun[1,2], Lei Zhang[2], Miao Dai[2], Xiaohui Cheng[2], Yunliang Zhao[2]

(1.Institute of Land Engineering and Technology, Shaanxi Provincial Land Engineering Construction Group Co., Xi'an; China; 2.Shaanxi Provincial Land Engineering Construction Group Co., Ltd., Xi'an, China; 3.School of Management, Xi'an Jiaotong University, Xi'an, China)

【Abstract】 To explore the formation mechanism of farmers' willingness to adopt eco-friendly fertilization technology (EFFT), which is important to promote the application of farmland conservation technologies and enhance the effect of farmland quality protection. Based on the extended stimulus-organism-response (S-O-R) theoretical analysis framework, this paper selects 295 field interview data of farmers in the high standard farmland grain main production area in Shaanxi Province, and uses the Partial least square structural equation model (PLS-SEM) of formative indicators to analyze the mechanism of external incentives on farmers' eco-friendly fertilization technology adoption behavior and the mediating effect of intrinsic perception and moderating effect of the family endowment. The results show that external incentives can effectively improve farmers' technology adoptive behavior; internal perception has a significant positive effect on adoptive behavior, and it plays an intermediary role between external incentives and eco-friendly fertilization technology adoption behavior; family endowment has a significant positive effect on farmers' technology adoption behavior, but the moderating effect of family endowment in external incentive-technology adoption behavior relationship is not significant. Therefore, we should choose appropriate and flexible government regulations, and give full play to the role of premium incentives, so as to improve the motivation of farmers to adopt eco-friendly fertilization technology.

【Keywords】 Eco-friendly Fertilization Technology (EFFT); Stimulus-Organism-Response (S-O-R); Intention to adopt; PLS-SEM; China

1 Introduction

Reform and opening up for more than 40 years, China's total agricultural economy can continue to increase, relying mainly on fertilizers, pesticides and other material factors such as a large number of inputs, but the high output behind the face of increasingly serious soil consolidation, pollution, agro-ecological degradation and other problems, soil productivity decline, the quality and safety of agricultural products under threat (Yang and Li, 2000; Chen, 2007). Soil testing and formulated fertilization technology is an eco-friendly fertilization technology that can improve the green productivity and output quality of land (Wu et al., 2017). In recent years, the country has implemented a number of soil improvement measures and completed the top-level policy design (Liu et al., 2017). The report of the 19th National Congress clearly points out that "we will implement an action plan for soil pollution prevention, control and restoration, and promote green development in the countryside" (Li et al., 2020a), and the "No. 1 Document" of the Central Government has also repeatedly mentioned the promotion of soil testing and formulated fertilization technology and the implementation of organic fertilizer replacement projects (Stewart et al., 2005; Meng et al., 2013). However, some research results in theoretical circles show that farmers' eco-friendly fertilization technology adoption behavior is not optimistic, and there is a phenomenon of high willingness but low behavior (Adesemoye and Kloepper, 2009; Xin, 2022), the main crux of which is the obvious difference between farmers' goals and government goals (Smith et al., 1999). The government, aiming to maximize the sustainable use of land resources and social welfare, tends to promote organic fertilizers with slow effect and low environmental impact to farmers. Rational farmers, aiming at increasing returns or reducing costs, focus more on short-term benefits and tend to use fast-acting chemical fertilizers, ignoring the agricultural surface pollution caused by over-application (Liu et al., 2019a; Qi et al., 2021). Organic fertilizer is a labor-intensive production factor, characterized by slow fertilizer efficiency, large volume, and high dosage (Godfray and Garnett, 2014), and its application by farmers increases input costs and does not significantly improve agricultural yields (Jordan-Meille et al., 2012). In addition, soil testing and formulated fertilization technology is a typical public welfare technology with positive environmental externalities, but the lack of quality and price mechanism makes it difficult to compensate the overflowing social benefits through the market price of agricultural products (Marenya and Barrett, 2007), and farmers' benefits are not effectively improved. The small-scale decentralized operation model will exist for a long time in the future, and it is important to explore how to improve the behavioral initiative of farmers to adopt conservation farming technologies, and whether external incentives can form an endogenous and long-term mechanism for farmers to adopt new technologies, i.e., to achieve cost savings and efficiency gains for farmers (Yan et al., 2008).

The current research on pro-environmental fertilizer application technology adoption behavior by scholars is mainly summarized into two views, one is that most of them use the planned behavior and its expansion model to analyze the influencing factors of farmers' technology adoption behavior, forming a research paradigm with endogenous factors such as farmers' individual characteristics, household characteristics, cognitive characteristics, and willingness to participate as the logical main line (Tey and Brindal, 2012; Daxini et al., 2018; Faruque-As-Sunny et al., 2018; Qi et al., 2021). Farmers' endowments such as the education level of decision makers, the number of household laborers, the degree of part-time employment and the scale of land operation all have significant effects on the adoption of pro-environmental fertilizer application technology (Onwezen et al., 2013). Some studies have extended to the level of internal subjective factors such as individual cognition and psychological factors, and then penetrated into the fields of social psychology and organizational behavior. Dong et al. (2022) proposed a theoretical framework including "economic rationality" and "ecological rationality", trying to find the behavioral logic of farmers' conservation farming technology adoption from the internal psychological mechanism. Some scholars clearly distinguished the differences in psychological cognition and behavior of economic individuals, and the limited rationality caused by incomplete cognition and uncertainty in the external environment, and farmers' motivation to adopt pro-environmental fertilizer application technology was insufficient. The difference in farmers' green perceptions is one of the important reasons for the paradox of soil testing and formulated fertilization technology adoption intention and behavior (Liu et al., 2019b; Li et al., 2020b). Second, from the perspective of exogenous dynamics such as policy incentives and constraints, social networks and market resource allocation, Elmustapha et al. (2018) point out that government incentives have a greater impact on adoption intentions and decision behavior than farmers' endowment characteristics. Mazvimavi (2009) and Bravo-Monroy et al. (2016) found that government subsidies, technology training, and agricultural returns were the key factors influencing the adoption decisions of surveyed farmers in choosing eco-friendly farmland soil management technologies. Government subsidies were more effective than controls in inducing proactive farmers' adoption decisions, and information induction was more applicable to farmers with lower environmental awareness (Atari et al., 2009). Based on social capital theory, interpersonal relationships and social networks can significantly increase the adoption level of green production technologies by farmers (Yang et al., 2020). The market environment has a direct influence on farmers' decision on soil testing and formulated fertilization technology application behavior mainly through information collection, quality testing, sales channels and prices (Smith and Siciliano, 2015).

The findings of existing studies provide important literature support for this paper. Many scholars believe that external variables such as farmers' endowment resources (personal characteristics, family characteristics), external environment (market incentives and policy incentives), and internal psychological variables such as cognitive level and willingness to adopt are significant factors affecting farmers' technology adoption. Bandura's interaction model of human behavior and motivation theory suggest that behavior is not constrained by internal factors or external stimuli alone, but is determined by the interaction between the individual's internal needs such as emotions and cognition and the environment, and is a process in which the external environment acts as an engine through internal factors (De Fano et al., 2019). As a rational economic person, farmers' behavioral decision of technology adoption must be the result of coupled constraints of internal and external factors, and internal psychological motivation and external environmental changes jointly affect farmers' behavioral decision, and the behavioral decision effect may also differ (Clark et al., 2003; Truelove et al., 2014). Based on the previous research results, this paper tries to conduct additional research in the following aspects: First, the existing literature mostly focuses on analyzing the influence of a single intrinsic factor or external environment on farmers' behavior, and less incorporates internal factors such as family endowment, intrinsic perception and external incentives into a theoretical analysis framework to explore the path and logical mechanism of external incentives affecting farmers' technology adoption behavior process. Second, most studies have used Logit regression model, stepwise regression, Bootstrap test and hierarchical regression to test the mediating and moderating effects of a variable step by step (Song et al., 2017), and structural equation model can deal with both the mediating effect of intrinsic perception and the moderating effect of family endowment. In short, this study aims to accomplish two main objectives:

· To explain the transmission mechanism through which external incentives influence farmers' technology adoption behavior through intrinsic perceptions.

· To test whether there is a moderating effect of family endowments on external incentives to influence technology adoption behavior.

Based on the research objectives, the research questions underpin this quantitative study:

· Does each dimension of Stimulus-Organism-Response (S-O-R) (i.e., external incentives, intrinsic perception, and family endowment) have a positive impact on technology adoption Behavior?

· Does the family endowment moderate the relationship between external incentives and technology adoption behavior?

2 Theoretical analysis and hypothesis

2.1 Theoretical model

The Stimulus-Organism-Response (S-O-R) theoretical model belongs to the domain of cognitive psychology, which specifically explains the influence of external environmental characteristics on individuals' mental cognitive and emotional responses and later behavioral decisions, with emphasis on mental cognition as a mediating element, and provides an effective research paradigm for analyzing individual behavioral mechanisms (Kamboj et al., 2018). The stimulus (S) represents the characteristics of the external environment in which the individual lives, which has a stimulating effect on the organism (O) 's cognition and emotions, and the organism undergoes a series of psychological reaction processes that eventually manifest themselves as response (R) to internal or external behavior patterns (Vieira, 2013). The S-O-R model does not necessarily follow a strict "S-O" and "O-R" path of action, as external stimulus (S) can act directly on the mental processes of the organism (O) and can also directly produce behavioral responses (R). The S-O-R model has been applied to information systems and e-commerce, and the research involves the user experience and consumer purchase behavior of online social platforms. The paper is based on the analytical framework of S-O-R model, but considering that the S-O-R model cannot better reflect the influence path of farmers' ability level on individual behavior, we borrow some ideas from the Motivation-Opportunity-Ability (MOA) model of behavioral organization theory. Ability (A) in the MOA model reflects the intrinsic possibility that an individual can engage in a certain behavior, and emphasizes that individual ability works together with Motivation (M) and Opportunity (O) to influence behavioral decisions (Ahmad et al., 2021). This paper incorporates the Ability (A) factor from MOA theory into the S-O-R theoretical model to construct an analytical framework for farmers' adoption behavior of organic matter reclamation technology and further enhance the explanatory power of the model to reality.

Among the external environmental factors, external incentives are an important component, and changing the incentive structure and incentive level of agricultural production is a fundamental way to improve farmers' fertilizer application technology and agricultural surface source pollution. Drawing on relevant research results (Pretty et al., 2001; Swinton et al., 2007; Bopp et al., 2019), external incentives mainly refer to incentives from three levels: government, market and society, and considering the principles of data availability, science and operability, external incentives are defined in this paper as two dimensions: policy incentives and market incentives. The policy incentives refer to the government-led technology promotion training and factor subsidies, while the market incentives refer to the market premium incentives brought by the "three products and one standard" and the brand effect to farmers. Among the internal factors, the adoption of technology by farmers is generally influenced by family endowment characteristics (Mendola, 2008), and the ability of family members largely determines whether farmers can overcome the operational challenges of green production technologies and cope with the high cost and riskiness of new technologies. Rogers' innovation diffusion theory shows that an important influencing factor of technology diffusion is technology attributes, and there may be significant differences in the intrinsic perception of the same technology by different individual farmers, which significantly affects their new technology adoption behavior (Wonglimpiyarat and Yuberk, 2005). Therefore, this paper focuses on the perspective of farmers and constructs a conceptual model of farmers' behavioral decision to adopt organic soil reclamation technology based on S-O-R extension theory, starting from family endowment, external incentives, and intrinsic motivation mechanisms. The specific theoretical analysis framework is shown in Fig.1.

2.2 Research hypothesis

2.2.1 Extrinsic incentives and technology adoption behavior

The impact of external incentives on farmers' soil testing and formulated fertilization technology adoption behavior. While improving soil fertility, soil testing and formulated fertilization technology reduces agricultural non-point source pollution caused by chemical fertilizer application, which is typically characterized

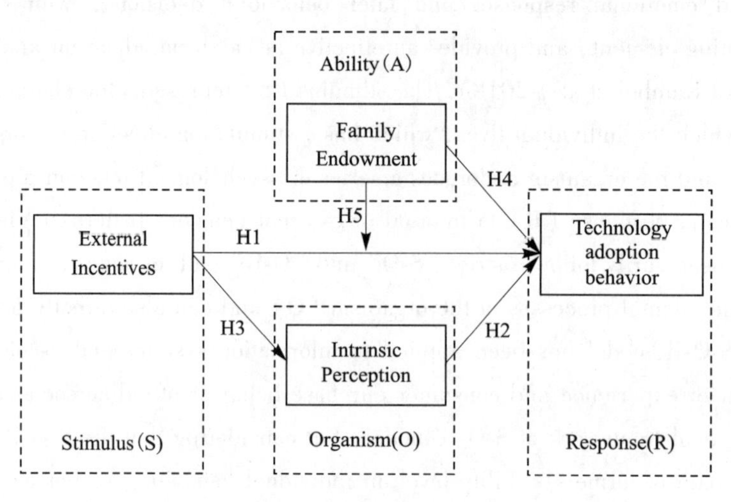

Fig.1 Research model

by high cost, high risk, and slow returns under positive spillovers over time. Small farmers generally have a risk aversion preference, lack the behavioral initiative to apply organic fertilizers that are slow to take effect and consume a lot of energy, and prefer to use chemical fertilizers to avoid the risk of crop yield reduction. External incentives can influence farmers' soil testing and formulated fertilization technology adoption behavior by generating endogenous long-term mechanisms through two paths: loss aversion and benefit drive. On the one hand, the government compensates farmers appropriately by using technology promotion and implementation factor subsidies to reduce the input costs of green production technologies, which increases the potential benefits of farmers and enhances their endogenous motivation to adopt soil testing and formulated fertilization technology. On the other hand, according to the externality theory, the marginal private gain of farmers adopting the technology is smaller than the marginal social gain, resulting in a low level of economic incentives. The role of market incentive is to solve the problem of farmers' distrust in market mechanism. "Three products and one standard" quality certification and brand building can make the premium effect appear, improve the expectation of the income of agricultural products, drive the farmers around the radiation to adopt soil testing and formulated fertilization technology, and promote the internalization of external effects to achieve the optimal allocation of resources. The higher the level of external incentives, the stronger the motivation of farmers to adopt soil testing and formulated fertilization technology. Thus, we propose the following hypothesis:

Hypothesis(H1): External incentives have a positive impact on farmers' technology adoption behavior.

2.2.2 Intrinsic perception and technology adoption behavior

The influence of intrinsic perception on farmers' soil testing and formulated fertilization technology adoption behavior. According to the theory of farmers' behavior, under the competitive market mechanism, farmers' behavioral decisions are fully rational and satisfy the principle of optimality. Whether a rational farmer accepts a new technology or decision depends on its technical characteristics and effects. The process of farmers' perception and awareness of soil testing and formulated fertilization technology characteristics generally includes the assessment of technology usefulness, ease of use, and expected adoption costs (Yu, Wei-Zhen et al., 2019), and intrinsic perception is a key psychological element for farmers to generate endogenous motivation to adopt green production technologies. In general, the stronger the intrinsic perception of the technology by farmers, the higher the likelihood of adopting soil testing and formulated fertilization technology. Thus, we propose the following hypothesis:

Hypothesis(H2): Farmers' internal perception has a positive impact on technology adoption behavior.

2.2.3 Intrinsic perception, external incentives and formulated fertilization technology adoption behavior

The theory of behavioral economics believes that human decision-making behavior is uncertain, and the introduction of human psychological factors can scientifically and effectively analyze decision-making behavior. The basic point of the S-O-R theoretical model is that individuals will make corresponding behavioral responses to certain characteristics of the external environment, and the individual's psychological cognition is the mediating factor that affects this process. Policy incentive tools or market incentives need to stimulate farmers' subjective initiative by meeting the actual needs of farmers to increase income and efficiency, and internalize them into positive psychological perceptions, in order to further influence their technology adoption behavior. Thus, we propose the following hypothesis:

Hypothesis(H3): Intrinsic perception mediates between external incentives and soil testing and formulated fertilization technology adoption behavior.

2.2.4 Family endowment and technology adoption behavior

Family endowment is the factor resources and capabilities possessed by the whole household, which plays a significant role in individual behavioral choices and decisions. mmm technology will increase material input costs and labor costs in the short term, but the cost-benefit ratio cannot increase faster at the same time, and rational farmers will arrange their factor allocation behavior rationally according to their resource endowments such as labor, capital, and land. Therefore, individual and family characteristics and other endowment capabilities are the basic influencing factors of MMM technology adoption, which determine farmers' attitude and ability to adopt new technologies and ultimately affect the technology diffusion effect (Kong et al., 2004). Thus, we propose the following hypothesis:

Hypothesis(H4): The family endowment of farmers has a significant impact on technology adoption behavior.

2.2.5 Family endowment, external incentives and technology adoption behavior

Faced with the same external incentive, differences in family endowments lead to different responses from farmers and affect their soil testing and formulated fertilization technology adoption behavior. External incentives reflect the government's long-term goal of improving land utilization and achieving green agricultural development, and do not directly improve farmers' economic efficiency in a short period of time; farmers' acceptance of external incentives depends on internal cognitive factors and preference factors. On the one hand, rich endowment resources mean that farmers have accumulated a large amount of production experience and knowledge, and have strong information acquisition and cognitive ability for policy incentives such as factor subsidies and technical training, which can reduce the learning cost of farmers and promote the effective transformation of farmers' behavioral cognition into technical adoption. On the other hand, farmers with high family endowment have certain market risk tolerance and are willing to invest more costs to achieve high quality and good price for agricultural products, and are more likely to choose soil testing and formulated fertilization technology under market incentives (Luo Xiaofeng et al., 2020). Therefore, the heterogeneity of family endowments can lead to a significant difference in the role of external incentives on soil testing and formulated fertilization technology adoption behavior. Thus, we propose the following hypothesis:

Hypothesis(H5): Family endowments play a moderating role in the relationship between external incentives and soil testing and formulated fertilization technology adoption behavior.

3 Methodology

3.1 Sample

The data in this paper come from a questionnaire survey conducted by the research team in Shaanxi Province from 2020 to 2021, and the research protocol uses a combination of stratified and random sampling to select the sample farmers. The first stage targeted grain-producing areas, i.e., sample counties were selected in the Guanzhong Plain and the Loess Plateau in northern Shaanxi based on factors such as rice grain and geographical location; the second stage aimed to find townships with large grain producers with the help of relevant agricultural departments in the sample counties; and the third stage was to find typical farmers in the selected townships. Finally, Yan'an, Yulin, Baoji, and Weinan cities in Shaanxi Province were selected as the sample areas. The group chose these study areas because, first, Shaanxi Province is located in the Guanzhong Plain region, backed by the Qinling Mountains, rich in water and fertile soil, and is an important grain-producing area in China. High-standard farmland projects are being built in recent years, which provides space for the development of soil testing and formulated fertilization technology. Second, according to previous research results, soil testing and formulated fertilization technology is more mature in the region and the sample data is more representative. The content of this survey mainly focused on the research of technology sources and communication, agricultural inputs and outputs, and ecological production behavior of environment-friendly arable land conservation technologies. A total of 325 questionnaires were obtained, and after eliminating invalid samples, 295 valid samples were obtained, with an effective rate of 90.8%.

3.2 Survey design

A questionnaire was designed, including (1) demographic characteristics; (2) evaluation of soil testing and formulated fertilization technology adoption predictors; and (3) evaluation of soil testing and formulated fertilization technology adoption implications. Each construct was measured using a validated research instrument developed by previous studies (modified to fit the research context) and based on the literature review findings and the theoretical foundation. A seven-point Likert scale was used to overcome measurement errors. The Likert scale ranges from "strongly disagree" (i.e., 1) to "strongly agree" (i.e., 7). Table 1 describes the used survey items.

3.3 Data analysis technique

Potential biases were considered in the survey, protocol design, and data analysis. Several approaches (e.g. direct contact by phone and assurance to share the results) were adopted to ensure the highest response rate and avoid a non-response bias (Frohlich and Westbrook, 2002). We used a partial least squares structural equation modelling (PLS-SEM) technique to analyse the data. This technique has been adopted because this process gives better results in the analysis of this type of exploratory study. This process can also analyse those data that are not normally distributed (Hair et al., 2012). This technique does not impose any sample restriction to conduct the survey. This process involves quantification of responses on a specific scale.

Table 1　Items used to measure each survey construct and loadings

Construct	Items	Loadings	VIF
Family endowment, adapted from Zeweld et al. (2017)	FE1: I think the scale of farming is beneficial to technology adoption behavior	0.875	3.352
	FE2: I think the number of years of farming is beneficial for technology adoption behavior	0.904	4.044
	FE3: I believe that farming family income is beneficial to technology adoption behavior	0.883	2.924
	FE4: I think the grain marketing channel is beneficial to technology adoption behavior	0.867	2.413
External incentives, adapted from Zeweld et al. (2017)	EI1: I think technology subsidies are beneficial to technology adoption behavior	0.857	2.893
	EI2: I think technology training is beneficial to technology adoption behavior	0.866	3.054
	EI3: I believe that product or corporate branding is beneficial to technology adoption behavior	0.790	2.001
	EI4: I think whether the product obtains the "three products and one standard" quality certification is beneficial to the technology adoption behavior	0.864	2.648
	EI5: I think the region is easy for new technology introduction	0.745	1.612
Intrinsic perception, adapted from Rezaei et al. (2020)	IP1: I think the soil testing and formulated fertilization technology is very useful	0.927	3.184
	IP2: I think the soil testing and formulated fertilization technology is easy to use	0.943	3.740
	IP3: I think the cost of soil testing and formulated fertilization technology is worth considering	0.878	2.470
Technology adoption behavior, adapted from Zeweld et al. (2017)	TAB1: I would like to use soil testing and formulated fertilization technology instead of the production technology used now	0.799	2.275
	TAB2: I would like to learn soil testing and formulated fertilization technology	0.878	3.955
	TAB3: I would like to apply soil testing and formulated fertilization technology	0.870	3.593
	TAB4: I would like to promote soil testing and formulated fertilization technology	0.891	3.932
	TAB5: I am willing to give feedback on the effectiveness of soil testing and formulated fertilization technology	0.880	4.193
	TAB6: I would like the village to develop eco-agricultural techniques on my land	0.852	3.118

4　Analysis result

4.1　Model evaluation

The reflective constructs were validated by testing internal consistency, composite reliability, convergent, and discriminant validity (Table 2 and Fig.2). To verify the internal consistency and composite reliability of the constructs, we verified that the value of Cronbach's alpha and composite reliability indices exceeded 0.7 (Hair et al., 2011a). This condition was valid for all the constructs. To test convergent validity, we verified that the average variance extracted (AVE) index was greater than 0.5. The lowest observed value (0.579) was substantially higher than this threshold. The discriminant validity of the reflective constructs was tested in three ways (Fornell and Larcker, 1981). The correlation matrix proved that the AVE was greater than the square correlation between each pair of latent constructs. These results suggest the validity of the reflective constructs used in our analysis and the adequacy of the items used as construct indicators. To determine the redundancy, the variance inflation factor (VIF) was used to test for multicollinearity among the different measurement variables, and the VIF values of the measurement models are shown in Table 1.the VIF values of all 18 observed variables were less than 5 (Hair et al., 2011b), indicating that there was no multicollinearity among the measurement variables.

Table 2 Construct consistency, reliability, convergent and discriminant validity squared value of the AVE reported on the main diagonal of the correlation matrix

Constructs	Composite reliability	Cronbach's alpha	Average variance extracted	External incentives	Family endowment	Intrinsic perception	Technology adoption behavior
External incentives	0.915	0.882	0.682	0.826	0.270	0.443	0.446
Family endowment	0.934	0.905	0.779	0.247**	0.882	0.343	0.637
Intrinsic perception	0.940	0.905	0.839	0.399**	0.320**	0.916	0.511
Technology adoption behavior	0.945	0.931	0.743	0.404**	0.590**	0.475**	0.862

Note: Significant level: $p<0.10$; * $p<0.05$; ** $p<0.01$; *** $p<0.001$, Bold diagonal entries are square root of AVEs, Heterotrait-Montrait ratios (HTMT) (Underlined) are below 0.85.

4.2 Analysis of results

Fig.2 plots the results of the structural equation model calculations, and Table 3 shows the direct, indirect and total effect coefficients of external incentives, family endowments affecting intrinsic perceptions and farmers' soil testing and formulated fertilization technology adoption behavior. These values can be considered as the predictive accuracy of the models among low, medium, and large (Hair et al., 2019). The analysis of the composite-based standardized root mean square residual (SRMR) yielded a value of 0.065, below the 0.10 threshold, which confirms the robustness of the model (Henseler et al., 2015).

The direct effect of external incentives on farmers' technology adoption behavior. External incentives have a statistically significant positive effect on farmers' technology adoption behavior at the 1% level with a path coefficient of 0.184, i.e., the more positive incentives farmers obtain, the more likely they are to adopt soil testing and formulated fertilization technology, and hypothesis H1 is verified. The incentive effect of government subsidies is to make up the difference between the extra cost of using the new technology and the expected benefit for farmers, and the more government subsidies, the more farmers are willing to adopt the new technology. The greater the number of technical trainings attended, the greater the chance for high-standard farmland grain farmers to receive direct technical services and support, and the lower the information cost of learning new technologies, which plays an important role in correcting farmers' behavioral habits of excessive fertilizer application and raising their awareness of scientific fertilizer application. The "lemon effect" may prevent farmers from adding value to the high-quality agricultural products they produce, resulting in uncertain net returns. Market incentives can achieve market premiums for agricultural products, which is the interest driver for farmers to choose soil testing and formulated fertilization technology.

The effect of intrinsic perception on farmers' technology adoption behavior. In Table 3, the path coefficient of intrinsic perception on adoption behavior was significant, and every 1 standard deviation increase in intrinsic perception would increase farmers' technology adoption behavior by 0.244 standard deviations, and hypothesis H2 was verified. In the study, it was found that growers generally believed that the application of chemical fertilizers was fast and stable, while organic fertilizers and bacterial fertilizers were not as "powerful" as chemical fertilizers. Driven by the goal of profit maximization, some growers formed a preference for chemical fertilizers, and the stronger the perceived usefulness of the technology, the easier it was to master the new technology, and the lower the cost of technology adoption, the more motivated farmers were to adopt soil testing and formulated fertilization technology.

Intrinsic transmission mechanism of external incentives: the mediating role of intrinsic perceptions. The hypothesis H3 holds that external incentives have a significant positive effect on intrinsic perceptions and intrinsic perceptions have a significant positive effect on adoption behavior, and that intrinsic perceptions play a mediating role between external incentives and soil testing and formulated fertilization technology adoption behavior. This is fully consistent with the expected hypothesis of S-O-R theory. The coefficient of direct effect of external incentives on intrinsic perception is 0.399, the coefficient of direct effect of intrinsic perception on adoption behavior is 0.244, and the coefficient of indirect effect of external incentives on adoption behavior through intrinsic perception is 0.097. External incentives reduce the marginal cost of technology adoption to a certain extent, and make farmers fully realize that soil testing and formulated fertilization technology can bring long-term economic benefits to high-standard farming. The level of farmers' intrinsic perceptions of technology adoption increased significantly, which eventually led to technology adoption behavior.

The effect of family endowment on farmers' technology adoption behavior. The path coefficient of family endowment on adoption behavior is statistically significant at the 1% level, indicating that family endowment directly affects farmers' technology adoption behavior, and hypothesis H4 is tested. Each 1 standard deviation increase in family endowment increased the adoption behavior of farmers by 0.459 standard deviations. Longer farming years and more experience of farmers imply more asset specificity of food on high standard farmland, more difficulty in switching to other crops, and more inclination to accept new technologies. Stable marketing channels make farmers pay more attention to the quality of high-standard farmland grain, creating conditions for the adoption of green production technologies. The application of organic fertilizer is a labor-intensive technology, and the lower the proportion of food income from high-standard farmland to total household income, the greater the opportunity cost for farmers to apply organic fertilizer and the less motivated they are to adopt soil testing and formulated fertilization technology. The more the scale effect generated by the expansion of planting scale can promote farmers' choice of organic fertilizer application.

Inter-farmer variation: moderating role of family endowment. The interaction coefficient of external incentives and family endowment on adoption behavior is insignificant, indicating that family endowment does not play a moderating role in external incentives affecting farmers' technology adoption behavior, i.e., family endowment heterogeneity does not lead to differential effects of external incentives on adoption behavior. The possible explanation for this is that in the historical process of rural revitalization and poverty alleviation, the government-led external incentives are universal policies, and the targets of implementation include high quality farmers with superior family endowments and poor households with poor resource endowments, and homogeneous and non-differentiated policy incentives cannot fundamentally stimulate different behavioral responses of farmers.

Table 3　Results of hypothesis testing

Hypothesis	Effect	Path	Path coefficient	Lower (2.5%)	Upper (95%)	*t*-statistics	*p*-value	Decision
Direct relationships								
H3	Direct	External Incentives→ Intrinsic Perception	0.399	0.272	0.512	6.546	0.000***	Accept
H1	Direct	External Incentives→ Technology adoption behavior	0.184	0.073	0.330	2.741	0.006***	Accept
H4	Direct	Family endowment → Technology adoption behavior	0.459	0.330	0.551	8.050	0.000***	Accept
H2	Direct	Intrinsic Perception→ Technology adoption behavior	0.244	0.139	0.332	4.925	0.000***	Accept
Mediating relationships								
H2 · H3	Indirect	External Incentives→ Intrinsic Perception→ Technology adoption behavior	0.097	0.054	0.141	4.399	0.000***	Accept
Moderating relationships								
H5	Direct	Family endowment* External Incentives→ Technology adoption behavior	0.040	−0.128	0.121	0.532	0.595^{NS}	Rejection
SRMR composite model = 0.059								
$R^2_{IP}=0.159$; $Q^2_{IP}=0.118$								
$R^2_{TAB}=0.474$; $Q^2_{TAB}=0.346$								

Note: *** $=p<0.001$; ** $=p<0.01$; * $=p<0.05$; NS $=p>0.05$.

5　Conclusions and implication

5.1　Conclusions

In this study, family endowment, external motivation, internal perception and adoption behavior were incorporated into the extended S-O-R theoretical framework, and the data of 295 major high-standard farmland grain-producing areas in Shaanxi Province were used. The structural equation model of formative indicators was constructed based on partial least square method to estimate the direct influence of external incentives on farmers' soil testing and formulated fertilization technology adoption behavior, as well as the mediating role of internal perception and the moderating role of family endowment. The research conclusions are as follows:

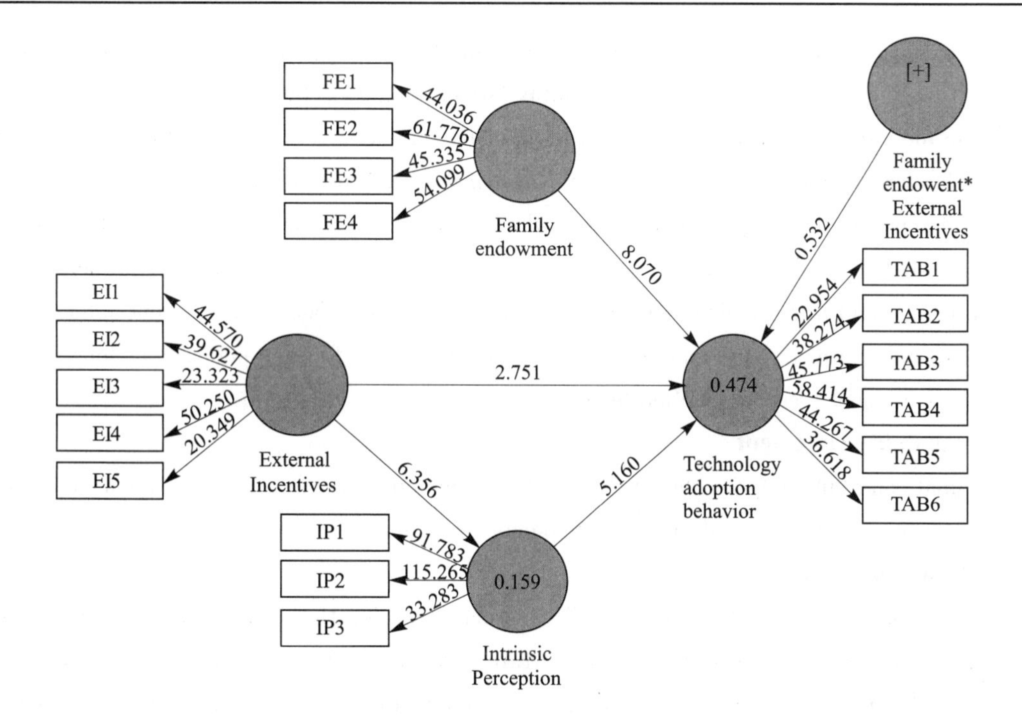

Fig.2 Model results calculation

First, external incentives have a significant positive impact on farmers' technology adoption behavior at the statistical level of 10%, that is, the more positive incentives farmers get, the more likely they are to adopt soil testing and formulated fertilization technology.

Second, the path coefficient of intrinsic perception on adoption behavior is significant, indicating that the latent variables of intrinsic perception, usability perception and technology adoption cost are the key factors determining adoption behavior. Internal perception plays a mediating role between external incentive and soil testing and formulated fertilization technology adoption behavior, which further confirms that in S-O-R theory, external environmental characteristics will act on internal perception and finally appear as farmers' technology adoption behavior.

Third, the characteristics of farmers' family endowment have a direct effect on soil testing and formulated fertilization technology adoption behavior. The interaction term between external incentives and family endowment has no significant path coefficient on soil testing and formulated fertilization technology adoption behavior, indicating that family endowment does not play a moderating role in the influence of external incentives on soil testing and formulated fertilization technology adoption behavior.

5.2 Implication

This paper explains the mechanism of soil testing and formulated fertilization technology adoption behavior of grain growers in high-standard farmland to a certain extent, and mainly obtains the following three policy implications. One is to choose suitable and flexible incentive means. The government's "top-down" system supply will largely ignore the actual needs of farmers, so it is necessary to provide targeted policy supply, such as providing technical training at the initial stage of technology promotion, replacing classroom training with field experiments as far as possible, and accelerating the process of farmers' mastery of new technology through "learning through work". Secondly, the farmers of high management ability and stability of the economic benefits is the premise and guarantee to accept new technology, the government in the implementation of the general policy of motivation at the same time, can't ignore the audience heterogeneity characteristics of object, should focus on high-quality farmers tilt, to achieve the optimal allocation of factors, stimulate the radiating and driving function of the "new farmers". The third is to improve the internal motivation of farmers to adopt technology. Farmers' response to soil testing and formulated fertilization technology is indifferent and the adoption rate is low, which is due to their concern about the technology effect and adoption cost. We should give full play to the incentive effect of premium, motivate farmers to carry out product certification and brand registration, raise product prices, guarantee the economic interests and production enthusiasm of farmers participating in green production, and form an endogenous long-term mechanism for farmers to adopt green production technology.

Data availability statement

The original contributions presented in the study are included in the article/Supplementary Material, and further inquiries can be directed to the corresponding authors.

Author contributions

Methodology and software, H.D. and H.Y.W.; formal analysis, H.D. and J.C.H.; resources and data curation, B.W. and Z.H.S.; investigation, H.D.; writing-original draft preparation, J.C.H.; writing-review and editing, H.D. and J.C.H.; supervision and project administration, H.D.; All authors have read and agreed to the published version of the manuscript.

Funding

This research was funded by Shaanxi Province Key R&D Program Project- Research, Development and Application of Dynamic Monitoring and Early Warning Technology for "Non-agriculturalization and Non-foodization" of Grain Fields (2022ZDLNY02-10).

Reference

Adesemoye, A. O., and Kloepper, J. W. (2009). Plant-microbes interactions in enhanced fertilizer-use efficiency. Appl Microbiol Biotechnol 85, 1-12.

Ahmad, B., Da, L., Asif, M. H., Irfan, M., Ali, S., and Akbar, M. I. U. D. (2021). Understanding the Antecedents and Consequences of Service-Sales Ambidexterity: A Motivation-Opportunity-Ability (MOA) Framework. Sustainability 13, 9675.

Atari, D. O. A., Yiridoe, E. K., Smale, S., and Duinker, P. N. (2009). What motivates farmers to participate in the Nova Scotia environmental farm plan program? Evidence and environmental policy implications. Journal of Environmental Management 90, 1269-1279.

Bopp, C., Engler, A., Poortvliet, P. M., and Jara-Rojas, R. (2019). The role of farmers' intrinsic motivation in the effectiveness of policy incentives to promote sustainable agricultural practices. Journal of Environmental Management 244, 320-327.

Bravo-Monroy, L., Potts, S. G., and Tzanopoulos, J. (2016). Drivers influencing farmer decisions for adopting organic or conventional coffee management practices. Food Policy 58, 49-61.

Chen, J. (2007). Rapid urbanization in China: A real challenge to soil protection and food security. CATENA 69, 1-15.

Clark, C. F., Kotchen, M. J., and Moore, M. R. (2003). Internal and external influences on pro-environmental behavior: Participation in a green electricity program. Journal of Environmental Psychology 23, 237-246.

Daxini, A., O'Donoghue, C., Ryan, M., Buckley, C., Barnes, A. P., and Daly, K. (2018). Which factors influence farmers' intentions to adopt nutrient management planning? Journal of Environmental Management 224, 350-360.

De Fano, A., Leshem, R., and Ben-Soussan, T. D. (2019). Creating an Internal Environment of Cognitive and Psycho-Emotional Well-Being through an External Movement-Based Environment: An Overview of Quadrato Motor Training. International Journal of Environmental Research and Public Health 16, 2160.

Dong, H., Wang, H., and Han, J. (2022). Understanding Ecological Agricultural Technology Adoption in China Using an Integrated Technology Acceptance Model—Theory of Planned Behavior Model. Front. Environ. Sci. 10, 927668.

Elmustapha, H., Hoppe, T., and Bressers, H. (2018). Consumer renewable energy technology adoption decision-making; comparing models on perceived attributes and attitudinal constructs in the case of solar water heaters in Lebanon. Journal of Cleaner Production 172, 347-357.

Faruque-As-Sunny, Huang, Z., and Karimanzira, T. T. P. (2018). Investigating Key Factors Influencing Farming Decisions Based on Soil Testing and Fertilizer Recommendation Facilities (STFRF)—A Case Study on Rural Bangladesh. Sustainability 10, 4331.

Fornell, C., and Larcker, D. F. (1981). Structural Equation Models with Unobservable Variables and Measurement Error: Algebra and Statistics. Journal of Marketing Research 18, 382-388.

Frohlich, M. T., and Westbrook, R. (2002). Demand chain management in manufacturing and services: web-based integration, drivers and performance. Journal of Operations Management 20, 729-745.

Godfray, H. C. J., and Garnett, T. (2014). Food security and sustainable intensification. Phil. Trans. R. Soc. B 369, 20120273.

Hair, J. F., Ringle, C. M., and Sarstedt, M. (2011a). PLS-SEM: Indeed a Silver Bullet. Journal of Marketing Theory and Practice 19, 139-152.

Hair, J. F., Ringle, C. M., and Sarstedt, M. (2011b). PLS-SEM: Indeed a Silver Bullet. null 19, 139-152.

Hair, J. F., Sarstedt, M., Ringle, C. M., and Mena, J. A. (2012). An assessment of the use of partial least squares structural equation modeling in marketing research. J. of the Acad. Mark. Sci. 40, 414-433.

Jordan-Meille, L., Rubaek, G. H., Ehlert, P. A. I., Genot, V., Hofman, G., Goulding, K., et al. (2012). An overview of fertilizer-P recommendations in Europe: soil testing, calibration and fertilizer recommendations: P fertilizer methods in Europe. Soil Use Manage 28, 419-435.

Kamboj, S., Sarmah, B., Gupta, S., and Dwivedi, Y. (2018). Examining branding co-creation in brand communities on social media: Applying the paradigm of Stimulus-Organism-Response. International Journal of Information Management 39, 169-185.

Li, B., Li, P., Zeng, X. C., Yu, W., Huang, Y. F., Wang, G. Q., et al. (2020a). Assessing the sustainability of phosphorus use in China: Flow patterns from 1980 to 2015. Science of The Total Environment 704, 135305.

Li, J., Feng, S., Luo, T., and Guan, Z. (2020b). What drives the adoption of sustainable production technology? Evidence from the large scale farming sector in East China. Journal of Cleaner Production 257, 120611.

Liu, X., Zhao, C., and Song, W. (2017). Review of the evolution of cultivated land protection policies in the period following China's reform and liberalization. Land Use Policy 67, 660-669.

Liu, Y., Ruiz-Menjivar, J., Zhang, L., Zhang, J., and Swisher, M. E. (2019a). Technical training and rice farmers' adoption of low-carbon management practices: The case of soil testing and formulated fertilization technologies in Hubei, China. Journal of Cleaner Production 226, 454-462.

Liu, Y., Ruiz-Menjivar, J., Zhang, L., Zhang, J., and Swisher, M. E. (2019b). Technical training and rice farmers' adoption of low-carbon management practices: The case of soil testing and formulated fertilization technologies in Hubei, China. Journal of Cleaner Production 226, 454-462.

Marenya, P. P., and Barrett, C. B. (2007). Household-level determinants of adoption of improved natural resources management practices among smallholder farmers in western Kenya. Food Policy 32, 515-536.

Mazvimavi, K., and Twomlow, S. (2009). Socioeconomic and institutional factors influencing adoption of conservation farming by vulnerable households in Zimbabwe. Agricultural Systems 101, 20-29.

Mendola, M. (2008). Migration and technological change in rural households: Complements or substitutes? Journal of Development Economics 85, 150-175.

Meng, Q., Hou, P., Wu, L., Chen, X., Cui, Z., and Zhang, F. (2013). Understanding production potentials and yield gaps in intensive maize production in China. Field Crops Research 143, 91-97.

Onwezen, M. C., Antonides, G., and Bartels, J. (2013). The Norm Activation Model: An exploration of the functions of anticipated pride and guilt in pro-environmental behaviour. Journal of Economic Psychology 39, 141-153.

Pretty, J., Brett, C., Gee, D., Hine, R., Mason, C., Morison, J., et al. (2001). Policy Challenges and Priorities for Internalizing the Externalities of Modern Agriculture. null 44, 263-283.

Qi, X., Liang, F., Yuan, W., Zhang, T., and Li, J. (2021). Factors influencing farmers' adoption of eco-friendly fertilization technology in grain production: An integrated spatial-econometric analysis in China. Journal of Cleaner Production 310, 127536.

Rezaei, R., Safa, L., and Ganjkhanloo, M. M. (2020). Understanding farmers' ecological conservation behavior regarding the use of integrated pest management- an application of the technology acceptance model. Global Ecology and Conservation 22, e00941.

Smith, L. E. D., and Siciliano, G. (2015). A comprehensive review of constraints to improved management of fertilizers in China and mitigation of diffuse water pollution from agriculture. Agriculture, Ecosystems & Environment 209, 15-25.

Smith, V. H., Tilman, G. D., and Nekola, J. C. (1999). Eutrophication: impacts of excess nutrient inputs on freshwater, marine, and terrestrial ecosystems. Environmental Pollution 100, 179-196.

Song, G., Min, S., Lee, S., and Seo, Y. (2017). The effects of network reliance on opportunity recognition: A moderated mediation model of knowledge acquisition and entrepreneurial orientation. Technological Forecasting and Social Change 117, 98-107.

Stewart, W. M., Dibb, D. W., Johnston, A. E., and Smyth, T. J. (2005). The Contribution of Commercial Fertilizer Nutrients to Food Production. Agronomy Journal 97, 1-6.

Swinton, S. M., Lupi, F., Robertson, G. P., and Hamilton, S. K. (2007). Ecosystem services and agriculture: Cultivating agricultural ecosystems for diverse benefits. Ecological Economics 64, 245-252.

Tey, Y. S., and Brindal, M. (2012). Factors influencing the adoption of precision agricultural technologies: a review for policy implications. Precision Agric 13, 713-730.

Truelove, H. B., Carrico, A. R., Weber, E. U., Raimi, K. T., and Vandenbergh, M. P. (2014). Positive and negative spillover of pro-environmental behavior: An integrative review and theoretical framework. Global Environmental Change 29, 127-138.

Vieira, V. A. (2013). Stimuli-organism-response framework: A meta-analytic review in the store environment. Journal of Business Research 66, 1420-1426.

Wonglimpiyarat, J., and Yuberk, N. (2005). In support of innovation management and Roger's Innovation Diffusion theory. Government Information Quarterly 22, 411-422.

Wu, Y., Shan, L., Guo, Z., and Peng, Y. (2017). Cultivated land protection policies in China facing 2030: Dynamic balance system versus basic farmland zoning. Habitat International 69, 126-138.

Xin, L. (2022). Chemical fertilizer rate, use efficiency and reduction of cereal crops in China, 1998-2018. J. Geogr. Sci. 32, 65-78.

Yan, X., Jin, J., He, P., and Liang, M. (2008). Recent Advances on the Technologies to Increase Fertilizer Use Efficiency. Agricultural Sciences in China 7, 469-479.

Yang, H., and Li, X. (2000). Cultivated land and food supply in China. Land Use Policy 17, 73-88.

Yang, Y., He, Y., and Li, Z. (2020). Social capital and the use of organic fertilizer: an empirical analysis of Hubei Province in China. Environ Sci Pollut Res 27, 15211-15222.

Zeweld, W., Huylenbroeck, G. V., Tesfay, G., and Speelman, S. (2017). Smallholder farmers' behavioural intentions towards sustainable agricultural practices. Journal of Environmental Management 187, 71-81.

本文曾发表于2022年《Frontiers in Environmental Science》第10卷

Analysis of Spatio-temporal Changes and Driving Forces of Cultivated Land in China from 1996 to 2019

Jianfeng Li[1,2,3], Jichang Han[1,2,3], Yang Zhang[1,2,3], Yingying Sun[1,2,3], Biao Peng[1,2,3], Xiao Xie[1,2,3], Chao Guo[1,2,3], Huping Ye[4]

(1.Institute of Land Engineering and Technology, Shaanxi Provincial Land Engineering Construction Group Co., Ltd., Xi'an, China;2.Technology Innovation Center for Land Engineering and Human Settlements, Shaanxi Land Engineering Construction Group Co., Ltd., and Xi'an Jiaotong University, Xi'an, China; 3.Shaanxi Provincial Land Engineering Construction Group Co., Ltd., Xi'an, China; 4.State Key Laboratory of Resources and Environmental Information System, Institute of Geographic Sciences and Natural Resources Research, Chinese Academy of Sciences, Beijing, China)

【**Abstract**】Cultivated land is an important prerequisite and guarantee for food production and security, and the change of cultivated land resources in China has always been concerned. National land survey is an effective way to accurately grasp the area and distribution of cultivated land resources. However, due to the differences in technical means and statistical standards at different stages, there are obvious breakpoints among the cultivated land area data of the three land surveys in China, which hinders the in-depth study of the spatio-temporal distribution of cultivated land resources in long-time series. The Autoregressive Integrated Moving Average (ARIMA) model is used to reconstruct and mine the cultivated land area data from 1996 to 2019 based on the data of the third land survey in China. The spatio-temporal variation characteristics of cultivated land area are explored by using Geographic Information System (GIS) spatial analysis, and the driving factors of cultivated land change are analyzed based on Geographical Detector (GeoDetector) from the perspective of social, economic, agricultural and natural. The results show that the area of cultivated land in China decreased continuously from 1996 to 2019, with a sharp decrease from 1996 to 2004 and a slow decrease from 2005 to 2019. From 1996 to 2019, there were obvious spatial differences in the change of cultivated land area in 31 provincial units. From 1996 to 2008, the cultivated land area in 29 provinces showed a downward trend, especially in the central and northern regions such as Shaanxi, Sichuan and Inner Mongolia. From 2008 to 2019, the cultivated land area in the underdeveloped areas of Heilongjiang, Jilin, Liaoning, Xinjiang, Gansu and Tibet increased significantly, while the rest showed a downward trend. Factor detection found that the q values of population, regional gross domestic product (GDP), grain output, the proportion of the added value of the primary industry and average slope were all more than 0.5, which had an important impact on the change of cultivated land area. The explanatory power of the interaction between factors on the change of cultivated land area is enhanced in different degrees compared with the single factor effect, which is manifested in the enhancement of bilinear or nonlinear enhancement, and the interaction of different factors promotes the change of cultivated land area. The change of cultivated land area is the result of complex interaction between factors, and is closely related to the land policy in the same period.

【**Keywords**】 Cultivated land; Spatio-temporal change; Driving force; ARIMA model; GeoDetector

1 Introduction

Cultivated land is an important prerequisite and guarantee for food production, and food security is an important part of national security (Liu et al., 2018; Wang et al., 2020; Yang et al., 2021). Cultivated land provides the main living guarantee for the rural population and is the main source of living materials for urban residents (Zou et al., 2020). China is a large agricultural country, and cultivated land is the main carrier to maintain population growth (Xu et al., 2021). However, China only has about 0.2 acres of cultivated land per citizen, accounting for less than 40% of cultivated land per citizen in the world (Cui and Shoemaker, 2018). With the development of social economy, the process of urbanization and industrialization in China is constantly advancing, the construction land is expanding rapidly, the cultivated land resources are constantly changing, and the protection of cultivated land is facing great challenges (Zhao et al., 2006; Kong, 2014; Wang et al., 2020). Therefore, much attention has been paid to the change of cultivated land resources in China (Anderson and Strutt, 2014; Liu et al., 2017; Tan et al., 2017). The change of cultivated land is the result of the comprehensive action of multiple factors, which is closely related to economic, social, ecological, political and other factors (Lin and Ho, 2003; Song et al., 2012; Cao et al., 2013b). Studying the spatio-temporal change law and influencing factors of cultivated land resources is conducive to comprehensively grasping the current situation of cultivated land use, revealing the driving mechanism of cultivated land change, and providing a scientific basis for rational use of land resources, policy formulation and trend prediction.

Since the 1990s, scholars have begun to study and discuss the changes of cultivated land resources from different perspectives (Deng et al., 2006; Ge et al., 2018; Ramankutty et al., 2018). Relevant research involves cultivated land change (Valbuena et al., 2010), influencing factor (Chen and Wang, 2021), influencing factor model and driving mechanism (He et al., 2005), etc. There are two main sources of cultivated land resources data: the data obtained by remote sensing image interpretation or model classification and the official land survey data. The rapid development of remote sensing technology provides long-time dynamic data for the study of cultivated land change. Currently, the widely used remote sensing land use datasets include Global Land Cover Characterization Database (GLCC) (Loveland et al., 2000)、Global Land Cover 2000 project data (GLC2000) (Bartholome and Belward, 2005)、University of Maryland land cover product (UMd) (Hansen et al., 2000)、Global Land Cover Product (GlobCover) (Arino et al., 2007) and GlobeLand30 (Jun et al., 2014). Scholars have studied the spatio-temporal change of cultivated land in China by using remote sensing data (Liu et al., 2003; Xu et al., 2017; Wang et al., 2020). Xu et al. found that from 1990 to 2010, the net increase of cultivated land in China was 1.30×10^6 hm^2 (Xu et al., 2017). The research results of Wang et al. show that from 1990 to 2000, the area of cultivated land increased by 1.62%, and then continuously decreased during 2000-2015, resulting in a national total growth rate of 0.80% from 1990 to 2015 (Wang et al., 2020). Due to the influence of the spatial-temporal resolution of remote sensing images, the phenomenon of different spectrum of the same object and the same spectrum of foreign objects, as well as the limitations of the algorithm, the accuracy of remote sensing classification in large-scale is generally not high, and there is still a certain gap with the accuracy of the official land survey data (Liu and Xia, 2010; Shao and Lunetta, 2012; Gómez-Chova et al., 2015; Cheng et al., 2020). Since 1996, China has completed three national land surveys. Based on the land survey data, the natural resources bureaus of each province will organize the land change survey every year, and the final summary will be released by the National Bureau of Statistics (NBS). In August 2021, the NBS of China released the data of the third national land survey, which is mainly based on remote sensing images or Unmanned Aerial Vehicle (UAV) images with a resolution better than 1m, combined with professional manual interpretation and field sampling verification to obtain high-precision land use data (Jiang et al., 2022). After statistical analysis, the difference between the cultivated land area of China in the commonly used remote sensing datasets and the data published in the third land survey is more than 10%. Therefore, the official land survey data can more accurately and truly reflect the status of cultivated land resources. However, there are obvious breakpoints in China's cultivated land area data in 2008 and 2018 due to differences in technical means and statistical standards adopted in different stages of the national land survey. Most scholars avoid the transition years with obvious changes in data, and focus on the changes of cultivated land area from 1996–2008 and 2009–2018 (Jin, 2014; Tan et al., 2017). At present, there are few studies on the spatio-temporal changes and driving forces of cultivated land area in the whole cycle of three land surveys. The ARIMA model does not depend on external variables and can effectively overcome the problem of insufficient model accuracy caused by external parameters. The performance of the model has been better than that of the complex structure model in short-term prediction, and has been widely used in the prediction of crop yield, climate change, economic trend and so on (Fattah et al., 2018; Singh and Mohapatra, 2019; Zheng et al., 2020).

Based on the data of the third land survey in China, the ARIMA model is used in this study to reconstruct and mine the cultivated land area data from 1996 to 2019 in order to scientifically grasp the current situation of cultivated land use in different periods. The spatio-temporal change characteristics of cultivated land in China are studied by using GIS spatial analysis. The driving factors of cultivated land change are explored from the social, economic, agricultural and natural dimensions based on GeoDetector to provide scientific support for the intelligent management and precise protection of cultivated land.

2 Materials and methods

2.1 Study area

China (Fig.1, omit) is located in eastern Asia, on the west coast of the Pacific Ocean. It starts from the center of the Heilongjiang River near Mohe River in the north and the Zengmu Reef in the Nansha Islands in the south. The land area is 9.6 million km^2, and the land boundary is more than 20000 km (Wang et al., 2021). In this study, 31 provinces, cities and autonomous regions except Taiwan Province, Hong Kong and Macao Special Administrative region were selected as the study area.

2.2 Data

Comprehensively considering the development status of China and the availability of data, and referring to previous literature (Wang et al., 2015; Arowolo and Deng, 2018), 12 indicators related to the change of cultivated land area are selected: (1) social factors: population (I_1, unit: $\times 10^4$ people); (2) Economic factors: regional GDP (II_1, unit: $\times 10^8$ yuan), per capita GDP (II_2, unit: yuan/person), proportion of added value of primary industry (II_3, unit: %), proportion of added value of secondary industry (II_4, unit: %), proportion of added value of tertiary industry (II_5, unit: %), added value of agriculture, forestry, animal husbandry and fishery (II_6, unit: $\times 10^8$ yuan); (3) Agricultural factors: grain output (III_1, unit: $\times 10^4$ tons), total power of agricultural machinery (III_2, unit: $\times 10^4$ kW); (4) Natural factors: average altitude (IV_1, unit: m), terrain relief (IV_2, unit: m); average slope (IV_3, unit: degrees). The cultivated land resources data come from the website of the Ministry of Natural Resources (http://www.mnr.gov.cn), and the data related to population and socio-economic development come from the website of the NBS of China (http://www.stats.gov.cn). Table 1 shows the official statistics of China's cultivated land area from 1996 to 2019. Table 2 shows the descriptive statistical characteristics of social, economic and agricultural factors from 1996 to 2019. Based on the 30-meter resolution SRTM DEM data (http://gdex.cr.usgs.gov/gdex), the average elevation, terrain relief and average slope of each province are calculated by Google Earth Engine cloud computing platform (Gorelick et al., 2017).

Table 1 Official statistics of cultivated land area in China from 1996 to 2019

Year	Area/×10^4 hm^2	Year	Area/×10^4 hm^2	Year	Area/×10^4 hm^2
1996	13003.92	2004	12244.43	2012	13515.85
1997	12990.31	2005	12208.27	2013	13516.34
1998	12964.21	2006	12177.59	2014	13505.73
1999	12920.55	2007	12173.52	2015	13499.87
2000	12824.31	2008	12177.68	2016	13492.10
2001	12761.58	2009	13538.46	2017	13488.12
2002	12593.00	2010	13526.83	2018	13480.00
2003	12339.22	2011	13523.86	2019	12786.19

Table 2 Descriptive statistical characteristics of social, economic and agricultural factors from 1996 to 2019

Factor	Minimum	Maximum	Mean	Standard deviation
I_1/×10^4	122389.00	141212.00	132752.64	5748.05
II_1/×10^8 yuan	71813.60	1015986.20	406730.43	315650.85
II_2/yuan/person	5898.00	72000.00	29833.60	22231.92
II_3/%	7.00	19.30	11.23	3.57
II_4/%	37.80	47.60	44.33	3.04
II_5/%	33.60	54.50	44.44	5.92
II_6/×10^8 yuan	14014.70	81103.90	37792.23	22005.20
III_1/×10^4 tons	43069.53	66949.15	55420.84	8156.62
III_2/×10^4 kW	38546.92	111728.07	78943.96	23853.29

2.3 ARIMA model

ARIMA is a time series data analysis and prediction model proposed by Box and Jenkins (Gilbert, 2005). Its main principle is to establish a corresponding model to describe or simulate its past behavior from the time series itself, so as to predict and infer the future value. The model can combine the dynamic and persistent characteristics of time series to reveal the relationship between past and present, future and present of time series (Lai and Dzombak, 2020).

The ARIMA(p, d, q) model consists of three parts: AR(p) represents the autoregressive process, that is, the current value of a time series can be expressed as a linear combination of delayed p-period observations; I(d) represents the difference, d is the number of differences required when the time series becomes stationary; MA(q) represents the moving average process, that is, the model value can be expressed as a linear function of the q-order residual term. The expression of the model is as follows:

$$x_t = \varphi_0 + \varphi_1 x_{t-1} + \cdots + \varphi_p x_{t-p} + \varepsilon_t - \theta_1 \varepsilon_{t-1} - \cdots - \theta_q \varepsilon_{t-q} (\varphi_p \neq 0, \theta_q \neq 0) \tag{1}$$

where, x_t is the actual value, φ_i and θ_j are coefficients, p is the order of the autoregressive model, q is the moving average order, ε_t represents the random error at t.

Fig.2 shows the modeling process of ARIMA model. Firstly, the stationarity of the time series data is tested. If it is a non-stationary time series, the d-order difference operation is required to convert it into a stationary time series. Then, white noise test is carried out on the data. If it is non-white noise, the best level p and order q are determined by autocorrelation function (ACF) and partial autocorrelation function (PACF) analysis. Finally, the white noise test is performed on the residual. Through the test, the modeling can be established to predict the future trend.

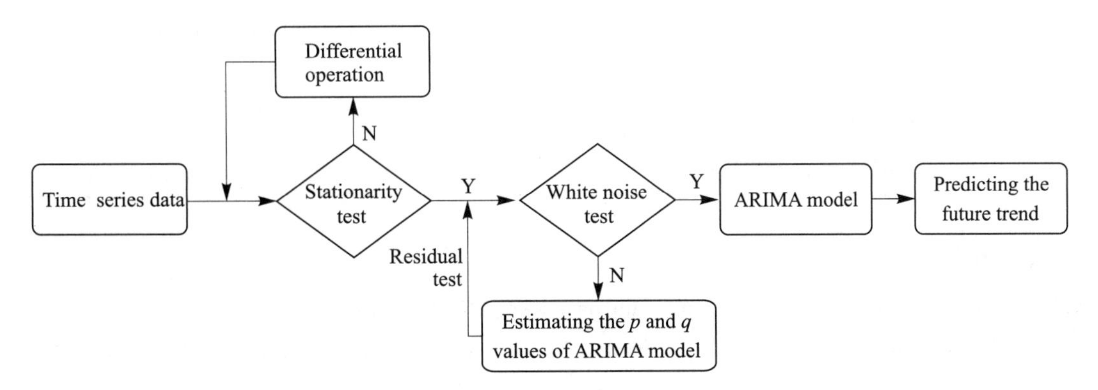

Fig. 2 Modeling process of ARIMA

2.4 GeoDetector

GeoDetector is a group of statistical methods to detect spatial differentiation and explain its driving force, including the factor detector, the interaction detector, the risk detector and the ecological detector (Wang et al., 2010). The main principle of the GeoDetector is to assume that the study is divided into several sub-regions. If the sum of the variance of the sub-region is less than the total variance of the region, there is a spatial difference; if the spatial distribution of the two variables tends to be consistent, there is a statistical correlation between the two variables. Geographic detector can evaluate spatial differentiation, detect explanatory factors and analyze the interaction between variables, and have been widely used in nature, environmental science, human health and other fields (Wang et al., 2010; Cao et al., 2013a; Liu et al., 2020).

The factor detector is used to detect the spatial differentiation of the dependent variable Y and the explanatory power of a factor X to the dependent variable Y, which is measured by the q value. The formula of q is:

$$q = 1 - \frac{\sum_{h=1}^{L} N_h \sigma_h^2}{N\sigma^2} = 1 - \frac{SSW}{SST} \qquad (h = 1,2,\cdots) \tag{2}$$

where: L is the stratification of variable Y or factor X, i.e. classification or partition; N_h and N are the number of units in layer h and the whole area respectively; $\sigma^2{}_h$ and σ^2 are the variances of the Y values of the layer h and the whole region respectively, and SSW and SST are the sum of the intra layer variance and the sum of the whole region variance respectively.

The interaction detector is used to evaluate the interaction between factors X_i and X_j, reflecting that the explanatory power of the two factors on variable Y is enhanced, weakened or independent. The interaction detector usually first calculates the interpretation force $q(X_1)$ and $q(X_2)$ of the two influence factors X_i and X_j on attribute Y, then calculates the value $q(X_1 \cap X_2)$ when they interact, and finally compares $q(X_1)$, $q(X_2)$ and $q(X_1 \cap X_2)$. There are five cases:

Table 3 Judgment basis of the interaction detector

Comparison	Interaction
$q(X_1 \cap X_2) < \mathrm{Min}[q(X_1), q(X_2)]$	Nonlinear weakening
$\mathrm{Min}[q(X_1), q(X_2)] < q(X_1 \cap X_2) < \mathrm{Max}[q(X_1), q(X_2)]$	Single factor nonlinear weakening
$q(X_1 \cap X_2) > \mathrm{Max}[q(X_1), q(X_2)]$	Bilinear enhancement
$q(X_1 \cap X_2) = q(X_1) + q(X_2)$	Independent
$q(X_1 \cap X_2) > q(X_1) + q(X_2)$	Nonlinear enhancement

3 Results

3.1 Reconstruction of cultivated land data based on ARIMA model

Fig.3(a) shows the change trend of cultivated land area in China according to official data. It can be seen from the figure that there are obvious breakpoints in China's cultivated land area data in 2008 and 2018, which is mainly due to the differences in the technical means adopted in different stages of the national land survey. 2008 is the dividing point between the first national land survey and the second national land survey, and 2018 is the dividing point between the second national land survey and the third national land survey. The technical means adopted in the first survey are backward, rely on manual operation, and the result data are stored in paper form, which is not conducive to land change investigation. However, the second survey is completed based on 3S (GIS, RS, GPS) technology, which greatly improves the accuracy and efficiency of land survey, and replenishes the previously uncounted cultivated land area due to underreporting or omission. Therefore, there was a sharp increase in the area of cultivated land in 2009. Compared with the second survey, the third survey is based on satellite images or UAV images with a resolution better than 1 m, combined with professional manual interpretation and machine learning, big data and cloud computing technology to further improve the accuracy of the data. The sharp decline in the area of cultivated land in 2019 is mainly due to the elimination of the false positives in the second survey. The two obvious breakpoints make the original cultivated land data lack of coherence and accuracy.

To study the dynamic characteristics of cultivated land area in China from 1996 to 2019, it is necessary to correct the data to improve the rationality and comparability. The cultivated land area data published in the third survey in 2019 were used as the basis for sectional correction based on the ARIMA model. Firstly, the 2009-2018 data were used to predict the 2019 data. After the augmented Dickey-Fuller (ADF) test, it is determined that the data remains stable after the second-order difference. The model was determined to be ARIMA (1,2,0) ($R^2 = 0.922$) after ACF and PACF analysis. The ratio of predicted data to actual data was determined to be 0.9488, and then the data from 2009 to 2018 were corrected. Secondly, the 1996-2008 data were used to predict the 2009 data. After analysis, ARIMA (1, 2, 0) ($R^2 = 0.908$) was selected as the model, and the ratio between the predicted data and the corrected data was determined to be 1.0543, and then the data from 1996 to 2008 were corrected. Fig.2(b) shows the change trend of cultivated land area after correction. It can be seen from the Figure that the corrected cultivated land data has stronger continuity and integrity, which is conducive to the dynamic analysis of long-time series.

3.2 Characteristics of spatio-temporal variation of cultivated land area in long-time series

As can be seen from Fig.2(b), China's cultivated land area showed a continuous decrease from 1996 to 2019. From 1996 to 2004, the cultivated land area showed a sharp decrease trend, with a total decrease

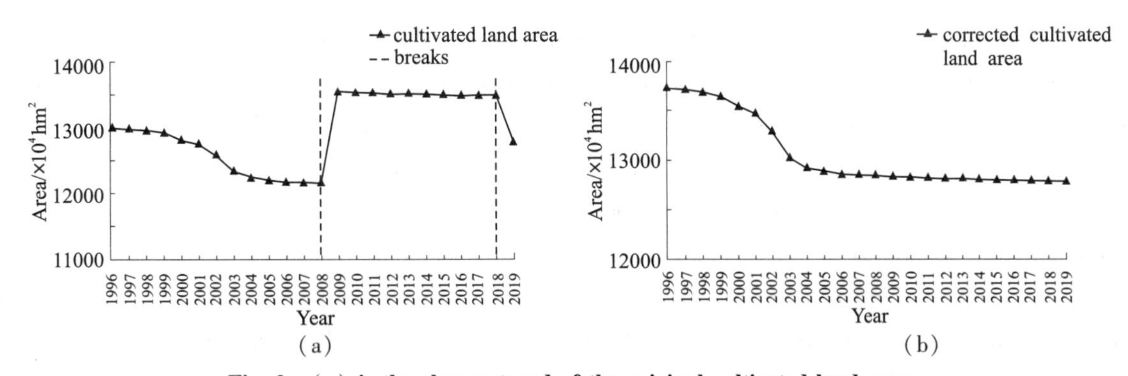

Fig. 3 (a) is the change trend of the original cultivated land area.
(b) is the change trend of cultivated land area after correction by ARIMA model

of 8.01×10^6 hm^2 and a reduction rate of 5.84%. From 2005 to 2019, the cultivated land area decreased slowly by 8.50×10^5 hm^2, with a reduction rate of 0.66%.

To further analyze the spatial distribution differences of cultivated land from the perspective of provincial units, the ARIMA model was used to reconstruct the cultivated land data of 31 provinces, cities and autonomous regions. Fig.4(omit) shows the amount and rate of change in cultivated land in China from 1996 to 2008 and 2008 to 2019. The change rate is the ratio of the difference between the cultivated land area at the end of the study and the initial stage of the study to the cultivated land area at the initial stage of the study. It can be seen from the figure that there are obvious spatial differences in the change of cultivated land area in 31 provincial units from 1996 to 2019. From 1996 to 2008, the cultivated land area of 29 provincial units showed a downward trend, accounting for more than 90%, especially in Shaanxi, Sichuan, Inner Mongolia and other central and northern regions, with a reduction of more than 1×10^7 hm^2, with a reduction rate of more than 10%. The change rate of Beijing is the highest, reaching −33%, and the change rate in most areas is maintained at −10%~0%. Only Heilongjiang and Xinjiang have seen a small increase in cultivated land area. It can be seen that the rapid progress of industrialization and urbanization has undoubtedly led to the occupation of a large amount of cultivated land. From 2008 to 2019, there are significant differences in cultivated land area changes in different regions. The cultivated land area in underdeveloped areas such as Heilongjiang, Jilin, Liaoning, Inner Mongolia, Xinjiang, Gansu and Tibet has increased significantly, especially in Xinjiang and Inner Mongolia, where the growth rate has exceeded 50%. The other regions showed a downward trend, and the decrease in central and southern provinces exceeded 1×10^7 hm^2. In developed regions such as Beijing, Shanghai, Guangdong and Zhejiang, the decline rate exceeded 30%. During this period, China began to implement the cultivated land occupation-compensation balance and cultivated land protection system. Although the decline rate of cultivated land area slowed down compared with 1996–2008, some provinces still face a severe situation of cultivated land reduction.

3.3 Driving force analysis of cultivated land area change

The factor detector results reflect the explanatory power of each factor on the change of cultivated land area in China, and the results are shown in Fig.5(a) According to the analysis, the order of explanatory power of each factor on the change of cultivated land area in China is as follows: population (I_1) > regional GDP (II_1) > grain output (III_1) > proportion of added value of the primary industry (II_3) > average slope (IV_3) > added value of agriculture, forestry, animal husbandry and fishery (II_6) > per capita GDP (II_2) > proportion of added value of the tertiary industry (II_5) > proportion of added value of the secondary industry (II_4) > average altitude (IV_1) > total power of agricultural machinery (III_2) > terrain relief (IV_2). The q values of the five factors, namely population, regional GDP, grain output, the proportion of the added value of the primary industry and average slope, exceeded 0.5, which had an important impact on the change of cultivated land area. With the continuous increase of population and the rapid rise of urbanization rate in China since 1996, the demand for residential land, living land and other construction land has increased sharply, resulting in the pressure of cultivated land being occupied to a certain extent. The regional GDP reflects the level of economic development of the region, and the increase in demand for construction land brought about by economic development is an important reason for the non-agriculturalization of cultivated land. Cultivated land is the most basic material condition of agricultural production, and the change of its quantity and quality will directly affect the grain yield. The qualitative and quantitative changes in the process of economic development are both characterized by the evolution and advancement of the industrial structure, which are mainly manifested in the continuous decline of the proportion of the primary industry and the increase in the proportion of the secondary and tertiary industries. The primary industry generally includes agriculture, forestry, fishing, animal husbandry and gathering. When the economic development enters the industrialization stage, the dominant factors of land use change are the market supply and demand of land products or services and the comparative benefits of land use. The land flows to the more efficient secondary and tertiary industries, and the agricultural land is rapidly non-agricultural. Slope is an important factor affecting the quality of cultivated land and the safety of cultivation. There is a positive correlation between slope and soil and water loss. With the increase of the slope, the runoff and scouring amount will increase accordingly. In order to control soil erosion and improve the ecological environment, China has implemented the conversion of cultivated land with large slopes to forests. Therefore, the slope has strong explanatory power to the change of cultivated land area.

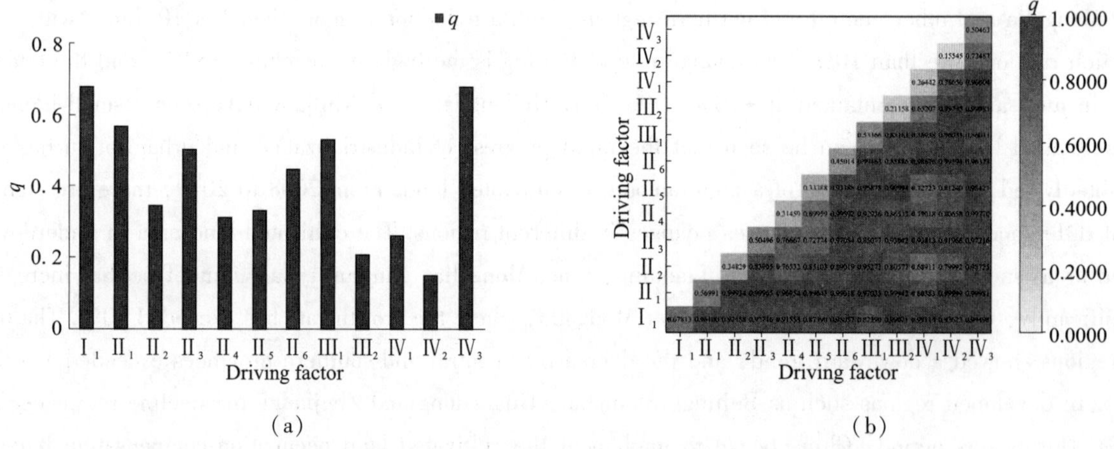

Fig.5 (a) is the result of factor detector. (b) is the result of the interaction detector

The interaction detector can reflect the interaction between different types of factors, which is helpful to further study the driving mechanism of cultivated land area change. Fig.5(b) shows the results of the interaction detector. It can be seen from the figure that the interaction detecor results between factors are both bilinear enhancement or nonlinear enhancement, and there is no independence or weakening. The q value of interaction between different types of factors are significantly larger than that of single factor. The q values of most of the interactions between factors are more than 0.8, accounting for more than 84%. The interaction q value of the average altitude and the total power of agricultural machinery is the smallest, which is close to the maximum value of the single factor q value. The interaction between regional GDP and other factors has a strong explanatory power to the change of cultivated land area, with an average value of more than 0.97. The experimental results show that the explanatory power of the interaction between factors on the change of cultivated land area is enhanced to varying degrees compared with the single factor effect, and the interaction between different factors will have varying degrees of impact on the change of cultivated land area. At the same time, it confirms that the change of cultivated land area is the result of complex interaction between factors.

4 Discussion

4.1 Analysis of driving forces in different geographical regions

China is generally divided into seven geographical regions: North China (Beijing, Tianjin, Hebei, Shanxi, Inner Mongolia), Northeast China (Liaoning, Jilin, Heilongjiang), East China (Shanghai, Jiangsu, Zhejiang, Anhui, Fujian, Jiangxi, Shandong), South China (Guangdong, Guangxi, Hainan), Central China (Henan, Hubei, Hunan), Southwest China (Chongqing, Sichuan, Guizhou, Yunnan, Tibet) and Northwest China (Shaanxi, Gansu, Qinghai, Ningxia, Xinjiang) (He et al., 2008). Fig.6 (omit) shows the amount and rate of change in cultivated land in seven geographical regions of China from 1996 to 2019. As can be seen from the figure, the area of cultivated land decreased in all areas except Northeast and North China from 1996 to 2009. The area of cultivated land in East China and Southwest China decreased greatly, both exceeding 5×10^7 hm^2. The decline rate of cultivated land area in South China is the largest, exceeding 30%. On the other hand, there has been a substantial increase in the Northeast China, with an increase rate of more than 30%.

Fig.7 (a) shows the factor detector results of cultivated land area change in seven geographical regions of China. It can be seen from the Fig.that the order of the explanatory power of each factor on the change of cultivated land area in the seven geographical regions of China is: proportion of added value of the primary industry ($Ⅱ_3$) > population ($Ⅰ_1$) > regional GDP ($Ⅱ_1$) > per capita GDP ($Ⅱ_2$) > grain output ($Ⅲ_1$) > average slope ($Ⅳ_3$) > proportion of added value of the secondary industry ($Ⅱ_4$) > added value of agriculture, forestry, animal husbandry and fishery ($Ⅱ_6$) > terrain relief ($Ⅳ_2$) > average altitude ($Ⅳ_1$) > proportion of added value of the tertiary industry ($Ⅱ_5$) > total power of agricultural machinery ($Ⅲ_2$). Except for terrain relief, average altitude, the proportion of added value of the tertiary industry and the total power of agricultural machinery, the q value of other factors exceeded 0.7, which have an important impact on the change of cultivated land area in seven geographical regions of China. The q value of the proportion of added value of the primary industry and the population are both more than 0.9, which has a strong explanatory power to the change of cultivated land area in the seven geographical areas. The q value of the total power of agricultural machinery is the lowest, only 0.289. The total power of agricultural machinery reflects the level of agricultural modernization. The progress of agricultural science and technology has improved the total power of agricultural machinery, promoted the increase of grain yield per unit area and the output rate of cultivated land, thus alleviating the pressure of cultivated land production and food security, but did not directly affect the change of cultivated land area.

Fig.7 (b) shows the interaction detector results of cultivated land area change in seven geographical regions of China. It can be seen from the figure that the interaction detecor results between factors are both bilinear enhancement or nonlinear enhancement, and there is no independence or weakening, which is consistent with the results of provincial cultivated land area. The q values of most of the interactions between factors are 1, accounting for more than 85%. The interaction q value between the proportion of added value of the tertiary industry and the terrain relief is the smallest, which is 0.6935. The average q value of interactive detection of all factors is 0.9839. The results show that the interaction between different factors has a strong explanatory power for the change of cultivated land area in seven geographical regions of China.

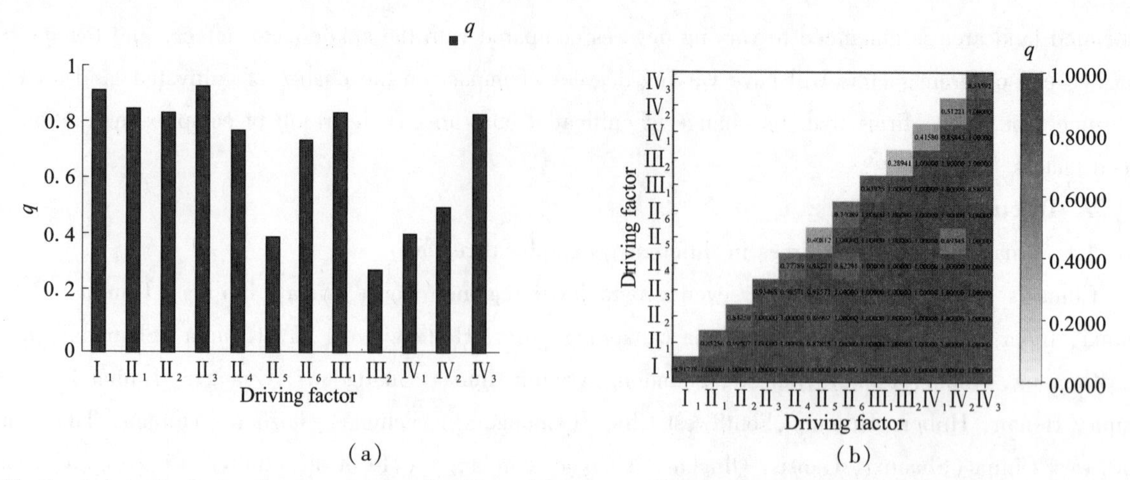

Fig.7 (a) is the factor detector results of cultivated land area change in seven geographical regions of China. (b) is the interaction detector results of cultivated land area change in seven geographical regions of China

4.2 Policy factors of cultivated land area change

Land not only has natural attribute, but also has social-economic attribute (Verburg et al., 2015). For a long time, due to the influence of social, economic, policy and technical means, it is difficult to carry out long-time series analysis of cultivated land resource changes from the perspective of land survey statistics. The existing studies on cultivated land change based on land survey statistics are mainly concentrated in the two periods of 1996–2008 and 2009–2018 (Jin, 2014; Tan et al., 2017). However, there is an obvious gap in accuracy between large-scale cultivated land area data obtained by remote sensing data and land survey data (Manandhar et al., 2009; Liu et al., 2015). In this study, the ARIMA model is used to reconstruct and mine the cultivated land area data from 1996 to 2019, which eliminates the problem of time series data fracture caused by differences in technical means, and improves the rationality and integrity of the data. GIS spatial analysis is used to explore the spatio-temporal change characteristics of cultivated land area based on the corrected data, and the GeoDetector is used to analyze the driving mechanism of cultivated land change from the perspective of single factor and factor interaction. GeoDetector has obvious advantages in explaining the spatial heterogeneity of geographical phenomena, and can make up for the weakness that conventional methods cannot explain the interaction mechanism (Hu et al., 2020; Li et al., 2021; Xiang et al., 2021).

The cultivated land area is not only affected by social, economic, agricultural and natural factors, but also closely related to national policies in the same period (Wang et al., 2012; Wang et al., 2018). Since the 1990s, China has entered a stage of rapid urbanization and industrialization (Liu et al., 2010). Driven by the national macro policy, the demand for construction land increases sharply, which leads to the occupation of a large number of cultivated land resources. Since 1999, in order to control soil erosion and improve the ecological environment, China has implemented the policy of returning farmland to forests, grasslands and lakes, resulting in a further rapid reduction in the area of cultivated land (Bi et al., 2021). In 2004, the government revised the land management law, emphasizing the need to strengthen the protection of cultivated land and implement the balance policy of occupation and compensation, which has played a positive role in the change of cultivated land area (Liu et al., 2014). Under the condition that the balance between occupation and compensation has been strictly implemented for the cultivated land occupied by non-agricultural construction, the main reasons for the reduction of cultivated land area are the adjustment of agricultural structure and land greening. In 2008, China promoted rural reform, improved the strict and standardized rural land management system, and adhered to the strictest cultivated land protection system (Liu et al., 2017). In the new urbanization and Rural Revitalization stage after 2014, China implemented the strictest cultivated land protection system and intensive and economical land use system to optimize the land use structure and improve the land use efficiency (Liu et al., 2017). The implementation of these policies has effectively slowed down the decline of cultivated land.

At present, China is in the "three peaks" period of population, industrialization and urbanization. Social development is faced with major problems, such as maintaining high-quality green economic development, ensuring ecological construction, protecting cultivated land resources and food security, and achieving the goal of carbon peak and carbon neutralization. At the same time, the international situation is complex and changeable, the epidemic situation of COVID-19 is intertwined with the war, and the food crisis in many countries has intensified. Therefore, the protection of cultivated land resources is facing unprecedented pressure. Looking forward to the future, the protection of cultivated land resources in China should turn to quantity, quality and ecology. On the basis of ensuring the red line of 0.3 billion acres of cultivated land, we should ensure the safety of China's cultivated land resources and food production by increasing land use control and balanced management of cultivated land, improving the incentive mechanism for cultivated land use, adjusting agricultural production methods, increasing investment in agricultural production, optimizing agricultural policies according to local conditions, changing land use patterns and reserving flexible cultivated land.

5 Conclusions

Based on the data of three land surveys in China, this study deeply analyzes the reasons for the obvious fracture of cultivated land area data in 2008 and 2018, and uses ARIMA model to reconstruct the cultivated land area data from 1996 to 2019. Combined with GIS spatial analysis and GeoDetector, the spatio-temporal change characteristics and driving factors of cultivated land area are analyzed. The results show that the area of cultivated land in China decreased continuously from 1996 to 2019, with a sharp decrease from 1996 to 2004 and a slow decrease from 2005 to 2019. From 1996 to 2019, there were obvious spatial differences in the change of cultivated land area of 31 provincial units. From 1996 to 2008, the cultivated land area of 29 provincial units showed a downward trend, accounting for more than 90%, especially in the central and northern regions such as Shaanxi, Sichuan and Inner Mongolia. From 2008 to 2019, the cultivated land area in the underdeveloped areas of Heilongjiang, Jilin, Liaoning, Xinjiang, Gansu and Tibet increased significantly, while the rest showed a downward trend. The factor detector found that the q value of the five factors, namely population, regional GDP, grain output, the proportion of the added value of the primary industry and average slope, exceeded 0.5, which had an important impact on the change of cultivated land area. The explanatory power of the interaction between factors to the change of cultivated land area is enhanced in varying degrees compared with that of single factor, which is characterized by bilinear enhancement or nonlinear enhancement. The proportion of interaction q greater than 0.8 is more than 84%, and the interaction of various factors promotes the change of cultivated land area. It can be inferred from the results that the change of cultivated land area is the result of complex interaction between factors, and is closely related to the land policy in the same period.

Data availability statement

The cultivated land resources data can be downloaded free of charge through t the website of the Ministry of Natural Resources (http://www.mnr.gov.cn), and population and socio-economic development data can be downloaded free of charge through the website of the National Bureau of Statistics of China (http://www.stats.gov.cn).

Author contributions

JL contributed by processing the data and wrote the main part of the manuscript. HY contributed to the research design, and gave constructive comments and suggestions. JH, YZ, YS and BP put forward detailed suggestions for the revision of the article, and assisted JL to complete all the experiments. XX and CG contributed to writing and collecting data.

Funding

This research was supported by the National Key Research and Development Program of China (No. 2019YFE0126500), National Science and Technology Major Project of China's High Resolution Earth Observation System (No. 21-Y20B01-9001-19/22), the Scientific Instrument Developing Project of the Chinese Academy of Sciences (No. YJKYYQ20200010) and Technology Innovation Center for Land Engineering and Human Settlements, Shaanxi Land Engineering Construction Group Co., Ltd and Xi'an Jiaotong University (No. 2021WHZ0090).

Acknowledgments

We are very grateful for the free data provided by China's Natural Resources Bureau and the National Bureau of Statistics.

References

Anderson, K., and Strutt, A. (2014). Food security policy options for China: lessons from other countries. Food Policy 49, 50-58.

Arino, O., Gross, D., Ranera, F., Leroy, M., Bicheron, P., Brockman, C., et al. (Year). "GlobCover: ESA service for global land cover from MERIS", in: 2007 IEEE international geoscience and remote sensing symposium: IEEE), 2412-2415.

Arowolo, A.O., and Deng, X. (2018). Land use/land cover change and statistical modelling of cultivated land change drivers in Nigeria. Regional Environmental Change 18(1), 247-259.

Bartholome, E., and Belward, A.S. (2005). GLC2000: a new approach to global land cover mapping from Earth observation data. International Journal of Remote Sensing 26(9), 1959-1977.

Bi, W., Wang, K., Weng, B., Yan, D., and Liu, S. (2021). Does the Returning Farmland to Forest Program improve the ecosystem stability of rhizosphere in winter in alpine regions? Applied Soil Ecology 165, 104011.

Cao, F., Ge, Y., and Wang, J.-F. (2013a). Optimal discretization for geographical detectors-based risk assessment. GIScience & Remote Sensing 50(1), 78-92.

Cao, Y., Bai, Z., Zhou, W., and Wang, J. (2013b). Forces Driving Changes in Cultivated Land and Management Countermeasures in the Three Gorges Reservoir Area,China. Journal of Mountain Science 10(1), 149-162.

Chen, L., and Wang, Q. (2021). Spatio-temporal evolution and influencing factors of land use in Tibetan region: 1995-2025. Earth Science Informatics 14(4), 1-12.

Cheng, G., Xie, X., Han, J., Guo, L., and Xia, G.-S. (2020). Remote sensing image scene classification meets deep learning: Challenges, methods, benchmarks, and opportunities. IEEE Journal of Selected Topics in Applied Earth Observations and Remote Sensing 13, 3735-3756.

Cui, K., and Shoemaker, S.P. (2018). "A look at food security in China". Nature Publishing Group).

Deng, X., Huang, J., Rozelle, S., and Uchida, E. (2006). Cultivated land conversion and potential agricultural productivity in China. Land use policy 23(4), 372-384.

Ge, D., Long, H., Zhang, Y., Ma, L., and Li, T. (2018). Farmland transition and its influences on grain production in China. Land Use Policy 70, 94-105.

Gilbert, K. (2005). An ARIMA supply chain model. Management Science 51(2), 305-310.

Gómez-Chova, L., Tuia, D., Moser, G., and Camps-Valls, G. (2015). Multimodal classification of remote sensing images: A review and future directions. Proceedings of the IEEE 103(9), 1560-1584.

Gorelick, N., Hancher, M., Dixon, M., Ilyushchenko, S., Thau, D., and Moore, R. (2017). Google Earth Engine: Planetary-scale geospatial analysis for everyone. Remote sensing of Environment 202, 18-27.

Hansen, M.C., DeFries, R.S., Townshend, J.R., and Sohlberg, R. (2000). Global land cover classification at 1 km spatial resolution using a classification tree approach. International journal of remote sensing 21(6-7), 1331-1364.

He, C., Li, J., Wang, Y., Shi, P., Chen, J., and Pan, Y. (2005). Understanding cultivated land dynamics and its driving forces in northern China during 1983-2001. Journal of Geographical Sciences (4), 387-395.

He, F., Ge, Q., Dai, J., and Rao, Y. (2008). Forest change of China in recent 300 years. Journal of Geographical Sciences 18(1), 59-72.

Hu, D., Meng, Q., Zhang, L., and Zhang, Y. (2020). Spatial quantitative analysis of the potential driving factors of land surface temperature in different "Centers" of polycentric cities: A case study in Tianjin, China. Science of the Total Environment 706, 135244.

Jiang, Y., Tang, Y.-T., Long, H., and Deng, W. (2022). Land consolidation: A comparative research between Europe and China. Land Use Policy 112, 105790.

Jin, T. (2014). Effects of cultivated land use on temporal-spatial variation of grain production in China. Journal of Natural Resources 29(6), 911-919.

Jun, C., Ban, Y., and Li, S. (2014). Open access to Earth land-cover map. Nature 514(7523), 434-434.

Kong, X. (2014). China must protect high-quality arable land. Nature 506(7486), 7-7.

Lai, Y., and Dzombak, D.A. (2020). Use of the autoregressive integrated moving average (ARIMA) model to forecast near-term regional temperature and precipitation. Weather and Forecasting 35(3), 959-976.

Li, D., Yang, Y., Du, G., and Huang, S. (2021). Understanding the contradiction between rural poverty and rich cultivated land resources: A case study of Heilongjiang Province in Northeast China. Land Use Policy 108, 105673.

Lin, G.C., and Ho, S.P. (2003). China's land resources and land-use change: insights from the 1996 land survey. Land use Police 20(20), 87-107.

Liu, C., Li, W., Zhu, G., Zhou, H., Yan, H., and Xue, P. (2020). Land Use/Land Cover Changes and Their Driving Factors in the Northeastern Tibetan Plateau Based on Geographical Detectors and Google Earth Engine: A Case Study in Gannan Prefecture. Remote Sensing 12(19), 3139.

Liu, D., Gong, Q., and Yang, W. (2018). The evolution of farmland protection policy and optimization path from 1978 to 2018. Chinese Rural Economy 12, 37-51.

Liu, D., and Xia, F. (2010). Assessing object-based classification: advantages and limitations. Remote sensing letters 1(4), 187-194.

Liu, J., Liu, M., Zhuang, D., Zhang, Z., and Deng, X. (2003). Study on spatial pattern of land-use change in China during 1995-2000. Science in China Series D: Earth Sciences 46(4), 373-384.

Liu, T., Liu, H., and Qi, Y. (2015). Construction land expansion and cultivated land protection in urbanizing China: Insights from national land surveys, 1996–2006. Habitat International 46, 13-22.

Liu, X., Zhao, C., and Song, W. (2017). Review of the evolution of cultivated land protection policies in the period following China's reform and liberalization. Land Use Policy 67, 660-669.

Liu, Y., Fang, F., and Li, Y. (2014). Key issues of land use in China and implications for policy making. Land Use Policy 40, 6-12.

Liu, Y., Liu, Y., Chen, Y., and Long, H. (2010). The process and driving forces of rural hollowing in China under rapid urbanization. Journal of Geographical Sciences 20(6), 876-888.

Loveland, T.R., Reed, B.C., Brown, J.F., Ohlen, D.O., Zhu, Z., Yang, L., et al. (2000). Development of a global land cover characteristics database and IGBP DISCover from 1 km AVHRR data. International journal of remote sensing 21(6-7), 1303-1330.

Manandhar, R., Odeh, I.O., and Ancev, T. (2009). Improving the accuracy of land use and land cover classification of Landsat data using post-classification enhancement. Remote Sensing 1(3), 330-344.

Ramankutty, N., Mehrabi, Z., Waha, K., Jarvis, L., Kremen, C., Herrero, M., et al. (2018). Trends in global agricultural land use: implications for environmental health and food security. Annual review of plant biology 69, 789-815.

Shao, Y., and Lunetta, R.S. (2012). Comparison of support vector machine, neural network, and CART algorithms for the land-cover classification using limited training data points. ISPRS Journal of Photogrammetry and Remote Sensing 70, 78-87.

Song, X., Ouyang, Z., Li, Y., and Li, F. (2012). Cultivated land use change in China, 1999-2007: Policy development perspectives. Journal of Geographical Sciences 22(06), 1061-1078.

Tan, Y., He, J., Yue, W., Zhang, L., and Wang, Q. (2017). Spatial pattern change of the cultivated land before and after the second national land survey in China. Journal of natural resources 32(02), 186-197.

Valbuena, D., Verburg, P.H., Bregt, A.K., and Ligtenberg, A. (2010). An agent-based approach to model land-use change at a regional scale. Landscape Ecology 25(2), 185-199.

Verburg, P.H., Crossman, N., Ellis, E.C., Heinimann, A., Hostert, P., Mertz, O., et al. (2015). Land system science and sustainable development of the earth system: A global land project perspective. Anthropocene 12, 29-41.

Wang, G., Liu, Y., Li, Y., and Chen, Y. (2015). Dynamic trends and driving forces of land use intensification of cultivated land in China. Journal of Geographical Sciences 25(1), 45-57.

Wang, J., Chen, Y., Shao, X., Zhang, Y., and Cao, Y. (2012). Land-use changes and policy dimension driving forces in China: Present, trend and future. Land use policy 29(4), 737-749.

Wang, J., Lin, Y., Glendinning, A., and Xu, Y. (2018). Land-use changes and land policies evolution in China's urbanization processes. Land use policy 75, 375-387.

Wang, J.F., Li, X.H., Christakos, G., Liao, Y.L., Zhang, T., Gu, X., et al. (2010). Geographical detectors - based health risk assessment and its application in the neural tube defects study of the Heshun Region, China. International Journal of Geographical Information Science 24(1), 107-127.

Wang, S., Xu, L., Zhuang, Q., and He, N. (2021). Investigating the spatio-temporal variability of soil organic carbon stocks in different ecosystems of China. Science of The Total Environment 758, 143644.

Wang, X., Xin, L., Tan, M., Li, X., and Wang, J. (2020). Impact of spatiotemporal change of cultivated land on food-water relations in China during 1990-2015. Science of the Total Environment 716, 137119.

Xiang, Y., Huang, C., Huang, X., Zhou, Z., and Wang, X. (2021). Seasonal variations of the dominant factors for spatial heterogeneity and time inconsistency of land surface temperature in an urban agglomeration of central China. Sustainable Cities and Society 75, 103285.

Xu, W., Jin, X., Liu, J., and Zhou, Y. (2021). Analysis of influencing factors of cultivated land fragmentation based on hierarchical linear model: A case study of Jiangsu Province, China. Land Use Policy 101, 105119.

Xu, X., Wang, L., Cai, H., Wang, L., Liu, L., and Wang, H. (2017). The influences of spatiotemporal change of cultivated land on food crop production potential in China. Food Security 9(3), 485-495.

Yang, Z., Xunhuan, L., and Yansui, L. (2021). Cultivated land protection and rational use in China. Land Use Policy 106, 105454.

Zhao, Q., Zhou, S., Wu, S., and Ren, K. (2006). Cultivated land resources and strategies for its sustainable utilization and protection in China. Acta Pedologica Sinica 43(04), 662-672.

Zou, L., Liu, Y., Yang, J., Yang, S., Wang, Y., and Hu, X. (2020). Quantitative identification and spatial analysis of land use ecological-production-living functions in rural areas on China's southeast coast. Habitat International 100, 102182.

本文曾发表于2022年《Frontiers in Environmental Science》第10卷

Ensemble Streamflow Forecasting Based on Variational Mode Decomposition and Long Short Term Memory

Xiaomei Sun[1,2,3,4], Haiou Zhang[1,2,3,4], Jian Wang[1,2,3,4],
Chendi Shi[1,2,3,4], Dongwen Hua[1,2,3,4], Juan Li[1,2,3,4]

(1.Shaanxi Provincial Land Engineering Construction Group Co., Ltd., Xi'an 710075, China;
2.Institute of Land Engineering and Technology, Shaanxi Provincial Land Engineering Construction Group Co., Ltd., Xi'an 710021, China;3.Shaanxi Provincial Land Consolidation Engineering Technology Research Center, Xi'an 710075, China;4.Key Laboratory of Degraded and Unused Land Consolidation Engineering, Ministry of Natural Resources, Xi'an 710075, China)

【**Abstract**】Reliable and accurate streamflow forecasting plays a vital role in the optimal management of water resources. To improve the stability and accuracy of streamflow forecasting, a hybrid decomposition-ensemble model named VMD-LSTM-GBRT, which is sensitive to sampling, noise and long historical changes of streamflow, was established. The variational mode decomposition (VMD) algorithm was first applied to extract features, which were then learned by several long short-term memory (LSTM) networks. Simultaneously, an ensemble tree, a gradient boosting tree for regression (GBRT), was trained to model the relationships between the extracted features and the original streamflow. The outputs of these LSTMs were finally reconstructed by the GBRT model to obtain the forecasting streamflow results. A historical daily streamflow series (from 1/1/1997 to 31/12/2014) for Yangxian station, Han River, China, was investigated by the proposed model. VMD-LSTM-GBRT was compared with respect to three aspects: (1) Feature extraction algorithm; ensemble empirical mode decomposition (EEMD) was used. (2) Feature learning techniques; deep neural networks (DNNs) and support vector machines for regression (SVRs) were exploited. (3) Ensemble strategy; the summation strategy was used. The results indicate that the VMD-LSTM-GBRT model overwhelms all other peer models in terms of the root mean square error (RMSE = 36.3692), determination coefficient ($R^2 = 0.9890$), mean absolute error (MAE = 9.5246) and peak percentage threshold statistics (PPTS(5) = 0.0391%). The addressed approach based on the memory of long historical changes with deep feature representations had good stability and high prediction precision.

【**Keywords**】Streamflow forecasting; Variational mode decomposition; Long short-term memory; Ensemble tree

1 Introduction

Streamflow forecasting is of great significance for the optimal management and effective operation of a water resources system. Therefore, it has been investigated by many researchers, and numerous forecasting models have been developed in the past decades. Among these models, forecasting techniques based on statistical modeling, data-driven models, seem to be in fashion for their simplicity and robustness[1-3].Regularization has played an important role in forecasting[4-6].

These data-driven models can be classified into time series models and artificial intelligence (AI) models. Many previous researchers have applied time-series models to forecast streamflow, including autoregression (AR), moving average (MA), autoregressive moving average (ARMA), and autoregressive integrated moving average (ARIMA) models[7-9]. However, due to the linear hypothesis of these models, they are not suited to forecast streamflow with non-linear and non-stationary characteristics. Therefore, AI models that have the ability of non-linear mapping are applied to streamflow forecasting, i.e., support vector machines for regression (SVRs)[1,10], fuzzy inference systems (FIS)[11-12], Bayesian regression (BR)[13-14], and artificial neural networks (ANNs)[15-16]. However, most of the AI models, which belong to the "shallow" learning category, cannot sufficiently represent instinctual information[17]. The deep learning models, e.g., the deep belief network (DBN) and recurrent neural networks (RNNs), can overcome this drawback due to their deeper representation ability[17]. However, these deep learning models completely rely on historical observed data, and some of the earlier changes of streamflow may or may not influence future streamflow. It is entirely possible for the gap between the streamflow information from further back in time and the current point where it is needed to become large. Therefore, using a deep learning model that can automatically "remember" or "forget" previous information should be able to enhance the accuracy of streamflow forecasting. Fortunately, LSTM[18], one of the deep learning models, has the ability to tackle this task. LSTM has been successfully used in some fields, e.g., accident diagnosis[19], electricity price prediction[20], water table depth forecasting[21], and others.

Unfortunately, due to the complicated non-linearity, extreme irregularity and multiscale variability of natural streamflow, the models directly built on original streamflow cannot appropriately identify streamflow change patterns[1]. For this reason, the processes of transformation, data pre-processing and feature extraction have attracted the attention of many researchers. In addition, feature extraction can efficiently improve the capability of these models[17,1,22-23]. Huang et al. (2014)[1] applied a modified empirical model decomposition (EMD) method to remove the most nonlinear and disorderly noise from the original series and then established one SVR-based model that computed a summation of all prediction results of all sub-series as an ensemble result. Wang et al. (2015)[24] used the EEMD technique to develop insight into the runoff characteristics and forecast each characteristic by the ARIMA model; the forecast results were summed to formulate an ensemble forecast for the raw runoff. Bai et al. (2016)[17] used EEMD to extract multi-scale features and reconstructed three deep neural networks (DNNs) by a summing strategy to forecast reservoir inflow. Yu et al. (2018)[10] exploited both Fourier transform (FT) and SVR models to extract and learn features and to forecast monthly reservoir inflow by adding all feature learning results. Obviously, feature representation of original data can contribute to the performance improvement of streamflow forecasting because of the advantage of feature extraction, which removes noise components and detect the hidden structures inside the raw time-series.

However, some recent commonly-used feature extraction methods, e.g., EMD, EEMD, and wavelet transforms (WT), suffer from drawbacks with respect to different aspects of actual signal decomposition. For example, EMD has limitations of sensitivity to noise and sampling[25], EEMD is not theoretically well-founded[26], and the effectiveness of WT heavily relies on the choice of the basic wavelet function[27]. Recently, a theoretically well-founded and robust VMD model[25] has been successfully applied to container throughput forecasting[28], vibro-acoustic feature extraction[29], chatter detection in milling processors[30], and other applications. This model is much more sensitive to noise and sampling than existing decomposition algorithms, such as EMD and EEMD[25].

Moreover, the ensemble strategy plays a vital role for integrating feature forecasting results to predict original streamflow. A straightforward and frequently used ensemble technique is summation, although it may cause error accumulation problems due to summation errors of the sub-results of streamflow prediction. Sometimes, the gap between the summation of extracted features and the original values may not be small.

Even a model that can achieve great performance in the prediction of sub-features may still not be able to accurately forecast the original time series. Therefore, building another supervised model[3], such as GBRT, for ensembles seems to be a good choice to avoid an accumulation of errors and obtain better performance.

Based on the above outline, this study addresses a decomposition-ensemble-based multiscale feature deep learning method to forecast streamflow. Our goal is to plug a memory framework into the process of deep feature learning that is robust to noise and sampling as well as long historical changes of streamflow, integrate an ensemble tree model with the capability to remove impacts caused by error accumulation and model the relationship between decomposition results and the original series to exploit the sophisticated nature of streamflow with a long history. To this end, VMD was used to extract smooth features, LSTM was applied to learn features sensitive to long historical streamflow changes, and GBRT was used to obtain an ensemble model to forecast the streamflow. This approach was evaluated by observed daily streamflow of the Yangxian station, Han River, China.

2 Methodologies

2.1 Variational mode decomposition (VMD)

The VMD algorithm, an entirely non-recursive variational mode decomposition model proposed by Dragomiretskiy and Zosso[25], is used to concurrently decompose a sophisticated signal into several band-limited intrinsic modes.

The VMD model assumes each mode u_k to be mainly compact around a center pulsation ω_k calculated with the decomposition. The following scheme proposed by Dragomiretskiy and Zosso (2014)[25] is applied to assess the bandwidth of a mode. The related analytic signal of each mode u_k is first computed by the Hilbert transform to acquire a unilateral frequency spectrum. Then, the frequency spectrum of each model is shifted to "baseband" by mixing with an exponential tuned to the respective evaluated center frequency. The bandwidth is finally assessed by the H^1 Gaussian smoothness of the demodulated signal, i.e., the squared L^2 -norm of the gradient[25]. To solve the decomposition problem of time series f, the constrained variational problem can be equivalently solved by the following equation:

$$\begin{cases} \min\limits_{\{u_k\},\{\omega_k\}} \left\{ \sum\limits_k \left\| \partial_t \left[\left(\delta(t) + \frac{j}{\pi t} \right) \cdot u_k(t) \right] e^{-j\omega_k t} \right\|_2^2 \right\} \\ s.t. \quad \sum\limits_k u_k(t) = f(t) \end{cases} \tag{1}$$

where $\{u_k\} := \{u_1, \cdots, u_k\}$ and $\{\omega_k\} := \{\omega_1, \cdots, \omega_k\}$ denote the set of modes and their center frequencies, respectively. To solve this variational problem, a Lagrangian multiplier λ and a quadratic penalty term are introduced to render the problem unconstrained. The augmented Lagrangian ℓ is defined as follows:

$$\begin{aligned} \ell(\{u_k\},\{\omega_k\},\lambda) := \alpha \sum_k \left\| \partial_t \left[\left(\delta(t) + \frac{j}{\pi t} \right) \cdot u_k(t) \right] e^{-j\omega_k t} \right\|_2^2 + \left\| f(t) - \sum_k u_k(t) \right\|_2^2 + \\ \left\langle \lambda(t), f(t) - \sum_k u_k(t) \right\rangle \end{aligned} \tag{2}$$

in which α indicates the balancing parameter of the data-fidelity constraint. In VMD, the Alternate Direction Method of Multipliers (ANMM) is used to solve Eq. (2). Eq. (3) is used to update the mode $u_k(\omega)$ in the frequency domain. The center frequencies ω_k are updated by Eq. (4), and λ is simultaneously updated by Eq. (5). In the time domain, the mode $u_k(t)$ can be obtained as the real part of the inverse Fourier transform of $u_k(\omega)$ expressed by Eq.(3).

$$\hat{u}_k^{n+1}(\omega) = \frac{\hat{f}(\omega) - \sum_{i \neq k} \hat{u}_i + \frac{\hat{\lambda}(w)}{2}}{1 + 2\alpha(\omega - \omega_k)^2} \tag{3}$$

$$\hat{\omega}_k^{n+1} = \frac{\int_0^{\infty} \omega \, |\hat{u}_k(\omega)|^2 \mathrm{d}\omega}{\int_0^{\infty} |\hat{u}_k(\omega)|^2 \mathrm{d}\omega} \tag{4}$$

$$\hat{\lambda}^{n+1}(\omega) = \hat{\lambda}^n(\omega) + \tau\left[\hat{f}(\omega) - \sum_k \hat{u}_k^{n+1}(\omega)\right] \tag{5}$$

The implementation process of the VMD model is summarized as Algorithm 1.

Algorithm 1 The process of VMD

Initialize $\{\hat{u}_k^1\}$, $\{\hat{\omega}_k^1\}$, $\hat{\lambda}^1$, $n \leftarrow 0$

Repeat

$n \leftarrow n+1$

for $k = 1:K$ do

Update $\hat{u}_k$ for all $\omega \geqslant 0$ using Eq. (3).

Update $\hat{\omega}_k$ using Eq. (4).

end for

Update $\hat{\lambda}^n(\omega)$ for all $\omega \geqslant 0$ using Eq. (5)

until convergence: $\sum_k \| \hat{u}_k^{n+1} - \hat{u}_k^n \|_2^2 / \| \hat{u}_k^n \|_2^2 < \varepsilon$.

Obtain $u_k^{n+1}(t)$ by the fast Fourier transform of $\hat{u}_k^{n+1}(\omega)$.

2.2 Long short-term memory (LSTM)

Long Short-Term Memory Networks (LSTMs) are a very specific kind of Recurrent Neural Networks (RNNs) for modeling sequential data. Therefore, it is essential to first introduce a normal version of an RNN. RNNs have chain-like structures of repeating modules that produce an output at each time step and have recurrent connections from the output at one time step to the hidden units at the next time step, illustrated in Fig. 1(a). The chain-like structure with self-connected hidden units can help RNNs to "remember" the previous information, which allows the RNNs to build a model for an arbitrarily-long time sequence.

The forward propagation algorithm is used to calculate the output for the RNN pictured in Fig.1(a). Begin with a specification of the initial state $h_0 = 0$ for each time step from $t=1$ to $t=\tau$; the following equations are used[31]:

$$h_t = \tanh(W h_{t-1} + U X_t + b_h) \tag{6}$$

$$o_t = V h_t + b_o \tag{7}$$

where the parameters are the bias vectors b_h and b_o along with the weight matrices U, V and W for input-to-hidden, hidden-to-output and hidden-to-hidden connections, respectively. X_t represents the input vector at time step t and h_{t-1} denotes the hidden cell state at time step $t-1$.

Back-Propagation Through Time (BPTT) is used to compute the gradients of the RNNs[32]. However, owing to the gradient vanishing or exploding problem, it is difficult and inefficient for BPTT to learn long-term dependencies in RNNs[33-34]. LSTMs are explicitly designed by Hochreiter and Schmidhuber (1997)[18] to avoid this long-term dependency problem. LSTMs also have chain-like repeating modules, although with complicated structures. Each repeating module of LSTMs includes a memory block called a "cell". This memory block help LSTMs to store or remove information over a long duration.

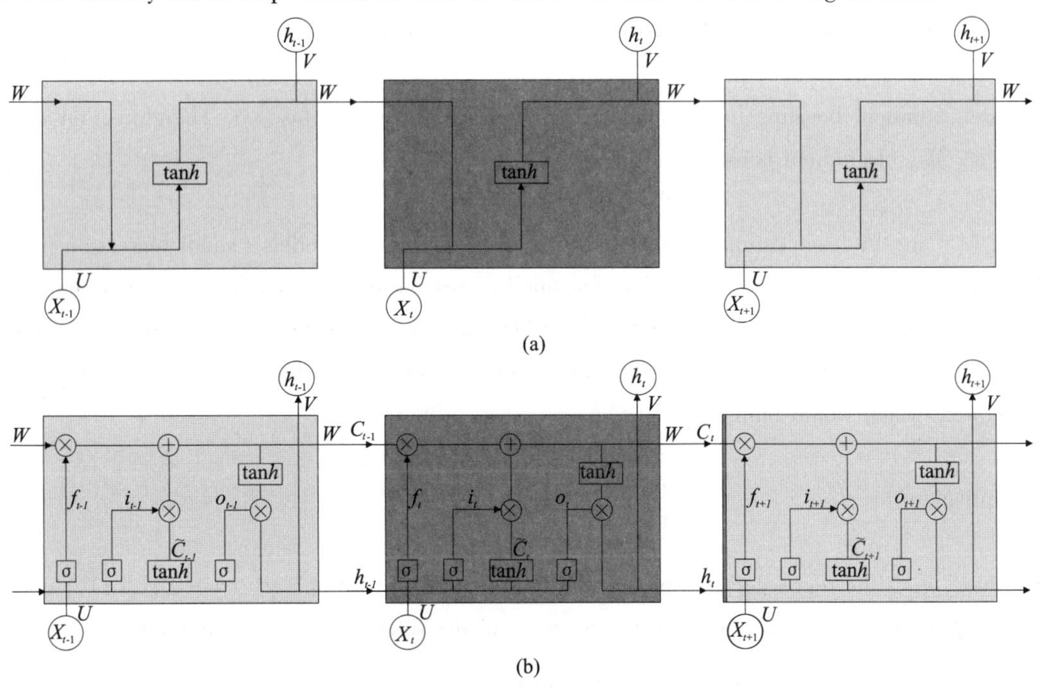

Fig.1 (a) Chain-like structure of the RNN. Because of the connections between hidden units, information can be passed from one time step to the next. (b) A graphical representation of the LSTM recurrent network with the memory cell block

The LSTM memory block diagram is illustrated in Fig. 1 (b). The LSTM memory block contains four parts, a cell state in addition to three special structures called gates. The horizontal line running through the top of the diagram is the cell state, which runs straight down the entire chain without any activation function; it is very easy for information to just flow along it unchanged. Therefore, the gradient does not vanish or explode when training an LSTM by BPTT. Moreover, the LSTM does have the ability to add or remove information to the cell state, regulated by the input, forget and output gates. Each gate is composed of a sigmoid unit and a pointwise multiplication operation, which can optionally pass information.

The corresponding forward propagation equations of LSTM are expressed for time steps from $t = 1$ to $t = \tau$ with initial state $C_0 = 0$ and $H_0 = 0$ as:

$$i_t = \sigma(W_i X_t + U_i h_{t-1} + b_i) \tag{8}$$

$$f_t = \sigma(W_f X_t + U_f h_{t-1} + b_f) \tag{9}$$

$$o_t = \sigma(W_o X_t + U_o h_{t-1} + b_o) \tag{10}$$

$$\widetilde{C}_t = \tanh(W_c X_t + U_c h_{t-1} + b_c) \tag{11}$$

$$C_t = f_t \otimes C_{t-1} + i_t \otimes \widetilde{C}_t \tag{12}$$

$$h_t = o_t \cdot \tanh(C_t) \tag{13}$$

where W, U and b are input weights, recurrent weights and biases, respectively, and the subscripts i, f and o represent the input, forget and output gates, respectively. The activation function *logistic sigmoid* is indicated by σ . i_t , f_t , o_t and C_t are the input, forget, output gates and cell state vectors at time step t , respectively. h_t is the output vector of the memory cell block, and $\otimes$ denotes element-wise multiplication.

2.3 Gradient boosting regression trees (GBRTs)

Gradient boosting is a powerful machine learning strategy to efficiently produce highly robust, competitive, interpretable procedures for both regression and classification[35,36]. The key to boosting is to combine the output of many weak prediction models ("learners"), typically decision trees, into a single strong ensemble model. Gradient boosting builds models in a forward stage-wise fashion. Therefore, for each stage m , $1 \leqslant m \leqslant M$, of gradient boosting:

$$F_m(x) = F_{m-1}(x) + h_m(x) \tag{14}$$

in which $h_m(x)$ are the basic estimators referred to as weak prediction models (small regression trees in the case of GBRT) and $F_m(x)$ is the summation of m small regression trees for GBRT. For iterations from $m = 1$ to $m = M$, the GBRT algorithm improves on F_m by adding a new regression tree h_m to its predecessor to provide a better model. Simultaneously, the procedure estimates the target value y based on the perfect h_m from the training set, which would imply:

$$F_m(x) = F_{m-1}(x) + h_m(x) = y \tag{15}$$

which is equivalent to

$$h_m(x) = y - F_{m-1}(x) \tag{16}$$

Therefore, h_m is the regression tree model that fits the current residuals $\gamma_m = y - F_{m-1}(x)$, and the residuals $y - F_{m-1}(x)$ for a given model are the negative of the squared error loss function, i.e.:

$$-\frac{\partial \frac{1}{2}[y - F_{m-1}(x)]^2}{\partial F_{m-1}(x)} = y - F_{m-1}(x) \tag{17}$$

Gradient boosting is thus a gradient descent algorithm obviously proved by Eq. (17), and generalizing it entails substituting the squared error with a different loss function and its gradient. For a more detailed description, see Friedman (2001)[35] and Hastie et al (2009)[37].

Moreover, the implicit idea behind gradient boosting is to apply a steepest-descent step to minimize the loss values between the response values and estimates to find an optimal approximation $\hat{F}(x)$. Therefore, for a training set $\{(x_1, y_1), \cdots, (x_n, y_n)\}$, the ensemble model would be updated in accordance with the following equations[35,37]:

$$F_m(x) = F_{m-1}(x) - \gamma_m \sum_{i=1}^{n} \frac{\partial L(y_i, F_{m-1}(x_i))}{\partial F_{m-1}(x_i)} \tag{18}$$

$$\gamma_m = \arg\min_{\gamma} \sum_{i=1}^{n} L\left(y_i, F_{m-1}(x_i) - \gamma \frac{\partial L(y_i, F_{m-1}(x_i))}{\partial F_{m-1}(x_i)}\right) \tag{19}$$

where the derivatives are obtained with respect to the function F_i for $i \in \{1, 2, \cdots, m\}$. In the m -th iteration of the GBRT model, the gradient boosting algorithm fits a regression tree $h_m(x)$ to the pseudo-residuals. Let J_m be the number of tree leaves; the regression tree splits the input space into J_m disjoint regions $R_{1m}, \cdots$, $R_{J_m m}$ and obtains a constant value for each region. The output of $h_m(x)$ for input x can thus be written as the sum[35] (Friedman, 2001):

$$h_m(x) = \sum_{j=1}^{J_m} b_{jm} 1_{R_{jm}}(x), (x \in R_{jm}) \tag{20}$$

where b_{jm} is the constant value predicted for the region R_{jm} and $1(\cdot)$ is an indicator function that has the value 1 if its argument is true and zero otherwise. Then, each coefficient b_{jm} is multiplied by an optimal value

γ_{jm} [35] (Friedman, 2001), and the model is then updated by the following rules:

$$F_m(x) = F_{m-1}(x) + \sum_{j=1}^{J_m} \gamma_{jm} 1_{R_{jm}}(x), (x \in R_{jm}) \tag{21}$$

$$\gamma_{jm} = \arg\min_{\gamma} \sum_{i=1}^{n} L(y_i, F_{m-1}(x_i) + \gamma) \tag{22}$$

The implementation process of the generic gradient boosting tree is summarized as Algorithm 2.

Algorithm 2　The process of GBRT

Input the loss function $L[y, F(x)]$, iteration number M and training set $\{(x_i, y_i)\}_{i=1}^{n}$.

1. Initialize model with a constant value:

$$F_0(x) = \arg\min_{\gamma} \sum_{i=1}^{n} L(y_i, \gamma)$$

2. For $m = 1$ to M:

2.1. For $i = 1, 2, \cdots, n$, compute pseudo-residuals:

$$\gamma_{im} = -\left[\frac{\partial L(y_i, F(x_i))}{\partial F(x_i)}\right]_{F(x) = F_{m-1}(x)}$$

2.2. Calculate the multiplier γ_{jm} by solving the following problem:

$$\gamma_{jm} = \arg\min_{\gamma} \sum_{i=1}^{n} L(y_i, F_{m-1}(x_i) + \gamma)$$

2.3. Fit a weak learner $h_m(x)$ to the pseudo-residuals using the training set $\{(x_i, y_i)\}_{i=1}^{n}$.

$$h_m(x) = \sum_{j=1}^{J_m} \gamma_{jm} 1_{R_{jm}}(x), (x \in R_{jm})$$

2.4. Update the model:

$$F_m(x) = F_{m-1}(x) + h_m(x)$$

3. Output $F_M(x)$.

2.4　The decomposition-ensemble model VMD-LSTM-GBRT

After discussing each key constituent separately, the approach of the proposed model VMD-LSTM-GBRT can be concluded as follows and is diagrammed in Fig. 2.

Step 1. Collect raw daily streamflow data $X = \{x_1, x_2, \cdots, x_N\}$.

Step 2. Use VMD to decompose the raw series X into several components.

Step 3. Plot the partial autocorrelation coefficient Figure of each component obtained in step 2 to select optimal numbers of inputs for it. Divide each of the components into three sub-sets: the training set (80%) for training multiple LSTM structures, the development set (10%) for searching optimal structure, and the test set (10%) for validating the ensemble model VMD-LSTM-GBRT.

Step 4. Given the test set, predict each component based on the optimal LSTM structure of each mode obtained in step 3.

Step 5. Build the ensemble tree model GBRT using the components obtained in step 2 as input and the original series obtained in step 1 as output. Use GBRT to reconstruct the predictions given by step 4.

Step 6. Output the forecasting streamflow results and perform error analysis.

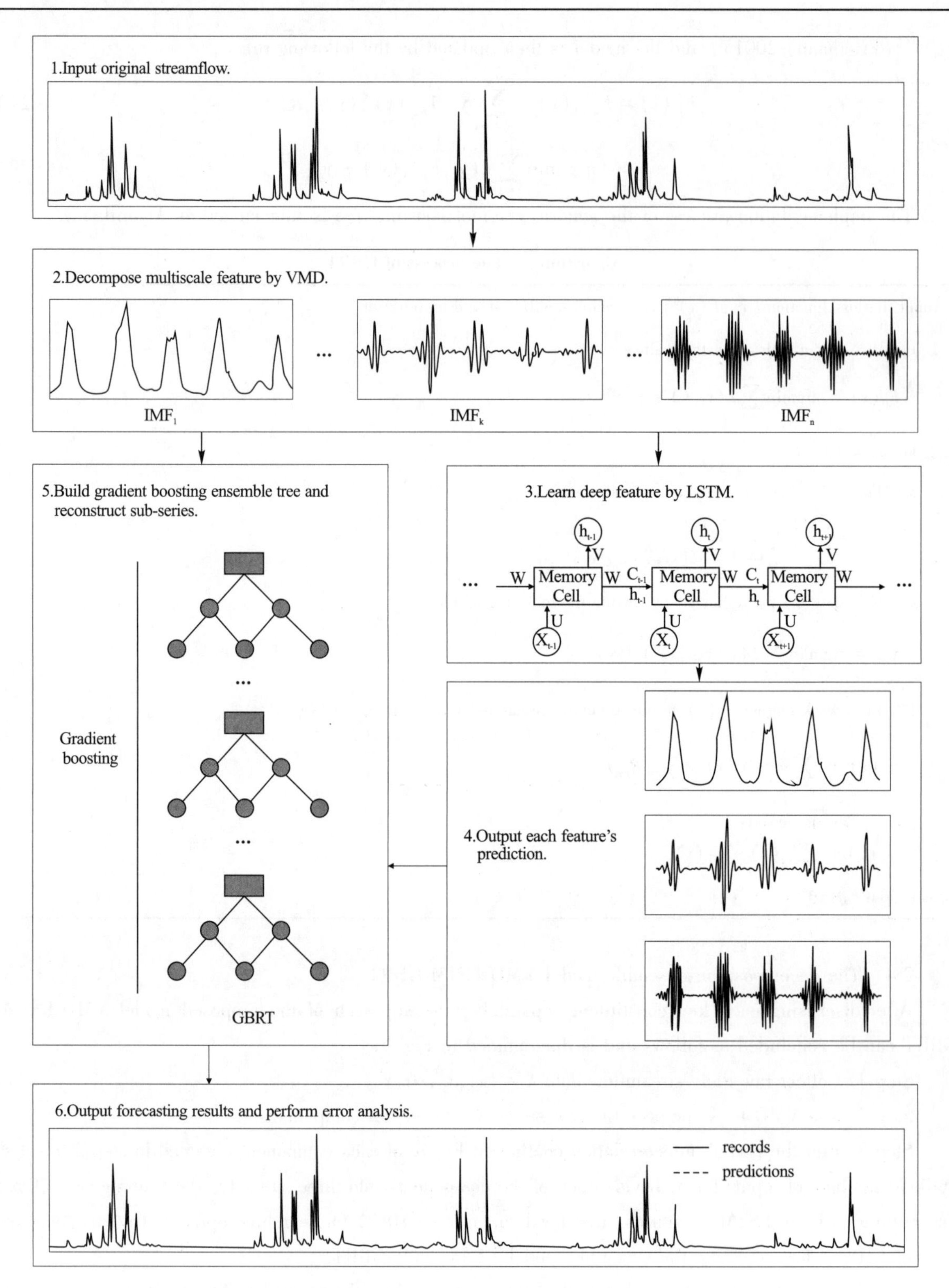

Fig. 2 Application of the proposed model VMD-LSTM-GBRT

3 Case study

3.1 Study site and observational data

In this paper, historical daily streamflow of the Yangxian hydrological station on the upstream of Han River were collected to assess the proposed model. The location of Yangxian station is illustrated in Fig. 3. The Han River, the biggest tributary of the Yangtze River, lies within 30°~34.5°N and 107°~114.5°E in the middle of China and has a total drainage area of 151000 km^2. The location of this basin is in a subtropical monsoon zone that has a humid climate and differentiated seasons. The water resources in this area are rich, with an annual average precipitation of more than 900 mm/a. The precipitation in the rainy season (July to September) accounts for 75% of the annual total, and the runoff has a similar seasonality. The study area of this paper is the upper source area of the Han River, which lies between BoZhong Mountain located in the south of the Qinling Mountains and the Yangxian hydrological station. The drainage area controlled by the Yangxian hydrological stations is 14484 km^2. As the main stream control station, forecasting the daily runoff of the Han River at Yangxian hydrological station can evaluate the short-term water production at the source of this basin.

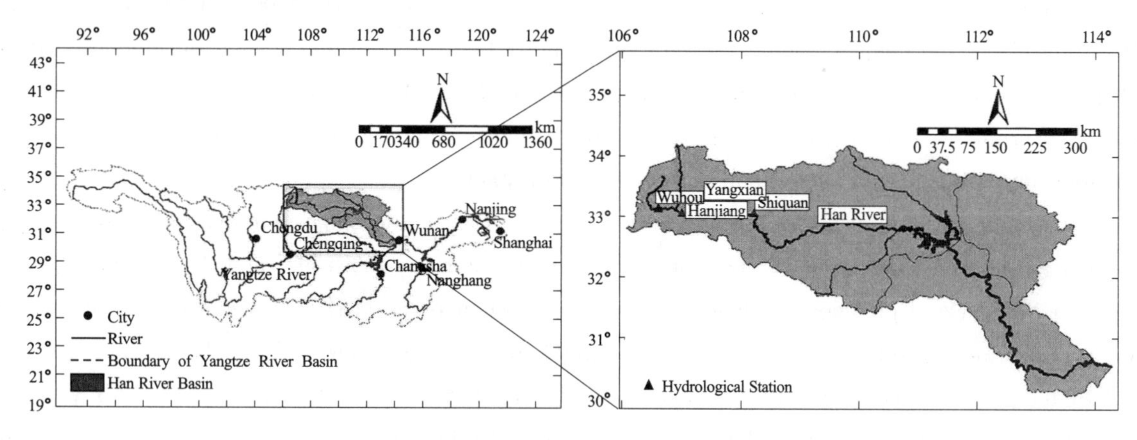

Fig. 3 **Location of the Yangxian hydrological station**

As shown in Fig. 4, daily streamflow records (total of 6574 samples) of the Yangxian hydrological station from 1/1/1997 to 31/12/2014 were used to develop the present model. For simplicity, the observation dates on the horizontal axis have been replaced with series numbers. These records were collected from the hydrological information datacenter of Shaanxi Hydrographic and the Water Resources Survey Bureau. The instantaneous value (m^3/s) observed at 8 a.m. was selected as the average daily streamflow.

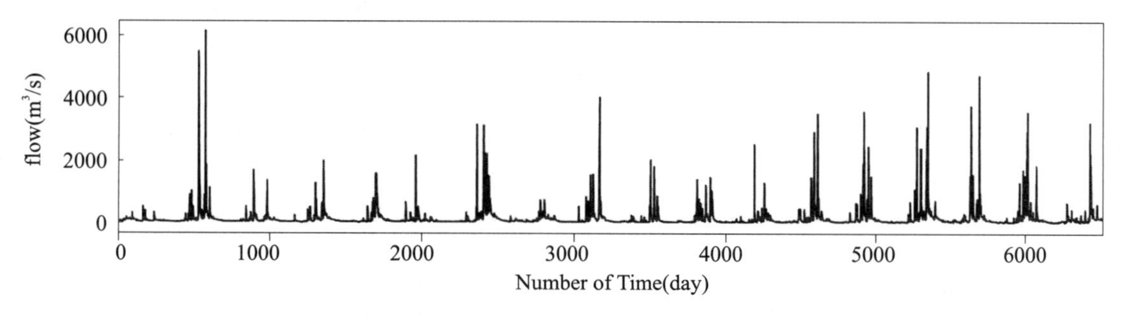

Fig. 4 **Daily Streamflow of the Yangxian station from** 1/1/1997 **to** 31/12/2014

3.2 Data pre-processing

Since the range of the streamflow time series and its decomposition sequences vary widely, in some cases of building machine learning models, the optimization algorithms applied for the loss function will not work well without feature normalization. Therefore, all the variables used for the model developed in the present study were normalized to the same scale. This pre-processing strategy can ensure that the optimization algorithm converges much faster than without normalization[38]. The normalization formula is as follows:

$$X_{normalized} = 2 \times \frac{X - X_{min}}{X_{max} - X_{min}} - 1 \tag{23}$$

where $X_{normalized}$ is the normalized vector and X is the raw vector. X_{max} and X_{min} are the maximum value and minimum value of X, respectively, and $X_{normalized}$ is calculated by element-wise mathematical operations. Once we have finished the simulation, the predictions can be re-normalized following the inverse procedure of Eq. (23).

3.3 Model evaluation criteria

The hidden layer is set with 1, 2, 3, 4 and 5; The learning rate is set to 0.001, 0.003, 0.007, 0.01, 0.03, 0.07, 0.1; The number of hidden layer neurons is set to 1~25, and the other parameters use the default parameters used by TensorFlow (the activation function of each layer is Rectified Linear Unit, the optimization algorithm is Adam algorithm, the loss function is mean squared error, the kernel initializer is Xavier uniform initialization, the bias initializer is zero initialization).

To evaluate the performance of the proposed model based on the decomposition-ensemble strategy, four error analysis criteria were applied. The expression of these criteria is shown in Table 1. The RMSE evaluates the performance of predicting high streamflow values, whereas the MAE accesses the average performance of the entire data set. The coefficient of determination R^2 indicates how well the observations are replicated by the proposed model. The peak percentage of threshold statistics, PPTS, denotes the ability to forecast peak flow[17,39]. The lower the PPTS, the better the capability to forecast peak flow. Note that the records are arranged in descending order to compute the PPTS and that the threshold level γ denotes the percentage of bottom data removed from this order; the parameter G is the number of top data at the threshold level γ.

Table 1 Formulas for error analysis criteria

Error analysis criteria	Definition
Root mean square error	$RMSE = \sqrt{\frac{\sum_{t=1}^{N} [x(t) - \hat{x}(t)]^2}{N}}$
Mean absolute error	$MAE = \frac{\sum_{t=1}^{N} \lvert x(t) - \hat{x}(t) \rvert}{N}$
Determination coefficient	$R^2 = 1 - \frac{\sum_{t=1}^{N} [x(t) - \hat{x}(t)]^2}{\sum_{t=1}^{N} [x(t) - \bar{x}(t)]^2}$
Peak percentage threshold statistics (%)	$PPTS(\gamma) = \frac{1}{100 - \gamma} \frac{1}{N} \sum_{t=1}^{G} \left\lvert \frac{x(t) - \hat{x}(t)}{x(t)} \times 100 \right\rvert$

Note: N is the number of samples, $x(t)$ is the original series, $\bar{x}(t)$ is the average of the original series and $\hat{x}(t)$ is the predicted series.

4 Results and discussion

4.1 Data decomposition with VMD

As mentioned in Section 2.4, when building the decomposition-ensemble based model, we first decomposed the raw daily streamflow data of the Yangxian hydrological station via VMD. The raw series and its decomposition results and the frequency spectra are shown in Fig. 5. However, it is hard to tell how many components the original series should be decomposed into. Too few components may not properly extract features inside raw data, whereas too many may be computationally expensive for training the model. By experiments, we found that the optimal decomposition mode number can be determined by the obvious aliasing phenomenon of the center frequency for the last component. It was first found in this study that when $k=10$, the frequency spectrum of the 10th mode had obvious aliasing phenomena (area surrounded by a red rectangular border shown in Fig. 6). To make the decomposition result satisfy orthogonality and avoid the spurious components as much as possible, the number of components was chosen to be 9.

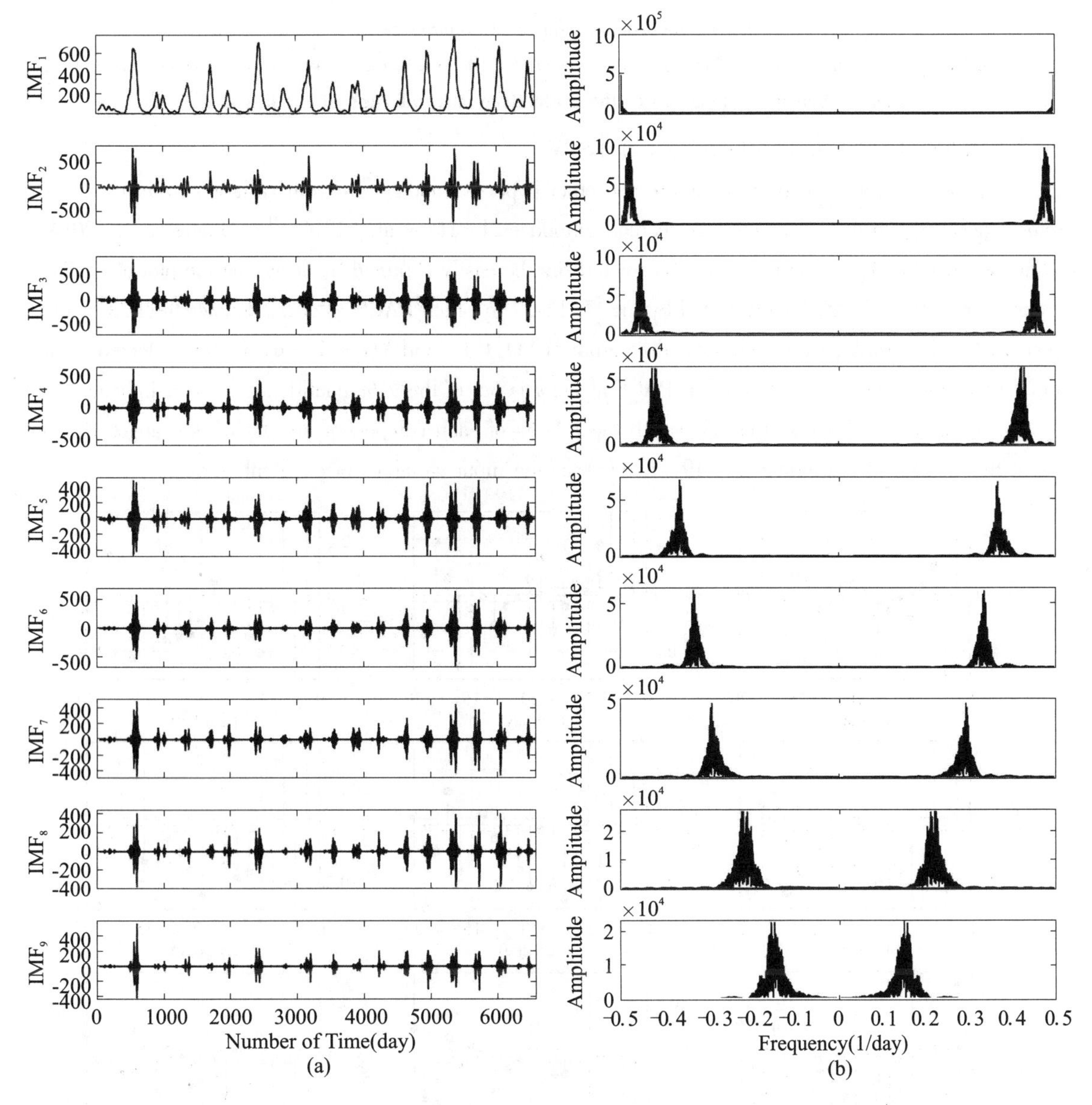

(a) the decomposition sequence waveform and (b) the frequency spectrum representation.

Fig. 5 VMD decomposition results

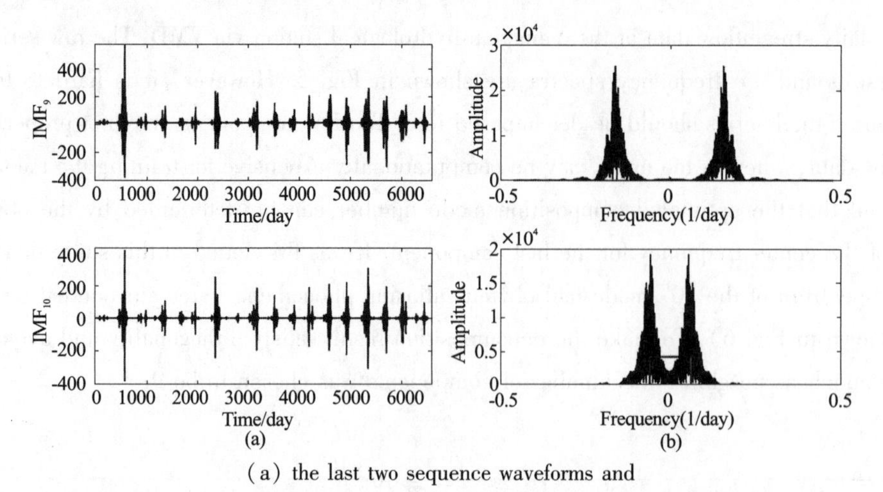

(a) the last two sequence waveforms and

(b) the frequency spectrum representations. The area surrounded by the red rectangular border indicates the aliasing.

Fig. 6 Schematic diagram of the center frequency aliasing of the last IMF

4.2 Multiscale deep feature learning with LSTM

The numbers of inputs, network structure and other hyper-parameters are vital variables in an LSTM model. Therefore, according to the input selection method introduced by He et al. (2014)[11], Huang et al. (2014)[1], and Wang et al. (2009)[16], the input variables could first be easily obtained by observing the plot of Partial Autocorrelation Functions (PACFs) illustrated by Fig. 7, Fig. 7 in which $PACF_1$–$PACF_9$ denotes the PACFs of each component. In other words, we assume that the output is $X(t+1)$ and $X(t+1-k)$ is then selected as one of the input variables under the condition that PACF at lag k is out of the 95% confidence interval indicated by the blue lines in Fig. 7. Fig. 7 shows that almost all the PACFs of each component are out of the range. Therefore, we select the 20 days of lag form $X(t-19)$ to $X(t)$ as the input variables of each sub-series.

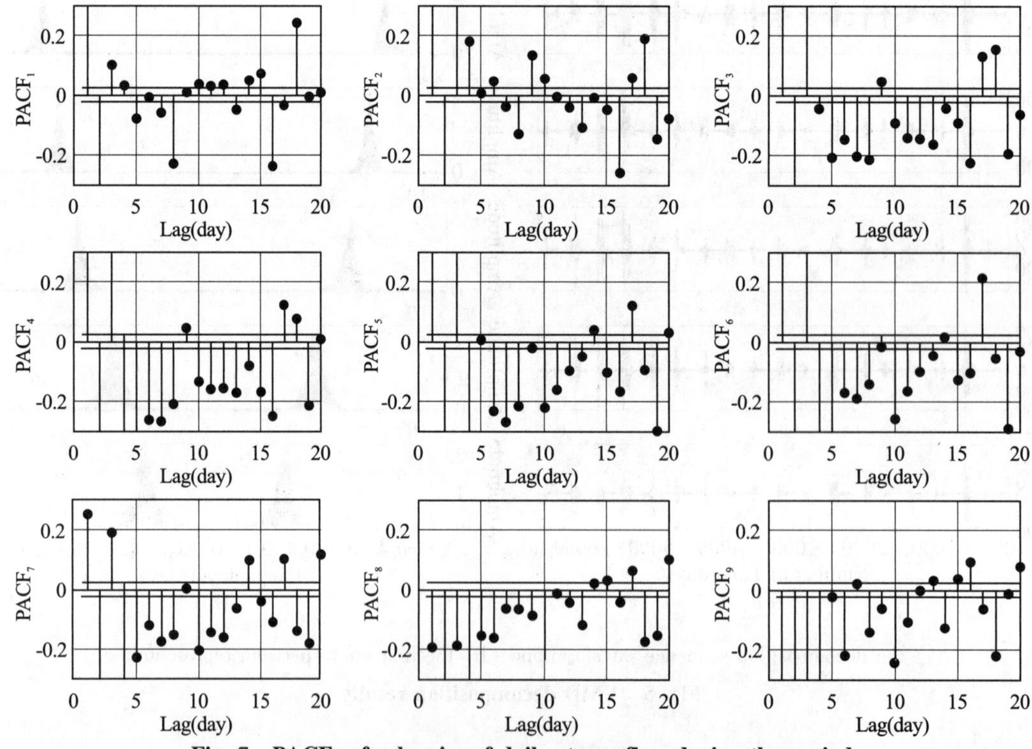

Fig. 7 PACFs of subseries of daily streamflow during the period 1997-01-01 to 2014-12-31 for the Yangxian hydrological station

Since there is no mature approach in theory to determine the number of hidden units and hidden layers, an experimental method can be used to select these two parameters for the LSTM network structure. In the experiment of this study, the sub-series shown in Fig. 5, first normalized by Eq. (23), were split into three parts: the training data set (1/1/1997–27/5/2011), the development set (28/5/2011–14/3/2013) and the testing set (15/3/2013–31/12/2014). The training set is used to train the model, that is, to train the parameters such as weight matrix and threshold in LSTM model; The development set is used to optimize model super parameters, such as learning rate, number of network neurons, number of network layers, etc.; The testing set is used to validate the accuracy of the model and to show the confidence level. The number of hidden units for each hidden layer was designed for 11 levels ranging from 15 to 25, and the number of input variables equaled the median of this interval. The hidden layers of LSTM were initialized from 1 to 5. Therefore, for each component of the original series, there were 55 LSTM network structures in this experiment. Training and developing the LSTM model to predict the first component, IMF_1, is an example used to describe the experimental process.

For each network structure mentioned in the previous paragraph, we first initialized an LSTM model with 20 input units and then trained these 55 structures using the training set. Then, the streamflow during the development period was forecast based on the trained structures. To find the optimal model structure, PPTS(5), illustrated in Fig. 8, was calculated for the training and development set. Fig. 8 (a) shows the changes of PPTS(5) of the training and development set for different levels of the hidden layers. According to the rule of bias and variance tradeoff[40], when the PPTS(5) of the development set is close to the PPTS(5) of the training set and both of these values are small, the model structure will obtain a great generalization and predicting ability. Therefore, we selected 20-15-1, 20-19-19-1, 20-19-19-19-1, 20-17-17-17-17-1 and 20-21-21-21-21-21-1 as the optimal structure of the 1 to 5 hidden layers, respectively. 20-19-19-1 means that the structure has 20 input features, 1 output target, and two hidden layers, with 19 hidden units for each layer. Fig. 8 (b) shows the boxplots of optimal structure for each level of the hidden layers, where the upper and lower quartiles are determined by the PPTS(5) of the training and development set. The range between the upper and lower quartiles indicates the degree of bias and variance tradeoff; the smaller the range, the better the tradeoff. From Fig. 8 (b), one can find that the structure 20-15-1 has the best bias and variance tradeoff. Therefore, the model structure 20-15-1 was selected as the optimal model to predict IMF_1.

To validate the optimal model for forecasting each sub-series, the predictions during the test period were renormalized to the original scale and are plotted in Fig. 9. The PPTS(5) and R^2 of the whole components during the training and development period are listed in Table 2. From Fig. 9 and Table 2, we can see that all the trained LSTM structures of all the sub-series have good accuracy.

4.3 Training ensemble tree model GBRT

We can simultaneously build an ensemble tree, the GBRT model, to represent the relationship between the original series and the sub-series decomposed by VMD; the GBRT algorithm can learn an ensemble function to reconstruct all the sub-series into a streamflow series. To find the optimal hyper-parameters, Bayesian optimization based on Gaussian processes was applied[41,42]. The entire data set for building the GBRT mode, consisting of 9 sub-series as input and raw streamflow as the output target, was divided into two parts: the training-validating set (1/1/1997 – 14/3/2013) and the testing set (15/3/2013 – 31/12/2014). The training-validating set is used to training the GBRT model and select the optimal parameters, while the testing set is used to validate the prediction performance of the models and to show the confidence level. The famous machine learning toolkit scikit-learn[43] was applied to train the GBRT model by use of the

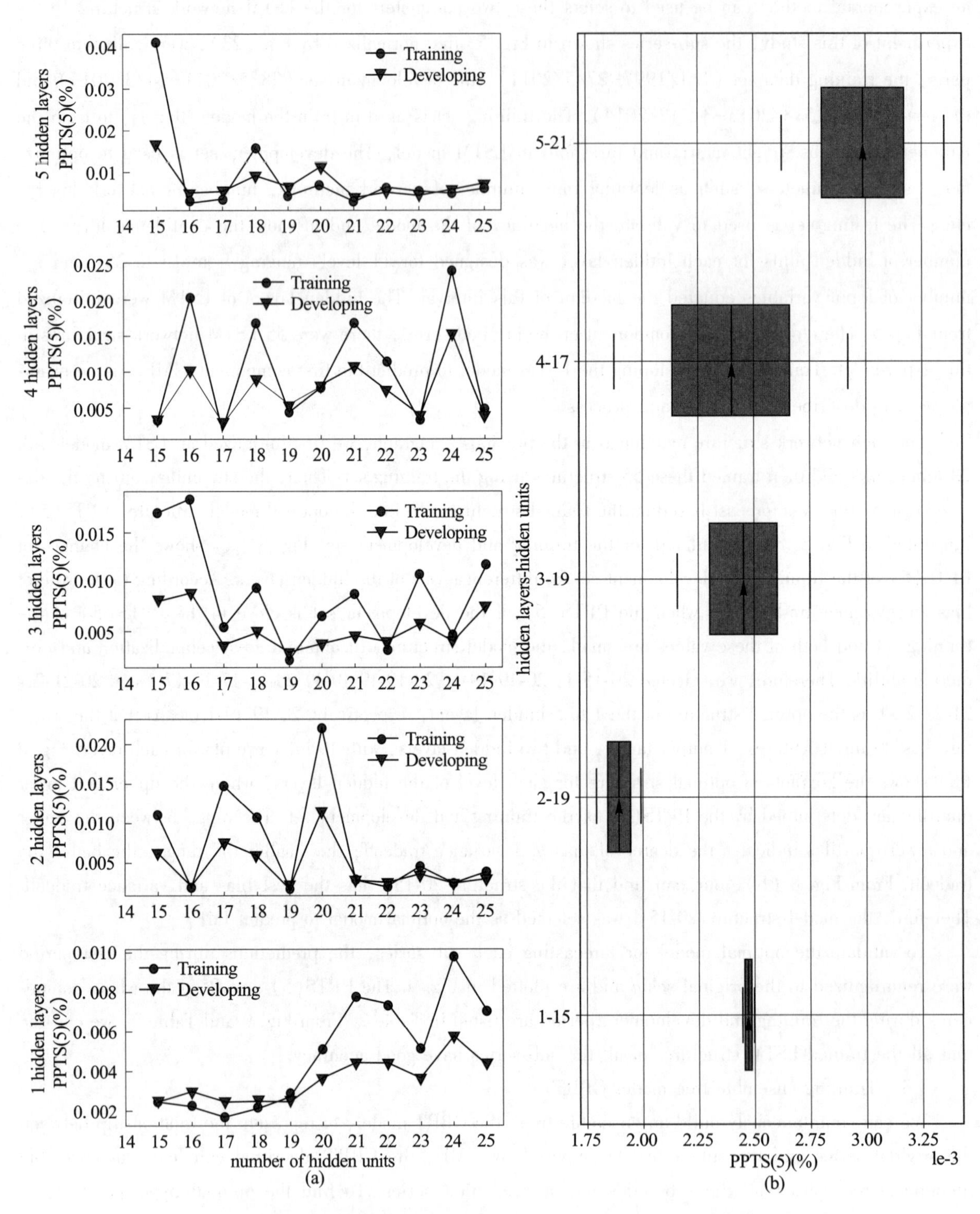

Fig. 8 PPTS(5) of different LSTM structures for predicting streamflow of IMF_1 during the training and development period: (a) line chart plots of PPTS(5) for different hidden layers and (b) boxplots of the optimal structure of each hidden layer

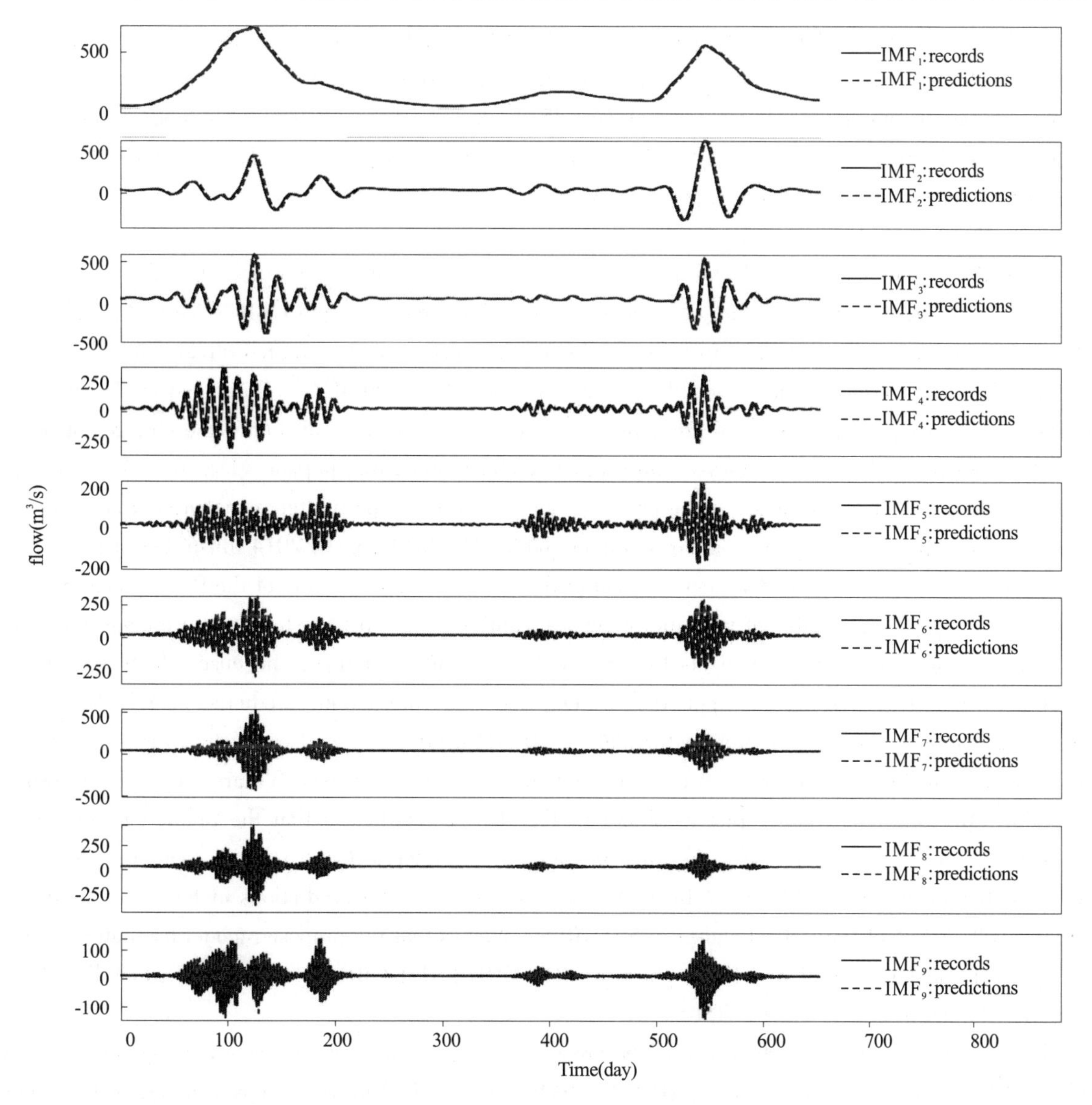

Fig. 9 Forecasting result of sub-series during the testing period

Table 2 Results of evaluation criteria with different hidden layers and hidden units for sub-sequences

Sequence	Hidden layers	Hidden units	Training		Developing	
			PPTS(5)(%)	R^2	PPTS(5)(%)	R^2
IMF1	1	15	0.0025	1.0000	0.0024	0.9999
IMF2	2	22	0.0396	0.9999	0.0227	0.9999
IMF3	1	16	0.2684	0.9999	0.2163	0.9997
IMF4	3	20	0.7646	0.9997	0.7156	0.9966
IMF5	4	20	0.8728	0.9993	0.2523	0.9983
IMF6	2	21	0.5031	0.9986	1.9367	0.9960
IMF7	2	23	1.2368	0.9973	0.4885	0.9943
IMF8	5	20	1.8371	0.9952	4.4562	0.9832
IMF9	1	25	1.5380	0.9972	2.1116	0.9897

training-validating set. To improve the performance of GBRT, 6-fold cross-validation was used. The optimal value of the hyper-parameters for GBRT, i.e., learning rate, maximum depth, maximum features, minimum sample split and minimum sample leaf, were 0.08, 25, 9, 9 and 10, respectively. The performance evaluation results of RMSE = 45.9766, R^2 = 0.9825, MAE = 11.4565 and PPTS(5) = 0.0438% indicate that GBRT had a good precision for multiscale feature ensembles.

4.4 Predicting results of VMD-LSTM-GBRT

Once the feature learning models based on LSTM were built, we could ensemble the sub-results to forecast the original streamflow. There are two ensemble methods: summation and GBRT. We first summed all 9 sub-results obtained in Section 4.2 to obtain the forecasting results of the original streamflow. The streamflows predicted by the model with ensemble summation, VMD-LSTM-SUM, are illustrated in Fig. 10. It can be observed from Fig. 10(a) that the predictions of a summation method can follow the changes of records during the test period but the accuracy of the peak streamflow forecasting is poor. Moreover, the scatter plot as shown in Fig. 10 (b) indicates that the peak predictions are not appropriately concentrated near the ideal fit. The detailed plot of the predictions during the period 07/09/2014-28/09/2014 illustrated in Fig. 10 (c) can also prove that point. By experiment, we found that the summation results of the 9 components decomposed by VMD were quite different from the original streamflow at the peak. Therefore, we could not simply sum the predictions of the 9 components to forecast the original streamflow; the ensemble tree model obtained in Section 4.3, GBRT, was applied to ensemble the sub-predictions predicted by LSTM. The final forecasting results forecast by the proposed model, VMD-LSTM-GBRT, are also shown in Fig. 10. From Fig. 10 (a), one can find that the peak predictions obtained by VMD-LSTM-GBRT are closer to the original streamflow. Moreover, the scatter plot as shown in Fig. 10 (b) indicates that the predictions forecast by VMD-LSTM-GBRT concentrated near the ideal fit and agreed better with the records, which could also be proved by the small angle between the linear fit and the ideal fit. The predictions of the sub-set (07/09/2014-28/09/2014) of the test set shown in Fig. 10(c) denotes that the proposed model has better performance at the peak flow. Therefore, the proposed model, VMD-LSTM-GBRT, has a better capability of peak streamflow forecasting than VMD-LSTM-SUM.

To assess the forecasting performance of the proposed model, a different decomposition algorithm, EEMD, and two different feature learning models, DNN and SVR, were applied for the comparisons using the identical dataset shown in Fig. 4. The building process of these models is the same as the approach mentioned in Section 2.4, except that the decomposition algorithm and the feature learning model are replaced, respectively.

Fig. 11 plots the streamflow predictions of Yangxian station by the proposed model VMD-LSTM-GBRT and the decomposition method-substituted model EEMD-LSTM-GBRT. As shown in Fig. 11(a), the proposed model performed better for peak flow forecasting than the traditional decomposition method EEMD, which can be validated by Fig. 11 (c). From the scatter plot illustrated by Fig. 11 (b), one can find that the recorded predicted values of the proposed model are much more concentrated than the model using EEMD. Moreover, the comparison of prediction performance between the proposed feature learning model LSTM and the other two machine learning models, DNN and SVR, were conducted and are indicated by Fig. 12. However, from the forecasting results shown in Fig. 12(a) and the scatters shown in Fig. 12 (b), one can observe that the difference between the three feature learning models is not that obvious. However, we can still determine that the best feature learning model is the LSTM from the quantitative evaluations given in Table 3 and the detailed predictions shown in Fig. 12(c). As shown in Table 3, the proposed model VMD-LSTM-GBRT has the lowest RMSE, MAE, PPTS(5) and the highest R^2 among these decomposition-ensemble-based models, which illustrates the proposed model superiority for both peak streamflow forecasting and global changes.

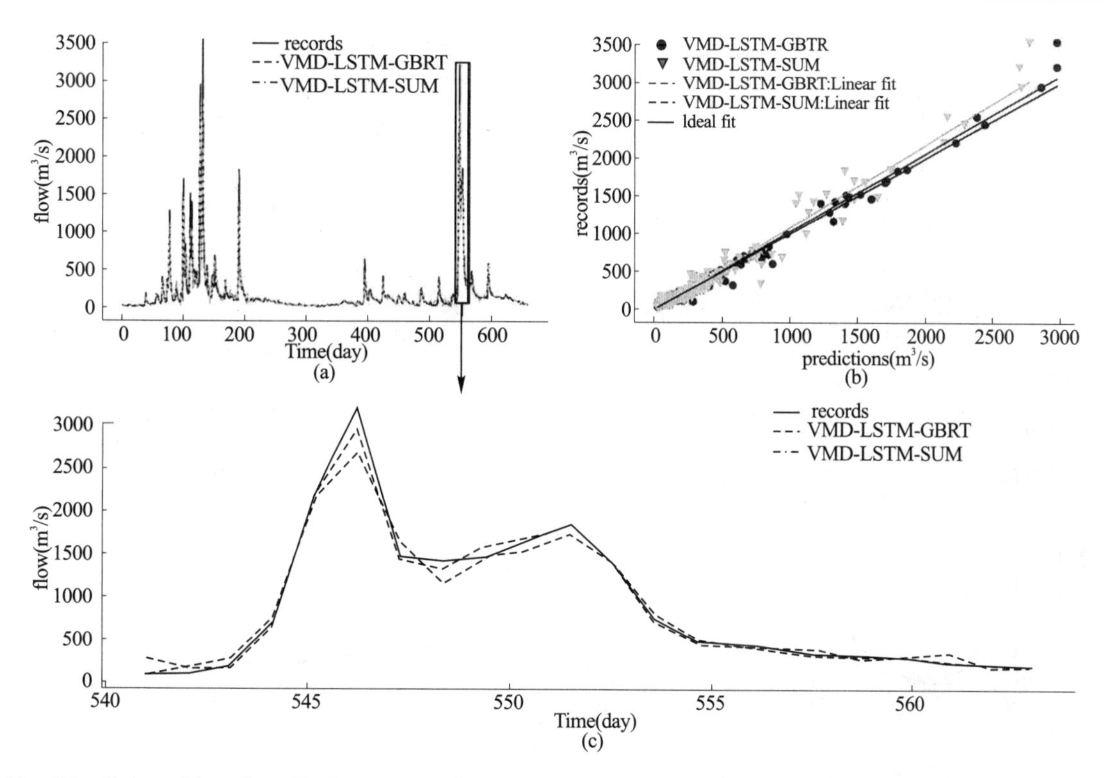

Fig. 10 Comparison of prediction results of the test dataset using different ensemble techniques, GBRT and summation: (a) plots of the prediction results and records, (b) scatters for the test set (15/3/2013-31/12/2014), and (c) plots of the prediction results for the period 07/09/2014-28/09/2014

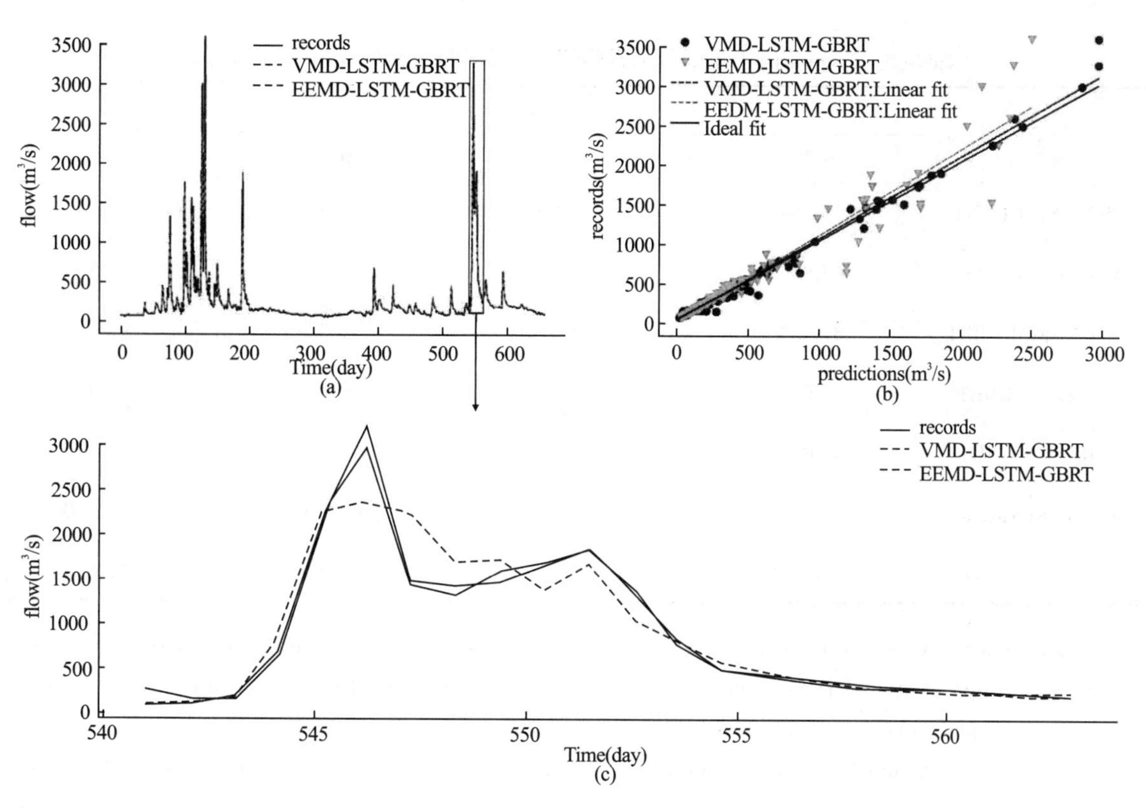

Fig. 11 Comparison of prediction results of the test dataset using the different decomposition algorithms VMD and EEMD: (a) forecasting results, (b) scatters for the testing data (15/3/2013-31/12/2014), and (c) plots of the prediction results for the period 07/09/2014-28/09/2014

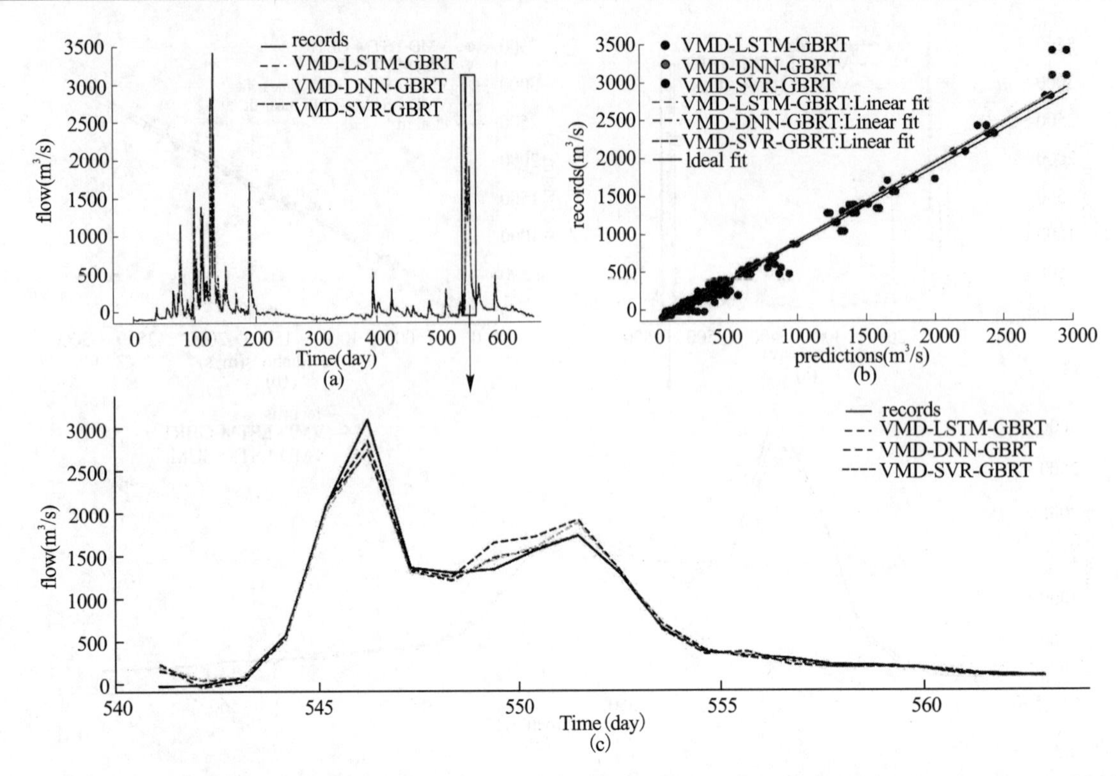

Fig.12 Comparison of forecasting results of the test dataset using the different feature learning models LSTM, DNN and SVR: (a) forecasting results, (b) scatters for the testing data (15/3/2013-31/12/2014), **and (c) plots of the forecasting results during the period** 07/09/2014-28/09/2014

Table 3 Comparison of the forecasting performances using different models

Model	Performance criteria			
	RMSE	R^2	MAE	PPTS(5)(%)
VMD-LSTM-GBRT	36.3692	0.9890	9.5246	0.0391
VMD-LSTM-SUM	67.8297	0.9619	27.8412	0.1500
EEMD-LSTM-GBRT	87.4506	0.9366	22.0321	0.0883
VMD-DNN-GBRT	44.9735	0.9832	12.1853	0.0451
VMD-SVR-GBRT	47.0555	0.9816	12.8919	0.0472
Linear regression	224.5310	0.5820	69.0201	0.2740
Multilayer perceptron	225.1806	0.5796	64.2869	0.1858

In the light of the above comparisons, all results fully indicate that the proposed model based on the decomposition algorithm, VMD, the multiscale deep feature learning model, LSTM, and the ensemble tree model, GBRT, performs very well for streamflow forecasting.

In order to verify that the proposed method will obtain similar performance on other flow data sets, the model is applied to Huaxian hydrological station which is located at the Wei River Basin in China, the result is as shown in Fig. 13 and Table 4. The results show that the performance of the proposed method is consistent with that in Yangxian hydrological station.

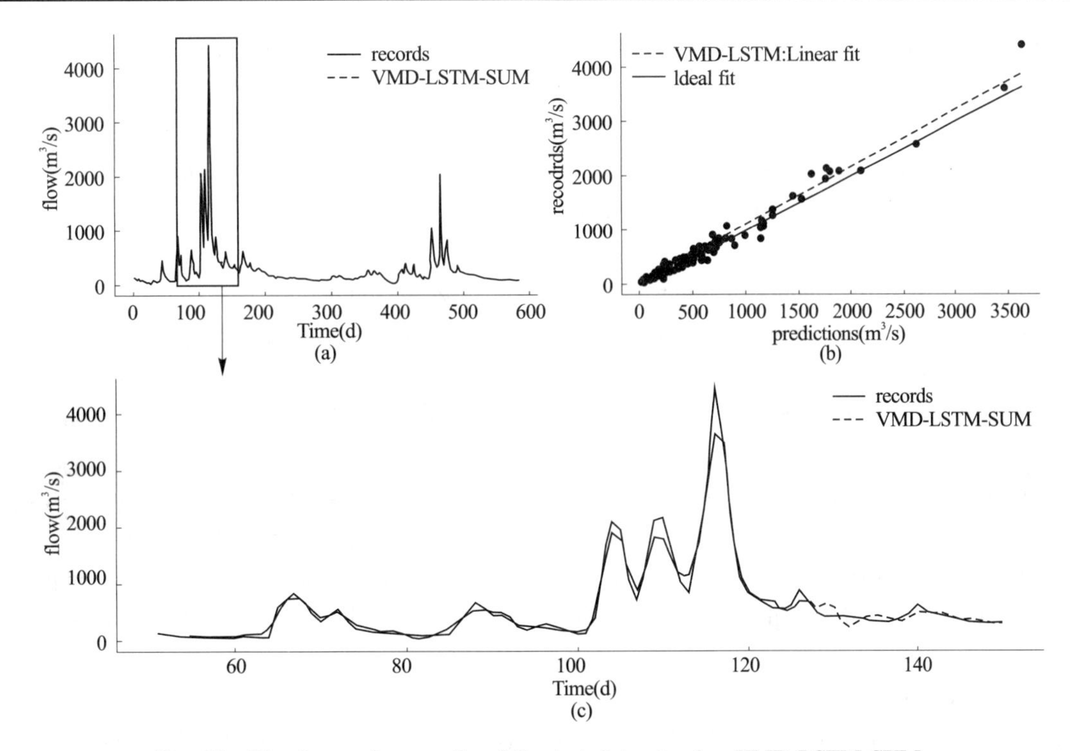

Fig. 13 **The forecasting results of the test dataset using VMD-LSTM-SUM: (a) forecasting results, (b) scatters for the testing data** (15/3/2013-31/12/2014), **and (c) plots of the forecasting results during the period** 07/09/2014-28/09/2014

Table 4 **The forecasting performances using the proposed model VMD-LSTM-SUM on the Huaxian hydrological station**

Model	Performance criteria			
	RMSE	R^2	MAE	PPTS(5)(%)
VMD-LSTM-SUM	57.5052	0.9754	22.5159	0.0725
Linear regression	148.0834	0.8363	43.1553	0.0951
Multilayer perceptron	141.5619	0.8505	40.7676	0.0931

5 Conclusions

In this paper, a decomposition-ensemble-based multiscale feature learning approach with hybrid models was developed for forecasting daily streamflow, and the approach was evaluated based on a historical river streamflow dataset for Yangxian station, Han River, China. To improve the accuracy and stability of forecasting, three aspects were considered in a comprehensive way: (1) Multiscale feature extraction by the algorithm with much more robustness with respect to sampling and noise; nine feature terms were extracted by VMD in this paper. (2) Deep feature learning with a model that can predict streamflow depending on the long historical changes of river flow; nine LSTMs were applied to sufficiently learn each feature. (3) An ensemble model with supervised learning; an ensemble tree GBRT was used to reconstruct the sub-results to obtain the final forecasting results. The daily streamflow forecasting capability of this decomposition-ensemble-based approach was compared with respect to three aspects: comparison with an approach that has the same feature learning model and ensemble technique but uses the traditional decomposition algorithm EEMD, comparison with an approach with the same decomposition algorithm and feature learning model but using the summation ensemble strategy, and comparison with an approach that has the same decomposition algorithm and ensemble technique but uses two different models, DNN and SVR. The results denote that the proposed model VMD-LSTM-GBRT exhibits the best forecasting performance among all the peer approaches for both global changes and peak streamflow forecasting.

This study proposes an approach to gain insight into the sophisticated features of natural river streamflow by designing a decomposition-ensemble framework. The three segments of this approach, i.e., using a robust model, VMD, to extract features that adequately represent natural river flow; using a long dependency model, LSTM, to remember or forget the historical changes of river flow; and reconstructing the extracted features by a tree model to remove the effect of error accumulation, are combined to determine what the values of the future streamflow should be. Note that the three segments of this framework can be replaced by other models, e.g., VMD can be replaced by an algorithm that is much more robust to sampling and noise but that still uses a regression strategy and replaces GBRT with DNN. Therefore, the present approach has value for river flow forecasting.

Streamflow forecasting is worthy of in-depth study. In the future, we will continue to study streamflow forecasting models. For instance, we could apply dynamic selection approaches[44,45] to improve the ensemble's performance in streamflow forecasting, and residual series modeling could be used to improve the accuracy of statistical and machine learning models[46,47]. It can make streamflow forecasting more and more accurate.

Acknowledgments

This work was supported by the Fund Project of Shaanxi Key Laboratory of Land Consolidation (2019-JC08) and Shaanxi Provincial Land Engineering Construction Group Internal Research Project (DJNY 2020-15).Sincere gratitude is extended to the editor and anonymous reviewers for their professional comments and corrections.

Authors' contributions

Conceptualization, X.S. and J.L.; Data curation, X.S., J.W., C.S. and D.H.; Formal analysis, X.S. and H.Z.; Funding acquisition, D.H.; Investigation, J.W., D.H. and C.S.; Methodology, X.S.; Project administration, J.L.; Resources, J.L. and X.S.; Supervision, J.L.; Validation, H. Z. and J.W.; Writing - original draft, X.S.. All authors reviewed the manuscript.

References

[1]Huang, S., Chang, J., Huang, Q., Chen, Y. Monthly streamflow prediction using modified EMD-based support vector machine. J. Hydrol., 511, 764-775.

[2]Lima, A.R., Cannon, A.J., Hsieh, W.W. Forecasting daily streamflow using online sequential extreme learning machines. J. Hydrol., 537, 431-443.

[3]Shiri, J., Kisi, O. Short-term and long-term streamflow forecasting using a wavelet and neuro-fuzzy conjunction model. J. Hydrol., 394(3-4), 486-493.

[4]Jiang, H., Zheng, W., Dong, Y. Sparse and robust estimation with ridge minimax concave penalty. Inf. Sci., 571, 154-174.

[5]Jiang, H., Tao, C., Dong, Y., Xiong, R. Robust Low-rank Multiple Kernel Learning with Compound Regularization. Eur. J. Oper. Res., 295(2), 634-647.

[6]Jiang, H., Luo, S., Dong, Y. Simultaneous feature selection and clustering based on square root optimization. Eur. J. Oper. Res., 289(1): 214-231.

[7] Castellano-Méndez, M. a., González-Manteiga, W., Febrero-Bande, M., Manuel Prada-Sánchez, J., Lozano-Calderón, R. Modelling of the monthly and daily behaviour of the runoff of the Xallas river using Box-Jenkins and neural networks methods. J. Hydrol., 296(1-4),38-58.

[8]Mohammadi, K., Eslami, H.R., Kahawita, R. Parameter estimation of an ARMA model for river flow forecasting using goal programming. J. Hydrol., 331(1-2), 293-299.

[9]Valipour, M., Banihabib, M.E., Behbahani, S.M.R. Comparison of the ARMA, ARIMA, and the autoregressive ar-

tificial neural network models in forecasting the monthly inflow of Dez dam reservoir. J. Hydrol., 476, 433-441.

[10]Yu, X., Zhang, X., Qin, H. A data-driven model based on Fourier transform and support vector regression for monthly reservoir inflow forecasting. J. Hydro-environ. Res., 18, 12-24.

[11]He, Z., Wen, X., Liu, H., Du, J. A comparative study of artificial neural network, adaptive neuro fuzzy inference system and support vector machine for forecasting river flow in the semiarid mountain region. J. Hydrol., 509,379-386.

[12]Yaseen, Z.M. et al.. Novel approach for streamflow forecasting using a hybrid ANFIS-FFA model. J. Hydrol., 554, 263-276.

[13]Humphrey, G.B., Gibbs, M.S., Dandy, G.C., Maier, H.R. A hybrid approach to monthly streamflow forecasting: Integrating hydrological model outputs into a Bayesian artificial neural network. J. Hydrol., 540, 623-640.

[14]Wang, H., Wang, C., Wang, Y., Gao, X., Yu, C. Bayesian forecasting and uncertainty quantifying of stream flows using Metropolis-Hastings Markov Chain Monte Carlo algorithm. J. Hydrol., 549, 476-483.

[15]Tan, Q.-F. et al. An adaptive middle and long-term runoff forecast model using EEMD-ANN hybrid approach. J. Hydrol., 567,767-780.

[16]Wang, W.-C., Chau, K.-W., Cheng, C.-T., Qiu, L. A comparison of performance of several artificial intelligence methods for forecasting monthly discharge time series. J. Hydrol., 374(3-4), 294-306.

[17]Bai, Y., Chen, Z., Xie, J., Li, C. Daily reservoir inflow forecasting using multiscale deep feature learning with hybrid models. J. Hydrol., 532,193-206.

[18]Hochreiter, S., Schmidhuber, J. Long Short-Term Memory. Neural Computation, 9(8): 1735-1780,(1997).

[19]Yang, J., Kim, J. An accident diagnosis algorithm using long short-term memory. Nucl. Eng. Technol., 50(4), 582-588,(2018).

[20]Peng, L., Liu, S., Liu, R., Wang, L. Effective long short-term memory with differential evolution algorithm for electricity price prediction. Energy, 162,1301-1314.

[21]Zhang, J., Zhu, Y., Zhang, X., Ye, M., Yang, J. Developing a Long Short-Term Memory (LSTM) based model for predicting water table depth in agricultural areas. J. Hydrol., 561, 918-929.

[22]Karthikeyan, L., Nagesh Kumar, D. Predictability of nonstationary time series using wavelet and EMD based ARMA models. J. Hydrol., 502,103-119.

[23]Seo, Y., Kim, S., Kisi, O., Singh, V.P. Daily water level forecasting using wavelet decomposition and artificial intelligence techniques. J. Hydrol., 520, 224-243.

[24]Wang, W.-c., Chau, K.-w., Xu, D.-m., Chen, X.-Y. Improving Forecasting Accuracy of Annual Runoff Time Series Using ARIMA Based on EEMD Decomposition. Water Resour. Manag., 29(8), 2655-2675.

[25]Dragomiretskiy, K., Zosso, D. Variational Mode Decomposition. IEEE Trans. Signal Process., 62(3), 531-544.

[26]Liu, H., Mi, X., Li, Y. Smart multi-step deep learning model for wind speed forecasting based on variational mode decomposition, singular spectrum analysis, LSTM network and ELM. Energy Conv. Manag., 159, 54-64.

[27]Naik, J., Dash, S., Dash, P.K., Bisoi, R. Short term wind power forecasting using hybrid variational mode decomposition and multi-kernel regularized pseudo inverse neural network. Renew. Energy, 118, 180-212.

[28]Niu, M., Hu, Y., Sun, S., Liu, Y. A novel hybrid decomposition-ensemble model based on VMD and HGWO for container throughput forecasting. Appl. Math. Model., 57,163-178.

[29]Mohanty, S., Gupta, K.K., Raju, K.S. Hurst based vibro-acoustic feature extraction of bearing using EMD and VMD. Measurement, 117, 200-220.

[30]Liu, C., Zhu, L., Ni, C. Chatter detection in milling process based on VMD and energy entropy. Mech. Syst. Signal Proc., 105,169-182.

[31]Goodfellow, I., Bengio, Y., Courville, A. Deep Learning. Adaptive Computation and Machine Learning. The MIT Press(2018).

[32]Werbos, P.J. Backpropagation through time: what it does and how to do it. Proc. IEEE, 78(10), 1550-1560.

[33]Bengio, Y., Simard, P., Frasconi, P. Learning Long-Term Dependencies with Gradient Descent is Difficult. IEEE Trans. Neural Netw., 5(2), 157-166.

[34]Hochreiter, S. The Vanishing Gradient Problem During Learning Recurrent Neural Nets and Problem Solutions. Int. J. Uncertainty Fuzziness Knowl.-Based Syst., 6(2),107-116.

[35]Friedman, J.H. Greedy Function Approximation: A Gradient Boosting Machine. Ann. Stat., 29(5), 1189-1232.

[36]Friedman, J.H. Stochastic gradient boosting. Computational Statistics & Data Analysis, 38(4), 367-378(2002).

[37]Hastie, T., Tibshirani, R., Friedman, J. The Elements of Statistical Learning. Springer, 745(2009).

[38]Ioffe, S., Szegedy, C. Batch normalization: accelerating deep network training by reducing internal covariate shift, Proceedings of the 32nd International Conference on International Conference on Machine Learning - Volume 37. JMLR.org, Lille, France :448-456,(2015).

[39]Stojković, M., Kostić, S., Plavšić, J., Prohaska, S. A joint stochastic-deterministic approach for long-term and short-term modelling of monthly flow rates. J. Hydrol., 544, 555-566.

[40]Valentini, G., Dietterich, T.G. Bias-variance analysis of support vector machines for the development of svm-based ensemble methods. J. Mach. Learn. Res., 5, 725-775,(2004).

[41]Bergstra, J., Bardenet, R., Bengio, Y., Kégl, B. Algorithms for Hyper-Parameter Optimization, 25th Annual Conference on Neural Information Processing Systems (NIPS 2011). Neural Information Processing Systems Foundation, Granada, Spain(2011).

[42]Shahriari, B., Swersky, K., Wang, Z., Adams, R.P., Freitas, N.d. Taking the Human Out of the Loop: A Review of Bayesian Optimization. Proc. IEEE, 104(1),148-175.

[43]Pedregosa, F. et al.. Scikit-learn: Machine Learning in Python. J. Mach. Learn. Res., 12, 2825-2830,(2011).

[44]De Oliveira, J. F. L., Silva, E. G., de Mattos Neto, P. S. G. A Hybrid System Based on Dynamic Selection for Time Series Forecasting. IEEE T. Neur. Net. Lear., 1-13.

[45]Silva, E. G., De Mattos Neto, P. S. G., Cavalcanti, G. D. C. A Dynamic Predictor Selection Method Based on Recent Temporal Windows for Time Series Forecasting. IEEE Access, 9, 108466-108479.

[46]de Oliveira, João Fausto Lorenzato, Pacífico, Luciano Demetrio Santos, de Mattos Neto, Paulo Salgado Gomes, etc.. A hybrid optimized error correction system for time series forecasting. Appl. Soft Comput., 87, 105970.

[47]de Mattos Neto, Paulo S.G.; Ferreira, Tiago A.E.; Lima, Aranildo R.; Vasconcelos, Germano C.; Cavalcanti, George D.C. A perturbative approach for enhancing the performance of time series forecasting. Neural Networks, 88, 114-124.

本文曾发表于2022年《Nature Portfolio》第12卷

Evaluation of Different Machine Learning Models and Novel Deep Learning-based Algorithm for Landslide Susceptibility Mapping

Tingyu Zhang[1,2], Yanan Li[1,2], Tao Wang[3], Huanyuan Wang[4], Tianqing Chen[1,2], Zenghui Sun[1,2], Dan Luo[3], Chao Li[3], Ling Han[4]

(1.Key Laboratory of Degraded and Unused Land Consolidation Engineering, the Ministry of Natural Resources, Xi'an, Shaanxi, China; 2.Institute of Land Engineering and Technology, Shaanxi Provincial Land Engineering Construction Group Co., Ltd, Xi'an, Shaanxi, China; 3.Shaanxi Provincial Land Engineering Construction Group Land Survey Planning and Design Institute Co., Ltd, Xi'an, Shaanxi, China; 4.School of Land Engineering, Chang'an University, Xi'an, Shaanxi, China)

【**Abstract**】The losses and damage caused by landslide are countless in the world every year. However, the existing approaches of landslide susceptibility mapping cannot fully meet the requirement of landslide prevention, and further excava-tion and innovation are also needed. Therefore, the main aim of this study is to develop a novel deep learning model namely landslide net (LSNet) to assess the landslide susceptibility in Hanyin County, China, meanwhile, support vector machine model (SVM) and kernel logistic regression model (KLR) were employed as reference model. The inventory map was generated based on 259 landslides, the training dataset and validation dataset were, respectively, prepared using 70% landslides and the remaining 30% landslides. The variance inflation factor (VIF) was applied to optimize each landslide predisposing factor. Three benchmark indices were used to evaluate the result of susceptibility map-ping and area under receiver operating characteristics curve (AUROC) was used to compare the models. Result demonstrated that although the processing speed of LSNet model is the slowest, it still significantly outperformed its corresponding benchmark models with validation dataset, and has the highest accuracy (0.950), precision (0.951), F1 (0.951) and AUROC (0.941), which reflected excellent predictive ability in some degree. The achievements obtained in this study can improve the rapid response capability of landslide prevention for Hanyin County.

【**Keyword**】Landslide susceptibility; Deep learning; Kernel logistic regression; Support vector machine; Evaluation

Instruction

Landslide is defined as the special geological phenomenon that is threatening to mankind triggered by human activities or natural factors. Under the dual background of human activities and natural transmutations, the occurrence rate of landslides in the world increased rapidly (Sun et al. 2020). Depending on the latest statistical report of Ministry of Natural Resources of China (http://www.mnr.gov.cn/), in total 6, 181 geological hazards occurred in 2019, during that year, 211 people were killed and direct economic loss valued at 397 million dollars. Due to complex terrain, tectonic development and human activities, there are more than 230000 potential geological hazards in China and landslides account for 53.50% of it. Therefore, the prevention of landslide development in China is crucial. In the face of increasingly serious landslide threats, the development of disaster prevention and mitigation work can effectively reduce the threat posed by landslides. In order to plan and construct the city safely and effectively, and to carry out the work of disaster prevention and mitigation successfully, it is necessary to quantitatively assess the landslide susceptibility on the regional scale.

Over the past years, various techniques and methods for forecasting the landslides have been applied to landslide susceptibility assessment (LSA). At first, with the help of historical landslide data and the geological environment background, the scope of landslide prediction was directly delineated, but this method relies heavily on experience, resulting in low reliability of the results (Guzzetti et al. 2012; Youssef and Pourghasemi 2021). With the development of geographic information system (GIS) and satellite remote sensing technology within each subject area, the more commonly used solutions can be summarized as a few steps. The first step of LSA is to collect the development characteristics and spatial distribution features of historical and hidden danger landslides (Pradhan and Lee 2010). Then the predisposing factors of landslide occurrence are selected from the geological and environment background. Subsequently, the linear or nonlinear mapping relationship between predisposing factors and the degree of landslide susceptibility is analyzed by using evaluation model (qualitative or quantitative), and the contribution rate of each landslide predisposing factor is determined. In the end, some techniques of analysis and comparison are used to choose the suitable model for the study area (Carrara et al. 1995).

In principle, the evaluation models used in LSA could be crude categorized as two classes: statistical model and machine learning model. Usually, the statistical models were conjugated to GIS for spatial variety prediction of landslide disaster. The frequently used models are index of entropy (IOE) (Constantin et al. 2011; Youssef et al. 2015), analytical hierarchy process (AHP) (Kayastha et al. 2013), frequency ratio (FR) (Umar et al. 2014; Razavizadeh et al. 2017), certainty factor (CF) (Fan et al. 2017; Li and Zhang 2017), logistic regression (LR) (Pourghasemi et al. 2013; Aditian et al. 2018) and evidential belief function (EBF) (Carranza 2015; Li and Wang 2019). As the predictive ability of statistical model is still deficient and with the development of information technology, machine learning models could be a better alternative to solve the susceptibility assessment problem, such as artificial neural networks (ANN) (Aditian et al. 2018; Polykretis and Chalkias 2018; SOMA et al. 2019), support vector machines (SVM) (Bui et al. 2016; Pandey and Pourghasemi 2020), adaptive neuro-fuzzy inference systems (ANFIS) (Aghdam et al. 2017), decision trees (DT) (Pham et al. 2016), fuzzy logic (FL) (Saadoud et al. 2018) and multivariate adaptive regression spines (MARS) (Conoscenti et al. 2014). Due to data quality, factor selection, model parameter adjustment and other factors, some low accuracy, over fitting, and under fitting problems often appear (Bui et al. 2018). In order to solve these problems, hybrid model was developed in recent years, such as reduced error pruning trees (REPT) (Pham et al. 2019), kernel logistic regression model integrated with fractal dimension ($KLR_{box\text{-}counting}$) (Zhang et al. 2019), support vector regression model integrated with gray wolf optimization algorithm (SVR-GWO) (Balogun et al. 2021), adaptive neuro-fuzzy inference system model integrated with satin bowerbird optimizer algorithms (ANFIS-SBO) (Chen et al. 2021). However, these machine learning models still cannot avoid some disadvantages, for example, (1) It still requires a lot of prior knowledge and erection; (2) the existing network cannot fully extract potential landslide features; (3) the models are sensitive to missing data, and prone to fall into local optima.

Compared with machine learning, deep learning (DL) does not need to manually construct and select feature layers when dealing with object features, and at the same time, deep learning accepts a larger sample size, which is gradually applied in various fields. For example, Panahi (2020) used convolutional neural networks and recurrent neural networks to predict the probability of flash flood (Panahi et al. 2020); Kumar (2020) used deep learning model to complete the prediction of ground water depth (Kumar et al. 2020); Benzekri (2020) employed the deep learning model to construct an early forest fire detection system (Benzekri et al. 2020). In general, DL model performed a satisfactory ability of classification and regression. The main reason is that DL is completely a data-driven feature learning method, and has multi-level non-linear operations, which can abstractly represent classification features from a large amount of data, and combines gradient transfer method to optimize its end-to-end network structure (Zhu et al. 2020). Recently, more and more DL models have been successively applied in the field of LSA (Xiao et al. 2018; Huang et al. 2020; Li et al. 2021), among the different DL models, convolutional neural network (CNN) plays a significant role in landslide recognition and prediction, for example, Wang (2019) applied three novel CNN architectures in LSA, which achieved higher prediction accuracy than conventional methods (Wang et al. 2019); Sameen (2020) developed a DL-based technique for LSA through a 1-dimensional CNN, which performed better than ANN and SVM (Sameen et al. 2020); Fang et al. (2021) constructed four heterogeneous ensemble-learning techniques combining with CNN, RNN, SVM, and LR, the final results also showed that the DL model performed best (Fang et al. 2021).

However, these previous studies have some shortcomings, for example, in the process of building DL model, each convolutional layer will bring different prediction effects, the construction of multi-channel networks is ignored for landslide predisposing factor. Moreover, due to China's vast territory and diverse geological environment, there is still a lack of databases and studies related to landslides, and more reasonable, reliable and more accurate DL models still need to be explored.

Therefore, this study proposed a novel deep learning network named LSNet composed of multiple convolutional layers to predict the landslide susceptibility in Hanyin County, Shaanxi Province, China. The patches of landslide predisposing factor maps were used as the input data to train the LSNet, meanwhile the LSI was regarded as the output to predict the landslide susceptibility. In addition, the support vector machine model (SVM) and kernel logistic regression model (KLR) were employed to compare with LSNet. The primary difference here between this study and the literature mentioned is that approaches existed in this paper are seldom used and compared in landslide susceptibility assessment, especially LSNet and KLR. Another point is that the three models were first applied in Hanyin County and LSNet with multiple channels was established combined with the data structure representing the landslide, with the aim to improve the accuracy of LSA in the study area. Finally, all the results may help the government to make efficient decisions about landslide prevention and provide prevention references for landslide risk.

Sample description of study area

Hanyin County belongs to the hilly area in southern Shaanxi Province, the geographical coordinates are 32°68′~33°09′ north latitude and 108°11′~108°44′ east longitude (Fig. 1). The study area is about 51 km wide from east to west, 58 km long from north to south, and covers an area of about 1 347 km^2. The climate type of study area is continental tropical monsoon climate and the temperature varies greatly. According to the local meteorological statistics, the mean annual precipitation in the past 50 years is about 920 mm, and the rainfall in the northern region is significantly less than that in the southern region. The water resources in the study area are very abundant, and there are 4 rivers in total, all of which belong to Yangtze River system. There are three types of groundwater in the study area, including loose rock pore water, carbonate fissure water, and bedrock fissure water.

The geomorphology of study area is dominated by low and middle mountains, with valleys, hills and basins, and the area of mountains accounts for 87%. The exposed strata and main lithology in the study area are shown in Table 1. Since the geotectonic location of the study area is located in the core zone of the Qinling microplate, there are many faults and folds in this area. In fact, there are a total of 5 faults that have been proven. Besides, according to the historical records, there have been 16 earthquakes in the study area, with an average magnitude of 4, but these earthquakes did not cause major damage.

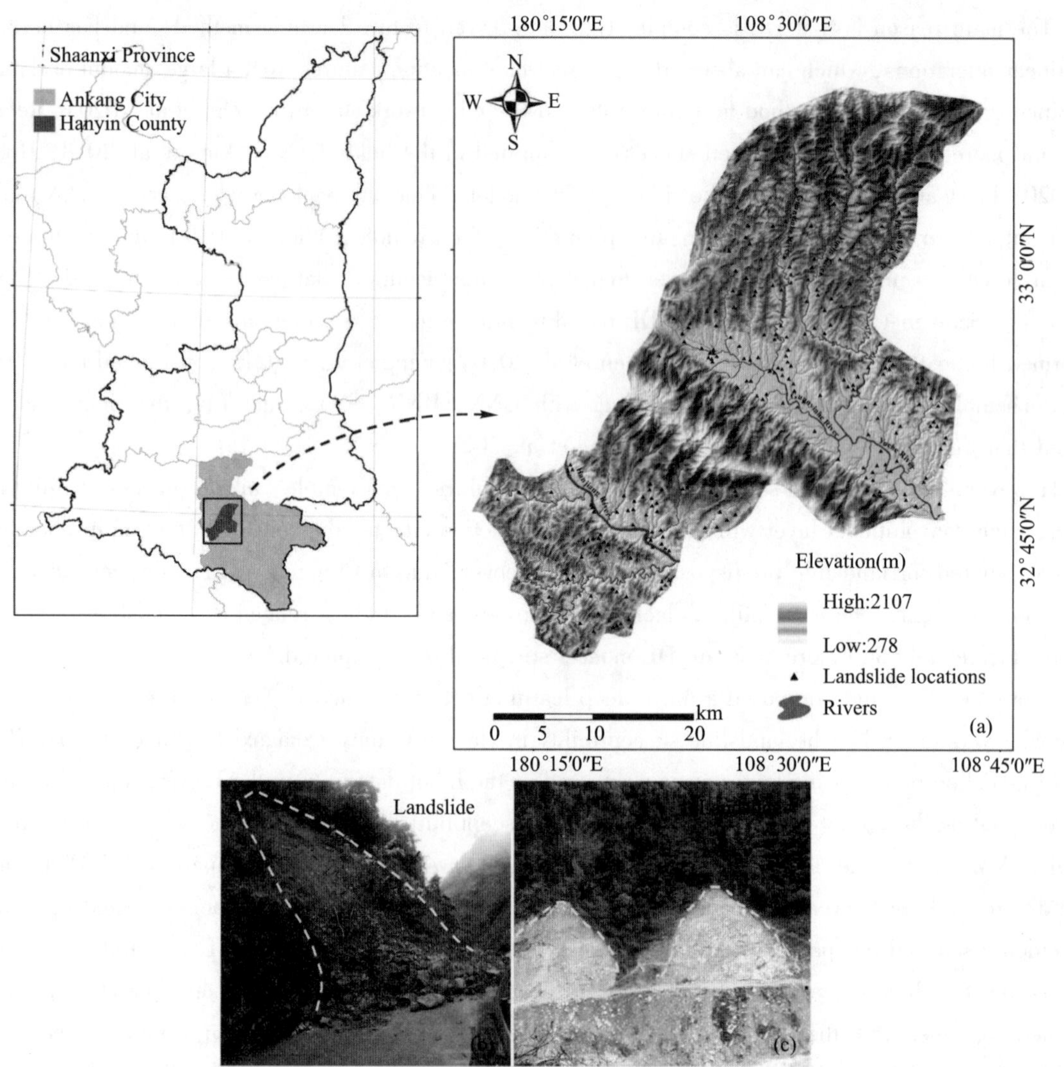

Fig.1 The study area (a) location and landslide inventory map; (b) forest destruction caused by landslide; (c) road damage caused by landslide

Table 1 The main lithology information of the study area

Geological age	Symbol	Main lithology
Quaternary	Q	Sandy clay, clay rock
Tertiary	E	Clay rock, siltstone, glutenite
Middle Devonian	D_2	Limestone, calcium schist
Lower Devonian	D_1	Calcium schist, calcium sandstone, granite
Silurian	S	Phyllite and siliceous rock, sandstone
Ordovician	O	Argillaceous limestone, carbonaceous schist, quartzite
Cambrian	ϵ	Limestone, slate, phyllite
Senian	Z	Limestone, quartzite, schist

Data preparation

Landslide inventory

Before carrying out the LSA, it is critical to verify about the information of landslides in the study area. Landslide inventory is to integration of landslide boundaries, locations, types and so on, which is the subsequent basis of data analysis and model construction. Based on the historical landslide data (SBGMR 1989; PRC 2020), remote sensing image (Cloud 2020), literatures (Liu and Huang 2006) and field survey, a total of 267 landslide locations were identified from 1989-2020 as the reference. These landslide locations were then imported into the GF-2 remote sensing images to delineated landslide boundary in ArcGIS software. In order to generate the landslide inventory map of study area (Fig. 1), all landslide boundaries were converted into polygons and resampled with the resolution of 30 m × 30 m. Fig.2 shows the preparation process of landslide inventory map.

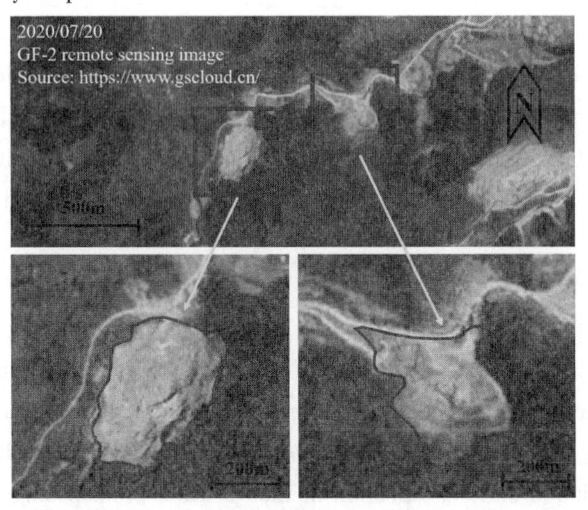

Fig.2 The preparation process of landsllde inventory map

Data preparation

In order to prepare the input dataset for model construction, 267 landslide samples were separated into two parts according to the ratio of 7/3 (Zhao and Chen 2020). Among them, 187 landslide samples were used as the training dataset to train the model, and the remaining 80 landslide samples were applied as the validation dataset to finish the validation purpose.

Analysis and quantification of landslide predisposing factors

There are no fixed guidelines for selecting the predisposing factors for LSA. After the comprehensive analysis of regional geo-environment characteristics and previous researches (Zhou and Fang 2015; Wang et al. 2020; Wu et al. 2020), we proposed altitude, slope angle, slope aspect, normalized difference vegetation index (NDVI), distance to rivers, distance to roads, distance to faults, mean annual precipitation (MAP) and lithology as the landslide predisposing factors.

Among them, altitude, slope angle, slope aspect are commonly used terrain factors, which can describe the influence of terrain on landslide with multi-dimension. For instance, altitude can affect the topographic properties, vegetation distribution and temperature differences at different heights. Slope angle is regarded as one major impact factor affecting instability and deformation of landslides, for example, the larger the slope angle, the less stable the slope is.

NDVI is a commonly used to indicate the vegetation coverage condition and it ranges from -1 to 1. Negative value expresses that there is water, snow and cloud coverage. Positive value expresses that the earth coverage is vegetation, and the larger the coverage extent of the vegetation lead to higher NDVI positive value.

At present, it is generally believed that the mechanism of landslide occurrence is the increase of pore water pressure in slope compositions by using the rainwater infiltration and accumulation during the rainy period. The high pore water pressure will lead to the increase of the effective load stress and the decrease of the shear strength of slope components, which are main triggering factors caused landslide. Therefore, MAP also occupies vital status of the landslide occurrence.

Rivers can affect the hydrogeology characteristics of slopes, and rivers usually corrode the toe of slope, which may decrease the anti-slide force. In mountainous area, it is common that numerous of landslide hazards are triggered by road constructions. Besides, in areas with frequent tectonic movements, the dislocation of faults can also cause landslide occurrence. Hence, distance to rivers, distance to roads, and distance to faults were regarded as predisposing factors in this study.

The lithology of the slope is the material founda-tion of the landslide. Some slopes are made up of hard rocks, some of which are made up of soft rocks, and some of them are made up of soil. Because of the difference between the lithologies, their shear strength varies. Often, slopes made of hard rocks do not easily fall, while slopes material of soft rocks or soil are easier to be desta-bilization deformation.

Since the original attribute data of each predisposing factor are very different, the frequency ratio (FR) is introduced to unify the dimension of each predisposing factor. The calculation process of FR value is shown in Eq. (1):

$$FR=\frac{Sam_{ij}}{Are_{ij}} \tag{1}$$

where Sam_{ij} stands for the percentage of landslides in each landslide predisposing factor class, and Are_{ij} is the area percentage of each landslide predisposing factor class (Siahkamari et al. 2017).

Additionally, in order to calculate the FR value, it is necessary to classify the predisposing factors, and the data sources, resolution and classification method of each predisposing factor are listed in Table 2.

Methodologies

The main research contents include 4 parts: (1) using the data that are already available to complete the landslide inventory; (2) using FR value to quantify the landslide predisposing factor maps, and partitioning dataset; (3) using the factor maps that are already quantified by FR to train the SVM model and KLR model, moreover using the original factor maps to train the LSNet model; (4) producing LSM corresponding to each model, assessing the result accuracy, and comparing the prediction performance of each model. The flowchart of this study is shown in Fig. 3. The techniques used in this study are described as follows.

Table 2 The information and data source of landslide predisposing factors

Landslide predisposing factors	Original format	Resolution	Classification method	Data source
Altitude/m	Grid	30 m×30 m	natural break (Jenks)	Extracting from DEM image (http://www.gscloud.cn/)
Slope angle/°	Grid	30 m×30 m	natural break (Jenks)	Extracting from DEM image (http://www.gscloud.cn/)
Slope aspect	Grid	30 m×30 m	natural break (Jenks)	Extracting from DEM image (http://www.gscloud.cn/)
NDVI	Grid	30 m×30 m	natural break (Jenks)	Generating by GF-2 remote sensing images obtained from Xi'an Satellite Measurement and Control Center
Distance to rivers/m	Vector	30 m×30 m	Equal interval	Generating by regional water system obtained from the local government
Distance to roads/m	Vector	30 m×30 m	Equal interval	Generating by regional traffic maps obtained from the local government
Distance to faults/m	Vector	30 m×30 m	Equal interval	Extracting from geological maps with 1:500000 scale obtained from the local government
MAP (mm/year)	Vector	30 m×30 m	Equal interval	Extracting from rainfall observation data from 2010 to 2021 obtained from the local government
Lithology	Vector	30 m×30 m	Custom interval	Extracting from geological maps with 1:500000 scale obtained from the local government

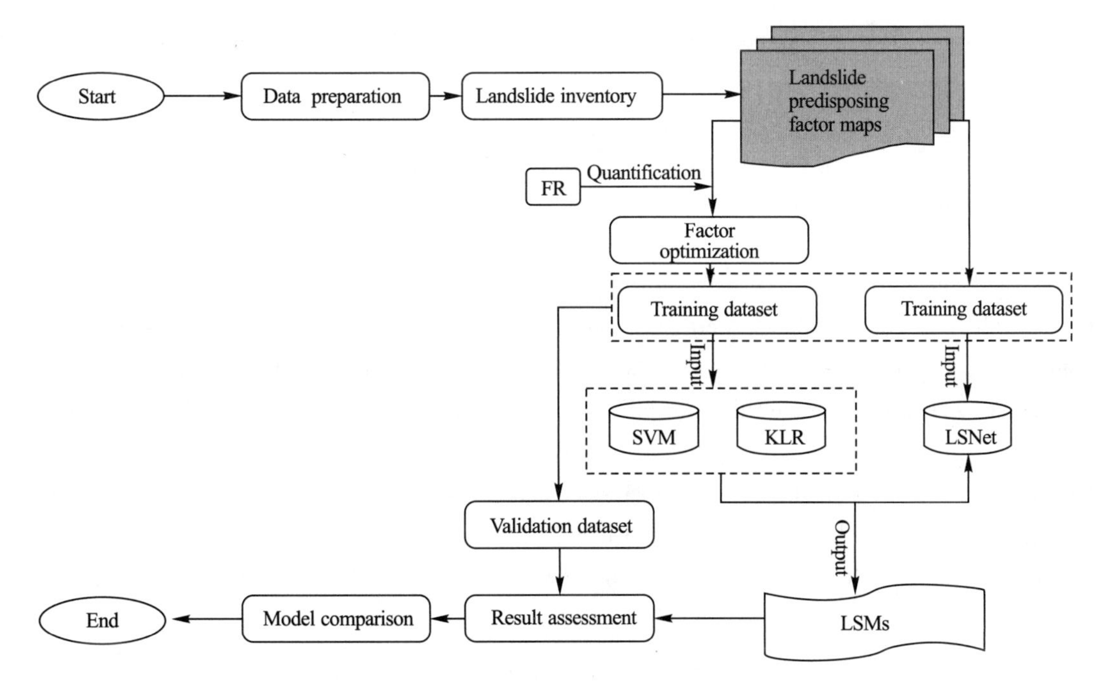

Fig.3 The flowchart of the study

Factor optimization method

Since the assumption of machine learning modeling is that the variables are independent of each other, it needs to detect whether there is strong correlation between the factors. This strong correlation relationship is called multicollinearity which may cause the over-fitting or under-fitting problems (Hong et al. 2018). In this study, the variance inflation factor (VIF) and tolerances (TOL) were applied to reflect the multicollinearity problem, which can be calculated by constructing a linear regression model based on the training dataset. When VIF > 10 and TOL < 0.1, it indicates that the predisposing factor has a multicollinearity problem and needs to be eliminated, vice versa (Pham et al. 2019).

Support vector machine model (SVM)

The basic principle of SVM is to search the optimal separating hyperplane that can maximize the interval between positive and negative samples in training dataset (Wang and Brenning 2021). Initially, SVM model was used as the supervised learning algorithm to solve binary classification problem, while the nonlinear classification problem can be solved after introducing the kernel function. Therefore, the SVM model was applied in many researches about landside susceptibility assessment. In addition, there are three parameters namely penalty factor (C_0), non-sensitive loss function (ε), and kernel function parameter (γ) that need to be adjusted appropriately in the process of constructing the SVM model (Xie et al. 2021). The main steps of SVM model construction can be described as below.

At first, the landslide predisposing factors are defined as the dataset of instance label pairs (s_i, t_i, $i = 1, 2, \cdots, n$), where s_i stands for the input data, t_i is the output classes (landslide and non-landslide), and n is the number of training samples (Kumar et al. 2017). The training samples are mapped in to a n-dimensional hyperplane by using the RBF kernel function which can be defined as:

$$K(s_i, s_j) = (-\gamma (s_i - s_j)), \gamma > 0 \tag{2}$$

Then mathematical expression of the n-dimensional hyperplane L needs to satisfy the following condition:

$$t_j(w \cdot s_j + b) + \varepsilon \geqslant 1 \tag{3}$$

where w denotes for the norm of normal hyperplane, and b is the constant. The maximum interval between vector and hyperplane can be derived by applying the Lagrangian multiplier (Abedini et al. 2019), and cost function can be expressed as:

$$L = 1/2w^2 - C_0 \sum_{i=1}^{n} \varepsilon \tag{4}$$

Kernel logistic regression model (KLR)

In statistical learning, when there are phenomena such as non-linear estimation, non-normal estimation, and uneven variance, it may cause invalid estimation by using the ordinary regression method (Chen et al. 2018). These problems were overcome after the introduction of logistic regression, and logistic regression is widely used to solve binary classification problem. However, the structure of original logistic regression model is relatively simple, the flexibility is relatively low, and it still has defects in dealing with non-linear classification problems (Chen et al. 2019). While the kernel function can help to solve these problems effectively in constructing logistic regression model. Therefore, the hybrid model, namely kernel logistic regression is created. In order to be consistent with the SVM model above, the RBF kernel function is determined to build KLR model. The expression of KLR model is as follows:

$$p_i(t=1|k_i)=\frac{1}{1+e^{-(k_i+\alpha)}} \tag{5}$$

where p_i is the probability of landslide occurrence, k_i stands for the i th row of $K(s_i, s_j)$, and is a constant for the intercept (Thai and Indra 2018).

Landslide net model (LSNet)

The deep learning has been widely used in the field of remote sensing image processing, including change detection, land use classification, image registration and so on. The deep belief networks, convolutional neural network (CNN), and auto coder are the three most commonly used network models in deep learning. The operating principle of these networks is to stack multiple layers within the model, and use the output of the previous item as the input of the next item, so that the features of each layer in the network can be converted into higher-dimensional features (Bui et al. 2020). Among them, the CNN has robust feature extraction capabilities and has been successfully applied in the field of image processing.

LSNet is a multi-layer feedforward neural network, the advantage of which is that it can process largescale data in the form of multiple arrays from the local and global input data. The structure of LSNet is consist of multiple layers, which are related to each other through a set of learnable weights and biases. The local and global-scale features can be captured by these convolutional blocks using scanning of the entire image. Meanwhile, the pooling layer and rectified linear unit (ReLU) layer are used for generalization to improve the non-linear fitting ability of the network (Li et al. 2021). Additionally, each convolutional layer contains feature maps obtained by multiple convolution kernels, and these feature maps share the node weights of the convolution kernels, so features can be extracted from different parts (Fig. 4). Specifically, the main operation performing in CNN can be generalized as follows:

$$O^l = pool_p(\sigma(O^{l-1} \cdot W^l + b^l)) \tag{6}$$

where O^{l-1} denotes the input feature map in lth layer, W^l and b^l, respectively, represent the weight and deviation of input feature layer convoluting by linear convolution, and is a non-linear function outside the convolution layer.

Assessment and comparison method

Result assessment method

In order to assess the accuracy of classification result and compare the performance of each model, statistical indexes are purposed to finish this work. A matrix (Table 3) is constructed by true positive (TP), false positive (FP), true negative (TN), and false negative (FN) calculating from training dataset (Pham et al. 2021). The accuracy and precision are calculated according to Eqs. (7) and (8) for accuracy assessment, meanwhile, the consistency of the results is verified with F1. The calculation process is as follows:

$$\text{Accuracy} = \frac{TP + TN}{TP + TN + FP + FN} \tag{7}$$

$$\text{Precision} = \frac{TP}{TP + FP} \tag{8}$$

$$1 = \frac{2 \times \text{Precision} \times \frac{TP}{TP+FN}}{\text{Precision} + \frac{TP}{TP+FN}} \tag{9}$$

Model comparison method

In this study, the work of model comparison is purposed to carry out from three indicators including the running speed of the model, the classification ability for landslide and non-landslide, and the generalized performance of the model. Among them, based on the validation dataset, the running speed of the model is quantitative expressed by time, and the sensitivity and specificity are, respectively, used to reflect the classification ability for landslide and non-landslide (Eqs. (9) and (10)) (Yanar et al. 2020). Additionally, the receiver operating characteristics curve (ROC) is used for assessing the generalized performance, and in general, the larger the area under ROC curve (AUROC), the stronger the generalization ability of the model (Dang et al. 2020):

$$\text{Sensitivtiy} = \frac{TP}{TP + FN} \tag{10}$$

$$\text{Specificity} = \frac{TN}{TN + FP} \tag{11}$$

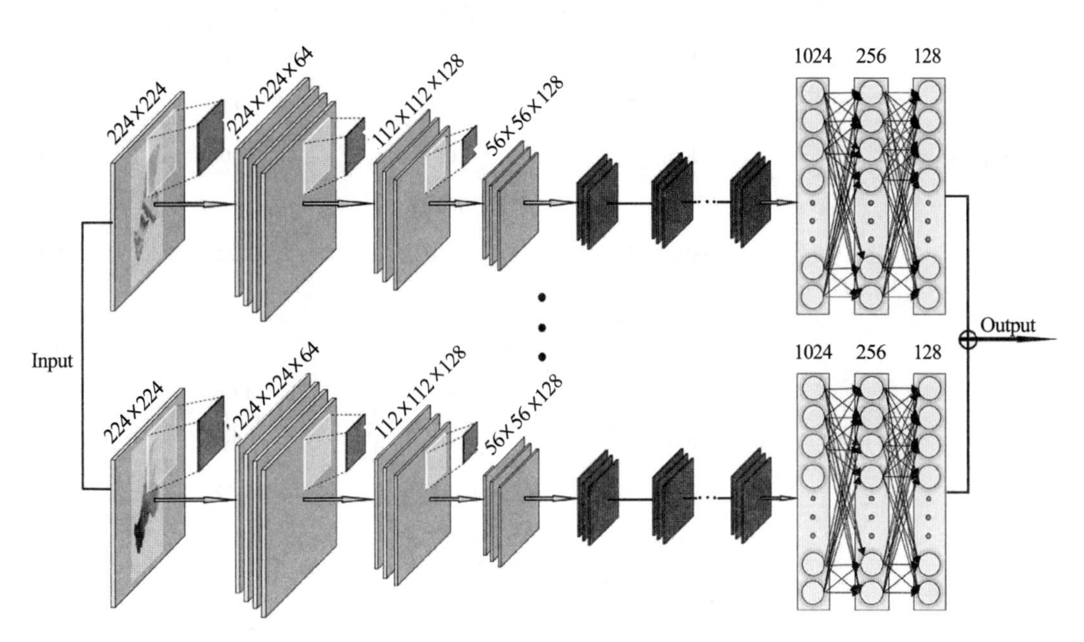

Fig.4 The structure schematic diagram of landslide net (LSNet)

Table 3 Discriminant matrix of statistical indexes

Samples		Predicted label	
		Landslide	Non-landslide
True label	Landslide	True positive (TP)	True negative (TN)
	Non-landslide	False positive (FP)	False negative (FN)

Results

The quantification results of FR for landslide predisposing factors

In this study, the FR value was employed to quantify each landslide predisposing factor according to the classification result. It can be observed from Table 4 that the interval of Tertiary from the lithology factor has the highest FR value (FR=2.32), followed by the range of < 100 from the distance to roads factor (FR=1.85), and the range of 278~548 from the altitude factor (FR=1.81). On the contrary, the lowest FR value appears in both the 1432~2107 interval of the altitude factor (FR=0.00) and the flat interval of the slope factor (FR=0.00).

Table 4 The *FR* calculation result for each class of landslide predisposing factors

Landslide predisposing factors	Classes	Are_{ij}/%	Sam_{ij}/%	FR
Altitude/m	278~548	32.00	57.98	1.81
	548~781	31.06	28.02	0.90
	781~1075	17.68	11.67	0.66
	1075~1432	12.33	2.33	0.19
	1432~2107	6.93	0.00	0.00
Slope angle/(°)	0.0000~9.6093	19.68	30.35	1.54
	9.6093~17.3502	27.24	31.91	1.17
	17.3502~24.8241	26.64	24.12	0.91
	24.8241~33.6327	18.95	10.51	0.55
	33.6327~67.7992	7.49	3.11	0.42
Slope aspect	Flat	0.06	0.00	0.00
	North	10.71	8.17	0.76
	Northeast	11.81	11.28	0.96
	East	15.32	19.84	1.30
	Southeast	12.82	12.84	1.00
	South	11.01	14.01	1.27
	Southwest	11.43	7.78	0.68
	West	14.37	16.34	1.14
	Northwest	12.47	9.73	0.78
NDVI	-0.0983~0.1717	7.26	5.45	0.75
	0.1717~0.2410	19.94	12.06	0.60
	0.2410~0.3030	28.86	31.13	1.08

Continued Table 4

Landslide predisposing factors	Classes	Are_{ij}/%	Sam_{ij}/%	FR
	0.3030～0.3698	28.81	35.02	1.22
	0.3698～0.5308	15.13	16.34	1.08
Distance to rivers/m	<100	7.19	8.56	1.19
	100～200	5.48	6.23	1.14
	200～300	5.16	7.39	1.43
	300～400	4.94	7.78	1.57
	>400	77.22	70.04	0.91
Distance to roads/m	<100	6.30	11.67	1.85
	100～200	5.39	8.17	1.51
	200～300	4.82	7.78	1.62
	300～400	4.50	6.23	1.38
	>400	79.00	66.15	0.84
Distance to faults/m	<1000	11.60	19.46	1.68
	1000～2000	10.87	12.84	1.18
	2000～3000	9.77	11.67	1.19
	3000～4000	8.88	9.73	1.10
	>4000	58.87	46.30	0.79
Lithology	Quaternary	18.02	28.40	1.58
	Tertiary	3.69	8.56	2.32
	Middle Devonian	9.46	1.95	0.21
	Lower Devonian	2.40	1.95	0.81
	Silurian	4.49	1.95	0.43
	Ordovician	2.38	1.17	0.49
	Cambrian	25.54	14.40	0.56
	Senian	33.99	36.19	1.06
MAP/(mm/year)	< 800	12.03	7.39	0.61
	800～850	22.39	19.84	0.89
	850～900	31.01	39.69	1.28
	900～950	14.82	10.12	0.68
	950～1000	3.37	5.06	1.50
	> 1000	16.39	17.90	1.09

The optimization result of landslide predisposing factors

The VIF and TOL values of each landslide predisposing factor were calculated based on the quantified landslide predisposing factors, and the calculation results are shown in Table 5. As can be seen from the results, the largest VIF value and the smallest TOL value appear in NDVI (VIF = 1.433, TOL = 0.698), followed by the altitude (VIF = 1.293, TOL = 0.773) and the aspect (VIF = 1.268, TOL = 0.789). By contrast, the distance to roads has the smallest VIF value and the largest TOL value (VIF = 1.019, TOL = 0.981). Since the VIF and TOL values of all landslide predisposing factors are not inside the critical range (VIF>10 and TOL<0.1), all factors are retained and used to prepare the dataset.

Based on the optimized landslide predisposing factors, the training and validation datasets were prepared according to aforementioned partition principle. Subsequently, the training dataset was used as the input data to implement the following three models.

Implementation of SVM model

In this study, the training dataset was used to construct the SVM model. Since the parameters of RBF kernel function are significant for model construction, the tenfold cross-validation method was used to search the most suitable parameter set (C_0, γ). The optimized parameter set is (241, 0.02). Then run the trained SVM model in the python platform, and adjust the output range of the model to 0.000 ~ 1.000 which also represents the LSI. In the end, the natural break (Jenks) method was used to divide the LSI into five ranges which, respectively, represent the very low susceptibility area (0.0899 ~ 0.2084), low susceptibility area (0.2085 ~ 0.4646), moderate susceptibil-ity area (0.4647 ~ 0.6228), high susceptibility area (0.6229 ~ 0.7893) and very high susceptibility area (0.7894 ~ 0.9224), furthermore the LSM was generated by converting these areas to image in ArcGIS software (Fig. 5).

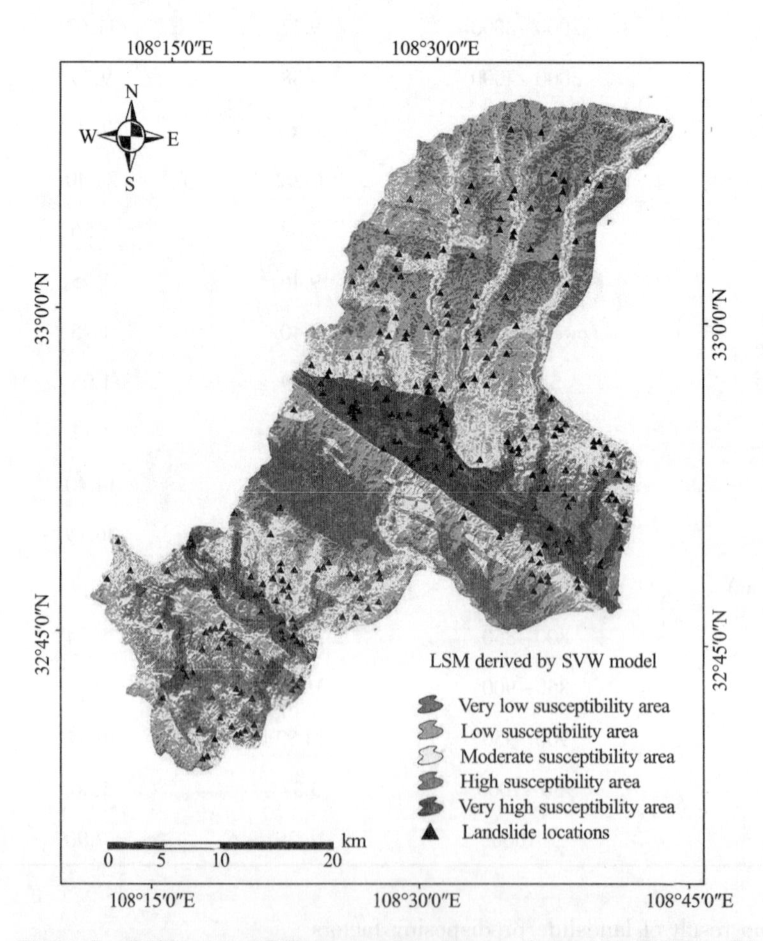

Fig.5 Landslide susceptibillty map of study area derived by SVM model

Table 5 The VIF and TOL values of each landslide predisposing factor

Landslide predisposing factors	VIF	Tolerances (TOL)
Altitude	1.293	0.773
Slope angle	1.032	0.969
Aspect	1.268	0.789
MAP	1.044	0.958
Lithology	1.103	0.907
Distance to rivers	1.148	0.871
Distance to faults	1.078	0.928
Distance to roads	1.019	0.981
NDVI	1.433	0.698

Implementation of KLR model

The construction progress of KLR model is similar to the SVM model. For the purpose of comparison, the parameter set (C_0, γ) was consistent with that of the SVM model. Subsequently, the training dataset was used as the input data for KLR model construction in the python platform, and the output range of the LSI was adjusted to 0.000~1.000. Finally, the LSI was divided into five ranges by using the natural break (Jenks) method. These five ranges, respectively, represent the very low susceptibility area (0.0145~0.2459), low susceptibility area (0.2460~0.3695), moderate susceptibility area (0.3696~0.5161), high susceptibility area (0.5162~0.6974) and very high susceptibility area (0.6975~0.9983), moreover the LSM corresponding to KLR model was generated in ArcGIS software (Fig. 6).

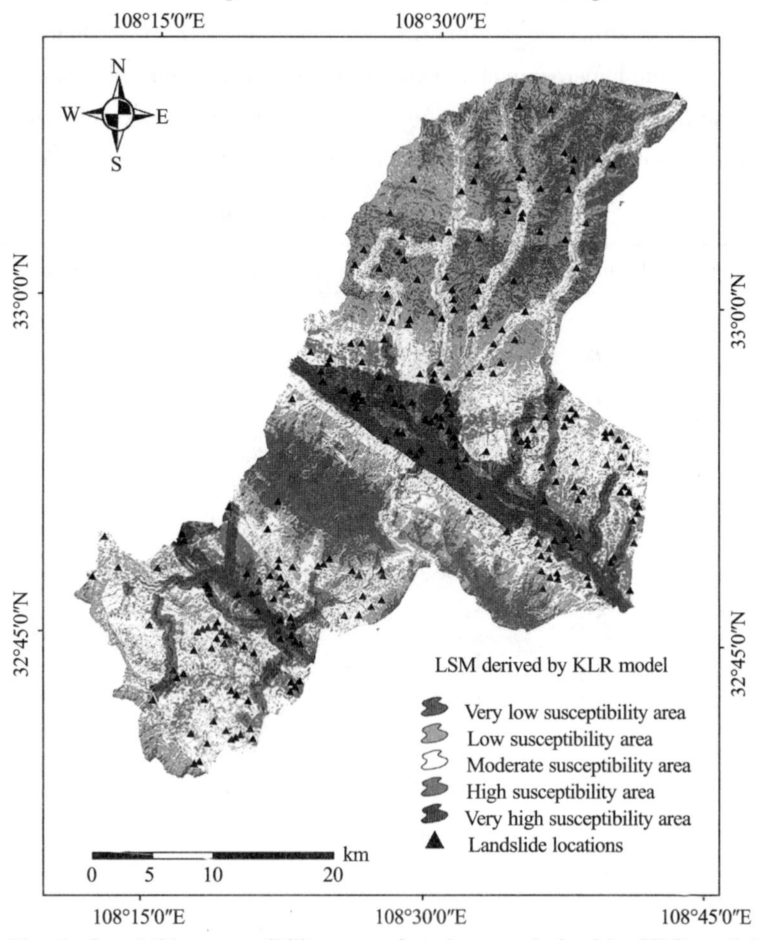

Fig. 6 Landslide susceptibility map of study area derived by KLR model

Implementation of LSNet model

The LSNet was coded using tensorflow 2.0 under the python environment, and running on a personal computer with Intel(R) Core(TM) i7-7700 k CPU, RTX 3080Ti GPU, 32 GB RAM, and the Windows 10 operating system. The LSNet had multi-layer structure, the size of input window was designed as 224×224. In fact, many researchers in image processing have demonstrated that the size of the data input has little effect on the task objective. Based on the data of multiple channels we input, the data size of 224 has already covered the entire area of Hanyin County. In theory, the context relationship of this range can fully meet the needs of LSA. In addition, we also used 96, 128, 196, 224, 256 image blocks to conduct experiments, respectively. The results show that the size of the input data has no obvious effect on the test accuracy.

AlexNet can implement more than 1000 categories of classification. In contrast, the landslide susceptibility mapping is a binary classification problem, which does not require deep network design. For this reason, in this study, the size of input layer in convolution kernel of LSNet was set to 5 × 5, the size of the convolution kernel for the other layers was set to 3 × 3, the number of feature maps for each layer was set to 64, 128, 128, 256, 256, respectively. At the same time, a pooling layer, non-linear activation function ReLU and batch normalized BN were set after each convolutional layer. Based on computer graphics vision, all other parameters of the LSNet were empirically optimized, for instance, the learning rate and epoch are set as 0.0001 and 600 to learn the depth features through back propagation. Subsequently, the number of neurons in the fully connected layers was set to 1024, 256, 128, 2, respectively (Table 6). Since the classical network, VGG and UNet have proved the validity of 3 × 3 convolution kernel and 64,128,256,1024 feature image parameters in various regression and classification tasks. Therefore, on the basis of these parameters, we design a network structure with multiple channel inputs and fuse them at the end of the network. In particular, we use the hyperparameter search method to obtain the fully connected layers and training parameters of the multi-channel fusion structure, such as learning rate, batch size and epoch. Finally, softmax was used to estimate the probability of landslide occurrence to output confidence, namely LSI.

Table 6 The architecture of LSNet

Layer	Feature map	Feature map	Size	Kernel Size	Stride	Activate
Input			224 × 224 × 3			ReLU
1	2×CNN	64	224 × 224 × 64	3 × 3	1	ReLU
	Max pooling	64	112 × 112 × 64	3 × 3	2	ReLU
2	2 × CNN	128	112 × 112 × 128	3 × 3	1	ReLU
	Max pooling	128	56 × 56 × 128	3 × 3	2	ReLU
3	2 × CNN	256	56 × 56 × 256	3 × 3	1	ReLU
	Max pooling	256	28 × 28 × 256	3 × 3	2	ReLU
4	2 × CNN	512	28 × 28 × 512	3 × 3	1	ReLU
	Max pooling	512	14 × 14 × 512	3 × 3	2	ReLU
5	2 × CNN	512	14 × 14 × 512	3 × 3	1	ReLU
	Max pooling	512	7 × 7 × 512	3 × 3	2	ReLU
6	FC	—	25088	—	—	ReLU
7	FC	—	1024	—	—	ReLU
8	FC	—	256	—	—	ReLU
9	FC	—	128	—	—	SoftMax

Similarly, the output range of LSI for LSNet was adjusted to 0.000 ~ 1.000, and, respectively, represents the very low susceptibility area (0.0045 ~ 0.2021), low susceptibility area (0.2022 ~ 0.3458), moderate susceptibility area (0.3459 ~ 0.4814), high susceptibility area (0.4815 ~ 0.8033) and very high susceptibility area (0.8034 ~ 0.9972) (Fig. 7).

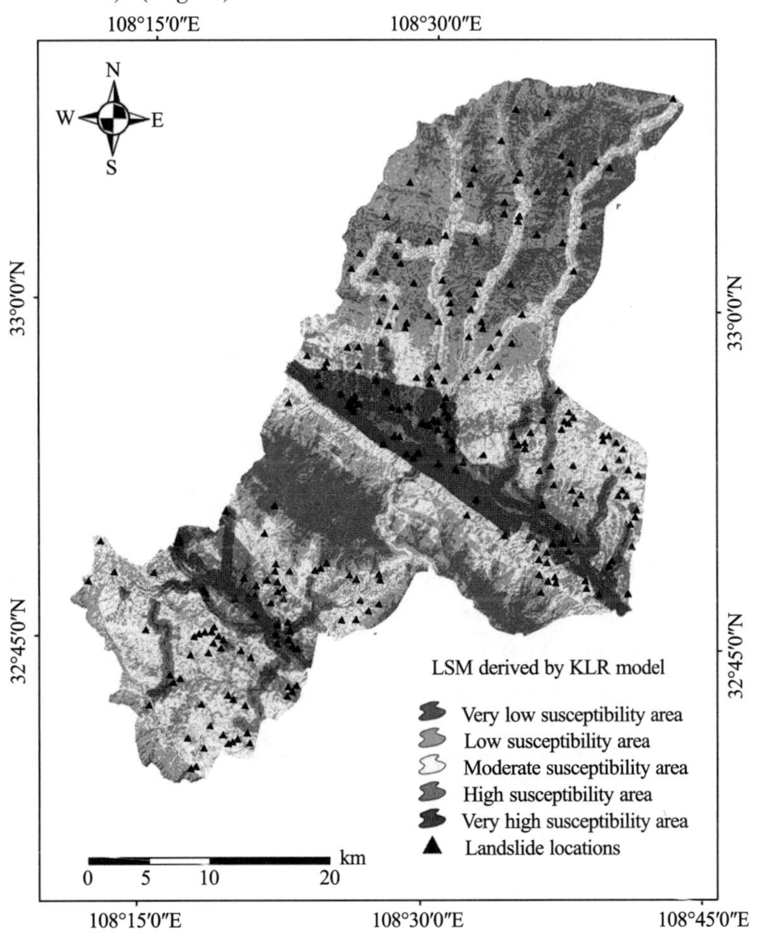

Fig.7 Landslide susceptibillty map of study area derived by LSNet model

Assessment of the results

The result of accuracy assessment

After mapping the LSMs of these three models, it is necessary to assess the quality of results. In this study, the matrix has been organized based on the validation dataset, then the accuracy, precision, and F1 values for each LSM were calculated (Table 7). As shown in Table 7, the LSNet model gets the highest accuracy value and precision value (accuracy = 0.950, precision = 0.951), by contrast, the SVM model gets the lowest accuracy value and precision value (accuracy = 0.825, precision = 0.850), while the performance of the KLR model is moderate. From the value of F1, the LSNet also gets the highest value (F1 = 0.951), followed by the KLR model and SVM model, which is also consistent with the ordering of accuracy and precision values.

Table 7 Calculation results of statistical indexes for landslide susceptibility mapping

Parameters	SVM	KLR	LSNet
TP	34	36	39
TN	32	36	37

Continued to Table 7

Parameters	SVM	KLR	LSNet
FP	6	6	2
FN	8	2	2
Accuracy	0.825	0.900	0.950
Precision	0.850	0.857	0.951
F1	0.829	0.900	0.951
Sensitivity	0.810	0.947	0.951
Specificity	0.842	0.857	0.949

The result of model comparison

In order to compare the running speed, classification and generalization performance, the run time, sensitivity, specificity and AUROC values were introduced to finish this work. As the results shown in Table 7, the largest sensitivity and specificity values belong to the LSNet model, indicating that the LSNet model has the best landslide and non-landslide classification abilities among these three models. On the contrary, the smallest sensitivity and specificity values belong to the SVM model, indicating that the landslide and non-landslide classification abilities of SVM model are the weakest among these three models.

For AUROC values (Fig. 8), the LSNet model also obtains the largest AUROC value (AUROC = 0.941), followed by the KLR model (AUROC = 0.899) and SVM model (AUROC = 0.835), and the results show that the LSNet model has the best generalization ability.

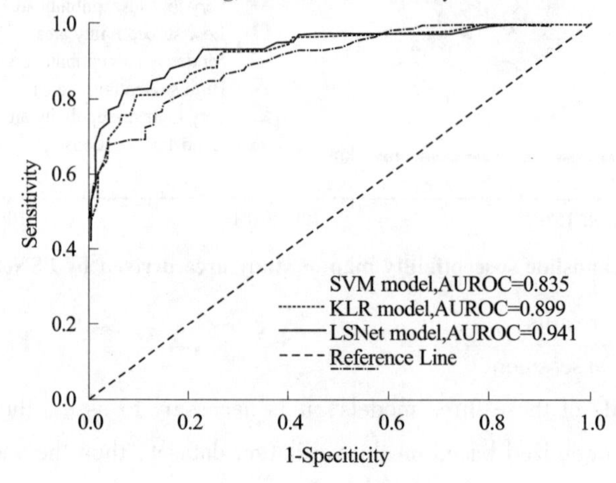

Fig.8 The ROC curves of each landslide susceptibillty model based on valldation dataset

Lastly, we measured the running speed of each model, the results show that the running speed of the SVM model (32 s) and the KLR model (27 s) are relatively close, while the running speed of the LSNet model (107 s) is significantly slower than the first models.

Discussion

In this paper, we show the progress and results of landslide susceptibility mapping based on SVM model, KLR model, and LSNet model in Hanyin County, Shaanxi Province, China. In terms of the model performance, although the classification accuracy of the three models is higher, the accuracy of LSNet and other statistical indexes are higher than that of SVM and KLR, which fully shows that the LSNet performs best in the study area.

Since both SVM and KLR are developed based on statistical theory, the quality of input data and the adjustment of model parameters in the process of model construction may affect the final result. Before preparation of input datasets, three classification methods, i. e., natural break (Jenks), equal interval, and custom interval were all used to grade FR-quantified landslide predisposing factors. However, the classification methods and results of landslide predisposing factors are inevitably affected by human factors, which may lead to over-fitting or under-fitting (Yacine and Pourghasemi 2019). For this reason, it is necessary to deeply analyze the impact of classification methods on data quality. Besides, this study only used two machine learning models for comparison, therefore, more models should be added for reference in subsequent research, so that the advantages and disadvantages of deep learning and machine learning in landslide susceptibility mapping can be more comprehensively compared.

In contrast, as a deep learning model, the input data of LSNet is a complete remote sensing image containing all the information. In order to distinguish landslide and non-landslide from image data, not only the objects in the image patch need to be characterized as landslides, but also need to accurately and reliably represent the contextual information of the landslide space background. The advantage of LSNet is to derive the category of the object at the image block level, and learn the spatial distribution through the CNN network with hierarchical representation, and finally obtain the probability of each object's category through multiple fully connected layers and softmax. It is different from machine learning in principle, and its specific advantages include: (1) LSNet can classify based on object blocks in a deep learning network of convolutional structure, and output the category probability; (2) LSNet uses the CNN model to learn the internal and overall spatial information of the object block to represent the contextual spatial semantic information of the category. LSNet represents the probability of the category at the object block level, which can avoid pixel-level misfits and improve the accuracy of classification (Dimililer et al. 2021); (3) LSNet can directly read remote sensing images without destroying the data structure, and can obtain richer and multi-source data sets, making the trained model robust. Interestingly, the running time of LSNet is significantly longer than that of SVM and KLR, which may be limited by the hardware performance of the computer, resulting in slower calculations. Nevertheless, this does not mean that the LSNet is not a state-of-the-art model and other studies have reached similar conclusions in their researches. However, since the occurrence of landslides is affected by multi-source factors, the characteristics of the landslides themselves are also very complex and cannot be described solely by human-represented features. Therefore, we try our best to use deep learning methods that are completely data-driven. In addition, in this study, the final network architecture of LSNet was determined by gradually increasing the ablation experiments, and at the same time, an LSNet with multiple channels was established combined with the data structure representing the landslide, which could well fulfill the requirements of LSA. Furthermore, because other network structures are generated based on their respective research objects and purposes, although we cannot conduct experiments on these networks one by one, we have designed a network structure that matches the confidence of the current landslide research. We have also demonstrated the effectiveness and accuracy of LSNet in our experiments.

On the other hand, as a black box model, DL cannot intuitively reflect the spatial distribution features of landslides in the study area during data preparation. On the contrary, in machine learning modeling, because FR is used to quantify the graded landslide predisposing factors, the spatial distribution of the landslide under the conditions of each predisposing factor can be intuitively reflected from the quantified results (Zhang et al. 2020). For instance, from the view of distance to rivers and roads, as the distance from roads and rivers increase, the FR value decreases, indicating that the closer to the river and the road, the more landslides are distributed. This is because the exposed rock and soil in study area have low mechanical strength, the surface is easily weathered and eroded, and the joints and fissures are very developed. Moreover, due to the scouring action from the river and excavation of the slope toe during road construction, the original stress structure of the slope was destroyed, which resulted in the instability of the slope and generated a large number of potential landslides. This consistent with the phenomenon we observed in the field, and is similar to the results of geological hazard studies in similar areas of the study area (Wang et al. 2016; Liu et al. 2020).

Conclusion

Landslide susceptibility mapping is a key step for landslide prevention work. This study used Hanyin County, Shaanxi Province, China as the study area to complete the work of landslide susceptibility mapping by building the LSNet model, SVM model, and KLR model, and generated the LSM. Then various statistical indexes were applied for the accuracy assessment, and the ROC curves was employed to compare the performance and classification ability of the models. In summary, the main conclusions are as follows: (1) in the process of dataset preparation and parameter adjustment, the machine learning model will inevitably be affected by human factors, resulting in unstable classification results. However, LSNet can overcome human interference and generate objective classification results. (2) LSNet can avoid the problems of over-fitting and under-fitting. The classification accuracy in the study area is high, moreover the generalization is stronger than the SVM model and the KLR model. The LSNet can be promoted and used in the study area.

This study introduced the construction method of LSNet model in detail, and compared the performance of LSNet model (deep learning), SVM model (machine learning), and KLR model (hybrid model), which enriched the landslide database of the study area and can provide reference for the application of deep learning model in landslide prevention in the future. Furthermore, the results of this study can improve the efficiency of landslide prevention for government decision-making in similar study areas, which is conducive to rapid response of landslide warning. Comprehensive risk assessors and land use planner can benefit from our study findings. Additionally, the proposed approach is an innovative method that may also help other scientists to develop landslide susceptibility maps in other areas, but also as an approach that could be used in geo-environmental problems besides natural hazard assessments.

Acknowledgements

The authors wish to express their sincere thanks to Heng Zhang (Chang'an University) for useful information provided.

Author contributions

Conceptualization, TZ and HW; methodology, TC; software, DL; validation, TZ; formal analysis, ZS; investigation, LH; resources, CL; data curation, TW; writing—original draft preparation, TZ; writing—review and editing, TZ and HW; visualization, YL; supervision, LH; project administration, TZ; funding acquisition, TZ. All authors have read and agreed to the published version of the manuscript.

Funding

This study is financially supported by Shaanxi Province Natural Science Basic Research Program (2022JQ-457), National Natural Science Foundation of China (211035210511), Fundamental Research Funds for the Central Universities (300102351502), Shaanxi Province Enterprises Talent Innovation Striving to Support the Plan (2021-1-2-1), and Inner scientific research project of Shaanxi Land Engineering Construction Group (SXDJ2021-10, SXDJ2021-30, SXDJ2022-16).

Availability of data and materials

Restrictions apply to the availability of these data. Data were obtained from Hanyin County People's Government and are available http://www.hanyin.gov.cn/ with the permission of Hanyin People's Government.

References

Abedini M, Ghasemian B, Shirzadi A et al (2019) A comparative study of support vector machine and logistic model tree classifiers for shallow landslide susceptibility modeling. Environ Earth Sci 78:560-577.

Aditian A, Kubotab T, Shinoharab Y (2018) Comparison of GIS-based landslide susceptibility models using frequency ratio, logistic regression, and artificial neural network in a tertiary region of Ambon, Indonesia. Geomorphology 318:101-111.

Aghdam IN, Pradhan B, Panahi M (2017) Landslide susceptibility assessment using a novel hybrid model of statistical bivariate methods (FR and WOE) and adaptive neuro-fuzzy inference system (ANFIS) at southern Zagros Mountains in Iran. Environ Earth Sci 76:237-255.

Balogun A-L, Rezaie F, Pham QB et al (2021) Spatial prediction of landslide susceptibility in western Serbia using hybrid support vector regression (SVR) with with GWO, BAT and COA algorithms. Geosci Front 12:101-104.

Benzekri W, Moussati AE, Moussaoui O et al (2020) Early forest fire detection system using wireless sensor network and deep learning. Int J Adv Comput Sci Appl 11:496-502.

Bui DT, Tuan TA, Klempe H et al (2016) Spatial prediction models for shallow landslide hazards: a comparative assessment of the efficacy of support vector machines, artificial neural networks, kernel logistic regression, and logistic model tree. Landslides 13:361-378.

Bui DT, Shahabi H, Shirzadi A et al (2018) Landslide detection and susceptibility mapping by AIRSAR data using support vector machine and index of entropy models in cameron highlands, Malaysia. Remote Sens 10:1527-1533.

Bui T-A, Lee P-J, Lum K-Y et al (2020) Deep learning for landslide recognition in satellite architecture. IEEE Access PP.

Carranza EJM (2015) Data-driven evidential belief modeling of mineral potential using few prospects and evidence with missing values. Nat Resour Res 24:291-304.

Carrara A, Cardinali M, Guzzetti F et al (1995) Gis Technology in Mapping Landslide Hazard. Geogr Inform Sys Assess Nat Hazards 8:135-175.

Chen W, Shahabi H, Shirzadi A et al (2018) Novel hybrid artificial intelligence approach of bivariate statistical-methods-based kernel logistic regression classifier for landslide susceptibility modeling. Bull Eng Geol Env 78:4397-4419.

Chen W, Yan X, Zhao Z et al (2019) Spatial prediction of landslide susceptibility using data mining-based kernel logistic regression, naive Bayes and RBFNetwork models for the Long County area (China). Bull Eng Geol Env 78:247-266.

Chen W, Chen X, Peng J et al (2021) Landslide susceptibility modeling based on ANFIS with teaching-learning-based optimization and Satin bowerbird optimizer. Geosci Front 12:93-107.

Cloud GD (2020) (GF-2) PMS sub-meter high resolution data products, 2020, Retrieved August 10, 2020.

Conoscenti C, Ciaccio M, Caraballo-Arias NA et al (2014) Assessment of susceptibility to earth-flow landslide using logistic regression and multivariate adaptive regression splines: a case of the Bence River basin (western Sicily, Italy). Geomorphology 242:49-64.

Constantin M, Bednarik M, Jurchescu MC et al (2011) Landslide susceptibility assessment using the bivariate statistical analysis and the index of entropy in the Sibiciu Basin (Romania). Environ Earth Sci 63:397-406.

Dang V-H, Hoang N-D, Nguyen L-M-D et al (2020) A novel GIS-based random forest machine algorithm for spatial prediction of shallow landslide susceptibility. Forests.

Dimililer K, Dindar H, Al-Turjman F (2021) Deep learning, machine learning and internet of things in geophysical engineering applications: an overview. Microprocess Microsyst 80:103-613.

Fan W, Wei XS, Cao YB et al (2017) Landslide susceptibility assessment using the certainty factor and analytic hierarchy process. J Mt Sci 21:100-119.

Fang Z, Wang Y, Peng L et al (2021) A comparative study of heterogeneous ensemble learning techniques for landslide susceptibility mapping. Int J Geogr Inf Sci 35:321-347.

Guzzetti F, Mondini AC, Cardinali M et al (2012) Landslide inventory maps: new tools for an old problem. Earth Sci Rev 112:42-66.

Hong H, Liu J, Bui DT et al (2018) Landslide susceptibility mapping using J48 decision tree with adaboost, bagging and rotation forest ensembles in the Guangchang area (China). CATENA 163:399-413.

Huang F, Zhang J, Zhou C et al (2020) A deep learning algorithm using a fully connected sparse autoencoder neural network for landslide susceptibility prediction. Landslides 17:217-229.

Kayastha P, Dhital MR, Smedt FD (2013) Application of the analytical hierarchy process (AHP) for landslide susceptibility mapping: a case study from the Tinau watershed, west Nepal. Comput Geosci 52:398-408.

Kumar D, Thakur M, S. Dubey C, et al (2017) Landslide susceptibility mapping & prediction using support vector machine for Mandakini River Basin, Garhwal Himalaya, India. Geomorphology 295:115-125.

Kumar D, Roshni T, Singh A et al (2020) Predicting groundwater depth fluctuations using deep learning, extreme learning machine and gaussian process: a comparative study. Earth Sci Inf 13:1-14.

Li R, Wang N (2019) Landslide susceptibility mapping for the Muchuan County (China): a comparison between bivariate statistical models (WoE, EBF, and IoE) and their ensembles with logistic regression. Symmetry 11:762-781.

Li J, Zhang Y (2017) GIS-supported certainty factor (CF) models for assessment of geothermal potential: a case study of Tengchong County, southwest China. Energy 140:552-565.

Li W, Fang Z, Wang Y (2021) Stacking ensemble of deep learning methods for landslide susceptibility mapping in the three gorges reservoir area China. Environmental Res Risk Assess.

Liu Y, Huang Q (2006) The formation and mechanism of an expansive soil highway landslide. Coal Geol Explor 13: 41-44.

Liu H, Li X, Meng T et al (2020) Susceptibility mapping of damming landslide based on slope unit using frequency ratio model. Arab J Geosci 13:178-192.

Panahi M, Jaafari A, Shirzadi A et al (2020) Deep Learning Neural Networks for Spatially Explicit Prediction of Flash Flood Probability. Geosci Front 12:370-383.

Pandey VK, Pourghasemi HR (2020) Landslide susceptibility mapping using maximum entropy and support vector machine models along the highway corridor, Garhwal Himalaya. Geocarto Int 35:168-187.

Pham BT, Bui DT, Dholakia MB et al (2016) A Comparative Study of Least Square Support Vector Machines and Multiclass Alternating Decision Trees for Spatial Prediction of Rainfall-Induced Landslides in a Tropical Cyclones Area. Geotech Geol Eng 34:1807-1864.

Pham BT, Prakash I, K. Singh S, et al (2019) Landslide susceptibility modeling using reduced error pruning trees and different ensemble techniques: Hybrid machine learning approaches. CATENA 175:203-218.

Pham QB, Yacine A, Ali SA et al (2021) A comparison among fuzzy multi-criteria decision making, bivariate, multivariate and machine learning models in landslide susceptibility mapping. Geomat Nat Haz Risk 12:1741-1777.

Polykretis C, Chalkias C (2018) Comparison and evaluation of landslide susceptibility maps obtained from weight of evidence, logistic regression, and artificial neural network models. Nat Hazards 93:249-274.

Pourghasemi HR, Moradi HR, Aghda SMF (2013) Landslide susceptibility mapping by binary logistic regression, analytical hierarchy process, and sta-tistical index models and assessment of their performances. Nat Hazards 69:605-609.

Pradhan B, Lee S (2010) Delineation of landslide hazard areas on Penang Island, Malaysia, by using frequency ratio, logistic regression, and artificial neural network models. Environ Earth Sci 60:1037-1054.

PRC, 2020. The Ministry of emergency management released the basic situation of natural disasters nationwide in 2019. In.

Razavizadeh S, Solaimani K, Massironi M et al (2017) Mapping landslide susceptibility with frequency ratio, statistical index, and weights of evidence models: a case study in northern Iran. Environ Earth Sci 76:499-512.

Saadoud D, Hassani M, Peinado FJM et al (2018) Application of fuzzy logic approach for wind erosion hazard mapping in Laghouat region (Algeria) using remote sensing and GIS. Aeol Res 32:23-24.

Sameen MI, Pradhan B, Lee S (2020) Application of convolutional neural networks featuring Bayesian optimization for landslide susceptibility assessment. CATENA 186:104249.

SBGMR, 1989. Regional geology of shaanxi province. geological publishing house. (In Chinese), Beijing, China.

Siahkamari S, Haghizadeh A, Zeinivand H et al (2017) Spatial prediction of flood-susceptible areas using frequency ratio and maximum entropy models. Geocarto Int 33:927-941.

Soma AS, Kubota T, Mizuno H (2019) Optimization of causative factors using logistic regression and artificial neural network models for landslide susceptibility assessment in Ujung Loe Watershed, South Sulawesi Indonesia. J Mt Sci 16: 144-162.

Sun X, Chen J, Han X et al (2020) Application of a GIS-based slope unit method for landslide susceptibility mapping along the rapidly uplifting section of the upper Jinsha River, South-Western China. Bull Eng Geol Env 79:533-549.

Thai PB, Indra P (2018) Machine learning methods of kernel logistic regression and classification and regression trees for landslide susceptibility assessment at part of Himalayan Area, India. Indian J Sci Technol 11:1-10.

Umar Z, Pradhan B, Ahmad A et al (2014) Earthquake induced landslide susceptibility mapping using an integrated ensemble frequency ratio and logistic regression models in West Sumatera Province, Indonesia. CATENA 118:124-135.

Wang Z, Brenning A (2021) Active-learning approaches for landslide mapping using support vector machines. Remote Sens 13:2588-2607.

Wang L, Guo M, Sawada K et al (2016) A comparative study of landslide susceptibility maps using logistic regression, frequency ratio, decision tree, weights of evidence and artificial neural network. Geosci J 20:117-136.

Wang Y, Fang Z, Hong H (2019) Comparison of convolutional neural networks for landslide susceptibility mapping in Yanshan County, China. Sci Total Environ 666:953-975.

Wang W, He Z, Han Z et al (2020) Mapping the susceptibility to landslides based on the deep belief network: a case study in Sichuan Prov- ince, China. Nat Hazards 103:3239-3261.

Wu R, Zhang Y, Guo C et al (2020) Landslide susceptibility assessment in mountainous area: a case study of Sichuan-Tibet railway, China. Environ Earth Sci 79:157-177.

Xiao L, Zhang Y, Peng G (2018) Landslide susceptibility assessment using integrated deep learning algorithm along the China-Nepal highway. Sensors 18:4436-4472.

Xie W, Nie W, Saffari P et al (2021) Landslide hazard assessment based on bayesian optimization-support vector machine in Nanping City, China. Nat Hazards 26:18-31.

Yacine A, Pourghasemi HR (2019) How do machine learning techniques help in increasing accuracy of landslide susceptibility maps? Geosci Front 11:328-345.

Yanar T, Kocaman S, Gokceoglu C (2020) Use of mamdani fuzzy algorithm for multi-hazard susceptibility assessment in a developing Urban Settlement (Mamak, Ankara, Turkey). Int J Geo-Inf 9:114-128.

Youssef AM, Pourghasemi HR (2021) Landslide susceptibility mapping using machine learning algorithms and comparison of their performance at Abha Basin, Asir Region, Saudi Arabia. Geosci Front 12:639-655.

Youssef AM, Mohamed A-K, Biswajeet P (2015) Landslide susceptibility mapping at Al-Hasher area, Jizan (Saudi Arabia) using GIS-based frequency ratio and index of entropy models. Geosci J 19:113-134.

Zhang T, Han L, Han J et al (2019) Assessment of landslide susceptibility using integrated ensemble fractal dimension with kernel logistic regression model. Entropy 21:218-234.

Zhang Y, Lan H, Li L et al (2020) Optimizing the frequency ratio method for landslide susceptibility assessment: a case study of the caiyuan basin in the Southeast mountainous area of China. J Mt Sci 17:340-357.

Zhao X, Chen W (2020) Optimization of computational intelligence models for landslide susceptibility evaluation. Remote Sensing 12:2180-2200.

Zhou S, Fang L (2015) Support vector machine modeling of earthquake- induced landslides susceptibility in central part of Sichuan prov- ince, China. Geoen Disasters 2:303-315.

Zhu L, Huang L, Fan L et al (2020) Landslide Susceptibility prediction modeling based on remote sensing and a novel deep learning algorithm of a cascade-parallel recurrent neural network. Sensors 20:1576-1591.

本文曾发表于 2022 年《Geoscience Letters》第 9 卷

Monitoring and Assessing Land Use/Cover Change and Ecosystem Service Value Using Multi-Resolution Remote Sensing Data at Urban Ecological Zone

Siqi Liu[1,2,3], Guanqi Huang[4], Yulu Wei[1,2] and Zhi Qu[3,5]

(1.Institute of Land Engineering and Technology, Shaanxi Provincial Land Engineering Construction Group Co., Ltd., Xi'an 710075, China; 2.Shaanxi Provincial Land Engineering Construction Group Co., Ltd., Xi'an 710075, China; 3.Key Laboratory of Degraded and Unused Land Consolidation Engineering, the Ministry of Natural Resources, Xi'an 710075, China; 4.Guangzhou Urban Planning & Design Survey Research Institute, Guangzhou 510060, China; 5.School of Land Engineering, Chang'an University, Xi'an 710000, China)

【**Abstract**】An urban ecological zone (UEZ) is an important part of a city, focusing on environmental conservation and ecological economic development simultaneously. During the past decade, the urban scale of Xi'an city in China has been expanding, and the population has been increasing rapidly. This dramatic change is a huge challenge to urban sustainability. It puts forward higher requirements for the construction of an UEZ. Under different spatial resolution scales, this study adopted Landsat8-OLI and gaofen-2 (GF-2) satellite high-resolution remote sensing data to interpret the land use/cover change (LUCC) of the Weihe River UEZ. The ecosystem service value (ESV) was assessed, and the ecological effect was analyzed based on LUCC. The results showed that the spatial distribution of land types in the Weihe River UEZ changed significantly from 2014 to 2020. The construction land gathered to the southeast. Especially, the vegetative land (i.e., forestland, grassland and other green land) and water body showed a slightly increasing trend since the official establishment of the UEZ in 2018. The cultivated land area gradually reduced, and the vegetative land area tended to be concentrated as well as expanded. Through the interpretation of GF-2 remote sensing data, the ESV at the Weihe River UEZ showed a downward trend in general. The high-value areas were mainly distributed in the Weihe River and its surrounding beach areas, which were greatly affected by river water scope. Construction land normally had low ESV, and it was affected by human activities obviously. Therefore, the development of urban construction had significant impacts on the Weihe River UEZ.

【**Keywords**】Land use/cover change (LUCC); Ecosystem service value (ESV); Multi-resolution remote sensing; Landscape pattern; Urban ecological zone (UEZ); Urban construction

1 Introduction

Land is the basis of human production and living, as well as being the carrier of all kinds of resources[1]. Due to the interference from human activities, the land is facing increasing pressure[2]. According to the natural characteristics of land itself and its certain economic and social purposes, human adopts biological, physical, chemical and other technical means to launch long-term and periodic management on land[3]. Land use/cover change (LUCC) is an important way in which humans act on ecosystems. LUCC directly affects the landscape pattern, biodiversity and ecosystem[4-7]. With the acceleration of urbanization and industrialization, especially in developing countries, LUCC driven by human activities has led to the deterioration of the ecological environment and the degradation of ecosystem service function[8-9].

Different land use types have different effects on ecosystem services in terms of extent and pattern. Similarly, different land use intensity has different effects on the ecosystem[10]. Ecosystem services refer to all ecosystem products and services that contribute to human survival and living quality[11-12]. Ecosystem service value (ESV) is the benefits achieved from the ecosystem for human beings, directly and indirectly. It mainly includes the input of useful materials and energy to the economic and social system, the acceptance and transformation of waste from the system, and the direct provision of services to human society[13-14].

The research object of ESV can be a single landscape unit including forest[15], island[16], basin[17], and even an administrative scope, such as a specific city[18] or regions[19-21] of different scales. In 1997, Costanza[22] systematically assessed the functional value of global ecosystem services. Through quantitative evaluation of ESV, the variation trend of ecosystem structure and function had been revealed; moreover, the impact mode and degree of human activities on the ecosystem had been figured out[23]. The analysis of ESV is mainly realized by calculating the values of different ecosystem service functions in the whole region[24]. Subsequently, some researchers improved the valuation method of ecosystem services and proposed the equivalent factor method to calculate the ESV[25], as well as the derived static ESV[26].

In recent years, studies combining ESV with urban expansion and landscape pattern evolution have gradually emerged[27-29]. This research is of great significance for promoting land ecology and urban sustainable development. Landscape pattern, namely the spatial arrangement pattern of landscape elements, is an important approach to studying ecosystem quality and function. The temporal and spatial change of landscape pattern significantly affects regional biodiversity and other ecological indicators. It is an important method to analyze ecological quality change, which has direct impacts on ESV[30]. The main reason leading to landscape pattern change is the external interference, whose action mechanism is comprehensive, involving the interaction among human activities, nature and various organisms[31]. At present, the study on landscape pattern combining LUCC is commonly accepted. In addition, the selection of landscape index, its particle size and scale effects have also been a concern for a long time[32].

Many researchers focused on the prediction of ecosystem services and landscape pattern based on LUCC. Cellular Automata (CA)—Markov prediction model has been commonly used[33-35]. Meanwhile, evaluating landscape ecological risk and analyzing various ecosystem service functions based on LUCC, such as calculating habitat quality and carbon storage based on the integrated valuation of ecosystem services and trade-offs (InVEST) model[36-37], have become hot issues in this field.

A city is a complicated and integrated organization. The environment and development conditions of different urban regions inside the city vary wildly. It is insufficient to study only large-scale objects such as regions and cities. More importantly, it is necessary to study the small and medium-sized ecological areas in the city. However, such research studies are still relatively few.

The research on LUCC and ESV of urban ecological areas based on remote sensing is an important method to assess urban ecological environment quality scientifically. In order to ensure research efficiency and accuracy, remote sensing data with different resolutions should be adopted. Different types of multi-resolution remote sensing image data for research are complementary. Although the obtained remote sensing information has a certain redundancy, it is more accurate and comprehensive. Higher resolution images can make up for the shortcomings of lower resolution images in fine structure extraction, while lower resolution images have advantages in large-scale observation and rapid information extraction.

The different imaging principles of remote sensing images result in their various target characteristics[38]. Therefore, combined with Landsat8-OLI and high-resolution gaofen-2 (GF-2) satellite remote sensing data, this study aims at (1) analyzing LUCC and comparing the scale effect of the Weihe River UEZ in Xi'an on different spatial resolution scales, (2) based on high-resolution remote sensing data, assessing the ESV and landscape pattern index, and analyzing the ecological effects, (3) putting forward some suggestions for the ecological development of Xi'an and the construction of the Weihe River UEZ.

2 Research area and data

2.1 Research area

The Weihe River is the largest tributary of the Yellow River. It is one of the most important rivers in the Guanzhong region. It originates from Niaoshu Mountain, Weiyuan County, Gansu Province. It is 502.4 km long in Shaanxi Province, with a drainage area of 67100 km^2, accounting for 32.6% of the total areas of Shaanxi Province[39].

The Weihe River flows through the loess area and carries great quantities of silt and sand. Most of the Weihe Basin belongs to the temperate continental monsoon climate zone, and the precipitation is concentrated in summer. The Weihe River mainly flows through the Weiyang district, Baqiao district, Gaoling district and Lintong district within Xi'an city[40]. The Weihe River UEZ is located in the north of Weiyang district, close to the south bank of the Weihe River, with a length of approximately 27 km and a maximum width of approximately 8.4 km. The total areas reach 120.78 km^2 (Fig.1).

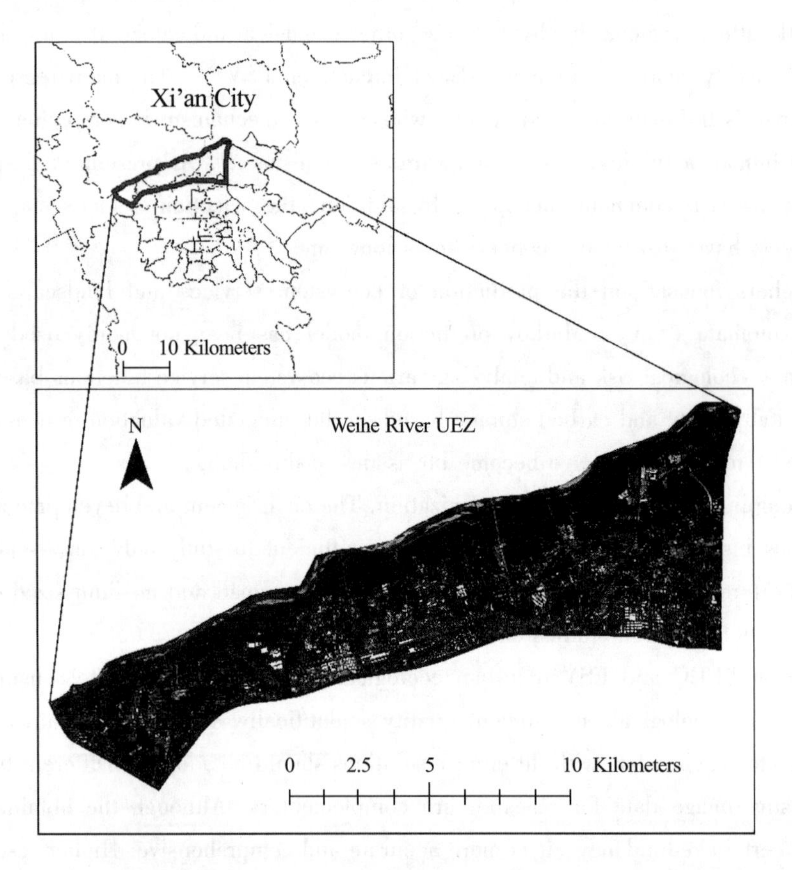

Fig. 1 The location and satellite image of research area

The Weihe River UEZ was built on the basis of a comprehensive treatment project for the Weihe River and was officially established by the government in 2018. In taking ecological protection as the main goal, this ecological zone is an important part of Xi'an sustainable development planning, as well as an important urban ecological barrier. It provides ecological services such as water conservation and eco-tourism for the city.

2.2 Data sources

In this study, Landsat8-OLI data in 2014, 2016, 2018 and 2020 were obtained from the geospatial data cloud platform (http://www.gscloud.cn, accessed on 1 July 2022), one Landsat 8-OLI image data of the bottom cloud volume was used each year to cover the study area, and these data were all imaged between April and July. Landsat8-OLI can achieve global coverage every 16 days, with a spatial resolution of 30 m; origin data were preprocessed in Envi 5.3 (Radiometric Calibration and FLAASH Atmospheric Correction). The 2018 and 2020 GF-2 high-resolution data were obtained from Shaanxi Satellite Application Technology Center, one GF-2 image data was used in 2018, and two GF-2 image data were used in 2020 to cover the study area. Due to the difficulty of acquiring GF-2 data, the imaging times for the data in 2018 and 2020 were October and March, respectively, and the data used were low cloud volumes. GF-2 global revisit frequency not greater than 5 days, with a spatial resolution reaching 0.8 m (after preprocessing and fusion). The maximum likelihood classification was used to complete the interpretation. The kappa coefficient of Landsat8-OLI data interpretation was higher than 0.8, and the GF-2 data interpretation was higher than 0.85, which met the research requirements.

Due to the difference in spatial resolution between the two remote sensing data, the Landsat8-OLI data was divided into five categories, including unused land, water body, vegetative land, cultivated land and construction land, in this study. At the same time, the GF-2 data was divided into six categories, including unused land, water body, forestland, grassland, cultivated land and construction land. The grassland and forestland types in the GF-2 classification are uniformly covered by vegetative land types in the Landsat8-OLI data classification. The boundary data of the Weihe River UEZ was obtained from the planning map of Xi'an city and the vectorization of the planning scope of the Weihe River UEZ. Other vector boundary data were from the resource and environmental science and data center of the Chinese Academy of Sciences (https://www.resdc.cn, accessed on 4 July 2022).

3 Methodology

3.1 Land use transfer matrix

The analysis of LUCC transfer is an important method to reveal regional human activities' impact on land use. Based on the land use data of the Weihe River UEZ interpreted by Landsat8-OLI and GF-2 remote sensing images, this study established the phase-by-phase transfer matrix in 2014, 2016, 2018 and 2020, respectively. Additionally, it compared the scale difference in the interpretation effect between Landsat8-OLI data with 30 m spatial resolution and GF-2 data with 0.8 m spatial resolution. The land use transfer matrix can scientifically reveal the overall land use structure and change characteristics, reflecting the change direction of land use types[41,42]. The formula is shown as follows:

$$S_{ij} = \begin{matrix} S_{11} & \cdots & S_{1n} \\ \vdots & & \vdots \\ S_{n1} & \cdots & S_{nn} \end{matrix} \tag{1}$$

where S represents the areas of various land use types. i and j refer to the land use types at the beginning and end of the study period, respectively. n is the total number of land use types. S_{ij} is the area transferred from land use i to land use j during the research period.

3.2 Landscape pattern index

Landscape pattern reflects the spatial differences of patches in size, shape and attributes. Landscape pattern is usually reflected by various landscape pattern indexes. In this study, separation degree, fragmentation degree and dominance degree were selected to analyze the landscape pattern of the Weihe River UEZ. Fragmentation statistics (Fragstats) is widely used to calculate landscape metrics for categorical map patterns, of which Fragstats 4.2 is the most authoritative. Based on this platform, the landscape pattern index of the Weihe River UEZ was calculated pixel by pixel through the moving-window method based on the GF-2 data in 2018 and 2020. Specifically, the selected window size was 100 m. The equations applied for calculation are shown as follows:

$$SPLIT = \frac{A^2}{\sum_{i=1}^{m}\sum_{j=1}^{n} a_{ij}^2} \tag{2}$$

$$F_i = \frac{np_{ij}}{ta_{ij}} \tag{3}$$

$$L_i = 1 - shei_{ij} \tag{4}$$

where Splitting index ($SPLIT$) is the splitting degree of landscape. a_{ij} is the area (m^2) of patch ij. A is the total landscape area. F_i is landscape fragmentation, np_{ij} is the number of patches of landscape type i, ta_{ij} is the area of landscape type i. F_i represents the fragmentation degree of landscape segmentation, reflecting the complexity of landscape spatial structure and the degree of human interference to the landscape to a certain extent. L_i represents landscape dominance and $shei_{ij}$ is the Simpson evenness of landscape type i. L_i is used to measure the deviation of landscape diversity from the maximum diversity, indicating the degree to which the landscape is controlled by several main landscape types.

3.3 Ecosystem service value (ESV)

ESV reflects the benefits that human beings can directly obtain from the ecosystem, which is closely related to the health and stability of the ecosystem and environmental policies. This study was based on the Arcgis 10.2 platform, using 300 m grid size, the ESV of the Weihe River UEZ was calculated in 2018 and 2020 using the GF-2 data by referring to the equivalent factor method and equivalent factor table proposed. Xie et al. considered that the equivalent ESV coefficient was 1/7 of the unit area value of market grain[25,42]. First, according to the grain yield, sowing area and market price (japonica rice, wheat and corn, obtained from Xi'an statistical yearbook and Shaanxi statistical yearbook) of prefecture-level cities in the study area, the unit area value of grain in the Weihe River UEZ was obtained. Second, the ESV of each category in Table 1 was calculated by the 1/7 multi-year from 2018 to 2020 average grain unit area value (equivalent ESV coefficient) multiplied by the equivalence factor. The data of equivalence factor referred to the studies of Xie et al. and Guo et al.[43-44]. Eventually, the ESV of the Weihe River UEZ in Xi'an can be calculated as Equation (5). The ESV calculations for both 2018 and 2020 are based on the same unit area ESV data (Table 1) since the same equivalence factor and multi-year average grain unit area values are used in the calculations.

$$\mathrm{ESV}_i = \sum (A_i \times V_i) \tag{5}$$

where ESV_i represents the ESV value of land use type i. A_i is the area of land use type i. V_i is the ESV per unit area of land use type i.

Table 1 Per unit ESV of various land use types at Weihe River UEZ (¥/hm²/year)

Primary type	Secondary type	Cultivated land	Forestland	Grassland	Water body	Construction land	Unused land
Supply service	Food production	1539.08	507.90	661.80	815.71	0.00	30.78
	Material production	600.24	4586.46	554.07	538.68	0.00	61.56
Regulation service	Gas regulation	1108.14	6648.83	2308.62	784.93	0.00	92.34
	Climate regulation	1492.91	6264.06	2400.96	3170.50	0.00	200.08
	Hydrological regulation	1185.09	6294.84	2339.40	28888.53	0.00	107.74
	Waste disposal	2139.32	2647.22	2031.59	22855.34	0.00	400.16
Support service	Soil conservation	2262.45	6187.10	3447.54	631.02	0.00	26.16
	Biodiversity	30.78	6941.25	2878.08	5279.04	0.00	615.63
Cultural service	Aesthetic Landscape	261.64	3201.29	1339.00	6833.52	0.00	369.38

3.4 Spatial analysis of ESV

3.4.1 Hotspot analysis

Based on ArcMap 10.2 platform, the hotspot analysis tool was used to test whether there existed significant spatial aggregation of high and low values of ESV in 2018 and 2020, using the GF-2 data in the Weihe River UEZ. The spatial distribution of ESV can be analyzed as a result.

$$G_i^* = \frac{\sum_{j=1}^{n} \omega_{i,j} x_j - \bar{X}\sum_{j=1}^{n} \omega_{i,j}}{S\sqrt{\left[n\sum_{j=1}^{n} \omega_{i,j}^2 - \left(\sum_{j=1}^{n} \omega_{i,j}\right)^2\right]/(n-1)}} \tag{6}$$

$$\bar{X} = \frac{1}{n}\sum_{j=1}^{n} x_j \tag{7}$$

$$S = \sqrt{\frac{\sum_{j=1}^{n} x_j^2}{n} - (\bar{X})^2} \tag{8}$$

where X_j is the attribute value of spatial unit j. $\omega_{i,j}$ represents the spatial weight between the spatial units i and j (value is 1 when adjacent, 0 when not adjacent). n is the number of spatial units. $\bar{X}$ is the mean value, and S is the standard deviation. The statistical result of G_i^* is Z points. Statistically significant positive Z-score indicates hot spots, and the higher the Z score, the closer the hot spots gather. At the same time, negative values indicate cold spots. The lower the Z score, the closer the cold spots gather.

3.4.2 Global spatial autocorrelation of ESV

Based on GeoDa platform, the spatial autocorrelation analysis of data is carried out to describe its spatial dependence and aggregation degree, and the global Moran's I index is selected for analysis in 2018 and 2020 using the GF-2 data. The specific formula is shown as follows:

$$I = \frac{n}{S_0}\frac{\sum_{i=1}^{n}\sum_{j=1}^{n} \omega_{i,j} z_i z_j}{\sum_{i=1}^{n} z_i^2} \tag{9}$$

where z_i is the deviation between the attribute of element i and its average value ($x_i - \bar{X}$), $\omega_{i,j}$ is the spatial weight between elements i and j. n is equal to the total number of elements, and S_0 is the aggregation of all spatial weights:

$$S_0 = \sum_{i=1}^{n} \sum_{j=1}^{n} \omega_{i,j} \tag{10}$$

Z_I -score is calculated in the following form:

$$Z_I = \frac{I - E[I]}{\sqrt{V[I]}} \tag{11}$$

where

$$E[I] = -1/(n-1) \tag{12}$$

$$V[I] = E[I^2] - E[I]^2 \tag{13}$$

3.4.3 Local spatial autocorrelation of ESV

The local indicators of spatial association (LISA) clustering map is used to evaluate the spatial clustering of land deformation in the study area in 2018 and 2020 based on the GF-2 data. When the LISA coefficient value is larger than 0, it indicates that there is a spatial positive correlation between local spatial units and adjacent spatial units, which is expressed as high or low. Otherwise, it indicates low-high or high-low, and there is a spatial negative correlation between local spatial units and adjacent spatial units. LISA was calculated by Equation (14).

$$I_i = \frac{(x_i - \bar{x})}{\frac{1}{n}\sum_{i=1}^{n}(x_i - \bar{x})^2} \sum_{i,j=1}^{n} \omega_{ij}(x_j - \bar{x}) \tag{14}$$

where I_i is LISA of Moran's I, and n is the number of spatial units participating in the analysis. $\omega_{i,j}$ is the space weight matrix. LISA statistics of Moran's I index test is the same as the global Moran's I[45].

4 Results

4.1 Land use transfer and scale effect analysis

This study constructed the land use transfer status of the Weihe River UEZ from different spatial resolution scales. Through the phase-by-phase land use distribution map and the construction of a land use transfer matrix, this paper analyzed the land use transfer from the spatial and temporal dimensions.

As shown in Fig.2, the construction land in the Weihe River UEZ was mainly distributed in the south and southeast. Additionally, with the construction of the ecological area, the aggregation degree in the southeast became higher. Green space was mainly distributed on both sides of the main stream of the Weihe River. Cultivated land was mainly distributed in a small part of the west, with concentrated spatial distribution characteristics.

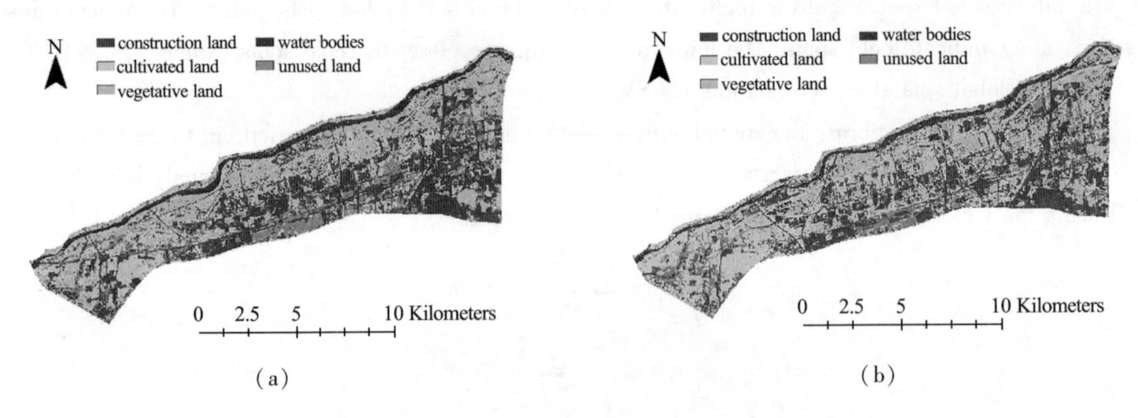

Fig. 2 Landsat8-OLI land use classification map of Weihe River UEZ in 2014 (a), 2016 (b), 2018 (c), 2020 (d), respectively

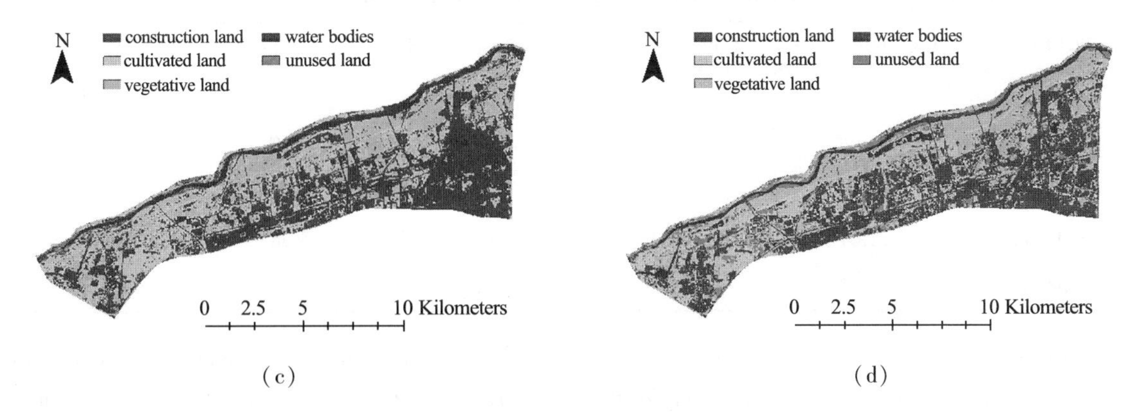

Continued to Fig.2

As shown in Fig.3, the construction land areas were all larger than 40 km^2 in four different years. This indicated that construction land was dominant in terms of land use types. In 2018 it reached its peak, 53.51 km^2, and then it decreased. By comparison, the areas of unused land and water body were relatively small in this UEZ. The water body areas were steady while the variation of unused land dramatically fluctuated during this period. In 2018, the areas of unused land reached the bottom, approximately 2.94 km^2. Moreover, it showed a remarkably increasing trend after 2018. Additionally, water body areas slightly decreased and cultivated land areas obviously decreased from 2016 to 2020.

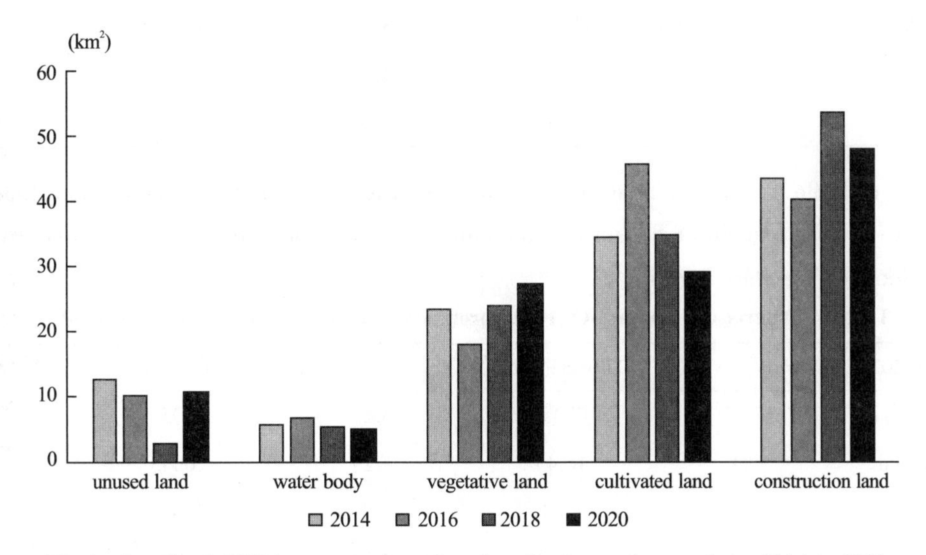

Fig.3 Landsat8-OLI interpretation of various land use changes from 2014 to 2020

Table 2 indicates that the change of construction land converted to cultivated land was remarkable from 2014 to 2016, reaching 11.58 km^2. At the same time, the areas of vegetative land transferred into cultivated land were relatively large. This period was in the early stage of the development of the Weihe River UEZ. The construction land was still dominant in the research area, a part of which was converted into vegetative land. This variation implemented the official planning of the Weihe River UEZ. A large area of vegetative land transferred into cultivated land, indicating that the impact of locally cultivated land on ecology was significant from 2014 to 2016.

Table 2 Matrix of Landsat8-OLI interpretation of land use data from 2014 to 2016

From 2014 to 2016	Construction land	Cultivated land	Vegetative land	Water body	Unused land
Construction land	26.12	11.58	1.72	0.78	3.49
Cultivated land	4.68	22.34	5.15	0.50	2.29
Vegetative land	4.12	6.11	10.38	2.08	0.91
Water body	1.61	0.33	0.37	3.38	0.15
Unused land	3.85	5.18	0.26	0.03	3.40

According to Table 3, the transformation of cultivated land into construction land was obvious from 2016 to 2018. By comparing the land use map of the Weihe River UEZ, the cultivated land area was mainly concentrated in the southwest, and the transformation from cultivated land to construction land occurred in this region concurrently. In addition, the transformation from cultivated land to vegetative land cannot be ignored, and it reached 9.54 km^2. The increase of vegetative land reflected the development of the Weihe River UEZ and the improvement of local ecological quality from 2016 to 2018.

Table 3 Matrix of Landsat8-OLI interpretation of land use data from 2016 to 2018

From 2016 to 2018	Construction land	Cultivated land	Vegetative land	Water body	Unused land
Construction land	30.86	3.48	3.86	1.53	0.65
Cultivated land	11.72	22.79	9.54	0.30	1.17
Vegetative land	3.48	5.20	8.30	0.37	0.55
Water body	1.68	0.33	1.32	3.13	0.30
Unused land	5.90	2.80	1.13	0.14	0.28

According to Table 4, the significant change from 2018 to 2020 was the transformation from cultivated land to vegetative land, reaching 8.29 km^2. In addition, the area from cultivated land to construction land was also remarkable, reaching 5.91 km^2.

Table 4 Matrix of Landsat8-OLI interpretation of land use data from 2018 to 2020

From 2018 to 2020	Construction land	Cultivated land	Vegetative land	Water body	Unused land
Construction land	37.35	7.45	3.69	2.05	3.10
Cultivated land	5.91	16.45	8.29	0.0045	3.94
Vegetative land	3.12	4.64	14.59	0.07	1.71
Water body	1.29	0.131	0.10	3.05	0.91
Unused land	0.41	0.57	0.78	0.09	1.09

Landsat8-OLI data with a spatial resolution of 30 m was not accurate enough to interpret land use, especially in small-scale areas. It would be confused in the interpretation of features with similar pixel spectra, such as small and scattered water bodies and buildings. However, it was good in the interpretation of features with large continuous areas, such as vegetative land and construction land. Generally, it reflected the trend of construction land gathering to the southeast. The land use spatial distribution map (Fig.4) and transfer matrix (Table 5) of the Weihe River UEZ were further built based on GF-2 remote sensing images in 2018 and 2020, respectively, and the resolution and accuracy were improved as a result.

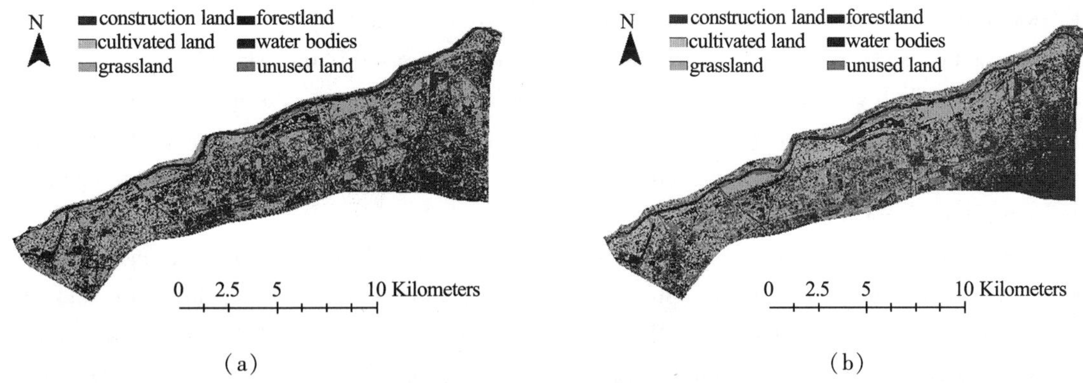

Fig. 4 GF-2 land use classification results of Weihe River UEZ in 2018 (a), 2020 (b)

Table 5 Matrix of GF-2 interpretation of land use data from 2018 to 2020

From 2018 to 2020	Construction land	Cultivated land	Grass land	Forest land	Water body	Unused land
Construction land	12.67	2.89	7.45	1.55	1.15	5.86
Cultivated land	6.37	3.73	7.03	2.03	1.59	4.83
Grass land	6.26	3.96	9.88	2.20	2.94	4.98
Forest land	2.85	1.93	4.74	1.34	1.31	2.12
Water body	3.24	1.66	3.12	0.64	1.13	2.23
Unused land	1.78	0.89	1.88	0.42	0.51	1.61

GF-2 land use classification results showed that the spatial distribution of construction land in 2018 and 2020 were concentrated in the southeast of the Weihe River UEZ. In terms of spatial agglomeration, the construction land in the research area is more obvious in 2020, while it was relatively scattered in 2018. Cultivated land was mainly concentrated in parts of western regions, and the largest land type was grassland.

Compared with Landsat8-OLI, GF-2 interpretation data showed an obvious scale effect in the proportion of various land use types (Fig.5). It was difficult to deal with different classes in a single pixel when interpreting small-scale regional land use data using 30 m resolution remote sensing data. The classifier will select a representative object as the pixel land classification according to its mechanism, ignoring the fragmented and scattered land classification. The dominant, concentrated and continuous land classification will expand, eventually resulting in the error in the total area of each category.

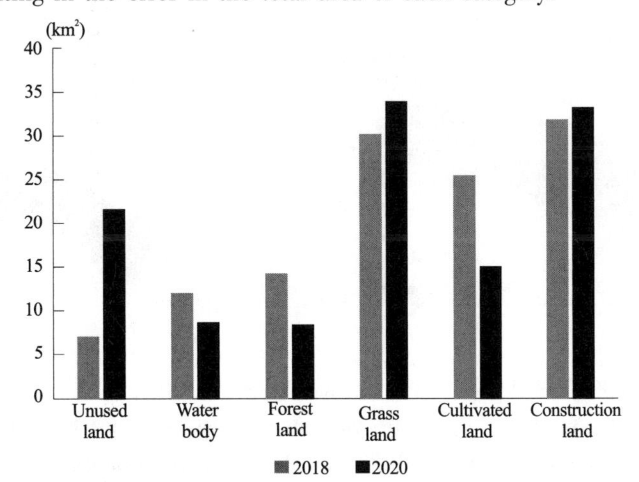

Fig. 5 GF-2 interpretation of various land use in 2018 and 2020

According to the interpretation of GF-2 data, the mutual conversion of land use types in the Weihe River UEZ was remarkable among construction land, grassland and cultivated land from 2018 to 2020 (Table 5). To be specific, there were 7.45 km^2 of construction land transferred into grassland, 7.03 km^2 of cultivated land transferred into grassland, 6.37 km^2 of cultivated land transferred into construction land and 6.26 km^2 of grassland transferred into construction land. The dramatic land use transformation reflected the instability of the current construction of the Weihe River UEZ. Meanwhile, it corresponded to the change of the spatial aggregation degree.

The transformation of land use in the Weihe River UEZ reflected the continuous strengthening of local ecological construction and the gradual improvement of ecological quality. However, due to the lack of interpretation of small area scale by Landsat8-OLI, the green space had not been classified in more detail, and the content covered by a single pixel and the information of similar spectral pixels were difficult to extract and classify. It was necessary to further utilize higher resolution GF-2 data for verification and analysis.

4.2 Landscape pattern analysis

According to the calculation of the landscape pattern index with the size of a 100 m window, the high-resolution indexes of spatial continuity of the Weihe River UEZ in 2018 and 2020 were obtained, as shown in Fig.6.

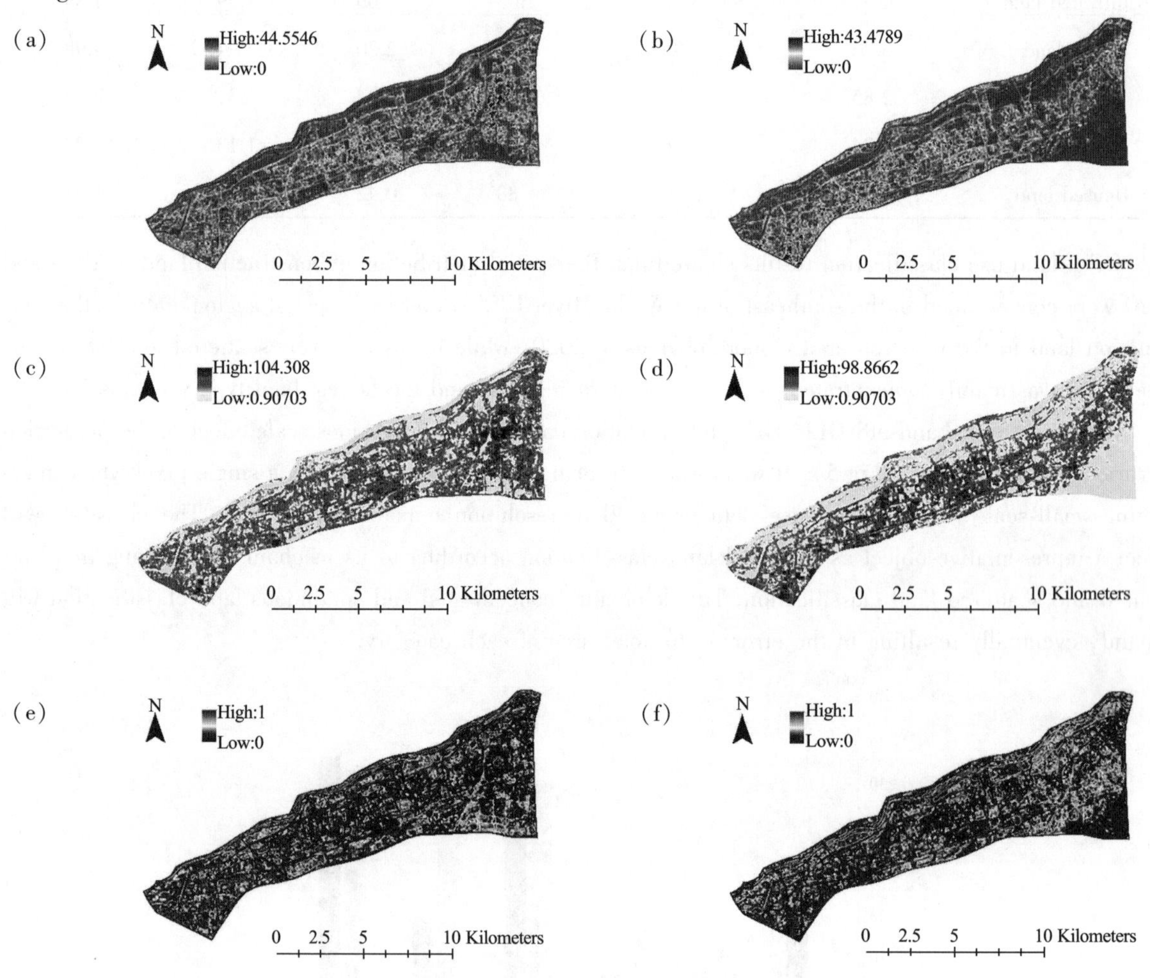

Fig.6 (a) The separation degree of Weihe River UEZ in 2018; (b) the separation degree of Weihe River UEZ in 2020; (c) the fragmentation degree of Weihe River UEZ in 2018; (d) the fragmentation degree of Weihe River UEZ in 2020; (e) the dominance of Weihe River UEZ in 2018; (f) the dominance of Weihe River UEZ in 2020. (GF-2 satellite based)

For different patches of a specific landscape pattern, the separation degree of individual distribution showed the spatial aggregation of landscape patches to a certain extent. From 2018 to 2020, areas with low separation degree concentrated in urban agglomeration areas, reflecting that the spatial agglomeration of construction land had a significant impact on the spatial distribution of separation degree. The regions with high separation had a trend of spatial dispersion from 2018 to 2020. Although the construction land showed a trend of spatial aggregation, the landscape separation degree of patches was relatively large within the construction land. It indicated that the spatial aggregation of construction land still has a large potential for improvement and optimization. Fragmentation was an important indicator reflecting the degree of the ecological health of the landscape. From 2018 to 2020, the low-value areas increased significantly. The low-value areas were concentrated in the southeast of the Weihe River UEZ and in part of the river's south beach. Generally, the ecosystem in this area was relatively stable. From 2018 to 2020, the areas with high dominance increased, which were mainly distributed in the concentrated construction land. This result can be mutually verified with land use interpretation data. The construction land is dominant in the study area. Since construction land has a great impact on ESV and landscape pattern, the dominance of construction land needs to be controlled to a certain extent.

4.3 ESV calculation based on high-resolution images

ESV calculation results showed that the total value of ESV in the Weihe River UEZ was 227.46 million yuan in 2018 and 175.27 million yuan in 2020. It decreased by 52.19 million yuan. The average value of ESV in the Weihe River UEZ was 0.17 million yuan in 2018 and 0.13 million yuan in 2020. It decreased by 0.04 million yuan. According to Fig.7, the ESV value along the river was generally at a higher level and the highest level. The ESV in 2018 and 2020 showed an outward attenuation trend along the river. By comparing the land use data interpreted by GF-2 in 2018 and 2020, the ESV in the construction area was at the lowest level and lower level as a whole. The ESV with significant human interference showed a lower trend than that with less interference. It showed a trend of gathering to the southeast from 2018 to 2020 with the change of spatial distribution of construction and other land types. The phenomenon that high-grade ESV existing in the river areas showed was that the ESV generated by water conservation service was the main ecological service in the Weihe River UEZ. Compared with 2018, the ESV in 2020 showed a higher degree of spatial aggregation in low-value areas, which was conducive to environmental management and ecological governance.

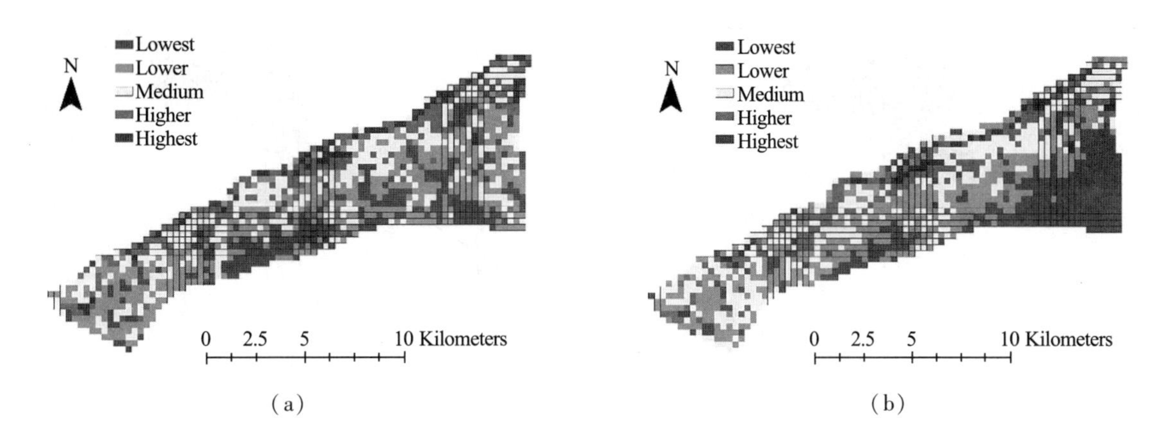

Fig.7 The ESV of Weihe River UEZ in 2018 (a) and 2020 (b). (GF-2 satellite based)

4.4 Spatial analysis of ESV

In 2018 and 2020, the hot spots of ESV in the Weihe River UEZ were spatially distributed around the main stream of the Weihe River (Fig.8). Generally, the spatial relation of the cold-hot spot is stable. However, the variation trend in the cold spot area was relatively conspicuous. It was mainly determined by the change in construction density. The areas with insignificant cold and hot spots correspond to the grassland, forestland and cultivated land based on GF-2 interpretation data, in which the building density was low. It reflected that the distribution of construction land had an obvious density center. Especially in 2020, it showed a decreasing trend from southeast to outside.

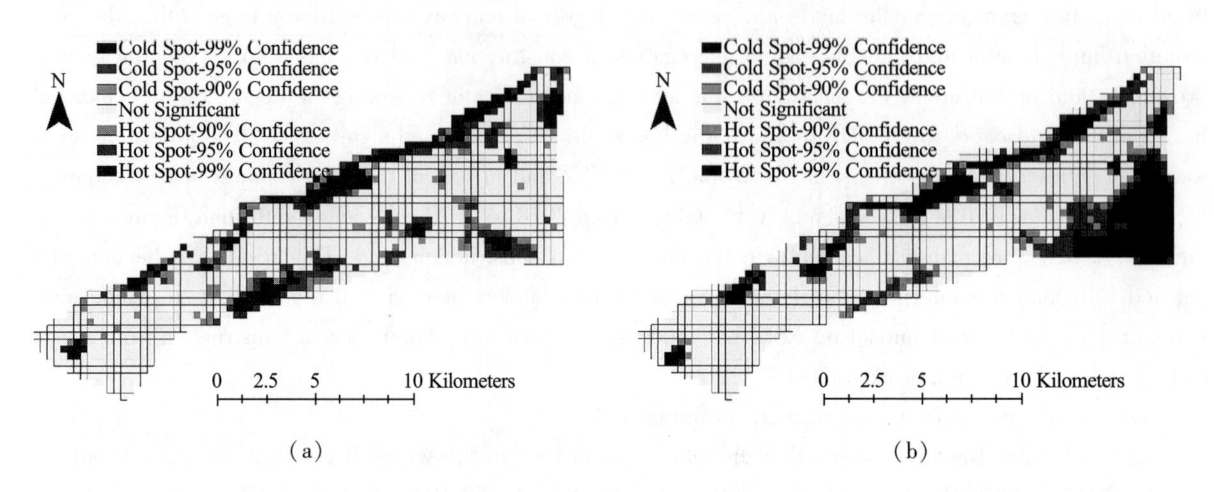

Fig. 8 The ESV cold-hot spot of Weihe River UEZ in 2018 (a) and 2020 (b).
The blue areas refer to cold spot in the 90%, 95% and 99% confidence interval (GF-2 satellite based)

Based on the GeoDa platform, the global spatial autocorrelation calculation of ESV was carried out. The Moran's I index of ESV in 2018 and 2020 was 0.527 and 0.613, respectively, which was in an upward trend (Fig.9). Under 999 randomized displacements of the GeoDa platform, the p value was 0.001, which passed the significance test. Local spatial autocorrelation and its LISA map were used to analyze further the spatial autocorrelation of ESV (Fig.10).

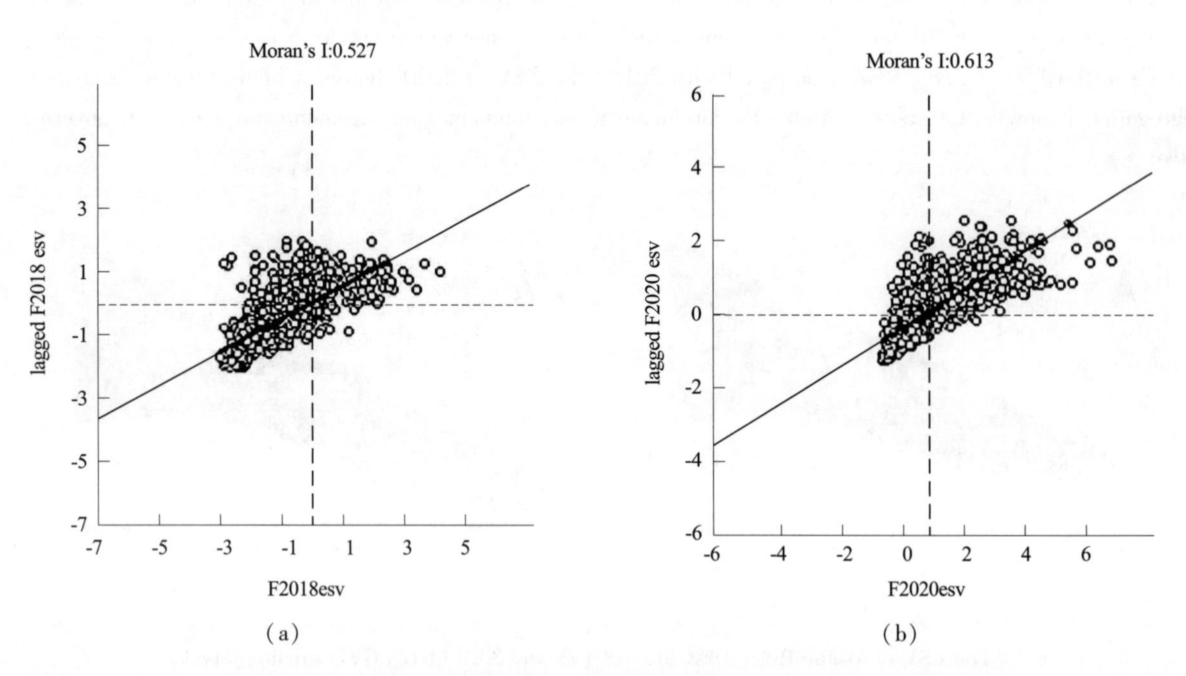

Fig. 9 The Moran's I scatter diagram of global spatial autocorrelation of ESV
at Weihe River UEZ in 2018 (a) and 2020 (b). (GF-2 satellite based)

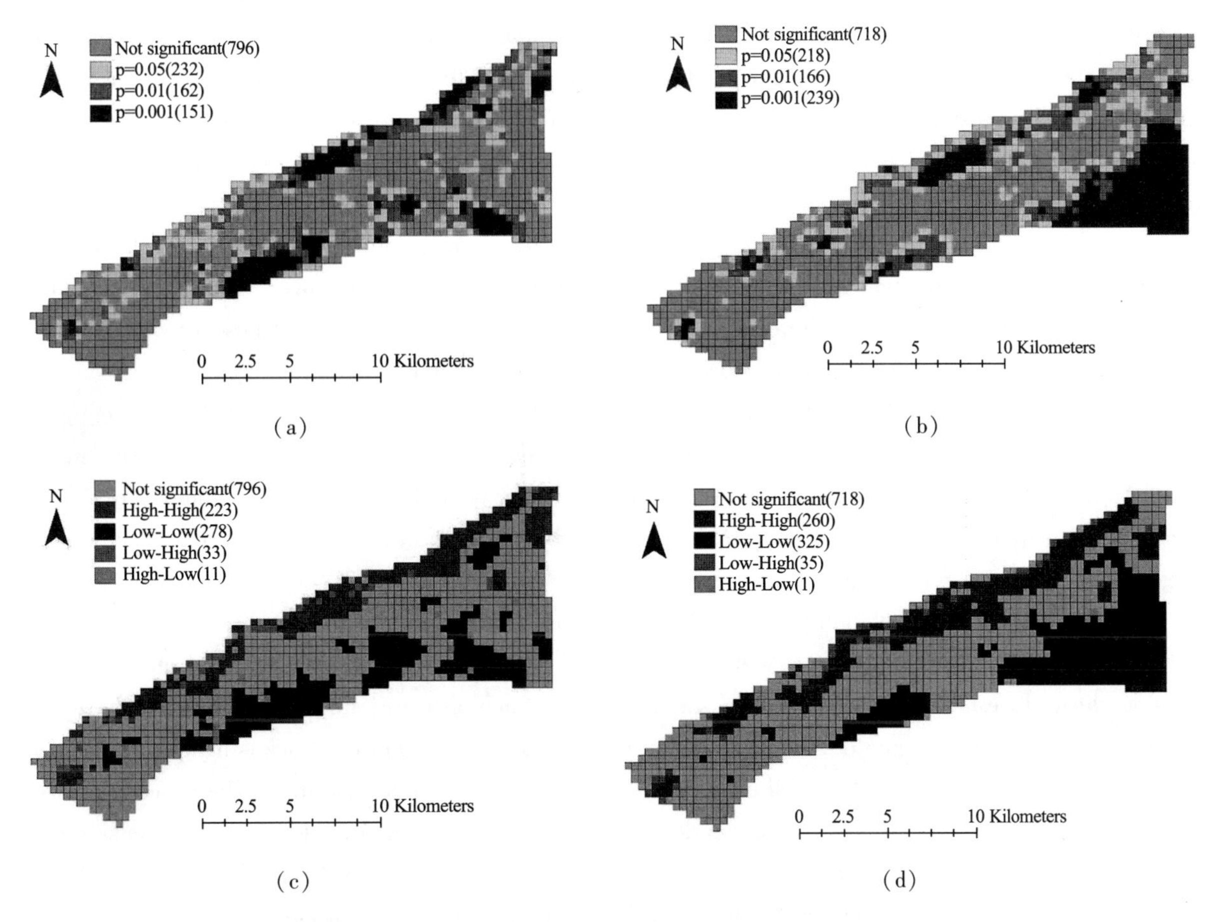

Fig. 10 LISA significance map of local spatial autocorrelation of ESV in 2018 (a) and 2020 (b); local spatial autocorrelation LISA clustering map of ESV in 2018 (c) and 2020 (d). The blue areas refer to low-low aggregation type

According to Fig.10, the regions with obvious significance were similar to the hot spot analysis diagram, indicating that there was a significant spatial autocorrelation phenomenon in the cold-hot spot significant region of the Weihe River UEZ. The region occupied by insignificant areas included 796 grids in 2018 and 718 grids in 2020, which showed a decreasing trend. In 2018 and 2020, the positive spatial autocorrelation regions of high-high aggregation and low-low aggregation corresponded to cold and hot spots, respectively. Between the regional distribution of low-high aggregation and high-high aggregation along the river, there were few high-low aggregation regions and decreased continuously with time. It was mainly distributed in low-low aggregation areas.

5 Discussion

5.1 Multi-resolution remote sensing method

The GF-2 high-resolution image can present more details, which can provide a reference for the selection of training samples when interpreting the Landsat8-OLI image in the same area. However, GF-2 is significantly affected by redundant information, such as building shadows during interpretation, causing a certain amount of blur and error in some information. It can be solved by post correction or adding training samples, but the efficiency is low. When using 30 m resolution remote sensing data to interpret small-scale regional land use data, it is difficult to deal with different classifications in a single pixel. The classifier will select a representative class as the pixel class according to its mechanism, resulting in the total area error of each class. The scale effect of remote sensing images with different spatial resolutions in a unified region has a great impact on the interpretation results. If the spatial resolution is particularly small, it will increase the difficulty and efficiency of training samples and interpretation, and the workload will increase exponentially. Therefore, it is suggested that the appropriate spatial resolution scale should be selected according to the research area and research targets. Moreover, the method and mechanism of fast processing high-resolution remote sensing images should be explored[46-47].

According to LUCC analysis by GF-2 remote sensing data interpretation, there existed a large-scale spatial aggregation phenomenon in the southeast of the Weihe River UEZ. Moreover, it transmitted to the ESV and its cold-hot spot results. However, this phenomenon did not appear in the landscape pattern index results calculated by using the moving window method. The reason probably is that the moving window has a certain fuzzy effect on the data. That was, the calculation process was not completely based on a single grid unit but took into account the values within a certain range around the central grid. Therefore, in the transitional boundary of the building aggregation area, the calculation of its landscape pattern index was significantly affected by the grid values of non-building aggregation areas. If the size of the moving window were larger, the calculation results would be more blurred.

5.2 Planning implementation and policy interpretation

According to the planning of the Weihe River UEZ, the government planned to build an education-housing cluster by establishing schools in the southeast areas. Such measures attracted more people to live in this region and drove economic development. From the point of view of land use, it was mainly reflected in the increase of construction land as well as the building density in the southeast areas. The concentration of urban settlement increases regional population density as well as human activity intensity. At the same time, the traffic efficiency and infrastructure utilization efficiency can be improved. Although the regional ESV decreases, it has positive impacts on reducing energy consumption and carbon emissions. This view corresponds to Song et al., who opined that high density, mixed land use and good connectivity of urban built environments are healthy and sustainable modes[48].

As shown in Fig.11, a large area of an ecological buffer zone, including protective green land and urban ecological parks, had been planned between the Weihe River and Xi'an downtown areas. Given the result of the research, scattered landscape patches gathered into a coherent large landscape belt, and the ESV in the north part had been improved from 2014 to 2020. Xu et al. pointed out that land resource was the main factor limiting urban sprawl. Urban construction had negative impacts on urban ESV[49]. Zhang et al. considered that urban parks could alleviate the impact of human activities on the ecological environment, especially in high-density cities[50]. The growth of urban parks is of great significance to urban sustainability. In the west part of the Weihe River UEZ, the continuous reduction of cultivated land indicated that urban expansion encroached on farmland. This phenomenon was spontaneous since these suburban areas are out of the planning scope. This spontaneous transformation of land use has potential threats, which should unify urban and rural planning, as well as strictly manage the non-agricultural conversion of farmland. In addition, the water of the Weihe River is obviously affected by local precipitation. The water flow varies greatly between different years and months. With many industrial factories relocated from downtown to suburbs, the water consumption, including agricultural, industrial and domestic water, use became quite large. This was consistent with the result that the water areas had decreased slightly. Therefore, the change in land use and ESV is just a symptom, and the key is to understand its causes and collective management behavior.

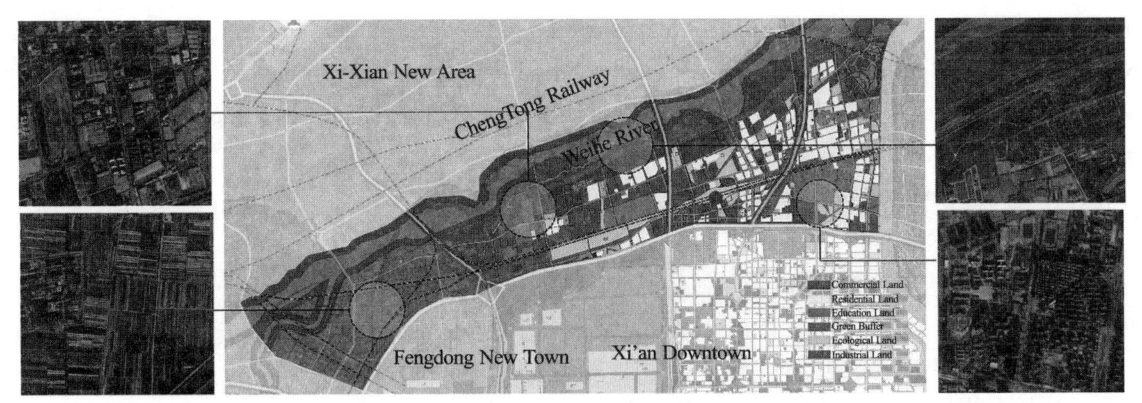

Fig.11 Land use planning and different land features including typical industrial area, farmland, ecological buffer zone and residential area (Source: Google Earth)

6 Conclusions

From 2014 to 2020, the spatial differentiation of land types in the Weihe River UEZ was significant. The construction land gradually gathered in the southeast part of the research area. The bounds between construction land and other land types, such as forestland and grassland, had become clear. With the development of the Weihe River UEZ, the vegetative land, including forestland, grassland and water body, showed a rising trend as well as non-scattered features. The cultivated land area was gradually reduced. Additionally, the grassland area was obviously expanded based on GF-2 data interpretation. The construction land area expanded, and it was mainly concentrated in the southeast. In general, the ecological quality of non-construction land areas has been optimized. The zoning regulation made the boundary of different regions clear.

The total ESV decreased by approximately 23% from 2018 to 2020. In terms of spatial distribution, the ESV of the ecological area represented the high value was mainly distributed in the Weihe River surrounding beach areas. It was greatly affected by river water and riverside ecological landscape. The distribution of low-value areas was largely affected by construction land. From 2018 to 2020, the ESV high-value areas remained comparatively stable while the medium-value areas increased significantly. The landscape pattern showed that the landscape separation degree, fragmentation degree and dominance degree were greatly affected by the changes in the construction area, especially the landscape fragmentation degree. The construction land was greatly disturbed by human activities and the spatial structure was complex. The Weihe River UEZ in 2018 and 2020 had obvious global spatial autocorrelation. Moran's I was in an upward trend. The distribution of ESV cold-hot spot significant area was similar to local spatial autocorrelation significant area. The high-high aggregation of the hot spot area and LISA cluster was distributed between the main stream of the Weihe River, and the low-low aggregation of the cold spot area and LISA cluster was greatly affected by construction land. With the development of the land system in China, land use changes under the guidance of planning, staying a benign and sustainable state. Significantly, spontaneous land use and landscape change still exist, especially in urban-rural transition areas. This issue needs to be brought to the forefront of public discussion.

Author contributions: Conceptualization, S.L.; methodology, S.L. and G.H.; software, Z.Q.; validation, S.L., G.H. and Y.W.; formal analysis, S.L.; investigation, Z.Q.; resources, Y.W.; data curation, G.H.; writing—original draft preparation, S.L.; writing—review and editing, G.H. and Y.W.; visualization, Z.Q.; supervision, G.H.; project administration, S.L.; funding acquisition, S.L. All authors have read and agreed to the published version of the manuscript.

Funding: This research was funded by Technology Innovation Center Funds for Land Engineering and

Human Settlements, Shaanxi Land Engineering Construction Group Co., Ltd. and Xi'an Jiaotong University (No. 2021WHZ0090); Enterprise Innovation and Youth Talent Support Program of Shaanxi Association for Science and Technology (No. 2021-1-2); The Project of Shaanxi Provincial Land Engineering Construction Group (DJNY2022-29).

Institutional review board statement: Not applicable.

Informed consent statement: Not applicable.

Data availability statement: The data presented in this study are available on request from the corresponding author.

Conflicts of interest: The authors declare no conflict of interest.

References

[1]Hu, B.Y., Zhang, P.T., Bai N.,et al. Land use scenario simulation in Qinglong Manchu Autonomous County based on CLUE-S and GMOP model. Chin. J. Agric. Resour. Reg. Plan.,2020, 41, 173-182.

[2]Jónsson, J.Ö.G., Davíðsdóttir, B. Classification and valuation of soil ecosystem services. Agr. Syst. 2016, 145, 24-38.

[3]Fu, B.J., Zhang, L.W. Land-use change and ecosystem services: Concepts, methods and progress. Prog. Geogr. 2014, 33, 441-446.

[4]Tang, J.X., Li, Y.M., Cui, S.H., et al. Linking land-use change, landscape patterns, and ecosystem services in a coastal watershed of southeastern China. Glob. Ecol. Conserv. 2020, 23, e01177.

[5]Zhao, G.S., Liu, J.Y., Kuang, W.H., et al. Disturbance impacts of land use change on biodiversity conservation priority areas across China: 1990-2010. J. Geogr. Sci. 2015, 25, 515-529.

[6]Dadashpoor, H., Azizi, P., Moghadasi, M. Land use change, urbanization, and change in landscape pattern in a metropolitan area. Sci. Total. Environ. 2019, 655, 707-719.

[7]Bryan, B.A., Ye, Y.Q., Zhang, J.E., et al. Land-use change impacts on ecosystem services value: Incorporating the scarcity effects of supply and demand dynamics. Ecosyst. Serv. 2018, 32, 144-157.

[8]Jing, Y.C., Chen, L.D., Sun, R.H. A theoretical research framework for ecological security pattern construction based on ecosystem services supply and demand. Acta Ecol. Sin. 2018, 38, 4121-4131.

[9]Qiu, J.Q., Huang, T., Yu, D.Y. Evaluation and optimization of ecosystem services under different land use scenarios in a semiarid landscape mosaic. Ecol. Indic. 2022, 135, 108516.

[10]Jiang, W., Wu, T., Fu, B.J. The value of ecosystem services in China: A systematic review for twenty years. Ecosyst. Serv. 2021, 52, 101365.

[11]Fu, B.J. Ecosystem Service and Ecological Security; Higher Education Press: Beijing, China, 2013:5-20.

[12]Hasan, S.S., Zhen, L., Miah, M.G., et al. Impact of land use change on ecosystem services: A review. Environ. Dev. 2020, 34, 100527.

[13]Zhang, Z.M., Liu, J.G. Progress in the valuation of ecosystem services. Acta Sci. Circumstantiae 2011, 31, 1835-1942.

[14]Sannigrahi, S., Chakraborti, S., Joshi, P.K., et al. Ecosystem service value assessment of a natural reserve region for strengthening protection and conservation. J. Environ. Manag. 2019, 244, 208-227.

[15]Grammatikopoulou, I., Vačkářová, D. The value of forest ecosystem services: A meta-analysis at the European scale and application to national ecosystem accounting. Ecosyst. Serv. 2021, 48, 101262.

[16]Xie, Z.L., Li, X.Z., Chi, Y., et al. Ecosystem service value decreases more rapidly under the dual pressures of land use change and ecological vulnerability: A case study in Zhujiajian Island. Ocean. Coast. Manag. 2021, 201, 105493.

[17]Rimal, B.,Sharma, R., Kunwar, R., et al. Effects of land use and land cover change on ecosystem services in the Koshi River Basin, Eastern Nepal. Ecosyst. Serv. 2019, 38, 100963.

[18]Akhtar, M., Zhao, Y.Y., Gao, G.L., et al. Assessment of ecosystem services value in response to prevailing and future land use/cover changes in Lahore, Pakistan. Reg. Sustain. 2020, 1, 37-47.

[19]Xiao, R., Lin, M., Fei, X.F., et al. Exploring the interactive coercing relationship between urbanization and ecosystem service value in the Shanghai-Hangzhou Bay Metropolitan Region. J. Clean. Prod. 2020, 253, 119803.

[20]Shao, X.Y., Jing, C.W., Qi, J.G., et al. Impacts of land use and planning on island ecosystem service values: A case study of Dinghai District on Zhoushan Archipelago, China. Ecol. Process. 2017, 6, 285-295.

[21]Su, S.L., Xiao, R., Jiang, Z.L., et al. Characterizing landscape pattern and ecosystem service value changes for urbanization impacts at an eco-regional scale. Appl. Geogr. 2012, 34, 295-305.

[22]Costanza, R., d'Arge, R., Groot, R.D., et al. The value of the world's ecosystem services and natural capital. Nature 1997, 387, 253-260.

[23]Qin, X.C., Fu, B.H. Assessing and predicting changes of the ecosystem service values based on land use/land cover changes with a random forest-cellular automata model in Qingdao metropolitan region, China. IEEE J. Sel. Top. Appl. Earth Obs. Remote Sens. 2020, 13, 6484-6494.

[24]Shi, Y., Wang, R.S., Huang, J.L., et al. An analysis of the spatial and temporal changes in Chinese terrestrial ecosystem service functions. Chin. Sci Bull. 2012, 57, 720-731.

[25]Xie, G.D., Zhang, C.X., Zhang, L.M., et al. Improvement of the evaluation method for ecosystem service value based on per unit area. J. Nat. Resour. 2015, 30, 1243-1254.

[26]Wu, J.S., Yue, X.X., Qin, W. The establishment of ecological security patterns based on the redistribution of ecosystem service value: A case study in the Liangjiang New Area, Chongqing. Geogr. Res. 2017, 36, 429-440.

[27]Yao, X.W., Zeng, J., Li, W.J. Spatial correlation characteristics of urbanization and land ecosystem service value in Wuhan Urban Agglomeration. Trans. Chin. Soc. Agric. Eng. 2015, 31, 249-256.

[28]Maimaiti, B., Chen, S.S., Kasimu, A., et al. Urban spatial expansion and its impacts on ecosystem service value of typical oasis cities around Tarim Basin, northwest China. Int. J. Appl. Earth Obs. 2021, 104, 102554.

[29]Haas, J., Ban, Y.F. Mapping and monitoring urban Ecosystem services using multitemporal high-resolution satellite data. IEEE J. Sel. Top. Appl. Earth Obs. Remote Sens. 2017, 10, 669-680.

[30]Hu, Z.N., Yang, X., Yang, J.J., et al. Linking landscape pattern, ecosystem service value, and human well-being in Xishuangbanna, southwest China: Insights from a coupling coordination model. Glob. Ecol. Conserv. 2021, 27, e01583.

[31]Xiao, D.N., Zhao, Y., Sun, Z.W., et al. Study on the variation of landscape pattern in the west suburbs of Shenyang. J. Appl. Ecol. 1990, 1, 75-84.

[32]Lam, N.S.N., Cheng, W.J., Zou, L., et al. Effects of landscape fragmentation on land loss. Remote Sens. Environ. 2018, 209, 253-262.

[33]Memarian, H., Balasundram, S.K., Talib, J.B., et al. Validation of CA-Markov for simulation of land use and cover change in the Langat Basin, Malaysia. J. Geogr. Inf. Syst. 2012, 4, 542-554.

[34]Zhou, L., Dang, X.W., Sun, Q.K., et al. Multi-scenario simulation of urban land change in Shanghai by random forest and CA-Markov model. Sustain. Cities Soc. 2020, 55, 102045.

[35]Firozjaei, M.K., Sedighi, A., Argany, M., et al. A geographical direction-based approach for capturing the local variation of urban expansion in the application of CA-Markov model. Cities 2019, 93, 120-135.

[36]Chu, L., Zhang, X.R., Wang, T.W., et al. Spatial-temporal evolution and prediction of urban landscape pattern and habitat quality based on CA-Markov and InVEST model. Chin. J. Appl. Ecol. 2018, 29, 4106-4118.

[37]Zhao, M.M., He, Z.B., Du, J., et al. Assessing the effects of ecological engineering on carbon storage by linking the CA-Markov and InVEST models. Ecol. Indic. 2019, 98, 29-38.

[38]Li, S.T., Li, C.Y., Kang, X.D. Development status and future prospects of multi-source remote sensing image fusion. Natl. Remote Sens. Bull. 2021, 25, 148-166.

[39]Guo, R. The Study of the Weihe River Ecological Corridor in the Xi'an Metropolis. Master's Thesis, Xi'an University of Architecture and Technology, Xi'an, China, 2013.

[40]Wu, L.N., Yang, S.T., Liu, X.Y., et al. Response analysis of land use change to the degree of human activities in Beiluo River basin since 1976. Acta Geogr. Sin. 2014, 69, 54-63.

[41]Wang, C., Chang, Y., Hou, X.Y., et al. Temporal and spatial evolution characteristics of habitat quality in Jia-

odong peninsula based on changes of land use pattern. J. Geo Inf. Sci. 2021, 23, 1809-1822.

[42]Xie, G.D., Zhang, C.X., Zhen, L., et al. Dynamic changes in the value of China' s ecosystem services. Ecosyst. Serv. 2017, 26, 146-154.

[43]Xie, G.D., Zhen, L., Lu, C.X., et al. Expert knowledge based valuation method of ecosystem services in China. J. Nat. Resour. 2008, 23, 911-919.

[44]Guo, M., Shu, S., Ma, S., et al. Using high-resolution remote sensing images to explore the spatial relationship between landscape patterns and ecosystem service values in regions of urbanization. Environ. Sci Pollut Res. 2021, 28, 56139-56151.

[45]Song, W., Pei, T., Chen, Y., et al. Evaluation of the modifiable unit problem in land expropriation compensation standards based on spatial autocorrelation. Inst. Agric. Resour. Reg. Plan. 2017, 38, 101-106.

[46]Han, P., Zhang, Q., Zhao, Y.Y., et al. High-resolution remote sensing data can predict household poverty in pastoral areas, Inner Mongolia, China. Geogr. Sustain. 2021, 2, 254-263.

[47]Tong, X.Y., Xia, G.S., Lu, Q.K., et al. Land-cover classification with high-resolution remote sensing images using transferable deep models. Remote Sens. Environ. 2020, 237,111322.

[48]Song, S.Q., Diao, M., Feng, C.C. Individual transport emissions and the built environment: A structural equation modelling approach. Transport. Res. A Pol. 2016, 92, 206-219.

[49]Xu, C., Jiang, W.Y., Huang, Q.Y., et al. Ecosystem services response to rural-urban transitions in coastal and island cities: A comparison between Shenzhen and Hong Kong, China. J. Clean Prod. 2020, 260, 121033.

[50]Zhang, S.Y., Liu, J.M., Song, C., et al. Spatial-temporal distribution characteristics and evolution mechanism of urban parks in Beijing, China. Urban. For. Urban. Gree. 2021, 64, 127265.

本文曾发表于 2022 年《Sustainability》第 14 卷

Daily Actual Evapotranspiration Estimation of Different Land Use Types Based on SEBAL Model in the Agro-pastoral Ecotone of Northwest China

Liangyan Yang[1,2], Jianfeng Li[1,2], Zenghui Sun[1,2], Jinbao Liu[1,3], Yuanyuan Yang[4], Tong Li[4]

(1.Shaanxi Provincial Land Engineering Construction Group Co., Ltd., Xi'an, Shaanxi, China; 2.Institute of Land Engineering and Technology, Shaanxi Provincial Land Engineering Construction Group Co., Ltd., Xi'an, Shaanxi, China; 3.Xi'an University of Technology, Xi'an, Shaanxi, China; 4.School of Earth Science and Resources, Chang'an University, Xi'an, Shaanxi, China)

【**Abstract**】Evapotranspiration (*ET*) plays a crucial role in hydrological and energy cycles, as well as in the assessments of water resources and irrigation demands. On a regional scale, particularly in the agro-pastoral ecotone, clarification of the distribution of surface *ET* and its influencing factors is critical for the rational use of water resources, restoration of the ecological environment, and protection of ecological water sources. The SEBAL model was used to invert the regional *ET* based on Landsat8 images in the agro-pastoral ecotone of northwest China. The results were indirectly verified by monitoring data from meteorological stations. The correlation between *ET* and surface parameters was analyzed. Thus, the main factors that affect the surface *ET* were identified. The results show that the SEBAL model determines an accurate inversion, with a correlation coefficient of 0.81 and an average root mean square error of 0.9 mm/d, which is highly suitable for research on water resources. The correlation coefficients of normalized vegetation index, surface temperature, land surface albedo, net radiation flux with daily *ET* were 0.5830, 0.8425, 0.3428 and 0.9111, respectively. The normalized vegetation index and the net radiation flux positively correlated with the daily *ET*, while the surface temperature and land surface albedo negatively correlated with the daily *ET*. The correlation from strong to weak is the net radiation flux > surface temperature > normalized vegetation index > surface albedo. In terms of spatial distribution, the daily *ET* of water was the highest, followed by woodland, wetland, cropland, built-up land, shrub land, grassland and bare land. However, the SEBAL model overestimates the inversion of daily *ET* of built-up land.

Introduction

Evapotranspiration (*ET*) is an important part of the global water cycle[1]. It is the channel that transforms surface and atmospheric water, which directly affects the spatial distribution of global precipitation and vegetation[2,3]. Studies have shown that > 60% of the precipitation that reaches the surface through condensation returns to the atmosphere through the process of *ET*, and this proportion can be as high as 90% owing to a reduction in precipitation and a higher *ET* in arid areas[4-6]. Solar short-wave radiation energy returns to the atmosphere in the form of latent heat during the process of *ET* on the surface, thus, forming an energy cycle process. Therefore, the surface *ET* plays an important role in the global water and energy cycles[7]. Simultaneously, the actual surface *ET* is a key parameter for research on environmental changes[8], hydrological processes[9], drought trends[10-11], and the potential evaluation of agricultural development[12,13]. Therefore, the study of surface *ET* is highly important scientifically to ensure the rational use of water resources and the protection of ecosystems in the study area[14-15].

Owing to the complexity of the surface, current *ET* observation methods include lysimeters[16], eddy correlator large aperture scintillation flux meters[17] and other methods. Such *ET* monitoring methods can obtain the *ET* more accurately at a specific location, but simultaneously, it is time-consuming, labor-intensive and expensive, and only a single point of *ET* data can be obtained. Therefore, the acquisition of regional surface *ET* data that is highly accurate over a continuous time-scale is always the key to solving difficult issues of research in the fields of meteorological science, hydrology, agriculture and geography. The development of remote sensing technology provides a new platform and way to obtain the actual surface *ET*, and it solves the problem of a lack of measured data owing to geographical reasons. In recent years, researchers have made substantial progress in utilizing remote sensing *ET* models for research, which successively established the surface energy balance index (SEBI) model[18], simplified surface energy balance index (S-SEBI) model[19], surface energy balance algorithm for land (SEBAL) model[20] and surface energy balance system (SEBS) model[21]. Among them, the SEBAL model benefits from a clearer mechanism, fewer model input parameters, simple data acquisition, high inversion accuracy, and a wide application range. It has become one of the most commonly used remote sensing methods to determine the inversion of surface *ET*[22].

However, the accuracy of SEBAL model varies in the estimation of *ET* for different products and environmental conditions. Liu and Hu investigated the effects of land use/cover change and climate change on wetland *ET* using SEBAL, and the results showed that the average relative error of regional *ET* estimated by the SEBAL model was 9.01%[23]. Rahimzadegan et al. studied the efficiency of SEBAL algorithm in estimating the *ET* of pistachio crops. The results of this study indicated that the determination coefficient and Root Mean Square Error of the estimated actual *ET* for pistachio plants was 0.8 and 2.5 mm, respectively[24]. The agro-pastoral ecotone of northwest China receives little rainfall, which is distributed unevenly, and has a strong surface *ET* and an extremely fragile ecological environment. A shortage of water is the most serious problem in this area. In addition, the research on the influence factors of *ET* mostly focuses on meteorological factors[25-26], and there is a lack of in-depth research on the influence of surface parameters on *ET*.

The objective of this study was to explore the applicability of SEBAL model and analyze the influencing factors in a wind and sandy grass beach area of northwest China. Therefore, in this study, (1) Landsat8 imagery and meteorological data were used as the source of data to estimate daily *ET* based on the SEBAL model; (2) the accuracy of estimation of daily *ET* was verified and used to analyze its spatiotemporal distribution, and (3) the influence of surface parameters on the daily *ET* was studied, providing an analysis of the response of daily *ET* to different land use types. These conclusions will enhance our understanding of the factors that affect *ET* and thus, provide a basis for the sustainable development of water resources and the restoration of the ecological environment in arid and semiarid regions in northwest China.

Materials and methods

Study area

The study area (107°58′~109°15′E, 37°31′~38°49′N) is located at the junction of Yulin City and Ordos City, China (Fig.1) in a sandy and grassy beach area southeast of Mu Us Sandy Land. It has a typical farming-pastoral transition zone with a composite underlying surface of bare land-grassland-farmland in China. The study area has a typical continental monsoon climate. The northwest is arid and lacks water, which results in infrequent precipitation. Consequently, the ecological environment is fragile. The land use types are primarily grassland and bare land. The southeast is low-lying, with rivers that pass through, and the land use types are primarily grassland and cropland. The average annual temperature is approximately 7 ℃, and the average annual precipitation is between 350~450 mm, which decreases from southeast to northwest in the study area. The interannual precipitation is highly uneven with a large seasonal variation.

The rainy season is from July to September, and it comprises more than 50% of the rain that falls during a whole year. The potential *ET* is relatively large, and the annual average *ET* is 2000 ~ 3000 mm in the study area. The sparse precipitation and strong *ET* activities are the main reasons for the drought in study area.

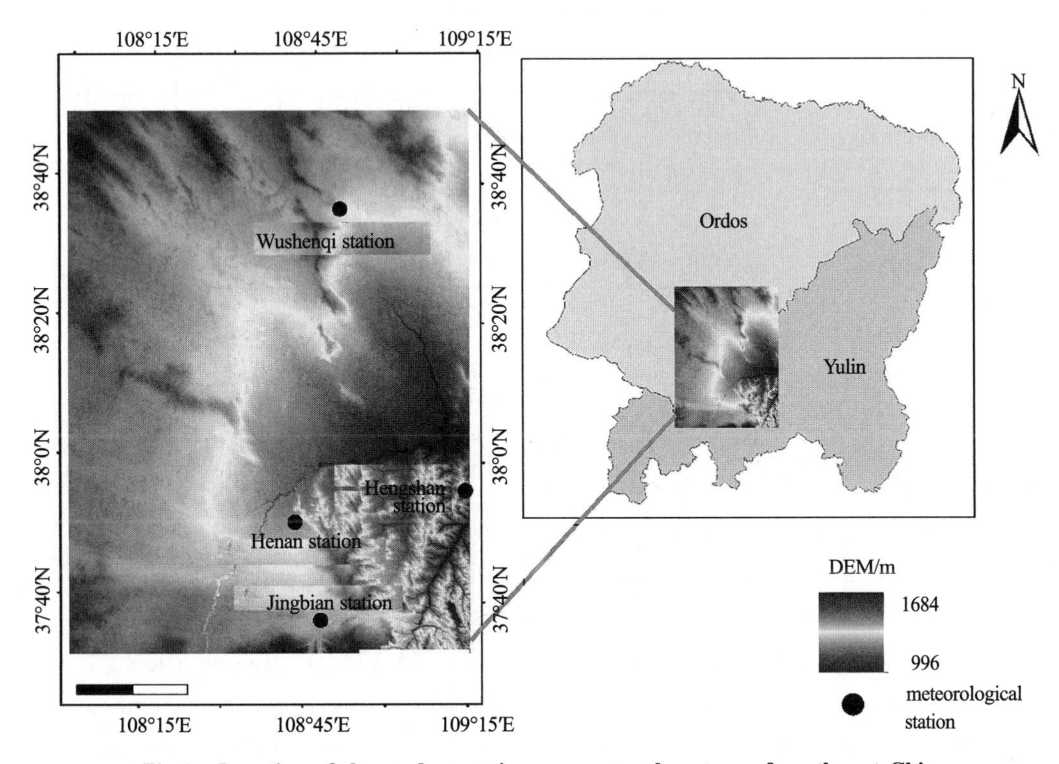

Fig.1 Location of the study area in agro-pastoral ecotone of northwest China

Data sources

In this study, the agro-pastoral ecotone of northwest China was used as the study area, and the Landsat8 images that cover the study area were numbered 128033 and 128034. They are all freely available at the Geospatial Data Cloud (http://www.gscloud.cn). Since the data is affected by cloud coverage, and parts of it are missing, it is impossible to obtain the image data for each period of the same year. To ensure that the data images can cover the study area, this study screened out 2017-01-02, 2016-04-21, 2017-05-26, 2017-07-13, 2016-09-28, 2016-12-17 and six other Landsat8 original remote sensing images. The selected data covered the four seasons of the year and are suitable for the study of the spatiotemporal distribution of surface *ET*. The Landsat8 remote sensing data requires preprocessing, such as radiometric calibration, atmospheric correction and image mosaics, before it can be substituted into the model. ENVI software (L3Harris Geospatial, Boulder, CA, USA) was used to process the images in this research.

Principle of the SEBAL model

In 1995, Bastiaanssen proposed the SEBAL inversion model, which was verified and optimized in 1998. In recent years, the surface *ET* model has been effectively developed and improved. The SEBAL model has been utilized for its advantages and widely used throughout the world. It has produced good inversion results and has become a high-precision regional *ET* remote sensing inversion model. The research foundation of the SEBAL model is the energy balance equation, which is composed of four parts, including net radiance, soil heat, sensible heat and latent heat fluxes, which form a closed-loop energy transmission. The formula is shown as follows:

$$R_n = H + G_n + \lambda ET \quad (1)$$

where R_n is the net solar radiation flux (W/m^2) ; G_n is the soil heat flux (W/m^2) ; H is the turbulent sensible heat flux (W/m^2) ; λET is the turbulent latent heat flux (W/m^2) , and λ is the coefficient of latent heat of vaporization of water, which is a function of temperature.

Net radiant flux (R_n). The net surface radiant flux is expressed per unit area, which includes the difference between total energy received from the solar short-wave radiation and atmospheric long-wave radiation on the surface and the long-wave radiation emitted by the surface, reflected solar short-wave radiation and atmospheric long-wave radiation[27]. The process of calculation is as follows:

$$R_n = (1 - \alpha) \times R_{swd\downarrow} + R_{lwd\downarrow} - R_{lwd\uparrow} - (1 - \varepsilon) \times R_{lwd\downarrow} \quad (2)$$

$$R_{swd\downarrow} = G_{sc} \times \cos\theta \times d \times \tau_{sw} \quad (3)$$

$$d = 1 + 0.033\cos(DAY \times 2\pi/365) \quad (4)$$

$$R_{lwd\downarrow} = \varepsilon_a \times \sigma \times T_a^4 \quad (5)$$

$$\varepsilon_a = 9.2 \times 10^\wedge[- 6 \times (T_a - 273.15)^2] \quad (6)$$

$$R_{lwd\uparrow} = \varepsilon \times \sigma \times LST^4 \quad (7)$$

where $R_{swd\downarrow}$ is the short-wave radiation energy that the sun reaches the surface (W/m^2) ; $R_{lwd\downarrow}$ is the long-wave radiation from the atmosphere to the surface (W/m^2) ; $R_{lwd\uparrow}$ is the ground upgoing long-wave radiation (W/m^2) ; α is the surface albedo; ε is the surface emissivity; G_{sc} is the solar constant, with a value of 1367 W/m^2; θ is the zenith angle of sun; d is the distance factor between the sun and the earth; τ_{sw} is the atmospheric transmittance; DAY is the number of days of the image in the current year; ε_a is the specific emissivity of the atmosphere; σ is a constant with a value of 5.67×10^{-8} $W/(m^2 \cdot K^4)$; T_a represents the temperature at the ground reference altitude, and LST represents the surface temperature.

Soil heat flux (G_n). The soil heat flux is the difference between surface and soil temperatures that enables the transmission of energy. A small part of the energy lost in the surface soil during this process is primarily affected by topography and land use types. The soil heat flux generally occupies a relatively small proportion in the energy balance equation. Factoring in the greater difficulty in its direct calculation, Bastiaanssen used calculations of net radiant flux, vegetation index, land surface temperature and land surface reflectance in the SEBAL model[28], and the calculations are shown as follows:

$$G_n = R_n \times \frac{T_s - 273.15}{\alpha}(0.0038\alpha + 0.0074\alpha^2)(1 - 0.98\times NDVI^4) \quad (8)$$

where $NDVI$ is the normalized vegetation index.

Sensible heat flux (H). Sensible heat flux is the temperature difference between the air and surface temperatures, and the energy that flows into the air from the surface is in the form of vertical transmission. The sensible heat flux is primarily related to the ground temperature difference, density, aerodynamic impedance and other parameters, and it is calculated is as follows:

$$H = \frac{\rho_{air} \times C_P \times d_T}{r_{ah}} \quad (9)$$

where C_P is the specific heat capacity of the atmosphere; ρ_{air} is the density of the atmosphere; r_{ah} is the aerodynamic impedance ($s \cdot m^{-1}$) , and d_T is the ground temperature difference (K). r_{ah} and d_T are unknown parameters. The specific process for calculation has been previously described[29].

Latent heat flux (λET). Latent heat flux is the heat exchange per unit area at constant temperature, which is the most important parameter in the inversion study of surface ET. It is the energy absorbed during the vaporization of water. Based on the surface energy balance formula in the remote sensing inversion

ET model, the latent heat flux is obtained as follows:

$$\lambda ET = R_n - H - G_n \tag{10}$$

λET represents the instantaneous latent heat flux (W/m^2).

Daily ET. The surface radiation flux is substantially affected by factors, such as surface temperature, wind speed, and surface coverage, and cannot be obtained directly. The difference in ratio of latent heat flux to the net radiant and soil heat fluxes in a day always remain basically stable, i.e., the theory of constant evaporation ratio and the instantaneous ET can be calculated by the theory of constant evaporation ratio, which can be extended to the daily ET[30-31]. The formula is calculated as follows:

$$EF = \frac{R_n - G_n - H}{R_n - G_n} = \frac{\lambda ET}{R_n - G_n} = EF_{24} \tag{11}$$

$$ET_{24} = \frac{86400 \times EF_{24}}{\lambda} \times (R_{n24} - G_{n24}) \tag{12}$$

where EF represents the instantaneous evaporation ratio; EF_{24} represents the entire day evaporation ratio; R_{n24} represents the daily net radiant flux (W/m^2), and G_{n24} represents the daily soil heat flux (W/m^2).

Results

Verification of the accuracy of daily ET

The verification and evaluation of the inversion accuracy of the surface ET model is the focus and difficulty of the research on ET. To verify the reliability of the daily ET retrieved by the SEBAL model, the water surface evaporation monitored by four weather stations in the study area was selected for relative verification. However, the evaporation on the water surface that was measured was significantly higher than the actual evaporation on the surface, and the water surface evaporator of the weather station has two types that include E-601 and 20 cm-caliber. There are differences in the installation plan, structure and observational methods of the two instruments. There are large deviations in the verification of data when the data of weather stations is used directly. Therefore, to more accurately explore the inversion accuracy of the SEBAL model and maintain the continuity and consistency of the data at each station, it is necessary to explore the conversion coefficients of the two surface evaporators and unify the data. E-601 and 20 cm-caliber meteorological stations, such as those at the Hengshan, Jingbian, Shenmu, Dingbian, Yanchi and Etuoke Banner Stations in China were compared in the study and surrounding areas. The monitoring data were analyzed and indicated that the average conversion coefficient of the two surface evaporators was 0.63, and the determination coefficient R^2 was 0.95, indicating that the conversion coefficients of the E-601 type and the 20 cm-caliber small evaporator are highly reliable, respectively, and can meet the requirements of the two types of surface evaporators.

The monitoring data of 20 cm-caliber small evaporator was converted into the E-601 monitoring data using the conversion coefficients of the two surface evaporators. This enabled the use of data to verify the SEBAL model inversion result. Since the meteorological sites were small meteorological stations that were established by research teams in Wushenqi and Henan in China, they did not have evaporation dishes to obtain the data on water surface evaporation. The only data involved in the validation were the monitoring data from the meteorological stations in Hengshan and Jingbian, China. Fig.2 is a scatter diagram of the water surface evaporation of the meteorological station and the inversion results. This figure. indicates that the actual surface ET based on the SEBAL model is highly correlated with the surface water surface ET. The coefficient of determination was 0.81, and the average root mean square error was 0.9 mm/d. These values indicate that the SEBAL model can be used to retrieve the surface ET in agro-pastoral ecotone of northwest China with a high degree of accuracy.

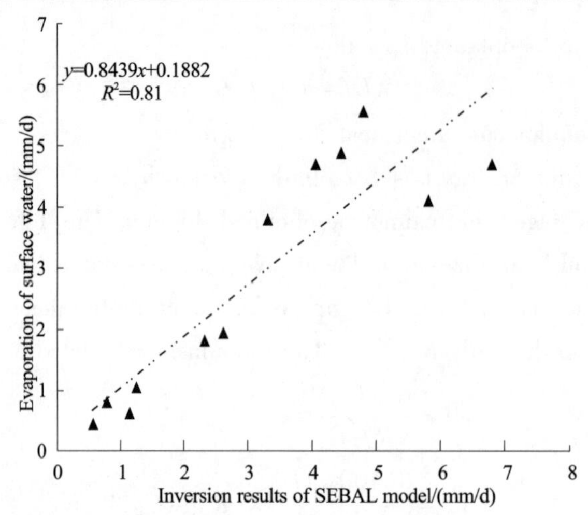

Fig.2 Validation scatter plot of the SEBAL model inversion results and meteorological data

Spatiotemporal distribution of the daily *ET*

In this study, the daily *ET* of the study area was obtained based on the theory of constant *ET* ratio. Table 1 shows the data on daily *ET* in different periods in the study area. An examination of the distribution revealed a pattern in the daily *ET* in the northwest agro-pastoral ecotone area. The pattern included July (5.17 mm/d)>May (3.51 mm/d)>September (3.31 mm/d)>April (2.02 mm/d)>December (0.72 mm/d)> January (0.56 mm/d). The possible reason is related to the local climate change during the four seasons. The temperature and precipitation were the lowest during the winter in January and December, and insufficient hydrothermal conditions limited the *ET* process. In April and May, the air temperature began to increase; the vegetation gradually turned green, and the crops were just beginning to be cultivated and sprouting. These factors indicate that the *ET* gradually increased. Since July is in the summer in this study area, the monthly average temperature reaches the highest value of the year, and this is the rainy season. The vegetation grows vigorously, and there is sufficient soil moisture. The hydrothermal conditions were the highest of the year. Therefore, the study area had its highest daily *ET* in the summer. In September, the vegetation in the sand of study area began to senesce; the temperature decreased; the rainfall was significantly reduced; the hydrothermal conditions were all lower than those in the summer, and the daily *ET* was less than in the summer. In summary, hydrothermal conditions are the main factor that affect the changes in evapotranspiration during the year.

Fig.3 shows the spatial distribution of the daily *ET* in different periods in the study area. The evapotranspiration varied with regions within the study area. It was higher overall in the southeast and lower in the northwest, and such a pattern of distribution corresponds to the terrain of the study area. There are rivers that pass through the southeast, and there are many areas of crops, woodland and grassland. The type of land use in northwest China primarily included denuded dunes or bare land. Thus, there is less vegetation coverage, and the soil is dry and lacks water. From a local perspective, the daily *ET* of different land use types differs significantly.

Table 1 Daily average evapotranspiration in the study area

Date	Min/(mm/d)	Max/(mm/d)	Daily average *ET*/(mm/d)	SD
2017-01-02	0.02	3.44	0.56	0.3477
2016-04-21	0.15	12.58	2.02	1.4135
2017-05-26	0.35	15.13	3.51	2.1273
2017-07-13	0.51	14.56	5.17	2.0476
2016-09-28	0.22	8.07	3.31	1.1839
2016-12-17	0.07	3.36	0.72	0.3363

ET, evapotranspiration; Min, minimum; Max, maximum; SD, standard deviation.

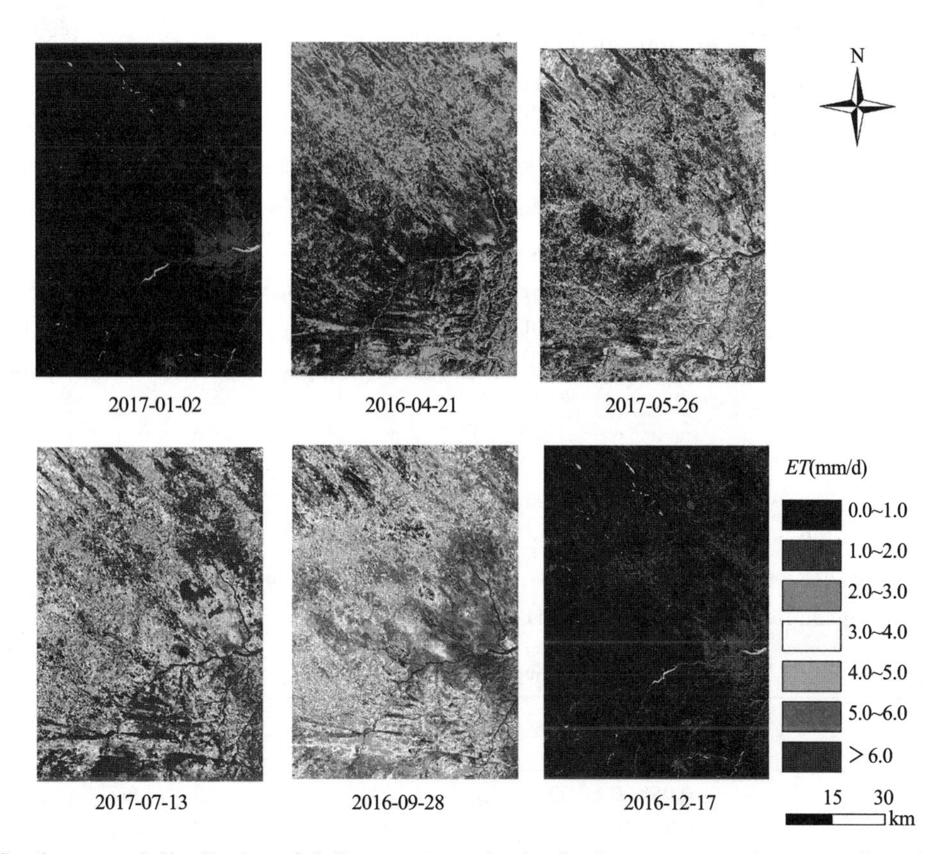

Fig. 3 Spatiotemporal distribution of daily evapotranspiration in the agro-pastoral ecotone of northwest China

Discussion

A correlation analysis of *ET* and meteorological factors

In this study, the annual daily data of the Hengshan station in 2016 were selected, including four meteorological factors: *ET* (mm/day), air temperature (TEM, ℃), relative humidity (RHU, %), and wind speed (WIN, m/s). A correlation analysis between meteorological factors and the surface *ET* was performed using SPSS, and the strength of correlation between the meteorological factors and *ET* was evaluated using the correlation coefficient R^2. Fig.4 shows the linear relationship between *ET* and the meteorological factors. *ET* significantly correlated with the air temperature with an R^2 of 0.6274. The wind speed weakly correlated with an R^2 of 0.3307. The RHU had a negative R^2 of 0.3374. In areas with higher air humidity, the surface moisture faces more resistance when entering the air, so the *ET* was relatively small.

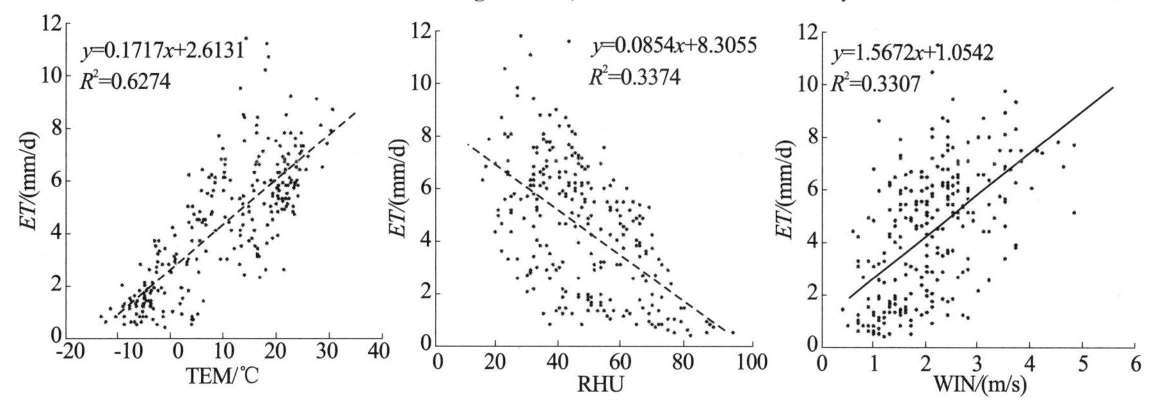

Fig.4 The linear relationship between evapotranspiration and meteorological factors

Sensitivity analysis of model parameters

To better understand the extent to which each model parameter affects the model inversion results, this study performed a sensitivity analysis of the model parameters to derive the amount of change in the final *ET* owing to changes in each parameter. The sensitivity analysis will be performed using the one-parameter variable method, i.e., the interaction of the parameters will not be considered. Only the parameters to be analyzed will be treated for variation. Since hydrothermal conditions have a large influence on *ET*, this study explored the sensitivity of atmospheric water vapor (wv) content, hot spot surface temperature (LST_h) and cold pixels to *ET*. The sensitivity index *I* was used to reflect the sensitivity of the evapotranspiration model to water vapor content. Larger values of *I* indicated that the model is highly sensitivity to this parameter. The formula is calculated as follows[32]:

$$I=\frac{\Delta ET}{\Delta X}\cdot\frac{X}{ET} \tag{13}$$

where ΔX represents the amount of change in the parameter, and ΔET represents the amount of change in the model output results owing to the change in the parameter. The sensitivity indices were classified into four classes by the magnitude of the value of *I* as shown in Table 2.

Table 2 Sensitivity classification

Grade	Sensitivity index	Sensitivity classification
Ⅰ	$0.0<\|I\|<0.05$	poor
Ⅱ	$0.05<\|I\|<0.20$	fair
Ⅲ	$0.20<\|I\|<1.00$	moderate
Ⅳ	$\|I\|>1.00$	excellent

The inversion accuracy of wv and LST_h were considered to be approximately 0.5 g/cm² and 2 ℃, respectively. $\Delta wv=0.5$ g/cm² and $\Delta LST=2$ ℃ were established. Table 3 and Table 4 show the mean values of the model outputs, as well as the sensitivity indices.

Table 3 Evapotranspiration changes and sensitivity indices from the changes in water vapor

Daily *ET* / wv	Average value (mm/d)	Average difference (mm/d)	$\|I\|$	Sensitivity classification
wv+1.5 g/cm²	2.73	−2.44	0.83	moderate
wv+1.0 g/cm²	3.24	−1.93	0.98	moderate
wv+0.5 g/cm²	4.04	−1.13	1.15	excellent
wv−0.5 g/cm²	6.05	0.88	0.90	moderate
wv−1.0 g/cm²	6.73	1.56	0.79	moderate
wv−1.5 g/cm²	7.24	2.07	0.70	moderate

ET, evapotranspiration; wv, atmospheric water vapor.

Table 4 Changes in evapotranspiration and sensitivity indices from the changes in hot surface temperature

Daily *ET* / LST_h	Average value (mm/d)	Average difference (mm/d)	$\|I\|$	Sensitivity classification
LST_h+6 ℃	5.54	0.37	0.56	moderate
LST_h+4 ℃	5.42	0.25	0.57	moderate
LST_h+2 ℃	5.30	0.13	0.59	moderate
LST_h−2 ℃	5.09	−0.08	0.36	moderate
LST_h−4 ℃	5.05	−0.13	0.30	moderate
LST_h−6 ℃	4.97	−0.20	0.30	moderate

ET, evapotranspiration; LST_h, hot surface temperature.

Table 3 shows the daily changes in *ET* and the sensitivity indices owing to changes in the wv. Increasing the wv by 0.5 g/cm^2 caused a daily change of 1.13 mm/d in the *ET*, and a decrease of 0.5 g/cm^2 wv caused a daily change of 0.88 mm/d in the *ET*. The sensitivity indices for increasing and decreasing the wv by Δ, 2Δ, and 3Δ were all > 0.2, indicating that the output of the SEBAL model-based inversion of daily *ET* is sensitive to the wv. The main reason is that the magnitude of wv directly affects the atmospheric transmittance, which causes a change in the net radiation flux. The effect on daily *ET* is weakened by the large changes in wv, which is owing to the large number of parameters in the SEBAL model. The changes in wv affect other parameters, thus, weakening the effect on daily *ET*.

Table 4 shows the daily changes in the *ET* and sensitivity indices owing to changes in the LST_h. A 2 ℃ increase in LST_h caused a daily change of 0.15 mm/d in the *ET*. A 2 ℃ decrease in LST_h caused an *ET* change of 0.08 mm/d. The sensitivity indices for increasing and decreasing the LST_h by Δ, 2Δ, and 3Δ were all > 0.2, indicating that the inversion results are sensitive to the LST_h based on the SEBAL model inversion of daily *ET*. The model output results on the LST_h are less sensitive compared with those of the wv. The main reason for this is that the LST_h is used to calculate the sensible heat flux, and the result of the sensible heat flux requires calculations that are continuously iterative. Multiple iterations weaken the influence of the LST_h. Thus, the change of a single parameter of the LST_h does not have a large influence on the daily *ET*.

Correlation analysis of the daily *ET* and surface parameters

An analysis of the spatiotemporal changes of *ET* in the study area indicated that the underlying surface characteristics, such as hydrothermal conditions and land use types, are important factors that affect energy transmission, and their interactions in concert affect the spatial distribution pattern of the surface *ET*[22]. To analyze the correlation between surface parameters and *ET* in more detail, this study obtained the spatial distribution map of the NDVI, LST, surface albedo and net radiant flux (Fig. 5) and their correlation with the daily *ET* (Fig. 6). The figures indicate terms of spatial distribution. For example, the high-value areas of NDVI are primarily concentrated in the southern part of the study area and distributed in strips and points. This pattern is consistent with the spatial distribution of daily *ET*. The spatial distribution of surface temperature is opposite to that of the daily *ET*. The high values are distributed in the west and north of the study area on primarily bare and sandy lands. The high value areas of surface albedo are primarily distributed in the west and north, similar to the distribution of surface temperature. The main reason for this is that crops are grown in the south and east, and the northwest is primarily bare and sandy lands. The surface is not very rough, and there is strong reflection from solar energy. The pattern of spatial distribution of the net radiation flux and daily evaporation remains highly consistent. The correlation coefficients between the NDVI, LST, surface albedo, net radiant flux and daily *ET* were 0.5830, 0.8425, 0.3428, and 0.9111, respectively. Among them, the NDVI and net radiant flux positively correlated with the daily *ET*, and the LST and surface albedo negatively correlated with the daily *ET*. Overall, the correlation between daily *ET* and the underlying surface parameters from strong to weak is the net surface radiation > LST > NDVI > surface albedo. The correlation of surface parameters and daily *ET* as shown in this study has also been widely reported from many studies[6,33-34].

Analysis of the daily *ET* with different land use types

Fig.7 shows the spatial distribution of land use types and daily *ET* in the agro-pastoral ecotone of northwest China. The amount of evapotranspiration varies from one land use type to another owing to differences in the physicochemical properties of the subsurface. The spatial distribution of the larger areas of *ET* was significantly similar to that of water and cropland . The spatial distribution of the smaller areas matches that

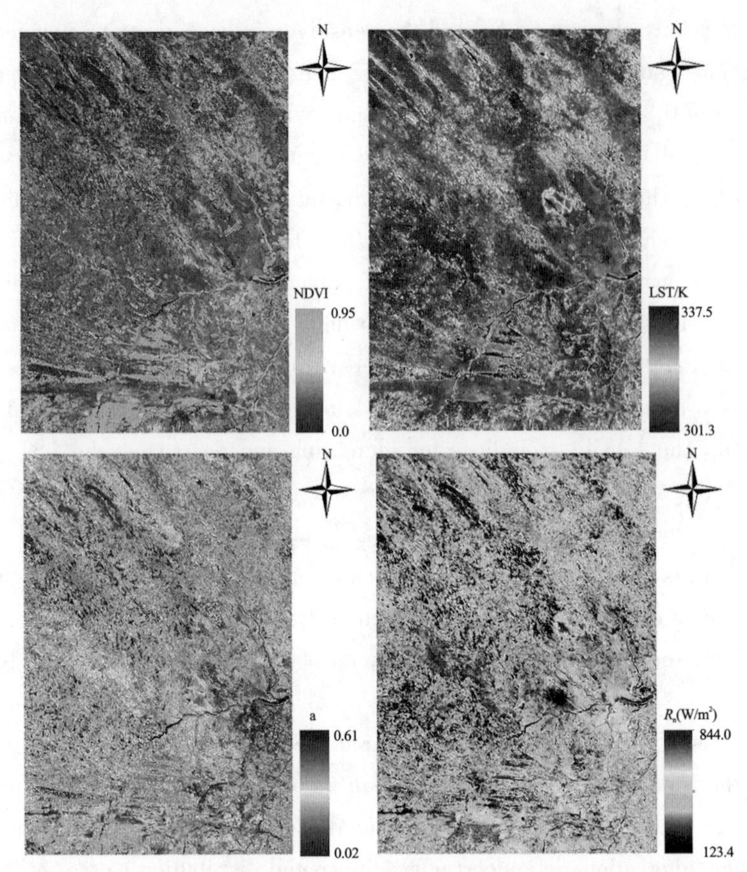

Fig. 5 Spatial distribution of the normalized vegetation index, surface temperature, surface albedo and the net radiation flux in the agro-pastoral ecotone of northwest China

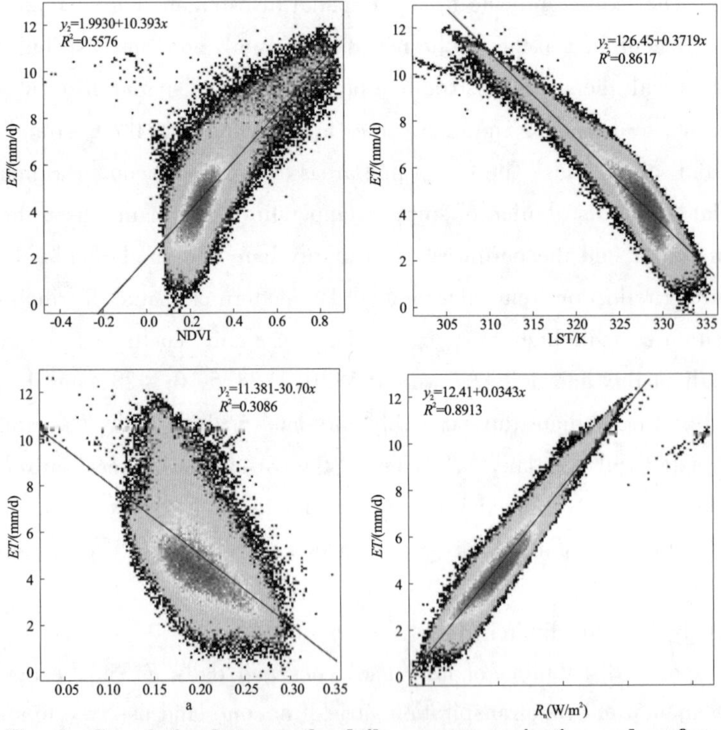

Fig. 6 Correlation between the daily evapotranspiration and surface parameters in the agro-pastoral ecotone of northwest China

of the bare land, and the daily ET of the built-up land did not differ significantly from that of the other surrounding land use types. Table 5 shows the data of daily ET values for different land use types. The daily ET of water and woodland were the highest in study area, with average daily ET values of 9.90 mm/d and 8.69 mm/d, respectively, which is because water provides sufficient water resources. Woodland have a dual role of water connotation and ET and can provide good moisture conditions for ET, which should be relatively high. This result is consistent with a study by Jeremiah et al.[35] The average daily ET of wetlands and croplands was 7.70 mm/d and 7.59 mm/d, respectively, which are relatively high categories of evapotranspiration. The combination of plant emissions in the results and evaporation from the soil and water surfaces results in high evapotranspiration. The croplands are predominantly dryland, and crop irrigation provides sufficient water for ET during warmer temperatures. The daily ET of built-up land, shrub land, grassland, and bare land were 6.39 mm/d, 5.74 mm/d, 4.92 mm/d and 3.78 mm/d, respectively. Among them, shrub lands are dominated by plantation woodland that are young and have a low density and few species of trees. Therefore, its ET is relatively low. Grassland and bare land comprise approximately 67.21% of the study area, which is the dominant land use type in the study area. The grasslands are primarily desert and sandy vegetation, which are sparsely distributed. Owing to the unique mosaic substratum characteristics of the region, grassland and bare lands are distributed in intervals. Transpiration is low. Therefore, it has a small value for its daily ET. There is only a small area of built-up land, and the surrounding land use types are mostly cropland. As a result, the daily ET data for the built-up land are higher overall than the daily ET of grassland and shrub land. In summary, water and woodland are the two land use types that had the highest daily ET, indicating that water resources and vegetation cover are the main factors that determined the daily ET in the study area. Wetlands have a higher ET owing to the dual effect of plant ET and evaporation from soil and water surfaces, and cropland has a higher ET owing to irrigation and other integrated human management measures. Shrub land, grassland, and bare land have a lower daily ET owing to the sparse vegetation and insufficient soil moisture. The daily ET of built-up land is clearly overestimated. The exact reasons for this merit further analysis and research.

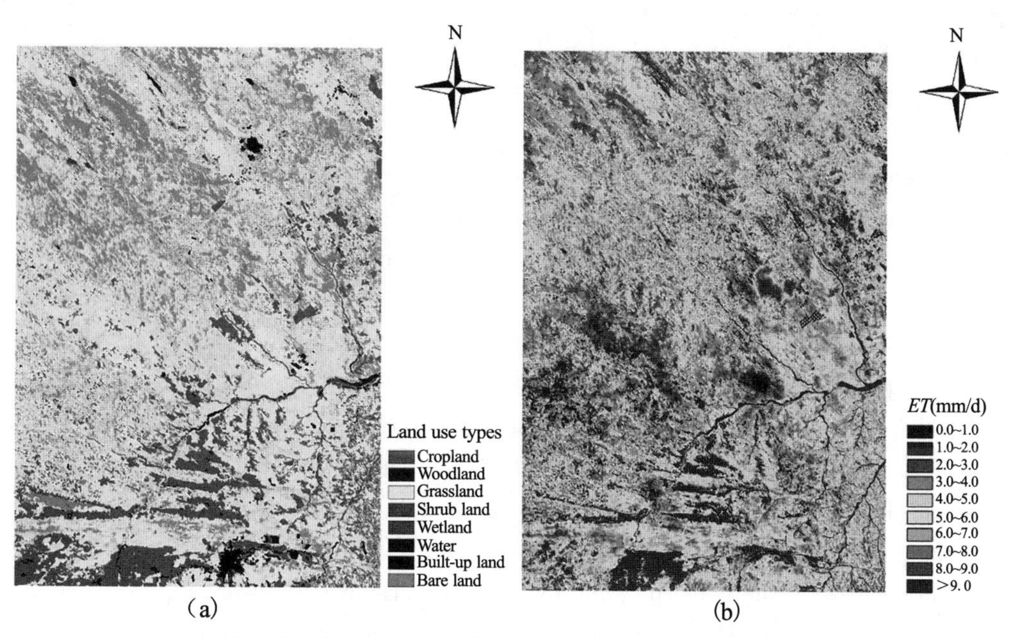

Fig.7 Spatial distribution of (a) land use types and (b) daily evapotranspiration in agro-pastoral ecotone of northwest China

Table 5 Daily evapotranspiration data for different land use types

Land use types	Area/km^2	Percentage of total area/%	Daily average *ET*/(mm/d)
Cropland	1982.86	13.92	7.59
Woodland	9.40	0.07	8.69
Grassland	9571.70	67.21	4.92
Shrub land	312.56	2.19	5.74
Wetland	14.88	0.10	7.70
Water	55.89	0.39	9.90
Built-up land	126.57	0.89	6.39
Bare land	2167.58	15.22	3.78

ET, evapotranspiration.

The verification of the *ET* model is an important part of the remote sensing inversion of *ET*. The use of remote sensing to estimate the surface flux in areas < 1 km^2 may have major problems, while the remote sensing inversion of large-scale areas is problematic owing to the problems of inadequate field observation data and errors in the observation instrument itself[14]. To more effectively vary the accuracy of the *ET* model, this study used the data of *ET* from the water surface of the weather station for relative verification and considered the problem of inconsistent water surface evaporators. The conversion coefficients of the E-601 type and 20 cm-caliber small evaporator in the study area were explored using multiple sets of meteorological data, and the errors caused by the inconsistent evaporators were reduced. The conversion coefficients that were obtained are basically consistent with those in China provided by Mao Yue[36]. When the SEBAL model was used to retrieve the *ET*, the absolute error < 1 mm/d, indicating that the SEBAL model is highly applicable to retrieve the *ET* in agro-pastoral ecotone of northwest China. Similar results can also be found in other studies of different areas[6,37]. There is no doubt that the improvement in inversion accuracy of the SEBAL model merits celebration. However, surface parameters are important ones that affect the accuracy of the *ET* model. Norman et al. delineated the error possible from the ground temperature, i.e., with a speed of 5 m/s for a canopy that is 10 m high and an error of the ground temperature difference that exceeds 1 K, the error of *H* can reach 87 W/m^2[38]. A study by Li et al.[22] found that the parameters with the strongest correlation between *ET* and surface characteristic parameters were surface temperature and net radiant flux, which are consistent with the results of this study. The uncertainty of current radiation temperature inversion is usually 1~3 K[39]. While the remote sensing *ET* model is widely used, it is also necessary to realize that there is still substantial uncertainty in the inversion of the surface parameters in the remote sensing *ET* model. Therefore, the future research of remote sensing *ET* models should ensure that each step of the inversion result and each surface parameter can be verified, so that the inversion result is accurate. The analysis of daily *ET* of different land use types found that the daily *ET* of built-up land was higher than that of the grasslands and shrub land. This may be owing to two reasons. First, the drought and water shortage in the study area can be owing to the sparseness of grasslands and shrub lands and their lower daily *ET*. The built-up land is mostly surrounded by cropland, which results in inaccurate data on the daily *ET* of built-up land. Secondly, the SEBAL model overestimates the inversion of the *ET* of built-up land, which is consistent with the results of Jin et al.[40].

Conclusions

The SEBAL model was used to retrieve the daily surface *ET* based on Landsat8 remote sensing images in the northwestern windy sand and grassland area. This data was combined with that of the water surface evaporation from a weather station to verify the accuracy of the inversion. In addition, the correlation between the *ET* and surface parameters was analyzed to clarify the impact factor of surface *ET*. This study produced the following conclusions:

(1) An analysis of the monitoring data of the E-601 type and 20 cm-caliber small evaporator in the same period indicated that the average conversion coefficient of the two surface evaporators was 0.63, and the correlation coefficient R^2 was 0.95, indicating that the conversion coefficients of E-601 type and the 20 cm-caliber small evaporator are highly reliable and can meet the conversion needs of the monitoring data of the two surface evaporators.

(2) The actual surface *ET* retrieved based on the SEBAL model correlated highly with the surface water *ET*. The correlation coefficient was 0.81, and the average root mean square error was 0.9 mm/d, indicating that the SEBAL model that is used to retrieve surface *ET* is highly accurate in the agro-pastoral ecotone of northwest China.

(3) The correlation coefficients between NDVI, LST, surface albedo and net radiant flux and daily *ET* were 0.5830, 0.8425, 0.3428, and 0.9111, respectively. Among them, the NDVI and net radiant flux positively correlated with the daily *ET*. The LST and surface albedo negatively correlated with the daily *ET*. The overall correlation between the daily *ET* and underlying surface parameters from strong to weak were the net radiant flux > LST > NDVI > surface albedo.

(4) The amount of daily *ET* varied substantially in terms of spatial distribution. The daily *ET* of water was the highest, followed by woodland, wetland, cropland, built-up land, and shrubland. The area of grassland and bare land were the dominant land use types in the study area, but they had the lowest value of daily *ET*. The inversion of daily *ET* of built-up land based on the SEBAL model was overestimated.

Acknowledgments

The data set is provided by Geospatial Data Cloud site, Computer Network Information Center, Chinese Academy of Sciences (http://www.gscloud.cn). And the authors gratefully acknowledge researchers at the Institute of Land Engineering and Technology, Shaanxi Provincial Land Engineering Construction Group, for their help with the field experiments. We wish to thank the editor of this journal and the anonymous reviewers during the revision process.

Author contributions

Conceptualization: Liangyan Yang, Jianfeng Li, Zenghui Sun. Data curation: Liangyan Yang, Yuanyuan Yang, Tong Li. Formal analysis: Liangyan Yang, Zenghui Sun, Jinbao Liu. Funding acquisition: Zenghui Sun, Jinbao Liu. Methodology: Liangyan Yang. Resources: Liangyan Yang, Jianfeng Li. Software: Liangyan Yang, Jianfeng Li. Validation: Liangyan Yang, Jianfeng Li. Writing-original draft: Liangyan Yang. Writing-review & editing: Liangyan Yang, Jianfeng Li, Zenghui Sun, Jinbao Liu.

References

[1]Yang FL, Zhang Q, Wang RY, Zhou J. Evapotranspiration measurement and crop coefficient estimation over a spring wheat Farmland ecosystem in the Loess Plateau. PLoS ONE. 2014; 9(6): e100031.

[2]Tang RL, Li ZL, Chen K, Jia Y, Li C, Sun, X. Spatial-scale effect on the SEBAL model for evapotranspiration estimation using remote sensing data (in Chinese). Agricultural and Forest Meteorology. 2013; 174: 28-42.

[3]Zhang RH. Characteristics of evapotranspiration based on remote sensing and analysis of water changes in Weihe

River basin, China (in Chinese). Beijing Normal University. 2013.

[4] Resenberg NJ, Blad BL, Verma SB. Microclimate: The biological environment of plants. New York: John Wiley & Sons. 1983; 495.

[5] L' Vovich MI, White GF, Belyaev AV, Kindler J, Koronkevic NI, Lee TR, et al. Use and Transformation of Terrestrial Water Systems. Cambridge: Cambridge University Pressure. 1990; 235-252.

[6] Kong JL, Hu YX, Yang LY, Shan ZB, Wang YT. Estimation of evapotranspiration for the blown-sand region in the Ordos basin based on the SEBAL model. International Journal of Remote Sensing. 2018; 40: 1945-1965.

[7] Brutsaert W. Hydrology: an introduction. Cambridge: Cambridge University Press. 2005.

[8] Bernier Troy P. The use of water vapor isotopes to determine evapotranspiration source contributions in the natural environment. Water. 2020; 12: 3203.

[9] Wang JJ, Liu H, Xu ZX, Wang SJ. Estimation of groundwater evapotranspiration with measured diurnal groundwater variations: A case study of typical oasis in Hexi Corridor (in Chinese). Arid Zone Research. 2021; 38: 59-67.

[10] Jiang TL, Su XL, Guo SM, Wu HJ. Spatiotemporal variation of vegetation water consumption and its response to meteorological drought in Northwest China (in Chinese). Shuili Xuebao. 2021; 52: 229-240.

[11] Huang J, Cui W, Zhong M, Li XN. Research on the characteristics of evapotranspiration and climate aridity based on remote sensing data in Yunnan-Guizhou region (in Chinese). China Rural Water and Hydropower. 2021; 1: 98-104.

[12] Xia JZ, Liang SL, Chen JQ, Yuan WP, Liu SG, Li LH, et al. Satellite-based analysis of evapotranspiration and water balance in the grassland ecosystems of Dryland East Asia. PLoS ONE. 2014; 9(5): e97295.

[13] Ning YZ, Zhang FP, Feng Q, Wei YF, Li L, Liu JY, et al. Estimation of evapotranspiration in Shule River Basin based on SEBAL model and evaluation on irrigation efficiency (in Chinese). ARID LAND GEOGRAPHY. 2020; 43: 928-938.

[14] Feng JZ, Wang ZJ. A review on evapotranspiration estimation models using remotely sensed data (in Chinese). SHUILI XUEBAO. 2012; 43: 914-925.

[15] Li F, Shen YJ. Progress in remote sensing-based models for surface heat and water fluxes (in Chinese). Resources Science. 2014; 36: 1478-1488.

[16] Wang WN, Zhang X, Zhang LF, Liu XQ, Zhao L, Li Q, et al. A comparison study of the evapotranspiration measured by lysimeter and eddy covariance (in Chinese). Chinese Journal of Ecology. 2019; 38: 3551-3559.

[17] Zhang JS, Meng P, Zheng N, Huang H, Gao J. The feasibility of using las measurements of the sensible heat flux from a mixed plantation in the hilly zone of the north China. Advances in Earth Science. 2010; 25: 1283-1290.

[18] Menenti M, Choudhury B. Parameterization of land surface evaporation by means of location dependent potential evaporation and surface temperature range. IAHS Publ. 1993; 212: 561-568.

[19] Roerink GJ, Su Z, Menenti M. S-SEBI: A simple remote sensing algorithm to estimate the surface energy balance. Physics and Chemistry of the Earth, Part B: Hydrology, Oceans and Atmosphere. 2000; 25: 147-157.

[20] Bastiaanssen WGM, Menenti M, Feddes RA, Holtslag AAM. A remote sensing surface energy balance algorithm for land (SEBAL): 1. Formulation. Journal of Hydrology. 1998; 212: 198-212.

[21] Su Z. The surface energy balance system (SEBS) for estimation of turbulent heat fluxes. Hydrology and Earth System Sciences. 2002; 6: 85-99.

[22] Li XL, Yang LX, Xu XF, Tian W, He CS. Analysis of evapotranspiration pattern by SEBAL model during the growing season in the agro-pastoral ecotone in Northwest China (in Chinese). Acta Ecologica Sinica. 2020; 40: 2175-2185.

[23] Liu MQ, Hu DY. Response of wetland evapotranspiration to land use/cover change and climate change in Liaohe river delta, China. Water. 2019; 11: 955.

[24] Rahimzadegan M, Janani AS. Estimating evapotranspiration of pistachio crop based on SEBAL algorithm using Landsat 8 satellite imagery. Agricultural Water Management, 2019; 217: 383-390.

[25] Zhang FJ, Liu ZH, Zhangzhong, LL, Yu JX, Shi KL, Yao L. Spatiotemporal distribution characteristics of reference evapotranspiration in Shandong province from 1980 to 2019. Water. 2020; 12: 3495.

[26] Luo Y, Gao P, Mu XM. Influence of meteorological factors on the potential evapotranspiration in Yanhe river ba-

sin, China. Water. 2021; 13: 1222.

[27] Bao PY. Estimation of the day evapotranspiration in semiarid area using remote sensing (in Chinese). Hohai University. 2007.

[28] Bastiaanssen WGM. SEBAL-based sensible and latent heat fluxes in the irrigated Gediz Basin, Turkey. Journal of Hydrology. 2000; 229: 87-100.

[29] Yang LY. Estimation of evapotranspiration in an arid area based on optimized SEBAL model using remote sensing (in Chinese). Chang'an University. 2019.

[30] Li BF, Chen YN, Li WH, Cao ZC. Remote sensing and the SEBAL model for estimating evapotranspiration in the Tarim River. Acta. Geographica. Sinica. 2011; 66: 1230-1238.

[31] Wang T, Tang RL, Li ZL, Jiang YZ, Liu M, Tang BH, et al. Temporal upscaling methods for daily evapotranspiration estimation from remotely sensed instantaneous observations (in Chinese). Journal of Remote Sensing. 2019; 23: 813-830.

[32] Huang QH, Zhang WC. Application and parameters sensitivity analysis of SWAT model (in Chinese). Arid Land Geography. 2010; 33(1):8-15.

[33] Tang, RL, Li ZL. Estimating daily evapotranspiration from remote sensed instantaneous observations with simplified derivations of a theoretical model. Journal of Geophysical Research: Atmospheres. 2017; 122: 10177-10190.

[34] Cheng MH, Jiao XY, Jin XL, Li BB, Liu KH, Shi L. Satellite time series data reveal interannual and seasonal spatiotemporal evapotranspiration patterns in China in response to effect factors. Agricultural Water Management. 2021; 55: 107046.

[35] Jeremiah K, Mul M, Mohamed Y, Bastiaanssen W, Pieter Z. Mapping ecological production and benefits from water consumed in agricultural and natural landscapes: a case study of the pangani basin. Remote Sensing. 2018; 10:1-24.

[36] Mao Y, Mao JX, Zhou HM. Study on evaporation correlation and conversion method for E-601B evaporator and small-scale evaporator (in Chinese). Meteorological, Hydrological and Marine Instruments. 2020; 37: 98-103.

[37] Mohammad IK, Rouzbeh N. Determination of maize water requirement using remote sensing data and SEBAL algorithm. Agricultural Water Management. 2018; 209: 197-205.

[38] Norman JM, Divakarla M, Goel NS. Algorithms for extracting information from remote thermal-IR observations of the earth's surface. Remote Sensing of Environment. 1995; 51: 157-168.

[39] Courault D, Seguin B, Olioso A. Review on estimation of evapotranspiration from remote sensing data: From empirical to numerical modeling approaches. Irrigation and Drainage Systems. 2005; 19: 223-249.

[40] Jin KL, Hao L. Evapotranspiration estimation in the Jiangsu-Zhejiang-Shanghai Area based on remote sensing data and SEBAL model (in Chinese). Remote Sensing for Land and Resources. 2020; 32: 204-212.

本文曾发表于2022年《PLOS One》第17卷第3期

Modeling Landslide Susceptibility Using Data Mining Techniques of Kernel Logistic Regression, Fuzzy Unordered Rule Induction Algorithm, SysFor and Random Forest

Tingyu Zhang[1,2], Quan Fu[3], Chao Li[4], Fangfang Liu[3], Huanyuan Wang[5], Ling Han[6], Renata Pacheco Quevedo[7], Tianqing Chen[1,2], Na Lei[1,2]

(1.Key Laboratory of Degraded and Unused Land Consolidation Engineering, The Ministry of Natural Resources, Xi'an, Shaanxi, China;2.Institute of Land Engineering and Technology, Shaanxi Provincial Land Engineering Construction Group Co., Ltd, Xi'an, Shaanxi, China;3.Shaanxi Provincial Land Engineering Construction Group Land Survey Planning and Design Institute Co., Ltd., Xi'an, Shaanxi, China;4.Shaanxi Land Engineering Construction Group Co., Ltd., Xi'an, Shaanxi, China;5.Provincial Key Laboratory of Chang'an University, School of Land Engineering, Chang'an University, Xi'an, Shaanxi, China;6.School of Land Engineering, Chang'an University, Xi'an, Shaanxi, China;7.Earth Observation and Geoinformatics Division, National Institute for Space Research (INPE), Sao Jose dos Campos, São Paulo, Brazil)

【Abstract】 This paper introduces four advanced intelligent algorithms, namely kernel logistic regres-sion, fuzzy unordered rule induction algorithm, systematically developed forest of multiple decision trees and random forest (RF), to perform the landslide susceptibility mapping in Jian'ge County, China, as well as well study of the connection between landslide occurrence and regional geo-environment characteristics. To start with, 262 landslide events were determined, and the proportion of randomly generated training data is 70%, while the proportion of randomly generated validation data is 30%, respectively. Then, through the comprehensive consideration of local geo-environment characteristics and relevant studies, fifteen conditioning factors were prepared, such as slope angle, slope aspect, altitude, profile curvature, plan curvature, sediment transport index, topographic wetness index, stream power index, distance to rivers, distance to roads, distance to lineaments, soil, land use, lithology and NDVI. Next, frequency ratio model was utilized to identify the corresponding relations for conditioning factors and landslides distribution. In addition, four data mining techniques were conducted to implement the landslide susceptibility research and generated landslide susceptibility maps. In order to examine and compare model performance, receiver operating characteristic curve was brought for judging accuracy of those four models. Finally, the results indicated that a traditional model, namely RF model, acquired the highest AUC value (0.859). Last but gained a lot of attention, the results can provide references for land use management and landslide prevention.

【Keywords】 Landslide susceptibility; Kernel logistic regression; Fuzzy unordered rule induction algorithm; Systematically developed forest of multiple trees; Random forest

1 Introduction

Landslide, as the most frequently occurred geological hazard in China, has caused casualties, economic losses and a serious of geo-environment problems (Liu et al. 2012; Zhao et al. 2015). Depending on the latest statistical report of Ministry of Natural Resources of China, in total 6181 geological hazards occurred in 2019, during that year, 211 people were killed and direct economic loss valued at 397 million dollars. Due to complex terrain, tectonic development and human activities, there are more than 230000 potential geological hazards in China and landslides account for 53.50% of it (Huang and Li 2011; Peng et al. 2019). Therefore, the prevention of landslide development in China is crucial. As it is all known that prediction of the area where landslide occurrence probabilities are higher is the first step in landslide prevention. Over the past years, various techniques and methods for forecasting the landslides have been applied to landslide susceptibility assessment. However, even as evaluation models have become more and more advanced, their predictive ability still is a critical point.

In principle, the evaluation models used in landslide susceptibility could be categorized as physical-based models, opinion-driven models, statistical models and machine learning models (Merghadi et al. 2020). Physical-based models involving detailed site characterization are currently not practical for large area risk zonation exercises. Opinion-driven models are structured based on limited information, which can be hard to quantify or evaluate a result objectively (Merghadi et al. 2020). Usually, the statistical models were conjugated to GIS for spatial variety prediction of landslide disaster. The frequently used methods are index of entropy (IOE) (Constantin et al. 2011; Demir 2019; Youssef et al. 2015), frequency ratio (FR) (Chen and Chen 2021; Hong et al. 2015b; Umar et al. 2014), certainty factor (CF) (Li and Zhang 2017; Xiao et al. 2020), logistic regression (LR) (Das and Lepcha 2019; Moon 2016; Talaei 2014) and evidential belief function (EBF) (Carranza 2015; Li and Wang 2019; Wang et al. 2016). As the predictive ability of statistical model is still deficient and with the development of information technology, machine learning models could be a better alternative to solve the susceptibility assessment problem, such as artificial neural networks (ANN) (Aditian et al. 2018; Lei et al. 2020a; Li et al. 2021; Polykretis et al. 2015), support vector machines (SVM) (Hong et al. 2015b; Pham et al. 2017; Yu and Chen 2020), adaptive neuro-fuzzy inference systems (ANFIS) (Chen et al. 2021a; Nasiri Aghdam et al. 2017; Polykretis et al. 2019), rotation forest (RF) (Hong et al. 2018; Pham et al. 2016; Sahin et al. 2020; Youssef and Pourghasemi 2021), support vector regression (SVR) (Liu et al. 2021; Panahi et al. 2020) and multivariate adaptive regression spines (MARS) (Conoscenti et al. 2015; Martinello et al. 2021; Rotigliano et al. 2019; Youssef and Pourghasemi 2021). For example, Wang et al. (2019b) used the CF method to conduct landslide hazard susceptibility analysis in southeastern Gansu, China. Nhu et al. (2020) applied twenty landslide conditioning factors and based on five machine learning methods (logistic model tree, LR, naïve Bayes tree, ANN and SVM) created corresponding landslide susceptibility maps for 111 landslides in Bijar city, Iran. Although these landslide modeling methods have good stability and flexibility, they still inevitably have some defects. For instance, the commonly used CF model does not take into account the difference in the impact of each landslide evaluation factor on the sensitivity of landslide hazards. The AHP method cannot analyze the influence of different characteristic variables in each conditioning factor on it. The LR model cannot solve the problem of merging multi-source data types. The SVM algorithm is difficult to implement for large-scale training samples and has difficulties in solving multi-classification problems. Therefore, it is crucial to discover new landslide modeling methods. Since there is currently no consensus on which models are most suitable for specific areas, it is necessary to explore various methods for different study areas to identify models with the best predictive power.

This paper introduces three less-applied and compared machine learning models (kernel logistic regression (KLR), fuzzy unordered rule induction algorithm (FURIA), multi-tree system development forest (SysFor)) and the commonly used RF model to evaluate the landslide susceptibility of Jian'ge County, China. In this study, combined with local geological environment characteristics and previous studies, 15 landslide conditioning factors were empirically identified, and the spatial correlation between landslides and conditioning factors was calculated based on the FR model. The obtained landslide modeling results are verified by some statistical methods ultimately. The results of the study can provide some reference for local landslide risk prevention.

2 Methodology

The methodology that run through in article can contain the following significant steps which are individually displayed in following text and Fig. 1.(1) data preparation; (2) data pretreatment for training dataset; (3) generating landslide susceptibility maps through application of KLR model, FURIA model, SysFor model and RF model; (4) validation of landslide susceptibility models.

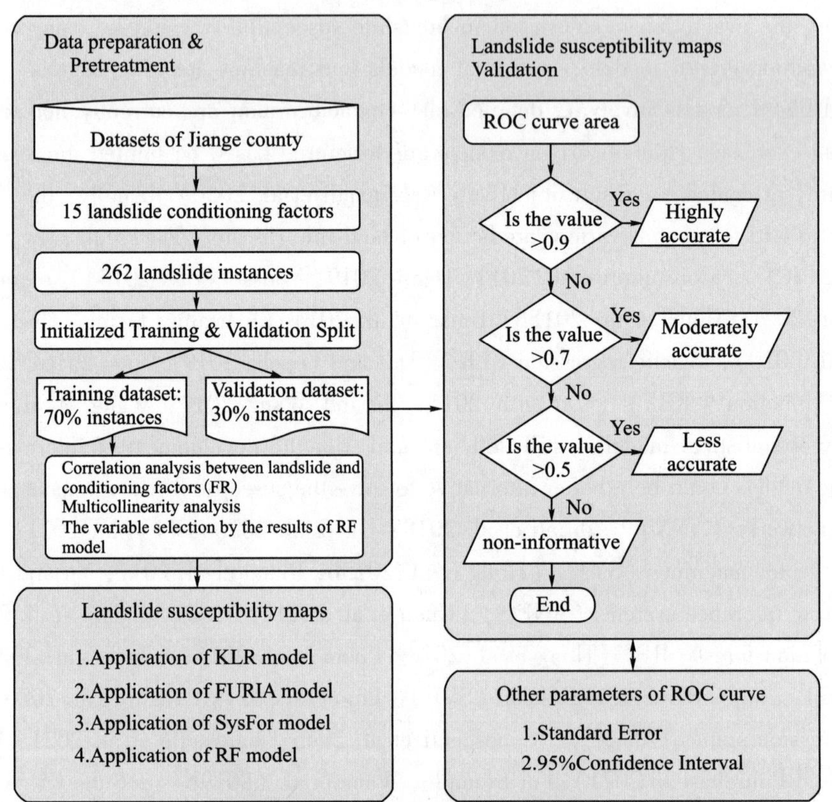

Fig. 1 The proposed technical roadmap

2.1 Frequency ratio

The frequency ratio (FR) model is a commonly used probabilistic model where major concept is relevance judgments for landslides and every landslide conditioning factors (Yalcin et al. 2011). The FR was defined, as in earlier research, by the ratio of the region where historical landslide distributed. In this paper, the frequency ratio of each class in each conditioning factor was determined across dividing the landslide occurrence probability through its area probability (Razavizadeh et al. 2017). The landslide susceptibility index (LSI) for FR was derived via summing across each factor ratio value, shown in the following equation according to the formulation method:

$$LSI = FR \tag{1}$$

where FR is the frequency ratio of per conditioning factor's class.

2.2 Kernel logistic regression

As a kind of machine learning method, kernel logistic regression (KLR) is a higher variant of normal logistic regression and commonly applied to binary prediction (Hong et al. 2015a). Since this model has capacity to classify vast amounts of data with high dimensional space, the landslide susceptibility of this paper was studied through the posterior probability that calculated by this method. Suppose D is the landslide training dataset and $D=(x_i, y_i)$, $x_i \in R^n$, $y_i \in$ (Aditian et al. 2018), x_i is the landslide conditioning factors in this research, such as attitude, elevation, slope angle, NDVI, land use and so on. y_i indicate two categories: landslide and non-landslide. The results of KLR were displayed as expression of Eq. (2) as following:

$$\log it\{y(x)\} = w \cdot \varphi(x) + b \tag{2}$$

where w parameter is the vector involving landslide conditioning factors, $\varphi(x)$ parameter represents a non-linear transformation to the original input vectors, and b parameter is a bias term.

After the logit transformation, Eq.(2) can be showed:

$$y(x) = \frac{1}{1+\exp\{-w \cdot \varphi(x) - b\}} \tag{3}$$

The landslide susceptibility index is generated by Eq. (3), and this index is normalized to the range of [0, 1] using the logit function. The nonlinear transformation is also known as the kernel function that comes from the inner product between the images of the vectors within the feature space.

$$K(x, x') = \varphi(x) \cdot \varphi(x') \tag{4}$$

There exist many kernel functions such as linear kernel, normalized polynomial kernel, radial basis function (RBF) kernel and kernel that have been applied in many areas. Here, the RBF kernel was chosen as its wide applications (Chen et al. 2019a).

$$K(x_i, x_j) = \exp((-\|x_i - x_j\|^2)/2\delta^2) \tag{5}$$

where δ is a turning parameter. The w-vector can minimize a cost function to generate. It was then expressed as following formula:

$$w = \sum_{i=1}^{n} a_i \varphi(x_i) \tag{6}$$

Where vector $a_i (i=1, 2, 3, \cdots, n)$ is called each landslide conditioning factors. Therefore, the form of the kernel logistic regression can finally be written as:

$$\log it\{y(x)\} = \sum_{i=1}^{n} a_i K(x_i, x) + b \tag{7}$$

2.3 Fuzzy unordered rule induction algorithm

Fuzzy unordered rule induction algorithm (FURIA) is a classification method that based on fuzzy rule (Hühn and Hüllermeier 2009). It is modified and expanded from a rule learner, which is simply referred as repeated incremental pruning to produce error reduction (RIP-PER). In this algorithm, the interval was generated by the fuzzy interval with trapezoidal membership function replacement. To utilize the benefits of RIPPER and fuzzy logic two approaches, a FURIA algorithm is proposed that indicates the rule order within the rule list is unconsidered and this was not available default rules here (Hühn and Hüllermeier 2010; Tien Bui et al. 2014). The rules of every class label are obtained, respectively, through the one-versus-the-rest strategy (Kumari and Kumar 2013; Yuan et al. 2011). Via the trapezoidal membership function, the FURIA model relies on the conversion of the clear rules into fuzzy rules of RIPPER (Palacios et al. 2013). Every fuzzy interval during this function is given by the following equation:

$$I(x) = \begin{cases} 1 & T_2 \leqslant v \leqslant T_3 \\ \dfrac{v - T_1}{T_2 - T_1} & T_1 < v < T_2 \\ \dfrac{T_4 - v}{T_4 - T_2} & T_3 < v < T_4 \\ 0 & \text{else} \end{cases} \tag{8}$$

where T_2 is the upper bound of the core fuzzy set while T_3 is the lower. Similarly, the T_1 is the lower bound of support while the T_4 is the upper. The method learns the antecedents in the form of a list (α_1, α_2, α_3, ⋯, α_m) which was presented in the order of the rules with importance of antecedents (Palacios et al. 2016). For the generalization rules, the importance is calculated by Eq. (9), as shown in the following:

$$\text{Importance} = \frac{p+1}{p+n+2} \times \frac{k+1}{m+2} \tag{9}$$

where p and n are the positive and negative samples with the identical rule, k is given by $j+1$that based on antecedent α_j.

2.4 SysFor

Systematically developed forest of multiple decision trees (SysFor) is initially proposed by Islam and Giggins in 2011 (Veenadhari et al. 2014). The principle of this new data mining method is based on the concept described by gain ratio. The main purpose of establishing the multiple trees is to obtain better knowledge by extracting a number of patterns (Bibri 2018). According to the framework of SysFor, it can be roughly introduced by three steps.

At the beginning, for identifying their segmentation points and a group of good attributes, according to separation values and gain ratio defined by the user. A numeric attribute can be selected multiple times in a good group of attributes when different segmentation points had a certain distance from each other and gain ratio had a higher value. After identification of the good attributes set, if the number of tree defined by the user is bigger than the size of the good attributes, SysFor will construct the tree by making each good attribute serve as the root attribute of the tree. The tree was then built as much as number of good attributes to the second step of SysFor. Otherwise, it will build the number of user defined trees from root attribute transformed by a group of good attributes. If the number of trees built at this step is still lower than the number of trees defined by users, SysFor will build more trees in the third step until the number of trees defined by user is reached.

The voting system proposed by Islam and Giggins is followed after generating the multitreeforest of the system to predict the class values that have not been recorded. In this voting system, the leaves that come from all trees which carry the record are all collected. The highest accuracy leaves were then identified from a maximum number of records to total number of records ratio. Finally, most class values of leaves are selected as the predicted class values of records.

2.5 Random forest

The random forest (RF) method, raised by Breiman, is a new and powerful machine learning method (Breiman 2001). RF belongs to one of flexible and useful machine learning algorithm that can achieve good results in most cases even without hyper-parametric tuning (Genuer et al. 2017; Song 2015). Because of its simplicity, it is convenient to use in both classification and regression tasks (Guio Blanco et al. 2018; Kamińska 2019). The process of data processing by RF algorithm has approximately four sessions. First of all, the method adopts decision tree as the base classifier. Then, use the Bagging method to generate the training datasets that are different from each other and establish the decision trees by random subspace method. Finally, select the partial attributes randomly from all the attributes and choose the optimal attribute from it when the tree is splitting. According to the previous studies (Pourghasemi and Kerle 2016; Youssef et al. 2016b), the RF method presents excellent performances in classification task compared to other methods and it has the following advantages: (1)the model is simple to operate and the accuracy will not be affected even if partial data is missing; (2) the model has a strong generalization ability and can detect the interaction of the conditioning factors in the training process; (3) after the training process, it can visually present the weight from each conditioning factor.

3 Description of the study area

The study area (Jian'ge County), with an area of 3204 km^2, is within latitude 31°31′ to 32°21′ N, and longitude 105°10′ to 105°49′E (Fig. 2). The attitudes vary in the range of 358～1284 m. Climatically, its climatic zone belongs to the subtropical humid monsoon. The annual mean temperature is 15.40 ℃. The terrain, road, lineament, river and landslide position in Jian'ge county are displayed in Fig. 2.

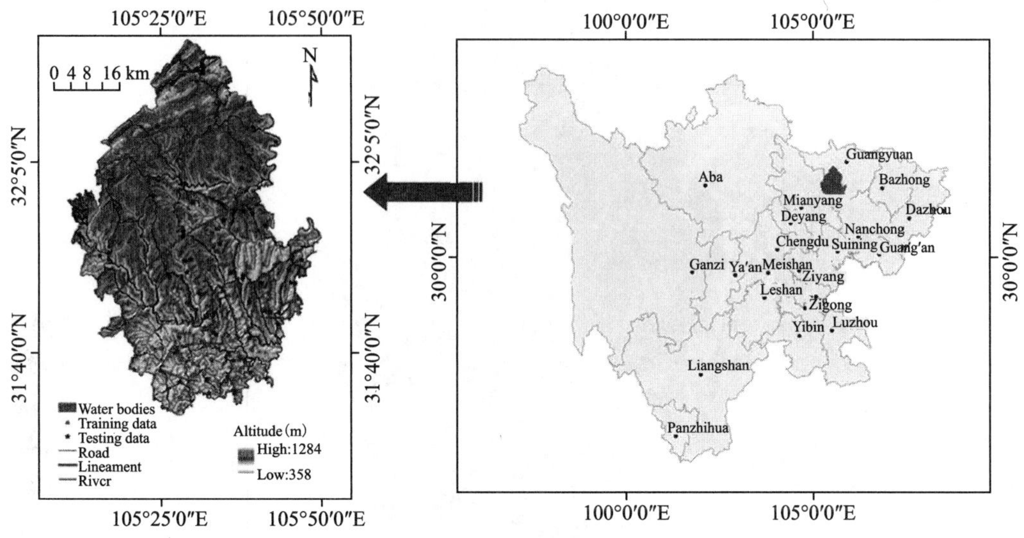

Fig. 2 Study area

4 Data preparation

It is known that a generated landslide inventory map containing varieties of information about the landslide, such as scale, state, threat objects and so on, is the most basic and significant material in identification for landslide risk research. Prior to plot the landslide inventory map, previous study of field investigation and the interpretation of remote sensing images have been conducted. A total of 262 landslides were identified and divided into training and validation datasets at a ratio of 70 ∶ 30 using a random sampling method. After the comprehensive analysis of regional geo-environment charac-teristics and previous researches (Wang et al. 2020b; Wu et al. 2020; Zhou and Fang 2015), we chose fifteen suitable landslide conditioning factors, and the sources of these factors are shown in Table 1. To represent the landslide conditioning factors, five layers have been drawn, namely altitude, NDVI, land use, soil and lithology (Figs. 3, 4, 5, 6, 7), and the detailed parameters of all conditioning factors can be seen in Table 2.

Table 1 Data sources for landslide conditioning factors

Conditioning factors	Type	Source of data
Slope angle	Terrain factors	Extracted from the digital elevation model (DEM) with a resolution of 20 m
Slope aspect		
Altitude		
Profile curvature		
Plan curvature		
STI	Hydrological factors	
TWI		
SPI		
Distance to rivers		
Distance to roads	Anthropogenic factors	Extracted from the topographic map with a scale of 1∶50000
Distance to lineaments	Geological factors	Extracted from the geological map with a scale of 1∶200000
Lithology		
Land use	Environmental factors	Extract from land use maps at 1∶100000 scale, soil maps at 1∶1 000000scale and Landsat-8 OLI images
Soil NDVI		

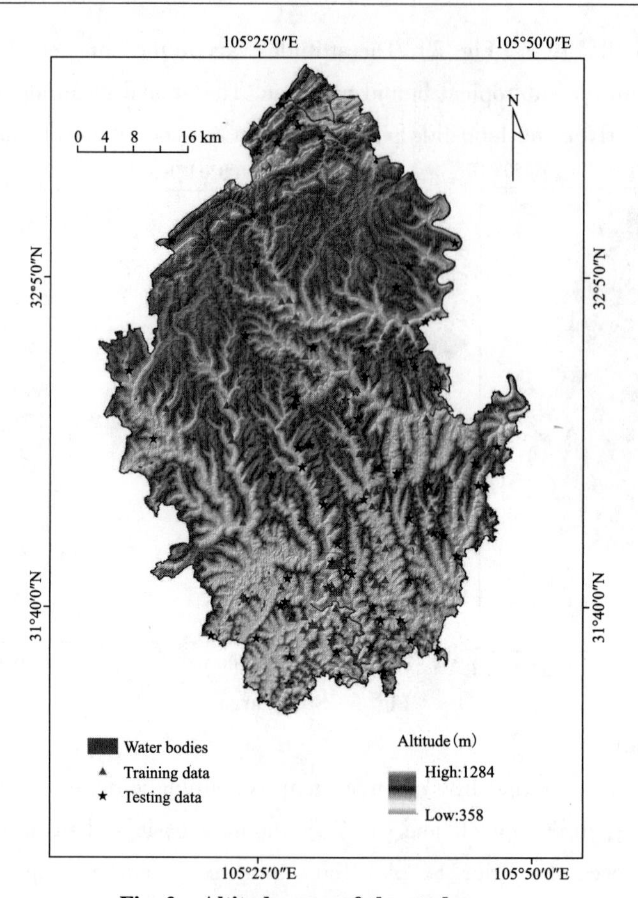

Fig. 3 Altitude map of the study area

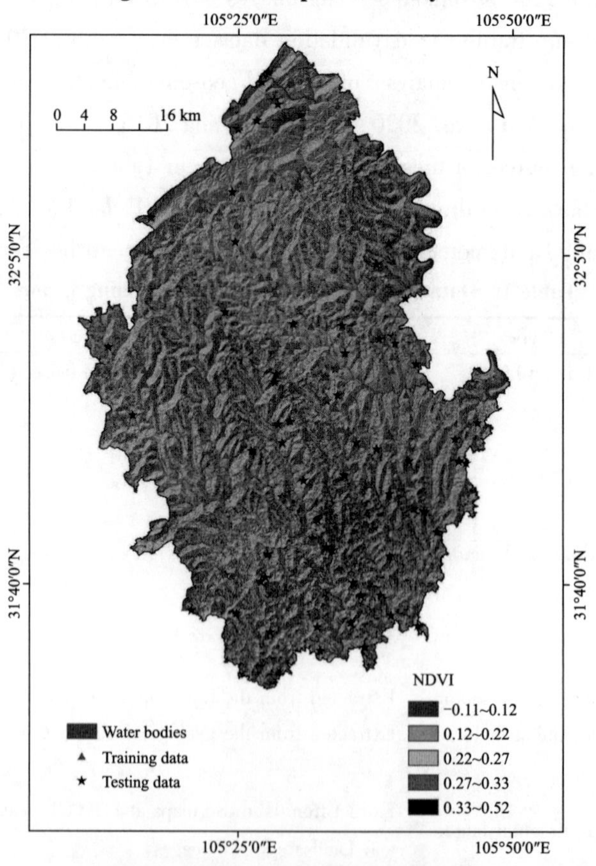

Fig. 4 NDVI map of the study area

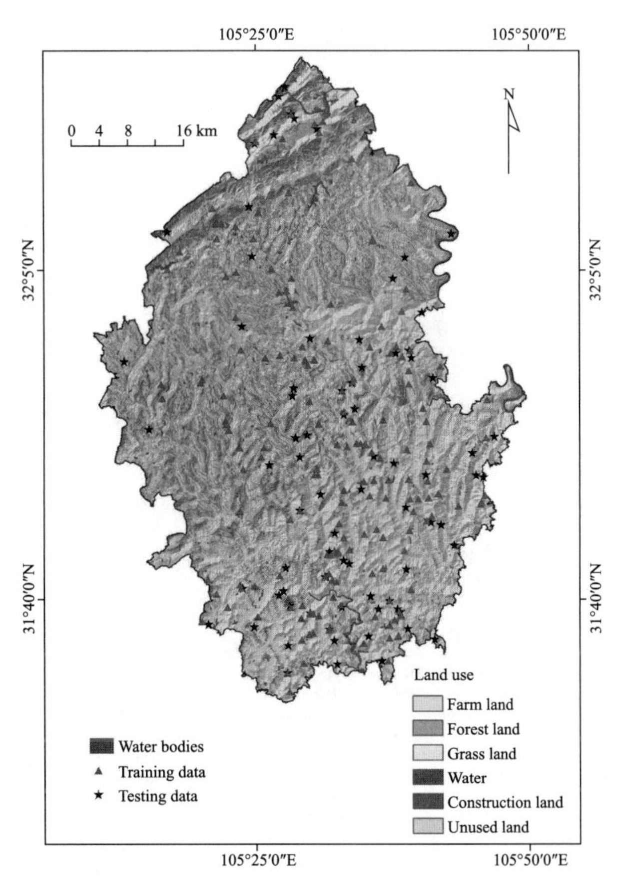

Fig. 5 Land use map of the study area

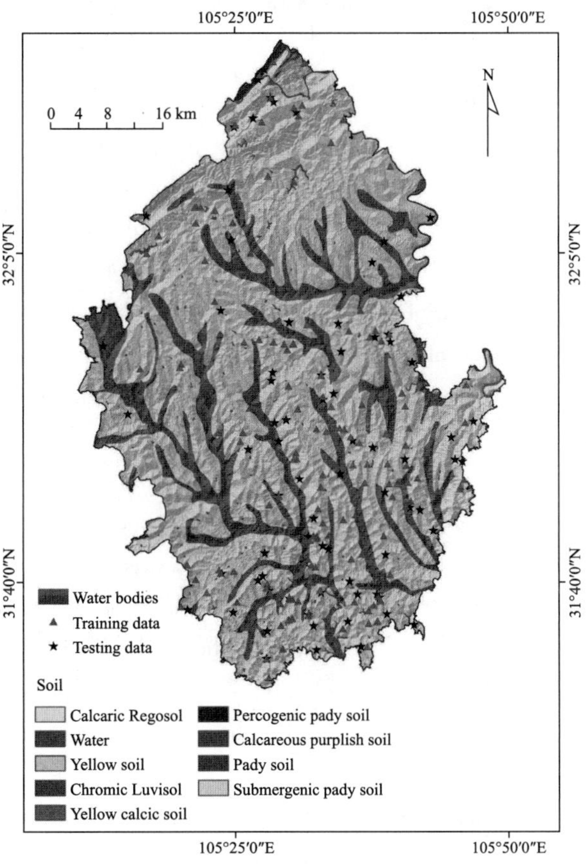

Fig. 6 Soil map of the study area

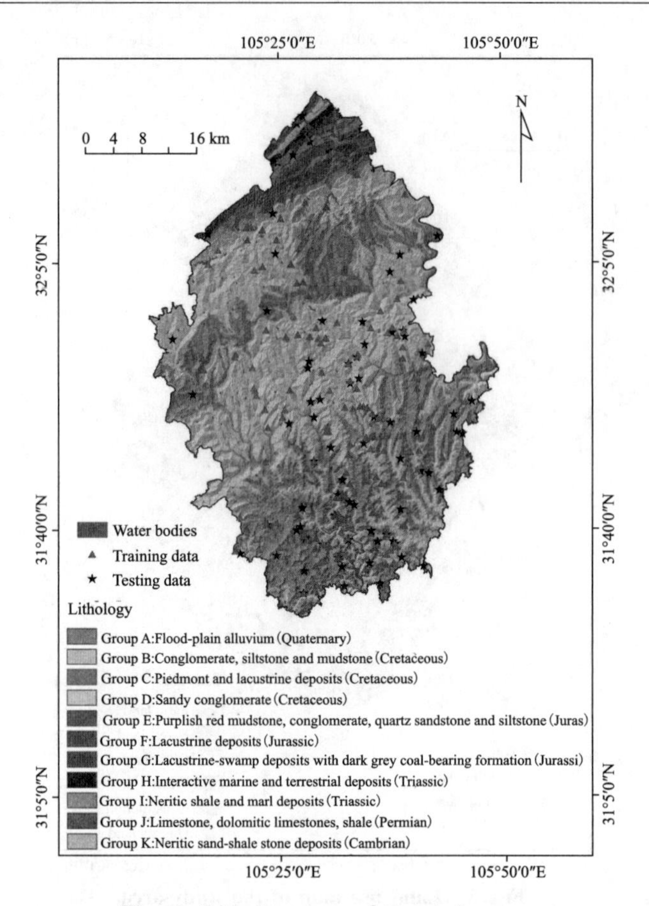

Fig. 7 Lithology map of the study area

5 Results

5.1 Landslide conditioning factors analysis

After the identification of landslide conditioning factors for predictor variables, FR model was introduced to analyze the corresponding relations for conditioning factors and landslides distribution. We present results of standard metrics (FR value) in Table 2. The geological environmental conditions in this region are benefit to develop the landslide when FR value is bigger than 0. For the slope angle, it changes from 0 to 78.59°, and more thaneighty percentage of landslides concentrate in the classes of 10°~20° and 20°~30°. Slope aspect influences the occurrence of landslide by controlling rainfall and sunshine. In this study, the aspects of east and west have a tendency to promote landslides. That is mainly because many slopes in study area are near the north-south trend. In the case of altitude, as human activities mainly concentrate in the area with the elevation of 500~600 m, it became the most concentrated area of landslides. Profile and plan curvatures describe the fluctua-tion of terrain from different aspects. It can be seen from Table 2 that landslides point is more opportunistic to be concentrated in an area with a convex slope and easy water catch-ment. For STI, more than seventy percentage of landslides are in areas with an index ofless than 10. However, landslides are more prone to appear in an area where STI value is greater than 40. For TWI, a factor usually applied to represent the spatial distribution state of soil moisture, more than sixty percentage of landslides clustered in sites where TWI values are less than 2. In the case of STI, the index range of 10~30 is most favorable for landslide occurrence.

Distance to rivers, roads and lineaments are three linear factors. Normally, the distance from linear factor is inversely proportional to the number of landslides, that is, the close the distance, the more landslides. In the present study, landslides concentrated within 400 m from rivers, 200 m from roads and 4000 m from lineaments. As for soil, study area has discriminated the eight diverse types of soil. Because of the inherent differences in physic-mechanical properties between different soils, their effects on landslide occurrence are also different. It can be seen from Table 2, landslides mainly distribute in yellow soil, chromic luvisol and pady soil. The main reason is that these kinds of soils are cultivated soils with frequent human activities and higher water content. For land use, farm land and construction land are two kinds of land use that are most prone to landslides. It also reflects that landslides are closely related to human activities. In the case of lithology, type J (limestone, dolomitic and shale) is prone to landslides, which mainly because study area lies in karst area, where water erodes rocks and is prone to landslide. For NDVI, the class of 0.33~0.52 obtained the maximum FR (1.232), while these areas are mainly covered by crops.

Before we proceed to evaluate landslide-prone area, one critical step is to analyze the connections among the factors. The accuracy of this research will be significantly affected if correlation analysis among factors is lacking. Hence, a frequently applied collinearity analysis method, multicollinearity analysis, was introduce to check the relationship among factors (Chen et al. 2021d). Tolerance (TOL) and variance inflation factor (VIF) are two statistical indicators which implement the identification, and they are a pair of reciprocals (Chen et al. 2021b). The results show that the highest VIF values (4.392) and the smallest TOL (0.228) belong to SPI (Fig. 8). The VIF and TOL values did not exceed the critical values of 10 and 0.1, respectively; that is, it can be considered that the factors are mutually independent.

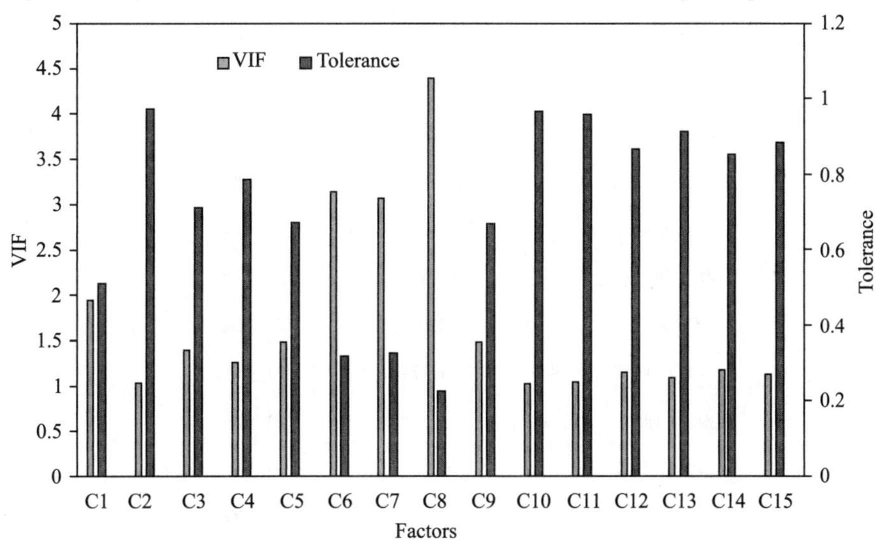

Fig. 8 Multicollinearity analysis: C1: slope angle, C2: slope aspect, C3: altitude, C4: profile curvature, C5: plan curvature, C6: STI, C7: TWI, C8: SPI C9: distance to rivers, C10: distance to roads, C11: distance to lineaments, C12: soil, C13: land use, C14: lithology, C15: NDVI

5.2 Application of KLR model

The prepared data was imported into Weka software for establishing the KLR model. In the modeling process, the optimal parameters of the KLR model were obtained by using the training data and tenfold cross-validation, so as to carry out heuristic test, and the selected parameters are shown in Table 3. After that, the model was applied to manage all the data in study area and calculate landslide susceptibility index (LSI). In this paper, in order to better exhibit the information contained in the data itself, the top compartmental method: natural break method was adopted to divide LSI. So, susceptibility map with LSI yielded five categories, that is, very high, high, moderate, low and very low. Generally, the larger the LSI gained the greater probability of landslide can be inferred (Fig. 9).

Table 2 Correlation between landslides and conditioning factors using FR model

Conditioning factors	Classes	Percentage of domain(a)	Percentage of landslides(b)	FR(b/a)
Slope angle(°)	0~10	32.607	0.000	0.000
	10~20	39.870	56.284	1.412
	20~30	18.930	25.137	1.328
	30~40	6.607	13.115	1.985
	40~50	1.667	5.464	3.279
	50~60	0.275	0.000	0.000
	60~70	0.041	0.000	0.000
	70~78.59	0.003	0.000	0.000
Slope aspect	Flat	0.171	0.000	0.000
	North	11.659	10.929	0.937
	Northeast	12.051	13.115	1.088
	East	14.915	18.033	1.209
	Southeast	11.967	12.022	1.005
	South	13.357	10.383	0.777
	Southwest	12.037	10.383	0.863
	West	13.574	15.301	1.127
	Northwest	10.267	9.836	0.958
Altitude (m)	358~500	8.154	0.000	0.000
	500~600	25.246	40.984	1.623
	600~700	26.566	26.776	1.008
	700~800	23.210	22.404	0.965
	800~900	12.627	9.290	0.736
	900~1 000	2.865	0.546	0.191
	1 000~1 100	1.023	0.000	0.000
	1 100~1 200	0.291	0.000	0.000
	1 200~1 284	0.017	0.000	0.000
Profile curvature	−42.11~−3.38	4.742	8.743	1.844
	−3.38~−1.20	22.051	24.590	1.115
	−1.20~0.68	35.780	24.044	0.672
	0.68~2.86	30.135	33.333	1.106
	2.86~37.53	7.291	9.290	1.274
Plan curvature	−36.05~−2.40	3.602	3.825	1.062
	−2.40~−0.94	14.274	18.033	1.263
	−0.94~0.23	43.177	32.240	0.747
	0.23~1.69	30.603	32.787	1.071
	1.69~38.55	8.345	13.115	1.572

continued to Table 2

Conditioning factors	Classes	Percentage of domain(a)	Percentage of landslides(b)	FR(b/a)
STI	0~10	79.058	73.224	0.926
	10~20	11.514	17.486	1.519
	20~30	4.193	3.279	0.782
	30~40	2.023	1.093	0.540
	>40	3.212	4.918	1.531
TWI	<1	0.658	2.186	3.323
	1~2	53.105	60.656	1.142
	2~3	33.987	29.508	0.868
	3~4	10.366	7.650	0.738
	>4	1.884	0.000	0.000
SPI	0~10	64.974	58.470	0.900
	10~20	11.199	15.847	1.415
	20~30	5.026	7.104	1.413
	30~40	3.261	2.186	0.670
	>40	15.540	16.393	1.055
Distance to rivers(m)	0~200	21.597	15.301	0.708
	200~400	27.527	34.426	1.251
	400~600	17.323	14.754	0.852
	600~800	14.667	12.568	0.857
	>800	18.887	22.951	1.215
Distance to roads (m)	0~200	13.370	20.219	1.512
	200~400	10.374	7.650	0.737
	400~600	9.285	9.290	1.001
	600~800	8.486	8.743	1.030
	>800	58.486	54.098	0.925
Distance to lineaments (m)	0~2 000	20.932	18.033	0.861
	2 000~4 000	18.181	24.590	1.353
	4 000~6 000	17.504	16.940	0.968
	6 000~8 000	16.710	14.208	0.850
	>8 000	26.673	26.230	0.983

continued to Table 2

Conditioning factors	Classes	Percentage of domain(a)	Percentage of landslides(b)	*FR*(b/a)
Soil	Calcaric regosol	78.381	74.863	0.955
	Water	0.975	0.000	0.000
	Yellow soil	0.138	0.546	3.966
	Chromic luvisol	0.454	1.093	2.406
	Yellow calcic soil	0.000	0.000	0.000
	Percogenic pady soil	0.024	0.000	0.000
	Calcareous purplish soil	0.613	0.000	0.000
	Pady soil	19.260	23.497	1.220
	Submergenic pady soil	0.154	0.000	0.000
Land use	Farm land	58.619	72.131	1.231
	Forest land	32.256	19.126	0.593
	Grass land	7.605	7.650	1.006
	Water	0.975	0.000	0.000
	Construction land	0.542	1.093	2.017
	Unused land	0.004	0.000	0.000
Lithology	Group A	0.294	0.546	0.773
	Group B	30.269	30.055	0.413
	Group C	38.840	44.809	0.480
	Group D	22.494	18.033	0.333
	Group E	4.612	3.279	0.296
	Group F	1.804	1.093	0.252
	Group G	0.368	0.546	0.619
	Group H	0.429	0.546	0.530
	Group I	0.618	0.000	0.000
	Group J	0.267	1.093	1.705
	Group K	0.006	0.000	0.000
NDVI	-0.11~0.12	2.435	1.639	0.673
	0.12~0.22	19.288	21.311	1.105
	0.22~0.27	40.882	36.612	0.896
	0.27~0.33	25.865	26.230	1.014
	0.33~0.52	11.530	14.208	1.232

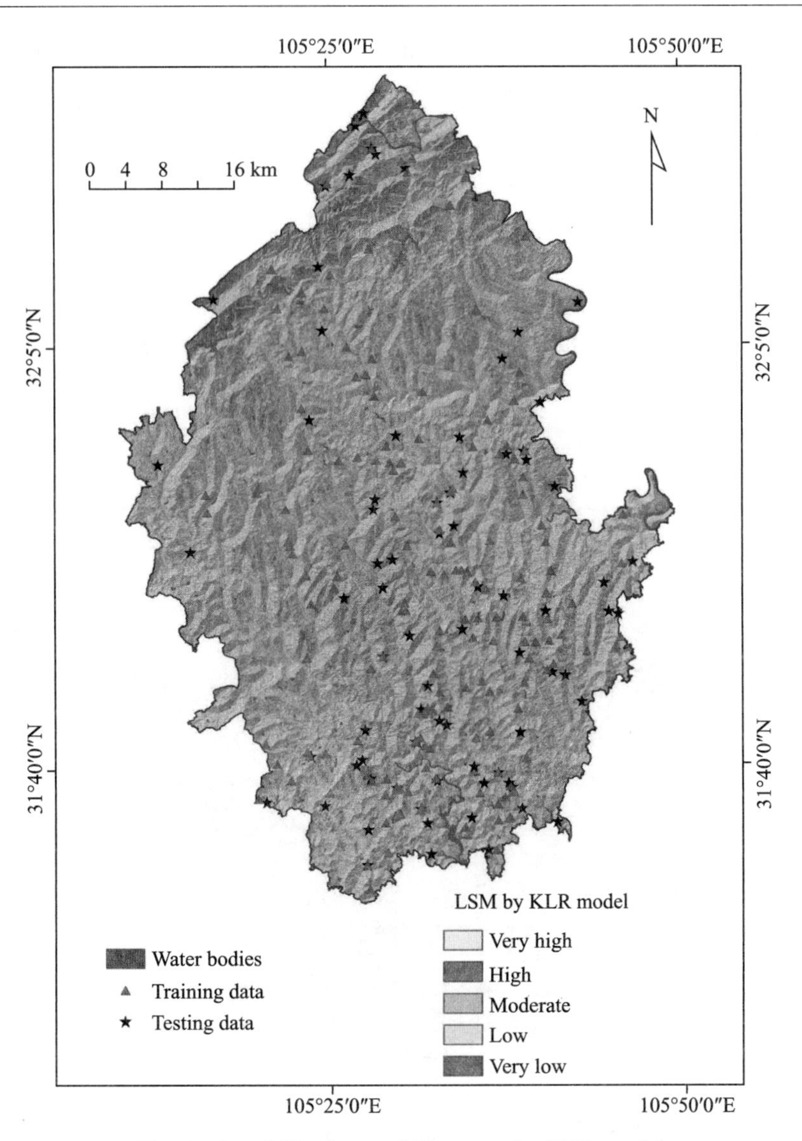

Fig. 9　Landslide susceptibility map by KLR model

Table 3　Parameters of the algorithm applied in this study

Model	Software	Parameter
KLR	Weka	seed, 1; Iambda, 0.01; kernel, RBF kernel
FURIA		seed, 1; optimizatons, 2
SysFor		confidence, 0.25; goodness, 0.3; minRe cLeaf, 10; Number of trees, 60; separation, 0.3
RF	R	mtry, 5; ntree, 1000

5.3　Application of FURIA model

The main reason for choosing the FURIA model is that even through the rule interpretation is more complex than other fuzzy rule induction methods, it can still provide compact rules in terms of rule length in a high precision. In the present study, the FURIA model was also built in Weka software (Table 3). FURIA replaces the rules and rule lists provided by traditional RIPPER algorithms with fuzzy rules and unordered rule sets. In addition, regular stretching method provided by FURIA model can also handle uncovered examples. In a nutshell, the LSM generation process applying FURIA model approximately contains two steps: (1) calculating LSI by its probability distribution function, (2) using natural break method to reclassify LSI, similarly for application of KLR model. Finally, LSM generated by FURIA model is displayed in Fig. 10.

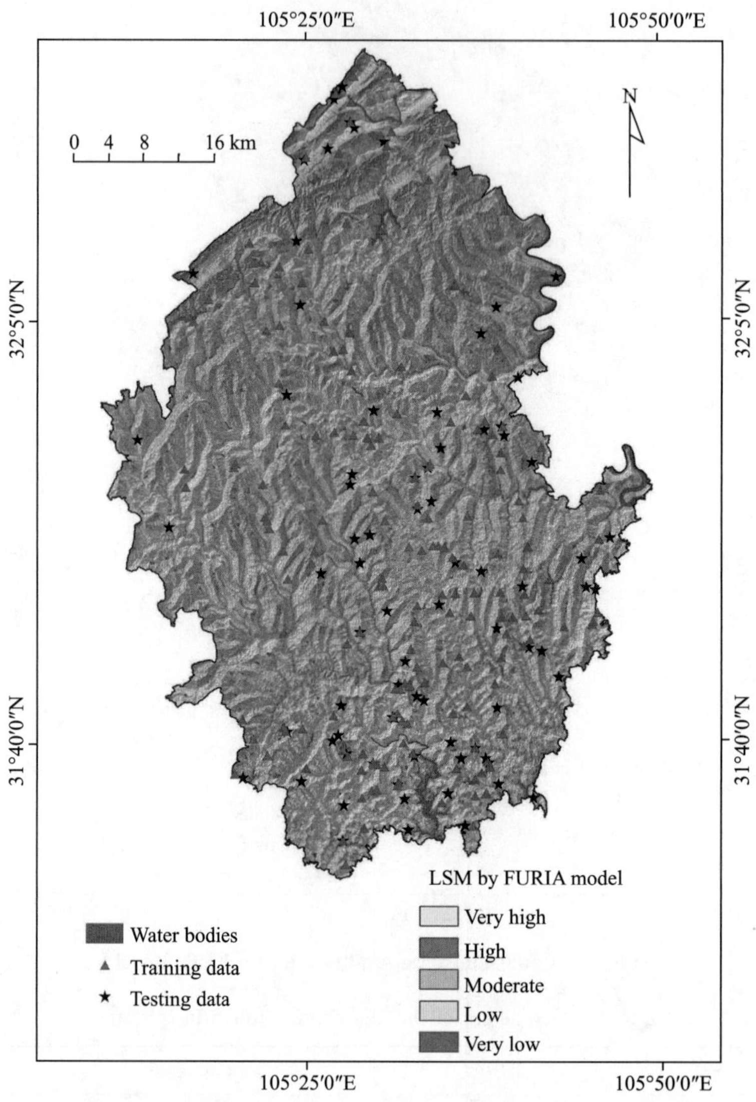

Fig. 10 Landslide susceptibility map by FURIA model

5.4 Application of SysFor model

SysFor is a new algorithm that based on information gain ratio, and this model's main purpose is to achieve better knowledge by extracting a number of patterns. In this work, Weka software was also used to build the SysFor model, and the process of landslide susceptibility research can be briefly divided into four steps. For the first, building the tree and checking whether the tree graph whether or not meets the expectation. Then, calculate the entropy function of the given dataset and delimit the dataset according to the given characteristics. Next, choose the best way to divide the dataset and train the algorithm. Finally, use the established tree to classify the results (landslide or non-landslide). After above four steps, LSI calculated by SysFor model was obtained. To compare against results from above two models, the LSI was also divided into four grades by natural break method, as with the two previous models. SysFor model result is shown in Fig. 11.

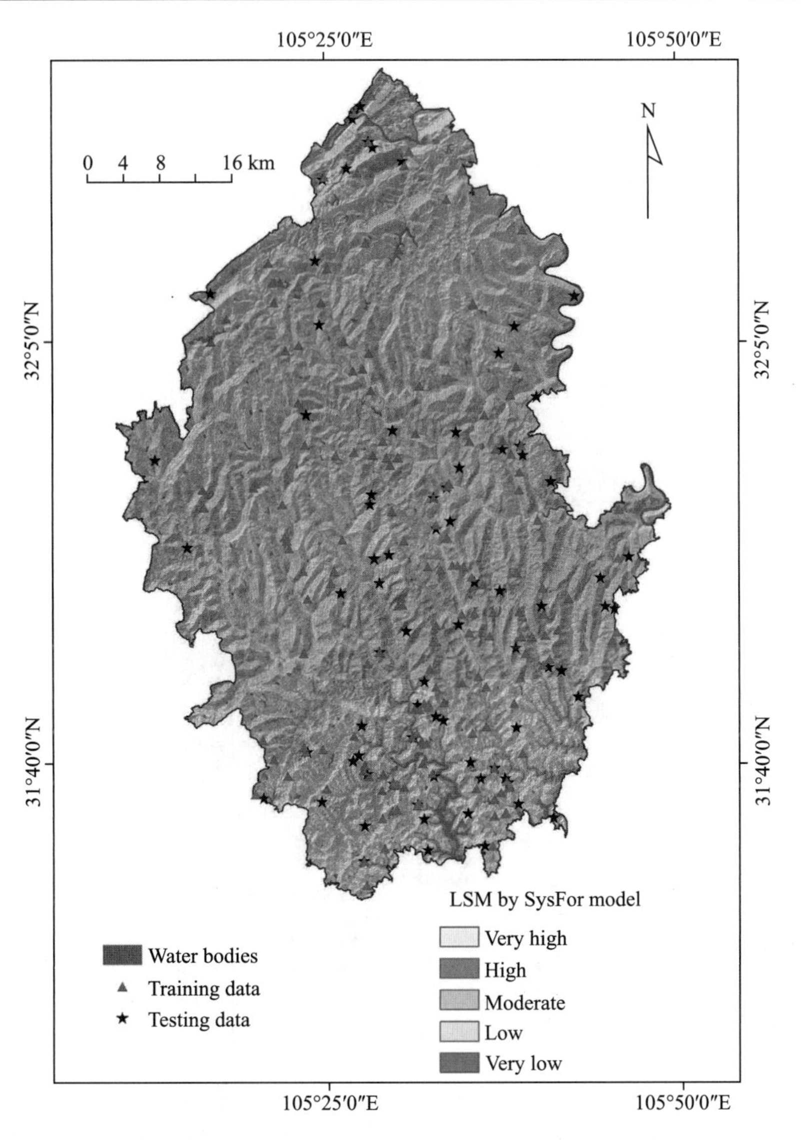

Fig. 11 Landslide susceptibility map by SysFor model

5.5 Application of RF model

In recent years, RF, a modeling approach, has been already extensively studied in assessing susceptibility outcomes with some advantages of perfect construction and very low computation. (Pourghasemi and Kerle 2016; Wang et al. 2019a). In this work, RF approach served to operate LSI of each evaluation unit. Practically speaking, RF is a kind of collective model that consists of bagging and decision tree two models. In this study, the RF model was run in the R programming environment, and the process of using this model to conduct landslide susceptibility research can be simply described as follows. For the first, the trees were constructed by training dataset which was formed by randomly selecting 70% data from dataset. Then, m feature subsets were randomly selected from M features, and the best feature was selected when the tree is splitting. Next, let each tree grew as much as possible without pruning. Besides, use generated trees to vote on the classification results (landslide or non-landslide) of each evaluation unit. Finally, the LSIs were calculated by RF model, and were divided like KLR model into five categories. Thus, the landslide susceptibility map was drawn according to 5 divided levels (Fig. 12).

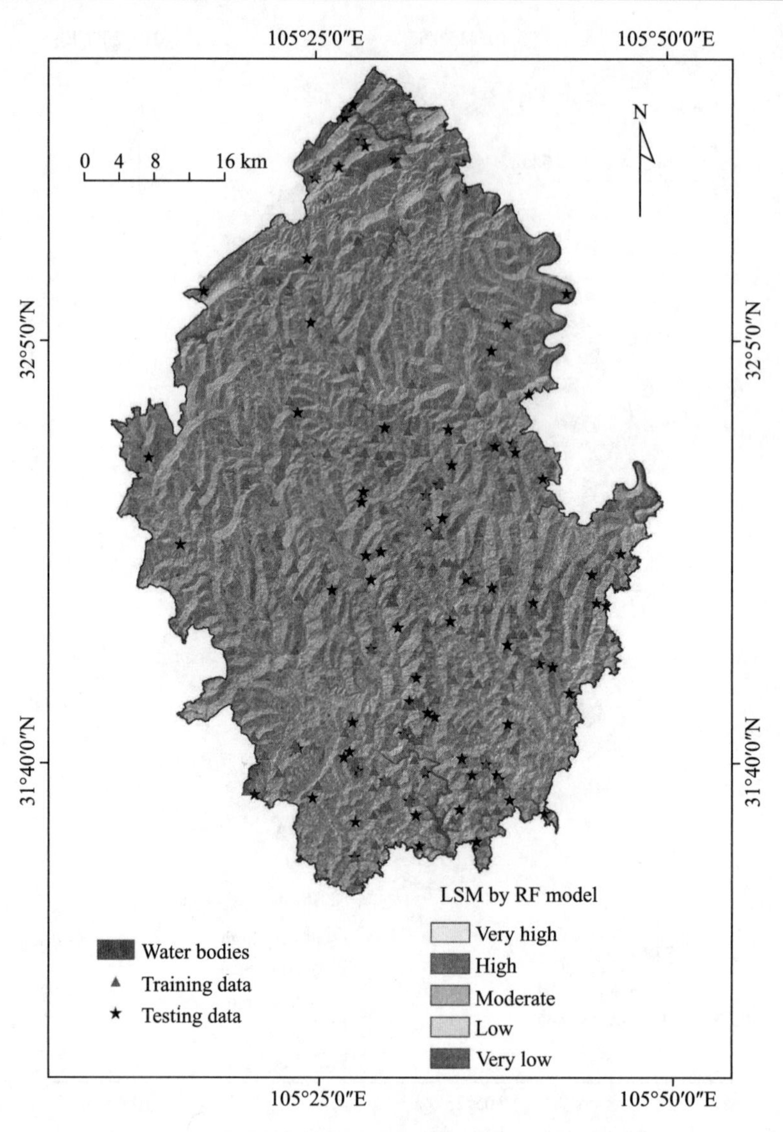

Fig. 12 Landslide susceptibility map by RF model

The variable selection was guided by the results of RF model shown in Fig. 13. As shown in the Fig., fifteen landslide condition factors are arranged on the rules basis of mean decrease accuracy and mean decrease Gini. First of all, the result of permutation of landslide conditioning factors based on mean decrease accuracy shows that slope, altitude, TWI, distance to rivers and land use have a great function of landslide occurrence with the importance indexes of 103.516, 42.95, 29.086, 18.7 and 24.832, respectively. In generating random forest, mean decrease Gini was introduced to measure contribution of each conditioning variable for the homogeneity of nodes and leaves. It should be noticed that slope, altitude, aspect, distance to rivers and distance to lineaments contributed a lot during the course of model building with the contribution indexes of 38.276, 18.847, 15.764, 13.301 and 11.868, respectively.

From the four final mapping divided into five groups perspective, percentage of each group area conform to realistic situation and experience (Wang et al. 2020a). These models, KLR, SysFor and RF, signify that the region's susceptibility level was very low maximumly occupied by 18.63%, 29.15% and 33.44% of the Jian'ge county, respectively. In the meantime, the application of these three models revealed that the proportions for the very high classes include the minimal research territory, throughout approximately 12% of the study area. Area with other groups (high, moderate and low) occupies a middle region of the study area. Correspondingly, as seen in the FURIA model, high, moderate and low groups occupy a small region of 6.17%, 12.1% and 6.79% study area. And the proportion (more than 44%) of the Jian'ge county covered with very low susceptibility group for landslide, while very high susceptibility also has larger region for 30.29% study area (Fig. 14).

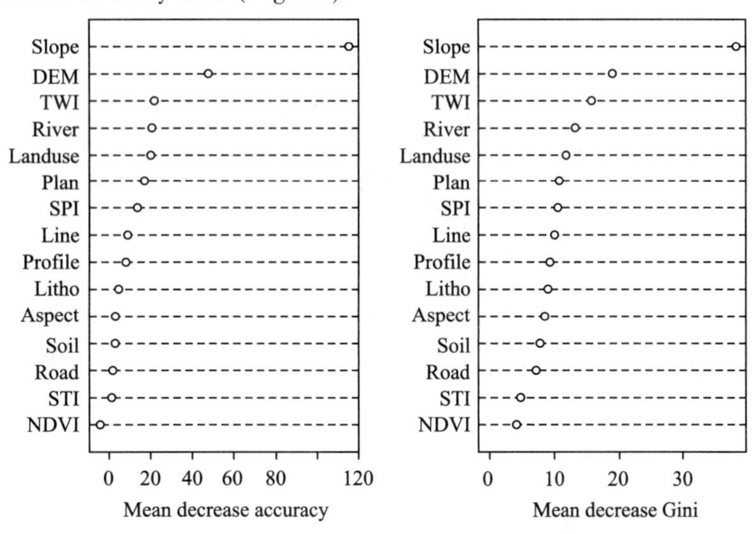

Fig. 13 **Mean decrease accuracy and mean decrease Gini**

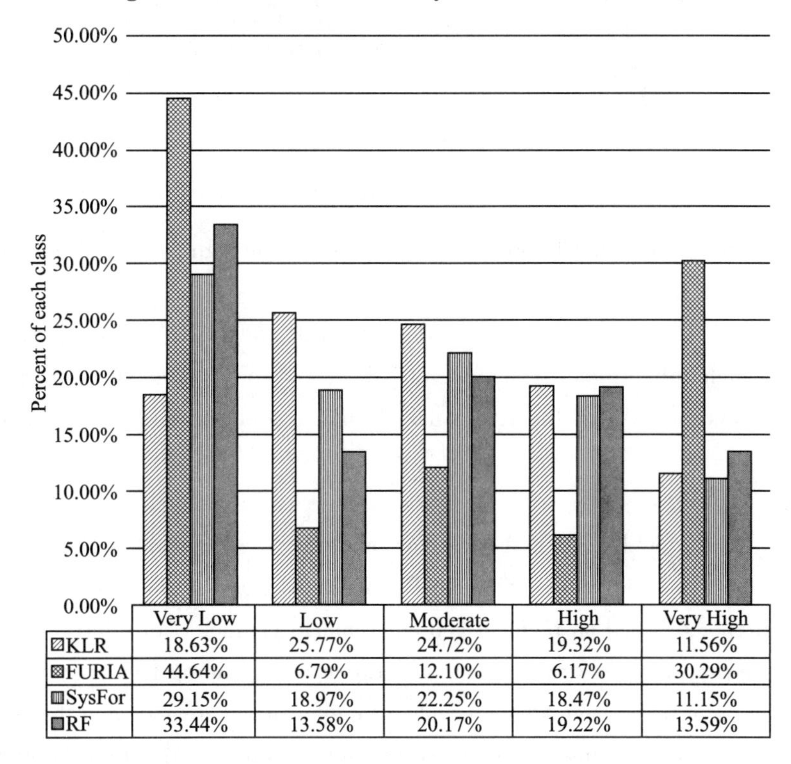

	Very Low	Low	Moderate	High	Very High
KLR	18.63%	25.77%	24.72%	19.32%	11.56%
FURIA	44.64%	6.79%	12.10%	6.17%	30.29%
SysFor	29.15%	18.97%	22.25%	18.47%	11.15%
RF	33.44%	13.58%	20.17%	19.22%	13.59%

Fig. 14 **Percentages of landslide susceptibility classes**

5.6 Validation of landslide susceptibility maps

Model validation is a pivotally final step among the study in landslide susceptibility. Without it, the accuracy of the research will be hard to determined, and it will exist no scientific meaning (Demir 2018; Guzzetti et al. 2006). Therefore, ROC curve was introduced to validate the precision of above models, and compared their performance (Chen et al. 2021c).

It can be seen from Fig. 15 and Table 4 that the validation result part was consisted of ROC curve and parameters of ROC curves. All this indicates that RF model showed its excellent performance, which had highest ROC curve area (0.859), followed by SysFor, KLR and FURIA models with ROC curve area values of 0.818, 0.806 and 0.795. Meanwhile, a similar result also appears in other two parameters of ROC curve, namely standard error (smallest for RF model) and 95% confidence interval (highest for RF model). Although RF model acquired promising ability in terms of predicting susceptibility, other models also performed well.

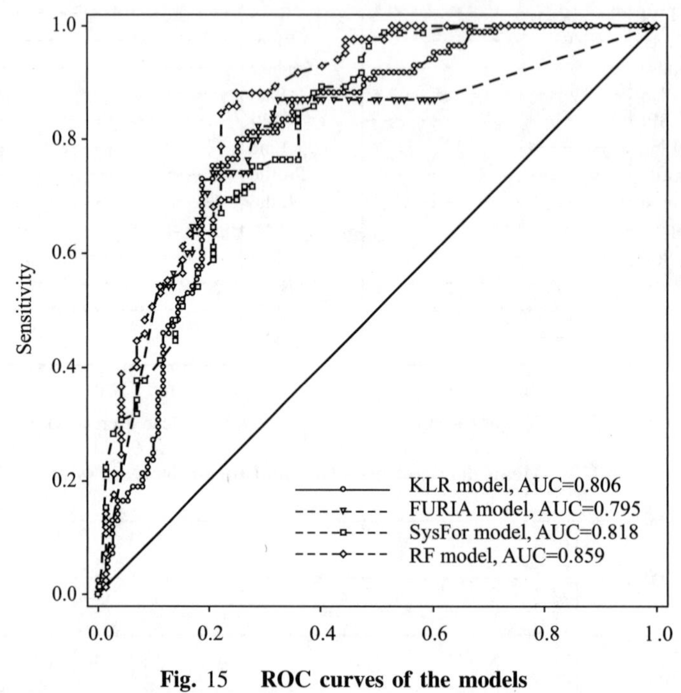

Fig. 15 ROC curves of the models

In addition, this study used the Chi-square test to compare the significance of the four models in pairwise comparisons (Table 5).

In the comparison of RF model and SysFor model with KLR and FURIA models, respectively, the Chi-square value is much higher than the critical value of 3.841, and the P value is also higher than the critical value of 0.05. The results showed that the RF model, SysFor model and KLR, FURIA models are significantly different.

6 Discussion

In the present study, KLR, FURIA, SysFor and RF were conducted to carry out landslide susceptibility research in Jian'ge County, Sichuan Province, China. For the first, fifteen landslide conditioning factors were prepared based on regional geo-environment characteristics and previous studies. Then, CF model was employed to well study the correction between one grouping factors and distribution of landslide. For example, the number of landslides presents a trend of first increasing and then decreasing with the increase in slope angle. More than ninety percentage of landslides occurred within the region of slope angle with 10°~40°degree. The relationship between landslide and linear factors (distance to rivers, roads and lineaments) can be briefly expressed as the closer to factor, the more landslide distribution. After the correction analysis, multicollinearity analysis was applied to inspect the relationship among factors, so as to avoid possible effects additions of correction factors on the research results.

Generally, the four models used in this study are relatively advanced, among which FURIA and SysFor belong to rare applications in the field of landslide susceptibility research. This implicates that the purpose of this study is not only to draw an accurate LSM for study area, but also to apply new mathematical models in this field. KLR is an improvement on logistic regression (LR) model and it can be regarded as either a linear model in multidimensional space or a linear model in which the kernel transforms the original features to produce new one. Chen et al. applied KLR, naive Bayes and RBFNetwork models to conduct and compare the performance of those three models in landslide susceptibility research for Long County Area, China. The results show that KLR model was superior to the rest and will give both highest success rate (0.847) and prediction rate (0.772), that is, compared with traditional naive Bayes and RBFNetwork models, KLR model has obvious advantages (Chen et al. 2019b). Similarly, Zhang et al. proposed landslide susceptibility research by KLR model and two improved models based on KLR model, namely $KLR_{box\text{-}counting}$ and $KLR_{correlation}$. Their results put forward that these used three models perform relatively good, among which $KLR_{correlation}$ obtained the highest success rate (0.878) and prediction rate (0.867), followed by $KLR_{box\text{-}counting}$ and KLR models (Zhang et al. 2019).

Table 4 Parameters of ROC curves

Test variables	KLR model	FURIA model	SysFor model	RF model
ROC curve area	0.806	0.795	0.818	0.859
Accuracy	0.720	0.720	0.726	0.758
Precision	0.773	0.741	0.763	0.797
Sensitivity	0.682	0.741	0.718	0.741
Kappa statistic	0.442	0.436	0.451	0.516
F-measure	0.725	0.741	0.739	0.768
Standard error	0.0314	0.0332	0.0308	0.0307
95% confidence Interval	0.744 to 0.867	0.729 to 0.860	0.752 to 0.884	0.798 to 0.919
P value	<0.0001	<0.0001	<0.0001	<0.0001

Table 5 Pairwise comparison for the four models using Chi-square test

Pair	KLR model, FURIA model	KLR model, SysFor model	KLR model, RF model	FURIA model, SysFor model	FURIA model, RF model	SysFor model, RF model
95% Con-fidence interval	−0.054 to 0.036	−0.102 to −0.022	−0.125 to −0.037	−0.080 to −0.026	−0.105 to −0.038	−0.044 to 0.007
Chi-square	0.161	9.303	12.980	14.565	17.794	2.079
P value	0.689	0.002	0.000	0.000	< 0.000	0.149

RF is a commonly used machine learning model, which improves prediction accuracy without significantly increasing computational complexity (Chang et al. 2019; Hong et al.2016; Youssef et al. 2016a). RF model is not sensitive to multicollinearity, and the result is relatively robust to missing and non-equilibrium data. It is considered as one of the best algorithms at present (Chang et al. 2019). Oliveira et al. carried out a landslide susceptibility research based on RF and artificial neural network (ANN) models, which are two of the most representative deep learning models. Their conclusion demonstrates that the RF model was superior to that of the ANN model, with the predictive rate of 0.949 and 0.925, respectively (de Oliveira et al. 2019). Dang et al. prepared a comparative landslide susceptibility study between RF model and other models based on five different benchmark approaches for landslide disaster along mountain roads. After the processes of data preprocess and model establishing, the results implied RF model performed outstanding of all models with the success rate and predictive rate of 0.988 and 0.917, followed by SVM, RMV, BPNN, DA and LR models (Dang et al. 2019).

This paper used Weka software and R to model the four selected machine learning methods, but did not consider the role of model parameters. However, according to the research of other scholars, the optimization of model parameters has a certain influence on the results of landslide spatial prediction. Xie et al. (2021) used the Bayesian optimization (BO) method to optimize the hyperparameters of the SVM, and the final results showed that the landslide susceptibility map generated by the parameter optimization method was better than the ordinary SVM model. In addition, some researchers have optimized the hyperparameters of the SVM model using random search, grid search or genetic algorithm, and obtained better landslide susceptibility evaluation results (Luo et al. 2019; Tang et al. 2019). Pradhan et al. (2021) proposed an optimization approach based on meta-learning for landslide spatial prediction and applied it to the tropical forest area of Cameron Highlands, Malaysia. The scholars tested the proposed meta-learning method based on Bayesian optimization as a hyperparameter tuning algorithm and random forest (RF) as a prediction model. Moreover, these researchers compared three different parameter optimization methods: (1) default parameters suggested by the sklearn library; (2) parameters suggested by the Bayesian optimization (BO); (3) parameters suggested by the proposed meta-learning approach. The accuracy of these three tuning methods was compared using the fivefold cross-validation method, and it was found that models with parameters optimized by Bayesian or meta-learning methods produced more accurate landslide susceptibility assessments than models with default parameters. However, this study only used the default parameters in the software, which needs to be improved in future study, applying different parameter optimization methods for landslide susceptibility modeling.

Although all models performed well observed on this study LSMs, in order to further distinguish their accuracy, validation and comparison each model's outcomes were implemented by introducing ROC curve method. For predicting outcomes, the biggest ROC curve area (0.859) with the best accuracy was achieved by RF model, followed by SysFor, KLR and FURIA models. Several parameters of ROC curve, such as standard error and 95% confidence interval, also indicated the same result that the RF yields best and the FURIA is worst. Additionally, the accuracy of LSMs can be tested through the statistical analysis of relative distribution in each class area, in common with other researchers (Chu et al. 2019; Lei et al. 2020b; Zhao and Chen 2020). In this process, relative distribution in the class area amply indicates the rationality of the result. Because Fig. 14, visualization of all LSMs, reveals that low susceptibility class takes up the largest region of Jian'ge county. If the very high susceptibility class was designated as the smallest area, this would demonstrate that the classification model for LSMs is appropriate. In the classification models, it can be legibly found out that the SysFor models' susceptibility distribution map is the optimal and the FURIA models' susceptibility distribution map is the worst. Last but by no means the least, LSMs as scientific reference produced in this paper can support the Jian'ge county government of establishing effective risk management mechanisms and useful for land use planning.

7 Conclusions

In the present study, four advanced intelligent algorithms, that is, KLR, FURIA, SysFor and RF models, along with prepared data, reduction of potential landslide have hope to execute in Jian'ge County, Sichuan Province, China. First of all, fifteen landslide conditioning factors were prepared through comparative analysis of local geo-environment characteristics and previous studies. Then, CF model was employed to well study the correction between conditioning factors and distribution of landslide. Next, multicollinearity analysis was introduced to verify the independence among all factors. After that, above four models were conducted to generated LSMs, respectively. The final process is the accuracy of inspection for those models by brought ROC curve, and the performance indicates that RF model obtained promising results. Compre-hensive risk assessors and land use planarians can benefit from our study findings.

Author contributions: TZ and HW were involved in conceptualization, methodology, writing—review and editing, funding acquisition; QF helped in resources, software, validation; FL and CL contributed to formal analysis, data curation; CL, RPQ, TC and NL were involved in visualization; TZ and HW helped in writ-ing—original draft preparation; LH helped in supervision.

Funding: This study is financially supported by Fundamental Research Funds for the Central Universities (300102351502), National Natural Science Foundation of China (211035210511), Shaanxi Province Natural Science Basic Research Program (2022JQ-457), Shaanxi Province Enterprises Talent Innovation Striving to Support the Plan (2021-1-2-1) and Inner scientific research project of Shaanxi Land Engineering Construction Group (DJNY2021-10, DJNY2022-16).

Availability of data and material: Yes.

Code availability: Yes.

Declarations

Conflict of interest: The authors declare that they have no known competing financial interests or personal relationships that could have appeared to influence the work reported in this paper.

Ethical approval: Not applicable.

Consent to participate: Yes.

Consent for publication: Yes.

References

Aditian A, Kubota T, Shinohara Y (2018) Comparison of gis-based landslide susceptibility models using frequency ratio, logistic regression, and artificial neural network in a tertiary region of Ambon, Indo-nesia. Geomo 318:101-111.

Bibri S E(2018)Data science for urban sustainability: data mining and data-analytic thinking in the next wave of city analytics. In: Bibri S E (ed) Smart sustainable cities of the future: the untapped potential of big data analytics and context-aware computing for advancing sustainability. Springer International Publishing, Cham, pp 189-246.

Breiman L (2001) Random forests. Mach Learn 45:5-32.

Carranza E J M (2015) Data-driven evidential belief modeling of mineral potential using few prospects and evidence with missing values. Nat Resour Res 24:291-304.

Chang K T, Merghadi A, Yunus A P, Pham B T, Dou J (2019) Evaluating scale effects of topographic variables in landslide susceptibility models using gis-based machine learning techniques. Sci Rep 9:12296.

Chen X, Chen W (2021) Gis-based landslide susceptibility assessment using optimized hybrid machine learning methods. CATENA.

Chen W, Shahabi H, Shirzadi A, Hong H, Akgun A, Tian Y, Liu J, Zhu A X, Li S (2019a) Novel hybrid artificial intelligence approach of bivariate statistical-methods-based kernel logistic regression classi-fier for landslide susceptibility modeling. Bull Eng Geol Env 78:4397-4419.

Chen W, Yan X, Zhao Z, Hong H, Bui D T, Pradhan B (2019b) Spatial prediction of landslide susceptibility using data mining-based kernel logistic regression, Naive Bayes and Rbfnetwork Models for the Long County Area (China). Bull Eng Geol Env 78:247-266.

Chen W, Chen X, Peng J, Panahi M, Lee S (2021a) Landslide susceptibility modeling based on anfis with teaching-learning-based optimization and satin bowerbird optimizer. Geosci Front 12:93-107.

Chen W, Lei X, Chakrabortty R, Chandra Pal S, Sahana M, Janizadeh S (2021b) Evaluation of different boosting ensemble machine learning models and novel deep learning and boosting framework for head-cut gully erosion susceptibility. J Environ Manage 284:112015.

Chen Y, Chen W, Chandra Pal S, Saha A, Chowdhuri I, Adeli B, Janizadeh S, Dineva A A, Wang X, Mosavi A (2021c) Evaluation efficiency of hybrid deep learning algorithms with neural network, decision tree and boosting methods for predicting groundwater potential. GeoIn.

Chen Y, Chen W, Janizadeh S, Bhunia G S, Bera A, Pham Q B, Linh N T T, Balogun A L, Wang X (2021d) Deep learning and boosting framework for piping erosion susceptibility modeling: spatial evaluation of agricultural areas in the semi-arid region. GeoIn.

Chu L, Wang L J, Jiang J, Liu X, Sawada K, Zhang J C (2019) Comparison of landslide susceptibility maps using random forest and multivariate adaptive regression spline models in combination with catchment map units. Geosci J 23:341-355.

Conoscenti C, Ciaccio M, Caraballo-Arias N A, Gómez-Gutiérrez Á, Rotigliano E, Agnesi V (2015) Assessment of susceptibility to earth-flow landslide using logistic regression and multivariate adaptive regression splines: a case of the Belice River Basin (Western Sicily, Italy). Geomorphology 242:49-64.

Constantin M, Bednarik M, Jurchescu M C, Vlaicu M (2011) Landslide susceptibility assessment using the bivariate statistical analysis and the index of entropy in the Sibiciu Basin (Romania). Environ Earth Sci 63:397-406.

Dang V H, Dieu T B, Tran X L, Hoang N D (2019) Enhancing the accuracy of rainfall-induced land-slide prediction along mountain roads with a Gis-based random forest classifier. Bull Eng Geol Env 78:2835-2849.

Das G, Lepcha K (2019) Application of logistic regression (Lr) and frequency ratio (Fr) models for landslide susceptibility mapping in Relli Khola River Basin of Darjeeling Himalaya, India. SN Appl Sci 1:1453.

De Oliveira G G, Ruiz L F C, Guasselli L A, Haetinger C (2019) Random forest and artificial neural networks in landslide susceptibility modeling: a case study of the Fão River Basin, Southern Brazil. Nat Hazards 99:1049-1073.

Demir G (2018) Landslide susceptibility mapping by using statistical analysis in the North Anatolian Fault Zone (Nafz) on the Northern Part of Suşehri Town, Turkey. Nat Hazards 92:133-154.

Demir G (2019) Gis-based landslide susceptibility mapping for a part of the North Anatolian Fault Zone between Reşadiye and Koyulhisar (Turkey). CATENA 183:104211.

Genuer R, Poggi J M, Tuleau-Malot C, Villa-Vialaneix N (2017) Random forests for big data. Big Data Res 9:28-46.

Guio Blanco CM, Brito Gomez V M, Crespo P, Ließ M (2018) Spatial prediction of soil water retention in a Páramo Landscape: methodological insight into machine learning using random forest. Geoderma 316:100-114.

Guzzetti F, Reichenbach P, Ardizzone F, Cardinali M, Galli M (2006) Estimating the quality of landslide susceptibility models. Geomorphology 81:166-184.

Hong H, Pradhan B, Xu C, Tien Bui D (2015a) Spatial prediction of landslide hazard at the Yihuang Area (China) using two-class kernel logistic regression, alternating decision tree and support vector machines. CATENA 133:266-281.

Hong H, Xu C, Bui D T (2015b) Landslide susceptibility assessment at the Xiushui Area (China) using frequency ratio model. Procedia Earth Planet Sci 15:513-517.

Hong H, Pourghasemi H R, Pourtaghi Z S (2016) Landslide susceptibility assessment in Lianhua County (China): a comparison between a random forest data mining technique and bivariate and multivariate statistical models. Geomo 259: 105-118.

Hong H, Liu J, Bui D T, Pradhan B, Acharya T D, Pham B T, Zhu A X, Chen W, Ahmad B B (2018) Landslide susceptibility mapping using J48 decision tree with Adaboost, Bagging and rotation forest ensembles in the Guangchang Area (China). CATENA 163:399-413.

Huang R, Li W (2011) Formation, distribution and risk control of landslides in China. J Rock Mech Geotech Eng 3: 97-116.

Hühn J, Hüllermeier E (2009) Furia: an algorithm for unordered fuzzy rule induction. Data Min Knowl Disc 19: 293-319.

Hühn J C, Hüllermeier E (2010) An Analysis of the Furia Algorithm for Fuzzy Rule Induction. In: Koronacki J, Raś Z W, Wierzchoń S T, Kacprzyk J (eds) Advances in machine learning i: dedicated to the memory of professor Ryszard Smichalski. Springer, Berlin Heidelberg, Berlin, Heidelberg, pp 321−344 Kamińska JA (2019) A random forest partition model for predicting No2 concentrations from traffic flow and meteorological conditions. Sci Total Environ 651:475-483.

Kumari V S R, Kumar P R (2013) Fuzzy unordered rule induction for evaluating cardiac arrhythmia. Biomed Eng Lett

3:74-79.

Lei X, Chen W, Pham BT (2020a) Performance evaluation of gis-based artificial intelligence approaches for landslide susceptibility modeling and spatial patterns analysis. ISPRS Int J Geo Inf 9:443.

Lei X X, Chen W, Avand M, Janizadeh S, Kariminejad N, Shahabi H, Costache R, Shahabi H, Shirzadi A, Mosavi A (2020b) Gis-based machine learning algorithms for gully erosion susceptibility mapping in a semi-arid region of Iran. Remote Sens.

Li R, Wang N (2019) Landslide susceptibility mapping for the Muchuan County (China): A comparison between bivariate statistical models (Woe, Ebf, and Ioe) and their ensembles with logistic regression. Symmetry 11:762.

Li J, Zhang Y (2017) Gis-supported certainty factor (Cf) models for assessment of geothermal potential: A case study of Tengchong County, Southwest China. Energy 140:552-565.

Li Y, Chen W, Rezaie F, Rahmati O, Davoudi Moghaddam D, Tiefenbacher J, Panahi M, Lee M J, Kulakowski D, Tien Bui D, Lee S (2021) Debris flows modeling using geo-environmental factors: developing hybridized deep-learning algorithms. GeoIn.

Liu X, Yu C, Shi P, Fang W (2012) Debris flow and landslide hazard mapping and risk analysis in China. Front Earth Sci 6:306-313.

Liu R, Peng J, Leng Y, Lee S, Panahi M, Chen W, Zhao X (2021) Hybrids of support vector regression with grey wolf optimizer and firefly algorithm for spatial prediction of landslide susceptibility. Remote Sens 13:4966.

Luo X, Lin F, Zhu S, Yu M, Zhang Z, Meng L, Peng J J P O (2019) Mine landslide susceptibility assessment using Ivm, Ann and Svm models considering the contribution of affecting factors. PLoS ONE 14:e0215134.

Martinello C, Cappadonia C, Conoscenti C, Agnesi V, Rotigliano E (2021) Optimal slope units partitioning in landslide susceptibility mapping. J Maps 17:152-162.

Merghadi A, Yunus A P, Dou J, Whiteley J, ThaiPham B, Bui DT, Avtar R, Abderrahmane B (2020) Machine learning methods for landslide susceptibility studies: a comparative overview of algorithm performance. Earth-Sci Rev 207:103225.

Moon K W (2016) Logistic regression. In: Moon K W (ed) Learn Ggplot2 Using Shiny App. Springer International Publishing, Cham, pp 51-54.

Nasiri Aghdam I, Pradhan B, Panahi M (2017) Landslide susceptibility assessment using a novel hybrid model of statistical bivariate methods (Fr and Woe) and adaptive neuro-fuzzy inference system (Anfis) at Southern Zagros Mountains in Iran. Environ Earth Sci 76:237.

Nhu V H, Shirzadi A, Shahabi H, Singh S K, Al-Ansari N, Clague J J, Jaafari A, Chen W, Miraki S, Dou J, Luu C, Górski K, Thai Pham B, Nguyen H D, Ahmad B B (2020) Shallow landslide susceptibility mapping: a comparison between logistic model tree, logistic regression, naïve bayes tree, artificial neural network, and support vector machine algorithms. Int J Environ Res Public Health 17:2749.

Palacios A M, Sanchez L, Couso I (2013) An Extension of the Furia Classification Algorithm to Low Quality Data. In: Pan J S, Polycarpou M M, Wo ź niak M, De Carvalho ACPLF, Quintián H, Corchado E (eds) Hybrid artificial intelligent systems. Springer, Berlin Heidelberg, Berlin, Heidelberg, pp 679-688.

Palacios A, Sánchez L, Couso I, Destercke S (2016) An extension of the furia classification algorithm to low quality data through fuzzy rankings and its application to the early diagnosis of dyslexia. Neurocom-puting 176:60-71.

Panahi M, Gayen A, Pourghasemi H R, Rezaie F, Lee S (2020) Spatial prediction of landslide susceptibility using hybrid support vector regression (Svr) and the adaptive neuro-fuzzy inference system (Anfis) with various metaheuristic algorithms. ScTEn 741.

Peng J, Wang S, Wang Q, Zhuang J, Huang W, Zhu X, Leng Y, Ma P (2019) Distribution and genetic types of loess landslides in China. J Asian Earth Sci 170:329-350.

Pham B T, Tien Bui D, Prakash I, Dholakia M B (2016) Rotation forest fuzzy rule-based classifier ensemble for spatial prediction of landslides using Gis. Nat Hazards 83:97-127.

Pham B T, Tien Bui D, Prakash I, Nguyen L H, Dholakia M B (2017) A comparative study of sequential minimal optimization-based support vector machines, vote feature intervals, and logistic regression in landslide susceptibility assessment using Gis. Environ Earth Sci 76:371.

Polykretis C, Ferentinou M, Chalkias C (2015) A comparative study of landslide susceptibility mapping using landslide susceptibility index and artificial neural networks in the Krios River and Krathis River catchments (Northern Peloponnesus, Greece). Bull Eng Geol Env 74:27-45.

Polykretis C, Chalkias C, Ferentinou M (2019) Adaptive neuro-fuzzy inference system (Anfis) modeling for landslide susceptibility assessment in a Mediterranean Hilly Area. B Eng Geol Environ 78:1173-1187.

Pourghasemi H R, Kerle N (2016) Random forests and evidential belief function-based landslide susceptibility assessment in Western Mazandaran Province, Iran. Environ Earth Sci 75:185.

Pradhan B, Sameen M I, Al-Najjar H A H, Sheng D, Alamri A M, Park H J (2021) A meta-learning approach of optimisation for spatial prediction of landslides. Remote Sens 13:4521.

Razavizadeh S, Solaimani K, Massironi M, Kavian A (2017) Mapping landslide susceptibility with fre-quency ratio, statistical index, and weights of evidence models: a case study in Northern Iran. Environ Earth Sci 76:499.

Rotigliano E, Martinello C, Hernandéz M A, Agnesi V, Conoscenti C (2019) Predicting the landslides triggered by the 2009 96e/Ida tropical storms in the Ilopango Caldera Area (El Salvador, Ca): Optimizing mars-based model building and validation strategies. Environ Earth Sci 78:210.

Sahin E K, Colkesen I, Kavzoglu T (2020) A comparative assessment of canonical correlation forest, random forest, rotation forest and logistic regression methods for landslide susceptibility mapping. Geo-carto Int 35:341-363.

Song J (2015) Bias corrections for random forest in regression using residual rotation. J Korean Stat Soc 44:321-326.

Talaei R (2014) Landslide susceptibility zonation mapping using logistic regression and its validation in Hashtchin Region, Northwest of Iran. J Geol Soc India 84:68-86.

Tang X, Hong H, Shu Y, Tang H, Li J, Liu W (2019) Urban waterlogging susceptibility assessment based on a Pso-Svm method using a novel repeatedly random sampling idea to select negative samples. J Hydrol 576:583-595.

Tien Bui D, Pradhan B, Revhaug I, Trung Tran C (2014) A Comparative assessment between the application of fuzzy unordered rules induction algorithm and J48 decision tree models in spatial prediction of shallow landslides at Lang Son City, Vietnam. In: Srivastava P K, Mukherjee S, Gupta M, Islam T (eds) Remote sensing applications in environmental research. Springer International Publishing, Cham, pp 87-111.

Umar Z, Pradhan B, Ahmad A, Jebur M N, Tehrany M S (2014) Earthquake induced landslide susceptibility mapping using an integrated ensemble frequency ratio and logistic regression models in West Sumatera Province, Indonesia. CATENA 118:124-135.

Veenadhari S, Misra B, Singh C (2014) Machine Learning Approach for Forecasting Crop Yield Based on Climatic Parameters. 2014 International Conference on Computer Communication and Infor-matics. pp 1-5.

Wang Q, Li W, Wu Y, Pei Y, Xing M, Yang D (2016) A comparative study on the landslide susceptibility mapping using evidential belief function and weights of evidence models. J Earth Syst Sci 125:645-662.

Wang H J, Xiao T, Li X Y, Zhang L L, Zhang L M (2019a) A novel physically-based model for updating landslide susceptibility. Eng Geol 251:71-80.

Wang Q, Guo Y, Li W, He J, Wu Z (2019b) Predictive modeling of landslide hazards in Wen County, Northwestern China Based on information value, weights-of-evidence, and certainty factor. Geo-mat Nat Haz Risk 10:820-835.

Wang G R, Chen X, Chen W (2020a) Spatial prediction of landslide susceptibility based on gis and discriminant functions. ISPRS Int J Geo Inf.

Wang W, He Z, Han Z, Li Y, Dou J, Huang J (2020b) Mapping the susceptibility to landslides based on the deep belief network: a case study in Sichuan Province, China. Nat Hazards 103:3239-3261.

Wu R, Zhang Y, Guo C, Yang Z, Tang J, Su F (2020) Landslide susceptibility assessment in mountainous area: a case study of sichuan-tibet railway, China. Environ Earth Sci 79:157.

Xiao T, Segoni S, Chen L, Yin K, Casagli N (2020) A step beyond landslide susceptibility maps: a simple method to investigate and explain the different outcomes obtained by different approaches. Landslides 17:627-640.

Xie W, Nie W, Saffari P, Robledo L F, Descote P Y, Jian W (2021) Landslide hazard assessment based on bayesian optimization-support vector machine in Nanping City, China. Nat Hazards 109:931-948.

Yalcin A, Reis S, Aydinoglu A C, Yomralioglu T (2011) A Gis-based comparative study of frequency ratio, analytical hierarchy process, bivariate statistics and logistics regression methods for landslide susceptibility mapping in Trabzon, Ne Turkey. CATENA 85:274-287.

Youssef A M, Pourghasemi H R (2021) Landslide susceptibility mapping using machine learning algo-rithms and comparison of their performance at Abha Basin, Asir Region, Saudi Arabia. Geosci Front 12:639-655.

Youssef A M, Al-Kathery M, Pradhan B (2015) Landslide susceptibility mapping at Al-Hasher Area, Jizan (Saudi Arabia) using Gis-based frequency ratio and index of entropy models. Geosci J 19:113-134.

Youssef A M, Pourghasemi H R, Pourtaghi Z S, Al-Katheeri M M (2016a) Erratum to: landslide susceptibility mapping using random forest, boosted regression tree, classification and regression tree, and general linear models and comparison of their performance at Wadi Tayyah Basin, Asir Region, Saudi Arabia. Landslides 13:1315-1318.

Youssef A M, Pourghasemi H R, Pourtaghi Z S, Al-Katheeri M M (2016b) Landslide susceptibility map-ping using random forest, boosted regression tree, classification and regression tree, and general linear models and comparison of their performance at Wadi Tayyah Basin, Asir Region, Saudi Arabia. Landslides 13:839-856.

Yu C, Chen J (2020) Landslide susceptibility mapping using the slope unit for Southeastern Helong City, Jilin Province, China: a comparison of ANN and SVM. Symmetry 12:1047.

Yuan F, Li X, Li-ming W, Le-ping P, Ying S (2011) Knowledge discovery of energy management system based on Prism, Furia and J48. In: Ma M (ed) Communication systems and information technology. Springer, Berlin Heidelberg, Berlin, Heidelberg, pp 593-600.

Zhang T, Han L, Han J, Li X, Zhang H, Wang H (2019) Assessment of landslide susceptibility using integrated ensemble fractal dimension with kernel logistic regression model. Entropy.

Zhao X, Chen W (2020) Optimization of computational intelligence models for landslide susceptibility evaluation. Remote Sens.

Zhao Y, Xu M, Guo J, Zhang Q, Zhao H, Kang X, Xia Q (2015) Accumulation characteristics, mechanism, and identification of an ancient translational landslide in China. Landslides 12:1119-1130.

Zhou S, Fang L (2015) Support vector machine modeling of earthquake-induced landslides susceptibility in central part of Sichuan Province, China. Geoenviron Disasters 2:2.

本文曾发表于2022年《Natural Hazards》第113卷第1期

Effects of Biochar on Soil Chemical Properties: a Global Meta-analysis of Agricultural Soil

Zenghui Sun[1,2,3], Ya Hu[1,3], Lei Shi[1,3], Gang Li[1,3], Zhe Pang[1,3], Siqi Liu[1,3], Yamiao Chen[1,3], Baobao Jia[4]

(1.Shaanxi Provincial Land Engineering Construction Group Co., Xi'an, China; 2.College of Life Sciences, Yulin University, Yulin, China; 3.Key Laboratory of Degraded and Unused Land Consolidation Engineering, Ministry of Natural and Resources of China, Xi'an, China; 4.Shaanxi Tourism Group Co., Ltd, Xi'an, China)

【Abstract】 Improved soil properties are commonly reported benefits of adding biochar to agriculture soils. To investigate the range of biochar's effects on soil chemical properties (e.g., soil pH, electrical conductivity (EC), cation exchange capacity (CEC), soil organic carbon (SOC), soil total carbon (TC), and soil carbon-nitrogen ratio (C:N ratio)) in response to varied experimental conditions, a meta-analysis was conducted on previously published results. The results showed that the effect of biochar on soil chemical properties varied depending on management conditi ons, soil properties, biochar pyrolysis conditions, and biochar properties. The effect size (Hedges' *d*) of the biochar was greatest for SOC (0.50), the C:N ratio of soil (0.44), soil pH (0.39), TC (0.35), EC (0.21), and CEC (0.20). Among the various factors examined by aggregated boosted tree analysis, the effects of biochar on soil chemical properties were largely explained by the biochar application rate, initial soil pH, and soil sand content. In conclusion, our study suggests that improving soil chemical properties by adding biochar not only requires consideration of biochar appli cation rates and chemical properties but also the local soil environmental factors, especially soil initial pH and sand content of the soil, should be considered.

【Keywords】 Charcoal; Organic material; Agricultural condition; Soil chemistry; Soil fertility

Biochar is a solid product formed by high-temperature pyrolysis of organic matter, such as straw, woody material, or livestock manure, under low anoxic conditions. Biochar is widely used as an amendment to improve soil quality for plant growth (Pandey et al. 2016, Kätterer et al. 2019), resource use efficiency (Stefaniuk et al. 2017), to reduce or mitigate environmental pollution (Meng et al. 2018, Lebrun et al. 2019, Norini et al. 2019), and as an effective measure for reducing greenhouse gas emissions (Case et al. 2012, Verhoeven and Six 2014, Lin et al.2017). Among its potential applications, agriculture is the most widely cited sector for biochar use, and biochar products have broad application prospects in ecological restoration and rehabilitation of barren lands such as degraded cultivated land, degraded grassland, and degraded orchards, as well as in new farmland (Deal et al. 2012, Jien and Wang 2013, Peake et al. 2014, Li et al. 2018).

The addition of biochar can stimulate soil carbon sequestration (Sandhu et al. 2017, Wang et al. 2017), improve soil structure (Mukherjee et al. 2014, Pranagal et al. 2017), physical and chemical properties (Herath et al. 2013, De la Rosa et al. 2014, Jeffery et al. 2015), enhance soil nutrient availability,

Supported by the Fund for Less Developed Regions of the National Natural Science Foundation of China, Grant No. 42167039, and by the Technology Innovation Center for Land Engineering and Human Settlement Environment, Shaanxi Land Engineering Construction Group Co., Ltd. and Xi'an Jiaotong University, Project No. 2021WHZ0091.

and aggregate stability (Bayabil et al. 2015, Molnár et al. 2016). By enhancing and improving soil microbial activity, biochar also indirectly affects crop growth and development, collectively improving soil quality and fertility and ultimately improving crop yields (Cheng et al. 2017, Yao et al. 2017, Wong et al. 2019). Especially in the context of increasingly widespread biochar application, in-depth studies of biochar char-acteristics, application management, and evaluation of its value in large-scale agricultural production have important theoretical and practical significance for the development of sustainable agriculture (Tan et al. 2017, Yu et al. 2019). Furthermore, in the context of the diversification of biochar feedstocks, the development of production technology, and the refinement of application techniques, biochars have very heterogeneous properties differing in resources and pyrolysis conditions (Mukherjee and Zimmerman 2013). Therefore, the analysis of the impacts of individual factors no longer fully informs applications across the diverse environmental conditions of farmlands.

In recent years, the impact of biochar application on crop growth and soil properties has become a dynamic area of research focused on the obtaining of increased agricultural yields and improving degraded soils (Borchard et al. 2014, Agegnehu et al. 2016). Biochar application can improve soil properties, particularly physical properties, and increase crop production. Biochar addition to soil could alter soil microbial composition and structure and their key soil processes such as carbon mineralisation and nutrient transformation. Additionally, the wide C: N ratio of biochar may decrease soil N availability in N deficient soils, therefore reducing crop yields. Other negative aspects of biochar on soil include decreased nutrient immobilisation and accelerated resolve of native soil organic carbon (SOC) by higher soil microbial activity (Jindo et al. 2012, Dai et al. 2020). Studies have shown that biochar addition also enhances soil chemical properties, such as soil pH, cation exchange capacity (CEC), and soil total carbon (TC), ensuring nutrient retention and soil fertility (Bera et al. 2016, Giagnoni et al. 2019). Furthermore, some studies have reported inconsistent findings on the effects of biochar on soil chemical properties due to various factors, including climate, soil properties, planting systems, and biochar properties in the experimental areas (Abujabhah et al. 2016a, Ajayi and Horn 2016, Sandhu et al. 2017). This has rendered the promotion and application of biochar technology unfavourable (Verhoeven and Six 2014, Hall and Bell 2015, Vaccari et al. 2015, Zheng et al. 2017). Biochar has complex and wide-scale effects on soil chemistry, and there is still an incomplete understanding of how the biochar generated from different feedstocks and experiment conditions affect the chemical properties of different soils (Unger et al. 2011). Based on the inconsistencies in the findings from previous studies, analysing the changes in soil chemical characteristics caused by the application of biochar with different characteristics and applied under different agricultural conditions is important for identifying key drivers that have positive effects on soil chemical properties. The differences in biochar properties and application rates, soil fertility, soil texture, climatic conditions, and experimental setup and duration could limit comparison between different studies and extrapolation of results to other conditions. Despite the numerous studies on biochar-soil interaction, there is still uncertainty on how biochar modifies soil chemical properties. Hence, developing a comprehensive biochar application policy based on experimental conditions is essential.

A meta-analysis is a statistical analysis that collates and combines data from existing research on a given subject matter (Jeffery et al. 2016, Omondi et al. 2016, Gurevitch et al. 2018, Jiang et al. 2019). This method has been applied to quantitatively analyse the effects of biochar on crop yield increases (Zhang et al. 2019), soil water properties (Omondi et al. 2016, Edeh et al. 2020, Razzaghi et al. 2020), plant uptake of heavy metals (Chen et al. 2018), and soil greenhouse gas emissions (Jeffery et al. 2016, Borchard et al. 2019, Liu et al. 2019b). In these studies, biochar parameters ((a) production conditions:

feedstock and pyrolysis temperature; (b) properties: pH, CEC, and C:N ratio, and (c) application rates), soil conditions (texture, pH, SOC, C: N ratio, and CEC), tested plant types, and experimental types (pot or field) were the important factors. A few meta-analysis studies have been conducted on the effects of biochar on soil chemical properties, and no study has comprehensively examined the influence of experimental conditions on soil chemical properties.

To fill this knowledge gap, we conducted a meta-analysis to compare the effects of different management conditions (experiment duration, type of crop, and biochar application rate), soil properties (soil texture and initial soil pH), pyrolysis conditions (feedstock and pyrolysis temperature), and biochar properties (biochar pH, carbon, nitrogen, oxygen, hydrogen, and C: N ratio of biochar) on soil chemical properties in order to provide a scientific basis for the selective application of biochar. In addition, this study aimed to determine the long-term effects of biochar on the improvement of soil environments.

Material and Methods

Data collection. The ISI Web of Science, Science Direct, and Google Scholar databases were searched for reports on the effects of biochar addition to agricultural soils on soil chemical properties published before May 2022. The keywords used for the literature search were biochar and soil chemical properties and/or soil pH and/or electrical conductivity (EC) and/or cation exchange capacity and/or soil organic carbon and/or soil total carbon and/or soil carbon-nitrogen ratio (C:N ratio). We only included studies that compared the changes between control (i.e., without biochar) and biochar-amended soils to form the literature database used for this meta-analysis. The information or data retrieved from the published articles included the measured soil chemical properties mentioned above, management conditions (cropping system, experiment duration, and biochar application rate), soil properties (soil texture and soil pH), pyrolysis conditions (feedstock and pyrolysis temperature) and biochar properties (biochar pH, carbon, nitrogen, oxygen, hydrogen, and C:N ratio of biochar). Ultimately, 682 comparisons with soil pH data, 326 comparisons with soil EC data, 208 comparisons with soil CEC data, 293 comparisons with SOC, 269 comparisons with total carbon, and 142 comparisons with the C:N ratio of soil were obtained from 138 previously published studies. The studies included in our dataset were from the continents of Asia, Europe, America, Africa, and Australia (Fig. 1, omit).

Data categorisation and treatment. For each of the investigated soil chemical properties, mean effect sizes were also compared when data were pooled into different categories according to management conditions, soil properties, pyrolysis conditions and biochar properties. For the meta-analysis, the duration of the experiment (the time since biochar was applied) was categorised into four levels(<3 months, or 3~6 months, or 6~12 months, or >12 months). Biochar application rates were grouped into low (≤20 t/hm^2), medium (21~40 t/hm^2), high (41~80 t/hm^2), and very high(>80 t/hm^2) application rates (Omondi et al. 2016). Soil acidity was categorised into three different levels (acidic soil with pH≤6.5, neutral soil with pH values from 6.5 to 7.5, and alkaline soil with pH>7.5). Soil texture was classified according to the USDA Soil Classification System into fine (clay, clay loam, silty clay loam, and silty clay), medium (loam, silt loam, and silt), and coarse (sandy loam, sandy clay loam, loamy sand, and sand) texture classes (Cayuela et al. 2017). Biochar was categorised based on feedstock as crop residue biochar, wood biochar, manure biochar, and sludge biochar. Biochar pyrolysis temperatures were grouped into four categories (≤400 ℃, 401~500 ℃, 501~600 ℃, and>600 ℃).Furthermore, studies were grouped according to biochar acidity at pH<6.0, pH 6.0-8.0, and pH>8.0; biochar carbon of<50, 50~75, and>75%; biochar nitrogen of<0.5, 0.5~1, and>1%; biochar oxygen<10, 10~20, and>20%; biochar hydrogen of≤3 and>3; and C:N of biochar at<30, 30≤50, 50≤100, 100≤500 and>500(Cayuela et al. 2014, Gao et al.

2019). The soil pH values measured with $CaCl_2$ were converted to values measured in distilled water using a method described by Minasny et al. (2011).

Data calculations and statistical analyses. For each standard pair-wise comparison (control and biochar treatment), the standardised mean difference Hedges' *d*(Hedges and Olkin 1985, Hedges et al. 1999) was calculated between the control and biochar treatment group to determine the effect size metric. The mean effect sizes for each grouping and 95% bootstrap confidence interval (CI) were calculated using Stata 15 software (Stata, College Station, USA) with Hedges-Olkin random-effects models (Adams et al. 1997). The effect size was considered to be significant if the 95% CI did not overlap zero. While the magnitude of the mean effect d is difficult to interpret, Cohen's benchmark gives a rough estimation with mean effect sizes of $d>0.8$ indicating a large effect, $0.2<d<0.8$ a moderate effect, and $0<d<0.2$ a small effect (Arft et al. 1999, Fedrowitz et al. 2014). Rosenthal's fail-safe number method was used to test the publication bias (Li et al. 2019). A Pearson correlation analysis was used to explore the relationship among the effect sizes of soil chemical properties and environmental variables using the R package "corrplot." An aggregated boosted tree (ABT) analysis was carried out using the "dismo" package in R (Hijmans et al. 2017) to quantitatively and visually evaluate the relative effect of treatment effects on soil chemical properties.

Results

The overall effect of biochar on soil chemical properties. Averaged across the entire dataset, the addition of biochar significantly increased all soil chemical properties (e.g., soil pH, EC, CEC, SOC, TC, and C:N ratio of soil) that were considered in our analysis (Fig. 2). When separating the dataset into different categories, we observed that biochar significantly increased soil pH, SOC, and C:N ratio in all of the investigated variable groups, except for the soil pH and SOC in the oxygen content of the biochar variable (Fig 3, Table 1). Moreover, type of crop, sand, silt, clay content, initial soil pH, feedstock materials, pyrolysis temperature, carbon, nitrogen, the hydrogen content of biochar, and the C:N ratio of biochar had positive effects on soil EC in the biochar addition treatments (Fig. 3, Table 1). The experimental duration and carbon of biochar had a positive effect on soil CEC in the biochar ad-dition treatments (Fig. 3, Table 1).

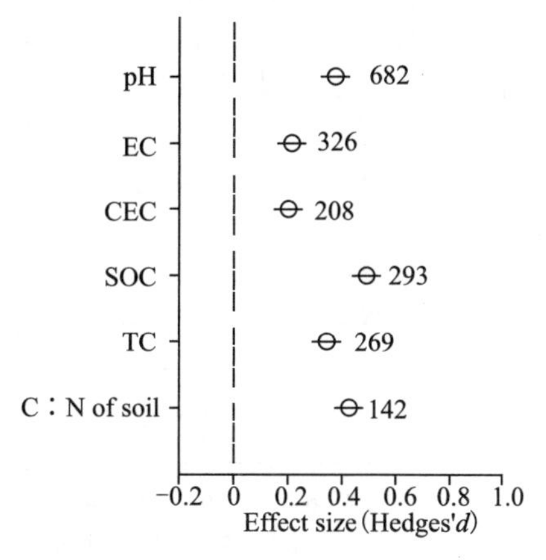

Fig. 2 Total effect of biochar on soil pH, electrical conductivity (EC), cation exchange capacity (CEC), soil organic carbon (SOC), soil total carbon (TC) and soil carbon nitrogen ratio (C:N ratio). The effect size was considered statistically significant if the 95% bootstrap confidence interval (CI) did not include zero. The numbers next to the bars are sample sizes for each variable

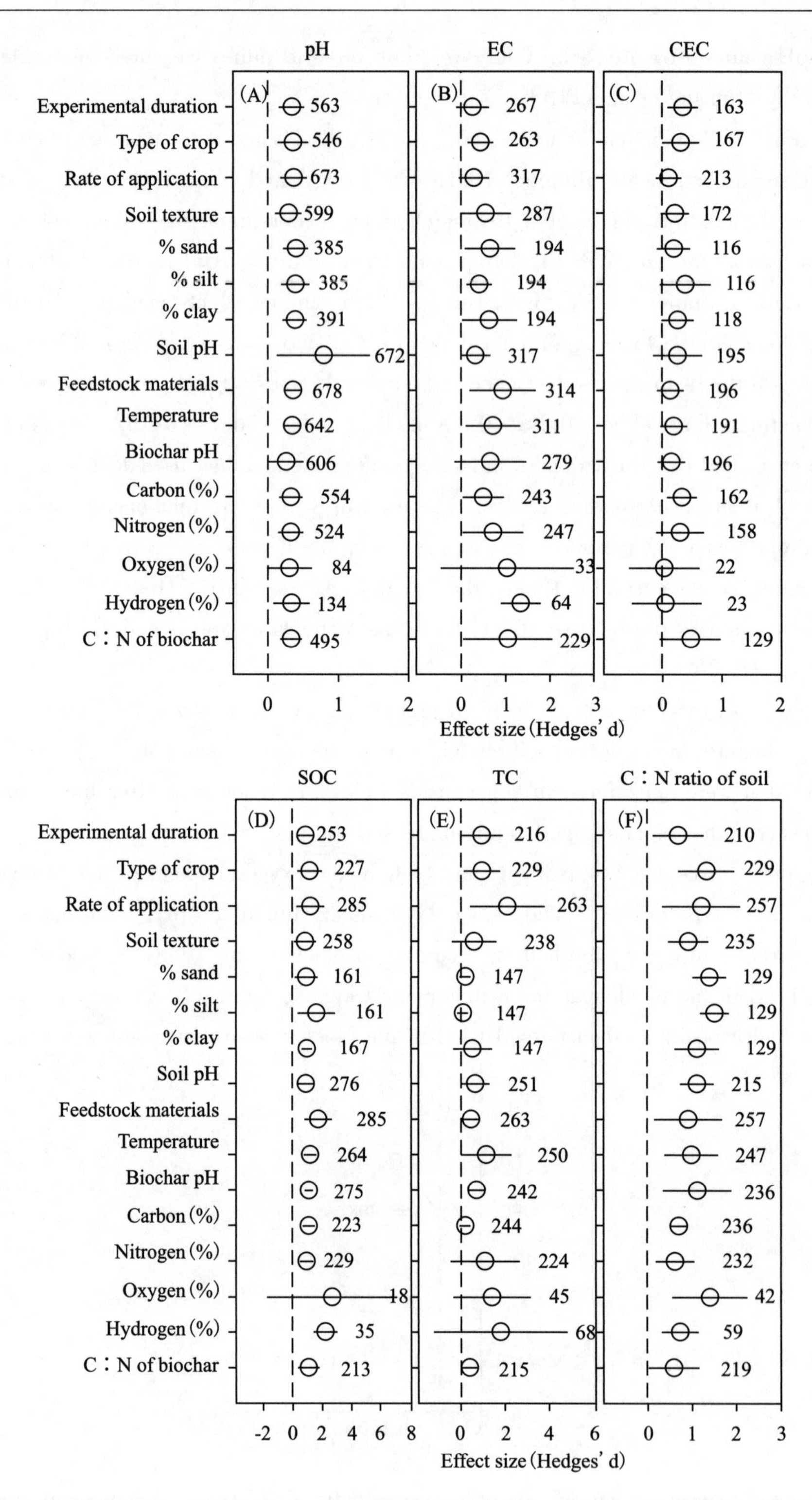

Fig. 3 Total effect of biochar on soil chemical properties (A) soil pH; (B) electrical conductivity (EC); (C) cation exchange capacity (CEC); (D) soil organic carbon (SOC); (E) soil total carbon (TC), and (F) soil carbon nitrogen ratio (C:N ratio) of soil to biochar addition in agriculture soil under different condi-tions. The effect size was considered statistically significant if the 95% bootstrap confidence interval (CI) did not include zero. The numbers next to the bars are sample sizes for each variable

Table 1 Summary of the effect size Hedges' *d* of soil chemical properties under different conditions

Variable	pH	EC	CEC	SOC	TC	C:N ratio
Experimental duration	0.38 (0.19~0.56)	0.13 (-0.04~0.30)	0.31 (0.08~0.53)	0.46 (0.29~0.64)	0.42 (0.23~0.62)	0.44 (0.19~0.69)
Type of crop	0.35 (0.16~0.53)	0.27 (0.10~0.44)	0.26 (0.03~0.49)	0.50 (0.31~0.69)	0.26 (0.07~0.46)	0.53 (0.31~0.74)
Rate of application	0.41 (0.22~0.61)	0.15 (-0.01~0.30)	0.08 (-0.1~0.27)	0.49 (0.28~0.69)	0.51 (-0.0~1.07)	0.36 (0.19~0.54)
Soil texture	0.31 (0.15~0.47)	0.19 (0.02~0.35)	0.19 (-0.0~0.40)	0.47 (0.30~0.65)	0.37 (-0.1~0.90)	0.37 (0.07~0.68)
Sand(%)	0.41 (0.27~0.55)	0.27 (0.07~0.47)	0.17 (-0.0~0.43)	0.48 (0.26~0.71)	0.17 (-0.0~0.40)	0.45 (0.21~0.70)
Silt(%)	0.41 (0.23~0.58)	0.30 (0.10~0.50)	0.29 (-0.0~0.59)	0.56 (0.25~0.87)	0.09 (-0.1~0.32)	0.45 (0.20~0.70)
Clay(%)	0.39 (0.23~0.55)	0.25 (0.05~0.45)	0.23 (-0.0~0.49)	0.49 (0.27~0.70)	0.23 (-0.0~0.50)	0.36 (0.11~0.60)
Initial soil pH	0.60 (0.11~1.08)	0.17 (0.02~0.33)	0.12 (-0.0~0.32)	0.46 (0.30~0.63)	0.46 (0.10~0.82)	0.47 (0.28~0.66)
Feedstock materials	0.36 (0.23~0.50)	0.14 (-0.02~0.30)	0.10 (-0.1~0.30)	0.48 (0.31~0.64)	0.39 (0.06~0.71)	0.33 (0.16~0.50)
Temperature	0.37 (0.25~0.48)	0.25 (0.09~0.41)	0.17 (-0.0~0.37)	0.50 (0.33~0.68)	0.31 (0.08~0.54)	0.44 (0.22~0.66)
Biochar pH	0.27 (0.01~0.52)	0.16 (-0.01~0.32)	0.12 (-0.0~0.32)	0.50 (0.33~0.67)	0.53 (0.26~0.80)	0.60 (0.27~0.94)
Carbon(%)	0.35 (0.18~0.52)	0.25 (0.08~0.43)	0.28 (0.06~0.51)	0.52 (0.33~0.71)	0.19 (0.01~0.37)	0.45 (0.26~0.64)
Nitrogen(%)	0.37 (0.17~0.57)	0.30 (0.04~0.55)	0.27 (-0.0~0.61)	0.49 (0.30~0.67)	0.37 (-0.1~0.85)	0.42 (0.20~0.64)
Oxygen(%)	0.31 (-0.01~0.62)	0.37 (-0.12~0.86)	0.07 (-0.5~0.69)	0.86 (0.00~1.72)	0.64 (0.22~1.07)	0.45 (0.02~0.89)
Hydrogen(%)	0.36 (0.12~0.60)	0.35 (0.00~0.70)	0.05 (-0.5~0.63)	0.64 (0.14~1.14)	0.90 (-0.3~2.14)	0.44 (0.07~0.80)
C:N of biochar	0.36 (0.23~0.49)	0.44 (0.030~0.85)	0.45 (-0.0~0.91)	0.51 (0.31~0.71)	0.32 (0.12~0.51)	0.38 (0.03~0.73)
Overall	0.39 (0.33~0.42)	0.21 (0.17~0.26)	0.20 (0.14~0.26)	0.50 (0.44~0.54)	0.35 (0.27~0.42)	0.44 (0.38~0.48)

Impacts of management conditions on the effects of biochar addition. Soil pH, SOC, and CEC showed the strongest responses to biochar application in experiments lasting 3~6 months. Meanwhile, experiments lasting<3 months, 6~12 months and>12 months produced the most significant effect on soil TC, EC, and C:N ratio, re-

Bold numbers represented significant changes when the 95% confidence interval of the effect size did not overlap with zero. pH-soil pH; EC-electrical conductivity; CEC-cation exchange capacity; SOC-soil organic carbon; TC-soil total carbon; C:N ratio-soil carbon nitrogen ratio.

spectively (Fig. 4). Soil pH, SOC, and the C∶N ratio showed consistent positive responses to horticulture, maize, and rice, while soil EC, CEC, and TC had the highest responses to maize, perennials, and rice, respectively (Fig. 4). On average, biochar effects on soil pH, SOC, TC, and C∶N ratio were significant when its application rate was higher than 40 t/ha, while its effects on soil EC were higher when the rate was between 21 and 40 t/ha. However, the effects on soil CEC were not significant at any of the application rates (Fig. 4).

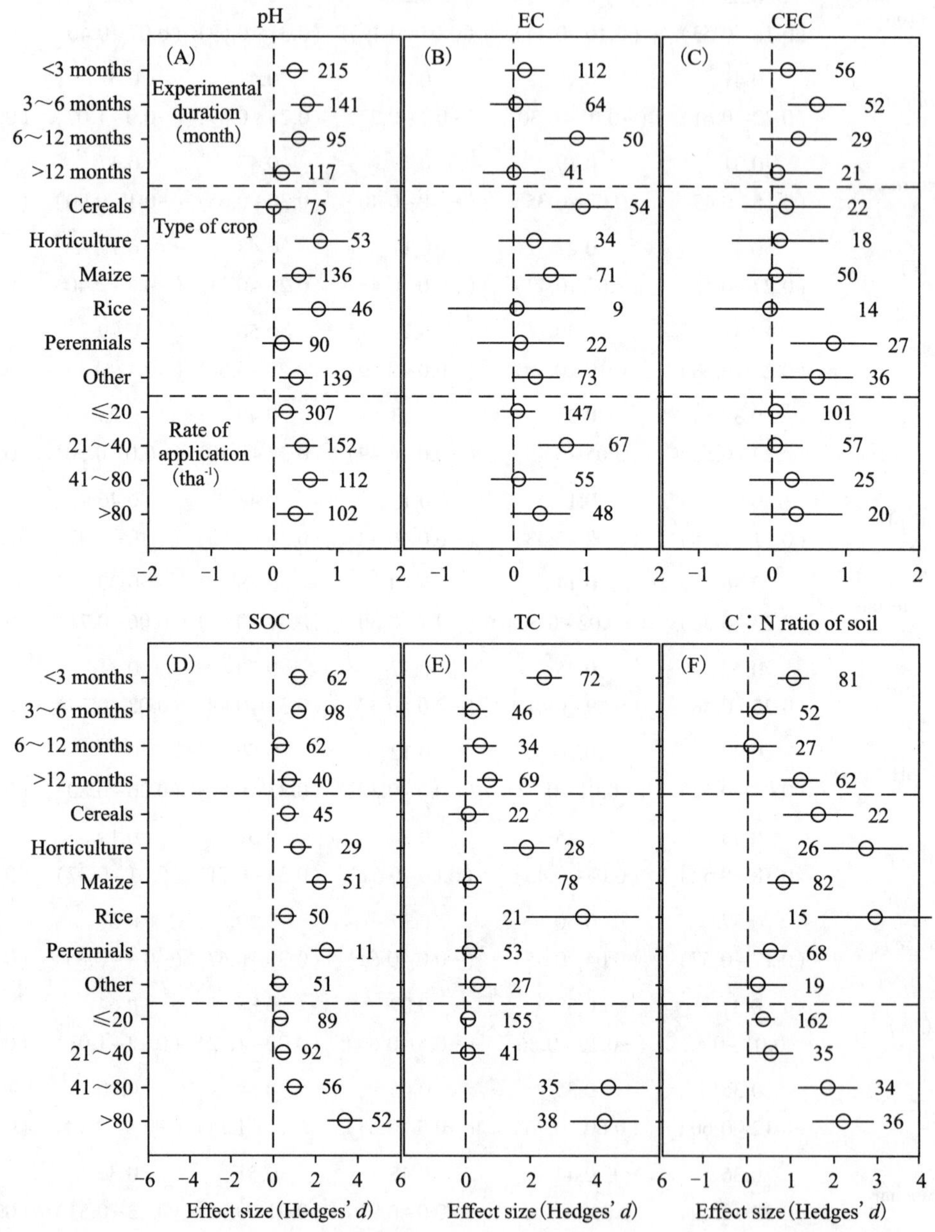

Fig. 4 Effect size Hedges' *d* for the soil chemical properties (A) soil pH; (B) electrical conductivity (EC); (C) cation exchange capacity (CEC); (D) soil organic carbon (SOC); (E) soil total carbon (TC), and (F) soil carbon nitrogen ratio (C∶N ratio) of soil under different management conditions (experiment duration, type of crop, and biochar application rate). The effect size was considered statistically significant if the 95% bootstrap confidence interval did not include zero. The numbers next to the bars are sample sizes for each variable

Impacts of soil properties on the effects of biochar addition. In general, treatment effects on soil pH, EC, SOC, TC, and the C∶N ratio increased with soil texture and initial soil pH. The effect size of SOC, TC, and the C∶N ratio was generally greater in medium texture soil than in soil with other textures, while soil pH and CEC had the highest responses to coarse texture. In contrast soil, EC had the highest response to fine texture soils (Fig. 5). On average, soil pH, CEC, SOC, TC, and the C:N ratio showed consistently positive responses to biochar amendment at a lower initial soil pH (acidic), whereas soil EC showed a positive response when the initial soil pH was alkaline (Fig. 5).

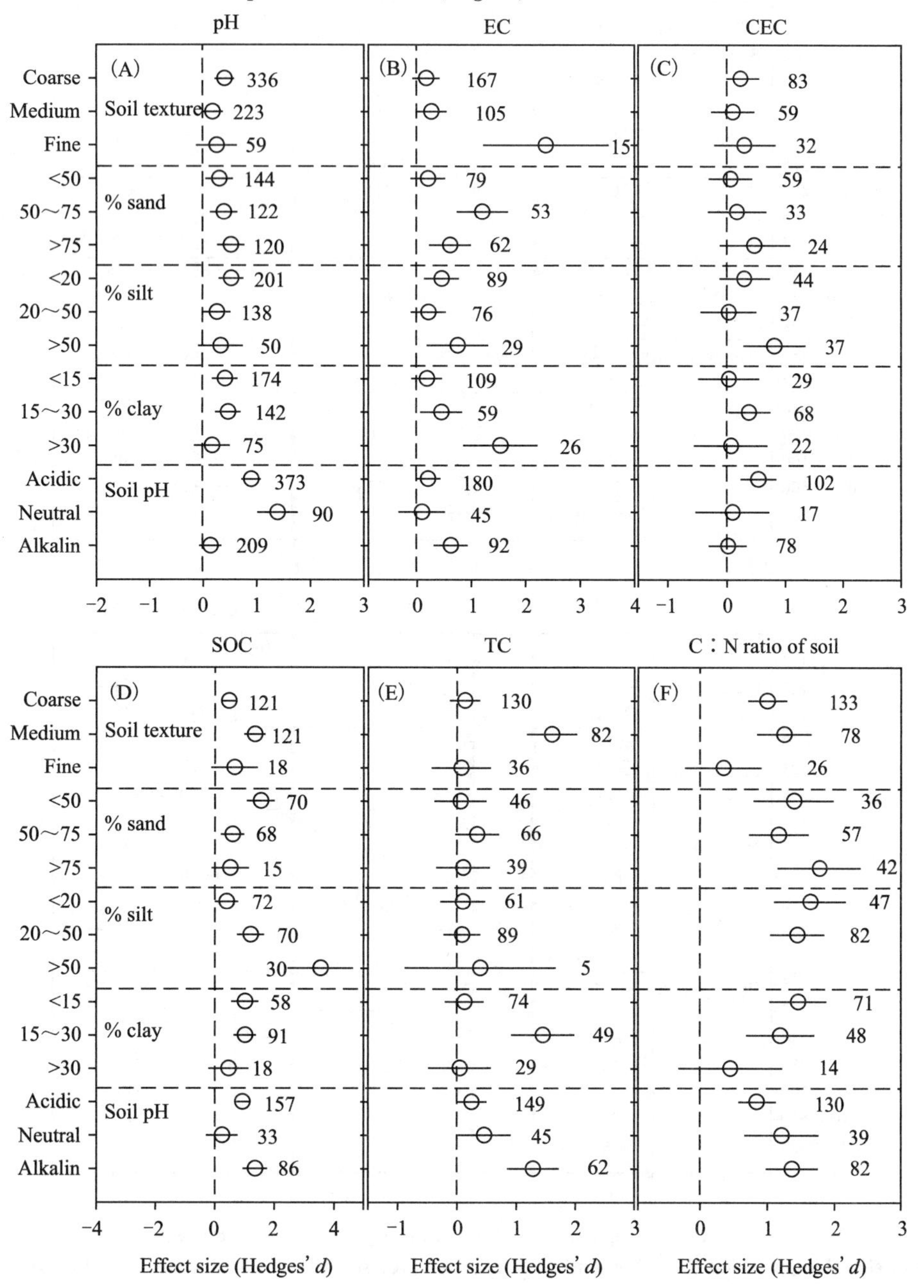

Fig. 5 Effect size Hedges' *d* for the soil chemical properties (A) soil pH; (B) electrical conductivity (EC); (C) cation exchange capacity (CEC); (D) soil organic carbon (SOC); (E) soil total carbon (TC), and (F) soil carbon nitrogen ratio (C:N ratio) under different soil properties (soil texture and initial soil pH). The effect size was considered statistically significant if the 95% bootstrap confidence interval did not include zero. The numbers next to the bars are sample sizes for each variable

Impacts of biochar pyrolysis conditions on the effects of biochar application. Biochar from different feedstocks significantly effects on soil chemical properties. Wood biochar and manure biochar produced the strongest effects on soil pH and EC, respectively, while sewage sludge biochar had the most significant effect on SOC, TC, and C:N ratio (Fig. 6). Overall, the effects on soil pH, SOC, and the C:N ratio were generally positive regardless of all the biochar pyrolysis temperatures. The highest effects on TC and soil EC were observed at when biochar pyrolysis temperatures were ≤400 ℃ and 401 ~ 500 ℃, respectively. Meanwhile, biochar pyrolysis temperature of 501 ~ 600 ℃ generated the strongest effects for SOC and the soil C∶N ratio (Fig. 6).

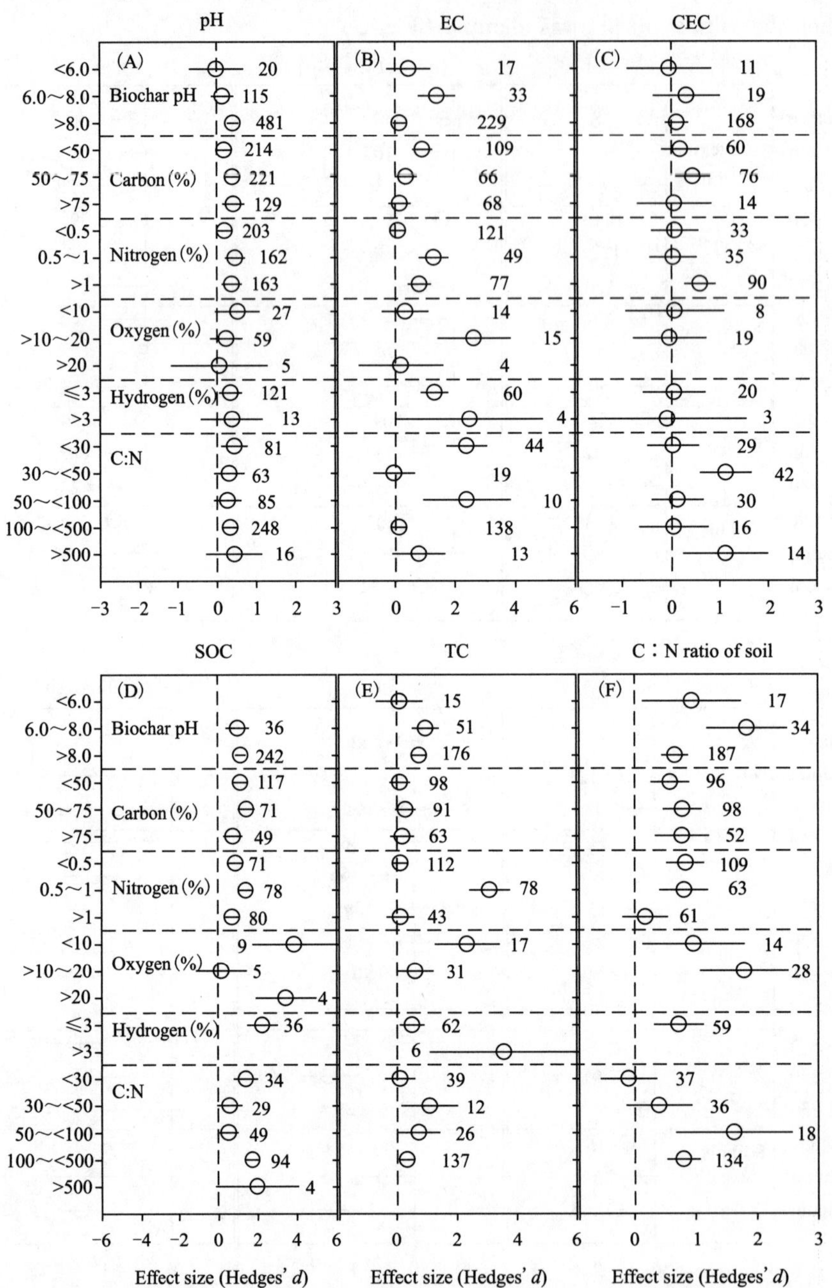

Fig. 6 Effect size Hedges' *d* for the soil chemical properties (A) soil pH; (B) electrical conductivity (EC); (C) cation exchange capacity (CEC); (D) soil organic carbon (SOC); (E) soil total carbon (TC), and (F) soil carbon nitrogen ratio (C∶N ratio) to biochar addition in agriculture soil under different pyrolysis conditions (feedstock and pyrolysis temperature). The effect size was considered statistically significant if the 95% bootstrap confidence interval did not include zero. The numbers next to the bars are sample sizes for each variable

Impacts of biochar properties on the effects of biochar application. A biochar pH value > 8.0, carbon content > 50%, nitrogen content > 0.5%, hydrogen content < 3%, and C∶N ratio < 30, significantly increased the soil pH, respectively (Fig. 7(A)). The magnitude of increase in soil EC was significant when biochar pH was 6.0 ~ 8.0, carbon content was < 75%, nitrogen was > 0.5%, oxygen was 10% ~ 20%, all hydrogen content conditions, and the C∶N ratio of biochar was < 30 and between 50 ~ 100 (Fig. 7(B)). In comparison, soil CEC was not noticeably affected by biochar properties, with the exception of the carbon content of biochar between 50%~75%, nitrogen>1%, and the C:N ratio of biochar 30~50 and>500(Fig. 6(C)). The effect of biochar application on SOC was significantly positive under all biochar pH, carbon, nitrogen, oxygen (except for 10%~20%), hydrogen, and C:N ratios (except for>500) (Fig. 7(D)). The effect of biochar application on TC was significantly positive when the biochar pH value was higher than 6.0, nitrogen was between 0.5%~1%, all oxygen and hydrogen content groups, and the biochar C:N ratio was >30 (Fig. 7(E)). The biochar pH, carbon, nitrogen (except for>1%), oxygen, hydrogen, and biochar C:N ratio (except for>50) exhibited an overall effect on the soil C:N ratio (Fig. 7(F)).

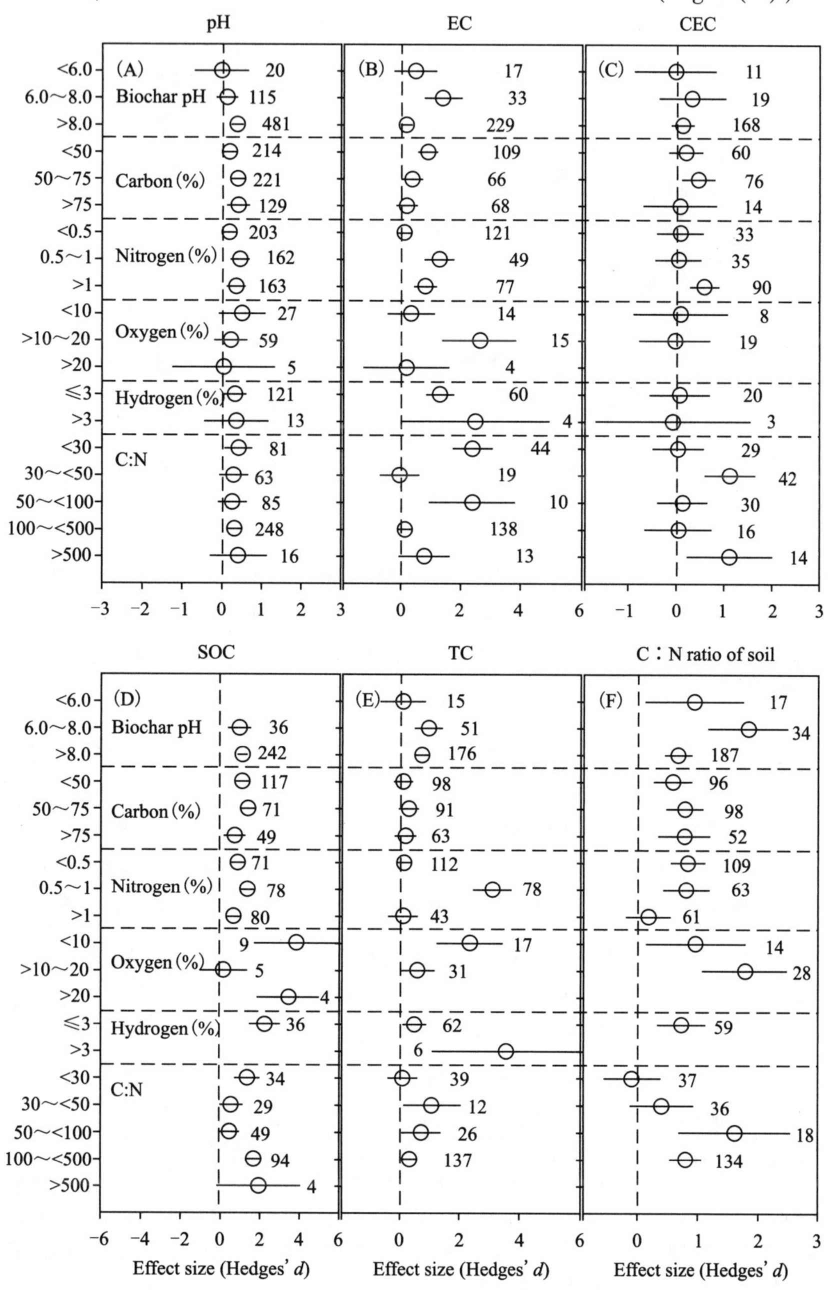

Fig. 7 Effect size Hedges' *d* for the soil chemical properties (A) soil pH; (B) electrical conductivity (EC); (C) cation exchange capacity (CEC); (D) soil organic carbon (SOC); (E) soil total carbon (TC), and (F) soil carbon nitrogen ratio (C:N ratio) to biochar addition in agriculture soil under different biochar properties (biochar pH, carbon, nitrogen, oxygen, hydrogen, and C:N ratio of biochar). The effect size was considered statistically significant if the 95% bootstrap confidence interval did not include zero. The numbers next to the bars are sample sizes for each variable

Correlation analysis among mean effect sizes of differences in soil chemical properties. Significant positive correlations were found among any two effect sizes of soil pH, TC, and SOC (Fig. 8). A significant positive correlation was also found between effect sizes of TC and the soil C∶N ratio, as well as the effect sizes of soil pH and CEC. The effect sizes of the soil C∶N ratio was negatively related to EC (Fig. 8). However, there were no significant relationships between any two effect sizes of TC, EC, and CEC. In addition, no significant relationships were found for the soil C∶N ratio with soil pH and SOC (Fig. 8).

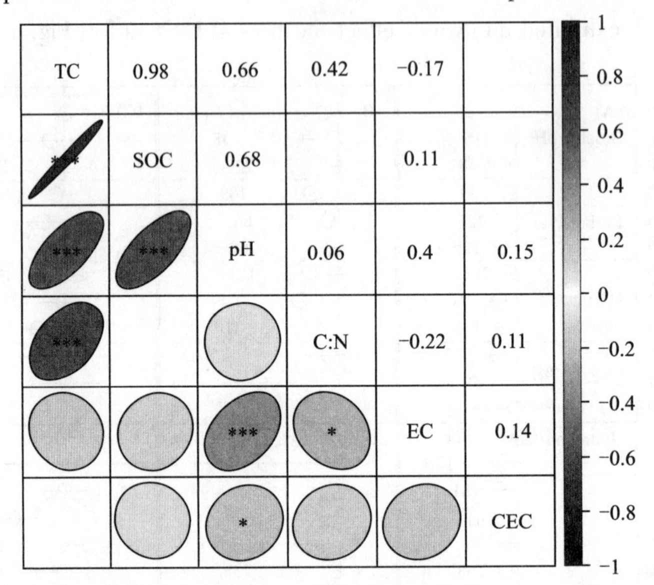

Fig. 8 Coefficients and *P*-values of the correlations among mean effect sizes of soil pH, electrical conductivity (EC), cation exchange capacity (CEC), soil organic carbon (SOC), soil total carbon (TC) and soil carbon nitrogen ratio (C∶N ratio) measurements for the tested studies. Asterisks highlight significant *P*-values(* P<0.05; * * P<0.01; * * * P<0.001)

Key drivers of changes in soil chemical proper−ties. We conducted an aggregated boosted tree (ABT) analysis to compare the relative importance of management conditions, soil properties, biochar pyrolysis conditions, and biochar properties on the soil chemical properties. Overall, 71.6%~81.6% of the variance in soil pH, EC, CEC, SOC, TC, and the C∶N ratio could be explained by the first four factors that were different for each soil chemical property (Fig. 9). Moreover, the initial soil pH (31.6%), biochar application rate (27.0%), and biochar pH (12.1%) were important in explaining the variation of soil pH (Fig. 9(A)); the biochar application rate (25.5%), soil pH (16.4%), biochar pH (15.4%), and biochar carbon (14.0%) were important for soil EC (Fig. 9(B)); the sand content (31.6%), silt content (15.7%), and experimental duration (15.5%) were important for soil CEC (Fig. 9(C)); the initial soil pH (26.2%), pyrolysis temperature (19.3%), biochar application rate (19.0%), and biochar carbon (14.2%) were important for SOC (Fig. 9(D)); the biochar application rate (48.0%) and biochar carbon (15.4%) were important for TC (Fig. 9(E)); and the biochar application rate (36.3%) and biochar carbon (26.2%) were also important for the C∶N ratio of soil (Fig. 9(F)). Together, these results suggest that the biochar application rate, initial soil pH, and sand content in the soil were the major drivers of the change in soil chemical properties in biochar amended soils.

Discussions

In this meta-analysis that used data from published experiments conducted worldwide, we found that biochar amendment can significantly improve soil chemical properties, and biochar addition can increase pH, EC, CEC, SOC, TC, and the C∶N ratio of agricultural soil (Fig. 2 and Table 1). Moreover, when biochar was added to soils, the soil chemical properties changed in a way that was related to experimental conditions, soil properties, pyrolysis conditions, and biochar properties (Fig. 3).

Effects of biochar on soil pH. Soil pH is an important indicator of soil properties, and it has a great influence on soil fertility and crop growth (Šimek and Cooper 2002). Our meta-analysis showed that the use of biochar as a soil amendment significantly ($P<0.01$) increased soil pH (Fig. 2, Table 1), depending on experimental duration, soil texture, initialsoil pH, biochar pyrolysis conditions, biochar pH, or the biochar application rate (Fig. 3, Figs 4(A) ~ 7(A)). Biochar pH depends on feedstock materials and pyrolysis conditions and determines its effect on soil pH (Lehmann and Joseph 2009, Shaaban et al. 2018). In general, biochar pH ranges from 5.9 to 12.3, with an average of 8.9 (Ahmad et al. 2014). The potential of biochar ameliorating soil acidity is related to its alkalinity (Shaaban et al. 2018) and application rate (Purakayastha et al. 2019). Biochar generally has a higher pH than the soil to which it was applied (Liang et al. 2016, El-Naggar et al. 2019); thus, as the application rate increases, the soil pH also increases (Molnár et al. 2016, Laird et al. 2017). This is likely because calcium, magnesium, potassium, and sodium base ions exist in biochar in the form of oxides and soluble carbonates, which dissolve in water and become alkaline, thus neutralising soil acidity (Tan et al. 2017). Moreover, the initial soil pH is important in determining the effect of biochar on soil pH.

Due to biochar's alkaline nature, acidic soils responded better to biochar applications compared to neutral and alkaline soils (Farhangi-Abriz et al. 2021). Since various biochar field trials are performed on acidic soils, applying an acid-neutralising material such as biochar is the most effective practice to in-crease yield and nutrient availability in such soils. Moreover, applying acidic biochar to alkaline soils may improve soil pH, while neutral biochar might decrease the soil pH of alkaline soils (Hailegnaw et al. 2019). Because most biochars are alkaline, limited information is available on biochar application to alkaline soils (Yu et al. 2019). Therefore, the effects of soil pH changes on plant growth in alkaline soils caused by biochar application need to be examined. Furthermore, it should be noted that the effect of biochar on improving soil pH progressively diminishes as time passes (Molnár et al. 2016). Within 12 months, biochar application increases soil pH, yet after more than 12 months, the effect on soil pH is not significant (Fig. 4). Oxidation and leaching processes may reduce the biochar effects and pH with time. This was demonstrated by Slavich et al. (2013) in a study showing decreasing effects of biochar made from animal manure over a few cropping seasons. As the effect of biochar on improving soil pH will weaken over time, the application period of biochar should be considered when using biochar to improve acidic soil.

Effects of biochar on soil EC and CEC. Soil soluble salt is directly proportional to EC, so changes in soil soluble salt can be inferred from changes in the EC value of soil solutions (Corwin and Lesch 2005). When biochar is applied, the soil EC value significantly increases (Table 1, Fig. 4), which may be caused by the soluble ash it contains, thus improving the soil base saturation (Liang et al. 2006). Our results indicated that the biochar application rate (25.5%) is important in explaining the variation of EC (Fig. 9(B)). The biochar made from different raw materials exhibits different EC, between 0.4 and 3.2, which is higher than soil's. However, some studies showed that soil EC was negatively correlated (Alotaibi and Schoenau 2016) or not correlated (Abujabhah et al. 2016b) with the amount of biochar applied. This is mainly because biochar increases the soil porosity and increases the leaching of water-soluble nutrient ions to the deep soil, and thus reduces the content of soluble ions in the soil.

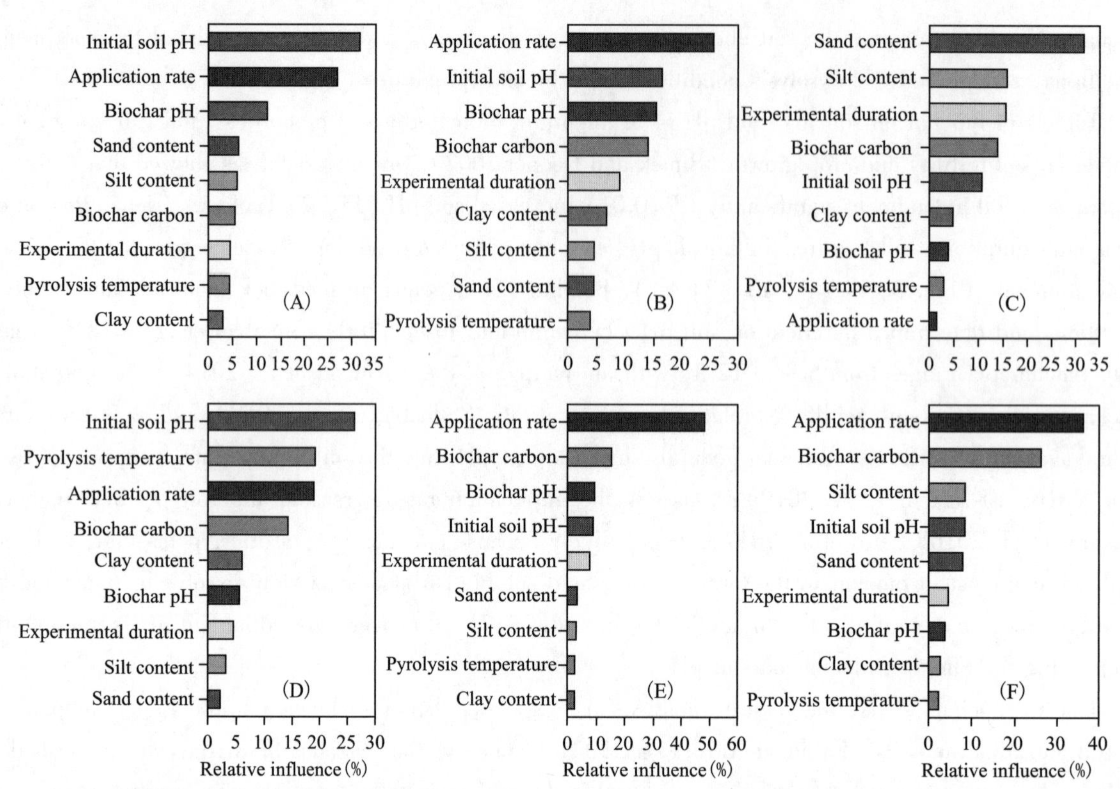

Fig. 9 The relative influence of experimental condition, soil properties, biochar pyrolysis condition and bio-char properties factors on the soil chemical properties ((A) soil pH; (B) electrical conductivity (EC); (C) cation exchange capacity (CEC); (D) soil organic carbon (SOC); (E) soil total carbon (TC), and (F) soil carbon nitrogen ratio (C:N ratio)) based on aggregated boosted tree model by analysis using the "dismo" package of R 4.0.2

CEC is used to estimate the ability of soil to adsorb, retain, and exchange cations and is, therefore, an important indicator of soil quality. A higher CEC indicates a high nutrient fixation capacity, which is beneficial for plant growth. When in contact with the soil, the active groups on the biochar surface, such as-COOH or-OH, react with metal cations in the soil and form metal ion complexes, resulting in electrostatic adsorption (Tan et al. 2017). These functional groups are negatively charged; thus, biochar has a high CEC, which increases soil CEC (Fig. 2). Tan et al. (2017) indicated that soil CEC increased by 0.92 cmol/kg when the biochar addition ratio was 1 : 100 and continues to increase with biochar addition compared to the no biochar addition soil. However, the effect of biochar on soil CEC is related to biochar feedstock materials, biochar production conditions, and soil type (Fig. 3(C)). The probable mechanisms have been extensively discussed (Liang et al. 2006, Lehmann and Joseph 2009, Hagner et al. 2016, Shaaban et al. 2018). Our metaanalysis indi-cated that biochar significantly improves the CEC of low CEC and acidic soils, but has no significant effect on CEC for alkaline soils with high CEC or soils with high organic matter content (Fig. 5(C)). This may be because biochar is rich in functional groups, it has a high adsorption capacity and can absorb mineral elements (Lehmann and Joseph 2009, Ahmad et al. 2014, Laghari et al. 2015).

It should be noted that biochar can increase soil CEC, particularly in sandy-textured soils (Fig. 9). In addition, the duration of biochar application and type of crop also affects the effect of biochar on soil CEC. It is acknowledged that a positive correlation between soil clay content and CEC exists (Ersahin et al. 2006, Gao et al. 2021). With biochar amendment, the CEC of coarse soils increased compared to noncoarse textured soils (Fig. 4(C)). This may be due to the loose structure of sandy soil, which makes it easy for biomass charcoal to pass. Therefore, the abundant functional groups on biochar surfaces and their larger surface areas can be used to increase soil cation exchange capacity, absorb more nutrient ions, avoid nutrient loss, and effectively improve soil fertility and fertiliser utilisation efficiency. Consistently, Burrell et al. (2016) and Razzaghi et al. (2020) showed in their experimental and statistical data models that coarse-textured soils could be improved structurally by biochar application. The CEC of agricultural soils gives an overall idea of their fertility status and water holding capacity (Obalum et al. 2013), thus, low-CEC soils lose nutrient ions with rainfall or irrigation water due to its weak adsorption capacity (Gregory and Nortcliff 2013).

Effects of biochar on SOC, TC, and the C:N ratio. Our results indicated that the effect of biochar on SOC, TC, and the C:N ratio of soil depended on soil pH, biochar properties, application amounts, experimental duration, and type of crop (Table 1, Fig. 3). Several other studies have confirmed an increase in soil carbon and the C:N ratio by the addition of biochar (Agegnehu et al. 2017, Tan et al. 2017, Shaaban et al. 2018). Moreover, our study indicated that soil pH is important for explaining the variation of SOC (26.2%), whereas the biochar application rate is important for explaining the variation of soil TC (48.0%) or the C:N ratio (36.3%) (Fig. 9).

SOC, the main energy source and key trigger for nutrient availability, is a major soil factor affecting plant growth. Its formation is a long-term process involving soil retention of exogenous organic materials, affected by climate, management practices, and soil properties. Biochar has high chemical and microbiological stability in soils due to its high carbon content, complex aromatised structure, and inherent chemical inertness. Therefore, it may change the composition of soil organic matter to increase the total SOC content. Applying biochar with a high C:N ratio results in microbial nitrogen immobilisation and carbon substrate input in soil (Kirkby et al. 2014); therefore, decreased soil microbial activities due to the high biochar C:N ratio would reduce soil greenhouse gas fluxes and increase SOC (Cleveland and Liptzin 2007, Kirkby et al. 2014). Biochar with moderate alkalinity would enhance SOC mineralisation, while that with extremely high or low pH values would cause macronutrient deficiencies. Applying biochar in acidic soils enhances its consumption by microorganisms for balanced soil pH conditions, which may trigger a vigorous priming effect on native SOC mineralisation (Foereid et al. 2011, Jones et al. 2011). Otherwise, adding biochar to neutral or alkaline soils would inhibit soil carbon mineralisation due to enhanced soil pH (Liu et al. 2019a).

Our study showed that the biochar application is the most influential variable of biochar effect on SOC among all variables. Biochar comprises highly concentrated aromatic ring structures, with hydroxyl, carboxyl, carbonyl groups, and lactone structures on their surfaces, thereby increasing soil carbon content (Lyu et al. 2018, Faloye et al. 2019). Moreover, the porous biochar structure can increase the stability of soil organic carbon against biodegradation, thereby reducing its mineralisation rate (Liu et al. 2019b). Due to abundant surface morphological structures, biochar can adsorb soil carbon onto its outer surfaces, reducing the availability of soil organic carbon and inhibiting the degradation of adsorbed carbon, thus indirectly increasing the carbon content of soil (Hartley et al. 2016, Tan et al. 2017, Yu et al. 2019).

Overall, our results showed that the effect size (Hedges' d) of the biochar was greatest for SOC (0.50), the C:N ratio of soil (0.44), soil pH (0.39), TC (0.35), EC (0.21), and CEC (0.20). Among the various factors examined using the ABT analysis, 71.6%~81.6% of the variance in soil pH, EC, CEC, SOC, TC, and the C:N ratio could be explained by the first four factors that were different for each soil chemical properties. In conclusion, our study suggests that improving soil chemical properties by adding biochar is affected not only by biochar application and biochar properties but also by local soil environmental factors, espe-cially soil pH and soil texture. Improving agricultural soil properties by adding biochar requires not only a particular type and implementation of management practices, but also the local environmental factors that should be considered when proposing a land management plan. Furthermore, there is an urgent need for long-term agriculture studies examining changes in soil chemical properties in order to improve our understanding of the potential of applications of biochar in global agriculture and soil environments.

Acknowledgement

We would like to convey our special thanks to various authors and organisations for their various contributions to the data used in this meta-analysis review. The authors gratefully acknowledge researchers at the Shaanxi Provincial Land Engineering Construction Group, for their help with the field experiments.

References

Abujabhah I.S., Bound S.A., Doyle R., Bowman J.P. (2016a): Effects of biochar and compost amendments on soil physico-chemical properties and the total community within a temperate agricultural soil. Applied Soil Ecology, 98: 243-253.

Abujabhah I.S., Doyle R., Bound S.A., Bowman J.P. (2016b): The effect of biochar loading rates on soil fertility, soil biomass, potential nitrification, and soil community metabolic profiles in three different soils. Journal of Soils and Sediments, 16: 2211-2222.

Adams C., Soares K. (1997): The Cochrane Collaboration and the Process of Systematic Reviewing. Advances in Psychiatric Treat-ment, 3: 240-246.

Agegnehu G., Nelson P.N., Bird M.I. (2016): The effects of biochar, compost and their mixture and nitrogen fertilizer on yield and nitrogen use efficiency of barley grown on a Nitisol in the highlands of Ethiopia. Science of The Total Environment, 569-570: 869-879.

Agegnehu G., Srivastava A.K., Bird M.I. (2017): The role of biochar and biochar-compost in improving soil quality and crop performance: a review. Applied Soil Ecology, 119: 156-170.

Ahmad M., Rajapaksha A.U., Lim J.E., Ming Z., Bolan N., Mohan D., Vithanage M., Lee S.S., Ok Y.S. (2014): Biochar as a sorbent for contaminant management in soil and water: a review. Che-mosphere, 99: 19-33.

Ajayi A.E., Horn R. (2016): Modification of chemical and hydrophysical properties of two texturally differentiated soils due to varying magnitudes of added biochar. Soil and Tillage Research, 164: 34-44.

Alotaibi K.D., Schoenau J.J. (2016): Application of two bioenergy byproducts with contrasting carbon availability to a prairie soil: three year crop response and changes in soil biological and chemical properties. Agronomy, 6: 13.

Arft A.M., Walker M.D., Gurevitch J., Alatalo J.M., Bret-Harte M.S., Dale M., Diemer M., Gugerl F., Henry G.H.R., Jones M.H., Hollister R.D., Jónsdóttir I.S., Laine K., Lévesque E., Marion G.M., Molau U., Molgaard P., Nordenhäll U., Raszhivin V., Robinson C.H., Starr G., Stenström A., Stenström M., Totland O., Turner P.L., Walker L.J., Webber P.J., Welker J.M., Wookey P.A. (1999): Responses of tundra plants to experimental warming: meta-analysis of the international tundra experiment. Ecological Monographs, 69: 491-511.

Bayabil H.K., Stoof C.R., Lehmann J.C., Yitaferu B., Steenhuis T.S. (2015): Assessing the potential of biochar and charcoal to improve soil hydraulic properties in the humid Ethiopian Highlands: the Anjeni watershed. Geoderma, 243-244: 115-123.

Bera T., Collins H.P., Alva A.K., Purakayastha T.J., Patra A.K. (2016): Biochar and manure effluent effects on soil biochemical properties under corn production. Applied Soil Ecology, 107: 360-367.

Borchard N., Schirrmann M., Cayuela M.L., Kammann C., Wrage-Mönnig N., Estavillo J.M., Fuertes-Mendizúbal T., Sigua G., Spokas K., Ippolito J.A., Novak J. (2019): Biochar, soil and land-use interactions that reduce nitrate leaching and N_2Oemissions: a meta-analysis. Science of The Total Environment, 651: 2354-2364.

Borchard N., Siemens J., Ladd B., Möller A., Amelung W. (2014): Application of biochars to sandy and silty soil failed to increase maize yield under common agricultural practice. Soil and Tillage Research, 144: 184-194.

Burrell L.D., Zehetner F., Rampazzo N., Wimmer B., Soja G. (2016): Long-term effects of biochar on soil physical properties. Geoderma, 282: 96-102.

Case S.D.C., McNamara N.P., Reay D.S., Whitaker J. (2012): The effect of biochar addition on N_2O and CO_2 emissions from a sandy loam soil-therole of soil aeration. Soil Biology and Biochemistry, 51: 125-134.

Cayuela M.L., Aguilera E., Sanz-Cobena A., Adams D.C., Abalos D., Barton L., Ryals R., Silver W.L., Alfaro M.A., Pappa V.A., Smith P., Garnier J., Billen G., Bouwman L., Bondeau A., Lassaletta L. (2017): Direct nitrous oxide emissions in Mediterranean climate cropping systems: emission factors based on a meta-analysis of available measurement data. Agriculture, Ecosystems and Environment, 238: 25-35.

Cayuela M.L., van Zwieten L., Singh B.P., Jeffery S., Roig A., Sánchez-Monedero M.A. (2014): Biochar's role in mitigating soil nitrous oxide emissions: a review and meta-analysis. Agricul-ture, Ecosystems and Environment, 191: 5-16.

Chen D., Liu X.Y., Bian R.J., Cheng K., Zhang X.H., Zheng J.F., Joseph S., Crowley D., Pan G.X., Li L.Q. (2018): Effects of biochar on availability and plant uptake of heavy metals-a meta-analysis. Journal of Environmental Management, 222: 76-85.

Cheng J.Z., Lee X.Q., Gao W.C., Chen Y., Pan W., Tang Y. (2017): Effect of biochar on the bioavailability of difenoconazole and microbial community composition in a pesticide-contaminated soil. Applied Soil Ecology, 121: 185-192.

Cleveland C.C., Liptzin D. (2007): C:N:P stoichiometry in soil: is there a "redfield ratio" for the microbial biomass? Biogeochemistry, 85: 235-252.

Corwin D.L., Lesch S.M. (2005): Apparent soil electrical conductivity measurements in agriculture. Computers and Electronics in Agriculture, 46: 11-43.

Dai Y.H., Zheng H., Jiang Z.X., Xing B.S. (2020): Combined effects of biochar properties and soil conditions on plant growth: a meta-analysis. Science of The Total Environment, 713: 136635.

De la Rosa J.M., Paneque M., Miller A.Z., Knicker H. (2014): Relating physical and chemical properties of four different biochars and their application rate to biomass production of *Lolium perenne* on a Calcic Cambisol during a pot experiment of 79 days. Science of The Total Environment, 499: 175-184.

Deal C., Brewer C.E., Brown R.C., Okure M.A.E., Amoding A. (2012): Comparison of kiln-derived and gasifier-derived biochars as soil amendments in the humid tropics. Biomass and Bioenergy, 37: 161-168.

Edeh I.G., Mašek O., Buss W. (2020): A meta-analysis on biochar's effects on soil water properties-new insights and future research challenges. Science of The Total Environment, 714: 136857.

El-Naggar A., Lee S.S., Rinklebe J., Farooq M., Song H., Sarmah A.K., Zimmerman A.R., Ahmad M., Shaheen S.M., Ok Y.S. (2019): Biochar application to low fertility soils: a review of cur-rent status, and future prospects. Geoderma, 337: 536-554.

Ersahin S., Gunal H., Kutlu T., Yetgin B., Coban S. (2006): Estimating specific surface area and cation exchange capacity in soils using fractal dimension of particle-size distribution. Geoderma, 136: 588-597.

Faloye O.T., Alatise M.O., Ajayi A.E., Ewulo B.S. (2019): Effects of biochar and inorganic fertilizer applications on growth, yield and water use efficiency of maize under deficit irrigation. Agricul-tural Water Management, 217: 165-178.

Farhangi-Abriz S., Torabian S., Qin R., Noulas C., Lu Y., Gao S. (2021): Biochar effects on yield of cereal and legume crops using meta-analysis. Science of the Total Environment, 775: 145869.

Fedrowitz K., Koricheva J., Baker S.C., Lindenmayer D.B., Palik B., Rosenvald R., Beese W., Franklin J.F., Kouki J., Macdonald E., Messier C., Sverdrup-Thygeson A., Gustafsson L. (2014): Can retention forestry help conserve biodiversity? A meta-analysis. Journal of Applied Ecology, 51: 1669-1679.

Foereid B., Lehmann J., Major J. (2011): Modeling black carbon degradation and movement in soil. Plant and Soil, 345: 223-236.

Gao S., DeLuca T.H., Cleveland C.C. (2019): Biochar additions alter phosphorus and nitrogen availability in agricultural ecosystems: a meta-analysis. Science of The Total Environment, 654: 463-472.

Gao Y., Shao G.C., Yang Z., Zhang K., Lu J., Wang Z.Y., Wu S.Q., Xu D. (2021): Influences of soil and biochar properties and amount of biochar and fertilizer on the performance of biochar in improving plant photosynthetic rate: a meta-analysis. European Journal of Agronomy, 130: 126345.

Giagnoni L., Maienza A., Baronti S., Vaccari F.P., Genesio L., Taiti C., Martellini T., Scodellini R., Cincinelli A., Costa C., Mancuso S., Renella G. (2019): Long-term soil biological fertility, volatile organic compounds and chemical properties in a vineyard soil after biochar amendment. Geoderma, 344: 127-136.

Gregory P.J., Nortcliff S. (2013): Soil Conditions and Plant Growth. 1st Edition. Oxford, Blackwell Publishing Ltd. IS-

BN: 9781405197700

Gurevitch J., Koricheva J., Nakagawa S., Stewart G. (2018): Meta-analysis and the science of research synthesis. Nature, 555: 175-182.

Hagner M., Kemppainen R., Jauhiainen L., Tiilikkala K., Setälä H. (2016): The effects of birch (*Betula* spp.) biochar and pyrolysis temperature on soil properties and plant growth. Soil and Tillage Research, 163: 224-234.

Hailegnaw N.S., Mercl F., Pračke K., Száková J., Tlustoš P. (2019): Mutual relationships of biochar and soil pH, CEC, and exchangeable base cations in a model laboratory experiment. Journal of Soils and Sediments, 19: 2405-2416.

Hall D.J.M., Bell R.W. (2015): Biochar and compost increase crop yields but the effect is short term on sandplain soils of Western Australia. Pedosphere, 25: 720-728.

Hartley W., Riby P., Waterson J. (2016): Effects of three different biochars on aggregate stability, organic carbon mobility and micronutrient bioavailability. Journal of Environmental Management, 181: 770-778.

Hedges L.V., Olkin I. (1985): Statistical Methods for Meta-Analysis. New York, Academic Press.

Hedges L.V., Gurevitch J., Curtis P.S. (1999): The meta-analysis of response ratios in experimental ecology. Ecology, 80: 1150-1156.

Herath H. M. S. K., Camps-Arbestain M., Hedley M. (2013): Effect of biochar on soil physical properties in two contrasting soils: an Alfisol and an Andisol. Geoderma, 209-210: 188-197.

Hijmans R.J., Phillips S., Leathwick J., Elith J. (2017): Species distribution modelling with R. R package "Dismo". Available at: cran.r-project.org/web/packages/dismo/vignettes/sdm. pdf (accessed March 13, 2018)

Jeffery S., Meinders M.B.J., Stoof C.R., Bezemer T.M., van de Voorde T.F.J., Mommer L., van Groenigen J.W. (2015): Biochar application does not improve the soil hydrological function of a sandy soil. Geoderma, 251-252: 47-54.

Jeffery S., Verheijen F.G.A., Kammann C., Abalos D. (2016): Biochar effects on methane emissions from soils: a meta-analysis. Soil Biology and Biochemistry, 101: 251-258.

Jiang Y., Carrijo D., Huang S., Chen J., Balaine N., Zhang W.J., van Groenigen K.J., Linquist B. (2019): Water management to mitigate the global warming potential of rice systems: a global meta-analysis. Field Crops Research, 234: 47-54.

Jien S., Wang C. (2013): Effects of biochar on soil properties and erosion potential in a highly weathered soil. Catena, 110: 225-233.

Jindo K., Sánchez-Monedero M.A., Hernández T., García C., Furukawa T., Matsumoto K. (2012): Biochar influences the microbial community structure during manure composting with agricul-tural wastes. Science of The Total Environment, 416: 476-481.

Jones D. L., Murphy D. V., Khalid M., Ahmad W., Edwards-Jones G., DeLuca T. H. (2011): Short-term biocharinduced increase in soil CO_2 release is both biotically and abiotically mediated. Soil Biology and Biochemistry, 43: 1723-1731.

Kätterer T., Roobroeck D., Andrén O., Kimutai G., Karltun E., Kirchmann H., Nyberg G., Vanlauwe B., Röing De Nowina K. (2019): Biochar addition persistently increased soil fertility and yields in maize-soybean rotations over 10 years in sub-humid regions of Kenya. Field Crops Research, 235: 18-26.

Kirkby C.A., Richardson A.E., Wade L.J., Passioura J.B., Batten G.D., Blanchard C., Kirkegaard J.A. (2014): Nutrient availability limits carbon sequestration in arable soils. Soil Biology and Biochemistry, 68: 402-409.

Laghari M., Mirjat M.S., Hu Z.Q., Fazal S., Xiao B., Hu M., Chen Z., Guo D. (2015): Effects of biochar application rate on sandy desert soil properties and sorghum growth. Catena, 135: 313-320.

Laird D.A., Novak J.M., Collins H.P., Ippolito J.A., Karlen D.L., Lentz R.D., Sistani K.R., Spokas K., Van Pelt R. S. (2017): Multi-year and multi-location soil quality and crop biomass yield responses to hardwood fast pyrolysis biochar. Geoderma, 289: 46-53.

Lebrun M., Miard F., Nandillon R., Scippa G.S., Bourgerie S., Morabito D. (2019): Biochar effect associated with compost and iron to promote Pb and As soil stabilization and *Salix viminalis* L. growth. Chemosphere, 222: 810-822.

Lehmann J., Joseph S. (2009): Biochar for environmental management: an introduction. In: Lehmann J., Joseph S. (eds.): Biochar for Environmental Management Science and Technology. UK, Earthscans, 1-12. ISBN: 9780367779184

Li M.F., Wang J., Guo D., Yang R.R., Fu H. (2019): Effect of land management practices on the concentration of dissolved organic matter in soil: a meta-analysis. Geoderma, 344: 74-81.

Li X.X., Chen X.B., Weber-Siwirska M., Cao J.J., Wang Z.L. (2018): Effects of rice-husk biochar on sand-based rootzone amendment and creeping bentgrass growth. Urban Forestry and Urban Greening, 35: 165-173.

Liang B., Lehmann J., Solomon D., Kinyangi J., Grossman J., O'Neill B., Skjemstad J.O., Thies J., Luizão F.J., Petersen J., Neves E.G. (2006): Black carbon increases cation exchange capacity in soils. Soil Science Society of America Journal, 70: 1719-1730.

Liang C.F., Gascó G., Fu S.L., Méndez A., Paz-Ferreiro J. (2016): Biochar from pruning residues as a soil amendment: effects of pyrolysis temperature and particle size. Soil and Tillage Research, 164: 3-10.

Lin Z.B., Liu Q., Liu G., Cowie A.L., Bei Q.C., Liu B.J., Wang X.J., Ma J., Zhu J.G., Xie Z.B. (2017): Effects of different biochars on *Pinus elliottii* growth, N use efficiency, soil N_2O and CH_4 emissions and C storage in a subtropical area of China. Pedosphere, Liu C., Wang H.L., Li P.H., Xian Q.S., Tang X.G. (2019a): Biochar's impact on dissolved organic matter (DOM) export from a cropland soil during natural rainfalls. Science of The Total Environment, 650: 1988-1995.

Liu X., Mao P.N., Li L.H., Ma J. (2019b): Impact of biochar application on yield-scaled greenhouse gas intensity: a meta-analysis. Science of The Total Environment, 656: 969-976.

Lyu H.H., Gao B., He F., Zimmerman A.R., Ding C., Huang H., Tang J.C. (2018): Effects of ball milling on the physicochemical and sorptive properties of biochar: experimental observations and governing mechanisms. Environmental Pollution, 233: 54-63.

Meng J., Tao M.M., Wang L.L., Liu X.M., Xu J.M. (2018): Changes in heavy metal bioavailability and speciation from a Pb-Zn mining soil amended with biochars from copyrolysis of rice straw and swine manure. Science of the Total Environment, 633: 300-307.

Minasny B., McBratney A.B., Brough D.M., Jacquier D. (2011): Models relating soil pH measurements in water and calcium chloride that incorporate electrolyte concentration. European Journal of Soil Science, 62: 728-732.

Molnár M., Vaszita E., Farkas É., Ujaczki É., Fekete-Kertész I., Tolner M., Klebercz O., Kirchkeszner C., Gruiz K., Uzinger N., Feigl V. (2016): Acidic sandy soil improvement with biochar-amicrocosm study. Science of The Total Environment, 563-564: 855-865.

Mukherjee A., Lal R., Zimmerman A.R. (2014): Effects of biochar and other amendments on the physical properties and greenhouse gas emissions of an artificially degraded soil. Science of the Total Environment, 487: 26-36.

Mukherjee A., Zimmerman A.R. (2013): Organic carbon and nutrient release from a range of laboratory-produced biochars and biochar-soil mixtures. Geoderma, 193-194: 122-130.

Norini M.-P., Thouin H., Miard F., Battaglia-Brunet F., Gautret P., Guégan R., Le Forestier L., Morabito D., Bourgerie S., Motelica-Heino M. (2019): Mobility of Pb, Zn, Ba, As and Cd toward soil pore water and plants (willow and ryegrass) from a mine soil amended with biochar. Journal of Environmental Management, 232: 117-130.

Obalum S.E., Watanabe Y., Igwe C.A., Obi M.E., Wakatsuki T. (2013): Improving on the prediction of cation exchange capacity for highly weathered and structurally contrasting tropical soils from their fine-earth fractions. Communications in Soil Science and Plant Analysis, 44: 1831-1848.

Omondi M.O., Xia X., Nahayo A., Liu X.Y., Korai P.K., Pan G.X. (2016): Quantification of biochar effects on soil hydrological properties using meta-analysis of literature data. Geoderma, 274: 28-34.

Pandey V., Patel A., Patra D.D. (2016): Biochar ameliorates crop productivity, soil fertility, essential oil yield and aroma profiling in basil (*Ocimum basilicum* L.). Ecological Engineering, 90: 361-366.

Peake L.R., Reid B.J., Tang X.G. (2014): Quantifying the influence of biochar on the physical and hydrological properties of dissimilar soils. Geoderma, 235-236: 182-190.

Pranagal J., Oleszczuk P., Tomaszewska-Krojańska D., Kraska P., Ró żyło K. (2017): Effect of biochar application on the physical properties of Haplic Podzol. Soil and Tillage Research, 174: 92-103.

Purakayastha T.J., Bera T., Bhaduri D., Sarkar B., Mandal S., Wade P., Kumari S., Biswas S., Menon M., Pathak H., Tsang D.C.W. (2019): A review on biochar modulated soil condition improve-ments and nutrient dynamics concerning crop yields: pathways to climate change mitigation and global food security. Chemosphere, 227: 345-365.

Razzaghi F., Obour P.B., Arthur E. (2020): Does biochar improve soil water retention? A systematic review and meta-analysis. Geoderma, 361: 114055.

Sandhu S.S., Ussiri D.A.N., Kumar S., Chintala R., Papiernik S.K., Malo D.D., Schumacher T.E. (2017): Analyzing the impacts of three types of biochar on soil carbon fractions and physiochemical properties in a corn-soybean rotation. Chemosphere, 184: 473-481.

Shaaban M., Van Zwieten L., Bashir S., Younas A., Núñez-Delgado A., Chhajro M.A., Kubar K.A., Ali U., Rana M.S., Mehmood M.A., Hu R.G. (2018): A concise review of biochar application to agricultural soils to improve soil conditions and fight pollution. Journal of Environmental Management, 228: 429-440.

Šimek M., Cooper J.E. (2002): The influence of soil pH on denitrification: progress towards the understanding of this interaction over the last 50 years. European Journal of Soil Science, 53: 345-354.

Slavich P.G., Sinclair K., Morris S.G., Kimber S.W.L., Downie A., Van Zwieten L. (2013): Contrasting effects of manure and green waste biochars on the properties of an acidic ferralsol and productivity of a subtropical pasture. Plant and Soil, 366: 213-227.

Stefaniuk M., Oleszczuk P., Ró żyło K. (2017): Co-application of sewage sludge with biochar increases disappearance of polycyclic aromatic hydrocarbons from fertilized soil in long term field ex-periment. Science of The Total Environment, 599-600: 854-862.

Tan Z.X., Lin C.S.K., Ji X.Y., Rainey T.J. (2017): Returning biochar to fields: a review. Applied Soil Ecology, 116: 1-11.

Unger R., Killorn R., Brewer C.E. (2011): Effects of soil application of different biochars on selected soil chemical properties. Communications in Soil Science and Plant Analysis, 42: 2310-2321.

Vaccari F.P., Maienza A., Miglietta F., Baronti S., Di Lonardo S., Giagnoni L., Lagomarsino A., Pozzi A., Pusceddu E., Ranieri R., Valboa G., Genesio L. (2015): Biochar stimulates plant growth but not fruit yield of processing tomato in a fertile soil. Agriculture, Ecosystems and Environment, 207: 163-170.

Verhoeven E., Six J. (2014): Biochar does not mitigate field-scale N_2O emissions in a Northern California vineyard: an assessment across two years. Agriculture, Ecosystems and Environment, 191: 27-38.

Wang D.Y., Fonte S.J., Parikh S.J., Six J., Scow K.M. (2017): Biochar additions can enhance soil structure and the physical stabilization of C in aggregates. Geoderma, 303: 110-117.

Wong J.T.F., Chen X.W., Deng W.J., Chai Y.M., Ng C.W.W., Wong M.H. (2019): Effects of biochar on bacterial communities in a newly established landfill cover topsoil. Journal of Environmental Management, 236: 667-673.

Yao Q., Liu J.J., Yu Z.H., Li Y.S., Jin J., Liu X.B., Wang G.H. (2017): Changes of bacterial community compositions after three years of biochar application in a black soil of northeast China. Applied Soil Ecology, 113: 11-21.

Yu H.W., Zou W.X., Chen J.J., Chen H., Yu Z., Huang J., Tang H.R., Wei X.Y., Gao B. (2019): Biochar amendment improves crop production in problem soils: a review. Journal of Environmental Management, 232: 8-21.

Zhang W.S., Liang Z.Y., He X.M., Wang X.Z., Shi X., Zou C., Chen X. (2019): The effects of controlled release urea on maize productivity and reactive nitrogen losses: a meta-analysis. Environ-mental Pollution, 246: 559-565.

Zheng J.F., Han J.M., Liu Z.W., Xia W.B., Zhang X.H., Li L.Q., Liu X.Y., Bian R.J., Cheng K., Zheng J.W., Pan G.X. (2017): Biochar compound fertilizer increases nitrogen productivity and economic benefits but decreases carbon emission of maize production. Agriculture, Ecosystems and Environment, 241: 70-78.

本文曾发表于2022年《Plant Soil and Environment》第6卷

Landslide Susceptibility Mapping Using Novel Hybrid Model Based on Different Mapping Units

Tingyu Zhang[a], Quan Fu[b], Renata Pacheco Quevedo[c], Tianqing Chen[a],
Dan Luo[b], Fangfang Liu[b] and Hui Kong[a]

(a.Shaanxi Provincial Key Laboratory of Land Rehabilitation, Xi'an 710064, China; b.Shaanxi Provincial Land Engineering Construction Group Land Survey Planning and Design Institute, Xi'an 710064, China; c.Earth Observation and Geoinformatics Division, National Institute for Space Research (INPE), Sao Jose dos Campos 12227010, Brazil)

【Abstract】 Landslide is a most widespread geohazards around the world. Reasonable landslide susceptibility mapping can aid decision-makers in landslide prevention. Therefore, drawing regional landslide susceptibility map is of great significance to landslide prevention. This research mainly aims to explore a new method and carry out the landslide susceptibility mapping based on terrain mapping unit (TMU) and grid cells mapping unit (GCMU) in Wuqi County, Yan'an City, China. Firstly, the landslide inventory map was prepared based on 717 landslides that were extracted. Secondly, the index of entropy model (IOE) was applied to quantify landslide predisposing factors based on TMU and GCMU, respectively. Finally, the systematically developed forest of multiple trees (SysFor) was integrated with IOE to construct a novel hybrid model (IOE-SysFor) and the landslide susceptibility maps based on TMU and GCMU were obtained. Statistical indices and receiver operating characteristic curve (ROC) were applied to evaluate the results and compare the approaches. From the results, it indicated that the accuracy of landslide susceptibility maps generated by the IOE-SysFor based on TMU and GCMU was satisfactory, furthermore, the performance of IOE-SysFor model running on the TMU was better than GCMU. Therefore, the TMU was considered as a more suitable landslide mapping unit for similar regions and the approach developed by this study can provide a reference for future researches.

【Keywords】 Geology; Systematically developed forest of multiple trees; Entropy; Hybrid model; Spatial analysis

1 Introduction

The definition of landslide refers to the large-scale movements of rock masses, soil, and debris caused by some influencing factors such as rainfall, earthquakes, or human activities (Chen et al., 2019a). As the most frequently occurred geological hazard in China, landslide has caused casualties, economic losses and a serious of geo-environment problems. According to reports, 7840 geological disasters were identified in China in 2020, of which 4810 were landslides, causing 139 deaths, 58 injuries and direct economic losses of approximately 770 million US dollars (China MoNRotPsRo, 2020), and from a certain point of view, the effect use of land resources and economic development in mountainous area are hindered by the occurrence of landslides. Since landslides occur frequently and extensively, and the occurrence of landslides is challenging to predict, the prediction is essential for landslide prevention, moreover, one of the hotspots in current research is to explore methods that can reasonably predict the spatial distribution of landslides (Nsengiyumva et al., 2018). Currently, landslide susceptibility mapping (LSM) is considered as one of the effective methods to know where landslide occurrence probabilities are higher (Kadavi et al., 2018). With the growing popularization of Geographic Information System (GIS) since 2000, the qualitative and semi-quantitative methods were commonly applied in LSM. However, these approaches are often influenced by subjective acknowledgement and expertise (Broeckx et al., 2018). For this reason, the objective and quantitative methods of LSM were introduced in recent years, which mainly consist of statistical models and machine learning models. The commonly applied statistical models comprise: the statistical indices (SI) (Razavizadeh et al., 2017; Liu and Duan, 2018), logistic regression (LR) (Lombardo and Mai, 2018; Chen et al., 2019b), the weight of evidence (WOE) (Ilia and Tsangaratos, 2016; Razavizadeh et al., 2017; Polykretis and Chalkias, 2018), certainty factors (CF) (Wu et al., 2016; Fan et al., 2017; Reichenbach et al., 2018), index of entropy (IOE) (Youssef et al., 2015; Bui et al., 2018), etc. Therefore, in order to ascertain the susceptibility of regional landslides, it can be achieved by calculating the weight of predisposing factors for landslide occurrence. However, it is undeniable that statistical methods have some presuppositions, which limit their development. Nevertheless, with the continuous breakthoughs in machine learning and data mining techniques in recent years, a large number of new models have been introduced in the field of landslide susceptibility mapping. The most common machine learning models involve decision tree (DT) (Pham et al., 2017; Park et al., 2019), random forest (RF) (Chen et al., 2017; Behnia and Blais-Stevens, 2018), support vector machine (SVM) (Ballabio and Sterlacchini, 2012; Kalantar et al., 2017; Kumar et al., 2017; Lee et al., 2017a), and kernel logistic regression model (KLR) (Bui et al., 2016; Chen et al., 2019c). Although complex rules are hidden in large amounts of data, these machine learning models can efficiently reveal the internal connections between them, and the results produced in this way are can more reliable. However, in different usage environment, due to differences in databases, the performance of machine learning models will dramatically change, which indicates that there is still a possibility for the improvement of generalization capabilities. Hence, many strategies for model integration have emerged, and a lot of hybrid models were applied for improving comprehensive performance of LSM, for instance, the reduced error pruning trees integral with rotation forest (Pham et al., 2019), the support vector machines integral with random subspace (Hong et al., 2017), multiboost integral with naïve Bayes trees (Nguyen et al., 2019). However, the operation of the above-mentioned models is based on the grid cells mapping unit, whereas researches and comparisons of the effects of the landslide susceptibility models running on other mapping units are still relatively rare.

The mapping unit, as the basic unit for LSM, has a significant impact on the prediction ability (Erener and Düzgün, 2012). Since the mapping unit is the base of landslide predisposing factors quantification and the establishment of LSM models, its suitability is crucial for LSM (Carrara et al., 1995). In current research of LSM, two types of mapping units were mainly applied: terrain mapping unit (TMU) and the grid cells mapping unit (GCMU). GCMU is a model based on the mosaic spatial data which divides the region into several regular grids, while TMU belongs to a part of a land surface that contains a set of ground conditions that are different from adjacent units on a definable boundary (Steger et al., 2016). Although previous studies and researches showed that mapping units and classification models impact the predictability of landslides, it is not yet clear which type is more suitable for LSM in terms of the effectiveness of mapping unit. How to improve the forecast accuracy of LSM is still the focus of current researches.

Therefore, in current research, a novel hybrid model combining the IOE model and systematically developed forest of multiple trees (SysFor) namely IOE-SysFor model is presented to generate the landslide susceptibility map based on TMU and GCMU in a selected study area located in Yan'an City, Shaanxi Province, China, where, each research technique was respectively evaluated and compared. The results of this study can not only reduce the burden of local landslide prevention and control work, and improve the efficiency of response, but also provide reference for future researches in the study area.

The structure of this study mainly contains five parts: 1) introduction; 2) describe the basic situation of the study area, explain the preparation of landslide inventory map and mapping units, and analyze the impact factor; 3) introduce the methods used in the study, such as evaluation methods of landslide predisposing factors, IOE model, SysFor model, and the method of validation and comparison; 4) quantify landslide predisposing factors, construct a hybrid model (IOE-SysFor), analyze the result of landslide susceptibility mapping; 5) discussion and conclusion.

2 Study Area and Data Used

2.1 The Study Area

The study area is located in the northwest of Yan'an City, Shaanxi Province, China, and extends between longitudes of 107°38′57″E and 108°32′49″E and latitudes of 36°33′33″N and 37°24′27″N, covering an area of 3 791.5 km^2 with a population of 127000 (Fig. 1). The landforms of the study area were generated after the uplift occurred in the Early Cretaceous, which covers an area of hilly gullies and broken terrain. With an elevation range of 1233 m to 1809 m, the gully region is generally greater in the northwest and smaller in the southeast, and it spreads all over the loess pitfalls. The study area has a semi-humid and semi-arid climate with temperate and continental monsoon characteristics. The mean annual temperature is 7.8 ℃ and the average annual precipitation is 483.4 mm; furthermore, their spatial and temporal distribution vary considerably. Besides, the duration of precipitation is generally concentrated from July to September, accounting for 62% of the total annual precipitation. The primary vegetation type is forest shrub grassland, which is in the transition from warm temperate deciduous broad-leaved forest belt to temperate forest shrub grassland. Moreover, the overall vegetation coverage is low, and the degree of soil erosion is quite severe. All these geological environments furnish convenient conditions for the development of landslides (Government WCPs, 2020).

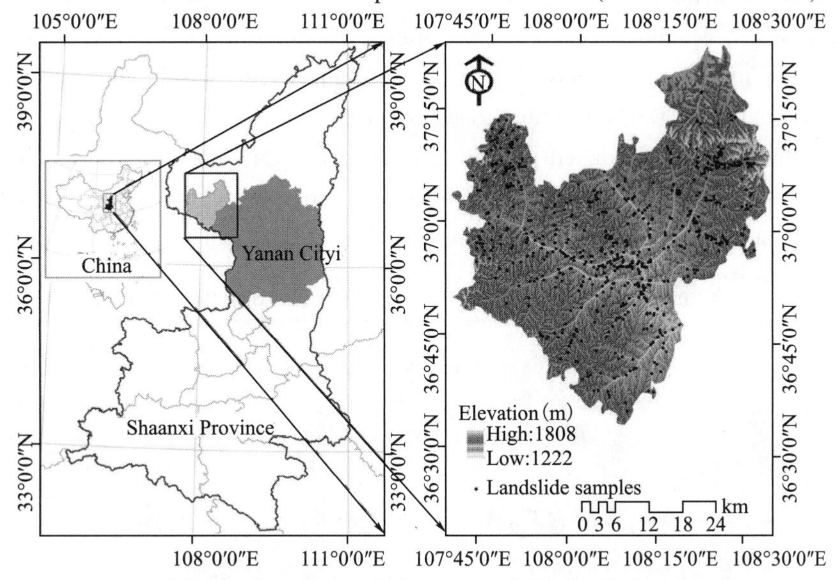

Fig. 1 Location and Inventory of the Study Area

The study area is widely covered by the loess of the Quaternary, and the middle and upper Pleistocene. Due to the large thickness of the loess deposit layer, the former Quaternary strata are only exposed in the deep cut valleys and the steep slopes. According to the sequence of geologic age, the main lithology of the study area is classified into five groups as follows: Group (A) the Lower Cretaceous Huanhe -Huachi Formation (K_1h), including mudstone, sandstone, siltstone, conglomerate, and arkose; Group(B) the Neogene Pliocene(N_2), including clay, sandy clay, conglomerate, glutenite, clay, and marlstone; Group(C) the Quaternary Middle Pleistocene (Q_2^{eol}), including Loess and silty clay; Group(D) The Quaternary New System (Q_4), including sand and gravel. The stratum of the study area is above a structure that is gently tilted towards the west as a whole with an inclination angle ranges between 1° and 3°; in addition to that, there is no fault structure in the study area, and the intensity of seismic activities is relatively low. The study area contains nine rivers, that covering a drainage area greater than 100 km^2. The types of groundwater mainly include the Quaternary loose layer pore water, pores, fissure water, and Cretaceous bedrock fissure water (Bullock and King, 2011).

2.2 Preparation of Landslide Inventory

The relationship between landslide conditioning factors and spatial distribution can be analyzed with the landslide inventory map. In present study, a total of 717 landslides pattern spots with the period of development from 1995 to 2018 were extracted based on historical data, field survey, and interpretation of GF-2 aerial photography, including 17 bedrock slides, 681 accumulated layer slides, 19 falls, and five debris flows generated on Loess. The mean areas of landslides including slides, falls, and debris flows were 7.7×10^3 m^2, 2.6×10^3 m^2, and 2.7×10^4 m^2, respectively. Since the total area of landslides only accounts for 0.04% in the study area, thus, in order to improve the efficiency of calculation, all landslides pattern spots were transformed into points by centroid method. This method has been used in many preceding studies and achieved satisfactory results (Pourghasemi et al., 2012; Moayedi et al., 2018). Besides, the theory of LSM is to solve the classification problems; thus, the positive and negative samples are necessary conditions. Hence a total number of 717 non-landslide points with the label of 0 were randomly generated in an area without landslides, whereas the label of 717 landslide points into two parts according to the ratio of 7/3, and also, the 717 non-landslide points were divided as per the same ratio. Then 70% (501 landslide points and 501 non-landslide points) of the points was used for training models, and the 30% remained (216 landslide points and 216 non-landslide points) for the validation. Finally, the landslide inventory map was generated according to all these landslides information, as shown in Fig. 1.

2.3 Preparation of Mapping Units

In this study, GCMU contains 8 331 296 cells with a resolution of 30 m (Fig. 2(a)). On the other hand, the terrain mapping unit was computed through three steps as follows: First, based on the digital elevation model(DEM) data with 30 m resolution, downloaded from Geospatial Data Cloud(2020), the drainage network(the valley line) was extracted by using hydrological analysis module of ArcGIS. Secondly, the original DEM data was then inverted, and the ridgeline was extracted by the same method as the previous step. Eventually, the study area was divided by the ridgeline and the valley line to obtain the terrain mapping unit(TMU), shown in Fig. 2(b).

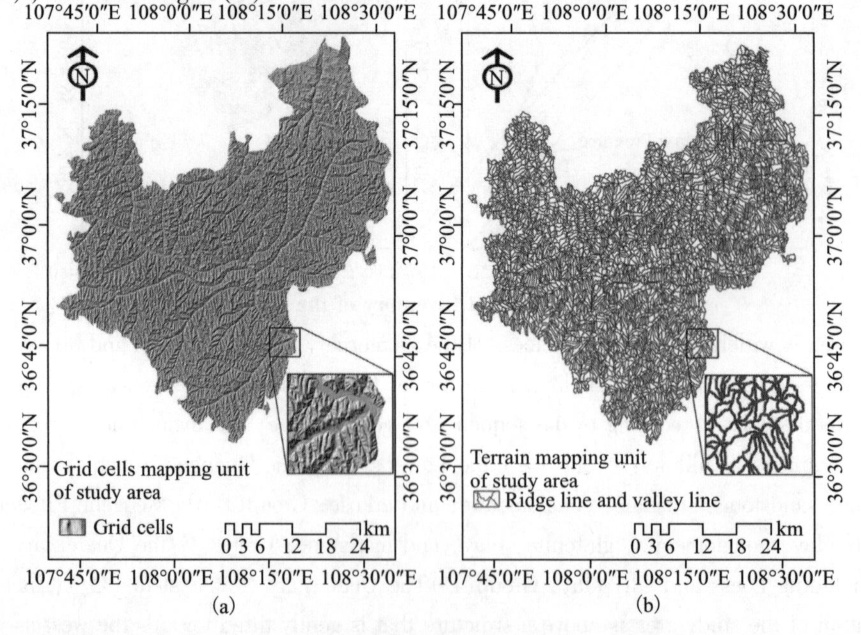

Fig. 2 The Grid Cells Mapping Units and the Terrain Mapping Unit of the Study Area: (a) GCMU, (b) TMU

2.4 Analysis of Landslide Predisposing Factors

According to the analyses of the available information and the field observation, the landslide predisposing factors includes 10 conditioning factors, which comprises: the altitude, lithology, slope angle, distance to roads, slope aspect, rainfall, plan curvature, distance to rivers, profile curvature, normalized differential vegetation index (NDVI) and distance to resident points.

Values of the aspect, altitude, slope, plan curvature, and profile curvature were all computed based on DEM using the digital terrain analysis method. One of the important factors affecting the gravitational potential energy inside the landside is altitude, which was divided into eight classes according to an equal interval of 300 m (Fig. 3(a)). The steepness of the terrain can be reflectedby slope angle Generally, under the same conditions, the probability of landslide occurrence will increase with the increase of slope angle (Arabameri et al., 2018). In this study, the slope angle was divided into eight classes with an equal interval of 10°, as shown in Fig. 3(b). The aspect value can affect the time of exposure to light, influencing the soil water content and causing the slope to become unstable (Ahmed and Dewan, 2017). Therefore, the aspect was grouped into nine categories, with an equal interval of 45° (Fig. 3(c)). The curvatures describe the slope shape and measure the convergence, and the concavity of the terrain, which can affect the soil moisture. Based on natural break method, the plan curvature (Fig. 3(d)) and profile curvature (Fig. 3(e)) were respectively divided into five classes.

The rainfall can have a large impact on landsides, where rainwater penetrates the slope through fissures, reducing the cohesion between the soil particles and decrease the anti-sliding force between the slope surfaces (Lee et al., 2017b). By collecting the rainfall observation data of the study area from 2007 to 2017, the mean annual precipitation map was generated and divided into five sections according to an interval of 100 mm (Fig. 3(f)). Similarly, river erosion makes the slope unstable by making the soil loose. By establishing a five-level buffer zone with an interval of 300 m, that is, the distance to rivers, it reflects the impact of the river of road construction will impact the slope stability (Ciurleo et al., 2017); thus, the distance to roads was used to reflect the influence of road construction on the landslides occurrence by making a five-level buffer zone with an interval of 500 m (Fig. 3(h)).

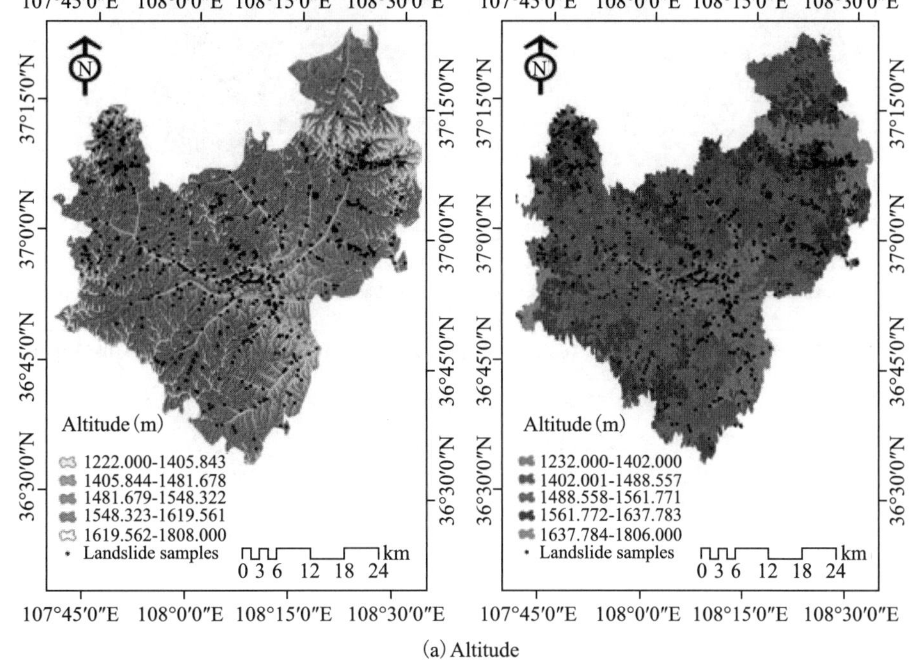

(a) Altitude

Fig. 3 Landslide Predisposing Factor Maps Based on GCMU and TMU of the Study Area

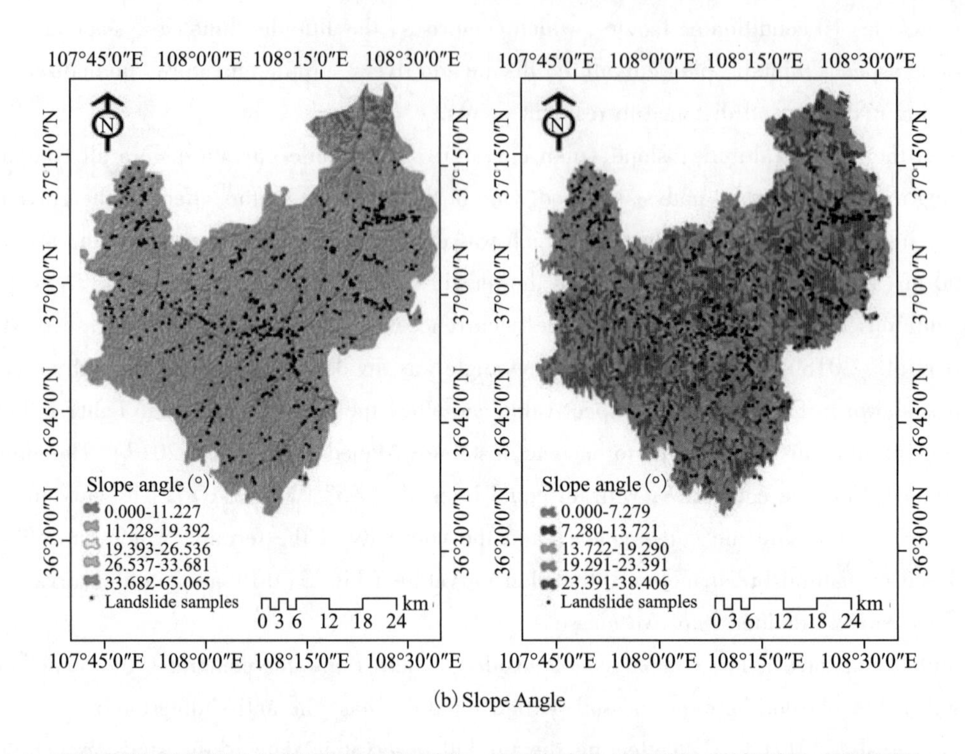

(b) Slope Angle

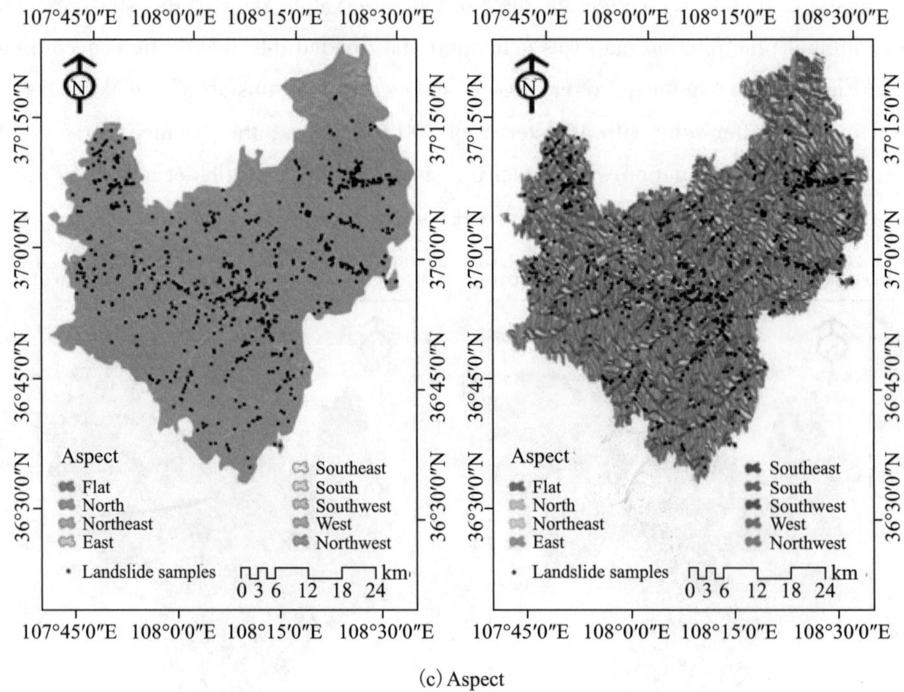

(c) Aspect

Fig. 3 (continued)

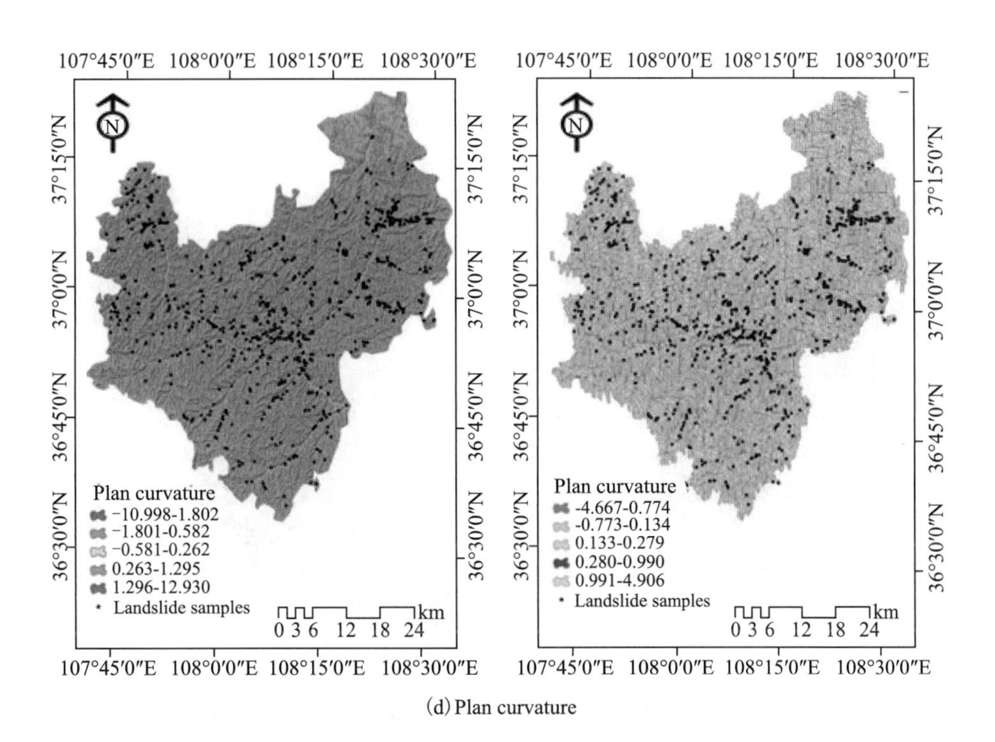

(d) Plan curvature

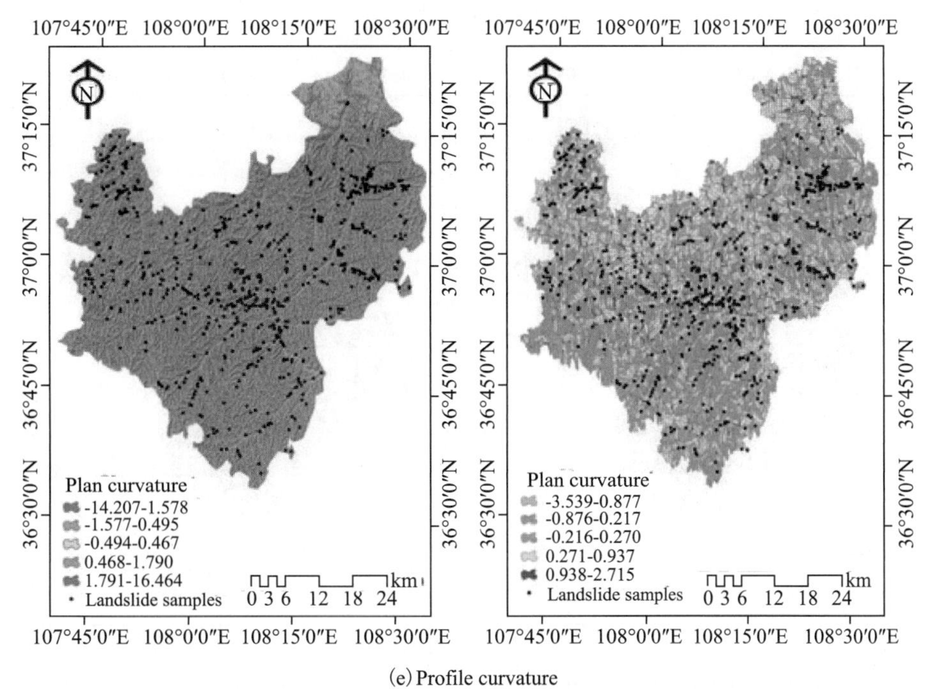

(e) Profile curvature

Fig. 3 (**continued**)

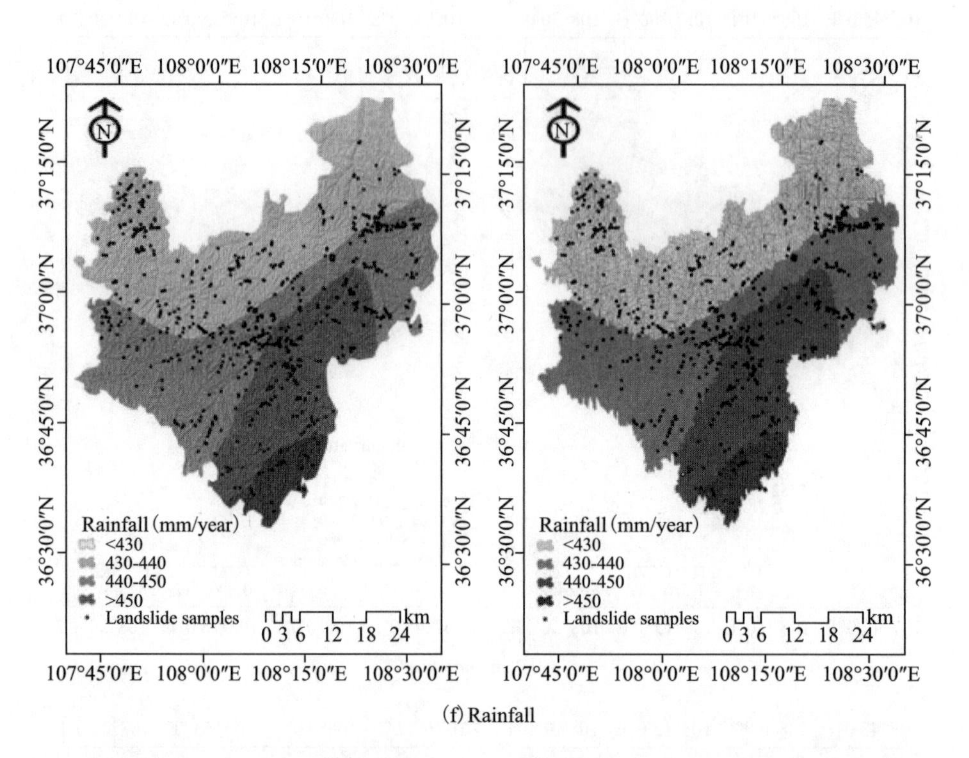

(f) Rainfall

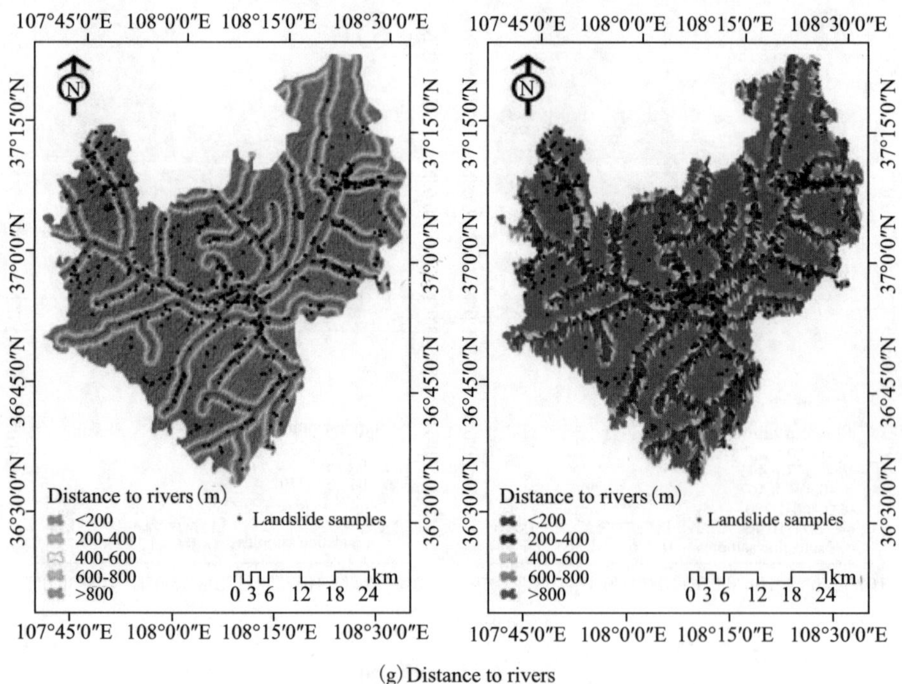

(g) Distance to rivers

Fig. 3 (**continued**)

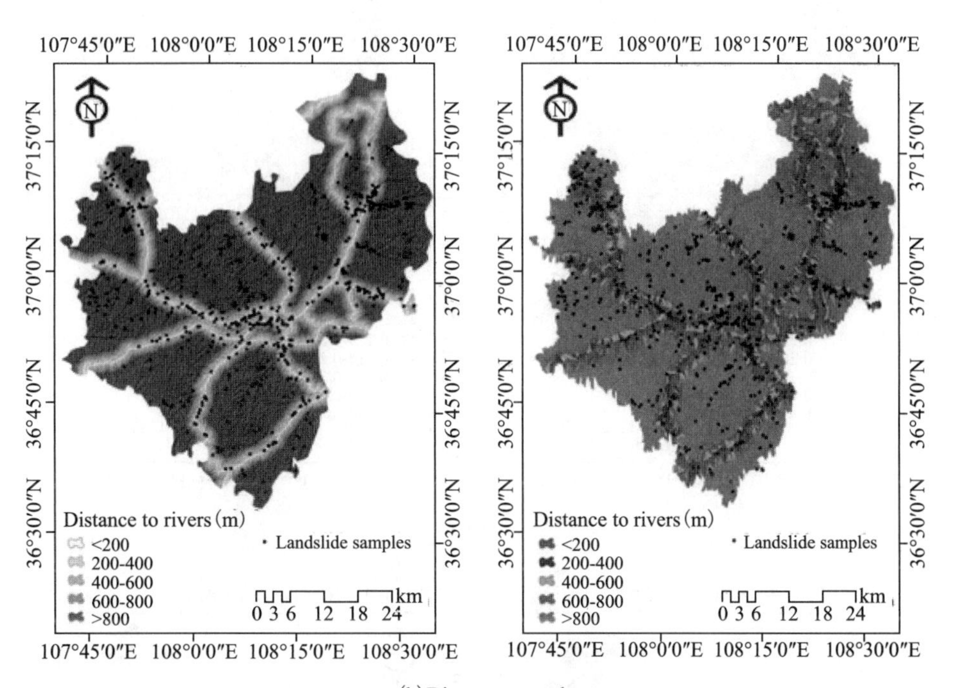

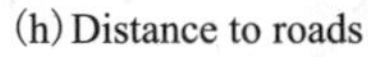

(h) Distance to roads

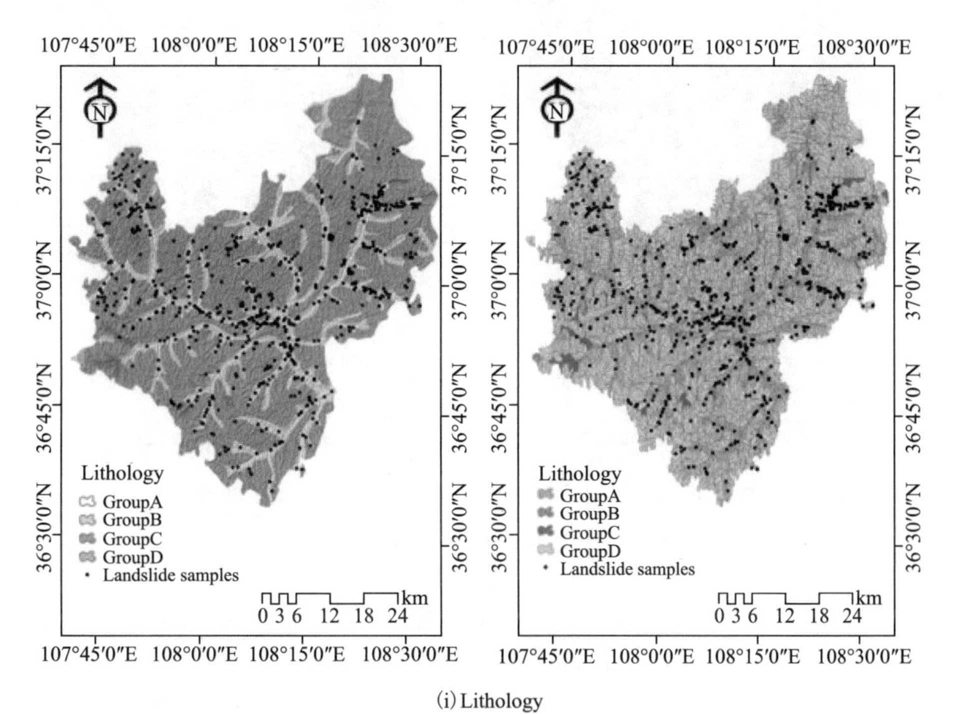

(i) Lithology

Fig. 3 **(continued)**

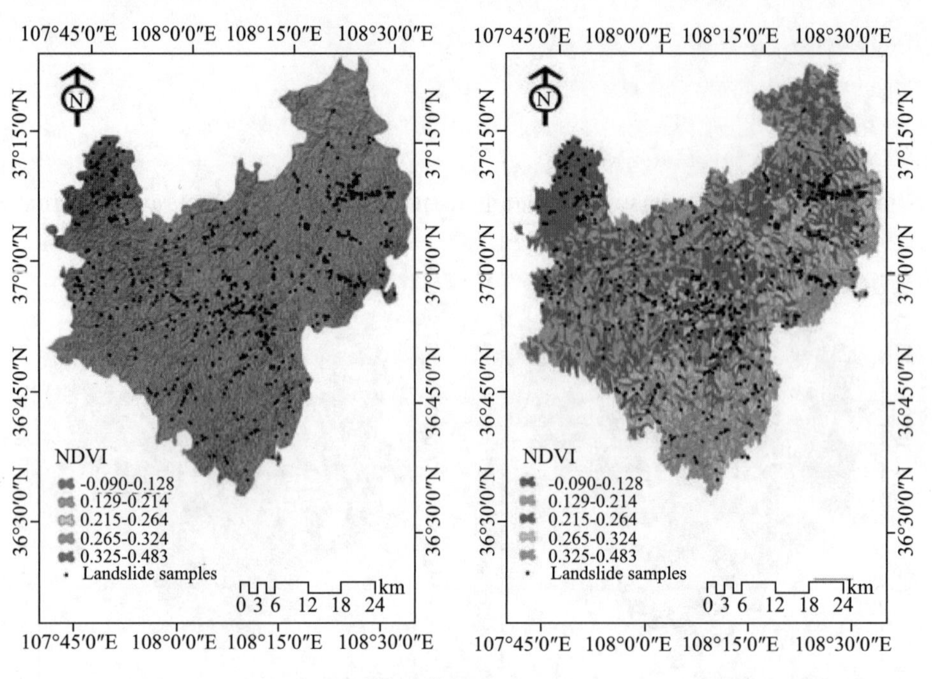

(j) NDVI

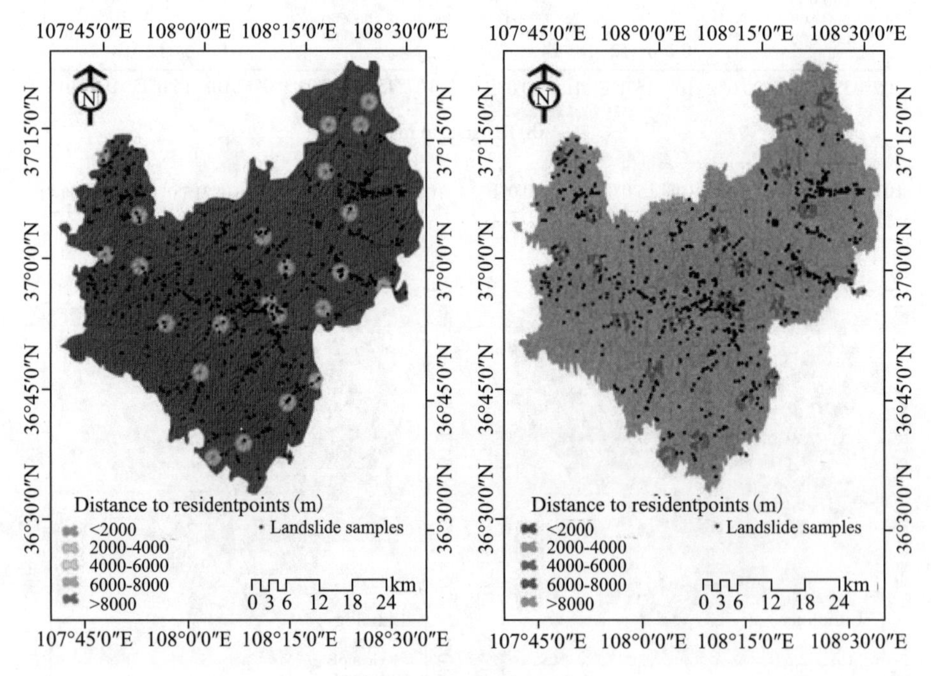

(k) Distance to resident points

Fig. 3 (continued)

Lithology is considered as an internal factor for landslide occurrence. In general, the higher the shale composition contained in the stratum, the worse its ability to resist destruction (Vakhshoori and Zare, 2017). The lithology information of the study area was generated from a 1:200 000 geological map and was grouped into five classes (Fig. 3(i)). The NDVI was extracted from GF-2 aerial photography and was then divided into five classes though the natural break method (Fig. 3(j)). The distance to resident points was generated based on the density map of the residential area offered by the local government and was separated into a five-level buffer zone with an interval of 2000 m (Fig. 3(k)). Finally, the resolution of all landslide predisposing factor maps was specified as 30 m.

3 Methodologies

The methodology followed in this research mainly comprised the following steps:1) according to the collected environmental and geological data of the study area, the mapping units (GCMU and TMU) were respectively prepared; 2) the IOE model wasused to quantify each landslide predisposing factors based onTMU and GCMU. The optimization of landslide predisposing factors was completed by calculating the information gain ratio,and then dataset1 based on TMU and $dataset_2$ based on GCMU were constructed; 3) $dataset_1$ and $dataset_2$ were respectively used as the input data for the SysFor model to construct a hybrid model, hereinafter referred to as the IOE-SysFor model. Subsequently, the IOE-SysFor model was employed to form the landslide susceptibility maps based on TMU and GCMU; 4) the results were then validated using statistical indices. Furthermore, the results were compared using the receiver operating characteristic curve (ROC) (Fig. 4).

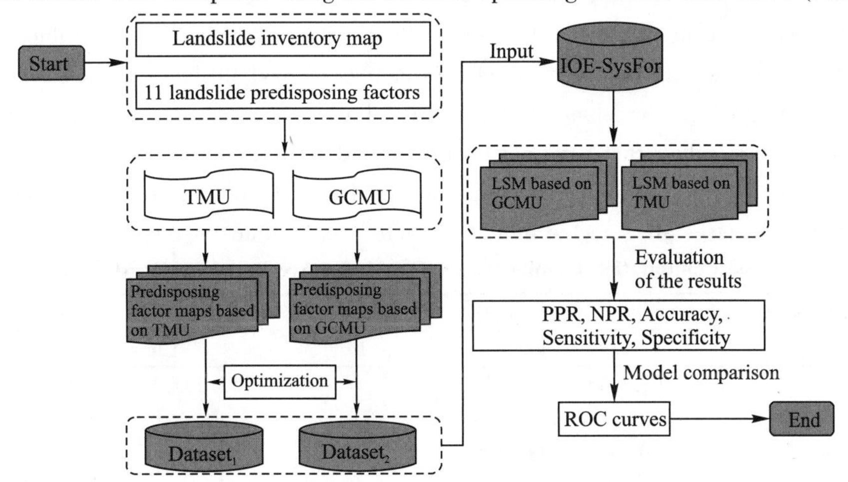

Fig. 4 The Workflow of the Present Study

3.1 Evaluation Methods of Landslide Predisposing Factors

The multicollinearity problem is caused by the highly correlated relationship between landslide predisposing factors, which may cause the landslide susceptibility model to be out of alignment (Mahdadi et al., 2018). In this study, by calculating the variance inflation factors (VIFs) and tolerances between the landslide predisposing factors, the potential multicollinearity can be detected. In general, if VIF>10 and corresponding tolerances<0.1, this indicates that there is potential multicollinearity among the predisposing factors (Moosavi and Niazi, 2016). Consequently, the important predisposing factors are retained, and the secondary predisposing factors are abandoned. On the other hand, the contribution of the landslide predisposing factors to the IOE-SysFor model was measured by computing the information gain ratio (IGR), whose main purpose was to optimize the predisposing factors. Similarly, the predisposing factors are excluded from landslide susceptibility modeling when IGR<0.5.

3.2 Description of the Index of Entropy Model

In current research, as a statistical model, the IOE model was employed to product the input data of the presented hybrid model. The following formulas present the pivotal computation processes of the IOE model:

$$FR_{ij} = \frac{b_{ij}}{a_{ij}} \tag{1}$$

$$P_{ij} = \frac{FR_{ij}}{\sum_{j-1}^{N_j} FR_{ij}} \tag{2}$$

$$H_j = -\sum_{j-1}^{N_j} p_{ij} log_2 p_{ij}, j = 1,2,3,\cdots,n \tag{3}$$

Where a_{ij} and b_{ij} respectively stand for the percentage of domain and landslide, and FR_{ij} is the frequency ratio value of landslides. N_j presents the classification number of landslides predisposing factors, and P_{ij} means the probability density. The entropy of the jth predisposing factor is expressed as H_j(Chen et al., 2018).

3.3 Description of the Systematically Developed Forest of Multiple Decision Trees

Systematically developed forest of multiple decision trees (SysFor) is proposed by Islam & Giggins in 2011 (Islam and Giggins, 2011). The principle of this new data mining method is based on the concept described by gain ratio, and the main purpose of this method is to establish the mlutiple tree to obtain better knowledge by extracting a number of patterns. According to the framework of SysFor, it can be roughly introduced by three steps.

At the beginning, for identifying their segmentation points and a group of good attributes, according to separation values and gain ratio defined by the user. A numeric attribute can be selected multiple times in a good group of attributes when different segmentation points had a certain distance from each other and gain ratio had a higher value. After identification of the good attributes set, if the number of tree defined by the user is bigger than the size of the good attributes, SysFor will construct the tree by making each good attribute serve as the root character of the tree. The tree was then built as much as number of good attributes to the second step of SysFor.

The voting system proposed by Islam and Giggins is followed after generating the multi-tree forest of the system to predict the class values that have not been recorded. In this voting system, the leaves come from all trees which carry the record are all a maximum number of records to total number of records ratio. Finally, most class values of leaves are selected as the predicted class values of records.

Table 1 The Computation Results of the IOE Model Are Based on GCMU and TMU

Landside predisposing factors	GCMU		TMU	
	Classes	H_j	Classes	H_j
Altitude(m)	1222.000~1405.843	1.841	1232.000~1402.000	2.059
	1405.844~1481.678		1402.001~1488.557	
	1481.679~1548.322		1488.558~1561.771	
	1548.323~1619.561		1561.772~1637.783	
	1619.562~1808.000		1637.784~1806.000	
Slope angle (°)	0.000~11.227	2.307	0.000~7.279	1.906
	11.228~19.392		7.280~13.721	
	19.393~26.536		13.722~19.290	
	26.537~33.681		19.291~23.391	
	33.682~65.065		23.392~38.406	
Aspect	Flat	2.942	Flat	2.833
	North		North	
	Northeast		Northeast	
	East		East	
	Southeast		Southeast	
	South		South	
	Southwest		Southwest	
	West		West	
	Northwest		Northwest	
Plan curvature	-10.998~-1.802	2.312	-4.667~-0.774	1.309
	-1.801~-0.582		-0.773~-0.134	
	-0.581~0.262		0.133~0.279	
	0.263~1.295		0.280~0.990	
	1.296~12.930		0.991~4.906	
Profile curvature	-14.207~-1.578	2.291	-3.539~-0.877	1.321
	-1.577~-0.495		-0.876~-0.217	
	-0.494~0.467		-0.216~0.27	
	0.468~1.790		0.271~0.937	
	1.791~16.464		0.938~2.715	

continued to Table 1

Landside predisposing factors	GCMU		TMU	
	Classes	H_j	Classes	H_j
Distance to rivers (m)	< 200	1.964	<200	2.144
	200~400		200~400	
	400~600		400~600	
	600~800		600~800	
	>800		>800	
Distance to roads (m)	<200	3.375	<200	3.641
	200~400		200~400	
	400~600		400~600	
	600~800		600~800	
	>800		>800	
Distance to resident points (m)	<2000	2.227	<2000	.180
	2000~4000		2000~4000	
	4000~6000		4000~6000	
	6000~8000		6000~8000	
	>8000		>8000	
NDVI	-0.203~0.178	2.308	-0.090~0.128	2.272
	0.179~0.229		0.129~0.214	
	0.230~0.289		0.215~0.264	
	0.290~0.364		0.265~0.324	
	0.365~0.562		0.325~0.483	
Lithology	Group (A)	1.678	Group (A)	1.350
	Group (B)		Group (B)	
	Group (C)		Group (C)	
	Group (D)		Group (D)	
Rainfall (mm/year)	<430	1.958	<430	1.555
	430~440		430~440	
	440~450		440~450	
	>450		>450	

3.4 Validation and Comparison Method of Results

In current research, five statistical indices, namely accuracy, sensitivity, positive prediction rate (PPR), specificity, and negative prediction rate (NPR), were applied to validate the results of LSM. These statistical indices were obtained by true positive (TP), true negative (TN), false positive (FP), and false negative (FN), where TP is the number of landslides that are correctly classified, TN is the number of non-landslides that are correctly classified, FP is the number of non-landslides that are incorrectly classified, and FN is the number of landslides that are incorrectly classified (Pham et al., 2018):

$$\text{Positive prediction rate}=\frac{\text{TP}}{\text{TP+FP}} \tag{4}$$

$$\text{Negative prediction rate}=\frac{\text{TN}}{\text{TN+FN}} \tag{5}$$

$$\text{Sensitivity}=\frac{\text{TP}}{\text{TP-FN}} \tag{6}$$

$$\text{Specificity}=\frac{\text{TN}}{\text{TN+FP}} \tag{7}$$

$$\text{Accuracy}=\frac{\text{TP+TN}}{\text{TP+TN+FP+FN}} \tag{8}$$

The ROC curve was then adopted to compare the classification ability of the landslide susceptibility model in this study. The ROC curve is a curve drawn with 1-specificity as the x-axis and sensitivity as the y-axis based on obtaining multiple pairs of sensitivity and specificity values. If the ROC curve is closer to the upper left corner, that is to say, the area under the curve (AUC) is closer to 1, it indicates that the landslide susceptibility model has better classification ability (Pourghasemi and Rossi, 2016). The following is the calculation method of AUC: where *P* stands for the total number of landslides and *N* is the total number of non-landslides.

$$\text{AUC}=\frac{\sum \text{TP}+\sum \text{TN}}{\text{P+N}}. \tag{9}$$

Where *P* stands for the total number of landslides and *N* is the total number of non-landslides.

4 Results

4.1 Quantification of Landslide Predisposing Factors

In current research, the landslide predisposing factors were respectively quantified using IOE model on TMU and GCMU. Table 1 shows the quantitative results of the IOE model. In the case of GCMU, the distance to roads had the highest entropy value ($H_j=3.375$). The entropy value of lithology was the lowest ($H_j=1.678$), indicating that the lithology had the weakest correlation with landslides. The biggest entropy value ($H_j=3.641$) of landslide predisposing factors based on TMU belonged to the distance to roads. In contrast, the smallest entropy value ($H_j=1.309$) belonged to plan curvature. Subsequently, all landslide predisposing factors quantified by the IOE model on GCMU and TMU were used as the input data to complete the next calculation.

4.2 Optimization Results of Landslide Predisposing Factors

According to optimization techniques mentioned before, the calculation results of VIF and tolerances of each landslide predisposing factors based on GCMU and TMU were exhibited in Table 2. The results clearly showed that the maximum VIF and the minimum VIF based on GCMU were distance to resident points and plan curvature respectively. Besides, the maximum VIF and the minimum VIF based on TMU were outside the specified section(VIF>10, tolerances<0.1), therefore the calculation of IGR involves all landslide predisposing factors.

Table 2　Calculation Results of VIF and Tolerances of Each Landslide Predisposing Factors Based on GCMU and TMU

Landslide predisposing factors	GCMU ($dataset_1$)		TMU ($dataset_2$)	
	VIF	Tolerances	VIF	Tolerances
Altitude	1.293	0.773	1.167	0.857
Plan curvature	1.028	0.973	1.232	0.812
Profile curvature	1.126	0.888	1.080	0.926
Slope angle	1.032	0.969	1.141	0.876
Aspect	1.268	0.789	1.084	0.923
Rainfall	1.044	0.958	1.096	0.912
Lithology	1.103	0.906	1.048	0.954
Distance to rivers	1.148	0.871	1.327	0.754
NDVI	1.078	0.928	1.112	0.899
Distance to resident points	1.323	0.756	1.297	0.771
Distance to roads	1.174	0.852	1.086	0.921

The IGR of landslide predisposing factors based on GCMU and TMU were shown in Fig. 5. For TMU, the distance to roads had the highest IGR value (IGR=0.897), followed by the slope angle (IGR=0.881) and the NDVI (IGR = 0.780), and the lowest IGR value belonged to the plan curvature (IGR=0.508). For GCMU, the largest IGR value belonged to the slope angle (IGR=0.873), followed by the distance to roads (IGR = 0.862) and the altitude (IGR = 0.786). In contrast, the distance to resident points had the smallest IGR value (IGR = 0.408). In other words, the factor map of distance to resident points generated from GCMU had no contribution to landslide susceptibility modeling, which need to be abandoned. Finally, all landslide predisposing factors generated from TMU were used to construct the $dataset_1$, and the remaining ten landslide predisposing factors generated from GCMU, excepting distance to resident points, were used to construct the $dataset_2$.

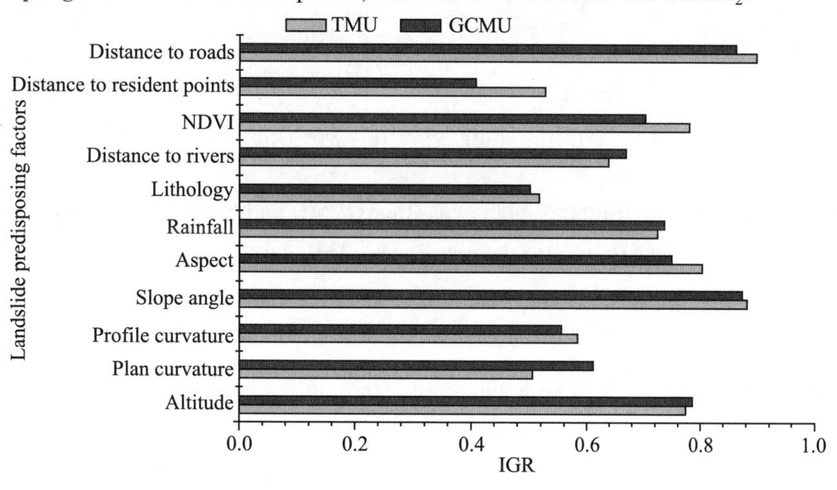

Fig. 5　The IGR of landslide Predisposing Factors Based on GCMU and TMU

4.3 Implementation of Hybrid Model

In this study, the process of using IOE-SysFor model to implement landslide sceptility research can be briefly divided into foursteps. For the first, building the tree and checking whether the tree graph whether or not meets the expectation. Then, calculating the entropy function of the given dataset ($dataset_1$ and $dataset_2$) and delimit the dataset according to the given characteristics. Next, choosing the best way to divide the dataset and training the algorithm. Finally, using the established tree to classify the results (landslide or non-landslide). After above four steps, the landslide susceptibility index (LSI) calculating by IOE-SysFor model based on GCMU and TMU was respectively obtained.

The LSI represented the probability degree of landslide occurrence, which the output range was from 0.0001 to 0.9999. The closer the value of LSI was to 0.9999, the greater the probability that a landslide will occur in that area and vice versa. By using natural break method, the LSI obtained based on GCMU were divided into five ranges as follows: 0.0015~0.2404, 0.2405~0.3931, 0.3932~0.5615, 0.5616~0.7494, and 0.7495~0.9674, which were then visualized to generate the landslide susceptibility map with corresponding five regions, such as very low area, low area, moderate area, high area and very high area (Fig. 6 (a)). Similarly, The LSI obtained based on TMU were divided into five ranges: 0.0145~0.2459, 0.2460~0.3695, 0.3696~0.5161, 0.5162~0.6974, and 0.6975~0.9983, which were also visualized to generate the landslide susceptibility map with corresponding five regions, such as very low area, low area, moderate area, high area and very high area (Fig. 6(b)).

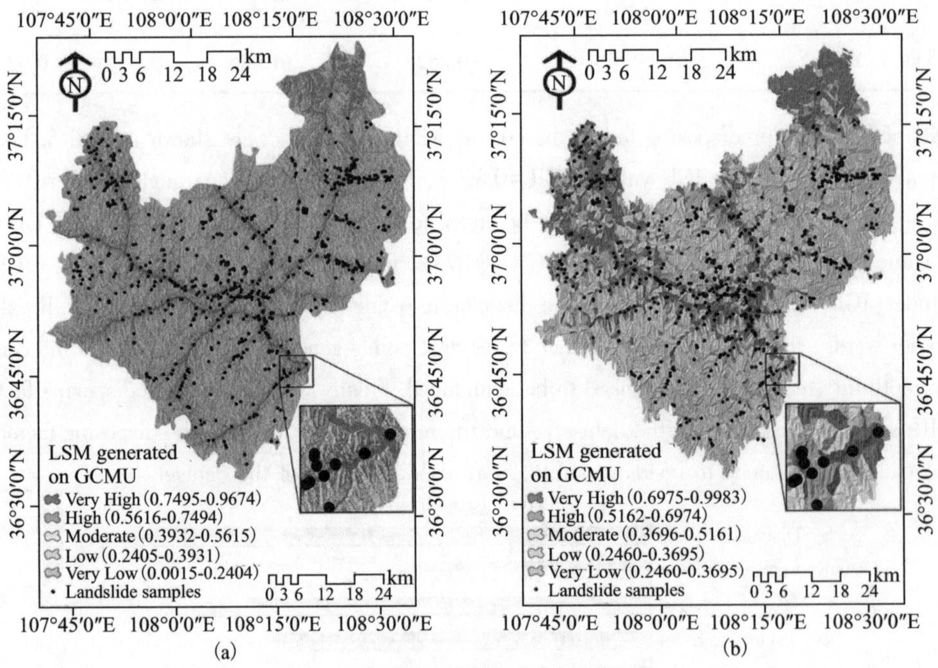

Fig. 6 Landslide Susceptibility Maps Based on GCMU and TMU of the Study Area: (a) GCMU ($dataset_1$), (b) TMU ($dataset_2$)

4.4 Model Evaluation and Comparison

4.4.1 Model Evaluation

Model evaluation is a pivotally final step among the study in landslide susceptibility. Without it, the accuracy of the research will be hard to determined, and it will exist no scientific meaning. Therefore, the performance of the hybrid model on two mapping units was evaluated by calculating the five aforementioned statistical indices based on the training datasets from $dataset_1$ and $dataset_2$ (Table 3). In the view of the rsuts on the TMU, it was found that the PPR, NPR, and the accuracy were 0.829, 0.901, and 0.862, respectively, furthermore the sensitivity and specificity were 0.906 and 0.820, respectively, which demonstrates a possibility of high sensitivity for landslides and non-landslides. While, the PPR, NPR, and the accuracy achieved from GCMU were 0.841, 0.835, and 0.838, respectively, and the sensitivity and specificity were 0.841 and 0.835, respectively.

Table 4 shows the results of the model validation of datasets from $dataset_1$ and $dataset_2$. The models on two mapping units were found to have an accuracy of more than 0.700, whichindicates a high degree of confidence (Chen and Li, 2020; Huang et al., 2020). Moreover, the values of PPR, NPR, sensitivity, and specificity on the TMU were higher than those on the GCMU.

Table 3 The Classification Results of the Training Data

Statistical indices	$Dataset_1$ prepared on GCMU	$Dataset_2$ prepared on TMU
TP	430	445
TN	410	419
FP	81	92
FN	81	46
PPR	0.841	0.829
NPR	0.835	0.901
ACC	0.838	0.862
Sensitivity	0.841	0.906
Specificity	0.835	0.820

Table 4 The Classification Results of the Validation Data

Statistical indices	$Dataset_1$ prepared on GCMU	$Dataset_2$ prepared on TMU
TP	159	167
TN	147	153
FP	67	54
FN	59	58
PPR	0.704	0.756
NPR	0.714	0.725
ACC	0.708	0.741
Sensitivity	0.729	0.742
Specificity	0.687	0.739

4.4.2 Result Comparison

The results of the model comparison were concluded into two categories, where the ROC curves, shown in Fig. 7(a), that were produced using the training datasets based on TMU and GCMU showed that the AUC values were 0.876 and 0.827, respectively. In addition, the AUC value obtained using the validation datasets based on TMU (AUC = 0.741) was found significantly higher than that obtained based on GCMU

(AUC=0.708), as shown in Fig. 7(b). In other words, the IOE-SysFor performed better on the TMU than on the GCMU.

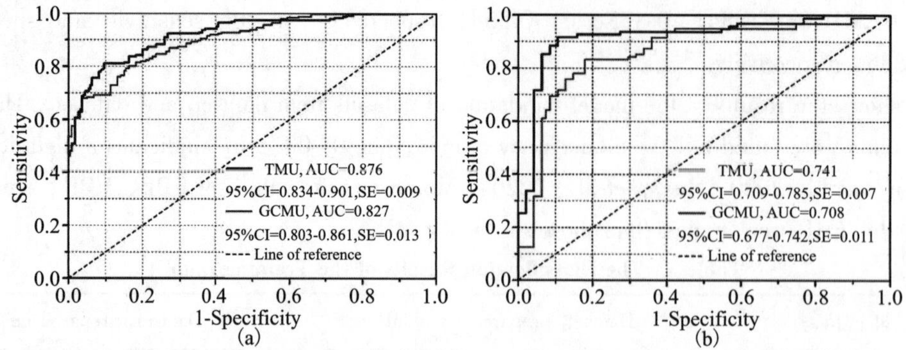

Fig.7 The ROC Curves and the Results of Model Comparison Based on GCMU and TMU: (a) Training Datasets, (b) Validation datesets

5 Discussion

In current research, a hybrid classification model was constructed by combining the IOE model and SysFor model, which was then used to generate the landslide susceptibility maps of the selected study area based on the TMU and GCMU. The results of this study showed a high degree of acceptance according to analysis results of statistical indices and ROC curves. Furthermore, the results generated based on the TMU were significantly more accurate than GCMU.

The relationship between landslide occurrence and landslide predisposing factors can be deeply revealed though entropy, as carried out in many previous studies (Mondal and Mandal, 2017; Siahkamari et al., 2017; Perera et al., 2019). From the computation results of the entropy value and IGR, it is clear that the entropy value and IGR of distance to roads based on GCMU and TMU were the highest, where this phenomenon could be reasonably explained as caused by human activities and the specific structure of the Loess. In recent years, lots of roads have been built because of the transportation of coal resources, which resulted in a large amount of excavation at the foot of the slope. Besides, the structure of the Loess in the study area is relatively loose and has certain collapsibility, which can improve the probability of landslide occurrence. These results are consistent with the results of other previous studies that took place in other study areas of similar conditions (Lei et al., 2020; Zhao and Chen, 2020).

Generally, hybrid models are increasingly applied in landslide susceptibility researches, where the results of LSM using the hybrid models were found effective and reliable (Aghdam et al., 2017; Bui et al., 2017). Literature reviews showed that a single statistical model or a single machine learning model has some inevitable drawbacks. For example, the statistical model, such as the statistical index model (SI), certainty factors model, and index of entropy model (IOE), can evaluate the influence degree of landslide predisposing factors on the landslide occurrence, but they cannot reflect the correlation between various predisposing factors (Reichenbach et al., 2018). On the other hand, the machine learning model can consider the correlation between various predisposing factors, but cannot reveal the predictive ability of various predisposing factors for LSM (Micheletti et al., 2014). Therefore, the IOE model was used to combine with SysFor model to generate the IOE-SysFor model for the accuracy improvement of LSM. On the other hand, SysFor belong to rare applications in the research field of LSM, which implicates that the purpose of this study is not only to draw an accurate LSM for study area, but also to apply new mathematical models in this field.

Although GCMU is a more common and easier to understand form in LSM and has been widely used to construct landslide susceptibility models, the results of LSM based on GCMU are susceptible to input data accuracy (landslide inventory map and landslide predisposing factors). In addition, each cell of GCMU has a uniform shape resulting in a lack of spatial connection between landslide predisposing factors and GCMU, which is not conducive to the comprehensive performance of conditions that lead to the occurrence of landslides. On the contrary, TMU is a hydrological unit that divided areas by drainage lines and divide lines, and there are some researches on LSM using the TMU (Guzzetti et al., 2006; Camilo et al., 2017). TMU performs block statistics on the data based on GCMU, the pixel values contained in the evaluation unit of TMU will be unified in the statistics. Thus, it can better remove the abnormalities in the data and can more accurately present the terrain characteristics. Moreover, LSM based on the TMU can avoid the disadvantages of large computational complexity.

The landslide susceptibility map contains 5 levels of areas, which are dividing using the natural break method in ArcGIS. By analyzing relevant literatures and researches, it can be seen that the main methods for classification of landslide susceptibility map are quantile method, natural break method, equal interval, and standard deviation method. Since the natural break method has the function of automatically searching for breakpoints, the difference between different categories can be maximized, so it is widely used in landslide research. Under this background, the natural break method can be used to achieve reasonable results, and these results also have a positive effect on the control and prevention of landslide in the study area.

6 Conclusions

This study integrated the index of entropy model (IOE) with the systematically developed forest of multiple trees (SysFor) and constructed a novel hybrid model (IOE-SysFor) to generate the landslide susceptibility maps based on terrain mapping unit (TMU) and grid cell mapping unit (GCMU) in Yan'an City, Shaanxi province, China. By computing the information gain ratio (IGR) of the landslide predisposing factors, two datasets ($dataset_1$ and $dataset_2$) based on the TMU and GCMU were established. Then the $dataset_1$ and $dataset_2$ were used as the input data to construct IOE-SysFor model, and finally, the landslide susceptibility maps based on TMU and GCMU were obtained. Five statistical indices: sensitivity, specificity, positive prediction rate (PPR), negative prediction rate (NPR), and accuracy were employed for the assessment of the results, while the area under the ROC curve (AUC) was used to compare the performance of the presented hybrid model on different mapping units. In conclusion, the accuracy of landslide susceptibility maps based on GCMU and TMU was greater than 0.700, which has high credibility and can provide references for local decision-makers. Besides, the performance of the novel hybrid model on the TMU (AUC=0.741) was found better than its performance on the GCMU (AUC=0.708), indicating that TMU may be more suitable than GCMU for the IOE-SysFor model as a basic mapping unit in Loess areas.

All in all, the methods and ideas used in this study are instructive for future relevant research. In addition, the results obtained in this paper can not only improve the response speed of landslide prevention in Wuqi County, but also serve as technical support for the construction of geological disaster prevention in the region.

Acknowledgments

This study is funded by Fundamental Research Funds for the Central Universities (300102351502) and Coordenação de Aperfeiçoamento de Pessoal de Nível Superior Brasil (CAPES) (Finance Code 001).

Nomenclature

a_{ij} = Percentage of domain

b_{ij} = Percentage of the landslide

H_j = The entropy of the jth predisposing factor

N_j = The classification number of landslides predisposing factors

P_{ij} = Probability density

References

Aghdam I N, Pradhan B, Ranahi M(2017) Landslide susceptibilityassessment using a novel hybrid model of statistical bivariate methods (FR and WOE) and adaptive neuro-fuzzy inference system (ANFIS) at southern Zagros Mountains in Iran. *Environmental Earth Sciences* 76:237-255.

Ahmed B, Dewan A (2017) Application of bivariate and multivariate statistical techniques in landslide susceptibility modeling in Chittagong city corporation, Bangladesh. *Remote Sensing* 9:304-311.

Arabameri A, Pradhan B, Rezaei K, Sohrabi M, Kalantari Z (2018) GIS-based landslide susceptibility mapping using numerical risk factor bivariate model and its ensemble with linear multivariate regression and boosted regression tree algorithms. *Journal of Mountain Science* 16:595-618.

Ballabio C, Sterlacchini S (2012) Support vector machines for landslide susceptibility mapping: The Staffora river basin case study, Italy. *Mathematical Geosciences* 44:47-70.

Behnia P, Blais-Stevens A (2018) Landslide susceptibility modelling using the quantitative random forest method along the northern portion of the Yukon Alaska Highway Corridor, Canada. *Natural Hazards* 90:1407-1426.

Broeckx J, Vanmaercke M, Duchateau R, Poesen J (2018) A data-based landslide susceptibility map of Africa. *Earth-Science Reviews* 185: 102-121.

Bui D T, Shahabi H, Shirzadi A, Chapi K (2018) Landslide detection and susceptibility mapping by AIRSAR data using support vector machine and index of entropy models in Cameron Highlands, Malaysia. *Remote Sensing* 10:1527-1533.

Bui D T, Tuan T A, Hoang N D, Thanh N Q, Nguyen D B (2017) Spatial prediction of rainfall-induced landslides for the Lao Cai area (Vietnam) using a hybrid intelligent approach of least squares support vector machines inference model and artificial bee colony optimization. *Landslides* 14:447-458.

Bui D T, Tuan T A, Klempe H, Pradhan B, Revhaug I (2016) Spatial prediction models for shallow landslide hazards: A comparative assessment of the efficacy of support vector machines, artificial neural networks, kernel logistic regression, and logistic model tree. *Landslides* 13:361-378.

Bullock A, King B (2011) Evaluating China's Slope land conversion program as sustainable management in Tianquan and Wuqi Counties. *Journal of Environmental Management* 92:1916-1922.

Camilo D C, Lombardo L, Mai P M, Dou J, Huser R (2017) Handling high predictor dimensionality in slope-unit-based landslide susceptibility models through LASSO-penalized generalized linear model. *Environmental Modelling & Software* 97: 145-156.

Carrara A, Cardinali M, Guzzetti F, Reichenbach P (1995) GIS technology in mapping landslide hazard. *Geographical Information Systems in Assessing Natural Hazards* 5:135-175.

Chen W, Li Y (2020) GIS-based evaluation of landslide susceptibility using hybrid computational intelligence models. *Catena* 195:104-107.

Chen W, Panahi M, Tsangaratos P, Shahabi H, Ilia I (2019a) Applying population-based evolutionary algorithms and a neuro-fuzzy system for modeling landslide susceptibility. *Catena* 172:212-231.

Chen W, Shahabi H, Shirzadi A, Hong H (2018) Novel hybrid artificial intelligence approach of bivariate statistical-methods-based kernel logistic regression classifier for landslide susceptibility modeling. *Bulletin of Engineering Geology and the Environment* 78:4397-4419.

Chen W, Sun Z, Han J (2019b) Landslide susceptibility modeling using integrated ensemble weights of evidence with logistic regression and random forest models. *Applied Siences* 9:171-192.

Chen W, Xie X, Wang J, Pradhan B, Hong H (2017) A comparative study of logistic model tree, random forest, and classification and regression tree models for spatial prediction of landslide susceptibility. *Catena* 151:147-160.

Chen W, Zhao X, Shahabi H, Shirzadi A (2019c) Spatial prediction of landslide susceptibility by combining evidential belief function, logistic regression and logistic model tree. *Geocarto International* 34: 1177-1201.

China MoNRotPsRo (2020) Statistics on natural disasters in China in 2020. China MoNRotPsRo, Retrieved September 15, 2020.

Ciurleo M, Cascini L, Calvello M (2017) A comparison of statistical and deterministic methods for shallow landslide susceptibility zoning in clayey soils. *Engineering Geology* 223:71-81.

Cloud G D (2020) GDEMV2 30M resolution digital elevation data. Geospatial Data Cloud, August 10, 2020.

Erener A, Düzgün H S B (2012) Landslide susceptibility assessment: What are the effects of mapping unit and mapping method? *Environmental Earth Sciences* 66:859-877.

Fan W, Wei X, Cao Y, Zheng B (2017) Landslide susceptibility assessment using the certainty factor and analytic hierarchy process. *Journal of Mountain Science* 14:906-925.

Government WCPs (2020) Basic information of Wuqi County. Government WCPs, Retrieved September 22, 2020.

Guzzetti F, Reichenbach P, Ardizzone F, Cardinali M, Galli M (2006) Estimating the quality of landslide susceptibility models. *Geomorphology* 81:166-184.

Hong H, Liu J, Zhu A X, Shahabi H (2017) A novel hybrid integration model using support vector machines and random subspace for weather-triggered landslide susceptibility assessment in the Wuning area (China). *Environmental Earth Sciences* 76:652-667.

Huang F, Cao Z, Guo J, Jiang S H, Li S, Guo Z (2020) Comparisons of heuristic, general statistical and machine learning models for landslide susceptibility prediction and mapping. *Catena* 191:104580.

Ilia I, Tsangaratos P (2016) Applying weight of evidence method and sensitivity analysis to produce a landslide susceptibility map. *Landslides* 13:379-397.

Islam M Z, Giggins H (2011) Knowledge discovery through SysFor -A systematically developed forest of multiple decision trees. Proceedings of the 9th Australasian data mining conference, December 15-16, Ballarat, Australia.

Kadavi P R, Lee C W, Lee S (2018) Application of ensemble-based machine learning models to landslide susceptibility mapping. *Remote Sensing* 10:1252-1269.

Kalantar B, Pradhan B, Naghibi S A, Motevalli A, Mansor S (2017) Assessment of the effects of training data selection on the landslide susceptibility mapping: A comparison between support vector machine (SVM), logistic regression (LR) and artificial neural networks (ANN). *Geomatics, Natural Hazards and Risk* 9:49-69.

Kumar D, Thakur M, Dubey S C, Shukla P D (2017) Landslide susceptibility mapping & prediction using support vector machine for Mandakini River basin, Garhwal Himalaya, India. *Geomorphology* 295:115-125.

Lee S, Hong S M, Jung H S (2017a) A support vector machine for landslide susceptibility mapping in Gangwon province, Korea. *Sustainability* 9:48-55.

Lee J H, Sameen M I, Pradhan B (2017b) Modeling landslide susceptibility in data-scarce environments using optimized data mining and statistical methods. *Geomorphology* 303:284-298.

Lei X, Chen W, Pham B T (2020) Performance evaluation of GIS-based artificial intelligence approaches for landslide susceptibility modeling and spatial patterns analysis. *International Journal of Geo-Information* 9:443-451.

Liu J, Duan Z (2018) Quantitative assessment of landslide susceptibility comparing statistical index, index of entropy, and weights of evidence in the Shangnan Area, China. *Entropy* 20:868-884.

Lombardo L, Mai P M (2018) Presenting logistic regression-based landslide susceptibility results. *Engineering Geology* 244:14-24.

Mahdadi F, Boumezbeur A, Hadji R, Kanungo D P, Zahri F (2018) GIS-based landslide susceptibility assessment using statistical models: A case study from Souk Ahras province, N-E Algeria. *Arabian Journal of Geosciences* 11:476-488.

Micheletti N, Foresti L, Robert S, Leuenberger M, Pedrazzini A (2014) Machine learning feature selection methods for landslide susceptibility mapping. *Mathematical Geosciences* 46:33-57.

Moayedi H, Mehrabi M, Mosallanezhad M, Rashid A S A (2018) Modification of landslide susceptibility mapping using optimized PSO-ANN technique. *Engineering with Computers* 35:967-984.

Mondal S, Mandal S (2017) Landslide susceptibility mapping of Darjeeling Himalaya, India using index of entropy (IOE) model. *Applied Geomatics* 11:129-146.

Moosavi V, Niazi Y (2016) Development of hybrid wavelet packet-statistical models (WP-SM) for landslide susceptibility mapping. *Landslides* 13:97-114.

Nguyen P T, Tuyen T T, Shirzadi A, Pham B T (2019) Development of a novel hybrid intelligence approach for landslide spatial prediction. *Applied Siences* 9:2824-2833.

Nsengiyumva J B, Luo G, Nahayo L, Huang X, Cai P (2018) Landslide susceptibility assessment using spatial multi-criteria evaluation model in Rwanda. *International Journal of Environmental Research and Public Health* 15:243-255.

Park S, Hamm S Y, Kim J (2019) Performance evaluation of the GIS-based data-mining techniques decision tree, random forest, and rotation forest for landslide susceptibility modeling. *Sustainability* 11:5659-5702.

Perera E N C, Jayawardana D T, Jayasinghe P, Ranagalage M (2019) Landslide vulnerability assessment based on entropy method: A case study from Kegalle district, Sri Lanka. *Modeling Earth Systems and Environment* 5:1635-1649.

Pham B T, Bui D T, Prakash I (2018) Landslide susceptibility modelling using different advanced decision trees methods. *Civil Engineering and Environmental Systems* 35:139-157.

Pham B T, Khosravi K, Prakash I (2017) Application and comparison of decision tree-based machine learning methods in landside susceptibility assessment at Pauri Garhwal Area, Uttarakhand, India. *Environmental Processes* 4:711-730.

Pham B T, Prakash I, Singh S K, Shirzadi A, Shahabi H (2019) Landslide susceptibility modeling using reduced error pruning trees and different ensemble techniques: Hybrid machine learning approaches. *Catena* 175:203-218.

Polykretis C, Chalkias C (2018) Comparison and evaluation of landslide susceptibility maps obtained from weight of evidence, logistic regression, and artificial neural network models. *Natural Hazards* 93:249-274.

Pourghasemi H R, Kornejady A, Kerle N, Shabani F (2012) Investigating the effects of different landslide positioning techniques, landslide partitioning approaches, and presence-absence balances on landslide susceptibility mapping. *Catena* 187:104364.

Pourghasemi H R, Rahmati O (2018) Prediction of the landslide susceptibility: Which algorithm, which precision? *Catena* 162:177-192.

Pourghasemi H R, Rossi M (2016) Landslide susceptibility modeling in a landslide prone area in Mazandarn Province, north of Iran: A comparison between GLM, GAM, MARS, and M-AHP methods. *Theoretical and Applied Climatology* 130:609-633.

Razavizadeh S, Solaimani K, Massironi M, Kavian A (2017) Mapping landslide susceptibility with frequency ratio, statistical index, and weights of evidence models: A case study in northern Iran. *Environmental Earth Sciences* 76:499-512.

Reichenbach P, Rossi M, Malamud B D, Mihir M, Guzzetti F (2018) A review of statistically-based landslide susceptibility models. *Remote Sensing* 180:60-91.

Siahkamari S, Haghizadeh A, Zeinivand H, Tahmasebipour N (2017) Spatial prediction of flood-susceptible areas using frequency ratio and maximum entropy models. *Geocarto International* 33:927-941.

Steger S, Brenning A, Bell R, Glade T (2016) The propagation of inventory-based positional errors into statisticalland-slide susceptibility models. *Natural Hazards & Earth System Sciences Discussions* 16:2729-2745.

Vakhshoori V, Zare M (2017) Landslide susceptibility mapping by comparing weight of evidence, fuzzy logic, and frequency ratio methods. *Geomatics, Natural Hazards and Risk* 7:1731-1752.

Wu Y, Li W, Wang Q, Liu Q, Yang D (2016) Landslide susceptibility assessment using frequency ratio, statistical index and certainty factor models for the Gangu County, China. *Arabian Journal of Geosciences* 9:84-99.

Youssef AM, Al-Kathery M, Pradhan B (2015) Landslide susceptibility mapping at Al-Hasher area, Jizan (Saudi Arabia) using GIS-based frequency ratio and index of entropy models. *Geosciences Journal* 19:113-134.

Zhao X, Chen W (2020) Optimization of computational intelligence models for landslide susceptibility evaluation. *Remote Sensing* 12: 2180-2200.

本文曾发表于2022年《KSCE Journal of Civil Engineering》第18卷第3期

Quantification of Soil Erosion in Small Watersheds on the Loess Plateau Based on a Modified Soil Loss Model

Hui Kong[a], Dan Wu[b], Liangyan Yang[a]

(a.Institute of Land Engineering and Technology Shaanxi Provincial Land Engineering Construction Group Co., Ltd., Xi'an,710075, China;b.Key Laboratory of Degraded and Unused Land Consolidation Engineering, Ministry of Natural Resources., Xi'an 710075, China)

【Abstract】The technology of slope vegetation system stability enhancement is an important part of the comprehensive ecological security improvement in small watersheds area of Loess Plateau. The results of the comprehensive soil erosion improvement in the sub-basin are used to give an evaluation of the effectiveness of the gully slope vegetation restoration project. Soil erosion quantification distribution in the Sheep Sap Gully sub-basin of the Loess Plateau hilly gully area was simulated and explored by combining the modified universal soil loss equation RUSLE model with GIS and RS spatial information technology. The quantitative values of LS factor were extracted using DEM data, the R factor of rainfall erosion force was calculated using meteorological monitoring station data around this region, the K factor of erosion resistance was obtained based on the soil survey database, and the C factor and P factor of soil and water conservation measures were obtained by combining MODIS image data and previous research experience. The study concluded that (1)The erosion area ratio within the study area is 36.33%; (2)the quantitative grading standard of the pattern, the size of the erosion distribution area at all levels is ranked as light>moderate>very strong>strong>intense; (3)Under the conditions of landuse and vegetation cover, strong erosion is mostly found in farming areas with sparse vegetation, while weak erosion is found in areas with lush vegetation such as forests and grasslands; (4)In terms of spatial distribution, erosion is greater in the south-western part of the basin than in the north-eastern part, and there is also strong erosion in the south-western part. The results of the study provide a reference for the research on "integration and synthesis of ecological security technologies for gully and slope management projects". Research content provides the basis and support for watershed governance and soil and water resource management and conservation.

【Keywords】RUSLE model ; Remote sensing images; Watershed; Loess plateau; Soil erosion amount

1 Introduction

One of the modern measures to combat soil erosion, one of the biggest ecological problems in the world, is the implementation of reforestation and grass restoration projects[1]. By changing the pattern of the ecosystem in the area, the project has a positive effect on the restoration of vegetation, improving soil erosion and controlling soil erosion in the area. Since the project was implemented in August 1999, significant results have been achieved. In the 21st century, soil erosion research in China is facing new opportunities and challenges. Therefore, how to accurately reveal the principles, processes and laws of soil erosion, investigate the ways in which natural and human factors act on soil erosion, establish soil erosion models, make objective and systematic evaluations of the environmental benefits of soil erosion on a regional and global scale, propose targeted strategic solutions and technical approaches to prevent and control soil erosion in conjunction with the use of water and soil resources, and create a benign ecological environment. It is of far-reaching significance to achieve national ecological security and healthy economic development, and to promote scientific and technological research and development in the field of soil and water conservation[2].

In this paper, with the help of RUSLE model, the factors and calculation results and raster data results related to soil erosion such as R rainfall erosion force, K soil erosion resistance, LS terrain factor, C vegetation cover factor were obtained respectively, and the calculation of soil erosion A was constructed with the help of ArcGIS 10.2 platform using Moder Builder modeling work The model is used to discuss the use of RUSLE model and the application of GIS Geographical statistics for this study, which gives a reference for the research on soil erosion in small watersheds of Loess Plateau and has far-reaching significance to promote soil and water conservation disciplines.

2 Data and Methods

2.1 Study Area Scope

The areas studied in this paper are in the Sheep Circle Gully sub-basin, which be part of the central area of the Loess Plateau region, located from 109°31′~109°71′ and 36°42′~36°82′ north of Liqu Town, Baota District, Yan' an City, northern Shaanxi Province. This small watershed is a secondary tributary of the Yan River Basin, a first-order tributary of the Yellow River and is a typical model of small watershed synthesis with a watershed area of 2.02 km^2[3] (Fig. 1).

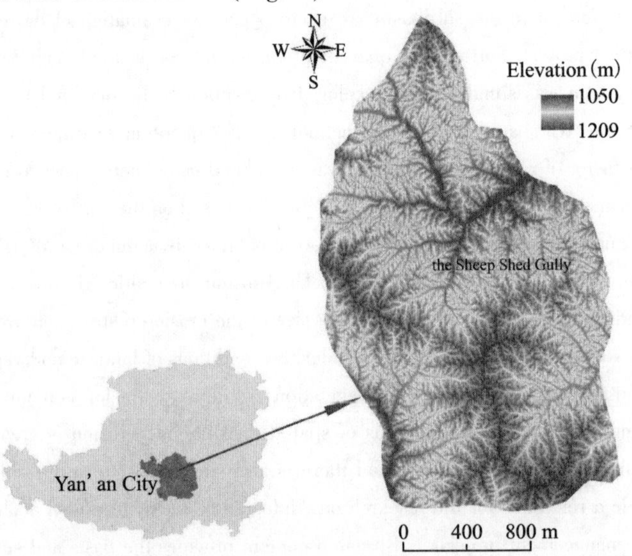

Fig. 1 Study area location diagram

The study area is located in a geomorphic gully interspersed with loess beams and loess ditches, with a density of 2.74 km/km^2[3] and an altitude of about 1050~1295 m. It is a semi-arid continental monsoon climateand an average annual precipitation of 535 mm, and the rainy season is mostly concentrated in July-September each year, accounting for about 79% of the annual precipitation, with a large interannual variability of precipitation. The annual sunshine hours are 2528.8 h, the annual average temperature is 9.40 ℃, and the annual difference in temperature is 29.40 ℃. The soils are mainly yellow cotton soil with loose structure, poor erosion resistance, and severe soil erosion[20]. The vegetation of the watershed belongs to the forest-steppe transition zone in terms of zoning. The natural vegetation in the watershed has been destroyed due to the disturbance of human activities, and nowadays it is mostly planted with artificial secondary vegetation, and the artificial species of plant species mainly include acacia, oleander, sea buckthorn, lemon grass, alfalfa, etc.[17] The average multi-year rainfall erosion force was 1755.83 (MJ · mm)/(hm^2 · h) before reforestation (before 1999) and 2155.71(MJ · mm)/(hm^2 · h) after reforestation. At present, the land types in the watershed are mainly sloping farmland, terraced farmland, forest, scrub, grassland, residential land, and water.

2.2 Experimental data

Soil erosion conditions in small watersheds were simulated with the modified universal soil loss equation (RUSLE), involving arithmetic parameters including precipitation, soil, topography, vegetation, soil and water conservation measures and other factors in the study target area. Daily rainfall data from 1997 to 2016 from the Yan'an Climate Observatory from the China Meteorological Data Network were used in this study. Soil texture data were provided through the Chinese soil texture base data from the Resource and Environment Science Data Center of the Chinese Academy of Sciences, which has small variability due to the characteristics of loess and is based on the current year in terms of temporal resolution. Topographic data were generated using DEM data with 5m spatial resolution by using interpolation of actual elevation points in the watershed. Vegetation cover data were obtained from MODIS vegetation index data through the Geospatial Data Cloud. The vegetation cover and soil conservation data were taken depending on the empirical values of the researchers.

2.3 Methodology

In this study, the study area was quantitatively calculatedusing RUSLE model, which was modified on the USLE model to establish a model for calculating soil erosion with a broader range of applicability. The calculation model consists of five input parameters: rainfall, soil erodibility, slope break length, Vegetation coverage, anthropogenic measures, and its quantitative calculation equation is:

$$A = R \times K \times \mathrm{LS} \times C \times P \tag{1}$$

Where each parameter is: A denotes the average annual soil erosion in the study area ; R denotes the rainfall erosivity factor; K denotes the soil erodibility factor; LS is the slope length-slope factor, dimensionless, and where L is the slope length factor, defined as a power function of slope length, S is the slope factor, and LS is the slope length slope factor, dimensionless, where L is the slope length factor, defined as a power function of slope length, S is the slope factor, and LS represents the ratio of soil loss on a slope for a given slope length and slope to soil loss on a typical slope in a standard runoff plot, all else being equal; C represents the vegetation cove, dimensionless, which refers to the ratio of soil loss under a particular crop or vegetation cover, all other factors being equal, to the ratio of soil loss under a particular crop or vegetation cover to that of continuous recreational land after cultivation; P refers to the ratio of soil loss after soil and water conservation measures to soil loss from downhill planting, dimensionless.

(1) R factor

This indicator of the ability of rainfall to strip and transport soil and is usually estimated as a model input parameter in terms of rainfall. The rainfall erosion force is calculated as

$$R = \alpha \cdot F^{\beta} \tag{2}$$

and according to the formula for calculating the average annual rainfall in the study by Zhang Wenbo[4] et al.

$$F = \sum_{i=1}^{12} \frac{r_i}{P} \tag{3}$$

In equation (2), F is the average annual rainfall (mm) and R the average annual rainfall erosion force; α and β are the model inputs; and in equation (3) P is the total annual rainfall (mm) and r_i is the monthly rainfall (mm) in that year. Based on the results of literature studies, the values of parameters α and β in the rainfall model of this study area were taken as $\alpha=0.184$ and $\beta=1.96$, respectively[9], and the average rainfall erosion force of the last 20 years in the Sheep Shed Gully sub-basin was derived from the daily rainfall data and equations from 1997 to 2016.

(2) K factor

The calculation of soil erodibility factor K is based on the participation of soil particle composition and organic matter parameters in soil texture properties, including content of sand, clay grain, chalk and organic matter. The factor K in the RUSLE model is a value obtained through experiments. It is quantitative and the K factor is usually obtained by direct measurement of soil loss per unit of precipitation erosion force in standard natural plots, but it is difficult to deploy natural plots in this watershed on a large scale, and in this paper we use Williams' calculation of K-value in the EPIC model[5].

$$K = 0.1317 \times \left\{0.2 + 0.3e^{\left[-0.0256\times SAN\left(1-\frac{SIL}{100}\right)\right]}\right\} \times \left\{\frac{SIL}{CLA + SIL}\right\}^{0.3} \times \left\{1.0 - \frac{0.25C}{C + e^{3.72-2.95c}}\right\} \times \left\{1.0 - \frac{0.7 \times \left(1 - \frac{SAN}{100}\right)}{\left(1 - \frac{SAN}{100}\right) + e^{\left[-551+22.9\left(1-\frac{SAN}{100}\right)\right]}}\right\} \tag{4}$$

Where: 0.1317 is the conversion factor between American and international units; SAN (0.1~2 mm) is the sand content (%); SIL (0.002~0.1 mm) is the powder content (%); CLA (<0.002 mm) is the clay content (%); and C is the organic carbon content (%). Using the raster computer in ArcGIS10.2 for the above parameters combined with equation (4), the results were calculated between 0.022 and 0.040 $t \cdot hm^2 \cdot h \cdot MJ^{-1} \cdot mm^{-1} \cdot hm^{-2}$, which is close to the results of Li Kui[16] and others, with a K value of 0.047 $t \cdot hm^2 \cdot h \cdot MJ^{-1} \cdot mm^{-1} \cdot hm^{-2}$ for yellow cotton soil.

(3) L, S factor

Slope and slope length, as a basic geomorphological indicator, affect the surface soil structure, surface runoff and soil erosion by influencing gravity, which in turn affects the deployment of soil and water conservation measures, the layout of agriculture, forestry and livestock production, and land resource development[23]. From this paper, the SL factor proposed by Wischmeier and Smith was calculated using the formula.

$$L = \left(\frac{\lambda}{22.13}\right)^{\alpha} \tag{5}$$

$$\alpha = \frac{\beta}{\beta + 1}, \beta = \frac{\frac{\sin\theta}{0.0896}}{3.0(\sin\theta)^{0.8} + 0.56} \tag{6}$$

Where: λ is the horizontal slope length, α is the slope length index, 22.13 is the slope length of the standard plot (m), and θ is the slope extracted using DEM data. S factor is calculated using the formula of McCool[7], and the results are more satisfactory.

$$S = \begin{cases} 10.8 \times \sin\theta + 0.03, \theta < 9° \\ 16.8 \times \sin\theta - 0.5, 9° \leqslant \theta < 14° \\ 10.8 \times \sin\theta + 0.03, \theta \geqslant 14° \end{cases} \tag{7}$$

and using the steep slope formula of Liu Baoyuan[6].

$$S = 21.9 \times \sin\theta - 0.96, \theta \geqslant 10° \tag{8}$$

(4) C factor

The vegetation cover and management C factor is highly sensitive parameters to soil erosion changes, and it has this good correlation with vegetation cover[19], which is calculated from the NDVI value calculated from remote sensing images as a dimensionless parameter. In this study, vegetation cover was obtained under the ENVI platform with the help of MYD13A1 synthetic vegetation index data from MODIS. Using the Tsai-Chung method to calculate C [11] as follows.

$$C = \begin{cases} 1, VFC = 0 \\ 0.650\,8 - 0.343\,6 \times \lg^{(VFC)}, 0 < VFC < 0.783 \\ 0, VFC \geqslant 0.783 \end{cases} \tag{9}$$

Where VFC is the formula shows that the vegetation cover is greater than 0.783, the erosion of the surface is extremely weak and the erosion is negligible. And when the vegetation cover is less than 0.1%, its erosion reduction effect is basically not reflected. The distribution of its C value ranges from 0 to 1 (no vegetation cover area such as bare rock and bare soil).

(5) P factor

The P factor is related to the type of land use in the subsurface, and according to Wischmier and Smith[12] showed that the P-value is between 0 and 1 and is a dimensionless number[22]. The soil conservation measure factor P is the most difficult parameter to obtain in the model, and theit can be measured by runoff plot tests, but this method is costly and time-consuming, and the measured results have regional limitations. In this study, we combined the research results of Deyan Zhong[13] and Baoyuan Liu[14] on the Sheep Circle Ditch sub-basin and combined with Landsat8 data to superimpose the results of land use classification decoding to get the P value, and other contained classifications as dams, roads, industrial and mining land, and oil fields.

3 Discussion and analysis

3.1 Factor calculation results

(1) K factor calculation

The obtained soil texture data were assigned and rasterized separately to obtain the range values of soil sand, powder, clay, and organic matter[15], where the soil sand was 0～12.6%, powder was 0～68.6%, clay was 0～18.7%, and organic carbon was 0.66%～12.2% (Fig. 2). And combined with equation (4), the K factor's calculation results which takes values between 0.022 and 0.04 (Fig. 4).

a.Quantification value of clay particles　b.Quantification value of sand particles　c.Quantification value of powder particles　d.Quantification of organic carbon values

Fig. 2 Soil mass content distribution

(2) R factor calculation

The Loess Plateau area's erosion type is water erosion, and rainfall is positively correlated with it, i.e., the greater the rainfall, the greater the intensity of soil erosion[18]. In this study, the average annual rainfall data of the region for the past 10 years were obtained, and the rainfall from 2006 to 2015 was obtained after the removal of abnormal values (Fig. 3). The total annual rainfall during this 10-year period was basically stable with small fluctuations, with a maximum peak of 959.1 mm in 2013, and the results were close to the R value of 2155.71 ($MJ \cdot mm \cdot hm^{-2} \cdot h^{-1} \cdot a^{-1}$) after reforestation according to equation (2)(3)(Fig. 5).

(3) L, S factor calculation

Small watersheds and regions to obtain the slope length L factor usually use DEM to extract the watershed slope S and the D8 algorithm in the hydrological analysis module to obtain the flow direction, and after repeated filling of the DEM depressions to obtain the depression-free DEM, the sink accumulation is finally calculated[21]. The product of the sink accumulation and the raster image element is used as the valuation of λ The quantified values of LS were calculated in Raster Calculator using equations (5)、(6)、(8), respectively (Fig. 6 and Fig. 7).

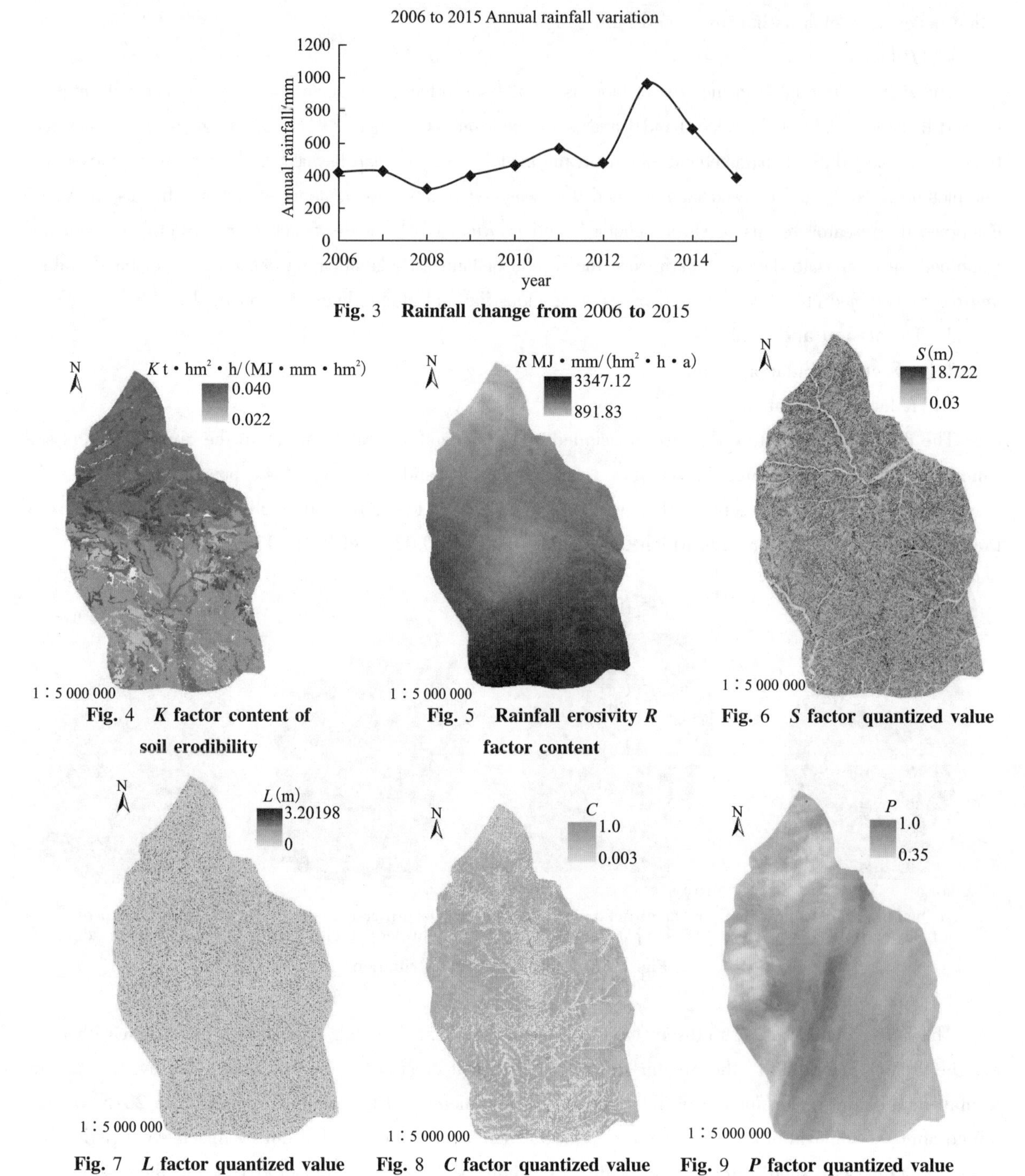

Fig. 3 Rainfall change from 2006 to 2015

Fig. 4 *K* factor content of soil erodibility

Fig. 5 Rainfall erosivity *R* factor content

Fig. 6 *S* factor quantized value

Fig. 7 *L* factor quantized value

Fig. 8 *C* factor quantized value

Fig. 9 *P* factor quantized value

(4) *C* factor calculation

This factor is a model parameter related to vegetation cover, and the calculation combines the Land Use Data with the MODIS16 synthetic vegetation normalized index NDVI raster data for overlaying, and the quantitative values of *C* factor under different vegetation types are calculated by equation (9) (Fig. 8 and Table 1). The value is inversely proportional to the amount of soil erosion, i.e., the smaller the *C*-value, the lighter the soil erosion condition.

Table 1 Quantized value of factor C under different land use classification (vegetation)

Land use type	Grassland	Forest	Cropland	Orchard	Terraces	bare ground	Other	Residential
C	0.23	0.03	0.35	0.05	0.31	1.0	0	0.9

(5) *P* factor calculation

The *P* factor in the general soil erosion equation has been a difficult problem to confirm, usually obtained by runoff plot test, and its value is from 0 to 1. The smaller the value, the lower the degree of soil erosion represents the perfect soil and water conservation measures deployed in the area, and vice versa, the higher the degree of soil erosion. This factor calculation is similar to the *C* factor, which Use of GIS platforms to obtain from land use data, and the following *P* factor quantification values are obtained according to Liu Baoyuan's research results (Table 2), and the raster quantification data are obtained by overlaying the land use data (Fig. 9).

Table 2 Quantified value of P factor under different land use classification

Land use type	Grassland	Forest	Cropland	Orchard	Terraces	water	bare ground	Other	Residential
P	0.9	0.7	1	0.8	0.35	0	0	0	1

3.2 Calculation results and analysis

Based on the raster data of each factor, the raster data were processed in the ARCGIS platform to construct a unified coordinate reference, and the raster image elements were converted into 5×5 size GRID format, and the raster calculator was used to bring in the linear calculation formula (1) for soil erosion to obtain the values of soil erosion and spatial distribution of the sheep-ring ditch sub-basin, and the soil erosion was classified symbolically Based on soil erosion classification standards (SL 196-2007). (SL 196-2007) for symbolic classification(Fig. 10, Table 3)

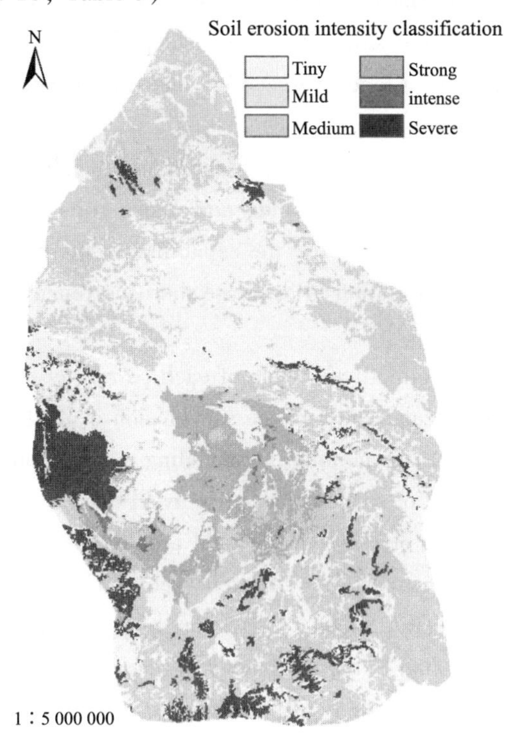

Fig. 10 P Soil erosion intensity classification mapping

Table 3　soil erosion grade classification table

	Soil erosion amount	Tiny	Mild	Medium	Strong	intense	Severe
Area(km^2)	71.75	125.75	43.86	18.74	6.26	9.07	0.59
Proportion(%)	36.33	63.67	22.21	9.49	3.17	4.59	0.3

The calculated soil erosion in this study area is 3175.16 t/(km^2 · a), which is close to 3270.19 t/(km^2 · a) as calculated by Zhao Wenqi[10] for the 2014 Sheep Sap Ditch sub-basin, concluding that the Loess Plateau area has experienced many years of reforestation and comprehensive small watershed management projects have contributed to soil and water conservation, which is a non-intense erosion area. From Table 3, From this it is clear that the soil erosion area of the whole watershed is 71.75 km^2, accounting for 36.33% of the total area of the study area. The area and proportion of different levels of soil erosion are: 27.25 km^2 for slight erosion, accounting for 63.67% of the area ; 9.27 km^2 for light erosion, area share of 22.21% ; 19.53 km^2 for moderate erosion, accounting for 9.49% of the area ; 75.65 km^2 for strong erosion, accounting for 3.17% of the total area of the study area; and 3.17% for very strong erosion. 3.17%; very strong erosion area 4.59 km^2, accounting for 1.16% of this; severe erosion's area is 2.71 km^2, accounting for 0.3% of the total area of the study area. Slight and mild erosion is the mainly soil erosion, accounting for 85.88% of the occurring soil erosion area. The largest area of slight erosion and the smallest area of severe erosion, and the overall distribution is contiguous. The soil erosion intensity of each level is arranged in the order of area: mild>medium>intense>strong>severe.

(1) In the model calculation, the perennial rainfall in the Loess Plateau area is basically stable, and the changes of soil texture, which is mainly yellow sheep soil, are not significant, resulting in constant rainfall erosion force R and soil erodibility factor K. Only the topographic factor LS and vegetation index-related C and P factors become strongly correlated to the soil erosion amount. The steeper the slope and the longer the slope length, both will bring the soil erosion intensity becomes larger and the erosion volume increases. Therefore, we can consider reducing the slope by cutting steep slope into gentle slope when carrying out comprehensive watershed management in land engineering, and truncating the length of slope gully management to reduce the runoff pooling time.

(2) From the perspective of vegetation cover, the planting of protective forests should be vigorously carried out within the red line of guaranteed arable land, and plants suitable for the survival of loess plateau should be planted in the gully and slope to improve the capability of soil and water conservation and increase the area covered by various types of vegetation to decrease the occurrence of soil erosion.

4　Conclusion

GIS and RS were used to obtain quantitative data of each parameter of the modified soil erosion equation (RUSLE), and the quantitative calculation of the soil erosion evaluation was carried out in the Sheep Sap Ditch sub-basin of Loess Plateau by combining the modeling method, and the influence of the differentiation of different factors on the change of soil erosion intensity in the sub-basin was analyzed. The soil erosion area is 71.75 km^2, accounting for 36.33% of the total area of the study area. The study area has the largest area of slight erosion and the smallest area of intense erosion, and the soil erosion intensity of each class in the order of area is light>moderate>strong>very strong>intense. In terms of spatial distribution, The watershed is larger to the south-west than to the north-east and is accompanied by highly erosive conditions, so appropriate protection measures and comprehensive watershed management are needed to stabilize the ecological environment. The erosion covers the whole small watershed, and the soil erosion varies significantly under different land use types and vegetation cover, mostly occurring in arable land, terraces and orchards, where the erosion condition is more severe; woodland, grassland and shrubs are mainly slightly and lightly eroded. Another influencing factor of PTU is that the soil erosion intensity also varies greatly under different slope topography, where the erosion area accounts for the largest 15%~25% slope zone, and the erosion intensity is mainly slight and mild.

Acknowledgements

This research was fund by Technology Innovation Center for Land Engineering and Human Settlements, Shaanxi Land Engineering Construction Group Co., Ltd. and Xi'an Jiaotong University(2021WHZ0088) and (2021WHZ0091). and Scientific Research Item of Shaanxi Provincial Land Engineering Construction Group (DJNY2021-32) and (DJNY2019-29).

References

[1] YUE Dapeng, LI Kui, LIU Peng, et al. Study on soil erosion and soil conservation in a small watershed of loess pits based on RUSLE[J]. Journal of Shaanxi Normal University: Natural Science Edition.2015,(2):85-91.

[2] Wang Xiaofeng, Xiao Feiyan, Yin Lixing, et al. Evaluation of soil erosion at different watershed scales in the middle reaches of the Yellow River[J]. Journal of Anhui Agricultural University.2017,(6):1070-1077.

[3] Wang Yafeng, Fu Bojie, Hou Prosperity, et al. Estimation of sediment siltation in silt dams based on differential GPS technology[J]. Journal of Agricultural Engineering. 2009,(9):79-83.

[4] Zhang Wenbo, Xie Yun, Liu Baoyuan. Research on the method of calculating rainfall erosion force using daily rainfall[J]. Geoscience. 2002,(6):705-711.

[5] Williams J R . EPIC-erosion/productivity impact calculator: 1. Model documentation.[J]. Technical Bulletin -United States Department of Agriculture, 1990, 4(4):206-207.

[6] Baoyuan L , Keli Z , Yun X . An Empirical Soil Loss Equation[C]// Sustainable Utilization of Global Soil and Water Resources vol.2: Process of Soil Erosion and Its Environment Effect. Department of Resources and Environmental Sciences, Beijing Normal University, Key Laboratory of Environmental Change and Natural Disaster, Department of Resources and Environmental Sciences, Beijing Normal University, Key Laboratory of Environmental Change and Natural Disaster, the Ministry of Education of China, Beijing 100875, PR China, 2002.

[7] Mccool D K , Brown L C , Foster G R , et al. Revised Slope Length Factor for the Universal Soil Loss Equation[J]. Transactions of the ASAE, 1989, 30(5):1387-1396.

[8] Wang Wanzhong. Study on the relationship between rainfall characteristics and soil loss in loess areas Ⅲ--Criteria for erosive rainfall[J]. Soil and Water Conservation Bulletin. 1984,(2):58-63.

[9] Wang Wanzhong. Research on the R-index of erosive force of rainfall in loess areas[J]. China Soil and Water Conservation, 1987(12).

[10] Zhao WQ, Liu Y, Luo ML, et al. Soil erosion effect assessment of vegetation restoration in small watersheds on Loess Plateau[J]. Journal of soil and water conservation.2016,(5):89-94.

[11] Cai, Chongfa, Ding, Shuwen, Shi, Zhihua, et al. Application of USLE model and GIS IDRISI to predict soil erosion in small watersheds[J]. Journal of Soil and Water Conservation. 2000,(2):19-24.

[12] Wischmeier, W.H, D.D. Smith. Predicting Rainfall Erosion Losses. Agriculture Handbook 537[J]. 1978.

[13] Zhong, D.Y.. Research on soil erosion in loess hills and gullies based on USLE model [D]. Northwest Agriculture and Forestry University, 2012.

[14] Liu, Baoyuan, Zhang, Keli, Jiao, Juying. Soil erodibility and its application in erosion forecasting[J]. Journal of Natural Resources. 1999,(4):345-350.

[15] Qin Zhijia. A comparative study of soil erosion in Guizhou Province based on remote sensing survey and RUSLE model[D]. Guizhou Normal University,2017.

[16] Li Kui. Research on soil erosion in Huangtuwa sub-basin based on polyquad sediment deposition analysis and RUSLE simulation[D]. Shaanxi Normal University,2014.

[17] Zhang Liwei, Fu Bojie, Lü Yihe, et al. Balancing multiple ecosystem services in conservation priority setting. landscape Ecology, 2015, 30(3): 535 -546.

[18]Xie Yun, Zhang Wenbo, Liu Baoyuan. Calculation of rainfall erosion force by daily rainfall and rain intensity[J]. Soil and Water Conservation Bulletin. 2001,(6):53-56.

[19]Li X.X., Guo Q.X., Yang X.F., et al. A quantitative study of soil erosion in a small watershed based on GIS and RUSLE model fork [J]. Journal of Shanxi Agricultural University: Natural Science Edition. 2013,(5):413-419.

[20]Fu Bojie, Liu Yu, Lü Yihe, et al. 2011. Assessing the soil erosion control service of ecosystems change in the Loess Plateau of China[J]. Ecological Complexity, 8(4): 284-293.

[21]Qi JY. Research on soil erosion and its influencing factors in the Weixiang River Basin based on GIS and RS[D]. Gansu Agricultural University,2017.

[22]Qin W, Zhu QK, Zhang Y. Soil erosion assessment of small watersheds in Loess Plateau based on GIS and RUSLE [J]. Journal of Agricultural Engineering.2009,(8):157-163.

[23]Zhang X.W., Zhou Y.M., Li X.S., et al. Progress of remote sensing research on soil erosion evaluation[J]. Soil Bulletin. 2010,(4):1010-1017.

本文曾发表于2022年《Water Supply》第2卷第7期

Spatiotemporal Evolution of Ecological Environment Quality in Arid Areas Based on the Remote Sensing Ecological Distance Index: a Case Study of Yuyang District in Yulin City, China

Liangyan Yang[1], Lei Shi[2], Jing Wei[3] and Yating Wang[4]

(1.Shaanxi Provincial Land Engineering Construction Group Co., Ltd, Xi'an 710075, China;2.Institute of Land Engineering and Technology, Shaanxi Provincial Land Engineering Construction Group Co., Ltd., Xi'an 710021, China; 3.Key Laboratory of Degraded and Unused Land Consolidation Engineering, the Ministry of Natural Resources, Xi'an 710021, China;4.Shandong urban and rural planning Design Institute Co., Ltd, Jinan 250013, China)

【Abstract】 The ecological environment in arid areas of Northwest China has undergone considerable changes under the combined effects of climate change and human factors Therefore, exploring the spatial and temporal evolution of ecological environment quality is of great significance for the protection and management of the ecological environment in arid areas of Northwest China. This study adopted Yuyang District as the study area. Landsat TM/OLI images from 1993 to 2018 were selected as the data source for the retrieval of important surface indicators and the construction of the remote sensing distance ecological index (RSEDI). The spatial distribution, trend, and grade classification of ecological environment quality were monitored and analyzed. The results showed that: (1) Ecological environment quality of Yuyang District from 1993 to 2018 showed an overall upward trend, mainly manifested as a sharp decline in the area of poor ecological environment from 84.81% to 53.36%. (2) The spatiotemporal changes in ecological environment quality showed a downward trend in the central urban area and an upward trend in the non-central urban area. (3) In general, rainfall and temperature had limited impacts on ecological environment quality. Urbanization seriously affected the local ecological environment quality and the implementation of ecological restoration policies, regulations, and measures were the main drivers of improvement to ecological environment quality in other surrounding areas.

【Keywords】 Yuyang District; Ecological environment quality; Remote sensing ecological distance index; Spatiotemporal evolution

1 Introduction

The rapid development of urbanization and over-exploitation of resources since the 1980s have led to a series of ecological and environmental challenges, including vegetation degradation, land desertification, water shortages, and frequent extreme weather and natural disasters. These challenges have greatly hindered the attainment of the sustainable development of society[1-2]. Since the arid and semi-arid regions of northwest China are characterized by scarce rainfall and low vegetation cover, the ecological environment of these regions shows obvious vulnerability[3]. Therefore, an improved understanding of the ecological health of ecologically fragile areas and clarifying the impacts of natural and social factors on ecological environment quality can assist the management and restoration of the ecological environment in northwest China.

There are various advantages associated with remote sensing technology, including a wide spatial monitoring range, a short revisit period, and low data cost. Remote-sensed data have been widely applied in ecology since these data provide an effective means of quantification, visualization, and evaluating the ecological environment[4]. The application of remote sensed data in previous studies include in the analysis of the response mechanisms of vegetation indices to human activities and climate change[5]; studying the spatiotemporal expansion and drivers of urban and rural construction land and analyzing the impact of urbanization on the ecological environment[6-8]; and assessing the urban heat island effect by analyzing changes in urban landscape patterns[9-11]. The study of changes in single indicators and their relationship with ecological response in the region has many limitations and is not able to provide a comprehensive evaluation of the regional ecological environment quality and a clarification of the ecological environment status of ecologically fragile areas[12]. Scholars have proposed a series of ecological environment comprehensive evaluation models to address this problem, including the "pressure-state-response" model[13], the remote-sensed ecological index model (RSEI) based on principle component analysis (PCA)[14], and the ecological vulnerability evaluation model based on "cause-effect" indicators[15]. The use of a comprehensive evaluation model overcomes the challenge of the one-sidedness of single factor evaluation and provides a more comprehensive evaluation result. However, the identification of methods for determining the weights of multiple indicators has become a key and difficult challenge facing ecological environmental evaluation research. In 2016, Zhang et al.[16] adopted the Guazhou-Dunhuang Basin as a study area and proposed a new system to evaluate ecological environment quality based on the four-dimensional space remote-sensed ecological distance index (RSEDI). Shi et al.[17] affirmed the applicability and objectivity of RSEDI by its application to assess and monitor the ecological environmental quality of the oasis area in the Shiyang River Basin. Huang et al.[12] used RSEDI to clarify the spatial and temporal changes in ecological environment quality in Urumqi. The RSEDI has the advantages of objective weighting, simplicity, and speed. It also overcomes the challenge of one-sidedness of single-factor evaluation and subjectivity in determining the weights of comprehensive evaluation indices. Therefore, the results of the RSEDI can objectively reflect the changes in the ecological environment in arid and semi-arid areas over the long term.

The Yuyang District of Yulin City is the center of modern agriculture in the agro-pastoral transition zone of northern China and an example of sustainable development within the Loess Plateau. At the same time, Yuyang District, as an influential city falling on the border between Shaanxi, Gansu, Ningxia, Mongolia, and Shanxi, acts as a base of national energy and chemical production in northern Shaanxi. Natural factors such as climate change and human activities such as agricultural development, energy development and utilization, and implementation of ecological projects have had certain impacts on the local ecological environment in recent decades. Therefore, the study of the process of ecological change and factors influencing Yuyang District is of practical significance for the ecological management of the area. The present study used Landsat TM/OLI image data, integrated the topography and geomorphology of the study area, combined the remote sensing ecological index, and constructed a four-dimensional remote-sensed ecological distance index based on the distance function. This index was used to analyze the ecological environment change process in Yuyang District and to explore the driving factors. The results of the present study can provide a scientific basis for achieving sustainable development of the Yuyang District.

2 Study area, materials, and methods

2.1 Study area

Yuyang District falls within the municipal district of Yulin City, Shaanxi Province, and is the political, economic, and cultural center of Yulin City (Fig.1). The elevation of Yuyang District generally decreases from the northeast to the south-central area. The study area is roughly bounded by the Great Wall of Ming, and consists of two major types of landforms: (1) wind-sand and grassland area in the north, accounting for 75% of the total area; (2) hilly and gully area in the south, accounting for 25% of the total area[18]. Yuyang District falls in a typical continental marginal monsoon climate zone characterized by four distinct seasons. The average annual rainfall and temperature of the study area are 412.2 mm and 8.8 ℃, respectively[19].

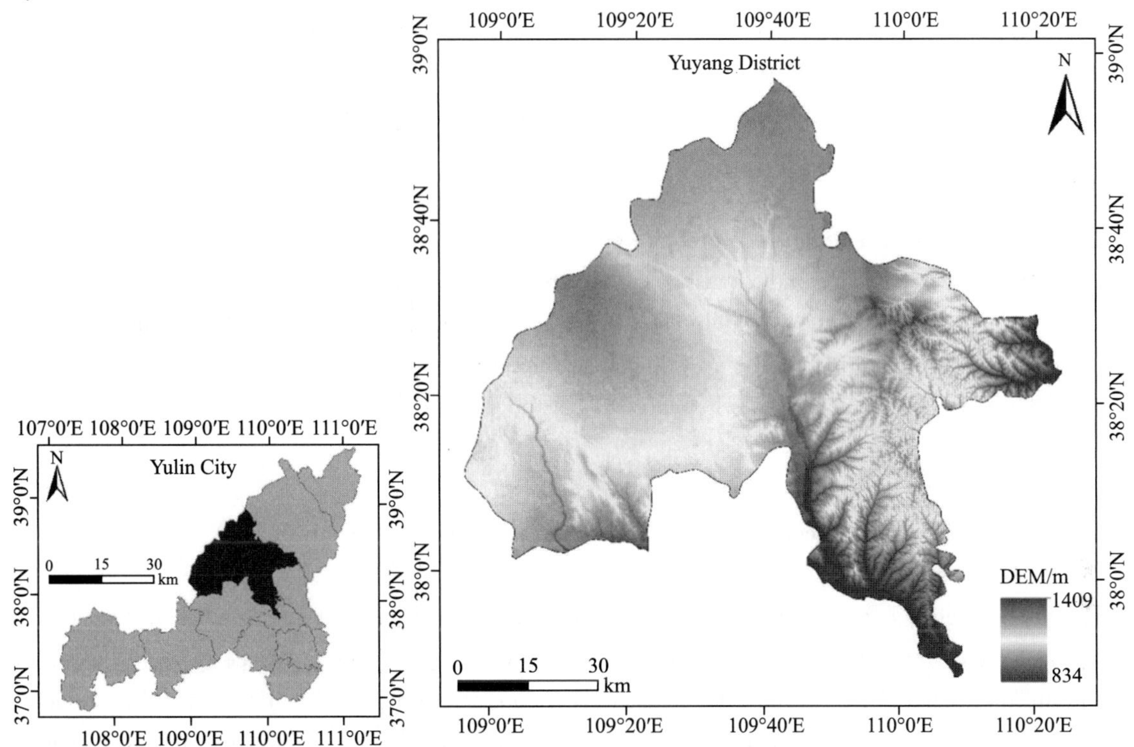

Fig. 1 Location and digital elevation map (DEM) of Yuyang District in Yulin City, China

2.2 Data sources

The Landsat program of the United States National Aeronautics and Space Administration (NASA) has launched eight satellites since the 23rd July, 1972. The data collected by these satellites can be downloaded from the United States Geological Survey (USGS) website (earthexplorer.usgs.gov/). Landsat data are long-term and large-scale ecological environmental monitoring data[20]. The spatial resolution of the Landsat data was 30 m and they underwent system radiation and geometric correction. The dataset used in the present study was provided by the Geospatial Data Cloud site, Computer Network Information Center, Chinese Academy of Sciences. (http://www.gscloud.cn). The present study maximized the accuracy of the surface information of the study area and ensured that image cloud cover was less than 3% by selecting three phases of satellite images with strip numbers (127, 33) and (127, 34) for the 18 June, 1993, 16 June, 2004, and 23 June, 2018. Remote sensing images need to be subjected to a series of preprocessing steps before they can be used for quantitative research. The ENVI 5.3 software was used to preprocess the images, including strip repair, image registration, radiometric calibration, atmospheric correction, cloud detection, image mosaic, and cropping. The temperature and rainfall data of Yuyang District were obtained from the China Meteorological Data Network (http://data.cma.cn/). The land use type data were obtained from the Global Geo-information Public Product (http://www.globallandcover.com/).

2.3 Research method

2.3.1 Humidity index

The humidity index indicates the degree of wetness of soil. The present study used the wetness index (WI) of the tassel cap transformation to reveal soil drought variability. The wetness component equation is:

$$WI = c_1B_1 + c_2B_2 + c_3B_3 + c_4B_4 + c_5B_5 + c_6B_6 \tag{1}$$

In Eq.(1), B_1-B_6 represent blue, green, red, near red, mid-infrared 1, and mid-infrared 2 bands, respectively, and c_1-c_6 are the sensor parameters. The humidity component calculation parameters for the Thematic Mapper (TM) and Operational Land Imager (OLI) sensors differ due to the different satellite sensor types. In the case of the TM sensor, the values of c_1-c_6 are 0.0315, 0.2021, 0.3012, 0.1594, -0.6806, and -0.6109, respectively[21], whereas they are 0.1511, 0.1973, 0.3283, 0.3407, -0.7117, and -0.4559, respectively for the TM sensor[22].

2.3.2 Greenness index

The greenness index is expressed through vegetation indices, among which the remote sensing-based normalized difference vegetation index (NDVI) better reflects the vegetation growth status, vegetation biomass, and other parameters, and is widely used in the study of the spatial and temporal changes of vegetation cover and in the monitoring of crop growth status. Therefore, the present study used NDVI as the greenness index based on the environmental characteristics of the study area. The NDVI can be calculated as[23]:

$$\mathrm{NDVI} = \frac{B_{\mathrm{NIR}} - B_{\mathrm{R}}}{B_{\mathrm{NIR}} + B_{\mathrm{R}}} \tag{2}$$

In Eq. (2), B_{NIR} and B_{R} represent the near red and red bands, respectively.

2.3.3 Dryness index

The dryness index is used to reflect the degree of dryness of the land surface, with both bare land and built-up land increasing land surface "dryness". The brightness values of the surface types of built-up and bare soil are higher than those of other land types. Therefore, the brightness value of the land surface can act as an indicator of bare surfaces and can represent the spatial distribution of built-up land. The ecological environment is influenced by the spatial distributions of both bare surfaces and built-up land[14]. Therefore, the index-based built-up index (IBI) and the soil index (SI) can be used to construct a normalized differential build-up and bare soil index (NDSI) to represent the dryness of the remote-sensed ecological index (RSEI). This is calculated as[24]:

$$\mathrm{SI} = \frac{[(B_5 + B_3) - (B_4 + B_1)]}{[(B_5 + B_3) + (B_4 + B_1)]} \tag{3}$$

$$\mathrm{IBI} = \frac{\left\{\dfrac{2B_5}{(B_5 + B_3)} - \left[\dfrac{B_4}{(B_4 + B_3)} + \dfrac{B_2}{(B_2 + B_5)}\right]\right\}}{\left\{\dfrac{2B_5}{(B_5 + B_3)} + \left[\dfrac{B_4}{(B_4 + B_3)} + \dfrac{B_2}{(B_2 + B_5)}\right]\right\}} \tag{4}$$

$$\mathrm{NDSI} = \frac{\mathrm{SI} + \mathrm{IBI}}{2} \tag{5}$$

2.3.4 Heat index

The heat index is reflected through the land surface temperature (LST). Since the Landsat TM data only includes one thermal infrared band, the present study selected a single-channel algorithm to invert the LST[25]:

$$\mathrm{LST} = \frac{T}{\left[1 + \left(\dfrac{\lambda T}{\rho}\right)\ln\varepsilon\right]} \tag{6}$$

$$T = \frac{K_2}{\ln\left(\dfrac{K_1}{B_{\mathrm{TIR}} + 1}\right)} \tag{7}$$

In Eq.(6) and Eq. (7), B_{TIR} is the thermal infrared band radiation, obtained from the sixth and tenth bands of the TM and OLI sensors, respectively, T is the brightness temperature at the sensor, and K_1 and K_2 are the sensor calibration parameters. For the sixth band of the Landsat-5 TM, K_1 = 607.76 W/(m^2 · sr · μm) and K_2 = 1,260.56 K. For the 10th band of Landsat-8 OLI/TIR, K_1 = 774.89 W/(m^2 · sr · μm) and K_2 = 1 321.08 K. λ is the central wavelength of the thermal infrared band, $\rho = 1.438 \times 10^{-2}$(m · K), and ε is the specific emissivity of the feature.

2.3.5 Construction of the remote-sensed ecological distance index

Four remote sensing ecological indicators, namely WI, NDVI, NDSI and LST, were selected to form a four-dimensional space (Fig. 2). Here, the minimum values of WI and NDVI and the maximum values of NDSI and LST represented the least desired ecological environment quality. The distance from all pixel points in the space to the worst point of ecological environment quality was used as the RSEDI to assess ecological environment quality in the study area[16]. The value of the RSEDI was proportional to ecological environment quality.

$$RSEDI = \sqrt{(WI - WI_{min})^2 + (NDVI - NDVI_{min})^2 + (NDSI - NDSI_{max})^2 + (LST - LST_{max})^2} \tag{8}$$

In Eq. (8), WI_{min} and $NDVI_{min}$ are the minimum values of the humidity index and the NDVI in the study area, respectively, $NDVI_{min}$ and LST_{max} represents the maximum values of the dryness index and surface temperature, respectively. The calculated RSEDI was normalized to [0, 1] according to a 99.5% confidence interval, after which the RSEDI was classified into five ecological environmental quality classes based on the equal spacing method: (1) poor (0.0~0.2); (2) fair (0.2~0.4); (3) moderate (0.4~0.6); (4) good (0.6~0.8); and (5) excellent (0.8~1.0).

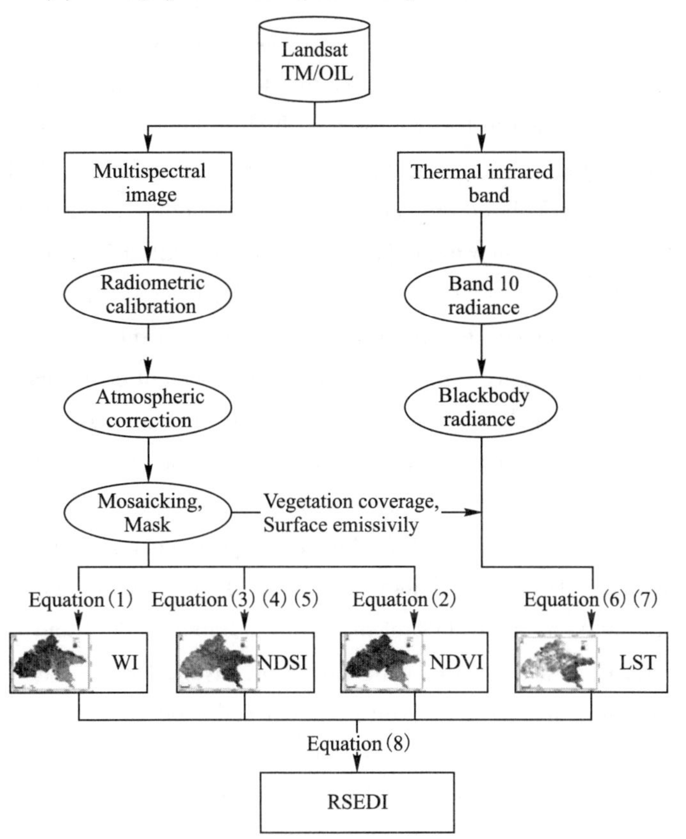

Fig. 2 Technique flow chart

3 Results

3.1 Overall evaluation of ecological environment quality of Yuyang District

Table 1 shows the mean values of each indicator and RSEDI in Yuyang District of Yulin City for 1993, 2004, and 2018. The trends of each indicator show that the mean LST decreased between 1993 and 2018, indicating a decreasing trend in heat in the study area. However, the rate of decrease during 2004-2018 was significantly lower than that during 1993-2004. Both the NDVI and WI showed increasing trends and larger increases during 1993-2018, indicating that the vegetation cover and soil moisture in Yuyang District have increased. Ecological environment quality of Yuyang District improved significantly from 1993to 2018, with the mean RSEDI increasing from 0.27 in 1993 to 0.33 in 2004 and further to 0.42 in 2018, an overall increase of 55.03%. Considered together, ecological environment quality of Yuyang District showed an overall upward trend during the 25-year study period.

Table 1 Average values of the remote-sensed ecological distance index (RSEDI) and indicators in Yuyang District, Shaanxi Province, China in 1993, 2004 **and** 2018

Year	Index				
	LST	NDSI	NDVI	WI	RSEDI
1993	0.85	0.52	0.18	0.27	0.27
2004	0.76	0.42	0.23	0.39	0.33
2018	0.75	0.40	0.35	0.50	0.42

3.2 Spatiotemporal distribution of ecological environment quality

The present study graded the RSEDI values according to the equal interval method using an interval of 0.2. The RSEDI values were assigned to five grades from high to low: (1) excellent; (2) good; (3) medium; (4) fair; and; (5) poor. Fig. 3 shows the spatiotemporal distribution of RSEDI in Yuyang District in 1993, 2004, and 2018 and Table 2 shows the percentages corresponding to each grade. The ecological environment of Yuyang District generally showed a spatial distribution pattern of ecological environment quality decreasing from the east to the west. Areas with excellent ecological environment quality in the study area from 1993 to 2018 showed a V-shaped change in ecological environment quality over time, whereas those with good ecological environment quality experienced an almost constant ecological environment quality during 1993-2004, following which ecological environment quality increased dramatically during 2004-2018 from 2.65% to 14.07%. Areas with medium ecological environment quality showed a continuous upward trend in ecological environment quality from 10.89% in 1993 to 30.03% in 2018. The regions with fair ecological environment quality showed an inverted V-shaped change in ecological environment quality. Regions with poor ecological environment quality showed significant decreasing trends in ecological environment quality, and accounted for 39.63% and 8.03% of the total land area in 1993 and 2018, respectively, a decrease of 31.60%. Overall, ecological environment quality of Yuyang District from 1993 to 2018 showed an increasing trend, and the proportions of areas with excellent and good grades in the study area continued to increase from 4.30% in 1993 to 16.62% in 2018. The areas with poor and fair ecological environment quality continued to decrease from 84.81% in 1993 to 53.36% in 2018.

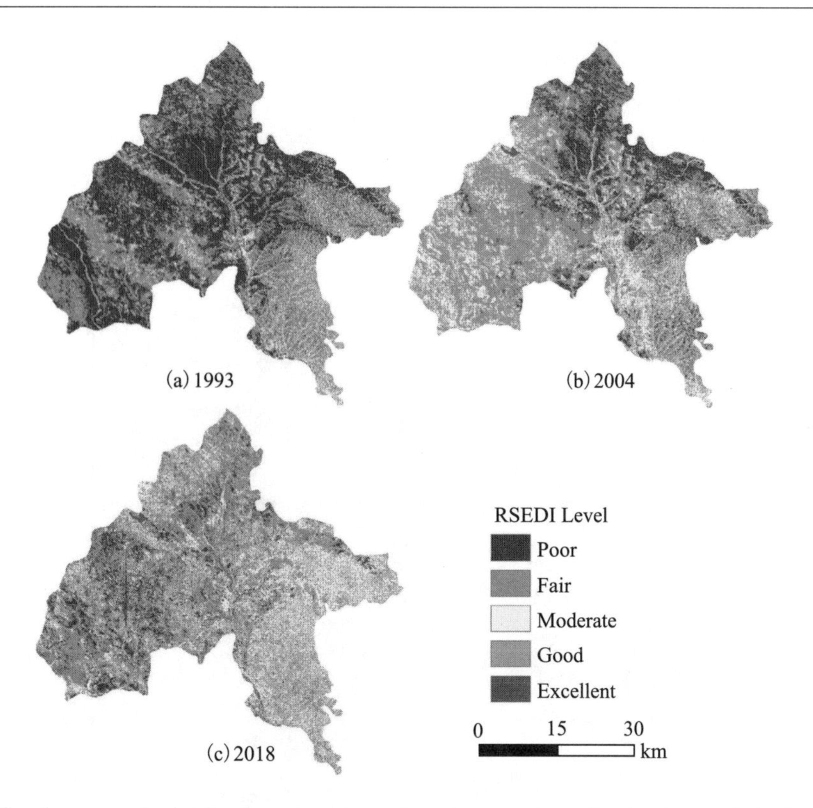

Fig. 3 Spatiotemporal distribution of ecological environment quality in Yuyang District, China

Table 2 Proportions of total land area (%) of Yuyang District, Shaanxi Province, China showing different grades of remote-sensed ecological distance index (RSEDI) from 1993 to 2018

Level	Grading standards	1993	2004	2018
		Percentage/%	Percentage/%	Percentage/%
poor	[0.0,0.2]	39.63	15.93	8.03
fair	(0.2,0.4]	45.18	57.02	45.33
moderate	(0.4,0.6]	10.89	23.96	30.03
good	(0.6,0.8]	2.79	2.65	14.07
excellent	(0.8,1.0]	1.51	0.44	2.55

3.3 Spatiotemporal evolution of ecological environment quality

This study used the difference method to detect changes in the RSDEI in Yuyang District in 1993 and 2018 to identify the spatiotemporal changes to ecological environment quality in Yuyang District from 1993 to 2018. The areas experiencing change were classified into five classes: (1) drastically degraded (≤ -2); (2) mildly degraded (-1); (3) largely unchanged (0); (4) mildly improved (1); and (5) drastically improved (≥ 2). Fig. 4 and Table 3 show the areas of the different classes as a proportion of total area. As shown in Table 3, the area of improved ecological environment quality between 1993 and 2018 accounted for 63.81% of Yuyang District. Over the 25-year study period, ecological environment quality in Yuyang District mainly improved or remained unchanged. Of the total area of Yuyang District, 31.75%, 48.29%, and 15.52% fell into the largely unchanged, mildly improved, and drastically improved classes, respectively. The area showing a degraded ecological environment quality in Yuyang District was relatively small, accounting for only 4.44% of the total area. As shown in Fig. 4, areas showing improved ecological environment quality in Yuyang District had a scattered distribution. Areas falling into the largely unchanged class showed a scattered distribution in the western part of the study area. Areas showing a degraded ecological environment quality were mainly distributed in urban areas and in the western and northeastern parts, distributed in a strip, point, and block pattern, respectively. The spatiotemporal change in ecological environment quality in Yuyang District during 1993-2018 generally showed a declining trend in urban areas and an increasing trend in the surrounding areas.

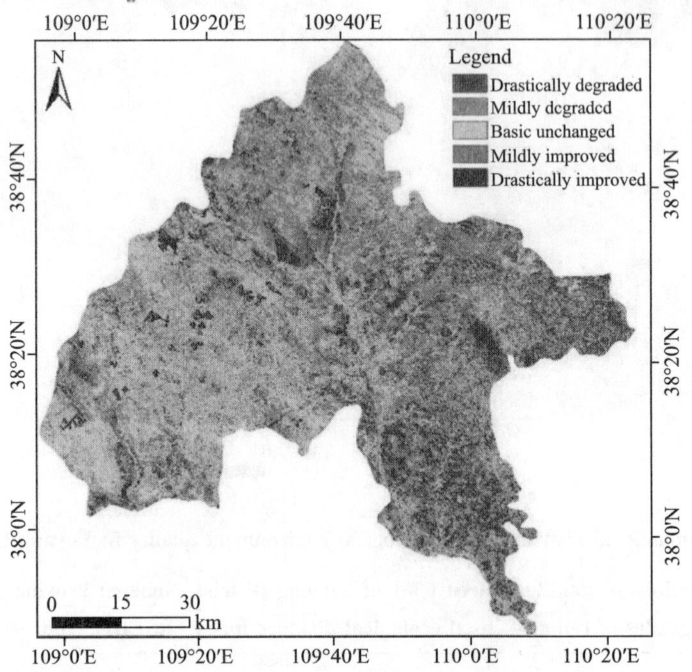

Fig. 4 Spatial distribution in changes to ecological environment quality in Yuyang District, Shaanxi Province, China from 1993 to 2018

Table 3 Percentage change in remote-sensed ecological distance index (RSEDI) in Yuyang District, Shaanxi Province, China from 1993 to 2018

Type	Level	Variation type	Level Percent/%
Degraded	≤-2	drastically degraded	0.79
	-1	mildly degraded	3.65
Unchanged	0	basic unchanged	31.75
Improved	1	mildly improved	48.29
	≥2	drastically improved	15.52

3.4 Responsive relationship between ecological environment quality and land use type

Fig. 5 shows the spatial distribution of land use types in Yuyang District. Superposition analysis was used to identify the RSEDI of different land use types in Yuyang District in 2018 (Table 4). The results showed that the land use types in Yuyang District are mainly grassland, cropland, and bare land, accounting for 62.56%, 19.91%, and 12.44% of the area of Yuyang District, respectively, among which grassland was spread all over Yuyang District and cropland and bare land were distributed between grassland in blocks. There was clear variability in the RSEDI among different land use types. Water and wetland were the two land use types with the highest ecological environment quality in the study area, with

RSEDI values of 0.91 and 0.70, falling in the excellent and good ecological grades, respectively. Cropland, woodland, and grassland followed with RSEDI values of 0.55, 0.53, and 0.50, respectively and all falling in the medium ecological grade. Built-up land and bare land showed the lowest RSEDI values of 0.37 and 0.26, respectively.

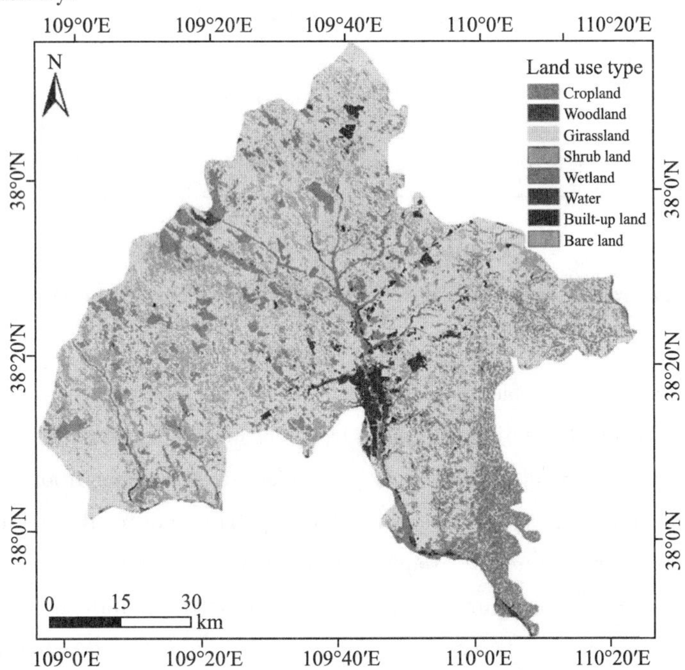

Fig. 5 Spatial distribution of land use types in Yuyang District, Shaanxi Province, China

Table 4 Remote-sensed ecological distance index RSDEI of different land use types in Yuyang District, Shaanxi Province, China in 2018

Land use type	Area/km^2	Percent/%	RSEDI	ecological rating
Cropland	1402.44	19.91	0.55	moderate
Woodland	20.22	0.29	0.53	moderate
Grassland	4406.12	62.56	0.41	moderate
Shrub land	98.50	1.40	0.50	moderate
Wetland	0.69	0.01	0.70	good
Water	32.16	0.46	0.91	excellent
Built-up land	206.68	2.93	0.37	fair
Bare land	875.87	12.44	0.26	fair

3.5 Responsive relationship between ecological environment quality and hydrothermal conditions

Fig. 6 shows the annual rainfall and average annual temperature of Yuyang District from 1991 to 2018. As shown in Fig. 5, there was an increasing trend in average temperature during the study period, with a rate of 0.0434 ℃/a, indicating a pronounced warming trend in Yuyang District. The study area showing a fluctuating increasing trend in annual rainfall at a rate of 9.213 mm/a, indicating a rising trend in air humidity in Yuyang District from 1991 to 2018.

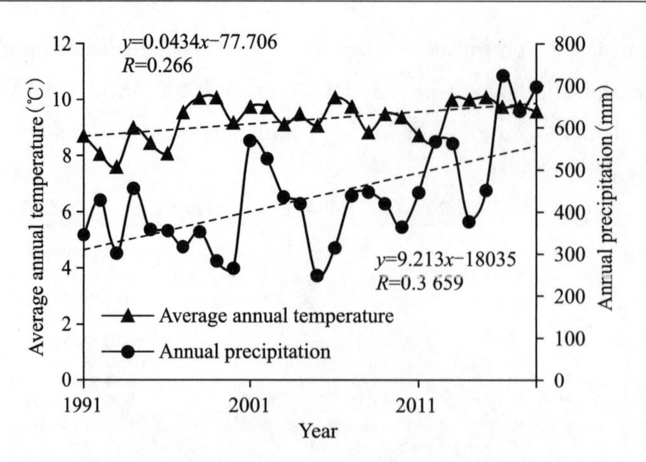

Fig. 6 Variations in annual rainfall and temperature in Yuyang District, Shaanxi Province, China from 1991 to 2018

4 Discussion

The present study exploring the factors influencing ecological environment quality in Yuyang District. The results showed that climate change is resulting in trends of increasing temperature and rainfall in Yuyang District, consistent with the results of other studies[26-27]. Water and heat conditions are the two most important natural conditions for promoting vegetation growth and vegetation cover, thereby improving the quality of the local ecological environment. Therefore, water and heat conditions play an important role in the process of improving the ecological environment in Yuyang District. However, significant variability in the trends of the heat and dryness indices, as indicators of temperature and rainfall, respectively, were evident in the study area. The heat and dryness indices continued to decline, despite rising temperature and rainfall. This result indicates the existence of other factors contributing to the trends in the heat and dryness indices.

There is an obvious correlation between the ecological environment index and land use types in Yuyang District, with the two factors showing coinciding spatial distributions. This result indicates that the greenness and dryness indices contribute the most to the evaluation of ecological environment quality, consistent with the findings of Wang et al.[28]. The effects of changes in land use type on ecological environment quality can be both positive or negative, with water bodies, wetland, woodland, grassland, and shrubland contributing to increased ecological environment quality. Therefore, these land use types can be collectively called "ecological land". An increase in vegetation cover has a positive effect on ecological environment quality. Built-up land and bare land decrease ecological environment quality. These results are consistent with those of Qiu et al.[29].

The "Land-to-Household" policy implemented in China in 1983 greatly encouraged the reclaiming of arable land by local people and ultimately resulted in the areas of cropland and bare land increasing and decreasing, respectively between 1980 and 1998[30]. The policy "Taking Farmland Returning to Forest or Grassland" was implemented in 1998. Yuyang District took the lead in implementing the pilot project of "Return Cropland to Woodland", resulting in 1.11 M mu of other land use types being converted to woodland between 1999 to 2008, including 272300 mu of cropland, 827700 mu of bare land suitable for afforestation, and 10000 mu of mountain area. The area of afforestation and conservation over the entire district reached 4.517 M mu, including 374000 mu of arbor woodland and 121000 mu of economic woodland[31]. The area of bare land continued to decrease from 1999 to 2018, thereby explaining the difference between the heat index and dryness index and hydrothermal conditions. The increase in woodland area and the trends of increasing temperature and rainfall explain the increasing greenness and humidity indices. The above analysis confirms that the implementation of ecological projects in China has promoted changes in land use types, thereby improving ecological environment quality in Yuyang District, Yulin City.

The evaluation of the ecological environment is currently a popular research topic. The dynamic monitoring of the ecological environment based on long-term remote sensing data provides a new platform and perspective for regional ecological environment evaluation[32-33]. The present study used the humidity (WI), dryness (NDSI), greenness (NDVI), and heat (LST) indices to represent ecological environment quality of Yuyang District. The minimum values of the WI and NDVI and the maximum values of sandiness and salinity were taken to represent the least desirable ecological environment quality in the study area to construct the RSEDI. The RSEDI was then used to monitor ecological environment quality. The results showed an increasing trend in ecological environment quality of Yuyang District from 1993 to 2018, mainly influenced by the combination of national reforestation, wind and sand control and other ecological projects and changes in hydrothermal conditions, consistent with the results of existing related studies[34-36]. However, the present study only analyzed changes in land use types and hydrothermal conditions relating to changes in ecological and environmental quality. Other factors contribute to changes to environmental quality, such as population changes and industrial structure. Future research on ecological and environment quality changes should integrate natural conditions and social and economic factors.

5 Conclusions

The present study constructed the RSEDI by integrating various indicators to enhance its applicability and to simplify the calculation method. The spatiotemporal evolution of ecological environment quality of Yuyang District from 1993 to 2018 was quantitatively and objectively analyzed. The driving forces behind the spatiotemporal evolution of ecological environment quality were also described from two perspectives, namely water and heat conditions and ecological policies implemented in China. The result showed that ecological environment quality in Yuyang District from 1993 to 2018 fell in the fair or moderate categories and showed an improving trend over time. The evolution of ecological environment quality of Yuyang District is the result of the combined effects of changes in water and heat conditions and implementation of ecological policies in China, with the latter being the dominant driving force. Urbanization had negative impacts on the ecological environment of Yuyang District, whereas the increase in rainfall and implementation of ecological policies had positive impacts on the ecological environment. Therefore, future restoration of the ecological environment of Yuyang District should focus on implementation of ecological policies and further greening of the city. These approaches can increase vegetation coverage and control the speed of expansion and development of built-up area.

Acknowledgments

This study was supported by the Scientific Research Item of Shaanxi Provincial Land Engineering Built-up Group (DJNY2021-33). And the authors gratefully acknowledge researchers at the Institute of Land Engineering and Technology, Shaanxi Provincial Land Engineering Construction Group, for their help with the field experiments. We wish to thank the editor of this journal and the anonymous reviewers during the revision process.

Author contributions

L. Y. managed the entire research project and also analyzed and considered the research materials. L. S. collected and analyzed the meteorological data; J. W. and Y. W. drew the figures for this paper; L. Y. and J. W. reviewed and edited the paper. All authors have read and agreed to the published version of the manuscript.

Conflicts of Interest

The authors declare no conflict of interest.

References

[1]Peng J, Wu J S, Pan Y J, Han Y N. Evaluation for regional ecological sustainability based on PSR Model: conceptual framework. PROGRESS IN GEOGRAPHY. 2012;31(7):933-940.

[2]Yao J Q, Yang Q, Chen Y N, Hu W F, Liu Z H, Zhao L. Climate change in arid areas of Northwest China in past 50 years and its effects on the local ecological environment. Chinese Journal of Ecology. 2013;32(5):1283-1291.

[3]Guo Z C, Wei W, Shi O J, Zhou L, Wang X F, Li Z Y, et al. Spatiotemporal changes of land desertification sensitivity in the arid region of Northwest China. ACTA GEOGRAPHICA SINICA. 2020;75(9):1948-1965.

[4]Liu R, Wang S X, Zhou Y, Yao Y, Han XD. Ecological environment condition evaluation mode of county region based on remote sensing techniques. China Environmental Science. 2012;32(1):181-186.

[5]Yang L Y, Cheng J, Li Y N. Spatial and temporal variation of NDVI and its response to hydrothermal conditions in northern shaanxi from 2000 to 2018. Journal of Irrigation and Drainage. 2020;39(5):110-118.

[6]Yang L J, Zhang X H, Pan J H, Yang Y C. Coupling coordination and interaction between urbanization and eco-environment in Cheng-Yu urban agglomeration, China. Chinese Journal of Applied Ecology. 2021;32(3):993-1004.

[7]Huang D M, Liu X Y, Zheng Q C, Liu J. Efects of polycentric mode on the coupling and coordinated development between urbanization and ecological environment: A case study of two metropolitan areas in Fujian Province. Acta Ecologica Sinica. 2020;40(21):7886-7896.

[8] Feng Y X, Li G D. Interaction between urbanization and eco-environment in Tibetan Plateau. ACTA GEOGRAPHICA SINICA. 2020;75(7):1386-1405.

[9]Yu Y H,Zhang W T, Wang J B, Yang S. Prediction of Wuhan Urban Agglomeration urban heat island and its response to land use. Environmental Science & Technology. 2018;41(12):158-168.

[10]Lu H M, Li F, Zhang M L, Yang G, Sun W W. Effects of Landscape Pattern on Annual Variation of Thermal Environment in Hangzhou. Remote Sensing Technology and Application. 2018;33(3):398-407.

[11]Ma Y G, Tashpolat T, Huang Y, Yang J L. Effects of landscape pattern change on urban heat-island effect in arid areas —a case study in Urumqi. ARIDZONE RESEARCH. 2006;23(1):172-176.

[12]Huang Y H, Yan H W, Li X J, Wu X S, Wang Z. Monitoring and evaluation of remote sensing ecologicar distance index in Urumqi city. Remote Sensing Information. 2019;34(6):72-77.

[13]Peter C, Schnlze. Overview: Measures of Enviromental Performance and Ecosystem Condition. Washington DC:National Academy Press, 1999.

[14]Xu H Q. A remote sensing index for assessment of regional ecological changes. China Environmental Science. 2013;33(5):889-897.

[15]Shang L Z, Zhang L S. Quantitative evaluation of ecological vulnerability in gansu counties based on "cause-result" index. Soil and Water Conservation in China. 2010;2010(6):11-13,23.

[16]Zhang J. The impact of the oasis development on ecological environment—take Guazhou-Dunhuang basin as an example. Lanzhou University. 2016.

[17]Shi S E, Wei W, Yang D, Hu X, Zhou J J, Zhang Q. Spatial and temporal evolution of eco-environmental quality in the oasis of Shiyang River Basin based on RSED. Chinese Journal of Ecology. 2018;37(4):1152-1163.

[18]Zhang C F. Optimization of water resources allocation and harmony of allocation system in Yuyang District of Yulin City considering uncertainty. Northwest A&F University. 2017.

[19]Feng J M, Guo L X, Li X H. Analysis of ecological vulnerability in Yuyang District based on landscape pattern. Research of Soil and Water Conservation. 2016;23(6):179-184.

[20]Nie X R, Hu Z Q, Zhu Q, Ruan M Y. Research on Temporal and Spatial Resolution and the Driving Forces of Ecological Environment Quality in Coal Mining Areas Considering Topographic Correction. Remote Sens. 2021;13,2815.

[21]Crist E P. A T M Tasseled Cap equivalent transformation for reflectance factor data. Remote Sensing of Environment. 1985;17(3):301-306.

[22]Baig M H A, Zhang L F, Shuai T, et al. Derivation of a tasselled cap transformation based on Landsat 8 at-satellite reflectance. Remote Sensing Letters, 2014, 5(5): 423-431.

[23]Gherardo C, Francesca G, Erica M, Saverio F, Davide T, Raffaello P, Joanne C. Monitoring clearcutting and subsequent rapid recovery in Mediterranean coppice forests with Landsat time series. Annals of Forest Science: A journal of the French National Institute for Agriculture, Food and Environment (INRAE), 2020, 77(1): 453-472.

[24]Zha Y, Gao J, Ni S. Use of normalized difference built-up index in automatically mapping urban areas from TM imagery. International Journal of Remote Sensing, 2003, 24(3): 583-594.

[25]XU H Q, Lin Z L, Pan W H. Some issues in land surface temperature retrieval of Landsat thermal data with the single-channel algorithm. Geomatics and Information Science of Wuhan University. 2015;40(4):487-492.

[26]Chen L X, Zhou X J, Li W L, Luo Y F, Zhu W Q. The characteristics and formation mechanism of climate change in China in the past 80 years. Acta Meteorologica Sinica. 2004;2004(5): 634-646.

[27]Liu X Q, Chen Y S, Liu Y S, Li T S, Lei, M, Rui Y, et al. Statistical Characteristics of Climate Change during 1974—2012 in Yulin, Shaanxi, China. Journal of Desert Research. 2017;37(2): 355-360.

[28]Wang Z J, Dai L. Assessment of land use/cover changes and its ecological effect in karst mountainous cities in central Guizhou Province: taking Huaxi District of Guiyang city as a case. Acta Ecologica Sinica. 2021;41(9):3429-3440.

[29]Qiu Q. Study on land use change and its impact on eco-environmental quality of Lin xia zhou. Xi'an University of Science and Technology. 2020.

[30]Bi G H, Yang Q Y, Zhang J Y, Cheng X Y. China's Rural Land Institutional Reform in the Past 40 Years since the Reform and Opening Up and Its Future Directions. China Land Science. 2018,32(10):1-7.

[31]Dang J J, Fu S P. Effects of the Grain for Green Project in the Development of Rural Area in the Northern Shannxi Province—A Case Study in Yuyang District. Research of Soil and Water Conservation, 2010, 17(6): 275-277,282.

[32]Xinghua Li, Hongyi Zhang, Junbo Yu, Yuting Gong, Xiaobin Guan, and Shuang Li "Spatial-temporal analysis of urban ecological comfort index derived from remote sensing data: A case study of Hefei, China," Journal of Applied Remote Sensing 15(4), 042403.

[33]Wang J, Liu D W, Ma Jiali, Cheng Y N, Wang L X. Development of a large-scale remote sensing ecological index in arid areas and its application in the Aral Sea Basin. Journal of Arid Land, 2021, 13(1): 40-55.

[34]Sun C J, Zhang W Q, Li X G, Sun J L. Evaluation of ecological effect of gully region of loess plateau based on remote sensing image. Transactions of the Chinese Society of Agricultural Engineering. 2019;35(12):165-172.

[35]Ma F, Zhuo J, He H J, Han S S. Ecological evolution and driving mechanism of vegetation in Yulin City, Shaanxi Province. Bulletin of Soil and Water Conservation. 2020;40(5):257-261.

[36]Wang Z Y, Chen X Y, Ma C S. Changes of soil erosion and ecological service value before and after implementing the project of returning farmland to forest in Yulin of Northern Shaanxi. Journal of Northwest Forestry University. 2021;36(3): 59-67.

本文曾发表于2022年《Ope Geosciences》第13卷

Study on the Spatiotemporal Variability of Soil Nutrients and the Factors Affecting Them: Ecologically Fragile Areas of the Loess Plateau, China

Liheng Xia[1], Ling Li[2], Jing Liu[3]

(1.Institute of Land Engineering and Technology, Shaanxi Provincial Land Engineering Construction Group Co., Ltd., Xi'an 710075, China; 2.Agricultural Inspection and Testing Center of Shaanxi Province, Xi'an 710014, China; 3.College of Natural Resources and Environment, Northwest Agriculture and Forestry University, Yangling 712100, China)

【Abstract】 The spatial variability of soil nutrients can improve the yield per unit area of food crops and protect the agricultural ecological environment. Geostatistics and geographic information system (GIS) technology were applied to analyse the spatiotemporal variability and main influencing factors in soil organic matter (SOM) and soil total nitrogen (STN) in ecologically fragile regions of the Loess Plateau, China. The results showed that the mean SOM and STN contents significantly increased in the past 40 years. Compared to the 1980 data, both SOM and STN global Moran's I indices are lower, with less spatial structure and an increased role for stochastic factors. The centre of gravity of soil nutrients in the study area mainly to the south-east, a reduction in the area of the ellipse and a tendency to concentrate the spatial distribution of nutrients. And it was concluded that the variation in nutrients was mainly influenced by fertiliser management practices. Therefore, a long time series of soil nutrient study is of great research value to clarify the soil nutrient status of the study area and to improve the efficiency of arable land use, as well as providing a theoretical basis for precision agriculture.

【Keywords】 Cultivated land; Geostatistics; Geographic information system; Centre of gravity

1 Introduction

Cultivated land soil nutrients are essential nutrients for plant growth and development and are important for ensuring the quality of cultivated land and grain yield (Keskinen et al., 2019). Soil organic matter (SOM) and soil total nitrogen (STN) are important factors that determine soil fertility, agricultural product yield and quality. Due to the combined effects of soil type, topography and human activities, soil nutrients have a high degree of spatial variability (Huang et al., 2012; Wang et al., 2009). Studying the temporal and spatial variability laws of soil nutrients has important theoretical and practical guiding significance for precise nutrient management and the promotion of farmland ecological and economic benefits.

In the early 2000s, domestic scholars began to use geostatistics combined with GIS technology to explore the spatiotemporal variability laws of soil characteristics (Xu et al., 2004). Some studies have been conducted at different spatial scales, such as the scale of farmland or paddy field (Duan et al., 2020; Liu et al., 2004, 2014), the watershed scale (Wei et al., 2008), and county or larger scales (Chen et al., 2016; Hu et al., 2014; Osat et al., 2016). Recently, an increasing number of studies have focused on the spatiotemporal variability of soil nutrients, e.g., for an agricultural county located in the southern Loess Plateau, China (Chen et al., 2016), five subcatchments in the Ping Gu intermontane basin in Beijing (Zhuo et al., 2019), and the hilly area of the Taihu Lake basin of China (Liao et al., 2017). These studies have provided more in-depth descriptions of the spatiotemporal change process of soil nutrients.

With its advantage of interpolation, the geostatistics method has been widely used in the study of soil spatial variability (Aghasi et al., 2017; Foroughifar et al., 2013). Most scholars have successfully used spatial autocorrelation analysis (Blanchet et al., 2017) and semivariogram analysis (Liu et al., 2013) to describe the spatial variability of soils. However, to date, most studies on the spatiotemporal variability of soil nutrients have used only two phases of data (Guo et al., 2001; Chuai et al., 2012; Wang et al., 2012), and there have been few reports comparing three phases of data with a longer time span. In addition, current studies generally use semivariograms to quantitatively describe spatial characteristics (Hoffmann et al., 2014), and few studies have analysed the spatiotemporal variability laws of soil nutrients from different perspectives. This limited approach will result in researchers not fully understanding the laws of spatiotemporal variability of soil nutrients, which will introduce uncertainty to the formulation of regional soil management measures.

Therefore, this study selects Baishui County, which is a typical region ecologically fragile regions of the Loess Plateau, China, as the research area. The SOM and STN contents in 1980, 2007 and 2020 are employed as the research objects. This study is based on geostatistics and Geographic Information System (GIS) technology, using spatial autocorrelation and Centre of gravity models to reveal the dynamics of the spatial distribution of SOM and STN, and to explore the main factors of nutrient content variation.

The main objectives of this study are (1) to characterise and compare the spatial variability of farmland SOM and STN in the dry plateau area of Weibei; (2) to quantify their spatial distributions and temporal changes; and (3) to clarify the main factors influencing the spatial variability of SOM and STN. Under the current situation of uneven distribution of nutrients in cultivated land, improve the use efficiency of low-nutrient cultivated land, protect the use efficiency of high-nutrient cultivated land, and provide a realistic basis for precision agriculture.

2 Materials and methods

2.1 Description of the study area

Baishui County is located in the transition zone between the Guanzhong Plain and the North Shaanxi Plateau, China, between 109°16′~109°45′ E and 35°4′~35°27′ N (Fig. 1). The total area of the district is 986.6 km^2, of which the cultivated area is 525.4 km^2, accounting for 53.2% of the total area. The area is high in the northwest and low in the southeast, with an altitude between 440 m and 1500 m. Baishui County has a temperate continental monsoon climate, an average annual temperature of 11.4 ℃, and an average annual precipitation of 577.8 mm. Due to the cutting of each branch gully of the Luo River and Baishui River, the study area has criss-crossed gullies and broken topography. The dominant soil types include loessial soil and cumulic cinnamon soil.

2.2 Data collection

Soil data, including the SOM and STN utilised in this study, were obtained from three soil sampling surveys: The second national soil survey in 1980, the 2007 survey of cultivated land quality in Shaanxi Province, and the 2020 survey of cultivated land quality in Shaanxi Province. The three datasets were named "Soil attribute-1980", "Soil attribute-2007" and "Soil attribute-2020". The average temperature and climate data were obtained from the Resource and Environment Science Data Center of CAS with a spatial resolution of 1 km×1 km and time taken from 2020.

In the 1980 soil survey, 2007 soil survey and 2020 soil survey, 154 farmland topsoil samples, 156 farmland topsoil samples and 73 farmland topsoil samples (0~20 cm), respectively, were collected from May to June (dry season) after the summer grain harvest. The sampling location was determined by the main topography, soil type and distribution of the sample plot, and the Global Positioning System (GPS) was employed to record the points while avoiding roads, residential areas and other easily disturbed areas when sampling. At each sampling point, 6~8 points were randomly selected and mixed into a soil sample. The collected soil samples were ventilated and dried, and the impurities were removed and finely ground for determination of the SOM and STN content.

In the 1980 soil survey, the SOM content of all 154 soil samples was obtained by the $K_2Cr_2O_7-H_2SO_4$ oxidation method, and the STN of 152 samples was measured by the Kjeldahl method (Agricultural Chemistry Committee of China, 1983). According to this method, the SOM and STN of the soil samples from the 2007 and 2020 surveys were analysed. The sampling point distribution is shown in Fig. 1.

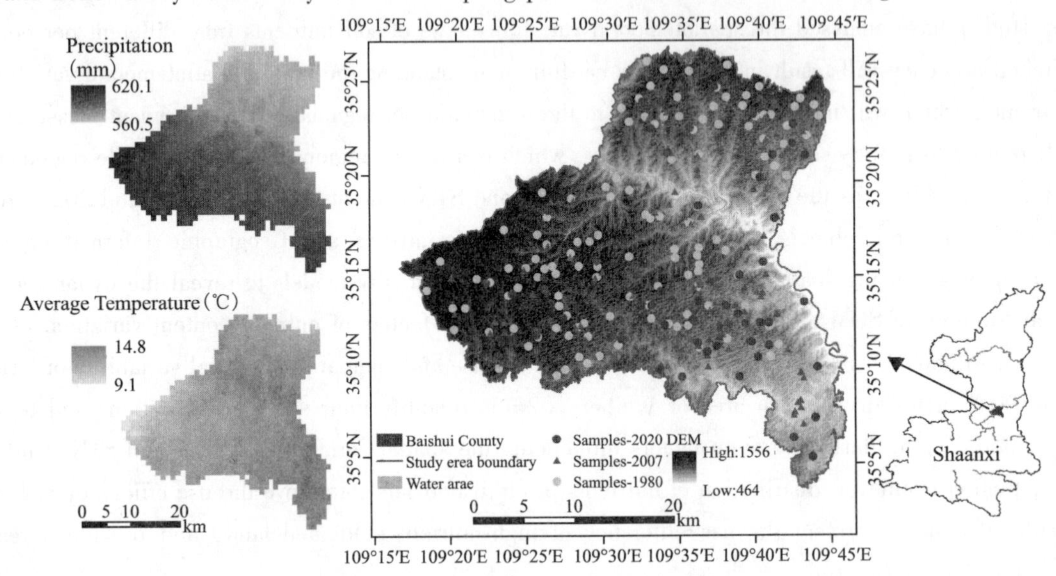

Fig. 1 Map of the soil sample distribution

2.3 Data processing and analysis

2.3.1 Data processing

The existence of outliers and the non-normal distribution of data can easily cause the proportional effect of the variogram and increase the estimation error. In this study, we applied the $A\pm3$ s method to eliminate outliers, where A denotes the average value for each variable and s is its standard deviation (Liu et al., 2009). Data that exceed the value ($A\pm3$ s) were obtained from the raw dataset; we replaced them with the maximum or minimal value of the dataset without outliers. In SPSS 20.0, a one-sample normality test (K-S test) was performed on the sample data after removing outliers. The R-language was applied to perform Box-Cox conversion of the data that did not conform to a normal distribution. The converted data passed the K-S test. The specific parameters are shown in Table 1.

Table 1 Soil nutrient description statistics of cultivated land in three periods (g/kg)

Variables	Samples	Mean ± standard deviation	Minimum	Maximum	Coefficient of variation (%)	P_{k-s}
SOM-1980	154	10.53 ± 2.36	4.71	17.74	22.4	0.14 *
STN-1980	152	0.60 ± 0.09	0.32	1.05	15.6	0.40 *
SOM-2007	156	12.49 ± 5.01	2.60	26.40	38.47	0.78 *
STN-2007	156	0.64 ± 0.37	0.18	1.11	37.16	0.74 *
SOM-2020	73	15.42 ± 4.23	7.96	27.53	27.23	0.43 *
STN-2020	73	0.73 ± 0.19	0.38	1.32	27.26	0.74 *

Level of significance: * $P<0.05$.

In this study, we compiled descriptive statistics of soil nutrients using Excel 2016 and performed Kriging interpolation to estimate the spatial distribution and spatial changes of SOM and STN in different years. We calculated the global Moran's I values using ArcGIS 10.6. Sample-independent t tests and a correlation analysis were carried out using SPSS 16.0. A geostatistics analysis was performed by GS +9.0.

2.3.2 Spatial autocorrelation analysis

Spatial autocorrelation refers to the potential dependence of the same variable at different spatial positions and is a statistical method used to test the correlation between adjacent positions of the studied variables in space (Gelaw et al., 2014). When judging the spatial autocorrelation, the global Moran's I index is often employed to reflect the spatial autocorrelation of the study area. The calculation formula is as follows:

$$I = \frac{n}{S_0} \times \frac{\sum_{i=1}^{n}\sum_{j=1}^{n} w_{ij}(X_i - \overline{X})(X_j - \overline{X})}{\sum_{i=1}^{n}(X_i - \overline{X})^2} \quad \text{(Eq.1)}$$

Where n is the total number of samples, w_{ij} is the symmetric binomial spatial weight matrix, and X_i and X_j are the measured values of spatial variable X at different positions i and j, respectively. The value of I ranges from -1 to 1. When $I>0$, it means that there is a positive spatial correlation, and larger values indicate stronger spatial correlation. When $I<0$, it means there is a negative spatial correlation, and smaller values indicate more obvious spatial differences (Darand et al., 2017).

2.3.3 Semivariogram analysis

The semivariogram is a commonly used function in geostatistical analysis. It can accurately describe the characteristics of the spatial variability of a study area based on known sample points. The semivariogram can reveal the internal relations of variables, and the relationship between spatial points in different distances and different directions can be used to obtain the spatial distribution law of variables, which makes the spatial interpolation more accurate (Awais et al., 2017); its calculation formula is as follows:

$$r(h) = \frac{1}{2N(h)}\sum_{i=1}^{N(h)}[Z(x_i) - Z(x_i + h)]^2 \quad \text{(Eq.2)}$$

Where h is the spatial interval of the sampling points, $N(h)$ is the number of samples with interval distance h, and $Z(x_i)$ and $Z(x_i+h)$ are the measured values of spatial variable $Z(x)$ at the different positions of xi and x_i+h, respectively (Balaguer-Beser et al., 2013). There are three important parameters in the semivariogram: the nugget value (C_0), the sill value (C_0+C) and the range (A). When the step size of h is 0, then $r(h)$ is the nugget value. As the step size of h increases, the semivariogram tends to be in a stable state. At this time, $r(h)$ is the sill value and h is the range (Onyejekwe et al., 2016).

2.3.4 Centre of gravity models and standard deviational ellipse

The centre of gravity model is mostly used to study the process of change in the spatial location of a geographical element in the process of regional development. The model reflects the changing trend of the spatial element through the direction, distance and speed of the centre of gravity migration. This paper uses a centre of gravity model to reveal the spatial aggregation characteristics and trends of soil nutrients. Its calculation formula is as follows:

$$\overline{X} = \sum_{i=1}^{n}\frac{M_i X_i}{\sum_{i=1}^{n} M_i} \quad \text{(Eq.3)}$$

$$\overline{Y} = \sum_{i=1}^{n}\frac{M_i Y_i}{\sum_{i=1}^{n} M_i} \quad \text{(Eq.4)}$$

where $\overline{X}$, $\overline{Y}$ are the soil nutrient coordinates at the beginning of the study; M_i is the nutrient content of the ith sample point, g/kg; X_i, Y_i denote the coordinates of the ith sample point.

The standard deviation ellipse is a visual representation of the state of aggregation of soil nutrients and their tendency to shift, and consists mainly of the angle of rotation θ, the standard deviation along the major axis (long axis) and the standard deviation along the minor axis (short axis). The long half-axis of the ellipse indicates the direction of the soil nutrient distribution and the short half-axis indicates the extent of the soil nutrient distribution. In this study, standard deviation ellipses were constructed to reflect the spatial pattern of soil nutrients on the basis of the centre of gravity model.

3 Results

3.1 Descriptive statistics

The descriptive statistical results of the data after eliminating the outliers in the three periods are summarised in Table 1. It can be seen that the average contents of SOM and STN of the cultivated soil in the study area in 1980 were 10.53 g/kg and 0.60 g/kg, respectively. In 2007, the average SOM content and STN content in the cultivated soil in the study area were 12.49 g/kg and 0.64 g/kg, respectively; in 2020, the corresponding values were 15.42 and 0.73 g/kg, respectively. Compared with 1980, the average SOM content and STN content in 2007 increased by 2.48 g/kg and 0.04 g/kg, respectively. From 2007 to 2020, the contents of the two nutrients increased by 2.41 g/kg and 0.09 g/kg, respectively. Although the data do not depend on the sample t test, the mean value presents a significant difference (Table 2). The coefficient of variability of the soil nutrients in the cultivated land in the study area in the three phases was between 15.6% and 38.47%, which indicated moderate variability.

Table 2 Significance test of soil nutrients in 1980, 2007 and 2020

Variables	P	Variables	P
SOM (1980–2007)	0.005**	SOM (2007–2020)	0.002**
STN (1980–2007)	0.03*	STN (2007–2020)	0.021*

Level of significance: * $P<0.05$; ** $P<0.01$.

3.2 Spatial autocorrelation analysis

The spatial statistical results of the soil nutrients of the cultivated land in the three phases are shown in Table 3. It can be seen from the table that the global Moran's I index of SOM and STN decreased from 0.41 and 0.44 to 0.26 and 0.17, respectively, from 1980 to 2007. In 2020, the global Moran's I index of SOM and STN were 0.25 and 0.21, respectively. By normalizing the global Moran' s I index, all Z values were greater than 2.56, which indicates that at the current sampling density, the three-stage SOM and STN show a very significant positive spatial correlation at the 0.01 statistic level, and their spatial distribution is characterised by agglomeration. By further comparing the global Moran's I index, the global Moran's I index of SOM and STN in 2007 and 2020 is higher than that of SOM and STN in 1980. This finding shows that SOM and STN in 1980 have stronger spatial dependence and better spatial structure, with weaker random variability.

3.3 Geostatistics analysis

The optimal fitting of the theoretical model revealed that SOM-1980, STN-1980, SOM-2020 and STN-2020 fit the exponential model, while SOM-2007 and STN-2007 conform to the linear model. The coefficient of determination (r^2) is between 0.785 and 0.995; the proximity of these values to 1 indicates that the excellent fit of the semivariogram (Fig. 2). The parameters of the model are listed in Table 3 (i.e., nugget, sill and nugget/sill ratio). The nugget/sill ratio of SOM and STN in the three periods ranges from 42.27% to 76.14%, which indicates that the spatial correlation is moderate and the spatial continuity is average. The nugget/sill ratios of SOM and STN in 2007 were 76.14% and 66.56%, respectively, which is higher than in 1980 (SOM with 42.95% and STN with 42.27%). The nugget/sill ratio of SOM and STN in 2020 has not changed substantially from 2007 and is also higher than that in 1980. The strong spatial variability in 2007 and 2020 indicates that they may be affected by more external factors, such as fertilization and irrigation. These results are consistent with the analysis results of the global Moran's I values.

The spatial range of soil SOM and STN in the three periods is 5050 to 17350 m, which is considerably larger than the average sampling interval (2500~4000 m). This finding shows that the number of sampling points in the three periods is sufficient for analysing the spatial distribution of nutrients on the county scale.

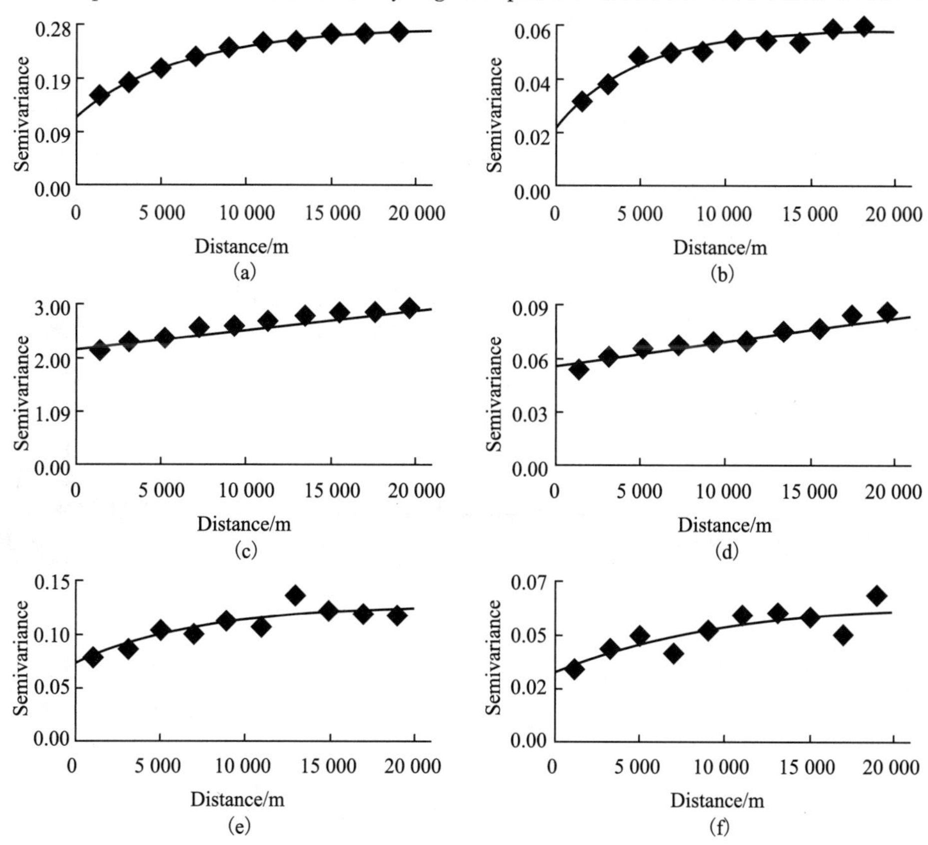

Fig. 2 Fitted models of SOM and STN in 1980 (a-b), 2007 (c-d), and 2020 (e-f). SOM: soil organic matter; STN: soil total nitrogen

Table 3 Semivariance model of soil nutrients in three periods and Moran's I index

Variables	Model	Nugget (C_0)	Sill (C_0+ C)	Nugget/sill (%)	Range (m)	R^2	Moran's I	Z
SOM-1980	E	0.1188	0.2766	42.95	6470	0.995	0.41 * *	32.36
STN-1980	E	0.0238	0.0563	42.27	5050	0.957	0.44 * *	35.48
SOM-2007	L	2.129	2.796	76.14	17350	0.867	0.26 * *	24.89
STN-2007	L	0.0547	0.0823	66.56	16280	0.816	0.17 * *	19.06
SOM-2020	E	0.0748	0.1315	56.88	9950	0.824	0.25 * *	24.96
STN-2020	E	0.0304	0.0638	47.64	11800	0.785	0.21 * *	17.56

E is the exponential model, L is the linear model. Level of significance: * * $P < 0.01$.

3.4 Spatial distribution characteristics

The spatial interpolation results of soil SOM and STN in the three periods are shown in Fig. 3. It can be seen from the Fig. that the spatial distribution characteristics of soil SOM content in the three phases are similar: the distribution characteristics are low in the northwest and high in the southeast (Fig. 3(a) ~ (c)). In 1980, the SOM content in the study area was at a low-medium level, while the content in the central region was relatively low. The area with a content greater than 13 g/kg was only 1.47 km^2. In 2007, the soil SOM content in the study area increased steadily, and the area with a content greater than 13 g/kg increased to 228.06 km^2. In 2020, the SOM content increased significantly, which shows high spatial distribution characteristics in the east and low spatial distribution characteristics in the west. Only 78.98 km^2 remained in the study area where the SOM content was less than 13 g/kg (Table A1 in the *Appendix*).

As SOM and STN have a significant positive correlation, they show similar spatial distribution characteristics (Fig. 3(d) ~ (f)). In 1980, the content of STN was low and mostly concentrated from 0.55 to 0.65 g/kg. In 2007, the STN content was distributed in blocks, and the content in some areas increased significantly, mainly in the southern part of the study area. In 2020, the STN content was at a relatively high level, with a content greater than 0.75 g/kg concentrated in the southeast part of the study area, with an area of 165.37 km^2, which accounts for 30.88% of the total study area (Table A2).

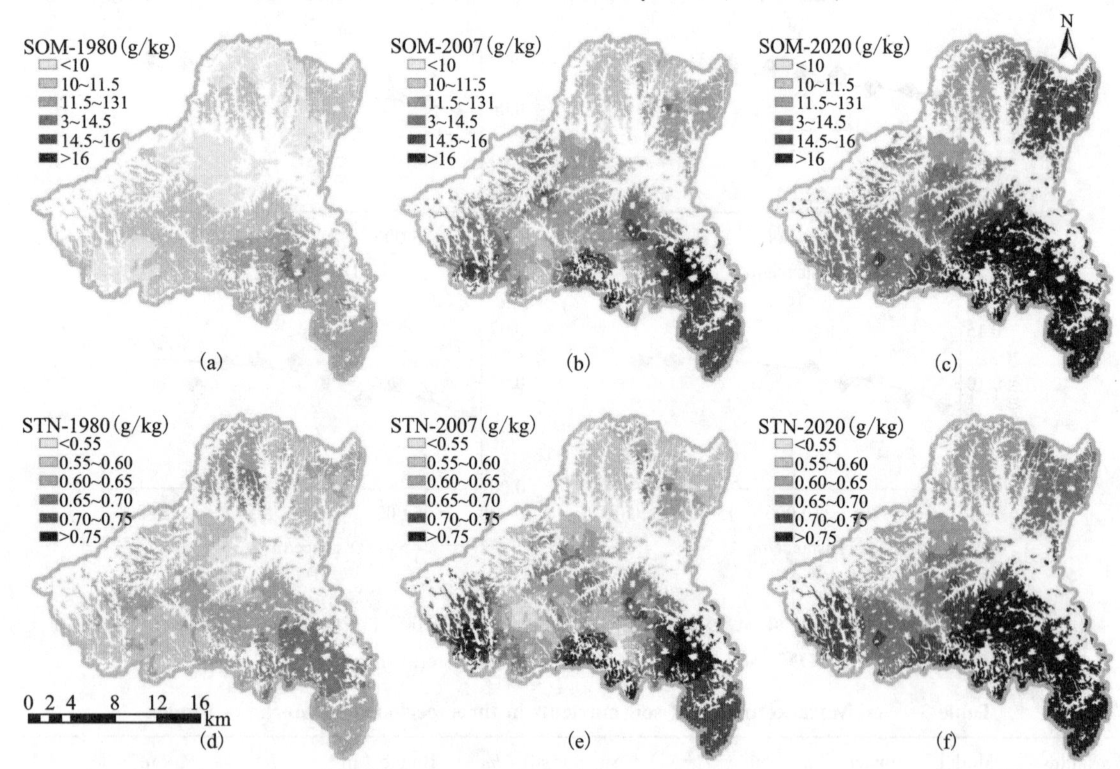

Fig. 3 Distribution map of SOM and STN in three different periods (a) ~ (f). SOM: soil organic matter; STN: soil total nitrogen

3.5 Temporal and spatial characteristics

We employed the ArcGIS vector mask extraction tool and raster calculator to calculate the rate of change according to the formula "(x2007−x1980)/x1980 and (x2020−x2007)/x2007". The results are shown in Fig. 4.

From 1980 to 2007, the soil SOM and STN content in the study area showed an overall increasing trend, but the degree of change was different in different regions. The area where the SOM increased accounted for 94.97% of the study area, with an area of 499.07 km^2. Compared with SOM, the increase in STN was relatively weak, and the increase was mainly concentrated in the range 0 ~ 15%, which accounts for 44.55% of the total area. The data reveals that 31.14% of the cultivated land exhibits a decrease in STN content (Table A3).

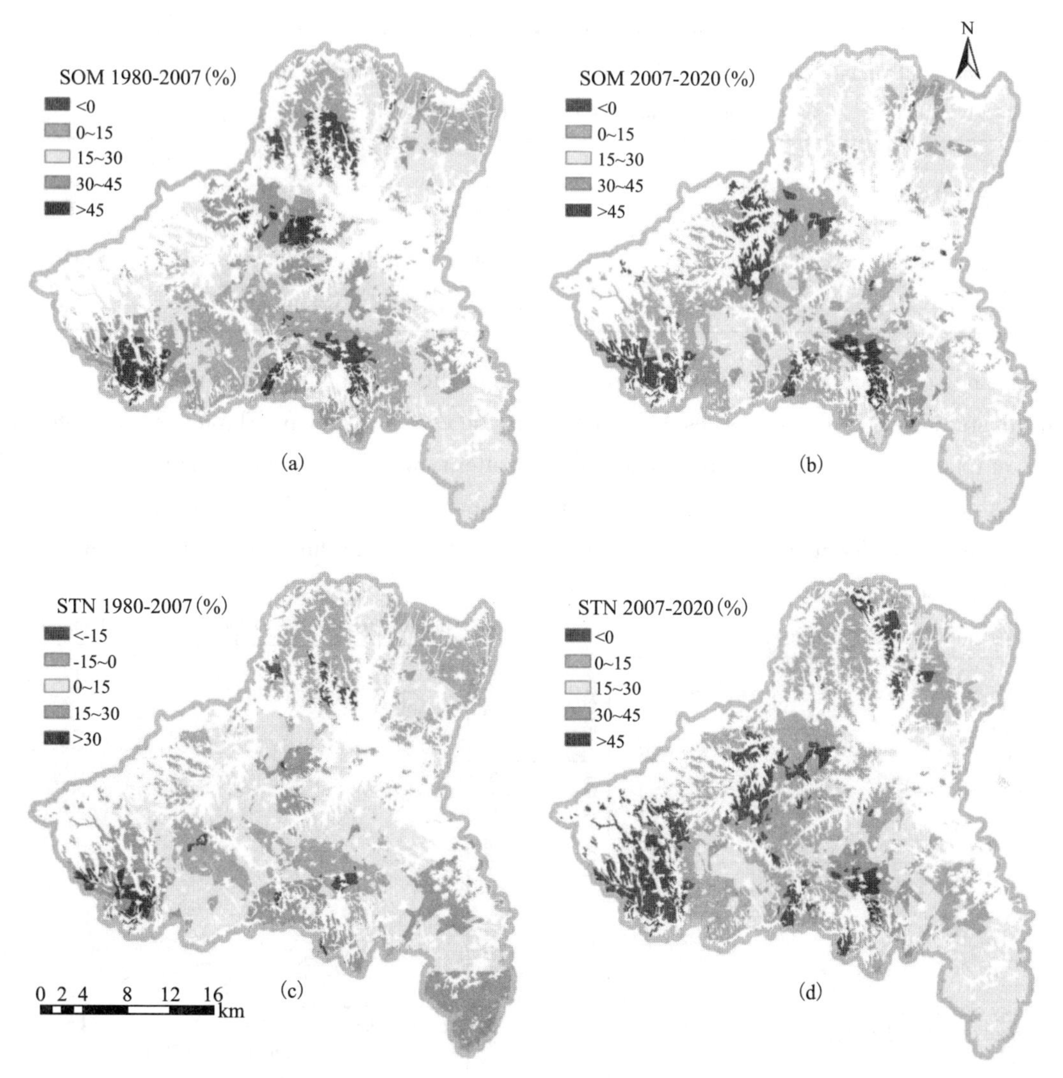

Fig. 4 Spatial distribution map of SOM (a) and STN (b) changes in Baishui County from 1980 to 2007, and the spatial distribution map of the changes in soil SOM (c) and STN (d) Baishui County from 2007 to 2020. SOM: soil organic matter; STN: soil total nitrogen

From 2007 to 2020, the content of SOM and STN in the study area increased steadily, with a relatively large increase in the east and a relatively small increase in the west. The SOM and STN content decreased in only a few areas. The growth rates of SOM and STN are similar, and the growth rates are mainly concentrated from 0 to 30%, which accounts for 79.52% of the study area and 79.75% of the study area, respectively (Table A4).

3.6 Centre of gravity models and standard deviational ellipse

The centre of gravity model was used to obtain the direction and distance of soil nutrient centre of gravity migration for each period respectively, and the results are shown in Tables 4 and 5 and Fig. 5, which show that the nutrient changes in the study area are mainly divided into the following two stages:

From 1980 to 2007, the centre of gravity of soil nutrients shifted towards southeast. Since the 1980s, farmers have been practising intensive farming and have begun to focus on land management, so the centre of gravity of soil nutrients has moved a long way. From 2007 to 2020, the centre of gravity of soil nutrients as a whole shifted 922.96 m and 1030.95 m to the southeast, respectively, a relatively short distance (Table 4). The centre of gravity as a whole shifted to the south-east as the lower ground made it easier for people to cultivate. In summary, it can be seen that the centre of gravity of soil nutrients in the study area has generally shifted to the south-east over the past 40 years, with a decreasing trend in the distance shifted.

Table 4 Soil nutrient gravity shift 1980–2020

Shift of gravity centre	SOM		STN	
	1980–2007	2007–2020	1980–2007	2007–2020
Moving direction	South-east	South-east	South-east	South-east
Movement distance (m)	3579.31	922.76	3903.31	1030.95

As can be seen from Fig. 5, there is a certain directionality to the change in the standard deviation ellipse in the study area, which correlates with the shift in the centre of gravity. Throughout the study period, the angle of rotation shows an "increasing-decreasing" pattern, with an increasing spatial distribution in the southeast. As can be seen from Table 5, the area of the standard deviation ellipse gradually decreases, indicating that the spatial distribution of soil nutrients in the study area gradually tends to concentrate and the nutrient content becomes more stable.

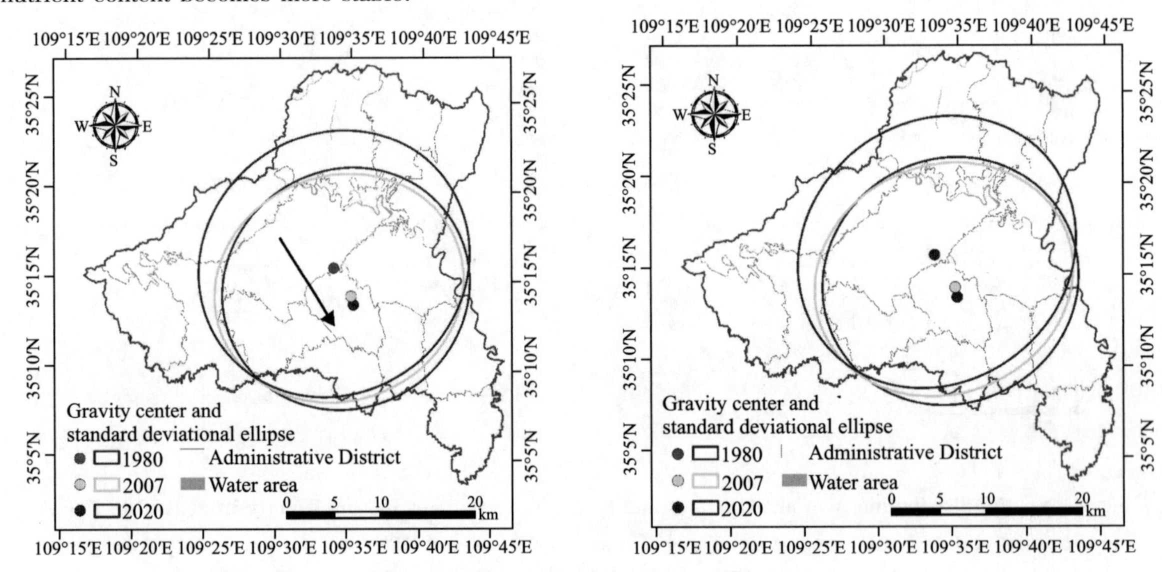

Fig. 5 Soil nutrient gravity shift 1980–2020

Table 5 Changes in standard deviational ellipse parameters for soil nutrients 1980–2020

Year	SOM			STN		
	1980	2007	2020	1980	2007	2020
Rotation (°)	56.32	72.63	61.98	51.49	70.71	63.66
Standard deviation along the x-axis (km)	14.72	13.38	13.29	14.58	13.49	13.31
Standard deviation along the y-axis (km)	13.48	11.63	12.32	13.25	11.59	12.37
Ellipse area (km^2)	623.69	514.61	489.24	618.76	514.61	491.54

4 Discussion

4.1 Temporal changes in SOM and STN

The distribution patterns of soil SOM and STN in the study area are similar. During the 38-year period from 1980 to 2020, the content of soil SOM and STN in Baishui County increased significantly. Generally, some agricultural practices can reduce soil carbon by soil disturbance and mineralization (Ma et al., 2016a). Some studies have focused on the impact of land use changes on SOM and STN content. For example, the conversion from cultivated land to orchard land increases the risk of nutrient loss in the watershed (Chen et al., 2019). Compared with farmland, the organic matter and total nitrogen content in the surface soil of orchard land increases (Lu et al., 2016). When the original grassland is changed to vegetable land (Kong et al., 2006), both the soil organic carbon (SOC) content and STN content increase. More research focuses on the impact of agricultural activities on SOM and STN content. In Kansas, the application of animal droppings (Schlegel et al., 2017) and the presence of plateau pikas in the Qinghai-Tibet Plateau significantly (Yu et al., 2017) increased the STN and SOC content. The average SOC content and STN content in the Tai Lake Basin increased during a 20-year period (1980–2000) (Liu et al., 2014a); and the STN and SOM content in the ecologically fragile area of the Loess Plateau increased (Guan et al., 2020). These findings show that fertilization measures and planting management methods may cause changes in soil nutrients. Therefore, it can be seen that the transformation of land use patterns and various agricultural practices will affect the soil SOM content and STN content to varying degrees. In this study, the general increase in SOM and STN content may be attributed to the implementation of the household contract responsibility system, the distribution of land to households, the widespread use of fertilizers, the improvements in irrigation and drainage facilities, and the intensive cultivation of farmers.

Among the major apple-producing areas in the country, Baishui County is the only county that meets the seven indicators used to identify the most suitable apple production areas, and the region has very superior natural conditions. Therefore, apples have become the main agricultural industry in the county. For apples, the content of organic matter in the soil is very important. It can not only improve the quality of apples but also fertilize the soil and improve the soil environment. The SOM content in the study area showed an increasing trend, which provided a strong guarantee for increasing apple output, rural residents' incomes and sustainable agricultural development.

4.2 Temporal changes of their spatial variability

Judging from the variability of the soil nutrients in the study area, the coefficient of variability of the nutrients in the three phases is between 15.6% and 38.47%, and the spatial heterogeneity is weak. From the spatial autocorrelation and semivariogram analysis, we obtain the same analysis results for the three-phase SOM and STN. The nugget/sill ratios of SOM and STN in 2007 and 2020 are higher, and the global Moran's I index is lower, which indicates that the spatial autocorrelation of SOM and STN is weakened, the distribution tends to be fragmented, and the proportion of random variability increases.

The global Moran's I index describes the spatial aggregation characteristics of the research variables from the perspective of correlation and uses the standard deviation of the approximate normal distribution hypothesis in random conditions to standardise it to determine whether the spatial autocorrelation is significant (or extremely significant). However, the global Moran's I index cannot provide a basis for spatial interpolation and is unable to adequately describe the spatial patterns of variables (Martin et al., 2014). A semivariogram can better compensate for the lack of interpolation of a spatial autocorrelation analysis. A semivariogram can not only quantitatively reveal the spatial correlation degree of regional variables and the scale range of spatial variability using indicators such as the block base ratio and variable range but also perform Kriging interpolation on a parameter basis; however, it cannot provide a statistical test for positive and negative spatial correlation significance, such as Moran's I standardised Z value (Ma et al., 2016).

4.3 Factors that influence an increase in SOM and STN

4.3.1 Topographic influence

To explore the influence of topography, the average SOM and STN content of different elevations and topography types were calculated (Tables 6 and 7). From 1980 to 2007, additional SOM and STN were accumulated in low-altitude areas. For example, SOM and STN content increased by 2.422 g/kg and 0.051 g/kg(<600 m), and high-altitude areas only increased by 1.899 g/kg and 0.034 g/kg (>900 m). This pattern also appeared from 2007 to 2020. This phenomenon may be caused by flat terrain in low-altitude areas, excessive agricultural production activities, and excessive fertilization, which produced greater biological residues and nitrogen accumulation in the soil (Zhu et al., 2019).

Table 6　Average SOM and STN content (g/kg) at different elevations

SOM: soil organic matter; STN: soil total nitrogen

Variables	Elevation			
	<700 m	700~800 m	800~900 m	> 900 m
SOM-1980	11.511 ± 2.374	11.230 ± 2.446	9.986 ± 2.309	10.429 ± 1.975
SOM-2007	13.933 ± 4.132	13.503 ± 5.779	12.148 ± 4.315	12.328 ± 4.537
SOM-2020	17.091 ± 2.895	17.717 ± 4.936	14.419 ± 3.671	13.281 ± 2.487
Increment (1980-2007)	2.422	2.273	2.162	1.899
Increment (2007-2020)	3.158	4.214	2.271	0.953
STN-1980	0.618 ± 0.087	0.635 ± 0.091	0.592 ± 0.090	0.601 ± 0.081
STN-2007	0.669 ± 0.217	0.663 ± 0.261	0.629 ± 0.203	0.635 ± 0.249
STN-2020	0.811 ± 0.131	0.821 ± 0.223	0.683 ± 0.135	0.653 ± 0.145
Increment (1980-2007)	0.051	0.028	0.037	0.034
Increment (2007-2020)	0.142	0.158	0.054	0.018

Table 7 shows the comparison results of the changes in the SOM and STN content for different landform types. From 1980 to 2007, the SOM and STN content in the valley terrace increased by 2.747 g/kg and 0.031 g/kg, respectively, while the increase in SOM and STN content in the middle mountain were negative. The changes in the SOM and STN content from 2007 to 2020 also showed this pattern.

Table 7　Average SOM and STN content (g/kg) for different terrains.

SOM: soil organic matter; STN: soil total nitrogen

Variables	Topography		
	Valley terrace	Loess tableland	Middle mountain
SOM-1980	10.881 ± 2.394	10.496 ± 2.109	10.982 ± 2.949
SOM-2007	13.628 ± 3.754	12.881 ± 4.142	10.400 ± 4.537
SOM-2020	17.354 ± 4.152	15.627 ± 4.210	11.278 ± 3.285
Increment (1980-2007)	2.747	2.385	-0.582
Increment (2007-2020)	3.726	2.746	0.878
STN-1980	0.606 ± 0.077	0.605 ± 0.086	0.590 ± 0.103
STN-2007	0.637 ± 0.193	0.632 ± 0.247	0.540 ± 0.249
STN-2020	0.750 ± 0.008	0.721 ± 0.185	0.574 ± 0.213
Increment (1980-2007)	0.031	0.027	-0.05
Increment (2007-2020)	0.113	0.089	0.034

The study revealed that the valleys and rivers located in the valley terrace of the study area have flat terrain with relatively satisfactory farming performance, which greatly improves the supply and utilization of carbon and nitrogen in the soil (Weihrauch and Opp, 2018). According to the survey results, the use of chemical fertilizers has increased steadily in recent years. The combined application of phosphate fertilizers and organic fertilizers, the increase in agricultural production input and the high degree of cultivation and maturation facilitate the accumulation of organic matter, which causes a steady increase in the SOM content in the study area.

4.3.2 Soil-type influence

The main soil types in Baishui County mainly include cumulic cinnamon soil and loessial soil. The effects of two soil types on the SOM and STN content were calculated separately (Fig. 6). It can be seen from the figure. that there are significant differences in the content of SOM and STN in the two soil types. For example, in 1980, the average SOM content and STN content of Cumulic cinnamon soil were 10.704 g/kg and 0.612 g/kg, respectively, while the average SOM content and STN content of loessial soil were only 9.985 g and 0.558 g/kg. From 1980 to 2007, the SOM content and STN content of cinnamon soil increased to 13.266 g/kg and 0.648 g/kg, respectively, and the SOM content and STN content of loess soil increased to 12.916 g/kg and 0.642 g/kg, respectively. From 2007 to 2020, the SOM and STN content in the two soil types also showed an increasing trend. These differences could be attributed to the different textures, parent materials, and soil formation processes that are associated with these soil types.

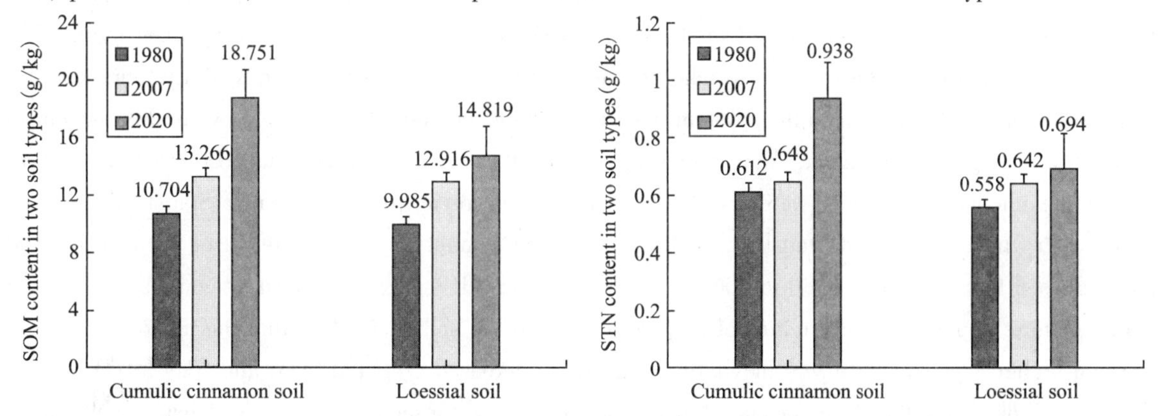

Fig. 6 Bar graphs for average SOM and STN content for two different soil types

SOM: soil organic matter; STN: soil total nitrogen

Cumulic cinnamon soil is a kind of anthropogenic soil that is neutral to slightly alkaline, has a deep plough layer, and has excellent water and fertility retention (Yan et al., 2019). Loessial soil is loose and soft with a light soil colour. Due to the lack of obvious profile development and serious soil erosion, loessial soil has weak water and fertilizer retention capabilities (Xin et al., 2016). Therefore, cumulic cinnamon soil has a higher SOM and STN content. Therefore, it can be concluded that soil type also has a substantial influence on the spatial changes of farmland SOM and STN.

4.3.3 Impact of land use practices

In addition to the influence of soil type, changes in land use patterns should not be ignored. In the 1980s, the land use pattern in Baishui county was mainly dry land. With the construction and improvement of farmland water conservancy facilities, dry land gradually decreased, watered land increased, and the effective irrigated area of arable land increased significantly. As can be seen from Fig. 7, from 1980 to 2007, except for the soil STN decreased under dryland-watered conditions, all the others showed an increase; from 2007 to 2020, the soil nutrient content increased regardless of the change in cropland use, with the highest increase in cropland maintained as watered land for a long time, with SOM and STN increasing by 3.74 g/kg and 0.148 g/kg respectively. Due to the favourable irrigation conditions on the watered land and the fine tillage management, the soil nutrient content increased significantly. It shows that the change in the use of arable land leads to a different pattern of change in soil nutrient content.

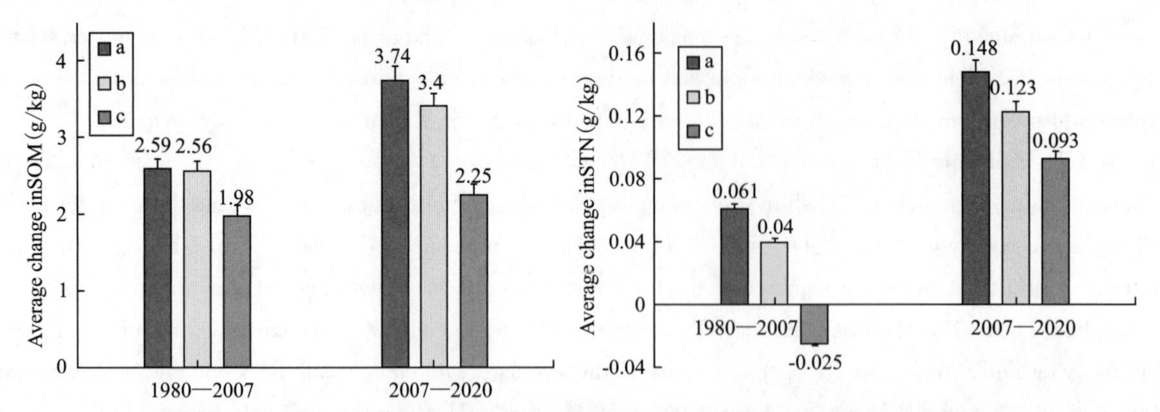

Fig. 7 Changes in soil nutrients in arable land under different land use practices.

(a) Watered land-Watered land; (b) Dryland-Dryland; (c) Dryland-Watered land

4.3.4 Farming management practice influence

In agricultural production, various farming practices, such as fertilizer application, irrigation, and crop residue returned to the field can significantly affect the SOM and STN dynamic change. In 1980, the lower SOM and STN contents on farmland reflected a long cultivation history with little or no fertilizer input (2542 t in 1980). Most crop residues were also taken off and used as fuel for cooking and heating. Since the early 1980s the Household Responsibility System has been implemented. The farmers were then given the authority to manage the contracted land, including all decisions regarding production. In order to get higher yield, more chemical fertilizers were used by farmers. After reviewing the Shaanxi Statistical Yearbook (Shaanxi Provincial Bureau of Statistics, 1980-2020), the amount of chemical fertilizer used in the study area increased from 2542 t in 1980 to 70643 t in 2020. Meanwhile, there has been a significant increase of organic manure applications and return of crop residues into the soil. It is clear that the use of organic manure and higher chemical fertilizers inevitably resulted in increased cropland SOM and STN levels.

In addition, with the development of irrigation and water conservation activities, many non-irrigated farmlands were transformed into irrigated lands, especially in the low-elevation areas. To reveal the irrigation effect, the average SOM and STN contents for different cropland types are calculated and their increment is shown in Fig. 8. During the period from 1980 to 2020, the highest increment of SOM was in the cropland that changed from dry land to irrigated land, whereas for STN the largest increase occurred for the irrigated land (over the whole period). Generally speaking, SOM and STN in irrigated cropland increased relatively most. Thus, it can be seen that irrigation has a clear influence on the change of farmland SOM and STN.

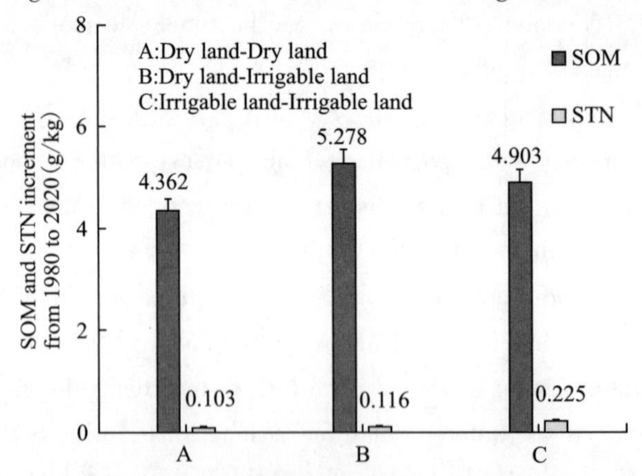

Fig. 8 Average increase of SOM and STN content between 1980 and 2020 for different cropland types.

SOM: soil organic matter; STN: soil total nitrogen

5 Conclusions

In this study, a combination of geostatistical and GIS techniques was used to quantitatively study and explore the spatial and temporal variability characteristics of the SOM and STN of cultivated soils in Baishui County, a typical region in ecologically fragile regions of the Loess Plateau, China. The main findings are as follows:

1. The average SOM content increased by 2.48 g/kg and 2.41 g/kg during 1980-2007 and 2007-2020, respectively, and the average STN content increased by 0.04 g/kg and 0.09 g/kg, respectively.

2. The coefficients of variability of soil nutrients on cultivated land in three years ranged between 15.6% and 38.5%, which represented moderate variability. Compared with 1980, the SOM and STN contents in 2007 and 2020 showed a higher nugget/sill ratio and a lower global Moran's I index, which indicates the spatial variability and weak spatial structure of SOM and STN.

3. The centre of gravity of soil nutrients generally shifted towards southeast, moving 922.96 m and 1030.95 m respectively. The spatial distribution pattern of the soil nutrient standard deviation ellipse is consistent with the direction of distribution in the study area, shifting to the southeast. The Rotation show an "increasing-decreasing" pattern of change, with the oval area decreasing and the spatial distribution of soil nutrients tending to concentrate.

4. We also found moderate spatial variation in both SOM and STN content as a result of a combination of intrinsic factors (topography, soil type, and land use patterns) and extrinsic factors (agricultural management practices). However, the main factors causing nutrient variation was agricultural management practices.

Acknowledgements

This research was supported by the National Natural Science Foundation (grant number 42071240) and Analysis of the Evolution of Spatial and Temporal Patterns of Arable Land in the Northern Weibei Dry Plateau Area and its Driving Forces(DJNY2022-36).

References

Aghasi, B., Jalalian, A., Khademi, H., Toomanian, N. (2017): Sub-basin scale spatial variability of soil properties in Central Iran. -Arabian Journal of Geosciences 10.

Agricultural Chemistry Committee of China (1983): Conventional Methods of Soil and Agricultural Chemistry Analysis. -Science Press, Beijing (in Chinese).

Awais, M., Arshad, M., Shah, S. H. H., Anwar-ul-Haq, M. (2017): Evaluating groundwater quality for irrigated agriculture: spatio-temporal investigations using GIS and geostatistics in Punjab, Pakistan. -Arabian Journal of Geosciences 10.

Balaguer-Beser, A., Ruiz, L. A., Hermosilla, T., Recio, J. A. (2013): Using semivariogram indices to analyse heterogeneity in spatial patterns in remotely sensed images. -Computers & Geosciences 50: 115-127.

Blanchet, G., Libohova, Z., Joost, S., Rossier, N., Schneider, A., Jeangros, B., Sinaj, S. (2017): Spatial variability of potassium in agricultural soils of the canton of Fribourg, Switzerland. -Geoderma 290: 107-121.

Chen, T., Chang, Q., Liu, J., Clevers JGPW (2016): Spatio-temporal variability of farmland soil organic matter and total nitrogen in the southern Loess Plateau, China: a case study in Heyang County. -Environmental Earth Sciences 75.

Chen, Z., Wang, L., Wei, A., Gao, J., Lu, Y., Zhou, J. (2019): Land-use change from arable lands to orchards reduced soil erosion and increased nutrient loss in a small catchment. -Science of the Total Environment 648: 1097-1104.

Chuai, X. W., Huang, X. J., Wang, W. J., Zhang, M., Lai, L., Liao, Q. L. (2012): Spatial variability of soil organic carbon and related factors in Jiangsu Province, China. -Pedosphere 22: 404-414.

Darand, M., Dostkamyan, M., Rehmanic, M. I. A. (2017): Spatial autocorrelation analysis of extreme precipitation in Iran. -Russian Meteorology and Hydrology 42: 415-424.

Duan, L. X., Li, Z. W., Xie, H. X., Li, Z. M., Zhang, L., Zhou, Q. (2020): Large-scale spatial variability of eight soil chemical properties within paddy fields. -Catena 188.

Foroughifar, H., Jafarzadeh, A. A., Torabi, H., Pakpour, A., Miransari, M. (2013): Using geostatistics and geographic information system techniques to characterize spatial variability of soil properties, including micronutrients. -Communications in Soil Science and Plant Analysis 44: 1273-1281.

Gelaw, A. M., Singh, B. R., Lal, R. (2014): Soil organic carbon and total nitrogen stocks under different land uses in a semi-arid watershed in Tigray, Northern Ethiopia. -Agric. Ecosyst. Environ. 188(15): 256-263.

Guan, Y., Zhou, W., Bai, Z., Cao, Y., Huang, Y., Huang, H. (2020): Soil nutrient variations among different land use types after reclamation in the Pingshuo opencast coal mine on the Loess Plateau, China. -Catena 188.

Guo, X. D., Fu, B. J., Ma, K. M., Chen, L. D., Wang, J. (2001): Spatio-temporal variability of soil nutrients in the Zunhua Plain, Northern China. -Physical Geography 22: 343-360.

Hoffmann, U., Hoffmann, T., Jurasinski, G., Glatzel, S., Kuhn, N. J. (2014): Assessing the spatial variability of soil organic carbon stocks in an alpine setting (Grindelwald, Swiss Alps). -Geoderma 232: 270-283.

Hu, K. L., Wang, S. Y., Li, H., Huang, F., Li, B. G. (2014): Spatial scaling effects on variability of soil organic matter and total nitrogen in suburban Beijing. -Geoderma 226: 54-63.

Huang, Y. L., Chen, L. D., Fu, B. J., Huang, Z. L., Gong, J., Lu, X. X. (2012): Effect of land use and topography on spatial variability of soil moisture in a gully catchment of the Loess Plateau, China. -Ecohydrology 5: 826-833.

Keskinen, R., Nyambura, M., Heikkinen, J., Sila, A., Eurola, M., Towett, E., Shepherd, K., Esala, M. (2019): Readily available concentrations of selected micronutrients and harmful metals in soils of Sub-Saharan Africa. -Geoderma 347: 203-209.

Kong, X., Zhang, F., Wei, Q., Xu, Y., Hui, J. (2006): Influence of land use change on soil nutrients in an intensive agricultural region of North China. -Soil & Tillage Research 88: 85-94.

Liao, K. H., Lai, X. M., Zhou, Z. W., Zhu, Q. (2017): Applying fractal analysis to detect spatio-temporal variability of soil moisture content on two contrasting land use hillslopes. -Catena 157: 163-172.

Liu, X. M., Xu, J. M., Zhang, M. K., Zhou, B. (2004): Effects of land management change on spatial variability of organic matter and nutrients in paddy field: a case study of Pinghu, China. -Environmental Management 34: 691-700.

Liu, X. M., Zhang, W. W., Zhang, M. H., Ficklin, D. L., Wang, F. (2009): Spatio-temporal variations of soil nutrients influenced by an altered land tenure system in China. -Geoderma 152: 23-34.

Liu, Q., Xie, W. J., Xia, J. B. (2013): Using semivariogram and Moran's i techniques to evaluate spatial distribution of soil micronutrients. -Communications in Soil Science and Plant Analysis 44: 1182-1192.

Liu, L.-L., Zhu, Y., Liu. X.-J., Cao W.-X., Xu, M., Wang, X.-K., Wang, E.-L. (2014a): Spatiotemporal changes in soil nutrients: a case study in Taihu region of China. -Journal of Integrative Agriculture 13.

Liu, Z. J., Zhou, W., Shen, J. B., He, P., Lei, Q. L., Liang, G. Q. (2014b): A simple assessment on spatial variability of rice yield and selected soil chemical properties of paddy fields in South China. -Geoderma 235: 39-47.

Lu, Y., Chen, Z., Kang, T., Zhang, X., Bellarby, J., Zhou, J. (2016): Land-use changes from arable crop to kiwi-orchard increased nutrient surpluses and accumulation in soils. -Agriculture Ecosystems & Environment 223: 270-277.

Ma, J. C., He, P., Xu, X. P., He, W. T., Liu, Y. X., Yang, F. Q., Chen, F., Li, S. T., Tu, S. H., Jin, J. Y., Johnston, A. M., Zhou, W. (2016a): Temporal and spatial changes in soil available phosphorus in China (1990-2012). -Field Crops Research 192: 13-20.

Ma, K., Zhang, Y., Tang, S., Liu, J. (2016b): Spatial distribution of soil organic carbon in the Zoige alpine wetland, northeastern Qinghai-Tibet Plateau. -Catena 144: 102-108.

Martin, M. P., Orton, T. G., Lacarce, E., Meersmans, J., Saby, N. P. A., Paroissien, J. B., Jolivet, C., Boulonne, L., Arrouays, D. (2014): Evaluation of modelling approaches for predicting the spatial distribution of soil organic carbon stocks at the national scale. -Geoderma 223-225(1): 97-107.

Onyejekwe, S., Kang, X., Ge, L. (2016): Evaluation of the scale of fluctuation of geotechnical parameters by autocorrelation function and semivariogram function. -Engineering Geology 214: 43-49.

Osat, M., Heidari, A., Karimian Eghbal, M., Mahmoodi, S. (2016): Spatial variability of soil development indices and their compatibility with soil taxonomic classes in a hilly landscape: a case study at Bandar village, Northern Iran. -Journal of Mountain Science 13: 1746-1759.

Schlegel, A. J., Assefa, Y., Bond, H. D., Haag, L. A., Stone, L. R. (2017): Changes in soil nutrients after 10 years of cattle manure and swine effluent application. -Soil & Tillage Research 172: 48-58.

Shaanxi Provincial Bureau of Statistics (1980-2017): Statistical Yearbook of Baishui. -China Statistics Press, Beijing (in Chinese).

Wang, Y. Q., Zhang, X. C., Zhang, J. L., Li, S. J. (2009): Spatial Variability of soil organic carbon in a watershed on the Loess Plateau. -Pedosphere 19: 486-495.

Wang, Z., Liu, G. B., Xu, M. X., Zhang, J., Wang, Y., Tang, L. (2012): Temporal and spatial variations in soil organic carbon sequestration following revegetation in the hilly Loess Plateau, China. -Catena 99: 26-33.

Wei, J. B., Xiao, D. N., Zeng, H., Fu, Y. K. (2008): Spatial variability of soil properties in relation to land use and topography in a typical small watershed of the black soil region, northeastern China. -Environmental Geology 53: 1663-1672.

Weihrauch, C., Opp, C. (2018): Ecologically relevant phosphorus pools in soils and their dynamics: the story so far. -Geoderma 325: 183-194.

Xin, Z. B., Qin, Y. B., Yu, X. X. (2016): Spatial variability in soil organic carbon and its influencing factors in a hilly watershed of the Loess Plateau, China. -Catena 137: 660-669.

Xu, X., Zhang, H., Zhang, O. (2004): Development of check-dam systems in gullies on the Loess Plateau, China. -Environ Sci Policy 7: 79-86.

Yan, P., Peng, H., Yan, L. B., Zhang, S. Y., Chen, A. M., Lin, K. R. (2019): Spatial variability in soil pH and land use as the main influential factor in the red beds of the Nanxiong Basin, China. -Peerj 7.

Yu, C., Pang, X. P., Wang, Q., Jin, S. H., Shu, C. C., Guo, Z. G. (2017): Soil nutrient changes induced by the presence and intensity of plateaupika (Ochotona curzoniae) disturbances in the Qinghai-Tibet Plateau, China. -Ecological Engineering 106: 1-9.

Zhu, M., Feng, Q., Qin, Y., Cao, J., Zhang, M., Liu, W., Deo, R. C., Zhang, C., Li, R., Li, B. (2019): The role of topography in shaping the spatial patterns of soil organic carbon. -Catena 176: 296-305.

Zhuo, Z. Q., Xing, A., Li, Y., Huang, Y. F., Nie, C. J. (2019): Spatio-temporal variability and the factors influencing soil-available heavy metal micronutrients in different agricultural sub-catchments. -Sustainability 11.

Appendix

Table A1 Statistical table of SOM in three periods

SOM (g/kg)	1980		2007		2020	
	Area (km^2)	Percentage (%)	Area (km^2)	Percentage (%)	Area (km^2)	Percentage (%)
<10	144.64	27.53	4.27	0.8	-	-
10~11.5	236.81	45.06	102.54	19.51	-	-
11.5~13	135.02	25.69	190.59	36.27	78.98	15.03
13~14.5	9.02	1.72	105.69	20.11	180.61	34.37
14.5~16	-	-	102.57	19.52	105.04	19.99
>16	-	-	19.83	3.77	160.86	30.61

SOM: soil organic matter.

Table A2 Statistical table of total nitrogen content in three periods

SOM (g/kg)	1980		2007		2020	
	Area (km²)	Percentage (%)	Area (km²)	Percentage (%)	Area (km²)	Percentage (%)
< 0.55	15.74	3.00	22.60	4.30	-	-
0.55~0.6	201.05	38.26	158.49	30.16	5.90	1.10
0.6~0.65	228.90	43.56	124.34	23.66	128.66	24.03
0.65~0.7	79.80	15.19	71.54	13.61	151.34	28.26
0.7~0.75	-	-	93.58	17.81	74.23	13.86
>0.75	-	-	54.96	10.46	165.37	30.88

Table A3 Area and percentage of soils with increased SOM and STN contents between 1980 and 2007

Variables and item	Different categories of content increase(%)					
	<-15	-15~0	0~15	15~30	30~45	>45
SOM						
Area(km²)	-	26.42	157.56	218.32	98.98	24.21
Percentage(%)	-	5.03	29.98	41.55	18.84	4.61
STN						
Area(km²)	8.22	155.42	234.12	117.13	10.61	-
Percentage (%)	1.56	29.58	44.55	22.29	2.02	-

SOM: soil organic matter; STN: soil total nitrogen.

Table A4 Area and percentage of soils with increased SOM and STN contents between 2007 and 2020

Variables and item	Different categories of content increase(%)					
	<-15	-15~0	0~15	15~30	30~45	>45
SOM						
Area(km²)		43.28	150.73	267.18	53.02	11.29
Percentage(%)		8.24	28.68	50.84	10.09	2.15
STN						
Area (km²)		75.79	248.36	170.71	23.87	6.75
Percentage(%)		14.42	47.26	32.49	4.54	1.28

SOM: soil organic matter; STN: soil total nitrogen.

本文曾发表于2022年《Applied Ecology and Environmental Researoh》第20卷第5期

Ecological Risk Assessment Management Methods of Heavy Metal Polluted Urban Green Lands

Tingting Meng[1,2,3], Jinbao Liu[1,2,3], Shaodong Qu[4]

(1.Shaanxi Provincial Land Engineering Construction Group Co., Ltd., Xi'an 710075, China;
2.Institute of Shaanxi Land Engineering and Technology Co., Ltd., Xi'an 710075, China;
3.Key Laboratory of Degraded and Unused Land Consolidation Engineering, Ministry of Land and Resources, Xi'an 710075, China;4.Shaanxi Ecological Industry Co., Ltd., Xi'an, 710075, China)

【Abstract】 Soil heavy metal pollution is a key problem that threatens the ecological environment of urban lands. In the present research, the differences in soil phytoremediation effects under fertilizing, irrigation, and natural growth management methods in urban green land were explored. Poplus plant (Populus tremula L.) grown experimental urban green plots were established. Soils were collected to determine the concentrations of Cr, Ni, Cu, Zn, As, Cd, and Pb. The soil heavy metal pollution and ecological risk degree under different management methods were quantitatively analyzed by calculating single factor pollution index, Nemerow comprehensive pollution index, and geo-accumulation index, and potential ecological risk evaluation method. The results showed that there were two trends of heavy metal content with the increase in soil depth: one was gradually decreasing, the other was first decreasing and then increasing, and the minimum value was obtained at the depth of 20~40 cm. The Cd concentration in the test area showed light pollution, which was the main source of ecological risk. The order of management measures for soil heavy metal pollution remediation effect was fertilization, irrigation and natural growth. Fertilization and irrigation measures in urban green space were beneficial to the improvement of urban ecological environment quality and remediation of heavy metal pollution in urban green space soil.

【Keywords】 Urban green land; Soil phytoremediation; Heavy metal pollution; Ecological risk; Regulative measure

1 Introduction

Soil heavy metal pollution refers to the phenomenon that human activities introduce heavy metals into the soil, resulting in the concentration of harmful elements exceeding the background value[1]. Related studies have shown that about 16.1% of soil samples in China are polluted by heavy metals and are above the national secondary quality standard[2]. After heavy metal elements enter into an ecosystem through dust deposition, storm runoff, sewage irrigation, etc. They not only affect the quality of land and the growth of food crops[3-5], but also disrupt soil structure, change soil properties, and destroy the stability of soil environment[6-7]. Heavy metals are non-biodegradable, highly active and cumulative elements and may have a serious impact on animal and human health at very low concentrations. Soil heavy metal pollution has become an environmental problem that cannot be ignored in the process of urbanization and industrial and agricultural development[8-9]. At present, researchers use methods of single factor-and soil accumulation-index, and enrichment factor, potential ecological risk assessment, etc. to describe the status quo of regional heavy metal pollution and its spatial distribution characteristics. As an important part of the ecosystem[10-11], urban green spaces can not only beautify the environment, purify the air, but also repair the polluted soil. Phytoremediation is a kind of green cleansing technology for soil heavy metal elements. The process generally depends upon the sowing of high capacity super-enrichment plants in the contaminated soil, thereby absorbing heavy metals from soil over time. Followed by harvesting the plants, so as to reduce the heavy metal content in the soil[11]. Under the influence of human agricultural, industrial, and mining activities, heavy metal pollution of urban green land soil will have a great impact on urban environment and human health[12]. In recent years, researchers from China have carried out a large number of studies on the element speciation[13-14], source analysis, spatial distribution characteristics of heavy metals and other aspects of urban heavy metal pollution[15]. But there are few studies on the remediation effect of soil heavy metal pollution under different management methods.

The goal of the present study is targeted to the reclamation of the urban green space from heavy metal pollution through phytoremediation. For the purpose different management practices at field levels will be tested. A quantitative analysis of soil on the differences under different management methods will be made and be considered as restoration effects. The study will determine the heavy metal pollution status and ecological risk degree of urban green spaces, so as to urban green space provides the basis for soil heavy metal pollution control and repair.

2 Materials And Methods

Overview of the study area. The test plot is located in the Project Department of Lanchi Road (108°50′38″,34°23′26″), Weicheng District, Xianyang City, Shaanxi Province, China (Fig. 1). The average temperature, wind speed, and relative humidity of the study area are 14.7 ℃, 2.1 m/s, and 63%, respectively. The area enjoys 228 days of frost-free period and the average annual precipitation is 443 mm. In April 2018, vegetation restoration of abandoned homestead in the project area began. After clearing and leveling the ground of the abandoned residential sites, the common fast-growing landscape poplar plant (Populus tremula L.) was used as the restoration vegetation. The planting density was 1 m row spacing and 0.5 m plant spacing, and the total planting area was about 3.07 ha. The management measures of poplar forests were three types namely, applying farm fertilizer + irrigating area (hereinafter referred to as "fertilizing"), only irrigation, and natural growth. Farm manure mainly consists of 70% pig manure, 18% human urine manure and 12% weeds. The organic matter, nitrogen, phosphorus, and potassium contents were 14.8, 0.47, 0.52, and 0.45% of the farm manure, respectively.

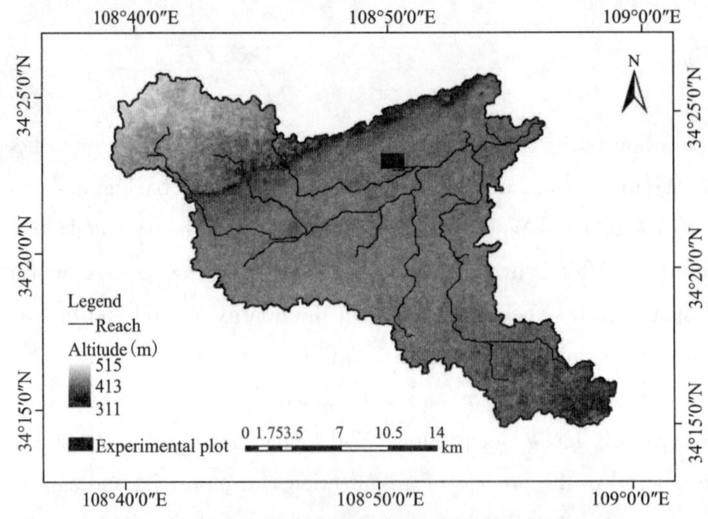

Fig. 1 Overview of experimental plot location

Experimental design. In May 2018, poplar trees basically with same growth trend and having 2.5 cm in diameter at breast height and 1.8 m in height were selected in the same direction of the study area to delimit the test plot. Each test area covers an area of 60 m^2 (15 m×4 m), and there are 12 experimental plots. The plots are designated as T0, undisturbed soil; T1, fertilization; T2, irrigation; and T3, natural growth. T1, T2, T3 treatment in the same plot, each experimental plot length and width interval is 3 m. The undisturbed soil plot is about 1000 m away from the experimental plot, and the interval length and width of the three treatments are also 3 m. The plots were fertilized according to three different management methods as described before. At the same time, undisturbed soil in the study area was selected as control, and three replicates were set for each treatment. The irrigation method was diffuse irrigation, and the frequency of fertilization and irrigation was once a month.

Sample collection and measurement of heavy metals. In July 2020, soil samples were collected in each test plot according to the diagonal method. In each test area with different treatments five sampling points each extending vertically from 0~100 cm soil depth was selected. At each sampling point soil samples were collected at 0~20, 20~40, 40~60, 60~80, and 80~100 cm, respectively. A 40 mm diameter soil drill was used for soil sampling in the field. After removing the unwanted identifiable debris from the collected soil samples, those were put into a zipped bag and brought back to the laboratory for further analysis. Accurately weighed 0.1~0.2 g of air-dried, ground soil sample with a particle size of less than 0.149 mm was put in a Teflon digestion tank adding 6 mL aqua regia, and the digestion tank was put into the microwave digestion device setting program. So that the sample is digested in a set microwave digestion parameter program. After the digestion was complete the solution was filtered in a 50 mL volumetric flask with a slow fixed-volume filter paper. After the digested solution is filtered out, the inner wall of the lid and the tank and the filter residue of the PTFE digestion tank were rinsed with a small amount of deionized water for at least 3 times into the volumetric flask and filled up to the mark accurately. The supernatant was taken and measured with an inductively coupled plasma mass spectrometer (ICP-MS, Agilent7700e, USA).

Index calculation method. The single-factor index method and the Nemerow comprehensive pollution index method[16] were used to evaluate the soil heavy metal pollution degree under different management measures in the test plots. The calculation formulas and grading standards of the single-factor index method and the Nemerow comprehensive pollution index method[6]. The calculation formula is: (Eq. 1) and (Eq. 2), the grading standards is Table 1.

$$P_i = \frac{C_i}{C_s^i} \tag{1}$$

$$P_n = \sqrt{\frac{P_{imax}^2 + P_{iave}^2}{2}} \tag{2}$$

P_i is the single-factor pollution index of heavy metal I, C_i is the measured value of heavy metal content, and C_s^i is the screening value of soil risk with heavy metal content. Regional background values are taken in this study, referring to the background values of soil elements in Shaanxi Province. The background soil element values used for Cr, Ni, Cu, Zn, As, Cd, and Pb were 62.5, 28.8, 21.4, 69.4, 11.0, 0.094, and 21.4 mg/kg, respectively. P_n is the comprehensive heavy metal pollution index, P_{imax} is the maximum value of the single-factor pollution index, and P_{iave} is the average value of all single-factor pollution indexes.

The pollution level of heavy metals in soil was evaluated by geological accumulative index method[6]. The calculation formula is:

$$I_{geo} = \log_2\left(\frac{C_i}{k \times S_i}\right) \tag{3}$$

I_{geo} is the geological accumulation index of heavy metal I, C_i is the measured value of heavy metal content, S_i is the background value of heavy metal, k is the correction coefficient, which is 1.5.

Table 1 Classification of pollution level, geological cumulative index and ecological risk degree

Pollution levels	I_{geo}	The degree of pollution	P_i	P_n	The degree of pollution	E_i^r	RI	Degree of hazard
0	<0	No pollution	<0.7	<0.7	No pollution	<40	<150	Slight harm
1	0~1	Light pollution	0.7~1 .0	0.7~1 .0	Alert value	40~80	150~300	Secondary damage
2	1~2	Moderate pollution	1 .0~2.0	1 .0~2.0	Light pollution	80~160	300~600	Strong harm
3	2~3	Medium pollution-heavy pollution	2.0~3.0	2.0~3.0	Moderate pollution	160~320	600~1200	Very Strong harm
4	3~4	Heavy pollution	>3.0	>3.0	Heavy pollution	≥320	≥1200	pole-strength harm
5	4~5	Heavy-very heavy pollution						
6	>5	Very heavy pollution						

Potential ecological risk index RI, proposed by Swedish geochemist Hakanson in 1980[17], is a method to evaluate the degree of heavy metal pollution. The calculation formula is: (Eq. 4)

$$RI = \sum_{i=1}^{n} E_r^i = \sum_{i=1}^{n} (T_r^i \times C_f^i) = \sum_{i=1}^{n} \left(T_r^i \times \frac{C^i}{C_n^i} \right) \tag{4}$$

C_i is the measured value of heavy metal content in soil, C_i^n is the regional background value of heavy metal, E_r^i is the potential ecological risk index of heavy metal I, and T_r^i is the toxicity response coefficient of heavy metal I. According to the existing results, the toxicity response coefficients of Cr, Ni, Cu, Zn, As, Cd and Pb are 2, 5, 5, 1, 10, 30 and 5, respectively. RI is the comprehensive potential ecological risk index, and the classification of ecological risk index is shown in Table 1.

Data processing. Office Excel 2016 was used for parameter calculation, heavy metal content distribution map and calculation result map were drawn, SPSS20 was used for heavy metal content correlation analysis, Arcgis10.2 was used to draw the location map of test area.

3 Results

Distribution characteristics of soil heavy metals under different management measures. The concentration of heavy metals and pH along the vertical depth gradient of the studied soil under different management methods has been presented in Table 2.

Table 2 Distribution of heavy metal content and pH

Treatment	Soil depth (cm)	Cr (mg/kg)	Ni (mg/kg)	Cu (mg/kg)	Zn (mg/kg)	As (mg/kg)	Cd (mg/kg)	Pb (mg/kg)	pH
Natural growth	0~20	73.44± 4.05Ac	36.87± 1.98Ac	37.29± 8.99Ac	96.52± 7.10Ab	13.26± 0.14Ab	0.27± 0.02Aa	27.77± 1.49Aa	8.67± 0.11Ac
	20~40	70.04± 0.46Ac	34.36± 1.16Ac	36.90± 7.99Ac	78.78± 3.10Ac	12.9± 0.05Ab	0.18± 0.00Ab	24.24± 1.12Ac	8.7± 0.02Ac
	40~60	82.09± 11.44Ab	42.17± 5.67Ab	42.37± 2.75Ab	89.66± 8.06Ab	16.09± 1.74Aa	0.19± 0.01Ab	26.09± 0.93Ab	8.68± 0.06Ac
	60~80	88.39± 7.63Aa	48.02± 4.89Aa	46.66± 10.9Ab	96.13± 3.52Ab	15.88± 1.07Aa	0.2± 0.00Ab	26.09± 1.02Ab	8.87± 0.01Ab
	80~100	89.58± 1.73Aa	48.67± 2.40Aa	56.23± 20.16Aa	101.85± 0.47Aa	16.57± 2.30Aa	0.19± 0.01Ab	27.24± 0.41Aa	8.98± 0.28Aa

continued to Table 2

Treatment	Soil depth (cm)	Cr (mg/kg)	Ni (mg/kg)	Cu (mg/kg)	Zn (mg/kg)	As (mg/kg)	Cd (mg/kg)	Pb (mg/kg)	pH
Irrigation	0~20	64.70± 8.31Bc	31.97± 4.95Bc	29.11± 2.58Bc	84.78± 9.51Bc	13.05± 0.16Ab	0.23± 0.05Bd	34.82± 11.46Ac	8.75± 0.23Bc
	20~40	68.45± 2.71Bc	33.58± 0.05Bc	30.96± 0.40Bc	75.19± 1.96Bd	13.01± 0.21Ab	0.17± 0.02Bc	23.24± 0.30Ab	8.66± 0.07Bc
	40~60	81.04± 12.92Bb	40.85± 7.54Bb	37.93± 9.03Bc	86.02± 13.2Bc	15.87± 2.05Ab	0.17± 0.01Bb	25.36± 1.96Ab	8.65± 0.01Bc
	60~80	89.44± 6.14Bb	47.56± 5.54Bb	39.13± 0.25Bb	92.83± 1.14Bb	17.21± 0.82Ab	0.19± 0.00Ba	26.04± 0.94Ab	8.9± 0.05Bb
	80~100	93.81± 4.27Ba	50.07± 0.42Ba	43.71± 2.45Ba	107.15± 7.02Ba	17.31± 3.34Aa	0.19± 0.01Ba	27.81± 0.40Aa	9.18± 0.00Ba
Fertilization	0~20	62.40± 5.05Bc	30.35± 2.66Bc	28.27± 1.39Cc	82.08± 5.70Cc	12.74± 0.28Bc	0.22± 0.04Bb	34.35± 12.13Bb	8.61± 0.43Ab
	20~40	66.58± 0.08Bc	33.03± 0.82Bc	29.02± 2.34Cc	73.41± 0.55Cc	13.28± 0.18Bc	0.16± 0.01Bc	22.21± 1.15Bb	8.55± 0.08Aa
	40~60	72.93± 1.45Bb	36.56± 1.48Bb	32.15± 0.85Cb	82.79± 8.63Ca	14.61± 0.27Bb	0.17± 0.01Ba	24.01± 0.05Bb	8.60± 0.08Ab
	60~80	80.11± 7.05Ba	41.33± 3.28Ba	37.26± 2.89Ca	90.25± 2.51Cb	16.21± 2.24Ba	0.18± 0.01Ba	25.27± 2.04Ba	8.79± 0.21Ab
	80~100	92.32± 6.38Ba	47.22± 3.60Ba	43.50± 2.75Ca	104.34± 11Ca	18.97± 0.99Ba	0.19± 0.01Ba	27.05± 1.47Ba	8.9± 0.40Ab

Note: Capital letters indicate significant differences between different underlying surfaces ($P<0.05$), lowercase letters indicate significant differences between different depths ($P<0.05$).

The studied soil was alkaline showing the range of pH from 8.55 to 9.18. However, the mean value of soil pH under different management methods showed variations in an order of irrigation (8.69) < natural growth (8.78), but fertilization (8.78) is equal to natural growth. On the other hand, the mean values of Cr, Ni, Cu, Zn, As, Cd, and Pb were 80.71, 42.02, 43.89, 92.59, 14.94, 0.21, and 26.29 mg/kg, respectively in soil under natural growth method. But, compared with natural growth, under irrigation, the mean contents of Cr, Ni, Cu, Zn, As, Cd, and Pb in soil decreased by 7.23%, 10.28%, 22.44%, 6.49%, −1.49%, 9.78% and −1.11%, respectively. A reduction in the concentrations of Cr, Ni, Cu, Zn, As, Cd, and Pb under the fertilization condition were observed by 9.13%, 12.65%, 30.95%, 10.76%, 6.78%, 14.85%, and 9.75%, respectively. With an increase in soil depth under natural growth method, the contents of other heavy metals in the soil decreased in the beginning, followed by an increase, it showed a minimum concentration at the depth of 20~40 cm. Under irrigation method, the contents of Cr, Ni, Cu and As gradually decreased with the increase of soil depth. While the contents of Zn, Cd and Pb decreased first and then increased, and reached the minimum value at the depth of 20~40 cm. Under fertilization method, the contents of Cr, Ni, Cu, As and Pb decreased first and then increased with the increase of soil depth, and the minimum values were obtained at 20~40 cm depth, while the contents of Zn and Cd did not show a uniform pattern of distribution with the change of soil depth.

Analysis of soil heavy metal pollution degree under different management methods. Combined with the results of single factor pollution index and Nemero comprehensive pollution index as shown Table 3.

It is shown that under natural growth conditions, the single factor pollution index of heavy metals in soil is in the range of 1.13 (Pb) ~2.92 (Cd). From the average value, except for Cu (2.05) and Pb (2.19), which showed moderate pollution, other heavy metals are in the range of 1.0~2.0, which belongs to light pollution. Under irrigation and fertilization conditions, soil heavy metal pollution was mild, and the single factor pollution index was in the range of 1.00 (Cr) ~2.37 (Cd) and 0.92 (Cr, Ni) ~1.93 (Cd), respectively. It can be seen that the individual heavy metal pollution degree of soil under different management methods was different. However, the heaviest heavy metal pollution in the three management methods was Cd, in the order of natural growth (2.19)>irrigation (1.98)>fertilization (1.86). Nemerow comprehensive pollution index showed that the comprehensive pollution status of heavy metals in the three management methods was moderate. The nemerow index of natural growth was the highest (1.98), followed by irrigation (1.73) and fertilization (1.62).

The geological accumulation index of heavy metals in soil is shown in Table 4. Combined with the classification standard of the geological accumulation index (Table 1), it can be known that Cu in the soil under the conditions of natural growth and irrigation is in the category of light pollution, and the size relationship is natural growth (0.43)>irrigation (0.07). The heavy metal Cd in the soil under the three management measures showed light pollution category. The size relationship was natural growth (0.53)>irrigation (0.39)>fertilization (0.31), and other heavy metal elements were pollution-free.

Analysis of ecological risk degree of soil heavy metals. The calculation results of single factor potential ecological risk index and comprehensive potential ecological risk index are shown in Table 5.

Table 3 P_i and P_n distribution

	Natural growth				Irrigation				Fertilization			
	Max.	Min.	Av.	CV	Max.	Min.	Av.	CV	Max.	Min.	Av.	CV
Cr	1.43	1.15	1.29	0.11	1.48	1.00	1.20	0.16	1.39	0.92	1.17	0.17
Ni	1.69	1.19	1.46	0.15	1.64	1.05	1.31	0.18	1.55	0.92	1.27	0.23
Cu	2.63	1.72	2.05	0.18	2.03	1.32	1.59	0.19	1.82	0.96	1.42	0.25
Zn	1.47	1.14	1.33	0.10	1.50	1.06	1.25	0.13	1.42	0.84	1.19	0.23
As	1.51	1.17	1.36	0.12	1.72	1.16	1.38	0.17	1.49	0.99	1.27	0.17
Cd	2.92	1.92	2.19	0.19	2.37	1.74	1.98	0.12	1.93	1.71	1.86	0.05
Pb	1.30	1.13	1.23	0.05	1.61	1.04	1.24	0.18	1.21	1.00	1.11	0.08
Pn	2.35	1.66	1.98	0.15	1.94	1.51	1.73	0.10	1.74	1.42	1.62	0.09

Table 4 I_{geo} numerical results

	Natural growth				Irrigation				Fertilization			
	Max.	Min.	Av.	CV	Max.	Min.	Av.	CV	Max.	Min.	Av.	CV
Cr	−0.07	−0.42	−0.22	−0.71	−0.02	−0.59	−0.34	−0.66	−0.11	−0.71	−0.37	−0.68
Ni	0.17	−0.33	−0.05	−4.16	0.13	−0.51	−0.21	−1.18	0.04	−0.71	−0.27	−1.28
Cu	0.81	0.20	0.43	0.58	0.44	−0.18	0.07	3.98	0.28	−0.64	−0.12	−3.11
Zn	−0.03	−0.40	−0.17	−0.82	0.00	−0.50	−0.28	−0.68	−0.08	−0.84	−0.37	−0.96
As	0.01	−0.36	−0.15	−1.13	0.20	−0.37	−0.14	−1.69	−0.01	−0.59	−0.26	−0.98
Cd	0.96	0.35	0.53	0.47	0.66	0.21	0.39	0.44	0.37	0.19	0.31	0.25
Pb	−0.21	−0.41	−0.29	−0.26	0.10	−0.53	−0.29	−0.83	−0.31	−0.58	−0.44	−0.25

Table 5 Ecological risk index

	Natural growth				Irrigation				Fertilization			
	Max.	Min.	Av.	CV	Max.	Min.	Av.	CV	Max.	Min.	Av.	CV
Cr	2.87	2.24	2.58	0.11	2.95	2.00	2.40	0.16	2.78	1.84	2.35	0.17
Ni	8.45	5.97	7.29	0.15	8.20	5.27	6.54	0.18	7.73	4.60	6.37	0.23
Cu	13.14	8.62	10.25	0.18	10.16	6.61	7.95	0.19	9.08	4.80	7.08	0.25
Zn	1.47	1.14	1.33	0.10	1.50	1.06	1.25	0.13	1.42	0.84	1.19	0.23
As	15.07	11.73	13.58	0.12	17.25	11.58	13.78	0.17	14.85	9.95	12.66	0.17
Cd	87.65	57.52	65.70	0.19	71.01	52.12	59.27	0.12	58.01	51.29	55.94	0.05
Pb	6.49	5.66	6.14	0.05	8.03	5.19	6.21	0.18	6.03	5.02	5.54	0.08
Pn	125.04	92.88	106.89	0.11	106.01	85.09	97.41	0.09	99.47	78.34	91.14	0.10

Combining with the classification of potential ecological risk degree (Table 1), it can be seen that Cd is medium hazard under the three management methods, and the order of size is natural growth (65.70) >irrigation (59.27) > fertilization (55.94). According to this order, the potential ecological risks of different heavy metals are in the order of Cd>As>Cu>Ni>Pb>Cr>Zn, and all of them are slight hazards. The ecological risk index of different management measures was natural growth (106.89) > irrigation (97.41) > fertilization (91.14), and the distribution range was 92.88~125.04, 85.09~106.01, and 78.34~99.47, respectively.

Correlation analysis of soil heavy metals. The correlation analysis of heavy metal elements showed (Table 6) that the concentrations of Cr, Ni, Cu, Zn and As were significantly correlated with each other, and the correlation coefficient was greater than 0.772 ($P<0.01$). The concentration of Cd and Pb were significantly correlated with each other, and the correlation coefficient was 0.675 ($P<0.01$). There was no significant correlation among the concentrations of the other five heavy metals. The results show that the sources and migration routes of different heavy metals are varied and/or there is a compound pollution phenomenon. Cr, Ni, Cu, Zn and As have a common pollution source, which is different from Cd and Pb. This may be related to agricultural activities and road traffic before and after the establishment of the pilot plot.

Table 6 Correlation analysis of heavy metal elements

Element	Cr	Ni	Cu	Zn	As	Cd	Pb
Cr	1						
Ni	0.991**	1					
Cu	0.877**	0.895**	1				
Zn	0.879**	0.877**	0.785**	1			
As	0.948**	0.925**	0.772**	0.815**	1		
Cd	0.123	0.134	0.274	0.439	0.029	1	
Pb	0.191	0.209	0.319	0.441	0.236	0.675**	1

4 Discussion

The present study shows that compared with the natural growth, under the condition of fertilization, soil heavy metal element concentration reduced. Under irrigation conditions, in addition to As, Pb concentration was higher than that of natural growth. The rest of the heavy metal concentrations were lower than the natural growth and increased along the depth gradient. Under the different management approaches, two kinds of trends in the change of heavy metal concentrations of soil were found. The first one of those, showed a decreasing trend followed by an increasing one in the concentrations. In this trend, the minimum value was obtained at the depth of 20~40 cm. The second trend showed a gradual reduction in the concentration of heavy metals along the depth gradient. This may be due to the dual influence of plant roots and moisture. Related studies have shown that under the influence of the rains or artificial irrigation, washed out heavy metals can follow the migration settlement occurring in the soil water infiltration process [18-19]. And the plants use their roots for the repair of soil, polluted by heavy metals[7]. The heavy metal concentration in soil can be reduced through the enrichment and absorption by plant roots[20], which transfer the same to the above-ground parts, and the chelation between plant tissues and heavy metals, thus occur. In addition, plant root-exudates not only regulate the physical and chemical properties of the rhizosphere environment, such as soil properties, pH, potential conditions, etc., but also promote a variety of microorganisms in the soil. Some microorganisms and their metabolites can change the available concentrations of heavy metals in the soil, which is more conducive to plant absorption[21]. In the soil of the test plot, the roots of plants at the depth of 20~40 cm were the densest, and the enrichment and absorption effect led to the low content of heavy metals at the depth of 20~40 cm.

Single factor pollution index analysis showed that Cd was the most polluted heavy metal in soil under different management methods, and showed the following trends of variation: natural growth (2.19)>irrigation (1.98)>fertilization (1.86). Nemerow index analysis shows that the comprehensive pollution status of heavy metals in the three management methods is moderate, in order of magnitude: natural growth (1.98)> (1.73) > irrigation fertilization (1.62). One study shows that in the urban areas, Cd concentration in the soil mainly originates from vehicle exhaust emissions, oil spill, rubber tire wear, etc[22-23]. Experimental plots, adjacent to the city roads showed higher intensity of Cd accumulation. However, among the different management measures, the intensity of Cd pollution under natural condition management system was the most serious. In the process of urban soil restoration, attention should be paid to the treatment of heavy metal Cd, and measures such as limiting vehicle traffic can be taken to reduce the Cd input of urban green space soil. Geological cumulative index and potential ecological risk index analysis showed that Cu in the soil under the conditions of natural growth and irrigation showed light pollution. Soil heavy metal Cd under three kinds of management measures showed light pollution status with medium hazard degree. The rest of the heavy metals showed no pollution, comprehensive damage size order natural growth, irrigation, and fertilization, respectively, were slight harm. This is not only related to the high content of Cd in the soil, but also to the far higher toxicity coefficient of Cd than other heavy metals[24].

In conclusion, although the development state of plant roots under natural growth conditions is better than that under fertilization, the frequent water input is not conducive to the accumulation of heavy metals in roots. Although there is natural growth and irrigation, the input of external nutrients will lead to the development of plant roots better than that of natural growth. Finally, the order of heavy metal pollution remediation capacity of soil was: fertilization>irrigation>natural growth.

5 Conclusions

This study established experimental plots to analyze and compare the pollution and ecological risk status of seven heavy metals (Cr, Ni, Cu, Zn, As, Cd and Pb) in the greenbelt soil under different management measures in a city in Xianyang, Shaanxi Province, and drew the following conclusions:

(1) With the increase of soil depth, the content of each heavy metal gradually decreased, or first decreased and then increased, and the minimum value was obtained at the depth of 20~40 cm.

(2)Cd in the test plot is light pollution and in medium hazard, which is the main source of ecological risk.

(3)There are differences in the effects of urban green space management measures on heavy metal pollution control. The remediation effect of soil heavy metal pollution under fertilization condition is the best, followed by irrigation condition, and only natural growth condition is the most unfavorable to plant remediation of soil heavy metal pollution.

6 Acknowledgements

Funding: This research was funded by Technology Innovation Leading Program of Shaanxi (Program No.2021CGBX-03).

References

[1]Chen, Y., Li, S., Zhang, Y., Zhang, Q. (2011) Assessing soil heavy metal pollution in the water-level-fluctuation zone of the three gorges reservoir, china. Journal of Hazardous Materials. 191(1-3), 366-372.

[2]Yang, J., Meng, X.Z., Duan, Y.P., Liu, L.Z., Cheng, H. (2014) Spatial distributions and sources of heavy metals in sediment from public park in shanghai, the yangtze river delta. Applied Geochemistry. 44, 54-60.

[3]Atafar, Z., Mesdaghinia, A., Nouri, J., Homaee, M., Yunesian, M., Ahmadimoghaddam, M., Amir, H.M. (2010) Effect of fertilizer application on soil heavy metal concentration. Environmental Monitoring & Assessment. 160(1-4), 83.

[4]Kupka, D., Kania, M., Pietrzykowski, M., Ukasik, A., Gruba, P. (2021) Multiple factors influence the accumulation of heavy metals (cu, pb, ni, zn) in forest soils in the vicinity of roadways. Water, Air, & Soil Pollution. 232(5), 1-13.

[5]Ewen, C., Anagnostopoulou, M.A., Ward, N.I. (2009) Monitoring of heavy metal levels in roadside dusts of thessaloniki, greece in relation to motor vehicle traffic density and flow. Environmental Monitoring and Assessment. 157(1-4), 483-498.

[6]Wang, H., Zhang, H., Tang, H., Wen, J. (2021) Heavy metal pollution characteristics and health risk evaluation of soil around a tungsten-molybdenum mine in luoyang, china. Environmental Earth Sciences. 80(7), 1-12.

[7]Ying, H., Chen, Q., Deng, M., Japenga, J., Li, T., Yang, X. (2018) Heavy metal pollution and health risk assessment of agricultural soils in a typical peri-urban area in southeast china. Journal of Environmental Management. 207(FEB. 1), 159-168.

[8]Sakram, G., Machender, G., Dhakate, R., Saxena, P.R., Prasad, M.D. (2015) Assessment of trace elements in soils around zaheerabad town, medak district, andhra pradesh, india. Environmental Earth Sciences. 73(8), 4511-4524.

[9]Tian, K., Huang, B., Xing, Z., Hu, W. (2017) Geochemical baseline establishment and ecological risk evaluation of heavy metals in greenhouse soils from dongtai, china. Ecological Indicators.72, 510-520.

[10]Cai, L., Xu, Z., Bao, P., He, M., Dou, L., Chen, L., Zhu, Y.G. (2015) Multivariate and geostatistical analyses of the spatial distribution and source of arsenic and heavy metals in the agricultural soils in shunde, southeast china. Journal of Geochemical Exploration. 148, 189-195.

[11]Borgese, L., Federici, S., Zacco, A., Gianoncelli, A., Bontempi, E. (2013) Metal fractionation in soils and assessment of environmental contamination in vallecamonica, italy. Environmental Science & Pollution Research International. 20(7), 5067-5075.

[12]Liu, Y., Chao, S., Hong, Z., Li, X., Pei, J. (2014) Interaction of soil heavy metal pollution with industrialisation and the landscape pattern in taiyuan city, china. Plos One. 9(9), e105798.

[13]Mandal, A., Voutchkov, M. (2011) Heavy metals in soils around the cement factory inrockfort, kingston, jamaica. International Journal of Geosciences. 2(1), 48-54.

[14]Zhang, H., Zhang, M., Wu, Y., Tang, J., Cheng, S., Wei, Y. (2021) Risk sources quantitative appointment of ecological environment and human health in farmland soils: a case study on jiuyuan district in china. Environmental Geochemistry and Health. 1-15.

[15]Haque, E., Thorne, P.S., Nghiem, A.A., Yip, C.S., Bostick, B.C. (2021) Lead (Pb) concentrations and speciation in residential soils from an urban community impacted by multiple legacy sources. Journal of Hazardous Materials. (416), 125886.

[16]El-Sherbiny, M.M., Ismail, A.I., El-Hefnawy, M.E. (2019) A preliminary assessment of potential ecological risk and soil contamination by heavy metals around a cement factory, western Saudi Arabia. Open Chemistry. 17, 671-684.

[17]Hakanson, L. (1980) An ecological risk index for aquatic pollution control.a sedimentological approach. Water Research. 14(8), 975-1001.

[18]Carreras, H.A., Pignata, M.L. (2002) Biomonitoring of heavy metals and air quality in Cordoba City, Argentina, using transplanted lichens. Environmental Pollution. 117(1), 77-87.

[19]Xiang, M., Li, Y., Yang, J., Lei, K., Cao, Y. (2021) Heavy metal contamination risk assessment and correlation analysis of heavy metal contents in soil and crops. Environmental Pollution. 278(2), 116911.

[20]Jiang, R., Huang, S., Wang, W., Liu, Y., Lin, C. (2020) Heavy metal pollution and ecological risk assessment in the maowei sea mangrove, china. Marine Pollution Bulletin. 161(Part B), 111816.

[21]Farmaki, E.G., Thomaidis, N.S. (2008) Current status of the metal pollution of the environment of Greece-a review. Global nest. The International Journal. 10(3), 366-375.

[22]Mandal, A., Voutchkov, M. (2011) Heavy metals in soils around the cement factory in Rockfort, Kingston, Jamaica. International Journal of Geosciences. 2(1), 48.

[23]Xu, X., Wang, T., Sun, M., Bai, Y., Fu, C., Zhang, L. (2019) Management principles for heavy metal contaminated farmland based on ecological risk-a case study in the pilot area of hunan province, china. The Science of the Total Environment. 684(SEP.20), 537-547.

[24]Mha, B., Td, B., Va, B. (2021) Heavy metal pollution of road dust in a city and its highly polluted suburb; quantitative source apportionment and source-specific ecological and health risk assessment. Chemosphere. 273, 129656.

本文曾发表于2022年《Fresenius Environmental Bulletin》第31卷第3A期

Study on Vegetation Growth Characteristics and Soil Physical and Chemical Properties in Coal Mine Reclamation Area in Northern Shaanxi

Na Wang [1,2,3,4], Zhe Liu [1,2,3,4]

(1.Institute of Land Engineering and Technology, Shaanxi Provincial Land Engineering Construction Group Co., Ltd., Xi'an 710075, China; 2.Shaanxi Provincial Land Engineering Construction Group Co., Ltd., Xi'an 710075, China; 3.Key Laboratory of Degraded and Unused Land Consolidation Engineering, the Ministry of Natural Resources, Xi'an 710075, China; 4.Shaanxi Provincial Land Consolidation Engineering Technology Research Center, Xi'an 710075, China)

【Abstract】To study the relationship between the characteristics of natural plant community and soil physical and chemical properties in the open-pit mining area of Northern Shaanxi, so as to provide a scientific basis for the selection of plant species in the process of vegetation reconstruction in this area. Taking four typical natural plants and soils in Northern Shaanxi mining area as the research object, the plant growth characteristics and soil physical and chemical properties under different management measures were compared and analyzed by using the method of indoor pot experiment. The results showed that there were great differences in soil physical and chemical properties under each treatment. The content of soil total nitrogen under A3H3C4 treatment was significantly higher than that of other treatments ($P<0.01$), and the corresponding content was 0.53 g/kg; The content of soil organic matter in A2H2C3 treatment was significantly higher than that in other treatments ($P<0.01$), and the corresponding content was 0.42 g/kg; The contents of soil total phosphorus and alkali hydrolyzable nitrogen in A4H4C1 treatment were significantly higher than those in other treatments ($P<0.01$), and the corresponding contents were 25.6 g/kg and 220.37 mg/kg respectively; The contents of soil available potassium and phosphorus in A4H4C4 treatment were significantly higher than those in other treatments ($P<0.01$), and the corresponding contents were 16.88 mg/kg and 216.3 mg/kg respectively. The variation ranges of soil microbial biomass carbon, nitrogen and phosphorus under each treatment were 83.19~589.00, 11.43~82.81 and 5.29~17.55 mg/kg respectively, and the average values were 309.86, 30.78 and 11.60 mg/kg respectively. From the perspective of vegetation growth characteristics, the plant height of vegetation increased most significantly under A2H2C3 treatment, and the chlorophyll content and specific surface area of vegetation roots changed significantly compared with other treatments. In general, when the pot field water capacity is 60%~65%, that is, the soil water content is 14%~16%, the nitrogen content is 0.75 g/kg, the phosphorus content is 4 mg/kg and the potassium content is 100 mg/kg, the rise trend of Caragana korshinskii is the best.

【Keywords】Coal mine reclamation area; Vegetation characteristics; Physical and chemical properties

1 Introduction

Coal resources, known as "black gold" and "industrial food", are one of the main energy sources supporting China's social and economic development. In the past 30 years, China has been the country with the highest coal production in the world. According to the data of the National Bureau of statistics, China's raw coal output reached 3.75 billion tons in 2019, with an annual growth rate of 4.2% [1]. The development of coal resources not only promotes the rapid development of China's economy, but also inevitably destroys the ecological environment of the mining area, which is particularly obvious in the open-pit mining area, such as the change of microbial community structure, the disorder of soil structure, the change of terrain and so on[2-3].

The ecological environment of the Loess Plateau is relatively fragile, but it carries half of the country's coal production capacity. The risk of land degradation is high, and ecological restoration and reconstruction is imminent. The destruction of coal mining is characterized by large quantity and many types. The contradiction between mining and land has seriously hindered the sustainable economic development of the mining area. At present, the transformation of the development concept of "green water and green mountains are golden mountains and silver mountains" also requires local governments, academia and relevant stakeholders to pay attention to the ecological restoration and governance of mining areas. As an effective measure to restore the post mining environment, guide the ecosystem to the ideal track and restore the post mining land use value, land reclamation has become an important means to coordinate coal resource mining and resource protection [4-5]. Among them, vegetation restoration is an effective way to change the bad ecological environment in the mining area. Vegetation restoration can not only build the initial community of the degraded mining area ecosystem, but also effectively improve the soil structure in a short time, increase soil fertility, restore the species and number of soil microorganisms, and promote the positive development of the whole ecosystem in the mining area [6-7]. However, the extreme habitats in the mining area make the natural recovery of vegetation extremely slow. Generally, it takes 50~100 a to achieve satisfactory coverage, while it takes 100~10000 a to recover the soil [8]. Therefore, it is particularly important to take artificial vegetation reconstruction in the process of ecological restoration in mining areas. For the screening of vegetation restoration species, species should have the characteristics of drought resistance, barren resistance, drought tolerance, certain improvement of soil, stress resistance and so on. Relevant studies show that the vegetation restoration of abandoned land in the mining area needs to screen relevant pioneer species according to local regionality and ecological succession. Plant species with nitrogen fixation function can be used as pioneer species for vegetation restoration, improve soil nitrogen content and increase soil fertility [9-10].

Many scholars put forward the concept of near natural ecological restoration in the process of ecological restoration in the mining area, learn from the ecosystem structure, composition and function of the undisturbed area around the mining area, and gradually approach or achieve the structure and function of the undisturbed ecosystem around the mining area on the basis of respecting the natural pattern and effectively combining the artificial induction method. The original natural plants are the preferred species. Therefore, the research on the natural plants in the mining area is particularly important.

2 Experimental methods and principles

2.1 Experimental design

The test soil is collected from the Goaf Collapse Area of coal mine in Northern Shaanxi. The collected soil sample is 0~20 cm topsoil, which is ground after air drying and sieved by 20 mesh for standby.

(1) Vegetation selection. Shrubs: Artemisia annua (A1), Caragana korshinskii (A2);

Arbor: Astragalus adsurgens (A3), oleander (A4);

(2) Experimental design of soil moisture gradient. Suitable moisture (H1, 75%~80% field capacity, soil moisture 18%~20%); Mild drought (H2, 60%~65% field water capacity, soil water content is 14%~16%); Moderate drought (H3, 45%~50% field water capacity, soil water content is 11%~12.5%); Severe drought (H4, 30%~35% field water capacity, soil water content is 7%~9%).

(3) Experimental design of soil nutrient gradient. According to the quality standard for newly added cultivated land of Shaanxi land improvement project (N≥0.3 g/kg, P≥2 mg/kg, K≥50 mg/kg), the following four gradients are set. The fertilizers used are urea (N46%), calcium superphosphate (P_2O_5 38%)

and potassium chloride(K_2O 57%).

Table 1 Soil nutrient gradient

Test number	Nutrient species		
	N(g/kg)	P(mg/kg)	K(mg/kg)
C1	0.3	2	50
C2	0.45	4	75
C3	0.6	6	100
C4	0.75	8	125

The specific fertilization amount is as follows:

Table 2 Fertilizer application rate

Test number	Nutrient species		
	Urea(g/kg)	calcium superphosphate(mg/kg)	Potassium oxide(mg/kg)
C1	0.65	5.26	87.72
C2	0.98	10.53	131.58
C3	1.30	15.79	175.44
C4	1.63	21.05	219.30

Select the dominant native species, and set up orthogonal tests with different drought degrees and different nutrient levels. The orthogonal test table is as follows:

Table 3 Orthogonal test table

Test number	Vegetation type	Drought degree	Nutrient status	Horizontal combination
1	A1	H1	C1	A1H1C1
2	A2	H2	C2	A2H2C2
3	A3	H3	C3	A3H3C3
4	A4	H4	C4	A4H4C4
5	A1	H1	C2	A1H1C2
6	A2	H2	C1	A2H2C1
7	A3	H3	C4	A3H3C4
8	A4	H4	C3	A4H4C3
9	A1	H1	C3	A1H1C3
10	A2	H2	C4	A2H2C4
11	A3	H3	C1	A3H3C1
12	A4	H4	C2	A4H4C2
13	A1	H1	C4	A1H1C4
14	A2	H2	C3	A2H2C3
15	A3	H3	C2	A3H3C2
16	A4	H4	C1	A4H4C1
17	Control	Do not add any measures		

2.2 Sample Detection

Table 4 Measurement methods of soil physical and chemical properties

Soil physical and chemical properties	Determination method
Total nitrogen	Kjeldahl method
Organic matter	Potassium dichromate colorimetry
Total phosphorus	Alkali fusion flame photometer method
Alkali hydrolyzed nitrogen	Alkaline hydrolysis diffusion method
Available phosphorus	$NaHCO_3$-Molybdenum antimony anti colorimetry
Available potassium	NH_4OAc extraction flame spectrophotometry
Microbial biomass carbon	
Microbial biomass nitrogen	Chloroform fumigation extraction
Microbial biomass phosphorus	

2.3 Data Processing And Analysis

Spss 20.0 software for statistical analysis of data. The differences of soil physical and chemical properties between different treatments were analyzed by one-way ANOVA and Pearson correlation analysis. Origin 9.1 was adopted to drawing.

3 Results And Analysis

(1) Effests OF Different Vegetation Restoration And Management Measures ON Soil Quality Restoration. Soil physical and chemical properties refer to the physical and chemical conditions of soil. The study of soil physical and chemical properties plays a decisive role in the ecological restoration of reclaimed soil [11-12]. In the field of mining area reclamation, the changes of soil physical and chemical properties will have a significant impact on the fertility of reclaimed soil, soil improvement and other adjustment measures [13]。

Soil total nitrogen is a necessary nutrient element for plant growth and development. It can maintain the balance of soil ecological nutrients, stimulate the growth of underground roots of vegetation and improve soil quality. It is one of the important indicators of soil fertility[14].It can be seen from Fig. 1 that under different management levels, the content of soil total nitrogen under A3H3C4 treatment is significantly higher than that of other treatments ($P<0.01$), and the corresponding content is 0.53 g/kg. The content of soil total nitrogen under A4H4C1 treatment is 0.20 g/kg, which is significantly lower than that of A2H2C3, A2H2C4 and A3H3C4 ($P<0.01$). Soil organic matter is one of the important components of soil. Organic matter can not only improve the effectiveness of soil nutrients, but also promote the formation of aggregate structure, so as to improve soil physical properties. It is one of the main evaluation indexes of ecological restoration [15-16]. The content of soil organic matter in A2H2C3 treatment was significantly higher than that in other treatments ($P<0.01$), and the corresponding content was 0.42 g/kg. In terms of soil total phosphorus, the content of A4H4C1 treatment was significantly higher than that of other treatments ($P<0.01$), and the corresponding content was 25.6 g/kg. Soil available nutrients are easily absorbed and utilized by plants. In this study, the contents of soil available potassium and phosphorus under A4H4C4 treatment are significantly higher than those of other treatments ($P<0.01$), and the corresponding contents are 16.88 mg/kg and 216.3 mg/kg respectively. Alkali hydrolyzable nitrogen (an) is a nutrient element that is necessary for plant growth and easy to be absorbed and utilized by plants. It plays an important role in crop growth and yield and is an important indicator of soil nitrogen availability[17]. In this study, the content of soil alkali hydrolyzable nitrogen under A4H4C1 treatment was significantly higher than that of other treatments ($P<0.01$), and the corresponding content was 220.37 mg/kg. Compared with the new cultivated land quality standards of the land remediation project in Shaanxi Province, the soil available phosphorus and available potassium are surplus after the harvest of each treatment vegetation, and the contents of organic matter and total nitrogen are in deficit. It shows that the demand for nitrogen for vegetation growth in the Loess Plateau is high.

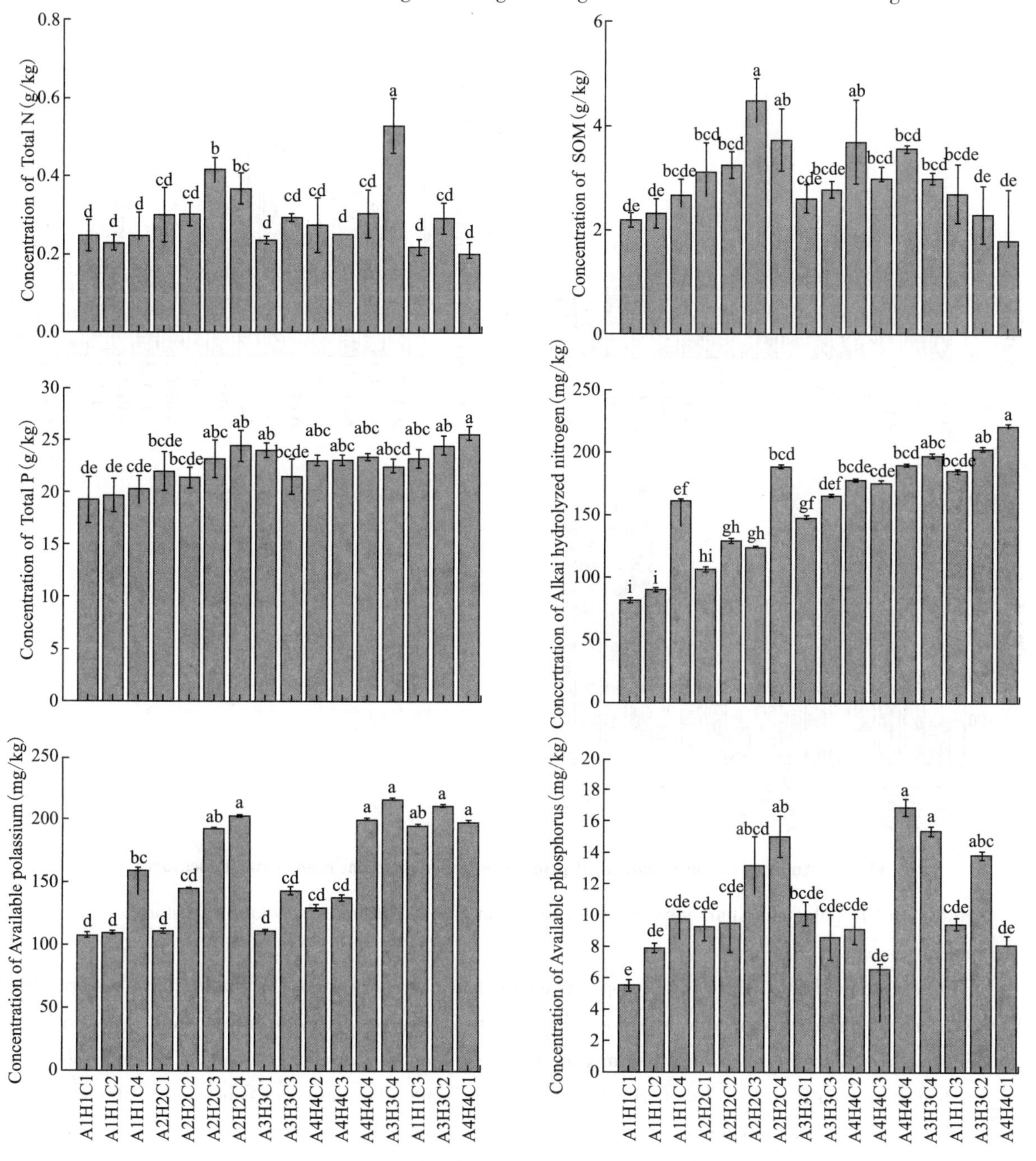

Fig. 1 Analysis of soil physical and chemical properties under different management measures

(2) Response of soil biochemical properties. Soil microbial biomass is an important index to measure the effective transformation of soil nutrients and is indispensable in the process of soil nutrient recycling and organic matter mineralization. As can be seen from Fig. 2, under different management levels, the variation

ranges of soil microbial biomass carbon, nitrogen and phosphorus are 83.19～589.00, 11.43～82.81 and 5.29～17.55 mg/kg respectively, with average values of 309.86, 30.78 and 11.60 mg/kg respectively. As for microbial biomass carbon, the content of A3H3C4and A4H4C1 was significantly higher than that of other treatments ($P<0.01$); In terms of microbial biomass nitrogen, the content of A3H3C1 was significantly higher than that of other treatments ($P<0.01$); In terms of microbial biomass phosphorus, the content of A2H2C3 treatment was significantly higher than that of other treatments ($P<0.01$).

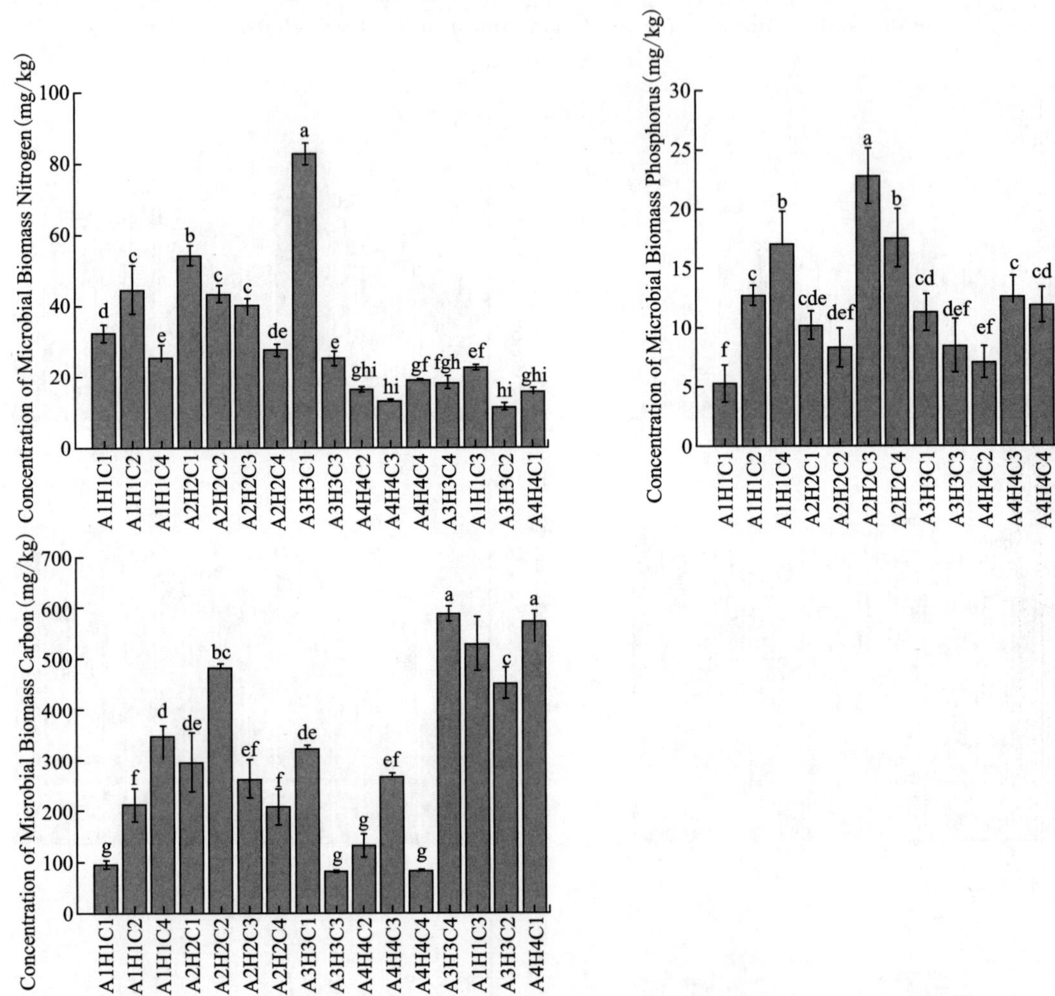

Fig. 2 Analysis of soil biochemical properties under different management measures

Study On Physiological Characteristics Of Vegetation Under Different Vegetation Restoration And Management Measures.

(1) Vegetation Height Change.The crop was conducted in the greenhouse of Fuping pilot base on April 1, 2020, and the indoor temperature was controlled at 20～25 degrees. The moisture content is controlled by weighing method, and the plant height is measured once a month. The change of plant height is shown in Fig. 6. It can be seen from Fig. 3 that with the extension of time, the plant height of each treatment increases greatly, and the most obvious change is A2H2C3 treatment, that is, when the water is controlled at 60%～65% of the field capacity, the nitrogen content is 0.6 g/kg, the phosphorus content is 4 mg/kg, and the potassium content is 100 mg/kg, the growth of Caragana korshinskii is the best.

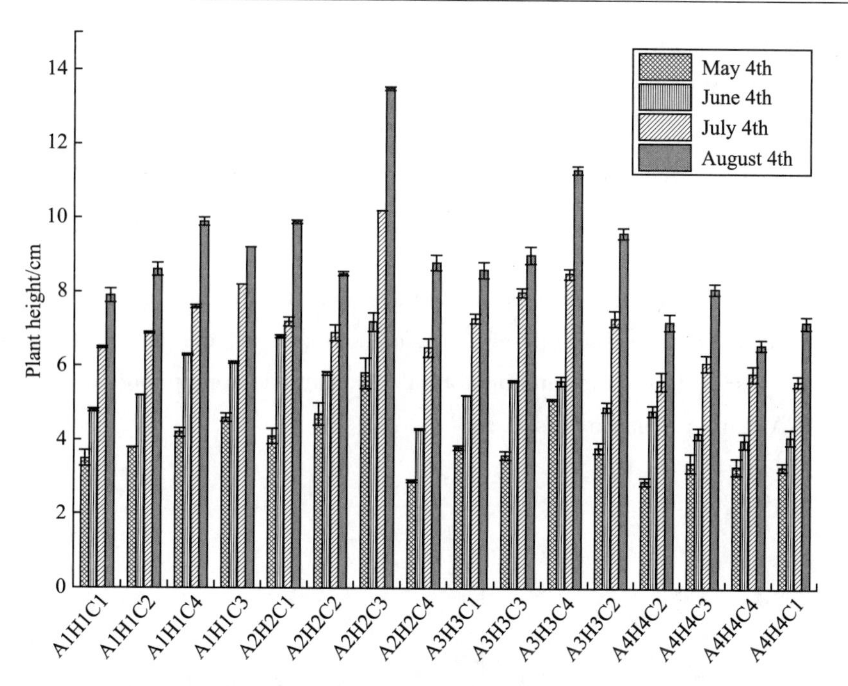

Fig. 3 Plant height change of vegetation

(2) Chlorophyll Change. It can be seen from Fig. 4 that the chlorophyll content of vegetation in A2H2C3 (60%~65% field capacity, soil water content of 14%~16%, vegetation of Caragana korshinskii, soil nutrients of 0.6 g/kg nitrogen, 4 mg/kg phosphorus and 100 mg/kg potassium) treatment is significantly higher than that in other treatments, followed by A3H3C4 (60%~65% field capacity, soil water content of 11% ~ 12.5%, vegetation of Astragalus adsurgens, soil nutrients of 0.6 g/kg nitrogen, 4 mg/kg phosphorus and 100 mg/kg potassium).

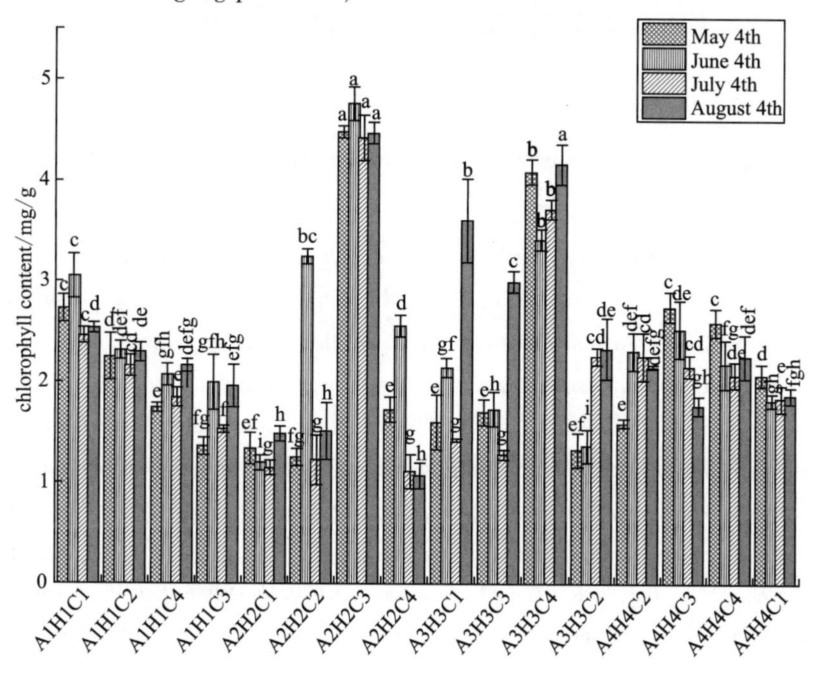

Fig. 4 Vegetation Chlorophyll Change

(3) Changes Of Root Specific Surface Area. As shown in Fig. 5, under suitable water conditions (75%~80% field water capacity and 18%~20% soil water content), there is little difference among treatments when the vegetation is Caragana korshinskii. In contrast, when the soil nutrient content is 0.45 g/kg

N, 3 mg/kg P and 75 mg/kg K, the root system of Caragana korshinskii is relatively developed and its specific surface area is large; When the water gradient was mild drought (60%~65% field water capacity, soil water content was 14%~16%), and the soil nutrients were 0.6 g/kg N, 4 mg/kg P and 100 mg/kg K, the root specific surface area of Artemisia annua was larger, and the root specific surface area of vegetation at this level was significantly higher than that of other treatments; When the water gradient is moderate drought (45%~50% field water capacity, soil water content is 11%~12.5%), and the soil nutrients are 0.75 g/kg N, 5 mg/kg P and 125 mg/kg K, the root specific surface area of Astragalus adsurgens is large, indicating that the demand for nutrients for vegetation growth is large; In general, the demand for water and nutrients for vegetation growth is large. Under the condition of mild drought and appropriate nutrient content, the specific surface area of Artemisia annua root is large.

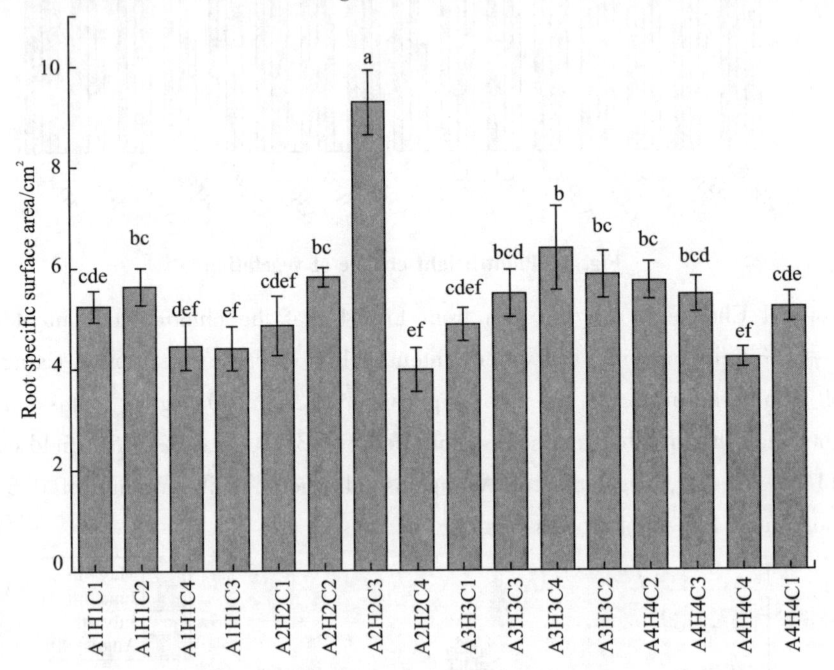

Fig. 5　Specific surface area of roots among different treatments

4　Conclusion

Vegetation restoration is the basis and key of ecological restoration in mining area. Vegetation restoration can not only beautify the environment, but also improve the soil, which can promote the restoration and reconstruction of the ecological environment in the mining area. In the vegetation reconstruction of the open-pit coal mine area in Northern Shaanxi, Caragana korshinskii is preferred as the best vegetation, and the soil water content is controlled not less than 14%. At the same time, the contents of nitrogen, phosphorus and potassium are not less than 0.75 g/kg, 4 mg/kg and 100 mg/kg respectively.

Reference

[1] Guan, Y. J., Zhou, W., Bai, Z. K., Cao, Y. G., Wang, J.(2021) Delimitation of supervision zones based on the soil property characteristics in a reclaimed opencast coal mine dump on the loess plateau, China. Science of The Total Environment.772(10),145006 .

[2] Feng, Y., Wang, J., Bai, Z.K., Reading, L.(2019) Effects of surface coal mining and land reclamation on soil properties: A review. Earth-Science Reviews. 191, 12-25.

[3]Li, P.F., Zhang, X.C., Hao, M.D., Cui, Y.X., Zhu, S. L.,Zhang, Y. J.(2019) Effects of Vegetation Restoration on Soil Bacterial Communities, Enzyme Activities, and Nutrients of Reconstructed Soil in a Mining Area on the Loess Plateau,

China. Sustainability. 11(8):2295-2310.

[4] Mukhopadhyay, S., Maiti, S. K., & Masto, R. E. (2014). Development of mine soil quality index (MSQI) for evaluation of reclamation success: A chronosequence study. Ecological Engineering, 71, 10-20.

[5] Zhong, K, B., Jing, K. Z.(1995) On the land reclamation benefit of China. Ecol. Econ. 2, 35-39. (In Chinese)

[6] Alam, M., Barbour, S. L., Huang, M., & Li, Y. (2020). Using Statistical and Dynamical Downscaling to Assess Climate Change Impacts on Mine Reclamation Cover Water Balances. Mine Water and the Environment, 39(4), 699-715.

[7] Hui, J., Bai, Z., Ye, B., & Wang, Z. (2021). Remote Sensing Monitoring and Evaluation of Vegetation Restoration in Grassland Mining Areas—A Case Study of the Shengli Mining Area in Xilinhot City, China. Land, 10(7), 743.

[8] Li, S. Q. (2019). Dominant species of herb community on the reclaimed coal gob pile in Yangquan mining area of Shanxi under different vegetations: interspecific relationship and niche. Chinese Agricultural Science Bulletin, 35(1), 80-87.

[9] Jochimsen, M. E. (2001). Vegetation development and species assemblages in a long-term reclamation project on mine spoil. Ecological Engineering, 17(2-3), 187-198.

[10] Alegre, J., Alonso-Blázquez, N., De Andrés, E. F., Tenorio, J. L., & Ayerbe, L. (2004). Revegetation and reclamation of soils using wild leguminous shrubs in cold semiarid Mediterranean conditions: litterfall and carbon and nitrogen returns under two aridity regimes. Plant and Soil, 263(1), 203-212.

[11] Sydnor, M. W., & Redente, E. F. (2002). Reclamation of high - elevation, acidic mine waste with organic amendments and topsoil. Journal of environmental quality, 31(5), 1528-1537.

[12] Dere, A. L., Stehouwer, R. C., & McDonald, K. E. (2011). Nutrient leaching and switchgrass growth in mine soil columns amended with poultry manure. Soil science, 176(2), 84-90.

[13] Sydnor M. E. W. and Redente E. F.. Reclamation of high-elevation, acidic mine waste with organic amendments and topsoil[J]. Journal of Environment Quality, 2002, 31(5): 1528-1537.

[14] Angel, H. Z., Priest, J. S., Stovall, J. P., Oswald, B. P., Weng, Y., & Williams, H. M. (2019). Individual tree and stand-level carbon and nutrient contents across one rotation of loblolly pine plantations on a reclaimed surface mine. New Forests, 50(5), 733-753.

[15] Rodríguez-Vila, A., Asensio, V., Forján, R., & Covelo, E. F. (2016). Carbon fractionation in a mine soil amended with compost and biochar and vegetated with Brassica juncea L. Journal of Geochemical Exploration, 169, 137-143.

[16] Jakab, G., Filep, T., Király, C., Madarász, B., Zacháry, D., Ringer, M., ... & Szalai, Z. (2019). Differences in mineral phase associated soil organic matter composition due to varying tillage intensity. Agronomy, 9(11), 700.

[17] Pietrzykowski, M., Socha, J., & van Doorn, N. S. (2015). Scots pine (Pinus sylvestris L.) site index in relation to physico-chemical and biological properties in reclaimed mine soils.New Forests,46(2), 247-266.

本文曾发表于2022年《Fresenius Environmental Bulletin》第31卷第8期

Prediction of Farmland Soil Organic Matter Content Based on Different Modeling Methods

Jinbao Liu[1, 2, 5], Shaodong Qu[3], Jing He[2,4,5], Zhongan Mao[2], Jiancang Xie[1]

(1.State Key Laboratory of Eco-hydraulics in Northwest Arid Region, Xi'an university of technology, Xi'an, 710048, China;2.Shaanxi Provincial Land Engineering Construction Group Co., Ltd., Xi'an, 710075, China;3.Shaanxi Ecological Industry Co., Ltd., Xi'an, 710075, China;4.Key Laboratory of Degraded and Unused Land Consolidation Engineering, Ministry of Natural Resource, Xi'an, 710075, China;5.Institute of Land Engineering and Technology, Shaanxi Provincial Land Engineering Construction Group Co., Ltd., Xi'an, 710075, China)

【**Abstract**】Soil Organic Matter (SOM) is an important source of crop growth. Its content can reflect the status of soil fertility, has a significant impact on the growth and development of crops, and is one of the indicators of land quality. In this study, 190 soil samples were collected in the Weihe Plain as the research area. ASD Fieldspec 4 was used to obtain the spectral data of the soil sample in the 350~2500 nm band, and the relationship between the SOM content and the soil reflectance spectrum was analyzed, and the effect of FD, SD, Log, FDL, SDl, CR combined with CARS in the characteristic band extraction was compared and analyzed. Cubist is used to establish a SOM estimation model to provide a basis for hyperspectral estimation of SOM content. The results show that the number of feature bands extracted based on CR-CARS is the least, and the accuracy is the best. Compare six kinds of transformations, CR has the least number of characteristic bands, which are 428 nm, 454 nm, 466 nm, 468 nm, 477 nm, 535 nm, 1406 nm, 1500 nm, 1609 nm, 1686 nm, 2172 nm, 2398 nm, 2399 nm, 2438 nm, 2449 nm. In the Cubist model, the coefficient of determination (Rv^2) in the validation ranges from 0.60 to 0.97, and the ranges for RPD is 1.19 to 3.02. Comparing the model calibration set Rc^2, RMSEC and the validation set Rv^2, RMSEP and scatter plot distribution, the CR-Cubist model has the highest prediction accuracy, the calibration set Rc^2 is 0.96, the RMSE is 0.49, the validation set Rv^2 is 0.97, the RMSEP is 0.51, and RPD = 3.02, which is stable, and the estimated model has better potential.

【**Keywords**】Soil organic matter; Vis-NIR spectroscopy; Spectral feature selection; Cubist

1 Introduction

Soil organic matter (SOM) is a key indicator to measure soil fertility, state and degradation degree[1-2]. It is also an important source, sink, and sink of carbon cycle in terrestrial ecosystems, which is of great significance to the sustainable development of precision agriculture[3-4]. Therefore, the rapid and accurate estimation of soil organic matter content in farmland is of great significance for soil fertility assessment and precision agriculture production. Although traditional methods have high accuracy in obtaining soil organic matter content, they are time-consuming, labor-intensive, and costly. The sample pretreatment takes a long time, and chemical waste is prone to pollute the environment[5]. Therefore, it is not suitable for large-scale farmland soil organic matter surveys. Recent years, with its high spectral resolution, strong band continuity, and ability to obtain fine spectral information that cannot be obtained by multispectral sensors, hyperspectral remote sensing has become more and more widely used in agriculture, forestry, and water bodies[6-8]. It is used for quantitative analysis of the earth. Surface biophysical and biochemical processes provide parameters and basis[9]. With the continuous development of remote sensing technology and machine learning technology, scholars at home and abroad have carried out a lot of research in soil near-Earth remote sensing[10-11], especially the feature information extraction and model construction based on improved machine learning methods have received widespread attention[12]. Based on indoor visible near-infrared reflectance (Vis-NIR) spectroscopy measurement, it mainly includes soil organic carbon (SOC)[13-14], soil pH[15], soil electrical conductivity (EC), iron oxide and other soil properties[16]. Among them, Savitzky-Golay smoothing denoising, multivariate scattering correction (MSC), standard normal variation (SNV), differential transformation of different orders, continuum removal[17-18]. Methods such as correlation analysis and morphological analysis to construct the spectral index reduce the spectral noise and enhance the signal-to-noise ratio. Genetic algorithm, interval partial least squares method. obtain spectral feature information, combined with BP neural network (Back Propagation Neural Network, BPNN), random forest (Random Forest, RF), support vector machine Non-linear machine learning models such as Support Vector Machine (SVM) and Convolutional Neural Network (CNN) have improved the accuracy of SOC estimation[39-43]. This paper selects the Weihe Plain as the study area, uses the indoor hyperspectral data of 190 surface soil samples combined with the measured SOM to preprocess the original hyperspectral data, and uses the competitive adaptive weighting method (CARS) to optimize the SOM sensitivity of different pretreatments. Band variables, using cubist to construct a spectral inversion model of soil organic matter content, in order to optimize the best modeling method, and provide a certain theoretical basis for the estimation of farmland soil organic matter content.

2 Materials and methods

Sample collection and data processing. The sampling time was July 2019. According to the typical farmland distribution characteristics in the study area, 190 representative sampling points were selected for field sampling, and GPS was used to record the location information of the sampling points. The five-point mixing method was adopted for sampling at each sampling point, and five soil samples were uniformly mixed as one sample, and the sampling depth was 0~20 cm (about 1kg). The collected soil samples were brought back to the laboratory for natural air-drying. After removing grass roots and gravel, they were ground through a 2 mm sieve and divided into two parts for spectrum collection and SOM determination. the concentrations of all soil samples were measured using the $K_2Cr_2O_7-H_2SO_4$ oxidization approach. Soil spectra were measured using an ASD FieldSpec © 4 high-resolution spectroradiometer.

Competitive Adaptive Re-weighting Algorithm (CARS) is based on Darwin's theory of biological evolution. The principle of "survival of the fittest" uses adaptive reweight sampling (ARS) technology to select the wavelength points with large absolute regression coefficients in the PLS model, remove the wavelength points with small weights, and use interactive verification to select the lowest subset of RMSECV. Can effectively find the optimal combination of variables.

Cubist is a rule-based model, an extension of Quinlan's M5 model tree. Grow a tree where the terminal leaves contain a linear regression model. Based on the predictor variables used in the previous split. In addition, each step of the tree has an intermediate linear model. The prediction is made using the linear regression model at the terminal node of the tree, but is "smoothed" by considering the prediction from the linear model in the previous node of the tree. The tree is reduced to a set of rules, which are initially the path from the top to the bottom of the tree. Simplify the rules by trimming or combining.

The coefficient of determination (Rc^2), the coefficient of verification determination (Rv^2), the root mean squared error (RMSE), and the relative percent deviation (RPD) comprehensively evaluate the performance of the model. When $0.60 \leqslant R_2 \leqslant 0.80$, the model fitting effect is better, when $0.80 \leqslant R_2 \leqslant 0.90$, the model fitting result is very good, when $R_2 \geqslant 0.90$, the model fitting effect is excellent. RMSE represents the predictive ability of the model. The closer to 0, the higher the predictive accuracy of the model and the stronger the predictive ability. When RPD≥2.5, the model has strong predictive ability, and the model can be used for inversion prediction of soil organic carbon content; when 2.0≤RPD≤2.5, it shows that the model has good quantitative predictive ability.

Data processing. R software (R core team, 2016) and MATLAB 2021a (MathWorks, Natick, MA, USA) were used for spectral data processing, modelling and plotting.

3 Results

Analysis of Spectral Characteristics of Soil Samples with Different SOM Content. The descriptive statistics table of soil organic matter content is shown in Table 1. Statistical analysis of organic matter content, the maximum, minimum, average, median, standard deviation and coefficient of variation were 14.86 g/kg, 3.16 g/kg, 8.77 g/kg, 1.84, 21%, respectively. Divide the modeling set and the verification set according to the soil spectral characteristics. Kennard-Stone algorithm is selected for the algorithm of representative modeling set and verification set. By comparison, we found that the validation values within the calibration range, indicating that the sample classification is correct and the established model is effective. For calibration and validation set, the maximum and minimum interval is 4.34~14.86 g/kg and 3.16~14.26 g/kg, the average, SD, and CV is 8.74~8.79, 1.79~1.98, and 20%~23%, respectively.

Select 190 soil samples with different contents, and then analyze the soil indoor spectral curves of different organic matter average contents (Fig. 1). It can be seen from Fig. 1 that the spectral curves of SOM with different contents are basically the same. For soil spectra with different soil organic matter content, the greater the soil organic matter content, the smaller the spectral reflectance, and the trough of the reflectance spectral curve shifts to the shortwave direction. In the range of 400~900 nm, the spectral reflectance of the soil increases; at 900~1400 nm, the rate of increase in reflectivity gradually decreases. A trough appears at 1400 nm, and then the reflectivity gradually increases; at around 1900nm, it decreases rapidly, and a trough appears; in the 1900~2200 nm band, the reflectivity increases rapidly, and the 2400 nm band slowly decreases.

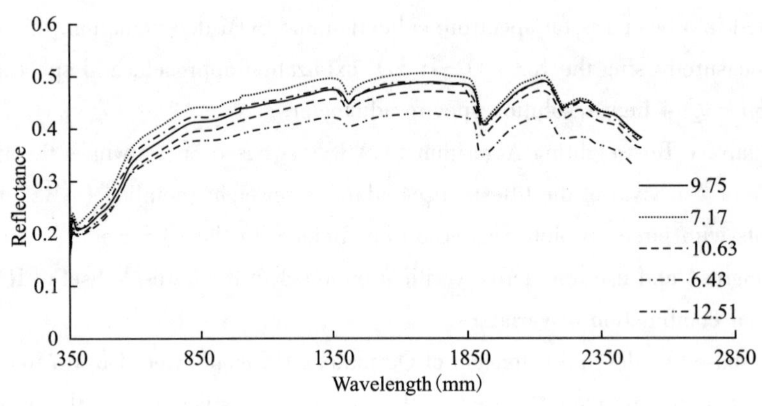

Fig. 1 Soil spectral reflectance curves with different SOM content

1420 nm, 1920 nm, 2340 nm have three absorption characteristics of water absorption bands, which are caused by OH-in the soil. In the 640~830 nm band, as the organic carbon content increases, the continuum removal value decreases, and the absorption area and absorption depth become larger. Since SOC organic molecules are related to the wavelengths absorbed by C—O, C=O and N—H, electromagnetic waves affect soil reflectivity in the visible light (400~700 nm), near-infrared (700~1400 nm), and short-wave infrared (1400~2500 nm) regions.

Table 1 Descriptive statistics of SOM content

Sample	Sample	Maximum (g/kg)	Minimum (g/kg)	Mean (g/kg)	SDv (g/kg)	CV
Total	190	14.86	3.16	8.77	1.84	0.21
Calibration	133	14.86	4.34	8.79	1.79	0.20
Validation	57	14.26	3.16	8.74	1.98	0.23

Note: SDv, Standard error; CV, Coefficient of Variation.

CARS feature band screening. In order to improve the accuracy and speed of the prediction model, the CARS algorithm selects characteristic variables to reduce the high degree of collinearity between spectral bands, treats each band variable as a single individual, and retains individuals with strong adaptability in the process of individual selection. First, randomly select a fixed ratio of samples as a calibration set to establish a PLS model, calculate the absolute value of the regression coefficient and the weight corresponding to each band point, and then use Exponentially Decreasing Function (EDP) and Adaptive Reweighted Sampling (ARS) to select variables. Calculate the Root Mean Square Error of Cross-Validation (RMSECV) through the cross-validation method, select N subsets after N Monte Carlo sampling to obtain N RMSECVs, and select the band subset with the smallest RMSECV. The variables contained in this subset are the optimal combination of variables. Among them, set the Monte Carlo sampling times to 100, and repeat the sampling times. By comparing the RMSECV value of each sampling, when the value is the smallest, the variable corresponding to the sampling times is selected as the optimal variable subset. First-order differential (FD), second-order differential (SD), reciprocal logarithm (Log), reciprocal logarithm first-order differential (FDL) and second-order differential transformation (SDL), combined with CARS to extract characteristic bands (Table 2, Fig. 2).

Table 2 Selected wavelength (nm) based on CARS

Method	Transforms	Visible region		Near infrared	
		Bands	Numbers	Bands	Numbers
CARS	FD	442,453,456,467,469,480,487,490,515,534,544	11	951,978,979,999,1423,2195,2261,2262,2345,2346,2351,2352,2405,2426,2434,2441,2480,2481,2486	19
	SD	451,457,468,481,489,516,535,540,545	9	953,972,978,980,998,2265,2370,2432,2433,2449,2462,2466,2482,2487	14
	Log	396,404,405,416,598	5	1067,1831,1832,1833,2045,2046,2054,2056,2073,2105,2219,2220,2487,2488,2500	15
	FDL	447,453,456,467,469,482,487,488,490,498,521,534	12	999,2420,2426,2433,2436,2441,2442,2446,2450,2455,2461,2481,2484,2486	14
	SDL	451,466,470,481,489,498,503,535,540,545	10	978,2420,2440,2449,2451,2454,2462,2466,2480,2482,2485,2487,2488	13
	CR	428,454,466,468,477,535	6	1406,1500,1609,1686,2172,2398,2399,2438,2449	9

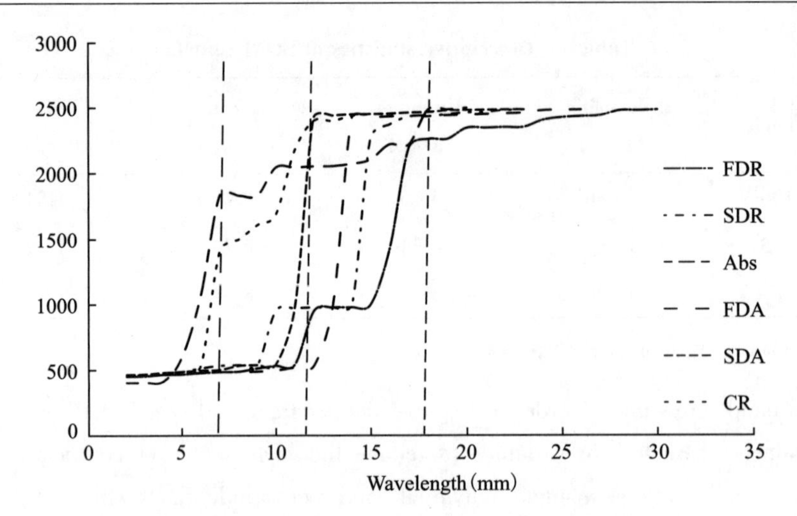

Fig. 2 Dynamics of migration distance of wetting front

It can be seen from Table 2 that in the visible light region, the number of feature bands extracted by FD, SD, Log, FDL, SDL, CR are 11, 9, 5, 12, 10, 6, respectively. Due to limited space, this article only takes the optimal variable process of FD as an example to illustrate, the feature bands extracted by FD+CARS are 442nm, 453 nm, 456 nm, 467 nm, 469 nm, 480 nm, 487 nm, 490 nm, 515 nm, 534 nm, 544 nm, respectively. In the near infrared region, the 19 preferred bands are 951 nm, 978 nm, 979 nm, 999 nm, 1423 nm, 2195 nm, 2261 nm, 2262 nm, 2345 nm, 2346 nm, 2351 nm, 2352 nm, 2405 nm, 2426 nm, 2434 nm, 2441 nm, 2480 nm, 2481 nm, 2486 nm, respectively. Compare 6 kinds of transformations, CR has the least number of characteristic bands, which are 428 nm, 454 nm, 466 nm, 468 nm, 477 nm, 535 nm, 1406 nm, 1500 nm, 1609 nm, 1686 nm, 2172 nm, 2398 nm, 2399 nm, 2438 nm, 2449 nm.

Cubist model establishment and verification. In this study, a variety of methods such as FD, SD, Log, FDL, SDL, CR, etc. are used for spectral preprocessing, and all preprocessing is performed on a smooth basis. After preprocessing, Cubist models were established for model comparison and accuracy verification. Characteristic bands extracted by CARS are used to establish Cubist models with different spectral variables, and the corresponding model effects are different, as shown in Table 3.

In the visible-near infrared region, SOM has multiple spectral response bands, which is the basis for predicting SOM using visible-near infrared spectroscopy. From Table 3, the model accuracy of the calibration set is generally higher than the model accuracy of the validation set. Without pretreatment, the Rc^2, RMSEC, RPD of the model accuracy based on FD are 0.61, 1.16g/kg, respectively. For the validation, the results of Rv^2, RMSEP, RPD are 0.60, 1.24, 1.19, respectively. In addition, by comparing SD and FD, SDL and FDL, whether it is calibration or validation, R^2 has been improved, which shows that the second-order differential can effectively extract features and the established model has better effects.

In the Cubist model, the best spectral processing method is CR combined with CARS. The coefficient of determination (Rv^2) in the validation ranges from 0.60 to 0.97, and the ranges for RPD is 1.19 to 3.02. Therefore, in the CR-Cubist model, the largest RPD value is 3.02, which is stable, and the estimated model has better potential.

In order to visually compare the predictive capabilities of different models, Fig. 3 shows the scatter plots of different preprocessing. Comparing the model calibration set Rc^2, RMSEC and the validation set Rv^2, RMSEP and scatter plot distribution, the CR-Cubist model has the highest prediction accuracy, the calibration set Rc^2 is 0.96, the RMSE is 0.49, the validation set Rv^2 is 0.97, the RMSEP is 0.51, and RPD=3.02.

Table 3 Cubist-based soil organic matter prediction statistics

	Pretreatments	Calibration		Validation		
		Rc^2	RMSEC	Rv^2	RMSEP	RPD
Cubist	FD	0.58	1.16	0.60	1.24	1.19
	SD	0.65	1.06	0.68	1.10	1.54
	Log	0.55	1.26	0.63	1.35	1.69
	FDL	0.68	1.11	0.66	1.33	1.73
	SDL	0.73	0.93	0.76	0.96	1.71
	CR	0.96	0.49	0.97	0.51	3.02

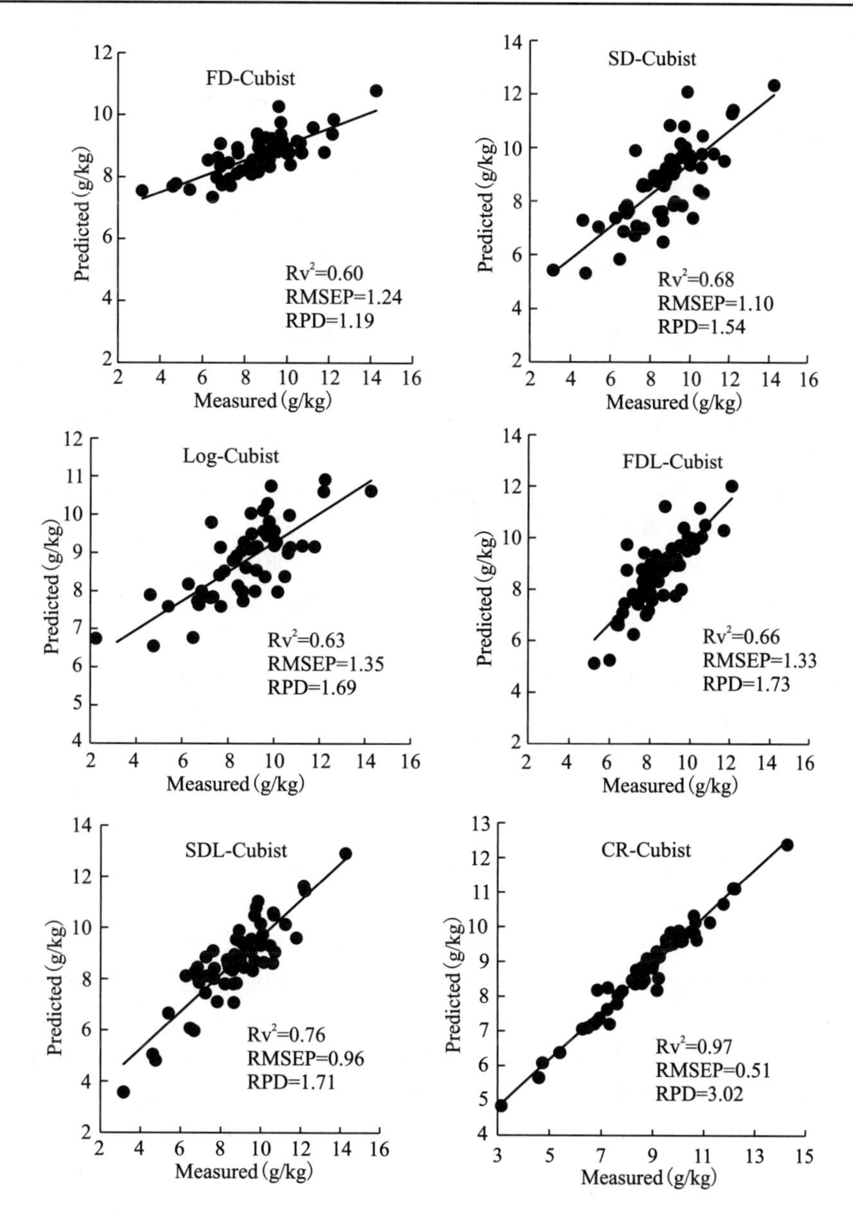

Fig. 3 Scatter plot of SOM content prediction based on different preprocessing

4 Conclusions

In this study, 190 soil samples were collected in the Weihe Plain as the research area. FD, SD, Log,

FDL, SDl, CR combined with CARS in the characteristic band extraction was compared and analyzed. Cubist is used to establish a SOM estimation model to provide a basis for hyperspectral estimation of SOM content. The results show that the number of feature bands extracted based on CR-CARS is the least, and the accuracy is the best. Compare six kinds of transformations, CR has the least number of characteristic bands, which are 428 nm, 454 nm, 466 nm, 468 nm, 477 nm, 535 nm, 1406 nm, 1500 nm, 1609 nm, 1686 nm, 2172 nm, 2398 nm, 2399 nm, 2438 nm, 2449 nm. In the Cubist model, the coefficient of determination (Rv^2) in the validation ranges from 0.60 to 0.97, and the ranges for RPD is 1.19 to 3.02. Comparing the model calibration set Rc^2, RMSEC and the validation set Rv^2, RMSEP and scatter plot distribution, the CR-Cubist model has the highest prediction accuracy, the calibration set Rc^2 is 0.96, the RMSE is 0.49, the validation set Rv^2 is 0.97, the RMSEP is 0.51, and RPD = 3.02, which is stable, and the estimated model has better potential.

Acknowledgements

Funding: This research was funded by Xi'an University of Technology Doctoral Dissertation Innovation Fund (310-252072018).

References

[1]Marchetti, A., Piccini, C., Francaviglia, R., Mabit, L. (2020) Spatial distribution of soil organic matter using geostatistics: a key indicator to assess soil degradation status in central italy. Pedosphere, 22(2), 230-242.

[2]Karhu, K., Fritze, H., Tuomi, M., Vanhala, P., Spetz, P., Kitunen, V., Liskia, J. (2010) Temperature sensitivity of organic matter decomposition in two boreal forest soil profiles. Soil Biology & Biochemistry, 42(1), 72-82.

[3]Kemmitt, S. J., Wright, D., Murphy, D.V., Jones, D. L. (2008). Regulation of amino acid biodegradation in soil as affected by depth. Biology and Fertility of Soils, 44(7), p.933-941.

[4]Laishram, J., Saxena, K. G., Maikhuri, R. K., Rao, K. S. (2012). Soil quality and soil health: a review. International Journal of Ecology & Environmental Sciences, 38(1).

[5]Rodrigues, A. F., Novotny, E. H., Knicker, H., Oliveira, R. D. (2019). Humic acid composition and soil fertility of soils near an ancient charcoal kiln: are they similar to terra preta de indios soils?. Journal of soil & sediments, 19(3), 1374-1381.

[6]Song, K., Lin, L., Li, S., Tedesco, L., Hall, B., Li, L. (2012). Hyperspectral remote sensing of total phosphorus (tp) in three central indiana water supply reservoirs. Water Air & Soil Pollution, 223(4), 1481-1502.

[7]Lausch, A., Zacharias, S., Dierke, C., Pause, M., I Kühn, Doktor, D., Dietrich, P., Werban, U. (2013). Analysis of vegetation and soil patterns using hyperspectral remote sensing, emi, and gamma-ray measurements. Vadose Zone Journal, 12(4), 108-112.

[8]Fabio, C., Sabine, C., Arwyn, J., Kristin, V., Bart, B., Bas, V. W. (2018). Soil organic carbon estimation in croplands by hyperspectral remote apex data using the lucas topsoil database. Remote Sensing, 10(2), 153.

[9]Numata, I., DA Roberts, Chadwick, O. A., Schimel, J. Sampaio, F. R., Leonidas, F. C., Soares, J.V. (2007). Characterization of pasture biophysical properties and the impact of grazing intensity using remotely sensed data. Remote Sensing of Environment, 109(3), 314-327.

[10]Khanal, S., Klopfenstein, A., Kushal, K. C., Ramarao, V., Shearer, S. A. (2021). Assessing the impact of agricultural field traffic on corn grain yield using remote sensing and machine learning. Soil and Tillage Research, 208, 104880.

[11]Le, B. T., Xiao, D., Mao, Y., He, D., Xu, J., Song, L. (2019). Coal quality exploration technology based on an incremental multilayer extreme learning machine and remote sensing images. IEEE Transactions on Geoence and Remote Sensing, 4192-4201.

[12]Shang, X., Chisholm, L. A. (2014). Classification of australian native forest species using hyperspectral remote sensing and machine-learning classification algorithms. IEEE Journal of Selected Topics in Applied Earth Observations & Remote Sensing, 7(6), 2481-2489.

[13]Gomez, C., Rossel, R., Mcbratney, A. B. (2008). Soil organic carbon prediction by hyperspectral remote sensing

and field vis-nir spectroscopy: an australian case study. Geoderma, 146(3-4), 403-411.

[14]Doetterl, S., Stevens, A., Oost, K. V., Wesemael, B. V. (2013). Soil organic carbon assessment at high vertical resolution using closed-tube sampling and vis-nir spectroscopy. Soilence Society of America Journal, 77(4), 1430-1435.

[15]Zhang, Y., Biswas, A., Ji, W., Adamchuk, V. I. (2017). Depth-specific prediction of soil properties in situ using vis-nir spectroscopy. Soil Science Society of America Journal, 81(5), 993.

[16][1] Mouazen, A., Baer De Maeker, J., Ramon, H. (2006). Effect of wavelength range on the measurement accuracy of some selected soil constituents using visual-near infrared spectroscopy. Journal of Near Infrared Spectroscopy, 14(1), 189-199.

[17]Masaya, A., Masamoto, A., Kimito, F. (2009). Prediction models for soil properties using vis-nir spectroscopy. Journal of Computer Aided Chemistry, 10, 53-62.

[18]Davari, M., Karimi, S. A., Bahrami, H. A., Hossaini, S., Fahmideh, S. (2021). Simultaneous prediction of several soil properties related to engineering uses based on laboratory vis-nir reflectance spectroscopy. Catena, 197, 104987.

[19]Nawar, S., Mouazen, A. M. (2019). On-line vis-nir spectroscopy prediction of soil organic carbon using machine learning. Soil and Tillage Research, 190, 120-127.

[20]Li, S., Ji, W., Chen, S., Jie, P., Yin, Z., Zhou, S. (2015). Potential of vis-nir-swir spectroscopy from the chinese soil spectral library for assessment of nitrogen fertilization rates in the paddy-rice region, china. Remote Sensing, 7, 7029-7043.

本文曾发表于2022年《Fresenius Environmental Bulletin》第31卷第2期

2000—2019年毛乌素沙地地表温度演变规律及影响因素分析

杨亮彦[1,2,3],石　磊[1,2,3],孔　辉[1,2,3]

(1.陕西省土地工程建设集团有限责任公司,西安 710075;2.陕西地建土地工程技术研究院有限责任公司,西安 710021;3.自然资源部退化及未利用土地整治工程重点实验室,西安 710021)

【摘要】【目的】明晰全球气候变化和退耕还林还草工程实施背景下毛乌素沙地地表温度的演变规律及其影响因素。【方法】基于地理空间数据云提供的2000—2015年MOSLT1M中国1 km地表温度和在NASA下载的2016—2019年MODIS/TERRA卫星的MOD11A2地表温度,利用线性回归斜率法对2000—2019年毛乌素沙地地表温度时空特征进行分析,并探究了地表温度与气候、土地利用类型和植被指数的相关性。【结果】毛乌素沙地地表温度在空间分布上存在明显差异性,呈从西北到东南逐渐降低、西部高于东部的总体变化趋势,在时间序列上,总体呈下降趋势,下降速率为0.59 ℃/(10 a);毛乌素沙地年均地表温度与气温波动趋势较为一致,二者在2000—2005、2005—2009、2009—2013年时间段间均出现先降低后升高的规律,并在2003、2012年同为极小值点;地表温度受土地利用类型变化影响较大,不同土地利用类型的地表温度差异性明显,建设用地和未利用地地表温度最高,其次是草地和林地,最后是耕地和水域;地表温度与*NDVI*的变化趋势负相关,决定系数为0.513 1;毛乌素沙地地表温度的时空变化趋势与全球气候变化逐渐变暖的趋势不同,其主要原因是人类活动改变了土地利用类型,增加了当地的植被覆盖度。【结论】毛乌素沙地地表温度时空变化是人类活动和气候变化共同影响的结果,且在2000—2019年人类活动是主导因素。

【关键词】毛乌素沙地;地表温度;气候变化;人类活动

Analysis of the Evolution Law and Influencing Factors of the Land Surface Temperature in Mu Us Sandy Land from 2000—2019

Liangyan Yang[1,2,3], Lei Shi[1,2,3], Hui Kong[1,2,3]

(1. Shaanxi Provincial Land Engineering Construction Group Co., Ltd, Xi'an 710075, China; 2. Institute of Land Engineering and Technology, Shaanxi Provincial Land Engineering Construction Group Co., Ltd., Xi'an 710021, China; 3. Key Laboratory of Degraded and Unused Land Consolidation Engineering, the Ministry of Natural Resources, Xi'an 710021, China)

【Abstract】【Objective】In order to clarify the evolution pattern of surface temperature and its influencing factors in Mu Us Sandy Land under the background of global climate change and the implementation of the project of

基金项目:陕西省土地工程建设集团内部项目(DJNY2021-33);陕西省技术创新引导专项基金(2021CGBX-03)。

returning farmland to forest and grass.【Methods】Based on the monthly synthetic products of MOSLT1M China 1km surface temperature from 2000—2015 provided by Geospatial Data Cloud and the surface temperature of MOD11A2 from MODIS/TERRA satellite downloaded at NASA from 2016—2019, we used linear regression slope method was used to analyze the spatial and temporal characteristics of surface temperature in the Mu Us Sandy Land from 2000—2019, and to explore the correlation between surface temperature and climate, land use type and vegetation index.【Results】The spatial distribution of surface temperature in the Mu Us Sandy Land showed significant variability, with a general trend of gradually decreasing from northwest to southeast and higher in the west than in the east, and an overall decreasing trend in the time series, with a decline rate of 0.59 ℃/(10 a); the annual average surface temperature and temperature fluctuation trends in the Mu Us Sandy Land were relatively consistent, and the two fluctuated in the time periods of 2000—2005, 2005—2009 and 2009—2013. The surface temperature is influenced by the change of land use types, and the difference of surface temperature of different land use types is obvious, the surface temperature of construction land and unused land is the highest, followed by grassland and forest land, and finally cultivated land and water; the trend of surface temperature and The trend of *NDVI* showed a negative correlation with a coefficient of determination of 0.5131; the spatial and temporal trends of surface temperature in the Mu Us Sandy Land were different from the gradual warming trend of global climate change, and the main reason was that human activities changed the land use types and increased the local vegetation cover.【Conclusion】The spatial and temporal variation of surface temperature in the Mu Us Sandy Land is the result of the joint influence of human activities and climate change, and human activities is the dominant factor in the period 2000—2019.

【Keywords】 Mu Us Sandy Land; Land surface temperature; Climate change; Human activities

0　前　言

【研究意义】地表温度(Land surface temperature,LST)是研究地表生态系统的重要参数,在地表与大气能量交换的过程中扮演着重要的角色[1-2]。地表温度的精确估算有助于评估区域水文与能量平衡、热通量和土壤水分[3]以及了解全球气候的长期变化动态,被广泛应用于城市热岛效应[4-6]、生态环境评价[7-8]、气候变化[9]、植被监测[10]等诸多研究方向。近些年,受全球气候变暖和多种人为活动的影响,全球生物多样性、生态系统及其服务发生重大改变[11],干旱半干旱区生态环境质量也有所起伏[12]。地表温度作为评价生态环境质量的重要地表参数,成为国内外学者的研究热点。

【研究进展】遥感技术快速发展和卫星数据的广泛应用,为获取大范围的地表温度提供了丰富的数据源。目前针对不同的遥感热红外数据,国内外学者提出了单窗算法[13]、劈窗算法[14-15]等地表温度遥感算法,在中小尺度水平获取了高精度的研究成果。地表温度数据产品的出现,使得基于遥感数据的地表温度研究取得了进一步发展,在天气晴朗和已知发射率的条件下,地表温度反演精度达到1K以内[12],在分析城市热岛效应和大尺度区域地表温度时空变化上有广泛的应用。陈彬辉等[16]基于中分辨率成像光谱仪(Moderate-resolution Imaging Spectroradiometer,MODIS)地表温度产品讨论了京津冀城区土地利用对城市热岛效应的影响;贺丽琴等[17]利用MODIS影像和不透水面积分析了珠江三角洲的热岛效应,得出热岛效应的大小与植被指数成负相关,与城市经济发展程度成正相关;管延龙等[12]基于2001—2013年MODIS地表温度产品分析了天山区域地表温度时空特征,结果表明地表温度呈逐年缓慢增加趋势,增加率为0.147 ℃/a。李琴等[18]基于MODIS数据利用劈窗算法反演了干旱半干旱地区地表温度,并结合气象数据论证了MODIS地表温度产品在地形复杂地区仍具有较好的应用精度。屈创等[19]利用GIS空间分析方法对石羊河流域地表温度空间分布进行研究,印证了MODIS地表温度产品与土地利用类型和植被指数的相关性。FENG等[20]利用2005—2018年LST和归一化植被指数(Normalized Difference Vegetation Index,NDVI)产品分析了我国的气候舒适区,并得出云南是我国气候最舒适的省份。【切入点】目前,地表温度的研究进展取得一些成果,但地表温度时空变化研究的时间跨度较短,研究时间缺乏连贯性[21];研究热点内容主要为北京[16,22]、珠江[17]等城市化进程的热岛效应和藏北高原[1]、天山[12]等受气候变化影响较大区域的地表温度时空变化分布,对我国实施生态工程的、受人类活动影响较大的西北干旱半干旱区研究较少,特别是生态环境脆弱的毛乌素沙地的地表温度演变规律有待研究。

毛乌素沙地气候干旱,生态环境脆弱,受国家政策和人为活动的影响,毛乌素沙地土地覆盖发生巨大变化。地表温度的持续监测,可为毛乌素沙地水资源的合理利用和生态环境的恢复提供理论支持。【拟解决的问题】本文基于2000—2019年MODIS地表温度和NDVI产品及中国土地利用类型遥感监测数据,采用线性回归斜率法和相关分析等方法,对毛乌素沙地区域地表温度的时空分布特征进行分析,并探究地表温度与气温变化、土地利用类型、归一化植被指数之间的联系,以期为毛乌素沙地的环境监测、生态规划等提供科学参考。

1 材料与方法

1.1 研究区概况

毛乌素沙地地处黄土高原腹部,横亘榆林市北部、鄂尔多斯市南部及盐池县东北部(37.45°~39.37°N,107.67°~110.67°E)(见图1),占地面积约为4.22万km^2,平均海拔为1 300 m左右,由东到西逐渐递增,南部最高达1900 m。毛乌素沙地处于干旱与半干旱过渡区,是我国典型的农牧交错带,也是我国荒漠草原-草原-森林草原的过渡地带,生态环境十分脆弱[23]。研究区以温度大陆性气候为主,干旱少雨且分布不均,降雨多集中在6~8月,年降水量250~440 mm,由西向东南方向递增。毛乌素沙地水体空间分布有差异性,西北部干旱缺水,东南部地表水与地下水都较为充足,河流众多,其中无定河、秃尾河、窟野河等河流贯穿沙地的东南部,为毛乌素沙地东南的生态环境恢复和农业发展提供了重要保障。

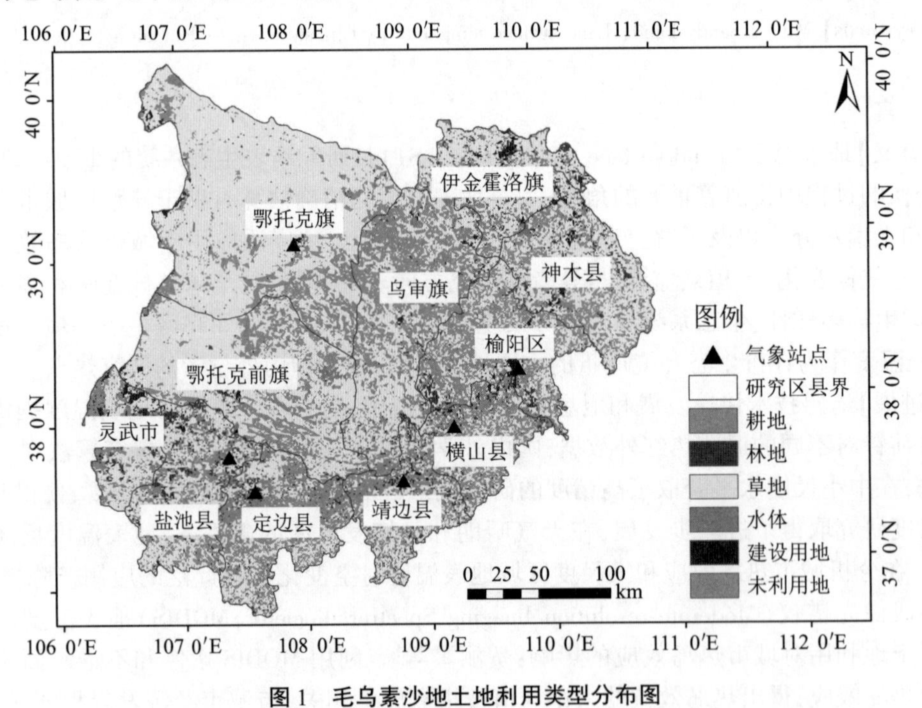

图1 毛乌素沙地土地利用类型分布图

Fig. 1 Distribution map of land use types in Mu Us Sandy Land

1.2 数据源及预处理

本研究采用数据为MODIS地表温度产品,其来源分为两个部分,第一部分由地理空间数据云提供的2000—2015年MOSLT1M中国1 km地表温度月合成产品;第二部分为在NASA下载的2016—2019年MODIS/TERRA卫星的MOD11A2地表温度8 d合成产品,空间分辨率为1 km。覆盖研究区的影像行列号为H26V04和H26V05,共计368幅影像。对第一部分数据通过拼接、裁剪获取毛乌素沙地2000—2015年月尺度和年尺度地表温度数据。针对第二部分地表温度8 d合成产品,通过格式转换、投影变换、拼接、裁剪获取研究区8 d合成地表温度数据,再利用加权平均法,获取2016—2019年毛乌素沙地月尺度和年尺度地表温度数据。遥感数据计算均在ENVI软件下完成。

气象数据来源于中国气象数据网,研究区内共有 7 个国家气象站点,分别为横山站(53740)、靖边站(53735)、定边站(53725)、盐池站(53723)、神木站(53651)、榆林站(53646)和鄂托克旗站(53529)。气象数据包含地表温度、气温、降水量等 8 种日值数据,通过统计分析获取每个站点的月平均和年平均气温数据,并计算毛乌素沙地 7 个气象站点的平均气温作为研究区的气温。土地利用/覆被栅格数据由中国科学院资源环境科学数据中心提供。数据生产制作以 Landsat 系列卫星数据为数据源,经过图像分类、人工目视解译完成[24]。本研究使用 2000 年、2005 年、2010 年、2015 年和 2019 年 5 期土地利用/覆被数据,空间分辨率为 1 km,土地利用类型包括耕地、林地、草地、水域、建设用地和未利用土地 6 个一级类型。

1.3 研究方法

1.3.1 地表温度时空分布分析方法

线性回归斜率法是一种基于像元尺度的变化趋势分析方法,在大尺度地表参数时空变化分析中应用广泛。本文利用线性回归斜率法对研究区像元尺度的地表温度变化趋势模拟,分析毛乌素沙地不同时期地表温度的空间变化规律,并获取每个像元在研究时间段内的变化斜率,利用斜率大小判断该像元年际变化趋势。其计算式参考文献[25]。

1.3.2 地表温度影响因素分析方法

(1)气温与地表温度的关系。年均气温和地表温度变化较大,两者无明显的线性关系。本研究利用定性分析的方法,分析气温与地表温度的变化规律。

(2)不同土地利用类型的地表温度。土地利用类型是影响地表温度的重要参数,本研究利用 ArcGIS 软件对不同土地利用类型的地表温度进行提取和统计,其提取步骤为 Spatial Analyst tools→zonal→zonal statistics as Table。具体参数设置和说明见表 1。

表 1 不同土地利用类型提取地表温度参数设置及说明

Table 1 Setting and description of surface temperature parameters for extraction of different land use types

参数	说明	数据类型
Input Zone data	定义区域的数据集,可通过整型栅格数据或者要素图层来定义区域(本研究选取的是土地利用类型栅格数据)	Raster or Feature Layer
Zone field	选取代表不同土地利用类型的字段	Field
Input value raster	含有要计算统计数据的值的栅格;本研究输入为地表温度栅格数据	Raster Layer
Output table	包含每个区域中值的汇总的输出表	Table
Statistics type(可选)	要计算的统计类型:ALL、MEAN、MAX、MIN、STD 等	String

(3)植被指数与地表温度的关系。本研究利用 ArcGIS 软件基于点要素,提取同一时间的植被指数与地表温度进行相关性分析,相关系数计算式参考文献[26]。

2 结果与分析

2.1 地表温度时空分布特征

图 2、表 2 分别为毛乌素沙地 2000 年、2005 年、2010 年、2015 年和 2019 年地表温度空间分布图和不同地表温度的占比。由图 2 和表 2 可知,在空间分布规律上 2000—2019 年毛乌素沙地地表温度无明显变化,均呈西部高于东部的规律。在时间分布上毛乌素沙地地表温度分布规律有所不同,其中 2000 年和 2005 年地表温度主要集中在 32~36 ℃,占比分别为 76.01%和 62.61%;2010 年和 2015 年地表温度集中分布在 30~34 ℃,占比分别为 67.37%和 58.22%;2019 年地表温度在 28~36 ℃分布相对均匀。2000—2019 年低温区不断增多,2000 年低于 28 ℃的区域在研究区内分布较为散落,之后面积不断扩张,2015 年、2019 年,低温区遍布在研究区东部和南部。地表温度高值分布在西部的主要原

因是该区域人类密度较小,土地利用类型为未利用地,受自然条件的影响,地表温度无明显下降。低值分布在东部和南部,该区域水资源相对丰富,人类活动较多,农业、林业发展相对较好,主要的土地利用类型是草地、耕地和林地,地表温度呈不断下降趋势。

为了进一步分析 2000—2019 年毛乌素沙地地表温度的变化规律,本研究根据趋势分析法将毛乌素沙地 2000—2019 年地表温度变化趋势分为 5 个等级(图 3):明显下降区(Slope<-0.2)占研究区 5.14%,下降区(-0.2<Slope<-0.05)占研究区 53.80%,稳定区(-0.05<Slope<0.05)占研究区 35.09%,上升区(0.05<Slope<0.2)占研究区 4.13%,明显上升区(Slope>0.2)占研究区 1.84%。地表温度明显上升区与上升区域集中分布在研究区的西部,占比较小。地表温度稳定区域分布在研究区的西部和散落在中部。下降区占研究区的比例高于 50%,主要分布在研究区的中部和东部,面积分布较广,明显下降区主要集中在东部,分布在榆阳市和神木县境内。毛乌素沙地地表温度的整体变化趋势与研究区内的高低温分布规律较为相似,其主要原因是人类活动对中东部土地利用类型的影响。

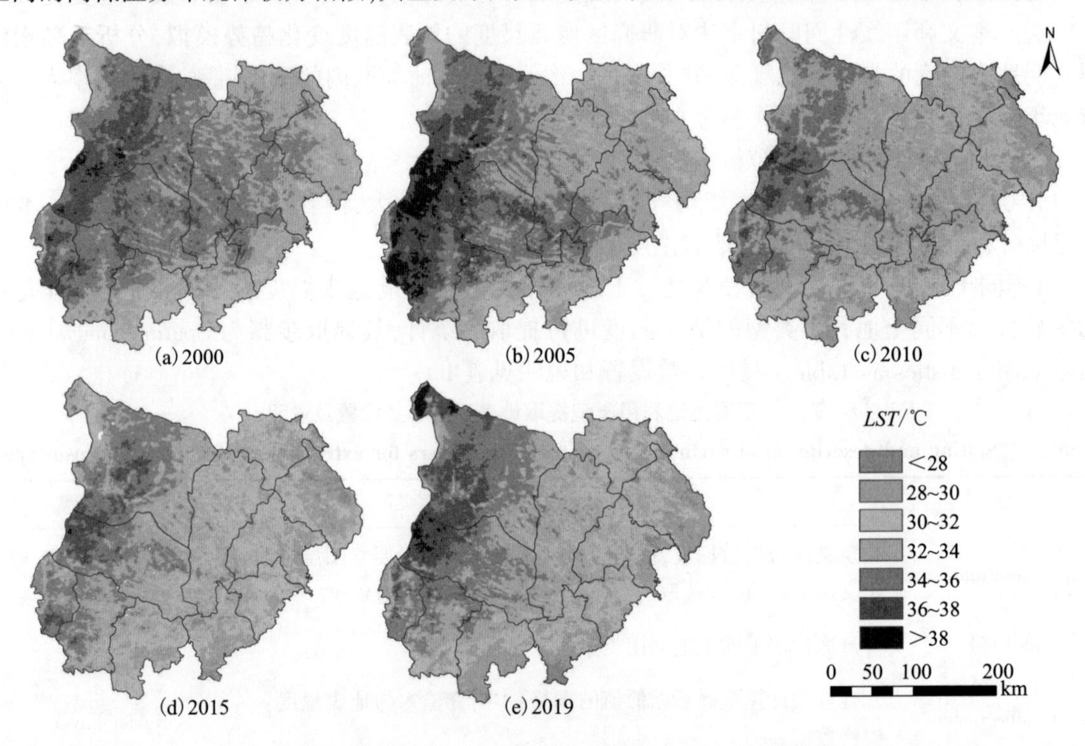

图 2 2000—2019 年毛乌素沙地年均 LST 空间分布

Fig.2 Spatial distribution map of annual LST in Mu Us Sandy Land from 2000 to 2019

表 2 2000—2019 年毛乌素沙地不同 LST 占比(%)

Table 2 Proportion of different LST in Mu Us Sandy Land from 2000 to 2019

LST/℃	2000	2005	2010	2015	2019
<28	0.54	0.51	1.23	1.37	7.86
28~30	0.72	1.66	11.53	10.77	24.44
30~32	9.05	13.92	27.94	27.62	23.65
32~34	31.07	35.37	39.43	30.60	23.91
34~36	44.94	27.24	17.71	22.92	16.49
36~38	13.14	16.47	2.13	6.55	3.27
>38	0.54	4.83	0.04	0.16	0.37

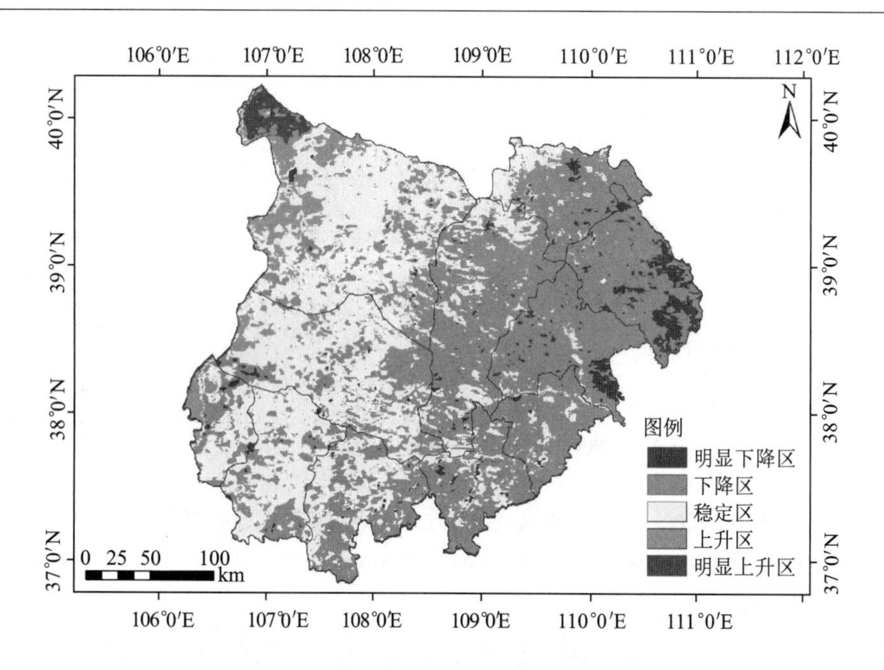

图 3　毛乌素沙地地表温度年际变化趋势

Fig.3　Interannual variation trend of LST in Mu Us Sandy Land

毛乌素沙地 LST 的年际变化趋势和相对变化率的波动情况见图 4。2000—2019 年毛乌素沙地地表温度在 30.94~34.23 ℃波动，平均地表温度为 32.38 ℃，总体呈下降趋势，下降速率为 0.59 ℃/(10 a)。地表温度的变化规律与全球气候变化逐渐变暖的趋势不同，其主要原因是毛乌素沙地是我国植树造林工程和退耕还林还草工程的主要区域，人类活动改变了研究区的土地利用类型，减少了未利用地面积，增加了空气湿度和地表蒸散发，从而降低了地表接收的太阳短波辐射，使地表温度保持下降趋势。研究区地表温度的相对变化率波动较大，但无明显的增加或降低的趋势，且与气温的波动变化规律基本一致，见图 5，二者在 2000—2005 年、2005—2009 年、2009—2013 年时间段内均出现先降低后升高的规律，并在 2003 年、2012 年同为极小值点。以上结果说明地表温度的时空变化是人类活动和气候变化共同影响的结果，且在 2000—2019 年人类活动占主导地位。

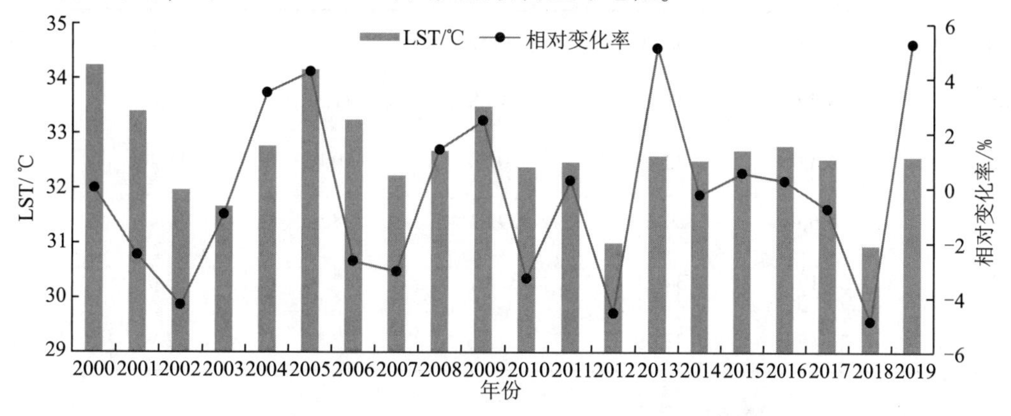

图 4　毛乌素沙地地表温度年际变化及相对变化率

Fig. 4　Interannual variation and relative change rate of surface temperature in Mu Us Sandy Land

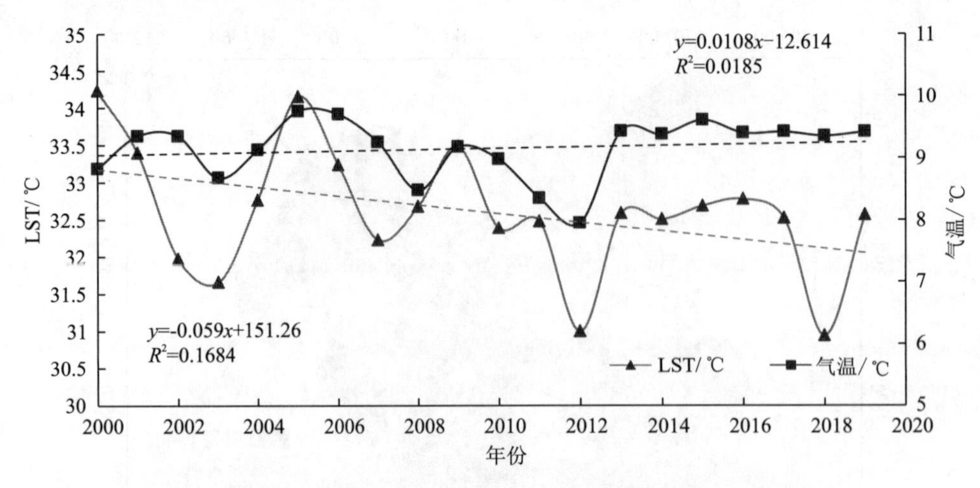

图 5　2000—2019 年毛乌素沙地年均地表温度与气温的变化趋势

Fig. 5　Variation trend of annual surface temperature and air temperature in Mu Us Sandy Land from 2000 to 2019

2.2　不同土地利用类型的年均 LST 变化特征

利用叠置分析统计毛乌素沙地 2000—2019 年不同土地利用类型的年均地表温度(表 3)。由表 3 可知,不同土地利用类型的年均地表温度差异性明显,建设用地和未利用地是研究区地表温度最高的 2 种地类,分别为 33.25 ℃和 33.20 ℃,其次是林地和草地,其地表温度分别为 32.53 ℃和 31.31 ℃,地表温度最低的为耕地和水域,分别为 30.99 ℃和 29.76 ℃。在耕地和水体区域,水资源较为丰富,较高的蒸散发量吸收了部分太阳辐射能量,因此地表温度较低。建设用地和未利用地表面无植被覆盖,在太阳辐射的作用下地表温度上升较快。各土地利用类型的平均地表温度均呈先降低后增加的趋势,2000—2015 年均保持逐年降低,与国家政策有关,1999—2010 年,国家实施退耕还林还草工程,该工程有效地阻止了土地沙化的进程,增加了研究区的植被覆盖度,使毛乌素沙地生态环境质量得到显著提升;2015—2019 年地表温度又有所回升,与全球气候不断升温有关。

表 3　2000—2019 年毛乌素沙地不同土地利用类型平均地表温度

Table 3　Average LST of different land use types in Mu Us Sandy Land from 2000 to 2019

LST/℃	2000	2005	2010	2015	2019	均值
耕地	32.18	32.20	30.33	29.94	30.29	30.99
林地	33.80	33.52	31.88	31.57	31.87	32.53
草地	32.43	32.08	30.50	30.08	31.46	31.31
水域	30.60	29.69	29.41	28.66	30.45	29.76
建设用地	34.19	34.21	32.47	32.32	33.05	33.25
未利用地	34.45	33.99	32.56	32.10	32.90	33.20

2.3　地表温度与植被指数的关系

图 6 反映了研究区年均地表温度和 NDVI 之间的关系。2000—2019 年,毛乌素沙地植被指数持续增加,波动范围为 0.25~0.43,而地表温度呈下降趋势,且两者在 2000—2002 年、2011—2013 年、2017—2019 年具有明显相反的变化趋势,表明植被指数与地表温度之间关系密切。图 7 为 2019 年 7 月地表温度与 NDVI 的关系图。由图 7 可知,地表温度与 NDVI 之间拟合的决定系数为 0.5131,说明人类活动对地表温度产生了一定的影响,也反映了绿色植被对调节区域气候地表温度的重要性。

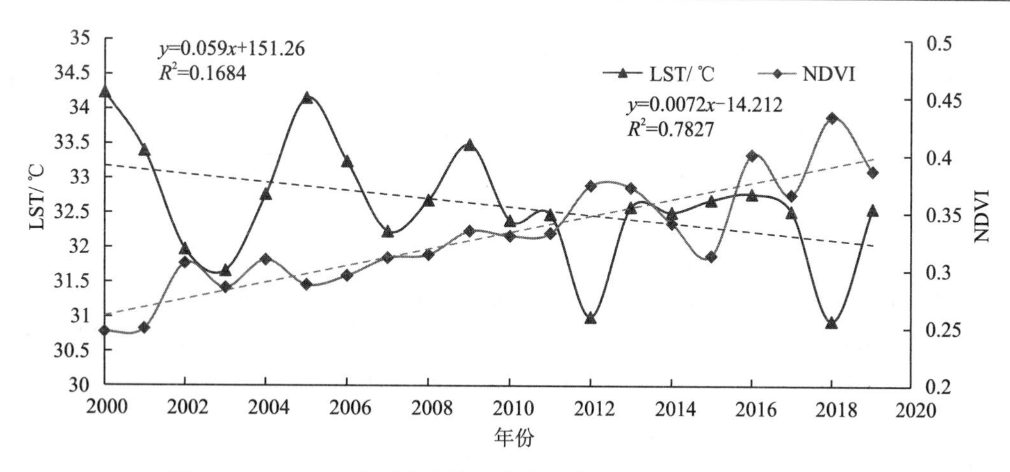

图 6 2000—2019 年毛乌素沙地年均地表温度与 NDVI 的变化趋势

Fig. 6 Variation trend of annual LST and NDVI in Mu Us Sandy Land from 2000 to 2019

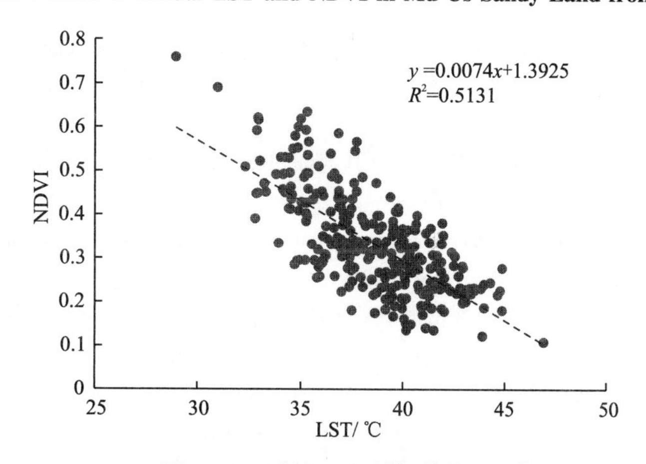

图 7 LST 与 NDVI 的相关关系

Fig. 7 Correlation between LST and NDVI

3 讨论

地表温度是研究水循环和能量循环的重要输入参数,明晰毛乌素沙地地表温度的演变规律,有助于研究区水资源的合理配置和生态环境的可持续发展。毛乌素沙地地表温度与气温的变化曲线具有相似的波动性,表明两者具有较为明显的正相关响应关系,与扎西欧珠的研究结果相似[1]。但是在全球气温升高的过程中,研究区地表温度出现下降趋势,与大兴安岭、广西等地变化规律有所差异[21,26],其原因是退耕还林还草和植树造林等生态政策的实施,改变了毛乌素沙地的土地利用类型,增加了研究区植被覆盖度,从而降低了地表温度。

地表温度的变化是地理因子、气候因子和社会经济等因素综合影响的结果,本研究仅分析了气温、土地利用类型和植被指数 3 种因素对地表温度的影响,研究局限性较大。毛乌素沙地面积较大,横跨多个行政区域,不同行政区域的土地利用政策有明显差异,导致毛乌素沙地南北区域土地利用类型变化的方向有所不同,这也是毛乌素沙地地表温度变化趋势呈现西部升高、东部下降的主要原因。土地政策影响机制的量化一直以来是土地科学和生态科学的研究热点和难点,也是日后重点研究的方向,因此对政策的解读仍需加深。在研究数据方面,缺乏详细的社会经济统计数据,是本研究的不足之处。此外,本研究所采用的地表温度数据分辨率为 1 km,空间分辨率较粗,且数据来自于不同的数据源,对本研究造成了一定的局限性。

4 结论

(1)毛乌素沙地地表温度在空间分布上存在明显差异性,呈从西北到东南逐渐降低、西部高于东部的总体变化趋势。在时间序列上,总体呈下降趋势,下降速率为 0.59 ℃/(10 a)。

(2)毛乌素沙地年均地表温度与气温波动趋势较为一致,两者在2000—2005年、2005—2009年、2009—2013年时间段内均出现先降低后升高的规律,并在2003年、2012年同为极小值点;地表温度受土地利用类型变化影响较大,不同土地利用类型的地表温度差异性明显,建设用地和未利用地地表温度最高,其次是草地和林地,最后是耕地和水域;地表温度与NDVI的变化趋势负相关,决定系数为0.5131。

(3)毛乌素沙地地表温度的时空变化趋势与全球气候变化逐渐变暖的趋势不同,其主要原因是人类活动改变了土地利用类型,增加了当地的植被覆盖度。人类活动对地表温度产生了明显的影响,也反映了绿色植被对调节区域地表温度的重要性,因此毛乌素沙地地表温度时空变化是人类活动和气候变化共同影响的结果,且在2000—2019年人类活动占主导地位。

参考文献

[1]扎西欧珠, 边多, 次珍, 等. 基于MODIS遥感数据和气象观测数据的藏北高原地表温度变化特征[J]. 中国农学通报, 2020, 36(20): 136-142.

ZHA Xiouzhu, BIAO Duo, CI Zhen, et al. Surface Temperature in the Northern Tibetan Plateau: Variation Characteristics Based on MODIS Remote Sensing Data and Meteorological Observed Data[J]. Chinese Agricultural Science Bulletin, 2020, 36(20): 136-142.

[2]张爱因, 张晓丽. Landsat-8地表温度反演及其与MODIS温度产品的对比分析[J]. 北京林业大学学报, 2019, 41(3): 1-13.

ZHANG Aiyin, ZHANG Xiaoli. Land surface temperature retrieved from Landsat-8 and comparison with MODIS temperature product[J]. Journal of Beijing Forestry University, 2019, 41(3): 1-13.

[3]马春芽, 王景雷, 陈震, 等. 基于温度植被干旱指数的土壤水分空间变异性分析[J]. 灌溉排水学报, 2019, 38(3): 28-34.

MA Chunya, WANG Jinglei, CHEN Zhen, et al. Spatial Distribution of Soil Moisture Estimated Using Thermal Vegetation Drought Indices[J]. Journal of Irrigation and Drainage, 2019, 38(3): 28-34.

[4]姚远, 陈曦, 钱静. 城市地表热环境研究进展[J]. 生态学报, 2018, 38(3): 1134-1147.

YAO Yuan, CHEN Xi, QIAN Jing. Research progress on the thermal environment of the urban surfaces[J]. Acta Ecologica Sinica, 2018, 38(3): 1134-1147.

[5]康文敏, 蔡芫镔, 郑慧祯. 福州城市地表温度时空变化与贡献度研究[J]. 地球科学进展, 2020, 35(1): 88-100.

KANG Wenmin, CAI Yuanbin, ZHENG Huizhen. Spatial-temporal patterns of land surface temperature and influencing factors contribution in Fuzhou City[J]. Advances in Earth Science, 2020, 35(1): 88-100.

[6]陈康林, 龚建周, 陈晓越, 等. 广州城市绿色空间与地表温度的格局关系研究[J]. 生态环境学报, 2016, 25(5): 842-849.

CHEN Kanglin, GONG Jianzhou, CHEN Xiaoyue, et al. The Pattern Relationship Research of Green Space and Surface Temperature in Guangzhou City[J]. Ecology and Environmental Sciences, 2016, 25(5): 842-849.

[7]刘思怡, 丁建丽, 张钧泳, 等. 艾比湖流域草地生态系统环境健康遥感诊断[J]. 草业学报, 2020, 29(10): 1-13.

Liu Siyi, Ding Jingli, Zhang Junyong, et al. Remote sensing diagnosis of grassland ecosystem environmental health in the Ebinur Lake Basin[J]. Acta Prataculturae Sinica, 2020, 29(10): 1-13.

[8]赵永华, 贾夏, 王晓峰. 泾河流域土地利用及其生态系统服务变化[J]. 陕西师范大学学报(自然科学版), 2011, 39(4):79-85.

ZHAO Yonghua, JIA Xia, WANG Xiaofeng. Land use change and its ecosystem services in Jinghe River basin[J]. Journal of Shaanxi Normal University (Natural Science Edition), 2011, 39(4):79-85.

[9]任鑫, 石正国. 中全新世全球极端温度响应-基于MPI-ESM-P模拟试验[J]. 地球环境学报, 2019, 10(5): 465-478.

REN Xin, SHI Zhengguo. Response of global temperature extremes in mid-Holocene: results from MPI-ESM-P experiments[J]. Journal of Earth Environment, 2019, 10(5): 465-478.

[10]高凡, 何兵, 闫正龙, 等. 2000—2016年叶尔羌河中下游植被覆盖动态变化遥感分析[J].灌溉排水学报, 2019, 38(4): 92-99.

GAO Fan, HE Bing, YAN Zhenglong, et al. Using Remote Sensing to Unravel Spatiotemporal Change in Vegetation Coverage in Middle-low Reaches of the Yarkant Basin[J]. Journal of Irrigation and Drainage, 2019, 38(4): 92-99.

[11]付景保. 黄河流域生态环境多主体协同治理研究[J].灌溉排水学报, 2020, 39(10): 130-137.

FU Jingbao. Coordinating Management of the Eco-environmental Systems in the Yellow River Basin[J]. Journal of Irrigation and Drainage, 2020, 39(10): 130-138.

[12]管延龙, 王让会, 李成, 等. 基于MODIS数据的天山区域地表温度时空特征[J]. 应用生态学报, 2015, 26(3): 681-688.

GUAN Yanlong, WANG Ranghui, LI Cheng, et al. Spatial-temporal characteristics of land surface temperature in Tianshan Mountains area based on MODIS data[J]. Chinese Journal of Applied Ecology, 2015, 26(3): 681-688.

[13]ROZENSTEIN O, QIN Z H, DERIMIAN K A. Derivation of Land Surface Temperature for Landsat-8 TIRS Using a Split Window Algorithm[J]. Sensors, 2014, 14(4): 5768-5780.

[14]JIMÉNEZ M J C, SOBRINO J A, SKOKOVIC D, et al. Land surface temperature retrieval methods from landsat-8 thermal infrared sensor data[J]. IEEE Geoscience and remote Sensing Letters, 2014, 11(10): 1840-1843.

[15]宋挺, 段峥, 刘军志, 等. Landsat8数据地表温度反演算法对比[J]. 遥感学报, 2015, 19(3): 451-464.

SONG Ting, DUAN Zheng, LIU Junzhi, et al. Comparison of four algorithms to retrieve land surface temperature using Landsat 8 satellite[J]. Journal of Remote Sensing, 2015, 19(3): 451-464.

[16]陈彬辉, 冯瑶, 袁建国, 等. 基于MODIS地表温度的京津冀地区城市热岛时空差异研究[J]. 北京大学学报(自然科学版), 2016, 52(6): 1134-1140.

CHEN Binhui, FENG Yao, YUAN Jianguo, et al. Spatiotemporal Difference of Urban Heat Island in Jing-Jin-Ji Area Based on MODIS Land Surface Temperature[J]. Acta Scientiarum Naturalium Universitatis Pekinensis, 2016, 52(6): 1134-1140.

[17]贺丽琴, 杨鹏, 景欣,等. 基于MODIS影像及不透水面积的珠江三角洲热岛效应时空分析[J]. 国土资源遥感, 2017, 29(4): 140-146.

He L Q,Yang P, Jing X, et al. Analysis of temporal-spatial variation of heat island effect in Pearl River Deltausing MODIS images and impermeable surface area[J]. Remote Sensing for Land and Resources, 2017, 29(4): 140-146.

[18]李琴, 陈曦, 包安明, 等. 干旱/半干旱区MODIS地表温度反演与验证研究[J]. 遥感技术与应用, 2008, 23(6): 602-603,643-647.

LI Qin, CHEN Xi, BAO Anming, et al. Validation of Land Surface Temperatures Retrieving in Arid/Semi-Arid Regions[J]. Remote Sensing Technology and Application, 2008, 23(6): 602-603,643-647.

[19]屈创, 马金辉, 夏燕秋, 等. 基于MODIS数据的石羊河流域地表温度空间分布[J]. 干旱区地理, 2014, 37(1): 125-133.

QU Chuang, MA Jinhui, XIA Yanqiu, et al. Spatial distribution of land surface temperature retrieved from MODIS data in Shiyang River Basin[J]. Arid Land Geography, 2014, 37(1): 125-133.

[20]FENG L, LIU Y X, FENG Z Z, et al. Analysing the spatiotemporal characteristics of climate comfort in China based on 2005-2018 MODIS data[J]. Theoretical and Applied Climatology, 2021, (11):1-15.

[21]熊小菊, 廖春贵, 胡宝清. 2007—2016年广西地表温度时空分异规律及其影响因素[J]. 科学技术与工程, 2019, 19(17): 44-52.

XIONG Xiaoju, LIAO Chungui, Hu Baoqing. Spatial and temporal variation of surface temperature in guangxi from 2007 to 2016 and its influencing factors[J]. Science Technology and Engineering, 2019, 19(17): 44-52.

[22]杨敏, 杨贵军, 王艳杰, 等. 北京城市热岛效应时空变化遥感分析[J]. 国土资源遥感, 2018, 30(3): 213-223.

YANG Min, YANG Guijun, WANG Yanjie, et al. Remote sensing analysis of temporal-spatial variations of urban heat island effect over Beijing[J]. Remote Sensing for Land and Resources, 2018, 30(3):213-223.

[23]刘娟, 刘华民, 卓义, 等. 毛乌素沙地1990—2014年景观格局变化及驱动力[J]. 草业科学, 2017, 34(2):

255-263.

LIU Juan, LIU Huamin, ZHUO Yi, et al. Dynamics and driving forces of landscape patterns in Mu Us Sandy Land, from 1990 to 2014[J]. Pratacultural Science, 2017 ,34(2) :255-263.

[24]LIU Jiyuan, ZHANG Zengxiang, Xu Xinling, et al. Spatial patterns and driving forces of land use change in China during the early 21st century[J]. Journal of Geographical Sciences, 2010, 20(4):483-494.

[25]程杰, 杨亮彦, 黎雅楠. 2000—2018 年陕北地区 NDVI 时空变化及其对水热条件的响应[J]. 灌溉排水学报, 2020, 39(5):111 119.

CHENG Jie, YANG Liangyan, LI Yanan. Spatiotemporal Variation in NDVI and Its Response to Hydrothermal Change from 2000 to 2018 in Northern Shaanxi Province [J]. Journal of Irrigation and Drainage, 2020, 39(5): 111-119.

[26]都彦廷, 张冬有. 大兴安岭地区 2001—2019 年地表温度时空分布及影响因素分析[J]. 森林工程, 2020, 36(6): 9-18.

DU Yanting, ZHANG Dongyou. Spatio-temporal distribution of surface temperature and its influencing factors in Daxing'an Mountains from 2001 to 2019[J]. Forest Engineering, 2020, 36(6): 9-18.

本文曾发表于2022年《灌溉排水学报》第41卷第1期

毛乌素沙地蒸散发时空分布及影响因素分析

杨亮彦[1,2],黎雅楠[1,2],范鸿建[1,2],王雅婷[3]

(1. 陕西省土地工程建设集团有限责任公司,西安 710075;2. 陕西地建土地工程技术研究院有限责任公司,西安 710021;3. 山东省城乡规划设计研究院有限公司,济南 250013)

【摘要】毛乌素沙地存在水资源短缺和生态环境脆弱的问题,探究其蒸散发时空格局可为区域水资源开发利用和生态环境保护提供依据。2000—2019 年基于 MOD16 数据分析毛乌素沙地 ET(Evapotranspiration)年际变化及其空间分布特征,选取气象数据、植被指数和土地利用/土地覆盖数据,通过趋势分析法和相关系数分析其对 ET 的影响。结果表明:(1)毛乌素沙地 ET 在空间上表现出显著的区域差异性,整体呈西北低、东南高的分布规律,2000—2019 年的均值为 258.8 mm;(2)2000—2019 年毛乌素沙地 ET 呈快速上升趋势,平均变化速率为 6.87 $mm \cdot a^{-1}$;(3)ET 与降水量相关系数为 0.74,与气温无明显相关性,不同土地利用/覆被的 ET 存在明显差异性,ET 与 NDVI 空间分布和年际变化曲线一致,两者全区域决定系数为 0.6288。在降水量、土地利用/覆被、NDVI 等因素共同影响下,2000—2019 年毛乌素沙地 ET 明显增加,其中降水量是毛乌素沙地气候条件的主要因素,NDVI 是影响同一时期 ET 的重要因素。该研究为毛乌素沙地生态水源涵养与保护提供理论基础和科学依据,对研究区落实水资源管理制度和生态环境保护政策具有重要的指导意义。

【关键词】毛乌素沙地;蒸散发;时空分布;影响因素;气候条件

Temporal and Spatial Distribution and Influencing Factors of Evapotranspiration in Mu Us Sandy Land

Liangyan Yang[1,2], Yanan Li[1,2], Hongjian Fan[1,2], Yating Wang[3]

(1. Shaanxi Provincial Land Engineering Construction Group Co., Ltd., Xi'an 710075, China; 2. Institute of Land Engineering and Technology, Shaanxi Provincial Land Engineering Construction Group Co., Ltd., Xi'an 710021, China; 3. Shandong urban and rural planning Design Institute Co., Ltd, Jinan 250013, China)

【Abstract】Mu Us Sandy Land has the problems of water shortage and fragile ecological environment, and exploring its spatial and temporal patterns of evapotranspiration can provide a basis for regional water resources development and utilization and ecological environmental protection. In this study, the MOD16 data was used to analyze the interannual variability of ET and its spatial distribution characteristics in Mu Us Sandy Land, China, and then the meteorological data, vegetation index and land use/land cover data were used to analyze the factors affecting the change of ET with trend analysis and correlation coefficient. The results can be summarized as follows:

基金项目:国家自然科学基金地区科学基金项目(42167039);陕西省自然科学基础研究计划项目(2021JM-447);陕西省土地工程建设集团内部项目(DJNY2021-33)。

(1)The ET of Mu Us Sandy Land showed significant regional variability in space, with an overall distribution pattern of low in the northwest and high in the southeast, and the average value of 258.8 mm from 2000 to 2019; (2)The ET of Mu Us Sandy Land showed a rapid increasing trend from 2000 to 2019, with an average change rate of 6.87 mm · a^{-1}; (3)The correlation coefficient between ET and rainfall was 0.74, with no significant correlation with temperature, and there was significant variability in ET for different LULC, and the spatial distribution and interannual variation curves of ET and NDVI were consistent, with a region-wide determination coefficient of 0.6288 for both. Under the combined influence of rainfall, LULC and NDVI, ET increased significantly in the Mu Us Sandy Land, where rainfall was the main factor of climatic conditions in the Mu Us Sandy Land, NDVI was an important factor affecting ET in the same period. The study provides a theoretical basis and scientific basis for ecological water conservation and protection in the Mu Us Sandy Land, and is an important guide to the implementation of water resources management system and ecological environmental protection policies in the study area.

【**Keywords**】Mu Us Sandy Land; Evapotranspiration; Spatial and temporal distribution; Influencing factors; Climatic conditions

蒸散发(Evapotranspiration,ET)主要包括水面和土壤蒸发、植被蒸腾[1],在地表水循环和能量平衡中扮演重要的角色。大约60%的全球地表降水通过蒸散发的方式返回大气,这一比例在干旱区达90%[2]。同时蒸散发直接影响着土壤湿度、地表温度[3]、地下水深度和植被生长状况[4],是气候变化[5]、水文过程[6]及农作物估产[7]等方面的重要参数。随着生态环境退化、水资源短缺等问题的出现,明晰蒸散发的时空变化和影响因素具有重要的研究意义。

由于地表的异质性,传统的地表蒸散发量获取方法以气象站提供的水面蒸发量为主,但该数据不能有效的反映区域尺度的地表蒸散发量,无法满足区域水资源合理开发利用的需要[8]。遥感技术的发展为地表蒸散发量的获取提供了新的方法,利用遥感数据可估算出净辐射通量、土壤热通量和显热通量,结合能量平衡公式得到潜热通量。国内外学者在此基础上相继发布了 SEBAL(Surface Energy Balance Algorithm for Land)、SEBS(Surface Energy Balance System)、SEBI(Surface Energy Balance Index)等[9]蒸散发模型,以此获取地表实际蒸散发量。基于区域蒸散发反演模型,众多学者及团队研发并发布了地表蒸散发遥感数据产品,如 Miralles 等[10]基于 Priestley-Taylor 方程发布的 GLEAM 产品(0.25°,Daily);NASA 基于 MODIS 卫星遥感数据,利用彭曼(Penman Monteith,P-M)公式考虑多种阻力算法制作的全球陆地蒸散量产品 MOD16 等(500 m,8 d)[11];美国地质调查局(USGS)利用陆面能量平衡模型,获取实际蒸散量的 SSEBop 产品(1 km,30 d)[12];Chen 等[13]基于 SEBS 模型获得了 ET-EB 产品;NASA 基于陆面模型和数据同化技术生成的全球陆面通量数据 GLDAS(0.25°,Monthly)[14]。其中,MOD16 产品因时空分辨率高、监测精度高、数据容易获取等优势,被广泛应用于典型地表蒸散发时空分布研究。张特等[15]利用 P-M 公式在漯河流域验证了 MOD16 数据精度,发现均相对误差 MRE 为 7.5%,可满足区域蒸散发时空变化研究需求。Ma 等[16]基于 MODIS 数据探究了 2000—2015 年中国黄土高原 ET 变化,结果表明 ET 在 2000—2010 年有明显增加,但 2011—2015 年没有明显变化趋势。郭晓彤等[17]基于 MOD16 数据采用复相关关系探究了淮河流域 ET 时空变化的驱动力,发现在淮河流域,人类活动是 ET 时空变化的主要驱动力。

毛乌素沙地是中国西北典型的农牧交错区,生态环境十分脆弱。由于该地区降水稀少、气候干燥、太阳辐射强、蒸发强烈等自然因素,毛乌素沙地长期存在水资源短缺问题,土地荒漠化进程迅速,对西部和京津冀地区的生态环境造成严重危害,因此毛乌素沙地一直是中国干旱研究学者的热点区域。但前人的研究多集中在毛乌素沙地植被覆盖变化、荒漠化、土地利用变化等方面,在蒸散发方面以蒸散发反演模型估算局部地区日蒸散发量为主,鲜有针对毛乌素沙地降水量和实际蒸散发时空变化的宏观研究。本研究基于 MOD16 ET 数据集,利用趋势分析法明晰毛乌素沙地 2001—2019 年蒸散发(Evapotranspiration,ET)时空变化特征,结合气象数据探究 ET 与水热条件的驱动力关系,并分析土地利用/覆被(Land use/land cover,LULC)与植被指数对 ET 时空变化的影响,为落实毛乌素沙地水资源管理及开发利用制度和生态环境保护政策提供参考。

1　材料与方法

1.1　研究区概况

毛乌素沙地地处中国西北部，黄土高原腹部，横亘榆林市北部、鄂尔多斯市南部及盐池县和灵武市(37.45°~39.37°N，107.67°~110.67°E)(见图1)。毛乌素沙地处于干旱与半干旱过渡区，是我国典型的农牧交错带，其西北部以沙地和草地为主，东南部以耕地和草地为主。研究区以温度大陆性气候为主，干旱少雨且分布不均，降水多集中在6—8月，年降水量250~440 mm，由西向东南方向递增。毛乌素沙地属于西北典型的生态脆弱区，是我国沙漠化最严重的地区之一，也是中国防风固沙重要的生态功能区之一[18-19]。

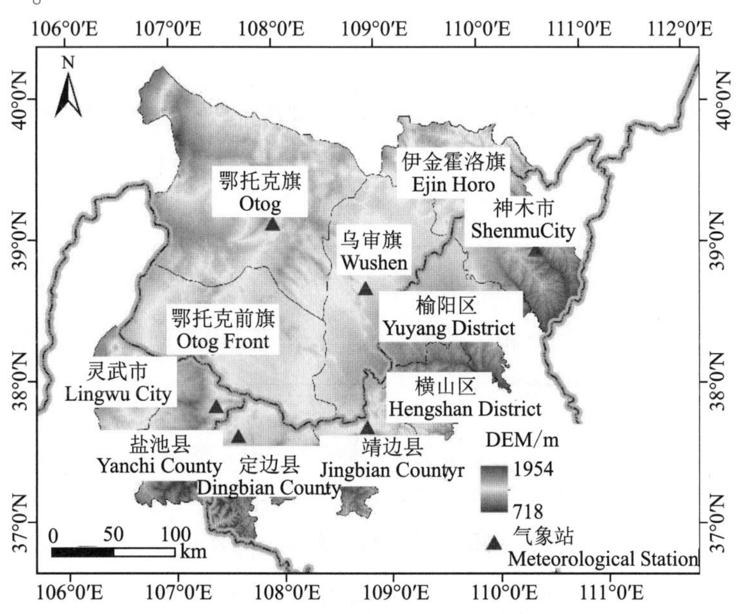

图1　毛乌素沙地区位

Fig.1　Distribution map of land use types in Mu Us Sandy Land

1.2　数据来源及处理

本研究采用的数据主要包括ET数据、气象数据、土地利用/覆被数据和植被指数数据，其中ET数据为MOD16产品，该数据由NASA基于P-M模型计算的数据集，空间分辨率为500 m，时间分辨率为8 d。选用2000—2019年MOD16A3GF数据集，该数据由MOD16产品全年统计合成的L4级产品，时间分辨率为年。本研究利用MODIS Reprojection Tool(MRT)工具对HDF类型数据进行格式转换、投影转换、裁剪、镶嵌等预处理。NDVI为MOD13数据，其空间分辨率为500 m，时间分辨率为16 d，每月可获取两期影像，取两者平均值作为当月植被指数，并利用最大值合成法获取每年NDVI数据。

气象数据来源于中国气象数据网，研究区内共有8个国家气象站点，分别为横山站(站点编号：53740)、靖边站(站点编号：53735)、定边站(站点编号：53725)、盐池站(站点编号：53723)、神木站(站点编号：53651)、榆林站(站点编号：53646)、鄂托克旗站(站点编号：53529)和乌审旗站(站点编号：53644)(见图1)。气象数据包含地表温度、气温、降水量等8种日值数据，本研究通过统计分析获取每个站点的年平均气温和降水量数据，并计算毛乌素沙地7个气象站点的年均气温和降水量。土地利用/覆被栅格数据由中国研制的30 m空间分辨率的GlobeLand30，该数据采用WGS-84坐标系，包括耕地、林地、草地、灌木地、建设用地、未利用地等10个一级土地利用/覆被。目前已发布2000、2010和2020版。

1.3 研究方法

1.3.1 ET 变化趋势分析

趋势分析法可模拟每个像元多年 ET 的变化趋势,分析 ET 在研究时段的时空变化规律。计算整个研究区 ET 长时间序列的像元回归斜率 K,若斜率为正则表示 ET 随时间变化呈上升趋势,为负则表示 ET 随时间呈下降趋势,且斜率的绝对值越大,说明 ET 变化趋势越明显。其计算公式为:

$$K = \frac{n \times \sum_{i=1}^{n} i \times \mathrm{ET}_i - \sum_{i=1}^{n} i \sum_{i=1}^{n} \mathrm{ET}_i}{n \times \sum_{i=1}^{n} i^2 - \left(\sum_{i=1}^{n} i\right)^2} \tag{1}$$

式中:K 为 ET 变化趋势斜率;n 为监测时间段的年数; ET_i 表示第 i 年的 ET。

1.3.2 相对变化率

相对变化率是相对值对时间的导数,主要用于探究研究时间段内因素的波动变化状况。

$$r = \frac{\mathrm{ET}_i - \overline{\mathrm{ET}}}{\mathrm{ET}_i} \tag{2}$$

式中:r 为相对变化率;ET_i 为第 i 年 ET;$\overline{\mathrm{ET}}$为研究时段 ET 多年平均值。

1.3.3 相关性分析

相关性分析是对两个或多个具有相关性的参数进行分析,用于衡量参数之间的相关程度。常用分析软件为 SPSS 和 EXCEL,其计算公式如下:

$$R = \frac{\sum_{i=1}^{n} (Y_i - \overline{Y})(X_i - \overline{X})}{\sqrt{\sum_{i=1}^{n} (Y_i - \overline{Y})^2 \sum_{i=1}^{n} (X_i - \overline{X})^2}} \tag{3}$$

式中:R 为线性相关系数;X_i、Y_i 分别为 X、Y 因子第 i 年的值;$\overline{X}$、$\overline{Y}$ 分别为 X、Y 因子的研究时段的均值。相关系数 R 取值范围-1~1,$R>0$ 表示参数为正相关关系,$R<0$ 表示参数呈负相关关系,且绝对值越大,表明相关程度越高。

2 结果与分析

2.1 毛乌素沙地蒸散发时空分布

2.1.1 毛乌素沙地 ET 空间变化特征

图 2 为毛乌素沙地 2000—2019 年 ET 逐年空间变化过程,可以看出,毛乌素沙地 ET 分布具有较强的空间差异性,整体呈西北低、东南高的空间规律,并且不同行政区域内 ET 分层明显,陕西省明显高于内蒙古自治区,这一特征可能与土地利用/覆被和地表植被覆盖度有关。毛乌素沙地 2000—2019 年 ET 有明显变化,除研究区西部的灵武市农业区 ET 较高外,20 年内研究区 ET 在不断上升,其中在 2000—2009 年,ET 在 300 mm/a 以下,2010 年后高于 300 mm/a 的区域明显增多,主要分布在陕西省境内的定边县、靖边县、神木市、榆阳区和横山区。

2.1.2 毛乌素沙地 ET 时间变化特征

为进一步表明不同时期 ET 的分布规律,统计了研究区 2000 年、2005 年、2010 年、2015 年和 2019 年 ET 分布区间及占比(见表 1),由表 1 可知,2000 年和 2005 年,研究区 ET 主要集中在 250 mm · a^{-1} 以下,分别占了研究区的 99.0%和 89.5%,比例略微下降;到了 2010 年逐渐增加至 250~300 和 300~350 mm · a^{-1},分别占了研究区 55.2%和 23.4%;2015 年,研究区 ET 出现回落,主要集中在 200-300 之间,占研究区的74.6%;在 2019 年毛乌素沙地 ET 主要区间恢复到 250~350 mm · a^{-1},其中 300~350 mm · a^{-1}首次成为研究区 ET 分布最多的区间。综上所述,毛乌素沙地年 ET 有上升趋势,且存在地域性差异,毛乌素沙地 ET 变化与行政区划生态环境政策和自然因素共同影响。

注:200~225代表200<ET≤225,下同。

图 2　毛乌素沙地 2000—2019 年 ET 逐年空间变化过程

Fig. 2　Annual spatial change process of ET in the Mu Us Sandy Land from 2000 to 2019

表 1　毛乌素沙地 2000—2019 年 ET 分布区间及占比

Table 1　Distribution range and proportion of ET in the Mu Us Sandy Land in 2000,2005,2010,2015 and 2019

ET/mm	比例 Proportion/%				
	2000 年	2005 年	2010 年	2015 年	2019 年
< 200	86.3	51.2	1.9	5.7	0.3
200~250	12.7	38.3	14.5	44.7	8.3
250~300	0.6	9.2	55.2	29.9	35.8
300~350	0.3	0.9	23.4	14.3	36.0
350~400	0.1	0.3	4.2	4.0	14.6
> 400	0.0	0.1	0.8	1.4	4.9

图 3 为 2001—2020 年毛乌素沙地 ET 年际变化及距平相对变化率,2000—2019 年研究区年均 ET 在 179.7~321.4 mm,多年平均 ET 为 258.8 mm。研究时段内,高于和低于平均水平的年份分别有 9 个和 11 个,其中高于平均水平的年份集中分布在 2010—2019 年,而低于平均水平的年份分布在 2000—2009 年,距平相对变化率最大的年份为 2000 和 2016 年 ET 分别为 179.7 mm · a^{-1}和 321.4 mm · a^{-1}。以上数据表明,2000—2019 年毛乌素沙地 ET 在波动中呈明显增加的趋势,且 2010 年之后波动范围变小。

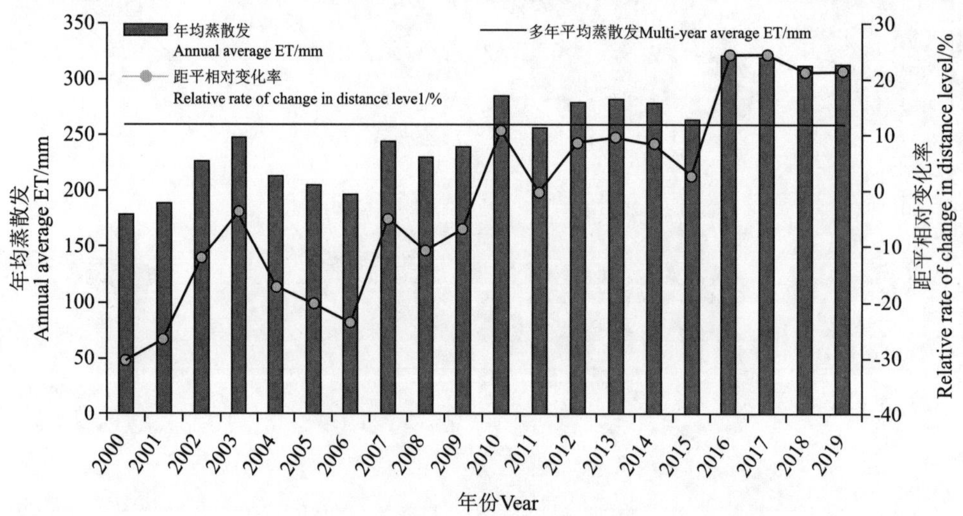

图 3　2001—2019 年毛乌素沙地 ET 年际变化及距平相对变化率

Fig. 3　The interannual change and relative change rate of ET in the Mu Us Sandy Land from 2001 to 2019

2.2　毛乌素沙地 ET 时空变化趋势

本研究根据毛乌素沙地 ET 年际变化将研究时段分为两个阶段:2000—2009 和 2010—2019 年,并将 ET 的趋势变化率分为 7 个等级(图 4),表 2 为 2000—2019 年毛乌素沙地 ET 的趋势变化率不同等级的比例,由图表可知,毛乌素沙地各阶段 ET 存在一定的差异性。在 2000—2009 年,毛乌素沙地 ET 的平均变化速率为 4.60 mm · a^{-1},呈快速上升趋势,ET 的增长速率集中在 2~10 mm · a^{-1},占研究区的 96.86%,其中 $K>5$ mm · a^{-1}a 的区域主要分布在研究区的周边,下降的区域仅占 0.04%,主要位于灵武市的农业区;2010—2018 年,毛乌素沙地 ET 的平均变化速率为 5.71 mm · a^{-1},ET 继续保持高速增加趋势的同时,速率有所加快,与 2000—2009 年相似,ET 的增长速率集中在 2~10 mm · a^{-1},但$K>5$ mm · a^{-1}区域成为毛乌素沙地最主要区域,主要分布在研究区的中部和东部。整体分析,2000—2019 年毛乌素沙地 ET 以上升为主,平均变化速率为 6.87 mm · a^{-1},其中 2~5 mm · a^{-1}的区域占研究区的 14.89%,分布在西北部;5~10 mm · a^{-1}的区域占研究区的 79.75%,分布在中部、南部和东部的大部分区域;$K>10$ mm · a^{-1}的区域占研究区的 5.25%,散落在研究区的东部和南部;基本不变和下降的区域仅占 0.10%。

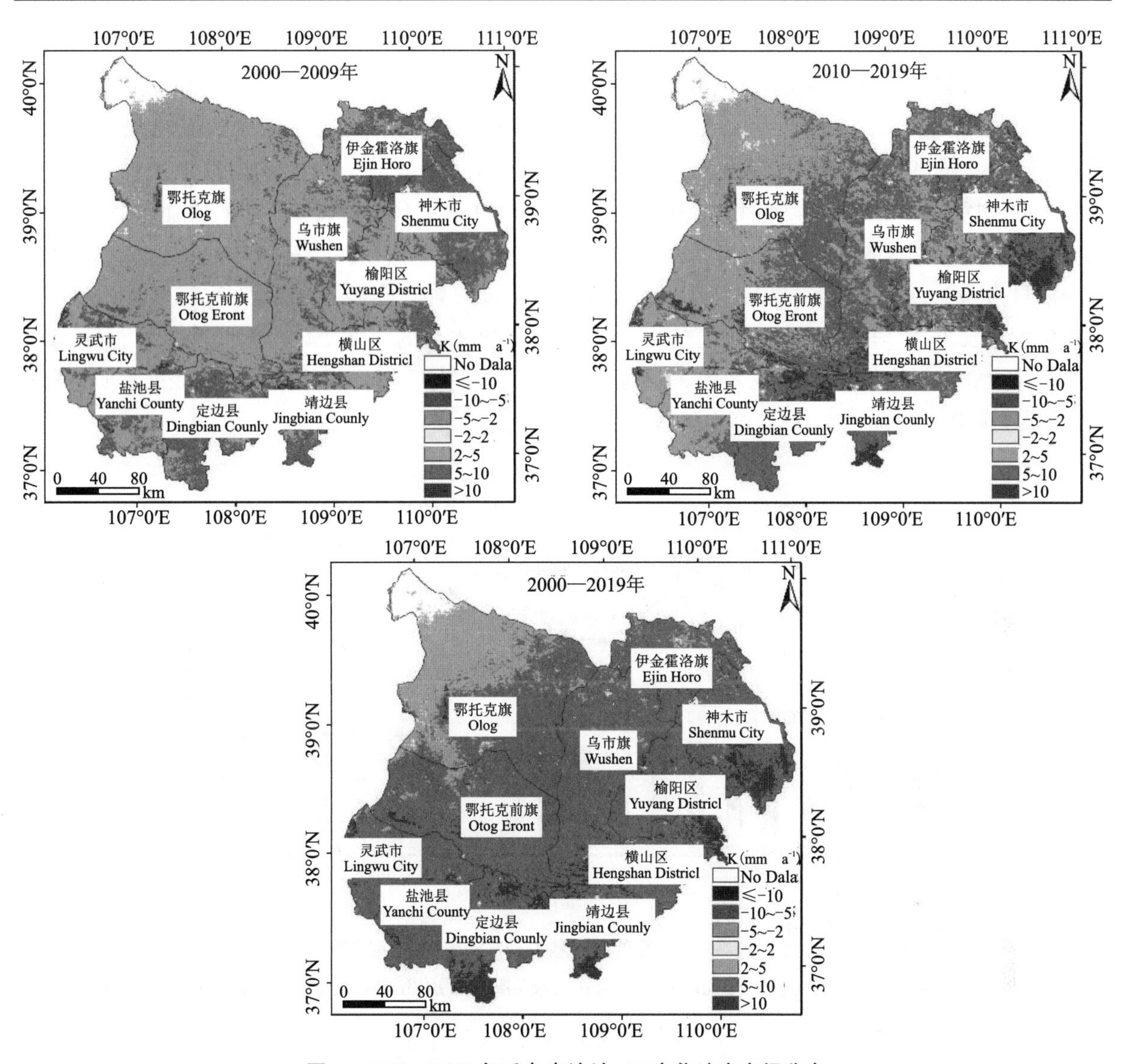

图 4　2000—2019 年毛乌素沙地 ET 变化速率空间分布

Fig. 4　Spatial distribution of ET change rate in the Mu Us Sandy Land from 2000 to 2019

表 2　2000—2019 年毛乌素沙地 ET 的趋势变化率不同等级的比例

Table 2　The proportion of different grades of ET trend change rate in the Mu Us Sandy Land from 2000 to 2019

变化趋势 Trend	变化速率 rate of change ($mm \cdot a^{-1}$)	2000—2009 年 比例 Proportion/%	2010—2019 年 比例 Proportion/%	2000—2019 年 比例 Proportion/%
下降 Falling	< -10	0.01	0.00	0.00
	-10 ~ -5	0.01	0.02	0.00
	-5 ~ -2	0.02	0.05	0.01
基本不变 Unchanged	-2 ~ 2	2.68	2.04	0.09
上升 Rising	2 ~ 5	72.62	44.99	14.89
	5 ~ 10	24.14	47.96	79.75
	> 10	0.53	4.94	5.25

2.3 ET影响因素分析

2.3.1 气候变化对毛乌素沙地ET时空变化的影响

气候变化是影响区域水热条件空间分布的重要因素,ET的变化与区域水热变化规律密切相关[20]。在区域研究中,年均气温表征区域热力条件,降水量表征水分条件,气温和降水量的年际变化研究有助于探究区域ET的影响因素。图5为2000—2019年毛乌素沙地ET与气温、降水量年际变化。研究区年均降水为337.8 mm,降水年际波动较大,上升趋势不明显,其中乌审旗站点2001—2010年降水量为383.1 mm,2011—2019年年降水量为430.1 mm,具有明显上升趋势;毛乌素沙地气温年际变化相对较小,年均气温在9.3 ℃左右。研究区ET与降水量的年际变化趋势基本一致,呈波动上升趋势,除2019年外,两者的相关系数为0.74;与气温变化趋势无明显相关性。以上数据表明毛乌素沙地ET受水热条件的影响,其中降水量为主要驱动因素。

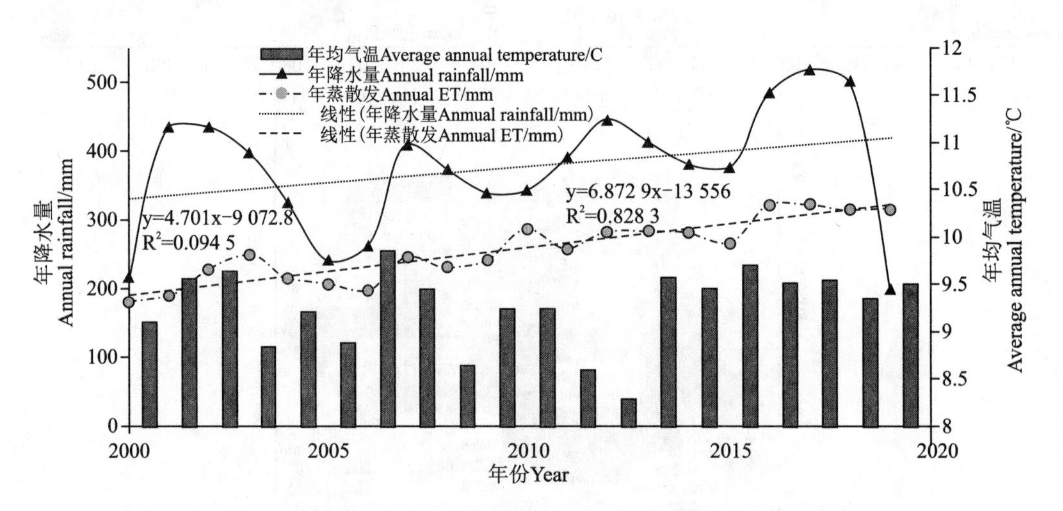

图5 2000—2019年毛乌素沙地ET与气温、降水量年际变化

Fig. 5 ET and interannual changes of temperature and precipitation in the Mu Us Sandy Land from 2000 to 2019

2.3.2 LULC对毛乌素沙地ET时空变化的影响

毛乌素沙地土地利用/覆被以草地和耕地为主,其次裸地、灌木地、林地、建设用地和水体,以2010年为例(图6),草地、耕地、裸地、灌木地、建设用地、林地和水体及湿地占比依次为73.05%、14.63%、9.59%、1.73%、0.50%、0.26%和0.26%,草地遍布整个研究区,耕地分布在东部和南部,未利用地主要分布为研究区中部,成条带状分布,其他土地利用/覆被散落分布在研究区内。由于MOD16数据缺乏水体蒸散发数据,本研究不再分析水体ET变化。

图7为2000—2019年毛乌素沙地不同土地利用/覆被面积及ET变化图。2000—2019年间,建设用地和耕地面积增加,草地、裸地面积减少,未利用地面积变化不大。6种土地利用/覆被的ET均呈显著增加的趋势,与2000年相比,2019年耕地、林地、草地、灌木地、建设用地和未利用地ET分别增加了70.33%、81.43%、64.52%、57.50%、43.51%和58.28%。不同土地利用/覆被平均ET的从大到小依次为耕地>林地>灌木地>建设用地>草地>裸地,耕地和建设用地面积的增加,裸地、草地面积的减少显著增加了毛乌素沙地ET,与研究区ET增加的趋势相符。综上所述,土地利用/覆被是影响毛乌素沙地ET变化的重要因素。但同时发现2019年的各土地利用/覆被的ET普遍高于2000年,说明土地利用/覆被不是毛乌素沙地ET变化的主导因素。

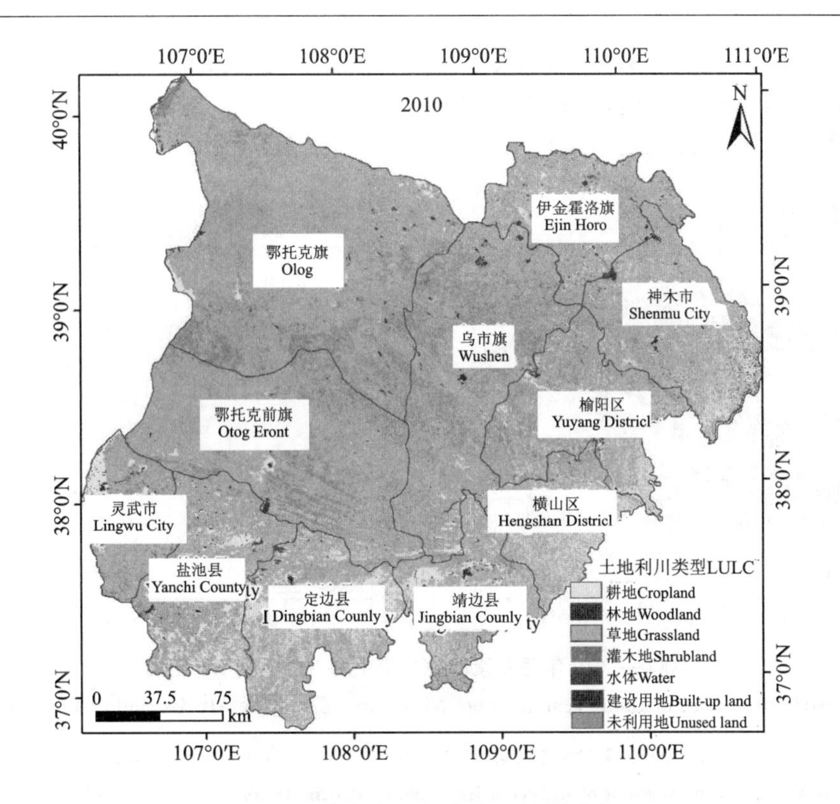

图 6　毛乌素沙地土地利用/覆被分布

Fig. 6　Distribution map of LULC in the Mu Us Sandy Land

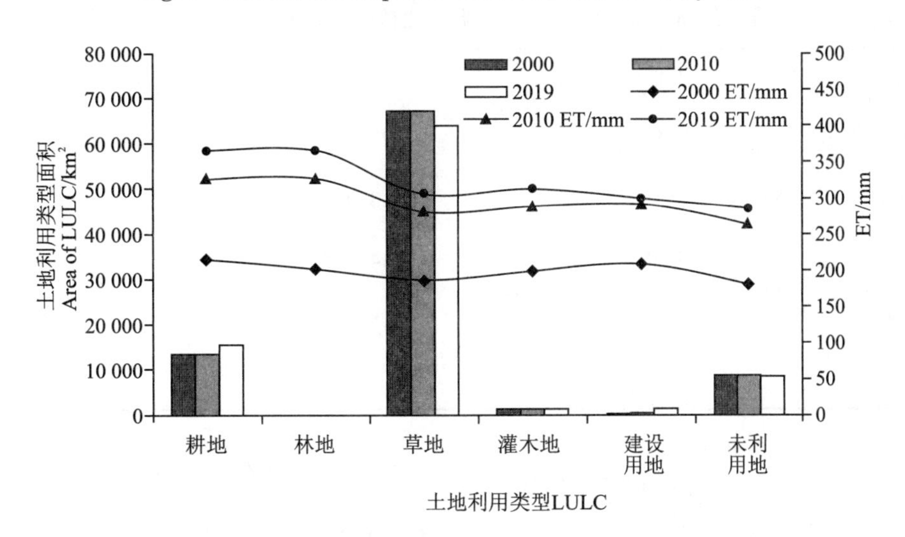

图 7　2000—2019 年毛乌素沙地不同土地利用/覆被面积及 ET 变化

Fig. 7　Changes in different LULC areas and ET in the Mu Us Sandy Land from 2000 to 2019

2.3.3　NDVI 对蒸散发时空变化的影响

为了更好地探究归一化植被指数(Normalized Difference Vegetation Index,NDVI)对 ET 的影响,本研究将毛乌素沙地多年平均 NDVI 和 ET 分为 6 级(见图 8)。毛乌素沙地多年平均 NDVI 为 0.33,8.56%的区域 NDVI<0.2,主要分布在西北部和腹部;仅有 1.25%的区域 NDVI>0.6,分布在灵武市农业区和靖边、定边境内;NDVI 介于 0.2~0.6 的区域占研究区的 90.19%,遍布整个研究区。毛乌素沙地 NDVI 整体呈西北到东南逐渐升高的趋势,与研究区 ET 空间分布较为一致,表明 ET 与 NDVI 具有较好的相关性。为了定量化分析两者的关系,图 9 为 NDVI 与 ET 年际变化曲线图,两者变化曲线具有高度的一致性。图 10 为 ET 与 NDVI 多年均值的全区域散点图(红色代表点密度大),两者决定系数为0.6288,进一步证明了ET与NDVI的相关性。整体来说,ET的变化情况与植被生长状态具有高度的相关性,植被的生长趋势是影响ET的重要因素。

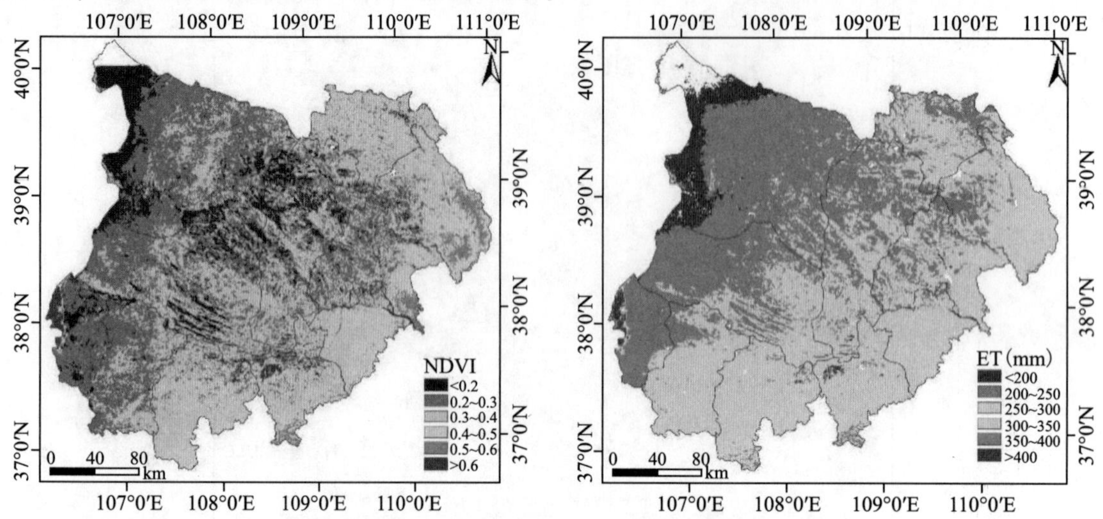

图8 2000—2019年毛乌素沙地多年平均NDVI和ET空间分布

Fig. 8 Spatial distribution map of multi-year average NDVI and ET in the Mu Us Sandy Land from 2000 to 2019

表3 2000—2019年毛乌素沙地多年平均NDVI和ET不同区间占比

Table 3 The proportion of different intervals of the multi-year average NDVI and ET in the Mu Us Sandy Land from 2000 to 2019

ET/mm	比例 Proportion /%	归一化植被指数 NDVI	比例 Proportion /%
< 200	4.47	< 0.2	8.56
200~250	40.89	0.2~0.3	37.26
250~300	42.15	0.3~0.4	29.89
300~350	10.86	0.4~0.5	18.63
350~400	1.13	0.5~0.6	4.41
>400	0.50	> 0.6	1.25

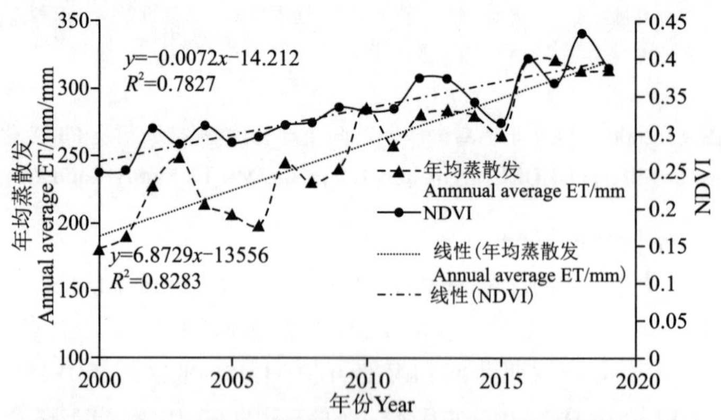

图9 2000—2019年毛乌素沙地NDVI与ET年际变化曲线

Fig. 9 Interannual variation curve of NDVI and ET in the Mu Us Sandy Land from 2000 to 2019

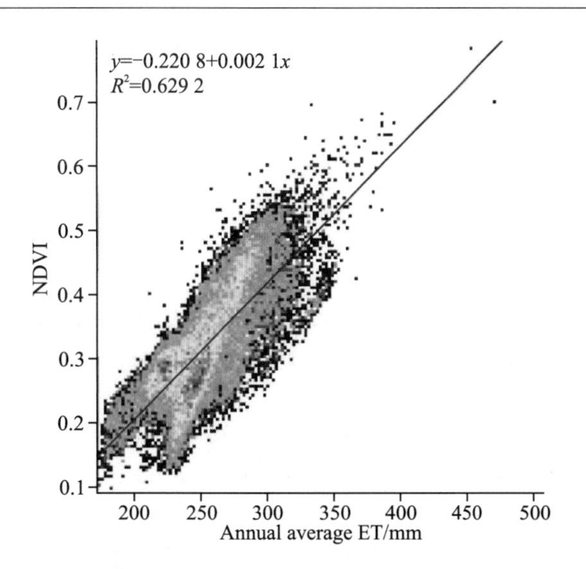

图 10 2019 年 ET 与 NDVI 多年均值的全区域热点

Fig. 10 Region-wide heat map of the multi-year average of ET and NDVI in 2019

3 讨论

ET 作为干旱区水循环的重要途径,明晰其时空变化规律对毛乌素沙地水资源合理开发及可持续发展具有重要的作用。毛乌素沙地 ET 在时空变化上存在一定的规律性,在空间分布上具有明显的分异性,呈西北向东南逐渐增加的趋势,其中陕西行政区内的 ET 明显高于内蒙古自治区的,与刘静等[18]研究结果一致。ET 与降水量年际变化趋势较为一致,与气温无明显相关性,表明在干旱半干旱区,降水量是 ET 的主要驱动因素,这个结论与毛乌素沙地缺水的自然特征相符。在不考虑水体的情况下,耕地和林地是毛乌素沙地 ET 最大的两种土地利用/覆被,与王卓月[8]的研究结果一致。多年平均 ET 与 NDVI 空间分布一致,年际变化曲线相似,表明毛乌素沙地植被覆盖及长势是 ET 的主要因素,与谷佳贺在黄河流域的研究结果相符[1]。

2000—2019 年毛乌素沙地 ET 增加明显,其增加速率略高于降水量的变化速率,考虑其原因是耕地、林地面积的增加加快了 ET 的增加,过度的植树造林存在恶化当地生态环境的可能性,因此如何平衡植树造林工程对研究区的影响是我国未来需要考虑的问题。该研究发现在不同土地利用/覆被中,灌木地和草地的 ET 明显低于林地和耕地的,因此增加灌木地和草地面积可有效减少水分流失,调整林木类型是改善植树造林政策合理性的重要选择。

本文针对 2000—2019 年毛乌素沙地时空特征及其与水热条件、土地利用/覆被和 NDVI 的关系进行了深入研究,但研究仍存在一定的局限性:(1) MOD16 数据空间分辨率为 500 m,且水体部分的 ET 缺失,较粗的空间分辨率和缺失值的问题导致部分研究信息无法获取,研究缺乏完整性。(2) 由于研究区内气象站点数量较少,且气象站周边缺乏 ET 数据,因此在时空立方体和新兴热点分析等研究方法[21]上受到一定的局限性。(3) 在探究相关驱动机制时,不同行政区域的土地政策不同,间接影响着研究区 ET 的变化,本文仅从宏观角度对 ET 进行分析,并未区分行政区域解读相关土地利用政策,政策影响机制的量化一直以来是科学研究的热点和难点,是日后工作重点研究的方向。

参考文献

[1] 谷佳贺, 薛华柱, 董国涛, 等. 黄河流域 NDVI/土地利用对蒸散发时空变化的影响[J]. 干旱区地理, 2021,44(1): 158-167.

GU J H, XUE H Z, DONG G T, *et al*.. Effect of NDVI/land use on spatiotemporal changes of evapotranspiration in the Yellow River Basin [J]. Arid land Geography, 2021,44(1): 158-167.

[2] KONG J L, HU Y X, YANG L Y, *et al*. Estimation of evapotranspiration for the blown-sand region in the Ordos basin based on the SEBAL model [J]. Int. J. Remote Sens., 2019,40(5): 1945-1965.

[3]何慧娟, 王钊, 董金芳, 等. 基于MODIS产品的秦岭地区NDVI、地表温度和蒸散变化关系分析[J]. 西北林学院学报, 2019, 34(4): 179-184,191.

HE H J, WANG Z, DONG J F, *et al*. NDVI, LST and ET Variation Analysis Based on MODIS Datasets in the Qinling Mountains [J]. Journal of Northwest Forestry University, 2019, 34(4): 179-184,191.

[4]YANG L S, FENG Q, ADAMOWSKI J F, *et al*. The role of climate change and vegetation greening on the variation of terrestrial evapotranspiration in northwest China's Qilian Mountains [J/OL]. Science of the Total Environment, 2021,759, 143532 [2020-11-16].

[5]ZHANG L F, SCHLAEPFER D R, CHEN Z G, *et al*.. Precipitation and evapotranspiration partitioning on the Three-River Source Region: A comparison between water balance and energy balance models [J/OL]. J. Hydrol. : Regional Studies,2021,38: 100936 [2021-10-07].

[6]WANG Y S, ZHANG Y E, YU X X, *et al*. Grassland soil moisture fluctuation and its relationship with evapotranspiration [J/OL]. Ecological Indicators, 2021, 131: 108196 [2021-09-13].

[7]HU X Y, LEI H M. Evapotranspiration partitioning and its interannual variability over a winter wheat-summer maize rotation system in the North China Plain [J/OL]. Agricultural and Forest Meteorology, 2021, 310: 108635 [2021-09-10].

[8]王卓月, 孔金玲, 李英, 等. 基于MOD16的银川平原地表蒸散量时空特征及影响因素分析[J]. 水文地质工程地质, 2021, 48(3): 53-61.

WANG Z Y, KONG J L, LI Y, *et al*. An analysis of spatio-temporal characteristics and influencing factors of surface evapotranspiration in the Yinchuan Plain based on MOD16 data [J]. Hydrogeology & Engineering Geology, 2021, 48(3): 53-61.

[9]李旭亮, 杨礼箫, 胥学峰, 等. 基于SEBAL模型的西北农牧交错带生长季蒸散发估算及变化特征分析[J]. 生态学报, 2020, 40(7): 2175-2185.

LI X L, YANG L X, XU X F, *et al*. Analysis of evapotranspiration pattern by SEBAL model during the growing season in the agro-pastoral ecotone in Northwest China [J]. Acta Ecologica Sinica, 2020, 40(7): 2175-2185.

[10]MARTENS B, MIRALLES D G, LIEVENS H, *et al*. GLEAM v3: satellite-based land evaporation and root-zone soil moisture [J]. Geosci. Model Dev, 2017, 10(5): 1903-1925.

[11]MU Q, ZHAO M, RUNNING S W. Improvements to a MODIS global terrestrial evapotranspiration algorithm [J/OL]. Remote Sens. Environ. 2011, 115(8): 1781-1800 [2011-04-08].

[12]VELPURI N M, SENAY G B, SINGH, R K, *et al*. A comprehensive evaluation of two MODIS evapotranspiration products over the conterminous United States: using point and gridded FLUXNET and water balance ET [J]. Remote Sens. Environ. 2013, 139: 35-49 [2013-08-24].

[13]CHEN X, SU Z, MA Y, *et al*. Development of a 10-year (2001—2010) 0.1° data set of land-surface energy balance for mainland China [J/OL]. Atmos. Chem. Phys. 2014, 14(10): 14471-14518 [2014-06-05].

[14]WANG W, CUI W, WANG X, *et al*. Evaluation of GLDAS-1 and GLDAS-2 forcing data and Noah model simulations over China at the monthly scale. Hydrometeorology. 2016, 17(11): 2815-2833.

[15]张特, 刘冀, 董晓华, 等. 基于MOD16的漳河流域蒸散发时空分布特征[J]. 灌溉排水学报, 2018, 37(8): 121-128.

ZHANG T, LIU X H, DONG X H, *et al*. Changing Water Price to Regulate Groundwater Extraction for Irrigating Winter Wheat in North China Plain [J]. Journal of Irrigation and Drainage, 2018, 37(8): 121-128.

[16]MA Z, YAN N, WU B, *et al*. Variation in actual evapotranspiration following changes in climate and vegetation cover during an ecological restoration period (2000—2015) in the Loess Plateau, China [J]. Sci. Total Environ. 2019, 689: 534-545.

[17]郭晓彤, 孟丹, 蒋博武, 等. 基于MODIS蒸散量数据的淮河流域蒸散发时空变化及影响因素分析[J]. 水文地质工程地质, 2021, 48(3): 45-52.

GUO X T, MENG F, JIANG B W, *et al*. Spatio-temporal change and influencing factors of evapotranspiration in the Huaihe River Basin based on MODIS evapotranspiration data [J]. Hydrogeology & Engineering Geology, 2021, 48(3): 45-52.

[18]刘静, 刘铁军, 杜晓峰, 等. 基于MOD16A2的毛乌素沙地实际蒸散量时空稳定性模拟[J]. 干旱地区农业研究, 2020, 38(2): 243-250.

LIU J, LIU T J, DU X F, *et al*. Simulation on spatio-temporal stability of ET based on MOD16A2 in Mu Us sandy land [J]. Agricultural Research in the Arid Areas, 2020, 38(2): 243-250.

[19]刘娟, 刘华民, 卓义, 等. 毛乌素沙地 1990—2014 年景观格局变化及驱动力. 草业科学, 2017, 34(2): 255-263.

LIU J, LIU H M, ZHUO Y, *et al*. Dynamics and driving forces of landscape patterns in Mu Us Sandy Land, from 1990 to 2014 [J]. Pratacultural Science, 2017, 34(2): 255-263.

[20]马建琴, 陈阳, 郝秀平, 等. 2001—2019 年河南省地表蒸散发时空变化及其影响因素[J]. 水土保持研究, 2021, 28(5): 134-141,151.

MA J Q, CHEN Y, HAO X P. Temporal and spatial changes of surface evapotranspiration and its influencing factors in Henan province from 2001 to 2019 [J]. Research of Soil and Water Conservation, 2021, 28(5): 134-141,151.

[21]张棋, 许德合, 丁严. 基于 SPEI 和时空立方体的中国近 40 年干旱时空模式挖掘[J]. 干旱地区农业研究, 2021, 39(3): 194-201.

ZHANG Q, XU D H, DING Y, *et al*. Spatio-temporal pattern mining of the last 40 years of drought in China based on SPEI index and spatio-temporal cube [J]. Agricultural Research in the Arid Areas, 2021, 39(3): 194-201.

本文曾发表于2022年《中国农业科技导报》第24卷第10期

耦合 SVM 和 Cloud-Score 算法的 Sentinel-2 影像云检测模型研究

李健锋[1,3,4,5]，刘思琪[1,3,4,5]，李劲彬[1,3,4,5]，彭　飚[1,3,4,5]，叶虎平[2*]

(1.陕西地建土地工程技术研究院有限责任公司,西安 710021;2.中国科学院地理科学与资源研究所资源与环境信息系统国家重点实验室,北京 100101;3.陕西省土地工程建设集团有限责任公司,西安 710075;4.自然资源部退化及未利用土地整治工程重点实验室,西安,710075;5.陕西省土地整治工程技术研究中心,西安,710075)

【摘要】云覆盖阻碍了光学遥感卫星对地观测的有效范围,快速、准确地云检测是遥感应用产品生成过程中的重要一步。针对 Google Earth Engine 云平台中缺乏适用且高质量的云检测模型,以热带多云的斯里兰卡为研究区,构建了耦合 SVM 和 Cloud-Score 算法的 Sentinel-2 影像云检测模型,通过实验从目视判读与定量分析两个角度对比了其与 QA60 法、Cloud-Score 算法以及 Fmask 的云检测精度,并在海南岛和亚马逊森林两个地区进行了云检测测试。研究结果表明:Fmask 模型的云检测性能最低,总体精度仅为 63.45%,存在严重的水体误分为云的现象,但其漏提率极低;QA60 法对卷云识别不足,漏提率较高,同时存在一定的误分现象,并且低空间分辨率影响了云体边界提取结果的细节性;Cloud-Score 算法的云检测性能明显好于 QA60 法,总体精度达到了 89.83%,误提率仅为 2.17%,但仍存在部分卷云漏提的现象;相比于其他三种云检测方法,本文提出的云检测模型总体精度最高,达到了 98.21%,并且拥有极低的漏提率和误提率,能比较精准的识别出云体的边界,可满足 Sentinel-2 遥感产品的云检测预处理需求。

【关键词】云检测;SVM;Cloud-Score 算法;Sentinel-2;Fmask

Research on Cloud Detection Model of Sentinel-2 Image Coupled with SVM and Cloud-Score Algorithm

Jianfeng Li[1,3,4,5], Siqi Liu[1,3,4,5], Jinbin Li[1,3,4,5], Biao Peng[1,3,4,5], Huping Ye[2]

(1. Institute of Land Engineering and Technology, Shaanxi Provincial Land Engineering Construction Group Co., Ltd., Xi'an 710021, China;2. State Key Laboratory of Resources and Environmental Information System, Institute of Geographic Sciences and Natural Resources Research, Chinese Academy of Sciences, Beijing 100101, China; 3. Shaanxi Provincial Land Engineering Construction Group Co., Ltd., Xi'an 710075, China; 4.Key Laboratory of Degraded and Unused Land Consolidation Engineering, the Ministry of Natural Resources, Xi'an 710075, China; 5.Shaanxi Provincial Land Consolidation Engineering Technology Research Center, Xi'an, 710075)

【Abstract】Cloud coverage hinders the effective range of earth observation by optical remote sensing satellite.

基金项目:国家重点研发项目(2019YFE0126500);高分辨率对地观测系统国家重大专项(21-Y20B01-9001-19/22);陕西地建-西安交大土地工程与人居环境技术创新中心开放基金资助项目(2021WHZ0090);陕西省土地工程建设集团内部科研项目(DJTD2022-4)。

Rapid and accurate cloud detection is an important step in the product generation process of remote sensing application. In view of the lack of suitable and high-quality cloud detection model in Google Earth Engine cloud platform, this study takes tropical cloudy Sri Lanka as the study area, constructs a Sentinel-2 image cloud detection model coupled with SVM and Cloud-Score algorithm. Through experiments, the cloud detection accuracy of this method is compared with that of QA60 method, Cloud-Score algorithm and Fmask from the point of view of visual interpretation and quantitative analysis. The results show that the cloud detection performance of Fmask model is the lowest, and the overall accuracy is only 63.45%. It has a serious phenomenon that water body is mistakenly divided into clouds, but its omission error is very low. The QA60 method has insufficient recognition of cirrus clouds, and the omission error is high. At the same time, it has a certain phenomenon of misclassification, and the low spatial resolution affects the details of cloud boundary extraction results. The cloud detection performance of the Cloud-Score algorithm is obviously better than that of the QA60 method, the overall accuracy is 89.83%, and the commission error is only 2.17%, but there is still a phenomenon that some cirrus clouds are missed. Compared with the other three cloud detection methods, the cloud detection model proposed in this study has the highest overall accuracy, reaching 98.21%, and has extremely low omission error and commission error. The model can accurately identify the boundary of the cloud, and can meet the cloud detection preprocessing requirements of Sentinel-2 remote sensing products.

【Keywords】 cloud detection; SVM; Cloud-Score algorithm; Sentinel-2; Fmask

1 引言

光学卫星影像中的云覆盖是影响其进行全球范围应用的关键因素,快速准确的云检测是遥感产品生成过程中的关键一步[1-3]。近年来,随着云计算、大数据技术的发展,Google Earth Engine[3]、PIE-Engine[4]等遥感云平台逐渐发布,遥感应用研究已突破了原先的“影像下载-预处理-模型构建-结果分析”模式,借助云平台强大的计算和存储能力,通过在线调用 API 接口对影像进行批量分析计算,极大地提高了成果转化的速度与效率,加速了遥感卫星的全球化应用[4-6]。

针对 Sentinel-2 影像,Google Earth Engine 提供了三种云检测方法:QA60 波段法[7]、Cloud-Score 算法[8]以及 S2cloudless[9]。QA60 波段由蓝色波段和两个短波红外波段生成,不同于其他地物,云和雪在蓝色波段反射率极高,而在短波红外波段云的反射率明显高于雪,但 QA60 波段法整体的云检测效果不稳定,通常会导致低估或高估的现象,并且其分辨率为 60 m,识别出的云体边界细节性不足[10-11]。Cloud-Score 算法主要原理是利用亮度、温度和归一化雪盖指数(Normalized Difference Snow Index, NDSI)组合计算出云的可能性分数,单个像素的分数越高,属于云的可能性就越大,最终通过设置阈值进行云掩膜处理。复杂的大气条件和 Sentinel-2 影像中缺少热红外波段等因素会导致 Cloud-Score 算法低估云的范围[11-12]。与目前最先进的基于阈值的云检测算法相比,机器学习算法具有更好的云检测性能,如 Fmask[13]、Sen2Cor[14]以及 MAJA[15]等[16]。S2cloudless[9]是 Sentinel-Hub 提供的专门针对 Sentinel-2 影像的一种基于机器学习的云检测模型,主要基于 LightGBM 算法[17],通过世界各地的多个云和非云样本上进行训练,最终通过分类获得云的范围。Google Earth Engine 提供了基于 S2cloudless 模型计算的 Sentinel-2 影像的云概率图像数据集合(“COPERNICUS/S2_CLOUD_PROBABILITY”),但该数据仍在生产中,尚未覆盖所有影像,并且在亚马孙热带雨林地区的云检测测试中,S2cloudless 的效果并不理想,仅为 52%[18]。因此,迫切需要一种更加精确且适用于 Google Earth Engine 云平台上 Sentinel-2 影像的云检测模型。

斯里兰卡是“21 世纪海上丝绸之路”的重要参与国,作为连接亚-非-欧海上航路的枢纽,具有极重要的经济及地理意义[19]。斯里兰卡属于热带地区,大气环境复杂,近中午前后常会形成各种云覆盖,致使对地观测卫星过境影像常年受云雾覆盖影响。通过 Google Earth Engine 统计,覆盖斯里兰卡的无云光学影像(Landsat-8、Sentinel-2)占比不到 1%。本文以斯里兰卡为研究区,基于 Google Earth Engine 云平台,提出了一种耦合 SVM 和 Cloud-Score 算法的 Sentinel-2 影像云检测模型,以期实现准确、可靠的云范围识别。

2 数据与研究区

2.1 研究区概况

斯里兰卡位于南亚次大陆的南端,印度的东南方向,整个岛屿在印度洋之中,东西宽 224 km,南北长 432 km,国土面积达 65610 km^2[20]。斯里兰卡地处热带地区,属于热带季风性气候,无四季之分,年平均气温达 28℃,终年如夏。除中南部山区外,一年中的气候变化很小,气候变化主要与降雨有关。本文选取了 4 个站点作为 Sentinel-2 影像云检测测试点,具体位置如图 1 所示。

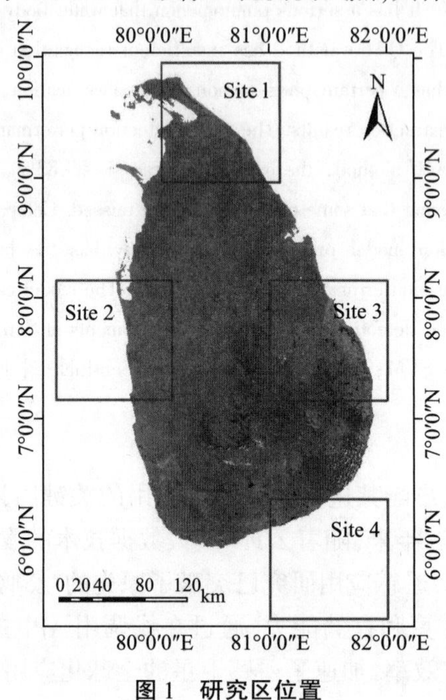

图 1 研究区位置

Fig. 1 Location of study area

2.2 数据来源及处理

Sentinel-2 属于高分辨率多光谱成像卫星,主要由 2A 和 2B 两颗卫星组成[21]。Sentinel-2A 发射于 2015 年 6 月 23 日,Sentinel-2B 发射于 2017 年 3 月 7 日,双星的重访周期大约为 5 d。Sentinel-2 携带有多光谱仪,其 13 个光谱带跨越可见光、近红外和短波红外,幅宽为 290km,高度为 756km。本研究所需的 Sentinel-2 数据来自 Google Earth Engine 云平台所提供的"COPERNICUS/S2_SR"数据集。针对每个站点,选取两景影像进行实验,影像的具体信息如表 1 所示。

表 1 影像具体信息

Table 1 Specific information of images

站点	条带号	日期(年/月/日)	传感器	含云量
站点 1	T44PMR	2019/07/25	Sentinel-2B	1.66%
		2020/12/11	Sentinel-2A	17.51%
站点 2	T44NLP	2019/04/06	Sentinel-2B	6.59%
		2020/02/15	Sentinel-2A	12.22%
站点 3	T44NNP	2019/04/03	Sentinel-2B	8.88%
		2020/04/07	Sentinel-2B	8.10%
站点 4	T44NNM	2019/04/08	Sentinel-2A	6.37%
		2020/11/08	Sentinel-2A	25.60%

3 研究方法

3.1 Google Earth Engine

Google Earth Engine 是 Google 提供的对大量全球尺度地球科学资料数据进行在线可视化计算和分析处理的云平台[5]。该平台主要存储了卫星影像和其他地球观测数据数据库中的资料,同时提供了足够的运算能力对存储数据进行调用和处理。相比于 ENVI、Erdas 等传统的影像处理软件,Google Earth Engine 云平台可以快速、批量地处理影像,并且不需要下载影像数据,同时提供多种格式数据的导出。Google Earth Engine 提供了在线的 JavaScript API 以及离线的 Python API,用户通过调用 API 可快速搭建基于 Google 云的 Web 端服务。

3.2 Cloud-Score 算法

Cloud-Score 算法主要利用云的光谱和热属性来识别并从遥感影像中进行剔除[10]。为了简化该过程,Cloud-Score 函数使用最小/最大归一化函数在 0 和 1 之间重新调整预期反射率(公式(1))。该算法主要原理是将归一化函数应用于计算 4 种云的光谱特性,最终获得的 Cloud-Score 值是这些归一化值中的最小值(公式(2)),以此寻求明亮和寒冷的像素,但同时不具有雪的光谱特性。具体公式为:

$$\mathrm{Norm}(r, r_{\min}, r_{\max}) = \frac{r - r_{\min}}{r_{\max} - r_{\min}} \tag{1}$$

$$\mathrm{CloudScore} = \min\left\{\begin{array}{l} 1.0, \mathrm{Norm}(\rho_{\mathrm{Blue}}, 0.1, 0.3), \mathrm{Norm}(\rho_{\mathrm{Blue}} + \rho_{\mathrm{Green}} + \rho_{\mathrm{Red}}, 0.2, 0.8), \\ \mathrm{Norm}(\rho_{\mathrm{NIR}} + \rho_{\mathrm{SWIR1}} + \rho_{\mathrm{SWIR2}}, 0.3, 0.8), \mathrm{Norm}\left(\dfrac{\rho_{\mathrm{Green}} - \rho_{\mathrm{SWIR1}}}{\rho_{\mathrm{Green}} + \rho_{\mathrm{SWIR2}}}, 0.8, 0.6\right) \end{array}\right\} \tag{2}$$

式中,Norm 表示归一化函数, r 为反射率, $r_{\max}$ 、$r_{\min}$ 分别为反射率的最大、最小值, ρ_{Red} 、ρ_{Green} 、ρ_{Blue} 、ρ_{NIR} 、ρ_{SWIR1} 、ρ_{SWIR2} 分别为遥感影像红、绿、蓝、近红外、短波红外 1 以及短波红外 2 波段的反射率。

3.3 支持向量机(SVM)

SVM 算法是一种建立在统计学习理论的 VC 维理论和结构风险最小化原理基础上的分类算法[24]。相比于神经网络或传统的基于统计的分类方法,SVM 通过向量的个数来控制模型的复杂度,不需要通过降维处理削减特征变量来控制模型的复杂度,因此在分类的过程中,SVM 分类器不会损失地物目标的特征信息,减少了一些过拟合现象的发生。SVM 在精度、泛化性以及高维数据处理等方面均具有明显的优势,在遥感影像分类应用中获得了广泛应用[25-26]。

SVM 的基本原理是将原始的特征向量变换到高维特征空间,在高维空间中求解最优分类超平面,分类精度主要取决于超平面与超平面两边的边界平面之间的距离,距离越大,分类器的精度就越高,误差就越小。给定一个样本集(x_i, y_i), i 为 1, 2, 3, …, N, $x_i \in R^n$, $y_i \in \{-1, 1\}$,分类线 $H = \omega \cdot x + b = 0$,最优分类线需要有最大的分类间隔,即两类样本之间的距离 $\frac{2}{\|\omega\|}$ 最大,等同于求函数的最小值问题。

$$\varphi(\omega) = \frac{1}{2}\|\omega\|^2 \tag{3}$$

本研究中云体的识别属于线性不可分问题,为此需要在上式中引入松弛变量 ξ_i 以及惩罚参数 c ,则 SVM 的可以表示为:

$$\min\left[\frac{1}{2}\|\omega\|2 + c\sum_{i=1}^{l}(\xi_i + \xi_i^*)\right]$$

$$\text{s.t.}\begin{cases} [(\omega \cdot x_i) + b - y_i] \leqslant \varepsilon + \xi_i \\ y_i - [(\omega \cdot x_i) + b] \leqslant \varepsilon + \xi_i^* \\ \xi_i, \xi_i^* \geqslant 0 \end{cases} \tag{4}$$

对于优化问题,需要变换为拉格朗日算子的对偶问题,最终 SVM 的表达式为:

$$f(x) = \sum_{i=1}^{n} (\alpha_i - \alpha_i^*) k(x_i, x) + b \tag{5}$$

其中, x_i 为训练样本的向量, b 为线性函数阈值。

3.4 耦合 SVM 和 Cloud-Score 算法的云检测模型

为了更加精确的对 Sentinel-2 影像进行云检测,本文基于 Google Earth Engine 云平台,借助其强大的分析与处理能力,通过耦合 SVM 和 Cloud-Score 算法,构建云检测模型。模型主要原理是在 Cloud-Score 算法的云检测结果基础上,利用分层随机抽样获得 SVM 的训练样本,最终通过训练和分类进行云检测。QA60 波段法和 Cloud-Score 算法具有阈值法高效率的特点,而 Cloud-Score 算法的云检测结果分辨率和精度更高,因此更适合作为样本的选取图层。图 2 为耦合 SVM 和 Cloud-Score 算法的云检测模型实现过程。首先以时间、含云量、位置为条件,通过编码筛选(ee.ImageCollection.filter())所需云检测的 Sentinel-2 影像("COPERNICUS/S2_SR");其次对筛选后的影像进行重投影(ee.Image.reproject())、裁剪(ee.Image.clip())等预处理操作;然后利用 Cloud-Score 算法识别云体范围,并设置 SVM 分类器(ee.Classifier.svm())的相关参数,核函数选择径向基核函数(Radial basis function, RBF),gamma、cost 参数默认设置为 0.5 和 10,具体可参考 Hus 等[27]提出的 SVM 参数设置指南;接着利用分层抽样函数(ee.Image.stratifiedSample())进行云和非云样本选取,采样比例设置为 10m,样本数量默认设置为 50 个;最终通过 SVM 训练(svm.train())和分类(svm.classify())确定云体范围,并利用 3×3 栅格窗口进行中值滤波(ee.Image.focal_median())去除噪声后得到云检测结果。

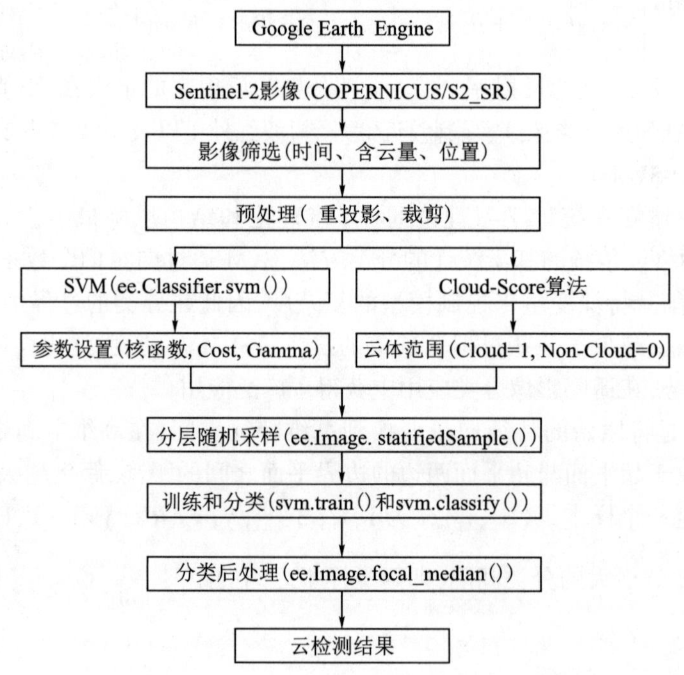

图 2 耦合 SVM 和 Cloud-Score 算法的云检测模型

Fig. 2 Cloud Detection Model coupled with SVM and Cloud-Score algorithm

4 结果与分析

针对斯里兰卡 4 个站点所选择的 8 景影像,分别利用 QA60 波段法、Cloud-Score 算法、Fmask 以及本文提出的模型进行云检测,每个站点首幅影像的云检测结果如图 3 所示。通过详尽的视觉分析可以发现,4 种方法的云检测结果存在显著差异,总体性能呈现出 Fmask<QA60<Cloud-Score<本文模型。QA60 波段法表现得不稳定,4 个站点的结果均出现了不同程度的漏提现象,尤其是对卷云识别效果较差;同时还存在误提现象,站点 1 中将海上部分清晰像素误识别为云;并且其分辨率为 60 m,而 Sentinel-2 影像的最高分辨率为 10 m,因此云边界的识别效果较差。相比于 QA60 波段法,由于更多

的波段参与云可能性分数的计算，Cloud-Score 算法在云体边界提取的完整性方面有了明显的提升，并且对密云的识别效果较好，但仍存在部分卷云的漏提现象，尤其是站点 4 南部海域存在较多的卷云未识别。Baetens[28]、Sanchez[22]等在不同地区通过实验均证明了 Fmask 模型的云检测精度在 90%以上，然而其在斯里兰卡 4 个站点的云检测测试中很不理想。Fmask 模型的云检测结果呈现出不同程度的误提现象，站点 1 沿海地区将部分水体识别为云，站点 2 和站点 4 的海域中存在大量的误提现象。虽然 Fmask 模型对密云和卷云均表现出了较好的识别效果，但严重的误提现象影响了云检测的整体精度。相比于其他三种模型的云检测结果，本文提出的云检测方法性能最好，能完整地识别出云体边界，对卷云的识别呈现出了良好的效果，视觉上仅在站点 2 北部海域出现了部分卷云的漏提现象。

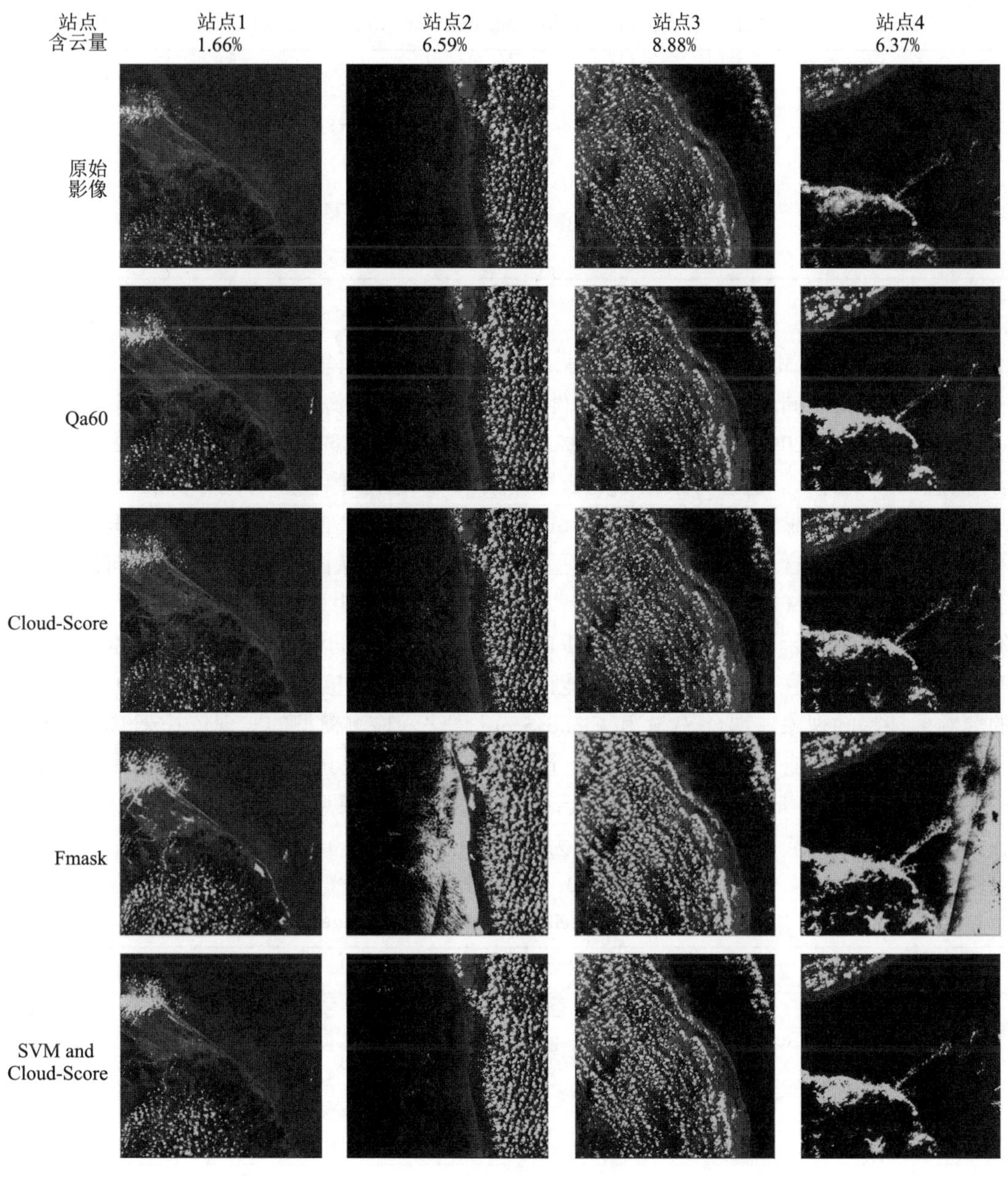

图 3　不同模型云检测结果

Fig. 3　Cloud detection results of different models

以每景影像的目视云检测结果为基准,分别选取200个云体样本和非云体样本,通过计算混淆矩阵[29]分别从漏提率、误提率和总体精度3个方面进一步反映不同模型云检测精度,其统计结果如表2所示。从表中可以看出,4种模型中Fmask模型的总体精度最低,仅为63.45%,误提率极高,达到了38.74%,但其漏提率极低;QA60法的漏提率最高,为25.35%,同时存在较高的误提率,总体精度为78.16%;Cloud-Score算法的云检测性能明显好于QA60法,总体精度达到了89.83%,误提率仅为2.17%,但仍存在较高的漏提率;本文提出的耦合SVM和Cloud-Score算法的云检测模型总体精度最高,达到了98.21%,并且拥有极低的漏提率和误提率,整体云检测性能较好,可满足Sentinel-2遥感产品的云检测预处理需求。

表2　不同模型云检测精度

Table 2　Cloud detection accuracy of different models

模型	漏提率/%	误提率/%	总体精度/%
QA60	25.35	7.21	78.16
Cloud-Score	8.64	2.17	89.83
Fmask	2.27	38.74	63.45
SVM and Cloud-Score	1.06	0.15	98.21

5　讨论

本文构建的耦合SVM和Cloud-Score算法的云检测模型在斯里兰卡地区表现出了较高的精度,为了进一步验证其在不同地区和不同云量条件的适用性,分别在海南岛和亚马孙热带森林按照含云量0~10%、10%~20%、20%~30%、30%~40%和40%~50%分别选取一景Sentinel-2影像进行云检测,具体信息如表3所示。通过每景影像的云检测结果与原始影像叠加对比分析发现,本文构建的模型在两个地区不同含云量条件下的云检测测试中均表现出了较高的精度,能准确地识别出云体的边界。图4为两个地区高含云量条件下不同模型的云检测结果。从图中可以看出,不同模型在高云量条件下的云检测结果同样差异显著,QA60波段法和Fmask存在明显的误提现象,将大量存在于云体中间的清晰像元识别为云;QA60波段法和Cloud-Score算对卷云的识别效果有限,但本文模型对海南岛和亚马孙热带森林的卷云识别效果较好,能比较精准的提取出卷云与其他地物的边界,仅存在部分少量的薄卷云漏提的现象。耦合SVM和Cloud-Score算法云检测模型的实现依托于Google Earth Engine云平台强大的计算能力,确保了模型的运行效率,并且针对单景影像的样本训练和分类满足了精度和通用性要求。同时Google Earth Engine云平台中易于实现模型的扩展和批量影像的处理操作,因此在大规模的云检测应用方面具有一定潜力。

表3　影像具体信息

Table 3　Specific information of images

地点	条带号	日期(年/月/日)	传感器	含云量
海南岛	T49QCB	2018/11/04	Sentinel-2B	8.79%
		2020/01/08	Sentinel-2B	13.23%
		2018/05/23	Sentinel-2A	21.22%
		2019/11/09	Sentinel-2B	30.92%
		2018/01/13	Sentinel-2A	42.60%
亚马孙热带森林	T18LXP	2020/04/22	Sentinel-2A	4.50 %
		2019/10/25	Sentinel-2A	19.74%
		2018/07/27	Sentinel-2B	27.86%
		2019/06/22	Sentinel-2B	36.21%
		2021/01/17	Sentinel-2A	47.09%

6 结论

本文以热带多云的斯里兰卡为研究区,通过耦合 SVM 和 Cloud-Score 算法,构建了适用于 Google Earth Engine 云平台的 Sentinel-2 云检测模型;对比了该模型与 QA60 法、Cloud-Score 算法以及 Fmask 的云检测精度,并在海南岛和亚马孙热带森林验证了该模型云检测的通用性和鲁棒性。研究结果表明:Fmask 模型的云检测性能最低,总体精度仅为 63.45%,存在严重的水体误分为云的现象,但其漏提率极低;QA60 法对卷云识别不足,漏提率较高,同时存在一定的误分现象,并且云体边界结果的细节性较差;Cloud-Score 算法的云检测性能明显好于 QA60 法,总体精度达到了 89.83%,误提率仅为 2.17%,但仍存在部分卷云漏提的现象;相比于其他三种云检测方法,本文提出的耦合 SVM 和 Cloud-Score 算法的云检测模型总体精度最高,达到了 98.21%,并且拥有极低的漏提率和误提率,能比较精准的识别出云体的边界,可满足 Sentinel-2 遥感产品的云检测预处理需求。在后续的研究中,将探究模型在 Landsat 系列影像的云检测潜力,同时聚焦大批量影像的云检测。

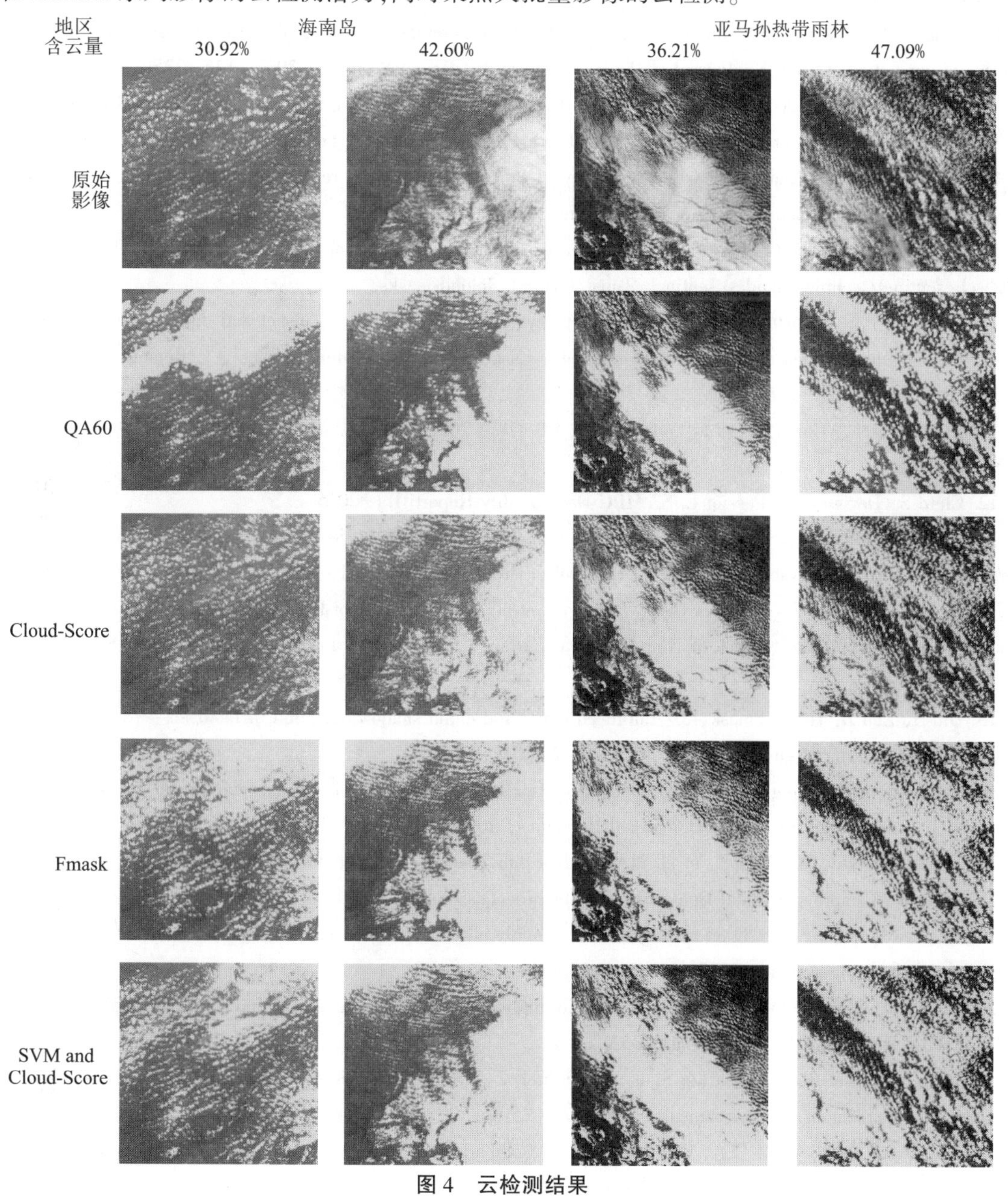

图 4 云检测结果

Fig. 4 Cloud detection results

参考文献

[1]Foga S, Scaramuzza P L, Guo S, et al. Cloud detection algorithm comparison and validation for operational Landsat data products[J]. Remote sensing of environment, 2017, 194: 379-390.

[2]Li Z, Shen H, Cheng Q, et al. Deep learning based cloud detection for medium and high resolution remote sensing images of different sensors[J]. ISPRS Journal of Photogrammetry and Remote Sensing, 2019, 150: 197-212.

[3]Sun L, Yang X, Jia S, et al. Satellite data cloud detection using deep learning supported by hyperspectral data[J]. International Journal of Remote Sensing, 2020, 41(4): 1349-1371.

[4]Wang C, Liu Y, Guo L, et al. A Smart Agricultural Service Platform for Crop Planting, Monitoring and Management-PIE-Engine Landscape[C]//2021 9th International Conference on Agro-Geoinformatics (Agro-Geoinformatics). IEEE, 2021: 1-6.

[5]Gorelick N, Hancher M, Dixon M, et al. Google Earth Engine: Planetary-scale geospatial analysis for everyone[J]. Remote sensing of Environment, 2017, 202: 18-27.

[6]Xiong J, Thenkabail P S, Gumma M K, et al. Automated cropland mapping of continental Africa using Google Earth Engine cloud computing[J]. ISPRS Journal of Photogrammetry and Remote Sensing, 2017, 126: 225-244.

[7]Carrasco L, O'Neil A W, Morton R D, et al. Evaluating combinations of temporally aggregated Sentinel-1, Sentinel-2 and Landsat 8 for land cover mapping with Google Earth Engine[J]. Remote Sensing, 2019, 11(3): 288.

[8]Wang L, Diao C, Xian G, et al. A summary of the special issue on remote sensing of land change science with Google earth engine[J]. Remote Sensing of Environment, 2020, 248, 112002.

[9]ESA, Level-1C Cloud Masks-Sentinel-2 MSI Technical Guide-Sentinel Online[EB/OL]. 2020. sentinels.copernicus.eu/web/sentinel/technical-guides/sentinel-2-msi/level-1c/cloud-masks.

[10]Housman I W, Chastain R A, Finco M V. An evaluation of forest health insect and disease survey data and satellite-based remote sensing forest change detection methods: Case studies in the United States[J]. Remote Sensing, 2018, 10(8): 1184.

[11]EO Research., Cloud Masks at Your Service[EB/OL]. 2020. medium.com/sentinel-hub/cloud-masks-at-your-service-6e5b2cb2ce8a.

[12]Clerc S, Devignot O, Pessiot L. S2 MPC Data Quality Report[R]. 2015.

[13]Coluzzi R, Imbrenda V, Lanfredi M, et al. A first assessment of the Sentinel-2 Level 1-C cloud mask product to support informed surface analyses[J]. Remote sensing of environment, 2018, 217: 426-443.

[14]Nguyen M D, Baez-Villanueva O M, Bui D D, et al. Harmonization of Landsat and Sentinel 2 for Crop Monitoring in Drought Prone Areas: Case Studies of Ninh Thuan (Vietnam) and Bekaa (Lebanon)[J]. Remote Sensing, 2020, 12(2): 281.

[15]Qiu S, Zhu Z, He B. Fmask 4.0: Improved cloud and cloud shadow detection in Landsats 4-8 and Sentinel-2 imagery[J]. Remote sensing of environment, 2019, 231: 111205.

[16]Main-Knorn M, Pflug B, Louis J, et al. Sen2Cor for sentinel-2[C]//Image and Signal Processing for Remote Sensing XXIII. International Society for Optics and Photonics, 2017, 10427: 1042704.

[17]Hagolle O, Huc M, Auer S, et al. MAJA Algorithm Theoretical Basis Document[R]. 2017.

[18]Zupanc A. Improving Cloud Detection with Machine Learning[R]. 2017.

[19]Ke G, Meng Q, Finley T, et al. Lightgbm: A highly efficient gradient boosting decision tree[J]. Advances in neural information processing systems, 2017, 30: 3146-3154.

[20]Chung C. What are the strategic and economic implications for South Asia of China's Maritime Silk Road initiative?[J]. The Pacific Review, 2018, 31(3): 315-332.

[21]Drusch M, Del Bello U, Carlier S, et al. Sentinel-2: ESA's optical high-resolution mission for GMES operational services[J]. Remote sensing of Environment, 2012, 120: 25-36.

[22]Sanchez A H, Picoli M C A, Camara G, et al. Comparison of Cloud cover detection algorithms on sentinel-2 images of the amazon tropical forest[J]. Remote Sensing, 2020, 12(8): 1284.

[23]Li Jianfeng, Ye Huping, Zhang Zongke, et al. Spatiotemporal Change Analysis of Sri Lanka Inland Water based on Landsat Imagery[J]. Geo-Information Science, 2019,21 (5):781-788. [李健锋，叶虎平，张宗科，等. 基于Landsat影像

的斯里兰卡内陆湖库水体时空变化分析[J]. 地球信息科学学报, 2019, 21(05):781-788.]

[24]Deng Naiyang, Tian Yingjie. A new method of data mining-support vector machine[M]. Science Publication, 2004. [邓乃扬,田英杰. 数据挖掘中的新方法:支持向量机[M]. 科学出版社, 2004.]

[25]Guo Y, Jia X, Paull D. Effective sequential classifier training for SVM-based multitemporal remote sensing image classification[J]. IEEE Transactions on Image Processing, 2018, 27(6): 3036-3048.

[26]Shaharum N S N, Shafri H Z M, Ghani W A W A K, et al. Oil palm mapping over Peninsular Malaysia using Google Earth Engine and machine learning algorithms[J]. Remote Sensing Applications: Society and Environment, 2020, 17: 100287.

[27]Hsu C W, Chang C C, Lin C J. A practical guide to support vector classification[EB/OL]. 2016. www.csie.ntu.edu.tw/~cjlin/papers/guide.

[28]Baetens L, Desjardins C, Hagolle O. Validation of copernicus Sentinel-2 cloud masks obtained from MAJA, Sen2Cor, and FMask processors using reference cloud masks generated with a supervised active learning procedure[J]. Remote Sensing, 2019, 11(4): 433.

[29]Congalton R G. A review of assessing the accuracy of classifications of remotely sensed data[J]. Remote sensing of environment, 1991, 37(1): 35-46.

本文曾发表于2022年《遥感技术与应用》第37卷第3期

陕北风沙草滩区新垦地风蚀输沙特征

张腾飞[1,2]

(1.陕西省土地工程建设集团有限责任公司,陕西 西安 710075;
2.自然资源部退化及未利用土地整治工程重点实验室,陕西 西安 710075)

【摘要】为探究新垦地野外风蚀输沙特征,本文以单次起沙风为研究对象,采用集沙仪定点监测方法对陕西北部风沙草滩区新垦田进行了为期一年的野外观测试验,并逐层对距离地表 0~50 cm 高度范围内风蚀物进行土壤粒度和电镜分析。结果显示:(1)集沙量和集沙粒径组成均与集沙高度呈负相关,随着集沙高度的增加,集沙量明显减少,0~10 cm 高度范围内集沙量占总集沙量的 82.16%,10~20 cm 高度范围内集沙量占总集沙量的 11.62%,20~50 cm 高度范围内集沙量占总集沙量的 6.22%。集沙高度范围内,集沙样品中黏粒和粉粒含量均为零,随着集沙高度的增加,细沙和粗沙质量比例逐渐由 1:1趋近于 6:4。(2)扫描电镜图像显示,新垦地风蚀物颗粒之间未能形成土壤胶结和团聚体,建议在后期耕种过程中注重土壤熟化措施的运用。

【关键词】新垦地;风力侵蚀;粒径;风沙草滩区;定边县

Characteristics of Wind Erosion and Sand Transport in Newly Reclaimed Land of Wind-sand Grassy Plain in Northern Shaanxi

Tengfei Zhang[1,2]

(1.Shaanxi Land Engineering Construction Group Co., Ltd., Xi'an 710075, China;2.Key Laboratory of Degraded and Unused Land Consolidation Engineering, Ministry of Natural Resources, Xi'an 710075, China)

【Abstract】In order to explore the wind erosion and particle size characteristics of newly reclaimed land in wind erosion area. Taking the Newly reclaimed land in wind-sand grassy beach area of northern Shaanxi as the research object, the single sand-driving wind was taken as the research unit, and the structure of wind-sand flow was observed by sand accumulation instrument, the composition of mass and particle size of wind erosion at different sites with a height of 0~50 cm were analyzed by laser particle size analyzer and electron microscope scanner. The results show that:(1) the sediment mass decreases from the bottom to the top layer, the amount of sediment collection in the height range of 0~10 cm accounts for 82.16% of the total amount of sediment collection, in the height range of 10~20 cm accounts for 11.62% of the total amount of sediment collection, and in the height range of 20~50 cm accounts for 6.22% of the total amount of sediment collection. The particle size of sand collecting material becomed finer layer by layer from bottom to top. with the range of five sand collecting heights, the content of clay and silt was zero. with the increase of sand collecting height, the mass ratio of fine sand and coarse sand grad-

基金项目:陕西省土地工程建设集团内部科研项目(DJNY2018-8)、陕西省重点科技创新团队计划项目(2016KCT-23)和长安大学中央高校基金项目(300102270503)资助。

ually increased from 1∶1 to 6∶4. (2) laser particle size analyzer and scanning electron microscope analysis showed that neither soil cementation nor aggregates was formed between the particles of aeolian sand erosion. These results suggest that the soil maturation measures should be taken into account in the later cultivation process of newly reclaimed land.

【**Keywords**】Newly reclaimed land; Wind erosion; Grain size; Sandy grassland area; DingBian county

风力侵蚀是我国北方干旱、半干旱区域最活跃的一种陆表过程[1-2]，强烈的土壤风蚀会加剧区域农田土地荒漠化程度，造成耕地资源浪费，影响生态环境和谐[3-4]。我国陕北长城沿线风沙草滩区地处毛乌素沙地南源，为典型的荒漠化气候环境，生态环境较为脆弱，土壤构成以沙土居多，风蚀现象严重[5]。近年来，随着快速城市化和工矿交通建设占用耕地现象增多[6]，在地区政府耕地占补平衡政策[7-8]影响下，陕北定边县将部分盐碱地开垦为耕地[9]，这些新垦地在风力侵蚀作用下易形成风沙流[10]。因此，研究长城沿线风沙草滩区新垦地土壤风蚀问题，对防治风力侵蚀、提高新垦土地农业生产力具有重要意义。

土壤颗粒受气流作用[11-12]形成风沙流的运动过程会造成大量扬尘影响空气质量[13]。风沙流结构是风沙流研究的一个重要内容[14]，李得禄等[15]研究认为在0~80 cm范围内，输沙量与风速呈指数函数关系，输沙率与高度呈幂函数关系。包岩峰等[16]研究认为在0~30 cm高度，输沙率均与风速成正比，与高度成反比，幂函数拟合关系最佳。郭树江等[10]研究青土湖干涸湖底不同立地输沙通量主要集中在近地表0~20 cm高度范围内，土壤颗粒组成是影响风蚀的主要因素之一[17]。苑依笑等[18]研究了农田土壤细颗粒物的损失特征，认为土壤细粒物质的损失粒径集中在0~0.040 mm，其中0.001 8~0.024 mm的细颗粒损失最为严重。董治宝[19]通过风洞模拟实验发现，0.075~0.400 mm为易蚀颗粒类型。Chepil[20]通过风洞实验研究认为，小于0.420 mm的颗粒物质为高度可蚀。王仁德等[21]通过野外风蚀观测发现，风蚀物粒度组成以粒径小于0.100 mm的悬移质为主。李晓娜等[22]研究认为0.050~0.002 mm的粉粒是延怀盆地葡萄种植区风蚀过程中损失的主要颗粒。王瑞东等[23]实验发现希拉穆仁围封区域荒漠草原近自然恢复状态下土壤风蚀颗粒范围为0.011~0.0250 mm。丁延龙[24]以吉兰泰盐湖北部沙垄内白刺灌丛沙堆为研究对象，发现白刺灌丛沙堆表层沉积物粒径多在0.0100~0.0300 mm范围内。以上研究成果对揭示风沙流结构特征和防治风沙灾害提供了重要支撑[10]，但通过野外试验观测研究新垦地土壤风蚀成果较少。

陕西省定边县开发实施盐碱地为农耕地，主要技术措施为通过工程方法降低地下水位，并在平整后的田面土层上摊铺20 cm厚的黄沙，搅拌形成30 cm厚的新土壤，经过逐年耕种后，新土壤逐渐具备种植耐盐碱作物的特性[25]。本文以定边县2018年底开发实施完成的新垦地田面为研究对象，课题组于2019年3月至2020年3月进行了野外观测试验，共收集到14场有效起沙风和风力侵蚀物，通过研究新垦地风蚀的输沙特征，以期为陕北风沙侵蚀区新垦地耕作层保护提供技术支撑。

1　研究区概况及研究方法

1.1　研究区概况

研究区处于毛乌素沙漠与黄土丘陵区过渡地带，隶属于定边县白泥井镇，地理坐标在东经107°50′0″~107°51′30″，北纬37°45′30″~37°46′30″。区内海拔高度介于1374~1387 m，地表无乔、灌木植物。区内多年平均气温7.9 ℃，多年平均降水量316.9 mm，其中7月、8月、9月3个月降水量150.7 mm，占全年降水量的62.16%。多年平均蒸发量为2 490.2 mm，为降水量的7.9倍。无霜期141 d，最长年无霜期为164 d。初开始解冻在次年3月，结冻天数95~105 d，冻结深度1.09~1.20 m。

风灾是研究区常见的自然灾害，以风蚀为主的土壤侵蚀普遍发生，危害较大，大风主要出现在春夏两季，春季多以冷空气影响形成的大风为主，夏季则以雷雨大风为主。起沙风最大瞬时风速多出现在4月和5月，此时正是农耕时节和作物幼苗生长阶段，大风易造成农作物减产。农作物种植期为5月至10月，休耕期为11月至次年4月，主要种植作物有玉米、荞麦、马铃薯。

1.2　研究方法

试验前首先将积沙仪安装在新垦地田面下风口位置,集沙样地面积为 30 hm^2,田面平坦且周围为草地,无其他起沙源。试验观测期间,试验区田面做裸露、无遮挡处理。试验采用 HY.JSY-A 型阶梯式集沙仪收集风沙物,集沙仪高 80 cm,集沙通道为 5 排,每排 2 个并行通道,集沙通道为 5 cm 的方形入口,每排集沙通道上沿距离地面高度分别是 10 cm、20 cm、30 cm、40 cm 和 50 cm,导向器装置可以使得集沙仪入口始终正对侵蚀风向。集沙布袋为白丝布,容量>1 000 g。采用 FC-16025 手持式风速风向仪测量起沙风速,主要风向等参数。

当田面遇风起尘时,立即开始气流和集沙的观测,风蚀物从 10 个进沙孔分别进入集沙布袋中,同步观测瞬时风速、平均风速、风向等数据。当田面不起沙尘时,暂停试验观测。试验期内共收集 14 场次有效风力侵蚀风沙土样品。每场起沙风后,收集量测各个集沙布袋中的集沙量,称量后留存以备试验分析。

1.3　数据处理

风沙土样品粒度检测使用 Mastersizer 3000 型激光粒度分析仪进行分析,根据土壤质地分类标准[26]将样品分为黏粒、粉粒、细沙、粗沙。电镜图像使用环境扫描电镜图像(FEIQ45)。测试工作在自然资源部退化及未利用土地整治工程重点实验室内进行,粒度分析结果以重量百分数表示。数据分析采用 SPSS 软件完成。

2　结果与分析

表 1 是 2019 年 3 月至 2020 年 3 月观测期间收集到的 14 场有效起沙风的风力特征。起沙风风向集中于西北和西南两个方向,且绝大多数集中在西北方向。起沙风平均风速为 6.28 m/s,起沙风的瞬时风速在 5.88~6.67 m/s。

表 1　试验期收集到的 14 场起沙风风力特征

风场	风场发生时间(y-m-d)	扬尘持续时间(h)	主要风向	起沙风速($m\cdot s^{-1}$)
风场 1	2019-03-08	12	SW	5.88
风场 2	2019-03-10	12	NW	5.75
风场 3	2019-03-20	12	NW	5.83
风场 4	2019-03-28	24	NW	6.02
风场 5	2019-04-06	24	SW	6.55
风场 6	2019-04-13	24	NW	6.67
风场 7	2019-04-15	12	NW	6.89
风场 8	2019-04-19	12	NW	6.63
风场 9	2019-04-26	12	SW	6.31
风场 10	2019-05-11	24	SW	6.96
风场 11	2019-12-14	12	SE	6.04
风场 12	2019-12-26	12	NW	5.93
风场 13	2020-02-21	24	NW	6.12
风场 14	2020-03-09	24	NW	5.89

2.1　输沙量的垂直变化

表 2 是收集到的 14 场有效风沙土样品质量,随着集沙高度的增加,集沙量逐层减少,其中 0~10 cm高度收集的集沙量占总集沙量的 82.16%,0~10 cm 高度范围内的集沙量是 10~20 cm 高度集沙量的 7.07 倍,10~20 cm 高度范围内的集沙量是 20~30 cm 高度集沙量的 3.11 倍,20~30 cm 高度范围

内的集沙量是30~40 cm高度集沙量的2.15倍,30~40 cm高度范围内的集沙量是40~50 cm高度范围的2.36倍。在距离地面每10 cm高度范围内,收集到的风蚀物质量呈现出不同的递减趋势。

表2 各风场内收集到不同高度处的风沙量 单位:g

风场	0~10 cm	10~20 cm	20~30 cm	30~40 cm	40~50 cm
风场1	29.70	8.50	7.78	6.44	5.02
风场2	65.58	0.00	0.00	0.00	0.00
风场3	62.28	9.25	0.00	0.00	0.00
风场4	181.61	49.75	7.83	5.53	0.00
风场5	74.35	12.57	0.00	0.00	0.00
风场6	41.86	26.99	20.34	7.14	4.87
风场7	330.30	5.01	0.00	0.00	0.00
风场8	116.58	0.00	0.00	0.00	0.00
风场9	208.88	14.65	0.00	0.00	0.00
风场10	241.43	15.39	3.58	0.00	0.00
风场11	83.20	16.42	5.69	0.00	0.00
风场12	117.48	35.31	12.56	6.27	0.00
风场13	90.77	29.06	14.73	8.23	2.45
风场14	151.19	31.03	9.22	4.43	3.79
合计	1795.21	253.93	81.73	38.04	16.13

2.2 集沙粒径组成分析

如表3所示为14场风沙样品的平均粒度试验数据,试验分析发现,集沙样品中黏粒和粉粒含量几乎为零,随着集沙高度的增加,细沙含量逐渐增加,粗沙含量逐层较少,二者的质量比由0~10 cm高度的1:1逐渐转变为40~50 cm高度的6:4。样品粒度组成与集沙高度的关系是,随着集沙高度的增加,粗沙含量逐渐降低,由0~10 cm的47.99%下降到40~50 cm的38.07%;细沙含量逐渐升高,由0~10 cm的52.02%上升到40~50 cm的61.93%。

表3 集沙高度与集沙物颗粒组成之间的关系

集沙高度(cm)	集沙物粒径分布(%)			
	黏粒(0~0.002 mm)	粉粒(0.002~0.020 mm)	细沙(0.020~0.200 mm)	粗沙(0.200~2 mm)
0~10	0.00	0.00	52.02	47.99
10~20	0.00	0.00	54.80	45.20
20~30	0.09	0.70	54.16	45.04
30~40	0.00	0.00	59.17	40.83
40~50	0.00	0.00	61.93	38.07

注:粒径数据是14场集沙物的平均数据。

2.3 集沙样品扫描电镜图像

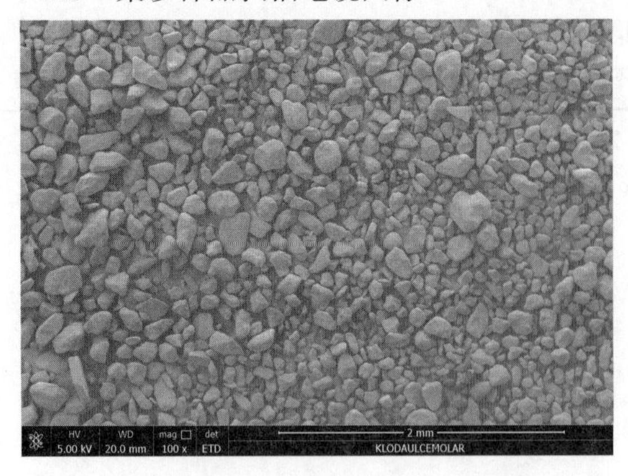

A:集沙样品 100 倍扫描电镜图像

B:集沙样品 800 倍扫描电镜图像

图 1 集沙样品扫描电镜图像

集沙样品 100 倍扫描电镜图像显示,风沙土质地均匀,质地以沙粒为主,存在极少的粉粒物质,几乎无法观测到黏粒(见图 1A)。颗粒之间不黏结,图像中没有团聚体,颗粒间隙较大,难以有效保水保肥,不利于颗粒形成稳定的连接关系[27]。从 800 倍颗粒外观来看,颗粒存在尖锐棱角和明显的边面(见图 1B)。由上述微形态结构可判断,集沙物颗粒粗骨化,土质疏松导致新垦田面遇风易起沙,与高倍扫描电镜下成熟农田土壤颗粒表面分布大量黏粒[28]形成反差,若不施加合理有效的防治措施,会逐渐带动田面土壤黏结物的流失,加剧田面沙化。

3 讨论

(1)本文通过分析阶梯式集沙仪自下而上 50 cm 不同高度收集的风沙样品粒径组成和质量关系发现,距离地面 0~10 cm 高度收集的风沙物质量占整个试验期风沙质量的 82.16%,说明近地面10 cm 范围内是新垦地风蚀输沙的主要阶层;距离地面 10~20 cm 范围内收集的风沙物质量占整个实验期风沙质量的 11.62%,距离地面 20~50 cm 范围收集的风沙物质量占整个试验期风沙质量的 6.22%,以上研究结果与刘铁军等[29]研究风沙流中土壤颗粒运动对风速的影响主要集中在 80 mm 高度以内较为接近。周炎广等[6]研究发现输沙率与高度的关系具有分段性,0~10 cm 高度区间内服从指数函数分布,10~20 cm 高度区间服从幂函数规律。但本研究观测的野外风沙量是以单次起沙风为单位进行观测,各风场中量测的风向、风速、持续时间不同,14 场风蚀物之间输沙率不具备可比性,各高度范围内的输沙函数关系不能拟合,这是本研究的不足之处。

(2)粒度分析试验显示,集沙物黏粒和粉粒含量几乎为零,试验结果与王仁德等[21]研究成熟农田野外风蚀物粒度组成以粒径小于 100 μm 的悬移质为主存在偏差,但与王陇等[30]研究毛乌素沙地风沙土优势粒径组为细沙较为接近,原因可能是新垦地土壤特性和成熟农田土壤结构特性不同造成的。同时扫描电镜图像发现风沙物颗粒棱角鲜明,颗粒之间无土壤胶结和团聚体,所以建议新垦地须在后期通过农艺等措施加强土壤熟化,使新造土壤逐渐向适宜盐碱作物生长的方向发展。

4 结论

本研究在同一种下垫面条件下进行了一年的野外风沙观测试验,试验期内共收集到 14 场有效起沙风,根据测量不同集沙高度范围内的集沙量发现:

(1)集沙量和集沙高度呈负相关,随着集沙高度的增加,集沙量明显减少;试验高度内集沙量呈较为明显的分层,0~10 cm 高度范围内收集的集沙量占总集沙量的 82.16%,10~20 cm 高度范围内收集的集沙量占总集沙量的 11.62%,20~50 cm 高度范围内收集的集沙量占总集沙量的 6.22%。

(2)集沙粒径组成与集沙高度呈负相关,风沙物中黏粒和粉粒含量几乎为零,随着集沙高度的增加,细沙和粗沙含量比例逐渐从 1:1向 6:4转变。

(3)激光粒度分析仪和扫描电镜图像显示,新垦地风蚀物几乎观测不到黏粒和粉粒,土壤颗粒粗骨化、不黏结,土壤结构不稳定,持水性差,土粒之间未能形成有效土壤胶结和团聚体。建议通过工程措施增加土壤中黏、粉粒含量,使新垦地土壤快速形成复配土[31],同时在后期耕种过程中注重土壤熟化措施的运用。

参考文献

[1]郑晓静.风沙运动的力学机制研究[J].科技导报,2007(14):22-27.

[2]高君亮,高永,罗凤敏,等.表土粒度特征对风蚀荒漠化的响应[J].科技导报,2014,32(25):20-25.

[3]南岭,董治宝,肖锋军.农牧交错带农田土壤风蚀 PM_(10)释放特征[J].中国沙漠,2017,37(6):1079-1084.

[4]哈斯,陈渭南.耕作方式对土壤风蚀的影响:以河北坝上地区为例[J].土壤侵蚀与水土保持学报,1996(1):10-16.

[5]张冰,高照良,王冬.干旱沙化生态脆弱区生态经济发展模式的探讨[J].生态经济(学术版),2012(2):67-75.

[6]周炎广,武子丰,胡日娜,等.毛乌素沙地新垦地土壤风蚀特征[J].农业工程学报,2020, 36(1):128-137.

[7]Liu L,Liu Z,Gong J,et al. Quantifying the amount, heterogeneity, and pattern of farmland: Implications for China's requisition-compensation balance of farmland policy[J]. Land Use Policy, 2019, 81:256-266.

[8]Liu Y,Fang F,Li Y.Key issues of land use in China and implications for policy making[J]. Land Use Policy, 2014, 40:6-12.

[9]韩霁昌,王曙光,李娟,等.浅析定边县盐碱地开发模式[J].中国土地,2013(5):40.

[10]郭树江,杨自辉,王强强,等.青土湖干涸湖底风沙流结构及输沙粒径特征[J].生态学杂志,2021,40(4):1166-1176.

[11]张卫华,韩霁昌,王欢元,等.砒砂岩对毛乌素沙地风成沙的改良应用研究[J].干旱区资源与环境,2015,(10):122-127.

[12]王金花,张荣刚,姚文艺,等.风蚀水蚀共同作用下沙漠河流泥沙级配特征[J].中国沙漠,2016, 36(6):1695-1700.

[13]曹翠红,蒋远营.基于 TVP-VAR-SV 模型的经济发展和 PM_(2.5)的联动关系研究:以北京市为例[J].生态经济,2020,36(8):185-193.

[14]武建军,何丽红,郑晓静.跃移层中沙粒浓度分布特征的研究[J].兰州大学学报,2002(3):15-21.

[15]李得禄,满多清,朱国庆,等.丘间低地不同部位风沙流结构特征[J].中国沙漠,2012,32(5):1210-1215.

[16]包岩峰,丁国栋,吴斌,等.毛乌素沙地风沙流结构的研究[J].干旱区资源与环境,2013,27(2):118-123.

[17]张春来,宋长青,王振亭,等.土壤风蚀过程研究回顾与展望[J].地球科学进展,2018,33(1):27-41.

[18]苑依笑,王仁德,常春平,等.风蚀作用下农田土壤细颗粒的粒度损失特征及其对土壤性质影响[J].水土保持学报,2018,32(02):104-109,119.

[19]董治宝,李振山.风成沙粒度特征对其风蚀可蚀性的影响[J].土壤侵蚀与水土保持学报,1998(4):2-6,13.

[20]Chepil W S.Dynamics of wind erosion. Nature of movement of soil by wind[J]. Soil Science, 1945, 60(4):305-320.

[21]王仁德,邹学勇,赵婧妍.北京市农田风蚀的野外观测研究[J].中国沙漠,2011,31(2):400-406.

[22]李晓娜,宋进库,张微微,等.延怀盆地葡萄种植区不同土地利用方式表土风蚀特征分析[J].水土保持学报,2019,33(6):86-91.

[23]王瑞东,高永,党晓宏,等.近自然恢复状态下荒漠草原不同群落表土粒度特征研究[J].草地学报,2019,27(5):1309-1316.

[24]丁延龙.白刺灌丛沙堆演化对地表蚀积的影响及其作用机制[D].呼和浩特:内蒙古农业大学,2019.3-5.

[25]张腾飞,盛晓磊,王波.风力侵蚀下沙堆吹蚀物土壤粒度特征研究[J]. 西部大开发(土地开发工程研究),2020,5(5):59-63.

[26]李天杰,赵烨,张科利,等.土壤地理学[M].北京:高等教育出版社,2003:36-37.

[27]李裕瑞,范朋灿,曹智,等.毛乌素沙地砒砂岩与沙复配农田的固沙效应及其微观机制[J].中国沙漠,2017,37(3):421-430.

[28]李裕瑞,范朋灿,曹智,等.基于扫描电镜解析毛乌素沙地砒砂岩与沙复配成土的微观结构特征[J].应用基础

与工程科学学报,2019,27(4):707-719.

[29]刘铁军,赵显波,赵爱国,等.东北黑土地土壤风蚀风洞模拟试验研究[J].水土保持学报,2013,27(2):67-70.

[30]王陇,高广磊,张英,等.毛乌素沙地风沙土粒径分布特征及其影响因素[J].干旱区地理,2019,42(5):1003-1010.

[31]韩霁昌,刘彦随,罗林涛.毛乌素沙地砒砂岩与沙快速复配成土核心技术研究[J].中国土地科学,2012,26(8):87-94.

本文曾发表于2022年《亚热带水土保持》第34卷第3期

Study on Plant Diversity and Community Quantitative Characteristics of Wetland on the North Bank of Hanjiang River in Chenggu, Shaanxi Province

Zongwu Li, Yi Zhang, Fujing Li, Mengge Duan

(Hanzhong Branch of Shaanxi Land Engineering Construction Group Co., Ltd., Hanzhong 723200, China)

【Abstract】 Hanjiang River is the longest tributary of the Yangtze River and an important research basin for ecological protection and high-quality development of the Yangtze River Basin. In order to master the plant diversity and quantitative characteristics of community in Hanjiang River. From 2020 to 2021, four water ecological surveys were carried out on the north bank of Hanjiang River in two seasons. Based on the field wetland quadrat survey and literature data, the species diversity index was selected. The research shows that there are 360 species of wetland plants in 108 families and 264 genera on the North Bank of Hanjiang River, and the geographical composition of seed plants is at the genus level, mainly temperate distribution types, with 160 species, In the community species diversity index, the mean value of Patrick index was 18.80 ± 5.19; The mean Shannon Wiener diversity index was 2.15 ± 0.39; The mean value of Simpson dominance index is 0.81 ± 0.10; The average Pielou evenness index is 0.75 ± 0.10; The mean value of Margalef index is 3.44 ± 0.87. The Patrick index and Margalef index of the *Bidens pilosa* form were the lowest, the Shannon-Wiener diversity index, the Simpson index, and the Pielou index of the *Potamogeton malaianus* form were the lowest; the Patrick index of the *Colocasia esculenta* form was the highest, and the Pielou of the *Sapium sebiferum* form was the highest. The Shannon-Wiener diversity index and Simpson index of the *Populus canadensis* form group were the highest, and the Margalef index of the *Flueggea virosa* form was the highest. The research results provide basic data and scientific basis for the protection and restoration of wetland ecosystems and management of wetland resources in Yuan'an and similar areas.

【Keywords】 Species diversity; Flora; Wetland plant community; North bank of Hanjiang River

Introduction

According to the definition of the "Wetland Convention", wetland refers to natural or artificial, perennial or seasonal, static or flowing fresh water, semi-aqueous or aquatic swamp, peatland or water area, including water depth of not more than 6 m at low tide Sea area (Liu et al., 2020). Wetlands are closely related to the survival, reproduction and development of human beings. They are the most biologically diverse ecosystems in nature and one of the most important living environments for human beings (Parreño et al., 2021; Su et al., 2021). They are used in flood storage and drought prevention, climate regulation, soil erosion control, land siltation, and environmental degradation. Pollution and other aspects reflect important ecological functions. As a natural body in the transitional form between water and land, wetland has the characteristics of fragility and instability (Harrison, 2020). With the rapid development of society and economy, about 80% of the world's wetland resources have been lost or degraded, which has seriously affected the sustainable development of the ecology, economy and society of the wetland area. Therefore, understanding the structure and function of the wetland ecosystem is of great significance to the management of the wetland ecosystem.

Within a certain time and space, biodiversity is an important link that maintains the energy flow, material circulation, and information transmission of the ecosystem, and it is also the material basis for human survival. Among them, species diversity is the most important structural and functional unit of biological diversity (Fu et al., 2021). Plants are an important part of the wetland ecosystem, and they are also the main primary producers in the wetland ecosystem. In addition to energy supply, plants also have ecological functions such as environmental instructions, biological habitat, soil improvement, etc., and play an important supporting role in the maintenance of the wetland ecosystem structure and the performance of ecological service functions (Padhy et al., 2021; Wang et al., 2021). There is a complex interaction between wetland plant species diversity and wetland ecosystem services, which is also the key basis for studying ecosystem services. Understanding the species diversity of wetland vegetation helps reveal the response of wetland ecosystems to climate change and human activities, and is a key prerequisite for the development of wetland ecosystem management.

The north bank of the Han River is an important part of the wetland in the middle reaches of the Yangtze River (Zhao, 2021). Its topography and geomorphic conditions are complex, with sufficient rain and heat conditions, diverse local microclimates, rivers, reservoirs and other types of wetlands are widely distributed, and the wetlands are rich in biodiversity (Han et al., 2021; Yabe et al., 2021). However, there are few related studies at present, and the diversity of wetland vegetation is still unclear. The results of Zhang's research on plant diversity on the north bank of the Han River focused on the analysis of species composition and flora types, but did not provide in-depth analysis of community species diversity (Zhang et al., 2021). This study intends to combine literature data and wetland sample surveys to study the plant species, flora, typical communities and diversity characteristics of the wetland on the north bank of the Hanjiang River, in order to provide basic data for the protection and restoration of wetland ecosystems on the north bank of the Hanjiang River and similar areas and wetland resource management And scientific basis.

Materials and Methods

Chenggu County is located in the central area of the Hanzhong Basin in southern Shaanxi, between 107°03′~107°30′ east longitude and 32°45′~33°40′ north latitude. The county is 101 kilometers long from north to south and 42 kilometers wide from east to west. The highest elevation is 2602.2 meters, the lowest is 467.0 meters, and the total county area is 2265 square kilometers. The topography of the city is high in the north and low in the south. It has a subtropical humid monsoon climate, warm and humid, with four distinct seasons, with an average annual temperature of 14.3 ℃, and an annual rainfall of 800~900 mm. It is known as the "most suitable city in the Northwest".

Based on the results of the second national wetland resource survey on the north bank of the Hanjiang River in Chenggu County, this study conducted a more comprehensive survey of the plant communities in the wetland patches on the north bank of the Hanjiang River in Chenggu County using the sample survey method. In accordance with the "Main Contents, Methods and Technical Specifications of the Inventory of Plant Communities", typical plant communities were randomly selected in each wetland patch, and standard squares of trees 5 m×5 m, shrubs 2 m×2 m, herbaceous and aquatic 1 m×1 m were set up respectively(Correa-Araneda et al., 2021; Ohore et al., 2021). Set up 40 plots, and record the wetland type, geographic location, altitude, all species, number of plants, and coverage of the plant community where the plots are located.

Through on-site investigation of plant communities in wetland patches on the north bank of the Han River, site identification and identification, investigation and recording of plant species names, photographs and records of suspected species, and samples taken back to the laboratory for identification (D. Li, Chu, et al., 2021; Quanz et al., 2021). Combining previous research data, we summarized the list of wetland plants in Chenggu County, and carried out statistical analysis on plant families, genera and flora distribution types (Hong et al., 2021). Please refer to the classification system of Flora of China for statistics of plant families, genera and species, as well as the classification of species flora and geographical components (Niu et al., 2021). The importance value is used to represent the degree of dominance of each species in the wetland plant community (Li, Tian, et al., 2021; Li et al., 2021), and the calculation formula is:

Fig. 1 Survey sample area

$$\text{Important species value(ISV)} = \frac{\text{Relative coverage} + \text{Relative density}}{2}$$

$$\text{Relative coverage} = \frac{\text{Coverage of a species}}{\text{The sum of all kinds of coverage}}$$

$$\text{Relative density} = \frac{\text{The number of individuals of a species}}{\text{Individuals of all species}}$$

Patrick Index (R), Margalef index (m), Simpson index (D), Shannon-wiener index (H) and Pielou evenness index (E) were used to measure plant species diversity and evenness, the calculation formula for each indicator is as follows: The Patrick index uses the number of species to represent the degree of species diversity (Hu et al., 2021; Zhang et al., 2021), focusing on the total number of different species existing in the plot:

$$R=S$$

The Margalef index is a species richness index based on the ratio of the number of species in the sample to the total number of individuals:

$$M = \frac{S-1}{\ln N}$$

The Simpson dominance index is the sum of the ratio of the number of individuals of each species to the total number of individuals (H. Wang, Zhang, et al., 2021; Zhan et al., 2021), which comprehensively reflects the species richness of the sample community:

$$D = \sum_{i=1}^{S} \left(\frac{N_i}{N}\right)^2$$

Shannon-Wiener diversity index applies the calculation formula of entropy in information theory (Hong et al., 2021), and the result reflects the degree of uncertainty of the individual appearance of the species:

$$H = -\sum_{i=1}^{S} (P_i \ln P_i)$$

The Pielou evenness index is the ratio of the Shannon-Wiener diversity index to the total number of species (Wang, Xu, et al., 2021), reflecting the uniformity of the number of species in the plot:

$$E = \frac{H}{\ln S}$$

In the above formula, S is the total number of plant species in the plot; N is the total number of plant individuals in the plot; N_i is the number of individuals of the i-th species in the plot; P_i is the important value of the i-th species and the species in the plot the ratio of the sum of important values.

Result and Discussion

Results showed that there are 360 species of wetland plants on the north bank of the Hanjiang River in Chenggu County, belonging to 108 families and 264 genera, including Charaphyta, 1 family, 1 genus, 1 species, 10 families, 11 genera, and 12 species of ferns, 3 families, 3 genera, and 3 species of gymnosperms. Angiosperms have 344 species in 251 genera and 94 families. The number of families, genera and species of vascular plants accounted for 30.31%, 8.29% and 1.20% of the total number of vascular plants in China, respectively. Including 5 species of national key protected plants, 1 species of first-level key protected wild plants, *Ginkgo biloba*, and 4 species of cultivated species of second-level key protected wild plants, namely *Cinnamomum cam phora*, *Nelumbo nuci fera*, *Glycine soja*, and *Juglans regia.*

Fig. 2 shows that among the 107 families of vascular plants in the wetland on the north bank of the Hanjiang River in Chenggu County, there are 6 families containing 8 genera and above, and a total of 88 genera, accounting for 33.46% of all species. The number of families containing less than 8 genera is 101, accounting for 66.54% of the total number of species and genera. The families containing more than 8 genera are Asteraceae 24 genera, Poaceae 24 genera, Fabaceae 13 genera, Rosaceae 10 genera, Lamiaceae 9 genera, and Euphorbiaceae 8 genera.

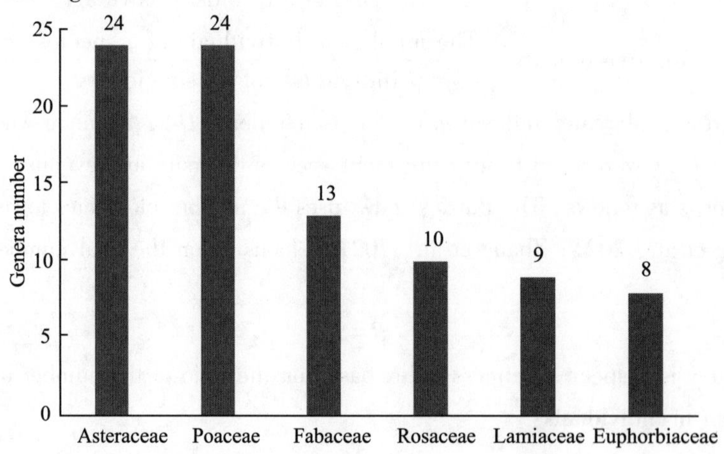

Fig.2 Number of dominant families and genera of wetland plants on the north bank of Hanjiang River in Chenggu County

Fig. 3 shows that in Chenggu County, the number of families containing 10 species or more of vascular plants in the wetland on the north bank of the Hanjiang River in Chenggu County is 8, including 147 species, accounting for 40.95% of the total number of species; the number of families containing less than 10 species There are 99, including 212 species, accounting for 59.05% of the total number of species; the number of families containing only one species is 60, accounting for 16.71% of the total number of species. Compositae, Gramineae, Rosaceae, Grandaceae, Leguminosae, Lamiaceae and Polygonaceae are families containing more than 10 species in order.

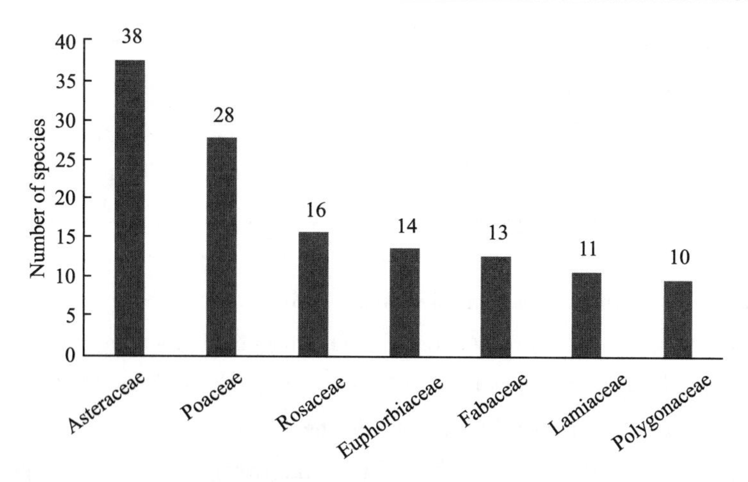

Fig. 3 Number of dominant family species of wetland plants on the north bank of Hanjiang River in Chenggu County

The total number of vascular plant genera in the wetland on the north bank of the Hanjiang River in Chenggu County is 263207 genera containing only one species, accounting for 78.41% of the total number of genera, and 57.50% of the total number of species; containing more than one species of genera There are 56 in total, accounting for 21.21% of the total number of genera, and the number of species accounting for 41.67% of the total number of species. Fig. 4 shows the top 8 genera with the largest number of species, namely Polygonum, Artemisia, Rubus, Euphorbia, Potamogeton, Cyperus, Sedum, Lysimachia.

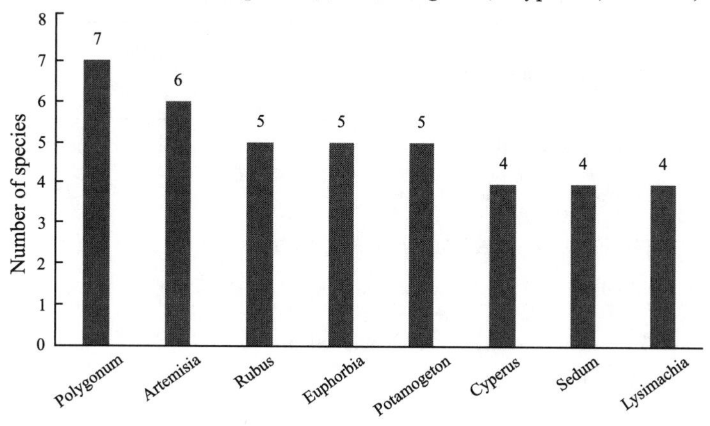

Fig. 4 Species of dominant genus of wetland plants on the north bank of Hanjiang River in Chenggu County

According to Flora of China, there are 107 families of vascular plants in the wetland on the north bank of the Hanjiang River in Chenggu County, which can be divided into 7 distribution types and 5 variants (Table 1). Among them, the world-wide type has 43 families, accounting for 40.19% of the total number of families; it contains a total of 164 genera, accounting for 62.36% of the total number of genera. The families with a large number of species are *Compositae*, *Gramineae*, *Rosaceae*, *Leguminosae*, *Lamiaceae*, *Asclepiaceae*, *Amaranthaceae*, *Cyperaceae*, *Ranunculaceae*, *Cruciferae*, *Rubiaceae*, *Umbelliferae*, *Crassulaceae*, *Primulaceae*, *Convolvulaceae*, *Pentoniaceae*, *Moraceae*, *Caryophyllaceae*, *Solanaceae*.

There are 27 families of tropical plants, accounting for 25.23% of the total number of families; 55 genera, accounting for 20.91% of the total number of genera. The families with a large number of species are *Euphorbiaceae*, *Urticaceae*, *Araceae*, *Vitis*, *Apocynaceae*, *Pteridaceae*, *Commelinaceae*, *Cucurbitaceae*, *Phyllanthus Sapindaceae*, *Malvaceae*. In addition, it also includes two tropical distribution variants, namely the tropical Asia-tropical Africa-tropical America distribution type and the pan-tropical distribution dominated by the southern hemisphere. The former includes *Iridaceae*, and the latter includes *Amaryllidaceae*, *Phytolacaceae*, and *Millinaceae*.

Both the old world tropical distribution type and the Chinese endemic distribution type only contain 1 family and 1 genus. The former includes *Musaceae* and the latter includes *Ginkgo biloba*. The proportions of both the total number of families and the total number of genera are 0.93% and 0.38%.

There are 8 families in the northern temperate zone, accounting for 7.48% of the total families; including 9 genera, accounting for 3.42% of the total genera. They are *Loniceraceae*, *Pinaceae*, *Stemonaceae*, *Liliaceae*, *Asparagaceae*, *Cannabisaceae*, *Hypericaceae*, and *Wufuhuaceae*. In addition, there are 3 variations of the distribution types in the northern temperate zone, namely the discontinuous distribution types in the north temperate zone and the south temperate zone, 9 families and 14 genera, accounting for 8.41% and 5.32% of the total number of families and genera; the discontinuous distribution types in the Eurasian and South American temperate zones 1 family and 1 genera, accounting for 0.93% and 0.38% of the total number of families and genera; Mediterranean, East Asia, New Zealand and Mexico-Chile discontinuous distribution type 1 family and 1 genera, accounting for 0.93% and 0.38 of the total number of families and genera %.

The number of families with discontinuous distribution types in East Asia and North America is 7, accounting for 6.54% of the total number of families; including 7 genera, accounting for 2.66% of the total number of genera. They are *Pteriaceae*, *Schisandraceae*, *Sauruaceae*, *Magnoliaceae*, *Acoraceae*, *Nelumboraceae*, and *Rootsaceae*.

Table 1 Typical statistical table of species and genera distribution of wetland seed on the north bank of Hanjiang River in Chenggu County

Regional character of distribution	Distribution type and its variant	Genera number	%	Number of species	%
World zone	1.world-wide distribution type	43	40.19	164	62.36
Tropical zone	2.Pantropical distribution type	27	25.23	55	20.91
	3.Tropical Asia-tropical Africa-tropical America distribution	1	0.93	1	0.38
	4. Pan-tropical distribution dominated by the southern hemisphere	3	2.8	4	1.52
	5.Discontinuous distribution in East Asia (tropical, subtropical) and tropical South America	5	4.67	5	1.9
	6.Old World Tropical Distribution	1	0.93	1	0.38
Temperate zone	7.North Temperate Distribution	8	7.48	9	3.42
	8. Intermittent distribution in the north temperate zone and the south temperate zone	9	8.41	14	5.32
	9.Temperate discontinuous distribution in Europe, Asia and South America	1	0.93	1	0.38
Paleo-mediterranean zone	10. Mediterranean, East Asia, New Zealand and Mexico-Chile intermittently distributed	1	0.93	1	0.38
	11.Discontinuous distribution in East Asia and North America	7	6.54	7	2.66
Specific zone	12.China-specific distribution	1	0.93	1	0.38
Sum		108		264	

The results of the species diversity index of 54 typical wetland plant communities on the north bank of the Hanjiang River in Chenggu County showed that they included Patrick index (R) , Shannon-Wiener diversity index (H) , Simpson index (D) , Pielou evenness index (E) and Margalef Index (M) . The Patrick index is between 9 and 30, with a mean of 18.80±5.19; the Shannon-Wiener diversity index is between 0.62 and 2.82, with a mean of 2.15±0.39; the Simpson index is between 0.22±0.92, with a mean of 0.81±0.10; Pielou the uniformity index is between 0.27 and 0.88, and the average is 0.75 ± 0.10; the Margalef index is between 1.72 and 5.06, and the average is 3.44 ± 0.87.

In Patrick ' s index, Bidens pilosa has the lowest diversity at 9, and taro community is the highest at 30; the Shannon-Wiener diversity index, Simpson index, and Pielou evenness index of the Pondweed community are the lowest, respectively at 0.616, 0.218, 0.267, the Shannon index and Simpson index of the Populus euphratica community were the highest at 2.817 and 0.925, respectively; the Pielou evenness index was the highest at 0.882; in the Margalef index, Flueggea virosa was the highest at 5.059, and Bidens pilosa was the lowest at 1.719. After being standardized, different wetland plant community diversity indexes have similarities in the changing trends of different community types.

In the wetland patch on the north bank of the Hanjiang River in Chenggu County, a total of 360 species of plants were investigated in 108 families, 264 genera, and charaphyta, including 1 family, 1 genera and 1 species, pteridophytes, 10 families, 11 genera, and 12 species, gymnosperms, 3 families, 3 genera, and 3 species. Angiosperms have 344 species in 251 genera and 94 families. The number of families, genera and species of vascular plants accounted for 30.31%, 8.29% and 1.20% of the total number of vascular plants in China, respectively. There are 8 families containing more than 10 species, including a total of 147 species, accounting for 40.95% of all wetland species on the north bank of the Han River in Chenggu County.

The 107 families of vascular plants in the wetland on the north bank of the Hanjiang River in Chenggu County can be divided into 7 distribution types and 5 variants. Among them, the world ' s widespread families account for 40.19% of the total number of families, accounting for the highest proportion. Followed by tropical distribution families and their variants, accounting for 34.28% of the total number of families. Among them, the pan-tropical distribution type subjects are the main ones, accounting for 25.23% of the total number of subjects. The temperate distribution type and the Chinese endemic type families are relatively few. The temperate distribution type families and their variants have a total of 26 families, accounting for 24.30% of the total number of families; the Chinese endemic distribution is only 1 family, 1 genus and 1 species in the Ginkgo family, accounting for the total number of families of 0.93%. The results show that the flora in this area has significant tropical characteristics, and at the same time has certain temperate characteristics. The north bank of Hanjiang River in Chenggu County is located in the northern subtropical region, with the subtropical continental monsoon climate as the main climate type. The characteristics of the wetland plants distributed in this area are consistent with the regional climate and environment at the family level.

Among 54 typical wetland plant community types on the north bank of Hanjiang River in Chenggu County, the average Patrick index is 18.80±5.19; the average Shannon-Wiener diversity index is 2.15±0.39; the average Simpson dominance index is 0.81±0.10; the average Pielou uniformity index is 0.75±0.10; the average value of Margalef index is 3.44±0.87. Ass.B.pilosa has the lowest Patrick and Margalef indexes; Ass.P.malaianus has the lowest Shannon diversity index, Simpson dominance index, and Pielou evenness index; Ass.C.esculenta has the highest Patrick index, and Ass.C.sebiferum The Pielou uniformity index of Ass.C.canadensis is the highest, the Shannon diversity index and Simpson dominance index of Ass. C.canadensis are the highest, and the Margalef index of ASS.F.virosa is the highest. The results show that the species diversity of aquatic plant communities or wet herb communities is generally low, while the wet shrub communities or arbor communities generally have higher species diversity. The reason may be that in the aquatic environment or wet environment with large changes in water level, flooding conditions have played a greater role in restricting and screening plant colonization, and herbaceous plants with shorter life cycles occupy most of the ecological niches. Among them, a few highly adaptable species have formed obvious advantages in the community. The community hierarchical structure is relatively simple and the species diversity is low. In wet shrub communities or tree communities, plants need sufficient water conditions, but are less affected by flooding(Fang et al., 2021). The community environment can support the colonization and survival of more species, and the wetland plant community can succeed to a higher level. In this stage, a richer community hierarchy and species diversity have been formed.

From the research results on the plant species, flora, typical communities and diversity characteristics of the wetland on the north bank of the Hanjiang River in Chenggu County, it can be seen that the wetland on the north bank of the Hanjiang River in Chenggu County is widely distributed and has more wetland plant species. In addition to the world's widespread type of wetland flora, the tropical distribution type occupies the largest part, which is consistent with the main climate types of the region(Chi et al., 2021). Wetland plant communities are rich in types, in which aquatic or wet herb communities are the main ones, and tree or shrub communities are few. However, compared with herb communities, arbor or shrub communities have more diverse species composition and usually have a higher species diversity index(C. Wang et al., 2021). In recent years, the wetland protection and restoration work implemented on the north bank of the Han River in Chenggu County has shown results, but there are still some problems, such as the invasion of alien species such as Eclipta prostrata and Eichhornia crassipes; water pollution and soil erosion caused by agricultural or industrial and mining production. And so on, it needs to be focused on in the future wetland protection and restoration work.

Acknowledgements

This work was financially supported by Shaanxi Provincial Land Engineering Construction Group fund (DJNY-2021-41).

References

Chi, Z., Wang, W., Li, H., Wu, H., & Yan, B. (2021). Soil organic matter and salinity as critical factors affecting the bacterial community and function of Phragmites australis dominated riparian and coastal wetlands. *Science of The Total Environment*, 10,1-7.

Correa-Araneda, F., Núñez, D., Díaz, M. E., Gómez-Capponi, F., Figueroa, R., Acuña, J., Boyero, L., & Esse, C. (2021). Comparison of sampling methods for benthic macroinvertebrates in forested wetlands. *Ecological Indicators*, 125,1-10.

Fang, J., Dong, J., Li, C., Chen, H., Wang, L., Lyu, T., He, H., & Liu, J. (2021). Response of microbial community composition and function to emergent plant rhizosphere of a constructed wetland in northern China. *Applied Soil Ecology*, 168,104-141.

Fu, X.-M., Tang, H.-Y., Liu, Y., Zhang, M.-Q., Jiang, S.-S., Yang, F., Li, X.-Y., & Wang, C.-Y. (2021). Resource status and protection strategies of mangroves in China. *Journal of Coastal Conservation*, 25,1-14.

Han, D., Kim, J., Choi, C., Han, H., Necesito, I. V., & Kim, H. S. (2021). Case study: On hydrological function improvement for an endemic plant habitat in Gangcheon wetland, Korea. *Ecological Engineering*, 160,443-454.

Hong, M. G., Nam, B. E., & Kim, J. G. (2021). Effects of microtopography and nutrients on biomass production and plant species diversity in experimental wetland communities. *Ecological Engineering*, 159, 10-25.

Hong, Z., Ding, S., Zhao, Q., Qiu, P., Chang, J., Peng, L., Wang, S., Hong, Y., & Liu, G.-J. (2021). Plant trait-environment trends and their conservation implications for riparian wetlands in the Yellow River. *Science of The Total Environment*, 767,323-337.

Hu, M., Le, Y., Sardans, J., & Peñuelas, J. (2021). Low-level saltwater intrusion alters soil diazotrophic community structure in a subtropical estuarine wetland. *Applied Soil Ecology*, 164,24-33.

Li, D., Tian, P., Luo, Y., Dong, B., Cui, Y., & Khan, S. (2021). Importance of stopping groundwater irrigation for balancing agriculture and wetland ecosystem. *Ecological Indicators*, 127,13-25.

Li, J., Chen, Q., Li, Q., Zhao, C., & Feng, Y. (2021). Influence of plants and environmental variables on the diversity of soil microbial communities in the Yellow River Delta Wetland, China. *Chemosphere*, 274,12-25.

Liu, L., Wang, H., & Yue, Q. (2020). China's coastal wetlands: Ecological challenges, restoration, and management suggestions. *Regional Studies in Marine Science*, 37,10-25.

Ohore, O. E., Zhang, S., Guo, S., Addo, F. G., Manirakiza, B., & Zhang, W. (2021). Ciprofloxacin increased abundance of antibiotic resistance genes and shaped microbial community in epiphytic biofilm on Vallisneria spiralis in mesocosmic wetland. *Bioresource Technology*, 323,124-136.

Parreño, M. A., Schmid, B., & Petchey, O. L. (2021). Comparative study of the most tested hypotheses on relationships between biodiversity, productivity, light and nutrients. *Basic and Applied Ecology*, 53, 175-190.

Su, X., Zhou, Y., & Li, Q. (2021). Designing Ecological Security Patterns Based on the Framework of Ecological Quality and Ecological Sensitivity: A Case Study of Jianghan Plain, China. *International Journal of Environmental Research and Public Health*, 18,83-92.

Wang, H., Xu, D., Han, J., Xu, R., & Han, D. (2021). Reshaped structure of microbial community within a subsurface flow constructed wetland response to the increased water temperature: Improving low-temperature performance by coupling of water-source heat pump. *Science of The Total Environment*, 781,334-352.

Wang, H., Zhang, M., Xue, J., Lv, Q., Yang, J., & Han, X. (2021). Performance and microbial response in a multi-stage constructed wetland microcosm co-treating acid mine drainage and domestic wastewater. *Journal of Environmental Chemical Engineering*, 9,10-26.

Yabe, K., Nakatani, N., & Yazaki, T. (2021). Cause of decade's stagnation of plant communities through16-years successional trajectory toward fens at a created wetland in northern Japan. *Global Ecology and Conservation*, 25,14-24.

Zhan, P., Liu, Y., Wang, H., Wang, C., Xia, M., Wang, N., Cui, W., Xiao, D., & Wang, H. (2021). Plant litter decomposition in wetlands is closely associated with phyllospheric fungi as revealed by microbial community dynamics and co-occurrence network. *Science of The Total Environment*, 753,142-165.

Zhang, K., Wang, J., Liu, X., Fu, X., Luo, H., Li, M., Jiang, B., Chen, J., Chen, W., Huang, B., Fan, L., Cheng, L., An, X., Chen, F., & Zhang, X. (2021). Methane emissions and methanogenic community investigation from constructed wetlands in Chengdu City. *Urban Climate*, 39,100-116.

Zhang, M., Xu, D., Bai, G., Cao, T., Liu, W., Hu, Z., Chen, D., Qiu, D., & Wu, Z. (2021). Changes of microbial community structure during the initial stage of biological clogging in horizontal subsurface flow constructed wetlands. *Bioresource Technology, in arid areas. ATENA*, 199,105-113.

本文曾发表于2022年《Bangladesh J.Bot.》第51卷第4期

4 房地产及建筑工程

基于产城融合视域下对城市开发运营商产业发展模式的思考

秦 悦

(陕西省土地工程建设集团,陕西 西安 710075)

【摘要】在党的十九大之后,我国针对区域协调发展和产业升级提出了全新的标准及要求,要求须在城市发展和产业升级上,和新型城镇化建设以及产业转移的需求互相适配。自此,各级地方政府将城市的综合开发和资源的全面运用提高到了进行城市化建设的战略发展高度,并提出需将产城融合作为重点,作为后续进行城镇化发展的重要形式。由此,城市开发模式已经不再简单地局限在城市建筑物的开发和运用上,逐步演变为推动产业发展的全新诉求。在这一趋势的影响之下,城市综合开发领域中的重要主体——开发商和房地产企业,也开始加强对产业发展的关注和重视。基于此,文章站在产城融合背景之下,针对城市开发运营商的产业发展作出相应的探讨和研究,并提出产业发展的全新模式,以供参考。

【关键词】产城融合;城市开发;开发运营;产业发展

引言

伴随着近些年我国城市化建设政策的不断转变以及市场需求的不断升级,我国城市化建设速度日渐加快,由此也带来了全新一轮的城市开发浪潮。再加上人们对美好生活需求的日渐提升,在我国经济发展迈入到全新常态的刺激和影响下,新型城镇化建设已经成为未来经济发展的重要引擎之一,城市的综合开发也面临着时代为其提供的全新发展机遇,但其中带来的挑战和压力也不容忽视。传统的房地产开发商需要积极地做出改变,将重点放在城市综合开发和产业发展等重点内容上。因此,地方政府针对城市综合开发商提出了较高的标准和要求,如何适应新型城镇化建设的发展常态,直击政府在城市综合开发上的痛点,需要对企业综合开发商业模式展开更为深入的探讨和研究,并提出配套新发展的新的应对模式,以推动我国城市发展迈入到全新的发展之路。

一、产城融合为城市开发运营商产业发展模式提供机遇

(一)业务空间拓展

基于产城融合这一背景之下,产业运营将成为传统城市资源开发的主旋律,从单一粗放式的运营发展模式转变为以下三种模式:一是政府主导+企业运营模式。目前在我国现代化城市的发展过程中,比较具有代表性的如郑州市所推出的华夏幸福地产产业园,2016年华夏幸福地产从我国的一线城市经济地区逐步向其他城市进行拓展,将重点放在了以郑州作为核心的中原地区。该模式主要是将地方政府作为主导,华夏幸福地产进行统一投资、开发、运营,最后政府以财政税收作为补贴等,降低企业在土地资源开发上的成本投入,形成了更为现代化的政府主导、企业运作、商业经营开发管理理念。相较于传统的房地产行业单一开发收入,政府在其中制定的土地优惠,帮助企业降低了在后期运营上的资源投入,将传统的城市开发短期盈利转变为长期盈利。二是企业主导+企业运营+政府支持模式。目前在我国各城市的发展过程中,比较具有代表性的是恒大在全国各地的文旅发展模式,主要是由恒大集团作为主导,和各地政府部门形成合作,以寻求在产业合作上的全新发展之路,换取在土地资源上的优惠,以吸引来自本土和外地的人口到此地进行旅游,进而对住宅进行销售,以通过政府部门的现金流,对后期运营进行补贴,创造出了更为全方位的生产、生活配套空间,除了推动各地区在传统房地产行业上的转型速度得到加快,对于本地旅游产业的升级、带动城市成为旅游城市方面也发挥了一定作用。三是企业主导+产业巨头运营+政府支持模式。目前比较具备代表性的如各地区所建立的科技小镇,房地产开发企业和本地产业巨头进行合作,形成长期战略合作的模式,最后共同

寻求本地政府对经营的支持,并由当地的商业巨头来进行运营,以弥补房企在产业运营上专业度的不足之处。例如可通过与中关村、华为公司的联手,形成房地产和科技领域的跨界合作,依托于云平台、人工智能技术等现代化科学技术的运用,吸引人才,聚焦产城融合,建设为以科技小镇为核心的产业园区。这一商业运行开发模式除了成为房地产行业发展的全新利润增长点,未来也会掀起一场全新的人居革命。

(二)政策环境利好

目前我国在推进产城融合发展的过程中,各地政府部门也在持续颁布相关落实和支持政策,为传统房地产行业的转型和城市开发运营商向产业发展转型提供了全新的道路,同时也提供了更为完善的外部环境,克服了企业在产业运营上存在的能力与资金等问题。一方面,可通过政策的颁布和落实,弥补企业在运营能力上存在的短板,将税收补贴和土地使用排他性作为主导,为城市开发运营商产业发展提供有力支持。另一方面,我国通过战略规划的持续推进,补齐企业在创新能力上存在的短板。尤其是伴随着国家各类战略规划平台在各地的持续推进,也可为传统房地产行业的转型和升级提供全新的发展之路,例如高新技术产业聚集区的持续建设以及经济技术开发区等聚集性板块的高速发展,再加上各类周边技术设施配套和相关交通枢纽功能的不断完善,吸引大批量的高端型产业人才、创新人才聚集在此,为城市开发、传统行业的转型和升级注入活力,开辟全新的发展空间。

(三)产业化有效赋能

通过持续推进产城融合,带动城市产业升级,城市发展也逐步走向智能化的发展之路,信息实现全面覆盖,解决了传统城市开发经营、房地产行业转型过程中产业化不足和信息化缺失的问题。尤其是伴随着产城融合的日渐深入,带动城市人工智能领域、大数据领域在城市中获得全面推广,为产业发展提供更多有价值的信息,掌握市场需求,降低由于信息孤岛问题所带来的发展风险。不管是前期的调研还是后期的运行,包括在资本金融运行等各个环节,都为其提供信息支持,以确保充分掌控市场变化。

二、基于产城融合视域下的城市开发运营商产业发展模式

综合城市的不同资源条件和产业经济发展水平,可使用到的城市综合开发运营模式有所不同,主要包括六种模式:

一是企业综合自身在城市开发建设上的能力,展开城市开发的一级开发和二级开发,获得的收益主要为一级开发和房地产所带来的销售效益,在这一发展模式之下,整体的发展能力相对较弱。

二是企业进行产业地产升级,加强产业园区的开发和建设,并以产业招商出售为主,实现对资金的全面回笼。其获得的收益包括在产业园区建设上的投资建设收益和地产资源出售所带来的收益。在这一模式之下,产业发展能力主要包括两类,分别为产业的招商以及后续产业园区规划能力,可以为后续的项目投资打下有利的基础。而产业招商能力主要是指在产业园区建设之后,资源销售的具体能力。

三是企业自持部分物业,通过物业管理来获得收益,包括项目建设所带来的收益、土地资源出售所带来的收益,还包括物业管理所带来的物业收益。在这一发展模式之下,产业发展能力涉及三方面,分别为产业的招商、研究规划和物业管理。在产业招商上,主要包括两方面,分别为产业的出租和出售。

四是企业直接参与到资产管理。除涉及物业管理外,还可以提供一定的地产配套服务,促使地产资源的价值得到增值、保值、升值。在获得的收益上,涉及的内容更多,除投资、建设、出售和物业管理外,还包括增值服务收益。在这一发展模式之下,产业的发展能力可通过产业进行招商、资产管理、研究规划。产业招商不再简单地局限在出售和出租上,还包括在产业转型升级过程中融入产业链中的优质企业。

五是企业参与到产业投资领域中去。通过更为成熟、已有经验的产业运营发展模式,进一步降低产业投资风险投入。与此同时,通过产业投资,也可以为当前项目的资源扩展和后续的产业招商奠定有利基础。涉及的收益除出租出售、投资建设、物业管理、增值服务外,还加入了产业投资收益。在这一模式中,产业能力涉及内容较多,除基础的研究规划、招商、资产管理之外,还包括产业投资。值得一提的是,这一模式中,随着园区发展到成熟阶段,企业在产业园区的入驻率上可望高达90%,若是有企业从园区中退出,依然会吸引大批量的新企业申请入驻。此时整个项目的管理团队和招商团队,只需要从这些申请的企业中选择更加优质的企业补充入驻。

六是上级企业以其强大的投资网络和顶尖的资本运作水平,为城市综合开发运营商的产业发展提供强有力的支持,以保障在产业发展上形成一体化发展模式。这种能力较强的开发运营商,是通过其背后的强大行业地位和资本力量,创造出城市发展的全新平台,形成在城市产业转型升级上的重要一环。其丰富和深厚的产业资源,对于产业招商和产业出售都有着非常重要的现实作用,更加关注产城融合发展过程中的资源配置水平。

不同的产业发展模式及其带来的产业资源也有所不同。在第一种和第二种发展模式中,产业资源多为自有资源,是推动房地产企业融入产业发展、走向产业地产转型之路的门槛。第三种模式除了需要围绕已有资源,实现对产业园的发展,同时还需要在产业园的发展过程中加强核心产业的培育,将产业链上下游产业链资源进行全面积累,为后续的资源开发提供保障。第四种产业发展模式除已有的产业资源外,还包括城市公共基础配套,产业多元化,如周边医疗配套、教育配套和商业配套等,还包括各类市政基础设施配套资源。第五、六种发展模式则扩充了产业发展的生活、生态和生产水平,实现了产业资源库的综合建设,既包括其中城市发展的核心产业,又包括产业发展过程中的上下游资源,有着更为丰富的教育资源、医疗资源和城市商业配套资源,可为人们提供更为完善的生活配套服务。

三、基于产城融合视域下对城市开发运营商产业发展模式保障

(一)企业层面

对于企业来说,需要加入本土产业发展的战略合作联盟中去,为传统企业的转型和升级,提供坚实有力的资源保障、人力资源保障。同时以企业合作平台为载体,优化企业间的合作水平,以推动资源的整合。在我国各城市持续推进产城融合的社会背景下,已经催生出了大量的协同空间组织,实现了各类资源的有机整合。因此,企业在转型升级时,需要发挥出其主观能动性,在产业链中积极寻找上中下游的合作节点,吸收其中的优质技术、人力资源和资金资源等,以对自身进行发展。同时,对这些资源进行合理运用,创造出全新的发展价值。除此之外,企业还需要在资源整合上进行不断的创新,强化对环境的适应水平。面对竞争越来越激烈的市场,需形成合作共赢、适当竞争的发展理念,在全产业链中寻求合作先机,构建企业战略发展联盟,确保产业和企业之间可形成优势互补。例如,可以通过建设大型的商业生态系统,创建出将分享和共享作为载体的商家联盟、投资联盟及合作联盟。

(二)政府层面

政府作为产城融合的重要推进者,需加强对传统产业转型和升级的政策指导,引导产业、传统的房地产企业,制定出符合自身发展需求的全新规划,有效规避产业发展的非特色和同质化。例如可以通过定期组织专业论坛,对传统企业转型和产业发展过程中的需求问题进行答疑解惑,并邀请其他城市的行业专家开设论坛,对其予以指导。还可通过制定出针对性的激励措施,对于符合产城融合需求的城市开发运营商,予以一定的奖励,以税收政策、产业发展政策、节约土地政策为代表,特别要勇于创新和突破,多采用"一事一议""一企一策"专项定制方案的形式,鼓励、吸引、支持更多更强大的企业参与到城市开发产业运行中去,为城市的发展、区域的协调、人们的美好生活贡献坚实的力量。

结论

综上所述,城市的综合开发运营为未来我国城市化建设过程中的必然发展趋势,因此企业需对其他城市的成功经验进行积极借鉴,同时自我积极开拓创新,综合自身在业务和产业以及资源上的核心优势,选择科学合理的发展模式,并综合城市的发展定位与需求,对其产业资源进行有机匹配,形成更为完善的产业发展能力,提升企业在产城融合发展过程中的核心竞争力。而整个过程中,除需要企业

做出努力外,更需要政府下足功夫,为企业提供更为完善的外部环境,给予更强力度、更合理的政策保障,以推动我国城市开发运营商产业发展模式走向更现代化的发展之路,为城市经济的提升打下有利基础。

参考文献

[1]龙海波.中小城市新区产城融合发展研究:以邯郸冀南新区为例[J].开发研究,2021(6):51-57.

[2]罗达,李长江,徐航,刘家财.产城融合视角下工业园建设对周边住房价格影响[J].中国储运,2022(3):114-115.

[3]王沛,倪敏东,王辉."十四五"时期鄞州区城市开发的若干思考[J].宁波经济(三江论坛),2022(2):26-29.

[4]刘康平. 产城融合战略背景下皇庆湖公司科技小镇建设项目研究[D].兰州:兰州理工大学,2021.

[5]何媛媛. 产城融合背景下郑州市传统房地产企业转型问题研究[D].郑州:郑州大学,2020.

[6]刘宇.产城融合区域综合开发思路与实现路径分析[J].住宅与房地产,2020(5):4.

本文曾发表于2022年《科学与技术》

Based on the CT Image Rebuilding the Micromechanics Hierarchical Model of Concrete

Guangyu Lei [1,2,,3], Jichang Han[1,2,3]

(1.Shaanxi Land Engineering Construction group, xi'an 710048, China; 2.Institute of Land Engineering and Technology, Shaanxi Provincial Land Engineering Construction Group Co. Ltd, xi'an 710048, China; 3.Key Laboratory of Degraded and Unused Land Consolidation Engineering, the Ministry of Natural Resources, Xi'an 710048, China)

【Abstract】 Establishing a mesoscopic numerical model to investigate the mechanical properties of concrete has very important significance. This paper considers that the random distribution of aggregate in concrete. Aggregate assumed to be spherical, respectively to simulate the interface layer as the entity unit or the contact elements. The random aggregate model and the interface model of random aggregate were established. Based on the CT image, the application of MATLAB and MIMICS software, the different characteristics of concrete model of 3D reconstruction was set up. Comparative analysis the advantages and disadvantages of different models, consider the CT number included in the CT images, this paper establishes the reconstruction model which include the shape of concrete aggregates, gradation, holes, etc. The analysis results have shown that the model can infer realistic concrete behavior, providing a new approach for studying concrete properties at the mesoscale.

【Keywords】 Meso concrete; Numerical model; Random aggregate model; Interface model of random aggregate; CT image; Reconstruction model

1 Introduction

Concrete is a composite material consisting of water, cement, and coarse aggregates. The special property of the composite material determines its macroscopic mechanical properties[Qu et al.,2022]. The destruction process involves the interaction of various scales from the microscopic to the macroscopic[Afroz et al.,2022; Nguyen et al.,2022]. Currently, the mesoscale is considered the linking bridge between the microscopic and macroscopic scales. The mesomechanical numerical method, based on the finite element method, uses the concrete micromechanical model to study its failure process and obtain the intrinsic relationship between the mesoscopic failure and macro-destruction, thus investigating the failure mechanism of concrete materials [Carol et al., 2001; Dupray et al., 2009; Kim et al., 2011; Shahbeyk et al., 2011; Du et al., 2017; Wang et al.,2021].

With the improvement of theoretical methods and computer hardware components, graphics processing software and meshing tools have been continuously upgraded, resulting in easier simulation of concrete mesostructures. Nowadays, using numerical methods to investigate the mechanical properties of concrete has received extensive attention from researchers. Indeed, numerical models' development has an important influence on the simulation results. Among the most representative models in the literature are the random particle model [Schutter et al.,1993], lattice model [Schlangen et al., 1997; Leite et al., 2004], stochastic mechanical property model [Tang et al., 2003], and statistical damage constitutive model [Mohamed et al., 1999; Bai et al., 2010; Ma et al., 2012].

The mesomechanical method is typically utilized to analyze the concrete characteristics by numerical experiments. Under the condition that the calculation model is reasonable and the material parameters of each concrete phase are sufficiently accurate, some tests can be realistically replicated to overcome the objective limitations of the test conditions and the human factors[Zhang et al.,2021]. This has greatly promoted the study of the mechanical properties of concrete [Du et al., 2012]. Nevertheless, previous research has outlined some limitations and drawbacks in the above models. At present, digital image processing technology is becoming significantly important for developing numerical models. Recently concrete computed tomography (CT) scan image reconstruction model has become a hotspot and frontier of research on the numerical simulation of construction materials [Gopalakrishnan et al., 2006; Dai, 2011; Onifade et al., 2013; Skarżyński et al., 2015].

The real shape and location of aggregate particles in concrete were assumed in numerical analyses based on X-ray CT scans. Trawiński et al. [2018] found that satisfactory agreement regarding the vertical force versus crack mouth opening displacement evolution and crack geometry was achieved between analyses and laboratory tests. Two-dimensional mesoscale finite element models with realistic aggregates, cement paste, and concrete voids are developed using microscale X-ray CT images. Cohesive elements with traction-separation laws are pre-embedded within cement paste and aggregate-cement interfaces. Ren et al. [2015] simulated complex nonlinear fracture and tension tests using a large number of images. Qin et al. [2012] proposed an objective judgment criterion of unit attribute recognition based on the method of mesh mapping and established the 3D finite element model with the same aggregate proportion as the specimen in the experiment.

This study combines the literature sources' achievements over the years to develop different microscopic concrete numerical models and analyze their respective deficiencies. After that, it establishes a concrete reconstruction model based on CT images. The model considers the concrete mesostructure to simulate its behavior realistically. This study is expected to provide a reference for future investigations of the mesoconcrete's mechanical properties.

2 Mesoscopic Concrete Stochastic Model

Generally, the concrete random model is used to replicate testing samples numerically using the knowledge of the random sampling method and statistics along with the actual mixture proportions. The advantage of this approach is the ability to simulate the influence of different factors on concrete behavior. At present, the method is considered mature and independent of the data size, and its development is relatively perfect. Perform calculations on different types of specimens.

2.1 Random aggregate model

The random aggregate model is a three-phase heterogeneous composite consisting of aggregates, mortar, and a bond zone between the two. Previously, the Monte Carlo method was used to generate random numbers [Metroplis et al., 1949]. Based on the self-programmed model generation procedure, the projected grid method is used to assign the corresponding material properties according to the unit type. Due to the involvement of different material parameters in each phase, the load-deformation relationship of the concrete specimen is nonlinear. It can be used to simulate the specimen's crack propagation process and damage patterns [Zheng, 2005].

This approach to determining the position of aggregates with different particle sizes in the sample involves calculating the number of aggregates using the actual mix ratio of concrete and assuming spherical shape aggregates with four variables each. The diameter of the sphere is used to distinguish the size of the aggregate, and the position of the sphere center (XYZ) is used to determine the location of the aggregate. According to the self-programming procedure, the spherical center coordinates are generated randomly using the Monte Carlo method. Generally, different spheres meet objective conditions by setting constraints and adopting a cyclical comparison method. A 3D random distribution geometric model of concrete aggregates can be obtained by compiling the generated random variables into corresponding FORTRAN programs and reading them into ANSYS in the command flow mode.

Once the concrete geometric model is generated, the aggregate, mortar, and interface are given corresponding parameters. Moreover, the material is meshed using the finite element method to obtain the 3D random distribution model of concrete aggregates. The overall model diagram is shown in Fig. 1, and the section diagram is shown in Fig. 2.

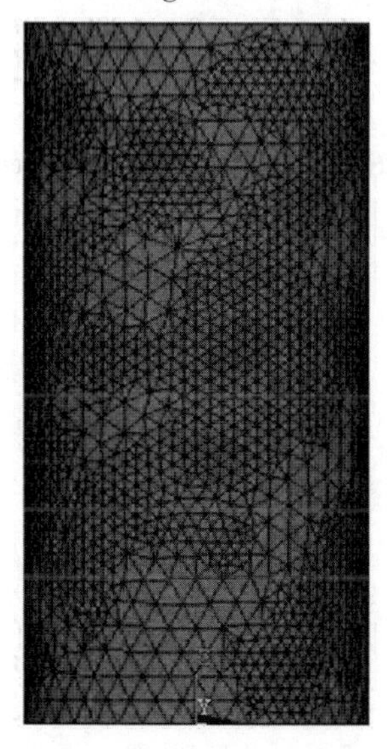
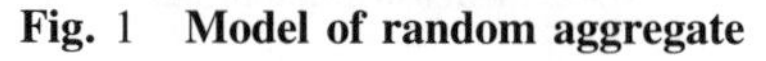

Fig. 1 Model of random aggregate

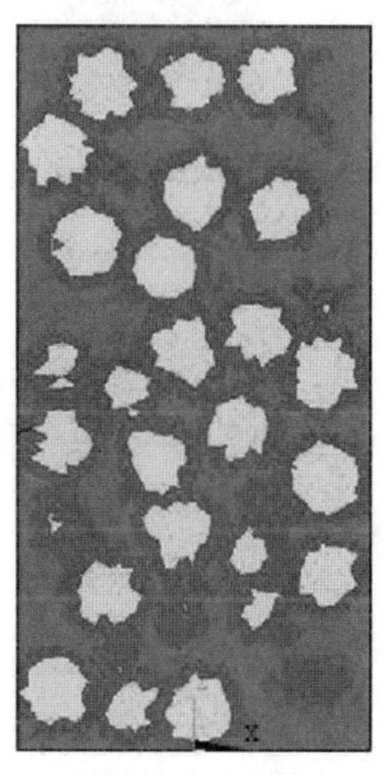

Fig. 2 Section of random aggregate

2.2 Random aggregate contact surface model

The concrete's random aggregate contact surface model is a two-phase material that consists of mortar and aggregate and uses an ANSYS contact surface unit to simulate their interface joint. This method starts by establishing a concrete cylinder body in ANSYS, then randomly generating aggregate spheres in the entity using the self-programming procedure to separate the cylinder and aggregate spheres for the BOOLEANS calculations. The position of the aggregate formed at this time is a hollow concrete specimen. Use the program again to read the position coordinates of the aggregate and generate its entity. The overall mesh is typically selected using the SOLID45 element. Besides, the aggregate projection grid method is used to distinguish the fundamental material properties. Moreover, the contact elements between the aggregate and the mortar are defined using the TARGE170 and CONTA174 unit types [Liang et al., 2011]. The model contact element is shown in Fig. 3, and the cross-section of the random aggregate contact surface model is shown in Fig. 4.

Compared with the random aggregate model, this technique uses a contact element instead of an interface element, resulting in considerably lower computational time and cost. The results of the two calculations are shown in Fig. 5. It can be seen that their destruction rules are the same, indicating that the random aggregate contact surface model can be a good substitute for the random aggregate model.

Fig. 3 Contact element of model **Fig. 4 Section of the contact element**

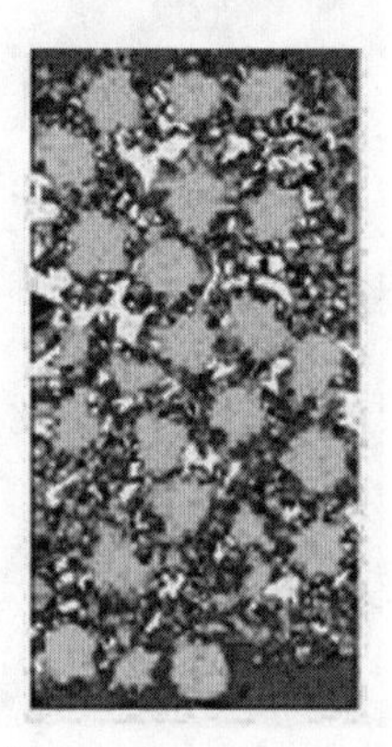

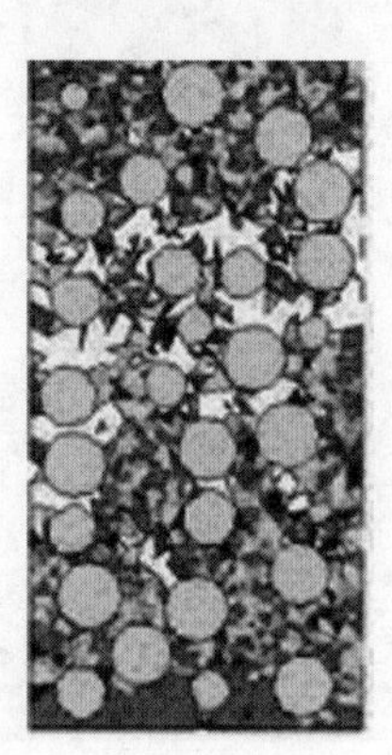

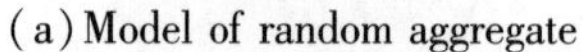

(a) Model of random aggregate (b) Interface model of random aggregate

Fig. 5 Different damaged sections of the model

3 Mechanical Model Based on Three-dimensional Reconstruction of Concrete Images

Due to the difference between the random model and the actual concrete internal structure, the research and analysis cannot lead to a complete correspondence with the test. Therefore, studying the mechanical model based on concrete images for 3D reconstruction has gradually gained attention and progress. The 3D reconstruction model is a model that is closer to the mesostructure of the actual concrete and is an important foundation for future research on the mechanical properties of meso-concrete.

3.1 Three-dimensional reconstruction model of concrete based on the joint unit method

Unlike previously explained techniques, the three-dimensional reconstruction model is closer to the real mesostructure of concrete and can be compared with the physical experiment. The 3D finite element model of concrete based on the joint-element method abandons the modeling order of points, lines, surfaces, and bodies by discretely separating the concrete into many nodes in the space, which are then connected to generate the unit. Then the material properties of the specimen are determined from the elements containing the node attributes, thereby establishing a 3D finite element model of the concrete mesostructure. The modeling procedure in this method is described as follows:

(1) Each CT image is processed to save its CT number and spatial position as a cross-section file.

(2) The image quality is enhanced using the threshold segmentation method to distinguish the aggregate from the mortar.

(3) The image information is saved as a two-dimensional matrix using MATLAB.

(4) The pixel attribute value matrix in MATLAB is read into the ANSYS array using a self-programming

code to generate a model.

The model is shown in Fig. 6.

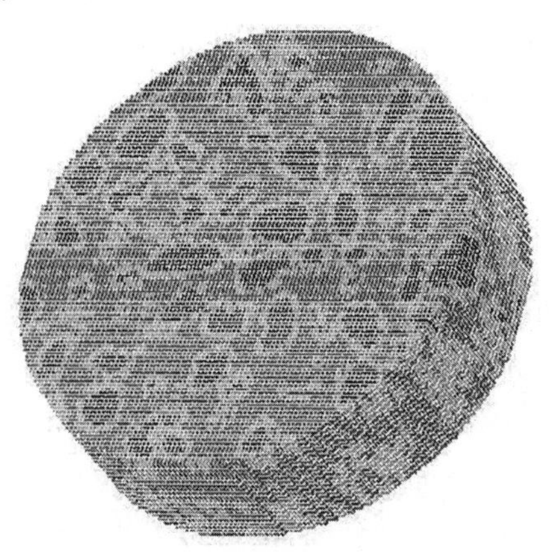

Fig. 6 The specimen FEM elements

On the other hand, this method cannot yet rebuild a complete concrete test piece due to using many elements. In order to reduce the amount of data, the influence of holes, small aggregate particles, etc., is not taken into account in the reconstruction process, resulting in relatively low simulation accuracy.

3.2 Three-dimensional reconstruction model of CT images based on MIMICS software

Some large-scale image creation and editing software, such as MIMICS has a 3D entity reconstruction function. MIMICS software can directly read Dicom format CT scan images through the following procedure:

(1) Interpolate the picture using image positioning.

(2) Set different thresholds for different materials.

(3) Modify the pixels of the image by different pixel modification methods.

(4) Once each layer of the image is processed to remove redundant data, a 3D geometric concrete model can be established upon the calculation.

(5) Use ANSYS software to read the list file of the 3D concrete built in the MIMICS software and generate the finite element model directly.

The overall model is shown in Fig. 8, and the profile is shown in Fig. 7.

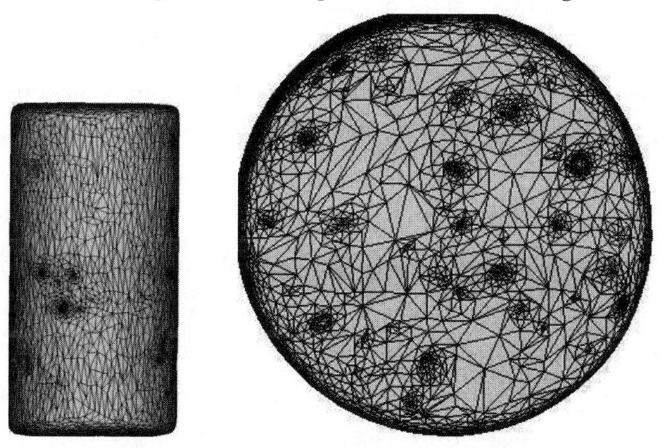

Fig. 7 3D-FEM elements of concrete

Based on the concrete reconstruction model obtained from the above CT images, a complete calculation cannot be performed at present due to the model's large number of elements and relatively low accuracy.

4 Key Issues and Technical Approaches to Concrete Micromechanical Models

The numerical model of mesoconcrete is the basis for studying the material's mechanical properties.

The diversity in the concrete's constituent materials and the randomness in the preparation process increase the complexity and non-uniformity of the model compared to other materials. Therefore, there are many key problems in the numerical model of the mesoconcrete. At present, this method cannot perform one-to-one comparative analysis with physical tests because the simulated aggregate shape and position are different from that of real concrete specimens, and the model rarely considers the influence of holes in the specimen. The 3D reconstruction model of concrete relies on CT images. Based on the digital image processing method and large-scale image processing software for finite element modeling, the mesostructure of the real specimen is well simulated. The advantage of this method is that according to the actual concrete shape, a consistent finite element model can be generated to perform concrete mechanical tests better. Accordingly, the development of the concrete reconstruction models is the direction of future concrete properties investigations. This research topic is still facing some deficiencies. For instance, image information data compression results in large data that causes prevent the completeness of a full model computation. Moreover, the model's accuracy is low due to the unit selection of feature points. Besides, the model cannot still consider the impact of small aggregates, holes, and other factors during the reconstruction process. These factors have an important influence on the mechanical properties of concrete. Combining the above problems with the concrete random model and concrete reconstruction model, the common limitation of the two is that the influence of small volume components such as holes and fine aggregates has not been taken into account, along with the influence of different aggregate parameters. These are problems that cannot be ignored in the study of concrete mechanics. Accordingly, a concrete mesoscopic model that reflects actual conditions is the basis for investigating concrete mechanical properties and obtaining reasonable and correct results. Therefore, based on the problems existing in the above models, starting from concrete CT images, reconstructing the concrete mechanical model, taking into account the influence of different aggregate parameters and concrete holes, reflecting these characteristics in the reconstruction model is the trend to study the numerical calculation model of the microscopic concrete.

5 Microscopic Concrete Reconstruction Model Based on CT Number

The CT machine scans the concrete sample layer by layer to obtain CT slices of any scanned surface inside the concrete. Each slice contains the material information and density of the layer. Finally, the material properties can be represented by the CT values of each pixel (Lei et al., 2016).

5.1 Image processing

In a single test specimen, the number of CT digital regions and the number of scanning CT slices varies due to the difference in the resolution of the CT machine and the scanning thickness. Before processing an image, it is first necessary to determine the resolution of the image and the geometric location of the CT slice in the concrete specimen and then select the CT slice that needs to be reconstructed. At the same time, since the CT machine scans the concrete specimens and the parts around the specimens, CT slice images also contain this information. The image is first divided into study areas, and the areas that need to be investigated for reconstruction are selected. A raw CT slice map is shown in Fig. 8. The Figure contains the scanned parts of the test specimen and the information on the surrounding blank parts. In the analysis, if the CT image is not processed, it will cause errors when analyzing the image data. Therefore, the image study area is extracted, as shown in Fig. 9.

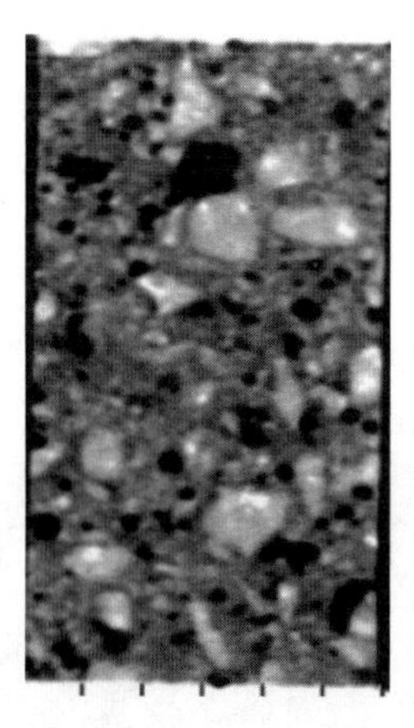

Fig. 8 Initial graph of the CT section

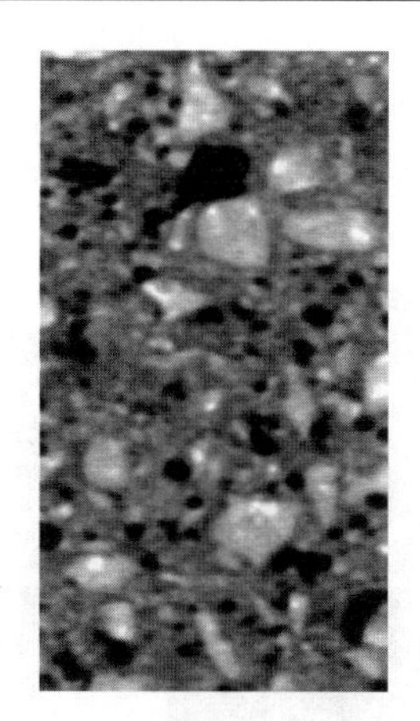

Fig. 9 After extraction of the CT section

Once the extraction of the CT slice image is completed, the size of the image should be determined to obtain the pixel matrix of the image. Then use the ENVI software to extract the CT value at each pixel. At this point, the coordinate value at the pixel point and the CT value is in a one-to-one correspondence and represent the material information at this position. For the convenience of calculation and statistics, CT values are normalized. Thereafter, the position information and material quantity are determined. Finally, the data, including the position coordinates of the aggregate, mortar, and holes and the size of the CT value, are saved.

5.2 Model reconstruction

The ABAQUS software generated a two-dimensional geometric model with the same dimensions as the CT slices. The two-dimensional plane geometry model was meshed. Each unit represented a pixel, and the ".inp" file was proposed once the partition was completed. According to the Fortran language, the corresponding program is compiled, and the node information in the ".inp" file is replaced with the information of the node coordinates of the aggregate, mortar, and hole in the CT slice, and the file is saved and read again. At this point, the information of the model's elements and nodes are the aggregates, mortar, and hole information. In this study, a new unit was added between the aggregate and the mortar, considered a transitional layer, giving it corresponding material properties. The reconstruction of the model is complete. Each material section is shown in Fig. 10, and the whole is shown in Fig. 11.

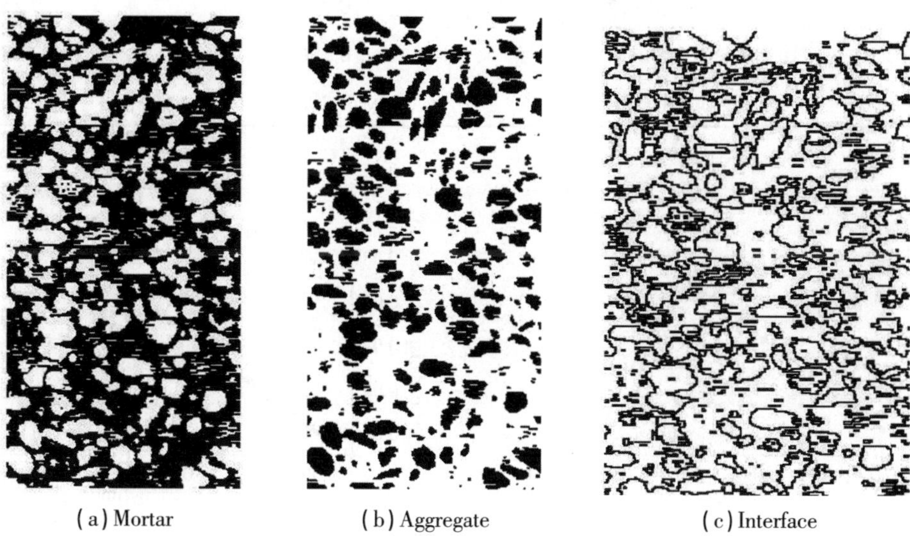

(a) Mortar (b) Aggregate (c) Interface

Fig. 10 Profile of every material in the CT section after reconstruction

After selecting the image and processing the data, the CT image is finally converted into a numerical model. It can be seen from the partial enlargement of Fig. 12 that the reconstructed model contains aggregates, interfaces, mortar, and holes at all levels.

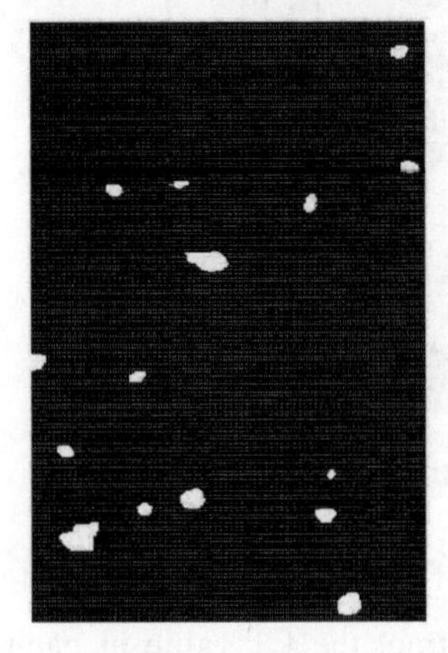

Fig. 11 Whole reconstruction model

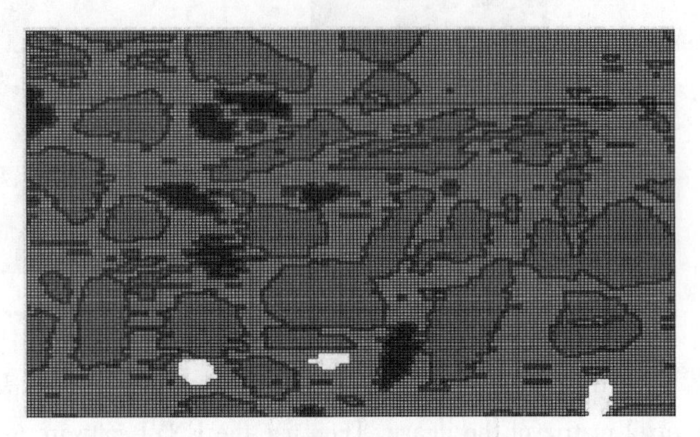

Fig. 12 Reconstruction partial enlarged

This model integrates the characteristics of the random models and CT reconstruction techniques. The interface influence in the random model and the aggregate shape and gradation effects in the CT reconstruction model were taken into account, and the influence of holes that both models did not consider was increased.

In order to verify the rationality of the model, a mechanical analysis was performed under pressure and tensile loads. This paper used the plastic damage model of concrete. The damage factor D is calculated separately according to the compression and tension conditions. The mechanical parameters of each material are shown in Table 1. This paper applied dynamic compression and tension forces using a displacement-controlled protocol. The loading curve is shown in Fig. 13 and Fig. 14. The dynamic compression loading time was 0.1 s, and the loading displacement was 1 mm. Besides, the loading time for the dynamic tensile case was 0.05 s, and the loading displacement was 0.12 mm.

Table 1 Material parameters

Material	Elastic modulus / Pa	Poisson's ratio	tensile strength /Pa
Aggregate	5.8731×10^{10}	0.2407	9.25e6
Mortar	1.7458×10^{10}	0.1960	2.78e6
Interface	1.3967×10^{10}	0.2000	1.56e6

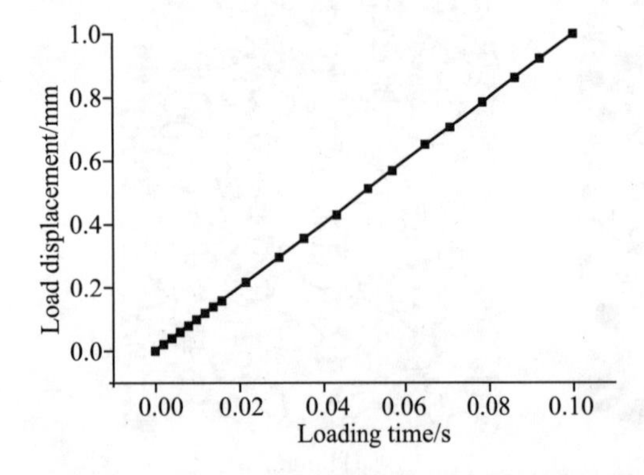

Fig. 13 The load curve of dynamic pressure load

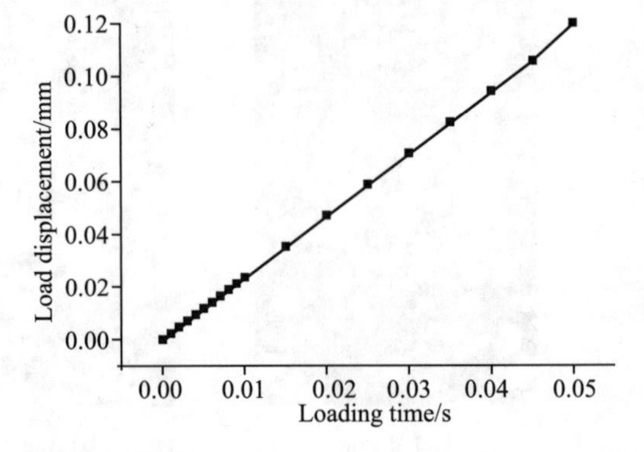

Fig. 14 The load curve of dynamic tensile load

5.3 Pressure load

The dynamic pressure load is applied to the model top, and the bottom surface of the model is restrained. The computation results are shown in Fig. 15 and Fig. 16. In the initial loading stage, when the specimen is not damaged, the displacement of the material changes regularly, where the top displacement is the largest, the bottom is the smallest, and the layering changes sequentially. When the specimen is destroyed, the displacement varies locally. Under the dynamic pressure load, the damage starts at the top ends of the specimen. As the load increases, the damage range gradually expands simultaneously over the whole section. When the stress reaches a certain value, damage in some areas begins to increase and gradually penetrates, and finally, several macroscopic cracks penetrating the entire specimen are formed. It can be seen from the figure that the hole has a certain impact on the concrete damage where the stress concentration is easy to occur first around the hole, causing the damage to initiate. Moreover, the hole influences the development of the damage path and affects the final failure surface.

Fig. 17 shows the analysis of different materials' damage. It can be seen that the crack is mainly developed along the intersection of the mortar and the aggregate. Two major failure cracks can be observed in the mortar, indicating that the crack mainly occurs. The damage diagram of the interface is slightly amplified in Fig. 18, where it suffered damage, and some interfaces collapsed. However, because the interface is relatively small, thin, and isolated, it cannot form a unified damage surface. Therefore, from an overall point of view, the damage mainly occurs in the mortar. However, in the initial damage stage, cracks initiation starts at the interface, and when the damage accumulates to a certain extent, due to the interaction of stress between the interface, aggregate, mortar, and holes, the cracks begin to develop and penetrate. In addition to the mortar, some aggregates have experienced damage. The partial damage enlargement shows that most cracks propagate around the aggregate, but some of them pass through the aggregate particles. Therefore, under the dynamic pressure load, the concrete as a composite material has a complicated failure process, and each material's stress interaction determines the cracks' development.

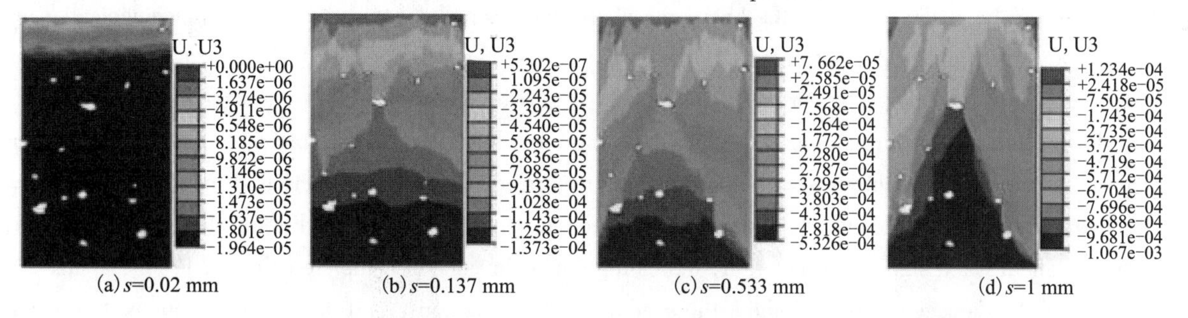

Fig. 15 Vertical displacement graphs under different loading displacement

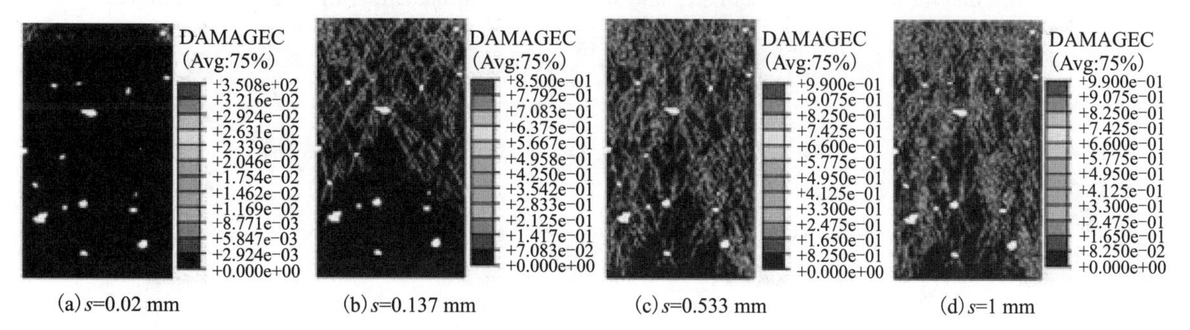

Fig. 16 Damage failure graphs under different loading displacement

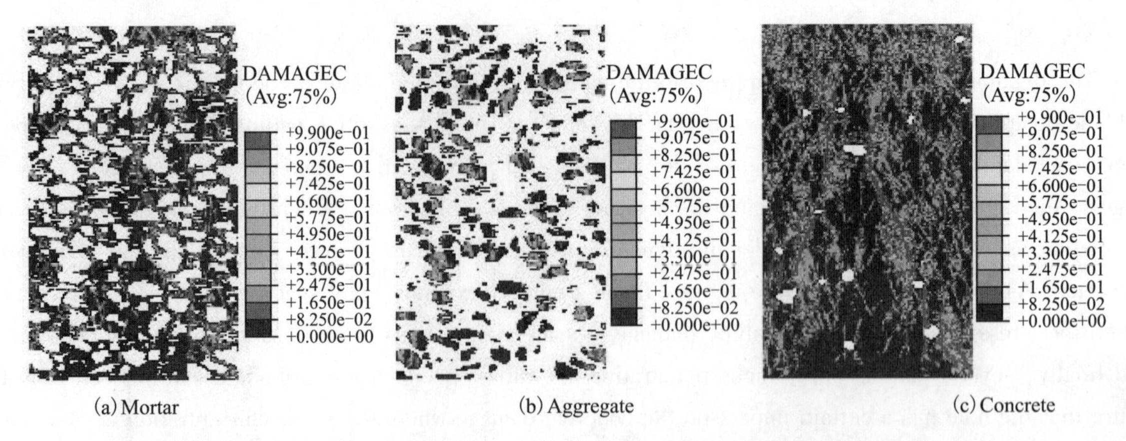

(a) Mortar (b) Aggregate (c) Concrete

Fig. 17 Damage graphs of different materials when the loading displacement is 1mm

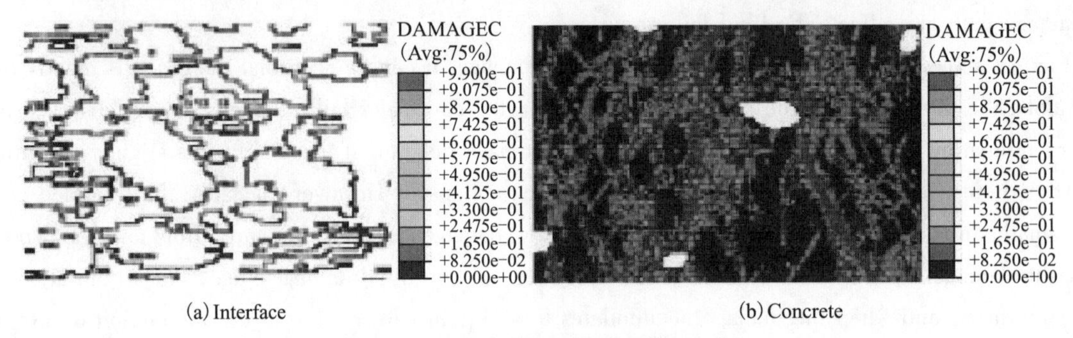

(a) Interface (b) Concrete

Fig. 18 Local failure graph of material

5.4 Tensile load

This study applies the dynamic tensile load to the model top while its bottom surface is restrained. The analysis results are shown in Fig. 19 and Fig. 20. It can be seen that when a concrete specimen is subjected to a dynamic tensile load, its failure mode is completely different from that of a dynamic compression load. For instance, it can be observed that at the beginning of the loading, the specimen's top displacement changed the most, and the damage did not occur. As the load increases, the change in displacement fluctuates, and the damage slowly develops. The vertical displacement begins to change unevenly, and the damage begins to occur slowly. Initially appearing around the hole, continuing to load, the damage begins to develop and gradually forms a damaged area throughout the specimen and perpendicular to the direction of loading. At the same time, there are intermittent damages in other parts of the test piece. When the loading reaches a certain value, the displacement has a significant mutation, indicating the occurrence of complete collapse. From the Fig. 20, the degree of the damaged area throughout the specimen is more serious, and the damage accumulation eventually leads to the destruction of the specimen, and one fracture surface runs through the specimen.

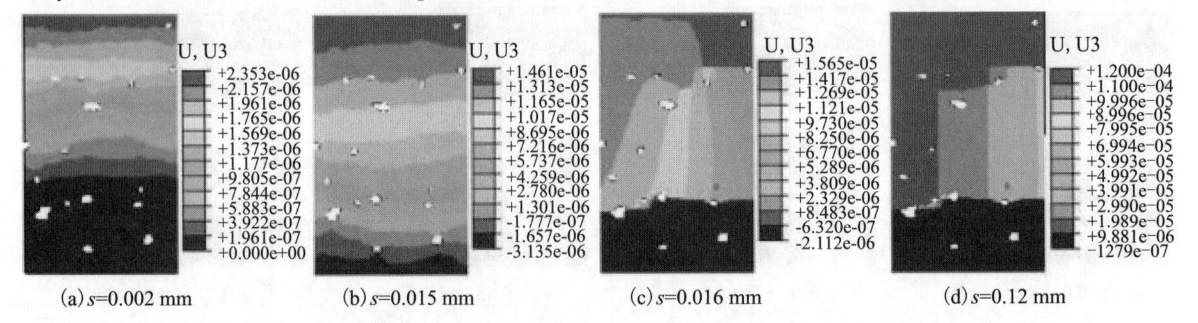

(a) s=0.002 mm (b) s=0.015 mm (c) s=0.016 mm (d) s=0.12 mm

Fig. 19 Vertical displacement graphs under different loading displacement

Further analysis of the mechanical changes of various materials during the stress process. As shown in Fig. 21, the damage mainly occurs in the mortar, and there are many cracks under the dynamic pressure load, but there is only one main crack under the dynamic tensile load where the aggregate has slight damage. The failure surface from the overall damage figure is relatively straight and passes through part of the aggregate and most of the mortar. The interface has also suffered damage, mainly concentrated at the location of the failure surface, as shown in Fig. 22. In the partially enlarged view of the overall damage, it can be clearly seen that the crack passes through small aggregate, mortar, and their interface, stops at the hole, and forms on its other side.

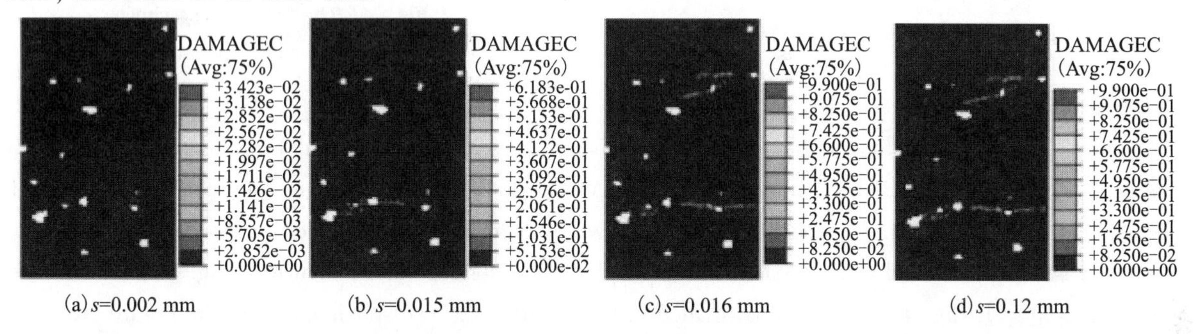

(a) s=0.002 mm (b) s=0.015 mm (c) s=0.016 mm (d) s=0.12 mm

Fig. 20 Damage failure graphs under different loading displacement

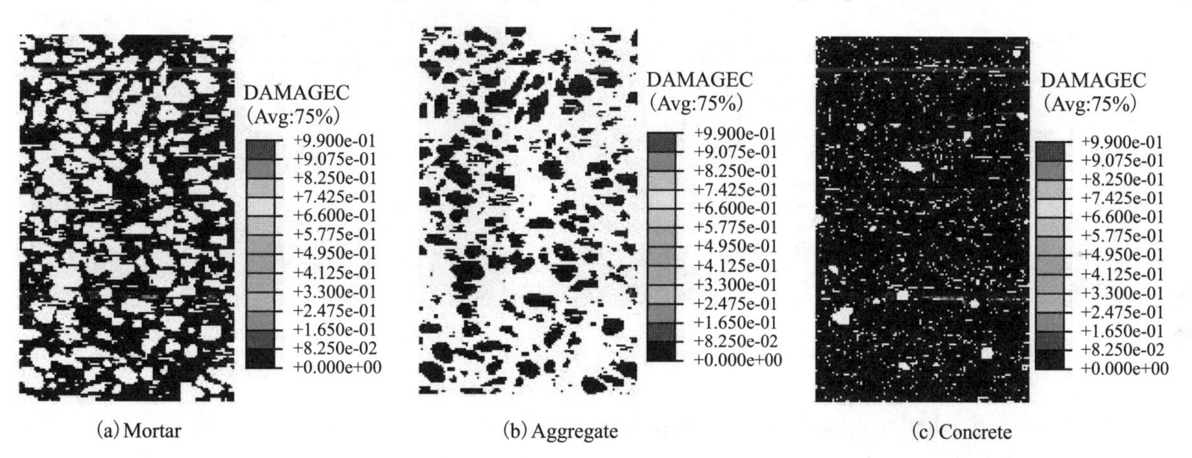

(a) Mortar (b) Aggregate (c) Concrete

Fig. 21 Damage graphs of different materials when the loading displacement is 0.12 mm

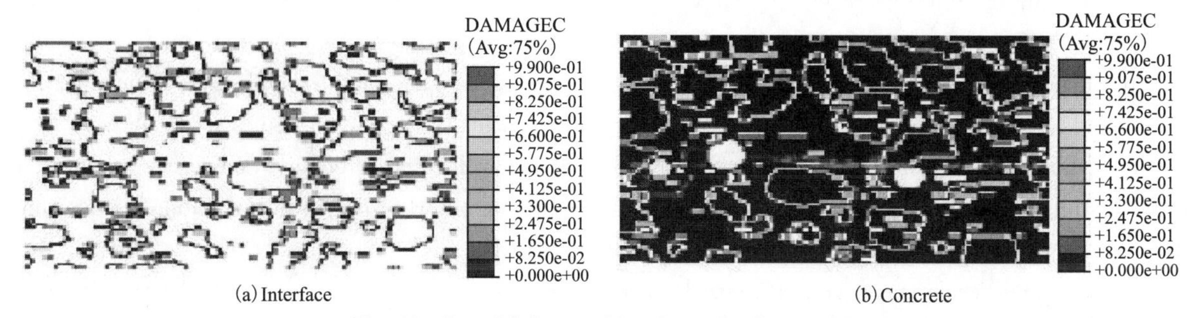

(a) Interface (b) Concrete

Fig. 22 Local failure enlarged graph of material

6 Model Verification

6.1 Dynamic pressure test results

The CT scanned sections under dynamic compressive load show that the concrete specimens' damage occurs abruptly. As a result, unlike damage characteristics under static pressure, it is difficult to capture the microcrack initiation process, propagation, and final development using CT. Fig. 23 and Fig. 24 show the final failure modes. It can be seen that under dynamic pressure, the damaged area is large, the damage is relatively complete, the aggregate is destroyed, and the damage is "double cone" extrusion damage.

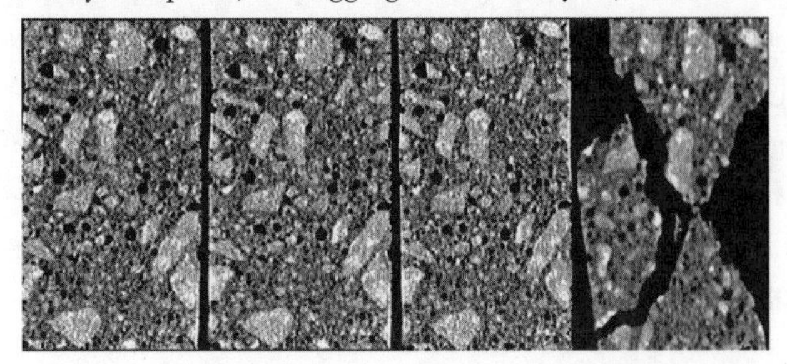

Fig. 23 CT scan sectional drawings

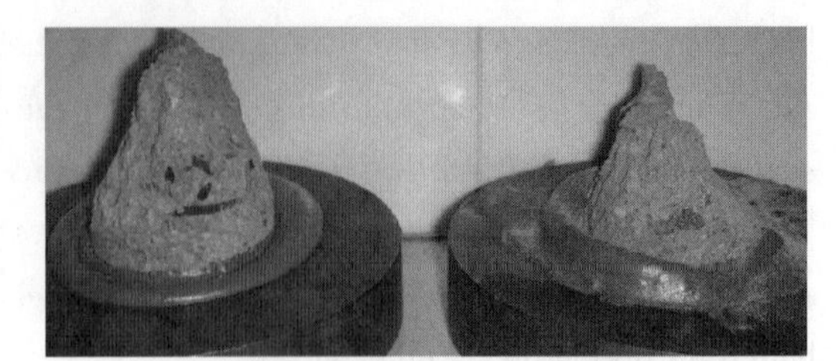

Fig. 24 The failure of specimens under dynamic pressure

6.2 Dynamic tensile test results

The CT scanning section of the concrete specimen (Fig. 25) depicts that the concrete specimen's damage specimen occurs suddenly under dynamic tensile load. Before the damage, each loading stage's CT image remained unchanged. After the damage, a horizontal main crack was formed in the weakest section of the whole specimen, and the crack passed through part of the aggregate. As shown in Fig. 26, the fracture surface is relatively flat, the aggregate is divided into two parts, the two specimens after the failure are completed, and the concrete surface is intact.

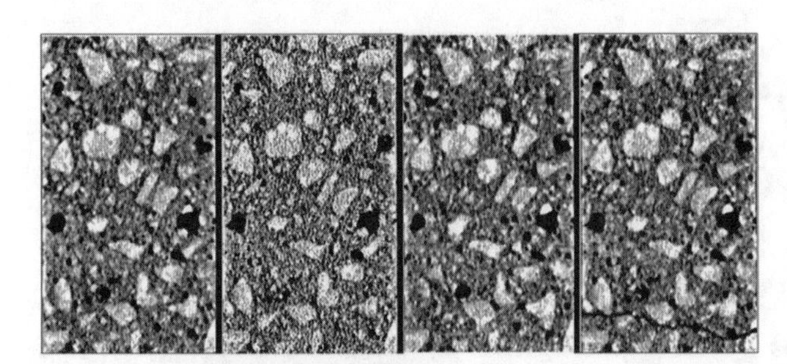

Fig. 25 CT scan sectional drawings

Fig. 26 The failure of specimens under dynamic tensile

The numerical calculation of the CT reconstruction model shows that under the dynamic load, the model's simulation results and the experimental ones from the CT test have high similarity. The crack is mainly developed along the intersection of the mortar with the aggregate. Additionally, some aggregates are damaged, and the hole has a certain influence on the development path of the crack. There are many cracks under dynamic pressure load with a high failure degree, a large failure area, and a double cone failure. On the other hand, there is one crack only under the dynamic tensile load, the fracture surface is relatively flat, and the failure mode is the fracture surface perpendicular to the loading direction. Compared with the CT test, in which the development of cracks before failure cannot be observed due to the limitations of test conditions, the numerical tests can solve this problem by its calculation characteristics. It can be observed that the initiation and development path of internal cracks in the concrete specimens occurs under the action of force in real-time.

Numerical experiments and CT physics tests are complementary. By establishing a numerical reconstruction model of mesoconcrete based on CT images, the two are well combined to provide effective help for further study of the mesomechanical mechanism of concrete.

7 Conclusion

As the basis of numerical experiments, numerical models have a crucial influence on the simulation test results. Therefore, choosing a suitable numerical model is critical for obtaining accurate results.

This paper develops a reconstruction model based on CT images to perform the mechanical analysis. The results of CT experiments are compared and verified, and the main conclusions are as follows:

(1) Both the random aggregate model and the random aggregate interface model can simulate the force characteristics of the mesoconcrete. When simulating the bond between the aggregate and the mortar in the concrete, the interface elements in the random aggregate model are relatively thick, and the contact parameters in the random aggregate contact surface model are not easily determined. Besides, both models do not consider aggregate shape and hole influence.

(2) The 3D reconstruction models based on the joint element method and the MIMICS CT image are closer to the actual concrete behavior than the random aggregate model. This observation is mainly attributed to the low precision and no consideration of the unit compression. As a result, a complete primary mechanical analysis cannot be performed at present.

(3) The CT image reconstruction model based on CT number considers the aggregate gradation influence, including the shape, hole, etc., to provide the advantages of the random aggregate model, thus achieving realistic concrete behavior compared to other models.

(4) The damage to concrete is affected by the hole, and the location of the hole is prone to crack. Moreover, the initiation of cracks often occurs at the interface and eventually extends through the mortar. The strength of its weak specimen parts mainly determines the strength of concrete.

Acknowledgement

The study is supported by Shaanxi Land Engineering Construction Group Scientific Research Project (DJNY2022-39).

References

Afroz S, Zhang Y D, Nguyen Q D, et al(2022). Effect of limestone in General Purpose cement on autogenous shrinkage of high strength GGBFS concrete and pastes[J]. Construction and Building Materials,327, Article ID 126949.

Bai W F, Chen J Y, Fan SL,et al (2010). Statistical damage constitutive model for concrete material under uniaxial compression[J].Journal of Harbin Institute of Technology, 17(3):338-344.

Carol I , Carlos M. Lόpez, Roa O(2001) . Micromechanical analysis of quasi-brittle materials using fracture-based interface elements[J]. International Journal for Numerical Methods in Engineering, 52(1-2):193-215.

Dupray F , Malecot Y , Daudeville L , et al(2009). A mesoscopic model for the behaviour of concrete under high confinement[J]. International Journal for Numerical and Analytical Methods in Geomechanics, 33(11):1407-1423.

Du M, Jin L, Li D, et al(2017). Mesoscopic simulation study of the influence of aggregate size on mechanical properties and specimen size effect of concrete subjected to splitting tensile loading[J].Engineering Mechanics, 34(9):54-62.

Du X L, Jin L(2012). Meso-element equivalent model for macro-scopic mechanical properties analysis of concrete materials[J].Chinese Journal of Computational Mechanics, 29(5):654-661.

Dai Q(2011). Two-and three-dimensional micromechanical viscoelastic finite element modeling of stone-based materials with X-ray computed tomography images[J]. Construction and Building Materials, 25(2):1102-1114.

Gopalakrishnan K, Ceylan H, Inanc F, and Heitzman, M. (2006). Characterization of Asphalt Materials using X-ray High-Resolution Computed Tomography Imaging Techniques, Proceedings of 2006 T&DI Airfield and Highway Pavement Specialty Conference, pp. 437-454, Atlanta, Georgia, April 30-May 3, 2006.

Onifade I, Jelagin D, Guarin A, Birgisson B , Kringos N(2013). Asphalt Internal Structure Characterization with X-Ray Computed Tomography and Digital Image Processing. Springer, Netherlands , 8 :139-158.

Kim S M , Alrub R K A(2011). Meso-scale computational modeling of the plastic-damage response of cementitious composites[J]. Cement & Concrete Research, 41(3):339-358.

Leite J P B , Slowik V , Mihashi H (2004). Computer simulation of fracture processes of concrete using mesolevel models of lattice structures[J]. Cement & Concrete Research, 34(6):1025-1033.

L. Skar żyński, Nitka M , Tejchman J (2015). Modelling of concrete fracture at aggregate level using FEM and DEM based on X-ray μCT images of internal structure[J]. Engineering Fracture Mechanics, 147:13-35.

Liang X Y, Dang F N, Tian W, et al(2011). Study on numerical model of concrete based on 3D meso-mechanics[J]. Chinese Journal of Applied Mechanics, 28(2):129-134.

Lei G Y, Han J C, Zhang Y, et,al(2016). Based on the CT image rebuilding the micromechanics hierarchical model of concrete[J]. Journal of Hydroelectric Engineering,35(3):105-112.

Mohamed A R, Hansen W(1999). Micormechanical modeling of crack-aggregate interaction in concrete materials[J]. Cement & Concrete Composites, 21(5/6):349-359.

Ma H F, Chen H Q, Yang C L(2012). Numerical tests of meso-scale damage mechanism for full graded concrete under complicated dynamic loads[J].China Civil Engineering Journal, 45(7):175-182.

Metroplis N, Lam S U(1949). Monte Carlo method [J].Ameriean: J. Amer.states, (44):335-341.

Nguyen Q D , Afroz S, Zhang Y D, et al(2022). Autogenous and total shrinkage of limestone calcined clay cement (LC3) concretes[J]. Construction and Building Materials, 314(12), Article ID 125720.

Qin W, Du C B(2012). Meso-level model of three-dimensional concrete based on the CT slices[J].Engineering Mechanics, 29(7):186-193.

Qu Z W, Liu Z H, Si R Z, et al (2022). Effect of Various Fly Ash and Ground Granulated Blast Furnace Slag Content on Concrete Properties: Experiments and Modelling[J].Materials, 15(9):3016.

Ren W Y, Yang Z, Sharma R, et al(2015). Two-dimensional X-ray CT image based meso-scale fracture modelling of concrete[J]. Engineering Fracture Mechanics, 133:24-39.

Shahbeyk S , Hosseini M , Yaghoobi M(2011). Mesoscale finite element prediction of concrete failure[J]. Computational Materials Science, 50(7).

Schlangen E, Garboczi E J(1997). Fracture simulations of concrete using lattice models: Computational aspects[J]. Engineering Fracture Mechanics, 57(2-3):319-332.

Schutter G D , Taerwe L (1993). Random particle model for concrete based on delaunay triangulation[J]. Materials & Structures, 26(2):67-73.Tang CA, Zhu WC(2003). Numerical Simulation Tests of Concrete Damnification and Fracture [M]. Beijing:Science Press.

Trawiński W, Tejchman J , Bobiński J(2018). A three-dimensional meso-scale modelling of concrete fracture, based on cohesive elements and X-ray μCT images[J]. Engineering Fracture Mechanics, 189.

Wang W, Yi Z, Tian B,et al (2021). Nonlinear finite element analysis of pbl shear connectors in hybrid structures [J]. Structures, 33(10):4642-4654.

Zheng Y N(2005). Three-Dimensional Numerical Simulation on Mesolevel and CT Experiment Verification of Failure Process in Concrete[D]. Xi'an: Xi'an University of Technology.

Zhang Y D, Afroz S, Nguyen Q D, et al(2021). Analytical model predicting the concrete tensile stress development in the restrained shrinkage ring test[J]. Construction and Building Materials, 307, Article ID 124930.

本文曾发表于2022年《Advances in Civil Engineering》

Measurement Method of Civil Engineering Complexity Structure Based on Logical Equivalent Model

Qian Li[1], Yan Gong[1], Jubao Zang[1,2], Yan Zhang[1], and Yiting Sun[1]

(1.Shaanxi Dijian Real Estate Development Group Co., Ltd, Xi'an Shaanxi, 710000, China;
2.Shaanxi Provincial Land Engineering Construction Group Co., Ltd, Xi'an Shaanxi, 710000, China)

【Abstract】 Civil engineering structures are generally large in volume and scale, especially the most commonly used reinforced concrete structures. There are many defects such as micro-cracks and air bubbles inside, and the concrete is easy to crack under tension. The short-gauge length (point) sensor will be affected by local damage and cannot accurately reflect the overall working state of the structure. The long-gauge strain sensing technology can reflect the average strain of the area to be measured when there are local cracks in the structure to be measured and can also be fully deployed on the structure to form a distributed sensor network to comprehensively monitor the entire structure. Therefore, long-gauge length sensing is more suitable for civil engineering structure monitoring. This study mainly focuses on fuzzy systems and introduces a generalized probability decision process model to describe the behavior of such fuzzy systems. The maximum-likelihood scheduling and the minimum-likelihood scheduling are discussed. The corresponding model checking methods for probabilistic linear time properties are final reachability, always reachability, persistent reachability, and repeated reachability, and the advantage of this method is that the verification process of their model checking is transformed into a fuzzy matrix. The model checking problems of generalized likelihood-regular security properties and generalized likelihood-correcting-regular properties are studied, respectively, and their model checking problems are transformed into generalized likelihood lineartime properties that are always possible. The failure state of the beam and the displacement of the beam bottom are obtained through the data of each unit sensor. The experimental results show that the sensor can effectively capture the cracks. The measured data of the sensor are relatively accurate. The dynamic test performance of the sensor is studied by the damage monitoring test of the vertical cantilever flat beam. After spectrum analysis, the sensor can accurately obtain the participation coefficient of each mode shape, the recognition result is close to that of the FBG sensor, the frequency recognition error does not exceed 0.2%, and the long-gauge length strain mode of the structure can also be accurately recognized. At the same time, the long-gauge strain mode obtained by the sensor monitoring can accurately locate the damage and can quantitatively identify the damage more accurately.

【Keywords】 Complexity Structural Measurement Model; Equivalence Verification; Logical Synthesis; Project Construction Phase

1 Introduction

Engineering quality is the guarantee for the survival and development of civil engineering enterprises. The extensive quality management model of civil engineering makes it extremely difficult for civil engineering enterprises to turn to quality and benefit-oriented intensive development. The application of information technology in the civil engineering industry has become an important way to transform the development mode and improve the quality of civil engineering[1]. Logic equivalent technology is the key technology of civil engineering informatization, creating conditions for "data-based" quality management and big data analysis. However, the research and application of logical equivalence and big data technology integration in quality management has just started, and the research on application methods is still in a "blank". Logic synthesis is a complicated process, and reasonable constraints need to be imposed on the design to produce good synthesis results. After the synthesis is completed, it is necessary to check whether the timing and various constraints are satisfied[2-5]. This study takes civil engineering quality management as the research object, innovatively applies logical equivalence and big data technology to civil engineering quality management, and systematically studies the realization path and application method of logical equivalence and big data in quality management. We provide reference for civil engineering quality management based on logical equivalence and big data.

However, current researches on civil engineering networks based on complex networks are mostly limited to the discussion of morphological complexity and structural common features, and lack quantitative indicators to describe the diversity, heterogeneity, and completeness of civil engineering cross-connections from a structural perspective. Entropy theory and fractal theory are representative methods of complexity research [6-8]. Scholars have proposed topological information entropy and several structural fractal dimension measurement methods. However, the current topological information entropy has the disadvantages of inconsistent methods and models and models and can only describe certain aspects of the network. Research on the fractal characteristics of civil engineering network structures is mostly limited to calculations. The box dimension, volume and degree volume dimension of civil engineering network are rarely studied, and the explanation of its geological mechanism needs to be in-depth. Existing map topology information measurement indicators pay more attention to the diversity of map symbol connectivity methods or the closeness of connectivity, and lack methods to quantitatively describe the spatial pattern and structural characteristics of map symbols, especially the compactness and compactness of map symbol topology. The paper defines the compactness of map symbol topology and heterogeneous information entropy, proposes its calculation method, and applies them to part of the civil engineering network. The results show that the method can describe the topology of the civil engineering network more accurately[9-11].

This article first takes the status quo of quality management as the breakthrough point, and through the analysis of total quality management theory, logical equivalent technology and big data theory expounds the feasibility and necessity of applying logical equivalent and big data technology in civil engineering quality management. We put forward the realization path of logical equivalence and big data in quality management, which provides ideas for civil engineering quality management based on logical equivalence and big data. Through systematic analysis of the whole cycle of logically equivalent quality management data, a framework for the big data analysis process of logically equivalent quality management was designed. Under the guidance of this framework, the structure of logically equivalent quality management data, the extraction and storage methods of logically equivalent quality management data, and the quality text big data analysis method based on text mining are proposed, thereby realizing logically equivalent research on quality management big data analysis methods and providing data support for total quality management based on logical equivalence. Integrating logical equivalence technology and big data analysis methods into traditional quality management workflow and information flow, a comprehensive quality management process based on logical equivalence and big data has been constructed. Guided by this process, the system analyzes the integration, storage, sharing and application methods of logically equivalent quality information, as well as the logically equivalent application points and implementation steps in total quality management, and proposes a logically equivalent-based approach in a big data environment.

2 Related Work

There have been a large number of researches on the characteristics of civil engineering network structures based on complex network theory. Qing et al.[12] used duality to characterize the civil engineering network of the regional city, revealing its small-world characteristics, selecting 6 urban civil engineering networks with different geometric forms and historical origins, and analyzing their complex network structure characteristics. Clark and Watson[13] analyzed the betweenness centrality of six domestic civil engineering networks as examples and found that the betweenness centrality values are distributed in a power law, and the betweenness centrality values can effectively reflect the civil engineering grade. Qi et al.[14] found the homology and heterogeneity of the civil engineering network were tested, and it was found that homogeneity and heterogeneity coexist in the network. These empirical studies reveal the potential structural characteristics of the civil engineering network, but most studies are limited to the description of certain aspects of the civil engineering network and lack a systematic description of the structural complexity of the civil engineering network. In the research of complex networks, the network structure entropy is used to describe the disorder and heterogeneity of the network, and it makes up for the shortcomings of the power law coefficients that are not easy to fit accurately and have good versatility.

Map information entropy is an extension of entropy theory in the field of cartography. It mainly describes the amount of information contained in map symbols from the perspectives of geometry, topology, and topics to evaluate its complexity. Map topology information entropy can effectively describe the complexity of the adjacency relationship of map symbols, and it has great potential in the measurement of the complexity of civil engineering network structure. It has been initially used in civil engineering selection and mapping synthesis[15-17]. However, the existing measurement models of map topology information entropy are not uniform, and the definitions are different, and the description of the complexity of the civil engineering network structure is still not accurate enough. The first issue that should be considered in the synthesis of reversible logic is the cascade of reversible logic gates, that is, the order in which those reversible logic gates can be used to realize the given reversible logic function; secondly, it is hoped that the realization cost of the reversible network is as small as possible to realize the reversible logic. Each technology of the network requires a reasonable price; the input bit and output bit must be equal in the process of reversible logic design, so in many cases useless information bits will be added, which is also an important factor to be considered in reversible logic synthesis. Yan et al.[18] used a variety of logic equivalent software to deepen the design of the prestressed steel structure, and realized construction simulation and dynamic monitoring. The Xsteel software was used in the deepening design of the project and the construction process, which demonstrated that the logic equivalent technology is in the large span. Special-shaped steel structure design construction have advantages in improving efficiency and sustainable development. Matheran et al.[19] carried out the Sports Center Cycling Hall Grid project, based on ProSteel software and using VB language for secondary development, a system for the rapid generation of spatial structure models was established.

Subsequently, scholars continue to propose new structural fractal dimension measurement methods. So far, small-world characteristics, scale-freeness, and structural fractal have been summarized as the three major characteristics of complex networks. Scholars tested the fractal characteristics and self-similarity of the civil engineering network structure of the top 50 cities in the United States using the box covering method and found that the civil engineering network also has fractal characteristics in its structure. However, compared with the research on the small world and scale-free characteristics of civil engineering network, the exploration of the fractal characteristics of civil engineering network structure is only in its infancy. This is mainly because structural fractal are not as intuitive and easy to understand as small world and scale-free. It is more abstract, and it is still difficult to explain its mechanism. Although studies have shown that the emergence and emergence of structural fractals are largely due to the mutual exclusion of hub nodes[20-23], this explanation is still partial to theory, and it is urgent to describe the fractal characteristics of complex network structures in reality. The researchers applied logical equivalence technology to prefabricated prefabricated houses, used Tekla to achieve deepening design, collision detection, etc., and proposed the establishment of a prefabricated apartment library and a prefabricated component product library to standardize prefabricated civil structures. They conducted research on the storage, protection and arrangement procedures of scaffolding engineering and construction site components of prefabricated civil construction. They believe that the application of logic equivalent technology is of great help in solving the current problems of prefabricated civil engineering. In the research, they also used the unique advantages of logically equivalent technologies such as collaborative design, 4D construction simulation, and collision detection.[24-26].

3 Logic Equivalence Model Space Architecture

3.1 Logical Level Distribution

The core of logical equivalence technology is information, which uses software as a carrier to realize model construction, model application and information management. The ultimate goal is to realize civil engineering informatization. According to the connotation of logical equivalence, a complete information model should be able to integrate engineering data and business data at different stages of the entire life cycle, dynamically realize the creation, management, and sharing of information, and always maintain the consistency and completeness of information.

Logical equivalent technology changes the way of information sharing and improves management efficiency with its own characteristics as follows. In different stages of the life cycle of civil engineering, the information remains consistent. The same information only needs to be created once, and it can automatically evolve on its basis. The model objects are modified and expanded in different stages without re-creation, which reduces the errors of inconsistent information.

$$M[t \mid (c,t) \to (1,1)] = f(u + mt) + Z(t), t = 0,1,2,\cdots,n \tag{1}$$

Agglomeration coefficient and characteristic path length are the two main indicators to determine whether a network has small-world characteristics. If the network has both a larger agglomeration coefficient and a smaller characteristic path length, then the network is a small-world network. Degree distribution, characteristic path length and clustering coefficient are the three most basic concepts in the statistical characteristics of complex network structures. Hierarchy is a common law and phenomenon in the objective world, and it is one of the basic organizational forms widely existing in various complex systems in nature and human society.

$$\sum (u + mt) + \sum_{i=1}^{t} (Y_i + u - mt) - \sum mt = 0 \tag{2}$$

The hierarchy in the civil engineering network is a hierarchical structure artificially organized or automatically emerging according to the function or importance of the civil engineering. Streets at the same level have similar properties and relatively close connections, while streets at different levels are progressively progressive in terms of features and functions. Hierarchical structure is not possessed by random network and regular network. It profoundly affects the distribution of traffic flow of the network and causes the difference in network complexity.

$$\begin{bmatrix} \sum_{i=1}^{t} Y_i & \sum_{i=1}^{t} X_i \end{bmatrix} \begin{bmatrix} \sum u + mt \\ \sum u - mt \end{bmatrix} - \begin{bmatrix} 1 & 0 \\ 0 & 1 \end{bmatrix} = 0 \tag{3}$$

The nodes are connected to each other to form a network. Their connection relationship directly affects the complexity of the network structure. The complexity of the connection relationship can be reflected by the diversity and difference of the connection methods. In the network, the connection relationship of nodes largely determines the network structure and the properties emerging from the structure. In a regular network, the connection relationship of nodes is very single, and by reconnecting edges with a certain probability on the basis of the regular network, the connection relationship of nodes has changed from single to rich, and the network structure has also changed from simple rules to complex and diverse.

3.2 Equivalent Data Indicators

Civil engineering quality management is a coordinated activity of command and control organization in terms of quality, to economically and efficiently construct civil engineering of qualified quality that meets design requirements and standards and user needs. Quality management is to implement and realize all activities of all quality management functions by determining and establishing quality policies, quality objectives and responsibilities, and in the quality management system through quality planning, quality control, quality assurance, and quality improvement.

$$\forall t \in (t, t-1, t+1 \mid t > x), \exists \lim_{x \to \infty} \sum_{i=1}^{t} [X_i + Y + \mathrm{cov}(x,y)] - 1 = 0 \tag{4}$$

The compactness of the topology and the heterogeneity of the topology comprehensively consider the pattern and structural characteristics of the nodes in the network. By comparing the importance of each node in the network (reflected by the degree value) and the difference between a node and its neighboring nodes in terms of adjacency, the spatial adjacency information of nodes at the global and local levels. is revealed After extracting the feature description parameters of the topological structure in Table 1, the improved information entropy calculation model is used to define the information entropy of the entire network.

Table 1 Network topology information

	Network calculation	Network adjacency	Network information
UPF 1	60.56	43.70	10.94
UPF 2	46.24	11.77	48.93
UPF 3	52.65	32.08	42.91
UPF 4	2.26	48.34	62.10
UPF 5	76.90	39.40	9.74

From the design flow point of view, starting from the RTL code and the initial UPF, RTL describes the logic function of the design, and UPF describes the power consumption intent of the design. RTL and UPF exist in different files, so that it is easy to adjust RTL and UPF. Read in the RTL code and the initial UPF, and comprehensively complete the output of the door-level netlist and the UPF', corresponding to the integrated door-level netlist. The newly generated UPF', contains the information of the original UPF, and also explains the power connections of the special units inserted during synthesis (such as isolation units, level conversion units, etc.).

$$\iint Z(t) - Z[t + \sigma w(t)] \mathrm{d}t\mathrm{d}w - \iint Z(t) - Z(w) \mathrm{d}t\mathrm{d}w = 0 \tag{5}$$

Based on logically equivalent civil engineering design, the objects facing are not unrelated points, lines, and surfaces, but civil components that contain rich attribute information and have bidirectional relationships, such as columns, beams, walls, slabs, and doors, etc. Each component object inherits the attributes of its own class, expressed and calculated by parameters and parameter values. Component entity objects include geometric attributes such as length, width, and height that describe their own characteristics; physical structure attributes such as material; basic data such as functional attributes; and extended attributes such as technical parameters, cost data, schedule data, construction information, and maintenance information. In the entire life cycle of a civil engineering project, it is only necessary to continuously update and improve the parameter information of the building information model.

3.3 Model Weight Iteration

Each link of the PDCA cycle in civil engineering quality management, the setting of quality control points, the quality management of inspection batches, subitem projects, unit projects, individual projects, and the completion and acceptance of the entire project, all require accurate and reliable information and data for support. In the existing quality management, most of the information is archived on paper, or stored separately in the enterprise management information system of each participant, and the integrated sharing of information cannot be realized. There "information gaps" and "information islands" make it difficult to communicate between the participants and reduce the predictability and controllability of the quality of civil engineering by the participants.

$$\begin{cases} Z[t+\sigma w(t)] = \sqrt{t^2+t+w^2}, t > w \\ Z[t-\sigma w(t)] = \sqrt{t}, t \leqslant w \end{cases} \tag{6}$$

Static timing analysis tools (such as PrimeTtme) and power analysis tools (such as PrimeTime PX) can read gate-level netlists and UPFs after synthesis or physical implementation to perform timing analysis and power analysis for low-power designs. PrimeTtme uses the information contained in UF to construct a virtual power supply network model and back-marks the voltage to the power supply pins of each gate-level netlist instance. Through corresponding constraints during synthesis, the synchronization circuit in Fig. 1 meets the requirements of setup time and hold time.

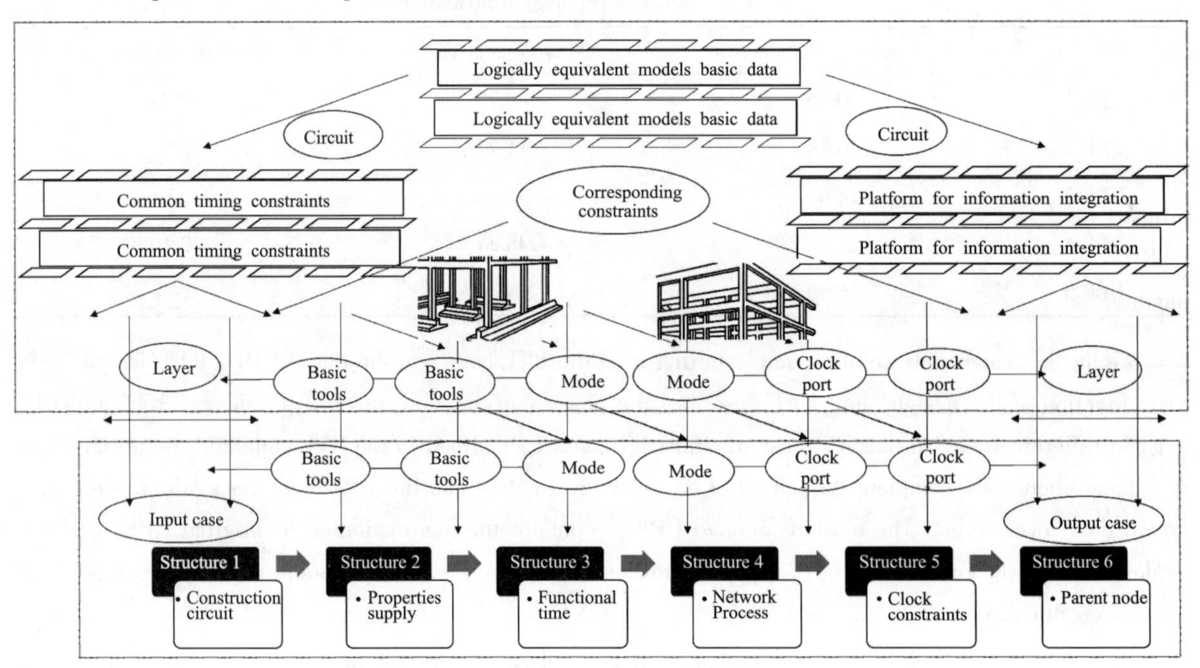

Fig. 1 Topology of civil structure model layer

Logic equivalent technology provides a platform for information integration with a powerful back-end storage system, including data layer, model layer and information application layer. The process of constructing logically equivalent models defines basic data such as geometric properties, physical structure properties, and functional properties of components to form a 3D model. With the progress of the project and the in-depth application of the 3D model, the extended information in the model is continuously enriched and improved. The information in the design phase, construction preparation phase, construction phase, completion phase, and operation and maintenance phase is continuously integrated on the basis of the 3D model to ensure the continuity and consistency of the information in each phase, and finally form the information of the project product and business process.

3.4 Spatial Factor Recursion

In the current quality management process, the formal communication between the parties is usually carried out in the form of emails, contact sheets, meetings, etc. Although it can solve communication problems, it is often manifested as a "point-to-point" communication mode, which is prone to untimely communication. The phenomenon of information asymmetry and information islands of all parties seriously affects the efficiency of quality management. There is a lack of a unified "point-to-face" information platform between all parties to improve work efficiency.

$$\frac{\partial f[t-\sigma w(t)]}{f(t)}+\frac{\partial f[t+\sigma w(t)]}{f(t)}=1,\mathrm{for}\{t,w\rightarrow 0,0\} \tag{7}$$

The R node can be used as both a premise and a conclusion, so we know that the model structure of the generative rule is an irregular directed graph structure. The specific irregularity is manifested as follows: in the topological graph of production rules, not only the nodes are related, but also there are certain connections within the nodes. The relationship is actually the rule. The relationship between the nodes, and the internal connection relationship of the node represents the relationship between the premise and the premise in a rule. There are many differences between these two relationships.

Because the quality planning and assurance work in Table 2 is not sufficient, the quality of quality control entities is prone to uneven quality. Information in the entire quality management process cannot be effectively integrated, making quality improvement work without a data basis, forming a vicious circle. The quality management based on logical equivalent technology realizes the collection and storage of information through its front-end collection equipment and back-end database, creating conditions for the formation of logically equivalent big data. Logical equivalent big data provides a reference for quality planning of similar projects, and can simulate the planning scheme to ensure that the planning scheme is economically feasible; real-time tracking of the process through information collection equipment, real-time control of project dynamics; through product information and process information.

Table 2　Standard civil engineering data structure

	Standard conditions	Structure quality	Structure entities	Structure assurance
Logical order 1	[800, 1200]	0	2.30	2.44
	[800, 1200]	1	1.77	1.87
Logical order 2	[800, 1200]	0	1.24	1.30
	[1000, 2100]	1	0.71	0.73
Logical order 3	[800, 1200]	0	0.18	0.16
	[1000, 2100]	1	0.35	0.41
Logical order 4	[800, 1200]	0	0.88	0.98
	[1000, 2100]	1	1.41	1.55

With the help of any of the 16 main roads in Fig. 2, the horizontal roads can easily reach each other. In the same way, vertical road markings can easily reach each other by means of horizontal road markings. Although the grid-like civil engineering network is not as dense as the sample civil engineering network, with the help of two circular civil engineering projects, any two roads can reach each other through a small number of transfers. The lowest compactness is the sample civil engineering network of Gavle City, which is on the road network delineated on the outer ring road.

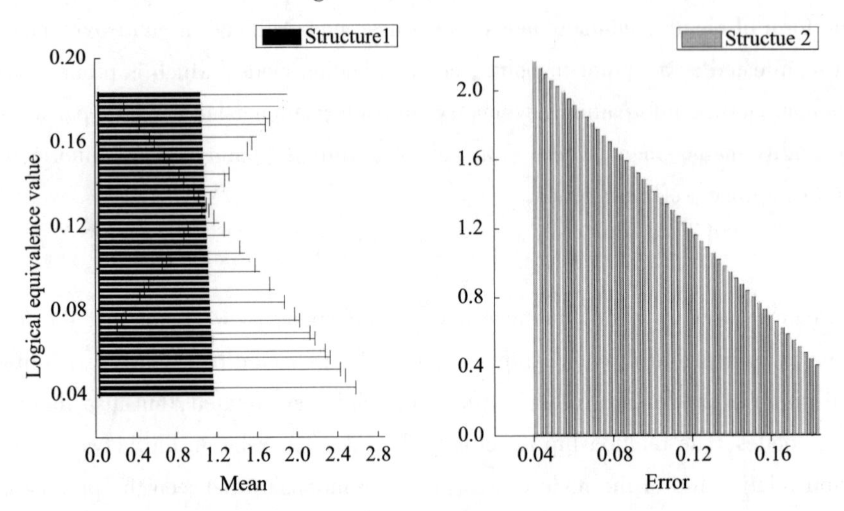

Fig. 2 Process information distribution of civil construction structure

The proposed topological information entropy calculation method only measures the uniformity of the road demarcation value distribution in each civil engineering network from the perspective of probability, and does not consider the degree value, that is, the compactness of the topology and the difference in the degree value of the connected roads. Therefore, the complexity of the topology of the sample road network is not well described.

$$\prod f[t-\sigma w(t)]+\cdots+f[t+\sigma w(t)]/\prod f(t)f[\sigma w(t)]=1/2,\text{if}\{t+1\in R\} \qquad (8)$$

The security rule system is an expandable database, so inputting security rules in a uniform format is the basis for the establishment of the database. When applying safety rules to natural language specifications that have been processed through knowledge engineering, first determine the accident subject of the safety accident, and continue to apply safety rules to conduct rule inspections within the scope of the identified accident subject. Therefore, the first node K11 is the subject of the accident. By searching the specification results of the first node K11, continue to search for safety information based on attributes, parameters, safety measures, etc., and extract the Ck conclusion node at the end of the search data, processing measures and other information in the clauses can be used for safety inspection.

4 Construction of a Civil Engineering Complexity Structure Measurement Model Based on Logical Equivalence

4.1 Civil Engineering Quality Management

The quality of civil engineering products refers to the use value of the product, that is, the natural attributes of the product that can meet the needs of the country and the people, such as applicability, safety (reliability), durability, aesthetics, economy, and environmental coordination. The operation quality of the quality management system can be reflected by its operational effectiveness. The effectiveness of the quality management system operation can be summarized as comprehensive implementation, behavior in place, timely management, moderate control, effective identification, and advancing with the times. It is the enterprise's purpose to achieve project quality. The efficiency and level of management, organization and technical work done by the standard.

$$\begin{cases} [1 + q(t)]/F(q,t) \to f[(q,t),(q,t-1)] \\ F(q,t)/F(q-1,t) \to 1 \end{cases} \tag{9}$$

The volume structure fractal dimension value of all twelve sample civil engineering network data is less than the degree volume structure fractal dimension value. This is because, in the dual graph of civil engineering network based on road planning, the heterogeneity of the network nodes is very strong, and the degree value mostly obeys the power law distribution. In addition, civil engineering networks have "hubs" serving local structures at different scales. So that there is a certain degree of heterogeneity between the global and local node degree values. Therefore, as the measurement radius increases, the speed at which the average number of nodes that can be reached by any node increases will be lower than the speed at which the node degree value that can be reached in Table 3 increases.

Table 3 Description of node degree value

Node number	Description of node	Degree cases
1	Local structures	Set_input_delay 3
2	Measurement radius	Set_input_delay 4
3	Certain degree of heterogeneity	Set_input_delay 5
4	Logical relationship	Get_ports IN 2
5	Production rules	Get_ports IN 3
6	The node degree value	Get_ports IN 4
7	Expression methods	Make_counts fn 2
8	Knowledge representation	Make_counts fn 3
9	Alter relationship	Make_counts fn 4

By analyzing and comparing the above knowledge representation methods, we can divide them into two categories. The first type is knowledge modeling and expression methods based on artificial intelligence AI technology. This type of method focuses more on knowledge reasoning, among which is based on production rules. Knowledge representation, Bayesian network knowledge representation, logic-based knowledge representation, and neural network knowledge representation are more representative knowledge representation methods. The second category is a knowledge modeling method based on Semantic Web technology.

Among them, knowledge representation based on Semantic Web, ontology, based on Framework and object-oriented knowledge representation methods are more typical methods. Among the above two types of knowledge representation methods, the knowledge representation method based on AI technology is more inclined to the logical relationship between knowledge and knowledge, and the knowledge represented can be derived through logical reasoning. The description of knowledge is convenient for knowledge classification and later query and retrieval.

The reference design and implementation design shown in Fig. 3 are two designs that use Formality to check whether they are equivalent. The reference design is used as a standard for comparison. It should be a functionally correct design, a form of design at a certain stage before the design is realized. The implementation design is the changed design, and it is the design that needs to be verified whether it is equivalent to the previous design. For example, the general design after synthesis is the realization design, and the reference design is the RTL design.

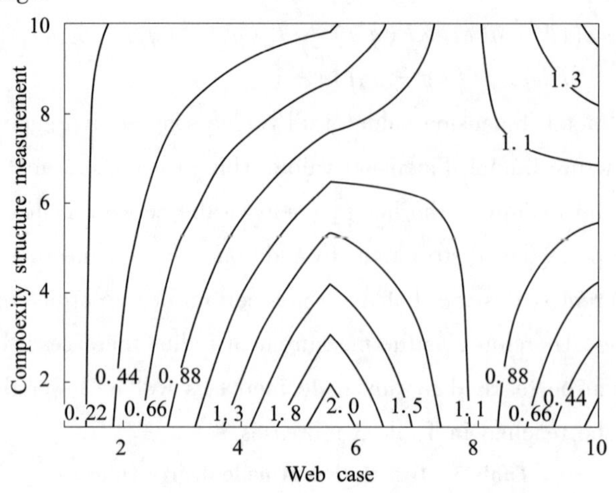

Fig. 3 Complexity structure measurement distribution

4.2 Complexity Quantification Post-Processing

The complexity quantification framework provides the possibility for the rapid development of IFC-based applications to support the complex business processes of the civil engineering industry. During the development process, software developers only need to add the binary files of IFC Java Toolbox to the path of the software project to realize the integration of Toolbox and the software, and then achieve full access to the logical equivalent model based on IFC. If you bind Toolbox in the initial stage of logical equivalence model creation, you can also read, write, modify, and create a java-based IFC model. We use IFC Java Toolbox to achieve complete access to the IFC-based logical equivalent model, as well as to read, write, modify and create IFC files, mainly through the following modules of IFC Java Toolbox.

$$\lim_{x\to\infty}\delta[q(t,x)] - \lim_{x\to\infty}[1 + q(x,t)]h = \begin{cases} q(x,t) - 1, q > x \\ 1, q \leqslant x \end{cases} \tag{10}$$

When the pedestrian updates the position, he will select the target grid point at the next moment according to the probability in the preference matrix. If the target grid is empty, and no other pedestrians move to this grid, the move is executed. If the target grid point is occupied, the pedestrian stands still. If multiple pedestrians choose the same target grid point, there will be a position conflict, you need to calculate the probability of moving to the target grid point, and the pedestrian with a higher relative probability in Table 4 occupies the target grid point.

Table 4 Grid moving algorithm of civil engineering target

Steps	Algorithm tables	Code texts
1	The pedestrian updates dm	Public static void main(String[ ] args)
2	Target grid is empty $\delta(q(t))$	Int abccount=0;
3	A higher relative probability in $q(t)$	Int spacecount=0;
4	The same target grid $f(\sigma)+f(h)$	Int numcount=0;
5	Pedestrian stands $\sigma q(t)$ some	Ylabel('Amplitude');
6	A position conflict $M^q(t)$	Title('Output Due to Weighted Input');
7	Implementation design $hq(t)$	Subplot(3,1,2)
8	The static field S is $dM^q(t)$	Stem(n,yt);

Table 4(**continued**)

Steps	Algorithm tables	Code texts
9	A correspondence between $f(h,t)$	Ylabel('Amplitude');
10	The $hq(t)dt$ move is executed	Title('Weighted Output: a \cdot y_{1}[n]');
11	During the matching and verification	Subplot(3,1,3)
12	Complete verification $\sigma q(t)$	Stem(n,d);
13	In some $dW(t)$ cases	Xlabel('Time index n');ylabel('Amplitude');
14	Achieve complete $q(t-1)$ comparison	Title('Difference Signal');
15	This $f(\sigma)$ change will be recorded	System.out.println(abccount);
16	Setting $hq(t-1)$ file is read	System.out.println(spacecount);
17	Initial comparison $dM^q(t) - dm$	A \cdot x_{1}[n] + b \cdot x_{2}[n]');
18	All comparison $-\sigma q(t)$ points	System.out.println(othercount);

In order for formality to perform complete verification, all comparison points should be verifiable. There must be a one-to-one correspondence between the reference design and the implementation design. However, in some cases, it is not necessary to achieve complete comparison when verifying consistency. For example, the implementation design has additional output ports, or the implementation logic or reference logic has additional nodes. The initial comparison point match will be matched according to the design object name, if the design object name is different. The phase of a node may be inverted, and this change will be recorded in the automatic setting file. When the automatic setting file is read in Formality, the tool can recognize the phase reversal during the matching and verification of the comparison points. The SVF file is helpful for successful verification, but it can also be verified when there is no SVF file.

$$\oiint \delta[q(t,x)] \cdot \delta[q(t-1,x)] \cdot \delta[q(t,x-1)]\mathrm{d}x\mathrm{d}t - \sqrt{x^2+t^2} = 0 \tag{11}$$

In order to better simulate the aggregation phenomenon, the "background field" can be divided into two types: static field and dynamic field. Among them, the static field S is only related to the pedestrian's movement environment, and has nothing to do with the evolution time of the system and the pedestrian's movement state. The size of the distance between each cell and the exit is used to indicate the size of the static field S value, the larger S is, the closer the cell is to the exit. We expand on the Burstedde cellular automata model, adopt the calculation method of "static field" in model, introduce local density, consider the dynamic changes of pedestrian local density in surrounding areas, and establish a kind of consideration localization.

Numerical simulation studies the evacuation process in single and multiple exits, as well as the dynamic behavior of high-density crowds in turning and circular passages. The results show that the new model can well reproduce the macroscopic phenomena of group movement, arching, following, and stop-and-go during pedestrian movement in different scenes.

4.3 Structural Level Measurement

Traditional data mining is aimed at structured data, that is, a pre-defined data model, which is described based on the IFC standard, and can be stored in a relational database through exchange or analysis, and can be logically expressed in a two-dimensional table structure. And models are classified as structured data. However, data is not all structured. With the development of network technology, organizations are full of unstructured data such as text documents, WEB pages, and e-mails. According to statistics, 80% of an organization's information is stored in the form of text, which is the most common form of information.

$$s.t.g(x) \cup g(t) = \varnothing, \overline{g[\min(t-1,x),\min(x),\min(t),\min(x-t)]}^{t,x\to 1,1} - \frac{\delta[q(t,x-1)]}{\delta(q(t))} = 0 \tag{12}$$

Compared with structured data, unstructured data is data that is inconvenient to use the two-dimensional table structure of the database for logical expression. Generally, it is text data, pictures, audio, video, etc. described by unstructured natural language. Standard format, the content cannot be directly parsed by the computer, and it will be difficult to store and retrieve it in the database. Regarding the storage of unstructured data in the database, the general approach is to create a data table that contains three fields: number, content keyword, and content summary. The content can be referenced through the number field, and its retrieval can be achieved through content keywords.

The object-oriented knowledge representation method mainly expresses knowledge through objects, classes, inheritance, encapsulation, interfaces, messages and methods. This method believes that the most basic condition of knowledge is the object, and the object can decompose any form of knowledge. Through data abstraction and information shielding, the method in Fig. 4 can not only deduct knowledge from general to special, but also generalize knowledge from special to general.

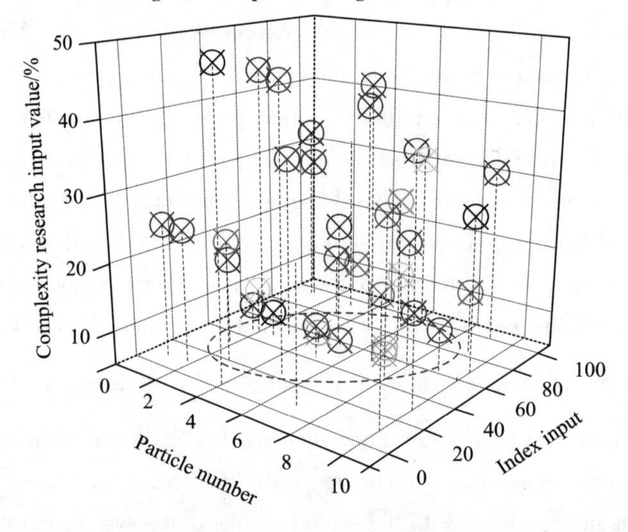

Fig. 4 Distribution of the level measurement of the civil structure

Dividing the signal into many small time intervals, and analyze each time interval with a Fourier transform to determine the frequencies present in the corresponding time interval. Equivalence verification requires related preparation steps and the support of library units, such as library files for standard cell libraries, level conversion units, and isolation units. When running FormMido, you need to read in all the relevant logic library files. The library files should contain the definitions of the corresponding logic functions, power pins and power shutdown functions. Before reading the reference design and implementing the design, you should read the SVF file generated by Compilef. The SVF file contains information such as design optimization and rename during synthesis, which can improve the verification performance and make the verification successful.

5 Application and Analysis of Civil Engineering Complexity Structure Measurement Model Based on Logical Equivalence

5.1 Logical Equivalent Data Preprocessing

The experiment divides the constructed logical equivalence rule knowledge base into eight "tables" to store data, which are the main table, the secondary main table, the direct parameter table, the indirect parameter table, the basis specification table, the accident type table, the handling measure table, and the risk level table has a certain logical relationship between the table and the table. When we extract the main and sub-subjects in the database, their corresponding direct and indirect parameters are also extracted at the same time, and calculated by comparing the parameters. We can judge whether the main body (or sub-subject) is a source of danger. Once it is proved to be a source of danger, the follow-up norms, accident types and handling measures will be output in this logical order.

$$\begin{pmatrix} \sqrt{a^2 - b^2} & a^2 + b^2 \\ \sqrt{a^2 + b^2} & a^2 - b^2 \end{pmatrix} - \begin{pmatrix} \delta[q(a,b)] & a + b \\ a - b & \delta[q(a,b)] \end{pmatrix} = 0 \tag{13}$$

As the scope of the sample data shrinks, the scale of the civil engineering network also becomes smaller. Generally, it is easier to find high-level transportation hubs or core nodes in large-scale or large-scale civil engineering networks, and some sub-hubs or centers are also locally arranged. As part of the large-scale civil engineering network data, the core node level in the medium-scale civil engineering network is slightly lower, the number of secondary central nodes is limited and the connectivity value will not be particularly high.

In other words, the heterogeneity of nodes in the large-scale civil engineering network is higher (that is, the difference in the connectivity value of the road in the civil engineering network is greater), the middle-scale is second, and the small-scale road network has the lowest degree of heterogeneity. From the basic principle and implementation process of the box covering method, it can be seen that the higher the degree of heterogeneity of the network, the faster the network convergence may be. In other words, as the box size increases, for networks with high heterogeneity, the minimum number of boxes required by the overlay network of Fig. 5 decreases faster, and thus the fractal dimension of the box overlay structure becomes higher.

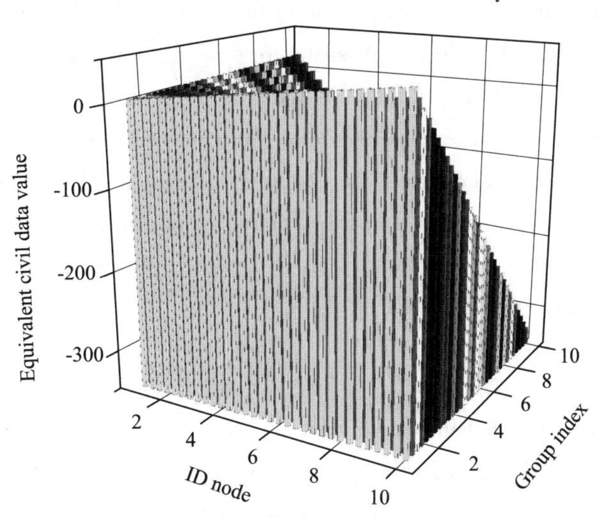

Fig. 5 Three-dimensional distribution of logically equivalent data

The first is the acquisition of target knowledge. For the knowledge that this research needs to acquire, that is, the current construction safety regulations, the norms are explicit knowledge expressed through text, images and symbols, so the method of acquisition is relatively simple. The second step is knowledge expression, that is, to transform the acquired knowledge and store the knowledge in the established knowledge base using computer language. The third step is the application of knowledge, that is, knowledge is extracted from the established knowledge base, and the problems encountered are solved through the extracted knowledge. The totality of these spectra represents how the spectrum changes over time.

In fact, in this process, the computer plays the role of an expert to solve practical problems. Using the knowledge engineering method, we translate the current construction safety specifications extracted above into language and translate them into a knowledge representation form that can be read by a computer to build a knowledge base of current construction safety specifications for subsequent rule inspections.

$$\begin{cases} m + \delta[q(t)] + q(t)h \geqslant 0 \\ m + q(t)h \geqslant 0 \\ \delta[q(t)] + q(t)h < 0 \end{cases} \tag{14}$$

Each component has its own unique identification ID, namely Element ID. Element ID exists as a condition item in the ".csv" file when operating Revit in Dynamo. Geometric dimensions, elevation parameters, etc., among which information such as component names, geometric dimensions, elevation parameters, etc. are used for rule inspection in the safety rule knowledge base, and Element ID is also output as the condition value regardless of ".csv" In the file or in the logical equivalence model, it is always with the component information.

Therefore, Element ID can be used as a bridge between the security rule check result and the visual rule check platform. The selected rule check platform needs to meet the requirements of being able to link to an external database. We import the original logical equivalent model into the visual rule check platform, and at the same time link the platform to the security rule knowledge base that has completed the rule check, so that the knowledge base with Element ID information and specific rule check results can be matched to the import through Element ID matching on the concrete logically equivalent model components in the platform.

5.2 Simulation of Structural Measurement of Civil Engineering Complexity

In the simulation process, compared with the frame structure system, the shear wall structure system is a medium rigid structure, and the lateral displacement of the house under the action of horizontal load is significantly reduced; compared with the shear wall structure system, the frame is very small. The plan layout of the shear wall structure system is free and flexible. When adding a load, you can use the 3D analysis view or the analysis view of each layer. In the analysis view, you can easily select the components to be loaded. Before structural analysis, load information can be counted in the schedule. If you need to modify the load value at this time, you only need to modify it in the schedule, and the system will automatically reflect this modification to the model instead of in the model. This makes the entire design process simple and efficient, while saving engineers more time and focusing on design.

$$\frac{\partial q(t)h\partial q(t-1)h}{\partial q(t)\partial h}\frac{\partial q(t)\partial h}{\partial q(t)\partial h\partial q(t-1)\partial h} = 1 \tag{15}$$

The internal space can be divided into aisle area and seating area, but different layouts have different numbers of horizontal and vertical aisles. In this paper, the movie hall is divided into two-dimensional grids of $W \times L$, and the exit widths of the movie hall are all equal, where the shape represents the number of grids in the width direction, and L represents the number of grids in the length direction. Each grid takes 0.4 m× 0.4 m to represent the area occupied by a single pedestrian, and each grid is either occupied by pedestrians or empty. Among the twelve sample data, the box-covered structure fractal dimension value of nine sample data is close to the degree volume structure fractal dimension value, while the volume structure fractal dimension value is significantly smaller than the two. Suppose x is an M-dimensional parameter vector, and its density $p(x)$ is the prior probability density.

This is determined by the definition of structural fractal dimension. As mentioned above, the fractal dimension of volume structure does not consider the difference of nodes, and the fractal dimension of volume structure is equivalent to giving each node a weight, which is the degree value of the node. Since the civil engineering network is a scale-free network, this weight may be very large, as high as hundreds or even thousands, which directly affects the calculation results of the fractal dimension of the volumetric structure in Fig. 6.

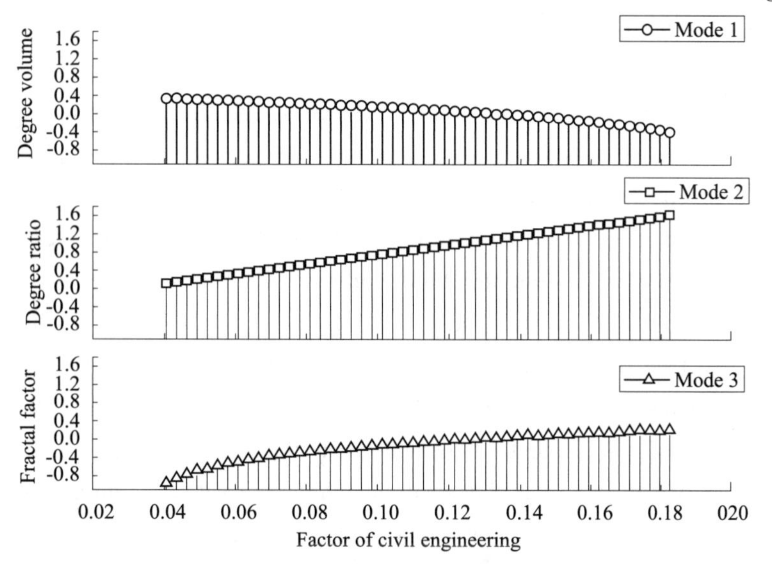

Fig. 6 Fractal comparison of volume structure of civil engineering

The civil engineering project is a multi-discipline complex with wide professional coverage and many subjects of work. Therefore, simply relying on the main body table sometimes cannot correspond to the specific information and location, such as the hole in the main body, the form that exists in the civil engineering can be divided into vertical openings and non-vertical openings; the scaffolding in the main body can be divided into portal scaffolding, bowl-buckle steel pipe scaffolding, bamboo scaffolding, etc. We regard this different existence form of the same main body as the main body. The sub-subject, that is, the specific manifestation of the main body, so the sub-subject table is designed in the knowledge base to reflect the specific existence form of the main body.

$$\sqrt{\nabla f(x,t)/f(x,h)} + \sqrt{\nabla f(x,t)/f(x-1,h-1)} + \sqrt{\nabla f(x,t)/f(x+1,h+1)} = 0 \tag{16}$$

In the box covering method, although the heterogeneity of nodes is considered in the box covering, that is, the difference in node degree value, the difference only affects the number of nodes contained in the box, and does not affect the minimum number of boxes required by the overlay network. In other words, the difference of the box itself is not directly mapped to the calculation process of the box covering method. The degree of connectivity diversity of the civil engineering network is different, and the degree of influence of the node heterogeneity on the fractal dimension of the corresponding structure is also different.

$$\text{if}\begin{cases} x > 0 \\ m > 0, \oint \dfrac{f(x,t) + g(x,t)}{f(m,n)}\mathrm{d}x\mathrm{d}t \xrightleftharpoons{x,t\to m,n} \oint \dfrac{f(x,t) - g(x,t)}{g(m,n)}\mathrm{d}x\mathrm{d}t \end{cases} \tag{17}$$

The results of the fractal dimension measurement of the three structures in different regions are very similar. Among them, the value ranges of the box covering structure fractal dimension, the volume structure fractal dimension and the volume structure fractal dimension of the measured twelve data are: [3.0987, 4.7685], [3.1460, 4.3539] and [3.9573, 5.2727]. The mean values are: 3.9464, 3.7410 and 4.5958 respectively. These results show that the topological fractal dimension of civil engineering network is greater than its geometric fractal dimension. As we all know, the geometric fractal dimension of civil engineering network is between 1 and 2, and the topological fractal dimension is all above 3. This is consistent with the definition of fractal dimension. The geometric fractal dimension describes the two-dimensional space filling capacity of geographic objects, while the topological structure can exceed the limit of the geometric space and fill the high-dimensional space arbitrarily.

5.3 Case Application and Analysis

After the calculation of the civil engineering model is completed, the calculation results can be analyzed to adjust the cross-section or structural layout of the unreasonable components. After multiple calculations, the modified structural model data can be fed back to the Revit model to obtain a reasonable and effective model. After entering the load transfer interface, we can check the basic structure analysis model settings such as the geometry, constraints, load conditions and combinations, and loads of the structure. You can choose to display the load settings under different load conditions layer by layer or as a whole to avoid serious errors when imported into the structural analysis software.

$$\begin{cases} dM^q(t) - d[m + \delta(q(t))]dt = M^q(t) + M^q(t-1) \\ dM^q(t) + d[m + \delta(q(t))]dt = M^q(t) - M^q(t-1) \end{cases} \tag{18}$$

After the rule is executed, the result of the rule check is output. According to the check result, it can be judged whether the logically equivalent component meets the safety standard. If the logically equivalent component meets the safety standard, the above rule matching and rule execution will be performed on the next component of the ".csv" file. Through a sensor, an IV-dimensional observation vector z is generated. The task of parameter estimation is to use z to restore the original parameter vector x.

If the result of the rule inspection judges that the logically equivalent component does not meet the safety standard, the safety inspection report will be output in the safety rule knowledge base, and the report content will include "accident type", "safety specification", "processing measures", and "risk level" items are embedded in the content of the condition result table of the safety rule knowledge base.

$$\begin{vmatrix} hq(t) & \sigma q(t) & 0 \\ -\sigma q(t) & hq(t-1) & \sigma q(t) \\ 0 & -\sigma q(t) & hq(t+1) \end{vmatrix} = \begin{vmatrix} hq(t) & 1 & 0 \\ -1 & hq(t-1) & 1 \\ 0 & -1 & hq(t+1) \end{vmatrix} \times \begin{vmatrix} h & \sigma & 0 \\ -\sigma & h & \sigma \\ 0 & -\sigma & h \end{vmatrix} \tag{19}$$

The rule check for direct parameters is relatively simple. It is judged whether the logically equivalent components meet the safety standards by comparing the explicit parameter information such as the geometric size and elevation of the accident subject to the "direct parameter table" in the safety rule knowledge base; for the rules for indirect parameters check, we first extract the calculation condition values of indirect parameters from the ".csv" file, such as the length and fineness of the template support column, and calculate whether the slenderness ratio of the support column is safe through the calculation formula embedded in the "indirect parameter table" rules.

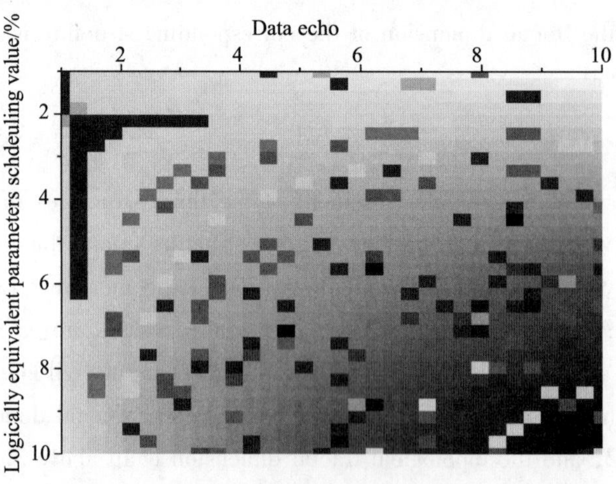

Fig. 7 Regular distribution of logically equivalent parameters

We import the ".csv" file into the structured security rule knowledge base in Fig. 7. The first step is to match the rules of the knowledge base. The matching condition is the "component name" in the ".csv" file, and the matching scope is the security rule knowledge base. For example, if the component name is "dongkou", it matches the subject whose subjectID is "4" in the main body table of the knowledge base; the component name is "wood scaffolding", which matches the sub-subject "Wood" whose subjectID is "10" in the sub-subject table of the knowledge base. "Scaffolding" and associate the "scaffolding" subject whose subject ID is "7" in the subject table, thereby completing the identification of the accident subject. The second stage of rule matching is to obtain the attribute parameter information of the ".csv" file.

This can be done by estimating the function. In addition, the conditional probability density p(zI, x) gives the relationship between the parameter vector and the observation vector. At this stage, the security rule knowledge base will identify the parameter information as a "direct parameter" or "indirect parameter condition value" according to the parameter structure, that is, extract and identify the subject of the accident after the parameter information and classify it into parameters for parameter verification.

6 Conclusion

In this paper, the civil engineering structure model is sent to different structural analysis software, and the model complexity of different analysis software is compared and analyzed. In particular, the use of analysis software, which is closely integrated with the Revit structural core modeling software, is detailed. At the same time, the application of logic equivalent technology assists and optimizes quality management work, it promotes the informatization and standardization of quality management, and improves quality management efficiency. By analyzing the structural fractal characteristics of different forms, different scales, and multi-scale civil engineering networks, and calculating the fractal dimensions of the box covering structure, the volume and the volume structure fractal dimensions, the comparative analysis of the fractal dimension values obtained shows that the civil engineering network is typical structural fractal body. The fractal dimension of the box covering structure decreases as the scale decreases, and the volume fractal dimension is closely related to the compactness and heterogeneity of the network. The paper further confirms that morphology affects structure. The shape of the civil engineering network affects the integrity, heterogeneity, accessibility, and compactness of the civil engineering network structure. The three fractal dimensions of the structure jointly describe this influence from different angles. Finally, the application of big data analysis to realize the value mining of logically equivalent quality management big data will help solidify the quality management experience and training of civil engineering and improve the quality.

Acknowledgements

This work was supported by Shaanxi Dijian Real Estate Development Group Co.,Ltd.

References

[1]Bao Y, Tang Z, Li H, et al. Computer vision and deep learning-based data anomaly detection method for structural health monitoring[J]. Structural Health Monitoring, 2019, 18(2): 401-421.

[2]Pepe M, Costantino D. Uav photogrammetry and 3d modelling of complex architecture for maintenance purposes: The case study of the masonry bridge on the sele river, italy[J]. Periodica Polytechnica Civil Engineering, 2021, 65(1): 191-203.

[3]Wang D, Fan J, Fu H, et al. Research on optimization of big data construction engineering quality management based on RNN-LSTM[J]. Complexity, 2018.

[4]Lee C Y, Chong H Y, Liao P C, et al. Critical review of social network analysis applications in complex project management[J]. Journal of Management in Engineering, 2018, 34(2).

[5]Yang L, Cheng J C P, Wang Q. Semi-automated generation of parametric for steel structures based on terrestrial laser

scanning data[J]. Automation in Construction, 2020, 112.

[6]Salehi H, Burgueño R. Emerging artificial intelligence methods in structural engineering[J]. Engineering structures, 2018, 171: 170-189.

[7]Lemenkov V, Lemenkova P. Measuring Equivalent Cohesion Ceq of the Frozen Soils by Compression Strength Using Kriolab Equipment[J]. Civil and Environmental Engineering Reports, 2021, 31(2): 63-84.

[8]Zhou C, Chase J G. A new pinched nonlinear hysteretic structural model for automated creation of digital clones in structural health monitoring[J]. Structural Health Monitoring, 2021, 20(1): 101-117.

[9]Belliazzi S, Lignola G P, Di Ludovico M, et al. Preliminary tsunami analytical fragility functions proposal for Italian coastal residential masonry buildings[C]//Structures. Elsevier, 2021, 31: 68-79.

[10]Grix T, Ruppelt A, Thomas A, et al. Bioprinting perfusion-enabled liver equivalents for advanced organ-on-a-chip applications[J]. Genes, 2018, 9(4): 176.

[11]Chen X, Fan H, Deng X, et al. Scaffold structural microenvironmental cues to guide tissue regeneration in bone tissue applications[J]. Nanomaterials, 2018, 8(11): 960.

[12]Qing X, Li W, Wang Y, et al. Piezoelectric transducer-based structural health monitoring for aircraft applications[J]. Sensors, 2019, 19(3): 545.

[13]Clark L A, Watson D. Constructing validity: New developments in creating objective measuring instruments[J]. Psychological assessment, 2019, 31(12): 1412-1427.

[14]Qi Q, Tao F, Hu T, et al. Enabling technologies and tools for digital twin[J]. Journal of Manufacturing Systems, 2021, 58: 3-21.

[15]Flah M, Nunez I, Chaabene W B, et al. Machine learning algorithms in civil structural health monitoring: A systematic review[J]. Archives of Computational Methods in Engineering, 2021, 28(4): 2621-2643.

[16]Harifi S, Mohammadzadeh J, Khalilian M, et al. Giza Pyramids Construction: an ancient-inspired metaheuristic algorithm for optimization[J]. Evolutionary Intelligence, 2021, 14(4): 1743-1761.

[17]Zhang X, Nguyen H, Bui X N, et al. Novel soft computing model for predicting blast-induced ground vibration in open-pit mines based on particle swarm optimization and XGBoost[J]. Natural Resources Research, 2020, 29(2): 711-721.

[18]Yan Y, Wang L, Wang T, et al. Application of soft computing techniques to multiphase flow measurement: A review[J]. Flow Measurement and Instrumentation, 2018, 60: 30-43.

[19]Mathern A, Steinholtz O S, Sjöberg A, et al. Multi-objective constrained Bayesian optimization for structural design[J]. Structural and Multidisciplinary Optimization, 2021, 63(2): 689-701.

[20]Lu G, Shang G. Impact of supply base structural complexity on financial performance: Roles of visible and not-so-visible characteristics[J]. Journal of Operations Management, 2017, 53: 23-44.

[21]Chernyshenko O S, Kankaraš M, Drasgow F. Social and emotional skills for student success and well-being: Conceptual framework for the OECD study on social and emotional skills[J]. 2018, 11: 241-245.

[22]Xu X, Augello R, Yang H. The generation and validation of a CUF-based FEA model with laser-based experiments[J]. Mechanics of Advanced Materials and Structures, 2021, 28(16): 1648-1655.

[23]Sihombing F, Torbol M. Parallel fault tree analysis for accurate reliability of complex systems[J]. Structural Safety, 2018, 72: 41-53.

[24]Nasiri S, Khosravani M R, Weinberg K. Fracture mechanics and mechanical fault detection by artificial intelligence methods: A review[J]. Engineering Failure Analysis, 2017, 81: 270-293.

[25]Yao G, Li J, Li Q, et al. Programming nanoparticle valence bonds with single-stranded DNA encoders[J]. Nature materials, 2020, 19(7): 781-788.

[26]Meixedo A, Ribeiro D, Santos J, et al. Progressive numerical model validation of a bowstring-arch railway bridge based on a structural health monitoring system[J]. Journal of Civil Structural Health Monitoring, 2021, 11(2): 421-449.

本文曾发表于2022年《Mathematical Problems in Engineering》

冻融循环对 U 型黄土渠道的稳定性分析

王秦泽

（陕西省土地工程建设集团有限责任公司，陕西 西安 710075）

【摘要】季节性冻土区的黄土输水渠道在冻胀融沉的反复作用下，冻结温度发生变化，渠道面板下部黄土冻结深度改变，导致输水渠道及衬砌面板会发生不同程度的损坏，影响灌溉输水渠道的正常使用。为了研究冻融循环对黄土冻结温度的影响规律及 U 型黄土渠道在冻融循环作用下的稳定性，本文以陕西省宝鸡市黄土为研究对象，通过室内冻结试验，进行了不同冻融次数下的冻结温度试验，研究了冻结温度随冻融次数的变化规律，分析了冻结温度的变化机制，并对冻结温度影响范围内黄土渠道的冻结深度及渠道稳定性进行了分析。结果表明：随冻融循环次数的增加，黄土试样冻结温度先降低后趋于平稳；冻融循环改变了土的宏细观结构，宏细观参数的变化导致孔隙分布指数随冻融次数增加先负向增大后趋于稳定，而冻结温度降低值只与孔隙分布指数有关；一定温度范围内，冻结温度越低，土层冻结深度越大；冻融循环降低了 U 型黄土渠道的稳定性。

【关键词】U 型黄土渠道；冻融循环；冻结温度；冻结深度；稳定性

0　引言

季节性冻融区的输水工程容易发生冻融病害，威胁农田水利设施的供水能力[1]。近年来，黄土高原地区渠道工程的病害频发，反复冻融作用下，冻胀和融沉引起黄土结构发生变化，黄土内部温度场和水盐重新分布，再次冻结时水热参数改变，影响冻结过程，渠道等岩土体的稳定性受到影响。渠道边坡黄土受季节性冻融循环影响出现了土层内盐分随水分向边坡表层持续迁移并汇聚的现象，表层含盐量逐渐增大，表层黄土逐渐剥落并逐步向边坡内部拓展，影响了渠道稳定性[2]。渠道的冻胀破坏一类现象频发于黄土高原灌区，由于冬季环境气温降低，渠道下部土层中原有水分冻结，并伴随温度梯度作用下水盐向基土表层迁移，该过程中产生的冻胀力导致衬砌面板出现冻胀裂缝[3]。进入春季，已冻土层逐渐融化，表层汇集大量水分，融沉变形出现，渠道衬砌发生滑塌[4]，影响输水渠道的稳定性。解决这些问题，首先要准确地评估渠基内的冻结范围，也就是需要准确地确定土的冻结温度，即界定土是否处于冻结状态的关键指标[5]。

目前，在土的冻结温度方面已经探讨了水分、盐分和压力等因素对土冻结温度的影响。然而，冻融作用作为一种影响土的结构的强风化作用仍然较少见诸报道，冻融作用对土结构的细观影响机制及其与土的冻结温度的联系仍然缺少相关研究。因此，本文通过试验分析了冻融次数对 U 型渠道内部黄土冻结温度的影响规律及其变化机制，研究了冻融作用对黄土渠道稳定性的影响，对工程实践具有指导意义。

1　试验方案

本文试验用土取自陕西省宝鸡市千阳县高崖镇一处农田输水渠道内，取土深度 0.3～1.0 m。黄土试样天然含水率 $w=15.0\%$，干密度 $\rho=1.42\ g/cm^3$，液限为 32.3%，塑限为 18.7%，塑性指数为 13.6，根据室内基本物性试验得出所取黄土的基本物理指标如表 1 所示。

对所取黄土土样进行烘干碾碎，并过 2 mm 土工标准筛。本文试验控制含盐量为 0.0%、0.2%、0.5%、1.0%、2.0%和 5.0%六个等级，控制含水量为 11.0%、14.0%、17.0%、20.0%和 23.0%五个等级，

基金项目：国家自然科学基金项目（51778528）。

为了防止其他离子对试验结果的影响,本文选用蒸馏水作为试验用水进行配土。将蒸馏水均匀地拌入过筛黄土中,充分搅拌并覆盖保鲜膜保湿,密封静置 24 h 以保证水分分布均匀。

将配好的土样制成宽 50 cm、高 30 cm 的渠道模型进行冻融循环试验,控制模型试样密度为 1.42 $g \cdot cm^{-3}$,冻融循环试验在控温环境箱内进行,控温精度为±0.3 ℃。为了模拟宝鸡市千阳县的真实自然气候条件变化状况,控制冻融循环控温箱的冻结环境温度为-20 ℃,冻结 8 h,之后将试样在平均室温为 20 ℃的环境中融化 10 h,即为 1 次冻融循环。为了保证试样在 8 h 内完全冻结,这里采用多向快速冻结方式。本次试验控制冻融次数分别为 0、2、5、10 次和 20 次。

表 1 陕西白鹿原黄土基本物理指标

物理指标	数据
相对密度, d_s	2.70
干密度, $\rho_d/(g \cdot cm^{-3})$	1.42
天然含水率, w/%	15.0
Atterberg 界限含水率	
液限, w_L/%	32.3
塑限, w_P/%	18.7
塑性指数, I_P	13.6
颗粒级配特征	
>0.05 mm	4.4%
0.01~0.05 mm	50.0%
0.005~0.01 mm	24.6%
<0.005 mm	21.0%

2 结果分析

2.1 冻融循环对黄土渠道冻结温度的影响分析

为了分析冻融循环对黄土冻结温度的影响,对黄土试样分别进行 0、2、5、10、20 次的冻融循环试验,得出含盐量一定条件下,黄土冻结温度随冻融次数的变化关系,如图 1 所示。

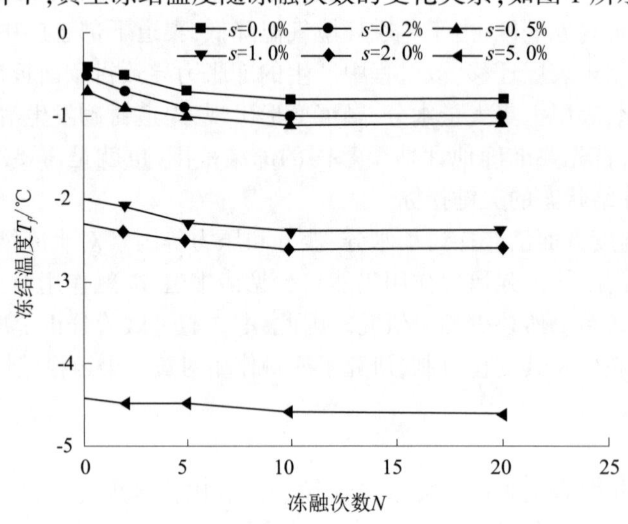

图 1 不同冻融次数条件下含盐黄土渠道冻结温度曲线

从图 1 中可以看出,相同未冻水含量和含盐量条件下,随冻融循环次数的增加,黄土试样冻结温度先降低后趋于平稳。$N<10$ 时,冻结温度随冻融循环次数的增加而降低;$10<N<20$ 时,冻结温度随冻融循环次数的增加几乎不变,$N<5$ 时,冻结温度随冻融循环降低最快,且冻结温度降幅约为 70%。

冻融循环后,土的宏细观结构发生变化,通过宏细观结构的变化来改变土体的热物理性质,最终改变土的冻结温度。

2.2 冻结温度变化机制分析

土的冻结实际为土中水分的冻结，水分达到冻结点后，冰晶开始生成。水分在冻结成冰时体积增大约9%[6]，土体发生冻胀，冰晶的体积膨胀对周围土颗粒产生挤压作用。

由图2可知，土颗粒间的连接方式发生变化，部分由原先的面接触变为点接触。冰晶的生长使土体颗粒受到挤压，增加了土颗粒的团聚性，土中孔隙率增大[7]。随着冻融次数的增加，团粒会破碎为均一的颗粒[8]，即冻融循环后，黄土中等效孔隙数量增多，等效孔隙直径减小，随冻融次数的增加，土的宏细观结构逐渐趋于稳定，土的孔隙率、等效孔隙数量随冻融次数的增加先增大后趋于稳定，而等效孔隙直径随冻融循环次数增多而减小，最后趋于稳定。

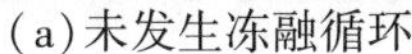
(a)未发生冻融循环

(b)冻融循环20次

图2 冻融循环对土颗粒连接方式的影响

为了描述冻结过程中孔隙的分布状况，袁俊平等[9]在冻胀变形规律的基础上提出了孔隙分布指数的概念，并将孔隙分布指数用式(1)表示：

$$\lambda = \sum \lg\left(\frac{d_0}{d}\right)\lg n \tag{1}$$

式中：λ 为孔隙分布指数，且 $\lambda<-1$；d_0为临界等效孔隙直径，为定值；d 为等效孔隙直径；n 为粒组等效孔隙数量。

由式(1)可知，临界孔隙直径一定时，随冻融循环次数增多，等效孔隙数量增大，等效孔隙直径减小，孔隙分布指数随冻融循环增加先负向增大，而后趋于稳定。冻融循环后，土的宏细观结构发生变化，各相物质在土中占比发生改变。由于土中各相物质的热物理性质不同，所以冻融循环通过宏细观结构的变化来改变土体的热物理性质，影响土在冻结过程中温度场的变化，最终改变土的冻结温度。

吴谋松[10]研究了冻结温度降低值与土的结构的关系，并建立了冻结温度降低值与土中孔隙分布指数的关系式，如式(2)所示：

$$\Delta T = \left(1 - \frac{E}{E_f}\right)^{d_2(-\lambda)+d_3} \tag{2}$$

式中：ΔT 为冻结温度降低值；E 为土样在温度 T 时的热量；E_f为土样在温度 T_f 时的热量；d_2，d_3 为经验常数，本文中设定 $d_2=d_3=1$。

由式(2)可得，E 和 E_f分别为土样温度为 T 和 T_f 时的热量，且 $E_f>E$，设 E 和 E_f 为定值，则式(2)中，冻结点(冻结温度)降低值只与孔隙分布指数有关。由式(1)可知，孔隙分布指数随冻融循环次数增多先负向增大后趋于稳定。结合式(1)和式(2)可得，冻结温度降低值随冻融循环次数增多而逐渐减小，最后趋于稳定。由此可得，黄土冻结温度随着冻融循环次数的增多先降低后趋于稳定。

2.3 U型黄土渠道冻融稳定性分析

由2.1可知，U型渠道内含盐黄土冻结温度随冻融循环次数的增多先降低后趋于稳定。冻融循环改变了土的内部结构,由于土的各相水热性质均不同,冻融作用改变了各相物质在土中所占比例,土的水热状态发生变化,最终影响土的冻结温度。

以冻结温度的变化为基础,对U型黄土渠道冻结深度进行数值模拟分析。发现冻结过程中,温度梯度不断增大,渠道冻结深度在一定范围内不断增大;土层冻结深度随冻结温度的降低而加深,即一定温度范围内,冻结温度越低,土层冻结深度越大。

由此可知,冻结温度影响范围内,U型黄土渠道衬砌下的黄土冻结深度随冻结温度的降低而逐渐加深,改变了黄土内部细观结构,导致黄土结构的整体性降低,力学性质降低,外力作用下,U型渠道的稳定性降低,出现衬砌面板出现冻胀裂缝、渠道衬砌发生滑塌等病害,影响灌溉输水渠道的正常使用。

3 结论

本文通过室内冻融循环试验,研究了不同冻融次数条件下U型渠道黄土冻结温度的变化,渠道冻结深度及稳定性的变化,主要结论如下:

(1)随冻融循环次数的增加,黄土试样冻结温度先降低后趋于平稳。

(2)冻融循环改变了土的宏细观结构,黄土冻融循环过程中,随冻融次数的增加,土中的孔隙体积先增大后减小,等效孔隙直径先增大后减小,而孔隙率会随着冻融循环次数的增大而增大,等效孔隙数量随冻融次数的增加而增多。

(3)临界孔隙直径一定时,随冻融循环次数增多,等效孔隙数量增大,等效孔隙直径减小,孔隙分布指数随冻融循环增加先负向增大,而后趋于稳定。

(4)冻结温度降低值只与孔隙分布指数有关,孔隙分布指数随冻融循环次数增多先负向增大,后趋于稳定,故冻结温度降低值随冻融循环次数增多而逐渐减小,最后趋于稳定。所以,黄土冻结温度随着冻融循环次数的增多先降低,后趋于稳定。

(5)土层冻结深度随冻结温度的降低而加深,即一定温度范围内,冻结温度越低,土层冻结深度越大。

(6)冻融循环改变了黄土内部细观结构,使得黄土结构的整体性降低,力学性质降低,在同等外力作用下,冻融影响下的U型渠道的稳定性降低。

参考文献

[1]何鹏飞,马巍.我国寒区输水工程研究进展与展望[J].冰川冻土,2020,42(1):182-194.

[2]李萍,黄丽娟,李振江,李新生,李同录.甘肃黄土高边坡可靠度研究[J].岩土力学,2013,34(3):811-817.

[3]王羿,王正中,刘铨鸿,肖旻.寒区输水渠道衬砌与冻土相互作用的冻胀破坏试验研究[J].岩土工程学报,2018,40(10):1799-1808.

[4]张国军,陆立国.影响衬砌渠道冻胀破坏严重的关键因素[J].中国农村水利水电,2012(9):105-108.

[5]Kozlowski T. A comprehensive method of determining the soil unfrozen water curves : 2. Stages of the phase change process in frozen soil-water system[J]. Cold Regions Science and Technology, 2003,36(1-3) :81-92.

[6]岳丰田,翁家杰,张勇,石荣剑,陆路.液氮地层冻结的理论与实践[M].北京:中国矿业大学出版社,2015.

[7]倪万魁,师华强. 冻融循环作用对黄土微结构和强度的影响[J].冰川冻土,2014,36(4):922-927.

[8]穆彦虎,马巍,李国玉,毛云程.冻融作用对压实黄土结构影响的微观定量研究[J].岩土工程学报,2011,33(12):1919-1925.

[9]袁俊平,李康波,何建新,刘亮,詹斌.基于孔隙分布模型的垫层料冻胀变形规律探讨[J].岩土力学,2014,35(8):2179-2183,2190.

[10]吴谋松. 冻融土壤水热盐运移规律研究及数值模拟[D].武汉:武汉大学,2016.

本文曾发表于2022年《南方农机》第53卷第17期

高强度预应力管桩在湿陷性黄土富集钙质结核地区的应用研究

刘亚军，冯浩然，程　浩，姜青山，周　健

（陕西省土地工程建设集团有限责任公司，陕西 西安 710075）

【摘要】西安少陵塬黄土地区存在较厚钙质结核坚夹层，预制管桩穿越该土层存在较大困难，针对这一问题，本文结合实际工程案例，对普通高强度预应力管桩（C80）在该类地区的应用进行多方案试验，经多方案对比分析，提出采用静力压桩及超高强预应力管桩（C105）+长螺旋引孔的方案，在保证较低爆桩率的情形下使桩尖达到设计标高，为此方案在严重湿陷性多层深厚钙质结核地区的应用提供理论研究及计算依据。

【关键词】湿陷性黄土；钙质结核；长螺旋；超高强预应力混凝土管桩

Application of High Strength Prestressed Pipe Pile in Collapsible Loess Rich of Calcareous Nodule Area

Yajun Liu, Haoran Feng, Hao Cheng, Qingshan Jiang, Jian Zhou

(Shaanxi Provincial Land Engineering Construction Group Co., Ltd.,, Xi'an 710075, China)

【**Abstract**】 Xi'an ShaoLing tableland loess areas with thick calcareous tuberculosis established interlayer, prefabricated pipe pile through the soil there is a big difficulty, to make the precast pipe pile tip to bearing layer, combining with the practical engineering case, the general high strength prestressed concrete pipe pile (C80) in this region solution test, the application of comparison and analysis of many alternatives, The scheme of static pile pressing, ultra-high strength prestressed pipe pile (C105) and long spiral hole guiding is proposed to make the pile tip reach the design elevation under the condition of low pile explosion rate, which provides theoretical research and calculation basis for the application of the scheme in seriously collapsible multi-layer and deep calcareous concretion area.

【**Keywords**】 Collapsible loess; Calcareous tuberculosis; Long spiral; Ultra-high strength prestressed concrete pipe pile

0　引言

黄土主要分布在北半球的中纬度干旱及半干旱地带[1-2]。黄土中碳酸盐成分占10%左右，导致黄土在降雨和气候干湿交替过程中形成一些钙质结核颗粒，这些钙质结核主要产出于黄土中的古土壤层底部、古土壤淋溶层之下一定深度的黄土中或分散于黄土层中[3]，其中在 Q_2 古土壤层上部的钙质结核一般较小，稀疏分布。钙质结核直径通常大于土颗粒，由于钙质结核的存在，使得土层具有特

基金项目：陕西省土地工程建设集团科研项目 DJNY2020-10。

殊的物质形态、结构组成和显著的非均匀性特点[4]。

目前穿越钙质结核层,主要桩基形式为钻孔灌注桩,但随着环境压力增大及对成本管控要求的提高,高强度预制管桩开始应用于此类土层。但在湿陷性黄土地区运用时受到场地湿陷等级、湿陷深度的影响,需做特殊的地基处理及强度计算。在钙质结核地区运用时,会出现坚硬钙质结核层难以穿越的情况。因此,在严重湿陷性夹杂钙质结核的复杂地质地区运用 PHC 管桩时将受到极大的限制,会出现断桩、爆桩及桩底达不到设计标高等情况。为解决此类地区混凝土管桩应用问题,本文引入超高强混凝土管桩(以下简称 UHC 管桩)施工工艺,通过多种试验方案及静载荷试验,对 UHC 管桩穿透湿陷性黄土及钙质结核硬层进行分析研究。

1　工程概况

本工程位于陕西西安,地貌单元属少陵塬二级黄土塬,总体地势东高西低。选取 47# 地质钻孔,场地地层物理力学性能指标如表 1 所示。根据地质勘察结果,地下水埋深为 57 m 左右,埋深较深。拟建场地为自重湿陷性黄土,湿陷等级为Ⅲ级(严重),10 m 以下古土壤夹杂白色钙质沉积层,含大量钙质结核和钙质条纹,⑨层底部钙质结核富集,直径为 2~5 cm,性状如图 1 所示。拟建建筑为普通住宅,地上 32 层,地下 2 层。基础埋深-10.7 m。采用静压成孔素土挤密桩为地基处理形式。地基基础首先采用普通 PHC 混凝土管桩进行试压,断桩、爆桩率较高,后引入 UHC 管桩+长螺旋引孔方式施工成功。

表 1　土体物理力学性质参数表(47# 钻孔)

土层编号	土层名称	土层厚度/m	深度/m	土层描述	承载力特征值/kPa	侧阻力端阻力标准值(预制桩)/kN	
						q_{sik}	q_{pk}
②	黄土 Q_3^{eol}	8.6	9.1	含钙质结核及钙质条纹,具轻微—中等湿陷性	155	-20(70)	
③	古土壤 Q_3^{el}	3.1	12.2	含较多碳酸钙条纹及零星钙质结核,具轻微—中等湿陷性和自重湿陷性	180	-20(74)	
④	黄土 Q_2^{eol}	8.3	20.5	含钙质条纹,具轻微—中等湿陷性和自重湿陷性	180	-20(75)	
⑤	古土壤 Q_2^{el}	4.1	24.6	具轻微—中等湿陷性,含较多碳酸钙条纹及零星钙质结核	190	-20(76)	
⑥	黄土 Q_2^{eol}	4.8	29.4	可见白色钙质条纹,具轻微—中等湿陷性	200	-20(70)	
⑦	古土壤 Q_2^{el}	2.9	32.3	含较多碳酸钙条纹及零星钙质结核,具轻微—中等湿陷性	210	80	3000
⑧	黄土 Q_2^{eol}	4.3	36.6	含较多白色钙质条纹,具轻微湿陷性	210	78	2800
⑨	古土壤 Q_2^{el}	3.1	39.7	含大量钙质结核和钙质条纹,底部钙质结核富集,不具湿陷性	210	80	3000
⑩	黄土 Q_2^{el}	4.1	43.8	零星的钙质结核,不具湿陷性	200	75	2700
⑪	古土壤 Q_2^{el}	1.9	45.7	含大量钙质结核和钙质条纹,底部钙质结核富集,不具湿陷性	210	82	2800
⑫	黄土 Q_2^{el}	9.4	55.1	含零星的钙质结核,不具湿陷性	210	80	3300
⑬	古土壤 Q_2^{el}	2.6	57.7	含大量钙质结核和钙质条纹,底部钙质结核富集,不具湿陷性	210	68	2700

注:括号内为经地基处理后的侧阻力标准值。

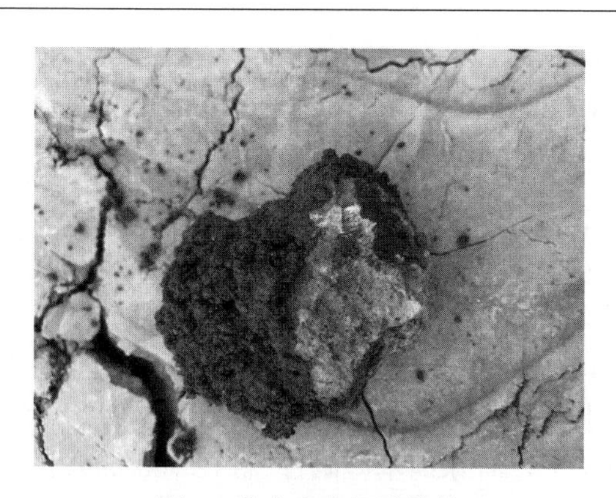

图 1 黄土夹杂钙质结核

2 湿陷性黄土对桩基承载力的影响

使用 PKPM 软件,采用荷载效应标准组合计算桩基结构的承载力,代入结构荷载,计算结果如图 2所示,桩基总反力为 527 746.188 kN。采用 PHC 管桩群桩基础,正三角形布桩,桩基间距 2100 mm,桩基根数为 253 根,单桩反力 $N_K = 2086$ kN。

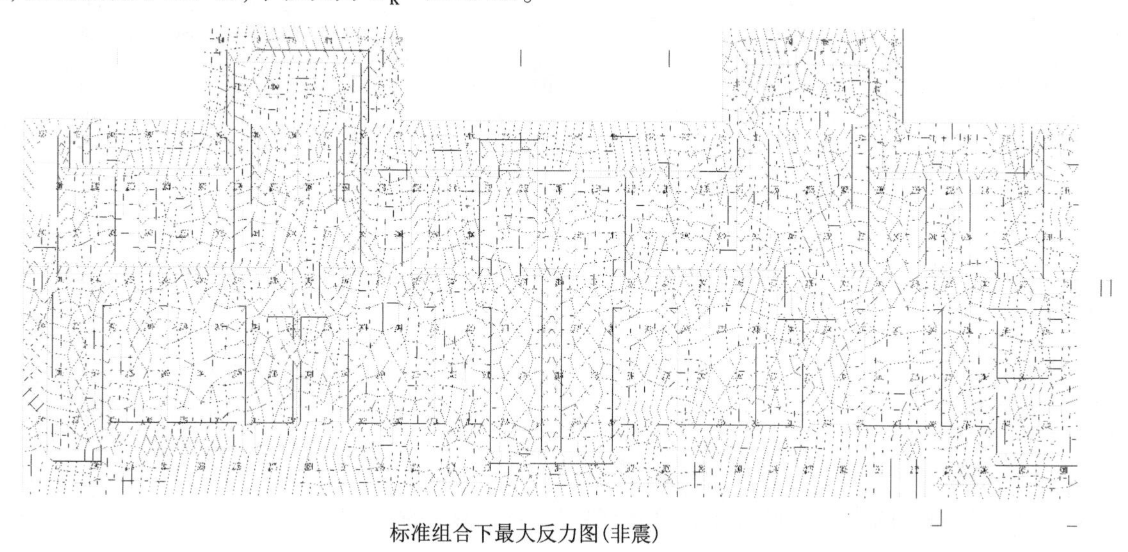

荷载=527977.000(kN) 桩反力=527746.188(kN) 桩承载力=632500.000(kN)(0.834)

图 2 标准组合下最大反力图

2.1 地基处理前基桩受力分析计算

因自重湿陷性黄土的存在,桩基承载力需考虑黄土湿陷的影响,按照建筑桩基技术规范要求,对于摩擦型基桩可取桩身中性点以上侧阻力为零[5],中性点深度计算如下:

$$l_n = l_0 \times 0.6 = 32.3 \times 0.6 = 19.38(\mathrm{m})$$

取⑫黄土层为桩端持力层,桩端进入持力层 8.32 m,桩基极限承载力计算如下:

$$\begin{aligned} Q_{uk} &= Q_{sk} + Q_{pk} = u \sum q_{sik} l_i + q_{pk}(A_j + \lambda_p A_p) \\ &= 3.14\times0.5\times(75\times1.12+76\times4.1+\times70\times4.8+80\times2.9+78\times4.3+80\times3.1+75\times4.1+82\times1.9+80\times5.32) \\ &\quad +3300\times(3.14\times(0.5^2-0.25^2)/4+0.8\times3.14\times0.25^2/4) \\ &= 3824.4+615 \\ &= 4439.4(\mathrm{kN}) \end{aligned}$$

承载力特征值 $R_a = \frac{1}{2} Q_{uk} = \frac{1}{2}\times4439.4 = 2219.7(\mathrm{kN})$

满足 $N_k = 2086 \leqslant R_a = 2\ 219.7$

此时计算桩长 $l = 1.12+4.1+4.8+2.9+4.3+3.1+4.1+1.9+5.32 = 45(\text{m})$

因 45 m 桩长过长,不经济且施工会有极大的困难,因此需进行地基处理。

2.2　湿陷性黄土的地基处理方案

为降低桩长,需对地基消除湿陷性,根据《湿陷性黄土地区建筑标准》(GB 50025—2018)6.1.2 条,对于大厚度湿陷性黄土地基上的甲类建筑,应将基础底面以下具自重湿陷性的黄土层全部处理,且应将附加压力与上覆土饱和自重压力之和大于湿陷起始压力的非自重湿陷性黄土层一并处理。根据地质勘察报告,并组织召开由《湿陷性黄土地区建筑标准》(GB 50025—2018)起草人参加的专家论证会,选取静压成孔素土挤密桩工艺消除地基湿陷性,桩径 560 mm,桩长 15 m,被处理土层处理前后土工参数变化如表 2 所示。由表 2 可知桩间土挤密系数及桩身土压实系数均满足规范要求。土体经过处理后含水率变化不大,土的干重度变大,孔隙比变小,土体因此更密实,土的自重湿陷性消除,湿陷系数为 0.01,基本消除湿陷性。

表 2　地基土处理前后相关参数变化对比

参数名称	处理前		处理后	
	范围值	平均值	范围值	平均值
含水率/%	21.5~19.0	20.07	25.0~15.8	20
干重度/(kN/m^3)	15.4~13.7	14.27	16.9~15.7	16.4
孔隙比	0.994~0.984	0.917	0.732~0.612	0.656
湿陷性系数	0.045~0.01	0.03	0.01~0.01	0.01
自重湿陷性系数	0.03~0.01	0.02	0~0.01	0
桩间土挤密系数			0.94~0.92	0.93
桩身土压实系数			0.98~0.94	0.96

2.3　地基处理后基桩受力计算分析

根据地基处理后的桩侧摩阻力标准值,按照建筑桩基技术规范,桩长计算如下:

$$
\begin{aligned}
Q_{uk} &= Q_{sk} + Q_{pk} = u\sum q_{sik} l_i + q_{pk}(A_j + \lambda_p A_{p1}) \\
&= 3.14\times0.5\times(70\times3.08+74\times3.1+75\times8.3+76\times4.1+\times70\times4.8+80\times2.9+78\times4.3+80\times2.42) \\
&= 3387.4+503 \\
&= 4390.4
\end{aligned}
$$

承载力特征值 $R_a = \dfrac{1}{2}Q_{uk} = \dfrac{1}{2}\times 4390.4 = 2195.2(\text{kN})$

满足 $N_k = 2086 \leqslant R_a = 2195.2$

此时计算桩长 $l = 3.08+3.1+8.3+4.1+4.8+2.9+4.3+2.42 = 33(\text{m})$

经地基湿陷性处理后桩长显著减小。

3　管桩施工工艺的研究

3.1　高强度预应力管桩试压桩试验

根据设计计算,桩长选用 PHC-500(125)AB 型管桩,桩长为 33 m,分为 3 节,分别为 10 m、11 m、12 m,选用 ZYJ860B-3 型液压静力压装机进行桩基施工。因进行湿陷性处理,场地土被挤密,且存在钙质结核,管桩压桩较为困难,选用 C80 管桩及 C80 管桩+长螺旋引孔进行试压,试桩成桩情况统计如表 3 所示。由表 3 可知,C80 管桩在进入⑨层古土壤 Q_2^{el} 存在压桩困难,达不到设计标高,当引孔 20 m 时桩底标高亦达不到设计标高或桩体提前爆桩。

表 3 C80 管桩压桩情况

C80 管桩压桩情况统计							
未引孔				引孔 20 m			
试桩编号	试桩入土桩长/m	终止压力值/kN	成桩情况	试桩编号	试桩入土桩长/m	终止压力值/kN	成桩情况
SZ1	33	3936	达到设计深度	SZ7	24	4700	爆桩
SZ2	25.6	5200	未达到设计深度	SZ8	31	5200	未达到设计深度
SZ3	23	5200	未达到设计深度	SZ9	33	5000	爆桩
SZ4	25.5	4900	未达到设计深度				
SZ5	26.1	5200	未达到设计深度				
SZ6	30.3	5200	爆桩				

因硬质钙质结核层的存在，普通混凝土管桩难以穿越，出现爆桩等现象，导致管桩压桩失败。本项目引入 C105 超高强预应力管桩工艺。

3.2 超高强度预应力管桩试压桩试验

试验采用 ZYJ860B-3 型液压静力压装机，共试压 6 根 33 m C105UHC 管桩，每节桩长为 10 m、11 m、12 m，采用长螺旋预引孔 20 m。试桩压入过程中每一行程(2 m)记录一次压力表读数，得出压桩力随桩底入土深度的变化规律，如图 3 所示。由图 3 中可知，在 10 m 深度范围内，压桩力随深度增长较为稳定，当超过 10 m 时，压力曲线变平缓，随入土深度增长，压力增加较大，是因为 10 m 左右需接桩，导致压桩停顿，此时孔隙水压力消散较快，桩端土体强度有一定恢复，导致再次压桩时压力增大。当入土深度 20 m 左右时，压桩力有一定回撤，是因为桩尖穿透钙质结核层，桩尖阻力降低导致压桩力降低。当桩尖穿透 20 m，随入土深度增加压桩力增长较快，与地勘报告地层中含大量钙质结核层的情况一致。

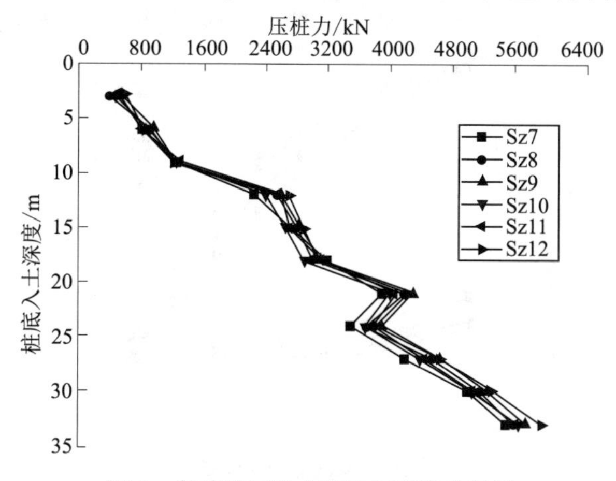

图 3 各试桩压桩力随桩身变化曲线图

3.3 超高强度预应力管桩单桩静载荷试验

试验装置由反力系统、加载系统、测量系统三部分组成。每组试验桩试验前，先对试验桩进行低应变动力测试，检测试验桩的桩身完整性是否符合规范要求。试验采用混凝土块堆载作为反力，总重量约为 5000 kN。将 2 台 6300 kN 油压千斤顶座于钢压板之上，千斤顶、荷重传感器、钢压板与试验桩中心严格对中。采用经系统标定的 CZLYB-4B(5000 kN)荷重传感器及其对应的数字测力仪，传感器编号为 01640704。采用 4 只百分表与强磁性表座及基准钢梁组成位移量测系统。

依据《建筑基桩检测技术规范》(JGJ 106—2014)，本次试验采用慢速维持荷载法。本次试验最大加载量为 5000 kN，分 9 级加载，第一级加载量为 1000 kN，以后每级加载增量为 500 kN。每级荷载施加后,应分别按第 5 min、15 min、30 min、45 min、60 min 测读桩顶沉降量,以后每隔 30 min 测读一次桩顶沉降量,每次测读值记入试验记录表。每小时内的桩顶沉降量不得超过 0.1 mm。当桩顶沉降速率达到相对稳定标准时,可施加下一级荷载。卸载时,每级荷载应维持 1 h,分别按第 15 min、30 min、60 min 测读桩顶沉降量后,即可卸下一级荷载;卸载至零后,应测读桩顶残余沉降量,维持时间不得少于 3 h,测读时间分别为第 15 min、30 min,以后每隔 30 min 测读一次桩顶残余沉降量。单桩静载试验结果如表 4 所示。

表 4　单桩静载荷试验结果

荷载	SZ1		SZ2		SZ3		SZ4		SZ5		SZ6		平均值	
/kN	沉降量	回弹量	沉降量	回弹量	沉降量	回弹量	沉降量	回弹量	沉降量	回弹量	沉降量	回弹量	沉降量	回弹量
0	0.00	11.78	0.00	12.28	0.00	13.02		11.41		13.44		12.48		8.39
1000	1.16	13.68	1.29	14.49		14.93		13.50		15.26		14.84		12.17
1500	1.89		2.02		2.15		2.33		2.08		2.47		2.16	
2000	2.70	14.65	2.86	15.48		16.07		14.74		16.33		16.03		13.11
2500	3.78		3.85		4.75		5.39		5.35		5.44		4.76	
3000	4.96	15.40	5.20	16.16		16.69		15.72		17.18		16.88		13.77
3500	6.33		7.01		8.20		9.16		9.48		9.26		8.24	
4000	8.24	15.71	9.29	16.55		17.03		16.30		17.78		17.36		14.17
4500	11.38		12.29		13.40		13.85		14.61		14.25		13.30	
5000	15.85		16.70		17.20		16.70		18.24		17.76		17.08	

由表 4 结果绘制相应的 Q-S 曲线图,如图 4 所示。由图 4 可知,沉降曲线为缓变型,没有陡降段,当达到最大加载量时沉降值均小于 40 mm,根据《建筑基桩检测技术规范》,对参加算术平均的试验桩检测结果,当极差不超过平均值的 30%时,可取其算术平均值为单桩竖向抗压极限承载力[6],取 $Q=5000$ kN,$R_a=\frac{1}{2}Q=2500$ kN 满足设计要求。

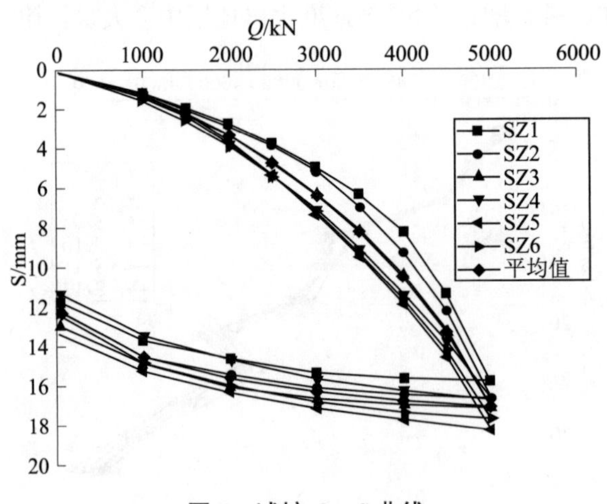

图 4　试桩 Q—S 曲线

4　结论

(1)湿陷性黄土经地基处理后,由于桩的挤密作用,桩侧阻力增加,有利于提高地基承载力,减小管桩桩长,降低工程造价。

(2)钙质结核层作为坚硬夹层,管桩穿越时,压桩力有突然变大的现象,穿过夹层后,压桩力有所下降。

(3)长螺旋引孔能够有效穿透钙质结核坚硬夹层,但由于长螺旋引孔施工时机械易晃动,造成引孔孔位出现偏差,施工完成后需复核引孔孔位及引孔的垂直度,必要时可复引。

(4)普通 C80 PHC 管桩因极限承载力较低,在用于穿越钙质结核类坚硬夹层时爆桩率较高,建议在此类地区谨慎采用。

(5)C105UHC 管桩+长螺旋引孔的方式能够成功穿透坚硬钙质夹层及深厚湿陷性黄土层,使桩尖置于非湿陷性的坚硬地层,能够在湿陷性黄土及富含钙质结核的地区推广运用。

参考文献

[1]刘祖典. 黄土力学与工程[M].西安:陕西科学技术出版社, 1997:1-4.

[2]Li Yanrong. A review of shear and tensile strengths of the Malan Loess in China[J]. Engineering Geology, 2018, 236: 4-10.

[3]孙建中. 黄土学 (上篇)[M].香港考古学会出版,2005: 206-208.

[4]黄宏翔, 陈育民, 王建平, 等. 钙质砂抗剪强度特性的环剪试验[J]. 岩土力学, 2018, 9(6): 2082-2088.

[5]中国建筑科学研究院.建筑桩基技术规范:JGJ94—2008[S].北京:中国建筑工业出版社,2004.

[6]中国建筑科学研究院. 建筑基桩检测技术规范:JGJ106—2014[S].北京:中国建筑工业出版社,2014.

本文曾发表于 2022 年《福建建筑》第 4 期

Deformation, Strength and Water Variation Characteristics of Unsaturated Compacted Loess

Xiao Xie[1,2,3], Li Qi[1,3], Xiaomeng Li[4]

(1.Shaanxi Provincial Land Engineering Construction Group Co. Ltd., Xi'an 710075, China; 2.Department of Geological Engineering, Chang'an University, Xi'an 710054, China; 3.Institute of Land Engineering and Technology, Shaanxi Provincial Land Engineering Construction Group Co. Ltd., Xi'an 710075, China; 4.Institute of Yunnan Provincial Highway Science and Technology, Kunming 650051, China)

【Abstract】The bearing capacity and safety of geotechnical structures are determined by the shear strength of soil. Soil is usually unsaturated in nature, and variation in unsaturated strength is very complicated. To investigate the unsaturated shear strength characteristics of compacted loess, a series of triaxial consolidated shear tests were conducted on compacted loess from the 6th loess layer in Yan'an, China, using GDS unsaturated soil triaxial apparatus. The saturation, volume and moisture change of compacted loess were analyzed, and the influence of matric suction (u_a-u_w) and net confining pressure (σ_3-u_a) on the deformation and strength characteristics of compacted loess were discussed. It was found that the deviatoric stress versus deviatoric strain curves exhibited a strain hardening state, and the volumetric strain transitions from shear contraction to shear dilatancy. In the testing pressure range ($u_a-u_w \leqslant 100$ kPa, $\sigma_3-u_a \leqslant 400$ kPa), both matric suction and net confining pressure significantly influence the deformation and strength characteristics of compacted loess. Under certain matric suction conditions, the shear strength of the compacted loess increased with increasing net confining pressure, the peak value of the shearing shrinkage increased, and the shear dilatancy decreased. Under a constant net confining pressure, the shear strength increased with the increasing matric suction. However, the peak values of shearing shrinkage and shear dilatancy decreased with increasing matric suction. In the unsaturated shear test, φ increased with the matric suction. The cohesion of unsaturated soil is affected by many factors and the data points from this research are limited owing to time constraints, hence, further analysis of cohesion strength is needed.

【Keywords】Unsaturated Compacted Loess; Strength; Water Variation; Deformation

1 Introduction

The shear streng is an important mechanical property of soil. The bearing capacity, earth pressure, slope stability, and many other issues related to geotechnical engineering in the limit state can be attributed to the soil shear strength problem. Shear strength depends on the stress state of the soil[1]. Saturated soil is a two-phase mixture, and its shear strength can be expressed by the Mohr-Coulomb failure criterion combined with the effective stress principle[2]. It can be observed from the expression that the shear strength of saturated soil depends on the change in the effective stress($\sigma-u_w$). Effective stress is the only state variable that controls the shear strength of saturated soil[3]. Compared with saturated soil, unsaturated soil has a different air phase and thus belongs to a three-phase mixture. The sides of the water-air interface bear different pressures, namely, pore air pressure and pore water pressure, the deviation of which is called matric suction. Therefore, the stress state of unsaturated soil is described by two variables: net normal stress ($\sigma-u_a$) and matric suction (u_a-u_w). The engineering properties of unsaturated soils are more complicated owing to the existence of matric suction. The related theory of saturated soil has been developed further, has matured in recent 80 years and is widely used in engineering. However, there are significant difficulties in studying related theoretical and resolving engineering problems of unsaturated soil because of the difficulty in obtaining relevant parameters and complex influencing factors.

The shear strength of unsaturated soils is affected by several factors [4-6]. Many studies have been conducted on unsaturated shear strength. Xing et al. [7] and Liao et al.[8] studied the changes in the shear strength parameters of original and remolded loess with different moisture contents through a shear strength test. Wang et al.[9] conducted a series of unsaturated consolidation drainage (CD) shear tests of soils with different compaction degrees under different loading rates and confining pressures to discuss the changes in shear strength parameters and deformation characteristics of this soil under different conditions. Patil et al. performed suction-controlled triaxial testing using axis translation and relative-humidity techniques to control matric suction at 0~750 kPa[10] and total suction at higher values, 20 MPa and 300 MPa[11]. Through these tests, an equation/model was proposed to predict unsaturated soil shear strength over a wide range of suction (0~300 MPa)[12]. Gao et al.[13] and Wang et al.[14] carried out CD tests under different suction and net confining pressures on loess through an unsaturated triaxial test that could control the suction. The influence of the dry density and matric suction on the strength and deformation characteristics of unsaturated loess was systematically studied. Previous studies have shown that matric suction has a significant influence on shear strength[15-17]; therefore, the contribution of matric suction must be considered in the study of the shear strength of unsaturated soil. Moisture content changes the shear strength by influencing the matric suction of the soil.

Loess has a large thickness and is widely distributed in north-western China. With further implementation of the Western Development Strategy, engineering construction has been propagating continuously in loess land[18], of which the "Moving Mountains and Building Cities" in Yan'an is the most typical. This project is located in the hilly loess area, with complex geological and hydrogeological engineering conditions and climate change[19]. Numerous engineering problems related to foundation bearing capacity and high fill slope stability will be encountered in the concrete implementation process, and the unsaturated shear strength of compacted loess is the key to solving these problems.

In this study, an effort was made to understand the shear strength characteristics and behavior of compacted loess. Loess from the 6th loess layer (L6) in Yan'an City, Shaanxi Province, was selected as the research object. Compacted loess samples were prepared using the static pressure method, and three different net confining pressures and matric suction were selected. Thus, nine groups of consolidated drained (CD) tests under different conditions were carried out using the GDS unsaturated triaxial test system, and the strength, moisture change, and deformation characteristics of the unsaturated compacted loess were analyzed.

2 Material and specimen preparation

Loess samples were obtained from Yan'an New District of Shaanxi Province, China, to investigate the unsaturated shear strength of the compacted loess used in this study. The sampling spot is receiving much concern[20] because "the campaign to bulldoze mountains to build cities." The sampled loess belonged to L6 according to strata age. The basic geotechnical properties of the tested materials are presented in Table 1. The loess is classified as lean clay according to the Unified Soil Classification System (USCS; ASTM 2011)[21].

The specimens were completely remolded and prepared in the laboratory for unsaturated shear strength testing. The block samples were air-dried, completely crushed, passed through a 2 mm sieve, and uniformly mixed with de-aired distilled water at an average gravimetric moisture content of 10% to prepare compacted loess specimens. The mixture was then placed in a sealed plastic bag in a moist chamber for at least 48 h to equilibrate the water content. Subsequently, the soil was split into five approximately equal parts and placed in a cylindrical specimen mold with a diameter of 39.1 mm and a height of 80 mm in five separate lifts. After each lift was placed in the mold, the static compaction approach was used to compact the soil to a predetermined dry density of 1.50 g/cm^3(similar to the intact loess). During each layer of compaction, the interfaces of the last layer were sufficiently scarified to ensure the good integrity of the compacted samples.

Table 1 Geotechnical properties of the tested material

Properties	Values	Properties	Values
Water Content (%)	13.3	Maximum Dry Density (g/cm^3)	1.79
Specific Gravity	2.71	Optimum Water Content (%)	16.0
Dry Density (g/cm^3)	1.54	Grain-size distribution (%)	
Natural Density (g/cm^3)	1.74	Gravel content (≥2 mm)	0
Plastic Limit (%)	18.7	Sand content (2~0.063 mm)	13.8
Liquid Limit (%)	28.5	Silt content (0.063~0.002 mm)	80.0
Plastic Index	9.8	Clay content (≤0.002 mm)	6.2

3 Testing device and the test program

3.1 Testing apparatus

This study used a GDS fully automatic unsaturated triaxial shear test device to obtain the unsaturated shear strength. It has remarkable advantages compared with traditional triaxial instruments; for example, it simultaneously controls the pore water pressure, pore air pressure, radial pressure, and axial pressure; measures and controls the matric suction through the axis translation technique[22]; and the volume change of the unsaturated specimens in the testing procedure can be accurately measured through the double-cell total volume change measuring system.

3.2 Test procedure

The test procedure used in this study was designed to provide information regarding the unsaturated shear strength of compacted loess at different matric suction and confining pressures. Nine consolidated drained (CD) tests were performed. Each test was performed under different matric suctions (25, 50, and 100 kPa) and net confining pressures (200, 300, and 400 kPa). The test procedure was mainly carried out in three stages: soil-water equalization, consolidation, and shearing. The desired matric suction was achieved during soil-water equalization by controlling the pore air and applying pore water pressure. Achieving the desired matric suction takes a relatively long time. It can be considered that the equalization state is achieved when the volume change of the sample is less than 0.05% of the total volume within 24 hours. After reaching the equilibrium state, the average water contents of the specimens corresponding to matric suction were 19.83%, 13.18%, and 12.23%, respectively.

The specimen was then allowed to undergo the next stage of consolidation. Matric suction was maintained at the target value during the consolidation stage. The confining pressure was set to increase linearly, and the rate of increase was 50 kPa/h. The soil sample was consolidated, and the back volume change was within 10 ml every 2 h. Subsequently, the matric suction and net confining pressure were maintained at the target value. The shearing rate should be sufficiently low to ensure dissipation of the pore water pressure[23]. Based on the current findings[10, 24], the specimen was sheared at a constant shearing rate of 0.005%/min (strain rate) by increasing the axial force.

4 Results and interpretation

4.1 Stress-strain relationship Analysis of compacted loess

The description of the soil stress-strain relationship is a key problem in studying various mechanical properties. The deviator stress-strain curve (Fig. 1, Fig. 2) of the GDS unsaturated triaxial shear test was obtained by processing the test data. These figures show that the deviator stress-strain curves are similar for both compacted loess samples and represent a weak hardening type.

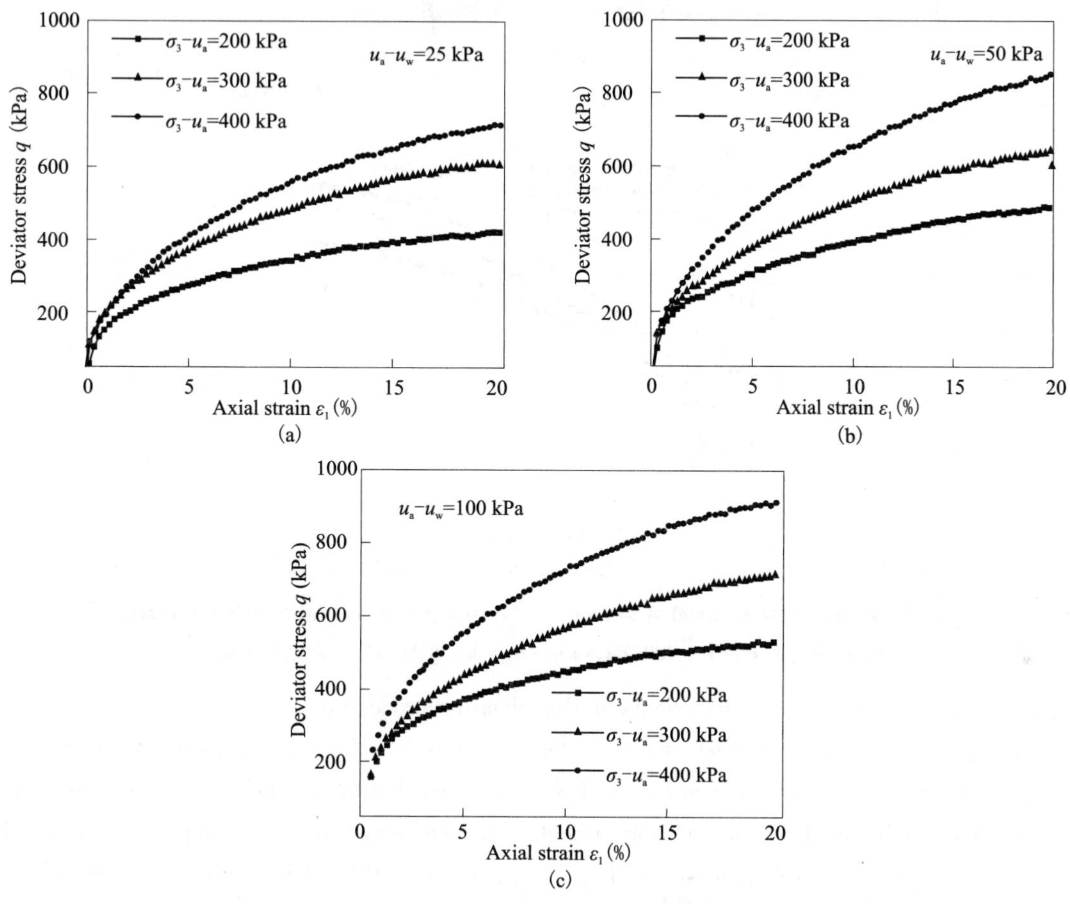

Fig. 1 Deviator stress-axial strain curve of compacted loess subjected to three different net confining stress conditions (a. $u_a - u_w = 25$ kPa, b. $u_a - u_w = 50$ kPa, c. $u_a - u_w = 100$ kPa)

The effect of the confining pressure on the stress-strain relationship is shown in Fig. 1. Under the same matric suction, the greater the net confining pressure, the higher the strength of the compacted loess. The effect of confining pressure on shear strength is reflected in two aspects: (i) at higher net confining pressures, the volume shrinkage of the specimens is larger, and the soil particles become denser; and (ii) the higher confining pressure is restricted more in the lateral direction; thus, the resistance that needs to be overcome when there is dislocation and rotation between soil particles increases. Therefore, the shear strength improved with net confining pressure. In addition, the specimens exhibited obvious hardening characteristics at higher confining pressures; however, the hardening characteristics were not evident under low confining pressures.

At a certain net confining stress, owing to the contribution of matric suction to the strength of the soil, the greater the matric suction, the higher the shear strength (Fig. 2). The effect of matric suction on the change in compacted loess specimens was not obvious in the suction control range in this study, which shows that matric suction contributes little to the shear strength of compacted loess when it is relatively small.

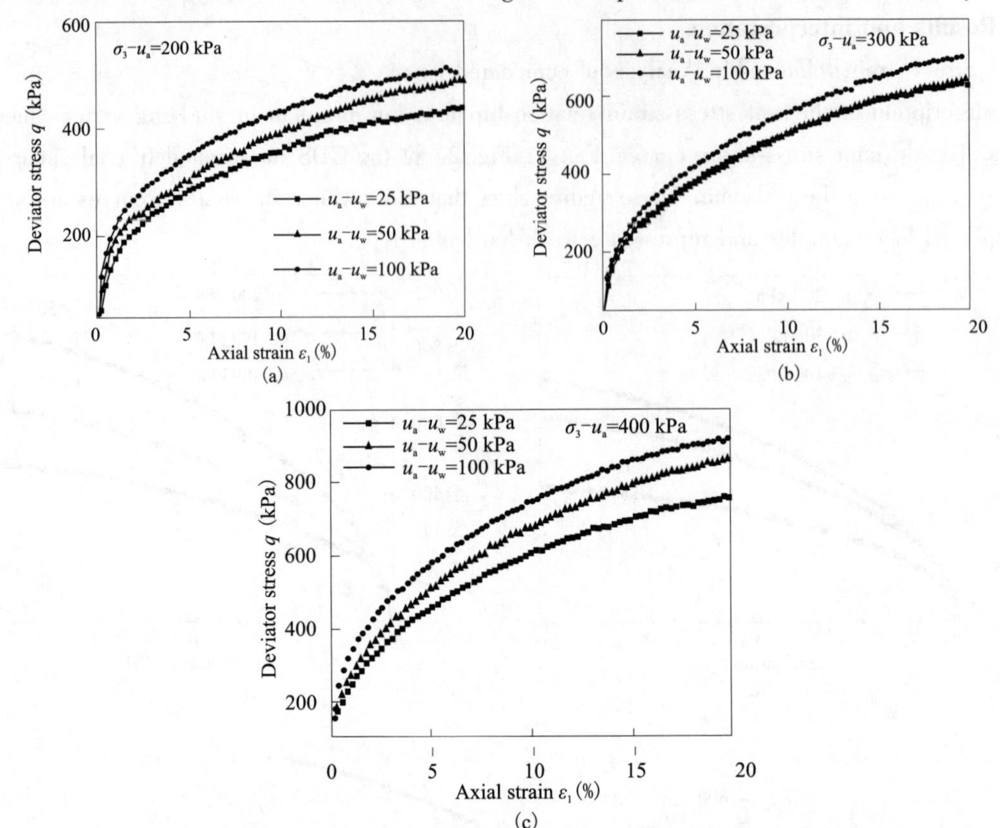

Fig. 2 Deviator stress-axial strain curve of compacted loess under different matric suctions (a. $\sigma_3-u_a=200$kPa, b.$\sigma_3-u_a=300$ kPa, c.$\sigma_3-u_a=400$ kPa)

4.2 Saturation and water volume change during shearing procedure

The saturation of the test specimens was calculated according to the test data, and the saturation strain curves of typical specimens are shown in Fig. 3. It was observed that during the shearing process, the saturation of the specimens initially increased until a peak value was achieved and subsequently decreased. The specimens with lower net confining pressures (Fig. 3(a), $\sigma_3-u_a=200$ and 300 kPa) and those with greater matric suction (Fig. 3(b), s = 50 and 100 kPa) showed little saturation change after shearing. However, the saturation of the other two specimens changed significantly.

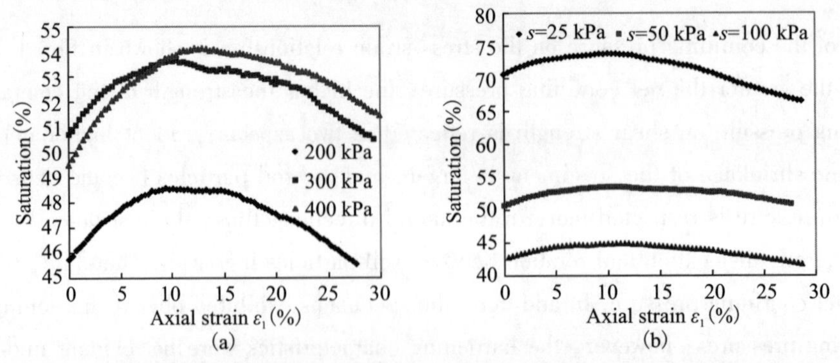

Fig. 3 Variation of saturation during the shearing process for typical specimens

(a) Variation of saturation during the shearing process for specimens with different net confining pressure.

(b) Variation of saturation during the shearing process for specimens with different matric suction

The saturation in the consolidation drained (CD) test was obtained by considering the overall volume change and in/out of the water volume of the specimens. The saturation calculation equation is as follows:

$$S_r = \frac{\omega G_s}{e} \tag{1}$$

Where S_r = saturation, ω = water content, G_s is the specific gravity, and e is the void ratio of the specimens. As illustrated in Eq. (1), the water content of the specimens was changed by the water in and out, and the volume change of the specimen changed the void ratio. In other words, the water inflow and volume shrinkage in the specimen can increase the soil saturation.

To better analyze the volume change and water volume of the specimens in the test process, we plotted the volume change and in/out of the water volume change in one picture (Fig. 4). The volume change trend of each specimen in Fig. 4 is essentially consistent: the specimens first shrink and reach a peak value, and then the volume increases. The total volume change of the sample after failure is represented by volume expansion. This phenomenon occurs in the initial stage, and the compression of the larger pores mainly causes the volume shrinkage of the specimens. As this process was carried out, the arrangement between soil particles became closer. As the deviator stress increased, the occlusal friction between the particles started to work, the particles moved, rotated, and rearranged, and the specimens began to swell.

By comparing Fig. 4 and Fig. 2, it can be seen that: (i) The net confining pressure significantly impacts the strength of the compacted loess. Under the same matric suction conditions, the greater the net confining pressure, the higher the strength of the compacted loess sample, the greater the peak value of shear shrinkage, and the smaller the final shear dilatancy. This is because when the specimen begins to compress under a high net confining pressure, the net confining pressure will also cause the specimen to shrink. In addition, the larger net confining pressure also causes the specimen to be subject to greater constraints, and the resistance to particle dislocation and rotation also increases. (ii) Under a certain net confining pressure, the matric suction is greater because of its contribution to soil strength. The larger the matric suction, the lower the moisture content, and the friction between soil particles increases. With an increase in matric suction, the peak shear shrinkage is smaller, and the dilatancy is more obvious. This is similar to the unsaturated triaxial test results for expansive soils conducted by Zhan and Wu[25]. However, the contribution of matric suction to the bulk deformation of the compacted loess was not significant.

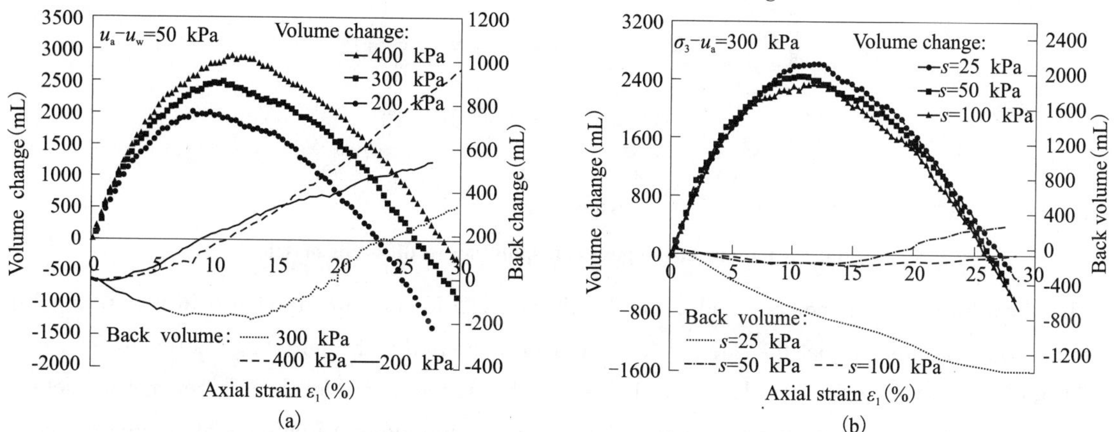

Fig. 4 Volume change and back volume change during shearing for typical specimens
(a) Specimens with different net confining pressure. (b) Specimens with different matric suction

In Fig. 4, the positive volume change indicates volume shrinkage, and the positive in/out of the water volume change shows the water inflow in the specimen. The same strain variation trend of the samples was observed: first shrinkage and then expansion. It can be seen from Fig. 3 and Fig. 4 that the trends of saturation and volume change are consistent, which indicates that during the shear process, the change in sample saturation is mainly affected by the change in acceptor, while the water inflow and outflow have little effect on the change in saturation. In Fig. 4 (a), the two specimens with matric suction of 50 kPa and net confining pressures of 200 kPa and 300 kPa show obvious volume expansion after shearing. However, the saturation of these two specimens changed slightly because the water entering the specimens offset the influence of the volume change on saturation. For the specimen with matric suction of 50 kPa and net confining pressure of 400 kPa, the volume expansion was smaller, and the water inflow was the maximum; thus, its saturation increased after shear. As shown in Fig. 4 (b), volume expansion and continuous drainage occurred in the specimen with a matric suction of 25 kPa and net confining pressure of 300 kPa, resulting in a greater saturation reduction. For the specimen with matric suction of 100 kPa and net confining pressure of 300 kPa, there was little change in water volume, and the saturation was mainly affected by volume change. From the water volume change of the specimens, it was observed that the influence of matric suction and net confining stress on the change in the water content of the specimens during the shear process had no significant regularity.

4.3 Effect of matric suction on shear strength

The shear strength of unsaturated soil is the total strength of saturated soil and the contribution of suction to the strength[26]. As can be seen from the stress-strain relationship curve of the unsaturated specimens (Fig. 1 and Fig. 2), there is no obvious peak point on this curve. Therefore, the stress corresponding to 15% of the axial strain was considered the failure stress. The failure stresses of the tested specimens are presented in Table 2. The strength envelopes are shown in Fig. 5. The strength parameters were calculated according to the methods described by Guo et al.[15] and Chen[27], and the calculated parameters are listed in Table 2.

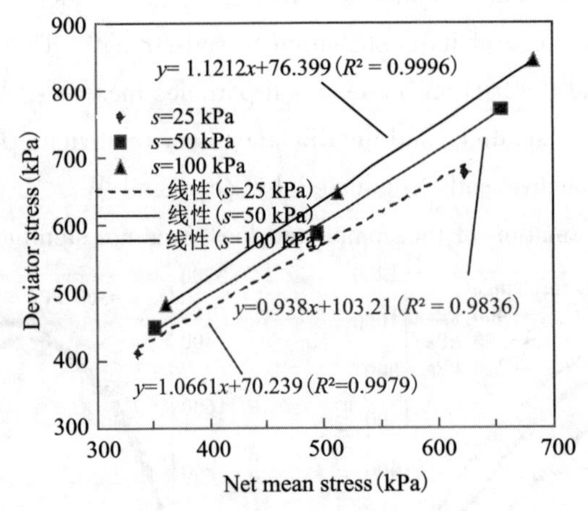

Fig. 5 Failure envelopes corresponding to different matric

The shear strength parameters c and φ under different matric suctions are shown in Fig. 6 and Fig. 7. Fig. 6 and Table 2 indicate that the internal friction angle increased with increasing matric suction, consistent with the observations of Cai et al.[28] and Lin et al.[29]. At the macro level, a larger matric suction implies a larger negative pore pressure and effective stress; thus, the internal friction angle increases according to the Mohr-Coulomb strength criterion. Zhai et al.[17] indicated that a larger suction stress results in additional compression stress on the soil structure and capillary bonding between the soil particles. Therefore, the curved air-water interface strengthens the unsaturated soil structure[17]. However, the cohesion first decreased and then increased, and the compacted loess exhibited peak cohesion. Based on the observations of many other scholars [30-32] who studied the strength of unsaturated undisturbed loess, cohesion increases with increasing matric suction, which is inconsistent with the results of this study.

Table 2 Parameters of shear strength on compacted loess

ρ_d(g/cm^3)	u_a-u_w(kPa)	σ_3-u_a(kPa)	p_f(kPa)	q_f(kPa)	tan w	φ'(°)	b(kPa)	c(kPa)
		200	336.71	410.12				
	25	300	496.12	588.66	0.94	24.0	103.21	48.83
		400	623.54	676.93				
		200	351.2	449.12				
1.5	50	300	494.68	589.13	1.07	27.0	70.24	33.45
		400	654.56	772.07				
		200	361.26	483.78				
	100	300	512.95	647.1	1.12	28.2	76.4	36.51
		400	683.35	844.65				

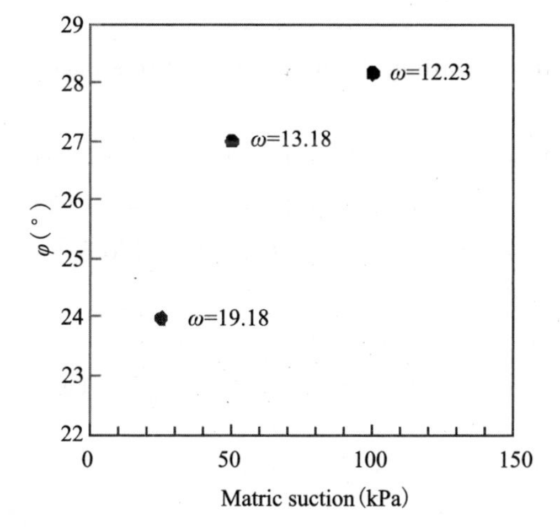

Fig. 6 The behavior of φ with respect to matric suction

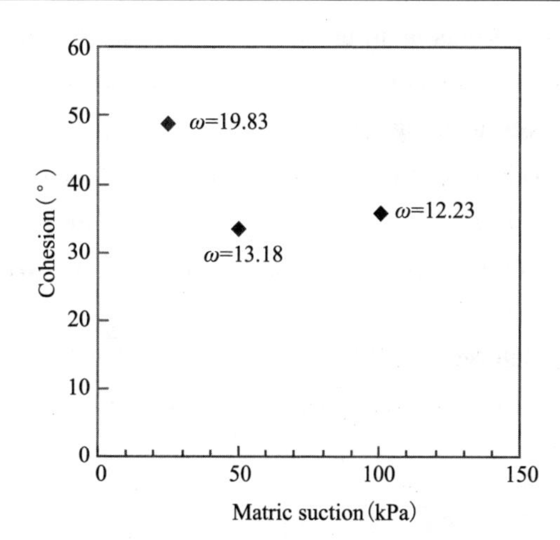

Fig. 7 The behavior of c with respect to matric suction

Many studies have been conducted on the change in the cohesion of unsaturated cohesive soil. Through a CU triaxial test on loess, Xing et al.[4] found that when the water content is higher than the plastic limit, the cohesion is essentially constant with decreasing water content (or increasing matric suction). When the water content was lower than the plastic limit, the cohesion increased gradually with decreasing water content. When the soil water content is between the plastic limit and saturated water content, the cohesion changes slightly. The transformation of cohesion is explained by the state of the soil water structure under the plastic limit. However, Lin et al.[29] showed that the cohesion of remolded soil with lower water content decreased with decreasing water content. They believed that in the boundary effect zone and transition zone of soil-water characteristic curve (SWCC), the cohesion of cohesive soil increased with the increasing matric suction. However, in the unsaturated residual zone, the cohesion changes with the soil state paths, and for remolded soil, the cohesion decreases with increasing matric suction. This may be related to the change in-bound water between the clay soil particles; the soil water in the residual volumetric water content belongs to bound water, which is not easy to remove. The recompacted soil structure was re-adjusted, and it was not easy to form a better cementing force. Therefore, cohesion decreases. It follows that the cohesion of unsaturated soil is affected by factors such as stress path, soil-water structure state, and soil type, and the study results of scholars have some differences. Owing to time limitations, the data points in this study were limited, and the variation characteristics of cohesion were not analyzed.

5 Conclusions

This paper presents test data and interpretation of the unsaturated shear strength characteristics of compacted loess. Unsaturated consolidated triaxial shear tests were performed under different net confining stresses, and matric suctions. The conclusions are as follows.

(1) The deviatoric stress versus deviatoric strain curves exhibit a strain-hardening state and volumetric strain transitions from shear contraction to shear dilatancy. Both matric suction and net confining pressure significantly influenced the deformation and strength characteristics of the compacted loess.

(2) The shear strength of the compacted loess increased with increasing net confining pressure, the peak value of the shearing shrinkage was larger, and the shear dilatancy decreased. The shear strength increased with the increase in matric suction, and the change in the peak value of the shearing shrinkage and shear dilatancy showed the opposite trend.

(3) In the unsaturated shear test, φ increased with the matric suction. The cohesion of unsaturated soil is affected by many factors, and several major findings have been listed.

Conflicts of interest

The authors declare that they have no conflicts of interest regarding the publication of this paper.

Acknowledgments

This work was supported by the National Natural Science Foundation of China (grant numbers 42072311); Technology Innovation Center for Land Engineering and Human Settlements, Shaanxi Land Engineering Construction Group Co.,Ltd and Xi' an Jiaotong University (Program No.2021WHZ0090).

References

[1] D. G. Fredlund, and N. R. Morgenstern "Stress state variables for unsaturated soils", *Journal of the Geotechnical Engineering Division*, vol.103, 1978.

[2] K.Terzaghi. "The shearing resistance of saturated soils and the angle between planes of shear", *Proc.of Int.conf.on Smfe*, vol.1, 1936.

[3] D. G. Fredlund and N. Yang. "Mechanical Properties and Engineering Application of Unsaturated Soil", *Chinese Journal of Geotechnical Engineering*, vol. 13, pp. 24-35, 1991.

[4] G.D. Hadžinikovi ć. "The influence of the grain-size distribution and soil structure on the unsaturated shear strength of loess sediments in Belgrade, Central Serbia", *Geoloski anali Balkanskog poluostrva*, no.70, pp. 83-91,2009.

[5] N.F. Zhao, W.M. Ye, Q. Wang, B. Chen, and Y.-J. Cui. "Influence of initial water content on unsaturated shear strength of compacted GaoMiaoZi (GMZ) bentonite", *Canadian Geotechnical Journal*. vol. 58, no.1, pp. 142-146, 2019.

[6] Kayadelen, C., Tekinsoy, M. A., & Taşkıran, T. "Influence of matric suction on shear strength behavior of a residual clayey soil", *Environmental Geology*, vol. 53, no.4, pp. 891-901, 2007.

[7] X. L. Xing, T. L. Li, P. Li et al. "Variation regularities of loess shear strength with the moisture content", *Hydrogeology & Engineering Geology*, vol. 41, no.3, pp. 53-59,97, 2014.

[8] H. J. Liao, T. Li, and J. B. Peng. "Study of strength characteristics of high and steep slope landslide mass loess", *Rock and Soil Mechanics*, vol. 32, no.7, pp. 1939-1944, 2011.

[9] Z. C. Wang, G. Jin, X. F. Wu et al. "Rate Dependent deformation characteristics and time dependent constitutive model of unsaturated compacted clay", *Rock and Soil Mechanics*, vol. 37, no.3, pp. 719-727, 2016.

[10] U. D. Patil, L. R. Hoyos, A. J. Puppala. "Modeling essential elastoplastic features of compacted silty sand via suction-controlled triaxial testing," *International Journal of Geomechanics*, vol. 16, no. 6, D4016012, pp.1-22,2016.

[11] U. D. Patil, A. J. Puppala, L. R. Hoyos, "Characterization of compacted silty sand using a double-walled triaxial cell with fully automated relative humidity control," *Geotechnical Testing Journal*, *ASTM*, vol. 39, no.5, pp. 742-756, 2016.

[12]U. D. Patil, A. J. Puppala, L. R. Hoyos et al., "Modeling critical-state shear strength behavior of compacted silty sand via suction-controlled triaxial testing," *Engineering Geology*, vol. 231, pp. 21-33, 2017.

[13]D.H. Gao, Z. H. Chen, N. Guo et al., "The influence of dry density and matric suction on the deformation and the strength characteristics of the remolded unsaturated loess soils," *Chinese Journal of Rock Mechanics and Engineering*, vol. 36, no.3, pp. 736-744, 2017.

[14]T. Wang, M. Zhou, J. W. Wang et al., "Experimental study on strength properties of unsaturated intact loess in loess tableland regions," *Chinese Journal of Geotechnical Engineering*, vol. 40, no.31, pp. 189-197, 2018.

[15]N. Guo, X. H. Yang, Z. H. Chen et al., "Influence of matricial suction on strength and deformation characteristic of unsaturated loess in intact form," *Journal of Lanzhou University of technology*, vol. 43, no. 6, pp. 120-25, 2017.

[16]F. R. Yan, X. Y. Hu, and F. Cui, "Experimental Study on Shear Strength Characteristics of Unsaturated Loess," *Journal of Water Resources and Architectural Engineering*, vol. 15, no. 5, pp. 100-104, 227, 2017.

[17]Q. Zhai, H. Rahardjo, A. Satyanaga et al., "Estimation of unsaturated shear strength from soil-water characteristic curve," *Acta Geotechnica*, vol. 14, pp. 1977-1990, 2019.

[18]D.Y. Xie, "Exploration of some new tendencies in research of loess soil mechanics," *Chinese Journal of Geotechnical Engineering*, vol. 1, pp. 3-13, 2001.

[19] Y. H. Wang, W. K. Ni, B. Y. Shi et al., "Stability Analysis of High-filled Loess Slope in Yan' an New District," *Journal of Water Resources and Architectural Engineering*, vol. 15, no. 5, pp. 52-56, 2014.

[20]P. Y. Li, Q. Hui, J. H. Wu, "Accelerate research on land creation," *Nature*, vol. 510, pp. 29-31, 2014.

[21]ASTM. Standard practice for classification of soils for engineering purposes (Unified Soil Classification System). ASTM standard D2487. American Society for Testing and Materials, West Conshohocken, Pa. 2011.

[22]C. W. W. Ng an R. Chen, "Advanced suction control techniques for testing unsaturated soils," *Chinese Journal of Geotechnical Engineering*, vol. 28, no. 2, pp.123-128, 2006.

[23]D. G. Fredlund, and H. Rahardjo. Soil mechanics for unsaturated soils (Wiley), 1993.

[24]U. D. Patil, A. J. Puppala, L. R. Hoyos, "Assessment of Suitable Loading Rate for Suction-Controlled Triaxial Testing on Compacted Silty Sand via Axis-Translation Technique," *Geo-Congress* 2014 *Technical Papers*, *Geotechnical special publication*, M. Abu-Farsakh, X. Yu, and L. Hoyos, eds. ASCE, Reston, VA, pp. 1307-1316, 2014.

[25]L. T. Zhan, C. W. W. Ng. "Experimental study on mechanical behavior of recompacted unsaturated expansive clay," *Chinese Journal of Geotechnical Engineering*, vol. 28, no. 2, pp.196-201, 2006.

[26]S. K.Vanapalli, D. G. Fredlund, D. E. Pufahl et al., "Model for the prediction of shear strength with respect to soil suction," *Canadian Geotechnical Journal*, vol. 33, no. 3, pp. 379-92, 1996.

[27]Z. H. Chen, "Deformation, strength, yield and moisture change of a remolded unsaturated loess," *Chinese Journal of Geotechnical Engineering*, vol. 21, no. 1, pp. 82-90, 1999.

[28]G. Q. Cai, C. Zhang, Z. W. Huang et al., "Experimental study on influences of moisture content on shear strength of unsaturated loess," *Chinese Journal of Geotechnical Engineering*, vol. 42, no. S2, pp. 32-36, 2020.

[29]H. Z. Lin, G. X. Li, Yuzhe Yu et al., "Influence of matric suction on shear strength behavior of unsaturated soils," *Rock and Soil Mechanics*, vol. 28, no. 9, pp.1931-1936, 2007.

[30]T. F. Gu, J. D. Wang, C. X. Wang et al., "Experimental study of the shear strength of soil from the Heifangtai Platform of the Loess Plateau of China," *Journal of Soils and Sediments*, vol. 19, no. 10, pp. 3463-3475, 2019.

[31]R. Schnellmann, H. Rahardjo, and H. R. Schneider, "Unsaturated shear strength of a silty sand," *Engineering Geology*, vol. 162, pp. 88-96, 2013.

[32]M. Y. Fattah, M. D. Ahmed, & H. A. Mohammed, "Determination of the Shear Strength, Permeability and Soil Water Characteristic Curve of Unsaturated Soils from Iraq," *Journal of Earth Sciences and Geotechnical Engineering*, vol. 3, no. 1, pp. 97-118, 2013.

本文曾发表于《Case Studies in Construction Materials》第16卷

Influence of Constructed Rapid Infiltration System on Groundwater Recharge and Quality

Chao Guo [1,2,3], Jiake Li[4], Yingying Sun[1,2], Wang Gao[1], Zhongan Mao[2], Shenglan Ye[3]

(1.Shaanxi Provincial Land Engineering Construction Group, Key Laboratory of Degraded and Unused Land Consolidation Engineering, Ministry of Natural Resources., Xi'an 710071, China;2.Shaanxi Engineering Research Center of Land Consolidation, Shaanxi Provincial Land Consolidation Engineering Technology Research Center, Xi'an 710071, China;3.Shaanxi Provincial Land Engineering Construction Group, Land Engineering Technology Innovation Center, Ministry of Natural Resources., Xi'an 710071, China;4.State Key Laboratory Base of Eco-Hydraulic Engineering in Arid Area, Xi'an University of Technology, Xi'an 710048, China)

【Abstract】 Infiltration of stormwater through green surfaces is an important way of groundwater recharge, however, the increase in constructed impervious area caused by intensive urbanization has led to a reduction of rainwater infiltration over the past decade. The constructed rapid infiltration system (CRIs) with an ample open space plays an important role in groundwater recharge. The objective of this study is to explore the influence of stormwater (roof runoff) concentration infiltration on groundwater table and quality in CIRs. More than 2 years (Ootober 2017—December 2019) groundwater table monitoring is conducted by the method of the continuous on-line monitor combined with manual sampling. The results show that the addition of zeolite to CRIs has good removal effect on rainfall runoff pollutants, and the influence of stormwater concentration infiltration on groundwater quality is small when the CRIs enters the stable running stage. The increased proportion of COD, N, and P in J1 are all less than 10% from 2018 to 2019, and they are less than 20% for heavy metals. The stormwater concentration infiltration can recharge groundwater and increase the groundwater depth, while the groundwater depth varies from 0.5 to 1.5 m during the monitoring period. The influence scope of the concentrated infiltration on groundwater table and quality is within 45 m, but it is more than 25 m. And the response of the groundwater table and quality to the stormwater concentration infiltration of J1, 25 m away from the CRIs, exhibits hysteresis, and the lag time is about 3 ~ 4 months. Conclusions from related research can provide important theoretical support for further study of groundwater recharge by CRIs.

【Keywords】 Concentration Infiltration; Groundwater Table; Groundwater Quality; Recharge; Constructed Rapid Infiltration System

1 Introduction

Urbanization has become a global trend with the rapid development of the economy and vast growth of the population, it is difficult for the urban storm drainage system to cope with the extreme climate. A series of serious ecological problems are emerged for various activities, such as urban flooding, waterlogging, no-point source pollution, and water shortage. This has caused the greatest challenge for stormwater management (Guo et al., 2019). Infiltrating stormwater locally into the ground instead of discharging to conventional pipe sewers is increasingly considered as a means of controlling urban tormwater runoff. The utilization of stormwater is an effective measure to solve the problem of the overexploitation of groundwater and to prevent urban disasters (Bach et al., 2010; Zhou et al., 2017). As an important source, stormwater can alleviate the crisis of urban water shortage. At present, the comprehensive regulation of surface water, groundwater and stormwater appears particularly important. Stormwater is frequently used to recharge groundwater, and it can increase groundwater level and reduce rainfall runoff volume. Importantly, the purpose of flood control and fully utilization of the rainwater resource has achieved (Zhai et al., 2017). Some industry developed countries, such as Germany, Japan, Australia and the United States, etc. have formed the mature rainwater utilization measures. The complete security systems have been established, and the research on low-impact development has gradually been carried out (Eisapour et al., 2011). China began to establish sponge city in 2014, and to achieve the optimal allocation of urban rainwater resources (Shao et al., 2016; Nguyen et al., 2019). At present, the stormwater recharge measures mainly include surface infiltration, which allows rainwater to infiltrate the soil through permeable surfaces. Rain gardens, lawn, permeable pavements, river beaches and the constructed rapid infiltration system are the permeable surfaces with good infiltration conditions (Trowsdale and Simcock, 2011; Brown et al., 2012).

The constructed rapid infiltration system (CRIs) is a sewage treatment technology developed from the sewage land treatment system(Yang et al., 2017; Su et al., 2020). It is usually filled the material with good permeability, such as natural river sand, ceramsite, gravel, coal gangue, etc. (Zhao et al., 2019; Duan and Su et al, 2014). The purpose is to replace the natural soil layer and to increase the hydraulic load of the system. CRIs is operated in a dry and wet alternate way. The pollutants in the CRIs are removed through the comprehensive action of filtration, adsorption and retention, physical and chemical reactions, and degradation of aerobic and anaerobic microorganisms (Kang, 2011).

The sewage treatment by CIRs to recharge groundwater has the risk of groundwater contamination (Vystavna et al., 2019; Al-Yamani et al., 2019). However, in terms of sewage infiltration, it is a most economical and safer way to use the stormwater to recharge groundwater. Recharge is what permits aquifers to remain active. It can make full use of the limited water resources to achieve the purpose of raising the groundwater table and preventing land subsidence. Meanwhile it takes into account the social, economic and ecological benefits (Mensah et al., 2022). Substrate and depth-to-water-table determine how quickly surface water will enter the underground watercourse. Shallow, sandy aquifers close to Earth's surface recharge much more quickly than aquifers nestled hundreds of meters below Earth's surface (Gallagher, 2017). At present, Sub-surface infiltration structures, CRIs, are the most common types of infiltration systems. CRIs can lead stormwater into aquifers through wells directly, and it eliminates the process of the interception and adsorption of rainwater by plants and soil. Since most of the fillers in CIRs are medium-coarse sand, and the water infiltration rate is relatively big, so it is easy to cause groundwater pollution(Lee et al., 2012). Therefore, 5% of zeolite, as the improvement material, is added to the CIRs. And the project aimed (1) explore the influence of stormwater concentration infiltration on groundwater quality in the CRIs; and (2) quantify the influence of stormwater concentration infiltration on groundwater table; and (3) estimate the rainfall runoff volume reduction rate and the pollutant removal rate with 5% of zeolite in CIRs. Conclusions from the related research can provide important theoretical support for further study of groundwater recharge by CRIs. And it can provide scientific and reasonable evaluation for the development of stormwater recharge measures.

2 Materials and methods

2.1 Description of the case study site and introduction of CIRs

The study area is located in a campus in Xianyang City, Shaanxi Province. The soil is mainly composed of silt loam. The bulk density is generally 1.35 g/cm^3, comprising 9% clay, 80% sandy silt, and 10% sand. The city is situated on widely distributed loess soil that generally has deep profile of more than 50 m. It has better geological conditions for the storage and migration of groundwater. The climate is semi-arid and humid. The annual average temperature in Xianyang is 13.3 ℃, and the rainfall is 580.2 mm, but the evaporation is

990 mm. The precipitation is unevenly throughout the year, and more than 80% of rainfall occur from May to October. This study involves a constructed rapid infiltration system(CIRs), the study area is seen in Fig.1.

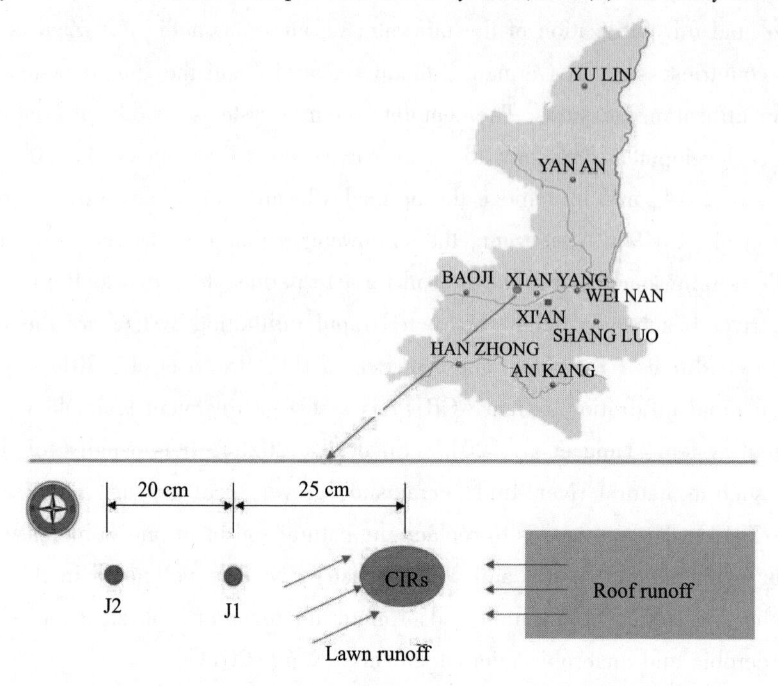

Fig.1 The study area

The CRIs was monitored to investigate stormwater concentration infiltration on groundwater table and quality from October 2017 to December 2019. It was installed in July 2017 and collects and treats stormwater runoff from 1100 m^2 of concrete roof, roads, and lawn. It is circular (diameter = 2.5 m, height = 3 m; area = 4.91 m^2). The CRIs is including two parts. The upper part, made of reinforced concrete, is a percolation tank, and the other part is the double glass-steel pipe installed at the bottom of the tank for water draining (Fig.2). The inner diameter of the pipe is 600 mm, and the length is 6 m. The height of the pipes inserted the medium sand of stratum is 0.5m, but it is more than 1 m beyond the highest groundwater table. The effective thickness of the filler layer is 2.4 m. The filter in this study is the mixed materials of medium-coarse sand and the zeolite, and the proportion of the zeolite 5%(mass ratio). The particle size of the sand varies from 0.10 to 1.0 mm. However, the size of the zeolite is 1~3 mm. The percolation tank area of CRIs is 4.91 m^2, and the confluence area is 1100 m^2(950 m^2 concrete impervious surface and 150 m^2 lawn). The confluence ratio is 220:1 (confluence ratio = the confluence area/tank area). A horizontal water sample collection pipe is set at the lower 1/3 of the filler layer, and a Φ350 PVC pipe is inserted vertically in the middle of the filler to collect the outflow (Fig.2 (a), (b) and(c)). The scene drawing of CRIs and groundwater monitoring well is seen in Fig.3.

2.2 Monitoring and analysis methodology

The groundwater flow in the study area is from east to west. In order to study the effluence of stormwater concentration infiltration on groundwater table and quality in CRIs, a monitoring well (J1) is drilled 25 m away from the CRIs, and a background well (J2) is drilled 45m away from the CRIs (Fig.2(c)and Fig.3). The groundwater table online monitor is installed in J1, and the data is recorded every 5 minutes. The water samples in J1 are collected manually to test the concentration of COD, N, P and the heavy metals. The groundwater depth and samples in J2 are measured manually, and they are collected 2~3 times a month. A total of 56 groundwater table values and water samples are collected in J2 from October 2017 to December 2019, The rainfall infiltration recharge is calculated by formula (1).

$$V = \alpha \cdot P \cdot F \tag{1}$$

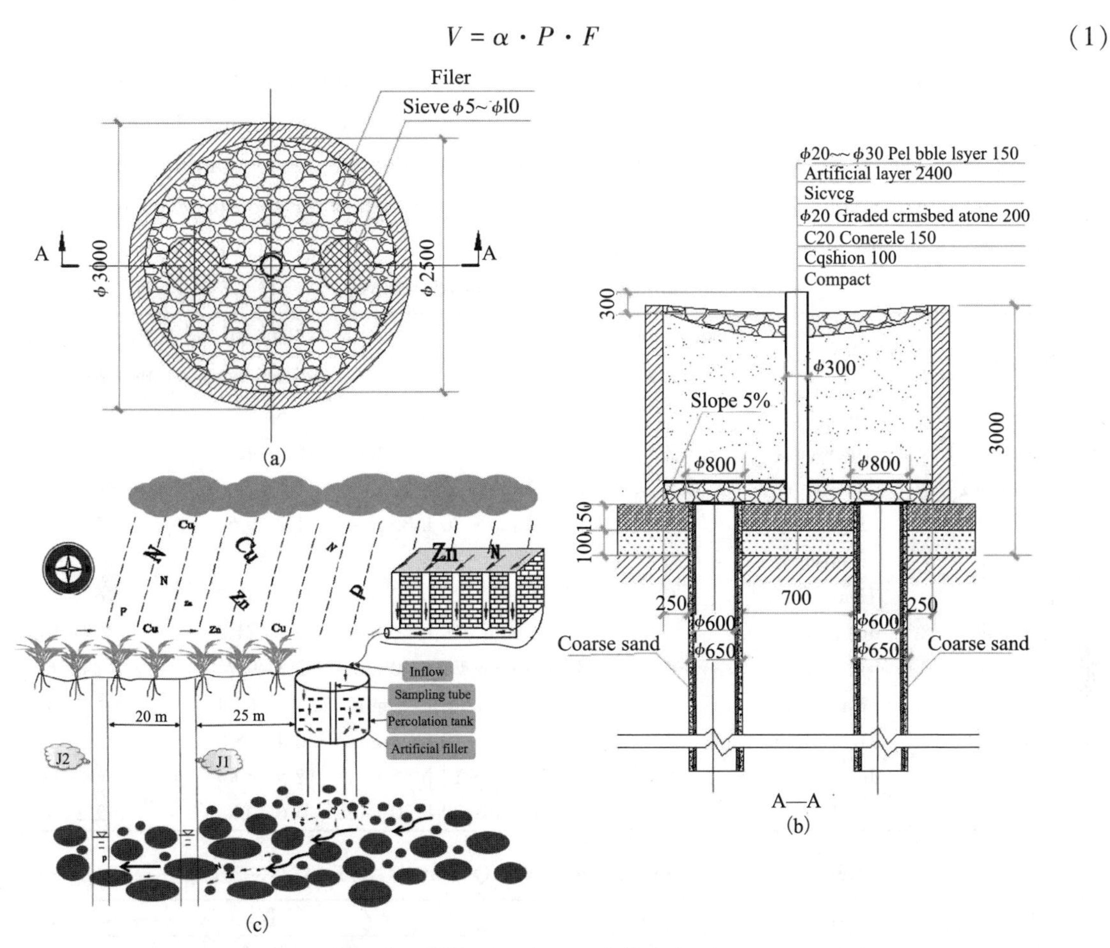

Fig. 2 CRIs structure and flowchart

Note: (a) the bottom structure of the percolation tank; (b) CRIs structure; (c) flowchart.

(a) (b)

Fig. 3 The scene drawing of CRIs and groundwater monitoring well (J1)

Note: (a) CRIs; (b) J1.

V is Recharge volume, m^3; α is Precipitation infiltration recharge coefficient, the rainfall volume varied from 348.0~726.2 mm, $\alpha = 0.15 \sim 0.24$ (Shu et al., 2008). Rainfall volume kept between 503.6 and 788.4 mm during the monitoring period, α is 0.15 in this article; P is Rainfall volume, mm; F is The confluence area, km^2.

Pollutant reduction can be evaluated through the indicator of the concentration removal rate (R_c), as follows:

$$R_c = \frac{C_{in} - C_{out}}{C_{in}} \times 100\% \tag{2}$$

Where R_c is the pollutant concentration removal rate, in %; C_{in}、C_{out} is the inflow and outflow concentration in a single rainfall event, mg/L.

Define the rainfall volume within 24 h as the criteria used to distinguish rainfall types. When the volume is more than 50 mm, it is defined as rainstorm, and if it kept between 25 and 50 mm, it is defined as heavy rain. However, the volume between 10 and 25 mm, it is defined as moderate rain, and it is defined as light rain when the volume is less than 10mm. Rainfall parameters for the storm events was shown in Table 1. The "Standard for Groundwater Quality" (GB/T 14848—2017) in China is shown in Table 2.

Table 1 Rainfall characteristics

Monitoring period	Rainfall type	Monitoring times	Rainfall duration	Rainfall volume/mm	Maximum rain intensity within 60 min(mm/h)
13 Ootober–18 November 2017	Rainstorm	1	1 h 20 min	54.5	45.8
	Heavy rain	3	3 h 25 min~16 h 30 min	25.6~42.5	16.3~35.6
	Moderate rain	5	9 h 35 min~21 h 20 min	12.6~24.5	8.6~15.3
	Light rain	7	8 h 10 min~20 h 35 min	1.2~9.5	1.1~7.5
12 April–15 September 2018	Rainstorm	0	—	—	—
	Heavy Rain	9	5 h 40 min~20 h 30 min	28.6~40.2	16.2~23.2
	Moderate rain	15	8 h 35 min~18 h 50 min	15.7~21.2	8.3~16.8
	Light rain	20	8 h 20 min~21 h 40 min	2.3~9.7	1.8~6.5
20 April–17 December 2019	Rainstorm	2	1 h 15 min~3 h 45 min	59.8~66.4	35.9~48.7
	Heavy Rain	7	4 h 40 min~10 h 25 min	30.4~45.8	15.2~22.8
	Moderate rain	8	6 h 20 min~17 h 55 min	12.5~20.4	6.8~18.2
	Light rain	9	9 h 10 min~20 h 30 min	5.6~9.8	2.2~7.5

Table 2 Standard for Groundwater Quality (GB/T 14848—2017) Unit: mg/L

Category	Ⅰ	Ⅱ	Ⅲ	Ⅳ	Ⅴ
NH_4^+-N	≤0.02	≤0.1	≤0.5	≤1.5	>1.5
NO_3^--N	≤2.0	≤5.0	≤20.0	≤30.0	>30.0
Cu≤	≤0.01	≤0.05	≤1.0	≤1.5	>1.5
Zn≤	≤0.05	≤0.5	≤1.0	≤5.0	>5.0

2.3 Samples collection analysis

The concentration of COD, N and P in water were tested and analyzed using standard analysis methods. The COD is measured by potassium dichromate rapid digestion spectrophotometry (DR5000). The NH_4^+-N and NO_3-N in water were measured by continuous flowing analysis (SKALAR, Holland), and TN and TP were measured by ultraviolet spectrophotometer (DR5000). SRP samples were first filtered through a 0.22 μm membrane filter, then analyzed using methods identical to TP analysis, except without the digestion process for SRP. TSS was processed gravimetrically by Standard Method 2540 D.

3 Results

3.1 Rainfall runoff reduction rate

A total of 86 rainfall events are monitored from October 2017 to December 2019. The characteristics of rainfall is shown in Table 1, and the infiltration volume and the runoff reduction rate are shown in Fig. 4. Most of the rainfall events are occurred in April to September every year. There are a total of 3 rainstorm events and 19 heavy rains, and the remaining are all moderate or light rain. The rainstorm in 2017 with rainfall volume is 54.5 mm, and the maximum rainfall intensity within 60 min reaches 45.8 mm. There is no rainstorm in 2018. 2 rainstorm events occur in 2019, and the rainfall volume varies from 59.8 to 66.4 mm, and the maximum rain intensity within 60 min ranges from 35.9 to 45.7 mm. Fortunately, there is no overflow during the monitoring period, and the runoff reduction rate and the flood peak reduction rate of CRIs all reach 100% (Fig.4). It indicates that even if the CRIs has a large confluence area, the rainfall runoff reduction effect is good. The main reason is that the filler in CRIs is medium-coarse sand, and the permeability coefficient is about 2.38×10^{-3} m/s, and it has good water-conducting effect on rainfall runoff. Relevant studies have shown that the reduction of rainfall runoff mainly depends on the moisture content of the filler (Crespo et al., 2011). The moisture content of CRIs is consistently small, and the permeability is good. Therefore, CRIs has better reduction effect on rainfall runoff.

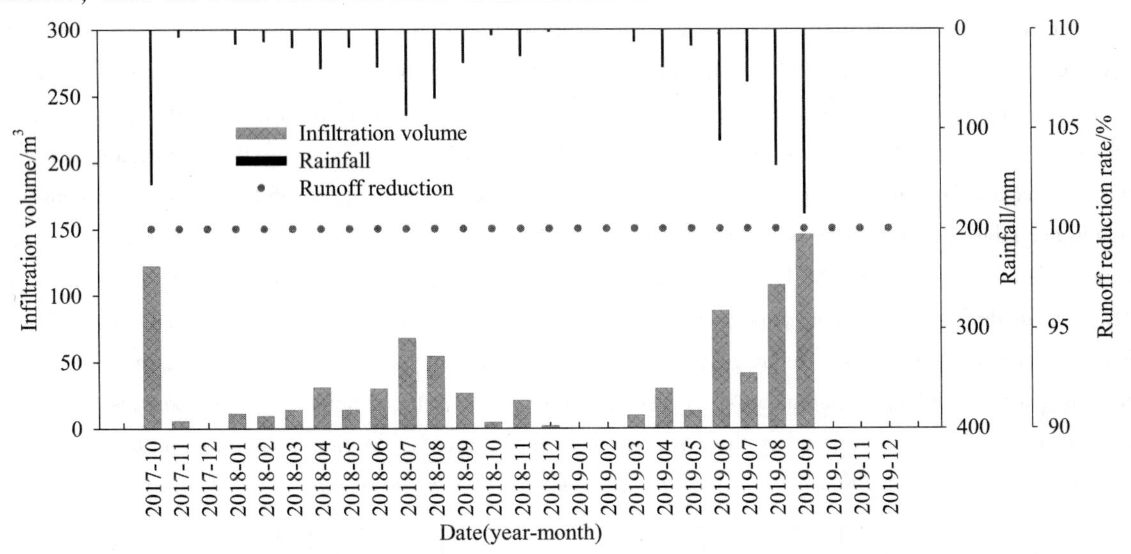

Fig. 4 Rainfall infiltration and runoff reduction of CIRs

3.2 Pollutants purification effect of CRIs

A total of 21 rainfall events are monitored from October 2017 to September 2019. The inflow and outflow water samples are collected to study the pollutants purification effect by the CRIs. The COD, TSS, N and P removal rate is calculated by formula (2), and the effect of runoff pollutant removal rate is shown in Table 3. The pollutants removal rate gradually increased with the monitoring time. The purification effect is poor for the first three rainfall events from October to November in 2017, and the removal rates are all negative except for COD, NH_4^+-N and TSS. It indicates that the pollutants in the CRIs fillers are leached with the infiltration of water. However, the mean removal rates of all the pollutants are positive in 2018, and the purification effect is very good for NH_4^+-N and TSS, and the removal rates reach 80.45% and 76.85%, respectively. But the removal rates of TN and TP are 70.49% and 64.75%. The main reason is that CRIs are completed in August 2017 in an unstable operation stage. At the same time, the CRIs mainly receives roof rainfall runoff with good water quality, and the dissolved pollutants in the CIRs are leached as a result of erosion by rainwater infiltration, so the removal rate of pollutants is mostly negative. However, the CRIs gradually reach a stable operation stage for rainfall runoff pollutants purification. The better removal purification of NH_4^+-N mainly depends on the zeolite. The natural zeolite was rich in Al^{3+}, and the Al^{3+} was hydrolyzed to form an $Al(OH)_3$ colloid, which plays an important role in the adsorption of NH_4^+ in a water body. Researches show that the zeolite has a good adsorption effect on NH_4^+-N (Hong et al., 2008). The removal rates of NH_4^+-N and TSS reach 85.64% and 80.16% in 2019, respectively, and they are 72.48% and 70.40% for TN and TP. The purification effect of NO_3^--N and SRP are poor during the monitoring period, and the removal rates remain at −40.15%~26.79% and −35.42%~32.48%, respectively.

3.3 Influence of stormwater concentration infiltration on groundwater table

The groundwater table depth is affected by rainfall infiltration recharge, and the groundwater in this study area is mainly recharged by rainfall runoff through the CRIs. Generally, the whole process of rainfall enters the soil and then reaches the groundwater is called rainfall infiltration recharge (Grismer et al., 2015). However, rainfall infiltration is different from the recharge, and when the rainfall reaches the ground, part of water is intercepted by soil or vegetation roots. Thus the amount of available water reaching the groundwater is reduced.

Fig. 5 shows the relationship among the daily average groundwater table depth, rainfall volume and rainfall infiltration recharge in J1. The groundwater depth in the study area is greatly affected by the concentration infiltration recharge of the CRIs. There is minor change for groundwater depth in the rainy season (March–September), and it is basically in the period of pollutant purification stability. While it fluctuates greatly in the non-rainy season (October–February of the following year). The change process of groundwater depth is inconsistent with the rainfall recharge. It indicates that the groundwater recharge by rainfall infiltration in the rainy season has obvious hysteresis, and the response of J1, 25 m far away from CRIs, to the concentration infiltration recharge is delayed, and the lag time is about 3 ~ 4 month. The stormwater concentration infiltration in CRIs causes the groundwater table to rise year by year, and the water depth becomes significantly shallow. The variation range of the depth is increased within 0.5 ~ 1.5 m. The groundwater depth of J2 is deeper than that of J1, and the variation of J2 is small during the monitoring period, and the range remains between 0 and 0.47 m. Hocking and Kelly (2016) studied the groundwater recharge and time lag measurement through Vertosols using impulse response functions, they found long-term groundwater level trend rose 1 m over the simulation period. Conclusions from the related research show that if it takes the CRIs as the concentrated infiltration point, the influence scope on groundwater table is greater than 25 m, but the impact is small when the distance exceeds 45 m.

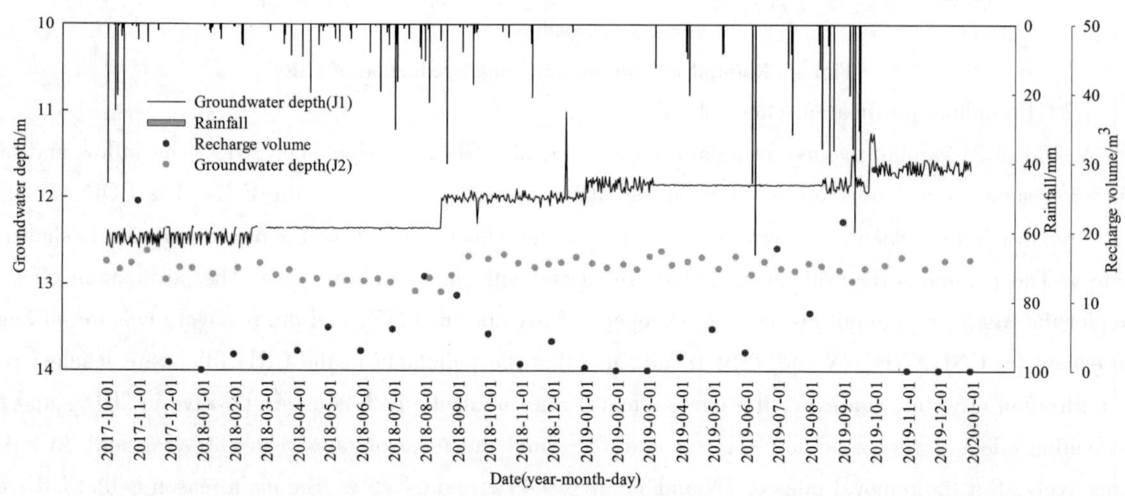

Fig. 5 Groundwater depth

3.4 Influence of stormwater concentration infiltration on groundwater quality

A total of 64 and 65 groundwater samples in J1 and J2 are collected from July 2017 to December 2019, respectively. It can be seen from the Table 4 that the pollutants concentration in J1 and J2 are all less than those of the inflow during the monitoring period, however, they are bigger in J1 than those in J2. The pollutants concentration in J1 reach the maximum value in the early operation stage of the CRIs from July to December 2017. And they increase slightly at the stable operation stage from 2018 to 2019, and the mean values of COD, TP, TN, NH_4^+-N, NO_3^--N, Cu, Zn, Cd in 2019 are increased by 5.32%, 8.00%, 8.25%, 3.13%, 9.49%, 17.39%, 19.67% and 15.38% respectively. The increased proportion of COD, N, and P is less than 10%, and it is less than 20% for heavy metals (the background value is small). The NH_4^+-N, NO_3^--N, Cu, and Zn concentrations at the stable operation stage reach the class Ⅱ or Ⅲ standard values specified in the "Standard for Groundwater Quality" (GB/T 14848—2017) (Table 2). Meanwhile, there is minor change of COD, N、P and heavy metals in J2 over the monitoring time. Thus it is revealed that the influence scope of the concentrated infiltration on groundwater quality is within 45 m, but it is more than 25 m.

Table 3 Runoff pollutant removal rate

Date	Times	Items	COD/%	TP/%	SRP/%	TN/%	NH_4^--N /%	NO_3^--N/%	TSS/%
13 Ootober–18 November 2017	3	Range	2.35~26.45	−35.42~−12.24	−45.67~−22.65	−13.89~20.45	−3.78~60.67	−28.88~−56.79	−8.09~55.78
		Avg±Sd	8.46±25.16	−22.46±36.05	−35.42±28.49	−20.28±32.46	34.45±42.78	−40.15±26.54	25.46±37.89
12 April–15 September 2018	7	Range	38.32~56.78	55.25~72.41	28.67~47.46	58.75~78.61	58.35~88.75	−35.46~48.62	50.46~85.06
		Avg±Sd	45.64±27.49	64.72±39.06	30.48±39.78	70.49±102.78	80.45±87.46	26.79±112.64	76.85±59.45
20 April–17 September 2019	11	Range	21.64~72.45	46.98~85.42	26.12~51.62	55.05~80.45	66.34~85.74	−25.45~50.48	56.75~88.64
		Avg±Sd	54.12±84.51	70.42±110.15	32.48±98.74	72.48±124.45	85.64±88.98	22.78±122.17	80.16±88.42

Table 4 Groundwater quality

Date		Times	Items	COD/(mg/L)	TP/(mg/L)	TN/(mg/L)	NH_4^+-N/(mg/L)	NO_3^--N/(mg/L)	Cu/(mg/L)	Zn/(mg/L)	Cd/(mg/L)
5 July 2017–18 December 2019	Inflow	21	Range	12.354~55.785	0.037~0.412	0.354~4.236	0.356~1.465	0.112~2.687	0.034~0.534	0.0346~1.264	0.015~0.253
			Avg±Sd	35.889±25.214	0.057±0.076	1.698±0.887	0.885±0.779	0.085±0.775	0.096±0.243	0.534±0.856	0.036±0.052
5 July–18 December 2017	J1	16	Range	9.432~39.24	0.021~0.241	0.222~2.542	0.217~0.930	0.054~1.933	0.003~0.183	0.019~0.864	0.005~0.086
			Avg±Sd	24.429±21.734	0.030±0.066	1.128±0.704	0.489±0.354	0.505±0.566	0.034±0.041	0.191±0.171	0.014±0.029
	J2	16	Range	7.709~22.545	0.009~0.083	0.228~2.025	0.058~0.989	0.025~1.805	0.005~0.250	0.020~0.708	0.007~0.075
			Avg±Sd	18.573±22.733	0.027±0.060	1.036±0.669	0.348±0.223	0.256±0.473	0.018±0.053	0.148±0.188	0.011±0.051
2 February–7 December 2018	J1	23	Range	15.839~39.544	0.013~0.312	0.398~2.542	0.471~1.075	0.004~1.815	0.011~0.212	0.022~0.858	0.006~0.094
			Avg±Sd	22.032±13.286	0.025±0.075	1.043±1.346	0.415±0.338	0.316±0.538	0.023±0.028	0.122±0.053	0.013±0.025
	J2	24	Range	16.240~30.953	0.008~0.083	0.232~2.145	0.155~0.809	0.025~1.240	0.009~0.221	0.023~0.812	0.004~0.092
			Avg±Sd	20.757±20.486	0.028±0.070	0.961±0.650	0.354±0.208	0.260±0.064	0.023±0.033	0.102±0.036	0.009±0.046
12 January–16 December 2019	J1	25	Range	9.432~39.24	0.019~0.128	0.222~2.742	0.217~1.230	0.041~1.753	0.012~0.125	0.021~0.564	0.008~0.102
			Avg±Sd	23.205±26.022	0.027±0.031	1.129±0.528	0.428±0.434	0.346±0.477	0.027±0.056	0.146±0.211	0.015±0.031
	J2	25	Range	7.709~32.545	0.011~0.172	0.128~2.257	0.042~0.975	0.208~1.805	0.005~0.250	0.020~0.708	0.006~0.046
			Avg±Sd	20.508±22.131	0.026±0.046	1.001±0.709	0.359±0.236	0.252±0.624	0.021±0.053	0.142±0.188	0.011±0.048

4 Discussion

The CRIs is widely used as rapid rainwater treatment facilities. The confluence area of the CRIs is large. Relevant studies have shown that the confluence ratio of CRIs can reach 50∶1～250∶1(Zhao et al., 2019). Since the CRIs can absorb rainfall runoff 50～250 times of their own area, and most of the runoff volume infiltrates to recharge the groundwater. Thus, the CRIs needs to treat roof runoff with good water quality, and it can't be used to treat the road runoff to prevent groundwater pollution (Ma et al., 2018).

The purification effect of pollutants by the CRIs mainly depends on the filler. Generally, the CRIs are filled with medium or coarse sand, and the well pollutants adsorption materials such as zeolite, granite and muscovite are added to improve the purification effect. Hydraulic retention time(HRT) is an important factor affecting the pollutants purification effect in the CRIs(Aris et al., 2016). However, the duration of the HRT should not be too long, otherwise it will affect the water infiltration rate of the CRIs. Zhao et al. (2019) showed that the water infiltration rate of the CRIs should be bigger than 2.6×10^{-4}m/s. And it is 2.38×10^{-3} m/s in this study, which can meet the requirements of rapid water infiltration. However, the purification effect on pollutants mainly depends on the filtration, adsorption and ion exchange of the fillers in CRIs. If the water infiltration rate of the CRIs is big, the purification effect on pollutants is poor (Zhao et al., 2011). Compared with bioretention facilities, the purification effect of CRIs on pollutants is relatively poor, mainly because the organic content of the fillers in bioretention facilities is big, which provides favorable living conditions for microorganisms. The microorganisms play an important role in degrading pollutants in bioretention facilities after rainfall (Li and Davis, 2008). The fillers of the CRIs are mainly medium-coarse sand with low organic matter content, which is not suitable for the survival of microorganisms, so the possibility of degradation of pollutants by microorganisms in CRIs is small (Zuo et al., 2019). The removal rate of NO_3^--N and SRP is small in this study. Firstly, the CRIs mainly receives roof runoff with good water quality, and it has a high water infiltration rate. Under the scouring effect of rainfall runoff, the soluble nitrate and phosphate is prone to leaching(Li and Davis, 2015). Secondly, due to the negative charge of NO_3^--N, and the filler is also negatively charged. NO_3^--N is extremely active in the filler, so it is easy to migrate downward with the water infiltration (Blecken et al., 2010). Finally, the CRIs are filled with the mixed fillers of medium-coarse sand and zeolite, and they have a low content of iron and aluminum salts, so they are less effective for P removal (Liu and Davis, 2014). Kamali et al. (2017) found dissolved phosphate in the groundwater was higher in concentration than the stormwater influent indicating that it was produced or added in the infiltration bed. Contaminants, such as nitrates and pathogens, are not completely removed before runoff enters the underground watercourse(Gallagher, 2017).

The groundwater recharge by rainfall infiltration is affected by the groundwater table depth. Relevant studies have shown that when the groundwater depth exceeds 25 m, the effect of groundwater recharge by rainfall infiltration is weak (Xia et al., 2016). It is because that when the rainfall has potential energy in the soil, it can flow and penetrate, and the infiltration depth is limited, rather than endless. That is to say, when the infiltration reaches a certain depth, the driving potential energy will be exhaust, and the water will not penetrate downward due to the resistance of the soil. The groundwater depth in the study area is about 12 m. The stomwater concentration infiltration by the CRIs can recharge the groundwater and increase the water depth. The increased range of groundwater depth varies from 0.5 to 1.5 m. The researchers found that groundwater at a depth of 1 meter below the water table consisted almost entirely of stormwater and that stormwater did not penetrate to depths greater than 3 meters below the water table (Watson et al., 2018). Boisson et al.(2014) found that the groundwater table in the irrigation area rose by an average of 2 m every year when surface water was used for irrigation. Some scholars also asserted that the groundwater recharge by irrigation water was negatively correlated with the groundwater depth (Massuel et al., 2014).

The response of groundwater table to rainfall runoff recharge has a hysteresis due to the impact of the groundwater flow field (Hocking and Kelly, 2016), and the lag time is about 3~4 month. Liu (2021) showed that with the increase of the groundwater depth, the lag duration time of infiltration recharge is more obvious. When the groundwater depth reach 8 m, the lag time of the peak value of irrigation infiltration recharge may reach more than 2~3 months. However, analysis of hydrograph response times with depth to water table (KCB, 2010) showed no clear correlation. The influence scope of the concentrated infiltration on groundwater table and quality is within 45 m, but it is more than 25 m. The impact of stomwater concentration infiltration on groundwater quality is small. The increased proportion of COD, N, and P in J1 are all less than 10% from 2018 to 2019, and they are less than 20% for heavy metals. The NH_4^+-N, NO_3^--N, Cu, and Zn concentrations at the stable operation stage reach the class Ⅱ or Ⅲ standard values specified in the "Standard for Groundwater Quality" (GB/T 14848—2017) (Table 2). This conclusion is consistent with Eshtawi et al. (2016). He declared that no evidence so far points at a high risk of groundwater contamination but the quality of surface soils will decrease due to long-term infiltration of polluted stormwater runoff.

5 Conclusions

The CRIs is an important facility for groundwater recharge and has good applicability for the treatment of roof rainfall runoff. In the early operation stage of the CRIs, the pollutants concentration in groundwater are relatively big. After the CRIs enters in a stable operation stage, the pollutants concentration in groundwater increase slightly over the monitoring time, but the increased proportion of COD, N, and P is less than 10%, and it is less than 20% for heavy metals. The stomwater concentration infiltration of the CRIs has little impact on groundwater quality, however, it can recharge groundwater and increase the groundwater table depth. The groundwater table has an obvious response to the rainfall runoff infiltration recharge. The increased of groundwater depth ranges from 0.5 to 1.5 me, but the response of groundwater table to rainfall runoff recharge in monitoring well, 25 m far away from the CRIs, has obvious hysteresis, and the lag time is about 3~4 months. The influence scope of the stomwater concentration infiltration on groundwater table and quality in CRIs is bigger than 25 m, while the impact is small when the distance exceeds 45 m.

Funding

This research was financially supported by the Key Research and Development Program of Shaanxi (2022ZDLSF-06-04), the Technology Innovation Center for Land Engineering and Human Settlements, Shaanxi Provincial Land Engineering Construction Group Co., Ltd and Xi' an Jiaotong University (2021WHZ0090), the project of Shaanxi Provence Land Engineering Construction Group (DJTD2022-4 and DJNY2022-26).

Refference

Al-Yamani, W., Green, S., Pangilinan, R., Dixon, S., Shahid, S. A., Kemp, P., Clothier, B. (2019). Water use of Al Samr (Acacia tortilis) forests irrigated with saline groundwater and treated sewage effluent in the hyper-arid deserts of Abu Dhabi. Agricultural Water Management, 216, 361-364.

Aris, M. A. M., Chelliapan, S., Din, M. F. M., Anwar, A. N., Shahperi, R., Selvam, S. B. (2016). Effect of effluent circulation and hydraulic retention time (HRT) on the performance of a modified anaerobic baffled reactor (MABR) during start-up period. Desalination and Water Treatment, 57(40), 18597-18605.

Bach, P. M., Mccarthy, D. T., Deletic, A. (2010). Redefining the stormwater first flush phenomenon. Water Resesrch, 44, 2487-2492.

Blecken, G.T., Zinger, Y., Deletic, A., Fletcher, T.D., Hedstrom, A., Viklander, M. (2010). Laboratory study on stormwater biofiltration nutrient and sediment removal in cold temperatures. Journal of Hydrology, 394 (3-4), 507-514.

Boisson, A., Baisset, M., Alazard, M., (2014). Comparison of surface and groundwater balance approaches in the evaluation of managed aquifer recharge structures: case of a percolation tank in a crystalline aquifer in India. Journal of Hydrology, 519, 1620-1633.

Brown, R. A., Line, D. E., Hunt, W. F. (2012). LID treatment train: Pervious concrete with subsurface storage in series with bioretention and care with seasonal high water tables. Journal of Environmental Engineering, 138 (6), 689-697.

Crespo, P. J., Feyen, P., Buytaert, W., Bücker, A., Breuer, L., Frede, H.G., Ramírez, M. (2011). Identifying controls of the rainfall-runoff response of small catchments in the tropical Andes (Ecuador). Journal of Hydrology, 407, 164-174.

Duan, J. M., Su, B. (2014). Removal characteristics of Cd(II) from acidic aqueous solution by modified steel-making slag. Chemical Engineering Journal, 246, 160-167.

Eisapour, A. H., Eisapour, M., Sardari, P. T., Walker, G. S. (2010). An innovative multi-zone configuration to enhance the charging process of magnesium based metal hydride hydrogen storage tank. Journal of Energy Storage, 36, 102443-102450.

Eshtawi, T., Evers, M., Tischbein, B., Diekkrüger, B. (2016). Integrated hydrologic modeling as a key for sustainable urban water resources planning. Water Research, 101, 411-428.

Grismer, M. E., Bachman, S., Powers, T. (2015). A comparison of groundwater recharge estimation methods in a semi-arid, coastal avocado and citrus orchard (Ventura County, California). Hydrological Processes, 14(14),2527-2543.

Gallagher, K. C. (2017). Stormwater Infiltration and Groundwater Integrity: An Analysis of BMP Siting Tools and Groundwater Vulnerability. University of South Florida, 3-16.

Guo, C., Li, J. K., Li, H. E., Li, Y. J. (2019). Influences of stormwater concentration infiltration on soil nitrogen, phosphorus, TOC and their relations with enzyme activity in rain garden. Chemosphere, 233,207, 207-215.

Hocking, M., Kelly, B. F. J. (2016). Groundwater recharge and time lag measurement through Vertosols using impulse response functions. Journal of Hydrology, 535,22-35.

Zheng, H., Han, L. J., Ma, H. W., Zheng, Y., Zhang, H. M., Liu, D. H., Liang, S. P. (2008). Adsorption characteristics of ammonium ion by zeolite 13X. Journal of Hazardous Materials, 158(2-3),577-584.

Kamali, M., Delkash, M., Tajrishy, M. (2017). Evaluation of permeable pavement responses to urban surface runoff. Jouranl of Environmental Management, 187,43-53.

Kang, A. (2011). The Treatment of High Ammonia Nitrogen Domestic Sewage by Three-Step Series Constructed Rapid Infiltration System. China University of Geosciences, Beijing, 38-56.

KCB, 2010. Central Condamine alluvium, stage II-Conceptual hydrogeology study. Prepared by Klohn Crippen Berger for the Department of Environment and Resource Management, Queensland.

Lee, B. S., Hong, S. W., Kang, H. J., Lee, J. S., Nam, K. P. (2012). Groundwater Recharge and Discharge in the Urban-rural Composite Area. Journal of Soil and Groundwater Environment, 17(2),37-46.

Li, H., Davis, A. P. (2008). Urban particle capture in bioretention media. II: Theory and model development. Journal of Environmental Engineering, 134 (6), 419-432.

Li, L.Q., Davis, A.P. (2015). Urban Storm-water Runoff Nitrogen Composition and Fate in Bioretention Systems. Environmental Science Technology, 48 (6), 3403-3410.

Liu, J. Y., Davis, A.P. (2014). Phosphorus speciation and treatment using enhanced phosphorus removal bioretention. Environmental Science Technology, 48(1), 607-614.

Liu, X. (2021). Research on groundwater hydrochemical evolution and recharge mechanism in the Chinese Loess Plateau on catchment scale. Northwest Agriculture and Forestry Technology, 35-43.

Massuel, S., Perrin, J., Mascre, C. (2014). Managed aquifer recharge in South India: what to expect from small percolation tanks in hard rock. Journal of Hydrology, 512, 157-167.

Ma, Y., Hao, S., Zhao, H., Fang, J., Zhao, J., Li, X. (2018). Pollutant transport analysis and source apportionment of the entire non-point source pollution process in separate sewer systems. Chemosphere, 211, 557.

Mensah, J. K., Ofosu, E. A., Yidana, S. M., Akpoti, K., Kabo-bah, A. T. (2022). Integrated modeling of hydrological processes and groundwater recharge based on land use land cover, and climate changes: A systematic review. Environmental Advances, 8,10224-10240.

Mikkelsen, P. S., Jacobson, P., Fujita, S.(2012). Infiltration practice for control of urban stormwater. Journal of Hydraulic Research,6, 827-840.

Nguyen, T. T., Ngo, H. H., Guo, W. S., Wang, X. C., Ren, N. Q., Li, G. B., Ding. J., Liang, H. (2019). Implementation of a specific urban water management-Sponge City. Science of Total Environment, 652,147-162.

Shao, W. W., Zhang, H. X., Liu, J. H., Yang, G. Y., Chen, X. D., Yang, Z. Y., Huang, H. (2016). Data Integration and its Application in the Sponge City Construction of CHINA. Procedia Engineering, 154,779-786.

Shu L C., Tao Y F., Liu P G. (2008). Reliability calculation method for groundwater recharge in consideration of uncertainty of hydrogeological parameters. SHUILI XUEBAO, 39(3),346-350.

Su, C. Y., Zhu, X. W., Shi, X. W., Xie, Y., Fang, Y. T., Zhou, X. B., Huang, Z., Xin, X. F., Chen, M. L. (2020). Removal efficiency and pathways of phosphorus from wastewater in a modified constructed rapid infiltration system. Journal of Cleaner Production, 267, 122063-122588.

Trowsdale, S. A., Simcock, R. (2011). Urban stormwater treatment using bioretention. Jouranl of Hydrology, 397 (3-4), 167-174.

Vystavna, Y., Schmidt, S. I., Diadin, D., Rossi, P. M., Vadillo, I. (2019). Multi-tracing of recharge seasonality and contamination in groundwater: A tool for urban water resource management, 161,413-422.

Watson, A., Miller, J., Fleischer, M., Clercq, D. W. (2018). Estimation of groundwater recharge via percolation outputs from a rainfall/runoff model for the Verlorenvlei estuarine system, west coast, South Africa. Journal of Hydrology, 558, 238-254.

Xia, J., Zhao, X., Chen, Y., Ying, F., Zhao, Z. (2016). Responses of Water and Salt Parameters to Groundwater Levels for Soil Columns Planted with Tamarix chinensis. Plos One, 2016, 11(1),0145828.

Yang, L., Kong, F. L., Xi, M.,Li, X., Wang S. (2017). Environmental economic value calculation and sustainability assessment for constructed rapid infiltration system based on emergy analysis. Journal of Cleaner Production, 167, 582-588.

Zhai, Y., Zhao, X., Teng, Y., Li, X., Zhang,J. J., Wu, J., Zuo, R. (2017). Groundwater nitrate pollution and human health risk assessment by using HHRA model in an agricultural area, NE China. Ecotoxicology and Environmental Safety, 137,130-142.

Zhao, F., Zhang, Y., Huang, M. S., Wu, X. H., Zhang, Y. F., He, Y. (2011). Study on the purification effects of aquatic plant floating-beds for urban polluted water. Journal of East China Normal University (Natural Science), 6, 63-70.

Zhao, R. S., Li, J. K. Guo, C., Li, F., Li, H. E. (2019). Filler improvement and purification effects of constructed rapid infiltration facility. Environmental science and pollution research international, 26(32), 33654-33669.

Zhou, X. L., Huang, K.Y., Wang, J. H. (2017). Numerical simulation of groundwater flow and land deformation due to groundwater pumping in cross-anisotropic layered aquifer system. Journal of Hydro-environment Research, 14,19-33.

Zuo, X. J., Guo, Z. Y., Wu, X., Yu, J. H. (2019). Diversity and metabolism effects of microorganisms in bioretention systems with sand, soil and fly ash. Science of The Total Environment, 676, 447-454.

本文曾发表于 2022 年《Frontiers in Environmental Science》第 10 卷

海绵城市 LID 设施运行效能衰减机理及寿命分析研究综述

郭　超[1,2,3],王璐瑶[1,2,3],谢　潇[1,2,3],牛　岩[1,2,3],王益权[4]

(1.陕西省土地工程建设集团 自然资源部退化及未利用土地整治工程重点实验室,陕西 西安 710075;
2.陕西省土地工程建设集团 陕西省土地整治工程技术研究中心,陕西 西安 710075;
3.陕西省土地工程建设集团 自然资源部土地工程技术创新中心,陕西 西安 710075;
4.西北农林科技大学,陕西 杨凌 712100)

【摘要】为了研究海绵城市 LID 设施长期受纳降雨径流后填料理化性质的演变规律,揭示设施运行效率的衰减过程及机理,探索设施合理的使用寿命,本文以海绵城市 LID 设施为研究对象,从设施填料理化性质与水分入渗和污染净化能力的关系、入渗性能改变及污染物吸附饱和、LID 设施运行效能衰减及运行寿命几个方面,通过文献资料综合评述国内外研究进展与不足。结果表明:现有研究大多围绕这类设施结构、效果与影响机制展开,对其填料理化性质的演变规律、入渗性能改变及污染物吸附饱和等缺乏系统研究,对设施运行效率的阶段性、动态化衰减过程研究鲜见报道,制约设施运行寿命的机制不明确,影响其合理使用。建议今后从以下 3 个方面开展进一步研究:①明确 LID 设施填料在运行期间理化性质的动态演变过程,发展基于 LID 设施水量调控和污染物净化效果的新方法;②揭示填料入渗性能和污染物吸附量随运行时间尺度的变化规律,从径流水质、材料两个方面探索海绵城市集中入渗条件下,LID 设施不同填料发生堵塞和污染物吸附饱和的内在机制及驱动因子;③阐明 LID 设施运行效能衰减机理,建立 LID 设施运行效能衰减评价方法,进一步研究并确定 LID 设施合理的使用寿命。

【关键词】海绵城市;LID 设施;理化性质;水分入渗;污染物净化;衰减机理;寿命分析

Research Progress on the Decay Mechanism of Operational Performance and Life Analysis of Sponge City LID Facilities

Chao Guo[1,2,3], Luyao Wang[1,2,3], Xiao Xie[1,2,3], Yan Niu[1,2,3], Yiquan Wang[4]

(1.Shaanxi Provincial Land Engineering Construction Group, Ministry of Natural Resources, Key Laboratory of Degraded and Unused Land Consolidation Engineering, Xi'an 710075; 2.Shaanxi Provincial Land Engineering Construction Group, Shaanxi Provincial Land Consolidation Engineering Technology Research Center, Xi'an 710075; 3.Shaanxi Provincial Land Engineering Construction Group, Ministry of Natural Resources, Land Engineering Technology Innovation Center, Xi'an 710075; 4.Northwest A&F University, Shaanxi Yangling 712100)

【Abstract】 Abstract: In order to study the evolution of fillers physical and chemical properties of LID facili-

基金项目:陕西省重点产业创新链项目(2022ZDLSF06-04);陕西省土地工程建设集团内部科研项目(DJNY2022-26、DJNY2021-23)。

ties in sponge city after long-term accepting rainfall runoff, and to reveal decay process of facility operation efficiency, then to explore its reasonable service life. Therefore, based on the LID facility in sponge city, this study reviews the progress and deficiency of investigations in China and beyond on the relations of fillers physical, chemical properties and water infiltration and pollutant purification capability, infiltration performance changes and pollutant adsorption saturation. The decay process of facility operation efficiency and its service life are also examined. The results showed that the existing researches mostly focus on the structure, effect and influence mechanism of such measures. There is a lack of systematic research of fillers physical and chemical properties, infiltration performance changes, and pollutant adsorption saturation. There are few reports on the phased and dynamic decay process of facility operation efficiency, and the mechanism that restricts the service life of such facilities is unclear. Which affects the reasonable use of this concentrated infiltration measures. We suggest that future studies should examine the following three aspects: ① to clarify the dynamic evolution of fillers physical and chemical properties during the concentrated infiltration of rainfall runoff in sponge cities, and to Develop a new method of water volume reduction and pollutants purification by the LID facilities. ② to reveal the infiltration performance of fillers and pollutant adsorption capacity with the running time. From the aspects of rainfall runoff water quality and materials to explore the intrinsic mechanism and driving factors of clogging and adsorption saturation of different fillers. ③ to clarify the operating efficiency decay mechanism of LID facilities, and establish a method for evaluating the efficiency decay process. Further to analyze and determine the reasonable service life of the LID facilities.

【**Keywords**】Sponge city; LID facilities; Physical and chemical properties; Water infiltration; Pollutant purification; Decay mechanism; Life analysis

1 研究背景

城市化进程加快，地面硬化带来的城市内涝、水环境污染、地下水补给匮乏、诱发城市地质灾害和建筑物安全等严峻的世界性难题，已经得到世界各国的重视。面对城市化存在的此类问题，国内外提出了许多应对之策，如美国的低影响开发（LID）、英国的可持续城市排水系统（SUDS）、澳大利亚的水敏感城市设计（WSUD）等[1]，我国提出建设自然积存、自然渗透、自然净化的“海绵城市”，均取得了良好效果。

在我国海绵城市建设过程中，对雨水花园、生物滞留设施、植草沟等 LID 设施的应用较多。这类设施也称为海绵设施、海绵体或雨水径流集中入渗设施，是人为设计的用于处理降雨径流的绿色基础设施[2-3]，具有导水[4]、渗水[5-6]、去污[7-8]、蓄水[9-10]等综合功能。LID 设施结构简单，在一定纵向深度范围内（≤1 m）填充天然填料或人工合成填料，利用填料的导水性能、可渗透性能、污染物净化性能以及蓄积水量的性能，将降雨径流导入地下或蓄积起来，并且将径流挟带的污染物经填料吸附、过滤、离子交换、微生物降解等物理、化学、生物作用得以自然净化。LID 设施的应用如同给硬化地面打开了一扇扇“窗户”，增加了可透水区域的面积，可有效缓减城市内涝、净化水质和涵养地下水资源[11-12]，是目前城市雨水处理较为有效的工程技术措施。由于城市雨水径流存在严重的面源污染，并且 LID 设施集中入渗有别于大规模城市化以前的面源入渗[13]，其入渗水量和污染负荷强度大，入渗水量通常为自身面积的 5~20 倍[14]。因此，LID 设施长期受纳降雨径流集中入渗时，其填料的机械组成、孔隙度、团聚体状态和饱和导水率等理化性质会发生改变[15-16]，造成填料堵塞、污染物吸附饱和等众多难题，降低设施削减水量和去除污染物的运行效率，缩短其使用寿命等，这些科学与技术难题已引起了众多专家的极大关注。目前，国内外主要针对 LID 设施结构设计与优化、不同填料对水量和污染物削减效果、影响机制、关键技术与模型模拟等开展了大量研究[17-19]，而对 LID 设施填料理化性质的演变规律、入渗性能改变及污染物吸附饱和等问题缺乏系统研究，对海绵城市 LID 设施运行效率的阶段性、动态化衰减过程研究鲜见报道，严重影响了这类技术的合理使用和海绵城市的纵深发展。而海绵城市建设既存在共性，也存在特殊性和复杂性，需对不同城市进行规划和设计[20]。我国分布广泛的黄土有其特殊性——质地均一、结构疏松、孔隙发育、下渗能力强，富含 SiO_2、Al_2O_3 和 CaO 等矿物，对调控雨洪和净化污染物起着重要作用。因此，对于资源型和水质型缺水、地下水补给不足的黄土地区,海绵城市 LID 设施具有良好的适用性。但黄土遇水具有湿陷性,且黄土地区径流雨水中含有大量泥沙、黏土颗粒等,理化性状易改变[21]。此外,黄土地区的水中富含游离的可溶性重质碳酸钙,降雨径流经过 LID 设施不同填料界面时,一般情况下流速会发生改变,引起重质碳酸钙中的CO_2溢出,形成碳酸钙沉淀,从而沉积在设施的不同填料界面,引起填料堵塞。故开展黄土地区海绵城市 LID 设施填料理化性质演变规律、入渗堵塞及污染物吸附饱和等方面的研究具有重要意义,以满足黄土地区海绵城市建设实践的迫切需要。

本文主要通过文献总结和分析目前国内外雨水径流集中入渗对 LID 设施填料影响的研究成果,分别从 LID 设施填料理化性质的变化规律,集中入渗对填料入渗性能及污染物吸附性能的影响研究,集中入渗条件下 LID 设施运行效能衰减机理及其运行寿命等方面,总结国内外研究进展,对我国现有海绵城市建设过程中理论研究存在的问题与不足进行探索,提出未来海绵城市建设模式下雨水径流集中入渗的发展方向,为城市雨水径流集中入渗工程技术的合理配置与推广应用提供科学依据,为我国海绵城市健康发展助一臂之力。

2 LID 设施填料理化性质研究现状

填料的理化性状是维系“海绵体”各项功能的重要指标,在 LID 设施中填料的物理结构左右着地表径流的下渗、土壤水的补给以及污染物净化等。填料物理性质退化主要表现在饱和导水率、容重、孔隙、质地等方面的变异[22],这些物理指标都与海绵城市建设中雨水滞留、渗透、蓄积等有着密切的相关性。已有研究表明,通过土壤结构性改善、孔隙结构调整等措施可以优化土壤的理化性质,调节土壤水分的入渗能力,提高土壤持水、保水性能[23]。土壤容重是指示 LID 设施填料物理性质的重要指标,对土壤水分入渗能力有较大影响,土壤入渗能力随容重增大而递减[23]。城市绿地土壤的持水性能和雨水的渗透、滞留之间的相关性是海绵城市建设成效的关键,土壤结构、容重在海绵城市雨水渗透和滞留方面起着重要作用[25]。一些绿地土壤结构性差,土体紧实,容重较大,孔隙较少,水分难以入渗,持水力差[26]。土壤容重还对土壤饱和导水率起着很大的影响[27]。土壤有机碳对团聚体的形成和稳定具有重要作用[14]。万松华等[28]指出,土壤中有机质含量与团聚体呈正相关关系,有机质通过调节土壤孔隙的大小以及分布情况来影响水分在土体内的传导能力。因此,有机质含量对土壤的物理结构有很大的影响,并且通过改变土壤入渗率、饱和导水率等理化指标影响着城市雨水的径流量。

海绵城市 LID 设施中通常添加粉煤灰、沸石、高炉渣等吸附性能强的材料,研究表明添加粉煤灰后,沙土的持水和保水能力提高了近 10%,并且显著提高了风沙土粉粒和黏粒含量,增加了沙土的稳定性,减小了孔隙结构空间尺度变异性[29]。海绵城市 LID 设施长期处于干湿交替状态,大量研究表明,干湿交替影响土壤团聚体与颗粒有机物和微生物群落之间的关系[30],改变土壤碳氮、磷等养分循环[31]。土壤黏土矿物比表面积巨大、富电荷,具有较强的吸附性、复水性及离子交换性能[32]。黄土区土壤黏土颗粒含量较高,占土壤组成的 10%~25%,土壤黏土矿物富含硅铝酸盐矿物,具有硅氧四面体和铝氧八面体相互叠加的层状结构,广泛存在于各类地质体中,具有较强的吸附性、脱水、复水性能和膨胀、收缩性能等[33]。唐双成[34]研究表明,雨水花园运行 4 年后,土壤颗粒中黏粒含量变化不明显,而提高了砂粒含量的比例,研究区雨水径流特性对雨水花园填料颗粒组分具有重要影响。LID 设施填料中黏土矿物的比例影响水分的持留和有机污染物的吸附,对削减洪峰流量、净化径流污染极为重要[35]。孟凡旭等[36]指出,土体结构变密实,水分入渗能力就会下降,土壤的孔隙结构也是水分入渗能力的重要指示。“植物、土壤和透水铺装”是海绵城市规划设计的三要素,植物可截留降雨,土壤对水分起传导作用,透水铺装结构及透水能力是水分持留、渗透的重要载体[37]。土壤“滞、蓄”水分主要依靠土壤的持水能力,并且土壤持留的水分为植物生长提供水源,也为土壤中养分的溶解和转移提供天然的溶剂[38]。

显然,国内外研究大多集中在土壤的某一属性对径流的作用强度,或对水文条件、土壤属性等某一区域特性进行分析研究,对海绵城市 LID 设施填料理化性质的演变规律研究较少。LID 设施填料是一个复杂的生态系统,需就多个属性对城市雨洪和径流污染影响进行对比研究,将填料理化性质研究由定态向动态演变规律研究转变。

3 LID 设施水分入渗性能及污染物净化能力研究进展

3.1 LID 设施水分入渗性能研究进展

海绵城市建设模式下,对于 LID 设施的应用较多,雨水径流主要以点状集中入渗。一般情况下 LID 设施的汇流比为 5:1~20:1,这说明 LID 设施要汇集其自身面积 5~20 倍的径流总量,入渗的水量明显增加[13],并且城市雨水径流面源污染严重。因此,LID 设施集中入渗的水量负荷与污染负荷强度均很大,长期集中入渗将改变填料的机械组成、孔隙度、团聚体、饱和导水率等理化性质,造成填料堵塞、污染物吸附饱和等。不同填料的表面电场不同,而土壤颗粒的表面电场对团聚体稳定性具有很大的影响,从而会改变土壤孔隙通道,影响水分入渗过程[39-40]。

通过国内外文献分析发现,农业领域水分入渗试验与实践研究较多,如王艳阳等[41]依据农田试验,通过施加研究不同生物炭,研究土壤水分入渗及其分布特性,结果表明生物炭添加能够改善黑土区土壤持水能力和水分入渗特性。曹崇文等[42]基于大田土壤水分入渗试验与相关参数研究,探讨了耕作层土壤水分入渗能力的衰减过程,揭示了耕作层土壤水分入渗能力衰减机理,发现在农业生产周期内,耕作层土壤水分入渗能力变化较大。吴丹[43]研究表明随着灌溉水中的悬浮固体浓度的增大,受灌土壤的入渗性能就会降低,而土壤容重增加、孔隙率降低,优先流的非均匀特征明显增大。冯锦萍[44]研究表明影响大田土壤水分入渗特性的直接影响因素有土壤结构、质地、含水率、表土结皮和板结等,农业耕作措施和土壤水的相变等间接影响水分入渗性能。Wang 等[45]根据美国伍德兰兹社区土壤的水分入渗特征,提出了以雨洪管理为目标的生态规划方法,为我国海绵城市建设提供了依据。通过调节土壤粉粒、黏粒比例,可提高水分入渗率,构建"渗透型"土壤结构[46]。杨珊基于海绵型生态绿地建设理念,以陕西龙湖湾别墅区为例,提出对于住宅绿地,应构建基于现状条件的水循环体系[47]。

综上所述,国内外对 LID 或海绵城市入渗性能的研究还是初步的,缺乏系统研究,迫切需要加强分析 LID 设施入渗性能改变的内在机制及驱动因子。有关常规土壤入渗性能研究较多,常规土壤入渗堵塞成因及机制分析将为 LID 设施水分入渗研究提供借鉴。

3.2 LID 设施污染物净性能研究进展

LID 设施对降雨径流污染物起净化作用,当 LID 设施长期受纳降雨径流时,将会由污染物受纳型变为污染物输出型[48]。降雨径流中大部分颗粒态污染物通过设施表层植被和填料拦截、过滤,降低了出水中污染物浓度[49]。在降雨期间,设施去除溶解态污染物以填料吸附、离子交换为主,不同填料类型对溶解态污染物的吸附能力不同[50]。对于溶解态污染物,主要通过填料吸附、植物吸收、微生物降解以及生化反应转化为难溶的形态而得到有效去除[51]。一般来讲,降雨期间,LID 设施对污染负荷量的去除主要受填料官能团、表面电荷、孔隙结构等因素的影响[52-54],植物吸收、微生物降解占比较少[55]。Gurung 等[56]研究得出雨水花园对重金属有较好的去除效果。Li 等[57]和 Liu 等[49]研究表明生物滞留设施对总氮负荷去除率大于 41%,对 TP 去除率达 90%以上。因此,为了提高设施对污染物的去除效果,许多专家对 LID 设施中填料进行了改良,砂基质填料的生物滞留剂对 TSS 和 TP 的去除效果较好,去除率均在 95%以上。由于受多种材料交叉吸附能力的影响,若填料中添加无机组分,对 TSS 的去除明显提高[34]。改性生物滞留设施对总氮(TN)、氨氮(NH_3-N)和硝态氮(NO_3-N)的平均去除率分别达到 59.45%、80.27%和 41.48%[58]。也有研究表明 LID 设施对降雨径流中硝酸盐氮(NO_3^--N)和溶解态磷(DP)的净化效果不理想[59]。

可见,海绵城市模式下 LID 设施对污染物净化能力分析也是初步的,缺乏系统研究,LID 设施人工合成填料或天然填料对污染物的净化效果不是一成不变的,需进一步进行深入研究,明确 LID 设施填料净化能力的动态演变过程。

4 LID 设施运行效能衰减过程研究现状

海绵城市基于 LID 开发理念,服务于缓解城市内涝、减轻面源污染、保护城市生态系统、恢复和修复生态功能等人居环境的目标。目前,国内外主要工作是针对 LID 或源头措施的本身结构设计、效果、影响机制与模型模拟展开,较深入地研究了不同海绵设施对径流水量、水质的调控效果。伍静[60]基于雨水花园基本构造和工作原理提出了基于水量平衡分析的设计方法,并探讨了其关键设计参数及影响因素。唐双成等[61]通过监测 14 场降雨事件,研究雨水花园对暴雨径流削减效果,发现分层填料雨水花园对水量的削减范围为 12.0%~85.9%,平均为 44.3%;对洪峰流量的削减范围为 11.2%~93.3%,平均为 55.8%;均质黄土填料雨水花园对水量削减范围为 9.8%~79.8%,平均为 39.2%;洪峰流量削减范围为 20.3%~89.8%,平均为 50.5%,并且随着运行时间其水量削减效果有下降的趋势。Hunt 等[62]等研究季节性变化对生物滞留设施水文效应的影响,结果表明夏天土壤入渗和蒸发蒸腾作用较强,不同地点的生物滞留设施出流量与入流量之比从冬天的 0.54 降低到夏天的 0.07。

目前,国内外学者初步开展了有关 LID 设施污染物累积和对污染物净化能力衰减等方面的研究。张兆鑫[63]研究了生物滞留系统中污染物的累积特征,结果表明,雨水花园中碳氮磷含量不稳定,而重金属含量均呈现增加趋势,受纳道路雨水的植生滞留槽均存在一定程度的 PAHs 累积,并且非汛期 PAHs 含量明显高于汛期,LID 设施长期运行存在一定程度效能衰减的风险。Guo 等[64]指出雨水花园土壤表层下 0~50 cm 范围内,NH_3-N、NO_2-N+TON 和 TN 含量较高,而土壤可溶性硝酸盐和磷酸盐随水分入渗发生了明显的迁移,对于重金属,大多集中在表层下 30 cm 范围内,具有显著的表层富集现象。Jiang 等[65]通过研究改良型生物滞留设施对径流中溶解性污染物的修复与累积特征,发现设施中污染含量呈现上层>中层>下层的趋势。Knappenberger 等[66]通过建立贝叶斯模型,预测生物滞留设污染物累积效果,结果表明随着运行时间,磷和氮的去除效果会受到极大影响,填料中污染物滞留量逐渐增加,净化效果具有衰减的风险。Xiong 等[67]在低温环境下明确了木屑填料的硝酸盐氮去除率由秋季的 70%~90 %下降到冬季的-23%~35%,总氮去除率由秋季的 75%~90%下降到冬季的 20%~50%。Guo 等[68]通过监测的 54 场降雨事件发现雨水花园对 NH_4^+-N 和 TP 的年均浓度去除率相对较好,但总体上随着监测时间的延长呈下降趋势。

可见,目前国内外有关海绵设施运行效能衰减过程的研究还是初步的,对海绵城市 LID 设施运行效率的阶段性、动态化衰减过程研究鲜见报道,需加强对海绵设施入渗性能衰减及污染物吸附饱和等方面的深入研究,明确 LID 设施入渗堵塞和填料吸附饱和的内在机制及驱动因子,使海绵城市由定态评价向阶段性的、动态化的过程评价转变。

5 LID 设施运行寿命研究现状

LID 设施的运行寿命与填料的入渗性能密切相关。设施填料容重小、孔隙结构性好,其导水能力相对较好,入渗率较大,对降雨径流的疏导作用明显。降雨径流通过填料入渗补给地下水,涵养地下水资源,对降雨径流削减率也越大。若 LID 设施长期受纳降雨径流,过流水量大,污染负荷高,径流挟带的颗粒态污染物以及泥沙在填料中得到累积,在填料表层逐渐形成淤积层,造成填料堵塞,影响其合理的使用寿命[69]。Casas-Mulet 等[70]研究表明填料拦截的固体颗粒和植物表面的生物膜由有机、无机的多聚物、腐殖质组成,可形成低密度的凝胶结构,具有很高的堵塞潜能,会加剧设施内填料孔隙堵塞,导致其导水能力逐渐减弱,入渗率减小,对降雨径流的削减逐渐降低,LID 设施的运行寿命逐渐缩短。有关研究表明人工湿地堵塞后,其出水流量将逐渐减小,对于小试试验系统,水力停留时间由堵塞前21.3 h延长至 32.5 h,对于中试试验系统,堵塞前水力停留时间为 19.4 h,堵塞后延长至 26.8 h[71]。Casas-Mulet 等[70]指出对于砾石床人工湿地,由于凝胶体的累积使得前 1/4 段床体填料的孔隙率减少了约 50%,且 80%累积的固体颗粒为无机物。Li 等[72]通过滤柱模拟和场地实验研究 LID 设施的过滤机制发现,当颗粒物质被 LID 设施截留后,填料内部会出现分层现象,细小土粒聚集在填料上层,而较大颗粒会滞留在基质下层,导致填料堵塞,渗透性降低,这不仅影响系统的水质净化效果,同时也会缩短填料的使用寿命。张瑞斌等[73]研究表明铝污泥与沸石配比为 4∶1的生物滞留系统其渗透性较强,且随着运行周期的增长,其渗透性下降较为缓慢。

制约LID设施运行寿命的另一因素是填料吸附饱和,有关LID设施填料吸附饱和寿命分析研究较少。不同填料对污染物的吸附能力不同,吴建[74]根据Langmuir模型,拟合得出煤质活性炭对COD的最大吸附量为62.86 mg/g。张彬鸿[75]通过动态吸附实验发现在传统LID设施填料中添加10%的粉煤灰、绿沸石或麦饭石,在进水量约相当于5年总降雨量时,设施对磷的吸附达到了饱和,吸附量为22~30 mg/kg;对于纯种植土壤,在进水量约为11年总降雨量时,设施对磷的吸附达到了饱和,吸附量约为59.44 mg/kg;而土壤中添加10%的给水厂污泥进行填料改良后,在进水量相当于15年降雨量时,设施对磷的吸附仍未达到饱和,吸附量达为94.29 mg/kg。仇付国明确了给水厂污泥对重金属Cr的吸附量随污泥粒径的减小而增加,35 ℃时饱和吸附量达到45.59 mg/g[76]。砾石地下流人工湿地在没有外部碳源的情况下,对重金属金属(Fe和Cr除外)的吸附迅速达到饱和[77]。绍兴华[78]通过人工合成的铁水合物,研究其对磷的吸附性能,结果表明人工合成的铁水合物对磷具有非常大的吸附潜力,磷的最大吸附容量为45.45 μg/g。郭超等[9]通过分析雨水花园运行寿命,提出了雨水花园"三阶段净化能力"的概念,包括净化增长期、净化稳定期和净化衰弱期,初步预测雨水花园运行10~15年后填料的吸附位将会处于饱和状态。国内外对人工湿地中填料使用寿命的研究较多。运行1~2年后的砾石床人工湿地,对磷的净化效果逐渐下降[79];美国某人工湿地运行4~5年后,基质中磷的积累量较大,对磷的净化效果逐渐丧失[80];通过探索填料类型和植物类型对人工湿地使用寿命的影响,表明种植芦苇的人工湿地使用寿命约22.423年;而种植黄花鸢尾的人工湿地使用寿命为约22.424[81];以给水厂污泥为填料的人工湿地,处理普通生活污水时,其使用寿命为4~17年[82]。而世界水协会(IWA)建议人工湿地参考使用寿命为15年[83]。LID设施有别于人工湿地,常年处于干湿交替状态,降雨过程及雨后一段时间,设施填料或处于饱和状态,污染物净化机制与人工湿地较为相似。在下一场雨来临之前处于干燥状态,其污染物净化机制类似于常规土壤系统。因此,亟待进一步加强研究,明确LID设施在干湿交替状态下对污染物净化的动态变化过程,确定合理的使用寿命。

6 结论与展望

6.1 结论

目前,国内外主要针对LID设施结构设计与优化、设施削减水量和污染物效果及其影响机制、关键技术与模型模拟等方面开展了大量的研究。对雨水径流集中入渗条件下,LID设施填料容重、孔隙、质地、饱和导水率等理化性质的演变规律研究较少。对LID设施不同填料的水分入渗能力和持水能力做了初步探索,针对某一土壤属性对降雨径流的作用强度进行了探索性研究。初步分析了海绵城市模式下LID设施不同填料对污染物净化能力,有关填料对污染物吸附饱和的研究较为单一,主要针对特定填料对某种污染物吸附容量进行了定量化研究,缺乏LID设施对污染物净化能力的系统性研究。初步开展了有关LID设施污染物累积效应和对污染物净化能力衰减等方面的研究,定性分析了并预测了LID设施的运行寿命,缺乏LID设施阶段性、动态化衰减过程研究,对LID设施合理的使用寿命有待进一步明确,而国内外对人工湿地使用寿命的研究较多,结论较为明确,具有可借鉴价值。

6.2 展望

为保障海绵城市LID设施的合理应用与纵深发展,探明其运行效能衰减机理及合理使用寿命,建议对以下问题加大研究力度:

(1)LID设施填料理化性质及演变过程。针对LID设施土壤理化性质的研究,国内外大多集中在某一土壤属性对径流的作用,对海绵城市LID设施土壤理化性质的演变规律研究较少。LID是一个复杂的生态系统,需针对多个属性对城市雨洪和径流污染影响进行对比研究,明确LID设施在雨水径流集中入渗过程填料理化性质随运行时间的变化过程,揭示不同填料理化性质的动态演变规律;研究雨水径流长期集中入渗对不同填料理化性质的作用强度;建立填料理化性质与水量和污染物削减效果之间的关系,将土壤理化性质研究由定态向动态演变规律研究转变具有极其重要的意义。

(2)LID设施填料入渗性能及污染物吸附容量。国内外对LID设施入渗性能的研究还是初步的,而对常规土壤入渗性能研究较多,常规土壤入渗堵塞成因及机制分析将为集中入渗提供借鉴。后期需进一步加强 LID 设施填料入渗性能和污染物吸附性能方面的研究,揭示填料入渗性能和污染物吸附容量随运行时间尺度的变化规律;从径流水质、材料两个方面明确探索海绵城市集中入渗条件下,LID 设施填料入渗堵塞成因及机制;探索不同填料发生堵塞和污染物吸附饱和的内在机制及驱动因子。

(3)LID 设施运行效能衰减及寿命分析。目前,有关海绵城市 LID 设施运行效率的阶段性、动态化衰减过程研究鲜见报道,后期可通过定量分析海绵设施不同填料入渗性能和污染物吸附容量随运行时间、进水水量和水质之间的关系,发展基于海绵城市水量调控和污染物去除效果的新方法。以填料堵塞、污染物吸附饱和、水量削减、水质净化等为参数,建立海绵设施运行效能衰减评价机制;用数值计算模型建立一套海绵设施运行效能衰减过程评价方法,实现黄土地区海绵设施入渗性能和水质净化过程由定态评价向阶段性的、动态化的过程评价转变。进一步分析并确定 LID 设施合理的使用寿命。

在长期运行条件下,LID 设施存在堵塞、通量下降等问题,探明海绵城市建设模式下 LID 设施运行效能衰减过程及合理使用寿命,能够确保海绵设施长期运行可靠,提高工程效益,破解我国城市内涝、控制城市面源污染、提高雨水贮存蓄积等面临的重要难题。

参考文献

[1]FLETCHER T D, SHUSTER W, HUNT W F, et al. SUDS, LID, BMPs, WSUD and more-The evolution and application of terminology surrounding urban drainage[J]. Urban Water Journal, 2015, 12(7):525-542.

[2]段小龙,李家科,蒋春博.雨水生物滞留系统关键设计参数研究进展[J].安全与环境工程, 2022,29(2):111-119.

[3]MENG Yike, WANG Yuan, WANG Chuanyue, et al. Nutrients adsorption characteristics and water retention capacity of polyurethane-biochar crosslinked material modified filler soil in stormwater treatment[J]. Journal of Water Process Engineering,2021,44: 102347-102357.

[4]KAYKHOSRAVI S, KHAN U T, JADIDI M A. A simplified geospatial model to rank LID solutions for urban runoff management[J]. Science of the Total Environment,2022,831: 154937-154937.

[5]欧阳友,潘兴瑶,张书函,等.典型海绵设施水循环过程综合实验场构建[J].中国给水排水, 2022,38(2):132-138.

[6]樊金璐,徐红,陆益忠,等.基于 LID 组合措施的张家港中心城区雨洪控制效果模拟研究[J].水电能源科学,2022,40(3):75-78.

[7]ZHANG Zhaoxin, LI Jiake, JIANG Chunbo, et al. Impact of nutrient removal on microbial community in bioretention facilities with different underlying types/built times at field scale[J]. Ecological Engineering,2022,176: 106542-106550.

[8]WANG Shumin, LIN Xiuyong, YU Hui, et al. Nitrogen removal from urban stormwater runoff by stepped bioretention systems[J]. Ecological Engineering, 2017, 106:340-348.

[9]徐宗学,李鹏,程涛.基于海绵城市理念的 LID 措施优化布局:以济南市黄台桥流域为例[J].南水北调与水利科技,2022,4:1-14

[10]YANG Dongdong, ZHAO Xin, ANDERSON B C. Integrating Sponge City Requirements into the Management of Urban Development Land: An Improved Methodology for Sponge City Implementation[J]. Water, 2022,14(7):1156-1156.

[11]纪亚星,同玉,侯精明,等.西咸新区海绵城市建设对沣河洪水特性影响模拟研究[J].水资源与水工程学报, 2021,32(02):50-57.

[12]徐宗学,李鹏.城市化水文效应研究进展:机制、方法与应对措施[J].水资源保护, 2022,38(1):7-17.

[13]李怀恩,贾斌凯,成波,等.海绵城市雨水径流集中入渗对土壤和地下水影响研究进展[J]. 水科学进展, 2019, (4): 589-600.

[14]蒋春博,李家科,李怀恩.生物滞留系统处理径流营养物研究进展[J].水力发电学报, 2017,36(8):65-77.

[15]许目瑶. 雨水花园集中入渗补给地下水受土壤非饱和区影响的试验研究[D].扬州:扬州大学, 2020.

[16]肖恒婧.西安地区雨水花园土壤中污染物的时空分布研究[D].西安：西安理工大学，2020.

[17]JIA Zhonghua, TANG Shuangcheng, LUO Wan, et al. Small scale green infrastructure design to meet different urban hydrological criteria[J]. Journal of Environmental Management, 2016, 117: 92-100.

[18]郭效琛，杜鹏飞，李萌，等.基于监测与模拟的海绵城市典型项目效果评估[J]. 中国给水排水，2019，(11)：130-134.

[19]李家科，张兆鑫，蒋春博，等. 海绵城市生物滞留设施关键技术研究进展[J].水资源保护，2020，36(1)：1-8.

[20]北京建筑大学.海绵城市建设技术指南. 北京：中国建筑工业出版社，2014.

[21]吕新波. 湿陷性黄土地区市政道路植草沟雨水入渗与路基沉降研究[D].西安：长安大学，2020.

[22]张毅川. 海绵城市导向下绿地典型下垫面的雨水特征及优化：以新乡市为例[D].武汉：武汉大学，2017.

[23]孙层层.土壤改良剂对土壤孔隙结构及其水分过程影响的定位研究[D].杨凌：西北农林科技大学，2019.

[24]PAN Limin, CHEN Yingyi, XU Yan, et al. A model for soil moisture content prediction based on the change in ultrasonic velocity and bulk density of tillage soil under alternating drying and wetting conditions[J]. Measurement, 2022, 189: 110411-110504.

[25]刘义存.绿地土壤物理性质在海绵城市规划建设中的影响[J]. 农村经济与科技，2016，27 (12)：21-22.

[26]刘澄静，角媛梅，高璇，等.哈尼梯田水源区不同景观类型土壤水分的入渗特性及影响因子[J].水土保持通报，2018，38(3)：99-105，111.

[27]刘家宏，王东，李泽锦，等.厦门海绵化改造区入渗性能测定及特征[J].水资源保护，2019，(6)：9-14.

[28]万松华，胡厚军，邓桂芹.浅析土壤有机质含量与土壤物理性能参数的相关性[J]. 农业与技术，2013，(8)：20-28.

[29]朱世雷. 晋陕蒙能源区新构土体孔隙结构及土壤水分入渗特征研究[D].杨凌：西北农林科技大学，2020.

[30]包丽君，贾仲君.模拟干湿交替对水稻土古菌群落结构的影响[J].土壤学报，2017，54(1)：191-203.

[31]赵金花，张丛志，张佳宝.农田生态系统中土壤有机碳与团聚体相互作用机制的研究进展[J].中国农学通报，2015，398(35)：160-165.

[32]ARIF M, LIU G, YOUSAF B, et al. Synthesis, characteristics and mechanistic insight into the clays and clay minerals-biochar surface interactions for contaminants removal-A review [J]. Journal of Cleaner Production, 2021, 310: 127548-127558.

[33]ZHU Ruiliang, CHEN Qingze, ZHOU Qing, et al. Adsorbents based on montmorillonite for contaminant removal from water: A review[J]. Applied Clay Science, 2016, 123: 239-258.

[34]唐双成.海绵城市建设中小型绿色基础设施对雨水径流的调控作用研究[D].西安：西安理工大学，2016.

[35]章林伟.海绵城市建设概论[J]. 给水排水，2015，(6)：1-7.

[36]孟凡旭，王树森，马迎梅，等.不同果农复合种植模式土壤入渗能力及其影响因素[J].干旱区研究，2020，37(6)：1469-1477.

[37]常青，刘晓文，孙艺.海绵城市"绿地规划设计三要素研究进展[J]. 中国农业大学学报，2017，(1)：139-150.

[38]任利东，黄明斌，樊军.不同类型层状土壤持水能力的研究[J].农业工程学报，2013，(19)：113-119.

[39]李喆，胡斐南，杨志花，等.土粒表面电场对典型黑土团聚体稳定性及水分入渗特性的影响[J].水土保持研究，2021，28(5)：61-66，75.

[40]杨志花.土壤内力作用对黄土母质发育土壤水力学性质的影响及模拟[D].北京：中国科学院大学，2019.

[41]王艳阳，魏永霞，孙继鹏，等. 不同生物炭施加量的土壤水分入渗及其分布特性[J].农业工程学报，2016，32(8)：113-119.

[42]曹崇文，樊贵盛.耕作土壤入渗能力衰减机制[J]. 太原理工大学学报，2007，(2)：25-27，38.

[43]吴丹.灌溉水中悬浮固体对土壤结构性质和水流运动特征的影响研究[D].长沙：长沙理工大学，2017.

[44]冯锦萍.用常规土壤物理参数确定入渗参数的方法研究[D]. 太原：太原理工大学，2003.

[45]WANG Zhifang, REN Zhongshen, ZHANG Min. Ecological planning based on soil factors—the planning and evaluation of the community planning process of in Texas, USA[J]. Urban Planning International, 2015, (4): 88-94.

[46]刘增超，李家科，蒋春博，等.4 种生物滞留填料对径流污染净化效果对比[J].水资源保护，2018，34(4)：71-79.

[47]杨珊.海绵型住宅绿地中水元素的设计研究：以山西龙湖湾为例[D].杨凌：西北农林科技大学，2016.

[48]MEHRING A S, HATT B E, KRAIKITTIKUN D, et al. Soil invertebrates in Australian rain gardens and their potential roles in storage and processing of nitrogen[J]. Ecological Engineering, 2016, 97: 138-143.

[49]LIU Jiayu, DAVIS A P. Phosphorus speciation and treatment using enhanced phosphorus removal bioretention[J]. Environmental Science & Technology, 2014, 48: 607-614.

[50]赵亚乾,杨永哲, AKINTUNDE B,等.以给水厂铝污泥为基质的人工湿地研发概述[J].中国给水排水, 2015,(11): 124-130.

[51]王鑫,刘雅慧,宁梓洁,等. 海绵城市下垫面基质优化及其对模拟雨水处理效果[J]. 环境工程学报, 2018, 12(7): 105-112.

[52]陈全. 改性纤维素对典型污染物的吸附性能和作用机制研究[D].广州:华南理工大学,2019.

[53]廖云.多孔纳米复合吸附剂的制备及其对铀(Ⅵ)离子吸附性能研究[D].上海:东华大学, 2020.

[54]林丽丹.沸石/水合金属氧化物的表面电荷特征以及吸附性能研究[D].上海:上海交通大学,2016.

[55]LI Jiake, DAVIS A P. A unified look at phosphorus treatment using bioretention[J]. Water Research, 2016, 90: 141-155.

[56]GURUNG S B, GERONIMO F K, HONG J, et al. Application of indices to evaluate LID facilities for sediment and heavy metal removal[J]. Chemosphere. 2018, 206: 693-700.

[57]LI Liqing, DAVIS A P. Urban Storm-water runoff nitrogen composition and fate in bioretention systems[J]. Environment Science & Technology, 2015, 48 (6): 3403-3410.

[58]CHENG Junrui, BI Junpeng, GONG Yuemin, et al. Processes of nitrogen removal from rainwater runoff in bioretention filters modified with ceramsite and activated carbon[J]. Environmental technology, 2022: 21-27.

[59]GUO Chao, LI Jiake, LI huaien, et al. Seven-Year Running Effect Evaluation and Fate Analysis of Rain Gardens in Xi'an, Northwest China[J]. Water,2018,10(7):944-959.

[60]伍静.基于海绵城市视域下居住社区雨水花园生态设计的研究进展[J].给水排水, 2019, 41: 105-107.

[61]唐双成,罗纨,贾忠华,等.填料及降雨特征对雨水花园削减径流及实现海绵城市建设目标的影响[J].水土保持学报, 2016, 30(1): 73-78.

[62]HUNT W F, JARRETT A, SMITH J, et al. Evaluating bioretention hydrology and nutrient removal at three field sites in North Carolina[J]. Journal of Irrigation and Drainage Engineering, 2006,132(6):600-608.

[63]张兆鑫. 生物滞留系统污染物累积特征及对微生态系统的影响研究[D].西安:西安理工大学, 2021.

[64]GUO Chao, LI Jiake, LI Huaien, et al. Influences of stormwater concentration infiltration on soil nitrogen, phosphorus, TOC and their relations with enzyme activity in rain garden[J]. Chemosphere, 2019, 233: 207-215.

[65]JIANG Chunbo, LI Jiake, LI Huaien, et al. Remediation and accumulation characteristics of dissolved pollutants for stormwater in improved bioretention basins[J]. Science of the Total Environment,2019,685:763-771.

[66]KNAPPENBERGER T, JAYAKARAN A D, STARK J D. A Bayesian modeling framework to predict stormwater pollutant reduction in bioretention media[J]. Ecological Engineering, 2022,178:106582-106590.

[67]XIONG Jiaqing, ZHU Junguo, LI Guohao, et al. Purification effect of bioretention with improved filler on runoff pollution under low temperature conditions[J]. Journal of environmental management,2021,295: 113065-113074.

[68]GUO Chao, LI Jiake, LI huaien, et al. Influences of Stormwater Concentration Infiltration in Rain Garden on Soil Heavy Metals at Field Scale. Water Science & Technology, 2020,81(5):1039-1051.

[69]郭超,李家科,马越,等.雨水花园运行寿命分析与价值估算[J]. 环境科学学报, 2018, 38(11): 171-179.

[70]CASAS-MULET R, ALFREDSEN K T, MCCLUSKEY A H, et al. Key hydraulic drivers and patterns of fine sediment accumulation in gravel streambeds: A conceptual framework illustrated with a case study from the Kiewa River, Australia[J]. Geomorphology, 2017, 299:152-164.

[71]詹德昊,吴振斌,张晟,等. 堵塞对复合垂直流湿地水力特征的影响[J].中国给水排水, 2003, 19: 1-4.

[72]LI HOUNG, DAVIS A P. Urban particle capture in bioretention media. I: Laboratory and field studies[J]. Journal of Environmental Engineering, 2008, 134(6): 409-418.

[73]张瑞斌, 潘卓兮, 奚道国, 等. 铝污泥填料改良生物滞留池对径流污染的削减效果[J]. 环境工程技术学报, 2021,11(4): 756-762.

[74]吴建.道路下沉式绿地截污功能层净化路面污染径流的性能与设计研究[D].南京:东南大学,2018.

[75]张彬鸿. 生物滞留系统填料改良及对磷素的净化性能试验研究[D]. 西安:西安理工大学,2018.

[76]仇付国,童诗雨,吕华东.给水厂污泥对Cr(VI)的吸附特性研究[J].水处理技术,2022,4:1-6.

[77]CHEN Jinquan, LI Xuan, JIA Wei, et al. Promotion of bioremediation performance in constructed wetland microcosms for acid mine drainage treatment by using organic substrates and supplementing domestic wastewater and plant litter broth[J]. Journal of Hazardous Materials,2021,404, Part A, 124125.

[78]绍兴华.水稻土淹水过程铁氧化物转化对磷饱和度和磷、氮释放的影响[D]. 杭州:浙江大学, 2005.

[79]MANN R A. Phosphorus adsorption and desorption characteristics of constructed wetland gravels and steelworks by-products[J]. Australian Journal of Soil Research, 1997, 35(2): 375-384.

[80]KADLEC R H, KNIGHT R L. Treatment Wetlands[M]. Chelsea, MI: Lewis Publishers, 1996.

[81]徐德福,李映雪,方华,等. 4 种湿地植物的生理性状对人工湿地床设计的影响[J]. 农业环境科学学报, 2013, 28(3): 587-591.

[82]赵晓红,赵亚乾,王文科,等.人工湿地系统以铝污泥为基质的几个关键问题[J].中国给水排水, 2015, (11): 131-136.

[83]KADLEC R H, KNIGHT R L, VYMAZAL J, et al. IWA Scientific and Technical Report[M]. London: IWA Publishing, 2000.

本文曾发表于2022年《水资源与水工程学报》第33卷第5期

Nonlinear Stability Characteristics of Porous Graded Composite Microplates Including Various Microstructural-Dependent Strain Gradient Tensors

Junjie Wang, Biao Ma, Jing Gao, Haitao Liu

(Shaanxi Provincial Land Engineering Construction Group Co., Ltd., Xi'an 710075, Shaanxi, China)

【Abstract】 In this study, the nonlinear buckling and postbuckling characteristics of composite microplates made of a porous functionally graded (PFG) material are scrutinized in the presence of different size-dependent strain gradient tensors as microscale. Accordingly, for the first time, the effect of each microstructural tensor is analyzed separately on the nonlinear stability of PFG microplates with and without a central cutout. In order to fulfill this goal, the isogeometric computation approach is engaged to integrate the finite element approach into the nonuniform B-spline-based computer aided design tool. Accordingly, the geometry of the microplate with a central cutout is modeled smoothly to verify C^{-1} continuity based upon a refined higher-order plate formulations. In this regard, the microstructural-dependent load-deflection paths associated with the nonlinear stability of axially compressed microplates are traced.

【Keywords】 Micromechanics; Size dependency; Isogeometric approach; Postbuckling; Geometric approximation; Porous graded composite materials

1 Introduction

In recent years, by taking high technology into the field of advanced materials, functionally graded (FG) and porous functionally graded (PFG) composite materials have been put to use for manufacturing multifunctional materials having a wide range of applications. Accordingly, it is required to predict mechanical behaviors of devices and systems made of these advanced materials with the aid of nonclassical continuum theories taking more accurate into consideration. Sahmani et al. [2014a, b] predicted the effect of surface stress size dependency in nonlinear free and forced oscillations of nanobeams. Sedighi et al. [2015] anticipated the stability behavior of asymmetric FG composite systems at nanoscale in the presence of surface stress size effect. Li et al. [2016] constructed a nonlocal strain gradient Timoshenko beam model for free oscillations of FG composite microbeams. Simsek [2016] reported nonlinear frequencies of FG composite nanobeams including nonlocal stress and strain gradient size dependencies. Sahmani and Aghdam [2017a, b, c] employed the nonlocal strain gradient continuum shell model for nonlinear mechanics of microtubules resting on cytoplasm of a living cell. Khakalo et al. [2018] used a simplified first strain gradient theory of elasticity for linear mechanical behaviors of microbeams having two-dimensional triangular lattices. Sahmani and Fattahi [2018] constructed nonlocal strain gradient shell model for size-dependent nonlinear buckling behavior of axially compressed FG composite cylindrical microshells. Thanh et al. [2019a, b] studied nonlinear thermal bending and buckling responses of PFG microplates using isogeometric couple stress-based analysis. Sahmani et al. [2019a] introduced an analytical solution framework for free oscillation response of FG laminated microbeams based upon the nonlocal strain gradient elasticity. Mercan et al. [2019] explored the influence of SiO_2 substrate on the stability behavior of zinc oxide nanotubes in the presence of various size effects. Sarafraz et al. [2019] analyzed nonlinear secondary resonance of nanobeams subjected to hard excitation in the presence of surface stress type of small scale effect. Tang et al. [2019] proposed a unified nonlocal strain gradient beam formulation for size-dependent bending behavior of PFG microbeams including thickness and shear deformations. Sahmani and Safaei [2019a,b, 2020] took nonlocal strain gradient beam model for different nonclassical nonlinear mechanics of bi-directional FG composite microbeams.

To mention some more recent investigations, Fang et al. [2020] put a new nonlocal model to use for free oscillations and thermal stability of rotating FG composite nanobeams. Li et al. [2020] analyzed the mechanical and thermal buckling characteristics of organic solar microcells on the basis of the modified strain gradient continuum theory. Yuan et al. [2020a,b, 2021] carried out size-dependent analyses based upon different continuum elasticity theories for nonlinear mechanics of FG composite conical microshells. Karamanli and Vo [2020] studied buckling and bending characteristics of FG sandwich microbeams on the basis of the modified strain gradient continuum model. Lin et al. [2020] incorporated the surface stress effect in the buckling and oscillation responses of single crystalline Reddy's nanobeams. Fan et al. [2020, 2021a, b] examined nonlinear oscillation and postbuckling behaviors of PFG microplates based on the nonlocal strain gradient isogeometric analysis (IGA).

Tang and Qing [2021] utilized the Laplace transform together with the integral nonlocal strain gradient model for elastic stability of FG composite Timoshenko microbeams. Belarbi et al. [2021] proposed an efficient nonlocal finite element formulation for bending behavior of FG composite nanobeams having a new distribution of the shear deformation. Yin et al. [2021] introduced a new numerical approach using IGA for Euler-Bernoulli beams at microscale on the basis of a reformulated strain gradient theory of elasticity. Wang et al. [2021] carried out a meshfree Galerkin approach for free oscillation and stability behaviors of strain gradient thin microplates. Bacciocchi and Tarantino [2021] utilized the nonlocal strain gradient continuum model for analytical solutions associated with mechanical responses of laminated composite nanoplates. Hou et al. [2021] analyzed sizedependent nonlinear free oscillations of elliptical and sector nanoplates made of a FG building material. Li and Xiao [2021] investigated the modified couple stressbased free oscillation response of one-dimensional piezoelectric quasicrystal beams at microscale. Tao and Dai [2021] explored the large-amplitude oscillations of FG laminated composite annular sector microplates on the basis of the modified couple stress theory of elasticity. Sahmani and Safaei [2021] and Zhang et al. [2021] employed the meshfree numerical method for nonlinear postbuckling behavior of axially compressed and lateral pressurized composite microshells reinforced randomly by graphene nanoplatelets.

The main scope of this exploration is to construct for the first time a microstructural-dependent plate model based upon refined higher-order shear deformable formulations to capture separately the effects of different microscaled strain gradient tensors on the nonlinear stability behavior of PFG inhomogeneous microplates with and without cutouts having different geometries.

2 Microstructural-Dependent PFG Plate Formulations

Consider PFG composite rectangular microplates having a thickness of h in the presence and absence of square and circular cutouts as shown in Fig.1. The porosity distribution along plate thickness is supposed based on a mixture rule satisfying the partition of unity for a material property as follows from Phung-Van et al. [2019]:

$$p(z)=p_m+(p_c-p_m)\left[\left(\frac{1}{2}+\frac{z}{h}\right)^k-\frac{\Gamma}{2}\right] \tag{1}$$

Where the subscripts m and c stand for the metal and ceramic phases of a PFG microplate, respectively. Also, k is the property gradient index and Γ represents the porosity index.

Therefore, corresponding to three different porosity distribution patterns illustrated in Fig. 1, the Young's modulus $E(z)$ and Poisson's ratio $v(z)$ of a PFG microplate can be given as follows:

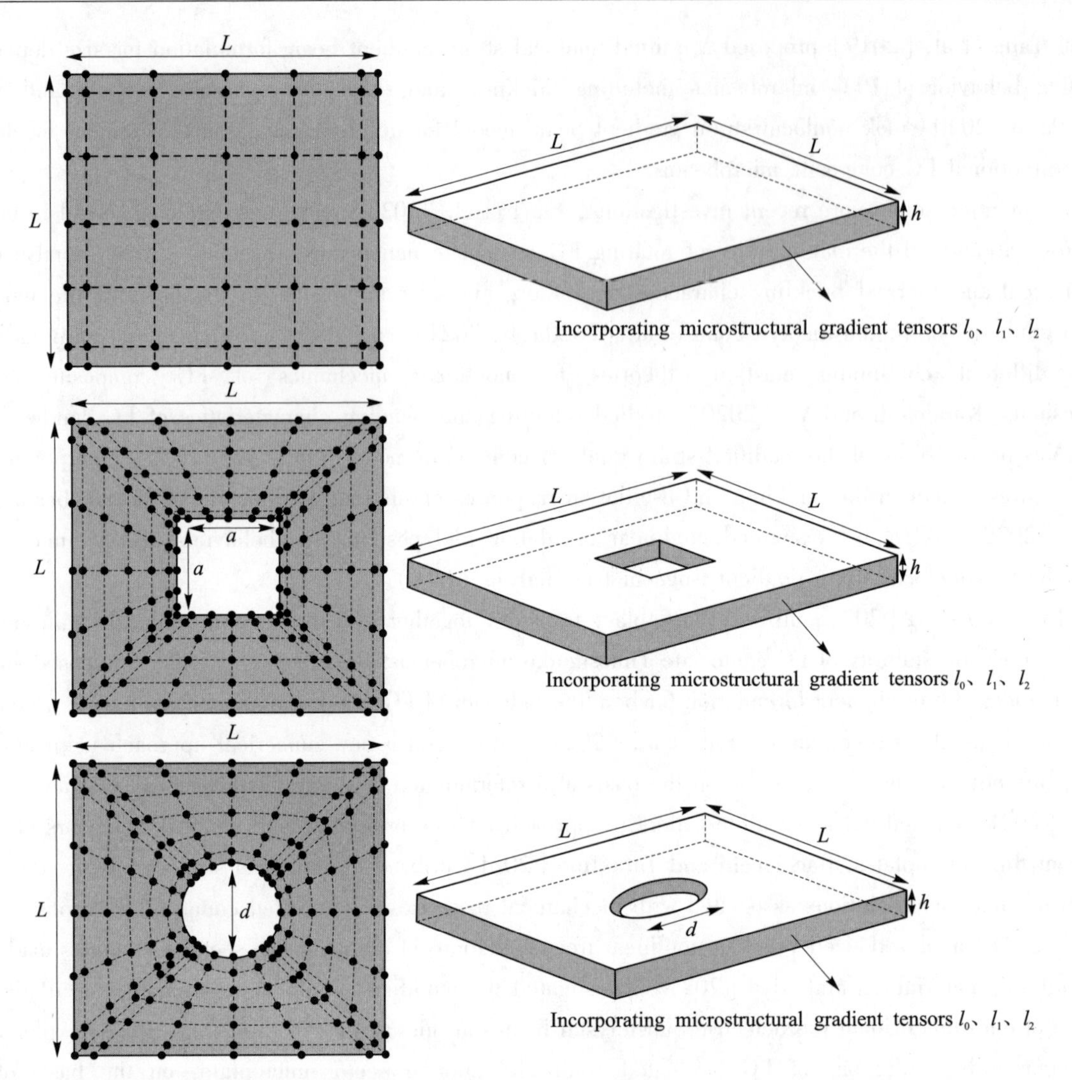

Fig.1 Discretization patterns of PFG microsized plates with and without central cutout using cubic elements

$$\text{for } U\text{-PFG}:\begin{cases} E(z)=(E_c-E_m)\left(\dfrac{1}{2}+\dfrac{z}{h}\right)^k+E_m-(E_c+E_m)\dfrac{\Gamma}{2} \\ v(z)=(v_c-v_m)\left(\dfrac{1}{2}+\dfrac{z}{h}\right)^k+v_m-(v_c+v_m)\dfrac{\Gamma}{2} \end{cases} \tag{2a}$$

$$\text{for } X\text{-PFG}:\begin{cases} E(z)=(E_c-E_m)\left(\dfrac{1}{2}+\dfrac{z}{h}\right)^k+E_m-(E_c+E_m)\dfrac{\Gamma|z|}{h} \\ v(z)=(v_c-v_m)\left(\dfrac{1}{2}+\dfrac{z}{h}\right)^k+v_m-(v_c+v_m)\dfrac{\Gamma|z|}{h} \end{cases} \tag{2b}$$

$$\text{for } O\text{-PFG}:\begin{cases} E(z)=(E_c-E_m)\left(\dfrac{1}{2}+\dfrac{z}{h}\right)^k+E_m-(E_c+E_m)\left(\dfrac{\Gamma}{2}-\dfrac{\Gamma|z|}{h}\right) \\ v(z)=(v_c-v_m)\left(\dfrac{1}{2}+\dfrac{z}{h}\right)^k+v_m-(v_c+v_m)\left(\dfrac{\Gamma}{2}-\dfrac{\Gamma|z|}{h}\right) \end{cases} \tag{2c}$$

Because in this study, the value of porosity index in considered maximum to $\Gamma=0.4$, the linear formulation is taken into account to capture effective material properties. For more large porosity, it is required to utilize a nonlinear evaluation as proposed by Guo et al. [2020].

Based upon the refined higher-order elastic model of plate as presented by Senthilnathan et al.

[1987], the transverse displacement can be separated into two parts of bending $w_b(x,y)$ and shear $w_s(x,y)$ ones. Accordingly, the strain components of a refined higher-order shear deformable PFG microplate are as

$$\boldsymbol{\varepsilon}_b=\begin{pmatrix}\varepsilon_{xx}\\ \varepsilon_{yy}\\ \gamma_{xy}\end{pmatrix}=\begin{pmatrix}\dfrac{\partial u}{\partial x}+\dfrac{1}{2}\left(\dfrac{\partial w_b}{\partial x}+\dfrac{\partial w_s}{\partial x}\right)^2\\ \dfrac{\partial u}{\partial y}+\dfrac{1}{2}\left(\dfrac{\partial w_b}{\partial y}+\dfrac{\partial w_s}{\partial y}\right)^2\\ \dfrac{\partial u}{\partial y}+\dfrac{\partial u}{\partial x}+\left(\dfrac{\partial w_b}{\partial x}+\dfrac{\partial w_s}{\partial x}\right)\left(\dfrac{\partial w_b}{\partial y}+\dfrac{\partial w_s}{\partial y}\right)\end{pmatrix}+z\begin{pmatrix}-\dfrac{\partial^2 w_b}{\partial x^2}\\ -\dfrac{\partial^2 w_b}{\partial y^2}\\ -2\dfrac{\partial^2 w_b}{\partial x\partial y}\end{pmatrix}+F(z)\begin{pmatrix}-\dfrac{\partial^2 w_s}{\partial x^2}\\ -\dfrac{\partial^2 w_s}{\partial y^2}\\ -2\dfrac{\partial^2 w_s}{\partial x\partial y}\end{pmatrix}=\boldsymbol{\varepsilon}_b^{(0)}+z\boldsymbol{\varepsilon}_b^{(1)}+F(z)\boldsymbol{\varepsilon}_b^{(2)}\tag{3a}$$

$$\gamma=\begin{Bmatrix}\gamma_{xz}\\ \gamma_{yz}\end{Bmatrix}=\left(1+\frac{\mathrm{d}F(z)}{\mathrm{d}z}\right)\begin{Bmatrix}\dfrac{\partial w_s}{\partial x}\\ \dfrac{\partial w_s}{\partial y}\end{Bmatrix}=\left(1+\frac{\mathrm{d}F(z)}{\mathrm{d}z}\right)\varepsilon_s\tag{3b}$$

By putting the modified strain gradient theory of continuum elasticity introduced by Yang et al. [2002] to use, the associated microstructural gradient tensors can be presented as follows:

$$\boldsymbol{\varsigma}_i=\frac{\partial\varepsilon_{mm}}{\partial x_i},\quad \boldsymbol{\chi}_{ij}=\frac{1}{4}\left(e_{imn}\frac{\partial^2\boldsymbol{u}_n}{\partial x_m\partial x_i}+e_{jmn}\frac{\partial^2\boldsymbol{u}_n}{\partial x_m\partial x_j}\right)$$
$$\boldsymbol{\eta}_{ijk}=\frac{1}{3}\left(\frac{\partial\varepsilon_{jk}}{\partial x_i}+\frac{\partial\varepsilon_{ki}}{\partial x_j}+\frac{\partial\varepsilon_{ij}}{\partial x_k}\right)-\frac{1}{15}\left[\delta_{ij}\left(\frac{\partial\varepsilon_{mm}}{\partial x_k}+2\frac{\partial\varepsilon_{mk}}{\partial x_m}\right)+\delta_{ik}\left(\frac{\partial\varepsilon_{mm}}{\partial x_i}+2\frac{\partial\varepsilon_{mi}}{\partial x_m}\right)+\delta_{ki}\left(\frac{\partial\varepsilon_{mm}}{\partial x_j}+2\frac{\partial\varepsilon_{mj}}{\partial x_m}\right)\right]\tag{4}$$

Accordingly, the components of the symmetric rotation gradient tensor for a refined higher-order PFG microplate can be obtained as follows:

$$\boldsymbol{\chi}_b=\begin{Bmatrix}\chi_{xx}\\ \chi_{yy}\\ \chi_{xy}\end{Bmatrix}=\begin{Bmatrix}\dfrac{\partial^2 w_b}{\partial x\partial y}+\dfrac{1}{2}\dfrac{\partial^2 w_s}{\partial x\partial y}\\ -\dfrac{\partial^2 w_b}{\partial x\partial y}-\dfrac{1}{2}\dfrac{\partial^2 w_s}{\partial x\partial y}\\ \dfrac{1}{2}\left(\dfrac{\partial^2 w_b}{\partial y^2}-\dfrac{\partial^2 w_b}{\partial x^2}\right)+\dfrac{1}{4}\left(\dfrac{\partial^2 w_s}{\partial y^2}-\dfrac{\partial^2 w_s}{\partial x^2}\right)\end{Bmatrix}+\frac{\mathrm{d}F(z)}{\mathrm{d}z}\begin{Bmatrix}-\dfrac{1}{2}\dfrac{\partial^2 w_s}{\partial x\partial y}\\ \dfrac{1}{2}\dfrac{\partial^2 w_s}{\partial x\partial y}\\ \dfrac{1}{4}\left(\dfrac{\partial^2 w_s}{\partial x^2}-\dfrac{\partial^2 w_s}{\partial y^2}\right)\end{Bmatrix}=\boldsymbol{\chi}_b^0+\frac{\mathrm{d}F(z)}{\mathrm{d}z}\boldsymbol{\chi}_b^{(1)}$$
$$\boldsymbol{\chi}_s=\begin{Bmatrix}\chi_{xz}\\ \chi_{yz}\end{Bmatrix}=\begin{Bmatrix}\dfrac{1}{4}\left(\dfrac{\partial^2 v}{\partial x^2}-\dfrac{\partial^2 u}{\partial x\partial y}\right)\\ \dfrac{1}{4}\left(\dfrac{\partial^2 v}{\partial x\partial y}-\dfrac{\partial^2 u}{\partial y^2}\right)\end{Bmatrix}+\frac{\mathrm{d}^2F(z)}{\mathrm{d}z^2}\begin{Bmatrix}-\dfrac{1}{4}\dfrac{\partial w_s}{\partial y}\\ -\dfrac{1}{4}\dfrac{\partial w_s}{\partial x}\end{Bmatrix}=\boldsymbol{\chi}_s^0+\frac{\mathrm{d}^2F(z)}{\mathrm{d}z^2}\boldsymbol{\chi}_s^{(1)}\tag{5}$$

In addition, the microstructural dilatation gradient tensor for a refined higher-order PFG microplate contains the following components:

$$\boldsymbol{\varsigma}=\begin{Bmatrix}\dfrac{\partial^2 u}{\partial x^2}+\dfrac{\partial^2 v}{\partial x\partial y}+\left(\dfrac{\partial w_b}{\partial x}+\dfrac{\partial w_s}{\partial x}\right)\left(\dfrac{\partial^2 w_b}{\partial x^2}+\dfrac{\partial^2 w_s}{\partial x^2}\right)+\left(\dfrac{\partial w_b}{\partial y}+\dfrac{\partial w_s}{\partial y}\right)\\ \left(\dfrac{\partial^2 w_b}{\partial x\partial y}+\dfrac{\partial^2 w_s}{\partial x\partial y}\right)\dfrac{\partial^2 u}{\partial x\partial y}+\dfrac{\partial^2 v}{\partial y^2}+\left(\dfrac{\partial w_b}{\partial y}+\dfrac{\partial w_s}{\partial y}\right)\left(\dfrac{\partial^2 w_b}{\partial y^2}+\dfrac{\partial^2 w_s}{\partial y^2}\right)+\left(\dfrac{\partial w_b}{\partial x}+\dfrac{\partial w_s}{\partial x}\right)\left(\dfrac{\partial^2 w_b}{\partial x\partial y}+\dfrac{\partial^2 w_s}{\partial x\partial y}\right)-\dfrac{\partial^2 w_b}{\partial x^2}-\dfrac{\partial^2 w_b}{\partial y^2}\end{Bmatrix}+$$
$$z\begin{Bmatrix}-\dfrac{\partial^3 w_b}{\partial x^3}-\dfrac{\partial^3 w_b}{\partial x\partial y^2}\\ -\dfrac{\partial^3 w_b}{\partial x^2\partial y}-\dfrac{\partial^3 w_b}{\partial y^3}\\ 0\end{Bmatrix}+F(z)\begin{Bmatrix}-\dfrac{\partial^3 w_s}{\partial x^3}-\dfrac{\partial^3 w_s}{\partial x\partial y^2}\\ -\dfrac{\partial^3 w_s}{\partial x^2\partial y}-\dfrac{\partial^3 w_s}{\partial y^3}\\ 0\end{Bmatrix}+\frac{\mathrm{d}F(z)}{\mathrm{d}z}\begin{Bmatrix}0\\ 0\\ \dfrac{\partial^2 w_s}{\partial x^2}-\dfrac{\partial^2 w_s}{\partial y^2}\end{Bmatrix}=\boldsymbol{\varsigma}_b^{(0)}+z\boldsymbol{\varsigma}_b^{(1)}+F(z)\boldsymbol{\varsigma}_b^{(2)}+\frac{\mathrm{d}F(z)}{\mathrm{d}z}\boldsymbol{\varsigma}_b^{(3)}\tag{6}$$

Also, one can obtain the components of the deviatoric stretch gradient tensor for a refined higher-order PFG microplate as follows:

$$
\bar{\boldsymbol{\eta}}=\left\{\begin{array}{l}
\frac{2}{5}\frac{\partial^2 u}{\partial x^2}-\frac{1}{5}\frac{\partial^2 u}{\partial y^2}-\frac{2}{5}\frac{\partial^2 v}{\partial x\partial y}+\frac{2}{5}\left(\frac{\partial w_b}{\partial x}+\frac{\partial w_s}{\partial x}\right)\left(\frac{\partial^2 w_b}{\partial x^2}+\frac{\partial^2 w_s}{\partial x^2}\right)\\
-\frac{1}{5}\left(\frac{\partial w_b}{\partial x}+\frac{\partial w_s}{\partial x}\right)\left(\frac{\partial^2 w_b}{\partial y^2}+\frac{\partial^2 w_s}{\partial y^2}\right)-\frac{2}{5}\left(\frac{\partial w_b}{\partial y}+\frac{\partial w_s}{\partial y}\right)\left(\frac{\partial^2 w_b}{\partial x\partial y}+\frac{\partial^2 w_s}{\partial x\partial y}\right)\\
-\frac{2}{5}\frac{\partial^2 u}{\partial x\partial y}+\frac{2}{5}\frac{\partial^2 v}{\partial y^2}-\frac{1}{5}\frac{\partial^2 v}{\partial x^2}+\frac{2}{5}\left(\frac{\partial w_b}{\partial y}+\frac{\partial w_s}{\partial y}\right)\left(\frac{\partial^2 w_b}{\partial y^2}+\frac{\partial^2 w_s}{\partial y^2}\right)\\
-\frac{1}{5}\left(\frac{\partial w_b}{\partial y}+\frac{\partial ws}{\partial y}\right)\left(\frac{\partial^2 w_b}{\partial x^2}+\frac{\partial^2 w_s}{\partial x^2}\right)-\frac{2}{5}\left(\frac{\partial w_b}{\partial x}+\frac{\partial w_s}{\partial x}\right)\left(\frac{\partial^2 w_b}{\partial x\partial y}+\frac{\partial^2 w_s}{\partial x\partial y}\right)\\
-\frac{1}{5}\frac{\partial^2 u}{\partial x^2}+\frac{4}{15}\frac{\partial^2 u}{\partial y^2}+\frac{8}{15}\frac{\partial^2 v}{\partial y\partial y}-\frac{1}{5}\left(\frac{\partial w_b}{\partial y}+\frac{\partial w_s}{\partial y}\right)\left(\frac{\partial^2 w_b}{\partial y^2}+\frac{\partial^2 w_s}{\partial y^2}\right)\\
+\frac{4}{15}\left(\frac{\partial w_b}{\partial y}+\frac{\partial w_s}{\partial y}\right)\left(\frac{\partial^2 w_b}{\partial x^2}+\frac{\partial^2 w_s}{\partial x^2}\right)+\frac{8}{15}\left(\frac{\partial w_b}{\partial x}+\frac{\partial w_s}{\partial x}\right)\left(\frac{\partial^2 w_b}{\partial x\partial y}+\frac{\partial^2 w_s}{\partial x\partial y}\right)\\
\frac{8}{15}\frac{\partial^2 u}{\partial x\partial y}-\frac{1}{5}\frac{\partial^2 v}{\partial y^2}-\frac{4}{5}\frac{\partial^2 v}{\partial x^2}-\frac{1}{5}\left(\frac{\partial w_b}{\partial x}+\frac{\partial w_s}{\partial x}\right)\left(\frac{\partial^2 w_b}{\partial x^2}+\frac{\partial^2 w_s}{\partial x^2}\right)\\
+\frac{4}{15}\left(\frac{\partial w_b}{\partial x}+\frac{\partial w_s}{\partial x}\right)\left(\frac{\partial^2 w_b}{\partial y^2}+\frac{\partial^2 w_s}{\partial y^2}\right)+\frac{8}{15}\left(\frac{\partial w_b}{\partial y}+\frac{\partial w_s}{\partial y}\right)\left(\frac{\partial^2 w_b}{\partial x\partial y}+\frac{\partial^2 w_s}{\partial x\partial y}\right)\\
-\frac{1}{5}\frac{\partial^2 u}{\partial x^2}-\frac{1}{15}\frac{\partial^2 u}{\partial y^2}-\frac{2}{15}\frac{\partial^2 v}{\partial x\partial y}-\frac{1}{5}\left(\frac{\partial w_b}{\partial x}+\frac{\partial w_s}{\partial x}\right)\left(\frac{\partial^2 w_b}{\partial x^2}+\frac{\partial^2 w_s}{\partial x^2}\right)\\
-\frac{1}{15}\left(\frac{\partial w_b}{\partial x}+\frac{\partial w_s}{\partial x}\right)\left(\frac{\partial^2 w_b}{\partial y^2}+\frac{\partial^2 w_s}{\partial y^2}\right)-\frac{2}{15}\left(\frac{\partial w_b}{\partial y}+\frac{\partial w_s}{\partial y}\right)\left(\frac{\partial^2 w_b}{\partial x\partial y}+\frac{\partial^2 w_s}{\partial x\partial y}\right)\\
-\frac{2}{15}\frac{\partial^2 u}{\partial x\partial y}-\frac{1}{15}\frac{\partial^2 v}{\partial x^2}-\frac{1}{5}\frac{\partial^2 v}{\partial y^2}-\frac{1}{5}\left(\frac{\partial w_b}{\partial y}+\frac{\partial w_s}{\partial y}\right)\left(\frac{\partial^2 w_b}{\partial y^2}+\frac{\partial^2 w_s}{\partial y^2}\right)\\
-\frac{1}{15}\left(\frac{\partial w_b}{\partial y}+\frac{\partial w_s}{\partial y}\right)\left(\frac{\partial^2 w_b}{\partial x^2}+\frac{\partial^2 w_s}{\partial x^2}\right)-\frac{2}{15}\left(\frac{\partial w_b}{\partial x}+\frac{\partial w_s}{\partial x}\right)\left(\frac{\partial^2 w_b}{\partial x\partial y}+\frac{\partial^2 w_s}{\partial x\partial y}\right)
\end{array}\right\}+z\left\{\begin{array}{l}
-\frac{2}{5}\frac{\partial^3 w_b}{\partial x^3}+\frac{3}{5}\frac{\partial^3 w_b}{\partial x\partial y^2}\\
-\frac{2}{5}\frac{\partial^3 w_b}{\partial y^3}+\frac{3}{5}\frac{\partial^3 w_b}{\partial x^3\partial y}\\
\frac{1}{5}\frac{\partial^3 w_b}{\partial x^3}+\frac{4}{5}\frac{\partial^2 w_b}{\partial x\partial y^2}\\
\frac{1}{5}\frac{\partial^3 w_b}{\partial y^3}-\frac{4}{5}\frac{\partial^3 w_b}{\partial x^2\partial y}\\
\frac{1}{5}\frac{\partial^2 w_b}{\partial x^2}+\frac{1}{5}\frac{\partial^2 w_b}{\partial x\partial y^2}\\
\frac{1}{5}\frac{\partial^3 w_b}{\partial y^3}+\frac{1}{5}\frac{\partial^3 w_b}{\partial x^2\partial y}
\end{array}\right\}+
$$

$$
F(z)\left\{\begin{array}{l}
\frac{2}{5}\frac{\partial^2 w_s}{\partial x^3}-\frac{3}{5}\frac{\partial^2 w_s}{\partial x\partial y^2}\\
\frac{2}{5}\frac{\partial^3 w_s}{\partial y^3}-\frac{3}{5}\frac{\partial^3 w_s}{\partial x^2\partial y}\\
-\frac{1}{5}\frac{\partial^3 w_s}{\partial x^3}+\frac{4}{5}\frac{\partial^2 w_s}{\partial x\partial y^2}\\
-\frac{1}{5}\frac{\partial^3 w_s}{\partial y^3}+\frac{4}{5}\frac{\partial^3 w_s}{\partial x^2\partial y}\\
-\frac{1}{5}\frac{\partial^2 w_s}{\partial x^2}-\frac{1}{5}\frac{\partial^2 w_s}{\partial x\partial y^2}\\
-\frac{1}{5}\frac{\partial^3 w_s}{\partial y^3}-\frac{1}{5}\frac{\partial^2 w_b}{\partial x^2\partial y}
\end{array}\right\}+\frac{\mathrm{d}^2F(z)}{\mathrm{d}z^2}\left\{\begin{array}{l}
-\frac{1}{5}\frac{\partial w_s}{\partial x}\\
-\frac{1}{5}\frac{\partial w_s}{\partial y}\\
-\frac{1}{15}\frac{\partial w_s}{\partial x}\\
-\frac{1}{15}\frac{\partial w_s}{\partial y}\\
-\frac{4}{15}\frac{\partial w_s}{\partial x}\\
-\frac{4}{15}\frac{\partial w_s}{\partial y}
\end{array}\right\}=\widetilde{\boldsymbol{\eta}}^{(0)}+z\widetilde{\boldsymbol{\eta}}^{(1)}+F(z)\widetilde{\boldsymbol{\eta}}^{(2)}+\frac{\mathrm{d}^2F(z)}{\mathrm{d}z^2}\widetilde{\boldsymbol{\eta}}^{(3)}
$$

$$\hat{\boldsymbol{\eta}}=\begin{Bmatrix}\frac{1}{5}\left(\frac{\partial^2 w_b}{\partial x^2}+\frac{\partial^2 w_b}{\partial y^2}\right)-\frac{1}{5}\left(\frac{\partial^2 w_s}{\partial x^2}+\frac{\partial^2 w_s}{\partial y^2}\right)\\ -\frac{4}{15}\frac{\partial^2 w_b}{\partial x^2}+\frac{1}{15}\frac{\partial^2 w_b}{\partial y^2}+\frac{4}{15}\frac{\partial^2 w_s}{\partial x^2}-\frac{1}{15}\frac{\partial^2 w_s}{\partial y^2}\\ \frac{1}{15}\frac{\partial^2 w_b}{\partial x^2}-\frac{4}{15}\frac{\partial^2 w_b}{\partial y^2}+\frac{4}{15}\frac{\partial^2 w_s}{\partial y^2}-\frac{1}{15}\frac{\partial^2 w_s}{\partial x^2}\\ -\frac{1}{3}\frac{\partial^2 w_b}{\partial x\partial y}+\frac{1}{3}\frac{\partial^2 w_s}{\partial x\partial y}\end{Bmatrix}+\frac{\mathrm{d}^2F(z)}{\mathrm{d}z}\begin{Bmatrix}-\frac{2}{5}\frac{\partial^2 w_s}{\partial x^2}-\frac{2}{5}\frac{\partial^2 w_s}{\partial y^2}\\ \frac{8}{15}\frac{\partial^2 w_s}{\partial x^2}-\frac{2}{15}\frac{\partial^2 w_s}{\partial y^2}\\ \frac{8}{15}\frac{\partial^2 w_s}{\partial y^2}-\frac{2}{15}\frac{\partial^2 w_s}{\partial x^2}\\ \frac{2}{3}\frac{\partial^2 w_s}{\partial x\partial y}\end{Bmatrix}=\tilde{\boldsymbol{\eta}}^{(0)}+\frac{\mathrm{d}^2F(z)}{\mathrm{d}z}\tilde{\boldsymbol{\eta}}^{(1)} \tag{7}$$

where

$$\boldsymbol{\eta}=\begin{Bmatrix}\tilde{\boldsymbol{\eta}}\\ \hat{\boldsymbol{\eta}}\end{Bmatrix}=\begin{Bmatrix}\{\eta_{xxx}\ \eta_{yyy}\ \eta_{yyx}\ \eta_{xxy}\ \eta_{zzx}\ \eta_{zzy}\}^{\mathrm{T}}\\ \{\eta_{zzz}\ \eta_{xxz}\ \eta_{yyz}\ \eta_{xyz}\}^{\mathrm{T}}\end{Bmatrix} \tag{8}$$

Consequently, the constructive equations can be expressed as

$$\begin{Bmatrix}\sigma_{xx}\\ \sigma_{yy}\\ \tau_{xz}\\ \tau_{yz}\\ \tau_{xy}\end{Bmatrix}=\begin{bmatrix}Q_{11}(z) & Q_{12}(z) & 0 & 0 & 0\\ Q_{21}(z) & Q_{22}(z) & 0 & 0 & 0\\ 0 & 0 & Q_{44}(z) & 0 & 0\\ 0 & 0 & 0 & Q_{55}(z) & 0\\ 0 & 0 & 0 & 0 & Q_{66}(z)\end{bmatrix}\begin{Bmatrix}\varepsilon_{xx}\\ \varepsilon_{yy}\\ \gamma_{xz}\\ \gamma_{yz}\\ \gamma_{xy}\end{Bmatrix}$$

$$\{m_{xx},m_{yy},m_{xy},m_{xz},m_{yz}\}=\frac{l_0^2E(z)}{1+v(z)}\{\chi_{xx},\chi_{yy},\chi_{xy},\chi_{xz},\chi_{yz}\}$$

$$\{p_x,p_y,p_z\}=\frac{l_1^2E(z)}{1+v(z)}\{\varsigma_x,\varsigma_y,\varsigma_z\}$$

$$\{\tau_{xxx},\tau_{yyy},\tau_{yyx},\tau_{xxy},\tau_{zzx},\tau_{zzy}\}=\frac{l_2^2E(z)}{1+v(z)}\{\eta_{xxx},\eta_{yyy},\eta_{yyx},\eta_{xxy},\eta_{zzx},\eta_{zzy}\}$$

$$\{\tau_{zzz},\tau_{xxz},\tau_{yyz},\tau_{xyz}\}=\frac{l_2^2E(z)}{1+v(z)}\{\eta_{zzz},\eta_{xxz},\eta_{yyz},\eta_{xyz}\}$$

in which

$$\begin{aligned}Q_{11}(z)=Q_{22}(z)&=\frac{E(z)}{[1+v(z)][1-2v(z)]}\\ Q_{12}(z)=Q_{21}(z)&=\frac{v(z)E(z)}{[1+v(z)][1-2v(z)]}\\ Q_{44}(z)=Q_{55}(z)=Q_{66}(z)&=\frac{E(z)}{2[1+v(z)]}\end{aligned} \tag{9}$$

and l_0, l_1 and l_2 are the internal length scale parameters associated with the rotation gradient, dilatation gradient, and deviatoric stretch gradient tensors, respectively. So, the variation of the strain energy is

$$\begin{aligned}\delta\Pi_S=&\iint_S\int_{-\frac{h}{2}}^{\frac{h}{2}}\{\sigma_{xx}\delta\varepsilon_{xx}+\sigma_{yy}\delta\varepsilon_{yy}+\sigma_{zz}\delta\varepsilon_{zz}+\tau_{xy}\delta\gamma_{xy}+\tau_{yz}\delta\gamma_{yz}+\tau_{xz}\delta\gamma_{xz}\}\mathrm{d}z\mathrm{d}S+\\ &\iint_S\int_{-\frac{h}{2}}^{\frac{h}{2}}\{m_{xx}\delta\chi_{xx}+m_{yy}\delta x_{yy}+m_{xy}\delta\chi_{xy}+m_{yz}\delta\chi_{yz}+m_{xz}\delta\chi_{xz}\}\mathrm{d}z\mathrm{d}S+\\ &\iint_S\int_{-\frac{h}{2}}^{\frac{h}{2}}\{p_x\delta_{\varsigma x}+p_y\delta_{\varsigma y}+p_z\delta_{\varsigma z}\mathrm{d}z\mathrm{d}S+\\ &\iint_S\int_{-\frac{h}{2}}^{\frac{h}{2}}\{\tau_{xxx}\delta\eta_{xxx}\tau_{yyy}\delta\eta_{yyy}+\tau_{yyx}\delta\eta_{yyx}+\tau_{xxy}\delta\eta_{xxy}+\tau_{zzx}\delta\eta_{zzx}+\tau_{xxy}\delta\eta_{xxy}+\tau_{zzz}\delta\eta_{zzz}+\tau_{xxz}\delta\eta_{xxz}+\end{aligned}$$

$$\tau_{yyz}\delta\eta_{yyz} + \tau_{xyz}\delta\eta_{xyz}\}\,dzdS \tag{10}$$

Furthermore, the external virtual work applied by the axial compression of P can be given as

$$\delta\Pi_W = \int_S P\left(\frac{\partial w_b}{\partial x} + \frac{\partial w_s}{\partial x}\right)\delta\left(\frac{\partial w_b}{\partial x} + \frac{\partial w_s}{\partial x}\right)dS \tag{11}$$

So, the Galerkin weak type of the governing differential equations in variational form are as follows:

$$\int_S P\{\delta(\bar{\bar{\varepsilon}}_b^{\mathrm{T}})\Re_b\bar{\bar{\varepsilon}}_b + \delta\bar{\bar{\varepsilon}}_s^{\mathrm{T}})\Re_s\varepsilon_s + \delta(\bar{\bar{\chi}}_b^{\mathrm{T}})\vartheta_b\Phi_1\bar{\bar{\chi}}_b + \delta(\bar{\bar{\chi}}_s^{\mathrm{T}})\vartheta_s\Phi_2\bar{\bar{\chi}}_s + \delta(\bar{\bar{\varsigma}}^{\mathrm{T}})\mathfrak{S}\,\bar{\bar{\varsigma}} + \delta(\tilde{\tilde{\eta}}^{\mathrm{T}})\Re_b\Phi_{3\eta}^{\cong} +$$

$$\delta(\overset{\triangle}{\eta}^{\mathrm{T}})\Re_s\Phi_4\overset{\triangle}{\eta} + P\left(\frac{\partial^2 w_b}{\partial x^2} + \frac{\partial^2 w_s}{\partial x^2}\right)\Bigg\}\,dS = 0 \tag{12}$$

in which

$$\bar{\bar{\varepsilon}}_b = \begin{Bmatrix}\varepsilon_b^{(0)}\\ \varepsilon_b^{(1)}\\ \varepsilon_b^{(2)}\end{Bmatrix},\quad \bar{\bar{\chi}}_b = \begin{Bmatrix}\chi_b^{(0)}\\ \chi_b^{(1)}\end{Bmatrix},\quad \bar{\bar{\chi}}_s = \begin{Bmatrix}\chi_s^{(0)}\\ \chi_s^{(1)}\end{Bmatrix},\ \bar{\bar{\varsigma}}_s = \begin{Bmatrix}\varsigma_b^{(0)}\\ \varsigma_b^{(1)}\\ \varsigma_b^{(2)}\\ \varsigma_b^{(3)}\end{Bmatrix},\ \tilde{\tilde{\eta}} = \begin{Bmatrix}\tilde{\eta}^{(0)}\\ \tilde{\eta}^{(1)}\\ \tilde{\eta}^{(2)}\\ \tilde{\eta}^{(3)}\end{Bmatrix},\ \hat{\hat{\eta}} = \begin{Bmatrix}\hat{\eta}^{(0)}\\ \hat{\eta}^{(1)}\end{Bmatrix} \tag{13}$$

3 Isogeometric-Based Numerical Solving Strategy

In the last decade, IGA has been employed as one of the efficient numerical techniques to construct a bridge between computer aided design methodology and finite element strategy in order to describe accurately an arbitrary geometry for a structure. Accordingly, by taking a two-dimensional regime into consideration, the associated knot vectors in two directions of ξ and ς are based upon a nonreducing scheme as follows:

$$K(\xi)=\{\xi_1,\xi_2,\xi_3,\cdots,\xi_{m+p+1}\},\qquad \xi_{i+1}\geqslant\xi_i\qquad \text{and}\qquad 1<i<m+p+1 \tag{14a}$$

$$M(\varsigma)=\{\zeta_1,\zeta_2,\zeta_3,\cdots,\zeta_{n+q+1}\},\qquad \zeta_{i+1}\geqslant\zeta_i\qquad \text{and}\qquad 1<i<n+q+1 \tag{14b}$$

in which m and n indicate the number of B-spline basis functions utilized for each direction, and p and q stand for their orders. In addition, each ith knot employed in IGA is selected within the considered ranges of $0 \leqslant \xi_i \leqslant 1$ and $0 \leqslant \zeta_i \leqslant 1$. As a result, on the basis of Cox-de Boor algorithm, the associated basis functions can be built mathematically in the following forms:

$$\begin{aligned}
x_{i,0}(\xi) &= \begin{cases}1, \xi_i \leqslant \xi < \xi_{i+1}\\ 0, \text{else}\end{cases}\\
y_{i,0}(\zeta) &= \begin{cases}1, \zeta_i \leqslant \zeta < \zeta_{i+1}\\ 0, \text{else}\end{cases}\\
x_{i,p}(\xi) &= \frac{\xi - \xi_i}{\xi_{i+p} - \xi_i}x_{i,p-1}(\xi) + \frac{\xi_{i+p+1} - \xi}{\xi_{i+p+1} - \xi_{i+1}}x_{i+1,p-1}(\xi)\\
y_{i,q}(\zeta) &= \frac{\zeta - \zeta_i}{\zeta_{i+q} - \zeta_i}y_{i,q-1}(\zeta) + \frac{\zeta_{i+q+1} - \zeta}{\zeta_{i+q+1} - \zeta_{i+1}}y_{i+1,q-1}(\zeta)
\end{aligned} \tag{15}$$

Therefore, for a plate-type microstructure, the appropriate B-spline basis functions can be derived using proper tensor product as

$$\mathscr{F}_{i,j}^{p,q}(\xi,\zeta) = \sum_{i=1}^{m} \mathfrak{z}_i(x,y)p_i \tag{16}$$

where P_i represents the nodal degrees of freedom related to a control net including two separate directions, and

$$\mathfrak{z}_i(\xi,\zeta) = (x_{i,p}(\xi)y_{j,q}(\zeta)W_{i,j})/\left(\sum_{i=1}^{m}\sum_{j=1}^{n}(x_{i,p}(\xi)y_{j,q}(\zeta)W_{i,j})\right) \tag{17}$$

in which Wi,j represents the weight coefficient for the related B-spline basis function.

Now, by putting the introduced basis functions into use, the correspondence displacement field for a refined higher-order PFG composite microplate satisfying the C^{-1}-requirment can be estimated as follows:

$$\{\tilde{u}^i,\tilde{v}^i,\tilde{w}_b^i,\tilde{w}_s^i\}^{\mathrm{T}}=\sum_{i=1}^{m\times n}\begin{bmatrix}3i(x,y) & 0 & 0 & 0\\ 0 & 3i(x,y) & 0 & 0\\ 0 & 0 & 3i(x,y) & 0\\ 0 & 0 & 0 & 3i(x,y)\end{bmatrix}\begin{Bmatrix}u^i\\ v^i\\ w_b^i\\ w_s^i\end{Bmatrix} \tag{18}$$

By combining Eqs. (18) and (20), and then putting it in the conventional and microstructural-dependent strain tensors, the discretized form of the governing equations becomes

$$\sum_{i=1}^{m\times n}K^i(X)X^i=P \tag{19}$$

in which $K(X)$ refers to the general stiffness matrix of a refined higher-order PFG microplate incorporating two parts of linear and nonlinear ones as below.

In order to trace the nonlinear stability equilibrium paths, the incrementaliterative Newton method together with the arc-length technique are put to use to achieve the nonlinear solution of Eq. (21). It is supposed that the external axial compression can be formulated in terms of a fixed load of P_0 with a proportional form as follows:

$$\Big(\sum_{i=1}^{m\times n}K_L^i(X)+\sum_{i=1}^{m\times n}K_{NL}^i(X)\Big)X^i=\lambda P_0 \tag{20}$$

After that, through an increment in the value of the applied axial compression from λP_0 to $(\lambda+\Delta\lambda)P_0$, the new equilibrium configuration near to the previous one can be written as

$$\Big(\sum_{i=1}^{m\times n}K_L^i(X)+\sum_{i=1}^{m\times n}K_{NL}^i(X)\Big)(X^i-\Delta X^i)=(\lambda+\Delta\lambda)P_0 \tag{21}$$

By taking the expansion of Taylor series into account, it yields

$$\Big(\sum_{i=1}^{m\times n}K_L^i(X)+\sum_{i=1}^{m\times n}K_{NL}^i(X)\Big)X^i-\lambda P_0+\Big(\sum_{i=1}^{m\times n}K_L^i(X)+\sum_{i=1}^{m\times n}K_{NL}^i(X)\Big)\Delta X^i\Delta\lambda P_0=0 \tag{22}$$

As a consequence, the incremental expressions related to the stability equation and the associated displacement can be read as

$$\Delta X_j^i=\Big(\sum_{i=1}^{m\times n}K_L^i(X)+\sum_{i=1}^{m\times n}K_{NL}^i(X)\Big)^{-1}\times\Big[(X_j^i+\Delta X_j^i)P_0-\Big(\sum_{i=1}^{m\times n}K_L^i(X)+\sum_{i=1}^{m\times n}K_{NL}^i(X)\Big)X_j^i\Big] \tag{23a}$$

$$X_{ij+1}=X_{ij}+\Delta X_{ij} \tag{23b}$$

where j stands for the load step number.

Another constrain is added to the solving process due to the addition of a new variable of ΔX^i to the problem related to each increment step. As a result, by taking the arc-length continuation method into account, the incremental-iterative equations can be written in a generalized form as follows:

$$\begin{aligned}\Delta X_j^i&=\Big(\sum_{i=1}^{m\times n}K_L^i(X)+\sum_{i=1}^{m\times n}K_{NL}^i(X)\Big)^{-1}\times\Big[(\lambda_j^{i-1}+\Delta\lambda_j^i)P_0-\Big(\sum_{i=1}^{m\times n}K_L^i(X)+\sum_{i=1}^{m\times n}K_{NL}^i(X)\Big)X_j^{i-1}\Big]\\&=\Delta\lambda_j^i[\overline{X}]_j+[\overline{\overline{\Delta X}}]_j^i\end{aligned} \tag{24}$$

in which, $[\overline{\overline{X}}]_j$ denotes the displacement increment related to the increment of applied compression, and $[\overline{\overline{\Delta X}}]_j^i$ refers to the displacement increment associated with the residual force. Therefore, ΔX_j^i is captured using an iterative arc-length strategy defined by Crisfield given by Crisfield [1991].

The tolerance using to check the convergence of the presented iteration procedure relevant to each step is selected as

$$\left\| \left(\sum_{i=1}^{m \times n} K_L^i(X) + \sum_{i=1}^{m \times n} K_{NL}^i(X) \right) \times X_j^i - \lambda_j^i P_0 \right\| \leqslant \vartheta \left\| \lambda_j^i P_0 \right\| \tag{25}$$

where ϑ stands for the tolerance parameter assumed equal to 0.0001.

4 Numerical Results and Discussion

In this section, by performing the developed numerical solution strategy, the microstructural-dependent nonlinear stability characteristics of the refined higher-order axially compressed PFG composite microplate are studied incorporating separately the various types of microsized strain gradient tensor. The material gradient of PFG microsized plates is undertaken as there is a ceramic-rich surface at the top and a metal-rich surface at the bottom of microplates, the properties of which are taken into account as $E_c = 210$ GPa, $\nu = 0.24$ corresponding to the ceramic phase and $E_m = 70$ GPa, $\nu = 0.35$ corresponding to the metal phase, as presented by Miller and Shenoy [2000]. Also, the applied axial compressive load and the associated induced deflection are made dimensionless as $\overline{P} = PL_1^2/E_m h^3$, $W = (w_b + w_s)/h$. Additionally, the geometric of rectangular microplates is selected as $h = 40$ μm, $L_1 = 50h$, $L_1 = L_2$.

In the commencement of this section, the validity of the employed solution tactic is surveyed. In accordance to this motivation, the dimensionless strain gradientbased critical buckling loads of square FG composite microplates are predicted corresponding to various property gradient indices and compared with those reported formerly by Thai et al. [2018] taking IGA into account. As presented in Table 1, the correction of the established refined higher-order PFG plate formulations in conjunction with the taken on solving process are confirmed.

Also, by ignoring the terms associated with the microstructural gradient tensors, the critical buckling loads of axially compressed PFG composite plates at macroscale are obtained corresponding to various ratios of L_1/L_2 and are compared with those reported by Chen et al. [2019] using Chebyshev-Ritz technique. As given in Table 2, an excellent agreement is found which confirms again the validity of this analysis.

Table 1 Comparison exploration for dimensionlessstrain gradient-based critical buckling loads of FG composite microplates corresponding to various propertygradient indices

k	$l_0 = l_1 = l_2$	Thai et al. 2018	Present study
0.5	01h	12.4241	12.4233
	02h	17.7378	17.7359
	05h	53.8834	53.8811
	h	181.0887	181.0862
1	01h	9.7001	9.6992
	02h	14.0645	14.0637
	05h	43.8203	43.8191
	h	148.6460	148.6444
2	01h	7.4530	7.4525
	02h	10.7648	10.7640
	05h	33.7503	33.7491
	h	115.5208	115.5189

Table 2 Comparison exploration for dimensionless critical buckling loads of simply supported PFG composite plates correspondingto various ratios of $\frac{l_1}{l_2}$

l_1/l_2	Chen et al.2019	Present study
1.2	0.00745	0.00743
1.4	0.00807	0.00806
1.6	0.00752	0.00749
1.8	0.00726	0.00723
2	0.00719	0.00717

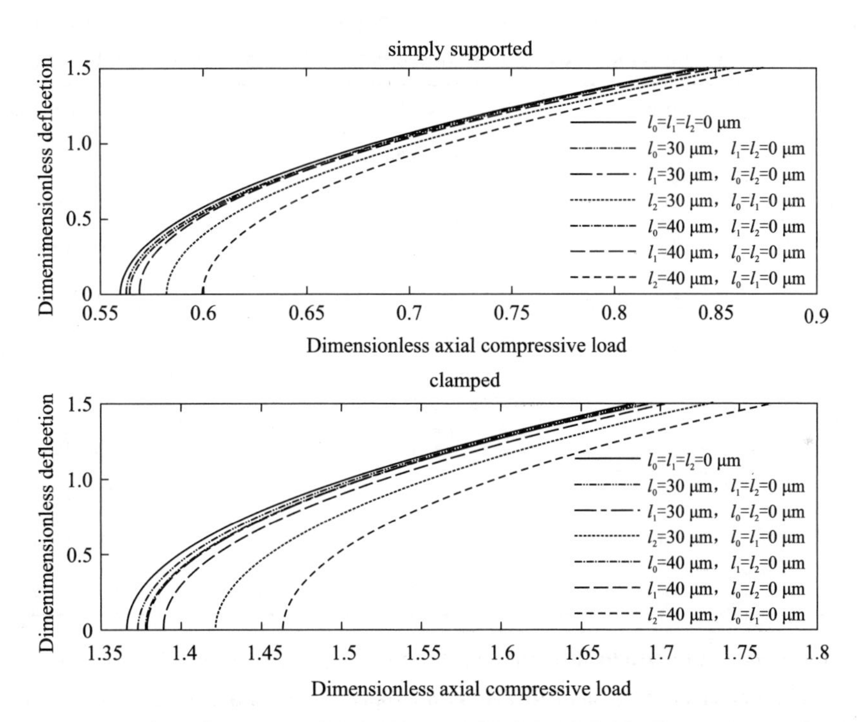

Fig.2 Microstructural-dependent nonlinear stability paths of U-PFG microplates in the presence of various strain gradient tensors($k = 0.5, \Gamma = 0.4, a = d = 0$)

Displayed in Fig.2 are the conventional and microstructural-dependent nonlinear stability paths of a U-PFG microplate without any central cutout corresponding to taking various types of strain gradient tensors into consideration. Through comparison of the traced load-deflection paths, it is reached to this point that the strengthening role of the deviatoric stretch type of gradient tensor is more than. the symmetric rotation and dilatation ones.Accordingly, it is observed that by taking all microstructural strain gradient tensors into account, the PFG microplate becomes more stable, so a higher axial compression is required to induce a specific lateral deflection in the PFG microplate.

In order to demonstrate the significance of stiffening character associated with each microstructural strain gradient tensor, the dimensionless critical buckling loads and two nonlinear postbuckling loads associated with two different values of the induced microplate deflection are given in Tables 3 and 4 corresponding to, respectively, the simply supported and clamped boundary conditions corresponding to various types of the microstructural strain gradient tensors. The percentages presented in parentheses reveal the difference between the obtained microsize-dependent axial compressions with their conventional counterpart. In the case of simply supported boundary conditions, it is found that the critical buckling load is enhanced about 1.36% by considering only the symmetric rotation gradient tensor, about 2.56% by taking only the dilatation gradient tensor, and 11.10% by considering only the deviatoric stretch gradient tensor. These percentages for the nonlinear postbuckling loads associated with the dimensionless induced deflections of $(w_b+w_s)/h$ = 0.5 and $(w_b+w_s)/h$ = 1 are, respectively, for each microstructural strain gradient tensors as 1.28%, 2.42%, 10.51%, and 1.02%, 1.95%, 8.76%. In the case of a clamped U-PFG microplate, it is revealed that the buckling stiffness is strengthened by the same percentages for the simply supported microplate, but within the nonlinear postbuckling regimes, the strengthening roles of microstructural strain gradient tensors become more significant; as for the dimensionless induced deflection of $(w_b+w_s)/h$ = 0.5, they are, respectively, 1.33%, 2.49%, and 10.82%, and for the dimensionless induced deflection of $(w_b+w_s)/h$ = 1, they are obtained as 1.19%, 2.25%, and 9.91%, respectively.

Table 3 Stiffening roles of different microstructural strain gradienttensors on the size dependent nonlinear stability behavior of a simplysupported U-PFG microplate without a central cutout (k=0.5, T=0.4)

l_0(μm)	l_1(μm)	l_2(μm)	Dimensionless axial compression
		$(w_b+w_s)/h=0$	
0	0	0	0.5594
50	0	0	0.5670 (1.36%)
0	50	0	0.5737 (2.56%)
0	0	50	0.6215 (11.10%)
		$(w_b+w_s)/h=1$	
0	0	0	0.6803
50	0	0	0.6872 (1.02%)
0	50	0	0.6935 (1.95%)
0	0	50	0.7399 (8.76%)

Table 4 Roles of different strain gradient tensors on the stiffening character of microstructural size dependency in nonlinearstability behavior of a clamped U-PFG microplate without a central cutout (k=0.5, T=0.4)

l_0(μm)	l_1(μm)	l_2(μm)	Dimensionless axial compression
		$(w_b+w_s')/h=0$	
0	0	0	1.3661
50	0	0	1.3847 (1.36%)
0	50	C	1.4011 (2.56%)
0	0	50	1.5178 (11.10%)
		$(w_b+w_s)/h=1$	
0	0	0	1.5017
0	0	0	1.5195 (1.19%)
0	50	0	1.5354 (2.25%)
0	0	50	1.6506 (9.91%)

Fig.3 illustrates the microsize-dependent nonlinear stability feature of PFG microplates without any central cutout relevant to diverse patterns of the throughthickness porosity dispersion. It is observed that the X-PFG and O-PFG porosity patterns result in, respectively, a higher and a lower nonlinear stiffness in comparison with a U-PFG microplate under axial compression. These findings can be deduced in the presence of microstructural strain gradient tensors with different values of length scale parameters as well as various types of boundary conditions.

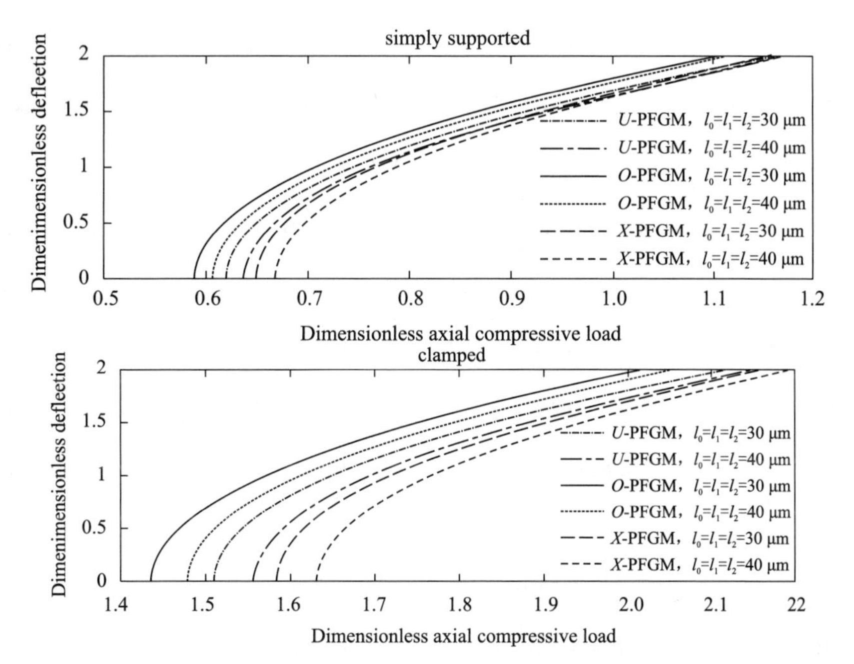

Fig.3 Microstructural-dependent nonlinear stability paths of PFG microplates corresponding to different patterns of porosity dispersion ($k=0.5, \Gamma=0.3, a=d=0$)

Tabulated in Table 5 are the conventional and microsize-dependent dimensionless axial compressive loads associated with specific induced deflections relevant to different material variation indices. The percentages presented in parentheses reveal the difference between obtained microsize-dependent axial compressions with their conventional counterpart. It can be deduced that the strengthening character of microstructural gradient tensors reduces a bit by moving to deeper part of the nonlinear stability regime resulting higher induced deflection. This expectation is reshown for different material variation indices. Moreover, it is stipulated that for various induced microplate deformations, the significance of stiffening impact associated with the microstructural size dependency in the nonlinear stability of a simply supported PFG microplate is somehow less than a clamped one. This consequence is repeated corresponding to all considered patterns for the through thickness dispersion of porosity.

The influence of a square or circular cutout having various sizes on the microsizedependent nonlinear stability paths of axially compressed U-PFG microplates is displayed in Fig.4 and Fig.5, respectively. It is demonstrated that existence of a central cutout leads to change the tendency of the traced load-deflection postbuckling paths causing various slopes. Accordingly, by existing a central cutout, the nonlinear stiffness of microplate at initial and deeper part of the postbuckling domain becomes different in comparison with the PFG microplate without any cutout. This prediction is more significant by taking the microstructural gradient tensors into account having lower values of the length scale parameters.

Table 5　Dimensionless conventional and microstructural strain gradient-based axial compressions of postbuckled PFG microsized plates for various material gradient indices ($\Gamma=0.4$)

k	$l_0=l_1=l_2$(μm)	U-PFG	O-PFG	X-PFG
Simply supported boundary conditions				
		$(w_b+w_s)/h=0.5$		
0.5	0	0.5893	0.5486	0.6259
	50	0.7971(35.26%)	0.7407 (35.01%)	0.8481 (35.50%)
		$(w_b+w_s)/h=1$		
	0	0.6803	0.6333	0.7101
	50	0.8847 (30.05%)	0.8181 (29.18%)	0.9296 (30.91%)
		$(w_b+w_s)/h=0.5$		
2	0	0.4619	0.4300	0.4908
	50	0.6256(35.45%)	0.5814(35.22%)	0.6659(35.67%)
		$(w_b+w_s)/h=1$		
	0	0.5259	0.4896	0.5502
	50	0.6878(30.79%)	0.6366(30.02%)	0.7238(31.55%)
Clamped boundary conditions				
		$(w_b+w_s)/h=0.5$		
0.5	0	1.3996	1.3029	1.4917
	50	1.9073(36.27%)	1.7738(36.14%)	2.0346(36.39%)
		$(w_b+w_s)/h=1$		
	0	1.5017	1.3980	1.5862
	50	2.0057(33.56%)	1.8605(33.08%)	2.1261(34.03%)
		$(w_b+w_s)/h=0.5$		
2	0	1.0999	1.0239	1.1726
	50	1.4999(36.37%)	1.3952(36.26%)	1.6003(36.47%)
		$(w_b+w_s)/h=1$		
	0	1.1717	1.0908	1.2399
	50	1.5697(33.96%)	1.4565(33.53%)	1.6653(34.38%)

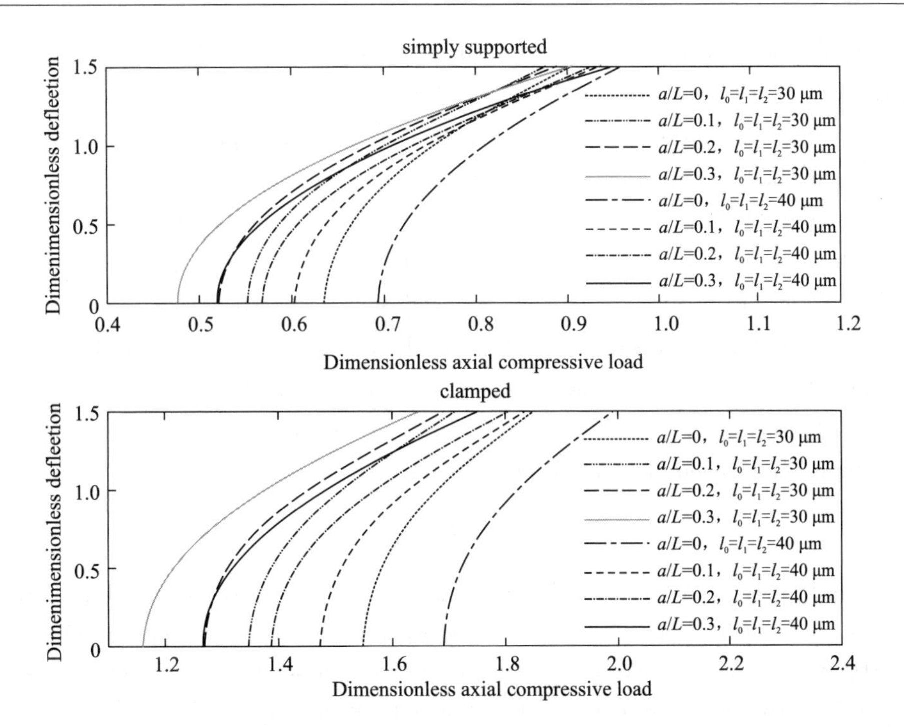

Fig.4 Conventional and microstructural-dependent nonlinear stability paths of U-PFG microplates with and without a central square cutout (k =0.5, Γ=0.4)

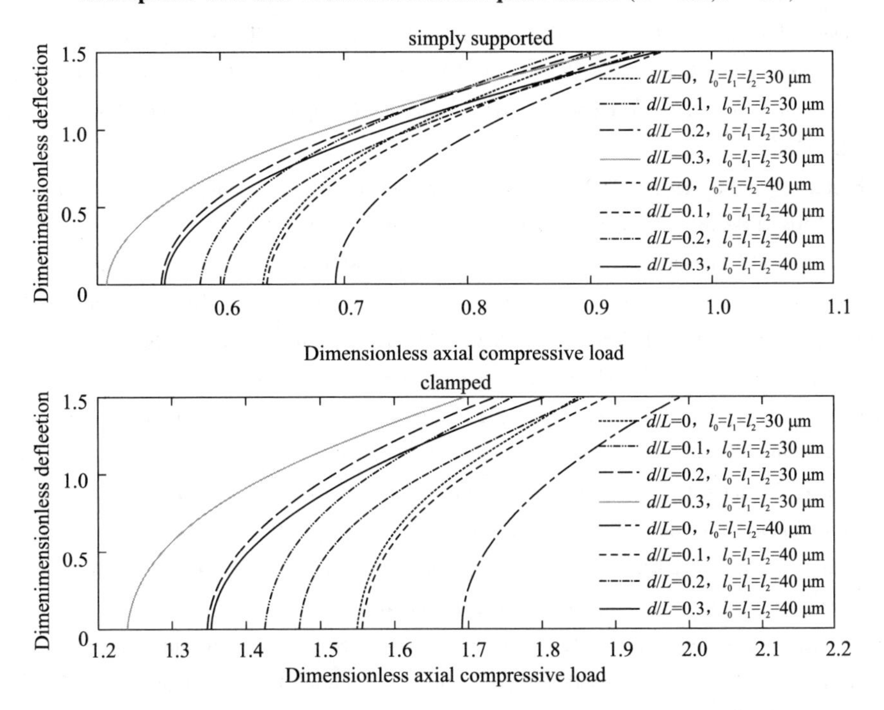

Fig.5 Conventional and microstructural-dependent nonlinear stability paths of U-PFG microplates with and without a central circular cutout (k=0.5, Γ=0.4)

5 Concluding Remarks

In the performed investigation, the microstructural-dependent nonlinear stability response associated with the axially compressed PFG rectangular microplates with and without a central cutout was studied incorporating various kinds of strain gradient tensors. A refined higher-order shear deformation plate model together with the IGA were put to use.

It was found that the strengthening role of the deviatoric stretch type of gradient tensor in the nonlinear stability characteristics of PFG microplates is more than the symmetric rotation and dilatation ones. In this regard, by taking all microstructural strain gradient tensors into account, a higher axial compression is required to buckle the PFG microplate. It was indicated that for a simply supported PFG microplate, the critical buckling load is enhanced about 1.36% by considering only the symmetric rotation gradient tensor, about 2.56% by taking only the dilatation gradient tensor, and 11.10% by considering only the deviatoric stretch gradient tensor. These percentages for the nonlinear postbuckling loads associated with the dimensionless induced deflections of $(w_b+w_s)/h=0.5$ and $(w_b+w_s)/h=1$ are, respectively, for each microstructural strain gradient tensors as 1.28%, 2.42%, 10.51%, and 1.02%, 1.95%, 8.76%. In the case of a clamped U-PFG microplate, it was deduced that the buckling stiffness is strengthened by the same percentages for the simply supported microplate, but within the nonlinear postbuckling regimes, the strengthening roles of microstructural strain gradient tensors become more significant.

References

Bacciocchi, M. and Tarantino, A. M. [2021] "Analytical solutions for vibrations and buckling analysis of laminated composite nanoplates based on third-order theory and strain gradient approach," *Composite Structures* 272, 114083.

Belarbi, M.-O. et al. [2021] "Nonlocal finite element model for the bending and buckling analysis of functionally graded nanobeams using a novel shear deformation theory," *Composite Structures* 264, 113712.

Chen, D., Yang, J. and Kitipornchai, S. [2019] "Buckling and bending analyses of a novel functionally graded porous plate using Chebyshev-Ritz method," Archives of Civil and Mechanical Engineering 19, 157-170. Crisfield, M. A. [1991] *Nonlinear Finite Element Analysis of Solids and Structures* (John Wiley & Sons, Chichester, UK).

Fan, F., Xu, Y., Sahmani, S. and Safaei B. [2020] "Modified couple stress-based geometrically nonlinear oscillations of porous functionally graded microplates using NURBSbased isogeometric approach," *Computer Methods in Applied Mechanics and Engineering* 372, 113400.

Fan, F., Sahmani, S. and Safaei B. [2021a] "Isogeometric nonlinear oscillations of nonlocal strain gradient PFGM micro/nano-plates via NURBS-based formulation," *Composite Structures* 255, 112969.

Fan, F., Safaei, B. and Sahmani, S. [2021b] "Buckling and postbuckling response of nonlocal strain gradient porous functionally graded micro/nano-plates via NURBS-based isogeometric analysis," *Thin-Walled Structures* 159, 107231.

Fang, J., Zheng, S., Xiao, J. and Zhang, X. [2020] "Vibration and thermal buckling analysis of rotating nonlocal functionally graded nanobeams in thermal environment," *Aerospace Science and Technology* 106, 106146.

Guo, Z., Wang, L., Guo, X., Chen, Y., and Dong, L. [2020] "On effective material properties of two-dimensional porous materials," *International Journal of Applied Mechanics* 12, 2050040.

Hou, R., Sahmani, S. and Safaei, B. [2021] "Nonlinear oscillations of elliptical and sector prefabricated nanoplate-type structures made of functionally graded building material," *Physica Scripta* 96, 115704.

Karamanli, A. and Vo, T. P. [2020] "Size-dependent behaviour of functionally graded sandwich microbeams based on the modified strain gradient theory," *Composite Structures* 246, 112401.

Khakalo, S., Balobanov, V. and Niiranen J. [2018] "Modelling size-dependent bending, buckling and vibrations of 2D triangular lattices by strain gradient elasticity models: Applications to sandwich beams and auxetics," *International Journal of Engineering Science* 127, 33-52.

Li, L., Li, X. and Hu, Y. [2016] "Free vibration analysis of nonlocal strain gradient beams made of functionally graded material," *International Journal of Engineering Science* 102, 77-92.

Li, Q., Wu, D., Gao, W. and Tin-Loi, F. [2020] "Size-dependent instability of organic solar cell resting on Winkler-Pasternak elastic foundation based on the modified strain gradient theory," *International Journal of Mechanical Sciences* 177, 105306.

Li, Y. S. and Xiao, T. [2021] "Free vibration of the one-dimensional piezoelectric quasicrystal microbeams based on modified couple stress theory," *Applied Mathematical Modelling* 96, 733-750.

Lin, F., Tong, L. H., Shen, H.-S., Lim, C. W. and Xiang, Y. [2020] "Assessment of first and third order shear deformation beam theories for the buckling and vibration analysis of nanobeams incorporating surface stress effects," *International Journal of Mechanical Sciences* 186, 105873.

Mercan, K., Emsen, E. and Civalek, O. [2019] "Effect of silicon dioxide substrate on buckling behavior of Zinc Oxide nanotubes via size-dependent continuum theories," *Composite Structures* 218, 130-141.

Miller, R. E. and Shenoy, V. B. [2000] "Size-dependent elastic properties of nanosized structural elements," *Nanotechnology* 11, 139-147.

Phung-Van, P., Thai, C. H., Nguyen-Xuan, H. and Abdel-Wahab, M. [2019] "An isogeometric approach of static and free vibration analyses for porous FG nanoplates," *European Journal of Mechanics - A/Solids* 78, 103851.

Sahmani, S., Bahrami, M., Aghdam, M. M. and Ansari, R. [2014a] "Surface effects on the nonlinear forced vibration response of third-order shear deformable nanobeams," *Composite Structures* 118, 149-158.

Sahmani, S., Bahrami, M. and Ansari, R. [2014b] "Surface energy effects on the free vibration characteristics of post-buckled third-order shear deformable nanobeams," *Composite Structures* 116, 552-561.

Sahmani, S. and Aghdam, M. M. [2017a] "Nonlinear vibrations of pre-and post-buckled lipid supramolecular micro/nano-tubules via nonlocal strain gradient elasticity theory," *Journal of Biomechanics* 65, 49-60.

Sahmani, S. and Aghdam, M. M. [2017b] "Size-dependent axial instability of microtubules surrounded by cytoplasm of a living cell based on nonlocal strain gradient elasticity theory," *Journal of Theoretical Biology* 422, 59-71.

Sahmani, S. and Aghdam, M. M. [2017c] "Size-dependent nonlinear bending of micro/nano-beams made of nanoporous biomaterials including a refined truncated cube cell," *Physics Letters A* 381, 3818-3830.

Sahmani S. and Fattahi, A. M. [2018] "Small scale effects on buckling and postbuckling behaviors of axially loaded FGM nanoshells based on nonlocal strain gradient elasticity theory," *Applied Mathematics and Mechanics* 39, 561-580.

Sahmani, S. and Safaei, B. [2019a] "Nonlinear free vibrations of bi-directional functionally graded micro/nano-beams including nonlocal stress and microstructural strain gradient size effects," *Thin-Walled Structures* 140, 342-356.

Sahmani, S. and Safaei, B. [2019b] "Nonlocal strain gradient nonlinear resonance of bidirectional functionally graded composite micro/nano-beams under periodic soft excitation," *Thin-Walled Structures* 143, 106226.

Sahmani, S. and Safaei, B. [2020] "Influence of homogenization models on sizedependent nonlinear bending and post-buckling of bi-directional functionally graded micro/nano-beams," *Applied Mathematical Modelling* 82, 336-358.

Sahmani, S. and Safaei, B. [2021] "Microstructural-dependent nonlinear stability analysis of random checkerboard reinforced composite micropanels via moving Kriging meshfree approach," *The European Physical Journal Plus* 136, 806.

Sahmani, S., Fattahi, A. M. and Ahmed, N. A. [2019] "Analytical mathematical solution for vibrational response of postbuckled laminated FG-GPLRC nonlocal strain gradient micro-/nanobeams," *Engineering with Computers* 35, 1173-1189.

Sarafraz, A., Sahmani, S. and Aghdam, M. M. [2019] "Nonlinear secondary resonance of nanobeams under subharmonic and superharmonic excitations including surface free energy effects," *Applied Mathematical Modelling* 66, 195-226.

Sedighi, H. M., Keivani, M. and Abadyan, M. [2015] "Modified continuum model for stability analysis of asymmetric FGM double-sided NEMS: Corrections due to finite conductivity, surface energy and nonlocal effect," *Composites Part B: Engineering* 83, 117-133.

Senthilnathan, N. R., Lim, S. P., Lee, K. H. and Chow, S. T. [1987] "Buckling of sheardeformable plates," *AIAA Journal* 25, 1268-1271.

Simsek, M. [2016] "Nonlinear free vibration of a functionally graded nanobeam using nonlocal strain gradient theory and a novel Hamiltonian approach," *International Journal of Engineering Science* 105, 12-27.

Tang, Y. and Qing, H. [2021] "Elastic buckling and free vibration analysis of functionally graded Timoshenko beam with nonlocal strain gradient integral model," *Applied Mathematical Modelling* 96, 657-677.

Tang, H., Li, L. and Hu, Y. [2019] "Coupling effect of thickness and shear deformation on size-dependent bending of micro/nano-scale porous beams," *Applied Mathematical Modelling* 66, 527-547.

Tao, C. and Dai, T. [2021] "Isogeometric analysis for size-dependent nonlinear free vibration of graphene platelet reinforced laminated annular sector microplates," *European Journal of Mechanics — A/Solids* 86, 104171.

Thai, C. H., Ferreira, A. J. M. and Nguyen-Xuan, H. [2018] "Isogeometric analysis of sizedependent isotropic and sandwich functionally graded microplates based on modified strain gradient elasticity theory," *Composite Structures* 192,

274-288.

Thanh, C.-L., Tran, L. V., Vu-Hu, T. and Abdel-Wahab, M. [2019a] "The size-dependent thermal bending and buckling analyses of composite laminate microplate based on new modified couple stress theory and isogeometric analysis," *Computer Methods in Applied Mechanics and Engineering* 350, 337-361.

Thanh, C.-L., Tran, L. V., Bui, T. Q., Nguyen, H. X. and Abdel-Wahab, M. [2019b] "Isogeometric analysis for size-dependent nonlinear thermal stability of porous FG microplates," *Composite Structures* 221, 110838.

Wang, B. B., Lu, C., Fan, C. Y. and Zhao, M. H. [2021] "A meshfree method with gradient smoothing for free vibration and buckling analysis of a strain gradient thin plate," *Engineering Analysis with Boundary Elements* 132, 159-167.

Yang, F.*et al.* [2002] "Couple stress based strain gradient theory for elasticity," *International Journal of Solids and Structures* 39, 2731-2743.

Yin, S., Xiao, Z., Deng, Y., Zhang, G., Liu, J. and Gu, S. [2021] "Isogeometric analysis of size-dependent Bernoulli-Euler beam based on a reformulated strain gradient elasticity theory," *Computers & Structures* 253, 106577.

Yuan, Y., Zhao, K., Han, Y., Sahmani, S. and Safaei, B. [2020a] "Nonlinear oscillations of composite conical microshells with in-plane heterogeneity based upon a couple stress-based shell model," *Thin-Walled Structures* 154, 106857.

Yuan, Y., Zhao, K., Zhao, Y., Sahmani, S. and Safaei, B. [2020b] "Couple stress-based nonlinear buckling analysis of hydrostatic pressurized functionally graded composite conical microshells," *Mechanics of Materials* 148, 103507.

Yuan, Y., Zhao, X., Zhao, Y., Sahmani, S. and Safaei, B. [2021] "Dynamic stability of nonlocal strain gradient FGM truncated conical microshells integrated with magnetostrictive facesheets resting on a nonlinear viscoelastic foundation," *Thin-Walled Structures* 159, 107249.

Zhang, Y., Sahmani, S. and Safaei, B. [2021] "Meshfree-based applied mathematical modeling for nonlinear stability analysis of couple stress-based lateral pressurized randomly reinforced microshells," *Engineering with Computers*.

本文曾发表于2022年《International Journal of Applied Mechanics》第14卷第1期

高层建筑施工质量管理问题及优化对策

王旭辉

（陕西省土地工程建设集团有限责任公司咸阳分公司，陕西 咸阳 712000）

【摘要】目前，由于土地资源的紧缺，我国建筑行业当下更侧重于发展高层建筑模式，这对于解决城市土地资源紧缺的问题有着很大的作用，与此同时，建筑工程的质量关乎到用户的安危，关系着整个社会的和谐，也关系着施工单位的名誉。因此，针对上述问题本文提出了高层建筑施工质量管理中存在的问题以及优化对策。

【关键词】高层建筑施工；质量管理；问题探究

引言

随着我国城市化进程的不断加快，人们对建筑的需求也不断提高，特别是在目前的城市规划中，可供使用的土地日益减少，为满足人们生活、工作和居住等需求，高层建筑应运而生，然而由于高层建筑结构复杂、建设周期长、工程风险高，因此采取有效的控制措施是必要的。

1 高层建筑施工质量管理

1.1 质量管理的必要性

（1）部分建筑工程质量较低。从当前的情况来看，施工单位对高层建筑有严格的管理制度，只有施工单位加强对建筑工程的管理，才有希望从根源上消除建筑施工中可能出现的隐患，从而降低施工中的安全风险。近年来，我国建筑工程规模不断扩大，不仅提升了施工和管理的难度，同时也对质量管理和控制工作提出了挑战。在这种情况下，必须对建筑工程施工质量通病进行全面的掌握，只有对建筑施工的全过程进行质量管理，才能够从根本上提升建筑工程的质量，为提升企业的综合竞争力奠定良好的基础。

（2）部分施工单位的法律意识淡薄。《中华人民共和国建筑法》的颁布，使建筑施工各方的职责与义务得到了明确，并对工程技术、质量管理的各个环节和标准进行了规定。但是，一些施工企业在施工中存在着法律意识薄弱、法律观念不足等问题，在施工中违规操作，进而出现了施工质量低下、工程事故频发的现象。

1.2 高层建筑的施工特点

高层建筑的优点在于高度集中，纵向发展，形象突出，可以成为标志性建筑。尤其是在人口稠密的区域，高层建筑更能发挥其自身的长处和特色。

（1）建筑规模大、成本高

与其他的建筑不同，高层建筑一般都是10层以上，所以在建造之前，必须要做好充分的预算，合理地购买质量可靠的建筑材料；高层建筑造价高，其建设周期也很长，因此，施工单位要统筹规划建设进度，科学安排工期，而高层建筑的高度越高，所耗费的材料也就越多。

（2）施工技术难度较大

高层建筑由于楼层较多，在实际施工过程中，涉及大量的高空作业，不仅施工技术复杂，对电力设施、钢筋配置等都有着更高的要求，材料运输也变得更为困难。此外，与普通建筑不同的是，高层建筑对于各类设备的要求也相当高，设备需要有较强的稳定性和可靠性才能够适应高层作业的高压环境。因此，高层建筑的施工方需要配置经验丰富人员和可靠设备，这无疑提高了建筑的经济性指标。此外,与地面施工相比,在高空施工中,运输方式较为复杂,操作难度较大,这会造成严重的影响。项目进度越慢,工期越长,受到自然因素的影响越多,无论是对项目本身,还是对工人来说,都是一种资源上的浪费。

2 案例分析

某项目混凝土结构设计强度等级为 C40,在完成地面二楼现浇后,出现了多个混凝土桩基松动、振捣不够、强度偏低的现象;在剪力墙拐角处的暗柱、钢筋密集部位出现裂缝、露筋、孔洞等质量问题,墙壁上出现了浮浆或浮石,按建筑工程质量评价标准,有很多指标不满足要求。因此,在试验过程中,对混凝土的强度进行了测试,结果表明,在施工过程中,所有的柱子、墙体、梁、楼板的混凝土强度等级都达不到设计指标,测试结果如表 1 所示。

表 1 检测的钢筋混凝土强度等级

楼层	原设计要求	实际检测强度
地下 2 层	C40	C27.5
地下 1 层	C40	C31.6
1 层	C40	C29.5
2 层	C40	C30.0

3 高层建筑施工管理中存在的问题

本文从理论上探讨了高层建筑及其施工管理的特点,以及如何有效地寻找影响高层建筑管理发展的根源,作者通过总结多年的施工管理经验,并结合大量国内外学术资料,对高层建筑施工管理中的不足进行了论述。

3.1 高层建筑施工技术管理不到位

高层建筑施工关注的重点是高层建筑的结构和功能,所以综合素质、综合水平、综合能力是对高层建筑技术管理的基本要求。因此,建筑企业必须严格落实工地的安全施工责任制,确保工地的施工质量。但就目前的情况来看,大部分的管理者在实施现场质量监控的过程中缺乏合理性,有些质量监控措施并没有得到很好的执行。比如,在审核高层建筑材料时,采购人员和材料核查人员的态度比较宽松,对部分强度较大的材料检查不够严格,从而导致不合格的材料进入工地,影响到工程的整体质量。因此,要确保施工单位明确质量监管的重要性,把施工现场的质量监督工作落实到位。

3.2 高层建筑施工分包管理不到位

(1)随着社会和经济的发展,我国的建筑工程建设技术得到了极大的提升,然而转包现象的出现,导致施工质量参差不齐。原因可能有:分包合同中的一些问题和职责并不明确,这就造成了高层建筑的承包商之间的推卸责任,疏忽工作[2]。

(2)分包的技术能力有限。虽然分包单位具有一定的施工资质,但是人员质量参差不齐,如业务知识、技术水平、管理能力等,都是制约工作推进顺利的因素,从而导致施工效率降低。

3.3 高层建筑施工安全管理不到位

安全生产,重如泰山。安全生产是我们国家生产建设中一贯坚持的指导思想,随着我国安全生产事业的不断发展,严守安全底线、严格依法监管、保障人民权益、生命安全至上已成为全社会共识。但仍然有部分企业存在如下缺点:一是缺乏以人为本的安全意识和管理机制,二是缺乏对员工安全意识和安全作业技能的培训,三是现场的安全管理不充分,安全生产责任制没有真正实行。在高层建筑施工中,如何达到安全生产的目的是十分必要的。由于高层建筑工程中,施工技术复杂,影响因素较多,因此,施工安全问题对工程的成功实施及工程建设的经济效益起着十分关键的作用。

3.4 高层建筑施工进度管理不到位

在建筑行业发展的进程中,由于建设的效率越来越高,施工的进度常常表现为工期的不断缩短,工期的缩短为建设单位带来了巨大的经济利益,但也催生了大量的施工企业为了利益,盲目地加快建设速度,缩短建设周期,从而造成了工程质量问题的恶化。一些建筑公司,为了缩短工期,获取一定的经济效益,往往会因为工程质量不过关,而再次返工,从而造成成本的浪费。

4 高层建筑施工管理中优化策略

“纵观人类发展历史,创新始终是推动一个国家、一个民族向前发展的重要力量,也是推动整个人类社会向前发展的重要力量”。工程建设行业是国民经济的支柱产业之一,对经济社会发展有着重要的意义。高层建筑在我国现阶段发展中仍然是提高城市空间内有限土地资源利用效率的一个重要途径,但经过长期的发展,在今后的施工过程中我们应该更多地着眼于如何科学有效地克服技术难点、堵点。

4.1 强化高层建筑施工方案的科学性

科学性、合理性、有效性是建筑工程建设中确保建筑工程质量的决定性条件,因此要确保高层建筑的施工质量,首先必须加强其科学性。完善高层建筑施工管理措施,使其具有科学性,需要从加强工程技术管理入手[3]。要完善高层建筑施工技术管理,必须加强技术管理队伍的建设,只有具有完善的人力资源、丰富的知识、高超的技术水平、综合实力、施工水平和责任意识,才能实现高层建筑施工技术管理的技术要求。其次,加强对高层建筑技术管理队伍的建设,加强对工程技术管理队伍在技术创新上的探索。探索建筑技术创新,既要充分认识目前工程建设的新材料、新技术,又要把工程建设中的关键技术牢记于心,探索改进管理方案、提升管理效率等工作方法,以达到我国高层建筑的不断完善和发展。

4.2 做好施工材料的质量管理及控制

在工程质量管理实施工作中,建材是工程建设的基本保障,因此在选用施工材料时最好选用三种不同的建材,既要达到施工要求,又要有一个合理的价位,避免施工单位因控制工程造价而忽视工程材料的质量。高层建筑工程对材料的需求很大,需要大批量地采购工程材料,并按规定运输到工地。由于建筑材料的型号、规格、性能、使用期限不同,因此必须分类储存,不仅便于以后使用,更便于登记在册,利于施工现场管理。另外,在购买设备的时候,要根据不同的施工需求,合理地选用设备,降低设备的搬运和存放费用,及时维修,延长设备的使用寿命,提高设备的利用率[4]。

4.3 强化高层施工过程管理监督工作

加强工程建设的全过程监控,对提高工程质量和缩短工期具有重要意义。在施工过程中,施工人员的技术、人员的配置都是非常严格的,当作业人员出现技术失误时,必须要及时纠正,以免造成不必要的损失。因此,高层建筑企业必须聘请高素质且具有熟练管理技能和管理经验的管理与监督人员,这对于高层建筑施工过程的整体把控尤为重要,能够促使高层建筑高质量施工的实现。

4.4 有效控制施工周期

每一项大型建筑工程都有一个建设周期,这也是高层建筑工程管理的一个重要因素。特别是施工方和设计方应认真编制切实可行的工程进度计划,同时建立完善的计划保证体系,以全生命周期为目标,以阶段控制计划为保证,采取工期动态管理,确保工程顺利完工。因此,在高层建筑的建设中,必须要对项目的工期进行科学的管理与控制,以免过早地造成项目的质量不达标,从而影响到高层建筑的经济效益。

4.5 钢筋笼上浮的防治

钢筋笼在灌浆的过程中,必须要将管道固定住,最好埋管深度应在 0.8 m 以上,在串动钢管时不能挂钢筋笼子。比如,在进行高层建筑施工质量管理时,要仔细辨别钢筋笼是不是悬笼,若为悬笼,则要留意钢筋笼在混凝土表面的位置有无位移,并要控制灌注速度,使其与混凝土的上升速度保持一致,通常采用这种方法来降低混凝土对钢筋笼的黏性[1]。在钢筋笼的埋设深度下,拆除管道时,应将钢管底部埋入钢筋笼;清理工作也很重要,要清理干净钻孔中的尘土和碎石;在选用管道时,要注意管道与钢筋笼的内径互相对应,且钢筋笼内径和管道外直径的间隙必须大于 100 mm。

4.6　加强建筑施工过程中的技术管理

在高层建筑施工中,加强技术管理,可以促进建筑施工的顺利进行,从而使建筑的整体质量得到改善。施工工艺管理的重点是施工进度、造价控制、施工质量控制、技术交底等。通过对施工进度和造价的控制,可以有效地减少不必要的费用,使项目按计划的时间完成;通过对施工质量的控制,可以确保工程按有关规范进行施工,保障人民的生命和财产的安全;由于高层建筑往往存在一些特殊和隐蔽的项目,因此,技术交底是确保工程进度和质量的重要措施。

在高层建筑中,由于其与竖向施工关系密切,所以对其垂直度的控制非常重要。施工单位应依据工程实际,依据高层建筑柱网的布置,结合图纸的要求,实施有效的施工组织和管理,科学地进行角柱模板的安装。为保证垂直度满足工程100%的要求,采用吊索的方法,使竖直度更加准确,从而实现模板的加固。四角柱卸料完毕后,应以四根柱子为基准,以提高其实际平整程度。为确保工程的施工质量,可采用激光铅锤进行垂直测量,并对其进行质量控制。在对坐标系进行控制时,必须保证坐标系的传输,以免参考坐标不能引测,从而影响后续的工程质量。所以,在进行高层建筑的施工时,必须做到三线管理,确定基准楼层,在地面上设方孔,并对轴线进行校正,以达到施工质量的目的。

结束语

由于高层建筑自身的特性,使得其施工技术难度大、施工工艺复杂、风险因素多,加之其施工技术的特殊性,使得高层建筑施工管理工作面临较大的困难。因此,探讨高层建筑施工管理中的问题和对策,是目前我国建设行业发展进程中,解决高层建筑工程质量管理问题的有效方法。

参考文献

[1]孙晓慧.高层建筑施工管理存在问题及优化对策[J].建筑工程技术与设计,2018(33):1971.

[2]贾栓勤,贾佳佳.关于高层建筑工程施工质量优化管理策略探析[J].建筑工程技术与设计,2016(26):1115.

[3]王中文.关于高层建筑工程施工质量管理控制分析[J].建筑工程技术与设计,2017(8):1341.

[4]顾建平.加强高层建筑工程施工质量管理措施探讨[J].乡村科技,2014(24):351-352.

本文曾发表于2022年《建筑实践》第11期

工程项目造价动态管理及控制策略

周 婉

(陕西省土地工程建设集团有限责任公司,陕西 西安 710075)

在工程项目施工环节中,工程项目造价管理与项目成本存在重要关联,属于企业关注点,针对成本管理问题,我国提出众多工程造价管理方法,从基础工作方面有效降低造价成本。伴随着市场经济不断变化,工程造价管理方法逐渐发生变化。在先进管理技术和管理方法支持下,经济市场出现工程造价动态管理方法,有效解决工程项目中成本和利润之间的问题,提高企业的经济效益,推进企业可持续发展。本文就目前工程项目造价管理的现状进行分析,并针对当前的情况提出相应的控制策略,以期为工程项目造价动态管理工作提供参考依据。

随着信息技术的快速发展,工程项目造价的管理方式也在逐渐发生变化,与传统的工程造价管理方式相比,先进的工程造价管理体系能够更加有效地适应当前的建设项目的施工要求。在工程项目中,企业需要根据企业发展实际状况确定工程项目造价管理方式,使企业合理开销,保证企业维持正常运营状态。在施工过程中,施工计划与工程项目造价管理方式要处于一致方向,根本目的是降低工程项目施工成本,为企业争取更高利益,前提条件是保证工程质量符合国家规定。确保在进行建设施工环节中工程造价管理能够充分地发挥自身的作用,加快工程项目的施工工期,提高企业的经济效益。

1 工程项目造价管理的重要性

在工程项目建设中,工程施工的各个流程都与工程造价之间存在着密切的联系,从工程开始施工一直到工程竣工以后,在这一环节中的任何一个施工环节中如果存在失误,企业的经济效益都会受到影响。因此,企业在进行工程施工时要对工程造价实施动态的管理。在市场发展模式不断变化背景下,企业愈加重视经济效益,旨在通过工程项目获得经济利益,探究工程项目造价管理方法对企业可持续发展十分重要。随着建设资金投入的力度不断加大,企业能否对工程项目中使用的资金进行有效的管理,直接影响企业的经济效益。随着经济水平的快速发展,工程项目建设逐渐受到人们的重视,并得到了快速的发展,为了更好地提高企业的经济效益,企业需要合理选择工程造价动态管理方式,在考虑自身经济发展水平基础上,创新适合企业发展的针对性工程造价动态管理体系。企业要结合工程造价动态管理的优势,来提高企业的综合实力,为发展奠定基础。工程造价动态管理模式与传统的工程管理模式相比,无论对企业自身发展还是对企业提高市场竞争力都有着极其重要的作用。

2 现状分析

目前,随着建筑工程的快速发展,工程项目造价动态管理工作已经成为企业工作的重要组成部分。在工程造价管理工作中,企业管理状态和工作人员整体素质水平与造价管理效果存在重要关联性,针对企业管理状态,企业管理内容中需要融入造价管理效果,体现造价管理重要性,使工作人员了解造价管理重要性,合理管理工程项目成本。工程项目施工环节具有复杂性,涉及材料种类较多,造价人员整体素质水平还需进一步提高,在项目管理中经常出现问题,影响造价管理效果。

2.1 缺乏相应的管理方法

在工程项目施工环节中,影响造价管理效果因素较多,主要包括造价人员整体素质水平、施工类型、施工计划、施工位置等内容,要综合考虑这些因素,使造价管理效果得到有效控制。对于不同工程项目施工来讲,工程的施工环境、施工背景、施工方案、工作技术人员的施工水平以及项目施工难度等都均存在较大的差异。在进行工程项目实际施工中,大多数参与工程造价核算的工作人员在对工程造价动态管理工作进行核算时,无法创新管理方法,缺乏工作经验,从而影响造价管理实施效果,造价人员难以高效完成造价工作,影响工程施工效果。企业对工程造价管理缺少正确认识,影响造价人员对工作内容的正确认知,使工程造价整体效果水平难以得到提高。造价人员对造价管理工作存在不正确认知,无法主动提升工作能力,为造价管理效果带来不良影响,造成在进行工程项目建设时不能更好地对工程施工以及其他方面制定出科学有效的管理方案,不能对工程造价实施动态管理。

2.2 缺少完善的监管体系

在工程项目施工中工程造价动态管理工作是工程施工环节中的主要工作环节,工程项目造价动态管理工作不仅涉及工程项目施工企业,还涉及工程项目投资者、工程项目承包商以及工程项目代理商之间的经济效益,正是基于工程造价动态管理工作中存在的这些问题,为造价管理工作带来消极影响,无法保证工程项目工作有效展开,工程造价管理难以发挥主要作用。企业容易忽视监管作用,缺乏高效监管体系,无法关注施工人员整体工作状态,无法保证施工质量,无法保证正常施工流程,影响工程施工进程。而且工程造价的监管不够全面,在进行工程造价动态管理工作不能够有效开展,最终增加工程的施工成本。

2.3 相关工作人员的工作能力不足

工程造价管理效果与造价人员工作能力存在重要关联性,造价人员工作能力较弱,工程项目施工效果差,影响企业项目利益。企业不重视造价管理工作内容,无法帮助造价人员提升工作能力,无法为他们提供培训机会,给工程项目带来不良影响。造价人员缺乏主动提升工作能力意识,当他们在工作过程中出现错误时,将影响整个工程项目成本,给企业经济效益带来不利影响。所以工程造价工作人员的工作能力直接影响工程造价动态管理效果以及对工程造价成本控制的准确性。在进行工程造价动态管理工作时,由于大多数的企业没有对相关的工作人员进行专业的培训,导致相关的工作人员工作能力不足,没有认识到工程造价在工程施工中的重要性,缺乏对工程造价实施动态管理的工作能力。

2.4 动态管理及控制力度不足

在工程造价管理方法中,动态管理属于重要方式,在日常管理工作中,存在动态管理及控制力度不足问题,影响造价动态管理效果。在施工过程中经常出现不可控影响因素,部分因素影响工程造价核算内容,造价动态管理主要针对这些不可控影响因素,将其融入造价核算内容中。在实际造价管理活动中,造价人员为节省核算时间,忽视核算步骤,无法保证造价核算结果准确性,企业容易出现实际经济效益与预测经济效益不相符问题。企业不关注造价管理方式和管理效果,使造价动态管理工作出现偏差,影响造价动态管理实际效果。

3 策略分析

3.1 加强对工程造价动态管理和控制力度

在工程项目施工环节中企业要想对工程造价进行动态管理及控制,便要加强对不同的的施工阶段进行管理和控制,对企业工程施工来讲,主要对工程每一阶段的施工环节进行细致的造价管控。在施工之前,企业需要关注成本问题,引导造价人员提前核算成本,保证企业可以获得最大化经济效益,在企业经济效益得到保障基础上开展工程项目施工活动。在工程施工环节中工程造价的动态管理是最为关键的施工环节,在这一环节中,相关工作人员要加强对工程预算费用的数据进行分析整理,确保工程造价动态管理中工程造价信息数据的精准度。在工程竣工后的工作任务便是加强对工程的施工质量以及投入资金的核算,对其工程施工中产生的各项费用进行统一核算,保障工程造价信息数据的准确性,提高工程造价动态管理工作的有效性。

3.2 提高相关工作人员的工作能力

造价人员工作能力与企业工程项目造价管理工作存在紧密联系,通过提高造价人员整体工作能力,可以改善工程造价管理效果。企业需要关注工程造价管理状态,斟酌工程造价招聘门槛,从基础工作中提高造价人员整体素质水平。企业需要为造价人员提供学习机会和发展平台,使他们有机会提升工作能力,保证工程造价管理有效性。企业需要为造价人员提供发展空间,使造价人员有信心完成工作内容,主动提高工作能力。造价人员需要充分考虑动态管理知识,学习先进造价管理知识,改善企业造价管理现状。为了工程造价动态管理能够取得良好的效果,企业要提高相关工作人员的工作能力,加大对工程造价工作人员的培养力度,企业需要为造价人员提供培训机会,并通过竞赛方式了解造价人员学习状态和学习成果,提高造价人员工作能力,使他们拥有良性竞争意识,还可以提高工作人员的技术能力,为项目造价动态管理工作培养更多的专业型人才。

3.3 借助 BIM 技术进行工程造价动态管理

工程项目造价动态管理工作主要的特点便是工程造价管理中所涉及的数据信息较多,工作人员在进行数据信息统计时工作力度较大,不能对信息数据的准确性进行保证。所以在这一环节中相关工作人员在进行工程造价管理时要结合信息技术方便快捷的优势,对工程造价管理进行创新优化,将所涉及的工作人员、施工工种、施工程序等借助信息技术进行有效结合,不仅能够在最大程度上满足对工程造价实施动态化管理,还可以有效地减轻相关工作人员的工作压力,提高企业的经济效益。例如 BIM 技术在工程项目造价管理中的运用,BIM 技术可以通过自身具备的优势为工程项目施工提供大数据计算和分析的能力,利用 BIM 技术的云平台对工程项目施工的工程量、材料价格、施工指标、工程造价等进行云造价管理、云材价、云指标、云计价、云算量、云碰撞检查等方式来提高工程的施工进度、施工质量以及工程造价管理水平。其中最有效的便是在项目造价动态管理及控制方面,利用 BIM 技术自身的优势构建信息数据共享平台,将工程造价的信息、工程施工进度等方面进行实时的信息数据共享,提高工程造价动态管理的工作效率,如图 1 所示。

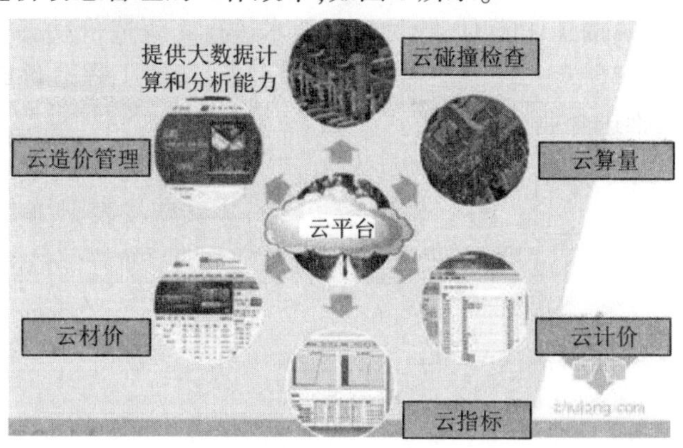

图 1 BIM 技术在工程项目造价动态管理中的运用

以某地铁工程 1 号线的工程施工为例,该地铁工程 1 号线施工项目的总长度为 70 km,建设施工车站共计 21 座。由于该项目在进行施工建设环节中需要投入大量的建设资金,为了能够有效地控制工程造价,节约工程成本,企业在进行工程施工时广泛地运用 BIM 技术,利用项目采用工厂化的施工方式,减少建筑材料消耗量,控制建筑企业成本,合理管理资本投入工作内容。在应用 BIM 技术基础上,可以发现工程项目中造价管理问题,还可以自动形成处理报告。在工程造价动态管理内容中,主要包括工程项目计划管理、工程项目设计管理、工程项目质量管理和工程项目成本管理等内容。针对计划管理,系统以工程项目实际工程量为基础,以保证质量和控制时间为目标,为施工项目提供计划;针对设计管理,为工程项目施工提供系统设计方法,节约施工时间;针对质量管理,以施工质量为核心,开启监督功能,提高建筑施工效率;针对成本管理,在保证施工质量基础上,为工程项目控制成本,提高企业经济效益水平,如图 2 所示。

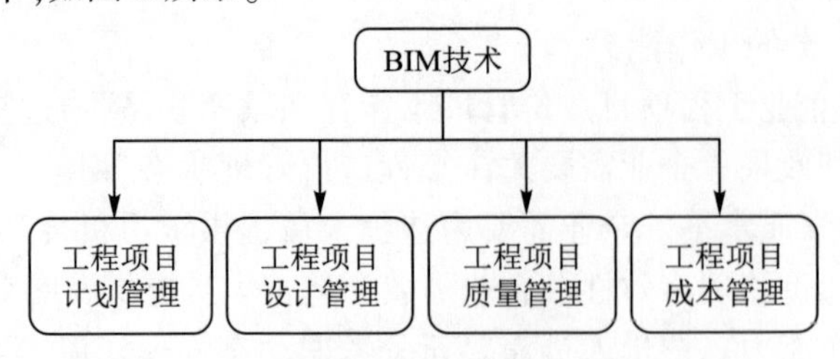

图 2　BIM 技术在工程项目造价的动态化管理

3.4　加大工程施工费用管理

在工程项目施工环节中,施工材料的使用性能也在不断发生着改变,同时也存在多样化的特点,在建筑材料市场上,随处可以看见不同性能、不同类型的建筑施工材料。工作人员在进行材料选择的同时,一定要严格地按照工程施工的标准和要求进行采买,在保障工程施工质量的前提下,还要对其企业的经济效益进行考虑,在不违背施工要求的情况下,本着节约施工成本的理念进行施工材料的采买。在材料采买之后,还要将施工材料进行统一的堆放与保管,避免施工材料受到外界因素的影响,导致材料变形、受潮、腐蚀等造成经济损失。针对于节约施工成本理念,企业可以加大对施工费用的管理,对工程施工中的每一项费用进行集中的管理,制定合理的管控措施,由专人进行负责对施工材料以及其他设备的采买,在确保施工质量的前提下,节约施工成本的投入,为企业带来更大的经济效益。

4　结束语

综上所述,随着市场经济水平的快速发展,企业要想在激烈的市场竞争中能够更好地发展,便要在工程项目施工环节中提高对工程造价动态管理及控制水平,在进行工程施工时加强工程造价动态管理的控制力度是保证企业在市场竞争中快速发展的关键环节。在对工程造价进行动态管理时,面对目前在工程造价动态管控工作中存在的问题,企业要充分重视,通过加强对工程造价的动态管理和控制力度、提高相关工作人员的工作能力、借助 BIM 技术进行工程造价动态管理,将工程造价动态管理工作落实到工程项目施工环节中,提高工程造价管理水平,节约工程造价成本,提高企业的经济效益,提高企业市场竞争力,推进企业持续发展。

本文曾发表于 2022 年《地产经济》第 18 期

5　综合管理

国企人力资源管理中绩效考核问题与对策研究

崔天宇

（陕西省土地工程建设集团有限责任公司，陕西 西安 710075）

【摘要】随着我国社会经济的快速发展，国有企业在经济方面做出的贡献得到了各行各业的认可。人力资源管理工作是国有企业管理工作的重要组成部分。绩效考核是人力资源管理工作的重要内容。做好人力资源管理工作，明确绩效考核的意义，创新绩效考核的方法，才能让国有企业当前存在的人力资源管理问题得以解决。积极探讨人力资源管理中的绩效考核现存问题，并开发有效的对策，是保障国有企业人力资源管理工作顺利进行的重要举措。

【关键词】国有企业；人力资源；绩效考核；方法；完善；对策

引言

在国有企业的发展过程中，人力资源管理工作对其影响直接且重大。人力资源管理工作的有效性，在一定程度上决定了国有企业发展的可持续性。重视人力资源管理工作，合理调配企业内的人力资源，并对绩效考核工作进行统一管理，才能提升每一位工作人员的工作热情与质量，促进企业经济效益提升。

一、国有企业人力资源之绩效考核管理问题

（一）绩效考核标准有待统一

绩效考核标准不统一，意味着绩效考核工作丧失标准，最终的考核结果也不会得到员工的认可，进而影响企业内的工作氛围，影响企业前进动力。在员工绩效考核工作实践中，基础的考核内容应当包括员工的实际出勤情况、岗位工作完成情况、个人工作纪律情况等，结合此类信息进行分析，打好员工绩效考核的基础。关于工作质量的绩效考核，应当重视实际工作内容与完成的数量，确保绩效考核的公平性与科学性。但在实际人力资源绩效考核工作中，部分企业的人力资源工作者未核实信息的真实性，未关注员工的个人工作态度等细节，使得最终考核结果引发争议，使企业内的绩效考核标准丧失公信力。

（二）绩效考核目标有待明确

绩效考核目标不明确，便会直接影响的成果。一些管理人员只重视企业的经济效益，忽视了绩效考核工作，未全面了解过绩效考核工作的实际意义，对其内涵与性质了解不够深入，甚至存在错误的认知。在这些管理思想中，所谓绩效考核，与工资分配无异，致使绩效考核难以在国有企业的发展过程中发挥积极作用。明确绩效考核的目标，才能使其在提高员工工作效率和质量方面发挥作用。给予绩效考核工作精准的定位，明确绩效考核的实质，优化考核的手段，才能让企业的发展与当前的时代进程相符。

（三）绩效考核方法有待完善

方法问题是实施方面的问题。绩效考核工作成果不佳，很大程度上受工作方法影响。很多国有企业并没有与企业发展相匹配的绩效考核工作方法，缺少科学的分配手段和考核方式，使得企业内部整体绩效考核评估工作不完善。考核人员会根据自己的主观经验做出评判，使考核的结果有失公

允,难以服众。主观臆断式的绩效考核,难以激起工作人员的热情,甚至激发愤懑情绪。

二、国有企业人力资源之绩效考核管理对策

(一)完善绩效考核的工作标准

国有企业的发展,需要人力资源绩效考核工作的大力支持。因此,每一位管理者都要明确绩效考核工作的地位,提高重视,在绩效考核工作制度、标准、目标、方法的完善与创新过程中给予多方面支持。用科学的绩效考核工作,打造公平、向上的企业文化与工作氛围,促进绩效考核工作在国有企业发展过程中发挥更大的积极作用。绩效考核标准的统一与完善,是促进绩效考核工作创新发展的第一步。只有标准明确、全面、公平,才能让绩效考核工作的落实更加容易,让绩效考核的最终结果为企业员工所接受。

企业内的岗位应当是绩效考核的基础。每一个工作岗位都有不同的工作职责与具体内容,以岗位为基础的绩效考核标准,能够让考核的结果具有实际性和针对性,不仅对工作人员的工作内容给予肯定,还应关注岗位之间的差异,推进绩效考核工作的公平化。人力资源工作者要结合企业内的岗位信息,打造岗位说明书,明确每一个岗位的实际要求,针对岗位建立绩效考核的标准,将平衡记分卡制度引入进来,通过定量与定性评价相结合的方式,使绩效考核标准更加全面,促进绩效考核工作的有效落实。人力资源的绩效考核工作,不仅是员工管理的一种有效方式,也是推进员工与企业共同成长的策略。在考核之时,每一位或者每一类工作人员的考核标准应当具有独立性。这样一来,人力资源管理者能够更好地调动员工的优势,员工也能更好地在企业发展过程中发挥个人价值,获得个人成长,实现个人职业目标。全面且明确的绩效考核标准,能够给予员工、企业双重回报,促进企业经济效益提升,促进员工专业技能水平提高。

(二)明确绩效考核的工作目标

明确绩效考核工作的最终目标,才能保证日常绩效考核工作的全面落实。方向正确,才能让过程中的每一步都具有明确的指向。

首先,利用绩效考核,提高员工的工作积极性。在绩效考核工作完成后,人力资源部门要及时反馈绩效考核的结果。通过结果分析,了解员工当前的工作方式的效率,明确下一步人员部署方式,做好工资调整与人员晋升等工作。绩效考核,不是企业"做减法"的手段,而是协手员工,共同"做加法"的工作环节。用绩效考核去提高员工的工作热情,营造更好的工作氛围,才能实现双赢目标。

其次,利用绩效考核,发现员工的工作问题。人无完人,在企业管理中,应当积极发现员工的优势、企业的长处,同时也要面对现实,从当前的发展中找到问题,及时修补漏洞。国有企业要结合企业的发展需求,对工作人员的工作方式加以调整,或寻找创新的工作路径,使以往国有企业人力资源管理中存在的问题得以解决。企业管理者定期利用绩效考核结果,对员工进行阶段性的工作评价。对工作表现突出的员工给予奖励,同时提醒工作中问题频出的员工,严于律己,反思己过。

(三)创新绩效考核的工作方法

合理选择绩效考核的方法,运用好多样化的评估方式,有利于国有企业人力资源绩效考核工作效率的提高与成果的优化。工作评估功能是绩效考核的首要功能。在实际工作中加强方法的创新,有利于人力资源合理分配,推进企业内各项工作的有效开展。

首先,对员工进行多维度的客观分析。在绩效考核过程中,要避免一切主观臆断式工作现象的出现。通过实际的工作数据,对员工进行多维度评价,关注其工作态度、工作成果与个人工作体验,使评估成果更加合理,绩效考核工作更加高效。

其次,结合绩效考核向上提供可用信息。一直以来,绩效考核工作呈"向下"趋势,即绩效考核工作在更多情况下是面对企业员工的,其在企业发展决策上的作用被忽视。人力资源管理者要通过绩效考核数据分析,发现企业的自身发展不足与发展需求,向上级领导人员提供数据,明晰其反映出的问题,帮助企业及时调整发展战略。

结语

综上所述,全新的经济环境下,国有企业要全面分析内外发展环境的变化,重视人力资源绩效考核工作。提升绩效考核工作板块的地位,在思想与方法上加大创新。明确绩效考核工作的目标,以统一的标准、共同的目标为指引,提高员工的活力与企业的创造力,让人力资源的优势发挥出来,推进国有企业的稳定发展。

参考文献

[1]张静明.国企人力资源管理中绩效考核问题与对策[J].现代企业,2022(9):16-17.
[2]李晓谋. A国有企业绩效管理问题及对策研究[D].石家庄:河北经贸大学,2022.
[3]屈艳梅.国企人力资源管理绩效考核中存在的问题及对策分析[J].活力,2022(6):112-114.
[4]骆小萍.国企人力资源管理中的绩效考核问题及对策分析[J].营销界,2022(3):64-66.
[5]林衎.关于国企人力资源管理中的绩效考核问题及对策研究[J].中国市场,2021(7):93-94.
[6]赵汝凤,赵芮,徐敬城.绩效考核在国企人力资源管理中的问题分析和策略探究[J].财经界,2021(6):191-192.
[7]陈景然.绩效考核在国企人力资源管理中的问题分析和策略探究[J].商讯,2020(25):183-184.
[8]姜宝山. 国有企业人力资源管理中能岗匹配问题研究[D].呼和浩特:内蒙古大学,2011.

本文曾发表于2022年《中国科技人才》第16期

新形势下企业资金管理的难点及应对策略

何雪亮[1],冯卓西[2]

(1.陕西省土地工程建设集团有限责任公司,陕西 西安 710075;
2.西安大兴新区综合改造管委会,陕西 西安 710077)

【摘要】资金要素直接影响到企业的日常生产经营,因此企业管理体系中资金管理是一个非常重要的组成环节。当前,企业在资金管理方面普遍存在资金集中管理执行不到位、资金预算管理体系薄弱、资金管理制度不健全、资金筹集来源单一等问题,针对这些共性问题深入分析研究,并提出加强资金集中管理、强化资金的预算管理、努力开拓企业融资渠道等切实可行的应对策略,有效防范化解相关财务风险。

【关键词】企业;资金管理;管理难点;应对策略

近年来,受国内外新冠疫情以及不断变化的国际环境的影响,企业作为市场主体,生产经营受到不同程度的影响,企业的日常生产经营发生了很大的变化。针对当前还在持续变化的国内外环境,特别是在有些企业融资规模和负债规模越来越大,资金链安全风险依然存在的情况下,企业要清醒地认识到要素资源配置在市场中的重要性,特别是资金要素在保障企业持续发展中的重要意义。资金作为企业发展离不开的重要生产要素之一,如何将这种资源配置做到最优,每个企业的经营者都应给予关注。通过调查发现,目前众多企业在资金管理工作中仍存在众多难点,需要不断改进才能达到资源配置最大化,确保资产保值增值。

一、企业在资金管理中存在的难点

(一)资金集中管理执行不到位,资金管理基础性工作有待提升

通过调研发现,有些企业由于项目多、分布广,项目资金集中管理不规范,资金管理流程不清晰,资金集中度较低,资金存在集中管理系统外循环的情况。资金集中管理必需的网络手段不健全、网络覆盖面不全,网银建设仍需进一步推进。资金管理岗位人员职责不清晰,业务流程不合理,银行账户管理混乱,随意开户现象较为普遍,部分企业在同一银行多个支行开户或者同一银行多头开立结算账户,不利于统一规范管理。

(二)资金预算管理体系薄弱,尚未形成完整的资金预算体系

由于在财务管理体系中,有些企业缺乏资金预算管理方面的经验,单位资金预算编制单纯依靠财务部门,没有形成计划经营、人力资源、生产经营等部门共同参与预算编制的有效工作机制。这就导致资金预算编制依据不充分、考虑因素不全面、编制方法不科学,预算与实际严重偏离,不仅难以有效发挥预算的引领作用,还可能对生产经营产生误导。一些企业未将资金预算目标层层分解落实到各生产及管理环节,对预算的执行结果不进行对照分析,对预算偏差没有补救或纠偏措施,对预算执行的奖惩制度没有贯彻到生产经营的各个环节,资金预算管理的作用尚未充分发挥,资金预算执行不规范、刚性约束较差。

(三)资金筹资来源单一,担保管理混乱

有些企业的资产负债率仍然偏高,银行贷款依旧是企业融资的主要来源,资产负债率随着经营呈快速上升趋势,在严峻的金融环境下蕴藏很大的财务风险。同时,企业为保证现金流,采取的融资渠道单一,销售回款率不高,银行借款仍是最主要的融资渠道,企业对银行的依赖度较强,融资成本较

高。有些企业的担保管理仍需进一步规范,有些企业未经上级单位批准,擅自为参股或者下属企业提供担保;有些企业对资金风险的关注程度和敏感性不够,资金风险预警机制不健全,对本单位蕴藏的经营风险及财务风险无法及时预警,特别是对外贸易企业对外币汇率变动风险缺乏认知。

(四)企业资金管理制度不健全,资金管理职能不明确

有些企业财务管理部门或者资金部门只负责日常的往来结算,忽视资金系统的规范性管理。由于企业经营者不能深刻认识到国际国内经济形势的严峻性和复杂性,无法及时把握所面临的经济、金融形势发展变化,缺乏基本的危机意识和忧患意识,无法应对各种挑战。一些企业的经营者也忽视制度建设的重要性,不能深刻理解资金管理制度体系同样是内部控制建设重要部分,没有做到用制度管权、用制度管事、用制度管人。

(五)企业资金管理队伍建设不足,管理人员业务素质不高

有些管理人员对管理制度和政策的理解不到位、领会不深刻;有些管理人员思想观念陈旧,被动应付,对工作缺乏激情,不能适应信息快速更新的工作方式,工作质量及效率低下;有些管理人员工作创新乏力,墨守成规,对资金管理中的新情况、新问题不求甚解,拿不出行之有效的解决办法;有些管理人员的专业素质和职业操守有待提高,学习动力不足、工作能力不强,难以维护资金安全、降低资产成本;有些管理人员不能多角度、多方位地发现和解决企业资金管理中存在的风险和问题。

二、企业在资金管理中应采取的策略

企业要围绕战略目标,以加强现金流监控、巩固资金链稳定为重点,以控制资产负债率、提高资金使用效益为目标,以强化资金监管、融资管控、担保管理为重要举措,切实提高资金管理水平,增强防范金融风险的实力,确保融资渠道畅通。

(一)加强资金集中管理,不断强化资金基础管理工作

一是进一步加强资金集中管理力度,着力提高企业总部和各单位的资金集中度,建立合理机制让各单位享受到资金集中管理带来的成果,进一步提高企业资金整体使用效率。二是规范企业银行账户管理,企业资金管理部门要对各单位的银行账户管理实行"核准制",对不合规的银行账户进行全面清理。三是着力加强资金现代网络管理手段建设,企业要加快搭建信息化网络资金管理平台,推广建立资金管理信息系统,实现数据信息的统一管理、有序传递、资源共享、高效处理、有效监控,较大程度上提高企业内部日常结算效率和资金使用效益,满足资金集中管理的需要。四是企业以资金管理信息化建设为契机,对资金管理岗位职责进行梳理,进行资金业务流程再造,理顺内部业务处理流程,建立健全科学、协调、高效的资金业务处理体系,提高资金集中管理力度,通过现代网络管理手段实现高效、准确、实时的资金监管。

(二)强化资金的预算管理,不断完善资金预算管理体系

一是从抓资金预算着手,建立资金计划报批制度,要求下属企业按要求统一按月编报资金计划表,细化资金预算,关注现金流的异常变动,及时掌握和控制未来的现金流量,增强对现金流的监控力度,杜绝超付款,力争做到一切资金支出均有计划。二是加强对工程款、大宗材料采购、设备购置等大额资金支付的监管,对于大额资金原则上应统一支付。三是进一步完善相关制度,资金管理部门要重点监管预算外资金、异常支付款项及非生产性资金等支出,严格控制资金风险。四是强化企业预算的基础管理,建立规范、有效的预算管控制度,抓好项目的月度预算,提高预算的编报质量,凸显预算的统领作用。

(三)不断加强银企关系,努力开拓企业融资渠道

企业要不断强化融资统筹管控力度和担保管理,严控融资风险。一是充分发挥优势,与大型金融机构积极联动,建立长期稳定的业务合作关系,利用好目前政府给予企业的"红利",积极争取更多的优惠政策。二是创新融资方式,关注资本金贷款的相关政策,把资本金贷款、信托融资、发行债券作为改变企业对外融资结构的主要方式,根据自身情况开拓多种融资方式,在当前宽松信贷形势下,抓住有利时机,抓紧落实项目融资,切实降低财务费用。三是切实根据企业实际需求从紧控制各单位的融

资规模和担保规模,严控资产负债率上升,加大融资和担保对"造血"能力强的投资项目的支持力度,确保重要投资项目资金通畅。四是更加严格控制担保管理,要及时撤回银行到期保函,同时对存续的对外担保尽快采取措施解除。

(四)加强资金管理制度建设,不断规范资金日常管理

一是结合企业自身实际情况,要制定出台资金监督管理办法和一系列配套制度,基本建立资金管理制度体系。二是进一步完善企业的结算流程,以适应目前新的管理需求,保障企业资金正常运营,不断规范资金调拨和担保办理流程,完善内部控制程序,加内部管理,适应新形势。三是对资金业务岗位的流程再造,对岗位职责进行梳理,明确职责分工,提高工作速度和效率,保持信息传递渠道畅通。

(五)加强资金管理队伍建设,不断强化专业学习培训

企业需加大资金管理人才培养力度,资金管理人才是企业资金管理工作向前发展的根本动力,要做好资金管理工作就需要培养一批懂理论、有能力、会实践的高素质复合型人才。人才培养是企业资金管理工作中的一个重要环节,企业需要求各单位资金管理部门加强队伍能力建设,加强资金管理人员的理论学习和业务培训,注重政策业务水平、政治思想觉悟、操作实践能力和沟通协调能力的培养,提高管理人员综合管理水平,打造具有较强现代化资金管理能力的专业团队,增强企业整体抗风险能力。

资金是企业核心竞争力的重要体现,企业的资金管理部门的主要职能就是管理调配好资金,优化财务资源配置,创造资金的大价值,发挥资金管理的最大效益,规避资金风险和财务危机。在做好统筹规划的同时,企业内部仍然要加大应收账项的回收力度,加大资金集中力度,减少沉淀,加快资金流转速度,提高资金使用效率。同时,企业要抓住机遇,创新资金管理思路,提高敏锐性和洞察力,有效化解资金管理风险,为企业的持续稳步发展做好要素保障。

参考文献

[1] 侯杰.简析新形势下建筑施工企业财务管理现状[J].财会学习,2017(3):66.

[2] 何雪亮,荆佩.国际工程项目资金管理的瓶颈及应对策略[J].西部财会,2017(4).

[3] 何雪亮. 施工企业境外项目资金集中管理探讨[J].国际商务财会,2018(9).

[4] 何雪亮. 施工企业日常财务管理中的难点及应对思路[J].会计师,2021(8).

[5] 刘芳.海外工程项目资金管理[J].经营者,2015(2).

[6] 许丽丽.上市公司财务管理中存在的常见问题及解决思路[J].时代金融,2018(35).

本文曾发表于2022年《西部财会》第12期

建设工程价款优先受偿权问题研究

沈珊珊

（陕西省土地工程建设集团有限责任公司，陕西 西安 710075）

【摘要】关于建设工程价款的优先受偿问题，长期以来一直是学术界和实践中的争议点，而这种争议又造成了我国现行法律制度中存在着不同程度的争议。最高人民法院《关于审理建设工程施工合同纠纷案件适用法律问题的解释（二）》中，特别设立了七条条款，明确了对建设项目价款优先受偿的重大问题，从而切实反映了实践的需要；《中华人民共和国民法典》（简称《民法典》）第八百零七条对《中华人民共和国合同法》（简称《合同法》）第二百八十六条中有关建设项目价款优先受偿的原则进行了修改。

【关键词】建设工程价款；优先受偿权；问题研究

前言

建设工程价款优先受偿权是以法律上的创造为基础，旨在保障施工单位和施工人员的权益，是一种保护社会弱势群体的制度。这一优先权体系的一个重要内容就是要权衡生存权利与社会运作、市场秩序之间的关系。

1　建设工程价款优先受偿权的适用

1.1　建设工程承包人

我国现行法律对优先受偿权的主体范围进行了严格的限定。在施工项目中，“承包人”享有施工价款优先受偿权。也就是说，只有“承包人”与发包人订立了施工合同，才能享受到优先支付的权利。

1.2　装饰装修工程的承包人

《建设工程施工合同纠纷司法解释（一）》（以下简称《司法解释（一）》）在2020年颁布，新法规于2021年正式生效，明确了装饰装修工程的承包商享有优先权。由于装修工程具有附着物的性质，因此，关于是否符合折价或拍卖的认定，在司法实践中，有许多意见分歧，目前司法解释尚未对其进行进一步的界定[1]。

1.3　工程勘察、设计的承包人

工程勘察设计的承包人不能享受工程价款的优先受偿权。建筑工程由土木工程、建筑工程、管道、设备安装、装饰工程等组成，但不包含工程勘察、设计，工程勘察、设计工作不能直接提供工程和建筑物的人力、材料支持。为工程勘察设计的承包人设立优先权，与保障施工人员的生存权相违背。因此，在我国的司法实践中，工程勘察设计的承包人并不享有工程价款的优先受偿权[1]。

1.4　分包人、转包人及实际施工人

我国现行司法实践中普遍存在的意见是：建设工程价款优先受偿权的主体应当严格控制，合法分包人、非法分包人和其他实际施工人员，均不是“承包人”，不在建设工程价款优先受偿权的主体范围内，对工程价款的优先受偿权也不具有优先受偿权。但是，这一看法在司法实践中却有争论，一些学者和法学家认为，为了保证施工人员的劳动报酬和生存权，“承包人”应该更宽泛一些，也就是赋予分包人、转包人和实际施工人员的工程价款优先受偿权[2]。

2　建设工程价款优先受偿权背后的价值衡量

2.1　建设工程价款优先受偿权优先保护的利益

从建筑工程价款优先受偿权产生之日起，“留置权”、“法定按揭”、“优先权”这三种学说都有其

特殊性。在实践中,过分纠结于工程价款优先受偿的定义与本质问题,在实践中尚无必然,但对工程建设工程价款优先受偿的权益计量却是必须的。从表面上来看,其问题在于其在法律上的应用,而不在于其理论上和立法上。然而,在实践中,我们无法摆脱对法的理解,而在对法的理解中,也无法与价值的度量相分离。建筑物的所有权是业主的合法权益,承包商是债权人,为什么承包人可以在诸多的利益纠纷中独树一帜,并获得合法的优先受偿权利呢?当前,承包商总体上处于劣势,施工服务市场呈现出明显的供大于求的局面,激烈的竞争使得承包商的谈判能力不断降低,低价中标现象时有发生。此外,承包商的费用也在持续上涨,近年来,人工费用居高不下,沙土等建材的成本也在持续攀升,导致承包商获利的机会越来越少。在这种形势下,建筑企业面临着来自上游和下游的巨大压力,若无法及时付款,不但会出现工人的薪酬支付问题,还会导致工程的品质问题,例如长沙市房地产交易中心,就出现过一次品质问题。工程质量是影响到无数家庭生活和财产的重要因素,其严重程度甚至超过了拖欠农民工的工资。所以,施工费用优先受偿权不仅要保障施工人员的权益,还要保障承包商的权益。同时,由于承包商的利益来自施工人员的工资和工程质量,所以与普通的债权人和抵押权人相比,应当享有更高的权利。另外,《民法典》明文规定,承揽人人对定制物具有优先保留的权利,该权利优于一般的权利,因而该法律的目的在于保护承租人的劳动权益。建筑工程承包作为一种特殊的承包行为,其性质与普通的承揽人行为没有本质上的差别,只不过合同的标的是房地产,建设项目的优先受偿权利和承包合同的留置权在法定上是相似的。据此,作者提出,在保证所有者利益的情况下,建筑工程的价款优先受偿权应该高于抵押权人和其他债权人[2]。

2.2 建设工程价款优先受偿权的范围

所谓的承包人权利应当反映在整个项目的价格中,利润是项目价格的一部分,这一点毋庸置疑。但是,有些学者认为,在建筑物的诸多利益冲突中,利润并不应该是优先的。然而,在现实生活中,由于利润关系到承包商的基本利益,如果不能保障其收益,势必会对其经营和财产造成严重的影响,在实际工作中无法得到应有的保障,最终无法达到立法目标。承包商的权益全部体现在工程总价中,若在项目费用中包含了质量保证金,则也应优先受偿。因此,任何一项属于工程的费用,都应作为承包商的优先受偿权,其效力要高于普通债权人和抵押权人,不能将其单独区分,也不能一概而论。然而,由于利息、违约金等并非项目成本的主要组成部分,与其他权利人的权益相比,也没有优先权,因此,不能被承包人的优先受偿权所涵盖[3]。

2.3 建设工程价款优先受偿权有无公示之必要

许多意见认为,承包人的优先受偿权是法律上的优先权利,具有自然的优先效力,无需公示,从权利的本质上来说,这是有道理的。从实践的角度来看,在许多国家立法中,都有对工程价款优先受偿权进行登记的制度。由于建设项目自身的众多利害关系,若能将项目建设价格的优先受偿权强加给其他利益攸关方,将有助于其他各方进行市场的决策,并能有效地化解权利的冲突,进而促进项目的价格优先受偿。这是一种强大的力量,但由于没有适当的公示手段,在实践中产生了很多问题。对建筑工程价款优先受偿权的强制登记不会削弱其效力,对其实施更为有利,希望今后的立法对这一问题进行探讨[3]。

2.4 建设工程价款优先受偿权事前放弃的效力

关于这一问题,历来有两种不同的看法:一种主张,施工价款优先受偿权系合同当事人的民事权益,依照私法自治的基本原理,当然可以不要;另一种主张,工程价款优先受偿权的背后,是施工人员的工资、工程质量和公共安全,而承包人却不能因自己的利益而放弃工程价款优先受偿权。《司法解释二》第二十三条,基于上述两种意见,提出了在不影响施工人员权益的情况下,可以对工程价款的优先受偿权进行约定。在条款文意方面,法院承认事先放弃或者限制的效果,但是如果侵害了施工人员的权益,就不能成立。从表面上来看,该条款具有一定的合理性,但实际上却缺乏操作性,这种条款在客观上很难避免对施工企业的权益造成损害。个人认为,在当前的实践中,对优先受偿权的预先放弃或限制的效力是不合适的。虽然承包人的放弃是对自己的实际权益进行处分,但大部分的放弃是否真的代表了承包人的真实意图？恐怕绝大部分都是由于市场竞争激烈，承包人或承包人的债权人施压造成的。另一些人则争辩说，承包商放弃这项权利，也许是出于自愿，想要在工程费用或工期上获得更多的好处，但也有其他方法可以选择，并不一定要放弃优先受偿权来获得，对于处境不利的承包商来说，除非有强大的优先受偿权保护，否则，其他的利益就会变得更加脆弱。而且，从表面上来看，发包方和承包商的合同是完全由市场决定的，但施工人员的利益和工程质量关系到整个社会的利益，不是承包商的契约所能改变的，所以法律需要进行特殊的干涉，而施工价款优先受偿权就是其中的一种。因此，在当前的市场条件下，不能允许承包人任意地放弃工程价款的优先受偿权[4]。

3 建设工程价款优先受偿权对企业的实用意义

3.1 对承包人企业的意义

3.1.1 法律创设的权利

施工项目价款优先受偿权是法律规定的优先权利，在不经合同当事人同意的情况下，不需要事先约定，也不需要进行注册。这是一种对承包人和建筑工人的特殊保护，它的设立目的就是为了维护承包人和建筑工人的合法利益。

3.1.2 优先顺位的权利

建设项目的优先受偿权，是对大多数抵押和非抵押债权的优先权。在不经协商的情况下，承包人获得了优先顺序，从而保证了最大限度地实现工程价款的清偿。

3.1.3 便于行使的权利

通过多年的实践，我国现行的建设工程价款优先受偿权制度已基本明确、规范、法定要件明确。以“发包人应支付施工费用，但不付款”为其权利行使的关键点，与承包人所要求的项目价格相一致。以工程质量标准作为权利行使的先决条件，对改善工程质量的要求是有益的。

3.1.4 合理期限的权利

施工项目价款优先受偿权的行使期限可以延长到18个月，赋予承包人更多的权利要求和充分的谈判和协商时间。

3.2 承包人企业应该如何重视

3.2.1 律师的作用

建筑工程案件一般都具有专业性强、标的高、诉讼难度大、诉讼过程长等特点，对此类案件的专业性要求较高。律师可以从项目招投标、项目实施、谈判、诉前、诉讼、执行等多个角度，全方位地参与施工项目。在上述过程中，律师可以有效地规范文件，完善诉讼程序，保存证据，协助谈判调解，制定诉讼策略和程序，从而减少诉讼费用和法律风险[5]。

3.2.2 应当保存的证据

第一，招标文件、合同及其他有关施工项目的主要文件；第二，在执行本协议时，双方签署的备忘录、修正案、工程量签证等；第三，双方在执行合约期间的电子邮件、通知、答复、指令等相关的往来文书；第四，工程请款函、催款函、查询函、结算书等结算资料；第五，竣工证明、交付证明等竣工资料；第六，如有需要，可以由专业的鉴定人出具工程量和工程价款的鉴定函。

4 结语

本文通过对建设项目价款优先受偿权的价值计量，以及在建设项目价款优先受偿权中的主次关系等问题上的实践思考，以期对我国现行的建设工程价款优先受偿权制度的实施起到一定的借鉴作用。

参考文献

[1]荆长坤. 建设工程价款优先受偿权问题研究[D].昆明:云南财经大学,2022.
[2]陈相荣. 实际施工人建设工程价款优先受偿权研究[D].邯郸:河北工程大学,2022.
[3]谢乐安,谢赟.建设工程价款优先受偿权的适用与意义[J].中国律师,2022(05):85-87.
[4]路梦瑶. 建设工程价款优先受偿权规则研究[D].石家庄:河北经贸大学,2022.
[5]张玉萍. 论建设工程价款优先受偿权的行使[D].兰州:甘肃政法大学,2022.

本文曾发表于2022年《工程管理前沿》第15期

大数据在企业纪检监察工作中的应用探究

刘兴姝

（陕西地建房地产开发集团有限责任公司）

【摘要】在从严治党的前提下，有必要强化高素质纪检监察队伍建设，不仅可以推动企业高质量发展，也可以强化纪检监察队伍战斗力、凝聚力以及监督力。在日常工作过程中，纪检监察人员不仅要强化思想建设，也要提升凝聚力，才能突出纪检监察队伍的先进性和纯洁性。基于此，在企业纪检监察工作中，将大数据应用其中，既要确保其准确性，也要确保其合法性。本文主要探讨在企业纪检监察工作中，大数据如何应用其中。

【关键词】大数据；企业纪检监察工作；应用

在大数据的背景下，纪检监察工作中新技术的广泛应用，使得纪检监察工作更加规范化、高效化，使其既在业务融合方面取得突破，也在模式创新方面有所发展，还在服务质量方面有所提升。为了促使纪检监察工作高效规范发展，应科学、合理地应用大数据技术，并对此进行深入研究。在此过程中，企业应根据大数据的特点，制定相关制度，规范大数据应用，使其在企业纪检监察工作中发挥重要作用，促使纪检监察工作更加深入人心。比如分析数据更加具有准确，分析结果更加令人信服，为纪检监察人员提供有力依据，并将其应用于实际工作中，在做好纪检监察工作的同时，确保纪检监察对象的合法权益，进而发挥大数据在企业纪检监察工作中的作用。可见，为了确保纪检监察工作更加顺利，大数据在纪检监察工作中应用，应对其进行规范，使其更加具有准确性和合法性，才能发挥其在纪检监察工作中的作用，促使纪检监察工作相关问题得到妥善解决，并为该领域的发展作出积极贡献。

一、大数据具有五大特点（5V）

（一）多样（Variety）

大数据的多样性是指不管是数据的种类，还是数据的来源，均较为多样化。数据既有结构化，也有半结构化，还有非结构化。数据的表现形式既有文本和图像，也有视频和 HTML 页面等。因为大数据具有多样性，才促使大数据能够准确地反映相关领域的数据以及信息，并在该领域发挥重要作用，进而解决该领域相关问题。

（二）大量（Volume）

大数据的大量性是指数据量的大小，即大数据可以容纳大量的数据，相对于传统的数据分析而言，大数据可以分析更多的数据，不仅具有更高的准确性，也具有更好的高效性。相关领域应用这些大数据，不仅可以提升工作人员的工作效率，也可以促使相关数据的分析更加准确，并为该领域的发展作出积极贡献，进而发挥大数据在该领域的作用。

（三）高速（Velocity）

大数据的高速性不仅指数据增长快速，也指数据处理快速。在日常工作中，不管任何行业，每天的数据都呈指数性爆炸增长。在大部分情况下，数据具有时效性，比如在几秒钟内，将用户的数据搜索出来。在面对持续增长的数据时，企业或者系统不仅要高速处理，也要快速响应。正因为大数据具有高速性，才能提升数据分析工作的效率，减少工作人员的工作量，促使数据分析更快地反馈给相关人员，并促使分析结果被广泛应用和认可，进而发挥大数据在该领域的作用。

(四)低价值密度(Value)

大数据的低价值密度性是指在大量的数据源中,有价值的数据相对较少,绝大多数数据存在错误的问题,不仅不够完整,也无法利用。从整体上来讲,数据总量中有价值的数据的密度极低,貌似海底捞针那样提炼数据。

(五)真实性(Veracity)

大数据的真实性不仅是指数据的准确度,也是指数据的可信赖度,即称之为数据的质量。因为大数据采用高科技采集数据,相对于人工采集数据而言,可以避免人工误差,数据的真实性相对较高。正因为大数据具有真实性,才能提升数据分析工作的准确性,使得数据分析结果更加具有可信赖度,并可以广泛应用于相关领域,进而发挥大数据的重要作用。

二、运用大数据开展纪检监察工作的具体实践

企业纪检监察大数据监督是指采用大数据技术,纪检监察人员监督各类公权力行使主体的行为,同时,借助大数据“自动化决策”等方式,发现问题,并作出监督决策。现阶段,这项工作是国家监察法律法规概括性赋予纪检监察人员的权力,结合大数据技术,创新和探索对各种公权力履行监督制约职能。在企业纪检监察工作中,采用大数据技术,需要更多措施进行推进。开展大数据监督工作,既要将大数据资源结合监督行为,也要将数据平台融合监督行为。先收集数据,后建立数据平台,为大数据监督提供有力支持,既要将权力行使行为标准化,也要将其数据化,并在大数据监督系统接入权力行使数据,方便纪检监察人员对比相关数据,并从中发现、查证相关问题。纪检监察人员借助线上公开平台,监督公权力运行流程;基层群众也可以借助网络公开平台,既可以监督公权力行使不规范行为,也可以监督其不正当行为,还可以监督其违法违规行为。可见,在企业纪检监察工作中,运用大数据技术,既能确保纪检监察数据更加合理,也能促使纪检监察工作进展顺利。

三、大数据应用在纪检监察工作中的问题

(一)缺乏应用规范

在企业纪检监察工作中,每个单位都在采用大数据技术,但没有建立统一的纪检监察信息网络平台,导致纪检监察工作无章可循。比如在企业纪检监察工作中,企业的总公司使用一个纪检监察信息网络平台,而企业的子公司使用不同的纪检监察信息网络平台,导致企业相关纪检监察数据存在差异,不利于为纪检监察工作提供有力依据,进而可能造成纪检监察工作存在失误。这样不仅不利于企业的发展,而且还可能引起员工的不满,非常不利于企业正常运转。因此,在企业纪检监察工作中,企业如果没有统一的纪检监察网络信息平台,那么,企业将无法通过大数据开展纪检监察工作,进而影响企业的发展。

(二)缺乏保障机制

在企业纪检监察工作中,采用大数据技术,需要建立健全保障机制,才能确保纪检监察工作进展顺利。比如在企业纪检监察工作中,企业只是根据企业发展的需要,将大数据技术引入企业,然后让相关技术人员使用这项技术,但没有为此提供保障机制,导致相关技术人员在使用的过程中发生问题,也无法找到相关保障机制解决问题,最终影响大数据技术在纪检监察工作中的应用,使其失去了原来的意义。因此,在企业纪检监察工作中,企业如果缺乏保障机制,那么,企业将无法通过大数据开展纪检监察工作,进而影响企业的发展。

(三)无视合法权益

在企业纪检监察工作中,因为现阶段企业不够重视大数据技术,导致企业纪检监察工作难免存在漏洞,非常不利于企业开展纪检监察工作,还可能损害纪检监察对象的合法权益,影响企业的形象以及运转。比如企业员工偷盗仓库的材料,给企业带来巨大的损失,导致企业开除员工,甚至将其起诉,后果较为严重。如果企业采用大数据技术,那么,员工进入仓库之后,企业发挥预警机制,杜绝偷盗事件发生,最多教育员工或者开除员工,杜绝侵犯员工合法权益。因此,在企业纪检监察工作中,企业如果无视合法权益,那么,企业将无法通过大数据开展纪检监察工作,进而影响企业的发展。

四、大数据应用在纪检监察工作中的相关支持

（一）企业领导高度重视

在企业纪检监察工作中，企业领导应高度重视大数据技术，才能促使纪检监察工作进展顺利。通常情况下，在企业纪检监察工作中，采用大数据技术，需要领导牵线或者指导，才能得到顺利落实和执行。如果领导较为重视这项技术，那么，企业才能将大数据技术贯彻执行，发挥其在纪检监察工作中的作用，使得纪检监察工作更具合理性。比如企业领导安排专业技术人员负责使用大数据技术，并赋予他们相关执行权，促使他们与其他部门取得良好沟通，积极执行相关工作，进而促成企业纪检监察工作。因此，在企业纪检监察工作中，如果领导重视大数据技术，那么，企业才能利用大数据技术，进而推动企业发展。

（二）技术应用有所规范

在企业纪检监察工作中，企业应规范应用大数据技术，才能推动纪检监察工作进展顺利。通常情况下，在企业纪检监察工作中，采用大数据技术，需要规范应用的范围以及行为，才能促使大数据技术发挥重要作用。如果企业在应用过程中足够规范，那么，采集的数据才能具有准确性，也才能确保纪检监察工作具有准确性。比如在企业的纪检监察工作中，根据企业相关规定，结合企业需求，对大数据技术进行运用，促使企业纪检监察工作进展顺利，既可以维护企业的形象，也能够推动企业的发展，进而发挥大数据技术在纪检监察工作中的意义。因此，在企业纪检监察工作中，如果企业规范应用大数据技术，那么，企业才能利用大数据技术，进而推动企业发展。

（三）技术应用有所创新

在企业纪检监察工作中，企业应创新大数据技术，才能促使纪检监察工作更加顺利。通常情况下，在企业纪检监察工作中，采用大数据技术，应对大数据技术持续创新，一方面保持与时俱进，另一方面满足企业需求，使其在企业纪检监察工作中发挥重要作用。如果大数据技术足够创新，企业纪检监察工作基本不会存在问题，企业纪检监察工作才能具备准确性和合法性，并且维护纪检监察对象的合法权益，进而发挥大数据技术在纪检监察工作中的作用。比如企业使用的大数据技术已经更新换代，企业应引进先进的大数据技术，不再沿用原有的大数据技术，避免数据采集与真实情况不符，确保数据采集更具全面性，才能确保数据分析更具准确性，发挥大数据技术在纪检监察中的重要作用，进而推动企业纪检监察工作进展顺利。因此，在企业纪检监察工作中，如果企业创新应用大数据技术，那么，企业才能利用大数据技术，进而推动企业发展。

五、大数据应用在纪检监察工作中的实施策略

（一）注重基础设计

在将大数据应用于纪检监察工作中，每个单位的大数据技术存在重合性。为了杜绝类似情况发生，可以采用大数据技术，建立统一的纪检监察信息网络平台，不管在大数据监督平台，还是在大数据监督模型，将相关单位和部门的数据壁垒打通，将数据进行共享，促使各纪检工作人员不管在任何时间，还是在任何地点，借助平台，既可以及时了解纪检监察工作信息，也可以及时了解其进程。比如企业总公司与子公司都建立统一的纪检监察信息网络平台，这样不仅方便纪检监察人员行使权力，也方便企业领导行使监督权力，使得纪检监察工作得到企业内部人员的全力支持以及监督。可见，在大数据应用的初级阶段，应事先强化顶层设计，为企业构建统一的纪检监察信息网络平台，数字化处理大量的监察信息，进而模块化对接各区域。

（二）建立保障机制

为了促使纪检监察相关数据具有安全性，避免数据监管存在漏洞，可以建立健全法制体系。不管是建立数据应用的监管制度，还是建立数据应用的问责机制，对清单管理进行强化，对业务流程进行优化，对数据管理权限进行明确，既可以规范数据采集和整理，也可以规范数据分析和使用，使得纪检监察相关数据更加合理。比如企业总公司与子公司都要规范数据应用，尤其是子公司，应该根据总公司相关规定，开展相关纪检监察工作，促使企业内部统一规范纪检监察行为，进而推动纪检监察工作

顺利开展。可见,企业建立保障机制,规范大数据应用,不仅可以促使纪检监察工作更具安全性,也可以促使纪检监察工作更具法制性。

(三)保障合法权益

合理应用大数据技术,不仅要落实技术威慑,也要体现人文关怀。不管在防微杜渐方面,还是在提前预警方面,既要建立有效的工作制度,也要建立良好的挽救机制。为了在企业推进纪检监察工作,不仅要建立完善的数据保护制度,也要建立健全数据使用机制,进而维护个人正当权益。纪检监察工作毕竟涉及个人合法权益,应该确保相关数据的准确性以及合法性,才能将其应用在纪检监察工作中,在不损害相关人员的合法权益的同时,尽最大优势发挥大数据技术的重要作用,进而确保大数据技术为该领域作出积极贡献。比如企业总公司与子公司都要合理应用大数据技术,在纪检监察中维护员工的合法权益,使得员工愿意配合纪检监察工作,一方面维护企业的形象,另一方面推动企业的发展,进而促使企业纪检监察工作进展顺利。可见,企业保障合法权益,规范大数据应用,不仅可以促使纪检监察工作更具法治性,也可以维护纪检监察对象的合法权益。总之,在大数据时代,不管任何行业,将大数据应用其中,相对于传统数据分析而言,一方面可以提升数据分析的效率,另一方面可以提升数据分析的准确性,并在相关行业发挥重要作用。在企业纪检监察工作中,将大数据应用其中,不仅可以促使纪检监察结果更具准确性,也可以促使纪检监察结果更具合法性,发挥大数据在纪检监察工作中的重要作用,促使纪检监察工作进展顺利,进而为该领域作出积极贡献。对纪检监察工作持续用大数据监督,应对相关理念进行更新,既要制定相关制度,也要细化相关制度,还要完善相关制度,促使大数据监督具有正当性,并在法治轨道上正常运转。

参考文献

[1] 钟梁. 关于大数据在企业纪检监察工作中应用的思考[J].海外文摘,2021(1):54-55.

[2] 刘雨丰. 关于提升企业纪检监察工作水平的思考[J]. 企业文化(下旬刊),2016(4):54.

[3] 靳玉山. 创新和深化企业纪检监察工作的几点思考[J]. 博鳌观察,2020(4):83.

[4] 李文峰, 齐道峰. 创新和深化企业纪检监察工作的几点思考[J].石油政工研究,2004(1):47-48.

本文曾发表于 2022 年《智库时代》

某国有建筑企业精细化管理案例分析

陈科皓,丁亚楠

(陕西省土地工程建设集团有限责任公司宝鸡分公司,陕西 宝鸡 721000)

【摘要】随着经济社会的快速发展,建筑行业面临激烈的市场竞争。实施精细化管理是提升企业竞争力、实现可持续发展的重要途径。本文以某国有建筑企业精细化管理为例,提出了以“三项建设、四项控制、五项管理”为核心的精细化管理方案,为加强企业管理、提升行业竞争力提供借鉴和参考。

【关键词】国有企业;精细化管理;案例分析

随着经济社会的快速发展,社会专业化分工越来越细,建筑行业面临激烈的市场竞争。精细化管理是一种精益求精的文化,是一种认真的工作态度。实施精细化管理,可以建立起公司强大的执行能力,为企业实现做强做优做大目标筑牢坚实、科学的管理基础。本文以某国有建筑企业精细化管理为例,提出了以“三项建设、四项控制、五项管理”为核心的精细化管理方案,有效促进了公司管理、项目管理的规范化、科学化和程序化,不断提高管理水平,确保工程安全生产及质量,打造精品工程。

1 精细化管理的原则

1.1 转变观念,坚持长期推动

精细化管理工作的核心是“精、准、细、实”,必须运用精细化管理理念和管理技术,通过提升员工素质,加强企业内部控制,强化协作管理,坚持长期推动,持续改进。

1.2 关注细节,把握核心内涵

精细化管理的本质在于对战略和目标进行分解细化和责任落实,通过精细化管理工作的开展最终实现公司的战略规划有效地贯彻到每个工作环节并发挥作用,从而提升公司整体执行力。

1.3 夯实基础,保障效益提升

不断梳理管理流程,夯实管理基础,通过制度建设的系统化和细致化,坚持标准化和信息化的原则,使内部组织机构能够精确、高效、协同和持续运行,最终确保公司效益得到稳步提升。

1.4 优化指标,兼顾过程改善

分解量化精细化管理目标体系,确保每项工作内容都有专人负责,各项指标数字、程序、责任得到持续改进,并能够得到及时固化,形成公司长效管理机制。

1.5 抓住重点,实现全面精细

建立以“月”为考核周期的战略目标管理体系,倒排工期、分解目标、细化落实到岗位上。抓住工作中的难点、痛点,通过以点带面、逐层推进的方法,把精细化逐步引入生产经营和企业管理的各个环节。

2 精细化管理的主要内容

精细化管理的实施包括“三项建设、四项控制、五项管理”。三项建设包括规章制度建设、人员组织建设和岗位责任建设。四项控制包括进度控制、质量控制、安全控制和成本控制。五项管理包括行政管理、协调管理、廉政建设管理、施工现场管理和科技创新管理。

2.1 三项建设

2.1.1 规章制度建设

按照部门日常工作的内容,梳理工作流程,完善公司行政管理、工程管理等制度。结合公司实际,制定年度学习培训计划,完善月考核制度、奖罚制度,严格落实三项机制。

2.1.2　人员组织建设

完善公司组织架构,根据工作开展情况适当调整部门业务、岗位分工,合理设置各机构部门。通过综合考查合理安排每位员工的岗位,尽量做到"人尽其才,才尽其用"。加强思想政治教育,加强团队凝聚力与战斗力,通过形式丰富的活动增进公司员工间的交流与沟通。加强培训与引导,通过"传帮带"的模式提高员工个人的专业技术水平。完善考核机制及激励机制,不吝啬表扬、不放纵错误,畅通员工晋升通道,培养员工承担责任的意识,营造良好的工作氛围。

2.1.3　岗位责任建设

分解年度目标任务,公司与各部门签订目标责任承诺书,抓好责任落实。各部门细分岗位职责,目标责任到人,从严从细抓好落实,在内部实行管理责任登记及追究制度。

2.2　四项控制

2.2.1　进度控制

充分做好施工准备工作。根据工程施工合同工期要求,在熟悉施工特点和工程量的情况下,要进行资金、材料、技术、劳动力和机械准备等工作。科学编制全年工程施工进度计划,加强计划的执行,落实倒排工期、挂图作战机制。按月编制月进度计划,按周编制周进度计划。每周在监理例会上落实周进度计划的完成情况,如进度滞后,要分析原因并及时采取补救措施。

2.2.2　质量控制

坚持组织月质量检查、召开月质量会议。针对项目质量管理的关键环节、重要部位,责任部门联合施工单位每月对现场质量全面检查,查找隐患,消除隐患,确保施工质量达标。在月质量会上对工程中出现的质量问题进行探讨和总结。坚持"样板引路,样板先行"的原则,在后期装饰、安装工程质量控制中,先进行样板施工,经监理、甲方验收合格后才开始大面积施工。对不符合质量要求的工程坚决要求返工,不按施工工艺要求进行施工的责令停工整改,确保工程质量。

2.2.3　安全控制

坚持组织月安全检查、召开月安全会议。针对现场施工起重机械、脚手架、深基坑、"三宝四口五临边"等重点环节、领域,按照公司《施工现场安全隐患排查清单》进行检查,做好隐患排查整改记录。在月安全会议上与各参建单位重点研究讨论工程中出现的安全隐患评定,做好登记。公司全员严格执行"三级培训",安全培训合格才能上岗,严格执行安全纪律、工艺要求、劳动纪律,各种原始记录做到标准化、规范化。安全工作人员在生产现场要做好各种信息的收集、传递、分析、处理工作,及时了解安全生产状况,及时处理生产中反映出的问题。

2.2.4　成本控制

加强项目劳务、物资、设备等管理,确保整个施工工程中推行成本精细化管理。合理安排工期与施工组织,节约因工期延长造成的施工成本浪费。加强后勤管理,实行生活物资采购领用登记制度,倡导节约意识,合理用支出,避免不必要的日常成本浪费。

2.3　五项管理

2.3.1　行政管理

认真做好各类会议的通知、会场安排、文件印发、会议记录等,保证及时、准确,不出差错,严格按照责任分工,扎实做好接待等各项服务保障工作。运用OA系统实现信息化传达管理科学化,及时完成领导督办事项,以及督办完成情况的反馈。严密各种手续,加强执行监督,确保各项工作在规范的程序下操作。

2.3.2　协调管理

内外部协调要做到及时性、畅通性、紧密性。及时性保证重要紧急事务可第一时间传达。畅通性保证全面掌握内外各项信息,合理部署各项工作。紧密性保证各个流程有督导,各个岗位有责任。安排工作要做到计划性、超前性、周密性。计划性保证工作安排无遗漏,轻重分配恰当。超前性保证人员量才使用,量力而行。周密性保证各项工作顺利开展、施工现场有序进行。

2.3.3 廉政建设管理

树立精细化管理理念,丰富廉政教育形式。依靠精细化管理标准,创新廉政工作模式。选准精细化管理角度,提升监督检查水平。把握精细化管理措施,确保反腐工作实效。

2.3.4 施工现场管理

围绕项目安全质量管理目标,实施全过程、全方位、全员化管控和监督,严格执行落实有关安全质量的法律法规,建立健全安全生产各项制度和应急预案。提高对现场的管理效率,争取工作做到日清日结,发现问题及时纠正,及时处理。

2.3.5 科技创新管理

充分利用集团科研平台,通过组织讲座、科普讲解等一系列活动,在公司营造人人讲学习的良好氛围。分解年度科研目标任务,将各项创新工作落实至各部门、每个人。组织学习先进企业创新管理案例及经验,组织员工开展创新思想研讨会,集思广益,为创新管理提供多样化思路,为公司精细化管理提供科学指引。

3 精细化管理实施步骤

3.1 学习宣传阶段

组织召开精细化管理工作动员会,让全体干部职工提高认识,转变思想观念,树立精细化管理理念。组织精细化管理知识培训会,组织学习精细化管理方案、简报等相关文件,提升全员精细化管理专业能力,并就如何提升公司精细化管理水平组织专题研讨会。拓展宣传平台,通过公司宣传栏、电子显示屏、活动展板、内刊等对精细化管理进行宣传,营造良好氛围。

3.2 自查自纠整改阶段

各部门按照职责分工,全面梳理内部流程,查找制度流程及岗位职责是否存在空白缺失或无法明确的问题,寻找管理短板并向公司递交自查报告。各部门对自查中发现的问题要及时纠正,按照目标任务对精细化管理工作进行分解,抓好工作实施和责任落实并向公司递交整改报告。

3.3 总结并持续推广阶段

加强制度建设,增强执行力度。根据各部门工作分工,修订完善公司内部管理制度及规范化流程,进一步细化岗位职责,强化执行、狠抓落实。强化评价考核,公司对各部门进行考核,综合评价各部门在精细化管理工作中取得的成果,并侧重于过程性考核,将精细化管理的阶段成果和先进经验进行固化,形成长效机制。

4 结语

结合建筑企业特点,为确保精细化管理措施落实落细,发挥应有效果,要持续加强精细化管理的组织领导,加大监督检查力度,督促整改。要紧密围绕年度目标任务,以加强项目管理、提升专业能力、增强服务保障为核心,明确责任分工,强化协作管理,建立科学的工程项目管理体系,实现工程管理的制度化、流程化和规范化。要持续巩固优化精细化工作成果,制定优化措施,建立精细化管理的长效机制,实现事事有流程、处处有标准、人人讲规范,努力打造标杆项目。

参考文献

[1]刘衡.精细化管理工作在公司的推进与实施[J].河北企业,2022(2):108-110.

[2]秦海涛. 精细化管理模式在民用建筑施工管理中应用[J]. 建筑与装饰,2022(1):28-30.

[3]王琼.浅谈精细化管理在国企绩效管理中的应用[J].中小企业管理与科技,2021(36):4-6.

[4]苏岳松.房建施工技术精细化管理实践和创新思路的结合[J].价值工程,2019,38(35):96-98.

本文曾发表于2022年《科协论坛》第1期

国企纪检干部教育培训提质增效的路径思考

夏　莺

(陕西省土地工程建设集团有限责任公司,陕西 西安 710075)

【**摘要**】教育培训是强化党的理论教育和党性教育的有效手段之一,国有企业党风廉政建设和反腐败工作任重而道远,国有企业纪检监察工作要想扎实开展就必须提升干部教育培训的管理水平,才能确保坚定不移的履行好监督责任。

【**关键词**】国有企业;教育培训;提质增效

引言

习近平总书记2016年在全国国企党建工作会上指出,我国国有企业的独特优势是,坚持党的领导、加强党的建设,是国有企业的“根”和“魂”。党的十九大强调全面从严治党永远在路上。而打铁必须自身硬,执纪者必先守纪,律人者必先律己,这对纪检队伍自身建设提出了更高的要求。想要建设一支素质高、专业能力水平高的纪检队伍,不断加大学习培训力度就尤为重要,这是切实帮助国有企业纪检干部快速进入角色、提高业务能力,更好地满足新时代纪检工作要求的必经之路。

一、高质量发展的新形势对加强国有企业纪检干部培训工作提出明确要求

(一)干部教育培训与党的发展壮大密不可分

我们中国共产党善于学习,重视干部的教育、培训,是历年以来保持的优势和传统,建党初期,我们党就大量成立地方党校、夜校和讲习所,加强对革命理论的宣讲培训工作。在抗战时期,我们党克服重重困难,成立三十余所干部学校加强战时干部教育培训工作,为培养优秀人才和干部提供坚强保障。改革开放后,干部培训又根据发展需要,培养了大批解放思想和创新、踏实、进取的干部。特别是进入新时代以后,我们党担负着进行伟大斗争、建设伟大工程、推进伟大事业、实现伟大梦想的历史使命。要实现新作为,完成新使命,就必须继续坚持和发扬好干部教育培训这个优良传统,教育引导干部树牢理想信念宗旨,增强业务能力,为党的事业发展提供强大的思想政治保证、干部人才和智力支持。

(二)加强党风廉政建设是国有企业高质量发展的重要纪律保障

我国国有企业是中国特色社会主义的重要物质基础和政治基础,是我们党执政兴国的重要支柱和依靠力量。党的十九大的一系列新部署新要求充分说明了党中央持续深入推进全面从严治党的态度、决心和信心。国有企业党风廉政建设和反腐败工作任重而道远,作为国有企业纪检监察干部既要理性认识当前反腐败工作形势,针对新形势新任务新挑战,要把干部教育培训的认识提升到更加紧要的任务来抓,从思想上更加引起重视,从措施上更加有效求实,从保障上更加周密健全,实现教育培训工作水平的整体提升,实现干部教育培训工作的整体加强和改进,引导干部结合自身工作实际,结合新时代我们党和国家带领民族复兴伟大实践,结合努力实现作风向好的使命,分类分级、精准有力地进行教育、引导和培训,督促广大干部做到真学、真用、知行合一。

(三)新形势下国有企业纪检干部培训的新要求新趋势

长期以来,纪检机关为守护党的初心使命、纯洁党的组织和队伍作出了重要贡献。但也应当看到,随着全面从严治党进入新阶段,“四大考验”“四种危险”长期存在,政治信念、思想认识、组织作风等方面一些问题依然突出,国有企业纪检干部培训工作与形势要求相比仍然存在一些短板。一是政

治教育不够细致深入，纪检机关作为管党治党政治机关，新时代纪检干部培训工作更要围绕党的旗帜、党的使命统一谋划，在培训的规划、任务、内容上毫不动摇地坚持政治统领，把提高纪检干部政治三力作为第一位的要求。二是培训内容不够全面提炼，新时代纪检工作总要求就是要实现高质量发展，围绕高质量发展这个主题，对纪检干部的培训就不能只仅仅局限于纪法贯通、法法衔接等业务知识，更要强化对经济意识、服务意识、成本意识等方式方法的灌输。三是业务指导还需深入，党的决策部署和工作重心在哪里，干部的教育培训就要跟进到哪里，随着"三转"和"三项改革"深入推进，监督的内容、方式、边界都产生了深刻变化，对纪检干部监督工作的职责、工作内容和监督对象的培训要更加务实，以免出现权限不清、工作发散，纪检干部承担大量职责以外的工作的情形。

二、加强国有企业纪检干部教育培训的途径

（一）七种能力，首要强化政治学习

在中央党校中青班"开学第一课"上，习近平总书记重点围绕提高年轻干部解决实际问题的能力，强调了培训学习的目的、重点和方法。他讲到在一名干部具有的全部能力中，要想圆满完成工作，政治能力是首位的、核心的能力。只有有了过硬的政治能力，才能自觉主动地做到不仅在思想上政治上，更关键是要在行动上同党中央保持高度一致，夯实党的领导，增强纪律意识，坚持把依规治党、从严治党要求从始至终贯穿到监督执纪执法工作中。

要通过干部培训自觉养成遇事首先考虑政治要求，办事考虑政治规矩、处事考虑政治效果，自觉养成从政治上解决问题的习惯。目的是要把习近平新时代中国特色社会主义思想学懂弄通做实，善于从政治上开展监督。作为纪检干部，更需要增强政治意识、磨砺政治眼光、强化干事担当、严守党的纪律、锤炼政治品格，不断提升对党忠诚的能力、风险防范的能力、履职尽责的能力、拒腐防变的能力、做"明白人"的能力，为祛邪扶正、维护社会稳定和人民安居乐业提供过硬政治能力保证。

对领导干部而言，学习不仅是一种兴趣爱好，更是一种政治责任。纪检干部要持续锤炼七种能力，练就过硬本领，不断提高运用党的创新理论指导我们应对重大挑战、抵御重大风险、克服重大阻力、化解重大矛盾、解决重大问题的能力，率先示范，争创一流。

（二）线上线下，模式创新系统全面

纪检教育培训工作要紧盯培训需求，创新培训形式，着力增强工与学、供与需、学与用之间的协调性，持续优化培训"供给侧"，提升业务培训与工作实践的契合度，不断推动纪检干部专业能力、实践本领跟上时代发展步伐。

线上充分运用信息化手段建立有效教育培训平台，可结合干部在线学习、学习强国等 APP 的同时，运用纪检监察内网系统组织干部学习纪检业务。制作优秀课件，整合系统师资力量，围绕纪检监察业务和工作实践中的重点、热点、难点问题，钻研年度课程项目，利用平台从不同角度讲解，从而引发实践思考，使线上学习也受益匪浅。

线下集中培训时可以采取错峰、分段、分业务的模式，进行"超市式教学"，给予纪检监察干部选择的权利，在监察法实务、政治监督、监督检查、审查调查、案件审理、信访工作等十多项供选课程中自主决定学习课程，自行选择上课时间。同时辅以辅导讲座、专题研讨、知识竞赛等多种方式，使得纪检监察干部培训更有针对性和方向性，快速掌握业务知识，并能将其很好地运用到工作实践中，纪检干部履职能力也得到有效提升。

（三）遴选内容，扎实锤炼培训基本功

众所周知，聚焦企业主责主业、服务企业发展大局才是干部教育培训的核心要义，要始终为党的政治路线服务，培训的内容要紧紧围绕党和国家的发展来开展。紧扣时代设定培训计划、培训内容，做到党的中心工作推进到哪里，纪检监察干部的培训就跟进到哪里。

着力解决"本领恐慌"的问题，按照缺什么补什么的原则，但一个干部真正所缺的东西和他自己认为缺的东西，不一定就一致。这就需要培训时在遴选内容方面考虑可否制度化，以锤炼党性和本领增强为目的，严把专业关，将所需要掌握的内容相对集中，必要时也可制度化培训。最终将培训内容定期报送学员所在单位,做好沟通联结,尽量与领导评价、年终考核等现实结合起来,充分运用考核结果。

三、把握好干部教育培训工作和干部能力提升之间的几个关键点

干部教育培训的任务是帮助干部提升能力,而终极目的是要打造一支忠诚干净担当、专业强素质高,拥有如钢铁一般纪律的队伍。因此教育培训工作更加不能脱离干部个人需要和个人成长,还需要通过几个关键点,来使之更加容易被干部吸收掌握,为干部个人所用,从具体可行、便于实施的层面讲,建议可以从四个方面着手。

加强返程课。为了把课上学到的道理、新意、高度和个人的学习思考进行全面的总结、提炼和概括。建议学员在培训结束后,都要围绕培训期间的内容、自己的感悟、工作的努力方向等方面,给所在单位讲一次党课,一方面放大培训的受众,加强了培训效应,也检验了培训成效。

抓住精气神。干部培训的目的,最重要就是推动作风转变,提振干部的精气神。要抓住干部的精气神来判断一次培训的效果如何,如果培训后"涛声依旧",没有任何变化,说明培训的实际效果不明显;反之,如果精神面貌焕然一新,说明培训成效明显。

做好关键事。就是通过关键事件,全面地检验干部在培训以后的表现,包括精神状态、应对措施、处置效果等。如果干部在处置突发事件时,能够临阵不乱、妥善应对,说明通过培训学到了本事。

关注身边人。可以通过对受训干部的领导、同事、下属,开展问卷调查、座谈访谈等方式,了解学员在接受培训后的表现。如果群众反响普遍较好,那么就可以大体认定培训效果明显。

由此可见,在新时代全面从严治党的新形势下,我们每一名从事纪检监察工作的干部都必须去认真思考、研究探索,正因为纪检监察工作对于国有企业健康稳步、高质量发展至关重要,意义重大,所以我们纪检监察干部都要进一步找准纪委的职责定位,适应反腐败新形势,最重要的是要有效精准履行监督执纪问责职责。

参考文献

[1]徐相锋.如何加强党员干部教育培训工作[N].学习时报,2017-04-19.

[2]张同锋.纪检干部提升思想政治教育的途径[J].长江丛刊,2019(15):160-161.

[3]师玮.浅析纪检监察工作在企业思想政治工作中的作用[J].才智,2017(33):207.

本文曾发表于2022年《魅力中国》第33期

浅析企业工会在职工思想政治工作中的作用

胡一萱

（陕西省土地工程建设集团有限责任公司，陕西 西安 710075）

【摘要】企业是推动经济发展的关键所在，企业所面临的环境日益复杂，同时也为其提供了多项全新的发展机遇，需要做好充分的内部管控工作。在现代企业发展过程中，必须加强思想政治强化工作，其中工会具有重要的作用，是对职工进行思想教育的基础组织，需要将其作用充分发挥，全面强化工会的教育、培训等基础性职能。但是结合部分企业的实际情况来看，针对职工的思政工作还存在着一些问题，工会的职责没有充分落实，不利于企业建设与发展，为此需要认识到工会在思政工作的作用。

【关键词】企业发展；工会组织；思想政治；重要作用；优化措施

工会作为维护企业基本权益的重要组织，是我们党和国家为了保证职工合法权益而设立的重要组织，在企业中发挥着重要的作用。工会不仅需要做好员工权益维护，同时需要做好思想政治教育工作，以此方式提升员工思想素质水平，强化职工政治觉悟。职工思政素质水平提升对于企业建设与发展具有重要的意义，为此需要全面落实工会组织的思政工作职能，在企业中开展多样化的思政教育活动，确保思政工作质量与效果全面提高。

1　工会组织与思政工作的关系概述

1.1　工会组织

工会代表着广大员工的基本权益，其职能对于员工与企业发展具有重要作用，企业工会的所有职能体现必须要以保障企业员工基本利益为中心，这是党和国家赋予工会的权利。所以企业工会必须履行好相应的职责，工作开展要站在企业员工合法权益的层面上，积极帮助企业员工解决生活和工作上存在的困难，对于企业员工提出的合理诉求要按照我国法律法规给予帮助，确保员工能够在健康、稳定的环境中工作，实现自我价值与企业价值。同时，工会具有企业宣传教育与文化组织的功能，需要结合企业发展以及员工需求，做好各项文化服务活动。工会在维护企业员工基本权益的同时，具有重要的教育、培训以及管理等职能，需要对员工进行思政培训与教育工作，将党的先进思想和优秀理论贯彻到教育培训过程中，从而提高员工的思想道德水平，提高员工的文化素质，这对于提高发展能力具有重要作用，能够推动企业发展目标实现[1]。

1.2　思政工作

思政工作一直都是企业工作中的一项基础工作，党和国家高度关注此项工作开展的效果，科学完善的思政体系建设，能够提高员工思想素质水平，满足职工职业发展与自我发展需求。思想政治包括员工的思想素质与政治觉悟，在复杂的国际形势以及市场环境下，企业不仅需要做好基础性的生产与经营工作，为了确保企业长远可持续发展，必须充分落实思政工作，通过对员工思想素质与政治意识的教育，能够全面提高员工思想与政治水平，把握社会与国家发展方向，为企业以及群众提供更好的服务，避免出现思想政治问题，对于社会、企业以及员工个人都具有重要的意义。

2　工会在企业思政工作中的作用分析

工会思政工作在企业中具有重要的作用，是保障企业建设发展的关键所在，但是由于工会思政工作模式缺乏创新，导致工会工作的实际效果不足，难以发挥出其具体作用，为此需要加强工会思政工作创新发展，确保工会组织的优势能够全面发挥出来，其重要作用体现在以下几个方面：①通过对工会思政工作的创新,能够提升工会思政工作模式质量,使得传统思政工作模式中存在的弊端问题得以解决,确保工会思政工作模式更加符合企业工作的实际需求,结合相关业务开展思政教育相关工作,能够全面保障企业各项工作开展。②通过强化思政培训工作,有利于促进工会教育职能更好地实现。在企业发展过程中,工会的思政教育职能逐渐凸显出来,当前工会的思政教育职能更加深化,工会的思政教育职能越来越受到各方的重视,因此工会需要采用更加主动的教育、培训态度,通多主动的工作模式,能够为员工提供更好的教育服务,主动调查了解当前的思想素质与政治意识状态,通过教育培训、管理等方式进行优化,确保员工能够学习优秀的思想与精神,从而能够促进员工思想水平提升,强化员工责任心与责任感,使其在工作中保持高度认真负责的意识,能够提高企业员工的工作积极性,从而保障企业各项工作更好地开展[2]。③工会作为企业思政工作的主体,承担着重要的教育职能,是党和国家赋予工会人员的权利与义务,需要构建完善的思政工作体系,确保思政工作能够全面落实,为员工提供丰富的思政学习资源与学习机会,积极转换传统的教育培训模式,提升工作实际效果,利用多样化的活动方式,提升其参与相关活动学习积极性,建设良好的思政学习氛围,从而能够全面提升企业员工的思想素质,对于企业生产、经营等基础性工作开展具有重要的现实意义,是工会在思政工作中的重要责任。在我国社会经济高速发展的过程中,企业需要结合时代以及市场变化,对思政工作模式进行全面创新,确保思政工作能够落实到企业的各个方面。企业思政工作不仅需要与企业的经营管理相结合,同时需要获得员工的广泛认同与支持,工会作为企业员工的代表,承担着重要的思政工作职能。

3　企业工会在思政工作中存在的主要问题分析

结合当前企业工会在此项工作中的实际情况来看,依然具有许多不足之处,从而使得其职能难以充分发挥,对企业各项工作开展造成了不利干扰,主要问题包括:①理念方面的滞后性问题。在相关工作开展过程中,部分工会人员、管理人员以及员工没有明确思政工作的内涵,导致相关理念没有创新,依然采用传统的思想理念,导致工会思政工作缺乏实效性,也就难以提高员工的综合水平,在落后的理念引影响下,企业思政工作整体质量不足,不符合党对企业思政工作的基本要求。②思政工作队伍存在问题。工会建设一直以来都是党和国家高度关注的内容,但是当前部分企业的工会成员在专业能力、综合素质等方面还存在着一些问题,整体队伍建设水平不足,从而导致工会思想政治工作的职能难以充分发挥,是影响工会工作质量的主要因素,需要全面强化队伍、团队综合水平。③工作形式缺乏创新。思政工作必须紧追党与时代发展,需要按照当前社会和市场发展实际情况,准确把握党的前进方向,依据党的基本要求对工作模式进行优化,但是部分企业在思想政治工作模式中较为落后,采用的工作模式缺乏创新,整体工作较为被动,无法有效提升员工政治意识,是影响思想政治工作开展的重要因素,为此需要进一步强化工会的思想政治工作模式,加强实施方式创新,确保此项工作能够高质量开展[3]。

4　企业工会做好员工思想政治工作的优化策略分析

结合上文叙述可以明确,工会与企业发展具有密切的关系,覆盖到企业发展的各个层面与环节,尤其是在企业思想政治工作方面,承载着重要的教育、培训以及管理责任。但是在多种因素的影响下,部分企业工会的思想政治工作存在着问题,导致思政工作质量受到很大负面影响,为此需要进一步强化思想政治工作,通过对工会工作理念、模式以及方法的创新,结合队伍建设工作,能够全面提升企业员工思政工作质量。因此,本文结合相关实践经验,总结如下多项科学有效的应用措施。

4.1　加强员工思政工作模式创新

企业员工思政工作是一项系统性、长期性工作,是企业全体成员发展过程中形成的价值认同,现代企业在经过多项改革调整后,其内部企业文化发生很大变化,但是相应的工作模式却没有得到创新,所以必须明确企业思政工作创新路径,采用科学的方法强化工作模式创新,使其能够与企业发展方向与内容相一致,才能够确保企业各项工作质量。为此,需要将工会组织与企业思政工作相连接,在现代企业思政工作过程中,结合实际工作内容对工作模式进行调整,确保工作模式符合思政工作实际情况与实际需求，同时思政工作具有明显的实践性，能够充分展现出企业发展中存在的问题，为此需要通过对工作模式的创新，促进工会思政工作模式完善，从而能够有效解决企业发展中存在的问题。与此同时，工会需要积极转变工作理念，工会不能仅仅完成上级派发的思想教育任务，要主动定期对员工、管理人员开展思想政治教育，并加强工会职能的宣传，让每一位员工都能明确工会的基本职能和权利，从而能够保证员工自觉地接受监督和教育，不断提高员工的思想政治水平，帮助员工树立高度的职业道德观念，进而能够拉近工会与员工之间的距离，在企业中营造良好的工作氛围，提高员工对企业的价值认同感和归属感，使得员工能够在良好的内部环境中开展各项工作，还能够促进员工学习与发展。

4.2　提升工作内容丰富性

在现代企业内涵逐渐丰富的背景下，为了提高现代企业文化建设工作实效性，工会组织需要采用多种不同的建设模式，构建多元化建设方式，从而满足现代企业不同文化领域的建设需求。例如，在现代企业的员工思想素质建设中，工会组织可以采用以诚信为核心的建设模式，强调生产经营诚信化，现代企业生产的产品、提供的服务必须保证质量，不能存在虚假宣传、质量较差等产品与服务，工会组织通过采用以诚信为核心的建设模式，提升员工思想道德水平，能够提高现代企业社会形象，获得社会群众的广泛认可；在现代企业思政工作中，工会组织可以采用以人为本的管理模式，深入到员工生活与工作中，了解员工的生活、工作以及学习需求，明确其当前的思想政治水平，通过对员工的思想政治教育，使其精神世界得到充实，从而能够促进企业员工个人发展。在新时代背景下，对员工思政工作提出了更多的要求，为此工会需要积极承担其思政教育职责，采用科学的教育方式，对思政工作内容进行扩充与拓宽，确保思政工作能够覆盖到员工的各项工作与生活中，从而能够提升思政工作质量，确保思政工作高质量完成，实现党和国家对企业思政工作的要求目标，是促进企业建设与发展的关键所在。

4.3　强化思政政治文化建设

在现代企业发展过程中，思想政治建设是一项重要内容，与企业文化建设具有密切的关系，所以工会组织需要将文化建设与思想政治工作进行融合，是现代企业文化建设创新的必然方向。首先，工会组织需要明确现代思政文化建设的重要意义，构建良好的思政文化体系，采用科学的方式将其进行结合，依据思想政治的基本内涵实现连接。现代企业文化所倡导的奉献精神、进取精神、创新精神，与思想政治理念具有一致性，通过将其进行融合，构建丰富多彩的现代企业文化建设活动，通过现代企业文化构建特有的表现形式，能够拉近思想政治工作与企业员工的距离，更容易被企业内部员工认同和接受，是工会组织在企业文化建设中需要履行的基础职责。其次，思政与文化具有重要的关系，都会对员工的精神产生影响，通过将思政与文化建设工作相结合，能够构建更好的氛围，从而提升员工参与思政工作积极性，确保各项工作能够充分落实，有利于全面提升思政工作质量，为此工会需要加强此项工作，优化文化与思想两项不同工作的衔接方式，推动相关工作质量提升。

4.4　加强工会思政工作队伍建设

首先，企业工会要优化内部成员结构，吸收更多优秀的党员，以党员的高素质、高思想政治觉悟来为工会的思政工作指明方向，在优秀党员的影响下，能够潜移默化地提高工会队伍整体的思想道德素质水平，从而促进思政工作质量提高。其次，工会还要积极吸收优秀员工，作为预备党员，提升党组织的力量，从而能够有效提高工会队伍的思想政治水平，促进工会思政工作更好的开展。再次，工会要加强工会队伍思政教育工作，定期对工会队伍开展思想政治教育课程，可以结合现代信息技术的方式，建设网络化的教育平台，依靠现代技术的优势，通过网络教育平台，工会人员可以不受限制地进行党的先进思想学习，使得常规培训模式中受到时间和地域限制的束缚被打破，从而提高思政教育的灵活性，能够有效提高工会人员参与思想政治学习的积极性，进而能够提高工会队伍整体的思想政治水平，推动进此项工作全面地开展与落实。最后，工会队伍需要提高综合能力，工会工作人员需要具备复合型的知识和能力，例如法律知识、经济知识、教育知识等，一方面可以为企业的发展提供建议，促进企业经济更好的发展,另一方面可以为企业员工提供更高质量的思政教育。加强思政教育丰富性是我国新形势下对工会思政工作所提出的更高要求,工会人员必须拥有自觉、主动的思政教育意识,才能够更好地发挥出工会的思政教育职能,为企业发展和员工发展提供双重保障,是新时期背景下企业工会思政工作创新发展的主要方向。

4.5 提供优质的思政教育服务

服务职能是新时期下企业工会职能定位出现的主要变化,因此工会要转变工作理念,在思政教育工作中做好服务工作,为企业的发展指明前进方向,保证企业发展方向能够始终符合党和国家的要求和发展理念,对于企业长远可持续发展具有重要的意义。为此需要不断深化工会的思政教育服务质量,确保工会能够为企业员工提供完善的思政教育服务,从而能够促进员工参与思政学习积极性的提高。通过对员工思政教育的服务,能够更好地完成思政教育工作,要求工会人员加强服务职能建设,结合思政工作的主要内容,全面强化思政教育服务能力,满足员工思政学习的主要需求,从而能够提高员工学习主动性,还能够提升思政教育培训效果,将工会的基本职能全面体现,是当前企业工会需要加强的重点工作,能够促进企业思政工作体系与模式创新发展。

5 结束语

综上所述,本文简要介绍了工会组织及思想政治工作的基本内涵,并对工会在思政工作中的作用进行分析,同时提出了当前工会思政工作存在的问题,总结了多项科学有效的思政工作开展优化措施,希望能够对企业工会思政工作起到一定的借鉴与帮助作用,不断提升思政工作水平。

参考文献

[1]马昕. 浅析工会组织在国有企业基层单位思想政治工作中的作用[J]. 环渤海经济瞭望, 2022(1):123-125.
[2]强敏荣. 工会在加强企业职工思想政治工作中的作用分析[J]. 中外企业文化, 2020(32):97-98.
[3]岑佳佳. 思想政治工作在企业工会工作中的作用[J]. 经济与社会发展研究, 2020(19).

本文曾发表于2022年《中文科技期刊数据库(全文版)社会科学》

全面预算管理在酒店经营管理中的应用

贾　娜

（陕西地建酒店管理集团有限责任公司，陕西 西安 710000）

【摘要】随着社会经济不断发展，各行各业市场竞争越发激烈，酒店行业也不例外，酒店利润逐渐下降，酒店经营管理在新时代背景下面临着严峻考验。在现代企业管理中，全面预算管理是提高企业财务管理水平的必要途径，也是促进企业稳定可持续发展的重要手段。酒店企业也应加强全面预算管理，以此提高酒店经营管理水平。本文从全面预算管理概述出发，简单分析了全面预算管理在酒店经营管理中存在的问题，并提出了相应的优化应用策略。

【关键词】全面预算管理；酒店经营管理；应用

引言

自改革开放以来，我国社会经济飞速发展，各行业也迎来了很好的发展机会，酒店行业也不例外。比如经济型连锁酒店在近年来发展迅速，各种经济型酒店越来越多，供给过剩问题逐渐凸显，这就给曾经非常辉煌的酒店带来了较大冲击。一方面，酒店平均价格很难得到提高，但人工成本、房子租金成本等却逐渐增加，导致酒店利润空间不断缩减。另一方面，人们对生活品质的追求越来越高，实际需求逐渐增多。酒店为了提高市场竞争能力，必须要在确保利润率的同时有效控制各方面成本，开展全面预算管理是新时代酒店经营发展的必要手段。

一、全面预算管理概述

企业根据自身经营情况与发展情况进行阶段性发展目标和规划的制定，并以以往各项财务数据为依据，统筹规划企业所有可量化的指标，对每项经营内容制定预算目标，这就是全面预算管理。全面预算管理是基于市场需求与科学预算，将销售预算作为主要切入点，逐渐延伸到各个层面，包括成本、投资以及收支等经济活动。根据企业各项业务活动，将预算管理目标分摊到酒店各个部门以及每个员工身上，将预算管理与酒店管控结合起来，方便评估与调整预算目标实施情况。将预算目标与实际经营情况相比较，及时发现酒店经营管理过程中与预算目标之间的偏差，采取相应举措进行完善。

对酒店企业而言，全面预算管理有助于内部控制水平、经营管理效率的有效提升，进而让企业在激烈的市场竞争中更具竞争力，为企业创造更大的经济价值，促进企业可持续长远发展。在新时代背景下，社会经济不断发展，经济管理相关理论也不断进步，全面预算管理从最初对资金的计划与协调逐渐发展到内部控制的各个方面，甚至与企业的总体规划关系密切。全面预算管理因其诸多优势受到诸多企业的认可并广泛应用，不管是制造型企业还是服务型企业，都纷纷开展了全面预算管理工作，酒店企业将全面预算管理体系引入到经营管理工作中，是在顺应时代发展的同时提高酒店经营管理水平的重要举措。

二、全面预算管理在酒店经营管理应用中存在的问题

（一）全面预算管理意识不强

在酒店经营管理中开展全面预算管理工作十分重要，部分酒店企业虽然意识到全面预算管理与酒店全局性和综合性管理目标关系密切，但在实际经营管理中将全面预算管理工作全权交给财务部门负责，各运营部门过于注重预算目标的完成情况，在业务预算收入与费用的编制过程中并未从全局

出发对酒店整体利益进行考虑,而只是考虑到部门利益,导致全面预算管理流于形式,不能将全面预算管理在实际运营管理中的积极作用充分发挥出来。

(二)欠缺完善的预算管理组织结构

很多酒店企业的预算管理组织结构都并不完善,甚至没有进行全面预算管理部门的构建。公司法中明确规定由董事会制订公司的年度财务预算方案,且由股东大会负责审批。相关数据表明,我国很多酒店企业经营预算管理决策都是由总经理制定的,由专门的预算机构制定最终决策的酒店企业不多。并且,酒店在制定全面预算过程中很少有其他部门参与进来,并未结合各方意见,预算的合理性受到限制,也在一定程度上影响了全面预算执行的可行性。与此同时,很多酒店企业对业务部门的重视程度较高,对职能部门较为忽视,导致酒店企业对财务管理、预算管理等工作并不够重视。酒店各项经营活动虽然都离不开财务部门的支持,但财务部管理人员却没有表决权,只能无条件执行企业高管的决策,在企业制定重大决策时财务部门不能发挥其作用,针对管理层制定的不具有财务可行性的决策,财务部门也很难否决。

(三)全面预算编制方式不够科学合理

很多酒店虽然进行了全面预算管理,但预算编制方法欠缺科学合理性,影响了编制质量,全面预算管理的执行依据欠缺可靠性,导致实际支出费用与预算之间的偏差较大。固定预算法是我国很多酒店最常用的预算编制方式,这种编制方法操作简单方便,且编制时间较短,有着较高的编制效率,但它是基于过去的数据,根据一定的变化量来进行预算编制,有一些不合理因素存在,科学性、先进性不足。固定预算法也没有对影响预算项目的内外部环境因素变化进行考虑,没有预测分析酒店经营活动,经营成本预算编制不够准确。酒店在预算编制程序上也没有进行上下结合、逐层汇总,只是财务部门将预算指标传达到预算执行部门与相关工作人员,预算执行部门从部门角度出发进行预算编制与执行,费用预算过高,再加上监督检查机制不够严格,导致预算数据真实性不足,与实际经营数据之间的差距过大。

(四)全面预算管理执行力度有待提高

酒店企业在全面预算管理执行中,预算编制工作更受重视,预算执行过程由于欠缺完善的预算执行机制,导致预算稳定性不能得到保证。部分酒店企业虽然意识到全面预算管理的积极作用,但一味照搬其他企业的预算管理机制,并未根据自身实际情况进行调整,导致企业预算管理体系可行性不高,执行效果不佳。部分酒店企业更加注重会计数据核算,认为会计数据分析能力较高就能让预算编制更加科学合理,并未对企业内部真实业务能力、实际经营效果等进行综合考虑,导致预算管理与战略发展目标不相符,不能充分发挥全面预算管理的真正价值。

(五)预算考核与激励机制不健全

酒店企业应做好全面预算管理的事后控制工作,严格考核预算执行效果,并根据激励机制采取相应的措施。但实际上,很多企业并未安排专业人员记录与及时反馈预算管理效果,预算执行效果考核较为随意。很多酒店并未针对预算执行制定相应的奖惩激励机制,只是考核了管理层人员,表现优秀者可以拿到一定的奖励,不达标的人并不会受到惩罚。针对普通员工,部分酒店企业并未进行考核,有的酒店企业虽然对员工进行了考核,但并未对不合格的员工进行惩罚,这样的考核方式影响了员工对预算执行的积极性,不利于全面预算管理工作的有序开展。

(六)欠缺专业的全面预算管理人才

一般而言,酒店预算编制工作都是由业务部、财务部负责的,总经理审核预算编制结果。这些预算人员只是对会计知识有所了解,预算管理水平很难得到保证,且他们在预算编制方面欠缺经验,很容易在全面预算管理过程中抓不住关键,不能全面分析酒店经营情况,影响了预算编制结果的准确性。很多酒店预算管理人才严重不足,也没有进行专业预算管理机构的设置,为全面预算管理在酒店经营管理中的顺利开展带来了严重负面影响。

三、全面预算管理在酒店经营管理中的应用策略

（一）增强全员预算管理意识

酒店管理人员应对全面预算管理更加重视，安排专门的全面预算管理人员，注重企业内部全面预算管理意识的培养。酒店企业应保证预算信息沟通渠道的畅通性与预算管理的相对独立性，创造良好的内部环境，为全面预算管理在酒店经营管理中的应用打下良好的基础。酒店企业管理人员应充分发挥引领带头作用，加强全面预算管理，并帮助企业员工认识到全面预算管理的重要性，提高全员预算管理知识与技能，增强全员全面预算管理意识，让各部门人员积极配合完成全面预算管理工作，确保预算执行效果。

（二）建立健全全面预算管理体系

酒店企业应根据酒店整体战略规划与经营目标建立健全全面预算管理体系，确保酒店预算的合理性。全面预算管理方案通过审批后必须要严格执行，为酒店战略规划与经营目标的良好实现提供助力。对此，酒店企业可以进行预算管理决策机构、工作机构以及执行单位的合理设置。其中，酒店负责人、财务负责人以及主要职能负责人应纳入到决策机构中，负责预算管理制度的制定、预算目标的拟定、考核预算执行情况等。一般情况下，工作机构会在财会部门设立，主要成员包括财务负责人、人力资源、业务等相关人员，主要负责预算管理制度的拟定、预算执行情况分析等。执行单位则将内部各职能部门包含在内，主要工作是为预算编制提供基础资料信息、本部门预算上报等。酒店企业应在经营管理的基础上对经营预算、资本预算以及财务预算进行深入研究。在这个过程中，应结合企业管理模式进行带有企业特色的报表、员工岗位职责分工等内容的子系统编制工作，才能将管理重点在全面预算管理体系中凸显，让相关工作人员了解实际管理需求。

（三）对全面预算编制流程进行规范

预算编制往往需要经过很多环节，明确经营目标后要逐步进行预算编制、调整、审批等各项工作。酒店管理决策人员应充分考虑到酒店经营环境、市场变化、竞争对手动态、消费者实际需求等各方面因素，通过定量与定性分析展开市场预测工作。酒店在确定经营目标时要进行预算方法的合理选择，还要保证预算目标具有可操作性。在这个经营目标的基础上对预算指标进行合理分解，将指标责任落实到各部门以及员工身上。除此之外，还要合理合规地开展全面预算编制后的审核与调整工作，确保预算书的合理性、可靠性，为酒店企业全面预算工作的开展提供重要保障。

（四）制定完善的预算执行机制

酒店企业应结合实际情况进行预算执行机制的优化与完善，提高酒店全面预算管理水平。部分酒店企业由于预算管理体系不健全导致很多全面预算管理措施得不到真正落实，甚至一些酒店企业照搬其他企业现有的预算管理体系，并未根据自身情况进行调整，执行性较低，全面预算的效果也很难得到充分发挥。针对这一问题，酒店企业必须要更加重视预算执行工作，深入落实各项全面预算管理措施，尤其是一些需要大额资金的经济活动，必须要对申请与审批流程有严格的要求，杜绝资金滥用等问题给酒店资金流带来不良影响，需要对预算执行过程中出现的问题进行深入分析与研究，及时采取有效的优化方案。同时，要对信息反馈与预警体系进行完善，降低酒店企业外部风险。酒店企业可以进行双向预算信息反馈机制的合理构建，为管理层获得的预算执行信息真实性提供保障，也能保证预算执行信息能及时传递到管理人员手中，避免出现信息孤岛、信息不对称等问题，也有助于管理层对预算执行情况的全面掌握，进而对全面预算管理工作进行适当调整。

（五）完善预算考核与奖惩机制

针对全面预算管理工作，应制定严格的预算考核机制与奖惩机制，确保全面预算各项工作真正落实到位。酒店企业应定期考核与评价预算执行情况，这就需要预算评价考核体系具有科学合理性，统一考核标准，根据主体不同适当变化考核方式。各部门负责人应进行目标责任书的签订，并将其作为考核依据。预算考核结果应与员工的升职晋升、经济利益等联系起来，酒店可以根据具体情况制定相应的激励奖惩机制，在预算管理中表现较好的员工可以得到一些奖励，充分调动员工在全面预算管理

中的积极性。针对预算执行效果不佳的员工,应适当进行绩效扣分处理。值得注意的是,预算考核必须要具有公平公开公正性,才能更好完成预算工作,促进酒店长远发展。

(六)注重相关工作人员专业素质的提升

全面预算管理效率与质量的提升离不开专业预算管理人才的支持。酒店企业应注重相关工作人员专业水平与职业素养的提升,在招聘时看重专业技能,且要求应聘人员具备较强的预算能力与管理能力。酒店企业还可以组织现有工作人员进行全面预算管理知识与技能培训,鼓励工作人员多多学习预算管理知识与技能。尤其是财务人员,必须要有长远的眼光,多学习预算管理方面的知识,了解相关政策。酒店企业还可以邀请全面预算管理方面的专家到酒店内部开展专业知识交流大会,提高酒店相关工作人员综合素质,更好地开展全面预算管理工作。

四、结束语

在当前市场环境下,酒店企业应充分意识到全面预算管理在经营管理中的重要作用,在酒店内部积极开展全面预算管理工作,增强员工全面预算管理意识,制定完善的预算管理体系、考核与激励奖惩机制,提高全面预算管理的执行力度,通过这些举措有效提高酒店企业经营管理水平,促使酒店企业在市场上的竞争能力得到提升,促进酒店企业可持续健康发展。

参考文献

[1]顾莹.全面预算管理在酒店经营中的应用问题探讨[J].企业改革与管理,2021(18):167-168.

[2]孙慧贞.全面预算管理在酒店经营管理中的应用问题探讨[J].企业改革与管理,2021(12):128-129.

[3]杨慧.全面预算管理在酒店经营管理中的应用[J].中国中小企业,2021(3):144-145.

本文曾发表于2022年《销售与管理》第5期